GEOCHIMICA ET COSMOCHIMICA ACTA

SUPPLEMENT 5

PROCEEDINGS
OF THE
FIFTH LUNAR SCIENCE CONFERENCE
Houston, Texas, March 18–22, 1974

COVER ILLUSTRATIONS

PAPERS IN THESE THREE VOLUMES deal primarily with Earth and its Moon, but in a very real sense they are discussing, speculating, and evolving new ideas concerning natural processes that may have governed origins and subsequent histories of planetary bodies throughout the solar system, if not the universe. It is in this rather philosophical vein that we have chosen to portray Earth and Moon in company with their fellow inner planets and Jupiter.

Within the past year, remote sensing vehicles have extended our horizons to four new planets and a second moon. Mariner 10 provided us with the mosaic of Mercury (Volume 1, front cover) on 29 March 1974. The observer is suspended approximately 210,000 kilometers over the equator in this view, with the north pole on the right (frame P-14580). Mariner 10 sent an image of Venus (Volume 1, back cover) with north pole at top left from a distance of 720,000 kilometers on 6 February 1974 (frame P-14400). Ronald Evans took the picture of Earth's southern hemisphere (Volume 2, front cover) from Apollo 17 on the way to the Moon (frame AS17-148-22725), while Lick Observatory provided the composite of the Moon (Volume 2, back cover) (frame L9).

Pioneer 10 took the view of Jupiter (Volume 3, front cover) from approximately 2,500,000 kilometers on 1 December 1973. Jupiter's Red Spot and the shadow of the moon, Io, are clearly visible (frame 73-HC-964). A composite view of Mars (Volume 3, back cover) shows the north polar cap (Jet Propulsion Lab Mosaic, based upon Mariner IX photography). The inset is a Mariner IX photograph of a Martian moon, Deimos (frame 12694).

GEOCHIMICA ET COSMOCHIMICA ACTA

Journal of The Geochemical Society and The Meteoritical Society

SUPPLEMENT 5

PROCEEDINGS
OF THE
FIFTH LUNAR SCIENCE CONFERENCE
Houston, Texas, March 18–22, 1974

Sponsored by

The NASA Johnson Space Center

and

The Lunar Science Institute

VOLUME 2

CHEMICAL AND ISOTOPE ANALYSES

ORGANIC CHEMISTRY

PERGAMON PRESS

NEW YORK · OXFORD · TORONTO · SYDNEY · BRAUNSCHWEIG

Pergamon Press, Inc., Maxwell House, Fairview Park
New York, N.Y. 10523 USA Teletype 137328

Pergamon Press, Ltd. Headington Hill Hall
Oxford OX3 OBW England

Pergamon of Canada Ltd., 207 Queen's Quay West
Toronto 1 Canada

Pergamon Gmbh Burgplatz 1, 33 Braunschweig, Germany

Pergamon House, 19a Boundary Street, Rushcutters Bay NSW2011, Australia

Type set by The European Printing Corporation, Ltd.
Dublin, Ireland

Printed by Publishers Production International
and bound by Arnold's Bindery
in the United States of America

Library of Congress Cataloging in Publication Data

Lunar Science Conference, 5th, Houston, Tex., 1974.
  Proceedings of the Fifth Lunar Science Conference.

  (Geochimica et cosmochimica acta. Supplement ; 5)
  Includes indexes.
  CONTENTS: v. 1. Mineralogy and petrology.--v. 2.
Chemical and isotope analyses. Organic chemistry.--
v. 3. Physical properties.
  1. Lunar geology--Congresses. 2. Cosmochemistry--
Congresses. I. Gose, Wulf Achim, ed. II. Lyndon
B. Johnson Space Center. III. Lunar Science Institute.
IV. Series.
QB592.L85 1974   559.9'1   74-23095
ISBN 0-08-018318-2

Supplement 5
GEOCHIMICA ET COSMOCHIMICA ACTA
Journal of The Geochemical Society and The Meteoritical Society

H. K. Hills, Department of Space Physics and Astronomy, Rice University, Houston, Texas 77001.

J. H. Hoffman, Institute of Physical Sciences, University of Texas at Dallas, Richardson, Texas 75080.

F. Hörz, NASA Johnson Space Center, Houston, Texas 77058.

K. A. Howard, Branch of Astrogeolgic Studies, U.S. Geological Survey, Menlo Park, California 94025.

R. Manka, Solar-Terrestrial Research Program, Atmospheric Sciences Section, National Science Foundation, Washington, D.C. 20550.

U. B. Marvin, Smithsonian Astrophysical Observatory, Cambridge, Massachusetts 02138.

R. Morris, NASA Johnson Space Center, Houston, Texas 77058.

W. R. Muehlberger, Department of Geological Sciences, University of Texas at Austin, Austin, Texas 78712.

G. D. O'Kelley, Oak Ridge National Laboratory, Oak Ridge, Tennessee 37830.

C. W. Parkin, NASA Ames Research Center, Moffett Field, California 94035.

J. A. Philpotts, NASA Goddard Spaceflight Center, Greenbelt, Maryland 20771.

G. W. Reed, Jr., Chemistry Division, Argonne National Laboratory, Argonne, Illinois 60439.

J. H. Reynolds, Physics Department, University of California, Berkeley, California 94720.

J. M. Rhodes, Lockheed Electronics Corporation, Houston, Texas 77058.

R. A. Schmitt, Department of Chemistry, Oregon State University, Corvallis, Oregon 97331.

J. R. Smyth, The Lunar Science Institute, Houston, Texas 77058.

D. F. Weill, Center for Volcanology, University of Oregon, Eugene, Oregon 97403.

R. Zartmann, U. S. Geological Survey, Denver, Colorado 80225.

EDITORIAL ASSISTANT

J. M. Shack, The Lunar Science Institute, Houston, Texas 77058.

EDITORIAL STAFF

P. Atkinson, The Lunar Science Institute, Houston, Texas 77058.

E. Ault, The Lunar Science Institute, Houston, Texas 77058.

J. H. Heiken, The Lunar Science Institute, Houston, Texas 77058.

Contents
(Volume 2)

Geochemical Evolution of Lunar Rocks and of the Moon as a Whole

Contents, Volume 2

Lunar Chronology

Contents, Volume 2

Contents, Volume 2

Noble Gases: Origins and Implications for Lunar History

Contents, Volume 2

Regolith Evolution Recorded by Cosmic-Ray Induced Nuclear Reaction Products

Contents, Volume 2

Search for Organic Compounds in Lunar Materials

GEOCHIMICA ET COSMOCHIMICA ACTA

SUPPLEMENT 5

PROCEEDINGS
OF THE
FIFTH LUNAR SCIENCE CONFERENCE

Houston, Texas, March 18–22, 1974

Proceedings of the Fifth Lunar Conference
(Supplement 5, Geochimica et Cosmochimica Acta)
Vol. 2 pp. 975–979 (1974)
Printed in the United States of America

The role of horizontal transport as evaluated from the Apollo 15 and 16 orbital experiments

I. Adler, M. Podwysocki, and C. Andre

University of Maryland, College Park, Md. 20742

J. Trombka, E. Eller, R. Schmadebeck, and L. Yin

Goddard Space Flight Center, Greenbelt, Md.

Abstract—Despite the large amount of accumulated data there is still considerable debate about the role of various processes in the formation of the lunar surface. One of the processes frequently discussed is that of horizontal transport and how it affected surface chemistry and surface features. The orbital experiments flown on Apollo 15 and 16 enable us to consider the extent of horizontal transport. In particular, the X-ray experiment because of its low penetration depth permits us to make at least crude estimates of the extent of horizontal transport.

Introduction

Among the various processes involved in the evolution of the lunar surface are horizontal transport, vertical and lateral mixing, erosion, volatile transport and mass wasting. How much of the moon's surface is primary, formed by accumulation processes and how much is secondary, generated by magmatic processes modified by subsequent infalls is still the subject of debate (Gold, 1972).

The existence of lateral transport as a consequence of major collisions such as the Imbrian event or the formation of Copernicus is well accepted. A large number of papers have in fact been published on Mare Imbrium and its relationship to lunar materials found elsewhere, even at very large distances. The earliest indication of such transport mechanisms was cited by Wood (1970) after finding anorthositic fragments in the Apollo 11 material. Other mechanisms which have been proposed include transport by ash flows and volatile transport.

Both Gold (1972) and Criswell (1972) have suggested lateral transport as a consequence of electrostatic charging. Criswell has proposed that electrostatic forces raise swarms of dust grains to levels from 3 to 30 cm above the surface and that this phenomenon explains the horizon glow seen by Surveyor 6 and 7 on the western horizon up to one hour after local sunset. Further, Criswell finds that the levitation mechanism preferentially removes dust from the rocks and smooths the dust around the rocks.

In this paper we do not argue that the various mechanisms described do not occur. The evidence that such phenomena occur is strong. We do however present semi-quantitative arguments, based on our geochemistry measurements which put limits on the extent of horizontal transport mechanisms and their effect on the surface composition.

RESULTS OF THE ORBITAL GEOCHEMICAL MEASUREMENTS

The various components of the orbital geochemistry experiments have shown us that there are distinct extensive regional variations in chemical composition. The X-ray experiment has clearly demonstrated the existence of very marked differences in chemistry between the highlands and the mare basin fill, the highlands being clearly richer in aluminum (inferentially anorthositic) while the mare basins are low in aluminum (basaltic). On a large scale there is a good correlation between topography and the aluminum content. The gamma-ray experiment which has shown unusual regional variations in radioactivity is now providing information of the regional distribution of a large number of chemical elements such as (Th and U), K, Fe, Si, Mg, and O (Metzger *et al.*, 1974). A good correlation for various lunar sites has been found between the X-ray and gamma-ray results for magnesium (Trombka *et al.*, 1974). This is particularly important because the X-ray experiment analyzes the topmost surface while the gamma-ray measurements involve a depth of at least a foot. Thus the good correlation between these experiments and the results of the lunar sample studies can be taken as an illustration that the surface material is representative of a substantially greater depth.

The X-ray fluorescence experiment is particularly well suited to look at the question of horizontal transport, in particular the relationship between the mare fill and the highland materials. This follows from the fact that the orbital X-ray experimental measurements involve secondary X-rays coming from very shallow depths. Simple calculations based on the assumption of either a basalt or feldspar show that a 10 micron layer represents infinite thickness. This is equivalent to about 3×10^{-4} g/cm^2 of material.

Gold (1973) proposed that horizontal transport by such mechanisms as electrostatic charging have played a large role in the formation of the flat mare basin fill. In a subsequent paper (Gold and Williams, 1974), it is stated that one can expect effects in the environment of the moon where the charge in neighboring grains is dissimilar and where electrons fields are set up on a micron scale. Under such circumstances the forces on a grain may exceed both gravitational forces on neighboring particles and as a consequence, produce surface transportation of a geologically significant scale. The authors conclude that magnetoshield electron bombardment provides an explanation for the great difference in the appearance and the surface topography between the back and front of the moon, if surface erosion by such a phenomenon has played a major part in shaping the moon.

It is our contention that if highland material has drifted into the basins to any marked extent, then the X-ray experiment would not have detected the substantial differences that were observed between the mare and the surrounding highlands. In point of fact, very marked differences have been found. There are a number of outstanding examples, one of which is the crater Tsiolkovsky, where X-ray observations showed the ratio of aluminum in the rim areas to that of the basin to be about 2:1. Further, recent studies (Podwysocki *et al.*, 1974) of the X-ray data using trend surface analysis have shown chemical deviations on a much finer

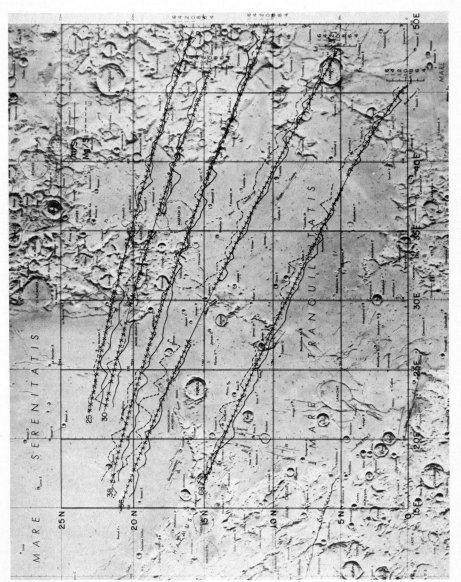

Fig. 1. Several ground tracks from the Apollo 15 orbital flight. The Al/Si variation is shown by the solid line and the Mg/Si values are given by the dashed lines. The range of values are indicated by the scales shown on the right. For 16 second data the calculated errors are ±25%.

regional scale. The crater Plinius can be distinguished by its higher Al/Si intensity ratio compared to the surrounding maria. Transitions between mare and highlands were found to be relatively sharp. In addition, areas of the mare floor were found where the Al/Si ratios deviated substantially from the local mean. The nature of these variations is shown in Fig. 1 for a number of projected ground tracks during the Apollo 15 mission. These particular tracks were chosen because they afforded a large number of data points. The plots show the variation of Al/Si and Mg/Si values along the tracks and their relationship to the underlying lunar features.

Additional substantiation of the chemical difference between the highlands and the mare basins even down to microscopic depths can be found in a recent paper on the results obtained by Lunakhod 2 in the crater Le Monnier (Kocharov and Victarov, 1974). The investigators were able to make an actual study of the transitional zone between Mare Serenitatis and the highland massif of the Taurus Mountains using a portable X-ray spectrometer on the roving Lunakhod 2 vehicle. They found that as Lunakhod moved to the hilly regions southwest of Le Monnier, the iron content began to fall while the aluminum percentage rose. In comparison to measurements in the landing area, the Si/Fe ratio rose 1.5 times while the Al/Fe ratio rose by a factor of 2. These increases are suggestive of rocks analogous to terrestrial rocks of anorthositic gabbro or gabbroic anorthosite, apparently widely distributed in the highlands.

Conclusion

Thus while electrostatic transport may occur, orbital experiments, particularly the X-ray show that this process is sufficiently limited so that the mare and highlands show a distinctly different chemistry. Further ground truth measurements previously determined (Adler *et al.*, 1972) show that the X-ray values for the Al and Mg in the mare agree well with the widespread basalts and the values for the highland sites are in close accord with the anorthositic rocks found there. If we accept the fact that the rate of impact gardening is slow at present but was much more rapid in the periods of highland formation, we can still state that the integrated effect has nevertheless been insufficient to wipe out the differences between the mare and the highlands as observed by the X-ray measurements.

References

Adler I., Trombka J., Gerard J., Lowman P., Schmadebeck R., Blodget H., Eller E., Yin L., Lamothe R., Osswald G., Gorenstein P., Bjorkholm P., Gursky H., Harris B., Golub L., and Harden R., Jr. (1972) X-ray fluorescence experiment. *Apollo 16 Preliminary Science Report*, NASA SP-315.
Criswell D. R. (1972) Lunar dust motion. *Proc. Third Lunar Sci. Conf., Geochim. Cosmochim. Acta*, Suppl. 3, Vol. 3, pp. 2671–2680. MIT Press.
Gold T. (1972) Conjectures about the evolution of the moon. *The Moon* 7(3/4), 293–306.
Gold T. and Williams G. J. (1974) Electrostatic effects on the lunar surface. *Proc. Fifth Lunar Sci. Conf., Geochim. Cosmochim. Acta*. This volume.
Kocharov G. E. and Victorov S. V. (1974) Koklady, Akad. Nauk. U.S.S.R. In press.
Metzger A. E., Trombka J. I., Reedy R. C., and Arnold J. R. (1974) Element concentrations from lunar

orbital gamma-ray measurements. *Proc. Fifth Lunar Sci. Conf., Geochim. Cosmochim. Acta.* This volume.

Podwysocki M. H., Weidner J. R., Andre C. G., Bickel A. L., Lum R. S., Adler I., and Trombka J. (1974) An analysis of the Apollo 15 X-ray fluorescence experiment for detailed lunar morphological and geochemical parameters. *Proc. Fifth Lunar Sci. Conf., Geochim. Cosmochim. Acta.* This volume.

Trombka J. (1974) Private communication.

Wood J., Dickey J. S., Marvin U. B., and Powell B. N. (1970) Lunar anorthosites and a geophysical model of the moon. *Proc. Apollo 11 Lunar Sci. Conf., Geochim. Cosmochim. Acta,* Suppl. 1, Vol. 1, pp. 965–988. Pergamon.

Proceedings of the Fifth Lunar Conference
(Supplement 5, Geochimica et Cosmochimica Acta)
Vol. 2 pp. 981–990 (1974)
Printed in the United States of America

Elemental composition of Apollo 17 fines and rocks

A. O. Brunfelt,[1] K. S. Heier,[1] B. Nilssen,[1] E. Steinnes,[2] and B. Sundvoll[1]

[1]Mineralogical-Geological Museum, University of Oslo, Oslo, Norway

[2]Institutt for Atomenergi, Kjeller, Norway

Abstract—Neutron activation analysis data for 33 elements in eight soils, five mare basalts, and one breccia from the Apollo 17 mission are discussed. A number of elements determined in the soils give linear plots versus the Al content, consistent with a simple two-component mixing of a basaltic and a plagioclase-rich end-member. The composition of each end-member is estimated assuming 4.8% Al in the mare basalt and 0.45% Ti in the "non-mare component." Elements associated with KREEP are low in soils from Stations 6 and 8 as compared to those from stations close to the South Massif, indicating a lower content of KREEP type material in the noritic rocks of the North Massif and/or Sculptured Hills. Soil 74261 is enriched in volatiles (Cu, Ga, Zn, alkali elements) relative to the other soils, probably due to a high content of "orange soil." Based on Cu and Ga data, the other soils probably contain <3% "orange soil." The Apollo 17 mare basalts are very similar to the Apollo 11 type B basalts. Two types of Apollo 17 basalt can be distinguished on the basis of the Ti content. The Apollo 17 "non-mare component" is lower in Al, Ca and higher in Fe, Mg, Mn, Cr than the typical Apollo 16 soils. Its composition corresponds to a 50%-50% contribution of noritic breccias and anorthositic rocks. Breccia 73235 is similar in composition to the "non-mare component."

INTRODUCTION

Samples received

The materials received from the Apollo 17 mission include eight samples of <1 mm fines (71501,19; 72501,25; 74121,9; 74261,6; 75061,20; 75081,23; 76321,9; 78501,23), five mare basalts (70017,21; 70215,57; 71055,36; 74275,68; 75035,39), and one breccia (73235,52).

Experimental work

The samples were analyzed with respect to 33 elements by neutron activation analysis following the procedure of Brunfelt and Steinnes (1972). In addition, analyses by X-ray fluorescence spectrometry are being made to obtain data for major elements not adequately determined by neutron activation. Furthermore, fines 78501 and basalts 70017, 71055, and 75035 were subjected to fragmental and mineral separation following the separation and identification methods of Brunfelt *et al.* (1972). The separated fractions were analyzed for the content of 24 elements by neutron activation analysis. The mineral fractions were also analyzed by electron microprobe for major elemental distribution. The results from the mineral investigations will be presented and discussed in a separate paper.

RESULTS AND DISCUSSION

Presentation of data

Data from bulk analyses of the soil samples have been published previously in Brunfelt *et al.* (1973) (75081) and Brunfelt *et al.* (1974) (71501, 72501, 74121, 74261,

75061, 76321, 78501). The chemistry of the main components contributing to the Apollo 17 soils is discussed here on the basis of our own data and data published for major and minor elements by other workers (LSPET, 1973; Duncan *et al.*, 1974; Ehmann *et al.*, 1974; Mason *et al.*, 1974; Nava, 1974; Rhodes *et al.*, 1974; Rose *et al.*, 1974; Scoon, 1974). Our major element data are in satisfactory agreement with those reported by other investigators, except for Ca and Mg, in which cases our data do not appear to be satisfactory. Concerning trace elements, the agreement with other workers (Barnes *et al.*, 1974; Chyi and Ehmann, 1974; LSPET, 1973; Philpotts *et al.*, 1974; Rhodes *et al.*, 1974) is good in most cases. However, our Sr values in 72501 and 76321 appear to be low. The same applies to Th in 75081, while our Rb values for that rock may be somewhat high. Analytical data for the five basalts are given in Table 1. They are in general in good agreement with results reported by other workers. Some of our Sr values seem to be somewhat low. The slight disagreement with Philpotts *et al.* (1974) for some rare earths in 70017 may be explained by different composition of the aliquots distributed. This is indicated by the major element data of Rose *et al.* (1974) for two different portions of this rock.

Values from the analysis of breccia 73235 are listed in Table 2, and are in particularly good agreement with those of Taylor *et al.* (1974), who investigated essentially the same elements.

Chemical characteristics of Apollo 17 soils

The soil samples from the Apollo 17 mission show a great compositional variation ranging from essentially mare basaltic (e.g. 75061) to almost entirely non-mare composition (e.g. 72501). It was shown by LSPET (1973) that Fe and Ca give linear plots versus the Al content of the soils, indicating a simple two-component mixing of a basaltic and an aluminium-rich end-member. Our data indicate the same trend for Na, Mg, Ti, Cr, Mn (Fig. 1) and the trace elements Sc, V, and Sr. On the basis of available literature data on Apollo 17 rocks, we may assume 4.8% Al in the average Apollo 17 basalt and 0.45% Ti in the non-mare derived fraction. From the Ti–Al plot in Fig. 2, which is based on all soil data available, the Al content of the high-Al end-member is fixed at 11.5%. On this basis, the composition of the end-members can be estimated from the binary plots. The composition derived from such plots is given in Table 1 and Table 2 for the mare basalt end-member and the non-mare component respectively.

As shown by LSPET (1973) soils sampled near the North Massif and Sculptured Hills (Stations 6, 8, 9) are depleted in P, Zr, and Y with respect to a simple two-component mixture. A similar trend was observed in the present work for the elements K, Rb, Cs, Ba, REE, Hf, Ta, W, Th (Fig. 3), and U, which are all known to some extent to be associated with the KREEP component. This trend is verified for K in Fig. 4, which is based on all available soil data. A possible explanation was proposed by LSPET (1973), who suggested that anorthositic gabbro may be more abundant in the North Massif or Sculptured Hills, or both, than it is in the South Massif. The MgO content of the Apollo 17 anorthositic rocks appears to be

Table 1. Elemental composition of Apollo 17 mare basalts.

	Average composition derived from soil data	Apollo 11 type B basalt (Mainly from Wakita *et al.*, 1970)	Analyzed basalts					Average, high-Ti type (70017, 70215, 71055, 74275)
			70017,21	70215,57	71055,36	74275,68	75035,39	
Na %	0.28	0.30	0.30	0.32	0.31	0.25	0.30	0.30
Mg %	5.7	4.2	5.5	4.5	5.3	6.5	4.1	5.5
Al %	4.8	5.4	5.01	4.82	4.93	4.57	5.12	4.8
K %	0.055	0.063	0.037	0.040	0.036	0.058	0.057	0.043
Ca %	7.4	8.2	8.2	7.8	8.1	8.1	7.6	8.1
Sc ppm	72	94	87	89	95	83	82	89
Ti %	6.5	6.3	8.23	7.93	8.61	7.85	5.51	8.2
V ppm	130	75	156	117	129	146	30	137
Cr %	0.31	0.20	0.355	0.251	0.279	0.410	0.156	0.32
Mn %	0.197	0.223	0.197	0.212	0.200	0.196	0.192	0.201
Fe %	14.9	14.4	14.2	14.8	14.8	14.0	14.3	14.5
Co ppm	—	19	20.6	20.4	21.6	21.3	13.7	21
Ni ppm	—	—	<10	<10	<10	<10	<10	—
Cu ppm	—	—	2.8	4.2	4.4	3.3	3.8	3.7
Zn ppm	—	—	2	2	3	2	2	2
Ga ppm	—	—	3.1	3.1	3.0	4.1	4.5	3.3
Rb ppm	0.8	0.8	0.4	0.3	0.9	0.9	0.6	0.6
Sr ppm	160	174	127	127	104	115	195	118
Cs ppm	0.055	—	0.03	0.02	0.07	0.06	0.04	0.05
Ba ppm	58	—	55	48	39	65	81	52
La ppm	5.0	8.9	4.11	4.96	4.67	5.98	7.60	5.0
Ce ppm	—	—	13.5	11.3	13.4	16.5	20.4	14
Sm ppm	8.7	15.5	7.53	6.79	7.05	9.70	12.9	7.8
Eu ppm	1.6	2.3	1.77	1.40	1.49	1.93	2.25	1.65
Tb ppm	—	—	1.77	1.66	1.74	2.05	2.81	1.80
Dy ppm	—	—	13.8	12.5	14.3	17.9	22.9	15
Yb ppm	9.4	14.2	6.3	5.9	5.4	7.5	10.7	6.3
Lu ppm	—	2.0	1.15	1.11	1.10	1.34	1.82	1.2
Hf ppm	8.5	14	8.0	8.3	6.6	8.3	10.0	7.8
Ta ppm	1.4	—	1.55	1.60	1.54	1.57	1.81	1.6
W ppm	0.11	—	0.075	0.075	0.089	0.099	0.120	0.09
Th ppm	0.4	1.6	0.17	0.21	0.32	0.38	0.35	0.3
U ppm	0.15	—	0.088	0.072	0.132	0.131	0.113	0.11
La/Sm	0.57	0.58	0.55	0.73	0.66	0.62	0.59	0.64
Sm/Eu	5.4	6.7	4.3	4.9	4.7	5.0	5.7	4.7
Sm/Yb	0.92	1.09	1.19	1.15	1.31	1.29	1.21	1.24

6–8%, while the average content of the noritic breccias, forming the other important class of non-mare rocks in this area, is about 12%. One might therefore expect the non-mare component in soils sampled near the South Massif to be higher in magnesium than the Stations 6, 8, and 9 soils. Examination of the available data from the literature suggests that the Mg content is not different in soils from the two areas. We therefore offer the alternative explanation that the observed trace element distribution trend is due to a lower percentage of KREEP in the noritic rocks derived from North Massif and/or Sculptured Hills.

The individual rare-earth elements show considerable scatter in the respect to

Table 2. Elemental composition of Apollo 17 "non-mare component."

	Average composition derived from soil data	Apollo 16 average fines (Brunfelt *et al.*, 1973)	50% noritic breccias* 50% anorthositic rocks† (PET, 1973)	VHAB average (Hubbard *et al.*, 1973)	Breccia 73235,52 present work
Na %	0.35	0.37	0.33	0.31	0.35
Mg %	5.9	3.8	5.8	7.0	6.4
Al %	11.5	14.5	11.7	11.8	11.76
K %	0.13	0.10	0.14	0.12	0.144
Ca %	9.3	11.0	9.6	8.9	9.2
Sc ppm	13	8.4			12.1
Ti %	0.48	0.32	0.68	0.44	0.45
V ppm	45	36			51
Cr %	0.130	0.073	0.112	0.085	0.125
Mn %	0.083	0.056	0.072		0.0706
Fe %	5.9	4.0	5.7	5.0	5.08
Co ppm	24	26			23.7
Ni ppm	—	—			150
Cu ppm	7	8			4.3
Zn ppm	25	20			3
Ga ppm	3.5	4.8			3.4
Rb ppm	3.6	2.5		3.9	4.1
Sr ppm	—	130			137
Cs ppm	0.16	0.16			0.17
Ba ppm	150	120		210	238
La ppm	12	11.5		21	19.7
Ce ppm	—	—			51.5
Sm ppm	7.2	5.2		9.3	9.43
Eu ppm	1.3	1.3		1.3	1.43
Tb ppm	—	1.0			1.58
Dy ppm	—	6.2			11.9
Yb ppm	5.7	4.2		6.5	5.9
Lu ppm	1.0	0.62			0.98
Hf ppm	6	4.0			7.7
Ta ppm	0.9	0.47			0.87
W ppm	0.4	—			0.26
Th ppm	2.0	1.3			2.85
U ppm	0.7	0.5		0.9	0.75
La/Sm	1.7	2.2		2.2	0.64
Sm/Eu	5.4	4.0		7.0	4.7
Sm/Yb	1.27	1.20		1.42	1.24

*Average of rocks 72435, 76315, 77135.
†Average of rocks 76230, 77017, 78155.

Al. The rare-earth distribution pattern, however, varies smoothly, as shown in Fig. 5, where plots of the La/Sm, Sm/Eu, and Sm/Yb ratios versus the Al content are given.

Soil 74261 from Station 4 was sampled adjacent to the orange soil 74220, which is characterized by a high content of ferromagnesian minerals and very high concentration of certain volatile elements such as Zn and Cl (LSPET, 1973). The plot in Fig. 6 shows a comparison of 74261 with the average of soils 75061 and 78501, which are assumed to contain only minor amounts of the orange soil component. The reason for selecting these soils as a reference is that their average Al content is

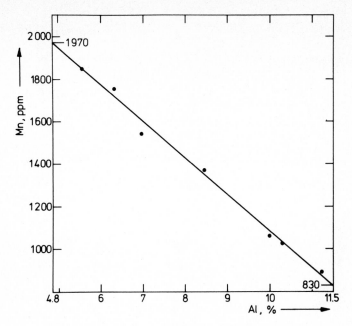

Fig. 1. Mn versus Al in Apollo 17 soils. Data from this work.

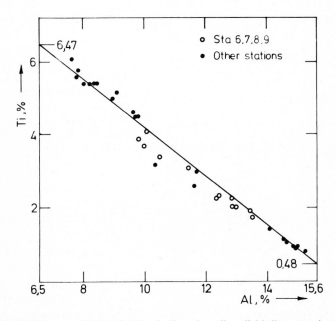

Fig. 2. Ti versus Al in Apollo 17 soils. Based on all available literature data.

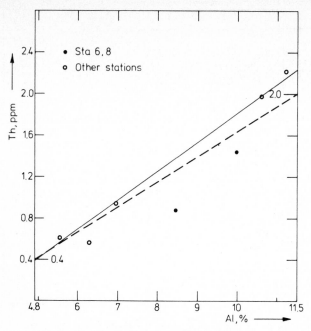

Fig. 3. Th versus Al in Apollo 17 soils. Data from this work. - - - - - - - All stations. ————
South Massif and valley stations.

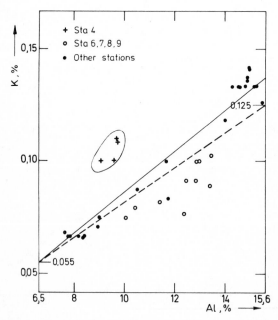

Fig. 4. K versus Al in Apollo 17 soils. Based on all available literature data. - - - - - - - All
stations except Station 4. ———— South Massif and valley stations.

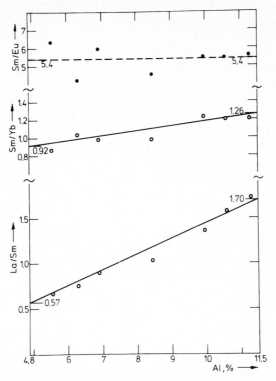

Fig. 5. Change in rare-earth pattern as a function of Al content in Apollo 17 soils, represented by the ratios La/Sm, Sm/Eu, and Sm/Yb.

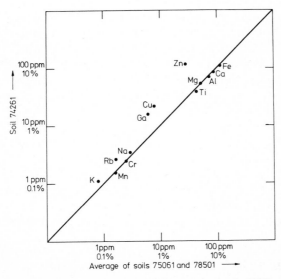

Fig. 6. Comparison plot of soil 74261 and the average of soils 75061 and 78501.

similar to that of 74261 (7.00, resp. 6.90%). The plot demonstrates a very high content in 74261 not only of Zn, but also of Cu and Ga, as compared with most other lunar materials. Enrichment of K and Rb relative to Al is also evident. When compared with the data for 74220 (Duncan *et al.*, 1974; Morgan *et al.*, 1974; LSPET, 1973) the Zn value of 120 ppm indicates about 40% orange soil in 74261.

The Zn content of the soils studied by us is in the range 18–30 ppm, while that of the mare basalts appears to be only about 2 ppm (Table 1). If the meteoritic component in the Apollo 17 soil is about 1.7% (Baedecker *et al.*, 1974), and the composition is similar to that of C_1 chondrites, the high Zn level in the soils is not explained by meteoritic admixture alone. If the excess Zn is assumed to originate from addition of orange soil (250 ppm Zn), the contribution of that component would be 3–8% in the soils investigated (except 74261). LSPET (1973), on the basis of their Zn data, indicated an average of 14% orange soil in dark mantle soils and about 5% in light mantle soils. If, instead of Zn, our data for Cu and Ga are considered, a somewhat different conclusion is reached. After subtraction of the estimated meteoritic contribution for Cu and Ga in the soils, the remaining concentrations are very similar to those of the rocks. Assuming that the high Cu and Ga content in 74261 is associated with orange soil, the concentration of those elements in the orange soil component is about 45 ppm and 40 ppm, respectively, provided that 74261 contains about 45% orange soil as indicated from the Zn data of 74240 and 74261. If that is so, the orange soil contribution to the other soils studied by us is probably 3% or less, which might indicate that some of the excess Zn in the soils is due to some unknown source. Unfortunately very few Cu and no Ga data are as yet available for other Apollo 17 soils.

Apollo 17 mare basalt

The composition of the average Apollo 17 mare basalt, as derived from the binary plots of soil data, is given in Table 1 with respect to 24 elements. With the exception of a higher Mg content and a lower level of REE and Th, it is very similar to the Apollo 11 type B basalt. The RE-distribution, characterized by a very significant depletion of the light rare earths as compared with a chondritic distribution, is also similar for these two types of basalt.

Analytical data for the five samples of basalt investigated are also listed in Table 1. For most elements studied the data show moderate individual variation, and are similar to those of the "average basalt." The titanium content, however, seems to define two distinct classes, one with about 8% Ti (70017, 70215, 71055, 74275), and another with about 5.5% Ti (75035). The lower Ti basalt also appears to contain less Mg, Cr, V, and Co, if 75035 is representative. Data by other investigators (LSPET, 1973; Rhodes *et al.*, 1974) seem to support this classification, placing rocks 75035 and 75075 in the high-Ti and 75055 in the lower Ti group.

Non-mare component

In Table 2, the composition of the "non-mare component" derived from the soil data is given with respect to 28 elements. The values for KREEP-associated elements are intended to be average values for the components derived from the North and South Massifs (cf. Fig. 3). Compared with the soils from the Apollo 16 landing site, the Apollo 17 non-mare soil has a lower content of Al and Ca and a correspondingly higher content of Fe, Mg, Cr, and Mn, reflecting the lower content of plagioclase in the highlands surrounding the Taurus-Littrow valley. It was indicated by LSPET (1973) that the highland component of the soil consists of about equal portions of noritic and anorthositic rocks. Using their data for rock samples, it is shown that a 50%-50% composition of noritic breccias/anorthositic rocks fits excellently with our "non-mare component" for all the major elements reported by us. The composition of breccia 73235, the analytical data of which are given in Table 2, also shows great similarity with the "non-mare component." It might also be mentioned that the Apollo 17 "non-mare component" is close in composition to the "very high Al_2O_3 basalts" by Hubbard *et al.* (1973), which they suggest to be a common type of rock in the lunar highlands.

Acknowledgment—Financial support by the Royal Norwegian Council for Scientific and Industrial Research (Contract B 1206 3070) is gratefully acknowledged.

REFERENCES

Baedecker P. A., Chou C. L., Grudewicz E. B., Sundberg L. L., and Wasson J. T. (1974) Extralunar materials in lunar soils and rocks (abstract). In *Lunar Science—V*, pp. 28–30. The Lunar Science Institute, Houston.

Barnes I. L., Garner E. L., Gramlich J. W., Machlan L. A., Moody J. R., Moore L. J., Murphy T. J., and Shields W. R. (1974) Isotopic abundance ratios and concentrations of selected elements in Apollo 17 samples (abstract). In *Lunar Science—V*, pp. 38–40. The Lunar Science Institute, Houston.

Brunfelt A. O. and Steinnes E. (1972) A neutron activation scheme developed for the determination of 42 elements in lunar material. *Talanta* 18, 1197–1208.

Brunfelt A. O., Heier K. S., Nilssen B., Steinnes E., and Sundvoll B. (1972) Distribution of elements between different phases of Apollo 14 rocks and soils. *Proc. Third Lunar Sci. Conf., Geochim. Cosmochim. Acta*, Suppl. 3, Vol. 2, pp. 1133–1147. MIT Press.

Brunfelt A. O., Heier K. S., Nilssen B., Steinnes E., and Sundvoll B. (1973) Geochemistry of Apollo 15 and 16 materials. *Proc. Fourth Lunar Sci. Conf., Geochim. Cosmochim. Acta*, Suppl. 4, Vol. 2, pp. 1209–1218. Pergamon.

Brunfelt A. O., Heier K. S., Nilssen B., Steinnes E., and Sundvoll B. (1974) Elemental composition of Apollo 17 fines (abstract). In *Lunar Science—V*, pp. 92–94. The Lunar Science Institute, Houston.

Chyi L. L. and Ehmann W. D. (1974) Implications of Zr and Hf abundances and their ratios in lunar materials (abstract). In *Lunar Science—V*, pp. 118–120. The Lunar Science Institute, Houston.

Duncan A. R., Erlank A. J., Willis J. P., and Sher M. K. (1974) Compositional characteristics of the Apollo 17 regolith (abstract). In *Lunar Science—V*, pp. 184–186. The Lunar Science Institute, Houston.

Ehmann W. D., Miller M. M. Ma M.-S., and Pacer R. A. (1974) Compositional studies of the lunar regolith at the Apollo 17 site (abstract). In *Lunar Science—V*, pp. 203–205. The Lunar Science Institute, Houston.

Hubbard N. J., Rhodes J. M., Gast P. W., Bansal B. M., Shih C.-Y., Wiesmann H., and Nyquist L. E. (1973) Lunar rock types: The role of plagioclase in non-mare and highland rock types. *Proc. Fourth Lunar Sci. Conf., Geochim. Cosmochim. Acta*, Suppl. 4, Vol. 2, pp. 1297–1312. Pergamon.

Jovanovic S. and Reed G. W. (1974) Labile trace elements in Apollo 17 samples (abstract). In *Lunar Science—V*, pp. 391–393. The Lunar Science Institute, Houston.

Keith J. E., Clark R. S., and Bennett L. J. (1974) Determination of natural and cosmic ray induced

radionuclides in Apollo 17 lunar samples (abstract). In *Lunar Science—V*, pp. 402–404. The Lunar
Science Institute, Houston.

LSPET (1973) Apollo 17 lunar samples: Chemical and petrographic description. *Science* **182**, 659–680.

Mason B., Jacobson S., Nelen J. A., Melson W. G., and Simkin T. (1974) Regolith compositions from the
Apollo 17 mission (abstract). In *Lunar Science—V*, pp. 493–495. The Lunar Science Institute,
Houston.

Morgan J. W., Ganapathy R., Higuchi H., Krähenbühl U., and Anders E. (1974) Lunar basins: Tentative
characterization of % projectiles from meteoric elements in Apollo 17 boulders (abstract). In *Lunar
Science—V*, pp. 526–528. The Lunar Science Institute, Houston.

Nava D. F. (1974) Chemistry of some rock types and soils from Apollo 15, 16, and 17 lunar sites
(abstract). In *Lunar Science—V*, pp. 547–549. The Lunar Science Institute, Houston.

Philpotts J. A., Schümann S., Kouns C. W., and Lum R. K. L. (1974) Lithophile trace elements in Apollo
17 soils (abstract). In *Lunar Science—V*, pp. 599–601. The Lunar Science Institute, Houston.

Rhodes J. M., Rodgers K. V., Shih C., Bansal B. M., Nyquist L. E., and Wiesmann H. (1974) The
relationship between geology and soil chemistry at the Apollo 17 landing site (abstract). In *Lunar
Science—V*, pp. 630–632. The Lunar Science Institute, Houston.

Rose H. J., Brown F. W., Carron M. K., Christian R. P., Cuttitta F., Dwornik E. J., and Ligon D. T. (1974)
Composition of some Apollo 17 samples (abstract). In *Lunar Science—V*, pp. 645–647. The Lunar
Science Institute, Houston.

Scoon J. H. (1974) Chemical analysis of lunar samples from the Apollo 16 and 17 collections (abstract). In
Lunar Science—V, pp. 690–692. The Lunar Science Institute, Houston.

Taylor S. R., Gorton M., Muir P., Nance W., Rudowski R., and Ware N. (1974) Lunar highland
composition (abstract). In *Lunar Science—V*, pp. 789–791. The Lunar Science Institute, Houston.

Wakita H., Schmitt R. A., and Rey P. (1970) Elemental abundances of major, minor and trace elements in
Apollo 11 lunar rocks, soil and core samples. *Proc. Apollo 11 Lunar Sci. Conf., Geochim. Cosmochim.
Acta*, Suppl. 1, Vol. 2, pp. 1685–1717. Pergamon.

Proceedings of the Fifth Lunar Conference
(Supplement 5, Geochimica et Cosmochimica Acta)
Vol. 2 pp. 991–1008 (1974)
Printed in the United States of America

Descartes Mountains and Cayley Plains: Composition and provenance

Michael J. Drake* and G. Jeffrey Taylor†

Smithsonian Astrophysical Observatory, Cambridge, Massachusetts 02138

Gordon G. Goles

Center for Volcanology, University of Oregon, Eugene, Oregon 97403

Abstract—Trace element compositions of petrographically characterized 2–4 mm lithic fragments from Apollo 16 soil samples are used to calculate initial REE concentrations in liquids in equilibrium with lunar anorthosites and to discuss the provenance of the Cayley Formation. Lithic fragments may be subdivided into four groups: (1) ANT rocks, (2) K- and SiO_2-rich mesostasis-bearing rocks, (3) poikiloblastic rocks, and (4) (spinel) troctolites. Model liquids in equilibrium with essentially monomineralic anorthosites have initial REE concentrations 5–8 times those of chondrites. The REE contents of K- and SiO_2-rich mesostasis-bearing rocks and poikiloblastic rocks are dominated by the mesostasis phases. ANT rocks appear to be more abundant in the Descartes Mountains, while poikiloblastic rocks appear to be more abundant in the Cayley Plains. Poikiloblastic rocks have intermediate to high LIL-element concentrations yet the low gamma-ray activity of Mare Orientale implies low LIL-element concentrations. Consequently, it is unlikely that the Cayley Formation is Orientale ejecta. A local origin as ejecta from smaller impacts is a more plausible model for the deposition of the Cayley Formation.

INTRODUCTION

THE LUNAR TERRAE are composed of a variety of rocks, the original characteristics of which have been partly or wholly masked by brecciation, mixing, metamorphism, or melting associated with intense bombardment early in lunar history. Nevertheless, studies of large rocks, lithic fragments, and glass from soil samples have led to the identification of several major suites of lunar materials comprising the bulk of the non-mare regions of the moon. In this paper we (1) report compositions of twenty-eight petrographically characterized 2–4 mm lithic fragments separated from Apollo 16 soil samples 60053,1, 60503,4, 61283,1, 68823,1, and 68843,2; (2) use these data in conjunction with experimentally determined values of distribution coefficients to calculate REE concentrations in liquids in equilibrium with lunar anorthosites; and (3) discuss the provenance of the Cayley Formation.

---

*Present address: Department of Planetary Sciences, University of Arizona, Tucson, Arizona 85721.

†Present address: Department of Earth Sciences, Washington University, St. Louis, Missouri 63136.

ANALYTICAL PROCEDURES

Analytical procedures employed in this study have been described previously (Drake *et al.*, 1973). Sixteen trace elements (Sc, Cr, Co, Zr, Hf, Ta, Ba, La, Ce, Sm, Eu, Tb, Yb, Lu, Th, and U) together with Na and Fe in twenty-eight 2–4 mm lithic fragments from Apollo 16 soil samples were determined by instrumental neutron activation analysis (INAA). Following INAA the fragments were sectioned and microprobe analyses of selected phases were performed. In electing to perform nondestructive INAA of very small solid chips one accepts less accuracy in the trace element concentrations than would be obtained using conventional INAA or the isotope dilution technique. This loss is offset by the gain of information obtained from subsequent petrographic and microprobe analysis of thin sections cut from the same chips. Thus in Figs. 1–4, only the general enrichments of the REE relative to chondrites and the overall slope of the patterns are considered important. With the exception of the Eu anomalies, no significance is attached to individual element anomalies.

In the interest of brevity, only selected analytical data are reported here. More complete data may be obtained from the authors upon request.

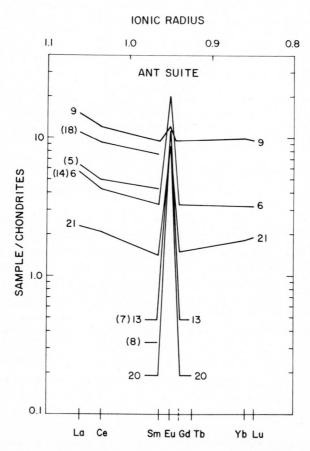

Fig. 1. Chondrite-normalized REE plots for the ANT suite. Samples in parentheses were plotted partially to avoid unneccessarily cluttering of the diagram.

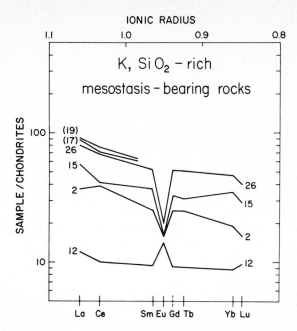

Fig. 2. Chondrite-normalized REE plots for K- and SiO$_2$-rich mesostasis-bearing rocks. Samples in parentheses were plotted partially to avoid unneccesarily cluttering the diagram.

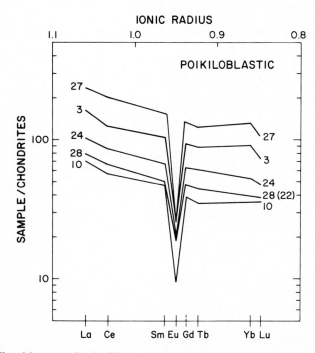

Fig. 3. Chondrite-normalized REE plots for poikiloblastic rocks. The sample in parentheses was not plotted to avoid unnecessarily cluttering of the diagram.

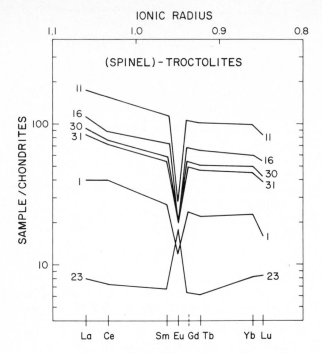

Fig. 4. Chondrite-normalized REE plots for (spinel)-troctolites.

RESULTS

Classification

The twenty-eight lithic fragments described in this paper were selected from a larger suite from which all obviously mixed (e.g., agglutinates, polymict breccias) or glassy particles had been excluded. They do not represent a statistical sampling of the larger suite; rather a deliberate effort was made to ensure that each group of lithic fragments identified by "macroscopic" binocular examination was studied. Consequently, the number of samples in each group below does not reflect their true frequency of occurrence. The lithic fragments may be separated using petrographical and geochemical criteria into four groups:

(1) The ANT suite. This group of anorthositic, noritic, and troctolitic rocks is characterized by greater than 60% plagioclase, no mesostasis, low K, P, and REE concentrations (La < 20× chondrites) and positive Eu anomalies.

(2) K- and SiO_2-rich mesostasis-bearing rocks. This group is similar to the ANT suite but is characterized by the presence of mesostasis which may be correlated with REE concentration (La ~ 10–100× chondrites). These samples are discussed separately from the ANT suite because they have higher K, P, and REE concentrations, and usually display negative Eu anomalies.

(3) Poikiloblastic rocks. This group is characterized by oikocrysts of a mafic mineral, usually low-Ca pyroxene. All samples contain K- and SiO_2-rich mesos-

tasis which may also be correlated with REE concentrations (La ~ 70–250×
chondrites). All samples display negative Eu anomalies.

(4) (Spinel)–troctolites. This group technically is a member of the ANT suite.
These rocks, defined by Prinz *et al.* (1973a), consist mainly of olivine and
plagioclase, and usually (but not always) contain minor (Mg, Al) spinel. They
always contain mesostasis characterized by low SiO_2, Al_2O_3, and K_2O, and high
FeO, MgO, and CaO concentrations. They are discussed as a separate group
because, with one exception, they have higher concentrations of K, P, and REE
(La ~ 40–200× chondrites) than is usual for ANT rocks, and they display negative
Eu anomalies.

Modal analyses of the twenty-eight samples are listed in Table 1, the
compositions of plagioclase and olivine are given in Table 2, compositions of
pyroxenes are displayed in Fig. 5, and photomicrographs of representative thin
sections are shown in Fig. 6.

Table 1. Modal analyses (%). Samples listed in order of increasing Sm concentration within each group. Abbreviations: CAT, cataclastic; REX, recrystallized; FRAG, fragment-laden; FSP, feldspathic; ANOR, anorthosite; BAS, basalt; NOR, norite; (SP), spinel; TROC, troctolite; p, present.

SAO #	NASA #	Rock type	# Points counted	Mafic	Plag	Glass or mesostasis	Oxide or sulfide	Metal	Other
20	60503	CAT ANOR	—	0	100	0	0		—
8	61283	REX ANOR	621	3.1	96.8	0.0	0.0	p	silica
13	60053	CAT ANOR	617	2.9	97.1	0.0	p	p	silica
7	61283	CAT ANOR	299	4.0	96.0	0.0	0.0	p	silica
21	60503	REX ANOR	973	19.0	80.9	0.0	0.1	p	—
6	61283	CAT ANOR	467	20.8	79.2	0.0	p	p	—
14	60053	REX ANOR	772	18.8	80.7	0.0	0.1	0.4	—
5	61283	FSP BAS	194	0.0	66.0	34.0	0.0	p	—
18	60053	REX ANOR	231	19.5	80.1	0.0	0.4	p	—
9	61283	REX NOR	389	32.1	67.9	0.0	p	p	—
12	61283	REX NOR	335	26.3	73.1	p*	0.6	p	phosphate
2	68843	FSP BAS	497	15.5	84.3	0.2*	p	0.0	phosphate
15	60053	REX NOR	586	24.9	74.2	0.2	0.5	0.2	phosphate
26	68823	FRAG BAS	643	37.2	61.4	0.7*	0.5	0.3	phosphate
17	60053	REX TROC	411	41.6	56.0	0.5*	0.7	1.2	phosphate
19	60053	REX NOR	429	35.7	62.2	1.0*	0.7	0.5	phosphate
10	61283	POIK	667	43.9	53.7	0.3*	0.6	1.5	phosphate
28	68823	POIK	848	39.4	59.2	0.2	0.8	0.4	phosphate
22	60503	POIK	758	39.7	59.2	0.5*	0.1	0.4	phosphate
24	60503	POIK	442	40.7	58.1	p*	0.9	0.2	spinel
3	68843	POIK	796	29.3	66.0	4.0*	0.6	0.1	phosphate
27	68823	POIK	828	35.0	59.3	3.6*	1.9	0.1	phosphate
23	60503	(SP) TROC	694	23.3	56.6	17.4	1.9	p	spinel
1	68843	(SP) TROC	459	42.3	56.6	0.9	p	p	spinel
31	68843	(SP) TROC	366	43.2	53.8	1.4	1.1	0.3	spinel
30	68843	(SP) TROC	411	17.8	53.3	28.0	0.2	0.5	spinel, phosphate
16	60053	(SP) TROC	574	40.9	52.3	5.2	p	1.6	—
11	61283	(SP) TROC	634	43.1	44.0	12.0	p	0.8	phosphate

*K- and SiO_2-rich interstitial mesostasis.

Table 2. Compositions of plagioclase and olivine. Samples listed in same order as Table 1.

SAO #	Plagioclase	Olivine
20	$An_{94-98}, Or_{0.0-0.09}$	none
8	$An_{96-98}, Or_{0.0-0.07}$	none
13	$An_{96-98}, Or_{0.0-0.03}$	none
7	$An_{94-97}, Or_{0.03-0.18}$	Fo_{64-66}
21	$An_{96-98}, Or_{0.0-0.17}$	Fo_{55}
6	$An_{94-97}, Or_{0.04-0.20}$	Fo_{88}
14	$An_{95-98}, Or_{0.03-0.24}$	Fo_{78}
5	$An_{94-97}, Or_{0.01-0.18}$	none
18	$An_{93-97}, Or_{0.0-0.58}$	Fo_{75-77}
9	$An_{94-97}, Or_{0.13-0.72}$	Fo_{70-76}
12	$An_{87-98}, Or_{0.0-0.77}$	none
2	$An_{90-97}, Or_{0.24-1.00}$	Fo_{74-76}
15	$An_{89-98}, Or_{0.33-1.66}$	Fo_{67-68}
26	$An_{91-97}, Or_{0.41-2.06}$	Fo_{77-80}
17	$An_{89-98}, Or_{0.14-2.29}$	Fo_{76-77}
19	$An_{88-97} Or_{0.30-1.82}$	Fo_{70-76}
10	$An_{92-97}, Or_{0.20-1.21}$	Fo_{79-80}
28	$An_{87-98}, Or_{0.26-5.74}$	Fo_{76-78}
22	$An_{90-95}, Or_{0.57-2.91}$	Fo_{75}
24	$An_{92-96}, Or_{0.24-1.52}$	Fo_{76-77}
3	$An_{91-98}, Or_{0.13-0.68}$	Fo_{68-70}
27	$An_{91-97}, Or_{0.17-1.63}$	Fo_{56-61}
23	$An_{86-94}, Or_{0.20-0.75}$	Fo_{77-80}
1	$An_{89-96}, Or_{0.23-1.34}$	Fo_{79-86}*
31	$An_{90-98}, Or_{0.17-1.11}$	Fo_{72-73}
30	$An_{90-98}, Or_{0.26-0.98}$	Fo_{75-76}
16	$An_{93-97}, Or_{0.04-0.55}$	Fo_{74-78}
11	$An_{88-97}, Or_{0.30-0.86}$	Fo_{75-83}

*Groundmass. Clasts range Fo_{87-91}.

The ANT suite

The ANT suite is composed of rocks which are dominated by plagioclase feldspar and contain from zero to approximately 40% of other phases, usually low-Ca pyroxene and olivine. A photomicrograph of a typical section is shown in Fig. 6a. Chondrite-normalized REE plots for ANT samples are shown in Fig. 1. All ANT rocks display positive Eu anomalies, and REE^{3+} concentrations increase as the modal percentage of phases other than plagioclase increases (Fig. 1 and Table 1). These samples have REE patterns similar to the proposed low-K anorthositic series (LKAS) of Hubbard et al. (1974). Comparison of Fig. 1 with Fig. 5a shows that the most plagioclase-rich ANT fragments (Nos. 20, 8, 7, 13, and 21) have the lowest REE and K concentrations and largest positive Eu anomalies,

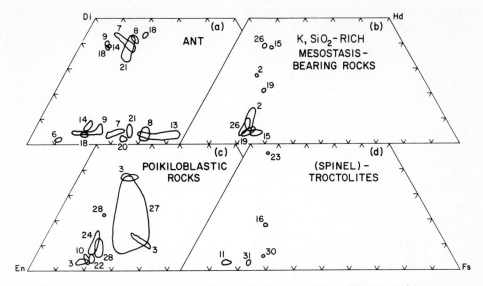

Fig. 5. Compositional fields in pyroxenes in various terra rock types. Sample numbers are given.

suggesting the crystallized from the least fractionated melt. Yet they contain the most Fe-rich mafic phases suggesting that they crystallized from the most fractionated melt. In several of these samples a silica mineral (also indicative of advanced fractionation) is present. Steele and Smith (1973) point out that this paradox has been confirmed for ANT rocks from all missions and appears to be a major, near surface lunar phenomenon, but as yet a plausible explanation for it has not been offered.

ANT sample 6, a cataclastic noritic anorthosite (Table 1) appears to be unusual in that it has very Mg-rich mafic phases and does not contain Ca-rich clinopyroxene, (Fig. 5a, Table 2). Its Eu concentration also appears to be higher than those of other ANT samples (Fig. 1) and it is possible that this sample is not directly related to the other members of the ANT suite.

Sample 5 is a quenched feldspathic melt consisting of plagioclase laths and abundant glass (Table 1). The glass is unusual in that its reflectivity is higher than that of plagioclase (for K- and SiO_2-rich glass the reverse is true). Average compositions of both glass and plagioclase are given in Table 3. Using a recent calibration (Drake, 1972) of the igneous plagioclase thermometer (Kudo and Weill, 1970) a quench temperature of 1395($\pm$50)°C was calculated. This temperature is consistent with that expected for a quenched, plagioclase-rich (normatively 91% plagioclase) melt, and indicates that this sample is probably a quenched impact melt of a preexisting plagioclase-rich rock.

K- and SiO_2-rich mesostasis-bearing rocks

These samples are frequently petrographically indistinguishable from ANT rocks except for the presence of mesostasis and phosphates (Fig. 6b). Phosphates,

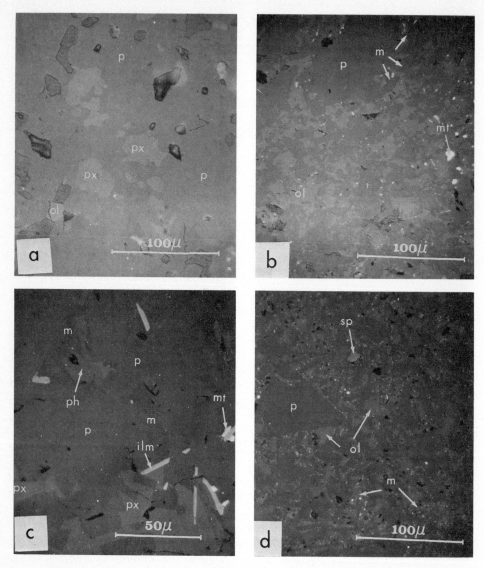

Fig. 6. Photomicrographs (40×, reflected light) of representative samples of each major sample subdivision. (a) ANT sample 21, (b), K- and SiO₂-rich, mesostasis-bearing sample 17, (c) poikiloblastic sample 27, (d) (spinel) troctolite 30. ilm, ilmenite; m, mesostasis; mt, metal; ol, olivine; p, plagioclase; ph, phosphate; px, pyroxene; sp, spinel.

especially whitlockite, usually contain high concentrations of the REE (Fig. 7) and thus exert a disproportionate influence on the REE geochemistry of a sample relative to their abundance. The low abundance and generally small dimensions of phosphates subject the percentage reported in a modal analysis to a very large uncertainty. But since phosphates are almost always associated with more abundant K- and SiO₂-rich mesostasis, it is reasonable to argue that the more

Table 3. Glass and residual phase compositions for selected samples.

SAO #	5	27	3	27	3	23	30
	Average glass	Average K-rich mesostasis	Average K-rich mesostasis	Average K feldspar	Average K feldspar	Average mesostasis	Average mesostasis
SiO_2	44.90	75.30	70.53	59.39	60.60	47.37	48.10
TiO_2	1.17	0.95	0.83	0.36	0.33	4.57	2.10
Al_2O_3	18.78	13.71	14.82	22.01	20.62	8.25	11.60
FeO	10.26	0.65	1.19	0.24	0.14	9.01	8.61
MgO	11.84	0.11	0.22	0.02	0.00	18.89	17.47
CaO	13.08	0.80	0.82	1.03	1.01	11.64	11.05
Na_2O	0.56	1.14	1.03	1.49	1.17	0.25	0.20
K_2O	0.09	6.72	7.66	11.93	12.81	0.04	0.12
P_2O_5	0.09	0.25	0.18	0.13	0.05	0.08	0.63
BaO	0.04	0.26	0.28	4.46	1.86	0.09	0.08
Total	100.81 An_{96}*	99.89	97.56	101.06	98.59	100.19	99.96

*Average composition of plagioclase in Sample 5.

accurately determined percentage of mesostasis is directly related to the percentage of phosphates. Mesostasis values of less than one percent often reflect only one or two points out of a total of many hundred (Table 1) so that considerable uncertainty is still associated with such values. Nevertheless, a plot of mesostasis (including phosphates) versus REE concentration (Fig. 8) should display a positive correlation. In Fig. 8 we have included poikiloblastic rocks (described below) because they also contain K- and SiO_2-rich mesostasis and Ca phosphates. In addition, samples from Apollos 14 and 15 together with an "average" Apollo 14 point (see figure caption for explanation) are plotted. A rough correlation between Sm concentration and percent mesostasis is observed.

Figure 5b shows pyroxene compositions in these samples. They correspond to the most Mg-rich pyroxenes of both the ANT suite and of the poikiloblastic rocks. Plagioclase feldspar compositions spread to more albitic values than are found in the ANT suite, and they are richer in Or component (Table 2).

Chondrite-normalized REE plots for these samples are shown in Fig. 2. The sample with the lowest REE concentrations (No. 12) displays a positive Eu anomaly and overlaps the range of concentrations observed for the ANT suite. It is transitional between the two groups and was excluded from the ANT suite for consistency because it contains K- and SiO_2-rich mesostasis. All other samples display negative Eu anomalies and have substantially higher REE concentrations than ANT rocks. The presence of samples which *could* be classified as ANT on the basis of petrography alone, but which have higher REE concentrations and negative Eu anomalies requires comment. It is possible that ANT rocks and K- and SiO_2-rich mesostasis-bearing rocks are members of a continuum of REE

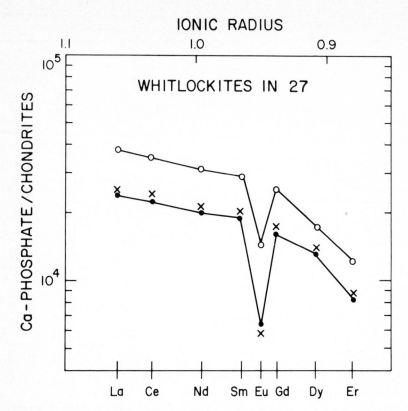

Fig. 7. Chondrite-normalized REE plots of three whitlockite crystals in sample 27, analyzed using the electron microprobe. The crosses are not linked in order to avoid confusing the diagram.

abundances. In this case, the mesostasis may be residuum left after crystallization. Alternatively, these samples may be the product of mechanical mixing of ANT and KREEP rocks, with KREEP dominating the trace element geochemistry. We have elected to discuss these K- and SiO_2-rich mesostasis-bearing samples as a separate group because we cannot distinguish between these or other hypotheses.

Poikiloblastic rocks

Poikiloblastic (or poikilitic) rocks have been described by numerous authors (Bence *et al.*, 1973; Simonds *et al.*, 1973; Taylor *et al.*, 1973b; Warner *et al.*, 1973). They are characterized by oikocrysts of a mafic mineral, usually Ca-poor pyroxene, enclosing smaller chadacrysts, usually olivine and plagioclase. The large areas interstitial to the oikocrysts have textures typical of recrystallized or partially melted breccias. These interstitial areas (Fig. 6c) are dominated by plagioclase and are severely depleted in mafic phases. Phases containing

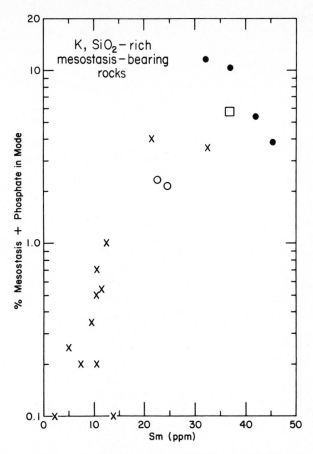

Fig. 8. Sm concentration versus residual phase content for K- and SiO$_2$-rich mesostasis-bearing rocks (including poikiloblastic samples), showing that the trivalent REE are predominantly dissolved in these phases. ×, this study; ○, Apollo 15; ●, Apollo 14; □, "average" Apollo 14 based on modal analyses by G. J. Taylor of 2–4 mm lithic fragments, and average Sm concentrations for Apollo 14 KREEP reported in the *Proceedings of the Third Lunar Science Conference.*

"excluded" elements (such as REE-rich Ca phosphates [Fig. 7], K feldspar, K- and SiO$_2$-rich mesostasis, ilmenite, and other opaque minerals) are relatively abundant. Typical analyses of some K-rich phases are given in Table 3.

Figure 5c shows the compositions of pyroxenes and Fig. 3 the chondrite-normalized REE patterns for the poikiloblastic samples. Most pyroxenes fall within the compositional range reported by other authors and are characteristically Mg-rich and Ca-poor. The two samples (Nos. 3 and 27) with the highest REE concentrations, which can justifiably be described as KREEP-rich poikiloblastic rocks, have unusual mafic phase compositions (Fig. 3, Table 2). Coexisting pigeonites, augites, and olivine are found together in the same oikocryst. The Mg-rich low-Ca pyroxene in sample 3 represents a single oikocryst. Although a single compositional envelope is drawn for sample 27, there is a concentration of compositions at both the low-Ca and high-Ca end, hence samples 3 and 27 are quite similar. Plagioclase in poikiloblastic rocks displays a range of Or contents similar to those of the K- and SiO_2-rich mesostasis-bearing rocks, but in general they are less albitic.

(*Spinel*) *troctolites*

Rocks of this type are present at all terra sites and are particularly abundant at Luna 20 and Apollo 16. A variety of textures is represented in the samples discussed here, ranging from recrystallized breccia rock (Nos. 1 and 11) through partially melted breccia rock (Nos. 16, 30, and 31) to diabasic texture rock (No. 23). All are characterized by the presence of olivine, plagioclase, usually (Mg, Al)-rich spinel, SiO_2-poor mesostasis (Table 3) and, rarely, pyroxene. A typical photomicrograph is shown in Fig. 6d. Rocks of this type have been called "(spinel)–troctolites" by many authors (Prinz *et al.*, 1973a; Taylor *et al.*, 1973a, 1973b; Drake *et al.*, 1974), "type A FIIR" (Delano *et al.*, 1973), and "olivine-rich, true spinel-bearing anorthosites" (Reid, 1972). None of the samples described here display cumulate textures, although a spinel troctolite with an unequivocally cumulate texture has been reported in the literature (Prinz *et al.*, 1973b). This sample is quite possibly a "primitive" lunar cumulate and hence has a different origin to the samples described in this report.

Compositions of olivine and plagioclase are given in Table 2, and compositions of rare pyroxenes are displayed in Fig. 5d. Chondrite-normalized REE plots are given in Fig. 4. Concentration patterns vary from those typical of ANT rocks (No. 23) to those typical of KREEP-rich samples (No. 11). This wide variety of REE concentrations suggests that these samples are unlikely to be directly related to one another. It is probable that they were produced by impact melting of a variety of rocks, hence a given sample is not necessarily simply related to a single rock type.

It is interesting to note that neither of the two most REE-rich samples (Nos. 11 and 16) contain spinel. It is not clear if this is coincidence or if this feature is of genetic significance.

DISCUSSION

Parent liquids of the ANT suite

Four of the ANT rocks (Nos. 7, 8, 13, and 20) are anorthosites composed of 96% or more plagioclase (Table 1). The monominerallic nature of these samples permits plagioclase/liquid distribution coefficients to be used to calculate REE concentrations in liquids with which they were in equilibrium (e.g. Hubbard *et al.*, 1971; Drake, 1972). For the purposes of the following calculations, it is assumed that the contribution of minor phases to the REE concentration in each fragment is negligible. As the most important minor phase is usually low-Ca pyroxene, this is a reasonable assumption. Although the REE concentrations in these anorthosites are so low that only Sm and Eu could be determined with meaningful precision, "Eu^{3+}" (which is normally obtained by interpolating a REE^{3+} versus ionic radius plot at the Eu position) may be estimated by assuming its normalized value to be identical to normalized Sm. Using plagioclase–liquid distribution coefficients for "Eu^{3+}" and "total Eu" (Weill and Drake, 1973) valid for lunar oxidation potentials ($f_{O_2} \sim 10^{-14}$ atm at 1100°C, $\sim 10^{-12.5}$ atm at 1200°C, $\sim 10^{-11.5}$ atm at 1300°C—see Nash and Hausel, 1973; Sato and Hickling, 1973; Weill and Drake, 1973) the concentrations of both Eu^{3+} and total Eu and hence the magnitude and direction of Eu anomalies in liquids that were in equilibrium with the anorthosites may be calculated.

These calculations are summarized in Table 4. Eu is enriched in these liquids by factors of 4–10 over chondritic abundances. Several of the model liquids

Table 4. Total Eu and "Eu^{3+}" (ppm) calculated for liquids in equilibrium with anorthosites. Distribution coefficients from Weill and Drake (1973).

		Eu^{tot}	"Eu^{3+}"	Anomaly
		Anorthosites		
	7	0.7	0.036	+ve
	8	0.9	0.024	+ve
	13	0.9	0.036	+ve
	20	0.8	0.014	+ve
1300°C	$D^{p/l}$	1.2	0.06	
1200°C	$D^{p/l}$	1.8	0.06	
1100°C	$D^{p/l}$	2.4	0.06	
		Liquids 1300°C		
	7	0.58	0.60	−ve
	8	0.75	0.40	+ve
	13	0.75	0.60	+ve
	20	0.67	0.23	+ve
		1200°C		
	7	0.39	0.60	−ve
	8	0.50	0.40	+ve
	13	0.50	0.60	−ve
	20	0.44	0.23	+ve
		1100°C		
	7	0.29	0.60	−ve
	8	0.38	0.40	−ve
	13	0.38	0.60	−ve
	20	0.33	0.23	+ve

display positive Eu anomalies above certain $T-f_{O_2}$ conditions which, *assuming an initial unfractionated REE distribution,* indicate that the anorthosites crystallized at lower temperatures and oxygen fugacities. However, liquids calculated for sample 20 have positive anomalies at all conditions considered here. This situation could arise if the values of distribution coefficients used in the calculations are not valid because of incorrectly chosen $T-f_{O_2}$ conditions or because of unexpected compositional effects (e.g. Morris and Haskin, 1974). Alternatively, one must consider the possibility that the assumption of an unfractionated REE distribution in the parent liquid may not be correct. For example, prior crystallization of Ca-rich clinopyroxene would generate a positive Eu anomaly in the liquid from which anorthosite No. 20 may subsequently have crystallized. It is also interesting to note that if the moon accreted inhomogeneously with an outer shell enriched in Ca, Al, and other refractory elements (Gast and McConnell, 1972), if the entire moon were made of such material (Anderson, 1972), or if the moon accreted with an excess of refractory condensates relative to cosmic abundances (Ganapathy and Anders, 1974), an initial positive Eu anomaly in the parent liquid might be a natural consequence (Grossman, 1973).

The same plagioclase–liquid distribution coefficients as above may be used in a perfect fractionation model (see Haskin *et al.,* 1970, for a comprehensive discussion) to calculate (1) the fraction of liquid which must already have crystallized as plagioclase from an originally unfractionated liquid prior to the formation of each sample and, (2) the initial concentration of Eu in that liquid. Results of these calculations are shown in Tables 5 and 6. Initial concentrations of Eu are approximately 5–8 times greater than those of chondrites, a result consistent with the conclusions of Taylor and Jakes (1974). In each case for which a calculation could be performed, at geologically reasonable temperatures a relatively small amount of the initial liquid need have crystallized as plagioclase to

Table 5. Fraction of calculated parent liquid which crystallized as plagioclase.

	1300°C	1200°C	1100°C
7	0.03	0.22	0.27
8	—	—	0.02
13	—	0.10	0.18

Table 6. Initial concentration of Eu (ppm) in calculated anorthosite parent liquids.

	1300°C	1200°C	1100°C
7	0.59	0.48	0.45
8	—	—	0.39
13	—	0.55	0.50

account for the Eu anomalies observed in the anorthosites. This conclusion is consistent with the observation that the anorthosites crystallized from an unfractionated melt and thus reemphasizes the paradox presented by the coexistence of Fe-rich mafic and accessory (e.g. silica) phases indicative of more advanced fractionation.

Provenance of the Cayley Formation

The surface expression of the light plains-forming units, of which the Cayley Formation is an example, suggests that they fill preexisting terra depressions. With the demise of the hypothesis of a volcanic origin for these materials (AFGIT, 1973) a reasonable alternative is that they were formed by ejecta from one or more impact events. Chao *et al.* (1973) date the Cayley Formation as younger than Imbrium ejecta but older than mare ferrobasalts. These authors single out the most recent cataclysmic impact, the Orientale event, as the source of most of the light plains deposits. Hodges *et al.* (1973) concur with this hypothesis. However, the composition and distribution of rock-types represented in Apollo 16 soil samples suggest that this hypothesis is not tenable.

A census of 1–2 mm lithic fragments in Apollo 16 soil samples (Taylor *et al.*, 1973b) is shown diagrammatically in Fig. 9. These data may be interpreted in

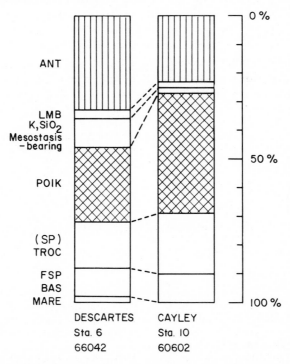

Fig. 9. Census of 1–2 mm lithic fragments in our best sample of the Descartes Mountains (66042) and Cayley Plains (60602). LMB = light matrix breccias, FSP BAS = feldspathic basalt.

terms of the stratigraphy of the Apollo 16 landing site. Although the continual gardening of the lunar surface has tended to produce a relatively homogeneous regolith, it is plausible to suggest that a bias toward a certain rock type at a given sample location reflects the constitution of nearby bedrock. Thus, although not compelling, the data in Fig. 9 suggest that ANT rocks are most abundant in our best sample of the Descartes Mountains (see also Aitken et al., 1973) while poikiloblastic rocks are most abundant in our best sample of the Cayley Plains (see also Bence et al., 1973). Figure 3 indicates that poikiloblastic rocks have intermediate to high LIL-element concentrations. The Cayley Plains are unlikely to be composed solely of poikiloblastic rocks, but, even if diluted with other materials, they should still contain higher concentrations of gamma-ray emitting elements than ANT rocks. Although the orbital gamma-ray instruments did not pass directly over Mare Orientale, the outer rings of the basin were grazed. Metzger et al. (1973) report very low gamma-ray activity and hence LIL-element concentrations (e.g. Th ~ 1 ppm) in that region which suggests the outer rings are composed largely of ANT rocks. The gamma-ray activity in the Apollo 16 landing region (including both Descartes Mountains and Cayley Plains) is higher by a factor of two, e.g., Th ~ 2 ppm. If the Cayley Plains could be independently resolved, they would presumably show still higher gamma-ray activity. In view of this observation, it appears unlikely that the Cayley Formation can be Orientale ejecta. It is more plausible that the Cayley Formation has a local provenance and was formed by one or more less important impacts. Oberbeck et al. (1973) and Grudewicz (1974) independently reached the same conclusion using different reasoning to that employed here. A corollary of a local provenance is that we might expect light plains deposits to exhibit different compositional characteristics at different locations (Horz et al., 1974).

Acknowledgments—We are indebted to Janice Bower who performed all of the microprobe analyses, and to Mary Robyn and Judy Stoeser for reducing INAA data. Discussions with John Wood are gratefully acknowledged. This work was supported by NASA grants NGL 09-015-150 to John Wood and NGL 38-003-024 to Gordon Goles.

REFERENCES

AFGIT (Apollo Field Geology Investigation Team) (1973) Apollo 16 exploration of Descartes: a geologic summary. *Science* **179**, 62–69.

Aitken F. K., Powell B. N., and Chichgar M. G. (1973) Apollo 16 lithic types: distribution and implications regarding the Descartes Formation. *EOS Trans. Amer. Geophys. Union* **54**, 1130.

Anderson D. L. (1972) The origin of the Moon. *Nature* **239**, 263–265.

Bence A. E., Papike J. J., Sueno S., and Delano J. W. (1973) Pyroxene poikiloblastic rocks from Apollo 16 (abstract). In *Lunar Science—IV*, pp. 60–62. The Lunar Science Institute, Houston.

Chao E. C. T., Soderblom L. A., Boyce J. M., Wilhelms D. E., and Hodges C. A. (1973) Lunar light plains deposits (Cayley Formation)—reinterpretation of origin (abstract). In *Lunar Science—IV*, pp. 127–128. The Lunar Science Institute, Houston.

Delano J. W., Bence A. E., Papike J. J., and Cameron D. D. (1973) Petrology of the 2–4 mm soil fraction from the Descartes region of the Moon and stratigraphic implications. *Proc. Fourth Lunar Sci. Conf.*, *Geochim. Cosmochim. Acta*, Suppl. 4, Vol. 1, pp. 537–551. Pergamon.

Drake M. J. (1972) The distribution of major and trace elements between plagioclase feldspar and magmatic silicate liquid: an experimental study. Ph.D. dissertation, University of Oregon.

Drake M. J., Stoeser J. W., and Goles G. G. (1973) A unified approach to a fragmental problem: petrological and geochemical studies of lithic fragments from Apollo 15 soils. *Earth Planet. Sci. Lett.* **20**, 425–439.

Drake M. J., Taylor G. J., and Goles G. G. (1974) Petrology and geochemistry of lunar crustal rocks (abstract). In *Lunar Science—V*, pp. 180–182. The Lunar Science Institute, Houston.

Ganapathy R. and Anders E. (1974) Bulk composition of the Moon and Earth, estimated from meteorites. In *Lunar Science—V*, pp. 254–256. The Lunar Science Institute, Houston.

Gast P. W. and McConnell R. K., Jr. (1972) Evidence for initial chemical layering of the Moon. *Revised Abstracts of the Third Lunar Science Conference*, pp. 289.

Grossman L. (1973) Refractory trace elements in Ca–Al-rich inclusions in the Allende meteorite. *Geochim. Cosmochim. Acta* **37**, 1119–1140.

Grudewicz E. B. (1974) Lunar upland plains relative age determinations and their bearing on the provenance of the Cayley Formation (abstract). In *Lunar Science—V*., pp. 301–303. The Lunar Science Institute, Houston.

Haskin L. A., Allen R. O., Helmke P. A., Paster T. O., Anderson M. R., Korotev R. A., and Zweifel K. A. (1970) Rare earths and other trace elements in Apollo 11 lunar samples. *Proc. Apollo 11 Lunar Sci. Conf., Geochim. Cosmochim. Acta*, Suppl. 1, Vol. 2, pp. 1213–1231. Pergamon.

Hodges C. A., Muehlberger W. R., and Ulrich G. E. (1973) Geologic setting of Apollo 16. *Proc. Fourth Lunar Sci. Conf., Geochim. Cosmochim. Acta*, Suppl. 4, Vol. 1, pp. 1–25. Pergamon.

Hörz F., Oberbeck V. R., and Morrison R. H. (1974) Remote sensing of the Cayley Plains and Imbrium basin deposits. In *Lunar Science—V*, pp. 357–359. The Lunar Science Institute, Houston.

Hubbard N. J., Gast P. W., Meyer C., Nyquist L. E., Shih C., and Weismann J. (1971) Chemical composition of lunar anorthosites and their parent liquids. *Earth Planet. Sci. Lett.* **13**, 71–75.

Hubbard N. J., Rhodes J. M., Nyquist L. E., Shih C., Bansal B. M., and Weismann H. (1974) Non-mare and highland rock-types: Chemical groups and their internal variations (abstract). In *Lunar Science—V*, pp. 366–368. The Lunar Science Institute, Houston.

Kudo A. M. and Weill D. F. (1970) An igneous plagioclase thermometer. *Contr. Mineral. and Petrol.* **25**, 52–65.

Metzger A. E., Trombka J. I., Peterson L. E., Reedy R. C., and Arnold J. R. (1973) Lunar surface radioactivity: preliminary results of the Apollo 15 and 16 gamma ray spectrometer experiments. *Science* **179**, 800–803.

Morris R. V. and Haskin L. A. (1974) EPR measurement of the effect of glass composition on the oxidation states of europium. *Geochim. Cosmochim. Acta.* In press.

Nash W. P. and Hausel W. D. (1973) Partial pressures of oxygen, phosphorus and fluorine in some lunar lavas. *Earth Planet. Sci. Lett.* **20**, 13–27.

Oberbeck V. R., Horz F., Morrison R. H., and Quaide W. L. (1973) Emplacement of the Cayley Formation. NASA Technical Memorandum **TM X-62**, 302.

Prinz M., Dowty E., Keil K., and Bunch T. E. (1973a) Mineralogy, petrology, and chemistry of lithic fragments from Luna 20 fines: Origin of the cumulate ANT suite and its relationship to high-alumina and mare basalts. *Geochim. Cosmochim. Acta* **37**, 979–1006.

Prinz M., Dowty E., Keil K., and Bunch T. E. (1973b) Spinel troctolite and anorthosite in Apollo 16 samples. *Science* **179**, 74–76.

Reid J. B., Jr. (1972) Olivine-rich, true spinel-bearing anorthosites from Apollo 15 and Luna 20 soils—possible fragments of the earliest formed lunar crust. *The Apollo 15 Lunar Samples*, pp. 154–156. The Lunar Science Institute, Houston.

Sato M., Hickling N. L., and McLane J. E. (1973) Oxygen fugacity values of Apollo 12, 14, and 15 lunar samples and reduced state of lunar magmas. *Proc. Fourth Lunar Sci. Conf., Geochim. Cosmochim. Acta*, Suppl. 4, Vol. 1, pp. 1061–1079. Pergamon.

Simonds C. H., Warner J. L., and Phinney W. C. (1973) Petrology of poikilitic rocks. *Proc. Fourth Lunar Sci. Conf., Geochim. Cosmochim. Acta*, Suppl. 4, Vol. 1, pp. 613–632. Pergamon.

Steele I. M. and Smith J. V. (1973) Mineralogy and petrology of some Apollo 16 rocks and fines:

General petrologic model for the Moon. *Proc. Fourth Lunar Sci. Conf., Geochim. Cosmochim. Acta,* Suppl. 4, Vol. 1, pp. 519–536. Pergamon.

Taylor G. J., Drake M. J., Wood J. A., and Marvin U. B. (1973a) The Luna 20 lithic fragments, and the composition and origin of the lunar highlands. *Geochim. Cosmochim. Acta* **37**, 1087–1106.

Taylor G. J., Drake M. J., Hallam M. E., Marvin U. B., and Wood J. A. (1973b) Apollo 16 stratigraphy: the ANT hills, the Cayley plains, and a pre-Imbrian regolith. *Proc. Fourth Lunar Sci. Conf., Geochim. Cosmochim. Acta,* Suppl. 4, Vol. 1, pp. 553–568. Pergamon.

Taylor S. R. and Jakeš P. (1974) Geochemical zoning in the Moon (abstract). In *Lunar Science—V*, pp. 786–788. The Lunar Science Institute, Houston.

Warner J. L., Simonds C. H., Phinney W. C., and Gooley R. (1973) Petrology and genesis of two "igneous" rocks from Apollo 17. *EOS Trans. Amer. Geophys. Union* **54**, 620–622.

Weill D. F. and Drake M. J. (1973) Europium anomaly in plagioclase feldspar: Experimental results and semiquantitative model. *Science* **180**, 1059–1060.

Proceedings of the Fifth Lunar Conference
(Supplement 5, Geochimica et Cosmochimica Acta)
Vol. 2 pp. 1009–1014 (1974)
Printed in the United States of America

Chemical and mineralogical composition of Surveyor 3 scoop sample 12029,9

E. J. Dwornik, C. S. Annell, R. P. Christian, Frank Cuttitta,
R. B. Finkelman, D. T. Ligon, Jr., and H. J. Rose, Jr.

U.S. Geological Survey, Reston, Virginia 22092

Abstract—Lunar sample 12029,9 consists of soil particles collected from the inner surfaces of the anodized aluminum scoop removed from the Surveyor 3 spacecraft by the Apollo 12 astronauts. The sample was examined by microscopic, semi-micro chemical, X-ray fluorescence, optical emission, and electron-probe methods. Compositional data for the bulk sample and various lithic and mineral components are compared to data from other lunar and Apollo 12 sites. Sample 12029,9, taken from within Surveyor Crater, is similar in composition, mineralogy, and meteoritic component to sample 12070 located outside the crater northwest of 12029,9, suggesting a common source and similar age for both.

Introduction

The return by Apollo 12 astronauts of portions of the Surveyor 3 spacecraft, softlanded at Oceanus Procellarum 31 months before (Fig. 1) provided a further sample for study of the lunar regolith. Surveyor 3's anodized aluminum metal scoop, used to determine soil compaction and consistency, was removed and brought back to the Lunar Receiving Laboratory as part of the Apollo 12 sample payload. This report is concerned with the study of 370 mg of soil particles found within the returned metal scoop; the elemental composition of the soil appears in Table 1.

Careful microscopic examination of the sample revealed an assortment of whole and partly fused rock, mineral and glass fragments of which the major components were fine- and medium-grained basalts and white to dark gray breccias. Lesser amounts of clear green glasses, tan to dark brown spherical and oblate glass particles, and ultra fine-grained material were present.

Some of the breccia fragments measured 6 mm long and the smallest fines were less than 100 Å, as determined from electron micrographs. Due to the relatively small amount of material allocated for mineralogical analysis and to the wide range in grain size, no attempt was made to further quantify the proportion of the various components.

Characterization of Sample 12029,9 Fragments

Grains of several of the more abundant particle types were analyzed with the electron microprobe using the procedure described by Finkelman (1973). The semiquantitative microprobe data and optical properties were used to characterize the particles.

Mineral clasts analyzed included three plagioclase feldspars with An values between 80 and

Table 1. Comparison of the composition of Surveyor 3 scoop sample 12029,9[a] with soil 12070 and other Apollo 12 soils (oxides in weight percent, elements in ppm).

Constituent	12029,9	12070,89[b]	Other Apollo 12 soils[c] Range	Average
SiO_2	46.60	45.8	45.7–48.2	46.4
Al_2O_3	11.90	12.9	12.7–15.1	13.5
FeO	16.32	16.3	12.9–16.7	15.5
MgO	10.25	10.2	8.45–10.5	9.73
CaO	10.35	10.5	10.4–10.6	10.5
Na_2O	0.50	0.50	0.47–0.87	0.59
K_2O	0.23	0.25	0.23–0.54	0.32
TiO_2	2.81	2.83	2.33–2.81	2.66
P_2O_5	0.29	0.33	0.32–0.55	0.40
MnO	0.19	0.22	0.18–0.24	0.21
Cr_2O_3	0.34	0.44	0.32–0.44	0.40
TOTAL	99.58	100.27		100.21
ΔRC	+2.35	+1.6	+0.2–1.6	+1.1
Zn	4	8	<4–8.2	6.7
Cu	13	12	8.0–14	10.6
Ga	5.2	4.9	4.4–5.2	4.9
Rb	6.6	6.2	5.2–16	8.2
Li	18	17	15–24	18
Co	44	52	36–106	58
Ni	190	222	137–235	195
Ba	420	423	420–990	563
Sr	140	123	110–178	131
V	130	121	80–121	107
Be	3.1	3.3	3.2–8.0	5.2
Nb	30	29	26–74	38
Sc	44	42	38–44	40
La	42	40	39–88	51
Y	140	145	128–245	164
Yb	13	14	12–23	16
Zr	500	498	410–790	548

[a]The following elements were looked for but not detected in the analyzed sample. If present, they would be in concentrations below those (in ppm) indicated in parenthesis: Ag(0.2), As(4), Au(0.2), B(10), Bi(1), Cd(8), Ce(100), Cs(1), Ge(1), Hf(20), Hg(8), In(1), Mo(2), Nd(100), Pb(1), Pt(3), Re(30), Sb(100), Sn(10), Ta(100), Te(300), Th(100), Tl(1), U(500), W(200), and Zn(4).

[b]Sample 12070,89 (Cuttitta *et al.*, 1970, Tables 2 and 5, pp. 1218 and 1220).

[c]Apollo 12 soils (Cuttitta *et al.*, 1970, Tables 3 and 6, pp. 1219 and 1220).

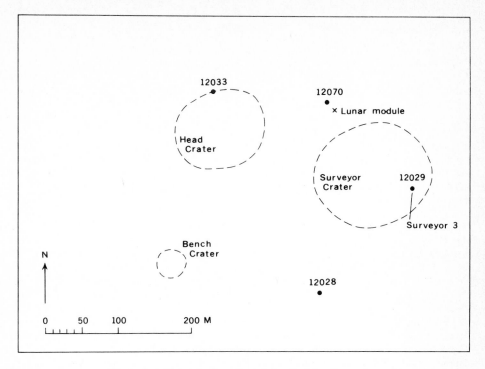

Fig. 1. Location map for selected Apollo 12 soil samples.

95 mole%; five olivines with Fo values between 60 and 77 mole%; three clino- and one orthopyroxene; two typical ilmenites; and a pyroxferroite grain.

Most of the glasses, including fragments, beads, and splatter glass, are high in iron (16.5–23.5 wt.% FeO) and titanium (2.5–4.0 wt.% TiO_2) contents. These values are typical of the analyzed rock and regolith samples from the Apollo 12 area (Cuttitta *et al.*, 1971). A number of clear pale green glasses were analyzed. The major element results are compared to data of Bunch *et al.* (1972) in Table 2. These glasses appear to form a coherent group with a limited range in composition; they probably are derivatives of anorthositic norites or troctolites, the ANT group as defined by Prinz *et al.* (1973). The clear color and uniform texture and composition of these glasses suggest that they were derived by impacts on a large uniform body of ANT rocks exposed at the lunar surface.

Several fragments of a dark, frothy, banded glass contained numerous metallic Fe–Ni blebs and spheres (diameters as much as 20 μm). Most of these inclusions contain about 6 weight percent nickel, except one, which contains more than 20 wt.% Cu. We suspect that this grain (diameter <5 μm) consists of a metallic copper phase adjacent to a metallic Fe–Ni phase. El Goresy *et al.* (1971) observed minute veins ($\sim$2 μm) of metallic copper within ulvospinel and ilmenite from two Apollo 12 rocks. They note that troilite and Fe–Ni always occur in direct contact with copper. Reid *et al.* (1970) also observed a copper-rich phase in association with troilite and iron in an Apollo 12 rock. However, the copper phase they reported had 22% nickel and 17% zinc, whereas the grain reported here has about 5 wt.% Ni (probably associated with the iron) and less than 1 wt.% Zn. The only other reports of native copper in lunar samples are by Simpson and Bowie (1970) who found copper in association with iron in small segregations in troilite at a troilite–ilmenite contact in an Apollo 11 sample, Ramdohr and El Goresy (1970) who observed native copper in a coarse grained Apollo 11 crystalline rock, and by the Lunar Science Preliminary Examination Team (LSPET, 1971) who reported extremely rare grains of copper in the Apollo 14 soil samples.

Data from Cuttitta *et al.* (1973) indicate that there is no significant difference in copper content

Table 2. Electron probe analyses of the major element composition of green glasses in Surveyor 3 scoop sample 12029,9 (in weight percent).

	Average[a]	Average[b]
SiO_2	46.2	42.6
Al_2O_3	23.5	25.6
FeO^c	5.9	5.9
MgO	9.1	9.9
CaO	15.2	14.3
Na_2O	<0.5	0.2
K_2O	<0.5	<0.1
TiO_2	<0.5	0.3

[a]Average of five semiquantitative analyses. Accuracy is within ±10% of the amount present.
[b]Average of first eight analyses from Table 5 of Bunch *et al.* (1972).
[c]Total iron expressed as FeO.

between Apollo 11, 12, 14, and 15 rocks and regolith. The data of Rose *et al.* (1972) suggest that Apollo 12 breccias may be depleted in copper relative to Apollo 11 and 14 samples. Furthermore, the reducing capacity suggests that Apollo 12 samples may be in a somewhat less reduced state than the samples returned by other Apollo missions. In view of these data, there does not appear to be any significant reason for the predominance of copper phases in Apollo 12 samples, other than the fact that data on such Cu grains is exceedingly meager.

The association of the copper with metallic iron in impact generated glasses leads us to suspect an extralunar origin for the copper.

A single metallic burr, identified as aluminum, was also found in the 12029,9 soil sample. It was probably derived from the anodized aluminum scoop by soil abrasion.

COMPARISON OF SURVEYOR 3 SCOOP SAMPLE 12029,9 WITH SOIL 12070 AND DISCUSSION

Sample 12029,9 is similar to 12070 for all major and trace components (Table 1). Consideration of various trace element ratios have led Schonfeld and Meyer (1972) to suggest that soil 12070 is a mixture of Apollo 12 mare basalts with 30% KREEP component and minor amounts of meteoritic, anorthositic, ultramafic and granitic constituents. Our compositional and mineralogic studies suggest a similar assemblage for Surveyor 3 scoop sample 12029,9.

Sample 12070 appears to typify most of the lunar soil found on the surface of the Apollo 12 site (LRL, Apollo 12 Catalog, 1970; LSPET, 1970): its color, texture, and amount of meteoritic component is similar to most portions of the drive-tube core material 12028 (Fig. 1; Ganapathy *et al.*, 1970). In examining our data on 12029 for evidence of a meteoritic component, it was evident that the concentration of one meteoritic element is very similar to that of sample 12070 (i.e. Ni ratio of 12070 to 12029 is ~1). We then assume that other meteoritic elements in 12029 have similar concentration to those found in 12070 (Keays *et al.*, 1970). This would suggest a similar age for both soils. It is to be noted also that amounts of meteoritic component in most portions (to a depth of 37.2 cm) of the drive-tube core material 12028 are not only very similar (Ganapathy *et al.*, 1970) but are also quite like those of surface soils 12029 and 12070. Lunar soil 12070 is part of the total contingency sample collected at the rim of a small crater 15 m northwest of the Apollo 12 lunar module (Fig. 1). Scoop sample 12029,9 is the lunar material covering the surface of the southern interior wall of Surveyor Crater which faces the LM 183 m to the southeast. Although soils 12070 and 12029,9 were collected about 165 m apart on the rims of two different craters, similarity in their composition and mineralogic assemblages either reflects a common parent material or suggests material transport.

The marked mineralogic and compositional similarities between samples 12029,9, 12070, and core-tube material 12028 suggest the following conclusions:

(1) The green glass components of 12029,9 appear to be derivatives of anorthositic norites or troctolites, perhaps the ANT suite suggested by Prinz *et al.* (1973).

(2) The relatively high nickel content and the presence of numerous metallic Fe–Ni blebs and spheres (diameters <20 μm) and of a copper-rich phase indicate a significant meteoritic component.

(3) There is a genetic and historical similarity for scoop sample 12029,9 collected on the southern interior wall of Surveyor Crater and soil sample 12070 collected on the rim of a small crater some 165 m to the northwest. In view of the 12029,9 studies, we concur with Ganapathy *et al.* (1970) that soils like 12070 can be considered as typifying the lunar regolith material at the Apollo 12 site.

REFERENCES

Bunch T. E., Prinz M., and Keil K. (1972) Electron microprobe analyses of lithic fragments and glasses from Apollo 12 lunar samples. Univ. of New Mexico Inst. of Meteorites, Spec. Publ. No. 4, 14 pp.
Cuttitta F., Rose H. J., Jr., Annell C. S., Carron M. K., Christian R. P., Dwornik E. J., Greenland L. P., Helz A. W., and Ligon D. T., Jr. (1971) Elemental composition of some Apollo 12 Lunar rocks and soils. *Proc. Second Lunar Sci. Conf., Geochim. Cosmochim. Acta,* Suppl. 2, Vol. 2, pp. 1217–1229. MIT Press.
Cuttitta F., Rose H. J., Jr., Annell C. S., Carron M. K., Christian R. P., Ligon D. T., Jr., Dwornik E. J., Wright T. L., and Greenland L. P. (1973) Chemistry of twenty-one igneous rocks and soils returned by the Apollo 15 Mission. *Proc. Fourth Lunar Sci. Conf. Geochim. Cosmochim. Acta,* Suppl. 4, Vol. 2, pp. 1081–1096. Pergamon.
El Goresy A., Ramdohr P., and Taylor L. A. (1971) The opaque minerals in the lunar rocks from Oceanus Procellarum. *Proc. Second Lunar Sci. Conf., Geochim. Cosmochim. Acta,* Suppl. 2, Vol. 1, pp. 219–235. MIT Press.
Finkelman R. B. (1973) Analysis of the ultrafine fraction of the Apollo 14 regolith. *Proc. Fourth Lunar Sci. Conf., Geochim. Cosmochim. Acta,* Suppl. 4, Vol. 1, pp. 179–189. MIT Press.
Ganapathy R., Keays R. R., and Anders E. (1970) Apollo 12 lunar samples: Trace element analysis of a core and the uniformity of the regolith. *Science* 170, 533–535.
Gault D. E., Collins R. J., Gold T., Green J., Kuiper G. P., Masursky H., O'Keefe J., Phinney R., and Shoemaker E. M. (1968) Lunar theory and processes. *J. Geophys. Res.* 73, 4115–4131.

Keays R. R., Ganapathy R., Laul J. C., Anders E., Herzog G. F., and Jeffrey P. M. (1970) Trace elements and radioactivity in lunar rocks: Implications for meteorite infall, solar-wind flux, and formation conditions of moon. *Science* **167**, 490–493.

LSPET (Lunar Sample Preliminary Examination Team) (1970) Preliminary examination of lunar samples from Apollo 12. *Science* **167**, 1325–1329.

LSPET (Lunar Sample Preliminary Examination Team) (1971) Preliminary examination of lunar samples from Apollo 14. *Science* **173**, 681–693.

Lunar Receiving Laboratory, Lunar Sample Information Catalog, Apollo 12, MSC-01512 (1970), 317 pp.

Prinz M., Dowty E., Keil K., and Bunch T. E. (1973) Mineralogy, petrology and chemistry of lithic fragments from Luna 20 fines: Origin of the cumulate ANT suite and its relationship to high-alumina and mare basalts. *Geochim. Cosmochim. Acta* **37**, 979–1006.

Ramdohr P. R. and El Goresy A. (1970) Opaque minerals of the lunar rocks and dust from Mare Tranquillitatis. *Science* **167**, 615–618.

Reid A. M., Meyer C., Jr., Harmon R. S., and Brett R. (1970) Metal grains in Apollo 12 igneous rocks. *Earth Planet. Sci. Lett.* **9**, 1–5.

Rose H. J., Jr., Cuttitta F., Annell C. S., Carron M. K., Christian R. P., Dwornik E. J., Greenland L. P., and Ligon D. T., Jr. (1972) Compositional data for twenty-one Fra Mauro lunar materials. *Proc. Third Lunar Sci. Conf., Geochim. Cosmochim. Acta*, Suppl. 3, Vol. 2, pp. 1215–1229. MIT Press.

Schonfeld E. and Meyer C., Jr. (1972) The abundances of components of the lunar soils by a least-squares mixing model and the formation age of KREEP. *Proc. Third Lunar Sci. Conf., Geochim. Cosmochim. Acta*, Suppl. 3, Vol. 2, pp. 1397–1420. MIT Press.

Simpson P. R. and Bowie S. H. U. (1970) Quantitative optical and electron-probe studies of the opaque phases. *Science* **167**, 619–621.

Proceedings of the Fifth Lunar Conference
(Supplement 5, Geochimica et Cosmochimica Acta)
Vol. 2 pp. 1015–1024 (1974)
Printed in the United States of America

Abundances of the group IVB elements, Ti, Zr, and Hf and implications of their ratios in lunar materials

W. D. EHMANN and L. L. CHYI

Department of Chemistry, University of Kentucky, Lexington, Kentucky 40506

Abstract—New data on Zr and Hf abundances in samples from recent Apollo missions and a few from earlier missions are presented. The data confirm our previous suggestion of a modest Zr–Hf fractionation among lunar rock types. The four major lunar rock types can be readily distinguished, based on Zr and Hf abundances and Zr/Hf mass ratios. Mare basalts exhibit mean Zr abundances ranging from 111 ppm for Apollo 15 to 534 ppm for Apollo 11 and low Zr/Hf mass ratios ranging from 34 to 39. KREEP basalts are characterized by very high Zr contents (mean = 1380 ppm Zr) and high Zr/Hf ratio of 47. Anorthosites exhibit extremely low Zr and Hf contents and possibly very low Zr/Hf ratios. Arguments that the fractionation may be partly due to the presence of Zr^{3+} in lunar magmas while Hf remains as Hf^{4+} are presented. These data for Zr and Hf together with data for Ti recently obtained on the same samples in our laboratory are discussed in support of a modified two-stage lunar crustal evolution model.

INTRODUCTION

DUE TO the strong geochemical coherence of Zr and Hf and the improbability of their forming volatile complexes in the lunar environment, we previously suggested that the modest Zr–Hf fractionation that we have observed among lunar rock types may be due in part to a Zr–Hf charge disparity under highly reducing conditions (Chyi and Ehmann, 1973). Under these conditions at least part of Zr could exist as Zr^{3+}, while Hf remains as Hf^{4+}. Trivalent Zr has a larger ionic radius (0.85 Å, calculated) than Zr^{4+} and Hf^{4+} (0.79 and 0.78 Å, respectively), which may prevent it from substituting for Ti^{4+} (0.69 Å) and being stabilized in the early crystallized Ti minerals. Hence, low Zr/Hf ratios would be expected in the early Ti minerals and higher Zr/Hf ratios would be expected in the late crystallized Zr oxides and silicates.

It should be noted that certain REE pairs such as Nd^{3+}–Sm^{3+}, Yb^{3+}–Lu^{3+}, and Tb^{3+}–Dy^{3+} also have similar ionic radii. Fractionation of these pairs has been explained by different solidus/liquidus partition coefficients and different degrees of partial melting. However, the abundance levels of REE are much closer to each other than is the case for the Zr/Hf pair (atomic abundance difference of a factor of ~100). The small amount of Hf that would have to camouflage Zr would not result in any appreciable lattice strain. Moreover, variable oxidation states for certain REE pairs are also known. The fractionation of Zr and Hf is less likely to be explained by the above processes involved to explain the REE fractionation in lunar materials, unless a Zr–Hf charge disparity is also involved.

New data for Zr and Hf in samples from the most recent Apollo missions and in a few samples from earlier missions have been obtained which tend to support

1015

our earlier suggestions. These new data together with the data presented by Chyi and Ehmann (1973) indicate that mare basalts, KREEP basalts, VHA basalts and anorthosites may be readily distinguished on the basis of their Zr and Hf abundances and Zr/Hf mass ratios. Ti/Zr abundance ratios for these rock types are also quite characteristic and exhibit a rather wide range, as compared to those found in terrestrial igneous rocks and chondrites. Our analyses in conjunction with other published information are used to delineate a modified two-stage lunar crustal evolution model in which the major rock types are genetically related.

Experimental

The radiochemical separation procedures for the simultaneous determination of Zr and Hf were developed by Kumar (1972) in our laboratory. A brief discussion of this precise radiochemical neutron activation analysis procedure and the standards used in this work is given in our Fourth Lunar Science Conference paper (Chyi and Ehmann, 1973). The error limits for the Zr and Hf determinations are $\pm 1.8\%$ and $\pm 2.9\%$, respectively, based on the computed standard deviations for replicate analyses of U.S.G.S. standard rock, W-1. Agreement of our data from analyses of various U.S.G.S. standard rocks with compilation values of Flanagan (1973) is excellent.

All Zr and Hf data and some of the Ti data discussed in this paper were obtained from analyses of single sample, in order that computed abundance ratios would be meaningful. In some cases, literature values for Ti on different subsamples are listed and the ratios derived in these cases are less certain. Ti abundances in Apollo 16 and 17 samples are reported by Miller *et al.* (1974) and Zr–Hf analyses on these samples are currently in progress.

Results and Discussion

New data for Zr and Hf in lunar materials are presented in Table 1. These data, in addition to our previously published values (Chyi and Ehmann, 1973), are the basis of the discussion which follows.

Fractionation among lunar rock types

Four major lunar rock types which include mare basalts, KREEP basalts, very high aluminum (VHA) basalts, and anorthosites (Bansal *et al.*, 1973) can be readily distinguished, based on their Zr and Hf abundances and Zr/Hf ratios (Fig. 1). Mare basalts have moderate but variable Zr contents, grouped according to mission, from an average of 111 ppm Zr for Apollo 15 up to 534 ppm Zr for Apollo 11. The mare basalts have low Zr/Hf mass ratios, ranging from 34 for Apollo 11 samples up to 39 for Apollo 15 samples.

KREEP basalts are clearly characterized by high Zr contents (Zr mean = 1380 ppm) and high Zr/Hf mass ratios (mean = 48). The VHA basalts have a moderate mean Zr content of 296 ppm, but their mean Zr/Hf ratio of 47 is essentially the same as that observed for the KREEP basalts. Hence, our Zr/Hf data do not provide any additional evidence that KREEP basalts and VHA basalts may be representative of independent magmas (Bansal *et al.*, 1973; Gast, 1973).

Anorthosites have extremely low Zr and Hf contents. At this writing we are reluctant to report an average Zr/Hf ratio for this rock type, due to the rather high

Table 1. Additional Zr and Hf abundances in lunar materials.

Sample	Description	Zr (ppm)	Hf (ppm)	Zr/Hf (weight)
10003,36	Mare basalt	336	10.1	33.3 [34.3]
		367	10.4	35.3
10022,32	Mare basalt	596	17.6	33.8 [33.8]
		617	18.3	33.7
10024,20	Mare basalt	598	18.0	33.3 [33.5]
		689	20.5	33.6
12051,46	Mare basalt	141	4.17	33.8 [34.0]
		147	4.30	34.2
12063,60	Mare basalt	137	4.23	32.4 [32.2]
		141	4.41	32.0
15016,31	Mare basalt	93.9	2.48	37.9 [38.1]
		98.5	2.58	38.2
60015,65B	Anorthosite	0.5	0.020	25.0 [25.0]
		<1	0.015	—
60025,72	Anorthosite	<1	0.015	—
		<1	0.012	—
60335,33	VHA basalt?	359	7.62	47.1 [47.0]
		343	7.32	46.9
61016,133B	VHA basalt	235	5.06	46.4 [46.5]
		244	5.24	46.6
61501,11	<1 mm fines	194	4.27	45.4 [45.0]
		193	4.34	44.5
64421,20A	<1 mm fines	183	4.14	44.2 [44.0]
		198	4.52	43.8
64501,10B	<1 mm fines	173	3.84	45.1 [46.1]
		187	3.93	47.6
65015,54	Metaclastic rock with high KREEP	920	19.8	46.5 [46.8]
		1010	21.4	47.1
71500,5	Unsieved fines	212	6.95	30.5 [31.0]
		229	7.28	31.5
73235,54	Blue gray breccia	n.d.*	0.21*	—
		365†	8.03†	45.5†
74220,90	Orange soil	203	6.07	33.4 [32.6]
		193	6.08	31.7

*Separated anorthositic clast.
†Basaltic clast.
[ ] = mean value.
Note: Other data discussed in the text have been published previously (Chyi and Ehmann, 1973). Replicate analyses are on homogenized splits, unless otherwise noted.

uncertainties in the Zr determination. A new Ge(Li) detector has just been received in our laboratory and we anticipate reporting a reliable Zr/Hf ratio for them at a later time. One of our anorthosite analyses yields a Zr content of 0.5 ± 0.3 ppm and a Hf content of 0.020 ± 0.002 ppm, which corresponds to a ratio of 25.

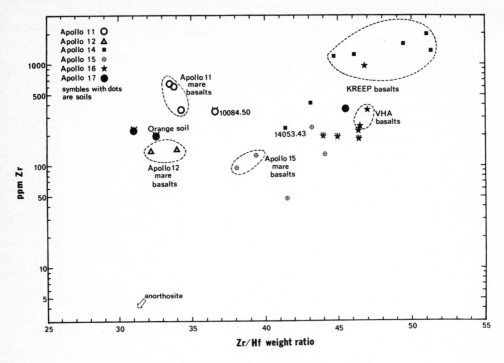

Fig. 1. Zr and Hf variation diagram for lunar materials.

It is worth noting that Apollo 14 rock 14053 with Al_2O_3, FeO and MgO contents intermediate between those of typical mare basalts and those of VHA basalts (Hubbard *et al.*, 1973) has a Zr/Hf ratio which is also intermediate between these two rock types.

Orange soil 74220 and unsieved soil 71500 are observed to have extremely low Zr/Hf ratios of approximately 32. These ratios are lower than the ratios in any rocks we have analyzed to date, with the exception of mare basalt 12063. It does not appear to be possible to produce the Zr/Hf ratios in these soils by simple mixing of rock types we have analyzed from other mission sites. It is also rather unlikely that impact melting or other secondary local effects could cause substantial Zr loss with respect to Hf. It would appear that these soils may contain a contribution of a material (Apollo 17 high-Ti basalts?) with a Zr/Hf ratio less than those found in mare basalts from other sites. Analyses of Apollo 17 mare-type basalts are currently in progress and will be reported elsewhere.

Differentiation of a homogeneous moon in light of group IVB elemental abundances

The Zr/Hf ratios in lunar rocks should be primary features of lunar crystalline rocks which are independent of cooling rate and enrichment or depletion of volatiles. Similarly, for lunar soils, the ratios should be independent of surface process such as the meteorite impact contribution and solar wind reduction. Zr and Hf fractionation as a result of impact thermal metamorphism has not as yet been demonstrated, and is likely to be negligible. Zr contents in the minerals ilmenite and ulvöspinel are a function of temperature and may change as a result of subsolidus redistribution under conditions of slow cooling (Taylor and McCallister, 1972). However, the Zr/Hf ratios we have determined by *bulk* sample analyses should reflect the pristine ratios at the time the rock was first crystallized. It was suggested by Butler and Thompson (1965) that diffusion of gaseous Zr and Hf compounds in an aqueous fluoride system could bring about fractionation. However, the lunar magma is essentially dry as indicated by the scarcity of hydrous minerals. The presence of whitlockite reflects very low fugacities of fluorine (Nash and Hausel, 1973). Hence, the formation of Zr and Hf gaseous complexes is hard to visualize. Since lunar soils have Zr contents ranging from 180 to 344 ppm and Zr contents in chondrites are generally less than 10 ppm (Ehmann and Chyi, 1974), a few percent of meteorite contribution would not measurably affect the Zr/Hf ratio in the soils. Schnetzler *et al.* (1972) have argued that rock 15555 may represent one of the least differentiated lunar materials, based on its low REE contents and small Eu anomaly. The fact that the Zr/Hf ratio for this rock is essentially at the midpoint in the Zr/Hf ratio distribution we observe which would support the primitive characterization given to this rock.

As new results from direct oxygen fugacity measurements in lunar rocks have become available (Sato *et al.*, 1973), it seems evident that lunar magmas were in highly reduced states throughout their cooling history. According to the calculations of Nash and Hausel (1973), the range of oxygen fugacities in lunar magma is rather narrow. They further conclude that the narrow range is due to geochemical evolution which has restricted the diversity of genesis and emplacement of magmas. These findings favor our belief that Zr^{3+} was present in the primitive lunar magmas, while Hf was present as Hf^{4+} (Chyi and Ehmann, 1973). The larger ionic radius Zr^{3+} would be incompatible with respect to substitution for Ti^{4+} in the early Fe–Ti minerals and would be enriched in the KREEP-rich residual melts, resulting in higher Zr/Hf ratios in rocks with higher percentages of KREEP. As was mentioned in the Introduction of this paper, the very large difference in the Zr and Hf atomic abundances, makes the explanation of their fractionation more difficult than in the case of REE pairs with similar small differences in ionic radii.

If the moon was accreted homogeneously and underwent early global differentiation down to a depth of 300–400 km in order to form the lunar highlands (Ringwood, 1974a) with accretional energy as the source of heat (Toksöz *et al.*, 1972; Wood, 1972), fractionation of this melt would yield a variety of gravitational differentiates. The lighter plagioclase cumulate would rise through the melt to form a highland anorthositic crust with very low Zr and Hf contents and low Zr/Hf ratios. The heavier pyroxene, olivine, and Fe–Ti mineral cumulates would sink and form a pyroxene-rich layer complementary to the highland anorthosites. The pyroxene-rich layer would have an intermediate Zr and Hf content, due to the limited solid solution of these two elements in Fe–Ti minerals. The layer would also exhibit low Zr/Hf ratios, due to the presence of Zr^{3+}. Partial melting of this pyroxene-rich layer is unlikely because of its depletion of radioactive elements which would provide the heat source. However, partial melting of underlying primitive mixtures of high-temperature condensates, as suggested by Ringwood (1974b), could provide the bulk of the material to form the mare basalts.

According to Ringwood (1974a), as partial melting commenced due to radiogenic heating, perovskite, which would contain most of the incompatible elements would react with pyroxene and enter the first liquid with other low-melting phases. On further melting, this liquid would be diluted by liquids from equilibrium melting. In this model, the first liquid would contain essentially all of the Zr and Hf and with further partial melting, Zr and Hf contents in the melts would increase because of dilution. The Zr/Hf ratio would remain relatively unchanged by this dilution and should have essentially the chondritic ratio of ~44, by weight. The partial melt liquids rising through the overlying pyroxene-rich layer may be contaminated, resulting in a lowered Zr/Hf ratio. The ratio could again be lowered by crystal fractionation on cooling, due to the presence of Zr^{3+}. The preceding mechanism may explain the over fivefold Zr and Hf content variation and the relatively low Zr/Hf ratio of ~33 we find in the mare basalts.

The gravitational differentiation of primitive lunar melts would eventually produce a residual liquid with a composition similar to the KREEP basalts with high Zr and Hf contents and high Zr/Hf ratios. As a consequence of prevailing impact events in the early lunar history, the KREEP-like residual melt could be transported into the plagioclase cumulate zone, where this zone was thin and unstable. As a result, rocks could be formed in these regions with quite variable Zr and Hf contents and Zr/Hf ratios, depending on the degree of mixing. VHA-type rocks may be formed in this manner. After the plagioclase cumulate becomes thick and rigid, impact events could be expected to produce cracks in the rigid crust and trigger extrusion of a nearly pure KREEP residual melt to form small localized igneous bodies of KREEP basalt. The X-ray fluorescence data of Adler et al. (1972) suggest that highland Al/Si concentration ratios vary geographically from 0.5 to 0.8. This would support the concept of a variable composition of the early crust, although it must be noted that the X-ray fluorescence experiment yields analyses of only a thin layer on the lunar surface.

Table 2 illustrates the much higher variability of Ti/Zr ratios in lunar materials. The ratios vary over a factor of twenty, in contrast to less than ten in terrestrial igneous rocks (Flanagan, 1973) and less than two in chondrites (Mason, 1971; Ehmann and Rebagay, 1971). This indicates that Ti and Zr are strongly fractionated during lunar crustal evolution. Low and relatively constant Ti/Zr ratios in KREEP and VHA basalts imply they both behave as incompatible lithophile elements. Ti/Zr ratios in mare basalts vary between 116 and 216. Since the lowest ratio in mare basalts is larger than the lowest ratio in lunar ilmenite, we believe most of Zr in the mare basalts is being incorporated in the crystal structure of ilmenite. The large variations of the Ti/Zr ratios in mare basalts support our subsequent contamination suggestion.

VHA basalts, originally suggested as a specific rock type, but actually polymict breccias, are apparently genetically related to KREEP basalts and anorthosite.

Table 2. Ti/Zr weight ratios in lunar materials.

Rock Type	Ti %*	Zr ppm	Ti/Zr	Ave.
Mare basalt				
10003,36	7.1 (1)	352	202	
10022,32	7.1 (1)	607	117	
10024,20	7.5 (1)	644	116	170
12051	2.9 (2)	144	201	
12063	3.0 (2)	139	216	
KREEP basalt				
14047,29	1.15 (3)	1340	8.6	
14310,113	0.64 (3)	1230	5.2	
14321,171A	0.83 (3)	1960	4.2	
14321,171B	0.83 (3)	1190	7.0	6.8
14321,225A	1.24 (3)	1570	7.9	
65015	0.76 (4)	965	7.9	
VHA basalt				
60335	0.39 (5)	351	11.1	
61016	0.46 (4)	240	19.2	15.2
Anorthosite				
60015,65B	—	<1		
60025	0.12 (6)	<1	>1200	
Ilmenite	31.6 (7)	3000–200 (7)	105–1580	

(1) Goles et al. (1970).
(2) Wakita and Schmitt (1971).
(3) Ehmann et al. (1972).
(4) Duncan et al. (1973).
(5) Walker et al. (1973).
(6) Rose et al. (1973).
(7) Calculation based on the stoichiometric Ti content and the Zr values of Arrhenius et al. (1971).
*Ti data from references (1) and (3) are based on analyses of the same exact splits used in our Zr and Hf analyses. Other data for Ti were obtained from the literature and are derived from different subsamples.

The REE, Ba, and U contents in VHA basalts are 1/3 and 1/4 those of KREEP. VHA basalts also have lower FeO/MgO ratios and higher Al contents than KREEP basalts. However, Bansal et al. (1973) pointed out that the relative abundances of REE, Ba, and U in the VHA basalts are essentially identical to those found in KREEP basalts. Our data indicate that Zr and Hf abundances in VHA basalts are approximately 1/5 those found in KREEP basalts, but the two materials do have essentially identical Zr/Hf ratios. Hence, our data are consistent with a model in which VHA basalts are derived by mixing (induced by an impact) of a high KREEP content residual melt with a cumulate rich in plagioclase. When we plot Zr content versus Zr/Hf mass ratio as shown in Fig. 1, we find that the majority of the lunar rocks and the soils derived from them fall along a main trend extending diagonally through the diagram with anorthosite at one end and KREEP basalt at the other. However, the Apollo 11 and 12 mare basalts and the Apollo 17 soils define a secondary trend with a Zr/Hf mass ratio centered around 33. This is compatible with the modified two-stage lunar crustal evolution model presented here. Rocks produced from the early stage melting fall on the main trend, while rocks from partial melting of the primitive condensates and subsequently contaminated by a pyroxene-rich layer fall on the secondary trend.

Condensation behavior of Zr and Hf from a gas of solar composition

The refractory elements Zr and Hf would be enriched in high-temperature condensates (Kurat, 1970; Ehmann and Rebagay, 1970). According to the calculations of Grossman (1973), ZrO_2 would condense at 1840°K and HfO_2 at 1719°K under a total pressure of 10^{-3} atm. The condensation temperatures change slightly at lower values for the total pressure, but the condensation sequence would remain the same. Grossman suggests that solid solution effects dominate the condensation behavior of many trace elements. Zr could condense at 1840°K as ZrO_2 which might later be physically trapped in perovskite as inclusions, or could be directly incorporated as a trace element into the crystal structure of a perovskite condensate at 1647°K due to crystal chemistry similarities.

Due to the lanthanide contraction, Zr and Hf have nearly identical atomic and ionic radii. The solar system *atomic* abundance of Hf is only approximately 1% that of Zr. A pure HfO_2 condensate from the solar nebula is therefore, quite unlikely and the condensation of Hf should occur with Zr. Ehmann and Chyi (1974) have noted that the Zr/Hf mass ratios for carbonaceous chondrites and for all stone meteorites analyzed in our laboratory are essentially identical (Zr/Hf = 44). This value is close to the mean value of 41 for all terrestrial igneous rocks, as calculated by Horn and Adams (1966). Ehmann and Chyi also note that there is insufficient evidence to identify any significant fractionation of Zr and Hf due to metamorphism among the various classes of chondrites. We would expect, therefore, the primordial Zr/Hf ratios in the various primary bodies in the solar system would reflect primarily their atomic abundances in the vicinity of the condensation and would be relatively independent of temperature and pressure conditions.

Gast (1972) and Ringwood (1972) originally suggested a heterogeneous accretion mechanism for the moon. In this model the outer portion of the moon would contain higher proportions of refractory elements, while the inner portion would contain higher proportions of Fe–Mg silicates. The anorthosites, VHA basalts, and KREEP basalts would then represent the outer portion, while the mare basalts would be representative of the inner portion of the moon. Both the outer and inner portions are assumed to condense directly from a massive, hot, primordial terrestrial atmosphere. Since it is difficult to rationalize the fractionation of Zr and Hf during a condensation process for the reasons presented previously, it would be expected that the Zr/Hf ratios in the various lunar rock types would be essentially identical in the heterogeneous accretion model. Since we find the Zr/Hf ratios for the highland rocks (~48) differ from those for the mare basalts (~36), we feel the homogeneous accretion models discussed by Philpotts *et al.* (1972), Taylor (1973) and others are preferred to the heterogeneous accretion models proposed. Magmatic differentiation of a homogeneously accreted moon to yield highland basalts and mare basalts with differing Zr/Hf ratios appears to be the most feasible approach to explaining the Zr and Hf data obtained to date.

Acknowledgments—This work has been supported in part by NASA grant NGR-18-001-058 and the University of Kentucky Research Foundation. We are grateful to the crew of Georgia Tech Research Reactor for the neutron irradiations. We are indebted to R. A. Schmitt and J. C. Laul for their comments and suggestions.

REFERENCES

Adler I., Trombka J., Gerard J., Lowman P., Schmadebeck R., Blodget H., Eller E., Yin L., Lamothe R., Osswald G., Gorenstein P., Bjorkholm P., Gursky H., and Harris B. (1972) Apollo 16 geochemical X-ray fluorescence experiment; Preliminary report. *Science* 177, 256–259.

Arrhenius G., Everson J. E., Fitzgerald R. W., and Fujita H. (1971) Zirconium fractionation in Apollo 11 and Apollo 12 rocks. *Proc. Second Lunar Sci. Conf., Geochim. Cosmochim. Acta*, Suppl. 2, Vol. 1, pp. 169–176. MIT Press.

Bansal B. M., Gast P. W., Hubbard N. J., Nyquist L. E., Rhodes J. M., Shih C.-Y., and Wiesmann H. (1973) Lunar rock types (abstract). In *Lunar Science—IV*, pp. 48–50. The Lunar Science Institute, Houston.

Butler J. R. and Thomson A. J. (1965) Zirconium: Hafnium ratios in some igneous rocks. *Geochim. Cosmochim. Acta* 29, 167–175.

Chyi L. L. and Ehmann W. D. (1973) Zirconium and hafnium abundances in some lunar materials and implications of their ratios. *Proc. Fourth Lunar Sci. Conf., Geochim. Cosmochim. Acta*, Suppl. 4, Vol. 2, pp. 1219–1226. Pergamon.

Duncan A. R., Erlank A. J., Willis J. P., and Ahrens L. H. (1973) Composition and inter-relationship of some Apollo 16 samples. *Proc. Fourth Lunar Sci. Conf., Geochim. Cosmochim. Acta*, Suppl. 4, Vol. 2, pp. 1097–1113. Pergamon.

Ehmann W. D. and Chyi L. L. (1974) Zirconium and hafnium in meteorites. *Earth Planet. Sci. Lett.* 21, 230–234.

Ehmann W. D. and Rebagay T. V. (1970) Zirconium and hafnium in meteorites by activation analysis. *Geochim. Cosmochim. Acta* 34, 649–658.

Ehmann W. D. and Rebagay T. V. (1971) Zirconium and hafnium. In *Handbook of Elemental Abundances in Meteorites*, pp. 307–318. Gordon and Breach, New York.

Ehmann W. D., Gillum D. E., and Morgan J. W. (1972) Oxygen and bulk element composition studies of Apollo 14 and other lunar rocks and soils. *Proc. Third Lunar Sci. Conf., Geochim. Cosmochim. Acta*, Suppl. 3, Vol. 2, pp. 1149–1160. MIT Press.

Flanagan F. J. (1973) 1972 values for international geochemical reference standards. *Geochim. Cosmochim. Acta* 37, 1189–1200.

Gast P. W. (1972) The chemical composition and structure of the moon. *The Moon* 5, 121–148.

Gast P. W. (1973) Lunar magmatism in time and space (abstract). In *Lunar Science—IV*, pp. 276–277. The Lunar Science Institute, Houston.

Goles G. G., Randle K., Osama M., Schmitt R. A., Wakita H., Ehmann W. D., and Morgan J. W. (1970) Elemental abundances by instrumental activation analyses in chips from 27 lunar rocks. *Proc. Apollo 11 Lunar Sci. Conf., Geochim. Cosmochim. Acta*, Suppl. 1, Vol. 2, pp. 1165–1176. Pergamon.

Grossman L. (1973) Refractory trace elements in Ca–Al-rich inclusions in the Allende meteorite. *Geochim. Cosmochim. Acta* **37**, 1119–1140.

Horn M. K. and Adams J. A. S. (1966) Computer-derived geochemical balances and elemental abundances. *Geochim. Cosmochim. Acta* **30**, 279–297.

Hubbard N. J., Rhodes J. M., and Gast P. W. (1973) Chemistry of lunar basalts with very high alumina contents. *Science* **181**, 339–342.

Kumar P. A. (1972) The rapid determination of zirconium and hafnium in standard rocks by neutron activation analysis. Master's thesis, University of Kentucky.

Kurat G. (1970) Zur Genese der Ca–Al-reichen Einschlusse in Chondriten von Lancé. *Earth Planet. Sci. Lett.* **9**, 225–231.

Mason B. (1971) Titantum. In *Handbook of Elemental Abundances in Meteorites*, pp. 181–184. Gordon and Breach, New York.

Miller M. D., Pacer R. A., Ma M-S., Hawke B. R., Lookhart G. L., and Ehmann W. D. (1974). Compositional studies of the lunar regolith at the Apollo 17 site. *Proc. Fifth Lunar Sci. Conf., Geochim. Cosmochim. Acta.* This volume.

Nash W. P. and Hausel W. D. (1973) Partial pressures of oxygen, phosphorus and fluorine in some lunar lavas. *Earth Planet. Sci. Lett.* **20**, 13–27.

Philpotts J. A., Schnetzler C. C., Nava D. F., Bottino M. L., Fullagar P. D., Thomas H. H., Schuhmann S., and Kouns C. W. (1972) Apollo 14: Some geochemical aspects. *Proc. Third Lunar Sci. Conf., Geochim. Cosmochim. Acta*, Suppl. 3, Vol. 2, pp. 1293–1305. MIT Press.

Ringwood A. E. (1972) Zonal structure and origin of the moon (abstract). In *Lunar Science—III*, pp. 651–653. The Lunar Science Institute, Houston.

Ringwood A. E. (1974a) Maria basalts and composition of lunar interior (abstract). In *Lunar Science—V*, pp. 636–638. The Lunar Science Institute, Houston.

Ringwood A. E. (1974b) Minor element chemistry of maria basalts (abstract). In *Lunar Science—V*, pp. 633–635. The Lunar Science Institute, Houston.

Rose H. J., Jr., Cuttita F., Berman S., Carron M. K., Christian R. P., Dwornik E. J., Greenland L. P., and Ligon D. T., Jr. (1973) Compositional data for twenty-two Apollo 16 samples. *Proc. Fourth Lunar Sci. Conf., Geochim. Cosmochim. Acta*, Suppl. 4, Vol. 2, pp. 1149–1158. Pergamon.

Sato M., Hickling N. L., and McLane J. E. (1973) Oxygen fugacity and values of Apollo 12, 14, and 15 lunar samples and related state of lunar magmas. *Proc. Fourth Lunar Sci. Conf., Geochim. Cosmochim. Acta*, Suppl. 4, Vol. 1, pp. 1061–1079. Pergamon.

Schnetzler C. C., Philpotts J. A., Nava D. F., Schuhmann S., and Thomas H. H. (1972) Geochemistry of Apollo 15 basalt 15555 and soil 15531. *Science* **175**, 426–428.

Taylor L. A. and McCallister R. H. (1972) An experimental investigation of the significance of zirconium partitioning in lunar ilmenite and ulvöspinel. *Earth Planet. Sci. Lett.* **17**, 105–109.

Taylor S. R. (1973) Chemical evidence for lunar melting and differentiation. *Nature Phys. Sci.* **245**, 203–205.

Toksöz M. N., Solomon S. C., Minear J. W., and Johnston D. H. (1972) Thermal evolution of the moon. *The Moon* **4**, 19–213.

Wakita H. and Schmitt R. A. (1971) Bulk elemental composition of Apollo 12 samples. *Proc. Second Lunar Sci. Conf., Geochim. Cosmochim. Acta*, Suppl. 2, Vol. 2, pp. 1231–1236. MIT Press.

Walker D., Longhi J., and Hays J. F. (1973) Petrology of Apollo 16 metaigneous rocks (abstract). In *Lunar Science—IV*, pp. 752–754. The Lunar Science Institute, Houston.

Wood J. A. (1972) Thermal history and early magmatism in the moon. *Icarus* **16**, 229–246.

Proceedings of the Fifth Lunar Conference
(Supplement 5, Geochimica et Cosmochimica Acta)
Vol. 2 pp. 1025–1033 (1974)
Printed in the United States of America

Primordial radioelement concentrations in rocks and soils from Taurus-Littrow*

James S. Eldridge, G. Davis O'Kelley, and K. J. Northcutt

Oak Ridge National Laboratory, Oak Ridge, Tennessee 37830

Abstract—Primordial radioelement (K, Th, and U) concentrations were determined nondestructively by gamma-ray spectrometry in thirteen soil and thirteen rock samples from the returned sample collection from Taurus-Littrow. Soil samples investigated were: 71131, 71501, 73121, 73131, 73141, 73221, 73241, 73261, 73281, 76501, 78501, 79221, and 79261. Rock samples were: 70135, 70185, 70215,4, 71135, 71136, 71175, 71566, 73215, 73255, 73275, 76295, 78597, and 79155.

Concentrations of K, Th, and U in the soil samples were found to have a narrow range of values for seven light-mantle soils that were distinctly different from those values for the other six soils in this study. The Th content of the light soils is the same as that found from the orbiting gamma-ray spectrometer determination of Trombka *et al.* (1973). Little or no variation of primordial radioelement content was observed as a function of depth, color, or texture in trench samples.

Variations of Th/U and K/U were observed in three types of basaltic rocks. These observations lend geochemical evidence to the speculation that two or three separate subfloor basalt units were sampled at the Taurus-Littrow site. Correlation of K/U mass ratios with potassium content of the Apollo 17 materials indicate a possible two component mixing of subfloor basalt and KREEP as end members to produce soil and breccias at Taurus-Littrow.

INTRODUCTION

THE RETURNED SAMPLE COLLECTION from Taurus-Littrow contained materials from three of five expected geologic units. Soil and rock samples were obtained that characterize the subfloor unit and its associated regolith, the light mantling material, and the highlands material from the North and South Massifs. Nondestructive gamma-ray spectrometry was used on a suite of rock and soil samples from each of the geologic units at Taurus-Littrow to determine the content of primordial radioelements (K, Th, and U) along with radionuclides produced by galactic and solar proton activation. Results and interrelationships of the activation product content of the same group of samples reported here are presented in a companion report in this volume (O'Kelley *et al.*, 1974).

A measure of the total radioactivity of the moon can yield important information concerning its thermal history. Schonfeld (1974a) estimated that the moon's upper mantle (or that part of the moon where most lunar samples were derived) has an average U concentration of 0.085 ppm. Wänke *et al.* (1973, 1974) calculated an average U concentration of 0.086 and 0.077 ppm for the part of the moon that underwent magmatic differentiation. Both Schonfeld (1974a) and Wänke *et al.* (1973) indicated that their calculated U average concentrations were consistent

*Research carried out under Union Carbide's contract with the U.S. Atomic Energy Commission through interagency agreements with the National Aeronautics and Space Administration.

with experimentally determined heat flow values from the Apollo 15 and 17 missions. In addition, the Th values reported here served as an important correlation with orbiting gamma-ray spectrometer measurements (Trombka *et al.*, 1973).

Part of the results from this study were obtained during the preliminary examination period of the Apollo 17 mission and are included in the LSPET (1973a) report. The present report includes additional results and further characterization of Th–U and K–U correlations that had been observed in samples from other Apollo landing sites.

Experimental Methods

The gamma-ray spectrometer system used in these studies has been previously described (O'Kelley *et al.*, 1970) and (Eldridge *et al.*, 1973). Resolution of the complex gamma-ray spectra from both the coincidence matrix and from the summed output of the single detectors was accomplished with ALPHA-M, a program for quantitative radionuclide determination (Schonfeld, 1967). Accurate replicas of all rock samples in this study were made from rock models provided by the Curator's Office, Johnson space Center. Standard gamma-ray spectrum libraries were acquired by measurement of accurate replicas (containing standards for each nuclide sought) under the same experimental conditions as the lunar samples. Error values listed for all determinations in this series are overall estimates including counting statistics and system calibrations. Sample weights ranged from 24 to 3600 g for the rock samples with the median weight near 400 g. Soil samples were nominally 50 or 100 g aliquots of <1 mm fines. Density determinations for rock samples were made by determining the volume of thin-shelled aluminum replicas by pressing aluminum foil around the rock models to make hollow shells. From the weight of the lunar sample and the volume of its replica shell, density values were calculated.

Results and Discussion

Rock and soil samples from a number of sampling stations were studied in order to characterize the primordial radioelement content of each of the geologic units at Taurus-Littrow. Figure 1 shows a diagram of the Apollo 17 landing site with the traverses, sampling stations, and major physiographic features indicated. Sampling stations are indicated by the bold-face Arabic numerals. Samples measured in this study are listed near the corresponding sampling station. Primordial radioelement concentrations for the 26 samples in this study are shown in Table 1.

The Apollo Field Geology Investigation Team speculated that sampled subfloor basalts were derived from 20 to 130 m depths, and that the stratigraphically lowest basalt unit was vesicular and coarse-grained. Other stratigraphic units graded upward to coarse-grained porphyritic, vesicular fine-grained, and finally aphanitic basalts in the shallowest recognizable type (AFGIT, 1973). Table 2 contains average primordial radioelement concentrations for 16 basalt samples, from this study and LSPET (1973a), grouped according to fine-, medium-, and coarse-grained classifications of LSPET (1973a). The average value for the K/U of fine-grained basalts is significantly different from that ratio in the medium and coarse basalts. This geochemical difference observed here in the discrepancy of

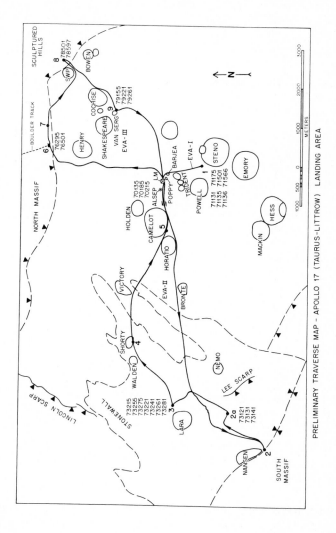

PRELIMINARY TRAVERSE MAP - APOLLO 17 (TAURUS-LITTROW) LANDING AREA

Fig. 1. Apollo 17 landing area in the Taurus-Littrow region. Bold-faced Arabic numerals indicate sampling stations and five digit numerals indicate sample numbers with soils, having a 1 as the final digit. The coordinates for the lunar module (LM) are: latitude 20°10′N and longitude 30°46′E.

J. S. Eldridge *et al.*

Table 1. Primordial radioelement concentrations in Apollo 17 samples.

Sample number	Density (g/cm³)	Type*	K (ppm)	Th (ppm)	U (ppm)	Th/U	K/U
			Rocks				
70135	3.0	CB	500 ± 30	0.31 ± .02	0.12 ± .01	2.6	4170
70185	3.0	CB	420 ± 35	0.30 ± .03	0.10 ± .02	3.0	4200
70215,4	3.3	FB	320 ± 64	0.36 ± .03	0.13 ± .03	2.8	2500
71135	1.9	FB	350 ± 40	0.60 ± .05	0.14 ± .03	4.3	2214
71136	2.4	FB	370 ± 100	0.46 ± .06	0.22 ± .05	2.1	1680
71175	2.4	MB	560 ± 28	0.39 ± .02	0.11 ± .01	3.5	5090
71566		CB	450 ± 20	0.31 ± .01	0.092 ± .008	3.4	4890
73215	2.5	BR	1665 ± 85	4.05 ± .20	1.10 ± .05	3.7	1514
73255	2.4	BR	1590 ± 80	3.47 ± .17	1.00 ± .05	3.5	1590
73275	2.2	BR	2240 ± 112	4.53 ± .23	1.20 ± .06	3.8	1867
76295	2.4	BR	2270 ± 114	5.30 ± .27	1.50 ± .08	3.5	1510
78597	2.6	FB	380 ± 20	0.38 ± .02	0.11 ± .01	3.4	3454
79155	2.6	CB	440 ± 30	0.32 ± .02	0.11 ± .01	2.9	4000
			Soils (<1 mm fines)				
71131		SS	625 ± 30	0.67 ± .05	0.23 ± .02	2.9	2720
71501		RS	585 ± 30	0.75 ± .04	0.23 ± .02	3.3	2540
73121		SS	1160 ± 60	2.63 ± .13	0.72 ± .04	3.7	1610
73131		SS	1160 ± 60	2.24 ± .11	0.63 ± .03	3.6	1840
73141		TB	1130 ± 60	2.25 ± .11	0.63 ± .03	3.6	1790
73221		TT	1180 ± 60	2.13 ± .11	0.63 ± .03	3.4	1870
73241		TM	1220 ± 60	2.25 ± .11	0.64 ± .03	3.5	1910
73261		TB	1090 ± 60	2.40 ± .12	0.67 ± .04	3.6	1630
73281		TB	1180 ± 150	2.33 ± .11	0.58 ± .04	4.0	2030
76501		RS	900 ± 50	1.39 ± .14	0.38 ± .04	3.6	2370
78501		RS	770 ± 40	1.11 ± .11	0.28 ± .03	4.0	2750
79221		TT	700 ± 40	1.12 ± .06	0.36 ± .03	3.1	1940
79261		TB	700 ± 40	1.08 ± .05	0.31 ± .02	3.5	2260

*CB = coarse basalt, FB = fine basalt, MB = medium basalt, BR = breccia, TT = trench top, TM = trench middle, TB = trench bottom, RS = rake soil, SS = surface soil.

K/U grading from coarse-grained to the fine-grained basalts would tend to verify the speculation of separate flow units in the mare basalts from the subfloor at Taurus-Littrow.

Basalt samples have Th/U ratios distinctly lower than the usual lunar and terrestrial value of ~3.8. Medium and coarse basalt have K/U ratios of ~4300; differing greatly from the previously observed lunar average of ~2500 (Schonfeld, 1974a). These geochemical differences indicate not only that the basalt flows filling the Taurus-Littrow Valley are different in themselves, but also different from previously sampled basalts from other Apollo sites. Duncan *et al.* (1974a) have noticed differences in K/Zr ratios of Apollo 17 samples compared to a near constant ratio they observed in Apollo 12, 14, and 15 samples. They correlated

Table 2. Average primordial radioelement content of Apollo 17 rocks.

Rock type	No.	K (ppm)	Th (ppm)	U (ppm)	Th/U	K/U
Fine basalt	6	382	0.40	0.14	$3.00 \pm .24$	2930 ± 260
Medium basalt	4	444	0.38	0.11	$3.48 \pm .24$	4015 ± 240
Coarse basalt	6	495	0.34	0.11	$3.12 \pm .16$	4590 ± 252
Breccia	4	1940	4.34	1.20	$3.62 \pm .12$	1620 ± 57

K/Zr ratios with TiO_2 contents and invoked a two-stage origin for high-titanium mare basalts.

Primordial radioelement concentrations in the four breccia samples of this study are approximately an order of magnitude greater than any of the three basalt types. In addition, the Th/U and K/U values are near normal and overlap those from Apollo 15 and 16 breccia samples (see Fig. 2 and Clark and Keith, 1973). Since the primordial radioelement content of the breccias is approximately twice that of even the light soils it appears that the breccias could not have been formed from simple impact induration of the upper regolith.

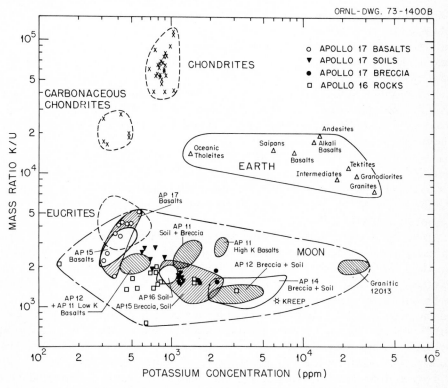

Fig. 2. Values of the mass ratio K/U as a function of the concentration of K. Apollo 17 basalts, soils, and breccias appear to fall on a line that intersects KREEP.

The U average value of 0.11 ppm shown in Table 2 for basalts is not greatly different from the predicted upper mantle average of 0.085 ppm as given by Schonfeld (1974a), and from the whole moon (or the part that underwent magmatic differentiation) average of 0.077 ppm calculated by Wänke *et al.* (1974).

Figure 2 shows a comparison of the mass ratio K/U and potassium concentration of Apollo 17 samples with earth, meteorite, and other lunar samples. It can be seen that the "moon" values fall in a region distinctly different from "earth" and most meteorite ratios. Apollo 17 basalts occupy a region which overlaps that of Apollo 15 basalts.

Grouping of the Apollo 17 basalts, soils, and breccias in the K–U systematic diagram of Fig. 2 indicates a possible two component mixing model for the soils and breccias; with coarse basalts and KREEP as the end members. In spite of the limited number of samples, this observation would lead to the speculation that breccias and soils at the Apollo 17 site may be indurated products of KREEP and basaltic materials with the KREEP component less abundant than that of the Fra Mauro breccias. This speculation is a very limited interpretation and is not as complete as the more extensive Apollo 17 regolith studies by several other groups (Duncan *et al.*, 1974b; Rhodes *et al.*, 1974; Schonfeld, 1974b).

The radioelement content of soil samples from the light-mantle unit differs greatly from that of soils collected from other points in the valley floor. Seven light-mantle soils in this study yielded average concentrations (ppm) of 1160, 2.32, and 0.64 for K, Th, and U, respectively. These concentrations are 2–3 times as great as those of the other six soils in the valley floor regolith samples from four different sampling stations.

The thorium concentration of 2.32 ppm for the light soil average compares favorably with the value of 2.2 ppm determined for a $5° \times 5°$ segment at Littrow by Trombka *et al.* (1973) in their orbital experiments. The orbital measurement covers a large area at a mare–highland interface; but Trombka *et al.* (1973) stipulate that the light soils are more typical of the general region covered by the orbital spectrometer. Since the orbital measurement averages the Th content of the entire $5° \times 5°$ segment, it is obvious that the dark valley soils cannot be a substantial fraction of the $5° \times 5°$ area because of the low Th content of the six dark- and intermediate-albedo soils.

Three light-mantle soils collected at Station 2a may be examined for near surface vertical variations. Samples 73121 and 73131 are surface soils at Station 2a, while 73141 was collected from the bottom of a 15 cm trench. There is no difference in potassium concentrations in the three soils and only slight elevation of Th and U in 73121 compared with 73131 and 73141. Field geology studies indicate that sample 73140 (the parent soil from which <1 mm fines 73141 were sifted) might be the most representative sample of light-mantle fines returned by the Apollo 17 mission (ALGIT, 1973). For this reason, analytical results for 73141 are especially useful as being representative of South Massif regolith.

Four samples from a trench in the light mantle at Station 3 were studied to determine any primordial radioelement variations with depth and with color variations within the 10 cm trench. It can be seen that there are no significant differences in K, Th, or U concentrations in 73221, 73241, 73261, or 73281 even though these four samples cover several depth variations from the upper 0.5 cm of soil to 10 cm below the surface. The trench samples also include color variations from white to gray along with variable agglutinate content (LSPET, 1973a).

Another pair of trench samples shown in Table 1 are 79221 and 79261 from the top and bottom of a 17 cm trench exposed in the southeast flank of Van Serg Crater ejecta blanket. The surface sample 79221 contains twice as much agglutinate (ubiquitous soil component consisting of mineral and lithic detritus bonded by grapelike clusters of brown glass) as the trench bottom, 79261, along with very different breccias (AFGIT, 1973). Again, this pair of samples shows no primordial radioelement differences in spite of major color, texture, and depth variations. This Station 9 trench pair is greatly different in chemical composition from the Stations 2a and 3 soils; and their K, Th, and U concentrations are similar to the dark soils at Taurus-Littrow.

A surface soil and a rake soil collected about 15 m apart at Station 1 (71131 and 71501) contain K, Th, and U concentrations that are similar to those of the basalts collected at the same location (71135, 71136, and 71175). In addition, these Station 1 soils are significantly lower in primordial radioelements than the other dark- or intermediate-albedo soils from Stations 6, 8, and 9 listed in Table 1. The Th/U ratio for the Station 1 soils ($2.91 \pm .33$ and $3.26 \pm .33$ for 71131 and 71501, respectively) is much lower than the ~ 3.8 value observed for previous Apollo soils. The low Th/U ratio together with the low overall primordial radioelement concentrations is a reflection of the high basaltic content of these soils. These dark soils from the valley floor contain the lowest primordial radioelement concentrations of all the other Apollo sampling sites (Eldridge *et al.*, 1974).

Thorium to uranium ratios discussed in this paper are not as precise as those measured by some other analytical techniques, such as isotope dilution mass spectrometry. The accuracy is limited by uncertainties in both Th and U due to statistics of counting. However, the nondestructive nature of gamma-ray spectrometry measurements allows one to use samples large enough to average out sampling inhomogeneities. Nunes *et al.* (1974) reported U and Th concentrations of 0.313 and 1.134 ppm, respectively for soil sample 79221; whereas the corresponding values from Table 1 are 0.36 ± 0.03 and 1.12 ± 0.06 ppm. These comparisons are for the same size fraction (<1 mm fines) of soil 79221. Silver (1974) reported U and Th concentrations of 0.406 and 1.337 ppm, respectively for an unsieved soil, 78500,6. Uranium and thorium concentrations reported in Table 1 for 78501, the <1 mm size fraction from soil 78500, are 0.28 ± 0.03 and 1.11 ± 0.11 ppm, respectively. It is obvious that the differences are outside the reported errors.

Duncan *et al.* (1974b) studied both major and trace element concentrations in Apollo 17 regolith and noted considerable compositional variations related to selenographic position. In addition, those authors separated a dark-mantle soil, 75081, into four size fractions and noted a significant increase in a KREEP-rich component and an "orange soil" component in the finest fraction. The <300 mesh fraction of the <1 mm fines contained 40% more K, 54% more P and 20% less Ti than the coarsest fraction of the <1 mm fines. In addition, Zn, Cu and Ni were present in a threefold increase over the coarse fractions.

Due to the compositional differences in a sieved sample in this work compared to the unsieved sample of Silver (1974) and to the significant correlation of a KREEP-rich component with the finest fraction of a dark-mantle soil by Duncan *et al.* (1974b), it can be postulated that analytical results for Apollo 17 regolith samples should be compared with caution. Large sample populations should be studied in order to average sampling defects. Silver (1974) reported average uranium and thorium levels in surface soils of the valley floor as 0.344 and 1.09 ppm, respectively. He reported that the average Th/U ratio of 3.14 was distinguishably lower than all Apollo mission sites except Apollo 16.

CONCLUSIONS

Primordial radioelement determinations in this study have shown geochemical evidence in support of field geology speculation concerning layering in the subfloor basalt flows along with a possible correlation of magmatic fractionation of K/U as a function of depth. Elevated values of primordial radioelement concentrations in breccia samples compared to the soils at Taurus-Littrow indicate that the breccias could not be formed from simple impact induration of the local regolith. On the other hand, similarities of the K, Th, and U content of the soils and basaltic rocks (as well as the similar Th/U ratio of ~3.1) indicate that the dark regolith in the Taurus-Littrow Valley may be the comminuted product of the underlying subfloor unit mixed with other components.

Correlation of a KREEP-rich component and an "orange soil" component with the finest fraction of <1 mm soil by Duncan *et al.* (1974b), coupled with an unresolved analytical uncertainty between an unsieved soil value of Silver (1974) and a <1 mm fraction of the same soil in this work, leads to the conclusion that Apollo 17 regolith samples are subject to sampling inhomogeneities. For this reason, care should be exercised in comparing analytical results of Apollo 17 regolith samples obtained with different size fractions.

Acknowledgments—The authors gratefully acknowledge the assistance provided by the Lunar Sample Analysis Planning Team in granting suitable sample allocations for this study. In addition, we wish to thank the Curator's staff at Johnson Space Center, especially J. O. Annexstad, R. S. Clark, and M. A. Reynolds, for the expeditious processing of the numerous samples in this work.

REFERENCES

AFGIT (Apollo Field Geology Investigation Team) (1973) Geologic exploration of Taurus-Littrow: Apollo 17 landing site. *Science* **182**, 672–680.

ALGIT (Apollo Lunar Geology Investigation Team) (1973) Documentation and environment of the Apollo 17 samples: a preliminary report. *U.S. Geological Survey Interagency Report-Astrogeology* **71**.

Clark R. S. and Keith J. E. (1973) Determination of natural and cosmic ray induced radionuclides in Apollo 16 lunar samples. *Proc. Fourth Lunar Sci. Conf., Geochim. Cosmochim. Acta,* Suppl. 4, Vol. 2, pp. 2105–2113. Pergamon.

Duncan A. R., Erlank A. J., Willis J. P., Sher M. K., and Ahrens L. H. (1974a) Trace element evidence for a two stage origin of high-titanium mare basalts (abstract). In *Lunar Science—V*, pp. 187–189. The Lunar Science Institute, Houston.

Duncan A. R., Erlank A. J., Willis J. P., and Sher M. K. (1974b) Compositional characteristics of the Apollo 17 regolith (abstract). In *Lunar Science—V*, pp. 184–186. The Lunar Science Institute, Houston.

Eldridge J. S., O'Kelley G. D., Northcutt K. J., and Schonfeld E. (1973) Nondestructive determination of radionuclides in lunar samples using a large low-background gamma-ray spectrometer and a novel application of least-squares fitting. *Nucl. Instrum. Methods* 112, 319–322.

Eldridge J. S., O'Kelley G. D., and Northcutt K. J. (1974) Primordial radioelement concentrations in rocks and soils from Taurus-Littrow (abstract). In *Lunar Science—V*, pp. 206–208. The Lunar Science Institute, Houston.

LSPET (Lunar Sample Preliminary Examination Team) (1973a) Preliminary examination of lunar samples. In *Apollo 17 Preliminary Science Report*, NASA SP-330, pp. 7-1 to 7-46.

LSPET (Lunar Sample Preliminary Examination Team) (1973b) Apollo 17 lunar samples: Chemical and petrographic description. *Science* 182, 659–672.

Nunes P. D., Tatsumoto M., and Unruh D. M. (1974) U–Th–Pb systematics of some Apollo 17 samples (abstract). In *Lunar Science—V*, pp. 562–564. The Lunar Science Institute, Houston.

O'Kelley G. D., Eldridge J. S., Schonfeld E., and Bell P. R. (1970) Primordial radionuclide abundance, solar proton and cosmic-ray effects and ages of Apollo 11 lunar samples by nondestructive gamma-ray spectrometry. *Proc. Apollo 11 Lunar Sci. Conf., Geochim. Cosmochim. Acta*, Suppl. 1, Vol. 2, pp. 1407–1423. Pergamon.

O'Kelley G. D., Eldridge J. S., and Northcutt K. J. (1974) Cosmogenic radionuclides in samples from Taurus-Littrow: Effects of the solar flare of August, 1972. *Proc. Fifth Lunar Sci. Conf., Geochim. Cosmochim. Acta*. This volume.

Rhodes J. M., Rodgers K. V., Shih C., Bansal B. M., Nyquist L. E., and Wiesmann H. (1974) The relationship between geology and soil chemistry at the Apollo 17 landing site (abstract). In *Lunar Science—V*, pp. 630–632. The Lunar Science Institute, Houston.

Schonfeld E. (1967) ALPHA-M-an improved computer program for determining radioisotopes by least-squares resolution of the gamma-ray spectra. *Nucl. Instrum. Methods* 52, 177–178.

Schonfeld E. (1974a) K and U systematics and average concentrations on the moon (abstract). In *Lunar Science—V*, pp. 672–674. The Lunar Science Institute, Houston.

Schonfeld E. (1974b) Component abundance and evolution of regoliths and breccias: Interpretation by mixing models (abstract). In *Lunar Science—V*, pp. 669–671. The Lunar Science Institute, Houston.

Silver L. T. (1974) Patterns of U–Th–Pb distributions and isotope relations in Apollo 17 soils (abstract). In *Lunar Science—V*, pp. 706–708. The Lunar Science Institute, Houston.

Trombka J. I., Arnold J. R., Reedy R. C., Peterson L. E., and Metzger A. E. (1973) Some correlations between measurements by the Apollo gamma-ray spectrometer and other lunar observations. *Proc. Fourth Lunar Sci. Conf., Geochim. Cosmochim. Acta*, Suppl. 4, Vol. 3, pp. 2847–2853. Pergamon.

Wänke H., Baddenhausen H., Dreibus G., Jagoutz E., Kruse H., Palme H., Spettel B., and Teschke F. (1973) Multielement analyses of Apollo 15, 16, and 17 samples and the bulk composition of the moon. *Proc. Fourth Lunar Sci. Conf., Geochim. Cosmochim. Acta*, Suppl. 4, Vol. 2, pp. 1461–1481. Pergamon.

Wänke H., Palme H., Baddenhausen H., Dreibus G., Jagoutz E., Kruse H., Spettel G., and Teschke F. (1974) Composition of the moon and major lunar differentiation processes (abstract). In *Lunar Science—V*, pp. 820–822. The Lunar Science Institute, Houston.

Proceedings of the Fifth Lunar Conference
(Supplement 5, Geochimica et Cosmochimica Acta)
Vol. 2 pp. 1035–1046 (1974)
Printed in the United States of America

Breccia 66055 and related clastic materials from the Descartes region, Apollo 16

J. S. Fruchter, S. J. Kridelbaugh, M. A. Robyn, and G. G. Goles

Center for Volcanology, University of Oregon, Eugene, Oregon 97403

Abstract—Trace and major element contents obtained by instrumental neutron activation are reported for a number of Apollo 16 soil samples and miscellaneous breccia fragments. In addition, data obtained by instrumental neutron activation and electron microprobe techniques along with petrographic descriptions are presented for selected subsamples of breccia 66055. The compositions of our soil samples can be modeled by mixtures of various amounts of anorthosite, anorthositic gabbro and low-K Fra Mauro basalt components. These mixtures are typical of those found in a number of petrographic surveys of the fines. Breccia 66055 is a complex regolith breccia which consists of at least four distinct types of microbreccias. No systematic relation with respect to stratigraphic age among the various microbreccia types was observed. Compositionally and texturally, the clasts which compose breccia 66055 are similar to a number of previously reported rock types from the Apollo 16 area. The entire breccia appears to have undergone a complex history of thermal metamorphism. We conclude from the study of these samples that the Cayley Formation is probably homogeneous in its gross compositional and petrographic aspects.

Introduction

Of all the Apollo missions, the Apollo 16 mission to the Descartes region of the moon was the only visit to a true lunar highlands site. As might be expected from an area which has apparently suffered a history of heavy and repeated bombardment, the vast majority of the returned samples are of a clastic nature. Here we report the results of two studies on some of these clastic materials. The first is an analysis and interpretation of a number of trace and major element abundances in bulk and sieved fractions of the lunar fines from the Apollo 16 landing area and in several miscellaneous breccia fragments allocated to us. The second is a thorough study of a complex regolith breccia, 66055, combining the techniques of instrumental neutron activation and electron microprobe analyses along with standard petrographic techniques. The latter study was undertaken as our contribution to a consortium study of breccia 66055.

Analytical Data

Eleven samples of <1 and 1–2 mm fines, five samples of miscellaneous breccias, and 39 clast and matrix subsamples of breccia 66055 were analyzed for 19 trace and major elements by instrumental neutron-activation analysis (INAA) (Lindstrom *et al.*, 1972). In addition five thin sections of breccia 66055 were studied by petrographic microscope and analyzed with an ARL-EMX microprobe.

1035

Soils

Abundances of 19 trace and major elements in our soil samples, determined by INAA, are reported in Table 1. The chondrite-normalized rare-earth plots are shown in Fig. 1. Although there are some significant compositional variations among these soils, the Apollo 16 soils appear in general to be considerably more homogeneous in composition than are the soils collected at other Apollo landing sites. This observation is somewhat surprising in view of the relatively large area sampled by the Apollo 16 astronauts. The compositional variations among our samples from a given station, and even the variations among different size fractions of the same sample, are as large as the variations from station to station.

An exception to this generalization is our sample from Station 11 on the rim of North Ray Crater, which seems to be significantly poorer in excluded trace elements than are the soil samples from other stations. Other samples from Station 11 analyzed by Wänke *et al.* (1973) and Haskin *et al.* (1973) show similar characteristics.

The majority of our soil samples come from only two stations, LM-ALSEP and Station 8, so that it is difficult for us to make general statements about the relationship of soil compositions to selenographic features. The unusual compositions at Station 11 seem to be closely related to North Ray Crater (Heiken *et al.*, 1973). The soils from Stations 1, 8, and the LM-ALSEP site are generally similar, although the soils at Station 8 are somewhat enriched in excluded trace elements such as the rare earths. The enrichment is probably due to the admixture of a small amount of low-K Fra Mauro basalt, as discussed below.

Table 1. Apollo 16 soils.

Sample	60051,6 LM	60501,19 LM	68821,5 Sta. 8	68841,26 Sta. 8	60052,2 LM	60502,1 LM	61282,2 Sta. 1	68822,2 Sta. 8	68842,3 Sta. 8	67481,20 Sta. 11	68121,4 Sta. 8	Relative % error
	<1 mm				1–2 mm					Unsieved fines		
Fe	3.2	4.0	3.9	4.0	3.4	5.1	4.1	3.4	4.0	3.2	4.0	3
Sc	8.1	10.1	9.8	10.1	8.5	8.7	9.4	8.4	9.6	7.9	9.8	2
Cr	611	774	749	770	684	734	758	695	831	526	732	3
Co	23.6	29.0	28.5	30.7	23.5	92.0	29.8	23.8	28.8	21.1	30.0	2
Hf	3.3	4.2	4.4	4.6	4.5	4.6	3.7	4.8	5.2	2.1	4.4	3
Ta	0.4	0.5	0.5	0.5	0.5	0.5	0.4	0.5	0.6	0.3	0.5	3
Na	3500	2250	3470	3470	3620	3560	3660	3650	3520	3806	3460	1
Ba	130	140	160	160	140	170	120	—	160	87	140	5
La	9.9	12.4	13.3	13.4	12.6	14.3	11.1	14.4	16.0	6.1	13.1	3
Ce	25.2	31.5	33.2	34.3	31.1	36.3	29.3	37.6	40.0	15.8	33.0	5
Nd	16	19	26	23	19	28	21	23	28	12	25	10
Sm	5.1	6.2	6.7	6.9	6.2	7.0	6.0	6.9	7.7	3.1	6.6	3
Eu	1.2	1.3	1.2	1.3	1.3	1.3	1.3	1.3	1.3	1.2	1.3	3
Tb	1.0	1.2	1.2	1.3	1.2	1.3	1.1	1.4	1.5	0.6	1.3	10
Yb	3.5	4.3	4.9	4.9	4.2	4.9	3.9	5.1	5.5	2.2	4.5	10
Lu	0.5	0.7	0.7	0.7	0.7	0.7	0.6	0.8	0.8	15.1	07	3
Al	14.9	14.5	14.8	—	—	15.2	14.8	16.3	—	15.1	—	5
Th	2.0	2.3	2.5	2.6	2.4	2.8	2.1	2.9	3.0	—	2.5	5

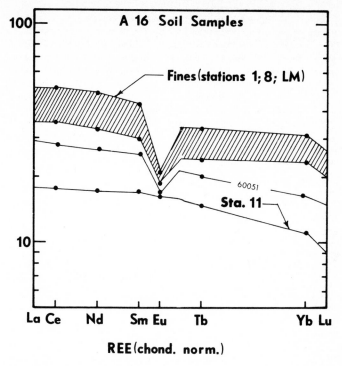

Fig. 1. Chondrite-normalized rare-earth plots for samples of fines from the Apollo 16 mission.

MISCELLANEOUS BRECCIA FRAGMENTS

Trace and major element data for several miscellaneous breccia fragments are presented in Table 2. The compositional data indicate that all are probably rock types previously recognized at the Apollo 16 landing site. Lithologically they seem to range from anorthosite to anorthositic gabbro. Data for these samples are helpful for constructing compositional models for the soils, because as will be discussed below, they provide additional examples of local rock types.

SOIL COMPONENTS

Ridley *et al.* (1973) report that the Apollo 16 soils they analyzed contained significant proportions of four components, anorthositic gabbro, gabroic anorthosite, anorthosite, and low-K Fra Mauro basalt. Although Ridley *et al.* defined these components on the basis of composition, not all of these components can be defined strictly enough on the basis of composition alone to be useful as mixing model end-members, as they compositionally grade into one another. It is, therefore, difficult to obtain precise estimates of the amounts of these components using compositional mixing models as we have in previous papers. Nevertheless,

Table 2. Apollo 16 miscellaneous breccias.

Sample	60335,4 LM	67455,14 Sta. 11	67455,13 Sta. 8	68815,w Sta. 8	68815,61 Sta. 8	61016,180 Sta. 1	60625,3 LM	Relative % error
			Breccia				Crystalline (rake)	
Fe	2.7	3.4	2.5	0.6	4.0	0.2	4.2	3
Sc	6.7	10.0	5.2	1.6	7.3	0.5	9.7	2
Cr	700	540	630	110	750	21	840	3
Co	15.2	10.2	19.4	2.6	50.9	0.5	27.0	2
Hf	5.1	0.3	3.4	0.4	5.3	—	6.3	3
Ta	0.6	—	0.4	—	0.5	—	0.7	3
Na	4300	2440	3360	3120	3700	3050	3660	1
Ba	170	—	120	160	160	—	190	5
La	16.2	0.9	9.8	1.4	15.4	—	20.7	3
Ce	39	—	26	3.9	37	0.3	49	5
Nd	29	—	18	—	26	—	34	10
Sm	8.2	0.6	4.8	0.7	7.6	0.1	10.5	3
Eu	1.2	0.8	1.1	—	1.2	0.9	1.4	3
Tb	1.5	0.1	1.0	0.1	1.4	—	1.4	10
Yb	5.1	0.6	3.6	—	5.1	—	6.7	10
Lu	0.8	0.1	0.5	0.1	0.8	—	1.0	3
Al	13.6	17.4	16.7	—	—	18.2	13.7	3
Th	2.8	0.1	1.9	0.3	2.8	—	4.1	5

linear least-squares mixing models are still valuable in obtaining estimates of the amounts and compositional types required to form the abundance patterns in each soil type, especially if locally collected examples of a given rock type are used as mixing end members. For example, anorthositic breccia fragment 67455 from Station 11 proves to be the best anorthositic component for soil sample 67481, also from Station 11. Thus, the conclusion that the lunar regolith has not been completely homogenized on a kilometer scale appears valid for the Cayley Plains as well as for previous areas. The composition of the soil from Station 11 can be modeled by approximately 70% anorthosite and 30% anorthositic gabbro, the soils from Station 8 consist of approximately 50% anorthositic gabbro, 35% anorthosite, and 15% low-K Fra Mauro basalt, while the soils from Station 1 and the LM site consist of approximately equal proportions of anorthosite and anorthositic gabbro compositional types. Our mixing calculations indicate that a mare-type basalt composition is not an important component of the Apollo 16 soils. This conclusion is in agreement with that reached by Ridley *et al.* (1973) in their petrographic study of two soil samples. Duncan *et al.* (1973) did find a basaltic type compositional component in Apollo 16 soils using a similar mixing model approach, but since their end-member components are defined somewhat differently, our results and their results cannot be directly compared.

Breccia 66055

Breccia 66055 was collected at the foot of Stone Mountain from a regolith developed on the Cayley Formation, but because of its angular shape, it is thought to represent ejecta from South Ray Crater and therefore presumably originated at least some meters below the present surface. A number of relatively simple stratigraphic models for the regolith which was sampled by the South Ray Crater event have been proposed (Head, 1973; Ulrich, 1973). These schemes may well have considerable validity in a general sense, but the complex relationships developed among the various chemical and petrologic clast types in breccia 66055 indicate that, as might be expected, the detailed history of the area must be quite complicated.

Petrography

Breccia 66055 consists of lithic and mineral clasts set in a fine- to coarse-grained moderately coherent matrix. Clast size varies from less than 1 mm to greater than 4 cm. Four major types of lithic clasts, three of them microbreccias and distinguishable on the basis of petrography and bulk chemical composition, are present. These same lithic clasts are well represented in the soil samples from the Apollo 16 site (Delano *et al.*, 1973; Steele and Smith, 1973; and Taylor *et al.*, 1973). No systematic relationships have been discovered between the various types of microbreccia clasts. In some instances, a particular lithologic type may be the host to a clastic inclusion of another lithologic type, while in other instances, the host-inclusion relationship between the same two lithologic types is reversed. The four major lithic clast types in microbreccia 66055 are designated (in order of abundance) as "partially molten microbreccias," nonrecrystallized microbreccias, clasts of the ANT suite (Prinz *et al.*, 1973), and a microbreccia that has brown glass as its matrix.

The partially molten microbreccias contain mineral clasts (olivine, orthopyroxene, clinopyroxene, and plagioclase ranging up to several millimeters in size) and lithic clasts set in a cryptocrystalline to medium-grained (50 μm) matrix composed of plagioclase laths (An_{93-95}), anhedral olivine (Fo_{75-82}), and mesostasis. The proportion of clastic fragments is low, commonly less than 20%. The texture of the matrix indicates that these microbreccia clasts were at one time partially or completely molten. The matrix of these partially molten microbreccias is enriched in aluminum and calcium and depleted in titanium and iron (Table 3) relative to the glass-rich and nonrecrystallized microbreccias. In terms of bulk composition, the matrix of the partially molten microbreccias is similar to the low-K Fra Mauro glass type described by Ridley *et al.* (1973) from the Apollo 16 site. The partially molten microbreccias are further distinguished by commonly containing rounded to irregular metallic blebs which are composed of the minerals kamacite, schreibersite, and troilite. The experimental and textural evidence (McKay *et al.*, 1973) indicates that these metallic particles have formed *in situ* under reducing conditions while the rock was in a partially or completely molten state. The partially molten microbreccias correspond to the olivine–plagioclase–mesostasis

Table 3. Broad beam analyses of 66055 clasts.

	Wt.%	S.D.*	Wt.%	S.D.*	Wt.%	S.D.*
	Brown† glass		Nonrecrystallized‡ clast matrix		Partially molten§ microbreccia	
SiO$_2$	46.9	1.21	45.4	0.60	45.1	0.46
Al$_2$O$_3$	19.06	1.14	20.9	1.04	23.2	0.56
TiO$_2$	1.2	0.18	1.12	0.13	0.9	0.07
FeO	7.79	1.26	7.45	0.77	5.76	0.24
MgO	10.6	0.51	9.06	0.57	9.05	0.54
CaO	11.9	0.80	12.2	0.51	13.0	0.25
K$_2$O	0.35	0.08	0.35	0.11	0.27	0.05
Na$_2$O	0.65	0.10	0.69	0.10	0.04	0.03
Cr$_2$O$_3$	0.19	0.02	0.16	0.01	0.16	0.01
P$_2$O$_5$	0.36	0.02	0.30	0.12	0.20	0.06
Total	99.5		97.6		98.2	

*Standard deviation of the mean of the number of analyses indicated in the subsequent footnotes.

†Brown glass, average of 11 clasts.

‡Average of 5 clasts.

§Average of 8 clasts.

basalts and mesotasis-rich basalts of Warner *et al.* (1973) and the Type A FIIR lithology of Delano *et al.* (1973).

Trace and major element abundances determined by INAA for selected partially molten microbreccia clasts are given in Table 4. The chondrite-

Table 4. Breccia 66055 clasts.

Sample	4-7	5-7	6-4	6-3	6-10	6-6	6-7	6-1	6-2	Relative % error
	Partially molten microbreccia clasts					Unrecrystallized microbreccia clasts				
Fe	6.5	7.0	7.4	6.2	5.7	3.4	4.9	4.7	3.8	3
Sc	11.5	11.8	11.5	13.4	11.1	7.3	9.7	9.3	8.6	2
Cr	1160	1100	1040	1200	1050	640	880	790	780	3
Co	55	85	113	11	46	28	42	38	22	2
Hf	9.6	10.3	9.3	8.0	9.3	5.7	7.4	7.5	2.8	3
Ta	1.2	1.2	1.1	—	1.1	0.7	0.9	1.0	—	3
Na	3880	3520	3670	4120	3710	3210	3470	3290	3580	1
K	1800	2600	1700	2200	2100	—	1700	1300	2500	10
Ba	280	300	250	—	260	160	180	240	—	5
La	27.3	29.5	27.8	32.0	—	15.3	22.5	20.3	9.3	3
Ce	76.9	79.9	73.4	85.5	63.8	42.1	57.8	73.4	38.0	5
Nd	51	56	55	—	45	32	42	46	—	10
Sm	13.6	14.6	13.2	15.6	12.8	6.9	10.6	9.5	5.9	3
Eu	1.6	1.7	1.5	—	1.6	1.4	1.4	1.5	—	3
Tb	2.5	2.7	9.9	4.2	2.6	2.1	2.1	2.1	—	10
Yb	9.1	10.4	9.9	10.8	9.6	5.5	8.1	7.0	5.2	10
Nu	1.3	1.4	1.2	1.6	1.1	0.7	1.1	1.0	0.6	3
Al	—	12.2	11.4	13.7	12.8	14.1	12.2	13.4	15.2	3
Th	4.4	4.4	4.6	5.7	4.7	3.0	4.0	3.7	6.2	5

normalized plots for the extremes of the range of rare-earth abundances found in this clast type are presented in Fig. 2. The partially molten microbreccias have relatively constant trace element abundance patterns with the exception of Co which varies almost over a factor of ten from clast to clast. This variability in Co contents is undoubtedly due to heterogeneous distribution of the characteristic metallic blebs mentioned above. The rare-earth patterns are all fairly similar to one another, being about 90 times chondrite abundances and having substantial negative Eu anomalies. Although texturally somewhat different, the trace and major element compositions of these clasts are quite similar to those reported by Drake *et al.* (1973) for poikiloblastic clasts found in the 2–4 mm lithic fragments collected during the Apollo 16 mission. The partially molten microbreccias are quite distinct in trace element compositions from an analysis of a poikiloblastic rock reported by Haskin *et al.* (1973).

The second major lithic clast type, the noncrystallized microbreccias, consist of abundant mineral clasts and (occasionally) lithic clasts set in a fine-grained,

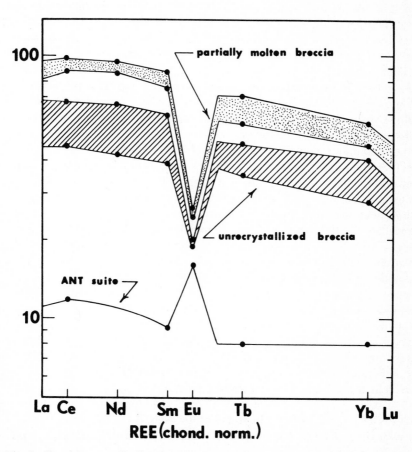

Fig. 2. Chondrite-normalized rare-earth plots for three clast types found in breccia 66055.

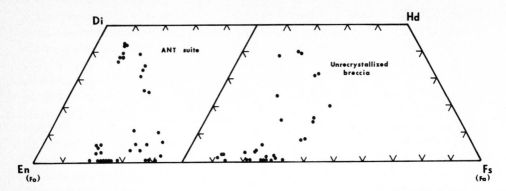

Fig. 3. Pyroxene and olivine compositions for two clast types found in breccia 66055.

dark-brown to dark-gray matrix, which has not undergone substantial recrystallization. The leucocratic members of this suite resemble the light-matrix breccia
type of Warner *et al.* (1973) and the dark-brown lithology corresponds to the
dark-matrix breccia of Delano *et al.* (1973). The mineral clasts within this
microbreccia include plagioclase, low-calcium pyroxene, clinopyroxene, olivine,
iron-rich metal, and troilite. Compositions of the mafic minerals are presented in
Fig. 3, and are representative of the mafic minerals found in the clasts of other
lithologies in 66055. The bulk composition of the fine-grained matrix (Table 3) is
indistinguishable from the composition of the brown glass, suggesting that these
two lithologies are closely related.

Trace and major element analyses of selected members of the noncrystallized
microbreccia group are given in Table 4. Their range of chondrite-normalized
rare-earth abundances is illustrated in Fig. 2. Because they are generally intermediate in trace element contents relative to other clast types, it is difficult to
identify them with previously recognized rock types on this basis. Textural
evidence shows that they are not simple mechanical mixtures of previously
existing clast types, an observation confirmed by linear least-squares mixing
calculations. The nonrecrystallized breccia clasts are compositionally similar to
two Apollo 16 basalts analyzed by Haskin *et al.* (1973).

Approximately one-quarter of the clasts in 66055 are members of the ANT
(anorthosite–norite–troctolite) suite, ranging in composition from nearly pure
anorthosite to norites and troctolites. The textures of the ANT clasts vary from
primary igneous (including cumulus), to cataclastic, to poikiloblastic types. The
compositions of the mafic phases found in the ANT clasts are presented in Fig. 3.
Trace and major element compositions of selected ANT suite microbreccia clasts
are given in Table 5 and a chondrite-normalized rare-earth pattern for ANT suite
clast 5-6a is shown in Fig. 2. The trace element contents of these clasts are typical
of low-KREEP ANT suite rocks found at the Apollo 16 and other Apollo landing
sites, with generally low abundance of excluded trace elements and pronounced
positive Eu anomalies.

Table 5. Breccia 66055 clasts.

Sample*	4-6	5-2	5-6a	4-4	6-9	4-12	6-8
		ANT suite clasts			Petrographically unassigned clasts		
Fe	0.1	3.0	3.3	3.2	5.3	4.2	7.9
Sc	0.4	4.9	6.2	6.5	9.7	6.3	7.7
Cr	21	987	755	825	911	530	680
Co	1	18	22	13	50	80	199
Hf	—	0.4	1.3	—	5.7	5.3	4.3
Ta	—	—	0.2	—	1.1	—	—
Na	2740	2020	3020	3010	3890	3260	3320
K	—	—	630	640	1920	1250	810
Ba	—	—	—	70	230	180	—
La	—	—	3.7	—	24.1	15.5	16.8
Ce	—	5.4	10.5	5.9	64.0	42.1	48.0
Sm	0.2	0.8	1.5	0.8	12.2	6.6	8.4
Eu	1.5	—	1.1	1.4	1.6	1.2	2.2
Tb	—	0.2	0.3	—	3.1	1.6	2.2
Yb	—	0.8	1.8	1.8	7.5	5.9	5.7
Lu	—	—	—	0.2	1.2	0.7	0.8
Al	19.4	14.5	15.9	14.7	6.2	14.9	15.2
Th	—	0.3	0.9	—	3.9	3.0	3.1

*For estimated relative error, see Table 6.

The clasts composed of lithic and mineral clasts set in a matrix of brown glass are least abundant. Plagioclase (An_{93-95}) is the predominant mineral fragment, but orthopyroxene, clinopyroxene, and olivine are also present in very minor amounts. Some of these clasts also contain "primary" plagioclase crystals, as large as 200 μm in size, that probably crystallized while the brown glass was molten. In clasts with abundant "primary" euhedral plagioclase crystals, clastic mineral fragments are rare. The proportions of clastic materials and of "primary" euhedral plagioclase crystals to brown glass vary over a wide range, with some clasts containing more than 90% glass, while other clasts of this type contain less than 10% glass. The bulk composition (from defocused beam microprobe analyses) of the brown glass (Table 3) is very similar to the major element composition of Apollo 16 KREEP basalts (LSPET, 1973). Its bulk composition is also similar to the "partial-melt" of rock 64455,35 (Grieve and Plant, 1973) except for a slight enrichment in titanium and depletion in aluminum. The clasts of brown glass occasionally contain spherules of metallic iron, schreibersite, and troilite identical to the metallic particles found in the "partially molten microbreccia." No example of the brown glass breccia large enough to separate for INAA analyses could be recognized in hand specimens. The compositions of four lithologically unassigned clasts are presented in Table 5.

MATRIX

The matrix of 66055 consists essentially of plagioclase-rich material varying in size from crystals less than 10 or 20 μm to crystals larger than 1 mm in size with the vast majority near 0.25 μm and less. It is often difficult, if not impossible, to discern the contact between the plagioclase-rich clast types and the matrix of this rock. Grains of pink pleonaste spinel rarely are found in the matrix, and they are not associated with any particular rock or clast types. Other minerals in the matrix are low-calcium pyroxene, clinopyroxene, ilmenite, iron-rich metal, and troilite.

As might be expected, the matrix has a composition intermediate between those of the various clast types, although the composition varies considerably because of the heterogenous distribution of clasts. The compositions of selected samples of the matrix are given in Table 6. A linear least-squares mixing calculation shows that the average matrix composition can be adequately approximated by a combination of essentially equal proportions of the three clast types. Thus, it appears that no one clast type was favored in whatever abrasion process led to the formation of the matrix.

Based upon the petrographic and chemical evidences, breccia 66055 is a soil breccia having as its major components rock fragments that are identical to those described in the soil samples collected at the Apollo 16 site (Delano *et al.*, 1973; Taylor *et al.*, 1973; Steele and Smith, 1973). The diversity of the textures of the various rock types, ranging from primary igneous (including cumulus) to low- and high-grade metamorphic, to cataclastic, to breccias, at the Apollo 16 and Luna 20

Table 6. Breccia 66055 matrix.

Sample	4-11	4-13	5-5	6-11	Relative % error
Fe	4.1	2.6	3.4	3.5	3
Sc	7.7	5.3	6.8	6.2	2
Cr	742	568	723	620	3
Co	30	19	20	32	2
Hf	5.1	2.0	3.2	2.9	3
Ta	0.6	0.3	0.4	0.4	3
Na	3270	3230	2300	3160	1
K	1050	—	940	810	10
Ba	130	89	120	84	5
La	14.4	6.0	9.3	7.3	3
Ce	37.3	15.2	24.3	21.3	5
Nd	24.7	—	—	—	3
Sm	6.9	2.8	4.3	3.2	3
Eu	1.2	1.1	1.1	1.2	3
Tb	1.4	0.6	0.8	0.7	10
Yb	5.1	2.6	3.7	3.2	10
Lu	0.7	0.4	—	0.5	3
Al	14.8	16.1	17.1	14.3	3
Th	3.0	1.2	1.8	1.9	5

sites (Kridelbaugh and Weill, 1973) indicates a very complex geological history for the highland areas.

CONCLUSION

We have analyzed two types of clastic materials from the Descartes region of the moon—miscellaneous samples of fines and breccias from the Cayley Formation, and a large sample of complex regolith breccia 66055. Breccia 66055 is presumed to be ejecta from South Ray Crater, and is, therefore, probably also a sample of the Cayley Formation from some distance away from the landing site. Although there are certainly recognizable differences in chemical composition and lithology among these various samples of the Cayley Formation, in gross aspect they are quite similar. Thus, it appears that the Cayley Formation is largely clastic in nature and is probably derived almost entirely from either the same source or at least compositionally related sources.

Acknowledgments—We would like to thank D. F. Weill, A. M. Reid and D. P. Blanchard for critical reviews of this paper. Cecily Bonham and Jeanne Heer performed the typing chores. This work was supported in large part by NASA grants NGL 38-003-024 and NGL 38-003-022.

REFERENCES

Delano J. W., Bence A. E., Papike J. J., and Cameron K. L. (1973) Petrology of the 2–4 mm soil fraction from the Descartes region of the moon and stratigraphic implications. *Proc. Fourth Lunar Sci. Conf.*, *Geochim. Cosmochim. Acta*, Suppl. 4, Vol. 1, pp. 537–551. Pergamon.
Drake M. J., Taylor G. J., and Goles G. G. (1974) Descartes Mountains and Cayley Plains: Composition and provenance. *Proc. Fifth Lunar Sci. Conf.*, *Geochim. Cosmochim. Acta*. This volume.
Duncan A. R., Erlank A. J., Willis J. P., and Ahrens L. H. (1973) Composition and inter-relationships of some Apollo 16 samples. *Proc. Fourth Lunar Sci. Conf.*, *Geochim. Cosmochim. Acta*, Suppl. 4, Vol. 2, pp. 1097–1113. Pergamon.
Grieve R. A. F. and Plant A. G. (1973) Partial melting on the lunar surface, as observed in glass coated Apollo 16 samples. *Proc. Fourth Lunar Sci. Conf.*, *Geochim. Cosmochim. Acta*, Suppl. 4, Vol. 1, pp. 667–679. Pergamon.
Haskin L. A., Helmke P. A., Blanchard D. P., Jacobs J. W., and Telander K. (1973) Major and trace element abundances in samples from the lunar highlands. *Proc. Fourth Lunar Sci. Conf.*, *Geochim. Cosmochim. Acta*, Suppl. 4, Vol. 2, pp. 1275–1296. Pergamon.
Head J. W. (1973) Stratigraphy of the Descartes region Apollo 16: Implications for the origin of samples. Preprint, Brown University, Providence, R.I. Submitted to *The Moon*.
Heiken G. H., McKay D. S., and Fruland R. M. (1973) Apollo 16 soils: Grain size analyses and petrography. *Proc. Fourth Lunar Sci. Conf.*, *Geochim. Cosmochim. Acta*, Suppl. 4, Vol. 1, pp. 251–265. Pergamon.
Kridelbaugh S. J. and Weill D. F. (1973) The petrology and mineralogy of the Lunar-20 soil sample. *Geochim. Cosmochim. Acta* **37**, 915–926.
Lindstrom M. M., Duncan A. R., Fruchter J. S., McKay S. M., Stoeser J. W., Goles G. G., and Lindstrom D. J. (1972) Compositional characteristics of some Apollo 14 clastic materials. *Proc. Third Lunar Sci. Conf.*, *Geochim. Cosmochim. Acta*, Suppl. 3, Vol. 2, pp. 1201–1229. MIT Press.
LSPET (Lunar Sample Preliminary Examination Team) (1973) The Apollo 16 lunar samples: Petrographic and chemical description. *Science* **179**, 23–34.
McKay G. A., Kridelbaugh S. J., and Weill D. F. (1973) The occurrence and origin of

schreibersite–kamacite intergrowths in microbreccia 66055. *Proc. Fourth Lunar Sci. Conf., Geochim. Cosmochim. Acta*, Suppl. 4, Vol. 1, pp. 811–818. Pergamon.

Prinz M., Dowty E., Keil K., and Bunch T. E. (1973) Spinel troctolite and anorthosite in Apollo 16 samples. *Science* **179**, 74–76.

Ridley W. I., Reid A. M., Warner J. W., Brown R. W., Gooley R., and Donaldson C. (1973) Glass compositions in Apollo 16 soils 60501 and 61221. *Proc. Fourth Lunar Sci. Conf., Geochim. Cosmochim. Acta*, Suppl. 4, Vol. 1, pp. 309–321. Pergamon.

Steele I. M. and Smith J. V. (1973) Mineralogy and petrology of some Apollo 16 rocks and fines: General petrologic model of moon. *Proc. Fourth Lunar Sci. Conf., Geochim. Cosmochim. Acta*, Suppl. 4, Vol. 1, pp. 519–536. Pergamon.

Taylor G. J., Drake M. J., Hallam M. E., Marvin U. B., and Wood J. A. (1973) Apollo 16 stratigraphy: The ANT hills, the Cayley Plains, and a pre-Imbrian regolith. *Proc. Fourth Lunar Sci. Conf., Geochim. Cosmochim. Acta*, Suppl. 4, Vol. 1, pp. 553–568. Pergamon.

Ulrich G. E. (1973) A geologic model for North Ray Crater and stratigraphic implications for the Descartes region. *Proc. Fourth Lunar Sci. Conf., Geochim. Cosmochim. Acta*, Suppl. 4, Vol. 1, pp. 27–60. Pergamon.

Wänke H., Baddenhausen H., Dreibus G., Jagoutz E., Kruse H., Palme H., Spettel B., and Teschke F. (1973) Multi-element analyses of Apollo 15, 16, and 17 samples and the bulk composition of the moon. *Proc. Fourth Lunar Sci. Conf., Geochim. Cosmochim. Acta*, Suppl. 4, Vol. 2, pp. 1275–1296. Pergamon.

Warner J. L., Simonds C. H., and Phinney W. C. (1973) Apollo 16 rocks: Classification and petrogenetic model. *Proc. Fourth Lunar Sci. Conf., Geochim. Cosmochim. Acta*, Suppl. 4, Vol. 1, pp. 481–504. Pergamon.

Proceedings of the Fifth Lunar Conference
(Supplement 5, Geochimica et Cosmochimica Acta)
Vol. 2 pp. 1047–1066 (1974)
Printed in the United States of America

Chemical studies of Apollo 16 and 17 samples

J. C. Laul, D. W. Hill, and R. A. Schmitt

Department of Chemistry and the Radiation Center, Oregon State University, Corvallis, Oregon 97331

Abstract—The abundances of the bulk elements TiO_2, Al_2O_3, FeO, MgO, CaO, Na_2O, K_2O, MnO and Cr_2O_3, and the trace elements Sc, V, Co, Zr, Ba, nine REE, La, Ce, Nd, Sm, Eu, Tb, Dy, Yb and Lu, Hf, Ta, Th and U and the siderophile elements Ni, Ir and Au have been determined by sequential instrumental neutron activation analysis (INAA) in seven Apollo 16 breccias, three Apollo 17 basalts, seven Apollo 17 breccias including five rocks chipped from boulder-2, Station 2, and ten Apollo 17 soils. Abundances of the siderophile, volatile and lithophile trace elements Ir, Re, Au, Ni, Co, Sb, Se, Ag, In, Zn, Cd, Tl, Rb, Cs, U, Ba, Sr, and Ga were determined by radiochemical neutron activation analysis (RNAA) in five rocks from boulder-2, Station 2, and in four Apollo 17 soils at Station 2 and one valley soil at Station 5. In general, the Apollo 16 breccias form a continuum of rocks ranging from 50% to 88% plagioclase and show Ba, REE, Hf, Ta, and Th (Ba–Th) normalized KREEP-type patterns that range from ~5× to ~200× chondrites. The REE and LIL trace element contents are related inversely to the plagioclase content. For REE patterns that are <15× chondrites, positive Eu anomalies are found; for >15×, negative Eu anomalies are observed. REE patterns in Apollo 17 basalts parallel the REE patterns observed in Apollo 11 basalts. Although the bulk compositions of the Apollo 11 and 17 high ilmenite basalts show many similarities, measurable bulk differences exist between these basalt groups. At least two and possibly three Apollo 17 lava flows are present in the valley. Similar to Apollo 11 basalts, high-Sc abundances were also observed in the three Apollo 17 basalts. The Sc enrichment in Apollo 11 and 17 mare basalts relative to the Apollo 12 and 15 mare basalts may be attributed to higher partial melting temperatures by ≈100°C for the formation of the former magmas relative to the latter magmas; this interpretation assumes the same compositional source material was subjected to partial melting. The average bulk and trace element composition of Apollo 17 "high-alumina" boulder-2 rocks, Station 2, are medium-K KREEP and are similar to other noritic breccias sampled at the South and North Massifs and this suggests fairly uniform composition for these highland massifs. The Ba–Th patterns are KREEP-like (~80× chondrites) and fit into the Apollo 16 breccia spectrum for the relationship between the Ba–Th patterns and plagioclase content. South Massif rocks contain siderophiles at the 2–4% Cl level and show an ancient meteoritic pattern. Valley soils have low siderophiles (1% Cl).

Introduction

We have analyzed by INAA nine Apollo 16 samples and 23 Apollo 17 samples for 26–30 major, minor, and trace elements. Six Apollo 16 boulder-2 rocks, five Apollo 17 soils and one Apollo 16 soil were also analyzed by RNAA for 18 siderophile, volatile, and other trace elements. For completeness, we include our results of the RNAA studies of Apollo 17 boulder-2 and soils (Laul and Schmitt, 1974).

Sample preparation, activation conditions, inclusion of the U.S.G.S. standard rock BCR-1, and the detailed INAA procedure are essentially the same as reported by Laul and Schmitt (1973a, 1974). Additionally, an Apollo soil sample analyzed in a previous batch by us was subsequently analyzed in our next batch of lunar samples for an additional safety check for both precision and accuracy.

RESULTS

Results are shown in Tables 1 and 2. We have made a direct comparison of our INAA data for soils 73121 and 73141 with the corresponding data reported for these soils by Wänke et al. (1974) and our data for soils 71501, 75081, 72501, and 73141, the anorthositic gabbro rock 77017, and the gabbroic anorthosite 63335 with the data reported for these soils and rocks by LSPET (1973a, 1973b). The agreement is excellent. Our RNAA data for soils 72321 and 75081 agree with Morgan et al. (1974) within experimental error.

DISCUSSION

Apollo 16 samples

64435 (B. Mason consortium). Separated fragments of the matrix, anorthositic clast and glass coating were analyzed. The light-gray matrix is a firmly welded rock with a cataclastic texture. It consists largely of anorthite (An 95–100) with minor amounts of orthopyroxene and olivine in a groundmass of pale brown partly devitrified glass (Mason, 1973). Based on our chemical data (Table 1), the matrix is a low-K gabbroic (noritic?) anorthosite with 88% plagioclase. The REE pattern (Fig. 1) is flat like the anorthositic gabbro 15418 (Laul et al., 1972a) with a positive Eu anomaly (Sm/Eu = 0.92).

The 64435 glass coating is gray–black and lustrous and shows no signs of divitrification. The chemical data indicate that the glass coating is anorthositic gabbro (67% plagioclase based on the Al_2O_3 content converted to anorthite) in composition. The siderophiles Ni, Ir, and Au, which are indicators for meteoritic impact, and Co are more abundant in this glass than observed in any other lunar rock or soil to date. Based on the low Ir/Au normalized cosmic ratio, an ancient meteorite of the LN type is associated with the glass fraction. The REE distribution in the glass is about 25 times chondritic and has a negative Eu anomaly (Sm/Eu = 4.7). The high-K KREEP pattern as indicated by K, Ba, REE (Sm/Eu = 15), Hf, Ta, and Th abundances (Fig. 1) is present in both the glass and matrix patterns and indicates significant amounts of KREEP in these two samples. Significant differences for the major, minor, and trace element data for the 64435 glass coating and the Apollo 16 soils preclude a derivation of the glass coating by meteoritic impact and melting of Apollo 16 soil.

The 64435 anorthositic clast consists of 97% plagioclase with minor contents of olivine, orthopyroxene, and augite. Higher REE abundances (except Eu) in this clast (Fig. 1) relative to 15415 and 60015 (Laul and Schmitt, 1973b) are not attributed to contamination by the matrix or the glassy phases of 64435. Similarities in the absolute REE abundances (except Eu) in the anorthositic phases of rocks such as 61016 (Philpotts et al., 1973), 15415 and 60015 (Laul and Schmitt, 1973b), and 64435 may indicate that these four anorthositic phases crystallized from magmas of similar LIL trace element composition. A low FeO/MnO ratio of 55 in the 64435 anorthositic clast merely reflects preferential partitioning of Mn relative to Fe in pure plagioclase.

Table 1. Elemental abundances via INAA in Apollo 16 and 17 samples.*

Sample No.	Station	Wt. (mg)	TiO_2 (%)	Al_2O_3 (%)	FeO (%)	MgO (%)	CaO (%)	Na_2O (%)	K_2O (%)	MnO (%)	Cr_2O_3 (%)	Sc (ppm)	V (ppm)	Co (ppm)	Zr (ppm)	Ba (ppm)
Apollo 16 Breccias																
60315,157	LM	44	1.3	17.0	10.4	14	9.4	0.76	0.42	0.112	0.219	14	50	88	640	460
62235,56	2	52	1.3	18.6	9.5	12	12.0	0.62	0.39	0.106	0.190	16	50	49	890	600
64435,39	4															
Matrix		271	0.2	32.1	3.0	3	17.0	0.34	0.024	0.040	0.064	5.2	15	7.0	—	20
Glass, 39		38	0.5	24.5	8.0	8	13.3	0.55	0.086	0.105	0.170	6.9	20	100	100	90
Anorthosite, 44		53	<0.1	35.5	0.61	—	19.0	0.29	0.025	0.011	0.0083	0.90	<4	1.3	—	<9
67095,28	11	217	0.66	21.7	5.8	12	12.4	0.59	0.25	0.080	0.140	8.4	30	10	250	220
Shadow Rock																
60017,81	13	295	0.37	31.2	3.6	3	17.0	0.52	0.056	0.048	0.054	6.7	10	7.1	30	50
63335,18		115	0.34	31.5	2.6	2	17.6	0.69	0.049	0.035	0.035	4.4	10	5.0	—	40
63355,10		137	0.88	21.5	8.3	8	12.0	0.50	0.22	0.089	0.169	12	35	62	280	280
Apollo 17 Basalts																
70135,35	ALSEP	231	13.8	7.0	21.8	9	8.7	0.40	0.11	0.270	0.506	82	120	20	—	210
72155,29	LRV-3	124	12.1	8.0	18.6	9	10.4	0.40	0.072	0.234	0.440	80	100	20	—	90
75035,48	5	286	9.0	9.9	18.8	7	11.3	0.42	0.074	0.236	0.221	76	30	16	—	95
Breccias Boulder-2																
72315,3 mostly ext.	2	43	1.4	19.8	8.5	11	11.6	0.61	0.32	0.111	0.186	16	50	21	400	290
72315,4 int.		34	1.4	19.2	8.5	12	11.3	0.70	0.35	0.111	0.187	16	50	32	400	280
72335,2 mostly ext.		39	0.60	27.3	4.8	8	15.4	0.45	0.12	0.060	0.100	8.0	30	25	150	120
72355,7 one ext. side		34	1.6	18.8	8.7	12	11.1	0.70	0.33	0.114	0.193	16	50	37	500	280
72375,2 mostly ext.		31	1.5	18.2	8.8	12	10.8	0.67	0.27	0.112	0.178	15	50	34	450	300
72395,3 int.		30	1.7	18.7	9.2	12	11.0	0.67	0.32	0.116	0.210	17	50	35	400	350
77017,57	7	88	0.75	26.0	6.2	6	14.5	0.31	0.050	0.085	0.140	12	40	24	—	30
Dark matrix		40	5.3	18.9	12.1	8	11.7	0.39	0.10	0.155	0.290	36	70	27	200	70
Gray fragment		35	0.35	27.1	5.7	6	15.7	0.36	0.076	0.077	0.126	9.8	40	23	—	40
79035,25	9	244	6.5	13.5	15.2	11	11.2	0.41	0.098	0.196	0.366	46	90	35	—	110

Table 1 (Continued).

Sample No.	Station	Wt. (mg)	TiO$_2$ (%)	Al$_2$O$_3$ (%)	FeO (%)	MgO (%)	CaO (%)	Na$_2$O (%)	K$_2$O (%)	MnO (%)	Cr$_2$O$_3$ (%)	Sc (ppm)	V (ppm)	Co (ppm)	Zr (ppm)	Ba (ppm)
Soils < 1 mm																
70008,218	ALSEP	94	9.4	10.8	18.0	10	10.3	0.46	0.085	0.231	0.463	60	110	31	—	130
71501,37	1a	66	8.5	11.3	17.6	9	10.2	0.36	0.076	0.222	0.442	69	100	38	—	50
72321,9	2	136	1.6	21.4	8.7	10	11.8	0.50	0.15	0.111	0.215	18	50	30	200	190
		44	1.6	21.1	8.7	10	12.1	0.45	0.16	0.113	0.210	18	50	28	220	190
72441,11	2a	170	1.5	21.1	8.8	10	11.6	0.49	0.15	0.111	0.220	18	50	30	200	190
72461,8	2a	151	1.4	21.8	8.7	10	12.5	0.47	0.15	0.110	0.212	18	50	30	180	190
72501,31	2	171	1.5	20.7	8.6	10	12.6	0.47	0.16	0.112	0.210	18	45	31	220	190
		41	1.7	21.0	8.6	10	12.8	0.49	0.16	0.114	0.217	18	50	35	200	170
73121,17	3	164	1.4	21.3	8.5	10	12.7	0.45	0.14	0.110	0.207	18	50	31	200	150
73141,22	3	175	1.1	21.4	7.8	10	12.6	0.45	0.14	0.103	0.200	17	45	30	220	160
75081,21	5	230	9.4	11.3	17.3	9	10.6	0.40	0.082	0.227	0.430	61	100	27	230	100
79511,15	9	214	5.9	14.0	15.5	10	10.0	0.42	0.082	0.190	0.355	50	85	37	150	100
BCR-1			2.20	13.6	12.2	—	6.9	3.30	1.70	0.176	—	32	410	36	—	590

Table 1. (Continued).

Sample No.	Station	La (ppm)	Ce (ppm)	Nd (ppm)	Sm (ppm)	Eu (ppm)	Tb (ppm)	Dy (ppm)	Yb (ppm)	Lu (ppm)	Hf (ppm)	Ta (ppm)	Th (ppm)	U (ppm)	Ni (ppm)	Ir (ppb)	Au (ppb)
Apollo 16 Breccias																	
60315,157	LM	50	120	73	21.5	1.90	4.4	29	15	2.1	16	2.0	8.1	1.7	1400	35	30
62235,56	2	61	150	96	27.0	2.10	5.3	32	19	2.7	20	2.3	10	2.2	720	20	16
64435,39	4																
Matrix		1.5	4	3	0.70	0.76	<0.2	0.8	0.58	0.082	0.41	0.07	0.25	<0.1	—	—	—
Glass, 39		9.6	24	15	4.3	0.91	0.80	5.1	2.8	0.43	3.2	0.35	1.1	0.4	1800	50	30
Anorthosite, 44		0.16	—	<0.4	0.086	0.69	0.03	0.2	0.06	0.008	<0.03	<0.02	—	<0.02	—	—	—
67095,28	11	21.4	50	33	8.9	1.40	1.9	12	6.4	0.90	6.3	0.83	3.2	0.80	120	4	4
Shadow Rock																	
60017,81	13	3.1	8	5	1.4	1.24	0.3	1.7	1.2	0.16	1.0	0.14	0.5	<0.14	45	1.5	4
63335,18		2.6	6	4	1.2	1.32	0.2	1.5	0.90	0.13	0.60	0.10	0.25	<0.1	—	2	4
63355,10		30	74	47	12.0	1.51	2.5	16	8.8	1.3	8.9	1.2	4.2	1.2	940	24	16

Apollo 17																	
Basalts																	
70135,35	ALSEP	12.6	52	50	18.0	2.84	4.5	29	16	2.2	14	2.6	0.3	—	—	—	—
72155,29	LRV-3	7.2	26	32	10.2	2.00	3.0	18	10	1.5	8.7	1.6	0.3	—	—	—	—
75035,48	5	7.3	27	30	10.8	2.20	3.1	20	10	1.5	8.7	1.6	0.3	—	—	—	—
Breccias																	
Boulder-2																	
72315,3 mostly ext.	2	30	76	50	12.8	1.82	2.6	17	10	1.3	10	1.4	5.2	1.4	180	5	3
72315,4 int.		31	77	50	12.9	1.83	2.7	17	10	1.3	10	1.4	5.4	1.5	330	10	4
72335,2 mostly ext.		13.2	31	21	5.8	0.90	1.1	7.0	4.2	0.55	4.2	0.59	2.4	0.8	330	12	4
72355,7 one ext. side		34	95	54	15.0	1.92	3.1	19	12	1.6	12	1.6	6.1	1.8	340	10	3
72375,2 mostly ext.		37	91	57	16.6	1.82	3.1	20	12	1.6	11	1.4	5.7	2.0	320	10	4
72395,3 int.		36	87	55	15.2	1.81	3.0	20	11	1.5	12	1.6	5.5	1.6	320	10	5
77017,57	7	3.3	9	5	1.5	0.78	0.3	2.4	1.6	0.21	1.5	0.22	0.4	—	290	10	3
Dark matrix		6.4	22	18	5.9	1.42	1.3	9.0	5.1	0.66	4.9	0.85	0.6	—	290	9	3
Gray fragment		3.6	10	5	1.7	0.81	0.3	2.4	1.4	0.18	1.0	0.14	—	—	300	10	3
79035,25	9	8.6	27	23	6.7	1.42	1.6	10	6.2	0.83	5.5	1.0	1.0	0.4	140	—	3
Soils < 1 mm																	
70008,218	ALSEP	7.3	27	20	7.9	1.67	1.8	12	7.0	0.97	6.9	1.3	0.4	—	—	—	—
71501,37	1a	6.7	24	20	7.0	1.70	2.2	13	6.7	0.89	7.3	1.3	—	—	—	—	—
72321,9	2	17.7	45	30	8.3	1.31	1.7	11	6.1	0.85	6.2	0.84	2.9	1.0	250	10	5
		18.3	45	32	8.2	1.32	1.6	11	6.2	0.84	6.4	0.84	2.8	1.0	260	10	4
72441,11	2	17.8	46	30	8.3	1.31	1.7	10	6.0	0.86	6.1	0.86	2.8	1.0	270	9	4
72461,8	2	17.6	45	28	8.2	1.32	1.6	10	6.0	0.86	6.0	0.80	2.8	1.0	230	12	5
72501,31	2	18.0	47	30	8.2	1.33	1.7	11	6.0	0.87	6.1	0.84	2.9	1.0	250	8	4
		17.8	47	31	8.3	1.32	1.6	11	6.2	0.84	6.1	0.84	3.0	1.0	340	12	5
73121,17	2a	15.6	39	27	7.2	1.20	1.4	10	5.3	0.77	5.0	0.78	2.4	0.7	280	11	3
73141,22	2a	15.5	38	25	7.2	1.15	1.4	9.0	5.1	0.75	4.9	0.75	2.1	0.7	240	8	4
75081,21	5	7.2	30	25	7.6	1.70	2.0	12	7.3	1.0	7.0	1.3	0.6	—	100	5	3
79511,15	9	8.7	29	21	7.3	1.52	1.5	10	5.7	0.86	5.6	0.99	—	—	170	8	3
BCR-1		25.5	54	30	6.7	1.95	1.0	6.3	3.4	0.50	5.0	0.85	—	—	—	—	—

*Estimated errors are: Al_2O_3, Na_2O, MnO, Cr_2O_3, Sc, La, and Sm, ±1–3%; FeO, K_2O, Co, Yb, Lu, and Hf, ±5%; TiO_2, CaO, MgO, Ni, Eu, Dy, Tb, and Th, ±5–10%; V, Ce, Ta, Zr, Ba, Nd, U, Ir, and Au, ±10–30%.

J. C. LAUL *et al.*

Table 2. Trace elements in Apollo 17 soils and boulder-2 rocks (Station 2), (m = ppm; b = ppb).*

	Wt. (mg)	Ir^b	Re^b	Au^b	Ni^m	Co^m	Sb^b	Se^b	Ag^b	In^b	Zn^m	Cd^b	Tl^b	Rb^m	Cs^b	U^m	Ba^m	Sr^m	Ga^m
Boulder-2																			
72315,3 mostly exterior	43	4.3	0.43	2.8	180	20	1.3	110	1.1	0.4	2.6	(300)	1.3	8.5	450	1.58	320	157	—
72315,4 totally interior	33	9.0	0.98	6.1	340	33	2.0	120	0.84	0.5	2.5	8.1	0.66	9.6	530	1.53	340	165	—
72335,2 mostly exterior	36	15	1.4	5.3	360	28	1.5	67	0.70	0.8	1.7	(80)	0.58	2.0	95	0.71	120	145	—
72355,7 one ext. side	33	7.3	0.73	4.9	310	34	2.2	75	0.87	0.2	2.4	5.1	0.24	8.0	280	2.00	380	157	—
72375,2 mostly exterior	30	8.5	0.84	5.3	320	34	2.2	90	0.82	0.2	2.3	7.2	0.59	6.2	250	1.85	370	149	—
72395,3 interior	30	8.0	0.79	5.8	290	33	2.1	190	1.4	0.2	2.1	(170)	0.29	5.3	160	1.72	360	152	—
Soils < 1 mm																			
72321,9	117	10	—	3.7	250	28	2.2	240	7.2	3.0	18	36	—	3.9	180	0.91	180	155	4.7
boulder-2 shadow	87	8.1	0.83	3.7	260	28	4.8	240	6.4	3.0	18	37	—	4.7	220	0.83	170	145	—
72441,11 under 0.7 m boulder	117	11	—	4.0	280	32	—	240	6.7	2.9	16	40	—	4.2	190	0.98	200	150	4.6
under 0.7 m boulder	75	9.6	—	4.0	280	31	(8.8)	220	6.0	3.5	15	—	—	4.4	200	0.84	180	150	—
72461,8 under 0.7 m boulder	165	11	—	4.0	280	32	—	240	6.5	3.0	15	40	—	4.2	190	1.10	220	145	4.5
72501,31 5 m E. of boulder-2	124	14	—	5.3	360	37	—	240	6.4	3.2	18	37	—	3.6	160	0.95	200	145	4.8
	97	8.5	—	4.0	260	27	2.3	220	6.5	4.0	20	—	2.4	3.6	160	0.83	170	150	—
75081,21 15 m from Camelot C	156	5.4	—	1.7	120	31	1.3	280	9.6	2.7	31	33	1.5	1.1	47	0.26	95	160	—
66041,20	112	15	—	7.1	460	33	4.0	340	8.5	(47)	24	77	(32)	2.7	120	0.70	140	165	5.0
BCR-1	85	0.02	—	0.41	10	36	—	85	24	86	150	135	290	45	900	≡1.75	590	330	—
BCR-1	78	0.01	0.90	0.54	10	36	680	84	24	≡86	130	140	270	≡46	920	≡1.75	580	315	—

*Abundances were determined by RNAA. Errors are ~3–10% for all elements. Values in parentheses are probably high due to contamination.

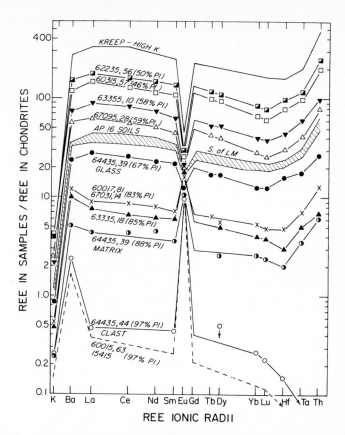

Fig. 1. Chondritic normalized K, Ba, REE, Hf, Ta and Th abundances in Apollo 16 samples. The normalized curves for soils south of the LM, gabbroic anorthosites 60017 and 67031 and the anorthosites 15415 and 60015 are taken from Laul and Schmitt (1973b). Plagioclase content was calculated from the anorthite equivalent of the Al_2O_3 content in the rock. Chondritic values (ppm) taken for normalization are K 830, Ba 3.8, La 0.34, Ce 0.91, Sm 0.195, Eu 0.073, Tb 0.047, Dy 0.30, Yb 0.22, Lu 0.034, Hf 0.20, Ta 0.020, and Th 0.040, respectively. The K point for KREEP-HIGH K should be at 10.

Shadow rock: Breccias 60017, 63335, and 63355. These three breccia rocks are chips from a 5 m boulder at Station 13. Rocks 60017 and 63335 are gabbroic anorthosites (85% plagioclase) and are similar in composition (Table 1). Breccia 60017 is a B4 dark matrix—light clast rock (Wilshire *et al.*, 1973) and is a partially molten, type IV rock and a product of partial melting of a clastic rock (LSPET, 1973a). Rock 67031 (Station 11), a fragment of breccia 67035, while gray in color and collected from the southeast rim of North Ray Crater, is chemically similar (Laul and Schmitt, 1973b) to breccia 60017 in our suite of 30 elements. This suggests that the 5 m Shadow Rock boulder, though 0.7 km distant from Station 11, was most likely ejected from the North Ray Crater.

Rock 63355 is a medium-K KREEP-type rock and is similar to Apollo 17 boulders and other noritic breccias at Station 2 on the South Massif (LSPET,

1973b; Laul and Schmitt, 1974). Similar Ni/Ir/Au ratios in rock 63355 and in the glass fraction of 64435 indicate association with similar types of ancient meteoritic projectiles, possibly group I Imbrium component (Morgan *et al.*, 1974). From the REE abundances and patterns, it is clear that these breccias are not compacted Apollo 16 soils.

The bulk and trace element data for the dark matrix breccias 60017, 63335, and 63355 indicate that their parent Shadow Rock is a heterogeneous rock with a complex history. Haskin *et al.* (1973) have found a slight positive Eu anomaly in the fines 67601 collected from the North Ray Crater and positive Eu anomalies in the dark matrix breccia 67915 and the light matrix breccia 67955 that were chipped from the Outhouse Rock boulder, ~40 m from the rim crest of North Ray Crater. Nakamura *et al.* (1973) also observed similar REE abundances and a positive Eu anomaly in 67915. Since both the bulk and REE abundances in the gabbroic anorthosites 60017 and 63335 from Shadow Rock are similar to the abundances in the gabbroic anorthosites from Outhouse Rock (LSPET, 1973a; Duncan *et al.*, 1973), the complex histories of these two large boulders, separated by ~0.7 km on the North Ray Crater apron, are probably very similar. Lower trace siderophile element abundances by a factor of ten or more were found in the Shadow Rock breccias 60017 and 63335 (Table 1) relative to breccias ejected from the South Ray Crater (Laul and Schmitt, 1973b). Similar Ni abundances in the above Outhouse Rock breccias and in breccias 67915 and 67015 (Duncan *et al.*) were found. These correlations support the hypothesis (Haskin *et al.*, 1973) that the impacting meteorite penetrated through the Cayley Plains and into Smoky Mountain and ejected Descartes Formation debris which is more anorthositic cumulate in origin.

60315 and 62235. Breccias 60315 (46% plagioclase) and 62235 (50% plagioclase) are C_2 crystalline metaclastic rocks (Wilshire *et al.*, 1973). Their chemical compositions are similar to the medium-K Fra Mauro-type basalts (Reid *et al.*, 1972) or KREEP basalts FeO/MgO = 1 (Gast, 1973). Judging from their high REE content relative to soils (Fig. 1), the breccias are not compacted soils. Both breccias have considerably high Ni, Ir, and Au contents suggesting significant ancient meteoritic contamination of LN type in them. These high siderophile contents are consistent with the petrographic study of Fe–Ni metal grains which are suggestive of extralunar origin (Taylor *et al.*, 1973).

67095. This breccia is a glassy rock (Wilshire *et al.*, 1973) with a bulk and lithophile trace element composition similar to other medium-K Apollo 16 breccias like 63355. However, considerably less ancient meteoritic component is present in our sample relative to 63355.

In general, the patterns of the LIL elements, K, Ba, REE, Hf, Ta, and Th clearly indicate a continuum of KREEP-type rocks ranging from 5× to 200× in Apollo 16 breccias (Fig. 1). The amount of KREEP in these breccias also seems to be inversely correlated with the plagioclase content. For ~75% plagioclase content, the Eu anomaly changes from positive to negative and the normalized Ba/La ratio also changes from a negative to positive slope. Several plausible explanations from the KREEP continuum have been suggested such as the variable degrees of partial melting of feldspar-rich solids (Haskin *et al.*, 1973), discrete spectrum of primary liquids by internal processes (Hubbard *et al.*, 1973) and partial fusion of feldspathic cumulates and primary aluminous "liquid" and subsequent mixing by impacts (Nava and Philpotts, 1973). In view of basin formation by large ancient meteoritic bodies and followed by multiple and local cratering events (Morgan *et al.*, 1974), we favor the former or the latter hypothesis for the KREEP continuum.

Weill *et al.* (1974) have proposed an intriguing model for the evolution of Sm and Eu abundances during lunar igneous processes. Their calculations are consistent with a model that generates KREEP abundances of Sm and Eu by either partial melting assemblages composed dominantly of olivine–ortho-pyroxene with <20% plagioclase and a few percent clinopyroxene, or assemblages with more plagioclase and more enrichment in both Sm and Eu (no negative anomaly) relative to the whole moon. Incipient partial melting of Sm and Eu abundances of 5 times chondrites in a 93% olivine plus orthopyroxene, 5% plagioclase, and 2% clinopyroxene will produce average Sm and Eu abundances in high-K KREEP material. Although it appears that plagioclase will be enriched relative to the mafic source material in the first liquid phase during partial melting, it is not clear whether the plagioclase content in the liquid phases will increase with increasing degrees of partial melting and thereby satisfy the observed inverse correlation of KREEP and plagioclase content (Fig. 1). Also it is not clear that the REE abundances and other LIL trace elements such as Ba, Hf, Ta, Th, U, etc. will decrease monotonically in the partial melt as the Al_2O_3 content or its plagioclase equivalent increases in the melt phase. For example, in Fig. 1, it is clear that the Sm content decreases from ≈250 to 20× as the plagioclase content increases from ≈52% to ≈76%. We conclude that currently no proposed model has been sufficiently tested whereby it will satisfy the KREEP continuum pattern (Fig. 1) and other trace element abundance distributions observed in a large number of highland breccia rocks.

Apollo 17 samples

Basalts 70135, 72155, and 75035. These three basalts are respectively coarse, medium, and medium low-K basalts, which approximate the bulk and trace composition of the low-K Apollo 11 basalts. The bulk compositions of 70135 and 72155 differ somewhat from the two basalts 70035 and 70215, reported by LSPET (1973b); however, the bulk compositions of 75035 of this work and 75055 by LSPET agree quite closely. The Y abundance of 113 ppm for 75035 interpolated from our REE abundances agrees well with the 112 ppm of Y by LSPET. This indicates that 75035 and 75055 originated from the same lava flow. However, the interpolated Y abundance of 164 ppm for 70135 differs considerably from the 75 ppm Y observed in 70035 and 70215 by LSPET. These comparisons suggest at least three lava flows at the Apollo 17 valley instead of two flows.

The REE and other trace element abundances (Table 1 and Fig. 2) fall within the range of the Apollo 11 low-K basalts. These similarities in trace element

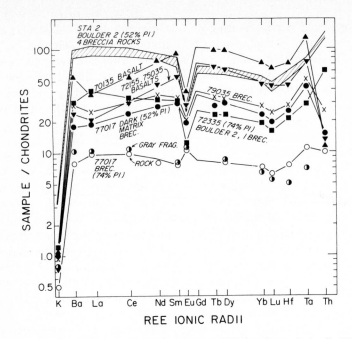

Fig. 2. Chondritic normalized K, Ba, REE, Hf, Ta, and Th abundances in Apollo 17 basalts and breccias.

content indicate similar partial melting mechanisms and source material for the formation of Apollo 11 and 17 lavas. Significant differences in the REE abundances and patterns for the Apollo 12 and 15 western lunar region and the Apollo 11 and 17 eastern lunar region could be attributed to either a partial melting of different source material or to larger degrees of partial melting in the western lunar region. Even varying combinations of the above two parameters may be responsible for the observed differences.

A recent model proposed by Taylor and Jakeš (1974) is a chemically zoned moon up to 1000 km and the source region of mare basalts is zoned by Ca–Al and Mg–Fe composition. The Apollo 11 and 17 mare basalts (high Ti) were extruded following intense meteoritic bombardment from Fe–Ti oxides zone from a depth of 150 km. This model is an attractive interpretation for many successive stages of differentiation and should be tested further.

A linear correlation between the lithophile trace element Sc and FeO (mafic indicator) was observed by Laul *et al.* (1972b). This correlation includes basalts, breccias and soils from the Apollo 12 and 15 maria, the Apollo 16 and Luna 20 highlands, the Apollo 14 Fra Mauro uplands and the Apollo 17 South Massif. However, basalts from the Apollo 11 and 17 sites deviate markedly from the linear correlation (Fig. 3). A number of Apollo 12 basalts (Laul *et al.*, 1972b; Wänke *et al.*, 1974) also deviate with higher Sc abundances. Soils from the Apollo 17 valley and the Apollo 11 and Luna 16 mare sites also deviate from the linear correlation; therefore, these soils obviously contain a large component of the high-ilmenite basalts.

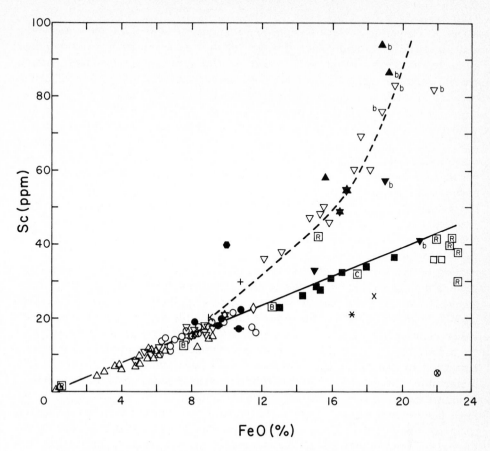

Fig. 3. Sc–FeO correlation in lunar rocks and soils, meteoritic and terrestrial matter. Symbols are identical to those found in Fig. 5. This correlation plot first shown by Laul *et al.* (1972b) has been brought up to date. High Sc points, deviating from the correlation line, are associated with high-ilmenite Apollo 11 and 17 basalts and soils and Luna 16 soils. The Sc basalt points (b) for both the Apollo 11 and 12 sites represent the average values for the Sc–FeO pairs for the two groups of basalts at both these two sites

The appreciable Sc deviation for the Luna 16 soil suggests that the average Sc content in Luna 16 basalts may be either closer to ~80 ppm rather than the 26 and 54 ppm observed in two tiny Luna 16 basaltic fragments (Helmke and Haskin, 1972), or the basaltic component in Luna 16 soil is much larger than that present in the Apollo 11 and 17 soils.

Goles *et al.* (1974) have recently measured the partition coefficient for Sc in clinopyroxene (augite), ilmenite, plagioclase, and magnetite minerals crystallizing from an olivine basaltic liquid. Scandium was found to be concentrated in augite relative to the other minerals. Goles *et al.* found that the following equation is valid for Sc distribution between augite and liquid:

$$\log D = \frac{5735}{T} - 3.703$$

where T is given in degrees Kelvin. For example, at 1100°C, $D \approx 3.0$. Assuming that (a) augite is an appreciable fraction of the source material a few hundred kilometers in depth, (b) the Sc and bulk element contents in the source material are the same for generation of all mare basalts, and (c) that bulk equilibrium between the melt and refractory assemblage is maintained during partial melting, we suggest two alternative interpretations will satisfy the Sc data. The first explanation suggests that the partial melting temperature for generation of the Apollo 11 and 17 basalts was higher by ≈ 100°C relative to the partial melting temperature for generation of Apollo 12 and 15 mare basalts. However, if the partition coefficient of Sc in olivine is appreciable relative to the Sc coefficient in augite, our conclusions are unchanged if the partition coefficient pattern for olivine parallels that observed in augite as a function of temperature. This prediction is contrary to the temperatures suggested by Ringwood and Green (1974) who predict partial melting temperatures of 1250°C for Apollo 11 and 17 to 1450°C for Apollo 15 basalt magmas. If mare basalts were generated by partial melting of pyroxene-rich cumulates (Schnetzler and Philpotts, 1971) and if the same temperature prevailed during the partial melting of the cumulates for generation of all mare basalts in the eastern (Apollo 12 and 15) and western (Apollo 11 and 17) halves of the moon's face, then the Sc data are consistent with lower temperatures for fractional crystallization and generation of the cumulates in the eastern half of the moon relative to corresponding temperatures for cumulate formation by fractional crystallization in the lunar western half. About 100° temperature differences would satisfy the Sc data. According to the Schnetzler–Philpotts model, if compositionally similar cumulates in the eastern and western lunar halves were formed at the same temperatures from identical magma, then the partial melting temperatures of the cumulates would be higher for the Apollo 11 and 17 basalts relative to the Apollo 12 and 15 basalts. At present, the temperature disagreement with the Ringwood–Green model favors the alternative model by Schnetzler–Philpotts.

Breccias 72315, 72335, 72355, 72375, and 72395 (Wasserburg consortium). The bulk and trace element abundances (Table 1 and Fig. 2) for five samples from four metaclastic rocks are remarkably similar with 2–10% dispersions observed for both bulk and trace elements. No compositional differences were found between "interior" or "exterior" (exposed to the lunar environment) specimens. The average rock composition for these four rocks (72335 excluded) corresponds to high-alumina rocks (52% plagioclase) and falls in the range of Apollo 14 clastic rocks and matches closely the Apollo 16 medium-K KREEP-type rocks (Fig. 2) with the exception of lower LIL trace element contents in the boulder relative to Apollo 16 medium-K KREEP rocks. Also, the average bulk composition of boulder-2 falls within the compositional range of five high-alumina noritic breccias returned from the South and North Massifs at Stations 2, 6, and 7 (LSPET, 1973b). The interpolated Y abundance of 102 ppm for boulder-2 agrees with the mean Y value of 106 ppm in the five noritic breccias analyzed by LSPET. Also average Rb, Sr, and Zr abundances of 7, 155, and 440 ppm, respectively, in these four rocks of boulder-2 (Tables 1 and 2) agree with average values of 6, 158, and 480 ppm, respectively, in the five noritic breccias analyzed by LSPET. Such close compositional agreement in a variety of breccias confirms LSPET's observation of a rather uniform composition for both the South and North Massifs and these may represent a single stratigraphic unit in the Serenitatis Basin event. Our high average Ni abundance of 310 ppm in these four rocks does not agree with the average abundance of 119 ppm in the five noritic breccias by LSPET; also our Ni value of 290 ppm does not agree with the 95 ppm of Ni reported by LSPET for the same rock 77017. We are confident of our Ni values because of a variety of cross-checks with other data, e.g. agreement with Ni abundances obtained by Anders and coworkers in a variety of lunar samples.

The chondritic REE patterns and other lithophile trace elements in these four boulder rocks are parallel to those observed in medium-K KREEP rocks. Assuming the validity of the arguments (Head, 1974) that the boulders are representatives of uplifted massifs during the Serenitatis basin forming event, our data indicate that medium-K KREEP may extend ~700 km ESE beyond the third Imbrian ring if the massifs originated from considerable depth. If the boulder originated from the top 200 m of the South Massif, then the boulders could be metamorphosed rocks of post-Serenitatis ejecta blanket material, as a result of ancient meteoritic bombardment.

Rock 72335 is a medium-K anorthositic gabbro (troctolite?) with 74% plagioclase. The REE and other lithophile element pattern (Fig. 2) is parallel to the KREEP pattern for medium-K KREEP-type rocks. The absolute REE and trace LIL abundances are about 0.4 of the corresponding abundances found in the other four rocks from boulder-2. The correlation observed between K, REE, Ba, Hf, Ta, and Th and the plagioclase content of Apollo 16 breccias (Fig. 1) also includes the two types of breccias (high-alumina and anorthositic gabbro) observed in the Apollo 17 boulder-2 rocks, i.e. the patterns observed in Fig. 2 for boulder-2 rocks fit in the expected locale of Fig. 1 for their respective plagioclase content. Such a correlation also includes the Luna 20 breccia rocks (Laul and Schmitt, 1973a). This observation underscores the ubiquitous nature of KREEP material in highland breccias and the relatively uniform spectrum of rocks in the lunar highlands facing the earth. It is not clear yet whether this uniformity may be attributed to either a restricted region of KREEP rocks (e.g. at the Imbrium Basin) and its subsequent distribution by the Imbrian basin event and subsequent homogenization by intermediate and localized smaller cratering events or whether KREEP source material was formed as a global layer or as a variety of pockets during differentiation of the outer lunar shell with subsequent partial melting for generation of KREEP-type rocks.

The breccia rocks of boulder-2 have a high content of siderophiles (2–4% Cl) and show fractionated patterns having an ancient meteoritic component. These patterns are similar to those first found by Morgan et al. (1973) in soil separates of highland soils, 14142 and 14146. We have analyzed the same suite of elements except Ge in these samples. Following the Chicago approach, we find that five samples of the four rocks have Ir/Au ratios of 0.44 which is typical for the LN group (Anders *et al.*, 1973; Ganapathy *et al.*, 1973). However, the 72335,2 sample has a high Ir/Au ratio of 0.84 and belongs to the DN group. Recently, Morgan *et al.* (1974) proposed six discrete ancient meteoritic groups for lunar basins based on ternary Au–Ir–Ge and Re–Ir–Sb diagrams. Based on their Re–Ir–Sb diagram, the four rocks of boulder-2, Station 2 (Ir/Au = 0.44) fall in group 2 and the sample 72335 with its high Ir/Au = 0.84 falls in group 3. It is noteworthy that rocks 72315 (medium-K KREEP rock) and 72335 (anorthositic gabbro) with two different ancient projectile components in them, are rocks separated by ~18 cm in the same clast of boulder-2. The difference in the Ir/Au ratios is attributed to a case of two-stage impact projectiles. Laul and Schmitt (1974) assigned the Ir/Au = 0.44 to the Serenitatis basin planetisemal which is concordant with group 2 assignment of Morgan *et al.* (1974). On the other hand, if sample 72335 is a distinct clast, it should be older than the matrix. Thus, younger Crisium Basin assignment of group 3 to the clast is contrary to the petrographic observations of this boulder. To circumvent this difficulty, Morgan *et al.* suggest remobilization of the matrix and incorporating clasts of older age during the Imbrium event. During such an event, the age of the clast may be reset. Alternatively, we suggest that the clast carries its own extralunar signature caused by a large local cratering event (not a basin-forming event); this clast was incorporated in the matrix by the subsequent Serenitatis Basin event.

Cesium varies over a fivefold range in content in the boulder-2 rocks and shows no correlation to U. Such variable ratios of Cs/U suggest selective volatilization of the heavier alkalies during cratering, brecciation, and metamorphic processes (Gibson and Hubbard, 1973). Variable Rb/Cs ratios of 19, 25, 29, and 33 found in the high-alumina rocks of boulder-2 indicate significant fractionation for the heavy alkalies. For these four rocks, the Na_2O and K_2O abundances were fairly constant. Such heavy alkali fractionation suggests variable time–high-temperature exposures for these four rocks. Further implications of the data shown in Tables 1 and 2 for boulder-2 rocks and Apollo 17 soils at Station 2 are discussed by Laul and Schmitt (1974).

77017. This breccia is a low-K anorthositic gabbro (norite?) with 71% plagioclase. The bulk composition (Table 1) of the whole rock is very similar to that of the breccia 72335 in boulder-2, Station 2; however, the abundances of the REE and other LIL trace elements are about four times lower relative to those found in 72335. A slight positive Eu anomaly was observed in the whole rock 77017.

A dark matrix and a gray fragment of rock 77017 were separated and analyzed. Based on the TiO_2 content, the bulk composition of the dark matrix is rich in ilmenite and is classed as a low-K, high-alumina, high-ilmenite rock. The REE, including a negative Eu anomaly, and other LIL trace elemental abundances and patterns (Table 1 and Fig. 2) in the dark matrix are similar to the patterns observed in the ilmenite-rich soils of Apollo 17, Stations LM, 1, 5, and 9 (Fig. 4). Much lower Th contents in these two respective patterns distinguish them from the KREEP-type pattern with enriched Th relative to the heavy REE, Hf, and Ta.

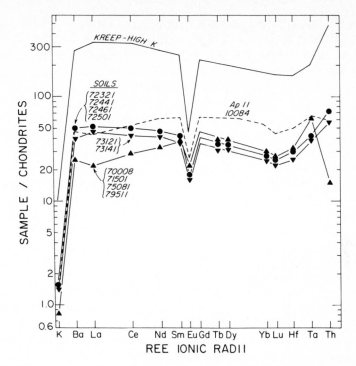

Fig. 4. Chondritic normalized K, Ba, REE, Hf, Ta, and Th abundances in Apollo 17 soils.
KREEP curve is average REE and other elements in high-K norite–KREEP (Laul and
Schmitt, 1973b).

Also, the overall composition of the dark matrix is similar to the high-ilmenite
breccia 79035, which was apparently ejected from the Van Serg Crater. Assuming
that the North Massif is similar to South Massif breccia rocks (see discussion
above), it seems that the distinctive chemical nature of the dark matrix is not
compatible with the suggestion (LSPET, 1973b) that breccia 77017 may be a
simple mass wasting product of the North Massif.

The overall composition of the gray fragment of 77017 is similar to the whole
rock. High-siderophile trace element abundances and similar ratios of Ni, Ir, and
Au in the whole rock, dark matrix and gray fragment of breccia 77017 compared to
rocks from boulder-2, Station 2, suggest similar ancient meteoritic components.

79035. The overall composition of this low-K high-ilmenite breccia is essentially identical with soils 79511 (Table 1) and soils 78501, 79221, and 79261 (Laul
and Schmitt, 1973b). If this breccia was ejected from Van Serg Crater, then the
soil on the rim of the crater very likely consists largely of comminuted breccias
like 79035. It is also possible that this breccia is an impact-indurated equivalent of
dark floor material such as 79511 (Schmitt, 1973).

Soils. Apollo 17 soils may be divided (LSPET, 1973b) into three groups; the
dark-mantle or valley soils at stations LM, 1, and 5 such as the 70008 drill steam

soil (unit 08-V; parent 140) of 5.7–6.2 cm depth, 71501 and 75081; intermediate
soils such as 79511 at Station 9; light-mantle soils such as 72321, 72441, 72461,
72501, 73121, and 73141 at Stations 2 and 3, derived from the South Massif
avalanche. Soils at Station 6 (North Massif) appear to be intermediate between the
light-mantle soils and the soils at Stations 8 and 9. The compositional data for four
common soils 71501, 75081, 72501 and 73141 by us and LSPET (1973b) agree well.
The interpolated Y abundances for these four soils also agree within 10% with the
Y abundances by LSPET. Our bulk data for some fourteen different Apollo 17
soils of this work and our previous studies (Laul and Schmitt, 1973b) support the
conclusions reached by LSPET.

Similar REE and other LIL trace element patterns, all with negative Eu
anomalies, but lower absolute REE abundances were found for the Apollo 17
dark-mantle and intermediate soils (Stations LM, 1, 5, and 9; see Fig. 4) relative to
Apollo 11 soil. Also the Th depletion was less severe in the Apollo 11 soil. These
general REE patterns reflect appreciable components of comminuted basalts in
these soils, especially at the LM, 1, and 5 sites. Significant KREEP contents may
be inferred in the light-mantle soils at Stations 2 and 3 from the KREEP-like
pattern (Fig. 4). Ten percent lower REE and LIL trace element abundances were
found in two Station 3 soils relative to Station 2 or South Massif soils. An
increased proportion of the ilmenite basaltic component in Station 3 soil is not
responsible for this distinct difference. No explanation is offered.

Four soils near boulder-2 on the South Massif have essentially identical
compositions (Table 1). Soil 72321 was in the boulder-2 shadow; soil 72441 was the
upper 4 cm of the soil from under the 0.7 m boulder-3, 30 m ENE of boulder-2; soil
72461 was the skim soil from under the same boulder-3; and soil 72501 was 5 m
east of boulder-2. Chou *et al.* (1973) characterized Cd, In, and Zn among other
volatiles as atmophile. Accordingly, it is expected that these elements should
show considerable enrichment in the boulder shadow soil relative to the exposed
soil. Our systematic study of shadowed and exposed soils (Table 2) rejects the
labile hypothesis. If these soils are relatively young; i.e. if shadowing is recent, a
test of the volatile transport hypothesis by our data is invalid. On the other hand,
the data of Chou *et al.* (1974) for shadowed soil 76240 and its control 76260, and
the various sieve fractions of soil 65500 show a considerable enrichment of Cd,
thereby supporting its labile character. Chou *et al.* further speculated that the
labile elements should also show evidence of movement from highlands to the
lower mare areas. Such volatile element movement should be observed at the
Apollo 17 interface between Station 2, as representative of highland avalanche
material, and the mare soils in the valley. A comparison in Table 2 between the
South Massif soils and the valley soil 75081 does not support such labile
processes.

The South Massif soils near the boulders have in general high Ir/Re/Ni/Ag/Au
ratios relative to the rocks. This is expected if the soils contained a mixture of
ancient meteoritic component and micrometeorites of Cl composition (Ganapathy
et al., 1973). Variable amounts of Ni and Au and a constant ratio of Ni/Au were
found in duplicate analyses of 72501 soil (Table 2); this seems to confirm the
extraneous addition of Cl material.

Interelement correlations

In Fig. 5, we have updated the FeO–MnO correlation by the inclusion of 20 Apollo 17 data points that span a factor of four in the FeO and MnO contents. Such a strong correlation (FeO/MnO = 80–85) found valid for rocks returned from all eight lunar sites implies no significant fractionation of this element pair during any lunar differentiation process in the generation of mare basalts or highland crustal rocks, and in turn strongly favors a homogeneous lunar accretion model for these elements.

Ehmann and Chyi (1974) have reported Zr/Hf ratios of 34 to 39 in mare basalts and a mean Zr/Hf ratio of 47 in KREEP rocks, which also have significantly higher Zr and Hf absolute abundances relative to mare basalts. Arguments have been advanced by them that the observed Zr/Hf fractionation may be attributed to the presence of Zr^{+3}, while the Hf^{+4} oxidation state is maintained in lunar magmas. In view of REE fractionation trends, we suggest that both Zr and Hf are present in the tetravalent state and that the trivalent Zr state need not be postulated. Significant REE fractionation (e.g. Nd–Sm, Tb–Dy, Yb–Lu pairs) has been

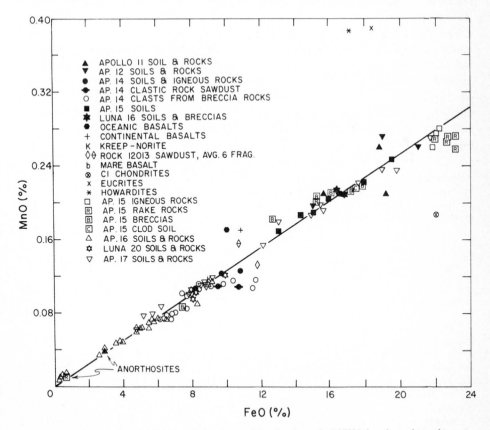

Fig. 5. MnO–FeO correlation first observed by Laul *et al.* (1972b) has been brought up-to-date and includes data from all eight lunar sample return missions. Terrestrial and meteoritic values are included for comparison.

observed in the REE patterns of mare basalts relative to KREEP-type rocks. In a series of isotope dilution analyses from 1970 to 1973 by a NASA team (e.g. Gast *et al.*, 1970; Hubbard *et al.*, 1973), the ratios of Nd/Sm abundances in Apollo 11, 12, and 15 basalts ranged from 2.40 to 3.14 with an average ratio of 2.86; for Apollo 12, 14, 15, and 16 KREEP-type rocks, the ratios of Nd/Sm abundances ranged from 3.32 to 3.76 with an average ratio of 3.54. The data were taken from the same laboratory in order to avoid any bias in absolute REE determinations. As observed for Zr and Hf, the REE abundances, and in particular Nd and Sm abundances, were higher in KREEP-type rocks relative to mare basalts; in addition, the degree of enrichment of Zr, Hf, Nd, and Sm in the former rocks relative to the latter rocks were very similar (see Fig. 1 for an appreciation of the uniformity of the Sm/Hf ratio as a function of plagioclase content in Apollo 16 KREEP rocks). The Nd^{+3}–Sm^{+3} pair, with hexacoordinated ionic radii of 1.06 and 1.04 Å, respectively, and octacoordinated ionic radii of 1.20 and 1.17 Å, respectively, is expected to be maintained in the trivalent state under lunar redox conditions. Hexacoordinated ionic radii of Zr^{+4} and Hf^{+4} are 0.80 and 0.79 Å, respectively, and octacoordinated radii are 0.92 and 0.91 Å, respectively. Since (a) the average ratio of the Nd–Sm pair has been increased by $\approx 24\%$ in average KREEP-type rocks relative to average mare basalts, (b) these two REE are maintained in the trivalent state, and (c) the general chemical properties of Nd^{+3}–Sm^{+3} are very similar, the comparable $\approx 30\%$ increase in the Zr/Hf ratio in average KREEP-type rocks relative to average mare basalt does not require the postulate that Zr^{+3} was present during lunar igneous processes. Significant changes in Zr/Hf ratios in terrestrial rocks (Taylor, 1965) observed in some gabbros (average ratio of 70) and granites (average ratio of 40) also do not support the Zr^{+3} postulate.

Acknowledgments—This work was supported by NASA grants NGL 38-002-039 and 38-002-020. We acknowledge the assistance of T. V. Anderson and W. T. Carpenter at the Oregon State University TRIGA reactor for neutron activations. We thank Dr. W. V. Boynton for helpful discussions. We also thank Jeanne Walsh for typing this manuscript.

REFERENCES

Anders E., Ganapathy R., Krähenbühl U., and Morgan J. W. (1973) Meteoritic material on the moon. *The Moon* **8**, 3–24.
Chou C. L., Baedecker P. A., and Wasson J. T. (1973) Atmophilic elements in lunar soils. *Proc. Fourth Lunar Sci. Conf., Geochim. Cosmochim. Acta*, Suppl. 4, Vol. 2, pp. 1523–1533. Pergamon.
Chou C. L., Baedecker P. A., Bild R. W., Robinson K. L., and Wasson J. T. (1974) Volatile elements in lunar soils (abstract). In *Lunar Science—V*, pp. 115–117. The Lunar Science Institute, Houston.
Duncan A. R., Erlank A. J., Willis J. P., and Ahrens L. H. (1973) Composition and inter-relationships of some Apollo 16 samples. *Proc. Fourth Lunar Sci. Conf., Geochim. Cosmochim. Acta*, Suppl. 4, Vol. 2, pp. 1097–1113. Pergamon.
Ehmann W. D. and Chyi L. L. (1974) Abundances of the group IV B elements Ti, Zr, and Hf and implications of their ratios in lunar materials. *Proc. Fifth Lunar Sci. Conf., Geochim. Cosmochim. Acta.* This volume.
Ganapathy R., Morgan J. W., Krähenbühl U., and Anders E. (1973) Ancient meteoritic components in

lunar highland rocks: Clues from trace elements in Apollo 15 and 16 samples. *Proc. Fourth Lunar Sci. Conf., Geochim. Cosmochim. Acta*, Suppl. 4, Vol. 2, pp. 1239–1261. Pergamon.

Gast P. W. (1973) Lunar magmatism in time and space (abstract). In *Lunar Science—IV*, pp. 275–277. The Lunar Science Institute, Houston.

Gast P. W., Hubbard N. J., and Wiesmann H. (1970) Chemical composition and petrogenesis of basalts from Tranquillity Base. *Proc. Apollo 11 Lunar Sci. Conf., Geochim. Cosmochim. Acta*, Suppl. 1, Vol. 2, pp. 1143–1163. Pergamon.

Gibson E. K. and Hubbard N. J. (1973) How to lose Rb, K and change the K/Rb ratio: An experimental study. *Proc. Fourth Lunar Sci. Conf., Geochim. Cosmochim. Acta*, Suppl. 4, Vol. 2, pp. 1263–1273. Pergamon.

Goles G. G., Hering C., and Leeman W. P. (1974) Private communication.

Haskin L. A., Helmke P. A., Blanchard D. P., Jacobs J. W., and Telander K. (1973) Major and trace element abundances in samples from the lunar highlands. *Proc. Fourth Lunar Sci. Conf., Geochim. Cosmochim. Acta*, Suppl. 4, Vol. 2, pp. 1275–1296. Pergamon.

Head J. W. (1974) Morphology and structure of the Taurus-Littrow highlands (Apollo 17): Evidence for their origin and evolution. *The Moon*. In press.

Helmke P. A. and Haskin L. A. (1972) Rare earths and other trace elements in Luna 16 soil. *Earth Planet. Sci. Lett.* **13**, 441–443.

Hubbard N. J., Rhodes J. M., Gast P. W., Bansal B. M., Shih C. Y., Wiesmann H., and Nyquist L. E. (1973) Lunar rock types: the role of plagioclase in non-mare and highland rock types. *Proc. Fourth Lunar Sci. Conf., Geochim. Cosmochim. Acta*, Suppl. 4, Vol. 2, pp. 1297–1312. Pergamon.

Laul J. C. and Schmitt R. A. (1973a) Chemical composition of Luna 20 rocks and soil and Apollo 16 soils. *Geochim. Cosmochim. Acta* **37**, 927–942.

Laul J. C. and Schmitt R. A. (1973b) Chemical composition of Apollo 15, 16, and 17 samples. *Proc. Fourth Lunar Sci. Conf., Geochim. Cosmochim. Acta*, Suppl. 4, Vol. 2, pp. 1349–1367. Pergamon.

Laul J. C. and Schmitt R. A. (1974) Chemical composition of boulder-2 rocks and soils, Apollo 17, station 2. *Earth and Planet Sci. Lett.* In press.

Laul J. C., Wakita H., and Schmitt R. A. (1972a) Bulk and REE abundances in anorthosites and noritic fragments. In *The Apollo 15 Samples*, pp. 221–224. The Lunar Science Institute, Houston.

Laul J. C., Wakita H., Showalter D. L., Boynton W. V., and Schmitt R. A. (1972b) Bulk, rare earth and other trace elements in Apollo 14 and 15 and Luna 16 samples. *Proc. Third Lunar Sci. Conf., Geochim. Cosmochim. Acta*, Suppl. 3, Vol. 2, pp. 1181–1200. MIT Press.

LSPET (Lunar Sample Preliminary Examination Team) (1973a) Preliminary examination of lunar samples from Apollo 16. *Science* **179**, 23–34.

LSPET (Lunar Sample Preliminary Examination Team) (1973b) Apollo 17 lunar samples: chemical and petrographic description. *Science* **182**, 659–672.

Mason B. (1973) Private communication.

Morgan J. W., Ganapathy R., Laul J. C., and Anders E. (1973) Lunar crater Copernicus: search for debris of impacting body at Apollo 12 site. *Geochim. Cosmochim. Acta* **37**, 141–154.

Morgan J. W., Ganapathy R., Higuchi H., Krähenbühl U., and Anders E. (1974) Lunar basins tentative characterization of projectiles, from meteoritic elements in Apollo 17 boulders (abstract). In *Lunar Science—V*, pp. 526–528. The Lunar Science Institute, Houston.

Nakamura N., Masuda A., Tanaka T., and Kurasawa H., (1973) Chemical composition and rare earth features of four Apollo 16 samples. *Proc. Fourth Lunar Sci. Conf., Geochim. Cosmochim. Acta*, Suppl. 4, Vol. 2, pp. 1407–1414. Pergamon.

Nava D. F. and Philpotts J. A. (1973) A lunar differentiation model in light of new chemical data on Luna 20 and Apollo 16 soils. *Geochim. Cosmochim. Acta* **37**, 963–973.

Philpotts J. A., Schuhmann S., Kouns C. W., Lum R. K. L., Bickel A. L., and Schnetzler C. C. (1973) Apollo 16 returned lunar samples: Lithophile trace element abundances. *Proc. Fourth Lunar Sci. Conf., Geochim. Cosmochim. Acta*, Suppl. 4, Vol. 2, pp. 1427–1436. Pergamon.

Reid A. M., Warner J., Ridley W. I., and Brown R. W. (1972) Major element composition of glasses in three Apollo 15 soils. *Meteoritics* **7**, 395–415.

Ringwood A. E. and Green D. H. (1974) Mare basalts and composition of lunar interior (abstract). In *Lunar Science—V*, pp. 636–638. The Lunar Science Institute, Houston.

Schmitt H. H. (1973) Apollo 17 report on the valley of Taurus-Littrow. *Science* **182**, 681–690.

Schnetzler C. C. and Philpotts J. A. (1971) Alkali, alkaline earth, and rare-earth element concentrations in some Apollo 12 soils, rocks and separated phases. *Proc. Second Lunar Sci. Conf., Geochim. Cosmochim. Acta*, Suppl. 2, Vol. 2, pp. 1101–1122. MIT Press.

Taylor L. A., McCallister R. H., and Sardi O. (1973) Cooling histories of lunar rocks based on opaque mineral geothermometers. *Proc. Fourth Lunar Sci. Conf., Geochim. Cosmochim. Acta*, Suppl. 4, Vol. 2, pp. 819–828. Pergamon.

Taylor S. R. (1965) The application of trace element data to problems in petrology. In *Physics and Chemistry of the Earth* (editors L. H. Ahrens, F. Press, S. K. Runcorn, and H. C. Urey), Vol. 6, pp. 133–214. Pergamon.

Taylor S. R. and Jakeš P. (1974) Geochemical zoning in the Moon (abstract). In *Lunar Science—V*, pp. 686–688. The Lunar Science Institute, Houston.

Wänke H., Palme H., Baddenhausen H., Dreibus G., Jagoutz E., Kruse H., Spettel B., and Teschke F. (1974) Composition of the moon and major lunar differentiation processes (abstract). In *Lunar Science—V*, pp. 820–822. The Lunar Science Institute, Houston.

Weill D. F., McKay G. A., Kridelbaugh S. J., and Grutzeck M. (1974) Modelling the evolution of Sm and Eu abundances during lunar igneous differentiation. *Proc. Fifth Lunar Sci. Conf., Geochim. Cosmochim. Acta*. This volume.

Wilshire H. G., Stuart-Alexander D. E., and Jackson E. D. (1973) Petrology and classification of the Apollo 16 samples (abstract). In *Lunar Science—IV*, pp. 784–786. The Lunar Science Institute, Houston.

Proceedings of the Fifth Lunar Conference
(Supplement 5, Geochimica et Cosmochimica Acta)
Vol. 2 pp. 1067–1078 (1974)
Printed in the United States of America

Element concentrations from lunar orbital gamma-ray measurements

A. E. Metzger

Jet Propulsion Laboratory, Pasadena, California

J. I. Trombka

Goddard Spaceflight Center, Greenbelt, Maryland

R. C. Reedy

Los Alamos Scientific Lab., Los Alamos, New Mexico

J. R. Arnold

University of California, San Diego, La Jolla, California

Abstract—We report the concentrations of Th, K, Fe, Mg, and Ti in 28 geographic regions overflown by the Apollo 15 and 16 spacecraft, as determined by gamma-ray spectrometry. The observed chemical compositions are consistent with ground truth, and the two missions give reasonable agreement in a region observed by both. The chemical compositions observed require a more complex model for interpretation than the previously reported radioactivity maps. Both highlands and maria show significant variations in composition. The van de Graaff region is unique; it may possibly be the source region for "granitic" materials such as rock 12013.

Introduction

The Apollo 15 and 16 service modules carried a gamma-ray spectrometer experiment in lunar orbit. Its prime purpose was the determination of the abundance of the radioactive elements K, Th, and U, and of several major elements, in the surface layers of the moon. The radioactive elements are determined using the gamma rays emitted as part of the decay process. The cosmic-ray bombardment of the lunar surface excites line spectra characteristic of the nuclei of each element bombarded (mainly by neutron capture and neutron inelastic scatter processes) (Reedy, Arnold, and Trombka, 1973). With the system employed we expected to be able to map the concentrations of a half dozen elements in the area of the moon (of order 20% of the total) overflown in these two missions.

We have earlier presented maps of the total radioactivity of the lunar surface (Metzger *et al.*, 1973; and frontispiece, 1973), and selected values for the content of radioactive elements (Trombka *et al.*, 1973). In this paper we present data for the abundances of Th, K, Fe, Mg, and Ti in a set of selected regions. We include a very preliminary discussion of the implications of these data for the composition and history of the lunar surface layers.

1067

EXPERIMENTAL

The experimental apparatus consisted of a 7 cm NaI(Tl) crystal with a plastic anticoincidence mantle to suppress the response to charged particles, and of the electronics required to power the instrument, monitor functions, and process the output pulses into digital form for transmission to earth stations. The details are described elsewhere (Harrington *et al.*, 1974; see also Arnold *et al.*, 1972).

The model on which the data analysis is based is that of Reedy, Arnold, and Trombka (1973) (designated RAT in the following). The experimental data as received from Johnson Space Center were processed for analysis by eliminating parity errors and other problematic material, and shifting the spectra to a constant level of gain (the overall instrument gain decreased over 40% on Apollo 15 and about 10% on Apollo 16 during the lunar phase of the mission). The rates were corrected for solid angle to a constant height of 110 km above the lunar surface. Then all data obtained over each region to be analyzed were accumulated into a single master spectrum from which the background and spacecraft contribution (derived from data obtained in trans-earth coast) was subtracted. Using present methods a minimum of 30–40 min of counting data are required for the quantitative elemental analyses presented; for smaller data blocks statistical errors produce unacceptable scatter.

The analyses were carried out at the Goddard Spaceflight Center using interactive matrix inversion procedures described in RAT. The most difficult part of the work is the derivation of the lunar continuum; that is, the portion of the flux in the 0.5–10 MeV region which does not contain characteristic lines. Figure 1 shows a sample spectrum, the derived continuum, and the net line spectrum after subtraction. The continuum produces about 85% of the counts in the detector. While its general shape is understood, it must be derived from the data. Below 3 MeV the lines due to the radioactive elements make an important contribution to this continuum by Compton scattering in the

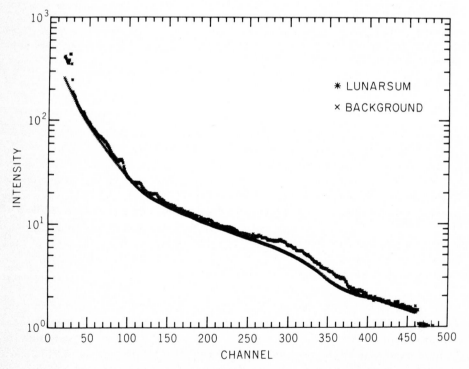

Fig. 1. (a) The processed sum spectrum for the whole ground track on Apollo 15. The scale is 19 KeV/channel. The smooth line below is the derived continuum (see text).

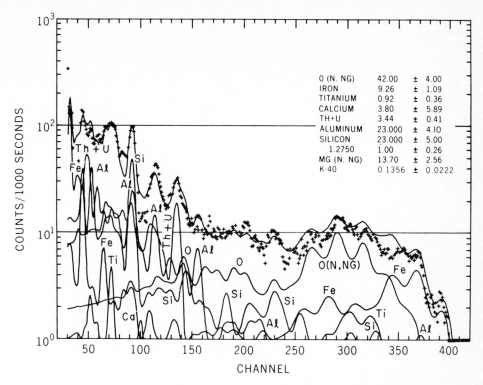

Fig. 1. (b) The net spectrum of (a) after continuum subtraction. Also shown are the component contributions as given by the RAT analysis.

lunar material. Hence we must break the continuum in practice into two parts. One is a count-rate-independent component (due to extranuclear processes and degraded gammas from nuclear interactions). The other varies with the count rate and, like the radioactive lines which are the main source of count rate variation, exists only at lower energies. These two continuum components were derived using the Apollo 15 data. The same derived continuum shapes were used directly for the Apollo 16 data; good agreement was achieved (see below) except in the region of the Ti lines. The radiation exposure was higher for Apollo 16 (more intense dose in the earth's radiation belts and a slightly higher cosmic ray flux), and the instrument resolution significantly better; we plan to repeat the full analysis with a continuum derived from Apollo 16 data.

   Some changes have been made in the procedure described in RAT. It has been found useful to break the analysis into two stages. In the first, following the removal of the continuum, only the spectrum from 5–9 MeV is analyzed; in this region only Fe (neutron capture), O, Si, and Ti contribute significant intensity. The component intensities derived by matrix inversion are then subtracted from the spectrum and the difference spectrum in the lower energy region is then analyzed using all the remaining components. Iron also produces a line spectrum in the 1 MeV region due to inelastic scatter of neutrons, the (n,n') process. Because of the good statistics for this element, the two iron components were analyzed separately, as a check for the presence of strong "thermal" neutron absorbers (such as Gd and other rare earths), which change the neutron spectrum (Lingenfelter et al., 1972), decrease the capture rate in other elements and thus the apparent abundances from (n,γ) data. These effects have been seen. For the other elements, such as Si, which have important lines of both types, we have combined the library spectra into a single component, to decrease the number of variables and increase accuracy. At this stage in our analysis we are also combining thorium and uranium into a single component (Th + U) using an abundance ratio of 3.8. Thorium values are obtained from this total component, using the same ratio. By far the strongest line of the combined component is the 2.61 MeV line due to Th.

Oxygen and silicon are well determined in the analysis, but their variation over broad areas of the moon is small, less than our error of about 4%. Hence in the analyses we report their values were fixed to minimize their interference with other components.

The elements which can be usefully analyzed fall into two groups. For the natural radioactivities, K and (Th + U), the elemental concentrations can be derived from the decay schemes and the properties of the detector, using theory and laboratory calibrations, with no ambiguity in principle. Our concentration values for these elements are derived in this way, without adjustable parameters. The elements which depend on cosmic-ray excitation, O, Si, Fe, Mg, and Ti, require input information from nuclear and cosmic-ray physics whose combined accuracy is not yet very high. The agreement to be expected is no better than that for the parent models—that of Reedy and Arnold (1972) for the production of cosmogenic nuclides in the moon, and that of Lingenfelter *et al.* (1972) for low energy neutrons. The depth variations of activity observed in the lunar soil are in rather good accord (Finkel *et al.*, 1974) with Reedy and Arnold's calculations, but the absolute amounts depart by ±25% or so from those calculated. The experiment of Woolum and Burnett (1974) on Apollo 17 confirms Lingenfelter *et al.* (1972), but suggests a somewhat harder neutron spectrum. Our observed deviations from the RAT calculations are also in this direction.

The best way to normalize the data for these elements is to use "ground truth," conforming the orbital data to the surface samples at one or more points. While the area mapped includes several sites from which samples have been returned, most of these are unsuitable because of local variability of the soil chemistry. The Apollo 15 site is a good example. The two best sites in our judgment are Mare Tranquillitatis, where the Apollo 11 analyses are supplemented by the results of the Surveyor V experiment (Turkevich *et al.*, 1969) at another point, and the Apollo 16 region near Descartes. We have chosen to normalize our Fe, Mg, and Ti data to the data for Mare Tranquillitatis, and to use the Descartes region as a check. Another available check is the comparison of regions overflown by both missions.

Table 1 gives the ground truth element concentrations used for the region of Mare Tranquillitatis. They are derived from the soil sample 10084, except for Ti which is taken to be somewhat lower. The count rates for the spectral region in which Ti plays an important role (5.97–6.37 MeV) show a variation over the Tranquillity region which is far outside the counting error. The Apollo and Surveyor sites are in the two regions of highest counting rate. Since O and Si are constant, and Fe appears to be in this region, we conclude that the true mean Ti concentration over the region is lower by a factor of

Table 1. Ground truth values, Mare Tranquillitatis.[a]

O	40.8%
Si	19.6%
Fe	12.1%
Mg	4.8%
Ti	2.9%[b]
Th	2.1 ppm
K	1100 ppm

[a]All data in this and later tables are in percent or ppm element by weight. The values of Th and K are given for convenience; they are not used for renormalization. The data are for soil 10084 (Wakita *et al.*, 1970; Schonfeld and Meyer, 1972).
[b]This is 0.65 times the soil 10084 concentration. See text for discussion.

0.65 than that observed on the ground; this correction of course increases the uncertainty of the Ti values.

In the following tables we do not quote errors. In complex interactive analyses of the present type the magnitude of random and systematic errors are difficult to evaluate. The main sources of error are: (1) statistical counting error, (2) the subtracted continuum, (3) interactions between components, and (4) where applicable, the ground truth normalization. At the present stage of our analysis, the ground truth comparisons and between-mission comparison give a fair measure of the uncertainties.

RESULTS

For this analysis we have divided the ground track of each mission into a series of regions, summing the data within each region as described above. Table 2 gives the number, assigned name, and lunar coordinates for each region. Figure 2 shows the actual regions in outline. In the case of Apollo 15 the data for early revolutions have not been used in this synthesis. Hence the western portion (southwestern or

Table 2. Lunar regions analyzed.

Number	Assigned name	Coordinates
	Apollo 15	
34	Van de Graaff Region	168°W–168°E
35	Highland East Farside	168°E–82°E (south of 10°S)
		168°E–88°E (north of 10°S)
36	Highland East Nearside	from (Region 35) to 60°E
37	Mare Fecunditatis	60°E–42°E
38	Mare Tranquillitatis	42°E–21°E
39	Mare Serentatis	21°E–6°E
40	Archimedes Region	6°E–15°W
41	Mare Imbrium	15°W–39°W
42	Aristarchus Region	39°W–54°W
43	Oceanus Procellarum (north)	54°W–81°W
44	Highland West Farside	81°W–168°W
	Apollo 16	
26	Highland East Farside	180°–142°E
22	Mendeleev	142°E–138°E
28	Highland East Limb	138°E–92°E
20	Mare Smythii	92°E–83°E
19	Highland East Nearside	83°E–55°E
17	Mare Fecunditatis	55°E–44°E
14	Intermediate Mare-Eastern Highlands	44°E–21°E
12	Highland Nearside Center	21°E–5°E
10	Ptolemaus-Albategnius	5°E–5°W
23	Fra Mauro Region	5°W–20°W
8	Mare Cognitum	20°W–30°W
5	Oceanus Procellarum (south)	30°W–58°W
3	Grimaldi Region	58°W–76°W
2	Orientale Rings	76°W–105°W
29	Highland West Limb	105°W–119°W
1	Hertzsprung Region	119°W–136°W
27	Highland West Farside	136°W–180°

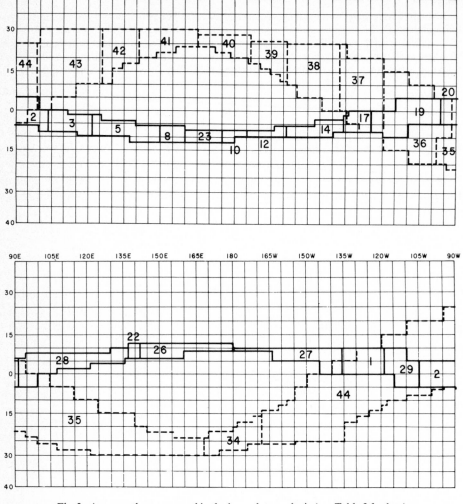

Fig. 2. Areas on the moon used in the lunar data analysis (see Table 2 for key).

northwestern in east and west longitudes respectively) of Apollo 15 regions (except those near 0° and 180°) are heavily weighted. The regions were chosen considering major topographic boundaries, the density of high quality data in a given region, and the contrasts observed in our earlier radioactivity maps. The smallest regions, like Mendeleev, were chosen to test the minimum area in which useful data can be derived; the errors in these regions are larger than normal because of poorer counting statistics. All the high quality data were used in this edition of our analysis; thus complex border regions are assigned to mare or highland. Doubtless this reduces true chemical contrasts in some regions.

Table 3 presents the results of the analysis for Apollo 15, normalizing Fe, Mg, and Ti as described above. The values for Fe are means for the two modes of

Table 3. Apollo 15: Element concentrations by regions.

Region	Name	Fe(%)	Mg(%)	Ti(%)	Th(ppm)	K(ppm)
34	V.d.Graaff	7.7	3.8	0.1	2.3	1600
35	Hi.E.Far.	6.5	4.5	1.3	1.0	940
36	Hi.E.Near.	9.3	5.7	0.8	1.4	1200
37	M.Fecund.	11.3	7.0	2.2	1.2	1400
38[a]	M.Tranq.	(12.1)	(4.8)	(2.9)	1.7	1200
39	M.Seren.	10.7	6.6	2.6	2.3	1700
40	Archim.R.	8.4[b]	6.3	0.8	6.8	3100
41	M.Imbr.	13.6	6.2	1.4	5.8	1700
42	Arist.R.	9.6[b]	4.9	2.2	6.9	2500
43	O.Proc.N.	10.5	4.6	2.0	3.9	1700
44	Hi.W.Far.	5.7	3.5	1.5	0.4	950
All data		8.7	4.8	1.45	2.2	1230

[a]Values in parentheses are ground truth data used to normalize concentrations for these elements in all regions.

[b]Region of apparent depressed thermal neutron flux (see text).

analysis, weighted by the nominal errors given by the computer analysis (usually giving the $[n,n']$ values about threefold greater weight). Values indicating reference to footnote b show regions where the values determined from neutron capture (low energy) are well below those from inelastic scatter (MeV neutrons). These regions appear to have high concentrations of KREEP, whose rare earth content should depress the thermal flux. In such places the true Fe content may be up to a few percent higher than shown here.

Table 4 gives the Apollo 16 results in the same format. The Apollo 15 continuum was used for this work, with very good consistency except for the Ti results, which are omitted. Here again Fra Mauro gives evidence of a depressed thermal neutron flux; the apparent depression in a few highland regions is not yet understood.

In Table 5 we show a number of instructive comparisons. In the crossover region near the east limb the regions analyzed on the two missions are roughly the same, and the comparison should give good agreement. The highland region from 5°E to 21°E, which includes Descartes, is compared with the soil analysis. The deviations seen are in the expected direction if more mare and KREEP material are mixed in near the margins of the region.

There are two ground stations not overflown where a plausible comparison is possible. Apollo 14 sampled the northern end of the Fra Mauro Formation; Apollo 16 overflew the southern end. Whatever the origin of this feature, similarities are to be expected. Apollo 12 sampled a part of Oceanus Procellarum not very far from, and topographically similar to, the part overflown by Apollo 16. These comparisons are shown in parts (c) and (d) of Table 5.

All these comparisons seem generally satisfactory, and serve to confirm the validity of the analysis. There is room for further work in a number of areas and this is under way. However, we expect no substantial modification of the values reported here.

Table 4. Apollo 16: Element concentrations by regions.

Region	Name	Fe(%)	Mg(%)	Th(ppm)	K(ppm)
26	Hi.E.Far.	6.2	3.4	0.6	920
22	Mendel.	4.3	3.0	0.5	960
28	Hi.E.Limb	7.2	2.9	0.5	840
20	M.Smyth.	8.8	2.8	1.3	1900
19	Hi.E.Near.	8.6	4.5	1.3	980
17	M.Fecund.	9.0	4.6	2.1	1100
14	Int.M.E.H.	9.0	3.5	1.5	1300
12	Hi.Near.C.	5.9	4.0	2.1	1400
10	Ptol.-Alb.	4.8[a]	5.2	5.0	2700
23	Fra M.R.	9.7[a]	5.7	10.5	3900
8	M.Cogn.	12.1	4.9	8.4	3600
5	O.Proc.S.	12.2	5.0	5.0	2300
3	Grim.R.	4.4[a]	3.6	2.4	1100
2	Ori.Rings	4.7	2.9	0.8	1000
29	Hi.W.Limb	3.5[a]	2.1	0.4	1200
1	Hertz.R.	3.6[a]	3.6	0.6	720
27	Hi.W.Far.	5.2	2.7	0.5	730
All data		7.2	3.6	2.1	1300

[a]Region of apparent depressed thermal neutron flux (see text).

DISCUSSION

In this paper we confine ourselves to a few of the more obvious conclusions to be drawn from these chemical data, deferring a more thorough analysis to a later time.

First, the contrasts between mare and highland regions are as expected nearly everywhere. Iron (and less strikingly magnesium) are relatively enriched in the maria. The radioactive elements show the same pattern except for one place.

The van de Graaff region (Table 3, Region 34) shows a chemical composition different from any we have seen on the moon. The major elements are highland-like, though Fe is a little high for this. The concentrations of K and Th are very similar to Mare Tranquillitatis, and typical for a mare. Such a composition could not be formed by mixing the major components we have seen elsewhere. Van de Graaff is also the deepest place on the moon (Sjogren and Wollenhaupt, 1973), and shows the largest magnetic anomaly (Russell *et al.*, 1973). From the detailed radioactivity maps (frontispiece, 1973), one can see that the radioactive feature extends over 30° in longitude. No trace of it is seen further north, along the Apollo 16 track.

The most exciting possibility is that van de Graaff or some chemically similar unmapped region is the source for the remarkable "granitic" rock 12013. The K/Th ratios are compatible (O'Kelley *et al.*, 1971) and about 5% of material of 12013 chemistry, added to highland soil, could produce the observed radioactivity. The major elements are also compatible, though this is not a useful check at the

Table 5.

Some comparisons

	(a) East Crossover		(b) Ground Truth	
	Apollo 15 Region 36	Apollo 16 Region 19	Apollo 16 Region 12	Descartes Soil[a]
Fe(%)	9.3	8.6	5.9	4.0
Mg(%)	5.7	4.5	4.0	3.3
Th(ppm)	1.4	1.3	2.1	2.0
K(ppm)	1200	980	1400	940

Ground analogies

	(c) Fra Mauro		(d) Oceanus Procellarum	
	Apollo 16 Region 23	Apollo 14 Soil[b]	Apollo 16 Region 5	Apollo 12 Soil[c]
Fe(%)	9.7	8.1	12.2	12.5
Mg(%)	5.7	5.6	5.0	6.2
Th(ppm)	10.5	11.6	5.0	7.6
K(ppm)	3900	4400	2300	2600

[a] Fe and Mg: Average of soil analysis, taken from five papers, Third Lunar Science Conference and PET report. K and Th from Eldridge *et al.* (1973).

[b] Fe and Mg: Average of soil analysis, taken from five papers, Third Lunar Science Conference and PET report. K and Th from Eldridge *et al.* (1972).

[c] Fe and Mg: Average of analyses of bulk soil 12070 and related samples in five papers, Second Lunar Science Conference. K and Th from O'Kelley *et al.* (1971).

5% level.* The van de Graaff region is very far from all the landing sites; it is plausible that such material would be uncommon in our samples.

The large highland regions are not entirely uniform. Most notable is an east-west asymmetry in the backside. The iron and thorium concentrations are higher in the eastern regions, an effect which extends to the limb and near to the mare edges. These eastern regions are lower topographically (Sjogren and Wollenhaupt, 1973). The radioactivity maps show more detail (frontispiece, 1973). The dry basins like Mendeleev give typical highland analyses within a rather large experimental error.

The values of Ti found in the backside highlands on Apollo 15 are unexpectedly high, and are another example of inhomogeneity in broad highland areas.

---

*We have used the whole rock Th and K values in this analysis. The rock is a very inhomogeneous breccia; our statements apply either to the whole rock or to any selected component containing most of the radioactive elements.

There are also differences among maria and KREEP-rich regions. The variability in Ti has been well documented in the mare samples, and is well displayed in the comparison of Mare Imbrium (Region 41) with Oceanus Procellarum (Region 43). The Archimedes and Aristarchus regions also differ in Ti and probably Fe.

The most interesting element ratio is K/Th (and its assumed close correlate K/U). In both missions this ratio rises from something like 400 in the KREEP-rich regions to the region of 1000–2000 in the highlands. Figure 3 shows plots of the two element concentrations for the two missions, as a function of longitude. The

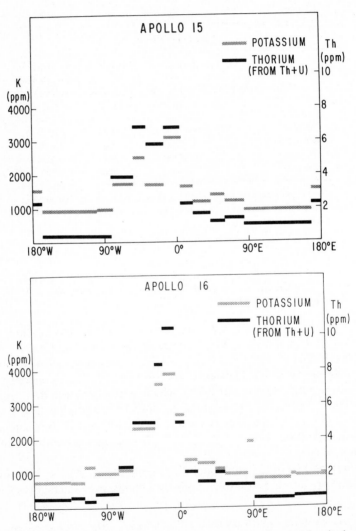

Fig. 3. (a) Potassium and thorium on the lunar surface, plotted against longitude on the Apollo 15 ground track. (b) The same plot for Apollo 16.

errors for both elements are of course larger at the lowest concentrations, but the trend is unmistakable.

Earlier we interpreted our radioactivity maps in terms of a three component model: mare basalt, KREEP, and a low-radioactivity highland component. Clearly this preliminary model is inadequate for the detailed information now available. At least at van de Graaff we are seeing material not prominently displayed in lunar samples so far described. However, no broad regions dominated by radically different chemistry (basic rocks or granites) have yet showed up in our analysis. If they exist, they lie in areas so far unmapped. At the same time, correlations previously observed with altitude and albedo seem to hold up rather well.

Acknowledgments—Since this paper represents a culmination of a long effort, we feel it necessary to acknowledge the help of the very large number of people who helped substantially in that effort, even though only a small fraction of them can be named here. At NASA headquarters Robert Bryson has supported our activities for many years, along with others in the Lunar Program Office and elsewhere. At Johnson Space Center the names of Anthony Calio and Eugene Krantz must stand not only for their generous help but for the assistance of many others in S. and A.D. and in Mission Control. Ed Stelly saw us successfully through the hardware phase. Charles Meyer and Frank Martin supported us throughout. The instrument was designed and constructed under the direction of Tim Harrington. The astronauts, especially Al Worden and Ken Mattingly, did the work at the moon.

The matrix inversion analyses described here were done at GSFC, with the participation of Michael Bielefeld and others, and using the full resources of the GSFC computer group. The data were prepared for analysis at JPL, under the guidance of Elliot Goldyn and Richard E. Parker.

The work described in this paper was carried out in part under NASA Contract NAS 7-100 at the Jet Propulsion Laboratory, California Institute of Technology, and in part under NASA Contrast NAS 9-10670 at the University of California, San Diego. It was entirely supported by NASA.

REFERENCES

Arnold J. R., Metzger A. E., Peterson L. E., Reedy R. C., and Trombka J. I. (1972) Gamma ray spectrometer experiment. *Apollo 16 Preliminary Science Report*, 18-1.

Eldridge J. S., O'Kelley G. D., and Northcutt K. J. (1972) Abundances of primordial and cosmogenic nuclides in Apollo 14 rocks and fines. *Proc. Third Lunar Sci. Conf., Geochim. Cosmochim. Acta*, Suppl. 3, Vol. 2, p. 1651. MIT Press.

Eldridge J. S., O'Kelley G. D., and Northcutt K. J. (1973) Radionuclide concentrations in Apollo 16 samples. *Proc. Fourth Lunar Sci. Conf., Geochim. Cosmochim. Acta*, Suppl. 4, Vol. 2, p. 2115. Pergamon.

Finkel R. C., Imamura M., Honda M., Nishiizumi K., Kohl C. P., Kocimski S. M., and Arnold J. R. (1974) Cosmic ray produced Mn and Be radionuclides in the lunar regolith (abstract). In *Lunar Science—V*, p. 228. The Lunar Science Institute, Houston.

Frontispiece (1973) *Proc. Fourth Lunar Sci. Conf., Geochim. Cosmochim. Acta*, Suppl. 4, Vol. 1. Pergamon.

Harrington T. M., Marshall J. H., Arnold J. R., Peterson L. P., Trombka J. I., and Metzger A. E. (1974) The Apollo gamma ray spectrometer. *Nucl. Instr. & Methods* **118**, 401.

Lingenfelter R. E., Canfield E. H., and Hampel V. E. (1972) The lunar flux revisited. *Earth Planet. Sci. Lett.* **16**, 355.

Metzger A. E., Trombka J. I., Peterson L. E., Reedy R. C., and Arnold J. R. (1973) Lunar surface radioactivity: Preliminary results of the Apollo 15 and Apollo 16 gamma ray spectrometer experiments. *Science* **179**, 800.

O'Kelley G. D., Eldridge J. S., Schonfeld E., and Bell P. R. (1971) Abundances of the primordial radionuclides K, Th, and U in Apollo 12 lunar samples. *Proc. Second Lunar Sci. Conf., Geochim. Cosmochim. Acta*, Suppl. 2, Vol. 2, p. 1159. MIT Press.

1078 A. E. METZGER *et al.*

Reedy R. C. and Arnold J. R. (1972) Interaction of solar and galactic cosmic ray particles with the moon. *J. Geophys. Res.* **77**, 537.

Reedy R. C., Arnold J. R., and Trombka J. I. (1973) Expected γ-ray emission spectra from the lunar surface as a function of chemical composition. *J. Geophys. Res.* **78**, 5847.

Russell C. T., Coleman P. J., Jr., Lichtenstein B. R., Schubert G., and Sharp L. R. (1973) Subsatellite measurements of the lunar magnetic field. *Proc. Fourth Lunar Sci. Conf., Geochim. Cosmochim. Acta*, Suppl. 4, Vol. 3, p. 2833. Pergamon.

Schonfeld E. and Meyer C. (1972) The abundances of components of the lunar soils by a least squares mixing model. *Proc. Third Lunar Sci. Conf., Geochim. Cosmochim. Acta*, Suppl. 3, Vol. 2, p. 1681. MIT Press.

Sjogren W. L. and Wollenhaupt W. R. (1973) Lunar shape via the Apollo laser altimeter. *Science* **179**, 275.

Trombka J. I., Arnold J. R., Reedy R. C., Peterson L. E., and Metzger A. E. (1973) Some correlations between measurements by the Apollo gamma-ray spectrometer and other lunar observations. *Proc. Fourth Lunar Sci. Conf., Geochim. Cosmochim. Acta*, Suppl. 4, Vol. 3, p. 2847. Pergamon.

Turkevich A. L., Franzgrote E. J., and Patterson J. H. (1969) Chemical composition of the lunar surface in Mare Tranquillitatis. *Science* **165**, 277.

Wakita H., Schmitt R. A., and Rey P. (1970) Elemental abundances of major, minor and trace elements in Apollo 11 lunar rocks. *Proc. First Lunar Sci. Conf., Geochim. Cosmochim. Acta*, Suppl. 1, Vol. 2, p. 1689. Pergamon.

Woolum D. S. and Burnett D. S. (1974) In situ measurement of the rate of ^{235}U fission induced by lunar neutrons. *Earth Planet. Sci. Lett.* **21**, 153.

Proceedings of the Fifth Lunar Conference
(Supplement 5, Geochimica et Cosmochimica Acta)
Vol. 2 pp. 1079–1086 (1974)
Printed in the United States of America

Compositional studies of the lunar regolith at the Apollo 17 site

M. D. Miller, R. A. Pacer, M.-S. Ma, B. R. Hawke,
G. L. Lookhart, and W. D. Ehmann

Department of Chemistry, University of Kentucky, Lexington, Kentucky 40506

Abstract—Major-, minor-, and trace-element abundance data are presented for twenty-two Apollo 17 samples, twelve Apollo 16 samples and one Apollo 15 basalt. The data were obtained by INAA employing a 14 MeV neutron generator, a ^{252}Cf isotopic neutron source and nuclear reactor thermal neutrons. Abundance summations for the twenty-four samples analyzed for major and minor abundance elements averaged $98.8 \pm 0.3\%$. Similarities between the Apollo 17 and the Apollo 11 mare basalts are discussed. Abundances in 13 Apollo 17 soils exhibit the greatest compositional range we have observed for soils from a specific landing site. A strong inverse Fe–Al correlation in all lunar materials we have analyzed with the exception of most Apollo 16 samples is defined by the equation: %Fe $= -1.36$ %Al $+ 21.5$. A strong direct Mn–Fe correlation is defined by the equation: %Fe $= 75.9$ %Mn $+ 0.08$. Both regression lines appear to have the mare basalts as one end-member and an unspecified Al-rich component at the other end. Comparisons of the data from direct O determinations and calculated O abundances based on normal stoichiometry are presented.

Introduction

Instrumental neutron activation analysis (INAA) employing 14 MeV, ^{252}Cf and reactor thermal neutrons has been applied to the determination of major-, minor-, and trace-element abundances in Apollo 15, 16, and 17 samples. Results of analysis for Zr/Hf on some of these samples and their relationship to a two stage lunar evolution model are presented elsewhere in these proceedings (Ehmann and Chyi, 1974).

The minor- and trace-element data on Apollo 14 and 15 samples supplement our earlier publications of major element abundances in these same samples (Ehmann et al., 1972). Major- and minor-element abundances in Apollo 16 samples supplement our earlier publication of similar data (Janghorbani et al., 1973). Results of major- and minor-element abundances on Apollo 17 samples represent our first extensive data compilation on this mission.

Experimental

Detailed procedures for the determination of O, Si, Al, Mg, and Fe by INAA have been reported in earlier publications by our group and will not be repeated here. A resumé of the procedures may be found in Morgan and Ehmann (1971). Major element data reported here are typically means of 3–5 replicate determinations.

The elements Ca, Ti, Mn, and Na are determined by ^{252}Cf neutron irradiation. The Ca and Ti abundances are determined by a tricyclic activation technique, while Mn and Na abundances are typically based on three replicate analyses.

Trace-element abundances were determined by instrumental reactor neutron activation. Irradiation times did not exceed 100 hours at a thermal neutron flux density of approximately 10^{13} n cm^{-2} sec^{-1}. Splits of standard rocks BCR-1 and W-1 were used as comparator standards for the trace element determinations. Abundances in the standard rocks were taken from Flanagan (1973).

Results and Discussion

Apollo 16 and 17 results

Major- and minor-element abundances in Apollo 16 and 17 rocks are presented in Table 1. The Apollo 17 basalts as a group exhibit rather uniform compositions, and closely resemble the major- and minor-element compositions of the Apollo 11 mare basalts. The obvious similarities are their high-Ti contents, their Al/Si ratios (Apollo 11 basalts = 0.255, Apollo 17 basalts = 0.255), and their Fe/Si ratios (Apollo 11 basalts = 0.80, Apollo 17 basalts = 0.82). Small dissimilarities include the slightly lower abundances of Si, Al, and Fe and the higher Mg abundance in the Apollo 17 basalts compared with the Apollo 11 basalts. The mean Apollo 17 basalt composition is intermediate between the type A and type B mare basalts, but is most similar to the type B material. Sample 75035,36 differs from the other Apollo 17 basalts we analyzed, having substantially higher abundances of Si and Ca and smaller abundances of Mg and Ti. These characteristics (also noted by LSPET on basalt 75055,6) cause this sample to resemble the elemental composition of the Apollo 11 basalts even more closely than do the other Apollo 17 basalts.

The compositions of the two Apollo 16 breccias 66075,30 and 67475,35 closely resemble the other breccias and basalts collected at this site which we reported

Table 1. Elemental abundances in Apollo 16 and 17 rocks (wt.%).*

Element	66075,30 Breccia	67475,35 Breccia	70017,28 Basalt	70215,46 Basalt	70215,59 Basalt (sur.)	71055,56 Basalt (int.)	73235,54 Breccia	74275,30 Basalt (sur.)	74275,63 Basalt (int.)	75035,36 Basalt	79035,32 Breccia
O	43.6	43.5	40.4	40.5	38.8	39.6	45.3	39.9	41.0	39.5	43.3
Si	21.3	20.8	18.0	17.8	17.3	17.5	22.3	17.9	18.5	19.1	20.3
Al	14.8	16.0	4.8	4.5	4.6	4.4	11.0	4.5	4.8	4.6	7.2
Mg	4.0	2.0	7.2	7.2	5.5	6.3	7.7	6.3	6.1	4.5	7.1
Fe	3.9	2.5	14.1	16.1	14.9	15.1	6.1	14.2	14.1	14.3	12.1
Ca	10.9	13.1	7.4	7.7	7.6	6.8	7.9	7.4	7.2	9.2	7.8
Ti	<0.5	<0.5	7.4	7.7	7.6	7.3	<0.5	7.1	7.2	5.3	3.4
Mn	0.048	0.036	0.181	0.203	0.194	0.200	0.080	0.192	0.197	0.193	0.157
Na	0.39	0.46	0.30	0.28	0.32	0.33	0.35	0.28	0.29	0.33	0.31
Total	99.2	98.6	99.8	102.0	96.8	97.5	101.0	97.8	99.4	97.0	101.7
Δ%O†	−2.1	−1.9	−1.2	−1.5	−1.2	−0.5	0.0	−0.6	−0.2	−0.8	+0.1

*Typical precisions of the replicate determinations presented here and in Table 2 are: ±0.3% O, 0.1% Si, 0.2% Al, 0.8% Mg, 0.3% Fe, 0.3% Ca, 0.3% Ti, 0.002% Mn and 0.01% Na.

Data with appreciably larger analytical uncertainties are italicized.

†Δ%O = Experimental % oxygen—Calculated stoichiometric % oxygen.

No correction has been applied for elements not determined by us nor for deviation of the abundance total from 100%.

earlier (Janghorbani *et al.*, 1973). The major element abundances in the two Apollo 17 breccias we analyzed exhibit a wide variation and do not resemble those in any other lunar materials we have analyzed. Their Si, Al, and Fe abundances are quite different from the basalts at this site and resemble the soils only because of the wide compositional ranges of the latter.

Major- and minor-element abundances for thirteen Apollo 17 soils are presented in Table 2. Apollo 17 soils exhibit the greatest compositional variation we have observed for soils from a specific mission site. The variability is pronounced for Al (3.5–11.3%), Fe (5.8–16.5%), Mn (0.087–0.207%) and Ti (<1–6%). Orange soil 74220 has the lowest Al content and the highest Mg, Fe, Ti, and Mn abundances of any lunar soil we have analyzed. Stations 2 and 2a soils 72501, 73121, and 73141 are virtually indistinguishable and may be characteristic of the "light-mantle" material distributed throughout this site. The blue–gray breccia from the Station 3 site, 73235, also appears to have this composition. The pattern of major- and minor-element abundances (higher Al and Ca, lower Fe, Ti, and Mn) in these samples appear to reflect a higher anorthositic component. However, a simple two-component mixing model based on anorthosite and mare basalt does not fit the general Fe–Al correlation which we will present later in this work. Station 1a soil, 71500, and Station 5 soil, 75081, have similar compositions, and may be representative of at least a portion of the "dark-mantle" material at this site. The Al, Fe, and Mn abundances, in particular, distinguish between the light- and dark-mantle materials. The four Station 8 soils have similar compositions and are intermediate between the light- and dark-mantle soils described above, resembling the light-mantle soils somewhat more closely. Soils 72141 and 72161 also have intermediate compositions, but resemble the composition of the dark-mantle materials. The intermediate composition of soil 74121 more closely resembles the light-mantle material.

The Fe–Al correlation

In our previous work (Janghorbani *et al.*, 1973) we noted the strong inverse Fe–Al correlation for all previously analyzed lunar rocks, breccias, and soils, excluding most of the Apollo 16 samples. The regression line (Fig. 1) is defined by the equation $\%Fe = -1.36 \%Al + 21.5$ excluding all of the Apollo 16 samples and sample 12013. The end-member at the Al-poor end appears to be the mare basalts. The basalts from the Apollo 11, 12, 15, and 17 sites all cluster around 4–5% Al and 14–16% Fe. The end member at the Al-rich end is not apparent, but should be a major component of the light-mantle material previously mentioned. The Apollo 16 breccias and soils lie on the Al-rich side of the Fe–Al regression line, which we interpret as reflecting an additional anorthositic component. The Apollo 16 samples lie on a straight line extending from the mare basalt composition to pure anorthite (60025,72). The Apollo 17 samples fall precisely along the primary Fe–Al regression line. Apollo 17 basalts lie exclusively at the Al-poor end of the regression, while the breccias and soils from this site cover the complete range of previously determined Al and Fe abundances (excluding the Apollo 16 samples).

Table 2. Elemental abundances in Apollo 17 soils (wt.%).†

Element	71500,5* Sta. la	72141,8 LRV2	72161,5 LRV3	72501,20 Sta. 2	73121,5 Sta. 2a	73141,6 Sta. 2	74121,7 LRV6	74220,90* Sta. 4	75081,55 Sta. 5	78421,23 Sta. 8	78441,9 Sta. 8	78461,7 Sta. 8	78481,21 Sta. 8
O	41.0	43.4	41.4	42.2	43.8	42.7	42.6	*40.2*	39.7	42.3	42.0	41.9	41.9
Si	18.8	20.7	20.0	20.8	21.2	20.9	20.7	*18.5*	18.1	20.9	20.6	19.9	20.2
Al	5.8	8.5	7.6	10.7	11.3	11.2	10.4	*3.5*	5.8	9.2	9.1	8.5	9.0
Mg	5.6	5.9	6.8	6.3	6.5	5.0	6.5	*9.1*	6.4	7.1	6.6	6.7	6.8
Fe	14.2	10.4	12.2	6.4	6.7	5.8	8.2	*16.5*	13.2	9.5	9.6	10.0	9.3
Ca	7.3	8.4	7.7	9.1	9.6	9.7	8.5	*6.6*	7.5	8.4	7.9	7.9	7.6
Ti	5.4	2.7	2.8	~1	~0.9	~1	1.4	*6*	5.6	2.3	1.9	2.1	1.8
Mn	0.180	0.131	0.149	0.089	0.092	0.087	0.101	*0.207*	0.172	0.127	0.131	0.122	0.124
Na	0.35	0.30	0.35	0.40	0.34	0.34	0.34	*0.35*	0.31	0.32	0.36	0.30	0.29
Total	98.6	100.4	99.0	97.0	100.4	96.7	98.7	101.0	96.8	100.1	98.2	97.4	97.0
Δ%O	0.0	+0.1	-1.2	-1.5	-1.2	-0.7	-1.3	-1.5	-1.0	-2.1	-1.2	-0.3	-0.6

*These samples are unsieved soils. All others are <1 mm fines.

†Additional data obtained by two Apollo 14 soils are: 14003,13 = 0.104% Mn, 0.47% Na and 14259,65 = 0.103% Mn, 0.54% Na. Other abundances for these two samples were reported previously. Data with appreciably larger analytical uncertainties are italicized.

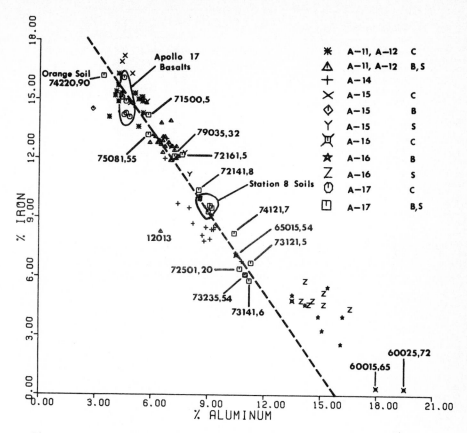

Fig. 1. The correlation of Fe and Al in lunar samples analyzed in this laboratory.
C = crystalline rocks, B = breccias, S = soils.

Basalt 65015,54 lies on the primary regression line and is much lower in Al and higher in Fe than the other Apollo 16 samples. A higher KREEP content for this rock is suggested by our trace element data (Table 3).

The Fe–Mn correlation

The excellent direct Fe–Mn correlation reported by other authors (Laul and Schmitt, 1973; Wänke *et al.*, 1973) is supported by our data (Fig. 2) and is consistent with the strong geochemical coherence of this element pair. Excluding the Apollo 16 samples, this regression line is defined by the equation %Fe = 75.9 %Mn + 0.08. As with the Fe–Al correlation, the mare basalts cluster at the Fe-rich end of the regression line. The two Apollo 16 anorthosites 60025 and 60015 lie close to the origin of the regression line. The Apollo 16 breccias and soils lie closer to the Fe–Mn regression line than in the case of the Fe–Al regression.

The Apollo 16 breccias and soils could be interpreted as being slightly displaced to the Fe-rich side of the Fe–Mn regression line at a Mn content of

Table 3. Element abundances (ppm) in some Apollo 15, 16, and 17 lunar samples.*

Lunar Sample	Fe(%)	Co	Sc	Cr	Eu	La
15016,31	17.6	60	40.8	6670	0.9	5
60015,65B	0.4	1	0.6	42	0.9	—
60025,72	0.3	1	0.4	—	1.1	—
60335,33	3.3	9	8.2	900	1.2	21
61016,133B	3.8	37	6.6	630	1.4	15
64421,20A	3.7	26	8.4	680	1.2	11
64501,10B	3.5	34	7.2	570	1.0	11
61501,11	4.3	34	9.0	840	1.2	12
65015,54	6.3	30	14.4	1230	2.0	56
73235,54A (anorthositic inclusion)	0.5	7	0.8	41	—	—
73235,54B (basaltic clast)	5.2	27	13.2	1350	1.6	23
71500,5	13.2	31	67.2	3120	1.6	7
74220,90	16.9	63	46.9	3790	1.9	7

*The values reported are the means of two replicate analyses on splits, except 73235,54 which had single analyses on one basaltic clast and one anorthositic inclusion.

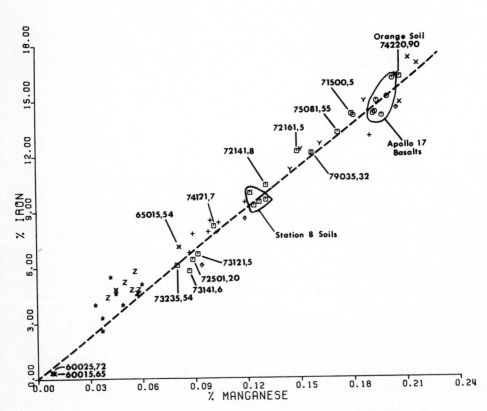

Fig. 2. The correlation of Fe and Mn in lunar samples analyzed in this laboratory.

approximately 0.05% Mn. At least a portion of this "displacement" could be explained by a higher "ancient" meteoritic component (high-Fe content and high Fe/Mn ratio) in these samples, as compared to similar materials from the other sites sampled. The Fe–Mn data of Laul and Schmitt (1973), however, show no apparent displacement of Apollo 16 materials.

Elemental abundance summations for all samples analyzed in this work average 98.8% and would be >99% if elements not determined by us were included. The distribution of the summations has a standard deviation of ±1.7%.

Trace element data

Trace element abundance data for some Apollo 15 and 16 samples (major- and minor-abundance elements were previously reported by Janghorbani et al. (1973)) and for several Apollo 17 samples are presented in Table 3. Fe abundances obtained by reactor thermal neutron irradiations are also included. For samples with low-Fe contents, such as the Apollo 16 anorthosites, these reactor irradiation data are preferred over the data we presented previously via 14 MeV neutron irradiations.

Anorthosites 60015,65 and 60025,72 exhibit extremely low Fe, Cr, Co, and Sc contents, as compared to other Apollo 16 materials. It is interesting to note that breccia 73235,54 is in effect a miniature sampling of the lunar highlands. Anorthositic inclusion 73235,54A closely resembles rocks 60015,65 and 60025,72, while basaltic clast 73235,54B has trace element abundances very similar to those in the metaclastic KREEP-rich rock, 65105,54. Orange soil 74220,90 is considerably enriched in Fe, Co, and Cr over soil 71500,5 from the "dark-mantle" region of the valley floor.

Oxygen abundances

In our previous papers we have pointed out that the O contents of lunar materials are governed by a number of factors including the basic mineralogy of the sample, the presence of reduced species such as Ti^{3+} resulting from the low f_{O_2} in the lunar melts, solar wind hydrogen reduction of lunar surface materials, and vaporization–redeposition effects resulting from impact events.

Comparisons of our direct O determinations with O contents required for simple stoichiometry (based on our determinations of other major- and minor-abundance elements but without corrections for elements we have not determined or deviations of the element summations from 100%) consistently indicate a 1–2% O deficiency with respect to stoichiometric requirements in lunar fines and many lunar breccias. If elements not determined by us were included in the stoichiometric O calculation, the magnitude of the depletion would be greater. Some lunar basalts also appear to exhibit a similar O deficiency (Janghorbani et al., 1973) which cannot be simply explained by the presence of metallic Fe.

Since the precision of our direct O analyses is of the order of 0.3–0.5% O and calculated values for stoichiometric O also have analytical uncertainties, quantification of the apparent O-depletion is extremely difficult. Comparisons of the O-depletion for interior and exterior samples of crystalline rocks, such as 70215 and 74275, are ambiguous. However, at all sites, the soils exhibit a greater average O-depletion than the crystalline rocks. We feel that the true magnitude of any O-depletion by solar wind interaction can be finally established only by use of analytical techniques capable of surface analyses. The bulk O analyses via 14 MeV neutron activation do, in our opinion, support an O-depletion of approximately 1–2% in lunar fines which may be related to solar wind exposure. While we are continuing to study this problem, we feel that surface analysis techniques ultimately must be used to quantify the phenomena.

Acknowledgments—This work has been supported in part by NASA grant NGR 18-001-058. The cooperation of the personnel at the Georgia Institute of Technology Research Reactor Facility is gratefully acknowledged. The assistance of Mr. John H. Jones in the computer processing of the data is also gratefully acknowledged.

REFERENCES

Ehmann W. D. and Chyi L. L. (1974) Abundances of the group IVB elements, Ti, Zr, and Hf and implications of their ratios in lunar materials. *Proc. Fifth Lunar Sci. Conf., Geochim. Cosmochim. Acta.* This volume.

Ehmann W. D., Gillum D. E., and Morgan J. W. (1972) Oxygen and bulk element composition studies of Apollo 14 and other lunar rocks and soils. *Proc. Third Lunar Sci. Conf., Geochim. Cosmochim. Acta*, Suppl. 3, Vol. 2, pp. 1149–1160. MIT Press.

Flanagan F. J. (1973) 1972 values for international geochemical reference standards. *Geochim. Cosmochim. Acta* **37**, 1189–1200.

Janghorbani M., Miller M. D., Ma M.-S., Chyi L. L., and Ehmann W. D. (1973) Oxygen and other elemental abundance data for Apollo 14, 15, 16 and 17 samples. *Proc. Fourth Lunar Sci. Conf., Geochim. Cosmochim. Acta*, Suppl. 4, Vol. 2, pp. 1115–1126. Pergamon.

Laul J. C. and Schmitt R. A. (1973) Chemical composition of Apollo 15, 16, and 17 samples. *Proc. Fourth Lunar Sci. Conf., Geochim. Cosmochim. Acta*, Suppl. 4, Vol. 2, pp. 1349–1367. Pergamon.

LSPET (Lunar Science Preliminary Examination Team) (1973) Apollo 17 lunar Samples: Chemical and Petrographic Description. *Science* **182**, 659–672.

Morgan J. W. and Ehmann W. D. (1971) 14 MeV neutron activation analysis of rocks and meteorites. *Activation Analysis in Geochemistry and Cosmochemistry* (editors A. O. Brunfelt and E. Steinnes), pp. 81–97. Universitetsforlaget, Oslo.

Wänke H., Baddenhausen H., Dreibus G., Jagoutz E., Kruse H., Palme H., Spettel B., and Teschke F. (1973). Multielement analyses of Apollo 15, 16, and 17 samples and the bulk composition of the moon. *Proc. Fourth Lunar Sci. Conf., Geochim. Cosmochim. Acta*, Suppl. 4, Vol. 2, pp. 1461–1481. Pergamon.

Proceedings of the Fifth Lunar Conference
(Supplement 5, Geochimica et Cosmochimica Acta)
Vol. 2 pp. 1087–1096 (1974)
Printed in the United States of America

Chemical compositions of some soils and rock types from the Apollo 15, 16, and 17 lunar sites

DAVID F. NAVA

Astrochemistry Branch, Code 691, NASA Goddard Space Flight Center, Greenbelt, Maryland 20771

Abstract—The major and minor element compositions of three Apollo 15 mare basalts (15065, 15555, and 15597), four Apollo 16 breccias (61016, 66095, 67559, and 68415), one soil (65500), three Apollo 17 soils (74121, 74220, and 74241), one mare basalt (70017), and one breccia (76055) have been determined by semi-micro combined atomic absorption and colorimetric spectrophotometries.

Material from two fragments, separated by some 10 cm and located in different domains of 15065, a coarse-grained mare basalt, were studied and found to possess distinctly different bulk chemical compositions. Apollo 15555 mare basalt, also coarse-grained, is another case of gross sample compositional heterogeneity. The chemical compositions of different small-sized fragments of these coarse-grained basalts could classify each, on the basis of parameters other than Si abundance and olivine norm, within both the olivine-normative and the quartz-normative fields of Apollo 15 mare basalt rock types. Conversely, the coarse-grained Apollo 15 mare basalts may constitute a third grouping in this classification scheme.

The composition determined for a hand-picked white separate of 61016 approximates 95% anorthite. Unsieved soil fines 65500, together with the very similar composition reported for one split of 69941, has higher Si and Mg and lower Al and Ca contents than all other Apollo 16 soils observed thus far. The suggestion is made that, in conjunction with the lunar locations of these two samples, the chemical compositional extremes may hint at some subtle chemical distinctions between the Cayley and the Descartes formations. 67559 from the southeast rim of North Ray Crater and 68415 from the northeast rim of South Ray Crater are extremely similar in major and minor element abundances. Both breccias may be hybrids.

The composition of coarse-grained sample 70017 is typical of the Apollo 17 subfloor basalts. Greenish gray breccia 76055 has a composition, except notably for higher Mg and lower Si, that is generally similar to Apollo 14 KREEP-type basalts.

A discussion regarding analysis of small subsamples of coarse-grained lunar materials, such as certain Apollo 15 and 17 basalts, is given in which is made the suggestion that analytical methods designed to determine the bulk "representative" chemical composition as well as the presence and extent of specimen heterogeneity are fundamentally necessary for a more complete understanding of genesis and history of lunar materials.

INTRODUCTION

THE MAJOR (>1 wt.%) and minor (0.01–1 wt.%) element compositions of three mare basalts from the Apollo 15 lunar mission, four breccias and one soil of Apollo 16 material, and three soils, one mare basalt, and one breccia returned by Apollo 17 have been determined by a semi-micro procedure combining atomic absorption spectrophotometry with colorimetry. Sample aliquants of 50 mg, in duplicate when available, were decomposed in a closed teflon pressure vessel with an HF-aqua regia acid media. The resultant solutions were complexed with boric acid and diluted to 100 ml final volumes, giving a solution concentration of 500 ppm total dissolved sample each. Aliquots were used for colorimetric determination of Ti and P. The other constituents, except K, were measured by flame atomic absorption. The K values reported for these Apollo samples are by mass-spectrometric isotope-dilution analysis of aliquants of our initial sample powders. All soil samples were analyzed as received from Houston; the breccia and basalt sample fragments and chips received were ground in boron carbide mortar to approximately <200 mesh powders. Additional analytical details of these procedures have been reported previously (Schnetzler and Nava, 1971; Philpotts *et al.*, 1972; Nava and Philpotts, 1973).

The relative precision for duplicates of the sample powders was ±1–2%, or better, for the major elements, and within ±5% for the minor elements. Analytical accuracy was monitored by concurrent analyses of USGS reference silicate rocks W-1 and BCR-1 and was found, as with our previously analyzed sample sets, to be well within the range of published analyses.

RESULTS

Analytical data for the Apollo 15 mare basalts studied are presented in Table 1. The chemical compositions determined for the Apollo 16 and 17 soils and rocks are given in Table 2. Data for rake soil sample 65500 and orange soil 74220 are bulk compositions of unsieved fines. The data listed for soil sample 74121 and gray reference soil 74241 are bulk compositions of the <1 mm fines. The chemical abundances presented for the breccia and basalt samples are bulk, whole-rock compositions, except for a heavy-liquid separation of black matrix material from an aliquant of basalt 15597 and a hand-picked 100 mesh size separate of the purest white material from breccia 61016.

Table 1. Chemical compositions of three Apollo 15 mare basalts.

Constituent	15065,6 Sta. 1	15065,42 Sta. 1	15555,18 Sta. 9	15555 Range of compositions reported by other analysts*	15597,19 Sta. 9 Matrix	15597,19 Sta. 9 Whole Rock
SiO_2	47.7	48.2	45.0	43.82–45.21	47.9	48.1
TiO_2	2.86	1.44	1.60	1.73–2.63	2.30	1.87
Al_2O_3	6.05	10.32	9.37	7.45–10.32	15.13	9.27
MgO	9.52	10.35	12.22	10.96–11.48	1.40	9.18
CaO	9.33	9.55	9.25	9.14–9.96	9.62	9.69
Na_2O	0.27	0.33	0.26	0.24–0.35	0.66	0.32
FeO	23.77	18.46	21.18	20.16–24.58	22.24	20.17
MnO	0.307	0.234	0.261	0.25–0.32	0.224	0.254
P_2O_5	0.119	0.104	0.066	0.05–0.07	0.151	0.107
Cr_2O_3	0.54	0.47	0.47	0.58–0.73	<0.005	0.49
K_2O	0.081	0.041	0.028	0.03–0.05	0.111	0.056
Total	100.55	99.50	99.71	—	99.74	99.51

*Sources for other Apollo 15555 reported data are: Maxwell *et al.* (1972), Mason *et al.* (1972), Cuttitta *et al.* (1973), Chappell and Green (1973), and LSPET (1972).

Table 2. Chemical compositions of some Apollo 16 and Apollo 17 soils and rocks.

Constituent	61016,184 STN 1 Anorthosite	65500,5 STN 5 Soil	66095,50 STN 6 Crystln*	67559,3 STN 11 Crystln	68415,79 STN 8 Crystln	70017,23 LM Basalt	74121,16 LRV 6 Soil	74220,40 STN 4 Orange	74241,20 STN 4 Gray soil	76055,3 STN 6 Breccia
SiO_2	45.0	46.2	44.9	45.3	45.9	38.8	44.9	38.9	42.3	45.7
TiO_2	0.02	0.62	0.60	0.26	0.28	12.44	2.47	8.96	7.33	1.38
Al_2O_3	34.85	25.17	23.00	27.42	28.19	9.73	18.75	6.38	13.69	15.84
MgO	<0.03	6.91	9.66	4.47	4.41	9.89	10.20	14.76	9.88	17.89
CaO	19.58	14.25	13.48	16.40	16.39	10.04	11.73	7.01	10.89	9.13
Na_2O	0.41	0.48	0.48	0.50	0.47	0.43	0.44	0.43	0.48	0.55
FeO	<0.05	5.65	6.86	4.31	4.01	17.60	10.43	22.34	14.66	9.27
MnO	<0.005	0.072	0.080	0.054	0.048	0.232	0.128	0.255	0.202	0.122
P_2O_5	0.047	0.137	0.240	0.113	0.072	0.048	0.120	0.097	0.124	0.220
Cr_2O_3	<0.002	0.12	0.13	0.08	0.07	0.45	0.23	0.68	0.38	0.19
K_2O	0.005	0.139	0.146	0.078	0.060	0.036	0.136	0.076	0.123	0.223
Total	99.99	99.75	99.58	98.99	99.90	99.70	99.53	99.89	100.06	100.52

*"Crystln" = Crystalline.

Ti and P colorimetrically, K by isotope dilution mass spectrometry, all others by flame atomic absorption spectrophotometry. Total iron expressed as FeO.

DISCUSSION

Apollo 15 mare basalts

Sample 15597 is a porphyritic clinopyroxene basalt from Station 9a near the Hadley Rille area of Mare Imbrium. Ti, Al, Na, Fe, P, and K are more abundant in the black matrix material than in the whole-rock portion analyzed. This difference in whole-rock and matrix composition is compatible with crystal fractionation from pyroxene. The whole-rock composition (determined on an aliquant from a 3.8 g sample) listed in Table 1 is in good agreement with that presented by Chappell and Green (1973).

Sample 15065 is a coarse-grained gabbroic rock collected from the rim of Elbow Crater at Station 1. Material from two separate fragments, separated by some 10 cm in the original specimen of this rock, were analyzed; one from a 1 g portion and the other from a 5 g portion designated, 6 and, 42 respectively. The Lunar Sample Information Catalog (1971) identifies two distinct domains in this rock, with boundaries being described as generally diffuse between them. The data in Table 1 for 15065 represent material from each of these domains and show that the bulk chemical composition of each of these fragments is distinctly different. Furthermore, these compositions differ from two other fragments found by Cuttitta et al. (1973) to also have markedly dissimilar compositions. This is a case of gross sample compositional heterogeneity similar to what we observed (Schnetzler et al., 1972) for 15555, a vuggy, porphyritic olivine mare basalt, also coarse-grained in nature, from Station 9a near the rim of Hadley Rille. Our data (Table 1) along with chemical analyses reported by others (LPSET, 1972; Mason et al., 1972; Maxwell et al., 1972; Chappell and Green, 1973; Cuttitta et al., 1973) show that 15555 has a very large compositional variation. It is difficult to document domains in the original 15555 specimen because the various fragments analyzed were taken from chips found loose in the Apollo 15555 sample bag. This rock is friable material and it was thus not possible to locate the original positions of these chips in the bulk rock (P. Butler, personal communication, 1973). Mason *et al.* (1972) first pointed out that these gross heterogeneities shown by the reported analyses for some Apollo 15 mare basalts are directly due to the relationship of subsample size taken for analysis to the average grain size of the original rock material. Studies have shown (e.g. Mason *et al.*, 1972; Philpotts *et al.*, 1972; Rhodes and Hubbard, 1973) that "representative" compositions can be obtained for small subsamples (50–500 mg) of fine-grained lunar specimens. For coarse-grained materials, however, "inadequate" sampling leads to discrepancies and precludes comparisons intended to establish analytical agreement of laboratory data obtained. An obvious point which appears to be most often overlooked is the fact that these small sample sizes *are* adequate enough to establish chemical heterogeneity. In turn, analysis data which establishes such heterogeneity usually give sufficient information to predict that the rock has a coarse-grained texture.

Appreciation of lunar analytical situations which involve small subsamples of coarse-grained material of heterogeneous composition (e.g. 15065 and 15555) is worth further discussion with respect to lunar rock types. This particularly has most important bearing, regarding the data in Table 1, upon the classification scheme (Mason *et al.*, 1972; Rhodes, 1972; Chappell and Green, 1973; Helmke *et al.*, 1973; Rhodes and Hubbard, 1973) for Apollo 15 mare basalts. This classification presents two distinct types based on their bulk chemical compositions, petrographic features, and petrogenetic aspects, and includes data from fine-grained as well as coarse-grained basalts. As depicted in Fig. 1, this classification scheme designates as separate, chemically defined groups an olivine-normative, low-Si, high-Ti series and a quartz-normative, high-Si, low-Ti series of basalts. It should be obvious in Fig. 1 that referring to these groups alone as chemically discrete depends upon which fragment of a particular coarse-grained basalt has been examined (and upon the subsumption criteria). The bulk chemical compositions of different small-sized fragments of heterogeneous rocks 15065 and 15555 places each in the fields of both groups. The interesting observation that Apollo 15 mare basalts constitute distinct chemical groupings may still be possible (1) if one considers a third grouping, namely a group populated by different analyzed small-sized fragments of coarse-grained basalts (as also suggested by Rhodes and Hubbard, 1973). This third grouping, on the basis of parameters other than Si abundance and olivine norm, shows overlap with both the olivine-normative and quartz-normative types in Fig. 1. Members of such a third grouping may be distinct on the very basis of their coarse-grained nature; or (2) pending resolution of the compositional variation data of these coarse-grained basalts via analytical studies of larger, more "representative" bulk subsamples. The caution, however, is that these groupings may be artificial due to limited laboratory analyses of rock subsamples and/or to limited sampling of the rock types at the Apollo 15 site.

If a "representative" chemical analysis of a particular coarse-grained basalt is fundamental to its detailed interpretive petrology (Rhodes and Hubbard, 1973), then better designed chemical studies are warranted. Perhaps an improved

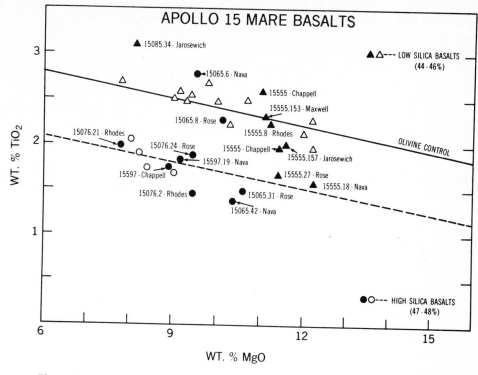

Fig. 1. Titanium and magnesium abundances in Apollo 15 mare basalts. Analysts' data for samples designated by filled symbols are from the following sources: Nava (this work), Jarosewich in Mason *et al.* (1972), Maxwell *et al.* (1972), Rose in Cuttitta *et al.* (1973), Chappell and Green (1973), and Rhodes in LSPET (1972).

approach would be via a judicious combination of classical wet chemical methods with high *accuracy* instrumental techniques applied to splits from larger powdered quantities of original parent samples, rather than undertaking chemical analyses solely via high *sensitivity* instrumental methods. From the standpoint of providing a single "representative" compositional analysis, resolution of the question of "adequate" sample size of any given material inherently becomes more of an enigmatic proposition as the degree, or scale, of chemical-petrologic heterogeneity increases. Since one cannot be in the position to fully understand the whole of a lunar sample (composition, history, genesis, etc.) without knowing the parts as well as the whole, chemically as well as petrologically, perhaps the more fundamental difficulty is that of "requiring" only general or singly "representative" compositional characteristics to be established by chemical analysis when evidence from other disciplines (e.g. petrology) indicates that certain specimens are heterogeneous (i.e. less homogeneous relative to materials that commonly are thought of as having a "norm" of homogeneity from subsample to subsample, as for example, Apollo 14 or Apollo 16 soils; fine-grained Apollo 15 mare basalts, etc). It should be noted that the petrologists encounter analogous difficulties in establishing agreeing descriptions of similar, but perhaps subtly distinct, specimen characterization features as discussed, for example, by Reid (1974).

Only Cuttitta *et al.* (1973) have even mentioned the observation that it is important also to have determined, via "unrepresentative" samples, that in fact these Apollo 15 mare basalts are chemically quite heterogeneous. For such specimen situations, it would appear fundamentally necessary that analytical efforts be designed and undertaken to provide information toward both aspects (namely heterogeneity characterization as well as bulk "representative" composition). In this way, more thorough specimen descriptions can be achieved, rather than from efforts which are intended to determine only one of these aspects (namely only the bulk "representative" composition). In terms of sample quantity, this combined approach would be expensive, but worthwhile certainly for some selected, particularly interesting lunar specimens.

Apollo 16 samples

Sample 61016 (Big Muley) is a coarse-grained nonvesicular anorthositic crystalline rock from Station 1 located on the Cayley Plains. Relative to the reported bulk composition of this rock (Rose *et al.*, 1973) the composition of this selected separate of whitest material (Table 2) is close to that of the anorthite end-member of the plagioclase feldspars (approximately 95% anorthite).

Unsieved soil fines 65500 is from Station 5 near South Ray Crater but onto the Stone Mountain part of the Descartes Formation. It is thought to be typical of the local soil with its probable origin being regolith possibly derived from underlying but reworked Descartes materials, as well as ejecta superposed on Descartes (ALGIT, Team, 1972). This soil, together with the very similar composition of soil 69941 ($a < 1$ mm fines soil sampled at Station 9 on the fringe of South Ray Crater ejecta) as determined by Haskin *et al.* (1973) for split 69941,25, exhibits the highest Si and Mg and the lowest Al and Ca contents compared to any other Apollo 16 soils reported thus far. Results reported by Duncan *et al.* (1973) for a split of the < 1 mm fines (i.e. 65501,3) show general similarity to parent 65500, but Si, Mg, Al, and Ca in 65501,3 are not at or near the compositional extremes defined by 65500,5 and 69941,25. Reported chemical studies (e.g. Haskin *et al.*, 1973; LSPET, 1973a; Philpotts *et al.*, 1973; Rose *et al.*, 1973) show that Apollo 16 soil fines exhibit a rather limited major, minor, and trace element compositional range. The chemical compositional extremes for Si, Mg, Al, and Ca in South Ray Crater area samples 65500 and 69941, and the generally high Al and Ca contents of soils from North Ray Crater (Duncan *et al.*, 1973; LSPET, 1973a), hint of some subtle chemical distinctions. In conjunction with the geological site descriptions (ALGIT, 1972), these chemical distinctions may relate to differences between the Descartes and the Cayley formations. The North Ray Crater soils probably contain more Descartes anorthositic cumulate. The present data for 65500 may imply a sampling of Cayley material (possibly South Ray Crater ejecta). This could lend support to the suggestion by Duncan *et al.* (1973) of a systematic but subtle compositional change in the traverse area being due to somewhat distinct Cayley and Descartes soil types. An apparent discrepancy regarding 69941 must be noted; however, in that analyses of two other splits, adjacent in split numbers to each other (69941,29 by Rose et al., 1973; 69941,30 by Laul and Schmitt, 1973), but not adjacent to that of 69941,25 by Haskin et al. (1973), gave bulk compositional results which agreed closely for these adjacent numbered splits, but which differed significantly from that of 69941,25, and hence also from 65500.

Sample 66095, a so-called "Rusty Rock" (Taylor et al., 1973) is an anorthositic breccia from Station 6 off the south rim of a subdued 10 m crater on the lowest observable bench of Stone Mountain and may possibly be ejected from South Ray Crater (ALGIT, 1972).

The southeast rim of North Ray Crater at Station 11 yielded sample 67559, a light-gray crystalline rock composed mostly of plagioclase. A boulder at Station 8 near the northeast rim of South Ray Crater yielded specimen 68415, an anorthositic gabbro crystalline rock. Aliquants from two separate 0.3 g interior fragments from our 4 g total sample chip weight of 68415 were analyzed and found to be, within the limits of analytical precision, virtually identical, although powdering during initial sample preparation led to a qualitative description of one fragment being "soft" and the other "hard" to powder. Petrographic description by Wilshire et al. (1973) terms this a fine-grained but texturally heterogeneous feldspathic rock containing large irregular inclusions, up to 1 cm, of plagioclase. The bulk chemical composition of 68415 presented in Table 2 agrees very well with that given by LSPET (1973a) and by Rose et al. (1973). Analyses of semi-micro sized quantities (i.e. 50 mg) of this rock appear to indicate sample homogeneity at this aliquant level. The extremely similar major and minor element chemical compositions (Table 2) of Apollo 16 rocks 67559 and 68415 are of particular interest when considered with their observed similar KREEP-like REE patterns (but at one-tenth the abundance level), Eu anomalies close to zero (Philpotts, unpublished data for 67559; Philpotts et al., 1973 for 68415), igneous aspects, and mineralogical, petrographical, and textural similarities described by Helz and Appleman (1973) and by Steele and Smith (1973). Following the argument of Helz and Appleman (1973) for 68415, rock 67559 may also be a hybrid, (even though the near-zero Eu anomaly alone might have suggested that both these rocks represent primitive liquids).

Apollo 17 samples

A vesicular, porphyritic coarse-grained basalt, sample 70017, was collected at the LM station near the center of the Taurus-Littrow Valley. The analysis reported in Table 2 shows it to be typical of the Apollo 17 subfloor basalts, which are characterized by low-Si and high-Ti concentrations and similar to Apollo 11 basalts (Rhodes, 1973; Rhodes et al., 1974). On the Fe versus Mg plot of Gast (1973) it lies in the field of intermediate iron-rich basalts. Due, however, to the coarse-grained nature of 70017, literature data for various aliquants indicate sample heterogeneity (e.g. Duncan et al., 1974a; Nava, 1974; Rhodes et al., 1974;

Rose *et al.*, 1974). This heterogeneity intimates similar analytical sampling difficulties as with the aforementioned Apollo 15 mare basalts.

Sample 74121 is the <1 mm soil fines collected as the LRV-6 grab sample of light-mantle material between Stations 3 and 4. It is somewhat high in Al and Ca contents and low in Ti, Fe, Cr, and Mn relative to a majority of the Apollo 17 soil analyses reported to date (e.g. LSPET, 1973b; Duncan *et al.*, 1974b; Mason *et al.*, 1974; Rhodes *et al.*, 1974; Rose *et al.*, 1974), indicating lower contents of mare basalt components. The chemistry of this light-mantle soil has been grouped compositionally with that of South Massif soils (e.g. Duncan *et al.*, 1974b; Rhodes *et al.*, 1974).

74220 is the unsieved bulk orange soil collected at Station 4 on the south rim of Shorty Crater. 74241 is the <1 mm fines portion of the adjacent 74240 gray reference soil.

Sample 76055 is a greenish gray breccia collected at Station 6 on the south slope of the North Massif and thought to be material that rolled off of North Massif. The chemical composition of 76055, except notably for higher Mg and lower Si, is generally similar to Apollo 14 KREEP-type basalts (e.g. Gast, 1973; Hubbard *et al.*, 1973).

Acknowledgments—The author is indebted to C. W. Kouns for powdering the breccia and basalt fragments and chips, for heavy-liquid separation of matrix from an aliquant of 15597, and for hand-picking of the anorthite-enriched separate of 61016. Deepest appreciation is expressed to the master analytical meteoriticist, Dr. H. B. Wiik, for continued interest in the research activities of this, his former student. And finally, in deference to Arch Reid for the author's probable opprobrious usage of geochemipetrographic terminology.

REFERENCES

ALGIT (Apollo Lunar Geology Investigation Team) (1972) Documentation and environment of the Apollo 16 samples: a preliminary report. U.S. Geol. Surv. Interagency Report: *Astrogeology* **51**, 252 pp. Washington, D.C.

Chappell B. W. and Green D. H. (1973) Chemical compositions and petrogenetic relationships in Apollo 15 mare basalts. *Earth Planet. Sci. Lett.* **18**, 237–246.

Chappell B. W., Compston W., Green D. H., and Ware N. G. (1972) Chemistry, geochronology, and petrogenesis of lunar sample 15555. *Science* **175**, 415–416.

Cuttitta F., Rose H. J., Jr., Annell C. S., Carron M. K., Christian R. P., Ligon D. T., Jr., Dwornik E. J., Wright T. L., and Greenland L. P. (1973) Chemistry of twenty-one igneous rocks and soils returned by the Apollo 15 mission. *Proc. Fourth Lunar Sci. Conf., Geochim. Cosmochim. Acta*, Suppl. 4, Vol. 2, pp. 1081–1096. Pergamon.

Duncan A. R., Erlank A. J., Willis J. P., and Ahrens L. H. (1973) Composition and interrelationships of some Apollo 16 samples. *Proc. Fourth Lunar Sci. Conf., Geochim. Cosmochim. Acta*, Suppl. 4, Vol. 2, pp. 1097–1113. Pergamon.

Duncan A. R., Erlank A. J., Willis J. P., Sher M. K., and Ahrens L. H. (1974a) Trace element evidence for a two stage origin of high titanium mare basalts (abstract). In *Lunar Science—V*, pp. 187–189. The Lunar Science Institute, Houston.

Duncan A. R., Erlank A. J., Willis J. P., and Sher M. K. (1974b) Compositional characteristics of the Apollo 17 regolith (abstract). In *Lunar Science—V*, pp. 184–186. The Lunar Science Institute, Houston.

Gast P. W. (1973) Lunar magmatism in time and space (abstract). In *Lunar Science—IV*, pp. 275–277. The Lunar Science Institute, Houston.

Haskin L. A., Helmke P. A., Blanchard D. P., Jacobs J. W., and Telander K. (1973) Major and trace element abundances in samples from the lunar highlands. *Proc. Fourth Lunar Sci. Conf., Geochim. Cosmochim. Acta*, Suppl. 4, Vol. 2, pp. 1275–1296. Pergamon.

Helmke P. A., Blanchard D. P., Haskin L. A., Telander K., Weiss C., and Jacobs J. W. (1973) Major and trace elements in igneous rocks from Apollo 15. *The Moon* **8**, 129–148.

Helz R. T. and Appleman D. E. (1973) Mineralogy, petrology, and crystallization history of Apollo 16 rock 68415. *Proc. Fourth Lunar Sci. Conf., Geochim. Cosmochim. Acta*, Suppl. 4, Vol. 1, pp. 643–659. Pergamon.

Hubbard N. J., Rhodes J. M., Gast P. W., Bansal B. M., Shih C., Wiesmann H., and Nyquist L. E. (1973) Lunar rock types: the role of plagioclase in non-mare and highland rock types. *Proc. Fourth Lunar Sci. Conf., Geochim. Cosmochim. Acta*, Suppl. 4, Vol. 2, pp. 1297–1312. Pergamon.

Laul J. C. and Schmitt R. A. (1973) Chemical composition of Apollo 15, 16, and 17 samples. *Proc. Fourth Lunar Sci. Conf., Geochim. Cosmochim. Acta*, Suppl. 4, Vol. 2, pp. 1349–1367. Pergamon.

LSPET (Lunar Sample Preliminary Examination Team) (1972) The Apollo 15 lunar samples: a preliminary description. *Science* **175**, 363–375.

LSPET (Lunar Sample Preliminary Examination Team) (1973a) The Apollo 16 lunar samples: petrographic and chemical description. *Science* **179**, 23–34.

LSPET (Lunar Sample Preliminary Examination Team) (1973b) Apollo 17 lunar samples: chemical and petrographic description. *Science* **182**, 659–672.

Lunar Sample Information Catalog Apollo 15 (1971) Lunar Receiving Laboratory MSC, Houston. MSC # 03209.

Mason B., Jarosewich E., and Melson W. G. (1972) Mineralogy, petrology, and chemical composition of lunar samples 15085, 15256, 15271, 15471, 15475, 15476, 15535, 15555, and 15556. *Proc. Third Lunar Sci. Conf., Geochim. Cosmochim. Acta*, Suppl. 3, Vol. 1, pp. 785–796. MIT Press.

Mason B., Jacobson S., Nelen J. A., Melson W. G., and Simkin T. (1974) Regolith compositions from the Apollo 17 mission (abstract). In *Lunar Science—V*, pp. 493–495. The Lunar Science Institute, Houston.

Maxwell J. A., Bouvier J. L., and Wiik H. B. (1972) Chemical composition of some Apollo 15 lunar samples. In *The Apollo 15 Lunar Samples*, pp. 233–238. The Lunar Science Institute, Houston.

Nava D. F. (1974) Chemistry of some rock types and soils from the Apollo 15, 16, and 17 lunar sites (abstract). In *Lunar Science—V*, pp. 547–549. The Lunar Science Institute, Houston.

Nava D. F. and Philpotts J. A. (1973) A lunar differentiation model in light of new chemical data on Luna 20 and Apollo 16 soils. *Geochim. Cosmochim. Acta* **37**, 963–973.

Philpotts J. A., Schnetzler C. C., Nava D. F., Bottino M. L., Fullagar P. D., Thomas H. H., Schuhmann S., and Kouns C. W. (1972) Apollo 14: some geochemical aspects. *Proc. Third Lunar Sci. Conf., Geochim. Cosmochim. Acta*, Suppl. 3, Vol. 2, pp. 1293–1305. MIT Press.

Philpotts J. A., Schuhmann S., Kouns C. W., Lum R. K. L., Bickel A. L., and Schnetzler C. C. (1973) Apollo 16 returned lunar samples: lithophile trace-element abundances. *Proc. Fourth Lunar Sci. Conf., Geochim. Cosmochim. Acta*. Suppl. 4, Vol. 2, pp. 1427–1436. Pergamon.

Reid A. (1974) Rock types present in lunar highland soils. *The Moon* **9**, 141–146.

Rhodes J. M. (1972) Major element chemistry of Apollo 15 mare basalts. In *The Apollo 15 Lunar Samples*, pp. 250–252. The Lunar Science Institute, Houston.

Rhodes J. M. (1973) Major and trace element chemistry of Apollo 17 samples. *EOS Trans. Amer. Geophys. Union* **54**, 609–610.

Rhodes J. M. and Hubbard N. J. (1973) Chemistry, classification, and petrogenesis of Apollo 15 mare basalts. *Proc. Fourth Lunar Sci. Conf., Geochim. Cosmochim. Acta*, Suppl. 4, Vol. 2, pp. 1127–1148, Pergamon.

Rhodes J. M., Rodgers K. V., Shih C., Bansal B. M., Nyquist L. E., and Wiesmann H. (1974) The relationship between geology and soil chemistry at the Apollo 17 landing site (abstract). In *Lunar Science—V*, pp. 630–632. The Lunar Science Institute, Houston.

Rose H. J., Jr., Cuttitta F., Berman S., Carron M. K., Christian R. P., Dwornik E. J., Greenland L. P., and Ligon D. T., Jr. (1973) Compositional data for twenty-two Apollo 16 samples. *Proc. Fourth Lunar Sci. Conf., Geochim. Cosmochim. Acta*, Suppl. 4, Vol. 2, pp. 1149–1158. Pergamon.

Schnetzler C. C. and Nava D. F. (1971) Chemical composition of Apollo 14 soils 14163 and 14259. *Earth Planet. Sci. Lett.* **11**, 345–350.

Schnetzler C. C., Philpotts J. A., Nava D. F., Schuhmann S., and Thomas H. H. (1972) Geochemistry of Apollo basalt 15555 and soil 15531. *Science* **175**, 426–428.

Steele I. M. and Smith J. V. (1973) Mineralogy and petrology of some Apollo 16 rocks and fines: general petrologic model of moon. *Proc. Fourth Lunar Sci. Conf., Geochim. Cosmochim. Acta,* Suppl. 4, Vol. 1, pp. 519–536. Pergamon.

Taylor L. A., Mao H. K., and Bell P. M. (1973) "Rust" in the Apollo 16 rocks. *Proc. Fourth Lunar Sci. Conf., Geochim. Cosmochim. Acta,* Suppl. 4, Vol. 1, pp. 829–839. Pergamon.

Wilshire H. G., Stuart-Alexander D. E., and Jackson E. D. (1973) Apollo 16 rocks: petrology and classification. *J. Geophys. Res.* **78**, 2379–2392.

Proceedings of the Fifth Lunar Conference
(Supplement 5, Geochimica et Cosmochimica Acta)
Vol. 2 pp. 1097–1117 (1974)
Printed in the United States of America

The relationships between geology and soil chemistry at the Apollo 17 landing site

J. M. RHODES,[1] K. V. RODGERS,[1] C. SHIH,[2] B. M. BANSAL,[1]
L. E. NYQUIST,[3] H. WIESMANN,[1] and N. J. HUBBARD[3]

[1]Lockheed Electronics Company, Houston, Texas

[2]Lunar Science Institute, Houston, Texas

[3]NASA Johnson Space Center, Houston, Texas

Abstract—Within the wide compositional range of the Apollo 17 soils, three distinct chemical groups have been recognized, each one corresponding broadly with a major geological and physiographic unit. These groups are: (a) Valley Floor type soils, (b) South Massif type soils, and (c) North Massif type soils. The observed chemical variations within and between these three groups is interpreted by means of mixing models in terms of lateral transport and mixing of prevailing local rock types, such as high-titanium basalts, KREEP-like noritic breccias, anorthositic gabbro breccias and orange glass.

The Valley Floor type soils are derived primarily by comminution of basalt and orange glass, and modified by mixing with varying amounts of previously homogenized massif soils. The extremely homogeneous South Massif type soils, which consist of approximately equal proportions of noritic breccia and anorthositic gabbro with minor basalt and orange glass, are believed to be derived by avalanching from the upper slopes of the South Massif. In contrast, the North Massif type soils evolved on the lower slopes of the massif and are extensively contaminated with basalt and orange glass. They also differ from the South Massif type soils in containing larger amounts of anorthositic gabbro relative to noritic breccia.

The contrasting composition of the massif type soils is interpreted as reflecting derivation from different levels of essentially stratigraphically similar massifs. According to this model, North Massif types evolved on the lower slopes of the North Massif and Sculptured Hills where anorthositic gabbro predominates over noritic breccia and where lateral mixing with basalt is effective, whereas the South Massif type soils originally developed on the upper slopes of the South Massif, where anorthositic breccia and noritic breccias are equally abundant, and where lateral mixing with basalt was minimal.

INTRODUCTION

THE APOLLO 17 MISSION had as one of its primary objectives the extensive sampling of lunar soils in order to characterize the main geological units in the valley of Taurus-Littrow. To this end, a total of 73 soil samples were collected, with particular attention to sampling the so-called dark and light mantles.

The chemical diversity of these soils is greater than that found at any previous landing site, ranging from soils that closely approach the high-titanium subfloor basalts in composition, to aluminous soils from the North and South Massifs. This paper outlines and discusses the compositional diversity of these soils with respect to their spatial distribution, the chemistry of prevailing rock types, and the inferred geological relationships at the site.

Thirty-seven analyses of 29 soil samples and two dark matrix or "soil" breccias collected from twelve of the major sampling stations are given in Tables 1A–C. Seventeen of the samples were analyzed during the Apollo 17 Preliminary

J. M. RHODES *et al.*

Table 1A. Chemical composition of Valley Floor type soils.

	70019,28* LM–ALSEP	70161,3 LM–ALSEP	70181,3 LM–ALSEP	71041,3 Sta. 1	71061,3 Sta. 1	71501,3 Sta. 1	74220,3† Sta. 4	74240,3 Sta. 4	74241,50 Sta. 4	74260,2 Sta. 4	75061,4 Sta. 5	75081,3 Sta. 5	75081,56 Sta. 5
Major element data (wt.%)													
SiO_2	40.66	40.34	40.87	39.74	40.09	39.82	38.57	40.78	41.55	41.22	39.32	40.27	40.00
TiO_2	8.26	8.99	8.11	9.57	9.32	9.52	8.81	8.61	7.45	7.68	10.31	9.41	9.40
Al_2O_3	12.38	11.60	12.30	10.80	10.70	11.13	6.32	12.54	13.35	13.25	10.42	11.31	11.18
FeO	16.38	17.01	16.37	17.73	17.85	17.41	22.04	15.84	14.89	15.31	18.19	17.20	17.30
MnO	0.24	0.23	0.24	0.24	0.24	0.25	0.30	0.24	0.22	0.23	0.25	0.25	0.25
MgO	9.50	9.79	9.82	9.72	9.92	9.51	14.44	9.15	9.19	9.47	9.53	9.59	9.42
CaO	11.03	10.98	11.05	10.72	10.59	10.85	7.68	11.36	11.54	11.37	10.72	10.97	10.87
Na_2O	0.47	0.32	0.35	0.35	0.36	0.32	0.36	0.38	0.48	0.38	0.33	0.33	0.38
K_2O	0.09	0.08	0.08	0.08	0.08	0.07	0.09	0.12	0.12	0.12	0.08	0.08	0.08
P_2O_5	0.07	0.08	0.06	0.07	0.07	0.06	0.04	0.09	0.10	0.09	0.06	0.07	0.07
S	0.10	0.12	0.11	0.13	0.13	0.12	0.07	0.14	0.12	0.12	0.13	0.12	0.11
Cr_2O_3	0.43	0.46	0.44	0.47	0.49	0.46	0.75	0.41	0.41	0.41	0.48	0.46	0.45
Total	99.61	100.00	99.80	99.62	99.84	99.52	99.47	99.66	99.42	99.65	99.82	100.06	99.51
Trace element data (ppm)													
Sr	—	168	169	165	174	157	205	163	154	167	166	165	159
Rb	—	1.4	1.9	1.1	1.1	1.2	1.2	2.3	2.5	2.0	1.6	1.1	1.3
Y	—	77	70	73	75	74	49	80	74	75	83	77	73
Zr	—	218	216	217	215	214	182	235	232	239	237	229	211
Nb	—	19	18	19	19	19	15	19	19	19	21	20	19
Zn	42	41	47	51	88	33	292	83	96	109	25	35	31
Ni	154	161	190	117	100	131	83	80	101	99	115	140	143

*Soil breccia.
†Orange soil.

Table 1B. Chemical composition of South Massif type soils.

	72321,5 Sta. 2	72441,3 Sta. 2	72461,3 Sta. 2	72501,2 Sta. 2	72501,22 Sta. 2	72701,2 Sta. 2	72701,21 Sta. 2	73121,6 Sta. 2A	73141,1 Sta. 2A	73141,7 Sta. 2A
Major element data (wt.%)										
SiO_2	44.91	45.03	44.98	45.12	45.17	44.87	44.96	44.60	45.06	44.91
TiO_2	1.56	1.53	1.50	1.56	1.55	1.52	1.53	1.42	1.29	1.24
Al_2O_3	20.57	20.51	20.87	20.64	20.63	20.60	20.55	20.83	21.52	21.42
FeO	8.65	8.85	8.58	8.77	8.74	8.65	8.94	8.59	8.10	8.14
MnO	0.13	0.13	0.12	0.11	0.13	0.12	0.13	0.13	0.11	0.12
MgO	9.84	9.89	9.69	10.08	9.87	9.97	9.98	10.00	10.04	9.94
CaO	12.82	12.83	12.97	12.86	12.84	12.80	12.83	12.87	13.04	13.06
Na_2O	0.47	0.46	0.47	0.40	0.46	0.40	0.51	0.44	0.38	0.44
K_2O	0.16	0.17	0.17	0.16	0.17	0.16	0.17	0.15	0.15	0.15
P_2O_5	0.15	0.17	0.16	0.13	0.15	0.15	0.14	0.13	0.12	0.12
S	0.06	0.07	0.06	0.09	0.06	0.07	0.07	0.07	0.06	0.06
Cr_2O_3	—	0.22	0.21	0.23	0.23	0.23	0.23	0.22	0.21	0.21
Total	99.32	99.88	99.78	100.15	100.00	99.54	100.04	99.45	100.08	99.81
Trace element data (ppm)										
Sr	—	155	155	153	153	155	153	150	148	151
Rb	—	4.3	4.2	4.2	4.5	3.9	3.9	3.5	3.5	3.8
Y	—	64	61	64	63	54	60	55	54	55
Zr	—	278	265	271	293	275	267	232	236	238
Nb	—	19	18	18	18	18	18	17	15	15
Zn	—	21	21	21	21	22	20	18	18	19
Ni	—	225	225	241	231	227	230	236	195	193

Examination (LSPET, 1973a), and are included here for the sake of completeness. The remainder, including duplicate analyses of different sub-samples, were analyzed subsequently. The data were obtained by X-ray fluorescence analysis following procedures used for the preliminary examination of the Apollo 15, 16 and 17 samples (LSPET, 1972, 1973b, 1973a); that is, major and minor elements were measured on a fused-glass disk, prepared by fusing a 280 mg aliquant of sample with a lanthanum-bearing lithium borate fusion mixture (Norrish and Hutton, 1969). Sodium was analyzed separately on a 10–20 mg portion by atomic absorption analysis. Trace elements were determined non-destructively using pressed powdered samples, and corrections made for matrix effects either by direct measurement of mass-absorption coefficients (Norrish and Chappell, 1967) or by calculating them from the major element data. Calibrations for both major and trace elements are based on primary synthetic standards, supplemented by previously analyzed U.S.G.S. and N.B.S. rock and mineral standards.

To supplement the data in Tables 1A–C, a single typical soil sample was selected, on the basis of the XRF data, from each of the major sampling stations and analyzed by stable isotope dilution analysis (Gast et al., 1970) for the large ion lithophile elements. These data are presented in Table 2.

J. M. Rhodes *et al.*

Table 1C. Chemical composition of North Massif type soils.

	76241,14 Sta. 6	76261,15 Sta. 6	76281,4 Sta. 6	76321,7 Sta. 6	76501,2 Sta. 6	76501,42 Sta. 6	77531,3 Sta. 7	78501,2 Sta. 8	79135,1* Sta. 9	79221,2 Sta. 9	79261,2 Sta. 9	72141,9 LRV 2	72161,6 LRV 3	70051,24 BSLSS-Residue
Major element data (wt.%)														
SiO_2	43.20	43.64	43.56	44.08	43.41	43.34	43.07	42.67	42.29	41.67	42.26	43.11	42.12	42.05
TiO_2	3.31	3.38	3.83	3.00	3.15	3.15	3.91	5.47	5.15	6.52	6.09	4.37	5.21	5.04
Al_2O_3	17.85	17.96	17.80	18.41	18.63	18.41	17.16	15.73	15.08	13.57	14.43	16.10	14.22	16.15
FeO	10.92	10.93	11.26	10.53	10.32	10.39	11.70	13.15	14.01	15.37	14.60	13.45	14.86	12.81
MnO	0.16	0.16	0.16	0.15	0.14	0.15	0.17	0.18	0.19	0.21	0.20	0.19	0.22	0.19
MgO	11.05	10.75	10.55	10.82	11.08	11.08	10.19	9.91	10.42	10.22	9.82	10.25	10.54	10.25
CaO	11.97	12.11	12.18	12.23	12.28	12.24	11.93	11.77	11.44	11.18	11.48	11.83	11.17	11.87
Na_2O	0.43	0.45	0.43	0.46	0.35	0.40	0.44	0.35	0.40	0.34	0.35	0.40	0.41	0.43
K_2O	0.12	0.12	0.11	0.13	0.10	0.11	0.11	0.09	0.10	0.09	0.11	0.12	0.11	0.10
P_2O_5	0.09	0.11	0.09	0.09	0.08	0.09	0.08	0.05	0.07	0.06	0.07	0.10	0.08	0.06
S	0.07	0.07	0.07	0.07	0.07	0.07	0.08	0.10	0.10	0.12	0.12	0.09	0.08	0.08
Cr_2O_3	—	0.28	0.29	0.26	0.26	0.27	0.31	0.37	0.39	0.42	0.40	0.37	0.42	0.33
Total	99.17	99.96	100.33	100.23	99.87	99.70	99.15	99.84	99.64	99.77	99.93	100.38	99.44	99.36
Trace element data (ppm)														
Sr	151	151	150	151	147	145	153	155	166	156	153	153	156	150
Rb	—	2.7	2.3	3.2	2.5	2.6	2.7	2.1	2.1	1.7	1.9	2.2	1.9	1.8
Y	52	52	48	54	46	43	52	58	55	61	59	53	55	49
Zr	197	197	174	204	158	168	198	189	185	193	183	197	207	169
Nb	15	15	13	15	13	13	15	15	14	16	16	15	16	14
Zn	26	26	33	26	29	32	31	40	72	51	48	50	58	34
Ni	182	182	169	210	206	200	231	194	218	236	177	271	273	169

*Soil breccia.

Table 2. Isotope dilution analyses of Apollo 17 soils.

		Valley Floor type soils					South Massif type soils			North Massif type soils					
		70181,3 LM–ALSEP	71501,3 Sta. 1	75061,4 Sta. 5	74241,50 Sta. 4	74220,44* Sta. 4	72501,22 Sta. 2	72701,21 Sta. 2	73141,7 Sta. 2A	76501,2 Sta. 6	77531,3 Sta. 7	78501,2 Sta. 8	79261,12 Sta. 9	72141,9 LRV 2	72161,6 LRV 3
Li	ppm	9.1	8.7	8.4	9.6	10.7	11.2	—	9.9	8.2	9.1	7.8	8.6	9.3	9.2
K	ppm	726	618	623	1000	701	1360	1330	1222	873	921	792	802	888	864
Rb	ppm	1.47	1.17	1.11	2.42	1.11	4.08	3.94	3.62	2.40	2.49	1.96	1.80	2.26	2.04
Sr	ppm	170	159	166	—	209	—	—	151	151	155	154	154	156	—
Ba	ppm	98.3	87.0	89.5	112	76.4	200	192	169	115	123	105	93.7	119	114
La	ppm	8.09	7.14	7.07	9.95	6.25	17.1	17.0	13.8	8.95	9.96	8.29	8.31	10.2	9.71
Ce	ppm	24.8	22.2	23.6	28.8	19.0	44.6	43.6	37.1	24.3	26.7	23.3	22.2	28.1	27.0
Nd	ppm	—	21.0	23.1	24.0	17.8	—	—	24.0	17.4	19.4	18.4	17.8	20.7	—
Sm	ppm	8.05	8.02	9.09	8.55	6.53	8.18	8.07	6.93	5.55	6.47	6.36	6.18	6.70	6.94
Eu	ppm	1.66	1.66	1.77	1.60	1.80	1.33	1.31	1.22	1.25	1.35	1.37	1.39	1.35	1.42
Gd	ppm	12.0	12.2	13.4	12.0	8.52	10.4	10.1	9.15	7.51	8.64	9.34	8.49	9.00	9.50
Dy	ppm	13.2	13.6	15.5	13.7	9.40	11.1	10.9	9.42	8.18	9.55	10.2	10.0	9.82	10.4
Er	ppm	7.63	7.95	9.02	8.07	5.10	6.58	6.46	5.60	4.89	5.72	6.02	5.99	5.72	6.01
Yb	ppm	7.02	7.37	8.36	7.45	4.43	6.15	5.99	5.26	4.53	5.26	5.54	5.53	5.28	5.51
Lu	ppm	—	—	—	—	0.61	—	—	—	—	—	—	0.71	—	—
Cr	ppm	2570	2900	3090	2680	4650	1480	—	1320	1740	1320	2300	2530	—	2670
Th	ppm	—	—	—	—	—	—	—	2.64	—	—	—	—	—	—
U	ppm	0.28	0.23	0.21	0.37	0.16	0.87	0.81	0.73	0.44	—	0.36	0.32	—	0.41
Zr	ppm	—	—	—	218	185	259	—	231	163	203	195	204	175	192
K/Rb	ppm	494	528	563	413	631	333	338	338	364	370	404	445	393	423

*Orange soil.

J. M. RHODES *et al.*

COMPOSITION OF MAJOR ROCK TYPES

A significant factor contributing to the wide compositional range of the Apollo 17 soils is the contrast in lithologies in the Taurus-Littrow Valley. Thus in order to evaluate the contribution of locally derived components to the chemical diversity of these soils, it is first necessary to survey briefly the chemistry of the returned rock samples and to identify prevailing compositional groups.

The data presented in Table 3 in conjunction with previously published whole rock analyses (LSPET, 1973a) indicate that, with few exceptions, three distinct, chemically defined rock types have been sampled. These are: subfloor basalts from the valley floor, and KREEP-like noritic breccias and anorthositic gabbro clasts and breccias from the North and South Massifs (LSPET, 1973a; Rhodes, 1973). These groups are shown in Figs. 1 and 2 in terms of Al_2O_3/FeO and Al_2O_3/Zr variation, these variables being three of the best discriminants for this particular data set. Averages or prevalent compositions for these three major rock types are given in Table 4.

The subfloor basalts appear to belong to two compositional groups. The majority of samples are olivine normative with $SiO_2 < 39\%$ and $TiO_2 > 12\%$, and are thus chemically distinct from Apollo 11 basalts. One sample (75055) appears to

Table 3. Composition of Apollo 17 rocks.

	Basalts			Noritic breccias					
	70017,35	70215,56	75075,58	73235,55	73275,30	76315,30M Matrix	76315,30,3 Clast	76315,35 Matrix	76315,52 Clast
Major element data (wt.%)									
SiO_2	38.07	38.46	37.64	46.20	46.16	45.64	46.45	46.21	48.57
TiO_2	13.10	12.48	13.45	0.67	1.43	1.50	1.43	1.50	0.32
Al_2O_3	8.79	9.01	8.20	21.28	18.49	17.53	18.18	18.14	17.91
FeO	18.07	19.40	18.78	7.32	9.05	9.53	8.83	8.95	7.66
MnO	0.27	0.29	0.28	0.11	0.13	0.13	0.13	0.12	0.13
MgO	9.81	7.91	9.49	11.05	11.54	12.50	12.34	12.02	13.84
CaO	10.30	10.94	10.29	12.55	11.30	10.97	11.30	11.32	10.36
Na_2O	0.40	0.42	0.40	0.48	0.67	0.70	0.64	0.60	0.47
K_2O	0.04	0.05	0.05	0.20	0.27	0.26	0.22	0.26	0.15
P_2O_5	0.05	0.10	0.05	0.20	0.26	0.30	0.29	0.29	0.12
S	0.15	0.17	0.10	0.04	0.08	0.08	0.07	0.07	0.00
Cr_2O_3	—	0.39	0.57	—	—	0.19	0.20	0.19	—
Total	99.05	99.62	99.36	100.10	99.38	99.33	100.08	99.67	99.53
Trace element data (ppm)									
Sr	—	123	166	—	—	174	172	177	—
Rb	—	0.9	0.5	—	—	6.7	3.6	6.2	—
Y	—	69	81	—	—	113	107	111	—
Zr	—	185	208	—	—	506	478	522	—
Nb	—	21	21	—	—	33	32	33	—
Zn	—	6	5	—	—	3	2	4	—
Ni	—	4	<1	—	—	77	82	74	—

Table 3. (*Continued*).

	Noritic breccias				Anorthositic gabbro	Troctolite
	76015,22 Matrix	76015,37 Matrix	76015,41 Matrix	76015,64 Matrix	77135,5	76315,62 Clast	76535,21
Major element data (wt.%)							
SiO_2	46.16	46.38	46.38	46.59	46.17	45.10	42.88
TiO_2	1.52	1.55	1.53	1.48	1.53	0.36	0.05
Al_2O_3	17.17	17.78	17.77	18.00	17.83	26.37	20.73
FeO	9.81	9.65	9.07	9.10	9.14	5.29	4.99
MnO	0.13	0.13	0.12	0.12	0.13	0.07	0.07
MgO	13.03	12.40	12.67	12.43	12.39	7.46	19.09
CaO	10.77	11.13	11.11	11.10	11.08	15.12	11.41
Na_2O	0.70	0.72	0.69	0.75	0.69	0.47	0.23
K_2O	0.26	0.26	0.26	0.29	0.27	0.10	0.03
P_2O_5	0.27	0.29	0.29	0.28	0.30	0.06	0.03
S	0.09	0.06	0.08	0.08	0.07	0.04	0.00
Cr_2O_3	—	—	—	—	0.21	—	0.11
Total	99.91	100.35	99.97	100.22	99.81	100.44	99.62
Trace element data (ppm)							
Sr	—	—	—	—	174	—	111
Rb	—	—	—	—	6.2	—	0.2
Y	—	—	—	—	111	—	4.4
Zr	—	—	—	—	508	—	12
Nb	—	—	—	—	33	—	1.2
Zn	—	—	—	—	4	—	1
Ni	—	—	—	—	62	—	25

represent a separate type that compares closely with Apollo 11 low-K basalts, having $SiO_2 > 39\%$ and $TiO_2 < 12\%$. Texturally, the low-silica basalts correspond to the type 2 and 3 basalts described by Papike and Bence (1974), and the high-silica basalts to their type 1 group. On the basis of the petrography of 2–4 mm soil fragments, these authors conclude that the type 1 basalts are a relatively minor component at the Apollo 17 site. Consequently, the average basalt composition given in Table 4 includes only the low-silica, high-titanium variants.

Although the massif rocks are petrographically and texturally complex (LSPET, 1973a), they are essentially bimodal compositionally, consisting of a large group of KREEP-like noritic breccias and a considerably smaller group of anorthositic gabbro clasts and breccias (Figs. 1 and 2). The matrix of the noritic breccias is remarkably uniform in composition, with no discernible distinction between those from the North and South Massifs (Table 3; LSPET, 1973a, Table 1). Averages for the noritic breccias and anorthositic gabbros are given in Table 4. The noritic breccia composition is comparable to the Apollo 16 KREEP-like breccias (LSPET, 1973b; Hubbard *et al.*, 1973), and both resemble the "low- to moderate-K Fra Mauro" composition proposed initially on the basis of glass compositions in the Apollo 15 soils (Reid *et al.*, 1972a). The anorthositic gabbro

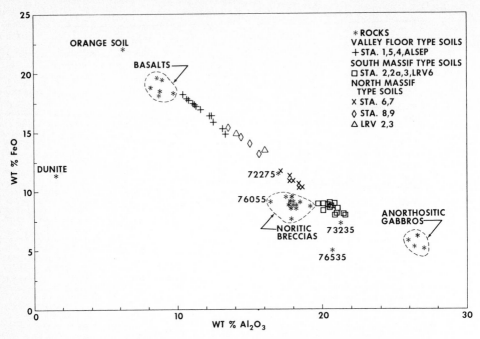

Fig. 1. Al₂O₃–FeO relationships in Apollo 17 soils and rocks. Data from Tables 1 and 3
and from LSPET (1973a). Data from LRV 6 and Station 3 from Duncan *et al.* (1974) and
Rose *et al.* (1974).

breccias, on the other hand, compare closely to the "highland basalt" composition
(Reid *et al.*, 1972b).

In addition to these two rock types, there are a small number of aluminum-,
iron-, and magnesium-rich variants of the noritic breccias (73235, 72275, 76055), a
dunite clast (72415), and a troctolite (76535). These have been omitted from the
compilation in Table 4, but are plotted in Figs. 1 and 2.

LSPET (1973a) and Heiken and McKay (1974a, 1974b) have shown that the
orange glass, and the black droplets that are inferred to be the quench-crystallized
chemical equivalent of the orange glass, form an additional major component in
the Apollo 17 soils. The bulk composition of the orange soil from Shorty Crater,
which is made up almost entirely of these droplets, is included in Table 4 as a
useful estimate of this component.

CHEMISTRY OF THE APOLLO 17 SOILS

In an earlier publication, Rhodes (1973) recognized, within the wide compositional range of Apollo 17 soils, three compositional groups, each one related to a
specific geological feature at the Apollo 17 landing site. The additional data
presented here confirms these three groups, both by inspection of the data and by
the more objective classificatory procedures of cluster analysis.

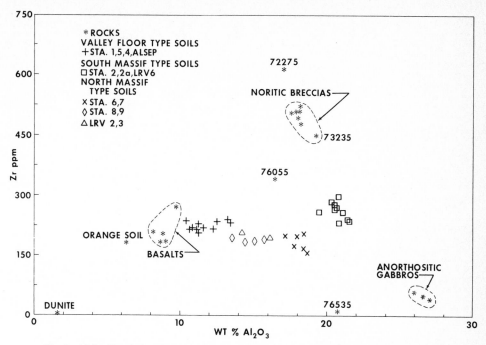

Fig. 2. Al₂O₃–Zr relationships in Apollo 17 soils and rocks. Data from Tables 1 and 3 and from LSPET (1973a). Data from LRV 6 from Duncan *et al.* (1974).

The clustering procedure outlined by Reid *et al.* (1972c) was used to assign soil samples into groups, such that the samples in a particular group are more similar to each other than they are to those in other groups, on the combined basis of all of the major, minor and trace elements (except Ni) given in Table 1A–C. The following three major chemical groups were distinguished:

I. Valley Floor Type Soils—soils from Stations 1, 5, 4, and LM-ALSEP
II. South Massif Type Soils—soils from Stations 2, 2A, and LRV 6
III. North Massif Type Soils—soils from Stations 6, 7, 8, 9, LRV 2, and LRV 3.

These groups are entirely consistent with a classification by visual inspection of the data, made largely with the help of Figs. 1 and 2. Similar conclusions were reached by Duncan *et al.* (1974), on the basis of principal component analysis. The rare-earth data (Table 2), which was not used in the cluster analysis, provide an independent test of this classification. Figure 3 illustrates the range in rare-earth concentrations, relative to chondrites, for each of the three major soil types, and clearly sustains the classification based on data from Table 1A–C. For example, the South Massif type soils are distinctly enriched in light rare earths relative to the other two soil types. Similarly, there is a marked hiatus in heavy rare-earth concentrations between the Valley Floor and North Massif type soils.

Table 4. Prevalent compositions for locally derived Apollo 17 soil
components.

	High-Ti subfloor basalt	Noritic breccia	Anorthositic gabbro	Orange soil
SiO$_2$	37.84	46.15	44.82	38.57
TiO$_2$	13.03	1.50	0.31	8.81
Al$_2$O$_3$	8.70	18.01	26.48	6.32
FeO	18.87	9.16	5.61	22.04
MnO	0.28	0.12	0.08	0.30
MgO	9.12	12.33	6.87	14.44
CaO	10.41	11.16	15.23	7.68
Na$_2$O	0.38	0.65	0.36	0.36
K$_2$O	0.05	0.26	0.07	0.09
P$_2$O$_5$	0.07	0.28	0.05	0.04
S	0.16	0.08	0.04	0.07
Cr$_2$O$_3$	0.50	0.20	0.13	0.75
Sr	147	173	144	205
Rb	0.6	5.7	1.3	1.2
Y	75	109	15	49
Zr	195	495	50	182
Nb	21	32	4	15
Zn	4.9	3.2	3.3	292
Ni	2	99	105	83

In order to demonstrate the close relationship between bulk chemistry and spatial and geological features, these groups have been given physiographically related names. It must be emphasized however, that the classification is entirely chemical and does not necessitate that a specific soil be located within the designated physiographic unit. Thus, soils at Station 9, for example, are located on the valley floor, yet they show closer chemical similarities to the soils from Stations 6 and 7 than to the valley floor soils from Station 1 or 5. Consequently, they are classified within the North Massif soil type. Similar arguments apply to the soils from LRV 2, LRV 3, and Station 4.

The overall correlation between soil chemistry and geographical location, together with substantial compositional changes in the soil from one geological unit to another, suggest that the bulk of these soils are derived locally by comminution and lateral mixing of ejecta from impact craters. If so, then the observed variations in soil chemistry can be usefully interpreted in terms of mixing locally derived rock types. This has been done in the following sections, where a least-squares compositional mixing model has been used to explain the composition of the soils in terms of the four prevailing rock types (Table 4) outlined in the previous section. The method used is basically the same as that described by Bryan *et al.*, (1969), with the exception that each of the 18 variables is given equal weight in order to allow for the use of both major and trace

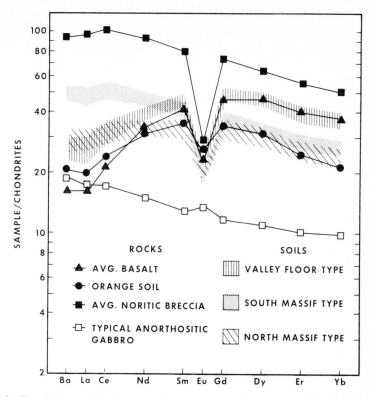

Fig. 3. Chondrite-normalized rare-earth patterns for Apollo 17 rocks and soils. Soils data from Table 2; data for average noritic breccias and anorthositic gabbro (76315,62) from Phinney *et al.* (1974). The basalt average is based on unpublished data.

elements in the model. Other weighting methods, such as weighting variables according to some measure of their precision (Schonfeld and Meyer, 1972), have also been tried. Although the results are essentially similar, we do not favor this approach since there is no *a priori* reason why elements that can be precisely determined should necessarily provide the most useful geological information. Nickel, which is assumed to be concentrated in a small meteoritic component, has been omitted from the calculations. The results of these calculations, normalized to 100%, are given in Table 5 and illustrated by means of pie diagrams in Fig. 4.

South Massif type soils

Soil samples from the South Massif (Station 2) and the light mantle (Stations 2A, 3, and LRV 6) are very uniform in composition (Tables 1A and 5; Figs. 1 and 2), and are characterized by high Al_2O_3 concentrations (19–21%), and low amounts of FeO (8–10%) and TiO_2 (1.3–2.6%). These soils have higher concentrations of P, K, Zr, and Rb, and are markedly enriched in the light rare-earth elements (Fig. 3) relative to the other soils.

Table 5. Average soil composition at Apollo 17 sample stations and the inferred proportions of locally derived components.

	Valley Floor type soils				South Massif type soils				North Massif type soils					
	LM-ALSEP	Sta. 1	Sta. 5	Sta. 4	Sta. 2	Sta. 2A	LRV 6*	Sta. 3*	Sta. 6	Sta. 7	Sta. 8	Sta. 9	LRV 2	LRV 3
Major element data (wt.%)														
SiO_2	40.62	39.88	39.86	41.18	45.01	44.86	44.51	44.84	43.54	43.07	42.67	42.07	43.11	42.12
TiO_2	8.45	9.47	9.71	7.91	1.54	1.32	2.56	1.81	3.30	3.91	5.47	5.92	4.37	5.21
Al_2O_3	12.09	10.88	10.97	13.05	20.62	21.26	19.36	20.29	18.18	17.16	15.73	14.36	16.10	14.22
FeO	16.59	17.66	17.56	15.35	8.74	8.28	10.24	8.75	10.73	11.70	13.15	14.66	13.45	14.86
MnO	0.24	0.24	0.25	0.23	0.12	0.12	0.13	0.11	0.15	0.17	0.18	0.20	0.19	0.22
MgO	9.70	9.72	9.51	9.27	9.90	9.99	9.93	10.25	10.89	10.19	9.91	10.15	10.25	10.54
CaO	11.02	10.72	10.85	11.42	12.85	12.99	12.44	12.89	12.17	11.93	11.77	11.37	11.83	11.17
Na_2O	0.38	0.34	0.35	0.41	0.45	0.42	0.40	0.42	0.42	0.44	0.35	0.36	0.40	0.41
K_2O	0.08	0.08	0.08	0.12	0.17	0.15	0.13	0.16	0.11	0.11	0.09	0.10	0.12	0.11
P_2O_5	0.07	0.07	0.07	0.09	0.15	0.12	0.14	0.15	0.09	0.08	0.05	0.07	0.10	0.08
S	0.11	0.13	0.12	0.13	0.07	0.06	0.08	—	0.07	0.08	0.10	0.11	0.09	0.08
Cr_2O_3	0.44	0.47	0.46	0.41	0.23	0.21	0.27	0.26	0.27	0.31	0.37	0.40	0.37	0.42
Total	99.79	99.66	99.79	99.57	99.85	99.78	100.19	99.93	99.92	99.15	99.84	99.77	100.38	99.44
Trace element data (ppm)														
Sr	169	165	164	162	154	150	151	—	149	153	155	158	153	156
Rb	1.7	1.1	1.3	2.3	4.2	3.6	3.6	—	2.7	2.7	2.1	1.9	2.2	1.9
Y	73	74	78	76	61	55	54	—	49	52	58	58	53	56
Zr	217	215	226	235	275	235	244	—	180	198	189	187	197	207
Nb	19	19	20	19	18	16	17	—	14	15	15	15	15	16
Zn	43	57	30	96	21	18	24	—	29	31	40	57	50	58
Ni	168	116	133	93	230	208	245	—	193	231	194	210	271	273
Content of major rock types														
Basalt	54.0	61.9	66.7	41.7	1.8	2.7	11.0	—	15.9	22.0	34.0	35.9	23.3	27.4
Orange glass	16.8	17.6	12.5	21.9	6.7	6.2	8.0	—	11.7	13.2	14.1	19.4	19.4	26.2
Noritic breccia	12.6	9.0	10.8	20.3	47.9	39.7	37.5	—	24.1	22.8	13.3	13.5	21.2	17.9
Anorthositic gabbro	16.7	11.5	10.0	16.1	42.9	51.3	43.5	—	48.3	42.1	38.5	31.2	36.1	28.5
% Noritic breccia in massif component	43.0	43.9	51.9	55.8	52.7	43.6	46.3	—	33.3	35.1	25.7	30.2	37.0	38.6
% Orange glass in mafic component	23.7	22.1	15.8	34.4	78.8	30.9	42.1	—	42.4	37.5	29.3	35.1	45.4	48.9

*Data for LRV 2 and Station 3 from Duncan et al. (1974) and Rose et al. (1974), respectively.

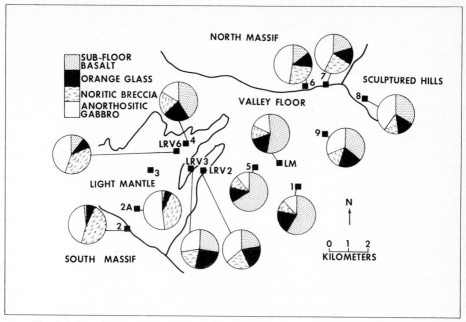

Fig. 4. Inferred proportions of locally derived components in the Apollo 17 soils. Each "pie" diagram is based on the average data given in Table 5.

Soils from Stations 2 and 2A have the lowest iron, titanium and chromium concentrations and are inferred, on the basis of mixing model calculations, to consist of roughly equal proportions of massif-derived noritic breccia and anorthositic gabbro (44–53% noritic breccia), with small amounts of orange glass (about 6%) and almost no basaltic component (Table 5; Fig. 4), these being derived laterally from the valley floor by impact processes. Heiken and McKay (1974b) suggest that the paucity of basaltic material in these soils is consistent with derivation from an avalanche deposit (Howard, 1973), originating high on the South Massif above the zone of lateral mixing with subfloor basalts. If this material is representative of the upper South Massif, it implies that it is composed of almost equal proportions of noritic breccias and anorthositic gabbros. This interpretation is in conflict with the preponderance of noritic breccias in the returned sample collection, and suggests that they may have been sampled preferentially to possibly less-coherent anorthositic material.

The uniformity of the South Massif soils, particularly in the relative proportions of noritic breccia and anorthositic gabbro, is surprising in view of the substantial compositional difference between the two major massif rock types and is in marked contrast to the wider compositional range generated by mixing massif material with subfloor basalt. Disparity between soil and rock variance is also observed at the Apollo 16 site, and to a lesser extent at the Apollo 14 site, and appears to be a characteristic of the lunar highlands. This feature may be attributed either to extensive impact mixing of soils early in lunar history, or to the derivation of the soils by degradation from rocks that, at the outcrop scale, have been intimately mixed during periods of extensive impact and brecciation. If the former applies, it follows that the South Massif soils were extensively mixed prior to basalt extrusion, and that subsequent to this event mixing was less efficient, permitting the development of a spectrum of soil compositions ranging between basalt and an already well-mixed massif soil component, intermediate in composition between the noritic breccias and anorthositic gabbros.

The basaltic and orange glass content of these soils increases from Stations 2 and 2A along the Light Mantle, away from the South Massif to LRV 6 (Duncan *et al.*, 1974), the most mafic soil in this group. Major element data for Station 3 soils (Rose *et al.*, 1974) indicate that these soils are intermediate in composition between those of LRV 6 and Stations 2 and 2A.

The soil at Station 2A is layered, the upper 15 cm of medium-gray soil being slightly more mafic, with a higher inferred basalt content (5%) than the underlying light-gray material. The basaltic component of the upper layer was probably added after the landslide that created the Light Mantle about 100 m.y. ago (AFGIT, 1973), and provides a measure of the small amount of lateral transport and mixing that has taken place since that time.

An interesting aspect of the mafic component in these soils, and also in the North Massif type soils, is the relative proportion of orange glass to basalt, which is higher than the proportions in the Valley Floor type soils, the presumed source of these components. This can be explained if the orange glass component and its devitrified derivatives are concentrated in the fine-sized fractions of the Valley Floor type soils, permitting them to be transported further relative to basalt during impact processes. The size fraction analysis of 70181 presented by LSPET (1973a) indicates that the orange soil and its equivalents are in greater abundance in the finer sized fraction, as does the Zn data presented by Duncan *et al.* (1974) for different size fractions of 75081.

Valley Floor type soils

Soils sampled at Stations 1, 5, and 4 and the LM-ALSEP site, as well as dark matrix "soil" breccia 70019, belong to a coherent, continuously variable compositional group, ranging from the more mafic soils found at Stations 5 and 1 to more aluminous soils from the LM–ALSEP site and Station 4 (Table 5). Chemically, they are the most mafic of the Apollo 17 site and are characterized by low aluminum and high iron, titanium, and chromium concentrations. Yttrium, niobium, zinc, and the heavy rare-earth elements are much more abundant in these than in any of the other soils from this site, and the chondrite-normalized rare-earth abundance patterns for these soils closely resemble those of the high-titanium subfloor basalts, particularly with respect to the depletion of the light rare earths relative to samarium or gadolinium (Fig. 3).

Most of the samples of this soil type were collected from the valley floor, where they overlie subfloor basalts. Although the gray soils from Shorty Crater (Station 4) are located on the light Mantle, their classification, on chemical grounds, with the valley floor soils is not inconsistent, since Shorty ejecta includes substantial contributions from underlying older valley floor soils and subfloor basalts (AFGIT, 1973).

Mixing model calculations using the prevalent compositions of locally derived components (Table 4) show that the chemistry of this group of soils is dominated by subfloor basalts (42–67%), with subordinate contributions of orange glass or its crystallized equivalents (12–22%), and aluminous massif components (21–36%) (Table 5; Fig. 4).

The orange glass and its devitrified equivalents are at their highest concentration in these Valley Floor type soils, and account for the high Zn concentrations. They are particularly abundant at Shorty Crater, where they constitute virtually the entire soil in 74220, and about 22% of the associated gray soils. The inferred wide distribution of the orange glass in the valley floor soils (Fig. 4) supports the suggestion of Heiken and McKay (1974a, 1974b) that it is not restricted to a local source but may be interlayered with the basalts and mixed into the soil by subsequent impacts. The work of Pieters et al. (1974) attributes the spectral reflectance of "dark-mantle" soils to the presence of black droplets of the type suggested by Heiken and McKay to be the crystallized equivalents of the orange glass, thus implying that these droplets and the orange glass are of regional extent, not only in the Taurus-Littrow Valley, but also elsewhere on the moon.

The calculated massif component in these soils is low, and is composed of approximately equal proportions of noritic breccia and anorthositic gabbro (Table 5; Fig. 4). The range (44–56%) in the proportions of noritic breccia to total massif component is similar to that found in the South Massif and Light Mantle type soils, suggesting that basalt and orange glass have been mixed with previously well-mixed massif soils, and not with the massif rocks. This point is emphasized in Fig. 1 by the well-developed linear trend shown by the soils, from the basalt field to the South Massif type soils.

Within this rather limited range in composition of massif component in the soils, there is a suggestion that the proportion of anorthositic gabbro increases slightly with respect to noritic breccia in soils that are furthest from the South Massif, and correspondingly closer to the North Massif (Table 5). This trend is further emphasized by soils at Station 9 which, although classified on chemical grounds as North Massif type soils, are intermediate between North Massif and Valley Floor type soils (Figs. 1 and 2). Can this trend be interpreted as being produced by varying contributions of the South and North Massifs to the Valley Floor type soils with distance across the Taurus-Littrow Valley? The inferred differences in noritic breccia and anorthositic gabbro components in the South and North Massif soils discussed in a later section certainly add support to this suggestion.

North Massif type soils

Soils collected at Stations 6, 7, 8, and 9 from the northern part of the Taurus-Littrow Valley encompass a wide compositional range, but are chemically distinct from the other two soil types. They are intermediate in composition, being more aluminous than the Valley Floor type soils and more mafic than the South Massif type soils (Fig. 1). Trace and minor element abundances are diagnostic for this group, particularly the concentrations of zirconium, niobium, and yttrium, which are lower than in any other soil type (Fig. 2). The light rare-earth elements are substantially lower than in the South Massif type soils and the chondrite-normalized pattern is flatter than for Valley Floor type soils (Fig. 3), with distinctly lower heavy rare earth abundances. Within the group, Station 6 and 7 soils, located on the lower slopes of the North Massif, are the most aluminous and are inferred to contain the highest massif component (Table 5; Fig. 4). However, unlike the soils from Stations 2 and 2A on the South Massif, they contain substantial contributions of basaltic and orange glass components (Fig. 4; Table 5), possibly thrown there by the central cluster event (AFGIT, 1973).

The Station 8 and 9 soils, located on the flanks of the Sculptured Hills and on the valley floor, respectively, are more mafic because of higher contributions of basalt and orange glass. The Station 9 soils are intermediate in chemistry between the North Massif and the Valley Floor type soils. They are placed within the North Massif group largely on the basis of low minor and trace element abundances (Fig. 2). The heavy rare-earth abundances in Station 9 soil (Table 2; Fig. 3) is consistent with this classification. The dark-gray surface soil (79221) collected from the upper 7 cm is more mafic than the underlying light-gray soil (79261), and is calculated to contain slightly more basalt and orange glass than the underlying soil. According to AFGIT (1973), the dark matrix "soil" breccias collected at Station 9 are samples of older regolith ejected and indurated by the impact creating the young Van Serg Crater. Chemically, this material (79135) is more aluminous than the overlying soils, and contains less basalt (26%) and more massif component (50%) and orange glass component (24%) than the overlying soils.

The calculated massif component in the North Massif type soils decreases in amount with distance from the North Massif from Station 6 to Station 9. Over this range the proportion of noritic breccia relative to anorthositic gabbro is roughly constant, ranging from a maximum 35% at Station 7 to a minimum of 26% at Station 8. If the massif component in the soils is derived by mass wasting from the lower slopes of the North Massif and Sculptured Hills, then these data suggest that anorthositic gabbro is substantially more abundant than noritic gabbro in those areas and that the lower slopes of the North Massif and Sculptured Hills contain a greater abundance of anorthositic gabbro than does the South Massif. Furthermore, the mixing model calculations result in lower calculated magnesium values for the Station 6 soils than the measured values, implying the presence in these soils of small amounts (less than 5%) of an additional high-magnesian component such as dunite or troctolite.

As was the case with the South Massif type soils, the calculated abundances for the anorthositic gabbro are much greater than would be inferred from the returned sample collection, suggesting that they are less coherent than the hornfels-like noritic breccias which dominate the sample collection. The Sculptured Hills have a more dissected morphology than the Massifs, with lower subdued slopes and a lack of boulders. These features are attributed by AFGIT (1973) to the presence in the Sculptured Hills of less-coherent breccias in contrast with the massifs, and may be correlated with a greater abundance of anorthositic gabbro. The soil data from Station 8, on the flanks of the Sculptured Hills (Table 5; Fig. 4) provide support for this interpretation, since the calculated anorthositic gabbro content of the massif component is much higher relative to noritic breccia than at Stations 6 or 7 at the North Massif or at any other sampling site. The Station 9 soils, which have probably received contributions of massif material from both the North Massif and Sculptured Hills, have an intermediate anorthositic gabbro/noritic breccia ratio consistent with a greater abundance of anorthositic gabbro in the Sculptured Hills.

Soils collected from LRV 2 and LRV 3 have been included, on chemical grounds, with the North Massif type soils, even though they are both situated on the valley floor close to the Light Mantle and are nearer to the South Massif rather than to the North Massif. They are similar in bulk chemistry to Station 8 and 9 soils (Figs. 1 and 2) and contain a substantial amount of basalt and orange glass (Table 5; Fig. 4). The calculated massif component in these soils, as in the North Massif soils proper, is dominated by anorthositic gabbro with subordinate noritic breccia, in contrast to South Massif and Valley Floor type soils that contain these two massif components in approximately equal amounts.

In an earlier section it was suggested that the low basalt content of the South Massif type soils can be explained if they were derived, by avalanching, from the upper slopes of the massif. On the other hand, the high content of basalt and orange glass in the North Massif soils implies derivation from the lower slopes of the massif, where lateral mixing of basalt and massif components is more effective. Thus, the differences in composition between the North and South Massif type soils, particularly their contrasting proportions of noritic breccia and anorthositic gabbro, may simply reflect derivation from different levels of a similar stratigraphic sequence in the two massifs. Similarly, the difference in composition between LRV 2 and 3 soils and the South Massif type soils may be due to the derivation of the massif components in these soils from a lower level of the South Massif than the Light Mantle type soils. The higher basalt and orange glass content is also consistent with this interpretation.

COMPARISON WITH APOLLO 11 SOIL

The Apollo 11 soil 10084 is remarkably similar in bulk chemistry to the more aluminous Apollo 17 Valley Floor type soils (e.g. LM–ALSEP, Station 4) and, like these soils, overlays and is presumed to be largely derived from high-titanium basalts of a similar age. Like the Apollo 17 soils again, it is more aluminous than the parental basalts and is inferred to contain substantial amounts of "exotic" highland-derived material (Wood, 1970).

Using chemical mixing models, Schonfeld and Meyer (1972) calculate that the highland component constitutes about 24% of the Apollo 11 soil. This is within the range calculated for massif components in the Apollo 17 Valley Floor type soils (Table 5). The major difference between these two soils is that the Apollo 17 soils are located in a valley surrounded by lunar highlands, and are less than 5 km from the nearest massif, whereas the Apollo 11 soils are situated on Mare Tranquillitatis, at least 40–50 km from the nearest highlands. To reconcile these differences assuming that the highland components have been introduced by impact related lateral transport and mixing, it is necessary to suggest that these processes have been extremely efficient at the Apollo 11 site and relatively inefficient at the Apollo 17 site.

Models of regolith evolution proposed by Shoemaker *et al.* (1970), and more recently by Oberbeck *et al.* (1973) do not favor large scale lateral transport of material, but imply that the bulk of the regolith is of local derivation. These requirements are consistent with the observations at the Apollo 17 site, but are in conflict with a distant highland source for the anorthositic components in the Apollo 11 soil (Wood, 1970). It may therefore be appropriate to reevaluate the source of this material at the Apollo 11 site, one possibility being that it is derived locally from beneath thin basalt flows, by cratering, as has been suggested by Wasson and Baedecker (1972) for the KREEP component in the Apollo 12 soils.

CONCLUSIONS

(1) Within the wide compositional range of the Apollo 17 soils, three distinct chemical groups have been recognized, each corresponding broadly to a major geological and physiographic unit.

(a) The Valley Floor type soils are derived primarily from subfloor basalts and orange glass, and modified by addition of varying amounts of massif components. The orange glass is an important, widespread constituent of these soils, and is probably interlayered with the underlying basalt, as suggested by Heiken and McKay (1974a, 1974b).

(b) The South Massif type soils are mixtures of noritic breccia and anorthositic gabbro in roughly equal proportions, with very minor basaltic or orange glass components. They are interpreted as being derived by avalanching from the upper slopes of the South Massif, where lateral contamination with valley floor assemblages was minimal.

(c) The North Massif type soils are also mixtures of noritic breccia and anorthositic gabbro, but with the latter dominant. Lateral mixing with basalt and orange glass has been considerable due to their derivation by mass-wasting of the lower slopes of the South Massif.

(2) The differences in composition for the North and South Massif type soils imply different proportions of the massif components, noritic breccia and anorthositic gabbro, in the source of these soils. This may imply that the lithologies of the North Massif and Sculptured Hills differ from those of the South Massif in that the former contain a greater proportion of anorthositic rocks, mainly anorthositic gabbro and minor dunite or troctolite, and lesser amounts of the KREEP-like noritic breccias. An alternative and preferred hypothesis is that

the lithologies and stratigraphic sequence of the two massifs are essentially the same, the differences in soil compositions being produced by derivation from different stratigraphic levels. Thus, the North Massif type soils are derived from the lower slopes where anorthositic gabbro predominates over noritic breccia, whereas the South Massif type soils derive from the upper slopes, where the massif components are considered to be in equal proportions. The compositional similarities of LRV 2 and LRV 3 soils and the North Massif type soils is clearly in conflict with the first hypothesis, but is not inconsistent with the second, providing the massif component in these soils derives from the lower slopes of the South Massif.

(3) The calculated abundance of anorthositic gabbro in the massif-derived soils is much larger than would be deduced from the returned sample collection. This discrepancy is attributed to preferential sampling of the hornfels-like noritic breccias with respect to less-coherent anorthositic gabbro. The friable nature of the anorthositic gabbro, together with its greater abundance in the Sculptured Hills soil than at any other sampling site, suggest that it may be the prevailing rock type in the Sculptured Hills.

(4) The wide range in soil composition, the persistence of three broad chemical types, and the gradational variation in composition across the valley floor (e.g. Station 6—Station 9—LM—ALSEP—Station 5), all indicate that lateral transport and mixing of the regolith was not an efficient process at the Apollo 17 site subsequent to basalt extrusion. Indications are that comminuted basalt and orange glass have been mixed to varying degrees with existing, relatively homogeneous, massif soils, and not with the prevailing massif rock types. It is not clear whether the massif soils attained their pre-basalt homogeneity by more drastic impact-related mixing, or whether they were derived by degradation of contrasting lithologies that were already well-mixed at the outcrop scale.

(5) The inefficient mixing of spatially close massif and basalt material at the Apollo 17 site is consistent with the regolith evolution models of Shoemaker et al. (1970) and Overbeck et al. (1973), which require that the bulk of the soil is of local derivation. On the other hand, the Apollo 11 soils, which are compositionally similar to the Valley Floor type soils, and contain comparable amounts of highland material, do not fit these models since they are located at least 40–50 km from the nearest lunar highlands. In view of the Apollo 17 observations, it is appropriate to re-evaluate the source of the anorthositic material in the Apollo 11 soils, one possibility being that they are derived locally, by cratering, from beneath thin basalt flows.

Acknowledgments—We wish to thank A. M. Reid, G. Heiken, R. Brett, and J. Wainwright for helpful comments. The assistance of J. Allred, M. Peil, C. Kime, and N. Brown is appreciated.

REFERENCES

AFGIT (Apollo Field Geology Investigation Team) (1973) Preliminary geologic investigation of the Apollo 17 landing site. NASA SP-330, pp. 6-1 to 6-91.

Bryan W. B., Finger L. W., and Chayes F. (1969) Estimating proportions in petrographic mixing equations by least-squares approximation. *Science* **163**, 926–927.

Duncan A. R., Erlank A. J., Willis J. P., and Sher M. K. (1974) Compositional characteristics of the Apollo 17 regolith (abstract). In *Lunar Science—V*, pp. 184–186. The Lunar Science Institute, Houston.

Gast P. W., Hubbard N. J., and Wiesmann H. (1970) Chemical composition and petrogenesis of basalts from Tranquillity Base. *Proc. Apollo 11 Lunar Sci. Conf., Geochim. Cosmochim. Acta*, Suppl. 1, Vol. 2, pp. 1143–1163. Pergamon.

Heiken G. and McKay D. S. (1974a) Petrography of Apollo 17 soils (abstract). In *Lunar Science—V*, pp. 319–321. The Lunar Science Institute, Houston.

Heiken, G. and McKay D. S. (1974b) Petrography of Apollo 17 soils. *Proc. Fifth Lunar Sci. Conf., Geochim. Cosmochim. Acta*. This volume.

Howard K. A. (1973) Lunar rock avalanches and Apollo 17. *Science* **180**, 1052–1055.

Hubbard N. J., Rhodes J. M., Gast P. W., Bansal B. M., Shih C., Wiesmann H., and Nyquist L. E. (1973) Lunar rock types: The role of plagioclase in non-mare and highland rock types. *Proc. Fourth Lunar Sci. Conf., Geochim. Cosmochim. Acta*, Suppl. 4, Vol. 2, pp. 1297–1312. Pergamon.

LSPET (Lunar Sample Preliminary Examination Team) (1972) The Apollo 15 lunar samples: A preliminary description. *Science* **175**, 363–375.

LSPET (Lunar Sample Preliminary Examination Team) (1973a) Apollo 17 lunar samples: Chemical and petrographic description. *Science* **182**, 659–672.

LSPET (Lunar Sample Preliminary Examination Team) (1973b) The Apollo 16 lunar samples: Petrographic and chemical description. *Science* **179**, 22–34.

Norrish K. and Chappell B. W. (1967) X-ray fluorescence spectrography. In *Physical Methods in Determinative Mineralogy*. (Editor J. Zussman). Academic Press, London.

Norrish K. and Hutton J. T. (1969) An accurate X-ray spectrographic method for the analysis of a wide range of geological samples. *Geochim. Cosmochim. Acta* **33**, 431–454.

Oberbeck V. R., Quaide W. L., Mahan M., and Paulson J. (1973) Monte Carlo calculations of lunar regolith thickness distributions. *Icarus* **19**, 87–107.

Papike J. J. and Bence A. E. (1974) Basalts from the Taurus-Littrow region of the moon (abstract). In *Lunar Science—V*, pp. 586–588. The Lunar Science Institute, Houston.

Phinney W. C., Anders E., Bogard D., Butler P., Gibson E., Gose W., Heiken G., Hohenberg C., Nyquist L., Pearce W., Rhodes M., Silver L., Simonds C., Strangeway D., Turner G., Walker R., Warner J., and Yuhas D. (1974) Progress Report: Apollo 17, Station 6 Boulder Consortium (abstract). In *Lunar Science—V*, Suppl. A, pp. 7–13. The Lunar Science Institute, Houston.

Pieters C., McCord T. B., and Adams, J. B. (1974) Evidence for regional occurrence of orange glass and related soils (abstract). In *Lunar Science—V*, pp. 605–607. The Lunar Science Institute, Houston.

Reid A. M., Warner J., Ridley W. I., and Brown R. W. (1972a) Major element composition of glasses in three Apollo 15 soils. *Meteoritics* **7**, 395–415.

Reid A. M., Ridley W. I., Harmon R. S., Warner J., Brett R., Jakes P., and Brown R. W. (1972b) Highly aluminous glasses in lunar soils and the nature of the lunar highlands. *Geochim. Cosmochim. Acta* **36**, 903–912.

Reid A. M., Warner J., Ridley W. I., Johnston D. A., Harmon R. S., Jakeš P., and Brown R. W. (1972c) The major element composition of lunar rocks as inferred from glass compositions in the lunar soil. *Proc. Third Lunar Sci. Conf., Geochim. Cosmochim. Acta*, Suppl. 3, Vol. 1, pp. 363–378. MIT Press.

Rhodes J. M. (1973) Major and trace element chemistry of Apollo 17 samples. *Trans. Amer. Geophys. Union* **54**, 609–610.

Rose H. J., Brown F. W., Carron M. K., Christian R. P., Cuttita F., Dwornik E. J., and Ligon D. T. (1974) Composition of some Apollo 17 samples (abstract). In *Lunar Science—V*, pp. 645–647. The Lunar Science Institute, Houston.

Schonfeld E. and Meyer J. (1972) The abundance of components of the lunar soil by a least-squares mixing model and the formation age of KREEP. *Proc. Third Lunar Sci. Conf., Geochim. Cosmochim. Acta*, Suppl. 3, Vol. 2, pp. 1397–1420. MIT Press.

Shoemaker E. M., Hait M. H., Swann G. A., Schleicher D. L., Schober R. L., Sutton R. L., Dahlem D.

H., Goddard E. N., and Waters A. C. (1970) Origin of the lunar regolith at Tranquillity Base. *Proc. Apollo 11 Lunar Sci. Conf., Geochim. Cosmochim. Acta*, Suppl. 1, Vol. 3, pp. 2399–2412. Pergamon.

Wasson J. T. and Baedecker P. A. (1972) Provenance of Apollo 12 KREEP. *Proc. Third Lunar Sci. Conf., Geochim. Cosmochim. Acta*, Suppl. 3, Vol. 2, pp. 1315–1326. MIT Press.

Wood J. A. (1970) Petrology of the lunar soil and geophysical implications. *J. Geophys. Res.* **75**, 6497–6513.

Proceedings of the Fifth Lunar Conference
(Supplement 5, Geochimica et Cosmochimica Acta)
Vol. 2 pp. 1119–1133 (1974)
Printed in the United States of America

Chemical composition of rocks and soils at Taurus-Littrow

HARRY J. ROSE, JR., FRANK CUTTITTA, SOL BERMAN, F. W. BROWN,
M. K. CARRON, R. P. CHRISTIAN, E. J. DWORNIK, and L. P. GREENLAND

U.S. Geological Survey, (923) National Center, Reston, Virginia 22092

Abstract—Seventeen soils and seven rock samples were analyzed for major elements, minor elements, and trace elements. Unlike the soils at previous Apollo sites, which showed little difference in composition at each collection area, the soils at Taurus-Littrow vary widely. Three soil types are evident, representative of (1) the light mantle at the South Massif, (2) the dark mantle in the valley, and (3) the surface material at the North Massif. The dark-mantle soils are chemically similar to those at Tranquillitatis. Basalt samples from the dark mantle are chemically similar although they range from fine to coarse grained. It is suggested that they originated from the same source but crystallized at varying depths from the surface.

INTRODUCTION

THE MAJOR-, MINOR-, AND TRACE-ELEMENT COMPOSITIONS of seven rocks and seventeen soils collected by the Apollo 17 Mission at the Taurus-Littrow Valley were determined by combined semimicro chemical, X-ray fluorescence and optical emission methods. Details of the procedures have been reported earlier (Annell and Helz, 1970; Cuttitta *et al.*, 1971; Rose *et al.*, 1970). Prior to analysis, each sample was photographed and then carefully examined microscopically. The rock samples were ground and mixed to ensure a representative split for each of the analytical methods.

Flanked by the North and South Massifs, the Taurus-Littrow Valley is in the Taurus Mountains–Littrow Region southeast of the Serenitatis Basin.

TAURUS-LITTROW SOILS

The lunar soils collected by earlier Apollo missions have a narrow range of composition (Rose *et al.*, 1971, 1972, 1973; Cuttitta *et al.*, 1971, 1973; LSPET, 1973). In contrast, the Apollo 17 soils have a much wider range of compositions (Tables 1 and 2). The 17 analyzed soils can be divided into two distinct groups on the basis of their TiO_2 and Cr_2O_3 content. One group of soils (the low-TiO_2 group—Table 1) is characterized by a TiO_2 content between 1.3% and 3.2% and a Cr_2O_3 content of less than 0.3%. The second group of Apollo 17 soils (the high-TiO_2 group—Table 2) is characterized by a TiO_2 content between 6.3% and 10.5% and Cr_2O_3 content greater than 0.4%. However, for soil 76501,30 collected at Station 6, an examination of major-, minor-, and particularly the trace-element compositions (Table 1) strongly suggests the presence of a third type of soil at the Taurus-Littrow site.

Table 1. Composition of some Apollo 17 soils having a relatively low titanium (1.3–3.2%) and chromium (<0.3%) contents (oxides in wt.%, elements in ppm).

Station	2	2	2a	2a	3	3	3	3	6
	72441,9	72461,7	73121,16	73141,21	73221,13	73241,14	73261,14	73281,12	76501,30
SiO_2	45.17	44.79	45.56	45.35	45.20	44.55	44.71	45.31	43.71
Al_2O_3	20.25	20.63	21.23	21.56	21.03	20.20	19.69	20.23	18.83
Fe_2O_3	0.00	0.00	0.00	0.00	0.00	0.00	0.00	0.00	0.00
FeO	8.68	8.61	8.45	8.02	8.85	8.45	8.86	8.82	10.35
MgO	10.78	10.52	9.73	10.28	8.97	11.11	10.95	9.95	10.71
CaO	12.75	12.87	12.82	12.91	12.86	12.90	12.90	12.91	12.06
Na_2O	0.40	0.43	0.39	0.38	0.41	0.46	0.40	0.41	0.38
K_2O	0.16	0.17	0.17	0.14	0.16	0.16	0.16	0.16	0.11
TiO_2	1.53	1.56	1.39	1.26	1.86	1.73	1.90	1.76	3.20
P_2O_5	0.15	0.16	0.15	0.12	0.15	0.15	0.14	0.14	0.08
MnO	0.11	0.11	0.11	0.11	0.11	0.11	0.11	0.11	0.13
Cr_2O_3	0.28	0.28	0.26	0.24	0.27	0.25	0.24	0.27	0.26
Total	100.26	100.13	100.26	100.37	99.87	100.07	100.06	100.07	99.82
ΔRC*	+1.82	+2.22	+1.70	+1.60	+0.75	+0.80	+2.14	+1.29	+2.40

	From beneath boulder upper 4 cm	From beneath boulder skim sample	Medium gray, from upper few cm	Light gray, from 15 cm below surface	Medium gray, upper 0.5 cm	Light gray, upper 5 cm	Medium gray, 5–10 cm below surface	Light gray, 5–10 cm below surface	Medium gray
Pb	<2	<2	<2	2	<2	<2	2	<2	<2
Zn	19	19	16	14	21	18	18	20	12
Cu	14	15	24	6.3	30	9.8	12	7.8	14
Ga	3.0	2.9	2.9	2.6	3.3	4.0	3.8	3.6	3.2
Li	8.7	9.0	10	8.7	9.5	10	8.7	9.5	7.5
Rb	2.4	2.6	3.2	2.8	3.1	2.8	1.3	2.8	2.3
Co	36	31	42	25	49	37	46	46	38
Ni	265	265	355	272	250	320	450	380	262
Ba	240	209	186	91	190	168	160	160	114
Sr	155	149	149	122	167	146	127	117	130
V	28	28	33	30	32	40	46	42	50
Be	1.9	1.8	2.2	1.5	2.2	1.8	2.0	1.8	<1
Nb	12	11	13	11	10	12	12	11	10
Sc	21	21	22	15	24	15	17	15	30
La	21	21	21	<10	19	<10	<10	<10	<10
Y	64	63	62	48	61	56	60	59	44
Yb	6.0	5.8	6.0	4.7	6.0	4.6	5.6	4.8	3.4
Zr	279	279	250	197	238	202	201	207	133

Trace elements listed in order of decreasing volatility in the DC arc.

*ARC—Total reducing capacity measured for the lunar samples less the reducing capacity attributable to the FeO content of the lunar samples. The following elements were looked for but not detected in the analyzed samples. If present they would be in concentrations below those (in ppm) indicated in parenthesis: Ag(0.2), As(4), Au(0.2), B(10), Bi(1), Cd(8), Ce(100), Cs(1), Ge(1), Hf(20), Hg(8), In(1), Mo(2), Nd(100), Pb(1), Pt(3), Re(30), Sb(100), Sn(10), Ta(100), Te(300), Th(100), Tl(1), U(500), W(200), and Zn(4).

Table 2. Composition of some Apollo 17 soils having a relatively high titanium (6.3–10.5%) and chromium (>0.4%) contents (oxides in wt.%, elements in ppm).

Station	LM	ALSEP	4	4	5	9	9	9
	70011,25	70181,18	74241,29	74261,16	75061,27	79221,30	79241,28	79261,29
SiO_2	41.03	40.90	42.00	42.08	39.70	41.63	41.73	42.58
Al_2O_3	11.98	12.40	13.19	13.70	10.60	13.48	13.90	14.51
Fe_2O_3	0.00	0.00	0.00	0.00	0.00	0.00	0.00	0.00
FeO	16.24	16.55	14.84	14.96	17.86	15.43	15.64	14.69
MgO	10.08	9.76	9.17	9.56	9.65	10.30	9.90	9.67
CaO	11.08	10.97	11.56	11.25	10.72	11.19	11.08	11.35
Na_2O	0.31	0.38	0.43	0.42	0.37	0.35	0.39	0.39
K_2O	0.08	0.09	0.14	0.13	0.08	0.11	0.09	0.10
TiO_2	8.30	8.40	7.90	7.45	10.46	6.48	6.79	6.28
P_2O_5	0.10	0.07	0.10	0.09	0.06	0.08	0.08	0.08
MnO	0.23	0.21	0.20	0.19	0.24	0.20	0.20	0.19
Cr_2O_3	0.41	0.46	0.42	0.48	0.48	0.44	0.46	0.41
Total	99.84	100.19	99.95	100.31	100.22	99.69	100.26	100.25
ΔRC*	+2.76	+1.70	+1.43	+0.94	+1.79	+2.72	+2.61	+0.81
Pb	<2	<2	<2	<2	<2	<2	<2	<2
Zn	28	13	37	35	12	21	22	15
Cu	23	28	28	32	27	20	24	19
Ga	7.2	4.1	10	10	4.3	4.6	5.1	4.4
Li	11	8.7	7.7	9.7	7.4	9.7	8.6	8.7
Rb	1.2	1.4	1.8	2.0	1.0	1.6	1.9	1.3
Co	52	42	28	38	36	48	43	48
Ni	240	220	126	133	154	430	275	300
Ba	290	107	84	140	135	115	117	100
Sr	180	144	144	151	157	135	147	131
V	84	60	75	70	95	74	81	63
Be	1.6	<1	1.3	1.5	1.5	1.4	1.5	1.4
Nb	26	18	14	23	20	20	19	19
Sc	57	61	59	57	78	59	56	50
La	<10	<10	<10	<10	<10	<10	<10	<10
Y	78	66	90	74	91	62	66	56
Yb	6.4	6.0	10	7.2	9.6	5.9	6.4	5.7
Zr	252	270	352	268	340	205	200	208
	Light gray 0–3 cm beneath LM	Med. to dark gray, 0–5 cm depth	Gray, from SW end of trench crossing orange soil 74420	Gray, from NE end of trench crossing orange soil 74420	Dark gray. Fills depression on boulder	Gray, 0–2 cm depth	Gray, 2–7 cm depth	White to light gray, 7–17 cm depth

Trace elements listed in order of decreasing volatility in the DC arc.

*ΔRC—Total reducing capacity measured for the lunar samples less the reducing capacity attributable to the FeO content of the lunar samples.

The following elements were looked for but not detected in the analyzed samples. If present they would be in concentrations below those (in ppm) indicated in parenthesis: Ag(0.2), As(4), Au(0.2), B(10), Bi(1), Cd(8), Ce(100), Cs(1), Ge(1), Hf(20), Hg(8), In(1), Mo2(2), Nd(100), Pb(1), Pt(3), Re(30), Sb(100), Sn(10), Ta(100), Te(300), Th(100), Tl(1), U(500), W(200), and Zn(4).

Elemental variations

The low-TiO$_2$ group soils (Table 1) were collected at the following locations: Station 2, at the foot of the South Massif at the southeast rim of Nansen Crater near the contact between the light mantle and materials of the South Massif; Station 2a, on the light mantle about 500 m northeast of Nansen Crater; Station 3, approximately 50 m east of the rim of Lara Crater on the light mantle; Station 6, on the south slope of the North Massif (Fig. 1).

The high-TiO$_2$ group soils (Table 2) were collected at Station 4 on the south rim of Shorty Crater near the north edge of the light mantle, at Station 5, within a block field on the southeast rim of Camelot Crater, and Station 9, on the southeast rim near the outer flank of Van Serg Crater. Although situated on the light mantle, Station 4 soils are grouped with the high-TiO$_2$ soils. These soils collected on the rim of Shorty Crater are probably dark-mantle material that was ejected through the thin veneer of white mantle. Station 4 soils, 74241,29 and 74261,16, were collected at the southwest and northeast ends, respectively, of the trench crossing orange soil 74220. The relatively high Zn content of these soils (36 ppm) reflects

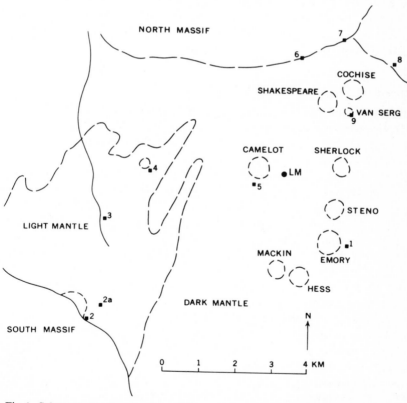

Fig. 1. Schematic drawing of valley of Taurus-Littrow showing sample collection sites.

the contribution of the orange soil which contains about 300 ppm Zn (LSPET, 1973).

A comparison of the data (Tables 1 and 2) shows that the low-TiO_2 group of soils is also characterized by significantly lower FeO and MnO contents, whereas the high-TiO_2 group is considerably lower in SiO_2, Al_2O_3, CaO, and P_2O_5. Although the two groups seem related geographically, the group collected around the South Massif, having low-TiO_2 and Cr_2O_3 contents, characterizes the light mantle material, whereas the other group collected in the Valley of Taurus-Littrow having high-TiO_2 and Cr_2O_3 contents, characterizes the dark-mantle material.

In addition to a higher FeO, TiO_2, and MnO content compared with the other samples presented in Table 1, the Station 6 North Massif soil 76501 contains a lower concentration not only of the volatile elements but also of refractory elements (Ba, Sr, Sc, Y, Yb, and Zr). This marked difference in the trace element composition of soil 76501 precludes it being a mixture of the light- and dark-mantle soils, and may imply that the soils of the North Massif have a distinctive composition. Results of a mixing model analysis (Schonfeld, 1974) also suggest that sample 76501 may comprise a separate soil type. The light-mantle soils at Stations 2, 2a, and 3 have similar concentrations of both major and trace elements. All of these soils are relatively high in K_2O and P_2O_5 content compared with the Station 6 soil 76501 and the dark-mantle soils (Stations 4, 5, and 9). This association may suggest the possible presence of a KREEP component in the South Massif soils.

Interelement correlations

Correlation coefficients between all possible pairs of elements in the high- and low-TiO_2 group of soils have been computed and tested for significance at the 95% confidence level. Soil 76501,30 was excluded from these groups. The correlations are shown in Table 3.

Figure 2 shows the correlation families obtained by linking elements that show positive correlations in the high-TiO_2 group of soils. The two distinct families shown in Fig. 2 reflect two major components of the soil: a feldspar–phosphate and a mafic component. The phosphate component is apparently not KREEP because Y, Yb, and Zr are associated with the mafic component. The mafic component may be only pyroxene, but it also could represent the simultaneous variation of a mixture of olivine, pyroxene, and oxides. The negative correlations in Table 3 show that the two components vary inversely thus suggesting that most of the chemical variation within the high-TiO_2 soils can be accounted for by random mixing of only two components.

Correlations in the low-TiO_2 soils are dominated by a strong KREEP component. The elements K, P, Zr, Yb, Y, Ba, Sr, Be, Sc, and Cr are too highly interlinked to be shown individually in correlation families in Fig. 2. A second family is also shown in Fig. 2 and includes Fe, Ti, Co, Zn, in addition to the weakly linked Ga, V, and Mn; this family is loosely linked to the KREEP component through Fe–Y and Co–Be correlations. It is not immediately obvious what mineral

Table 3. Interelement correlations in Apollo 17 soils (95% confidence level).

	Si	Al	Fe	Mg	Ca	Na	K	Ti	P	Mn	Cr	Zn	Ga	Li	Rb	Co	Ni	Ba	Sr	V	Be	Nb	Sc	Y	Yb	Zr
Si		+	—	0	+	0	0	—	0	—	0	0	0	0	0	0	0	0	0	0	0	—	0	0	0	0
Al	0		—	0	+	0	0	—	0	0	0	0	0	0	0	0	0	0	0	0	0	0	0	0	0	0
Fe	0	0		0	0	0	+	0	+	0	0	0	0	0	0	0	0	0	0	0	0	+	0	0	0	0
Mg	0	0	0		0	—	0	0	0	0	0	0	0	0	0	+	+	0	0	0	0	0	0	0	0	—
Ca	0	0	0	0		0	+	0	+	—	0	0	0	0	0	0	0	0	0	0	0	—	0	0	0	0
Na	—	0	0	0	0		+	0	0	0	0	0	0	0	0	0	—	0	—	0	0	0	0	0	0	0
K	0	0	0	0	0	0		0	0	0	0	+	+	0	+	0	0	0	0	0	0	0	0	0	0	0
Ti	0	—	+	0	0	0	0		0	+	0	0	0	0	0	0	0	0	0	0	0	+	+	0	0	+
P	0	0	0	0	0	0	+	0		0	0	+	+	0	0	0	0	0	0	0	0	0	0	0	0	0
Mn	0	0	0	0	0	0	0	0	0		0	0	0	0	—	0	0	0	+	+	0	+	0	0	0	0
Cr	0	0	0	0	0	0	0	0	0	0		0	0	0	0	0	0	0	0	0	0	0	0	0	0	0
Zn	0	0	+	0	0	0	0	+	0	0	0		+	0	0	0	0	0	0	0	0	0	0	0	0	0
Ga	0	—	0	0	0	0	0	+	0	0	0	0		0	0	0	0	0	0	0	0	0	0	0	0	0
Li	0	0	0	0	0	0	0	0	0	0	0	0	0		0	+	0	+	0	0	0	+	0	0	0	0
Rb	0	+	0	0	0	0	0	0	0	0	0	0	0	0		0	0	0	0	0	0	0	0	0	0	0
Co	0	0	+	0	0	0	0	+	0	0	0	0	0	0	0		+	0	0	0	0	0	0	0	—	—
Ni	0	0	0	0	0	0	0	0	0	0	0	0	0	0	0	0		0	0	0	0	0	0	—	—	—
Ba	0	0	0	0	—	0	+	0	+	0	+	0	0	0	0	0	0		+	0	0	+	0	0	0	0
Sr	0	0	0	0	0	0	0	0	+	0	0	0	0	0	0	0	0	0		0	0	0	0	0	0	0
V	0	0	0	0	0	0	0	0	0	0	0	+	0	0	0	+	0	0	0		0	0	0	0	0	0
Be	0	0	0	0	0	0	0	0	0	0	0	0	0	0	0	+	0	0	0	0		0	0	0	0	0
Nb	0	0	0	0	0	0	0	0	0	0	0	0	0	0	0	0	0	0	0	0	0		0	0	0	0
Sc	0	0	0	0	0	0	0	0	0	0	0	0	0	0	0	0	0	+	0	+	0	0		0	0	0
Y	0	0	+	0	0	0	+	0	+	0	+	0	0	0	0	0	0	+	0	0	0	0	0		+	+
Yb	0	0	0	0	—	0	0	0	0	0	0	0	0	0	0	0	+	0	0	+	0	+	+	0		+
Zr	0	0	0	0	—	0	0	0	+	0	+	0	0	0	0	0	+	0	0	—	0	0	+	+	+	

(Upper-right triangle: High-Titanium Soils; Lower-left triangle: Low-Titanium Soils)

reflects the TiO₂ correlation family, but the links with the KREEP component and the absence of negative correlations (except Zr–V) between the families show that the two components are associated and have been added together (or deleted). The negative correlations of Al with Ti and Ga, and of Ca with Ba, Yb, and Zr suggest that plagioclase varies inversely with the other two components. It is likely that the low-TiO₂ soils have arisen by addition of varying amounts of KREEP and, possibly, Fe–Ti oxides to an original plagioclase-rich soil.

The two groups of soils are distinguished by (among others) Cr content. It is notable that Cr is uncorrelated with any other element in the high-TiO₂ soils whereas it is, surprisingly, strongly associated with the KREEP component of the low-TiO₂ soils. Therefore, the differing Cr contents of the two soil types is unrelated to the mixing processes and reflects inherent differences of the original soils.

Comparison to soils from other Apollo sites

Table 4 lists the average compositions of soils returned by all the Apollo missions. In order to facilitate comparisons, all the data shown are those obtained in this laboratory. The three soil types at Taurus-Littrow are listed separately for comparison. All three groups of Apollo 17 soils show a lower Na₂O content when compared to those returned by prior Apollo missions. Notably, the high-TiO₂ group is lower in SiO₂, Al₂O₃, Na₂O, K₂O, P₂O₅, and Rb and higher in FeO, TiO₂,

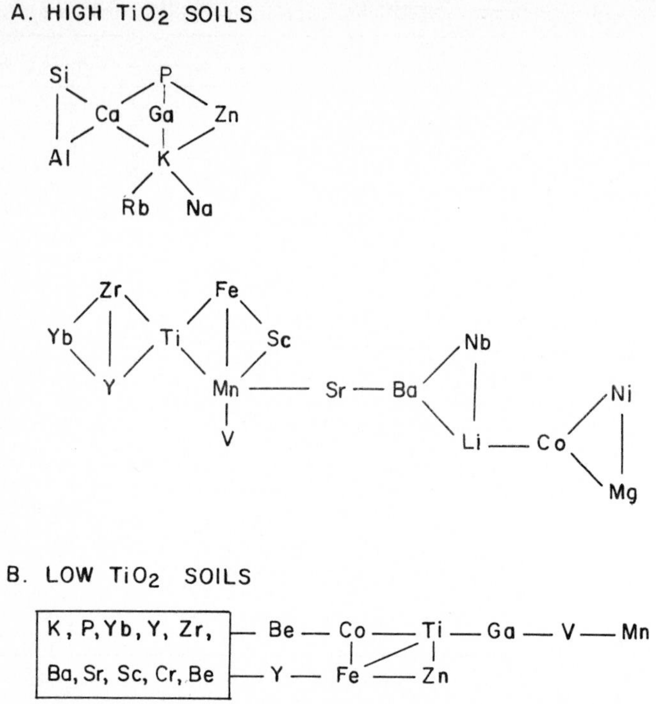

A. HIGH TiO$_2$ SOILS

B. LOW TiO$_2$ SOILS

Fig. 2. Correlation families for Apollo 17 soils. Elements which are positively correlated at 95% confidence level are linked by solid lines. Elements shown within the block are too strongly interlinked to enable individual representation.

Cr_2O_3, Cu and Sc compared with the other soils. Most of the other elements are very nearly in the same concentration as those found in the soils at the Apollo 11 site at Tranquillitatis. The one notable difference between the soils is the greater ΔRC value found at the Apollo 11 site when compared to the high-TiO$_2$ soils (dark mantle) at the Taurus-Littrow (+4.1% and 1.97%, respectively). This difference strongly suggests that the reducing environments at these sites were quite different and may reflect a greater solar wind contribution at Tranquillitatis at some time during lunar history.

Although represented by a single sample (76501,30), the North Massif soil is distinctly different from the dark- and light-mantle soils. These data and conclusions support those previously reported by other investigators (Rhodes *et al.*, 1974; Mason *et al.*, 1974; LSPET, 1973). The Ba, Be, and Zr contents of the North Massif soil (76501) are lower than the soils collected at other lunar sites.

Elemental ratios

Table 5 lists some of the calculated and averaged elemental ratios for the soils collected by each of the Apollo missions. Again, the three different soil types at Taurus-Littrow are shown separately. The differences between 76501 and the

Table 4. Comparison of average composition of lunar soils returned by the Apollo missions. (Oxides in wt.%, elements in ppm).

						Apollo 17		
Constituent	Apollo 11	Apollo 12	Apollo 14	Apollo 15	Apollo 16	low-TiO$_2$ group	high-TiO$_2$ group	Sample 76501
SiO$_2$	42.04	46.40	47.93	46.61	44.94	45.08	41.46	43.71
Al$_2$O$_3$	13.92	13.50	17.60	17.18	26.71	20.60	12.97	18.83
Fe$_2$O$_3$	0.00	0.000	0.00	0.00	0.00	0.00	0.00	0.00
FeO	15.74	15.50	10.37	11.62	5.49	8.59	15.78	10.35
MgO	7.90	9.73	9.24	10.46	5.96	10.29	9.76	10.71
CaO	12.01	10.50	11.19	11.64	15.57	12.86	11.15	12.06
Na$_2$O	0.44	0.59	0.68	0.46	0.48	0.41	0.38	0.38
K$_2$O	0.14	0.32	0.55	0.20	0.13	0.16	0.10	0.11
TiO$_2$	7.48	2.66	1.74	1.36	0.58	1.62	7.76	3.20
P$_2$O$_5$	0.12	0.40	0.53	0.19	0.12	0.14	0.08	0.08
MnO	0.21	0.21	0.14	0.16	0.07	0.11	0.21	0.13
Cr$_2$O$_3$	0.30	0.40	0.25	0.25	0.12	0.26	0.44	0.26
Total	100.30	100.21	100.22	100.13	100.17	100.12	100.10	100.25
ΔRC*	+4.1	+1.3	+2.8	+2.1	+2.1	+1.64	+1.97	+0.81
Pb	<2	<2	10	2.8	3.8	<2	<2	<2
Zn	19	6.7	25	18	27	18	23	12
Cu	10	11	18	9.3	8.8	15	25	14
Ga	3.8	4.9	5.5	3.3	3.7	3.3	4.0	3.2
Li	11	18	23	9.2	7.3	9.3	9.4	7.5
Rb	2.7	8.2	13	5.3	2.8	2.6	1.5	2.3
Co	24	58	35	43	24	39	42	38
Ni	185	195	370	332	344	320	235	262
Ba	210	563	1030	320	121	176	136	114
Sr	130	131	189	159	149	142	149	130
V	50	107	56	87	23	35	75	50
Be	1.6	5.2	6.6	3.1	1.2	1.9	1.4	<1
Nb	18	38	55	16	10	12	20	10
Sc	56	40	27	22	10	19	60	30
La	16	54	74	32	<10	14	<10	<10
Y	81	164	276	80	38	59	73	44
Yb	—	—	—	8.8	3.1	5.4	7.2	3.4
Zr	273	548	813	299	151	232	262	133

*ΔRC—Total reducing capacity less the reducing capacity attributable to the FeO content of the sample, in % FeO.

The following elements were looked for but not detected in the analyzed samples. If present, they would be in concentrations below those (in ppm) indicated in parenthesis: Ag(0.2), As(4), Au(0.2), B(10), Bi(1), Cd(8), Ce(100), Cs(1), Ge(1), Hf(20), Hg(8), In(1), Mo(2), Nd(100), Pb(1), Pt(3), Re(30), Sb(100), Sn(10), Ta(100), Te(300), Th(100), Tl(1), U(500), W(200), and Zn(4).

other two Apollo 17 groups can clearly be seen in the K/Rb, Ba/Sr, and Cr/V ratios. The high-TiO$_2$ group of dark-mantle Apollo 17 soils has ratios very similar to those found earlier at Tranquillitatis, except for variations in the K/Ba, Ba/V, Ba/Sr, and Fe/Ni ratios. Both the Apollo 11 and Apollo 17 high-TiO$_2$ soils have comparable FeO contents (Table 5), yet their Fe/Ni ratios vary more than twofold. In contrast, the Fe/Ni ratios for all the Apollo 17 soils are similar. The difference

Table 5. Comparison of some elemental ratios in soils returned by the Apollo 11–17 missions.

Ratio	Apollo 11	Apollo 12	Apollo 14	Apollo 15	Apollo 16	Apollo 17 low TiO$_2$	Apollo 17 high TiO$_2$	Apollo 17 76501
Si/Al	2.67	3.04	2.04	2.39	1.48	1.94	2.84	2.06
Ca/Si	0.43	0.35	0.31	0.38	0.53	0.44	0.41	0.42
Mg/Fe	0.50	0.41	0.89	0.70	0.84	0.91	0.48	0.80
Fe/Ni	661	617	218	272	128	208	291	307
Cr/Ni	9.4	15	4.7	5.2	2.4	5.6	13	6.8
Ni/Co	7.7	3.4	11	7.7	16	8.2	5.5	6.9
Al/Ti	1.6	4.5	9.0	11	41	11	1.5	5.5
10^{3}Nb/Ti	0.40	2.4	5.3	2.0	2.9	1.0	0.44	0.52
K/Rb	518	390	423	313	385	525	567	383
K/Ba	5.3	4.5	4.4	5.2	8.9	8.0	7.3	7.7
Ba/V	4.2	5.3	19	3.7	5.3	4.9	1.8	2.3
Ba/Sr	1.6	4.3	5.5	2.0	0.84	1.2	0.89	0.44
Rb/Sr	0.015	0.062	0.069	0.033	0.019	0.018	0.011	0.017
Cr/V	35	27	31	20	31	51	41	36

in the Fe/Ni ratios between the Tranquillitatis and the Taurus-Littrow dark-mantle soils reflect a greater meteoritic component in the Apollo 17 soils. The influx of extralunar material could have admixed with ejecta from the event that formed the Serenitatis Basin. The light mantle soils which constitute the low-TiO$_2$ group and the North Massif soil 76501 do not appear to be related to any of the soils from the other Apollo sites.

Trench samples

Several suites of soil samples collected at different depths in the lunar regolith were analyzed. Sample 73121,16 was collected at the surface, whereas 73141,21 was collected at a depth of 15 cm at Station 2a. Although their major element contents are very similar, the trace elements are depleted in 73141,21. The Ni content, for example, is about 25% less in the subsurface material, presumably because of a lower meteoritic contribution.

Four soils, 73221,13, 73241,14, 73261,14, and 73281,12, were collected at increasing depths from the surface to 10 cm. Except for Sr which becomes depleted with depth, little variation can be seen in either major-, minor-, or trace-element content among the samples. The flanks of the South Massif dive into the light-mantle area partially covering the southwest part of Nansen Crater (Fig. 1). The avalanche that spilled into the plain at Taurus-Littrow may have occurred in at least two stages, the earlier one that formed much of the light mantle and the later one that partly filled Nansen Crater (perhaps as the result of the cratering event). The relative uniformity of the soils at Stations 2, 2a, and 3 suggest that mixing was extensive during the avalanching and that the bulk composition of these soils may characterize the rocks composing the South Massif.

Three samples, 79221,30, 79241,28, and 79261,29 were collected from the surface to a depth of about 17 cm. Al_2O_3 apparently increases as MgO decreases with depth; this cannot be attributed to a simple admixture of any of the local soils. There is a considerable decrease in the ΔRC value, however, of the deepest soil sample, which may be due to a greater shielding of that soil from solar wind bombardment.

Samples 72441,9 and 72461,7 were both collected from beneath a boulder at Nansen Crater. There is little difference in chemical composition between the skim sample and that taken at a depth of 4 cm.

IGNEOUS ROCKS AND BRECCIAS

Seven rocks were analyzed: five basalts and two breccias. The results are reported in Table 6. Two of the samples were splits from Goodwill Rock (70017), a coarse-grained vesicular basalt collected close to the LM. Sample 70215, a fine-grained basalt having only a few vesicles, was collected about 50 m east–northeast of the LM. Sample 71055, a fine-grained basalt, is characterized by a high proportion of vugs and vesicles and was collected within a small crater near Station 1a. Sample 75075, a medium-grained vuggy basalt, was collected at Station 5 on the southwest rim of Camelot Crater. The light matrix breccia 72275 was collected on the lower slopes of the South Massif and the dark matrix breccia 79135 was collected at Station 9 on the southeast rim of Van Serg Crater.

Elemental variations

As shown in Table 6, the basalts are very similar in their major-, minor-, and trace-element contents. The samples were collected within a narrow area near the LM in the dark mantle and presumably are of the same origin. The variation in grain size suggests that the basalts may have crystallized at varying depths in the same flow.

The elemental composition of the two breccias differs considerably. Sample 79135,35 is notably higher in FeO, TiO_2, Cr_2O_3, Zn, Cu, and Ni and lower in SiO_2, Al_2O_3, CaO, K_2O, P_2O_5, Ba, La, and Zr than sample 72275,90.

A comparison of the composition of breccia 79135,35 with the soils collected at Van Serg Crater (Table 2) shows little variation in either major or trace elements. The similarity of chemical composition indicates that the breccia may be indurated soil material which may have formed at the time of the cratering event.

The breccia, sample 72275,90, collected on the slope of the South Massif cannot be related directly to the soils analyzed at that site. The soils are more anorthositic while the breccia has a higher SiO_2 and FeO content. The data show the presence of a KREEP component in sample 72275,90.

Comparison to rocks from other sites

The composition of the average igneous rocks returned by all Apollo missions (analyzed in this laboratory) are reported in Table 7. In comparison to other sites,

Table 6. Composition of some Apollo 17 basalts and breccias (oxides in wt.%, elements in ppm).

	Basalt 70017,30	Basalt 70017,50	Basalt 70215,73	Basalt 71055,51	Basalt 75075,72	Breccia 72275,90	Breccia 79135,35
SiO_2	38.80	38.68	37.62	38.14	38.51	47.31	42.57
Al_2O_3	8.54	7.40	8.79	8.62	8.29	16.90	14.74
Fe_2O_3	0.00	0.00	0.00	0.00	0.00	0.00	0.00
FeO	18.12	18.77	19.22	19.20	18.85	12.45	15.19
MgO	10.16	10.45	9.34	9.04	9.68	9.47	9.10
CaO	10.56	10.05	10.82	10.77	10.17	11.72	10.91
Na_2O	0.33	0.34	0.31	0.31	0.37	0.35	0.40
K_2O	0.07	0.07	0.08	0.06	0.11	0.22	0.11
TiO_2	12.84	13.75	13.20	13.41	13.33	0.94	6.33
P_2O_5	0.04	0.04	0.07	0.08	0.12	0.38	0.09
MnO	0.24	0.25	0.27	0.26	0.25	0.19	0.19
Cr_2O_3	0.49	0.49	0.41	0.41	0.55	0.34	0.45
Total	100.19	100.29	100.13	100.30	100.23	100.27	100.08
ΔRC*	+0.98	+1.26	+2.03	+1.75	—	+1.40	+2.41
Pb	<2	<2	<2	<2	<2	<2	<2
Zn	<4	<4	<4	<4	<4	<4	39
Cu	28	84	22	31	34	5.4	26
Ga	5.8	5.4	6.3	8.1	6.5	3.2	7.5
Li	7.8	7.8	11	9.6	8.9	13	8.9
Rb	0.9	0.7	1.0	0.9	1.0	4.6	1.5
Co	32	32	33	51	32	37	52
Ni	<1	24	<1	43	31	127	280
Ba	250	180	475	315	348	330	129
Sr	217	155	170	170	190	135	163
V	98	80	64	88	108	75	76
Be	<1	<1	<1	<1	<1	3.8	1.5
Nb	23	18	20	27	31	24	17
Sc	80	77	92	87	82	40	50
La	<10	<10	<10	<10	<10	35	<10
Y	94	100	73	69	98	88	64
Yb	7.7	8.3	5.0	5.4	7.4	9.2	6.4
Zr	254	250	223	223	296	545	260

*ΔRC—Total reducing capacity measured for the lunar samples less the reducing capacity attributable to the FeO content of the lunar samples.

The following elements were looked for but not detected in the analyzed samples. If present they would be in concentrations below those (in ppm) indicated in parenthesis: Ag(0.2), As(4), Au(0.2), B(10), Bi(1), Cd(8), Ce(100), Cs(1), Ge(1), Hf(20), Hg(8), In(1), Mo(2), Nd(100), Pb(1), Pt(3), Re(30), Sb(100), Sn(10), Ta(100), Te(300), Th(100), Tl(1), U(500), W(200), and Zn(4).

the basalts at Taurus-Littrow are highest in TiO_2, Cr_2O_3, Cu, and Ga and lowest in SiO_2 and Al_2O_3 content. As with the dark-mantle soils, the igneous rocks at Taurus-Littrow are chemically similar to those of Tranquillitatis. Although not as high as that found at the Apollo 11 site, the ΔRC value at the Apollo 17 site is at least twice that found at all other locations. The low-Ni content of basalts at both

Table 7. Comparison of average* composition of igneous rocks returned by Apollo 11–17 missions (oxides in wt.%, elements in ppm).

Const.	Apollo 11	Apollo 12	Apollo 14	Apollo 15	Apollo 16	Apollo 17
SiO_2	40.10	47.10	47.70	46.10	44.91	38.35
Al_2O_3	8.60	12.80	21.44	8.63	27.39	8.33
Fe_2O_3	0.00	0.00	0.00	0.00	0.00	0.00
FeO	18.90	17.40	7.78	21.67	4.44	18.83
MgO	7.74	6.80	7.29	10.46	6.49	9.73
CaO	10.70	11.40	13.05	9.85	15.63	10.45
Na_2O	0.46	0.64	0.70	0.31	0.43	0.33
K_2O	0.30	0.07	0.48	0.06	0.08	0.08
TiO_2	12.20	3.17	1.16	2.17	0.43	13.31
P_2O_5	<0.2	0.17	0.42	0.09	0.08	0.07
MnO	0.25	0.24	0.11	0.28	0.06	0.25
Cr_2O_3	0.37	0.31	0.25	0.48	0.13	0.47
Total	99.82	100.10	100.38	100.10	100.07	100.20
ΔRC†	+2.1	+0.37	+0.32	+0.24	+0.73	+1.50
Pb	<2	<2	11	<2	<1	<2
Cu	8.8	11	9.0	15	9.6	40
Ga	4.8	4.9	3.7	4.4	2.2	6.4
Li	17	5.9	22	5.9	6.2	9.0
Rb	5.1	1.4	14	<1	1.8	0.7
Co	31	64	13	61	18	36
Ni	6.6	70	116	76	253	20
Ba	440	63	740	58	83	314
Sr	135	64	170	115	127	180
V	73	160	38	187	19	88
Be	3.1	1.4	4.1	<1	nd	<1
Nb	24	13	13	<10	<10	24
Sc	97	40	23	38	8.0	84
La	26	<20	59	<20	nd	<10
Y	162	39	192	29	25	87
Yb	20	5.2	16	4.3	2.2	6.8
Zr	594	110	615	72	101	249

*Averages are only of the rocks analyzed in this laboratory.

†ΔRC—Total reducing capacity less the reducing capacity attributable to the FeO content of the sample in % FeO.

Average of crystalline rocks 61016, 68415, and 68416.

Tranquillitatis and Tauris-Littrow is characteristically less than that found at the sites visited by other missions.

Elemental ratios

Table 8 lists some elemental ratios found in igneous rocks returned by all the Apollo missions. The basalts at Taurus-Littrow have the lowest Al/Ti, Nb/Ti,

Table 8. Comparison of some elemental ratios in igneous rocks returned by the Apollo 11–17 missions.

	Apollo 11	Apollo 12	Apollo 14	Apollo 15	Apollo 16	Apollo 17
Si/Al	4.1	3.2	2.0	4.8	1.4	4.1
Ca/Si	0.41	0.37	0.43	0.33	0.53	0.42
Mg/Fe	0.63	0.39	0.92	0.37	1.1	0.40
Fe/Ni	21,000	2,520	540	2,200	136	7,030
Ni/Co	0.23	0.98	8.9	1.3	14	0.56
Cr/Ni	357	39	15	43	3.5	160
Al/Ti	0.86	3.6	16	3.6	56	0.55
$10^3 \times$ Nb/Ti	0.33	0.68	4.4	~0.7	~4.0	0.23
K/Rb	588	500	342	~500	367	720
Ba/V	6.0	0.48	20	0.32	4.4	3.6
Ba/Sr	3.2	0.95	4.4	0.49	0.65	1.7
Rb/Sr	0.037	0.022	0.082	~0.009	0.014	0.005
Cr/V	34	14	46	18	47	37
Zr/Y	3.7	2.8	3.2	2.5	4.0	3.0
K/Zr	4.2	5.3	6.5	6.9	6.5	2.1
Zr/Ni	90	1.6	5.3	0.95	0.40	12
Mg/Si	0.25	0.19	0.20	0.29	0.19	0.33

Rb/Sr, and K/Zr, and the highest K/Rb and Mg/Si ratios when compared to all other Apollo igneous rocks. The Apollo 16 crystalline rocks display the greatest number of high and low ratios among all the sites as a result of their greater anorthositic and lower ilmenite and ferromagnesian content. The Apollo 17 basalts seem to have elemental ratios more nearly like those found for the Apollo 11 igneous rocks (Table 7). Also, as mentioned previously, the high-TiO_2 soils from the dark mantle (Table 2) are chemically similar to the soils at Tranquillitatis (Table 4).

CONCLUSIONS

(1) Unlike those from previous missions, the soils collected at Taurus-Littrow are extremely varied in composition and have three distinguishable groups: light-mantle soils, representative of the South Massif surface material; dark-mantle soils in the valley area between the South and North Massifs; and the surface material at the North Massif.

(2) The composition of soil 76501 at North Massif cannot be explained by a simple mixing of the dark- and light-mantle soils because it is depleted in most trace elements with respect to the other lunar soils. The North Massif soil is distinctly different from the dark- and light-mantle material.

(3) Except for differences in excess reducing capacities and Fe/Ni ratios, the dark-mantle soils (the high-TiO_2 soils at Taurus-Littrow) are compositionally similar to the soils collected at Tranquillitatis. Compared to the high-TiO_2 soils at Taurus-Littrow, the excess reducing capacity and Fe/Ni ratios strongly suggest that the Mare Tranquillitatis soil has been exposed to a more intense solar wind radiation and contains a smaller meteoritic component.

(4) The dark-mantle basalts of Apollo 17 appear to have a common source and the differences in their grain size may reflect only different rates of crystallization. They are also compositionally very similar to those collected at the Apollo 11 site.

(5) The breccia and soils at Van Serg Crater are identical chemically. It is likely that the breccia may be indurated soil material which formed at the time of the cratering event.

Acknowledgments.—We are grateful to the National Aeronautics and Space Administration for making available the lunar samples for this investigation. The work was undertaken in part on behalf of NASA through MSC order No. T-75447 and 2630 A.

REFERENCES

Annell C. S. and Helz A. W. (1970) Emission spectrographic determination of trace elements in lunar samples from Apollo 11. *Proc. Apollo 11 Lunar Sci. Conf., Geochim. Cosmochim. Acta*, Suppl. 1, Vol. 2, pp. 991–994. Pergamon.

Cuttitta F., Rose H. J., Jr., Annell C. S., Carron M. K., Christian R. P., Dwornik E. J., Greenland L. P., Helz A. W., and Ligon D. T., Jr. (1971) Elemental composition of some Apollo 12 lunar rocks and soils. *Proc. Second Lunar Sci. Conf., Geochim. Cosmochim. Acta*, Suppl. 2, Vol. 2, pp. 1217–1229. MIT Press.

Cuttitta F., Rose H. J., Jr., Annell C. S., Carron M. K., Christian R. P., Ligon D. T., Jr., Dwornik E. J., Wright T. L., and Greenland L. P. (1973) Chemistry of twenty-one igneous rocks and soils returned by the Apollo 15 Mission. *Proc. Fourth Lunar Sci. Conf., Geochim. Cosmochim. Acta*, Suppl. 4, Vol. 2, pp. 1081–1096. Pergamon.

LSPET (Lunar Sample Preliminary Examination Team) (1973) Apollo 17 lunar samples: Chemical and petrographic description. *Science* **182**, 659–672.

Mason Brian, Jacobson Sara, Nelen J. A., Melson W. G., and Simkin Tom (1974) Regolith compositions from the Apollo 17 Mission (abstract). In *Lunar Science—V*, pp. 493–495. The Lunar Science Institute, Houston.

Rhodes J. M., Rodgers K. V., Shih C., Bansal B. M., Nyquist L. E., and Wiesman H. (1974) The relationship between geology and soil chemistry at the Apollo 17 landing site (abstract). In *Lunar Science—V*, pp. 630–632. The Lunar Science Institute, Houston.

Rose H. J., Jr., Cuttitta F., Dwornik E. J., Carron M. K., Christian R. P., Lindsay J. R., Ligon D. T., Jr., and Larson R. R. (1970) Semimicro X-Ray fluorescence analysis of lunar samples. *Proc. Apollo 11 Lunar Sci. Conf., Geochim. Cosmochim. Acta*, Suppl. 1, Vol. 2, pp. 1493–1497. Pergamon.

Rose H. J., Jr., Cuttitta Frank, Annell C. S., Carron M. K., Christian R. P., Dwornik E. J., Greenland L. P., and Ligon D. T., Jr. (1972) Compositional data for twenty-one Fra Mauro lunar materials. *Proc. Third Lunar Sci. Conf., Geochim. Cosmochim. Acta*, Suppl. 3, Vol. 2, pp. 1215–1229. MIT Press.

Rose H. J., Jr., Cuttitta F., Berman S., Carron M. K., Christian R. P., Dwornik E. J., Greenland L. P., and Ligon D. T., Jr. (1973) Composition data for twenty-two Apollo 16 samples. *Proc. Fourth Lunar Sci. Conf., Geochim. Cosmochim. Acta*, Suppl. 4, Vol. 2, pp. 1149–1158. Pergamon.

Schonfeld, Ernest (1974) Component abundance and evolution of regoliths and breccias: Interpretation by mixing models (abstract). In *Lunar Science—V*, pp. 669–671. The Lunar Science Institute, Houston.

Proceedings of the Fifth Lunar Conference
(Supplement 5, Geochimica et Cosmochimica Acta)
Vol. 2 pp. 1135–1145 (1974)
Printed in the United States of America

K and U systematics and average concentrations on the moon

ERNEST SCHONFELD

NASA Johnson Space Center, Houston, Texas 77058

Abstract—The K–U, Th–U, and Th–Al systematics for lunar samples from the Apollo 11–17 missions and the Luna 16 and 20 missions are summarized. With few exceptions (granitic portion of 12013 and the Apollo 17 mare basalts) the Th/U ratio is 3.8 ± 0.2. The K/U ratio for lunar samples is about 2600 and is distinctly different from the K/U ratio found in terrestial samples and in the majority of meteorites. The K–U systematics are similar to the K–La, K–Sm, K–Ba, and K–Zr systematics, suggesting that the moon accreted approximately homogeneously. Based on a model the moon contains an average abundance of about 80 ppb U, about 200 ppm K, and about 7 times chondritic abundances for all the refractory elements such as Al, Ca, Ti, Zr, Ba, Sr, and REE. The Th–Al systematics of many lunar highland rocks are dominated by the presence of the rock-type KREEP.

INTRODUCTION

THE K–U SYSTEMATICS of lunar samples, terrestrial samples and meteorites are important in understanding the processes by which the moon and the earth were formed and evolved (Gast, 1972). Also K–U systematics are useful to derive thermal models of the moon and understand the results of the heat-flow experiments. The purpose of this paper is to summarize K and U concentrations from the Apollo 11–17 missions and the Luna 16 and 20 missions and find interelemental relationships between K and U with other elements such as Th, Sm, La, Ba, Zr, and Al.

Lunar samples have a K/U ratio of about 2000 (Gast, 1972); this ratio is distinctly different from the K/U ratio found in terrestrial samples (Wasserburg *et al.*, 1964) and the majority of the meteorites (O'Kelley *et al.*, 1971; see also Fig. 1). The variation between the earth, moon, and the meteorites in the K/U ratio, along with the lack of K/U variation observed in igneous processes, suggests that the variations in K/U ratio were established by a process that dates from the formation of the moon, earth, and the meteorites (Gast, 1972). Since K is a volatile element and U is a refractory element, the variations of the K/U ratio between the moon, earth, and meteorites suggests a chemical fractionation process based on the condensation temperatures of these elements.

The chemical compositions of the lunar samples in this study are averages selected from many sources (e.g. *Proc. LSC*, 1970; *Proc. LSC*, 1971; *Proc. LSC*, 1972; *Proc. LSC*, 1973; *Lunar Science—V*, 1974; LSPET, 1973).

K–U SYSTEMATICS OF LUNAR SAMPLES

The concentrations of K, U, and Th from the Apollo 11–17 missions and Luna 16 and 20 missions were summarized and are shown (for U and K) in Figs. 1, 2,

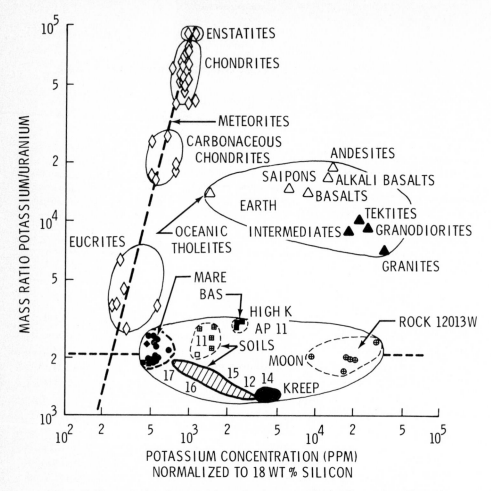

Fig. 1. K–U systematics of terrestrial samples, lunar samples, and meteorites. There are two trend lines. One is the meteoritic trend line and the other is a line representing an approximately constant K/U ratio of about 2600 for the moon. The intersection of both trend lines permits to estimate the average U and K concentrations of the source material where the lunar samples were derived. These concentrations are about 80 ppb U and 200 ppm K. A similar procedure for the earth gives concentrations of 31 ppb U and 400 ppm K (for a core free of radioactive elements).

and 3. Figure 1 represents an updated version of the K–U systematics (O'Kelley *et al.*, 1971; Gast, 1972). There are several observations one can make: (A) The K/U ratio for the lunar samples is about 2600 (O'Kelley *et al.*, 1971; Fig. 1). As shown in Fig. 3, the anorthosites are an exception with K/U ratios of 10^4 and higher. The anorthosites are not very abundant on the lunar surface and are made up of almost pure plagioclase. Since the mineral plagioclase does not concentrate U this exception to the constancy of the K/U ratio probably does not have any significance. (B) The Th/U ratio for the majority of the lunar samples is 3.8 ± 0.2

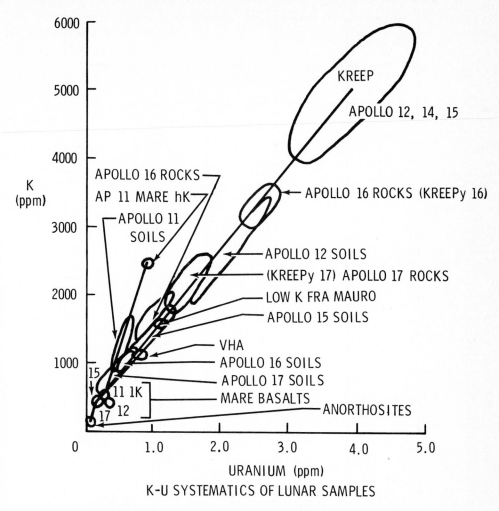

Fig. 2. K–U systematics of lunar samples. KREEPy 16 is the average of 60315, 65016, and 62235. KREEPy 17 is the average of the Apollo 17 noritic breccias (LSPET, 1973). "VHA" is the average of 61016, 61156, and 62295. Apollo 11 mare basalts hK and 1K are respectively the high-K and low-K basalts. Notice the trend line between KREEP and the mare basalts.

(O'Kelley *et al.*, 1971). The exceptions are the granitic portion of rock 12013 which has a Th/U ratio of 3.3 and the Apollo 17 mare basalts which have Th/U ratios of about 3.2 (Eldridge *et al.*, 1974). (C) The majority of the soils and breccias follow a K:U trend line (Fig. 2) between the mare basalts and KREEP suggesting simple mixtures between those two components. There are problems with this simple approach since there are other rock types such as low-K Fra Mauro basalt (Reid *et al.*, 1972), "VHA" (Hubbard *et al.*, 1973), KREEPy 16 (average of 65015, 60315, and 62235), and KREEPy 17 (noritic breccia from Apollo 17; LSPET, 1973) that have rather similar K/U ratios to KREEP and fall on the trend line shown in Fig. 2.

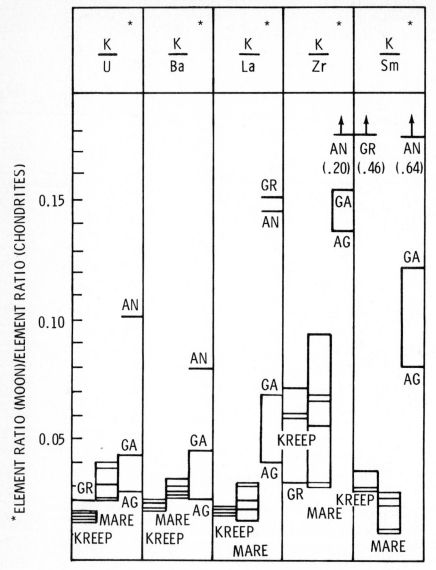

KREEP INCLUDES VHA, LOW K-FRA MAURO, KREEPy 16 AND 17.
GA = GABBROIC ANORTHOSITE. AG = ANORTHOSITIC GABBRO.
AN = ANORTHOSITE. GR = GRANITE. MARE = MARE BASALTS.

Fig. 3. K–U, K–Ba, K–La, K–Zr, and K–Sm systematics of lunar samples. The volatile (K)/refractory element ratio is normalized to chondrites. The horizontal lines in the mare basalts group represents mare basalts from different landing sites. In the case of KREEP the horizontal lines represent values for KREEP, "VHA," low-K Fra Mauro basalt, KREEPy 16, and KREEPy 17. For chondrites the values used for the K/U, K/Ba, K/La, K/Zr, and K/Sm ratio are respectively 75000, 250, 2900, 65, and 4100. The K/Ba* for granite is 0.022 (not shown).

Therefore, based only on the K–U systematics it is difficult to determine the component abundance in the soils and a multicomponent and multielement mixing model analysis is required (Schonfeld and Meyer, 1972; Schonfeld, 1974). The results of these mixing model calculations show that the relatively high K, Th, and U concentrations found in the majority of the lunar samples are associated with the presence of the lunar rock-type KREEP in these samples. (D) The Apollo 11 soils and breccias follow a different trend line (Fig. 2) between the high- and low-K mare basalts from Apollo 11. Mixing model calculations using many components and elements, show that there are other components present such as an "anorthositic" and a KREEP component in these samples (Schonfeld and Meyer, 1972). This trend line has a negative intercept in the K axis that does not have a geochemical significance since it is a mixing line.

COMPARISON OF THE K–U WITH THE K–LA, K–SM, K–BA, AND K–ZR SYSTEMATICS

The K/U ratio in lunar samples is rather constant when compared with the ratios found in terrestrial samples and meteorites (Fig. 1). Systematics of other similar pairs of volatile–refractory elements such as K/La (Wänke et al., 1973), K/Ba (Schnetzler and Philpotts, 1971; Duncan et al., 1973), and K/Zr (Duncan et al., 1973) have also been developed showing similar relative constancy of these ratios in lunar samples. In order to compare all these ratios in a single figure, these ratios were normalized by dividing each ratio by the respective value found in ordinary chondrites. The result of this comparison is shown in Fig. 3. This figure shows that in all cases the ratios for these volatile–refractory pairs in lunar samples is much smaller than the value found in chondrites suggesting enrichment of refractory elements and/or depletion of volatile elements on the moon (e.g. Ganapathy et al., 1970; Gast, 1972; Taylor, 1973). The normalized ratios for each system are rather similar but there is a "fine" structure or small degree of variability. For example, the relative position of the ratio for KREEP with respect to the ratio in mare basalts is variable. If these ratios were the same, they would suggest that the moon accreted homogeneously (Philpotts et al., 1972; Brett, 1973).

THORIUM–ALUMINUM SYSTEMATICS

The Th–Al systematics of lunar samples are summarized in Fig. 4. The orbital geochemical sensors of the Apollo 15 and 16 missions (Metzger et al., 1972; Adler et al., 1973) measured the Th and Al concentrations for about 20% of the moon's surface. In order to understand the results of these experiments it is necessary to summarize the Th–Al systematics of the lunar samples or "ground-truth." As can be seen in Fig. 4, the Th–Al systematics of many highland rocks that have composition similar to KREEP can be distinguished from other rock types. In cases where there are mixtures, the Th–Al systematics of lunar samples can be helpful but not sufficient to precisely identify the components and determine their respective abundances (Schonfeld, 1974).

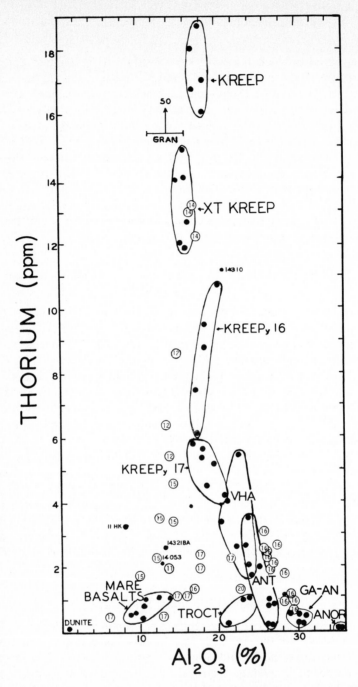

Fig. 4. Thorium–Al_2O_3 systematics of lunar samples. XT KREEP is crystalline KREEP from Apollo 15. The black circles are rocks and open circles are soils. The numbers in the open circles indicate the mission. One soil labelled 16 with an Al_2O_3 concentration of 16% is from Luna 16. The rest of the points labelled 16 are from the Apollo 16 mission. KREEP like rocks such as KREEPy 16, KREEPy 17, and "VHA" can be identified from the Th–Al_2O_3 systematics.

K AND U AVERAGE CONCENTRATION OF THE MOON

The average concentrations of K and U of the moon can be derived in the following way. In Fig. 1, two trend lines are shown. One is the meteoritic trend line (O'Kelley *et al.*, 1971) and the other is the approximately constant K/U line for the moon of about 2600. Assuming that the bulk composition of that part of the moon where most of the lunar samples were derived had K and U concentrations that follow the meteoritic trend line, then the intersection of those two trend lines would give the K and U concentrations of the upper mantle of the moon or that part of the moon where most lunar samples were derived. The intersection of those two trend lines corresponds to about 220 ppm K and 85 ppb U (K/U = 2600). In these calculations it was assumed that the average Si concentration for that part of the moon where most lunar samples were derived is 18% (average Si concentration in ordinary chondrites). If this last assumption is not correct, the K and U concentrations will change proportionally to the Si concentration change. For a K/U ratio of 1500 and 3500 the calculated values are 110 and 70 ppb, respectively for the U concentrations and 165 and 245 ppm, respectively for the K concentrations. Using a different model, Wänke *et al.* (1973) calculate concentrations of 86 ppb U and 250 ppm K for the whole moon. As discussed before, the ratio of volatile/refractory elements is not exactly constant and therefore it is possible to have a "fine" structure that could imply a small degree of original heterogeneity in the moon. Alternatively, the small variations of the volatile/refractory elemental ratios could be explained assuming a homogeneously accreted moon (Brett, 1973) and assuming slightly different crystal–liquid equilibria properties (Taylor, 1973) for the volatile and the refractory elements.

The calculated U and K concentrations for the moon have special significance when they are compared with the results of the heat-flow experiments of the Apollo 15 and 17 sites (Langseth *et al.*, 1973). The results of these experiments correspond to an average concentration of U of about 75 ppb (assuming a K/U ratio of about 2600 and a Th/U ratio of 3.8), suggesting that not only the upper mantle but probably the whole moon was originally enriched about 7 times in U with respect to chondritic abundances. This composition is quite similar in U concentration to the one proposed for the composition of the upper 150 km of the moon (Gast, 1972). Therefore it is possible that the moon accreted with a rather uniform composition that is enriched in all the refractory elements (e.g. Al, Ca, Ti, Zr, Ba, Sr, Th, U, and REE) about 7 times chondritic abundances.

A similar calculation as described before for the moon can be applied to the earth to estimate the average U and K concentrations. The solution of the intersection of the meteoritic trend line and the earth's K/U ratio trend line is equivalent to average concentrations of 31 ppb U and 400 ppm K (for K/U = 13000 and Si = 18%). The average heat flow of the earth is 63.9 erg cm^{-2} sec^{-1} (Lee and MacDonald, 1963) and for a K/U ratio of 13000 and a Th/U ratio of 3.8 this average heat-flow measurement is equivalent to average concentrations of 33 ppb U and 430 ppm K (for a core free of radioactive elements). The agreement between the two methods for estimating the average K and U concentrations is good.

Discussion

The model described before (intersection of two trend lines) to determine the average concentrations of U and K for the moon and the earth requires further discussion. First is the problem of deriving an average K/U ratio for the crust of the moon. At this stage we do not know the exact proportions of each rock type in the lunar crust and therefore it is difficult to determine an exact K/U ratio for the crust of the moon. An approximate method to estimate the abundance of rock types can be obtained from the results of the orbiting geochemical sensors (Metzger *et al.*, 1972; Adler *et al.*, 1973) and by the determination of rock type abundances in lunar soils using the mixing model method (Schonfeld, 1974). From these results it appears that the "anorthositic" materials (gabbroic anorthosites and anorthositic gabbros) are the most abundant materials on the surface of the moon, followed by mare basalts, and finally by KREEP. Based on these relative estimated abundances of rock types, one can derive an estimated value of 2600 for the K/U ratio of the lunar crust.

The second problem is the validity of the meteoritic trend line shown in Fig. 1. Meteorites are complicated bodies. In some cases there are chondrules and matrix present (e.g. carbonaceous chondrites type 2 and 3; Larimer and Anders, 1970), in other cases such as the howardites there are polymict breccias (Duke and Silver, 1967), that appear to be mixtures of eucrites and diogenites (McCarthy *et al.*, 1972) and in still other cases the meteorites appear to be differentiated bodies (Duke and Silver, 1967; Schnetzler and Philpotts, 1969). The majority of the eucrites are monomict breccias and when the chemical composition data (Duke and Silver, 1967; Schnetzler and Philpotts, 1969; McCarthy *et al.*, 1972) are averaged, then the relative concentration of Al, Ca, Ti, Sr, Eu, Ba, Zr, U, Th, and the rare earths with respect to ordinary chondrites are in the range of 6–8. It is difficult to obtain these concentrations by differentiating a source material similar to ordinary chondrites. Therefore, based on this argument, it is assumed that the average eucrite chemical composition represents a rather primitive body that underwent little or no differentiation. In the case of howardites, since they appear to be a rather clear case of mixtures of eucrites and diogenites they were excluded from Fig. 1. In the case of carbonaceous chondrites, the situation is not very clear at this time. Larimer and Anders (1970) postulated that the C-2 and C-3 type carbonaceous chondrites are mixtures of two components, but we do not have analysis of U and K for the chondrules. There are two other samples that were included in the derivation of this meteoritic trend line. One is the meteorite Angra dos Reis that has a K/U ratio of about 230 and fits the trend line. This meteorite is a single sample in its class and has a slight negative Eu anomaly (Schnetzler and Philpotts, 1969) and could be a differentiated body. In the case of the Ca-rich inclusions of the Pueblito de Allende meteorite, there are not enough data unless one combines data from two sources (Gast *et al.*, 1970; Wänke *et al.*, 1973). In general, combining meteoritic chemical data is not a good procedure because of possible sampling errors. The K/U ratio is about 900 and also fits approximately the trend line. It is significant that the Ca-rich inclusion of the Pueblito de Allende has a concentration of the rather volatile element K of 96.4 ppm, since according to

Larimer and Anders (1970) and Grossman (1973), it should have no K present. On the other hand, Arrhenius (1972) using a different model, where the ionization of the elements plays an important role, can account for the presence of K in the high Ca, Al, and U inclusions in Pueblito de Allende.

Even with some of the problems discussed before, the author believes that the use of a meteoritic trend line is a valid approach, and that more chemical analysis of meteorites are required to clarify this question.

CONCLUSIONS

(1) The K/U ratio for lunar samples is about 2600 and distinctly different from the K/U ratio found in terrestrial samples (about 13000) and the majority of the meteorites.

(2) The Th/U ratio in lunar samples is 3.8 ± 0.2. Exceptions are the granitic portion of rock 12013 and the Apollo 17 mare basalts.

(3) The K–U systematics of lunar samples are similar to the K–La, K–Sm, K–Ba, and K–Zr systematics suggesting that the moon has accreted approximately homogeneously.

(4) Based on a model presented in this paper, the average concentration of the moon is about 80 ppb U, about 200 ppm K, and about 7 times chondritic abundances for all the refractory elements such as Al, Ca, Ti, Zr, Ba, Sr, REE, etc. A similar procedure for the earth gives concentrations of 31 ppb U and 400 ppm K for a core free of radioactive elements.

(5) In general, the relatively high concentrations of Th, K, and U found in lunar samples are associated with the lunar rock-type KREEP. In many highland rocks such as "low-K Fra Mauro basalt," "VHA," KREEPy 16, and KREEPy 17, the Th–Al systematics is dominated by the rock-type KREEP.

Acknowledgments—The authors thank E. Anders, D. P. Blanchard, R. Brett, J. Philpotts, S. R. Taylor, and H. Wiesmann for their critical review and C. Hardy for typing this paper.

REFERENCES

Adler I., Trombka J. I., Schmadebeck R., Lowman P., Blodget H., Yin L., Eller E., Podwysocki M., Weidner J. R., Bickel A. L., Lum R. K. L., Gerard J., Gorenstein P., Bjorkholm P., and Harris B. (1973) Results of the Apollo 15 and 16 X-ray experiment. *Proc. Fourth Lunar Sci. Conf., Geochim. Cosmochim. Acta*, Suppl. 4, Vol. 3, pp. 2783–2791. Pergamon.

Arrhenius G. (1972) Chemical effects in plasma condensation. In *From Plasma to Planet* (editor A. Elvius), pp. 117–129. Almqvist and Wiksell, Stockholm.

Brett R. (1973) The lunar crust: A product of heterogeneous accretion or differentiation of a homogeneous Moon? *Geochim. Cosmochim. Acta* 37, 2697–2703.

Duke M. B. and Silver L. T. (1967) Petrology of eucrites, howardites and mesosiderites. *Geochim. Cosmochim. Acta* 21, 1637–1665.

Duncan A. R., Erlank A. J., Willis J. P., and Ahrens L. H. (1973) Composition and inter-relationships of some Apollo 16 samples. *Proc. Fourth Lunar Sci. Conf., Geochim. Cosmochim. Acta*, Suppl. 4, Vol. 2, pp. 1097–1113. Pergamon.

Eldridge J. S., O'Kelley G. D., and Northcutt K. J. (1974) Primordial radioelement concentrations in rocks and soils from Taurus-Littrow (abstract). In *Lunar Science—V*, pp. 206–208. The Lunar Science Institute, Houston.

Ganapathy R., Keays R. R., Laul J. C., and Anders E. (1970) Trace elements in Apollo 11 lunar rocks: Implications for meteoritic influx and origin of moon. *Proc. Apollo 11 Lunar Sci. Conf., Geochim. Cosmochim. Acta*, Suppl. 1, Vol. 2, pp. 1117–1142. Pergamon.

Gast P. W. (1972) The chemical composition and structure of the Moon. *The Moon* **5**, 121–148.

Gast P. W., Hubbard N. J., and Wiesmann H. (1970) Chemical composition and petrogenesis of basalts from Tranquillity Base. *Proc. Apollo 11 Lunar Science Conf., Geochim. Cosmochim. Acta*, Suppl. 1, Vol. 2, pp. 1143–1163. Pergamon.

Grossman L. (1973) Refractory trace elements in Ca–Al-rich inclusions in the Allende meteorite. *Geochim. Cosmochim. Acta* **37**, 1119–1140.

Hubbard N. J., Rhodes J. M., and Gast P. W. (1973) Chemistry of lunar basalts with very high alumina contents. *Science* **181**, 339–342.

Langseth M. G., Chute J. L., and Keihm S. (1973) Direct measurements of heat flow from the moon (abstract). In *Lunar Science—IV*, pp. 455–456. The Lunar Science Institute, Houston.

Larimer J. W. and Anders E. (1967) Chemical fractionations in meteorites—II. Abundance patterns and their interpretation. *Geochim. Cosmochim. Acta* **31**, 1239–1270.

Lee W. H. K. and MacDonald G. J. F. (1963) The global variation of terrestrial heat flow. *J. Geophys. Res.* **68**, 6481–6492.

LSPET (Lunar Sample Preliminary Examination Team) (1973) Apollo 17 Lunar samples: Chemical and petrographic description. *Science* **182**, 659–672.

Lunar Science—V (1974) Fifth Lunar Science Conference Abstracts. The Lunar Science Institute, Houston.

McCarthy T. S., Ahrens L. H., and Erlank A. J. (1972) Further evidence in support of the mixing model for howardite origin. *Earth Planet. Sci. Lett.* **15**, 86–93.

Metzger A. E., Trombka J. I., Peterson L. E., Reedy R. C., and Arnold J. R. (1973) Lunar surface radioactivity: Preliminary results of the Apollo 15 and Apollo 16 gamma-ray spectrometer experiments. *Science* **179**, 800–803.

O'Kelley G. D., Eldridge J. E., Schonfeld E., and Bell P. R. (1971) Abundances of the primordial radionuclides K, Th and U in Apollo 12 lunar samples by nondestructive gamma-ray spectrometry: Implications for origin of lunar soils. *Proc. Second Lunar Sci. Conf., Geochim. Cosmochim. Acta*, Suppl. 2, Vol. 2, pp. 1159–1168. MIT Press.

Philpotts J. A., Schnetzler C. C., Nava D. F., Bottino M. L., Fullagar P. D., Thomas H. H., Schuhmann S., and Kouns C. W. (1972) Apollo 14: Some geochemical aspects. *Proc. Third Lunar Sci. Conf., Geochim. Cosmochim. Acta*, Suppl. 3, Vol. 2, pp. 1293–1305. MIT Press.

Proceedings of the Apollo 11 Lunar Science Conference (1970) *Geochim. Cosmochim. Acta*, Supplement 1. Pergamon.

Proceedings of the Second Lunar Science Conference (1971) *Geochim. Cosmochim. Acta*, Supplement 2. MIT Press.

Proceedings of the Third Lunar Science Conference (1972) *Geochim. Cosmochim. Acta*, Supplement 3. MIT Press.

Proceedings of the Fourth Lunar Science Conference (1973) *Geochim. Cosmochim. Acta*, Supplement 4. Pergamon.

Reid A. M., Warner J., Ridley W. I., and Brown R. W. (1972) Major element composition of glasses in three Apollo 15 soils. *Meteoritics* **7**, 395–415.

Schnetzler C. C. and Philpotts J. A. (1969) Genesis of the calcium-rich achondrites in light of rare-earth and barium concentrations. In *Meteorite Research* (editor P. Millman), pp. 206–216. D. Reidel.

Schnetzler C. C. and Philpotts J. A. (1971) Alkali, alkaline earth, and rare-earth element concentrations in some Apollo 12 soils, rocks, and separated phases. *Proc. Second Lunar Sci. Conf., Geochim. Cosmochim. Acta*, Suppl. 2, Vol. 2, pp. 1101–1222. MIT Press.

Schonfeld E. (1974) Component abundance and evolution of regoliths and breccias: Interpretations by mixing models (abstract). In *Lunar Science—V*, pp. 669–671. The Lunar Science Institute, Houston.

Schonfeld E. and Meyer C., Jr. (1972) The abundances of components of the lunar soils by a least-squares mixing model and the formation age of KREEP. *Proc. Third Lunar Sci. Conf., Geochim. Cosmochim. Acta*, Suppl. 3, pp. 1397–1420. MIT Press.

Taylor S. R. (1973) Chemical evidence for lunar melting and differentiation. *Nature* **245**, 203–205.

Wänke H., Baddenhausen H., Breibus G., Jagoutz E., Kruse H., Palme H., Spettel B., and Teschke F. (1973) Multielement analyses of Apollo 15, 16 and 17 samples and the bulk composition of the moon. *Proc. Fourth Lunar Sci. Conf., Geochim. Cosmochim. Acta,* Suppl. 4, Vol. 2, pp. 1461–1481. Pergamon.

Wasserburg G. J., MacDonald G. J. F., Hoyle F., and Fowler W. A. (1964) Relative contributions of uranium, thorium, and potassium to heat production in the Earth. *Science* **143**, 465–467.

Proceedings of the Fifth Lunar Conference
(Supplement 5, Geochimica et Cosmochimica Acta)
Vol. 2 pp. 1147–1157 (1974)
Printed in the United States of America

Trace element evidence for a two-stage origin of some titaniferous mare basalts

A. R. Duncan, A. J. Erlank, J. P. Willis,
M. K. Sher, and L. H. Ahrens

Department of Geochemistry, University of Cape Town, Rondebosch 7700, Cape, South Africa

Abstract—Values of the K/Zr, K/Y, Ba/Zr, and Zr/Nb ratios in Apollo 17 mare basalts are significantly different from the near-constant values previously observed for these ratios in most lunar materials. Perturbation of these ratios from "normal" lunar values cannot be due to near surface or post-eruptive crystal fractionation processes and it is concluded that these perturbed interelement ratios are a compositional characteristic of Apollo 17 basaltic *liquids*. A model is proposed in which high-titanium basaltic liquids with these perturbed interelement ratios may be derived by melting of a preexisting cumulate in which ilmenite and a Zr–Y–Nb-rich phase(s) are cumulus phases. This two-stage genetic model is also applicable to the Apollo 11 low-potassium basalts and is compatible with major element data, experimental petrology, and models of lunar differentiation which involve early melting of the outer portion of the moon.

Introduction

Analytical data for 23 elements in fourteen Apollo 17 samples are presented in Table 1. The data were obtained by X-ray fluorescence spectrometry using essentially the same techniques described by Willis *et al.* (1971, 1972a). The four U.S.G.S. Standard rocks G-1, W-1, BCR-1, and GSP-1 were again used to standardize trace element data and to avoid systematic errors between this and previous data we have reported. The accuracy and precision of our data may be assessed from Willis *et al.* (1972a, Table 2).

We have recently discussed the relationship between composition and selenographic position of regolith samples at the Apollo 17 landing site (Duncan *et al.*, 1974). In this paper we concentrate our discussion on what we consider to be trace element evidence for a two-stage origin of some high-Ti mare basalts.

We have previously commented (Willis *et al.*, 1971; Erlank *et al.*, 1972; Willis *et al.*, 1972b; Duncan *et al.*, 1973) on interelement relationships between the "incompatible" elements K, Ba, Zr, Nb, and Y that we consider to be of particular importance in considering certain aspects of lunar origin and the petrogenesis of lunar lavas. For the Apollo 12, 14, and 15 samples analyzed by us (basalts, soils, and breccias) the relative constancy of the K/Zr, K/Ba, Ba/Zr, and Zr/Nb ratios (Duncan *et al.*, 1973, Figs. 3, 4, and 5) indicates that these ratios are very similar in the source regions from which both the basalts and the igneous rocks present as major components in the clastic materials (e.g. KREEP) were derived, and that they have not undergone major subsequent modification. This suggests that this group of five elements is essentially excluded from mineral phases which are left in the residuum after partial melting of the source material, and from mineral phases which may have

1147

A. R. DUNCAN *et al.*

Table 1. Chemical analyses of Apollo 17 soils, breccia, and basalts.

	72701,12 Soil	74121,2 Soil	74220,31 Soil	76321,3 Soil	78221,1 Soil	78501,12 Soil	75081,66 A Soil +60 #
Major elements (%)							
SiO$_2$	45.24	44.51	39.03	44.19	43.67	42.83	39.73
TiO$_2$	1.50	2.56	8.72	2.95	3.84	5.28	10.45
Al$_2$O$_3$	20.70	19.36	6.47	18.68	17.13	15.65	10.18
FeO	8.78	10.24	22.13	10.36	11.68	13.18	17.66
MnO	0.116	0.132	0.273	0.135	0.157	0.177	0.235
MgO	9.99	9.93	14.44	10.82	10.55	10.01	9.36
CaO	12.74	12.44	7.62	12.24	11.79	11.51	11.04
Na$_2$O	0.44	0.40	0.34	0.40	0.37	0.38	0.37
K$_2$O	0.154	0.134	0.077	0.124	0.092	0.090	0.067
P$_2$O$_5$	0.153	0.136	0.043	0.113	0.080	0.082	0.071
Cr$_2$O$_3$	0.229	0.269	0.684	0.272	0.321	0.355	0.491
S	0.074	0.083	0.073	0.080	0.088	0.109	0.125
Subtotal	100.116	100.194	99.900	100.364	99.768	99.653	99.779
O≡S	0.037	0.042	0.037	0.040	0.044	0.055	0.063
Total	100.079	100.152	99.863	100.324	99.724	99.598	99.716
Trace elements (ppm)							
Nb	16.9	16.5	13.6	15.8	13.3	15.2	20.2
Zr	264	244	186	210	173	186	241
Y	54.7	53.7	43.8	48.6	45.1	50.4	73.7
Sr	150	151	200	150	147	150	149
Rb	4.3	3.6	1.5	3.2	2.6	2.4	1.8
Zn	17.5	24.1	253	20.4	25.6	32.8	12.0
Cu	4.5	8.4	26.3	5.2	5.8	10.0	3.2
Ni	231	245	74.7	190	221	177	68.5
Co	30	33	62	30	34	32	25
V	47	58	132	54	68	82	111
Ba	190	164	82	139	109	102	95

been added to or removed from the derived liquids (by crystal fractionation or assimilation processes), i.e. that they are truly behaving as "incompatible" elements. The Apollo 16 samples that are relatively low in K (<700 ppm) and high in Al$_2$O$_3$ (>25%) have significantly higher K/Zr and K/Ba ratios than other Apollo 16 samples, or samples from Apollo 12, 14, and 15. We believe that these plagioclase-rich samples are cumulates, or are essentially monomict breccias derived from cumulates, and ascribe their higher K/Zr and K/Ba ratios to preferential incorporation of K into plagioclase relative to Zr and Ba. Although the K content of plagioclase is low, it would constitute a significant fraction of the total K content in plagioclase cumulates which contain relatively minor amounts of K-rich mesostasis.

Table 1. (*Continued*).

	75081,66 B Soil −60 + 120 #	75081,66 C Soil −120 + 305 #	75081,66 D Soil −305 #	73235,53 Breccia	70017,18 Basalt	70215,55 Basalt	74275,58 Basalt
Major elements (%)							
SiO_2	40.06	40.03	40.65	45.96	38.37	37.91	38.43
TiO_2	9.89	9.73	8.33	0.60	12.83	13.08	12.66
Al_2O_3	10.25	11.03	13.77	22.57	8.78	8.86	8.51
FeO	17.74	17.75	16.01	6.68	18.71	19.96	18.25
MnO	0.237	0.237	0.202	0.091	0.247	0.264	0.247
MgO	9.71	9.61	9.08	9.61	9.41	7.99	10.26
CaO	10.79	10.75	10.97	13.18	10.43	10.77	10.38
Na_2O	0.37	0.33	0.35	0.44	0.43	0.38	0.37
K_2O	0.065	0.071	0.093	0.200	0.047	0.041	0.075
P_2O_5	0.069	0.080	0.109	0.192	0.052	0.114	0.074
Cr_2O_3	0.477	0.479	0.434	0.196	0.577	0.431	0.639
S	0.120	0.100	0.189	0.027	0.175	0.188	0.141
Subtotal	99.778	100.197	100.187	99.746	100.058	99.988	100.036
O≡S	0.060	0.050	0.095	0.014	0.088	0.094	0.071
Total	99.718	100.147	100.092	99.732	99.970	99.894	99.965
Trace elements (ppm)							
Nb	19.8	20.2	20.6	19.7	18.5	20.8	22.1
Zr	235	224	238	315	218	192	248
Y	71.2	67.3	65.5	62.3	71.2	63.6	81.5
Sr	154	164	180	145	166	122	158
Rb	1.4	1.4	2.1	5.6	1.2	<1	1.9
Zn	16.8	21.8	49.1	<2	<2	<2	<2
Cu	4.1	5.0	11.5	<2	<3	<3	<3
Ni	87.5	116	198	118	<3	<3	<3
Co	26	30	37	22	18	23	24
V	103	82	75	40	146	50	79
Ba	89	91	112	252	83	77	89

INTERELEMENT RELATIONSHIPS IN APOLLO 17 SAMPLES

Our data for Apollo 17 materials (Table 1) show perturbation of the interelement relationships observed in Apollo 12, 14, 15, and some Apollo 16 samples. The ratios K/Zr, K/Y, Ba/Zr, and Zr/Nb in Apollo 17 samples vary systematically as a function of TiO_2 content as shown in Fig. 1. The majority of the samples for which data are plotted in Fig. 1 are soil samples. We thus consider that the trends shown represent mixing lines between a low TiO_2 component with what we have come to consider as "normal" lunar interelement ratios and a high TiO_2 component (Apollo 17 basalts) with severely perturbed interelement ratios. We stress that the variation *trends* shown by us do not represent magmatic variation trends. The

Table 2.

	No. of samples	K/Zr	K/Y	Zr/Nb	Ba/Zr
KREEP-rich basaltic materials	(4)	4.59	22.3	16.0	0.81
Apollo 12, 14, and 15 mare-type basalts Apollo 16 rocks with K > 1000 ppm	(10)	3.88	14.2	16.2	0.64
Apollo 17 basalts	(3)	2.02	6.15	10.7	0.38
Apollo 11 low-K basalts	(6)	2.14	5.42	13.3	—
Apollo 11 high-K basalts	(3)	5.56	15.1	17.5	—

average Apollo 17 basalt (Table 1, LSPET, 1973; Rhodes *et al.*, 1974) has K/Zr, Ba/Zr, and Zr/Nb ratios which are displaced approximately 5–6 standard deviations from the mean ratios for Apollo 12, 14, 15, and high-K Apollo 16 samples. These ratio perturbations should be explicable either by processes occurring in the basaltic magma before or after eruption, or by processes which have resulted in a source region compositionally unlike those regions which were melted to produce the igneous rocks collected from the Apollo 12, 14, 15, and 16 landing sites. In all of the following discussions we have made the basic assumption that

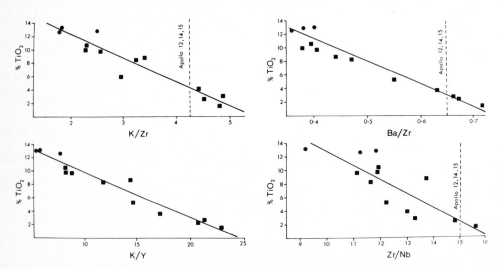

Fig. 1. Variation of selected interelement ratios with TiO$_2$ content in Apollo 17 samples. Data are from Table 1. Symbols: ● = basalt, ■ = soil. Dashed lines represent the mean values of K/Zr, Ba/Zr, and Zr/Nb ratios in *all* Apollo 12, 14, and 15 samples we have analyzed (data from Willis *et al.*, 1971; Willis *et al.*, 1972b; Erlank *et al.*, 1972). K/Y ratios are somewhat variable in Apollo 12, 14, and 15 samples (see Duncan *et al.*, 1973) and for this reason overall means have not been indicated on this figure.

the basaltic materials with perturbed interelement ratios have been generated from an original material with "normal" lunar interelement ratios. Our reasons for making this assumption are discussed in a later section.

We are not aware of any early-crystallizing mineral phases in the lunar basalts which concentrate K, Ba, and minor Nb and exclude Y and Zr. Consequently, if the perturbed ratios in the Apollo 17 basalts are due to fractional crystallization of the basaltic magma, then the basalts must be partial cumulates and not residual liquids whose ratios have been perturbed by separation of mineral phases prior to eruption. In examining whether or not these basalts are partial cumulates, one must consider textural evidence, experimental evidence, and the geochemical effects to be expected from cumulus concentration of liquidus or near-liquidus phases. Two of the Apollo 17 basalts we have analyzed are very fine grained (70215, 74275) and are texturally most unlike cumulates. This same pair of basalts has been studied by Green *et al.* (1974) who concluded from experimental and petrographic evidence that the lavas which these samples represent were saturated in olivine alone or olivine + Ti–Cr–spinel on extrusion, and that pyroxene, plagioclase, ilmenite and armalcolite were not present in the lavas prior to extrusion. Walker *et al.* (1974), however, state that armalcolite, not olivine, is the liquidus phase of 70215 although both 70215 and 70017 (analysis given in Table 1) become doubly saturated with olivine and armalcolite close to their liquidus temperatures, with subsequent reaction of armalcolite to form ilmenite. O'Hara *et al.* (1974a) claim that high-Ti basalts from both Apollo 17 and Apollo 11 are near-cotectic in character, and consider that the basalts have been derived from a high level magma chamber in which fractionation was well advanced. O'Hara *et al.* (1974a, 1974b) also suggest that the large rock samples of basalts from the Apollo 17 site (including 70017, 70215, and 74275) represent the basal portion of lava flows in which post-eruptive cumulus enrichment of olivine and Fe–Ti oxides has occurred, and that cumulus enrichment of microphenocrysts would not be texturally discernible.

In view of this considerable disagreement on both textural and experimental data and its interpretation, it is clear that neither of these lines of evidence can unequivocally determine whether or not the Apollo 17 basalts contain cumulus phases.

Let us instead consider whether cumulus concentration of any reasonable combination of phases into Apollo 17 basalts could explain the perturbation of interelement ratios from those values observed in the majority of lunar materials sampled. The hypothetical cumulus assemblage must be depleted in K and Ba, and enriched in Zr, Y, and Nb relative to the basaltic magma into which it is sinking. The only minerals present in the basalts at significant (>0.5%) modal concentration which possess some of the required characteristics are ilmenite and armalcolite. Lunar ilmenites exclude K and Ba and concentrate Zr with up to 2500 ppm Zr (Meyer and Boctor, 1974) reported from ilmenites in Apollo 17 basalts. Armalcolites in the Apollo 17 basalts are relatively low in Zr with less than 400 ppm being typical (e.g. El Goresy, 1973). Assuming Zr contents of 2000 ppm in ilmenite and 300 ppm in armalcolite, it would require 5% of the rock to be cumulus ilmenite, or

33% of the rock to be cumulus armalcolite to explain the observed perturbations of the K/Zr and Ba/Zr ratios in the Apollo 17 basalts. Since the calculated amount of armalcolite is far in excess of the modal concentrations of armalcolite reported in Apollo 17 basalts, we can conclude that cumulus armalcolite cannot be responsible for the observed ratio perturbations. In view of the high ilmenite content of Apollo 17 basalts, the possibility of 5% cumulus ilmenite in the rock does not seem unreasonable. However, as the Y content of lunar ilmenites is extremely low (Andersen *et al.*, 1970), it appears impossible to explain the perturbed K/Y ratio in Apollo 17 basalts by addition of cumulus ilmenite.

We conclude that the perturbed interelement ratios (K/Zr, K/Y, Ba/Zr, and Zr/Nb) cannot be characteristic of any assemblage of reasonable cumulus phases in the Apollo 17 basalts and thus infer that the Apollo 17 basaltic *liquids* have perturbed ratios.

TWO-STAGE ORIGIN OF APOLLO 17 BASALTS

Since it appears impossible to explain the perturbation of interelement ratios in Apollo 17 basalts by processes occurring in the basaltic magma before or after eruption, it is now necessary to examine whether these perturbations could have been caused by partial melting processes or a source material of unusual composition.

From what is now known or inferred concerning lunar mineralogy, there is no phase or combination of phases which would be partially retained in the residuum after partial melting and which concentrates K, Ba and minor Nb, while excluding Zr and Y. In the local equilibrium model of partial melting advocated by Graham and Ringwood (1971) and Ringwood (1974a) variation in the degree of partial melting can only vary the absolute abundances of the "incompatible" elements but cannot change their relative proportions. It thus appears most unlikely that a partial melting model with either bulk or local equilibrium melting of the same source material that was melted to form other mare basalt magmas could explain the ratio perturbations observed in Apollo 17 basalts.

We turn then to consideration of a source material of unusual composition. In previous papers (Erlank *et al.*, 1972; Duncan *et al.*, 1973) we have argued that the relative constancy of certain ratios of volatile and involatile elements in lunar materials other than high-Ti basalts is strong evidence against the proposal of inhomogeneous accretion of the outer portion of the moon on a simple volatility dependent model (e.g. Gast and McConnell, 1972; Ringwood, 1972). Other authors (e.g. Brett, 1973; Wänke *et al.*, 1973; Ringwood, 1974b) have come to similar conclusions. Since the volatile–involatile element ratios in Apollo 17 basalts differ from those we have observed in the majority of returned lunar samples, it is clear that we cannot exclude the possibility of Apollo 17 basalts being derived by partial melting of a portion of the moon that is of unique composition, and that this unique character may reflect initially inhomogeneous accretion. However, the involatile–involatile element ratios (e.g. Zr/Nb, Zr/Y) in Apollo 17 basalts also differ from other lunar samples with the exception of some Apollo 11 basalts (see later section), indicating that if these differences are a product of inhomogeneous accretion, the process controlling inhomogeneity is not volatility dependent. We believe therefore that the differences in the source material for the Apollo 17 basalts are best explained by igneous processes operating within the outer portion of the moon which was relatively homogeneous with respect to the elemental ratios with which we are concerned.

Schnetzler and Philpotts (1971) and Philpotts *et al.* (1973) suggested that Apollo 11 low-K basalts and Apollo 17 basalts could have been formed by extensive partial melting of a previously formed cumulate, which they inferred from their trace element data to be rich in a high-Ti, heavy rare-earth-bearing phase or phases. We consider that such a two-stage model for the origin of Apollo 17 basalts could explain the ratio perturbations we have observed, but we have had difficulty in precisely defining the assemblage of cumulus minerals required. Ilmenite as a cumulus phase in the source region would explain the low K/Zr and Ba/Zr ratios in the derived liquids. If lunar ilmenites have low Zr/Nb ratios, as do the Zr-rich ilmenite megacrysts in terrestrial kimberlites (Zr/Nb ratios of 0.5–2; A. J. Erlank, unpublished data), then ilmenite as a cumulus phase in the source region could also explain the relatively low Zr/Nb ratios in the derived liquids. However, since lunar ilmenite contains insignificant Y (~4 ppm atomic, Andersen *et al.*, 1970), it cannot be responsible for perturbations in the K/Y ratio. Most armalcolites are too low in Zr (~200–300 ppm, e.g. El Goresy *et al.*, 1973) to have any appreciable effect on K/Zr ratios, and those armalcolites which are high in Zr have low Nb and Y contents (Hinthorne, personal communication, 1974) and are thus unsuitable in this respect.

Kirsten and Horn (1974) have recently described a poikilitic troctolite (78503,13A) in which zirconolite is unusually abundant and appears to be an early-crystallizing primary phase. This raises the possibility of zirconolite being included in a cumulus assemblage formed in the lunar interior. Although analyzed lunar zirconolites (e.g. Wark *et al.*, 1973; Meyer and Boctor, 1974) have variable composition, many contain abundant Zr (24%), Y (10%), and Nb (3%) with relatively low Zr/Nb and Zr/Y ratios. Zirconolites are also relatively enriched in heavy rare earths (Lovering and Wark, 1974) and could prove a suitable cumulus phase for satisfying the constraints of Philpotts *et al.* (1973). However, in view of the normal lunar occurrence of zirconolite as a very late crystallizing mineral found in the basaltic mesostasis, the possibility of cumulus zirconolite (in line with Kirsten and Horn's observations) must be considered speculative.

An ilmenite-bearing cumulate with minor amounts of a cumulus phase rich in Y and heavy REE (such as zirconolite) would be a satisfactory source material to give rise to Apollo 17 basalts by extensive partial melting. The liquid produced by extensive partial melting of such a cumulate would have the following characteristics relative to the liquid from which the cumulus phases had separated: lower K/Zr, K/Y, Ba/Zr, Zr/Nb, and Zr/Y ratios. These are precisely the ratio perturbations observed in Apollo 17 basalts. Walker *et al.* (1974) have noted that experimental data is compatible with an origin of high-Ti Apollo 17 basalts by partial melting of a cumulus assemblage containing olivine, pigeonite, ilmenite, spinel, and minor plagioclase at depths of approximately 100 km.

We conclude that there is strong trace element evidence in favor of a two-stage origin of the Apollo 17 basalts by extensive partial melting of ilmenite-bearing cumulates containing a minor Y-rich cumulus phase. Such an origin is compatible with the major element chemistry of the samples and experimental investigations of their petrogenesis.

APOLLO 11 REVISITED

We have only analyzed one Apollo 11 basalt (10017) in this laboratory and have therefore had to extract data for K, Zr, Y, and Nb in Apollo 11 basalts from the Lunar Sample Database (a magnetic tape file of lunar analytical data compiled by Dr. J. L. Warner and co-workers at J.S.C.). The K/Zr, K/Y, and Zr/Nb ratios of both high-K and low-K Apollo 11 basalts are given in Table 2, with data for basalts from other landing sites for comparison. In spite of the problems inherent in comparing data from different laboratories, it seems clear that the Apollo 11 low-K basalts have perturbed K/Zr and K/Y ratios that are very similar to those observed in Apollo 17 basalts. It is not clear whether the slightly lower Zr/Nb ratio in Apollo 11 low-K basalts represents a significant perturbation. Apollo 11 high-K basalts do not show the same perturbation of K/Zr, K/Y, and Zr/Nb ratios as shown by Apollo 17 and Apollo 11 low-K basalts.

It seems very probable in view of the major and trace element similarities (including perturbed K/Zr and K/Y ratios) between Apollo 11 low-K basalts and Apollo 17 basalts that a two-stage model as developed for the Apollo 17 basalts could also be applied to the origin of the Apollo 11 low-K basalts.

THE ORIGIN OF MARE BASALTS AND MODELS OF THE MOON

Taylor and Jakeš (1974) have recently proposed a rather comprehensive model of the moon and a comparison of our data on interelement relationships in mare basalts with the implications of the Taylor–Jakeš model may be of interest. They have proposed that all mare basalts have been derived from remelted cumulus assemblages which were formed by fractional crystallization of an original 1000 km thick molten zone. This model is compatible with our data provided that the cumulus assemblages which are parental to the Apollo 12, 14, and 15 mare-type basalts contain no cumulus phases which differentially concentrate the incompatible elements, and provided that the zone which is parental to high-Ti basalts contains a minor Y-rich cumulus phase in addition to the Fe–Ti oxides suggested by Taylor and Jakeš. It remains an open question as to whether the overall distribution of K, U, and Th in the Taylor–Jakeš model will provide suitable thermal evolution to permit remelting of the cumulate layers.

In a more conventional model where the surface zone of melting is restricted to some 250 km depth, the low-Ti mare basalts could be produced by partial melting in previously unmelted (primitive) mantle material, heated by downward propagation of the surface heat "pulse," combined with radioactive heating (e.g. thermal models of Wood, 1972). As stated by Ringwood (1974a), the Ti-rich cumulates will form relatively late in the crystallization sequence of an initial "skin melt" and will thus be rather close to the zone containing residual liquids that are enriched in K, U, and Th. This position, as the "filling in a sandwich" between two zones of high temperature, with considerable radioactive heat production in the overlying zone, could lead to remelting of the Ti-rich cumulate layer.

In almost any model of the origin of a Ti-rich cumulate layer it is reasonable to infer that this cumulate layer and a residual liquid are complementary to one another. This residual liquid will of necessity be strongly enriched in what can be termed the "KREEP characteristic elements" (e.g. K, REE, P, Zr, Nb, Y) and the Taylor–Jakeš model suggests that this residual liquid is either directly tapped, or crystallized and subsequently remelted to form KREEP-basalts. If the production of a Ti-rich cumulate layer (with ilmenite and zirconolite (?) as cumulus phases) extracted a significant proportion of some but not other "incompatible" elements from the residual liquid, then KREEP material should show interelement ratio perturbations which are opposite in sense, but not necessarily equal in magnitude, to those observed in the Apollo 17 and Apollo 11 low-K basalts. Table 2 shows that the K/Zr, K/Y, and Ba/Zr ratios in KREEP are higher than those in typical low-Ti mare basalts, whereas the K/Zr, K/Y, and Ba/Zr ratios in Apollo 17 basalts are lower than those in low-Ti mare basalts. We had previously noted that KREEP materials had slightly different values for those ratios (Duncan *et al.*, 1973) but had ascribed the difference to the possible production of KREEP materials by a very low degree of partial melting.

The apparent complementary perturbations of certain interelement ratios in KREEP and high-Ti mare basalts is considered additional evidence for the two-stage origin of some high-Ti basalts by melting of ilmenite-rich cumulates. The occurrence of Apollo 17 basalts and Apollo 11 low-K basalts with extraordinary similar chemistry (including some unusual interelement ratios) at widely separated sites on the moon's surface suggests that the formation of the proposed cumulate source materials is very reasonably linked to the moonwide skin melting hypothesis of early lunar differentiation that has been proposed by many authors.

Acknowledgments—We acknowledge support from the University of Cape Town and the C.S.I.R., and the invaluable assistance of Dr. J. J. Gurney who exercised his customary dexterity in the preparation of fusion disks for major element analysis.

REFERENCES

Andersen C. A., Hinthorne J. R., and Fredriksson K. (1970) Ion microprobe analysis of lunar material from Apollo 11. *Proc. Apollo 11 Lunar Sci. Conf., Geochim. Cosmochim. Acta*, Suppl. 1, Vol. 1, pp. 159–167. Pergamon.

Brett R. (1973) The lunar crust: a product of heterogeneous accretion or differentiation of a homogeneous moon? *Geochim. Cosmochim. Acta* 37, 2697–2703.

Duncan A. R., Erlank A. J., Willis J. P., and Ahrens L. H. (1973) Composition and inter-relationships of some Apollo 16 samples. *Proc. Fourth Lunar Sci. Conf., Geochim. Cosmochim. Acta,* Suppl. 4, Vol. 2, pp. 1097–1113. Pergamon.

Duncan A. R., Erlank A. J., Willis J. P., and Sher M. K. (1974) Compositional characteristics of the Apollo 17 regolith (abstract). In *Lunar Science—V,* pp. 184–186. The Lunar Science Institute, Houston.

El Goresy A., Ramdohr P., and Medenbach O. (1973) The opaque minerals in Apollo 17 samples. *Trans. Amer. Geophys. Union* 54, 591–592.

Erlank A. J., Willis J. P., Ahrens L. H., Gurney J. J., and McCarthy T. S. (1972) Inter-element relationships between the moon and stony meteorites with particular reference to some refractory elements (abstract). In *Lunar Science—III,* pp. 239–241. The Lunar Science Institute, Houston.

Gast P. W. and McConnell R. K. (1972) Evidence for initial chemical layering of the moon (abstract). In *Lunar Science—III,* pp. 289–290. The Lunar Science Institute, Houston.

Graham A. L. and Ringwood A. E. (1971) Lunar basalt genesis: The origin of the Europium anomaly. *Earth Planet. Sci. Lett.* 13, 105–115.

Green D. H., Ringwood A. E., Ware N. G., and Hibberson W. O. (1974) Petrology and petrogenesis of Apollo 17 basalts and Apollo 17 orange glass (abstract). In *Lunar Science—V,* pp. 287–289. The Lunar Science Institute, Houston.

Kirsten T. and Horn P. (1974) ^{39}Ar–^{40}Ar chronology of the Taurus-Littrow region II: a 4.28 b.y. old troctolite and ages of basalts and highland breccias (abstract). In *Lunar Science—V,* pp. 419–421. The Lunar Science Institute, Houston.

Lovering J. F. and Wark D. A. (1974) Rare earth element fractionation in phases crystallizing from lunar late-stage magmatic liquids (abstract). In *Lunar Science—V,* pp. 463–465. The Lunar Science Institute, Houston.

LSPET (Lunar Sample Preliminary Examination Team) (1973) Apollo 17 Lunar Samples: Chemical and Petrographic Description. *Science* 182, 659–672.

Meyer H. O. A. and Boctor N. Z. (1974) Opaque minerals in basaltic rock 75035 (abstract). In *Lunar Science—V,* pp. 512–514. The Lunar Science Institute, Houston.

O'Hara M. J., Biggar G. M., Humphries D. J., and Saha P. (1974a) Experimental petrology of high titanium basalt (abstract). In *Lunar Science—V,* pp. 571–573. The Lunar Science Institute, Houston.

O'Hara M. J., Biggar G. M., Hill P. G., Jeffries B., and Humphries D. J. (1974b) Plagioclase saturation in lunar high-titanium basalt. *Earth Planet. Sci. Lett.* 21, 253–268.

Philpotts J. A., Schuhmann S., Schnetzler C. C., Kouns C. W., Doan A. S., Wood F. M., Bickel A. L., Lum Staab R. K. L. (1973) Apollo 17: Geochemical aspects of some soils, basalts and breccia. *Trans. Amer. Geophys. Union* 54, 603–604.

Rhodes J. M., Rodgers K. V., Chin C., Bansal B. M., Nyquist L. E., and Wiesmann H. (1974) The relationship between geology and soil chemistry at the Apollo 17 landing site (abstract). In *Lunar Science—V,* pp. 630–632. The Lunar Science Institute, Houston.

Ringwood A. E. (1972) Zonal structure and origin of the moon (abstract). In *Lunar Science—III,* pp. 651–653. The Lunar Science Institute, Houston.

Ringwood A. E. (1974a) Minor element chemistry of maria basalts (abstract). In *Lunar Science—V,* pp. 633–635. The Lunar Science Institute, Houston.

Ringwood A. E. (1974b) The lunar crust: Comments on a letter by R. Brett. *Geochim. Cosmochim. Acta.* 38, 983–984.

Schnetzler C. C. and Philpotts J. A. (1971) Alkali, alkaline earth, and rare-earth element concentrations in some Apollo 12 soils, rocks, and separated phases. *Proc. Second Lunar Sci. Conf., Geochim. Cosmochim. Acta,* Suppl. 2, Vol. 2, pp. 1101–1122. MIT Press.

Taylor S. R. and Jakeš P. (1974) Geochemical zoning in the moon (abstract). In *Lunar Science—V,* pp. 786–788. The Lunar Science Institute, Houston.

Walker D., Longhi J., Stolper E., Grove T., and Hays J. F. (1974) Experimental petrology and origin of titaniferous lunar basalts (abstract). In *Lunar Science—V,* pp. 814–816. The Lunar Science Institute, Houston.

Wänke H., Baddenhausen H., Dreibus G., Quijano-Rico M., Palme H., Spettel B., and Tesche F. (1973) Multi-element analysis of Apollo 16 samples and about the composition of the whole moon (abstract). In *Lunar Science—IV,* pp. 761–763. The Lunar Science Institute, Houston.

Wark D. A., Reid A. F., Lovering J. F., and El Goresy A. (1973) Zirconolite (versus zirkelite) in lunar rocks (abstract). In *Lunar Science—V*, pp. 764–766. The Lunar Science Institute, Houston.

Willis J. P., Ahrens L. H., Danchin R. V., Erlank A. J., Gurney J. J., Hofmeyr P. K., McCarthy T. S., and Orren M. J. (1971) Some inter-element relationships between lunar rocks and fines, and stony meteorites. *Proc. Second Lunar Sci. Conf., Geochim. Cosmochim. Acta*, Suppl. 2, Vol. 2, pp. 1123–1138. MIT Press.

Willis J. P., Erlank A. J., Gurney J. J., Theil R. H., and Ahrens L. H., (1972a) Major, minor, and trace element data for some Apollo 11, 12, 14, and 15 samples. *Proc. Third Lunar Sci. Conf., Geochim. Cosmochim. Acta*, Suppl. 3, Vol. 2, pp. 1269–1273. MIT Press.

Willis J. P., Erlank A. J., Gurney J. J., and Ahrens L. H. (1972b) Geochemical features of Apollo 15 materials. In *The Apollo 15 Lunar Samples*, pp. 268–271. The Lunar Science Institute, Houston.

Wood J. A. (1972) Thermal history and early magmatism in the moon. *Icarus* **16**, 229–240.

Proceedings of the Fifth Lunar Conference
(Supplement 5, Geochimica et Cosmochimica Acta)
Vol. 2 pp. 1159–1180 (1974)
Printed in the United States of America

The history of lunar breccia 14267*

G. Eglinton, B. J. Mays, and C. T. Pillinger

School of Chemistry, University of Bristol, England

S. O. Agrell and J. H. Scoon

Department of Mineralogy and Petrology, University of Cambridge, England

J. C. Dran and M. Maurette

Laboratoire René Bernas, C.N.R.S., 91406 Orsay, France

E. Bowell and A. Dollfus

Observatoire de Paris, 92190 Meudon, France

J. E. Geake

Department of Physics, U.M.I.S.T., Manchester, England

L. Schultz and P. Signer

Swiss Federal Institute of Technology, 8006 Zurich, Switzerland.

Abstract—Breccia 14267 has been studied for its bulk chemistry, mineralogy and petrology, cosmic ray tracks, noble gases, carbon chemistry, and optical polarization. It was probably formed 3.9 b.y. ago (although a more recent origin cannot be completely discounted) by a shock-induced lithification of soil grains, only a small portion of which had been exposed to solar wind irradiation. The temperature reached during compaction did not exceed 800°C and the matrix grade of the rock is Warner metamorphic group 2.

After formation, 14267 was buried at a depth greater than two meters until ejection, presumably by the Cone Crater event, on to the near surface of the Apollo 14 site about 30 m.y. ago.

Polarization curves obtained for dust-coated exterior surfaces of 14267 resemble the lunar fines and large-scale lunar surface. Carbon chemistry data and cosmic ray track density measurements obtained for the rock are discussed in relation to other lunar breccias.

INTRODUCTION

Rock 14267 was collected as part of the comprehensive sample taken from the ALSEP area at the Apollo 14 site in the Fra Mauro Formation. No photographic documentation of its surface location or orientation is available. The sample (54.77 g; 5 × 3 × 2 cm) is a polymict breccia having a dark-gray fine-grained matrix (LSPET, 1972). The exterior surface was 40% glass covered and glass-lined zap pits ranging from 0.4 to 1.2 mm occur with a frequency of about one every four square centimeters of surface (LSPET, 1972).

The studies reported herein refer to optical properties, chemical analysis,

---

*This paper constitutes a contribution by the European Consortium.

mineralogy and petrology, cosmic ray track measurements, rare gas analysis, and carbon chemistry. A separate section is devoted to each discipline. To avoid repetitive discussion the overall conclusions of each investigation have been combined into a single section.

EXPERIMENTAL

The rock was divided at Bristol into two halves by cutting along the minor axis; one half was retained by the consortium and the other returned to Houston. Throughout the cutting procedure the original surface of the rock was protected by aluminium foil. A slab 2–3 mm thick was removed from the face of the consortium half and interior chips 14267,3 and 14267,69 were taken for carbon chemistry and rare gas measurements, respectively. Portions of the exterior surface (14267,51 and 14267,59) were used for track studies. Because of the friability of the sample, which tended to break into small chips along an irregular system of penetrative fractures (see Fig. 1), it was impossible to locate the exact position of any of these chips. The saw cuttings (14267,85), obtained from subdividing the sample, were used for chemical analysis. A second slab, 0.5 mm thick from the entire cut face, adjacent to the first, was used to prepare a polished thin section for mineralogical and petrological studies. Optical measurements were performed using the exterior surface of the remaining total

Fig. 1. Cross section of 14267. The whole of the cut face shown was used to prepare a polished thin section. Samples used in other studies were taken from a 2 mm thick slice which had been previously cut from material adjacent to this face. Because of the extreme friability of the rock it was impossible to unambiguously identify sampling locations. The approximate areas sampled are indicated by their respective numbers.

specimen without further subdivision. Except for the preparation of the thin section, all sawing operations were performed, without water cooling, in a stainless steel, nitrogen-filled glove box (Pathfinder Limited, Havant), using a 6×0.040 in. diamond cutting disk (Impregnated Diamond Products, Gloucester). The rock was fed slowly into the saw with frequent stops to allow cooling. The glove box constituted the primary contamination barrier; it was situated in a small, clean air facility which constituted the secondary barrier (Abell *et al.*, 1970). All materials and instruments placed directly in contact with the rock were precleaned by ultrasonic extraction in redistilled Analar grade toluene/methanol (3:1). The interior of the glove box was extracted by successive washing with the same solvents, plus hexane. The sample allocated for bulk chemical analysis (14267,85) contained some small pieces of aluminium which derived from the foil used to protect the sample during cutting. The material was sieved into three fractions (a) <150 mesh, which was virtually free of aluminium foil particles, (b) 150–90 mesh which contained aluminium foil particles only some of which could be removed by handpicking, (c) >90 mesh which contained some foil particles all of which were removed by hand picking. Fraction (b) was rejected. Fraction (c) was then crushed to pass through a 150 mesh sieve and added to fraction (a) for the analysis. It is considered that there was <.2% aluminum contaminant in the analyzed product.

Classical methods of chemical analysis were used as described by Agrell *et al.* (1970) but, in this case, because amounts of material available were small, a semi-micro balance was employed. Lime was weighed as $CaCO_3$, and the total iron was determined colorimetrically using thioglycollic acid. There was insufficient material to carry out a satisfactory determination of total water. The composition of glass fragments and the minerals in 14267 were determined on polished thin sections with a Cambridge Geoscan electron-probe operating at an accelerating potential of 20 kV. The experimental procedure and treatment of data used here have been described fully by Sweatman and Long (1969) and all analyses have been referred, either directly or indirectly, to the primary standards listed by these authors.

The procedure used for track measurement (Borg *et al.*, 1970), noble gas analysis (Nyquist *et al.*, 1973) and carbon chemistry (Cadogan *et al.*, 1972) have been described in detail previously.

RESULTS AND DISCUSSION

Optical polarization

Three surface types have been identified on breccia 14267 by scanning electron microscopy: freshly chipped (interior), glass-coated exposed surfaces, and dust-covered regions. They polarize light in rather different ways. Detailed discussions of the optical polarization properties of 14267 and other lunar samples and meteorites are given elsewhere (Dollfus *et al.*, 1973). The polarization produced is partial, and the degree of polarization P is plotted as a function of the angle source-sample-observer (phase angle V). The curve for phase angles smaller than 30° displays a specific shape with a "negative branch" resulting from multiple scattering between the submicron particle facets.

Dust-covered surfaces exhibit complex structures built up by adhesion; most of the individual grains are rough on a micron scale. The negative branch of the polarization curve descends to -11×10^{-3} near $V = 9°$; inversion angle $V(0) = 20°$. The telescopic lunar negative branch shown $P_{min} = -12 \times 10^{-3}$ at $V = 10°$; $V(0) = 23°$. Such a region closely resembles both the large-scale lunar surface and the lunar fines.

Glass-coated regions show individual grains on smooth glassy surfaces. The negative branch observed for such regions is quite unlike that measured telescopically for the moon; it descends only to $P_{min} = -4 \times 10^{-3}$ at $V = 9°$ with $V(0) = 17°$.

The effect of multiple scattering is much reduced in this instance since complex particle structures are few or absent, and the polarization curve approaches that of a plane dielectric surface. The negative polarization branch observed for freshly chipped areas suggests that such surfaces cannot reproduce the overall lunar polarization characteristics, even though the albedo curve is representative of the lunar highlands in both color and albedo.

Mineralogy and petrology

Rock 14267, in hand specimen, is a fine-grained coherent polymict breccia with a dark matrix of vitreous appearance. Pale-gray to dark-gray, fine-grained lithic clasts predominate over mineral and vitric clasts; the latter are not readily seen on the natural or cut surfaces of the rock. The breccia falls into group F2 of Wilshire and Jackson (1972) and the matrix texture corresponds to group 2 of Warner (1972).

The outer surface is partially coated by a very thin skin (<0.1 mm) of dark glass, both the glass and the underlying body of the rock showing sparse, glass-lined zap pits from 0.03 to 0.30 mm in diameter.

The maximum clast size is 6 mm and the dominant size is in the range 0.3–1.5 mm. If any subparallelism of unequidimensional clasts exists it is very weak and varies over the area of the cut surface of the rock, Fig. 1.

In the dark vitreous matrix, minute vesicles, <0.1 mm in diameter, mark tenuous zones of glass, some of which penetrate from the exterior surface. The composition of one of these glass seams is quoted as analysis 12 in Table 1, and is comparable with the bulk composition of rock 14267. This can be interpreted as demonstrating that the glass coating and tenuous veins were formed at a late stage, subsequent to lithification of the matrix of the rock, and possibly associated with its excavation and transportation by the Cone Crater event to the site where it was collected.

The majority of the lithic clasts are angular to subangular in outline, whereas the vitric clasts are nearly all flow-banded forms, Fig. 2d. The latter can be related to the fragmentation of ropy or twisted masses of glass.

A clast count on fragments >0.5 mm in thin section gave:

Glass	fawn to pale brown	40
	brown to orange yellow	14
	colorless to pale yellow	8
	olivine-bearing chondrule	1
Metaclastic	fine-grained, medium gray	60
	medium-grained, pale gray	19
Igneous	intersertal to microporphyritic basalts	8
	microharzburgite	3
	orthoclase–tridymite rock	1
	poikilitic norite	1
Minerals	plagioclase, hypersthene, olivine	20

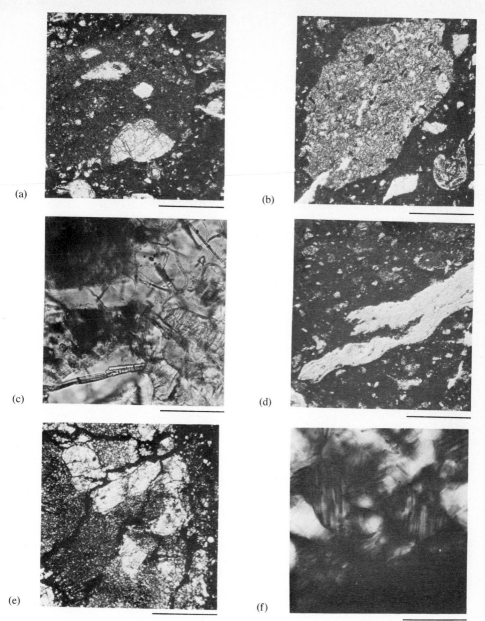

Fig. 2. *Clasts and minerals in 14267*: (a) Fine-grained, medium-gray metaclastic fragment with mosaicized olivine and plagioclase porphyroclasts. Scale bar: 0.5 mm. (b) Medium-grained, pale, metaclastic fragment. Scale bar: 0.5 mm. (c) Partially melted, granitic fragment enclosed in fine-grained metaclastic. Inverted protohypersthene crystallites in siliceous glass with partially-resorbed barian orthoclase. Scale bar: 2.5 μm. (d) Ropy fawn, glass fragment in microbreccia. Scale bar: 0.25 mm. (e) Microharzburgite enclosed in fine-grained, metaclastic fragment. Composed of polygonized olivine crystals set in a fine matrix of twinned clinoenstatite. Crossed polars. Scale bar: 0.5 mm. (f) Clinoenstatite showing lamellar twins. From Harzburgite fragment. Crossed polars. Scale bar: 25 μm.

All mineral and lithic fragments show some signs of shock metamorphism in either granulation, polygonization, kink-banding, undulose extinction or partial maskelynitization.

It is not clear how many shock overprints are to be seen. It is felt that development of clinoenstatite, kinkbanding of hypersthene and polygonization of olivine are early features characteristic of particular fragments and that granulation, partial maskelynitization, undulose extinction are late events which affect all fragments equally. If tenuous glass veins are generated at a late stage in the Cone Crater event, then the ubiquitous weak shock features may have been developed at an earlier stage of this event, even if only separated by seconds.

Below 0.1 mm, lithic fragments are essentially absent and mineral predominate over glass fragments. The mineral fragments, in decreasing order of abundance, consist of plagioclase, orthopyroxene, clinopyroxene, olivine, ilmenite, kamacite, troilite, and spinel. Under the microscope, the mineral fragments are recognizable down to the smallest sizes (1–2 μm) and are closely set in a matrix of fawn-colored glass or its cryptocrystalline devitrification products.

Small, rounded areas of kamacite and, more rarely, taenite, up to 25 μm in diameter occur sporadically both in the matrix of the breccia and within some of the lithic clasts (the fine-grained metaclastic type). Both are interpreted as being of meteoritic origin on the basis of Ni and Co contents. So far only single phase areas have been observed. The kamacite has Ni 4.0–7.0%, Co 0.35–0.50%, the rare taenite Ni 10.0%, Co 0.7%.

Chemical analysis

The composition of 14267,85 is given as analysis 1 on Table 1. The quantity of Fe_2O_3 reported (0.28%) is unusually high for a lunar rock. This is probably due to the difficulty in obtaining accurate figures for FeO and Fe_2O_3, especially FeO, using the rather small quantity of material available for the analysis. Alternatively, some oxidation may have occurred during the sawing operations. The composition is typical of Fra Mauro or KREEP basaltic glasses which have values of Na_2O, K_2O, and P_2O_5 that are high compared with mare-derived material. It may be compared with type C Fra Mauro glass (Reid *et al.*, 1972) and with the averages of dark and pale metaclastic fragments from the <1 mm fines of the Apollo 14 site (Carr and Meyer, 1972). These analyses are quoted in Table 1 as analyses 2, 3, and 4, respectively.

The vitric clasts. The majority of the vitric clasts are flow-banded forms that can be related to small ropy and twisted masses of glass, Fig. 2d, which have suffered a variable degree of fragmentation. Glass spheres are characteristically absent.

Most of the glass is relatively homogeneous, but marked inhomogeneities can occur, analyses 6 and 7, Table 1. These reflect complete melting with only partial mixing of an inhomogeneous protolith. Incomplete melting is represented by similar ropy fragments and microfladden-like masses in which bands and areas contain many micron-sized residual mineral fragments. These are predominantly

Table 1. Analyses of 14267 and representative glass fragments.

	1	2	3	4	5	6	7	8	9	10	11	12	13
SiO_2	48.35	48.20	49.20	48.80	44.45	48.02	52.14	48.34	48.48	49.22	77.19	48.27	51.39
TiO_2	2.11	2.20	2.30	1.67	7.23	4.90	1.65	4.07	3.47	1.53	0.34	1.77	3.12
Al_2O_3	16.87	17.10	16.80	17.90	10.32	13.92	15.13	13.87	15.64	16.52	11.69	16.32	11.06
Cr_2O_3	0.19	n.d.	n.d.	n.d.	0.11	0.13	0.13	0.13	0.31	0.25	0.00	0.17	0.12
Fe_2O_3	0.28	n.d.	n.d.	n.d.	n.d.	n.d.	n.d.	n.d.	n.d.	n.d.	n.d.	n.d.	n.d.
FeO	10.10	10.50	10.40	9.60	21.09	13.77	11.69	12.49	10.33	11.00	0.45	10.27	16.70
MnO	0.13	n.d.	n.d.	n.d.	0.23	0.17	0.15	0.28	0.08	0.13	0.16	0.10	0.19
MgO	7.92	7.60	6.90	10.40	5.38	5.09	5.56	8.20	9.03	9.02	0.00	7.39	2.09
CaO	10.29	10.90	10.20	9.90	11.01	9.27	9.40	9.01	9.99	9.35	0.47	11.18	7.88
BaO	n.d.	n.d.	n.d.	n.d.	0.08	0.25	0.26	0.23	n.d.	0.23	0.16	0.23	0.26
Na_2O	0.90	0.80	0.90	0.90	0.82	0.96	1.12	1.01	0.86	0.94	1.13	1.07	1.13
K_2O	0.78	0.60	0.60	0.60	0.37	1.28	1.34	0.99	0.35	1.12	6.82	1.08	2.18
P_2O_5	0.57	n.d.	n.d.	n.d.	0.38	0.64	0.74	0.32	0.53	0.77	0.03	0.59	1.72
S	0.08	n.d.	n.d.	n.d.	n.d.	n.d.	n.d.	n.d.	n.d.	n.d.	n.d.	n.d.	n.d.
	100.37	97.90	97.30	99.77	101.47	98.40	99.32	98.95	99.19	100.08	98.44	98.44	97.84

C.I.P.W. NORM

	1	2	3	4	5	6	7	8	9	10	11	12	13
Q2	0.06	2.20	4.84	—	1.55	6.44	7.47	3.02	4.56	0.83	43.21	0.41	11.48
COR	—	—	—	—	—	—	—	—	—	—	1.56	—	—
OR	4.61	3.55	3.55	3.55	2.19	7.56	7.92	5.88	2.23	6.62	40.30	6.40	12.88
AB	7.60	6.77	7.62	7.62	6.94	8.12	9.52	8.55	7.34	7.95	9.56	9.07	9.56
AN	36.69	41.30	40.02	43.03	23.39	28.89	32.28	30.37	37.54	37.55	2.43	36.53	18.67
DI	6.23	10.51	8.71	4.95	24.51	10.36	8.36	10.40	6.78	3.47	0.00	13.92	8.38
HY	36.09	29.41	28.18	35.05	28.09	24.88	28.57	31.92	32.25	38.46	0.56	28.24	26.62
OL	—	—	—	2.41	—	—	—	—	—	—	—	—	—
CM	0.30	—	—	—	0.17	0.19	0.20	0.20	0.45	0.37	0.00	—	0.18
IL	3.99	4.18	4.37	3.17	13.73	9.31	3.13	7.74	6.53	2.91	0.65	3.35	5.93
AP	1.34	—	—	—	0.89	1.52	1.75	0.76	1.34	1.82	0.07	1.41	4.07
PY	0.14	—	—	—	—	—	—	—	—	—	—	—	—
$\frac{FS \times 100}{FS + EN}$	31.95	38.62	40.39	30.41	60.63	51.09	50.95	38.23	30.65	37.48	100.00	40.0	79.02

Columns:

(1) 14267/85 Chemical analysis by J. H. Scoon. Contamination by aluminum foil estimated at 0.2%. Fe_2O_3 probably due to oxidation during or subsequent to cutting procedure from which analyzed powder was produced.

(2) Fra Mauro basalt glass, type C average. Reid et al. (1972).

(3) Dark metaclastic fragments, average from <1 mm fines from Apollo 14 — Carr and Meyer (1972).

(4) Pale metaclastic fragments, average from <1 mm fines from Apollo 14 — Carr and Meyer (1972).

(5) Orange–brown glass fragment.

(6) Orange–yellow schlieren in glass 7.

(7) Very pale yellow glass composing bulk of small "fladden"-like fragment.

(8) Dark-brown glass fragment.

(9) Red–brown glass enclosing barred olivine "chondrule."

(10) Pale-fawn fragment.

(11) Siliceous colorless glass produced by partial melting of "granitic" clast enclosed in finegrained dark metaclastic fragment.

(12) Fawn-colored glass seam.

(13) Pale-brown fluidal glass fragment.

Analyses 2–13 by electron probe.

resorbed crystals of plagioclase with subordinate pyroxene. Seams crowded with angular micron-sized mineral chips occur as bands and in convolutions in some of the glass fragments. These represent solid material that coexisted in the impact ejecta and was incorporated into or adhered to the still plastic glass fragments while in flight.

A few cored lapillae occur in which hypersthene crystals, coarser grained, pale metaclastic rock or a chondrule-like melt with bars of olivine (analysis 9, Table 2) are enclosed in a shell of brownish glass (analysis 9, Table 1). The finer grained, paler and darker gray metaclastic fragments never have a glass coating and so may be derived from a source different from that of the vitric clasts and the glass-coated fragments.

Analyses of typical glass fragments from 14267 are quoted in Table 1, analyses 5–12. The FeO and MgO contents for these and other glass fragments from 14267 are plotted on Fig. 3 along with representative glasses from 14259,27. All the analyses bar one correspond to quartz-normative basalt glasses of the Fra Mauro type. The wide range of FeO:MgO ratios they show is illustrated in Fig. 3.

One glass, analysis 5, Table 1, shows a basaltic composition with the high iron and titanium content of a mare basalt. If this is its real affinity we have evidence for mare basalt types existing prior to 3.9 b.y., the provisional age of rock 14267. Caution should be exercised here as one 0.2 mm glass fragment does not

Table 2. 14267 Porphyroclasts in fine grained metaclastic fragments.

	1	2	3	4	5	6	7	8	9
SiO_2	38.31	35.98	40.94	53.52	50.73	50.19	1.98	2.18	39.14
TiO_2	n.d.	n.d.	n.d.	(0.35)	(0.35)	2.02	53.27	1.81	0.22
Al_2O_3	n.d.	n.d.	n.d.	(0.75)	(0.74)	4.12	0.45	53.18	0.14
Cr_2O_3	0.54	0.52	0.12	(0.62)	(0.62)	0.85	0.64	9.70	0.11
FeO	22.02	31.72	11.43	15.70	19.44	16.20	40.30	22.60	18.14
MnO	n.d.	n.d.	0.08	n.d.	n.d.	n.d.	n.d.		0.08
MgO	39.17	31.22	47.40	26.07	20.30	13.72	3.20	11.36	41.71
CaO	0.11	0.19	0.92	1.64	5.00	15.33	0.03	1.22	0.18
Total	100.15	99.63	100.89	98.65	97.18	102.43	99.87	102.05	99.72
Cations per unit formulae									
Si	0.994	0.987	1.004	1.966	1.953	1.858	0.048	0.059	1.000
Ti	—	—	—	(0.010)	(0.011)	0.057	0.977	0.024	0.005
Al	—	—	—	(0.033)	(0.034)	0.186	0.016	1.686	0.005
Cr	0.012	0.012	0.003	(0.019)	(0.019)	0.025	0.012	0.207	0.003
Fe	0.478	0.728	0.235	0.483	0.626	0.502	0.814	0.509	0.388
Mn	—	—	0.002		n.d.	n.d.			0.002
Mg	1.515	1.277	1.731	1.427	1.165	0.757	0.115	0.455	1.588
Ca	0.003	0.006	0.025	0.065	0.207	0.608	0.002	0.036	0.006
				Wo 3.3	10.4	32.6			
Fo	76.0	63.7	88.0	En 72.2	58.3	40.5	Ge 12.4		Fo 80.4
Fa	34.0	36.3	12.0	Fs 24.5	31.3	26.9	Il 87.6		Fa 19.6

Columns: (1)–(3) Olivine.
         (4)   Hypersthene.
         (5)   Pigeonite.
         (6)   Augite.
         (7)   Ilmenite.
         (8)   Spinel.
         (9)   Olivine, scalariform crystal in chondrule-like body coated with brown glass.
                  Cr, Al, Ti conjectural in analyses 4 and 5.

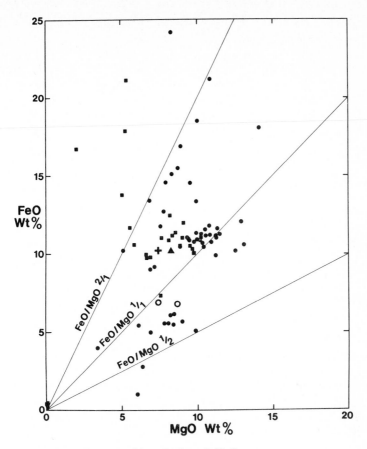

▲   14267 Bulk composition. Analyst, J. H. Scoon.
■   14267 Glass fragments.
○   14267 Fine- and medium-grained plagioclase-rich hornfelsic clasts.
●   14259,27 Glasses, spheres + fragments from <1 mm lunar fines.
+   14267 Fawn-colored glass seam.

Fig. 3. 14267—Bulk composition, lithic clasts, glass fragments, FeO–MgO.

constitute a rock type! It may represent the melting of a fortuitous association of mineral grains in a preexisting fragmental rock or soil. Or it may represent the product of melting of a small iron-enriched volume of a parental igneous rock which has not suffered gross homogenization in the glass-forming process.

Analysis 13, Table 2 illustrates a high iron glass fragment with high Na_2O, K_2O, and P_2O_5 which probably represents a glass derived from endstage products of crystallization of a Fra Mauro basalt.

As has been pointed out previously, the bulk compositions of 14267 and the veining glass are very similar. Both are consistently slightly richer in Al_2O_3 and CaO than the vitric clasts that occur in the rock. This would point to a slightly greater content of modal plagioclase in some of the lithic clasts as compared with the normative plagioclase of the vitric clasts. This again emphasizes a slight but significant difference in composition between their source regions.

The lithic clasts. Most of the lithic clasts are fine-grained medium-gray metaclastic fragments with a porphyroclastic texture, Fig. 2a. A variation in color from darker to lighter is directly correlated with the grain size of their matrix silicates and the degree of segregation of the dominant oxide phase, ilmenite.

The darker, finer grained varieties have a matrix of intergrown plagioclase and orthopyroxene outlined by droplike micron-sized crystals of ilmenite with subordinate kamacite and troilite. The texture is thought to arise from a short period of annealing of an original glass matrix. In some fragments there is a patchy increase in grain size and they pass into paler medium-gray metaclastics. In these the matrix is composed of hypidiomorphic rods of hypersthene, probably with some subordinate clinopyroxene, that are set in a matrix of allotriomorphic plagioclase, small composite areas being outlined by beaded plates and rods of ilmenite with accessory kamacite and troilite in the micron size range. Larger, ± 25 μm, pools of kamacite occur which are thought to represent material of meteoritic origin. In the more recrystallized clasts they may be penetrated by small crystals of pyroxene growing from the silicate matrix.

Porphyroclasts of plagioclase, An_{95-85}, make up 10–15% of the metaclastic fragments and all show features due to shock: undulose extinction, secondary twinning, anomalously low birefringence, granulation, etc. Against some of the larger plagioclase crystals a slight increase in grain size of the groundmass pyroxene is apparent. Olivine, hypersthene, augite, ilmenite, purple spinel also occur as shocked porphyroclasts. Typical compositions are quoted in Table 2 and pyroxene and olivine compositions are plotted on Fig. 4. Three distinct olivine compositions occur: Fo_{88} in fine mosaicized crystals, Fig. 2a, Fo_{76} in what looks like small fragments of dunite hornfels, and finally, Fo_{65} in isolated crystals. A similar wide range in composition of the pyroxene porphyroclasts is observed. The analyses are too few to allow more to be said than that the olivine and pyroxene cannot all be in equilibrium with each other. The compositions recorded are in the range of those found for the Fra Mauro basalts and their metaclastic equivalents and for members of the ANT suite.

In the finest grained metaclastics, sporadic small patches of partially melted rock of granitic composition occur, Fig. 2c. These are composed of siliceous glass (analysis 11, Table 2) associated with resorbed patches of potash feldspar ($Ab_{8.5}Or_{88.0}Cs_{3.5}$). In the glass occur isolated quench crystals of clinopyroxene as curved trichites <1.5 μm in width. These show a characteristic cross fracture, inclined extinction and occasional twinning. Their small size precludes analysis. It is tentatively suggested, on the basis of unpublished observations by S.O.A. on the terrestrial buchites of Ardmuckish, Scotland, (Kynaston and Hill, 1908), that the crystals were metastable quench protohypersthene that has inverted to clinohypersthene. It is probable, as they occur in a "granitic" glass, that they have a high Fe/Mg ratio.

These partially melted granitic patches were probably formed during the short

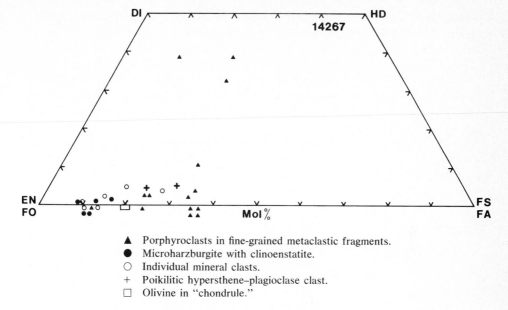

14267

▲ Porphyroclasts in fine-grained metaclastic fragments.
● Microharzburgite with clinoenstatite.
○ Individual mineral clasts.
+ Poikilitic hypersthene–plagioclase clast.
□ Olivine in "chondrule."

Fig. 4. 14267—Pyroxenes and olivines.

period of annealing during which the fine crystalline matrix of the metaclastic fragments was formed from the original glass matrix at some temperature around 900°C. This must have been prior to the event which brought the pale metaclastic fragments and vitric clasts into association in what is now rock 14267.

A few ultramafic fragments occur both as independent clasts and as inclusions within the fine-grained metaclastic rock type described previously. These are of interest as some contain shock-produced clinoenstatite and this has not been previously recorded from lunar rocks.

One 5 mm fragment of a poikilitic microharzburgite, Fig. 2e, occurs as an inclusion in a fine-grained, gray metaclastic rock. It is composed of strongly mosaicized rounded crystals of olivine (Fo_{89}) set in a "felsitic" matrix of twinned clinoenstatite (En_{91}), Fig. 2f. The individual crystals of clinoenstatite are 2–3 μm in size and show a marked preferred orientation (observable with the sensitive tint plate) that is controlled by the orientation of the three or four larger enstatite crystals originally present in the rock fragment. A sample of clinoenstatite was extracted from the hand specimen and its identity confirmed on an X-ray powder photograph (by M. G. Bown). Analyses of the olivine and clinoenstatite are quoted as 1 and 2 in Table 3. It is clear that one has here the shock-metamorphosed equivalent of an ultramafic igneous rock, probably a very fine-grained accumulate with a high Mg/Fe ratio.

Two half millimeter fragments of microharzburgite occur as isolated lithic clasts in the vitric matrix of the breccia. One shows residual patches of hypersthene in a "felsitic" matrix of what is either untwinned clinohypersthene or clinohypersthene reinverted to hypersthene. No material was available for X-ray confirmation. The compositions are quoted in Table 3 as analyses 5 and 6, respectively. Although the differences in composition may be real, it is suggested that too much significance should not be attached to them, as the polish on the fragment was poor and pitted and relatively large analytical errors may be present.

The presence of the two pyroxene polymorphs in one clast is not unexpected when we know that shock effects may be extremely localized. For example, even within a single crystal of plagioclase, maskelynite, and optically normal plagioclase can exist side by side.

If the "felsitic" material is clinohypersthene reinverted to hypersthene, then there must have been sufficient residual heat for subsequent reinversion to take place in the material of this clast. The temperature required is probably in excess of 850°C (Raleigh *et al.*, 1971).

A few independent single-crystal clasts of mosaicized olivine and enstatite occur in the breccia. These are up to one millimeter in diameter and are coarser in

Table 3. 14267 Olivines and pyroxenes associated with shocked microharzburgites with clinoenstatite.

	1	2	3	4	5	6
SiO_2	40.92	57.87	40.47	56.52	55.31	56.05
TiO_2	0.15	0.08	0.14	0.03	0.13	0.08
Al_2O_3	0.09	0.27	0.23	0.35	0.38	0.40
Cr_2O_3	0.09	0.37	0.00	0.44	0.17	0.34
FeO	10.24	6.23	11.31	6.75	11.05	8.28
MnO	0.09	0.10	0.13	0.16	0.12	0.12
MgO	47.00	34.03	47.98	34.41	30.98	32.59
CaO	0.21	0.33	0.07	0.37	0.66	0.35
Total	98.79	99.28	100.33	99.03	98.80	98.21
Cations per unit formulae						
Si	1.016	2.006	0.995	1.976	1.977	1.977
Ti	0.003	0.003	0.003	0.001	0.004	0.002
Al	0.003	0.072	0.007	0.015	0.016	0.017
Cr	0.002	0.011	0.000	0.013	0.005	0.070
Fe	0.213	0.181	0.233	0.198	0.331	0.245
Mn	0.002	0.003	0.003	0.005	0.004	0.004
Mg	1.738	1.758	1.759	1.793	1.650	1.713
Ca	0.006	0.013	0.002	0.014	0.026	0.014
$\dfrac{Fe \times 100}{Fe + Mg}$	10.9	9.3	11.7	9.9	16.5	12.4

Columns: (1) Olivine—mosaicized crystal in shocked microharzburgite.
(2) Clinoenstatite pseudomorphing enstatite in shocked microharzburgite.
(3) Olivine—mosaicized single-crystal clast.
(4) Enstatite—kink-banded single-crystal clast.
(5) Hypersthene—residual in microharzburgite clast.
(6) Clinohypersthene—matrix in microharzburgite clast.

grain size than the microharzburgite although their compositions, Table 3, analyses 3 and 4, are similar. The enstatite shows extensive kink-banding and some mosaicism. It is very similar in appearance to the hypersthene from the mylonitized harzburgite described by Trommsdorf and Wenk (1968) but in the lunar example, no clinoenstatite was observed in the kink bands.

The enstatite–clinoenstatite transformation has been studied experimentally by a number of authors: Turner *et al.* (1960), Carter (1971), Ahrens and Gaffney (1971), Raleigh *et al.* (1971). It was shown by Raleigh *et al.* (1971), that enstatite deforms either by slip to kink-banded or polygonized enstatite or deforms by a martensitic shear transformation to clinoenstatite. These were shown to be competing rate-controlled processes dependent on temperature, stress and strain rate: high strain rate favoring the shear transformation (production of clinoenstatite) and high temperature the slip process. It was established that deformation produced clinoenstatite does not reinvert to enstatite when heated to a temperature of 850°C.

The rarity of the enstatite–clinoenstatite transformation in terrestrial metamorphic rocks was interpreted by Trommsdorf and Wenk (1968) as due to the low strain rate normally operative in terrestrial geological processes. In meteorites, shock-induced transformations associated with high strain rates are not uncommon and in shock metamorphism of lunar rocks of appropriate mineralogy one could expect clinoenstatite or clinohypersthene to be produced and preserved unless the rocks were strongly annealed in a hot ejecta blanket. The rarity of shock-produced clinohypersthenes is puzzling, perhaps less so if one considers that the great bulk of the material that is ejected in crater-forming processes is not necessarily subjected to deformation at high strain rate. Any high strain and stress phenomena may be localized in shear zones within or at the boundary surfaces of ejected masses as they are produced from the original body of rock.

Cosmic ray tracks

Various radiation damage features have been studied in the constitutent grains of breccia 14267, by applying transmission and scanning electron microscope techniques. With a 1 MeV transmission electron microscope, micron-sized grains obtained by crushing a fragment of an external chip (14267,51) have been examined. The following observations have been made:

(i) No amorphous coating or high density of latent tracks characteristic of highly irradiated soil particles has been observed.

(ii) Tiny microcrystallites appear in a small proportion of grains heated 2 hours at 800°C, under vacuum (Fig. 5). Such crystallites look similar to those previously found, either in mature lunar dust grains heated under the same conditions, or in the matrices of mildly metamorphosed Apollo 11 breccias (Dran *et al.*, 1972a, 1972b).

Scanning electron microscopy has been used to examine etched tracks in polished thick sections of two exterior chips (14267,51 and 14267,59). The sections

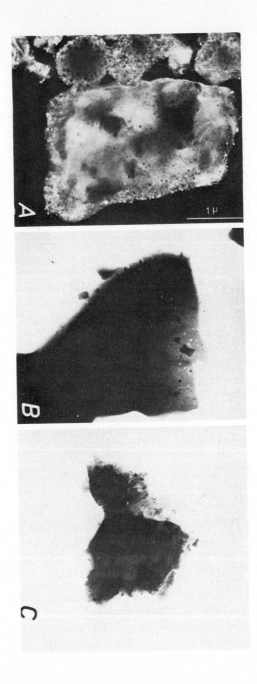

Fig. 5. 1 MeV electron micrograph of lunar grains showing microcrystallites (small black dots). A and C show grains of 10084 and 14267, respectively, in which microcrystallites are artificially produced by heating 2 hours at 800°C under vacuum. B shows microcrystallites naturally occuring in breccia 10046.

were first irradiated with a ^{252}Cf source to produce a low density of etchable fission fragment tracks in order to identify the various mineral phases for fossil track studies and to insure that such phases were capable of registering tracks even if no fossil tracks were observed in them. Irradiated sections were gradually etched in chemical reagents to sequentially reveal tracks in glassy grains, in feldspars and in pyroxenes. Each phase was further identified with the electron microprobe attachment of the SEM. The major results of this investigation are as follows:

(i) The cosmic ray track density measured in interior glassy clasts ($\geqslant 50\ \mu$m) ranges from $\sim 10^5$ to $\sim 10^7$ tracks/cm^2. The external glass coating has $\sim 2 \times 10^6$ tracks/cm^2, with no observable track gradient.

(ii) The cosmic ray track density in feldspar ranges from ~ 1 to 8×10^7 tracks/cm^2.

(iii) In Fig. 6, the track density distribution in rock 14267 is compared to those observed in various other breccias, including several glass-rich Apollo 15 breccias as well as two other breccias, 10046 and 14305, previously studied (Dran *et al.*, 1972a). In 10046, the high track density clearly shows that the constituent grains have retained the tracks stored during their irradiation as individual dust grains in the solar flare cosmic rays. On the other hand, there is no remnant of the prebreccia solar irradiation in the strongly metamorphosed 14305 breccia. Among the other breccias, only 14267 and 15459 have a track density distribution similar to 14305 and characteristic of a galactic cosmic ray origin. Furthermore the track densities in the various mineral phases are different and this complex effect cannot be quantitatively used at the present time as it depends on the following factors: different efficiency of track registration, different extent of track metamorphism as the tracks are generally more stable in feldspars than in the other phases, and admixture of grains with different irradiation histories, etc.

Track metamorphism evidence. The microcrystallites induced by strong heating suggest that partially faded solar flare tracks, which cannot be observed directly as latent tracks, exist in the constituent grains of breccia 14267. Indeed, Price and Walker (1962) have shown that fission fragment tracks in muscovite nucleate microcrystallites upon heating. Furthermore, the "phase" transformation of fossil tracks into crystallites in artificially heated lunar dust grains has already been reported (Dran *et al.*, 1972b).

In the case of the mildly metamorphosed Apollo 11 breccias both microcrystallites and latent tracks are observed before heating. On the other hand, the metamorphism that was involved in the formation of breccia 14267 has partially erased the latent tracks that were registered in the grains before their compaction into a breccia, without triggering the growth of the crystallites. This difference between the two types of breccias could be the result of a violent shock metamorphism, which has transformed the latent tracks in rock 14267 into "dotted" tracks that cannot be observed with the electron microscope, without producing the heat metamorphism required to grow crystallites. This conclusion concerning the brecciation history of rock 14267 is also supported by its high

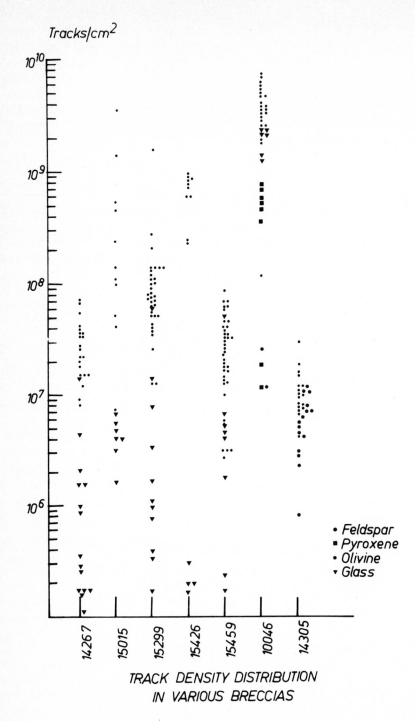

Fig. 6. Track density distribution in various breccias.

degree of "dust sintering" as estimated from scratch hardness measurements (Dran *et al.*, 1972b).

Noble gases

Concentrations and isotopic composition of He, Ne, and argon were determined in a 33 mg chip from 14267,69 of an imprecisely defined interior location (Fig. 1). It contained both, the fine-grained, dark vitreous matrix and gray, lithic clast material. The results of the mass spectrometric noble gas analyses are given in Table 4. The noble gases found are of three different origins:

Radiogenic ^{40}Ar is clearly present. Considerable amounts of solar Ne and Ar are indicated by the ^{20}Ne/^{22}Ne and ^{36}Ar/^{38}Ar ratio, respectively. A *spallogenic* component is suggested by the low ^{22}Ne/^{21}Ne ratio.

Spallogenic and trapped neon can be evaluated by assuming a ^{22}Ne/^{21}Ne ratio of 1.10 ± 0.5 in spallogenic Ne and one of 32 ± 1 in trapped Ne. Then, the concentration of spallogenic ^{21}Ne is found to be $10.1 \pm 1.3 \times 10^{-8}$ cc STP/g. From this concentration, together with the chemical composition given in Table 1 and individual elemental production rates, the exposure age (60 ± 10 m.y.) given in Table 4 is computed.

The predominance of trapped argon renders the calculation of the spallogenic Ar component more uncertain. Assuming a ^{36}Ar/^{38}Ar ratio for trapped Ar of 5.35 ± 0.05 and one of 0.65 ± 0.03 for spallogenic Ar, a concentration of spal-

Table 4. Measured He, Ne, and Ar concentrations, spallogenic, trapped and radiogenic components. Exposure and apparent gas retention ages. (Sample 14267,69 weight: 33.17 mg).

	^{3}He	^{4}He	^{20}Ne	^{21}Ne in 10^{-8} cc STP/g	^{22}Ne	^{36}Ar	^{38}Ar	^{40}Ar
	28.8	63800	2970	18.4	277	3190	621	40600
	±0.8	±1000	±70	±0.7	±8	±40	±7	±700
spallogenic component	≤28.8		10	0.1	11.1	18	28	
			±2	±1.6	±1.7	±10	±15	
trapped (3) component			2960	8.3	266	3170	593	(6400)
			±70	±1.3	±9	±50	±22	
radiogenic		≤63800						~34200
Exposure age [m.y.]								
(1, 2)	≤29			60			200	
				±10			±100	
(1, 4)							95	
							±100	
Gas retention age m.y. (1.3)		≤750						~3900

Columns: (1) Chemistry from Table 1, column 1.
(2) Production rates for ^{3}He and ^{21}Ne from Bogard *et al.* (1971). $P(^{38}$Ar$) = 1.4 \times 10^{-10}$ cc STP(gCa)$^{-1}$ m.y.$^{-1}$.
(3) ^{40}Ar/^{36}Ar in trapped gas assumed to be 2.
(4) Production rate from Weber (1973).

logenic ^{38}Ar of about 28×10^{-8} cc STP/g is found. With a Ca concentration of 7.4% and a production rate of 1.4×10^{-10} cc STP/g Ca^{-1} m.y.$^{-1}$ an exposure age of 200 ± 100 m.y. is calculated. However, using production rates given by Weber (1973) $(P_{38} = \{(29 \pm 11)[K] + (1.3 + 0.1)[Ca] + (0.95 \pm 0.34)[Ti]\}10^{-10}$ cc STP g^{-1} m.y.$^{-1}$) an exposure age of 95 ± 100 m.y. is observed. Exact evaluation of the radiogenic and trapped ^{40}Ar components is not possible due to the uncertainty of the ^{40}Ar/^{36}Ar ratio of the trapped gas. However, if a ^{40}Ar/^{36}Ar ratio in the trapped gas of about 2 is assumed, a K/Ar-age of 3.9 b.y. is obtained.

The determination of the spallogenic, radiogenic, and trapped He is impossible because of the severe alteration of the element abundances of the trapped gas. Therefore, only limiting cases can be considered: assuming that no trapped He is retained, the apparent U/Th–He age is only 0.7 b.y.; if a uranium concentration of 3.6 ppm is assumed together with a Th/U ratio of 3.6, the spallogenic ^{3}He is (assuming total loss of trapped ^{3}He and no loss of spallogenic ^{22}Ne) depleted by at least a factor of 2.

The occurrence of solar Ne and Ar, indicate that the matrix of the breccia 14267 contains constituents which were exposed to solar and cosmic irradiation as finely dispersed regolith material prior to the formation of the breccia. This observation is supported by the occurrence of carbide as well as developable microcrystallites. However, compared to solar He, Ne, and Ar in lunar fines, the breccia contains not only much lower concentrations but also severely altered abundance patterns. Linear heating experiments show clearly that He is more easily released from lunar fines than Ne and Ar (Frick *et al.*, 1973). Thus, the altered abundance pattern can be taken to indicate the occurrence of shock and thermal events. Such metamorphic events are also visible in the petrographic and track patterns.

Taking the exposure ages at face value, it appears that loss of spallogenic gases also occurred. But one may speculate that at least some constituents of the breccia 14267 had a cosmic ray irradiation in a pre-compaction state.

Carbon chemistry

The CH$_4$ and carbide content of two interior chips (14267,3,1, 0.0805 g and 14267,3,2, 0.0646 g) were measured by gas chromatographic analysis of the gases released by DCl (38% in D$_2$O) treatment (Cadogan *et al.*, 1972, 1973). Neither chip contained a high abundance of included clasts or a large single clast; both 14267,3,1 and 2 were representative of predominantly fine-grained matrix material. In neither case was CH$_4$ detected, and the CD$_4$ observed accounted for only 0.11 and 0.15 μg/g of carbon, respectively (Table 5). Since carbide has not yet been found in igneous rocks (Abell *et al.*, 1971), 14267 must have been formed from soil, a proportion of which had been exposed to solar wind.

The breccias collected at the Apollo 14 site have been classified by Warner into seven groups within three metamorphic grades (Warner, 1972). Table 5 shows the CH$_4$ and carbide content of other breccias in relation to 14267. The group 1 breccias (least metamorphic, e.g. 14047, 14313, and 10059) show CH$_4$ and CD$_4$ measurements similar to bulk soils collected at the same site (14148, CH$_4$ =

3.6 $\mu g/g$, $CD_4 = 11.6$ $\mu g/g$ and 10086, $CH_4 = 4.2$ $\mu g/g$, $CD_4 = 18.6$ $\mu g/g$). Breccias of higher grades than group 1, i.e. groups 2–7, contain little or no CH_4 and very small amounts of carbide (Table 5). However, the presence of at least some carbide in 14267 and, indeed, in other metamorphic breccias suggest that the fine grains comprising the matrix of these rocks have experienced some exposure to the solar wind. The data in Table 5 suggest that on the basis of carbon chemistry 14049, classified as group 2 by Warner, should be reclassified as group 1. This breccia also contains substantial amounts of solar wind rare gases.

Matrix texture suggests that 14267 is Warner group 2. Gas release patterns of CO and CO_2 lunar fines (Gibson and Moore, 1972) suggest that the group 1 and group 2 transition temperature is 800°C (Williams, 1972). Therefore, 14267 presumably has not been heated above this temperature and other consortium

Table 5. Carbon chemistry data, together with metamorphic group, for a number of lunar breccias.

Sample No.	CH_4 $\mu g/g$	CD_4 $\mu g/g$	Total C[†] ppm	Metamorphic Group[‡]
10059	5.0	20	n.m.	n.c.
	5.2	17		
14047	2.59	10.4	140	1
	1.2*	5.4*	210	
14313	3.3	9.9	130	1
14049	1.1	4.3	135	2§
			190	
14301	0.26*	0.22*	50	2
			70	
14267	n.d.	0.11	n.m.	2¶
	n.d.	0.15		
14063	<0.04	0.07	80	3
14311	n.d.	0.80	44	5
	n.d.	0.37		
14305	n.m.	n.m.	32	6
14321	n.d.	0.29	22	6
	n.d.*	0.05*	28	
14318	0.20*	0.05*	86	6
14066	<0.03	0.16	90	7

*Burlingame et al. (1972).
†Moore et al. (1972).
‡Warner (1972)
§Carbon and rare gas data suggest reclassification as group 1.
¶Classified by Agrell, this work.
n.d. not detected.
n.m. not measured.
n.c. not classified.

investigators substantiate this conclusion. Heating to 800°C, although sufficient to remove CH₄ from the fines constituting the breccia matrix, would not completely destroy carbide synthesized during exposure to the solar wind. Therefore, the amount of solar wind-exposed material incorporated into 14267 is probably small and this rock was formed from predominantly immature fines. The alternative conclusion that 14267 was formed from mature, well-exposed grains and that carbon species were lost by thermal or shock processes during the compaction is unlikely.

CONCLUSIONS

The consortium investigations allow the following tentative history for 14267 to be proposed:

At least part of the material which constitutes breccia 14267 must have once existed in a finely dispersed, regolith-like state on the lunar surface. This is a firm conclusion based on the presence of solar noble gases, carbides and the fact that artificial heating develops microcrystallites. The variety of clasts and the presence of material of presumed meteoritic origin point to a similar conclusion; the presence of metaclastic fragments in the breccia indicates that impact events in the soil were common prior to the formation of 14267. However, from the low abundance of etchable tracks, the low proportion of grains showing microcrystallites after heating and the low abundance of carbide it appears that the extent of irradiation before brecciation was small.

During, and/or after compaction of this portion of the regolith the constituent grains underwent a process which elevated the temperature to between 600°C and 800°C. At least 600°C was necessary to bring about sintering and the formation of the vitreous matrix. However, it is unlikely that 800°C was exceeded since the boundaries between the iron and the troilite show no evidence of melting. Other measurements also show the effects of increased temperature. The solar noble gases are fractionated compared to solar abundances and the volatile carbon species, methane, normally found in exposed soils, is absent. The appearance of microcrystallites when the rock is heated to 800°C in the laboratory indicates that latent tracks have been only partially erased and that individual dust grains have not been thoroughly thermally metamorphosed during brecciation. Thus, the temperature of the rock could not have exceeded 800°C during formation. The observed extent of track metamorphism (i.e. the formation of faded tracks rather than the direct production of microcrystallites) argues in favor of shock rather than thermal lithification. Indeed, there is ample petrographic evidence for shock processes. The matrix texture and thermal history of 14267 is consistent with this rock being categorized in Warner group 2.

Due to the partial erasure of tracks and the loss of noble gases, no firm conclusions seem possible with regard to the time and sequence of events which led to the formation of 14267. However, the following tentative history is in keeping with the generally accepted scheme for the formation of the Fra Mauro region: a regolith which contained a proportion of exposed material was compacted and metamorphosed approximately 3.9 b.y. ago, as indicated by the [40]Ar retention age. The track density in feldspar grains and the rare gas spallation ages suggest that 14267 was subsequently excavated, possibly by the Cone Crater event (25 m.y., Walker *et al.*, 1972). During this process He was partially lost, due to thermal and/or mechanical action as indicated by the low U/Th–He retention age. However, the temperature increase was insufficient to destory the clinoenstatite present in some of the lithic clasts. The low track density in the exterior glass coating indicates that either this coating was formed more recently (~2 m.y. ago) or that a substantial track-annealing took place. However, the density of tracks in feldspars and internal glassy clasts is inconsistent with a process which annealed the whole rock. The chemical analysis data and the nature of the glass coating is consistent with formation by melting of the bulk material at a late stage in the rock's evolution.

The tentative history outlined in the previous paragraph is based on the concordance of the ^{40}Ar age with the accepted age of the Fra Mauro Formation. However, if the original soil was indeed immature, as indicated by the carbide content and the track data, then an alternative history, with a younger lithification age may be deduced. The metamorphic temperature of ~800°C could induce small losses of solar wind He, Ne, and Ar, without substantial loss of radiogenic ^{40}Ar.

A single event of elevated temperature and/or shock could then account for the observed rare gas isotope patterns, track densities and carbon chemistry and the presence of a glass component of mare basalt composition.

Acknowledgments—We would like to thank the Science Research Council for a grant toward the installation and operation of the sample distribution facility at Bristol. We are particularly grateful to the staff of the Engineering Workshop at the School of Chemistry, University of Bristol, who helped design and construct the rock cutting equipment and to Mr. I. Manning and Mrs. A. P. Gowar for technical assistance. Transmission and scanning electron microscope studies of radiation damage were performed at the Institut d'Optique Electronique du C.N.R.S., Toulouse and the B.R.G.M., Orleans, respectively. The help received from the staff of these two institutions is gratefully acknowledged. The following agencies have provided funds to support the analyses: Science Research Council (SGS 128), Natural Environment Research Council GR3/1144, Swiss National Science Foundation (2.589.71), and C.N.R.S. Centre National de la Recherche Scientifique and Centre National d'Etude Spatiale.

REFERENCES

Abell P. I., Draffan G. H., Eglinton G., Hayes J. M., Maxwell J. R., and Pillinger C. T. (1970) Organic analysis of the returned Apollo 11 Lunar sample. *Proc. Apollo 11 Lunar Sci. Conf., Geochim. Cosmochim. Acta*, Suppl. 1, Vol. 2, pp. 1757–1773. Pergamon.

Agrell S. O., Scoon J. H., Muir I. D., Long J. V. P., McConnell J. D. C., and Peckett A. (1970) Observations on the chemistry, mineralogy and petrology of some Apollo 11 lunar samples. *Proc. Apollo 11 Lunar Sci. Conf., Geochim. Cosmochim. Acta*, Suppl. 1, Vol. 1, pp. 93–128. Pergamon.

Ahrens T. J. and Gaffney E. S. (1971) Dynamic compression of enstatite. *J. Geophys. Res.* **76**, 5504–5513.

Bogard D. D., Funkhouser J. G., Schaeffer O. A., and Zähringer J. (1971) Noble gas abundances in lunar material—Cosmic ray spallation products and radiation ages from the Sea of Tranquility and Oceans of Storms. *J. Geophys. Res.* **76**, 2757–2779.

Borg J., Dran J. C., Durrieu L., Jouret C., and Maurette M. (1970) High voltage electron microscope studies of fossil nuclear particle tracks in extra terrestrial matter. *Earth Planet. Sci. Lett.* **8**, 379–386.

Cadogan P. H., Eglinton G., Firth J. M. N., Maxwell J. R., Mays B. J., and Pillinger C. T. (1972) Survey of lunar carbon compounds II. The carbon chemistry of Apollo 11, 12, 14, and 15 samples. *Proc.*

Third Lunar Sci. Conf., Geochim. Cosmochim. Acta, Suppl. 3, Vol. 2, pp. 2069–2090. MIT Press.

Carr M. H. and Meyer C. E. (1972) Chemical and petrographic characterization of Fra Mauro Soils. *Proc. Third Lunar Sci. Conf., Geochim. Cosmochim. Acta,* Suppl. 3, Vol. 1, pp. 1015–1027. MIT Press.

Carter N. L. (1971) Static deformation of silica and silicates. *J. Geophys. Res.* **76,** 5514–5540.

Crozaz G., Drozd R., Hohenberg C. M., Hoyt H. P., Ragan D., Walker R. M., and Yuhas D. (1972) Solar flare and galactic cosmic ray studies of Apollo 14 and 15 samples. *Proc. Third Lunar Sci. Conf., Geochim. Cosmochim. Acta,* Suppl. 3, Vol. 3, pp. 2917–2931. MIT Press.

Dollfus A., Boswell E., Zellner B., and Geake J. E. (1973) Polarimetric properties of the lunar surface and its interpretation. Part 6: Albedo determinations from polarimetric characteristics. *Proc. Fourth Lunar Sci. Conf., Geochim. Cosmochim. Acta,* Suppl. 4, Vol. 3, pp. 3167–3174. Pergamon.

Dran J. C., Durand J. P., Maurette M., Durran L., Jouret C., and Legressus C. (1972a) Track metamorphism in extraterrestrial breccias. *Proc. Third Lunar Sci. Conf., Geochim. Cosmochim. Acta,* Suppl. 3, Vol. 3, pp. 2883–2903. MIT Press.

Dran J. C., Durand J. P., and Maurette M. (1972b) Low energy solar nuclear particle irradiation of lunar and meteorite breccias. In *The Moon* (editors H. C. Urey and S. K. Runcorn), p. 309.

Frick U., Baur H., Funk H., Phinney D., Schafer Chr., Schultz L., and Signer P. (1973) Diffusion properties of light noble gases in lunar fines. *Proc. Fourth Lunar Sci. Conf., Geochim. Cosmochim. Acta,* Suppl. 4, Vol. 2, pp. 1987–2002. Pergamon.

Gibson E. K. and Moore G. W. (1972) Inorganic gas release and thermal analysis study of Apollo 14 and 15 soils. *Proc. Third Lunar Sci. Conf., Geochim. Cosmochim. Acta,* Suppl. 3, Vol. 2, pp. 2029–2040. MIT Press.

Holland P. T., Simoneit B. R., Wszolek P. C., and Burlingame A. L. (1972) Compounds of carbon and other volatile elements in Apollo 14 and 15 samples. *Proc. Third Lunar Sci. Conf., Geochim. Cosmochim. Acta,* Suppl. 3, Vol. 2, pp. 2131–2147. MIT Press.

Kynaston H. and Hill J. B. (1908) The geology of the country near Oban and Dalmally. Mem. 45, Geol. Survey, Scotland, H.M.S.O. Glasgow.

LSPET (The Lunar Sample Preliminary Examination Team) (1972) *Apollo 14 Lunar Samples Information Catalogue.* NASA TM X-58P62.

Moore C. B., Lewis C. F., Cripe J., Delles F. M., Kelly W. R., and Gibson E. K. (1972) Total carbon, nitrogen and sulfur in Apollo 14 lunar samples. *Proc. Third Lunar Sci. Conf., Geochim. Cosmochim. Acta,* Suppl. 3, Vol. 2, pp. 2051–2058. MIT Press.

Nyquist L. E., Funk H., Schultz L., and Signer P. (1973) He, Ne, and Ar in chondritic Ni–Fe as irradiation hardness sensors. *Geochim. Cosmochim. Acta* **37,** 1655–1685.

Price P. B. and Walker R. M. (1962) Electron microscope observation of a radiation nucleated phase transformation in mica. *J. Appl. Phys.* **33,** 2625–2628.

Raleigh C. B., Kirby S. H., Carter N. K., and Avé Lallemont H. G. (1971) Slip and the clinoenstatite transformation as competing rate processes in enstatite. *J. Geophys. Res.* **76,** 4011–4022.

Reid A. M., Warner J., Ridley W. I., Johnston D. A., Harmon R. S., Jakeš P., and Brown R. W. (1972) The major element compositions of lunar rocks inferred from glass compositions in lunar soils. *Proc. Third Lunar Sci. Conf., Geochim. Cosmochim. Acta,* Suppl. 3, Vol. 1, pp. 363–378. MIT Press.

Sweatman T. R. and Long J. V. P. (1969) Quantitative electron-probe microanalysis of rock forming minerals. *J. Petrol.* **10,** 332–379.

Trommsdorf V. and Wenk H. R. (1968) Terrestrial metamorphic clinoenstatite in kinks in bronzite crystals (1968) *Contrib. Mineral. Petrol.* **19,** 158–168.

Turner F. J., Heard H., and Griggs D. T. (1960) Experimental deformation of enstatite and accompanying inversion to clinoenstatite. *Rep. Int. Geol. Congr. 21st session* **17,** 339–408.

Warner J. L. (1972) Metamorphism of the Apollo 14 breccias. *Proc. Third Lunar Sci. Conf., Geochim. Cosmochim. Acta,* Suppl. 3, Vol. 1, pp. 623–643. MIT Press.

Weber H. W. (1973) Bestrahlungsalter und radiogene Edelgasalter von Mondmaterial. Ph.D. thesis, University of Mainz.

Williams R. J. (1972) The lithification and metamorphism of lunar breccias. *Earth Planet. Sci. Lett.* **16,** 250–256.

Wilshire H. G. and Jackson E. D. (1972) The petrology and stratigraphy of the Fra Mauro formation of the Apollo 14 site. *U. S. Geol. Surv. Prof. Paper* 785.

Proceedings of the Fifth Lunar Conference
(Supplement 5, Geochimica et Cosmochimica Acta)
Vol. 2 pp. 1181–1206 (1974)
Printed in the United States of America

Bulk compositions of the moon and earth, estimated from meteorites

R. Ganapathy and Edward Anders

Enrico Fermi Institute and Department of Chemistry, University of Chicago,
Chicago, Illinois 60637

Abstract—The bulk composition of the earth and moon was calculated on the assumption that these planets had formed by the same processes as the chondrites. Three basic condensates (refractories, nickel–iron, magnesium silicates) were fractionated from each other in the solar nebula, and were transformed by condensation and remelting to a total of seven components: early condensate, FeS, remelted and unremelted FeNiCo and (Fe,Mg)-silicates, and a carbonaceous, volatile-rich silicate. The proportions of these components in a differentiated body may be estimated from geochemical constraints, e.g. K/U, bulk U and Fe abundances, etc. In terms of this model the earth contains 9.2% early condensate and 1.5% volatile-rich material. For the moon these percentages are 30.1 and 0.04. Complete abundance tables for 83 elements are constructed on the assumption that each component carried its cosmic complement of trace elements.

When lunar and terrestrial basalt data are normalized to these model compositions (thus correcting for differences in bulk composition), the abundance patterns become strikingly similar. This would seem to demonstrate the essential sameness of igneous processes on both planets. Many volatile elements (Cd, Ga, Pb, Tl, etc.) are enriched by the same factors as large-ion, lithophile elements such as Ba or U. The model correctly predicts the abundance ratios of certain volatile/refractory element pairs, e.g. Cd/Ba, Ga/La, Sn/Th, and Pb/U.

Some of the differences between lunar and terrestrial basalts (V, W, etc.) seem to reflect the more reducing environment of the moon. For Cr, on the other hand, it is the low terrestrial abundance, not the high lunar abundance, which is anomalous in terms of this model.

The high refractory element and low metal content of the moon are compatible with an origin either in heliocentric or in geocentric orbit. By the time the volatiles accreted, however, the moon appears to have been in a geocentric orbit.

All input data required by this model can be obtained by unmanned spacecraft or ground-based observations. Thus, if this model proves viable, it will make it possible to construct a detailed geochemical profile of a differentiated planet after a single visit by an unmanned spacecraft.

Introduction

The traditional approach to planetary modeling has been largely intuitive. Planets were constructed from a single meteorite type, or a mixture of such types, with the proportions adjusted so as to match the density and a few other properties. Though the results have looked reasonable (a tribute to the intuition of the modelists!), the arbitrary, *ad hoc* nature of this approach is esthetically objectionable.

We have therefore developed another approach, based on a highly restrictive set of rules. The fundamental assumption is that the inner planets were made by exactly the same processes as the chondrites. For the chondrites, it has been obvious for some time that they are mixtures of about half a dozen components that formed in a cooling solar nebula (Larimer and Anders, 1970; Anders, 1971; Grossman and Larimer, 1974). We assume that the earth and moon were made from the same components, but in different proportions. The problem then reduces to estimating these proportions, from appropriate geochemical constraints such as K/U ratios. And since each component carries a characteristic suite of trace elements in predictable (generally cosmic) proportions, this model can predict the abundance of all 83 elements in the planet.

A similar approach has been developed independently by Wänke et al. (1973, 1974). However, they use only two components: early condensate and a chondritic component of adjustable composition.

Our first attempt in this direction was published last year (Krähenbühl et al., 1973). We now present a second iteration, with improved constraints and a review of the results and their implications. We hope that the detailed predictions set forth in this paper will stimulate critical tests of our model, leading to further improvements or outright dismissal.

PLANETARY COMPONENTS

Condensation sequence of a solar gas

The basic framework of our model is the equilibrium condensation sequence of a solar gas (Fig. 1), as worked out by Larimer (1967), Grossman (1972), and Grossman and Larimer (1974). Three kinds of primordial dust condense from such a gas on cooling: (1) an early condensate, consisting of refractories, (2) metallic nickel–iron, and (3) magnesium silicates.

Together, these three materials comprise about 70% of the potentially condensable "rocky" material. They collect volatiles on cooling; metal reacts with H_2S to give FeS, and with H_2O to give FeO, which finds its way into the silicates (Larimer and Anders, 1967).

Fractionation processes

This is not the whole story, however. The study of chondrites has shown that three additional processes (italicized in Fig. 2) take place on cooling. They double the number of components and alter their proportions (Larimer and Anders, 1970).

1. *Fractionation of refractories.* Early condensate is gained or lost, presumably by settling to the median plane before the other two condensates have precipitated from the gas.

2. *Fractionation of metal.* Nickel–iron is gained or lost (Urey and Craig, 1953), presumably by ferromagnetic attraction (Wood, 1962; Harris and Tozer, 1967).

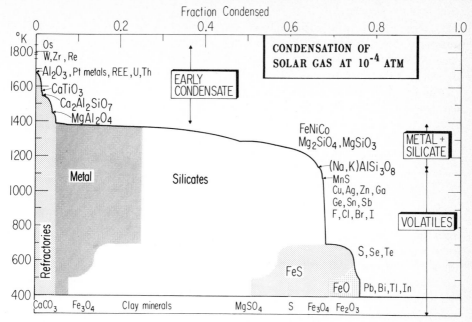

Fig. 1. Three types of dust condense from a solar gas: refractories, metallic nickel–iron, and magnesium silicates. On cooling, iron reacts with H_2S and H_2O to give FeS and FeO. Further major changes take place below 400°K.

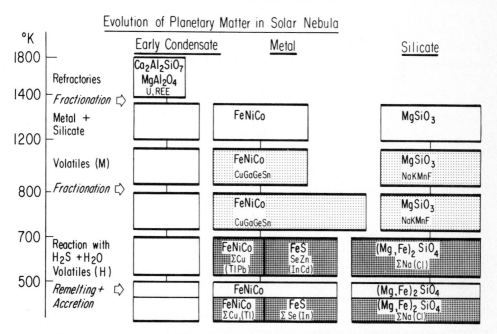

Fig. 2. Three additional processes, not predicted by the condensation sequence, take place on cooling (*cf.* arrows on left). They effectively double the number of components, from three to six. Presence of moderately (M) and highly (H) volatile elements is indicated by light and dark shading. Partial condensation is indicated by parentheses.

3. *Remelting* (= *chondrule formation*). Just before or during accretion, some portion of the micron-sized dust is briefly remelted and converted to millimeter-sized chondrules, probably by collisions (Whipple, 1972). Volatiles are lost, FeS reverts to metal, but time is too short to permit reduction of FeO to Fe.

The first two processes merely alter the proportions of the components. The last process doubles their number, to a total of six:

1 early condensate
2 metals (unremelted and remelted)
1 sulfide
2 silicates (unremelted and remelted)

The trace element content of these components can generally be predicted from the condensation sequence. Elements fully condensed at the temperature of fractionation will be present in cosmic proportions. Thus, if we know the abundance of one element from the early condensate, such as U, we can predict the abundance of 37 others. Problems arise for 16 elements that condense below 600°K: Cd, Hg, B, In, Tl, Pb, Bi, Cl, Br, I, H, C, N, Ar, Kr, and Xe. They will be only partially condensed at the temperature of accretion, to a degree that is hard to predict in advance. For the sake of definiteness, we have therefore assigned these elements to a seventh, volatile-rich component, containing the first ten elements in cosmic proportions (Cameron, 1973), and the last six, in the amounts found in C3 chondrites, relative to Tl.

This step may not be as arbitrary as it sounds. There are indications that the earth and moon may have acquired their volatiles mainly as a thin veneer of carbonaceous-chondrite-like material (Kokubu *et al.*, 1961; Anders, 1968; Turekian and Clark, 1969; Krähenbühl *et al.*, 1973).

The Model

Constraints

In terms of our model, the above seven components are the basic ingredients of the chondrites and planets. Their composition is substantially fixed; only their proportions vary from one body to the next. We must now try to estimate these proportions for the moon and earth.

As in the past, we have used various geochemical constraints to estimate the extent of the fractionation processes and hence the proportions of the components (Table 1). Whole-planet abundances of a single element are most useful, but they are known only for U, Fe, and a few other elements (Larimer, 1971). We have therefore relied mainly on abundance ratios of elements that belong to different components but are geochemically coherent in igneous processes (e.g. K/U, Tl/U, etc.). Let us discuss these constraints, one by one.

Refractories. The amount of early condensate is given by the bulk U content of the planet. The lunar value of 59 ppb was obtained from Apollo 15 and 17 heat flow data (Langseth *et al.*, 1973), which agree with earlier microwave measurements by Tikhonova and Troitskii (1969), and hence are representative for the whole moon. The terrestrial U content was taken to be 18 ppb, the "maximum" value of Larimer (1971). Contributions from Th and K were subtracted on the basis of Th/U = 3.7 and the K/U ratios in Table 1.

Table 1. Constraints on composition.

| | | Constraint | | Component affected* | | | | | | |
| | | Moon | Earth | EC | Metal R | Metal U | FeS | Silicate R | Silicate U | Volatile-rich |
Process										
Refractory fractionation	Bulk U, ppb	59	18	×						
Metal–silicate fractionation	Bulk Fe, %	9	36		×	×	×			
Oxidation of Fe	FeO/MnO	82	62		×	×	×	×	×	
Formation of FeS	Supply of:	Fe°	S		×	×	×			
Remelting	K/U	1625	9440		×	×	×	×	×	
Volatiles, <600 K	Tl/U	0.0023	0.27							×

*EC = early condensate; R = remelted; U = unremelted.

Metal–silicate fractionation. The extent of this fractionation is estimated from the bulk Fe content of the planet. The lunar value of $9 \pm 4\%$ Fe is based on magnetometer data (Parkin *et al.*, 1973). It is lower than our previous value of 13%, which was taken from Reynolds and Summers (1969). The terrestrial value of 36% Fe (or 38% FeNiCo) comes from Reynolds and Summers (1969).

Oxidation of Fe. In our 1973 paper, we had no suitable constraint for FeO, and therefore arbitrarily assumed FeO/(FeO + MgO) = 0.30 (Model 1). However, it appears that the FeO/MnO ratio may serve as an approximate constraint. These two elements do not separate readily from each other in igneous processes, judging from the constancy of their ratio in rocks from a given planet (Laul and Schmitt, 1973). But they differ in volatility (Fig. 1), and so in carbonaceous chondrites at least, about two-thirds of the Mn but no Fe is lost during chondrule formation (Schmitt *et al.*, 1965; Larimer and Anders, 1967).

The degree of loss will depend on the peak temperatures reached during chondrule formation, and may be greater or smaller than two-thirds for the moon. In the ideal case of 100% loss, the MnO/FeO ratio reflects *only* the degree of remelting and the FeO content of the silicate. Because the former is given by the K/U ratio (see below), the latter can readily be calculated.

We have made our basic set of calculations on the assumption of 100% loss of Mn (Models 2 and 3a). The resulting FeO/(FeO + MgO) ratio was uncomfortably low, and we have therefore repeated the calculation for 90% loss (Model 3b). Clearly, these numbers are wild guesses, and so FeO/(FeO + MgO) remains poorly determined in our model.

Formation of FeS. This reaction takes place after the metal–silicate fractionation, and so the amount of FeS formed may be limited by either Fe or S, whichever is in shorter supply. In reality, the reactions $Fe \rightarrow FeO$ and $Fe \rightarrow FeS$ take place more or less concurrently (Grossman, 1972). However, we have found it more convenient to treat them sequentially, first allowing the Fe to form the proper amount of FeO (as indicated by FeO/MnO), and then letting the remaining Fe react with H_2S until one or the other reactant was exhausted. This procedure differs from that in our previous paper where we assumed that the S/K ratio was equal to the cosmic ratio.

Remelting. With bulk U known, the K content of the planet can be obtained from the K/U ratio. Potassium is fairly volatile, and is therefore lost during chondrule formation if temperatures are high enough (e.g. carbonaceous chondrites, parent material of eucrites; Larimer and Anders (1967); Laul *et al.* (1972)). Assuming that this was also the case for the earth and moon, we can infer the fraction of unremelted material from the K-content. This fraction is the same for all three components (Fig. 2), because the remelting process is nonselective.

The lunar K/U ratio in Table 1 is an average for lunar soils. The terrestrial value is based on the "maximum" K and U values of Larimer (1971).

Volatiles, <600 K. Of the 16 highly volatile elements, Tl is the most tractable. It behaves like an alkali ion in igneous processes, correlating with U (see Fig. 1 of Krähenbühl *et al.*, 1973), but is far more volatile, condensing only at ~450°K rather than ~1100°K (Grossman and Larimer, 1974). We have therefore used Tl/U ratios to estimate the amount of the volatile-rich component. The ratio for the moon was the grand average of 33 mare and terra rocks (Krähenbühl *et al.*, 1973). The value for the earth was taken from deAlbuquerque *et al.* (1972).

Components

Composition. The composition of the individual components was based on the condensation sequence of Fig. 1 (Grossman, 1972). The boundary between early and main condensate was drawn just prior to condensation of diopside, $T = 1387°K$. We used Cameron's (1973) solar-system abundances, with some modifications. The abundance of Fe was raised from 8.3×10^5 (Schmitt *et al.*, 1972) to 8.9×10^5 (Mason, 1962), because the latter value was based on larger samples and a much more accurate analytical method. The Ca abundance was lowered from 7.21×10^4 to 6.25×10^4, in order to bring the Ca/Al mass ratio to 1.08, a very well established value in nearly all meteorites (Ahrens, 1970). The Th abundance was lowered from 0.058 to 0.0453, to bring the Th/U ratio into line with the chondritic value of 3.6 ± 0.2 (J. W. Morgan, private communication). The K abundance was lowered from 4200 to 3790, on the basis of recent data by Nichiporuk and Moore (1974).

Compositional data for a few key elements are given in Table 2. All others can be derived from Cameron's abundance table, with the help of Figs. 3 and 4 showing the distribution of the elements among the various components. We must

Table 2. Composition of planetary components (wt.%).*

	Early condensate	Silicates (Model 3a)			Silicates (Model 3b)		
		Remelted	Unremelted	Volatile-rich†	Remelted	Unremelted	Volatile-rich
Ca	21.13	—	—	—	—	—	—
Al	19.34	—	—	—	—	—	—
Ti	1.121	—	—	—	—	—	—
Mg	8.71	23.59	23.00	21.29	22.51	21.99	20.43
Si	8.21	25.83	25.19	23.32	24.66	24.08	22.37
O	41.48	46.21	45.66	42.28	45.12	44.64	41.47
Fe	—	4.38	4.27	3.95	7.66	7.48	6.95
Na	—	—	1.281	1.187	—	1.225	1.138
Mn (ppm)	—	—	4746	4395	465	4538	4216
K (ppm)	—	—	1376	1274	—	1316	1223
H_2O	—	—	—	4.88	—	—	4.68
C	—	—	—	2.46	—	—	2.36
N	—	—	—	0.063	—	—	0.060
Tl (ppb)	—	—	—	338	—	—	324
U (ppb)	197	—	—	—	—	—	—

*These values apply to the moon. Model 3a assumes 100% Mn loss from remelted fraction; model 3b assumes 90% loss. Values for the earth are lower by a factor of 0.968, except for Fe where the following values apply for the three silicates: 6.72%, 6.56%, and 6.09%.

†Also $Ar^{36} = 926 \times 10^{-8}$ cc STP/g. The remaining gases were calculated from their abundance ratios in C3's (Mazor et al., 1970): $Ne^{20}/Ar^{36} = 0.19$; $He^4/Ne^{20} = 292$; $Ar^{36}/Kr^{84} = 202$; $Ar^{36}/Xe^{132} = 278$.

stress that these compositions are "theoretical," being based on solar-system abundances and the equilibrium condensation sequence. They are similar but not necessarily identical to actual meteoritic components, such as Allende inclusions.

Calculations. Let us use the following symbols for the mass fraction of each component in the moon: c = early condensate, s = silicate, t = troilite, m = metal, w = wüstite (FeO). The remelted, unremelted, and volatile-rich components are designated by subscripts r, u, and v: $s = s_r + s_u + s_v$.

The early condensate was found directly from the U content (Table 2): $c = 59/197 = 0.301$. The other components required a more elaborate approach. Each silicate fraction was decomposed into a magnesium silicate part, σ, and the requisite number of complements (all normalized to $\sigma = 1$): α = mass fraction of the lithophiles condensing between 1300 and 600 K: MnO, Na_2O, K_2O, Rb_2O, Cs_2O, and F; β = mass fraction of FeO; γ = mass fraction of H_2O, C, N, Tl, and 12 volatiles condensing below 600 K (Fig. 3). Then:

$$s_r = \sigma_r(1 + \beta) \qquad s_u = \sigma_u(1 + \alpha + \beta) \qquad s_v = \sigma_v(1 + \alpha + \beta + \gamma)$$

$$\beta = w/(\sigma_u + \sigma_r + \sigma_v)$$

The remelted fraction f is defined as:

$$f = \sigma_r/(\sigma_u + \sigma_r + \sigma_v)$$

Two of the silicate fractions were found directly from their indicator elements:

$$\sigma_v = (Tl_{\leftmoon}/Si_\sigma)(Si/Tl)_{cosmic}$$
$$\sigma_u + \sigma_v = (K_{\leftmoon}/Si_\sigma)(Si/K)_{cosmic}$$

$Tl_{\leftmoon}$ and $K_{\leftmoon}$ were found from the Tl/U, K/U ratios and the bulk U content. Here Si_σ is the Si content of the magnesium silicate.

MnO was found from $K_{\leftmoon}$ and the cosmic Mn/K ratio.

The three iron compounds were written in terms of the mass fraction of iron, i:

$$t = 1.574 i_t$$
$$w = 1.286 i_w = (FeO/MnO)_{\leftmoon} MnO_{\leftmoon}$$
$$m = Fe_{\leftmoon}[1 + (Ni + Co)_{cos}/Fe_{cos}] - i_t - i_w$$

The two numerical coefficients are stoichiometric factors. Also, the unremelted fraction equals the ratio of troilitic Fe to troilitic + metallic Fe:

$$1 - f = i_t/(Fe_{\leftmoon} - i_w)$$

and $s + m + t + c = 1$.

RESULTS

Compositions in terms of the seven components are given in Table 3. They differ somewhat from those in our abstract (Ganapathy and Anders, 1974), which were based on Cameron's original Ca abundance and an erroneously high U abundance (0.013 rather than 0.0098 atoms/10^6 Si). They differ even more strongly from our 1973 values (Krähenbühl *et al.*, 1973), owing to changes in input parameters and the model itself. The composition and U content of the early

Table 3. Model composition of the moon and earth.

Component	Moon Model 3a	Moon Model 3b	Earth Model 3a
	Mass fraction*		
Early condensate	0.301	0.301	0.092
Metal, remelted	0.061	0.041	0.240
Metal, unremelted	—	—	0.071
Troilite	0.011	0.0071	0.050
Silicate, remelted	0.557	0.577	0.418
Silicate, unremelted	0.070	0.073	0.114
Volatile-rich material	0.00040	0.00042	0.015
FeO/(FeO + MgO) mole%	7.47	12.9	11.4
Fe/Si atomic	0.243	0.245	1.26
Fraction remelted	0.891	0.890	0.771

*Model 3a assumes 100% loss of Mn from the remelted fraction, while model 3b assumes 90% loss.

Table 4. Abundances of the elements in the moon and earth, according to Model 3a (ppm unless otherwise marked; noble gases 10^{-10} cc STP/g).

Element	Moon	Earth	Element	Moon	Earth
H	2.2	78	Ru	4.9	1.48
He^4	2100	74,000	Rh	1.05	0.32
Li	8.7	2.7	Pd	0.25	1.00
Be ppb	186	56	Ag ppb	9.6	80
B ppb	13	470	Cd ppb	0.58	21
C	9.9	350	In ppb	0.075	2.7
N	0.26	9.1	Sn	0.085	0.71
O %	41.42	28.50	Sb ppb	7.6	64
F	30	53	Te	0.20	0.94
Ne^{20}	7	250	I ppb	0.48	17
Na	900	1580	Xe^{132}	0.13	4.8
Mg %	17.37	13.21	Cs ppb	33	59
Al %	5.83	1.77	Ba	16.8	5.1
Si %	18.62	14.34	La	1.57	0.48
P	538	2150	Ce	4.2	1.28
S %	0.39	1.84	Pr	0.53	0.162
Cl	0.70	25	Nd	2.9	0.87
Ar^{36}	37	1330	Sm	0.86	0.26
K	96	170	Eu	0.33	0.100
Ca %	6.37	1.93	Gd	1.18	0.37
Sc	40	12.1	Tb	0.22	0.067
Ti	3380	1030	Dy	1.49	0.45
V	340	103	Ho	0.33	0.101
Cr	1200	4780	Er	0.96	0.29
Mn	330	590	Tm	0.145	0.044
Fe %	9.00	35.87	Yb	0.95	0.29
Co	240	940	Lu	0.160	0.049
Ni %	0.51	2.04	Hf	0.95	0.29
Cu	6.9	57	Ta ppb	96	29
Zn	19.9	93	W	0.75	0.250
Ga	0.66	5.5	Re ppb	250	76
Ge	1.66	13.8	Os	3.6	1.10
As	0.90	3.6	Ir	3.5	1.06
Se	1.30	6.1	Pt	6.9	2.1
Br ppb	3.8	134	Au	0.072	0.29
Kr^{84}	0.18	6.6	Hg ppb	0.28	9.9
Rb	0.33	0.58	Tl ppb	0.136	4.9
Sr	60	18.2	Pb^{204} ppb	0.055	1.97
Y	10.9	3.29	Bi ppb	0.104	3.7
Zr	65	19.7	Th ppb	210	65
Nb	3.3	1.00	U ppb	59	18
Mo	9.8	2.96			

condensate is now based on cosmic abundances and Grossman's condensation sequence, rather than the measured composition of Allende inclusions. This has raised U from 120 ppb (Wänke *et al.*, 1973) to 197 ppb. The bulk Fe content of the moon is now 9% rather than 13%, and the FeO/(FeO + MgO) ratio is based on FeO/MnO, rather than being arbitrarily fixed at 30 mole%.

From the proportions of these components, we have calculated whole-planet abundances of all stable elements (Table 4 and Figs. 3 and 4). The underlying assumption, best illustrated in the figures, is that each trace element associates exclusively with one component. Thus a whole group of cosmochemically similar elements is depleted or enriched by the same factor, represented by the plateaus in Figs. 3 and 4.

Judging from chondrites, this is not a bad assumption, except for a few elements from the fringes of each volatility group (Li, Cs, Zn, Te, Na, and Mn). The first four are probably less abundant, and the last two more abundant by perhaps a factor of <2–3. The highly volatile group (open symbols) is also uncertain, as discussed later on.

The gross trends are unsurprising, having been qualitatively stated by many authors. The earth and especially the moon are depleted in volatiles and enriched in refractories, as first noted by Gast (1960, 1968, 1970). The individual abundances agree fairly well with Larimer's (1971) estimates for the earth, but are considerably higher than his values for the moon.

We can compare the major element compositions predicted by various models (Table 5). Our model gives lower Ca, Al and higher Mg, Si contents than most others, because of the smaller amount of early condensate. Our current value from Table 3 is 0.30, compared to 0.42 from Krähenbühl *et al.* (1973); 0.50 for Wänke *et al.* (1974); and ~1 for Anderson (1973). Much of the difference is due to the higher U content used for the early condensate, Table 2. Interestingly, our composition 2 falls very close to that of Taylor and Jakeš (1974), which was derived by a rather different set of arguments.

Discussion

Composition of basalts

Gross patterns. It has been obvious ever since Apollo 11 that lunar basalts differ profoundly from terrestrial basalts. Refractories are enriched while volatiles and some siderophiles are depleted, relative to terrestrial basalts (see, for example, Fig. 10 in Ganapathy *et al.*, 1970). It has not been clear, however, whether this dissimilarity reflects mainly differences in bulk composition of earth and moon, or differences in igneous processes (e.g. absence of water, lower f_{O_2}, lower g, lack of an atmosphere).

Now that we have data for the bulk composition of the two planets (Table 4), we can separate these two factors. We divide the abundance of each element in basalt by its abundance in the bulk planet. This correction cancels out differences in bulk composition, permitting the pure igneous pattern to show through. The results are shown in Figs. 5 and 6. Several trends are apparent.

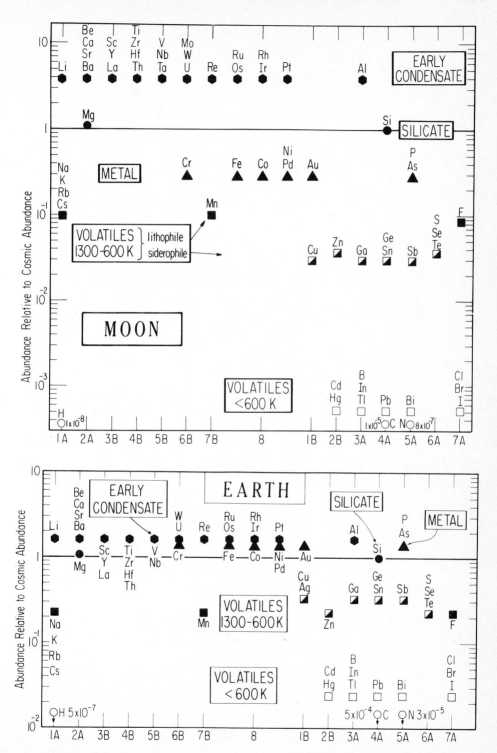

Figs. 3 and 4. Model compositions of the moon and earth. They are based on the assumption that each component contains its cosmic complement of trace elements.

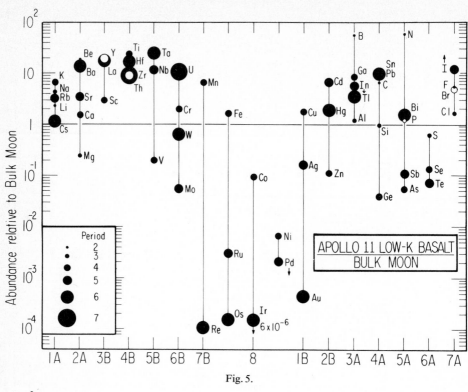

Fig. 5.

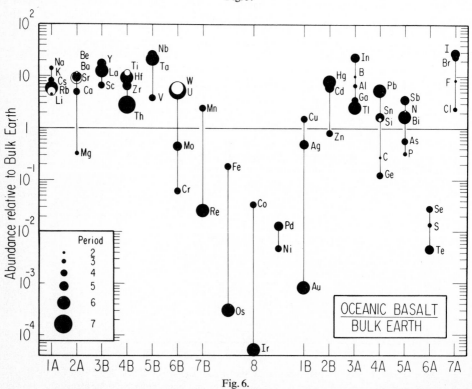

Fig. 6.

Table 5. Model compositions of the lunar interior (weight percent, without metal or troilite).

	SiO_2	TiO_2	Al_2O_3	FeO	Cr_2O_3	MgO	CaO	Na_2O	MnO	References
Interior	53.1	1.0	5.0	13.5	0.5	22.5	4.0	0.1	0.4	Ringwood and Essene (1970)
Lower mantle	49.1	0.45	6.0	18	0.5	19.6	5.3	0.1	0.5	Gast (1972)
Upper 150 km	48	0.8	16	9	—	14	12	—	0.4	Gast (1972)
Interior	30.8	1.6	27.6	0	—	17.8	22.2	—	—	Anderson (1973)
Bulk moon	30.4	1.0	20.2	7.8	—	15.4	15.9	—	—	Wänke et al. (1974)
Bulk moon	44.0	0.3	8.2	10.5	—	31.0	6.0	0.30	0.10	Taylor and Jakeš (1974)
Bulk moon (1)	35.9	0.63	12.8	12.7	0.39	20.9	10.6	0.58	0.24	Krähenbühl et al. (1973)
Bulk moon (2)	42.7	0.420	8.22	3.21	0.164	30.7	7.68	0.109	0.039	Ganapathy and Anders (1974)
Bulk moon (3a)	39.8	0.564	11.02	3.52	0.175	28.8	8.91	0.121	0.043	This work; 100% loss of Mn
Bulk moon (3b)	39.5	0.564	11.02	6.39	0.175	28.6	8.91	0.121	0.078	This work; 90% loss of Mn

(1) Both graphs have become symmetrical. Alkalis on the extreme left and volatiles on the right are enriched to nearly the same degree as refractory LIL elements on the left, Groups 2A to 6B. Apparently the volatility of these elements is important only in pre-accretional, condensation processes in the nebula, not in post-accretional, igneous processes on the planet. In the latter setting, most of these elements show "LIL" behavior.

(2) The gross dissimilarity between lunar and terrestrial basalts is greatly reduced. Apparently most of the difference reflected bulk composition. With this factor removed, the basic sameness of partial melting processes is becoming apparent. Of the remaining differences, some may represent real differences in igneous pattern (Cr, Mo, W, Re, Fe, S, Se, and Te), while others probably are due to analytical and sampling errors (the specific rocks analyzed were not the same for all elements).

(3) A number of trace elements often regarded as siderophile-chalcophile (Co, Cu, Ag, Zn, Cd, Ga, Ge, Pb, As, Sb) are not as strongly depleted as the out-and-out siderophiles (Ni, Pt metals) or chalcophiles (S, Se, Te). Apparently these elements were at least slightly lithophile during initial differentiation of the planet.

Moreover, most of these elements show substantially the same depletion factors on the earth and moon (Figs. 5, 6). Perhaps these depletion factors can be interpreted as distribution coefficients.

It may be useful to recapitulate what we have done in order to arrive at Figs. 5 and 6. We have scaled the abundances of each cosmochemical group to that of an

Figs. 5 and 6. Composition of lunar and terrestrial basalts normalized to bulk composition of each planet. Elements are arranged according to position in the Periodic Table. Similarity of patterns suggests that gross differences between lunar and terrestrial basalts reflect mainly differences in bulk composition, rather than the nature of igneous processes on the two planets. See text for discussion of remaining differences. Lunar basalt data from Apollo 11 and Second Lunar Science Conference Proceedings; oceanic basalt data from Larimer (unpublished compilation), Laul et al. (1972), and Kay et al. (1970).

"indicator" element from the group, as follows:

Refractories: U
Siderophiles: Fe
Moderately volatile elements: K (with additional small adjustments according to geochemical character, see Figs. 3 and 4)
Highly volatile elements: Tl

The underlying assumption is that the trace elements associated with each component were present in cosmic proportions. Thus, if the abundance of one element is known, that of all others follows.

Correlation with ionic radii and electronic structure. We can look for correlations between the abundance of lithophile elements and their crystal-chemical properties (Fig. 7). Ions with noble-gas structure are represented by solid symbols; all others, by open symbols. Ionic charge is indicated by the degree of the symmetry axis; thus ellipses, with a twofold axis, stand for +2; triangles, with a threefold axis, for +3, etc. Major elements are italicized.

The maximum enrichment relative to the bulk planet is tenfold (give or take a factor of two), in both lunar and terrestrial basalts. It is displayed both by the classical (refractory) LIL elements such as Ta, Zr, U, Ba, REE, etc., and by volatiles such as In, Cd, Pb. As noted above, the concept of LIL or incompatible elements can be broadened to include the latter. A limiting enrichment by a factor of ten means that these basalts were produced by ~10% partial melting, with the most highly incompatible elements (of distribution coefficients $\geqslant 100$) being almost entirely concentrated in the melt.

Because of the mixed origin of the data, not much can yet be made of minor trends (e.g. the variations in enrichment factor between 5 and 20). It will be interesting to look for such trends in a more homogeneous body of data. For the time being, we shall consider mainly the lunar graph as the better of the two, and limit ourselves to major trends, i.e. enrichment factors of less than five.

Five of the elements showing smaller enrichments (Mg, Fe, Si, Al, Ca) are major elements, whose fractionation follows a different set of rules. A few others (Sc, Sr, perhaps also Li and Bi) are probably retained in mantle minerals, by isomorphous substitution for Al^{3+}, Ca^{2+}, etc. (Frondel, 1968).

The alkalis show just the reverse trend from the alkaline earths, with depletion increasing with ionic radius. We therefore suspect that this trend reflects their abundance in the bulk planet rather than crystal-chemical factors. The volatilities of alkali metals differ greatly, and so our assumption that these elements condensed in cosmic proportions may not be justified. In chondrites, Rb and Cs often are more depleted than elements of the same nominal volatility group (Laul *et al.*, 1973). To complicate matters further, in terrestrial oceanic basalts Cs often is depleted by water leaching (Muehlenbachs and Clayton, 1972). An alternative possibility is that Rb and Cs were extracted into the core as sulfides (Murthy and Hall, 1970; Goettel, 1972).

Zinc is conspicuously underabundant, as are several other ions not plotted here: Co(II), Cu(II), Ag(I), Ge(IV), As(III), and Sb(III) (Figs. 5 and 6). As mentioned above, this probably reflects their chalcophile and siderophile tendencies.

Chromium is much more abundant in lunar basalts than in terrestrial basalts. Haggerty *et al.* (1970) have attributed this to incorporation of Cr^{2+} in olivine, but Burns *et al.* (1973) found no clear evidence for Cr^{2+} in the absorption spectra. The abundance trends place the problem in a different perspective. The Cr abundance in *lunar* basalts is compatible with either a +3 or a +2 oxidation state; in the former case it falls between Al^{3+} and Sc^{3+}; in the latter (R = 0.81 A) between Mg^{2+} and Mn^{2+}. But it is the Cr abundance in *terrestrial* basalts which is anomalous on our model. Here the oxidation state is indisputably +3, yet Cr falls far below all other +3 ions. Perhaps this reflects incorporation in high-pressure phases (spinels, garnets) which are rare or absent in the moon.

Vanadium shows the opposite trend, being more depleted on the moon. Reduction to V or V^{2+} is a possibility; in the latter case, V should be enriched in mantle olivines and perhaps other minerals (Ringwood and Essene, 1970). Tungsten is likewise depleted, probably due to reduction to the metal (Wänke *et al.*, 1971).

In summary, it appears that most trace elements are incompatible with the mantle mineralogy of the earth and moon. The majority are lithophile, and concentrate in the first liquid during partial melting. This includes several elements from the right-hand side of the Periodic Table, that were previously regarded as chalcophile.

Abundance ratios. One interesting test of our model is its ability to predict abundance ratios of trace elements belonging to different planetary components. To minimize the distorting effect of igneous fractionations, such tests must be limited to incompatible elements that are largely concentrated in the liquid during partial melting. From Fig. 7, seven such elements are available: Cd, Ga, Sn, In, Hg, Pb, and Tl. Three of these must be eliminated: In (contamination), Hg (mobile on lunar surface), and Tl (used as indicator element in our model). The remaining four elements are paired off with refractory ions, generally of the same charge: Ba^{2+}, La^{3+}, Th^{4+}, and U^{4+}. A survey of literature data (compiled by Dr. Jeffrey L. Warner) gave mean ratios of the right order (Table 6), though only two of the four

Table 6. Volatile/refractory element ratios.

	Cd/Ba	Ga/La	Sn/Th	Pb^{204}/U
Moon, model 3a	5.6×10^{-5}	0.43	0.32	9×10^{-4}
Moon, mare basalts, mean*	5.1×10^{-5} (23)	0.41 (42)	0.26 (5)	16×10^{-4} (20)
Earth, model 3	4.0×10^{-3}	11.6	10.9	0.11
Earth, oceanic basalts	3.4×10^{-3}	3.3	5†	0.11

*Number of samples is given in parentheses.
†Very poorly determined.

pairs (Cd/Ba, Pb/U) showed reasonable linear correlations. The remaining two pairs correlated poorly, with the volatile element (Ga, Sn) showing a much smaller abundance range than the refractory element. It is curious that the model nonetheless predicts the right ratios.

Petrology

Hodges and Kushiro (1974) have studied the melting relations of two of our earlier model compositions for the moon [bulk moon 1 and 2, Table 5]. The former was too poor in silica to yield plagioclase, and was hence disqualified. (Presumably other compositions of even lower SiO_2 content must also be ruled out for this reason). The latter did give plagioclase, preceded by olivine, orthopyroxene, and clinopyroxene. It also produced a reasonable sequence of rock types, in thicknesses consistent with seismic data: anorthositic gabbro with an ilmenite-rich zone, followed by plagioclase lherzolite, harzburgite, and dunite. The density of this model composition (including its complement of Fe and FeS) was 3.34 g/cm^3 at 20 kb and a temperature slightly below the solidus (Hodges and Kushiro, 1974 and private comm.). The agreement with the bulk density of the moon (3.34 g/cm^3) is of some interest, because our model included no prior constraint on density.

Two flaws of model 2 were the relatively high Fe/Mg ratios of the anorthositic rocks, and the inability to produce troctolites or spinel troctolites. Both could be avoided if plagioclase crystallized before orthopyroxene and clinopyroxene. This requires a higher alumina content.

Model 3 attempts to solve this problem by lowering the Ca/Al ratio from Cameron's value of 1.26 to Ahrens' 1.08. However, the absolute amount of both Ca and Al has been raised by correction of the U abundance, and since this change also lowers SiO_2, it remains to be seen whether model 3 is indeed an improvement over model 2, insofar as petrology is concerned.

Highly volatile elements and noble gases

Anders (1968) compared the abundances of volatile elements in the earth's crust with those in Cl and ordinary chondrites. These two meteorite classes, respectively, typified material that had condensed its volatiles at low temperatures, in cosmic proportions, or at higher temperatures, in proportions declining with volatility. (The temperatures in question are ~360°K and ~420–500°K; see Anders (1972) for references). The earth's pattern was intermediate between the two: it was flat in the initial portion, like the Cl pattern, but then declined, like the ordinary chondrite pattern.

For this reason we used an intermediate material for our model, containing H, C, N, and noble gases in C3 chondrite proportions, and all other highly volatile elements in cosmic proportions. It corresponds essentially to a 400–430°K condensate.

Figure 8 compares the crustal abundances of highly volatile elements with those predicted by our model 3a (solid circles) and a variant using Cl rather than

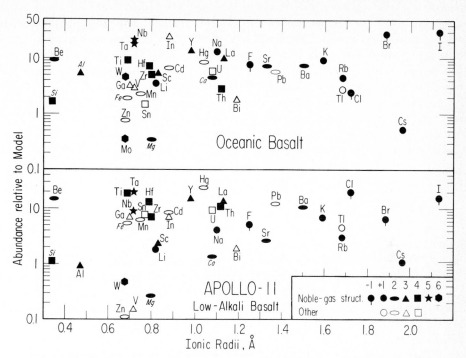

Fig. 7. Many volatiles, like the lithophiles, are enriched by factors of 5–20. Apparently they, too, concentrate largely in the liquid during partial melting. See text for discussion of exceptions.

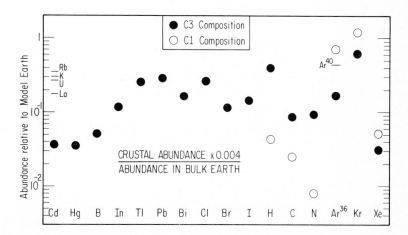

Fig. 8. Most highly volatile elements are present in the earth's crust in about the proportions assumed in the model: cosmic proportions for the first 10; C3 chondrite proportions for the last six (solid circles). A variant using Cl proportions for the last six elements (open circles) is also shown. In terms of this model, crust contains about 8–30% of earth's inventory of most volatiles.

C3 abundances for the last six elements (open circles). The mass of the crust is $0.004M_\oplus$, and so the ratio (abundance in crust) $\times 0.004$/(abundance in bulk earth) which is plotted here gives the fraction of each element released into the crust. Crustal abundances were taken from Larimer (1971 and unpublished) except Tl (de Albuquerque *et al.*, 1972) and Cl (Garrels and Mackenzie, 1971).

Taken at face value, the C3-based pattern suggests that most of these elements have been released from the mantle to a rather uniform extent of 8 to 40%.* Similar results are obtained for less volatile, LIL elements such as K, Rb, U, and La (horizontal lines on left). These release factors are not too different from earlier estimates by Gast (1968), Hurley (1968), Larimer (1971), and others. The agreement may be significant, because the present approach, in contrast to the earlier ones, involves no prior assumptions about outgassing of volatiles.

The Cl-based variant gives less uniform release factors for the last six elements. The factor for H_2O (0.04) is lower than that for most other volatiles, and since various authors, beginning with Rubey (1951), have shown that the water content of the mantle must be low, this argument alone would seem to rule out Cl composition. On the other hand, Cl composition does have one advantage over C3. The Ar^{40} content of the atmosphere, for a terrestrial K abundance of 170 ppm, corresponds to a release factor of 0.44. For comparison, the release factors of Ar^{36} are 0.12 on the C3 model and 0.53 on the Cl model. Now, it does not make any sense for the Ar^{36} factor to be smaller than the Ar^{40} factor, because primordial Ar^{36}, having been present in the earth from the beginning, should be outgassed more completely than radiogenic Ar^{40} which was produced gradually over geologic time. We conclude that the volatile-rich fraction of the earth must have had a lower Ar^{36}/K ratio than C3 chondrites, by a factor of four or more. Such meteorites exist (Cl or H3, L3, LL3 chondrites), but they have other shortcomings.

The model also gives reasonable results for the light noble gases, because the Ne^{20}/Ar^{36} ratio in the atmosphere (0.52) is similar to that in C3 chondrites used here (0.19). The release factor for Ne^{20} is 0.60, close to that for most other elements in Fig. 8. The He^4 value is based on the assumption that the He^4/Ne^{20} ratio in the volatile-rich material was equal to the mean ratio in C3V chondrites (292; Mazor *et al.*, 1970). Interestingly, Tolstikhin (1974) has obtained similar Ne^{20} release factors and He^4 abundances by an entirely different line of reasoning, based on the high He^3/He^4 ratios in some volcanic gases (Tolstikhin *et al.*, 1972).

Range of cosmochemical fractionations

Our model has been based on the premise that the earth and moon were made by the same processes as the chondrites. It is proper to ask whether we had to

---

*Values for Cd, Hg, B, and Xe are lower, but at least the last two may be in error. Cameron's cosmic abundance of B may have been overestimated (Reeves, 1973), and Larimer's crustal abundance of Xe is too low, being based on the atmosphere alone. Substantial amounts of Xe are locked up in shales (Canalas *et al.*, 1968).

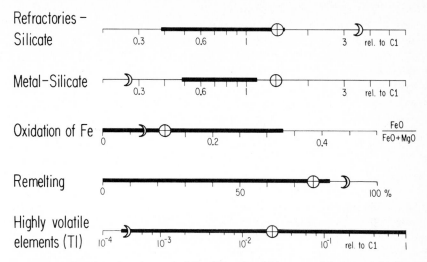

RANGE OF FRACTIONATIONS IN EARTH, MOON, AND CHONDRITES

Fig. 9. Fractionations for the earth fall within chondritic range (black bars) in four cases out of five. For the moon, they fall well outside the chondritic range in three cases out of five. Either the observed range in chondrites does not cover the full gamut of these processes, or the moon formed under special conditions (earth orbit?) where fractionation was enhanced.

stretch any of these processes beyond their observed range in chondrites, in order to account for the earth and moon.

Figure 9 attempts to answer this question. The chondritic range is indicated by heavy black bars. We see that the earth falls outside the chondritic range just once, and then only barely. The moon, on the other hand, falls outside this range in three cases out of five.

An obvious excuse is that we know only eight classes of chondrites. There may be hundreds of others not sampled by the earth (Anders, 1964), and it would be very provincial to assume that these eight classes cover the full gamut of these processes. Indeed, the range could be expanded greatly by including mesosiderites, a class of meteorites that may possibly represent another type of primitive condensate (J. A. Wood, private communication). Moreover, chondrites probably come from the asteroid belt at 2–3 a.u., whereas the moon presumably formed at ~1 a.u. The fractionation processes may have operated more efficiently at 1 a.u. Still, it is rather disturbing that the lunar point falls so far outside the chondritic range for the first two fractionations. No chondrite and no known planet has so low a metal/silicate ratio. Not even the extinct basin-forming objects seem to have had Fe/Si ratios as low as the lunar value (Morgan *et al.*, 1974).

It is rather suggestive that the only body showing such extreme behavior also is the only one which is a satellite. Perhaps its unusual composition reflects competition with the central planet. In metal/silicate ratio at least, it is complementary to the earth. We shall pursue this question in the next section.

Why is the moon enriched in refractories and depleted in volatiles?

Whatever the success of our model in describing a planet's composition in terms of seven components, it does not explain why the proportions of these components vary from one planet to the next. We can try to probe a little more deeply and see whether the chemical differences between earth and moon shed any light on their formation conditions.

Within the general framework of accretion from a cooling nebula, two scenarios may be envisioned for the moon: formation as a sunbound planet or as an earthbound satellite.

Formation in sun orbit. At first sight, heterogeneous accretion as two independent planets would seem capable of accounting for the chemical differences between earth and moon. Following Öpik (1971), we consider a planet of radius R and mass μ sweeping up dust and planetesimals, traveling at an encounter velocity U relative to circular velocity. Its target radius σ can be expressed in terms of the escape velocity $v_\infty = \sqrt{2\mu/R}$:

$$\sigma^2 = R^2[1 + v_\infty^2/U^2] \tag{1}$$

For low encounter velocities, the unity term can be disregarded, giving

$$\sigma^2 \approx 2\mu R/U^2 \tag{2}$$

If we neglect differences in density, $\mu \propto R^3$, and the mass accretion rate equals

$$dM/dt \propto \sigma^2 = \mu^{4/3}/U^2 \tag{3}$$

This equation can be integrated, and the result expressed as the variable mass ratio of two accreting bodies:

$$\mu_1/\mu_2 = (1 + C\mu_1^{1/3})^3 \tag{4}$$

In order to obtain the earth/moon mass ratio of 81.5 when $\mu_1 = 1$, the adjustable parameter C is set equal to 3.34. This equation gives the mass ratio as a function of μ_1, going backward in time.

Because the growth rate is a strong function of mass, larger bodies outgrow smaller ones at an ever-increasing pace. Thus "the rich get richer." Ganapathy *et al.* (1970) have drawn attention to an important chemical consequence of this fact. If the composition of the accretable material changes with time, the smaller body will be relatively richer in "early" material (refractories), and the larger body, in "late" material (volatiles).

Qualitatively, this is just the trend observed for the earth and moon. We can check whether the agreement is quantitative as well. Let us compare the mass ratio of the principal components predicted from model 3a (Table 3) and Eq. (4). We combine metal, silicate, and troilite because in our model, they remain in equilibrium with the gas and accrete concurrently to an intimate, chondrite-like mixture. For the volatile-rich material, we use a more exact relation that takes into account density differences between the two planets (Ganapathy *et al.*, 1970).

Heterogeneous accretion in heliocentric orbit.

Ratio $n_{\oplus}/n_{\mathrm{moon}}$[a]	Early condensate	Metal + Silicate + Troilite	Volatile-rich material
Model 3a	0.30	1.30	37
Eq. (4)	0.19	1.10	3.6

[a]n = mass fraction of the component.

The agreement for the first two stages of accretion is not bad.* The discrepancy for volatiles, on the other hand, is large and unbridgeable. The predicted ratio of 3.6 is a firm, asymptotic upper limit which is reached only at very low values of U. There is no way of stretching it to the observed ratio of 37.

Moreover, if the earth/moon differences were due to heterogeneous accretion in heliocentric orbit, then bodies smaller than the moon should be still richer in early condensate and poorer in volatiles. We have compositional data on two populations of such bodies: the parent bodies of the chondrites and the extinct objects that produced the lunar mare basins (Morgan et al., 1974). The latter are particularly germane, because they presumably formed in the moon's neighborhood (Morgan et al., 1974).

Compositional data for six such bodies show exactly the opposite trend. They are lower in refractories and higher in volatiles than the bulk Moon (Morgan et al., 1974). This is also true of all known chondrite classes. Some special factors, reversing the trend, must have been at work when the moon was built.

Formation in earth orbit. The depletion of volatiles can apparently be accounted for by accreting the moon in earth orbit. Owing to its orbital motion, an earthbound moon would encounter dust in heliocentric orbits at high velocities, and hence have a small capture cross section. The original treatment by

---

*It conceals the great difference in iron content; however, our model explains it not by heterogeneous accretion, but by a separation based on ferromagnetism or some other unique property of iron.

Interestingly, there exist circumstances where Öpik's equation correctly predicts the low iron content of the moon. In an extreme heterogeneous accretion model, where iron, moreover, is assumed to condense *after* silicate, the following ratios $n_{\oplus}/n_{\mathrm{moon}}$ are predicted:

	Early condensate	Silicate + Troilite	Metal	Volatiles
Model 3a	0.30	0.9	5	37
Eq. (4)	0.19	1.3	4	3.6

Delayed condensation of iron is not out of the question. Blander and Katz (1967) have in fact argued that iron will condense after silicates, owing to nucleation difficulties. However, the texture of chondrites shows that silicate and metal accreted simultaneously. Thus an extreme heterogeneous accretion model is incompatible with ours, and should perhaps be explored in its own right.

Ganapathy *et al.* (1970) was quantitatively incorrect, as noted by Singer and Bandermann (1970), but a recent, more thorough treatment by Whipple (1973), that takes into account the effects of gas, shows that the mechanism can indeed explain the observed compositional differences. The orbital motion of the moon produced a wind that aerodynamically prevented the capture of dust and smaller planetesimals. The same mechanism would also favor selective accretion of chondrules over dust (Whipple, 1972), thus explaining the larger amount of remelted material on the moon (Table 3).*

Though formation in earth orbit provides additional mechanisms for chemical fractionation, we believe that these mechanisms should be invoked only where the standard chondritic fractionation processes fail. Figure 9 shows clearly that all chemical fractionations seen in the moon are also seen in chondrites, though often to a lesser degree. Thus a treatment based entirely on chondritic fractionation rules is a valid first step. At the very least it points out problem areas where additional processes must be invoked.

Selenopoetry

We can try to assemble the different parts of this mosaic into a single, coherent picture. This entails a higher order of speculation, because many intermediate pieces are missing, and the relative order of even the well-reconstructed pieces often is uncertain. We have acknowledged these uncertainties in the heading of this section.

(1) At the time when early condensate was the principal solid available in the nebula, two nuclei accreted, of 0.092 and 0.0037 $M_\oplus$. No later than at the end of this stage, the smaller nucleus (proto-moon) was in orbit around the larger (proto-earth).

(2) When metal became available, the proto-earth collected it more efficiently, depriving the moon of its share.†

(3) Chondritic material was accreted next, both as dust and as larger bodies. Its *average* composition may have been close to that of lunar silicate (Table 3).

(4) Volatile-rich material accreted last, again both as dust and as larger bodies. (However, we cannot rule out the possibility that it accreted first, as proposed by Urey, 1972).

---

*Some authors have contended that accretion in geocentric orbit would lead to "impact differentiation," with selective loss of volatiles from the moon owing to the higher impact velocity. We do not believe that such loss will be significant. Studies of the micrometeorite component in lunar soils (Anders *et al.*, 1973) show that even highly volatile metals such as Bi are retained by the moon, although they are vaporized on impact.

†We have no clear preference for a mechanism. Acceptable alternatives are fractionation based on ferromagnetism, as in chondrites, or sorting by density in a circumterrestrial ring (Öpik, 1972; Smith, 1974). It seems preferable to put much of the metal into the earth at an early stage, rather than leaving it in an iron-rich, chondritic component to be accreted later. Though it is conceivable that two chondritic components of very different Fe/Si ratio (1.23 and 0.21) formed at some distance from each other (*cf.* L and H chondrites), the problem of keeping them separate becomes formidable at short distances.

Applications to other planets

Obviously, this model will have to pass a number of tests before it deserves to be taken seriously. However, if it does survive, it may become a useful planetological tool. Because of its conceptual rigidity, it can make very detailed predictions (e.g. abundances of 83 elements; Table 4) from a limited set of input data (six constraints in Table 1). All of these constraints can be determined by unmanned spacecraft, or by ground-based observations.

The U content can be determined by a heat flow measurement. The bulk Fe content can be estimated from the planet's density. The K/U ratio can be measured by a γ-ray spectrometer, as was recently done for Venus (Vinogradov et al., 1973). The FeO content can be measured remotely, as demonstrated by the Surveyor α-scattering instrument (Turkevich, 1973). The MnO and Tl contents have not been measured on past missions, but it seems that appropriate methods could be developed, if a strong scientific incentive were provided.

In this manner, it may be possible to construct a detailed cosmochemical profile of a differentiated planet, after a single visit by an unmanned spacecraft.

Acknowledgments—We are indebted to Dr. Jeffrey L. Warner for a computer printout of chemical data on lunar rocks. This work was supported in part by NASA Grant NGL 14-001-167. One of us (EA) thanks the John Simon Guggenheim Memorial Foundation for a fellowship.

REFERENCES

Ahrens L. H. (1970) The composition of stony meteorites—IX. Abundance trends of the refractory elements in chondrites, basaltic achondrites and Apollo 11 fines. Earth Planet. Sci. Lett. 10, 1–6.

de Albuquerque C. A. R., Muysson J. R., and Shaw D. M. (1972) Thallium in basalts and related rocks. Chem. Geol. 10, 41–58.

Anders E. (1964) Origin, age, and composition of meteorites. Space Sci. Rev. 3, 583–714.

Anders E. (1968) Chemical processes in the early solar system, as inferred from meteorites. Acc. Chem. Res. 1, 289–298.

Anders E. (1971) Meteorites and the early solar system. Ann. Rev. Astron. Astrophys. 9, 1–34.

Anders E., Ganapathy R., Krähenbühl U., and Morgan J. W. (1973) Meteoritic material on the Moon. The Moon 8, 3–24.

Anderson A. T. (1974) Chlorine and sulfur in and out of some basaltic magmas. Bull. Geol. Soc. Am. In press.

Anderson D. L. (1973) The composition and origin of the Moon. Earth Planet. Sci. Lett. 18, 301–316.

Blander M. and Katz J. L. (1967) Condensation of primordial dust. Geochim. Cosmochim. Acta 31, 1025–1034.

Burns R. G., Vaughan D. J., Abu-Eid R. M., Witner M., and Morawski A. (1973) Spectral evidence for Cr^{3+}, Ti^{3+}, and Fe^{2+} rather than Cr^{2+} and Fe^{3+} in lunar ferromagnesian silicates. Proc. Fourth Lunar Sci. Conf., Geochim. Cosmochim. Acta, Suppl. 4, Vol. 1, pp. 983–994. Pergamon.

Cameron A. G. W. (1973) A new table of abundances. In Explosive Nucleosynthesis (editors D. N. Schramm and W. D. Arnett), pp. 3–21. University of Texas Press, Austin.

Canalas R. A., Alexander E. C., Jr., and Manuel O. K. (1968) Terrestrial abundance of noble gases. J. Geophys. Res. 73, 3331–3334.

Frondel C. (1968) Crystal chemistry of scandium as a trace element in minerals. Z. Kristallogr. 127, 121–138.

Ganapathy R. and Anders E. (1974) Bulk compositions of the Moon and Earth, estimated from meteorites (abstract). In Lunar Science—V, pp. 254–256. The Lunar Science Institute, Houston.

Ganapathy R., Keays R. R., Laul J. C., and Anders E. (1970) Trace elements in Apollo 11 lunar rocks: Implications for meteorite influx and origin of Moon. *Proc. Apollo 11 Lunar Sci. Conf., Geochim. Cosmochim. Acta*, Suppl. 1, Vol. 2, pp. 1117–1142. Pergamon.

Garrels R. M. and Mackenzie F. T. (1971) *Evolution of Sedimentary Rocks*. Norton, New York. 397 pp.

Gast P. W. (1960) Limitations on the composition of the upper mantle. *J. Geophys. Res.* **65**, 1287–1297.

Gast P. W. (1968) Upper mantle chemistry and evolution of the Earth's crust. In *The History of the Earth's Crust* (editor R. A. Phinney), pp. 15–27. Princeton University Press, Princeton.

Gast P. W., Hubbard N. J., and Wiesmann H. (1970) Chemical composition and petrogenesis of basalts from Tranquillity Base. *Proc. Apollo 11 Lunar Sci. Conf., Geochim. Cosmochim. Acta*, Suppl. 1, Vol. 2, pp. 1143–1163. Pergamon.

Goettel K. A. (1972) Partitioning of potassium between silicates and sulphide melts: Experiments relevant to the Earth's core. *Phys. Earth Planet. Interiors* **6**, 161–166.

Grossman L. (1972) Condensation in the primitive solar nebula. *Geochim. Cosmochim. Acta* **36**, 597–619.

Grossman L. and Larimer J. W. (1974) Early chemical history of the solar system. *Rev. Geophys. Space Phys.* **12**, 71–101.

Haggerty S. E., Boyd F. R., Bell P. M., Finger L. W., and Bryan W. B. (1970) Opaque minerals and olivine in lavas and breccias from Mare Tranquillitatis. *Proc. Apollo 11 Lunar Sci. Conf., Geochim. Cosmochim. Acta*, Suppl. 1, Vol. 1, pp. 513–538. Pergamon.

Harris P. G. and Tozer D. C. (1967) Fractionation of iron in the solar system. *Nature* **215**, 1449–1451.

Hartmann W. K. (1968) Growth of asteroids and planetesimals by accretion. *Astrophys. J.* **152**, 337–342.

Hodges F. N. and Kushiro I. (1974) Apollo 17 petrology and experimental determination of differentiation sequences in model moon compositions. *Proc. Fifth Lunar Sci. Conf., Geochim. Cosmochim. Acta*, Suppl. 5. Vol. 1.

Hurley P. M. (1968) Absolute abundance and distribution of Rb, K, and Sr in the Earth. *Geochim. Cosmochim. Acta* **32**, 273–283 and 1025–1030.

Kay R., Hubbard N. J., and Gast P. W. (1970) Chemical characteristics and origin of oceanic ridge volcanic rocks. *J. Geophys. Res.* **75**, 1585–1613.

Kokubu N., Mayeda T., and Urey H. C. (1961) Deuterium content of minerals, rocks, and liquid inclusions from rocks. *Geochim. Cosmochim. Acta* **21**, 247–256.

Krähenbühl U., Ganapathy R., Morgan J. W., and Anders E. (1973) Volatile elements in Apollo 16 samples: Implications for highland volcanism and accretion history of the Moon. *Proc. Fourth Lunar Sci. Conf., Geochim. Cosmochim. Acta*, Suppl. 4, Vol. 2, pp. 1325–1348. Pergamon.

Langseth M. G., Keihm S. J., and Chute J. L., Jr. (1973) Heat flow experiment. In *Apollo 17 Preliminary Science Report*, NASA SP-330, 9-1 through 9-24.

Larimer J. W. (1967) Chemical fractionations in meteorites—I. Condensation of the elements. *Geochim. Cosmochim. Acta* **31**, 1215–1238.

Larimer J. W. (1971) Composition of the Earth: Chondritic or achondritic? *Geochim. Cosmochim. Acta* **35**, 769–786.

Larimer J. W. and Anders E. (1967) Chemical fractionations in meteorites—II. Abundance patterns and their interpretation. *Geochim. Cosmochim. Acta* **31**, 1239–1270.

Larimer J. W. and Anders E. (1970) Chemical fractionations in meteorites—III. Major element fractionations in chondrites. *Geochim. Cosmochim. Acta* **34**, 367–388.

Laul J. C., Keays R. R., Ganapathy R., Anders E., and Morgan J. W. (1972) Chemical fractionations in meteorites—V. Volatile and siderophile elements in achondrites and ocean ridge basalts. *Geochim. Cosmochim. Acta* **36**, 329–345.

Laul J. C., Ganapathy R., Anders E., and Morgan J. W. (1973) Chemical fractionations in meteorites— VI. Accretion temperatures of H-, LL-, and E-chondrites, from abundances of volatile trace elements. *Geochim. Cosmochim. Acta* **37**, 329–357.

Laul J. C. and Schmitt R. A. (1973) Chemical composition of Apollo 15, 16, and 17 samples *Proc. Fourth Lunar Sci. Conf., Geochim. Cosmochim. Acta*, Suppl. 4, Vol. 2, pp. 1349–1367. Pergamon.

Morgan J. W., Ganapathy R., Higuchi H., Krähenbühl U., and Anders E. (1974) Lunar basins: Tentative characterization of projectiles, from meteoritic elements in Apollo 17 boulders. *Proc. Fifth Lunar Sci. Conf., Geochim. Cosmochim. Acta*, Suppl. 5. This volume.

Muehlenbachs K. and Clayton R. N. (1972) Oxygen isotope studies of fresh and weathered submarine basalts. *Can. J. Earth Sci.* **9**, 172–184.

Murthy V. R. and Hall H. T. (1970) The chemical composition of the Earth's core: Possibility of sulphur in the core. *Phys. Earth Planet. Interiors* **2**, 276–282.

Nichiporuk W. and Moore C. B. (1974) Lithium, sodium, and potassium abundances in carbonaceous chondrites. *Geochim. Cosmochim. Acta* **38**, in press.

Öpik E. J. (1951) Collision probabilities with the planets and distribution of interplanetary matter. *Proc. Roy. Irish Acad.* **54[A]**, 165–199.

Öpik E. J. (1971) Cratering and the Moon's surface. In *Advan. Astron. & Astrophys.* (editor Z. Kopal), Vol. 8, pp. 107–337. Academic Press, New York.

Öpik E. J. (1972) Comments on lunar origin. *Irish Astron. J.* **10**, 190–238.

Parkin C. W., Dyal P., and Daily W. D. (1973) Iron abundance in the moon from magnetometer measurements. *Proc. Fourth Lunar Sci. Conf., Geochim. Cosmochim. Acta*, Suppl. 4, Vol. 3, pp. 2947–2961. Pergamon.

Reeves H. (1973) On the origin of the light elements. Report number SEP/SES/73-72R/HR/nnm. S. E. P. Saclay, France, 75 pp.

Reynolds R. T. and Summers A. L. (1969) Calculations on the composition of the terrestrial planets. *J. Geophys. Res.* **74**, 2494–2511.

Ringwood A. E. and Essene E. (1970) Petrogenesis of Apollo 11 basalts, internal constitution and origin of the Moon. *Proc. Apollo 11 Lunar Sci. Conf., Geochim. Cosmochim. Acta*, Suppl. 1, Vol. 1, pp. 769–799. Pergamon.

Rubey W. W. (1951) Geologic history of seawater: An attempt to state the problem. *Geol. Soc. Am. Bull.* **62**, 1111–1147.

Schmitt R. A., Smith R. H., and Goles G. G. (1965) Abundances of Na, Sc, Cr, Mn, Fe, Co, and Cu in 218 individual meteoritic chondrules via activation analysis, 1. *J. Geophys. Res.* **70**, 2419–2444.

Schmitt R. A., Goles G. G., Smith R. H., and Osborn T. W. (1972) Elemental abundances in stone meteorites. *Meteoritics* **7**, 131–213.

Singer S. F. and Bandermann L. W. (1970) Where was the Moon formed? *Science* **170**, 438–439.

Smith J. V. (1974) Origin of Moon by disintegrative capture with chemical differentiation followed by sequential accretion (abstract). In *Lunar Science—V*, pp. 718–720. The Lunar Science Institute, Houston.

Tatsumoto M. (1966) Genetic relations of oceanic basalts as indicated by lead isotopes. *Science* **153**, 1094–1101.

Taylor S. R. and Jakeš P. (1974) Geochemical zoning and early differentiation in the Moon. *Nature.* In press.

Tikhonova T. V. and Troitskii V. S. (1969) Effect of heat from within the Moon on its radio emission for the case of lunar properties which vary with depth. *Soviet Astron.* **13** [1], 120–128.

Tokstikhin I. N., Mamyrin B. A., and Khabarin L. V. (1972) Anomalous isotopic composition of He in several xenoliths. *Geokhimiya*, 629–631.

Tolstikhin I. N. (1974) Temperature condition of the earth accretion, as inferred from an abundance of the rare gas isotopes in the earth, in the atmosphere, and in meteorites. In *Proc. of US–USSR Conference on Cosmochemistry of the Moon and Planets*, Moscow, June 4–8, in press.

Turekian K. K. and Clark S. P., Jr. (1969) Inhomogeneous accumulation of the Earth from the primitive solar nebula. *Earth Planet. Sci. Lett.* **6**, 346–348.

Turkevich A. L. (1973) The chemical composition of the lunar surface. *Acc. Chem. Res.* **6**, 81–89.

Urey H. C. (1972) Evidence for objects of lunar mass in the early solar system. *The Moon* **4**, 383–389.

Urey H. C. and Craig H. (1953) The composition of the stone meteorites and the origin of the meteorites. *Geochim. Cosmochim. Acta* **4**, 36–82.

Vinogradov A. P., Surkov Yu. A., and Kirnozov F. F. (1973) The content of uranium, thorium, and potassium in the rocks of Venus as measured by Venera 8. *Icarus* **20**, 253–259.

Wänke H., Wlotzka F., Baddenhausen H., Balacescu A., Spettel B., Teschke F., Jagoutz E., Kruse H., Quijano-Rico M., and Rieder R. (1971) Apollo 12 samples: Chemical composition and its relation to sample locations and exposure ages, the two component origin of the various soil samples and studies on lunar metallic particles. *Proc. Second Lunar Sci. Conf., Geochim. Cosmochim. Acta*, Suppl. 2, Vol. 2, pp. 1187–1208. MIT Press.

Wänke H., Baddenhausen H., Dreibus G., Jagoutz E., Kruse H., Palme H., Spettel B., and Teschke F. (1973) Multielement analyses of Apollo 15, 16, and 17 samples and the bulk composition of the Moon. *Proc. Fourth Lunar Sci. Conf., Geochim. Cosmochim. Acta*, Suppl. 4, Vol. 2, pp. 1461–1481. Pergamon.

Wänke H., Palme H., Baddenhausen H., Dreibus G., Jagoutz E., Kruse H., Spettel P., Teschke F., and Thacker R. (1974) Chemistry of Apollo 16 and 17 samples: Bulk composition, late stage accumulation and early differentiation of the Moon. *Proc. Fifth Lunar Sci. Conf., Geochim. Cosmochim. Acta*, Suppl. 5. This volume.

Whipple F. L. (1972) On certain aerodynamic processes for asteroids and comets. In *Nobel Symposium 21, from Plasma to Planet* (editor A. Elvius), pp. 211–232. Almqvist and Wiksell, Stockholm.

Whipple F. L. (1973) On the growth of the Earth–Moon system (abstract). *Bull. Am. Astron. Soc.* 5, 292.

Wood J. A. (1962) Chondrules and the origin of the terrestrial planets. *Nature* 194, 127–130.

Zähringer J. (1962) Isotopie-Effekt und Häufigkeiten der Edelgase in Steinmeteoriten und auf der Erde. *Z. Naturforsch.* 17a, 460–471.

Proceedings of the Fifth Lunar Conference
(Supplement 5, Geochimica et Cosmochimica Acta)
Vol. 2 pp. 1207–1212 (1974)
Printed in the United States of America

Oxygen isotopic constraints on the composition of the moon

LAWRENCE GROSSMAN

Department of the Geophysical Sciences

ROBERT N. CLAYTON

Enrico Fermi Institute and Departments of Chemistry and Geophysical Sciences

TOSHIKO K. MAYEDA

Enrico Fermi Institute, The University of Chicago, Chicago, Illinois 60637

Abstract—The mean oxygen isotopic composition of 5 Apollo 17 soils, one Apollo 17 breccia and one Apollo 12 soil is $\delta O^{18} = 5.63 \pm 0.05$ and $\delta O^{17} = 3.8 \pm 0.2\%o$. These values are within several tenths of a part permil of the composition of a large fraction of the lunar interior. High-temperature condensate aggregates from Allende and other C2 and C3 chondrites are vastly enriched in O^{16} compared to this composition. The moon cannot be a mixture of ordinary chondrites and Allende inclusions, nor can it be derived from such a mixture by chemical fractionation processes. The moon's isotopic composition *is* consistent with a mixture of high- and low-temperature condensates but the refractory fraction would have to be free of the O^{16}-rich component so prevalent in the meteoritic aggregates, a fact which makes such models less attractive than they once seemed.

INTRODUCTION

THE HIGH CONCENTRATIONS of refractory elements and low abundances of volatiles in the lunar rocks have prompted some investigators to propose that part or all of the moon accreted from materials enriched, relative to the solar system abundances, in high-temperature condensates such as the Ca–Al-rich inclusions found in the Allende meteorite. See, for example, Gast *et al.* (1970), Anderson (1972, 1973), Krähenbühl *et al.* (1973), Wänke *et al.* (1973, 1974), Ganapathy and Anders (1974). These models have been criticized by Clayton *et al.* (1973b) and Grossman and Larimer (1974) because the oxygen isotopic compositions of lunar rocks differ enormously from those of Allende inclusions. The purpose of this paper is to compare the δO^{18} and δO^{17} of lunar, meteoritic, and terrestrial materials and to discuss the origin of the moon in the light of these data.

ANALYTICAL METHODS

Analytical procedures for lunar samples were the same as those used in this laboratory for previous Apollo missions (Onuma *et al.*, 1970). O^{17}/O^{16} variations were measured as outlined by Clayton *et al.* (1973a). Oxygen isotopic compositions are reported in the δ-terminology, as permil (‰) deviations from the SMOW standard. Uncertainties in δO^{18} and δO^{17} are estimated to be ± 0.1 and ± 1 permil, respectively.

RESULTS

Table 1 shows the oxygen isotopic compositions of five Apollo 17 soils, one Apollo 17 breccia, and one Apollo 12 soil. The mean of these analyses is $\delta O^{18} = 5.63 \pm 0.05$ and $\delta O^{17} = 3.8 \pm 0.2$.

Table 1. Oxygen isotopic compositions of lunar samples.

Sample No.	Description	δO^{18} (SMOW)	δO^{17} (SMOW)
70019,10	Breccia	5.53	3.7
73221,8	Soil	5.64	3.4
73281,7	Soil	5.66	4.3
76501,14	Soil	5.68	4.1
79221,25	Soil	5.61	3.3
79261,17	Soil	5.64	3.7
12033,55	Soil	5.62	4.0

DISCUSSION

Oxygen isotopic composition of the moon. The oxygen isotopic compositions of the common rock-forming minerals in lunar igneous rocks vary little from one sampled locality to another (Onuma *et al.*, 1970; Taylor and Epstein, 1970, 1973; Clayton *et al.*, 1973b). The δO^{18} of pyroxene is always between 5.3 and 5.8‰. Olivines are between 4.9 and 5.1‰ and plagioclases between 5.6 and 6.4‰. These minerals come from igneous rocks which are crystallization products of partial melts of source regions in the lunar interior. The small range of δO^{18} implies that the source regions of the rocks from all the sites are very similar in oxygen isotopic composition. Because isotope fractionations are very small at basalt melting temperatures, the δO^{18} of a mineral in a surface igneous rock is within several tenths of a part permil of the δO^{18} of the same mineral in the source region. Assuming that the source regions are composed largely of a mixture of olivine, pyroxene, and plagioclase, a large fraction of the lunar interior has a δO^{18} of $\sim 5.5 \pm 0.2$‰. Furthermore, the presence of large quantities of melilite (Anderson, 1973) in the lunar mantle would not alter this conclusion since unpublished data on terrestrial basalts show that coexisting melilite and pyroxene have the same isotopic composition at these temperatures.

A chemical isotope effect is known to be associated with the surfaces of lunar soil particles (Epstein and Taylor, 1971). This causes bulk soil samples to be enriched in δO^{18} by 0.3 to 0.5‰ (Taylor and Epstein, 1973) and in δO^{17} by 0.15 to 0.25‰ relative to igneous rocks. When the compositions of the soils in Table 1 are corrected for this effect, their δO^{18} remains very close to that of the lunar interior inferred above. The δO^{17} values for the soils must also be within a few tenths of a part permil of the δO^{17} of the mantle because the isotopic compositions of the igneous rocks are related to those of the source regions through chemical effects.

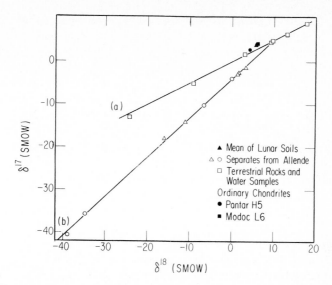

Fig. 1. Oxygen isotopic compositions of lunar samples, terrestrial rock and water samples, ordinary chondrites and separated inclusions from the Allende meteorite. Note that the lunar composition does not lie on any of the mixing lines which can be drawn between Allende inclusions and H-group chondrites.

Oxygen isotopic compositions of solar system materials. Figure 1 is a plot of δO^{17} versus δO^{18} for a variety of samples from different bodies in the solar system. The data for terrestrial rocks and water samples plot along line *a* which has a slope of $+\frac{1}{2}$. This is expected for all samples derived by chemical processes from the same batch of nucleosynthetic material. The mean composition of the lunar samples in Table 1 is seen to lie close to line *a*, but slightly above it. The refractory Ca–Al-rich aggregates in Allende, however, fall along line *b* which has a slope of $+1$. These objects are vastly lower in δO^{17} and δO^{18} than the moon. In particular, the lunar point does not fall on any of the family of mixing lines which can be drawn between the H-group ordinary chondrites and any of the Allende inclusions. The L-group chondrite data point is indistinguishable from the lunar point. The moon cannot be a mixture of Allende inclusions and ordinary chondrites, a model suggested by Wänke *et al.* (1973). Nor can the moon be derived by chemical fractionation processes from such a mixture since lines joining the lunar point to points on the mixing lines do not have slopes of $+\frac{1}{2}$.

There is a suggestion in our data that the lunar point is displaced slightly from the terrestrial chemical line *a*. This may indicate that the moon is depleted in the O^{16} component by about 2‰ relative to the earth. Although this is an extremely tenuous conclusion, it may have very important cosmochemical consequences if supported by future measurements.

Oxygen isotopic constraints on the origin of the moon. The fact that line *b* has a slope of unity indicates that nuclear processes, rather than chemical ones, dominated the establishment of the isotopic compositions of the Allende inclusions. Thus, these compositions are *not* the natural consequence of chemical isotope effects accompanying the condensation of high-temperature minerals from the solar nebula. Clayton *et al.* (1973a) suggested that line *b* is a mixing line between "ordinary" solar system oxygen, having an isotopic composition at the intersection of lines *a* and *b*, and pure, or nearly pure, O^{16}. They proposed that the O^{16} entered the solar nebula bound in preexisting interstellar grains which escaped vaporization and isotopic homogenization during the early high-temperature stage of the nebula. These grains were then incorporated by the minerals of the Allende inclusions as they condensed from the cooling nebula. Alternatively, O^{18} and O^{17} may have been consumed in nuclear reactions during an early stage of the sun, with the resulting O^{16}-enriched gas ejected into the nebula where it was implanted into the surfaces of condensate grains. Whatever the origin of the O^{16}-rich component, it is clearly a contaminant in the sense that its presence in the inclusions masks the isotopic compositions which they would have inherited purely from the chemical isotope effects of condensation.

By projecting their compositions alone line *b* onto line *a*, one can correct the oxygen isotopic compositions of the inclusions to the same O^{16} content as the moon. The average inclusion would then plot at the intersection of the two lines, $\delta O^{18} = 9.6‰$.

If this inclusion is assumed to have condensed at 1400°K (Grossman, 1972), then the mean equilibration temperature of the materials which accreted to form the moon can be inferred from the difference in the δO^{18} between the intersection and the lunar point. Using the principles behind the cosmothermometer of Onuma *et al.* (1972), and a nebular δO^{18} of $\sim 14‰$ (Onuma *et al.*, 1974), this temperature is $\sim 880°K$.

The oxygen isotopic composition of the moon is thus consistent with a mixture of high-temperature condensates with materials having lower equilibration temperatures. *The high-temperature condensates incorporated by the moon, however, must have had vastly different oxygen isotopic compositions than the Allende inclusions.* In fact, every chondritic high-temperature condensate so far measured (Clayton *et al.*, 1973a) is vastly enriched in O^{16} compared to the moon. The lunar equilibration temperature derived above is considerably below the condensation temperatures of nickel–iron and magnesium silicates in the pressure range 10^{-3}–10^{-5} atm. These phases could have made up a significant fraction of the lower temperature component.

Another important difference between the lunar rocks and the Allende inclusions is their radically different refractory siderophile element contents. As predicted by condensation calculations (Grossman, 1973), the Allende inclusions are enriched in refractory platinum metals, for example, by factors of 12–24 relative to the C1 chondrites (Grossman, 1973; Wänke *et al.*, 1973). The lunar rocks are depleted in these elements by factors of 10^2–10^4 compared to C1 chondrites (Ganapathy *et al.*, 1970), suggesting that a very efficient, post-accretional removal process was active. The possibility that nickel–iron is a significant fraction of the lower temperature component suggests that this process could have been the gravity separation of a siderophile element-rich Fe or Fe–S melt.

The validity of any lunar model in which the moon accretes from substantial quantities of high-temperature condensates rests critically on the assumption that these materials can form somewhere in the nebula in a way which does not lead to the incorporation of the O^{16}-rich component. This component has been found in every chondritic high-temperature condensate measured, including the Ca–Al-rich inclusions in Allende (Clayton *et al.*, 1973a).

If the source of the O^{16}-rich component was interstellar grains, its relative absence from the lunar rocks implies that the grains were not present in the part of the nebula where protolunar material condensed. This may mean that a thorough mechanism was available for the separation of preexisting grains from that part of the nebula before condensation began. Alternatively, it may mean that all preexisting grains were evaporated and their unique isotopic compositions homogenized with the rest of the nebula. This would imply that the moon formed from materials which condensed in a region of the nebula which had a higher initial temperature than the condensation site of the Allende inclusions. This, in turn, suggests that the moon condensed closer to the center of the nebula than the meteoritic condensates.

If the O^{16}-rich component was produced in the sun and then implanted into the condensate grains, the relative absence of it from the lunar rocks implies that protolunar material had a smaller mean surface/volume ratio than the Allende condensates at that stage of nebular evolution.

In summary, it is possible that the moon accreted from a super-chondritic proportion of high-temperature condensate, but such models seem less attractive than previously because this component with the required isotopic composition has never been found in meteorites.

Acknowledgments—This work was supported by the Research Corporation, the Louis Block Fund of the University of Chicago, and the National Aeronautics and Space Administration through grants NGL 14-001-169 and NGR 14-001-249.

REFERENCES

Anderson D. L. (1972) The origin of the moon. *Nature* **239**, 263–265.
Anderson D. L. (1973) The composition and origin of the moon. *Earth Planet. Sci. Lett.* **18**, 301–316.
Clayton R. N., Grossman L., and Mayeda T. K. (1973a) A component of primitive nuclear composition in carbonaceous meteorites. *Science* **182**, 485–488.
Clayton R. N., Hurd J. M., and Mayeda T. K. (1973b) Oxygen isotopic compositions of Apollo 15, 16, and 17 samples, and their bearing on lunar origin and petrogenesis. *Proc. Fourth Lunar Sci. Conf., Geochim. Cosmochim. Acta*, Suppl. 4, Vol. 2, pp. 1535–1542. Pergamon.
Epstein S. and Taylor H. P., Jr. (1971) O^{18}/O^{16}, Si^{30}/Si^{28}, D/H, and C^{13}/C^{12} ratios in lunar samples. *Proc. Second Lunar Sci. Conf., Geochim. Cosmochim. Acta*, Suppl. 2, Vol. 2, pp. 1421–1441. MIT Press.
Ganapathy R. and Anders E. (1974) Bulk compositions of the moon and earth, estimated from meteorites (abstract). In *Lunar Science—V*, pp. 254–256. The Lunar Science Institute, Houston.
Ganapathy R., Keays R. R., Laul J. C., and Anders E. (1970) Trace elements in Apollo 11 lunar rocks: Implications for meteorite influx and origin of moon. *Proc. Apollo 11 Lunar Sci. Conf., Geochim. Cosmochim. Acta*, Suppl. 1, Vol. 2, pp. 1117–1142. Pergamon.
Gast P. W., Hubbard N. J., and Wiesmann H. (1970) Chemical composition and petrogenesis of basalts from Tranquillity Base. *Proc. Apollo 11 Lunar Sci. Conf., Geochim. Cosmochim. Acta*, Suppl. 1, Vol. 2, pp. 1143–1163, Pergamon.

Grossman L. (1972) Condensation in the primitive solar nebula. *Geochim. Cosmochim. Acta* **36**, 597–619.

Grossman L. (1973) Refractory trace elements in Ca–Al-rich inclusions in the Allende meteorite. *Geochim. Cosmochim. Acta* **37**, 1119–1140.

Grossman L. and Larimer J. W. (1974) Early chemical history of the solar system. *Revs. Geophys. Space Phys.* **12**, 71–101.

Krähenbühl U., Ganapathy R., Morgan J. W., and Anders E. (1973) Volatile elements in Apollo 16 samples: Implications for highland volcanism and accretion history of the moon. *Proc. Fourth Lunar Sci. Conf., Geochim. Cosmochim. Acta*, Suppl. 4, Vol. 2, pp. 1325–1348. Pergamon.

Onuma N., Clayton R. N., and Mayeda T. K. (1970) Apollo 11 rocks: Oxygen isotope fractionation between minerals, and an estimate of the temperature of formation. *Proc. Apollo 11 Lunar Sci. Conf., Geochim. Cosmochim. Acta*, Suppl. 1, Vol. 2, pp. 1429–1434. Pergamon.

Onuma N., Clayton R. N., and Mayeda T. K. (1972) Oxygen isotope cosmothermometer. *Geochim. Cosmochim. Acta* **36**, 169–188.

Onuma N., Clayton R. N., and Mayeda T. K. (1974) Oxygen isotope cosmothermometer revisited. *Geochim. Cosmochim. Acta* **38**, 189–191.

Taylor H. P., Jr. and Epstein S. (1970) O^{18}/O^{16} ratios of Apollo 11 lunar rocks and minerals. *Proc. Apollo 11 Lunar Sci. Conf., Geochim. Cosmochim. Acta*, Suppl. 1, Vol. 2, pp. 1613–1626. Pergamon.

Taylor H. P., Jr. and Epstein S. (1973) O^{18}/O^{16} and Si^{30}/Si^{28} studies of some Apollo 15, 16, and 17 samples. *Proc. Fourth Lunar Sci. Conf., Geochim. Cosmochim. Acta*, Suppl. 4, Vol. 2, pp. 1657–1679. Pergamon.

Wänke H., Baddenhausen H., Dreibus G., Jagoutz E., Kruse H., Palme H., Spettel B., and Teschke F. (1973) Multielement analyses of Apollo 15, 16, and 17 samples and the bulk composition of the moon. *Proc. Fourth Lunar Sci. Conf., Geochim. Cosmochim. Acta*, Suppl. 4, Vol. 2, pp. 1461–1481. Pergamon.

Wänke H., Palme H., Baddenhausen H., Dreibus G., Jagoutz E., Kruse H., Spettel B., and Teschke F. (1974) Composition of the moon and major lunar differentiation processes (abstract). In *Lunar Science—V*, pp. 820–822. The Lunar Science Institute, Houston.

Proceedings of the Fifth Lunar Conference
(Supplement 5, Geochimica et Cosmochimica Acta)
Vol. 2 pp. 1213–1225 (1974)
Printed in the United States of America

Chemical evidence for the origin of 76535 as a cumulate

L. A. Haskin,[1] C.-Y. Shih,[2] B. M. Bansal,[3]

J. M. Rhodes,[3] H. Wiesmann,[3] and L. E. Nyquist[1]

[1]NASA Johnson Space Center, Houston, Texas 77058

[2]National Research Council, NASA Johnson Space Center, Houston, Texas 77058

[3]Lockheed Electronics Company, Houston, Texas 77058

Abstract—Lunar sample 76535 is a coarse-grained troctolitic granulite. It is characterized by low REE concentrations and a positive Eu anomaly. Its original petrographic character has been disturbed by metamorphic reequilibration. Its chemical characteristics are those of an olivine–plagioclase cumulate. The amount of trapped parent liquid in the rock is estimated to be in the range of 8–16%. The REE concentrations of the parent liquid, if 16%, range from 13 times the chondritic value for Lu to 27 times for La. The parent liquid had no appreciable Eu anomaly.

Introduction

THE NATURE OF the earliest igneous rocks that were produced in the lunar crust and the parent liquids of those rocks remains obscure. Simple models have been postulated for the early, extensive melting of the outer few hundred kilometers of the lunar surface, with crystallization of the molten material resulting in anorthositic rocks, which now constitute the bulk of the lunar highlands (e.g. Wood *et al.*, 1970; Keil *et al.*, 1970). The presence of cataclastic anorthosites such as 15415 or 60015 in the lunar samples is in qualitative agreement with this assumption, as is the generally plagioclase-rich nature of the highland surface as indicated by the orbital chemical mapping (Adler *et al.*, 1972; Metzger *et al.*, 1973). In the returned lunar samples, however, rocks consisting almost exclusively of plagioclase are not quantitatively as important as may have initially been anticipated. It is expected that any large segregations of plagioclase would have been broken up and mixed with other materials during the heavy bombardment by meteoroids that presumably occurred during crustal formation (e.g. Warner *et al.*, 1974). Most highland rocks have lost the evidence for their original petrographic textures. Surviving pieces of almost pure plagioclase, olivine, or mineral mixtures that might have formed by crystal accumulation may be samples of large segregations of crystals at considerable depth that were excavated relatively late so that they did not go through enough cycles of meteoroid bombardment to destroy their characteristics. Alternatively, those materials might be relics of smaller scale events of crystal fractionation similar to those that produced differentiated terrestrial intrusions. These would have had to occur late enough in lunar crustal formation that impacts did not demolish them.

In this paper we suggest that rock 76535 may be a relict of an early event of crystal separation from a parent liquid. It has characteristics similar to those of rocks that can be sampled in terrestrial layered intrusions. The arguments used here for an origin by crystal separation are based on the chemical composition of the rock. Strong additional support for this hypothesis can be inferred from the present metamorphic texture and mineralogy of the rock.

Description of 76535

Rock 76535 has been extensively examined by Gooley *et al.* (1974), who classify it as a troctolitic granulite. The rock consists mainly of large aggregates of plagioclase (up to 1 cm) and olivine (up to 0.8 cm). Based on studies of two thin sections plus point counting of a mosaic photograph, the following model composition was obtained: plagioclase 58% (An_{96}); olivine 37% (Fo_{88}); bronzite 4% (En_{86}); and approximately 1% accessory minerals. Study of a single thin section by Brown *et al.* (1974) produced similar results (plagioclase 56%, An_{95}; olivine 38%, Fo_{87}; and two pyroxenes, 5%, the more abundant being bronzite of composition En_{87}).

Gooley *et al.* (1974) argue on the basis of the overall petrographic character, that the specimen 76535 was originally a cumulate of plagioclase and olivine that cooled very slowly, with considerable recrystallization and internal equilibration, at a depth of 10–30 km beneath the lunar surface. Most of the evidence for the depth of origin stems from the presence of symplectic intergrowth between the crystals of olivine and plagioclase. For details of these arguments, the reader is referred to the paper by Gooley *et al.*

Bogard *et al.* (1974) report an exposure age of 200×10^6 yr for the rock. Crozaz *et al.* (1974) found a Kr/Kr exposure age of $195 \pm 16 \times 10^6$ yr and a track age of not more than $2 \pm 0.3 \times 10^6$ yr, which indicates significant complexity in the near-surface history of the rock. Studies by Morgan *et al.* (1974) indicate very low concentrations of volatile and siderophile elements, which is consistent with the absence of any significant meteoritic component and with the suggested origin and metamorphism of the rock at depth. Jovanovic and Reed (1974) report a high lunar ratio of F to Cl, analogous to the high ratios of F to Cl in terrestrial ultramafic rocks as compared to terrestrial basalts.

No firm crystallization or metamorphic age has been determined for 76535. Bogard *et al.* (1974) estimate a "corrected" K/Ar age for the rock of $4.34 \pm .08 \times 10^9$ yr. Data for isotopes for Rb and Sr did not yield an isochron but could be interpreted in a manner consistent with the K/Ar age. The isotopic ratio of $^{87}Sr/^{86}Sr$ in the plagioclase, corrected back to the inferred K/Ar age of 4.34 AE, shows an initial $^{87}Sr/^{86}Sr$ ratio equivalent to BABI. Tera *et al.* (1974) were unable to obtain an unambiguous age for the rock from U–Th–Pb systematics. They concluded, however, that there was no evidence for an origin by impact melting and that the rock was either intruded at a depth where it remained hot enough to permit the continuous isotopic equilibration of Pb until the time of its excavation or it was a near-surface rock that formed 3.9 AE ago.

In conclusion, published data on 76535 are consistent with, and offer some support for, the origin of the rock at depth. Formation of the rock was accompanied by slow cooling or was followed by reheating with slow cooling. This caused extensive equilibration among the constituent minerals. The rock was then exhumed from depth in a large but relatively late event so that it was not significantly degraded by subsequent impact processes. In this paper we offer chemical evidence for the origin of 76535 by partial separation of olivine and plagioclase from their parent liquid. This chemical evidence is independent of most of the above observations but is consistent with them.

EXPERIMENTAL

In order to provide the most nearly representative sample obtainable for this very coarse-grained lunar rock, a mini-consortium was formed and two 2-gram samples, 76535,21 and 76535,22, were pooled for analysis. Much of the work on the resulting sample is reported elsewhere (Bogard *et al.*, 1974). Figure 1 is a flow chart showing the separations and uses of the material in the composite sample. These will be discussed here only in so far as they pertain to the major and trace element analyses.

Both subsamples (76535,21 and 76535,22) were derived from parent 76535,19, which was in turn derived from 76535,9, a conical end piece of the original rock. It was intended that 76535,21 and 76535,22 be interior samples. However, the sample crumbled completely during curatorial processing, so there is a high probability that both subsamples contained some exterior surface, as well as some sawn surface (P. Butler, personal communication). The sample as received for analysis consisted of numerous chunks, several millimeters each in size.

The entire 4 g were recombined and crushed to ~1 mm size in an agate mortar. A 1.0 g aliquot was then further ground to fine grain size (~50 μ) in a boron carbide mortar. This sample became the parent for all subsequent "whole-rock" analyses. It was first analyzed nondestructively by X-ray fluorescence spectrometry for the trace elements Sr, Rb, Y, Zr, Cr, Nb, and Zn (Rhodes *et al.*, 1974). Aliquants of this sample were analyzed for major elements (XRF), REE, Li, K, Rb, Sr, Ba, U, Zr, and Hf (isotope dilution mass spectrometry), $^{87}Sr/^{86}Sr$, and noble gases (Bogard *et al.*, 1974). Na was determined by atomic absorption spectrometry on a 20 mg aliquant.

Another 0.7 g portion of the coarsely crushed fraction was taken for mineral separation. The fine material (<147 μ) was removed by sieving through a nylon sieve. The remaining material was separated into "magnetic" and "nonmagnetic" fractions with a Frantz magnetic separator. Mineral separations were carried out on a laminar flow bench. The stainless steel trough of the Frantz was covered with aluminum foil; the sample hopper was a cone of glassine powder paper (see also Nyquist *et al.*, 1974). A portion of the nonmagnetic fraction was further purified by handpicking and analyzed for the REE and other trace elements and for $^{87}Sr/^{86}Sr$. This plagioclase fraction was milky white in appearance, presumably due to numerous tiny metal-rich inclusions (Gooley *et al.*, 1974). No attempt was made to eliminate these inclusions.

The "magnetic" fraction was also further purified by handpicking as illustrated in Fig. 1 (OL 2). First, all grains visibly containing plagioclase intergrowths were removed. Second, all grains containing significant amounts of minor phases were removed. The latter were identified as dark brownish grains, presumably consisting of symplectites and mosaic assemblages (Gooley *et al.*, 1974). As a practical necessity, some grains containing vestiges of symplectite were retained in the OL 2 fraction. No attempt to discriminate between olivine and bronzite was made at this stage. It was subsequently learned that our "olivine" fractions commonly contained 5–10% bronzite. The plagioclase separate was estimated to contain ≥99% plagioclase, the Ol 2 to contain ≥99% (olivine + bronzite).

OL 2 was analyzed separately for Rb, Sr, $^{87}Sr/^{86}Sr$, and the REE. This was necessitated by the Rb and Sr blank contributed by the REE spikes. Major and minor elements were determined, by X-ray fluorescence spectrometry, using a glass disc prepared by fusing a 280 mg aliquant of the sample with a lanthanum-bearing lithium borate fusion mixture (Norrish and Hutton, 1969). Sodium (Na_2O) was analyzed by atomic absorption on a separate 20 mg portion of the sample. The accuracy of this

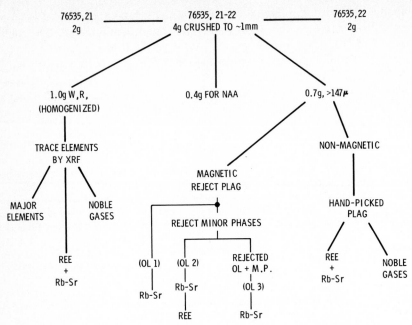

Fig. 1. Flow chart for treatment and separations on 76535,21-22.

Table 1. Blank contributions in percent of measured value for the analyses for trace elements
(76535,21–22).

	Whole Rock	Plagioclase	Olivine
La	0.3	0.2	—
Ce	1.3	0.8	15
Nd	1.1	0.7	10
Sm	0.5	0.3	2.5
Eu	0	0	0
Gd	—	—	—
Dy	0	0	0
Er	0.5	1.5	0.4
Yb	1.6	6.8	1.6
Lu	1.8	10	1.7
Li	14	14	—
K	7.2	6.7	—
Rb	8.8	6.8	22
Sr	0.2	0.2	2
Ba	2.6	1.5	17
U	10	10	—
Zr	10	3	—
Hf	16	4	—
Th	7.5	30	—

technique has been evaluated by Norrish and Hutton (1969), and is better than ±1% of the amount present for the major elements.

All of the trace elements reported here were determined by isotopic dilution mass spectrometry. Precision and accuracy are better than 2% and usually better than 1% (Gast *et al.*, 1970). About 50 mg of sample were used for each whole-rock and plagioclase analysis. However, due to the relatively low trace element content in olivine, 100 mg of olivine (OL 2) were used. Furthermore, because of the limited amount of pure olivine sample available and the significant blanks of Rb and Sr contributed by Ba and REE spike solutions, Ba and REE were spiked and determined on OL 2 after eluting from the Rb–Sr column. The Ba and REE yield of the columns was checked by a similar analysis of a separate plagioclase sample and found to be 99–100%. Blank contributions are generally insignificant (<1–2%) for REE, Ba, and Sr except for the light REE in olivine and heavy REE in plagioclase, as shown in Table 1.

RESULTS AND DISCUSSION

Analytical results obtained for the whole rock and for the separates of olivine and plagioclase are given in Tables 2 and 3. Data for the major elements obtained in our sample (Table 2, column 1) are contrasted with those of Gooley *et al.* (1974), which are based on microprobe analyses of the individual, homogeneous minerals, and the mode as determined by point counting of a photomicrograph mosaic of the surface of the rock. The results are similar, but different outside of analytical uncertainty for the analysis done in this work. This may indicate that our sample, 76535,21–22, was richer in plagioclase and poorer in mafic minerals than the mosaic would indicate. Calculations given below for modeling of the history of the rock are based on the analytical data for 76535,21–22, even though the modal

Table 2. Major element compositions of 76535.

	1	2
SiO_2	42.9	43.2
TiO_2	0.05	0.03
Al_2O_3	20.7	19.1
FeO	4.99	5.4
MnO	0.07	0.04
MgO	19.1	21.0
CaO	11.4	10.1
Na_2O	0.2	0.2
K_2O	0.03	0.05
P_2O_5	0.03	—
S	<.01	—
Cr_2O_3	0.11	0.3

(1) XRF analysis of 76535,21-22 (Na_2O by AA). This work.

(2) Gooley *et al.* (1974), composition of 76535 from mineral compositions plus modal analyses.

Table 3. Trace element concentrations (ppm) in
76535,21-22.

	Whole Rock	Plagioclase	Olivine
La	1.51	2.5	—
Ce	3.8	5.5	0.101
Nd	2.3	3.2	0.125
Sm	0.61	0.81	0.061
Eu	0.73	1.26	0.0044
Gd	0.73	0.84	0.148
Dy	0.80	0.57	0.34
Er	0.53	0.23	0.37
Yb	0.56	0.159	0.54
Lu	0.079	0.012	0.100
Li	3.0	3.2	—
K	233	390	—
Rb	0.24	0.41	0.010
Sr	114	198	0.82
Ba	33	56	1.15
U	0.056	0.055	—
Zr	23.6	108	—
Hf	0.52	2.54	—

composition assumed is that given by Gooley *et al.* (1974). The conclusions are insensitive to these assumptions over the range of variation observed in the modal and chemical composition.

The concentrations of the REE, K, Rb, Ba, and Sr in the three materials analyzed are compared with the average values for chondrites (Hubbard and Gast, 1971) in Fig. 2.

Qualitative arguments for a cumulate origin

The overall troctolitic mineralogy and composition of 76535 themselves suggest a cumulate origin for that rock. The plagioclase is highly clacic (An_{95-96}) and the olivine highly magnesian (Fo_{87-88}) (Gooley *et al.*, 1974; Brown *et al.*, 1974). A liquid with the composition of the whole rock could only correspond to a very extensive partial melt of its source region to have such strongly calcic and magnesian character. The predominance of plagioclase and olivine in the rock, together with an absence of spinel, suggests that 76535 may be a cumulate rock derived from a parental magma that simultaneously crystallized both plagioclase and olivine. Such a liquid would lie somewhere on the olivine–plagioclase cotectic of the silica–olivine–plagioclase phase diagram (Walker *et al.*, 1973). The position of the whole-rock composition on this diagram, close to the olivine–anorthite side, and with relative proportions of olivine to plagioclase of 37:58 modal percent, is entirely consistent with this suggestion. The proportion of bronzite in the rock is relatively low, suggesting its origin along with that of most of the accessory minerals by late-stage crystallization of trapped parent liquid.

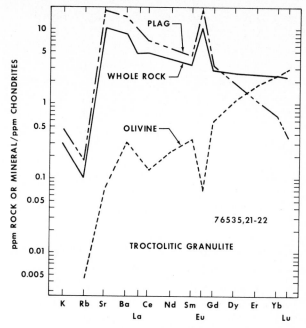

Fig. 2. Comparison diagram with chondrites for 76535,21-22 whole rock, olivine, and plagioclase. The olivine separate analyzed contained some ≥5% bronzite.

Concentrations of the large ion lithophile elements are relatively low for a lunar rock (for range for REE, see Haskin *et al.* [1973] or Hubbard *et al.* [1973]). The concentration of Ni (44 ppm) (Morgan *et al.*, 1974) is rather high for a sample showing no meteoritic component and suggests withdrawal of that element from the parent liquid into the olivine (for example, Henderson and Dale, 1969/1970). The whole rock has a positive Eu anomaly. This feature is an expected result of accumulation of plagioclase from a liquid that does not itself have a large relative depletion in Eu as compared with the other rare earths.

The positive Eu anomaly could, in principle, result from selective exclusion of Eu as a +2 ion from earlier crystallized mafic minerals such as olivine or pyroxene. It is difficult, however, to produce a liquid with a significantly positive Eu anomaly either by partial melting of an olivine/orthopyroxene source or by even extensive crystallization of olivine or orthopyroxene from a liquid. Under reducing lunar conditions, it may be possible to produce a significant positive Eu anomaly in a liquid relative to its source by small degrees of partial melting of a clinopyroxene-rich source or by extensive crystallization of clinopyroxene from a liquid. The presence of the required amount of clinopyroxene is unlikely in view of the modal composition of the rock and general considerations of the source regions for highland materials (e.g. Walker *et al.*, 1973). Finally, those rocks such as KREEP or mare basalts, whose major element compositions suggest that they formed by partial melting, are all characterized by significant negative Eu anomalies, and no evidence has yet been presented for the presence of lunar liquids with large positive Eu anomalies. Thus, we ascribe the large positive Eu anomaly of 76535 to the presence of cumulus plagioclase.

The coarse grain size of 76535 is consistent with the accumulation of olivine and plagioclase but could, in principle, also be an artifact of the slow cooling and long equilibration time. The Sr isotopic data are inconsistent with complete equilibration of the plagioclase with the rest of the rock during its slow cooling, and this is consistent with an original large grain size for the rock.

The parent liquid

A method commonly used to estimate the trace element concentrations of the parent liquid for a rock is to analyze a mineral from the rock for the trace elements of interest, then to divide the concentrations of the various elements by their respective distribution coefficients for that mineral (e.g. Hubbard *et al.*, 1971). Such an approach can, at best, provide only an upper limit to the concentrations in the parent liquid of those elements that are preferentially retained by the parent liquid as it crystallizes. Because of possible effects of crystallization as a closed system, this procedure can result in a gross overestimate (Helmke *et al.*, 1972). Finally, there is an additional, large uncertainty in the upper limit because of the wide range of reported values for distribution coefficients that can reasonably be used for the calculation (e.g. Philpotts and Schnetzler, 1970; Schnetzler and Philpotts, 1970).

Using the published range of values for the crystal/liquid distribution coefficients of plagioclase and olivine, we have estimated the maximum concentrations of REE, K, Rb, Sr, and Ba in the parent liquid by the above procedure. The result, labeled "extreme case," is shown in Fig. 3. The estimate yields REE concentrations, including a significant Eu depletion, in the range found for Apollo 17 noritic breccias (KREEP) and intermediate to those for Apollo 16 KREEP and proposed VHA basalts (e.g. Hubbard *et al.*, 1973, 1974). Thus, the "extreme" liquid does have REE concentrations similar to those actually observed in other materials analyzed, materials whose overall compositional character suggests that they are frozen products of equilibrium partial melting. In this sense, 76535 and Apollo 17 noritic breccias seem to form a complementary solid–liquid equilibrium pair. If indeed they are, additional evidence is provided for the systematic loss of Eu relative to the other REE from the source materials for the lunar crust (e.g. Nguyen *et al.*, 1973).

Such liquids are not reasonable parents, however, for other lunar cumulates such as the cataclastic anorthosites (e.g. Hubbard *et al.*, 1971; Helmke *et al.*, 1972). Furthermore, the overall mineralogical and chemical composition of 76535 is not consistent with the above, extreme interpretation. Paster *et al.* (1974) discuss in detail the modeling of cumulate rocks and the relative roles of cumulus, adcumulus, and orthocumulus (i.e. frozen trapped parent liquid) materials. The above, extreme case corresponds to the production of the entire rock or, at least, all of the plagioclase and olivine, by cumulus plus adcumulus growth. That is, either the rock has no trapped, frozen parent liquid, or the bronzite and accessory

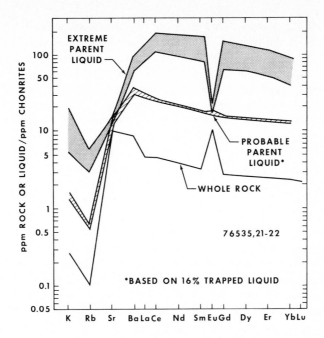

Fig. 3. Comparison diagram showing extreme and probable estimates of trace element concentrations in the liquid parent of 76535.

minerals represent the trapped liquid and there was no effect of the crystallization of that trapped liquid on the olivine or plagioclase. Neither situation is realistic.

Alternatively, Apollo 17 noritic material might be considered a product of equilibrium partial melting of a rock such as 76535. It is not possible to regard 76535 as a residue from partial melting to form such a liquid because the bronzite and accessory minerals would have been the first to disappear.

The most straightforward interpretation of 76535 is that of a (metamorphosed) cumulate consisting of primary olivine and plagioclase (cumulus plus adcumulus) that formed in equilibrium with a parent liquid, plus frozen, trapped parent liquid (orthocumulus material) that crystallized to form additional olivine and plagioclase as well as bronzite (partly by reaction of the liquid in its last stages of crystallization with solid olivine) and accessory minerals. Textural evidence, including most zoning, was annealed out during slow initial cooling or metamorphism. From this assumed mode of formation for the rock, information on the concentrations of trace elements in the parent liquid can be obtained.

The first step is to estimate the amount of parent liquid trapped in the rock. The minimum amount is that needed to produce the bronzite (4–5%) plus accessory minerals ($\sim 1\%$). From the diagram of Walker *et al.* (1973) (Fig. 3), the relative proportions (by moles) of olivine, anorthite, and silica at the peritectic are 30:25:45. On considering the reaction of silica with olivine to produce pyroxene, this composition corresponds to pyroxene, 0.3; anorthite, 0.25; silica, 0.3. The excess silica would, of course, react with previously crystallized olivine to produce additional pyroxene. That is, about two-thirds of the bronzite would have come directly from the trapped liquid and one-third by reaction with crystalline olivine. The ~4% bronzite in 76535 plus 1% accessory minerals correspond to about 8% by weight of trapped liquid for 76535. This is the minimum possible amount of trapped liquid, because it implies that the composition of the parent liquid for the rock was the peritectic composition.

More likely, the composition of the parent liquid corresponded to some less silica-rich point along the olivine–anorthite cotectic line. The least silica-rich composition possible is that of the olivine–spinel–anorthite triple point. Liquid of that composition, or any composition between that composition and the peritectic composition, would, on being trapped, continue to precipitate olivine and plagioclase until the peritectic composition was reached, at which point pyroxene would appear. The 4% of bronzite in the rock corresponds to a maximum amount of trapped liquid in the rock of 17%, which assumes that the composition of the parent liquid was that of the olivine–spinel–anorthite triple point.

The limits on the amount of parent liquid, while quite accurately defined in terms of the model used, are somewhat less tightly defined in reality. The diagram of Walker *et al.* (1973) does not exactly match the composition of 76535. The modal proportion of bronzite in our analyzed sample (76535,21–22) is not known to better than about ±20%. Modal and normative mineralogies seldom match exactly. Thus, the allowed range of 8–16% (wt.) for trapped parent liquid in 76535 is somewhat approximate, but reasonable in terms of observations on terrestrial layered intrusions (e.g. Paster *et al.*, 1974). The best overall fit to the major and trace element compositions for 76535 was obtained using a value of 16% for the trapped liquid component. Unless otherwise stated, the discussion below is based on this value.

A model for the formation of 76535

The logarithmic model for trace element partition between a silicate liquid and its crystallizing minerals as given by Paster *et al.* (1974) is now used to describe the formation of 76535. Concentrations of trace elements in the parent liquid are first obtained through the use of Eq. (1), which expresses the mass balance for the rock.

$$C_{E,L} = C_{E,R} / \left[\sum_i f_i D_i + (1 - Y) \right] \tag{1}$$

In Eq. (1), $C_{E,L}$ is the concentration of trace element E in the parent liquid L; $C_{E,R}$ is the concentration of trace element E in the whole rock R; f_i is the weight fraction of the whole rock that corresponds to the cumulus plus adcumulus portions of each mineral; D_i is the solid–liquid distribution coefficient for element E and mineral i; and $(1 - Y)$ is the weight fraction of orthocumulus material in the rock (including orthocumulus additions to the cumulus minerals). Results for 16% (wt.) trapped parent liquid $(1 - Y)$ are shown in Fig. 3 and labeled as "probable parent liquid." We assumed that the modal proportions of plagioclase and olivine in the cumulus plus adcumulus portion of the rock were the same as the modal proportions in the whole rock. The modal proportions were corrected for mineral density, the relative amounts by weight of plagioclase and olivine were calculated, and the values for f_i were obtained by multiplying the fractions by weight of olivine and plagioclase in the mixture of those two minerals by 0.84, the cumulus plus adcumulus portions of the whole rock. This presumes that the orthocumulus additions to those two minerals did not significantly alter the proportions in which they accumulated initially. This is a fairly good assumption because the proportions of those minerals do not change appreciably for crystallization along their cotectic and the addition of plagioclase plus loss of olivine during crystallization of the cotectic liquid do not greatly alter the relative amounts of the two minerals in the whole rock. Values for $C_{E,R}$ are those of Table 2 for the whole rock. Values of D_i for plagioclase and olivine for the trivalent REE, K, Rb, and Ba were varied to cover most of the range available from the literature (e.g. Philpotts and Schnetzler, 1970; Schnetzler and Philpotts, 1970). The variations in D_i have only a small effect on the concentrations of these elements in the parent liquid because all values used are much smaller than unity. Thus, the trapped parent liquid is responsible for nearly the entire quantity of those elements present in the whole rock. Values of D_{pl} for Eu were allowed to vary between 0.8 and 1.2 (Green *et al.*, 1971). Eu and Sr are the only elements studied whose distribution coefficients (for plagioclase) are high enough that a cumulate mineral contributes a significant fraction of their concentrations in the whole rock.

The parent liquid, as calculated for 16% trapped liquid in 76535, ranges in concentration from about 13 times the average for chondrites for Lu to about 27 times that of the chondrites for La. The regular increase in chondrite-normalized concentrations is reminiscent of the trend found at the surface of the earth (e.g. Haskin *et al.*, 1966) with a similar break in slope about Eu, but with less extreme relative enrichment in the lighter REE than commonly found in terrestrial materials. Judicious selection of values of D_{pl} for the trivalent REE from within the range found in the literature can eliminate most of the slope shown in Fig. 3 for the probable parent liquid. This is done by forcing the cumulus plagioclase to contain a sufficiently high concentration of REE to produce the slope of the distribution. However, this requires values for D_{pl} that are much too high (e.g. $D_{Sm,pl} \sim 0.2$) to describe most systems (e.g. Paster *et al.*, 1974; Shih, 1972). In addition, the required high values of the distribution coefficients significantly lower the calculated concentrations for the parent liquids and produce positive Eu anomalies in those liquids. There is no negative Eu anomaly in the liquid shown in Fig. 3, and thus no requirement for selective removal of that element from the source region of the liquid. Parent liquids for such plagioclase-rich rocks as 15415 appear not to have had significant relative depletions in Eu (Helmke *et al.*, 1972).

If the calculation is made using less trapped liquid, the concentrations of the REE, K, Rb, and Ba are increased almost in inverse proportion so that the REE concentrations range from about 20 to over 40 times the values for chondrites. Concentrations of Sr and Eu are hardly affected, so a significant negative Eu anomaly (~ 0.6 of the interpolated value) appears for the most extreme case (8% trapped liquid).

In principle, concentrations for major elements can be calculated from a mass balance for the system. It is necessary to infer from studies of phase equilibria values for the relative amounts of Na and Ca in the cumulus plagioclase and its parent liquid (e.g. Bowen, 1913) and for Fe and Mg in the cumulus olivine (Roeder and Emslie, 1970). These cannot be measured directly because cumulus and orthocumulus portions of the rock have at least partially annealed (Gooley *et al.*, 1974). In practice, data of sufficient accuracy cannot be obtained to produce an accurate calculation of the major element composition of the parent liquid. The problem is one of multiplying the proper (unknown) modal proportions for the cumulate parts of the olivine and plagioclase by the (unknown) compositions of those cumulate minerals, and subtracting from the measured major element composition for the sample whose mode was used. By trial and error, a number of compositions were tested.

From these it can only be said that liquids with compositions in the vicinity of the cotectic line for plagioclase and olivine (Walker *et al.*, 1973) or in the spinel field near the spinel–olivine–anorthite triple point can be accommodated within the uncertainties of the model.

Acknowledgments—We thank Miss K. V. Rodgers for assisting with the analyses for major elements. We thank Mrs. D. L. Coennen for assistance in preparing the manuscript. C.-Y. Shih was supported by an NRC Research Associateship.

REFERENCES

Adler I., Gerard J., Trombka J., Schmaderbeck R., Lowman P., Blodget H., Yin L., Eller E., Lamothe R., Gorenstein P., Bjorkholm P., Harris B., and Gursky H. (1972) The Apollo 15 X-ray fluorescence experiment. *Proc. Third Lunar Sci. Conf., Geochim. Cosmochim. Acta*, Suppl. 3, Vol. 3, pp. 2157–2178. MIT Press.

Bogard D. D., Nyquist L. E., Bansal B. M., and Wiesmann H. (1974) 76535: An old lunar rock? (abstract). In *Lunar Science—V*, pp. 70–72. The Lunar Science Institute, Houston.

Bowen N. L. (1913) Melting phenomena in plagioclase feldspars. *Am. J. Sci.* **35**, 577–599.

Brown G. M., Peckett A., Emeleus C. H., and Phillips R. (1974) Mineral-chemical properties of Apollo 17 mare basalts and terra fragments (abstract). In *Lunar Science—V*, pp. 89–91. The Lunar Science Institute, Houston.

Crozaz G., Drozd R., Hohenberg C., Morgan C., Ralston C., Walker R., and Yuhas D. (1974) Lunar surface dynamics: Some general conclusions and new results from Apollo 16 and 17 (abstract). In *Lunar Science—V*, pp. 157–159. The Lunar Science Institute, Houston.

Gast P. W., Hubbard N. J., and Wiesmann H. (1970) Chemical composition and petrogenesis of basalts from Tranquility Base. *Proc. Apollo 11 Lunar Sci. Conf., Geochim. Cosmochim. Acta*, Suppl. 1, Vol. 2, pp. 1143–1163. Pergamon.

Gooley R., Brett R., Warner J., and Smyth J. R. (1974) A lunar rock of deep crustal origin: Sample 76535. *Geochim. Cosmochim. Acta.* In press.

Green D. H., Ringwood A. E., Ware N. G., Hibberson W. O., Major A., and Kiss E. (1971) Experimental petrology and petrogenesis of Apollo 12 basalts. *Proc. Second Lunar Sci. Conf., Geochim. Cosmochim. Acta*, Suppl. 2, Vol. 1, pp. 601–615. MIT Press.

Haskin L. A., Frey F. A., Schmitt R. A., and Smith R. H. (1966) Meteoritic, solar and terrestrial rare-earth distributions. *Phys. Chem. Earth* **7**, 167–321.

Haskin L. A., Helmke P. A., Blanchard D. P., Jacobs J. W., and Telander K. (1973) Major and trace element abundances in samples from the lunar highlands. *Proc. Fourth Lunar Sci. Conf., Geochim. Cosmochim. Acta*, Suppl. 4, Vol. 2, pp. 1275–1296. Pergamon.

Helmke P. A., Haskin L. A., Korotev R. L., and Ziege K. (1972) Rare earths and other trace elements in Apollo 14 samples. *Proc. Third Lunar Sci. Conf., Geochim. Cosmochim. Acta*, Suppl. 2, Vol. 2, pp. 1275–1292. MIT Press.

Henderson P. and Dale I. M. (1969/1970) The partitioning of selected transition element ions between olivine and ground mass of oceanic basalts. *Chem. Geol.* **7**, 73–95.

Hubbard N. J. and Gast P. W. (1971) Chemical composition and origin of nonmare lunar basalts. *Proc. Second Lunar Sci. Conf., Geochim. Cosmochim. Acta*, Suppl. 2, Vol. 2, pp. 999–1020. MIT Press.

Hubbard N. J., Gast P. W., Meyer C., Nyquist L. E., Shih C., and Wiesmann H. (1971) Chemical composition of lunar anorthosites and their parent liquids. *Earth Planet. Sci. Lett.* **13**, 71–75.

Hubbard N. J., Rhodes J. M., Gast P. W., Bansal B. M., Shih C., Wiesmann H., and Nyquist L. E. (1973) Lunar rock types: The role of plagioclase in non-mare and highland rock types. *Proc. Fourth Lunar Sci. Conf., Geochim. Cosmochim. Acta*, Suppl. 4, Vol. 2, pp. 1297–1312. Pergamon.

Hubbard N. J., Rhodes J. M., Nyquist L. E., Shih C., Bansal B. M., and Wiesmann H. (1974) Non-mare and highland rock types: Chemical groups and their internal variations (abstract). In *Lunar Science—V*, pp. 366–368. The Lunar Science Institute, Houston.

Jovanovic S. and Reed G. W. (1974) Labile trace elements in Apollo 17 samples (abstract). In *Lunar Science—V*, pp. 391–393. The Lunar Science Institute, Houston.

Keil K., Bunch T. E., and Prinz M. (1970) Mineralogy and composition of Apollo 11 lunar samples. *Proc. Apollo 11 Lunar Sci. Conf., Geochim. Cosmochim. Acta*, Suppl. 1, Vol. 1, pp. 561–598. Pergamon.

Metzger A. E., Trombka J. I., Peterson L. E., Reedy R. C., and Arnold J. R. (1973) Lunar surface radioactivity: Preliminary results of the Apollo 15 and Apollo 16 gamma-ray spectrometer experiments. *Science* **179**, 800–803.

Morgan J. W., Ganapathy R., Higuchi H., Krähenbühl U., and Anders E. (1974) Lunar basins: Tentative characterization of projectiles, from meteoritic elements in Apollo 17 boulders (abstract). In *Lunar Science—V*, pp. 526–528. The Lunar Science Institute, Houston.

Nguyen L. D., de Saint Simon M., Puil G., and Yokoyama Y. (1973) Rare earth elements in Luna 20 soils and their implications for cosmochemistry. *Proc. Fourth Lunar Sci. Conf., Geochim. Cosmochim. Acta*, Suppl. 4, Vol. 2, pp. 1415–1426. Pergamon.

Norrish K. and Hutton J. T. (1969) An accurate X-ray spectrographic method for the analysis of a wide range of geological samples. *Geochim. Cosmochim. Acta* **33**, 431–454.

Nyquist L. E., Bansal B. M., Wiesmann H., and Jahn B.-M. (1974) Taurus Littrow chronology: Some constraints on early lunar crustal development. *Proc. Fifth Lunar Sci. Conf., Geochim. Cosmochim. Acta*. This volume.

Paster T. P., Schauwecker D. S., and Haskin L. A. (1974) A trace element study of the Skaergaard layered series. *Geochim. Cosmochim. Acta*. In press.

Philpotts J. A. and Schnetzler C. C. (1970) Phenocryst-matrix partition coefficients for K, Rb, Sr and Ba, with application to anorthosite and basalt genesis. *Geochim. Cosmochim. Acta* **34**, 307–322.

Rhodes J. M., Rodgers K. V., Shih C., Bansal B. M., Nyquist L. E., Wiesmann H., and Hubbard N. J. (1974) The relationship between geology and soil chemistry at the Apollo 17 landing site. *Proc. Fifth Lunar Sci. Conf., Geochim. Cosmochim. Acta*. This volume.

Roeder P. L. and Emslie R. F. (1970) Olivine-liquid equilibrium. *Contr. Mineral. Petrol.* **29**, 275–289.

Schnetzler C. C. and Philpotts J. A. (1970) Partition coefficients of rare-earth elements between igneous matrix material and rock forming mineral phenocrysts—II. *Geochim. Cosmochim. Acta* **34**, 331–340.

Shih C.-Y. (1972) The rare earth geochemistry of oceanic igneous rocks (Unpbl. Ph.D. thesis).

Tera F., Papanastassiou D. A., and Wasserburg G. J. (1974) The lunar time scale and a summary of isotopic evidence for a terminal lunar cataclysm (abstract). In *Lunar Science—V*, pp. 792–794. The Lunar Science Institute, Houston.

Walker D., Grove T. L., Longhi J., Stolper E. M., and Hays J. F. (1973) The origin of lunar feldspathic rocks. *Earth Planet. Sci. Lett.* **20**, 325–336.

Warner J. L., Simonds C. H., and Phinney W. C. (1974) Impact induced fractionation in the highlands (abstract). In *Lunar Science—V*, pp. 823–825. The Lunar Science Institute, Houston.

Wood J. A., Dickey J. S., Jr., Marvin U. B., and Powell B. N. (1970) Lunar anorthosites and a geophysical model of the moon. *Proc. Apollo 11 Lunar Sci. Conf., Geochim. Cosmochim. Acta*, Suppl. 1, Vol. 1, pp. 965–988. Pergamon.

Proceedings of the Fifth Lunar Conference
(Supplement 5, Geochimica et Cosmochimica Acta)
Vol. 2 pp. 1227–1246 (1974)
Printed in the United States of America

The chemical definition and interpretation of rock types returned from the non-mare regions of the moon

Norman J. Hubbard,[1] J. Michael Rhodes,[2] H. Wiesmann,[2]
C.-Y. Shih,[1] and B. M. Bansal[2]

[1]NASA Johnson Space Center, Houston, Texas 77058

[2]Lockheed Electronics Corporation, Houston, Texas 77058

Abstract—The chemical compositions of rocks returned from the non-mare regions of the moon fall into two major groups: (1) rocks with basaltic chemical compositions, high concentrations of rare earths and related lithophile trace elements, deep negative Eu and Sr anomalies, and a uniform slope in their trivalent rare-earth abundance patterns (KREEP-VHA), and (2) anorthositic rocks with low concentrations of rare earths and related lithophile trace elements and positive Eu and Sr anomalies (LKAS). Both groups 1 and 2 are interpreted to have been produced from the same suite of related precursors, with group 1 being produced by partial melting and group 2 by removal of olivine and/or orthopyroxene by fractional crystallization. During these igneous events, 10–20% of the Eu was in the trivalent oxidation state and the precursors to the KREEP-VHA group already had the distinctive slope to their trivalent REE abundance patterns. The precursors are proposed to be the original material that accreted to form the outer tens of kilometers of the moon and to be chemically indistinguishable from the bodies that impacted the moon in its early post accretional history.

Introduction

Our past efforts at defining rock groups among lunar non-mare samples have been limited to groups with basaltic chemical compositions for major elements, high concentrations of rare earths and related lithophile trace elements and deep negative Eu and Sr anomalies, such as KREEP of various varieties and VHA basalt (Hubbard *et al.*, 1971, 1972, 1973). Apollo 17, 16, and to a lesser extent Apollo 15 have provided enough samples with more than 24.0% Al_2O_3 so that it is now possible to study the chemical composition of anorthositic lunar samples with the same quality and quantity of chemical and Sr isotope data as for non-mare materials of basaltic composition. Sufficient samples of basaltic and anorthositic chemical compositions are now available so that we are probably seeing the effects of the major factors controlling the genesis and fractionation of these materials.

We do not know to what extent the chemical compositions of non-mare samples reflect the chemical composition of original lunar materials or the results of igneous differentiation after the moon's formation. At least one model for the moon's origin (Wood and Mitler, 1974) provides a mechanism for making the moon from materials that may have already experienced igneous differentiation as part of a prior planetary body. If the moon did originate in this way, then an initially heterogeneous moon is an expected result and some of the igneous

1227

differentiation inferred in this and other papers (Walker *et al.*, 1973, for example) may be pre-lunar.

One of the serious unknowns in the study of lunar samples, other than for mare basalts, is the extent to which the samples studied are the result of extensive mixing of non-cogenetic materials. With the exception of soil breccias, there are essentially no objective, model independent, means of directly identifying fine-grained rocks which have their present chemical composition as a result of mixing large fractions of non-cogenetic materials. A start has been made (Meyer *et al.*, 1974) which demonstrates that visually polymict breccias may contain anorthositic clasts that are cognate crystal cummulates. Specificly, one VHA sample, 61156, investigated by Meyer *et al.* (1974) appears to contain cognate anorthositic material and anorthositic clasts in Apollo 14 KREEP breccias were found to be cognate. Clearly, the visual appraisals sited by Dowty *et al.* (1973) as evidence that VHA basalts are mixtures may well be wrong. Many, perhaps most, of the visually polymict features of lunar samples are potentially the result of brecciating a cognate igneous complex.

Since the discovery of KREEP it has become fashionable to view all lunar breccias as polymict rocks and to interpret their chemical compositions as mixtures of a few rock types of extreme, end-member, compositions (Goles *et al.*, 1971; Wänke *et al.*, 1972; Schonfeld, 1974). To demonstrate the probability of any mixing model that matches the chemical composition of a rock, one must demonstrate the existence, and adequate abundance, of the end-members before the mixing event. Proof of a mixing model occurs when the sample in question can be shown to contain the end-members in the abundance calculated. Thus far *no* mixing model calculation, although mathematically satisfactory, has been proven correct and many are geologically improbable. Further, the demonstration that any rock can be a mixture of other rocks rapidly loses its validity as the number of components are increased beyond two. Our past usage of mixing models has been essentially limited to showing that Apollo 12 soils can be a mixture of KREEP and Apollo 12 mare basalts (Hubbard *et al.*, 1971b) and more recently to show that VHA basalts cannot simply be mixtures of KREEP and anorthositic rocks (Hubbard *et al.*, 1973). Dowty *et al.* (1974) have recently proposed a vaguely defined three-component mixing model to support their interpretation of the VHA basalt group of rocks. They use almost entirely data from Luna 20 soil particles to propose that the VHA basalt composition, which is primarily from Apollo 16, was somehow made by mixing KREEP, anorthositic rocks and high-magnesian non-mare rocks. They make no attempt to demonstrate that adequate volumes of high-magnesian rocks are, or were, present at the Apollo 16 site, where VHA basalts are common, or that the VHA samples listed by us contain adequate amounts of admixed high-magnesian material. Thus far the only chemically well-characterized high-magnesian non-mare rocks are from the Apollo 17 site. The VHA basalt composition clearly exists at the Apollo 16 site and was also described by Ridley *et al.* (1973), who broadened their low-K Fra Mauro group to include this composition. We have long been aware that one could imitate the VHA chemical composition (or almost any other!) by the kind of mixing model proposed by Dowty *et al.* (1974), but have considered it untenable because the required high-magnesian rocks are not common at the Apollo 16 site. At several points in this paper we will show that the VHA basalt composition is the logical extreme to a series of chemical compositions ranging from VHA basalt to Apollo 14 common KREEP. See Nyquist *et al.* (1974) for a discussion of Rb–Sr and Sr^{87} data for the VHA to KREEP series of rocks.

Given the extensive cratering record of the lunar highlands (Short and Foreman, 1972), one must conclude that mixing is an important process on the moon. However, impacts have not erased the strong evidence that igneous processes, lunar or extra-lunar, have produced an abundance of lunar rock types with a wide range of chemical compositions. If the rocks that we are studying are mechanical mixtures of igneous rocks produced on pre-lunar planetary bodies (Wood and Mitler, 1974), then mixing models are perhaps the best way to describe the results of their lunar experience. If instead, lunar rocks are the result of an igneous modification of lunar material, then it is appropriate to study the rocks, using relevant partial melting or fractional crystallization models, in an attempt to deduce the precursors. This is the ultimate goal of this paper.

The study of lunar rocks has thus far resulted in three major view points of lunar petrogenesis: (1) KREEP, anorthositic and high-magnesian rocks, as well as mare basalts are primary materials, while all others are mixtures of these; (2) nearly all types of lunar rocks could have been produced by normal igneous processes and are therefore suitable for studying these igneous processes, with the caveat that one must ultimately be able to directly estimate the nature and extent of any mixing processes; and (3) a recent proposal by Warner *et al.* (1974) that impacts themselves homogenize by mixing and then differentiate by partial fusion of the material involved in the impact. Each of these viewpoints has at least one major weakness. Success of the Warner *et al.* (1974) hypothesis depends on the possibility of proving that impact related differentiation is more effective in providing a range of chemical compositions than is impact mixing in reducing this range. The mixing model approach (viewpoint 1) is internally inconsistent and simplistic because igneous processes adequate to produce the primary materials are certainly adequate to produce intermediate compositions. The approach taken in this paper (viewpoint 2) is currently limited by the minimal amount of high quality chemical data for whole rock samples and by the extreme lack of modern, chemically based petrographic data that would allow direct determination of the rock types that make up a lunar breccia and the amounts of cognate and noncognate materials.

In this paper we will also introduce the concept that chemically related lunar igneous rocks, i.e. a chemically defined rock group, cannot be expected to show any lesser extent of internal chemical variation than a single large terrestrial volcano, such as a Hawaiian, or a large basaltic volcanic province such as, for example, the Columbia Plateau or a single volcanic province in Iceland.

PRESENTATION OF DATA

Major element data

Perhaps the best introduction to major element data is via a triangular diagram of silica, olivine, and anorthite (Fig. 1). This presentation allows one to compare the chemical data with the experimental data of Walker *et al.* (1973) when both the analytical and experimental data are recalculated in the same manner. The majority of lunar non-mare rocks plot in or near one of the three chemical groups plotted in Fig. 1. The area around the olivine–plagioclase cotectic line from point A to somewhat beyond point B is populated by a group of non-mare rocks of basaltic chemical composition ranging from VHA basalts at high Al_2O_3 values, through Apollo 16/17 KREEP at intermediate Al_2O_3 values to common Apollo 14 KREEP (not plotted) which plots in the general area of point B. The scatter of points on Fig. 1 suggests that the VHA and Apollo 16/17 KREEP groups may have obtained their chemical compositions by partial melting and then had them scattered somewhat by subsequent fractional crystallization and/or impact mixing. We suggest that the mineral clasts actually observed (Simonds *et al.*, 1973) may be derived from original phenocrysts. The third group shown on Fig. 1, anorthositic rocks, plots in a fashion suggestive of a different origin although some samples approach the spinel–plagioclase boundary.

An even broader chemical overview of lunar rocks is given in Fig. 2, where MgO is plotted versus Al_2O_3. The striking observation provided by this view of the chemical data is that the chemical rock groups that plot along the center cotectic of Fig. 1 have parallel internal Al_2O_3 versus MgO variations and that these variations parallel olivine–Mg-rich orthopyroxene fractionation trends. Further,

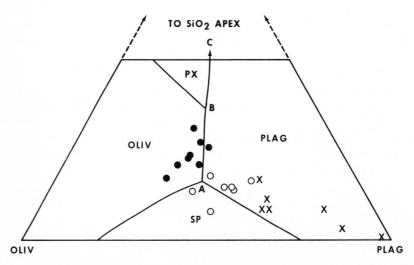

Fig. 1. A portion of the pseudoternary system used by Walker *et al.* (1973). The dots are Apollo 17/16 KREEP, the circles are VHA and the × are anorthositic samples. All data points and points A and B were processed by the same normative calculation.

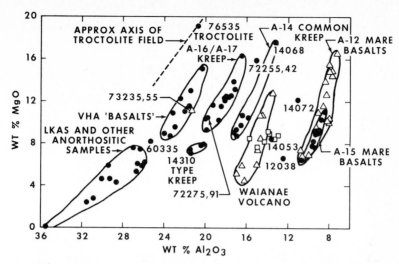

Fig. 2. MgO versus Al₂O₃ for a wide range of lunar rocks. Data for Waianae Volcano and Apollo 12/15 mare basalts are included with non-mare and highland samples in order to make comparisons with known igneous rocks. The Waianae Volcano data are from Macdonald and Katsura (1964) and the Apollo 15 mare basalt data are from Rhodes and Hubbard (1973). The other data are from numerous published sources and are mostly XRF data.

they parallel the internal MgO versus Al₂O₃ variations of mare basalts and a typical Hawaiian volcano, Waianae Volcano (MacDonald and Katsura, 1964). The range in MgO/FeO ratios for Waianae Volcano is 1.9 fold (not shown in Fig. 3) which is essentially as large as for the VHA basaltic composition, which is 2 fold, and the Apollo 16/17 KREEP composition. The mare basalts have a lesser range of MgO/FeO ratios, 1.7 fold (not shown in Fig. 3). In other words, the Al₂O₃, MgO, and MgO/FeO variations in the KREEP and VHA chemical rock groups are near or within the limits for known cogenetic volcanic rocks both on earth and on the moon.

The anorthositic rocks (Fig. 3) show an even broader range in MgO/FeO ratios than VHA and Apollo 16/17 KREEP rocks, with lower average MgO/FeO ratios than VHA and KREEP. The known extremes, 67075, and 15455 white (Taylor, 1973), are 3 fold apart although the majority of ratios fall within a lesser range of 1.6 fold. The anorthositic lunar rocks, like the non-mare basaltic lunar rocks, show a slight increase in MgO/FeO with decreasing Al₂O₃ concentration.

Trace element data

The data for one set of lithophilic trace elements, the REE and Ba, are summarized in Fig. 4 for a wide range of Apollo 15, 16, and 17 non-mare basaltic and anorthositic rocks. The data in this diagram are readily divided into two major groups: (1) rocks with deep negative Eu anomalies and high concentrations of REE and Ba, and (2) rocks with almost no Eu anomalies or positive Eu anomalies

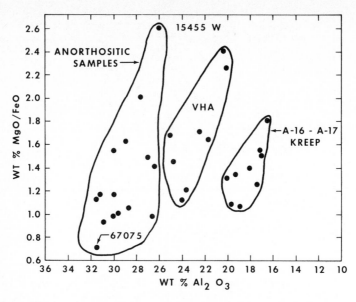

Fig. 3. MgO/FeO ratios versus Al_2O_3 for anorthositic, VHA and Apollo 16/Apollo 17
KREEP samples. The data base is as for Fig. 2.

and low concentrations of REE and Ba. Not surprisingly, the samples in group 2
have anorthositic major element compositions and plot in the anorthositic fields
on Figs. 1 and 2. The group 1 samples plot near the plagioclase–olivine or the
plagioclase–pyroxene cotectic in Fig. 1 and in the KREEP–VHA region of Fig. 2.
The chemical data in Figs. 1, 2, and 4 demonstrate that rocks of KREEP–VHA

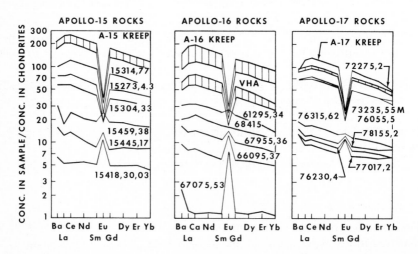

Fig. 4. Rare-earth and Ba data for non-mare and highland samples from Apollo 15, 16,
and 17. The data are all from our research group and are published or are in this paper in
Tables 1 and 2.

Table 1. Rocks with deep negative Eu anomalies. All analyses are by isotope dilution except Na, which is by atomic absorption.

		15304,33 53.1 mg	15314,77 54.4 mg	15386,1 37.9 mg	66095,36 49.5 mg	60335,77 64.9 mg	72275,2 52.8 mg	72435,1 53.7 mg	73235,55M 50.0 mg	76015,22M 53.2 mg	76055,5 48.7 mg	76315,2 52.4 mg	77135,2 52.6 mg
Na	%				0.29				0.38				
Li	ppm	7.3	22.1	27.2	10.7	10.7	13.8	17.5	11.9	18.3		14.6	19.3
K	ppm	1364	2484	5706	1279	2007	2404	1803	1746	2223	1758	2252	2452
Rb	ppm	4.06	8.11	18.5	3.61	6.18	8.97	3.93	5.13	6.41	5.17	5.88	7.32
Sr	ppm	218	379	187	233	162.5	122.7	171.6	146.9	171.8	156.6	179.5	172.2
Ba	ppm			837		193	350	334	263	348	253	359	337
La	ppm			83.5	24.9	19.9	41.0	31.7	23.3		22.6	30.1	32.1
Ce	ppm	46.4	97.3	211	63.2	51.5	106	80.6	60.6	83.3	56.3	84.6	81.2
Nd	ppm	27.8	61.3	131	40.0	32.4	67.4	51.3	37.0	52.8	35.8	53.5	51.6
Sm	ppm	7.9	17.4	37.5	11.2	9.05	18.8	14.5	10.4	14.9	10.1	15.1	14.6
Eu	ppm	1.33	1.70	2.72	1.43	1.28	1.49	1.88	1.37	1.94	1.71	2.00	1.99
Gd	ppm	9.6	21.0	45.4	14.0	10.6	23.4	18.3	12.8	18.7	12.7	18.9	18.5
Dy	ppm	11.0	22.5	46.3	14.4	11.5	23.2	18.6	13.8	19.5	13.5	19.9	19.1
Er	ppm	6.9	13.2	27.3	8.34	6.77	13.7	11.3	8.2	11.5	8.18	11.7	11.4
Yb	ppm			24.4	7.63	6.23	11.6	10.1	7.7	10.6	7.64	11.0	10.5
Lu	ppm			—		0.675	—	—	—	—		—	
Cr	ppm	800	1400	2124	823	850	2345	1291	1331	1316	1283	1228	1178
Th	ppm								4.19	5.44			
U	ppm	1.17	1.84	—	1.07	1.0	1.56	1.40	1.14	1.46		1.52	
Zr	ppm	266	492	—	—	292	605	473	341	490		477	
Hf	ppm	6.6	12.5	—	—	7.3	14.6	12.7		12.9		12.5	

Table 2. Anorthositic rocks. All data are by isotope dilution except Na, which is by atomic absorption.

		15445,17 51.0 mg	15459,38 24.8 mg	61016,3 53.3 mg	61016,79 81.5 mg	61016,79 gray plag. 88.0 mg	61016,84 48.4 mg	61295,34 73.9 mg	63335,36 70.0 mg	63549,2 47.1 mg	64435,59 49.5 mg
Na	%	0.24	0.54	0.25	0.32	0.24	0.24	0.35	—	0.32	
Li	ppm	—	—	2.2	1.2	1.2	1.0	7.1	—	4.8	2.6
K	ppm	582	872	184	729	40	40	818	437	594	216
Rb	ppm	1.43	1.69	0.446	0.038	0.017	0.040	2.31	1.146	1.764	0.638
Sr	ppm	130	205.3	177.9	180.4	179.0	181.7	186.0	222.1	170.2	
Ba	ppm	61.9	119.1	40.7	6.97	7.05	7.11	116.7	56.2	74.0	21.1
La	ppm	4.02	5.80	3.47	<0.3	0.143		10.4	3.15	6.39	1.32
Ce	ppm	11.1	19.9	8.61	<0.7	0.37	0.44	25.7	7.76	16.6	4.09
Nd	ppm	5.91	12.2	5.60	0.20	0.205		16.4	4.99	10.6	2.35
Sm	ppm	1.65	3.60	1.56	0.045	0.058		4.59	1.44	2.99	0.691
Eu	ppm	0.929	1.74	0.926	0.805	0.77		1.22	1.39	1.03	0.759
Gd	ppm	2.05		1.84	0.045	0.054		5.76	1.82	3.67	0.842
Dy	ppm	2.69	5.7	1.91	0.025	0.065		5.99	1.96	3.90	1.03
Er	ppm	1.72	3.5	1.16	0.067	0.040		3.55	1.22	2.40	0.660
Yb	ppm	1.78	3.47	1.01	0.020	0.045		3.25	1.14	2.23	0.611
Lu	ppm	0.268	0.522	0.149	0.005	0.024		—	0.175	—	0.0950
Cr	ppm	1,560	—	190	<20	375	<46	—	383	625	498
Th	ppm	—	—	0.486	—	—	—	—	0.49	—	0.218
U	ppm	—	0.35	0.101	0.002	0.0015	—	0.495	0.136	0.329	0.062
Zr	ppm	—	—	51.2	3	2.4	—	203	44.0	—	11.2
Hf	ppm	—	—	1.06	—	—	0.0002	—	1.15	—	
K/Rb		407	516	413	—	—	—	354	381	337	339
Th/U		—	—	4.8	—	—	—	—	3.6	—	3.5
Zr/Hf		—	—	48.3	—	—	—	—	38.3	—	

chemical compositions have many chemical similarities and are clearly different from the other major group of non-mare lunar rocks, the rocks of anorthositic chemical composition.

Anorthositic lunar rocks, as a group, are presented in Fig. 5. There is a high degree of regularity in the REE, Ba, and U abundance patterns of these rocks which in general can be correlated with Al_2O_3 concentrations, suggesting that we may be dealing with a series of anorthositic rocks where the major chemical variations are a function of plagioclase concentration. First, in order to subdivide these lunar anorthositic rocks we need to reject the rocks that deviate from the simple requirements of an internally consistent anorthositic series, in this case a correlation of REE abundance patterns with Al_2O_3 concentration. The purpose of this exercise is to remove samples which may be mixtures, or have suffered extensive chemical modification by impact related fusion, or simply to reject samples that formed in equilibrium with materials of different chemical composition than did the majority of samples. The primary rejection is accomplished by plots of Al_2O_3 versus Eu and Sm (Figs. 6a and b). Additional rejection, reduction to conditional membership, or simply further subdivision may be made on other chemical parameters, such as the slope of the trivalent rare-earth abundance pattern, MgO/FeO ratio, and abnormally low or high concentrations of elements like Si, K, Ti, Cr, etc. We can immediately reject 63335,36, 15459,38, 61295,34,

Table 2. (*Continued*).

		66095,37 24.3 mg	67075,53 64.2 mg	67955,56 54.7 mg	68415,10 86.4 mg	68416,33 54.1 mg	76315,62 52.5 mg	76230,4 78.1 mg	77017,2 46.0 mg	78155,2 51.6 mg
Na	%	0.26	—	0.27	0.35	0.30	0.35	0.29	0.32	0.26
Li	ppm	8.0	—	5.1	5.1	4.4	9.5	11.0	4.4	5.2
K	ppm	596	130	441	565	559	809	440	422	656
Rb	ppm	1.591	0.593	0.884	1.704	1.705	2.336	0.448	1.310	2.061
Sr	ppm	162.7	145.0	169.1	182.4	—	153.1	145.9	141.5	146.7
Ba	ppm	36.0	8.85	61.9	76.2	78.2	72.8	50.2	49.0	58.8
La	ppm	2.57	0.393	4.45	6.81	7.24	5.41	3.04	3.48	4.02
Ce	ppm	6.56	0.891	11.3	18.3	18.4	13.7	7.54	8.90	10.2
Nd	ppm	3.87	0.664	7.09	10.9	11.5	8.60	4.64	5.56	6.29
Sm	ppm	1.09	0.209	2.02	3.09	3.28	2.42	1.34	1.60	1.81
Eu	ppm	0.88	0.650	0.973	1.11	1.11	0.940	0.805	0.794	0.874
Gd	ppm		0.301	2.57	3.78	4.07	2.99	1.70	2.01	2.32
Dy	ppm	1.75	0.343	2.81	4.18	4.29	3.39	2.02	2.34	2.64
Er	ppm		0.255	1.79	2.57	2.86	2.14	1.31	1.50	1.69
Yb	ppm	1.04	0.251	1.74	2.29	2.42	2.07	1.37	1.50	1.73
Lu	ppm		0.038	0.25	0.34	—	0.30	0.20	—	0.259
Cr	ppm	578	372	659	599	683	770	685	881	1008
Th	ppm	—	0.023	1.03	1.26	—	1.234	0.72	—	1.013
U	ppm	0.138	0.013	0.38	0.32	0.342	0.343	0.198	0.22	0.284
Zr	ppm	—	7.6	124	97.5	—	95	—	59.1	—
Hf	ppm	—	0.12	3.1	2.4	—	5.3	—	1.6	—
K/Rb		375	219	499	332	328	346	982	322	318
Th/U		—	1.8	2.7	3.9	—	3.6	3.64	—	3.6
Zr/Hf		—	63	40	40.6	—	17.9	—	36.9	—

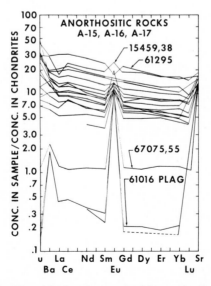

Fig. 5. Rare-earth, Ba, U, and Sr data for anorthositic samples from Apollo 15, 16, and 17. Data for most of these samples are listed in Table 2. All data are from our research group.

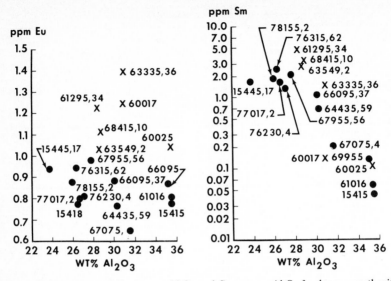

Fig. 6a and 6b. Plots of Eu versus Al_2O_3 and Sm versus Al_2O_3 for lunar anorthositic rocks. Samples that are retained as members of the LKAS are shown by large dots, ×s are for other samples.

68415,10, and 63549,2 because 63335,36 has abnormal Eu anomaly and high Eu and Sr concentrations, 15459,38 has too much Eu, Sr, and Sm for its Al_2O_3 concentrations, as do 61295,34, 68415,10 and to a lesser extent 63549,2 (see also Tables 2 and 3). Samples 15418, 67075, and 15445,17 are reduced to conditional membership and sample 67955 is marginal. Sample 15418 has low Sm for its Al_2O_3 value of 26% and also has a flat trivalent rare-earth pattern. Sample 67075,53 also has a flat trivalent rare-earth pattern and has the lowest MgO/FeO ratio of anorthositic samples included in this study. Sample 15445,17 has low Al_2O_3, high MgO/FeO ratio and high SiO_2 concentration. Sample 15455 white (Taylor *et al.*, 1973) has not been included because of its high MgO/FeO ratio, low TiO_2, very high Eu concentration and the steep slope of its trivalent rare-earth pattern. Samples 60017 and 60025 are rejected because of their high Eu concentrations (Laul and Schmitt, 1973). Sample 64422,13,20 is rejected because of its high Eu and Sr concentrations. The remaining samples include 15415, 15418, 15445,17, 61016 plagioclase, 64435,59, 66095,37, 67955,56, 76230,4, 76315,62, 77017,2, and 78155,2 and possibly includes samples 69955, parts of 60015, numerous coarse fine and rake fragments. Samples that are retained within the subset are plotted in Figs. 6a and 6b. This series is proposed to be a major subset of anorthositic lunar rocks and will be referred to as the Low-K Anorthositic Series (LKAS).

Lunar samples are also seen to form groups on a plot of Eu versus Sr (Fig. 7). This figure also provides fundamental information about the oxidation state of Eu in the lunar petrogenetic environment. If Eu was entirely divalent during lunar petrogenesis, then Eu should closely follow Sr if lunar samples were derived from

Table 3. Major element data for anorthositic rocks. All data are by XRF analyses done by either H. Rose or J. M. Rhodes, with Rhodes the analyst for those samples labeled PET.

	Rhodes 15418,30,07A	PET 15418	Rhodes 15418,51 Saw dust	Rhodes 15445,17	Rose 60017,62	Rose 60025,95	PET 61295,5	PET 63335,1	Rhodes 63549,2
SiO_2	45.53	44.97	44.21	48.67	44.43	43.94	45.19	45.20	45.68
FeO	6.66	5.37	6.65	3.88	2.97	0.67	4.52	3.23	4.27
Al_2O_3	25.98	26.73	26.57	23.76	30.90	35.21	28.29	30.86	28.59
CaO	15.63	16.10	15.98	13.26	17.72	18.92	16.16	17.25	15.20
MgO	6.09	5.38	5.08	9.94	2.77	0.27	4.72	2.81	4.33
TiO_2	0.29	0.27	0.27	0.15	0.30	0.02	0.56	0.42	0.30
Na_2O	0.31	0.31	0.27	0.33	0.58	0.49	0.45	0.57	—
K_2O	0.03	0.03	0.02	0.06	0.06	0.03	0.09	0.05	0.07
MnO	0.10	0.08	0.10	0.08	0.04	0.03	0.06	0.04	0.05
P_2O_5	0.03	0.03	0.03	0.03	0.02	0.00	0.10	0.03	0.07
Cr_2O_3	—	0.11	—	—	0.06	0.00	—	—	—
S	0.03	0.03	0.04	0.00	—	—	0.06	0.03	0.04

	Rhodes 64435,59	Rhodes 66095,37	PET 67075,4	Rose 67115,16	Rose 67455,17	PET 67955,8	Rose 68415,85	PET 68415,6	Rhodes 68416,33
SiO_2	44.55	44.07	44.80	44.75	44.87	45.01	45.30	45.40	45.04
FeO	3.42	3.03	3.41	2.60	3.41	3.84	4.12	4.25	4.27
Al_2O_3	30.25	30.02	31.54	31.15	30.42	27.68	28.70	28.63	28.75
CaO	17.16	16.65	18.09	17.76	18.30	15.54	16.24	16.39	16.31
MgO	3.83	4.72	2.42	3.03	2.30	7.69	4.35	4.38	4.49
TiO_2	0.19	0.18	0.09	0.24	0.30	0.27	0.29	0.32	0.33
Na_2O	—	0.35	0.26	0.51	0.41	0.40	0.50	0.41	0.34
K_2O	0.03	0.08	0.01	0.08	0.03	0.05	0.09	0.06	0.08
MnO	0.05	0.05	0.06	0.04	0.05	0.05	0.05	0.06	0.07
P_2O_5	0.03	0.06	0.00	0.02	0.02	0.03	0.06	0.07	0.08
Cr_2O_3	—	—	—	0.06	0.08	—	0.14	—	—
S	0.04	0.14	0.01	—	—	0.03	—	0.04	0.05

	Rose 68416,40	Rose 69935,24	Rose 69955,12	PET 76230,4	Rhodes 76315,62	PET 77017,2	PET 78155,2
SiO_2	45.60	44.69	44.10	44.52	45.10	44.09	45.57
FeO	4.22	2.34	0.36	5.14	5.29	6.19	5.82
Al_2O_3	28.40	31.47	35.15	27.01	26.37	26.59	25.94
CaO	16.32	17.97	19.30	15.17	15.12	15.43	15.18
MgO	4.64	2.63	0.23	7.63	7.46	6.06	6.33
TiO_2	0.30	0.22	0.01	0.20	0.36	0.41	0.27
Na_2O	0.44	0.43	0.42	0.35	0.47	0.30	0.33
K_2O	0.08	0.08	0.02	0.06	0.10	0.06	0.08
MnO	0.06	0.03	0.01	0.06	0.07	0.08	0.10
P_2O_5	0.07	0.15	0.01	0.05	0.06	0.03	0.04
Cr_2O_3	0.13	0.06	0.04	0.11	0.04	0.13	0.14
S	—	—	—	0.03	—	0.15	0.04

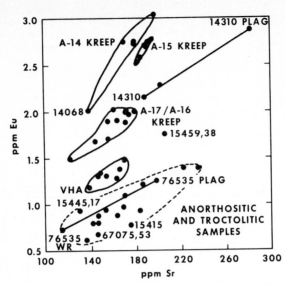

Fig. 7. Sr versus Eu for lunar non-mare and highland rocks. All data are from our research group or from Philpotts *et al.* (1972).

a precursor with a single Eu/Sr ratio. Instead the relative variation of Eu is much wider than that of Sr in lunar non-mare rocks. Coupled with this observation is the companion observation that within a chemically defined rock group Eu and Sr are closely correlated, suggesting that the Eu/Sr ratio was fractionated when the chemical compositions of the individual rock groups were formed.

The chemical linkage between Eu, Sr and plagioclase and the aluminous nature of lunar non-mare rocks provides a powerful means of unraveling the genesis of these complex rocks (Hubbard *et al.*, 1971, 1973). Unfortunately, there is no similarity potent combination of a ferromagnesian mineral, Fe or Mg concentrations, and one or more trace elements that allow an equal level of study of the nonplagioclase portion of these rocks. We have instead chosen to combine the advantages of the Eu–Sr pair with Mg data. A plot of MgO versus Sr (Fig. 8) is strikingly simple considering that Sr and MgO are not coupled by any mineralogical control. The simple variation in Sr and MgO seen for Apollo 14, 15, and 17 KREEP samples suggests a pervasive commonality, of precursors or generative mechanisms for these samples. The quantitatively different behavior of Sr versus Mg in the Apollo 16 KREEP and VHA samples, the general mass of anorthositic samples, and at least one rather Fe-rich and Sr-poor Apollo 17 KREEP like sample (72275,2) require either different precursors or a different generative mechanism because these differences cannot be caused simply by varying the ratio of plagioclase to ferromagnesian minerals. A line joining the whole rock and plagioclase points for 76535 troctolite forms an approximate boundary between rocks with basaltic chemical compositions and deep negative Eu anomalies and rocks with anorthositic chemical compositions and positive Eu anomalies. Note that some "non-normal" anorthositic samples, 63335 and 15459,38, plot far

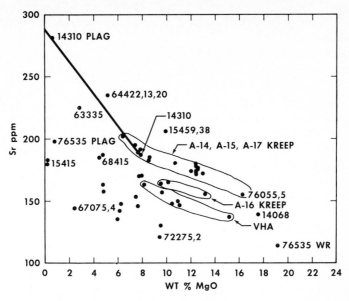

Fig. 8. Sr versus MgO for lunar non-mare and highland rocks. Compare this figure with Fig. 8. All data are from our research group or from Philpotts *et al.* (1972).

removed from the other anorthositic samples, reinforcing our rejection of these samples from the LKAS on the basis of other data.

INTERPRETATION OF DATA

The data presented in Figs. 1–8 document the chemical features of non-mare samples that appear to stem from lunar petrogenetic processes other than accidental, impact-related processes. These features are:

(1) The association of deep negative Eu anomalies and high levels of rare earths and related lithophile trace elements with rocks of basaltic chemical composition and the association of positive Eu anomalies and low levels of lithophilic trace elements with rocks of anorthositic chemical composition.

(2) The rather constant slope of the trivalent abundance pattern in lunar non-mare rock types of basaltic composition.

(3) The MgO–Al$_2$O$_3$ variation internal to individual chemical groups, such as VHA and Apollo 16/17 KREEP, and the parallel nature of this variation with the MgO–Al$_2$O$_3$ variation of terrestrial basaltic rocks and mare basalts.

(4) The greater range in Eu concentration relative to Sr concentrations.

(5) The inverse correlation of Eu and Al in the series LKAS → VHA → Apollo 16/17 KREEP.

The association of deep negative Eu anomalies with basaltic chemical composition (KREEP–VHA) is readily explained as the result of partially melting an aluminous source material that retains plagioclase in the residue. The experimen-

tal data of Walker *et al.* (1973) further describe the genesis of these chemical
compositions.

The rather constant slope of the trivalent REE abundance pattern in rocks
with deep negative Eu anomalies and basaltic chemical composition is almost
certainly a feature inherited from the source material for these rocks. Earlier
partial melting models (Hubbard *et al.*, 1971; Gast, 1972) generated this slope from
a flat, chondrite-normalized rare-earth abundance pattern in the source by using
clinopyroxene and very small percentages of melting to fractionate the La/Yb
ratio in liquid to double the value that existed in the source material prior to the
partial melting event. However, the absence or very minor role of clinopyroxene
in lunar non-mare rocks essentially invalidates partial melting models that depend
on clinopyroxene. The extraction of small percentages of liquid has always
appeared to be a problem. Partial melting models that use only plagioclase,
olivine, and orthopyroxene (Hubbard and Shih, 1973) are more consistent with
experimental and petrographic data for lunar non-mare rocks and can produce the
required enrichments. For example, models that have about 50% orthopyroxene
and low plagioclase (Fig. 9 and Table 4) can produce La/Yb ratios in the liquid that
are higher than in the source material by an amount approaching that observed in
the KREEP–VHA composition samples. However, this occurs only for the
production of very small percentages of liquid (1%), thus encountering a major
weakness of the earlier partial melting models, i.e. the extraction of small
percentages of liquid. If we allow 5% liquid, then numerous plagioclase–
orthopyroxene–olivine partial melting models still provide about a 20 fold
enrichment of rare earths and allow the production of even Apollo 14 common
KREEP by single-stage partial melting of a source material that had rare-earth
concentrations about 10 fold chondritic. On the other hand, 5% melting of either
clinopyroxene-bearing or clinopyroxene-free source materials cannot produce the
observed La/Yb ratio in the liquid by single-stage partial melting of a source

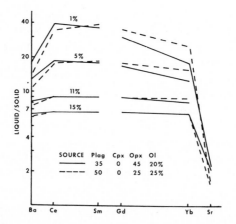

Fig. 9. Partial melting model diagram for two plagioclase, orthopyroxene, and olivine
assemblages. Distribution coefficients used are shown in Table 4.

Table 4. Distribution coefficients used for
Fig. 9. D = liquid/solid.

	Plag.	Opx	Oliv.
Ba	8	83	116
Ce	27	80	111
Sm	42	35	167
Gd	46	23	143
Yb	63	10	111
Sr	0.8	96	106

Plag. values are from Hubbard *et al.* (1971).

Opx values are from Gast (1973).

Oliv. values are from Schnetzler and Philpotts (1970) and Philpotts and Schnetzler (1970).

material that has absolutely chondritic relative abundances of the rare earths. If we make the extreme assumption that the observed La/Yb ratio already existed in the source material, then the deviation from absolutely chondritic relative abundances is not large, only requiring La to be 1.35× and Yb 0.675× chondritic. In view of the grossly nonchondritic composition of lunar samples this deviation is quite permissible.

The $MgO-Al_2O_3$ variation internal to the VHA and Apollo 16/17 KREEP and its parallelness with the $MgO-Al_2O_3$ variation in Apollo 12 and 15 mare basalts and in terrestrial basalts such as those from Waianae Volcano, Hawaii is considered to have the same basic cause in all four sets of samples, i.e. ordinary fractional crystallization in basaltic volcanic rocks that have olivine and secondarily pyroxene as the *mobile* early crystallizing minerals. Early crystallizing, but largely immobile, phases, such as plagioclase, are of minor importance in such fractional crystallization so long as they physically move very little relative to the liquid. In comparing lunar mare basalts and non-mare rocks of basaltic chemical composition with terrestrial basalts from a single volcano we assert that lunar rocks of basaltic composition cannot be expected to exhibit more systematic internal chemical variations than a single large basaltic volcano on earth. This is not meant to suggest that the moon once had large Hawaiian-type volcanoes, only that Hawaiian volcanoes are excellent examples of voluminous, but geographically restricted, production of basaltic magma. Probably only small, one eruption, volcanoes can be expected to show lesser internal chemical variations.

The greater range of Eu concentrations relative to Sr concentrations is readily explained if about 15% of the Eu is trivalent. The chemical data for KREEP-VHA samples can be used to estimate the $Eu^{+3/+2}$ ratio if one makes the assumption that Eu^{+2} should show nearly the same chemical behavior as Sr and that Eu^{+3} should be midway between Gd^{+3} and Sm^{+3} in its chemical behavior (Philpotts, 1970). It then follows via any relevant partial melting model, or even direct consideration of the REE and Sr data, that the approximately 2 fold greater range in relative Eu concentrations with respect to Sr and the increase in Eu concentration with increasing trivalent REE concentrations are the result of 10–20% of the Eu being in the trivalent oxidation state in the source material prior to melting. Experimental work by Morris and Haskin (1974) and Morris *et al.* (1974) give much the same $Eu^{+2/+3}$ ratio. Morris and Haskin found that for a fixed p_{O_2} the $Eu^{+3/+2}$ ratio is a steep function of bulk composition for glasses in the compositional range from anorthite to Ca–Mg pyroxene. Specifically, the nearer the glass composition to the pyroxene end member the higher is the $Eu^{+3/+2}$ ratio. Related research by Morris *et al.* (1974) found that even in glass of anorthite composition ($CaAl_2Si_2O_8$) at 1370–1600°C the Eu^{+3} was about 10% of the total Eu at $p_{O_2} = 10^{-12}$ and about 4% of the total at $p_{O_2} = 10^{-14}$. Thus, even a pure anorthite liquid at lunar p_{O_2} should contain a few percent of trivalent Eu. Glass of diopside composition ($CaMgSi_2O_6$) at 1450°C at $p_{O_2} = 10^{-14}$ was found to have $Eu^{+3/+2} \sim 0.5$. These experimental data suggest that silicate liquids on the moon of KREEP–VHA chemical composition, i.e. 50% or less pyroxene, will have about $\frac{1}{4}$ or less of their Eu in the trivalent oxidation state.

The inverse correlation of Al and Eu for the series of chemical compositions LKAS–VHA–Apollo 16/17 KREEP–Apollo 14 common KREEP is partly explained by coupling the pseudoternary diagram of Walker *et al.* (1973) (our Fig. 1) with partial melting model calculations. On the pseudoternary diagram (Fig. 1) the series Apollo 14 common KREEP → Apollo 16/17 KREEP → VHA is moving toward higher liquidus temperatures and more Al-rich compositions (see also Fig. 2). The Eu and trivalent REE are decreasing in this series which, in terms of partial melting models, implies increasing percentages of liquid, which in turn is consistent with the higher liquidus temperatures observed for the more Al-rich compositions shown on the pseudoternary diagram. Basically this is a description of the partial melting of an aluminous source material where Al is retained by the residual plagioclase and Eu is enriched in the liquid due to the presence of 10–20% of Eu^{+3}. Thus, as the degree of melting varies, the inverse relationship of Eu and Al results. Extrapolation of the above progression to include the LKAS is not valid because the LKAS does not have the negative Eu and Sr anomalies required for the production of Al-rich liquids by partially melting plagioclase-bearing source materials.

The LKAS was anticipated by Hubbard *et al.* (1971) when they calculated the REE, Ba, and Sr concentrations in a hypothetical silicate liquid in equilibrium with anorthosite 15415. We still accept the crystal plus equilibrium liquid model behind those calculations as the best chemical explanation of the internal, plagioclase-related chemical variations. However, the composition of the whole rock (i.e. liquid plus plagioclase), before fractionation of plagioclase, had to have a higher Al_2O_3 concentration than the equilibrium liquid, which is approximated by samples like 77017 ($\sim$26% Al_2O_3), although how much is unclear from present data. The positive Eu anomaly associated with the whole-rock composition, or of the LKAS members with 26% Al_2O_3, rules out the generation of these samples by removal of large amounts of plagioclase from their precursors unless these precursors once had much larger positive Eu anomalies and high Al_2O_3 concentrations. The direct origin of this series as a partial melt is also ruled out because of the required higher Al_2O_3 concentrations and by the small positive Eu anomaly. Although the plagioclase fractional crystallization model for the internal chemical variations of the LKAS requires that the LKAS itself was once partly molten, the LKAS was not made by plagioclase fractional crystallization or by partial melting, unless an even more Al_2O_3-rich precursor can be found. One might consider the LKAS, and lunar anorthositic rocks in general, to be the residue from a partial melting event. However, unless very large percentages of liquid were produced (30–50%) and removed the initial material had to be very Al-rich and we have not escaped from the basic problem of explaining the origin of highly aluminous lunar rocks. The LKAS and anorthositic samples in general could have been made by fractional crystallization of olivine, orthopyroxene, and/or spinel from a less Al-rich precursor, i.e. the denser ferromagnesian minerals sank leaving a residue enriched in plagioclase and liquid. The observed LKAS variations in plagioclase content could then postdate the ferromagnesian fractional crystallization and be the result of filter pressing. The extent of such ferromagnesian fractional crystallization cannot be specified from present data, but the troctolite sample 76535 cannot be a direct result of crystal accumulation from the LKAS because of the higher Sr and Eu concentrations of the plagioclase in 76535. The sinking of ferromagnesian minerals could be increasingly retarded by a mush of earlier crystallizing plagioclase crystals as the bulk composition of the liquid residue approached the Al_2O_3 concentrations of the anorthositic rocks. This is probably the most serious weakness of this model. Several fractional crystallization plus filter pressing schemes can be used to generate much of the chemical variations found in the anorthositic group of lunar rocks. However, the salient problem is how to produce such Al-rich compositions on the moon. The alternative is to assume a precursor that is already of anorthositic composition. Nonetheless, we currently consider the fractional crystallization of ferromagnesian minerals, perhaps aided by filter pressing, to be the most reasonable way to produce anorthositic rocks on the moon.

THE RELATIONSHIP BETWEEN BASALTIC AND ANORTHOSITIC COMPOSITIONS

The relationships suggested by Walker *et al.* (1973), i.e. the partial melting of a series of rocks of troctolite to anorthositic norite chemical composition, to give VHA–KREEP is accepted as the best current explanation of the major element chemistry although we prefer less feldspathic compositions. We add the constraint that the parent materials for non-mare rocks of basaltic chemical composition (VHA–KREEP) must have nearly the same trivalent REE pattern as VHA–KREEP itself and that if the degree of partial melting is to be small (~5%), then the concentrations in the parent material must be about 10 times chondritic and with little or no Eu and Sr anomalies. This parent material is probably not the observed LKAS because of the lower average MgO/FeO ratio of anorthositic rocks relative to VHA–KREEP. It cannot presently be determined if any analyzed rocks were, or represent, the parental materials for VHA–KREEP.

We have repeatedly suggested that lunar non-mare and highland rocks were made from material richer in Al and lithophilic trace elements (Hubbard and Gast, 1971; Hubbard *et al.*, 1971; Hubbard *et al.*, 1972) than the parent material for mare basalts and that this difference stems from original heterogeneities rather than large-scale differentiation of once homogeneous lunar material. Although it may be argued (Taylor and Jakes, 1974) that the source regions for all observed lunar rocks were produced from an initially homogeneous moon by extensive differentiation, this approach seems improbable to us because of the higher MgO/FeO ratio of non-mare rocks relative to mare basalts (Ringwood *et al.*, 1972), the high Cr concentration of non-mare basaltic and anorthositic rocks (Hubbard *et al.*, 1972) and the more aluminous nature of the non-mare rocks, especially so long as large-scale plagioclase flotation remains unproven. These data all argue for noninvolvement of the source regions for non-mare rocks in any differentiation that may have produced the source regions of mare basalts. We propose the following petrogenesis of non-mare rocks and their immediate precursors: (1) the KREEP–VHA series of chemical compositions was made by partially melting a suite of precursors that may have been related to one another; (2) extreme melting of these precursors produced conditions allowing generation of LKAS by removal of olivine and/or orthopyroxene by fractional crystallization; (3) the entire suite of non-mare materials was involved in a metamorphism and igneous differentiation separate from any that may have involved the source regions of mare basalts; (4) the precursors were the original material that accreted onto the outer tens of kilometers of the moon and are proposed to have been chemically indistinguishable from bodies that impacted the moon early in its post accretional history; and (5) thermal metamorphism of the precursors preceded the generation of KREEP–VHA or LKAS from the precursors and served to convert original mineral assemblages to the plagioclase, orthopyroxene, olivine ± spinel assemblages inferred by Walker *et al.* (1973) and the present interpretation. This metamorphism is the natural result of rising temperatures prior to melting. It may not be unreasonable to expect that some returned samples have experienced only this metamorphism. These precursors are predicted to have more than about 10% Al_2O_3, MgO/FeO ratios greater than 1.5, Ti and lithophile trace elements were at 5–15 fold chondritic concentrations.

No heat source is proposed in this paper but the chemical record itself is direct evidence of an adequate heat source because the only known way to produce the observed range of chemical compositions is by igneous processes.

REFERENCES

Agrell S. O., Agrell J. E., Arnold A. R., and Long J. V. P. (1973) Some observations on rock 62295 (abstract). In *Lunar Science—IV*, pp. 15–17. The Lunar Science Institute, Houston.

Gast P. W. (1972) The chemical composition and structure of the moon. *The Moon* 5, 121–148.

Goles G. G., Duncan A. R., Lindstrom D. J., Martin M. R., Beyer R. L., Osawa M., Randle K., Meek L. T., Steinborn T. L., and McKay S. M. (1971) Analyses of Apollo 12 specimens: Compositional variations, differentiation processes and lunar soil mixing models. *Proc. Second Lunar Sci. Conf., Geochim. Cosmochim. Acta*, Suppl. 2, Vol. 2, pp. 1063–1081. MIT Press.

Haskin L. A., Helmke P. A., Blanchard D. P., Jacobs J. W., and Telander K. (1973) Major and trace element abundances in samples from the lunar highlands. *Proc. Fourth Lunar Sci. Conf., Geochim. Cosmochim. Acta*, Suppl. 4, Vol. 2, pp. 1275–1295. Pergamon.

Helz R. T. and Appleman D. E. (1973) Mineralogy, petrology and crystallization history of Apollo 16 rock 68415. *Proc. Fourth Lunar Sci. Conf., Geochim. Cosmochim. Acta*, Suppl. 4, Vol. 1, pp. 643–660. Pergamon.

Hubbard N. J. and Gast P. W. (1971a) Chemical composition and origin of nonmare lunar basalts. *Proc. Second Lunar Sci. Conf., Geochim. Cosmochim. Acta*, Suppl. 2, Vol. 2, pp. 999–1020. MIT Press.

Hubbard N. J., Meyer C., Jr., Gast P. W., and Wiesmann H. (1971b) The composition and derivation of Apollo 12 soils. *Earth Planet. Sci. Letts.* 10, 341–350.

Hubbard N. J., Gast P. W., Rhodes J. M., Bansal B. M., Wiesmann H., and Church S. E. (1972) Nonmare basalts: Part II. *Proc. Third Lunar Sci. Conf., Geochim. Cosmochim. Acta*, Suppl. 3, Vol. 2, pp. 1161–1179. MIT Press.

Hubbard N. J. and Shih C. (1973) Opposing partial melting models and the genesis of some lunar magmas (abstract). *36th Annual Meeting Meteoritical Society*, pp. 72–73.

Hubbard N. J., Rhodes J. M., Gast P. W., Bansal B. M., Shih C.-Y., Wiesmann H., and Nyquist L. E. (1973) Lunar rock types: The role of plagioclase in non-mare and highland rock types. *Proc. Fourth Lunar Sci. Conf., Geochim. Cosmochim. Acta*, Suppl. 4, Vol. 2, pp. 1297–1312. Pergamon.

Hubbard N. J., Rhodes J. M., and Gast P. W. (1973) Chemistry of very high Al_2O_3 lunar basalts. *Science* 181, 339–342.

Hubbard N. J., Gast P. W., Meyer C., Nyquist L. E., Shih C., and Wiesmann H. (1974) Chemical composition of lunar anorthosites and their parent liquids. *Earth Planet. Sci. Lett.* 13, 71–75.

Longhi J., Walker D., and Hays J. F. (1972) Petrography and crystallization history of basalts 14310 and 14072. *Proc. Third Lunar Sci. Conf., Geochim. Cosmochim. Acta*, Suppl. 3, Vol. 1, pp. 131–140. MIT Press.

LSPET (Lunar Sample Preliminary Examination Team) (1972) The Apollo 15 lunar samples: A preliminary description. *Science* 175, 363–375.

LSPET (Lunar Sample Preliminary Examination Team) (1973) The Apollo 16 lunar samples: Petrographic and chemical description. *Science* 179, 23–34.

LSPET (Lunar Sample Preliminary Examination Team) (1973) Apollo 17 lunar samples: Chemical and petrographic description. *Science* 182, 659–674.

MacDonald G. A. and Katsura T. (1964) Chemical composition of Hawaiian lavas. *J. Petrology* 5, 82–133.

Meyer C., Jr., Anderson D. H., and Bradley J. G. (1974) Ion microprobe mass analyses of plagioclase from "non-mare" lunar samples. *Proc. Fifth Lunar Sci. Conf., Geochim. Cosmochim. Acta.* This volume.

Morris R. V., Haskin L. A., Biggar G. M., and O'Hara M. J. (1974) Measurement of the effects of temperature and partial pressure of oxygen on the oxidation states of europium in silicate glasses. *Geochim. Cosmochim. Acta.* In press.

Morris R. V. and Haskin L. A. (1974) EPR measurement of the effect of glass composition on the oxidation states of europium. *Geochim. Cosmochim. Acta.* In press.

Philpotts J. A. (1970) Redox estimation from a calculation of Eu^{+2} and Eu^{+3} concentrations in natural phases. *Earth Planet. Sci. Lett.* 9, 257–268.

Philpotts J. A. and Schnetzler C. C. (1970) Phenocryst-matrix partition coefficients for K, Rb, Sr, and Ba, with application to anorthosite and basalt genesis. *Geochim. Cosmochim. Acta* 34, 307–322.

Philpotts J. A., Schnetzler C. C., Nava D. F., Bottino M. L., Fullagar P. D., Thomas H. H., Schuhmann S., and Kouns C. W. (1972) Apollo 14: Some geochemical aspects. *Proc. Third Lunar Sci. Conf., Geochim. Cosmochim. Acta*, Suppl. 3, Vol. 2, pp. 1293–1306. MIT Press.

Rhodes J. M. and Hubbard N. J. (1973) Chemistry, classification and petrogenesis of Apollo 15 mare basalts. *Proc. Fourth Lunar Sci. Conf., Geochim. Cosmochim. Acta*, Suppl. 4, Vol. 2, pp. 1127–1148. Pergamon.

Ridley W. I., Reid A. M., Warner J. L., Brown R. W., Gooley R., and Donaldson C. (1973) Glass compositions in Apollo 16 soils 60501 and 61221. *Proc. Fourth Lunar Sci. Conf., Geochim. Cosmochim. Acta*, Suppl. 4, Vol. 1, pp. 309–322. Pergamon.

Ringwood A. E., Green D. H., and Ware N. G. (1972) Experimental petrology and petrogenesis of Apollo 14 basalts (abstract). In *Lunar Science—III*, pp. 654–656. The Lunar Science Institute, Houston.

Rose H. J., Jr., Cuttitta F., Berman S., Carron M. K., Christian R. P., Dwornik E. J., Greenland L. P., and Ligon D. T., Jr. (1973) Compositional data for twenty-two Apollo 16 samples. *Proc. Fourth Lunar Sci. Conf., Geochim. Cosmochim. Acta*, Suppl. 4, Vol. 2, pp. 1149–1158. Pergamon.

Schnetzler C. C. and Philpotts J. A. (1970) Partition coefficients of rare earth elements between igneous matrix material and rock forming mineral phenocrysts—II. *Geochim. Cosmochim. Acta* **34**, pp. 331–340.

Schonfeld E. (1974) Component abundance and evolution of regoliths and breccias: Interpretation by mixing models (abstract). In *Lunar Science—V*, pp. 669–1671. The Lunar Science Institute, Houston.

Shih C. (1972) The rare earth geochemistry of oceanic igneous rocks. Ph.D. thesis, Columbia University.

Simonds C. H., Warner J. L., and Phinney W. C. (1973) Petrology of Apollo 16 poikilitic rocks. *Proc. Fourth Lunar Sci. Conf., Geochim. Cosmochim. Acta*, Suppl. 4, Vol. 1, pp. 613–632. Pergamon.

Taylor S. R., Gorton M. P., Muir P., Nance W., Rudowski R., and Ware N. (1973) Lunar highlands composition: Apennine Front. *Proc. Fourth Lunar Sci. Conf., Geochim. Cosmochim. Acta*, Suppl. 4, Vol. 2, pp. 1445–1460. Pergamon.

Taylor S. R. and Jakes P. (1974) Geochemical zoning in the moon (abstract). In *Lunar Science—V*, pp. 786–788. The Lunar Science Institute, Houston.

Walker D., Grove T. L., Longhi J., Stolper E. M., and Hays J. F. (1973) Origin of lunar feldspathic rocks. *Earth Planet. Sci. Lett.* **20**, 325–336.

Wänke H., Baddenhausen H., Balacescu A., Teschke F., Spettel B., Dreibus G., Palme H., Quijano-Rico, Kruse H., Wlotzka F., and Begeman F. (1972) Multielement analyses of lunar samples and some implications of the results. *Proc. Third Lunar Sci. Conf., Geochim. Cosmochim. Acta*, Suppl. 3, Vol. 2, pp. 1251–1268. MIT Press.

Warner J. L., Simonds D. H., and Phinney W. C. (1974) Impact induced fractionation in the highlands (abstract). In *Lunar Science—V*, pp. 823–825. The Lunar Science Institute, Houston.

Weill D. and Drake M. (1973) Europium anomaly in plagioclase feldspar: An explanation. *Science* **180**, 1059–1060.

Proceedings of the Fifth Lunar Conference
(Supplement 5, Geochimica et Cosmochimica Acta)
Vol. 2 pp. 1247–1253 (1974)
Printed in the United States of America

Possible REE anomalies of Apollo 17 REE patterns

Akimasa Masuda,[1] Tsuyoshi Tanaka,[2] Noboru Nakamura,[1]
and Hajime Kurasawa[2]

[1]Department of Earth Sciences, Kobe University, Nada-ku, Kobe, Japan 657

[2]Geological Survey of Japan, Hisamoto, Takatsu-ku, Kawasaki, Japan 213

Abstract—Six Apollo 17 samples were analyzed for the rare-earth elements and Ba. It is pointed out that the "liquid-type" lunar materials show the positive Ce anomaly, while the "solid-type" lunar materials do not show the corresponding effect. These results from the Apollo 17 analyses are in agreement with Nakamura's (1974a) recent observation about the Ce anomaly.

Introduction

We (Masuda *et al.*, 1972a) are the first to point out the positive cerium anomaly in lunar materials. Laul *et al.* (1972) also admitted a similar anomaly, but seemed not confident of its being real. Nakamura (1974a) observed that the apparent positive Ce anomaly bears on the types of chondrite-normalized REE patterns. According to his investigation, the straight-type patterns show positive Ce anomaly, while the curved-type patterns show little if any anomaly for Ce. The data on six Apollo 17 samples studied here appear to be in accord with his suggestion.

Experimental

The RE elements and Ba were determined by stable isotope dilution with much caution (Masuda *et al.*, 1972b, 1973). The sample size for each analysis was about 150–200 mg. In general, the precisions in REE determination are better than 1% and the accuracies are considered to be better than 1.5%.

Results and Discussion

The analytical data are presented in Table 1. The samples with the curved-type ("solid-type") RE patterns are shown in Fig. 1, and those with the straight-type ("liquid-type") ones are shown in Fig. 2. Here the Leedey chondrite (Masuda *et al.*, 1973) is chosen for normalization. Even if the average values (Nakamura, 1974b) for common ordinary chondrites are employed instead, it makes no difference, except for the absolute magnitude level.

The solid-type group can be understood to represent the products of almost complete melting of the material system which was once separated from the primary liquid-type magma (Masuda, 1972; Philpotts *et al.*, 1973). Masuda's inference like this is based on a good fitting of "linear" bulk partition coefficients to the observed curved-type patterns.

As reported previously, when the results on 74220 (Tanaka *et al.*, 1973) are

Table 1. REE abundances (ppm) in the Apollo 17 samples and apparent anomalies for Ce and Yb.

	74220,52 Unsieved soil	74260,13 Unsieved soil	72441,6 Soil (<1 mm)	73235,59 Blue–gray breccia	70215,78 Fine basalt	75075,59 Medium basalt
La	6.54	9.81	17.0	22.8	5.35	5.67
Ce	19.7	28.2	47.2	60.4	17.3	19.5
Nd	18.4	23.2	29.9	36.7	17.0	21.0
Sm	6.82	8.35	8.73	10.4	6.98	8.90
Eu	1.91	1.73	1.40	1.42	1.45	2.00
Gd	8.99	11.0	10.9	12.6	10.3	12.9
Dy	9.54	12.8	11.6	13.7	12.7	15.7
Er	5.17	7.71	7.13	8.28	7.91	9.48
Yb	4.41	6.96	6.50	7.74	7.45	8.71
Lu	0.604	0.975	0.886	1.07	1.07	1.22
Ba	80.4	114	214	260	61.8	72.3
ΔCe(%)	~0	+3	+5 ~ +10	+8	+2	+1
ΔYb(%)	—	—	—	+7	—	+5
Type	Curved	Curved	Straight	Straight	Curved	Curved

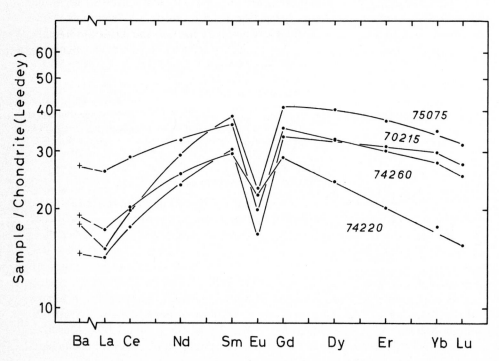

Fig. 1. The chondrite-normalized RE patterns of four Apollo 17 samples, 75075, 70215, 74260, and 74220.

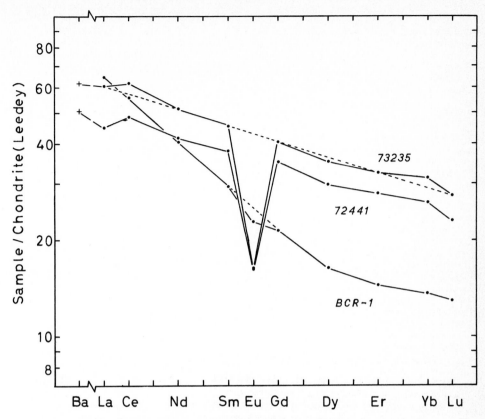

Fig. 2. The chondrite-normalized RE patterns of two Apollo 17 samples, 73235 and 72441, and a terrestrial standard basalt (BCR-1).

divided by the bulk partition coefficients estimated by Masuda and Nakamura (1970), the logarithmically rectilinear pattern is obtained (Tanaka *et al.*, 1973). The sample 10058 (Gast *et al.*, 1970) from Mare Tranquillitatis resembles 75075 (medium basalt) with respect to the chondrite-normalized RE pattern. When the chondrite-normalized RE values for 75075 are divided by the similar bulk partition coefficients but with some modification for heavier REE, a horizontal RE pattern is obtained (see Fig. 3). This suggests that the precursory material for 75075 was the solid material system (cumulate) separated from the melt with the same relative RE abundances as in chondrites. According to the notation presented by Masuda and Jibiki (1973), the set of bulk partition coefficients obtained by Masuda and Nakamura (1970) is expressed as $K(-3.3/14.0; 7h)$, while the modified set employed for 75075 is defined as $K(-3.3/14.0; 6d [0.4])$. Note that there is no difference for the lightest REE between two sets.

The applicability of the bulk partition coefficients advanced by us would mean either that the materials like 74220 and 75075 are the products of nearly complete

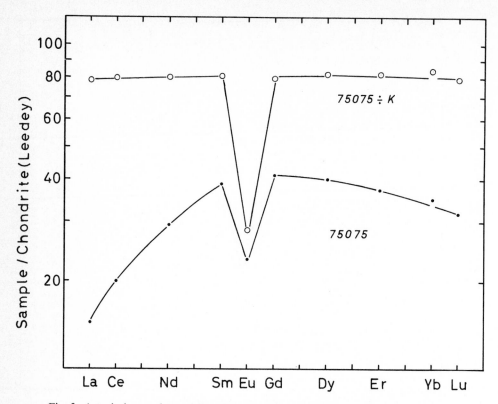

Fig. 3. A typical curved-type RE pattern for 75075 and a pattern resulting from the division by the bulk partition coefficients K. For the coefficients used, see text.

melting of cumulate or that substantially all amounts of REE in the cumulate moved in the product of the partial melting.

The curvatures of solid-type RE patterns of 70215 and 74260 (the latter in particular) are less than those mathematically anticipated from $K(-3.3)$. Theoretically, it is likely that this could be ascribed to some extent of inclusion of the liquid-type material. Based on the numerical treatment employing $K(-3.3/14.0)$, 74260 can be accounted for as involving 25% liquid-type material.

At the same time, the mathematical treatment of data in this fashion enables us to estimate the Ce anomaly. As presented in Table 1, the lunar materials with the curved-type (or solid-type) RE patterns have no recognizable Ce anomaly. However, the value of 3% for 74260 might be regarded as tangible. This is parallel with the aforementioned possible inclusion of some extent of the liquid-type (straight-type) material.

The samples 73235 and 72441 are similar in the RE patterns (Fig. 2), and can be considered as belonging to the liquid-type materials. For 73235, La, Nd, Sm, Gd, Er, and Lu fall closely on a straight line within the allowances of ±0.7%, with no gap between the light REE group and the heavy one. (A small gap between such two groups is observed sometimes (Masuda et al., 1972b).) Apart from the great anomaly of Eu, there are some deviations for Ce, Yb, and Dy. Meanwhile, the pattern of 72441 is somewhat less smooth. It may be worth mentioning that 72441 is the sieved material. Although the effect of sieving on the smoothness of the RE pattern has never been studied precisely, it may deserve further investigation. Anyway this sample is not very good for assessment of the degree of possible Ce anomaly.

The sample 73235 could be judged to have the positive Ce anomaly of 7.8%, based on a smooth straight line passing through La, Nd, and Sm. This value is comparable with those for Apollo 15 samples (Masuda *et al.*, 1972b). According to our interpretation, this breccia appears to have a positive Yb anomaly. Deviation of Dy is small ($\sim$3%). Though not accountably well, such a small negative "aberration" of Dy arises sometimes, for example, in Peace River, 15101 (Masuda *et al.*, 1972b), and an Allende inclusion (Tanaka and Masuda, 1973).

It is true that La^{3+} and Lu^{3+} display some peculiarity in certain chemical properties with respect to adjacent trivalent REE, which suggests that apparent La and Lu "anomalies" are not necessarily inconceivable. Generally, however, the possible variation of valency would give rise to relatively prominent anomaly. For additional reasons as mentioned below, we have preferred to consider that the rather small irregularities observed are due to Ce and Yb anomalies. (Of course, someone might suggest a compromise, for instance, that the particular unsmoothness for the heaviest REE is due to both Yb and Lu.)

Nakamura and Masuda (1973) have discovered a chondrite with undeniably high positive Ce anomaly. In their discussions they paid attention to the exceptional chemical state of cerium oxide in the gaseous state and the comparatively high vapor pressure of cerium oxide (Shchukarev and Semenov, 1961; Benezech and Foëx, 1969). It is evident that the behavior of Ce can be irregular under certain conditions at elevated temperatures when compared with other REE; in particular, the relatively high volatility of cerium oxide would be noteworthy. A marked anomalous behavior of Yb is also detected in meteoritic samples (Tanaka and Masuda, 1973; Nakamura and Masuda, 1973).

Emphasis should be placed on the fact that our analyses of terrestrial rocks seldom or never reveal such positive Ce anomalies as found in lunar samples. It is noted that our present statement rests on our results for meteoritic, lunar and terrestrial samples and that the same spike solutions are used in analyzing all of them. As an example, the pattern based on our data (Nakamura *et al.*, 1973) on the USGS standard basalt BCR-1 is presented in Fig. 2. It can be seen that this terrestrial rock has no tangible positive Ce anomaly (just 1.1% according to the least-square method).

Apart from these facts, a mathematically great significance can be given to a good fitting of linear bulk partition coefficients to the solid-type RE patterns as well as a logarithmical linearity of the liquid-type RE patterns, because these aspects covering La, Nd, and Sm are mathematically consistent to each other and imply that there is little mathematically sound reason to admit the La anomaly unless this consistency is a fortuitous one. From a similar consideration about the liquid-type and solid-type RE patterns, the Yb anomaly is preferred to the Lu anomaly.

Our data on the Apollo 17 samples are in general agreement with the observation by Nakamura (1974a) as mentioned above. But it is a puzzling problem why the liquid-type (straight-type) materials have the positive Ce anomaly, while the solid-type (curved-type) ones are free from such an effect. Would it be too foolhardy to speculate that excess cerium was added to the surface layer material immediately after the primary separation of liquid-type material and underlying solid-type material? Tatsumoto *et al.* (1972) suggested that one could not rule out the possibility that a lead greater in $^{207}Pb/^{206}Pb$ than that of primordial lead was added. Although it is premature to associate the positive Ce anomaly with the above problem of lead without the deliberate interchecking, this subject would deserve further consideration.

Acknowledgment—This work was supported by Grant-in-Aid for Scientific Research from the Ministry of Education, Japan.

REFERENCES

Benezech G. and Foëx M. (1969) Vapor pressure measurements of lanthanide oxides between 2000 and 2400°. *Compt. Rend. Acad. Sci., Ser. C* **263**, 2315–2318.

Gast P. W., Hubbard N. J., and Wiesmann H. (1970) Chemical composition and petrogenesis of basalts from Tranquility Base. *Proc. Apollo 11 Lunar Sci. Conf., Geochim. Cosmochim. Acta,* Suppl. 1, Vol. 2, pp. 1143–1163. Pergamon.

Laul J. C., Wakita H., Showalter D. L., Boynton W. V., and Schmitt R. A. (1972) Bulk, rare earth, and other trace elements in Apollo 14 and 15 and Luna 16 samples. *Proc. Third Lunar Sci. Conf., Geochim. Cosmochim. Acta,* Suppl. 3, Vol. 2, pp. 1181–1200. MIT Press.

Masuda A. (1972) Lunar solid-type and liquid-type materials and rare-earth abundances. *Nature (Phys. Sci.)* **235**, 132–133.

Masuda A. and Jibiki H. (1973) Rare-earth patterns of Mid-Atlantic Ridge gabbros: Continental nature? *Geochem. J.* **7**, 55–65.

Masuda A. and Nakamura N. (1970) Evolution of lunar materials in light of the abundance variation of oxyphile trace elements. *Contr. Mineral. and Petrol.* **29**, 11–27.

Masuda A., Nakamura N., Kurasawa H., and Tanaka T. (1972a) Chondrite-normalized rare earth patterns of the Apollo 14 samples (abstract). In *Lunar Science—III*, pp. 517–519. The Lunar Science Institute, Houston.

Masuda A., Nakamura N., Kurasawa H., and Tanaka T. (1972b) Precise determination of rare-earth elements in the Apollo 14 and 15 samples. *Proc. Third Lunar Sci. Conf., Geochim. Cosmochim. Acta,* Suppl. 3, Vol. 2, pp. 1307–1313. MIT Press.

Masuda A., Nakamura N., and Tanaka T. (1973) Fine structures of mutually normalized rare-earth patterns of chondrites. *Geochim. Cosmochim. Acta* **37**, 239–248.

Nakamura N. (1974a) Positive Ce anomaly and two types of lunar samples. *Geochem. J.* **8**, 67–74.

Nakamura N. (1974b) Determination of REE, Ba, Fe, Mg, Na and K in carbonaceous and ordinary chondrites. *Geochim. Cosmochim. Acta* **38**, 757–775.

Nakamura N. and Masuda A. (1973) Chondrites with peculiar rare-earth patterns. *Earth Planet. Sci. Lett.* **19**, 429–437.

Nakamura N., Masuda A., Tanaka T., and Kurasawa H. (1973) Chemical compositions and rare-earth features of four Apollo 16 samples. *Proc. Fourth Lunar Sci. Conf., Geochim. Cosmochim. Acta,* Suppl. 4, Vol. 2, pp. 1407–1414. Pergamon.

Philpotts J. A., Schuhmann S., Schnetzler C. C., Kouns C. W., Doan A. S., Jr., Wood F. M., Jr., Bickel A. L., and Lum Staab R. K. L. (1973) Apollo 17: Geochemical aspects of some soils, basalts, and breccia. *Trans. Amer. Geophys. Union* **54**, 603–604.

Shchukarev S. A. and Semenov G. A. (1961) A mass-spectrometric study of the vapor composition over oxides of rare-earth elements. *Dokl. Akad. Nauk SSSR* **141**, 652–654.

Tanaka T. and Masuda A. (1973) Rare-earth elements in matrix, inclusions, and chondrules of the Allende meteorite. *Icarus* **19**, 523–530.

Tanaka T., Kurasawa H., Nakamura N., and Masuda A. (1973) Rare-earth elements in fines 74220. *EOS Transactions* **54**, 614.

Tatsumoto M., Hedge C. E., Doe B. R., and Unruh D. M., (1972) U–Th–Pb and Rb–Sr measurements on some Apollo 14 lunar samples. *Proc. Third Lunar Sci. Conf., Geochim. Cosmochim. Acta*, Suppl. 3, Vol. 2, pp. 1531–1555. MIT Press.

Proceedings of the Fifth Lunar Conference
(Supplement 5, Geochimica et Cosmochimica Acta)
Vol. 2 pp. 1255–1267 (1974)
Printed in the United States of America

Origin of Apollo 17 rocks and soils

John A. Philpotts, S. Schuhmann, C. W. Kouns, R. K. L. Lum,
and S. Winzer

Astrochemistry Branch, Code 691, NASA–Goddard Space Flight Center,
Greenbelt, Maryland 20771

Abstract—Lithophile trace element abundances have been determined by mass spectrometric isotope dilution for a suite of Apollo 17 samples. The six mare basalts have generally similar relative trace element abundances; they are also similar to Apollo 11 trace element poor basalts. It is suggested that these basalts were derived by partial fusion of cumulates. The Apollo 17 highland breccias show an order of magnitude range in trace element abundances although there is a clustering of KREEP-rich samples which are interpreted as mixtures. The Apollo 17 soils show only a limited range of trace element abundances. They are mixtures of highland breccias, mare basalts, and orange–black "soil." There appear to be two groups of soils, Light Mantle and the rest. Both groups seem to have the same basalt component, which is similar to Station 4 basalt from Shorty Crater and probably is the uppermost basalt unit throughout the Taurus-Littrow valley. Dark Mantle within the valley seems to be largely comminuted basalt. The main group of soils appears to contain a highland component of uniform composition. The sources of both this component and the Light Mantle are uncertain at this time.

INTRODUCTION

THE APOLLO 17 MISSION to the moon returned samples from the Taurus-Littrow valley located at the southeastern edge of Mare Serenitatus. The site is geologically interesting and contains both highland and mare units (AFGIT, 1973; Schmitt, 1973). We have determined lithophile trace element abundances in a suite of Apollo 17 samples. The analytical technique utilized is mass-spectrometric stable-isotope dilution (Schnetzler *et al.*, 1967). The determined concentrations are believed to be good within several percent. Abundance values obtained for the standard rock powder BCR-1 are reported in the data table. Analytical blank (Philpotts *et al.*, 1972c) is generally less than 1%.

The purpose of this investigation was to provide high quality lithophile trace element abundance data for Apollo 17 samples, to examine variations in these abundances within the various types of samples, to relate the samples to those from other lunar sites in terms of a genetic model, and to consider the origin of the soils in terms of the local rocks and site geology.

MARE BASALTS

Lithophile trace element abundances in six mare basalts are reported in Table 1 and plotted in Fig. 1. Three of the basalts are fragments from the Station 4, 1–2 mm fines. The other samples are from Stations LM, 1A, and 5 (Fig. 2). The data indicate that all the basalts are similar and could be consanguinous. Sample

1255

Table 1. Lithophile trace element abundances in ppm by weight in Apollo 17 samples.

	MARE BASALT						Core Tube				SOILS			Trench		Drive Tube	
			74242,8				~45 cm <250 μm	~65 cm						~1 cm	~15 cm	~40 cm	
	70017,23	71055,38	f.g.	int.	c.g.	75035,44	70008,226	70008,249	70181,15	71501,31	72141,19	72161,16	72501,30	73121,14	73141,18	74001,11	74121,1
	LM	STN 1A	STN 4	STN 4	STN 4	STN5	LM	LM	LM	STN 1A	LRV 2	LRV 3	STN 2	STN 2A	STN 2A	STN 4	LRV 6
Li	8.57	9.32	8.83	9.59	9.26	11.0	10.2	10.5	9.85	8.63	9.33	9.17	11.9	9.91	9.82	11.8	10.3
K	307	342	534	609	525	549	605	720	696	602	843	835	1400	1160	1170	596	1130
Rb	0.280	0.362	0.750	0.944	(0.851)	0.679	1.26	1.68	1.42	1.14	2.19	2.01	4.60	3.51	3.56	0.814	3.24
Sr	168	121	161	168	148	192	167	195	167	156	154	155	155	144	147	203	150
Ba	43.0	62.4	76.6	78.7	65.4	92.9	95.0	—	104	86.0	120	115	211	171	171	73.8	167
Ce	10.7	15.6	21.5	23.3	—	23.6	23.5	24.6	24.4	21.5	27.7	25.7	—	38.1	37.7	18.4	39.0
Nd	12.1	17.0	23.8	25.8	23.5	27.3	23.1	20.7	22.0	20.7	20.7	20.5	27.8	24.8	24.8	17.8	25.6
Sm	5.13	6.72	9.55	10.6	9.04	11.2	8.87	7.32	8.18	8.02	6.69	6.84	8.18	7.14	7.00	6.61	7.55
Eu	1.62	1.36	1.87	1.97	1.85	2.52	1.60	1.57	1.71	1.67	1.41	1.45	1.38	1.26	1.24	1.86	1.33
Gd	—	—	13.6	14.9	13.8	17.1	—	10.0	11.0	10.7	8.66	9.13	9.74	8.75	8.15	8.52	—
Dy	10.2	13.0	16.6	19.2	16.8	19.7	14.0	11.5	13.1	13.3	9.73	9.99	11.0	9.65	9.39	9.01	10.4
Er	6.31	7.74	9.68	10.3	10.0	11.1	8.52	7.27	7.52	7.84	5.68	5.77	6.33	5.85	5.73	4.68	6.44
Yb	6.25	7.75	9.14	9.74	10.2	11.4	7.65	6.13	7.06	7.28	5.29	5.39	6.14	5.38	5.46	4.00	5.79
Lu	0.954	—	1.35	1.51	1.62	1.70	1.11	0.956	1.07	1.11	0.819	0.837	0.929	—	0.825	0.617	0.895
Zr	223	—	309	489	—	—	208	208	338	224	191	206	288	238	235	223	213
wt. in mg	117.34	88.29	48.63	41.73	9.6	87.29	42.17	37.30	120.40	91.28	122.89	132.14	98.75	203.90	96.61	73.80	90.43

Table 1 (*continued*).

| | SOILS | | | | | | | | | | | SOIL BRECCIA | BRECCIAS | | BCR-1 |
| | Trench | | | | | Trench | | | | | | | | | |
	74220,40 ~6 cm STN 4	74220,40 spheroids STN 4	74241,20 STN 4	76501,28 STN 6	78121,10 LRV 11	78421,33 ~20 cm STN 8	78441,14 ~10 cm STN 8	78461,14 ~5 cm STN 8	78481,30 ~1 cm STN 8	78501,30 STN 8	79511,13 STN 9	79135,34 STN 9	73235,88 STN 3	76055,3 STN 6	
Li	11.4	10.8	10.3	8.83	9.29	8.58	8.89	8.90	8.96	9.13	9.22	10.6	12.5	13.5	13.0
K	647	532	1030	831	831	774	806	774	791	748	763	865	1640	1850	14100
Rb	1.11	0.644	2.55	2.36	2.22	2.04	2.15	2.01	1.94	1.87	1.70	1.99	5.26	5.00	46.7
Sr	206	205	155	150	154	152	149	149	150	153	164	171	141	154	325
Ba	78.4	73.9	116	116	113	112	113	109	111	102	103	123	288	291	667
Ce	19.9	17.7	29.6	23.4	25.2	24.4	24.2	23.2	23.4	24.1	23.6	29.2	58.4	65.5	53.1
Nd	17.9	17.4	24.8	17.3	19.2	18.5	18.6	17.6	17.8	18.0	20.0	21.7	37.3	42.1	28.3
Sm	6.50	6.40	8.80	5.60	6.26	5.96	6.08	5.88	5.91	6.34	7.00	7.51	10.4	12.0	6.57
Eu	1.84	1.83	1.64	1.26	1.39	1.35	1.35	1.31	1.35	1.42	1.56	1.64	1.35	1.81	1.95
Gd	8.46	8.50	12.2	—	—	8.04	8.04	7.79	8.58	8.31	9.73	—	12.6	—	6.64
Dy	9.16	8.82	14.0	8.12	9.84	9.14	9.18	8.88	9.01	9.89	10.9	11.7	13.7	16.0	6.34
Er	4.82	4.59	7.85	4.73	5.74	5.32	5.43	5.19	5.30	5.47	6.27	—	8.27	9.66	3.54
Yb	4.20	3.96	7.60	4.55	5.27	5.01	5.17	4.88	4.99	5.48	5.89	5.85	7.69	8.84	3.39
Lu	0.627	0.608	1.14	0.717	0.830	0.758	0.806	0.780	0.762	0.847	0.901	0.792	1.17	1.37	0.54
Zr	184	194	565	180	188	171	196	166	202	179	224	—	366	399	195
wt.	83.87	67.20	103.58	119.84	130.39	110.13	93.41	134.79	199.89	129.58	206.25	77.55	95.09	93.15	

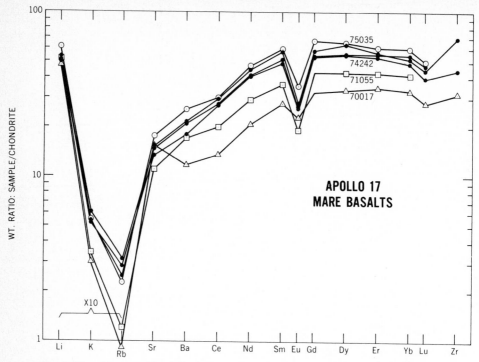

Fig. 1. Chondrite-normalized lithophile trace element abundances in Apollo 17 mare basalts.

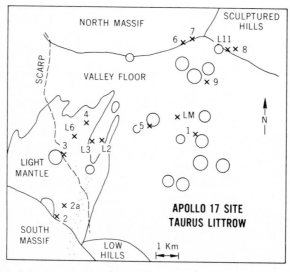

Fig. 2. Schematic map (after AFGIT, 1973) of the Apollo 17 site in the Taurus-Littrow valley showing major units, craters, and sample collection stations. The dots extending south from LRV2 and 3 represent a dark streak that continues over South Massif.

70017 is the most atypical of the group, being relatively enriched in Sr and Eu, probably owing to an excess of accumulative plagioclase. The other basalts differ mostly in terms of the absolute, rather than the relative, trace element abundances. As noted below in the discussion of the regolith, it appears probable that the uppermost basalt unit in the Taurus-Littrow valley is most nearly similar to the Station 4 samples. In general, the basalts are similar in petrography, chemistry, and age to the Apollo 11 basalts as noted in the Apollo 17 Preliminary Examination Teams Report (1973). In terms of the lithophile trace elements, the Apollo 17 basalts overlap the trace-element-poor Apollo 11 basalts and extend the abundance range to lower values. (Compare basalt 10062 with the Apollo 17 samples in Fig. 5.) A similar situation holds for the major elements.

A number of models attempting to explain the genesis of mare basalts, including those from Apollo 17, have been advanced. In the latest version of his model, Ringwood (1974) suggests that the Apollo 11 and 17 basalts were derived by very limited local-equilibrium partial fusion of (primary) perovskite–pyroxene in a source rock that also contains armoured (primary) melilite. The Apollo 12 and 15 basalts would be derived by larger degrees of melting. This is an interesting hypothesis with much to recommend it although it is somewhat complex and *ad hoc*. One possible objection is that many Apollo 11 and 17 basalts have Rb abundances as low as those in Apollo 12 and 15 basalts (e.g. Philpotts and Schnetzler, 1970; Schnetzler and Philpotts, 1971; Schnetzler *et al.*, 1972). Rb, as a large ion that strongly partitions into silicate melts, should be a good indicator of extent of partial fusion. We accept it as such and would suggest that trace-element-poor Apollo 11 basalts, Apollo 12, 15 and 17 and Luna 16 (e.g. Philpotts *et al.*, 1972a) basalts all represent substantial amounts of partial fusion; the trace-element-rich Apollo 11 basalts may be the only ones collected to date that represent limited partial fusion. These interpretations are in full accord with the most obvious interpretation of the Rb–Sr data for these basalts (e.g. Papanastassiou and Wasserburg, 1971; Schnetzler and Philpotts, 1971). Ringwood's model does not explain the trace element characteristics of the trace-element-rich Apollo 11 basalts or their young model Rb–Sr ages. To explain the similar Rb contents of Apollo 11 and 17 and 12 and 15 basalts in light of different Yb or Zr or Ti abundances, the model would probably have to involve accretionary heterogeneities. That there are heterogeneities in the source materials of mare basalts seems probable. What we have pointed out in our previous papers on mare basalts is that these apparent differences can be nicely explained in terms of igneous differentiation. That they can also be explained in terms of heterogeneous accretion is not at all obvious.

Another lunar model has been put forward recently by Taylor and Jakeš (1974). Although more comprehensive, this model has many similarities to the model (e.g. Nava and Philpotts, 1973) that we have been building since Apollo 11, including the hypotheses of frozen primary liquid in the crust, and derivation of mare basalts by melting of cumulates. Some of the aspects important to this model have been discussed recently by Brett (1974). Experimental support has been provided by Walker *et al.* (1973, 1974). The Taylor and Jakeš model proposes a moon that is regularly zoned to considerable depth (1000 km), and that mare material (green glass) has been derived from depths as great as 300 km. We consider these aspects of the model to be highly speculative. The anorthite content of the plagioclases, the Mg/Fe ratios of the mafics, and the Rb concentrations discussed above suggest that if mare basalts result from large degrees of partial fusion of cumulates, then the various parental cumulates could have been derived from similar primary liquids, not the considerably different liquids postulated in the Taylor–Jakes model. Also, the depth of origin of mare basalts is unknown. Also it is not obvious that a regularly zoned model moon has more to commend itself than a shallow differentiated layered igneous-complex model, for example. But in spite of these caveats, the model does fit fairly well our own hypotheses for the genesis of mare basalt.

Briefly, our model for mare basalt production is as follows. The old Rb–Sr model ages of most mare basalts suggest they (with a few exceptions such as the trace-element-rich Apollo 11 basalts) were derived by almost total extraction of the basaltic component of their source regions. These source materials were apparently of different compositions. It is unlikely that these differences were due to heterogeneous accretion. The differences can be explained in terms of igneous processes. The source materials might have been residues from a prior partial fusion event. However, the chemical differences are more readily explained if the source materials were cumulates. The high Fe content of mare basalts is due to these cumulates being fairly late-stage and/or their containing significant amounts of iron bearing oxides. The trace element characteristics of mare basalts appear to preclude their parental cumulates having trapped much liquid. We have therefore proposed that mare basalts were derived by partial fusion of cumulates formed during an early differentiation of at least the outer portions of the moon. Those with a prejudice against two-stage igneous processes should consider that the most abundant type of terrestrial basalt, oceanic tholeiite, also appears to be the product of two-stage melting (e.g. Gast, 1968).

Ringwood (1974) has criticized our model on a number of grounds. He believes that the model does not account satisfactorily for (a) the small amount of relative fractionation observed for mare basalt lithophile trace elements over an order of magnitude range in absolute abundances, (b) the correlation of the size of the Eu anomaly with trace element abundances, (c) the source of the heat of fusion, (d) the high Ti abundances, and (e) the fairly constant Sr abundance. However, we do not believe that any of these criticisms represent insurmountable difficulties and, indeed, believe most of them are better accounted for by our model than by Ringwood's model. Multi-stage genetic processes are expected, in general, to produce a wider range of trace element abundances than single-stage processes. The lack of relative fractionation of lithophile trace elements in mare basalts may be more apparent than real. Most mare basalts can be classified into two families, Apollos 11–17, and 12–15. It appears that within a family, source materials and type and extent of genetic processes are generally similar. However, it is not possible to relate all samples in both families in terms of a single-stage equilibrium process operating on the same source material. The problem is amplified if unusual mare basalts such as 12038 (Schnetzler and Philpotts, 1971) and Luna 16 (Philpotts *et al.*, 1972a) are included. The source materials of mare basalts apparently had various compositions. These differences are better explained in terms of a prior igneous event than in terms of accretionary heterogeneities. The general correspondence of the size of the Eu anomaly and abundances of other rare earths can be explained as well by fractional crystallization (preceding accumulation) as by partial fusion. Neither of these affects is totally adequate, however, as indicated by such phenomena as the similarity of Rb contents associated with differences in other trace element abundances mentioned above. It appears probable that in the source regions of some mare basalts there are phases that have both high heavy rare-earth abundances and large negative Eu anomalies. As far as heat sources are concerned, the limited amounts and types of mare basalt tend to favor localized rather than global fusion events; impact related phenomena are an obvious possible cause of fusion. Another of Ringwood's arguments is that Ti is not expected to be abundant in early cumulates. Our model, however, does not specify that the cumulates be early in the differentiation sequence. High Ti and Fe can be explained in terms of late-stage cumulation, or Fe–Ti cumulate phases, or both. Finally, the fairly constant abundance of Sr in mare basalts is surprising. It appears, however, to be better explained by models, such as ours, that involve feldspar crystallization. As noted above, the Rb–Sr coherence suggested by old model ages argues against hypotheses such as Ringwood's that involve leaving Sr in the residue.

Highland Samples

Lithophile trace element abundances for highland breccias 73235 and 76055 are reported in Table 1. These data are also plotted in Fig. 3 along with data we

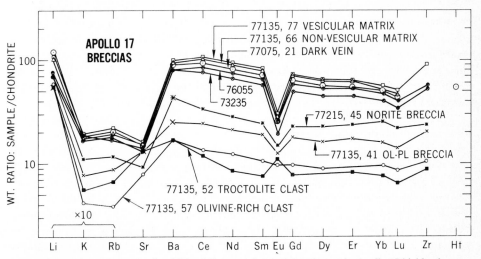

Fig. 3. Chondrite-normalized lithophile trace element abundances in Apollo 17 highland breccias.

have obtained for samples from the Station 7 consortium boulder, which will be reported and more fully discussed in a separate paper. Figure 3 and the Apollo 17 Preliminary Examination Team (1973) data show that lithophile trace element abundances in Apollo 17 highland samples range over at least an order of magnitude. Within this range, however, there is some evidence of clustering. It is apparent that the Station 3, Station 6, and three of the Station 7 samples have similar trace element composition. The relative abundances are KREEP-like although the absolute abundances are somewhat lower than those in Apollo 14 breccias (e.g. Philpotts *et al.*, 1972b).

Warner *et al.* (1974) have made the interesting suggestion that KREEP rocks are products of partial melting of surficial material at many locations resulting from meteorite impact. Apparently at odds with this hypothesis, however, is the rather uniform variation in the surficial abundances of the radioactive elements indicated by the orbital γ-ray data (Metzger *et al.*, 1973). These data have been used previously (Nava and Philpotts, 1973) as an argument against the existence of a large number of primary liquids having the relative trace element abundances of KREEP. The evidence indicates that KREEP was produced by partial fusion in a very limited number of events (one? two?). We favor an explanation of highland breccias with various abundances of trace elements in KREEP proportions in terms of surficial mixing (see Schonfeld, 1974). This will be discussed further below in the soil section. Our model (Nava and Philpotts, 1973) proposes that highland rocks are largely mixtures of primary "liquid," primary cumulates, and KREEP.

Soils

Lithophile trace element abundances for twenty-two Apollo 17 regolith samples are reported in Table 1. The "soil breccia" 79135 can also be treated as a regolith sample. All of the soil samples analyzed are < 1 mm sieved fractions except for 74220,40 which is a bulk soil, 74220,40 black spheroids which are > 570 μ, and the 70008 core tube samples which are < 250 μ. It should be noted that about 90% of the typical lunar soil consists of the < 1 mm fraction. The samples analyzed provide good areal coverage of the Apollo 17 site. Some vertical coverage is provided by trench and core samples from Stations LM, 2A, 4, and 8.

The regolith samples do not show as large a range in lithophile trace element absolute abundances as do the rocks. Nevertheless, there are considerable variations in the relative abundances of these elements in the soils. Figure 4 shows the most fractionated patterns. The other soils are intermediate to these extreme cases. In view of the geology of the site, it might be expected that variations in soil composition are due largely to different contents of highland material, mare basalt, and Station 4 type orange–black soil which for purposes of discussion is better considered a "rock-type." This expectation is further supported by petrographic studies of the soils (e.g. Heiken and McKay, 1974). And indeed the Apollo 17 regolith samples can be modeled very well in terms of these components (e.g. Schonfeld, 1974). The nature and proportions of the indicated components in the various soils, however, are not without surprises.

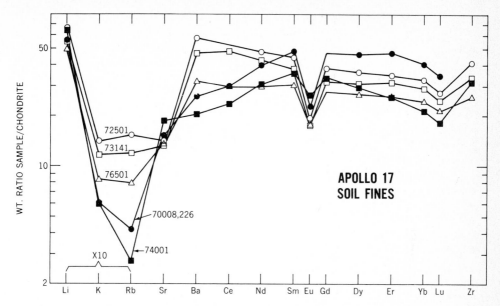

Fig. 4. Chondrite-normalized lithophile trace element abundances in Apollo 17 soils showing extremes of fractionation.

It is difficult to keep track of all the lithophile trace elements at the same time. We therefore narrow our field of view to Rb and Yb, two of the elements of most use in sorting out the various Apollo 17 materials. It should be recognized, however, that the discussion has wider validity not only in terms of the other lithophile trace elements but also the major elements and many other trace elements (see *Lunar Science—V*). In Fig. 5 are plotted Rb and Yb data for all of our Apollo 17 samples and some reference materials we have analyzed from previous missions. The highland samples show a range from troctolites 77135,52 and 57 to KREEP-rich samples such as 77135,77. The mare basalts all have low Rb contents. They exhibit a range of Yb concentrations suggestive of addition or subtraction of phases low in both Rb and Yb. The orange and black "soils" from Station 4 plot in their own distinctive position. The soils plot between these other materials.

Rhodes and co-workers (Preliminary Examination Team Report, 1973; Rhodes *et al.*, 1974) have made the widely accepted suggestion that there are three chemically distinct groups of Apollo 17 soils, namely South Massif and Light Mantle, Valley Floor, and North Massif and Sculptured Hills. We would suggest that (excluding 74220 and 74001) a better classification might involve only two groups, namely Light Mantle and all other soils. This latter group is well defined in Fig. 5. It extends from 76501 to 70008,226. Our Light Mantle group consists of three similar soils, 73121 and 73141 from Station 2A near South Massif (Fig. 2), and 74121 from LRV6 which is halfway between Lara and Shorty Craters toward the edge of the Light Mantle. It is suggested that 74241 from the rim of Shorty Crater should also be considered a member of this group inasmuch as its composition can be

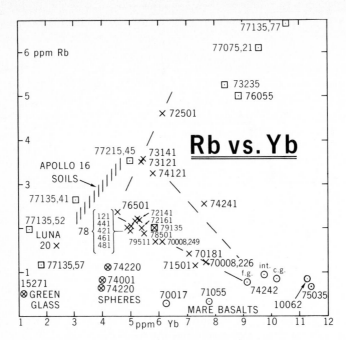

Fig. 5. Rb versus Yb abundances in Apollo 17 materials: mare basalts. #, orange and black "soils" ⊗, soils ×, soil breccia ⊠, and highland breccias ⊡. Also shown are data for an Apollo 11 trace element poor basalt 10062 (Philpotts and Schnetzler, 1970), Apollo 15 green glass (unpublished), Luna 20 soil (Philpotts *et al.*, 1972c), and the approximate range for Apollo 16 soils (Philpotts *et al.*, 1972c).

matched very well by a mixture of Light Mantle and Station 4 basalt. Soil 72501, collected at South Massif, is similar to the Light Mantle soils, although somewhat richer in lithophile trace elements (Fig. 4). It is uncertain whether this is a local aberration or reflects a more general compositional difference between South Massif and Light Mantle soils. In any case, the Light Mantle group is interpreted as a mixture of a highland component and local mare basalt. The other group is interpreted as a mixture of a distinct highland component and local mare basalt. The nature of this second highland component is not exactly defined but a good estimate can be made if, as the TiO_2 concentration for example appears to indicate, soil 76501 contains about 15% mare basalt. Both groups contain various but lesser amounts of orange–black "soils." This grouping of Apollo 17 soils and the proposed components are by no means the only ones possible but they do appear to fit the compositional data and are reasonable in view of the local geology. They also generate some interesting corollaries.

It appears that most of the normal valley floor soils have essentially the same kind of mare basalt component. This mare basalt component is most closely matched in composition, of the samples we have analyzed, by the Station 4 basalts. These basalts, of course, also occur in the Light Mantle soils at Shorty Crater. The Station 5 basalt is more nearly similar to these basalts than is the

Station 1 basalt; 70017 is most different. It appears likely that the Taurus-Littrow valley is underlain by basalt, of which the uppermost unit, flow or margin, resembles Station 4 basalt. The soils containing the largest mare basalt component are the core sample 70008,226 from about 45 cm depth, the surficial soil 70181 from the same station, LM, and the Station 1 soil 71501. All of these samples are from the center of the valley, and the high basalt content may reflect a relatively recent addition. Eberhardt et al. (1974) have interpreted the lower exposure ages of 70181 and a core sample from 26 cm, relative to those of core samples from 62 to 290 cm depth, as possibly due to Camelot ejecta ($\sim$ 100 m.y.?). Our other core sample 70008,249 has less basalt component, only slightly greater than those of the main cluster (Fig. 5) of Taurus-Littrow soils which includes samples from Stations LRV2, LRV3, LRV11, 8 and 9(Fig. 2). At Station 8 there appears to be little variation in basalt content down to 20 cm depth. Sample 76501 from North Massif has the lowest content of basalt of this group of soils. Of the Light Mantle group, only 74241 from Shorty Crater has much basalt, roughly 50%. Sample 74121 from LRV6 is almost as free of basalt as the Station 2A soils. Sample 72501 is also essentially basalt free. It appears that what has been called Dark Mantle at the Apollo 17 site is, in fact, soil rich in mare basalt component. It has been suggested (Pieters et al., 1974) that the term be restricted to dark areas northwest of the landing site believed rich in 74001-type "soil."

An unusual feature of the main group of Apollo 17 soils is the lack of variation in the composition of the highland component. This is somewhat surprising in view of the wide variety of highland rocks collected at the site. It is also in contrast to the situation at the Apollo 12, 15, and 16 sites where the soils show a wide range of KREEP component abundances. This homogeneity of the Apollo 17 main group soils might be due to little recent addition of highland material or to efficient mixing of highland materials perhaps owing to the long steep slopes of the massifs. It is also conceivable that the highland component composition reflects that of a homogeneous areal ejecta blanket that has been reworked and tended to accumulate in the valley. This explanation of the mass-wasting hypothesis would seem to fit with the observations (Eberhardt et al., 1974) of layering and fairly uniform exposure ages of the deep core samples. Of course, it is entirely possible that there are no large bodies of highland rock at the site with distinctive compositions. There is an indication from the study of the Apollo 17 boulders (e.g. Chao, 1974) that this might be the case; and rock heterogeneities may be relatively small scale, easily obliterated during regolith formation. Complicating the matter, however, is the fact that the Light Mantle soils have compositions distinct from that of the main group.

Rhode's classification of Apollo 17 soils implies that the Light Mantle soils are representative of South Massif. One problem with this is that the LRV2 and 3 soils, 72141 and 72161 respectively, which were collected almost equidistant from North and South Massifs, are similar to Station 8 soils collected at the foot of the Sculptured Hills (Fig. 2) and fit on the main trend ranging from 76501 to mare basalt. Why is there no South Massif component in these two soils? A possible explanation, fitting also with their relatively low basalt content, is that they might represent relatively recent ray material ejected from near or on North Massif. Some support for this hypothesis comes from the interfingered appearance of the Light Mantle and from the fact that a dark streak can be traced on orbital photographs from LRV2 and 3, across the Light Mantle and up over South Massif (Fig. 2). However, soils 72141 and 72161 are only part of a more general problem. There is no indicated change in the composition of the highland component in the soils in the sequence Stations 6, 8, LM, 1, LRV2, LRV3, from North Massif toward South Massif. Thus, too ready assignment of one highland component as representative of North Massif and the other, the Light Mantle, as representative of South Massif should be viewed with caution. Indeed, it is still an open question whether Light Mantle is avalanche, ray, or even subsurface material. Perhaps, studies of the Apollo 17 boulders will provide some definitive information on the nature of the massifs and hence on the highland components in the soils.

REFERENCES

AFGIT (Apollo Field Geology Investigation Team) (1973) Geologic exploration of Taurus-Littrow: Apollo 17 landing site. *Science* **182**, 672–680.

Apollo 17 Preliminary Examination Team (1973) Apollo 17 lunar samples: chemical and petrographic description. Science **182**, 659–672.

Brett R. (1973) The lunar crust: a product of heterogeneous accretion or differentiation of a homogeneous Moon? *Geochim. Cosmochim. Acta* **37**, 2697–2703.

Chao E. C. T. and the International Consortium (1974) First results of consortium study of Apollo 17 Station 7 boulder samples (abstract). In *Lunar Science—V*, pp. 1–3. The Lunar Science Institute, Houston.

Eberhardt P., Eugster O., Geiss J., Graf H., Grogler N., Guggisberg S., Jungck M., Maurer P., Morgeli M., and Stettler A. (1974) Solar wind and cosmic radiation history of Taurus Littrow regolith (abstract). In *Lunar Science—V*, pp. 197–199. The Lunar Science Institute, Houston.

Gast P. W. (1968) Trace element fractionation and origin of tholeiitic and alkali magma types. *Geochim. Cosmochim. Acta* **32**, 1057–1086.

Heiken G. and McKay D. S. (1974) Petrography of Apollo 17 soils (abstract). In *Lunar Science—V*, pp. 319–321. The Lunar Science Institute, Houston.

Metzger A. E., Trombka J. I., Peterson L. E., Reedy R. C., and Arnold J. R. (1973) Lunar surface radioactivity: preliminary results of the Apollo 15 and Apollo 16 gamma-ray spectrometer experiments. *Science* **179**, 800–803.

Nava D. F. and Philpotts J. A. (1973) A lunar differentiation model in light of new chemical data on Luna 20 and Apollo 16 soils. *Geochim. Cosmochim. Acta* **37**, 963–973.

Papanastassiou D. A. and Wasserburg G. J. (1971) Lunar chronology and evolution from Rb–Sr studies of Apollo 11 and 12 samples. *Earth Planet. Sci. Lett.* **11**, 37–62.

Philpotts J. A. and Schnetzler C. C. (1970) Apollo 11 lunar samples: K, Rb, Sr, Ba and rare-earth concentrations in some rocks and separated phases. *Proc. Apollo 11 Lunar Sci. Conf., Geochim. Cosmochim. Acta*, Suppl. 1, Vol. 2, pp. 1471–1486. Pergamon.

Philpotts J. A., Schnetzler C. C., Bottino M. L., Schuhmann S., and Thomas H. H. (1972a) Luna 16: some Li, K, Rb, Sr, Ba, rare-earth, Zr, and Hf concentrations. *Earth Planet. Sci. Lett.* **13**, 429–435.

Philpotts J. A., Schnetzler C. C., Nava D. F., Bottino M. L., Fullagar P. D., Thomas H. H., Schuhmann S., and Kouns C. W. (1972b) Apollo 14: some geochemical aspects. *Proc. Third Lunar Sci. Conf., Geochim. Cosmochim. Acta*, Suppl. 3, Vol. 2, pp. 1293–1305. MIT Press.

Philpotts J. A., Schuhmann S., Bickel A. L., and Lum R. K. L. (1972c) Luna 20 and Apollo 16 core fines: large-ion lithophile trace-element abundances. *Earth Planet. Sci. Lett.* **17**, 13–18.

Pieters C., McCord T. B., and Adams J. B. (1974) Evidence for regional occurrence of orange glass and related soils (abstract). In *Lunar Science—V*, pp. 605–607. The Lunar Science Institute, Houston.

Rhodes J. M., Rodgers K. V., Shih C., Bansal B. M., Nyquist L. E., and Wiesmann H. (1974) The relationship between geology and soil chemistry at the Apollo 17 landing site (abstract). In *Lunar Science—V*, pp. 630–632. The Lunar Science Institute, Houston.

Ringwood A. E. (1974) Minor element chemistry of maria basalts (abstract). In *Lunar Science—V*, pp. 633–635. The Lunar Science Institute, Houston.

Schmitt H. H. (1973) Apollo 17 report on the valley of Taurus-Littrow. *Science* **182**, 681–690.

Schnetzler C. C. and Philpotts J. A. (1971) Alkali, alkaline earth, and rare-earth element concentrations in some Apollo 12 soils, rocks, and separated phases. *Proc. Second Lunar Sci. Conf., Geochim. Cosmochim. Acta*, Suppl. 2, Vol. 2, pp. 1101–1122. MIT Press.

Schnetzler C. C., Philpotts J. A., Nava D. F., Schuhmann S., and Thomas H. H. (1972) Geochemistry of Apollo 15 basalt 15555 and soil 15531. *Science* **175**, 426–428.

Schnetzler C. C., Thomas H. H., and Philpotts J. A. (1967) Determination of rare-earth elements in rocks and minerals by mass-spectrometric, stable isotope dilution technique. *Anal. Chem.* **39**, 1888–1890.

Schonfeld E. (1974) Component abundance and evolution of regoliths and breccias: interpretation by mixing models (abstract). In *Lunar Science—V*, pp. 669–671. The Lunar Science Institute, Houston.

Taylor S. R. and Jakes P. (1974) Geochemical zoning in the moon (abstract). In *Lunar Science—V*, pp. 786–688. The Lunar Science Institute, Houston.

Walker D., Grove T. L., Longhi J., Stolper E. M., and Hays J. F. (1973) Origin of Lunar feldspathic rocks. *Earth Planet. Sci. Lett.* **20**, 325–336.

Walker D., Longhi J., Stolper E., Grove T., and Hays J. F. (1974) Experimental petrology and origin of titaniferous lunar basalts (abstract). In *Lunar Scienve—V*, pp. 814–816. The Lunar Science Institute, Houston.

Warner J. L., Simonds C. H., and Phinney W. C. (1974) Impact induced fractionation in the highlands (abstract). In *Lunar Science—V*, pp. 823–825. The Lunar Science Institute, Houston.

Proceedings of the Fifth Lunar Conference
(Supplement 5, Geochimica et Cosmochimica Acta)
Vol. 2 pp. 1269–1286 (1974)
Printed in the United States of America

The contamination of lunar highland rocks by KREEP:
Interpretation by mixing models

Ernest Schonfeld

NASA Johnson Space Center, Houston, Texas 77058

Abstract—A mixing model method was used to determine the component abundance in the Apollo 16 and 17 soils. This method uses the chemical composition of the soils for up to 30 elements and a weighted least-squares mixing model technique. Elements included in the calculations are: Si, Ti, Al, Ca, Fe, Mg, P, Cr, Mn, Na, K, Rb, Ba, U, Th, La, Ce, Sm, Eu, Sr, Yb, Y, Sc, V, Zr, Nb, Co, Ni, Li, Au, and Ir.

The method was used to examine the possibility that some of the highland rocks such as VHA and low-K Fra Mauro basalt are mixtures. The results of the mixing model calculations show that is is *possible* that these rocks are mixtures of KREEP, troctolite, "anorthosites," and a meteoritic component. The Rb–Sr systematics are consistent with such a model for the genesis of the highland rocks. KREEP has high relative concentrations of Rb, U, and radiogenic Sr and Pb and its model age of about 4.4 AE dominates the model age of all the "contaminated" highland rocks.

Introduction

ONE OF THE MOST SIGNIFICANT RESULTS of the lunar analysis program is that on a chemical classification basis there are three abundant rock types: "anorthosites" (anorthosites, gabbroic anorthosites, and anorthositic gabbros), mare basalts, and KREEP (Gast, 1972). In addition to these major rock types other rock types have been postulated on the basis of petrographic examination and chemical analysis of glasses, lithic fragments, and breccias. These rock types are low-K Fra Mauro basalt (LK-FM) (Reid *et al.*, 1972), very high alumina basalt VHA (Hubbard *et al.*, 1973), (spinel) troctolite (Prinz *et al.*, 1973), dunite (LSPET, 1973), etc. These other rock types are mainly concentrated close to or in the lunar highlands.

In this work the possibility that low-K Fra Mauro basalt, VHA, and other highland rocks such as 68415, 14068, etc. are mixtures was investigated using the chemical composition of these rocks and a weighted least-squares mixing model technique (Schonfeld and Meyer, 1972). The results of the mixing model calculations show (Schonfeld, 1973, 1974) that it is possible to generate low-K Fra Mauro basalt, VHA, 68415, 14068, etc. by mixing "anorthosites," KREEP and troctolite and/or dunite. The meteoritic and granitic components are only minor contributors in these rocks. Also Dowty *et al.* (1974a, b) have proposed that "VHA basalt" and highland melt rocks are admixtures of KREEP-like material with cumulate ANT rocks.

Mixing in the highlands

There is evidence that mixing is an important process in the lunar highlands. In the mare sites such as Mare Tranquillitatis (Apollo 11), Oceanus Procellarum

(Apollo 12), Palus Putredinis (Apollo 15), and Mare Fecunditatis (Luna 16), there are soils and breccias that are mixtures of several components. The lunar highlands have been exposed to a total planetesimal bombardment of about 20 to 30 times higher than the mare sites (Hartmann and Wood, 1971). Therefore the probability of mixing and resetting ages is much higher in the highlands than in the mare sites.

The high concentration of siderophile elements such as Ni, Au, and Ir has been interpreted (Baedecker *et al.*, 1973; Ganapathy *et al.*, 1973) as evidence of meteoritic contamination of the lunar highlands.

The majority of the highland rocks are breccias. Textural evidence suggests that at one time most of the highland rocks were polymict glassy breccias or soils (Warner *et al.*, 1973). At the Apollo 15 site the most abundant highland rock is the brown-glass matrix breccia (Phinney *et al.*, 1972; Cameron *et al.*, 1973). These breccias contain considerable amounts of solar wind gases (Husain, 1972) suggesting that they originated from soils. The Apollo 16 poikilitic rocks include up to 10% pore space suggesting a protolith made up of a polymict, gas-rich soil or glassy breccias (Simonds *et al.*, 1973). Also these highland rocks have relics of plagioclase, olivine, and spinel (Albee *et al.*, 1973; Simonds *et al.*, 1973; Dowty *et al.*, 1974; Meyer *et al.*, 1974). The presence of these relics has been interpreted by Dowty *et al.* (1974a,b) and by Meyer *et al.* (1974) as an indication that some highland rocks are mixtures.

Hörz *et al.* (1974) have shown that there is no positive correlation between the results of the orbiting geochemical experiments and the Cayley Plains or the "Fra Mauro," "Montes Alpes," and "Montes Apenninus" formations. This conclusion, according to Hörz *et al.*, is consistent with a localized origin for these geological units, and suggests that local mixing by secondary impacts (by incorporating local components into ejecta blankets) and primary impacts was a significant process.

The use of mixing models is one method to test whether or not rocks or soils are possible mixtures (see section on mixing models).

Difficulties with the partial melting model

A partial melting origin has been proposed for VHA and low-K Fra Mauro basalts (Hubbard *et al.*, 1973a; Walker *et al.*, 1973; Warner *et al.*, 1974). There are several difficulties with this model. One is that all these rocks have large ion lithophile (LIL) trace element patterns almost identical to KREEP, another is that they have high concentrations of meteoritic Au, Ir, and Ni. Also, these rocks have large variations in the Mg/Fe ratio.

The remarkably similar LIL trace element patterns for many highland rocks are shown in Fig. 1. In the same figure the trace element patterns for some mare basalts and highland soils are shown for comparison. Included in the same figure are the partial melting calculations of a model derived by Gast (1972) for KREEP. This model predicts very well the trace element pattern for KREEP. When the degree of partial melting is increased there is a considerable change in the trace element pattern. The source for KREEP in Gast's model has the same relative

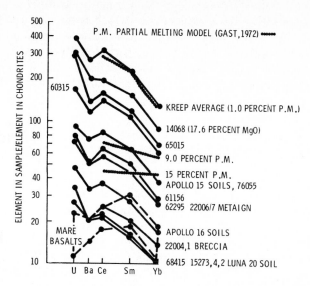

Fig. 1. Concentrations of U, Ba, Ce, Sm, and Yb normalized to chondrites in KREEP, highlands rocks, and highlands soils. Also included for comparison are data for mare basalts. Notice the remarkable similarity of these samples pattern with the KREEP pattern even when absolute concentrations change by a factor of 20. Included are the calculations of Gast (1972) (partial melting model for KREEP).

concentrations to chondrites for U, Ba, Ce, Sm, and Yb. Therefore, it is impossible to generate VHA and low-K Fra Mauro basalts by different degrees of partial melting from this source. To solve this problem Hubbard and Shih (1973) have proposed *ad hoc* models where the sources for VHA and LK-FM *do not* have equal concentrations relative to chondrites of LIL trace elements. The simplest explanation of this remarkable similarity of the trace element pattern is by diluting KREEP (by mixing) with other rock types with relatively low concentrations of LIL trace elements (Philpotts *et al.*, 1973).

VHA and LK-FM basalts have very high concentrations of meteoritic Au, Ir, and Ni, and these concentrations are similar to those found in highland soils (Baedecker *et al.*, 1973; Laul and Schmitt, 1973; Ganapathy *et al.*, 1973). In contrast, crystalline and near-crystalline rocks such as anorthosites, mare basalts, KREEP samples 15205, 15272-9-22, and troctolite 76535 (Baedecker *et al.*, 1973; Ganapathy *et al.*, 1973; Laul and Schmitt, 1973) have much lower relative concentrations of Au and Ir. The Au, Ir, and Ni are nearly all concentrated in the metal phase in soils and breccias (Wlotzka *et al.*, 1973) and the equilibrium metal-silicate melt highly favors the metal phase (Brett, 1971). Therefore, if VHA and low-K Fra Mauro basalt are made from highland soils and/or breccias by partial melting, it is difficult to understand why they have such high concentrations of Au, Ir, and Ni. It is easier to explain these high Au, Ir, and Ni concentrations in VHA and LK-FM basalts if there are mixtures contaminated with a meteoritic component.

Dowty *et al.* (1974a) pointed out that "VHA basalts" have a wide range of

Fe/Mg ratios with similar REE abundances, which is not consistent with their crystallization from a single liquid, but might be expected in cumulate rocks.

There is evidence that small amounts of partial melting occurred in surface highland rocks (Albee et al., 1973). The author believes that in general the very intense planetesimal bombardment on the lunar surface will tend to remix the small amounts of partial melts with the residues of the partial melts.

MIXING MODELS

Technique

The method used is a linear mixing model of chemical elements according to the weighted least-squares method of Gauss (Kendall and Stuart, 1961) and it is assumed that the chemical composition of a soil or breccia is the result of adding the contribution of each of the rock types present according to their abundance (Schonfeld and Meyer, 1972). This technique has been applied to lunar soils and breccia (Goles et al., 1971; Wänke et al., 1971; Meyer et al., 1971; Schonfeld, 1973; Duncan et al., 1973; Schonfeld, 1974). The chemical compositions for the soils, breccias, "rocks," and each of the components used in this mixing model technique include data from the published literature for about 30 elements (Si, Ti, Al, Ca, Fe, Mg, P, Cr, Mn, Na, K, Rb, Ba, U, Th, La, Ce, Sm, Eu, Sr, Yb, Y, Sc, V, Zr, Nb, Co, Ni, Li, Au, and Ir). The rock types used are: mare basalts, KREEP, "anorthositic" rocks (anorthositic gabbro, gabbroic anorthosite, anorthosites), granite, meteorite (CC-1), and ultramafic (Schonfeld and Meyer, 1972). In addition to these rock types, in this paper, troctolites and dunite have been added. For the troctolites the data from the Luna 20 troctolitic fragments (Prinz et al., 1973), sample 73235W (Taylor et al., 1974), 14321Tr (Taylor et al., 1972), and 76535 (Haskin et al., 1974) were used assuming that the LIL trace element concentration was similar to the rock 76535. For dunite the data for rock 72415 (LSPET, 1973) and 15445,71 (Ridley et al., 1973b) were used.

The chemical composition for the "rocks," soils, and breccia are averages of selected analytical data taken from many sources (Proc. Apollo 11 Lunar Sci. Conf., 1970; Proc. Second Lunar Sci. Conf., 1971; Proc. Third Lunar Sci. Conf., 1972; Proc. Fourth Lunar Sci. Conf., 1973; Lunar Science—V, 1974; LSPET, 1973).

Apollo 16 soils and 68415

Mixing model calculations were performed on 9 Apollo 16 soils and rock 68415 using the published chemical analysis of about 26 elements. The results are shown in Table 1.

Duncan et al. (1973) have performed mixing model calculations on the Apollo 16 soils using the analysis of 14 chemical elements. They found that if VHA rock 61016,139 was included as a component it did not give a good fit to the soils. In the analysis presented here similar conclusions are obtained using an average

Table 1. Results of mixing model for Apollo 16 samples.

	Solution I[a,b,c]		Solution II[a,c]	
	KREEPy 16	"Anorth"	KREEP	"Anorth"
60601	17.9 ± 3	71 ± 6	13 ± 1	85 ± 3
61221	12 ± 2	80 ± 4	6.7 ± 1	92 ± 5
61801	15 ± 3	75 ± 4	10 ± 1	86 ± 2
63501	10.7 ± 3	82 ± 6	6.9 ± 1	93 ± 3
64421	22 ± 3	65 ± 5	11.6 ± 1.5	87 ± 2
68501	18.5 ± 2	70 ± 4	11.7 ± 1	86 ± 2
67601	8 ± 3	88 ± 3	4.7 ± 1	95 ± 2
67461	5 ± 2	92 ± 3	2.5 ± .5	96 ± 2
68415	8 ± 1	87 ± 3	4.7 ± .5	94 ± 2

[a]All samples had a meteoritic component.

[b]VHA was included as a component and approximately correlated with KREEPy 16 (VHA ≈ 0.4 KREEPy 16).

[c]"Anorth" ≈60% Gab. Anorth + 20% Anorthosite + 20% An. Gabbro.

Solution I gave slightly better overall fit than Solution II, but much better fit for Mg.

Mare basalt is present about 1–2% and granite is less than 0.3%. (Exception: 68415 has no mare basalt).

composition for VHA (61156, 62295, and 61016,143) as a component. When they used rock 65015 as a component they did not find an improvement in the fit. In this analysis when the average composition for rocks similar to 65015 was used (65015, 60315, and 62235; KREEPy 16) there was no significant improvement in the overall fit, but the Mg fit improved considerably, suggesting that KREEPy 16 is more important than KREEP in the Apollo 16 soils. Both sets of results, using a KREEPy 16 (solution I) and KREEP (solution II) as components, are shown in Table 1.

KREEPy 16 (Fig. 3) can be modeled as a mixture of about 51% KREEP + 34% "anorthosite" + 15% dunite + 0.3% granite. The lowest amount of KREEPy 16 is in the soils from North Ray Crater (67601 and 67461) and rock 68415. Rock 68415 contains meteoritic contamination (Ganapathy et al., 1973) and gives a very good fit as a possible mixture. Simple mixtures of VHA + "anorthositic" + meteoritic and LK-FM + "anorthosites" + meteoritic gave poor fits for Mg in all cases. Therefore, there appear to be no significant amounts of VHA (Haskin et al., 1973) and/or LK-FM in the Apollo 16 soils and KREEPy 16 appears to be the main contributor of LIL elements. This interpretation is contrary to the conclusion of Hubbard et al. (1973a) based on a Sr–Al variation diagram, that more than half of some Apollo 16 soils may consist of VHA basalts.

In all the soils there is also about 1–2% mare basalt. The exact abundance depends on a good chemical definition of this mare basalt. This result agrees with the analyses of glasses in the Apollo 16 soil (Ridley et al., 1973a). The amount of granitic component was in all cases less than 0.3%. In all cases also there was a rather large amount of non-chondritic meteoritic Ni, Au, and Ir (Ganapathy *et al.*, 1973; Laul and Schmitt, 1973).

Preliminary analysis of the Apollo 17 soils

A preliminary mixing model analysis was performed on the chemical composition of 38 of the Apollo 17 soils (Duncan *et al.*, 1974; Eldridge *et al.*, 1974; LSPET, 1973; Philpotts *et al.*, 1974; Rhodes *et al.*, 1974; Rose *et al.*, 1974). The elements used in the mixing model calculations are those published (LSPET, 1973) with the exception of Zn. In 12 cases Li, Ba, Sr, Ce, Sm, Eu, Yb, K, and Rb measured by isotopic dilution (Philpotts *et al.*, 1974) were included with similar but better results. In 12 cases Th, U, and K measured by gamma-ray spectroscopy (LSPET, 1973; Eldridge *et al.*, 1974) were included with similar results. Samples from the trench in Station 3 had the shortest list of elements with the major and minor elements measured by Rose *et al.* (1974), and the trace elements Th, U, and K measured by Eldridge *et al.* (1974). In the case of samples 76321, 78221, and the size fractions of soil 75081 (data from Duncan *et al.*, 1974), Ba was also included in the list of elements.

Two sets of components were tested. In the first test KREEP, dunite, "anorthosites," mare basalt from Apollo 17, orange glass from Apollo 17, granite, and meteoritic components were used. In the second test KREEPy 17 (noritic breccia 17) and anorthositic gabbro from Apollo 17 were used instead of KREEP and "anorthosites." In both cases the fit was similar and quite good. Also, instead of KREEPy 17 other components were tested such as VHA and LK-FM (low-K Fra Mauro basalt). In most cases KREEPy 17 gave a better fit, suggesting that KREEPy 17 (that could itself be a mixture as shown in Fig. 3) and the anorthositic gabbro from Apollo 17 are significant components of the Apollo 17 soils.

The results of the mixing model calculations are given in Table 2 and are also shown for some of the soils in Fig. 2. In Fig. 2 the distance between stations is approximately proportional to the actual distance between stations. The Apollo 17 site has some similarities with the Apollo 15 site in the sense that in both places the abundance of mare basalts is inversely proportional to the abundance of "anorthositic" material (Schonfeld and Meyer, 1972). Rhodes *et al.* (1974) have classified the Apollo 17 soils in three distinct groups based largely on the minor and trace element data. These groups are the South Massif and light-mantle soils, the valley floor soils, and the North Massif and Sculptured Hills soils. Based on the results of the mixing model calculations presented in Table 2 it appears that the only distinct group of samples are from Stations 2, 2a, and possibly LRV-6, corresponding to the Light Mantle, or avalanche-generated on the South Massif slope (Muehlberger *et al.*, 1974). The results of the mixing model calculations show that the other two groups, as defined by Rhodes *et al.*, are not distinct groups and are only approximate groups. Philpotts *et al.* (1974) have questioned Rhodes *et al.* classification of the Apollo 17 soils, and have proposed only two groups of soils. One is the Light Mantle group and the other is all other soils. The results of

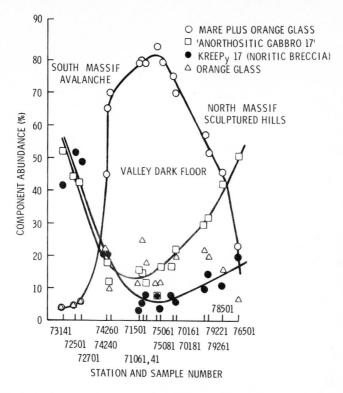

Fig. 2. Preliminary mixing model calculations results for the Apollo 17 soils.

the mixing model calculations are in agreement with Philpotts *et al.* classification of the Apollo 17 soils. Duncan *et al.* (1974) have also performed mixing model calculations of the Apollo 17 soils and in the cases where a comparison is possible the agreement is good.

There are several observations one can make. (1) Soils from Stations 6, 8, and possibly 7, have dunite. This dunite can also be interpreted as a KREEPy 17 component that has higher Mg concentration, such as sample 76055, or as the presence of a troctolite component (Prinz *et al.*, 1973) in these soils. (2) The ratio of anorthositic gabbro 17/KREEPy 17 (called R) and the ratio of mare basalt 17/orange glass 17 (called Q) can be used to characterize the Apollo 17 soils. All the light-mantle soils have R values close to one and very small relative values of Q. The majority of the other soils have R ratios ranging from about 1.7 to 3.5 with an average of about 2.4. Exceptions are soils 74241, 74260, 75061, 75081,D, 76321, and the trench soils from Station 3. These soils have R values found in the light-mantle soils. Soils 74241 and 74260 are from the rim of Shorty Crater and the highland component is from the Light Mantle (Muehlberger *et al.*, 1974). Soil 75061 was collected from the top of a boulder from Camelot Crater and soil 76321 was collected from the boulder in Station 6 (Muehlberger *et al.*, 1974). Soil 75081,D is the finest size fraction of soil 75081 and was collected near Camelot Crater

E. SCHONFELD

Table 2. Preliminary results of mixing model calculations of Apollo 17 soils.

Sample	Station	Type[a]	Anorthositic gabbro 17	KREEPy 17	Mare basalt 17	Orange glass 17	Dunite	Meteorite[b] CC-1	Anor. Gabbro KREEPy 17	Mare basalt Orange glass
72441	2	UB	39±3	55±3	3.1±4	4±3	0±2	—	0.71±0.1	0.8±1.2
72461	2	UB	42±4	52±4	0.7±5	4±3	0±2	1.3±0.2	0.81±0.0	0.2±1.1
72501	2		42±4	52±4	0±6	5±3	0±2	1.2±0.3	0.81±0.1	0.1±1.3
72701	2		44±3	49±3	0.7±3	6±3	0±1	1.2±0.2	0.89±0.1	0.1±0.5
73121	2A	TT	49±3	42±3	1.0±3	7±4	1.5±2		1.17±0.1	0.1±0.5
73141	2A	TB	52±3	42±3	0.8±3	4±2	1.7±2	0.84±0.2	1.25±0.1	0.2±0.8
73221	3	TT	52±5	37±5	3.5±9	9±5	0±5		1.40±0.3	0.4±1.0
73241	3	TM	48±7	40±7	0±10	12±8	0±5		1.20±0.3	0±0.8
73261	3	TB	50±6	37±6	0±9	13±9	0±5		1.35±0.3	0±0.7
73281	3	TB	51±5	27±5	0±9	12±5	0±5		1.35±0.3	0±1.0
74121	LRV-6		44±3	38±3	9.7±4	6±3	1±2	1.4±0.2	1.17±0.1	1.5±1.0
74240	4		13±5	20±5	56±10	10±5	0±3	0.3±0.4	0.63±0.3	5.6±3.0
74241	4		18±4	24±4	47±9	10±6	0±3	0.3±0.3	0.74±0.2	4.6±2.9
74260	4		18±5	20±4	48±7	14±5	0±2	0.4±0.2	0.90±0.3	3.4±1.3
72141	LRV-2		40±4	20±3	20±7	20±7	0±2		1.9±0.3	1.0±0.5
72161	LRV-3		30±4	16±3	20±7	30±5	1.7±2		1.8±0.4	0.7±0.3
75061	5	BT	8.2±2	6.6±2	74±6	10±4	0±2	0±0.2	1.2±0.4	6.9±2.6
75081	5		15±3	6.3±2	64±5	14±4	0±1	1.0±0.2	2.4±0.9	4.3±1.2
75081,A	5	SF	11±4	5.5±3	80±6	2±7	2.5±2	0.46±0.3	2.0±1.3	40±140

Sample	Station	Type								
75081,B	5	SF	12 ± 4	6.1 ± 3	71 ± 7	7 ± 6	2.4 ± 2	0.56 ± 0.3	2.0 ± 1.3	9.0 ± 7.0
75081,C	5	SF	14 ± 4	6.7 ± 2	64 ± 5	15 ± 4	0.3 ± 1.5	0.80 ± 0.2	2.0 ± 0.9	4.3 ± 1.2
75081,D	5	SF	18 ± 4	16 ± 4	52 ± 7	13 ± 7	0 ± 2	1.4 ± 0.3	1.1 ± 0.3	4.0 ± 2.2
70161	ALSEP		17 ± 3	7.8 ± 3	53 ± 6	22 ± 5	0 ± 2	0.52 ± 0.2	2.2 ± 1	2.4 ± 0.6
70181	ALSEP		20 ± 3	10 ± 2	51 ± 4	19 ± 4	0 ± 2	1.3 ± 0.2	2.0 ± 0.5	2.7 ± 0.6
71501	1		15 ± 2	4.9 ± 1	67 ± 4	12 ± 4	0.90 ± 1	0.97 ± 0.2	3.1 ± 0.7	5.8 ± 2.0
71041	1	TT	12 ± 3	7.4 ± 2	61 ± 5	18 ± 4	0 ± 2	0.80 ± 0.2	1.7 ± 0.6	3.4 ± 0.8
71061	1	TB	14 ± 3	7 ± 2	55 ± 5	25 ± 5	0 ± 2	0.55 ± 0.2	2.0 ± 0.7	2.2 ± 0.5
70051	EVA-3	SM	40 ± 5	14 ± 5	27 ± 8	16 ± 8	3.0 ± 2	—	3.0 ± 1.2	1.7 ± 1.0
79221	9		30 ± 4	10 ± 3	35 ± 7	22 ± 6	1.9 ± 2	1.7 ± 0.3	3.0 ± 1.0	1.6 ± 0.5
79261	9	BR	32 ± 4	15 ± 4	32 ± 7	19 ± 7	1.2 ± 2	1.1 ± 0.3	2.2 ± 0.6	1.6 ± 0.7
79135	9		33 ± 3	18 ± 2	20 ± 4	28 ± 5	0.0 ± 2	1.3 ± 0.2	1.8 ± 0.3	0.7 ± 0.2
76261	6	TT	42 ± 3	28 ± 3	15 ± 5	11 ± 5	3.4 ± 1.5	0.7 ± 0.2	1.5 ± 0.2	1.4 ± 0.8
76281	6	TB	45 ± 3	21 ± 2	18 ± 5	11 ± 5	4.0 ± 1.5	—	2.1 ± 0.2	1.7 ± 0.9
76321	6		44 ± 3	32 ± 2	11 ± 4	8 ± 4	3.8 ± 1.5	0.90 ± 0.2	1.4 ± 0.2	1.4 ± 0.9
76501	6		52 ± 2	17 ± 1	17 ± 2	6 ± 2	6.6 ± 1.0	1.1 ± 0.2	3.1 ± 0.2	2.8 ± 1.0
77531	7		41 ± 4	23 ± 3	16 ± 7	16 ± 7	1.9 ± 1.9	—	1.8 ± 0.3	1.0 ± 0.6
78221	8	UB	46 ± 4	18 ± 3	18 ± 6	12 ± 6	4.7 ± 1.7	1.3 ± 0.3	2.5 ± 0.5	1.4 ± 0.8
78501	8		41 ± 3	12 ± 3	31 ± 5	13 ± 5	2.7 ± 1.5	1.2 ± 0.2	3.5 ± 0.9	2.3 ± 0.9

[a]TT = top trench; TM = middle trench; TB = bottom trench; UB = under boulder; BT = top of boulder; BR = soil breccia; SF = size fractions: 75081,A = +60#; 75081,B = −60+120#; 75081,C = −120+305#; 75081,D = −305# (Duncan et al., 1974). SM = Residue from several stations in EVA-3.

[b]In excess of what is present in the end members.

(Duncan *et al.*, 1974). The fact that soils 75061, 76321, and 75081,D have an R similar to the R value found in the light-mantle soils might be coincidental. The R value for the soils of the trench in Station 3 are similar and slightly higher than average light-mantle soils. The highland component of these soils could be a mixture of light-mantle soil and a regolith excavated by a nearby 10 m crater (Muehlberger *et al.*, 1974). (3) Soils from the valley floor have in general higher values of Q when compared with the light-mantle soils and soils from Stations 6, 7, 8, and 9. (4) The meteoritic contamination of these soils is estimated mainly from the Ni concentration and in general is "equivalent" to about 1% (in excess of what might be present in the end members). (5) The abundance of granites is in all cases less than 0.3%. (6) The material balance of Zn can be tested using the results of the mixing model calculations. The material balance for Zn is in general good with the exception of soils 74240 and 74260 where there appear to be a large excess of Zn. These samples were collected next to the orange soil (sample 74220) which has a high concentration of Zn, but there appears not to be enough orange soil in samples 74240 and 74260 to account for the excess Zn. One possible explanation for the excess Zn is that some of the Zn from the orange soil (sample 74220) diffused out into soils 74240 and 74260. (7) In the case of sulfur it was assumed that the meteoritic component contribution is negligible (Moore *et al.*, 1972), and the results of the mixing model calculations show that with the exception of the finest fraction of soil 75081 (sample 75081,D) there is good material balance of sulfur consistent with the results of Gibson and Moore (1974). This good material balance for sulfur obtained by the mixing model calculations suggests, that within the experimental error, there is little or no bulk volatilization of sulfur in these soils. (8) The finest fraction of soil 75081 (sample 75081,D) when compared with the coarsest fraction of this soil (sample 75081,A) is enriched in anorthositic gabbro 17, KREEPy 17, orange glass 17, and meteoritic component. Also, the finest fraction 75081,D has excess sulfur. (9) The results of the mixing model calculations can be used to estimate the Th concentration of the *highland portion* of the Apollo 17 soils for comparison studies with the results of the orbiting gamma-ray experiment. A R value of 1.0 corresponds to a Th concentration of about 2.9 ppm and a R value of 2.4 corresponds to a Th concentration of about 2.1 ppm. The measured value for a 5×5 (degrees) segment area in the region of Littrow is 2.2 ppm Th (Trombka *et al.*, 1973).

Highland rocks

The possibility that VHA and low-K Fra Mauro basalt (LK-FM) are mixtures was investigated using the published chemical composition of these rocks and the weighted least-squares mixing model technique described in this paper. Samples studied were VHA basalts (61016-143, 61156, 15273-4-3, and 62295), KREEPy 16 rocks (65015, 60315, 62235), KREEPy 17 rocks also called noritic breccias (LSPET, 1973) (72435, 72275, 76315, 77135, and 76055), Luna 20 soil, sample 15273-4-2, sample 67955, sample 14310, sample 68415, sample 15445 dark, and sample 14068. The results of the mixing model calculations using up to 27 chemical elements show that it is not possible to generate these rock types by using only mare basalts, "anorthosites," KREEP, meteoritic component (met), and granite (gr). If dunite is included as another component, then the mixing model calculations show (Fig. 3) that it is *possible* to generate VHA, LK-FM, and the other samples mentioned before by mixing dunite, "anorthosite," and KREEP. In Fig. 3 there is an approximate trend line between KREEP and (spinel) troctolite (Prinz *et al.*, 1973). (Spinel) troctolite can be a mixture of "anorthosite" and dunite, but the trend line suggests that (spinel) troctolite is an important rock type and not a mixture. The deviations from the trend line could be caused by sampling errors and/or the variable composition of the (spinel) troctolites. The fact that these rock types (VHA, LK-FM) can be mixtures from the standpoint of chemical material balance is not an absolute proof that they are actual mixtures, but it is considered to be a significant argument in favor of this proposed model. In conclusion it appears that it is possible that highland rocks are "contaminated" with KREEP, and that (spinel) troctolite and possibly dunite are significant rock types. This conclusion extends the previous conclusion (Schonfeld and Meyer, 1972) that KREEP was the main "contaminant" of LIL trace elements and radiogenic isotopes in the Apollo 11, 12, and 15 mare sites.

Rb–Sr systematics of highland rocks

The results of the mixing model calculations on highland rocks permit interpretation of the Rb–Sr systematics of these rocks. The Rb–Sr systematics of some of these rocks is shown in Fig. 4 using the data of Nyquist *et al.* (1973). The

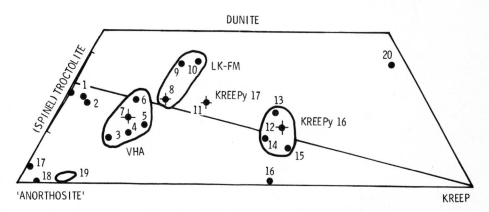

Fig. 3. Results of mixing model. Points labeled 1,2,3... correspond to samples Luna 20 soil, 15273-4-2, 61016-143, 61156-2, 15273-4-3, 62295, average VHA, average low-K Fra Mauro basalt, 15445 dark, 76055, average KREEPy 17, average KREEPy 16, 60315, 65015, 62235, 14310, 67955, 68415, Apollo 16 soils, and 14068. "Anorthosite" is a mixture of anorthosite, gabbroic anorthosite, and anorthositic gabbro. KREEPy 16 is the average of 60315, 65015, and 62235. Average VHA is the average of 61016, 61156, and 62295. Average KREEPy 17 is the average of 72435, 72275, 76315, 77135, and 76055.

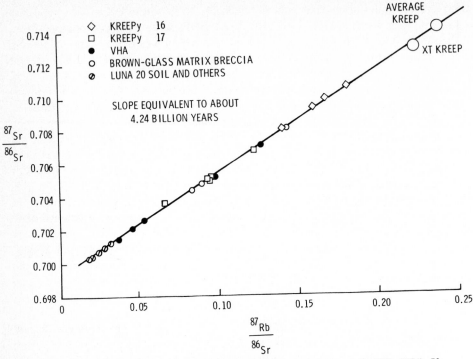

Fig. 4. Rb–Sr systematics of highland rocks. (Data by Nyquist *et al.*, 1973, 1974). If KREEPy 16, KREEPy 17, VHA, and KREEP are cogenetic then the line is an isochron with a corresponding age of 4.24 AE. If the mentioned samples are mixtures of KREEP and troctolite then the line is a mixing line with no age significance.

isotopic properties of KREEP are fairly well known (Nyquist *et al.*, 1973) but for other end members such as "anorthosites," troctolites, and dunites are not well known since as shown previously many highland rocks probably are contaminated with KREEP. In the following discussion it is assumed that "anorthosites," troctolites, and dunites have $^{87}Rb/^{86}Sr$ ratios less than about 0.05 and model ages about 4.6 AE (AE = 10^9 yr). If the VHA basalts, Apollo 17 noritic breccias (KREEPy 17), and KREEPy 16 breccias are mixtures of troctolite and KREEP the line in Fig. 4 is a mixing line and has no age significance. Slight deviations from the line are interpreted as variations in KREEP composition (see later). Adopting the hypothesis that rocks with KREEP-like relative REE abundance patterns (VHA, Apollo 17 noritic breccias, and crystalline KREEP) differentiated contemporaneously from the same or similar source materials leads to construction of the primary 4.22 AE isochron which is essentially the same as that determined from VHA and KREEP alone (Nyquist *et al.*, 1974). According to Nyquist *et al.* the validity of this construction rests on the final understanding of the petrogenesis of these rocks.

    In order to better understand Fig. 4, a brief review of the model ages (T_{BABI}) of the highland rocks and KREEP follows. Schonfeld and Meyer (1972) have shown that when the KREEP rocks from Apollo 12 and 14 site are averaged one obtains an average model age for KREEP of 4.4 AE. Since then many samples have been analyzed by Nyquist *et al.* (1973) that have KREEP-type trace element chemistry (Hubbard *et al.*, 1973b). The average model age for the Apollo 16 KREEPy rocks (65015, 62235, 60315, and 64815) is 4.44 AE, for the "VHA" basalt group rocks (62295, 64567, 63545, and 61016) is 4.40 AE, for the noritic breccias KREEPy 17 (72435, 76055, 76315, 77135, and 72275) is 4.44 AE, and for the Apollo 15 crystalline KREEP (15023, 15273, 15382, 15314, and 15304) is 4.30 AE. KREEP has much higher relative concentrations of Rb and radiogenic Sr than "anorthosites", troctolites, and dunites and therefore its average model age of about 4.4 AE dominates in these samples if they are mixtures. The individual model ages of each rock and the KREEP fragments deviates slightly from 4.40 AE. These deviations are interpreted as being due to the system being partially open on account of the very intense planetesimal bombardment, and that in each site we are able to collect enough samples so the average model age remains quite constant and represents the model age for this rock type. This average model age of 4.4 AE for KREEP has been interpreted by Schonfeld and Meyer (1972), by using a two stage evolution model, to be close to the original time of extrusion of KREEP. Based on the U–Pb systematics of several Apollo 16 samples (60315, 68415, and 65015), Nunes *et al.* (1973) have shown that an older event at about 4.4 AE occurred. This observation is consistent with the view presented in this work where many of the Apollo 16 rocks are contaminated with KREEP that dominates the Rb–Sr and U–Pb systematics of these samples.

Crystalline KREEP from Apollo 15 (Meyer, 1972) has higher Si and Cr concentration than the Apollo 12 and 14 KREEP (Powell *et al.*, 1973) and as mentioned before has an average model age of 4.30 AE (Nyquist *et al.*, 1973). This is not surprising considering that the mare basalts have slightly different compositions and extrusion times. Therefore the time of extrusion of KREEP could be in the range of about 4.3 to 4.4 AE. The average model age of Apollo 15 crystalline KREEP of 4.30 AE based on the two stage evolution model (Schonfeld and Meyer, 1972) represent the original extrusion time for KREEP, in agreement with the 4.3 AE track retention age found by Haines *et al.* (1974). Nyquist *et al.* (1973) proposed a Rb loss model (by volatilization) to account for the difference between the model ages of crystalline KREEP and the Apollo 14 soils. Some difficulties with this model are the chemical differences between crystalline KREEP and the Apollo 14 KREEP breccias and soils (as mentioned before) and also the assumption that there is only loss and no gain of Rb (by volatility) in the soils. The author believes that the loss of Rb by volatility as proposed by Nyquist *et al.* (1973) might be responsible, but in a small degree, for some of the variations of the model ages of the KREEP-rich soils (for example, soil 14259). At the same time probably other soils gained Rb.

DISCUSSION

Based on the results of the mixing model calculations (for the highland rocks) KREEP, "anorthosites" (anorthositic gabbros, gabbroic anorthosites, and anorthosites), troctolites, and possibly dunites are important rock types on the moon. Minor components are the meteoritic component and granites. As far as the LIL trace elements and radiogenic isotopes, KREEP is the main contaminant in highland rocks. According to the results of the gamma-ray experiment (Metzger *et al.*, 1973) it appears that the main source for KREEP are the Imbrium–Procellarum regions. The existence of other local sources of KREEP remains an open question. The lowest concentration of Th in the highlands measured by the gamma-ray orbiting experiment is about 0.5 ppm (Metzger *et al.*, 1974). This concentration is close to the Th concentration estimated in uncontaminated gabbroic anorthosite and anorthositic gabbro. Therefore, there seems to be large areas of the highlands with little or no KREEP. The low Th concentration areas include a large portion of the Orientale ejecta (120°W to 80°W). The Van de Graaff region is relatively low in altitude with respect to the rest of the highlands (Sjogren and Wollenhaupt, 1973), and the higher levels of radioactivity (Metzger *et al.*, 1974) might indicate local sources of KREEP.

With the exception of the Luna 20 site (Prinz *et al.*, 1973) the rather low abundance of troctolites appears to be a problem. Where should we find high Mg rocks? If troctolites and dunites represent deep cumulates during the formation of the lunar crust of about 60 km thickness (Toksöz *et al.*, 1973) it is expected that it would probably be difficult to find troctolites and dunites in large amounts on the surface of the moon. The most likely place to find them would be around the rims of the large basins. A preliminary analysis of the Al and Mg concentrations obtained by the orbiting X-ray fluorescence experiment (Adler *et al.*, 1973) was done. In this analysis it was assumed that the top portion of the highlands originally had only anorthositic rocks (anorthositic gabbro, gabbroic anorthosites, and anorthosites). From what we know from the lunar samples (ground truth) the Mg concentration for these rocks is approximately inversely proportional to the Al concentration. Therefore, with the assumption given before, for each Al–Mg concentration pair obtained by the orbiting X-ray fluorescence experiment, it is possible to estimate if there is an excess Mg. Preliminary analysis of these data shows that there appears to be an excess of Mg at the rims of Imbrium, Procellarum, Serenitatis, Tranquillitatis, and Fecunditatis. There is little or no Mg excess at the rim of Crisium and no excess at the rim of Smythii. The lack of excess of Mg in Smythii is consistent with the interpretation of the laser altimetry experiment (Kaula *et al.*, 1974), where the crust is thicker on the far side of the moon. Based on this preliminary interpretation of the X-ray fluorescence data it is possible to explain the rather low abundance of troctolites close to the Imbrium–Procellarum regions. In those regions there probably was originally troctolite and KREEP and the very intense bombardment made mixtures of those two rock types and the "anorthosites." VHA basalts would be troctolites contaminated with KREEP. In the rim of the Fecunditatis region the situation was different. There probably were troctolites and little or no local KREEP. Therefore we would expect to find troctolites at the rim of Fecunditatis with little contamination of KREEP as the Luna 20 troctolites (Prinz *et al.*, 1973) appear to be.

Conclusions

(1) Based on the results of mixing model calculations using up to 30 chemical elements, it is possible that VHA basalts and low-K Fra Mauro basalt, and other highland rocks (68415, 14068, 65015, 60315, 62235, etc.) are mixtures of KREEP, "anorthosites," troctolite, and dunite. Meteoritic and granitic components are present only in small amounts.

(2) The Rb–Sr model ages of these rocks are dominated by the high concentrations of ^{87}Rb and radiogenic ^{87}Sr in KREEP. The average model age of KREEP in all sites is about 4.4 AE and is interpreted as the original extrusion time of KREEP.

(3) The Apollo 16 soils are mixtures of "anorthositic" rocks and KREEPy 16. Minor contributors to these soils are "VHA", mare basalt, meteoritic and granitic components.

(4) The Apollo 17 soils are mixtures of local mare basalt, local orange basaltic composition glass, "anorthositic" rocks, and KREEPy 17 (noritic breccias).

Acknowledgments—The author thanks R. Brett, K. Keil, J. S. Fruchter, R. Merrill, C. Meyer, Jr., D. F. Nava, L. E. Nyquist, and J. A. Philpotts for their useful comments and C. Hardy for typing this paper.

References

Adler I., Trombka J. I., Schmadebeck R., Lowman P., Blodget H., Yin L., Eller E., Podwysocki M., Weidner J. R., Bickel A. L., Lum R. K. L., Gerard J., Gorenstein P., Bjorkholm P., and Harris B. (1973) Results of the Apollo 15 and 16 X-ray experiment. *Proc. Fourth Lunar Sci. Conf., Geochim. Cosmochim. Acta*, Suppl. 4, Vol. 3, pp. 2783–2791. Pergamon.

Albee A. L., Gancarz A. J., and Chodos A. A. (1973) Metamorphism of Apollo 16 and 17 and Luna 20 metaclastic rocks at about 3.95 AE: Samples 61156, 64423,14-2, 65015, 67483,15-2, 76055, 22006, and 22007. *Proc. Fourth Lunar Sci. Conf., Geochim. Cosmochim. Acta*, Suppl. 4, Vol. 1, pp. 569–595. Pergamon.

Baedecker P. A., Chou C.-L., Grudewicz E. B., and Wasson J. T. (1973) Volatile and siderophilic trace elements in Apollo 15 samples: Geochemical implications and characterization of the long-lived and short-lived extralunar materials. *Proc. Fourth Lunar Sci. Conf., Geochim. Cosmochim. Acta*, Suppl. 4, Vol. 2, pp. 1177–1195. Pergamon.

Brett R. (1971) The earth's core: Speculations on its chemical equilibrium with the mantle. *Geochim. Cosmochim. Acta* 35, 203–221.

Cameron K. L., Delano J. W., Bence A. E., and Papike J. J. (1973) Petrology of the 2–4 mm soil fraction from the Hadley–Apennine region of the moon. *Earth Planet. Sci. Lett.* 19, 9-21.

Dowty E., Prinz M., and Keil K. (1974a) "Very High Alumina Basalt": Mixture and not a magma type. *Science* 183, 1214–1215.

Dowty E., Keil K., and Prinz M. (1974b) Igneous rocks from Apollo 16 rake samples (abstract). In *Lunar Science—V*, pp. 174–176. The Lunar Science Institute, Houston.

Duncan A. R., Erlank A. J., Willis J. P., and Ahrens L. H. (1973) Composition and inter-relationships of some Apollo 16 samples. *Proc. Fourth Lunar Sci. Conf., Geochim. Cosmochim. Acta*, Suppl. 4, Vol. 2, pp. 1097–1113. Pergamon.

Eldridge J. S., O'Kelley G. D., and Northcutt K. J. (1974) Primordial radio-element concentrations in rocks and soils from Taurus-Littrow (abstract). In *Lunar Science—V*, pp. 206–208. The Lunar Science Institute, Houston.

Ganapathy R., Morgan J. W., Krähenbühl U., and Anders E. (1973) Ancient meteoritic components in lunar highland rocks: Clues from trace elements in Apollo 15 and 16 samples. *Proc. Fourth Lunar Sci. Conf., Geochim. Cosmochim. Acta*, Suppl. 4, Vol. 2, pp. 1239–1261. Pergamon.

Gast P. W. (1972) The chemical composition and structure of the moon. *The Moon* **5**, 121–148.

Gibson E. K., Jr. and Moore G. W. (1974) Total sulfur abundances and distributions in the valley of Taurus-Littrow: Evidence of mixing (abstract). In *Lunar Science—V*, pp. 267–269. The Lunar Science Institute, Houston.

Goles G. G., Duncan A. R., Lindstrom D. J., Martin M. R., Beyer R. L., Osawa M., Randle K., Meek L. T., Steinborn T. L., and McKay S. M. (1971) Analyses of Apollo 12 specimens: Compositional variations, differentiation processes and lunar soil mixing models. *Proc. Second Lunar Sci. Conf.*, *Geochim. Cosmochim. Acta*, Suppl. 2, Vol. 2, pp. 1063–1081. MIT Press.

Haines E. L., Hutcheon I. D., and Weiss J. R. (1974) Excess fission tracks in Apennine front KREEP basalts (abstracts). In *Lunar Science—V*, pp. 304–306. The Lunar Science Institute, Houston.

Hartmann W. K. and Wood C. A. (1971) Moon: Origin and evolution of multi-ring basins. *The Moon* **3**, 3–78.

Haskin L. A., Helmke P. A., Blanchard D. P., Jacobs J. W., and Telander K. (1973) Major and trace element abundances in samples from the lunar highlands. *Proc. Fourth Lunar Sci. Conf.*, *Geochim. Cosmochim. Acta*, Suppl. 4, Vol. 2, pp. 1275–1296. Pergamon.

Haskin L. A., Shih C.-Y., Bansal B. M., Rhodes J. M., Wiesmann H., and Nyquist L. E. (1974) Chemical evidence for the origin of 76535 as a cumulate (abstract). In *Lunar Science—V*, pp. 313–315. The Lunar Science Institute, Houston.

Hörz F., Oberbeck V. R., and Morrison R. H. (1974) Remote sensing of the Cayley plains and Imbrium basin deposits (abstract). In *Lunar Science—V*, pp. 357–359. The Lunar Science Institute, Houston.

Hubbard N. J. and Shih C.-Y. (1973) Opposing partial melting models and the genesis of some lunar magmas. *Meteoritics* **8**, 384–386.

Hubbard N. J., Rhodes J. M., and Gast P. W. (1973a) Chemistry of lunar basalts with very high alumina contents. *Science* **181**, 339–342.

Hubbard N. J., Rhodes J. M., Gast P. W., Bansal B. M., Shih C.-Y., Wiesmann H., and Nyquist L. E. (1973b) Lunar rock types: The role of plagioclase in non-mare and highland rock types. *Proc. Fourth Lunar Sci. Conf.*, *Geochim. Cosmochim. Acta*, Suppl. 4, Vol. 2, pp. 1297–1312. Pergamon.

Husain L. (1972) The ^{40}Ar–^{39}Ar and cosmic ray exposure ages of Apollo 15 crystalline rocks, breccias and glasses. In *The Apollo 15 Lunar Samples*, pp. 374–377. The Lunar Science Institute, Houston.

Kaula W. M., Lingenfelter R. E., Sjogren W. L., and Wollenhaupt W. R. (1974) Apollo laser altimetry and inferences as to lunar structure (abstract). In *Lunar Science—V*, pp. 399–401. The Lunar Science Institute, Houston.

Kendall M. G. and Stuart A. (1961) *The Advanced Theory of Statistics*. Hafner, New York.

Laul J. C. and Schmitt R. A. (1973) Chemical composition of Apollo 15, 16, and 17 samples. *Proc. Fourth Lunar Sci. Conf.*, *Geochim. Cosmochim. Acta*, Suppl. 4, Vol. 2, pp. 1349–1367. Pergamon.

LSPET (1973) Apollo 17 lunar samples: Chemical and petrographic description. *Science* **182**, 659–672.

Metzger A. E., Trombka J. I., Peterson L. E., Reedy R. C., and Arnold J. R. (1973) Lunar surface radioactivity: Preliminary results of the Apollo 15 and Apollo 16 gamma-ray spectrometer experiments. *Science* **179**, 800–803.

Metzger A. E., Trombka J. I., Reedy R. C., and Arnold J. R. (1974) Element concentrations from lunar orbital gamma-ray measurements (abstract). In *Lunar Science—V*, pp, 501–502. The Lunar Science Institute, Houston.

Meyer C., Jr., Brett R., Hubbard N. J., Morrison D. A., McKay D. S., Aitken F. K., Takeda H., and Schonfeld E. (1971) Mineralogy, chemistry, and origin of the KREEP component in soil samples from the Ocean of Storms. *Proc. Second Lunar Sci. Conf.*, *Geochim. Cosmochim. Acta*, Suppl. 2, Vol. 1, pp. 393–411. MIT Press.

Meyer C., Jr. (1972) Mineral assemblages and the origin of non-mare lunar rock types (abstract). In *Lunar Science—III*, pp. 542–544. The Lunar Science Institute, Houston.

Meyer C., Jr., Anderson D. H., and Bradley J. G. (1974) Ion microprobe mass analysis of plagioclase from "non-mare" lunar samples (abstract). In *Lunar Science—V*, pp. 506–508. The Lunar Science Institute, Houston.

Moore C. B., Lewis C. F., Cripe J., Delles F. M., Kelly W. R., and Gibson E. K., Jr. (1972) Total carbon, nitrogen and sulfur in Apollo 14 lunar samples. *Proc. Third Lunar Sci. Conf.*, *Geochim. Cosmochim. Acta*, Suppl. 3, Vol. 2, pp. 2051–2058. MIT Press.

Muehlberger W. R., Batson R. M., Cernan E. A., Freeman V. L., Hait M. H., Holt H. E., Howard K. A., Jackson E. D., Larson K. B., Reed V. S., Rennilson J. J., Schmitt H. H., Scott D. H., Sutton R. L., Stuart-Alexander D., Swann G. A., Trask N. J., Ulrich G. E., Wilshire H. G., and Wolfe E. W. (1973) Preliminary geologic investigation of the Apollo 17 landing site. NASA SP-330, pp. 6-1 to 6-60.

Nunes P. D., Tatsumoto M., Knight R. J., Unruh D. M., and Doe B. R. (1973) U–Th–Pb systematics of some Apollo 16 lunar samples. *Proc. Fourth Lunar Sci. Conf., Geochim. Cosmochim. Acta*, Suppl. 4, Vol. 2, pp. 1797–1822. Pergamon.

Nyquist L. E., Hubbard N. J., Gast P. W., Bansal B. M., Wiesmann H., and Jahn B. (1973) Rb–Sr systematics for chemically defined Apollo 15 and 16 materials. *Proc. Fourth Lunar Sci. Conf., Geochim. Cosmochim. Acta*, Suppl. 4, Vol. 2, pp. 1823–1846. Pergamon.

Nyquist L. E., Bansal B. M., Wiesmann H., and Jahn B. M. (1974) Taurus-Littrow chronology: Implications for early lunar crustal development (abstract). In *Lunar Science—V*, pp. 565–567. The Lunar Science Institute, Houston.

Philpotts J. A., Schuhmann S., Kouns C. W., Lum R. K. L., Bickel A. L., and Schnetzler C. C. (1973) Apollo 16 returned lunar samples: Lithophile trace-element abundances. *Proc. Fourth Lunar Sci. Conf., Geochim. Cosmochim. Acta*, Suppl. 4, Vol. 2, pp. 1427–1436. Pergamon.

Philpotts J. A., Schuhmann S., Kouns C. W., and Lum R. K. L. (1974) Lithophile trace elements in Apollo 17 soils (abstract). In *Lunar Science—V*, pp. 599–601. The Lunar Science Institute, Houston.

Philpotts J. A., Schuhmann S., Kouns C. W., Lum R. K. L., and Winzer S. (1974) Origin of Apollo 17 rocks and soils. *Proc. Fifth Lunar Sci. Conf., Geochim. Cosmochim. Acta*. This volume.

Phinney W. C., Warner J. L., Simonds C. H., and Lofgren G. E. (1972) Classification and distribution of rock types at Spur Crater. In *The Apollo 15 Lunar Samples*, pp. 149–152. The Lunar Science Institute, Houston.

Powell B. N., Aitken F. K., and Weiblen P. W. (1973) Classification, distribution and origin of lithic fragments from the Hadley-Apennine region. *Proc. Fourth Lunar Sci. Conf., Geochim. Cosmochim. Acta*, Suppl. 4, Vol. 1, pp. 445–460. Pergamon.

Prinz M., Dowty E., Keil K., and Bunch T. E. (1973) Mineralogy, petrology and chemistry of lithic fragments from Luna 20 fines: Origin of the cumulate ANT suite and its relationship to high-alumina and mare basalts. *Geochim. Cosmochim. Acta* 37, 979–1006.

Reid A. M., Warner J., Ridley W. I., and Brown R. W. (1972) Major element composition of glasses in three Apollo 15 soils. *Meteoritics* 7, 395–415.

Rhodes J. M., Rogers K. V., Shih C., Bansal B. M., Nyquist L. E., and Wiesmann H. (1974) The relationship between geology and soil chemistry at the Apollo 17 landing site (abstract). In *Lunar Science—V*, pp. 630–632. The Lunar Science Institute, Houston.

Ridley W. I., Reid A. M., Warner J. L., Brown R. W., Gooley R., and Donaldson C. (1973a) Glass compositions in Apollo 16 soils 60501 and 61221. *Proc. Fourth Lunar Sci. Conf., Geochim. Cosmochim. Acta*, Suppl. 4, Vol. 1, pp. 309–321. Pergamon.

Ridley W. I., Hubbard N. J., Rhodes J. M., Weismann H., and Bansal B. (1973b) The petrology of lunar breccia 15445 and petrogenetic implications. *Jour. Geology* 81, 621–631.

Rose H. J., Jr., Brown F. W., Carron M. K., Christian R. P., Cuttitta F., Dwornik E. J., and Ligon D. T., Jr. (1974) Composition of some Apollo 17 samples (abstract). In *Lunar Science—V*, pp. 645–647. The Lunar Science Institute, Houston.

Schonfeld E. and Meyer C., Jr. (1972) The abundances of components of the lunar soils by a least-squares mixing model and the formation age of KREEP. *Proc. Third Lunar Sci. Conf., Geochim. Cosmochim. Acta*, Suppl. 3, Vol. 2, pp. 1397–1420. MIT Press.

Schonfeld E. (1973) Lunar rock types and weighted least-squares mixing models. *Meteoritics* 8, 432–435.

Schonfeld E. (1974) Component abundance and evolution of regoliths and breccias: Interpretation by mixing models (abstract). In *Lunar Science—V*, pp. 669–671. The Lunar Science Institute, Houston.

Simonds C. H., Warner J. L., and Phinney W. C. (1973) Petrology of Apollo 16 poikilitic rocks. *Proc. Fourth Lunar Sci. Conf., Geochim. Cosmochim. Acta*, Suppl. 4, Vol. 1, pp. 613–632. Pergamon.

Sjogren W. L. and Wollenhaupt W. R. (1973) Lunar shape via the Apollo laser altimeter. *Science* 179, 275–278.

Taylor S. R., Kaye M., Muir P., Nance W., Rudowski R., and Ware N. (1972) Composition of the lunar uplands: Chemistry of Apollo 14 samples from Fra Mauro. *Proc. Third Lunar Sci. Conf., Geochim. Cosmochim. Acta*, Suppl. 3, Vol. 2, pp. 1231–1249. MIT Press.

Taylor S. R., Gorton M., Muir P., Nance W., Rudowski R., and Ware N. (1974) Lunar highland composition (abstract). In *Lunar Science—V*, pp. 789–791. The Lunar Science Institute, Houston.

Toksöz M. N., Dainty A. M., Solomon S. C., and Anderson K. R. (1973) Velocity structure and evolution of the moon. *Proc. Fourth Lunar Sci. Conf., Geochim. Cosmochim. Acta*, Suppl. 4, Vol. 3, pp. 2529–2547. Pergamon.

Trombka J. I., Arnold J. R., Reedy R. C., Peterson L. E., and Metzger A. E. (1973) Some correlations between measurements by the Apollo gamma-ray spectrometer and other lunar observations. *Proc. Fourth Lunar Sci. Conf., Geochim. Cosmochim. Acta*, Suppl. 4, Vol. 3, pp. 2847–2853. Pergamon.

Walker D., Grove T. L., Longhi J., Stolper E. M., and Hays J. F. (1973) Origin of lunar feldspathic rocks. *Earth Planet. Sci. Lett.* **20**, 325–336.

Wänke H., Wlotzka F., Baddenhausen H., Balacescu A., Spettel B., Teschke F., Jagoutz E., Kruse H., Quijano-Rico M., and Rieder R. (1971) Apollo 12 samples: Chemical composition and its relation to sample locations and exposure ages, the two component origin of the various soil samples and studies on lunar metallic particles. *Proc. Second Lunar Sci. Conf., Geochim. Cosmochim. Acta*, Suppl. 2, Vol. 2, pp. 1187–1208. MIT Press.

Warner J. L., Simonds C. H., and Phinney W. C. (1973) Apollo 16 rocks. Classification and petrogenetic model. *Proc. Fourth Lunar Sci. Conf., Geochim. Cosmochim. Acta*, Suppl. 4, Vol. 1, pp. 481–504. Pergamon.

Warner J. L., Simonds C. H., and Phinney W. C. (1974) Impact induced fractionation in the highlands (abstract). In *Lunar Science—V*, pp. 823–825. The Lunar Science Institute, Houston.

Wlotzka F., Spettel B., and Wänke H. (1973) On the composition of metal from Apollo 16 fines and the meteoritic component. *Proc. Fourth Lunar Sci. Conf., Geochim. Cosmochim. Acta*, Suppl. 4, Vol. 2, pp. 1483–1491. Pergamon.

Proceedings of the Fifth Lunar Conference
(Supplement 5, Geochimica et Cosmochimica Acta)
Vol. 2 pp. 1287–1305 (1974)
Printed in the United States of America

The geochemical evolution of the moon

S. R Taylor[1] and P. Jakeš[2]

The Lunar Science Institute, 3303 Nasa Road 1, Houston, Texas 77058

Abstract—It is assumed that the moon accreted from refractory material from which the volatile elements had already been depleted. The accretion was homogeneous since there is (a) a good correlation between volatile/involatile element ratios in both highland and maria samples, and (b) the element distribution in crustal rocks is not governed by volatility differences.

The overall composition of the highland crust is derived from the orbital Al/Si and Th data and the observed interelemental relationships. The highland REE abundances have a positive Eu anomaly. If the moon has the same relative pattern as chondrites, the interior has a negative Eu anomaly, with a pattern resembling those of maria basalts. The abundances of the refractory trace elements, are about five times those in Type 1 carbonaceous chondrites. In order to account for the high near-surface element abundances, very efficient large-scale element fractionation must occur, implying melting of most or all of the moon.

The following geochemical model is proposed. The center of the moon (below 1000 km) is primitive unfractionated material, now partially molten due to trapped K, U, and Th.

Following accretional melting, the first silicate phase to separate was Mg-rich olivine. As crystallization proceeded, orthopyroxene precipitated. In the low-pressure (<50 kbar) environment, most cations except Mg, Fe, Ni, Co, and Cr^{2+} were excluded from the olivine and orthopyroxene lattice sites, and migrated upward. A frozen crust quickly developed, although continually broken up by the declining meteorite bombardment. This frozen surface layer, analogous to a chilled margin retained high concentrations of Mg, Cr, etc. in near-surface regions. These elements are not derived from chondritic meteorites, which would have contributed high Ni, Ir, etc. Because of the refractory nature of the total lunar composition, a Ca–Al-rich residuum develops. Increasing crystallization at depth leads to a concentration of these elements, trapped under the frozen surface layer. When the concentration of Al reaches 12–17% Al_2O_3, An-rich plagioclase precipitates, and concentrates beneath the frozen surface, whereas the Mg–Fe phases sink. The Ca–Al-rich region incorporates Sr^{2+} and Eu^{2+}.

As crystallization proceeds in both top (crustal Ca–Al) and bottom (mantle Fe–Mg) regions, additional fractionation changes the Mg/Fe ratio, and produces zones of Fe–Ti oxide accumulation. Those elements unable to enter the plagioclase above or the Mg–Fe sites below are trapped between. In this zone, all the remaining elements concentrate. These include K, Ba, Rb, Cs, REE, Th, U, Zr, and Nb. Thus following the primordial fractionation, a chemically zoned moon is produced.

The crustal zonation established at about 4.5 aeons was changed very quickly. The declining stages of the meteoritic bombardment pulverized the chilled zone and larger impacts mixed in the underlying anorthosite. The high concentration of heat-producing elements K, U, and Th (and Zr, Hf, REE, etc.) beneath the plagioclase zone provide the high-element abundances for the Fra Mauro or KREEP basalts. Possibly this zone did not solidify but the residual liquids invaded the crust, where impact mixing produced the parent material for the anorthositic gabbro (highland basalt) and the Fra Mauro basalts. This stage continued to 3.9 aeons.

Partial melting next occurred in successively deeper layers and a succession of maria basalts were erupted. The first of these were the aluminous basalts from shallow depths beneath the KREEP source layer. These overlapped with the later stages of the bombardment, and predate the Imbrium collision in part. Next (3.8–3.6 aeons), the Ti-rich Apollo 11 and 17 basalts were erupted from a zone where Fe–Ti

---

[1]Research School of Earth Sciences, Australian National University, Canberra.
[2]Ustredni ustav geologicky, Hradebni 9, Praha, Czechoslovakia.

oxides accumulated. Finally (3.4–3.2 aeons), the Apollo 12 and 15 quartz and olivine normative basalts were extruded.

1. INTRODUCTION

THE LARGE VOLUME of geochemical, petrological, and geophysical data now available place many constraints on the composition and evolution of the moon. In this paper we consider the evidence for and the implications of homogeneous accretion of the moon. Inter-element ratios are used to calculate the composition of the highland crust. The rare-earth data for the highlands show a positive Eu anomaly allowing the deduction that the interior has a negative anomaly. The Cr/Ni ratios in the highlands are considered in regard to the overall lunar abundance of the siderophile elements. These abundance data, and the constraints from the heat flow measurements are next used to set limits on the bulk composition for the moon. The geochemical constraints are integrated with the petrological and geophysical data in an attempt to provide a consistent model for the geochemical evolution of the moon. Many of the points cannot be discussed in the detail which they deserve within the limitations of space imposed by these Conference Proceedings. A further and extended account will be given in a forthcoming book (Taylor, 1975).

2. HOMOGENEOUS ACCRETION OF THE MOON

One of the first-order facts which appears to be well established is that the volatile elements (e.g. Rb, Pb, Tl, Bi, Cs) had already been depleted at the time of accretion (Papanastassiou et al., 1970; Wetherill, 1971). Accordingly, it may be assumed that the moon initially accreted from material somewhat enriched in refractory constituents (Grossman and Larimer, 1974). Whether the siderophile elements (e.g. Ir, Au, Ni) were depleted before accretion is a current question.

Much insight has been gained from the study of elemental ratios, briefly summarized here. Analytical data of high quality for lunar samples have revealed a host of correlations often unexpected on the basis of terrestrial geochemical experience. Values for well-established element ratios in lunar samples are given in Table 1.

Three distinct reasons appear for the correlations:

(a) Similarity between elements of ionic radius, valency and bond type leading to conventional geochemical coherence. Good examples are Th/U, Zr/Hf, REE (except Eu), K/Rb, Ba/Rb, Rb/Cs, and Fe^{2+}/Mn^{2+}. In this context, the lack of water and other volatiles in the moon produces simple crystal chemical relationships. None of the ions are hydrated, for example.

(b) Concentration of elements of diverse chemistry in residual melts during fractional crystallization due to their inability to enter the major rock-forming minerals. For the same reasons, these elements may be concentrated in initial melts during partial melting (e.g. K/U, Zr/Nb, Cs/U, K/Zr, K/La). These ratios are very similar in highlands and mare samples. Such differences as exist (e.g. K/U)

Table 1. Lunar element average ratios (±20%).

(A) Geochemically associated elements		Reference
K/Ba	6.10	1
Rb/Cs	23	2
K/Tl	2×10^5	2
Th/U	3.8	3
Tl/Vs	4×10^{-2}	2 highlands
	1.2×10^{-2}	2 maria
Sr^{2+}/Eu^{2+}	100	4
Cr/V	28	6
Zr/Hf	45	5 highlands
	35	5 maria
Cr/Sc	70	4
FeO/Sc	5400	6
Ba/Rb	60	4
FeO/MnO	80	6

(B) Volatile/involatile element ratios		Reference
K/Zr	4.23	1
K/Nb	67	1
K/La	70	7
Cs/U	0.23	2
K/Hf	210	1, 5
K/Th	500	3
K/U	1500	3 highlands
	2500	3 maria

(C) Involatile elements		Reference
Ba/Zr	0.69	1
Zr/Ba	1.44	1
Zr/Nb	14	1

References for Table 1:
1. Duncan et al. (1973).
2. Krähenbühl et al. (1973).
3. Eldridge et al. (1973).
4. Taylor et al. (1973).
5. Chyi and Ehmann (1974).
6. Laul and Schmitt (1973).
7. Wänke et al. (1973).

may be explained by crystal chemical effects (Taylor, 1973). In other examples (Cs/U) the volatile element is somewhat enriched in the highlands samples, the converse of that predicted by the heterogeneous accretion hypotheses. The geochemical association of elements of such widely differing valency, radius, bond type and volatility as K^+, La^{3+}, and Zr^{4+}, through multistage processes is indeed striking. One immediate consequence may be noted. The similarity of those volatile/involatile element ratios place severe constraints on heterogeneous accretion models invoking volatility differences to explain the highland chemistry (Duncan *et al.*, 1973).

The range of correlations for some involatile elements in highland samples is illustrated in Fig. 1. A wide variety of data from Apollo 14–17 missions, all from this laboratory, show close correlations for many involatile elements. In this figure, total REE abundances are plotted. Individual elements of the group would show similar correlations. The significance of the correlations is that, once the value for one element, such as Th is known, those for other elements may be obtained.

(c) Correlations produced by mixing of different rock types during intense cratering of the highlands. This is a further process which will contribute to the observed close inter-element ratios in highland samples. Since the chemistry of the highland samples appears to be dominated by two major rock types (anorthositic gabbro, and low-K Fra Mauro basalt), many of the highland breccias will

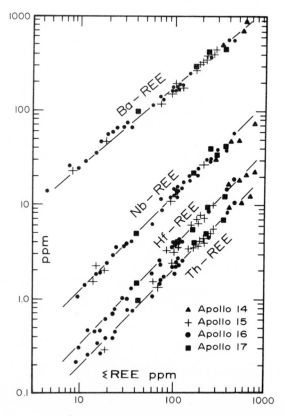

Fig. 1. Element correlations between ΣREE and Th, Hf, Nb, and Ba for highland samples. Data from Taylor *et al.* (1973).

show element abundances dominated by two component mixtures and hence exhibit simple element correlations.

A further relationship emerges from the abundances of the involatile elements. The assumption is made that there is no relative fractionation of the *involatile* elements between the moon and the primitive solar nebula abundances. For example, it is assumed that the rare-earth abundance patterns are parallel in both cases. If this assumption is correct, then the involatile elements in the highland rocks have been fractionated to a degree which depends not on volatility, but on the differences in ionic radii and/or valency from those of Fe^{2+} and Mg^{2+}. These observations have been interpreted as evidence for fractionation based on crystal-liquid equilibria, and hence imply large-scale melting in the outer layers, of the moon (Taylor, 1973).

These lines of evidence encourage the view that the moon was accreted homogeneously (Brett, 1973b; Duncan *et al.*, 1973; Taylor, 1973). Theories that the highlands represented a late addition of chemically distinct refractory material have been abandoned (Ringwood, 1974).

An important consequence of homogeneous accretion theories is that very efficient large-scale elemental fractionation is required to account both for the high near-surface concentrations of refractory elements (e.g. Th, U, REE, Zr, Ba, etc.), and for the Ca–Al-rich crust. This is because the observed concentrations in lunar samples commonly represent enrichments by two orders of magnitude over any estimates of primordial solar nebula abundances.

3. The Composition of the Highlands Crust

The observed inter-element correlations greatly facilitate the study of overall highland chemistry. The various ratios form an interlocking system so that, provided some are known, the rest may be calculated. The orbital XRF and gamma-ray data provide information on Al/Si and Mg/Si ratios and Th values across wide regions of the highlands (Adler *et al.*, 1973; Trombka *et al.*, 1973).

These data form a base on which to build up a multi-element compositional table, by using established inter-element relationships (Table 1). Such a table was first prepared by Taylor *et al.* (1973). In the present paper, the calculation is refined using more recent data. Most notably, the average abundance of Th in the highlands is somewhat lower than previously reported (J. R. Arnold, personal communication).

Average Al/Si and Mg/Si values are 0.62 ($\pm$0.10) and 0.24 ($\pm$0.05), respectively. SiO_2 concentrations are relatively uniform at 45% in highland samples, yielding values of 24.6% Al_2O_3 and 8.6% MgO. A value of 6.6% FeO is obtained from the MgO/FeO relationship observed in the highland soils and breccias. From the Fe/Cr relationship, a figure of 0.10% Cr_2O_3 results. No estimates are made for Ni and Co because of the random meteoritic component. Allowing a typical lunar Na_2O abundance of 0.45%, the remaining major constituent is CaO, which yields, by difference, a value of 14.2%. This value is consistent with that observed in samples of this approximate composition and with the Ca/Al relationship.

The Th average for much of the highlands is in the range 1.0–2.0 ppm (Metzger *et al.*, 1974). Adopting 1.5 ppm Th as an average, and using the well-established Th/U ratio of 3.8, the U abundance is 0.4 ppm. From the K/U value of 1500 (Table 1) the K value may be calculated as 600 ppm or 0.75% K_2O. From K/Rb = 350, Rb = 1.7 ppm and from Rb/Cs = 23, Cs = 0.07 ppm.

Further abundances can be calculated by utilizing the inter-element relations shown in Fig. 1. Thus for a Th value of 1.5 ppm, ΣREE = 65 ppm, and hence Hf = 2.3 ppm, Nb = 7.5 ppm and Ba = 110 ppm, from the relationships in Table 1. The Zr/Nb ratio of 14 is firmly established, yielding 105 ppm Zr from the Nb value. Individual values for the rare earths (with ΣREE = 65 ppm) can be obtained by reference to Fig. 2, which shows the chondrite normalized REE patterns. Assuming that the La/Yb ratio is the same (3.1) as in nearly all the Apollo 16 samples (Taylor *et al.*, 1973) (so that the patterns remain parallel), individual rare-earth element abundances are obtained. On this basis, La = 8.8 ppm, with appropriate values for the others listed in Table 2. For these REE values, the average highlands composition has a positive Eu anomaly. The consequences of a positive Eu anomaly in the highland rocks as discussed later are profound.

A value of 22.4 ppm Y is obtained from the observed relationship that chondrite normalized yttrium values are similar to those for the heavy REE (notably Ho, which has the same ionic radius). The Cr/V ratio of 28 yields a value

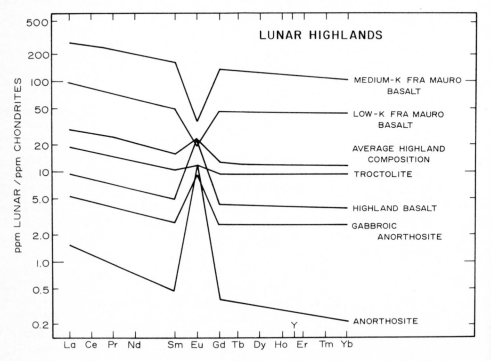

Fig. 2. Rare-earth patterns for average highland rock types. Data from Taylor *et al.* (1973).

Table 2. Trace element abundances in the total moon, in the highlands (0–60 km) and in the region 60–300 km. It is assumed that none of these elements are present in the zone 300–1000 km. The composition of the zone 1000–1738 km (7.6%) is assumed to be the same as the bulk moon.

	Total moon 5 × CCI ppm	Average highlands ppm	% in highlands	Residual conc. in upper 60–300 km ppm
Ba	17.5	110	63	18.7
Sr	43	200	47	67
Th	0.23	1.5	65	0.23
U	0.06	0.4	67	0.057
Zr	30	105	35	60
Hf	0.67	2.3	34	1.20
Nb	2.2	7.5	34	3.84
La	1.1	8.8	80	0.60
Ce	3.10	21.6	70	2.67
Pr	0.44	3.06	70	0.39
Nd	2.10	12.4	59	2.46
Sm	0.75	3.26	43	1.23
Eu	0.27	1.70	63	2.94
Gd	1.10	3.86	35	2.07
Tb	0.18	0.59	33	0.30
Dy	1.15	3.87	34	2.23
Ho	0.27	0.89	34	0.54
Er	0.75	2.51	34	0.72
Tm	0.11	0.37	34	0.21
Yb	0.73	2.25	31	1.50
Lu	0.11	0.35	32	0.22
ΣREE	12	66	55	13.5
Y	7.5	22.4	30	15.5
K	100	600	60	120
Rb	0.29	1.7	59	0.36
Cs	0.013	0.07	55	0.018

of 21 ppm V. The Cr/Sc ratio yields 7.6 ppm Sc. A Tl value of 4 ppb results from the K/Tl ratio of 2×10^5 (Table 1).

The overall highland composition so derived is given in Tables 2 and 3. This is specifically based on orbital Al/Si ratios of 0.62 and a Th value of 1.5 ppm. Compositions for individual highland areas can be obtained from this same approach if the Al/Si and Th values are specified. These data place a premium on the acquisition of further values from both XRF and gamma-ray orbital experiments.

Mixing program models have been employed to see if the values obtained by this empirical technique are realistic in terms of the various postulated individual highland crustal rock types. By testing against four components (anorthosite—highland basalt—low-K Fra Mauro basalt—medium-K Fra Mauro basalt), the closest match was obtained from a two-component mixture of 80% anorthositic gabbro (highland basalt) and 20% low-K Fra Mauro basalt.

Table 3. Major element compositions of various zones in the moon. The composition of the asthenosphere (1000–1738 km; 7.6% volume) is assumed to be undifferentiated, and equivalent to that of the whole moon.

	1	2	3	4	5	6
SiO_2	44.0	44.3	45.2	44.9	43.3	42.73
TiO_2	0.3	—	0.6	0.56	0.7	0.42
Al_2O_3	8.2	0.6	16.9	24.6	14.4	8.21
FeO	10.5	9.9	6.8	6.6	12.6	10.95
MgO	31.0	44.7	18.3	8.6	17.7	30.70
CaO	6.0	0.7	12.5	14.2	11.3	7.68
Volume %	100	49	43.4	10	33.4	100

(1) Whole moon composition.
(2) Composition of 300–1000 km of moon.
(3) Composition of upper 300 km of moon (including crust).
(4) Average highland crust (upper 60 km).
(5) Composition of 60–300 km zone of moon (source of maria basalts).
(6) Bulk moon (Ganapathy and Anders, 1974).

Previous calculations, using a value of 28% Al_2O_3 for highland basalt (Taylor *et al.*, 1973) gave a 70:30 ratio for the proportions of highland basalt to low-K Fra Mauro basalt in the overall average highland composition. A set of new revised estimates for highland rock types based on wider sampling however lowers the average Al_2O_3 value for anorthositic gabbro (highland basalt) to 26% (Reid *et al.*, 1973). Using this value the average highland composition is produced by mixing 80% of highland basalt and 20% of low-K Fra Mauro. Lowering the LKFM component from 30% to 20% (Taylor *et al.*, 1973) produces a positive Eu anomaly (normalized to chondrites) in the overall highland average, as was shown also by the abundances derived from the REE–Th plot, based on the average orbital gamma-ray data. As the Th value decreases, the positive Eu anomaly increases.

The lateral variations in composition of the highland crust are most clearly shown by both the orbital Al/Si data (Alder *et al.*, 1973) and the orbital gamma-ray values (Trombka *et al.*, 1973). Are there significant analogous variations in depth? Do the abundances derived in Table 1 refer only to a superficial layer, or are they representative of the 60 km thick highland crust indicated by the geophysical data (Toksöz *et al.*, 1973)? This is a question vital to geochemical models of total lunar composition.

The large craters have overturned crustal segments down to at least 10 km. The depth of excavation and overturning by the ringed basin collisions is much less certain. The estimates of very deep excavation (>200 km) are probably excessive (Dence *et al.*, 1974), since there is little evidence of the presence of material from these depths in the Imbrium ejecta at either the Apollo 14 or 15 sites. The emerald green glass from Apollo 15 (15426) is associated with the much later mare basalt filling, not the impact (Podosek and Huneke, 1973). The lower estimates envisage cratering depths of 40–60 km (Baldwin, 1972).

The 40 observed ringed basin forming events (Stuart-Alexander and Howard, 1970) must have effectively overturned large segments of the deep crust. Many lunar models predict somewhat enhanced KREEP-type components with depth, so that the abundance problem is likely to be exacerbated rather than alleviated by lack of thorough mixing. It seems likely that adequate deep excavation has occurred. If this has not done a very good job of homogenization, at least extensive overturning has occurred, and any primary crustal zonation has been destroyed. Accordingly, the composition of the crust is probably not grossly different at depth, and it is assumed here that the overall 60 km thick crust has similar element abundances to those of the surficial rocks. There does not seem to be any significant admixture of subcrustal material from the basin impacts. The existence of the 60 km discontinuity, which appears to be a compositional break, is in itself good evidence which limits the depth of excavation.

4. Europium Anomalies and the REE Pattern of the Lunar Interior

It is a reasonable assumption that the total REE patterns in the moon parallel those of chondrites, although the overall concentration levels are several times those of the Type 1 carbonaceous chondrites (Taylor and Jakeš, 1974). Given the assumption of a homogeneously accreted moon, in which the highland crust developed by differentiation, then the interior should display a pattern complementary to that of the crust.

This is not a trivial consequence on account of the great enrichment of rare-earth elements in the crust. As will be shown later, estimates that 70–80% of the total lunar abundances of the light REE are in the highland crust (10% of the whole moon) are not unreasonable. In contrast, the volume of the maria basalts is so small ($\sim\frac{1}{2}\%$ of the total lunar volume) that they may be neglected in these calculations.

Both crustal and interior residual patterns are shown in Fig. 3. These are based on a total lunar abundance of $5 \times CCI$ for the rare earths, as discussed in the next section. The patterns shown are not sensitive to the overall abundances. On this basis, crustal abundances are enriched about 10–20 times chondrites. It is probable that most of the REE and associated elements in the lunar interior are in the upper 300 km (as indicated in Fig. 3) and even within this zone the REE abundances are higher in shallower parts.

The resemblance of the residual REE pattern to that found in the maria basalts is notable. Eu is depleted. There is a progressive depletion of the light (La–Sm) REE, similar to that observed in many maria basalts, particularly the high-Ti varieties found by Apollo 11 and 17. The interior pattern is complementary to those exhibited by plagioclase and resembles that of pyroxenes.

Accordingly, the REE abundance patterns in the interior would be consistent with prior removal of a plagioclase component from source region of maria basalts (e.g. Philpotts and Schnetzler, 1970; Schnetzler and Philpotts, 1971; Taylor and Jakeš, 1974) explaining the Eu depletion observed in them. The subtler features such as depletion of the light REE (La–Sm), a common feature, are consistent with

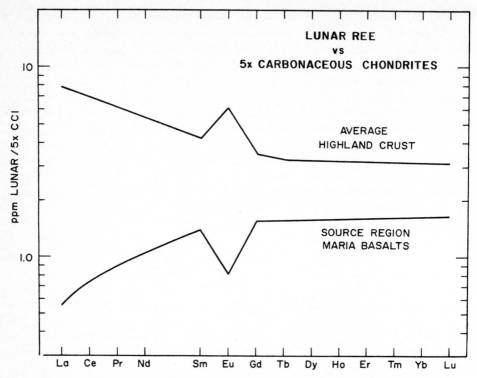

Fig. 3. REE patterns relative to $5 \times CCI$ abundances, taken here as representing the whole moon composition (see text). Note the positive Eu anomaly in the highland crust. All the remaining REE in the moon are assumed concentrated in the upper 60–300 km zone (33.4%) except for the deep interior (1000–1738 km: 7.6%) which is assumed to contain unfractionated material.

a pyroxene source. These conclusions follow from a simple differentiation model, following melting and homogeneous accretion, without recourse to more exotic mechanisms.

5. Chromium/Nickel Ratios in the Highlands

Important constraints on lunar origin and composition arise from the abundances of Cr and Ni (and other siderophile elements) in the lunar highlands. The highland compositions are characterized by three geochemically unusual element associations. These are (1) high Mg and Cr, (2) Ca–Al–Sr–Eu, and (3) high REE and other refractory trace elements.

In the geochemical model developed later, these three elemental associations are derived from different sources. The high-Mg and Cr (and related elements) abundances are thought to be derived from the initial chilled or frozen surface layer. The composition of this zone is essentially that of the accreted material. If this is a valid deduction, the material forming the moon was depleted in siderophile elements before accretion took place. The evidence for this comment comes from the Cr/Ni ratios.

In the primitive solar nebula composition (as represented by the Type 1 carbonaceous chondrite abundances) the Cr/Ni ratio is about 0.25. In contrast, in the lunar highlands Cr/Ni is about 5. (Chromium averages about 500–600 ppm while Ni is rarely more than 100 ppm.) Thus Ni is depleted relative to Cr by at least an order of magnitude. The initial depletion must have been much greater, for much of the Ni (and other siderophile elements) at present in the highland rocks, is derived from the later meteoritic bombardment. Hence, if our original assumption is correct, Ni (and the other siderophiles) were strongly depleted in the primordial lunar composition.

6. THE OVERALL COMPOSITION OF THE MOON

Several independent constraints are available. All these estimates refer to the *involatile* elements. However, once the abundances for these elements have been established it is possible for example to derive values for volatile elements such as K from K/La, K/Zr, or K/Ba ratios, etc.

(a) The heat flow data provide a critical compositional constraint. The high heat flow ($\sim$0.7 HFU) (HFU $= 10^{-6}$ cal cm^{-2} sec^{-1}) indicates that the total lunar abundance of uranium is about 60 ± 15 ppb (Toksöz *et al.*, 1973). The abundance of U in CCI is variable (Mason, 1970) but generally estimated at about 12 ppb. On this basis the moon is enriched at least 5 times over the CCI abundances. Even the extreme value of 17 ppb (Mason, 1970) still requires about $4 \times$ CCI levels.

(b) The orbital gamma-ray value for Th averages about 1.5 ppm (Trombka *et al.*, 1973). If the 60 km thick highland crust (10% of the moon) has this value, then $4 \times$ CCI levels are required just to provide for the crustal concentrations. In addition there must have been some Th in the lunar interior to provide radioactive heating during the period 3.8–3.2 aeons in order to produce the maria basalts.

(c) The highland trace element abundances (Table 1) set further limits. If the concentration levels are representative of the 60 km thick crust and not some thinner surficial zone, then $3–4 \times$ CCI abundances are required to provide enough Ba and light REE for the crustal abundances alone (Table 2). Maria basalts, derived from deeper levels by partial melting after the highland crust was formed, also contain high levels of Ba, U, Th, REE, Zr, Hf, Nb, etc. (relative to CCI abundances), increasing the abundance problem.

(d) The uniform inter-element ratios in lunar samples (e.g. K/La, K/Ba, K/Th, K/Zr, etc.) set limits for the lunar abundance of K. Assuming $5 \times$ CCI for the *involatile* elements, the K concentration is $\sim$100 ppm. On this basis, about 60% of the K (and associated Rb and Cs) are in the highland crust (Table 2).

In summary, some limits can be set. The lower limit of $4 \times$ CCI appears well established by the element abundance levels. The heat flow values set upper limits of perhaps $7 \times$ CCI. An overall lunar average for the involatile elements of about $5 \times$ CCI appears reasonable and is adopted here. The trace element data are given in Table 2.

The major element abundances in the moon may then be estimated as follows. The refractory elements Ca, Al, and Ti are taken as $5 \times$ CCI. The iron content is set at 10.5% FeO to accommodate density and magnetic requirements (Parkin *et al.*, 1974). The Si/Mg ratio in chondrites is used to obtain Si and Mg concentrations. This composition is very close to that proposed by Ganapathy and Anders (1974). The major element values are given in Table 3.

7. GEOPHYSICAL AND PETROLOGICAL CONSTRAINTS

The moon's moment of inertia ($.3953 \pm .0045$) suggests a nearly uniform density distribution with depth (Kaula *et al.*, 1974). This and the low bulk density of the moon ($3.34 \, \text{g/cm}^3$) rules out the existence of abundant dense (eclogitic) phases at depth (Wetherill, 1968; Ringwood and Essene, 1970). Thus neither the highland anorthosites nor the maria basalts are suitable compositions for the deep lunar interior. The lack of major seismic discontinuities between the base of the crust at 60 km and a depth of 1000 km suggests a rather uniform interior with gradational boundaries in mineralogy and chemical composition. The interior of the moon is divided into (a) crust (upper 60 km: 10% volume) (b) lithosphere (b) lithosphere (60–1000 km: 82.4% volume) and (c) asthenosphere (1000–1738 km: 7.6% volume) with at least the upper portion of this zone partially liquid (Toksöz *et al.*, 1973; Latham *et al.*, 1973). The presence of orthopyroxene (with some Al and Ca) below 300 km is inferred from *P*-wave velocities which fit an olivine–orthopyroxene mixture better than olivine alone. The gravity data (Kaula *et al.*, 1974) indicate that the outer 100 km of the moon was rigid and thus cool when the maria basalts were erupted (forming the mascons). A lower limit for the source region of the maria basalts from the experimental petrology data is about 300–400 km (Ringwood and Essene, 1970; Walker *et al.*, 1974). Accordingly, the source of the maria basalts lies between depths of 100–300 km. Several types of maria basalts are distinguished using major and trace element criteria; their common feature is a high Ca/Al ratio, relatively low Mg/Fe, and varying degrees of REE enrichment with Eu depletion, and varying but low contents of Ni. Although the maria basalts could be related by petrological schemes involving different degrees of partial melting, the TiO_2 and Al_2O_3 contents, accompanied by Eu anomalies, and varying nickel contents, suggest as well that partial melting of slightly different mineralogies involving a combination of pyroxene, olivine, Fe–Ti phases and plagioclase (and sulfides) has taken place. The REE abundances in maria basalts suggest that plagioclase was absent from their source region (Schnetzler and Philpotts, 1971).

8. A GEOCHEMICAL MODEL

We now attempt to reconcile this geochemical information with the geophysical constraints, derive the composition of the deep lunar interior and correlate the overall composition with the sequence of geological events, beginning with a homogeneously accreted moon, which underwent partial or total melting, due to accretional heating. Toksöz *et al.* (1973) have shown that early heating and melting of a large part of the moon is compatible with the later thermal history of the moon. Total melting of the moon can be achieved if short accretional times of the order of 100–1000 yr are considered (Mizutani *et al.*, 1973).

Three alternative models may account for the deep lunar interior (asthenosphere) (below 1000 km) (Brett, personal communication).

(a) An immiscible Fe–FeS liquid sinks to form a core effectively removing most siderophile and chalcophile elements (Brett, 1973a). The core radius is restricted to less than 700 km by the coefficient of moment of inertia (0.395). Enough sulfur (~0.5%) is retained in the whole moon to form FeS. The partially liquid zone as suggested by the seismic evidence below 1000 km is interpreted as due to dispersed Fe–FeS in an olivine–orthopyroxene matrix. The magnetic field appearing in remanently magnetized rocks results from a core dynamo mechanism. An important feature of this model is that temperature at the 1000 km discontinuity may be as low as 1000–1100°C (Brett, 1973a).

(b) The initial melting did not extend below 1000 km. The central part of the moon is formed of primitive unfractionated material, now in a partially molten state from heating due to trapped initial K, U, and Th. The seismic data are satisfied by $\frac{1}{2}$–1% partial melting. This model precludes core formation, since sinking Fe–FeS will drive these incompatible elements upward. In this model, the remanent magnetism of lunar rocks is caused by external magnetic fields (Strangway and Sharpe, 1974).

(c) The molten zone is a relic of early melting.

The choice between these models for the lunar interior depends critically on the amount of the early siderophile and chalcophile element depletion. If the siderophile elements were accreted, even in the amounts corresponding to the supposed low-bulk Fe content of the moon, a lunar core is required to remove them in a very efficient manner. If these elements were depleted before accretion (see section on Cr/Ni ratios) then this reason for postulating a core disappears. We assume in the following discussion that a core did not form and that an inner zone some 700 km thick (7.6% volume) was not melted initially (Model b).

Following accretional melting of 90% of the moon the first silicate phase to separate from the primitive molten moon was Mg-rich olivine. The early precipitation of olivine removes Co^{2+}, Cr^{2+}, and Ni^{2+}. If the oxygen fugacity is close to or below the Fe–FeO buffer, much of the nickel will be in the metallic state (Brett, personel communication). However, since the extreme reduction observed in maria basalts may be a near-surface phenomenon developed during crystallization (Sato *et al.*, 1973) much nickel in the moon may be present as Ni^{2+}.

The surface of the moon cooled rapidly forming a "frozen crust," although this was continually broken up by the declining meteoritic bombardment. This frozen surface layer, analogous to the chilled margins of terrestrial intrusions, retained high concentrations of Mg, Cr, etc. in near-surface regions. Thus, its composition is probably representative of the melt composition during the early stages of crystallization of the lower interior. This chilled early lunar crust is later incorporated into the overall highlands composition and contributes the Mg and

Cr "primitive" component in the crust. This material is now thoroughly incorporated in the highland crustal material, although the 4.6 b.y. old dunite (Albee *et al.*, 1974) may have been derived from such an early primitive crust.

As crystallization proceeded in the deep lunar interior, the Si/Mg ratio changed and orthopyroxene precipitated. This had a similar effect on the composition of the residual melt as olivine except for Si/Mg ratios. Most cations except Mg, Fe, Ni, Co, and Cr^{2+} were excluded from the olivine and orthopyroxene lattice sites, and migrated upward, concentrating in the still voluminous residual melt. These included Ca and Al. The high Cr^{3+} abundances in most accessible lunar materials indicate that separation of clinopyroxene if any was minor, and olivine and orthopyroxene were probably the major components.

The density and seismic properties, and the Si/Mg ratio in the deep lunar interior are satisfied by 75–80% olivine ($\sim Fo_{85}$) and 20–25% orthopyroxene ($\sim En_{85}$). The composition of the lower part of the moon (300–1000 km) (assuming 2.2% Al_2O_3 and 2.5% CaO in the orthopyroxene) is given in Table 3, col. 2. The model requires however that even the lower parts of the moon are compositionally and mineralogically zoned. Orthopyroxene with some Al_2O_3 and CaO content is present at shallower depths together with olivine, whereas Mg-rich olivine is present in deeper parts.

Increasing crystallization of Mg-rich olivine, later accompanied by orthopyroxene at depth leads to an increasing concentration of refractory elements (Ca–Al) trapped between the already crystallized lower (Ol–Opx) lunar interior and the chilled surface layer. The composition of this zone (upper 300 km of the moon) is given in Table 3, col. 3. When the concentration of Al reaches 12–17% Al_2O_3, An-rich plagioclase precipitates, and concentrates or remains suspended beneath the frozen surface, whereas the Mg–Fe phases continue to crystallize and sink (Wood, 1973). The Ca–Al-rich region (plagioclase) incorporates Sr^{2+} and Eu^{2+}, but most other elements are unable to enter the Ca^{2+} sites in significant amounts.

Experimental petrology provides constraints on the Al_2O_3 content of melts to precipitate plagioclase together with Fe–Mg silicates (olivine and pyroxene). Values above 12% Al_2O_3 are necessary (Walker *et al.*, 1973).

The absence of plagioclase in the source region of maria basalts explains the high Ca/Al ratios, negative Eu anomalies and the low content of Al in maria basalts (Schnetzler *et al.*, 1972). Plagioclase is not a liquidus phase in maria basalts (Green *et al.*, 1971), except for high-Al maria basalts such as 12038 (Reid and Jakeš, 1974). We suggest (following Schnetzler *et al.*, 1972) that the source region of maria basalts is zoned in respect of Ca–Al and also Mg/Fe and consequently in mineralogy.

The high-Al maria basalts come from plagioclase-bearing regions shallower than other maria basalts (Reid and Jakeš, 1974); our model suggests that the content of plagioclase decreases from the base of crust (60 km) downward, whereas the contents of Fe–Mg silicate increases. These silicates become more Mg-rich with depth. Pockets or zones rich in Fe–Ti oxides and FeS in shallower regions account for the high Ti and S contents of the Apollo 11 and 17 basalts

(LSPET, 1973). Small zones of FeS provide a convenient place to trap lead in a uranium- and thorium-free environment (Nunes and Tatsumoto, 1973).

The source region of 15555 (Great Scott) and the green glass may represent a lower boundary of plagioclase precipitation. We assume that plagioclase separation commenced when a large part ($\simeq$40%) of the moon was still molten, i.e. during formation of source region of maria basalts above 300 km depth. The composition of the region from the base of the crust at 60 km to a depth of 300 km, which encompasses the source region of the maria basalts is given in Table 3, col. 5. (For the composition of the highland crust, see Table 3, col. 4.)

As crystallization of source region of maria basalts and crust proceeds, elements unable to enter the Ca–Al sites in plagioclase (above) or the Mg–Fe sites (below) are trapped between. In this trapped or residual zone, all the remaining elements concentrate. These include K, Ba, Rb, Cs, REE, Th, U, Zr, and Nb. It is a geochemical characteristic of great importance that the principal lunar mineral phases do not readily accommodate the refractory trace elements. The evidence of high concentrations of these elements near the surface of the moon is a dramatic consequence of this crystal chemical fact. The behavior of ions in entering crystal lattice sites in lunar minerals should be simpler than in terrestrial geochemistry, due to the nearly total absence of water and other volatiles.

Following the primordial fractionation, a chemically zoned moon is produced, with residual phases enriched below chilled margin and plagioclase crust and above source region of maria basalts (Fig. 4). This crustal zonation established at about 4.5 aeons is changed very quickly. The declining stages of the meteoritic bombardment pulverize the chilled zone and larger impacts mix in the underlying anorthosite. The high concentration of heat-producing elements K, U, and Th (and Zr, Hf, REE, etc.) trapped beneath the plagioclase zone provide the high-element abundances for the Fra Mauro or KREEP basalts. Possibly this zone did not solidify but the residual liquids invaded the crust, where impact mixing of the primitive surface layer, the Ca–Al plagioclase-rich layer, and the residual liquids beneath produced the parent material for the anorthositic gabbro (highland basalt) and the Fra Mauro basalts. The activity continued to 3.9 aeons, culminating in the production of the ringed basins and the cessation of the intense highland cratering. The details of such processes and the role of the great basin-forming collisions in the petrogenesis of the highland rock types remain to be established (Walker *et al.*, 1973; Head, 1974).

Partial melting next occurs in successively deeper layers as the smaller amounts of the heat producing elements induce partial melting, and a succession of "maria-type" basalts are erupted. The progressive deepening of the source region of the maria basalts is attributed here to the decrease in K, U, and Th with depth. The higher near-surface concentrations of these heat producing elements result in early melting and removal of the heat sources (since K, U, and Th are concentrated in the melt). The lesser amounts of K, U, and Th at depth generate a smaller amount of heat and it takes a longer period for melting temperatures to be reached.

The high-Al maria basalts formed in this model at shallow depths beneath the

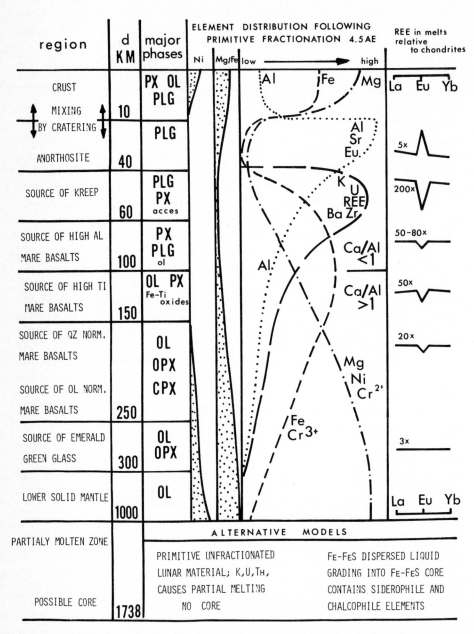

Fig. 4. Geochemical model of the lunar interior at about 4.4 aeons.

crust. This emplacement overlaps with the later stages of the bombardment, and predates the Imbrium collision in part, as shown by their presence in Fra Mauro breccias. Some of the high-Al maria basalts were emplaced later (≈ 3.4 aeons) suggesting that partial melting in shallower zones was not limited to early periods of maria formation (Luna 16 rocks, 12038).

Following these the Ti-rich Apollo 11 and 17 basalts were extruded during the period 3.8–3.6 aeons from a zone where Fe–Ti oxides and FeS accumulated. They have about 1 ppm Ni. Later (3.4–3.2 aeons) the Apollo 12 and 15 quartz and olivine normative basalts were extruded. These contain nickel, indicating extensive partial melting involving olivine, and many show evidence of near-surface fractionation. A negative Eu anomaly is characteristic of all maria basalts and contrasts with the deepest material erupted, the Apollo 15 emerald green glass (15426) (Ridley *et al.*, 1972; Green *et al.*, 1973) with 180 ppm Ni and primitive REE patterns, low total REE (3–5 times chondrites) and a small Eu anomaly (Taylor *et al.*, 1973). This material is the least fractionated lunar material which has been sampled. The thickness of the crust above this zone, as well as the residual high melting point material and lack of heat sources due to the upward concentration of K, U, and Th, causes cessation of lunar vulcanism at 3.2 aeons.

Acknowledgments—We wish to thank many colleagues at the Planetary and Earth Sciences Division, NASA–Johnson Space Center, Houston, for stimulating discussions, and Drs. Robin Brett, John Philpotts, R. J. Williams and an anonymous reviewer for constructive criticism of this manuscript. This paper was written at the Lunar Science Institute, which is operated by the Universities Space Research Association under Contract No. NSR 09-051-001 with the National Aeronautics and Space Administration. This paper constitutes the Lunar Science Institute Contribution No. 190.

REFERENCES

Albee A. L., Chodos A. A., Dymek R. F., Gancarz A. J., Goldman D. S., Papanastassiou D. A., and Wasserburg G. J. (1974) Dunite from the lunar highlands (abstract). In *Lunar Science—V*, pp. 3–5. The Lunar Science Institute, Houston.

Adler I., Trombka J. I., Schmadebeck R., Lowman P., Blodget H., Yin L., and Eller E. (1973) Results of the Apollo 15 and 16 X-ray experiment. *Proc. Fourth Lunar Sci. Conf., Geochim. Cosmochin. Acta*, Suppl. 4, Vol. 3, pp. 2783–2791. Pergamon.

Baldwin R. B. (1972) The Tsunami model of the origin of ring structures concentric with large lunar craters. *Phys. Earth Planet. Interiors* **6**, 327–339.

Brett R. (1973a) A lunar core of Fe–Ni–S. *Geochim. Cosmochim. Acta* **37**, 165–170.

Brett R. (1973b) The lunar crust: A product of heterogeneous accretion or differentiation of a homogeneous moon. *Geochim. Cosmochim. Acta* **37**, 2697–2703.

Chyi L. L. and Ehmann W. D. (1974) Implications of Zr and Hf abundances and their ratios in lunar minerals (abstract). In *Lunar Science—V*, pp. 118–120. The Lunar Science Institute, Houston.

Dence M. R., Grieve R. A. F., and Plant A. G. (1974) The Imbrium basin and its ejecta (abstract). In *Lunar Science—V*, pp. 165–167. The Lunar Science Institute, Houston.

Duncan A. R., Erlank A. J., Willis J. P., and Ahrens L. H. (1973) Compositions and inter-relationships of some Apollo 16 samples. *Proc. Fourth Lunar Sci. Conf., Geochim. Cosmochim. Acta*, Suppl. 4, Vol. 2, pp. 1097–1113. Pergamon.

Eldridge J. S., O'Kelley G. D., and Northcutt K. J. (1973) Radionuclide concentrations in Apollo 16 lunar samples determined by nondestructive gamma-ray spectrometry. *Proc. Fourth Lunar Sci. Conf., Geochim. Cosmochim. Acta*, Suppl. 4, Vol. 2, pp. 2115–2122. Pergamon.

Ganapathy R. and Anders E. (1974) Bulk compositions of the moon and earth, estimated from meteorites (abstract). In *Lunar Science—V*, 254–256. The Lunar Science Institute, Houston.

Green D. H. and Ringwood A. E. (1973) Significance of a primitive lunar basaltic composition. *Earth Planet. Sci. Lett.* **19**, 1–8.

Grossman L. and Larimer J. W. (1974) Early chemical history of the solar system. *Rev. Geophys. Space Phys.* **12**, 71–101.

Head J. W. (1974) Orientale multi-ringed basin interior, and implications for the petrogenesis of lunar highland samples. *The Moon*. In press.

Kaula W. M., Schubert G., Lingenfelter R. E., Sjogren W. L., and Wollenhaupt W. R. (1974) Apollo laser altimetry and inferences as to lunar structure (abstract). In *Lunar Science—V*, pp. 399–401. The Lunar Science Institute, Houston.

Krähenbühl U., Ganapathy R., Morgan J. W., and Anders E. (1973) Volatile elements in Apollo 16 samples: Implications for highland volcanism and accretion history of the moon. *Proc. Fourth Lunar Sci. Conf., Geochim. Cosmochim. Acta*, Suppl. 4, Vol. 2, pp. 1325–1348. Pergamon.

Laul J. C. and Schmidt R. A. (1973) Chemical composition of Apollo 15, 16, and 17 samples. *Proc. Fourth Lunar Sci. Conf., Geochim. Cosmochim. Acta*, Suppl. 4, Vol. 2, pp. 1349–1367. Pergamon.

Latham G., Dorman J., Duennebier F., Ewing E., Lammlein D., and Nakamura Y. (1973) Moonquakes, meteoroids, and the state of the lunar interior. *Proc. Fourth Lunar Sci. Conf., Geochim. Cosmochim. Acta*, Suppl. 4, Vol. 2, pp. 2515–2527. Pergamon.

Latham G., Nakamura Y., Lammlein D., Dorman J., and Duennebier F. (1974) Structure and state of the lunar interior based on seismic data (abstract). In *Lunar Science—V*, p. 434. The Lunar Science Institute, Houston.

LSPET (1973) Apollo 17 lunar samples: Chemical and petrographic description. *Science* **182**, 659–672.

Mason B. H. (1970) *Elemental Abundances in Meteorites*. Gordon and Breach, New York.

Metzger A. E., Trombka J. I., Reedy R. C., and Arnold J. R. (1974) Element concentrations from lunar orbital gamma-ray measurements (abstract). In *Lunar Science—V*, pp. 501–502. The Lunar Science Institute, Houston.

Mizutani H., Matsui T., and Takeuchi H. (1972) Accretion process of the moon. *The Moon* **4**, 476–489.

Nunes P. D. and Tatsumoto M. (1973) Excess lead in "Rusty Rock" 66095 and implications for early lunar differentiation. *Science* **182**, 916–920.

Papanastassiou D. A., Wasserburg G. J., and Burnett D. S. (1970) Rb–Sr ages of lunar rocks from the Sea of Tranquillity. *Earth Planet. Sci. Lett.* **8**, 1–19.

Parkin C. W., Dyal P., and Daily W. D. (1973) Iron abundance in the moon from magnetometer measurements. *Proc. Fourth Lunar Sci. Conf., Geochim. Cosmochim. Acta*, Suppl. 4, Vol. 3, pp. 2947–2961. Pergamon.

Philpotts J. A. and Schnetzler C. C. (1970) Apollo 11 lunar samples: K, Rb, Sr, Ba and rare-earth concentrations in some rocks and mineral phases. *Proc. Apollo 11 Lunar Sci. Conf., Geochim. Cosmochim. Acta*, Suppl. 1, Vol. 1, pp. 1471–1486. Pergamon.

Podosek F. A. and Huneke J. C. (1973) Argon in Apollo 15 green glass spherules (15426): $^{40}Ar–^{39}Ar$ age and trapped argon. *Earth Planet, Sci. Lett.* **19**, 413–421.

Reid A. M. and Jakeš P. (1974) Luna 16 revisited. The case for aluninuous mare basalts (abstract). In *Lunar Science—V*, pp. 627–629. The Lunar Science Institute, Houston.

Ridley W. I., Reid A. M., Warner J. L., and Brown R. W. (1972) Apollo 15 green glasses. *Phys. Earth Planet. Interior* **7**, 133–136.

Ringwood A. E. (1974) Heterogeneous accretion and the lunar crust. *Geochim. Cosmochim. Acta.* **38**, 983–984.

Ringwood A. E. and Essene E. (1970) Petrogenesis of Apollo 11 basalts, internal constitution and origin of the moon. *Proc. Apollo 11 Lunar Sci. Conf., Geochim. Cosmochim. Acta*, Suppl. 1, Vol. 1, pp. 769–799. Pergamon.

Sato M., Hickling N. L., and McLane J. E. (1973) Oxygen fugacity values of Apollo 12, 14 and 15 lunar samples and reduced state of lunar magmas. *Proc. Fourth Lunar Sci. Conf., Geochim. Cosmochim. Acta*, Suppl. 4, Vol. 1, pp. 1061–1079. Pergamon.

Schnetzler C. C. and Philpotts J. A. (1971) Alkali, alkaline earth and rare-earth element concentrations in some Apollo 12 soils, rocks and separated phases. *Proc. Second Lunar Sci. Conf., Geochim. Cosmochim Acta*, Suppl. 2, Vol. 2, pp. 1101–1122. MIT Press.

Schnetzler C. C., Philpotts J. A., Nava D. F., Schuhmann S., and Thomas H. H. (1972) Geochemistry of Apollo 15 basalt 15555 and soil 15531. *Science* **175**, 426.

Strangway D. W. and Sharpe H. A. (1974) Models of lunar evolution (abstract). In *Lunar Science—V*, pp. 755–757. The Lunar Science Institute, Houston.

Stuart-Alexander D. and Howard K. A. (1970) Lunar maria and circular basins—A review. *Icarus* **12**, 440–456.

Taylor S. R. (1973) Chemical evidence for lunar melting and differentiation. *Nature* **245**, 203–205.

Taylor S. R. (1975) *Lunar Science: A post-Apollo view*. Pergamon. In press.

Taylor S. R., Gorton M. P., Muir P., Nance W. B., Rudowski R., and Ware N. (1973) Composition of the Descartes region, lunar highlands. *Geochim. Cosmochim. Acta* **37**, 2665–2683.

Taylor S. R. and Jakeš P. (1974) Geochemical zoning in the moon (abstract). In *Lunar Science—V*, pp. 786–788. The Lunar Science Institute, Houston.

Toksöz M. N., Dainty A. M., Solomon S. C., and Anderson K. R. (1973) Velocity structure and evolution of the moon. *Proc. Fourth Lunar Sci. Conf., Geochim. Cosmochim. Acta*, Suppl. 4, Vol. 3, pp. 2529–2547. Pergamon.

Trombka J. I., Arnold J. R., Reedy R. C., Peterson L. E., and Metzger A. E. (1973) Some correlations between measurements by the Apollo gamma-ray spectrometer and other lunar observations. *Proc. Fourth Lunar Sci. Conf., Geochim. Cosmochim. Acta*, Suppl. 4, Vol. 3, pp. 2847–2853. Pergamon.

Walker D., Grove T. L., Longhi J., Stolper E. M., and Hays J. F. (1973) Origin of lunar feldspathic rocks. *Earth Planet. Sci. Lett.* **20**, 325–336.

Walker D., Longhi J., Stolper E., Grove T., and Hays J. F. (1974) Experimental petrology and the origin of titaniferous lunar basalts (abstract). In *Lunar Science—V*, pp. 814–817. The Lunar Science Institute, Houston.

Wänke H., Baddenhausen H., Dreibus G., Jagoutz E., Kruse H., Palme H., Spettel B., and Teschke F. (1973) Multielement analyses of Apollo 15, 16 and 17 samples and the bulk composition of the moon. *Proc. Fourth Lunar Sci. Conf., Geochim. Cosmochim. Acta*, Suppl. 4, Vol. 2, pp. 1461–1481. Pergamon.

Wetherill G. W. (1968) Lunar interior: Constraint on basaltic composition. *Science* **160**, 1256–1257.

Wetherill G. W. (1971) Of time and the moon. *Science* **173**, 383–392.

Wood J. A. (1973) The moon's feldspathic crust: Early differentiation or heterogeneous accretion. *Meteoritics* **8**, 467–468.

Proceedings of the Fifth Lunar Conference
(Supplement 5, Geochimica et Cosmochimica Acta)
Vol. 2 pp. 1307–1335 (1974)
Printed in the United States of America

Chemistry of Apollo 16 and 17 samples: Bulk composition, late stage accumulation and early differentiation of the moon

H. Wänke, H. Palme, H. Baddenhausen, G. Dreibus, E. Jagoutz, H. Kruse, B. Spettel, F. Teschke, and R. Thacker

Max-Planck-Institut für Chemie (Otto-Hahn-Institut), Abteilung Kosmochemie, Mainz, Germany

Abstract—Five Apollo 16 breccias and thirteen Apollo 17 samples (nine soil samples, three breccias, and one basalt) have been analyzed for all major and many minor and trace elements. Breccia 65315 consists of nearly 100% plagioclase. Its trace element abundances are very similar to those of other anorthosites: 60015, 61016 (light), and 15415.

The element correlations established previously are confirmed by the new data. Ratios of volatile/nonvolatile elements can be used to show a loss of alkali elements and Mn in some Apollo 16 breccias. This loss is ascribed to volatilization during the brecciation process. Later on some of the breccias were enriched in highly volatile elements.

Mixing diagrams for the highland samples proved the presence of a chondritic component, representing material, having accumulated at a time when the original feldspathic crust had already been formed. This chondritic component is responsible for most of the Fe and Mg present in the non-mare samples; it also helped to define the chondritic component in our two-component model of the bulk composition of the moon. Taking into account some other element correlations, we finally end up with a contribution of 50% of the high temperature component and 50% of the chondritic component for the bulk moon.

Parallel REE patterns and an analogous behavior of other LIL elements in all non-mare samples suggest that the KREEP fractionation of the LIL elements is typical for the whole crust. From heat flow data we conclude that most of these elements are concentrated in the crust. Partial melting processes in most parts of the moon produced this differentiation. From the residuum the LIL-element patterns of the mare basalts were generated in a second partial melting process. Under certain assumptions the resulting LIL-element distribution can be calculated.

1. Introduction, Analytical Methods, and Results

ALL MAJOR AND MANY TRACE ELEMENTS were determined in five Apollo 16, and thirteen Apollo 17 samples and in the eucrite Ibitira. Altogether data on 53 elements have been obtained. With the exception of fluorine, neutron activation methods were applied for all elements. Samples between 0.3 and 1 g were sequentially irradiated and counted according to the scheme described in our last Apollo paper (Wänke *et al.*, 1973). Besides INAA with 14 MeV and thermal neutrons, RNAA was used for a number of elements. The methods for the chemical separations were similar to those described previously (Rieder and Wänke, 1969; Wänke *et al.*, 1970).

Lithium, Cl, Br, and I were measured on aliquants (100 mg) using RNAA. Fluorine was determined in these aliquants, too, via a specific ion electrode. For many of the samples X-ray fluorescence measurements were carried out for the major elements and P using additional aliquants of 100 mg.

The results are given in Tables 1–4. For the major elements (Tables 1 and 3)

Table 1. Major and minor elements in Apollo 16 samples and in the eucrite Ibitira.

Element %	Method	61016-151 Breccia	61156-35 Breccia	65315-52 Breccia	66075-31 Breccia	68815-130 Breccia	Ibitira Eucrite	Accuracy %
O	IFNAA	44.5	44.1	46.3	44.9	44.8	42.6	1
Mg	IFNAA	6.27	5.94	0.15	3.96	3.87	4.3	4
	XRF	6.11	—	—	3.85	3.84	4.4	4
	av.	6.19	5.94	0.15	3.90	3.85	4.3	
Al	IFNAA	13.2	12.2	18.8	14.1	14.2	6.67	2
	XRF	12.8	—	18.1	14.1	—	6.68	2
	av.	13.0	12.2	18.45	14.1	14.2	6.67	
Si	IFNAA	20.3	21.2	20.7	21.1	21.8	22.9	1
	XRF	21.1	—	21.4	20.4	—	22.3	4
	av.	20.3	21.2	20.7	21.1	21.8	22.9	
Ca	IFNAA	9.8	9.5	14.3		11.1		7
	ITNAA	10.8	10.0	13.3	10.9	11.0		7
	XRF	10.10	—	13.54	11.20	10.6	7.79	4
	av.	10.16	9.75	13.62	11.12	10.77	7.79	
Ti	IFNAA	0.45	0.37	0.0072		0.31		5
	XRF	0.48	—	—	0.27	—	0.47	4
	av.	0.47	0.37	0.007	0.27	0.31	0.47	
Fe	IFNAA	3.97	5.92	0.27*	3.67	3.83	13.9	3
	ITNAA	4.02	6.09	0.22*	3.71	3.91		3
	XRF	4.05	—	0.23*	3.54	—	14.4	3
	av.	4.01	6.01	0.24	3.64	3.87	14.1	
ppm								
Na	ITNAA	2700	3270	2250	3650	3950	1440	3
P	XRF	524	—	40	460	—	205	10
K	ITNAA	670	940	58	740	1810	119	5
Cr	ITNAA	610	940	20.0	540	650	2460	3
	XRF	580	—	—	520	—	2490	3
	av.	600	940	20.0	530	650	2480	
Mn	ITNAA	385	660	59	440	470	3780	3
	XRF	381	—	—	432	—	3940	3
	av.	384	660	59	436	470	3860	3
Σ		100.1	100.4	99.8	99.8	100.4	99.7	

IFNAA = Instrumental fast neutron activation analysis.
ITNAA = Instrumental thermal neutron activation analysis.
XRF = X-ray fluorescence analysis.
*Accuracy reduced by a factor of 2.

the data obtained by up to three independent methods are listed separately together with the weighted averages. The mean accuracy for the elements analyzed is given in the last column of the tables. For the major elements the listed accuracies can be checked via the sums as we have determined all major and minor elements including oxygen. The values for the accuracy of the trace elements were obtained by critical comparison of our data with those of other authors for standard rocks and lunar fines where direct comparisons were possible. For three Apollo 17 soil samples a comparison of our results with those of Philpotts *et al.* (1974) obtained by mass spectrometric isotope dilution is given in Table 5. As can be seen, our data which, except for Li, were all obtained by INAA, are in most cases in excellent agreement with the isotope dilution data. The larger differences for Ba and

Table 2. Trace elements in Apollo 16 samples and in the eucrite Ibitira.

Element	61016-151 Breccia	61156-35 Breccia	65315-52 Breccia	66075-31 Breccia	68815-130 Breccia	Ibitira Eucrite	Accuracy %
F ppm	31		14		40	4	5
Cl ppm	270		104		260		5
K ppm	670	940	58	740	1810	119	5
Sc ppm	6.6	9.36	0.39	6.62	7.2	31.5	3
Co ppm	36.7	59.4	0.58	25.3	30.2	14.3	5
Ni ppm	510	1190	1.4*	350	500		7
Cu ppm	6.4	6.6	2.1	4.3	7.8	5.1	10
Zn ppm		5.0	93	7.6		2.7	10
Ga ppm	3.48	4.5	3.25	5.1	3.6	1.02	7
Ge ppm	0.85	1.75		1.7	1.4		10
As ppb	270	370	2*	94	740	330	10
Br ppm	0.88		0.93		0.48		5
Rb ppm	2.84	2.44	0.17*	2.01	8.8		10
Sr ppm	130	155	167	200	160		7
Y ppm	44	60.5	<0.3		64.4		10
Zr ppm	209	304	15		331		7
Nb ppm	13	15.6	<0.2		20		10
Pd ppb	25	59		17	36		15
I ppb	70		47				25
Cs ppb	120	150	15	120	460		15
Ba ppm	160	220	4.8*	106	300		7
La ppm	16.7	22.5	0.12*	11.8	22.3	2.89	3
Ce ppm	46	55		28	61		5
Pr ppm	5.8	8.0		4.3	6.8		10
Sm ppm	6.9	9.82	0.04	5.5	9.4	1.75	3
Eu ppm	1.38	1.34	0.74	1.20	1.84	0.67	3
Gd ppm	9.5	12.1			11.9		10
Tb ppm	1.5	2.1		1.0	2.0		10
Dy ppm	9.7	13.3	0.056*	6.9	14.1		5
Ho ppm	2.0	3.3		1.4	3.1		10
Er ppm	4.8	8.3		4.1	7.6		10
Yb ppm	4.4	6.8	0.026*	3.44	6.86	1.84	3
Lu ppm	0.61	0.90	0.004*	0.46	1.0	0.42	5
Hf ppm	4.9	7.25	0.49	3.80	7.5		5
Ta ppm	1.02	0.83		0.56	0.93		10
W ppm	0.25	0.56	0.019	0.21	0.45	0.14	10
Re ppb	14	2.6		1.0			10
Ir ppb	15	23		10	11		15
Au ppb	13	22	1.0	7.8	15	3.2	10
Th ppm	1.6	3.2		1.6	3.74		15
U ppm	0.46	0.92	<0.0006	0.41	1.09	0.081	7
Fe/Mn	103	91	40.7	84	83		
K/La	40	41	483	63	81		
Zr/Hf	42.6	41.9	30.6		44.1		
La/Hf	3.41	3.10	0.24	3.10	2.97		

*Accuracy reduced by a factor of 2.

Table 3. Major and minor elements in Apollo 17 samples.

Element %	Method	7001l-12 Fines Sta. LM	72141-22 Fines Sta. LRV-2	73121-18 Fines Sta. 2a	73141-23 Fines Sta. 2a	73221-16 Fines Sta. 3	73241-17 Fines Sta. 3	73261-17 Fines Sta. 3	73281-18 Fines Sta. 3	74121-18 Fines Sta. 4	73235-91 Breccia Sta. 3	79035-27 Breccia Sta. 9	79135-38 Breccia Sta. 9	74275-69 Basalt Sta. 4	Accuracy %
O	IFNAA	41.5	42.8	43.8	44.0	43.9	44.0	44.1	44.7	43.7	44.7	41.8	41.9	40.9	1
Mg	IFNAA	5.86	5.93	6.05	6.18	5.66	5.85	5.78	6.02	5.89	6.91	6.03	6.42	6.1	4
	XRF	6.12	6.00	6.23	6.10	—	5.80	5.85	—	6.00	7.01	5.93	6.62	6.15	4
	av.	5.99	5.96	6.14	6.09	5.66	5.82	5.82	6.02	5.95	6.96	5.98	6.52	6.12	
Al	IFNAA	6.50	8.30	10.9	11.3	10.9	10.8	10.8	11.0	10.1	10.8	6.47	7.32	4.40	2
	XRF	6.66	8.39	11.1	11.1	—	10.95	10.6	—	9.97	10.9	6.50	7.31	4.48	2
	av.	6.58	8.35	11.0	11.2	10.9	10.88	10.7	11.0	10.04	10.85	6.49	7.32	4.44	
Si	IFNAA	19.4	20.1	21.3	21.4	21.1	21.2	21.1	21.5	21.0	21.8	19.5	19.9	18.1	1
	XRF	19.1	19.7	21.0	20.6	—	21.0	20.2	—	20.2	21.7	18.7	19.8	17.4	4
	av.	19.4	20.1	21.3	21.4	21.1	21.2	21.1	21.5	21.0	21.8	19.5	19.9	18.1	
Ca	IFNAA	7.8	8.5	—	9.1	9.0	8.6	9.0	9.1	8.2	8.1	8.3	8.2	7.4	7
	ITNAA	7.9	8.4	9.4	9.0	8.7	9.2	9.1	7.6	8.9	8.0	7.9	7.5	6.7	7
	XRF	7.9	8.4	9.4	9.3	—	9.0	9.1	8.4	8.9	8.8	8.0	7.9	7.4	4
	av.	7.9	8.4	9.4	9.1	8.8	9.0	9.1	8.4	8.7	8.5	8.0	7.9	7.2	
Ti	IFNAA				0.72	1.09			1.05		0.37		3.19	7.05	5
	XRF	4.41	2.63	0.83	0.76	—	1.01	1.09	—	1.54	0.40	4.79	3.30	7.15	4
	av.	4.41	2.63	0.83	0.74	1.09	1.01	1.09	1.05	1.54	0.39	4.79	3.25	7.10	
Fe	IFNAA	12.5	10.1	6.73	6.28	6.66	6.46	6.78	6.72	7.96	5.94	12.6	11.8	14.1	3
	ITNAA	12.41	10.44	6.53	6.24	6.94	6.52	6.83	6.55	7.96	5.52	12.9	11.73	14.05	3
	XRF	12.33	10.59	6.74	6.40	—	6.58	7.01	—	8.19	5.77	12.98	11.4	14.4	3
	av.	12.4	10.38	6.67	6.31	6.80	6.52	6.87	6.64	8.04	5.74	12.83	11.64	14.18	
Na	ITNAA	2780	2980	3170	3200	3420	3320	3400	3300	3170	3380	3035	3480	2745	3
P	XRF	208	420	590	520	—	550	585	—	510	810	240	330	275	10
K	ITNAA	645	880	1170	1110	1180	1210	1250	1140	1120	1640	680	810	660	5
Cr	ITNAA	2680	2430	1430	1410	1500	1460	1460	1410	1730	1335	2830	2610	3780	3
	XRF	2570	2300	1440	1500	—	1370	1440	—	1660	1380	2670	2490	3600	
	av.	2630	2370	1440	1460	1500	1370	1450	1410	1700	1360	2750	2550	3690	
Mn	ITNAA	1630	1370	845	820	895	840	910	850	1020	750	1700	1500	1840	3
	XRF	1704	1355	852	852	895	851	910	850	1053	797	1665	1518	1889	
	av.	1670	1360	848	836	895	846	910	850	1040	774	1680	1510	1865	
Σ		99.2	99.5	99.9	99.8	99.2	99.3	99.6	100.1	99.7	99.4	100.3	99.3	99.1	

IFNAA = Instrumental fast neutron activation analysis.
ITNAA = Instrumental thermal neutron activation analysis.
XRF = X-ray fluorescence analysis.

Sr merely reflect the lower accuracy of the INAA for these elements. In cases where the samples are radiochemically processed the accuracy for these elements is better (see Table 2 and 4).

1.1. *Apollo 16 samples*

All Apollo 16 samples listed in Table 1 and 2 are breccias. In the scheme of Hubbard *et al.* (1973a) the samples 61016 and 61156 are classified as VHA basalts, 66075 and 68815 are anorthositic gabbros, while 65315 consists of nearly 100% plagioclase. In Table 6 this 65315 anorthosite is compared to other anorthosites from Apollo 16 and to 15415. There seems to exist a lower limit for the concentrations of LIL elements in these anorthosites. The constancy of the trace element pattern in the anorthosites gives us, with the help of partition coefficients, an important clue for the composition of the melt from which they crystallized (Hubbard *et al.*, 1971). The primitive Sr-isotopic composition for 60025 and 61016 (Papanastassiou *et al.*, 1972b; Nyquist *et al.*, 1973) indicates a very early time for this differentiation process.

The Zr and Hf abundances in our anorthosite are too high compared to the REE. Taylor *et al.* (1973) have found 1.1 ppm Zr in 60015 and Laul *et al.* (1973) 0.013 ppm Hf in 60015 and 0.014 ppm Hf in 60025. Our sample may contain Zr and Hf bearing mineral grains, e.g. zircon. Terrestrial contamination cannot be excluded. Gallium is not depleted in the anorthosite as it substitutes for Al. Tungsten is less depleted than other siderophile elements; it closely follows K (see Section 2).

In contrast to the anorthositic sample, the other breccias contain large amounts of siderophile elements, indicating a substantial meteoritic component (Krähenbühl *et al.*, 1973). The Pd/Au ratio seems to be fairly constant (Table 2).

The LIL-element patterns of these breccias are identical to the KREEP pattern in a relative sense; their abundance ratios are constant and equal to the ratios of Apollo 12, 14, and 15 KREEP-rich samples.

1.2. *Apollo 17 samples*

We have analyzed nine soil samples, three breccias, and one basalt. The soil samples seem to be mixtures of a KREEP-containing aluminous end member and the local-mare basalts. This can be seen by the trace element patterns varying from KREEP to a pattern resembling the Apollo 11 rocks. A convenient measure is the La/Hf ratio (Table 4), changing from 3.12 for KREEP to 0.80 for the mare basalt, while Ti is simultaneously increasing. This has been observed for other pairs of LIL elements by Duncan *et al.* (1974).

2. ELEMENT CORRELATIONS

Element correlations as observed by various authors (Laul *et al.*, 1972; Philpotts *et al.*, 1972; Willis *et al.*, 1971) including ourselves (Wänke *et al.*, 1973)

H. WÄNKE *et al.*

Table 4. Trace elements in Apollo 17 samples.

Element	70011-12 Fines Sta. LM	72141-22 Fines Sta. LRV-2	73121-18 Fines Sta. 2a	73141-23 Fines Sta. 2a	73221-16 Fines Sta. 3	73241-17 Fines Sta. 3	73261-17 Fines Sta. 3	73281-18 Fines Sta. 3	74121-18 Fines Sta. 4	73235-91 Breccia Sta. 3	79035-27 Breccia Sta. 9	79135-38 Breccia Sta. 9	74275-69 Basalt Sta. 4	Accuracy %
Li ppm	8.4	9.7		9.7	13.2					13.5	9.3	10.4		5
F ppm	58	59		31	71					27	61	90		5
Cl ppm	12.5	25		15	16.0					20.0	12.3	26		5
K ppm	645	880	1170	1110	1180	1210	1250	1140	1120	1640	680	810	660	5
Sc ppm	53.1	35.8	17.3	16.6	19.5	17.7	19.5	17.5	24.3	13.4	56.6	39.5	74.0	3
Co ppm	31.6	38.2	37.0	26.5	29.3	27.7	28.0	26.7	36.4	26.3	29.6	38.3	22.5	5
Ni ppm	110	230	315	250	275	170	240	280	327	205	160	170		7
Cu ppm	10.9					6.25				3.8	11.0	19.6	3.5	10
Zn ppm	38.9					20					32	97.6	1.7	10
Ga ppm	5.03					4.7				4.0	5.5	8.0	3.4	7
Ge ppm	0.29					0.41				0.36	0.19	0.44	<0.1	10
As ppb	23					41				130	14	24	4	10
Br ppm	0.12	0.2		0.11	0.13	4.80				0.11	0.090	0.28		5
Rb ppm										—	1.62		1.22	10
Sr ppm	210	180*	160*	130*	160*	170*	130*	150*	170*	150	170	200	195	7
Y ppm	61			46	48			52		69		47	79	10
Zr ppm	210			225	227			245		343		186	246	7
Nb ppm	15			14	14			16		20.4		12.6	19	10
Pd ppb	5									11	10		<2	15
I ppb	20	30		<20		10					<10			25

Cs ppb	102		170	195*	160*	194	180*	150*		170	60.5	50	53	15
Ba ppm			190			185	180*			260	108	120	83	7
La ppm	8.03	10.9	15.6	15.4	16.0	16.1	16.5	15.0	15.5	24.5	8.70	9.88	6.70	3
Ce ppm	23.5	28	38.4	37.3	37.8	38	39.3	35.0	37.0	58.5	24.1	25.8	22.1	5
Pr ppm	4.0	4.2				5.9				8.4	4.0	3.9	4.2	10
Sm ppm	8.0	6.78	7.34	6.93	7.57	7.8	7.85	7.58	8.06	9.4	8.43	7.26	9.76	3
Eu ppm	1.67	1.39	1.30	1.22	1.29	1.25	1.37	1.22	1.28	1.25	1.70	1.60	1.91	3
Gd ppm	10.9	9.9				9.1				12.5	11.3	9.5	14.2	10
Tb ppm	1.9	1.5	1.6	1.6	1.7	1.6	1.8	1.7	1.9	2.2	2.1	1.8	2.5	10
Dy ppm	12.5	11.0	10.1	9.5	9.6	10.5	9.8	10.5	11.3	14.3	12.7	10.5	15.8	5
Ho ppm	2.8	2.5	2.5	2.4	2.6	2.3	2.5	2.4		3.3	3.0	2.5	3.6	10
Er ppm	7.7	6.3				5.5				7.8	7.4	6.2	9.4	10
Yb ppm	7.04	5.62	5.63	5.50	5.87	5.84	5.78	5.52	6.02	7.9	7.37	5.71	9.02	3
Lu ppm	1.02	0.74	0.78	0.81	0.81	0.81	0.82	0.83	0.81	1.08	1.10	0.78	1.30	5
Hf ppm	6.5	5.64	5.52	5.55	5.82	5.72	5.74	5.78	6.02	7.85	7.20	5.70	8.33	5
Ta ppm	1.2	0.77	0.73	0.76	0.81	0.76	0.78	0.68	0.87	0.94	1.34	0.94	1.50	10
W ppm	0.12					0.30				0.58	0.12	0.18	0.060	10
Re ppb	0.63					0.6					0.5	0.88	<0.5	10
Ir ppb		25	15	17	9		15	11	11	3.2		10		15
Au ppb	5.7	10	6	8	13	4			13	3.75	4.5	3.1	0.19	10
Th ppm			2.8	2.6	2.7	2.3	2.8	2.5	2.6					15
U ppm	0.24	0.30				0.58				1.05	0.28	0.33	0.16	7
Fe/Mn	76.7	74.5	78.7	76.2	75.4	77.4	74	77.6	78	76	75.0	78	76	
K/La	80.3	80.7	75.0	72.1	73.8	75	75.8	76	72.3	66.9	78.2	82	98.5	
Zr/Hf	32.3			40.5	39			42.2		43.7		32.6	29.5	
La/Hf	1.24	1.93	2.83	2.77	2.75	2.81	2.87	2.59	2.57	3.12	1.21	1.73	0.80	

*Accuracy reduced by a factor of 2.

Table 5. Comparison of analytical data for three Apollo 17 soils between Philpotts *et al.* (1974) (Mass spectrometric isotope dilution) and our laboratory (Instrumental neutron activation analysis, except for Li).

Element ppm	Philpotts *et al.* 72141,19	This work 72141,22	Philpotts *et al.* 73121,14	This work 73121,18	Philpotts *et al.* 73141,18	This work 73141,23
Li	9.33	9.7			9.82	9.7
K	843	880	1160	1170	1170	1110
Sr	154	180	144	160	147	130
Ba	120		171	190	171	195
Ce	27.7	28.0	38.1	38.4	37.7	37.3
Sm	6.69	6.78	7.14	7.34	7.00	6.93
Eu	1.41	1.39	1.26	1.30	1.24	1.22
Dy	9.73	11.0	9.65	10.1	9.39	9.5
Yb	5.29	5.62	5.38	5.63	5.46	5.5
Lu	0.819	0.74	—	0.78	0.825	0.81

hold for the Apollo 16 and 17 samples as well as for the samples from previous missions. They are found for groups of elements with similar geochemical behavior, like Fe–Mn, or for the LIL elements (Fig. 1).

Among the more important is the W/La correlation as W has siderophile tendencies. Only that portion of W which is present in oxidized form can correlate with La; hence the W/La ratio must depend on the ratio of metallic iron to iron oxide. Reliable data of the W/La ratio for terrestrial samples are not available. From our few data on samples other than lunar it may be inferred that there is no clear difference in the Fe°/Fe^{++} ratio for the moon, the earth, and the parent body of the achondrites (Fig. 2).

A number of previously unrecognized element correlations was observed. Figure 3 proves the excellent correlation of yttrium versus ytterbium in all lunar samples. In some cases the data points indicate a bimodal distribution; the mare basalts having considerably different ratios than the samples containing KREEP (see Fig. 4 and Fig. 5). This will be discussed in detail in Section 5. For the elements Ga versus Al a reasonable correlation is observed for the mare basalts and for KREEP-rich samples as well as for achondrites (Fig. 6).

Table 6. Comparison of some trace element abundances in lunar anorthosites.

Element ppm	15415.11 Hubbard *et al.* (1971)	60015.63 Laul *et al.* (1973)	61016.184 Philpotts *et al.* (1973)	65315.52 This work
K	120	83	44.8	58
Ba	6.3	8 ± 2	6.01	4.8
La	0.12	0.13	—	0.12
Sm	0.046	0.051	0.036	0.038
Eu	0.806	0.81	0.671	0.74

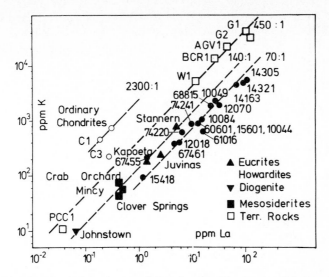

Fig. 1. K versus La in lunar, terrestrial, and meteoritic samples. All data from this laboratory.

The Sc/Fe correlation (Fig. 7) was first pointed out for lunar samples by Laul *et al.* (1972). In this case a very good correlation is found for eucrites. For the lunar samples the expected correlation may be obscured by the addition of material which is not in equilibrium with the material of the source region of the mare basalts. We will return to this question in Section 4.

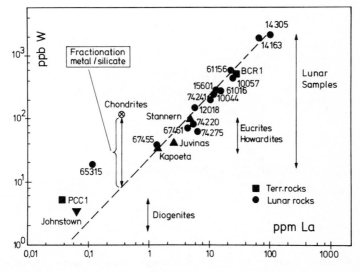

Fig. 2. W versus La in lunar, terrestrial, and meteoritic samples. All data from this laboratory.

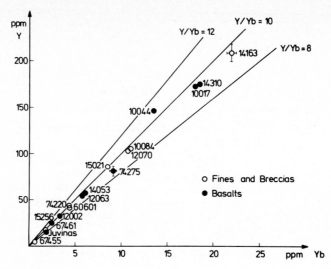

Fig. 3. Y versus Yb in lunar samples and in the eucrite Juvinas. All data from this laboratory.

In many Apollo 16 rocks and breccias the volatile to refractory element ratios deviate considerably from the general trends. For example the K/La ratios are more variable in these samples than in the Apollo 16 soil samples or in samples from other landing sites, where ratios of about 70 have been observed (Wänke *et al.*, 1973). To illustrate this, a number of Apollo 16 rocks and breccias has been

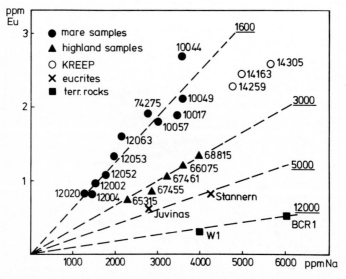

Fig. 4. Eu versus Na in lunar and terrestrial samples and in the eucrites Juvinas and Stannern. All data from this laboratory.

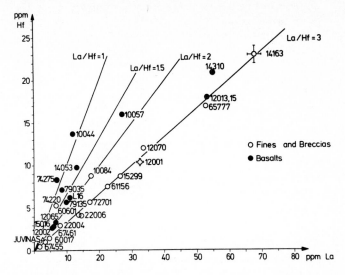

Fig. 5. Hf versus La in lunar samples and in the eucrite Juvinas. Data for Luna 16 and 20 samples from Laul and Schmitt (1973) and Gillum *et al.* (1972), for sample 12013 from Wakita and Schmitt (1970), all other data from this laboratory.

listed in Table 7. All these samples (mostly VHA basalts) have about the same concentration of La and the same KREEP-type REE pattern. Their main element abundances are rather similar. Yet their K/La ratios vary from 33 for 62295 to 105 for 60335 (Table 7). Rubidium and Cs are variable even within the same samples (66095, 68815), although their ratio is remarkably constant. The ratio of K/Rb is

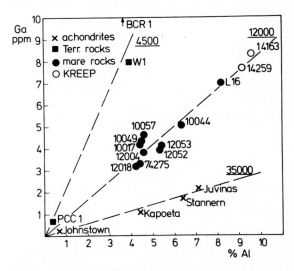

Fig. 6. Ga versus Al in lunar, terrestrial, and meteoritic samples. All data from this laboratory.

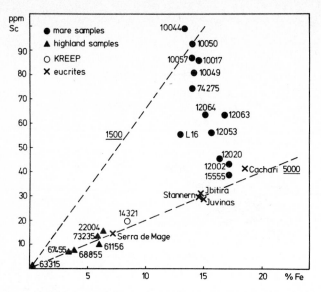

Fig. 7. Sc versus Fe in lunar samples and in eucrites. Data from Laul and Schmitt (1973), Schmitt *et al.* (1972), Gillum *et al.* (1972), and this laboratory.

Table 7. Comparison of some Apollo 16 samples with similar La contents. The samples are arranged in the order of increasing K/La ratios. Classification of rocks according to Wilshire *et al.* (1973).

	62295-34 C2[a]	66095-48 B4[b]	61016-151 C1[c]	61156-35 B4[c]	63545-2[a]	66095-52 B4[d]	64567-4[a]	68815-130 B5[e]	64455 C2[e]	60335 C2[e]
K	605	730	670	940	1030	1303	1540	1810	2034	2316
La	18.6	18.5	16.7	22.5	19.7	22.66	21.7*	22.3	21.1	22.0
K/La	32.5	39.5	40.1	41.8	52.3	57.5	71.0	81.2	96.4	105
Rb	4.59 5.8[f]	11.0 3.9[f]	2.84 1.3[f]	2.44	3.16	—	4.93	8.8 2.0[f]	6.0	7.1
Cs	0.53[f]	0.4 0.16[f]	0.12 0.056[f]	0.15				0.46 0.125[f]	0.27	0.28
K/Rb	132	83	235	385	326	—	312	206	339	326
Rb/Cs	10.9[f]	27.5 24.4[f]	23.6 23.2[f]	16.3				19.1 16.0[f]	22.2	25.4
Na	3412	3400	2700	3270	2820	3560	3120	3950	4230	4670
Na/K	5.64	4.66	4.03	3.48	2.74	2.73	2.03	2.18	2.08	2.02
Al	10.74	13.7	13.0	12.2	11.6	12.9	11.12	14.2	11.85	12.9
Mg	8.63	5.6	6.27	5.94	7.42	4.92	6.93	3.87	5.6	4.73

*La was obtained by multiplying Sm with 2.2.
References: [a]Hubbard *et al.* (1973b); [b]Brunfelt *et al.* (1973); [c]This work; [d]Nakamura *et al.* (1973); [e]Haskin *et al.* (1973); [f]Krähenbühl *et al.* (1973).

not constant and it does not systematically vary with increasing K/La ratios. The Na/K ratio is fairly constant for K-enriched samples (samples with a K/La ratio > 70). The variable ratios of K and Rb relative to the refractory REE has been observed by Mark *et al.* (1973) in microbreccias from Apollo 14 (14066 and 14321). Their conclusion was: "The lack of correlation (for the 14066 and 14321 microbreccia clasts with KREEP basalt patterns) of REE with K and Rb suggest that there is not a single KREEP component."

We suggest that the loss and enrichment of K, Rb, and Na is due to volatilization processes during the formation of these breccias:

(1) The soil samples of Apollo 16 and Apollo 14 have constant K/La ratios. (Anorthosite samples with very low La concentrations have high K/La ratios, because K enters the plagioclase lattice more readily than La, which is best seen in the pure anorthosite 65315 (Table 2) with a K/La ratio of 483.)

(2) In some samples the lower K/La ratios are accompanied by lower Mn/Fe and Na/Eu ratios, indicating parallel loss of K, Mn, and Na (Fig. 8).

(3) In Fig. 9 we have plotted the Rb/Sr ratios of some rocks against their initial Sr^{87}/Sr^{86} ratios, calculated to the same age of 3.9 b.y. Where no internal isochrons are determined, the whole rock Sr^{87}/Sr^{86} ratios have been calculated back to 3.9 b.y.

If these rocks and breccias formed about 3.9 b.y. ago, and if their Rb and La is essentially determined by the admixture of KREEP then a straight line should be expected. One can see that the lower the Rb/La ratio is, the more these samples deviate from the line connecting samples with the typical KREEP Rb/La ratios of 0.22. All these samples had 3.9 b.y. ago a Sr^{87}/Sr^{86} ratio, according to their amount of La (respectively KREEP). During the brecciation process they lost some Rb

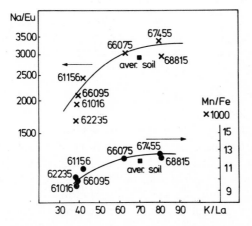

Fig. 8. Na/Eu and Mn/Fe versus K/La. The parallel depletion of the more volatile element in each pair indicates volatilization loss. Data from Brunfelt *et al.* (1973) and this laboratory. Data points for the average soil compositions from the Lunar Module station and from Stations 1, 4, and 5 are indicated (data from Wänke *et al.*, 1973).

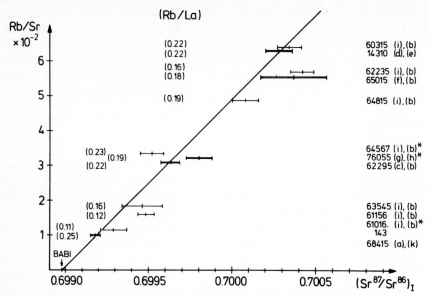

Fig. 9. Rb/Sr against (Sr⁸⁷/Sr⁸⁶) initial values for samples with KREEP-like REE patterns from Apollo 14, 16, and 17. Heavy lines indicate samples, where initial values have been corrected for a common age of 3.9 b.y. The other samples are total rock measurements, calculated back to 3.9 b.y. From 65015 the total rock-quintessence isochron has been used. A correction for interlaboratory bias has been applied to the Sr^{87}/Sr^{86} ratios as follows: C.I.T. group + 0.00012 = Nyquist *et al.* (1973); C.I.T. group + 0.00005 = Mark *et al.* (1974). The error bars indicate only errors in the Sr^{87}/Sr^{86} ratios. The first reference indicates the source for Rb/Sr measurements, the second for the La concentrations. Rb/La ratios are given in parentheses. The more the samples deviate from the straight line, the bigger is the difference of the Rb/La to the "KREEP-value" of about 0.22. This indicates loss or gain of Rb, during the formation of these breccias about 3.9 b.y. ago. References: (a) Papanastassiou and Wasserburg (1972a); (b) Hubbard *et al.* (1973b); (c) Mark *et al.* (1974); (d) Papanastassiou and Wasserburg (1971); (e) Hubbard *et al.* (1972); (f) Papanastassiou and Wasserburg (1972b); (g) Tera *et al.* (1974); (h) Philpotts *et al.* (1974); (i) Nyquist *et al.* (1973); (k) Bansal *et al.* (1972). *La obtained by multiplying Sm with 2.2.

(indicated by their observed Rb/La ratio), and the respective data points are shifted to lower Rb/Sr values.

(4) It is very difficult to explain the observed variation in volatile to refractory element ratios by magmatic processes. There is no systematic variation in the major elements with increasing K/La ratios (Table 7). Grieve and Plant (1973) found partial melts of 64455 to be enriched in Na and K. Addition or loss of such a melt could perhaps explain the observed variations. But one would expect the REE to enter this first melt, too, and consequently the K/La ratio would not change.

Some of the above mentioned rocks (e.g. rusty rock 66095) contain a large amount of very volatile elements like Pb or Tl. This enrichment probably occurred after the brecciation and crystallization, as already proposed by Krähenbühl *et al.* (1973).

3. Origin and Composition of the Lunar Highlands

The KREEP-rich soils from the Apollo 14 area and all highland samples seem to contain a component with a Mg/Fe ratio close to one. It may either be a mafic cumulate (Nava and Philpotts, 1973) or, neglecting metal and troilite, a nearly unfractionated chondritic component.

In Fig. 10 we plotted Fe/Mg versus Fe/Cr. A clear distinction between mare and highland samples is observed. In magmatic differentiations Mg/Cr varies less than Fe/Mg. Hence, magmatic processes could shift the samples along the dashed line of Fig. 10.

When normalized to C1 abundances, the patterns of the siderophile elements observed in soils from the Apollo 11, 12, 14, and 15 landing sites were always close to horizontal lines, indicating the meteoritic origin of these elements (Fig. 11). The patterns of the highland samples, especially of the breccias, look very different. It was already pointed out by Ganapathy *et al.* (1973) that the refractory siderophile elements appear depleted, as one would expect in an inhomogeneous accretion model. The material accumulating last would be poor in refractories. Some of the Apollo 16 breccias contain siderophile elements in amounts corresponding to an admixture of up to 15% C1 material. For lithophile elements an even higher chondritic contribution could be expected. Both a possible loss of metal during accretion as well as segregation of metal in small local melting processes would act in this direction.

The highland samples contain a feldspathic, i.e. aluminum-rich component and a magnesium-rich component. We have set up a mixing model assuming that only

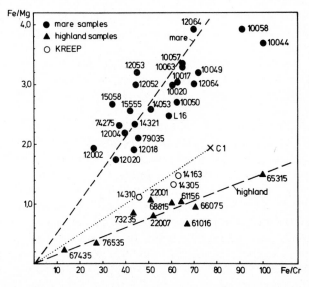

Fig. 10. Fe/Mg versus Fe/Cr. The mare samples are clearly separated from the highland samples. The KREEP-rich Apollo 14 samples fall close to the highland samples from the Apollo 16 and 17 and Luna 20 landing sites. Magmatic fractionations result in a shift approximately along the dashed lines.

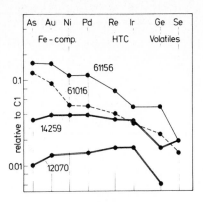

Fig. 11. Siderophile elements in lunar soils and breccias relative to C1 abundances. All
data from this laboratory.

these two components are of importance. The small concentrations of Ti in these
samples, which are not correlated with Mg, indicate that in fact at least one
additional component must be present, but in small amounts. In Fig. 12 we have
calculated the position of the individual samples along the x-axis from the data on
Al, Ca, and Mg alone, using a Tschebyscheff approximation. With the position of
the samples along the x-axis fixed via the Al, Ca, and Mg concentrations, the data
points for Fe, Cr, and Si were plotted as well as those for Na and Sc (Fig. 13) and
the element lines calculated from least square fits.

Among the highland samples there are a few which do not fit in the mixing
diagram, i.e. samples 61016, 67435 (Prinz *et al.*, 1973), and 76535 (Gooley *et al.*,
1973). These samples show clear evidence for magmatic differentiation (see Fig.
10); we therefore have not included them into our mixing diagram. An excellent fit
to the mixing lines is observed for all other samples. Only the Na points scatter
considerably. As discussed in the preceding section some samples have been
subjected to loss or gain of K and Na, probably due to volatilization. Samples with
a high K/La ratio are high in Na, too (Fig. 13), while samples with exceptionally
low K/La ratios are low in Na content. For the calculation of the Na line we have
therefore neglected all samples with a K/La ratio below 45 and above 95. All
samples with a La content higher than 40 ppm have been neglected, too.

One of the two main components of all these samples is clearly a feldspathic
component which is represented in pure form by the anorthositic rocks. Hence, as
one would expect, the concentration of Ca is proportional to Al for all highland
samples. The ratio of Ca/Al being about 0.78 is considerably lower than that
observed in chondrites (1.08) and much lower than in mare samples (about 1.4).
Magnesium and Cr which are both absent in the anorthosites correlate very well
with each other; the observed Mg/Cr ratio of 52 is not too far off the C1 value of
40, but highly different from the value of about 15–20 found in most of the mare
samples.

The observed concentrations of iron have been corrected. Considerable
amounts of metallic iron have been observed in the highland samples by Nagata *et*

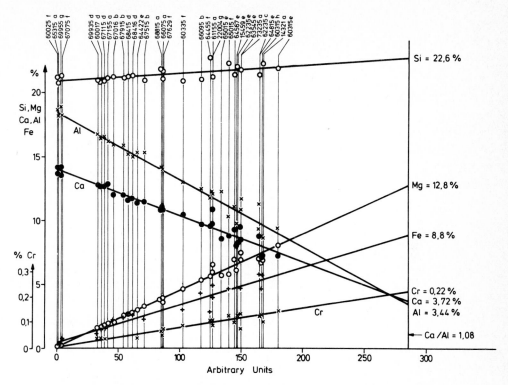

Fig. 12. Mixing diagram. The sample positions along the x-axis were calculated from the data of Al, Ca, and Mg alone. All the samples fit well into the mixing diagram. Together with the constancy of the Mg/Fe ratio this can be taken as strong indication that these samples are mechanical mixtures. Data are from the following sources: (a) this laboratory; (b) Duncan *et al.* (1973); (c) Brunfelt *et al.* (1973); (d) Rose *et al.* (1973); (e) Hubbard *et al.* (1973); (f) Haskin *et al.* (1973b); (g) Laul and Schmitt (1973); and (h) Morrison *et al.* (1973).

al. (1973), Pearce *et al.* (1973), and Tsay *et al.* (1973). Assuming a Fe/Ni ratio of 12 in this metal we have estimated the metal content of all samples for which Ni concentrations are known. This metallic Fe was then subtracted from the observed Fe concentrations. As already mentioned, the presence of Ti in the highland samples indicates the presence of a third component. The lowest Fe/Ti ratio observed in lunar samples is about two. To account for this third component we have further reduced the Fe values by the twofold amount of Ti. The contribution of the third component is only significant for Fe and probably can be neglected for the elements in Figs. 12 and 13.

Constant element ratios, especially Fe/Mg and a Mg/Na correlation are never observed in lunar, meteoritic (eucritic), and common terrestrial petrologic series. The close match of the C1 ratios for all elements belonging to the Mg-rich component argue against a mafic origin of this component. In the mixing diagrams in Figs. 12 and 13, beside 30 Apollo 16 breccias, the few samples from other

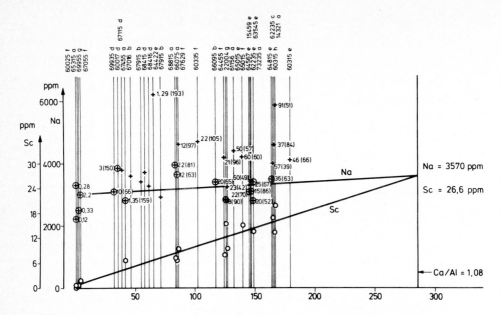

Fig. 13. Mixing diagram. Positions along the x-axis from Fig. 12. Only the encircled data points for Na were used for the calculation of the Na line (see text). The numbers in brackets refer to the K/La ratio of the samples; the numbers in front of the brackets refer to the observed La concentrations. For references see Fig. 12. In normal petrologic series Na should be anticorrelated with Mg. According to this diagram Na is supplied proportional to the Mg content from the Mg-rich component, if we consider the decrease of the feldspathic fraction, going from left to right.

landing sites, i.e. Apollo 14, Apollo 17, and Luna 20, fit equally well. An admixture of a mafic component with nearly identical composition especially with the same Mg/Fe ratio to the breccias found at localities up to 2000 km apart is highly improbable. We therefore suggest that the Mg-rich component represents material of the last accretion stage of the moon, which was never incorporated in the lunar magmatic differentiation processes. It was subjected to local melting processes only, which in most cases did not lead to substantial differentiations. Hence, the chemical compositions of the highland samples reflect mechanical mixing processes.

From the mixing diagram we can directly read the Mg/Fe, Mg/Cr, and Mg/Sc ratios of this Mg-rich component. Absolute values for the elemental composition of the Mg-rich component can also be obtained. Material not subjected to magmatic fractionations must have the cosmic Ca/Al ratio of 1.08. Thus the position along the x-axis for which Ca/Al equals 1.08 gives us the composition of the Mg-rich component.

Many of the Apollo 16 and most Apollo 14 samples have high concentrations of KREEP elements. However, no correlation between KREEP elements and the major elements Al, Ca, Fe, and Mg is observed. Hence, we have to conclude that the major elements connected with KREEP do not contribute significantly to the bulk composition of the KREEP-containing samples. The major elements in a microbreccia clast of 14321 (La = 91 ppm) and those in most Apollo 16 breccias can be explained as mixtures of the same two components (Fig. 12). If any of the major elements shows a trend to correlate with La it is Ti, at least for samples with higher (La > 20 ppm) La concentrations; here the Ti/La ratio is 170 ± 40.

4. BULK COMPOSITION OF THE MOON

The observed element correlations and the relations from the condensation sequence (Larimer, 1967; Anders, 1971; Grossman, 1972) give us a framework from which the bulk composition of the moon can be calculated (Fig. 14).

In our model we have only two major components: the high temperature condensates (HTC) and the "chondritic" component (ChC). The chondritic component as defined here, also contains the moderately volatile elements. Following Anders (1971) we assume constant depletion factors for the moderately volatile elements, except for the heavy alkali elements.

The two-component model for the bulk composition of the moon, as put forward in our last Apollo paper (Wänke *et al.*, 1973) is based on the K/La ratio. In that model the chondritic component was not very well defined. Cosmochemical groups of elements, distinguished by their condensation behavior vary in their relative proportions even among different classes of chondrites. In addition, the

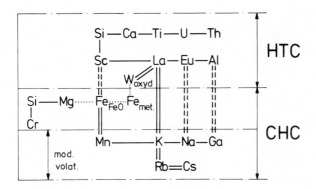

Connected by: ═══ strongly correlated elements
==== moderately correlated elements
—— group relations
·········· other relations

Fig. 14. Condensation sequence and element correlations as a framework for the estimation of the bulk chemistry of planets. Within the two major groups (HTC and ChC) only the elements of the HTC appear in C1 abundance ratios. In the ChC, the ratio $Fe_{FeO}/Fe_{met.}$ and Fe/Mg may vary considerably. As discussed in Section 2 the La/W correlation could be used to estimate the $FeO/Fe_{met.}$ ratio. For the moderately volatile elements fractionations according to their different volatilities might occur.

varying Fe to Mg ratios indicate a metal/silicate fractionation (Anders, 1971).

If one accepts our conclusion in Section 3 that the Mg-rich component found in the lunar highlands represents material of the last accretion stage of the moon, we can estimate the composition of the ChC. The concentrations of Al, Ca, and also Sc in the Mg-rich component indicate a mixture of HTC and ChC in this component, where the HTC amounts to 25%. This value may be a bit too high, because some Al may originate from differentiated lunar components (KREEP), not taken into account in our mixing model in Section 3. The presence of at least one further component is indicated by the fact that the sum of the elements of the Mg-rich component is only 94%. The remaining 6% are made up by Fe, Ti, and K, elements which together with Al and Ca probably are associated with KREEP. For the evaluation of the composition of the ChC given in Table 8, column 2, we have assumed that the third component has a SiO_2 content of 50%.

Metal and troilite were not considered so far and the amount of Fe in the Mg-rich component refers to iron in oxidized form only. We have also seen from our mixing diagram that the Mg/Cr ratio in this component is higher than the C1 ratio. If Cr condenses together with metallic iron (Grossman and Olsen, 1974) the lower Cr/Mg ratio indicates loss of metal relative to silicate.

Finally we can estimate the concentrations of the moderately volatile elements from the Fe/Mn correlation. In this way we obtain a value of 4100 ppm Na in the ChC of the bulk moon. This value seems more reliable than the one of 4700 ppm, resulting from the composition of the Mg-rich component (Fig. 13). Nevertheless the agreement is quite satisfactory.

From the observed lunar K/La ratio of 70 we then calculate the proportions of HTC and ChC in the bulk moon as described by Wänke *et al.* (1973) (see Table 9).

Table 8. Composition of the HTC and ChC used for the calculation and the resulting bulk composition of the moon.

Element	HTC	ChC	Moon	Element	HTC	ChC	Moon
O %	43.5	37.5	40.5	Mn ppm	—	1500	750
Mg	6.5	15.2	10.9	Cr	—	2900	1450
Al	17	—	8.5	Sr	130	—	65
Si	14	20.3	17.2	Ba	47	—	24
Ca	18	—	9	Sc	120	—	60
Ti	0.91	—	0.46	Y	33	—	17
Fe (total)	—	22.2	11.1	La	4.9	—	2.5
Fe (FeO)	—	11.6	5.8	Eu	1.14	—	0.57
Fe (metal) }	—	10.6	5.3	Zr	93	—	47
Fe (FeS) }				Hf	3.3	—	1.7
Ni	—	1.2	0.6	V	620	—	310
Na ppm	—	4100	2050	Nb	6.6	—	3.3
K	—	(415)	175	Ta	0.25	—	0.13
Rb	—	(2)	0.4	W	1.84	—	0.9
Cs	—	(0.16)	0.02	Th	0.44	—	0.22
				U	0.12	—	0.06

Table 9. Proportions of HTC and ChC in the bulk moon.

	HTC %	ChC %
Calculated from Sc/Fe–Fe/Mn	40	60
Calculated from K/La	54	46

The value of 46% ChC may be considered as lower limit only as a depletion of K larger than that of Mn would raise this number. Other pairs of volatile to nonvolatile elements can be used for similar calculations.

In Table 9 we also give the proportions of the two components obtained from the Sc/Fe–Fe/Mn correlations (Fig. 14). The lowest observed Fe/Sc ratio may be an upper limit of the Fe/Sc ratio of the bulk moon and, hence, the value of 60% ChC deduced from this ratio is an upper limit. As can be judged from Fig. 1 and Fig. 7, the lunar K/La ratio is more constant and, hence, more reliable than the Fe/Sc ratio. Therefore we adopted for the calculation of the bulk composition of the moon a ratio of HTC to ChC of 0.5:0.5. The result of this calculation is summarized in Table 8. All refractory elements are found to be enriched in the bulk moon by a factor of 13 relative to C1 abundances. In the case of the heavier alkali elements K, Rb, and Cs we did not take the C1 ratios relative to Na and Mn, but used the observed ratios from the element correlations K/La, K/Cs, and Rb/Cs (Wänke *et al.*, 1973). It seems that these elements are somewhat more depleted than Na and Mn. The new data for the bulk composition of the moon differ by up to 30% from those given in our last Apollo paper (Wänke *et al.*, 1973).

As discussed in Section 3, the fraction of HTC in the last accreting material is below 25%. In order to get a 50:50 ratio of the bulk moon, we must have had an inhomogeneous accretion in which the moon and probably also the earth and the other planetary objects started with a core consisting mainly of HTC. During the further accretion the amount of HTC decreased and that of ChC increased. The faster growing earth accreted more of the ChC compared to the moon.

A calculation for the bulk composition (silicates only) of the earth yields a value of 21% for the HTC component, not far from the 25% found as upper limit for the material added to the moon in the final accretional stage.

The Fe/Mn ratio for all non-highland samples is about 80. Nearly the same value is found for the highland samples. Thus in spite of the varying ratio of HTC/ChC, the composition of the ChC and especially the portion of its moderately volatile elements did not change during accretion.

Some of the element correlations yielded different ratios of mare and highland samples (Eu/Na in Fig. 4, Sc/Fe in Fig. 7, and Mg/Cr see Fig. 10). In all these cases the ratio of the HTC element to the ChC element is about a factor of 2 to 3 lower for the highland samples compared to mare samples. This factor corresponds to the difference of the ratio HTC/ChC of 1 (50:50) for the bulk moon and of 0.33

(25/75) for the old surface regions (highlands) and may be interpreted as further indication of an inhomogeneous accretion of the moon. The mare basalts probably were generated from material in equilibrium with the bulk moon. The K/La ratio is identical for mare samples and KREEP-containing samples. Thus KREEP also represents material in equilibrium with the bulk moon.

Independent estimates of the bulk composition of the moon based partly on different constraints were published by Anders and coworkers (Krähenbühl *et al.*, 1973; Ganapathy and Anders, 1974). Table 10 gives a comparison of the results of both their and our model. The main difference lies in the amount of HTC. Ganapathy and Anders have 30% HTC (early condensate) compared to our value of 50%. This difference arises mainly from the fact that we use the analytical values of the Ca-, Al-rich inclusions of Allende meteorite (Clarke *et al.*, 1970; Wänke *et al.*, 1974) as face values for the composition of the HTC, while Ganapathy and Anders use theoretical predictions, resulting from the condensation sequence as calculated by Grossman (1972).

The La/U ratio in the Allende inclusion is about 40, compared to a value of about 30 in ordinary chondrites. In their model Ganapathy and Anders (1974) assumed an abundance ratio La/U = 27. They define the amount of HTC by the bulk U content (59 ppb) of the moon as derived from heat flow data (Langseth *et al.*, 1973) and thermal calculations (Toksöz and Solomon, 1973). Langseth *et al.* give an estimated error of 20% for the heat flow. An additional error results from the thermal model. Assuming a moon in thermal equilibrium, the value for the U content of 59 ppb would rise to 81 ppb (see Section 5). The uncertainties in our model are in the same range, as partly illustrated in Table 9 with values between 40% and 54% for the contribution of the HTC.

Our model is exclusively based on geochemical and cosmochemical data and the uncertainties mainly come from the uncertainties of the composition of the ChC. In the comparison in Table 10 we would like to stress the conformity rather than the difference.

Hodges and Kushiro (1974), who studied the melting relations of model compositions for the moon found that a composition with $SiO_2 = 35.88\%$ is too poor in silica to yield plagioclase. Our value for the SiO_2 of 36.8% is only slightly higher. But we believe that this constraint can easily be removed with an inhomogeneous moon as described above. According to our observation of a ratio

Table 10. Bulk compositions of the moon (weight percent).

Authors	SiO_2	MgO	FeO	Al_2O_3	CaO	TiO_2	Cr_2O_3	Na_2O	MnO	Fe	FeS	Ni
Krähenbühl et al. (1973)	35.9	20.9	12.7	12.8	10.6	0.63	0.39	0.58	0.24	~2.5*	~2.5	—
Wänke et al. (1973)	32.1	14.8	7.2	22.6	17.9	1.17	0.50	0.24	0.09	1.9	1.7	0.53
Ganapathy and Anders (1974)	39.8	28.8	3.52	11.02	8.91	0.564	0.175	0.121	0.043	5.6	1.1	0.51
This work	36.8	18.1	7.5	16.1	12.6	0.77	0.21	0.28	0.097	~6.3†		0.6

*Fe + Ni + Co.
†Fe + FeS assuming for S an abundance of 1%.

of HTC/ChC of <0.33 for the outer regions of the moon the SiO_2 content would rise there to about 40%.

5. Lunar Differentiation Processes as Deduced from Trace Element Abundances

Nearly all non-mare samples have the same relative abundances of LIL elements. This is clearly demonstrated by the parallel REE patterns. Another example is, e.g. the constancy of the La/Hf ratio (Fig. 5). This ratio is constant for an absolute range of 1.3 to 70 ppm La in non-mare samples from Apollo 12, 14, 15, 16, and 17 and Luna 20 landing sites. Mare basalts have a different and variable ratio of La/Hf. As the trace element behavior of non-mare samples is determined by the admixture of the KREEP component, we call this specific LIL-element pattern KREEP fractionation, irrespective of the absolute concentration of the LIL elements. In Fig. 15 some LIL elements of the soil sample 14163 with typical KREEP fractionation are plotted relative to C1 chondritic abundances (x-axis). Potassium abundances are multiplied by 30 to account for the deficiency of the alkali elements on the moon. For W only the oxidized part behaves like a LIL element (see Section 2).

As one can see, U, Ba, and La have the highest enrichment factors. The upper half of Fig. 15 shows again how well these elements are correlated in KREEP-containing samples. Soil sample 12070 consists of only 30% KREEP (Wänke *et al.*, 1972), and it seems that the mare basalt part is responsible for the deviation of La, U, and Ba from the correlation line. In sample 67455 the anorthositic component changes the abundances of Sr, Eu, Ba, and K. If we assume a chondritic LIL-element pattern and a 13-fold enrichment (see preceding section) of refractory elements in the bulk moon relative to C1 chondrites, we have to postulate major magmatic differentiation processes in order to produce the KREEP pattern (see Taylor *et al.*, 1972).

If the heat flux through the surface of the moon at the Apollo 15 and 17 landing sites (Langseth *et al.*, 1973) is representative of the whole moon and if the moon is in thermal equilibrium, then a bulk content of U of 81 ppb can be calculated (with a heat flux $q = 0.7 \mu cal/cm^2$ sec). The U concentrations of the bulk moon, used in most thermal models is somewhat lower for two reasons:

(a) There is still initial heat contributing to the present day heat flux.
(b) The present day heat flux through the lunar surface refers to the decay of U at a time, when the concentration of U was higher than today.

If we do not neglect convection, we can forget the first point (Tozer, 1972). It will be shown that most of the lunar U is now concentrated in the crust. Therefore the relaxation time for conduction is about 0.5 b.y. ($\sim L^2/k$ for $L = 100$ km, k = thermal diffusivity). At this time the heat production by U and Th was 7% higher than today. A reasonable value for the bulk lunar U content would thus be 75 ppb, with essentially the error of the measured heat flux.

With a reasonable conductivity K of 5×10^{-3} cal/cm^2 sec °C (Robie and

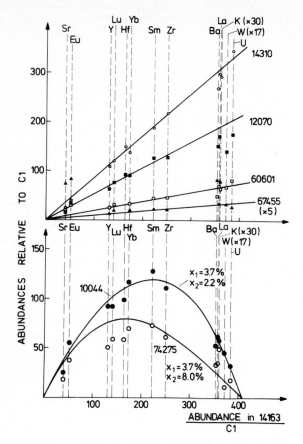

Fig. 15. Trace element abundances in KREEP-containing samples (14310, 12070, 60601, and 67455) and mare basalts (10044 and 74275) versus their abundances in 14163; all are normalized to C1. The curves in the lower half are calculated from a two step model, x_1 and x_2 are fractions of partial melting.

Hemingway, 1971; Mizutani *et al.*, 1972), and the above mentioned heat flux q, we can calculate the depth where the moon would melt.

$$q = \frac{K}{d} \Delta T$$

where $d =$ depth

$\Delta T =$ temperature difference.

At about a depth of 70 km we would reach the melting point. To avoid this kind of conclusion we have to assume that at least one half of the total U is concentrated in the first 70 km. Starting with a homogeneous moon or with a moon with increasing concentration of refractory elements towards the center, we would have to extract U and the other LIL elements from the major part of the moon. A refractory interior would begin to melt roughly 100 m.y. after accretion, if the accretion temperature is about 1200°K. A partial melting process would then form a melt very rich in LIL elements and the fractionation of these elements would depend upon the mineral assemblage of the interior. This melt is mixed to different degrees to the already existing highlands, which probably were formed in an earlier differentiation of the outer region of the moon, as a consequence of additional heat sources (accretion, external heat). The Mg and Fe concentrations and the ratio Mg/Fe would be low in this melt. Addition of the late accumulating chondritic material determines the Mg and Fe contents of non-mare samples.

It should be noted that the Eu anomaly of KREEP, reflecting this first melt, is not due to plagioclase separation on the surface of the moon, but is instead a consequence of Eu-rich minerals in the interior, e.g. melilite, similar to the model of Ringwood (1974).

In the lower part of Fig. 15 the abundances of some LIL elements of two mare basalts relative to C1 chondrites are plotted against the enrichments relative to C1 chondrites for KREEP. The resulting parabola-shaped curves can be explained by a second partial melting process, producing the mare basalt patterns of LIL elements. The curves in Figs. 15 and 16 demonstrate a model calculation assuming a moon 15-fold enriched in refractory elements relative to C1 chondrites.

The abundances of the LIL elements of soil sample 14163 have been assumed to be representative for this first partial melt. From the ratio C_1^1/C_b (concentration in 14163 to concentration in the bulk moon) and from an assumed fraction of partial melting x_1 we calculated an effective distribution coefficient D for each LIL element:

$$C_1^1/C_b = \frac{1}{x_1}(1-(1-x_1)^D) \qquad \text{(Shaw, 1970)}$$

We calculated the LIL-element distribution in the residual moon (see for example Fig. 16). If we then apply the above distribution coefficients to a second partial melting process (with a fraction x_2), we can obtain the resulting LIL-element pattern C_1^2 by the following equation:

$$C_1^2/C_b = \frac{1}{x_2(1-x_1)} \cdot \left[\left(1 - \frac{C_1^1}{C_b} \cdot x_1\right) - \left(1 - \frac{C_1^1}{C_b} \cdot x_1\right)^{\frac{\ln(1-x_2)}{\ln(1-x_1)}+1} \right]$$

The fractions x_1 and x_2 of partial melt can be used as parameters to fit the observed LIL-element patterns in mare basalts. Figures 15 and 16 are examples for such calculations. As one can see, we can quite well reproduce the observed element patterns, although the absolute values of x_1 and x_2 should not be taken too seriously, as we do not really know the absolute enrichment of the LIL elements in first melt (i.e. in KREEP).

The above calculations have been performed under the assumption that each fraction of liquid separated continuously from preceding liquids. It can be shown that the results are not very different, if the liquid remains at all times in equilibrium with the residual solid, as long as the fractions of partial melt x_1 and x_2 are small (<10%).

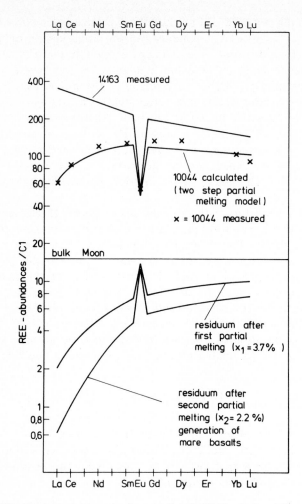

Fig. 16. REE-abundance patterns in the mare basalt 10044 calculated from a two step
partial melting model, x_1 and x_2 are fractions of partial melting.

Figure 15 is very useful in another sense. One can immediately see the
elements which are well correlated for the whole moon. Ratios like Y/Yb (Fig. 3)
or Yb/Hf are constant in all samples; La/Hf clearly is not constant. Potassium
should correlate well with La or Ba, but the K/Zr and K/U ratios are expected to
be more variable, being different for mare basalts and KREEP.

Acknowledgments—We are grateful to NASA for making available the lunar material for this
investigation. The samples were activated in TRIGA-research reactor of the Institut für Anorganische
Chemie und Kernchemie der Universität Mainz. We wish to thank the staff of the TRIGA-research
reactor and the members of the staff of our Institute who are not listed as authors, in particular Mr. P.
Deibele and Miss H. Prager.

REFERENCES

Anders E. (1971) Meteorites and the early solar system. *Ann. Rev. Astron. Astrophys.* **9**, 1–34.

Bansal B. M., Church S. E., Gast P. W., Hubbard N. J., Rhodes J. M., and Wiesmann H. (1972) Chemical composition of soil from Apollo 16 and Luna 20 sites. *Earth Planet. Sci. Lett.* **17**, 29–35.

Brunfelt A. O., Heier K. S., Nilssen B., and Sundvoll B. (1973) Geochemistry of Apollo 15 and 16 materials. *Proc. Fourth Lunar Sci. Conf., Geochim. Cosmochim. Acta*, Suppl. 4, Vol. 2, pp. 1209–1218. Pergamon.

Clarke R. S., Jr., Jarosewich E., Mason B., Nelen J., Gómez M., and Hyde J. R. (1970) The Allende, Mexico, Meteorite Shower. *Smithsonian Contributions to the Earth Sciences*, No. 5, pp. 1–53.

Duncan A. R., Erlank A. J., Willis J. P., and Ahrens L. H. (1973) Composition and inter-relationships of some Apollo 16 samples. *Proc. Fourth Lunar Sci. Conf., Geochim. Cosmochim. Acta*, Suppl. 4, Vol. 2, pp. 1097–1113. Pergamon.

Duncan A. R., Erlank A. J., Willis J. P., Sher M. K., and Ahrens L. H. (1974) Trace element evidence for a two stage origin of high titanium mare basalts (abstract). In *Lunar Science—V*, pp. 187–189. The Lunar Science Institute, Houston.

Ganapathy R., Morgan J. W., Krähenbühl U., and Anders E. (1973) Ancient meteoritic components in lunar highland rocks: Clues from trace elements in Apollo 15 and 16 samples. *Proc. Fourth Lunar Sci. Conf., Geochim. Cosmochim. Acta*, Suppl. 4, Vol. 2, pp. 1239–1261. Pergamon.

Ganapathy R. and Anders E. (1974) Bulk compositions of the moon and earth, estimated from meteorites. Preprint.

Gillum D. E., Ehmann W. D., Wakita H., and Schmitt R. A. (1972) Bulk and rare earth abundances in the Luna 16 soil levels A and D. *Earth Planet. Sci. Lett.* **13**, 444–449.

Gooley R., Brett R., Warner J., and Smyth J. R. (1973) Sample 76535, a deep lunar crustal rock. Preprint.

Grieve R. A. F. and Plant A. G. (1973) Partial melting on the lunar surface, as observed in glass coated Apollo 16 samples. *Proc. Fourth Lunar Sci. Conf., Geochim. Cosmochim. Acta*, Suppl. 4, Vol. 1, pp. 667–679. Pergamon.

Grossman L. (1972) Condensation in the primitive solar nebula. *Geochim. Cosmochim. Acta* **36**, 597–619.

Grossman L. and Olsen E. (1974) Origin of the high-temperature fraction of C2 chondrites. *Geochim. Cosmochim. Acta* **38**, 173–187.

Haskin L. A., Helmke P. A., Blanchford D. P., Jacobs J. W., and Telander K. (1973) Major and trace element abundances in samples from the lunar highlands. *Proc. Fourth Lunar Sci. Conf., Geochim. Cosmochim. Acta*, Suppl. 4, Vol. 2, pp. 1275–1296. Pergamon.

Hodges F. N. and Kushiro I. (1974) Apollo 17 petrology and experimental determination of differentiation sequences in model moon compositions (abstract). In *Lunar Science—V*, pp. 340–342. The Lunar Science Institute, Houston.

Hubbard N. J., Gast P. W., Meyer C., Nyquist L. E., Shih C., and Wiesmann H. (1971) Chemical composition of lunar anorthosites and their parent liquids. *Earth Planet. Sci. Lett.* **13**, 71–75.

Hubbard N. J., Gast P. W., Rhodes J. M., Bansal B. M., and Wiesmann H. (1972) Nonmare basalts: Part II. *Proc. Third Lunar Sci. Conf., Geochim. Cosmochim. Acta*, Suppl. 3, Vol. 2, pp. 1161–1179. MIT Press.

Hubbard N. J., Rhodes J. M., and Gast P. W. (1973a) Chemistry of lunar basalts with very high aluminum contents. *Science* **181**, 339–342.

Hubbard N. J., Rhodes J. M., Gast P. W., Bansal B. M., Shih C., Wiesmann H., and Nyquist L. E. (1973b) Lunar rock types: The role of plagioclase in non-mare and highland rock types. *Proc. Fourth Lunar Sci. Conf., Geochim. Cosmochim. Acta*, Suppl. 4, Vol. 2, pp. 1297–1312. Pergamon.

Krähenbühl U., Ganapathy R., Morgan J. W., and Anders E. (1973) Volatile elements in Apollo 16 samples: Implications for highland volcanism and accretion history of the moon. *Proc. Fourth Lunar Sci. Conf., Geochim. Cosmochim. Acta*, Suppl. 4, Vol. 2, pp. 1325–1348. Pergamon.

Langseth M. G., Keihm S. J., and Chute J. L., Jr. (1973) Heat flow experiment. In *Apollo 17 Preliminary Science Report*, NASA SP-330, 9-1 through 9-24.

Larimer J. W. (1967) Chemical fractionations in meteorites—I. Condensation of the elements. *Geochim. Cosmochim. Acta* **31**, 1215–1238.

Laul J. C. and Schmitt R. A. (1973) Chemical composition of Apollo 15, 16 and 17 samples. *Proc. Fourth Lunar Sci. Conf., Geochim. Cosmochim. Acta*, Suppl. 4, Vol. 2, pp. 1349–1367. Pergamon.

Laul J. C., Wakita H., Showalter D. L., Boynton W. V., and Schmitt R. A. (1972) Bulk, rare earth, and other trace elements in Apollo 14 and 15 and Luna 16 samples. *Proc. Third Lunar Sci. Conf., Geochim. Cosmochim. Acta*, Suppl. 3, Vol. 2, pp. 1181–1200. MIT Press.

Mark R. K., Cliff R. A., Lee-Hu C., and Wetherill G. W. (1973) Rb–Sr studies of lunar breccias and soils. *Proc. Fourth Lunar Sci. Conf., Geochim. Cosmochim. Acta*, Suppl. 4, Vol. 2, pp. 1785–1795. Pergamon.

Mark R. K., Lee-Hu C., and Wetherill G. W. (1974) Rb–Sr measurements on lunar igneous rocks and breccia clasts (abstract). In *Lunar Science—V*, pp. 490–492. The Lunar Science Institute, Houston.

Mizutani H., Fujii N., Hamano Y., and Osako M. (1972) Elastic wave velocities and thermal diffusivities of Apollo 14 rocks. *Proc. Third Lunar Sci. Conf., Geochim. Cosmochim. Acta*, Suppl. 3, Vol. 3, pp. 2557–2564. MIT Press.

Morrison G. H., Nadkarni R. A., Jaworski J., Botto R. I., and Roth J. R. (1973) Elemental abundances of Apollo 16 samples. *Proc. Fourth Lunar Sci. Conf., Geochim. Cosmochim. Acta*, Suppl. 4, Vol. 2, pp. 1399–1405. Pergamon.

Nagata T., Fisher R. M., Schwerer F. C., Fuller M. D., and Dunn J. R. (1973) Magnetic properties and natural remanent magnetization of Apollo 15 and 16 lunar materials. *Proc. Fourth Lunar Sci. Conf., Geochim. Cosmochim. Acta*, Suppl. 4, Vol. 3, pp. 3019–3043. Pergamon.

Nakamura N., Masuda A., Tanaka T., and Kurasawa H. (1973) Chemical compositions and rare-earth features of four Apollo 16 samples. *Proc. Fourth Lunar Sci. Conf., Geochim. Cosmochim. Acta*, Suppl. 4, Vol. 2, pp. 1407–1414. Pergamon.

Nava D. F. and Philpotts J. A. (1973) A lunar differentiation model in light of new chemical data on Luna 20 and Apollo 16 soils. *Geochim. Cosmochim. Acta* **37**, 963–973.

Nyquist L. E., Hubbard N. J., Gast P. W., Bansal B. M., and Wiesmann H. (1973) Rb–Sr systematics for chemically defined Apollo 15 and 16 materials. *Proc. Fourth Lunar Sci. Conf., Geochim. Cosmochim. Acta*, Suppl. 4, Vol. 2, pp. 1823–1846. Pergamon.

Papanastassiou D. A. and Wasserburg G. J. (1971) Rb–Sr ages of igneous rocks from the Apollo 14 mission and the age of the Fra Mauro formation. *Earth Planet. Sci. Lett.* **12**, 36–48.

Papanastassiou D. A. and Wasserburg G. J. (1972a) The Rb–Sr age of a crystalline rock from Apollo 16. *Earth Planet. Sci. Lett.* **16**, 289–298.

Papanastassiou D. A. and Wasserburg G. J. (1972b) Rb–Sr systematics of Luna 20 and Apollo 16 samples. *Earth Planet. Sci. Lett.* **17**, 52–63.

Pearce G. W., Gose W. A., and Strangway D. W. (1973) Magnetic studies on Apollo 15 and 16 samples. *Proc. Fourth Lunar Sci. Conf., Geochim. Cosmochim. Acta*, Suppl. 4, Vol. 3, pp. 3045–3076. Pergamon.

Philpotts J. A., Schnetzler C. C., Nava D. F., Bottino M. L., Fullager P. D., Thomas H. H., Schuhmann S., and Kouns C. W. (1972) Apollo 14: Some geochemical aspects. *Proc. Third Lunar Sci. Conf., Geochim. Cosmochim. Acta*, Suppl. 3, Vol. 2, pp. 1293–1305. MIT Press.

Philpotts J. A., Schuhmann S., Kouns C. W., and Lum R. K. L. (1974) Lithophile trace elements in Apollo 17 soils (abstract). In *Lunar Science—V*, pp. 599–601. The Lunar Science Institute, Houston.

Prinz M., Dowty E., Keil K., and Bunch T. E. (1973) Spinel troctolite and anorthosite in Apollo 16 samples. *Science* **179**, 74–76.

Rieder R. and Wänke H. (1969) Study of trace element abundance in meteorites by neutron activation. In *Meteorite Research* (editor P. Millman) pp. 75–86. D. Reidel, New York.

Ringwood A. E. (1974) Minor element chemistry of maria basalts (abstract). In *Lunar Science—V*, pp. 633–635. The Lunar Science Institute, Houston.

Robie R. A. and Hemingway B. S. (1971) Specific heats of the lunar breccia (10021) and olivine dolerite (12018) between 90° and 350° Kelvin. *Proc. Second Lunar Sci. Conf., Geochim. Cosmochim. Acta*, Suppl. 2, Vol. 3, pp. 2361–2365. MIT Press.

Rose H. J., Jr., Cuttitta F., Berman S., Carron M. K., Christian R. P., Dwornik E. J., Greenland L. P., and Ligon D. T., Jr. (1973) Compositional data for twenty-two Apollo 16 samples. *Proc. Fourth Lunar Sci. Conf., Geochim. Cosmochim. Acta*, Suppl. 4, Vol. 2, pp. 1149–1158. Pergamon.

Schmitt R. A., Goles G. G., Smith R. H., and Osborn T. W. (1972) Elemental abundances in stone meteorites. *Meteoritics* **7**, 131–214.

Shaw D. M. (1970) Trace element fractionation during anatexis. *Geochim. Cosmochim. Acta* **34**, 237–243.

Taylor S. R., Kaye M., Muir P., Nance W., Rudowski R., and Ware N. (1972) Composition of the lunar uplands: Chemistry of Apollo 14 samples from Fra Mauro. *Proc. Third Lunar Sci. Conf., Geochim. Cosmochim. Acta*, Suppl. 3, Vol. 2, pp. 1231–1249. MIT Press.

Taylor S. R., Gorton M. P., Muir P., Nance W. B., Rudowski R., and Ware N. (1973) Composition of the Descartes region, lunar highlands. *Geochim. Cosmochim. Acta* **37**, 2665–2683.

Tera F., Papanastassiou D. A., and Wasserburg G. J. (1974) Isotope evidence for a terminal lunar cataclysm. *Earth Planet. Sci. Lett.* **22**, 1–21.

Toksöz M. N. and Solomon S. C. (1973) Thermal history and evolution of the moon. *The Moon* **7**, 251–278.

Tozer D. C. (1972) The moon's thermal state and an interpretation of the lunar electrical conductivity distribution. *The Moon* **5**, 90–105.

Tsay F. D., Manatt S. L., Live D. H., and Chan S. I. (1973) Metallic Fe phases in Apollo 16 fines: Their origin and characteristics as revealed by electron spin resonance studies. *Proc. Fourth Lunar Sci. Conf., Geochim. Cosmochim. Acta*, Suppl. 4, Vol. 3, pp. 2751–2761. Pergamon.

Wakita H. and Schmitt R. A. (1970) Elemental abundances in seven fragments from lunar rock 12013. *Earth Planet. Sci. Lett.* **9**, 169–176.

Wänke H., Rieder R., Baddenhausen H., Spettel B., Teschke F., Quijano-Rico M., and Balacescu A. (1970) Major and trace elements in lunar material. *Proc. Apollo 11 Lunar Sci. Conf., Geochim. Cosmochim. Acta*, Suppl. 1, Vol. 2, pp. 1719–1727. Pergamon.

Wänke H., Baddenhausen H., Balacescu A., Teschke F., Spettel B., Dreibus G., Palme H., Quijano-Rico M., Kruse H., Wlotzka F., and Begemann F. (1972) Multielement analyses of lunar samples and some implications of the results. *Proc. Second Lunar Sci. Conf., Geochim. Cosmochim. Acta*, Suppl. 2, Vol. 2, pp. 1251–1268. MIT Press.

Wänke H., Baddenhausen H., Dreibus G., Jagoutz E., Kruse H., Palme H., Spettel B., and Teschke F. (1973) Multielement analyses of Apollo 15, 16 and 17 samples and the bulk composition of the moon. *Proc. Fourth Lunar Sci. Conf., Geochim. Cosmochim. Acta*, Suppl. 4, Vol. 2, pp. 1461–1481. Pergamon.

Wänke H., Baddenhausen H., Palme H., and Spettel B. (1974) On the chemistry of the Allende inclusions and their origin as high temperature condensates. To be published.

Willis J. P., Ahrens L. H., Danchin R. V., Erlank A. J., Gurney J. J., Hofmeyr P. K., McCarthy T. S., and Orren M. J. (1971) Some inter-element relationships between lunar rocks and fines, and stony meteorites. *Proc. Second Lunar Sci. Conf., Geochim. Cosmochim. Acta*, Suppl. 2, Vol. 2, pp. 1123–1138. MIT Press.

Wilshire H. G., Stuart-Alexander D. E., and Jackson E. D. (1973) Apollo 16 rocks: Petrology and classification. *J. Geophys. Res.* **78**, 2379–2392.

Proceedings of the Fifth Lunar Conference
(Supplement 5, Geochimica et Cosmochimica Acta)
Vol. 2 pp. 1337–1352 (1974)
Printed in the United States of America

Modeling the evolution of Sm and Eu abundances during lunar igneous differentiation

D. F. Weill, G. A. McKay, S. J. Kridelbaugh, and M. Grutzeck

Center for Volcanology, University of Oregon, Eugene, Oregon 97403

Abstract—Models are presented for the evolution of Eu and Sm abundances during lunar igneous processes. The effect of probable variations in lunar temperature and oxygen fugacity, mineral–liquid distribution coefficients, and the crystallization or melting progression are considered in the model calculations. Changes in the proportions of crystallizing phases strongly influence the evolution of trace element abundances during fractional crystallization, and models must include realistic estimates of the major phase equilibria during crystallization.

The results are applied to evaluate the possibility of generating KREEP-rich materials by lunar igneous processes. Fractional crystallization of a primordial liquid shell is a very unlikely origin because the characteristic Sm and Sm/Eu abundances of KREEP are achieved only after a very high degree of fractionation involving separation of mostly ferromagnesian minerals. The low Fe/Mg found in KREEP is incompatible with such a history. All fractional crystallization sequences involving appreciable separation of plagioclase produce negative Eu anomalies that are much too large.

Partial melting of olivine–orthopyroxene–clinopyroxene–plagioclase assemblages can produce KREEP abundances of Sm and Sm/Eu. The most probable parent assemblages are (1) dominantly olivine plus orthopyroxene (<20% plagioclase and a few percent clinopyroxene) with bulk Sm and Sm/Eu concentrations comparable to those estimated for the whole moon; i.e. enriched approximately tenfold or less relative to chondrites, with no Eu anomaly, or (2) a rock with more plagioclase, enriched in both Sm and Eu (no negative Eu anomaly) relative to the whole moon. An origin for either of these parental bulk compositions via fractional crystallization implies a low concentration of plagioclase constituent relative to ferromagnesian constituents in the moon's primordial liquid shell.

Although the range of Sm and Sm/Eu abundances in KREEP may be related to varying degrees of partial melting of these parental assemblages, it is also possible that the variation may simply reflect different amounts of olivine plus orthopyroxene relative to plagioclase plus clinopyroxene in the parent material.

Introduction

MUCH OF THE CHEMICAL VARIATION observed in returned lunar rocks and in lunar orbital surveys (Adler *et al.*, 1973; Metzger *et al.*, 1973) is generally compatible with the range and type of variation expected from igneous differentiation processes. Furthermore, both geophysical and geochemical models of the moon's early history involve large-scale melting and igneous differentiation of its outer portions (Taylor and Jakeš, 1974; Toksöz *et al.*, 1973). An interesting chemical feature of many of the lunar rocks is their high concentration of REE (e.g. Sm) relative to chondritic abundances coupled with a pronounced depletion in Eu relative to Sm. A systematic approach to quantitative modeling of possible igneous differentiation paths to the observed abundances of Sm and Eu in lunar rocks is clearly desirable. Valuable attempts along these lines have been made by

Gast *et al.* (1970), Haskin *et al.* (1970), Philpotts and Schnetzler (1970), Hubbard and Gast (1971), Helmke *et al.* (1972), and Hubbard *et al.* (1973).

The evolution of Sm and Eu abundances during crystallization or melting is generally a function of such interrelated variables as temperature and oxygen potential ($T-f_{O_2}$ path), the distribution coefficients for Sm and Eu between the precipitating (or melting) solid phases and the magmatic liquid, and the relative proportions of the solid and liquid phases during the crystallization (or melting) sequence. It may never be possible to treat rigorously all of these variables in a general quantitative way when trying to reconstruct past igneous processes, but we believe that there are now sufficient data from various sources to warrant a systematic discussion of the more important factors.

CONDITIONS OF LUNAR CRYSTALLIZATION

Temperature–oxygen fugacity paths

Sato *et al.* (1973) have made direct measurements of oxygen fugacities over lunar samples between 1000°C and 1200°C. The range of their results is shown in Fig. 1. An approximate theoretical prediction of a representative lunar $T-f_{O_2}$ path may be obtained by considering that lunar magmas are saturated (or only slightly undersaturated) with respect to the ferromagnesian solid solutions, olivine and pyroxene, and to metallic iron during lunar igneous processes. To the extent that this approximation is valid, lunar $T-f_{O_2}$ is determined by the equilibrium

$$Mg_2SiO_4(SS) + Fe_2SiO_4(SS) = 2Fe(S) + 2MgSiO_3(SS) + O_2(G)$$

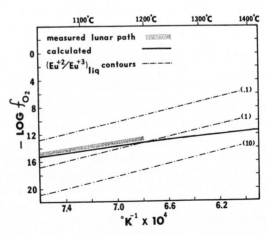

Fig. 1. Temperature and oxygen fugacity in lunar magmas. Stippled band from laboratory measurements of Sato *et al.* (1973). $T-f_{O_2}$ path and approximate Eu^{+2}/Eu^{+3} contours calculated as described in text.

and the oxygen fugacity over this system is given by

$$\log f_{O_2} = \frac{-\Delta G^\circ}{2.3\,RT} + 2[\log X_{fo} + \log (1 - X_{fo}) - \log X_{en}]$$

where $fo = Mg_2SiO_4$ and $en = MgSiO_3$. The activity of fo in olivine is approximated by X_{fo}^2 and that of en in pyroxene by X_{en}. We have considered a total range of magmatic temperature from 1700°K to 1300°K with $X_{fo} = X_{en}$ decreasing linearly with temperature from 0.9 to 0.3 in calculating the T–f_{O_2} plot shown in Fig. 1. All petrologically reasonable choices of $X_{fo}(T)$ and $X_{en}(T)$ result in similar curves with a total spread of no more than approximately one log unit in f_{O_2}. The theoretical approximations are in good agreement with the direct measurements of Sato et al. (1973) and we refer to the calculated curve in Fig. 1 in our subsequent discussions of lunar T–f_{O_2} paths.

Distribution coefficients

REE distribution coefficients (weight fraction in solid/weight fraction in liquid) for plagioclase have been experimentally determined by Drake (1972), Sun et al. (1974) and additional unpublished work in this laboratory. Where the experimental conditions overlap there is general agreement in the values of the distribution coefficients, but real differences persist which are most probably due to a dependence of the distribution coefficients on the bulk composition of the systems investigated. The values we have adopted for present purposes are shown in Fig. 2. No conclusions drawn in this paper are affected by varying these distribution coefficients within a factor of two.

Grutzeck et al. (1973) have measured REE distribution coefficients for clinopyroxene as a function of oxygen fugacity at 1265°C. Sun et al. (in press) experimentally investigated the distribution coefficient of Eu in clinopyroxene (10^{-8}–10^{-14} atmospheres of O_2) at 1150°C. The distribution coefficients for clinopyroxene shown in Fig. 2 are consistent with these two sets of data.

Sm and Eu distribution coefficients for olivine and orthopyroxene are known to be very small. Reconstructed values based on analyses of quenched natural rock systems indicate that D_{Eu}^{OL} and $D_{Sm}^{OL} \leqslant 0.01$ and D_{Eu}^{OPX} and $D_{Sm}^{OPX} \leqslant 0.08$ (Onuma et al., 1968; Schnetzler and Philpotts, 1968; Higuchi and Nagasawa, 1969; Schnetzler and Philpotts, 1970; Nagasawa and Schnetzler, 1971). These approximate maximum values are indicated in Fig. 2.

The distribution coefficient, D_{Eu}, is given to a first approximation by summing the contributions due to $Eu^{+2}(D_2)$ and $Eu^{+3}(D_3)$, weighted according to the relative amounts of these two ions present in the coexisting liquid (Weill and Drake, 1973).

$$D_{Eu} = D_2(Eu^{+2}) + D_3(Eu^{+3}) \tag{1}$$

where $(Eu^{+2}) + (Eu^{+3}) \equiv 1$. For magmatic melts saturated with plagioclase and/or clinopyroxene the Eu^{+2}/Eu^{+3} equilibrium conditions may be approximated by

$$\log K(T) = \log \frac{(Eu^{+2})}{(Eu^{+3})} f_{O_2}^{1/4} = \frac{C}{T} + A \tag{2}$$

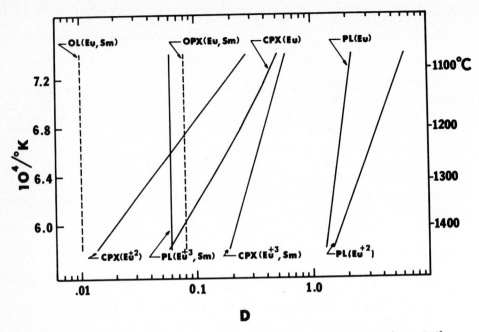

Fig. 2. Approximate distribution coefficients (solid/liquid) for Sm, Eu, Eu^{+2}, and Eu^{+3}. Temperature variation of D_{Eu} is drawn to correspond to $T-f_{O_2}$ path of Fig. 1. Sources discussed in text.

where C and A are constants over a restricted range of temperature and composition. Additional experimental work will be necessary to define more closely the compositional ranges over which C and A may be treated as constants (cf. Morris and Haskin, 1974). All experimental determinations of $D_{Eu}(T, f_{O_2})$ for plagioclase and clinopyroxene that we have made to date are compatible with approximate values of $C = -11,300°K^{-1}$ and $A = 4.4$, and we have adopted these values in our calculations. Equation (2) may be used to calculate the approximate oxidation state of Eu dissolved in magmatic melts, and we have plotted the resulting Eu^{+2}/Eu^{+3} contours in Fig. 1. Despite its necessarily approximate nature, Fig. 1 clearly shows that lunar $T-f_{O_2}$ paths involve a modest *decrease* of oxygen fugacity during crystallization accompanied by an *oxidation* of Eu dissolved in the melt. All processes that are dependent on Eu^{+2}/Eu^{+3} (e.g. liquid–solid partitioning) will be affected by $T-f_{O_2}$ changes during the course of lunar magmatism.

Equations (1) and (2) may be combined to obtain

$$D_{Eu}(T, f_{O_2}) = \frac{D_2 K + D_3 f_{O_2}^{1/4}}{K + f_{O_2}^{1/4}} \qquad (3)$$

which predicts that at constant temperature D_{Eu} approaches D_3 at high oxygen potential and varies along an S-shaped curve to approch D_2 at low oxygen potential. Plots of this model equation showing the contrasted behavior of

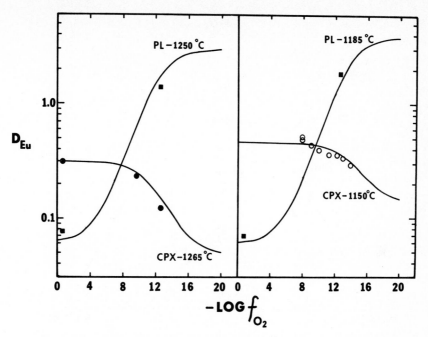

Fig. 3. Variation of D_{Eu} with oxygen fugacity. Curves are based on model Eq. (3) as described in text. Experimental data: solid circles (Grutzeck *et al.*, 1973); solid squares (Weill and Drake, 1973); open circles (Sun *et al.*, 1974).

plagioclase $(D_2 > D_3)$ and clinopyroxene $(D_2 < D_3)$ are shown in Fig. 3 along with experimental results. The values of D_{Eu}, D_2 and D_3 in Fig. 2 are compatible with Eqs. (1), (2), and (3) and with the calculated lunar igneous T–f_{O_2} path shown in Fig. 1.

Crystallization paths

If the moon accreted homogeneously and rapidly, then its large-scale chemical zoning is probably due to extensive melting and crystal–liquid fractionation at the time of formation, possibly altered by subsequent internal partial melting and surface impact processes. Many of the rocks collected at the lunar surface may have gone through several cycles of liquid–crystal fractionation. It therefore seems appropriate to investigate the effects of fractional crystallization on the Eu and Sm abundances of probable igneous materials.

Liquid–solid phase equilibria for many lunar rock compositions at low to modest pressure and strongly reducing conditions can be approximated by the pseudoternary $(Mg, Fe)_2 SiO_4$–SiO_2–$CaAl_2Si_2O_8$ liquidus diagram (Walker *et al.*, 1973). Such diagrams have obvious shortcomings in predicting crystallization or melting paths of actual lunar compositions. The most obvious of these is that they cannot convey any information about $Mg/(Mg + Fe)$ variations in olivine, pyroxene and liquid. They also assume that all the Ca is included in plagioclase and

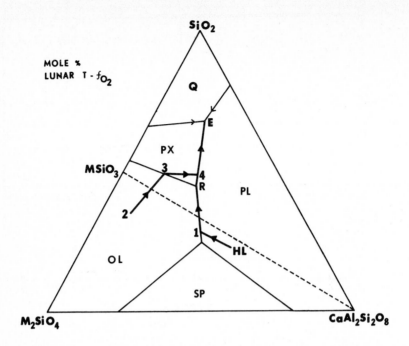

Fig. 4. Fractional crystallization paths on the pseudoternary liquidus surface for the system (Mg, Fe)$_2$ SiO$_4$–SiO$_2$–CaAl$_2$Si$_2$O$_8$. Adapted from Walker *et al.*, (1973) who are not responsible for its abuse in this paper.

therefore they cannot directly account for the crystallization of Ca-pyroxene. Nevertheless, they are very useful guides to general crystallization paths, and we will make use of Fig. 4 (taken from Fig. 6 in Walker *et al.*, 1973) in order to discuss the effect of contrasting crystallization paths on the evolution of Eu and Sm abundances.

In Fig. 4 we consider the fractional crystallization of two liquid compositions. Composition HL is a projection of the average lunar highland composition estimated by Taylor *et al.* (1973). It is also very similar to the average terra composition of Turkevich (1973). The present lunar highland surface composition is not necessarily significant as a parental liquid composition because it most probably represents a mixture of various igneous materials, further modified by impact processes. Nevertheless, it seems likely that if the moon has an anorthositic crust not far below the surface, then HL may be considered to represent an upper limit to the CaAl$_2$Si$_2$O$_8$ concentration in a potential global parental liquid shell. A more mafic possibility is typified by composition 2. The fractional crystallization paths for these two liquids are depicted in Fig. 5 in terms of the weight fraction of crystalline and liquid phases coexisting during the crystallization process.

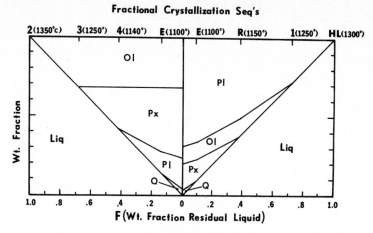

Fig. 5. Phase assemblages (weight fractions) during fractional crystallization sequences 2–3–4–E and HL–1–R–E. The figure is constructed from the phase assemblages at the reference points in Fig. 4. The crystallization of individual mineral phases is assumed to be linear with respect to F between the reference points. In order to convert from mole to weight fractions it is assumed that in Fig. 4 the mole fraction of Fe/(Fe + Mg) in the liquid varies linearly from 0.3 to 1 as the mole fraction of LIQ/(LIQ + OL + PX) decreases from 1 to 0 during fractional crystallization. Temperature estimates for the reference points are based on experimental phase equilibria reported in Walker *et al.* (1972 and 1973).

FRACTIONAL CRYSTALLIZATION

Composite distribution coefficients

At any given interval in a fractional crystallization sequence the composite solid–liquid distribution coefficient for a trace element depends on the distribution coefficients of the individual solid phases and the relative proportions of the solids precipitating during that interval. For example, in the crystallization interval 1–R (cf. Fig. 5), the composite distribution coefficient for Eu is given by $\bar{D}_{\mathrm{Eu}} = D_{\mathrm{Eu}}^{\mathrm{PL}} W^{\mathrm{PL}} + D_{\mathrm{Eu}}^{\mathrm{OL}} W^{\mathrm{OL}}$, where $D_{\mathrm{Eu}}^{\mathrm{PL}}$ and $D_{\mathrm{Eu}}^{\mathrm{OL}}$ are the temperature and oxygen fugacity dependent coefficients plotted on Fig. 2, and W^{PL} and W^{OL} are the weight fractions (normalized to a sum of one) of plagioclase and olivine precipitating during interval 1–R. The crystallization sequence shown in Fig. 5 has been combined with the temperature dependence of the individual distribution coefficients of Fig. 2 (assuming that temperature varies linearly with F between reference points) in order to plot the variation of the composite coefficients shown on Fig. 6. Several features are illustrated by this figure. The $\bar{D}(F)$ curves clearly reflect the contrasting influences of a crystallization sequence dominated by plagioclase and one dominated by the ferromagnesian minerals. The large discontinuities in the curves are caused by changes in the relative proportions of

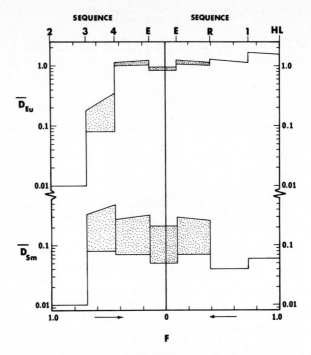

Fig. 6. Composite distribution coefficients during crystallization sequences 2–3–4–E and HL–1–R–E. Shaded areas correspond to intervals where pyroxene is crystallizing. Upper curve drawn for clinopyroxene and lower curve for orthopyroxene.

crystallizing phases encountered at multiple-phase saturation boundaries. Unfortunately, this effect is usually totally neglected in models for trace element differentiation. *A knowledge of the major phase equilibria is not incidental to modeling trace element behavior.* Smooth variations in $\bar{D}$ resulting from the effects of changes in temperature and oxygen fugacity on the individual mineral coefficients are important, but may in many cases be minor relative to the discontinuous variations resulting from changing mineral proportions. Uncertainties in $\bar{D}$ due to incomplete knowledge of the crystallization sequence may be at least as great as uncertainties due to inexact knowledge of the individual mineral distribution coefficients.

Finally, fundamentally different magma types, namely compositions 2 and HL in Fig. 4, will result in very different values of $\bar{D}$ in early differentiation but may converge on identical $\bar{D}$ values during later stages unless they are effectively separated by "thermal barriers" in the liquidus phase diagram.

Sm and Eu during fractional crystallization

The fractionation of Sm and Eu in the residual liquid during an interval of crystallization is given by

$$C/C_0 = F^{(\bar{D}-1)} \tag{4}$$

where C stands for concentration of the trace element in the liquid and F is the weight fraction of residual liquid remaining relative to the amount of liquid present at the start of the crystallization interval (where $C = C_0$). $\bar{D}$ is the composite distribution coefficient as previously defined. Equation (4) is valid only for constant $\bar{D}$. Consequently, in calculating the evolution of Sm and Eu over an extended range of fractional crystallization, it is necessary to divide the sequence into suitable stages such that $\bar{D}$ is approximately constant between successive stages. For example, the sequence between reference points 2 and 3 (Fig. 6) need not be further divided since $\bar{D}$ is constant, but the sequence between 3 and 4 must be subdivided to account for the variation in $\bar{D}$. In the general case, a crystallization sequence will be divided into stages at $F = F_0, F_1, F_2, F_3, \ldots, F_{n-1}$ and F_n such that $|\bar{D}_n - \bar{D}_{n-1}| \ll \bar{D}_n$. The concentration in the liquid at any stage n relative to the original bulk concentration is given by

$$C_n/C_0 = \prod_1^n (F_n/F_{n-1})^{[.5(\bar{D}_n + \bar{D}_{n-1})-1]}$$

The evolution of Eu and Sm abundances calculated for the two crystallization sequences are contrasted in Fig. 7. Enrichment in Sm progresses at approximately the same rate for the two sequences as illustrated by the nearly symmetric disposition of the two sets of Sm curves in Fig. 7. The differences in $\bar{D}_{Sm}$ shown in Fig. 6 are not sufficient to cause appreciably different trends. Enrichment factors of ten or more are reached with approximately 10% liquid remaining. The strong

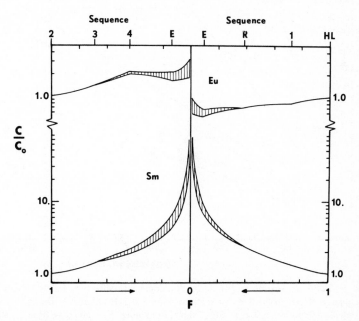

Fig. 7. Evolution of Eu and Sm concentrations during fractional crystallization. Hatched areas correspond to intervals where pyroxene is crystallizing. Upper curve drawn for orthopyroxene and lower curve for clinopyroxene.

influence of plagioclase may be seen in the curves for Eu evolution. The sequence HL–E, dominated by a total crystallization of nearly 70% plagioclase, shows an overall depletion in Eu. The initial Eu-richment trend in the ferromagnesian mineral dominated sequence, 2–E, is reversed when plagioclase saturation occurs at reference point 4. In a general way, the relative abundances of Sm and Eu in magmatic residua should be moderately sensitive indicators of the total amount of fractional crystallization and of the proportion of plagioclase involved.

REE enrichment and the Eu anomaly

The evolution of trivalent REE enrichment and the Eu anomaly is most easily seen in a plot of Sm versus Sm/Eu as shown in Fig. 8 where we have also indicated the trend defined by lunar KREEP-rich materials. The geochemical evidence indicates that the moon probably accreted as a homogeneous body (Duncan *et al.*, 1973; Brett, 1973) which was probably enriched in REE (including Eu) relative to chondritic abundances. Estimates of the REE enrichment of the whole moon range from 10× chondrites (Wänke *et al.*, 1973) to 2–5× chondrites (Taylor and Jakeš, 1974) with no Eu anomaly. With these starting compositions it is quite clear

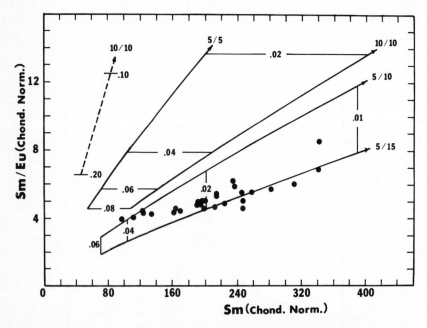

Fig. 8. Evolution of Sm and Sm/Eu during fractional crystallization. Dashed line is used for sequence HL–E; solid lines all refer to sequence 2–E (all curves drawn assuming pyroxene is orthopyroxene). Fraction of liquid remaining (F) is given as .01, .02, .04, etc. Composition of initial liquid is indicated by 10/10, 5/5, etc., where the first number signifies that Sm = 10× chondritic abundance and the second number applies to Eu. Plotted points are representative of KREEP-rich fragments returned from lunar surface.

that any fractional crystallization sequence involving appreciable plagioclase (e.g. HL–E) will not be capable of generating the abundances observed in REE-rich materials such as KREEP. All such sequences produce a very sharp rise in the Eu anomaly (Sm/Eu) relative to the overall REE enrichment (Sm), falling very wide of the KREEP trend. Even sequences such as 2–E, dominated by the crystallization of approximately 80% of ferromagnesian phases which reject Eu^{+2} effectively, produce fractionation trends with appreciably larger Sm/Eu than found in KREEP. Starting with an initial liquid of 10× chondrites and no Eu anomaly (liquid 10/10 in Fig. 8), it is possible to generate KREEP characteristics only with crystallization sequences that involve even less plagioclase than in sequence 2–E. The effect of starting with initial liquids that have inherited different abundances relative to chondrites is indicated in Fig. 8. Initial liquids with negative Eu anomalies and/or lower enrichments relative to chondrites (e.g. liquid 5/5 in Fig. 8) all result in trends even further removed from KREEP. Liquids with inherited positive Eu anomalies (e.g. liquid 5/15 in Fig. 8) could produce a trend approximating that of KREEP with a crystallization sequence like 2–E, but it is very difficult to see how the moon could have inherited such a composition or how subsequent igneous processes could have generated it without involving the addition of plagioclase. Bulk compositions rich in plagioclase component would yield crystallization sequences resembling HL–E rather than 2–E. In addition, it is apparent from Fig. 8 that such sequences will reproduce the KREEP trend only after extreme fractionation. While it is not unreasonable to expect such liquids to produce veins, dikes or sills in the lunar crust, the fractionation would also produce high Fe/Mg ratios not found in KREEP-rich rocks. The origin of KREEP-rich rocks via any reasonable sequence of fractional crystallization processes seems to be ruled out. Furthermore, any highly differentiated residual liquids produced by fractional crystallization would eventually solidify into accessory minerals and mesostasis. Subsequent remelting of these phases in toto would not get around the problem of too high Sm/Eu relative to Sm.

PARTIAL MELTING

General considerations

Having ruled out fractional crystallization processes for generating KREEP-rich lunar materials directly from a primordial liquid shell we now turn to an analysis of partial melting, restricting ourselves to the case of bulk equilibrium. The simple mass balance equation for a trace element in a partially molten system in equilibrium is

$$C/C_0 = [F + \bar{D}(1 - F)]^{-1} \tag{5}$$

where C is the concentration of the trace element in the partial melt, C_0 is the concentration in the bulk or total system, F is the fraction of melt, and the composite distribution coefficient, $\bar{D}$, is the sum of contributions from the individual mineral coefficients weighted according to their weight fraction in the refractory residue. As before, we note that although the distribution coefficients for the individual minerals depend on temperature and oxygen fugacity, the dominant factors determining $\bar{D}$ are the proportions of the minerals in the refractory assemblage. If we can assume that the refractory assemblage is composed primarily of ferromagnesian minerals and plagioclase, the partial melting can be considered graphically by means of Fig. 9. Assuming an approximate temperature and oxygen fugacity, the phase proportion triangle can be contoured for $\bar{D}_{Sm}$ and $\bar{D}_{Eu}$. Each point in the triangle (i.e. each refractory assemblage) uniquely defines a combination of $\bar{D}_{Sm}$ and $\bar{D}_{Eu}$ values. According to Eq. (5), any combination of Sm and Eu concentrations in the partial melt corresponds to a unique set of $\bar{D}_{Sm}$ and $\bar{D}_{Eu}$ values at a specified degree of partial melting (F) and initial concentration (C_0). Consequently, each point on the KREEP trend in Fig. 8 has a corresponding point on Fig. 9 that represents the refractory assemblage left behind after partial melting. For modest degrees of partial melting, the refractory assemblage and the parent assemblage are similar.

Derivation of KREEP-rich liquids

In Fig. 9 we have plotted the trend of refractory assemblages corresponding to the trend of KREEP abundances in Fig. 8 for $F = 0$ (incipient melting) and assuming initial abundances of $10\times$ chondrites for Sm and Eu (10/10). All the

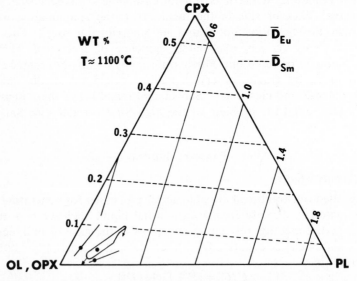

Fig. 9. D'Agostino plot of $\bar{D}_{Eu}$ and $\bar{D}_{Sm}$ for CPX + PL + (OL + OPX) refractory mineral assemblages. Olivine and orthopyroxene are treated as a single "phase" with a "low" distribution coefficient. Envelope corresponds to refractory assemblages equilibrated with KREEP incipient ($F = 0$) melts (cf. Fig. 8) assuming bulk abundances 10/10. Arrows indicate shift of position for envelope at $F = .04$. Point represents average of KREEP materials. Trend of parental assemblages for KREEP incipient melts also shown for bulk abundances 10/5 (upper line) and 5/10 (lower line).

individual KREEP points are concentrated in the OL + OPX corner, and to avoid overcrowding they are represented by an envelope and a single point for the average. The arrows indicate the effect of increasing the postulated degree of partial melting. The trends of mineral assemblages capable of producing incipient melts with KREEP concentrations of Sm and Sm/Eu for postulated initial bulk concentrations of 10/5 and 5/10 are also indicated in the figure by a line (representing the long axis of the envelope) and point (representing average KREEP). All such parental materials are predominantly composed of olivine and/or orthopyroxene. Of course, the initial concentration of Sm and Eu postulated for the parental material need not be restricted to the range described above. It is conceivable that earlier igneous differentiation had seriously altered the original lunar abundances so that KREEP parental materials could have inherited rather different abundances of Sm and Eu. In order to evaluate this effect without overcrowding Fig. 9 we have tabulated various refractory assemblages equilibrated with an average KREEP melt for various combinations of assumed starting abundances (Table 1). Only those parental compositions with large overall enrichments of both Sm and Eu and/or initial positive Eu anomalies are compatible with substantial amounts of plagioclase relative to olivine and orthopyroxene. A possible way of generating such concentrations from whole moon abundances is by means of fractional crystallization. Our previous analysis of fractional crystallization indicated that it was possible to achieve enrichment factors (C/C_0) of approximately two for both Sm and Eu in residual liquids after 50% crystallization of ferromagnesian minerals in a crystallization sequence such as 2–E. Further crystallization in this system involves plagioclase and a rapid development of a negative Eu anomaly. Plagioclase dominated sequences such as HL–E develop large negative Eu anomalies right away. The only systems that are capable of generating large enrichment factors via fractional crystallization without developing a negative Eu anomaly must be even richer in the ferromagnesian constituents than composition 2. Fractional crystallization of clinopyrox-

Table 1. Refractory phase assemblages in equilibrium with an incipient melt ($F = 0$) of average KREEP at 1100°C for various combinations of starting abundances of Sm and Eu. Phase assemblages are given in the order (OL + OPX):PL:CPX (wt.%). Sm and Eu are given relative to chondritic abundances.

Sm \ Eu	5×	10×	20×	40×
5×	93:5:2	87:12:1	75:25:0	—
10×	89:4:7	83:11:6	71:24:5	48:50:2
20×	82:2:16	76:9:15	64:22:14	41:48:11
40×	—	62:5:33	50:18:32	26:44:30

ene can introduce a positive Eu anomaly in the residual liquid, but a very large amount of clinopyroxene crystallization is required for a modest positive Eu anomaly. We conclude that either the parental phase assemblage for KREEP-rich partial melts was predominantly composed of olivine and orthopyroxene or, if it contained appreciable plagioclase, it was originally derived by fractional crystallization of a system that itself contained limited amounts of plagioclase constituent. Either way, an analysis of the derivation of KREEP-rich materials by igneous processes indicates *limited concentrations of plagioclase component* (relative to olivine and orthopyroxene) in the moon's primordial liquid shell. If so, the fact that plagioclase is so abundant in returned lunar samples lends further support to the idea of a zoned moon with plagioclase concentrated near the surface.

Regardless of the initial concentrations assumed for the parental material or of the degree of partial melting assumed, it is generally true that all the trend lines for KREEP-producing assemblages on the D'agostino plot (Fig. 9) are approximately radial to the OL + OPX corner. This suggests that the range of KREEP-rich materials (cf. Fig. 8) could have been generated by a relatively constant degree of partial melting of parent material with variable olivine plus orthopyroxene content. This is a result imposed directly by the relative values of the distribution coefficients and is distinct from the proposition that the range of REE enrichment seen in KREEP can be attributed to differing degrees of partial melting. The two statements are equivalent only if the variation of olivine plus orthopyroxene content calculated for the refractory assemblages is itself considered to be the result of the partial melting process. There is not enough detailed information about melting in such systems to comment further on this possibility. A simple alternative interpretation is that the KREEP sequence of Sm and Sm/Eu values was generated by approximately equivalent degrees of partial melting of parent assemblages with variable contents of olivine plus orthopyroxene relative to plagioclase plus clinopyroxene.

The derivation of KREEP-rich material by partial melting of lunar rocks, leaving behind a refractory residue of olivine, orthopyroxene, clinopyroxene, and plagioclase, is possible. The most likely source is either (1) a rock rich in olivine and orthopyroxene (>80%) relative to plagioclase and clinopyroxene that inherited Sm and Eu abundances not greatly different from those estimated for the whole moon, or (2) a rock with larger amounts of plagioclase (>20%) that was appreciably enriched relative to the whole moon in both Sm and Eu and did not have a negative Eu anomaly (cf. Hubbard and Gast, 1971). Such a source rock could have originated by fractional crystallization of an initial liquid poor in plagioclase constituent.

Plagioclase is the only common lunar mineral known to impart a strong negative Eu anomaly signature on coexisting liquids, and its presence is certainly necessary for the derivation of KREEP-rich materials, notable for their large negative Eu anomalies. Ironically, a systematic analysis of the possibilities for generating KREEP liquids seems to indicate a modest upper limit to the amount of plagioclase involved in the overall process.

Acknowledgments—Our research was supported by NASA grants NGL 38-003-020 and NGL 38-003-022. Helpful comments by F. Frey, G. Goles, N. Hubbard, D. Lindstrom, and R. Schmitt were much appreciated. Readers familiar with the work of the late Paul Gast will realize that we have derived much inspiration from it.

REFERENCES

Adler I., Trombka J. I., Schmadebeck R., Lowman P., Blodgett H., Yin L., and Eller E. (1973) Results of the Apollo 15 and 16 X-ray experiment. *Proc. Fourth Lunar Sci. Conf., Geochim. Cosmochim. Acta*, Suppl. 4, Vol. 3, pp. 2783–2791. Pergamon.

Brett R. (1973) The lunar crust: A product of the heterogeneous accretion or differentiation of a homogeneous Moon. *Geochim. Cosmochim. Acta* 37, 2697–2703.

Drake M. (1972) The distribution of major and trace elements between plagioclase feldspar and magmatic silicate liquid: An experimental study. Ph.D. thesis, University of Oregon.

Duncan A. R., Erlank A. J., Willis J. P., and Ahrens L. H. (1973) Composition and inter-relationships of some Apollo 16 samples. *Proc. Fourth Lunar Sci. Conf., Geochim. Cosmochim. Acta*, Suppl. 4, Vol. 2, pp. 1097–1113. Pergamon.

Gast P. W., Hubbard N. J., and Wiesmann H. (1970) Chemical composition and petrogenesis of basalts from Tranquility Base. *Proc. Apollo 11 Lunar Sci. Conf., Geochim. Cosmochim. Acta*. Supp. 1. Vol. 2, 1143–1163. Pergamon.

Grutzeck M., Kridelbaugh S., and Weill D. (1973) REE partitioning between diposide and silicate liquid (abstract). *Trans. Amer. Geophys. Union* 54, 1222.

Haskin L. A., Allen R. O., Helmke P. A., Paster T. P., Anderson M. R., Korotev R. L., and Zweifel K. A. (1970) Rare earths and other trace elements in Apollo 11 lunar samples. *Proc. Apollo 11 Lunar Sci. Conf., Geochim. Cosmochim. Acta*, Suppl. 1, Vol. 2, 1213–1231. Pergamon.

Helmke P. A., Haskin L. A., Korotev R. L., and Ziege K. E. (1972) Rare earths and other trace elements in Apollo 14 samples. *Proc. Third Lunar Sci. Conf., Geochim. Cosmochim. Acta*, Suppl. 3, Vol. 2, pp. 1275–1292. MIT Press.

Higuchi H. and Nagasawa H. (1969) Partition of trace elements between rock forming minerals and the host volcanic rocks. *Earth Planet. Sci. Lett.* 7, 281–287.

Hubbard N. J. and Gast P. W. (1971) Chemical composition and origin of nonmare lunar basalts. *Proc. Second Lunar Sci. Conf., Geochim. Cosmochim. Acta*, Suppl. 2, Vol. 2, pp. 999–1020. MIT Press.

Hubbard N. J., Rhodes J. M., Gast P. W., Bansal B. M., Shih C. Y., Wiesmann H., and Nyquist L. E. (1973) Lunar rock types: The role of plagioclase in non-mare and highland rock types. *Proc. Fourth Lunar Sci. Conf., Geochim. Cosmochim. Acta*, Suppl. 4, Vol. 2, pp. 1297–1312. Pergamon.

Metzger A. E., Trombka J. I., Peterson L. E., Reedy R. C., and Arnold J. R. (1973) Lunar surface radioactivity: Preliminary results of the Apollo 15 and Apollo 16 gamma-ray spectrometer experiments. *Science* 179, 800–803.

Morris R. V. and Haskin L. A. (1974) EPR measurement of the effect of glass composition on the oxidation states of europium. *Geochim. Cosmochim. Acta*. In Press.

Nagasawa H. and Schnetzler C. (1971) Partitioning of rare earth, alkali, and alkaline earth elements between phenocrysts and acidic igneous magma. *Geochim. Cosmochim Acta* 35, 953–968.

Onuma N., Higuchi H., Wakita H., and Nagasawa H. (1968) Trace element partitioning between two pyroxenes and the host lava. *Earth Planet. Sci. Lett.* 5, 44–51.

Philpotts J. A. and Schnetzler C. C. (1970) Apollo 11 lunar samples: K, Rb, Sr, Ba and rare-earth concentrations in some rocks and separated phases. *Proc. Apollo 11 Lunar Sci. Conf., Geochim. Cosmochim. Acta*, Suppl. 1, Vol. 2, pp. 1471–1486. Pergamon.

Sato M., Hicklin N. L., McLane J. E. (1973) Oxygen fugacity values of Apollo 12, 14 and 15 lunar samples and reduced state of lunar magmas. *Proc. Fourth Lunar Sci. Conf., Geochim. Cosmochim. Acta*, Suppl. 4, Vol. 1, pp. 1061–1079. Pergamon.

Schnetzler C. C. and Philpotts J. A. (1968) Partition coefficients of rare-earth elements and barium between igneous matrix material and rock-forming phenocyrsts—I. In *Origin and Distribution of the Elements* (editor L. H. Ahrens), pp. 929–938. Pergamon.

Schnetzler C. C. and Philpotts J. A. (1970) Partition coefficients of rare-earth elements between igneous matrix material and rock-forming mineral phenocrysts—II. *Geochim. Cosmochim. Acta* **34**, 331–340.

Sun C.-O., Williams R. J., and Sun S.-S. (1974) Distribution coefficients of Eu and Sr for plagioclase–liquid and clinopyroxene–liquid equilibria in oceanic ridge basalt: An experimental study. *Geochim. Cosmochim. Acta.* In press.

Taylor S. R., Gorton M. P., Muir P., Nance W., Rudowski R., and Ware N. (1973) Lunar highlands composition: Apennine Front. *Proc. Fourth Lunar Sci. Conf., Geochim. Cosmochim. Acta*, Suppl. 4, Vol. 2, pp. 1445–1459. Pergamon.

Taylor S. R. and Jakeš P. (1974) Geochemical zoning in the moon (abstract). In *Lunar Science—V*, Part II, pp. 786–788. The Lunar Science Institute, Houston.

Toksöz M. N., Dainty A. M., Solomon S. C., and Anderson K. R. (1973) Velocity structure and evolution of the moon. *Proc. Fourth Lunar Sci. Conf., Geochim. Cosmochim. Acta*, Suppl. 4, Vol. 3, pp. 2529–2547. Pergamon.

Turkevich A. L. (1973) The average chemical composition of the lunar surface. *Proc. Fourth Lunar Sci. Conf., Geochim. Cosmochim. Acta*, Suppl. 4, Vol. 2, pp. 1159–1168. Pergamon.

Walker D., Longhi J., and Hays J. F. (1972) Experimental petrology and origin of Fra Mauro rocks and soil. *Proc. Third Lunar Sci. Conf., Geochim. Cosmochim. Acta*, Suppl. 3, Vol. 1, pp. 797–817. MIT Press.

Walker D., Longhi J., Grove T. L., Stopler E., and Hays J. F. (1973) Experimental petrology and origin of rocks from the Deschutes Highlands. *Proc. Fourth Lunar Sci. Conf., Geochim. Cosmochim. Acta*, Suppl. 4, Vol. 1, pp. 1013–1032. Pergamon.

Wänke H., Baddenhausen H., Dreibus G., Jacoutz E., Kruse H., Palme H., Spettel B., and Teschke F. (1973) Multi-element analyses of Apollo 15, 16, and 17 samples and the bulk composition of the moon. *Proc. Fourth Lunar Sci. Conf., Geochim. Cosmochim. Acta*, Suppl. 4, Vol. 2, pp. 1461–1481. Pergamon.

Weill D. and Drake M. (1973) Europium anomaly in plagioclase feldspar: Experimental results and semiquantitative model. *Science* **180**, 1059–1060.

Proceedings of the Fifth Lunar Conference
(Supplement 5, Geochimica et Cosmochimica Acta)
Vol. 2 pp. 1353–1373 (1974)
Printed in the United States of America

^{40}Ar–^{39}Ar studies of lunar breccias

E. Calvin Alexander, Jr.* and S. B. Kahl

Physics Department, University of California, Berkeley, California 94720

Abstract—Ar isotopic data obtained from stepwise heating experiments on neutron irradiated lunar samples 14270, 15362, 15455, 63335, 14301, and 14313 are presented. ^{40}Ar–^{39}Ar gas retention ages are 3.89 ± 0.05 b.y. for 14270 and 3.92 ± 0.04 b.y. for 15455. Lower limits of 3.98 b.y. and 3.64 b.y. are set on the crystallization ages of 15362 and 63335, respectively. ^{38}Ar exposure ages are 244 ± 25 m.y. for 14270, 428 ± 43 m.y. for 15362, 205 ± 21 m.y. for 15455, and 41 ± 8 m.y. for 63335. Evidence is presented that 15455 received its cosmic ray exposure in a more shielded location than 14270.

14301 and 14313 were never isotopically homogenized and therefore do not yield ages. However, both samples contain large quantities of trapped gas which are apparently composed of at least two components. The component released at lower temperatures is enriched in ^{40}Ar relative to the component released at higher temperatures.

Introduction

The ^{40}Ar–^{39}Ar method (Sigurgeirsson, 1962; Merrihue, 1965; Merrihue and Turner, 1966; Turner, 1970) has proven very useful to lunar chronologists. This work is a continuation of the ^{40}Ar–^{39}Ar dating program in Berkeley (Davis *et al.*, 1971; Alexander *et al.*, 1972a; Alexander *et al.*, 1972b; Reynolds *et al.*, 1974; Alexander and Davis, 1974). We report here ^{40}Ar–^{39}Ar data from three Apollo 14 samples, two Apollo 15 samples, and one Apollo 16 sample. All of the samples except one are breccias.

Lunar breccias are an important but enigmatic portion of the returned lunar samples. They account for less than 10% of the rocks collected at the Apollo 12 site but completely dominate the Apollo 14 and 16 collections. They range in coherence from the "clods" of Apollo 14 to hard, strongly recrystallized breccias which grade into igneous textures. Some breccias record four generations of brecciation within a single hand specimen. The lunar breccias are extremely heterogeneous as a group and as individuals.

The chronologist, therefore, is faced with two fundamental questions which must be answered for each individual lunar breccia. First, was there any event in the history of the breccia which completely reset the radioisotope clocks, i.e. is the breccia datable in the conventional sense? Second, if the radioisotope clocks were reset, what precisely was the event recorded? A number of workers have shown that the radioisotope clocks in many of the more highly metamorphosed breccias were reset and that the event recorded is probably the excavation of the

---

*Present address: Geology and Geophysics Department, University of Minnesota, Minneapolis, Minnesota 55455.

major circular basins (cf. Turner *et al.*, 1971, 1972; Papanastassiou and Wasser-burg, 1971). However, many breccias did not completely equilibrate during the final assemblage event particularly in the Rb–Sr system (cf. Albee *et al.*, 1970; Mark *et al.*, 1973).

Samples

14270,1-7

14270 is a 25.59 g breccia chip collected as part of the comprehensive Apollo 14 sampling. It is a group 7 breccia on Warner's (1972) scale and is being analyzed by a consortium headed by G. Goles. Compositional data for 14270 are given by Lindstrom *et al.* (1972).

14301,110

14301 is a 1.35 kg breccia collected at Station G1. It is a group 2 breccia on Warner's (1972) scale and is one of three Apollo 14 breccias described by King *et al.* (1972) as containing "abundant chondrules." These three breccias have also been shown to contain fission Xe from the decay of ^{244}Pu (Drozd *et al.*, 1972; Behrmann *et al.*, 1973; Reynolds *et al.*, 1974).

14313,77

14313 is a 144 g breccia collected at Station G1. It is a group 1 breccia on Warner's (1972) scale and has been described by Floran *et al.* (1972). 14313, like 14301, contains chondrules and evidence of ^{244}Pu fission Xe.

15362,3

15362 is a 4.2 g anorthosite rake sample collected at Spur Crater, Station 7. Laul *et al.* (1972) report compositional data for 15362.

15455

15455 is the "black-and-white breccia" being analyzed by a consortium headed by L. Silver. It is a 937 g breccia collected at Spur Crater, Station 7. We have analyzed splits of homogenized samples of the dark phase, 15455,183 and light phase, 15455,70A. Silver (private communication) has described the light phase as a leucogabbro and indicates that in thin section the dark phase can be seen to intrude the light phase. Ganapathy *et al.* (1973) report trace element concentrations in splits of the same homogenized samples.

63335,3

63335 is a 65.4 g breccia collected at Station 13 near North Ray Crater. It is a B5 breccia in the Wilshire *et al.* (1973) classification.

Experimental

Samples 14270, 15455, 15363, and 63335 were irradiated together in one layer of an irradiation designated Val. 6 and 14301 and 14313 were irradiated in one layer of a separate irradiation designated Val. 8. Both irradiations were performed in the shuttle tube facility of the General Electric Test

Reactor, Vallecitos Nuclear Center, Pleasanton, California. The experimental method is described in Alexander and Davis (1974) and references listed therein. In the Val. 8 irradiation, high-purity Ni wires were used to monitor possible fluence variations within the irradiation container, via the ^{58}Ni (n, p) ^{58}Co reaction, instead of the Co-doped Al wires previously used.

St. Severin monitor

Table 1 listed a summary all of the data we have accumulated from our St. Severin monitor. All of the listed ^{40*}Ar/^{39*}Ar ratios are based on the >800°C fraction(s) except V501 which is based on the >700°C fractions. The reproducibility of ^{40*}Ar/^{39*}Ar ratios in replicate analyses of monitors from a single irradiation is limited by three factors: (1) the homogeneity of the appropriate neutron fluence within the irradiation container, (2) the homogeneity of the ^{40*}Ar/^{40}K ratio in replicate splits of the monitor, and (3) the instrumental precision associated with the measurement of a given Ar sample. In order to separate these three factors, starting with the Val. 6 samples, we split each sample of the monitor *after* the irradiation. The intra-sample comparisons are a measure of the sum of factors 2 and 3 above while the intersample comparisons are the sum of all three factors.

Table 1. St. Severin monitor data.

Sample	Date Analyzed	^{40*}Ar/^{39*}Ar	‰dev. from average	J†
V409	30/8/71	419.6 ± 2.8		0.02388
V411	30/6/71	410.1 ± 2.1	11.45	±0.00049
V501	4/6/72	424.8 ± 3.4		0.02335
V507	16/9/71	424.1 ± 2.0	0.82	±0.00029
V606.1	18/3/73	202.68 ± 0.49		0.04887
V606.2	19/3/73	202.77 ± 0.58	0.22	±0.00058
V612.1	25/1/73	197.22 ± 0.70		0.05024
V612.2	25/1/73	197.15 ± 0.50	0.18	±0.00060
V618.1	26/3/73	204.89 ± 0.55		0.04841
V618.2	26/3/73	204.42 ± 0.54	1.15	±0.00058
V801.1	24/8/73	229.47 ± 0.50		0.04316
V801.2	24/8/73	229.59 ± 0.58	0.26	±0.00051
V807.1	20/11/73	231.69 ± 1.89		0.04277
V807.2	5/2/74	231.64 ± 0.96	0.11	±0.00052

†$J = (e^{\lambda Tm} - 1)/^{40*}$Ar/^{39*}Ar, where $\lambda = 5.305 \times 10^{-10}$ yr^{-1} and $Tm = 4.504 \pm 0.020 \times 10^9$ yr (Alexander and Davis, 1974).

The maximum deviation in ^{40*}Ar/^{39*}Ar observed in five intra-sample comparisons from Val. 6 and Val. 8 is 1.15 per mil and the average deviation is 0.4 per mil. Evidently the >800°C ^{40*}Ar/^{39*}Ar ratios in St. Severin are reproducible to about 1 per mil. The 2% inter-sample spread in the Val. 4 samples (Alexander *et al.*, 1972b) is evidently due to fluence inhomogeneity. Note that while most of the pairs were analyzed on the same day, V501 and V507 were analyzed $8\frac{1}{2}$ months apart and V807.1 and V807.2 were analyzed $2\frac{1}{2}$ months apart. These data indicate the long-term reproducibility of the Ar ratio measurements.

V606 is the monitor for the layer containing the Val. 6 lunar samples and V801 is the monitor for 14301 and 14313.

Data and Results

Tables 2 and 3 list the argon data from Val. 6 and Val. 8, respectively. The measured isotope ratios for each temperature fraction were corrected in the following fashion to arrive at the tabulated data. (1) The measured ratios,

Table 2. Argon from neutron irradiated lunar samples.

Temp. °C	^{36}Ar	^{37}Ar	^{38}Ar	^{39}Ar	^{40}Ar
			in units of 10^{-8} cm^3 STP/g		
		14270,1–7		**113.1 mg**	
300	0.720 ± 0.012	7.56 ± 0.22	2.269 ± 0.014	6.490 ± 0.014	199.0 ± 3.1
400	4.154 ± 0.027	21.05 ± 0.38	1.495 ± 0.020	15.604 ± 0.041	1506.7 ± 3.1
500	8.09 ± 0.13	35.71 ± 0.92	2.422 ± 0.049	24.66 ± 0.13	3718.7 ± 3.1
600	18.19 ± 0.11	94.7 ± 1.1	5.471 ± 0.037	45.72 ± 0.14	6786.2 ± 3.1
700	18.61 ± 0.11	205.6 ± 2.2	7.313 ± 0.037	46.72 ± 0.14	6590.8 ± 3.1
800	7.218 ± 0.042	209.9 ± 2.4	5.008 ± 0.022	30.516 ± 0.075	4370.9 ± 3.1
900	4.678 ± 0.033	214.0 ± 2.3	4.516 ± 0.036	19.100 ± 0.064	2712.3 ± 3.1
1000	4.063 ± 0.052	135.4 ± 1.7	3.308 ± 0.045	10.580 ± 0.035	1357.5 ± 3.1
1200	9.301 ± 0.059	513.6 ± 5.6	12.730 ± 0.047	18.465 ± 0.063	2079.5 ± 3.1
1680	1.438 ± 0.019	114.7 ± 1.2	2.132 ± 0.014	3.082 ± 0.014	421.5 ± 3.1
Total	76.46 ± 0.23	1552.2 ± 7.3	46.66 ± 0.11	220.94 ± 0.27	$29{,}743.1 \pm 9.8$
		15362,3		**125.7 mg**	
300	0.1643 ± 0.0097	9.83 ± 0.36	0.2530 ± 0.0025	—	5.9 ± 1.6
500	2.686 ± 0.023	120.9 ± 2.1	3.429 ± 0.036	0.230 ± 0.014	24.1 ± 1.6
700	21.87 ± 0.37	942 ± 12	29.82 ± 0.15	1.076 ± 0.019	146.0 ± 1.6
800	11.99 ± 0.12	579.4 ± 7.7	18.32 ± 0.11	0.631 ± 0.011	92.9 ± 1.6
900	6.093 ± 0.042	289.3 ± 3.6	9.208 ± 0.047	0.3215 ± 0.0097	46.1 ± 1.6
1000	5.029 ± 0.032	239.8 ± 2.9	7.591 ± 0.035	0.2491 ± 0.0092	38.2 ± 1.6
1100	3.727 ± 0.028	178.6 ± 2.1	5.677 ± 0.024	0.1906 ± 0.0070	29.0 ± 1.6
1300	5.552 ± 0.058	269.6 ± 3.5	8.549 ± 0.066	0.2778 ± 0.0056	43.3 ± 1.6
1700	8.876 ± 0.063	435.7 ± 5.6	13.76 ± 0.10	0.4403 ± 0.0058	73.1 ± 1.6
Total	65.99 ± 0.40	3065 ± 17	96.61 ± 0.23	3.416 ± 0.031	498.6 ± 4.8
		15455,70A		**147.4 mg**	
300	0.2160 ± 0.0096	23.48 ± 0.41	0.374 ± 0.010	0.4092 ± 0.0026	34.1 ± 1.4
400	0.9075 ± 0.0091	85.9 ± 1.0	1.3165 ± 0.0070	1.2626 ± 0.0085	127.5 ± 1.4
500	1.217 ± 0.013	107.6 ± 1.3	1.733 ± 0.014	1.424 ± 0.013	170.4 ± 1.4
600	2.750 ± 0.019	235.8 ± 2.6	3.806 ± 0.013	2.8593 ± 0.0056	378.4 ± 1.4
700	3.876 ± 0.024	357.8 ± 4.0	5.715 ± 0.019	3.995 ± 0.013	562.1 ± 1.4
800	0.7996 ± 0.0088	77.08 ± 0.89	1.2165 ± 0.0058	0.8576 ± 0.0050	123.5 ± 1.4
900	0.7308 ± 0.0082	68.92 ± 0.86	1.1001 ± 0.0093	0.8108 ± 0.0067	112.3 ± 1.4
1100	1.930 ± 0.018	182.7 ± 2.2	2.885 ± 0.021	1.9783 ± 0.0068	251.8 ± 1.4
1300	2.020 ± 0.014	185.8 ± 2.2	3.071 ± 0.016	1.5948 ± 0.0092	226.4 ± 1.4
1650	0.552 ± 0.010	51.54 ± 0.66	0.8599 ± 0.0066	0.5201 ± 0.0071	76.4 ± 1.4
Total	14.999 ± 0.045	1376.6 ± 6.1	22.077 ± 0.042	15.712 ± 0.026	2062.9 ± 4.4

Table 2. (*Continued*).

Temp. °C	^{36}Ar	^{37}Ar	^{38}Ar	^{39}Ar	^{40}Ar
			in units of 10^{-8} cm^3 STP/g		
		15455,183		150.7 mg	
300	0.1215 ± 0.0098	7.27 ± 0.27	0.2129 ± 0.0051	1.737 ± 0.010	143.2 ± 1.4
400	0.3427 ± 0.0061	23.88 ± 0.36	0.4950 ± 0.0084	3.086 ± 0.011	343.6 ± 1.4
500	0.620 ± 0.019	46.75 ± 0.89	0.851 ± 0.013	3.233 ± 0.017	436.0 ± 1.4
600	1.454 ± 0.012	111.2 ± 1.3	2.010 ± 0.018	5.358 ± 0.017	769.9 ± 1.4
700	1.549 ± 0.012	130.8 ± 1.5	2.316 ± 0.012	5.1667 ± 0.0090	775.4 ± 1.4
800	1.642 ± 0.018	146.4 ± 1.7	2.572 ± 0.015	5.273 ± 0.011	806.7 ± 1.4
900	1.991 ± 0.023	178.2 ± 2.0	3.182 ± 0.018	6.108 ± 0.017	908.9 ± 1.4
1000	0.887 ± 0.011	73.95 ± 0.91	1.4139 ± 0.0097	2.275 ± 0.014	305.3 ± 1.4
1100	3.219 ± 0.021	263.9 ± 2.9	4.895 ± 0.021	4.980 ± 0.020	646.7 ± 1.4
1300	3.369 ± 0.023	268.0 ± 2.9	5.070 ± 0.020	5.512 ± 0.017	784.5 ± 1.4
1670	0.3003 ± 0.0065	21.94 ± 0.35	0.426 ± 0.010	0.4328 ± 0.0086	64.4 ± 1.4
Total	15.496 ± 0.053	1272.3 ± 5.4	23.444 ± 0.048	43.162 ± 0.047	5984.6 ± 4.6
		63335,3		187.4 mg	
300	0.032 ± 0.011	1.93 ± 0.25	0.0326 ± 0.0038	0.0831 ± 0.0035	4.4 ± 2.9
400	0.052 ± 0.020	5.584 ± 0.089	0.058 ± 0.018	0.189 ± 0.031	2.3 ± 2.9
500	0.203 ± 0.011	14.03 ± 0.33	0.0832 ± 0.0048	0.1908 ± 0.0063	< 2.9
600	0.565 ± 0.014	29.06 ± 0.39	0.1936 ± 0.0048	0.2803 ± 0.0075	3.9 ± 2.9
700	0.421 ± 0.013	36.44 ± 0.43	0.1902 ± 0.0038	0.2788 ± 0.0072	2.8 ± 2.9
800	0.200 ± 0.010	55.19 ± 0.60	0.1967 ± 0.0066	0.3535 ± 0.0097	2.9 ± 2.9
900	0.150 ± 0.010	59.70 ± 0.67	0.2167 ± 0.0069	0.4294 ± 0.0030	4.1 ± 2.9
1000	0.205 ± 0.016	99.1 ± 1.1	0.3578 ± 0.0054	0.7857 ± 0.0063	7.6 ± 2.9
1100	0.324 ± 0.013	193.9 ± 2.1	0.7579 ± 0.0079	1.5168 ± 0.0084	45.2 ± 2.9
1200	0.451 ± 0.014	281.4 ± 3.0	1.111 ± 0.011	2.046 ± 0.012	140.7 ± 2.9
1300	0.314 ± 0.012	189.6 ± 2.1	0.7465 ± 0.0081	1.372 ± 0.012	119.8 ± 2.9
1400	0.758 ± 0.018	454.5 ± 5.4	1.799 ± 0.016	3.829 ± 0.018	363.9 ± 2.9
1680	1.309 ± 0.021	864.6 ± 9.4	3.430 ± 0.017	6.768 ± 0.024	817.5 ± 2.9
Total	4.984 ± 0.052	2285 ± 12	9.173 ± 0.036	18.122 ± 0.050	1515 ± 10

The absolute amount of any isotope is uncertain by an untabulated error which is estimated to be $\pm 10\%$. The listed errors are the 1σ fractional errors to be used when calculating isotope ratios within a temperature fraction. See text for discussion of the meaning and use of the listed errors.

normalized to ^{40}Ar, were corrected for mass discrimination and then converted to amounts of each isotope in arbitrary units with ^{40}Ar $= 1.000$. All of the 1σ statistical error in a ratio, as defined by Srinivasan's (1973) Eq. (1), plus the error associated with the mass discrimination correction was assigned to the numerator isotope in the ratio. (2) ^{37}Ar and ^{39}Ar were corrected for radioactive decay during and after irradiation using $\lambda^{37} = (1.974 \pm 0.0056) \times 10^{-2}$ day^{-1} and $\lambda^{39} = (7.05 \pm 0.079) \times 10^{-6}$ day^{-1} (Stoenner *et al.*, 1965). (3) ^{36}Ar, ^{38}Ar, and ^{39}Ar were corrected for Ar produced by nuclear reactions in Ca by assuming that all of ^{37}Ar

Table 3. Argon from neutron irradiated 14301 and 14313.

Temp. °C	^{36}Ar	^{37}Ar	^{38}Ar	^{39}Ar	^{40}Ar
			in units of 10^{-6} cm^3 STP/g		
		14301,110		**61.2 mg**	
300	0.1627 ± 0.0031	0.0659 ± 0.0015	0.06368 ± 0.00080	0.04201 ± 0.00030	8.00 ± 0.44
400	1.450 ± 0.022	0.1708 ± 0.0028	0.2893 ± 0.0022	0.14285 ± 0.00048	56.84 ± 0.44
500	2.754 ± 0.020	0.1660 ± 0.0017	0.5150 ± 0.0024	0.1043 ± 0.00039	78.23 ± 0.44
600	16.94 ± 0.13	1.063 ± 0.011	3.181 ± 0.014	0.4737 ± 0.0013	299.36 ± 0.44
700	20.93 ± 0.15	2.126 ± 0.020	3.991 ± 0.015	0.4671 ± 0.0015	213.37 ± 0.44
800	7.176 ± 0.056	2.928 ± 0.026	1.4216 ± 0.0080	0.3853 ± 0.0024	101.28 ± 0.44
900	3.690 ± 0.029	2.564 ± 0.023	0.7544 ± 0.0029	0.21031 ± 0.00091	57.04 ± 0.44
1000	4.979 ± 0.037	1.686 ± 0.017	1.0065 ± 0.0067	0.12718 ± 0.00053	44.46 ± 0.44
1200	2.861 ± 0.048	5.37 ± 0.10	0.6684 ± 0.0044	0.1821 ± 0.0012	32.12 ± 0.44
1400	0.1978 ± 0.0023	0.7307 ± 0.0063	0.05002 ± 0.00046	0.01870 ± 0.00011	2.81 ± 0.44
1600	0.0026 ± 0.0015	0.0228 ± 0.0013	0.00108 ± 0.00036	0.00066 ± 0.00013	< 0.31
Total	61.14 ± 0.22	16.89 ± 0.11	11.942 ± 0.024	2.1542 ± 0.0036	893.5 ± 1.4
		14313,77		**67.4 mg**	
300	0.8576 ± 0.0064	0.0955 ± 0.0014	0.16720 ± 0.00070	0.03857 ± 0.00012	8.89 ± 0.32
400	3.285 ± 0.025	0.2261 ± 0.0022	0.6251 ± 0.0024	0.13275 ± 0.00039	45.46 ± 0.32
500	17.29 ± 0.12	0.5120 ± 0.0061	3.209 ± 0.015	0.27030 ± 0.00068	189.16 ± 0.32
600	53.32 ± 0.39	0.933 ± 0.010	9.847 ± 0.043	0.3350 ± 0.0012	309.33 ± 0.32
700	119.96 ± 0.87	1.954 ± 0.020	22.479 ± 0.089	0.33896 ± 0.00086	294.34 ± 0.32
800	92.53 ± 0.70	3.505 ± 0.033	17.882 ± 0.071	0.35515 ± 0.00079	202.09 ± 0.32
900	47.33 ± 0.34	4.049 ± 0.036	9.359 ± 0.037	0.2457 ± 0.0011	124.77 ± 0.32
1000	21.27 ± 0.17	3.014 ± 0.030	4.346 ± 0.027	0.11431 ± 0.00026	50.83 ± 0.32
1200	5.418 ± 0.044	2.372 ± 0.024	1.2274 ± 0.0066	0.05868 ± 0.00028	14.06 ± 0.32
1400	0.0895 ± 0.0013	0.0697 ± 0.0011	0.02166 ± 0.00026	0.00170 ± 0.00016	0.21 ± 0.32
1600	0.0072 ± 0.0012	—	0.00148 ± 0.00024	—	0.28 ± 0.32
Total	361.4 ± 1.2	16.730 ± 0.066	69.16 ± 0.13	1.8911 ± 0.0022	1239.4 ± 1.1

See notes to Table 2.

was produced in Ca and using production ratios of ^{36}Ar/^{37}Ar = $(3.147 \pm 0.092) \times 10^{-4}$, ^{38}Ar/^{37}Ar = $(4.768 \pm 0.093) \times 10^{-4}$, and ^{39}Ar/^{37}Ar = $(7.225 \pm 0.085) \times 10^{-4}$ (Alexander and Davis, 1974). (4) ^{38}Ar and ^{40}Ar were corrected for Ar produced in K assuming that all of the ^{39}Ar left from step (3) was produced in K and using production ratios of ^{38}Ar/^{39}Ar = $(1.463 \pm 0.014) \times 10^{-2}$ and ^{40}Ar/^{39}Ar = $(6.252 \pm 0.087) \times 10^{-2}$ (Alexander and Davis, 1974). (4) ^{40}Ar, ^{38}Ar, and ^{36}Ar were corrected for blanks using the average of hot blanks analyzed before and after each sample. Conservative errors equal to $\frac{1}{2}$ of the blank corrections were assigned to this step. (5). All of the data were converted to units of cm^3 STP/g by use of the sensitivity of the mass spectrometer as determined from calibration analyses using known amounts of atmospheric argon. (6) The "total" values for each sample were determined by mathematically summing the individual temperature fractions. This last step implicitly assumes that there was no variation in the sensitivity of the system during a complete temperature analysis.

The errors listed in Tables 2 and 3 are the quadratic sums of the errors associated with steps (1)–(4) of the preceeding paragraph. The listed errors are those germane to the age, K/Ca ratio and exposure history calculations in this paper. *The absolute concentration of any isotope is uncertain by an untabulated amount which is estimated to be ± 10%.* The untabulated error arises from uncertainties in the calibration of the pipette system and affects only the calculated concentration of K and Ca. The excellent agreement of the ^{40}*Ar/^{39}*Ar ratio in replicate analyses of our St. Severin monitor (see Table 1) indicates that the system of error analysis outlined above leads to errors which are reasonable if not overstated.

K and Ca contents and K/Ca ratios

The total ^{37}Ar and ^{39}*Ar in each sample are measures of the Ca and K contents of the samples. The conversion factors listed in Alexander and Davis (1974) were scaled for fluence differences between layers and irradiations using the ^{40}*Ar/^{39}*Ar ratios in the appropriate St. Severin monitors to obtain the following values: ^{39}*Ar/K = 3.56×10^{-4} and 3.14×10^{-4} cm^3/g and ^{37}Ar/Ca = 1.89×10^{-4} and 1.66×10^{-4} cm^3/g for the Val. 6 and Val. 8 samples, respectively. The K and Ca contents of the samples are summarized in Table 4.

Turner *et al.* (1972) have shown that the variation of the K/Ca ratio as a function of ^{39}*Ar release often is helpful in interpreting the age plots. Such K/Ca variation plots are shown at the top of Figs. 1, 3, 4, and 6.

Partitioning of stable Ar isotopes

The ^{36}Ar, ^{38}Ar, and ^{40}Ar data in Tables 2 and 3 are mixtures of trapped and spallogenic Ar plus varying amounts of ^{40}*Ar from the decay of ^{40}K (both *in situ*

Table 4. Age, exposure age and K and Ca contents.

Sample	K[a] (ppm)	Ca[a] (%)	Age[b] (b.y.)	Exposure Age[c] (m.y.)	P_K^{38}/P_{Ca}^{38}
14270,1–7	6200	8.2	3.89 ± 0.05	244 ± 25	1.34 ± 0.22
14301,110	6900	10.2	—	—	—
14313,77	6000	10.1	—	—	—
15362,3	96	16.2	>3.98	428 ± 43	—
15455,70A	440	7.3	>3.82	205 ± 21	7.1 ± 1.0
15455,183	1200	6.7	3.92 ± 0.04	205 ± 21	7.1 ± 1.0
63335,3	510	12.1	>3.64	41 ± 8	—

[a]Error estimated to be ±10%.
[b]The interpretation of this age varies from sample to sample, see text.
[c]Nominal, surface exposure ages. The error is formal only and is due mainly to the error in ^{37}Ar/Ca.

and external) and $^{38}Ar_{Cl}$ from $^{37}Cl\,(n, \gamma, \beta)\,^{38}Ar$. Trapped and spallogenic Ar were separated assuming $(^{36}Ar/^{38}Ar)_t = 5.32$ and $(^{36}Ar/^{38}Ar)_s = 0.65$. When the observed $^{36}Ar/^{38}Ar$ ratios were <0.65, it was assumed that $^{38}Ar_{Cl}$ was responsible and all of the observed ^{36}Ar was assumed to be spallogenic. In all of the Val. 6 samples the trapped and spallogenic ^{40}Ar components were very small compared to radiogenic ^{40}Ar. Nominal trapped and spallogenic ^{40}Ar corrections were made assuming $(^{40}Ar/^{36}Ar)_t = 1 \pm 1$ and $(^{40}Ar/^{36}Ar)_s = 0.23 \pm 0.23$. The trapped Ar in the Val. 8 samples is a major component and is discussed separately in a later section.

Cosmic ray exposure ages

Nominal surface cosmic ray exposure ages can be calculated from $^{38s}Ar_{Ca}/^{37}Ar$ ratios and an assumed ^{38}Ar production rate from cosmic rays via the equation:

$$t = (1/P_{Ca}^{38})(^{38s}Ar_{Ca}/^{37}Ar)(^{37}Ar/Ca)$$

where P_{Ca}^{38} is the production rate of ^{38}Ar from Ca by cosmic rays at the lunar surface and is taken to be $1.4 \times 10^{-8}\,cm^3$ STP $^{38}Ar/g$ Ca/m.y. in this work. Since ^{38}Ar is produced from spallation on K, Ti, and Fe and from neutrons on ^{37}Cl as well as from spallation on Ca, it is necessary to identify portions of the thermal release data where the ^{38s}Ar data is dominated by production from Ca or to correct for the production from other elements in some fashion. The calculation of $^{38s}Ar_{Ca}/^{37}Ar$ is discussed for individual rocks in the following sections. The resulting exposure ages are summarized in Table 4. These exposure ages may be reasonable estimates of the cosmic ray exposure times for samples with simple histories but are only lower limits for the exposure time of samples with complex histories.

$^{40}Ar-^{39}Ar$ retention ages

The apparent $^{40}Ar-^{39}Ar$ age of each temperature release fraction can be calculated from the $^{40*}Ar/^{39*}Ar$ ratio by

$$t = 1/\lambda\,\ln\,[1 + J(^{40*}Ar/^{39*}Ar)]$$

where $\lambda = 5.305 \times 10^{-10}\,yr^{-1}$ and

$$J = (e^{\lambda Tm} - 1)/(^{40*}Ar/^{39*}Ar)_m$$

with $Tm = 4.504 \pm 0.020 \times 10^9\,yr$ and where $(^{40*}Ar/^{39*}Ar)_m$ is the value determined from the monitor.

The apparent ages of the temperature fractions are shown in Figs. 1, 3, 4, and 6 and discussed below. The height of a data point corresponds to the $1\,\sigma$ error associated with the $^{40*}Ar/^{39*}Ar$ ratio. The width of the data point corresponds the fraction of $^{39*}Ar$ released. The small numbers by the data points are the release temperatures in hundreds of degrees centigrade. The errors on the numerical ages shown in the figures and summarized in Table 4 are the quadratic sum of contributions from: (1) the statistical error on the $^{40*}Ar/^{39*}Ar$ ratios, (2) the "spread error" associated with the scatter about the plateau (or the mean of the portion of the release used to calculate the age), (3) the 1% fluence uncertainty, and (4) the uncertainty in J. The quadratic sum of the latter two sources of error are shown to scale on Figs. 1, 3, 4, and 6.

DISCUSSION

14270

Figure 1 shows the K/Ca ratio and apparent age plots for 14270,1–7. The K/Ca ratio decreases monotonically from 0.46 at 300°C to 0.014 at 1680°C. The apparent age rises abruptly to a maximum of 4.00 b.y. at 500°C, decreases slowly to 3.90 b.y. at 900°C, drops sharply to 3.53 b.y. at 1200°C and then rises again in the final temperature fraction. Both the K/Ca and apparent age plots are typical of those displayed by the more metamorphosed Apollo 14 breccias (Turner *et al.*, 1971; York *et al.*, 1972; Husain *et al.*, 1972; Alexander and Davis, 1974). The age is commonly interpreted to be the date of a major metamorphic event which degassed all the constituent parts of Apollo 14 high-grade breccias. This metamorphic event was presumably the formation of the Fra Mauro Formation.

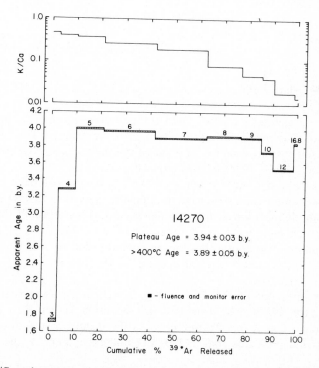

Fig. 1. K/Ca and apparent age data from 14270,1–7. In this and all the following figures, the numbers by the data points are the temperature of the fraction in hundreds of degrees C.

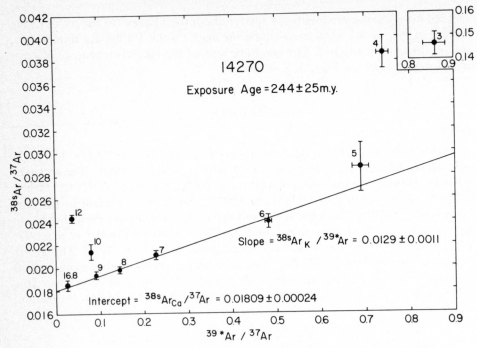

Fig. 2. ^{38s}Ar/^{37}Ar versus ^{39*}Ar/^{37}Ar for data from 14270,1–7. Error bars are shown where they are larger than the points.

The 500–900°C data define a rough plateau which contains 75% of the ^{39*}Ar released and which corresponds to an age of 3.94 ± 0.03 b.y. However, Turner *et al.* (1974) have presented evidence that the high-temperature drop-off in ages is an artifact due to ^{39*}Ar recoil during the ^{39}K(n, p) ^{39*}Ar reaction. If redistribution by recoil is a closed system, then the best age of the sample would be the >400°C age (Alexander and Davis, 1974) of 3.89 ± 0.05 b.y. from 14270.

Figure 2 is a plot of ^{38s}Ar/^{37}Ar versus ^{39*}Ar/^{37}Ar for the 14270 data. Turner *et al.* (1971) discuss in detail the calculation of exposure ages from this type of data. The 500–900°C data define a linear array indicating that this portion of the data can be interpreted as a simple mixture of ^{38s}Ar$_{Ca}$ and ^{38s}Ar$_K$. The 300°C, 400°C, 1000°C, and 1200°C points plot above the line presumably due to contributions to ^{38s}Ar from ^{38s}Ar$_{Fe}$, ^{38s}Ar$_{Ti}$, and ^{38}Ar$_{Cl}$. The slope of the line is ^{38s}Ar$_K$/^{39*}Ar and the intercept is ^{38s}Ar$_{Ca}$/^{37}Ar.

From the intercept of the line in Fig. 2 we calculate an exposure age of 244 ± 25 m.y. for 14270. The exposure age is much longer than the generally accepted age, $\sim$26 m.y. (cf. Turner *et al.*, 1971; Lugmair and Marti, 1972), for Cone Crater and indicates that 14270 was not excavated by the Cone Crater impact.

From the slope and intercept of the line in Fig. 2 and the relationship:

$$P_K^{38}/P_{Ca}^{38} = \frac{(^{38s}Ar_K/^{39*}Ar)(^{39*}Ar/K)}{(^{38s}Ar_{Ca}/^{37}Ar)(^{37}Ar/Ca)}$$

we calculate the ratio of the production of ^{38s}Ar from K and Ca to be $P_K^{38}/P_{Ca}^{38} = 1.34 \pm 0.22$. This value is indicative of a "harder" or more near surface exposure for 14270 than 15455 (see below) since P_K^{38}/P_{Ca}^{38} should increase with shielding. Turner *et al.* (1971) have observed a similar value of 1.0 ± 0.5 in 14001,7,1.

15362

Figure 3 gives the apparent age and K/Ca data for 15362. The K/Ca ratio is about 0.001 in the initial, 500°C, fraction, then remains constant around 0.0006 for the remaining temperature fractions. Such behavior is characteristic of an essentially pure mineral and is consistent with the identification of 15362 as an anorthosite.

It is not possible to define a line analogous to that in Fig. 2 using the 15362 data because the ^{39*}Ar/^{37}Ar ratios are so small and constant. The ^{38s}Ar/^{37}Ar ratios start at 0.0257 ± 0.0010 at 300°C, rise through 0.0276 ± 0.007 and 0.0311 ± 0.009 at 500°

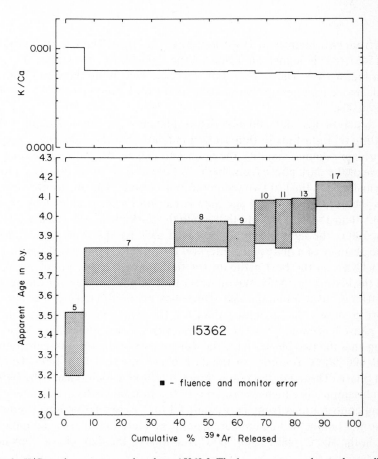

Fig. 3. K/Ca and apparent age data from 15362,3. The large errors are due to the small K content and low K/Ca ratio.

and 700°C and remain at 0.0317 ± 0.001 through the remaining temperatures. Assuming that all of the targets except Ca are negligible, we calculate an exposure age of 428 ± 43 m.y. for 15362. The lower $^{38s}Ar/^{37}Ar$ ratios in the 300°C, 500°C, and 700°C fractions indicate gas loss during the exposure history of 15362.

The apparent age pattern in Fig. 3 does not define a good plateau. (The large errors in the data are due to the low K content, 96 ppm, and very low K/Ca ratio, 0.0006.) The low ages in the 500°C and 700°C fractions are compatible with the evidence for recent gas loss mentioned in the preceeding paragraph. However, the remaining temperature fractions yield increasing ages, from 3.92 b.y. at 800°C to 4.12 b.y. at 1700°C. The simplest explanation of the release pattern is that material older than 4.1 b.y. was extensively but not completely outgassed around 3.92 b.y. However, this interpretation is not unique and the error bars on the $\geqslant 800$°C fractions overlap so we interpret the $\geqslant 800$°C total age, 3.98 ± 0.06 b.y., as a minimum age of crystallization.

15455

Data from two samples of 15455 are shown in Fig. 4. The K content of the dark phase, 15455,183, is higher than that of the light phase, 15455,70A, by approximately a factor of 3. The K/Ca ratios decrease with temperature in both phases but the decrease is larger, 0.129 to 0.0102, in the dark phase than in the light phase, 0.0094 to 0.0046.

The apparent age data do not define plateaus for either sample. Although crosscutting relationships in thin section indicate that the light phase is older (L. Silver, private communication), the apparent ages of the light phase are younger than those of the dark phase from 400°C to 1100°C. The ages of the two phases are indistinguishable in the last two temperature fractions. The apparent contradiction between the petrographic and age data is caused by the greater loss of Ar from 15455,70A than 15455,183. Note that while the dark phase has released 43% of its $^{39*}Ar$ by 700°C the light phase has released 63% by the same temperature.

In the absence of a plateau we interpret the >400°C total age of the dark phase, 3.92 ± 0.04 b.y., as the best estimate for the age of the last major metamorphic event in the history of 15455. We interpret the >400°C total age of the light phase, 3.82 ± 0.04 b.y., as a minimum age which has been affected by ^{40}Ar loss.

Figure 5 shows the spallation data for the light and dark phases of 15455. Neither phase has enough spread in $^{39*}Ar/^{37}Ar$ ratios to define a line. However, by *assuming* that the two phases have the same exposure history we can define a line using the 600–900°C fractions of the dark phase and the 500–1100°C fractions of the light phase. These are the temperature fractions which release the bulk of the Ar in each sample. The higher temperature fractions in each case lie above the line indicating contributions from Ti, Fe, and Cl to ^{38s}Ar while the lower temperature fractions plot below the line indicating gas loss. Note that there is no indication of gas loss in the 500°C, 600°C, and 700°C fractions of the light phase. This indicates

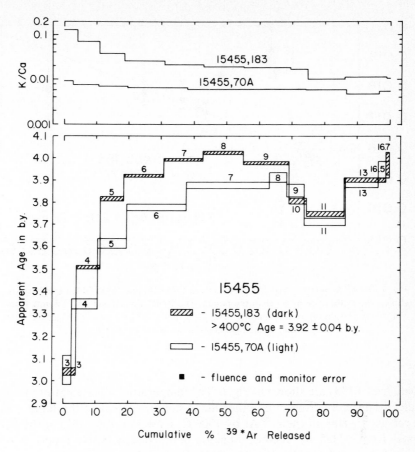

Fig. 4. K/Ca and apparent age data from 15455,183 and 15455,70A.

that the selective gas loss which affected the ^{40}Ar–^{39}Ar age of the light phase occurred before the start of the cosmic ray exposure.

From the ratio of the slope, 0.0577 ± 0.0021, to the intercept, 0.015206 ± 0.000049, of the line in Fig. 5 we calculate $P_K^{38}/P_{Ca}^{38} = 7.1 \pm 1.0$. (The value of ~ 70 listed in Alexander and Kahl (1974) is a misprint.) The high P_K^{38}/P_{Ca}^{38} value for 15455 is completely dependent on our *assumption* that both phases experienced the same exposure history. However, the high value indicates that 15455 accumulated most of its cosmic ray exposure in a more shielded location than did 14270. The intercept in Fig. 5 corresponds to an exposure age of 205 ± 21 m.y. but in view of the P_K^{38}/P_{Ca}^{38} value this is only a lower limit on the exposure of 15455 to cosmic rays.

The long exposure ages of 15455 and 15362 and the high P_K^{38}/P_{Ca}^{38} value of 15455 indicate that both rocks probably record long complicated exposure histories and that neither can be used to date the Spur Crater impact.

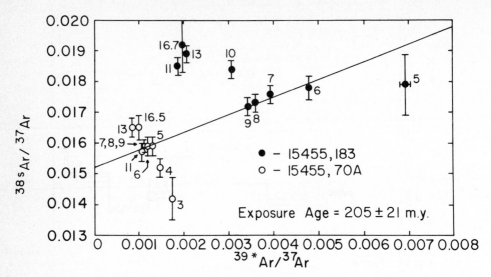

Fig. 5. $^{38s}Ar/^{37}Ar$ versus $^{39*}Ar/^{37}Ar$ for data from 15455,183 and 15455,70A. Error bars are shown where they are larger than the points. The 700°C, 800°C, and 900°C points from 15455,70A plot on top of one another. The 300°C and 400°C fractions of 15455,183 plot offscale to the right and slightly below the line.

63335,3

Data from 63335 are shown in Fig. 6. The temperature release pattern for 63335 is radically different from most other lunar samples. Most of the lunar samples have released ~80% of their $^{39*}Ar$ by 1000°C. However, only 15% of the $^{39*}Ar$ is released from 63335 by 1000°C. Shaeffer and Husain (1974) report a similar behavior for another Apollo 16 rock, 60015. Both rocks contain evidence of shock melting (cf. Sclar *et al.* (1973) for 60015 and the Apollo 16 Lunar Sample Information Catalog for 63335).

The K/Ca ratio in 63335 drops monotonically from 0.023 at 300°C to 0.0035 at 800°C and then varies irregularly between 0.0039 and 0.0045 for the remaining 90+% of the $^{39*}Ar$ release. The fairly constant K/Ca ratio indicates that most of the K in 63335 is in a single high-temperature phase.

Above 900°C the $^{36}Ar/^{38}Ar$ ratios are <0.65 indicating a contribution from $^{38}Ar_{Cl}$. Assuming all of the ^{36}Ar in the >900°C fractions to be spallogenic we calculate an average $^{38s}Ar/^{37}Ar$ in the >400°C fractions of 0.0030 ± 0.0005 which corresponds to an exposure age of 41 ± 8 m.y. This exposure age is the same within analytical error of Marti *et al.*'s (1973) more precise value of 48.9 ± 1.7 m.y. for the age of North Ray Crater.

The apparent age data do not define a plateau but rather indicate extensive gas loss. The 1680°C fraction defines a minimum crystallization age of >3.64 b.y. The <1100°C data give some indication that the gas loss occurred about 0.77 b.y. ago but these data are very uncertain due to large blank corrections to the ^{40}Ar. The data do not rule out the obvious possibility that the outgassing event was the North Ray Crater forming event.

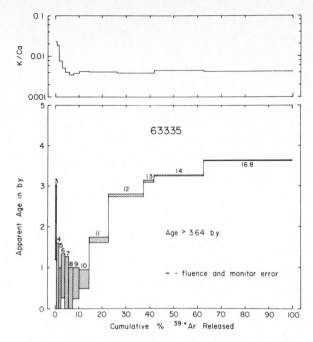

Fig. 6. K/Ca and apparent age data from 63335,3.

14301,110 and 14313,77

Three Apollo 14 breccias, 14301, 14313, and 14318 contain abundant chondrules (King *et al.*, 1972) and fissiogenic Xe attributable to the decay of extinct ^{244}Pu (Drozd *et al.*, 1972; Behrmann *et al.*, 1973; Reynolds *et al.*, 1974) but the latter effect is marginal in the case of 14313 (Basford *et al.*, 1973). Reynolds *et al.* (1974) have reported that 14318 has a reasonably well defined ^{40}Ar/^{36}Ar versus ^{39}Ar/^{36}Ar isochron with a slope corresponding to an age of 3.69 ± 0.09 b.y. and contains a large trapped component with an unusual isotopic composition, $(^{40}$Ar/^{36}Ar)$_t = 13.68 \pm 0.25$. Megrue (1973) using a laser probe technique has reported that 14301 contains clasts which can be interpreted to be 3.7 and 2.9 b.y. old and that the matrix of 14301 contains trapped Ar with $(^{40}$Ar/^{36}Ar)$_t = 14$. Therefore, we have extended Reynolds *et al.*'s (1974) study of 14318 to include 14301 and 14313.

The K/Ca ratios in 14301 and 14313 decrease monotonically with temperature—from 0.344 to 0.014 in 14301 and from 0.218 to 0.013 in 14313. Both samples contain so much trapped ^{38}Ar (^{38s}Ar is undetectable in many temperature fractions) that exposure ages could not be determined for either sample.

Since the trapped component is so significant in 14301 and 14313, the data must be plotted on isochron type diagrams in an effort to determine internally both the age and trapped Ar composition. Figure 7 is a plot of ^{40}Ar/^{36}Ar versus ^{39*}Ar/^{36}Ar for the 14301 data. A similar plot for the 14313 data (not shown) looks very similar. Data from neither sample define an isochron.

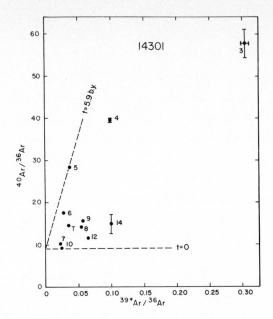

Fig. 7. ^{40}Ar/^{36}Ar versus ^{39*}Ar/^{36}Ar for data from 14301,110. The data do not define an isochron but lie in the region bounded by 0 and 5.9 b.y. reference isochrons. Error bars are shown where they are larger than the data points. The point labeled "T" is the total point.

The scatter of the data indicates that the assumptions of an isochron (namely complete isotopic homogenization at a definite point in the past, followed by no subsequent isotopic exchange) have not been met. If complete isotopic homogenization has not been reached, then 14301 and 14313 can be made up of three different assemblages of material. The breccias could have been assembled from material: (1) having different ages but the same isotopic composition of trapped argon, or (2) with trapped argon having different compositions but the same age, or (3) with trapped argon having different compositions and different ages. These different assemblages are discussed below as models 1, 2, and 3, respectively.

Megrue's (1973) results would seem to support model 1. However, the dashed reference isochrons in Fig. 7 demonstrate that our data are not compatible with a homogeneous trapped component in 14301. If one assumes that there is a homogeneous trapped component in 14301, then an *upper* limit on the trapped ^{40}Ar/^{36}Ar ratio is set by assigning the fraction with the lowest observed ^{40}Ar/^{36}Ar ratio to zero age material. This upper limit can then be used to calculate *lower* limits on the apparent ages of the other temperature fractions. Figure 7 shows that, for 14301, this leads to the absurd conclusion that the 500°C fraction represents material >5.9 b.y. old. (A similar procedure for 14313 data indicates that the 600°C fraction represents material >6.1 b.y. old.) Therefore, the trapped component in 14301 (and 14313) cannot be homogeneous.

Model 3 is probably the most realistic model to describe 14301 and 14313. However, we can treat model 2 quantatively and it will serve as a limiting case for model 3. If we *assume* that all of the material in 14301 and 14313 has the same age, we can calculate effective $(^{40}Ar/^{36}Ar)_t$ ratios for each temperature fraction. Figures 8 and 9 show the results of such calculations for 14301 and 14313, respectively. We assumed the age of the two breccias to be 3.7 ± 0.2 b.y. (Megrue's (1973) age for parts of 14301 and Reynolds *et al.*'s (1974) age for 14318.) 14313 contains a factor of 6 more trapped ^{36}Ar and has a $(^{40}Ar/^{36}Ar)_t$ ratio $3\frac{1}{2}$ times smaller than 14301. The shapes of the two curves are strikingly similar. The low-temperature fractions yield much higher $(^{40}Ar/^{36}Ar)_t$ ratios than do the higher temperature fractions. The presence of a low-temperature trapped component enriched in ^{40}Ar has been detected previously during analyses of lunar glass spherules (Podosek and Huneke, 1973; Huneke *et al.*, 1973). The trapped argon in 14301 and 14313 is not homogeneous but is made up of at least two components which fractionate in a stepwise heating experiment. The component released at lower temperatures is enriched in ^{40}Ar.

A low-temperature enrichment of the trapped argon in ^{40}Ar is compatible with ion-pumping of lunar atmospheric ^{40}Ar into lunar surface grains (Manka and Michel, 1970; Heymann and Yaniv, 1970); the "potassium-coated regolith" model of Bauer *et al.* (1972), and an "internal atmosphere" during the lithification of these breccias. However, one caveat is in order. Part of patterns in Figs. 8 and 9 may be $^{39*}Ar$ recoil artifacts of the type described by Turner *et al.* (1974).

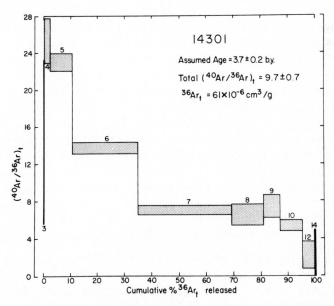

Fig. 8. Trapped $^{40}Ar/^{36}Ar$ versus the cumulative % trapped ^{36}Ar released for 14301. The trapped $^{40}Ar/^{36}Ar$ ratios are enriched in ^{40}Ar in the lower temperature data. The height of the data rectangle is the error—mainly due to the error in the assumed age.

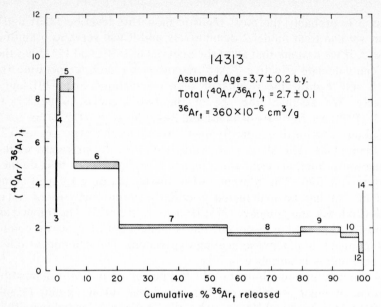

Fig. 9. Trapped $^{40}Ar/^{36}Ar$ versus the cumulative % trapped ^{36}Ar released for 14313. The higher trapped ^{36}Ar content of 14313 results in smaller relative radiogenic ^{40}Ar (from *in situ* decay of ^{40}K) corrections and smaller errors than in the 14301 data shown in Fig. 8.

SUMMARY

Our results from 14301 and 14313 indicate that isotopic homogenization of Ar has not been attained in Apollo 14 breccias classified as groups 1 and 2 in Warner's (1972) scale. Reynolds *et al.*'s (1974) results from 14318 indicate that the argon in a group 3 breccia was homogenized. Therefore, only breccias which have been metamorphosed to group 3 or higher are potentially datable.

The trapped argon in 14301, which appears in Megrue's (1973) work to be homogeneous, is made up of at least two components. These two trapped components fractionate during a heating experiment with a component rich in ^{40}Ar being released at lower temperatures. Our data do not necessarily contradict those of Megrue (1973). If the various components of the trapped argon occur in proportional amounts, then the trapped argon will appear to be homogeneous in a laser probe experiment. Laser probe and stepwise heating experiments provide information which is complimentary.

The following basic chronological information has been obtained:

(1) 14270 was completely outgassed at 3.89 ± 0.05 b.y. It has been exposed to cosmic rays, in a fairly unshielded environment, for 244 ± 25 m.y.

(2) 15362 is at least 3.98 ± 0.06 b.y. old and may consist of material >4.1 b.y. old which was partially outgassed ~3.9 b.y. It has been exposed to cosmic rays for at least 428 ± 43 m.y.

(3) The dark portion of 15455 was completely outgassed 3.92 ± 0.04 b.y. ago.

The light portion of 15455 has suffered greater gas loss than the dark portion. Both portions have been exposed to cosmic rays for a minimum of 205 ± 21 m.y. A major part of the exposure occurred in a shielded location.

(4) 63335,3 has been exposed to cosmic rays for 41 ± 8 m.y. It is comprised of material at least 3.6 b.y. old which has been extensively outgassed in a more recent event.

Acknowledgments—J. H. Reynolds has given indispensable aid, encouragement and critical evaluation to all phases of this work. G. A. McCrory gave constant technical and electronics support to the experimental phase. We thank both of these individuals for their assistance. The experimental phase of this work was supported by NASA under grant 05-003-409. One of us (ECA) acknowledges NASA support under grant 24-005-223 during the data evaluation and manuscript preparation phases of this work.

REFERENCES

Albee A. L., Burnett D. S., Chodos A. A., Haines E. L., Huneke J. C., Papanastassiou D. A., Podosek F. A., Russ G. P., III, and Wasserburg G. J. (1970) Mineralogic and isotopic investigations on lunar rock 12013. *Earth Planet. Sci. Lett.* **9**, 137–163.

Alexander E. C., Jr. and Davis P. K. (1974) ^{40}Ar–^{39}Ar ages and trace element contents of Apollo 14 breccias; an interlaboratory cross-calibration of ^{40}Ar–^{39}Ar standards. *Geochim. Cosmochim. Acta.* **38**, 911–928.

Alexander E. C., Jr. and Kahl S. B. (1974) ^{40}Ar–^{39}Ar studies of lunar breccias (abstract). In *Lunar Science—V*, pp. 9–11. The Lunar Science Institute, Houston.

Alexander E. C., Jr., Davis P. K., and Lewis R. S. (1972a) Argon 40–argon 39 dating of Apollo sample 15555. *Science* **175**, 417–419.

Alexander E. C. Jr., Davis P. K., and Reynolds J. H. (1972b) Rare-gas analyses on neutron irradiated Apollo 12 samples. *Proc. Third Lunar Sci. Conf., Geochim. Cosmochim. Acta*, Suppl. 3, Vol. 2, pp. 1787–1795. MIT Press.

Basford J. R., Dragon J. C., Pepin R. O., Coscio M. R., Jr., and Murthy V. R. (1973) Krypton and xenon in lunar fines. *Proc. Fourth Lunar Sci. Conf., Geochim. Cosmochim. Acta*, Suppl. 4, Vol. 2, pp. 1915–1955. Pergamon.

Baur H., Frick U., Funk H., Schultz L., and Signer P. (1972) Thermal release of helium, neon, and argon from lunar fines and minerals. *Proc. Third Lunar Sci. Conf., Geochim. Cosmochim. Acta*, Suppl. 3, Vol. 2, pp. 1947–1966. MIT Press.

Behrmann C. J., Drozd R. J., and Hohenberg C. M. (1973) Extinct lunar radioactivities: Xenon from ^{244}Pu and ^{129}I in Apollo 14 breccias. *Earth Planet. Sci. Lett.* **17**, 446–455.

Davis P. K., Lewis R. S., and Reynolds J. H. (1971) Stepwise heating analyses of rare gases from pile-irradiated rocks 10044 and 10057. *Proc. Second Lunar Sci. Conf., Geochim. Cosmochim. Acta*, Suppl. 2, Vol. 2, pp. 1693–1703. MIT Press.

Drozd R., Hohenberg C. M., and Ragan D. (1972) Fission xenon from extinct ^{244}Pu in 14301. *Earth Planet. Sci. Lett.* **15**, 338–346.

Floran R. J., Cameron K. L., Bence A. E., and Papike J. J. (1972) Apollo 14 breccia 14313: A mineralogic and petrologic report. *Proc. Third Lunar Sci. Conf., Geochim. Cosmochim. Acta*, Suppl. 3, Vol. 1, pp. 661–671. MIT Press.

Ganapathy R., Morgan J. W., Krähenbühl U., and Anders E. (1973) Ancient meteoritic components in lunar highland rocks: Clues from trace elements in Apollo 15 and 16 samples. *Proc. Fourth Lunar Sci. Conf., Geochim. Cosmochim Acta*, Suppl. 4, Vol. 2, pp. 1239–1261. Pergamon.

Heymann D. and Yaniv A. (1970) Ar^{40} anomaly in lunar samples from Apollo 11. *Proc. Apollo 11 Lunar Sci. Conf., Geochim. Cosmochim. Acta*, Suppl. 1, Vol. 2, pp. 1261–1267. Pergamon.

Huneke J. C., Jessberger E. K., Podosek F. A., and Wasserburg G. J. (1973) ^{40}Ar/^{39}Ar measurements in Apollo 16 and 17 samples and the chronology of metamorphic and volcanic activity in the

Taurus-Littrow region. *Proc. Fourth Lunar Sci. Conf., Geochim. Cosmochim. Acta*, Suppl. 4, Vol. 2, pp. 1725–1756. Pergamon.

Husain L., Schaeffer O. A., Funkhouser J., and Sutter J. (1972) The ages of lunar material from Fra Mauro, Hadley Rille, and Spur crater. *Proc. Third Lunar Sci. Conf., Geochim. Cosmochim. Acta*, Suppl. 3, Vol. 2, pp. 1557–1567. MIT Press.

King E. A., Jr., Carman M. F., and Butler J. C. (1972) Chondrules in Apollo 14 samples: Implications for the origin of chondritic meteorites. *Science* 175, 59–60.

Laul J. C., Wakita H., and Schmitt R. A. (1972) Bulk and REE abundances in anorthosites and noritic fragments. *The Apollo 15 Lunar Samples*. pp. 221–224. The Lunar Science Institute, Houston.

Lindstrom M. M., Duncan A. R., Fruchter J. S., McKay S. M., Stoeser J. W., Goles G. G., and Lindstrom D. J. (1972) Compositional characteristics of some Apollo 14 clastic materials. *Proc. Third Lunar Sci. Conf., Geochim. Cosmochim. Acta*, Suppl. 3, Vol. 2, pp. 1201–1214. MIT Press.

Lugmair G. W. and Marti K. (1972) Exposure ages and neutron capture record in lunar samples from Fra Mauro. *Proc. Third Lunar Sci. Conf., Geochim. Cosmochim. Acta*, Suppl. 3, Vol. 2, pp. 1891–1897. MIT Press.

Manka R. H. and Michel F. C. (1970) Lunar atmosphere as a source of argon-40 and other lunar surface elements. *Science* 169, 278.

Mark R. K., Cliff R. A., Lee-Hu C., Wetherill G. W. (1973) Rb–Sr studies of lunar breccias and soils. *Proc. Fourth Lunar Sci. Conf., Geochim. Cosmochim. Acta*, Suppl. 4, Vol. 2, pp. 1785–1795. Pergamon.

Marti K., Lightner B. D., and Osborn T. W. (1973) Krypton and xenon in some lunar samples and the age of North Ray crater. *Proc. Fourth Lunar Sci. Conf., Geochim. Cosmochim. Acta*, Suppl. 4, Vol. 2, pp. 2037–2048. Pergamon.

Megrue G. H. (1973) Spatial distribution of $^{40}Ar/^{39}Ar$ ages in lunar breccia 14301. *J. Geophys. Res.* 78, 3216–3221.

Merrihue C. M. (1965) Trace-element determinations and potassium–argon dating by mass spectroscopy of neutron-irradiated samples (abstract). *Trans. Amer. Geophys. Union* 46, 125.

Merrihue C. M. and Turner G. (1966) Potassium–argon dating by activation with fast neutrons. *J. Geophys. Res.* 71, 2852–2857.

Papanastassiou D. A. and Wasserburg G. J. (1971) Rb–Sr ages of igneous rocks from the Apollo 14 mission and the age of the Fra Mauro Formation. *Earth Planet. Sci. Lett.* 12, 36–48.

Podosek F. A. and Huneke J. C. (1973) Argon in Apollo 15 green glass spherules (15426): $^{40}Ar-^{39}Ar$ age and trapped argon. *Earth Planet. Sci. Lett.* 19, 413.

Reynolds J. H., Alexander E. C., Jr., Davis P. K., and Srinivasan B. (1974) Studies of K–Ar dating and xenon from extinct radioactivities in breccia 14318; implications for early lunar history. *Geochim. Cosmochim. Acta* 38, 401–418.

Schaeffer O. A. and Husain L. (1974) Chronology of lunar basin formation and ages of lunar anorthosites (abstract). In *Lunar Science—V*, pp. 663–665. The Lunar Science Institute, Houston.

Sclar C. B., Bauer J. F., Pickart S. J., and Alperin H. A. (1973) Shock effects in experimentally shocked terrestrial ilmenite, lunar ilmenite of rock fragments in 1–10 mm fines (10085,19), and lunar rock 60015,127. *Proc. Fourth Lunar Sci. Conf., Geochim. Cosmochim. Acta*, Suppl. 4, Vol. 1, pp. 841–859. Pergamon.

Sigurgeirsson T. (1962) Age dating of young basalts with the potassium argon method (in Icelandic) Report Physical Laboratory of the University of Iceland, Reykjavik. 9pp.

Srinivasan B. (1973) Variations in the isotopic composition of trapped rare gases in lunar sample 14318. *Proc. Fourth Lunar Sci. Conf., Geochim. Cosmochim. Acta*, Suppl. 4, Vol. 2, pp. 2049–2064. Pergamon.

Stoenner R. W., Schaeffer O. A., and Katcoff S. (1965) Half-lives of argon-37, argon-39, and argon-42. *Science* 148, 1325–1328.

Turner G. (1970) Argon-40/argon-39 dating of lunar rock samples. *Proc. Apollo 11 Lunar Sci. Conf., Geochim. Cosmochim. Acta*, Suppl. 1, Vol. 2, pp. 1665–1684. Pergamon.

Turner G., Huneke J. C., Podosek F. A., and Wasserburg G. J. (1971) $^{40}Ar-^{39}Ar$ ages and cosmic ray exposure ages of Apollo 14 samples. *Earth Planet. Sci. Lett.* 12, 19–35.

Turner G., Huneke J. C., Podosek F. A., and Wasserburg G. J. (1972) ^{40}Ar–^{39}Ar systematics in rocks and separated minerals from Apollo 14. *Proc. Third Lunar Sci. Conf., Geochim. Cosmochim. Acta,* Suppl. 3, Vol. 2, pp. 1589–1612. MIT Press.

Turner G., Cadogan P. H., and Yonge C. J. (1974) The early chronology of the moon and meteorites (abstract). In *Lunar Science—V,* pp. 807–808. The Lunar Science Institute, Houston.

Warner J. L. (1972) Metamorphism of Apollo 14 breccias. *Proc. Third Lunar Sci. Conf., Geochim. Cosmochim. Acta,* Suppl. 3, Vol. 1, pp. 623–643. MIT Press.

Wilshire H. G., Stuart-Alexander D. E., and Jackson E. D. (1973) Apollo 16 rocks: Petrology and classification. *J. Geophys. Res.* **78,** 2379–2392.

York D., Kenyon J. W., and Doyle R. J. (1972) ^{40}Ar–^{39}Ar ages of Apollo 14 and 15 samples. *Proc. Third Lunar Sci. Conf., Geochim. Cosmochim. Acta,* Suppl. 3, Vol. 2, pp. 1613–1622. MIT Press.

Proceedings of the Fifth Lunar Conference
(Supplement 5, Geochimica et Cosmochimica Acta)
Vol. 2 pp. 1375–1388 (1974)
Printed in the United States of America

K–Ar analysis of Apollo 11 fines 10084

JEFFREY R. BASFORD

School of Physics and Astronomy, University of Minnesota, Minneapolis, Minnesota 55455

Abstract—Regolithic history of Mare Tranquillitatus is studied with rare gas and mass spectrometric techniques. Grain size-ordinate intercept analysis is applied to bulk soil, ilmenite, and plagioclase separates obtained from ≤1 mm fines sample 10084,48. Although all the rare gases have been examined, only argon and the associated potassium analyses are discussed in this report.

K–Ar ages for ilmenite, plagioclase, and total soil separates are determined to be 2.80, 3.38, and 4.00 AE, respectively (1 AE = 10^9 yr). It appears that 10084,48 may consist primarily of two components. In this view, the first is locally derived, forms 65–75% of the present day regolith and may show that a local degassing event(s) occurred between 2.8 and ~3.5 AE ago. The second component is similar to material from the lunar highlands. It may be local or foreign in origin and contributes the remaining 25–35% of the soil. This component exhibits an age of 4.00 AE and is either slightly older than, or contemporaneous with, the oldest of the returned Apollo 11 basalts.

Within the limits of experimental error, the trapped $^{38}Ar/^{36}Ar$ ratio is constant at 0.1890 for all the soil components examined. Trapped $^{40}Ar/^{36}Ar$ ratios, on the other hand, vary. The ilmenite and soil values of 0.677 and 0.893 differ significantly from each other. In the plagioclase separate, trapped $^{40}Ar/^{36}Ar$ shows a large (~5%) scatter around 0.90. As would be expected, the proposed locally derived material, represented by ilmenite, has a well-defined ratio. The postulated older and possibly more heterogenous second component, represented by plagioclase, has a less precise value. Whether these differences are due to mineralogical, selenographical, or secular variations cannot at present be determined.

INTRODUCTION

LUNAR SOILS have been studied extensively with rare gas and grain size techniques but little is known beyond the inventories and chronologies of mineralogically undifferentiated soil samples. All models and studies of the regolith, however, show us a complex interaction of comminution, gardening, transport, and aggregation processes. One immediately wonders whether the different mineralogical components of soils contain information that is lost in analysis of "whole" soil samples and size separates. The present paper is therefore a study of the feasibility and value of examining the separate mineralogical components of lunar soils.

Fines from the Apollo 11 mission were selected because they, as mare regolith, may be representative of a major morphological feature of the moon. In addition, ilmenite, although distributed nonuniformly over the moon, is enriched in the region of the Sea of Tranquillity and may act as a tracer for local regolith producing events.

In particular, ≤1 mm fines sample 10084,48, returned with the first Apollo mission, was separated into ilmenite, plagioclase, and total soil size separates. In each fraction, all the rare gases were measured for isotopic compositions and abundances. Trace element concentrations of K, Rb, Sr, and Ba were determined on representative aliquots.

1375

Chronologies are a point of major interest and in rare gas work there exist only two inert radioactive daughters whose progenitors have half-lives on the order of the 3–4 AE ages commonly measured in lunar material. These daughters are ^{4}He and ^{40}Ar.

It is difficult to use ^{4}He as a geochronometer in lunar soils because of its high mobility and its overwhelming presence in the solar wind. As a result, even with the problem of parentless argon, the decay of ^{40}K to ^{40}Ar forms a much more usable radiometric clock and is used in this paper. The focus of this paper will be centered on argon analysis and the information that it yields about indigenous and external influences on the moon's surface.

Krypton and xenon measurements made on the total size separates have been previously published (Basford *et al.*, 1973). The lighter rare gases for all the separates, and krypton and xenon determinations in plagioclase and ilmenite, are presently being examined and will be published separately.

EXPERIMENTAL PROCEDURE

The mass spectrometric and grain sizing techniques used in this experiment are similar to those used previously. They are discussed in detail in Basford *et al.* (1973) and Evensen *et al.* (1973). The emphasis of the present study results in several modifications which are discussed below.

Bulk soil grain size fractions, once obtained by sieving and sedimentation in acetone, were separated into two aliquots. One aliquot of each size range was maintained as a bulk soil "reservoir." The other was further separated with heavy liquid and magnetic techniques into ilmenite and plagioclase concentrates.

Ilmenite separates were obtained by sedimentation in Clerici's solution (Thallium Malonate–Formate) with $\rho \simeq 4.1$ g/cc. Optical and electron microprobe analyses confirmed that this separate was ~98% ilmenite. Under microprobe analyses, a few high-Zr patches were seen in the >74 μ size fractions. (Arrhenius *et al.* (1971) have previously described patches of high-Zr concentration that they observed in lunar ilmenite.)

Plagioclase was magnetically separated from the total soil. Fractions coarser than 25 μ were obtained with a Frantz Isodynamic Separator; finer fractions were separated in acetone with jig and hand held magnetic wands. Optical, electron microprobe and X-ray diffraction measurements showed that these concentrates were 90–95% plagioclase with 1–2% opaque inclusions.

Losses in the initial size separation were ≤3%, and were concentrated in the finer size fractions. (A grain size-mass yield curve for this separation of 10084,48 is in Venkatesan *et al.* (1974).) Complete removal of plagioclase and ilmenite was not attempted. Thus mineral recovery (~1–2% of each fraction) was not representative and is not reported.

All the rare gas mass spectrometric measurements were done on the 6-in. double focusing machine described in Basford *et al.* (1973). Spectrometer sensitivity and mass discrimination were determined by analyzing atmospheric argon introduced into the sample processing system through a gas pipette. Mass discriminations were based on Nier's (1950) measurements. Calibrations were done before and after each sample, or each pair of samples.

Potassium analysis was done by isotope dilution mass spectrometry and is discussed in Evensen *et al.* (1973).

Gas extractions were all done at 1650°C for one hour heating periods. At mass 40 system blanks were typically <1.5% for total soil separates, and <3% for plagioclase except for the 16–25 μ fraction where the blank was 11%. In ilmenite the situation was complicated by low potassium levels and small sample sizes. Here the system residuals at ^{40}Ar ranged from 3% in the finer fractions to 22% in the coarsest fractions. Blank corrections of this magnitude are uncertain and the two ilmenite samples with blanks ~22% (74–105 μ, 147–250 μ) were tabulated in the tables but were not used in the determination of correlations.

Concentrations and isotopic compositions of ilmenite, plagioclase, and total soil argon contents are listed by grain size in Tables 1, 2, and 3. Listed in the same tables are the corresponding sample masses and potassium abundances. All rare gas and potassium determinations have been corrected for blank contributions. Errors listed for the rare gas isotopic measurements in the tables are $1\ \sigma$ statistical errors which include contributions from blank and mass discrimination uncertainties. Errors in the absolute concentrations of ^{40}Ar include systematic errors resulting from gas standard and pipette volume uncertainties as well as statistical and linearity uncertainties.

Table 1. Argon in grain size separates of Apollo 11 <1 mm bulk fines 10084,48.

Grain size fraction	Mass (mg)	K (ppm)	^{40}Ar ($\times 10^{-4}$ ccSTP/g)	$\dfrac{^{38}\text{Ar}}{^{36}\text{Ar}}$	$\dfrac{^{40}\text{Ar}}{^{36}\text{Ar}}$
590–1000 μ	61.16	1200	2.573 ±.064	0.19283 ±.00040	1.6818 ±.0047
250–590 μ	28.18	1580	1.975 ±.045	0.19338 ±.00037	1.5724 ±.0044
147–250 μ	28.49	1330	1.727 ±.040	0.19282 ±.00037	1.4100 ±.0039
105–147 μ	15.81	1170	1.582 ±.035	0.19351 ±.00037	1.3501 ±.0043
74–105 μ	12.85	1190	1.370 ±.032	0.19435 ±.00039	1.2904 ±.0085
37–74 μ	12.93	1080	2.210 ±.051	0.19200 ±.00038	1.1229 ±.0035
25–37 μ	10.31	1120	2.719 ±.052	0.19136 ±.00038	1.0526 ±.0033
16–25 μ	7.99	1130	3.429 ±.065	0.19072 ±.00038	1.0011 ±.0031
8–16 μ	7.21	1180	5.32 ±.10	0.19017 ±.00036	0.9722 ±.0022
4–8 μ	3.81	1200	7.49 ±.23	0.19182 ±.00067	0.9677 ±.0074
1–4 μ	1.36	1350	15.25 ±.36	0.19011 ±.00066	0.9631 ±.0073
≤1 μ	0.50	1630	19.08 ±.37	0.1887 ±.0013	1.005 ±.010
unsieved	12.56		4.054 ±.085	0.19040 ±.00061	1.0483 ±.0031

DISCUSSION AND RESULTS

There is abundant evidence that rare gases are present in lunar fines in separable surface (sc) and volume (vc) correlated components (e.g. Eberhardt *et al.*, 1970; Basford *et al.*, 1973). The surficially sited gases reside within a few thousand angstroms of the surface. Agreement is general that they result from solar wind implantation and possibly, to a lesser extent, from recycling in the lunar atmosphere. In the absence of admixed solar wind gases, volume sited gases result from high-energy spallation and *in situ* radioactive decay and fission. Although there is evidence for surface correlated radiogenic production in xenon

J. R. Basford

Table 2. Argon in plagioclase grain size separates of Apollo 11 <1 mm fines 10084,48.

Grain size fraction	Mass (mg)	K (ppm)	^{40}Ar ($\times 10^{-5}$ ccSTP/g)	$\dfrac{^{38}Ar}{^{36}Ar}$	$\dfrac{^{40}Ar}{^{36}Ar}$
250–590 μ	10.38	990 ±40	4.752 ±.071	0.21947 ±.00075	3.257 ±.023
147–250 μ	7.24	540 ±22	5.041 ±.081	0.21588 ±.00078	2.430 ±.017
105–147 μ	6.84	760 ±30	5.356 ±.086	0.21125 ±.00070	2.301 ±.016
74–105 μ	5.45	*600 ±24	6.25 ±.10	0.20494 ±.00068	2.227 ±.016
37–74 μ	10.54	*910 ±36	9.07 ±.15	0.19864 ±.00066	1.739 ±.012
25–37 μ	2.94	840 ±34	12.69 ±.20	0.19420 ±.00064	1.552 ±.011
16–25 μ	0.66	1320 ±53	19.52 ±.31	0.19134 ±.00063	1.423 ±.028
8–16 μ	3.93	*1110 ±44	22.33 ±.36	0.19117 ±.00059	1.1086 ±.0049
4–8 μ	1.74	1220 ±48	60.41 ±.97	0.19037 ±.00057	0.9957 ±.0040
1–4 μ	0.717	1330 ±53	73.4 ±1.2	0.18981 ±.00057	1.0559 ±.0055

*Duplicate K determinations.

Table 3. Argon in ilmenite grain size separates of Apollo 11 <1 mm fines 10084,48.

Grain size fraction	Mass (mg)	K (ppm)	^{40}Ar ($\times 10^{-5}$ ccSTP/g)	$\dfrac{^{38}Ar}{^{36}Ar}$	$\dfrac{^{40}Ar}{^{36}Ar}$
147–250 μ	3.34	600 ±36	1.94 ±.10	0.19740 ±.00066	*1.231 ±.039
105–147 μ	4.90	340 ±20	2.360 ±.071	0.19715 ±.00083	1.052 ±.022
74–105 μ	4.55	270 ±16	2.84 ±.12	0.19423 ±.00047	*0.691 ±.047
37–74 μ	8.14	226 ±14	2.891 ±.078	0.19264 ±.00044	0.844 ±.034
25–37 μ	4.52	220 ±13	6.83 ±.20	0.19061 ±.00046	0.730 ±.022
16–25 μ	8.38	290 ±17	11.14 ±.57	0.18990 ±.00057	0.7380 ±.0096
8–16 μ	6.50	230 ±14	16.22 ±.83	0.18875 ±.00085	0.7131 ±.0093
4–8 μ	9.64	210 ±13	28.7 ±1.5	0.18944 ±.00094	0.6782 ±.0081
1–4 μ	4.23	280 ±17	—	—	—

*Large blank correction (~22%).

(Basford *et al.*, 1973), ^{40}Ar resulting from *in situ* potassium decay is considered a volume component. Potassium concentrations do increase with decreasing grain size in the tota! soil ($<74 \mu$) and plagioclase ($<37 \mu$) (Tables 1 and 2), but as this effect is absent in the ilmenite fractions (Table 3) it is probably the result of mixing in a fine-grained, K-rich component rather than the result of surficial coatings.

In a simple model of spherical soil grains, the surface-to-volume ratios of the particles behave functionally as $1/r$. Grain sizing is therefore a relatively easy way to differentially fractionate volume and surface contributions and forms the basis for the ordinate intercept analysis of soil data. This approach was introduced by Eberhardt *et al.* (1970) and results from treating the various sources of an isotope as additive terms in a linear equation.

On this basis the equations for ^{36}Ar and ^{38}Ar may be manipulated to yield (with the convention in equations that $M \equiv {}^M\text{Ar}$)

$$\frac{38}{36} = \left(\frac{38}{36}\right)_{sc} + \frac{36_{vc}}{36}\left[\left(\frac{38}{36}\right)_{vc} - \left(\frac{38}{36}\right)_{sc}\right] \tag{1}$$

According to Eq. (1), a set of samples with common surface and volume compositions should plot as a straight line on a graph of ^{38}Ar/^{36}Ar versus $1/^{36}$Ar, if ^{36}Ar$_{vc}$ is also constant throughout the set. In this situation, it is apparent that the intercept gives the isotopic composition of the surficially trapped component. To the best of our knowledge, ^{38}Ar and ^{36}Ar result from the solar wind and spallation. The surface and volume components are identified as being, respectively, from these sources.

For ^{40}Ar the situation is more complex. First of all, although trapped ^{40}Ar/^{36}Ar ratios ~ 1 are commonly measured, the ^{40}Ar content of the solar wind is thought to be negligible. Cameron (1968) estimates $(^{40}\text{Ar}/^{36}\text{Ar})_{sw} \sim 10^{-4}$ and limits of <0.05 have been determined for this quantity in meteoritic studies (Pepin *et al.*, 1970).

Thus much of the trapped ^{40}Ar in the soil is anomalous in the sense that it is not supported by solar wind exposure. On the contrary, it is thought to be lunar in origin and to result from the production of ^{40}Ar in the lunar crust by potassium decay. A portion of this argon then escapes from the moon's surface and is energetically implanted in lunar material by atmospheric recycling (Heymann and Yaniv, 1970; Manka and Michel, 1971). Whatever the method of accumulation, ^{40}Ar is sited similarly to the solar wind-derived ^{36}Ar and is resolvable from volume sited gases.

A second complication in ^{40}Ar exists because ^{40}Ar$_{vc}$ is not produced only by spallation. Unlike the lighter isotopes, it is shielded from this process by Ca, and results almost entirely from the *in situ* decay of potassium. In most lunar soils, argon from this source (^{40}Ar$_{rad}$) is much more abundant than either ^{36}Ar$_{sp}$ or ^{38}Ar$_{sp}$. Since neither of these isotopes is shielded from spallation, it follows that ^{40}Ar$_{sp}$ is much smaller than ^{40}Ar$_{rad}$. An assumption used in this paper is that ^{40}Ar$_{sp} \ll {}^{40}$Ar$_{rad}$.

An equation similar to Eq. (1) could be written for the ^{40}Ar–^{36}Ar system. However, in chronologies the quantity of interest is the ratio of the radioactive parent and daughter. Equation (2) is an equivalent form, which emphasizes this ratio, and is used instead.

$$\frac{40}{(36)_{sc}} = \left(\frac{40}{36}\right)_{sc} + \frac{K}{(36)_{sc}} \left(\frac{40_{rad}}{K}\right) \tag{2}$$

This equation appears in Venkatesan *et al.* (1974) and has the advantage (for samples with common $(^{40}Ar/^{36}Ar)_{sc}$ and $^{40}Ar_{rad}/K$ values) that its slope yields the isochron necessary for K–Ar dating. Equation (2) is derived with the assumption that $^{40}Ar_{sp} \ll {}^{40}Ar_{rad}$.

Before isochrons may be determined with Eq. (2), it is necessary to calculate the values of $^{36}Ar_{sc}$. This calculation is most easily done on the basis that the $^{38}Ar/^{36}Ar$ ratios measured in a sample reflect the mixing of only solar wind and spallation components. The ratio of surface correlated ^{36}Ar to the total amount of ^{36}Ar in a sample is then simply determined from a mixing formula, and

$$\frac{(36)_{sc}}{36} = \frac{\dfrac{38}{36} - \left(\dfrac{38}{36}\right)_{sp}}{\left(\dfrac{38}{36}\right)_{sc} - \left(\dfrac{38}{36}\right)_{sp}} \tag{3}$$

Trapped and spallation compositions of $^{38}Ar/^{36}Ar$ are well defined and very uniform from sample to sample. Values of 0.189 (Table 6) and 1.52 (Huneke *et al.*, 1972) were used to represent $^{38}Ar/^{36}Ar$ surface and volume compositions. In all samples, variations as large as 30% in the assumed spallation ratio produced less than 0.5% change in the abundances determined for $^{36}Ar_{sc}$. The effect of the correction on Eq. (2) was small, and in all samples examined $(36)_{sc}/36$ was >0.97.

Ideally all the size separates of a mineral or soil component would be used in the correlations described by Eqs. (1) and (2). But in practice, the assumption of uniformity of isotopic composition and volume concentration in a set of samples can not be taken for granted. The conditions do appear to be met in the $^{38}Ar/^{36}Ar$ correlations. Except as noted below for the $\geqslant 105\ \mu$ fractions of the total soil, all size separates were used in Eq. (1). The ^{40}Ar correlations were more difficult and are discussed in the following paragraphs. Regardless of sample selection, all correlations were made with straight lines fit according to York's (1966) criteria.

In ilmenite, as already noted, two coarse fractions had 22% blanks and the measured $^{40}Ar/^{36}Ar$ ratios fell off the general trend. Although plotted, these samples were not used in the straight line fitting of the data. The coarsest plagioclase separate (250–590 μ) falls below the isochron of Fig. 1a in the direction of younger age. This point, although shown in the figure, was not considered in the determination of the isochron.

10084 is a mature soil (Venkatesan *et al.*, 1974) and, in common with other lunar soils (Butler and King, 1974), contains a significant proportion of agglutinates in the coarsest fractions. (These authors also find that the finer grain sizes of lunar soils consist primarily of mineral grains and glass.) Agglutinates contain fractionated gases of surficial origin in their interiors. These gases are more abundant than *in situ* produced radiogenic and spallation gases, and Table 1 shows that coarse fractions containing agglutinates are far more gas rich than would be expected from their grain size. Only for fractions $\leqslant 105\ \mu$ does this

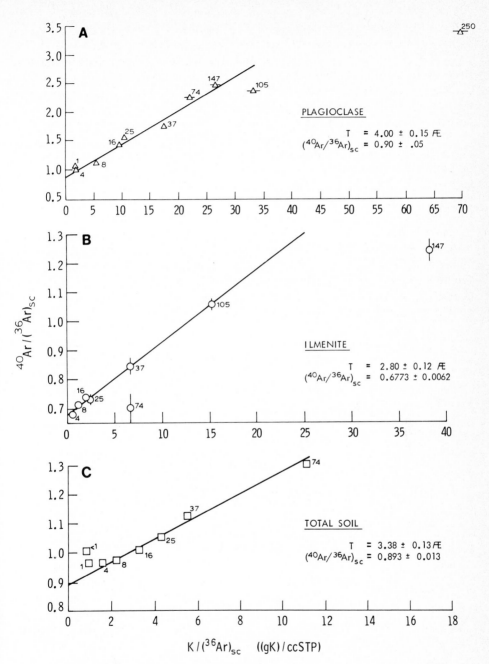

Figs. 1a–c. Plagioclase, ilmenite and total soil isochrons for <1 mm fines 10084. Intercepts are $(^{40}Ar/^{36}Ar)_{sc}$ and ages are calculated with Eqs. (2) and (4). As explained in the text, the total soil correlation is restricted to grain sizes finer than 105 μ. Numbers identify data points by minimum grain size of the fraction. Uncertainties in ages and intercepts are 1 σ. Error bars not shown are smaller than the symbols.

additional volume component seem to disappear. Here a simple ($\sim 1/r$) gas dependence exists and total soil correlations are restricted to this range.

A feature common to the total soil and plagioclase separates is the tendency of $^{40}Ar/^{36}Ar$ to increase in the finest grain sizes (Tables 1 and 2). This effect may be present in ilmenite also but could not be examined because of experimental limitations. Similar behavior in the finer grain sizes has been seen in other soils, such as 14149, Luna 16 and 75081 (Pepin *et al.*, 1972; Venkatesan *et al.*, 1974).

It is interesting that low-energy fluxes of ^{40}Ar are present on the moon (Vondrak *et al.*, 1974). It is also possible that finer particles in the soil would be preferentially exposed to the lunar atmosphere (McCoy and Criswell, 1974). However, a convincing explanation of this $^{40}Ar/^{36}Ar$ effect does not exist at present. Fractions of size $<1\,\mu$ and $1-4\,\mu$ have been plotted in ^{40}Ar isochrons, but have been excluded from any determinations of the correlations.

Immediately noticeable in the figures and in Table 4 is the constancy of the surficially trapped $^{38}Ar/^{36}Ar$ ratios and the variations of the slopes and intercepts of the $^{40}Ar/^{36}Ar$ isochrons. Other workers (Eberhardt *et al.*, 1970) have done a similar study on 10084 soil and ilmenite separates. Where comparable, the results of this paper are in agreement with their work.

The potassium isotope ^{40}K is radioactive and decays by both electron capture

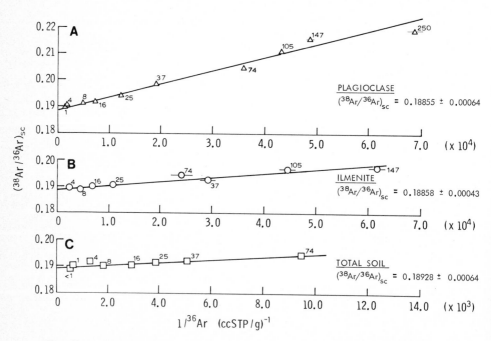

Figs. 2a–c. Plagioclase, ilmenite and total soil ordinate intercept determinations of $(^{38}Ar/^{36}Ar)_{sc}$. As explained in the text, the total soil correlation is restricted to grain sizes finer than 105 μ. Numbers identify data points by minimum grain size of the fraction. Uncertainties in the surface compositions are 1 σ. Error bars not shown are smaller than the symbols.

Table 4. K–Ar ages and Ar surface correlated ratios
in Apollo 11 <1 mm fines.

Component	K–Ar (AE)	$\frac{^{38}Ar}{^{36}Ar}$	$\frac{^{40}Ar}{^{36}Ar}$
Ilmenite	2.80	0.18858	0.6773
	±.12	±.00043	±.0062
Total soil	3.38	0.18928	0.893
	±.13	±.00064	±.013
Plagioclase	4.00	0.18855	0.90
	±.15	±.00064	±.05

and β^- emission with decay constants $\lambda_{ec} = 0.584 \times 10^{-10}$ yr^{-1} and $\lambda_{\beta^-} = 4.72 \times 10^{-10}$ yr^{-1}. Only the electron capture branch results in ^{40}Ar production and the use of these decay constants, along with a current $^{40}K/K$ ratio of 1.19×10^{-4}, allows calculation of the radiometric age equation. With the age (T) in AE, ^{40}Ar in units of 10^{-8} ccSTP/g and K in ppm, this equation is

$$T = 1.8854 \ln \left[1 + 1.3315 \left(\frac{40_{rad}}{K} \right) \right]$$ (4)

The ages obtained from applying Eq. (4) to the separates are shown in the isochron plots (Figs. 1a–c) and Table 4. Table 4 in particular shows that the ilmenite, total soil and plagioclase fail to have a common age or trapped $^{40}Ar/^{36}Ar$ value. Ilmenite has the youngest radiometric age at 2.80 AE, plagioclase has the oldest at 4.00 AE and the total soil is intermediate at 3.38 AE.

There is always the problem of what one is dating with a radiometric clock and the K–Ar system is no exception. The pristine sample is one that was thoroughly heated and outgassed at its formation and has, in addition, remained a closed system up to the moment of analysis. This ideal is seldom met in practice and the possibilities of the soil components being mixtures must be considered as well as the effects of incomplete sample outgassing, extraneous argon, and diffusion.

One might think that the ^{39}Ar–^{40}Ar dating technique would be an effective way to examine these possibilities. But, although its application to rocks has proved very fruitful, in soils its use has been blocked by the presence of surface correlated ^{40}Ar. This problem may soon be solved, but at present estimates of $^{40}Ar_{rad}$ retention in the soil from ^{39}Ar–^{40}Ar measurements must be based on rock determinations.

Stettler *et al.* (1973, 1974) have recently reported on their ^{39}Ar–^{40}Ar analysis of two Apollo 11 basalts. These rocks, 10003,39 and 10071,30, were crushed and several mineral fractions, including ilmenite and plagioclase, were examined.

The behavior in each rock was uniform, and all minerals from the same rock had a common intermediate or high-temperature plateau age (3.51 AE in 10071 and 3.91 AE in 10003). The low-temperature releases were more variable and showed depressed release of $^{40}Ar_{rad}$ with respect to ^{39}Ar.

As ^{39}Ar is a measure of the potassium concentration, the most direct explanation of this release pattern is that $^{40}Ar_{rad}$ is being lost from sites of low retentivity. A rough measure of this effect is obtained by comparing "total" ages (which utilize argon released at all temperatures) with plateau ages. In 10071 (Stettler *et al.*, 1973), this comparison shows that the plateau age was 3.51 AE and the total ages for the plagioclase, ilmenite, and total rock samples were ~3.32, 3.11, and 3.00 AE, respectively.

Thus, if gas loss is represented by age, plagioclase is the most retentive separate and shows a "loss" of about 0.2 AE. Ilmenite, on this basis, shows a larger loss (0.4 AE) and the total rock a still larger one (0.5 AE).

Such effects, present in the rocks, are undoubtedly also present in the individual grains of the soil. Whether loss of radiogenic argon from soil and rock mineral fractions is similar is unknown, but it seems prudent to assume that similar depletion may exist.

As a result of these considerations, the plagioclase separates from the soil have probably retained the most radiogenic argon and their ages are significant indicators of lunar events. Ilmenite is more suspect and its K–Ar age may be as much as ~0.5 AE low. The age of the total soil is not an independent quantity and reflects the effects of mixing of components as well as soil working processes.

The substantial loss of radiogenic argon in ilmenite is interesting because this mineral shows high retention of solar wind-derived gases (Eberhardt *et al.*, 1970; Basford, 1974). The reason for this difference may result from the different sitings of these gases. The solar wind gases are implanted directly into the surface layers of the ilmenite mineral matrix. It is likely, however, that $^{40}Ar_{rad}$ is sited primarily in small, non-ilmenite regions of high potassium concentrations. In fact, inclusions of high-K silica glass have been observed in lunar ilmenite (Roedder and Weiblen, 1970, 1971) and may well contain a significant proportion of the potassium found in separates of this mineral.

One of the results of this study is that different mineral fractions, in the same soil sample, record different K–Ar ages. Neither age is unique and each has been observed in other work.

The 2.8 AE age of ilmenite is younger than most radiometric ages in lunar soils. It has, however, been reported by Pepin *et al.* (1972, 1974).

The 4.0 AE age of the plagioclase separates has been observed much more frequently. It is roughly as old as, or slightly older than, the ages of the oldest Apollo 11 rocks (Turner, 1970; Stettler *et al.*, 1974). It is also observed in the mineralogically similar anorthositic material of the lunar highlands.

The caveats about K–Ar dating must be remembered, but it is felt that the soil ages determined in this experiment may be discussed in either of two models. One model views material transport as an important phenomena at this site, and the other considers that it is a relatively unimportant process.

In the view of the transport model, ilmenite and plagioclase ages record different events. This view grants that ilmenite may show the effects of large radiogenic argon losses, but feels that it still records a local event(s) that occurred between 2.8 and ~3.5 AE ago. The model also postulates that much of the plagioclase at this site originates in the highlands and is transported to the mare with minimum gas loss. (^{39}Ar–^{40}Ar experiments on plagioclase indicate that temperatures associated with the transport would have to be <500°C.) However, since highland regions exist within 50 km of the site, low-energy, lateral transport may well be an acceptable process.

Evidence for soil transport is found in the major and minor chemistry of 10084. This soil is poorer in ilmenite than are the rocks that were returned with the Apollo 11 mission (Gast *et al.*, 1970; Maxwell *et al.*, 1970; Wänke *et al.*, 1970). As highland material is rich in anorthosites and poor in ilmenite (LSPET, 1972), it is tempting to explain the discrepancy with the admixture of highland material. On a more quantitative level, it is straightforward to make chemical balances of representative oxides (e.g. Al_2O_3, TiO_2, and FeO) in Apollo 11 and Apollo 16 rocks and soils. When this is done, it is possible to consistently reproduce the soil in 10084 with a mixture derived ~70% from local rocks and ~30% from material with a highland composition.

While interesting, the evidence for this transport is not overwhelming. An obvious objection is that the rocks returned from the mare site failed to sample a major, anorthosite-rich, local contributor. Reid (1974) has recently suggested that this may be true for maria in general and that aluminous mare basalts may be a common rock type which has not been representively sampled in the rocks returned by the astronauts.

The second model notes these observations of Reid's and views the mineral components of 10084 as being locally derived. The local rocks, which have been sampled, range in age between 3.5 and 3.95 AE (Turner, 1970; Stettler *et al.*, 1974). The young age of ilmenite thus reflects either a later degassing event(s) or severe loss of radiogenic argon by diffusion. The old age of plagioclase reflects its high retentivity. And, as this mineral is by hypothesis local in origin, its source is in local material at least as old as the oldest Apollo 11 rocks that have so far been examined.

Both models explain the total soil age as the result of mixing and soil processes. The transport model mixes 30% material with the plagioclase isochron and 70% material with the ilmenite isochron. A soil with an age of 3.25 AE is derived. This is slightly lower than, but within error of, the 3.38 AE age of the total soil.

The local origin model also views the soil as a mixture. Here the mixture results from the comminution of local rocks only. As a result it would not be unreasonable to expect that the soil should exhibit an age similar to the "total" ages of whole rock samples dated with ^{39}Ar–^{40}Ar techniques. Earlier discussion showed that losses as large as ~0.5 AE are possible in rocks with 3.5 AE ages. The Apollo 11 rocks range in age between 3.5 and 3.95 AE. As they are present in the soil in unknown percentages, the 3.38 AE age of the total soil can be accounted for on the basis of only moderate gas loss by a soil formed from the local rocks.

A complication to the discussion of the origin of plagioclase has so far been ignored. Tables 2 and 3 show that, in contrast to the behavior in ilmenite, the K concentration rises in the finer fractions of plagioclase. It is possible that this is an artifact of the separation technique and represents increasing content of locally derived mesostasis with finer grain size. It may also represent the addition of foreign KREEP-like or Fra Mauro "exotic" components. Evensen et al. (1974) postulate that such a component may represent as much as 12% of the Apollo 11 soil and may be concentrated in the finer grain sizes due to the dynamics of long distance, lateral transport. The ages of Fra Mauro (Turner et al., 1972), the oldest Apollo 11 rocks and much of the highland material are within error of each other. As a result, no decision between these as sources of the plagioclase is possible on the basis of age alone.

Surficially sited gases also show variations between the separates. Table 4 displays these results and shows that, although $(^{38}Ar/^{36}Ar)_{sc}$ may be marginally higher in the total soil, $(^{38}Ar/^{36}Ar)_{sc}$ is within error of 0.1890 for all three determinations. This value has been seen in other lunar soils (e.g. Pepin et al., 1974; Venkatesan et al., 1974) and may be rather uniform on the moon. Although real variations undoubtedly exist, both these isotopes originate in the solar wind and constancy of $(^{38}Ar/^{36}Ar)_{sc}$ is not unexpected.

Unlike the behavior of surface correlated $^{38}Ar/^{36}Ar$, $(^{40}Ar/^{36}Ar)_{sc}$ has been observed to be quite variable with depth and location on the moon (e.g. Pepin et al., 1972; Bogard and Nyquist, 1973). Table 4 shows that this variability extends to separates from the same location. The respective values of $(^{40}Ar/^{36}Ar)_{sc}$ for ilmenite, total soil, and plagioclase are: 0.677, 0.893, and 0.90, and correlate roughly with their K–Ar ages.

In particular, the distinct variations of the ilmenite and plagioclase $^{40}Ar/^{36}Ar$ ratios are of interest. The differences arise in the local and solar wind origins of these isotopes and may show the effects of surface transport, secular decrease in the concentration of ^{40}Ar in the lunar atmosphere or different retentivity of these minerals to ^{36}Ar and ^{40}Ar.

The possibility of surface transport has already been discussed. Secular decrease of the $^{40}Ar/^{36}Ar$ trapped ratio with lunar activity seems qualitatively plausible and has been discussed by Yaniv and Heymann (1972). (If the variations in the ilmenite and plagioclase $^{40}Ar/^{36}Ar$ ratio are attributed to this cause, the different radiometric ages do reflect different times of formation.)

The third possibility, differential trapping and retention of argon by the soil constituents, is also a real possibility. According to the atmospheric model (Heymann and Yaniv, 1970; Manka and Michel, 1971), ^{40}Ar is cycled into the lunar surface with energies of 0–2 keV. Solar wind argon ions strike with approximately 30 keV. As a result, the solar wind argon ions are initially sited much more firmly than are the ^{40}Ar ions. Diffusion experiments, however, show that much of this distinction is lost (Ducati et al., 1972) due to the migration of gases in the soil particles. At the moment, it is not clear whether ilmenite and plagioclase, after prolonged exposures to the same ^{40}Ar and ^{36}Ar fluxes, would show different trapped ratio fractionations.

Analysis of the components of lunar soils with grain size, mineralogical and rare gas techniques is laborious but feasible and useful. The results of this work on 10084 show both that the ages of the components and their trapped $^{40}Ar/^{36}Ar$ ratios vary, and that this variation may presently be explained in terms of either a local or transport model. The same observations on $(^{38}Ar/^{36}Ar)_{sc}$ show that this quantity is very uniform from sample to sample and may be rather constant for lunar soils in general.

Acknowledgments—I am happy to acknowledge the encouragement and criticism that Professor R. O. Pepin supplied to this research. James Dragon shared spectrometer duties throughout most of this work and along with Michael Coscio, Norman Evensen, and Allan Bates of Professor V. R. Murthy's group furnished invaluable assistance. This research was supported by NASA grant NGL 24-005-225.

REFERENCES

Arrhenius G., Eversen J. E., Fitzgerald R. W., and Fujita H. (1971) Zirconium fractionation in Apollo 11 and Apollo 12 rocks. *Proc. Second Lunar Sci. Conf., Geochim. Cosmochim. Acta*, Suppl. 2, Vol. 1, pp. 169–176. MIT Press.

Basford J. R., Dragon J. C., Pepin R. O., Coscio M. R., and Murthy V. R. (1973) Krypton and xenon in lunar fines. *Proc. Fourth Lunar Sci. Conf., Geochim. Cosmochim. Acta*, Suppl. 4, Vol. 2, pp. 1915–1955. Pergamon.

Basford J. R. (1974) Rare gases in Apollo 11 regolith. In preparation.

Bogard D. D. and Nyquist L. E. (1973) $^{40}Ar/^{36}Ar$ variations in Apollo 15 and 16 regolith. *Proc. Fourth Lunar Sci. Conf., Geochim. Cosmochim. Acta*, Suppl. 4, Vol. 2, pp. 1975–1985. Pergamon.

Butler J. C. and King E. A. (1974) Analysis of the grain size frequency distributions of lunar fines (abstract). In *Lunar Science—V*, pp. 97–99. The Lunar Science Institute, Houston.

Cameron A. G. W. (1968) A new table of abundances of the elements in the solar system. In *Origin and Distribution of the Elements* (editor L. H. Ahrens), pp. 125–143. Pergamon.

Ducati H., Kalbitzer J., Kiko J., Kirsten T., and Müller H. W. (1972) Rare gas diffusion studies in individual lunar soil particles and in artificially implanted glasses. Max-Planck-Institut Für Kernphysik, Heidelberg, MPI H-V36.

Eberhardt P., Geiss J., Graf H., Grögler N., Krähenbühl U., Schwaller H., Schwarzmüller J., and Stettler A., (1970) Trapped solar wind noble gases, exposure age and K/Ar-age in Apollo 11 lunar fine material. *Proc. Apollo 11 Lunar Sci. Conf., Geochim. Cosmochim. Acta*, Suppl. 1, Vol. 2, pp. 1037–1070. Pergamon.

Evensen N. M., Murthy V. R., and Coscio M. R. (1973) Rb–Sr ages of some mare basalts and the isotopic and trace element systematics in lunar fines. *Proc. Fourth Lunar Sci. Conf., Geochim. Cosmochim. Acta*, Suppl. 4, Vol. 2, pp. 1707–1724. Pergamon.

Evensen N. M., Murthy V. R., and Coscio M. R. (1974) Episodic lunacy—V: Origin of the exotic component (abstract). In *Lunar Science—V*, pp. 220–221. The Lunar Science Institute, Houston.

Gast P. W., Hubbard N. J., and Wiesmann H. (1970) Chemical composition and petrogenesis of basalts from Tranquillity Base. *Proc. Apollo 11 Lunar Sci. Conf., Geochim. Cosmochim. Acta*, Suppl. 1, Vol. 2, pp. 1143–1163. Pergamon.

Heyman D. and Yaniv A. (1970) Ar^{40} anomaly in samples from Tranquillity Base. *Proc. Apollo 11 Lunar Sci. Conf., Geochim. Cosmochim. Acta*, Suppl. 1, Vol. 2, pp. 1261–1267. Pergamon.

Huneke J. C., Podosek F. A., Burnett D. S., and Wasserburg G. J. (1972) Rare gas studies of the galactic cosmic ray irradiation history of lunar rock. *Geochim. Cosmochim. Acta* 36, 269–301.

Huneke J. C., Jessberger E. K., Podosek F. A., and Wasserburg G. J. (1973) $^{40}Ar/^{39}Ar$ measurements in Apollo 16 and 17 samples and the chronology of metamorphic and volcanic activity in the Taurus-Littrow region. *Proc. Fourth Lunar Sci. Conf., Geochim. Cosmochim. Acta*, Suppl. 4, Vol. 2, pp. 1725–1756. Pergamon.

LSPET (Lunar Sample Preliminary Examination Team) (1972) Preliminary examination of lunar samples. *Apollo 16 Preliminary Science Report*, NASA.

Manka R. H. and Michel F. C. (1971) Lunar atmosphere as a source of lunar surface elements. *Proc. Second Lunar Sci. Conf., Geochim. Cosmochim. Acta*, Suppl. 2, Vol. 2, pp. 1717–1728. MIT Press.

Maxwell J. A., Peck L. C., and Wilk H. B. (1970) Chemical composition of Apollo 11 lunar samples

10017, 10020, 10072, and 10084. *Proc. Apollo 11 Sci. Conf., Geochim. Cosmochim. Acta*, Suppl. 1, Vol. 2, pp. 1369–1374. Pergamon.

McCoy J. E. and Criswell D. R. (1974) Evidence for a lunar dust atmosphere from Apollo orbital observations (abstract). In *Lunar Science—V*, pp. 475–477. The Lunar Science Institute, Houston.

Nier A. O. (1950) A redetermination of the relative abundances of the isotopes of carbon, nitrogen, oxygen, argon and potassium. *Phys. Rev.* 77, 789–793.

Pepin R. O., Nyquist L. E., Phinney D., and Black D. C. (1970) Rare gases in Apollo 11 lunar material. *Proc. Apollo 11 Lunar Sci. Conf., Geochim. Cosmochim. Acta*, Suppl. 1, Vol. 2, pp. 1435–1454. Pergamon.

Pepin R. O., Bradley J. G., Dragon J. C., and Nyquist L. E. (1972) K–Ar dating of lunar fines; Apollo 12, Apollo 14, and Lunar 16. *Proc. Third Lunar Sci. Conf., Geochim. Cosmochim. Acta*, Suppl. 3, Vol. 2, pp. 1569–1588. MIT Press.

Pepin R. O., Basford J. R., Dragon J. C., Coscio M. R., and Murthy V. R. (1974) K–Ar ages and depositional chronologies of Apollo 15 drill core fines. *Proc. Fifth Lunar Sci. Conf., Geochim. Cosmochim. Acta*. This volume.

Reid A. M. (1974) Luna 16 revisited: The case for aluminous mare basalts (abstract). In *Lunar Science—V*, pp. 627–629. The Lunar Science Institute, Houston.

Roedder E. and Weiblen P. W. (1970) Lunar petrology of silicate melt inclusions. *Proc. Apollo 11 Lunar Sci. Conf., Geochim. Cosmochim. Acta*, Suppl. 1, Vol. 1, pp. 801–837. Pergamon.

Roedder E. and Weiblen P. W. (1971) Petrology of silicate melt inclusions, Apollo 11 and Apollo 12 and terrestrial equivalents. *Proc. Second Lunar Sci. Conf., Geochim. Cosmochim. Acta*, Suppl. 2, Vol. 1, pp. 507–528, MIT Press.

Stettler A., Eberhardt P., Geiss J., Grögler N., and Maurer P. (1973) Ar^{39}–Ar^{40} ages and Ar^{37}–Ar^{38} exposure ages of lunar rocks. *Proc. Fourth Lunar Sci. Conf., Geochim. Cosmochim. Acta*, Suppl. 4, Vol. 2, pp. 1865–1888. Pergamon.

Stettler A., Eberhardt P., Geiss J., Grögler N., and Maurer P. (1974) On the duration of lava flow activity in Mare Tranquillitatus. *Proc. Fifth Lunar Sci. Conf., Geochim. Cosmochim. Acta*. This volume.

Turner G. (1970) Argon-40/Argon-39 dating of lunar rock samples. *Proc. Apollo 11 Lunar Sci. Conf., Geochim. Cosmochim. Acta*, Suppl. 1, Vol. 2, pp. 1665–1684. Pergamon.

Turner G., Huneke J. C., Podosek F. A., and Wasserberg G. J. (1972) ^{40}Ar–^{39}Ar systematics in rocks and separated minerals from Apollo 14. *Proc. Third Lunar Sci. Conf., Geochim. Cosmochim. Acta*, Suppl. 3, Vol. 2, pp. 1589–1612. MIT Press.

Turner G., Cadogan P. H., and Yonge C. (1973) Argon selenochronology. *Proc. Fourth Lunar Sci. Conf., Geochim. Cosmochim. Acta*, Suppl. 4, Vol. 2, pp. 1889–1914. Pergamon.

Venkatesan T. R., Johnson N. L., Pepin R. O., Evensen N., Coscio M. R., and Murthy V. R. (1974) K–Ar dating of lunar fines: The Apollo 17 dark mantle. *Earth Planet. Sci. Lett.* In press.

Vondrak R. R., Freeman J. W., and Lindeman R. A. (1974) Measurements of lunar atmosphere loss rate (abstract). In *Lunar Science—V*, pp. 812–813. The Lunar Science Institute, Houston.

Wänke H., Rieder R., Baddenhausen H., Spettel B., Teschke F., Quijano-Rico M., and Balacescu A. (1970) Major and trace elements in lunar material. *Proc. Apollo 11 Lunar Sci. Conf., Geochim. Cosmochim. Acta*, Suppl. 1, Vol. 2, pp. 1719–1727. Pergamon.

Yaniv A. and Heymann D. (1972) Atmospheric Ar^{40} in lunar fines. *Proc. Third Lunar Science Conf., Geochim. Cosmochim. Acta*, Suppl. 3, Vol. 2, pp. 1967–1986. MIT Press.

York D. (1966) Least squares fitting of a straight line. *Can. J. Phys.* 44, 1079–1086.

Proceedings of the Fifth Lunar Conference
(Supplement 5, Geochimica et Cosmochimica Acta)
Vol. 2 pp. 1389–1400 (1974)
Printed in the United States of America

Lead isotope systematics of some Apollo 17 soils and some separated components from 76501

S. E. CHURCH and G. R. TILTON

Geological Sciences, University of California, Santa Barbara, California 93106

Abstract—Uranium, thorium, and isotopic lead data have been measured on bulk samples of soils 71500,7, 76501,34, and 78500,7 and on agglutinate, plagioclase, and on the <74 micron fraction of 76501,34. The bulk soil analyses, when corrected for primordial lead and plotted in a concordia diagram, show the presence of a component of excess radiogenic lead having a $^{207}Pb/^{206}Pb$ ratio of 1.32.

All soil samples from the North Massif, including those from Station 9 analysed by Nunes *et al.* (1974), lie on a chord intersecting the concordia curve at ca. 3.75 and 4.35 AE. The lead isotopic systematics of these soils can be accounted for by mixtures of various highland components derived from the crystalline rocks and the local Taurus-Littrow area basalts. Data reported by other laboratories show that the lead isotope systematics for soils at Stations 2 and 4, both near the South Massif, plot along a 4.4–3.9 AE chord (Radiogenic $^{207}Pb/^{206}Pb = 1.45$). This is probably due to control of the lead isotope systematics at these stations by initial lead like that found in the orange soils [which also contain very high concentrations of lead].

Agglutinate from 76501,34 shows loss of lead compared to the bulk soil. The fine material of the soil (assumed source of the agglutinates) and the agglutinates define a chord in a concordia diagram having a lower intersection with the curve at 0.6 AE. This value probably has no direct significance since all of the loss would not have occurred at that particular point in time, but it does agree rather closely with the nominal "exposure ages" of many Apollo 17 soils. It is also possible to reconcile the data with models in which agglutinate formation has occurred over approximately the past 4 AE.

Plagioclase from 76501,34 contains radiogenic lead which is largely supported by the uranium in the sample. The acid washed sample yields nearly concordant model ages of 4.4 AE, a result that contrasts sharply with the large fractions of unsupported radiogenic lead in highland anorthosites 15415, 60015, 67075, and 66095. The model ages for the acid washed plagioclase agree closely with those for anorthositic gabbro 77017 (Nunes *et al.*, 1974). No evidence for undisturbed material with an age of 4.6 AE was found in any of the samples.

INTRODUCTION

IN A CONVENTIONAL Rb–Sr isochron diagram, the best fit line through soils from all missions corresponds to a model age of 4.35 ± 0.10 AE, with an initial $^{87}Sr/^{86}Sr$ of 0.69921 ± 16 (Papanastassiou and Wasserburg, 1972). These authors postulate that the soil data record Rb–Sr differentiation events that occurred 4.3–4.6 AE ago. The events, if real, predate the oldest directly measured ages for crystalline rocks, and merit further study.

Several studies have shown that soils contain materials having a variety of Rb–Sr model ages. For example, Evensen *et al.* (1973, 1974) have shown that the model ages of soils at the Apollo 14, 15, and 17 sites vary systematically with particle size. They postulate that an "exotic component" controls the ages of the finest particle sizes, yielding values that approach 4.6 AE.

Nyquist *et al.* (1973) demonstrated a correlation between the Rb–Sr model age and agglutinate content of the 90–150 micron fractions of Apollo 14 and 16 soils.

High agglutinate contents correlated with higher model ages. Nyquist *et al.* (1973) postulated that, in the absence of agglutinates, the model ages of soils are controlled by crystalline KREEP, with a model age of 4.4 AE. Higher model ages are presumably due to the influence of agglutinates, which have lost Rb by volatilization during their formation. In this case the authors note that the average Rb–Sr model age of ca. 4.5 AE for all soils may be an accidental result of the amount of Rb lost by the agglutinates.

U–Pb measurements on bulk soils generally yield single stage $^{207}Pb/^{206}Pb$ ages greater than 4.8 AE (Nunes *et al.*, 1973; Silver, 1972). These ages result from the transfer to the lunar crust of unsupported radiogenic lead with a high $^{207}Pb/^{206}Pb$ ratio that was produced during approximately the first 0.7 AE of the moon's history (Nunes *et al.*, 1973). Nunes *et al.* (1973) also cite evidence for episodic loss of lead from Apollo 12, 14, and 16 soils approximately 2.0 AE (so-called "third event"). The significance of this number is obscure, as it has not been detected by the other dating methods. Pepin *et al.* (1973), however, have noted K–Ar ages of 2.5–3.0 AE in lunar soils, but no correlation in these ages can be demonstrated.

The present studies use the U–Pb and Th–Pb systems to study soil evolution problems along the lines outlined above for the Rb–Sr method. Where possible, various fractions are separated from the soils instead of relating data for bulk samples to the percentage of a given component. Relatively few isotopic data on soil separates have been reported. Tatsumoto *et al.* (1971, 1972) give isotopic lead data on Apollo 11, 12, and 14 soils. Their separates were obtained with the aid of heavy liquids, which are difficult to purify from lead. The separations in this work were accomplished by sieving, magnetic separation, and hand picking.

EXPERIMENTAL PROCEDURES

Sample handling and mineral separations

Each of the bulk soils was first analysed by pouring a 100–200 mg aliquot from a weighing paper into a 5 ml beaker. The effect of this sampling procedure, as opposed to taking a scoop of bulk material from the soil, is to concentrate coarser fragments (e.g. 76501) in the analysed sample (see Fig. 1 and discussion). In this respect, our "bulk" soils are not good statistical splits of the unknowns.

Mineral and agglutinate separates were obtained by first sieving the sample (Spex nylon screen-lucite sieve set) and separating them into three fractions (coarse: >149 microns; intermediate: 74–149 microns; and fine: <74 microns). One gram of the fines (76501) was then washed in distilled acetone for 4 hours, the acetone removed, and the Pb present in the acetone wash analysed. About 2 ng of terrestrial Pb was observed; no evidence of leached or volatile lunar Pb was found. A split of this washed fines sample was then analyzed.

The coarse fraction contained fragments of the most interest, but all of the grains were coated with fines dust making identification and separation of pure end members impossible. This fraction was also washed in acetone to remove the adhering fines. Unfortunately, this acetone wash was not analyzed. Mineral fragments, rock fragments, and agglutinate particles were clearly visible and separates (>99% purity) of plagioclase and agglutinate components were quickly obtained by the use of the Franz isodynamic separator. These separates were then cleaned of impurities by hand picking using a teflon minivacuum sweeper. The two separates were again washed in acetone prior to analysis to remove any terrestrial contamination. This method of obtaining separates for Pb isotopic work is far preferable to the use of heavy liquids because of the presence of Pb compounds in heavy liquids and the difficulty of washing all the heavy liquids out of porous components, namely the agglutinate fraction of lunar soils. All mineral separations and handling were done in laminar flow work areas.

Chemistry

The chemical procedures were adapted from those described by Tilton (1973) for meteorite analyses. All U, Th, and Pb data reported were obtained on aliquants of digested lunar samples. These were placed in a 5 ml beaker with 100λ HNO_3, 150λ $HClO_4$, and 1 ml HF for each 100 mg of sample. Digestion was done in covered teflon beakers in a laminar flow environment. The sample was taken into clear solution in dilute HCl and aliquanted. Approximately 30–40% of the solution was separated for the Pb concentration analysis. U and Th concentrations were also determined on this same aliquant for all samples, except for the plagioclase determinations, where the entire sample was spiked for the U and Th analysis. Pb separation was achieved by anion chromotography in the bromide medium using a 500λ teflon column (0.32 cm $\times$ 6.5 cm). The Pb was further purified by anion chromotography in the chloride medium on a second 100λ teflon column (0.2 cm $\times$ 3.5 cm). Pb blanks, determined simultaneously using 1 ng ^{208}Pb, range from 0.9 to 2.0 ng but these values appear excessive because of poor yields in the blank chemistry during the precipitation procedure. This step involved the precipitation of Pb with ca. 20 mg of Fe carrier after the dissolution procedure. Total reagent blanks for an analysis are 0.5–0.6 ng and the column procedure and loading blanks are 0.1–0.15 ng. All data in Table 2 have been corrected for 1 ng blank for the total sample, proportioned between the two aliquants on the basis of their relative weights. The lack of agreement of our $^{204}Pb/^{206}Pb$ values in the two splits is undoubtedly due to the nonreproducibility of our handling after the sample is split. Further work (in progress) on the evaluation of our procedure blanks should clarify this problem and allow us to improve on the agreement of these data.

U and Th were separated using anion exchange in the nitrate medium (Tatsumoto, 1966). Blanks during the initial stages were in the tenths of ng range, but drastically improved to less than 0.01–0.02 ng by the introduction of a 2MHCl–1MHF elution and wash procedure. A 1 ml teflon column (0.3 cm $\times$ 12.5 cm) is used; all chemistry is carried out in class 100 laminar flow work areas.

The isotopic measurements were carried out on a 35 cm, 90° sector AVCO mass spectrometer equipped so that data can be taken either with a Faraday cup or with an electron multiplier. The signal is measured using a Cary 401 vibrating reed electrometer with remote ranging capability. Peaks are scanned by changing the magnetic field and data are automatically collected and reduced by an on-line small computer. Data collection is as follows: Reference Peak, baseline, Peak A, baseline, Reference Peak, baseline, Peak B, etc.; the program may be set up for 2, 3, or 4 isotopes and data sets normally consist of from 8 to 15 cycles. Each position is counted for eight seconds with the first two readings being discarded to allow the Cary 401 to settle. Monitoring the reference peak every other time allows for short integration time between peaks. Range scales have been intercalibrated relative to a 1.35 volt mercury cell using a voltage divider. The instrument was calibrated by replicate analyses of S.R.M. 981, 982, 983 for Pb and S.R.M., U-050, U-500, U-970 for U, (available from National Bureau of Standards, Washington, D.C.). The Pb analyses are made using the standard silica gel-H_3PO_4 emitter technique. F. Tera kindly provided the silica gel batch we are using for the lunar analyses and allowed us to decrease our loading blanks substantially (i.e. from 0.1 to 0.14 ng to 0.02–0.04 ng). One ng of ^{208}Pb routinely gives signals of 3×10^{-12} amps. Isotopic fractionation is monitored by running the S.R.M. Pb standards at the level of the analyses (i.e. 30–50 ng for this study). All reported data have been measured on the Faraday cup except for the $^{204}Pb/^{206}Pb$ values which are measured on the electron multiplier and appropriately corrected for mass discrimination.

Pb, U, and Th concentration data are measured using ^{208}Pb, ^{235}U, and ^{230}Th spikes, respectively. The Pb analyses are performed as mentioned above. U and Th analyses for all samples except the plagioclase analyses were made using ~120 ng ^{235}U and ~230 ng ^{230}Th spikes, by the triple filament method of ionization. The U and Th analyses for the plagioclase work were done using the Ta_2O_5 procedure described by Tera and Wasserburg (1972). High purity Ta_2O_5 ($^{238}U < 3$ picograms and $^{232}Th < 10$ picograms per analysis) was readily prepared using anion chromatography in the HF medium. Samples were spiked with ~5 ng ^{235}U and ~10 ng ^{230}Th.

The U, Th, and Pb concentrations given in Table 1 are believed accurate to ±1%. The observed $^{207}Pb/^{206}Pb$ and $^{208}Pb/^{206}Pb$ ratios in Table 2 are believed accurate to ±0.1%; the $^{204}Pb/^{206}Pb$ ratios to ±0.5%.

Results

The data and results are given in Tables 1–3 and summarized graphically in Fig. 1. We have used the newer values for the decay constants of ^{238}U and ^{235}U determined by Jaffey et al. (1971). This has the effect of lowering ^{207}Pb/^{206}Pb ages at 4.5 AE by ca. 0.06 AE compared to ages derived from the decay constants in common use since 1950.

Discussion

Bulk soils. The concentration of U, Th, and Pb in soils 71500, 76501, and 78500 are all in the general ranges found for the Apollo 11, 15, and 16 soils and are lower than those found in the Apollo 12 and 14 soils. The soils do appear to have substantially higher concentrations of these elements than do the local crystalline rocks (Nunes et al., 1974), a relationship typical of all sites, but they are markedly lower than the U, Th, and Pb concentrations found in the breccias from the South Massif (Tatsumoto et al., 1974).

Compositionally, the soils represent a wide range of mixtures of local rock types. Heiken and McKay (1974), Rhodes et al. (1974), and Philpotts et al. (1974) argue on the basis of petrographic, major element, and trace element chemistry that these soils are essentially mixtures of basaltic components from the Taurus-Littrow valley and the material from the surrounding anorthositic highlands. The soils however do represent a rather wide range of components as shown by Heiken and McKay (1973) and by the data on soils given by LSPET (1973, Table 7-III). Soil 71500 is predominantly basaltic in composition, particularly among the coarser grain size fractions, whereas soil 76501, from the base of the North Massif, contains a substantial anorthositic highland component. Soil 78500 is

Table 1. Concentrations of uranium, thorium, and lead in some Apollo 17 soils.

Sample	Fraction	Weight (mg)	Concentrations, ppm			Atom ratios	
			U	Th	Pb	^{232}Th/^{238}U	^{238}U/^{204}Pb
78500,7	Total	47.1	0.330	1.190	1.113	3.72	76.2
71500,7	Total	68.2	0.214	0.792	0.727	3.85	92.6
76501,34	Total	53.5	0.468	1.649	1.267	3.62	194.6
76501,34	Agglutinate	39.1	0.458	1.681	1.162	3.79	185.2
76501,34	Fines[a]	58.0	0.455	1.697	1.349	3.84	167.7
76501,34	Plagioclase, leached[b]	51.7[c]	0.214	—	0.410	—	206.5
		139.0					
76501,34	Plagioclase leach solution	—	0.0136[d]	0.0781[d]	0.110[d]	5.99	20.36
76501,34	Plagioclase, total	139.0	0.227	—	0.520	—	146.1

[a]Fines: less than 74 microns.
[b]Sample was leached 10 min with cold 1M HNO₃ before analysis.
[c]Total sample, 139.0 mg spiked for U, Th; 51.7 mg fraction taken for Pb determination.
[d]Concentrations are micrograms of element removed per gram of sample.

Table 2. Isotopic composition of lead in some Apollo 17 soils.

Sample	Fraction	Weight (mg)	Observed ratios[a] $\frac{^{204}Pb}{^{206}Pb}$	$\frac{^{207}Pb}{^{206}Pb}$	$\frac{^{208}Pb}{^{206}Pb}$	Corrected for analytical blank[b] $\frac{^{204}Pb}{^{206}Pb}$	$\frac{^{207}Pb}{^{206}Pb}$	$\frac{^{208}Pb}{^{206}Pb}$
78500,7	Total	73.3	0.01023	0.7557	1.1112	0.00996	0.7551	1.1056
71500,7	Total	47.1	0.00958	0.7540	—	0.00940	0.7529	—
		121.5	0.00806	0.7571	1.0612	0.00779	0.7567	1.0553
76501,34	Total	68.2	0.00795	0.7492	—	0.00774	0.7485	—
		81.2	0.00466	0.6792	1.0095	0.00444	0.6774	1.0046
76501,34	Agglutinate	53.5	0.00551	0.6781	—	0.00531	0.6771	—
		53.0	0.00540	0.7004	1.0268	0.00504	0.6995	1.0190
76501,34	Fines	39.1	0.00557	0.6933	—	0.00525	0.6899	—
		82.2	0.00497	0.7107	1.0130	0.00477	0.7102	1.0085
76501,34	Plagioclase, leached	58.0	0.00518	0.7077	—	0.00499	0.7061	—
		87.3	0.00524	0.5677	0.5942	0.00472	0.5648	0.5781
76501,34	Plagioclase leach solution	51.7	0.00382	0.5579	—	0.00323	0.5401	—
		—	0.01470	0.9546	1.3467	0.01301	0.9601	1.3152
76501,34	Plagioclase, total	—	0.01345	0.9581	—	0.01248	0.9149	—
		—	—	—	—	0.00596	0.6237	0.6880

[a]Entries with no ^{208}Pb data are from isotope dilution experiments.
[b]All samples corrected for blank of 1 ng, except plagioclase leach solution for which the blank was taken as 0.45 ng.

Table 3. Model ages of Apollo 17 soil samples.

Sample	Fraction	Atom ratios, corrected for primordial lead[a] $\frac{^{206}Pb}{^{238}U}$	$\frac{^{207}Pb}{^{235}U}$	$\frac{^{207}Pb}{^{206}Pb}$	$\frac{^{208}Pb}{^{232}Th}$	Model ages in million years $\frac{^{206}Pb}{^{238}U}$	$\frac{^{207}Pb}{^{235}U}$	$\frac{^{207}Pb}{^{206}Pb}$	$\frac{^{208}Pb}{^{232}Th}$
78500,7	Total	1.2423	123.23	0.7193	.2998	5205	4896	4770	5300
71500,7	Total	1.2963	130.37	0.7294	.3015	5359	4953	4790	5326
76501,34	Total	1.1182	101.60	0.6589	.2800	4838	4702	4644	4990
76501,34	Agglutinate	1.0279	96.29	0.6794	.2476	4557	4648	4688	4471
76501,34	Fines	1.2044	114.90	0.6918	.2840	5095	4826	4714	5053
76501,34	Plagioclase, leached	0.9766	72.67	0.5399	—	4392	4366	4355	—
76501,34	Bulk plagioclase	1.06825	87.70	0.5954	—	4684	4554	4498	—

[a]^{204}Pb: ^{206}Pb: ^{207}Pb: ^{208}Pb:: 1.00: 9.307: 10.294: 29.480 (Tatsumoto, Knight and Allegre, 1973; Tilton, 1973).

intermediate between the two extremes (e.g. Philpotts *et al.*, 1974, Fig. 1). Likewise, they also vary in soil maturity, as defined on the basis of their agglutinate content, in the same order. 76501, which contains 47% agglutinate material in the 90–150 micron fraction, is more mature and 78500 and 71500 containing 35% agglutinate particles are less mature (LSPET, 1973, Table 7-III).

There is also a compositional dependence on size fraction noted for these lunar soils and more components are present in the fine-grained fractions.

In Fig. 1 we present the results of this study on a conventional concordia diagram together with pertinent results from other workers (Nunes *et al.*, 1974; Silver, 1974; Tera *et al.*, 1974). The bulk soils from this study, those from site 9 (Nunes *et al.*, 1974), and the Apollo 17 basalts form a mixing line chord with a slope corresponding to inherited radiogenic lead with ^{207}Pb/^{206}Pb = 1.32. This chord is distinctly different from the cataclysm chord (dashed line of Fig. 1) for which this ratio is 1.45. The lower intersection of the chord was fixed at 3.75 AE (the Rb–Sr mineral isochron age of 70035, Nyquist *et al.*, 1974), to reflect the substantial basaltic component present in the soils. This line also passes through the three whole rock Apollo 17 basalt results reported by Nunes *et al.* (1974). The two soils from the South Massif (72701 and 75120) do not fall on this line and probably reflect the relatively sparse basaltic component in these soils. 74220 and 74240 (Nunes *et al.*, 1974), which are not shown in Fig. 1, also plot on the same

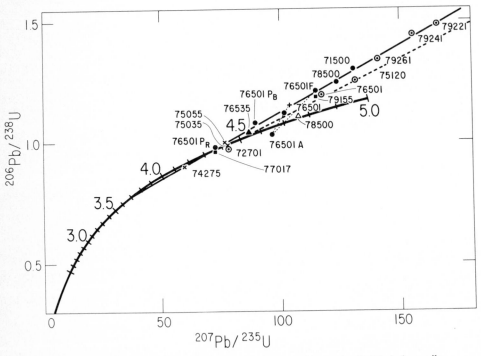

Fig. 1. Concordia diagram for Apollo 17 soils and crystalline rocks. Solid circles: soils, this work; A = agglutinates, P_R = leached plagioclase, P_B = bulk plagioclase, F = fines. Open circles: soils from Nunes *et al.* (1974). X: basalt from Nunes *et al.* (1974); solid squares: anorthositic gabbro from Nunes *et al.* (1974); open triangle: soil from Silver (1974); solid triangle: troctolite from Tera *et al.* (1974). Solid line is a chord intersecting concordia at 3.75 AE, with a slope corresponding to radiogenic lead with ^{207}Pb/^{206}Pb = 1.32 after correction for primordial lead; dashed chord intersects concordia at 3.90 AE and has a slope corresponding to radiogenic lead with ^{207}Pb/^{206}Pb = 1.45.

chord as 72701 and 75120. The lead isotope systematics in the soils from the stations near the South Massif seem to be controlled by initial lead with an isotopic composition similar to that found in the orange soils. At present we cannot characterize the chord for soils from Station 5 since 75081, as reported by Barnes *et al.* (1974), plots along the 3.75–4.35 AE chord in Fig. 1.

Comparison of our results with those of Nunes *et al.* (1974) for 76501 suggests that our bulk sample contains a higher plagioclase component—note that our bulk soil (76501) lies on a tie line between our plagioclase separate and their 76501 bulk soil analysis. The probable reason for this was explained in the section on experimental results and represents a biasing of our bulk sample in favor of plagioclase. A comparison of our bulk soil analysis for 78500 with that reported by Silver (1974) is not possible because of the lack of data on components in the soil. We are now working on separations of components in 78500.

Separates. As mentioned above, the Apollo 17 soils appear to represent mixtures of several local rock types. In attempting to understand the U, Th, and Pb systematics of lunar soils, it is necessary to analyse not only the rock components of the soils, but also to examine the effect of micrometeorite bombardment of the fines to produce the agglutinate component in the soils. To evaluate this process, we have extracted three components from 76501, based both on grain size and physical properties, in an attempt to better understand bulk lunar soil data. These three fractions form a triangle around the bulk soil analysis (Fig. 1). Both the plagioclase fraction and the fines lie above the cataclysm chord (dashed line in Fig. 1) and appear to represent the influence of local basalt components in the fines of the soils. We view the fines as essentially mixtures of the local rock types as supported by the arguments made from the physical and chemical data referenced above.

The plagioclase separate from soil 76501 was first leached in cold 1M HNO_3, following Tera *et al.* (1972), in an attempt to remove any loosely bound Pb. Differential leaching of Pb relative to U (and Th) is readily apparent from the data in Tables 1 and 2. The leached elements have been added to the residual plagioclase concentrations after correction for the blank to obtain the bulk sample. Only the bulk sample and the residual sample are plotted on Fig. 1. We have also included the data on the troctolite 76535 (Tera *et al.*, 1974), an anorthositic gabbro (77017), and a gabbro (79155) from the North Massif (Nunes *et al.*, 1974). Lead from the plagioclase separate is very similar to that obtained in these rocks and the composition of the leach lead closely parallels that reported by Tera *et al.* (1974) for their leach experiments on 76535. As in the case of 76535, a line through our bulk plagioclase sample and our residual plagioclase sample would be nearly tangent to the concordia curve at 4.0 AE. Since this sample does not contain an appreciable component of excess radiogenic Pb which is isotopically distinct from that in the local crystalline rocks, we conclude that the lead in the plagioclase component in the soil has not been altered by transport or inclusion processes in the soil. The model ages for the leached plagioclase are nearly concordant at 4.35 AE and agree closely with the model ages of anorthositic gabbro 77017 (Nunes *et al.*, 1974). This contrasts markedly with the virtually unsupported radiogenic lead found in highland anorthosites 15415 (Tatsumoto *et al.*, 1972; Tera *et al.*, 1972), 60015 and 66095 (Nunes *et al.*, 1973), and 67075 (Silver, 1973).

The analysis of the agglutinate component from 76501 represents the first U–Th–Pb work done on pure agglutinate separates from a lunar soil. Tatsumoto *et al.* (1971, 1972) report results on size and density fractions of 10084, 12070, and 14003 but no component separations were made on these fractions. The various chords in Fig. 1 connecting the bulk soil analysis and the agglutinate component show the effect of the several components on the bulk soil analysis. Furthermore, the data of LSPET (1973, Table 7-III) can be used to estimate the effect the agglutinate component has on the bulk soil and to correct for it. The corrected point (+ on the agglutinate-bulk soil vector, Fig. 1) lies on the tie line between the plagioclase and the fines, and on the 3.75 AE chord.

Agglutinate components have been identified from all the Apollo sites (Duke *et al.*, 1970; McKay *et al.*, 1972; Heiken *et al.*, 1973). McKay *et al.* (1972) have suggested that agglutinate particles are formed as a result of micrometeorite impact of the lunar soils, welding together lunar fines. (Some agglutinate particles may include coarser grained constituents, but such mixed grains were removed from the agglutinate sample analyzed here in an attempt to remove as many variables as possible.) If we can assume then that the analyzed agglutinate represents the lunar fines of this soil fused together as the result of micrometeorite impact, the isotopic data in Fig. 1 formally indicate loss of lead in relatively recent time—<0.6 AE ago on the average. This age agrees rather closely with the average "exposure ages" of many Apollo 17 soils (Eberhardt *et al.*, 1974), but the agreement is probably fortuitous. In fact, the age value probably has no real significance since the micrometeorite bombardment undoubtedly took place over an extended period of time. The lead loss pattern can be duplicated rather closely by a model in which bombardment has occurred at a constant rate in which case the instantaneous production of lead in the agglutinate fraction is expressed by:

$$\frac{d\ (^{206}\text{Pb})}{dt} = {}^{238}\lambda\ {}^{238}\text{U} - k\ (^{206}\text{Pb})$$

where ^{206}Pb and ^{238}U are the atom concentrations of ^{206}Pb and ^{238}U, respectively; $^{238}\lambda$ is the decay constant for ^{238}U; k is a constant related to the fraction of total lead lost by volatization during agglutinate formation. An analogous equation can be written for ^{207}Pb and ^{235}U. The trajectory for lead loss by such a model as a function of k will have nearly constant slope over much of its length and the slope will depend on duration assumed for the bombardment process. For example, if the rate of bombardment has been constant over the past 4.3 AE, the slope of the trajectory closely matches that of the line through 76501F and 76501A in Fig. 1. The slope of the trajectory can also be controlled by varying the time dependence of the rate of bombardment. Until more agglutinate results are obtained, it hardly seems worth pursuing the matter in greater detail, but we emphasize that lead data from agglutinates may ultimately yield significant information on the time scale of micrometeorite bombardment processes on the moon.

Nyquist *et al.* (1973) summarized the available data on agglutinate-rich and agglutinate-poor soils and related the results to the process involved in agglutinate formation. They noted the general correlation of the loss of volatile elements with agglutinate content in the soils. Heiken *et al.* (1973a) also show this effect very clearly on the Hg concentrations and the rare gas contents of agglutinate-rich and agglutinate-poor layers in the Apollo 15 core. There is also a general correlation with the Pb isotopic data on the cores reported by Silver (1972), but the effect may be partially masked by the Pb contamination introduced in obtaining and processing the core samples for distribution. Finally, Nyquist *et al.* (1973) pointed out the strong correlation between Rb–Sr model ages of bulk soils and the agglutinate content. Using the data of Gibson and Hubbard (1973), they also showed that the process of bombardment would cause Rb loss by volatilization and would account for small changes of the T_{BABI} model ages because of small loss of the parent isotope. These observations were essentially confirmed by the size fraction analysis of the agglutinate-rich soil 14259 by Murthy *et al.* (1973). Smaller scale effects are probably also present in 75081, but they were not prominent in the Rb–Sr system.

In the U, Th, Pb system, the daughter, rather than the parent isotope is volatile. Data on comparative volatile effects for both elements are sparse. Cliff *et al.* (1971) noted that terrestrial basalt samples heated in vacuum to the melting point experienced extensive loss of K, Rb, and Pb relative to Sr. Volatilization extraction techniques have been used by a number of workers (Huey *et al.*, 1971; Silver, 1971; Doe and Tatsumoto *et al.*, 1972), all indicating significant Pb loss at temperatures well below melting (600–800°C) whereas detectable loss of Rb from similar material did not occur until 1050°C (Gibson and Hubbard, 1973). Thus, the Pb isotope system will show larger effects on parent/daughter ratios than the Rb–Sr system in response to the same instantaneous heating event, and the direction of the effect will be reversed, tending to decrease the model ages because the daughter rather than the parent isotope is volatile. Comparisons of the model ages of the bulk soil 76501 with the agglutinate and fines separates (see Table 3 and Fig. 1) dramatically illustrates the effect of lead loss on the model ages. Although the agglutinate has a model age of near 4.6 AE, this value clearly has no age significance. Rather it reflects lead loss from a soil which has already inherited a significant amount of radiogenic Pb early in its lunar history (i.e. the first 0.7 AE).

CONCLUSIONS

(1) Isotopic lead data from bulk samples of Apollo 17 soils 71500,7, 76501,34, and 78500,7, plus data from the fine-sized fraction of 76501,34, define a chord in a concordia diagram showing the presence of a component or components containing excess radiogenic lead with $^{207}Pb/^{206}Pb$ equal to approximately 1.32. Three samples from Station 9, reported by Nunes *et al.* (1974), also lie along this chord. The chord is distinctly different from the cataclysm chord, for which $^{207}Pb/^{206}Pb$

equals approximately 1.45. The difference apparently is due to control of lead in these soils by components from local rocks having ages of about 3.75 AE. Other soils from Stations 2 and 4 near the South Massif yield leads that do plot along the cataclysm chord (Nunes *et al.*, 1974).

(2) Plagioclase separated from 76501,34, when leached with cold, dilute HNO_3, gives nearly concordant lead ages of around 4.35 AE, an age that agrees closely with the lead age measured on anorthositic gabbro 77017 by Nunes *et al.* (1974).

(3) Agglutinates from 76501,34 show loss of approximately 15% of lead, nominally at relatively recent time but long-term processes can also account for the data. This loss of lead correlates with other data suggesting loss of Rb, and may prove a useful mechanism to help explain some of the scatter in lead age data from soils. Further data from agglutinates (work in progress) will help to assess whether the time scale for micrometeorite bombardment can be ascertained from the lead isotope systematics.

(4) Inspection of the data points in Fig. 1 shows that there is no evidence for samples whose U–Pb ratios have not been disturbed since 4.6 AE ago. Instead the soil data give evidence for source materials with ages of around 4.3–4.4 AE.

Acknowledgments—We are indebted to Mr. Mark Stein for invaluable assistance in the design, fabrication, and maintenance of electronic components of the mass spectrometer. This project was supported by NASA Grant NGR 05-010-081.

References

Barnes I. L., Garner E. L., Gramlich J. W., Machlan L. A., Moody J. R., Moore L. J., Murphy T. J., and Shields W. R. (1974) Isotopic abundance ratios and concentrations of selected elements in Apollo 17 samples (abstract). In *Lunar Science—V*, pp. 38–40. The Lunar Science Institute, Houston.

Cliff R. A., Lee-Hu C., and Wetherill G. W. (1971) Rb–Sr and U, Th–Pb measurements on Apollo 12 material. *Proc. Second Lunar Sci. Conf., Geochim. Cosmochim. Acta*, Suppl. 2, Vol. 2, pp. 1493–1502. MIT Press.

Doe B. R. and Tatsumoto M. (1972) Volatilized lead from Apollo 12 and 14 soils. *Proc. Third Lunar Sci. Conf., Geochim. Cosmochim. Acta*, Suppl. 3, Vol. 2, pp. 1981–1988. MIT Press.

Duke M. B., Woo C. C., Sellers G. A., Bird M. L., and Finkleman R. B. (1970) Genesis of lunar soil at Tranquility Base. *Proc. Apollo 11 Lunar Sci. Conf., Geochim. Cosmochim. Acta*, Suppl. 1, Vol. 1, pp. 347–361. Pergamon.

Eberhardt P., Eugster O., Geiss J., Graf H., Grögler N., Guggisberg S., Jungck M., Maurer P., Mörgeli M., and Stettler A. (1974) Solar wind and cosmic radiation history of Taurus-Littrow regolith (abstract). In *Lunar Science—V*, pp. 197-199. The Lunar Science Institute, Houston.

Evensen N. M., Murthy V. R., and Coscio M. R., Jr. (1973) Rb–Sr ages of some mare basalts and the isotopic and trace element systematics in lunar fines. *Proc. Fourth Lunar Sci. Conf., Geochim. Cosmochim. Acta*, Suppl. 4, Vol. 2, pp. 1707–1724. Pergamon.

Evensen N. M., Murthy V. R., and Coscio M. R., Jr. (1974) Episodic lunacy—V: Origin of the exotic component (abstract). In *Lunar Science—V*, pp. 220–221. The Lunar Science Institute, Houston.

Gibson E. K. and Hubbard N. J. (1972) Volatile element depletion investigations on Apollo 11 and 12 lunar basalts by means of thermal volatilization. *Proc. Third Lunar Sci. Conf., Geochim. Cosmochim. Acta*, Suppl. 3, Vol. 2, pp. 2003–2014. MIT Press.

Heiken G., Duke M., McKay D. S., Clanton U. S., Fryxell R., Nogle J. S., Scott R., and Sellers G. A. (1973a) Preliminary stratigraphy of the Apollo 15 drill core. *Proc. Fourth Lunar Sci. Conf., Geochim. Cosmochim. Acta*, Suppl. 4, Vol. 1, pp. 191–213. Pergamon.

Heiken G. H., McKay D. S., and Fruland R. M. (1973b) Apollo 16 soils: Grain size analyses and petrography. *Proc. Fourth Lunar Sci. Conf.*, *Geochim. Cosmochim. Acta*, Suppl. 4, Vol. 1, pp. 251–265. Pergamon.

Heiken G. and McKay D. S. (1974) Petrography of Apollo 17 soils (abstract). In *Lunar Science—V*, pp. 319–321. The Lunar Science Institute, Houston.

Huey J. M., Ihochi H., Ostic R. G., and Kohman T. P. (1971) Lead isotopes and volatile transfer in the lunar soil. *Proc. Second Lunar Sci. Conf.*, *Geochim. Cosmochim. Acta*, Suppl. 2, Vol. 2, pp. 1547–1564. MIT Press.

Jaffey A. H., Flynn K. F., Glendenin L. E., Bentley W. C., and Essling A. M. (1971) Precision measurement of the half-lives and specific activities of ^{235}U and ^{238}U. *Phys. Rev.* **C4**, 1889.

LSPET (Lunar Sample Preliminary Examination Team) (1973) Preliminary examination of lunar samples from Apollo 17, 1–46. In *Apollo 17 Preliminary Science Report*. NASA SP 330. U.S. Government Printing Office.

McKay D. S., Heiken G. H., Taylor R. M., Clanton U. S., and Morrison D. A. (1972) Apollo 14 soils: Size distribution and particle types. *Proc. Third Lunar Sci. Conf.*, *Geochim. Cosmochim. Acta*, Suppl. 3, Vol. 2, pp. 983–994. MIT Press.

McKay D. S., Fruland R. M., and Heiken G. H. (1974) Grain size distribution as an indicator of the maturity of lunar soils (abstract). In *Lunar Science—V*, pp. 480–482. The Lunar Science Institute, Houston.

Murthy V. R., Evensen N. M., Jahn B., and Coscio M. R., Jr. (1972) Apollo 14 and 15 samples: Rb–Sr ages, trace elements, and lunar evolution. *Proc. Third Lunar Sci. Conf.*, *Geochim. Cosmochim. Acta*, Suppl. 3, Vol. 2, pp. 1503–1514. MIT Press.

Nunes P. D. and Tatsumoto M. (1973) Excess lead in "rusty rock" 66095 and implications for an early lunar differentiation. *Science* **182**, 916–920.

Nunes P. D., Tatsumoto M., Knight R. J., Unruh D. M., and Doe B. R. (1973) U–Th–Pb systematics of some Apollo 16 lunar samples. *Proc. Fourth Lunar Sci. Conf.*, *Geochim. Cosmochim. Acta*, Suppl. 4, Vol. 2, pp. 1797–1822. Pergamon.

Nunes P. D., Tatsumoto M., and Unruh D. M. (1974) U–Th–Pb systematics of some Apollo 17 samples (abstract). In *Lunar Science—V*, pp. 562–564. The Lunar Science Institute, Houston.

Nyquist L. E., Hubbard N. J., Gast P. W., Bansal B. M., Wiesmann H., and Jahn B. (1973) Rb–Sr systematics for chemically defined Apollo 15 and 16 materials. *Proc. Fourth Lunar Sci. Conf.*, *Geochim. Cosmochim. Acta*, Suppl. 4, Vol. 2, pp. 1823–1846. Pergamon.

Nyquist L. E., Bansal B. M., Wiesmann H., and Jahn B. (1974) Taurus-Littrow chronology: Implications for early lunar crustal development (abstract). In *Lunar Science—V*, pp. 565–567. The Lunar Science Institute, Houston.

Papanastassiou D. A. and Wasserburg G. J. (1972) Rb–Sr age of a Luna 16 basalt and the model age of lunar soils. *Earth Planet. Sci. Lett.* **13**, 368–374.

Pepin R. O., Bradley J. G., Dragon J. C., and Nyquist L. E. (1973) K–Ar dating of lunar fines: Apollo 12, Apollo 14 and Luna 16. *Proc. Third Lunar Sci. Conf.*, *Geochim. Cosmochim. Acta*, Suppl. 3, Vol. 2, pp. 1569–1588. MIT Press.

Philpotts J. A., Schumann S., Kouns C. W., and Lum R. K. L. (1974) Lithophile trace elements in Apollo 17 soils (abstract). In *Lunar Science—V*, pp. 599–601. The Lunar Science Institute, Houston.

Rhodes J. M., Rodgers K. V., Shih C., Bansal B. M., Nyquist L. E., and Wiesmann H. (1974) The relationship between geology and soil chemistry at the Apollo 17 landing site (abstract). In *Lunar Science—V*, pp. 630–632. The Lunar Science Institute, Houston.

Silver L. T. (1971) U–Th–Pb relations in Apollo 11 and Apollo 12 lunar samples. Second Lunar Science Conference (unpublished proceedings).

Silver L. T. (1972a) U–Th–Pb abundances and isotopic characteristics in some Apollo 14 rocks and in Apollo 15 soil (abstract). In *Lunar Science—III*, pp. 704–706. The Lunar Science Institute, Houston.

Silver L. T. (1972b) Uranium–Thorium–Lead isotopes and the nature of the mare surface debris at Hadley-Apennine. In *The Apollo 15 Lunar Samples*, pp. 388–390. The Lunar Science Institute, Houston.

Silver L. T. (1973) Uranium–Thorium–Lead isotopic characteristics in some regolith materials from the Descartes region (abstract). In *Lunar Science—IV*, pp. 672–674. The Lunar Science Institute, Houston.

Silver L. T. (1974) Patterns of U–Th–Pb distributions and isotope relations in Apollo 17 soils (abstract). In *Lunar Science—V*, pp. 706–708. The Lunar Science Institute, Houston.

Tatsumoto M. (1966) Isotopic composition of lead in volcanic rocks from Hawaii, Iwo Jima, and Japan. *J. Geophys. Res.* **71**, 1721–1733.

Tatsumoto M., Hedge C. E., Knight R. J., Unruh D. M., and Doe B. R. (1972a) U–Th–Pb, Rb–Sr and K measurements on some Apollo 15 and Apollo 16 samples. In *The Apollo 15 Lunar Samples*, pp. 391–395. The Lunar Science Institute, Houston.

Tatsumoto M., Hedge C. E., Doe B. R., and Unruh D. M. (1972b) U–Th–Pb and Rb–Sr measurements on some Apollo 14 lunar samples. *Proc. Third Lunar Sci. Conf., Geochim. Cosmochim. Acta*, Suppl. 3, Vol. 2, pp. 1531–1555. MIT Press.

Tatsumoto M., Knight R. J., and Allegre C. J. (1973) Time differences in the formation of meteorites as determined by ^{207}Pb/^{206}Pb. *Science* **180**, 1279–1283.

Tatsumoto M., Knight R. J., and Doe B. R. (1971) U–Th–Pb systematics of Apollo 12 lunar samples. *Proc. Second Lunar Sci. Conf., Geochim. Cosmochim. Acta*, Suppl. 2, Vol. 2, pp. 1521–1546. MIT Press.

Tatsumoto M., Nunes P. D., Knight R. J., and Unruh D. M. (1974) Rb–Sr and U–Th–Pb systematics of boulders 1 and 7 Apollo 17 (abstract). In *Lunar Science—V*, pp. 774–776. The Lunar Science Institute, Houston.

Tera F. and Wasserburg G. J. (1972) U–Th–Pb systematics in lunar highland samples from the Luna 20 and Apollo 16 missions. *Earth Planet. Sci. Lett.* **17**, 36–51.

Tera F., Ray L. A., and Wasserburg G. J. (1972) Distribution of Pb–U–Th in lunar anorthosite 15415 and inferences about its age. In *The Apollo 15 Lunar Samples*, pp. 396–401. The Lunar Science Institute, Houston.

Tera F., Papanastassiou D. A., and Wasserburg G. J. (1974) The lunar time scale and a summary of isotopic evidence for a terminal lunar cataclysm (abstract). In *Lunar Science—V*, pp. 792–794. The Lunar Science Institute, Houston.

Tilton G. R. (1973) Isotopic lead ages of chondritic meteorites. *Earth Planet. Sci. Lett.* **19**, 321–329.

Proceedings of the Fifth Lunar Conference
(Supplement 5, Geochimica et Cosmochimica Acta)
Vol. 2 pp. 1401–1417 (1974)
Printed in the United States of America

Provenance of KREEP and the exotic component: Elemental and isotopic studies of grain size fractions in lunar soils

N. M. Evensen, V. Rama Murthy, and M. R. Coscio, Jr.

Department of Geology and Geophysics, University of Minnesota, Minneapolis, Minnesota 55455

Abstract—We have determined the abundances of K, Rb, Sr and Ba in size fractions of an Apollo 16 soil 63321, two Apollo 17 soils 71501 and 79261, and samples from the Apollo 15 drill core, as a continuation of our program of chemical and isotopic studies of sieve fractions of lunar fines. The soils show enrichment of these elements in finer size fractions similar to those we have previously observed. These abundance patterns demonstrate the presence of a fine-grained exotic component enriched in these elements and in radiogenic Sr, in all soils. The probable source of this exotic component is the areas of high-surficial radioactivity observed by orbital gamma ray spectrometry, such as those at Fra Mauro and Archimedes. Meteorite impacts or other lunar transport processes have dispersed these materials widely across the lunar surface. A correlation exists between distance from the proposed source areas and proportion of exotic component in the soils. We suggest that the exotic component represents trace element enriched material located at some depth at the Imbrium area which was surficially deposited during Imbrium excavation. The association of this enriched component only with Imbrium may reflect either primary inhomogeneities in the moon or the greater depth to which the Imbrium Basin was excavated. If the exotic component is fine-grained KREEP, similar conclusions may apply to the origin and distribution of KREEP fragments in the soils. The Imbrium excavation may be the only event in lunar history which has directly sampled such highly differentiated materials. The widespread distribution of these materials on the lunar surface need not imply a similarly widespread distribution of primary source areas, nor does it unequivocally lead to the inference of a global radioactive crust.

Introduction

Our recent work on the lunar samples has focused on the characterization of components in the lunar soils by examining abundances of K, Rb, Sr and Ba, and Sr isotopic systematics in soil grain size fractions (Murthy *et al.*, 1972; Evensen *et al.*, 1973). In particular, most of the soils we have examined show higher abundances of these elements in the finest size fractions, accompanied by relatively radiogenic Sr isotopic compositions, in a pattern which is difficult to account for by simple mixing of locally sampled rock types. We have attributed this effect to the admixture of an "exotic" component which we identify with the "magic component" of Papanastassiou *et al.* (1970), and which may also be related to KREEP (Hubbard *et al.*, 1971).

In this paper we present sieve fraction analyses of an Apollo 16 soil, 63321, two Apollo 17 soils, 71501 and 79261, plagioclase and ilmenite separates from sieve fractions of the Apollo 11 soil 10084, and 6 samples from the Apollo 15 drill core (15001–15006). The discussion deals with data from these and previously analyzed soils, and incorporates them into a model of the origin and distribution of an exotic component, including possible relations with KREEP.

EXPERIMENTAL PROCEDURES

Procedures followed for sieving, chemical separations, and mass spectrometry are essentially those described earlier (Murthy *et al.*, 1971; Evensen *et al.*, 1973). Sieving is done in an acetone medium, using a combination of nylon mesh and electroformed nickel screens, and particle sedimentation in the finest size fractions (Basford and Coscio, 1973). Sieving and chemical procedures are carried out in a clean room environment. In many cases, as noted specifically below, the small amounts of sample available required combining several sieve fractions which normally are analyzed separately. In some cases, replicate analyses were made on separate aliquots of sieved material; these appear as separate determinations in Tables 1 and 2. All abundance determinations were performed by stable isotope dilution techniques, using a 30.5 cm radius 60° sector single focusing mass spectrometer with an automated data collection system (Murthy *et al.*, 1971). Blank corrections for total analytical procedures are K = 20 ng, Rb = 0.1 ng, Sr = 1 ng, Ba = 6 ng; as noted below, sample sizes on the order of 1 mg often produced appreciable blank corrections. Errors of ±5% are assigned to abundance determinations on these samples. Errors for other samples are ±2%.

ANALYTICAL RESULTS

Fines sample 63321

This is a sample of <1 mm fines collected by the Apollo 16 astronauts from Station 13, approximately 1 km south of North Ray Crater. Elemental abundances of the bulk soil are near the middle of the range observed for the 6 Apollo 16 soils we have measured (Evensen *et al.*, 1973); this is the only Apollo 16 soil on which we have done sieve analyses to date (Table 1). Because of a relatively small proportion of fine-grained material in this soil, the normal 25 μ and 8 μ sieving operations were omitted and, therefore, only two fractions <37 μ were determined. This may obscure to some extent the K and Rb enrichments in the smaller fractions, which are clearly seen only in the 4–16 μ range, but in general the enrichments appear rather small. This and other soil systematics will be discussed in more detail below.

Fines samples 71501 and 79261

We have previously reported elemental and isotopic analyses of grain size fractions of the dark-mantle soil 75081 (Evensen *et al.*, 1973). Sample 71501 consists of <1 mm fines collected at Station 1a, about a kilometer south of the Apollo 17 landing site and in an area of the dark mantle. Sample 79261 is <1 mm fines from Station 9 on the ejecta blanket at the comparatively young Van Serg Crater. The sample was collected from a trench ~17 cm deep and represents the lower 10 cm of light-colored material overlain by darker soil; thus it is presumably distinct from the dark-mantle material.

Elemental abundances in comparable grain size fractions of these two soils as well as the previously analyzed 75081 (Evensen *et al.*, 1973) are quite similar and show no appreciable difference between dark-mantle and trench soils (Table 1).

Table 1. Elemental abundance data on bulk fines and size fractions.

Sample	wt.[a] (mg)	K[b] (μg/g)	Rb[b] (μg/g)	Sr[b] (μg/g)	Ba[b] (μg/g)
63321,17					
147–10000 μ	9.18	775	2.12	172	120
74–147 μ	9.73	740	1.90	168	102
37–74 μ	10.04	755	1.97	171	105
16–37 μ	10.09	776	1.99	166	103
4–16 μ	10.79	913	2.24	172	110
71501,27					
Total sample	12.14	507	0.91	131	70.9
147–1000 μ	15.92	—	—	123	87.0
74–147 μ	15.59	589	0.99	152	—
74–147 μ	15.70	613	1.01	155	78·4
37–74 μ	16.08	588	1.06	155	73.9
25–37 μ	14.43	674	1.21	161	91.5
16–25 μ	16.52	—	—	166	95.2
16–25 μ	1.70	692	1.28	168	95.6
8-16 μ	15.49	852	1.88	186	119
4–8 μ	15.52	1020	2.28	188	134
79261,28					
147–1000 μ	10.63	584	1.40	133	73.5
74–147 μ	8.41	629	1.21	142	79.9
37–74 μ	10.58	668	1.46	144	84.3
16–37 μ	12.02	748	1.74	143	95.3
8–16 μ	8.49	995	2.56	169	119
4–8 μ	9.78	1063	2.91	163	128

[a]Weights refer to amounts analyzed and not to total amounts obtained in sieving.

[b]Errors of ±2% are assigned.

These three Apollo 17 soils show moderate enrichments of all four elements analyzed with decreasing grain size. The enrichments are less marked than in Apollo 15 and possibly Apollo 11 soils. The total sample analysis of 71501 appears to be unrepresentative; mass balance indicates that total sample abundances are somewhat higher in both this soil and 79261 and are comparable with those reported for 75081 (Evensen *et al.*, 1973).

Fines sample 10084 mineral separates

We have previously reported bulk and sieve analyses of this Apollo 11 contingency soil sample (Murthy *et al.*, 1970; Evensen *et al.*, 1973). In the present study, plagioclase and ilmenite mineral fractions were obtained by performing

Table 2. Elemental abundances in plagioclase and ilmenite size fractions
of soil 10084.

Sample	wt.[a] (mg)	K[b] (μg/g)	Rb[b] (μg/g)	Sr[b] (μg/g)	Ba[b] (μg/g)
10084,48 Plagioclase					
250–590 μ	3.07	—	—	249	80
147–250 μ	2.06	540	0.58	284	67
105–147 μ	2.05	761	1.11	309	76
74–105 μ	2.15	609	0.61	319	71
74–105 μ	0.74	601	0.73	314	61
37–74 μ	5.65	899	1.18	355	105
37–74 μ	1.41	924	1.09	—	—
25–37 μ	0.96	842	1.15	—	—
16–25 μ	1.66	1318	2.29	311	178
8–16 μ	2.88	1125	2.11	210	161
8–16 μ	0.77	1035	2.11	235	156
4–8 μ	1.31	1220	2.88	185	189
1–4 μ heavy[c]	1.83	1334	3.23	226	219
1–4 μ light[c]	1.51	1486	4.85	237	245
10084,48 Ilmenite					
105–147 μ	1.42	340	0.86	27.7	41
74–105 μ	1.33	268	0.60	19.6	38
37–74 μ	1.72	226	0.54	17.5	—
25–37 μ	1.76	224	0.68	16.9	28
16–25 μ	1.07	294	0.70	33.5	38
8–16 μ	1.15	232	0.49	20.6	30
4–8 μ	1.20	206	—	22.3	26
1–4 μ	0.93	275	0.73	32.0	40
10084,48 Black glass					
74–105 μ	7.30	1179	2.99	145	164

[a]Weights refer to amounts analyzed and not to total amounts obtained
in sieving.
[b]Error of ±5% are assigned.
[c]Heavy and light plagioclase fractions obtained. Heavy fraction corresponds to separation technique used for coarser fractions.

magnetic and density separations on each of the grain size fractions previously
obtained (Basford, 1974). In addition, a magnetic fraction was obtained from the
74–105 μ fraction using a hand magnet, and was found to consist of black glass
spheres. Elemental abundance data for these separates are shown in Table 2.
Because of the small amounts of sample available for analysis (generally 1–2 mg),
blank corrections are often nonnegligible and therefore larger errors of ±5% are
assigned to these determinations. In addition, although replicate samples yield
generally similar results, sample aliquoting may be unrepresentative, particularly
in the coarser fractions where relatively few grains are included in the analyses.

Apart from analytical difficulties, interpretation of the mineral data is complicated by the fact that mineral separations were performed individually on each of the sieve fractions. The purity of the separates and nature of the contaminants undoubtedly varies through the series, probably both randomly and as a function of grain size. Electron probe and X-ray studies indicate that the plagioclase separates are probably ~90% pure, while the ilmenites are >95% pure.

Despite the scatter introduced by these effects, there is a clear rising trend of K, Rb, and Ba with decreasing grain size, as was seen in the unseparated sieve fractions of 10084, but without the accompanying rise in the coarser fractions as well, which gave a U shape to the distribution. In our previous discussion of these data, we tentatively attributed this U shape to the effect of agglutinates in the Apollo 11 soil (Evensen *et al.*, 1973). The plagioclase separate, relatively free of agglutinates, does not show this trend. The ilmenite fractions, by contrast, show appreciable scatter, but no clear trend for any of the elements analyzed. This is in accord with the observation that in ilmenite these trace elements are characteristically found as small inclusions dispersed through the volume of the grains. Purely mechanical breakup of such material, unaccompanied by importation of exotic, trace element enriched ilmenite, would produce such a flat abundance-grain size relationship.

Apollo 15 drill core

Six samples, 15001–15006, from ~40 cm intervals in the Apollo 15 drill core were each sieved into four grain size fractions. The elemental abundances in these fractions are given in Table 3. Again, the small amounts of sample available for analysis (typically <1 mg) result in larger errors of ±5% being assigned to these determinations. There is also a definite possibility of inhomogeneous aliquoting with these small sample sizes.

These difficulties, together with the poor resolution afforded by only four sieve fractions, and the complex stratigraphy of the Apollo 15 core tube, which has 42 identifiable units (Heiken *et al.*, 1973), render the interpretation of the abundance data very complex, and we will not undertake an extensive discussion in this paper. We note that all elemental abundances are generally higher in the two deepest samples (2.0 and 2.4 m), as previously noted by Philpotts and Schnetzler (1972), though K abundances rise again in the shallowest samples (0.4 and 0.8 m). All abundances show an increase from the $16\,\mu$–$37\,\mu$ fraction to the $<16\,\mu$ fraction, except for K and Rb in the shallowest sample (0.4 m depth). Our previous analyses have generally shown the most marked rise within the individual size fractions $<16\,\mu$ which are not resolved here (i.e. 8–16, 4–8, 1–4 μ). The abundance pattern within the three coarsest fractions is variable from sample to sample and between elements analyzed in the same sample. It may be monotonically increasing, decreasing, V-shaped or A-shaped. We have mentioned that U-shaped abundance patterns (maximum in coarsest and finest size fractions) may be due to agglutinates (Evensen *et al.*, 1973). Some correlation is seen between U-shaped patterns for K and Rb and proportion of agglutinates in the 90–150 μ size range as determined for core tube samples by Heiken *et al.* (1973). Rare gas

N. M. Evensen *et al.*

Table 3. Elemental abundances in grain size fractions of Apollo 15 drill core samples.

Sample	wt.[a] (mg)	K[b] (μg/g)	Rb[b] (μg/g)	Sr[b] (μg/g)	Ba[b] (μg/g)
15001,30 (~241 cm depth)					
>74 μ	1.12	2390	6.63	139	377
37–74 μ	0.44	2540	10.1	128	344
16–37 μ	0.63	2330	6.86	—	340
<16 μ	0.68	3930	8.19	154	383
15002,26 (~202 cm depth)					
>74 μ	1.20	2400	7.38	120	337
37–74 μ	0.44	1990	6.50	116	280
16–37 μ	1.09	1890	5.44	130	291
<16 μ	0.76	2290	7.23	155	355
15003,28 (~161 cm depth)					
>74 μ	0.83	1820	5.62	114	249
37–74 μ	0.63	1890	5.60	116	245
16–37 μ	0.70	1990	6.00	130	263
<16 μ	1.30	2060	6.72	153	297
15004,26 (~122 cm depth)					
>74 μ	0.87	1850	—	130	—
37–74 μ	0.78	1420	5.58	117	248
16–37 μ	0.86	1700	5.21	127	240
<16 μ	0.89	1810	5.88	138	268
15005,25 (~82 cm depth)					
>105 μ	0.42	2120	5.30	119	340
37–105 μ	1.13	1780	5.21	—	253
16–37 μ	0.89	1880	5.47	133	279
<16 μ	0.75	2120	6.56	149	316
15006,26 (~39 cm depth)					
>105 μ	0.92	1900	5.86	129	273
37–105 μ	0.80	2010	6.12	—	278
16–37 μ	0.65	2440	6.07	133	270
<16 μ	0.41	1750	5.77	135	302

[a]Weights refer to amounts analyzed and not to total amounts obtained in sieving.

[b]Errors of ±5% are assigned.

studies on these sieved core tube samples are reported elsewhere in this volume (Pepin *et al.*, 1974).

DISCUSSION

Nature of the exotic component

The existence of "magic component" at the Apollo 11 landing site was originally inferred by Papanastassiou *et al.* (1970) from the Rb–Sr systematics of the rocks and soils, which indicated that despite the presence of both high- and low-K rocks at this site, the soil (with intermediate abundances of K and many other trace elements) could not be derived solely from a mixture of high- and low-K rocks. Instead, the soil appeared to be derived largely from the low-K rocks, with an admixture of material of high-K and Rb and radiogenic Sr composition, and of higher model age (~4.6 AE) than the high-K rocks (~3.9 AE).

When the Apollo 12 site was sampled, the necessity of the exotic component was even more clearly demonstrated. The basaltic rocks sampled were even less radiogenic than the Apollo 11 low-K rocks, while the soils were more radiogenic and K and Rb rich than those from Apollo 11; i.e. it was evident even from the bulk chemistry of rocks and soils that some additional component was present in the soils (e.g. Papanastassiou and Wasserburg, 1970; Compston *et al.*, 1971; Murthy *et al.*, 1971). The presence of the unusual rock 12013 at this site, portions of which had chemical characteristics permissible for the exotic component (e.g. Albee *et al.*, 1970; Schnetzler *et al.*, 1970), did not explain how such material could be widely disseminated through the soils at the Apollo 12 site, as well as those almost a thousand kilometers distant at the Apollo 11 site.

Examination of Apollo 12 materials also revealed fragments of material in the soils having high abundances of K, Rb, rare earths, and other incompatible elements; this material has been termed KREEP (Hubbard *et al.*, 1971). Chemical and isotopic analyses show that KREEP compositions are also suitable for the exotic component. However, since the origin and mechanism of distribution of KREEP materials is also unknown, and since KREEP was defined by examination of petrologically identifiable soil fragments rather than inferred from mixing calculations, the distinction between KREEP and exotic component will be retained through the part of this discussion dealing with the results of sieve analyses.

Characterization of the exotic components

Bulk compositions of rocks and soils from the Apollo 15 site, as with the Apollo 12 site, show clearly the presence of exotic component at this location. The analyses we have performed on grain size fractions of soil 15531 show marked increases (~3 ×) in K and Rb abundances in the finer size fractions relative to the coarser; these abundances rise essentially monotonically as grain size decreases (Fig. 1). Although Rb/Sr ratio increases monotonically, Sr and Ba abundances rise as well, though less steeply than K and Rb.

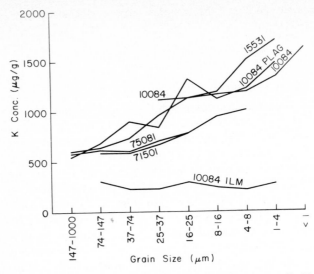

Fig. 1. Potassium abundances in grain size fractions of soils from Apollo 11, 15, and 17 missions and in plagioclase and ilmenite separates from Apollo 11 soil. The two Apollo 17 soils are indistinguishable in their finer grain size fractions.

These effects can be partially explained by a comminution model in which the trace element rich mesostasis of the rocks concentrated along grain boundaries and in inclusions, becomes preferentially located in the fine-grained fractions of the soil. If this were the only or dominant effect, Sr isotopic compositions of all soil fractions would plot along a mixing line on the Rb–Sr evolution diagram defined by the internal isochron of the comminuted rocks. Measured isotopic compositions of successively finer sieve fractions, however, deviate increasingly not only from the average basalt composition, but from the internal isochron, in a direction indicating addition of radiogenic material of ~4.6 AE model age (Evensen et al., 1973).

Volatilization effects have been experimentally observed for K and Rb in lunar materials (e.g. DeMaria et al., 1971; Naughton et al., 1971; Gibson and Hubbard, 1972), and thus volatilization could be invoked to explain some of the effects observed. Fine-grained materials, with a higher surface to volume ratio, should tend to lose volatiles more readily than coarser size fractions. This is the opposite effect from that observed. Alternatively, if volatile K and Rb are already present from some other source, the finer grained fractions would act as more efficient "getters." Chou et al. (1974) have observed marked Cd enrichment in finer grained fractions of an Apollo 17 soil; this, together with their observation of Cd abundances twice as high in shadowed as in unshadowed soil, suggests that such a getter effect may be present for highly volatile elements such as Cd. Such a model cannot, however, explain many features of our data, including the slope of the enrichment pattern, which is much less than the $1/r$ dependence of surface correlated effects; the enrichments of Sr and Ba, which are relatively refractory

(Gibson and Hubbard, 1972); the Sr isotopic data; and the lack of K/Rb fractionation between size fractions (Gibson *et al.*, 1973). Furthermore, to explain *bulk* soil abundances, massive amounts of these elements must be transported into the Apollo 15 site in the vapor phase. Some volatilization effects are almost surely present; however, their presence appears to be largely masked by other, more dominant processes which seem to exert primary control over the soil systematics.

By far the most probable explanation, therefore, is that some material enriched in K, Rb, Sr, and Ba has been added preferentially to finer size fractions, i.e. the effects seen are produced by a fine-grained exotic component. Since we are seeing the exotic component only in mixtures, it is impossible to fully characterize the composition of the pure end-member; nevertheless assuming that the enrichment effects are primarily due to addition of exotic component, some statements can be made on the basis of trends in elemental abundances and isotopic ratios. This component must have high enrichments of K and Rb relative to the locally derived component, lesser enrichment in Ba, and moderate enrichment in Sr. The K/Rb ratio must be similar to that of the 15 soil (~ 350), Rb/Sr appreciably greater than ~ 0.0025, and Ba/Sr significantly greater than 1. The dispersion of Sr isotopic data along a 4.6 AE reference isochron strongly suggests that the exotic component lies along this isochron, i.e. that it has a high Rb/Sr ratio with model age ~ 4.6 AE (Evensen *et al.*, 1973).

The fine-grained nature of the exotic component can perhaps be most readily explained as a consequence of its transport from some source area into the Apollo 15 site. Whether this transport is accomplished by dispersal resulting from meteorite impacts which will tend to pulverize the impacted material, or by electrostatic transport process which favor the transport of fine-grained particles, the end result is to produce an exotic component which is finer grained than the soil produced by *in situ* breakup of crystalline rocks.

Analyses of soils from the Apollo 11 and 17 landing sites yield conclusions essentially similar to the above, with some differences in detailed interpretation apparently resulting from the differences in local lithology. These are discussed in more detail by Evensen *et al.* (1973). The two additional Apollo 17 soils for which analyses are presented in Table 1 show no significant differences in elemental abundance from the previously analyzed 75081. The single Apollo 16 soil we have analyzed (63321, Table 1) shows some increase in K and Rb abundances with decreasing grain size, although the lack of more size fractions in the $<37\ \mu$ range does not permit a very clear picture of this trend. Detailed discussion of this soil will be deferred until Sr isotopic analyses are available. There appears thus far to be at least an approximate correlation between degree of enrichment of trace elements in soils from various sites and the proportion of exotic component in these sites as indicated by comparison of soils and rocks. Thus in Fig. 1, enrichments are highest in the Apollo 15 soils, intermediate in Apollo 11 soil, and lowest in the Apollo 17 soils.

The data on 10084 mineral separates (Table 2, Fig. 1) are relevant to the question of transport of exotic component. The ilmenite is probably almost entirely of local derivation, since rocks from the Apollo 11 site are the most ilmenite-rich rocks returned from the moon. The ilmenite fractions show a very flat slope, consistent with this interpretation, and indicating that even the finest fractions are locally derived. Plagioclase, on the other hand, occurs ubiquitously on the lunar surface and could well be a mineralogical constituent of the exotic component. As would be expected in this case, the plagioclase values show a rise paralleling that of the unseparated size fractions.

Provenance of the exotic component

If, then, the exotic component is a fine-grained, trace element enriched material transported into the various sites sampled, identification of the source of this material should provide insight into the nature and origin of such relatively differentiated material. It is possible that numerous localized sources could be scattered globally over the moon; such sources would have to be small enough to fall below the resolution of the orbital gamma-ray spectrometer. Depending on the efficiency of dispersal from the sources and the uniformity of their distribution over the lunar surface, the resulting distribution of exotic component in various sites might be uniform or varying as a function of distance from the most local source regions. Global patterns of exotic component distribution would necessarily result from systematic global variation of source characteristics which would in turn require explanation.

The data from the lunar orbital gamma-ray spectrometers reveal a definite pattern in the distribution of radioactivity over the lunar surface (see Metzger *et al.*, 1973 and Plate II from the *Proceedings of the Fourth Lunar Science Conference*, Vol. 1). High gamma-ray activity is seen almost exclusively in the Imbrium–Procellarum region, with highest activities in localized areas in the vicinity of Aristarchus, Archimedes, and Fra Mauro. Of these locations, Fra Mauro was directly sampled at the Apollo 14 landing site. The materials returned by Apollo 14 show uniformly high concentrations of the radioactive elements K, U, Th, as well as Rb, rare earths and other incompatible trace elements (LSPET, 1971). The K/Rb ratio is 250–300, Rb/Sr ~0.1 and Ba/Sr ~5; Sr compositions are rather radiogenic and yield ~4.5 AE model ages (Murthy *et al.*, 1972). These characteristics are all in good agreement with those inferred above for the exotic component. Furthermore, abundance and isotopic measurements of grain size fractions of the Fra Mauro soils (Murthy *et al.*, 1972) show only random variation rather than the systematic patterns that we have observed in all other soils and have attributed to addition of exotic component. This characteristic is to be expected of a source area, as is the close similarity in composition between rocks and soils at the Fra Mauro site.

Fra Mauro (and perhaps the other orbitally identified areas of high radioactivity) is thus a region of surficially exposed material which, if transported into other sites and mixed with locally derived soils, would produce the characteristics we have observed, at least to a first approximation. It seems inevitable that such transport must have occurred to some degree; the extent of transport over several

Table 4. Averaged Rb and Sr abundances in lunar rocks and soils, and inferred percentages of exotic component in soils.

Site	Rb (ppm)			Sr (ppm)			Sources
	rocks	soil	% exotic component[a]	observed in rocks	observed in soil	expected for soil[b]	
Apollo 14	—	15	100[c]	—	175	—	1
Apollo 11	1.2[d]	2.9	12	180[d]	160	179	2, 3, 4
Apollo 12	1.0	6.2	37	110	140	134	3, 5, 6
Apollo 15	0.65	2.6	14	90	100	98	1, 3
Apollo 16	1.3	2.5	9–18[e]	180	175	180	7, 8
Apollo 17	0.55	1.2	4.5	160	165	161	7
Luna 16	1.8[f]	1.9	0.8	350[f]	260	349	9
Luna 20	≥0.9[g]	1.6	≤5	—	140	—	10, 11

[a]% = [Rb(soil) − Rb(rocks)]/[Rb(Apollo 14 soil) − Rb(rocks)] × 100.

[b]Calculated by using inferred percentage of exotic component.

[c]Assumed to be the exotic component.

[d]Rock data assumes that low-K rocks are the major local contributors to soil (1).

[e]Rock values are highly dependent upon proportion of different rock types assumed. Also the regolith at this site is ~2× thicker than at other sites, so % exotic component is dependent upon mixing depth. Higher value corresponds to complete mixing of exotic component with full depth regolith.

[f]Rock values based on only three small chips and may not be representative.

[g]Rock data not available; Rb abundance calculated from K abundance at ~900 ppm (20) and K/Rb of 1000; probably a lower limit for Rb.

Sources: 1. Murthy *et al.* (1972), 2. Papanastassiou *et al.* (1970), 3. Schonfeld and Meyer (1972), 4. Murthy *et al.* (1970), 5. Papanastassiou and Wasserburg (1971), 6. Murthy *et al.* (1971), 7. Evensen *et al.* (1973), 8. LSPET (1973), 9. Philpotts *et al.* (1972), 10. Papanastassiou and Wasserburg (1972), 11. Prinz *et al.* (1973).

aeons is difficult to assess. Certainly appreciable amounts of presumably highland-derived anorthositic fragments have been transported into the mare sites (e.g. Wood *et al.*, 1970). If finer grained materials are transported in greater abundance, as our discussion above suggests, then significant amounts of material may be transported over distances of 100 km or more (Basford, 1974).

One test of the adequacy of a hypothesized source area is to seek a correlation between observed abundances of exotic component and distance from the source area. In Table 4 we have attempted to estimate proportion of exotic component in all of the Apollo and Luna landing sites. Since the original inference of the presence of exotic component was based on Rb–Sr systematics of the soils, we have used Rb abundances in rocks and soils as parameters in a simple mixing model. We assume that the soils are derived from a mixture of comminuted local rocks and Fra Mauro material, in proportions determined from their Rb abundances (Table 4, note a). Such a two component model is undoubtedly a simplification; however, other likely soil components (e.g. anorthositic materials in mare regions) are relatively Rb-poor and thus do not grossly affect the Rb balance unless present in large quantities. The estimation of average abundances in rocks and soils is somewhat difficult, particularly in complex regions such as the Apollo 16 site. Using, for example, the averages compiled by Rose *et al.* (1973) for the Apollo 11, 12, 14, 15, and 16 sites in our mixing model, we would obtain higher proportions of exotic component in the Apollo 12 and 15 sites, but the relative positions of the five sites are not altered. Our estimates are also in approximate agreement with those obtained using more refined mixing models (e.g. Schonfeld and Meyer, 1972; Wasson and Baedecker, 1972). A primary difference results from our assumption, on the basis of Rb–Sr systematics, that low-K rocks are the dominant local contributors to the Apollo 11 soil (Papanastassiou *et al.*, 1970); this leads to a somewhat higher proportion of exotic component in this soil. In Table 4 we have also estimated Sr abundances in the soil resulting from our mixing model. The general agreement with observation (except for the Luna 16 site where rock data are doubtful) suggests that the exotic component has a Rb/Sr ratio similar to that of Fra Mauro materials.

We realize that several of the landing sites are highly complex areas showing considerable variability in soil and rock compositions at a given site. However, rocks returned from all sites except Fra Mauro are very dominantly trace element poor compared to KREEP or exotic component. Soil enrichments within a site may be rather variable, reflecting different histories of derivation from local rocks, exposure on the surface, and mixing with exotic component. However, our use of average soil analyses, rather than individual analyses, minimizes the problem of local variability and should more closely reflect the average composition over a broad area. We believe that the abundances of exotic component in Table 4 are at least qualitatively correct in their assignments of relative enrichments, and probably fairly accurate quantitatively.

In Fig. 2 these estimated percentages are plotted against distance from the Fra Mauro site. The correlation is rather good, considering that Fra Mauro is almost

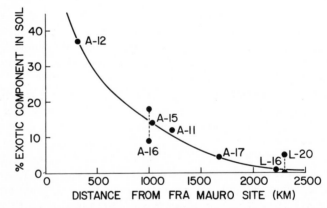

Fig. 2. Percentage of exotic component in soils from Apollo (A) and Luna (L) missions plotted against distance from the Apollo 14 landing site at Fra Mauro. Percentages taken from Table 4; uncertainties in Apollo 16 and Luna 20 values are discussed in the table and text.

certainly *not* the only source of exotic component. The dominant contribution of the Apollo 15 site, for example, is probably from the gamma-ray high at Archimedes, which is as close to the Apollo 15 site as the Apollo 12 site is to Fra Mauro. The lower abundance of exotic component in Apollo 15 as opposed to Apollo 12 then suggests that the Archimedes source may be smaller or less enriched than that at Fra Mauro. At the other Apollo and Luna sites, more distant from Imbrium, distinction between individual sources in the Imbrium region is impossible.

Relationships with the Imbrium Basin

Material sampled at Fra Mauro, and probably at the other regions of high gamma-ray activity in the vicinity at Mare Imbrium, is believed to be derived from excavation of the Imbrium Basin (Sutton *et al.*, 1972). Regardless of the detailed relationships between landing sites and source areas, we can then interpret the history of the exotic component in a fashion which is illustrated diagramatically in Fig. 3. The first stage is Imbrium excavation (Fig. 3-1). Trace element enriched material was excavated and deposited on the lunar surface in the Imbrium region. The fact that these enriched materials appear to be associated only with Imbrium and not with other major mare basins may result from primary inhomogeneity in lunar composition or differentiation processes. Alternatively, it may simply reflect the greater size of the Imbrium Basin, which must have brought up material from a greater depth than was sampled in other basin excavating events. This latter interpretation is supported by the observation that ejecta blankets show reverse stratification, i.e. materials from deepest in the crater are found most surficially on the ejecta (Shoemaker, 1963). Deep-seated materials would probably be given the most vertical trajectories, and would thus be even more sharply concentrated in the vicinity of the basin than the ejecta blanket as a whole. The presently observed gamma-ray highs at Fra Mauro and Aristarchus may reflect in-homogeneities in ejecta distribution, or they may be relics of an originally more disseminated layer of radioactive materials which has since been largely obliterated by subsequent events such as impacts and mare filling.

Following deposition of enriched materials on the surface, transport processes begin diffusing these materials outward (Fig. 3-2). Mare filling would bury material exposed on the surface of the basin itself, except where large cratering events such as Archimedes have excavated it from beneath the mare surface. The much later Copernican event probably also excavated Fra Mauro formation (Schmitt *et al.*, 1967), though not necessarily trace element rich material. In any event, diffusion from primary and secondary sources ultimately results in a broad pattern of exotic component distribution centered in the Imbrium region (Fig. 3-3). Other sources may well exist to the north of Imbrium; current orbital coverage does not extend to this area.

An alternative model which could be invoked is one in which discrete sources of exotic component are replaced by a distribution grading uniformly away from Mare Imbrium, resulting from uniform dispersal of excavated trace element rich

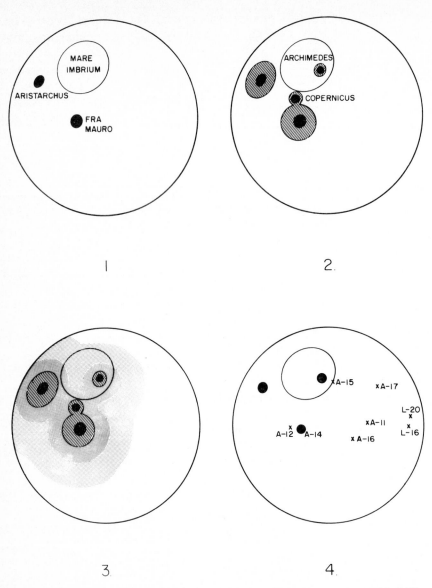

Fig. 3. Diagrammatic representation of dispersion of exotic component over the visible hemisphere of the moon. See text for a complete discussion. Upper left: Immediately following Imbrium basin excavation. Surrounding "hot spots" formed, typified by those of Aristarchus and Fra Mauro. Upper right: At subsequent time, diffusion outward from hot spots has begun, together with formation of new hot spots by subsequent cratering, typified by Archimedes and Copernicus. Lower left: As diffusion proceeds, material from hot spots becomes more widely disseminated over the surface. Lower right: Apollo (A) and Luna (L) landing sites shown together with Imbrium and known hot spots as revealed by gamma ray spectrometry. Apollo 14 landing site is within the Fra Mauro hot spot.

material. The present surficial distribution of exotic component, which emphatically does not fit this pattern, must then be explained by localized processes in the vicinity of each landing site (particularly Apollo 12, 14, and 15) which have buried or uncovered variable amounts of Imbrium ejecta. Wasson and Baedecker (1972) have presented a fairly detailed model of this type for the Apollo 12 site. Attractive features of this approach are that primary inhomogeneities of ejecta distribution are not required (although they are not absolutely necessary to our model either) and that extensive lateral transport is replaced by primarily vertical transport. However, obvious inhomogeneities in the distribution of Imbrium ejecta, as represented by the Fra Mauro Formation, do indeed exist (Mutch, 1970). Furthermore, examination of the gamma-ray spectrometry map shows that in the regions between areas of high and low radioactivity there are broad, smooth gradations extending over distances of 1000 km or more. These patterns are strongly suggestive of mixing by extensive lateral transport of surface materials.

Conclusions

Whichever model (or combination of them) is chosen, the conclusion remains that the trace element enriched material presently found as exotic component and disseminated widely over the lunar surface could have had its origin in the lunar interior at a depth such that it was sampled directly *only* by the excavation of the Imbrium Basin. It appears that no other primary source *need be* invoked. A deep layer of late-stage differentiate is a feature of several recent models of the lunar interior (e.g. Smith, 1974). It is important to note, however, that such a layer need *not* be uniformly present within the moon, but could be localized to the earth facing side, for example, or even to the Imbrium Basin region. Lateral inhomogeneity seems to be the rule rather than the exception in the case of the moon, as evidenced by mare distribution, center of mass and principal moments of inertia.

As stated at the beginning of our discussion, we have focused on the exotic component in the lunar soils. If Fra Mauro material is taken as representative of the exotic component in its pure form, then the relationship between exotic component and KREEP becomes apparent; Fra Mauro materials are considered to be essentially pure KREEP (Hubbard *et al.*, 1972; Schonfeld and Meyer, 1972). The dissemination of fine-grained exotic component must be accompanied by the dissemination of a lesser amount of coarse-grained KREEP fragments as well; the exotic component can be considered fine-grained KREEP (Schonfeld and Meyer, 1972). For example, our data for the 147–1000 μ fraction of 15531 (Evensen *et al.*, 1973) indicate that this fraction contains ~6% exotic component; we would therefore, expect a few percent of identifiable KREEP fragments larger than 1 mm to be present in this soil. If formed from late-stage liquids at the base of an anorthositic crust, KREEP materials should be associated with plagioclase, as is observed (Hubbard *et al.*, 1972).

Other highly differentiated lunar materials (e.g. the so-called lunar "granites") are extremely rare, and may well be derived from KREEP by localized differentiation or metamorphism (Schnetzler *et al.*, 1970; Meyer, 1972; Hubbard *et al.*, 1972). They could also be formed within the deep source area by immiscible liquid separation (Roedder and Weiblen, 1970). If it were not for the Imbrium collision, we might see only trace element poor mare basalts and anorthositic rocks on the lunar surface today.

Acknowledgments—This work was prepared under NASA grant NGR 24-005-223. We are greatly indebted to Professor R. O. Pepin and Mr. J. R. Basford for many stimulating discussions and collaborative research efforts. Mr. Allan T. Bates performed the mass spectrometric work for some samples.

REFERENCES

Albee A. L., Burnett D. S., Chodos A. A., Haines E. L., Huneke J. C., Papanastassiou D. A., Podosek F. A., Russ G. P., III, and Wasserburg G. J. (1970) *Earth Planet Sci. Lett.* **9**, 137–163.
Basford J. R. (1974) K–Ar analysis of Apollo 11 fines 10084. *Proc. Fifth Lunar Sci. Conf., Geochim. Cosmochim. Acta.* This volume.
Basford J. R. and Coscio M. R. (1973) An improved method for rapid low loss density separations with heavy liquids. *Amer. Mineral.* **58**, 1094–1095.
Chou C.-L., Baedecker P. A., Bild R. W., Robinson K. L., and Wasson J. T. (1974) Volatile elements in lunar soils (abstract). In *Lunar Science—V*, pp. 115–117. The Lunar Science Institute, Houston.
Compston W., Berry H., Vernon M. J., Chappell B. W., and Kaye M. J. (1971) Rubidium–strontium chronology and chemistry of lunar material from the Ocean of Storms. *Proc. Second Lunar Sci. Conf., Geochim. Cosmochim. Acta*, Suppl. 2, Vol. 2, pp. 1471–1485. MIT Press.
Evensen N. M., Murthy V. Rama, and Coscio M. R., Jr. (1973) Rb–Sr ages of some mare basalts and the isotopic and trace element systematics in lunar fines. *Proc. Fourth Lunar Sci. Conf., Geochim. Cosmochim. Acta*, Suppl. 4, Vol. 2, pp. 1707–1724. Pergamon.
Gibson E. K., Jr. and Hubbard N. J. (1972) Thermal volatilization studies on lunar samples. *Proc. Third Lunar Sci. Conf., Geochim. Cosmochim. Acta*, Suppl. 3, Vol. 2, pp. 2003–2014. MIT Press.
Gibson E. K., Jr., Hubbard N. J., Wiesmann H., Bansal B. M., and Moore G. W. (1973) How to lose Rb, K, and change the K/Rb ratio: An experimental study. *Proc. Fourth Lunar Sci. Conf., Geochim. Cosmochim. Acta*, Suppl. 4, Vol. 2, pp. 1263–1273. Pergamon.
Heiken G., Duke M., McKay D. S., Clanton U. S., Fryxell R., Nagle J. S., Scott R., and Sellers G. A. (1973) Preliminary stratigraphy of the Apollo 15 drill core. *Proc. Fourth Lunar Sci. Conf., Geochim. Cosmochim. Acta*, Suppl. 4, Vol. 1, pp. 191–213. Pergamon.
Hubbard N. J., Gast P. W., Rhodes J. M., Bansal B. M., Wiesmann H., and Church S. E. (1972) Nonmare basalts: Part II. *Proc. Third Lunar Sci. Conf., Geochim. Cosmochim. Acta*, Suppl. 3, Vol. 2, pp. 1161–1179. MIT Press.
Hubbard N. J., Meyer C., Jr., Gast P. W., and Wiesmann H. (1971) The composition and derivation of Apollo 12 soils. *Earth Planet. Sci. Lett.* **10**, 341–350.
LSPET (Lunar Sample Preliminary Examination Team) (1971) Preliminary examination of lunar samples from Apollo 14. *Science* **173**, 681–693.
LSPET (Lunar Sample Preliminary Examination Team) (1973) Preliminary examination of lunar samples from Apollo 16. *Science* **197**, 23–34.
Metzger A. E., Trombka J. I., Peterson L. E., Reedy R. C., and Arnold J. R. (1973) Lunar surface radioactivity: Preliminary results of the Apollo 15 and Apollo 16 gamma-ray spectrometer experiments. *Science* **179**, 800–803.
Meyer C., Jr. (1972) Mineral assemblages and the origin of non-mare lunar rock types (abstract). In *Lunar Science–III*, pp. 542–544. The Lunar Science Institute, Houston.
Murthy V. Rama, Evensen N. M., and Coscio M. R., Jr. (1970) Distribution of K, Rb, Sr and Ba and Rb–Sr isotopic relations in Apollo 11 lunar samples. *Proc. Apollo 11 Lunar Sci. Conf., Geochim. Cosmochim. Acta*, Suppl. 1, Vol. 2, pp. 1393–1406. Pergamon.

Murthy V. Rama, Evensen N. M., Jahn Bor-ming, and Coscio M. R., Jr. (1971) Rb–Sr ages and elemental abundances of K, Rb, Sr, and Ba in samples from the Ocean of Storms. *Geochim. Cosmochim. Acta* **35**, 1139–1153.

Murthy V. Rama, Evensen N. M., Jahn Bor-ming, and Coscio M. R., Jr. (1972) Apollo 14 and 15 samples: Rb–Sr ages, trace elements, and lunar evolution. *Proc. Third Lunar Sci. Conf., Geochim. Cosmochim. Acta*, Suppl. 3, Vol. 2, pp. 1503–1514. MIT Press.

Mutch T. A. (1970) *Geology of the Moon.* Princeton University Press. 324 pp.

Naughton J. J., Derby J. V., and Lewis V. A. (1971) Vaporization from heated lunar samples and the investigation of lunar erosion by volatilized alkalis. *Proc. Second Lunar Sci. Conf., Geochim. Cosmochim. Acta*, Suppl. 2, Vol. 1, pp. 449–457. MIT Press.

Papanastassiou D. A. and Wasserburg G. J. (1970) Rb–Sr ages from the Ocean of Storms. *Earth Planet. Sci. Lett.* **9**, 269–278.

Papanastassiou D. A. and Wasserburg G. J. (1971) Lunar chronology and evolution from Rb–Sr studies of Apollo 11 and 12 samples. *Earth Planet. Sci. Lett.* **11**, 37–62.

Papanastassiou D. A. and Wasserburg G. J. (1972) Rb–Sr systematics of Luna 20 and Apollo 16 samples. *Earth Planet. Sci. Lett.* **17**, 52–63.

Papanastassiou D. A., Wasserburg G. J., and Burnett D. S. (1970) Rb–Sr ages of lunar rocks from the Sea of Tranquillity. *Earth Planet. Sci. Lett.* **7**, 1–19.

Pepin R. O., Basford J. R., Dragon J. C., Coscio M. R., and Murthy V. Rama (1974) Rare gases and trace element chemistry in the Apollo 15 deep drill core: Depositional chronologies and K–Ar ages. *Proc. Fifth Lunar Sci. Conf., Geochim. Cosmochim. Acta.* This volume.

Philpotts J. A., Schnetzler C. C., Bottino M. L., Schuman S., and Thomas H. H. (1972) Luna 16: Some Li, K, Rb, Sr, Ba, rare-earth, Zr and Hf concentrations. *Earth Planet. Sci. Lett.* **13**, 429–435.

Prinz M., Dowty E., Keil K., and Bunch T. E. (1973) Mineralogy, petrology and chemistry of lithic fragments from Luna 20 fines: Origin of the cumulate ANT suite and its relationship to high-alumina and mare basalts. *Geochim. Cosmochim. Acta* **37**, 979–1006.

Roedder E. and Weiblen P. W. (1970) Lunar petrology of silicate melt inclusions, Apollo 11 rocks. *Proc. Apollo 11 Lunar Sci. Conf., Geochim. Cosmochim. Acta*, Suppl. 1, Vol. 1, pp. 801–837. Pergamon.

Rose H. J., Jr., Cuttitta F., Berman S., Carron M. K., Christian R. P., Dwornik E. J., Greenland L. P., and Ligon D. T., Jr. (1973) Compositional data for twenty-two Apollo 16 samples. *Proc. Fourth Lunar Sci. Conf., Geochim. Cosmochim. Acta*, Suppl. 4, Vol. 2, pp. 1149–1158. Pergamon.

Schmitt H. H., Trask N. J., and Shoemaker E. M. (1967) Geologic map of the Copernicus quadrangle of the moon. Map I-515, U.S. Geological Survey.

Schnetzler C. C., Philpotts J. A., and Bottino M. L. (1970) Li, K, Rb, Sr, Ba and rare-earth concentrations, and Rb–Sr age of lunar rock 12013. *Earth Planet. Sci. Lett.* **9**, 185–192.

Schonfeld E. and Meyer C., Jr. (1972) The abundances of components of the lunar soils by a least-squares mixing model and the formation age of KREEP. *Proc. Third Lunar Sci. Conf., Geochim. Cosmochim. Acta*, Suppl. 3, Vol. 2, pp. 1397–1420. MIT Press.

Shoemaker E. M. (1963) Impact mechanics at Meteor Crater, Arizona. In *The Moon, Meteorites and Comets* (editors B. M. Middlehurst and G. P. Kuiper), pp. 301–336. University of Chicago Press.

Smith J. V. (1974) Origin of moon by disintegrative capture with chemical differentiation followed by sequential accretion (abstract). In *Lunar Science—V*, pp. 718–720. The Lunar Science Institute, Houston.

Sutton R. L., Hart M. H., and Swann G. A. (1972) Geology of the Apollo 14 landing site. *Proc. Third Lunar Sci. Conf., Geochim. Cosmochim. Acta*, Suppl. 3, Vol. 1, pp. 27–38. MIT Press.

Wasson J. T. and Baedecker P. A. (1972) Provenance of Apollo 12 KREEP. *Proc. Third Lunar Sci. Conf., Geochim. Cosmochim. Acta*, Suppl. 3, Vol. 2, pp. 1315–1326. MIT Press.

Wood J. A., Dickey J. S., Jr., Marvin U. B., and Powell B. N. (1970) Lunar anorthosites and a geophysical model of the moon. *Proc. Apollo 11 Lunar Sci. Conf., Geochim. Cosmochim. Acta*, Suppl. 1, Vol. 1, pp. 965–988.

Proceedings of the Fifth Lunar Conference
(Supplement 5, Geochimica et Cosmochimica Acta)
Vol. 2 pp. 1419–1449 (1974)
Printed in the United States of America

High resolution argon analysis of neutron-irradiated Apollo 16 rocks and separated minerals

E. K. Jessberger,* J. C. Huneke, F. A. Podosek,† and G. J. Wasserburg

The Lunatic Asylum of the Charles Arms Laboratory, Division of Geological and Planetary Sciences,‡ California Institute of Technology, Pasadena, California 91109

Abstract—Ar analyses are reported for detailed thermal release studies on neutron-irradiated samples of mineral separates and a whole sample of breccia 65015 and rock fragments 67483,15-2 and 60503,3-3.

The ^{37}Ar release patterns show a single release peak at ~800°C for 65015 plagioclase and a distinctive release peak at ~1200°C for 65015 pyroxene. 60% of the ^{37}Ar is released from the pyroxene above 1000°C, whereas only 15% of the ^{39}Ar is released above this temperature. From a graph of ^{40}Ar/^{39}Ar ages versus ^{37}Ar, the anomalously low ages of high-temperature releases, which are a common feature in ^{40}Ar/^{39}Ar age spectra of lunar samples, can be associated with the unique high-temperature release of ^{37}Ar from the pyroxene.

The age spectrum of 65015 plagioclase plotted versus ^{37}Ar shows a plateau at 3.98 AE and then rises with a sharp ascent to a maximum of 4.47 AE for the last stages of gas release that correlates with the proportion of ancient relict plagioclase as determined by Rb–Sr analyses. This confirms the existence of ancient lithic fragments within this metaclastic rock and indicates that the protolith of 65015 contained plagioclase fragments older than 4.47 AE and the 65015 breccia was severely metamorphosed in an event at 3.98 AE. The time interval between this metamorphic event and the event dated by Luna 20 samples (neutron-irradiated with 65015) is 80 m.y. These events occurred at distinctly separate times and presumably represent the times of formation of the Nectaris and Crisium basins, respectively. The terminal lunar cataclysm involved at least two separate major impact events.

Introduction

The ^{40}Ar–^{39}Ar dating technique has proven a useful tool in deciphering the history of lunar and meteoritic material. However, little work has yet been done to fully understand the method and *all* effects which might be important in interpreting the results and to find the limitations of its applicability. The answers to such questions are important in the case that only a limited number and facies of samples are available for dating and if the age differences of a group of samples are small.

Model calculations by Turner (1969) demonstrated the general applicability of the ^{40}Ar–^{39}Ar technique in a simple system. However, natural samples are usually more complex in terms of diffusion characteristics, their thermal release patterns, and individual histories of the components. Turner *et al.* (1971, 1972) investigated separated minerals in an attempt to reduce the complexity of polymineralic systems and demonstrated that plagioclase generally yielded an obvious well-defined age

*Present address: Max Planck Institut für Kernphysik, Heidelberg, BRD.
†Present address: Washington University, St. Louis, Missouri 63130.
‡Contribution No. 2483.

plateau, while the total rock age spectrum reflected the complexities of the mineral assemblage. While this was true for lunar mare plagioclase, at least one highland rock plagioclase did not exhibit a similarly well-defined age plateau (Huneke *et al.*, 1973). The importance of understanding age spectra obtained on lunar highland samples in order to delineate early lunar history 3.9–4.6 AE ago has led to yet another approach. By releasing Ar in many small increments, a high resolution was obtained in the age spectra to obtain more information about the gas release and, more importantly, to isolate the gas release from different sites within a sample in order to be able to attribute $^{40}Ar/^{39}Ar$ ratios and corresponding ages to the various minerals.

Combining both of these approaches, we analyzed the composition of argon released in many steps from neutron-irradiated whole rock and mineral separates of Apollo 16 breccia 65015 and two fragments 60503,3-3 and 67483,15-2. A small portion of these results has already been published (Jessberger *et al.*, 1974; Huneke *et al.*, 1974).

Rock 65015 was the first rock in which Papanastassiou and Wasserburg (1972) demonstrated the existence of large plagioclase clasts containing Sr which was isotopically unequilibrated with matrix Sr in the last severe metamorphism at ~4.0 AE. The Sr in these clasts has a more primitive isotopic composition. Detailed Ar analyses of neutron-irradiated whole rock and mineral separates were made to establish if these clasts were also incompletely degassed of ^{40}Ar and to place limits on their earlier formation time.

Included in the mineral separates from rock 65015 were (1) a high purity plagioclase B separate containing an impurity of only ~1.5% quintessence, (2) a less pure plagioclase C separate with ~2.5% quintessence, (3) a separate of low-Ca pyroxene with 2.6% high-Ca pyroxene, 2.1% plagioclase and 1.3% olivine, and (4) a separate (Ph–B) with 4% phosphates, 54% low-Ca pyroxene, 7% high-Ca pyroxene, and 35% plagioclase. The mineral percentages were estimated from electron probe analysis of 100–200 grains of each separate. Characteristic compositions of each of the minerals have been reported by Albee *et al.* (1973).

We fully recognize the complexity that often characterizes the ^{40}Ar–^{39}Ar results on some lunar samples (see, for example, results on Apollo 14 breccias (Turner *et al.*, 1971)) and which makes the unraveling of an age pattern rather difficult. Our approach is to utilize the results from different techniques on the same sample to aid us in establishing a consistent and physically reasonable interpretation of what might otherwise remain as yet another example of a complex Ar release pattern.

EXPERIMENTAL

The samples were neutron-irradiated in an irradiation designated LAV 3 (Podosek *et al.*, 1973). For convenience, the ^{39}Ar and ^{37}Ar conversion factors are repeated here

$$C_{39}(K) = {}^{39}Ar^*/K = 8.02 \cdot 10^{-4} \text{ cc STP/g}$$
$$C_{37}(Ca) = {}^{37}Ar/Ca = 4.33 \cdot 10^{-4} \text{ cc STP/g}$$

The asterisk (*) denotes K-derived Ar.

Gas extraction and analysis procedures are substantially the same as previously used (cf. Podosek and Huneke, 1973). After loading into the extraction system, the samples were degassed at about 90°C. In order to further clean samples of adsorbed gas, several low-temperature extractions were usually made. High ^{39}Ar* and ^{37}Ar concentrations allowed a large number (30–49) of extraction steps, generally extracting enough gas in each step to be well above the blank amounts. In the temperature range ~500–1000°C, at least two extractions at each temperature were made. The extraction furnace generally was not cooled between extractions. The minimum extraction time for one step was ~1 hour, the time required for purification and separation from heavy rare gases, transfer to the mass spectrometer and measurement of the argon from the previous extraction. In some cases the second extraction time was increased to 2 hours in order to increase the gas amounts released in that extraction. Extraction temperatures were inferred from an empirically obtained power-law formula relating temperature to the power input into the resistance heated furnace, originally calibrated by optical pyrometer measurements on a fresh tantalum crucible (spectral emissivity 0.49 at 0.65 μ). Temperatures below 900°C were extrapolated. The reproducibility of temperatures is believed to be better than ±20°C. A series of blanks were performed at various temperatures before and after each sample. The ^{40}Ar amounts of all blanks are given in Fig. 1. Although the blanks were taken over a period of six months, they were remarkably stable in time. After each sample a high temperature ($\geqslant$1500°C) blank was taken to ensure the completeness of the sample extraction.

RESULTS

Measured ^{40}Ar amounts and isotopic ratios are extrapolated to gas inlet time and corrected for mass discrimination (0.5%/amu favoring heavier masses). In addition to the statistical error of the ratios, a relative uncertainty of 2.5% in ^{40}Ar amounts is included, corresponding to the maximum variation of all air argon calibrations in this series of analysis. Thereafter, the following corrections are made in the order listed (cf. Podosek and Huneke, 1973):

(a) Decay of ^{39}Ar and ^{37}Ar during and after irradiation.
(b) Contributions from neutrons on Ca to ^{36}Ar (^{36}Ar/^{37}Ar = 0.0003), ^{38}Ar (^{38}Ar/^{37}Ar = 0.0001) and ^{39}Ar (^{39}Ar/^{37}Ar = 0.000732).
(c) Contributions from neutrons on K to ^{38}Ar (^{38}Ar/^{39}Ar = 0.01) and ^{40}Ar (^{40}Ar/^{39}Ar = 0.01).
(d) Corrections for the relative fluxes received by the individual samples as monitored by Ni wires.
(e) Blank contributions on ^{40}Ar, ^{38}Ar, and ^{36}Ar, assuming air composition of the blank with the assigned error given by the shaded area in Fig. 1.

After these corrections, the remaining argon consists of spallogenic, radiogenic and trapped isotopes in addition to K-derived ^{39}Ar* and Ca-derived ^{37}Ar. The data in this stage of processing are listed as "measured" data in Tables 2 and 3. K and Ca concentrations of all the samples, calculated from ^{39}Ar* and ^{37}Ar, respectively, using the LAV 3 conversion coefficients, are listed in Table 1.

APOLLO 16 ROCK 65015

Ar release patterns

In the ^{40}Ar/^{39}Ar dating technique, an unambiguous relationship between ^{40}Ar*/K ratios, measured as ^{40}Ar*/^{39}Ar* ratios, and the ^{40}Ar* released from

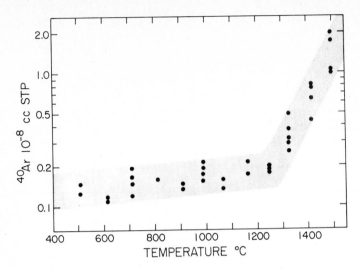

Fig. 1. Amounts of ^{40}Ar measured in system blanks at various temperatures. The shaded area gives the maximum variation. An error corresponding to the width of this area has been assigned to blank corrections to measurements.

Table 1. ^{36}Ar/^{38}Ar and ^{38}Ar$_s$/^{37}Ar plateau values, exposure ages, and calcium and potassium concentrations.

Sample	^{36}Ar/^{38}Ar plateau	^{38}Ar$_s$/^{37}Ar plateau	Exposure[a] age (m.y.)	Ca (%)	K (ppm)
65015					
Total rock	.595 ±.003	.01585 ±.00006	490 ± 20	7.8	3000
Plagioclase B	.613 ±.007	.01498 ±.00007	463 ± 19	12.7	1475
Plagioclase C	.605 ±.003	.01523 ±.00007	471 ± 20	12.7	2115
Pyroxene	.596 ±.007	.01482 ±.00015	458 ± 20	3.4	202
Ph B	.587 ±.003	.01494 ±.00008	463 ± 20	6.6	850
60503,3-3	3.19 ±.04	.00426[b] ±.00013	132 ± 7	14.7	1370
67483,15-2	—	.0033[b] ±.0002	102 ± 7	8.5	1675

[a]Exposure ages calculated with $P_{38} = 1.4 \times 10^{-8}$ cc STP ^{38}Ar$_s$/g Ca/m.y. The error is due to statistical errors in the measured ratios, a conservative error in (^{36}Ar/^{38}Ar)$_s$ of .05, and the uncertainty in C_{37}. The uncertainty in P_{38} is not included.

[b]Because of the high ^{36}Ar/^{38}Ar ratios, no correction for Cl-derived ^{38}Ar can be applied.

Table 2. Argon analyses of a whole sample and mineral separates of Apollo 16 breccia 65015.

Extraction Temp. [°C]	Time [min]	^{40}Ar [10^{-8} cc STP/g]	$\dfrac{^{36}Ar}{^{40}Ar}$	$\dfrac{^{37}Ar}{^{40}Ar}$	$\dfrac{^{38}Ar}{^{40}Ar}$	$\dfrac{^{39}Ar}{^{40}Ar}$	Age [m.y.]
Total rock (63.5 mg)			$\times 10^{-5}$	$\times 10^{-4}$	$\times 10^{-5}$	$\times 10^{-5}$	
220	65	2 ± 1	1870 ± 1070	—	1210 ± 760	780 ± 540	5040 ± 1230
345	65	4 ± 1	310 ± 120	540 ± 125	2620 ± 500	3050 ± 580	2830 ± 280
460	65	132 ± 3	130 ± 7	455 ± 14	506 ± 7	3648 ± 50	2578 ± 19
550	65	1050 ± 25	22 ± 2	219 ± 2	107 ± 2	2093 ± 7	3409 ± 5
550	68	605 ± 12	25 ± 3	139 ± 3	63 ± 3	1640 ± 3	3801 ± 3
550	65	381 ± 8	23 ± 5	125 ± 5	62 ± 6	1543 ± 4	3900 ± 4
550	67	307 ± 6	19 ± 6	126 ± 5	50 ± 5	1513 ± 5	3933 ± 5
550	65	249 ± 5	32 ± 6	119 ± 3	53 ± 5	1487 ± 6	3961 ± 7
580	65	291 ± 7†	15 ± 7	178 ± 4	58 ± 5	1484 ± 6	3966 ± 7
580	65	589 ± 13	23 ± 3	190 ± 2	53 ± 2	1465 ± 3	3987 ± 3
580	65	482 ± 10	25 ± 3	168 ± 3	53 ± 3	1464 ± 4	3988 ± 4
580	66	452 ± 10	29 ± 4	174 ± 3	57 ± 3	1458 ± 4	3996 ± 4
625	65	796 ± 14	28 ± 4	256 ± 2	70 ± 1	1473 ± 2	3977 ± 2
625	70	658 ± 14	28 ± 2	246 ± 1	65 ± 3	1482 ± 3	3968 ± 3
625	65	464 ± 10	36 ± 3	258 ± 2	71 ± 3	1490 ± 3	3959 ± 4
685	70	533 ± 11	37 ± 3	331 ± 4	81 ± 3	1503 ± 3	3945 ± 3
685	65	381 ± 8	34 ± 10	329 ± 6	71 ± 5	1507 ± 4	3940 ± 4
730	60	583 ± 13	62 ± 3	569 ± 2	117 ± 4	1528 ± 3	3917 ± 3
730	65	546 ± 12	66 ± 4	621 ± 3	121 ± 2	1525 ± 2	3919 ± 3
730	70	401 ± 9	69 ± 6	633 ± 3	125 ± 5	1533 ± 4	3911 ± 5
775	65	768 ± 15	117 ± 2	1131 ± 2	205 ± 2	1554 ± 2	3889 ± 2
775	65	573 ± 12	126 ± 2	1253 ± 3	221 ± 3	1549 ± 3	3896 ± 3
775	65	430 ± 9	131 ± 4	1426 ± 3	242 ± 2	1535 ± 5	3909 ± 5
775	65	316 ± 7	142 ± 5	1585 ± 8	269 ± 8	1536 ± 6	3908 ± 7
830	65	645 ± 14	228 ± 5	2430 ± 4	393 ± 2	1541 ± 3	3902 ± 3
830	65	495 ± 11	259 ± 5	2700 ± 6	441 ± 4	1523 ± 3	3922 ± 3
830	65	298 ± 7	276 ± 6	2850 ± 10	469 ± 2	1505 ± 6	3941 ± 6
870	65	261 ± 5†	375 ± 9	3698 ± 12	597 ± 7	1510 ± 6	3935 ± 7
870	65	336 ± 7	378 ± 4	3986 ± 14	639 ± 2	1499 ± 5	3948 ± 6
870	65	219 ± 5	435 ± 7	4429 ± 18	706 ± 5	1502 ± 7	3944 ± 8
910	66	254 ± 5	530 ± 7	5530 ± 18	882 ± 4	1502 ± 7	3943 ± 7
910	60	205 ± 5	629 ± 7	6753 ± 27	1057 ± 9	1494 ± 8	3952 ± 9
950	65	278 ± 6	878 ± 9	9314 ± 30	1481 ± 7	1514 ± 5	3929 ± 7
950	70	185 ± 4	1061 ± 8	11227 ± 48	1756 ± 12	1529 ± 9	3913 ± 10
1000	70	253 ± 6	1125 ± 8	11733 ± 37	1887 ± 13	1544 ± 7	3895 ± 8
1000	65	68 ± 2	881 ± 26	9229 ± 110	1565 ± 30	1606 ± 24	3832 ± 24
1050	67	126 ± 3	643 ± 13	6374 ± 41	1219 ± 12	1650 ± 13	3788 ± 13
1050	68	67 ± 2	600 ± 28	6178 ± 79	1197 ± 23	1621 ± 21	3818 ± 21
1145	72	140 ± 3	725 ± 10	6417 ± 38	1320 ± 19	1726 ± 13	3715 ± 12
1145	71	60 ± 2	1283 ± 85	11617 ± 155	2252 ± 30	1656 ± 27	3781 ± 26
1235	70	72 ± 2	2373 ± 39	21413 ± 261	4002 ± 52	1566 ± 27	3868 ± 28
1235	66	12 ± 1	2420 ± 210	20930 ± 1540	3900 ± 330	1640 ± 150	3790 ± 140
1290	65	170 ± 4	2741 ± 21	24970 ± 170	4575 ± 32	1651 ± 14	3780 ± 16
1290	65	13 ± 1	1800 ± 200	17100 ± 1550	2870 ± 280	1540 ± 150	3900 ± 150
1355	65	29 ± 2	1390 ± 120	13870 ± 950	2330 ± 170	1480 ± 110	3970 ± 120

†Unexpected low value due to mistake during manifold operation.

Table 2. (*Continued*).

Extraction Temp. [°C]	Time [min]	^{40}Ar [10^{-8} cc STP/g]	$\dfrac{^{36}\text{Ar}}{^{40}\text{Ar}}$	$\dfrac{^{37}\text{Ar}}{^{40}\text{Ar}}$	$\dfrac{^{38}\text{Ar}}{^{40}\text{Ar}}$	$\dfrac{^{39}\text{Ar}}{^{40}\text{Ar}}$	Age [m.y.]
1355	65	6 ± 2	1970 ± 630	13230 ± 4050	2345 ± 740	1490 ± 500	3940 ± 520
1420	68	21 ± 3	1270 ± 260	12080 ± 2250	2065 ± 390	1410 ± 280	4050 ± 315
1420	65	6 ± 4	1575 ± 1080	13800 ± 9200	2270 ± 1520	1630 ± 1160	3800 ± 1100
1502	65	29 ± 7	1700 ± 470	16000 ± 4300	2580 ± 700	1310 ± 380	4170 ± 460
Total		15254	225 ± 7	2210 ± 55	398 ± 5	1580 ± 7	3852 ± 5
Plagioclase B (47.4 mg)			$\times 10^{-5}$	$\times 10^{-3}$	$\times 10^{-5}$	$\times 10^{-5}$	
200	80	7 ± 1	912 ± 205	—	431 ± 64	223 ± 57	7316 ± 473
265	80	14 ± 1	1010 ± 87	5 ± 50	386 ± 48	140 ± 24	8179 ± 318
345	80	12 ± 1	1294 ± 74	64 ± 8	537 ± 43	590 ± 86	5539 ± 262
345	85	9 ± 1	1380 ± 104	80 ± 11	547 ± 95	1020 ± 131	4578 ± 223
410	70	29 ± 1	1373 ± 30	75 ± 3	511 ± 12	644 ± 76	5383 ± 212
410	85	37 ± 1	1223 ± 32	90 ± 4	515 ± 21	1247 ± 141	4237 ± 191
475	70	85 ± 2	976 ± 36	88 ± 4	452 ± 5	1456 ± 116	3984 ± 133
475	90	120 ± 3	281 ± 33	74 ± 6	256 ± 12	2062 ± 52	3431 ± 40
520	55	179 ± 4	240 ± 36	119 ± 17	275 ± 19	1798 ± 33	3650 ± 30
520	85	202 ± 5	110 ± 12	99 ± 3	182 ± 8	1683 ± 10	3758 ± 9
520	115	102 ± 3	89 ± 24	72 ± 2	133 ± 21	1528 ± 16	3917 ± 17
565	65	281 ± 7	124 ± 12	150 ± 1	199 ± 7	1487 ± 7	3962 ± 8
565	80	268 ± 7	114 ± 7	116 ± 3	182 ± 5	1454 ± 7	3999 ± 8
620	60	607 ± 15	133 ± 5	171 ± 2	234 ± 4	1463 ± 3	3988 ± 4
620	85	470 ± 12	136 ± 4	163 ± 1	247 ± 5	1492 ± 3	3956 ± 4
685	60	541 ± 14	231 ± 11	272 ± 2	390 ± 3	1546 ± 5	3896 ± 5
685	90	343 ± 9	253 ± 5	287 ± 3	440 ± 11	1544 ± 6	3900 ± 6
705	65	530 ± 13	438 ± 6	474 ± 2	694 ± 2	1580 ± 4	3860 ± 5
705	105	491 ± 12	511 ± 6	556 ± 2	831 ± 3	1553 ± 5	3888 ± 6
775	65	574 ± 14	769 ± 4	840 ± 2	1261 ± 4	1550 ± 3	3890 ± 5
775	125	470 ± 12	893 ± 5	954 ± 3	1443 ± 5	1510 ± 4	3932 ± 6
810	65	414 ± 10	1114 ± 4	1196 ± 3	1817 ± 6	1497 ± 7	3946 ± 9
810	70	230 ± 6	1231 ± 16	1317 ± 6	1988 ± 12	1482 ± 7	3962 ± 9
865	70	335 ± 8	1434 ± 9	1527 ± 4	2312 ± 8	1474 ± 7	3969 ± 10
865	75	226 ± 5	1508 ± 12	1591 ± 7	2425 ± 10	1470 ± 10	3974 ± 13
910	60	317 ± 8	1523 ± 8	1636 ± 5	2481 ± 6	1473 ± 5	3971 ± 9
910	90	103 ± 3	1517 ± 30	1592 ± 14	2474 ± 29	1451 ± 15	3996 ± 18
1000	75	269 ± 7	1421 ± 11	1509 ± 7	2372 ± 8	1464 ± 9	3983 ± 12
1085	75	295 ± 7	1183 ± 7	1208 ± 5	1932 ± 8	1360 ± 7	4106 ± 10
1170	85	189 ± 5	1206 ± 14	1194 ± 7	1892 ± 14	1211 ± 8	4300 ± 12
1290	75	103 ± 5	993 ± 36	895 ± 29	1507 ± 52	1100 ± 37	4465 ± 55
1430	65	71 ± 9	761 ± 108	747 ± 93	1146 ± 137	1088 ± 137	4484 ± 205
Total		7925 ± 45	683 ± 3	696 ± 3	1073 ± 5	1496 ± 3	3940 ± 3
Plagioclase C (33.2 mg)			$\times 10^{-5}$	$\times 10^{-3}$	$\times 10^{-5}$	$\times 10^{-5}$	
410	60	30 ± 1	1291 ± 39	92 ± 5	578 ± 23	938 ± 114	4724 ± 212
410	80	66 ± 2	891 ± 56	86 ± 5	480 ± 19	2102 ± 177	3391 ± 133
410	105	78 ± 2	460 ± 57	46 ± 6	343 ± 11	2811 ± 129	2952 ± 69

Table 2. (*Continued*).

Extraction Temp. [°C]	Time [min]	^{40}Ar [10^{-8} cc STP/g]	$\dfrac{^{36}\text{Ar}}{^{40}\text{Ar}}$	$\dfrac{^{37}\text{Ar}}{^{40}\text{Ar}}$	$\dfrac{^{38}\text{Ar}}{^{40}\text{Ar}}$	$\dfrac{^{39}\text{Ar}}{^{40}\text{Ar}}$	Age [m.y.]
465	65	226 ± 6	177 ± 34	42 ± 3	190 ± 13	2119 ± 34	3388 ± 26
465	80	162 ± 4	731 ± 47	70 ± 3	368 ± 12	1732 ± 102	3702 ± 95
510	65	294 ± 7	168 ± 25	58 ± 10	244 ± 10	1821 ± 30	3631 ± 26
510	80	132 ± 3	433 ± 32	52 ± 6	227 ± 17	1369 ± 53	4094 ± 65
545	60	404 ± 10	130 ± 9	84 ± 3	171 ± 6	1574 ± 12	3868 ± 13
570	80	579 ± 15	77 ± 6	77 ± 11	124 ± 6	1508 ± 5	3939 ± 5
630	65	892 ± 22	85 ± 4	98 ± 2	159 ± 2	1482 ± 3	3967 ± 3
630	80	604 ± 25	89 ± 3	99 ± 1	146 ± 2	1505 ± 3	3942 ± 4
670	65	1050 ± 26	139 ± 4	157 ± 2	249 ± 1	1580 ± 2	3863 ± 2
670	80	732 ± 18	164 ± 3	181 ± 3	278 ± 2	1588 ± 5	3856 ± 5
715	70	597 ± 15	252 ± 6	269 ± 3	414 ± 5	1600 ± 3	3841 ± 3
715	82	639 ± 16	304 ± 4	323 ± 3	502 ± 3	1604 ± 3	3836 ± 4
765	65	860 ± 21	484 ± 3	521 ± 2	803 ± 2	1607 ± 3	3833 ± 4
765	80	567 ± 14	584 ± 3	635 ± 2	971 ± 4	1565 ± 4	3875 ± 5
810	60	429 ± 11	788 ± 6	856 ± 3	1294 ± 5	1562 ± 6	3877 ± 7
810	80	430 ± 11	907 ± 6	977 ± 4	1480 ± 7	1534 ± 5	3908 ± 7
860	65	504 ± 13	1253 ± 6	1371 ± 6	2053 ± 9	1522 ± 5	3919 ± 8
860	80	319 ± 8	1394 ± 9	1505 ± 6	2265 ± 11	1509 ± 7	3932 ± 9
910	80	393 ± 10	1477 ± 7	1608 ± 7	2444 ± 9	1513 ± 6	3927 ± 9
995	85	316 ± 8	1462 ± 9	1571 ± 9	2401 ± 10	1489 ± 7	3954 ± 10
1035	80	132 ± 3	1320 ± 28	1421 ± 15	2235 ± 29	1467 ± 19	3979 ± 21
1075	81	118 ± 3	1336 ± 25	1441 ± 19	2240 ± 34	1456 ± 20	3992 ± 23
1110	75	76 ± 2	1454 ± 43	1506 ± 32	2380 ± 53	1387 ± 32	4072 ± 37
1155	60	47 ± 2	1585 ± 54	1583 ± 49	2387 ± 83	1421 ± 52	4029 ± 58
1200	60	31 ± 2	1066 ± 66	1430 ± 80	2046 ± 110	1480 ± 97	3965 ± 103
1245	62	30 ± 2	1335 ± 95	1218 ± 74	2217 ± 215	1372 ± 91	4090 ± 106
1330	63	13 ± 2	1178 ± 245	822 ± 161	1539 ± 276	1141 ± 207	4399 ± 296
1475	70	30 ± 12	809 ± 369	425 ± 179	691 ± 292	500 ± 224	5849 ± 763
Total		10777 ± 67	506 ± 3	509 ± 3	799 ± 4	1574 ± 3	3866 ± 3
Pyroxene (57.0 mg)		× 10^{-1}	× 10^{-4}	× 10^{-3}	× 10^{-4}	× 10^{-4}	
375	75	42 ± 5	65 ± 34	2 ± 0.2	92 ± 62	39 ± 16	
510	67	284 ± 8	6 ± 4	268 ± 32	140 ± 4	258 ± 5	3085 ± 28
510	76	179 ± 6	10 ± 7	144 ± 60	32 ± 3	192 ± 7	3540 ± 55
570	67	360 ± 10	8 ± 5	152 ± 38	42 ± 3	154 ± 3	3910 ± 28
570	65	212 ± 7	13 ± 4	254 ± 45	34 ± 6	149 ± 4	3950 ± 50
615	63	587 ± 15	16 ± 1	196 ± 19	44 ± 2	141 ± 2	4045 ± 17
615	91	531 ± 14	17 ± 2	173 ± 11	39 ± 2	141 ± 2	4050 ± 18
670	60	612 ± 16	22 ± 2	316 ± 17	53 ± 2	144 ± 2	4017 ± 18
670	66	451 ± 12	28 ± 2	299 ± 15	48 ± 2	146 ± 2	3993 ± 22
715	65	913 ± 23	43 ± 1	500 ± 17	74 ± 1	147 ± 1	3982 ± 12
715	92	531 ± 14	42 ± 2	532 ± 18	78 ± 1	146 ± 2	3995 ± 20
765	65	893 ± 23	63 ± 1	690 ± 11	101 ± 1	146 ± 1	3989 ± 14
765	90	670 ± 17	66 ± 2	683 ± 12	102 ± 1	145 ± 1	3999 ± 17
810	60	824 ± 21	72 ± 1	799 ± 18	123 ± 2	145 ± 1	3999 ± 13
810	91	476 ± 13	77 ± 2	812 ± 24	123 ± 2	141 ± 2	4050 ± 22

Table 2. (*Continued*).

Extraction Temp. [°C]	Time [min]	^{40}Ar [10^{-8} cc STP/g]	$\dfrac{^{36}\text{Ar}}{^{40}\text{Ar}}$	$\dfrac{^{37}\text{Ar}}{^{40}\text{Ar}}$	$\dfrac{^{38}\text{Ar}}{^{40}\text{Ar}}$	$\dfrac{^{39}\text{Ar}}{^{40}\text{Ar}}$	Age [m.y.]
860	65	483 ± 13	81 ± 2	982 ± 24	143 ± 2	144 ± 2	4011 ± 21
860	90	304 ± 9	90 ± 3	1015 ± 35	153 ± 3	140 ± 3	4061 ± 34
910	61	305 ± 9	104 ± 3	1085 ± 40	174 ± 4	147 ± 3	3975 ± 32
985	96	600 ± 16	133 ± 2	1440 ± 20	222 ± 2	159 ± 2	3849 ± 19
1070	92	290 ± 10	334 ± 9	3285 ± 100	516 ± 13	229 ± 7	3255 ± 45
1165	90	202 ± 11	1660 ± 80	14725 ± 710	2470 ± 120	237 ± 18	3125 ± 110
1245	65	187 ± 9	2490 ± 105	22040 ± 930	3750 ± 160	146 ± 15	3850 ±
1325	67	142 ± 10	522 ± 35	4400 ± 310	750 ± 50	615 ± 45	1880 ± 85
1495	63	655 ± 49	84 ± 7	470 ± 40	81 ± 6	7 ± 1	
Total		10733 ± 82	141 ± 4	1360 ± 35	225 ± 5	151 ± 2	3782 ± 9
Ph-B			× 10^{-5}	× 10^{-3}	× 10^{-5}	× 10^{-5}	
390	65	8 ± 1	440 ± 110		1545 ± 195	1835 ± 170	3620 ± 150
460	65	47 ± 1	84 ± 36	130 ± 35	1240 ± 35	4265 ± 55	2363 ± 18
490	65	139 ± 4	88 ± 10	97 ± 17	340 ± 14	2611 ± 14	3067 ± 8
490	80	114 ± 3	44 ± 18	81 ± 5	188 ± 7	1984 ± 13	3494 ± 10
565	65	187 ± 5	89 ± 13	99 ± 8	208 ± 7	1641 ± 8	3800 ± 8
565	90	167 ± 4	96 ± 18	89 ± 5	183 ± 7	1518 ± 8	3927 ± 9
615	70	287 ± 7	90 ± 5	121 ± 4	195 ± 4	1464 ± 5	3987 ± 5
615	90	232 ± 6	98 ± 7	106 ± 4	176 ± 4	1464 ± 6	3988 ± 7
660	65	313 ± 8	157 ± 4	168 ± 5	267 ± 5	1518 ± 6	3928 ± 6
660	95	306 ± 8	176 ± 5	205 ± 4	304 ± 6	1507 ± 5	3940 ± 6
730	70	317 ± 8	283 ± 7	318 ± 3	486 ± 6	1537 ± 6	3908 ± 6
730	90	233 ± 6	341 ± 6	385 ± 6	600 ± 9	1530 ± 6	3914 ± 6
780	65	376 ± 9	500 ± 4	557 ± 5	846 ± 4	1530 ± 4	3913 ± 5
780	90	284 ± 7	560 ± 5	635 ± 7	942 ± 6	1491 ± 5	3956 ± 6
820	65	260 ± 7	696 ± 6	770 ± 6	1176 ± 6	1493 ± 5	3953 ± 6
820	90	162 ± 4	772 ± 9	848 ± 11	1294 ± 10	1471 ± 8	3977 ± 10
865	65	165 ± 4	904 ± 8	1014 ± 8	1539 ± 11	1472 ± 8	3975 ± 9
865	90	110 ± 3	981 ± 15	1094 ± 16	1670 ± 14	1459 ± 12	3985 ± 14
910	65	133 ± 3	1124 ± 9	1278 ± 13	1930 ± 13	1492 ± 10	3951 ± 12
990	90	202 ± 5	1257 ± 6	1369 ± 7	2100 ± 10	1608 ± 6	3829 ± 9
1085	90	81 ± 2	2275 ± 30	2315 ± 30	3725 ± 50	2285 ± 30	3264 ± 22
1170	90	43 ± 2	7870 ± 240	7320 ± 220	12390 ± 370	2830 ± 100	2910 ± 60
1250	65	32 ± 1	14170 ± 470	12930 ± 430	22300 ± 750	1900 ± 100	3495 ± 85
1335	65	14 ± 1	3525 ± 325	3160 ± 300	5325 ± 485	510 ± 75	
1420	65	22 ± 2	600 ± 80	330 ± 60	480 ± 75	105 ± 20	
1500	65	44 ± 6	400 ± 80	45 ± 40	190 ± 30	60 ± 13	
Total		4276 ± 26	640 ± 7	666 ± 7	1080 ± 11	1597 ± 4	3789 ± 3

well-defined retentive phases in a sample must be demonstrated. In a simple system comprised of a single mineral, like mare basalt plagioclase (Turner *et al.*, 1972), this relationship is demonstrated by a broad plateau in the graph of ^{40}Ar*/^{39}Ar* versus Ar release. In a polymineralic system, the usual graph of age versus fraction of $^{39}Ar^*$ released relates the ages to unspecified K-containing phases. The $^{40}Ar/^{39}Ar$ technique provides, through the $^{39}Ar^*$ and ^{37}Ar release, a measure of the K and Ca contents of the minerals, which characterize the degassing phases. In a high resolution, fine-scaled temperature release such as performed on 65015, the release patterns of $^{40}Ar^*$ (an alternative measure for K) and ^{37}Ar can be used to identify the phases being degassed at a given temperature.

In Fig. 2 the differential releases of $^{40}Ar^*$ and ^{37}Ar are plotted versus the extraction temperature for 65015 whole rock and mineral separate samples. The $^{40}Ar^*$ release curve from 14310 plagioclase and K-rich quintessence separates (Turner *et al.*, 1972) are included for comparison. The neutron fluence for 14310

Table 3. Ar analyses for Apollo 16 rock fragments 67483,15-2 and 60503,3-3.

Extraction Temp. [°C]	^{40}Ar [10^{-8} cc STP/g]	$\dfrac{^{36}Ar}{^{40}Ar}$ $\times 10^{-4}$	$\dfrac{^{37}Ar}{^{40}Ar}$ $\times 10^{-3}$	$\dfrac{^{38}Ar}{^{40}Ar}$ $\times 10^{-5}$	$\dfrac{^{39}Ar}{^{40}Ar}$ $\times 10^{-5}$	Apparent age [m.y.]
67483,15-2 (29.0 mg)						
240	31 ± 2	35 ± 8	6 ± 5	175 ± 35	230 ± 45	
455	15 ± 1	156 ± 23	140 ± 21	1330 ± 250	3170 ± 360	2760 ± 170
455	12 ± 1	152 ± 24	132 ± 17	495 ± 165	4160 ± 650	2380 ± 210
455	11 ± 1	142 ± 32	118 ± 21	640 ± 135	3220 ± 460	2740 ± 210
555	148 ± 4	247 ± 4	150 ± 3	615 ± 20	2142 ± 21	3335 ± 40
555	67 ± 2	108 ± 5	77 ± 3	345 ± 55	1620 ± 45	3805 ± 50
555	63 ± 2	108 ± 5	77 ± 3	345 ± 55	1624 ± 50	3800 ± 52
630	327 ± 8	175 ± 2	109 ± 1	391 ± 9	1479 ± 9	3940 ± 30
630	319 ± 8	105 ± 1	91 ± 1	223 ± 10	1472 ± 9	3962 ± 20
630	214 ± 6	85 ± 1	91 ± 3	174 ± 24	1482 ± 14	3953 ± 21
755	560 ± 14	115 ± 1	161 ± 1	261 ± 5	1500 ± 5	3930 ± 20
755	351 ± 9	102 ± 2	176 ± 1	241 ± 13	1526 ± 12	3903 ± 21
755	325 ± 6	98 ± 2	193 ± 1	219 ± 25	1524 ± 14	3906 ± 22
830	800 ± 20	143 ± 1	353 ± 1	367 ± 5	1563 ± 4	3857 ± 23
830	480 ± 12	64 ± 1	385 ± 2	221 ± 10	1533 ± 9	3902 ± 25
830	366 ± 9	55 ± 2	429 ± 2	220 ± 18	1520 ± 8	3918 ± 12
925	844 ± 21	104 ± 1	542 ± 1	363 ± 5	1504 ± 4	3927 ± 16
925	643 ± 16	106 ± 1	547 ± 1	368 ± 4	1497 ± 6	3934 ± 17
925	265 ± 7	97 ± 2	486 ± 3	347 ± 13	1398 ± 11	4049 ± 19
1015	506 ± 13	142 ± 1	625 ± 2	462 ± 6	1484 ± 7	3943 ± 23
1015	251 ± 6	127 ± 2	578 ± 4	444 ± 12	1482 ± 20	3950 ± 30
1015	136 ± 4	133 ± 2	484 ± 5	490 ± 17	1500 ± 18	3928 ± 28
1110	357 ± 9	215 ± 2	446 ± 3	630 ± 15	1513 ± 8	3900 ± 35
1110	127 ± 4	114 ± 2	415 ± 6	509 ± 18	1517 ± 25	3915 ± 30
1110	84 ± 3	73 ± 3	399 ± 8	385 ± 20	1390 ± 30	4070 ± 40
1225	136 ± 4	76 ± 2	539 ± 6	402 ± 24	1450 ± 20	4000 ± 26
1225	39 ± 2	48 ± 7	533 ± 30	305 ± 30	1170 ± 75	4350 ± 105
1311	55 ± 2	45 ± 5	700 ± 25	260 ± 60	1150 ± 45	4390 ± 65
1400	1104 ± 28	10 ± 0.3	697 ± 2	280 ± 3	1538 ± 5	3907 ± 5
1500	214 ± 12	9 ± 2	475 ± 23	163 ± 14	1365 ± 70	4100 ± 80
Total	8866 ± 56	101 ± 1	415 ± 1	338 ± 5	1517 ± 4	3904 ± 5

Table 3. (*Continued*).

Extraction Temp. [°C]	Time [min]	^{40}Ar [10^{-8} cc STP/g]	$\frac{^{36}\text{Ar}}{^{40}\text{Ar}}$ $\times 10^{-4}$	$\frac{^{37}\text{Ar}}{^{40}\text{Ar}}$ $\times 10^{-3}$	$\frac{^{38}\text{Ar}}{^{40}\text{Ar}}$ $\times 10^{-4}$	$\frac{^{39}\text{Ar}}{^{40}\text{Ar}}$ $\times 10^{-4}$	Apparent age [m.y.]
60503,3-3 (21.8 mg)							
175	65	3 ± 2	150 ± 140	90 ± 75	12 ± 10	29 ± 28	
350	67	12 ± 2	31 ± 12	27 ± 9	75 ± 13	66 ± 14	5360 ± 390
455	65	105 ± 3	58 ± 5	57 ± 4	72 ± 4	119 ± 9	4325 ± 125
455	65	24 ± 2	130 ± 11	118 ± 10	79 ± 12	291 ± 24	2890 ± 120
550	65	145 ± 4	274 ± 10	266 ± 10	106 ± 4	298 ± 5	2830 ± 40
550	65	81 ± 3	200 ± 8	210 ± 7	55 ± 3	253 ± 9	3090 ± 60
640	65	251 ± 7	300 ± 3	397 ± 3	81 ± 2	202 ± 2	3420 ± 50
640	65	182 ± 5	170 ± 5	344 ± 4	59 ± 2	187 ± 3	3560 ± 33
640	65	127 ± 4	134 ± 5	357 ± 6	51 ± 5	179 ± 3	3635 ± 36
725	65	306 ± 8	234 ± 2	620 ± 4	70 ± 2	171 ± 2	3700 ± 42
725	65	173 ± 5	220 ± 4	574 ± 6	64 ± 3	161 ± 4	3800 ± 50
725	65	145 ± 4	215 ± 4	564 ± 7	63 ± 3	161 ± 3	3795 ± 42
830	65	815 ± 20	369 ± 1	915 ± 3	102 ± 1	156 ± 1	3825 ± 60
830	70	361 ± 9	223 ± 3	877 ± 5	73 ± 2	151 ± 1	3905 ± 37
830	68	227 ± 6	206 ± 2	937 ± 8	74 ± 1	148 ± 1	3935 ± 35
930	70	665 ± 17	393 ± 1	1469 ± 4	125 ± 1	149 ± 1	3900 ± 60
930	77	410 ± 10	251 ± 1	1262 ± 6	93 ± 1	150 ± 1	3910 ± 40
930	70	202 ± 5	229 ± 5	687 ± 6	85 ± 2	152 ± 2	3890 ± 40
995	65	388 ± 10	481 ± 2	849 ± 5	143 ± 1	154 ± 1	3830 ± 75
995	65	228 ± 6	506 ± 4	737 ± 6	154 ± 2	159 ± 2	3775 ± 85
995	65	150 ± 4	483 ± 7	685 ± 8	160 ± 3	154 ± 4	3825 ± 85
1105	65	288 ± 7	635 ± 9	812 ± 7	187 ± 3	169 ± 1	3650 ± 100
1105	67	440 ± 11	840 ± 22	1184 ± 22	203 ± 5	143 ± 5	3885 ± 150
1105	65	19 ± 2	920 ± 95	1265 ± 125	285 ± 35	164 ± 18	3650 ± 240
1205	65	115 ± 3	1013 ± 20	1344 ± 25	251 ± 5	161 ± 4	3670 ± 180
1205	70	31 ± 2	800 ± 50	1025 ± 60	208 ± 16	112 ± 10	4300 ± 200
1290	67	134 ± 4	444 ± 7	1143 ± 16	149 ± 3	148 ± 3	3900 ± 75
1290	65	24 ± 3	108 ± 25	940 ± 100	87 ± 17	107 ± 15	4500 ± 230
1390	70	620 ± 16	135 ± 2	1662 ± 12	104 ± 1	183 ± 2	3606 ± 20
1390	68	13 ± 6	120 ± 60	1380 ± 620	75 ± 40	125 ± 60	4250 ± 780
Total		6684 ± 43	359 ± 2	951 ± 8	114 ± 1	164 ± 1	3729 ± 16

was only about half that for 65015, and a shift in the temperature of the maximum ^{40}Ar* release is possible (Alexander *et al.*, 1973). There are maxima in the ^{37}Ar release from *all* samples between 780°C and 850°C and a peak at 950°C and a strong peak at ~1200°C in the whole-rock, Ph–B, and pyroxene samples. As discussed below, these features are in some way related to the degassing properties of the individual minerals and can serve to identify contributions from these minerals in more complex systems.

A single peak at relatively low temperature (~800°C) is evidently characteristic of 65015 plagioclase. A ^{37}Ar release peak around this temperature is evident in all separates. As plagioclase is a major constituent in the whole rock and Ph–B

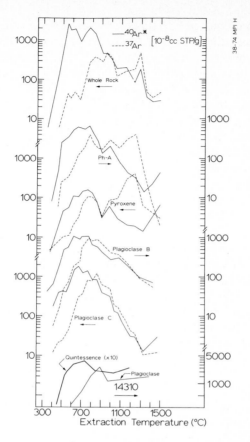

Fig. 2. The absolute amounts of ^{37}Ar and ^{40}Ar* released versus extraction temperature. When several extractions have been made at a given temperature the amounts have been added. The arrows below the sample names indicate left or right hand scales. The ^{40}Ar* release patterns of 14310 plagioclase and quintessence (Turner *et al.*, 1972) have been included for comparison with the 65015 samples.

samples and a minor constituent in the pyroxene separate, it is reasonable to interpret the 800°C peak as attributable to plagioclase. Release of ^{37}Ar from the plagioclase B separate does not drop off as rapidly as for the plagioclase C separate, suggesting a larger concentration of a more retentive plagioclase in plagioclase B. The occurrence of the distinct large, high-temperature peak in the ^{37}Ar release from the pyroxene separate, its reoccurrence in other samples with significant pyroxene contents and its absence from the plagioclase separates identify the high-temperature peak as a release characteristic of pyroxene in 65015. It also identifies pyroxene as one of the most retentive phases, at least in laboratory heating experiments.

The Ca-bearing minerals in the pyroxene separate are low-Ca pyroxene, high-Ca pyroxene and plagioclase. Using the chemical analyses of Albee *et al.* (1973) we estimate a total Ca content for the separate of only 2.2%, as compared to 3.4% derived from ^{37}Ar measurements. We have been unable to reconcile this difference, and thus cannot make assignments of portions of the ^{37}Ar release pattern to specific minerals. The high-temperature peak is unique to the 65015 pyroxene separate. It is difficult to ascertain how much of the low-temperature release is due to plagioclase or if a particular pyroxene is responsible. We note only that the shape of the low-temperature ^{37}Ar release from the pyroxene separate differs from that of the plagioclase in being sharper and occurring at slightly lower temperatures. ^{37}Ar release from a pyroxene could well constitute a significant portion of the low-temperature release.

At lower temperatures in all samples, there is a preferential release of ^{40}Ar* relative to ^{37}Ar. This is almost certainly due to an admixture of a K-rich contaminating phase. Comparison in Fig. 2 of the ^{40}Ar* release patterns of 14310 quintessence and plagioclase (Turner *et al.*, 1972) is a demonstration of this preferential release.

In some temperature regimes, the release of ^{37}Ar and ^{40}Ar* is similar. This is most clear in the release from pyroxene just below ~900°C and from the plagioclase above ~800°C and is interpreted to indicate that the Ca and K distributions are similar in sites contributing Ar at these temperatures. This is expected to be the case if the Ar is released primarily from a single mineral phase. In other cases, as above 900°C in Ph–B and pyroxene, it is evident that the K cannot be associated primarily with the Ca-rich mineral.

Plagioclase

The ^{40}Ar/^{39}Ar ages obtained for the plagioclase separates, particularly for plagioclase B, have been discussed in detail elsewhere (Jessberger *et al.*, 1974; Huneke *et al.*, 1974). The complete data on both are presented here in Table 2 and Figs. 3 and 4.

Both plagioclase separates display unusual structures in the graph of apparent age versus Ar release. In both Figs. 3 and 4, the ages are plotted as a function of ^{37}Ar release, to properly identify ages corresponding to Ar release from the plagioclase, which contains all of the Ca (^{37}Ar) but only 25–35% of the K (^{39}Ar*). The remainder of the K is contained in fine-grained quintessence. As seen in the lower parts of the Figs. 3 and 4, the quintessence contributions dominate the age plots versus ^{39}Ar* release in the lower temperature release and yield a broad undulating plateau extending to ~60% of the total ^{39}Ar* release. This is correlated with high K/Ca ratios. Plotted against ^{37}Ar release, the ^{40}Ar/^{39}Ar ages increase rapidly to a maximum and decrease sharply within the first 10% of the ^{37}Ar release. With increasing temperature, the ^{40}Ar* release which is identified with the plagioclase appears with K/Ca values compatible with the plagioclase. The resolution of a low-temperature (615°C) age peak rather than a monotonic increase in apparent age may reflect special differences in grain size and activation energies for the quintessence, or may be due to recoil effects as indicated in the preliminary experiment by Turner and Cadogan (1974), which suggests this pattern is due to recoil of ^{39}Ar* from fine-grained quintessence into more retentive phases during the neutron-irradiation.

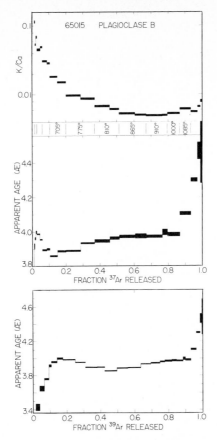

Fig. 3. Apparent age and K/Ca ratios versus fractional ^{37}Ar release (top) and fractional ^{39}Ar* release (bottom) from 65015 plagioclase separate B. The higher purity of plagioclase B compared to plagioclase C (Fig. 4) is demonstrated by the faster decrease in the K/Ca ratios down to the microprobe value for plagioclase in 65015 (Albee *et al.*, 1973). A high-K impurity is responsible for the age peak at ~600°C, as is clearly shown by plotting the ages with respect to the ^{37}Ar instead of the ^{39}Ar* release. At higher temperatures, plagioclase B shows a well-developed age plateau at 3.98 AE clearly separated from the effects of the isotopically unequilibrated clasts of ~4.5 AE age (Jessberger *et al.*, 1974). Note the expanded age scale.

In the plagioclase sample least contaminated by quintessence, the plot versus ^{37}Ar release also reveals an extensive intermediate age plateau at 3.98 AE associated with the plagioclase. At higher temperatures the apparent ages increase to values up to 4.47 AE, instead of forming well defined high-temperature plateaus. The Ar isotopic variations are complex and our understanding of the natural and experimental causes of Ar isotopic variations is incomplete. The interpretation of ^{40}Ar*/^{39}Ar* variations in terms of real ages consequently deserves cautious treatment. Nevertheless, we presently find no alternative to accepting the ages of up to 4.47 AE as real. The following discussion clarifies this

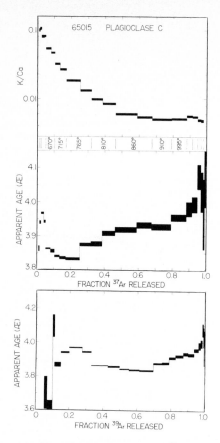

Fig. 4. Apparent age and K/Ca ratios versus fractional ^{37}Ar release (top) and fractional ^{39}Ar* release (bottom) from 65015 plagioclase separate C.

judgment, and illustrates the use of reasonable criteria forming the basis for such a decision. Rock 65015 will be contrasted with rock 68415, for which a roughly comparable ^{40}Ar*/^{39}Ar* spectrum was obtained on a plagioclase separate but not fully interpreted as measuring real ages.

An interpretation of the larger ^{40}Ar*/^{39}Ar* ratios as due to contributions from older relict clasts, incompletely degassed when 65015 formed, is strongly supported by petrographic, chemical and Sr isotopic measurements on 65015. In fact, these studies provided the impetus for the Ar analyses. Petrographic and electron microprobe data unequivocally demonstrate the presence of chemically unequilibrated plagioclase clasts in 65015. Recent ion microprobe trace element analyses by Meyer *et al.* (1974) lead to the same conclusion. In a diagram of ^{87}Sr/^{86}Sr versus ^{87}Rb/^{86}Sr, all plagioclase separates from 65015 lie on a straight line which intersects the ^{87}Sr/^{86}Sr axis close to the primitive BABI ratio and do not lie on the isochron defined by all other mineral separates and the total rock sample (Tera *et al.*, 1973). The Sr in some part of the plagioclase is not isotopically equilibrated with the remainder of the rock. Figure 5 shows the region of the diagram containing both the intersection of these two lines and the Rb/Sr data obtained for the plagioclase separates studied here (Papanastassiou and Wasserburg, 1972). A plagioclase separate comprised of large handpicked clasts contained very little Rb relative to Sr and had a Sr composition very close to BABI. Assuming the plagioclase line is a mixing line defined by admixing variable proportions of clast cores to matrix and clast rims which were recrystallized and isotopically equilibrated with the rest of the mineral assemblage in the last metamorphic event, and assuming the same Sr concentration in both types of plagioclase, we estimate that plagioclase B and C contain 14% and 8% admixed clasts, respectively. The clasts are chemically and isotopically unequilibrated, and there is thus reason to expect that these clasts may retain some $^{40}Ar^*$ generated prior to their inclusion in 65015.

There is no comparable body of evidence leading one to expect older ages in the high-temperature release from 68415 plagioclase. There are rare large plagioclase xenocrysts which are An-rich, not chemically zoned, and show neither sign of resorbtion in the 68415 melt nor visible overgrowths (Helz and Appleman, 1973). Trace element analyses by Meyer et al. (1974) demonstrate equilibrium partitioning of Ba, Li, and Sr between xenocrysts and a melt of the total rock composition, which was not true of 65015. Papanastassiou and Wasserburg (1972) find no

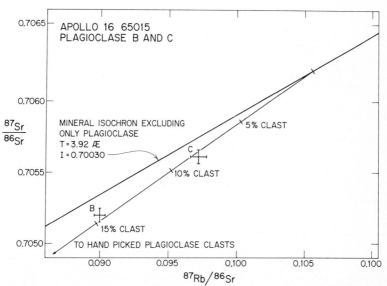

Fig. 5. The portion of the Rb–Sr evolution diagram for 65015 showing Rb–Sr data on the plagioclase B and C mineral separates (Papanastassiou and Wasserburg, 1972). Plagioclase B and C lie distinctly off the mineral isochron defined by all other minerals (pyroxene, phosphate enriched separate, quintessence, ilmenite, and total rock) due to the admixture of isotopically unequilibrated clasts with low Rb/Sr and primitive Sr composition. The estimated percentage of admixed clasts required to displace the points from the isochron are noted along the mixing line defined by all of the plagioclase data.

significant contributions to Sr in 68415 plagioclase from unequilibrated relict clasts with Sr of a more primitive composition. There is thus no evidence for isotopically unequilibrated clasts in the 68415 plagioclase analyzed.

There is no conflict in the chronological information obtained for 65015 using the ^{40}Ar–^{39}Ar and Rb–Sr techniques, but rather strong complimentary observations which lead to the same conclusion. The Rb–Sr age of 3.92 AE is well within the 3.90–4.00 AE range of variation of ages associated with the Ar release from quintessence and is not significantly different from the intermediate plagioclase plateau age of 3.98 AE. An older age for the relict clasts is consistent with the more primitive Sr isotopic composition but is not necessary. In the instance of 68415, an intermediate plateau age of 4.09 AE was obtained for plagioclase, considerably older than the 3.84 AE Rb–Sr isochron age and the 3.86 AE ^{40}Ar–^{39}Ar intermediate plateau age of a whole sample.

There are no independent and self-sufficient criteria from any "dating" method which prove the validity of a result. It is both the strict internal consistency within one method and between two methods which permit us to infer a specific history in the case that the results are complex.

All Ar isotopic data on plagioclase B are amenable to a self-consistent interpretation. The 33% increase in ^{40}Ar*/^{39}Ar* and 13% increase in ^{38}Ar$_s$/^{37}Ar (Fig. 10) are the only outstanding anomalies at the higher temperatures. The covariant increase of both ^{40}Ar*/^{39}Ar* and ^{38}Ar$_s$/^{37}Ar requires only that the older clasts have been exposed to cosmic rays prior to incorporation in 65015. The corresponding increase in K/Ca is well within the range of K/Ca measured in relict plagioclase clasts by electron microprobe (A. Albee, personal communication). ^{38}Ar/^{36}Ar varies within the range of values observed for spallogenic Ar$_s$ in meteorites and other lunar samples (Fig. 9). There is no systematic mass dependence in the Ar isotopic variations. A similar consistency was not observed in all isotopic variations in Ar released from 68415 plagioclase, as discussed by Huneke *et al.* (1973).

The maximum age of 4.47 AE in the 65015 plagioclase B spectrum is approached in a sequence of several consecutive releases with ages increasing monotonically until all Ar is extracted, a pattern naïvely expected if older, incompletely degassed relict clasts are present. In contrast, the oldest age in the 68415 plagioclase age spectrum occurs in the last Ar release, discontinuous from the well-defined intermediate plateau, and the oldest ages observed for sample 67483,15-2 (Fig. 11) occur in intermediate releases, with subsequent releases again having younger ages. The observation of a continuous approach to the oldest ages requires higher resolution data than previously acquired. About 15% of the ^{37}Ar released from plagioclase B, but only ~5% of the ^{37}Ar released from plagioclase C, is associated with the older ages above 3.98 AE. This is in striking agreement with the estimated percentages of admixed clasts using the Rb/Sr data (see Fig. 5), again providing evidence that these older ages are a property of these clasts.

In conclusion ^{40}Ar/^{39}Ar versus ^{37}Ar analysis of the plagioclase separates indicates that some of the plagioclase clasts in 65015 have ages of at least 4.47 AE. Thus rocks with large plagioclase crystals of high An content must have existed before 4.47 AE and formed part of the protolith of 65015 (cf. Albee *et al.*, 1973). The 65015 breccia was severely metamorphosed at 3.98 AE.

65015 Pyroxene

The $^{40}Ar/^{39}Ar$ ages of the pyroxene separate are plotted in Fig. 6, again versus ^{37}Ar. A $^{40}Ar*/^{39}Ar*$ versus ^{37}Ar analysis is particularly appropriate for the pyroxene since ~70% of the ^{37}Ar is given off in a high temperature release with a peak at ~1100°C. This peak is present only in the pyroxene separate and samples containing large amounts of pyroxene. This distinctive release pattern for the pyroxene is shown in Fig. 2, where the ^{37}Ar release is seen to be well resolved

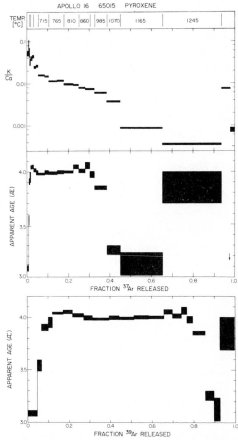

Fig. 6. Apparent age and K/Ca ratio of the 65015 pyroxene separate versus fractional ^{37}Ar release (top) and fractional $^{39}Ar*$ release (bottom). The first 30% of the ^{37}Ar release corresponds to almost the entire $^{39}Ar*$ release and gives an age plateau. The last 70% of ^{37}Ar release, which is separated by a release minimum from the first part (Fig. 2), and which is associated with release from pyroxene, does not yield meaningful ages. Note the expanded age scales.

from that of the ^{40}Ar* which has 70% of the total release below 810°C. The pyroxene shows a peak in apparent age in the first 5% of ^{37}Ar release corresponding to high K/Ca ratios, as observed in the plagioclase.

K/Ca does not decrease as rapidly between 715°C and 910°C as at higher and lower temperatures, indicating that Ar released from a single mineral phase provides a significant if not dominant portion of the total Ar release at these temperatures. This temperature interval encompasses the low-temperature Ar release maximum from the pyroxene. An intermediate age plateau at 4.00 AE is also relatively well defined over most of this interval. Again, it is unclear what specific mineral is responsible for the low-temperature maximum in the ^{37}Ar release. To the extent that a pyroxene contributes significant ^{40}Ar* at these temperatures, the spectrum indicates that age information can be obtained for pyroxene below ~900°C. Above 900°C the apparent ages decrease sharply to very low values over a range of ^{37}Ar release corresponding to the high-temperature ^{37}Ar release peak from pyroxene. This behavior in the apparent age spectrum is thus very clearly associated with the pyroxene, but the cause is not established. A change in the release characteristics of ^{40}Ar* and ^{39}Ar* at ~900°C is strongly indicated and may be associated with structural changes in the pyroxene at elevated temperatures.

The 4.00 AE intermediate plateau age of the pyroxene supports the 3.98 AE intermediate plateau age inferred from the plagioclase B measurements. Although relict pyroxene clasts have also been observed in 65015, the pyroxene separate shows no evidence for the admixture of isotopically unequilibrated pyroxene containing more primitive Sr (Tera *et al.*, 1973) nor does it show evidence in the ^{40}Ar*/^{39}Ar* age spectrum of older, incompletely degassed relict clasts (Fig. 6). This contrasts with the observation from Fig. 2 that some pyroxene is more retentive than plagioclase. However, the apparent age spectrum of pyroxene at higher temperatures does not provide a good background against which contributions from older clasts can be clearly observed.

Ph–B separate and whole rock

The most complex samples mineralogically are the whole rock and Ph–B samples, neither of which are significantly enriched in any one mineral. Characteristic age spectra and release patterns previously inferred for separated minerals provide limited guidelines to the interpretation of age spectra of these more complicated samples. Two features dominate the age spectra of these two samples (Figs. 7 and 8). The first is the high-temperature decrease in apparent age, which can be correlated with the release from pyroxene both by its association with low K/Ca values and the high-temperature release peak of ^{37}Ar (Fig. 2). The second is the apparent age peak at ~600°C which is related to Ar release from quintessence, as inferred from associated high K/Ca values.

The ~600°C age peak, the intermediate minimum, and the second age plateau or peak at ~900°C are features which occur in all 65015 samples. Usually, increases in apparent age with degassing temperature are interpreted as being the

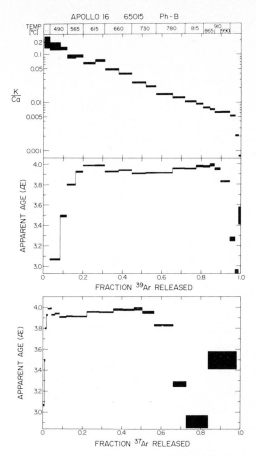

Fig. 7. Apparent age and K/Ca ratio versus fractional ^{39}Ar* release (top) and fractional ^{37}Ar release (bottom) from 65015 mineral separate Ph–B, which is a composite of plagioclase, pyroxene, and phosphates. The age spectrum is similar to that of the total rock (Fig. 8), while the K/Ca ratios show a different pattern due to the smaller K contributions from quintessence. Note the expanded age scale.

result of ^{40}Ar* loss (Turner, 1969). If this mechanism is responsible for the low ages at ~750°C, then phases which show an age plateau at ~900°C have lost a large part of their ^{40}Ar*, while the phase releasing its gas at ~600°C has retained most of the radiogenic argon. A more real possibility is that ^{39}Ar* has been redistributed in the neutron activation by recoil from fine-grained, nonretentive quintessence into more retentive phases. Preliminary experiments by Turner and Cadogan (1974) have demonstrated that this process can lower the apparent ages of sites which are degassed at ~750°C and increase the apparent ages of sites degassed at ~600°C. At ~800°C this ^{39}Ar component is quantitatively removed, so that the ^{40}Ar*/^{39}Ar* ratios show the actual values at higher temperatures. In no instance for 65015 has an age plateau been well defined when

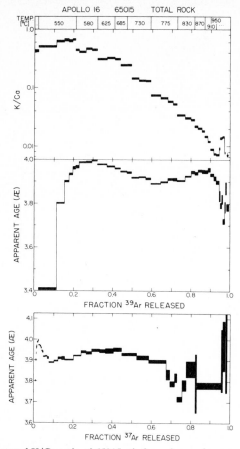

Fig. 8. Apparent age and K/Ca ratio of 65015 whole-rock sample versus fractional ^{39}Ar* release (top) and fractional ^{37}Ar release (bottom). The high K/Ca indicates that K-rich quintessence dominates the age spectrum. The low-temperature apparent ages indicate moderate ^{40}Ar* loss prior to measurement. The ~600° age peak is related to the argon release from quintessence. At higher temperatures no relationship between degassed phases and apparent ages can be obtained, except the high-temperature decrease associated with pyroxene (Fig. 6). Note the expanded age scales.

observed at this resolution and with small errors. This is a severe limitation to obtaining very precise ages from such samples.

No evidence is found in the extremely detailed age spectra of either the Ph–B or whole-rock sample for ages older than 4.0 AE. The data show that such ages would not be clearly observed unless the pyroxenes are removed from the sample.

^{36}Ar/^{38}Ar ratios and exposure ages

In order to calculate spallogenic ^{38}Ar concentrations, all other possible ^{38}Ar components (air contamination, trapped and Cl-derived ^{38}Ar) have to be subtracted. The actual spallogenic $^{36}Ar/^{38}Ar$ ratio must first be established. In Fig. 9 the measured $^{36}Ar/^{38}Ar$ ratios of all 65015 samples are shown versus fractional $^{39}Ar^*$ release. Two characteristic patterns of $^{36}Ar/^{38}Ar$ variations are evident. The total rock, pyroxene, and Ph–B have initial values below 0.6; the plagioclase separates have initial values above 0.6. At higher temperatures ($\geqslant$650–750°C) the $^{36}Ar/^{38}Ar$ ratios in all 65015 samples converge to a very constant plateau at the same value. Thus there is one common ^{36}Ar and ^{38}Ar component in the total rock and all mineral separates, with $^{36}Ar/^{38}Ar = 0.6$, which is characteristic of Ar produced by cosmic ray spallation. The additional ^{38}Ar released at lower temperatures in samples of the first group has to be regarded as Cl-derived. If this additional ^{38}Ar represents the total Cl, the Cl concentrations of 10, 4.3, and

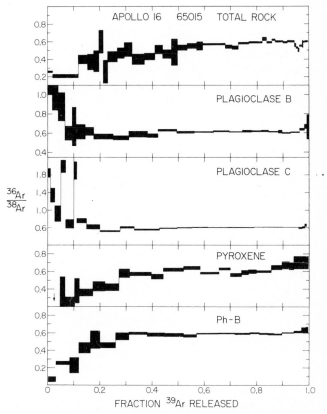

Fig. 9. Measured $(^{36}Ar/^{38}Ar)_m$ ratios versus fractional $^{39}Ar^*$ release for 65015 whole rock and mineral separates. Note the differing scales. After 20–50% of the $^{39}Ar^*$ release, $(^{36}Ar/^{38}Ar)_m$ in all samples shows a plateau at ~0.6. In the plagioclase separates this value is reached from higher values. In the other samples the presence of Cl produced ^{38}Ar is indicated at lower temperatures. Since $(^{36}Ar/^{38}Ar)_m$ has the same plateau value for a variety of separates with very different chemical compositions (Albee *et al.*, 1973), a Cl contribution over the whole extraction can be excluded and $^{36}Ar/^{38}Ar = 0.6$ is taken to be the spallogenic ratio.

2.7 ppm for total rock, Ph–B and pyroxene, respectively, can be calculated (Podosek, 1973). The total rock Cl concentration corresponds to the lower limit of Cl concentrations for Apollo 16 rocks given by Jovanovic and Reed (1973).

The high $^{36}Ar/^{38}Ar$ ratios at ~500°C in the plagioclase separates is probably caused by air contamination and indicate a lack of Cl in these separates. In the total-rock, pyroxene, and Ph–B samples, this air contamination is overprinted by the Cl component, since the amounts of Cl-derived ^{38}Ar are ~ 10 times the amounts of "trapped" ^{38}Ar, assuming $^{36}Ar/^{38}Ar = 5.35$ for trapped Ar. Taking 0.6 as the spallogenic $^{36}Ar/^{38}Ar$ ratio for 65015, the measured ^{38}Ar is formally decomposed into spallogenic $^{38}Ar_s$ and Cl-derived ^{38}Ar if $(^{36}Ar/^{38}Ar)_m < 0.6$ or cosmogenic and trapped or contaminant if $(^{36}Ar/^{38}Ar)_m \geq 0.6$. Ca is the predominant target element, and $^{38}Ar_s/^{37}Ar$ ratios are plotted in Fig. 10 as a measure of the cosmic ray exposure age. The $^{39}Ar*$ release is chosen as the abcissa only to permit comparison with Fig. 9. A plot versus the fractional ^{37}Ar release would be more relevant.

An appreciable variation in consecutive $^{38}Ar_s/^{37}Ar$ ratios is observed in the low-temperature fractions with small gas amounts released. At higher temperatures, however, large portions of the Ar release show plateaus at $^{38}Ar_s/^{37}Ar = 0.016$ for total rock and 0.015 for the plagioclase, pyroxene, and Ph–B separates. The uniformity of $^{38}Ar_s/^{37}Ar$ ratios in the mineral separates, in spite of large variations in minor target element concentrations (Fe, Ti: Albee *et al.*, 1973), indicates that contributions from Fe and Ti to $^{38}Ar_s$ are small and that the cosmic ray spectrum was relatively soft (cf. Huneke *et al.*, 1972).

The higher $^{38}Ar_s/^{37}Ar$ ratio in total rock on the other hand, can be explained by an $^{38}Ar_s$ contribution from Ti, which is ~1.4% in the total rock (Albee *et al.*, 1973). The total $^{38}Ar_s$ exposure age, calculated with

$$P_{38} = 1.4 \times 10^{-8} \left(1 + \frac{Ti}{5\,Ca} + \frac{Fe}{15\,Ca}\right) cc \; ^{38}Ar_s/g \; Ca/m.y.$$

(Huneke *et al.*, 1972) is 484 m.y. An $^{38}Ar_s/Ca$ cosmic ray exposure age of 465 m.y. can be calculated from the plateau $^{38}Ar_s/^{37}Ar$ ratio of the mineral separates and a production rate of 1.4×10^{-8} cc $^{38}Ar/g$ Ca/m.y. (Turner *et al.*, 1971). An upper limit for Fe and Ti contributions can be calculated from the small, high-temperature $^{38}Ar_s/^{37}Ar$ variations in pyroxene, giving a maximum spallation yield for iron of $(2.7 \pm 1) \times 10^{-10}$ cc $^{38}Ar_s/g$ Fe/m.y. and a minimum relative production rate ratio of $P_{38}(Ca)/P_{38}(Fe) = 52 \pm 10$, neglecting contributions from Ti (Fe and Ti concentrations from Albee *et al.*, 1973).

An increase in $^{38}Ar_s/^{37}Ar$ is observed in the plagioclase separate corresponding to the large increases in apparent age in the last, high-temperature portion of ^{37}Ar release. The increase cannot be explained in terms of variable target element chemistry, and is taken to reflect additional exposure of the relict clasts to cosmic rays prior to 3.98 AE.

Table 1 summarizes the $^{38}Ar_s/^{37}Ar$ cosmic-ray exposure ages and $(^{36}Ar/^{38}Ar)_m$ values for all samples.

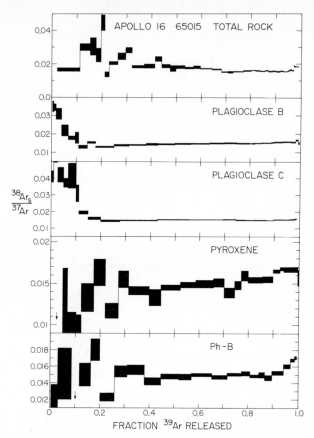

Fig. 10. ($^{38}Ar_s/^{37}Ar$) versus fractional $^{39}Ar^*$ release for 65015 whole rock and mineral separate samples. Note the differing scales. $^{38}Ar_s$ is calculated taking $(^{36}Ar/^{38}Ar)_t = 5.35$ and $(^{36}Ar/^{38}Ar)_s = 0.6$ (Fig. 9). In the absence of major spallogenic contributions from Fe and Ti, the $^{38}Ar_s/Ca$ ratios, or equivalently the $^{38}Ar_s/^{37}Ar$ ratios, are proportional to the cosmic ray exposure time. Taking the production rate $P_{38}(Ca) = 1.4 \times 10^{-8}$ cc $^{38}Ar/g$ Ca/m.y., the $^{38}Ar_s/^{37}Ar$ plateau for all samples corresponds to an exposure age of ~465 m.y.

APOLLO 16 ROCK FRAGMENTS 67483,15-2 AND 60503,3-3

Two rock fragments, one from the North Ray Crater rim (67483,15-2) and one from the ALSEP site (60503,3-3), have also been studied in detail, again in an attempt to isolate contributions from older phases within the rocks.

67483,15-2

67483,15-2 is a breccia fragment from the 2–4 mm fraction of soil sample 67480 taken on the North Ray Crater rim (Station 11). The sample consists primarily of

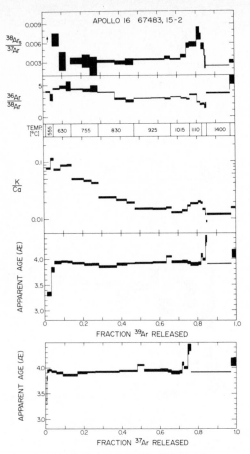

Fig. 11. $^{38}Ar_s/^{37}Ar$, $(^{36}Ar/^{38}Ar)_m$, K/Ca and the apparent age versus the fractional $^{39}Ar^*$ release from fragment 67483,15-2.

large plagioclase clasts in a fine-grained metaclastic matrix of plagioclase and pyroxene.

The K/Ca ratios (Fig. 11) show only minor variation and are almost constant over the last 50% of $^{39}Ar^*$ release. There are some exceptional points at 1100°C with corresponding variations in the $(^{36}Ar/^{38}Ar)_m$ and $^{38}Ar_s/^{37}Ar$ ratios, indicating the degassing at this temperature of either a high-K phase contributing a significant ^{38}Ar from neutron reactions on K, or a low-Ca phase where another target element (Fe or Ti) becomes dominant. The latter seems more reasonable since the variations in $(^{36}Ar/^{38}Ar)_m$ are not strictly correlated. However, it is not possible to describe this phase even qualitatively. The constant $^{38}Ar_s/^{37}Ar$ ratios at intermediate temperatures define a cosmic ray exposure age of 102 ± 7 m.y., assuming spallogenic $^{36}Ar/^{38}Ar = 0.6$.

Minor trapped ^{40}Ar contributions in rock samples are customarily subtracted using $(^{40}Ar/^{36}Ar)_t = 1 \pm 1$ (cf. Huneke *et al.*, 1973). The relatively large trapped component in 67483,15-2 allows a test of the validity of this correlation. The intercept of a least-squares fit (York, 1966) to the data on a plot of $(^{40}Ar/^{36}Ar)_m$ versus $(^{39}Ar/^{36}Ar)_m$ yields a trapped ratio of $(^{40}Ar/^{36}Ar)_t = 1.1 \pm 1.1$, which substantiates the standard $^{40}Ar_t$ correction.

The apparent age pattern of 67483,15-2 shows a very well defined plateau. The fragment has lost almost no radiogenic argon on the lunar surface. The $^{40}Ar/^{39}Ar$ age is 3.93 ± 0.04 AE. It is important to note that the rock fragment studied here shows no clear evidence for an older age for the rock or any phase within the rock. The very rapid fluctuation in age up to 4.4 AE occurs simultaneously with rapid variations in all other isotope ratios. The high ages are also not maintained for any significant span of Ar release. Together, these observations suggest the ages are probably experimental artifacts.

60503,3-3

60503,3-3 is a fine-grained, metaclastic breccia fragment from the 2–4 mm fraction of soil sample 60500 collected at Station 10 near the ALSEP site.

The Ar analyses of fragment 60503,3-3 (Fig. 12) is in some aspects similar to 67483,15-2. Again, there is a temperature region where the K/Ca ratio rises sharply from 0.006 to 0.013 and then suddenly drops back to 0.006. Below 930°C, the K/Ca ratios of subsequent degassing steps at the same temperature are constant, indicating the mineralogical control of the release process. Above 930°C fluctuations in the K/Ca pattern occur with constant extraction temperatures. The $^{38}Ar_s/Ca$ exposure age, as calculated from the plateau value, is 132 ± 7 m.y.

The apparent age pattern is of the type observed for 14053 (Turner et al., 1971): increasing low-temperature ages, a short maximum and subsequent decreasing ages at higher temperatures, presumably due to Ar release from pyroxene. The maximum apparent age given by five consecutive $^{40}Ar^*/^{39}Ar^*$ ratios, which must be interpreted as a minimum age of the fragment, is $3.91 \pm .06$ AE. Again, there is no evidence for substantially older ages.

K, Ca, exposure ages and $(^{36}Ar/^{38}Ar)_m$ values for both rock fragments are included in Table 1.

Schaeffer and Husain (1973) have measured $^{40}Ar/^{39}Ar$ gas retention ages as great as 4.26 AE for some light matrix breccia fragments from the North Ray Crater site. The relationship of the fragments studied here to the light matrix breccias is uncertain. In any case, the present results demonstrate that even at high resolution in the $^{40}Ar/^{39}Ar$ technique, some fragments have been totally reset in gas retention ages.

DISCUSSION

The appearance of the face of the moon is dominated by huge impact structures, with ejecta from each spread over a very large portion of the moon. The ejecta has contributed widely to the formation of the lunar highlands. Most Apollo 14, Apollo 17 and Luna 20 highland samples fall into the age bracket

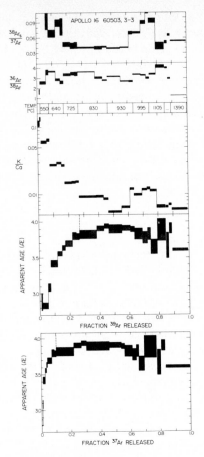

Fig. 12. $^{38}Ar_s/^{37}Ar$, $(^{36}Ar/^{38}Ar)_m$, K/Ca and the apparent age versus fractional $^{39}Ar^*$ release from fragment 60503,3-3.

3.9–4.0 AE (Podosek *et al.*, 1973; Turner *et al.*, 1971; Huneke *et al.*, 1973; and others). Since the associated errors in the ages are of the order of 0.05 AE, it has not been clearly established whether the observed ages represent a number of large events within this time range or a single event blanketing the lunar surface with debris which has been sampled at every site. The correct alternative depends on the resolution obtainable by the dating techniques.

In discussing the error of a $^{40}Ar–^{39}Ar$ age determination, one differentiates between the absolute and the relative error of the age. The absolute error must include the uncertainty in the $^{39}Ar^*$/K ratio from the monitor, which typically produces an error in a ~4 AE old sample of about 0.04 AE, neglecting uncertainty in the decay constants, which will produce the same systematic errors in all K–Ar ages. This error must be included if samples from different neutron irradiations are to be compared and is too large to allow resolution of ages within the interval 3.9–4.0 AE. If, however, samples from the same neutron irradiation are compared, age differences between samples do not depend on the monitor uncertainty, and these age differences may be determined with considerably better precision than 0.04 AE. In this case, the principal systematic uncertainty involved in comparing sample ages is due to uncertainty in the neutron dosage received by the samples. For irradiations performed in this laboratory, the uncertainty in relative neutron fluence is typically about 0.5%, corresponding to an age uncertainty on the order of 10 m.y. The remaining uncertainties are those due to random statistical effects in ratio measurements, blank variability, and the decomposition into different Ar components. These statistical errors form the basis on which the tabulated age uncertainties are computed. In favorable cases, these statistical uncertainties are small and an apparent age plateau well defined. In such cases, the dominant source of error is the neutron dosage uncertainty, and samples from the same irradiation may be compared with an age uncertainty of about 10 m.y.

To our knowledge there presently exists only one pair of lunar highland samples from different localities, which have been irradiated in the same capsule. These are the Luna 20 samples 22006, 22007 (Podosek *et al.*, 1973) and the Apollo 16 samples of this work, all irradiated in LAV 3. The *relative* uncertainties in the ages of these samples are ~0.01 AE. The ages of the Luna 20 samples are 3.91 and 3.90 AE, respectively, and of 65015, taking the plagioclase B plateau, 3.98 AE. These ages are interpreted as the times of metamorphism of these samples in impact events (Podosek *et al.*, 1973; Albee *et al.*, 1973; Jessberger *et al.*, 1974). The difference in these ages establishes the nonsimultaneity of the major basin-forming impacts in the terminal lunar cataclysm 3.9–4.0 AE ago, and sets a time scale of ~10^8 yr over which these events occurred. In general, it is not possible to assign an age of metamorphism on a few fragments with specific lunar cratering events. However, we will pursue a series of speculative arguments and attempt to assign ages to major basin-forming episodes.

McGetchin *et al.* (1973) have modeled the radial thickness variations of impact crater ejecta and have predicted average ejecta thickness at the Apollo landing sites for the major basin-forming events. Using their model, we have included the Luna 20 landing site. The results are shown in Fig. 13. The bottom of the Imbrium ejecta stratum is arbitrarily set to the same level at each site. The layers have been ordered using the stratigraphically determined sequence of ring basin forming events (Stuart-Alexander and Howard, 1970; Wilhelms, 1970). Although the model from McGetchin *et al.* (1973) has some uncertainty, we use it to decipher the relationship between sample ages and huge basin-forming impacts. We interpret the ejecta thicknesses as the probabilities of finding samples from these events at a given site. For example, it is highly improbable that material older than Imbrium and Crisium ejecta would be found at the Apollo 15 and Luna 20 sites, respectively. Further, the Apollo 16 site provides the highest probability of penetrating through the Imbrium debris to older samples. Figure 13 can also be interpreted to show that *all* highland samples with ages of 3.9–4.0 AE could be Imbrium ejecta. However, the chance of sampling pre-Imbrium material, either by small impacts or because of the local topographic situation, is good at several of the sites.

At the Apollo 16 and Luna 20 sites, the probability is highest for sampling

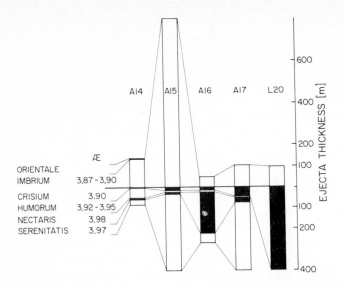

Fig. 13. Stratigraphy of ring basin ejecta at several landing sites. Ejecta thicknesses (McGetchin *et al.*, 1973) are normalized to the bottom of Imbrium ejecta. The relative morphologic age scale is from Stuart-Alexander and Howard, (1970). The assignment of ages giving the probable times when the basin forming events happened is discussed in the text. The depths are arbitrarily truncated at -400 m. Crisium ejecta at the L20 site is ~800 m thick. Serenitatis ejecta at the A15 and A17 sites is ~900 and ~1650 m thick, respectively.

material from the Nectaris and Crisium events, respectively. On this basis, we tentatively associate the 3.98 AE age obtained from rock 65015 with the Nectaris-forming event and the 3.90 age of the Luna 20 samples with the Crisium-forming event (Podosek *et al.*, 1973). The relative age scale being established by this association, we then tentatively ascribe all highland rock ages to basin-forming events, in spite of the uncertainty in the absolute ages which must now be included. Two groups of ages are identifiable at the Apollo 14 site (Turner *et al.*, 1971, 1972; Huneke *et al.*, 1972). The first contains samples of 3.87–3.90 AE age found in the smooth terrain away from the large Cone Crater and which we associate with Imbrium ejecta. The other group of ages of 3.92–3.95 AE is associated primarily with Cone Crater ejecta of underlying material, which, if not Imbrium, is most likely to be ejecta from the Humorum event. This identification is made in Fig. 13. The age previously inferred for the Serenitatis basin-forming impact and included in Fig. 13 is 3.97 AE (Huneke *et al.*, 1973), which is younger than the 3.98 AE age presently inferred for the Nectaris event. However, the contradiction with the stratigraphic sequence is not outside error limits and is not serious. We repeat that these assignments of ages to different basin forming events should be interpreted as an attempt to correlate observed ages of rocks with the lunar stratigraphic sequence and the likelihood of finding ejecta from a given event at a given site. Other possibly more valid assignments may be made.

Independent of the correctness of these assignments, the data do indicate that not all of these events were simultaneous. At least two events separated by about 80 m.y. seem to be required, but further resolution within this interval is not justified by presently available data, even though the ability to correlate ages and stratigraphic sequence is remarkable.

We may also consider whether these events are more or less singular in lunar history or whether they mark the end of an approximately exponentially decaying flux of large bodies onto the lunar surface. If the latter is the case and the flux decayed with a half-life of ~0.1 AE then a total of ~300 huge bodies impacted on the moon during the time interval 4.0–4.6 AE. Such a flux of large impact bodies would be accompanied by a much larger number of smaller bodies, producing medium sized craters (10–100 km) and obliterating the morphology of the large basins. Nonetheless, a large number of basins would still be observable but are not presently observed. Further, the amount of material deposited beyond the rim crest of 300 large basin-forming craters can be estimated at ~350×10^6 km^3, corresponding to the material of the upper ~10 km layer of the moon. If smaller craters are included, a much larger amount of material would have been thoroughly reworked. The existence of 2–4 mm size soil fragments with ages 4.3 AE (Schaeffer and Husain, 1973; Kirsten and Horn, 1974), breccia inclusions which have retained their argon since 4.5 AE (Jessberger et al., 1974, and this work) and large clasts with Rb–Sr ages of 4.6 AE (Albee et al., 1974) provide some argument against such a continuous, intense reworking of the lunar crust.

The other possibility is that the basin-forming events are singular highlights superimposed on a steadily decaying background of medium and small cratering events. This would be the case if the moon were in an environment together with other large bodies several times in its history. Such large bodies would have also probably been accompanied by a much larger number of smaller bodies, and when, for example, the Imbrium impact occurred, many other, smaller impacts might have happened at almost the same time. In this way, the crater-density age of the older basins like Nectaris will have been increased. This would obviate the requirement for a rapidly decaying flux of cratering bodies to account for the observed correlation of crater densities with radiometric ages of the different highland sites (Shoemaker, 1972). Two distinct *exponentially decaying* populations of extralunar bodies impacting the lunar surface (Shoemaker 1972; Baedecker et al., 1972) would not be necessary; it is at least equally likely that over a steady, slowly decaying background cratering flux, at least two, and maybe more flux peaks are superimposed.

It is also possible that the times of lunar basin-forming events are related to the gas retention ages of the achondrites, e.g. Pasamonte (4.1 AE, Podosek and Huneke, 1973), Stannern, (3.5–3.9 AE, Podosek and Huneke, 1973), Malvern (3.4–3.8 AE, Horn and Kirsten, 1974), and the Rb/Sr ages of clasts from Kapoeta (3.6 and 3.8 AE, Papanastassiou et al., 1974). These may represent relicts of other cataclysms at approximately the same time elsewhere in the solar system and provide evidence that the cataclysm was a system wide phenomenon.

Acknowledgments—It is our pleasure to thank Drs. A. Albee, A. Gancarz, P. Horn for stimulating discussions and T. Gay for Ni flux wire counting. One of us (EKJ) has been supported by the Bundesministerium für Forschung und Technologie (BRD) and the National Research Council (USA). This research was supported by NASA grant NGL-05-002-188 and NSF grant GP-28027.

REFERENCES

Albee A. L., Gancarz A. J., and Chodos A. A. (1973) Metamorphism of Apollo 16 and 17 and Luna 20 metaclastic rocks at about 3.95 AE: Samples 61156, 64424, 14-2, 65015, 67483, 15-2, 76055, 22006 and 22007. *Proc. Fourth Lunar Sci. Conf., Geochim. Cosmochim. Acta*, Suppl. 4, Vol. 1, pp. 569–595. Pergamon.

Albee A. L., Chodos A. A., Dymek R. F., Gancarz A. J., Goldman D. S., Papanastassiou D. A., and Wasserburg G. J. (1974) Dunite from the lunar highlands: Petrography, deformational history, Rb–Sr age (abstract). In *Lunar Science—V*, pp. 3–5. The Lunar Science Institute, Houston.

Alexander E. C., Jr., Davis P. K., Reynolds J. H., and Srinivasan B. (1973) Radiogenic xenon and argon in 14318 and implications (abstract). In *Lunar Science—IV*, pp. 30–32. The Lunar Science Institute, Houston.

Baedecker P. A., Chou C.-L. Grudewicz E. B., and Wasson J. T. (1973) The flux of extra lunar materials onto the lunar surface as a function of time (abstract). In *Lunar Science—IV*, pp. 45–47. The Lunar Science Institute, Houston.

Helz R. T. and Appleman D. E. (1973) Mineralogy, petrology and crystallization history of Apollo 16 rock 68415. *Proc. Fourth Lunar Sci. Conf., Geochim. Cosmochim. Acta.*, Suppl. 4, Vol. 1, pp. 643–660. Pergamon.

Horn P. and Kirsten T. (1974) ^{39}Ar–^{40}Ar dating of basalts and rock breccias from Apollo 17 and the Malvern achondrite. Submitted to Soviet-American conference on the cosmochemistry of the moon and the planets. Moscow, USSR June 4–8, 1974.

Huneke J. C., Podosek F. A., Burnett D. S. and Wasserburg G. J. (1972) Rare gas studies of the galactic cosmic ray irradiation history of lunar rocks. *Geochim. Cosmochim. Acta* **36**, 269–301.

Huneke J. C., Jessberger E. K., Podosek F. A., and Wasserburg G. J. (1973) ^{40}Ar/^{39}Ar measurements in Apollo 16 and 17 samples and the chronology of metamorphic and volcanic activity in the Taurus-Littrow region. *Proc. Fourth Lunar Sci. Conf., Geochim. Cosmochim. Acta*, Suppl. 4, Vol. 2, pp. 1725–1756. Pergamon.

Huneke J. C., Jessberger E. K., and Wasserburg G. J. (1974) The age of metamorphism of a highland breccia (65015) and a glimpse at the age of its protolith (abstract). In *Lunar Science—V*, pp. 375–377. The Lunar Science Institute, Houston.

Husain L. and Schaeffer O. A. (1973) Lunar Volcanism: Age of the glass in the Apollo 17 orange soil. *Science* **180**, 1358–1360.

Jessberger E. K., Huneke J. C., and Wasserburg G. J. (1974) Evidence for a ~4.5 AE age of plagioclase clasts in a lunar highland breccia. *Nature* **248**, 199–201.

Jovanovic S. and Reed G. W., Jr. (1973) Trace element studies in Apollo 16 samples (abstract). In *Lunar Science—IV*, pp. 418–420. The Lunar Science Institute, Houston.

Kirsten T. and Horn P. (1974) ^{39}Ar–^{40}Ar chronology of the Taurus-Littrow region II: A 4.28 b.y. old troctolite and ages of basalts and highland breccias (abstract). In *Lunar Science—V*, pp. 429–431. The Lunar Science Institute, Houston.

McGetchin T. R., Settle M., and Head W. (1973) Radial thickness variation in impact crater ejecta: implications for lunar basin deposits. *Earth Planet. Sci. Lett.* **20**, 226–236.

Meyer C., Jr., Anderson D. H., and Bradley J. G. (1974) Ion microprobe mass analysis of plagioclase from "non-mare" lunar samples (abstract). In *Lunar Science—V*, pp. 506–508. The Lunar Science Institute, Houston.

Papanastassiou D. A. and Wasserburg G. J. (1972a) The Rb–Sr age of a crystalline rock from Apollo 16. *Earth Planet. Sci. Lett.* **16**, 289–298.

Papanastassiou D. A. and Wasserburg G. J. (1972b) Rb–Sr systematics of Luna 20 and Apollo 16 samples. *Earth Planet. Sci. Lett.* **17**, 52–63.

Papanastassiou D. A., Rajan R. S., Huneke J. C., and Wasserburg G. J. (1974) Rb–Sr ages and lunar analogs in a basaltic achondrite; implications for early solar system chronologies (abstract). In *Lunar Science—V*, pp. 583–585. The Lunar Science Institute, Houston.

Podosek F. A. (1973) Thermal history of the nakhlites by the ^{40}Ar–^{39}Ar method. *Earth Planet. Sci. Lett.* **19**, 135–144.

Podosek F. A. and Huneke J. C. (1973) Argon 40–argon 39 chronology of four calcium-rich achondrites. *Geochim. Cosmochim. Acta* **37**, 667–684.

Podosek F. A., Huneke J. C., Gancarz A. J., and Wasserburg G. J. (1973) The age and petrography of two Luna 20 fragments and inferences for widespread lunar metamorphism. *Geochim. Cosmochim. Acta* **37**, 887–904.

Schaeffer O. A. and Husain L. (1973) Early lunar history; ages of 2 to 4 mm soil fragments from the lunar highlands. *Proc. Fourth Lunar Sci. Conf., Geochim. Cosmochim. Acta*, Suppl. 4, Vol. 2, pp. 1847–1863. Pergamon.

Shoemaker E. M. (1972) Cratering history and early evolution of the moon (abstract). In *Lunar Science—III*, pp. 696–698. The Lunar Science Institute, Houston.

Stuart-Alexander D. E. and Howard K. A. (1970) Lunar maria and circular basins—A review. *Icarus* **12**, 440.

Tera F., Papanastassiou D. A., and Wasserburg G. J. (1973) A lunar cataclysm at ~3.95 AE and the structure of the lunar crust (abstract). In *Lunar Science—IV*, pp. 723–725. The Lunar Science Institute, Houston.

Turner G. (1969) Thermal histories of meteorites by the ^{39}Ar–^{40}Ar method. In *Meteorite Research* (editor P. M. Millman), pp. 407–417. D. Reidel.

Turner G. and Cadogan P. H. (1974) Possible effects of ^{39}Ar recoil in ^{40}Ar–^{39}Ar dating of lunar samples. *Proc. Fifth Lunar Sci. Conf., Geochim. Cosmochim. Acta.* This volume.

Turner G., Huneke J. C., Podosek F. A., and Wasserburg G. J. (1971) ^{40}Ar–^{39}Ar ages and cosmic ray exposure ages of Apollo 14 samples. *Earth. Planet. Sci. Lett.* **12**, 19–35.

Turner G., Huneke J. C., Podosek F. A., and Wasserburg G. J. (1972) Ar^{40}–Ar^{39} systematics in rocks and separated minerals from Apollo 14. *Proc. Third Lunar Sci. Conf., Geochim. Cosmochim. Acta*, Suppl. 3, Vol. 2, pp. 1589–1612. MIT Press.

Wilhelms D. E. (1970) Summary of lunar stratigraphy-telescopic observations. U.S. Geological Survey Prof. paper 599-F.

York D. (1966) Least-squares fitting of a straight line. *Can. J. Phys.* **44**, 1079–1086.

Proceedings of the Fifth Lunar Conference
(Supplement 5, Geochimica et Cosmochimica Acta)
Vol. 2 pp. 1451–1475 (1974)
Printed in the United States of America

Chronology of the Taurus-Littrow region III:
Ages of mare basalts and highland breccias and some remarks about the interpretation of lunar highland rock ages

T. Kirsten and P. Horn

Max-Planck-Institut für Kernphysik, Heidelberg, Germany

Abstract—Results of ^{39}Ar–^{40}Ar dating of twelve Apollo 17 highland and mare rocks are reported. Six samples are from large rocks 70215, 79155, 75055, 77017, and 76055; the remaining being rock fragments from coarse fines 74243, 78503, and from the "picking pot" 70053.

The ages of the mare basalts indicate cessation of major volcanism in the Taurus-Littrow Valley at 3.78 ± 0.04 b.y. ago. The age of two of the highland breccias is 4.05 ± 0.07 b.y. whereas an unshocked poikilitic gabbro has an age of 4.28 ± 0.05 b.y. which is one of the highest ages found among lunar rocks. The ^{38}Ar–^{37}Ar exposure ages of four of the rocks group around 95 ± 15 m.y. and very probably date the Camelot cratering event.

The geological significance of the highland rock ages is discussed in detail. It is concluded that at present no unequivocal interpretation of lunar highland rock ages in relation to the ages of lunar formations and large circular multiring impact basins is possible. Geochronological arguments supporting this statement are presented.

INTRODUCTION

THE APOLLO 17 SPACECRAFT landed in a valley of the eastern mountains surrounding Mare Serenitatis. On the basis of pre-mission photogeological interpretations and crater counts, the Serentitatis Basin is one of the oldest nearside multiring impact basins of the moon. One of the objectives in exploring the vicinity of an old basin was to search for samples which represent the oldest lunar crust and which supposedly have not been metamorphosed by more distant subsequent large impacts like those which formed the Imbrium and Orientale basins. Another aim was to sample material from the stratigraphically young dark mantle, a unit which is prominent near the eastern margins of Mare Serenitatis and which was also expected to occur on top of the basalts in the Taurus-Littrow Valley (Scott *et al.*, 1972).

After having obtained now a set of age data for rocks from every highland expedition, we think that it is time for a critical review of our own interpretations of highland rock ages and of those given by the other investigators concerned with lunar rock dating by radiometric methods. Such a review will be given after the presentation of new results obtained by ^{39}Ar–^{40}Ar dating of twelve Apollo 17 mare and highland rocks by our laboratory.

EXPERIMENTAL PROCEDURES AND RESULTS

A general description of the samples is given in Table 1. Samples were either chipped from larger igneous rocks (70215, 79155) or manually selected under the binocular microscope from breccia 77017 and from coarse fines (70053, 74243, 78503).

Table 1. Sample description and location.

Sample number	Locality	Rock type	Probable origin
70215,21	50 m ENE of LM	Quenched mare basalt; fine grained	Camelot Crater*
79155,24	SE rim of Van Serg Crater	Heavily shocked mare basalt; coarse grained	Cochise, Shakespeare, Henry craters?
75055,11,2	SW rim of Camelot Crater	Mare basalt; coarse grained	Camelot Crater*
78503,13,B	Base of Sculptured Hills	Quenched mare basalt; medium grained	Camelot Crater*
74243,4,A	SW rim of Shorty Crater	Mare basalt; coarse grained	Shorty Crater
74243,4,B	SW rim of Shorty Crater	Quenched mare basalt; very fine grained	Shorty Crater
74243,4,C	SW rim of Shorty Crater	Poikilitic mare basalt; medium grained	Shorty Crater
77017,32,A	Base of North Massif	Anorthositic breccia with poikilitic clasts	North Massif
77017,32,B	Base of North Massif	Black glass vein penetrating A	North Massif
76055,4,1	Base of North Massif	Annealed polymict breccia; noritic matrix	North Massif
70053,8	Picking pot; very probably broken from 76335; Base of North Massif	Monomict anorthositic breccia	North Massif
78503,13,A	Base of Sculptured Hills	Poikilitic gabbro	Unknown

*This work, taking the exposure age as an indication.

    Samples which were expected to contain trapped solar wind gases were etched in a $HF + H_2SO_4 + H_2O$ mixture to remove the gas loaded surficial layers. Lunar samples and terrestrial hornblende monitors were ultrasonically cleaned in ethanol, wrapped in aluminum foil, stacked alternately in quartz ampoules and vacuum sealed together with a nickel wire to check the uniformity of the fast neutron flux from the induced ^{58}Co activity. Chips of extremely pure optical grade CaF_2 lenses were included to monitor the $^{37}Ar/^{39}Ar$ production ratio from calcium and to determine the production of ^{36}Ar and ^{38}Ar from calcium.

    We have carried out three different neutron irradiations behind Cd shielding: two in the BR-2 reactor at Mol, Belgium, one in the FR-2 reactor at Karlsruhe, Germany. Nominal fast neutron fluencies were 1.5×10^{19} and 3.5×10^{19} n/cm^2 at Mol, 0.8×10^{19} n/cm^2 at Karlsruhe. Each ampoule contained hornblende monitors in the top, middle, and bottom position with at most three samples placed between two respective monitors. J-values (defined in Table 2) were calculated for each monitor and for the individual sample positions (Table 2). The inferred neutron flux gradient along the total lengths of the ampoules was 1% at Karlsruhe, <1% and 3% at Mol, respectively. The further experimental procedure has been described in Kirsten et al. (1972).

    The following corrections have been applied:

(a) Blank corrections.
(b) Mass discrimination (0.9% per mass unit).

Table 2. Summary of ^{39}Ar–^{40}Ar results.

Sample number	J-value[a]	Total gas released (10^{-8} cc STP/g)					K (ppm)	Ca (%)	Exposure age (m.y.)	Total Ar age (b.y.)	Plateau age (b.y.)	Plateau range (% of ^{39}Ar released)
		^{40}Ar*	^{39}Ar$_K$	^{38}Ar$_C$	^{37}Ar$_{corr.}$	^{36}Ar$_T$						
70215,21	0.0616	1710	15.9	12.3	1725	1.31	345 ± 25	7.2 ± 0.5	100 ± 12	3.77 ± 0.10	3.84 ± 0.04	19–67(100)[c]
79155,24	0.0336	1865	10.0	69.6	1085	3.80	390 ± 30	7.1 ± 0.5	575 ± 60	3.73 ± 0.07	3.80 ± 0.04	20–55
75055,11,2	see Kirsten *et al.* (1973a)						480 ± 30	8.0 ± 0.3	85 ± 10	3.62 ± 0.07	3.82 ± 0.05	30–92(100)[c]
78503,13,B	0.1135	2515	46.9	14.6	3655	3.65	550 ± 30	8.1 ± 0.6	105 ± 15	3.68 ± 0.08	3.83 ± 0.04	46–90(100)[c]
74243,4,A	0.1135	3735	66.9	39.6	3325	6.95	790 ± 35	7.4 ± 0.6	315 ± 40	3.76 ± 0.08	3.78 ± 0.04	18–94
74243,4,B	0.1135	3295	49.6	6.5	3045	3.04	580 ± 30	6.75 ± 0.5	57 ± 10	4.04 ± 0.10	no plateau	—
74243,4,C	0.1135	2555	44.0	6.9	3280	1.33	515 ± 30	7.3 ± 0.6	58 ± 8	3.82 ± 0.08	3.93 ± 0.05[b]	46–64
77017,32,A	0.0628	1385	19.2	12.1	2345	1.43	410 ± 30	9.6 ± 0.8	80 ± 10	3.23 ± 0.07	4.05 ± 0.05	60–100
77017,32,B	0.0631	(500)	22.2	15.6	2450	346	475 ± 30	10 ± 0.8	90 ± 40	1.67 ± 0.3	1.5 ± 0.3	4–100
76055,4,1	see Kirsten *et al.* (1973a)						1780 ± 100	6.5 ± 0.2	120 ± 15	4.05 ± 0.08	4.05 ± 0.07	10–100
70053,8	0.0338	1485	6.6	23.7	1320	3.5	255 ± 25	8.65 ± 0.8	160 ± 20	4.06 ± 0.08	(4.3?)	see Fig. 7
78503,13,A	0.0619	5040	36.0	40.4	1975	8.74	790 ± 40	8.2 ± 0.7	290 ± 35	4.26 ± 0.08	4.28 ± 0.05	13–84(100)[c]

[a] $J = \dfrac{\exp(t_m/\tau - 1)}{(^{40}\text{Ar*}/^{39}\text{Ar})_{\text{monitor}}}$; t_m = monitor age; τ = mean life of ^{40}K: 1.885×10^9 yr.

[b] Maximum value in only one temperature fraction.

[c] Plateau possibly extending up to 100% release (see figures); nonetheless this part not included in calculation of plateau age.

(c) ^{37}Ar decay during and after irradiation.

(d) Reactor produced ^{36}Ar from Ca, with $(^{36}Ar/^{37}Ar)_{Ca} = (3 \pm 0.2) \times 10^{-4}$.

(e) ^{38}Ar produced from K and Ca, with $(^{38}Ar/^{39}Ar)_K = (11.4 \pm 3) \times 10^{-3}$ (Brereton, 1970) and $(^{38}Ar/^{37}Ar)_{Ca} = (6 \pm 1.5) \times 10^{-5}$.

(f) ^{39}Ar from Ca with $(^{37}Ar/^{39}Ar)_{Ca}$ as measured for the different irradiations $(1100 \pm 20$ and $1280 \pm 30)$.

(g) To correct for trapped ^{40}Ar in sample 77017,32,B we have used the $(^{40}Ar/^{36}Ar)_T$—ratios inferred from a $^{40}Ar/^{36}Ar$ versus $^{39}Ar/^{36}Ar$ plot (insert of Fig. 6). For the other samples we used $(^{40}Ar/^{36}Ar)_T = 1 \pm 1$. There is a possibility that a ratio as high as 10 could apply for coarse fines 74243 because of the exceptional nature of soil 74240 (Hübner *et al.*, 1974). If this is the case, ages for 74243,4,A,B,C would have to be reduced by 0.03, 0.015, and 0.01 b.y., respectively.

　　For this correction it is required to split ^{36}Ar and ^{38}Ar into the trapped and the cosmogenic components. We have used $(^{38}Ar/^{36}Ar)_T = 0.187 \pm 0.002$ and $(^{38}Ar/^{36}Ar)_C = 1.7 \pm 0.2$.

(h) For sample 79155, ^{40}Ar had to be corrected for cosmogenic ^{40}Ar. We have assumed a ratio $^{40}Ar_C/^{38}Ar_C = 0.2 \pm 0.1$ as deduced for Fe targets (Lämmerzahl and Zähringer, 1966). This assumes a recent irradiation since $(^{40}K/^{38}Ar)_C \approx 1$.

We have not corrected for:

(a) ^{38}Ar from chlorine since our samples were effectively Cd-shielded.

(b) Decay of ^{39}Ar (insignificant because of decay times < 3 months).

(c) Gas reduction during measurement (insignificant because of low emission current).

The corrected isotopes are listed for each temperature fraction in Tables 3–7. Note that ^{36}Ar and ^{38}Ar are the corrected total quantities as measured but not $^{36}Ar_T$ and $^{38}Ar_C$. This is different in Table 2 which summarizes the results for all samples.

　　Our monitor is a purified hornblende separate from the Northern Light Gneiss of Northern Minnesota. It is a split of a new preparation, NL-25-2, kindly supplied to us by Dr. O. A. Schaeffer of the State University of New York, Stony Brook. An elaborate determination of the K–Ar age of this standard at Stony Brook yields an age of 2.669 ± 0.018 b.y. (Husain, 1974) in agreement with our result of 2.68 ± 0.04 b.y. (Kirsten *et al.*, 1973b). Because of the higher precision of the Stony Brook value, we have adopted 2.67 ± 0.02 b.y. for the age of the monitor. The error of the J-values is $\pm 2\%$ for all samples. The resulting age data are given in Table 2. In addition, ^{39}Ar–^{40}Ar release patterns are shown in Figs. 1, 4, 5, 6, and 7. The applied temperature schedule was usually one hour each at 600°C, 700°C, 800°C, 900°C, 1000°C, 1200°C, and 1500°C. The apparent K/Ca ratios measured for each temperature step via $^{39}Ar_K/^{37}Ar$ ratios are also shown in Figs. 1, 4–7.

　　In calculating ^{38}Ar–Ca exposure ages (Table 2), we have adopted $(^{38}Ar/^{36}Ar_C) = 1.7 \pm 0.2$; $(^{38}Ar/^{36}Ar)_T = 0.187 \pm 0.002$ and a production rate of 1.7×10^{-8} cc ^{38}Ar/g Ca, m.y. (Bogard *et al.*, 1971). This production rate includes a flat 20% contribution of ^{38}Ar from Fe and Ti and is thus applicable for mare basalts. For

Table 3. Ar isotopes in neutron-activated samples 70215,21 and 79155,24.

Temperature (°C)	$^{40}Ar^*$ $(10^{-8}$ cc/g$)$	$\dfrac{^{40}Ar^*}{^{39}Ar_K}$	$\dfrac{^{39}Ar_K}{^{37}Ar_{corr.}}$ $(\times 10^{-3})$	$\dfrac{^{38}Ar_{lunar}}{^{37}Ar_{corr.}}$ $(\times 10^{-3})$	$\dfrac{^{36}Ar_{lunar}}{^{38}Ar_{lunar}}$	Apparent age $(10^9$ yr$)$
70215,21 (33.63 mg)						
600	44.9 ± 1.5	92.7 ± 1	98.8 ± 1.2	12.7 ± 0.3	1.81 ± 0.1	3.59 ± 0.05
700	69.8 ± 2	95.6 ± 1	52.0 ± 0.7	7.12 ± 0.2	0.79 ± 0.1	3.64 ± 0.05
800	194 ± 5	103.6 ± 1	37.3 ± 0.4	6.27 ± 0.2	0.78 ± 0.1	3.77 ± 0.05
900	266 ± 7	107.4 ± 1.2	29.1 ± 0.3	6.16 ± 0.1	0.70 ± 0.1	3.83 ± 0.05
1000	237 ± 6	108.2 ± 1.4	25.7 ± 0.3	6.35 ± 0.1	0.71 ± 0.1	3.84 ± 0.05
1200	316 ± 9	108.6 ± 2	10.6 ± 0.1	7.13 ± 0.2	0.66 ± 0.1	3.85 ± 0.07
1500	583 ± 30	112.4 ± 4.5	4.3 ± 0.1	7.44 ± 0.15	0.67 ± 0.1	3.90 ± 0.12
79155,24 (101.48 mg)						
600	364 ± 8	182.5 ± 2	327 ± 5	66.9 ± 1.5	0.71 ± 0.05	3.70 ± 0.05
700	403 ± 9	195 ± 2	181 ± 2	66.3 ± 1.5	0.71 ± 0.05	3.81 ± 0.05
800	287 ± 8	193.5 ± 2	57.9 ± 1	59.7 ± 1	0.74 ± 0.02	3.80 ± 0.05
900	201 ± 5	185.2 ± 2	20.0 ± 0.3	59.2 ± 1	0.67 ± 0.02	3.73 ± 0.05
1000	138 ± 3	179.5 ± 2	14.4 ± 0.2	68.2 ± 1	0.64 ± 0.02	3.67 ± 0.05
1200	286 ± 9	171.7 ± 3	6.2 ± 0.1	69.0 ± 1	0.62 ± 0.02	3.60 ± 0.07
1400	176 ± 8	188.2 ± 4	1.44 ± 0.05	63.6 ± 1	0.64 ± 0.02	3.75 ± 0.07
1500	10 ± 6	348 ± 220	2.6 ± 0.6	67.0 ± 6	0.64 ± 0.05	4.79 ± 1.9

Table 4. Ar isotopes in neutron-activated samples 78503,13,B and 74243,4,A.

Temperature (°C)	$^{40}Ar^*$ $(10^{-8}$ cc/g$)$	$\dfrac{^{40}Ar^*}{^{39}Ar_K}$	$\dfrac{^{39}Ar_K}{^{37}Ar_{corr.}}$ $(\times 10^{-3})$	$\dfrac{^{38}Ar_{lunar}}{^{37}Ar_{corr.}}$ $(\times 10^{-3})$	$\dfrac{^{36}Ar_{lunar}}{^{38}Ar_{lunar}}$	Apparent age $(10^9$ yr$)$
78503,13,B (28.93 mg)						
600	154 ± 3	28.1 ± 0.4	275 ± 5	5.22 ± 0.4	1.76 ± 0.1	2.70 ± 0.05
700	428 ± 10	49.0 ± 0.5	111 ± 2	4.54 ± 0.2	1.61 ± 0.1	3.54 ± 0.05
800	410 ± 10	56.8 ± 0.5	46.2 ± 1	4.20 ± 0.1	1.26 ± 0.05	3.78 ± 0.05
900	512 ± 12	58.7 ± 0.7	25.9 ± 0.5	3.78 ± 0.1	0.98 ± 0.05	3.84 ± 0.05
1000	402 ± 10	58.5 ± 0.7	19.8 ± 0.4	3.76 ± 0.1	0.92 ± 0.05	3.84 ± 0.05
1200	310 ± 8	57.8 ± 0.9	11.5 ± 0.3	4.38 ± 0.1	0.87 ± 0.05	3.81 ± 0.05
1500	301 ± 50	66.6 ± 13	2.0 ± 0.1	4.24 ± 0.1	0.67 ± 0.05	4.05 ± 0.30
74243,4,A (56.36 mg)						
600	533 ± 14	44.0 ± 0.6	598 ± 10	10.3 ± 0.5	2.39 ± 0.1	3.38 ± 0.05
700	559 ± 15	55.7 ± 0.6	278 ± 3	11.4 ± 0.4	1.25 ± 0.05	3.75 ± 0.05
800	660 ± 15	56.9 ± 0.6	107 ± 1	11.7 ± 0.3	0.92 ± 0.03	3.79 ± 0.05
900	648 ± 16	56.4 ± 0.6	43.2 ± 0.5	11.9 ± 0.3	0.89 ± 0.03	3.77 ± 0.05
1000	456 ± 12	56.0 ± 0.6	28.4 ± 0.4	11.8 ± 0.3	0.93 ± 0.03	3.76 ± 0.05
1200	555 ± 15	57.2 ± 0.8	16.8 ± 0.2	12.5 ± 0.3	0.83 ± 0.03	3.79 ± 0.05
1500	324 ± 60	84.9 ± 16	1.88 ± 0.06	12.4 ± 0.2	0.64 ± 0.02	4.46 ± 0.30

Table 5. Ar isotopes in neutron-activated samples 74243,4,B and 74243,4,C.

Temperature (°C)	$^{40}Ar^*$ $(10^{-8}$ cc/g)	$^{49}Ar^*$ $\overline{^{39}Ar_K}$	$\dfrac{^{39}Ar_K}{^{37}Ar_{corr.}}$ $(\times 10^{-3})$	$\dfrac{^{38}Ar_{lunar}}{^{37}Ar_{corr.}}$ $(\times 10^{-3})$	$\dfrac{^{36}Ar_{lunar}}{^{38}Ar_{lunar}}$	Apparent age $(10^9$ yr)
74243,4,B (41.47 mg)						
600	373 ± 10	43.3 ± 0.6	423 ± 7	4.67 ± 0.5	1.84 ± 0.2	3.35 ± 0.05
700	576 ± 12	55.2 ± 0.6	139 ± 2	2.44 ± 0.2	1.06 ± 0.1	3.74 ± 0.05
800	542 ± 11	56.7 ± 0.7	46.6 ± 0.6	1.92 ± 0.1	0.82 ± 0.05	3.78 ± 0.05
900	333 ± 8	64.8 ± 0.8	29.1 ± 0.4	2.10 ± 0.1	1.02 ± 0.05	4.00 ± 0.05
1000	326 ± 8	65.3 ± 0.8	24.2 ± 0.4	2.12 ± 0.1	1.04 ± 0.05	4.01 ± 0.05
1200	594 ± 30	74.1 ± 4	9.11 ± 0.15	2.29 ± 0.05	0.97 ± 0.05	4.23 ± 0.1
1500	548 ± 80	192 ± 30	1.93 ± 0.05	2.42 ± 0.05	0.95 ± 0.04	5.89 ± 0.28
74243,4,C (96.05 mg)						
600	92.5 ± 4	35.7 ± 0.7	477 ± 10	2.76 ± 0.6	1.47 ± 0.2	3.05 ± 0.06
700	364 ± 8	52.5 ± 0.6	353 ± 5	1.99 ± 0.4	1.52 ± 0.2	3.66 ± 0.05
800	634 ± 13	59.1 ± 0.6	73.7 ± 1	1.92 ± 0.2	1.06 ± 0.06	3.85 ± 0.05
900	475 ± 10	61.9 ± 0.7	26.5 ± 0.5	1.94 ± 0.1	0.84 ± 0.05	3.93 ± 0.05
1000	402 ± 10	58.8 ± 0.7	17.2 ± 0.3	1.88 ± 0.1	0.72 ± 0.05	3.84 ± 0.05
1200	261 ± 8	54.6 ± 0.8	12.5 ± 0.3	2.20 ± 0.1	0.84 ± 0.03	3.72 ± 0.06
1500	325 ± 40	73.7 ± 9	2.16 ± 0.06	2.28 ± 0.05	0.71 ± 0.03	4.22 ± 0.15

Table 6. Ar isotopes in neutron-activated samples 77017,32,A and 77017,32,B.

Temperature (°C)	$^{40}Ar^*$ $(10^{-8}$ cc/g)	$^{40}Ar^*$ $\overline{^{39}Ar_K}$	$\dfrac{^{39}Ar_k}{^{37}Ar_{corr.}}$ $(\times 10^{-3})$	$\dfrac{^{38}Ar_{lunar}}{^{37}Ar_{corr.}}$ $(\times 10^{-3})$	$\dfrac{^{36}Ar_{lunar}}{^{38}Ar_{lunar}}$	Apparent age $(10^9$ yr)
77017,32,A (99.97 mg)						
600	68.5 ± 2	38.1 ± 0.8	9.35 ± 0.2	3.57 ± 0.1	0.852 ± 0.02	2.30 ± 0.04
700	63.1 ± 2	21.7 ± 0.5	9.27 ± 0.2	4.52 ± 0.1	0.690 ± 0.02	1.62 ± 0.04
800	166 ± 4	33.3 ± 0.5	9.15 ± 0.2	5.26 ± 0.1	0.682 ± 0.02	2.13 ± 0.04
900	153 ± 4	89.0 ± 1.5	8.93 ± 0.2	5.49 ± 0.1	0.695 ± 0.02	3.55 ± 0.05
1000	200 ± 4	119.6 ± 2	8.62 ± 0.2	5.57 ± 0.1	0.701 ± 0.02	4.04 ± 0.05
1100	335 ± 4	120.5 ± 2	8.24 ± 0.2	5.56 ± 0.1	0.672 ± 0.02	4.05 ± 0.05
1300	229 ± 5	121.2 ± 2	5.05 ± 0.1	5.90 ± 0.1	0.684 ± 0.02	4.06 ± 0.06
1500	170 ± 10	123.1 ± 7	7.11 ± 0.1	5.81 ± 0.1	0.673 ± 0.02	4.08 ± 0.10
77017,32,B (32.60 mg)						
600	83.2 ± 2	267 ± 8	21.0 ± 0.3	21.8 ± 0.3	4.24 ± 0.05	5.43 ± 0.1
700	27.9 ± 1	47.6 ± 2	13.3 ± 0.2	11.5 ± 0.2	3.22 ± 0.05	2.61 ± 0.1
800	24.4 ± 1.3	19.0 ± 1.5	11.4 ± 0.2	11.7 ± 0.2	3.00 ± 0.05	1.49 ± 0.1
900	21.7 ± 1.5	23.7 ± 1.8	13.1 ± 0.2	20.8 ± 0.2	3.75 ± 0.05	1.72 ± 0.1
1000	25.7 ± 1.6	22.3 ± 2	13.7 ± 0.2	20.3 ± 0.2	3.77 ± 0.05	1.66 ± 0.1
1100	20.5 ± 2	15.2 ± 1.6	11.2 ± 0.2	20.4 ± 0.2	3.80 ± 0.05	1.27 ± 0.1
1300	94.7 ± 30	16.4 ± 6	8.84 ± 0.15	47.9 ± 0.4	4.63 ± 0.05	1.34 ± 0.4
1500	204 ± 35	18.8 ± 4	8.00 ± 0.15	30.4 ± 0.2	4.42 ± 0.05	1.48 ± 0.3

Table 7. Ar isotopes in neutron-activated samples 70053,8 and 78503,13,A.

Temperature (°C)	$^{40}Ar^*$ $(10^{-8} cc/g)$	$\dfrac{^{40}Ar^*}{^{39}Ar_K}$	$\dfrac{^{39}Ar_K}{^{37}Ar_{corr.}}$ $(\times 10^{-3})$	$\dfrac{^{38}Ar_{lunar}}{^{37}Ar_{corr.}}$ $(\times 10^{-3})$	$\dfrac{^{36}Ar_{lunar}}{^{38}Ar_{lunar}}$	Apparent age $(10^9 yr)$
70053,8 (54.58 mg)						
600	98.9 ± 2	158 ± 2	5.48 ± 0.1	16.2 ± 0.3	0.713 ± 0.01	3.48 ± 0.05
700	173 ± 4	193 ± 3	4.85 ± 0.1	18.1 ± 0.3	0.707 ± 0.01	3.80 ± 0.05
800	181 ± 4	219 ± 3	4.76 ± 0.1	18.6 ± 0.3	0.752 ± 0.01	4.01 ± 0.05
900	372 ± 8	235 ± 3	4.81 ± 0.1	18.4 ± 0.3	0.690 ± 0.01	4.13 ± 0.05
1000	330 ± 8	245 ± 4	5.04 ± 0.1	18.4 ± 0.3	0.714 ± 0.01	4.20 ± 0.05
1100	74.8 ± 2	236 ± 4	5.29 ± 0.1	19.3 ± 0.3	0.783 ± 0.015	4.13 ± 0.06
1400	218 ± 6	259 ± 5	5.23 ± 0.1	19.5 ± 0.3	0.737 ± 0.01	4.29 ± 0.06
1500	37.4 ± 2	265 ± 10	4.87 ± 0.1	19.6 ± 0.3	0.712 ± 0.015	4.33 ± 0.11
78503,13,A (25.68 mg)						
600	80.3 ± 2	147 ± 3	22.8 ± 0.3	16.0 ± 0.3	1.75 ± 0.04	4.36 ± 0.07
700	175 ± 5	127 ± 2	20.5 ± 0.2	20.2 ± 0.3	1.23 ± 0.03	4.11 ± 0.06
800	366 ± 9	137 ± 2	20.7 ± 0.2	22.1 ± 0.3	1.10 ± 0.03	4.24 ± 0.06
900	1188 ± 26	141 ± 2	19.2 ± 0.2	20.8 ± 0.3	0.724 ± 0.02	4.29 ± 0.05
1000	1486 ± 32	140 ± 2	19.0 ± 0.2	20.8 ± 0.3	0.725 ± 0.02	4.28 ± 0.06
1200	912 ± 25	139 ± 5	19.7 ± 0.2	21.4 ± 0.3	0.746 ± 0.02	4.27 ± 0.07
1500	834 ± 55	145 ± 10	13.5 ± 0.2	22.5 ± 0.3	0.706 ± 0.03	4.33 ± 0.14

highland rocks, the effective production rate may be lower by $\sim$ 10–15%, but since this depends on the degree of shielding during exposure, our adopted value is probably as good as any "refined" one.

The K and Ca contents which are listed in Table 2 were determined from $^{39}Ar_K$ and ^{37}Ar in sample and monitor and from the K and Ca contents of the monitor (3070 ± 8 ppm K, $8.04 \pm 0.2\%$ Ca).

Error discussion

The errors of the Ar quantities, ages and K and Ca concentrations are listed in Tables 2–7. Interlaboratory comparisons have shown that the absolute calibration of our argon gas standard is accurate within $\pm 2\%$ (Hübner *et al.*, 1974). Ar ratios are obtained from $\geqslant 5$ recordings at masses 36–38, but for ^{40}Ar and ^{39}Ar from $\geqslant 10$ recordings. The statistical error of isotopic ratios (2σ) is mostly close to 1%. Blank corrections were usually negligible except for the 1200°C and 1500°C fractions. In such cases the influence of their error on the apparent age is shown by error bars in the figures. Errors in isotopic ratios caused by the flux gradient correction and by corrections for interfering reactions as well as the error of the monitor age (see above) were considered in calculating the errors of the plateau ages as given in Tables 2–7 (but not included in the figures). In general, these errors exceed the total fluctuation of ages among those temperature fractions which define the plateau. Errors of the K, Ca contents and the total Ar ages are larger (Table 2) since they are affected by the cumulation of errors in seven subsequent measurements and by the larger blank corrections for the 1500°C fraction. Errors of the exposure ages assume a 10% uncertainty of the production rate and include the bias caused by the partitioning of trapped and cosmogenic Ar. Any larger deviation of the actual production rate would also change the exposure ages. Also a possible systematic error due to varying effective cosmic ray energy-spectra leading to different production rates is not taken into account.

As the fragments from coarse fines had been etched prior to irradiation to remove the trapped solar wind, one could suppose that leaching of K could affect the ages. However, it has been found that leaching of terrestrial whole-rock basalt samples to remove carbonates and zeolithes and to reduce the amount of air argon contamination does not change the measured K–Ar age (Horn and Lippolt, 1971, unpublished result). Acid leaching of basaltic rocks evidently removes K and Ar in proportional amounts.

Similar conclusions were drawn by Evernden and Curtis (1965) from experiments on HF-leached sanidines and plagioclases. Leaching effects on the $^{40}Ar-^{39}Ar$ release curves of lunar rocks are presently studied in our laboratory.

DISCUSSION OF APOLLO 17 RESULTS

Gas retention ages

Mare basalts. The mare basalts from the Taurus-Littrow Valley dated in this laboratory comprise three samples from large rocks. Four fragments were separated from 2–4 mm coarse fines. These samples are further identified by their sample numbers and by a short characterization in Table 1.

70215,21 This rock is fine grained with a quench texture and very likely is a devitrified vitrophyre. The argon release curve obtained (Fig. 1) reveals a reasonable plateau at temperatures from 900°C to 1200°C. The plateau defines an age of solidification of 3.84 ± 0.04 b.y. The steady decrease in the K/Ca ratio over the temperature range of the experiment (600–1500°C; Fig. 1) is atypical for Apollo 17 mare basalts in that the K/Ca ratio changes only by a factor of about 20, while variations by a factor of 130–300 are observed for the other basalt samples (Figs. 1, 4, and 5). The smaller variations may be explained by the quenching of the rock which prevented K and Ca to separate by differential solidification of Ca-bearing phases and late-stage K-rich phases (mesostasis) which release their gas at different temperatures in the experiment.

Highland rocks which do not contain a late-stage crystallization mesostasis show variations of the K/Ca ratio over the whole temperature range up to a factor of five only (under our heating conditions).

79155,24 Basalt 79155 is coarse grained and has suffered severe shock as indicated by planar elements in pyroxenes and plagioclases and by the occurrence of maskelynite and *in situ* produced rock glass with residual (but corroded) ilmenites (Fig. 2b). The glass has Schlieren and shows an orange color reminiscent of that of the orange spherules in 74220.

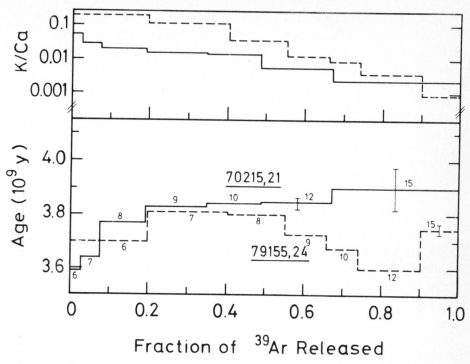

Fig. 1. ^{39}Ar–^{40}Ar release patterns and apparent K/Ca ratios for mare basalts 70215 and 79155. Only errors due to blank corrections are shown, otherwise see Tables 2 and 3.

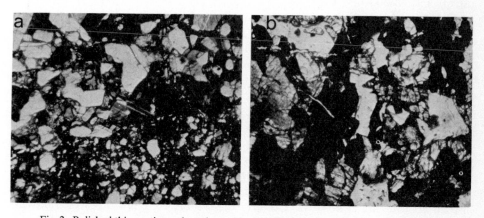

Fig. 2. Polished thin sections of portions of breccia 77017 (a) and mare basalt 79155 (b) in transmitted light: (a) Shows part of a coarse-grained relict of a plagioclase-rich poikilitic clast embedded in finely crushed matrix of similar mineralogical composition; width of picture 3.5 mm; partly crossed Nicols. (b) Heavily shocked and fractured pyroxenes (gray), ilmenites (black) and maskelynite (white). Light gray patches in maskelynite are plagioclase relicts; dark vein is shock-melted glass of orange color; width of picture 2 mm; slightly crossed Nicols.

Interestingly, it is this shocked mare basalt which shows a decrease in ^{40}Ar–^{39}Ar ratios of the high-temperature gas fractions (Fig. 1). High-temperature drop-offs in apparent ages are frequently observed in highland rocks but only rarely in mare basalts. The general opinion is that it cannot be explained by shock effects. This view might have to be reconsidered in the presence of the above-mentioned result.

At intermediate temperatures a plateau is defined at an age of 3.80 ± 0.04 b.y. It is considered to be a lower limit for the crystallization age of this rock but might well be equal or close to the age of primary crystallization if the ages of the other basalts in this area are considered. The high-temperature age of 3.6 b.y. for the 1200°C fraction could be an upper age limit for the shock event; but it is doubtful whether it has a geological significance at all. This basalt is an example for rocks that have not been completely degassed in an event involving very high temperatures (beginning of melting of ilmenite) locally.

75055,11,2   This coarse-grained variety of the subfloor basalts has been already described (Kirsten *et al.*, 1973a). Its crystallization age is 3.82 ± 0.05 b.y.

78503,13,B, 74243,4,A, B, and C   The textures of these basaltic fragments selected from coarse fines range from quench textures through poikilitic to ophitic (Figs. 3a, b, c, d) with large variations in grain size—very probably reflecting various cooling rates on or near the lunar surface. Sample 74243,4,C (Fig. 3c) appears to be crushed and subsequently metamorphosed (e.g., the ilmenites are the oldest phase but they are anhedral in this rock). Two of the samples reveal reasonable intermediate plateaus (Fig. 4) with ages of 3.78 ± 0.04 b.y. and 3.83 ± 0.04 b.y. at gas release temperatures between 800°C and 1200°C. Despite the large errors introduced into the ^{40}Ar–^{39}Ar ratio in the 1500°C fraction by high blank values in these particular measurements, it seems as if there is a true rise in the ^{40}Ar–^{39}Ar ratios. This high-temperature rise in apparent ages is observed in all our Apollo 17 basalt measurements (also in basalt 75055 which was irradiated and measured at a different time than the basalts under discussion; see Kirsten *et al.*, 1973a). Measurements by other investigators occasionally reveal a similar trend (e.g. Huneke *et al.*, 1973; Husain and Schaeffer, 1973; Schaeffer and Husain, 1973; Stettler *et al.*, 1973; Turner, 1970). It could be an experimental artifact but a more plausible tentative explanation for this phenomenon could be that in such cases inherited contamination ^{40}Ar is present. Either this had been trapped at depth by early formed phenocrysts of olivine, pyroxene or armalcolite/ilmenite or it is due to older xenocrysts from the overlying crust which were incorporated into the upwelling magma. Occasional displacements of data points from Rb/Sr internal isochrons (Murthy *et al.*, 1971; Compston, personal communication) are possibly due to the same effect. Continental terrestrial basalts very frequently contain xenoliths or xenocrysts from the country rocks (Horn *et al.*, 1972). Until now we have no microscopical indications for such contaminations in lunar mare basalts—but they would be very difficult to recognize.

The other two fragments of mare basalts from the gray control soil do not show age plateaus (Fig. 5). The release pattern of sample 74243,4,B is totally

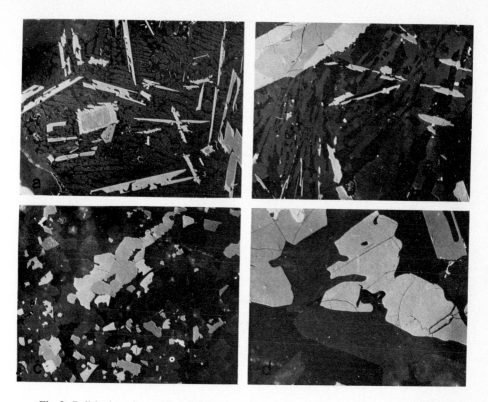

Fig. 3. Polished sections of basaltic fragments in interference contrast (a) 74243,4,B; (b) 78503,13,B; (c) 74243,4,C; (d) 74243,4,A; revealing different grain sizes and textures. The principle phases are glass and plagioclase (black), pyroxene and olivine (dark gray), ilmenite and armalcolite (brighter gray) and NiFe metal (white). Widths of pictures is 0.4 mm.

disturbed and indicates large amounts of excess argon (if an age of 3.8 b.y. is assumed). This excess Ar is either due to an uneven redistribution of argon on the occasion of some shock event (now texturally annealed) or to argon inheritance by the mechanisms considered above. Trapped solar wind argon is not the source of the excess argon since it is not accompanied by ^{36}Ar.

For sample 74243,4,C a tentative lower age limit of 3.93 b.y. may be inferred. This seems to be too high for Apollo 17 basalts; but an age of 3.92 ± 0.03 b.y. has already been found for Apollo 11 mare basalt 10003 (Stettler et al., 1974; Turner, 1970) and 3.93 b.y. is therefore a possible mare basalt age.

The mare basalts from Apollo 17 fall into at least two chemically, petrologically and probably genetically distinct groups (Brown et al., 1974; El Goresy et al., 1974; Papike and Bence, 1974). As our suite of dated rocks covers these different basalt types and as all basalt ages obtained agree within the limits of errors at 3.82 ± 0.04 b.y. it follows that genetical differences of the basalts do not necessarily imply large differences in the times of eruption.

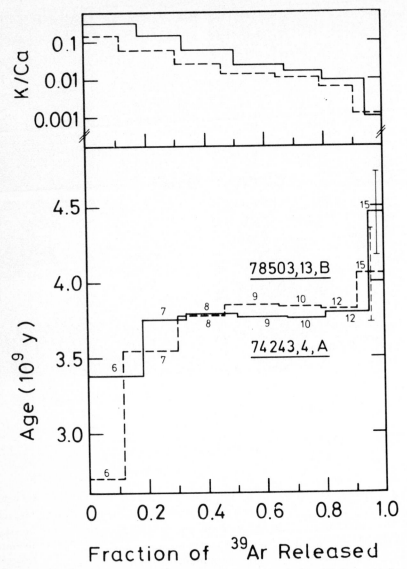

Fig. 4. ^{39}Ar–^{40}Ar release patterns and apparent K/Ca ratios for basaltic fragments from coarse fines 78503 and 74243. Only errors due to blank corrections are shown, otherwise see Tables 2 and 4.

Highland rocks. The highland rock samples investigated are from two large rocks and two are fragments from coarse fines (Table 1).

77017,32 Anorthositic breccia 77017 is one of the abundant breccias that contain texturally intact rock clasts within a crushed and annealed matrix. In Fig. 2a the difference in texture and grain size between a plagioclase rich clast and the fine-grained matrix is shown. A detailed description of this rock and an attempt to

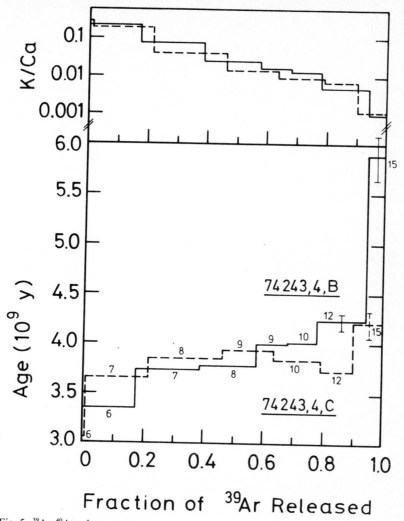

Fig. 5. $^{39}Ar-^{40}Ar$ release patterns and apparent K/Ca ratios for basaltic fragments form coarse fines 74243. Only errors due to blank corrections are shown, otherwise see Tables 2 and 5.

reveal its petrological history is given by Helz and Appleman (1974). Despite of the very complex history of this rock (*ibid.*) the argon release pattern of subsample A (Fig. 6) shows a very well defined plateau at temperatures from 1000°C to 1500°C. The age defined by the plateau is 4.05 ± 0.05 b.y. Subsample A is a whole-rock sample from which contaminations by the black glass (subsample B) have been carefully removed by handpicking.

The flat intermediate- to high-temperature plateau indicates that we do not deal with a mixture of different clast and matrix ages. The age very likely dates the time when the augite and pigeonite poikiloblasts crystallized (event 3 according to

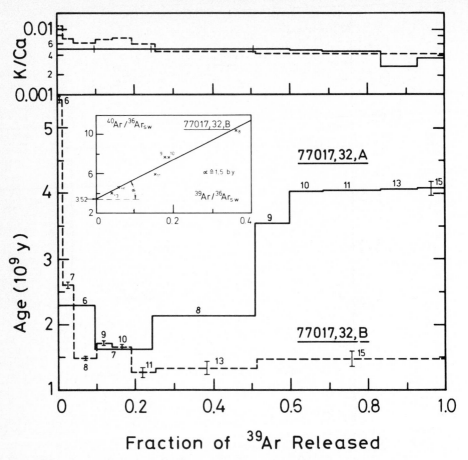

Fig. 6. ^{39}Ar–^{40}Ar release patterns and apparent K/Ca ratios for anorthositic breccia 77017,32,A and for a black glass vein penetrating the breccia (77017,32,B). The insert is a ^{40}Ar/^{36}Ar$_{s.w.}$–^{39}Ar/^{36}Ar$_{s.w.}$ isochron for the various temperature fractions released from the glassy sample. A least square fit yields an age of 1.5 ± 0.3 b.y. and a ratio of ^{40}Ar/^{36}Ar = 3.52 for the trapped component which occurs in large quantities in the glass (346×10^{-8} cc STP ^{36}Ar/g). Only errors due to this admixture and due to blank corrections are shown. Otherwise, see Tables 2 and 6.

Helz and Appleman). The large total gas loss the sample has suffered (~40%) is in our opinion mainly related to the injection of the black glass. As at release temperatures of 1000–1500°C a plateau is defined and the age does not steadily increase toward higher temperatures we consider the plateau age as being unaffected by the short-termed shock induced thermal metamorphism which formed the glass. This glass (subsample B; Fig. 6) releases its gas preferably at very high temperatures and contains large amounts of trapped solar wind Ar isotopes. After correction for trapped ^{40}Ar (insert in Fig. 6; see also chapter on experimental procedures and results) we obtain an age of 1.5 ± 0.3 b.y. for the glass. The trapped solar wind gases indicate that the glass is genetically unrelated to the host rock although the K and Ca contents of rock and glass are similar (Table 2). The glass certainly represents impact melted soil which was injected into cracks of the host rocks ~ 1.5 b.y. ago (very probably events 4 and/or 5 described by Helz and Appleman). The shock event, even though it caused appreciable changes in the mineralogy of the rock, was not effective in completely annihilating its older age record.

76055,4,2   The age of this annealed breccia (4.05 ± 0.07 b.y.) coincides with that of 77017,32A and is interpreted as the age of metamorphism of this rock (see also Kirsten *et al.*, 1973a; Chao, 1973).

70053,8 and 78503,13,A   Anorthositic fragment 70053,8 consists of slightly shocked plagioclases with interstitial pyroxenes, olivine and opaques; a single pink spinel has also been observed. The release pattern (Fig. 7) rises steadily to a maximum at 4.29 b.y. (or 4.33 b.y. at 1500°C).

It is rather doubtful whether one can infer a (minimum?) age of 4.3 b.y. from the release pattern. The K/Ca ratio remains extremely constant over the whole gas release, indicating that only a mineral phase (or mineral phases) with a constant K/Ca ratio and presumably similar gas retention characteristics is dated. This sample was taken from the "picking pot" but (judged from sample comparison) very probably represents a piece from large rock 76335 which was transported in the same sample bag.

Sample 78503,13,A described as a troctolite (Kirsten and Horn, 1974) turned out to be a poikilitic gabbro after thin section and microprobe analyses were available, while until then only a polished section had been investigated. The pyroxenes are of the low Ca type and enclose magnesian olivine, plagioclase and accessories. One grain of K feldspar has been observed.

The age of this rock is well defined by a broad plateau at 4.28 ± 0.05 b.y. (Fig. 7). The very constant K/Ca ratio over the same temperature range at which the plateau occurs indicates a homogeneous mineral phase releasing argon. The rise of the $^{40}Ar-^{39}Ar$ ratio (paralleled by a significant decrease in the K/Ca ratio) in the last (1500°C) temperature fraction indicates that a distinct phase (or lattice site) with an apparent age of ≥ 4.33 b.y. might probably be present in this rock.

All highland rocks described here have relatively constant K/Ca ratios over the whole temperature range of gas release and reflect the observation that a K-rich mesostasis is not present.

The ages of the highland rocks reported here are in the range of 4.05–4.28 (4.33) b.y. Similar and younger ages have been found by various authors for Apollo 17 highland rocks (Huneke *et al.*, 1973; International Consortium, 1974; Schaeffer and Husain, 1974; Tera *et al.*, 1974; Turner *et al.*, 1973).

Exposure ages of mare and highland rocks

The $^{38}Ar-^{37}Ar$ exposure ages are given in Table 2. The assumed origins are noted in Table 1 and imply a comparison of the spallogenic exposure ages with the relative ages as indicated in the Apollo 17 Preliminary Science Report (1974).

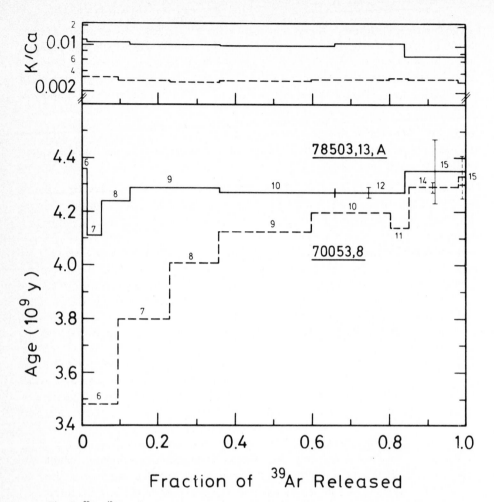

Fig. 7. ^{39}Ar–^{40}Ar release patterns and apparent K/Ca ratios for anorthositic breccia 70053,8 and poikilitic gabbro 78503,13,A from coarse fines. Only errors due to blank corrections are shown, otherwise see Tables 2 and 7.

The only exposure ages which can be attributed with some confidence to a specific cratering event group around 95 m.y. and are believed to date Camelot Crater. The exposure ages of the highland rocks have to be interpreted in terms of mass wasting from the mountains surrounding the Taurus-Littrow Valley—but no specific events (craters, boulder tracks, avalanches) are recognized so far.

Conclusions from the Apollo 17 results

For five out of seven mare basalts we obtain plateau ages within the narrow range of 3.78 ± 0.04 b.y. to 3.84 ± 0.04 b.y. These ages are comparable to those of Apollo 11 low-K basalts (Turner, 1971).

Since the five dated basalts represent at least two different types, subsequent lava flows must have occurred within ≤60 m.y. Three basalts have exposure ages of about 95 ± 15 m.y. and very probably date the Camelot Crater. Judged from its diameter Camelot Crater is only about 100 m deep. Therefore, the dated basalts derive from the uppermost stratigraphical horizons of the basalts in the Taurus-Littrow Valley. Because no ejecta debris from either one of the large impact basins has been deposited on top of the basalts, their ages define a lower age limit for the time of formation of all of the basins (it is very unlikely that, for example, Imbrium Basin ejecta is interbedded at greater depth within the basalts).

The ages obtained for the basaltic fragments from coarse fines are consistent with the ages of the larger rocks which more likely are derived from the local bedrock. Therefore, no indication for extraneous volcanic material (dark mantling material) is given by the age measurements. One of the mare basalts was influenced by strong shock metamorphism ≤ 3.6 b.y. ago, but nevertheless reveals a reasonable age for its primary solidification 3.80 ± 0.04 b.y. ago.

Highland breccias 77017,32,A and 76055,4,1 represent debris from the North Massif. Both rocks give well defined plateau ages of 4.05 ± 0.05 b.y. and 4.05 ± 0.07 b.y., respectively.

A shock and heating event which affected breccia 77017 about 2.5 b.y. after the beginning of radiogenic ^{40}Ar accumulation did not reset this rock's radiometric clock completely. This observation is again taken as proof that only extremely intense shock metamorphism or complete shock melting of a rock upon a cratering or basin-forming event can lead to its complete degassing. Coarse fine highland rock fragments from two different stations show ages up to 4.28 ± 0.05 b.y.

The question is where and how these old rocks survived the early intense bombardment of the lunar crust by meteorites, comets and asteroide-sized bodies, while the majority of the rocks from the same lunar formations or even from the same breccia have ages which are significantly lower. This leads us to a review of the interpretations and implications of lunar highland rock ages.

INTERPRETATION OF LUNAR HIGHLAND ROCK AGES

In the case of mare basalts, the age data obtained by isotopic age determinations are relatively straightforward to interpret in terms of crystallization ages and times of emplacement of the respective mare formations. Major uncertainties are whether a given mare rock collected at the moon's surface derives originally from the surface of the underlying bedrock, from more ancient lava flows beneath, or from younger or older surface flows at some distance from the sampling area. With the above-mentioned difficulties in mind, we may state that mare basalts can tentatively be correlated with more or less definite mare units. One reason is that almost all mare rocks are definitely igneous (basalts) which did not undergo secondary alteration by either thermal and/or shock metamorphism. Accordingly, only a very small number of the rare mare breccias are polymict in character.

A completely different situation is encountered for highland rocks which are usually brecciated and shocked—often multiply. Furthermore, they are frequently polymict. Until now, there is not one highland rock for which we know the locality of primary origin with some certainty. All the highland rocks collected so far have been embedded in pyroclastic sediments and highland bedrocks have not been detected, neither on the moon nor from orbit.

Evidently this is due to the fact that that at the time of highland rock formation the flux of extralunar solid material that collided with the moon's surface was very high (Neukum *et al.*, 1974). This intense bombardment resulted in lateral and vertical mixing of the existing rocks; rocks were metamorphosed and new types (impactites) were formed. Due to these perturbations in the rocks histories, it is difficult to ascribe the measured ages to specific rock forming and brecciation events—especially as they are not readily distinguishable by mineralogical or chemical methods.

An important question in this respect is whether ages measured on highland rocks bear information on the time of deposition of formations in which they are found today and which are commonly believed—on a large scale—to be related to large circular multiring impact basins.

Among the laboratories engaged in lunar rock dating there are various approaches in interpreting the age figures obtained for rocks or minerals.

One opinion is that the age of a highland rock implicitly gives the age of the formation in which it was found or from which it was derived by some smaller event (local impact, mass wasting, avalanche). The argument is that at the time when the material was blasted out of the lunar crust by the large impact, the temperature in the ejecta blanket was high enough to reset the radiometric clock (Huneke *et al.*, 1973; Jessberger *et al.*, 1974; Schaeffer and Husain, 1974; Stettler *et al.*, 1973; Turner *et al.*, 1973; York *et al.*, 1972). A complement of this idea is that the large impacts excavated material from depths of several tens of kilometers where the temperature was high as to maintain open isotopic systems in the rocks and minerals. Upon excavation these rocks cooled and started to accumulate radiogenic daughter isotopes (Kirsten *et al.*, 1973b; Kirsten and Horn, 1974; Podosek *et al.*, 1973; Turner *et al.*, 1973).

Faced with the problem that at a given landing site and presumably within one and the same stratigraphical unit a range of rock ages is found, it is argued that the formation has to be younger than (or as young as) the youngest rock found within that unit (Kirsten *et al.*, 1972, 1973b; Papanastassiou and Wasserburg, 1971; Mark *et al.*, 1973; Nunes *et al.*, 1973)—here one should not forget that the youngest rock is always still to be found! We think that all stratigraphical problems have to be treated in this manner when clastic sediments are involved that were not subjected to age-relevant metamorphism *after* deposition.

To explain the occurrence of rocks showing different ages at one landing site, it was argued that these materials are of different provenience as there are ejecta deposits with varying thicknesses from many impact basins beneath any landing site visited so far (McGetchin *et al.*, 1973). However, this does not explain the existence of many highland breccias with clasts or xenoliths of different ages— prominent examples being rocks 68415/68416 with well-defined ^{39}Ar–^{40}Ar ages of

3.85 ± 0.06 b.y. and 4.00 ± 0.05 b.y., respectively (Kirsten *et al.*, 1973b). Other examples of breccias which were not isotopically equilibrated or degassed at the occasion of their final assemblage are described elsewhere in the literature (e.g. International Consortium, 1974; Mark *et al.*, 1973; Albee *et al.*, 1974). As noted before (Kirsten *et al.*, 1973b), it seems as if impacts are not effective in completely extinguishing radiometric records from the time before. Resetting seems to us of low probability, especially if the rocks are found outside the crater or basin rim where material of the Bunte Brekzie-type will mainly be found (Oberbeck *et al.*, 1973; Dence *et al.*, 1974; Hörz *et al.*, 1974). This material consists of weakly to moderately shocked debris intermixed with appreciable or even dominant amounts of material of local origin (*ibid.*). According to Dence *et al.* (1974), the vast majority of impact melt and thermally highly affected rocks are to be found inside the basins—as it is the case in terrestrial impact craters. In the case of the lunar nearside basins these melt rocks are now mostly covered by mare basalts and therefore have not been sampled at the Apollo or Luna landing sites.

Chao *et al.* (1972) give mineralogical support to the idea that the effects of basin-forming impacts on the involved rocks have not been as catastrophic as supposed. These authors show that the various metamorphic effects recorded in single Apollo 14 breccias cannot be due to just one event (here, the Imbrium impact). Similarly, variations of metamorphic grades were observed in other lunar highland breccias, e.g. 77017 (Helz and Appleman, 1974) or Apollo 17, Boulder 1 (Consortium Indomitabile, 1974).

The above-mentioned arguments allow one to deduce only upper age limits for lunar highland formations. Then, in turn, if the rocks collected are basin ejecta (e.g. Apollo 14 rocks from Imbrium Basin, Sutton *et al.*, 1972; Apollo 16 rocks from Orientale Basin, Chao *et al.*, 1973; Apollo 17 rocks from Serenitatis Basin, ALGIT, 1973) upper age limits for these impacts are deduced.

There are several objections published against the notion that the highland rocks are basin ejecta (Schonfeld and Meyer, 1973; Dence *et al.*, 1974; Oberbeck *et al.*, 1974; Stöffler *et al.*, 1974). Due to these objections and especially because the considerations of Schonfeld and Meyer (1973) are in conflict with our own interpretation of lunar highland rock ages (Kirsten *et al.*, 1973b), we reconsidered the age data in the light of the new evidence.

We have argued that the large impacts did little in the way of defining the measured rock ages. Therefore, it could well be that one only measures "survival" ages. This means that at a time when the flux of impacting projectiles has decreased in intensity below a critical value, a rapidly increasing number of rocks escapes isotopic reequilibration or degassing. Resetting before or at that time occurred mostly within smaller craters (of sizes larger than approximately 200 m)—as it is inside of a crater where kinetic energy is converted into thermal energy. As indicated by crater statistics and other arguments (Hartmann and Wood, 1971), some time must have elapsed between the formation of the various nearside impact basins. If basin formation affected the rock ages that are now measured, the differences in the ages of the basins should show up in an age histogram of lunar highland rocks; if the large events are resolvable in time, then

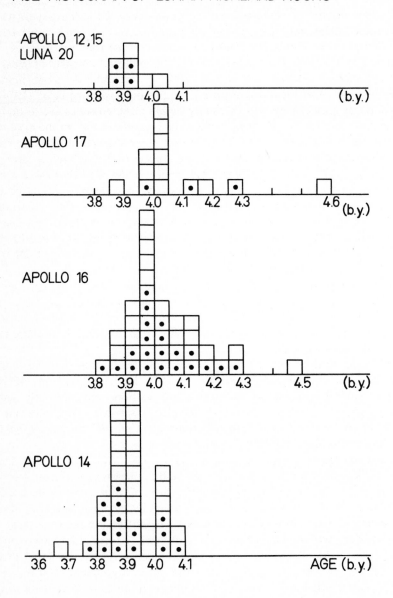

AGE HISTOGRAM OF LUNAR HIGHLAND ROCKS

the histogram should show distinct peaks; if they are not resolvable or if only survival ages are measured, one peak should be the result. The shape of the histogram would critically depend on the shape of the flux curve for the projectiles (Hartung and Neukum, private communications). Age histograms for highland rocks from various lunar landing sites are shown in Fig. 8. The way how the histogram is derived is explained in the figure caption. Although it appears as if the peaks in the graphs for the Apollo 14 (15), 16, and 17 rocks seem to lie at slightly different ages (3.9 b.y. for Apollo 14; 4.0 b.y. for Apollo 16 and 17) they are statistically not readily resolvable. Statistical significance tests show this very clearly—even when the intervals are spaced at 50 m.y. as in Fig. 8. If the actual errors of the ages considered and intervals of 100 m.y. are chosen, the age differences mentioned above would almost disappear.

Interlaboratory bias in measured ages or age differences due to different dating technics (Rb/Sr versus ^{39}Ar–^{40}Ar) are generally small and lie well within the errors given for the ages if one and the same rock is dated by different laboratories and technics (see also caption to Fig. 8). The histogram is not easy to interpret. Provided that the slight age differences among Apollo 14 (and 15), 16, and 17 rocks are significant one has to decide whether one has to take the maxima of the respective distributions to assign an age to the respective formations (Model A) or whether one has to take the younger end of the curves to infer *upper* age limits for the formations (Model B).

As a third possibility one may regard the higher ages of any of the distributions as *lower* limits for the time of deposition of the formations in question (Model C). This would mean that at the time given by the highest rock age, or before, the respective formation was deposited. Subsequent bombardment annealed the radiometric records almost completely and only very few rocks survived (terrestrial analogs are sedimentary formations which have been metamorphosed after deposition).

All three models would be consistent with the idea that at the times of the large impacts rocks were excavated from extreme depths where temperatures were high. Only after excavation they began to accumulate daughter isotopes because the rocks were then isotopically closed systems.

Fig. 8. Histograms of lunar highland rock ages. Included are almost all dated highland rocks for which the stated errors are ≤0.1 b.y. A mean error of approximately ± 50 m.y. for the ages has to be considered. The ages are ^{39}Ar–^{40}Ar gas retention ages, whole-rock Rb–Sr isochron ages for metamorphic breccias and internal Rb–Sr isochron ages and one Pb–Pb isochron age for homophaneous rocks or rock clasts. Petrologically different clasts from polymict breccias are plotted separately. If more than one age is available for a given sample, arithmetic means were used. Dots indicate fragments from coarse fines 2–4 mm. Data sources are not explicitly stated but all of the data are extracted from publications contained in the Reference list of this paper. The concordance of ^{39}Ar–^{40}Ar plateau ages and Rb–Sr internal isochron ages for a given rock is proved to be very good (see, for example, comparisons by Stettler *et al.* (1973), Tera *et al.* (1974), Kirsten and Horn, unpublished). When metamorphosed polymict breccias are considered, care has to be taken that only ages on material of *one* petrological type are compared in judging concordance.

The ages obtained for the formations would then be:

Formation	Model A (b.y.)	Model B (b.y.)	Model C (b.y.)
Apollo 14 (15) Fra Mauro	3.85–3.95	≤ 3.80	≥ 4.1
Apollo 16 Cayley	3.95–4.05	≤ 3.85	≥ 4.3
Apollo 17 Massifs and Sculptured Hills	4.00–4.05	≤ 3.95	≥ 4.3

As we have discussed in detail, we like to prefer the ages obtained for the formations by Model B. We are aware (see discussion above) that it is unclear whether the formations are related to the large multiring basins in that the material from within the formations is basin derived. If for example, on the basis of proximity we relate formations and basins (and use their relative ages; Hartmann and Wood, 1971) we arrive at ages of ≤ 3.8 b.y. for the Imbrium Basin, 3.8–3.85 b.y. for the Nectaris Basin and 3.85–3.95 b.y. for the Serenitatis Basin (but see the objections to such young ages for the Imbrium Basin by Schonfeld and Meyer (1973) and by Stettler *et al.* (1974), the latter pointing out that an 3.92 b.y. old near surface mare basalt (Apollo 11, rock 10003) conflicts with the Imbrium event as being younger).

The lower age limit for the Imbrium Basin (and Orientale Basin, too) is in any case given by the oldest mare surface which in turn is dated by the uppermost (youngest) mare rocks there. If rock fragment 74243,4,B (Table 2) is a sample of an upper basalt layer from the ancient mare in the Taurus-Littrow Valley, the lower age limit for the Imbrium impact would be 3.78 ± 0.04 b.y. The ages for the events given here are somewhat (~ 50 m.y.) different from the ages obtained if only data from our laboratory are used (Kirsten *et al.*, 1973b).

Summing up we conclude: (1) There are only *very slight* indications that various lunar formations differ in their radiometric ages. (2) These age differences might be due to local events. It is by no means clear that one implicitly has resolved the ages for various large multiring impact basins, not even those for the basins which have extreme relative ages (Orientale or Imbrium and Serenitatis basins); (3) the unorthodox "Old Imbrium Hypothesis" (Schonfeld and Meyer, 1973) is neither proved nor disproved by the available age data. (4) No decision can be made at present when the large basin-forming impacts occurred; it can only be stated that the basins are older than ~ 3.8 b.y. Whether they were formed almost contemporaneously or spread over the time interval of 4.6–3.8 b.y. cannot be decided by an unprejudiced interpretation of the available age data.

At the moment it is not possible to figure out a correlation of rock ages and rock types. This is in part due to the fact that the lunar rock nomenclature is in the same chaotic disorder as is the terrestrial rock nomenclature. Informations like, for example, chemical data which possibly allow to correlate basin ejecta with different projectiles (Morgan *et al.*, 1974) are needed for a sufficiently large number of dated rocks. This would help to find a more definite answer to the question what the inferences to be drawn from lunar rock ages really are.

Acknowledgments—We thank NASA and especially M. Duke for providing us with lunar samples. We appreciate the skillful operation of the mass spectrometer by H. Richter. J. Kiko helped in data processing and J. Janicke in microprobe analysis. Technical assistance was given by H. Lämmler, R. Schwan, O. Stadler, and H. Weber. H. Urmitzer helped in styling and typed the manuscript. Discussions with A. ElGoresy, J. Hartung, A. Hoffmann, E. Jessberger, O. Müller, G. Neukum, O. A. Schaeffer, and E. Schonfeld are acknowledged. We thank the reviewers J. Huneke and L. Husain for some valuable comments.

References

Albee A., Chodos A., Dynek R., Gancarz A., Goldman D., Papanastassiou D., and Wasserburg G. (1974) Dunite from the lunar highlands: Petrography, deformational history, Rb–Sr-age (abstract). In *Lunar Science—V*, pp. 3–5. The Lunar Science Institute, Houston.

ALGIT (Apollo Lunar Geology Investigation Team) U.S. Geological Survey (1973) Preliminary geologic analysis of the Apollo 17 site. *Interagency Report: Astrogeology* 72, U.S. Geol. Survey, March 17.

Apollo 17 Preliminary Sci. Report, NASA publication SP-330, pp. 6-1 to 6-91.

Bogard D. D., Funkhouser J. G., Schaeffer O. A., and Zähringer J. (1971) Noble gas abundances in lunar material—cosmic ray spallation products and radiation ages from the Sea of Tranquillity and the Ocean of Storms. *J. Geophys. Res.* 76, 2757–2779.

Brereton N. (1970) Corrections for interfering isotopes in the ^{40}Ar–^{39}Ar dating method. *Earth Planet. Sci. Lett.* 8, 427–433.

Brown G. M., Peckett A., Emeleus C. H., and Phillips R. (1974) Mineral-chemical properties of Apollo 17 mare basalts and terra fragments (abstract). In *Lunar Science—V*, pp. 89–91. The Lunar Science Institute, Houston.

Chao E. C. T. (1973) The petrology of 76055,10, a thermally metamorphosed fragment-laden olivine micronorite hornfels. *Proc. Fourth Lunar Sci. Conf., Geochim. Cosmochim. Acta*, Suppl. 4, Vol. 1, pp. 719–732. Pergamon.

Chao E. C. T., Minkin J. A., and Best J. B. (1972) Apollo 14 breccias: General characteristics and classification. *Proc. Third Lunar Sci. Conf., Geochim. Cosmochim. Acta*, Suppl. 3, Vol. 1, pp. 645–659. MIT Press.

Chao E. C. T., Soderblom L., Boyce J., Wilhelms Don E., and Hodges C. (1973) Lunar light plains deposits (Cayley Formation)—a reinterpretation of origin (abstract). In *Lunar Science—IV*, pp. 127–128. The Lunar Science Institute, Houston.

Consortium Indomitabile—J. A. Wood (1974) Investigations of a KREEPy stratified boulder from the South Massif. In *Lunar Science—V*, Suppl. A, pp. 4–6. The Lunar Science Institute, Houston.

Dence M., Grieve R., and Plant A. (1974) The Imbrium Basin and its ejecta (abstract). In *Lunar Science—V*, pp. 165–167. The Lunar Science Institute, Houston.

El Goresy A., Ramdohr P., Medenbach O., and Bernhardt H. J. (1974) Taurus-Littrow TiO_2-rich basalts: Opaque mineralogy and geochemistry. *Proc. Fifth Lunar Sci. Conf., Geochim. Cosmochim. Acta.* Volume 1.

Evernden J. and Curtis G. (1965) Potassium–argon dating of late Cenozoic rocks in East Africa and Italy. *Current Anthropology* 6, 343–385.

Hartmann W. K. and Wood C. A. (1971) Moon: Origin and evolution of multiring basins. *The Moon* 3, 3–78.

Helz R. and Appleman D. (1974) Poikilitic and cumulate textures in rock 77017, a crushed anorthositic gabbro (abstract). In *Lunar Science—V*, pp. 322–324. The Lunar Science Institute, Houston.

Hörz F., Oberbeck V., and Morrison R. (1974) Remote sensing of the Cayley Plains and Imbrium Basin deposits (abstract). In *Lunar Science—V*, pp. 357–359. The Lunar Science Institute, Houston.

Horn P., Lippolt H. J., and Todt W. (1972) Kalium–Argon–Altersbestimmungen an tertiären Vulkaniten des Oberrheingrabens I. Gesamtgesteinsalter. *Eclogae geol. Helv.* **65** (1), 131–156.

Hübner W., Kirsten T., and Kiko J. (1974a) Rare gases in Apollo 17 soils with emphasis of size and mineral fractions of soil 74241 (abstract). In *Lunar Science—V*, pp. 369–371. The Lunar Science Institute, Houston.

Hübner W., Kirsten T., and Kiko J. (1974b) Rare gases in Apollo 17 soils with emphasis on analysis of size and mineral fractions of soil 74241. To be published.

Huneke J., Jessberger E., Podosek F., and Wasserburg J. (1973) ^{40}Ar/^{39}Ar measurements in Apollo 16 and 17 samples and the chronology of metamorphic and volcanic activity in the Taurus-Littrow region. *Proc. Fourth Lunar Sci. Conf., Geochim. Cosmochim. Acta*, Suppl. 4, Vol. 2, pp. 1725–1756. Pergamon.

Husain L. (1974) ^{40}Ar–^{39}Ar chronology and cosmic ray exposure ages of the Apollo 15 samples. *J. Geophys. Res.* **79**, 2588–2606.

Husain L. and Schaeffer O. A. (1973) Lunar volcanism: Age of the glass in the Apollo 17 orange soil. *Science* **180** (4093), 1358–1360.

International Consortium—E.C.T. Chao (1974) First results of consortium study of Apollo 17 station 7 boulder samples (abstract). In *Lunar Science—V*, Suppl. A., pp. 1–3. The Lunar Science Institute, Houston.

Jessberger E., Huneke J., and Wasserburg G. J. (1974) Evidence for a ~4.5 AE age of plagioclase clasts in a lunar highland breccia. *Nature* **248**, 199–201.

Kirsten T. and Horn P. (1974) ^{39}Ar–^{40}Ar chronology of the Taurus-Littrow Region II: A 4.28 b.y. old troctolite (abstract). In *Lunar Science—V*, pp. 419–421. The Lunar Science Institute, Houston.

Kirsten T., Deubner J., Horn P., Kaneoka I., Kiko J., Schaeffer O. A., and Thio S. K. (1972) The rare gas record of Apollo 14 and 15 samples. *Proc. Third Lunar Sci. Conf., Geochim. Cosmochim. Acta*, Suppl. 3, Vol. 2, pp. 1865–1889. MIT Press.

Kirsten T., Horn P., and Heymann D. (1973a) Chronology of the Taurus-Littrow Region I: Ages of two major rock types from the Apollo 17 site. *Earth Planet. Sci. Lett.* **20**, 125–130.

Kirsten T., Horn P., and Kiko J. (1973b) ^{39}Ar–^{40}Ar dating and rare gas analysis of Apollo 16 rocks and soils. *Proc. Fourth Lunar Sci. Conf., Geochim. Cosmochim. Acta*, Suppl. 4, Vol. 2, pp. 1757–1784. Pergamon.

Lämmerzahl P. and Zähringer J. (1966) K–Ar Altersbestimmungen an Eisenmeteoriten II. Spallogenes Ar^{40} und Ar^{40}–Ar^{39} Bestrahlungsalter. *Geochim. Cosmochim. Acta* **30**, 1059–1074.

Mark R., Cliff B., Lee-Hu C., and Wetherill G. (1973) Rb–Sr studies of lunar breccias and soils. *Proc. Fourth Lunar Sci. Conf., Geochim. Cosmochim. Acta*, Suppl. 4, Vol. 2, pp. 1785–1796. Pergamon.

McGetchin T. R., Settle M., and Head J. W. (1973) Radial thickness variation in impact crater ejecta: Implications for lunar basin deposits. *Earth Planet. Sci. Lett.* **20**, 226–250.

Morgan J. W., Ganapathy R., Higuchi H., and Anders E. (1974) Lunar basins: Characterization of projectiles from meteoritic trace elements in highland samples (abstract). Submitted to: Proc. Soviet American Conference on Cosmochemistry of the Moon and Planets, Moscow.

Murthy V. Rama, Evensen N. M., Bor-Ming Jahn, and Coscio M. R. (1971) Rb–Sr ages and elemental abundances of K, Rb, Sr, and Ba in samples from the Ocean of Storms. *Geochim. Cosmochim. Acta* **35**, 1139–1153.

Neukum G., König B., and Arkani-Hamed J. (1974) Effects of local processes on lunar crater distributions. To be published.

Nunes P., Tatsumoto M., Knight R., Unruh D., and Doe B. (1973) U–Th–Pb systematics of some Apollo 16 lunar samples. *Proc. Fourth Lunar Sci. Conf., Geochim. Cosmochim. Acta*, Suppl. 4, Vol. 2, pp. 1797–1822. Pergamon.

Oberbeck V. R., Hörz F., Morrison R. H., Quaide W. L., and Gault D. E. (1974) Effects of formation of large craters and basins on emplacement of smooth plain materials (abstract). In *Lunar Science—V*, pp. 568–570. The Lunar Science Institute, Houston.

Papanastassiou D. A. and Wasserburg G. J. (1971) Rb–Sr ages of igneous rocks from the Apollo 14 mission and the age of the Fra Mauro formation. *Earth Planet. Sci. Lett.* **12**, 36–48.

Papike J. J. and Bence A. E. (1974) Basalts from the Taurus-Littrow region of the Moon (abstract). In *Lunar Science—V*, pp. 586–588. The Lunar Science Institute, Houston.

Podosek F., Huneke J., Gancarz A., and Wasserburg G. J. (1973) The age and petrography of two Luna 20 fragments and inferences for widespread lunar metamorphism. *Geochim. Cosmochim. Acta* **37**, 887–904.

Schaeffer O. A. and Husain L. (1973) Early lunar history: Ages of 2–4 mm soil fragments from the lunar highlands. *Proc. Fourth Lunar Sci. Conf., Geochim. Cosmochim. Acta,* Suppl. 4, Vol. 2, pp. 1847–1863. Pergamon.

Schaeffer O. A. and Husain L. (1974) Chronology of lunar basin formation and ages of lunar anorthositic rocks (abstract). In *Lunar Science—V*, pp. 663–665. The Lunar Science Institute, Houston.

Schonfeld E. and Meyer C., Jr. (1973) The old Imbrium hypothesis. *Proc. Fourth Lunar Sci. Conf., Geochim. Cosmochim. Acta,* Suppl. 4, Vol. 1, pp. 125–138. Pergamon.

Scott D. H., Lucchita B. K., and Carr M. H. (1972) Geologic maps of the Taurus-Littrow region of the moon. U.S. Geol. Survey Misc. Geol. Inv. Map I-800.

Stettler A., Eberhardt P., Geiss J., Grögler N., and Maurer P. (1973) ^{39}Ar–^{40}Ar ages and ^{37}Ar–^{38}Ar exposure ages of lunar rocks. *Proc. Fourth Lunar Sci. Conf., Geochim. Cosmochim. Acta,* Suppl. 4, Vol. 2, pp. 1865–1888. Pergamon.

Stettler A., Eberhardt P., Geiss J., Grögler N., and Maurer P. (1974) Sequence of rock formation and basaltic lava flows on the moon (abstract). In *Lunar Science—V*, pp. 738–740. The Lunar Science Institute, Houston.

Stöffler D., Dence M. R., Abadian M., and Graup G. (1974) Ejecta formations and pre-impact stratigraphy of lunar and terrestrial craters: Possible implications for the ancient lunar crust (abstract). In *Lunar Science—V*, pp. 746–748. The Lunar Science Institute, Houston.

Sutton R., Hait M., and Swann G. (1972) Geology of the Apollo 14 landing site. *Proc. Third Lunar Sci. Conf., Geochim. Cosmochim. Acta,* Suppl. 3, Vol. 1, pp. 27–38. MIT Press.

Tatsumoto M., Nunes P., Knight R., and Unruh D. (1974) Rb–Sr and U–Th–Pb systematics of Boulders 1 and 7, Apollo 17 (abstract). In *Lunar Science—V*, pp. 774–776. The Lunar Science Institute, Houston.

Tera F., Papanastassiou D., and Wasserburg G. (1974) Isotopic evidence for a terminal lunar cataclysm. *Earth Planet. Sci. Lett.* **22**, 1–21.

Turner G. (1970) Argon 40–Argon 39 dating of lunar rock samples. *Proc. Apollo 11 Lunar Sci. Conf., Geochim. Cosmochim. Acta,* Suppl. 1, Vol. 2, pp. 1665–1684. Pergamon.

Turner G. (1971) ^{40}Ar–^{39}Ar ages from the lunar maria. *Earth Planet. Sci. Lett.* **11**, 169–191.

Turner G., Cadogan P. H., and Yonge C. J. (1973) Argon selenochronology. *Proc. Fourth Lunar Sci. Conf., Geochim. Cosmochim. Acta,* Suppl. 4, Vol. 2, pp. 1889–1914. Pergamon.

York D., Kenyon W., and Doyle R. (1972) ^{40}Ar–^{39}Ar ages of Apollo 14 and 15 samples. *Proc. Third Lunar Sci. Conf., Geochim. Cosmochim. Acta,* Suppl. 3, Vol. 2, pp. 1613–1622. MIT Press.

Proceedings of the Fifth Lunar Conference
(Supplement 5, Geochimica et Cosmochimica Acta)
Vol. 2 pp. 1477–1485 (1974)
Printed in the United States of America

Equilibration and ages: Rb–Sr studies of breccias 14321 and 15265

ROBERT K. MARK, CHIN-NAN LEE-HU, and GEORGE W. WETHERILL

Department of Planetary and Space Science, University of California, Los Angeles,
California 90024

Abstract—We have previously demonstrated that the Rb–Sr system of Fra Mauro breccia 14321 did not completely equilibrate at the time of assembly. Basalt and microbreccia clasts, with approximately the same internal isochron ages (~4.0 b.y.) have distinctly different initial ratios. This result did not preclude the possibility that individual clast ages were reset by short range diffusion in a hot ejecta blanket. We now demonstrate that equilibration between basalt and adjacent microbreccia did not occur at ~4 b.y. on a 1 mm scale. The internal isochron ages of the clasts therefore probably date pre-assembly (pre-Imbrium?) magmatic and impact events. We also report the presence of a mare basalt clast (3.16 ± 0.11 b.y., 0.69945 ± 3) in relatively KREEP-rich breccia 15265, indicating formation of such breccias after the basin-forming events.

INTRODUCTION

WE HAVE CONTINUED our Rb–Sr studies of lunar breccias in order to further elucidate the Imbrium and pre-Imbrium bombardment and magmatic history of the moon.

The interpretation of Rb–Sr chronologic data from lunar breccias requires an evaluation of the effects of thermal metamorphism during the assembly of these rocks. The breccias are presumed to be the result of impact processes, produced either by shock welding or by deposition in a hot or cold ejecta blanket (e.g. Christie *et al.*, 1973; LSPET, 1971; Dence and Plant, 1972).

A study of Rb–Sr systematics can provide information about the conditions of assembly as well as temporal information. In a thoroughly thermally metamorphosed breccia the clast whole rock isochron would be identical with clast internal isochrons and would date the impact and metamorphism. In an unmetamorphosed breccia each clast internal isochron may date some pre-assembly igneous or metamorphic (impact-deposition) event. It is possible that some breccias are only partially equilibrated in a hot ejecta blanket by short-range diffusion (Compston *et al.*, 1972; Mark *et al.*, 1973). In such a case the clasts may each exhibit internal isochrons of the same age (i.e. of the metamorphism), but with different initial isotopic ratios. The clast whole rocks would not be expected to define an isochron. It is likely that breccias exhibiting the entire range of metamorphic gradation exist on the lunar surface.

Several igneous and microbreccia clasts from probable Fra Mauro breccia samples (Imbrium ejecta blanket) have similar Rb–Sr ages of ~ 4.0 b.y. (Pananastassiou and Wasserburg, 1971a; Compston *et al.*, 1971, 1972; Mark *et al.*, 1973). Compston *et al.* (1972) proposed that these are metamorphic ages, dating diffusion within the hot ejecta blanket. We have previously demonstrated that, in the case of polymict breccia 14321, complete Rb–Sr equilibration of the whole rock did not occur at ~4 b.y. (Mark *et al.*, 1973).

14321, the largest rock sample returned by Apollo 14, was collected on the rim of Cone Crater and is interpreted to be a Fra Mauro breccia ejected by the Cone Crater impact. It is composed of multigenerational microbreccia clasts of Warner's group ≤6 (Warner and Heiken, 1972; Grieve *et al.*, 1974) and igneous clasts, dominantly of an atypical mare affinity (Duncan *et al.*, 1974a), in a light colored, heterogeneous matrix.

An understanding of the assembly history of 14321 is tied to the nature of the light matrix. The extent to which this matrix has been recrystallized is not adequately determined. Christie (personal communication) reports observing unrecrystallized glass-bonded porous matrix material with an electron microscope, while Warner and Heiken (1972) report metamorphism to Warner's group 4 (medium grade metamorphism).

Rb–Sr ages of the igneous clasts in the range 3.9–4.0 b.y. have been measured by several laboratories and have been interpreted both as crystallization ages and as ages reset by diffusion in a hot, Imbrium ejecta blanket (Papanastassiou and Wasserburg, 1971a; Compston *et al.*, 1971, 1972; Mark *et al.*, 1973). We have reported a similar age for a microbreccia clast, but with an initial Sr isotopic ratio distinct from the igneous clasts. Thus 14321 could not have totally reequilibrated at that time. This result does not preclude the possibility that short-range diffusion reset internal isochrons within the clasts. In this paper we report further work indicating that this was not the case, and interpret these ages as predating the final assembly and possible metamorphism of 14321.

15265 is a relatively KREEP-rich breccia collected at the Apennine Front. We report data demonstrating that it was assembled after 3.16 b.y., significantly later than the basin-forming impacts.

ANALYTICAL TECHNIQUE

Rb, Sr, and K concentrations and Sr isotopic compositions were measured using a nine-inch, single filament mass spectrometer with automatic magnetic peak switching and on-line digital data acquisition system. Sr isotopic data are normalized to $^{86}Sr/^{88}Sr = 0.1194$ and adjusted to a value of 0.71014 for NBS SRM 987. Our currently measured value for this standard, during this study, (88 sets of 10 ratios) is 0.71014 with a standard deviation of 0.00023 and standard error of the mean of 0.000024. Errors reported for Sr isotopic measurements are the standard error of the mean among data sets of 10 ratios each. The ^{84}Sr tracer is NBS SRM 985. The ^{87}Rb tracer, previously calibrated against Matthey "SpecPure" RbCl, is now calibrated against NBS SRM 984, with no significant difference. The random thermal fractionation error in $^{87}Rb/^{86}Sr$, using a Re filament with silica gel for Rb measurements, is ≤1%. Sample blanks are typically 0.8 ng Sr and 0.1–0.2 ng Rb.

14321

Sample 14321,478 (parent 184, Goles Consortium sample) was a 295 mg fragment containing a basalt clast and adjacent microbreccia (Fig. 1). Rb–Sr measurements on the basalt and contiguous microbreccia material are reported in Table 1 and Fig. 2.

Figure 2 illustrates the basalt mineral isochron (4.0 ± 0.12 b.y., 0.69957 ± 13, 2 σ York errors) and microbreccia clast internal isochron (4.05 ± 0.06 b.y., 0.70104 ± 33) previously reported by Mark *et al.*

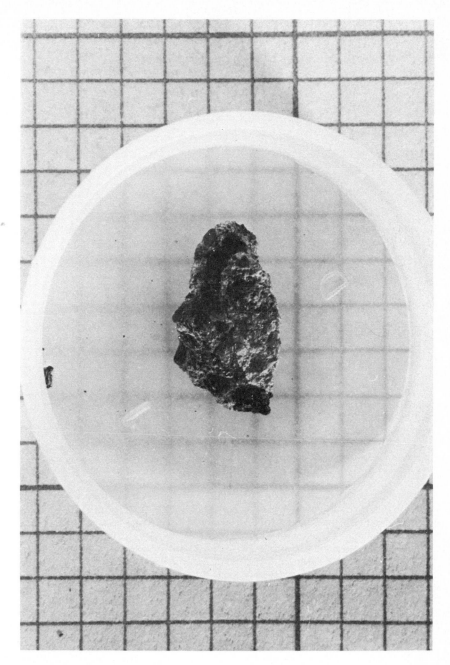

Fig. 1. Sample 14321,478. Note basalt clast on right side. Scale is 2.5 mm/div.

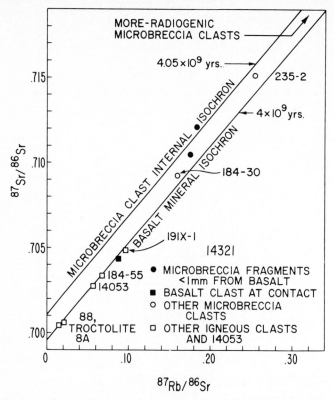

Fig. 2. Isochron diagram illustrating basalt and microbreccia internal isochrons from Mark *et al.* (1973). Additional igneous clast whole rock points (basalt and troctolite from Papanastassiou and Wasserburg (1971a), Compston *et al.* (1972), and Mark *et al.* (1973)) plot on the basalt internal isochron. Additional microbreccia clasts plot above it. The basalt from 14321,478 plots on this isochron while the adjacent microbreccia fragments plot distinctly above it, indicating that equilibration did not occur at the 1 mm scale 4 b.y. ago.

(1973). Other 14321 igneous clasts (basalt and troctolite) also plot on the basalt isochron (Papanastassiou and Wasserburg, 1971a; Compston *et al.*, 1972; Mark *et al.*, 1973). Other microbreccia clasts plot distinctly above the basalt isochron. Basalt 14053 is included for reference because of its chemical and age similarity to the basalt clasts (Duncan *et al.*, 1974a; Papanastassiou and Wasserburg, 1971a). If the basalt clast and adjacent microbreccia material had equilibrated at about 4 b.y., the three filled points (Fig. 2) would plot on a 4 b.y. isochron. This is not the case; the basalt shows no deviation from the basalt mineral isochron, and the microbreccia plot distinctly above it. The result indicates that equilibration did not occur on the 1 mm scale after the assembly of 14321. It thus seems unlikely that 14321 is a block of Fra Mauro breccia that underwent sufficient thermal metamorphism in a hot ejecta blanket to reset the internal isochron ages as proposed by Compston *et al.* (1972); it is more likely that the internal isochron ages of the clasts predate the final assembly of 14321.

15265

Sample 15265 is a relatively friable and moderately KREEP-rich breccia (LSPET, 1972) collected at the Apennine Front. The sample contains high siderophile element concentrations (Ganapathy *et al.*, 1973)

Table 1. Rb, Sr, K concentrations and Sr isotopic composition.

Sample	Weight (mg)	K (ppm)	Rb (ppm)	Sr (ppm)	K/Rb (weight)	$^{87}Rb/^{86}Sr$ (atomic)	$^{87}Sr/^{86}Sr \pm 2\,\sigma$
14321,478							
basalt calst							
frag. (F)	6	1225	2.93	95.97	418	0.0879	0.70433 ∓ 10
microbreccia							
frag. (O)	17	4546	11.68	173.3	389	0.194	0.71207 ± 8
microbreccia							
frag. (R)	5	4738	11.17	182.5	424	0.176	0.71046 ± 12
15265,9005							
basalt clast							
plagioclase (H)	9	487	0.461	289.8	1056	0.00458	0.69965 ± 6
"whole rock" (F)	22	337	0.743	98.06	507	0.0218	0.70042 ± 8
"clinopyroxene" (I)	31	338	0.762	63.79	444	0.0343	0.70102 ± 10
"clinopyroxene" (I')	27	359	0.809	65.88	444	0.0353	0.70101 ± 6
15265,9009							
breccia matrix (B)	32	2188	6.96	142.0	314	0.140	0.70771 ± 22

and the matrix contains unrecrystallized glass fragments. K concentration and Rb–Sr systematics of a 30 mg matrix sample (Table 1) fall within the range reported for Apollo 15 soils (Papanastassiou and Wasserburg, 1972b). A $\sim\frac{1}{2}$ gm basalt clast was removed from the breccia (Burlingame Consortium sample). Plagioclase and clinopyroxene separates were prepared using heavy liquids. A mineral isochron of 3.16 ± 0.11 b.y., $I = 0.69945 \pm 3$ was obtained (Fig. 3). These values are typical for a mare basalt, being somewhat closer to those reported for Apollo 12 rather than Apollo 15 basalts (Papanastassiou and Wasserburg, 1971b, 1972a). Stettler et al. (1973) report a similar K–Ar age from a basalt clast in 15459, also collected at the Apennine Front. Such breccias must have been assembled after the Imbrium event.

Discussion

The Fra Mauro Formation, underlying the Apollo 14 site, is composed of Imbrium ejecta, perhaps mixed with local material and modified by Orientale ejecta and subsequent cratering (e.g. McGetchin et al., 1973; Oberbeck et al., 1974). The unit may have been deposited as a hot ejecta blanket, undergoing thermal autometamorphism (e.g. LSPET, 1971; Compston et al., 1972; Warner, 1972; Williams, 1972; Wilshire and Jackson, 1972), or alternatively, may have been deposited as cool, poorly consolidated material, analogous to the Bunte breccia of the Ries Crater (Dence and Plant, 1972; Chao, 1973). In the former case the coherent Apollo 14 breccia samples may be blocks of the Fra Mauro Formation (Wilshire and Jackson, 1972) and in the latter, clasts within it. The degree of recrystallization of the breccia matrices should bear upon this question.

If 14321 is a Fra Mauro breccia, then it is likely that the ages retained by the clasts date earlier, pre-Imbrium igneous and impact events. The clasts did not equilibrate among themselves in a hot ejecta blanket ~ 4 b.y. ago. Several other

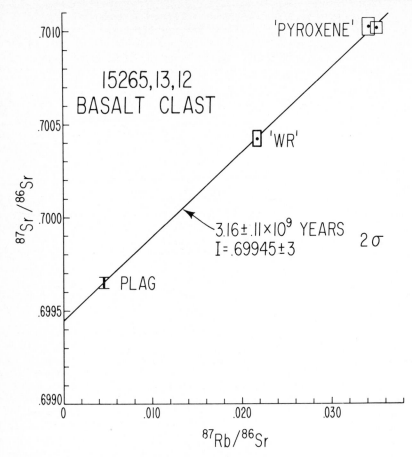

Fig. 3. Isochron diagram for a basaltic clast in relatively KREEP-rich breccia 15265, collected at the Apennine Front. The age and initial ratio is consistent with data from mare basalts.

alternatives for the history of 14321 warrant consideration in interpreting the geochronologic results.

One possibility is that the Fra Mauro Formation is analogous to the Bunte breccia of the Ries Crater; that is, was deposited cold and is only weakly lithified. The coherent Apollo 14 breccia samples are thus preexisting material, now clasts in the Fra Mauro Formation (Dence and Plant, 1972; Chao, 1973). Chao's hypothesis was based, in part, on the low seismic velocity reported for the Fra Mauro Formation and may underestimate the effect of fracturing and heterogeneity on reduction of such velocities. In this case, the internal isochron ages would certainly predate the Imbrium event.

A second alternative suggests that 14321 was assembled or reworked by a post-Imbrium impact at the Apollo 14 site. The light matrix of 14321 was apparently produced from the associated clasts, without addition of an additional extra-lunar siderophile element component (Duncan *et al.*, 1974a). Thus 14321 is apparently not a regolith breccia. Several craters older than Cone Crater and up to 1 km in diameter occur in the immediate vicinity of Cone Crater (Swann *et al.*, 1971). It is possible that 14321 was assembled by one of these minor post-Imbrium impacts that penetrated the Fra Mauro Formation (Duncan *et al.*, 1974b) or even by the Cone Crater event itself. A variant of this model is that rather than being assembled, 14321 was remobilized by this latter shock in a manner analogous to that proposed by Wilshire *et al.* (1973) for the Apollo 16 breccias.

Evidence has been presented (Gose *et al.*, 1972; Hargraves and Dorety, 1972; Pearce *et al.*, 1972) indicating that 14321 has a consistent thermoremanent magnetization suggesting that the entire rock was heated briefly (i.e. not long enough for Rb–Sr equilibration), to a relatively high temperature in the presence of a magnetic field. It is unlikely that such a field was present ~25–30 m.y. ago when Cone Crater was formed (e.g. Turner *et al.*, 1971; Lugmair and Marti, 1972). If 14321 is reworked Fra Mauro material, then the basalt clasts could be fragments of post-Imbrium basalt flows incorporated during assembly, or they could have been clasts in the Fra Mauro Formation (i.e. pre-Imbrium). The microbreccia clasts could be Fra Mauro matrix or Fra Mauro clasts. That is, the microbreccia clasts could date or predate the Imbrium event.

A third alternative proposes the deposition of 14321 at the future Cone Crater site by a post-Imbrium impact elsewhere on the lunar surface. The highest probability for such an event would be during the 3.9–4.0 b.y. "cataclysm," while the impacting flux was still high. The provenance is probably limited to KREEP-rich terrain, thereby suggesting a relatively local source, within the Fra Mauro Formation. If this is the case, the chronologic implications are similar to those for locally reworked Fra Mauro Formation. Otherwise, it indicates the existence of KREEP-rich, ~4 b.y. old breccia elsewhere on the lunar surface.

While it is generally accepted that the 3.9–4.0 b.y. "cataclysm" includes the Imbrium event, the actual age of this impact has not been unequivocally determined (e.g. Mark *et al.*, 1973). Although a preponderance of measured ages are in this interval, they could date earlier events, ages retained by clasts within the Imbrium ejecta (Dence and Plant, 1972; Chao, 1973; Mark *et al.*, 1973). We have demonstrated the likelihood that clasts within 14321, presumably a Fra Mauro breccia, retain ages of events which occurred prior to final assembly. If such clasts are commonly present in weakly metamorphosed distal facies of Imbrium ejecta, then the common occurrence of such material at the Apollo sites does not necessarily date the Imbrium event.

Ages reported for clasts in 14321 span the range 3.95–4.05 b.y. (Compston *et al.*, 1971, 1972; Papanastassiou and Wasserburg, 1972a; Mark *et al.*, 1973). We suggest that these ages probably predate the Imbrium impact. If 14310 was also a clast in the Fra Mauro Formation, this upper age limit can be set by the reported ages in the range 3.87–3.94 b.y. (Papanastassiou and Wasserburg, 1971a; Turner *et al.*, 1971; Compston *et al.*, 1972; Murthy *et al.*, 1972; York *et al.*, 1972; Mark *et al.*, 1974). A lower age limit can be set by the oldest mare basalts, which are stratigraphically younger than the Fra Mauro Formation. Ages of about 3.8 b.y.

have been measured for the subfloor basalt at Taurus Littrow (e.g. 3.82 ± 0.06 b.y., Evensen et al., 1973; 3.77 ± 0.06 b.y., Tera et al., 1974; $3.73 \pm .11$ b.y., Nyquist et al., 1974).

The occurrence of young mare basalt clasts in breccias such as 15265 and 15459 demonstrates that their formation occurred significantly later than the Imbrium event, and that relatively KREEP-rich breccias are associated with more minor impact events as well as with basin formation.

Acknowledgments—We are grateful for the helpful discussion with many of our colleagues in the lunar science program and the assistance of the curator's office at JSC. We thank M. Stein for construction and maintenance of the automated data acquisition system and A. D. Sharbaugh for assistance in the chemical preparation of samples. This work was supported under NASA Grant NGL 05-007-287.

REFERENCES

Chao E. C. T. (1973) Geologic implications of the Apollo 14 Fra Mauro breccias and comparison with ejecta from the Ries Crater, Germany. *J. Res. U.S. Geol. Survey* 1, 1–17.

Christie J. M., Griggs D. T., Hever A. H., Nord G. L., Jr., Radcliffe S. V., Lally J. S., and Fisher R. M. (1973) Electron petrography of Apollo 14 and 15 breccias and shock-produced analogs. *Proc. Fourth Lunar Sci. Conf., Geochim. Cosmochim. Acta*, Suppl. 4, Vol. 1, pp. 365–382. Pergamon.

Compston W., Vernon M. J., Berry H., and Rudowski R. (1971) The age of the Fra Mauro Formation: A radiometric older limit. *Earth Planet. Sci. Lett.* 12, 55–58.

Compston W., Vernon M. J., Berry J., Rudowski R., Gray C. M., Ware N., Chappell B. W., and Kaye M. (1972) Apollo 14 mineral ages and the thermal history of the Fra Mauro Formation. *Proc. Third Lunar Sci. Conf., Geochim. Cosmochim. Acta*, Suppl. 3, Vol. 2, pp. 1487–1501. MIT Press.

Dence M. R. and Plant A. G. (1972) Analysis of Fra Mauro samples and the origin of the Imbrium Basin. *Proc. Third Lunar Sci. Conf., Geochim. Cosmochim. Acta*, Suppl. 3, Vol. 1, pp. 379–399. MIT Press.

Duncan A. R., McKay S. M., Stoeser J. W., Lindstrom M. M., Lindstrom D. J., Fruchter J. S., and Goles G. G. (1974a) Lunar polymict breccia: A compositional study of principal components. *Geochim. Cosmochim. Acta*. In press.

Duncan A. R., Grieve R. A. F., and Weill D. F. (1974b) The life and times of Big Bertha: Lunar breccia 14321. *Geochim. Cosmochim. Acta*. In press.

Evensen N. M., Murthy V. R., and Coscio M. R., Jr. (1973) Rb–Sr ages of some mare basalts and the isotopic and trace element systematics in lunar fines. *Proc. Fourth Lunar Sci. Conf., Geochim. Cosmochim. Acta*, Suppl. 4, Vol. 2, pp. 1707–1724. Pergamon.

Ganapathy R., Morgan J. W., Krähenbühl V., and Anders E. (1973) Ancient meteoritic components in lunar highland rocks: Clues from trace elements in Apollo 15 and 16 samples. *Proc. Fourth Lunar Sci. Conf., Geochim. Cosmochim. Acta*, Suppl. 4, Vol. 2, pp. 1239–1261. Pergamon.

Gose W. A., Pearce G. W., Strangway D. W., and Larson E. E. (1972) Magnetic properties of Apollo 14 breccias and their correlation with metamorphism. *Proc. Third Lunar Sci. Conf., Geochim. Cosmochim. Acta*, Suppl. 3, Vol. 3, pp. 2387–2395. MIT Press.

Grieve R. A., McKay G. A., Smith H. D., and Weill D. F. (1974) Lunar polymict breccia 14321: A petrographic study. *Geochim. Cosmochim. Acta*. In press.

Hargraves R. B. and Dorety N. (1972) Natural remnent magnetization in lunar breccia 14321. *Proc. Third Lunar Sci. Conf., Geochim. Cosmochim. Acta*, Suppl. 3, Vol. 3, pp. 2417–2421. MIT Press.

LSPET (1971) Preliminary examination of lunar samples from Apollo 14. *Science* 173, 681–693.

LSPET (1972) The Apollo 15 lunar samples; a preliminary description. *Science* 175, 363–375.

Lugmair G. W. and Marti K. (1972) Exposure ages and neutron-capture record in lunar samples from Fra Mauro. *Proc. Third Lunar Sci. Conf., Geochim. Cosmochim. Acta*, Suppl. 3, Vol. 2, pp. 1891–1897. MIT Press.

Mark R. K., Cliff R. A., Lee-Hu C., and Wetherill G. W. (1973) Rb–Sr studies of lunar breccias and soils. *Proc. Fourth Lunar Sci. Conf., Geochim. Cosmochim. Acta*, Suppl. 4, Vol. 2, pp. 1785–1795. Pergamon.

Mark R. K., Lee-Hu C., and Wetherill G. W. (1974) Rb–Sr ages of lunar igneous rocks 62295 and 14310. *Geochim. Cosmochim. Acta*. In press.

McGetchin T. R., Settle M., and Head J. W. (1973) Radial thickness variations in impact crater ejecta: Implications for lunar basin deposits. *Earth Planet Sci. Lett.* **20**, 226–236.

Murthy V. R., Evensen N. M., Jahn B. M., and Coscio M. R., Jr. (1972) Apollo 14 and 15 samples: Rb–Sr ages, trace elements, and lunar evolution. *Proc. Third Lunar Sci. Conf., Geochim. Cosmochim. Acta*, Suppl. 3, Vol. 2, pp. 1503–1514. MIT Press.

Nyquist L. E., Bansal B. M., and Wiesmann H. (1974) Taurus-Littrow chronology: Implications for early lunar crustal development (abstract). In *Lunar Science—V*, pp. 565–567. The Lunar Science Institute, Houston.

Oberbeck V. R., Hörz F., Morrison R. H., Quaide W. L., and Gault D. E. (1974) Effects of formation of large craters and basins on emplacement of smooth plains material (abstract). In *Lunar Science—V*, pp. 568–570. The Lunar Science Institute, Houston.

Papanastassiou D. A. and Wasserburg G. J. (1971a) Rb–Sr ages of igneous rocks from the Apollo 14 mission and the age of the Fra Mauro Formation. *Earth Planet Sci. Lett.* **12**, 36–48.

Papanastassiou D. A. and Wasserburg G. J. (1971b) Lunar chronology and evolution from Rb–Sr studies of Apollo 11 and 12 samples. *Earth Planet. Sci. Lett.* **11**, 37–62.

Papanastassiou D. A. and Wasserburg G. J. (1972a) Rb–Sr ages and initial strontium in basalts from Apollo 15. *Earth Planet. Sci. Lett.* **17**, 324–337.

Papanastassiou D. A. and Wasserburg G. J. (1972b) Rb–Sr systematics of Luna 20 and Apollo 16 samples. *Earth Planet Sci. Lett.* **17**, 52–63.

Pearce G. W., Strangway D. W., and Gose W. A. (1972) Remnent magnetization of the lunar surface. *Proc. Third Lunar Sci. Conf., Geochim. Cosmochim. Acta*, Suppl. 3, Vol. 3, pp. 2449–2464. MIT Press.

Stettler A., Eberhardt P., Geiss J., Grögler N., and Maurer P. (1973) Ar^{39}–Ar^{40} ages and Ar^{37}–Ar^{38} exposure ages of lunar rocks. *Proc. Fourth Lunar Sci. Conf., Geochim. Cosmochim. Acta*, Suppl. 4, Vol. 2, pp. 1865–1888. Pergamon.

Swann G. A., Bailey N. G., Batson R. M., Eggleton R. E., Hait M. H., Holt H. E., Larson K. B., McEwen M. C., Mitchell E. D., Schaber G. G., Schaffer J. P., Shepard A. B., Sutton R. L., Trask N. J., Ulrich G. E., Wilshire H. G., and Wolfe E. W. (1971) Preliminary geologic investigations of the Apollo 14 landing site. In NASA SP-272 *Apollo 14 Prelim. Sci. Rept.*, 39–85. U.S. Govn. Printing Office.

Tera F., Papanastassiou D. A., and Wasserburg G. J. (1974) Isotopic evidence for a terminal lunar cataclysm. *Earth Planet. Sci. Lett.* **22**, 11–21.

Turner G., Huneke J. C., Podosek F. A., and Wasserburg G. J. (1971) ^{40}Ar–^{39}Ar ages and cosmic ray exposure ages of Apollo 14 samples. *Earth Planet. Sci. Lett.* **12**, 19–35.

Warner J. L. (1972) Metamorphism of Apollo 14 breccias. *Proc. Third Lunar Sci. Conf., Geochim. Cosmochim. Acta*, Suppl. 3, Vol. 1, pp. 623–643. MIT Press.

Warner J. and Heiken G. (1972) Metamorphism and surface mapping of lunar sample 14321. Unpublished manuscript.

Williams R. J. (1972) The lithification and metamorphism of lunar breccias. *Earth Planet. Sci. Lett.* **16**, 250–256.

Wilshire H. G. and Jackson E. D. (1972) Petrology and stratigraphy of the Fra Mauro Formation at the Apollo 14 site. *U.S. Geol. Survey Prof. Paper* **785**, 1–26.

Wilshire H. G., Stuart-Alexander D. E., and Jackson E. D. (1973) Apollo 16 rocks: Petrology and classification. *J. Geophys. Res.* **78**, 2379–2392.

York D., Kenyon J., and Doyle F. J. (1972) ^{40}Ar–^{39}Ar ages of Apollo 14 and 15 samples. *Proc. Third Lunar Sci. Conf., Geochim. Cosmochim. Acta*, Suppl. 3, Vol. 2, pp. 1613–1622. MIT Press.

Proceedings of the Fifth Lunar Conference
(Supplement 5, Geochimica et Cosmochimica Acta)
Vol. 2 pp. 1487–1514 (1974)
Printed in the United States of America

U–Th–Pb systematics of some Apollo 17 lunar samples and implications for a lunar basin excavation chronology*

P. D. NUNES, M. TATSUMOTO, and D. M. UNRUH

U.S. Geological Survey, Denver, Colorado 80225

Abstract—U, Th, and Pb concentrations and lead isotopic compositions of selected Apollo 17 soil and rock samples are presented. Concordia treatments of U–Pb whole samples of Apollo 17 mare basalts and highland rocks probably reflect several early thermal events ~4.5 b.y. old more consistently than do U–Pb ages of samples collected at other lunar sites.

We propose that all lunar U–Th–Pb data reflect a multistate U–Pb evolution history most easily understood as being related to a complex planetesimal bombardment history of the moon which apparently dominated lunar events from ~4.5 to ~3.9 b.y. ago. Semi-distinct events at ~4.0, ~4.2, and 4.4–4.5 b.y. are evident on whole-rock frequency versus $^{207}Pb/^{206}Pb$ age histograms. Each of these events may reflect multiple cratering episodes. For mare basalts, complete resetting of the source rock U–Pb systems owing to Pb loss relative to U was apparently often approached after a major planetesimal impact. It further appears that during later melting and extrusion of these basalts 500–800 m.y. after basin formation, the U–Pb total rock systems were negligibly disturbed while the $^{40}Ar/^{39}Ar$ whole-rock and U–Pb and Rb–Sr mineral systems were completely reset.

INTRODUCTION

THE ISOTOPIC COMPOSITIONS of Pb, and the concentrations of U, Th, and Pb were determined in selected samples collected during the Apollo 17 mission to the Taurus-Littrow region of the moon. Astronauts Eugene Cernan and Harrison Schmitt collected soils, mare igneous rocks from the valley floor, and brecciated highland-type rocks at the base of the massifs.

We undertook this U–Th–Pb study of Apollo 17 samples with the hope of clarifying our understanding of the early history of the moon. We consider to what extent new data presented here and published U–Th–Pb and other geochemical data suggests a chronology of events for the first 0.8 b.y. of lunar evolution. An attempt is made to correlate the early events which followed lunar accretion with the basin excavation stratigraphy outlined by McGetchin *et al.* (1973).

EXPERIMENTAL

Samples

Sample localities and descriptions are listed in the Appendix. Our analyses of soils 72701, 74001, 74220, 74240, 75120, 76501, 79221, 79241, and 79261 (except for 72701) contain excess Pb relative to U and appear to reflect at least 3 different mixing components.

Mare basalts analyzed included rake basalt 71569,17 and basalts 75055, 75035, 74275, 74255, 74235,

---

*Publication authorized by the Director, U.S. Geological Survey.

and 72155. Phase separates of 75035 were also analyzed. A glass concentrate and whole-rock sample of the glass-bearing (mare?) basalt 79155 were analyzed.

Highland rocks studied include brecciated anorthositic gabbros 78155 and 77017 as well as the following samples from Wood's Boulder 1 Station 2 consortium: 72275,73 (matrix); 72275,81 (Clast 1 black rind); 72275,117 (clast 1 white interior); 72255,67 (dark matrix); 72255,54 (light matrix); 72255,60 (dark and light matrix mixture); and 72255,49 ("Civet Cat" clast separated into separately analyzed plagioclase-enriched and plagioclase-deficient fractions). The Civet Cat clast, named by Harrison Schmitt, is a norite.

Analytical procedure

Soils were analyzed directly. Rock samples were crushed in a stainless-steel mortar. The crushed samples were generally washed in twice-distilled acetone two times with the aid of an ultrasonic vibrator. The acetone wash was adopted to remove possible terrestrial Pb contamination because large amounts of non-analytical terrestrial Pb contamination were detected in several Apollo 16 samples (Nunes *et al.*, 1974a)—particularly the less than 100 mesh 60018 sample. The large amount of Pb contamination found for this sample is, for some reason, graphically illustrated by Tera and Wasserburg (this issue, Fig. 3). The glass concentrate of 79155 was handpicked in a laminar flow work area. The plagioclase and pyroxene separates of 75035 were made with bromoform and handpicking techniques respectively. All phase separates were washed two times with twice-distilled acetone.

All samples were digested in teflon bombs using an HF–HNO_3 solution. Some samples were split from solution into concentration and composition portions prior to spiking. Most were approximately halved from a pile of crushed solid material using a stainless-steel spatula, and the concentration split totally spiked with a combined ^{235}U–^{230}Th–^{208}Pb-enriched solution. Totally spiked concentration runs yielded more reproducible U/Pb ratios than the samples which were divided from solution prior to spiking. Therefore, except for sample 74001,8 all U/Pb ratios of data plotted in this paper on U–Pb evolution diagrams were calculated from the concentration data—using only the $^{208}Pb/^{206}Pb$ ratio from the composition run. This is important because other authors may choose to plot our data by calculating the U/Pb ratios in a less accurate manner. The major drawback to the total spiking procedure was that it did not yield perfectly homogeneous splits. Such heterogeneity is reflected in slightly different Pb isotopic compositions obtained from the composition and concentration runs. Concentration and composition $^{207}Pb/^{206}Pb$ ratios of a single sample commonly varied by >0.3% (the approximate maximum error due to mass fractionation and mainly blank errors), owing to heterogeneity of the splits. For some reason, Tera and Wasserburg (this issue) have chosen to illustrate this graphically. Resulting uncertainties in the sample $^{208}Pb/^{206}Pb$ ratios generally affected the U/Pb uncertainties calculated from the concentration data by <0.5%. Except for 60315, all samples reported by Nunes *et al.* (1973) were checked with the total spike technique. Except for run 4 of 67701 and run 1 of 60315 all data in Table 1 of Nunes *et al.* (1973) were obtained by totally spiking samples. Consideration of all sources of error indicate U/Pb values of all our totally spiked data are precise to better than ±2%.

Lead was extracted from samples 75055 and 74220 with a combined Ba-coprecipitation and electrodeposition technique (Tatsumoto, 1970a). Lead was extracted from all other samples by precipitating the Pb (and U and Th) in a hydroxide precipitation ($pH \sim 8$). The precipitate was washed in ~ 2 ml 5 × distilled H_2O, and was dissolved in $1-1\frac{1}{2}$ ml 1.5 N HBr for loading on the first anion exchange column of the two column Pb isolation method (Nunes *et al.*, 1973). Lead blanks ranged from 0.59 to 1.96 ng for the two-column procedure and up to 2.9 ng for the 2 larger samples which were Ba-coprecipitated and electrodeposited.

Isotope abundance measurements of Pb were made with the silica gel technique (Cameron *et al.*, 1969). Replicate analyses of the N.B.S. equal atom standard revealed that our measured Pb isotope ratios were depleted in the heavy isotopes by about 0.1% per mass unit. This correction factor was used for calculations.

Uranium and thorium solutions which passed through the first Pb resin column were collected and evaporated; the U and Th were then isolated using a Dowex 1 × 8 NO_3^- form resin (Tatsumoto, 1966). Uranium and thorium blanks were <0.01 ng.

All reagents were purified with a sub-boiling technique. Water and HNO_3 were sub-boiled in French quartz stills and HF and HCl were prepared in a teflon still (Mattinson, 1972).

RESULTS AND DISCUSSION

Concentration data

U and Th concentrations of Apollo 17 soils varied considerably from ~0.14 ppm U and 0.47 ppm Th at Station 4 (sample 74001) to 0.81 ppm U and 2.96 ppm Th at Station 2 (Table 1). Uranium and thorium concentrations of soils from Stations 5, 6, and 9 are rather uniform and ranged from 0.31 to 0.41 ppm U and from 1.1 to 1.5 ppm Th. These U and Th concentration data appear to correlate with the three basic geologic units described by LSPET (1973) and Heiken and McKay (1974). The Station 4 soils with the lowest U and Th contents are almost entirely composed of orange and black droplets: the Station 2 soil with the highest U and Th concentrations was from the South Massif "avalanche deposit"; soils from Stations

Table 1. Concentrations of uranium, thorium, and lead in some Apollo 17 whole-soil and whole-rock samples.

| Sample | Weight (mg) | Concentrations (ppm) | | | $^{232}Th/^{238}U$ | $^{238}U/^{204}Pb$ |
		U	Th	Pb		
		Soil				
72701	141.8	0.8079	2.962	1.781	3.79	395
*74001,8 (core tube)	37.8	0.1391	0.4664	1.080	3.47	10.5
74220 (orange soil)	435.8	0.1610	0.5555	2.481	3.57	5.13
74240	208.2	0.3141	1.136	1.983	3.74	38.9
75120	170.1	0.3337	1.216	1.094	3.76	107
76501	87.4	0.4053	1.484	1.183	3.78	183
79221	205.3	0.3127	1.134	1.255	3.75	78.4
79241	213.9	0.3250	1.163	1.234	3.70	95.9
79261	144.9	0.3216	1.165	1.115	3.74	101
		Mare Basalt				
75055	533.8	0.1359	0.4472	0.3111	3.40	250
75035	209.4	0.1505	0.4879	0.3258	3.35	507
74275	181.0	0.1360	0.4654	0.2649	3.54	430
74255	207.4	0.1323	0.4451	0.2421	3.48	427
74235	235.9	0.1200	0.4004	0.2786	3.45	444
72155	187.8	0.1182	0.3879	0.2589	3.39	452
71569,17 rake basalt	158.6	0.1176	0.3881	0.2663	3.41	370
		Glass-bearing "basalt"				
79155	152.8	0.2198	0.7930	0.6517	3.73	138
		Anorthositic gabbro				
77017	94.7	0.2699	1.025	0.5733	3.92	643
	79.8	0.4147	1.489	0.8663	3.71	863
78155	112.1	0.2683	0.9352	0.8513	3.60	165

Table 1. (*Continued*).

	Weight (mg)	Concentrations (ppm)			$^{232}Th/^{238}U$	$^{238}U/^{204}Pb$
Sample		U	Th	Pb		
Boulder 1, Station 2						
72275,73 matrix	131.8	1.561	5.962	3.096	3.95	4,284
	150.0	1.672	6.285	3.451	3.89	4,712
72275,81 clast # 1						
black rind	31.7	3.500	13.21	7.878	3.90	2,493
72275,117 clast # 1						
white interior	50.7	0.670	—	1.410	—	2,445
72255,67 matrix						
(dark)	132.5	1.145	4.222	2.478	3.81	1,414
72255,54 matrix						
(light)	98.2	1.536	5.724	3.080	3.85	2,998
72255,60 matrix mixture						
(dark and light)	196.7	1.663	6.362	3.540	3.95	2,135
72255,49 Civet Cat clast plag.-						
deficient	51.6	0.3874	1.216	0.9448	3.24	195
72255,49 Civet Cat clast plag.-						
enriched	35.7	0.2151	—	0.6939	—	177

*Concentration run divided from solution; all other analyses were of splits from crushed solid material obtained prior to spiking.

5 and 9 were from the valley floor "dark-mantle" deposit; and Station 6 soil is from the North Massif unit.

Pb concentrations do not correlate so well with geologic units, a reflection of the mobility of Pb in the lunar regolith. Indeed, although Station 4 orange soil sample 74220 had the next-to-lowest U and Th concentrations measured, its Pb concentration was the highest measured (2.48 ppm). Pb contents of the other soils ranged from 1.1 ppm (74001 and 75120) to 1.8 ppm (72701).

The Apollo 17 mare basalts have U (0.11–0.15 ppm), Th (0.35–0.49 ppm), and Pb (0.24–0.49 ppm, Table 1) contents slightly lower than Apollo 11 low-K basalts and the Apollo 17 station 4 "glassy soils" 74220 and 74001. These are the lowest U, Th, and Pb contents yet reported for lunar mare basalts and they correlate with similarly low concentrations of other large ion lithophile elements (Philpotts *et al.*, 1974; Haskin *et al.*, 1974) suggesting a source material rather depleted in these elements.

Glass-bearing basalt 79155 and highland rock samples 77017, 78155, 72275, and 72255 have variable U (0.21–3.5 ppm), Th (0.94–13.2 ppm), and Pb (0.57–7.9 ppm) contents reflecting varying amounts of KREEP component. Two separate totally spiked analyses of anorthositic gabbro 77017 yielded markedly different U, Th, and Pb concentrations (Table 1) indicating gross chemical heterogeneity over several millimeters, in this sample, although large differences in calculated age parameters were not detected. The black rind of Clast 1 in 72275 has the highest U, Th, and Pb concentrations and therefore the highest KREEP content of all Apollo 17 samples so far analyzed for U, Th, and Pb. These concentrations are nevertheless lower than those of the most KREEP-rich rocks at the Apollo 14 and 12 sites.

^{232}Th/^{238}U ratios in the seven Apollo 17 mare basalts analyzed range from 3.35 to 3.48—consistently and significantly below the ^{232}Th/^{238}U ratios of other Apollo mare basalts which average about 3.8. With the exception of the Civet Cat clast in Boulder sample 72255 which has a ^{232}Th/^{238}U ratio of ~3.2, all Apollo 17 highland whole-rock samples have ^{232}Th/^{238}U ratios which range from 3.60 to 3.95 and average about 3.8 which is typical of most lunar mare and highland (excluding anorthosites) whole-rock samples. The Th/U ratios obtained for whole-rock samples (Table 1) generally agree well with the ratio of the stable end product isotopes, ^{208}Pb/^{206}Pb, found in these samples (Table 2). Apollo 17 soils analyzed by us have Th/U ratios of about 3.7 (Table 1)—in contrast to the 3.14 value reported by Silver (1974).

Lead isotope ratios

Perhaps the most significant discovery at the Apollo 17 site was the high ^{204}Pb content of the "orange soil" at Shorty Crater. The very nonradiogenic nature of lead in Station 4 orange and black "glassy droplet soils" (^{206}Pb/^{204}Pb = 23.37; ^{207}Pb/^{204}Pb = 26.00; ^{208}Pb/^{204}Pb = 41.20 for a whole soil analysis of 74220; Table 2) and the associated low ^{238}U/^{204}Pb ratio of 5.1 relative to all other lunar samples analyzed may be indicative of a deep source region *not* grossly depleted in volatile elements (Tatsumoto *et al.*, 1973a). Studies involving acid leaching, stepwise lead volatilization, and size-fraction analyses of "orange soil" 74220 (Tatsumoto *et al.*, 1973a, Nunes *et al.*, 1974b; Silver, 1974) show that the nonradiogenic Pb is concentrated on or in the outermost portions of the glass spherules. The data further suggest that the nonradiogenic Pb component evolved in a ^{238}U/^{204}Pb ($=\mu$) environment of about 35 from 4.65 to 3.63 b.y. ago when it was presumably concentrated on the glass spherules if a simple two-stage U–Pb evolution history is applicable for this sample. It is not within the scope of this work to further discuss the orange soil U–Pb data, which will be the subject of a separate report. Station 2 soil sample 72701 from the South Massif "avalanche deposit" (Heiken and McKay, 1974), is rather radiogenic (^{206}Pb/^{204}Pb ~ 500; Table 2) apparently owing to a distinguishable KREEP component—higher than that found at Apollo 11, 15, and 16 sites and similar to the less KREEPy Apollo 14 and 12 soils. The ^{206}Pb/^{204}Pb values of soils from the North Massif–Sculptured Hills unit and the valley floor "dark-mantle deposits" (Heiken and McKay, 1974), range from about 120–250 and are lower than almost all other Apollo soil analyses (excluding the Apollo 17 glass-droplet soils of course) but similar to Luna 16 and 20 soil analyses reported by Tera and Wasserburg (1972a, 1972b) and Tatsumoto (1973). These soils at Stations 5, 6, and 9 contain varying degrees of an orange and black droplet component and a plagioclase component. Thus, their rather nonradiogenic nature relative to most other lunar soils may be attributable to dilution by unradiogenic components such

Table 2. Isotopic composition of Pb in some Apollo 17 samples.

Sample	Weight (mg)	Run	Observed ratios[†]			Corrected for analytical blank[‡]				
			$\frac{^{206}Pb}{^{204}Pb}$	$\frac{^{207}Pb}{^{204}Pb}$	$\frac{^{208}Pb}{^{204}Pb}$	$\frac{^{206}Pb}{^{204}Pb}$	$\frac{^{207}Pb}{^{204}Pb}$	$\frac{^{208}Pb}{^{204}Pb}$	$\frac{^{207}Pb}{^{206}Pb}$	$\frac{^{208}Pb}{^{206}Pb}$
Soil										
72701	111.3	P	433.2	264.7	420.8	507.3	309.7	492.3	0.6104	0.9705
	90.7	C1	459.9	280.9	—	497.7	303.6	—	0.6100	—
	82.2	C2*	369.4	221.4	—	391.7	234.5	—	0.5987	—
74001 (core tube)	48.4	P	25.32	24.40	42.92	25.49	24.60	43.18	0.9654	1.695
	37.8	C	25.31	24.38	—	25.46	24.56	—	0.9649	—
74220	623.2	P	23.33	25.90	41.04	23.38	25.99	41.20	1.1116	1.762
	435.8	C*	23.34	25.97	—	23.36	26.00	—	1.1130	—
74240	93.6	P	86.26	88.43	95.72	87.29	89.52	96.84	1.0256	1.109
	77.3	C1	87.19	89.49	—	88.24	90.62	—	1.0269	—
	208.2	C2*	90.18	91.53	—	90.79	92.18	—	1.0153	—
75120	99.35	P	135.8	107.2	141.9	141.1	111.4	147.0	0.7893	1.042
	80.56	C1	134.3	106.2	—	140.1	110.7	—	0.7902	—
	170.1	C2*	138.9	108.9	—	143.1	112.1	—	0.7837	—
76501	107.8	P	236.0	162.0	233.4	248.5	170.4	245.2	0.6859	0.9866
	96.0	C1	220.7	151.7	—	232.4	159.6	—	0.6867	—
	87.4	C2*	215.8	157.6	—	226.7	165.5	—	0.7299	—
79221	117.3	P	120.2	100.8	127.9	123.0	103.1	130.6	0.8387	1.063
	102.8	C1	119.6	100.4	—	122.4	102.7	—	0.8391	—
	205.3	C2*	123.5	103.2	—	125.4	104.7	—	0.8351	—
79241	108.3	P	133.2	106.3	140.3	137.2	109.5	144.2	0.7981	1.051
	121.0	C1	133.7	106.8	—	137.9	110.1	—	0.7986	—
	213.9	C2*	144.6	116.6	—	147.1	118.6	—	0.8062	—
79261	97.4	P	154.9	118.3	157.5	160.5	122.5	162.8	0.7632	1.014
	118.5	C1	157.2	120.1	—	163.9	125.1	—	0.7635	—
	144.9	C2*	142.0	111.8	—	144.9	114.1	—	0.7871	—
Mare Basalt										
75055	621.4	P	179.7	106.9	179.5	236.4	139.2	229.0	0.5888	0.9725
	533.8	C*	231.3	134.0	—	260.1	150.0	—	0.5767	—
75035	120.7	P	384.6	228.1	352.3	579.2	341.0	520.5	0.5888	0.8987
	92.1	C1	350.8	208.6	—	582.2	342.9	—	0.5890	—
	209.4	C2*	397.5	232.4	—	509.1	296.3	—	0.5819	—
74275	105.1	P	321.7	161.1	304.7	449.9	226.6	418.3	0.4947	0.9298
	109.8	C1	360.9	180.0	—	519.9	256.3	—	0.4931	—
	181.0	C2*	305.6	152.9	—	397.4	196.9	—	0.4952	—
74255	207.4	P	478.1	227.7	441.9	680.7	321.5	621.1	0.4723	0.9125
	207.4	C*	431.0	206.7	—	586.7	278.8	—	0.4752	—
74235	191.3	P	178.9	111.3	174.7	215.0	133.0	205.9	0.6186	0.9579
	235.9	C*	339.7	208.6	—	464.8	283.7	—	0.6105	—
72155	219.4	P	343.7	205.5	313.9	433.6	258.3	391.3	0.5957	0.9025
	187.8	C*	352.7	274.9	—	459.6	274.9	—	0.5982	—
71569,17 rake basalt	213.7	P	180.4	109.8	176.2	220.3	133.7	210.7	0.6042	0.9564
	158.6	C1*	249.6	149.9	—	380.2	225.7	—	0.5936	—
Glass-bearing brecciated "basalt"										
79155	92.0	P	147.2	106.5	152.6	168.9	121.9	172.3	0.7217	1.020
	152.8	C1*	158.4	115.3	—	172.7	125.5	—	0.7266	—
Anorthositic gabbro										
77017	103.5	P	383.6	223.8	370.9	582.0	337.0	553.0	0.5790	0.9502
	87.2	C1	374.7	219.0	—	616.4	357.0	—	0.5791	—
	94.7	C2*	449.2	252.5	—	636.1	355.2	—	0.5584	—
	79.8	C3*	512.6	278.9	—	832.4	449.3	—	0.5398	—
78155	113.4	P	235.4	190.2	230.9	263.0	212.4	256.0	0.8076	0.9734
	94.5	C1	196.6	158.9	—	218.3	176.4	—	0.8078	—
	112.1	C2*	198.4	161.8	—	216.5	176.5	—	0.8154	—

Table 2. (*Continued*).

Sample	Weight (mg)	Run	Observed ratios†			Corrected for analytical blank‡				
			$\frac{^{206}Pb}{^{204}Pb}$	$\frac{^{207}Pb}{^{204}Pb}$	$\frac{^{208}Pb}{^{204}Pb}$	$\frac{^{206}Pb}{^{204}Pb}$	$\frac{^{207}Pb}{^{204}Pb}$	$\frac{^{208}Pb}{^{204}Pb}$	$\frac{^{207}Pb}{^{206}Pb}$	$\frac{^{208}Pb}{^{206}Pb}$
Boulder 1, Station 2										
72275,73 matrix	162.0	P	1,097	537.1	1,090	1,225	599.3	1,218	0.4893	0.9945
	131.8	C1*	2,715	1,308	—	3,961	1,905	—	0.4811	—
		C2*	3,220	1,545	—	4,556	2,183	—	0.4792	—
72275,81 clast #1 black rind	53.3	P	1,578	959.2	1,532	1,937	1,176	1,880	0.6072	0.9705
	31.7	C*	1,688	1,000	—	2,521	1,492	—	0.5918	—
72275,117 clast #1 white interior	83.3	P	902.2	520.8	860.4	1,423	818.2	1,347	0.5752	0.9472
	50.7	C*	920.4	533.3	—	2,361	1,360	—	0.5761	—
72255,67 matrix (dark)	100.8	P	1,596	908.5	1,526	2,815	1,601	2,682	0.5685	0.9527
	132.5	C*	1,092	624.7	—	1,405	802.5	—	0.5711	—
72255,54 matrix (light)	124.1	P	2,089	1,085	2,023	3,296	1,709	3,186	0.5187	0.9666
	98.2	C*	1,734	898.9	—	2,803	1,449	—	0.5171	—
72255,60 matrix mixture (dark and light)	190.3	P	1,987	1,107	1,909	2,212	1,233	2,128	0.5573	0.9619
	196.7	C*	1,897	1,060	—	2,090	1,166	—	0.5582	—
72255,49 Civet Cat clast plag.-deficient	66.2	P	185.7	120.9	163.5	198.2	128.9	173.3	0.6505	0.8743
	51.6	C*	199.2	127.2	—	217.6	138.6	—	0.6369	—
72255,49 Civet Cat clast plag.-enriched	32.9	P	204.9	153.2	148.8	245.4	183.3	173.3	0.7467	0.7062
	35.7	C*	160.0	138.5	—	180.5	156.4	—	0.8663	—

*Samples totally spiked prior to digestion.

†^{208}Pb spike contribution subtracted from Pb concentration data.

‡Analytical total Pb blanks ranged from 0.59 to 1.96 ng except for the 75055 composition blank (2.9 ng), and the 74220 concentration blank (2.8 ng).

P = composition data; C = concentration data.

as the glass droplets so conspicuous at Station 4 and plagioclase from anorthositic rocks. A smaller KREEP contribution to these soils could also be partially responsible for the relatively low ^{206}Pb/^{204}Pb ratios observed.

The radiogenic nature of lead in Apollo 17 mare basalts (Table 2) and their U, Th, and Pb concentrations (Table 1) are quite similar to those of Apollo 11 low-K basalts.

The ^{206}Pb/^{204}Pb values of Apollo 17 highland rocks vary over a wide range from about 200–3000, apparently reflecting varying amounts of KREEP component content.

Age relations

In this paper, we shall continue to use the ^{206r}Pb/^{238}U versus ^{207r}Pb/^{235}U concordia diagram (Wetherill, 1956) *and* the ^{206}Pb/^{207}Pb versus ^{238}U/^{207}Pb positive slope variant of Tera and Wasserburg's (1972a) ^{207}Pb/^{206}Pb versus ^{238}U/^{206}Pb plot to complement one another. Thus, we correct our data only for blanks when plotting on the Pb/Pb versus U/Pb diagram, but we subtract blanks and an *assumed* primordial Pb from the data prior to plotting on the Wetherill (1956) diagram—as previously discussed (Nunes *et al.*, 1973).

Single-stage age parameters for Apollo 17 samples corrected for blank and primordial Pb are shown in Table 3. Decay constants used in this paper are $\lambda_{238} = 0.15369 \times 10^{-9}$ yr^{-1}, $\lambda_{235} = 0.97216 \times 10^{-9}$ yr^{-1} (Fleming *et al.*, 1952), and $\lambda_{232} =$

Table 3. Age parameters and single-stage ages of some Apollo 17 samples.

Sample	Run	Corrected for blank and primordial Pb				Single-stage ages in m.y.			
		$\frac{^{206}Pb}{^{238}U}$	$\frac{^{207}Pb}{^{235}U}$	$\frac{^{207}Pb}{^{206}Pb}$	$\frac{^{208}Pb}{^{232}Th}$	$\frac{^{206}Pb}{^{238}U}$	$\frac{^{207}Pb}{^{235}U}$	$\frac{^{207}Pb}{^{206}Pb}$	$\frac{^{208}Pb}{^{232}Th}$
Soil									
72701	C2P	0.9739	80.76	0.6018	0.2389	4,425	4,530	4,578	4,389
	C2	0.9690	78.34	0.5864	—	4,408	4,498	4,540	—
74001	C1P*	1.543	188.0	0.8844	0.3772	6,072	5,392	5,137	6,557
	C1*	1.542	187.7	0.8836	—	6,070	5,390	5,136	—
74220	C1P	2.753	423.4	1.116	0.6433	8,605	6,224	5,469	10,176
	C1	2.738	421.8	1.118	—	8,580	6,220	5,471	—
74240	C2P	2.081	291.8	1.017	0.4813	7,322	5,842	5,337	8,049
	C2	2.092	289.7	1.005	—	7,346	5,835	5,319	—
75120	C2P	1.248	132.1	0.7678	0.2957	5,271	5,031	4,933	5,307
	C2	1.252	131.3	0.7613	—	5,282	5,025	4,921	—
76501	C2P	1.194	110.3	0.6703	0.2847	5,113	4,847	4,736	5,132
	C2	1.191	117.2	0.7138	—	5,102	4,908	4,827	—
79221	C2P	1.476	166.4	0.8177	0.3509	5,900	5,267	5,024	6,162
	C2	1.481	165.9	0.8133	—	5,911	5,264	5,017	—
79241	C2P	1.427	155.7	0.7766	0.3462	5,770	5,179	4,950	6,090
	C2	1.437	155.6	0.7859	—	5,796	5,199	4,967	—
79261	C2P	1.344	137.6	0.7430	0.3168	5,544	5,073	4,886	5,638
	C2	1.337	141.1	0.7652	—	5,524	5,098	4,928	—
Mare basalt									
75055	C1P	1.001	78.30	0.5675	0.2596	4,512	4,499	4,492	4,792
	C1	1.004	77.09	0.5572	—	4,523	4,483	4,464	—
75035	C2P	0.9833	78.73	0.5810	0.2529	4,456	4,504	4,526	4,619
	C2	0.9850	77.66	0.5721	—	4,461	4,490	4,503	—
74275	C2P	0.9005	59.85	0.4824	0.2247	4,178	4,226	4,249	4,152
	C2	0.9017	59.71	0.4805	—	4,182	4,224	4,244	—
74255	C1P	0.8764	55.98	0.4635	0.2222	4,095	4,159	4,190	4,110
	C1	0.8745	56.03	0.4650	—	4,088	4,159	4,194	—
74235	C1P	1.001	82.32	0.5966	0.2491	4,514	4,549	4,565	4,557
	C1	1.027	84.85	0.6003	—	4,593	4,580	4,574	—
72155	C1P	0.9929	79.98	0.5845	0.2497	4,487	4,520	4,535	4,567
	C1	0.9941	80.50	0.5877	—	4,491	4,527	4,543	—
71569 rake basalt	C1P	0.9844	78.95	0.5820	0.2479	4,459	4,506	4,529	4,538
	C1	1.003	80.24	0.5808	—	4,519	4,523	4,525	—
Glass-bearing brecciated "basalt"									
79155	C1P	1.180	113.8	0.6993	0.2835	5,074	4,879	4,798	5,114
	C1	1.183	114.9	0.7050	—	5,079	4,889	4,809	—

Table 3. (*Continued*).

Sample	Run	Corrected for blank and primordial Pb				Single-stage ages in m.y.			
		$\frac{^{206}Pb}{^{238}U}$	$\frac{^{207}Pb}{^{235}U}$	$\frac{^{207}Pb}{^{206}Pb}$	$\frac{^{208}Pb}{^{232}Th}$	$\frac{^{206}Pb}{^{238}U}$	$\frac{^{207}Pb}{^{235}U}$	$\frac{^{207}Pb}{^{206}Pb}$	$\frac{^{208}Pb}{^{232}Th}$
Anorthositic gabbro									
77017	C2P	0.9540	75.08	0.5711	0.2223	4,359	4,456	4,501	4,112
	C2	0.9596	72.77	0.5503	—	4,377	4,424	4,446	—
	C3	0.9541	70.13	0.5334	—	4,359	4,386	4,399	—
78155	C2P	1.265	139.0	0.7974	0.3136	5,320	5,083	4,988	5,589
	C2	1.256	138.9	0.8023	—	5,294	5,082	4,997	—
Boulder 1, Station 2									
72275,73 matrix	C1P	0.9175	61.26	0.4845	0.2274	4,236	4,250	4,256	4,198
	C1	0.9223	60.95	0.4796	—	4,252	4,245	4,241	—
72275,81 clast #1 black rind	C1P	1.006	83.88	0.6048	0.2478	4,531	4,568	4,585	4,535
	C1	1.008	81.90	0.5899	—	4,534	4,544	4,548	—
72275,117 clast #1 white interior	C1P	0.9595	75.60	0.5717	—	4,377	4,463	4,502	—
	C1	0.9620	76.09	0.5740	—	4,385	4,469	4,508	—
72255,67 matrix (dark)	C1P	0.9906	77.36	0.5667	0.2459	4,480	4,486	4,489	4,503
	C1	0.9873	77.22	0.5676	—	4,469	4,484	4,491	—
72255,54 matrix (light)	C1P	0.9321	66.41	0.5170	0.2324	4,285	4,331	4,353	4,281
	C1	0.9317	66.13	0.5151	—	4,284	4,327	4,348	—
72255,60 matrix mixture (dark and light)	C1P	0.9745	74.53	0.5550	0.2349	4,427	4,448	4,458	4,322
	C1	0.9743	74.61	0.5557	—	4,426	4,449	4,460	—
72255,49 Civet Cat clast plag.-deficient	C1P	1.065	92.15	0.6281	0.2499	4,717	4,664	4,641	4,570
	C1	1.069	90.76	0.6159	—	4,732	4,649	4,612	—
72255,49 Civet Cat clast plag.-enriched†	C1P	1.464	147.8	0.7326	—	5,382	5,050	4,916	—
	C1	1.443	169.7	0.8533	—	5,867	5,146	4,865	—

*Note: Concentration and composition splits were divided from solution prior to adding the ^{208}Pb enriched spike. All other analyses were of splits from crushed solid material and the concentration portions were totally spiked prior to dissolution.

†The gross difference between the CP and C only calculations must be because of an heterogeneous splitting of this sample prior to spiking—calculated U/Pb ratios from the concentration only data (i.e. where only the $^{208}Pb/^{206}Pb$ ratio from the composition run was utilized) are the most accurate.

0.048813×10^{-9} yr^{-1} (Senftle *et al.*, 1956). The $^{238}U/^{235}U$ atomic ratio used is 137.8 previously reported for the moon (Rosholt and Tatsumoto, 1970), and the primordial lead isotopic compositions used are $^{206}Pb/^{204}Pb = 9.307$, $^{207}Pb/^{204}Pb = 10.294$, and $^{208}Pb/^{204}Pb = 29.476$ (Tatsumoto *et al.*, 1973b). The blank lead isotopic composition used in our calculations is $^{206}Pb/^{204}Pb = 18.36$, $^{207}Pb/^{204}Pb = 15.58$, $^{208}Pb/^{204}Pb = 37.88$. If the U decay constants of Jaffey *et al.* (1971) were used, ages reported here would be reduced by about 1%.

Apollo 17 soils

Except for soil 72701, all Apollo 17 soils analyzed by us contain excess Pb relative to U. The scatter in these analyses as shown in Fig. 1 exceeds analytical error and apparently reflects mixing of at least three different components. This is consistent with Heiken and McKay's (1974) explanation that the Apollo 17 soils are mixtures of three basic geologic units.

Mare basalts

Whole-rock U–Pb analyses of mare basalts 75055, 75035, 74235, 72155, and 71569 plot within error on concordia in the 4.49–4.57 b.y. region. Mare basalts 74275 and 74255 plot slightly below concordia and have $^{207}Pb/^{206}Pb$ ages of 4.24 and 4.19 b.y., respectively (Fig. 2, Table 3). Rb–Sr internal isochron ages of mare basalts

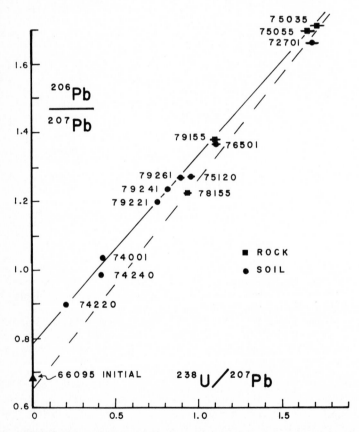

Fig. 1. $^{206}Pb/^{207}Pb$ versus $^{238}U/^{207}Pb$ diagram. All Apollo 17 soils for which we have data are plotted (circles). Several Apollo 17 rocks (squares) and the initial $^{206}Pb/^{207}Pb$ ratio of breccia 66095 (triangle, Nunes and Tatsumoto, 1973) and two lines are plotted for reference. Data are corrected for blanks only. U/Pb errors plotted are ±2%.

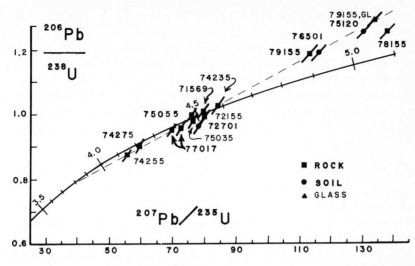

Fig. 2. Concordia diagram (Wetherill, 1956). Apollo 17 mare basalts (74275, 74255, 74235, 75055, 75035, 72155, and 71569), highland rocks (77017 and 78155), soils (72701, 75120, and 76501), and a whole-rock and glass separate of 79155 are plotted. U/Pb errors are ±2%. Data are corrected for blank and primordial lead.

75055 (Tatsumoto *et al.*, 1973a; Evensen *et al.*, 1973a), 70035 (Evensen *et al.*, 1973b; Nyquist *et al.*, 1974), and 71055 (Tera *et al.*, 1974) suggest the Apollo 17 mare basalts were extruded between 3.85 and 3.65 b.y. ago. ^{40}Ar/^{39}Ar Apollo 17 mare basalt ages of about 3.8 b.y. for 75055 (Huneke *et al.*, 1973; Turner *et al.*, 1973; Kirsten *et al.*, 1973), 3.84 b.y. for 70215 (Kirsten and Horn, 1974), 3.82 b.y. for 78503 (Kirsten and Horn, 1974), and 3.76 b.y. for 74243 (Kirsten and Horn, 1974) support the interpretation that the Apollo 17 basalts crystallized about 3.85–3.65 b.y. ago. The fact remains, however, that five out of seven Apollo 17 whole-rock U–Pb mare basalt analyses (Table 3) are nearly concordant at ~4.5 b.y. This is the first instance that approximately concordant whole-rock U–Pb ages of *mare basalts* have yielded the 4.5 b.y. age. We doubt this is coincidence and we suspect that although extrusion of these basalts 3.8–3.7 b.y. ago may have totally reset the Rb–Sr and U–Pb mineral and whole-rock ^{40}Ar/^{39}Ar systems, the whole-rock Rb–Sr systems (e.g. see Compston, 1974; Nyquist *et al.*, 1974), and U–Pb whole-rock systems were affected only slightly or not at all. The 4.5 b.y. age presumably reflects differentiation event(s) involving melting and gross partitioning of U and Pb and Rb and Sr. Slightly discordant mare basalts 74255 and 74275 have ^{207}Pb/^{206}Pb ages of 4.19 and 4.24 b.y., respectively. These ages may or may not reflect a discrete event ~4.2 b.y. old. Highland crystalline rock 68415 also yielded a whole-rock U–Pb concordant age of 4.47 (Nunes *et al.*, 1973; Tera *et al.*, 1974) and a 3.84 b.y. Rb–Sr internal isochron age and a 3.95 b.y. internal U–Pb isochron age (Tera *et al.*, 1973). U–Pb workers have repeatedly noted that nearly all lunar crystalline whole rock ^{207}Pb/^{206}Pb ages are significantly older than corresponding Rb–Sr internal isochron ages and $^{40}Ar/^{39}Ar$ ages (e.g. Silver, 1970; Tatsumoto *et al.*, 1972a; Tera and Wasserburg, 1972a).

Analyses of mare basalt 75035 plagioclase, pyroxene, and whole-rock samples (Table 4) yielded an internal U–Pb isochron age of 3.56 ± 0.40 b.y. (Fig. 3; maximum age uncertainty estimated graphically). The large uncertainty reflects the plagioclase Pb contamination correction assuming 0–100% of the plagioclase ^{204}Pb is of terrestrial origin. Nevertheless, it is clear that the U–Pb mineral systems were open well after the apparent 4.5 b.y.-old event reflected in the U–Pb whole-rock data.

Highland gabbros and glass-bearing basalt 79155

Brecciated anorthositic gabbro 78155 and coarse basalt 79155 whole rocks plot above concordia (Fig. 2) and contain excess Pb relative to U. The position of these rocks relative to concordia suggests that they contain a significant soil component. Black glass specks in basalt 79155 were hand picked and analyzed for U, Th, and Pb (Table 4). The glass contains even more excess Pb relative to U than does the whole rock (Fig. 2) and may be the site of our hypothesized soil component in this rock.

Two slightly discordant U–Pb analyses of brecciated gabbro 77017 (Fig. 2) have $^{207}Pb/^{206}Pb$ ages of 4.42 and 4.46 b.y. (Table 3). These ages may reflect a partially reset U–Pb system of ~4.5 b.y.-old material by one or more impact events 3.8–4.4 b.y. ago which caused the brecciation and injection of high-Ti glass (we did not observe this glass in our sample) described by Helz and Appleman (1974). Impacts may also have introduced excess Pb relative to U into brecciated anorthositic gabbro 78155 and glass-bearing basalt 79155. *If* both rocks represent old material of approximately the same age (say 4.5 b.y.) which later were injected with excess Pb by impacts, the relative positions in Fig. 2 require that the impact events affecting 78155 were older (~4.2–4.5 b.y. old) than those affecting 79155 (~3.8–4.0 b.y. old). That is, the distinct nonlinear scatter of data points above concordia in Fig. 2 requires a multistage U–Pb evolution for these rocks during the first 0.8 b.y. of lunar history.

Boulder 1 Station 2 rocks

Boulder 1 highland breccia samples 72275 and 72255 were described in detail by Wood's Consortium Indomitable (1974). We divided sample 72275 into matrix, Clast 1 white interior, and Clast 1 black rind samples for U–Th–Pb analyses. Three matrix samples of Boulder 1 sample 72255 were analyzed: a dark-gray to black dense sample, a light-gray less coherent sample, and a sample containing both matrix types. In sample 72255,60, we observed megascopically and under the microscope that the dark-gray to black material appeared to be included within the light-gray material. In addition to these matrix samples, a portion of the Civet Cat clast was divided into plagioclase-enriched (~85–95% plagioclase) and plagioclase-deficient (<25% white material) samples which were analyzed separately. Tables 1, 2, and 3 show analytical and age data.

Except for the two Civet Cat analyses, which contain excess Pb relative to U, all

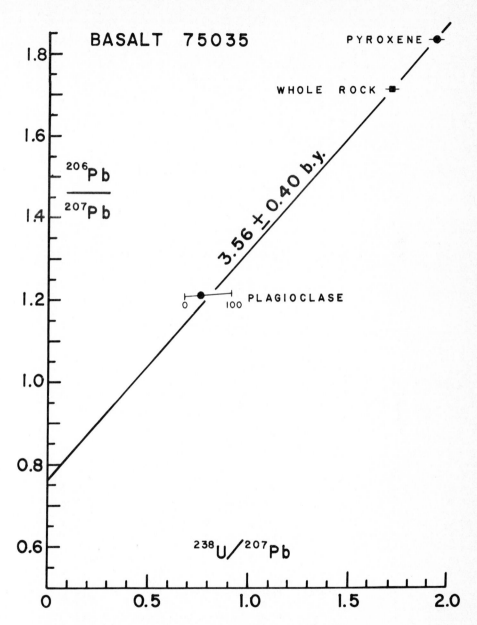

Fig. 3. $^{206}Pb/^{207}Pb$ versus $^{238}U/^{207}Pb$ evolution diagram. Internal isochron for Apollo 17 başalt 75035. Data are corrected for blanks only. Pyroxene and whole-rock U/Pb errors are ±2%. Plagioclase U/Pb errors are the maximum possible errors assuming from 0% (0) to 100% (100) of the measured ^{204}Pb is attributable to terrestrial contamination. The initial $^{206}Pb/^{207}Pb$ ratio and slope of the line drawn are 0.767 and 0.556, respectively. The slope corresponds to an age of 3.56 ± 0.40 b.y. The error in this age is a maximum error estimated graphically.

Table 4. Whole-rock and phase separate analyses of mare basalt 75035 and "basalt" 79155.

Analysis	Comp. Conc.	Weight (mg)	Concentrations (ppm) U	Th	Pb	Observed* 206Pb/204Pb	207Pb/204Pb	208Pb/204Pb	Corrected for blank† 206Pb/204Pb	207Pb/204Pb	208Pb/204Pb	232Th/238U	238U/204Pb	Corrected for blank and primordial Pb 206Pb/238U	207Pb/235U	207Pb/206Pb	208Pb/232Th
75035																	
Whole rock	P	120.7	0.1505	0.4879	0.3258	384.6	228.1	352.3	579.2	341.0	520.5	3.35	507	0.9850	77.66	0.5721	0.2529
	C	209.4				397.5	232.4	—	509.1	296.3	—						
Handpicked pyroxene	P	63.4	0.1316	0.4354	0.2721	146.8	83.32	146.7	203.1	113.1	194.8	3.42	538	0.9328	68.40	0.5321	0.2262
	C	35.7				185.8	104.5	—	511.0	277.3	—						
Plag. ρ < 2.8	P	145.5	0.0255	—	0.1011	83.07	68.56	95.33	114.8	94.67	123.9	—	65.1	1.436	160.6	0.8117	—
	C	92.1				68.33	57.36	—	102.7	86.09	—						
79155																	
Whole rock	P	92.0	0.2198	0.7930	0.6517	147.2	106.5	152.6	168.9	121.9	172.3	3.73	138	1.183	114.9	0.7050	0.2835
	C	152.8				158.4	115.3	—	172.7	125.5	—						
Handpicked –glass	P	100.8	0.2985	1.086	0.9941	137.4	105.1	144.6	151.4	115.8	157.5	3.76	109	1.287	134.6	0.7589	0.3087
	C	106.0				136.8	106.9	—	149.6	116.8	—						

Note: All samples were totally spiked prior to digestion.
*^{208}Pb spike contribution was subtracted from concentration data.
†Total Pb blanks ranged from 1.11 to 1.96 ng.

the Boulder 1 data shown in Fig. 4 plot within error of concordia in the
4.24–4.55 b.y. range. The excess Pb in the Civet Cat clast probably reflects transfer
of Pb from the relatively Pb rich surrounding material into the clast. The dark-gray
to black matrix material of 72255 and the white Clast 1 from 72275 both plot
essentially at the 4.5 b.y. concordia point. The ^{207}Pb/^{206}Pb ages of the dark-gray to
black and the light-gray matrix samples from 72255 (4.49 and 4.35 b.y., respectively)
appear to reflect the relative age difference observed under the microscope (i.e. the
dark-gray to black matrix material occurred as inclusions in, and is therefore older
than, the light-gray matrix material). Sample 72255,60, which under the microscope
appeared to be a mixture of dominantly dark-gray to black matrix material and
light-gray material has an intermediate ^{207}Pb/^{206}Pb age of 4.46 b.y. A best-fit line
drawn through all the Boulder 1 samples in Fig. 4 intersects concordia at ~4.5 and
~4.0 b.y. These data alone, therefore, may be explained with a simple two-stage
U–Pb evolution history whereby ~4.5 b.y.-old material was disturbed by a later
(impact?) event ~4.0 b.y. ago (Tera *et al.*'s, 1973, 1974, "Lunar Cataclysm"). As
discussed above, however, other Apollo 17 highland rocks require a multistage
U–Pb evolution history so that other events occurring between ~4.0 and ~4.5 b.y.
may be masked in the analytical uncertainty of the data plotted in Fig. 4.

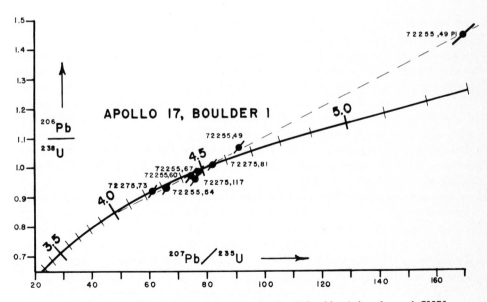

Fig. 4. Concordia diagram (Wetherill, 1956). Apollo 17 Boulder 1 data for rock 72275
(72275,73 = matrix; 72275,81 = black rind of Clast 1; 72275,117 = Clast 1 white interior),
and 72255 (72255,67 = dark matrix; 72255,54 = light matrix; 72255,60 = dark and light
matrix mixture; 72255,49 = plagioclase-deficient. Civet Cat clast; 72255,49 pl =
plagioclase enriched Civet Cat Clast) are plotted. All data are corrected for blank and
primordial lead. U/Pb errors are ±2%.

Summary of lunar whole-rock $^{207}Pb/^{206}Pb$ ages, and their possible relation to major basin excavation events.

Recognizing and characterizing isotopic "age events" in the early history of the moon is critical to understanding the moon's history. If a simple two-stage "4.5 b.y. differentiation—4.0 b.y. cataclysm" U–Pb evolution history characterized the early lunar history, a *random* array of $^{207}Pb/^{206}Pb$ whole-rock ages $\geqslant 4.0$ b.y. is expected, because the various U–Pb whole-rock systems should be affected to various (random) degrees by the 4.0 b.y. event. To investigate this further, we compiled a list of published whole-rock $^{207}Pb/^{206}Pb$ ages of all lunar rocks which within error plot on or below concordia (Table 5). We excluded data which falls outside of error above concordia (arbitrarily defined as when the $^{238}U/^{206}Pb$ age >2% higher than the $^{207}Pb/^{206}Pb$ age) because samples which have gained Pb relative to U clearly are in themselves unreliable single-stage age indicators. On the other hand, some samples which may have been totally reset by a major thermal event (i.e. all Pb lost relative to U) may withstand subsequent events which totally reset still other samples and yield approximately meaningful $^{207}Pb/^{206}Pb$ ages. Samples which appear to lie analytically below concordia are included with concordant samples as possibly having meaningful $^{207}Pb/^{206}Pb$ ages because we cannot rule out the possibility that very recent real or apparent Pb loss has occurred either on earth or on the moon. For example, nonequilibration of U and Pb spikes with sample U and Pb during dissolution, or imperfect splitting of samples prior to spiking *could* introduce somewhat larger analytical U/Pb precision errors than might normally be expected (we have no reason to suspect this is a serious problem, however). On the moon there is good evidence that Pb in lunar soils is labile possibly owing to surface heating (Silver, 1970, 1973) processes such as micro-meteoroid impacts. Possible leakage of ^{222}Rn (an intermediate member of the ^{238}U, ^{206}Pb decay scheme) from the lunar regolith is suggested by the anomalies documented by Bjorkholm *et al.* (1973). Conceivably, then, minor amounts of Pb were lost from surface rock samples in the last few million years—particularly for those samples with large cosmic ray exposure ages. The degree of discordance of samples which plot slightly below concordia is slightly reduced when the U decay constants of Jaffey *et al.* (1971) are used.

Let us first consider the non-mare whole-rock $^{207}Pb/^{206}Pb$ age versus frequency distribution. Figure 5 shows a *nonrandom* distribution with a broad hump in the 4.35–4.55 b.y. region, a semi-distinct hump at 4.2–4.3 b.y., and a dribbling of $^{207}Pb/^{206}Pb$ ages down to ~3.95 b.y. Each "hump" (Fig. 5) may reflect several events (major impacts?), the number of such events increasing with age. Regardless of what caused the apparent thermal events at ~4.0, 4.25, and 4.4–4.5 b.y. ago (Fig. 5), at least three semi-distinct events may have occurred at these times. If the moon formed ~4.65 b.y. ago as suggested by very primitive initial $^{87}Sr/^{86}Sr$ ratios (Nyquist *et al.*, 1973; Nunes *et al.*, 1973; Tatsumoto *et al.*, 1974), and by Albee *et al's.* (1974) 4.60 ± 0.09 b.y. dunite Rb–Sr isochron, a *minimum* of 4 stages of U–Pb evolution appears necessary to explain the U–Pb lunar highland rock data. Thus, stage 1 would last from ~4.65 to ~4.5 b.y., stage 2 from ~4.5 to 4.25 b.y., stage 3 from ~4.25 to 4.0 b.y., and stage 4 from ~4.0 b.y. ago to the present.

Table 5. Apollo 11, 12, 14, 15, 16, and 17 whole-rock ^{207}Pb/^{206}Pb ages. Rocks having excess Pb relative to U are not listed (see text for explanation). Only one age is shown for each sample. All ages were calculated by subtracting primordial Pb (Tatsumoto *et al.*, 1973b) from blank corrected data and using $\lambda_{238} = 0.15369 \times 10^{-9}$ yr^{-1}, and $\lambda_{235} = 0.97216 \times 10^{-9}$ yr^{-1}.

Non-mare rocks			Mare basalts		
Sample	^{207}Pb/^{206}Pb Age (m.y.)	Reference*	Sample	^{207}Pb/^{206}Pb Age (m.y.)	Reference*
12013,10	4,388	1	10003	4,154	11
12034,16	4,215	1	10017	4,156	11
14307 clast	4,373	2	10020	4,180	11
14310	4,272	2	10045	4,209	12
14063 clast	4,621	2	10050	4,233	11
14073 clast	4,373	2	10057	4,212	11
14318 clast	4,453	2	10071	4,216	11
14321 matrix	4,414	3	10072	4,114	12
60018	4,408	4	12009	4,002	1
60315	3,993	5	12021	3,937	1
60035	4,141	6	12022	3,980	1
60636	4,007	7	12035	3,564	1
61156	4,297	7	12038	3,924	1
62235	4,068	7	12052	3,996	1
65015	3,995	5	12063	3,994	1
67015 clast	4,418	5	12064	4,333	1
67559	4,478	7	15065	3,945	13
68415	4,470	5	15076	3,961	13
72255,57 matrix			15085	3,974	13
(light)	4,351	8	15476	4,087	13
72255,67 matrix					
(dark)	4,490	8	15555	3,965	13
72275,7 matrix	4,241	8	71569	4,526	8
72275,81 clast					
rind	4,548	8	72155	4,540	8
72275,117 clast	4,505	8	74235	4,570	8
72395	4,136	7	74255	4,192	8
76015,53	4,340	9	74275	4,225	8
76055	4,001	7	75035	4,500	8
76315,71 matrix	4,473	9	75055	4,490	8
76315,72 clast	4,487	9			
77017	4,446	8			
77075,22 dike	4,485	10			
77115,35 matrix	4,449	10			
77135,33 matrix					
(less vesicular)	4,420	10			
77135,34 matrix					
(more vesicular)	4,382	10			
77135,57A clast	4,368	10			
77215,37 clast	4,490	10			

*References:
1. Tatsumoto *et al.* (1971).
2. Tatsumoto *et al.* (1972a).
3. Barnes *et al.* (1972).
4. Nunes *et al.* (1974a).
5. Nunes *et al.* (1973).
6. Barnes *et al.* (1973).
7. Tera *et al.* (1974).
8. This Report.
9. Silver (1974) in Phinney *et al.* (1974).
10. Nunes, *et al.* (1974c).
11. Tatsumoto (1970b).
12. Silver (1970).
13. Tatsumoto *et al.* (1972b).

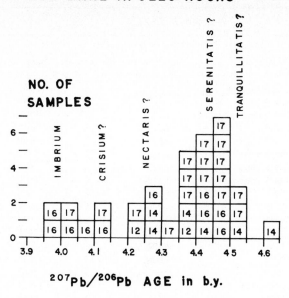

Fig. 5. Frequency versus ^{207}Pb/^{206}Pb whole-rock age of non-mare Apollo rocks. Rocks which contain excess Pb relative to U are not plotted. The Apollo mission number is listed for each age plotted.

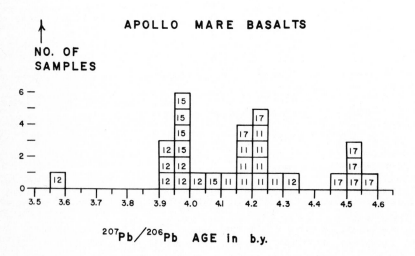

Fig. 6. Frequency versus ^{207}Pb/^{206}Pb whole-rock age of mare basalts. The Apollo mission number is listed for each age plotted.

Certainly major basin-forming events cause discrete and widespread periods of heating. Is there any relation, then, between the basin excavation ejecta stratigraphy which McGetchin *et al.* (1973) recently described and the pattern in Fig. 5? The large apparent 4.35–4.55 b.y. event in Fig. 5 is largely produced by Apollo 17 brecciated rocks. The Apollo 17 site is apparently covered by a thicker deposit of Serenitatis ejecta (1650 m; McGetchin *et al.*, 1973) than any other sample site and is also overlain by a relatively thin series of younger ejecta (~230 m; McGetchin *et al.*, 1973). We see no reason to ascribe this to coincidence and speculate that the Serenitatis excavation event occurred about 4.4–4.5 b.y. ago. Of course partially reset rocks from previous basin excavation events (e.g. the Tranquillitatis event) likely occur in the Serenitatis ejecta and may be responsible for the broadness observed in the 4.35–4.55 b.y. "hump" (Fig. 5). If Serenitatis was excavated ~4.45 b.y. ago, still older basins (Nubian, Tranquillitatis, Fecunditatis, Procellarum, and "South Imbrium") must then predate 4.45 b.y. Such a situation would require a reasonably solid lunar crust at least 4.5 b.y. ago in order for the oldest basins to be preserved. Using an arbitrary maximum age of 4.5 b.y., crater count data, and the observed rate of slumping of crater rims, Baldwin (1971) calculated that 33 (out of 83) planetesimals which formed craters with diameters > one hundred miles impacted the moon between 4.35 and 4.50 b.y. ago. The age agreement between his 4.35–4.50 b.y. episode and our 4.35–4.55 b.y. hump in Fig. 5 is remarkable. Had Baldwin (1971, Table 2) chosen 4.55 b.y. (rather than 4.50 b.y.) as a maximum age, the agreement would be even better. Because U–Pb whole-rock systems may or may not be reset by later excavation events, and because we have a limited amount of available, data, it is difficult to make further semiquantitative correlations between the pattern in Fig. 5 and the McGetchin *et al.* (1973) ejecta stratigraphy model. Assignments of the ~4.0 b.y. event to the Imbrium excavation event and the ~4.25 b.y. event to the Nectaris excavation, however, are consistent with the McGetchin *et al.* (1973) ejecta stratigraphy. An Imbrium event at ~4.0 b.y. continues to explain the numerous 3.9–4.0 b.y. ages of brecciated lunar rocks published by many $^{40}Ar/^{39}Ar$, U–Th–Pb, and Rb–Sr workers.

Now let us examine a frequency versus $^{207}Pb/^{206}Pb$ whole-rock age histogram of mare basalts (Fig. 6). Curiously enough, a similar but more distinct pattern emerges as that for non-mare rocks; three apparent events occurred at ~3.95, ~4.20, and ~4.5 b.y. But how can this be if the basalts were all extruded in the interval between 3.2 and 3.9 b.y. as suggested by their $^{40}Ar/^{39}Ar$ ages and Rb–Sr internal isochrons? Perhaps large excavation events profoundly altered the U–Pb systems of the *source* regions of the mare basalts due to associated large-scale partial melting, and the whole-rock U–Pb systems were scarcely affected by later eruption which grossly affected the Rb–Sr and U–Pb internal isochron and $^{40}Ar/^{39}Ar$ whole-rock systems. If this were so, some correlation would be expected between the mare basalt $^{207}Pb/^{206}Pb$ ages and their position relative to the large mare basins. The Apollo landing sites are plotted in Fig. 7 (adopted from Wilshire and Jackson, 1972) which depicts the relative age sequence of the mare basins (Stuart-Alexander, and Howard, 1970) and the approximate extent of their ejecta. Apollo 15 mare basalts which define the "Imbrium event" in Fig. 6 are closely associated with the Imbrium

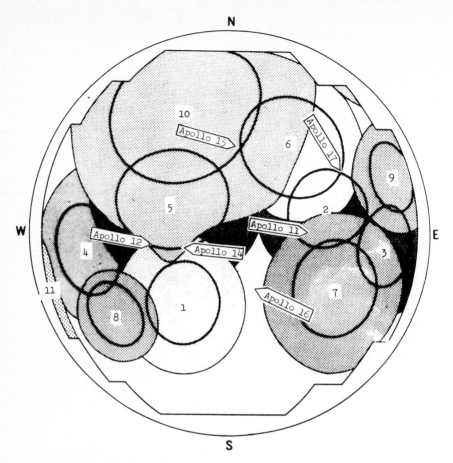

Fig. 7. Map adopted from Wilshire and Jackson (1972) showing main outer mountain rings of major lunar basins and the extent of their continuous ejecta blankets. Numbers 1–11 refer to the relative ages (11 = youngest) of the basins (primarily adopted from Stuart-Alexander and Howard, 1970). The numbered basins are: (1) Nubium, (2) Tranquillitatis, (3) Fecunditatis, (4) Procellarum, (5) "South Imbrium," (6) Serenitatis, (7) Nectaris, (8) Humorum, (9) Crisium, (10) Imbrium, and (11) Orientale. Apollo 11, 12, 14, 15, 16, and 17 sites are designated.

Basin. Five out of eight Apollo 12 mare basalts also enhance the "Imbrium" peak (Fig. 6). The Apollo 12 site is closer to the Imbrium Basin than any other mare basalt site except for the Apollo 15 site. Apollo 11 basalts which are primarily responsible for the ~4.2 b.y. event (Fig. 6) are closely related to the Nectaris Basin. Apollo 17 mare basalts, which define the ~4.5 b.y. event (Fig. 6) are most closely related to the Tranquillitatis and Serenitatis basins but also not far from Crisium which may have somehow affected two of the Apollo 17 mare basalt U–Pb systems. This striking general correlation between whole-rock mare basalt ^{207}Pb/^{206}Pb ages and location indeed suggests that the major mare excavation events grossly affected the whole-rock $^{207}Pb/^{206}Pb$ ages of the mare basalts. Exactly what mechanisms operated during the nearly complete resetting of these U–Pb clocks is not clearly understood, but partially or entirely molten rock at depth may have somehow lost Pb to the surface regions after a major impact event. This interpretation of Pb loss in the mare basalt source regions suggests that these regions were shallower (less than 100 km?) than is commonly accepted. Pre-Imbrium lunar soils, then, may have contained excess Pb relative to U 4.1 b.y. ago and earlier, not attributable to the Imbrium event. We agree with Wilshire and Jackson (1972), who state from their petrologic and stratigraphic study of the Fra Mauro formation that the Fra Mauro clasts reflect "a long history of impact and ejection that probably records a half billion years of planetary accretion, of which the Imbrium impact was one of the last large events on the moon." We fully agree with their conclusion which is supported by U–Pb whole-rock age determinations as already discussed.

The $^{207}Pb/^{206}Pb$ age pattern observed for mare rocks (Fig. 6) is even more enhanced when combined with the non-mare whole-rock data (compare Figs. 5 and 6). Many workers, too numerous to list, found evidence using Rb–Sr internal isochrons, $^{40}Ar/^{39}Ar$ data and U–Th–Pb data for an Imbrium event about 4.0 b.y. ago. One or more events about 4.4–4.5 b.y. old were later clearly documented by U–Pb and Rb–Sr workers (e.g. Murthy *et al.*, 1970; Tatsumoto, 1970a, Schonfeld and Meyer, 1972; Tera and Wasserburg, 1972a; Tera *et al.*, 1973, 1974; Nunes and Tatsumoto, 1973; Nunes *et al.*, 1973; Nyquist *et al.*, 1973). Silver (1970) and Tatsumoto *et al.* (1971) first called attention to the ~4.2 b.y. age. Then the significance of such an age was not clear, and it received little notice until Husain and Schaeffer (1973) began to see discrete $^{40}Ar/^{39}Ar$ plateaus of ~4.2 b.y. in Apollo 16 highland rocks. Interestingly, Doe and Tatsumoto (1972) found evidence for both ~4.2 and ~4.5 b.y.-old events in their stepwise volatilization Pb isotope study of Apollo 12 and 14 soils. Discrete $^{40}Ar/^{39}Ar$ ages between 4.0 and ~4.3 b.y. have since been documented by several workers (e.g. Huneke *et al.*, 1973; Kirsten and Horn, 1974; Stettler *et al.*, 1974; Schaeffer and Husain, 1974). Even the ~4.5 b.y. event(s) have apparently become visible to the $^{40}Ar/^{39}Ar$ clock as recently reported by Jessberger *et al.* (1974). We accept the >4.0 b.y. $^{40}Ar/^{39}Ar$ ages as reflecting discrete events and note that they are entirely consistent with our U–Pb observations. The ~4.2 event is apparent in Rb–Sr whole-rock data (e.g. Nyquist *et al.*, 1973) and a Rb–Sr internal isochron from the Civet Cat clast in Apollo 17 Boulder 1, Station 2 (Compston and Gray, 1974). Thus, the ~4.6–4.0 b.y. age "gap" emphasized by Papanastassiou and Wasserburg (1971, 1972) may not exist. Baldwin (1971) and Schaeffer and Husain (1974) deserve credit as the first workers to attempt to assign distinct ages to pre-Imbrium major basin excavation events. Their age sequence assignments are consistent with our interpretation of the U–Pb results. We cannot clearly resolve events between ~4.0 and ~4.2 b.y. using U–Pb data, such as Schaeffer and Husain (1974) possibly can do using $^{40}Ar/^{39}Ar$ ages. We can, however, see a broad smear of $^{207}Pb/^{206}Pb$ ages suggesting perhaps several unresolved events back in the 4.35–4.55 b.y. range not readily visible to the $^{40}Ar/^{39}Ar$ clock. Adopting Stuart-Alexander and Howard's (1970) relative age assignments of the major basin excavation events and Schaeffer and Husain's (1974) basin-age interpretation and extending it back using our U–Pb data, we propose a very tentative assignment of ages to large basin excavations (Table 6). This is largely guess work at this stage using the $^{207}Pb/^{206}Pb$ data in Figs. 5 and 6, knowledge of lunar ejecta stratigraphy (McGetchin *et al.*, 1973) and our knowledge of relative ages of the lunar basins (Stuart-Alexander and Howard, 1970). Because KREEP material was found mostly at the Apollo 12 and 14 sites (Fig. 7) and because γ-radiation is high just south of the Imbrium area (Metzger *et al.*, 1973), we speculate that at least some of the KREEP in this area was formed about 4.47 b.y. ago as a result of the "South Imbrium" excavation event. The 4.47 b.y. episode is the upper concordia intercept age defined by Tera and Wasserburg (1972a), for KREEP-rich rocks 14310 and 14053, and also the age Nunes and Tatsumoto (1973), calculated for the formation of a KREEP-rich source region inferred from the U–Th–Pb systematics of breccia 66095. "South Imbrium" Basin has no official name and its center is located immediately east of the crater Copernicus (Wilhelms and McCauley, 1971). This explanation of the formation of KREEP ~4.47 b.y. ago immediately following the excavation of "South Imbrium" weakens Schonfeld and Meyer's (1973) argument that the Imbrium Basin predates KREEP formation, but it also is consistent with their view of relative events if "South Imbrium" is substituted for Imbrium in their discussion. The relative age of "South Imbrium" is very uncertain although it appears definitely older than Serenitatis (D. E. Wilhelms, oral communication, April 1974). Four Apollo 17 whole-rock U–Pb analyses of mare basalts discussed earlier which are within error concordant at 4.49–4.57 b.y. suggest to us that the Tranquillitatis basin excavation event may have affected the source regions of these oldest concordant mare U–Pb basalt ages yet found. Thus, we suspect

Table 6. Proposed major basin excavation chronology for the moon.

Basin	Possible Age (b.y.) of excavation
Nubium	>4.55?
Tranquillitatis	~4.55?
Fecunditatis	4.47 – 4.55?
Procellarum	4.47 – 4.55?
"South Imbrium"*	~4.47
Serenitatis	~4.45?
Nectaris†	~4.2
Humorum‡	4.13 – 4.2
Crisium‡	~4.13
Imbrium†	~3.99
Orientale‡	~3.85

*"South Imbrium" refers to the large basin whose center lies just east of the crater Copernicus.

†From both our interpretation and that of Schaeffer and Husain (1974).

‡Age assignments from Schaeffer and Husain (1974)—consistent with our U–Pb data.

Tranquillitatis basin and the still older Nubian Basin predate the "South Imbrium" excavation. The relative age of "South Imbrium" with respect to Fecunditatis and Procellarum is not clear and was arbitrarily assigned (Table 6). Regardless of the exact relative age sequence of the pre-Serenitatis basins, if the stated basin age interpretation is approximately correct, the moon's outer crust must have been rather solid >4.55 b.y. ago in order to support the earliest basins. This reasoning supports an old age of about 4.65 b.y. for the moon because the moon was probably largely or entirely molten following accretion (e.g. Mitzutani *et al.*, 1972; Taylor and Jakeš, 1974; Tatsumoto *et al.*, 1974) and because some time (100 m.y.?) must be required for a reasonably thick crust to form. About 45% of the 29 major lunar basins Stuart-Alexander and Howard (1970) considered are as old or older than Serenitatis and may have impacted the lunar surface in a narrow time span about 4.55–4.45 b.y. ago with the remaining 55% of the planetesimals impacting over a much longer time interval between about 4.45 and 3.8 b.y. ago. Thus, the proposed basin excavation chronology in Table 6 seems to require an exponentially decreasing flux of impacting planetesimals such as that proposed by Baldwin (1971) and Hartmann (1972). Our proposed chronology, then, avoids contriving a 0.5–0.7 b.y. orbital storage interval for the large impacting planetesimals which is essential to the 4.0 b.y. cataclysm model (Papanastassiou and Wasserburg, 1971). The hypothesized chronology (Table 6) is consistent with Morgan *et al.*'s (1974) attempt to characterize the lunar basins with siderophile trace element ratios related to the impacting planetesimals. Our proposed basin excavation chronology (Table 6) closely resembles the independently derived impact-cratering chronology Baldwin (1971) published.

Admittedly, our chronological model (Table 6) leaves unresolved some major problems, such as the problem of explaining why mare basalts with internal Rb–Sr isochron ages (i.e. crystallization ages) greater than 3.85 b.y. have not been found. Possibly even these relatively young Rb–Sr internal isochron ages are related to young impact events (Birck and Allegre, 1974). Whether this is a valid mechanism for melting mare basalts or not, however, does not affect our interpretation of the U–Th–Pb and Rb–Sr systematics. Judging from the difference between mare basalt ^{207}Pb/^{206}Pb ages and Rb–Sr internal isochron ages (emphasized by Tatsumoto *et al.*, 1972a), a time span of about 500–800 m.y. appears to separate basin formation and later mare basalt extrusion. Possibly this time span is related to the heating of adjacent highland basement rocks which were thermally insulated by basin fractures and ejecta blankets, as Arkani-Hamed (1973) proposed. Regardless of the unresolved difficulties, the basin excavation chronology model in Table 6 deserves serious reflection when seeking to understand lunar U–Pb whole-rock data, and the early evolution of the moon.

SUMMARY

Although two-stage (e.g. Tatsumoto, 1970b; Tera and Wasserburg, 1972a) and three-stage (e.g. Nunes and Tatsumoto, 1973; Tera and Wasserburg, 1972a) U–Pb evolution histories may be applicable for certain individual rock systems, a complex multiple stage U–Pb evolution history more appropriately explains all the available U–Pb lunar rock data (i.e. simple two- and three-stage U–Pb evolution models are simply too simple for the moon as a whole). These multiple events reflected in the U–Pb lunar data, we speculate, are related to the complex planetesimal bombardment history which dominated lunar events from ~4.5 to ~3.9 b.y. ago. The ~4.0, ~4.2, and ~4.4 to ~4.5 b.y. "events" which are evident on whole rock frequency versus ^{207}Pb/^{206}Pb age histograms probably each reflect multiple cratering episodes. It appears that for mare basalts, complete resetting of the source rock U–Pb systems owing to Pb loss relative to U was often approached after a major planetesimal impact. It further appears that during later melting and extrusion of these basalts 500–800 m.y. after impact, the U–Pb total rock systems were negligibly disturbed.

APPENDIX: SAMPLE DESCRIPTIONS* AND LOCALITIES

Sample	Description	Station	Location
71569,17	Fine grained rake basalt	1A	East of Emory Crater
72155,26	Medium basalt	LRV3	Near boundary, dark and light mantle
72255,49	Boulder 1, layered light gray breccia	2	Large boulder, south massif
72255,54	Boulder 1, layered light gray breccia	2	Large boulder, south massif
72255,60	Boulder 1, layered light gray breccia	2	Large boulder, south massif
72255,67	Boulder 1, layered light gray breccia	2	Large boulder, south massif
72275,73	Boulder 1, layered light gray breccia	2	Large boulder, south massif
72275,81	Boulder 1, layered light gray breccia	2	Large boulder, south massif
72275,117	Boulder 1, layered light gray breccia	2	Large boulder, south massif
72701,35	<1 mm fines, rake	2	South massif, light mantle
74001,8	Drive tube (L44, lower)	4	Rim of Shorty Crater
74220,14	Unsieved fines, orange	4	Rim of Shorty Crater
74240,15	Unsieved fines	4	Rim of Shorty Crater
74255,25	Coarse basalt	4	Boulder on rim of Shorty Crater
74275,53	Fine basalt	4	Rim of Shorty Crater
75120,3	Unsieved fines	LRV7	Near Victory Crater
75035,45	Medium basalt	5	South rim of Camelot
75055,9	Coarse basalt	5	South rim of Camelot, small boulder
76501,32	<1 mm fines, rake	6	~10 m south of small crater, north massif
77017,38	Brecciated olivine gabbro	7	15–20 m northeast of Boulder 7, north massif
78155,38	Gabbroic breccia	8	Inside small crater, southwest edge of Sculptured Hills
79155,32	Coarse basalt	9	Southeast of Van Serg
79221,23	<1 mm fines	9	~60 m southeast of Van Serg; northeast of crater
79241,23	<1 mm fines	9	~60 m southeast of Van Serg; northeast of subdued crater
79261,32	<1 mm fines	9	~60 m southeast of Van Serg; northeast of subdued crater

*Detailed descriptions and sample localities may be obtained from NASA, MSC 03211, *Lunar Sample Information Catalog, Apollo 17*, prepared by the Lunar Receiving Laboratory; NASA SP-330, *Apollo 17 Preliminary Science Report*, prepared by the Manned Spacecraft Center; and U.S. Geological Survey, Interagency Report: *Astrogeology 71, Documentation and Environment of the Apollo 17 Samples: A Preliminary Report*, prepared by Apollo Field Geology Investigation Team, U.S. Geological Survey.

Acknowledgments—We benefited from manuscript reviews by B. R. Doe, R. E. Zartman, L. T. Silver, and R. S. Naylor. We thank A. P. Schwab for laboratory assistance, and Sandra Blanchard and Maybelle Kamrath for help with manuscript preparation. This work was supported primarily by NASA Interagency Transfer order T-2407A.

REFERENCES

Albee A. L., Chodos A. A., Dymek R. F., Gancarz A. J., Goldman D. S., Papanastassiou D. A., and Wasserburg G. J. (1974) Dunite from the lunar highlands: Petrography, deformational history, Rb–Sr age. (abstract). In *Lunar Science—V*, pp. 3–5. The Lunar Science Institute, Houston.

Arkani-Hamed J. (1973) On the formation of lunar mascons. *Proc. Fourth Lunar Sci. Conf., Geochim. Cosmochim. Acta*, Suppl. 4, Vol. 3, pp. 2673–2684. Pergamon.

Baldwin R. B. (1971) On the history of lunar impact cratering: The absolute time scale and the origin of planetesimals. *Icarus* **14**, 36–52.

Barnes I. L., Carpenter B. S., Garner E. L., Gramlich J. W., Kuehner E. C., Machlan L. A., Maienthal E. J., Moody J. R., Moore L. J., Murphy T. J., Paulsen P. J., Sappenfield K. M., and Shields W. R. (1972) Isotopic abundance ratios and concentrations of selected elements in Apollo 14 samples. *Proc. Third Lunar Sci. Conf., Geochim. Cosmochim. Acta*, Suppl. 3, Vol. 2, pp. 1465–1472. MIT Press.

Barnes I. L., Garner E. L., Gramlich J. W., Machlan L. A., Moody J. R., Moore L. J., Murphy T. J., and Shields W. R. (1973) Isotopic abundance ratios and concentrations of selected elements in some Apollo 15 and Apollo 16 samples. *Proc. Fourth Lunar Sci. Conf., Geochim. Cosmochim. Acta*, Suppl. 4, Vol. 2, pp. 1197–1207. Pergamon.

Birck J. L. and Allegre C. J. (1974) Constraints imposed by ^{87}Rb–^{87}Sr on lunar processes and on the composition of the lunar mantle (abstract). In *Lunar Science—V*, pp. 64–66. The Lunar Science Institute, Houston.

Bjorkholm P. J., Golub L., and Gorenstein P. (1973) Distribution of ^{222}Rn and ^{210}Po on the lunar surface as observed by the alpha particle spectrometer. *Proc. Fourth Lunar Sci. Conf., Geochim. Cosmochim. Acta*, Suppl. 4, Vol. 3, pp. 2793–2802. Pergamon.

Cameron A. E., Smith D. H., and Walker R. L. (1969) Mass spectrometry of nanogram-size samples of lead. *Anal. Chem.* **41**, 525–526.

Compston W. (1974) REE trends and Rb–Sr model ages in mare basalts (abstract). In *Lunar Science—V*, pp. 135–137. The Lunar Science Institute, Houston.

Compston W. and Gray C. M. (1974) Rb–Sr age of the Civet Cat clast, 72255,41. In interdisciplinary studies of samples from Boulder 1, Station 2, Apollo 17 (abstract). *Consortium Indomitabile* (editor John Wood), Vol. 1, pp. 139–144. Smithsonian Astrophysical Laboratory.

Doe B. R. and Tatsumoto M. (1972) Volatilized lead from Apollo 12 and 14 soils. *Proc. Third Lunar Sci. Conf., Geochim. Cosmochim. Acta*, Suppl. 3, Vol. 2, pp. 1981–1988. MIT Press.

Evensen N. M., Murthy V. R., and Coscio M. R., Jr. (1973a) Taurus-Littrow: Age of mare volcanism; chemical and Rb–Sr isotopic systematics of the dark mantle soil. *EOS* **54**, 587–588.

Evensen N. M., Murthy V. R., and Coscio M. R., Jr. (1973b) Rb–Sr ages of some mare basalts and the isotopic and trace element systematics in lunar fines. *Proc. Fourth Lunar Sci. Conf., Geochim. Cosmochim. Acta*, Suppl. 4, Vol. 2, pp. 1707–1724. Pergamon.

Fleming E. H., Jr., Ghiorso A., and Cunningham B. B. (1952) The specific alpha-activities and half-lives of U^{234}, U^{235}, and U^{236}. *Phys. Rev.* **88**, 642–652.

Hartmann W. K. (1972) Paleocratering of the moon: Review of post-Apollo data. *Astrophys. Space Sci.* **17**, 48–64.

Haskin L. A., Blanchard D. P., Jacobs J. W., Korotev R. L., Herrmann A. G., and Reid A. M. (1974) Rare earth and other trace elements in some individual 1–2 mm fines from Apollo 16 and 17 (abstract). In *Lunar Science—V*, pp. 310–312. The Lunar Science Institute, Houston.

Helz R. T. and Appleman D. E. (1974) Poikilitic and cumulate textures in Rock 77017, a crushed anorthositic gabbro (abstract). In *Lunar Science—V*, pp. 322–324. The Lunar Science Institute, Houston.

Heiken G. H. and McKay D. S. (1974) Petrography of Apollo 17 soils (abstract). In *Lunar Science—V*, pp. 319–321. The Lunar Science Institute, Houston.

Huneke J. C., Jessberger E. K., Podosek F. A., and Wasserburg G. J. (1973) ^{40}Ar/^{39}Ar measurements in Apollo 16 and 17 samples and chronology of metamorphic and volcanic activity in the Taurus-Littrow region. *Proc. Fourth Lunar Sci. Conf., Geochim. Cosmochim. Acta*, Suppl. 4, Vol. 2, pp. 1725–1756. Pergamon.

Husain L. and Schaeffer O. A. (1973) ^{40}Ar/^{39}Ar crystallization ages and ^{38}Ar–^{37}Ar cosmic ray exposure ages of samples from the vicinity of the Apollo 16 landing site (abstract). In *Lunar Science—V*, pp. 406–408. The Lunar Science Institute, Houston.

Jaffey A. H., Flynn K. F., Glendenin L. E., Bentley W. C., and Essling A. M. (1971) Precision measurement of half-lives and specific activities of ^{235}U and ^{238}U. *Phys. Rev.* **CH**, 1889–1906.

Jessberger E. K., Huneke J. C., and Wasserburg G. J. (1974) Evidence for a ~4.5 aeon age of plagioclase clasts in a lunar highland breccia. *Nature* **248**, 199–202.

Kirsten T. and Horn P. (1974) ^{39}Ar–^{40}Ar chronology of the Taurus-Littrow region II: A 4.28 b.y. old Troctolite and ages of basalts and highland breccis (abstract). In *Lunar Science—V*, pp. 419–421. The Lunar Science Institute, Houston.

Kirsten T., Horn P., and Heymann D. (1973) Chronology of the Taurus-Littrow region I: Ages of two major rock types from the Apollo 17 site. *Earth Planet Sci. Lett.* **20**, 125–130.

LSPET (1973) Apollo 17 lunar samples: Chemical and petrographic description. *Science* **182**, 659–672.

Mattinson J. M. (1972) Preparation of hydrofluoric, hydrochloric, and nitric acids at ultralow lead levels. *Anal. Chem.* **44**, 1715–1716.

McGetchin T. R., Settle M., and Head J. W. (1973) Radial thickness variation in impact crater ejecta: Implications for lunar basin deposits. *Earth Planet Sci. Lett.* **20**, 226–236.

Metzger A. E., Trombka J. I., Peterson L. E., Reedy R. C., and Arnold J. R. (1973) Lunar surface radioactivity: Preliminary results of the Apollo 15 and 16 gamma-ray spectrometer experiments. *Science* **179**, 800–803.

Mitzutani H., Matsui T., and Takeuchi H. (1972) Accretion process of the moon. *The Moon* **4**, 476–489.

Morgan J. W., Ganapathy R., Higuchi H., Krähenbühl U., and Anders E. (1974) Lunar basins: Tentative characterization of projectiles, from meteoritic elements in Apollo 17 boulders (abstract). In *Lunar Science—V*, pp. 526–528. The Lunar Science Institute, Houston.

Murthy V. R., Evensen N. M., and Coscio M. R. (1970) Distribution of K, Rb and Ba, and Rb–Sr isotopic relations in Apollo 11 lunar samples. *Proc. Apollo 11 Lunar Sci. Conf., Geochim. Cosmochim. Acta*, Suppl. 1, Vol. 2, pp. 1393–1406. Pergamon Press.

Nunes P. D. and Tatsumoto M. (1973) Excess lead in "Rusty Rock" 66095 and implications for an early lunar differentiation. *Science* **182**, 916–920.

Nunes P. D., Tatsumoto M., Knight R. J., Unruh D. M., and Doe B. R. (1973) U–Th–Pb systematics of some Apollo 16 lunar samples. *Proc. Fourth Lunar Sci. Conf., Geochim. Cosmochim. Acta*, Suppl. 4, Vol. 2, pp. 1797–1822. Pergamon.

Nunes P. D., Knight R. J., Unruh D. M., and Tatsumoto M. (1974a) The primitive nature of the lunar crust and the problem of initial Pb isotopic compositions of lunar rocks: A Rb–Sr and U–Th–Pb study of Apollo 16 samples (abstract). In *Lunar Science—V*, pp. 559–561. The Lunar Science Institute, Houston.

Nunes P. D., Tatsumoto M., and Unruh D. M. (1974b) U–Th–Pb systematics of some Apollo 17 samples (abstract). In *Lunar Science—V*, pp. 562–564. The Lunar Science Institute, Houston.

Nunes P. D., Tatsumoto M., and Unruh D. M. (1974c) U–Th–Pb and Rb–Sr systematics of Apollo 17 Boulder 7 from the North Massif of the Taurus-Littrow Valley. *Earth Planet. Sci. Lett.* (in press).

Nyquist L. E., Hubbard N. J., and Gast P. W. (1973) Rb–Sr systematics for chemically defined Apollo 15 and 16 materials. *Proc. Fourth Lunar Sci. Conf., Geochim. Cosmochim. Acta*, Suppl. 4, Vol. 2, pp. 1823–1846. Pergamon.

Nyquist L. E., Bansal B. M., Wiesmann H., and Jahn B. M. (1974) Taurus-Littrow chronology: Implications for early lunar crustal development (abstract). In *Lunar Science—V*, pp. 565–567. The Lunar Science Institute, Houston.

Papanastassiou D. A. and Wasserburg G. J. (1971) Rb–Sr ages of igneous rocks from the Apollo 14 mission and the age of the Fra Mauro Formation. *Earth Planet. Sci. Lett.* **12**, 36–48.

Papanastassiou D. A. and Wasserburg G. J. (1972) The Rb–Sr age of crystalline rock from Apollo 16. *Earth Planet Sci. Lett.* **16**, 289–298.

Philpotts J. A., Schuhmann S., Kouns C. W., and Lum R. K. L. (1974) Lithophile trace elements in Apollo 17 soils (abstract). In *Lunar Science—V*, pp. 599–601. The Lunar Science Institute, Houston.

Phinney W. C., Anders E., Bogard D., Butler P., Gibson E., Gose W., Heiken G., Hohenberg C., Nyquist L., Pearce W., Rhodes J. M., Silver L., Simonds C., Strangway D., Turner G., Walker R., Warner J., and Yuhas D. (1974) Progress Report: Apollo 17, Station 6 Boulder consortium (abstract). In *Lunar Science—V*, Suppl. A, pp. 7–13. The Lunar Science Institute, Houston.

Rosholt J. N. and Tatsumoto M. (1970) Isotopic composition of uranium and thorium in Apollo 11 samples. *Proc. Apollo 11 Lunar Sci. Conf., Geochim. Cosmochim. Acta*, Suppl. 1, Vol. 2, pp. 1499–1502. Pergamon.

Schaeffer O. A. and Husain L. (1974) Chronology of lunar basin formation and ages of lunar anorthositic rock (abstract). In *Lunar Science—V*, pp. 663–665. The Lunar Science Institute, Houston.

Schonfeld E. and Meyer C. (1972) The abundances of components of the lunar soils by a least squares mixing model and the formation age of KREEP *Proc. Third Lunar Sci. Conf., Geochim. Cosmochim. Acta*, Suppl. 2, Vol. 2, pp. 1397–1420. MIT Press.

Schonfeld E. and Meyer C., Jr. (1973) The old Imbrian hypothesis. *Proc. Fourth Lunar Sci. Conf., Geochim. Cosmochim. Acta*, Suppl. 4, Vol. 1, pp. 125–138. Pergamon.

Senftle F. E., Farley T. A., and Lazar N. (1956) Half-life of Th^{232} and the branching ratio of Bi^{212}. *Phys. Rev.* **104**, 1629.

Silver L. T. (1970) Uranium–thorium–lead isotopes in some Tranquillity Base samples and their implications for lunar history. *Proc. Apollo 11 Lunar Sci. Conf., Geochim. Cosmochim. Acta*, Suppl. 1, Vol. 2, pp. 1533–1574. Pergamon.

Silver L. T. (1970) Uranium–thorium–lead isotopic characteristics in some regolithic materials from the Descartes Region (abstract). In *Lunar Science—IV*, pp. 672–674. The Lunar Science Institute, Houston.

Silver L. T. (1974) Patterns of U–Th–Pb distributions and isotopic relations in Apollo 17 soils (abstract). In *Lunar Science—V*, pp. 706–708. The Lunar Science Institute, Houston.

Stettler A., Eberhardt P., Geiss J., Grögler N., and Maurer P. (1974) Sequence of Terra Rock Formation and basaltic lava flows on the moon (abstract). In *Lunar Science—V*, pp. 738–740. The Lunar Science Institute, Houston.

Stuart-Alexander D. E. and Howard K. A. (1970) Lunar maria and circular basins: A review. *Icarus* **12**, 440–456.

Tatsumoto M. (1966) Isotopic composition of lead in volcanic rocks from Hawaii, Iwo Jima, and Japan. *J. Geophys. Res.* **71**, 1721–1733.

Tatsumoto M. (1970a) U–Th–Pb age of Apollo 12 rock 12013. *Earth Planet. Sci. Lett.* **9**, 193–200.

Tatsumoto M. (1970b) Age of the moon: An isotopic study of U–Th–Pb systematics of Apollo 11 lunar samples—II. *Proc. Apollo 11 Lunar Sci. Conf., Geochim. Cosmochim. Acta*, Suppl. 1, Vol. 2, pp. 1595–1612. Pergamon.

Tatsumoto M. (1973) U–Th–Pb measurements of Luna 20 soil. *Geochim. Cosmochim. Acta*, **37**, 1079–1086.

Tatsumoto M., Knight R. J., and Doe B. R. (1971) U–Th–Pb systematics of Apollo 12 lunar samples. *Proc. Second Lunar Sci. Conf., Geochim. Cosmochim. Acta*, Suppl. 2, Vol. 2, pp. 1521–1546. MIT Press.

Tatsumoto M., Hedge C. E., Doe B. R., and Unruh D. M. (1972a) U–Th–Pb and Rb–Sr measurements on some Apollo 14 lunar samples. *Proc. Third Lunar Sci. Conf., Geochim. Cosmochim. Acta*, Suppl. 3, Vol. 2, pp. 1531–1555, MIT Press.

Tatsumoto M., Hedge C. E., Knight R. J., Unruh D. M., and Doe B. R. (1972b) U–Th–Pb, Rb–Sr, and K measurements on some Apollo 15 and Apollo 16 samples (abstract). In *The Apollo Lunar Samples*, pp. 391–395. The Lunar Science Institute, Houston.

Tatsumoto M., Nunes P. D., Knight R. J., Hedge C. E., and Unruh D. M. (1973a) U–Th–Pb, Rb–Sr, and K measurements of two Apollo 17 samples (abstract). *EOS* **54**, 614–615.

Tatsumoto M., Knight R. J., and Allegre C. J. (1973b) Time differences in the formation of meteorites as determined by $^{207}Pb/^{206}Pb$. *Science* **180**, 1279–1283.

Tatsumoto M., Nunes P. D., and Unruh D. M. (1974) Early history of the moon: Implications of

U–Th–Pb and Rb–Sr systematics. *Soviet-American Lunar Sci. Conf., Moscow.* In press.

Taylor S. R. and Jakeš P. (1974) Geochemical zoning in the moon (abstract). In *Lunar Science—V*, pp. 786–788. The Lunar Science Institute, Houston.

Tera F. and Wasserburg G. J. (1972a) U–Th–Pb systematics in three Apollo 14 basalts and the problem of initial Pb in lunar rocks. *Earth Planet. Sci. Lett.* **14**, 281–304.

Tera F. and Wasserburg G. J. (1972b) U–Th–Pb systematics in lunar highland samples from the Luna 20 and Apollo 16 missions. *Earth Planet. Sci. Lett.* **17**, 36–51.

Tera F., Papanastassiou D. A., and Wasserburg G. J. (1973) A lunar cataclysm at ~3.95 AE and the structure of the lunar crust (abstract). In *Lunar Science—IV*, pp. 723–725. The Lunar Science Institute, Houston.

Tera F., Papanastassiou D. A., and Wasserburg G. J. (1974) Isotopic evidence for a terminal lunar cataclysm. *Earth Planet. Sci. Lett.* **22**, 1–21.

Turner G., Cadogan P. H., and Yonge C. J. (1973) Argon selenochronology. *Proc. Fourth Lunar Sci. Conf., Geochim. Cosmochim. Acta*, Suppl. 4, Vol. 2, pp. 1889–1914. Pergamon.

Wetherill G. W. (1956) Discordant uranium-lead ages, 1. *Trans. Amer. Geophys. Union* **37**, 320–326.

Wilhelms D. E. and McCauley J. F. (1971) Geologic map of the near side of the moon. U.S. Geol. Surv. Misc. Geol. Inv. Map I-703.

Wilshire H. G. and Jackson E. D. (1972) Petrology and stratigraphy of the Fra Mauro Formation at the Apollo 14 site. U.S. Geol. Survey Prof. Paper 785, 26 pp.

Wood J. A. (representing the Consortium Indomitable) (1974) Investigations of a KREEP Stratified Boulder from the south massif (abstract). In *Lunar Science—V*, Suppl. A, pp. 4–6. The Lunar Science Institute, Houston.

Proceedings of the Fifth Lunar Conference
(Supplement 5, Geochimica et Cosmochimica Acta)
Vol. 2 pp. 1515–1539 (1974)
Printed in the United States of America

Taurus-Littrow chronology: Some constraints on early lunar crustal development

L. E. Nyquist,[1] B. M. Bansal,[2] H. Wiesmann,[2] and B.-M. Jahn[1]*

[1]NASA Johnson Space Center, Houston, Texas 77058

[2]Lockheed Electronics Corporation, Houston, Texas 77058

Abstract—A mineral isochron for Apollo 17 mare basalt 70035,6 yields an age $T = 3.73 \pm 0.11$ AE (2 σ, 1 AE = 10^9 yr), and initial Sr isotopic composition, $I = 0.69924 \pm 0.00005$ (2 σ). This age is interpreted as the time of mare flooding. The low I-value indicates that Rb/Sr in the source material (Rb/Sr $\approx$ 0.006) was similar to that of the source of low-K Apollo 11 basalts but lower than that of the source of Apollo 12 and 15 basalts.

A whole rock isochron for KREEP-rich Apollo 17 noritic breccias yields an age, $T = 4.02 \pm .10$ AE (2 σ), and initial Sr, $I = 0.69974 \pm .00015$ (2 σ). Alternatively, if an age of 4.0 AE, as determined by ^{39}Ar/^{40}Ar techniques, is *assumed* for the noritic breccias, then correction for radiogenic growth in this interval shows that they had very similar ^{87}Sr/^{86}Sr ratios 4.0 AE ago. Some material mixing seems to be called for to produce this effect, but it must also have been accompanied by a redistribution of Rb/Sr. The latter process is most easily attributable to volatilization during the Serenitatis event, and the whole rock age probably gives a good approximation to the time of its occurrence.

A whole-rock isochron for four Apollo 17 anorthositic gabbros yields the imprecise results $T = 4.09 \pm .33$ AE and $I = 0.69924 \pm 17$. Alternatively, assumption of a 4.0 AE age from published ^{39}Ar/^{40}Ar results yields I (4.0 AE) = 0.69932 ± 4 for three of the samples and I (4.0 AE) = 0.69919 ± 6 for the fourth, indicating that isotopic homogenization probably was not achieved among them 4.0 AE ago even though the Rb-Sr systems seem to have been disturbed.

A mineral isochron obtained for an igneous Apollo 15 Apennine Front KREEP fragment yields $T = 3.91 \pm .04$ AE and $I = 0.70070 \pm .00010$, similar to results for Apollo 14 igneous KREEP. If this represents a metamorphic resetting or a total remelting, whole-rock Rb-Sr systematics can be applied to estimate the differentiation time. Several models are considered, each contingent on a particular petrogenetic viewpoint. If, as has recently been suggested, anorthositic gabbros (norites) approximate the source material for KREEP, then a time of initial differentiation about 4.26 AE ago is most probable for at least the Apollo 15 igneous-textured KREEP. The precursor rocks for Apollo 17 noritic breccias and VHA basalts may also have differentiated at or near this time from the same or similar source materials. On the other hand, if initial differentiation of the Apollo 15 KREEP occurred at the time given by the mineral isochron, the source must have had a high Rb/Sr ratio (Rb/Sr ≥ 0.06), similar to that of Apollo 16 KREEP. In this latter case, materials similar to known anorthositic gabbros are not viable candidates as source materials.

Introduction

The Apollo 17 mission resulted in the return of an extraordinarily diverse and informative suite of lunar samples. These samples may be subdivided in several ways, one of the more informative of which is into groups of similar chemical composition. Chemical analyses of soils from the site yield estimates of the relative abundances of each of the major compositional groups. The major

---

*Also at National Research Council, Houston, Texas 77058.

components at the site appear to be the high-titanium mare basalts and the associated(?) orange glass, noritic breccias, and anorthositic gabbros (cf. Rhodes *et al.*, 1974). Two unique samples of great importance were also returned; dunite 72417 and troctolite 76535. These samples derive much of their importance from their probable formation as crystal cumulates early in lunar history. The dunite has been dated at 4.6 AE by the Rb–Sr technique (Albee *et al.*, 1974) and the troctolite has a K–Ar age of 4.3 AE (Bogard *et al.*, 1974). A lithic fragment from coarse fines 78503 has also been dated at 4.3 AE by the $^{39}Ar/^{40}Ar$ method (Kirsten and Horn, 1974). Thus, the foregoing samples give direct evidence of ancient (prior to 4 AE ago) lunar processes.

We believe the isotopic data on whole-rock samples also contain indirect evidence for early processes involved in the formation of the more abundant noritic breccias and anorthositic gabbros. This information is partially obscured by events at and/or prior to ~ 4.0 AE ago. Nevertheless, the study of these samples seems important because (a) they make up the major portion of the boulders derived from the North and South Massifs, (b) surveys of lithic fragments and glasses indicate compositional groupings with average characteristics similar to those of anorthositic gabbros and noritic breccias (e.g. Ridley *et al.*, 1973; Prinz *et al.*, 1973), (c) the orbiting geochemistry experiments (Adler *et al.*, 1972a, 1972b; Metzger *et al.*, 1973) indicate significant portions of these materials may be in the present lunar crust; for example, Taylor *et al.* (1974) conclude from the orbiting data that anorthositic gabbro and low-K Fra Mauro basalt, chemically similar to the noritic breccias, appear to be the most important components of the highland crust, (d) the noritic breccias contain high concentrations of the "KREEP" elements and thus complement data obtained from KREEP basalts found at other sites. (e) KREEP basalts may be one of the first rock types produced by partial melting of the lunar interior (e.g. McConnell and Gast, 1972), (f) the major and trace element concentrations allow a petrogenetic relation between anorthositic gabbro and the KREEP group (Walker *et al.*, 1973; Hubbard *et al.*, 1974). Here we will discuss constraints imposed on such a possible relationship by "whole-rock" Rb–Sr data. We also present mineral isochrons for an Apollo 17 mare basalt and for a 69 mg fragment of crystalline (= igneous-textured) KREEP from the Apennine Front. The former has obvious bearing on the chronology of the Taurus-Littrow site, the latter on the evolution of the ubiquitous KREEP components. Data on a number of Apollo 17 soils and on some Apollo 15 and 16 whole-rock samples are also presented, but are not discussed in detail.

ANALYTICAL PROCEDURES

Analytical results are reported in Table 1. Whole rock concentrations of Rb and Sr and $^{87}Sr/^{86}Sr$ ratios were obtained for the same samples for which mass spectrometric isotope dilution analyses of rare-earth and other trace elements are reported by Hubbard *et al.* (1974) and in references therein. These samples were taken from powders analyzed for major and trace elements by X-ray fluorescence spectrometry (Rhodes *et al.*, 1974; LSPET, 1974). Usually a total sample of ~ 1.3 g was finely ground (~ 325 mesh) prior to the XRF analyses. Our samples (~ 50 mg) were aliquants of this fine powder and should be good representatives of the whole rocks.

Table 1. Rb and Sr analytical results.

Sample	wt. (mg)	Rb (ppm)	Sr (ppm)	$\frac{^{87}Rb^a}{^{86}Sr}$	$\frac{^{87}Sr^b}{^{86}Sr}$	T_B^c	T_L^d
I. *Apollo 17 Noritic Breccias*							
72275,2	52.8	8.97	122.7	0.2115 ± 17	0.71188 ± 6	$4.22 \pm .05$	$4.24 \pm .05$
72435,1	53.7	3.93	171.6	0.0662 ± 6	0.70360 ± 5	$4.73 \pm .09$	$4.80 \pm .09$
76055,5	47.7	5.17	156.6	0.0955 ± 8	0.70511 ± 9	$4.39 \pm .10$	$4.44 \pm .10$
76315,2	52.4	5.88	179.5	0.0948 ± 8	0.70515 ± 5	$4.45 \pm .08$	$4.50 \pm .08$
76315,35M	49.2	5.78	174.4	0.0960 ± 8	0.70521 ± 7	$4.44 \pm .09$	$4.49 \pm .09$
76315,30C3	66.7	3.85	171.5	0.0650 ± 6	0.70351 ± 10	$4.72 \pm .14$	$4.80 \pm .14$
76315,30M	51.6	6.56	174.8	0.1086 ± 9	0.70595 ± 5	$4.40 \pm .08$	$4.46 \pm .08$
76315,52	38.9	3.73	115.2	0.0937 ± 9	0.70491 ± 6	$4.33 \pm .08$	$4.40 \pm .08$
76015,22M	52.3	6.41	171.8	0.1079 ± 9	0.70589 ± 5	$4.39 \pm .07$	$4.45 \pm .07$
76015,37M	53.5	6.67	177.5	0.1088 ± 9	0.70605 ± 5	$4.45 \pm .07$	$4.52 \pm .07$
76015,41M	63.6	6.57	176.6	0.1076 ± 9	0.70589 ± 11	$4.40 \pm .11$	$4.45 \pm .11$
76015,64M	51.5	7.46	173.8	0.1242 ± 10	0.70693 ± 6	$4.40 \pm .07$	$4.44 \pm .07$
73235,55M	50.0	5.13	146.9	0.1010 ± 9	0.70539 ± 6	$4.35 \pm .08$	$4.39 \pm .08$
73275,30	51.1	6.62	171.8	0.1112 ± 10	0.70619 ± 5	$4.45 \pm .07$	$4.49 \pm .07$
77135,2	52.6	7.32	172.2	0.1230 ± 10	0.70688 ± 7	$4.41 \pm .08$	$4.45 \pm .08$
II. *Apollo 17 Anorthositic Gabbros*							
76230,4	78.1	0.448	145.9	0.0089 ± 2	0.69982 ± 7	$5.60 \pm .65$	$6.12 \pm .66$
77017,2	68.4	1.310	141.5	0.0268 ± 3	0.70072 ± 6	$4.22 \pm .20$	$4.40 \pm .20$
78155,2	51.6	2.06	146.7	0.0406 ± 4	0.70164 ± 6	$4.37 \pm .14$	$4.48 \pm .14$
76315,62	52.5	2.34	153.1	0.0441 ± 5	0.70185 ± 5	$4.35 \pm .13$	$4.46 \pm .13$
III. *Apollo 17 Soils and Soil Breccia*							
71501,3	52.1	1.171	159.4	0.0213 ± 3	0.70058 ± 10	$4.83 \pm .32$	$5.05 \pm .32$
75061,4	58.3	1.106	165.8	0.0193 ± 3	0.70036 ± 5	$4.55 \pm .25$	$4.79 \pm .25$
76501,1	44.5	2.403	150.8	0.0461 ± 5	0.70217 ± 14	$4.64 \pm .20$	$4.74 \pm .20$
78501,2	53.8	1.959	153.7	0.0369 ± 4	0.70154 ± 9	$4.61 \pm .16$	$4.73 \pm .16$
79135,1	46.1	1.937	168.6	0.0332 ± 4	0.70119 ± 5	$4.39 \pm .10$	$4.53 \pm .10$
79261,12	47.6	1.80	154.3	0.0338 ± 4	0.70128 ± 6	$4.50 \pm .17$	$4.69 \pm .18$
70181,3	50.0	1.468	169.8	0.0250 ± 4	0.70083 ± 4	$4.81 \pm .19$	$5.00 \pm .19$
IV. *15434,73 Mineral Separates (69 mg crystalline KREEP fragment)*							
200–325 Mesh	6.1	15.77	140.7	0.3242 ± 26	0.71899 ± 4	$4.27 \pm .05$	$4.29 \pm .05$
Plag	4.6	2.826	314.4	0.0260 ± 3	0.70214 ± 6		
Mesostasis + Px	4.0	53.37	169.3	0.912 ± 7	0.75138 ± 6		
V. *70035 Mineral Separates (mare basalt)*							
70035,1	58.5	0.461	173.7	0.00772 ± 23	0.69967 ± 6	$5.1 \pm .7$	$6.0 \pm .7$
70035,6	53.2	0.628	161.3	0.01126 ± 29	0.69980 ± 6	$4.3 \pm .5$	$4.9 \pm .5$
Plag	4.4	0.0948	687.5	0.00040 ± 4	0.69924 ± 10		
Ilm 1	11.9	0.8345	47.79	0.05053 ± 37	0.70195 ± 8		
PX	22.7	0.3738	43.73	0.02473 ± 28	0.70059 ± 8		
PX + Ilm	26.9	0.6334	52.01	0.03524 ± 39	0.70112 ± 20		
Ilm 2	8.9	0.832	66.01	0.03647 ± 50	0.70116 ± 4		
VI. *Apollo 15 and 16 Whole Rocks*							
15386,1 (KREEP)	37.9	18.46	187.4	0.285 ± 2	0.71640 ± 7	$4.25 \pm .04$	$4.28 \pm .04$
60335,77 (VHA)	64.9	6.18	162.5	0.1101 ± 9	0.70570 ± 5	$4.19 \pm .06$	$4.23 \pm .06$
67955,56	54.7	0.885	169.1	0.0151 ± 3	0.70012 ± 8	$4.70 \pm .46$	$5.01 \pm .46$
63335,36	70.0	1.146	222.1	0.0149 ± 2	0.69997 ± 5	$4.08 \pm .29$	$4.40 \pm .29$
61295,34	73.9	2.308	186.0	0.0359 ± 3	0.70130 ± 6	$4.28 \pm .15$	$4.41 \pm .15$
67075,53	64.2	0.593	145.0	0.0118 ± 3	0.69984 ± 7	$4.38 \pm .52$	$4.78 \pm .52$

Table 1. (*Continued*).

Sample	wt. (mg)	Rb (ppm)	Sr (ppm)	$\frac{^{87}Rb^a}{^{86}Sr}$	$\frac{^{87}Sr^b}{^{86}Sr}$	T_B^c	T_L^d
Concurrent Analyses of NBS 987 SrCO₃							
1/30/73[e]					0.71022 ± 6		
5/7/73[e]					0.71027 ± 6		
5/8/73[e]					0.71032 ± 6		
9/18/73[e]					0.71025 ± 6		
9/19/73[f]					0.71026 ± 8		
10/22/73[f]					0.71027 ± 7		

[a] Uncertainties correspond to last figures.
[b] Uncertainties correspond to last figures and represent 2 σm. Normalized to $^{88}Sr/^{86}Sr = 8.37521$.
[c] Model age assuming $I = 0.69910$ (BABI plus our bias).
[d] Model age assuming $I = 0.69903$ (Apollo 16 anorthosites for $T = 4.6$ AE).
[e] Voltage scanning.
[f] Magnetic field scanning.

Mineral separates of 70035 were prepared by handpicking and by Frantz magnetic separation. About 200 mg of a coarsely crushed whole-rock sample was taken for mineral separation after completion of the whole-rock chemical and isotopic studies. The sample was ultrasonically cleaned in spectrophotometric grade acetone and sieved into $+ 100$, 100–200, 200–325, and $- 325$ mesh fractions with new nylon sieves in a previously cleaned Spex sieve. The $+ 100$ mesh fraction ($\sim 200\ \mu$m) was handpicked to yield plagioclase (plag), ilmenite (Ilm 1), pyroxene (PX) and a plagioclase free mixture of ilmenite and pyroxene (PX + Ilm). An additional ilmenite separate (Ilm 2) was prepared from the 100–200 mesh fraction by magnetic separations (Frantz) followed by handpicking. The purity of the handpicked fraction was visually estimated at $\geq 99\%$, that of the magnetic separate at $\geq 90\%$.

The 15434,73 sample was a 69 mg fragment of crystalline KREEP. The following petrographic data were supplied by W. C. Phinney (personal communication) from study of the corresponding thin section (15434,47).

In general, rocks of this type contain about 35–40% plagioclase, 20–30% pigeonite, 15–25% mesostasis, and 5–10% hyptersthene by volume. This sample was relatively coarse grained with plagioclase, pigeonite, and hypersthene laths up to 1.2 mm long and 0.2 mm wide. The opaque-rich matrix occurred in irregular patches ~ 100 to $\sim 500\ \mu$m across.

No attempt was made to obtain a representative "whole-rock" sample of this fragment. It was crushed in a boron carbide mortar and sieved in the same manner as the 70035 sample. About 20 mg of the 100–200 mesh fraction and 15 mg of the 200–325 mesh fraction were obtained. The 100–200 mesh fraction was divided into magnetic and non-magnetic fractions by several passes through a Frantz magnetic separator. The nonmagnetic fraction was further purified by handpicking to obtain a mesostasis free plagioclase separate. The magnetic portion is referred to as "mesostasis + PX" in Table 1, and yielded the most radiogenic sample analyzed. A 6 mg portion of the 200–325 mesh fraction was analyzed and is also quite radiogenic because of the relatively high-Rb content of this type of rock.

Mineral separations were carried out in laminar flow hoods. For handpicking the samples were displayed on a cleaned TFE teflon petri dish and sucked through a Pyrex tube into a clean plastic vial by a vacuum device. Some handpicking was also done with stainless steel tweezers. The Frantz magnetic separator has a stainless steel trough which is cleaned prior to use and covered by an aluminum foil during separations. The sample hopper is a replaceable cone of glassine powder paper. Samples are deposited into cleaned Pyrex beakers after passage through the Frantz. Sample transfer operations are done with glassine powder paper.

Chemical procedures for the whole-rock samples were as reported previously (Nyquist *et al.*, 1973a). Blanks for this procedure are about 1.3 ng Rb and 18 ng Sr, most of which is derived from cross contamination from other spikes used in our multielement analysis. Blank corrections were made assuming maximum variations of ± 100% in the Rb blank and ± 50% variations in Sr blanks. This is probably conservative as the amount of other spikes used does not vary significantly from sample to sample and thus the total cross contamination blank should be rather constant.

Abbreviated chemical procedures were used for the mineral separates. Blanks for these procedures ranged from 0.05 to 0.10 ng for Rb and from 1.6 to 3.4 ng for Sr. Procedures were slightly modified from time to time, but blanks have remained essentially constant for more than a year.

Mass spectrometry was similar to that reported previously (Nyquist et al., 1973a). However, we have changed from scanning the Sr spectrum via the accelerating voltage to scanning via the magnetic field. No difference in results has been detected, although the mass discrimination has been reduced from ⩽ 0.5%/mass to ⩽ 0.1%/mass. This conclusion was verified both by analyses of the SRM 987 $SrCO_3$ standard and by alternating "magnetic" and "voltage" data on several samples.

A nonrandom source of error in our previous analytical techniques was discovered during the early portion of these studies. Previously, a radioactive ^{89}Sr tracer had been routinely added to the multielement analyses in order to locate the Sr elution peak with certainty. The amount of the tracer added was increased with the decay of the ^{89}Sr activity until the ^{88}Sr present in the tracer became a significant contributor to the total ^{88}Sr peak, causing an error in the normalization factor used in calculating $^{87}Sr/^{86}Sr$. Fortunately, we had retained the Sr fractions of earlier analyses. The $^{88}Sr/^{90}Sr$ ratio in the tracer and the ^{90}Sr signal in samples to which the tracer had been added was determined and the appropriate correction made. This effect caused an error of about − 0.004% in analyses of Apollo 16 anorthosites and thus in our previously published LUNI value (Nyquist et al., 1973a). The BABI model ages of 76055 and 77135 referred to in Nyquist et al. (1973b) are also in error due to this effect. The corrected values are given in Table 1. The effect was negligible for other earlier reported analyses.

The radioactive tracer was not added to any of the mineral separates, but was added to the whole-rock samples of 70035. Uncorrected data for the latter were colinear with data for the mineral separates within the analytical uncertainty. However, correction for the tracer increased their $^{87}Sr/^{86}Sr$ ratios by an amount comparable to the analytical precision, improving the fit of the mineral isochron and reducing the calculated age by 0.09 AE. Our earlier result, orally reported at the Fourth Lunar Science Conf., is thus revised here. The revised value is well within the originally assigned uncertainty, however.

RESULTS

The isochron diagram for mare basalt 70035 is given in Fig. 1. The data define an age of $3.73 \pm .11$ AE (2σ) and initial $^{87}Sr/^{86}Sr = 0.69924 \pm .00005$ (2σ). The isochron parameters are in satisfactory agreement with the values obtained for 70035,9 by Evensen et al. (1973). Duplicate $^{39}Ar/^{40}Ar$ analyses of 70035,6 by Stettler et al. (1973) yielded ages of 3.72 ± 0.07 AE and 3.75 ± 0.07 AE, in excellent agreement with our Rb–Sr results. The possibility that the isochron is generated by the mixing of two phases is excluded because a plot of Rb versus Sr does not yield a straight line and because very pure separates of the major minerals were obtained.

Several groups have determined ages of about 3.78 AE for mare basalt 75055 by the Rb–Sr and $^{39}Ar/^{40}Ar$ techniques (Tera et al., 1973; Huneke et al., 1973; Kirsten et al., 1973; Turner et al., 1973). A Rb–Sr age of $3.64 \pm .09$ AE has been reported for 71055 (Tera et al., 1974). Kirsten and Horn (1974) report $^{39}Ar/^{40}Ar$ ages of fine-grained subfloor basalt 70215 and two fine-grained basalt fragments between $3.76 \pm .07$ AE and $3.84 \pm .03$ AE. These ages appear to restrict the period of mare flooding of the valley subfloor to the period of ~ 3.65 to ~ 3.85 AE. Nearly identical ages were determined for similar basalts at the Apollo 11 site (Papanastassiou et al., 1970; Turner, 1970) indicating nearly coincident periods of mare

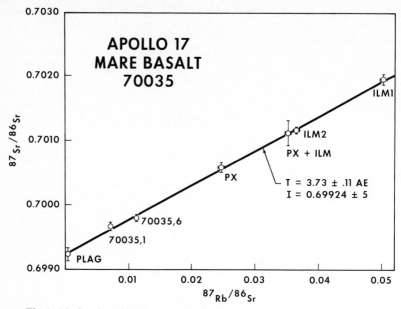

Fig. 1. Rb–Sr mineral isochron for Apollo 17 high-titanium mare basalt 70035.

flooding for the two sites. The I-value of 70035 is also similar to that of low-K Apollo 11 mare basalts (Papanastassiou *et al.*, 1970) indicating that Rb/Sr in the source materials was also similar. For LUNI = 0.69903 and an assumed 4.60 AE age for the moon a two-stage evolution model yields Rb/Sr = 0.006 in the source material. I-values calculated for our data on Apollo 15 mare basalts (Nyquist *et al.*, 1973a) assuming ages of 3.35 AE (e.g. Papanastassiou and Wasserburg, 1973) yields an average I (3.35 AE) = 0.69946 and first stage Rb/Sr = 0.009, slightly higher than for low-K Apollo 11 and 17 basalts, but similar to that for Apollo 12 basalts (Papanastassiou and Wasserburg, 1971a).

Figure 2 presents whole-rock data for the noritic (blue–gray, green–gray) breccias and the anorthositic gabbros. The noritic breccias lie along a line corresponding to an age of $4.02 \pm .10$ AE. A whole-rock age always has some ambiquity because of the inability to guarantee the same initial $^{87}Sr/^{86}Sr$ ratios in all samples. However, we have been reasonably successful in applying the whole-rock technique to high-grade metamorphic samples belonging to the same compositional group (Nyquist *et al.*, 1972, 1973a). Possible reasons for this success will be discussed in the next section. Here we simply note that the noritic breccia isochron whole-rock age is in good agreement with $^{39}Ar/^{40}Ar$ ages determined for some of these very highly recrystallized samples. Several investigators (Huneke *et al.*, 1973; Turner *et al.*, 1973; Kirsten *et al.*, 1973) have determined $^{39}Ar/^{40}Ar$ ages ranging from 3.97 to 4.05 AE for 76055. Turner (oral communication at Fifth Lunar Science Conference) reported ages of $3.98 \pm .04$ AE for boulder consortium samples from 76315 (corresponding matrix samples are included on our noritic breccia isochron). Apparently whole-rock Rb–Sr systematics for these rocks were reset by the event recorded in the $^{39}Ar/^{40}Ar$ ages.

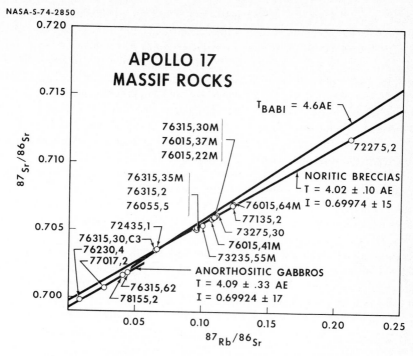

Fig. 2. Rb–Sr whole-rock isochrons for two types of Apollo 17 massif rocks.

Because of the metamorphic texture of these rocks and their probable derivation from Serenitatis ejecta (ALGIT, 1973), this age has been consistently identified with the Serenitatis event (Huneke *et al.*, 1973; Turner *et al.*, 1973; Kirsten *et al.*, 1973). We agree that this conclusion is the most probable one and note that analogous Rb–Sr systematics were also found for high-grade breccias at the Fra Mauro site (Nyquist *et al.*, 1972).

Whole-rock data for anorthositic gabbros do not allow a precise age determination but also strongly suggest that Rb and Sr were redistributed by the 4.0 AE event. Again the $^{39}Ar/^{40}Ar$ ages appear to reflect the same event. Kirsten *et al.* (1974) have determined a 4.05 ± .03 AE age for 77017 by $^{39}Ar/^{40}Ar$ and Turner (oral presentation at Fifth Lunar Science Conference) has reported 3.98 ± .04 AE for an anorthositic gabbro clast of 76315, the same result as for the noritic matrix portion of the rock.

Petrographic studies of the noritic breccias indicate that they were formed by a very high degree of metamorphism and recrystallization and the Ar retention ages are interpreted as defining the time of this event (Albee *et al.*, 1973; Chao, 1973). This has a simple but profound implication: materials enriched in the KREEP elements must have been present at or near the lunar surface prior to 4.0 AE ago. This conclusion follows independently of the final mode of formation of the rocks. If the breccias themselves are reasonable approximations to the precursor materials, containing only a small admixture of foreign material (Hubbard *et al.*, 1974), then the primary rock type approximated was present prior to 4.0 AE ago. If instead, as has been suggested as an alternative (Schonfeld, 1974), the rocks represent intricate mixtures of trace element poor materials with a KREEP basalt component, then the latter must have been present prior to 4.0 AE ago. Indeed, several workers (cf. Hubbard and Gast, 1971; Schonfeld and Meyer, 1972, Nyquist *et al.*, 1973a) have argued on geochemical grounds that KREEP basalts must have been formed by small degrees of partial melting with attendant large changes in Rb/Sr so that the single-stage model ages of about 4.3–4.4 AE must approximate the time of differentiation. Mineral isochron ages of 3.88 AE obtained on Apollo 14 KREEP basalts (Papanastassiou and Wasserburg, 1971b) seem at variance with this argument. If the mineral ages represent a metamorphic resetting or a total remelting, the dilemma would be solved. This has been the preferred interpretation of those wishing to ascribe significance to the various model age arguments, and seems to be supported by the age data on the Apollo 17 noritic breccias. Compston *et al.* (1972) have discussed the conditions necessary for resetting mineral ages.

Assuming that KREEP differentiation did occur prior to 3.9 AE ago, there would seem to be a chance that some of this material escaped the 3.9 AE "cataclysm" and would show an older mineral age. Some fragments and pebbles from the Apennine Front are another source of KREEP and, at least on the scale sampled, have igneous textures (Meyer, 1972). The isochron diagram shown in Fig. 3 is the first mineral isochron obtained for Apollo 15 KREEP. The age,

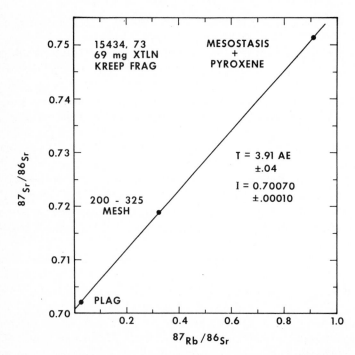

Fig. 3. Rb–Sr mineral isochron for Apollo 15 Apennine Front KREEP fragment 15434,73.

3.91 ± 0.04 AE, is the same as that found for KREEP rake sample 15382 by the ^{39}Ar/^{40}Ar method (Stettler *et al.*, 1973; Turner *et al.*, 1973) and within error of that of Apollo 14 crystalline KREEP (Papanastassiou and Wasserburg, 1971b; Wasserburg and Papanastassiou, 1971). The initial ^{87}Sr/^{86}Sr ratio, $I = 0.70070 \pm 10$, is slightly higher than that found for Apollo 14 crystalline KREEP, within error of that found for a whole-rock isochron for high-grade Apollo 14 breccias (Nyquist *et al.*, 1972), and perhaps slightly lower than that for breccia clast 14321,184-53 (Mark *et al.*, 1973). Both of the "radiogenic" samples analyzed are mixtures of phases, raising the possibility that the isochron is also a mixing line. However, a plot of Rb versus Sr for the three samples does not yield a straight line indicating that more than two Rb- and Sr-bearing phases are present and that the isochron is not simply a mixing line. Thus, this Apollo 15 fragment fails to confirm a pre-3.9 AE age for KREEP. This, however, may simply mean that KREEP at both the Apollo 14 and 15 sites was reset by the Imbrium impact. Intriguingly, Haines *et al.* (1974) find extremely high track excesses in whitlockites from two Apennine Front Kreep fragments. These authors point out that the excess suggests a substantial contribution from the spontaneous fission of ^{244}Pu and is consistent with a track retention age of 4.3 AE for an initial Pu/U $= 0.02$. The Rb–Sr model ages of Apennine Front KREEP is also 4.3 AE (Nyquist *et al.*, 1973a) and represents our preferred time of initial differentiation of this material.

DISCUSSION

Whole-rock isochrons

Several questions need to be addressed before the whole-rock data can be further interpreted. They are: (a) Under what conditions are whole-rock ages such as those shown in Fig. 2 valid? (b) What constraints can the isotopic data put on the alternative theories of petrogenesis? (c) What is the reason for the difference in Rb–Sr systematics between KREEP-rich breccias and the igneous textured variety?

Figure 4 aids in addressing these questions. In this figure we have corrected whole-rock ^{87}Sr/^{86}Sr and ^{87}Rb/^{86}Sr values for radiogenic ^{87}Sr growth in the past T years, where $T = 3.9$–4.0 AE, assuming closed system evolution. In cases where mineral isochrons have been determined we have plotted the I-values versus the decay-corrected ^{87}Rb/^{86}Sr ratios. We have done this in order to consider the effects of processes acting at or before ~ 4.0 AE ago. The data considered are: (a) Apollo 17 anorthositic gabbros ($T = 4.09$ AE from Fig. 2); (b) Apollo 16 VHA basalts ($T = 3.93$ AE from a whole-rock isochron using data of Nyquist *et al.*, 1973, and Table 1); (c) Apollo 17 noritic breccia ($T = 4.00$ AE from Fig. 2 and ^{39}Ar/^{40}Ar ages); (d) Apollo 16 meta-clastic KREEP, ($T = 3.93$ AE from ^{39}Ar/^{40}Ar (Husain and Schaeffer, 1973) and a Rb–Sr secondary isochron on 65015 (Papanastassiou and Wasserburg, 1972); (e) Apollo 15 crystalline KREEP ($T = 3.91$ AE from 15434,73); (f) Apollo 14 KREEP mineral isochron values ($T = 3.88$ AE), and (g) Apollo 14 breccias and soils for an assumed $T = 3.94$ AE. The shaded regions give I-values obtained from whole-rock isochrons were appropriate.

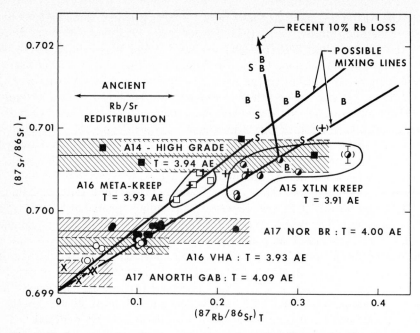

Fig. 4. Rb–Sr systematics of some non-mare rocks at time *T*, assuming closed system evolution during the last *T* years. Values of *T* are given in the figure. Symbols used are: ×: Apollo 17 anorthositic gabbro, ○: Apollo 16 VHA, ●: Apollo 17 noritic breccia, □: Apollo 16 KREEP, ◐: Apollo 15 KREEP, ■: Apollo 14 high grade KREEP breccias, + Apollo 14 igneous KREEP, B: Apollo 14 low grade breccias, S: Apollo 14 soils. The horizontal arrow labeled "ancient Rb/Sr redistribution" refers to redistribution at time *T*. The nearly vertical arrow represents the effects of recent Rb loss (i.e. lowering of Rb/Sr) on the calculation used to correct for radiogenic growth. Two possible mixing lines, through the Apollo 16 and 15 KREEP fields, respectively, are shown. The diagram is not completely rigorous for inferring mixing relations because slightly differing values of *T* have been used for the different rock types. Data are from this laboratory and Papanastassiou and Wasserburg (1971b) and Wasserburg and Papanastassiou (1971) adjusted for interlaboratory bias. The data for 15434,73 are shown enclosed in parentheses because of the probability that the 200–325 mesh fraction analyzed was unrepresentative of the whole rock.

The figure illustrates several points, but perhaps none more clearly than that some rocks with essentially the same $^{87}Sr/^{86}Sr$ ratios and chemical composition 3.9–4.0 AE ago had a wide variation in Rb/Sr. This argument may appear somewhat circular in that by choosing whole rock ages to make the radiogenic growth corrections we have guaranteed the best attainable horizontal alignment in Fig. 4 for data included on the respective isochrons. The circularity could be broken by using ages determined in an independent manner, e.g. $^{39}Ar/^{40}Ar$ ages. This is somewhat academic as the Ar ages agree with the whole-rock Rb–Sr values within the respective uncertainties. We have already commented on this for the Apollo 17 noritic breccias and anorthositic gabbros. The same is true for VHA basalts if the suggestion by Delano *et al.* (1973) that "type A F11R" and VHA basalts are equivalent is correct. The latter does indeed seem to be the case judging from the major element composition of "type A F11R" given by these authors. The average $^{39}Ar/^{40}Ar$ age found by Schaeffer and Husain (1973) for F11R rocks is 3.98 AE. Further, Mark et al. (1974) report a mineral isochron for VHA basalt 62295 corresponding to an age of $4.00 \pm .06$ AE and $I = 0.69956 \pm 6$. Our earlier reported VHA whole-rock isochron parameters (Nyquist et al., 1973a) were $T = 4.06 \pm .11$ AE and $I = 0.69944 \pm 13$. We are revising this value slightly due to the exclusion of sample 61016,143 which does not completely fulfill the VHA criteria (cf. Hubbard et al., 1974) and the inclusion of 60335,77 (Table 1) which does. Our revised values are $T = 3.93 \pm .08$ AE, and $I = 0.69958 \pm 10$. The aberrant behavior of 61016,143 is apparent from Fig. 4 where it falls below the other VHA data and is enclosed in parentheses for emphasis. In any case, all available age data on VHA basalts agree within the respective uncertainties. The case for Apollo 14 breccias is less clear, as their $^{39}Ar/^{40}Ar$ ages are not well defined. However, total Ar ages are consistent at 3.9 AE (Huneke et al., 1972) and agree with the whole-rock age for high-grade breccias (Nyquist et al., 1972) used in constructing Fig. 4.

The consequence of the horizontal alignments shown in Fig. 4 are twofold: (a) some mechanism must have operated to produce the essentially constant $^{87}Sr/^{86}Sr$ values 4.0 AE ago, (b) a further mechanism must have operated to redistribute Rb/Sr in rocks which otherwise have very similar compositions. It seems unlikely for these rocks that these effects could have been caused by atomistic diffusion and isotope exchange which operate to produce mineral isochrons. The Apollo 17 noritic breccias represent samples from both the North and South Massifs and from stations kilometers apart. Data on boulder rock 76315 show that the Sr isotopic composition in the trace element poor clast represented by 76315,62 was unaltered by the close proximity of trace element rich matrix material even though the clast showed a flow structure indicative of a molten or near-molten state. Moreover, isotopic equilibration is not even achieved between various mineral phases in these rocks (Tera et al., 1973). Thus, some other mechanism seems to be called for. Intimate mixing of "protolith" material prior to recrystallization of the rock seems a very probable mechanism. This would act to produce approximately constant $^{87}Sr/^{86}Sr$ on the "whole-rock" scale while leaving the $^{87}Sr/^{86}Sr$ of the granular constituents unaffected. Subsequent recrystallization apparently only caused partial intergranular melting and isotopic equilibration.

Most petrographic studies seems to agree that these rocks were produced from a protolith; however, its nature is open to some debate. The process described above would work most effectively for a protolith derived from essentially a single rock type; one with a wide diversity of rock types would require a more efficient mixing process. Schonfeld (1974) has suggested that VHA basalts and Apollo 17 noritic breccias (his "KREEPy 17") could be produced by mixing KREEP with trace element poor materials. This requires an extremely efficient mixing process to produce the uniform REE abundances and initial Sr observed in these rocks. Granting this efficiency, the process would be allowed by the isotopic data. In fact, there is a tendency for the majority of the KREEPy 17 in Fig. 4 to form a diagonal array approximately along a line which could be interpreted as a mixing line through KREEP and a material with low Rb/Sr. The VHA data do not show this tendency, whereas the Apollo 16 and 15 KREEP do. An alternative explanation to the "mixing" interpretation for lines through the lunar initial Sr value is that they represent approximate closed system evolution from an early differentiation. We see no means to distinguish the alternatives on the basis of Rb–Sr systematics alone.

The mixing mechanism cannot account for all of the characteristics of Figs. 2 and 4, however. For most compositional groups there are data with extreme values of Rb/Sr. We have previously argued that for Apollo 14 high-grade breccias this is due to Rb volatilization, or at least redistribution on the millimeter scale (Nyquist *et al.*, 1972). A similar argument can be made for the Apollo 17 noritic breccias.

Figure 5 shows Rb and Sm concentrations for anorthositic gabbros, VHA basalts, Apollo 17 noritic breccias, and Apollo 15 and 16 KREEP. According to the mixing hypothesis, the concentration of trace elements in a highland breccia is essentially determined by the amount of KREEP basalts admixed into the breccias. One would thus expect to see a linear correlation between Rb and Sm, as

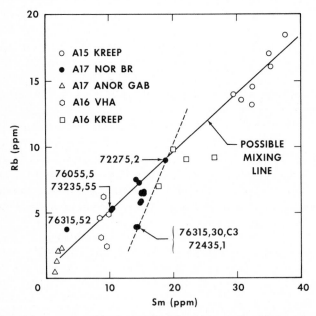

Fig. 5. Rb and Sm concentrations in anorthositic gabbros and noritic breccias studied in this investigation compared to those in Apollo 16 VHA and KREEP and Apollo 15 KREEP. An example of a mixing line with anorthositic gabbro and Apollo 15 KREEP as end components is shown. Cumulate rocks such as anorthosites, dunite and troctolites would plot near the origin and their inclusion as mixing components might change the slope of the line slightly. A tie line between samples at the extremities of the Fig. 2 noritic breccia isochron does not pass through the data field occupied by other samples of this type. Data from this laboratory as reported by Hubbard *et al.* (1973, 1974).

illustrated by the line in Fig. 5 drawn between anorthositic gabbro and Apollo 15 KREEP. Whereas it could be argued that the compositional groups roughly adhere to this trend, individual samples show considerable scatter. In particular, Fig. 5 shows that the noritic breccia isochron in Fig. 2 cannot be only a mixing line. The isochron is determined by samples 76315,30,C3, 72435,1, and 72275,2. The other samples might be considered as lying on the tie line between them. A tie line between corresponding data in Fig. 5 misses *all* of the other noritic breccia data and passes through neither the KREEP nor anorthositic gabbro fields. It is also worth noting that while ten of fourteen noritic breccia samples have Sm concentrations between 14 and 16 ppm, the Rb concentrations for these samples vary by a factor of two from 4 to 8 ppm. Thus, the Rb and Sm concentrations in these samples are not determined solely by sprinkling in variable amounts of KREEP basalt.

The relative abundances of K, Rb, and Ba in the Apollo 17 breccias can also be examined in the same manner as was previously done for Apollo 14 breccias (Nyquist *et al.*, 1972). It is found that most of the Apollo 17 noritic breccia data (Hubbard *et al.*, 1974), satisfy K/Ba = 6.2–7.2, Rb/Ba = 0.016–0.022, and K/Rb = 325–380. Again, exceptions are 72275 (K/Ba = 6.6, Rb/Ba = .026, K/Rb = 260) and 72435,1 and 76315,30,C3 (K/Ba = 5.4, Rb/Ba = .012, K/Rb = 470). A fourth exception is the uniformly aberrant 76315,52 which also has a very unique REE abundance pattern and was excluded from the isochron fit to the Fig. 2 data. A tie line between 72275,1 and 72435,2-76315,30,C3 on a K/Ba versus Rb/Ba diagram again misses *all* of the other data. Apollo 15 KREEP data form a tight group satisfying K/Ba = 6.0–6.8, Rb/Ba = .019–.022, and K/Rb = 308–327, nearly coincident with the majority of the Apollo 17 noritic breccia data. Thus, while Fig. 5 and a mixing hypothesis might suggest that 72275,2 is simply enriched in KREEP basalt component relative to other Apollo 17 samples, this is not supported by either the Sr isotopic data or the relative K, Rb, and Ba abundances. Even if one accepts the hypothesis that the gross features of the Apollo 17 noritic breccias are due to mixing, some special process must be invoked to explain variation of Rb with respect to more refractory elements. Rb volatilization and mobilization seems most likely.

In summary, our preferred explanation of the horizontal alignments shown by the shaded regions in Fig. 4 is $^{87}Sr/^{86}Sr$ averaging by mixing in a protolith prior to ~ 4.0 AE ago, followed by redistribution of Rb/Sr probably by volatilization during the 4.0 AE event(s). The mathematics of the Rb/Sr isochron are impartial to the physical processes acting to produce equality of $^{87}Sr/^{86}Sr$, so that the whole-rock age of the noritic breccias in Fig. 2 is at least a good approximation to the time of the event which stopped the averaging process for $^{87}Sr/^{86}Sr$. In fact, other methods of dating these rocks have their own problems. $^{39}Ar/^{40}Ar$ age spectra are commonly complex, particularly when the rock dated contains older remnants (Huneke *et al.*, 1974). Typically, isotopic equilibration between the minerals in these rocks cannot be demonstrated making the mineral age technique difficult to apply. Thus, we feel our whole-rock age for Apollo 17 noritic breccias to be valid within the stated uncertainties and to give the time at which these breccias were formed from the protolith. This time is most reasonably equated with the time of the Serenitatis impact.

Figure 4 leads to some additional comments. First, the Apollo 14, 15 and 16 KREEP data do not occupy a well-defined region of the diagram. In particular, the Apollo 14 soils and low-grade breccias scatter toward higher $(^{87}Sr/^{86}Sr)_T$ than for the other materials. This could be, and perhaps is, an artifact of the calculation used to correct for radiogenic ^{87}Sr growth. The calculation assumes closed system evolution for the past 3.9 AE. If this assumption is violated by a relatively recent loss of Rb, for example, the calculation will yield a $(^{87}Sr/^{86}Sr)_T$ higher than the actual value 3.9 AE ago. This is illustrated in the figure by an arrow arbitrarily starting at one of the Apollo 15 KREEP data points. Assume this sample was ground and subjected to a volatilization experiment resulting in a 10% loss of Rb. If then Rb, Sr, and $^{87}Sr/^{86}Sr$ are measured and corrected for closed system radiogenic growth in the last 3.9 AE, the resultant would plot at the tip of the arrow in Fig. 4. We have previously suggested (Nyquist *et al.*, 1973a) that some of the Apollo 14 soils have lost Rb by volatilization during exposure to micrometeorite bombardment at the lunar surface. If this exposure occurred in the "recent" past, at a time comparable to the "exposure age" of the soil, for example, one would expect the soils to plot at spuriously high locations in Fig. 4. Indeed, all of the Apollo 14 soils and most of the low-grade breccias plot above the data field for igneous KREEP. This does not mean that all of these data represent systems open to Rb loss in the past 3.9 AE, but certainly is consistent with the hypothesis that at least some of them have. Note that Fig. 4 allows a distinction between "ancient" and "recent" Rb mobilization, the former producing horizontal displacements in the diagram, the latter nearly vertical ones.

It has been shown that Apollo 16 KREEP contains at least some foreign material (Meyer *et al.*, 1974). The distribution of the data along a possible mixing line is consistent with variable amounts of the foreign component. If, however, the crystalline varieties of KREEP (Apollo 14 and 15) represent pure KREEP, the total quantity of foreign material of low Rb/Sr in these rocks is restricted by the near equality of the ancient Sr isotopic composition in these three types of KREEP material. Assuming pure KREEP had the isotopic composition given by the mineral isochron initial ratio for 15434,73 leads to a calculation of between 14% and 32% foreign material with assumed Rb/Sr = 0. Choice of an "average" value for Apollo 15 KREEP leads to calculation of between ~0% and ~20% of such foreign admixture. Note that it is not possible to make Apollo 16 KREEP from Apollo 15 KREEP by simple admixture of known lunar materials. Three possibilities are suggested: (a) the Apollo 16 KREEP or its precursor differentiated earlier than did the Apollo 15 variety, and Fig. 4 just reflects a longer period of radiogenic growth of ^{87}Sr. (b) Apollo 15 KREEP was produced from Apollo 16 KREEP by large degrees of partial melting about 3.9 AE ago. (c) Apollo 15 KREEP was present prior to 3.9 AE ago and Apollo 16 KREEP was produced from it by comminution and admixture of some foreign material accompanied by volatile loss of some Rb during its formation from protolith ~3.9 AE ago. Hypothesis (b) is weakened by the apparent absence of siderophiles of meteoritic origin in Apollo 15 KREEP (Morgan *et al.*, 1973; Baedecker *et al.*, 1973). Further studies of this nature could probably exclude (b). Hypothesis (c) is strengthened by the almost total loss of Pb from Apollo 16 KREEP ~ 4.0 AE ago (Tera *et al.*, 1973). Some loss of Rb at this time is not unreasonable, particularly in light of demonstrated losses for individual samples. This same consideration weakens hypothesis (a). We cannot uniquely choose between the alternatives but favor (c).

Differentiation of a KREEP suite

Various authors have argued for an early differentiation of KREEP since the discovery of its unique geochemical properties as has been discussed above. Mineral ages obtained to date do not support this interpretation. The latter can be explained if the mineral ages are interpreted as representing a metamorphic event. Such situations are not uncommon in the terrestrial case (cf. Hart *et al.*, 1968). As commented above, the Apollo 16 KREEP and the Apollo 17 noritic breccias (called Apollo 17 KREEP by Hubbard *et al.*, 1974) give evidence for the existence of a "KREEP component" prior to ~ 4.0 AE ago. The identification of this old KREEP component is difficult because it has become increasingly obvious that not all KREEPs are created equal, as illustrated by the preceding discussion. Because of the pervasiveness of the "KREEP component" in the Apollo 17 samples and the importance of the question of the time of KREEP differentiation, we continue previous discussions here (cf. Nyquist *et al.*, 1972, 1973).

We have previously adopted two approaches to the question of the KREEP differentiation time. First, following earlier suggestions, we have calculated average single-stage model ages for a KREEP component, assuming the lunar initial Sr isotopic composition to be near BABI. A refinement to this approach has been the two-stage model formalism developed by Schonfeld and Meyer (1972) which assumes KREEP differentiation from a low-Rb/Sr source so that in practice the two-stage solution approaches the single-stage one. Here we extend this approach to the Apollo 17 samples. In this paper we shall use LUNI for the lunar initial, causing the calculated model ages to be ~ 1% older than previous ones.

The identification of additional lunar crustal rock types opened the possibility of applying a second approach to the question of KREEP differentiation. This second approach relies on the identification of a "KREEP suite" of rocks and applying the whole-rock Rb/Sr technique to them, as might be done for cogenetic terrestrial lavas. The application of the technique is of course much more ambiguous in the lunar than in the terrestrial case. However, in the previous paper (Nyquist *et al.*, 1973) we obtained consistent results from a "KREEP + VHA isochron" and the two-stage model. In both cases, only crystalline KREEP from Apollo 14 and 15 were used. The recent paper by Walker *et al.* (1973) encourages us to continue this approach. In most respects we shall adopt the model of these authors and ask what isotopic constraints can be put on it.

Walker *et al.* (1973) postulate that KREEP (Fra Mauro, alkalic–high-alumina) basalt was produced by a small degree of partial melting of an "anorthositic norite" source rock and that low-K Fra Mauro (low-alkali, high-alumina) basalt was produced by more extensive partial melting of the same rock. The Apollo 17 rocks which we have called "anorthositic gabbro" and "noritic breccia" fulfill the chemical criteria of anorthositic norite and low-K Fra Mauro basalt, respectively. Thus, they and the various varieties of KREEP are candidates for a "KREEP suite." We shall also follow our previous choice of including VHA basalts in this suite. In this we appear to depart from Walker et al. (1973), but it is not at all clear the departure is real. Inspection of the composition of low-K Fra mauro basalt given by Ridley et al. (1973) for the Apollo 16 and Luna 20 site show a striking resemblance to the composition of VHA basalt. The main difference is in silica content. Bulk samples of VHA basalt tend to have lower silica values placing them into a different portion of the Walker diagram. This diagram is extremely sensitive to variation in silica content, small differences in silica can change the apparent petrogenetic interpretation drastically in the VHA region of the diagram. The average SiO_2 content of five VHA samples for which we have determined the Sr isotopic composition is $44.9 \pm 0.6\%$ (1 σ) (Hubbard et al., 1973 and J. M. Rhodes, private communication). The silica content of Apollo 16 low-K Fra Mauro basalt is $46.7 \pm 2.1\%$ (1 σ) (Ridley et al., 1973), that of Luna 20 high-alumina basalts is $46.6 \pm 1.5\%$ (1 σ) (Prinz et al., 1973). The average composition of "type A F11R" is 45.2 ± 0.8 (1 σ) (Delano et al., 1973). Readers wishing to pursue this point are referred to the work of Ridley et al. (1973), Prinz et al. (1973), Delano et al. (1973), Walker et al. (1973), and Hubbard et al. (1974) and references. In any case, the Sr isotopic characteristics of VHA basalts are consistent with production from a source very similar to that for low-K Fra Mauro and we continue to include it as a candidate for membership in a KREEP suite.

Table 2 summarizes results of the model age calculations for the various compositional groups. Samples which were obviously open systems 4.0 AE ago have been excluded. These are Apollo 17 noritic breccias 72275, 72435, 76315,30C3, and Apollo 14 breccia 14006 and 14161,35,3 (Nyquist et al., 1972). The two-stage model ages use the formalism of Schonfeld and Meyer (1972), assuming the age of the moon is 4.65 AE, the lunar initial Sr composition is LUNI or BABI and Rb/Sr = 0.01 in the source region. The "fractionation factor," F, is the ratio of Rb/Sr during the second stage to that during the first stage. Thus, F is the average Rb/Sr for the compositional group divided by 0.01. In our previous paper we used Rb/Sr = 0.008 for the source. In this paper we take it equal to that of the Apollo 17 anorthositic gabbros (norites), resulting in a minor change in F-values.

Single-stage model ages are upper limits to the differentiation time for these materials, as they assume no radiogenic [87]Sr growth in the source region. Two-stage model ages range from 4.27 AE for Apollo 15 crystalline KREEP to 4.46 AE for Apollo 14 breccias. As discussed in the previous section, we distrust the Apollo 14 breccia data because of the possibility of recent Rb loss from some of these samples. With the exception of the Apollo 14 and 15 crystalline KREEP, all other groups are subject to uncertainties of containing admixed foreign material and/or Rb volatilization ~ 4.0 AE ago. If the amount of foreign admixture is small and of a material like the low-Rb/Sr Apollo 16 anorthosites, a small

Table 2. Average model ages of chemically defined rock types.

Rock type and model	No. Samples	T (AE)
1. A 17 NB[a]—single-stage LUNI[b]	12	$4.46 \pm .04$[d]
2. A 17 NB—two-stage, $F = 3.7$[c]	12	4.39
3. A 16 VHA—single-stage LUNI	5	$4.46 \pm .22$
4. A 16 VHA—two-stage, $F = 2.8$	5	4.36
5. A 16 KREEP—single-stage LUNI	4	$4.48 \pm .04$
6. A 16 KREEP—two-stage $F = 5.6$	4	4.44
7. A 15 KREEP—single-stage LUNI	7	$4.32 \pm .04$
8. A 15 KREEP—two-stage, $F = 8.2$	7	4.27
9. A 14 XTLN—single-stage BABI (CIT)[e]	5	$4.33 \pm .04$
10. A 14 XTLN—two-stage, $F = 7.8$ (CIT)[e]	5	4.28
11. A 14 breccia*—single-stage BABI	9	$4.48 \pm .13$
12. A 14 breccia—two-stage, $F = 9.2$	9	4.46

[a]NB = noritic breccia. "XTLN" and "Breccia" for Apollo 14 refer to the crystalline and brecciated varieties of KREEP, respectively.
[b]LUNI = 0.69903 and BABI = 0.69910 on our scale.
[c]F = Rb/Sr during the second stage divided by that during the first stage.
[d]One standard deviation of the population of values averaged. No uncertainties have been given to the two-stage values because the F-values are model dependent.
[e]Papanastassiou and Wasserburg (1971b) and Wasserburg and Papanastassiou (1971).

percentage of foreign component will not significantly affect the model age. Rb volatilization would bias the ages to higher values. All data are consistent with differentiation $\leqslant 4.46$ AE ago, but for reasons given earlier in this and the previous paper, we favor the data for the crystalline varieties of KREEP, i.e. $T \sim 4.28$ AE, as the best choice for this material. The interpretation of the model age for each individual compositional group depends on the petrogenesis of that group and may need to be individually argued.

Table 3 gives the results of fitting a whole-rock "isochron" to various combinations of the compositional groups. As the age is controlled by the KREEP

Table 3. Whole-rock isochrons for combinations of chemically defined rock types.

Rock types combined	No. Samples	T (AE)	I
1. A 17 NB[a] + A 17 AG[b]	18	$4.32 \pm .11$	$.69922 \pm 14$
2. A 17 NB + A 16 VHA + A 15 KREEP	27	$4.19 \pm .08$	$.69940 \pm 13$
3. Above + A 17 AG	31	$4.25 \pm .07$	$.69927 \pm 10$
4. A 17 NB + A 16 VHA + A 15 KREEP + A 14 XTLN KREEP	32	$4.21 \pm .06$	$.69938 \pm 12$
5. Above + A 17 AG	36	$4.26 \pm .06$	$.69926 \pm 10$
6. A 17 NB + A 16 VHA + A 16 KREEP	24	$4.26 \pm .13$	$.69933 \pm 21$
7. Above + A 17 AG	28	$4.34 \pm .09$	$.69919 \pm 13$
8. A 17 NB + A 17 AG + A 16 KREEP	22	$4.37 \pm .09$	$.69918 \pm 13$

[a]NB = noritic breccia.
[b]AG = anorthositic gabbro.

group and as inclusion of the crystalline varieties yields results differing from those obtained by inclusion of the brecciated ones, we have considered the two cases separately. The method applied here is the same as that in our earlier paper, except that we have included the Apollo 17 noritic breccias as an example of "low-K" KREEP and the anorthositic gabbros as an example of a possible source rock. Ages obtained by this method range from 4.19 to 4.37 AE. We have repeatedly emphasized that we feel Rb volatilization losses in the formation of the breccias to be a problem. If it occurs, it obviously violates the closed system assumption upon which all of these models rest. We have no examples of truly igneous-textured VHA basalt or low-K KREEP. Using the brecciated variety of these rocks with the igneous-textured KREEP could lead to ages which are slightly low in contrast to the model ages of the brecciated rock which may be too high. Inclusion of a rock type of truly lower Rb/Sr (i.e. the anorthositic gabbro) will partly compensate for this effect. Thus, if this technique is valid, line (5) probably represents the best application of it. We have thus chosen this case for representation in Fig. 6.

The isochron in Fig. 6 is essentially identical to that given by Nyquist *et al.* (1973a) for "VHA + KREEP." It entails the following assumptions: (a) the alkali-poorer and alkali-richer varieties of KREEP differentiated approximately simultaneously by variation in the degree of partial melting of an anorthositic gabbro (norite) source rock (Walker *et al.*, 1973; hypothesis); (b) VHA basalts

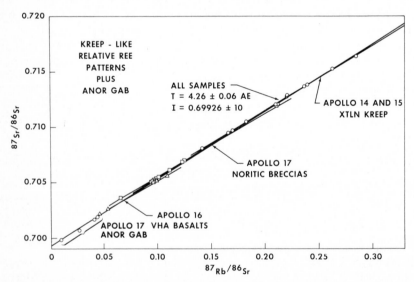

Fig. 6. A whole-rock "isochron" for Apollo 17 anorthositic gabbros (open circles in lower left corner) and samples with KREEP-like relative REE abundance patterns. The "isochron" has time significance only under the assumption of a specific petrogenetic model (see text). Triangles: VHA basalts; squares: Apollo 17 noritic breccias; open circles in upper right corner: Apollo 14 and 15 crystalline KREEP. Secondary isochrons for $T \sim 4.0$ AE are also shown.

were also produced at nearly the same time from a source of nearly the same composition; (c) none of the rocks are only simple mechanical mixtures; (d) the igneous-textured crystalline varieties of KREEP were closed systems on the whole-rock scale during a 3.9 AE metamorphism. We cannot guarantee any of these assumptions but feel that they are consistent with current isotopic and geochemical data and with the experimental petrology studies. It has been suggested that the entire petrogenesis of VHA basalts and Apollo 17 noritic breccias can be explained by mechanical mixing of KREEP and trace element poor rocks (Schonfeld, 1974). If these rocks are indeed mixtures, then the central portion of the Fig. 6 isochron will disappear, but the result will be little affected if condition (a) above remains (appropriately modified). If (a) also falls then one is driven back to the model age arguments previously given.

Figure 7 examines the various petrogenetic hypotheses with the aid of a (T, Δ) diagram. Here Δ is the increment in $^{87}Sr/^{86}Sr$ above the initial lunar value, assumed to be LUNI. An evolution line for Rb/Sr = 0.01, appropriate to Apollo 17 anorthositic gabbro, and starting at 4.60 AE is shown. The (T, Δ) values obtained from Fig. 3 for Apollo 15 crystalline KREEP sample 15434,73 is also shown. A

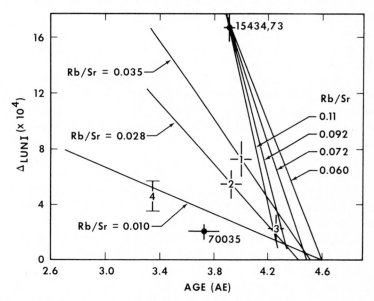

Fig. 7. Illustration of evolution of $^{87}Sr/^{86}Sr$ from an assumed lunar initial value of LUNI (0.69903 on our scale) under various conditions. This is, $\Delta_{LUNI} = I - 0.69903$. Points 1, 2, and 3 are values for whole-rock isochrons for Apollo 17 noritic breccias, Apollo 16 VHA basalts, and the combined rock type "isochron" of Fig. 5, respectively. Evolution lines through data points 1 and 2 correspond to a backward extrapolation in time assuming a Rb/Sr ratio equal to the average measured in the respective rock types. Various possible evolution lines for 15434,73 are shown (see text). The evolution line from 4.6 AE is that appropriate to a source region with Rb/Sr = 0.01, the average of four Apollo 17 anorthositic gabbros. (T, Δ) values of mare basalt 70035 and a range for Apollo 15 mare basalts assuming all have identical ages of 3.35 AE (point 4) are shown for comparison.

strong constraint on the petrogenesis of the KREEP sample is that it cannot have been produced from an anorthositic gabbro (norite) source rock 3.9 AE ago, or from a source of similar Rb/Sr. On the other hand, a line through the KREEP point with Rb/Sr = 0.11, the measured value in the 200–325 mesh fraction, intersects the anorthositic gabbro evolution line at 4.25 AE, within the error limits of point (3) obtained from the Fig. 6 isochron. It could be argued that the 200–325 mesh fraction is not a representative whole-rock sample. This is in fact strongly implied by Fig. 4 which indicates Rb/Sr = 0.09 is more appropriate to the initial Sr in this fragment. A corresponding evolution line intersects the "primitive" evolution line at 4.33 AE. Averaging Rb/Sr for all Apollo 15 KREEP analyzed to date gives Rb/Sr = 0.07, leading to intersection at 4.46 AE. Fig. 4 indicates that this is very probably too low a Rb/Sr for this sample, and thus too high an age.

The (T, Δ) values of the separate whole rock isochrons of VHA basalts and Apollo 17 noritic breccias are consistent with contemporaneous differentiation of the precursor (protolith?) material and KREEP.

Alternatively, from the mineral isochron data alone, one could argue that the immediate precursor to 15434,73 must have had a Rb/Sr ratio as high as Rb/Sr = 0.06. This ratio is typical of Apollo 16 KREEP, and again implies that some Rb-rich material (by lunar standards) was produced early in lunar history. An alternative hypothesis, advanced to explain the high I-values of Apollo 14 basalts (Papanastassiou and Wasserburg, 1971b) is that these rocks picked up their radiogenic ^{87}Sr, as well as other trace elements, by crustal contamination. Other trace element data do not support this hypothesis, and it seems further weakened by the clear implication of some kind of "KREEP component" present at or near the lunar surface prior to 4.0 AE ago.

Comparison of the compatibility of hypotheses discussed here with other isotopic data, in particular the U–Th–Pb data, is beyond the scope of this paper. However, it has been pointed out (Tera *et al.*, 1974) that almost all highland rocks lie along a chord intersecting concordia at ~ 4.0 and ~ 4.43 AE. This chord appears to be dominated by Pb evolved in KREEP-rich rocks. The two-stage model interpretation is that 4.43 AE represents the start of Pb evolution from the primitive isotopic value and that the system was disturbed (changed U/Pb) at 4.0 AE. The Rb–Sr models we have been discussing are really three-stage models: i.e. (1) Sr isotopic composition = LUNI at 4.6 AE ago, (2) differentiation of KREEP-rich rocks sometime later, and (3) resetting of the mineral isochron at ~ 3.9 AE. An obvious difference between the U–Th–Pb and Rb–Sr systems is that the former are evidently much more severely affected by the ~ 4.0 AE event. The two-stage U–Th–Pb models are really most analogous to the single-stage Rb–Sr model ages given in Table 2 in that both start out from a primitive isotopic composition. This is an oversimplification. U concentrations in anorthositic gabbros are not negligible, being greater than those of Apollo 11 and 17 mare basalts and of the order of those for Apollo 12 and 15 mare basalts, quite analogous to the Rb–Sr situation. If anorthositic gabbro, or rocks similar to it, is the source material for KREEP one might expect significant evolution from the primitive Pb values before differentiation of KREEP. Thus, the 4.43 AE two-stage concordia intersection seems to us to be best viewed as an upper limit to the KREEP differentiation, in analogy to the single-stage Rb–Sr age.

CONCLUSIONS AND SPECULATIONS

The age of 70035 and other Apollo 17 mare basalts indicates flooding of the Taurus-Littrow Valley about 3.7–3.8 AE ago by basaltic lavas similar to the low-K variety at the Apollo 11 site. The initial $^{87}Sr/^{86}Sr$ in these lavas indicates production from a source of very low Rb/Sr ratio; lower than that from which Apollo 12 and Apollo 15 mare basalts were produced (Papanastassiou and Wasserburg, 1973).

$^{87}Sr/^{86}Sr$ was nearly identical for Apollo 17 noritic breccias 4.0 AE ago. Material mixing very likely played some role in producing this near equality. The degree of mixing need not have been great if the starting material was relatively uniform in composition. A diversity of starting materials requires more efficient mixing. A similar conclusion can be reached by considering the narrow range of REE abundances in these breccias. The homogenizing process must have occurred at or prior to 4.0 AE ago.

Rb abundances for these rocks show wider variations than either the REE abundances or the back-calculated $^{87}Sr/^{86}Sr$ ratios and are not highly correlated to either. This suggests operation of a mechanism capable of decoupling Rb abundances from those of more refractory trace elements. Volatilization seems a probable process. Formal calculation of a whole-rock Rb–Sr age from these data yields a result in agreement with $^{39}Ar/^{40}Ar$ ages on individual rocks of this type, suggesting that the decoupling of Rb from refractory elements occurred during the 4.0 AE event.

Interpretation of early lunar history is by its nature speculative. The following considerations seem at least consistent with most of the observations and to leave the greatest possibility of unraveling lunar evolution.

The formation of the large multiringed basins must certainly be viewed as one of the dominant features of lunar history. The gigantic impacts responsible for them must have produced huge quantities of impact melt. Head (1974c) estimates the quantity of impact melt associated with the formation of the Orientale Basin to be comparable to the entire mare basalt fill of the Tranquilitatis Basin. Most of this impact melt falls back into the original excavation but a substantial covering extends to at least the second basin ring (Head, 1974c). It seems reasonable that a thinner covering might extend considerably further. "Igneous" impact melts from terrestrial craters are characterized by abundant inclusions but in general agree closely with surrounding country rocks in composition (Dence, 1971).

The Serenitatis, Nectaris, and Imbrium basin events (in order of occurrence (Stuart-Alexander and Howard, 1970)) must be considered in connection with the petrogenesis of Apollo 17 noritic breccias, KREEP, and VHA rocks. Following discussions by Head (1974a, 1974b, 1974c) we note that the Fra Mauro Formation sampled by Apollo 14 has been interpreted to be impact ejecta of the Imbrium event and that the Apollo 15 landing site is within the third ring of the Imbrium

Basin. Further, the Taurus-Littrow highlands are part of the second ring of the Serenitatis Basin, and the Apollo 16 landing site essentially lies at the base of the back slope of the Nectaris Basin outer ring. According to Head (1974b), a 60 km diameter crater ("Unnamed B") was formed on the Nectaris ejecta and dominates the geology and petrogenetic history of the site. Head (1974b) further suggests that Apollo 16 metaclastic crystalline rocks, in particular VHA basalts, may be impact melts associated with the Unnamed B crater. The latter is post-Nectarian but pre-Imbrian. The target material for this crater consisted of Nectaris ejecta underlain by older highland material. A significant amount of Imbrium ejecta may also occur at the Apollo 16 site, which may be the source of Apollo 16 KREEP. Thus, it is possible that the Apollo 17 noritic breccias, KREEP and VHA rocks can be traced in one or two steps of impact melting to the excavation of the Serenitatis, Nectaris and Imbrium basins. Their radiometric ages are consistent with the relative ages of the basins but require the latter to have formed in a short time span ~ 3.9–4.0 AE ago (Tera *et al.*, 1973). The extent to which one can see through the cataclysmic events 3.9–4.0 AE ago depends on the extent to which the chemical systems remained closed during them. If they remained at least approximately closed, as seems most consistent with the existence of preferred compositional groupings, then there is justification for viewing the basins as sampling previously differentiated magmas at some depth below the lunar crust, and in using the whole-rock Rb–Sr data in the manner discussed above to arrive at a time of primary differentiation of these magmas. Extensive mixing of diverse rock types to produce the present samples would invalidate this approach, but a small admixture of foreign material into a basically monolithic melt could be tolerated. If the Fra Mauro Formation truly represents Imbrium ejecta, then at least one case exists where compositional characteristics survived the ejection process. However, Schonfeld and Meyer (1974) have presented an alternative interpretation of the Fra Mauro Formation.

The apparent lack of a meteoritic component (Morgan *et al.*, 1973; Baedecker *et al.*, 1973) in Apollo 15 KREEP indicates that this material may not be as intimately associated with the actual Imbrium impact as the Apollo 14 KREEP is. Close agreement of the mineral isochron age of 15434,73 with that of Apollo 14 KREEP rocks would then indicate closely associated but distinct events, with the Apollo 15 KREEP perhaps representing an impact triggered lava flow instead of an impact melt. In either case, the immediate (pre-3.9 AE) precursor to Apollo 15 KREEP must have had a high Rb/Sr ratio unlike that of anorthositic gabbro. If the Rb/Sr ratio was unchanged by the 3.9 AE event, then production of Apollo 15 KREEP by partially melting of an anorthositic gabbro source material 4.26 AE ago is an attractive hypothesis.

Acknowledgments—W. C. Phinney arranged transfer of sample 15434,73 and supplied a petrographic description. Both light and heat were generated by discussions with N. J. Hubbard, C. Meyer, Jr., and E. Schonfeld.

REFERENCES

Adler I., Trombka J., Gerard J., Lowman P., Schmaldebeck R., Blodgett H., Eller E., Yin L., Lamonthe R., Gorenstein P., and Bjorkholm P. (1972a) Apollo 15 geochemical X-ray fluorescence experiment: Preliminary report. *Science* **175**, 436–440.

Adler I., Trombka J., Gerard J., Lowman P., Schmaldebeck R., Blodgett H., Eller E., Yin L., Lamonthe R., Osswald G., Gorenstein P., Bjorkholm P., Gursky H., and Harris B. (1972b) Apollo 16 geochemical X-ray fluorescence experiment: Preliminary report. *Science* **177**, 256–259.

Albee A. L., Gancarz A. J., and Chodos A. A. (1973) Metamorphism of Apollo 16 and 17 and Luna 20 metaclastic rocks at about 3.95 AE: Samples 61156, 64423,14-2, 65015, 67483,15-2, 76055, 22006 and 22007. *Proc. Fourth Lunar Sci. Conf., Geochim. Cosmochim. Acta*, Suppl. 4, Vol. 1, pp. 569–595. Pergamon.

Albee A. L., Chodos A. A., Dymek R. F., Gancarz A. J., Goldman D. S., Papanastassiou D. A., and Wasserburg G. J. (1974) Dunite from the lunar highlands: Petrography, deformational history, Rb–Sr age (abstract). In *Lunar Science—V*, pp. 3–5. The Lunar Science Institute, Houston.

ALGIT (Apollo Lunar Geology Investigation Team) (1973) Interagency Report: *Astrogeology* **72.**, United States Department of the Interior, Geological Survey, March 16, 1973.

Baedecker P. A., Chou C.-L., Grudewicz E. B., and Wasson J. T. (1973) Volatile and siderophilic trace elements in Apollo 15 samples: Geochemical implications and characterization of the long-lived and short-lived extralunar materials. *Proc. Fourth Lunar Sci. Conf., Geochim. Cosmochim. Acta*, Suppl. 4, Vol. 2, pp. 1177–1195. Pergamon.

Bogard D. D., Nyquist L. E., Bansal B. M., and Wiesmann H. (1974) 76535: An old lunar rock? (abstract). In *Lunar Science—V*, pp. 70–72. The Lunar Science Institute, Houston.

Chao E. C. T. (1973) The petrology of 76055,10, a thermally metamorphosed fragment-laden olivine micronorite hornfels. *Proc. Fourth Lunar Sci. Conf., Geochim. Cosmochim. Acta*, Suppl. 4, Vol. 1, pp. 719–732. Pergamon.

Compston W., Vernon M. J., Berry H., Rudowski R., Gray C. M., and Ware N. (1972) Apollo 14 mineral ages and the thermal history of the Fra Mauro formation. *Proc. Third Lunar Sci. Conf., Geochim. Cosmochim. Acta*, Suppl. 3, Vol. 2, pp. 1487–1501. MIT Press.

Delano J. W., Bence A. E., Papike J. J., and Cameron K. L. (1973) Petrology of the 2–4 mm soil fraction from the Descartes region of the moon and stratigraphic implications. *Proc. Fourth Lunar Sci. Conf., Geochim. Cosmochim. Acta*, Suppl. 4, Vol. 1, pp. 537–551. Pergamon.

Dence M. R. (1971) Impact melts. *J. Geophys. Res.* **76**, 5552–5565.

Evensen N. M., Murthy V. R., and Coscio M. R., Jr. (1973) Rb–Sr ages of some mare Basalts and the isotopic and trace element systematics in lunar fines. *Proc. Fourth Lunar Sci. Conf., Geochim. Cosmochim. Acta*, Suppl. 4, Vol. 2, pp. 1707–1724. Pergamon.

Haines E. L., Hutcheon I. D., and Weiss J. R. (1974) Excess fission tracks in Apennine Front KREEP basalts (abstract). In *Lunar Science—V*, pp. 304–306. The Lunar Science Institute, Houston.

Hart S. R., Davis G. L., Steiger R. H., and Tilton G. R. (1968) A comparison of the isotopic mineral age variations and petrologic changes induced by contact metamorphism. In *Radiometric Dating for Geologists.* pp. 73–110. Interscience, Publ., London–New York–Sydney.

Head J. W. (1974a) Morphology and structure of the Taurus-Littrow Highlands (Apollo 17): Evidence for their origin and evolution. *The Moon* **9**, 355–396.

Head J. W. (1974b) Stratigraphy of the Descartes region (Apollo 16): Implications for the origin of samples. *The Moon.* In press.

Head J. W. (1974c) Orientale multi-ringed basin interior and implications for the petrogenesis of lunar highland samples. *The Moon.* In press.

Hubbard N. J. and Gast P. W. (1971) Chemical composition and origin of nonmare lunar basalts. *Proc. Second Lunar Sci. Conf., Geochim. Cosmochim. Acta*, Suppl. 2, Vol. 2, pp. 999–1020. MIT Press.

Hubbard N. J., Rhodes J. M., Gast P. W., Bansal B. M., Shih C.-Y., Wiesmann H., and Nyquist L. E. (1973) Lunar rock types: The role of plagioclase in non-mare and highland rock types. *Proc. Fourth Lunar Sci. Conf., Geochim. Cosmochim. Acta*, Suppl. 4, Vol. 1, pp. 1297–1312. Pergamon.

Hubbard N. J., Wiesmann H., Shih C.-Y., and Bansal B. (1974) The chemical definition and interpretation of rock types returned from the non-mare regions of the moon. *Proc. Fifth Lunar Sci. Conf.*, *Geochim. Cosmochim. Acta.* This volume.

Huneke J. C., Podosek F. A., Turner G., and Wasserburg G. J. (1972) ^{40}Ar–^{39}Ar systematics in lunar rocks and separated minerals of lunar rocks from Apollo 14 and Apollo 15 (abstract). In *Lunar Science—III*, pp. 413–414. The Lunar Science Institute, Houston.

Huneke J. C., Jessberger E. K., Podosek F. A., and Wasserburg G. J. (1973) ^{40}Ar/^{39}Ar measurements in Apollo 16 and 17 samples and the chronology of metamorphic and volcanic activity in the Taurus-Littrow region. *Proc. Fourth Lunar Sci. Conf.*, *Geochim. Cosmochim. Acta*, Suppl. 4, Vol. 2, pp. 1725–1756. Pergamon.

Huneke J. C., Jessberger E. K., and Wasserburg G. J. (1974) The age of metamorphism of a highland breccia (65015) and a glimpse at the age of its protolith (abstract). In *Lunar Science—V*, pp. 375–376. The Lunar Science Institute, Houston.

Husain L. and Schaeffer O. A. (1973) ^{40}Ar–^{39}Ar crystallization ages and ^{38}Ar–^{37}Ar cosmic ray exposure ages of samples from the vicinity of the Apollo 16 landing site (abstract). In *Lunar Science—IV*, pp. 406–408. The Lunar Science Institute, Houston.

Kirsten T. and Horn P. (1974) ^{39}Ar–^{40}Ar-chronology of the Taurus-Littrow Region II: A 4.28 b.y. old troctolite and ages of basalts and highland breccias (abstract). In *Lunar Science—V*, pp. 419–421. The Lunar Science Institute, Houston.

Kirsten T., Horn P., and Heymann D. (1973) Chronology of the Taurus-Littrow region I: Ages of two major rock types from the Apollo 17-site. *Earth Planet. Sci. Lett.* **20**, 125–130.

LSPET (Lunar Sample Preliminary Examination Team) (1974) Apollo 17 lunar samples: Chemical petrographic description. *Science* **182**, 659–690.

Mark R. K., Cliff R. A., Lee-Hu C., and Wetherill G. W. (1973) Rb–Sr studies of lunar breccias and soils. *Proc. Fourth Lunar Sci. Conf.*, *Geochim. Cosmochim. Acta*, Suppl. 4, Vol. 2, pp. 1785–1795. Pergamon.

Mark R. K., Lee-Hu C., and Wetherill (1974) Rb–Sr measurements on lunar igneous rocks and breccia clasts (abstract). In *Lunar Science—V*, pp. 490–492. The Lunar Science Institute, Houston.

McConnell R. K. and Gast P. W. (1972) Lunar thermal history revisited. *The Moon* **5**, 171–181.

Metzger A. E., Trombka J. I., Peterson L. E., Reedy R. C., and Arnold J. R. (1973) Lunar surface radioactivity: Preliminary results of the Apollo 15 and Apollo 16 gamma-ray spectrometer experiments. *Science* **179**, 800–803.

Meyer C., Jr. (1972) Mineral assemblages and the origin of non-mare lunar rock types (abstract). In *Lunar Science—III*, pp. 542–544. The Lunar Science Institute, Houston.

Meyer C., Jr., Anderson D. H., and Bradley J. G. (1974) Ion microprobe mass analyses of plagioclase from "non-mare" lunar samples. *Proc. Fifth Lunar Sci. Conf.*, *Geochim. Cosmochim. Acta*, Suppl. 5, Vol. 1. Pergamon.

Morgan J. W., Krähenbühl U., Ganapathy R., and Anders E. (1973) Trace element abundances and petrology of separates from Apollo 15 soils. *Proc. Fourth Lunar Sci. Conf.*, *Geochim. Cosmochim. Acta*, Suppl. 4, Vol. 2, pp. 1379–1398. Pergamon.

Nyquist L. E., Hubbard N. J., Gast P. W., Church S. E., Bansal B. M., and Wiesmann H. (1972) Rb–Sr systematics for chemically defined Apollo 14 breccias. *Proc. Third Lunar Sci. Conf.*, *Geochim. Cosmochim. Acta*, Suppl. 3, Vol. 2, pp. 1515–1530. MIT Press.

Nyquist L. E., Hubbard N. J., Gast P. W., Bansal B. M., Wiesmann H., and Jahn B.-M. (1973a) Rb–Sr systematics for chemically defined Apollo 15 and 16 materials. *Proc. Fourth Lunar Sci. Conf.*, *Geochim. Cosmochim. Acta*, Suppl. 4, Vol. 2, pp. 1823–1846. Pergamon.

Nyquist L. E., Hubbard N. J., Bansal B. M., Wiesmann H., and Jahn B.-M. (1973b) On the differentiation time of KREEP and VHA basalts (abstract). *Meteoritics* **8**, 416–417.

Papanastassiou D. A. and Wasserburg G. J. (1971a) Lunar chronology and evolution from Rb–Sr studies of Apollo 11 and 12 samples. *Earth Planet. Sci. Lett.* **11**, 37–62.

Papanastassiou D. A. and Wasserburg G. J. (1971b) Rb–Sr ages of igneous rocks from the Apollo 14 mission and the age of the Fra Mauro Formation. *Earth Planet. Sci. Lett.* **12**, 36–48.

Papanastassiou D. A. and Wasserburg G. J. (1972) Rb–Sr systematics of Luna 20 and Apollo 16 samples. *Earth Planet. Sci. Lett.* **17**, 52–64.

Papanastassiou D. A. and Wasserburg G. J. (1973) Rb–Sr ages and initial strontium in basalts from Apollo 15. *Earth Planet. Sci. Lett.* **18**, 324–337.

Papanastassiou D. A., Wasserburg G. J., and Burnett D. S. (1970) Rb–Sr ages of lunar rocks from the Sea of Tranquility. *Earth Planet. Sci. Lett.* **8**, 1–19.

Prinz M., Dowty E., Keil K., and Bunch T. E. (1973) Mineralogy, petrology and chemistry of lithic fragments from Luna 20 fines: Origin of the cumulate ANT suite and its relationship to high-alumina and mare basalts. *Geochim. Cosmochim. Acta* **37**, 979–1006.

Rhodes J. M., Rodgers K. V., Shih C., Bansal B. M., Nyquist L. E., Wiesmann H., and Hubbard N. J. (1974) The relationship between geology and soil chemistry at the Apollo 17 landing site. *Proc. Fifth Lunar Sci. Conf., Geochim. Cosmochim. Acta.* This volume.

Ridley W. I., Reid A. M., Warner J. L., Brown R. W., Gooley R., and Donaldson C. (1973) Glass compositions in Apollo 16 soils 60501 and 61221. *Proc. Fourth Lunar Sci. Conf., Geochim. Cosmochim. Acta*, Suppl. 4, Vol. 1, pp. 309–321. Pergamon.

Schaeffer O. A. and Husain L. (1973) Early lunar history: Ages of 2 to 4 mm soil fragments from the lunar highlands. *Proc. Fourth Lunar Sci. Conf., Geochim. Cosmochim. Acta*, Suppl. 4, Vol. 2, pp. 1847–1863. Pergamon.

Schonfeld E. (1974) Component abundance and evolution of regoliths and breccias: Interpretation by mixing models (abstract). In *Lunar Science—V*, pp. 669–671. The Lunar Science Institute, Houston.

Schonfeld E. and Meyer C., Jr. (1972) The abundances of components of the lunar soils by a least-squares mixing model and the formation age of KREEP. *Proc. Third Lunar Sci. Conf., Geochim. Cosmochim. Acta*, Suppl. 3, Vol. 2, pp. 1397–1420. MIT Press.

Schonfeld E. and Meyer C., Jr. (1974) The old Imbrium hypothesis. *Proc. Fourth Lunar Sci. Conf., Geochim. Cosmochim. Acta*, Suppl. 4, Vol. 1, pp. 125–138. Pergamon.

Stettler A., Eberhardt P., Geiss J., Grögler N., and Maurer P. (1973) Ar^{39}–Ar^{40} ages and Ar^{37}–Ar^{38} exposure ages of lunar rocks. *Proc. Fourth Lunar Sci. Conf., Geochim. Cosmochim. Acta*, Suppl. 4, Vol. 2, pp. 1865–1888. Pergamon.

Stuart-Alexander D. and Howard K. (1970) Lunar maria and circular basins—A review. *Icarus* **12**, 440–456.

Taylor S. R., Gorton M., Muir P., Nance W., Rudowski R., and Ware N. (1974) Lunar highland composition. In *Lunar Science—V*, pp. 789–791. The Lunar Science Institute, Houston.

Tera F., Papanastassiou D. A., and Wasserburg G. J. (1973a) A lunar cataclysm at ~ 3.95 AE and the structure of the lunar crust (abstract). In *Lunar Science—IV*, pp. 723–725. The Lunar Science Institute, Houston, and oral communication at the Fourth Lunar Science Conference, March 1973, Houston.

Tera F., Papanastassiou D. A., and Wasserburg G. J. (1973b) The lunar time scale and a summary of isotopic evidence for a terminal lunar cataclysm (abstract). In *Lunar Science—V*, pp. 792–794. The Lunar Science Institute, Houston.

Turner G. (1970) Argon-40/argon-39 dating of lunar rock samples. *Proc. Apollo 11 Lunar Sci. Conf., Geochim. Cosmochim. Acta*, Suppl. 1, Vol. 2, pp. 1665–1684. Pergamon.

Turner G., Cadogan P. H., and Yonge C. Y. (1973) Argon selenochronology. *Proc. Fourth Lunar Sci. Conf., Geochim. Cosmochim. Acta*, Suppl. 4, Vol. 2, pp. 1889–1914. Pergamon.

Walker D., Grove T. L., Longhi J., Stolper E. M., and Hays J. F. (1973) Origin of lunar feldspathic rocks. *Earth Planet. Sci. Lett.* **20**, 325–336.

Wasserburg G. J. and Papanastassiou D. A. (1971) Age of an Apollo 15 mare basalt: Lunar crust and mantle evolution. *Earth Planet. Sci. Lett.* **13**, 97–104.

Proceedings of the Fifth Lunar Conference
(Supplement 5, Geochimica et Cosmochimica Acta)
Vol. 2 pp. 1541–1555 (1974)
Printed in the United States of America

Chronology of lunar basin formation

O. A. Schaeffer and Liaquat Husain

Department of Earth and Space Sciences, State University of New York at Stony Brook,
Stony Brook, New York 11794

Abstract—The lunar anorthositic rocks 60025,86 and 60015,22 give ^{40}Ar–^{39}Ar ages of $4.19 \pm .06$ and $3.50 \pm .05$ G.y., respectively. Two coarse fine anorthositic rock fragments 78503,7,1 and 72503,8,12 have ages of $4.13 \pm .03$ and $3.96 \pm .02$ G.y., respectively. The rock 60025 is the first large lunar rock with an age well in excess of 4.0 G.y. Evidence for the conclusion that the cratering events reset the radiometric ages is discussed. The chronology of the lunar basin forming events is deduced from the ages of the lunar highland impact breccias. Using the ages of lunar breccias, we have deduced the following chronology for the last five lunar basin forming events: Orientale $3.85 \pm .05$, Imbrium $3.95 \pm .05$, Crisium 4.05–4.20, Humorum 4.05–4.20, and Nectaris $4.25 \pm .05$. It is concluded that the "lunar cataclysm" is the Imbrium event.

Introduction

Impact generated breccias and cataclastic rocks have been found by all Apollo missions. The Apollo 14 and 16 samples are predominantly this rock type (LSPET, 1971, 1973a). At the Apollo 17, they also are a major part of the returned samples (LSPET, 1973b). These materials appear to form the morphological unit which is widespread over the moon. It appears that the impact generated breccias are for the most part related to major multiring impact basins (Hodges *et al.*, 1973). On the nearside of the moon, there are ten clearly visible multiring impact structures. Starting with the youngest, these are Orientale, Imbrium, Crisium, Humorum, Nectaris, Serenitatis, Fecunditatis, Tranquillitatis E, Tranquillitatis W, and Nubium. It may be possible to obtain a chronology for the lunar basin forming era on the moon by studying these breccias. At least, it should be possible to decide whether most of the basins were formed in a lunar cataclysm (Tera *et al.*, 1973) or whether the basins were formed over a period of hundreds of millions of years.

In this paper we report new radiometric and exposure ages for some Apollo 16 and 17 anorthositic rocks. Lunar impact breccias are discussed in terms of a basin forming chronology.

Experimental Procedure

The experimental setup for Ar isotopic analysis and neutron irradiation was the same as described in detail by Husain (1974). Two samples each of rocks 60015 and 60025, and 2–4 mm fragments 72503,8,12 and 78503,7,1 were analyzed. Owing to the very low K contents of the rocks 60015 and 60025, relatively large samples of $\sim$300 mg, were used. In all, three irradiations were performed. We have continued to use the Stony Brook hornblende mineral separate, NL-25-2 to monitor the neutron flux (Husain, 1974) during irradiation. The $^{40}Ar/^{39}Ar$ ratios measured for multiple samples from the three irradiations are given in Table 1.

Table 1. Argon isotopic data for Apollo 16 and 17 rocks.

Temp. (C)	36/38	38/37	39*/37	40*/39*	39* $\times 10^{-8}$ ccSTP/g	Exp. age (m.y.)	Age (G.y.)
60015,22	311.1 MG Cataclastic anorthosite						
1050	1.68 ± 20^a	0.00039 ± 0	0.0009 ± 0	30.6 ± 1.5	0.29 ± 1	5 ± 1	2.546 ± 0.069
1200	0.96 ± 10	0.00035 ± 0	0.0009 ± 0	53.9 ± 2.7	0.32 ± 2	5 ± 1	3.388 ± 0.080
1300	0.76 ± 7	0.00034 ± 0	0.0009 ± 0	56.6 ± 2.6^b	0.42 ± 2	6 ± 1	3.468 ± 0.073
1400	0.83 ± 6	0.00034 ± 0	0.0009 ± 1	60.3 ± 4.2^b	0.22 ± 1	5 ± 1	3.568 ± 0.112
1550	0.75 ± 3	0.00036 ± 0	0.0010 ± 0	58.0 ± 2.1^b	0.63 ± 2	6 ± 1	3.505 ± 0.058
1610	0.67 ± 3	0.00040 ± 0	0.0010 ± 0	56.2 ± 2.0^b	0.40 ± 1	7 ± 1	3.456 ± 0.057
1640	0.72 ± 7	0.00036 ± 0	0.0011 ± 1	56.4 ± 3.0^b	0.44 ± 2	6 ∓ 1	3.460 ± 0.085
1670	0.72 ± 4	0.00038 ± 0	0.0010 ± 0	58.2 ± 1.6^b	1.94 ± 5	6 ± 1	3.510 ± 0.045
1700	0.75 ± 7	0.00038 ± 0	0.0009 ± 0	63.2 ± 3.5	0.40 ± 2	6 ± 1	3.644 ± 0.090
1750	0.60 ± 27	0.00043 ± 1	0.0010 ± 2	71.8 ± 13.1	0.08 ± 1	8 ± 3	3.849 ± 0.299
$J_I^c = 0.09349 \pm 0.00045$		40/36 trapped $= 0.5 \pm 0.5$					
60015,69	310.5 MG cataclastic anorthosite						
650	1.53 ± 69	0.00132 ± 5	0.0011 ± 0	443.9 ± 9.9	0.01 ± 0	21 ± 16	7.077 ± 0.042
900	1.47 ± 37	0.00027 ± 1	0.0008 ± 1	20.3 ± 2.2	0.07 ± 1	3 ± 2	2.011 ± 0.132
1050	0.97 ± 5	0.00032 ± 0	0.0008 ± 0	13.4 ± 1.3	0.11 ± 1	5 ± 0	1.540 ± 0.101
1200	0.88 ± 5	0.00032 ± 0	0.0008 ± 0	42.1 ± 2.3	0.21 ± 1	5 ± 1	3.019 ± 0.084
1300	0.90 ± 5	0.00032 ± 0	0.0007 ± 0	54.7 ± 1.9	0.31 ± 1	5 ± 1	3.422 ± 0.056
1350	0.89 ± 4	0.00033 ± 0	0.0008 ± 0	56.4 ± 1.8	0.31 ± 1	5 ± 0	3.470 ± 0.051
1400	0.95 ± 8	0.00034 ± 0	0.0010 ± 1	56.5 ± 4.5	0.17 ± 1	5 ± 1	3.472 ± 0.125
1600	0.99 ± 8	0.00032 ± 0	0.0009 ± 0	57.3 ± 2.6	0.59 ± 2	5 ± 1	3.496 ± 0.073
1700	0.94 ± 11	0.00028 ± 0	0.0008 ± 0	62.0 ± 2.5	2.62 ± 9	4 ± 1	3.621 ± 0.067
1800	0.91 ± 51	0.00071 ± 1	0.0035 ± 1	58.8 ± 21.5	0.31 ± 1	12 ± 9	3.537 ± 0.584
$J_{II} = 0.09402 \pm 0.00056$		40/36 trapped $= 0.5 \pm 0.5$					
60025,86	364.8 MG Cataclastic anorthosite						
650	0.73 ± 2	0.00193 ± 0	0.0009 ± 0	51.4 ± 1.8	0.21 ± 1	39 ± 2	3.323 ± 0.056
800	2.37 ± 4	0.00119 ± 0	0.0009 ± 0	48.3 ± 2.0	0.43 ± 1	15 ± 1	3.229 ± 0.064
900	2.74 ± 3	0.00098 ± 0	0.0008 ± 0	64.5 ± 2.0	0.70 ± 1	10 ± 1	3.685 ± 0.052
950	2.67 ± 7	0.00080 ± 0	0.0008 ± 0	74.9 ± 2.0	0.51 ± 1	8 ± 1	3.930 ± 0.044
1000	2.11 ± 5	0.00066 ± 0	0.0008 ± 0	79.5 ± 2.2	0.49 ± 1	8 ± 1	4.028 ± 0.047
1050	1.55 ± 2	0.00063 ± 0	0.0008 ± 0	82.9 ± 3.5	0.36 ± 1	9 ± 1	4.097 ± 0.071
1100	1.62 ± 7	0.00062 ± 0	0.0009 ± 0	85.0 ± 4.3	0.30 ± 1	9 ± 1	4.141 ± 0.086
1150	1.47 ± 6	0.00062 ± 0	0.0008 ± 0	89.0 ± 3.3^b	0.28 ± 1	9 ± 1	4.217 ± 0.063
1200	1.31 ± 5	0.00068 ± 0	0.0008 ± 0	88.0 ± 3.5^b	0.30 ± 1	11 ± 1	4.198 ± 0.067
1250	1.13 ± 1	0.00084 ± 0	0.0008 ± 0	86.9 ± 2.8^b	0.31 ± 1	15 ± 1	4.176 ± 0.036
1300	0.84 ± 4	0.00093 ± 0	0.0008 ± 0	89.9 ± 3.9^b	0.18 ± 1	18 ± 1	4.234 ± 0.073
1350	0.62 ± 4	0.00093 ± 0	0.0008 ± 0	91.3 ± 5.6^b	0.10 ± 1	18 ± 1	4.260 ± 0.103
1600	0.94 ± 3	0.00088 ± 0	0.0009 ± 0	90.9 ± 4.7^b	0.17 ± 1	16 ± 1	4.252 ± 0.088
1700	0.52	0.00083 ± 0	0.0032 ± 0	94.8 ± 45.4	0.10 ± 0	16 ± 21	4.323 ± 0.811
$J_{III} = 0.09402 \pm 0.00056$		40/36 trapped $= 0.5 \pm 0.5$					
60025,86	345.1 MG Cataclastic anorthosite						
650	0.59 ± 2	0.00103 ± 0	0.0009 ± 0	55.3 ± 1.7	0.74 ± 2	21 ± 2	3.429 ± 0.050
800	1.34 ± 6	0.00045 ± 0	0.0008 ± 0	75.9 ± 3.0	1.08 ± 4	7 ± 1	3.941 ± 0.066
900	1.00 ± 6	0.00049 ± 0	0.0008 ± 0	87.4 ± 3.0	0.70 ± 2	8 ± 1	4.177 ± 0.059
950	1.05 ± 6	0.00043 ± 0	0.0008 ± 0	93.7 ± 4.3	0.24 ± 1	7 ± 1	4.295 ± 0.079
1000	0.83 ± 4	0.00044 ± 0	0.0009 ± 0	88.6 ± 4.9	0.26 ± 1	8 ± 1	4.199 ± 0.094
1050	0.72 ± 7	0.00051 ± 0	0.0009 ± 1	87.1 ± 9.4	0.25 ± 3	9 ± 1	4.171 ± 0.182
1100	0.54 ± 3	0.00062 ± 0	0.0008 ± 0	87.5 ± 3.9	0.27 ± 1	12 ± 1	4.178 ± 0.075
1150	0.45 ± 2	0.00073 ± 0	0.0008 ± 1	87.6 ± 5.8	0.27 ± 2	15 ± 1	4.182 ± 0.111
1200	0.41 ± 3	0.00085 ± 0	0.0009 ± 1	85.6 ± 5.5	0.26 ± 2	17 ± 1	4.142 ± 0.108
1250	0.39 ± 2	0.00090 ± 0	0.0008 ± 1	86.5 ± 11.0	0.19 ± 2	19 ± 1	4.159 ± 0.213

Table 1. Argon isotopic data for Apollo 16 and 17 rocks.

Temp. (C)	36/38	38/37	39*/37	40*/39*	39* $\times 10^{-8}$ ccSTP/g	Exp. age (m.y.)	Age (G.y.)
1300	0.28 ± 3	0.00090 ± 0	0.0008 ± 1	85.8 ± 7.5	0.12 ± 1	19 ± 2	4.146 ± 0.146
1350	0.30 ± 4	0.00094 ± 0	0.0008 ± 2	94.4 ± 19.9	0.07 ± 1	20 ± 2	4.307 ± 0.357
J_I = 0.09349 ± 0.00045		40/36 trapped = 0.5 ± 0.5					
78503,7,1	36.6 MG						
600	0.68 ± 4	0.00438 ± 2	0.0139 ± 8	14.3 ± 1.1	1.63 ± 10	90 ± 8	1.636 ± 0.085
700	1.67 ± 40	0.00522 ± 3	0.0181 ± 63	29.4 ± 24.5	0.04 ± 1	84 ± 52	2.538 ± 1.163
800	1.63 ± 12	0.00383 ± 3	0.0115 ± 4	36.2 ± 1.7	1.15 ± 4	62 ± 8	2.834 ± 0.071
900	1.24 ± 9	0.00418 ± 3	0.0090 ± 1	42.9 ± 0.7	3.91 ± 5	76 ± 9	3.089 ± 0.026
1000	0.93 ± 1	0.00410 ± 0	0.0082 ± 1	62.4 ± 1.1	6.59 ± 12	81 ± 5	3.677 ± 0.031
1100	0.96 ± 3	0.00416 ± 1	0.0080 ± 1	75.8 ± 0.8	4.16 ± 4	82 ± 6	3.996 ± 0.020
1200	0.78 ± 2	0.00411 ± 0	0.0081 ± 1	80.8 ± 0.8ᵇ	5.68 ± 5	84 ± 5	4.102 ± 0.020
1300	0.71 ± 1	0.00404 ± 0	0.0080 ± 1	81.4 ± 1.3ᵇ	6.20 ± 9	84 ± 5	4.115 ± 0.028
1400	0.69 ± 1	0.00402 ± 0	0.0078 ± 1	84.2 ± 1.2ᵇ	3.78 ± 4	84 ± 5	4.171 ± 0.025
1500	0.67 ± 2	0.00408 ± 0	0.0077 ± 1	81.5 ± 2.1ᵇ	3.66 ± 5	86 ± 6	4.117 ± 0.043
1600	0.68 ± 2	0.00397 ± 0	0.0079 ± 1	84.3 ± 1.6ᵇ	6.38 ± 11	83 ± 5	4.174 ± 0.033
J_{III} = 0.09672 ± 0.00056		40/36 trapped = 0.5 ± 0.5					
72503,8,12	21.8 MG						
600	3.06 ± 35	0.04232 ± 49	0.4902 ± 150	14.3 ± 0.4	5.75 ± 8	334 ± 124	1.656 ± 0.031
700	3.40 ± 11	0.04500 ± 40	0.3203 ± 271	40.9 ± 0.8	6.36 ± 10	334 ± 75	3.044 ± 0.031
800	2.99 ± 4	0.03654 ± 5	0.2445 ± 38	72.3 ± 1.1ᵇ	18.35 ± 26	345 ± 26	3.948 ± 0.026
900	2.78 ± 5	0.03344 ± 7	0.1533 ± 18	73.2 ± 0.7ᵇ	37.40 ± 28	367 ± 27	3.969 ± 0.016
950	2.37 ± 3	0.02697 ± 3	0.0862 ± 9	72.2 ± 0.8ᵇ	25.49 ± 22	357 ± 26	3.945 ± 0.019
1000	1.90 ± 2	0.02198 ± 3	0.0557 ± 6	72.9 ± 0.8ᵇ	24.49 ± 17	344 ± 21	3.962 ± 0.019
1050	1.50 ± 3	0.01919 ± 4	0.0432 ± 3	74.0 ± 0.7ᵇ	22.50 ± 16	338 ± 23	3.986 ± 0.017
1100	1.93 ± 2	0.02210 ± 3	0.0378 ± 6	73.8 ± 1.3ᵇ	13.28 ± 21	347 ± 21	3.983 ± 0.030
1200	1.81 ± 3	0.02176 ± 4	0.0400 ± 7	69.5 ± 1.1	14.54 ± 19	353 ± 25	3.884 ± 0.026
1300	1.39 ± 1	0.02077 ± 1	0.0304 ± 3	66.4 ± 0.8	13.05 ± 11	380 ± 24	3.808 ± 0.020
1400	0.79 ± 1	0.01765 ± 3	0.0219 ± 3	70.7 ± 0.9	18.43 ± 19	374 ± 23	3.912 ± 0.021
1500	0.76 ± 2	0.01796 ± 3	0.0221 ± 7	77.1 ± 6.3	3.49 ± 11	384 ± 25	4.055 ± 0.137
1600	0.83 ± 8	0.07517 ± 2215	0.0768	124.6 ± 59.6	0.62 ± 28		4.875 ± 0.832
J_{IV} = 0.09853 ± 0.00029		40/36 trapped = 0.5 ± 0.5					

ᵃCorrections have been made as follows: ^{36}Ar and ^{38}Ar for system blanks; ^{37}Ar for radioactive decay; ^{39}Ar for ^{39}Ar$_{ca}$ contribution; and ^{40}Ar for system blank and trapped ^{40}Ar.

ᵇTemperature fractions used to determine final ages.

ᶜThe ages determined here are based on the hornblende mineral separate, NL-25-2 which contains $0.3070 \pm 0.0008\%$ K and $(7.20 \pm 0.09) \times 10^{-5}$ ccSTP^{40} Ar/g (Husain, 1974). In each irradiation four standards were included. $J = [\exp(t/\tau) - 1]/^{40}$Ar*$/^{39}$Ar*, where $\tau = 1.885$ G.y., t = age of the standard = 2.6698 ± 0.018 G.y.

$J_1 = 0.09349 \pm 0.00045$ is based on the average of two measurements with ^{40}Ar*$/^{39}$Ar* = 33.32 ± 0.16 and 33.49 ± 0.36. The criterion of the selection of standards for age determination is the closeness of lunar samples to a given standard. The other splits of NL-25-2 in this irradiation yielded ^{40}Ar*$/^{39}$Ar* ratios of 33.38 ± 0.15 and 33.56 ± 0.10.

$J_{II} = 0.09402 \pm 0.00056$ is based on the average of four measurements on NL-25-2 with ^{40}Ar*$/^{39}$Ar* = 33.21 ± 0.28, 33.04 ± 0.20, 33.23 ± 0.14, and 33.34 ± 0.21.

J_{III} and J_{IV} are from the same irradiation which, as is also shown by the nickel wire, has a gradient in neutron flux near the very bottom of the quartz tube. Four hornblende splits gave ^{40}Ar*$/^{39}$Ar* ratios of 31.47 ± 0.22, 31.63 ± 0.10, 31.75 ± 0.10 and 32.28 ± 0.19. The standard used in each case was next to the lunar sample, hence the uncertainty due to neutron flux gradient is negligible.

The samples analyzed in this work are:

Rock 60015, which weighs 5.57 kg, is a coarsely crystalline cataclastic anorthosite consisting of highly strained plagioclase (Sclar *et al.*, 1973). The plagioclase contains myriads of bubble-like inclusions which appear to be glass. The rock was apparently severely shocked, raised almost to the melting point, and is coated with a dark grey vesicular glass.

Rock 60025, which weighs 1.8 kg, is also a cataclastic anorthosite. It shows well-developed shock lamellae and linear chains of low-relief isotropic inclusions (along with shock lamellae) that may represent incipient melting (Hodges and Kushiro, 1973). This rock is similar to the anorthositic fragments which give ages ~4.25 G.y. (Schaeffer and Husain, 1973).

Rock 72503,8,12 is a 2–4 mm fragment of the melt rock lithologic type (olivine–pyroxene–plagioclase) described by Bence *et al.* (1974). A detailed description is given in the Appendix.

Rock 78503,7,1 is a 2–4 mm gabbroic anorthosite fragment. The petrographic description is given in the Appendix.

Results

The Ar isotopic data are given in Table 1. All Ar isotopes have been corrected for system blanks and mass discrimination. In addition, the following corrections were also applied: (1) ^{37}Ar for radioactive decay, using a half-life of 35.1 days, (2) ^{39}Ar for contribution from ^{42}Ca (n, α) ^{39}Ar reaction, using ^{39}Ar/^{37}Ar = $(6.45 \pm 0.29) \times 10^{-4}$ measured for several Ca targets, (3) ^{40}Ar for trapped Ar using $(^{40}$Ar/^{36}Ar$)_{tr}$ ratios given in Table 1. The ^{40}Ar and ^{39}Ar produced from K alone are denoted by an asterisk. The uncertainties in the K–Ar ages of the individual temperature fractions, Table 1, are based on the uncertainties in the ^{40}Ar*/^{39}Ar* ratios of the sample and the hornblende standard. The uncertainty in the age of the hornblende standard NL-25-2 (2.668 ± 0.018 G.y.) is included only in the final age determination.

The ^{40}Ar–^{39}Ar formation age of a rock is inferred from the ^{40}Ar*/^{39}Ar* spectrum from K sites of varying retentivities as determined from stepwise-heating experiments. The thermal release patterns of the individual samples are discussed below.

Rock 60025

In Fig. 1 we have plotted the apparent ages against cumulative fractions of ^{39}Ar* for two analyses of this rock. In both analyses, the same temperatures were used to extract Ar. The Ar is somewhat more easily diffused from 60025,86,1 than 60025,86. However, the shapes of the age spectra are very similar. The low-temperature fractions (<900°C) yield relatively younger ages indicating Ar diffusion losses from low retentivity K phases. From the plateau ^{40}Ar–^{39}Ar and K–Ar ages, we estimate ^{40}Ar* losses of 8.0% for 60025,86,1 and 13.6% for 60025,86. For sample 60025,86, 58% ^{40}Ar* is released in 1000–1600°C fractions. A high resolution age spectrum is obtained as ^{40}Ar* is quite evenly distributed in nine fractions between 1000°C and 1600°C. An extraction at 1700°C contained only 2% of the total ^{40}Ar* and yielded an age with large uncertainty. The uncertainties in ages of other fractions are also larger than our usual measurements, due mainly

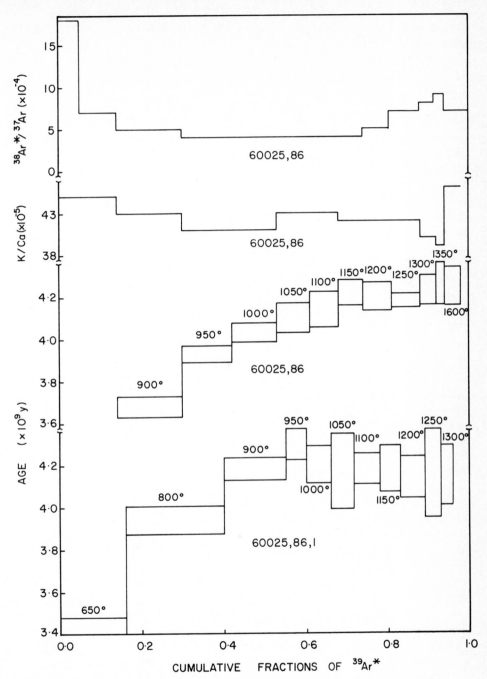

Fig. 1. The results of stepwise heating experiments on cataclastic anorthosite 60025. Temperatures shown are in degrees centigrade. The K/Ca and $^{38}Ar^*/^{37}Ar$ ratios are shown only for 60025,86. These ratios for 60025,86,1 are in excellent agreement with those for 60025,86.

to a large correction, about 40% for an ^{39}Ar contribution from Ca. The ^{40}Ar$_{trapped}$ correction was made using trapped $(^{40}$Ar/^{36}Ar$) = 0.5 \pm 0.5$. However, this correction has only a minimal effect as this rock contains very little trapped Ar. The 1150–1600°C fractions show a well-defined plateau. Even the 1050°C and 1100°C fractions yield ages which are within experimental uncertainties of those which define the plateau. This plateau corresponds to an age of 4.21 ± 0.06 G.y. for rock 60025,86. The uncertainty quoted is a 2σ error and includes the uncertainty in the age of the hornblende standard. If we were to include the 1000°C fraction in the plateau, the ^{40}Ar–^{39}Ar would decrease only by 0.01 G.y.

In comparison to 60025,86, the second analysis of this rock, 60025,86,1 suffers from relatively higher uncertainties due mainly to a temporary mass-spectrometer memory in ^{39}Ar left from the first analysis. Nevertheless, a reasonably well defined age plateau is observed for 1000–1300°C fractions. The latter corresponds to an ^{40}Ar–^{39}Ar age of 4.17 ± 0.06 G.y. (1σ error). This age is in excellent agreement with the one derived above for 60025,86. We believe the average of these two ages, 4.19 G.y., to be the best age for rock 60025.

Since a great many of the lunar highland breccias have suffered severe shocks and multiple brecciations, the possibility of relict or inherent Ar must be carefully considered. The two most direct ways to study the relict Ar are by studies of ^{40}Ar*/^{39}Ar* age spectrum and Rb–Sr systematics. A comparison of the ^{40}Ar*/^{39}Ar* systematics with those of Rb/Sr for a given rock provides a test of relative isotopic equilibration. An agreement between ^{40}Ar–^{39}Ar and Rb–Sr ages of breccias means that both isotopic systems were equilibrated at the same time. In case of partial resetting of K–Ar clock, no ^{40}Ar*/^{39}Ar* plateau would result. It is well to recall at this time that for lunar breccias the Rb–Sr isochron ages are in excellent agreement with the whole rock ^{40}Ar–^{39}Ar ages. The recent results of Huneke et al. (1974) for breccia 65015 clearly demonstrate that relict plagioclases, when present, are identifiable both in the ^{40}Ar*/^{39}Ar* age spectrum and Rb–Sr isochron. A constant age in high-temperature fractions of 60025 as opposed to increase in ages at high-temperature fractions for 65015 plagioclase clearly proves that no relict Ar is present in 60025.

In Fig. 1, we have also plotted the K/Ca and ^{38}Ar*/^{37}Ar ratios (the asterisk denotes spallation ^{38}Ar) for 60025,86. These ratios obtained for the second analysis, 60025,86,1 are in excellent agreement with the first and are not shown in Fig. 1. It is observed that the K/Ca ratios are essentially constant, within experimental uncertainties, at all temperatures. The ^{36}Ar/^{38}Ar ratios, column 2, Table 1, show that 60025,86 contains some solar wind ^{36}Ar and ^{38}Ar. These two components were resolved assuming $(^{36}$Ar/^{38}Ar$)_{sw} = 5.35$ and $(^{36}$Ar/^{38}Ar$)_{spall.} = 0.65$. The ^{38}Ar*/^{37}Ar ratios start out high for 650°C and 800°C fractions but then attain a plateau for 900–1150°C fraction. An increase in this ratio occurs subsequently. An explanation of this is provided by the sample 60015,86,1 which contains lower solar wind ^{36}Ar and ^{38}Ar. The temperature fractions 1100–1350°C yield ^{36}Ar/^{38}Ar ratios between 0.28 and 0.54. This is most certainly due to contribution to ^{38}Ar* from a source other than protons spallation of K, Ca, Fe, and Ti. This excess ^{38}Ar* could be from ^{37}Cl (n, γ) ^{38}Cl $\xrightarrow{\beta^-}$ ^{38}Ar reaction during neutron bombardment. A cosmic ray exposure age of 8.9 ± 0.5 m.y. from the plateau $^{38}Ar^*/^{37}Ar$ ratio was determined.

Figure 2 shows results for rock 78503,7,1. The $^{40}Ar^*/^{39}Ar^*$ thermal release diagram is relatively simple. The low-temperature fractions suggest $^{40}Ar^*$ diffusion losses from mineral phases of low retentivity. However, 1200–1600°C fractions containing 60% of the total $^{40}Ar^*$ all give the same ages within experimental uncertainties. The K/Ca ratios also shown in Fig. 2 clearly show that all mineral phases, releasing Ar above 1000°C, have the same K/Ca ratio. Thus, the discussion given above about relict plagioclases is valid also for the rock 78503,7,1. An $^{40}Ar–^{39}Ar$ formation age of 4.13 ± 0.03 G.y. is determined from the 1200–1600°C plateau. The $^{38}Ar^*/^{37}Ar$ release pattern is similar to the $^{40}Ar^*/^{39}Ar^*$ thermal release pattern. But $^{38}Ar^*$ losses from the low-retentivity sites are relatively quite small, less than 3%, compared to 13.5% for $^{40}Ar^*$. A mean exposure age, 84 ± 3 m.y., is determined from the plateau (1200–1600°C) $^{38}Ar^*/^{37}Ar$ ratios.

In Fig. 3, we have plotted the $^{40}Ar/^{39}Ar$ thermal release patterns from duplicate runs of rock 60015, which like rock 60025 contains only 0.0070%K, but 15.3% Ca, making the $^{40}Ar–^{39}Ar$ age determination more difficult. Large samples, ~310 mg, were analyzed to obtain an accurate age. The Ar isotopic data are given in Table 1. The K/Ca ratios are essentially constant in all temperature fractions of both analyses. Thus, $^{39}Ar_{ca}$ corrections amount to approximately 41.5% at all temperatures. In the first analysis, 60015,69, 55% of the total $^{40}Ar^*$ was released in a single extraction at 1700°C. This fraction yields a slightly older age, 3.62 ± 0.07 G.y. than the 1350–1600°C fractions with ages between 3.47 and 3.50 G.y. The 3.62 G.y. age, though higher, is still in agreement with the 3.50 G.y. age. However, the agreement is at the limit of experimental uncertainties. In the second analysis of this rock, four extractions were made between 1600°C and 1700°C. A very well defined $^{40}Ar/^{39}Ar$ plateau is obtained from the 1300–1670°C fractions, which yields an age of 3.50 ± 0.05 G.y. This age is taken to be the age of the rock 60015, in view of the better release pattern obtained for the second run.

An unusual feature of the release pattern is that about 50% of the total $^{40}Ar^*$ is released in a very narrow temperature range, 1610–1670°C. Chemically, rocks 60015 and 60025 are very similar, but the Ar isotopes are released in very different manners. Although rocks 60015 and 60025 are shocked and belong to the cataclastic anorthosites (type III) (LSPET, 1973a), both yield very well defined plateaus.

For rock 60015, K/Ca ratios are constant at all temperatures, Table 2. Thus, we have determined the $^{38}Ar^*/Ca$ exposure age from the total $^{38}Ar^*$ to total ^{37}Ar ratio. The duplicate analyses yield exposure ages of 4.6 ± 0.6 and 6.1 ± 0.5 m.y., with a mean of 5.3 ± 0.7 m.y.

The lithic fragment 72503,8,12, a gray–green recrystallized breccia with poikiloblastic texture (Bence *et al.*, 1974) gives a more conventional $^{40}Ar^*/^{39}Ar^*$ thermal release diagram, Fig. 4. This sample has almost quantitatively retained its $^{40}Ar^*$ since its last metamorphism. A plateau $^{40}Ar–^{39}Ar$ age of 3.96 ± 0.02 G.y. and cosmic ray exposure age of 360 ± 20 m.y. are determined.

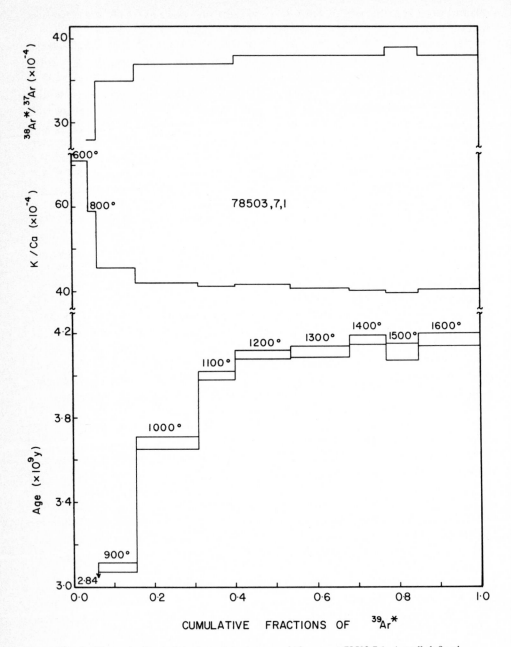

Fig. 2. Thermal release data for a 2 to 4 mm rock fragment 78503,7,1. A well-defined plateau ^{40}Ar–^{39}Ar age of 4.13 ± 0.03 G.y. is determined.

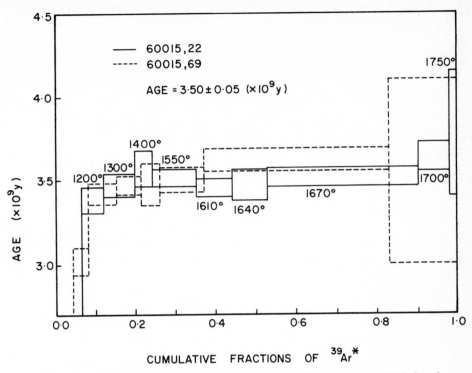

Fig. 3. Age data from duplicate runs of cataclastic anorthosite, 60015. About 50% of total Ar is released in a narrow temperature range, 1610–1670°C, indicating melting of a single phase. The 3.50 ± 0.05 G.y. age of this rock is the youngest for a lunar highland sample. See text.

Table 2. Summary of ages, and K and Ca concentrations of lunar rocks.

Sample	% K[a]	% Ca	Exposure age $\times 10^6$ yr	$^{40}Ar-^{39}Ar$ Age $\times 10^9$ yr
60015,22	0.0067	15.7	4.6 ± 0.6	3.50 ± 0.05
60015,69	0.0073	15.0	6.1 ± 0.5	3.51 ± 0.06[b]
60025,86	0.0063	14.5	8.9 ± 0.5	4.21 ± 0.06
60025,86,1	0.0065	14.7	8.3 ± 0.7	4.17 ± 0.06
72503,8,12	0.2754	9.9	360 ± 20	3.96 ± 0.02
78503,7,1	0.0594	14.0	8.4 ± 3	4.13 ± 0.03

[a]The uncertainties in K and Ca concentrations are estimated to be ±10%. The procedure for K and Ca determination has been given by Husain (1974).

[b]Average of 1350–1700°C fractions.

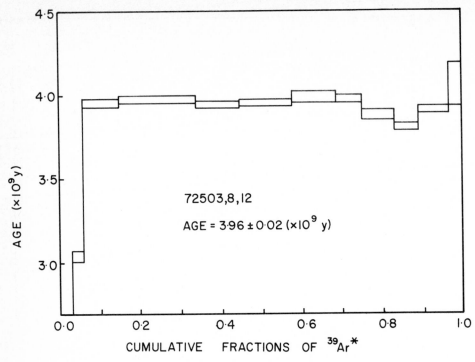

Fig. 4. Age data for an Apollo 17 2–4 mm fragmental rock, 72503,8,12.

DISCUSSION

The ages determined in this work are summarized in Table 2 along with the K and Ca concentrations of the samples.

Rocks 60015 and 60025 were collected at Station 10 and the LM landing site, respectively. We have determined exposure ages of 5.3 ± 0.7 and 8.6 ± 0.8 m.y. for 60015 and 60025, respectively. We have previously reported (Husain and Schaeffer, 1973a) on the age of the rock 60315, also from Station 10. This rock has an exposure age of 6 ± 2 m.y., very similar to those for rocks 60015 and 60025. There are several small craters in the vicinity of Station 10/LM landing site. But the most prominent and closest to the station are craters Buster and Spooky. On the basis of the position of recovery of the rocks 60015 and 60315, we suggest that the Buster cratering event occurred 6 m.y. ago and Spooky some 8.6 m.y. ago.

Station 2, Apollo 17

We report here an exposure age of 360 ± 20 m.y. for rock 72503,8,12. Our preliminary results on two other 2–4 mm fragments yield cosmic ray exposure ages of 140 ± 10 and 45 ± 5 m.y. Thus, for three samples from Station 2, we have a span of 45–360 m.y. in exposure ages. Apparently, the regolith at Station 2 has undergone frequent turnovers and mass wasting.

In view of the above conclusions for the samples from South Massif, the exposure age of 78503,7,1 from Station 8 in North Massiff cannot be reliably used to date prominent local craters.

$^{40}Ar-^{39}Ar$ ages

We have reported in this work the oldest age for a kilogram size rock. This age adds to our earlier work (Schaeffer and Husain, 1973) which provided the first evidence for lunar history during 4.0–4.6 G.y. The Ar isotopic studies discussed in detail earlier in this paper clearly lead us to the conclusion that the 4.19 G.y. age for this rock represents the time of last isotopic equilibration. Since that time this rock has been a "closed" Ar system.

Rock 60015 yields the "youngest" formation age for a highland breccia. The $^{40}Ar^*/^{39}Ar^*$ release patterns observed for two samples of this rock are almost "ideal" inasmuch as the reproducibility of the $^{40}Ar^*/^{39}Ar^*$ ratios at plateau are concerned. The 3.50 ± 0.05 G.y. age is puzzling as its shows the resetting of the Ar clock of a large rock after the end of the major basin forming era. Since this is a highly metamorphosed rock, its age cannot support any major highland volcanism at ~3.5 G.y. However, we cannot rule out burial at a depth where Ar will not be retained. A subsequent impact could bring the rock to lunar surface. This would be consistent with the shocked nature of the rock.

CHRONOLOGY OF THE BASIN FORMATION

It has been proposed from the first ages reported for highland type breccias (Sutter *et al.*, 1971) that these rocks represent the ages of the large impact forming events. A lunar rock can be at its present location either because it was removed from the local bedrock by a relatively small crater and placed in the regolith, or because it was part of the ejected material from a large basin forming event.

It appears that samples of basin ejecta, especially from Apollo 14, 15, 16, and 17 sites, date the time of the basin forming events. These materials are predominantly breccia with few samples which do not show extensive remelting or shock features. The rocks were apparently sufficiently metamorphosed during the basin forming event that both the $^{40}Ar-^{39}Ar$ and the Rb–Sr mineral isochron ages were reset.

The K–Ar clock is reset by the diffusion of argon. It has been estimated (Turner *et al.*, 1973) that a temperature of 400–700°C may be adequate to degas a rock of its argon in a geologically short time, approximately one year at 700°C and 10,000 yr at 400°C. Rb–Sr ages for the highland rocks are frequently defined by the analysis of plagioclases and a Rb-rich finely crystalline mesostasis variously called quintessence. The resetting of the Rb–Sr clock by the homogenization of such a mineral assemblage is quite likely for the pressures and temperatures accompanying the formation of an ejecta layer.

There are three principal mechanisms by which the rock can be heated to reset the radiometric clocks: (1) During the impact, itself, the high pressure and temperature may be sufficient at least for some of the materials to reset the clocks. Studies of the Brent Crater in Canada (Hartung et al., 1971) indicated that certain of the materials had completely lost their argon. (2) The thermal gradient on the moon, 4–4.5 G.y. ago, when most of the large basins were formed, may have been such that a good fraction of the materials excavated were at temperatures of several hundred degrees centigrade, and as a result the rocks before impact may have been an essentially open system for argon diffusion and for Sr isotopic equilibrium in the Rb-rich phases. (3) The ejecta blanket is heated by the impact but probably derives most of its heat from the thermal gradient in the moon at the time of impact. As a result, it is highly likely that all the breccia which petrologically appear to have a complex history are reset to the age of the basin forming event. It is of interest to note that Strangway et al., (1973) have suggested that the magnetic anomalies at the Apollo 16 site are caused by breccia flows which cooled in place from above 770°C, a temperature more than adequate to reset the radiometric clocks.

It has been estimated by McGetchin et al. (1973) that the ejecta blankets from the large lunar basins at the Apollo 14, 16, and 17 landing sites are hundreds of meters thick. At each site, there is only a thin layer, several meters at most, of Orientale ejecta. The striking feature is that the Imbrium ejecta predominates in the upper 100 m at all landing sites. The results of McGetchin et al. (1973) are shown in Fig. 5. Only at the Apollo 16 site is Imbrium ejecta less than 100 m thick.

If one assumes that the radiometric ages of the breccias have all been reset by the basin forming events, one is struck by the fact that the age 3.95–4.00 G.y. predominates in all the measurements. It appears that this age is related to the Imbrium event whose ejecta predominates in upper layers at all the sites sampled (Podosek et al., 1973).

There have been ages reported from breccias at the Apollo 16 and 17 landing sites (Schaeffer and Husain et al., 1973; Stettler et al., 1974; Kirsten et al., 1974) which are well in excess of 4 G.y., ranging up to 4.28 G.y. At Apollo 16, the Imbrium ejecta is estimated to be 50 m thick so that local craters with diameters of hundreds of meters are likely to excavate pre-Imbrium ejecta. At Apollo 17, the massifs were probably uplifted during the time of formation of the Serenitatis Basin. As a result, they should be covered with ejecta from all basins formed after Serenitatis. It is likely that at the Apollo 17 site, one can find samples of all basin forming events back to and including Serenitatis.

In addition to rocks 60025 and 78503,7,1 a number of samples have been found on the rim of North Ray Crater which give ages in the range of 4.04–4.26 G.y. (Schaeffer and Husain, 1973). From the estimates of McGetchin et al. (1973) (Fig. 5), it would appear that Nectaris ejecta would predominate underneath the Imbrium layer. The North Ray cratering event excavated deep enough, ~200 m, so it could bring the Nectaris ejecta to lunar surface. It is thus possible to associate the oldest of these ages with the Nectaris event. This would place the Nectaris event at 4.25 ± .05 G.y. The other old ages between 4.05 and 4.20 G.y. would represent Crisium and Humorum ejecta. There are at present not enough ages to allow a precise age assignment for these two basin forming events.

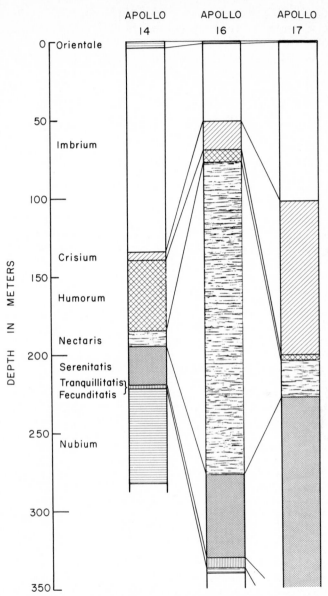

Fig. 5. Depth of ejecta, from basin forming lunar cratering events, at the Apollo 14, 16, and 17 sites (McGetchin *et al.*, 1973).

The regolith at the Apollo 17 site below 200 m is predominated by Serenitatis ejecta (Fig. 5). It should be possible to determine the age of the Serenitatis event by a study of the coarse fines near the North and South Massifs. However, the age of 78503,7,1 is significantly younger.

A few ages of approximately 3.85 G.y. have been measured at the Apollo 14 site (Husain *et al.*, 1972; Papanastassiou and Wasserburg, 1971). It is likely that these represent Orientale ejecta which would place the Orientale event at 3.85 ± .05 G.y.

As a result, we propose the following chronology for the five most recent lunar basin forming events:

Orientale	3.85 ± .05 (G.y.)
Imbrium	3.95 ± .05
Crisium ⎱ Humorum ⎰	4.05 – 4.20
Nectaris	4.25 ± .05

It seems quite clear that the era of lunar basin formation extended over many hundreds of millions of years of the lunar chronology up to the formation of the Orientale event. It also seems quite clear that the evidence for a lunar cataclysm at 3.95 G.y. (Tera *et al.*, 1973) is a reflection of the widespread occurrence of Imbrium ejecta at the Apollo landing sites and that the lunar cataclysm is really the Imbrium event.

REFERENCES

Bence A. E., Delano J. W., and Papike J. J. (1974) Nature of the Highland Massifs of Taurus Littrow: An Analysis of the 2 to 4 mm Soil Fraction. *Proc. Fifth Lunar Sci. Conf., Geochim. Cosmochim. Acta.* This volume.
Hartung J. B., Dence M. R., and Adams J. A. S. (1971) Potassium–argon dating of metamorphosed rocks from the Brent impact crater, Ontario, Canada. *J. Geophys. Res.* **76**, 5437–5448.
Hodges F. N. and Kushiro I. (1973) Petrology of Apollo 16 lunar highland rocks. *Proc. Fourth Lunar Sci. Conf., Geochim. Cosmochim. Acta*, Suppl. 4, Vol. 1, pp. 1033–1048. Pergamon.
Hodges C. A., Muehlberger W. R., and Ulrich G. E. (1973) Geologic setting of Apollo 16. *Proc. Fourth Lunar Sci. Conf., Geochim. Cosmochim. Acta*, Suppl. 4, Vol. 1, pp. 1–25. Pergamon.
Huneke J. C., Jessberger E. K., and Wasserburg G. J. (1974) The age of metamorphism of a highland breccia (65015) and a glimpse at the age of its protolith (abstract). In *Lunar Science—V*, pp. 375–377. The Lunar Science Institute, Houston.
Husain L. (1974) ^{40}Ar–^{39}Ar chronology and cosmic ray exposure ages of the Apollo 15 samples. *J. Geophys. Res.* **79**, 2588–2606.
Husain L., Schaeffer O. A., Funkhouser J., and Sutter J. (1972) The ages of lunar material from Fra Mauro, Hadley Rille and Spur Crater. *Proc. Third Lunar Sci. Conf., Geochim. Cosmochim. Acta*, Suppl. 3, Vol. 2, pp. 1557–1567. MIT Press.
Husain L. and Schaeffer O. A. (1973) Lunar volcanism: Age of the glass in Apollo 17 orange soil. *Science* **180**, 1358–1360.
Husain L. and Schaeffer O. A. (1973a) ^{40}Ar–^{39}Ar crystallization ages and ^{38}Ar–^{37}Ar cosmic ray exposure ages of samples from the vicinity of the Apollo 16 landing site (abstract). In *Lunar Science—IV*, pp. 406–408. The Lunar Science Institute, Houston.
Kirsten T. and Horn P. (1974) ^{39}Ar–^{40}Ar chronology of the Taurus-Littrow Region II: A 4.28 b.y. old troctolite and ages of basalts and highland breccias (abstract). In *Lunar Science—V*, pp. 419–421. The Lunar Science Institute, Houston.
LSPET (Lunar Sample Preliminary Examination Team) (1971) Preliminary examination of samples from Apollo 14. *Science* **173**, 681–693.
LSPET (Lunar Sample Preliminary Examination Team) (1973a) Preliminary examination of samples from Apollo 16. *Science* **179**, 23–34.
LSPET (Lunar Sample Preliminary Examination Team) (1973b) Preliminary examination of samples from Apollo 17. *Science* **182**, 659–671.

McGetchin T. R., Settle M., and Head J. W. (1973) Radial thickness variation in impact crater ejecta: Implications for lunar basin deposits. *Earth Planet. Sci. Lett.* **20**, 226–236.

Papanastassiou D. A. and Wasserburg G. J. (1971) Rb–Sr ages of igneous rocks from the Apollo 14 mission and the age of the Fra Mauro Formation. *Earth Planet. Sci. Lett.* **12**, 36–48.

Podosek F. A., Huneke J. C., Gancarz A. J., and Wasserburg G. J. (1973) The age and petrography of two Luna 20 fragments and inference for widespread lunar metamorphism. *Geochim. Cosmochim. Acta* **37**, 887–904.

Schaeffer O. A. and Husain L. (1973) Early lunar history: Ages of 2–4 mm soil fragments from the lunar highlands. *Proc. Fourth Lunar Sci. Conf., Geochim. Cosmochim. Acta,* Suppl. 4, Vol. 2, pp. 1847–1863. Pergamon.

Sclar C. B., Bauer J. F., Pickart S. J., and Alperin H. A., (1973) Shock effects in experimentally shocked terrestrial ilmenite, lunar ilmenite of rock fragments in 1–10 mm fines (10085,19) and lunar rock 60015,127. *Proc. Fourth Lunar Sci. Conf., Geochim. Cosmochim. Acta,* Suppl. 4, Vol. 1, pp. 841–859. Pergamon.

Stettler A., Eberhardt P., Geiss J., Grögler N., and Maurer P. (1974) Sequence of terra rock formation and basaltic lava flows on the moon. In *Lunar Science—V,* pp. 738–740. The Lunar Science Institute, Houston.

Strangway D. W., Gose W. A., Pearce G. W., and McConnell R. K. (1973) Lunar magnetic anomalies and the Cayley formation. *Nature, Phys. Sci.* **246**, 112–115.

Sutter J. F., Husain L., and Schaeffer O. A. (1971) $^{40}Ar/^{39}Ar$ ages from Fra Mauro. *Earth Planet. Science Lett.,* **11**, 249–253.

Tera F., Papanastassiou D. A., and Wasserburg G. J. (1973) A lunar cataclysm at ~3.95 AE and the structure of the lunar crust, (abstract). In *Lunar Science—IV,* pp. 723–725. The Lunar Science Institute, Houston.

Turner G., Cadogan P. H., and Yonge C. J. (1973) Argon selenochronology. *Proc. Fourth Lunar Sci. Conf., Geochim. Cosmochim. Acta,* Suppl. 4, Vol. 2, pp. 1889–1914. Pergamon.

Appendix

Petrographic Description of Fragments Dated by $^{40}Ar/^{39}Ar$ Techniques from Bence *et al.* (1974).

78503,7,1 Gabbroic Anorthosite (large rock equivalent 77017).
This fragment is a clast of gabbroic anorthosite surrounded by a clast-laden glassy rim. The glass accounts for ~$\frac{1}{3}$ the area of the total fragment in thin section.
The gabbroic anorthosite fragment (modal mineralogy, plagioclase 70%, clinopyroxene 20%, olivine < 5%) has been moderately to severely shocked and granulated.
$^{40}Ar/^{39}Ar$ plateau age 4.13 ± 0.03 G.y.

72503,8,12 Melt Rock
This fragment is a clast-laden melt rock with a fine-grained groundmass containing laths of plagioclase, poikilitic pigeonites, and an unusually high abundance of opaques (some NiFe metal but primarily Fe–Ti oxides). Clasts include single fragments of olivine and plagioclase and one lithic clast. Slight development of reaction rims is observed around the plagioclase clasts.
Zoned pigeonites have $Fe/(Fe + Mg)$ ratios of 0.20–0.30.
Olivine compositions are $Fe/(Fe + Mg) \sim 0.3$.
$^{40}Ar/^{39}Ar$ plateau age 3.96 ± 0.02 G.y.

Acknowledgments—We wish to thank: G. Barber and T. Ludkewycz for running the mass spectrometer and the crew of the Brookhaven High Flux Beam Reactor for neutron irradiation. This work was supported by NASA grant NGL 33-015-130.

Proceedings of the Fifth Lunar Conference
(Supplement 5, Geochimica et Cosmochimica Acta)
Vol. 2 pp. 1557–1570 (1974)
Printed in the United States of America

On the duration of lava flow activity in Mare Tranquillitatis

A. Stettler, P. Eberhardt, J. Geiss, N. Grögler, and P. Maurer

Physikalisches Institut, University of Bern, Sidlerstrasse 5, 3012 Bern, Switzerland

Abstract—Ar^{39}–Ar^{40} investigations on whole rock and separated mineral samples of Apollo 11 mare basalts have resulted in ages between 3.91 ± 0.03 and 3.51 ± 0.06 AE. Rock ages and exposure ages of Apollo 11 basalts are related to their petrologic classification and their potassium contents; the rocks with the lowest K concentrations give the highest Ar^{39}–Ar^{40} ages. These results strongly imply that Apollo 11 basalts with significantly different chemical and petrologic characteristics originated from different lava flows and that volcanic activity in Mare Tranquillitatis lasted for at least 400 m.y. Crater erosion observations indicate in fact that there are significant differences in the ages of surface areas surrounding Tranquillity Base. However, these estimates are not detailed enough so far to allow a reconstruction of the lava flow history in this area. Fra Mauro rock 14053 has been described as an aluminous mare basalt. Its age of 3.95 ± 0.04 AE indicates that partial melting processes of the type which have produced mare basalts were operating even before the excavation of the Imbrian basin.

Introduction

THE FORMATION of the maria is one of the most significant geological processes in lunar history. Although the chemical composition and ages of many mare rocks from five different locations have been determined and their petrology studied in detail, a complete understanding of the evolution of maria has not been reached (cf. Ridley *et al.*, 1973). There is little doubt that the basins of the quasi-circular maria were excavated by impacts about 3.95 AE ago or earlier ($1 \text{ AE} = 10^9 \text{ y}$). Subsequently the basins were filled by lava flows of iron-rich, basaltic composition. It is a widely held view (Ringwood and Essene, 1970; Gast and Hubbard, 1970; Gast, 1972) that these lavas originated by partial melting of a pyroxene-rich lunar mantle material at a depth of several hundred kilometers. However, other hypotheses have been advanced. For instance, Philpotts and Schnetzler (1970) have proposed that mare basalts were formed by partial melting of a mafic residuum which was produced during highland differentiation. As implied by the ages of mare basalts collected in the Palus Putredinis (Apollo 15), the process of basin filling only terminated about 0.5 AE after the excavation of the basin, at least in the case of Mare Imbrium. A possible explanation is that heating of the moon progressed inward very slowly and temperatures sufficient for partial melting of materials in the mantle were attained only many hundred million years after the main phase of lunar accretion.

In broad terms, two periods of mare basalt formation have been identified so far, 3.2–3.4 AE for the Apollo 12 and 15 and Luna 16 rocks, and 3.5–3.9 AE for rocks collected near the Apollo 11 and 17 landing sites. These age groups correspond to the low- and high-Ti grouping found for the mare basalts in the five regions visited so far.

As early as 1970 Turner noted a significant difference between the Ar^{39}–Ar^{40} ages of Apollo 11 high-K and low-K basalts. Both rock types were collected in Mare Tranquillitatis within a range of a few hundred meters. Subsequent investigations have supported this finding (Turner, 1971; Eberhardt et al., 1971; Stettler et al., 1973). Moreover, Husain (1974) has recently reported smaller but still significant differences in the ages of Apollo 15 mare basalts.

Clearly, the large differences found in the Ar^{39}–Ar^{40} ages of Apollo 11 basalts bear very directly on any model of mare basin filling. We have, therefore, embarked on a detailed study of Ar^{39}–Ar^{40} ages of Apollo 11 rocks and report here the data and conclusions obtained so far.

EXPERIMENTAL PROCEDURE

Sample description and preparation

Rock 10047 is an ophitic basalt type 3; rock 10003 an ophitic basalt type 1 (Warner, 1971). Both rocks are members of the low-K group of Apollo 11 rocks (Tera et al., 1970; Gast et al., 1970). After cleaning the surfaces by ultrasonic treatment with acetone, chips from samples 10047.38 and 10003.39 were crushed to <150 μm. In addition, we separated with heavy liquids four mineral fractions from the aliquot of sample 10003.39. The purity of the mineral fractions was estimated from X-ray diffraction patterns (cf. Table 1).

Neutron irradiation

The procedures and analytical results concerning the investigated whole rock sample 10071 and the three mineral fractions separated from this rock have been presented in an earlier publication (Stettler et al., 1973). In the work reported here four mineral fractions from rock 10003 and the whole rock sample 10047 were irradiated along with four meteorites and eight terrestrial irradiation monitors for 20.8 days. This neutron bombardment took place in two different Harwell cans in neighboring positions of the same isotope channel facility of the FR-2 reactor, Gesellschaft für Kernforschung, Karlsruhe. Each sample was wrapped in a high purity Al foil, packaged into an Al container and evacuated to a pressure of less than 10^{-2} torr. Then, after filling with slightly more than 1 atm N_2, the Al containers were sealed and a series of nine of them placed inside the 0.5 mm Cd shielding of the two Harwell capsules. Each capsule contained four of our terrestrial hornblende CC-27 and biotite CAA-12 standards previously described (Stettler et al., 1973). A small Ni wire was fixed outside each Al container for monitoring the fast neutron fluence ($E > 0.1$ MeV), using the reaction Ni^{58} (n, p) Co^{58}.

Table 1. Mineral fractions of samples 10003,39 and 10071,30.

Bern number	Mineral fraction	Grain size [μm]	Density [g/cm³]	Purity [%]
10003,39, LA 4	feldspar	33–128	2.96	>95
LA 9	feldspar	128–150	2.96	>95
LA 10	pyroxene	26–150	3.30–3.53	>90
LA 6	ilmenite	26–150	4.25	>95
10071,30, CA 4	feldspar	26–92	2.46–2.91	>90
CA 6	feldspar	26–92	2.91–2.99	>80
CA 2	ilmenite	26–92	4.03	>95

Based on these measurements the fluence gradients over the capsule dimensions in horizontal and vertical direction were found to be $\leqslant 2.4\%$ and were corrected for. The two capsules received integrated neutron fluences ($E > 0.1$ MeV) of $(8.00 \pm 0.13) \times 10^{18}$ cm^{-2} and $(8.60 \pm 0.17) \times 10^{18}$ cm^{-2}. The results obtained for the monitors included in this series are listed in Table 2.

From the Ar_*^{40}/Ar_*^{39} ratios we derived J-values of 0.03615 ± 0.00035 for CC-27 and of 0.03590 ± 0.00040 for CAA-12. For the purpose of calculating the K–Ar ages reported in this paper, we adopted a weighted mean value of $J = 0.03610 \pm 0.00035$. The conversion factors for K and Ca determined from decay corrected Ar_*^{39} and Ar^{37} were

$$C_{39}(\text{K}) = 2.55 \times 10^{-4} \text{ cm}^3 \text{ STP } Ar_*^{39}/\text{gK} \ (\lambda_{39} = 7.2 \times 10^{-6} \text{ day}^{-1})$$

$$C_{37}(\text{Ca}) = 1.315 \times 10^{-4} \text{ cm}^3 \text{ STP } Ar_*^{37}/\text{gCa} \ (\lambda_{37} = 0.0203 \text{ day}^{-1})$$

and thus Ca/K $= 1.94 \ (Ar^{37}/Ar_*^{39})$ assuming K $= 2600$ ppm and Ca $= 8.5\%$, respectively for CC-27. The observed fluctuations in the Ar^{37}/Ar_*^{39} ratio could reflect small variations in the Ca/K ratio of CC-27 or differences in neutron energy spectra. Since the samples of the present investigation and those studied earlier by us were irradiated in the same pile position, the same correction factors were applied for interfering neutron reactions, in both cases (Stettler et al. 1973). In order to evaluate a better correction for the Cl^{37} (n, $\beta\gamma$) Ar^{38} reaction which can be significant in lunar samples with low exposure ages, we activated one NaCl sample and obtained a production rate of $(165 \pm 15) \times 10^{-12}$ cm^3 Ar^{38}/ppm Cl for a fluence of 10^{19} cm^{-2} ($E > 0.1$ MeV) and our irradiation conditions. To determine Ar^{39}–Ar^{40} ages and Ar^{37}–Ar^{38} exposure ages the following corrections for trapped gases and spallation products have been used:

$$(Ar^{36}/Ar^{38})_{sp} = 0.65 \pm 0.1; \ (Ar^{40}/Ar^{38})_{sp} = 0.2 \pm 0.1$$

$$(Ar^{36}/Ar^{38})_{tr} = 5.32 \pm 0.15 \text{ and } (Ar^{40}/Ar^{36})_{tr} = 1 \pm 1$$

Table 2. Irradiation standard measurements.

Standard	$(Ar_*^{40}/Ar_*^{39})^{\text{a}}$	(Ar^{37}/Ar_*^{39})
CC-27	86.40 ± 0.80	16.65 ± 0.20
	85.15 ± 0.90	16.78 ± 0.25
	83.8 ± 1.0	16.95 ± 0.35
	85.30 ± 0.80	16.84 ± 0.20
	84.90 ± 0.90	16.77 ± 0.30
	85.40 ± 0.90	16.95 ± 0.25
CAA-12	40.20 ± 0.60	
	40.85 ± 0.70	

The following notations, constants and data were used throughout this paper:

Ar_*^{39}: pile produced Ar^{39} from potassium only
Ar_*^{40}: radiogenic Ar^{40}
Index M: monitor
$J = (e^{\lambda t_M} - 1)/(Ar_*^{40}/Ar_*^{39})_M$ with
$\lambda = 5.305 \times 10^{-10} \text{ y}^{-1}$, $\lambda_e = 0.585 \times 10^{-10} \text{ y}^{-1}$
$K^{40}/K = 1.19 \times 10^{-4}$
$t_M(\text{CC-27}) = 2.650 \pm 0.025$ AE
$t_M(\text{CAA-12}) = 1.695 \pm 0.020$ AE
    [a] Ar_*^{40}/Ar_*^{39} ratios have been corrected for fluence inhomogenity using the Ni wire measurements.

Argon analyses

The Ar extraction and analysis procedures for the irradiated lunar samples were performed generally as described by Stettler *et al.* (1973). Extraction temperature control was carried out by infrared pyrometry in the range 400–950°C and optical pyrometry above 800°C. The following extraction blanks for Ar^{40} (in 10^{-8} cm^3 STP) were measured: ≤ 0.03 for ≤ 1200°C and 0.2 at 1600°C.

Results

The results of the stepwise release of argon from the 10047 bulk sample and from four mineral separates of rock 10003 are presented in Table 3 and Figs. 1 to

Table 3. Argon results from stepwise heating of neutron activated samples. All isotopes are corrected for analytical blanks and interference contributions due to neutron irradiation.

Temp. °C	$\dfrac{Ar^{36}}{Ar^{38}}$	$\dfrac{Ar^{38}}{Ar^{37}}$ ($\times 10^{-2}$)	$\dfrac{Ar^{38}_{sp}}{Ar^{37}}$ ($\times 10^{-2}$)	$\dfrac{Ar^{39}_*}{Ar^{37}}$ ($\times 10^{-2}$)	$\dfrac{Ar^{40}_*}{Ar^{39}_*}$	Ar^{39}_* (10^{-8} cm^3 STP/g)	Apparent age (AE)
10003 (Plagioclase 33–128 μ, 17.0 mg)		$J = 0.03668$					
400	0.27 ± 0.10	4.84 ± 0.17	4.80 ± 0.50	13.0 ± 7.0	46.7 ± 6.0	1.70	1.88 ± 0.25
600	0.0560 ± 0.0020	20.45 ± 0.30	20.2 ± 1.2	3.505 ± 0.050	128.5 ± 2.0	2.23	3.285 ± 0.035
760	0.2150 ± 0.0070	4.685 ± 0.060	4.63 ± 0.25	2.810 ± 0.030	181.8 ± 1.9	2.24	3.840 ± 0.030
800	0.730 ± 0.030	1.246 ± 0.050	1.225 ± 0.090	2.780 ± 0.030	184.2 ± 2.0	1.66	3.860 ± 0.030
860	0.720 ± 0.012	1.238 ± 0.020	1.220 ± 0.070	1.690 ± 0.015	186.3 ± 1.6	4.08	3.880 ± 0.025
900	0.680 ± 0.012	1.238 ± 0.018	1.230 ± 0.070	1.450 ± 0.018	189.7 ± 2.5	2.98	3.910 ± 0.035
1050	0.677 ± 0.011	1.244 ± 0.019	1.235 ± 0.070	1.477 ± 0.014	186.8 ± 1.7	5.95	3.885 ± 0.025
1600	0.718 ± 0.011	1.250 ± 0.020	1.230 ± 0.070	1.149 ± 0.012	195.0 ± 4.0	3.19	3.955 ± 0.040
Total	0.615	2.382	2.355	1.787	172.3	24.05	3.750
10003 (Plagioclase 128–150 μ, 27.4 mg)		$J = 0.03710$					
400	0.0135 ± 0.0050	2.00 ± 0.50	1.95 ± 0.70	9.2 ± 2.0	52 ± 14	0.57	2.05 ± 0.50
600	0.730 ± 0.030	1.320 ± 0.060	1.30 ± 0.10	6.850 ± 0.090	119.3 ± 2.0	2.84	3.190 ± 0.035
750	0.773 ± 0.018	1.253 ± 0.030	1.220 ± 0.070	3.272 ± 0.025	184.1 ± 1.2	3.63	3.880 ± 0.025
810	0.715 ± 0.013	1.227 ± 0.020	1.210 ± 0.070	2.166 ± 0.014	188.60 ± 0.90	4.84	3.920 ± 0.020
860	0.700 ± 0.011	1.207 ± 0.018	1.195 ± 0.070	1.654 ± 0.010	188.30 ± 0.90	5.23	3.915 ± 0.020
910	0.667 ± 0.010	1.221 ± 0.018	1.215 ± 0.070	1.591 ± 0.010	187.4 ± 1.0	4.25	3.910 ± 0.020
980	0.687 ± 0.011	1.248 ± 0.019	1.240 ± 0.070	1.632 ± 0.011	185.4 ± 1.1	3.69	3.892 ± 0.025
1070	0.676 ± 0.010	1.308 ± 0.019	1.300 ± 0.070	1.202 ± 0.011	182.7 ± 1.6	2.65	3.868 ± 0.025
1600	0.726 ± 0.014	1.303 ± 0.020	1.280 ± 0.080	0.742 ± 0.010	192.0 ± 7.0	0.98	3.950 ± 0.070
Total	0.700	1.250	1.238	1.900	177.4	28.68	3.820
10003 (Ilmenite 54.2 mg)		$J = 0.03556$					
400	0.30 ± 0.10	4.70 ± 0.45	4.65 ± 0.65	13.0 ± 1.0	29.0 ± 7.0	0.135	1.35 ± 0.50
600	0.125 ± 0.035	78.4 ± 1.9	77.5 ± 5.0	17.6 ± 0.45	125.0 ± 3.0	0.266	3.195 ± 0.050
700	0.133 ± 0.020	153.5 ± 6.0	152 ± 12	7.80 ± 0.50	176.0 ± 9.0	0.0922	3.735 ± 0.090
760	0.635 ± 0.070	19.85 ± 0.40	19.6 ± 1.2	4.70 ± 0.13	214.0 ± 5.0	0.187	3.940 ± 0.050
810	0.630 ± 0.045	1.550 ± 0.080	1.53 ± 0.12	3.920 ± 0.070	190.7 ± 4.5	0.229	3.870 ± 0.050
900	0.628 ± 0.015	2.67 ± 0.30	2.64 ± 0.35	3.380 ± 0.040	195.2 ± 3.0	0.348	3.900 ± 0.030
1000	0.598 ± 0.014	6.93 ± 0.15	6.85 ± 0.45	2.685 ± 0.070	200.0 ± 7.0	0.175	3.940 ± 0.070
1600	0.566 ± 0.014	31.23 ± 0.80	31.0 ± 3.0	0.522 ± 0.017	250 ± 20	0.144	4.31 ± 0.17
Total	0.57	22.87	22.67	2.69	154.4	1.575	3.52

Table 3. (*continued*).

Temp. °C	$\dfrac{Ar^{36}}{Ar^{38}}$	$\dfrac{Ar^{38}}{Ar^{37}}$ ($\times 10^{-2}$)	$\dfrac{Ar^{38}_{sp}}{Ar^{37}}$ ($\times 10^{-2}$)	$\dfrac{Ar^{39}_*}{Ar^{37}}$ ($\times 10^{-2}$)	$\dfrac{Ar^{40}_*}{Ar^{39}_*}$	Ar^{39}_* (10^{-8} cm^3 STP/g)	Apparent age (AE)
10003 (Clinopyroxene 149.7 mg)		$J = 0.03394$					
400	0.547 ± 0.050	3.95 ± 0.35	3.92 ± 0.70	10.40 ± 0.80	51.4 ± 8.0	0.108	1.90 ± 0.30
600	0.785 ± 0.050	1.450 ± 0.070	1.41 ± 0.11	7.81 ± 0.11	126.3 ± 1.9	0.390	3.140 ± 0.035
700	0.545 ± 0.080	1.61 ± 0.17	1.60 ± 0.25	5.79 ± 0.10	179.5 ± 3.0	0.137	3.692 ± 0.035
810	0.808 ± 0.014	1.380 ± 0.020	1.335 ± 0.070	2.358 ± 0.014	208.30 ± 0.90	0.647	3.935 ± 0.020
900	0.741 ± 0.012	1.342 ± 0.025	1.316 ± 0.080	2.355 ± 0.014	207.35 ± 0.90	0.770	3.928 ± 0.020
1010	0.697 ± 0.011	1.490 ± 0.020	1.475 ± 0.080	1.374 ± 0.012	197.3 ± 1.5	0.398	3.847 ± 0.025
1600	0.685 ± 0.060	1.390 ± 0.020	1.380 ± 0.080	0.0235 ± 0.0020	141 ± 15.0	0.210	3.30 ± 0.20
Total	0.690	1.394	1.383	0.269	181.1	2.66	3.705
10047 (Whole rock 106.6 mg)		$J = 0.03740$					
400	0.730 ± 0.040	1.475 ± 0.070	1.45 ± 0.20	4.95 ± 0.30	45.0 ± 5.0	0.16	1.85 ± 0.20
600	1.05 ± 0.11	1.00 ± 0.10	0.90 ± 0.14	16.20 ± 0.60	73.5 ± 3.0	1.46	2.492 ± 0.020
700	1.19 ± 0.14	0.90 ± 0.10	0.80 ± 0.14	15.00 ± 0.10	114.95 ± 0.60	0.80	3.143 ± 0.020
810	1.270 ± 0.080	0.805 ± 0.050	0.700 ± 0.070	7.715 ± 0.040	167.45 ± 0.35	6.64	3.738 ± 0.020
900	0.822 ± 0.035	0.718 ± 0.030	0.690 ± 0.050	4.158 ± 0.025	169.70 ± 0.60	7.20	3.760 ± 0.020
990	0.740 ± 0.025	0.805 ± 0.030	0.790 ± 0.060	3.912 ± 0.020	165.45 ± 0.50	4.57	3.718 ± 0.020
1170	0.679 ± 0.011	0.853 ± 0.014	0.850 ± 0.050	0.3870 ± 0.0030	161.4 ± 1.2	5.34	3.677 ± 0.025
1600	0.672 ± 0.012	0.865 ± 0.0144	0.860 ± 0.050	0.3460 ± 0.0030	169.0 ± 2.5	0.52	3.752 ± 0.035
Total	0.750	0.830	0.815	2.040	158.8	26.70	3.650

3. Our data from a bulk sample and three mineral separates of rock 10071 have been given earlier (Stettler *et al.*, 1973). In Table 4 we have summarized the results of all our Apollo 11 basalt investigations.

Table 4. Ar^{39}–Ar^{40} ages, Ar^{37}–Ar^{38} exposure ages and K and Ca concentrations of Apollo 11 rocks and separated minerals investigated in our laboratory.

Sample	High-temp. plateau age (AE)	Intermediate temp. age (AE)	Ar^{37}–Ar^{38} Exposure age (10^6 y)	Ca (%)	K ppm	Ca/K
10071,30 Whole rock	$(3.47 \pm .11)$	—	425	7.5	2300	32.6
10071,30 CA 4 Plag.	$3.53 \pm .06$	—	360	9.1	3800	23.9
10071,30 CA 6 Plag.	$3.49 \pm .08$	—	380	8.1	3720	21.8
10071,30 CA 2 Ilm.	—	$(3.50 \pm .07)$	340	1.2	1020	11.8
10047,38 Whole rock	$3.74 \pm .03$	—	70	8.6	900	95.5
10003,39 Whole rock*		—	140	8.0	400	200
10003,39 LA 4 Plag.	$3.90 \pm .04$	—	115	10.2	940	108
10003,39 LA 9 Plag.	$3.91 \pm .03$	—	120	11.6	1110	104
10003,39 LA 10 CPX	—	$3.93 \pm .03$	125	8.0	110	730
10003,39LA 6 Ilm.	—	$3.91 \pm .05$	100	0.45	60	75.0

Values in parentheses: Corresponding plateau not very well defined (cf. Stettler *et al.*, 1973).
*Total extraction only.

DISCUSSION

The ages of Apollo 11 basalts obtained by other authors from Rb–Sr internal isochrons or from Ar^{39}–Ar^{40} high-temperature plateaus are tabulated in Table 5 along with the results of our laboratory. Significant differences between the Ar^{39}–Ar^{40} ages of low-K and high-K basalts are obtained by both the Sheffield and Bern groups, whereas no such difference has been found by the Rb–Sr method. To some extent, this can be explained as follows: (1) the mare basalt with the highest Ar^{39}–Ar^{40} age, 10003, was never dated by the Rb–Sr method, and (2) the Rb–Sr age of 10071 is appreciably higher than its Ar^{39}–Ar^{40} age, which is the lowest among the Apollo 11 rocks. Since the correlation between ages and alkali contents is based only on data obtained from the Ar^{39}–Ar^{40} method, it is important to analyze the meaning of the data in detail. Of particular significance are the investigations on different mineral separates of the same rock, because complications in the history of a rock which could lead to a misinterpretation of an Ar^{40}_*/Ar^{39}_* release curve should affect different minerals in a different way. We have analyzed a set of mineral phases separated from 10003 and 10071, the Apollo 11 basalts with the highest and lowest Ar^{39}–Ar^{40} age, respectively.

Table 5. Compilation of ages determined for Apollo 11 rocks. Included are ages with errors <0.12 AE for high-K basalts and all ages for low-K basalts. For the Ar^{39}–Ar^{40} ages of 10003 and 10071 the weighted averages of the feldspar fractions were taken.

Sample	Ar^{39}–Ar^{40} Age high-temp. plateau age (AE)	Rb–Sr–Age (AE)	K ppm
10003	3.92 ± .07 (1)		445 (4)
	3.91 ± .03 (2)		
10062	3.82 ± .06 (1)		645 (5)
10044	3.73 ± .05 (1)	3.71 ± .11 (6)	781 (9)
10058		3.63 ± .20 (6)	877 (5)
10047	3.74 ± .03 (2)		900 (2)
10022	3.59 ± .06 (1)		2290 (11)
10017		3.59 ± .05 (6)	2388 (9)
		3.80 ± .11 (7)	
10072	3.52 ± .05 (1)	3.78 ± .10 (7)	2410 (7)
10069		3.68 (6)	2438 (9)
10024		3.61 ± .07 (6)	2490 (7)
10057		3.63 ± .002 (6)	2588 (10)
10071	3.51 ± .06 (3)	3.68 ± .02 (6)	2740 (8)

(1) Turner, 1970.

(2) This work.

(3) Stettler et al. (1973).

(4) Eberhardt et al. (1971).

(5) Gast et al. (1970).

(6) Papanastassiou et al. (1970).

(7) Compston et al. (1970).

(8) Eberhardt et al. (1974).

(9) Tera et al. (1970).

(10) Wanless et al. (1970).

(11) Murthy et al. (1970).

Fine-grained, low alkali basalt 10003

The potassium–argon age of this rock is very well documented. Turner (1970) found an Ar^{39}–Ar^{40} age of 3.92 ± 0.07 AE for a whole rock sample. Even the conventional K–Ar age gave 3.74 ± 0.06 AE for a bulk sample and 3.82 ± 0.05 AE for the feldspar sample 10003,39 LA 4 (Eberhardt *et al.*, 1971). Two feldspar fractions differing in grain size gave well-defined and long high-temperature Ar_*^{40}/Ar_*^{39} plateaus (Fig. 1) at 3.90 ± 0.04 AE and 3.91 ± 0.03 AE, in agreement with the age of the whole rock (Turner, 1970). The ilmenite and pyroxene concentrates give less regular release patterns, but in both cases fairly long plateaus were observed at intermediate temperature (Fig. 2), corresponding to ages which are identical with the feldspar ages. The Ca/K ratios in the whole rock and in the mineral separates vary markedly, and so do the Ca/K thermal release curves (cf. Figs. 1 and 2, and Turner *et al.*, 1972). This indicates that our mineral-enrichment processing has succeeded at least to a certain degree in separating different potassium bearing phases. This is of course also reflected in the Ar_*^{40}/Ar_*^{39} release curves (Figs. 1 and 2). In spite of these differences, the Ar_*^{40}/Ar_*^{39} ratios are

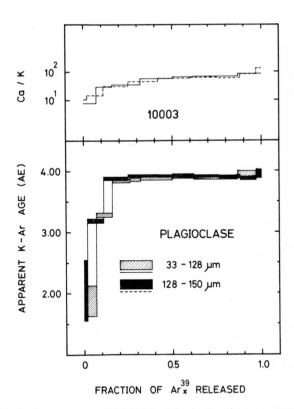

Fig. 1. Ar^{39}–Ar^{40} release curves of 2 feldspar grain size fractions, separated from mare basalt 10003. Ca/K release curves calculated from ratios are given in the upper part of the figure.

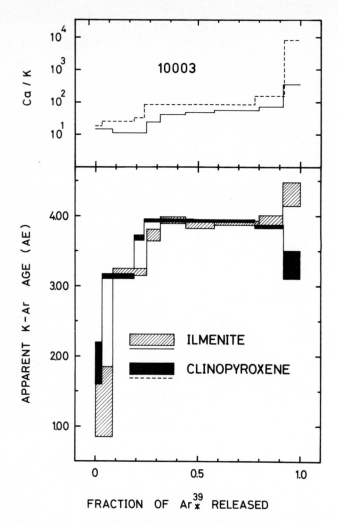

Fig. 2. Ar^{39}–Ar^{40} and Ca/K release curves of an ilmenite and a pyroxene fraction separated from mare basalt 10003.

virtually identical in the plateau regions of all four mineral concentrates. This observation and the good plateaus obtained for the feldspar samples suggest to us that the history of rock 10003 was in no way more complicated than that of other mare basalts and that its age is well represented by the 3.91 ± 0.03 AE obtained from its feldspar fractions.

Fine-grained, low alkali basalt 10047

The argon thermal release pattern of a whole rock sample is given in Fig. 3. The well-defined Ar^{40}_*/Ar^{39}_* plateau indicates a regular thermal history with

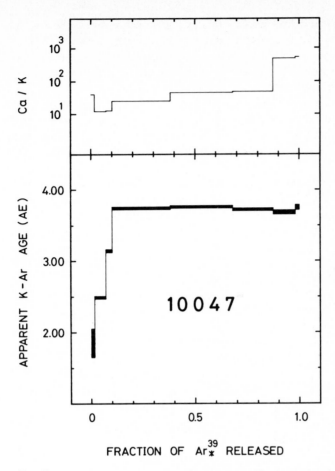

Fig. 3. Ar^{39}–Ar^{40} and Ca/K release curves of a whole rock sample of mare basalt 10047.

cristallization at 3.74 ± 0.03 AE and only small subsequent Ar_*^{40} diffusion loss. The agreement of the Ca/K curve in shape and absolute range with the corresponding curve obtained by superposition of the data from the 10003 components shows a similar temperature behavior of the Ca- and K-bearing mineral phases.

Coarse-grained, high alkali basalt 10071

As reported earlier (Stettler *et al.*, 1973) and summarized in Table 4, feldspar separates of this rock gave fairly good high-temperature plateaus at 3.49 ± 0.08 AE and 3.53 ± 0.06 AE. The Ar_*^{40}/Ar_*^{39} release curves of a whole rock sample and an ilmenite fraction give a maximum at intermediate temperatures which is very close to the plateau of the feldspar curves (cf. Stettler *et al.*, 1973, Fig. 5). Thus, there is nothing in the argon release observations which would offer an explanation for the difference with the Rb/Sr age. Only a fairly complete episodic or continuous argon loss about 100 m.y. after the minerals in the rock became a closed system with respect to Rb–Sr could result in such a difference. (It is not known whether a complete Ar loss can occur in lunar samples without affecting the Rb–Sr system.) The coarse-grained texture implies slow cooling, but we know of no petrographic observation which would suggest cooling over a period of 100 m.y. Since generally there is very good agreement between Rb–Sr and Ar^{39}–Ar^{40} ages of lunar rocks, even for petrologically more complicated rocks such as highland breccias or Fra Mauro rocks, it would be worthwhile if further studies, concerning the observed discrepancy in the ages of the 10071 basalt were made.

The internal consistency of the Ar_*^{40}/Ar_*^{39} release curves of the 10071 whole rock and mineral fractions as well as the agreement among the Ar^{39}–Ar^{40} ages of the high-K rocks (cf. Table 5) indicate to us that these ages represent the time of formation or at least the time of the last major cooling period of these rocks.

CONCLUSIONS

In Fig. 4 we have plotted all available Ar^{39}–Ar^{40} ages of Apollo 11 rocks versus their potassium contents. Turner (1970) had pointed out that the high-K rocks are systematically younger than the low-K rocks. Our data confirm Turner's conclusion. Figure 4 shows that there may actually be a whole sequence of ages; at least there are three significantly different ages or age groups (10003, 10044/10047, and

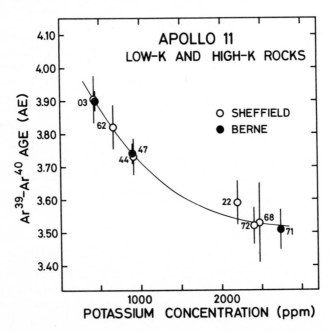

Fig. 4. Relation between age and K concentration of basalts collected at Tranquillity Base. Included are all Ar^{39}–Ar^{40} plateau ages determined so far.

10071/10072). The observed age groupings appear to be correlated with Warner's (1971) detailed classification of the Apollo 11 basalts (cf. Table 6). Only additional investigations on Apollo 11 basalts will decide whether there are these three age groupings or whether there is a quasi-continuous sequence of ages related to the K contents of the rocks.

It is likely that most sizable rocks ($\geqslant 100$ g) not showing major traces of shock were thrown onto the lunar surface by nearby craters. The Apollo 15 rock inventory may serve as a clue. Almost without exception mare basalts were collected at the stations in the plains, and highland-type rocks were found only on or near the slopes of Hadley Delta. A somewhat similar systematic is observed in the Apollo 17 area. Thus, in regions which are not geographically dominated by one or a few large craters at intermediate distances, most of the sizable rocks were apparently transported by impacts over distances not exceeding the order of 10 km. It is therefore even more remarkable that at the Apollo 11 landing site, the rocks differ in age by as much as 400 m.y. Soderblom and Lebofsky (1972) have derived relative ages of lunar surface areas from crater erosion observations. They found a relatively large span of ages for mare Tranquillitatis, the lowest ages corresponding to the highest in Mare Imbrium; their highest relative ages in Mare Tranquillitatis overlap with those obtained for the Cayley formation.

Thus, the high age found for 10003 as well as the range of ages of 400 m.y. is not inconsistent with the crater information. Also, Ronca (1973) has published relative ages of mare surfaces, using the "geomorphical index," another parameter based on crater observations. His contour map of these indices for mare Tranquillitatis suggests a particularly high gradient of relative ages near the

Table 6. Textural-mineralogical classification of Apollo 11 basalts (Warner, 1971).

	Group 2 Ophitic Basalts		Group 3 Intersertal Basalts	
	Sample number	Ar^{39}–Ar^{40} Age (AE)	Sample number	Ar^{39}–Ar^{40} Age (AE)
Type 1	10003 10062 10020 10045	$3.91 \pm .03^B$ $3.82 \pm .06^S$	10049 10069	
Type 2	—		10017 10022 10057	$\geqslant 3.23 \pm .06^S$ $3.59 \pm .06^S$
Type 3	10044 10047 10050 10058	$3.73 \pm .05^S$ $3.74 \pm .03^B$	10024 10071 10072	$\geqslant 3.48 \pm .05^S$ $3.51 \pm .06^B$ $3.52 \pm .05^S$

S: Sheffield (Turner 1970).
B: Bern (Stettler et al., 1973, this work).

Apollo 11 landing site. This again would be consistent with the large variability of Apollo 11 rock ages observed. It is emphasized that we consider these comparisons between rock ages and relative ages of surfaces based on crater observations to be quite tentative. They appear, however, interesting enough to warrant further studies of the relative ages.

Eberhardt *et al.* (1970) found a clear difference between the exposure ages and irradiation histories of high-K and low-K Apollo 11 basalts. They offered three models to explain these results. In any case, the differences in exposure history which they found, indicate separate locations for the origin of low-K and high-K rocks in a manner consistent with the idea that these two rock types originated from successive lava flows.

Based on rock ages in the Fra Mauro region and the Cayley Plains, it is generally believed that the Imbrium and Orientale impacts occurred at about 3.95 AE (cf. LSAPT, 1973; Schaeffer and Husain, 1974). Stratigraphy and relative ages based on crater observations indicate that the Tranquillitatis impact was earlier, but it is not known how much. Considering the large spread of Apollo 11 basalt ages, the fact that a rock with an age of 3.91 ± 0.03 AE was found devoid of shock features near the surface of the lava-filled basin suggests that this basin was excavated appreciably earlier than 3.91 AE.

The information contained in rock 14053 may perhaps help to shed light on the general question concerning the processes generating mare basalt-like lavas.

Based on chemical composition, Reid *et al.* (1974) have argued that 14053 and some other rocks and glasses in the Luna 16 and Apollo 11, 12, and 14 inventories are representatives of an aluminous type of mare basalt, which is somewhat different from the Al poorer basalts commonly found at the mare sites visited so far. They suggest that the aluminous mare basalts may have originated from a source region transitional between the aluminium-rich crust and a mafic mantle.

The Rb–Sr and Ar^{39}–Ar^{40} methods give ages of 3.95 ± 0.04 AE (cf. Papanastassiou and Wasserburg, 1971; Turner *et al.*, 1971; Stettler *et al.*, 1973) for rock 14053. Its place of collection near the rim of Cone Crater and its exposure age of 24 m.y. make it quite certain that it was excavated by Cone Crater. It appears therefore that this rock is a regular constituent of the Fra Mauro deposit, i.e. the Imbrian ejecta. This implies that the partial melting processes which have produced mare basalt-like lavas were already operating at the time of the Imbrian impact.

Further investigation of the ages of old mare basalts and of related rocks should lead to a more precise characterization of the partial melting processes which have formed these rocks and of the thermal history of the moon in general.

Acknowledgments—We should like to thank the National Aeronautics and Space Administration for generously supplying the lunar samples for this investigation. We are grateful to Eveline Giger, Elisabeth Bühler, E. Lenggenhager, A. Schaller, U. Schwab, F. Schweizer, and M. Zuber for their help in various aspects of this work. This research was supported by the Swiss National Science Foundation (grants NF 2.405.70, 2.592.71, and 2.768.72).

REFERENCES

Compston W., Chappell B. W., Arriens P. A., and Vernon M. W. (1970) The chemistry and age of Apollo 11 lunar material. *Proc. Apollo 11 Lunar Sci. Conf., Geochim. Cosmochim. Acta*, Suppl. 1, Vol. 2, pp. 1007–1027. Pergamon.

Eberhardt P., Geiss J., Graf H., Grögler N., Krähenbühl U., Schwaller H., Schwarzmüller J., and Stettler A. (1970) Correlation between rock type and irradiation history of Apollo 11 igneous rocks. *Earth Planet. Sci. Lett.* **10**, 67–72.

Eberhardt P., Geiss J., Grögler N., Krähenbühl U., Mörgeli M., and Stettler A. (1971) Potassium–argon age of Apollo 11 rock 10003. *Earth Planet. Sci. Lett.* **11**, 245–247.

Eberhardt P., Geiss J., Graf H., Grögler N., Krähenbühl U., Schwaller H., and Stettler A. (1974) Noble gas investigations of lunar rocks 10017 and 10071. *Geochim. Cosmochim. Acta* **38**, 97–120.

Gast P. W. (1972) The chemical composition and structure of the moon. *The Moon* **5**, 121–148.

Gast P. W. and Hubbard N. J. (1970) Abundance of alkali metals, alkaline and rare earths, and strontium-87/strontium-86 ratios in lunar samples. *Science* **167**, 485–487.

Gast P. W., Hubbard N. J., and Wiesmann H. (1970) Chemical composition and petrogenesis of basalts from Tranquillity Base. *Proc. Apollo 11 Lunar Sci. Conf., Geochim. Cosmochim. Acta*, Suppl. 1, Vol. 2, pp. 1143–1163. Pergamon.

Husain L. (1974) ^{40}Ar–^{39}Ar chronology and cosmic ray exposure ages of the Apollo 15 samples. *J. Geophys. Res.* In press.

LSAPT (1973) Fourth Lunar Science Conference. *Science* **181**, 615–622.

Murthy V. R., Evensen N. M., and Coscio M. R., Jr. (1970) Distribution of K, Rb, Sr and BA and Rb–Sr isotopic relations in Apollo 11 lunar samples. *Proc. Apollo 11 Lunar Sci. Conf., Geochim. Cosmochim. Acta*, Suppl. 1, Vol. 2, pp. 1393–1406. Pergamon.

Papanastassiou D. A., Wasserburg G. J., and Burnett D. S. (1970) Rb–Sr ages of lunar rocks from the Sea of Tranquillity. *Earth Planet. Sci. Lett.* **8**, 1–19.

Papanastassiou D. A. and Wasserburg G. J. (1971) Rb–Sr ages of igneous rocks from the Apollo 14 mission and the age of the Fra Mauro formation. *Earth Planet. Sci. Lett.* **12**, 36–48.

Philpotts J. A. and Schnetzler C. C. (1970) Apollo 11 lunar samples: K, Rb, Sr, Ba, and rare-earth concentrations in some rocks and separated phases. *Proc. Apollo 11 Lunar Sci. Conf., Geochim. Cosmochim. Acta*, Suppl. 1, Vol. 2, pp. 1471–1486. Pergamon.

Reid A. M. (1974) Luna 16 revisited: The case for aluminous mare basalts (abstract). In *Lunar Science—V*, p. 627. The Lunar Science Institute, Houston.

Ridley W. I., Reid A. M., and Brett P. R. (1973) Lunar Science Institute, contribution 156.

Ringwood A. E. and Essene E. (1970) Petrogenesis of lunar basalts, internal constitution and origin of the moon. *Science* **167**, 607–609.

Ronca L. B. (1973) The Filling of the Lunar Mare Basins. *The Moon* **7**, 239–248.

Schaeffer O. A. and Husain L. (1974) Chronology of lunar basin formation and ages of lunar anorthositic rocks (abstract). In *Lunar Science—V*, p. 663. The Lunar Science Institute, Houston.

Soderblom L. A. and Lebofsky L. A. (1972) Technique for rapid determination of relative ages of lunar areas from orbital photography. *J. Geophys. Res.* **77**, 279–296.

Stettler A., Eberhardt P., Geiss J., Grögler N., and Maurer P. (1973) Ar^{39}–Ar^{40} ages and Ar^{37}–Ar^{38} exposure ages of lunar rocks. *Proc. Fourth Lunar Sci. Conf., Geochim. Cosmochim. Acta*, Suppl. 4, Vol. 2, pp. 1865–1888. Pergamon.

Tera F., Eugster O., Burnett D. S., and Wasserburg G. J. (1970) Comparative study of Li, Na, K, Rb, Cs, Sr and Ba abundances in achondrites and in Apollo 11 lunar samples. *Proc. Apollo 11 Lunar Sci. Conf., Geochim. Cosmochim. Acta*, Suppl. 1, Vol. 2, pp. 1647–1657. Pergamon.

Turner G. (1970) Argon-40/Argon-39 dating of lunar rock samples. *Proc. Apollo 11 Lunar Sci. Conf., Geochim. Cosmochim. Acta*, Suppl. 1, Vol. 2, pp. 1665–1684. Pergamon.

Turner G. (1971) ^{40}Ar–^{39}Ar ages from the lunar maria. *Earth Planet. Sci. Lett.* **11**, 169–191.

Turner G., Huneke J. C., Podosek F. A., and Wasserburg G. J. (1971) ^{39}Ar–^{40}Ar ages and cosmic ray exposure ages of Apollo 14 samples. *Earth Planet. Sci. Lett.* **12**, 19–35.

Turner G., Huneke J. C., Rodosek F. A., and Wasserburg G. J. (1972) Ar^{40}–Ar^{39} systematics in rocks and separated minerals from Apollo 14. *Proc. Third Lunar Sci. Conf., Geochim. Cosmochim. Acta,* Suppl. 3, Vol. 2, pp. 1589–1612. MIT Press.

Wanless R. K., Loveridge W. D., and Stevens R. D. (1970) Age determinations and isotopic abundance measurements on lunar samples (Apollo 11). *Proc. Apollo 11 Lunar Sci. Conf., Geochim. Cosmochim. Acta,* Suppl. 1, Vol. 2, pp. 1729–1739. Pergamon.

Warner J. L. (1971) Lunar cristalline rocks: Petrology and geology. *Proc. Second Lunar Sci. Conf., Geochim. Cosmochim. Acta,* Suppl. 2, Vol. 1, pp. 469–480. MIT Press.

Proceedings of the Fifth Lunar Conference
(Supplement 5, Geochimica et Cosmochimica Acta)
Vol. 2 pp. 1571–1599 (1974)
Printed in the United States of America

U–Th–Pb systematics on lunar rocks and inferences about lunar evolution and the age of the moon

Fouad Tera and G. J. Wasserburg

The Lunatic Asylum of the Charles Arms Laboratory, Division of Geological and Planetary
Sciences,* California Institute of Technology, Pasadena, California 91109

Abstract—All of the available data on terra rocks define a striking linear array when plotted on the U–Pb evolution diagram with a generally limited scatter around a reference isochron which intersects the concordia curve at 4.42 and 3.95 AE and defines an initial radiogenic $^{207}Pb/^{206}Pb$ of $I_{Pb} = 1.45$. In addition to the observed wide range of $^{207}Pb/^{206}Pb$ (0.4–1.45) the samples exhibit a vast range for $\mu \equiv {}^{238}U/^{204}Pb$ from 0.9 to 10^4 which is generally correlated with the degree of discordance. On the basis of similar observations, on a limited number of terra rocks, we had previously inferred that a major impact had caused global metamorphism in the lunar crust. The more extensive results presented here demonstrate that the terra rocks in general fall on the linear array although it is not clear as to how much of the scatter reflects lunar processes and how much is due to insufficient precision in some of the existing data. The data array is taken as an evidence for the validity of the terminal lunar cataclysm hypothesis which ascribes large-scale metamorphism at $\sim$3.9 AE to a period of intense bombardment of an ancient lunar crust, effecting element redistribution and extensive mobilization of Pb. A time interval of up to 200 m.y. (3.8–4.0 AE) is assigned to the duration of the terminal cataclysmic bombardment during which several of the major ringed basins may have been formed. Some of the impacted materials lost almost all their Pb and consequently fall on the concordia curve at the lower intersection. The mobilized Pb was deposited and/or incorporated in other rocks which became discordant. Some rocks are dominated by this labile Pb and lie close to the intercept (e.g. 15415). While the target rocks may be of variable age and the mobilized Pb not have a unique isotopic composition the gross isotopic composition of the radiogenic component of the mobilized lead ($I_{Pb} \sim 1.45$) presumably represents the average value of $^{207}Pb/^{206}Pb$ in the lunar crust at $\sim$3.8–3.9 AE. This I_{Pb} is consistent with its production in a single stage in the time span 4.4–3.9 AE. The upper intersection of the linear array with the concordia curve appears well defined by the fact that two terra rocks and one mare basalt are concordant by the U–Th–Pb method at 4.42 AE indicating a very strong possibility that this is close to the actual time of differentiation of the lunar crust. A firm interpretation of the upper intersection is not yet possible.

It must be recognized, however, that from the point of view of the U–Th–Pb systematics that it is not possible to distinguish between differentiates produced by magmatic activity at $\sim$3.9 AE and those rocks produced in cataclysmic impact metamorphism. Thus, a sample might fall accidentally on the array without having a genetic relationship to terra rocks on the array. This might apply to the orange soil and basalt 75055 which plot on the reference isochron. New data on two lunar basalts show that the ^{204}Pb concentration levels are much lower than reported previously.

General Systematics for Terra Rocks

The lead–uranium data on terra rocks from widely separated areas have been found to have remarkably regular characteristics. The data are conveniently displayed on a U–Pb evolution diagram with coordinates ($^{238}U/^{206}Pb$, $^{207}Pb/^{206}Pb$). In this representation, all possible mixtures of two components lie on a straight-

---

*Contribution number 2492.

line segment between the two end-members. If one end-member is composed only of inherited Pb (initial lead) and the other is *in situ* radiogenic Pb produced subsequent to time τ_1 in a closed system without any initial Pb, then the two points lie on the y-axis (at I_{Pb}) and on the concordia curve (at time τ_1), respectively. This is shown schematically in Fig. 1. Such a line segment corresponds to an isochron defining a time τ_1. A second intersection of the isochron with the concordia curve

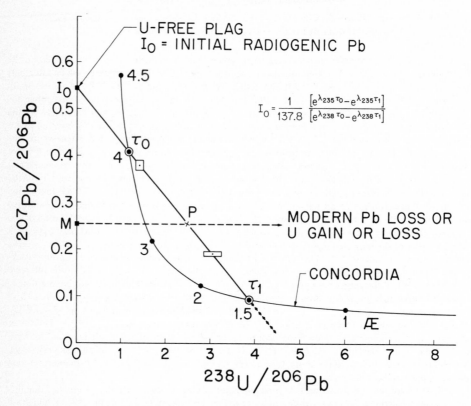

Fig. 1. Schematic representation of an arbitrary linear data array on the coupled U–Pb evolution diagram after Tera and Wasserburg (1972a, 1973a). In terms of a simple two stage model, the system was formed at $\tau_0 = 4$ AE with no initial Pb and was then recrystallized or differentiated at $\tau_1 = 1.5$ AE. Upon differentiation at 1.5 AE, the Pb produced in the time span 4–1.5 AE is considered the initial Pb of the system and its composition is given by the intercept I_0. This would be the Pb isotopic composition for a U free phase which was formed (or lost all its U) at the metamorphic event 1.5 AE ago. Alternatively, I_0 could represent an initial Pb produced by a complex evolution and τ_0 would not correspond to an actual time event. The concordant point τ_1 corresponds to a phase which at the time of metamorphism contained no Pb. All the lead in the phase concordant at τ_1 is produced by *in situ* decay for 1.5 AE. The isochron could be treated as a mixing line with the two end components being at I_0 and τ_1. Mineral phases with different U/Pb ratios will lie on the mixing line between the two end components. The dotted extension of the isochron is a forbidden region. The effects of modern loss of Pb or U or gain of U on an arbitrary point P are shown.

at time τ_0 may also have a specific time meaning (Tera and Wasserburg, 1972a, 1972c, 1973a) (see caption of Fig. 1).

The first internal isochrons on lunar rocks were determined on two igneous terra rocks and showed that the isochrons were indistinguishable within analytical errors and gave $I_{Pb} \sim 1.45$ (Tera and Wasserburg, 1972a). Some highland soils were then found to lie in the general region of this isochron and also showed the enrichment of radiogenic Pb relative to U (Tera and Wasserburg, 1972c; Nunes et al., 1973; Silver, 1973). This observation was then extended to several terra rocks including an Apollo 15 anorthosite whose *average* value lies very close to the end point with $I_{Pb} \sim 1.45$. Based on these observations, we proposed a two-stage evolution model in which a more ancient lunar crust was subject to a last cataclysmic bombardment at ~ 3.95 AE (Tera et al., 1973b). Since only a limited number of rocks had been analyzed, there was some question of the universal character of the U–Pb systematics upon which the hypothesis was in part based. A more extensive suite of samples was analyzed and the correlation firmly established using data from both this laboratory and the USGS laboratory in Denver. Further, the geometrical positions of the data points were found to be highly correlated with the observed value of $^{238}U/^{204}Pb$ ($\equiv \mu$) as measured today. An extensive discussion of these data is given by Tera et al. (1974a).

In the present paper, we will continue on the theme which was exposed in the article by Tera et al. (1974a) and attempt to summarize all of the data currently available in the literature on terra materials and to discuss the implications of the upper intersection (τ_0) with the concordia curve. Figure 2 shows the data on total rocks and some plagioclase separates. The reference isochron is the original one found for 14310 (Tera and Wasserburg, 1972a). Inspection of Fig. 2 shows that the data form a remarkable linear array in spite of the fact that the rocks comprise a wide spectrum of lithic types with highly variable Pb and U concentrations and with μ ranging from ~ 0.9 to 1.5×10^4. The μ values are correlated with the $^{238}U/^{206}Pb$ values, and the data points which fall near the lower intersection with concordia at ~ 3.9 AE typically have high μ values (over 5×10^3). This array of data is a strong confirmation of the correlation which we had previously proposed for terra rocks and appear to give further support to our model of a terminal lunar cataclysm associated with intense global bombardment at ~ 3.9 AE and which caused widespread metamorphism and element redistribution. In particular, lead was extensively mobilized, redeposited, and incorporated into some rocks as initial Pb. The gross isotopic composition of this initial lead of $I_{Pb} \sim 1.45$ is consistent with its production in a first stage between 4.4 and 3.9 AE prior to the terminal cataclysm.

INTERSECTIONS WITH THE CONCORDIA CURVE

Dispersion from the reference isochron

Because of the near tangency of the chord with the concordia curve as shown in Fig. 2, the times assigned to the two intersections are approximate. Some

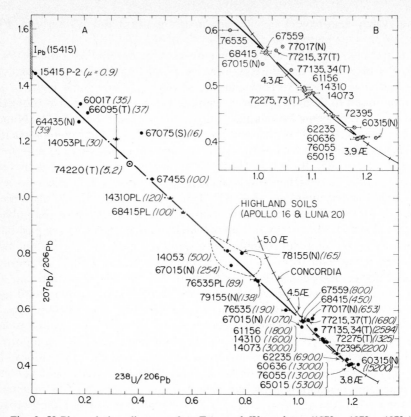

Fig. 2. U–Pb evolution diagram after Tera and Wasserburg (1972a, 1972c, 1973a) showing all the data currently available in the literature on terra rocks. Data are from this laboratory (Tera and Wasserburg 1972a, 1972b, 1972c; Tera et al., 1974), Tatsumoto et al. (labeled T) (1973a, 1973d), Nunes et al. (N) (1973, 1974a, 1974b), and Silver (S) (1973). The line intersecting the concordia at 4.4 and 3.95 AE is a reference isochron passing through the total rock 14310, and its plagioclase separate. The measured $\mu \equiv {}^{238}U/{}^{204}Pb$ for each sample is shown in parenthesis next to the sample number. Note the wide range of μ values (from $\mu = 0.9$ for 15415 to $\mu = 15200$ for 60315) and their general correlation along the discordia chord. Data are corrected for blank and primordial Pb (PAT). Those data uncorrected for PAT are shown as small dots close to the corresponding large data points for the data from this laboratory. Only data from this laboratory are shown with error bars. The majority of total rock data lie between the two concordia intersections, and are shown in inset B. The concordia curve was calculated using $\lambda_{238U} = 1.5525 \times 10^{-10}$ y^{-1}, and $\lambda_{235U} = 9.8485 \times 10^{-10}$ y^{-1}. In addition to terra rocks and soils, the orange soil 74220 (T) (Tatsumoto et al., 1973a) is also included in the figure.

dispersion of the data is to be expected if the rocks have different ages. Any spread due to analytical errors will, of course, obscure the real problems. The Rb–Sr crystallization ages of these non-mare rocks show ~200 m.y. spread (3.8–4.0 AE) (Papanastassiou and Wasserburg, 1971, 1972a, 1972b; Albee et al., 1970; Tatsumoto et al., 1974; Tera et al., 1974a) and hence the departure of total rocks from a single reference isochron is partly understandable. Inspection of the data in Fig. 2, however, shows that some of the deviations are extreme and cannot be explained in terms of *slight* variations in the time of metamorphism and/or the age of the parent material from which the rocks were derived. The extreme example of 67075 is discussed by Tera *et al.* (1974a). In particular, some of the recent results of Nunes *et al.* (1973, 1974) which are useful in defining the general trend also show considerable scatter that would imply metamorphic event(s) *much younger* than ~3.9 AE. This prompted us into examination of the published data. Some of the results of Nunes *et al.* (1973, 1974) shown on the Pb isotope correlation diagram (Fig. 3) would suggest serious internal inconsistencies. In this representation (Tera and Wasserburg, 1972c), the uncorrected data points for the composition and concentration *aliquots* of a given sample, as well as the two corresponding points corrected for the blank should all fall on a mixing line with

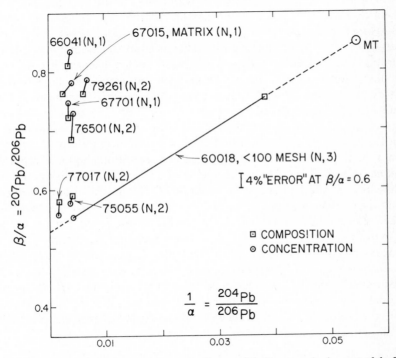

Fig. 3. Pb isotopes correlation diagram showing possible discrepancies in some of the Pb data of Nunes *et al.* (1973, 1974a, 1974b) which might be in part correlated to the dispersion of their data from the linear trend in Fig. 2. Note in particular 77017 and 67015. Critical sample 60315 does not appear to show this disparity. According to Pb isotopes systematics, all data points when corrected for spike contributions from *aliquots* of a given sample should fall on a mixing line. The end components of the mixing line are the composition of the blank (presumably near MT) and the characteristic Pb in the sample. For example, sample 60018 satisfies the requirements of the systematics. However, many other pairs of data points from Nunes *et al.* (labeled N1, N2, and N3) (1973, 1974a, 1974b) form lines which extrapolate into regions of completely unreasonable composition for the blank. This disparity may introduce errors in radiogenic $^{207}Pb/^{206}Pb$ values which affect the U–Pb systematics.

the blank (presumably a modern terrestrial Pb (MT)) being one end component and the other end component being the sample. This requirement is independent of blank fluctuation as long as the blank composition remains reasonably constant. As seen in Fig. 3, this condition is satisfied by sample 60018, <100 mesh, (Nunes *et al.*, 1973). All other pairs of data points, however, fall on lines which are at *distinct angles* to the corresponding mixing lines with MT. The effect cannot be readily explained in terms of slight changes in blank composition because in most of the cases shown, the line connecting the composition and concentration points extrapolates into regions of totally unreasonable values for the blank. This is suggestive of a third component or could be due to an artifact stemming from data processing and may be symptomatic of errors which could affect the present argument. We experienced such a discrepancy with a small plagioclase separate of rock 14053 (Tera and Wasserburg, 1972a). This apparent third component could be due to improper splitting of crushed sample so that the splits were not true aliquots (Nunes and Tatsumoto, written communication, 1974). This possible procedural difficulty would limit the full value of analyzing both composition and concentration aliquots. Irrespective of its source, such a disparity is quite serious because uncertainties in the isotopic abundances could introduce large uncertainties into the radiogenic $^{207}Pb/^{206}Pb$ ratio, or possibly into the concentration results. For example, 77017 shows a difference of 3.5% for $^{207}Pb/^{206}Pb$ which is far outside of the precision of $\pm 0.2\%$ usually obtainable for this ratio. We note that point 67015 in Fig. 2 would move upwards to within close proximity of the reference isochron if the higher $^{207}Pb/^{206}Pb$ value of the concentration aliquot as reported by Nunes *et al.* is used. This might be the preferred analysis if aliquoting is the source of the difficulty. Similarly, the displacement of 77017 from the chord is reduced by about half if the concentration value of $^{207}Pb/^{206}Pb$ is used; however, 78155 would be further displaced from the chord. Such discrepancies do not appear to be present in the data for 60315. While we are convinced that the data *do not form* an *exact* linear array, it is not at present clear whether the deviations are as extensive or as common as some of the published results indicate. Certainly the analyses should be repeated under controlled conditions to precisely ascertain the positions of these data points relative to the general trend. Any attempt to determine the "exact" intersections with concordia will require far more stringent efforts on the part of all of us.

The lower intersection

Considering the data array in Fig. 2, the first order question is whether one can assign a time meaning to the intersections in terms of a simple two-stage model as discussed in the schematic diagram. Using the results from this laboratory only, it is evident from the scatter that while certain data points lie on the concordia curve at 3.95 AE, others are displaced from the chord by 2 σ error bars and seem to indicate a lower intersection at a younger age of about 3.8 AE. Thus irrespective of any technical difficulty associated with the near tangency of the chord, it is not possible to assign a *precise* time to the lower intersection. This is correlated with the fact that these terra rocks do not have a common unique age as demonstrated by the precise Rb–Sr crystallization ages for those rocks which cover the span 3.8–4.0 AE as mentioned above. We note that for those samples which lie precisely on the concordia curve, the $^{208}Pb/^{232}Th$ ages are also in almost precise agreement (new decay constants). See, for example, sample 65015 (Tera *et al.*, 1974a). This observation suggests that those samples which are concordant for *all* methods may have a special significance and possibly represent simple closed systems for the ages determined. In general, the quality of our Th concentration data must be improved if this complete concordance is to be adequately tested.

If the data of Nunes *et al.* (1973, 1974) are included, then it follows that there is a much wider spread which would extend down to ~3.7 AE, well into the period of what we had interpreted as post-major impact mare flooding. If this should prove to be the case, our hypothesis regarding the terra rocks (Tera *et al.*, 1973b, 1974a, 1974b) would be seriously in doubt.

In our preferred model, the entire U–Pb distribution spectrum in Fig. 2 as well as the vast range of μ values is primarily the direct consequence of major impact metamorphism. Thus associating the lower intersection with this metamorphism, and considering the Rb–Sr crystallization ages, we assign a time interval of up to 0.2 AE (3.8–4.0 AE) to a terminal lunar cataclysm. In this time span, several of the major lunar basins must have been formed by the major impacts or else one or two major impacts are responsible for the widespread metamorphism over the whole-earth facing side of the moon.

The upper intersection

The upper intersection with concordia is not defined simply by a best fit line, but appears more precisely defined because of the fact that two terra rocks lie *on* the concordia curve at this intersection and have $^{232}Th–^{208}Pb$ ages that are in very good agreement with U–Pb ages (Tera *et al.*, 1974a). In spite of the actual concordance of two terra rocks at 4.42 AE (67559 and 68415), ascribing a strict genetic meaning to this upper intersection as being the "age of the moon" or the "age of the crust" is not self evident. If we ignore the scatter of the data, three basic possibilities are available to explain the upper intersection:

(a) The moon formed at 4.42 AE. This would mean that its formation was accompanied by instant differentiation and U–Pb fractionation resulting in the enrichment of the outer layer in U and Th and in the high μ values of surface materials.

(b) 4.42 AE represents the age of the crust. The moon itself formed earlier, say at 4.6 AE. The question then arises: Where is the radiogenic Pb formed in the earlier time span (4.6–4.4 AE) prior to crust formation?

(c) The upper intersection at 4.42 AE has no specific time meaning but is a consequence of an *average* initial lead value $I_{Pb} \simeq 1.45$. The initial Pb in this case is a multicomponent mixture, averaging a spectrum of I_{Pb} values on both sides of 1.45, corresponding to rocks of widely different ages which were isotopically homogenized at the terminal cataclysm ~ 3.9 AE.

Considering case (a), the 4.42 AE age could conceivably be the actual age of the moon; however, a major difficulty is that this would require instantaneous differentiation upon accretion with no significant addition of later crust or that there is no Pb–U fractionation. From this argument, an older age would be more plausible. An older age of ~ 4.65 AE has been alluded to repeatedly by Nunes *et al.* (1973, 1974) but this appears to be influenced by the more recent results from the Rb–Sr method on some meteorites (Burnett and Wasserburg, 1967; Wasserburg and Burnett, 1968; Wasserburg *et al.*, 1969; Sanz and Wasserburg, 1969; Kaushal and Wetherill, 1969) and is not based on rigid application of U–Pb systematics on lunar materials.

In their argument for a 4.65 AE old moon, Tatsumoto (1970) and Nunes *et al.* (1973) used as evidence the near concordancy at ~ 4.65 AE of soil 10084 and breccia 10061. Soils are very complex systems and cannot be taken as representative of the moon as a whole. Breccia 10061 is actually slightly discordant and its Pb–U and Pb–Th ages are close to 4.7 AE. This may indicate the presence of initial lead, in which case, the near concordancy at ~ 4.7 AE would be a deceptive accident. Nunes *et al.* (1973) have recently replied to criticism of this approach and indicated that they have used several whole-rock basalt analyses in addition to the soil and breccia data discussed above. We do not agree with their implication that there exists a well-defined set of U–Pb systematics on lunar basalt samples which clearly point to an age of 4.65 AE. In particular, Tatsumoto *et al.* (1972a) reported on an internal isochron for rock 14310 intersecting the concordia curve at 4.65 AE and with a corresponding $I_{Pb} = 1.63$. These workers used these data as further evidence of the validity of 4.65 AE for the age of the moon. These data are in disagreement with the $I_{Pb} = 1.45$ obtained earlier by Tera and Wasserburg (1972a) from an internal isochron on the same rock. Subsequently, however, it was stated by Tatsumoto *et al.* (1972b) that the higher I_{Pb} value obtained by them is in error due to an over correction which would affect all the data points they reported on 14310. Since the upper intersection is correlated to the I_{Pb} value, it follows that the discussion based on an upper intersection at 4.65 AE with $I_{Pb} = 1.63$ is invalid.

Injecting the Pb isotope ages of some meteorites into the argument (Nunes *et al.*, 1973) is not justified, for these are single stage *model ages* for Pb evolution (Tatsumoto *et al.*, 1973b; Tilton, 1973), and at least one of these samples is known to have a complex evolution history (Gray *et al.*, 1973). At present there are no reliable concordant U–Pb age determinations for meteorites and many of them give $^{207}Pb-^{206}Pb$ model ages of 4.50–4.57 AE (Tatsumoto *et al.*, 1973b). It is now firmly established that meteorites were formed or altered over a prolonged time span and there is no *a priori* justification for ignoring possible differences in formation ages (Papanastassiou *et al.*, 1974a, 1974b). In this regard, one might mention that "the" geochron obtained from the Pb isotope systematics for the earth is close to 4.45 AE which, by a parallel argument to that mentioned above, could be used as an indication of the fact that some planetary bodies are of approximately that age.

We do not believe that 4.65 AE is an unreasonable value for the age of the moon; however, no cogent experimental evidence for this value has yet been presented. While the approximate age of the solar system and the planets is reasonably well defined at ∼4.5 AE, the precise ages of the planetary objects are uncertain by ∼200 m.y. This fundamental problem of the reliable determination of the relative ages of the planetary bodies is certainly not yet resolved.

Considering case (b), if 4.42 AE corresponds to the age of the crust, then one must provide a plausable explanation for the absence of I_{Pb} produced in the preceding stage of evolution. This basic problem was first pointed out by Tera and Wasserburg (1972a) based on their study of Apollo 14 basalts. These workers stated "this (age) appears to be significantly younger than the 'age of the solar system.'" and concluded that the results "are suggestive of a time scale of ∼ 10^8 yr crustal formation with concomitant depletion of initial lead in the crust relative to U and Th." This initial Pb was either left behind in the moon's interior or it might have been lost from the moon's surface by an efficient outgassing mechanism. Irrespective of any mechanism, the initial μ value for the moon must have been much lower than the μ value of the crust upon differentiation, i.e. the crust formation was accompanied by strong U–Pb fractionation with the crust being highly enriched in U relative to Pb. If the Pb was left behind, then the interior of the moon would be enriched in both primordial Pb and in ancient radiogenic Pb produced between the nominal ages of 4.6 and 4.4 AE in an environment with a much lower μ than that of the outer layer of the moon. Tera et al. (1970) and Tera and Wasserburg (1972a) have suggested that Pb depletion in the lunar crust might be due to the segregation of calcophile and siderophile elements in pods (?) of Fe and FeS.

An alternative to the model of Pb burial at depth as discussed above, is the enrichment of the outer layer of the moon in both Pb and U possibly without fractionation, followed by the strong depletion of Pb from the lunar crust at 4.42 AE by some undetermined mechanism (possibly by volatile losses at the surface during the process of rapid crustal growth). This would result in an outer layer of 4.42 AE of age and an interior characterized by a much lower μ and much lower U and Pb content than the outer layer.

Considering case (c), a variety of models for crustal growth could be constructed to fit this condition. An extreme case would be uniform crustal growth with a constant rate of addition of U (but no Pb) over the time span from 4.60 to 3.91 AE (Tera et al., 1974a).

The possibility that the I_{Pb} ∼ 1.45 is not due to a simple single-stage process but is a grand average of the Pb in the crust of the moon at the time of the last cataclysmic bombardment has been discussed extensively by Tera et al. (1974a). In this case, the upper intersection τ_0 is a sort of average value. We may use the above parameters to estimate the proportion of early lunar crust if we assume an age for the moon. One simple crustal evolution model is that the crust grew over a prolonged time scale. For simplicity, assuming that only ^{238}U is added to the initial crust at a constant rate, then the value of $^{207}Pb/^{206}Pb$ for the total crust at a time t years after the formation of the moon is

$$I_{Pb} = \frac{e^{(\lambda_5 - \lambda_8)T}\left[\dfrac{\lambda_5(1 - e^{-(\lambda_5 - \lambda_8)t})}{\lambda_8(\lambda_5 - \lambda_8)} - \left(\dfrac{1}{\lambda_8} - Rt\right)(1 - e^{-\lambda_5 t})\right]}{137.8\left[t - \left(\dfrac{1}{\lambda_8} - Rt\right)(1 - e^{-\lambda_8 t})\right]}$$

Here T is the "age of the moon" and R is the ratio of the ^{238}U atoms in the original outer crust formed at T AE ago to the number of ^{238}U atoms added between T and $T - t$. That is $1/(1 + R)$ is the fraction of the total crust which grew uniformly after the formation of the initial crust. The limits $R = \infty$ and $R = 0$ correspond to an initial crust without subsequent growth and to a uniformly growing crust of zero initial size respectively.

Given an exact value for I_{Pb} at a time t, it is possible to estimate the rate of crustal growth if one knows T. It is clear in the simple limiting model stated above with infinite Pb–U fractionation that τ_0 is younger than the actual "age of the moon." Using an estimate for T of 4.60 AE and $\tau_1 = T - t = 3.86$ AE we obtain $\tau_0 = 4.43$ AE and would require that $\sim 15\%$ of the lunar crust have been formed very rapidly. According to this model the upper limit for the age of the moon (T) is 4.79 AE, assuming that $\tau_1 = 3.80$ AE and $R = 0$. A realistic crustal growth model must take into account the addition of Pb along with U, in which case τ_0 approaches T.

A schematic representation of different crustal evolution rates is shown in Fig. 4. In the upper figure the rate of formation is shown as a function of time. This represents the grams/cm^2 sec of material which is differentiated from the lunar interior and extruded to make the crust. The lower figure is complimentary and shows the percent of the total lunar crust which existed as a function of time. For the rapidly formed crust as shown, about 80% of the crust is formed in the first 150 m.y. Subsequent to 3.9 AE only small crustal addition takes place with the extrusion of mare basalt flows. The model showing a uniform rate of crustal formation shows two cases; one for which the mare basalts are a major fraction and have persisted from the earliest stages of lunar history; the other case is where the mare basalts are minor fraction. Our viewpoint is that mare basalts are only a minor fraction of the total crust in all areas and hence that almost all the major crustal growth took place prior to ~ 3.9 AE.

From a schematic point of view we prefer a model in which the rate of formation of lunar crust is intense in the first 0.2 AE in which, for example, about 60% of the crust was formed. About 30% more of the crust is formed by ~ 3.9 AE during a moderate rate of crustal addition. There appears to be a sharp break in the nature of the lithic materials added to the crust after 3.9 AE. The last few percent of crust being formed in mare basalt dribbles between ~ 3.9 and 3.0 AE. Note that the end point of 4.6 AE in Fig. 4 is a nominal value. It is quite possible that there was a more intermittent form of crustal growth in some later spurts (4.0–3.8 AE) triggered by major impacts on the moon.

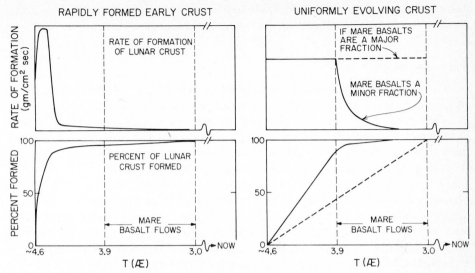

Fig. 4. Schematic diagrams showing the rate of formation of lunar crust as a function of time. This is also directly related to the rate of differentiation of the moon as a whole. The percent of the crust versus time is also shown. In the diagram on the left, is the case of a rapidly formed early crust. The diagram on the right is for a uniformly evolving crust down to ~3.9 AE. Note that the mare basalts are shown as a minor contribution to the crust and that almost no crustal growth took place after 3.0 AE. The age of ~4.6 AE is a *nominal* value. The actual age of the moon is not firmly established and could be different by 0.15 AE.

Summary

Considering the data available on terra rocks, no unique interpretation of the upper intersection τ_0 can be given and there is no conclusive evidence on the models of crustal growth, partly due to our ignorance of igneous events in the first 0.5 AE of lunar history, and in part due to imprecise knowledge of the age of the moon, and the degree of Pb–U fractionation in the lunar mantle in generating the "average" crust. The interpretation of the lower intersection, however, appears reasonably well defined as the period of a terminal cataclysmic bombardment.

The Animals in the Zoo

There are a wide variety of rock types which are part of the general systematics shown in Fig. 2. The coherence of the data appears to be independent of any specific petrologic characteristics. The categorization of these samples as "terra" rocks is at best obscure since they range from extrusive and intrusive igneous rocks to metaclastic rocks. If a data point lies in the linear array, there is no *a priori* evidence of a deep genetic significance. In order to explore this matter we will present some new data on a few samples of diverse origin ranging from an intrusive "terra" rock to a mare basalt and finally a soil. Each sample will be discussed and an attempt will be made to ascertain how and if the Pb–U isotopic characteristics can be related to the overall models discussed earlier.

Troctolite 76535

Troctolite 76535, which was considered by Gooley *et al.* (1974) to have been formed at depth, is close to but *not* precisely on the reference isochron (see Fig. 2). This rock is distinctive from any known mare rocks and is associated petrogenetically with the older (?) crust and may plausibly have been excavated during a major impact. The total rock point is somewhat discordant indicating that it is not a simple closed system, or else our sample is not representative because this rock is coarse grained and our sample size was small. As seen in Fig. 5a, the plagioclase from this rock has a Pb isotopic composition and a U/Pb ratio which are quite different from the total rock indicating the presence of initial radiogenic lead. There does not appear to be a significant amount of exotic labile Pb in/on the rock as found in the shocked lunar anorthosite 15415. The results in Fig. 5a indicate that this rock was last equilibrated (or formed) at ~4.0 AE (see caption of Figs. 5a and 5b and Table 1 for details).

With regard to the time of formation of this rock, we can only refer to the last time of equilibration presumably at ~4.0 AE although no precise value can be given. Whether this rock was formed from a magma at 4.0 AE or was an ancient

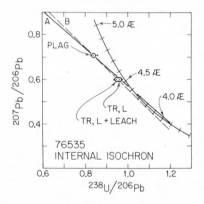

Fig. 5a. A segment of U–Pb evolution diagram showing the data points for 76535. Both total rock sample and the plagioclase separate were mildly leached (See Table 1 for details). The large dispersion in $^{207}Pb/^{206}Pb$ and $^{238}U/^{206}Pb$ between these two samples is evident and indicates the presence of radiogenic initial Pb. The leached total rock (TR, L) and the combined total (TR, L + Leach) plot very close to each other. The leached plag (Plag, L) and the combined plag (Plag, L + Leach) are almost superimposed. If our total rock sample is representative, then the discordancy of the TR point would indicate that it was an open system. A line through the TR and Plag points is almost tangent to concordia so that no precise age can be determined. Line A which intersects concordia at ~4.4 and 3.9 AE passes through the data points within the error bars. Line B is drawn through the center of the data points and misses concordia. These data indicate that troctolite 76535 must have been crystallized or the Pb isotopically homogenized at ~4.0 AE.

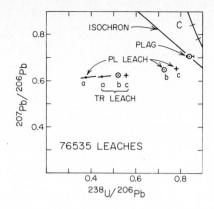

Fig. 5b. U–Pb evolution diagram showing a small segment of concordia (C) and a portion of the "isochron A" from Fig. 5a for 76535. The total rock and the plagioclase were mildly acid leached. The leaches are strongly displaced to the left of the "isochron." This mild leaching removed ~4% ^{206}Pb, 8.7% ^{208}Pb, and ~1.6% U from the total rock. Points a are corrected for nominal blank and points c have all ^{204}Pb removed as blank. Adjusted points b are calculated following a construction described earlier (Fig. 6, Tera et al. 1974b). Even the *extreme limit* of corrections are far off any isochron and prove that the leaching must represent differential element dissolution, i.e. both leaches show a deficit in ^{238}U relative to ^{206}Pb. In both cases, the 207/206 composition of the leach is different from the Pb composition of the corresponding leached sample. These data demonstrate the complexities involved in leaching.

4.4 AE rock which was reequilibrated at 4.0 AE cannot be decided. Furthermore, if 76535 was a rock originally formed at ~4.4 AE, it cannot be decided whether equilibration of the 4.4 AE rock occurred by shock metamorphism at the time of excavation at ~4.0 AE or if equilibration was due to continuous Pb isotopic exchange in the solid state in a reservoir which was kept sufficiently hot in the interval 4.4–4.0 AE due to its burial at depth in a steep lunar thermal gradient.

Troctolite 76535 shows an evolved initial Sr (Papanastassiou and Wasserburg, unpublished data, 1974) indicating equilibration at times younger than 4.4 AE. The attempt to establish an Rb–Sr isochron for this rock by Bogard et al. (1974) revealed that the Rb–Sr systematics are seriously disturbed. The K–Ar data on plagioclase from this rock indicate an age of 4.34 AE ± 0.08 (Bogard et al., 1974). However, the results also show the presence of a large trapped component (Bogard et al., 1974), and there could be redistribution of argon during the shock processes as has been inferred by Turner et al. (1971) for some Apollo 14 breccias. Other lunar troctolites (e.g. 78235) show signs of different degrees of shock and possible injection of extraneous materials.

There is no simple model which explains the complex and disparate behavior of the different isotopic systems. This is distinct from the behavior of well-recrystallized terra rocks such as 76055 (Tera et al., 1974a) and 65015 (Papanastassiou and Wasserburg, 1972b; Jessburger et al., 1974), which show well-defined recrystallization ages and only vestages of the much more ancient original "age." The complexities in the various isotopic systematics of 76535 are most simply

Table 1.

Sample[a]	Weight mg	Lunar lead[b,c]				α[d]		^{204}Pb[b] blank	$\Delta(^{204}$Pb$)$[b,s] blank	^{238}U[b]	^{232}Th[b]	$\dfrac{\text{Th}}{^{238}\text{U}}$[b]	Model ages (AE)[e,f]			
		^{208}Pb	^{207}Pb	^{206}Pb	^{204}Pb	comp.	conc.						$\dfrac{207}{206}$	$\dfrac{206}{238}$	$\dfrac{207}{235}$	$\dfrac{208}{232}$
76535 TR, L	522.4	73.73	52.53	84.16	0.4428	169.2	170.4	0.0571	0.0023	77.01	—	—	4.51	4.59	4.53	—
Leach[g]		7.013	2.329	3.342	0.0576	47.53	46.85	0.0223	0.0050	1.267	—	—	4.55	7.57	5.33	—
76535, PL, L[h]	216.1	26.49	16.08	21.68	0.1896	117.3	101.2	0.0300	0.0349	16.74	—	—	4.75	5.05	4.84	—
Leach[g]		5.618	2.045	2.596	0.0980	26.09	25.12	0.0187	0.0131	0.6296	—	—	4.55	8.38	3.21	—
75055 TR	192.6	116.0	73.70	127.8	0.2760	427.5	420.7	0.0289	0.0063	127.4	440.2	3.45	4.42	4.41	4.42	4.43
75055 TR[m]	533.8	305.5	184.1	311.1	1.3034	(236)[n]	(260)[n]	~0.16[o]	—	302.8	1028	3.40	4.44	4.42	4.43	4.67
15555 PL 2[i]	255.1	13.19	9.580	10.94	0.1725	54.98	55.55	0.0364	0.0024	4.060	—	—	4.99	7.69	8.00	—
Leach[j]		21.38	11.18	15.15	0.3821	38.26	38.42	0.0227	0.0040	7.844	—	—	4.57	5.84	4.93	—
15555 PL 1, L[k,l]	70.0	—	—	—	—	24.70	—	0.0200	—	—	—	—	—	—	—	—
15555 TR	583.9	278.1	113.0	278.9	0.4474	574.8	580.6	0.0341	0.0036	354.3	1334	3.76	3.89	3.70	3.82	3.66
15555 TR[p]	1000	423.3	173.0	412.1	1.1322	—	(360)[q]	~0.25[r]	—	527.3	1981	3.76	3.96	3.65	3.82	3.63

[a] Acid washed sample designated by L; material removed from sample in acid wash designated Leach. Total atoms in sample obtained by summing. [b] In picomoles. [c] Corrected for blank with $\alpha = 18.26$, $\beta = 15.46$, and $\gamma = 37.59$. [d] Uncorrected for blank, α values for concentration runs are corrected for cross contamination from spikes. [e] After correction for blank and PAT (after Tatsumoto, Knight and Allegre). [f] $\lambda_{238} = 1.5525 \times 10^{-10}$ y^{-1}, $\lambda_{235} = 9.8485 \times 10^{-10}$ y^{-1}, and $\lambda_{232} = 4.9475 \times 10^{-11}$ y^{-1}. [g] 1 N cold HNO$_3$ followed by 1 N cold HCl. [h] Hand picked. [i] By Frantz and hand picking. [j] 0.5 N cold HNO$_3$. [k] Heavy liquid and acetone used. [l] $\beta = 21.30$ for this sample. [m] Calculated from data by Nunes et al. (1974a). [n] Corrected values reported by Nunes et al. (1974a). [o] Estimated from the blank range reported by Nunes et al. (1974a) for this sample. [p] Calculated from data by Tatsumoto et al. (1972b). [q] Corrected value reported by Tatsumoto et al. (1972b). [r] Estimated on the basis of blank range reported by Tatsumoto et al. (1972b). [s] See Tera et al. (1974a, Section 2.2, p. 2).

explained by assuming it to be an old rock (~ 4.4 AE) which was partially shock metamorphosed at $\sim 3.9–4.0$ AE. We must assume that the Pb was more labile than the elements comprising the other isotopic systems. If the rock had been *completely* isotopically rehomogenized for all systems and with no inherited Ar, with, say, a recrystallization age of 3.9 AE, we would be unable to distinguish it from a plutonic or hypabyssal differentiate associated with early mare volcanism.

Concordant mare basalts

Mare basalts in general do not appear to show an intimate relationship to the terminal cataclysm. While mare-type volcanism may have taken place prior to 3.9 AE, there is no evidence from either petrographic studies of terra breccias or from isotopic age determinations that this type of igneous activity was abundant in the first 0.5 AE of lunar history.

In our previous paper (Tera *et al.*, 1974a), we noted that the only common characteristic of the rocks in the linear array is that "none of these rocks is a typical mare basalt," although it was recognized that "old" mare basalts (e.g. ~ 3.9 AE) may lie very close to the reference isochron and would be unresolvable from the general data array. Recently, Nunes *et al.* (1974a) reported data on two typical mare basalts from Apollo 17 (75055 and 75035) which showed them to be concordant at 4.44 AE (new decay constants). These data points are found to plot on the linear array and further they lie very close to the point of upper intersection of the reference isochron with the concordia curve. It follows that some mare basalts are members of the zoo and in particular that they define the same value of τ_0 as general terra rocks. Because of the importance of this relationship, we have carried out our own analyses of 75055 which confirm the general concordancy first reported by Nunes *et al.* Our results (Table 1) show that the total rock is concordant at 4.42 AE for *all* U–Th–Pb systems. The data for this sample are shown in Fig. 6, along with the original reference isochron and the two concordant terra rocks. Irrespective of any model, the concordancy of basalt 75055 at the precise location of two terra rocks at 4.42 AE (see Fig. 6) and the very good agreement of Pb–U and Pb–Th ages for these concordant rocks, is of great significance and cannot be taken as an accident. In a simple model, this concordant age would be taken to be the time of formation of a closed system without any initial radiogenic Pb.

The time of 4.42 AE which was first determined on "terra" rocks thus appears to be associated with some mare basalts which are the product of internal magmatism. These concordant U–Pb ages at 4.42 AE are much higher than the Rb–Sr internal isochron ages and the ^{40}Ar–^{39}Ar total rock ages determined on the Apollo 17 mare basalts. The ages obtained from these two methods are in the range of 3.7–3.9 AE (Tera *et al.*, 1974a, 1974b; Evensen *et al.*, 1973; Tatsumoto *et al.*, 1973a; Schaeffer and Husain, 1973). For 75055 in particular, the Rb–Sr internal isochron ages reported are 3.77 AE (Tera *et al.*, 1974a) and 3.83 AE (Tatsumoto *et al.*, 1973a) and are in agreement with precise ^{40}Ar–^{39}Ar ages (Huneke *et al.*, 1973). These ages are assumed to be associated with mare volcanism (see, for example,

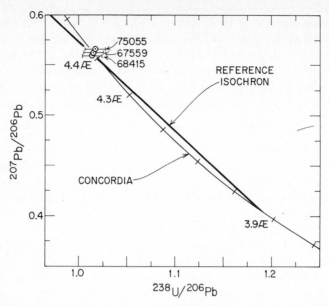

Fig. 6. U–Pb evolution diagram showing the exact position of mare basalt 75055. This sample is concordant at 4.42 AE. Two terra rocks (67559 and 68415) are also concordant at the same location. All three samples fall on the reference isochron within error bars close to the upper intersection at 4.4 AE. ^{208}Pb/^{232}Th ages for samples 67559 (Tera *et al.*, 1974a) and 75055 are in excellent agreement and that of 68415 (Tera *et al.*, 1974a; Nunes *et al.*, 1973) is in good agreement with the U–Pb ages. This indicates the strong possibility that these samples represent closed systems and suggests that the upper intersection of the reference isochron at 4.42 AE might have a specific time meaning.

Tera *et al.*, 1974a; Schaeffer and Husain, 1973; Huneke *et al.*, 1973). This mare volcanism took place several hundred million years after the formation of the moon during which time the isotopic composition of Pb in the magma source region was evolving. The coincidence of the age of extrusion with the neighborhood of the lower intersection would suggest that an internal melt might lie *near* the chord in Fig. 2. However, in order for the data point to lie precisely on the chord the initial Pb in the basalt would be required to have evolved in a Pb–U environment very similar to that found in the crustal rocks without any complicated history of multistage Pb–U fractionation which would tend to displace them from the isochron. In particular, if the intersection at 4.42 AE is a true closed system age, this must represent *no* Pb–U fractionation. This behavior is quite distinctive from the U–Pb results on basalts from Apollo 11, Apollo 12, and Apollo 15 which indicated a two stage evolution with $\mu \equiv {}^{238}$U/^{204}Pb of the source being much lower than the value measured today (μ_2) (Tatsumoto, 1970; Tatsumoto *et al.*, 1971). On that basis, we concluded that the normal formation of basaltic magmas is accompanied by U–Pb fractionation with U/Pb being enriched in the magma relative to the rock source (Tera *et al.*, 1974a). The results on 75055 and 75035 show that this assumption is not a universally valid rule if the magma was extruded at 3.8 AE from a 4.42 AE reservoir.

This apparent lack of U–Pb fractionation might be explained by one or more of the following models:

(a) These basalts are 4.42 AE old and were remelted by impact and recrystallized at 3.8 AE.

(b) A 4.42 AE deep-seated basaltic source was *totally melted* by an internal heat source at 3.8 AE resulting in the magma extrusion at that time. The lack of U–Th–Pb fractionation in spite of differences in their distribution coefficients would then be attributed to the fact that the source was totally melted.

(c) The magma generated at 3.8 AE was extremely depleted in U, Th, and Pb and upon extrusion it engulfed and totally melted part of the ancient crust. The 4.42 AE would then be a reflection of the age of the assimilated crust and not the source itself.

(d) Basalts are originated from a common ultrabasic mantle source, and the partial melting of the source always results in the fractionation of U, Th, and Pb. However, if the *distribution coefficients of these three elements are high*, then the change in the relative abundances of these elements in the melt relative to the source will be manifested only in the cases involving a few percent partial melting. This change in abundance pattern in the melt relative to the source will decrease with the increase of the degree of partial melting. For example: assuming distribution coefficients on the order of 200 and 1000 for two elements A and B respectively, one would calculate $A/B = 0.55$ in the melt relative to the source for 1% partial melting. If, however, one allows for 10% of partial melting, then the ratio becomes 0.98. That is to say in spite of a factor of five difference in the distribution coefficients, one would observe only a 2% disparity between the two elements in a rock formed from a 10% partial melt of the source for the arbitrary parameters used. The appealing aspect of this model is that it does not require a special mechanism to explain the basalts concordant at 4.42 AE. In such a model, the degree of displacement of a given rock from the 4.42 AE intersection (the assumed age of the source) is inversely proportional to the degree of partial melting (within a limited percentage range of partial melting depending on the magnitude of the distribution coefficients and their ratio).

Such a model must be more critically examined using correlations of Pb–U–Th discordance with REE systematics and Rb–Sr model ages in mare basalts. In contradistinction to the "equilibrium" partial melting mechanism proposed above, Graham and Ringwood (1971) and Albee and Gancarz (1974) have suggested disequilibrium partial melting to explain the REE and Rb–Sr data. If this were also applied to the Pb–U results, we would require that the Pb–U-rich phases (that would have concordant 4.4 AE ages) which are present in the source regions be part of the lowest melting fraction and did not equilibrate with the other mineral phases. This would presumably provide concordant Pb–U ages for the lowest melting (unequilibrated) fraction. Increased melting would presumably give more opportunity for detailed chemical and isotopic equilibration and produce more discordant rocks. These alternative models have an inverse relationship to the degree of partial melting.

Orange soil 74220

The orange soil 74220 taken from the work of Tatsumoto *et al.* (1973a) is included in Fig. 2. It is striking that the orange soil plots so close to the reference isochron (within less than 1%). The sample has a lower μ value than what would be expected from its location in the array. However, this would not have been considered an exceptional anomaly if it were a terra rock sample since we have observed other reversals in the general trend of μ values (Tera *et al.*, 1974a). The close association of this sample with the array of terra samples would seem to suggest that the orange glasses were produced in the terminal cataclysm. The glass could have been produced by impact on a low μ segment of the crust. The empirical fact that this soil fits on the linear array in Fig. 2 is in harmony with the results of a leaching experiment carried out by Tatsumoto *et al.* (1973a) who showed that the Pb isotope correlation for the data from the leaches, the residue and the whole soil define a mixing line corresponding to an age of 3.79 AE (new decay constants). This could mean that the glass represents material produced near the end of the terminal cataclysmic interval within the existing uncertainties. On the other hand, many workers associate this type of glass with magmatic activity produced by internal heat sources (lava fountains, explosive volcanism). As was pointed out previously (Tera *et al.*, 1974a), simple magmatic products formed in the general neighborhood of ~ 3.9 AE would be indistinguishable from the results of major impact metamorphism on the basis of U–Pb systematics. The 3.79 AE age obtained by Tatsumoto *et al.* (1973a) is close enough to ~ 3.9 AE so that the sample could lie on the reference line for a simple Pb–U evolution independent of any genetic correlation between the glasses and the terminal lunar cataclysm.

The picture is further complicated by the fact that the ^{40}Ar–^{39}Ar age on this soil ranges from 3.55 to 3.77 AE although there is some uncertainty in the trapped component (Schaeffer and Husain, 1973; Eberhardt *et al.*, 1973; Huneke *et al.*, 1973). If one accepts the younger age of 3.55 AE as an actual age of the ultra-basic orange glasses, then its location on the reference isochron (Fig. 2) is an accident. However, if we attribute the younger value for the age by ^{40}Ar–^{39}Ar method as due to gas loss, then it might still be possible to associate the Pb of orange soil with the terminal cataclysm.

Tatsumoto *et al.* (1973a) assumed an age of 3.63 AE for the orange soil and calculated a μ_1 value of 34.5 for the source of Pb in the orange soil using an age of 4.65 AE for the moon. If we take $\tau_1 = 3.79$ AE on the basis of the leaching experiment (Tatsumoto *et al.*, 1973a), then we calculate $\mu_1 = 52$ (assuming $\tau_0 = 4.45$ AE). In either case, the μ_1 value is below or at the very lowest end of the range calculated by Tatsumoto (1970) and by Tatsumoto *et al.* (1971) for the sources of mare basalts (60–300). Tatsumoto *et al.* (1973a) have suggested that the orange glasses from Apollo 17 originated in magma at a greater depth than the source of mare basalts. If these glasses are from the lunar interior, then we must concur with the conclusion of Tatsumoto *et al.* (1973a) that a significant portion of the lunar mantle has low μ values. This was evident earlier in the range of μ_1

(60–300) on some mare basalts (Tatsumoto, 1970; Tatsumoto *et al.*, 1971) and must be considered in the broader problem of the source regions of mare lavas and the average value of μ for the moon.

DISCORDANT MARE BASALTS

From the existing data, the main basaltic mare flooding covered a span of time from 3.8 to 3.2 AE (Papanastassiou and Wasserburg, 1973). The actual contribution of mare basalts to the lunar crust during this period is very small. We estimate that only a few percent of the total crustal thickness in mare regions is composed of mare basalts. These appear to be thin flows and intercalated sills in a thick debris field. As mare basalts are interpreted as partial melts from the lunar interior, the detailed interpretation of their evolution is critical to the understanding of the nature of the lunar mantle. If mare basalts are the result of simple two-stage evolution, it should be possible to obtain a consistent value for the "age" of the moon from U–Pb internal isochrons on these rocks of different ages. Such determinations would be independent of the estimate made from the linear array for "terra" rocks. In general, the discordant mare basalts plot far from the isochron shown in Fig. 2 and are from this purely geometrical point of view not part of the zoo. Mare basalt samples from Apollo 11, Apollo 12, and Apollo 15 give total rock U–Pb model ages younger than ~4.5 AE but older than their crystallization ages. This implies U gain and/or Pb loss associated with derivation from their sources. The measured μ_2 values of those mare basalts are in the range 300–1000 (Tatsumoto, 1970; Tatsumoto *et al.*, 1971) as compared to a calculated μ_1 in the range of 60–300, assuming a two-stage model for their evolution (Tatsumoto, 1970; Tatsumoto *et al.*, 1971). The question at hand is whether the sources of mare basalts are primary undifferentiated reservoirs isolated immediately after the formation of the moon or if the source regions themselves are second or higher generation differentiates produced over a wide time interval. This issue could be clarified through the establishment of U–Pb internal isochrons for mare basalts.

Figure 7 and Tables 1 and 2 show the results of an exploratory experiment on mare basalt 15555. This rock was chosen for study because it is a relatively young, coarse-grained mare basalt. The data for the total rock, the leached plagioclase (Plag, L), and the leach of the plagioclase (PL, Leach) are shown on a U–Pb evolution diagram (Fig. 7). Because of the large contamination encountered in mineral separation relative to the content of lunar Pb in the sample, the U–Pb isochron (Fig. 7) is imprecise. However, the results point clearly to the technical feasibility of establishing U–Pb internal isochrons for low-Pb mare basalts. Some element fractionation is induced by leaching. Points *a* are corrected for nominal blank and points *b* have all [204]Pb removed as MT. In Fig. 7, lines A and B are drawn to intersect concordia at 3.3 AE which is the age of the rock by the Rb–Sr (Papanastassiou and Wasserburg, 1973b) and the ^{40}Ar–^{39}Ar methods (Podosek *et al.*, 1972). We have used these other age determinations for τ_1 in lieu of a precise U–Pb internal isochron. All of the subsequent discussion is subject to the

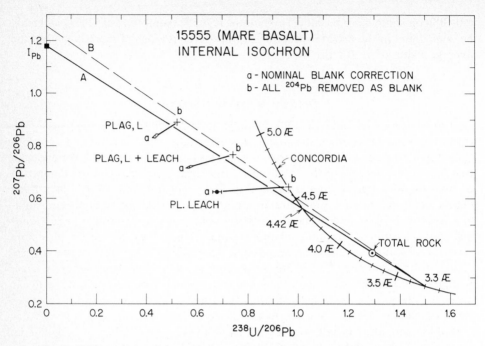

Fig. 7. U–Pb evolution diagram showing total rock-plagioclase internal "isochron" for mare basalt 15555. The plag separate (PL 2 in Fig. 8) was obtained by mechanical means and was leached in cold 0.5 N HNO₃. The leached plagioclase (Plag, L) and its leach (PL, Leach) are shown. A significant amount of Pb is contained in the leach (see Table 1). Some element fractionation is induced by leaching. Points *a* are corrected for nominal blank and the limiting end points *b* have all ²⁰⁴Pb removed as blank. Lines A and B are drawn to intersect concordia at 3.3 AE (⁴⁰Ar–³⁹Ar and Rb–Sr age for 15555). Line A is drawn through the total rock point taken from this work. Note that this line intersects concordia at the magic point of 4.42 AE. Line B is the *limiting line* determined by the combined plag with all ²⁰⁴Pb removed as blank. Note that the upper intersection of the limiting line B with concordia is *less* than 4.55 AE. Thus the upper intersection of 4.6 AE (new decay constants) assumed by Tatsumoto *et al.* (1972b) for this rock would appear to be subject to question.

uncertainties in the crystallization age. Line A is drawn through the total rock point taken from this work. Note that this line intersects the concorida curve at 4.42 AE. This value is in surprising agreement with the upper intersection (see Figs. 2 and 6) discussed earlier in this paper and is distinct from the value of 4.6 AE obtained by Tatsumoto *et al.* (1972b). Line B is a *limiting line* determined by the combined plag (Plag, L + Leach) with *all* ²⁰⁴Pb removed as blank. Thus, as indicated by line B, the upper intersection with concordia must be less than 4.55 AE. This upper limit value is 50 m.y. younger than the 4.6 AE upper intersection assumed by Tatsumoto *et al.* (1972b). Because of the large errors in the plagioclase due to contamination, it is not possible to determine the exact location of the upper intersection, although line A indicates an upper intersection at ~4.42 AE which appears to be a magic value and might suggest that the source of 15555 is a primary magma produced from undifferentiated lunar mantle or a lunar mantle which differentiated extremely rapidly at a time of formation very near 4.42 AE. The more detailed discussion of the analyses of the plagioclase separates and the leaches in the appendix points out some technical questions that are not yet resolved.

PRIMORDIAL LEAD

An important parameter in defining the evolution of a planet is its content of primordial and radiogenic Pb and the distribution pattern of those two types of Pb relative to each other in the planet. The μ value is another parameter which describes the degree of U–Pb fractionation associated with the events of the planetary evolution. The measured μ values for lunar basalts range from 300–1000 versus μ_1 ranging from 60 to 300 for the source of those mare basalts, assuming a two-stage model (Tatsumoto, 1970; Tatsumoto et al., 1971).

In this paper, we report results on two mare basalts analyzed earlier (Tatsumoto et al., 1972b; Nunes et al., 1974a). Our results show much higher μ values. We report μ values of 461 and 792 for 75055 and 15555, respectively versus 250 and 457 reported by these workers. This points to the possibility that the actual μ values of mare basalts might be higher than the data in the literature indicate. The measured μ value of 792 which we report here for 15555 moves it from the lower range of μ values for mare basalts (~ 300) to close to the assumed upper range (~ 1000). This raises the question of whether the low-range values are not similarly influenced by analytical factors. An immediate important consequence to the higher measured μ value for 15555 is that the μ_1 environment from which this rock was derived must be much higher (240) than the value calculated from the existing data (140).

By virtue of its concordance at 4.42 AE, the measured μ value for mare basalt 75055 is essentially equal to its μ_1 value. This is about a factor of 10 larger than the values for the lower range of μ_1 calculated by Tatsumoto (1970) for mare basalts. Insofar as anything like this range is due to lunar U–Pb fractionation, one must conclude that there exists several distinct source regions for mare basalts.

The technical difficulty inherent in the measurement of ^{204}Pb stems from the fact that the lunar samples are depleted in that isotope relative to radiogenic Pb. Lunar α, β, and γ values in general are very high relative to terrestrial Pb. This makes these values as well as lunar μ values very sensitive to terrestrial contamination and subject to large uncertainties unless the Pb blank is low and its fluctuation is limited. A concentrated effort has been directed to this problem in this laboratory with some success, although more stringent controls and improvement in sample handling and in mineral separations are required.

In using lunar Pb data from the literature, one reasonable criterion of choice from different results on the same sample is to use the results with the higher α, β, and γ values, provided that the results used show internal consistency. For example, analyses of aliquots (from the same solution) should show only limited range of fluctuation in α if the blank is under control. It must be recognized, however, that lunar samples (particularly breccias) are heterogeneous and thus some variability in the results from different laboratories is to be expected. Based on the above reasoning, we conclude that the μ values for 75055 and 15555 reported here are more reliable than the data existing in the literature.

The disagreement in the μ values underlines the difficulty in adopting a realistic range for the measured μ values for mare basalts and μ_1 for their sources. This compounds the difficulty in estimating an average μ_1 for the moon as a whole which is important in calculating ^{204}Pb content for the planet (see Tera *et al.*, 1974a). It would seem that second generation experiments with higher reliability for ^{204}Pb determination are required to supplement the earlier pioneering results of Tatsumoto (1970) on Apollo 11. Such a study may result in changes in the magnitude and the range of μ values presently in use for the characterization of mare basalts.

Conclusions

The majority of terra rocks define a rather coherent linear array on the U–Pb evolution diagram with the data clustering around a reference isochron intersecting the concordia at ~4.42 and ~3.95 AE. The lower intersection is not well defined due primarily to the fact that the terra rocks are not of the same age. This lower intersection is associated with global impact metamorphism with a duration of up to 200 m.y. (3.8–4.0 AE), which caused element redistribution and extensive Pb mobilization. Subsequent to this terminal cataclysm the rate of bombardment decreased sharply. Precise details of the history prior to ~4.0 AE is at present somewhat obscure. In spite of some scatter, the upper intersection with concordia seems to be rather well defined by the fact that two terra rocks are concordant at 4.42 AE.

There is a wide variety of rock types which comprises the linear array. Recent data show that in addition to "terra" rocks, some mare basalts also lie in the array at the upper intersection with concordia at 4.42 AE. The orange soil reported by Tatsumoto *et al.* was also found to lie on the array. At the present time, there does not appear to be a well-defined self-consistent criterion which explains the types of materials included in the array. Some of these lithic types are not obviously related by the terminal lunar cataclysm.

The presence of an old mare basalt on the data array of the terra rocks precisely at the magic point of 4.42 AE (Fig. 6) suggests that this time may represent the actual age of crustal formation and large-scale lunar differentiation. This possibility is further hinted at by the results on a younger mare basalt (Fig. 7). The 4.42 AE age could conceivably be a close approximation of the age of the moon itself. There is no justification for the *a priori* assumption that the moon is much older than 4.42 AE. Irrespective of the exact time of the crust formation, this process was accompanied by strong U–Pb fractionation resulting in the formation of an outer layer with higher U concentration and higher μ values than the interior of the moon.

The concordance of a mare basalt of Apollo 17 (75055) at 4.42 AE is in marked contrast with Apollo 11 and Apollo 12 basalts. The latter appear to show a two-stage evolution with the second stage being enriched in U and resulting in μ_2 values much higher than μ_1. This disparate behavior, further underlines the heterogeneity of the sources of mare basalts and may be indicative of differences in the mechanism of magma generation on the moon. The diversity in the magma sources of mare basalts combined with the earlier observation of 4.42 AE "age" (Fig. 7) appear to suggest that the moon differentiated rapidly after formation resulting in considerable heterogeneity in the trace elements' chemistry particularly in $^{238}U/^{204}Pb$ values. From this point of view, all lunar igneous rocks including mare basalts may be viewed as third generation differentiates derived from sources (second generation differentiates) formed shortly after the accretion of the moon. Such a model would require accretional energy as the heat source because of the postulated rapid differentiation. This question which is intimately tied up with the age of the moon and the internal U–Pb systematics of mare basalts is certainly not yet resolved.

Acknowledgments—We are indebted to our colleagues A. L. Albee, A. J. Gancarz, and D. A. Papanastassiou for discussions which contributed constructively to this paper. We wish to thank G. R. Tilton for passing water to us. This manuscript was much improved by the many careful comments and criticisms of P. Banks, B. R. Doe, and R. E. Zartman. One section title was derived from the witicisms of F. A. Podosek. This work was supported by a grant from the National Aeronautics and Space Administration, NGL 05-002-188.

REFERENCES

Albee A. L. and Gancarz A. J. (1974) Petrogenesis of lunar rocks: Rb–Sr constraints and lack of H_2O. *Proc. of Soviet–American conf. on Cosmochim. of Moon and Planet.* Moscow. In press.

Albee A. L., Burnett D. S., Chodos A. A., Haines E. L., Huneke J. C., Papanastassiou D. A., Podosek F. A., Russ G. P. III, and Wasserburg G. J. (1970) Mineralogic and isotopic investigation on lunar rock 12013. *Earth Planet. Sci. Lett.* **9**, 137–163.

Bogard D. D., Nyquist L. E., Bansal B. M., and Wiesmann H. (1974) 76535: An old lunar rock? (abstract). In *Lunar Science—V*, pp. 70–72. The Lunar Science Institute, Houston.

Burnett D. S. and Wasserburg G. J. (1967) $^{87}Rb–^{87}Sr$ ages of silicate inclusions in iron meteorites. *Earth Planet. Sci. Lett.* **2**, 397–408.

Eberhardt P., Geiss J., Grögler N., Maurer P., and Stettler A. (1973) $^{39}Ar–^{40}Ar$ ages of lunar material. *Meteortics* **8**, 360–361.

Evensen N. M., Murthy V. R., and Coscio M. R., Jr. (1973) Taurus-Littrow: Age of mare volcanism; chemical and Rb–Sr isotopic systematics of dark mantle soil. *EOS* **54**, 587.

Gooley R. E., Brett R., Warner J., and Smyth J. R. (1974) Sample 76535, A deep lunar crustal rock. *Geochim. Cosmochim. Acta.* In press.

Graham A. L. and Ringwood A. E. (1971) Lunar basalt genesis: the origin of the europium anomaly. *Earth. Planet. Sci. Lett.* **13**, 105–115.

Gray C. M., Papanastassiou D. A., and Wasserburg G. J. (1973) The identification of early condensates from the solar nebula. *Icarus* **20**, 213–239.

Jessberger E. K., Huneke J. C., Podosek F. A., and Wasserburg G. J. (1974) High resolution argon analysis of neutron-irradiated Apollo 16 rocks and separated minerals. *Proc. Fifth Lunar Sci. Conf., Geochim. Cosmochim. Acta.* This volume.

Kaushal S. K. and Wetherill G. W. (1969) $^{87}Rb–^{87}Sr$ age of bronzite (H group) chondrites. *J. Geophys. Res.* **74**, 2717–2726.

Huneke J. C., Jessberger E. K., Podosek F. A., and Wasserburg G. J. (1973) $^{40}Ar–^{39}Ar$ measurements in Apollo 16 and 17 samples and the chronology of metamorphic and volcanic activity in the

Taurus-Littrow region. *Proc. Fourth Lunar Sci. Conf., Geochim. Cosmochim. Acta*, Suppl. 4, Vol. 2, pp. 1725–1756. Pergamon.

Nunes P. D., Tatsumoto M., Knight R. J., Unruh D. M., and Doe B. R. (1973) U–Th–Pb systematics of some Apollo 16 lunar samples. *Proc. Fourth Lunar Sci. Conf., Geochim. Cosmochim. Acta*, Suppl. 4, Vol. 2, pp. 1797–1822. Pergamon.

Nunes P. D., Tatsumoto M., and Unruh D. M. (1974a) U–Th–Pb systematics of some Apollo 17 samples (abstract). In *Lunar Science—V*, pp. 562–564. The Lunar Science Institute, Houston.

Nunes P. D., Knight R. J., Unruh D. M., and Tatsumoto M. (1974b) The primitive nature of the lunar crust and the problem of initial Pb isotopic compositions of lunar rocks: A Rb–Sr and U–Th–Pb study of Apollo 16 samples (abstract). In *Lunar Science—V*, pp. 559–561. The Lunar Science Institute, Houston.

Papanastassiou D. A. and Wasserburg G. J. (1971) Rb–Sr ages of igneous rocks from the Apollo 14 mission and the age of the Fra Mauro Formation. *Earth Planet. Sci. Lett.* **12**, pp. 36–48.

Papanastassiou D. A. and Wasserburg G. J. (1972a) The Rb–Sr age of a crystalline rock from Apollo 16. *Earth Planet. Sci. Lett.* **16**, 289–298.

Papanastassiou D. A. and Wasserburg G. J. (1972b) Rb–Sr systematics of Luna 20 and Apollo 16 samples. *Earth Planet. Sci. Lett.* **17**, 52–63.

Papanastassiou D. A. and Wasserburg G. J. (1973) Rb–Sr ages and Initial strontium in basalts from Apollo 15. *Earth Planet. Sci. Lett.* **17**, 324–337.

Papanastassiou D. A. and Wasserburg G. J. (1974) Evidence for late formation and young metamorphism in the achondrite Nakhla. *Geophys. Res. Lett.* **1**, 23–26.

Papanastassiou D. A., Rajan R. S., Huneke J. C., and Wasserburg G. J. (1974) Rb–Sr ages and lunar analogs in a basaltic achondrite; Implications for early solar system chronologies (abstract). In *Lunar Science—V*, pp. 583–585. The Lunar Science Institute, Houston.

Podosek F. A., Huneke J. C., and Wasserburg G. J. (1972) Gas-retention and cosmic-ray exposure ages of lunar rock 15555. *Science* **175**, 423–425.

Sanz G. H. and Wasserburg G. J. (1969) Determination of an internal ^{87}Rb–^{87}Sr isochron for the olivenza chondrite. *Earth Planet. Sci. lett.* **6**, 335–345.

Schaffer O. A. and Husain L. (1973) Isotopic ages of Apollo 17 lunar material. *EOS* **54**, 614.

Silver L. T. (1973) U–Th–Pb isotopic characteristics in some regolithic materials from the Descartes region (abstract). In *Lunar Science—IV*, pp. 672–674. The Lunar Science Institute, Houston.

Silver L. T. (1974) Patterns of U–Th–Pb distributions and isotope relations in Apollo 17 soils (abstract). In *Lunar Science—V*, pp. 706–708. The Lunar Science Institute, Houston.

Tatsumoto M. (1970) Age of the moon: an isotopic study of U–Th–Pb systematics of Apollo 11 lunar samples, II. *Proc. Apollo 11 Lunar Sci. Conf., Geochim. Cosmochim. Acta*, Suppl. 2, Vol. 2, pp. 1595–1612. Pergamon.

Tatsumoto M., Knight R. J., and Doe B. R. (1971) U–Th–Pb systematics of Apollo 12 lunar samples. *Proc. Second Lunar Sci. Conf., Geochim. Cosmochim. Acta*, Suppl. 2, Vol. 2, pp. 1521–1546. MIT Press.

Tatsumoto M., Hedge C. E., Doe B. R., and Unruh D. M. (1972a) U–Th–Pb and Rb–Sr measurement on some Apollo 14 lunar samples. *Proc. Third Lunar Sci. Conf., Geochim. Cosmochim. Acta*, Suppl. 3, Vol. 2, pp. 1531–1555. MIT Press.

Tatsumoto M., Hedge C. E., Knight R. J., Unruh D. M., and Doe B. R. (1972b) U–Th–Pb, Rb–Sr and K measurements on some Apollo 15 and Apollo 16 samples. In *The Apollo 15 Lunar Samples*, pp. 391–395. The Lunar Science Institute, Houston.

Tatsumoto M., Nunes P. D., Knight R. J., Hedge C. E., and Unruh D. M. (1973a) U–Th–Pb measurements of two Apollo 17 samples. *EOS* **54**,(6), 614–615.

Tatsumoto M., Knight R. J., and Allegre C. J. (1973b) Time differences in the formation of meteorites as determined from the ratio of Lead-207 to Lead-206. *Science* **180**, 1279–1283.

Tatsumoto M., Nunes P. D., and Knight R. J. (1973d) U–Th–Pb systematics of some Apollo 16 samples (abstract). In *Lunar Science—IV*, pp. 705–707. The Lunar Science Institute, Houston.

Tatsumoto M., Nunes P. D., Knight R. J., and Unruh D. M. (1974) Rb–Sr and U–Th–Pb systematics of boulder 1 and boulder 7, Apollo 17 (abstract). In *Lunar Science—V*, pp. 774–776. The Lunar Science Institute, Houston.

Tera F., Eugster O., Burnett D. S., and Wasserburg G. J. (1970) Comparative study of Li, Na, K, Rb, Cs, Ca, Sr, and Ba abundances in achondrites and in Apollo 11 lunar samples. *Proc. Apollo 11 Lunar Sci. Conf., Geochim. Cosmochim. Acta,* Suppl. 1, Vol. 2, pp. 1637–1657. Pergamon.

Tera F. and Wasserburg G. J. (1971) U–Th–Pb analyses of soil from the Sea of Fertility. *Earth Planet. Sci. Lett.* **13**, 457–466.

Tera F. and Wasserburg G. J. (1972a) U–Th–Pb systematics in three Apollo 14 basalts and the problem of initial Pb in lunar rocks. *Earth Planet. Sci. Lett.* **14**, 281–304.

Tera F. and Wasserburg G. J. (1972b) U–Th–Pb systematics in lunar highland samples from Luna 20 and Apollo 16 missions. *Earth Planet. Sci. Lett.* **17**, 36–51.

Tera F. and Wasserburg G. J. (1973a) A response to a comment on U–Pb systematics in lunar basalts. *Earth Planet. Sci. Lett.* **19**, 213–217.

Tera F., Ray A. L., and Wasserburg G. J. (1972) Distribution of Pb–U–Th in lunar anorthosite 15415 and inferences about its age. In *The Apollo 15 Lunar Samples,* pp. 396–401. The Lunar Science Institute, Houston.

Tera F., Papanastassiou D., and Wasserburg G. J. (1973b) A lunar cataclysm at ~3.95 AE and the structure of the lunar crust (abstract). In *Lunar Science—IV,* pp. 723–725. The Lunar Science Institute, Houston.

Tera F., Papanastassiou D. A., and Wasserburg G. J. (1974a) Isotopic evidence for a terminal lunar cataclysm. *Earth Planet. Sci. Lett.* **22**, 1–21.

Tera F., Papanastassiou D. A., and Wasserburg G. J. (1974b) The lunar time scale and a summary of isotopic evidence for a terminal lunar cataclysm (abstract). In *Lunar Science—V,* pp. 792–794. The Lunar Science Institute, Houston.

Tilton G. R. (1973) Isotopic lead ages of chondritic meteorites. *Earth Planet. Sci. Lett.* **19**, 321–329.

Turner G., Huneke J. C., Podosek F. A., and Wasserburg G. J. (1971) ^{40}Ar–^{39}Ar ages and cosmic ray exposure ages of Apollo 14 samples. *Earth Planet. Sci. Lett.* **12**, 19–35.

Wasserburg G. J. and Burnett D. S. (1968) The status of isotopic age determinations on iron and stone meteorites. In *Meteorite Research* (editor Millman), pp. 467–479. D. Reidel.

Wasserburg G. J., Papanastassiou D. A., and Sanz H. G. (1969) Initial Sr for a chondrite and the determination of a metamorphism or formation interval. *Earth Planet. Sci. Lett.* **7**, 33–43.

APPENDIX

Leaching experiment on 15555 plagioclase

The establishment of a precise internal isochron for 15555 is our objective in the next generation of experiments. In preparation for that goal, we carried out a series of leaching experiments on plagioclase separates from this rock to explore the removal of surface contamination and to gain some information about the composition of Pb in the plagioclase itself.

Because of the serious Pb contamination problem encountered in sample handling and in mineral separations, we have carried out a variety of experiments to distinguish laboratory contamination from elements which are indigenous to the sample. Most total rock samples are prepared by removing the outer surface of the sample as received using a tungsten microchisel. A fragment is then pried loose and put into dissolution without crushing. This requires a longer time than for crushed samples but eliminates various types of contamination caused by crushing. If there is a question about the source of the lead, the fragment is subjected to a mild acid wash and both the acid wash and the washed fragment are analyzed. Such a procedure allows the reconstruction of the total sample and readily identifies surficial Pb contamination. Such procedures have been used by us previously and were applied in the present study to samples 76535. We first leached plagioclase separates under mild conditions to remove surficial Pb contamination; although we know that even this mild leaching might cause some differential dissolution as discussed in the case of troctolite 76535 mentioned earlier. For the purpose of Pb isotope systematics, however, this differential dissolution is desirable since it might allow the construction of an isotope mixing line for the sample under consideration. The two end components of the mixing line would be the radiogenic Pb produced by *in situ* decay and the Pb

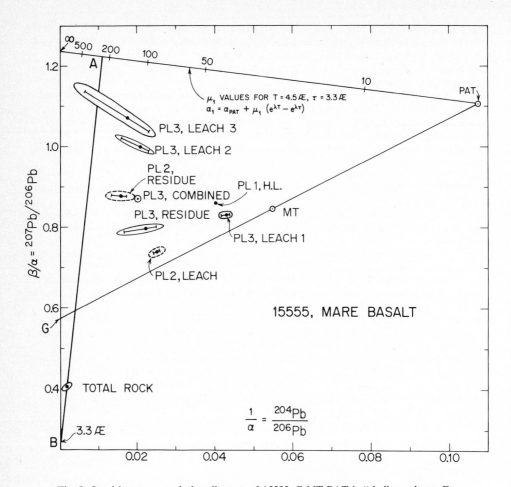

Fig. 8. Lead isotope correlation diagram of 15555. G-MT-PAT is "the" geochron. For visual aid data points and error bars are enclosed in ellipses. Error bars correspond to $+100\%$ increase and -50% decrease in the nominal blank for each experiment. Plagioclase "components" show a wide range of β/α. Line AB is a 3.3 AE reference isochron which could be treated as a mixing line with two end components: Points A and B. PL 1, H. L. is heavily contaminated as seen from its proximity to MT, but it clearly shows the high $^{207}Pb/^{206}Pb$ value for the plagioclase as compared with the total rock. PL 2 was obtained by Frantz and handpicking, and was leached. The leach of the plagioclase (PL 2, Leach) contained a large amount of contamination. PL 3 was obtained by Frantz and heavy liquid separations (PL 3). As can be seen, the first leach contains a large proportion of ^{204}Pb due to easily leached surface contamination. The next two subsequent leaches (PL 3, Leach 2 and PL 3, Leach 3) are far less contaminated but do not fall on the reference isochron. It is clearly seen that leaches show progressive increase in $^{207}Pb/^{206}Pb$ with the succession of leaching. Note that the Pb composition of the third leach (PL 3, Leach 3) is very close to that of the initial Pb as given approximately by A. The remaining residue (PL 3, Residue) contains Pb with low $^{207}Pb/^{206}Pb$ relative to the second and third leaches and shows that a portion of the *in situ* radiogenic Pb is not easily leached. Note that final residue (PL 3, Residue) and first leach (PL 3, Leach 1) are colinear with MT suggesting one component. PL 3, combined (Leach 1 (adjusted)) +

Table 2. Leaching experiment on 15555 plag 3[a] (Sample weight = 225 mg, total Lunar Pb ≃ 10 ng).

	α [b]	Lunar lead[c,d]			
		^{208}Pb	^{207}Pb	^{206}Pb	^{204}Pb
Leach, 1[e,f]	23.184	48.030	22.119	26.649	1.1420
		(13.235)	(7.809)	(9.710)	(0.2164)
Leach, 2[e,g,h]	39.271	5.276	4.117	4.121	0.0862
Leach, 3[e,g,h]	35.051	1.868	1.645	1.535	0.0270
Residue	36.122	4.350	2.539	3.189	0.0711
Total[i]		24.729	16.110	18.555	0.4007

[a]About 30% of plagioclase was obtained by heavy liquid. The rest by handpicking and by Frantz. [b]Uncorrected for blank. [c]Corrected for blank (0.035 picomoles of ^{204}Pb). Blank composition: $\alpha = 18.26$, $\beta = 15.45$, and $\gamma = 37.59$. [d]In picomoles. [e]For 40 min by 10 ml 1 N HNO$_3$, followed by 10 ml 1 N HCl, then 10 ml H$_2$O. [f]At room temperature. [g]At 100°C. [h]About 20% of sample was dissolved. ( ) Adjusted values obtained by increasing the blank correction so that the "adjusted" datum would have the same α as the residue. [i]Adjusted values of leach 1 were used.

inherited by the rock at the time of crystallization (initial Pb). It is obvious that to establish such a mixing line, one must correct for the blank which otherwise would be a third component and thus disturb the linearity of the isotope mixing line. If the data points, when *properly corrected for blank* do not fall on a line, then it would indicate that there was not a unique initial Pb.

The results for leaching experiments on plagioclase from mare basalt 15555 are shown in Table 2 and on a Pb isotope correlation diagram (Fig. 8) (Tera and Wasserburg, 1972c). Line PAT-MT-G is the geochron. It is seen that all leaches and residues (which are the samples remaining after leaching) plot far above the total rock point, indicating the presence of ancient radiogenic initial Pb. A prototype experiment was carried out on plagioclase separated by heavy liquid (PL 1, H.L.). This sample was not leached. While this sample shows excessive contamination as seen from its proximity to MT, it clearly shows the presence of initial Pb. A larger high-purity plagioclase separate was obtained by purely mechanical means (PL 2) and another separate (PL 3) was obtained by mixing an intermediate-purity mechanical separate with 30% of a high-purity heavy liquid separate. One of the major impediments is the contamination introduced during the grinding and mineral separation. For example, one milligram of an impurity with 1 ppm of Pb would add ~10% to the total Pb in the largest sample.

Let us now examine the different leaches. Sample PL 2 was subjected to a single leach (PL 2, Leach) and the remainder (PL 2, Residue) was then analyzed. Inspection of these two data points shows that these steps yield widely divergent ^{207}Pb/^{206}Pb ratios that are apparent in spite of the large contribution of processing blank in PL 2, Leach.

Sample PL 3 was subjected to three leachings after which the remaining material (PL 3, Residue) was then totally dissolved and analyzed. Ten percent of each solution was taken to determine the number of atoms. The first leach was carried out at room temperature and yielded a lead dominated by terrestrial contamination, but with a significant lunar contribution. Leaches 2 and 3 were carried out at

Leach 2 + Leach 3 + Residue) falls close to PL 2. This indicates that bulk plagioclase has Pb with reasonably well-defined isotopic composition. However, the leaching experiment shows that the Pb in the bulk plagioclase is resolvable into a wide spectrum of isotopic compositions. Note that the leached samples do not fall on line AB indicating possible technical problems.

higher temperature and yielded leads with very high $^{207}Pb/^{206}Pb$ ratios of 0.98 and 1.07. These leaches are clearly dominated by lunar leads and show that the plagioclase separates contain lead components which vary by 40% in their $^{207}Pb/^{206}Pb$ ratios and which are a factor of 3 greater than found for the total rock. The PL 3 residue lies on the MT-PL 3 leach 1 line indicating *in situ* decay Pb common to both "components." The total Pb in sample PL 2 (not shown) was found to plot very close to the total Pb in PL 3 (see PL 3 combined) if the adjusted points for the first leach are used (see footnotes of Table 2). In addition, it may be seen that PL 1, H.L. lies on the tie line connecting MT and PL 3 combined. This strongly suggests that the total Pb composition in the plagioclase separate is reasonably well defined, although its precise value is uncertain due to terrestrial Pb contamination. However, the total plagioclase Pb is a mixture of highly variable components. These variations may be a reflection of the presence of very small amounts of mesostasis or other phases rich in U and Th or may possibly be the result of the differential solubility of radiogenic atoms located in different lattice sites from the initial Pb atoms which were incorporated in the feldspar during the original crystallization process.

The results prove the presence of initial radiogenic (and primordial) Pb in a mare basalt as was first postulated by Tatsumoto (1970) to explain the discrepancy between the total rock Pb–U data and the internal Rb–Sr isochrons. It also appears possible to isolate Pb which is very close to the end-member initial Pb (see point A in Fig. 8). At present, we are unable to prove that the initial leads present in a mare basalt are characteristic of the pristine basaltic magma or if there is Pb incorporated by interaction of the magma with the radioactive crust. The difference between these two cases is of great importance since it is the indigeneous initial lead of the magma which gives information on the source regions for lunar basalts. The fact that our leached samples do not lie on the reference isochron AB but appear displaced to the right in the direction of increasing $^{204}Pb/^{206}Pb$ indicates that either we have some technical difficulty with terrestrial contamination or that there is in fact no internal isochron. It should be noted that the precise position of the line AB is uncertain due to the assumed age and the close proximity of the total rock to the y axis.

New analysis of 15555 TR

Our analysis of this rock shows that its α and μ values are much higher than the results of Tatsumoto *et al.* (1972b) indicate. Our uncorrected α values from the composition and the concentration aliquots are in good agreement at 575 and 581, respectively (Table 1). We calculate a corrected α value of 623 for this rock versus 360 reported by Tatsumoto *et al.* (1972b). We also report a μ value of 792 versus a value of ~460 reported by these workers. From this we calculate the blank level of Tatsumoto *et al.* (1972b) for this rock to be ~9 ng Pb/g sample. There is also a disagreement of ~2% on the location of the datum on the U–Pb evolution diagram which could be due to compounded analytical errors from both laboratories.

Basalt 75055

Recently, Nunes *et al.* (1974a) reported that two mare basalts from Apollo 17 are concordant at ~4.44 AE (new decay constants). The results of our analysis on 75055 shown in Table 1 and Fig. 6 confirm the concordancy of 75055 but at the younger age of ~4.42 AE. The source of this disagreement is not clear but might be due to the fact that Nunes *et al.* (1974a) underestimated the level of their blank. This is clearly seen from the lower $^{206}Pb/^{204}Pb$ value reported by those workers. As can be seen from Table 1, our uncorrected α values from both composition and the concentration runs are in good agreement (427 and 421, respectively). Our corrected α value of 463 is much higher than the corrected value of $\alpha \cong 250$ reported by Nunes *et al.* (1974a). Consequently, we calculate a μ value of 461 versus a value of 250 reported by those workers. The blank correction applied by Nunes *et al.* (1974a) is 2.9 ng. To bring their value into agreement with ours, a blank correction of 10 ng must be applied to their sample. It is of course possible that the sample of Nunes *et al.* was contaminated by handling prior to analysis. It is also possible, but unlikely, that 75055 is very heterogeneous. We conclude that the more reliable μ value for this important rock is 461 which is almost in the center of the range of measured μ_2 values (300–1000) for mare basalts (Tatsumoto, 1970; Tatsumoto *et al.*, 1971). Nunes *et al.* (1974a) reported a $\mu = 507$ for another concordant mare basalt (75035) from the same location as 75055. It should be pointed out, however, that for rock 75055, the $^{207}Pb/^{206}Pb$ ratios tabulated by these workers for the composition and concentration aliquots are disparate by $\sim 2\%$ (see Fig. 3). This could affect the parent–daughter ratios and change the exact position of the datum on the evolution diagram. Rock 75035 shows a similar disparity of 1% (see Table 1, Nunes *et al.* (1974a)).

Proceedings of the Fifth Lunar Conference
(Supplement 5, Geochimica et Cosmochimica Acta)
Vol. 2 pp. 1601–1615 (1974)
Printed in the United States of America

Possible effects of ^{39}Ar recoil in ^{40}Ar–^{39}Ar dating

G. TURNER and P. H. CADOGAN

University of Sheffield, Sheffield, England

Abstract—The extent of ^{39}Ar recoil as a result of the ^{39}K$(n, p)^{39}$Ar reaction has been calculated. Near the surface of a K-bearing mineral the ^{39}Ar concentration is shown to be depleted relative to ^{40}Ar. The relative concentration of ^{39}Ar, $C(x)$, at depth x, in this depletion layer can be approximated by the expression, $C(x) = 1 - 0.5 \exp(-x/x_0)$, where $x_0 = 0.082 \mu$m.

The possible effect of this depletion layer on ^{40}Ar–^{39}Ar release patterns is considered with reference to published data. An attempt was made to demonstrate how the argon release pattern from a finely crushed basalt can be modified by recoil. The experimental results were similar to other previously not understood release patterns, although a net loss of ^{39}Ar occurred leading to an overall increase in the apparent age.

Argon measurements on the uncrushed sample of the basalt, 75035, indicate a crystallization age of (3.76 ± 0.05) G.y.

INTRODUCTION

THE DETERMINATION of precise and meaningful ^{40}Ar–^{39}Ar ages has contributed significantly to our understanding of the moon's history. The method has several advantages which make it appropriate for the analysis of small, rare or in-homogeneous samples and, if stepwise degassing is used, for samples with complex histories. In the latter situation the measurement of (^{40}Ar/^{39}Ar) ratios, and hence (^{40}Ar/K) ratios, from lattice sites of differing argon retentivity can, in principle at least, provide information concerning the argon retention history of the sample.

However, the ^{40}Ar–^{39}Ar release patterns appear to be relatively well under-stood only in those cases where the (^{40}Ar/^{39}Ar) ratio is either constant, as in a number of lunar plagioclases (Turner *et al.*, 1972) or shows a monotonic increase to a so-called "plateau" or constant ratio, reflecting in the initial stages of gas release the effects of recent argon loss (Turner, 1970, 1971a).

The simple release patterns described above have, in the case of lunar samples, been confined to mare basalts or separated minerals. Following the early measurements on Apollo 14 samples (Turner *et al.*, 1971) an increasing number of cases have been documented where the release patterns are more complex and the relationship to the samples previous history less clear. Two general types of nonideal release pattern are common. The first type is similar to the ideal cases described above except that in the gas released at high temperatures the (^{40}Ar/^{39}Ar) ratio shows a sharp decrease. Experiments on separated minerals from Apollo 14 basalts have shown that the decrease is in some way associated with the pyroxene in these rocks. A comparison with ^{40}Ar–^{39}Ar ages from separated plagioclase and with Rb–Sr ages indicates that in several cases the "intermediate plateau" (^{40}Ar/^{39}Ar) ratio yields a meaningful age. The second type of complex pattern shows a monotonic decrease in (^{40}Ar/^{39}Ar) ratio over most of the gas release. In most cases of this type, the *total* gas (^{40}Ar/^{39}Ar) ratio appears to correspond to a meaningful age. This type of release pattern appears to be associated with fine-grained or glassy samples and seems to reflect local redistribution of argon without loss. Whether it is the radiogenic ^{40}Ar or the reactor-produced ^{39}Ar which is redistributed is not clear, nor is it clear whether the nonideal release patterns have distinct causes or whether the monotonically decreasing release pattern is merely an extreme form of the pattern which exhibits a sharp decrease at high temperatures. Both types of release have been observed in meteorites (Turner and Cadogan, 1973).

The recoil of ^{39}Ar, produced by the ^{39}K$(n, p)^{39}$Ar reaction is capable of producing small differences (Mitchell, 1968), of the order of 0.1 μm, in the spatial distribution of ^{39}Ar and K which may account for some of the effects described above. It is the purpose of this paper to present calculations of the extent of ^{39}Ar recoil, to consider possible effects on ^{40}Ar–^{39}Ar release patterns and finally to present the preliminary results of an experiment designed to demonstrate how the argon release pattern from a lunar basalt may be modified by recoil.

^{39}Ar Recoil Calculations

The cross section, $\sigma(E_n)$, for the ^{39}K$(n, p)^{39}$Ar reaction has been determined by Bass *et al.* (1961, 1964) in the neutron energy range 1.46–8.7 MeV and at 13.7 MeV. The total reaction cross section, which contains a large number of resonances, has been smoothed to facilitate computation and is plotted in Fig. 1 as a function of incident neutron energy, E_n. The ^{235}U neutron fission spectrum, $\phi(E_n)$, measured in the Herald reactor, AWRE, Aldermaston (Davies, 1965) is also plotted in Fig. 1.

The product $\phi(E_n) \cdot \sigma(E_n)$ is plotted in Fig. 1, in arbitrary units, and indicates the relative proportion of reactions induced by neutrons of energy E_n. It is clear that neutrons in the energy range 1.5–6 MeV dominate the ^{39}Ar production. The effective cross section, $\bar{\sigma}$, for ^{39}Ar production is given by

$$\bar{\sigma} = \frac{\displaystyle\int_0^\infty \phi(E_n)\sigma(E_n)\,dE_n}{\displaystyle\int_0^\infty \phi(E_n)\,dE_n} \tag{1}$$

Numerical integration of this expression indicates $\bar{\sigma} \simeq 70$ mb, which compares favorably with a directly measured value of (76 ± 4) mb (Turner, 1971b).

The energy, E, of the recoiling ^{39}Ar nucleus, measured in the laboratory reference frame, may be calculated from elementary mechanics as,

$$E = \frac{m}{(m + M)}\left\{ Q + 2E_n \cdot \frac{M}{(m + M)}\left[1 - \cos\theta\left(1 + \frac{(m + M)}{M} \cdot \frac{Q}{E_n}\right)^{1/2}\right]\right\} \tag{2}$$

where m is the mass of the neutron (or proton), M is the mass of ^{39}Ar (or ^{39}K), Q

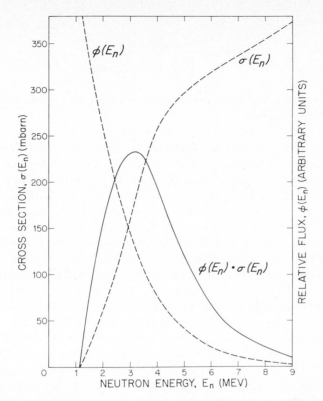

Fig. 1. Cross section, $\sigma(E_n)$, for the reaction ^{39}K$(n, p)^{39}$Ar smoothed and plotted as a function of incident neutron energy, E_n (Bass, 1961). Neutron flux, $\phi(E_n)$ from ^{235}U fission in the Herald reactor (Davies, 1965). The product $\phi(E_n) \cdot \sigma(E_n)$ indicates the relative proportion of reactions induced by neutrons of energy, E_n.

is the energy released by the reaction and θ is the angle between the incident neutron and emerging proton, measured in the center of mass reference frame.

Substituting $m = 1$ and $M = 39$ into expression (2) and considering the case where $Q \ll E_n$ we obtain,

$$E \simeq \frac{E_n}{20}(1 - \cos \theta) \tag{3}$$

indicating that the recoil energy will typically be of the order of $E_n/20$ or around 200 keV.

The proportion of recoils, $N(E)\,dE$, between energy E and $E + dE$ is given by,

$$N(E) = \int_0^\pi n(E_n)\left(\frac{\partial E_n}{\partial E}\right)f(\theta)\,d\theta \tag{4}$$

where $n(E_n)$ expresses the proportion of reactions induced by neutrons of energy

E_n and is given by,

$$n(E_n) = \frac{\phi(E_n)\sigma(E_n)}{\displaystyle\int_0^\infty \phi(E_n)\sigma(E_n)\,dE_n} \tag{5}$$

and $f(\theta)$ expresses the angular distribution of emerging protons relative to the incident neutron direction.

Expression (4) has been integrated numerically using expression (2) to determine E_n and $(\partial E_n/\partial E)$ in terms of E. Because of the dependence of expression (2) on the Q-value of the reaction the calculation made use of the partial cross sections and appropriate Q-values for reactions leading to the ground state $(Q = +0.22\text{ MeV})$ and first three excited states of ^{39}Ar (Bass et al., 1961, 1964) $(Q = -1.05, -1.30, -1.88\text{ MeV})$. The integration (4) has been performed for (a) an isotropic distribution of emergent proton directions, $f(\theta) = (\sin\theta)/2$, and (b) a distribution which is peaked in the forward and backward direction, $f(\theta) = (3\sin\theta\cos^2\theta)/2$. The actual angular distribution has not been determined, although it is likely that departures from isotropy are small (T. Tombrello, private communication). As will be seen at a later stage the effect of departures from isotropy on the spatial distribution of the recoiling ^{39}Ar nuclei is itself small. The calculated distribution of ^{39}Ar recoil energies is shown in Fig. 2.

The recoil energy distribution has been converted to a recoil distance distribution making use of the range-energy measurements of ^{41}Ar in aluminium by Davies et al. (1963). At energies in the 0–500 keV range straggling of the recoiling Ar ions is significant and in order to treat the range-energy data analytically we have attempted to fit it with a lognormal probability distribution using two energy dependent parameters tabulated by Davies et al.; R_m the median depth of penetration and W the straggling width at half maximum. The probability, $P(R)$, of a ^{39}Ar ion recoiling a distance R was then obtained by numerically integrating the expression,

$$P(R) = \int_0^\infty N(E)p(R, E)\,dE \tag{6}$$

where $p(R, E)$ is the probability of an ion of energy E recoiling a distance R. The graph of $P(R)$ against R is superficially similar to Fig. 2 and is not shown. The effect of straggling is to smear out the features of Fig. 2 particularly for the nonisotropic case.

The final stage of the recoil calculations was to determine the concentration of ^{39}Ar as a function of depth, x, below the plane surface of a semi-infinite K-bearing mineral bombarded by an isotropic flux of neutrons. Because the neutron flux is assumed to be isotropic the recoil distribution will also be isotropic regardless of any assumptions about $f(\theta)$. Thus, ^{39}Ar nuclei recoiling a distance R may arrive with equal probability from points on the surface of a sphere of radius R centered on the final position of the ^{39}Ar nucleus. It is a simple matter to deduce from this that the concentration, $c(x, R)$, at depth x below the surface would, if the recoil

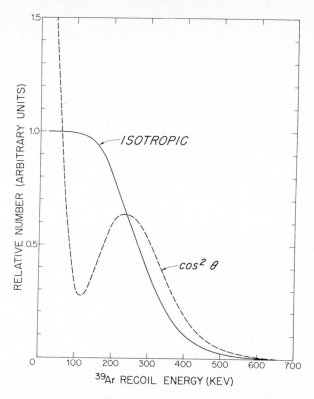

Fig. 2. Energy spectrum of recoiling ^{39}Ar nuclei produced by neutron energy spectrum shown in Fig. 1. The two curves correspond to different assumptions about the angular distribution of the reaction products with respect to the incident neutron direction. The curve labeled isotropic is probably the more realistic one.

range were constant, be given by

$$x \leq R \qquad c(x, R) = \frac{1}{2}\left(1 + \frac{x}{R}\right)$$
$$x > R \qquad c(x, R) = 1 \tag{7}$$

Taking account of the distribution of recoil distances, $P(R)$, the actual concentration, $C(x)$, as a function of x was calculated from the expression,

$$C(x) = \int_0^\infty c(x, R)P(R)\, dR \tag{8}$$

The results of this integration are shown in Fig. 3. The depth, x, is shown in micrometers, the conversion from the area densities presented by Davies *et al.* (1963) to thickness having been made assuming a nominal density of 2.7 g cm^{-3}. If the semi-infinite potassium-bearing mineral were adjacent to a semi-infinite potassium free mineral, of similar density, the distribution, $C'(x)$, of ^{39}Ar in that

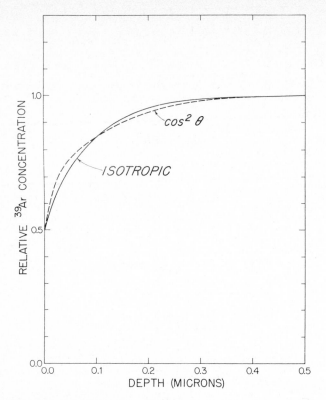

Fig. 3. The relative concentration, $C(x)$, of ^{39}Ar as a function of depth, x, below the plane surface of a semi-infinite K-bearing mineral bombarded by an isotropic flux of neutrons. The two curves correspond to different assumptions about the angular distribution of the reaction products (see text). The ^{39}Ar depletion layer is caused by ^{39}Ar recoiling out of the mineral. The mean depth of the depletion layer is 0.08 μm. The concentration in an adjacent K-poor mineral would be represented by $1 - C(x)$.

mineral would be given by

$$C'(x) = 1 - C(x) \tag{9}$$

It is clear from Fig. 3 that the distributions for isotropic and nonisotropic $f(\theta)$ are not very different. The nonisotropic case produces a higher proportion of low-energy recoils.

The effect of ^{39}Ar recoil from a typical K-bearing mineral is clearly to produce a depletion layer of ^{39}Ar. The mean depth of this layer is calculated to be 0.082 μm. The depletion layer for the isotropic case shown in Fig. 3 can be approximated quite well with an *empirical* expression of the form

$$C(x) = 1 - 0.5 \exp\left(-x/x_0\right) \tag{10}$$

where x_0 is the mean depth of the depletion layer, 0.082 μm (22.2 μg cm^{-2}). This approximate relationship agrees with the calculated distribution to $\frac{1}{2}\%$ or better for all values of x. The corresponding fit to $C'(x)$ (expression (9)) is better than 5% for $x \leqslant 0.2 \, \mu m$, which accounts for 91% of the recoiling ^{39}Ar. For $x = 0.3 \, \mu m$ expression (10) yields a value of $C'(x)$ of 0.0136 some 30% higher than the calculated value. At larger values of x the percentage error in $C'(x)$ increases rapidly but represents a relatively insignificant proportion of the recoiled ^{39}Ar.

^{39}Ar RECOIL AND ^{40}Ar–^{39}Ar RELEASE PATTERNS

Following electron capture in ^{40}K the resulting ^{40}Ar atom recoils with an energy of $\sim 23 \, eV$ and therefore may be expected to change its location in the crystal structure by only a few atomic spacings. In contrast we have seen in the preceding section that the ^{39}Ar produced by neutron activation is expected to recoil a distance of $0.08 \, \mu m$ on average. It is clear, therefore, that regions of variable (^{40}Ar/^{39}Ar) ratio of this width may be produced where the potassium concentration changes significantly.

The magnitude and nature of any recoil effects on the ^{40}Ar–^{39}Ar release pattern will depend on a number of factors. We would anticipate that the following factors would be the most significant:

(a) The fractional volume of K-bearing phases which are within $0.08 \, \mu m$ of a discontinuity in K concentration. This will presumably be a major factor in determining the *magnitude* of recoil effects on the argon release patterns. In rocks where the K-bearing phases are coarse grained we would expect recoil effects to be small or negligible since most ^{39}Ar is expected to recoil to sites equivalent to those originally containing the ^{39}K. In a situation where the K-bearing phase is concentrated in regions with at least one dimension of the order of $1 \, \mu m$, or so, we would expect recoil effects to be significant since much of the ^{39}Ar may recoil into sites of a quite different character to those where the potassium was situated.

(b) The diffusion characteristics of the K-bearing minerals and of adjacent minerals are expected to determine the *nature* of the ^{40}Ar–^{39}Ar release pattern. In the region of discontinuities in K concentration we have shown that there is expected to be an ^{39}Ar depletion layer in the K-bearing mineral and an ^{39}Ar concentration in adjacent K-poor minerals, relative to what would exist in the absence of recoil. Thus the (^{40}Ar/^{39}Ar) ratio will be anomalously high in the depletion layer of the K-rich mineral. For a semi infinite, (i.e. thickness $> 0.5 \, \mu m$) K-rich mineral the surface (^{40}Ar/^{39}Ar) ratio is expected to be enhanced by up to a factor of two. For a K-rich mineral with thickness $< 0.1 \, \mu m$ the (^{40}Ar/^{39}Ar) ratio may be enhanced by a greater factor. In contrast, the (^{40}Ar/^{39}Ar) ratio in the surface regions of adjacent K-poor minerals may be depressed to some value between zero and the nominal (no recoil) value.

We may estimate the value of the diffusion coefficient at which major release of argon will occur from the ^{39}Ar depletion (or ^{39}Ar enhancement) layers. For major argon release we may write

$$2Dt \simeq a^2 \tag{11}$$

where D is the diffusion coefficient, t the heating period and a the width of the ^{39}Ar depletion layer. Substituting $t = 1$ hour (for a typical stepwise heating experiment) and $a = 0.08 \mu$m we may calculate that $D \simeq 2.10^{-14} \, cm^2 \, sec^{-1}$ (assuming that the K discontinuity represents a crystallographic discontinuity from which argon may readily escape). Thus at temperatures where $D \simeq 2.10^{-14} \, cm^2 \, sec^{-1}$ in the K-rich mineral major argon release will occur from the ^{39}Ar depletion layer and the (^{40}Ar/^{39}Ar) ratio should be enhanced. At temperatures where $D \simeq 2.10^{-14} \, cm^2 \, sec^{-1}$ in the adjacent K-poor mineral major argon release will occur from the ^{39}Ar enhancement layer in these minerals and the (^{40}Ar/^{39}Ar) ratio should be depressed.

It is clear then that if the diffusion characteristics of K-rich phase and adjacent K-poor phase are different the spatial variations in (^{40}Ar/^{39}Ar) must be reflected in variations in (^{40}Ar/^{39}Ar) observed in a stepwise heating experiment. There appears to be no fundamental reason to associate any particular temperature range with argon release from K-rich or K-poor minerals. However in most, if not all, lunar rocks and mineral separates so far analyzed by the ^{40}Ar–^{39}Ar technique the measured (K/Ca) ratios of sites releasing argon has been found to decrease with increasing temperature implying that the K-poor minerals are more retentive (lower diffusion coefficients at low temperatures, higher activation energies) than the K-rich minerals. (To some extent this may be a grain size effect and misleading when one is concerned with argon diffusion from a fixed recoil depth.) The extensive literature of mineral diffusion studies also lends support to the association of low retentivity with high potassium content, although the effect of radiation damage on argon diffusion coefficients has so far received only limited attention from experimenters. With suitable reservations then, it would be anticipated from the preceding discussion that the tendency would be for recoil effects to be manifested by an enhancement of (^{40}Ar/^{39}Ar) in the low-temperature release of argon from a polymineralic assemblage and a depression of (^{40}Ar/^{39}Ar) in the high-temperature release. These features would be superimposed on the already established effects associated with ^{40}Ar loss from low retentivity sites.

This section will be concluded with a brief discussion of the possible relationship of observed release patterns to ^{39}Ar recoil.

The monotonically decreasing type of release pattern referred to earlier and exemplified by 14321(F) (Turner *et al.*, 1971) displays a number of features which may be *qualitatively* understood in terms of recoil. They are: argon redistribution without apparent loss, implied by the "meaningful" total K–Ar age; enhanced (^{40}Ar/^{39}Ar) associated with high (K/Ca), and vice versa; the apparent association of these extreme release patterns with very fine-grained samples. While alternative explanations of these release pattern are possible, there appears at the moment to be no clear evidence which conflicts with the recoil hypothesis.

At first sight the patterns with the sharp high-temperature decrease are less obviously the result of ^{39}Ar recoil for while the high-temperature decrease in (^{40}Ar/^{39}Ar), could be the result of ^{39}Ar recoil into pyroxene, there is in general no obvious (^{40}Ar/^{39}Ar) enhancement to compensate. A possible explanation of this would be that the surface regions of the K-rich phase have already lost a substantial proportion of radiogenic ^{40}Ar due to solar heating and the (^{40}Ar/^{39}Ar) enhancement is in consequence suppressed.

It is a simple matter to show that substantial ^{40}Ar loss will have occurred from the recoil region if the *activation energy* for argon diffusion in that region is less than some critical value. For a rock which has spent 20 m.y. on the lunar surface at an effective temperature of 355°K (see Turner (1971a) for details of the calculation) this critical value of the activation energy, ΔE, is $52/(2.82–10^3/T)$ kcal mole^{-1}, where T is the temperature at which the same loss would occur in a time of 1 hour. Thus for $T = 650°C$, $\Delta E \simeq 30$ kcal mole^{-1} and for $T = 1250°C$, $\Delta E \simeq 24$ kcal mole^{-1}. These calculations imply that recoil effects would be obscured due to solar heating on the lunar surface, in minerals with an activation energy for argon diffusion of less than around 30 kcal mole^{-1}. Such a low activation energy is not unreasonable for the K-rich phase of lunar basalts (Turner, 1971).

While evidence for a low-temperature enhancement in the (^{40}Ar/^{39}Ar) ratio is less apparent than for the high-temperature decrease it is not entirely lacking. There are a number of instances of published release patterns which show fine structure with a low-temperature enhancement of the (^{40}Ar/^{39}Ar) ratio which may be the result of recoil. The published results of Jessberger *et al.* (1974) on a plagioclase separate from 65015 show most clearly an initial enhancement of (^{40}Ar/^{39}Ar) associated with high K/Ca followed by a depressed ratio associated possibly with ^{39}Ar which has recoiled into the plagioclase from the K-rich phase.

One final comment we wish to make is in connection with the apparent association of the high-temperature decrease of (^{40}Ar/^{39}Ar) in Apollo 14 basalts with pyroxene (Turner *et al.*, 1972). These authors pointed out that part of the release pattern observed from the pyroxene separates may have been the result of very small amounts of contaminating K-rich phase (quintessence). The K/Ca ratios observed lend support to this statement. If this were the case, it is more than likely that the contamination was very fine-grained interstitial material, having survived the mineral separation procedure. It therefore seems eminently reasonable to explain the high-temperature decrease in (^{40}Ar/^{39}Ar) observed in the pyroxene in terms of ^{39}Ar recoil from quintessence into the (more retentive) pyroxene.

There are clearly a number of observations which can be qualitatively interpreted in terms of ^{39}Ar recoil. However, quantitative evidence is lacking. The recoil calculations make a number of predictions which could be tested by the use of published (^{40}Ar–^{39}Ar) data in conjunction with microprobe measurements of potassium distribution in lunar samples. These predictions are:

(1) (^{40}Ar/^{39}Ar) depressions should be associated with argon release from K-poor minerals, (^{40}Ar/^{39}Ar) enhancements with argon release from fine-grained K-rich minerals.

(2) In rocks or mineral concentrates which are otherwise similar mineralogically the extent of the high-temperature (^{40}Ar/^{39}Ar) decrease should correlate with the amount of K-rich interstitial phase and inversely with the grain size of this phase.

(3) The calculated thickness of the ^{39}Ar depletion layer implies that the parameters of this correlation are calculable.

(4) Rocks with retentive K-rich phases should show an enhancement of the (^{40}Ar/^{39}Ar) ratio at an early stage of the argon release from this phase.

(5) Finely crushed samples, in which there is a greater area of contact between K-rich and K-poor phases, should show larger (^{40}Ar/^{39}Ar) variations than uncrushed samples. The parameters relating the extent of the variations to grain size are in principle calculable.

Experimental Test of Recoil Predictions

An experiment has been performed to test the last prediction of the previous section. An attempt was made to modify the release pattern of a medium-grained Apollo 17 basalt by crushing to a very fine grain size.

The basalt used was 75035 and was chosen principally on the basis of it being the coarsest grained mare basalt in our present allocation. 75035 is a medium grained ferrobasalt containing plagioclase (33%), clinopyroxene (45%), cristobalite (6%), and opaque minerals, especially ilmenite (15%). A representative whole-rock sample was gently crushed to a very fine powder using a stainless steel pestle and mortar. The grain size was estimated from electron microscopy to be in the range 1–10 μm. A 55 mg sample of this powder and a 58 mg fragment of uncrushed rock were irradiated with a high fluence of fast neutrons ($2.10^{19} n$ cm^{-2}), together with terrestrial hornblende monitor, Hb3gr, in an irradiation designated SH31 (Boulder 6 Consortium, 1974). The argon from each sample was subsequently extracted in several heating steps, from 400°C to 1400°C, and analyzed.

The results of the analysis of the uncrushed rock fragment are tabulated in Table 1 and shown in Fig. 4. They are typical of the results obtained previously on mare basalts in that there is evidence, in the low-temperature release, of radiogenic ^{40}Ar loss ($\sim 8\%$). Over much of the argon release (extractions indicated by a double asterisk in Table 1) the (^{40}Ar/^{39}Ar) ratio remains essentially constant and corresponds to an age of (3.76 ± 0.05) G.y.* Above 950°C the last 20% of the release pattern shows some fine structure with significant departures from the "plateau" age, averaging 0.09 G.y. The plateau age is typical of ages obtained on other Apollo 17 mare basalts (Turner *et al.*, 1973; Evensen *et al.*, 1973; Huneke *et al.*, 1973; Kirsten and Horn, 1974; Schaeffer and Husain, 1973; Tatsumoto *et al.*, 1973; Stettler *et al.*, 1973; Tera *et al.*, 1973).

It was anticipated that recoil effects would be minimal in 75035. On account of the relatively coarse grain size we anticipated that only a minor fraction of the K-bearing phase would be within 0.1 μm of an adjacent low-K mineral. This appears to be confirmed by Fig. 4. The small departures from the plateau at high temperatures represent 0.7% of the total ^{39}Ar (or ^{40}Ar) release.

In attempting to crush the second sample of 75035 to a grain size of the order

---

*Uncertainty includes 2σ uncertainty of monitors.

Table 1. Argon release patterns from basalt 75035.

Temp. °C	$\dfrac{36}{38}$	$\dfrac{38}{37}$	$\dfrac{38c^a}{37}$	$\dfrac{39*}{37}$	$\dfrac{40*}{39*}$	$39*^b$	Apparent age (G.y.)^c
75035 Whole rock (58.5 mg) $J = 0.1295$							
530	0.9 ± 0.3	0.021 ± 6	0.020 ± 6	0.421 ± 11	7.7 ± 1.1	1.0	1.31 ± 0.14
580	2.5 ± 1.7	0.005 ± 3	0.003 ± 3	0.449 ± 6	12.75 ± 0.46	2.6	1.84 ± 0.04
630	2.04 ± 0.31	0.0034 ± 5	0.0024 ± 8	0.308 ± 3	26.66 ± 0.34	3.8	2.82 ± 0.02
700	1.03 ± 0.45	0.0022 ± 8	0.0020 ± 10	0.210 ± 2	39.39 ± 0.30	4.5	3.41 ± 0.01
720	0.69 ± 0.14	0.0022 ± 4	0.0022 ± 5	0.144 ± 1	46.14 ± 0.32	5.6	3.66 ± 0.01
780	0.92 ± 0.15	0.0020 ± 2	0.0018 ± 3	0.0987 ± 8	48.94 ± 0.27	6.4	$3.76 \pm 0.01**$
820	0.66 ± 0.04	0.00215 ± 9	0.00215 ± 9	0.0587 ± 4	49.61 ± 0.24	8.1	$3.78 \pm 0.01**$
840	0.61 ± 0.04	0.00207 ± 6	0.00207 ± 6	0.0312 ± 3	49.26 ± 0.20	8.1	$3.77 \pm 0.01**$
900	0.64 ± 0.02	0.00211 ± 4	0.00211 ± 4	0.0282 ± 2	49.13 ± 0.20	8.8	$3.76 \pm 0.01**$
950	0.57 ± 0.06	0.00237 ± 11	0.00237 ± 11	0.0316 ± 3	48.72 ± 0.35	5.3	$3.75 \pm 0.01**$
990	0.51 ± 0.07	0.00320 ± 17	0.00320 ± 17	0.0274 ± 3	47.03 ± 0.45	4.9	3.69 ± 0.02
1030	0.55 ± 0.02	0.00273 ± 5	0.00273 ± 5	0.0104 ± 1	49.23 ± 0.59	3.4	3.77 ± 0.02
1090	0.59 ± 0.01	0.00237 ± 3	0.00237 ± 3	0.00334 ± 5	51.17 ± 0.83	2.2	3.83 ± 0.03
1100	0.58 ± 0.02	0.00238 ± 3	0.00238 ± 3	0.00189 ± 5	54.0 ± 1.8	1.1	3.92 ± 0.05
1200	0.58 ± 0.01	0.00242 ± 4	0.00242 ± 4	0.00183 ± 3	51.8 ± 0.9	1.9	3.85 ± 0.03
1260	0.64 ± 0.01	0.00238 ± 2	0.00238 ± 2	0.00244 ± 7	52.6 ± 2.2	0.9	3.88 ± 0.07
1320	0.85 ± 0.15	0.00214 ± 8	0.00205 ± 10	0.00223 ± 23	47.3 ± 13.2	0.2	3.70 ± 0.45
Total	0.61	0.0024	0.0024	0.016	45.2	68.6	3.63
75035 Finely crushed whole rock (54.8 mg) $J = 0.1295$							
530	0.9 ± 0.4	0.0057 ± 17	0.0054 ± 18	0.0961 ± 26	31.5 ± 1.3	1.2	3.07 ± 0.06
580	1.03 ± 0.19	0.0038 ± 5	0.0035 ± 6	0.0756 ± 10	44.88 ± 0.70	1.2	3.62 ± 0.03
630	1.09 ± 0.18	0.0037 ± 3	0.0033 ± 4	0.0593 ± 10	62.09 ± 1.14	2.6	4.15 ± 0.03
720	1.20 ± 0.05	0.00251 ± 9	0.0022 ± 9	0.0474 ± 5	61.16 ± 0.31	9.3	4.10 ± 0.01
810	0.79 ± 0.03	0.00232 ± 4	0.00225 ± 4	0.0309 ± 2	53.15 ± 0.19	14.9	3.89 ± 0.01
900	0.74 ± 0.04	0.00237 ± 4	0.00233 ± 4	0.0223 ± 2	51.96 ± 0.26	9.9	3.86 ± 0.01
990	0.71 ± 0.03	0.00341 ± 4	0.00336 ± 4	0.0178 ± 2	51.09 ± 0.45	4.7	3.83 ± 0.01
1040	0.65 ± 0.02	0.00263 ± 5	0.00263 ± 5	0.00416 ± 5	47.28 ± 0.77	2.0	3.70 ± 0.03
1120	0.62 ± 0.01	0.00238 ± 2	0.00238 ± 2	0.00069 ± 2	64.8 ± 2.6	0.9	4.22 ± 0.07
1220	0.74 ± 0.06	0.00231 ± 8	0.00226 ± 8	0.00063 ± 8	82.3 ± 14.1	0.1	4.63 ± 0.29
Total	0.73	0.0025	0.0025	0.014	53.6	48.2	3.91

^{a}(^{38}Ar$_c$/^{37}Ar) was calculated assuming that ^{38}Ar originates solely from cosmogenic argon (^{36}Ar/^{38}Ar)$_c$ = 0.65, and trapped argon (^{36}Ar/^{38}Ar)$_t$ = 5.35.

^{b}Amounts in units of 10^{-8} cc STP/g, accurate to $\pm 10\%$.

^{c}Apparent age = 1.885 ln (1 + J(^{40}Ar*/^{39}Ar*)) G.y.

**Ages used to calculate "plateau" age.

of 1 μm, we estimated that by doing this roughly 10% of the K-bearing sites would be brought within 0.1 μm of a grain boundary, i.e. the ^{39}Ar depletion layer would constitute 10% of the K-bearing phase. Since the K-rich phase accounts for only a small proportion of the rock, the minerals brought into contact with the depletion

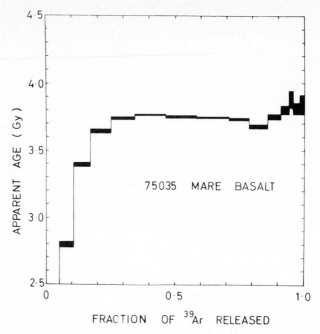

Fig. 4. ^{40}Ar–^{39}Ar release pattern from whole-rock chip of Apollo 17 mare basalt, 75035. The release pattern shows evidence of 8% radiogenic argon loss from low retentivity K-rich sites. Over the major part of the release the ($^{40}Ar/^{39}Ar$) ratio is constant indicating a crystallization age for this sample of (3.76 ± 0.05) G.y.

layer by the crushing were expected to be the major minerals, plagioclase, pyroxene, and ilmenite. These should act as catchers for the recoiled ^{39}Ar. It was anticipated that relative to the uncrushed basalt the low-temperature ($^{40}Ar/^{39}Ar$) ratio would be enhanced as argon is released from the depletion zone and the higher temperature ($^{40}Ar/^{39}Ar$) ratio depressed as recoiled ^{39}Ar is released from the plagioclase and pyroxene. If 10% of the ^{39}Ar had indeed recoiled from high- to low-K sites we expected that the average fluctuation in ($^{40}Ar/^{39}Ar$) would be of this order.

The results obtained from the finely crushed sample are given in Table 1 and displayed graphically in Fig. 5. Superficially it appears that a monotonically decreasing type of release pattern has been generated as anticipated. The ($^{40}Ar/^{39}Ar$) ratio is certainly enhanced in the initial stages of the gas release and the ratio decreases with increasing temperature. The release pattern for the un-crushed sample is included in Fig. 5 for comparison. Detailed comparison of the two release patterns reveals features which were not anticipated.

The most obvious feature is that, contrary to expectation, the ($^{40}Ar/^{39}Ar$) ratio in the crushed sample does not fall significantly *below* that of the uncrushed sample. Thus while ^{39}Ar appears to have recoiled out of the K-rich sites (notice the greatly reduced ($^{39}Ar*/^{37}Ar$) ratios in the low-temperature fractions), it does not appear to have recoiled *into* K poor sites and presumably it escaped into the

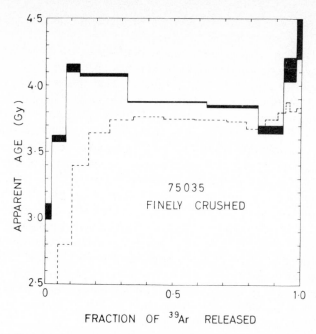

Fig. 5. ^{40}Ar–^{39}Ar release pattern from finely crushed whole-rock chip of 75035. The low temperature (^{40}Ar/^{39}Ar) ratios are enhanced relative to the corresponding ratios from the uncrushed sample (dashed line) in a manner which can be understood in terms of recoil of ^{39}Ar nuclei from the fine-grained K-rich phase during neutron activation.

quartz phial during irradiation. That this must be the case is implied by the observation that the overall (^{40}Ar/^{39}Ar) ratio of the crushed sample is 18% higher than the corresponding ratio for the uncrushed sample. A comparison of absolute concentrations of the other argon isotopes reveals a general depletion in the crushed sample, but with the ^{39}Ar depletion being by far the largest (30%). Relative to the uncrushed samples the percentage change in absolute concentration of the argon isotopes in the crushed sample are (^{36}Ar: ^{37}Ar: ^{38}Ar: ^{39}Ar: ^{40}Ar) = (+2: −17: −13: −30: −17). It is possible that the loss of these other isotopes may be the result of diffusive loss due to heating of the fine grained material in the reactor or possibly during the grinding process itself. This being the case, some diffusive loss of ^{39}Ar would also be expected but since ^{39}Ar is present, on average, for only half the neutron irradiation we would expect loss of ^{39}Ar due to thermal diffusion to be *less* than for ^{40}Ar. This process would, therefore, be expected to lower not raise (^{40}Ar/^{39}Ar).

A search for recoil effects in the (^{38}Ar$_c$/^{37}Ar) ratio reveals that if present they are much smaller than is the case for (^{40}Ar/^{39}Ar) and are difficult to separate unambiguously from the real fluctuations in (^{38}Ar$_c$/^{37}Ar) arising from the sample chemistry. This is at first sight surprising since the ^{37}Ar is produced by the ^{40}Ca$(n, \alpha)^{37}$Ar reaction and since the ^{37}Ar recoils against an α-particle rather than a proton the recoil energy of the ^{37}Ar would be expected to be roughly four times the recoil energy for ^{39}Ar in the ^{39}K$(n, p)^{39}$Ar reaction. Thus, recoil distances of the order of 0.3 μ are expected for ^{37}Ar.

We note also that, in contrast to radiogenic ^{40}Ar, cosmogenic ^{38}Ar is produced by high-energy reactions and it too would be expected to have recoiled even greater distances at the time it was produced. That this does not invalidate the (^{38}Ar–^{37}Ar) method of cosmic ray dating may be explained by the fact that Ca is a major element and does not occur in small localized regions to the extent that potassium does. The fact that ^{37}Ar does not appear to have been lost preferentially in the crushed sample may itself be due to the higher recoil energy.

One final aspect of Fig. 5 requiring some comment is the enhancement of (^{40}Ar/^{39}Ar) in the two final extractions, leading to apparent ages of 4.2 and 4.6 G.y. In seeking to understand these extremely high ages we note that the (^{39}Ar*/^{37}Ar) ratios for these extractions are extremely low (≈ 0.0007) and consequently a 50% correction for neutron induced ^{39}Ar from Ca has been applied to arrive at the tabulated figures. This correction is applied on the assumption of a constant (^{39}Ar/^{37}Ar) ratio for the Ca-derived argon equal to (0.00067 ± 0.00003). It seems likely that this assumption may not be valid in a finely crushed sample due to differential recoil of ^{37}Ar and ^{39}Ar. These two isotopes are both produced by (n, α) reactions but with Q-values differing by 1.4 MeV. For this reason, we are inclined to view the 50% Ca correction with suspicion.

While the results described above do not provide definitive proof of the importance of recoil to the understanding of (^{40}Ar–^{39}Ar) release patterns, they do suggest that this may be a fruitful avenue for future investigation. More definitive experiments are planned making use of (a) (impure) mineral concentrates to provide a well-defined ^{39}Ar catcher mineral and (b) a range of different and more well-defined grain sizes. Some attempt must also be made to understand why in a powdered sample ^{39}Ar does not appear to recoil efficiently into the catcher mineral.

SUMMARY

We have shown theoretically that near the surface of a K-bearing mineral the ^{39}Ar concentration should be depleted relative to ^{40}Ar and enhanced in adjacent K-poor minerals. The relative concentration of ^{39}Ar, $C(x)$, at depth x, in the depletion layer can be approximated by the expression, $C(x) = 1 - 0.5 \exp(-x/x_0)$, where $x_0 = 0.082$ μm.

The existence of (^{40}Ar/^{39}Ar) variations resulting from recoil can account qualitatively for many of the variations in (^{40}Ar/^{39}Ar) ratio observed in stepwise heating of lunar samples but other mechanisms which have been suggested are not precluded.

An attempt to modify, in a predictable way, the release pattern of a medium-grained Apollo 17 basalt by crushing was only partially successful. The experiment duplicated anomalously high (^{40}Ar/^{39}Ar) ratios observed in fine-grained lunar samples but produced an increase in the overall (^{40}Ar/^{39}Ar) ratio due to preferential loss of ^{39}Ar following recoil. The experiment provides some evidence in support of the suggestion that recoil is an important factor in understanding (^{40}Ar–^{39}Ar) release patterns and points the way to a number of more definitive tests.

Acknowledgments—The recoil calculations were made when one of us (GT) was a Visiting Associate at the Lunatic Asylum, California Institute of Technology. Our ideas on the possible relationship of recoil to argon release patterns have benefited greatly from discussions with Professor H. Wänke and J. C. Huneke. The sample analyses were made possible by a grant from the Natural Environment Research Council (GR/3/1715) and support for neutron irradiations came from the Science Research Council (B/RG/5624.8).

REFERENCES

Bass R., Häemi H. P., Bonner T. W., and Gabbard F. (1961) Disintegration of K^{39} by fast neutrons. *Nuclear Physics* **28**, 478–493.

Bass R., Fanger U., and Saleh Fatma M. (1964) Cross sections for the reactions ^{39}K(n, p) ^{39}Ar and ^{39}K(n, α) ^{36}Cl. *Nuclear Physics* **56**, 569–576.

Davies A. R. (1965) Private communication.

Davies J. A., Brown F., and McCargo M. (1963) Range of Xe^{133} and Ar^{41} ions of kiloelectron volt energies in aluminum. *Can. J. Phys.* **41**, 829–843.

Evensen N. M., Rama Murthy V., and Coscio M. R., Jr. (1973) Rb–Sr ages of some mare basalts and the isotopic and trace element systematics in lunar fines. *Proc. Fourth Lunar Sci. Conf., Geochim. Cosmochim. Acta*, Suppl. 4, Vol. 2, pp. 1707–1724. Pergamon.

Huneke J. C., Jessberger E. K., Podosek F. A., and Wasserburg G. J. (1973) ^{40}Ar/^{39}Ar measurements in Apollo 16 and 17 samples and the chronology of metamorphic and volcanic activity in the Taurus Littrow region. *Proc. Fourth Lunar Sci. Conf., Geochim. Cosmochim. Acta*, Suppl. 4, Vol. 2, pp. 1725–1756. Pergamon.

Jessberger E. K., Huneke J. C., and Wasserburg G. J. (1974) Evidence for a 4.5 AE age of plagioclase clasts in a lunar highland breccia. *Nature* **248**, 199–201.

Kirsten T. and Horn P. (1974) ^{39}Ar–^{40}Ar chronology of the Taurus Littrow region II. A 4.28 G.y. old troctolite and ages of basalts and highland breccias (abstract). In *Lunar Science—V*, pp. 419–421. The Lunar Science Institute, Houston.

Mitchell J. G. (1968) Ph.D. thesis, University of Cambridge.

Schaeffer O. A. and Husain L. (1973) Isotope ages of Apollo 17 Lunar material. *EOS, Trans. Amer. Geophys. Union* **54** (6), 614.

Stettler A., Eberhardt P., Geiss J., Grögler N., and Maurer P. (1973) Ar^{39}–Ar^{40} ages and Ar^{37}–Ar^{38} exposure ages of lunar rocks. *Proc. Fourth Lunar Sci. Conf., Geochim. Cosmochim. Acta*, Suppl. 4, Vol. 2, pp. 1865–1888. Pergamon.

Tera F., Papanastassiou D. A., and Wasserburg G. J. (1974) The lunar time scale and a summary of isotopic evidence for a terminal lunar cataclysm (abstract). In *Lunar Science—V*, pp. 792–794. The Lunar Science Institute, Houston.

Tatsumoto M., Nunes P. D., Knight R. J., Hedge C. E., and Unruh D. M. (1973) U–Th–Pb, Rb–Sr and K measurements of two Apollo 17 samples. *EOS, Trans. Amer. Geophys. Union* **54** (6), 614–615.

Turner G. (1970) Argon 40/Argon 39 dating of lunar rock samples. *Proc. Apollo 11 Lunar Sci. Conf., Geochim. Cosmochim. Acta*, Suppl. 1, Vol. 2, pp. 1665–1684. Pergamon.

Turner G. (1971b) Argon 40–Argon 39 dating: The optimization of irradiation parameters. *Earth Planet. Sci. Lett.* **10**, 227–234.

Turner G. (1971a) ^{40}Ar–^{39}Ar ages from the lunar maria. *Earth Planet. Sci. Lett.* **11**, 169–191.

Turner G. and Cadogan P. H. (1973) ^{40}Ar–^{39}Ar Chronology of Chondrites. *Meteoritics* **8**.

Turner G., Huneke J. C., Pososek F. A., and Wasserburg G. J. (1971) ^{40}Ar–^{39}Ar ages and cosmic ray exposure ages of Apollo 14 samples. *Earth Planet. Sci. Lett.* **12**, 19–35.

Turner G., Huneke J. C., Podosek F. A., and Wasserburg G. J. (1972) Ar^{40}–Ar^{39} systematics in rocks and separated minerals from Apollo 14. *Proc. Third Lunar Sci. Conf., Geochim. Cosmochim. Acta*, Suppl. 3, Vol. 2, pp. 1589–1612. MIT Press.

Proceedings of the Fifth Lunar Conference
(Supplement 5, Geochimica et Cosmochimica Acta)
Vol. 2 pp. 1617–1623 (1974)
Printed in the United States of America

A study of ^{204}Pb partition in lunar samples using terrestrial and meteoritic analogues*

R. O. ALLEN, JR.,† S. JOVANOVIC, and G. W. REED, JR.

Chemistry Division, Argonne National Laboratory, Argonne, Illinois 60439

Abstract—^{204}Pb, Bi, Tl, and Zn have been measured in Apollo 16 samples and Apollo 17 orange soil. These data in addition to already reported Apollo 14 and 15 ^{204}Pb results and literature data are used to support the ^{204}Pb-metallic phase coherence. The partition of ^{204}Pb into very fine metallic grains appears to be the result of an impact process rather than metal segregation in a magma.

INTRODUCTION

THE CONCENTRATIONS of primordial Pb as ^{204}Pb and of Bi, Tl, and Zn have been measured in Apollo 16 samples and in Apollo 17 orange soil by neutron activation techniques. In addition, in order to gain some insights into the possible geochemistry of the heavy metals in lunar samples, selected terrestrial and meteoritic samples were studied. These included Disko Island basalt and metal from this basalt; basaltic achondrites; and metallic spherules from Meteor Crater, Arizona.

Apollo 16 soils 64501 and 64801 from Stone Mountain, 63501 from the North Ray Crater ejecta blanket and 61221 from Station 1 on the rim of Plum Crater were measured. Breccias 61016 and 66095 which have unusually high-halogen contents were studied since labile ^{204}Pb and Bi fractions correlated with halogens were observed in Apollo 14 and 15 samples (Allen *et al.*, 1973). Apollo 17 orange soil 74220, which contains large amounts of halogens and no metallic Fe, was also measured.

In this work, as in previous studies, an effort was made to distinguish between surficial and/or readily soluble fractions of the cations studied and the fractions in more retentive sites. To accomplish this the samples were leached in the presence of carriers for ~10 min in a hot weakly acid solution, HNO$_3$ at pH 5–6. We have proposed that the nonleachable ^{204}Pb is associated with a metal-related phase. The data for Apollo 16 samples obtained in this work have been combined with those obtained for Apollo 14 and 15 samples and with literature data in support of this relationship. It should be noted that the phase referred to is *not* related to the coarse >100 μm metal particle which may be meteoritic, but to the submicron metal detected by magnetic resonance techniques. The correlation of labile ^{204}Pb and Bi with Br and the correlation of residual Bi with P$_2$O$_5$ reported by Allen *et al.* (1973) will also be examined for Apollo 16 samples.

*Work sponsored by USAEC and NASA.
†Permanent Address: Dept. of Chemistry, Univ. of Virginia, Charlottesville, Virginia.

RESULTS

^{204}*Pb, Bi, Tl, and Zn in Apollo 16 samples*

The data for ^{204}Pb, Bi, Tl, and Zn are summarized in Table 1. The ^{204}Pb contents of a few ppb in the Apollo 16 soils are similar to those in other lunar soil samples. Bi contents of soils are 1–2 ppb and Tl contents are 4–13 ppb. The Apollo 17 orange soil is distinct in having an order of magnitude more ^{204}Pb and

Table 1. Lead, Bismuth, Thallium, and Zinc in Apollo 16 samples and Apollo 17 orange soil.

Sample*	^{204}Pb r† (ppb)	^{204}Pb l† (ppb)	Bi r (ppb)	Bi l (ppb)	Tl r (ppb)	Tl l (ppb)	Zn r (ppm)	Zn l (ppm)
61221,17	<2.6	<0.65	0.67	0.81	11	2.2	27	1.5
63501,41	3.0 ± 0.3	0.4 ± 0.3	0.7	—	3.8	0.41	11	0.7
64501,18	3.2 ± 0.5	0.5 ± 0.2	0.62	1.1	7.2	1.4	19	1.5
64801,41	1.9 ± 0.9	0.4 ± 0.1	0.93	0.24	5.3	0.68	25	1.0
61016,131	3.0 ± 1.9	1.8 ± 0.9	9.7	2.5	57	2.7	2.9	2.3
66095,23	41 ± 6	4.8 ± 0.9	12	0.6	197	79	18	0.99
74220,111	36 ± 2	3.9 ± 0.4	0.54	0.18	16	6.2	135	5.6

*61016 and 66095 are breccias; all the other samples are soils.
†r = residue after leach; l = leach, HNO$_3$ solution at pH 5–6.

2–4 times more Tl and Zn than any other soil measured. The Bi concentration was erroneously transcribed for the Abstract for the Fifth Lunar Science Conference; it is 0.73 ppb, hence as low as any other soil measured. The leachable fractions of ^{204}Pb and Bi in the Apollo 16 soils are similar to those found for other soils. With regard to the Zn results we note only that except for 61016 the Zn in these samples is less labile than are the heavier metals.

Breccia 66095 exhibits an appreciable enhancement in ^{204}Pb, Bi and Tl over all other samples except for ^{204}Pb in 74220. These samples contain comparable amounts of ^{204}Pb, 41 and 36 ppb, respectively. Breccia 61016 has ^{204}Pb at the few ppb level but its Bi and Tl contents are at least an order of magnitude higher and its Zn is lower than that found in soils. The enrichment or depletion processes to which these elements were subjected must have been very selective.

The relation of Bi$_r$ (r = residual) to P$_2$O$_5$ for samples reported here is not the same as that observed previously. There is an apparent correlation including 74220, but not the two rocks with Bi$_r$/P$_2$O$_5$ ~7.3 in units of ppb/wt.%. The mean Bi$_r$/P$_2$O$_5$ ratio, in the same units, for 6 Apollo 14 and 15 samples was 2.9. Both means have average deviations of 20%. We make no effort here to present literature data that tend to support these two trends since only the toal concentrations of Bi are available. In fact the data of Morgan *et al.* (1974) on Apollo 17 samples indicate a third population with an even lower Bi/P$_2$O$_5$ ratio of 0.55.

The ^{204}Pb$_l$ and Br$_l$ (l = leached) correlation noted for Apollo 14 and

15 samples is extended by the orange soil Pb$_l$ using a Br$_l$ content of 300 ppb measured in the same specific sample. The Apollo 16 soils show no correlation between ^{204}Pb$_l$ and Br$_l$; they have an approximately constant ^{204}Pb$_l$ content of ~0.45 ppb but variable Br$_l$. Breccia 66095 contains a large amount of both leachable ^{204}Pb$_l$ and Br$_l$; breccia 61016 has lower ^{204}Pb$_l$ but comparable Br$_l$, neither fall along the correlation line.

This lack of coherence between the Apollo 16 samples and those from Apollo 14 and 15 lead us in our Fifth Lunar Science Conference Abstract to postulate a lack of homogenization on a regional scale coupled with local alteration to account for the differences. Such conjecture is premature.

The correlation we attempt to explore in the current investigation is that between the residual ^{204}Pb and metal or metal-related phases. The Apollo 16 soils measured scatter about the correlation line. These and the literature data are plotted in Fig. 1. An upper limit is indicated for the literature data since total ^{204}Pb was measured. Usually a 10–50% correction for the labile fraction might be required. Most of these data fall near the correlation line. This is reassuring and the discussion in the next section is concerned with an attempt to understand this trend. It should be noted again though that some samples have appreciable ^{204}Pb but no metal, 74220 has been added to this group.

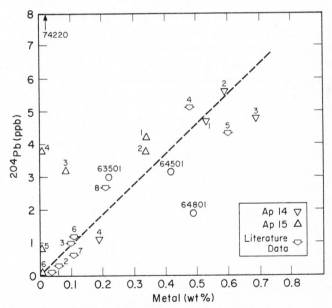

Fig. 1. ^{204}Pb versus metallic content of lunar samples. Our data are for nonleachable ^{204}Pb; Apollo 14, 1–4: 14049, 14163, 14259, 14321; Apollo 15, 1–6: 15031, 15101 (<75 μm), 15101 (>75 μm), 15205, 15418, 15555; literature data are for total ^{204}Pb, 1–8: 12021, 12063, 14310, 15301, 60501, 67701, 68415, 68501, are from: Nunes *et al.* (1973), Silver (1973), Tera *et al.* (1973), and Tatsumoto *et al.* (1972). Metal data are from: Allen *et al.* (1973), Griscom *et al.* (1973), Housley *et al.* (1974), Huffman *et al.* (1974), Nagata *et al.* (1973), Pearce *et al.* (1973), and Tsay *et al.* (1973).

^{204}Pb, Bi, Tl partition in terrestrial and meteoritic samples

The lunar chemistry of ^{204}Pb in Apollo 14 and 15 samples apparently involved extraction of some of this Pb into a metal-related phase and the association of the remainder with halogens. Possible parallels in the geochemistry of lunar samples with non-lunar systems were explored. Comparisons were sought between coexisting silicate- and metal-related phases in terrestrial and meteoritic samples. The metal was separated by repeated cycles of crushing, magnetic separation, and ultrasonic cleaning in absolute alcohol. The final metallic phase was better than 90% free of adhering silicates. No leach data were acquired since leaching was done primarily to remove terrestrial contamination.

Samples which were pulverized silicates or fine metallic grains were leached for 10 min with macro amounts of Pb, Bi, and Tl salts in pH 5–6 HNO$_3$ solution heated to near boiling. It is expected that trace amounts of contaminant metals deposited on surfaces either as volatiles or from aqueous solution (ground water) would be dissolved or exchanged during this treatment. The samples were rinsed with dilute acid, water, alcohol, and ether before dissolution.

The metal in basalt from Disko Island was formed *in situ* by the reaction and equilibration of basaltic magma with carbonaceous shales (Melson and Switzer, 1966). This highly reduced basalt may be compared with lunar samples. The ^{204}Pb and Tl are enriched in the silicate relative to the metal and the Bi is enriched in the metal, Table 2. On the basis of these results ^{204}Pb is not partitioned into Disko basalt metal as it appears to be in metal in lunar samples.

The achondrites may be magmatic differentiates from which a metal phase has been separated. For comparison with lunar samples the heavy metals in two basaltic achondrites were measured. The results, Table 2, suggest that ^{204}Pb is

Table 2. Distribution of Lead, Bismuth, and Thallium in shock melted and magmatic meteoritic and terrestrial samples (in ppb).*

Sample	^{204}Pb	Bi	Tl
Shergotty, eucrite	41 ± 6	—	13
Moore County, eucrite	58 ± 11	6.0	0.4
Canyon Diablo metal from spherules	122 ± 6	4.4	2.9
Canyon Diablo spherules ($\sim$36% metal)	<62†	—	—
Disko basalt metal	6.1 ± 1.9	15	23
Disko basalt silicate	138 ± 30	7.5	1.6
Canyon Diablo metal‡	2.76	—	0.25
	4.33		
Canyon Diablo FeS‡	72.8	40	7.8, 13, 1.5

*All samples were leached at pH 5–6 primarily to remove assumed terrestrial contamination. The leach was discarded.

†A limit only is set because of excessive decay of the sample after the end of the irradiation.

‡The concentrations are from data compiled in Mason (1971).

highly partitioned into the silicate phase if the iron meteorites are representative of the metal phase. This is also true for Bi on the basis of its concentration in iron meteorite troilite and possibly also for Tl.

The impact of iron meteorites causes the ejection of spherules of metallic Fe containing eutectics with Fe, FeS, and Fe$_3$P components. The structures are similar to some found in lunar metal (Blau *et al.*, 1973). We have separated the metal phase from associated oxides and silicates in spherules from Meteor Crater, Arizona. The metal-bearing phase constituted about 36% by weight of the spherule mass and after separation was 90–95% pure. The ^{204}Pb content in the metallic phase is 122 ppb. The spherules themselves contain <62 ppb ^{204}Pb and this limit is consistent with most or all of the ^{204}Pb being in the metallic phase. Probably, the high concentration of ^{204}Pb accommodated in this phase is primarily due to the presence of the eutectic. Troilite from the Canyon Diablo meteorite contains up to ~600 ppb ^{204}Pb (Oversby, 1973), although concentrations of ~100 ppb are more common (Mason, 1971).

A possible lunar ^{204}Pb extraction mechanism

The ^{204}Pb contents of lunar samples are too high to be attributed to Pb in the metal only. The ^{204}Pb contents of iron meteorites (metal + troilite) are too low to account for the ^{204}Pb in lunar samples. A mechanism then has been operative on the moon that concentrates ^{204}Pb in a metal-associated phase. The results discussed above suggest that the process of spherule formation is an efficient way to accomplish this. Most of the metal in lunar samples is endogenous according to a number of investigators. Housley *et al.* (1972) point out that the bulk of the metal fines resulted from *in situ* reduction. It is associated with agglutinates which are the result of impact processes.

A constraint on the mechanism by which the metal is formed is that it must cause the extraction of ^{204}Pb into the metal or metal-related phase at least in basaltic samples. A Canyon Diablo spherule-type process would require that S be involved. Lunar fines contain from 0.05 to 0.1 wt.% S, which if associated with the 0.1–0.5 wt.% metal could provide the troilite and/or eutectic. The amount of S and its isotopic composition varies with the apparent maturity of the soil (Rees and Thode 1972). Whereas the metal can be formed *in situ*, the incremental S must be introduced from an external source. Since most of the ^{204}Pb is associated with the metal, increments of ^{204}Pb with increasing metallic content of the soil must also be supplied from an external source. An amount of ^{204}Pb of ~10^{-6} g/g of metal present in lunar soils is probably too high to be attributed even to meteoritic troilite; therefore another source is necessary. A lunar source would require transporting the ^{204}Pb and probably S to surfaces from which it may become incorporated into splatter material involved in agglutinate formation. Clayton *et al.* (1974) have in fact suggested a vapor transport mechanism to account for isotopically fractionated S in lunar fines. According to McKay *et al.* (1972), agglutinates form in finer grained, more highly comminuted soils. These would supply the surface for deposition of Pb and S. The heating and melting caused by

micrometeorite impacts and required for agglutinate formation could also pro-
mote metal-eutectic formation and Pb extraction. This is illustrated in Fig. 2. Such a
mechanism may account for the behavior of ^{204}Pb in Apollo 14, 15, and 16 soils.
We have already noted, however, that some ^{204}Pb may be incorporated in sites not
related to metal. Both 74220 and 66065 contain much more nonleachable ^{204}Pb
than can be accounted for by the metal contents.

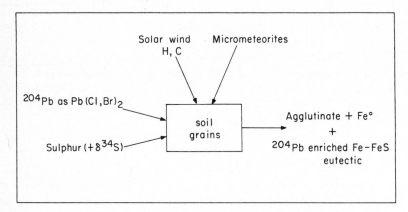

Fig. 2. Model for concentration of ^{204}Pb into metallic phase in lunar soils.

CONCLUSIONS

(1) Apollo 16 soils have ^{204}Pb, Bi, and Tl contents similar to those in soils from
other sites.

(2) Apollo 16 breccias 66095 and 61016 and Apollo 17 orange soil 74220 have
highly fractionated heavy-metal contents. Only in 66095 do very large enrichments of all these heavy metals occur.

(3) In general, Apollo 16 samples do not follow the correlation patterns of
^{204}Pb$_l$–Br$_l$ and Bi$_r$–P$_2$O$_5$ noted for Apollo 14 and 15 samples.

(4) The ^{204}Pb$_r$-metallic phase coherence may be due to the extraction of
surface deposited Pb and S into a metallic phase produced during agglutinate
formation by micrometeorite impact.

Acknowledgments—We thank C. Moore for providing the Canyon Diablo spherules and L. Fuchs for
preparing the spherules and Disko basalt metal for metallographic examination. We acknowledge the
assistance of D. Showalter during the initial stage of this work. We thank the crew of the BNL High Flux
Reactor for irradiations.

REFERENCES

Allen R. O., Jr., Jovanovic S., and Reed G. W., Jr. (1973) Geochemistry of primordial Pb, Bi and Zn in
    Apollo 15 samples. *Proc. Fourth Lunar Sci. Conf., Geochim. Cosmochim. Acta*, Suppl. 4, Vol. 2, pp.
    1169–1175. Pergamon.
Blau P. J., Axon H. J., and Goldstein J. I. (1973) Investigation of the Canyon Diablo metallic spheroids
    and their relationship to the breakup of the Canyon Diablo meteorite. *J. Geophy. Res.* **78**, 363–374.

Clayton R. N., Mayeda T. K., and Hurd J. M. (1974) Loss of O, Si, S and K from the lunar regolith (abstract). In *Lunar Science—V*, pp. 129–131. The Lunar Science Institute, Houston.

Griscom O. L., Friebele E. J., and Marquardt C. L. (1973) Evidence for a ubiquitous, sub-microscopic "magnetite-like" constituent in the lunar soils. *Proc. Fourth Lunar Sci. Conf., Geochim. Cosmochim. Acta*, Suppl. 4, Vol. 3, pp. 2709–2727. Pergamon.

Heiken G. and McKay D. S. (1974) Petrography of Apollo 17 Soils (abstract). In *Lunar Science—V*, pp. 319–321. The Lunar Science Institute, Houston.

Housley R. M., Cirlin E. H., and Grant R. W. (1974) Solar wind and micrometeorite alteration of the lunar regolith (abstract). In *Lunar Science—V*, pp. 360–362. The Lunar Science Institute, Houston.

Housley R. M., Grant R. W., and Abdel-Gawad M. (1972) Study of excess Fe metal in the lunar fines by magnetic separation, Mossbauer spectroscopy, and microscopic examination. *Proc. Third Lunar Sci. Conf., Geochim. Cosmochim. Acta*, Suppl. 3, Vol. 1, pp. 1065–1076. MIT Press.

Huffman G. P., Schwerer F. C., Fisher R. M., and Nagata T. (1974) Iron distributions and metallic-ferrous ratios for Apollo lunar samples: Mössbauer and magnetic analysis (abstract). In *Lunar Science—V*, pp. 372–374. The Lunar Science Institute, Houston.

Mason B. (1971) *Handbook of Elemental Abundances in Meteorite*. Gordon and Breach, New York.

McKay D. S., Harkin G. H., Taylor R. M., Clanton U. S., Morrison D. A., and Ladle G. H. (1972) Apollo 14 soils: Size distribution and particle types. *Proc. Third Lunar Sci. Conf., Geochim. Cosmochim. Acta*, Suppl. 3, Vol. 1, pp. 983–994. MIT Press.

Melson W. G. and Switzer G. (1966) Plagioclase-spinel-graphite xenoliths in metallic iron-bearing basalts, Disko Island, Greenland. *Am. Min.* **51**, 664–676.

Morgan J. W., Ganapathy R., Higuchi H., Krähenbuhl U., and Anders E. (1974) Lunar basins: Tentative characterization of projectiles from meteoritic elements in Apollo 17 boulders (abstract). In *Lunar Science—V*, pp. 526–528. The Lunar Science Institute, Houston.

Nagata T., Fisher R. M., Schwerer F. C., Fuller M. D., and Dunn J. R. (1973) Magnetic properties and natural remanent magnetization of Apollo 15 and 16 lunar materials. *Proc. Fourth Lunar Sci. Conf., Geochim. Cosmochim. Acta*, Suppl. 4, Vol. 3, pp. 3019–3043. Pergamon.

Nunes P. D., Tatsumoto M., Knight R. J., Unruh D. M., and Doe B. R. (1973) U-Th-Pb systematics of some Apollo 16 lunar samples. *Proc. Fourth Lunar Sci. Conf., Geochim. Cosmochim. Acta*, Suppl. 4, Vol. 2, pp. 1797–1822. Pergamon.

Oversby V. M. (1973) Redetermination of lead isotopic composition in Canyon Diablo troilite. *Geochim. Cosmochim. Acta*, **37**, 2693–2645.

Pearce G. W., Gose W. A., and Strangway D. W. (1973) Magnetic studies on Apollo 15 and 16 lunar samples. *Proc. Fourth Lunar Sci. Conf., Geochim. Cosmochim. Acta*, Suppl. 4, Vol. 3, pp. 3045–3076. Pergamon.

Rees C. E. and Thode H. G. (1972) Sulfur concentrations and isotope ratios in lunar samples. *Proc. Third Lunar Sci. Conf., Geochim. Cosmochim. Acta*, Suppl. 3, Vol. 2, pp. 1479–1485. MIT Press.

Silver L. T. (1973) Uranium–thorium–lead isotopic characteristics in some regolith materials from the Descartes region (abstract). In *Lunar Science—IV*, pp. 672–674. The Lunar Science Institute, Houston.

Tatsumoto M., Hedge C. E., Doe B. R., and Unruh D. M. (1972) Uranium–thorium–lead and Rb–Sr measurements on some Apollo 14 lunar samples. *Proc. Third Lunar Sci. Conf., Geochim. Cosmochim. Acta*, Suppl. 3, Vol. 2, pp. 1531–1555. MIT Press.

Tera F., Papanastassiou D. A., and Wasserburg G. J. (1973) A lunar cataclysm at ~3.95 AE and the structure of the lunar crust (abstract). In *Lunar Science—IV*, pp. 723–725. The Lunar Science Institute, Houston.

Tsay F. D., Manatt S. L., Live D. H., and Chan S. I. (1973) Metallic Fe phases in Apollo 16 fines: Their origin and characteristics as revealed by electron spin resonance studies. *Proc. Fourth Lunar Sci. Conf., Geochim. Cosmochim. Acta*, Suppl. 4, Vol. 3, pp. 2751–2761. Pergamon.

Proceedings of the Fifth Lunar Conference
(Supplement 5, Geochimica et Cosmochimica Acta)
Vol. 2 pp. 1625–1643 (1974)
Printed in the United States of America

Volatile and siderophilic trace elements in the soils and rocks of Taurus-Littrow

P. A. Baedecker,* C.-L. Chou, L. L. Sundberg, and J. T. Wasson

Departments of Chemistry and Planetary and Space Science,
Institute of Geophysics and Planetary Physics, University of California,
Los Angeles, California 90024

Abstract—The extralunar component in mature Apollo 17 soils is about 1.8%; mean Ni/Ir, Ge/Ir, and Au/Ir ratios are intermediate between low values observed at the Apollo 11 and 12 mare-type sites and high values at the Apollo 14 and 16 highlands-type sites. Variations in these ratios at different Apollo 17 stations can be related to the fractional contribution of highlands–regolith materials to the extralunar component; trends in the estimated contribution are consistent with those expected from the geography of the Taurus-Littrow site.

The orange-glass content of Apollo 17 soils and breccias can be estimated from Zn, Ga, and Cd concentrations after correction for highlands and mare basalt contributions. The highest orange-glass components in our suite of samples were found in breccia 79135 (34%) and soil 72161 (16%).

Analyses of our suite of elements in shaded soil 76240 and its control 76260 show enhancement in the former of only one element, Cd. We tentatively assign this enhancement to the highly labile character of Cd. Trench soils from Station 8 differ only in terms of a small enhancement of siderophiles in the surficial materials. The subboulder soils 72441 and 72461 show no significant differences in composition and are very similar to the other Station 2 soil 72701. Basaltic rocks 71055 and 79155 have very low concentrations of many trace elements, and are similar in this regard to Apollo 11 low-K basalts.

Introduction

Apollo 17 landed in a lava-filled valley surrounded by highlands. The geochemistry of the Taurus-Littrow landing site is marked by four different types of materials: mare basalts, noritic breccias, anorthositic materials, and orange glass (Apollo 17 PET, 1973). All Apollo 17 soils and soil breccias include components from these materials, from short-lived and long-lived extralunar materials, and from exotic lunar sources such as ancient magmatic volatiles.

We report data on 22 elements determined on 11 soils, one soil breccia, and two crystalline rocks from the Apollo 17 site. Among these 22 elements are four siderophilic (Ni, Ge, Ir, Au), four volatile (Na, Zn, Cd, In), two mafic (Mn, Fe), and eight large-ion lithophilic (Zr, Ce, Eu, Tb, Yb, Hf, Ta, and Th) elements. Together these provide useful diagnostic indicators of the soil components discussed above, and in some cases allow the discrimination of alternate sources for similar geochemical components (e.g. mare- versus highlands-type extralunar materials).

---

*Now at Analytical Laboratories Branch, U.S. Geological Survey, Reston, Virginia 22092.

EXPERIMENTAL

Samples and sample preparation

Sample preparation, packaging, flux monitor preparations, and irradiation procedures for radiochemical neutron activation analysis (RNAA) method have been described in reports of our previous work (Wasson and Baedecker, 1970; Baedecker *et al.*, 1971, 1972, 1973). Samples for instrumental neutron activation analysis (INAA) were powders, packaged for irradiation in 2.5 dram polyethylene vials. Rock samples were prepared by grinding 0.2–0.5 g chips in a steel percussion mortar. Flux monitors for INAA were samples of the USGS standard rock W-1. Since our data-reduction program (Baedecker, 1974) could not always find Zr in the W-1 spectra, we have estimated Zr monitor activities to maximize agreement with previously published results while allowing for differences in neutron fluence between different irradiations as estimated by other flux monitors.

Analytical and radiometric procedures

We determine eight elements by RNAA: Ni, Zn, Ga, Ge, Cd, In, Ir, and Au. Analytical and radiometric procedures for these elements were described in Baedecker *et al.* (1973) and our previous reports cited there.

INAA irradiations were carried out in the UCLA reactor at a flux of $2 \times 10^{12} \, \mathrm{cm^{-2} \, sec^{-1}}$. A 10 min irradiation was used for the determination of Mn and Na. The samples were irradiated for 3 hour prior to measuring the longer lived activities. All samples were counted on a Nuclear Diodes Ge(Li) detector having 11% efficiency relative to a 3×3 in. NaI detector, 1.9 keV resolution for the 1333 keV ^{60}Co photopeak, and a peak-to-Compton ratio of 40:1. Spectral data were recorded in a 4096-channel analyzer and processed using our SPECTRA computer program (Baedecker, 1974).

Precision and accuracy

We estimate 95% confidence limits on the mean of two determinations to be Ni, ±6%; Zn, ±5%; Ga, ±4%; Ge, ±4%; Cd, ±9%; In, ±4%; based on replicate analyses of BCR-1; these are essentially as estimated in Baedecker *et al.* (1973). Our data for Ir and Au in BCR-1 (Wasson *et al.*, 1973) show considerable scatter. Based on the observed agreement of replicates for lunar soils, we estimate 95% confidence limits on our means to be about ±10% for Ir and ±15% for Au.

We have assessed the precision of our INAA procedure by the analysis of 20 aliquots of a single powdered obsidian sample. Estimates of the standard deviation of a single analysis for 10 elements and their observed concentrations are listed in Table 1. For comparison, standard deviations based on Poisson-distributed counting statistics alone are also presented. In these samples, sample inhomogeneity seems to contribute very little (perhaps 2%) to the scatter of the data. In lunar samples, inhomogeneities play a greater role, and we suggest that relative standard deviations from sampling would be roughly 5%, and that the larger of 5% or the "observed" values tabulated in Table 1 are

Table 1. Estimates of precision for the determination of 10 elements by INAA based on analyses of 20 aliquots of obsidian powder.

Element		Sc	Cr	Fe	Ce	Eu	Tb	Yb	Hf	Ta	Th
Concentration (μg/g)		1.8	6.1	0.8*	90	0.24	1.14	5.1	6.5	3.2	22
Observed	σ (%)	1.7	8	1.6	1.8	5	7	3	6	4	1.5
Poisson	σ (%)	0.7	10	1.1	0.5	8	7	3	2	3	5

*Fe concentration in %.

reasonable precision estimates for lunar samples. We have not adequately tested the precision of our INAA method for Na, Mn, Cr, and Zr. However, estimates of 1 σ based on counting statistics are: Na, ±2%; Mn, ±1%; Co, ±1%; Zr, ±8%, and for the first three it is probable that errors resulting from sample inhomogeneities are greater than those from the INAA procedure itself.

In order to test the accuracy of our INAA procedure we present a comparison of some of our data with that of other analysts in Table 2. The agreement is generally satisfactory, although there is an indication that our Mn data are systematically high by 6–10%. We find no clear evidence for systematic errors for the remaining elements.

RESULTS

Our RNAA data for 8 elements are listed in Table 3. Duplicate determinations were made on each sample except the two crystalline rocks, for which the duplicate runs have not as yet been carried out.

Table 2. A comparison of our INAA data for 14 elements with those determined by other analysts (units of $\mu g/g$ except Fe and Na, %).

Sample	Ref.	Na	Sc	Cr	Mn	Fe	Co	Zr	Ce	Eu	Tb	Yb	Hf	Ta	Th
10084		0.34	61	1950	1700	12.5	32		48	2.0	3.2	8.9	9.8	1.35	3.2
	a	0.32		1850	1600	12.35	34		58						
	b								47.7	1.70		11.5			
	c	0.323	59.5	1920	1600	11.3	20.5	290	51.6	1.78	2.6	10.9	10.5	1.3	
	d		61.7	1824	1530	12.2	26.8		47	1.75	3.1	11.3	7.6		
	e	0.33		1900	1600	12.3									
	f								46.1	1.77		10.6			
	g	0.315	61	1830	1560	12.0			47	1.67	2.8	8.3	10.2	1.3	1.6
	h	0.317		1900	1630	12.1			47	2.0	3.0	10.8			2.6
65015		0.46	17.0	1270	980	6.9	30	851	180	2.3	6.1	16.1	22	0.14	12
	j	0.41			6.67			909							
	k	0.42	14.9		850	6.60			153	2.12	5.2	19.1	19.3		
	l	0.33			930	6.14			125	1.91		15.3			
	q							970							
66041		0.32	10.4	760	620	4.7	38	187	39	1.3	1.6	4.3	4.8	0.52	2.1
	m	0.336	9.8	766	560	4.6	32	155	36	1.23	1.3	4.6	4.6	0.62	2.6
	n	0.30		820		4.5									
72701		0.39	19	1430	1060	6.9	32	243	41	1.3	1.7	4.4	7.1	0.72	3.4
	o	0.323	17.3	1370	850	6.7	28.3	256	47	1.4	1.7	5.9	5.8	0.70	2.4
	q	0.33		1570	900	6.8	30	264							
	s	0.36		1200	900	7.1									
75081		0.31	67	3000	2000	14.1	33	251	23	1.8	2.2	6.4	8.3	1.4	<0.60
	i	0.16	65	3090	930	12.8	30.6		23	1.2	1.97	9.0	6.9	1.2	0.3
	m	0.272	60	2740	1700	13.8	29		32	1.8	1.9	7.1	7.5	1.3	0.7
	r	0.27		3300	1820	13.8	~27	235							
	s	0.31			1720	13.2									

References:
a—Compston et al. (1970), 1007;
b—Gast et al. (1970), 1143;
c—Goles et al. (1970), 1177;
d—Haskin et al. (1970), 1213;
e—Maxwell et al. (1970), 1369;
f—Philpotts and Schnetzler (1970), 1471.
g—Wänke et al. (1970), 931;
h—Wakita et al. (1970), 1685;
i—Brunfelt et al. (1973), 1209;
j—Duncan et al. (1973), 1097;
k—Haskin et al. (1973), 1275;
l—Hubbard et al. (1973), 1297;
m—Laul and Schmitt (1973), 1349;
n—Rose et al. (1970), 1149;
o—Wänke et al. (1973), 1461;
p—Chyi and Ehmann (1974), 118;
q—Duncan et al. (1974), 184;
r—Ehmann et al. (1974), 203;
s—Scoon (1974), 690.
1970—Proc. Apollo 11 Lunar Sci. Conf.; 1973—Proc. Fourth Lunar Sci. Conf.; 1974—Lunar Science—V.

Table 3. Replicate and mean concentrations of eight trace elements in Apollo 17 samples. Italicized values considered less accurate and given one-half weight in the determination of means. Values in parentheses are of poor quality and disregarded in the determination of means.

Sample (description)	Station	Ni (μg/g) Repl.	Mean	Zn (μg/g) Repl.	Mean	Ga (μg/g) Repl.	Mean	Ge (ng/g) Repl.	Mean	Cd (ng/g) Repl.	Mean	In (ng/g) Repl.	Mean	Ir (ng/g) Repl.	Mean	Au (ng/g) Repl.	Mean
*Rocks**																	
7055,31 (basalt)	1A	2.0		1.9		4.27		3.3		≤1.5		4.7		(1.1)		0.082	
79155,34 (gabbro)	9	2.7		1.9		4.34		2.0		≤6.5		0.226		0.069		0.26	
Breccia																	
79135,39	9	166, 155	161	87, 108	98	8.60, 8.53	8.57	281, 290	286	93, 130	112	6.6, 7.2	6.9	6.4, 5.1	5.8	(6.7), 2.6	2.6
Soils																	
72161,19	LRV-3	290, 282	286	56.1, 50.0	53.0	6.57, 6.37	6.47	452, 441	446	58	58	3.8, 5.4	4.6	10.8, 10.6	10.7	3.7, (5.8)	3.7
72441,13 (under 0.7 m boulder)	2	288, 278	283	19.3, 18.8	19.0	4.54, 4.91	4.72	471, 537	493	36, 39	38	3.1, 4.1	3.6	8.8, 6.9	7.8	3.7, 3.8	3.8
72461,9 (under 0.7 m boulder)	2	290, 313	302	19.2, 21.1	20.2	4.76, 4.63	4.70	415, 447	431	35, 55	42	2.7, 3.0	2.9	9.6, (15)	9.6	3.6, 4.1	3.8
72701,39	2	349, 285	317	19.3, 17.4	18.4	4.47, 4.43	4.45	514, (142)	514	36, 36	36	2.1, 0.8	1.4	11.6, 9.5	10.6	4.4, 3.5	4.0
75081,20	5	130, 120	125	25.6, 26.5	26.0	4.93, 5.09	5.01	210, 204	207	29, 35	32	2.1, 2.0	2.0	4.4, 5.6	5.0	1.6, 1.1	1.4
76240,19 (shadowed)	6	242, 188	215	24.8, 24.6	24.7	5.75, 4.89	5.32	560, 370	465	76, 67	72	2.5, 2.4	2.5	7.9, 6.4	7.2	3.8, 2.7	3.3
76260,8 (control)	6	191, 197	194	23.1, 23.9	23.5	5.29, 5.13	5.21	(440), 408	408	36, 37	37	3.7, 4.2	3.9	7.0, 7.2	7.1	3.0, 4.3	3.6
Trench soils (depth)																	
78421,35 (15–25 cm)	8	251, 258	254	29.4, 29.0	29.2	4.99, 4.91	4.95	497, 404	442	41, 50	46	2.7, 3.0	2.8	9.7, 9.7	9.7	3.3, 3.6	3.4
78441,15 (ca. 10 cm)	8	306, 274	290	28.6, 28.9	28.8	5.25, 4.94	5.10	534, 400	467	35, 42	38	2.4, 2.7	2.5	9.7, 10.1	9.9	2.8, 3.8	3.5
78461,15 (ca. 5 cm)	8	246, 265	256	27.7, 28.4	28.0	4.94, 4.98	4.96	463, 428	446	44, 36	40	2.7, 2.8	2.8	8.7, 9.4	9.0	3.3, 3.3	3.3
78481,32 (0–1 cm)	8	262, 440†	262	27.6, 30.0	28.8	5.24, 5.14	5.19	509, 660†	509	43, 36	39	4.5, 4.4	4.5	10.4, 14.2†	10.4	3.6, 7.5†	3.6

*Replicate analyses of rocks in progress.

†Aliquot of sample showing enrichment of siderophiles. Values not included in means.

Table 4. Concentrations of 14 elements determined by instrumental neutron activation analysis in 3 rock and 11 soil samples from the Apollo 17 site (μg/g except Na and Fe, %).

Sample	Type	Station	Na	Sc	Cr	Mn	Fe	Co	Zr	Ce	Eu	Tb	Yb	Hf	Ta	Th
71055	Basalt	1a	0.31	94	2800	2100	16.4	26	—	23	1.5	2.1	5.7	7.0	1.3	<0.65
79155	Gabbro	9	0.29	86	3600	2100	15.9	24	255	23	1.6	2.3	6.8	8.5	1.1	<0.64
79135	Breccia	9	0.38	43	2700	1580	12.5	41	178	36	1.8	1.7	4.8	6.2	0.79	<0.46
72161	Soil	LRV-3	0.34	41	2700	1700	11.8	44	—	27	1.5	1.9	3.7	5.8	0.83	1.5
72441	Soil	2	0.40	22	1640	1050	7.8	37	256	47	1.6	1.8	5.5	7.5	0.92	3.7
72461	Soil	2	0.40	21	1530	1010	7.5	33	251	45	1.6	3.1	5.0	7.5	0.86	3.7
72701	Soil	2	0.39	19	1430	1060	6.9	32	243	41	1.3	1.7	4.4	7.1	0.72	3.4
75081	Soil	5	0.31	67	3000	2000	14.1	33	251	23	1.8	2.2	6.4	8.3	1.4	<0.60
76240	Soil	6	0.36	31	1980	1210	9.6	37	210	34	1.6	1.8	4.8	6.7	0.88	2.7
76260	Soil	6	0.33	30	1840	1160	8.7	32	—	1.5	—	—	5.6	—	—
78421	Soil	8	0.32	37	2100	1420	9.8	36	152	25	1.4	1.6	5.4	5.1	0.79	1.3
78441	Soil	8	0.32	35	2100	1460	9.6	35	—	22	1.4	1.5	4.0	5.1	0.57	2.1
78461	Soil	8	0.31	36	2100	1420	9.8	37	165	29	1.3	1.6	5.0	5.0	0.74	1.3
78481	Soil	8	0.32	36	2100	1390	9.8	38	—	24	1.4	1.7	5.3	5.6	0.68	1.6

Our INAA data for 14 elements are listed in Table 4. Only one determination was made on each sample. Data on Zr are missing in several instances because our computer program failed to find a peak of the proper energy. Some of these may still be recovered by hand reduction of the data. Several elements were not determined in 76260 because our sample was needed for RNAA study before the nuclides with long half-lives could be counted.

Extralunar Component in Soils and Breccias

Concentration of the extralunar component

The siderophilic elements found in lunar soils are mainly of extralunar origin, as can be seen by comparison of their concentrations in mature soils with those in major lunar rock types (Ganapathy et al., 1970; Baedecker et al., 1973). In our work we have relied exclusively on Ni, Ge, Ir, and Au to estimate the amount of extralunar materials in lunar samples. The ratios of these elements to each other are moderately variable in the primitive meteorite groups (such as the chondrites) which are considered to be representative of the most abundant types of solid materials falling onto the moon (the solar wind is only a weak source of siderophiles) The most primitive (i.e. most sunlike in composition) meteorites are CI chondrites; there is some evidence that they are the most abundant meteorites in earth-crossing orbits (see, for example, Wasson, 1974). Although the siderophilic element ratios in mare soils superficially resemble those in CI chondrites, they differ in detail. In order to estimate the amount of extralunar materials in lunar soils one needs to divide the net siderophile concentrations (after subtraction of the indigenous contribution) by the mean concentrations in the extralunar materials. Baedecker et al. (1972) estimated the latter for the mare soils to be 1.35% Ni, 28.3 μg/g Ge, 634 ng/g Ir, and 184 ng/g Au by normalizing the average concentrations found in soils from Apollo 11, 12, and 15 missions to the Ir concentration of about 634 ng/g observed in water-free CI chondrites. We now estimate higher net Ni values in these soils, and, as a result, must revise the Ni concentration up to 1.45%.

Mean concentrations of the four siderophiles in mature (based on high-siderophile concentrations) Apollo 17 soils (72161, 72441, 72461, 72701, 78421, 78441, 78461, 78481) are given in Table 5. After subtracting assumed indigenous components (Table 5) we estimate mean extralunar components (ELC) in Apollo 17 mature soils to be 1.87 ± 0.06, 1.60 ± 0.05, 1.51 ± 0.07, and $1.94 \pm 0.05\%$, from the concentrations of Ni, Ge, Ir, and Au, respectively. Weighting these by their inverse variance gives a grand mean ELC of 1.76%. If we followed the practice of Ganapathy et al. (1970) and calculated the ELC concentration based on a hydration state similar to that of CI chondrites at the earth's surface, the ELC value would be 25% larger, 2.20%.

Figure 1 shows the ELC abundance ratios of Ni, Ge, Ir, and Au in mature soils from the six Apollo landing sites. Mean abundance ratios are obtained by weighting the elements by their inverse variances. In order to reduce the effect of systematic interlaboratory errors we have relied exclusively on our data where available. The exceptions are 12070 Ni (Compston et al., 1971; Cuttitta et al., 1971; Taylor et al., 1971; Willis et al., 1971) and Au (Laul et al., 1971) and 14259 Au (Morgan et al., 1972). The mean concentrations are low at the mare sites (11, 12), intermediate at the mare–highlands interface sites (15, 17) and high at the highlands sites (14, 16). The high concentrations in the latter provide a record of early, intense bombardment at these old sites (cf. Baedecker et al., 1972). Because of more extensive sample handling operations, small sizes, and the absence of data on certain siderophiles, the ELC abundance ratios in the Luna 16 and 20 soils are not well defined, and for this reason not included in Fig. 1. However, the published data (Laul et al., 1972; Morgan et al., 1973) can be interpreted in terms of abundance ratios of $1.4 \pm 0.2\%$ and $1.6 \pm 0.2\%$, respectively.

The Apollo 17 ELC of 1.76% is about 20% higher than the value of 1.33% estimated for the Apollo 11 site. Radiogenic ages of mare basalts from the two sites are very similar. Most $^{39}Ar-^{40}Ar$ ages of Apollo 17 basalts fall in the range 3.7–3.8 G.y. (Huneke et al., 1973; Stettler et al., 1973); those of Apollo 11 basalts

Table 5. Mean siderophilic element concentrations in mature soils and in lunar source rocks at the Apollo landing sites. Data are from our papers except for Ni at Apollo 12 and Au at Apollo 12 and 14 sites, where literature values were used.

Apollo site	Ni (μg/g)		Ge (ng/g)		Ir (ng/g)		Au (ng/g)	
	soil	rock	soil	rock	soil	rock	soil	rock
11	214	10	385	10	9.0	0.1	2.1	0.1
12	198	35	355	15	8.3	0.1	2.2	0.1
14	414	40	750	20	12.5	0.1	6.6	0.2
15	266	40	437	20	8.5	0.1	3.3	0.2
16	516	30	1120	20	16.2	0.1	8.8	0.2
17	281	10	462	5	9.7	0.1	3.7	0.1

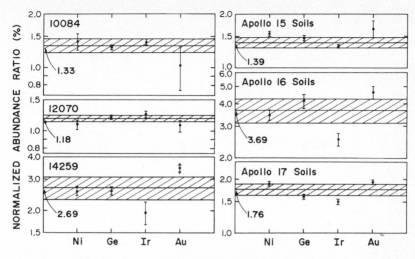

Fig. 1. Concentration of extralunar component in lunar soils obtained by dividing net soil contents of Ni, Ge, Ir, and Au by normalized concentrations of the ELC in mare-type soils of 1.45%, 28.3 µg/g, 634 ng/g, and 184 ng/g, respectively. The error bars give 70% confidence limits on values for each element, the shaded bands give 70% confidence limits on mean ELC concentrations.

fall in two groups; 3.6 G.y. (high·K), and 3.8 G.y. (low-K) (Turner, 1971; Stettler *et al.*, 1973; Turner *et al.*, 1973). Perhaps the mean regolith age is slightly higher at the Apollo 17 site, in which case this accounts for part of the difference in ELC. The difference appears to be too great to be attributable entirely to differences in age, however, and it seems likely that an appreciable contamination by the regolith from the nearby highlands has also occurred. Compositional evidence of this contamination will be discussed in the following subsection.

Compositional variation of the extralunar component

Morgan *et al.* (1972) pointed out the existence of significant variations in the Ir/Au ratio in lunar soils and breccias. Baedecker *et al.* (1973) showed that their data showed remarkable systematic trends in siderophilic element ratios, with the ratios of Ni, Ge, and Au to Ir at each lunar landing site increasing as a function of the mean extralunar component at that site. The addition of Apollo 17 data from this paper confirms the general trend found by Baedecker *et al.* (1973), though the ratios no longer show a strict monotonic increase.

The simplest interpretation of the trends in the ratios is that there have existed two distinct populations of extralunar materials—one associated with the early, intense bombardment, the other with the more recent, relatively constant influx (Baedecker *et al.*, 1972, 1973). If only two populations were involved, linear plots of siderophilic element/Ir ratios against each other should give straight lines indicative of the mixing process. If this model is correct it is not the mean ELC which controls the ratios, but rather the relative contributions of the two

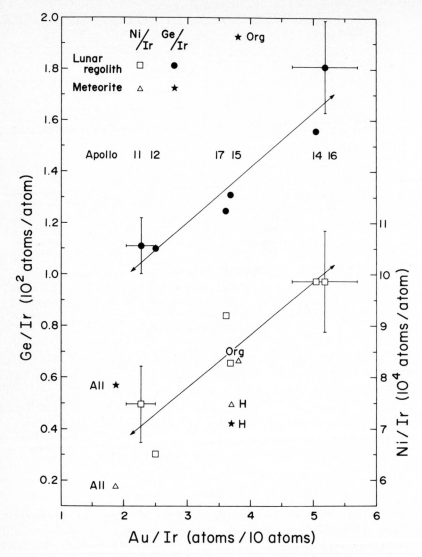

Fig. 2. Variation of siderophilic element ratios of short-lived and long-lived extralunar components, and meteorites. The highland regoliths (Apollo 14 and 16) are dominated by the short-lived extralunar population, characterized by higher Ni/Ir, Ge/Ir, and Au/Ir ratios, the mare regoliths by the long-lived population, characterized by lower ratios. Apollo 15 and 17 regoliths have intermediate ratios as results of mixing of highland and mare regoliths. Mixing lines are indicated.

extralunar populations, which need not be strictly dependent on the magnitude of the ELC.

In Fig. 2 mean soil Ge/Ir and Ni/Ir ratios are plotted against Au/Ir for the six Apollo sites. All data are from our research group with the exception of 12070 Ni and Au and 14259 Au values. Also shown are the loci of our unpublished data for the Orgueil (CI1), Allende (CV3), and mean H-group chondritic meteorites. Representative error bars (amounting to 10%, roughly 95% confidence limits) are shown on points at the extremes of each distribution.

On the Ni/Ir versus Au/Ir diagram the soil data give a relatively linear array, and the meteorites fall within experimental error of the same least-squares line. Thus, the data are consistent with the two-population hypothesis.

The relationship between Ge/Ir and Au/Ir is not obviously linear, but a least-squares line passes within the error limits on each soil point. The chondrite data fall well away from this line, and are themselves not linearly distributed.

One can draw several conclusions from Fig. 2 and considerations regarding the treatment of the data: (1) That diverse chondrite types fall near the Ni/Ir versus Au/Ir line indicates that there is no significant lunar source of the high Au concentrations found in highlands soils. Oxidized forms of Au and Ni are not geochemically coherent, and it is very unlikely that Au enrichment by processes such as hydrothermal transport (Hughes et al., 1973) could have supplied an appreciable fraction of the Au found in the highlands regoliths without destroying the chondrite-related linear array. (2) The Ge/Ir versus Au/Ir plot could be interpreted either in terms of two or multiple components (e.g. in addition to two extralunar sources there might be a local Ge-rich source which contributed 10% of the Ge at the Apollo 16 site, but negligible amounts of Ni, Ir, and Au). In fact no Apollo 16 material with Ge concentration greater than 500 ng/g (half the mean soil concentration) which has low relative contents of Ni, Au, and Ir has been discovered (Krähenbühl et al., 1973; Ganapathy et al., 1974; Wasson et al., 1974). For this reason we tentatively conclude that the two-component interpretation is more appropriate.

In order to simplify the following discussion of Fig. 2 we have marked with arrows points which we hold to be lower limits on the element/Ir ratios in the short-lived extralunar population and upper limits on the long-lived population. These were calculated by tentatively accepting the lines as two-component mixing lines, and assuming that the short-lived population contributed $\leqslant 95\%$ of the extralunar material at the Apollo 16 site, and that the long-lived population contributed $\leqslant 95\%$ of the extralunar material at the Apollo 11 site. (3) It appears that Orgueil CI-type material has not contributed an appreciable fraction of the early, short-lived population of extralunar materials. The less likely alternative is that these materials consisted of mixtures of CI material with another substance having an extremely high Au/Ir ratio ($\geqslant 0.8$). (4) The Ge/Ir ratios in the long-lived population are about a factor of two less than those in Orgueil, (or in other CI chondrites—see Krähenbühl et al., 1973) and, in fact, are more like those in Allende. Note that this is in conflict with the statement regarding this material by Anders et al. (1973) that "the dominant component has CI composition." The difference cannot be attributed to lunar sources of Ge. If we overestimated the Ge contribution of local rocks the extralunar Ge/Ir ratio in the Apollo 11 and 12 soils could at most be raised by 5%.

It would be very valuable if we could use the relationships illustrated in Fig. 2 to determine the relative contributions of the two extralunar populations to individual soils. We have attempted this by plotting all our Apollo 17 soil data on the same plot, but the scatter between points is large, and the relationships between the soils obscured. However, upon averaging the soil data for individual stations some relationships emerge (Fig. 3). Bars showing the total range of the analyses are attached where data from several soils are averaged. The points on the Ge/Ir versus Au/Ir diagram fall near the mixing line defined in Fig. 2. Those on

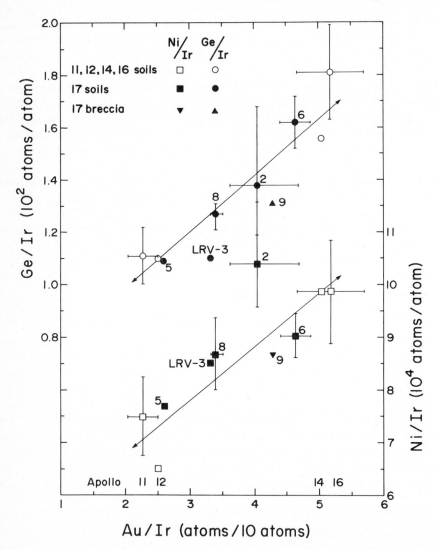

Fig. 3. Variation of Ni/Ir, Ge/Ir, and Au/Ir ratios among Apollo 17 sample stations. Mean ratios of each station are more or less along the mixing lines of short-lived and long-lived extralunar populations (see Fig. 1). The amount of highlands–regolith in soils decreases in the order: North Massif (Station 6), South Massif (Station 2), Sculptured Hills (Station 8), and Dark Mantle (Stations LRV-3 and 5).

the Ni/Ir versus Au/Ir plot fall near the line with the exception of those for Station 2. It appears possible to order the stations in order of decreasing contribution of the short-lived extralunar materials (i.e. of highlands regolith) as follows: 6, 2, 8, LRV-3, and 5. This is as expected from the local topography; the highest concentrations are found near North and South Massifs; the Sculptured Hills and LRV-3 (near the interface between white and dark mantling materials) areas have intermediate contamination levels; and Station 5, in the dark-mantled area well away from the massifs has siderophilic ratios nearly as low as those at the Apollo 11 and 12 sites. The siderophiles in soil breccia 79135 are mainly of highlands origin.

As will be discussed in the following section, one can also use the siderophile data to estimate the fraction of highlands regolith material added to individual soils.

ORANGE-GLASS COMPONENT IN SOILS AND BRECCIAS

All Apollo 17 soils contain orange-glass spherules (and spherule fragments), and sample 74220 recovered near Shorty Crater consists entirely of this material. The orange glass is closely related to the Apollo 15 green glass, differing chiefly in terms of a much higher Fe_2TiO_4 content in the former (Reid et al., 1973). The orange glass is compositionally very similar to the Apollo 17 mare basalts; it approximates a mixture of basalt plus olivine (Reid et al., 1973). Because of the close compositional resemblance to the local basalts, Green and Ringwood (1973) have proposed that it formed from the same ultramafic source by a higher degree of partial melting. Other workers have proposed alternative origins of the orange glass from impacts on local materials, including lava lakes. Numerous authors have shown that 74220 has an extremely high content of volatile elements (such as Zn, Cd, Pb). Following a suggestion of L. T. Silver (private communication), we have argued in our companion paper by Chou et al. (1974) that the high-volatile content of the orange glass is only consistent with origin in a lava fountain having a high ratio of gaseous to condensed materials, and is not consistent with an origin by impact.

A review of published data shows that volatile-element concentrations in 74220 are often an order of magnitude higher than those in typical local soils (those without high orange-glass contents), and two orders of magnitude higher than those in mare basalts. In particular, elements with 74220/mature-soil ratios $\geqslant 4$ include Cl (Jovanovic and Reed, 1974); Zn (Wänke et al., 1973; Duncan et al., 1974; Morgan et al., 1974); Cd (Morgan et al., 1974); Tl (Allen et al., 1974; Morgan et al., 1974); and Pb[204] (Allen et al., 1974; Silver, 1974). A number of other elements show more modest enrichments in the orange glass.

Concentrations of these elements in mare basalts and anorthositic materials are much lower than those in mature Apollo 17 soils. Concentrations in KREEP-rich materials are generally comparable to those in the soils, but most of these are breccias, and concentrations in pristine KREEP are probably much smaller. In fact, the only other important source for these volatiles in Apollo 17 soils appears to be the regolith of the surrounding highlands.

We can estimate the relative importance of these sources in favorable cases where the magnitude of the highlands–regolith component can be estimated from siderophilic element data, as discussed in the previous section. Our studies of Apollo 16 soils shows that volatiles such as Zn and Cd are strongly correlated with siderophiles such as Ir, and that subregolith materials collected near North Ray Crater have low concentrations of both types of elements.

The variations in Zn and Ga concentration in Apollo 17 soils and soil breccias are illustrated in Fig. 4. The two elements are strongly correlated; the total range in Zn values is about a factor of 15, that in Ga about a factor of 4. The highest concentrations are observed in bulk 74220 orange soil (Wänke *et al.*, 1973; Morgan *et al.*, 1974).

The relationship between Cd and Zn is shown in Fig. 5. The orange soil is highly enriched in both elements (Morgan *et al.*, 1974). With the exception of the shaded soil 76240, the Cd/Zn ratios are in the range $(1.1–2.0) \times 10^{-3}$, near or below the CI-chondrite ratio of 1.9×10^{-3}. Most other lunar samples plot above the CI ratio (Wasson *et al.*, 1974). The enrichment of Cd in 76240 is attributed to lability, as discussed in the following section.

The highlands–regolith contributions of Zn, Ga, and Cd to Apollo 17 samples with high concentrations of these elements are not negligible. We shall assume that the concentrations of these elements and Ir are those reported by Wasson *et al.* (1974) for mature Apollo 16 soils: 26 µg/g Zn, 5.4 µg/g Ga, 98 ng/g Cd, and 16.2 ng/g Ir. One can derive a relationship giving the fraction of mature

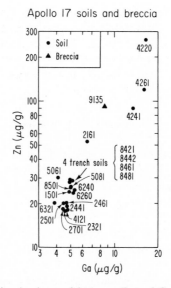

Fig. 4. Positive correlation is observed between Zn and Ga in Apollo 17 soils and a breccia based on all available data (this paper; Wänke *et al.*, 1973; Brunfelt *et al.*, 1974; and Laul and Schmitt, 1974). Soils 72161, 74241, and 74261 and breccia 79135 are enriched in orange glass (74220) relative to other soils.

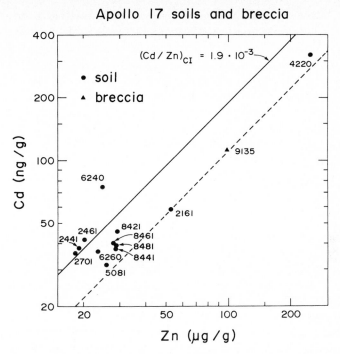

Fig. 5. Relationship between Cd and Zn in Apollo 17 soils and a breccia. Data are ours except for 74220, from published literature. The Cd/Zn ratios are between 1.1×10^{-3} (the dashed line) and 2.0×10^{-3} with the exception of the shaded soil 76240 which is enriched in Cd, probably from volatile transport. Samples 72161 and 79135 are significantly enriched in orange-glass components. The solid line shows the Cd/Zn ratio in the Orgueil (CI) chondrite.

highlands–regolith x in a soil sample

$$x = \frac{(R_s - R_m)[\text{Ir}]_s}{(R_h - R_m)[\text{Ir}]_h}$$

where R is the ratio of Ni, Ge or Au to Ir and the subscripts s, m, and h designate sample, mare and highlands quantities. We will arbitrarily define the mare and highlands Au/Ir ratios to be 0.23 and 0.52 (see Fig. 2) and the Ge/Ir and Ni/Ir ratios those defined by the lines in Fig. 2 at these Au/Ir values. Highlands–regolith contributions estimated from the three sets of siderophilic element ratios and the net Ir concentrations in Apollo 17 samples are listed in Table 6. Soils from individual stations are combined. Also listed in Table 6 are the Zn, Ga and Cd concentrations in these samples resulting from the highlands–regolith contamination.

Two samples with very high Zn, Ga and Cd contents are soil 72161 and breccia 79135. The largest highlands contribution to these volatiles is for Cd in 72161, and amounts to 35% of the gross Cd content. It is necessary to correct for mare basalt

Table 6. Highlands regolith contribution of Zn, Ga, and Cd to Apollo 17 soils, and orange-glass components estimated as described in the text.

Sample	Highlands component (%)	Highlands contribution			Orange-glass component (%)
		Zn (μg/g)	Ga (μg/g)	Cd (ng/g)	
72161	20	5.2	1.09	19.7	16 ± 3
79135	19	5.0	1.04	18.9	34 ± 4
75081	4	0.96	0.20	3.6	8 ± 2
Station 6	35	9.1	1.88	34.2	6 ± 2
Station 8	25	6.6	1.37	24.8	6 ± 2
Station 2	43	11.1	2.31	41.8	1 ± 2

contributions to Ga, but this source contributes amounts of Zn and Cd which are negligible for present purposes.

After correcting for the highlands–regolith and basalt contributions, we can estimate the orange-glass components in these samples from the published concentrations in 74220: 16.5 μg/g Ga (Wänke *et al.*, 1973); 250 ± 20 μg/g Zn (numerous authors); and 320 ng/g Cd (Morgan *et al.*, 1974). This exercise yields orange-glass concentrations in 72161 of 16 ± 3% and in 79135 of 34 ± 4% (Table 6) where the limits correspond to the ranges of estimates from the three elements, and appear to be reasonable limits on our estimates. Meyer and Marvin (1973) noted as much as 40–50% glass in the matrix (80% of total) seen in thin sections of 79135. It would appear that the bulk of this glass is orange glass. We are not aware of petrographic investigations of soil 72161. The elevated [204]Pb content of 72160 by Silver (1974) appears to be consistent with our estimated orange-glass content.

We have also estimated orange-glass contents of our other Apollo 17 soils (Table 6): about 1% at Station 2 to about 6–8% at Stations, 5, 6, and 8. The listed errors are again based on the ranges in estimates obtained for the three elements.

It is of interest to recall that typical mature Apollo 11 soils (e.g. 10084) contain about 25 μg/g Zn and 44 ng/g Cd, similar to those in most Apollo 17 soils. Reid *et al.* (1973) note that glasses very similar in composition to the orange glass are abundant in Apollo 11 soils. Perhaps the Apollo 11 volatiles are genetically associated with this glass.

There is some evidence (Ganapathy *et al.*, 1973; Morgan *et al.*, 1974) that the volatiles are present as surficial coatings on the green- and orange-glass spherules. If so, it would be remarkable if the ratio of volatiles to glass were always the same over relatively wide areas. It will be interesting to test this coherence by comparing petrographic estimates of the orange-glass contents of Apollo 17 soils with our compositional values.

Labile Elements in Lunar Soils

In Tables 3 and 4 we report data on permanently shaded soil 76240 (collected 50 cm behind the edge of an overhanging boulder) and its control 76260. Measured concentrations in both soils are similar for all elements with the exception of Cd, which is a factor of two more abundant in the shaded soil. Morgan *et al.* (1974) report very similar data on the two soils, including the factor of two enhancement in Cd. Their sample of 76241 is slightly enriched in siderophiles relative to 76261.

Soils 72441 and 72461 were originally located under a small (70 cm) boulder; the latter is a skim sample, the former was material from the upper 4 cm. Laul and Schmitt (1974) designate these "shadowed soils," and argue that the absence of appreciable enrichments in Zn, Cd, and In (as compared to exposed mare soils) is evidence against labile behavior on the part of these elements. Our data (Table 3) confirm the absence of appreciable enrichments in these elements. However, these do not appear to be shadowed soils, since it does not appear that there was an opening through which labile elements could effuse in and condense. These soils are more comparable to trench samples.

Laul and Schmitt (1974) and Morgan *et al.* (1974) also studied "shaded" soil 72321, and found no enrichment in potentially labile elements such as Zn, Br, Cd, In, Tl, and Bi. This sample was collected nearer (20 cm behind) the edge of the overhang than was 76240 (ALGIT, 1973) and as a result may have experienced more extensive mixing with nearby unshadowed materials.

There are two possible interpretations of the observed enrichment of Cd in 76240: (1) The soil has been gardened during recent times, and as a result, is not enriched in labile elements; the high Cd value is a fluke. (2) The soil has remained undisturbed for a period long compared to the hopping lifetime of labile elements; Cd is the only labile element among those determined by Morgan *et al.* (1974) or by us. There is no way to conclusively choose between these alternatives. However, this shaded soil was collected from a location a greater distance behind the overhang of the boulder than was 72321, and we have additional (though tentative because of evidence of contamination) evidence based on sieving of soil 65500 that Cd is surface-correlated (Wasson *et al.*, 1974). The more likely explanation appears to be the second, that Cd is labile and the other volatiles surveyed by Morgan *et al.* (1974) and by us are not.

We have looked into the question of the volatility of Cd compounds. The most volatile form under lunar conditions appears to be the monatomic metal. At 400°K, the temperature of the sunlit lunar surface, the vapor pressure of the metal is 2.5×10^{-6} torr, much higher than those of other heavy ($Z > 25$) metals with the exception of Hg, the alkalies, Se and Te. The latter are probably not as easily reduced as are Cd and Hg. A pressure of 2.5×10^{-6} torr is much greater than the minimum necessary to allow all the Cd on the lunar surface to remain in the atmosphere without undergoing homogeneous condensation.

DISCUSSION OF INTERSAMPLE RELATIONSHIPS

Trench soils 78421–78481

The four trench soils were collected at depths up to 25 cm at Station 8 (Tables 3 and 4). The only difference we can resolve is a slightly higher siderophilic content in the surficial sample (which would be still higher if we were to include the high-siderophile aliquot in our mean). Clearly, the idea of Chou *et al.* (1973) that labile elements might fall through interstices and concentrate at subsurficial levels in the regolith is not supported by these data.

The siderophilic element data indicate a 25% highlands–regolith component at Station 8 (Table 6). The low Fe and Mn contents relative to mare basalts indicate additional dilution by materials with low contents of these elements—perhaps subregolith material from North Massif or the Sculptured Hills.

Under-boulder soils 72441–72461 and soil 72701

As with the trench soils, these two samples show no significant differences in composition, with the possible exception of Ir. As stressed by Laul and Schmitt (1974), there is no evidence that labile elements are enriched in these subsurficial samples.

The data in Tables 3 and 4 show that 72701 is also very similar to soils 72441 and 72461 in composition. The only significant difference appears to be a 20% lower content of large-ion lithophiles in 72701. Thorium concentrations are high in all samples, which is possibly related to a high noritic component at Station 2 (Papike *et al.*, 1973). The siderophilic element data indicate a large (43%) highlands–regolith component in these soils. There is a large error associated with this value, however, since the Ni/Ir ratio gives much higher estimates than the other two ratios. The low Fe and Mn contents show that these soils have the lowest mare-basalt components among our suite of soil samples.

Soil 72161

Soil 72161 was collected at LRV-3, in the dark-mantled area adjacent to the light-mantling detritus from the South Massif. As we discussed in the section on orange glass, its orange-glass content estimated from Zn, Ga, and Cd data is about 16%, much higher than in any other soil in our suite. The high Fe and Mn values indicate a high content of mare basalt as well. The moderately low content of large-ion lithophiles shows that the South Massif contribution is small despite the nearness of the light-mantling materials to the collection site.

Soil 75081

Soil 75081 collected near the rim of Camelot Crater has an extralunar component only about 40% of that observed in mature soils, and a very low highlands–regolith component. By these standards it is an immature soil, and probably includes an appreciable content of subregolith materials ejected in a recent cratering event. Its contents of elements other than volatiles and siderophiles are nearly identical to those of the two mare-basalt rocks we have studied; its mare-basalt component must be about 90%.

Soil breccia 79135

We designate 79135 a soil breccia on the basis of its moderately large extralunar component, which is about 60% as large as those in mature Apollo 17 soils. Its high apparent orange-glass content (about 34%) has been discussed earlier. The very low contents of large-ion lithophiles indicate a composition intermediate between mare basalts and orange glass.

Igneous rocks 71055 and 79155

Medium-grained basalt 71055 and coarse-grained gabbro 79155 are very similar in composition despite differences in texture and collection sites linearly separated by 2 km. These basaltic rocks are marked by very low concentrations of Ni, Zn, Ge, Cd, and probably In, Ir, and Au (we suspect contamination of the higher values listed in Table 3). Concentrations of Sc, Cr, Mn, and Fe are higher than observed in our soil samples, indicating lower concentrations of these elements in other major soil components such as noritic and anorthositic materials. The large-ion lithophiles are lower in these samples than in all other samples except the orange-glass breccia 79135.

Acknowledgments—We are indebted to LSAPT and the JSC curatorial staff for the lunar samples, and to R. W. Bild, E. B. Grudewicz, S. Jones, J. F. Kaufman, G. Pusavat, and K. L. Robinson for technical assistance. We are grateful to L. T. Silver for very informative discussions. Neutron irradiations at the UCLA and Ames Laboratory reactors were capably handled by J. Brower, B. Link, A. F. Voigt, and their associates. This research was supported in large part by NASA grant NGL 05-007-367.

REFERENCES

ALGIT (Apollo Lunar Geology Investigation Team) (1973) Documentation and environment of the Apollo 17 samples: A preliminary report. *USGS Interagency Report: Astrogeology* **71**, 322 pp.

Allen R. O., Jr., Jovanovic S., Showalter D., and Reed G. W., Jr. (1974) ^{204}Pb, Bi, Tl and Zn in Apollo 16 samples and inferences on the lunar geochemistry of ^{204}Pb based on meteoritic and terrestrial sample studies (abstract). In *Lunar Science—V*, pp. 12–14. The Lunar Science Institute, Houston.

Anders E., Ganapathy R., Krähenbühl U., and Morgan J. W. (1973) Meteoritic material on the moon. *The Moon* **8**, 3–24.

Apollo 17 Preliminary Examination Team (1973) Apollo 17 lunar samples: Chemical and petrographic description. *Science* **182**, 659–672.

Baedecker P. A. (1974) SPECTRA: A computer code for the evaluation of high resolution gamma-ray spectra from instrumental activation analyses experiments. In *Advances in Obsidian Glass Studies: Archaeological and Geochemical Perspectives* (editor R. E. Taylor). Noyes Press. In press.

Baedecker P. A., Schaudy R., Elzie J. L., Kimberlin J., and Wasson J. T. (1971) Trace element studies of rocks and soils from Oceanus Procellarum and Mare Tranquillitatis. *Proc. Second Lunar Sci. Conf., Geochim. Cosmochim. Acta*, Suppl. 2, Vol. 2, pp. 1037–1061. MIT Press.

Baedecker P. A., Chou C.-L., Sundberg L. L., and Wasson J. T. (1972) Extralunar materials in Apollo-16 soils and the decay rate of the extralunar flux 4.0 G.y. ago. *Earth Planet. Sci. Lett.* **17**, 79–83.

Baedecker P. A., Chou C.-L., Grudewicz E. B., and Wasson J. T. (1973) Volatile and siderophilic trace elements in Apollo 15 samples: Geochemical implications and characterization of the long-lived and short-lived extralunar materials. *Proc. Fourth Lunar Sci. Conf., Geochim. Cosmochim. Acta*, Suppl. 4, Vol. 2, pp. 1177–1195. Pergamon.

Brunfelt A. O., Heier K. S., Nilssen B., Steinnes E., and Sundvoll B. (1974) Elemental composition of Apollo 17 fines (abstract). In *Lunar Science—V*, pp. 92–94. The Lunar Science Institute, Houston.

Chou C.-L., Baedecker P. A., and Wasson J. T. (1973) Atmophilic elements in lunar soils. *Proc. Fourth Lunar Sci. Conf., Geochim. Cosmochim. Acta*, Suppl. 4, Vol. 2, pp. 1523–1533. Pergamon.

Chou C.-L., Baedecker P. A., Bild R. W., and Wasson J. T. (1974) Volatile-element systematics and green glass in Apollo-15 lunar soils. *Proc. Fifth Lunar Sci. Conf., Geochim. Cosmochim. Acta*. This volume.

Compston W., Berry H., Vernon M. T., Chappell B. W., and Kaye M. J. (1971) Rubidium–strontium chronology and chemistry of lunar material from the Ocean of Storms. *Proc. Second Lunar Sci. Conf., Geochim. Cosmochim. Acta*, Suppl. 2, Vol. 2, pp. 1471–1485. MIT Press.

Cuttitta F., Rose, H. J., Jr., Annell C. S., Carron M. K., Christian R. P., Dwornik E. T., Greenland L. P., Helz A. W., and Ligon D. T., Jr. (1971) Elemental composition of some Apollo 12 lunar rocks and soils. *Proc. Second Lunar Sci. Conf., Geochim. Cosmochim. Acta*, Suppl. 2, Vol. 2, pp. 1217–1229. MIT Press.

Duncan A. R., Erlank A. J., Willis J. P., and Sher M. K. (1974) Compositional characteristics of the Apollo 17 regolith (abstract). In *Lunar Science—V*, pp. 184–186. The Lunar Science Institute, Houston.

Ganapathy R., Keays R. R., Laul J. C., and Anders E. (1970) Trace elements in Apollo 11 lunar rocks: Implications for meteorite influx and origin of moon. *Proc. Apollo 11 Lunar Sci. Conf., Geochim. Cosmochim. Acta*, Suppl. 1, Vol. 2, pp. 1117–1142. Pergamon.

Ganapathy R., Morgan J. W., Krähenbühl U., and Anders E. (1973) Ancient meteoritic components in lunar highlands rocks: Clues from trace elements in Apollo 15 and 16 samples. *Proc. Fourth Lunar Sci. Conf., Geochim. Cosmochim. Acta*, Suppl. 4, Vol. 2, pp. 1239–1261. Pergamon.

Ganapathy R., Morgan J. W., Higuchi H., and Anders E. (1974) Meteoritic and volatile elements in Apollo 16 rocks and in separated phases from 14306. *Proc. Fifth Lunar Sci. Conf., Geochim. Cosmochim. Acta*. This volume.

Green D. H. and Ringwood A. E. (1973) Significance of a primitive lunar basaltic composition present in Apollo 15 soils and breccias. *Earth Planet. Sci. Lett.* **19**, 1–8.

Hughes T. C., Keays R. R., and Lovering J. F. (1973) Siderophile and volatile trace elements in Apollo 14, 15, and 16 rocks and fines: Evidence for extralunar component and Tl-, Au-, and Ag-enriched rocks in the ancient lunar crust (abstract). In *Lunar Science—IV*, pp. 400–402. The Lunar Science Institute, Houston.

Huneke J. C., Jessberger E. K., Podosek F. A., and Wasserburg G. J. (1973) $^{40}Ar/^{39}Ar$ measurements in Apollo 16 and 17 samples and the chronology of metamorphic and volcanic activity in the Taurus-Littrow region. *Proc. Fourth Lunar Sci. Conf., Geochim. Cosmochim. Acta*, Suppl. 4, Vol. 2, pp. 1725–1756. Pergamon.

Jovanovic S. and Reed G. W., Jr. (1974) Labile trace elements in Apollo 17 samples (abstract). In *Lunar Science—V*, pp. 391–393. The Lunar Science Institute, Houston.

Krähenbühl U., Ganapathy R., Morgan J. W., and Anders E. (1973) Volatile elements in Apollo 16 samples: implications for highland volcanism and accretion history of the moon. *Proc. Fourth Lunar Sci. Conf., Geochim. Cosmochim. Acta*, Suppl. 4, Vol. 2, pp. 1325–1348. Pergamon.

Laul J. C. and Schmitt R. A. (1974) Chemical composition of Apollo 17 boulder-2 rocks and soils (abstract). In *Lunar Science—V*, pp. 438–440. The Lunar Science Institute, Houston.

Laul J. C., Morgan J. W., Ganapathy R., and Anders E. (1971) Meteoritic material in lunar samples: Characterization from trace elements. *Proc. Second Lunar Sci. Conf., Geochim. Cosmochim. Acta*, Suppl. 2, Vol. 2, pp. 1139–1158. MIT Press.

Laul J. C., Ganapathy R., Morgan J. W., and Anders E. (1972) Meteoritic and non-meteoritic trace elements in Luna 16 samples. *Earth Planet. Sci. Lett.* **13**, 450–454.

Meyer C., Jr. and Marvin U. B. (1973) 79135. In *Lunar Sample Information Catalogue, Apollo 17*, pp. 422–424. NASA Johnson Space Center (MSC 03211).

Morgan J. W., Laul J. C., Krähenbühl U., Ganapathy R., and Anders E. (1972) Major impacts on the moon: Characterization from trace elements in Apollo 12 and 14 samples. *Proc. Third Lunar Sci. Conf., Geochim. Cosmochim. Acta*, Suppl. 3, Vol. 2, pp. 1377–1395. MIT Press.

Morgan J. W., Krähenbühl U., Ganapathy R., and Anders E. (1973) Luna 20 soils: Abundance of 17 trace elements. *Geochim. Cosmochim. Acta* **37**, 953–961.

Morgan J. W., Ganapathy R., Higuchi H., Krähenbühl U., and Anders E. (1974) Lunar basins: Tentative characterizations of projectiles, from meteoritic elements in Apollo 17 boulders (abstract). In *Lunar Science—V*, pp. 526–528. The Lunar Science Institute, Houston.

Papike J. J., Bence A. E., Cameron K., and Delano J. (1973) Petrology of the 2–4 mm soil fragments from Apollo 17. *Trans. Amer. Geophys. Union* **54**, 601–603.

Reid A. M., Lofgren G. E., Heiken G. H., Brown R. W., and Moreland G. (1973) Apollo 17 orange glass, Apollo 15 green glass and Hawaiian lava fountain glass. *Trans. Amer. Geophys. Union* **54**, 606–607.

Silver L. T. (1974) Patterns of U–Th–Pb distributions and isotope relations in Apollo 17 soils (abstract). In *Lunar Science—V*, pp. 706–708. The Lunar Science Institute, Houston.

Stettler A., Eberhardt P., Geiss J., Grögler N., and Maurer P. (1973) Ar^{39}–Ar^{40} ages and Ar^{37}–Ar^{38} exposure ages of lunar rocks. *Proc. Fourth Lunar Sci. Conf., Geochim. Cosmochim. Acta*, Suppl. 4, Vol. 2, pp. 1865–1888. Pergamon.

Taylor S. R., Rudowski R., Muir P., Graham A., and Kaye M. (1971) Trace element chemistry of lunar samples from the Ocean of Storms. *Proc. Second Lunar Sci. Conf., Geochim. Cosmochim. Acta*, Suppl. 2, Vol. 2, pp. 1083–1099. MIT Press.

Turner G. (1971) ^{40}Ar–^{39}Ar ages from the lunar maria. *Earth Planet. Sci. Lett.* **11**, 169–191.

Turner G., Cadogan P. H., and Yonge C. J. (1973) Argon selenochronology. *Proc. Fourth Lunar Sci. Conf., Geochim. Cosmochim. Acta*, Suppl. 4, Vol. 2, pp. 1889–1914. Pergamon.

Wänke H., Baddenhausen H., Dreibus G., Jagoutz E., Kruse H., Palme H., Spettel B., and Teschke F. (1973) Multielement analyses of Apollo 15, 16, and 17 samples and the bulk composition of the moon. *Proc. Fourth Lunar Sci. Conf., Geochim. Cosmochim. Acta*, Suppl. 4, Vol. 2, pp. 1461–1481. Pergamon.

Wasson J. T. (1974) *Meteorites—Classification and Properties.* Springer. In press.

Wasson J. T. and Baedecker P. A. (1970) Ga, Ge, In, Ir, and Au in lunar, terrestrial, and meteoritic basalts. *Proc. Apollo 11 Lunar Sci. Conf., Geochim. Cosmochim. Acta*, Suppl. 1, Vol. 2, pp. 1741–1750. Pergamon.

Wasson J. T., Chou C.-L., Bild R. W., and Baedecker P. A. (1973) Extralunar materials in Cone-Crater soil 14141. *Geochim. Cosmochim. Acta* **37**, 2349–2353.

Wasson J. T., Chou C.-L., Robinson K. L., and Baedecker P. A. (1974) Trace elements in Apollo-16 rocks and soils. *Geochim. Cosmochim. Acta*, submitted.

Willis J. P., Ahrens L. H., Danchin R. V., Erlank A. J., Gurney J. J., Hofmeyr P. K., McCarthy T. S., and Orren M. J. (1971) Some interelement relationships between lunar rocks and fines, and stony meteorites. *Proc. Second Lunar Sci. Conf., Geochim. Cosmochim. Acta*, Suppl. 2, Vol. 2, pp. 1123–1138. MIT Press.

Proceedings of the Fifth Lunar Conference
(Supplement 5, Geochimica et Cosmochimica Acta)
Vol. 2 pp. 1645–1657 (1974)
Printed in the United States of America

Volatile-element systematics and green glass in Apollo 15 lunar soils

C.-L. Chou, P. A. Baedecker,* R. W. Bild, and J. T. Wasson

Departments of Chemistry and Planetary and Space Science, Institute of Geophysics and
Planetary Physics, University of California, Los Angeles, California 90024

Abstract—A number of volatiles (Zn, Cd, In, ^{204}Pb, excess ^{40}Ar, excess radiogenic Pb) in Apollo 15 soils correlate with each other and with decreasing distance from the Apennine Front. These correlations can be understood in terms of a deposition of green glass and cogenetic volatiles on the flanks of the Front followed by addition of secondary magmatic volatiles including excess ^{40}Ar. Mixing of these materials with mare-type regolith resulted in the observed volatile-element patterns.

The absence of an increase in siderophilic element content with decreasing distance from the Front probably resulted from the failure of the steeply sloping front to develop a highlands-type mature regolith. Breccia 15015 has a siderophilic element signature typical of a mature Apollo 15 soil, which tends to rule out its transport to the site as part of an Autolycus or Aristillus ray.

Introduction

THE APOLLO 15 AND APOLLO 17 SITES consist of mare plains adjacent to highlands, with substantial altitude differences between the two types of regions. Both sites are also distinguished by major deposits of igneous glass spherules: the Apollo 15 green glass which is abundant in several samples collected near Station 7 on the slope of the Apennine Front, and the Apollo 17 orange glass which is abundant near Shorty Crater. The glassy spherules are found in soils from all sampled areas at both sites.

In our previous paper (Chou *et al.*, 1973), we summarized evidence on the distribution of atmophilic volatiles (i.e. volatiles concentrated in the regolith as a result of transport through the lunar atmosphere) in the Apollo 15 soils, and found positive correlations between Zn, Cd, and excess ^{40}Ar, with higher concentrations in samples collected near the Front. We suggested that this reflected a greater admixture of one or more ancient components from the Front, since Heymann and Yaniv (1970) had pointed out that the rate of liberation of ^{40}Ar was greater early in the moon's history, and it seemed likely that Zn and Cd were enriched in old surficial materials as a result of their production as volatiles during the period of intense magmatic activity 3.9–3.2 Gyr ago.

The siderophilic elements (almost exclusively of extralunar origin) in the Apollo 15 soils show no correlation with Zn, Cd, and ^{40}Ar, and do not increase in concentration toward the Front. This absence of higher siderophilic-element concentrations near the Front was unexpected, since the concentrations of siderophiles in mature soils collected at the Apollo 14 and 16 sites were,

---

*Now at Analytical Laboratories Branch, U.S. Geological Survey, Reston, Virginia 22091.

respectively, two and three times higher than those in the mature soils collected at the Apollo 11 and 12 mare landing sites. Baedecker *et al.* (1973) assumed that the Zn, Cd, and siderophiles were all more or less uniformly mixed in the Front regolith, and suggested that the absence of a correlation between the volatiles and siderophiles indicated that regolith materials ejected from the Front had experienced more reheating (and volatile loss) the greater the distance of ejection during impact events.

Since the preparation of the previous paper (Chou *et al.*, 1973) we have received two soil samples with excess ^{40}Ar contents (Jordan *et al.*, 1974) greater than and less than those in the suite of samples previously assigned us. Our data base is increased by the inclusion of these results and by data for additional volatile elements recently published by other investigative teams. We also present results on soil breccia 15015, which was a late Apollo 15 allocation to our group.

The most important new development, however, is not in terms of these additional soil data but rather the data (Ganapathy *et al.*, 1973; Wänke *et al.*, 1973; Duncan *et al.*, 1974; Morgan *et al.*, 1974) for relatively pure samples of the Apollo 15 green glass and Apollo 17 orange glass, which show that these have volatile contents far in excess of those previously found in lunar samples, and indicate that they play a dominant role in the observed distribution of volatile elements such as Zn, Cd, and In at the Apollo 15 and 17 sites.

Volatiles in the Green and Orange Glasses

The Apollo 15 green glass and Apollo 17 orange glass are compositionally closely related to the basalts recovered from these sites, although the glasses contain more normative pyroxene and olivine (Prinz *et al.*, 1973). Their more ultramafic composition and lower content of elements of large ionic radius led Green and Ringwood (1973) to propose that the glasses originated at depth from the same source materials as the local basalts, but by higher degrees of partial melting. We will discuss other ideas regarding their origin in the following section.

Radiometric age determinations by the ^{39}Ar–^{40}Ar method have yielded values of 3.38 ± 0.06 Gyr (Podosek and Huneke, 1973) and 3.79 ± 0.08 Gyr (Husain, 1972) for green-glass clod 15426. The former workers give a more detailed description of their study, which was carried out on a narrow size fraction (250–750 μm); we tend to attach more significance to it. In any case there is strong evidence that the green glass was formed after the Imbrium event (at $\geqslant 3.9$ Gyr) and before the end of basalt formation at the Apollo 15 site at about 3.2 Gyr (Husain, 1974).

Huneke *et al.* (1973) reported an age of 3.54 ± 0.05 Gyr for orange-glass fines 74220; Schaeffer and Husain (1973) reported an age of 3.71 ± 0.06 Gyr. We find no means for determining which of these is more nearly correct. The former is considerably later than the time of formation of the bulk of the Apollo 17 basalt samples at about 3.7–3.8 Gyr (Huneke *et al.*, 1973) but well within the period during which basalt formation was occurring at other lunar sites.

The most nearly pure sample of green glass surveyed for volatile trace elements is green clod 15426 (Ganapathy *et al.*, 1973). About 35% of the sample was green glass and about 65% matrix, with minor amounts of brown glass (Ganapathy, private communication). Thus, by appropriate weighting, Zn, Cd, In, and Tl are an order of magnitude more abundant in green glass than in typical Apollo 15 basalts, and 1.5–5 times more abundant than in mature soils. The only soil with comparable elemental abundances is 15401, an immature soil very rich in green glass (Carusi *et al.*, 1972).

Studies of Apollo 17 orange glass 74220 (Wänke *et al.*, 1973; Duncan *et al.*, 1974; Morgan *et al.*, 1974) show a similar though still more striking pattern. Zinc, Cd, and Tl are two orders of magnitude more abundant than in Apollo 17 mare basalts, and about one order of magnitude more abundant than in mature soils.

The significance of these volatile enrichments in the green-glass and orange-glass samples is discussed in the following section.

APOLLO 15 SOILS REVISITED

Chou *et al.* (1973) reported correlations between Zn, Cd, and excess ^{40}Ar (^{40x}Ar) at the Apollo 15 site. We analyzed two additional soils having ^{40x}Ar contents (Jordan *et al.*, 1974) above and below the range for which our Zn and Cd data were available. The results are listed in Table 1.

Figure 1 shows a revised Zn–^{40x}Ar plot. Soil 15421 with high ^{40x}Ar does indeed have a very high Zn content. Zinc data on soil 15601 (with low ^{40x}Ar content) by Brunfelt *et al.* (1973) and Haskin *et al.* (1973) disagree by almost a factor of 2. Our replicates also disagree by a similar factor, and our mean Zn value plots above the correlation trend defined by the soils with higher ^{40x}Ar contents. In Fig. 1 we also

Table 1. Replicate and mean concentrations of eight trace elements in two Apollo 15 soils and an Apollo 15 breccia. Values in parentheses considered unreliable and not included in means.

	Station	Ni (μg/g) Repl.	Mean	Zn (μg/g) Repl.	Mean	Ga (μg/g) Repl.	Mean	Ge (ng/g) Repl.	Mean
Soils									
15421,29	7	215, 215	215	52.6, 57.3	55.0	4.52, 4.37	4.44	281, 245	263
15601,183	9a	160, 154	157	25.0, 14.0	19.5	3.66, 3.16	3.41	179, 188	184
Breccia									
15015,40	LM	242, 238, 229	236	13.3, 12.6, 13.5	13.1	4.16, 4.67, 3.78	4.20	321, 352, 462	378

	Station	Cd (ng/g) Repl.	Mean	In (ng/g) Repl.	Mean	Ir (ng/g) Repl.	Mean	Au (ng/g) Repl.	Mean
Soils									
15421,29	7	234, 246	240	44, 49	46	2.5, 2.5	2.5	3.7, 2.3	3.0
15601,183	9a	121, 83	102	4.3, 2.4	3.4	3.1, 2.9	3.0	1.2, 1.1	1.2
Breccia									
15015,40	LM	40, 46, 33	40	3.5, 2.5, (5.4)	3.0	(0.2), 7.1, 7.6	7.3	5.5, 2.5, 1.4	3.1

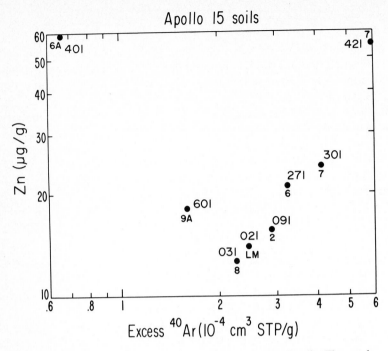

Fig. 1. Relationship between Zn and excess ^{40}Ar in Apollo 15 soils. The numbers in larger print give the last three digits of the sample numbers. The numbers or letters in smaller print give the station designation.

plot soil 15401 (Zn data by Ganapathy *et al.*, 1973) which has an extremely low ^{40x}Ar content. The high Zn content of this soil places it far from the trend observed for soils with ^{40x}Ar $> 10^{-4}$ cm^3/g. The correlation observed in Fig. 1 for soils other than 15401 is significant at the 95% confidence level.

In Fig. 2, In and Cd are plotted against ^{40x}Ar (In and Cd determined by us except 15401 by Ganapathy *et al.*, 1973). Whereas no correlation between In and ^{40x}Ar was observed by Chou *et al.* (1973), the addition of 15601 and 15421 data results in a good correlation for samples with ^{40x}Ar contents $> 10^{-4}$ cm^3/g (significant correlation at the 90% confidence level). Some caution must be exercised, however, because the rather high In content of 15421 may partially reflect contamination. A weak correlation between Cd and ^{40x}Ar is observed for soils with ^{40x}Ar $> 10^{-4}$ cm^3/g.

Chou *et al.* (1973) noted that Zn and ^{40x}Ar values increase as one proceeds from the mare (LM, 2, 8, 9) to the Front stations (6, 7), and that the correlations may reflect either enrichment of ^{40x}Ar and Zn in the same or separate components which are more abundant in Front materials than in mare basalts. They observed no correlation between Zn and (solar wind implanted) ^{36}Ar, and suggested that Zn and ^{40x}Ar are enriched in one or more ancient components. Atmophilic Zn was probably released in large amounts during the period of mare-type volcanism 3.9–3.2 Gyr ago; release of ^{40}Ar was also greater earlier in lunar history (Yaniv and

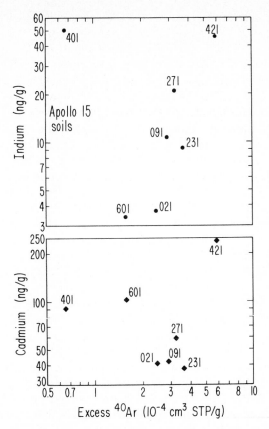

Fig. 2. Plots of In and Cd versus excess ^{40}Ar for Apollo 15 soils. The numbers by each
point are the last three digits of the sample number.

Heymann, 1972), both because of magmatic enhancement and the relatively short
half-life of ^{40}K. It is possible that outgassing of the interior reached a maximum
during the period of mare formation, and that the maximum ^{40}Ar release was
contemporaneous with that of Zn and other magmatic volatiles.

As noted in the previous section, volatiles are enriched in the Apollo 15 green
glass and in the Apollo 17 orange and brown glass. Ganapathy *et al.* (1973) suggest
that the high contents of Zn, Cd, and In in soil 15401 reflect its high green glass
content (Carusi *et al.*, 1972), and that the four-times lower volatile contents of
coarse green-glass and brown-glass hand separates from 15426 suggest that some
fraction of these elements are surface correlated. As discussed in the previous
section, volatiles are also abundant in Apollo 17 soil 74220 (and in 74240 and
74260), which contains large amounts of the red- and brown-glass spherules.

The ^{40}Ar content of relatively coarse ($>105\ \mu$m) separates of green-glass
spherules is very small, generally in the range $(0.6–2) \times 10^{-5}$ cm^3/g (Lakatos *et al.*,
1973). Unfortunately, no bulk determination of ^{40}Ar in green-glass clod 15426 is

available, although data on the coarse separates by Lakatos *et al.* (1973) fall in the indicated concentration range. Jordan *et al.* (1974) report ^{40}Ar data on soil 15421 which was collected near 15426 at Station 7, and which appears to be rich in green glass; the ^{40x}Ar content of bulk 15421 is one of the highest reported in an Apollo 15 soil, and is 100 times greater than that in the 250–354 μm size fraction analyzed by Lakatos *et al.* (1973). The ^{40x}Ar/^{36}Ar ratio in 15421 is about 3.4, much higher than in other Apollo 15 soils (Jordan *et al.*, 1974); according to Yaniv and Heymann (1972) such high ^{40x}Ar/^{36}Ar ratios are higher in older samples, and may be attributable to Ar implantation during a period when the lunar atmosphere had a high ^{40}Ar content. It would be very useful to have bulk data on 15426 and to know which component of 15421 is responsible for its high ^{40x}Ar content.

The ^{40x}Ar content of the orange glass is not well defined. Heymann *et al.* (1974) report moderately high bulk ^{40}Ar contents of 14 and 3.3×10^{-4} cm^3/g in 74241 and 74220. However Hübner *et al.* (1974) find only 0.45×10^{-4} cm^3/g ^{40}Ar in orange glass separated from 74241. Taking the latter as an upper limit for ^{40x}Ar, we tentatively conclude that the ^{40x}Ar content of the orange glass is also very low.

In Chou *et al.* (1973), we considered the possibility that the increase in Zn and Cd toward the Front resulted from the higher anorthosite or KREEP components in Front materials, and showed this to be unlikely on the basis of the relatively low contents of these elements in Apollo 15 samples rich in anorthosite and KREEP. Thus, concentrations of Cd and In in certain Apollo 16 rocks (Krähenbühl *et al.*, 1973; Wasson *et al.*, 1974) are 2–3 times higher than those observed in the 15421 soil with highest volatile concentrations, but Zn is not enriched in the Apollo 16 rocks (66095 is an exception) and there is no evidence for a high anorthosite content of 15421. Similarly, KREEP-rich materials from the Apollo 14 site have Zn, Cd, and In concentrations comparable to the levels observed in 15421, but low concentrations of other KREEP-associated elements (Taylor *et al.*, 1973) indicate a very low KREEP content for this soil. It thus appears that green glass is the only viable candidate for a petrographically recognizable soil component which could be responsible for the bulk of the variation in Zn, Cd, and In concentration at the Apollo 15 site.

The alternative possibility presented by Chou *et al.* (1973) was that Zn, Cd, In, and ^{40x}Ar are correlated because they are all atmophilic. We assumed that the condensing volatiles accumulated on exposed high sites during the period of intense lunar magmatic activity, and that those which escaped subsequent burial remained as sources of volatile-rich materials.

We are confident that the magmatic volatilization model has validity, but the immediate question regarding the Apollo 15 site is the relative importance of a strictly atmophilic source vis-à-vis one associated with the green glass. We have noted the association of the high Zn, Cd, and In contents of 15401 with a high green-glass content; our 15421, another sample with high Zn, Cd, and In, also appears to have a high green-glass content. It may be that the entire variation in Zn, Cd, and In at the Apollo 15 site can be understood in terms of a green-glass associated component, and that a strictly atmophilic component need not be invoked.

The Zn, Cd, and In may be incorporated in the bulk glass or it may be a cogenetic condensate which nucleated on the surface of or is otherwise associated with the glass. The latter possibility was suggested by Ganapathy *et al.* (1973) for 15426 on the basis of the much higher volatile abundances in the fine matrix than in a hand-picked green-glass separate.

If the bulk of the material is present as a condensate, it should be attacked by a dilute acid leach, or vaporized by heating to moderate temperatures. To our knowledge, no such experiment has been carried out for Zn, Cd, and In in either green glass-rich Apollo 15 samples or orange glass-rich Apollo 17 samples.

Some evidence regarding volatile siting is provided by studies of the Pb isotopes. Allen *et al.* (1973) reported both leaching and vaporization (at 450°C) studies of ^{204}Pb in Apollo 15 soils; however, none of their samples appears to have appreciable contents of green glass. Allen *et al.* (1974) carried out a leach experiment involving about 10^{-5} N HNO$_3$ on the Apollo 17 orange glass 74220, and found only about 10% of the ^{204}Pb in the leachate. Tatsumoto *et al.* (1973) showed that 74220 is rich in ^{204}Pb, and that various leachates had lower ^{207}Pb/^{204}Pb and ^{206}Pb/^{204}Pb ratios than the whole sample. Silver (1974) reported that 60–80% of the total Pb was released from 74220 by heating to 600°C, and that the volatilized Pb had an even higher ^{204}Pb content than did the leachates of Tatsumoto *et al.* (1973). It appears that the Allen *et al.* (1974) leach was too weak to attack the Pb-bearing materials, and that the bulk of the Pb associated with 74220 is either near or on the surfaces of insoluble particles. On the basis of similar geochemical behaviors one would predict that Zn, Cd, and In are similarly sited on the orange glass and, by extrapolation, on the Apollo 15 green glass as well. This should, of course, be checked by experiment.

Evidence of the geochemical coherency of Pb and Zn in Apollo 15 soils is shown in Fig. 3, where excess ^{206}Pb, ^{207}Pb, and ^{208}Pb are plotted against Zn. The excess concentrations of these isotopes are calculated from published Pb isotopic data and U concentrations by assuming a nominal age of U-bearing phases of 4.4 Gyr. This arbitrary age is obtained by assuming that the bulk of the U is concentrated in the KREEP component of the soil; Schonfeld and Meyer (1973) summarized evidence that the formation age of KREEP is 4.3–4.4 Gyr. The strength of the correlations between the excess Pb isotopes (significant at 98–99% confidence levels) and Zn is only slightly affected by varying the assumed age over the range 4.0–4.6 Gyr. The correlation is also not related to the U or Th concentration. Figure 4 shows that ^{208}Pb is uncorrelated with Th if the 15426 point is not included (including 15426 yields a negative correlation significant at the 90% confidence level). Because of the relatively long ^{232}Th half-life, the calculation of excess ^{208}Pb is relatively insensitive to the assumed age, and the absence of a positive correlation in Fig. 4 supports the interpretation that the correlations observed in Fig. 3 are unrelated to the KREEP content of the soil. On the other hand, we can infer from the studies of 74220 by Tatsumoto *et al.* (1973) and Silver (1974) that the excess radiogenic Pb is probably not cogenetic with the green glass.

The relationship between ^{204}Pb and Zn is shown in Fig. 5. Because of the very low concentrations of ^{204}Pb in lunar samples, variable blanks introduce substantial

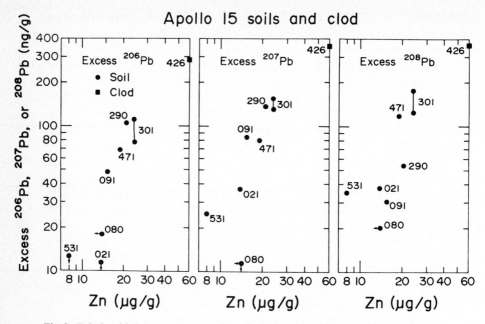

Fig. 3. Relationship between excess radiogenic Pb isotopes and Zn in Apollo 15 soils and a green-glass clod. Concentrations of Pb isotopes, U and Th are from Silver (1972, 1973), Tatsumoto *et al.* (1972), and Barnes *et al.* (1973). An age of 4.4 Gyr was assumed in order to calculate the excess amounts. Numbers by each point are sample numbers.

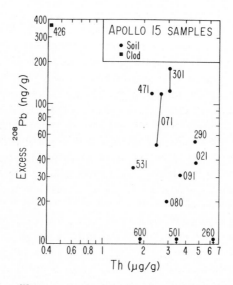

Fig. 4. Plot of excess ^{208}Pb versus Th in Apollo 15 soils and a green-glass clod. Data sources are those listed in Fig. 3 caption. No correlation is observed. Numbers by each point give sample numbers.

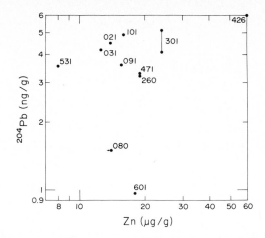

Fig. 5. Relationship between ^{204}Pb and Zn in Apollo 15 soils and a green-glass clod. Data sources are those listed in Fig. 3 caption and Allen *et al.* (1973). Sample numbers are listed near each point.

errors. Despite this, the bulk of the points plotted in Fig. 5 show a weakly positive correlation; only 15080 and 15601 fall far away from the trend. A 90% significance level of the correlation is calculated when these latter points are ignored.

The volatile-element story at the Apollo 15 site is fraught with fascinating complexity. The concentrations of the volatiles Zn, Cd, In, ^{204}Pb, excess radiogenic Pb and ^{40x}Ar all show relatively well-defined interelement correlations. Some points on each plot are exceptions to the general trends; most often these are samples very rich in green glass. Most of the above elements (or isotopes) show no correlation with siderophilic elements such as Ni, Ge, Ir or Au or with solar wind isotopes such as ^{36}Ar.

Baedecker *et al.* (1973) explained the absence of an increase in siderophiles in the direction of the Front in terms of a model involving equal degrees of contamination of the Apollo 15 landing site by ancient regolith materials from the Front, but with preferential outgassing of materials ejected farther. We now believe that it is more likely that the steepness of the Front prevented the buildup of a mature regolith with high siderophile content such as those found at the Apollo 14 and 16 sites. Although siderophilic-element ratios indicate that the Apollo 15 soils *are* contaminated with materials from an ancient, mature, highlands-type regolith (Baedecker *et al.*, 1973), this material must have come from more distant areas of lower relief. The relatively uniform amount of highlands contamination across the site can be readily understood if the mean distance was appreciably greater than the dimensions of the Apollo 15 site.

The abundance of solar wind isotopes appears to be mainly related to the recent residence near the surface. There is no particular reason why the mean solar wind age should increase toward the Front; if (as discussed below) erosion is more important than gardening near the Front, precisely the opposite trend would be expected. Figure 6, a plot of ^{36}Ar versus Zn, shows a negative correlation (at a

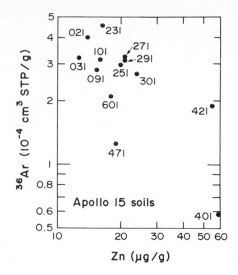

Fig. 6. Plot showing negative correlation of ^{36}Ar and Zn in Apollo 15 soils. The ^{36}Ar data are from Kirsten *et al.* (1972), Bogard and Nyquist (1973), Hintenberger and Weber (1973), and Jordan *et al.* (1974).

99% confidence level), and thus supports this interpretation. If one ignores the two green glass-rich soils 15401 and 15421, the confidence level is 70%.

Before we attempt to model these observations, it is pertinent to consider models for the origin of the green glass. These fall into two chief categories: (1) impact, either on a lava lake (Prinz *et al.*, 1973), the nearby highlands (Carusi *et al.*, 1972) or during the creation of the Imbrium basin (Dence *et al.*, 1974); or (2) formation at depth and ejection as a lava fountain (Green and Ringwood, 1973; Reid *et al.*, 1973; Silver, private communication). We find that the similarity in age between the green glass and the local basalts and their close compositional resemblance provide strong evidence in favor of the latter endogenous origin. As pointed out to us by Silver (private communication), the high content of volatiles provides perhaps the most cogent evidence of all. The contents of Zn, Cd, and ^{204}Pb are far in excess of those in mare basalts, and several times higher even than those found in mature highlands soils. There seems to be no straightforward way to cause such an enrichment of volatiles in association with impact ejecta. Although a small enrichment of volatiles might occur in the plume, the higher mean velocity of gas relative to liquid droplets should result in a separation of volatiles before the spherules return to the surface of the moon. On the other hand, Silver notes that a lava fountain may have associated with it a large amount of a volatile-rich vapor phase such that the high observed concentration of condensed volatiles could result even though a sizable additional fraction of the vapor phase failed to undergo condensational deposition together with the glass.

A model which appears to fit most of the volatile element data at the Apollo 15 site is the following: (1) Some time after the Imbrium event the lava flooding of the Palus Putredinus site began; (2) green glass and associated primary volatiles were produced in one or more lava-fountaining events; (3) additional lava generation covered most of the green-glass deposits, but accumulations on the Apennine Front which were higher than the highest lava flows remained on the lunar surface; (4) the upper layers of these deposits as well as those of other Front materials were gardened and accumulated, as secondary volatiles, excess ^{40}Ar and lunar magmatic volatiles including radiogenic Pb isotopes; (5) because of the slope of the Front a thick regolith layer did not accumulate on top of these materials; rather, they remained near the surface and were occasionally excavated by small impacts; (6) mixing of these materials into the adjacent mare-type regolith produced the observed volatile-element correlations. In some cases, large amounts of green glass from below the ancient gardened layer were added, yielding high contents of primary volatiles but only minor contents of secondary volatiles.

Soil Breccia 15015

Eglinton *et al.* (1974) note that breccia 15015 has a bulk composition very similar to Apollo 15 soils. Our data (Table 1) show that the siderophilic element content is very similar to that of relatively mature soils (e.g. about 20% lower than the high values reported by Baedecker *et al.* (1973) for 15021, which was collected about 10 m away, and quite similar to those in 15080 collected near Elbow Crater). We were particularly interested in 15015 because it appeared to offer the possibility of determining the true In content of a mature soil. Handling operations had resulted in serious contamination of most soils from earlier missions. In fact, the observed In concentration of 3.0 ng/g confirms the tentative conclusion of Baedecker *et al.* (1973) that the values of 3–5 ng/g they observed in numerous Apollo 15 soils did not include appreciable contamination contributions.

Interest attaches to sample 15015 because its situation on the Aristillus–Autolycus ray which crosses the Apollo 15 site suggests that it might be ejecta from one of these craters. The remarkable resemblance to local mature soils effectively rules out such a possibility, however. It is both (1) unlikely that the regolith at the sites of these craters had so nearly the same composition as the present Apollo 15 regolith; and (2) very unlikely that a breccia produced during the formation of a crater with a diameter of 35–50 km would manage to preserve the composition of the uppermost few meters of surficial materials. It is possible that the petrification of 15015 was produced by a projectile ejected from the Aristillus or Autolycus events, but the record of that object is not obvious.

Acknowledgments—We are indebted to LSAPT and the JSC curatorial staff for the lunar samples, and to E. B. Grudewicz, S. Jones, J. F. Kaufman, G. Pusavat, K. L. Robinson, and L. L. Sundberg for technical assistance. We are grateful to L. T. Silver for very informative discussions. Neutron irradiations at the UCLA and Ames Laboratory reactors were capably handled by J. Brower, A. F. Voigt, and their associates. This research was supported in large part by NASA grant NGL 05-007-367.

References

Allen R. O., Jr., Jovanovic S., and Reed G. W., Jr. (1973) Geochemistry of primordial Pb, Bi, and Zn in Apollo 15 Samples. *Proc. Fourth Lunar Sci. Conf., Geochim. Cosmochim. Acta*, Suppl. 4, Vol. 2, pp. 1159–1175. Pergamon.

Allen R. O., Jr., Jovanovic S., Showalter D., and Reed G. W., Jr. (1974) ^{204}Pb, Bi, Tl and Zn in Apollo 16 samples and inferences on the lunar geochemistry of ^{204}Pb based on meteoritic and terrestrial sample studies (abstract). In *Lunar Science—V*, pp. 12–14. The Lunar Science Institute, Houston.

Baedecker P. A., Chou C.-L., Grudewicz E. B., and Wasson J. T. (1973) Volatile and siderophile trace elements in Apollo 15 samples: Geochemical implications and characterization of the long-lived and short-lived extralunar materials. *Proc. Fourth Lunar Sci. Conf., Geochim. Cosmochim. Acta*, Suppl. 4, Vol. 2, pp. 1177–1195. Pergamon.

Barnes I. L., Garner E. L., Gramlich J. W., Machlan L. A., Moody J. R., Moore L. J., Murphy T. J., and Shields W. R. (1973) Isotopic abundance ratios and concentrations of selected elements in some Apollo 15 and Apollo 16 samples. *Proc. Fourth Lunar Sci. Conf., Geochim. Cosmochim. Acta*, Suppl. 4, Vol. 2, pp. 1197–1207. Pergamon.

Bogard D. D. and Nyquist L. E. (1973) ^{40}Ar/^{36}Ar variations in Apollo 15 and 16 regolith. *Proc. Fourth Lunar Sci. Conf., Geochim. Cosmochim. Acta*, Suppl. 4, Vol. 2, pp. 1975–1985. Pergamon.

Brunfelt A. O., Heier K. S., Nilssen B., and Sundvoll B. (1973) Geochemistry of Apollo 15 and 16 materials. *Proc. Fourth Lunar Sci. Conf., Geochim. Cosmochim. Acta*, Suppl. 4, Vol. 2, pp. 1209–1218. Pergamon.

Carusi A., Cavarretta G., Cinotti F., Civitelli G., Coradini A., Funiciello R., Fulchignoni M., and Taddeucci A. (1972) Lunar glasses as an index of the impacted sites lithology: The source area of Apollo 15 "green glasses." *Geol. Rom.* **11**, 137–151.

Chou C.-L., Baedecker P. A., and Wasson J. T. (1973) Atmophilic elements in lunar soils. *Proc. Fourth Lunar Sci. Conf., Geochim. Cosmochim. Acta*, Suppl. 4, Vol. 2, pp. 1523–1533. Pergamon.

Dence M. R., Grieve R. A. F., and Plant A. G. (1974) The Imbrium basin and its ejecta (abstract). In *Lunar Science—V*, pp. 165–167. The Lunar Science Institute, Houston.

Duncan A. R., Erlank A. J., Willis J. P., and Sher M. K. (1974) Compositional characteristics of the Apollo 17 regolith (abstract). In *Lunar Science—V*, pp. 184–186. The Lunar Science Institute, Houston.

Eglinton G., Mays B. J., Pillinger C. T., Agrell S. O., Scoon J. H., Agrell J. E., Arnold A. A., Kaplan I. R., Petrowski C., Smith J. W., Gast P. W., Nyquist L. E., Dran J. C., Maurette M., Bowell E., Dollfus A., Geake J. E., Cadogan P. H., Turner G., Schultz L., and Signer P. (1974) The history of lunar breccia 15015 (abstract). In *Lunar Science—V*, pp. 217–219. The Lunar Science Institute, Houston.

Ganapathy R., Morgan J. W., Krähenbühl U., and Anders E. (1973) Ancient meteoritic components in lunar highlands rocks: Clues from trace elements in Apollo 15 and 16 samples. *Proc. Fourth Lunar Sci. Conf., Geochim. Cosmochim. Acta*, Suppl. 4, Vol. 2, pp. 1239–1261. Pergamon.

Green D. H. and Ringwood A. E. (1973) Significance of a primitive lunar basaltic composition present in Apollo 15 soils and breccias. *Earth Planet. Sci. Lett.* **19**, 1–8.

Haskin L. A., Helmke P. A., Blanchard D. P., Jacobs J. W., and Telander K. (1973) Major and trace element abundances in samples from the lunar highlands. *Proc. Fourth Lunar Sci. Conf., Geochim. Cosmochim. Acta*, Suppl. 4, Vol. 2, pp. 1275–1296. Pergamon.

Heymann D. and Yaniv A. (1970) Ar^{40} anomaly in samples from Tranquillity base. *Proc. Apollo 11 Lunar Sci. Conf., Geochim. Cosmochim. Acta*, Suppl. 1, Vol. 2, pp. 1261–1267. Pergamon.

Heymann D., Jordan J. L., Walton J. R., and Lakatos S. (1974) An inert gas "borscht" from the Taurus-Littrow site (abstract). In *Lunar Science—V*, pp. 331–333. The Lunar Science Institute, Houston.

Hintenberger H. and Weber H. W. (1973) Trapped rare gases in lunar fines and breccias. *Proc. Fourth Lunar Sci. Conf., Geochim. Cosmochim. Acta*, Suppl. 4, Vol. 2, pp. 2003–2019. Pergamon.

Hübner W., Kirsten T., and Kiko J. (1974) Rare gases in Apollo 17 soils with emphasis on analysis of size and mineral fractions of soil 74241 (abstract). In *Lunar Science—V*, pp. 369–371. The Lunar Science Institute, Houston.

Huneke J. C., Jessberger E. K., Podosek F. A., and Wasserburg G. J. (1973) ^{40}Ar/^{39}Ar measurements in Apollo 16 and 17 samples and the chronology of metamorphic and volcanic activity in the Taurus-Littrow region. *Proc. Fourth Lunar Sci. Conf., Geochim. Cosmochim. Acta*, Suppl. 4, Vol. 2, pp. 1725–1756. Pergamon.

Husain, L. (1972) The ^{40}Ar–^{39}Ar and cosmic ray exposure ages of Apollo 15 crystalline rocks, breccias and glasses. *The Apollo 15 Lunar Samples* (editors J. W. Chamberlain and C. Watkins), pp. 374–377. The Lunar Science Institute, Houston.

Husain L. (1974) ^{40}Ar–^{39}Ar chronology and cosmic ray exposure ages of the Apollo 15 samples. *J. Geophys. Res.* **79**, 2588–2606.

Jordan J. L., Heymann D. and Lakatos S. (1974) Inert gas patterns in the regolith at the Apollo 15 landing site. *Geochim. Cosmochim. Acta*, **38**, 65–78.

Kirsten T., Duebner J., Horn P., Kaneska I., Kiko J., Schaeffer P., and Thio S. K. (1972) The rare gas record of Apollo 14 and 15 samples. *Proc. Third Lunar Sci. Conf., Geochim. Cosmochim. Acta*, Suppl. 3, Vol. 2, pp. 1865–1889. MIT Press.

Krähenbühl U., Ganapathy R., Morgan J. W., and Anders E. (1973) Volatile elements in Apollo 16 samples: Implications for highland volcanism and accretion history of the moon. *Proc. Fourth Lunar Sci. Conf., Geochim. Cosmochim. Acta*, Suppl. 4, Vol. 2, pp. 1325–1348. Pergamon.

Lakatos S., Heymann D., and Yaniv A. (1973) Green spherules from Apollo 15: Inferences about their origin from inert gas measurements. *The Moon*, **7**, 132–148.

Morgan J. W., Ganapathy R., Higuchi H., Krähenbühl U., and Anders E. (1974) Lunar basins: Tentative characterizations of projectiles, from meteoritic elements in Apollo 17 boulders (abstract). In *Lunar Science—V*, pp. 526–528. The Lunar Science Institute, Houston.

Podosek F. A. and Huneke J. C. (1973) Argon in Apollo 15 green glass spherules (15426): ^{40}Ar–^{39}Ar age and trapped argon. *Earth Planet. Sci. Lett.* **19**, 413–421.

Prinz M., Dowty E., and Keil K. (1973) A model for the origin of orange and green glasses and the filling of mare basins. *Trans. Amer. Geophys. Union* **54**, 605–606.

Reid A. M., Lofgren G. E., Heiken G. H., Brown R. W., and Moreland G. (1973) Apollo 17 orange glass, Apollo 15 green glass and Hawaiian lava fountain glass. *Trans. Amer. Geophys. Union* **54**, 606–607.

Schaeffer O. A. and Husain L. (1973) Isotopic ages of Apollo 17 lunar material. *Trans. Amer. Geophys. Union* **54**, 614.

Schonfeld E. and Meyer C., Jr. (1973) The Old Imbrium hypothesis. *Proc. Fourth Lunar Sci. Conf., Geochim. Cosmochim. Acta*, Suppl. 4, Vol. 1, pp. 125–138. Pergamon.

Silver L. T. (1972) Uranium–thorium–lead isotopes and the nature of the mare surface debris at Hadley-Apennine. In *The Apollo 15 Lunar Samples* (editors J. W. Chamberlain and C. Watkins), pp. 388–390. The Lunar Science Institute, Houston.

Silver L. T. (1973) Uranium–thorium–lead isotope relations in the remarkable debris blanket at Hadley-Apennine (abstract). In *Lunar Science—IV*, pp. 669–674. The Lunar Science Institute, Houston.

Silver L. T. (1974) Patterns of U–Th–Pb distributions and isotope relations in Apollo 17 soils (abstract). In *Lunar Science—V*, pp. 706–708. The Lunar Science Institute, Houston.

Tatsumoto M., Hedge C. E., Knight R. J., Unruh D. M., and Doe B. R. (1972) U–Th–Pb, Rb–Sr, and K measurements on some Apollo 15 and Apollo 16 samples. In *The Apollo 15 Lunar Samples* (editors J. W. Chamberlain and C. Watkins), pp. 391–395. The Lunar Science Institute, Houston.

Tatsumoto M., Nunes P. D., Knight R. J., Hedge C. E., and Unruh D. M. (1973) U–Th–Pb, Rb–Sr, and K measurements of two Apollo 17 samples. *Trans. Amer. Geophys. Union* **54**, 614–615.

Taylor S. R., Gorton M. P., Muir P., Nance W., Rudowski R., and Ware N. (1973) Lunar highlands composition: Apennine Front. *Proc. Fourth Lunar Sci. Conf., Geochim. Cosmochim. Acta*, Suppl. 4, Vol. 2, pp. 1445–1459. Pergamon.

Wänke H., Baddenhausen H., Dreibus G., Jagoutz E., Kruse H., Palme H., Spettel B., and Teschke F. (1973) Multielement analyses of Apollo 15, 16, and 17 samples and the bulk composition of the moon. *Proc. Fourth Lunar Sci. Conf., Geochim. Cosmochim. Acta*, Suppl. 4, Vol. 2, pp. 1461–1481. Pergamon.

Wasson J. T., Chou C.-L., Robinson K. L., and Baedecker P. A. (1974) Trace elements in Apollo 16 rocks and soils. Preprint.

Yaniv A. and Heymann D. (1972) Atmospheric Ar^{40} in lunar fines. *Proc. Third Lunar Sci. Conf., Geochim. Cosmochim. Acta*, Suppl. 3, Vol. 2, pp. 1967–1980. MIT Press.

Proceedings of the Fifth Lunar Conference
(Supplement 5, Geochimica et Cosmochimica Acta)
Vol. 2 pp. 1659–1683 (1974)
Printed in the United States of America

Meteoritic and volatile elements in Apollo 16 rocks and in separated phases from 14306

R. Ganapathy, John W. Morgan, Hideo Higuchi,
and Edward Anders

Enrico Fermi Institute and Department of Chemistry

Alfred T. Anderson

Department of Geophysical Sciences, University of Chicago, Chicago, Illinois 60637

Abstract—The trace elements Ag, Au, Bi, Br, Cd, Cs, Ge, Ir, Ni, Rb, Re, Sb, Se, Te, Tl, U, and Zn were analyzed by radiochemical neutron activation in 13 Apollo 16 samples (11 rocks and 2 soil separates) and 3 Apollo 14 samples (magnetic and nonmagnetic separates from 14306 and a metal fragment from 14258). Thin section examination of typical soil separates (12 fractions from 67602,14 and 5 from 67702,16) suggests that coarse crystalline breccias, in which larger grains of metal and sulfide tend to associate with vugs and oikocryst boundaries, are richer in Fe metal, siderophiles and alkalis than fine-grained types. In fine breccias, dense types are higher in these elements than more porous varieties. At North Ray Crater at least 3 types of ancient meteoritic component are recognized by abundance ratios of Ir, Re, Au, Ge, and Sb. These have been tentatively assigned to basin ejecta from Imbrium, Nectaris, and Humorum (?) in increasing order of age and inferred depth, with lower strata being more heavily represented at the crater rim. Rocks from the upper strata are more prominent near South Ray where Imbrium-type material predominates and meteorite-free cataclastic anorthosites may represent (inverted) deep Nectaris ejecta. Rocks from the southern part of the Apollo 16 site are often enriched in Tl, Br, and Cd by up to a factor of 10^2. A smaller group shows Tl depletion by a factor of 6 relative to Cs. In Apollo 16 < 1 mm soils Bi, Te, and Se correlate with Ne^{21} exposure age, indicating an addition of $(2.3 \pm 0.1) \times 10^{-5}$ g Cl chondrite-type material per gram of soil per million years, and an equilibrium level of 0.91% Cl material assuming a saturation age of 400 m.y. Three soils (61221, 63321, 64501) and rock 60315 apparently contain an ancient meteoritic component not previously recognized which is characterized by very low Ir/Au (~0.2 Cl-normalized).

The fine crystalline soil separate 67602,14-3 contains a unique meteoritic component characterized by a very high ratio of refractory siderophiles to Au and Ni (Ir/Au = 6.4; Re/Au = 7.0 relative to Cl), indicative of a high-temperature condensate.

Siderophile elements in 14306 metal are enriched ~ 100 fold over the bulk rock. The metal, bulk rock and a rhyolitic clast have similar Ir/Au (~ 0.5) but different Ge/Au, suggesting Ge redistribution in this breccia.

Introduction

The relationship between the stratigraphy at Apollo 16 and the trace element distribution (Krähenbühl et al., 1973c) may be profitably reexamined in the light of recent evidence from two sources. First, surface exposure ages are now available, allowing many of our samples to be related to specific local impact events (Behrmann et al., 1973; Drozd et al., 1973; Huneke et al., 1973; Kirsten et al., 1973; Marti et al., 1973; Stettler et al., 1973; Turner et al., 1973). Second, trace element analyses of Apollo 17 rocks and especially boulders have enabled at least 5 ancient meteoritic components to be tentatively assigned to individual basin-forming impacts (Morgan *et al.*, 1974). A petrographic examination has been made of separates from 2 soils from Station 11 (North Ray Crater), some of which have also been analyzed for trace elements (Krähenbühl *et al.*, 1973c; present work).

Materials and Methods

Samples

Coarse 1–2 mm soils were cleaned ultrasonically in redistilled acetone. After separating a magnetic fraction, we handpicked texturally distinct fractions under a binocular microscope, using stainless steel tools.

Fragments of breccia 14306 weighing 2.3 g were finely ground and separated into a magnetic and a nonmagnetic fraction with a small hand magnet. Each fraction was subjected to at least 10 cycles of ultrasonic disaggregation in redistilled acetone followed by further magnetic separation. Intermediate partially magnetic fractions and residue from the acetone treatment were set aside, and the final products were 1 mg of a pure magnetic separate and 0.87 g of relatively nonmagnetic material.

Neutron activation

Samples were sealed in ultrapure silica vials and irradiated in a thermal flux of $\sim 10^{14} \, n \, cm^{-2} \, sec^{-1}$ for 10 days at the National Bureau of Standards reactor at Gaithersburg, Maryland. The analytical procedure has recently been described in detail by Keays *et al.* (1974).

The transfer of 1 mg of finely divided magnetic material from 14306 after irradiation required special care. After opening the vial the larger particles were collected on soft iron wire attached to a magnet. The whole wire was dropped into the fusion mixture. The vial was "rinsed" with iron powder, mixing magnetically, and the powder all transferred to the fusion mixture. The vial was then scavenged using magnetized wire. The procedure was repeated several times until the activity of the vial was down to background levels, and then repeated twice more.

Trace Elements

The new analytical data are given in Table 1. The accuracy and precision of our method is generally 10% or better, except where governed by counting statistics at very low levels (Keays *et al.*, 1974).

It is interesting to seek resemblances between Apollo 16 soil separates and large rocks using the trace element data. The regolith at the Apollo 16 site is apparently made up of ejecta from several basin-forming events, plus a much smaller contribution from local craters. At least 5 types of ancient meteoritic material may be recognized from the relative proportions of Ir, Au, and Ge; or Re, Au, and Sb (Morgan *et al.*, 1974). Figure 1 illustrates the distribution of Apollo 16 rocks and soil separates between these groups on the basis of Ir, Au, and Ge.

Table 1. Abundances of trace elements in Apollo 16 and 14 samples (ppb; Ni, Zn, and Rb, ppm).

	Class.*	Ir	Re	Au†	Ni	Sb†	Ge	Se	Te	Ag	Br	Bi	Zn†	Cd	Tl	Rb	Cs	U
Rocks																		
60017,8A‡	B4 DG-MMSB	1.24	0.103	4.25	47	0.37	9.4	21	6.8	3.4	230	0.36	3.3	5.0	1.76	0.70	41	135
60017,8B	B4 DG-MMSB	1.75	0.161	0.41	35	1.01	20	17.7	7.2	0.59	560	0.24	5.4	4.1	3.09	0.78	49	117
60095,5	Glass	25.4	2.17	7.11	560	2.62	306	167	26	1.2	136	0.59	1.55	1.8	1.66	1.67	64	670
60315,79	C2 Poik	11.0	1.36	18.3	798	11.0	625	520	4.7	0.94	23	0.13	0.30	5.0	0.20	10.8	540	2280
63335,17	B5	1.32	0.136	0.81	70	3.19	28	24	6.1	4.9	310	2.02	16.3	12.4	2.03	1.20	67	159
63355,7	B4 DG-MRB	16.6	2.27	18.4	800	5.87	1910	340	38	2.3	230	0.44	5.2	5.7	6.0	6.5	300	980
64455,25	Glass	40.6	4.11	12.7	905	3.16	500	390	12.8	1.6	930	0.63	2.4	5.2	83.2	3.9	144	860
64455,27	C2 Basalt	2.25	0.284	1.56	80	0.45	62	190	2.5	1.2	1200	0.08	2.2	5.3	109	6.6	280	1430
65016,7	Glass	26.3	2.29	7.19	532	1.66	225	96	12.8	0.59	42	0.23	0.52	1.5	0.60	1.44	62	650
67915,63a‡	B4 matrix	7.33‖	0.670	1.90	160	0.41	180	23	18	2.8	123	0.42	5.10	2.60	0.54	0.80	45	180
67955,20	C2(B1?) LMB	5.56	0.572	1.60	231	0.23	59	26	9.7	1.2	199	0.34	6.7	4.3	0.66	1.20	64	360
68115,77	B5 DG	0.040	0.005	2.28	≤7	0.19	6.7	3.4	0.4	0.19	700	0.22	0.47	81	130	0.043	8.1	1.8
69935,8	B4	12.7	1.55	11.9	583	3.63	325	190	2.8	1.3	220	0.19	0.88	6.6	0.25	5.9	260	870
Soil Separates																		
64502,4a	Magnetic	76.7	10.4	73.7	4070	31.1	6995	804	14	8.4	200	1.8	5.3	199	50.4	7.1	300	1720
67702,16-5	Breccia	2.62	0.279	1.45	120	0.44	123	34	5.0	0.61	13	0.12	1.35	2.8	0.44	1.1	54	180
67702,16-6	Breccia	1.63	0.144	0.34	60	0.16	17	19.0	0.7	0.40	6.3	0.15	1.35	2.8	0.21	0.34	18	57
Apollo 15																		
15265,13,6	B matrix	0.023	0.006	0.09	55	0.14	6.2	117	28	6.4	30	0.20	0.97	0.66	0.25	0.84	36	167
Apollo 14																		
14306,35,8§	B Rhyolite	3.6	0.28	2.2		11.8	39	50	23	0.92	550	1.2	2.9	10	28	114	4500	7200
14306,35,9§	B Bulk	8.14	0.64	5.3		1.41	390	99	5.3	2.5	270	0.28	2.7	31	6.0	18.6	1030	4800
14306,35,10	B Metal	950	90.2	553	45600	525	64600	98	53	73	17600	32	41600	1430	24.0	2.7	440	630
14306,35,11	B Nonmagn.	3.71	0.361	3.05	151	22.8	195	75	3.9	1.6	220	0.56	4.0	36	5.87	23	1100	5200
14258,36,14	S Metal	36.2	4.75	45.2	3570	163	129000	430	16	1.3	7.8	0.37	0.65	17	0.45	2.1	61	820

*Classified according to Wilshire et al. (1973) and Warner et al. (1973):

B = breccia
B1 = light matrix, light clast
B4 = dark matrix, light clast
B5 = dark matrix, dark clast
C2 = crystalline, metaclastic
DG = devitrified glass with
  plagioclase and/or lithic relics

LMB = light matrix breccia
MMSB = melted matrix shocked
  breccia
MRB = mesostasis-rich basalt
poik = poikilitic rock with plagioclase,
  olivine, and/or lithic relics

†Italicized values are high, owing to contamination.
‡Krähenbühl et al. (1973c).
§Ganapathy et al. (1973).
‖Corrected value.

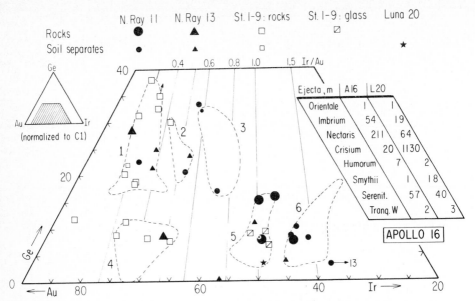

Fig. 1. Ancient meteoritic material appears to fall into 6 chemical groups (Morgan *et al.*, 1974) on the basis of Ir, Au, Ge abundances, normalized to Cl chondrites (Krähenbühl *et al.*, 1973a). Five of these groups have been tentatively associated with basin-forming impacts: 1, Imbrium; 2, Serenitatis; 3, Crisium; 5, Nectaris; 6. Humorum (?). Group 4 may be spurious, being populated by members of Groups 1 and 2 which have lost Ge. Groups 5 and 6 predominate at North Ray, Group 1 (and 4) at South Ray Crater.

Clearly, if we are to seek possible parent rocks of the soil separates, the ancient meteoritic component provides a good negative criterion; a rock and separates with the same meteoritic component may be related, but a different component strongly suggests a mismatch. All six groups are represented at Apollo 16, although Groups 2 and 3 only sparsely. The large rocks from North Ray fall almost exclusively into Groups 5 and 6, whereas the rocks from other locations chiefly occupy Groups 1 and 4. Group 4 may not be a real grouping at all but appears to be populated by members of Groups 1 and 2 (depending on Ir/Au) which have lost Ge.

The abundance of Rb, representing alkalis, and U, representing other large ion lithophile elements, serve as a link between trace elements and rock type (Fig. 2). Soil separates and rocks of similar lithophile content also tend to belong to the same meteoritic group. Most North Ray samples lie in a rather restricted range of Rb and U content. The cataclastic anorthosites of very low U and Rb content (not plotted in Fig. 2) on the one hand, and the KREEP-rich samples on the other, largely come from other areas of the Apollo 16 site.

Although there is no compelling reason for a correlation, siderophile-rich Apollo 16 material tends also to be high in alkalis (Krähenbühl *et al.*, 1973c). The variation of Au with Rb content illustrates this trend well (Fig. 3). Soil separates tend to group according to Rb, Au content and meteoritic component. The general

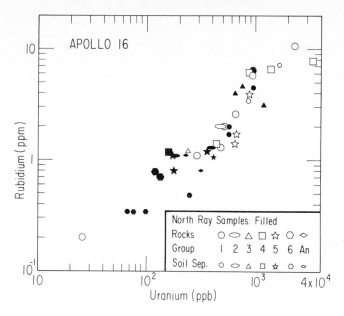

Fig. 2. The lithophile elements Rb and U give a rough indication of rock type. Most North Ray samples lie in a rather restricted range of Rb and U content. Cataclastic anorthosites (too low in Rb, U to be shown), and KREEP-rich material largely come from other areas of the Apollo 16 site. Soil separates with the same meteoritic component are similar in Rb, U content, and in many cases are likewise related to large rocks. An = anomalous.

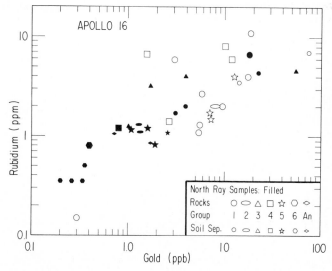

Fig. 3. Apollo 16 samples high in siderophiles tend also to be high in alkalis, and generally group according to ancient meteoritic component. North Ray samples are distinctly separated from those of other stations. An = anomalous.

Table 2. Summary of Petrographic and Selected Trace Element Data for Apollo 16 Rocks and Soil Separates.

Sample	Rb, ppm	U, ppb	Au, ppb	Group	Exposure Age, m.y.	Description*†	Petrography†	pl %	Metal % Visible	Metal % Magnetic	Metal/Sulfide ratio
60025,84	0.017	0.9	0.007	—	2–3[i]	B1 CA	maskelynite	99	<0.001		
61016,156	0.018	0.9	0.020	—	≤7[d]	C1 CA					
68115,77	0.043	1.8	—	—	2[j]	B5 DG					
61222,1b	0.10	12	0.020	—		glass colorless CA					
61222,1c	0.14	5	0.011	—		C1	breccia an	98	<0.001		<0.1
69955,11	0.15	26	0.31	1	2[g]		with glass rim		0.05		large
63502,13-1	0.34	76	0.20	6(?)	42[c]	bc wh chalky very light norite					
63502,13-8	0.34	25	0.27	6	42[c]	bc m–lt gray	breccia d f gr	80	0.006		>5
67702,16-6	0.34	57	0.34	6		bc wh mx dk clasts	breccia p pmct	75	0.005		>1
67602,14-7	0.48	29	0.36	6		B4 DG-MMSB					
60017,8	0.70; 0.78	135; 117	—; 0.41	6			bst w/an clast	70	0.02		1
63502,13-9	0.81	315	1.7	An(5?)		white coated aph light norite					
63502,13-3	1.1	410	2.6	5		bc lt–m gray f–m gr	bc f–m grained	75	0.01		1
67602,14-4	1.2	230	1.0	3(5?)							
67915,63b	0.80	180	1.9	5	30–50[bgi]	B4 FB–GB	breccia polymict	80	<0.05		>5
67915,63a	1.1	175	1.1	5	30–50[bgi]	B4 FB–GB	breccia polymict	80	<0.05		>5
67955,20	1.2	360	1.6	5	50[j]	C2(B1?) LMB					
67602,14-3	1.1	230	0.7	An		bc v f gr m gray	breccia dense f gr	70	0.002		?>1
67702,16-5	1.1	180	1.4	2		breccia m gray f gr	breccia dense f gr	70	0.0005		?>1
63502,13-2	1.3	405	1.3	2		breccia dark f gr		80	0.03	0.3[l], 0.9[m]	1
63335,17	1.2	160	0.81	4		B5	basalt pl				
68415,67	1.4	445	2.7	4	87–106[abdj]	C1 Di					

Sample					Classification (1)		Classification (2)			
65016,7	1.4	650	7.2	5	glass sphere					
60095,5	1.7	670	7.1	5	glass sphere					
64455,25	3.9	860	13	5	glass coating					0.7[t]
63502,13-11	1.7	565	3.1	1	breccia light gray					
63502,13-10	2.0	575	4.0	1	breccia gray					
61222,1a	3.4	865	14	1	bc dk coherent					
67602,14-2	4.3	955	23	1	magnetic dark		metabc ves m gr	50	0.3	5
64502,4a	7.1	1720	74	1	metal-rich					
65095,30	1.1	290	5.4	1	B2 MN		breccia p pmct	50	0.2	3
61016,132,b	1.3	485	5.6	1	C1 MRD	≤7[d]	oph di w/ an cl	65	0.3	10
61016,132,a	2.0	515	9.6	1	C1 MRD	≤7[d]	oph di w/ an cl	65	0.3	10
68815,124	2.0	570	8.3	2(1?)	B5 DG	2[g]	metabc w/ g pmct	70	0.1	0.5 / 0.6[i]
60016,6	2.6	655	5.9	1	B2 FB	1.1[e]	metabc mso	55	1	>10
62095,55	3.9	1020	18	1	C2 Di		metabreccia	50	0.2	3 / 1.2[l], 1.4[m]
62295,40	5.8	935	3.1	1	C2 MRD	310[f]	basalt int ol sp	60	0.1	1 / 0.3[i]
69935,8	5.9	870	12	1	B4		breccia pmct m gr	55	0.1	5
63355,7	6.5	980	18	1	B4 DG-MRB	3.3[g]				
67602,14-8	3.1	1200	1.7	3	metabreccia light		metabreccia poik	50	~0.5	5
67702,16-4	4.0	650	3.9	3	metabc lt-m gray		metabreccia c di	60	~0.05	1
67702,16-1	4.6	755	55	3	magnetic dk f gr					
64455,27	6.6	1430	1.6	4	C2 basalt	365[b]	metabc c gr poik	40	0.1	2
65015,51	7.8	3420	10.2	4	C2 Poik	4.5[b]	metabc c gr poik	45	10	2
60315,79	10.8	2280	18	An	C2 Poik		metabc c gr poik			1.4[k], 4.4[l]

*Rock classifications are given according to two systems.

(1) Wilshire et al. (1973): B1 = light matrix, light clast breccia; B2 = light matrix, dark clast breccia; B4 = dark matrix, light clast breccia; B5 = dark matrix, dark clast breccia; C1 = crystalline, igneous; C2 = crystalline, metaclastic.

(2) Warner et al. (1973): CA = cataclastic anorthosites; DG = devitrified glass with plagioclase and/or lithic relics; Di = diabase, "igneous", generally subophitic with plagioclase, olivine, and/or lithic relics; FB = feldspathic polymict breccia with some tan glass; GB = brown to tan glassy polymict breccia; LMB = light matrix breccia; MMSB = melted matrix shocked breccia; MN = meta-norite; MRB = mesostasis-rich diabase with plagioclase and/or lithic relics; Poik = poikilitic rock with plagioclase, olivine, and/or lithic relics.

†Abbreviations: An = anomalous; an = anorthositic; aph = aphanitic; bc = breccia; bst = basalt; c = coarse; cl = clast; d = dense; di = diabase; dk = dark; f = fine; g = glass; gr = grained; int = intersertal; lt = light; m = medium; mso = microsubophitic; mx = matrix; ol = olivine; oph = ophitic; p = porous; poik = poikilitic; pl = plagioclase; pmct = polymict; sp = spinel; v = very; ves = vesicular; wh = white.

References: (a) Huneke et al., 1973; (b) Kirsten et al., 1973; (c) Schaeffer and Husain, 1973; (d) Stettler et al., 1973; (e) Heymann et al., 1973; (f) Turner et al., 1973; (g) Behrmann et al., 1973; (h) Marti et al., 1973; (i) Lightner and Marti, 1974; (j) Drozd et al., 1974; (k) Brecher et al., 1973; (l) Nagata et al., 1973; (m) Pearce et al., 1973.

separation of North Ray samples from those of other sites, as pointed out previously, is even more marked here. The data plotted in Figs. 1, 2, and 3 are summarized in Table 2. Samples are generally arranged in order of increasing Rb, and grouped according to ancient meteoritic component. Very brief petrographic data are given for each sample. In these groupings, we have attempted to relate the soil separates to large rocks of similar trace element composition and petrology. Detailed petrographic descriptions are given below.

PETROGRAPHY OF 1–2 mm SOIL FRACTIONS

General comments

Much of the material examined is fairly dense crystalline breccia in which the abundance of metallic iron crudely correlates with grain size: less than 0.01% metal for fine-grained varieties and up to a few percent metal in some coarsely crystalline breccias with sieve textured pyroxene grains (oikocrysts) up to 1 mm long. (A similar conclusion was made recently on the basis of magnetic measurements; Pearce and Simonds, 1974.) Some multigeneration breccias with a fine, porous or glassy matrix contain fragments of both metal-rich and metal-poor breccia. In coarsely crystalline metal-rich breccias, the coarse metal particularly occurs at boundaries between pyroxene oikocrysts, and, accompanied by practically all of the sulfide, is associated with vugs and porous regions. It appears to have moved as vapor into the voids, or, alternatively, to have emanated the vapor which caused the cavities.

In the following petrographic descriptions, the macroscopic appearance is briefly given first (in italics). Not all fractions described have yet been analyzed, but all are included here for ease of reference. Estimates of metallic iron determined optically are given in *volume* percent and are accurate perhaps to a factor of 5. Metal loss during thin section preparation may lead to systematically low estimates. There exists a close correlation ($r = 0.852$) between Au and metal content in lunar rocks, determined magnetically. For the sample size used for our activation analyses (0.01–0.10 g), sampling errors are generally not serious.

Soil fractions which have been analyzed (Krähenbühl *et al.*, 1973c and this work) are marked with an asterisk (*).

Medium to coarse metabreccias

These are common in the soil and among the rocks (60016,88, 60315,65, 65015,90, and 66095,79). Fractions 67602,14-1, 67602,14-2, and 67602,14-8 have metal contents approaching those of some rocks, but 67702,16-4 and metabreccia in the polymict breccia 67602,14-9 are only duplicated by some metal-poor clasts in 65095,53.

Fraction 67602,14-2 closely resembles rock 66095. Both contain similar amounts of visible metal and siderophile elements, and have a Group 1 ancient meteoritic component. A dark coherent breccia, 61222,1a, similar to those described by Steele and Smith (1973), resembles these two samples in siderophile and alkali trace elements. Neither soil separate has undergone the volatile (fumarolic?) enrichment seen in 66095 (Krähenbühl *et al.*, 1973b, 1973c).

Breccias 67602,14-8 and 67702,16-4 are quite similar to each other in nonmeteoritic trace elements (except for very high Br in the latter), but do not resemble any of the large rocks studied. Siderophile element concentrations differ by more than a factor of two, clearly a reflection of the irregular metal distribution observed in thin section. Both breccias contain a Group 3 meteoritic component, which has not been found in any large rock at Apollo 16. Magnetic fraction 67702,16-1 has a prominent Group 3 component (55 ppb Au) and resembles 67702,16-4 in Se, Te, Bi, Zn, Cd, Rb, Cs, and U.

67602,14-1: Magnetic fraction, mainly recrystallized breccia. Two medium-grained fragments with granular to subophitic textures. These are probably high-temperature recrystallized breccias. They are vesicular and comparatively metal-rich. Metal and sulfide occur together in one vesicle. A large metal grain (nital etched) has an inclusion of another phase (Fe_3P?). Perhaps 0.02–0.5% metal is present.

*67602,14-2: Dark magnetic particles (similar to 67602,14-1). Vesicular breccia with large metal grains. The granular to microsubophitic texture suggests high-temperature recrystallization. There may be $0.3 \pm 0.1\%$ metal.

*67602,14-8: Plagioclase-rich crystalline fragments. Medium-grained, with poikilitic texture. The grain size, texture and vesicles attest to recrystallization in a partially melted regime. Metal is between 0.2 and 1.0%, irregularly distributed.

*67702,16-4: Light- to medium-gray crystalline fragments. Medium-grained fraction. One fragment is holocrystalline (no glass) and granular, suggestive of recrystallization. It has about 70% plagioclase, 20% pyroxene, 5% olivine, traces of ilmenite, and ~0.05% metal mostly as one large irregular grain. A second fragment is coarser grained. It contains cristobalite and minor interstititial glass in addition to dominant plagioclase and intergranular pyroxene and/or olivine. About 0.0005% metal occurs as tiny grains associated with residual glass and sulfide.

Fine crystalline breccias

The soils have abundant, very finely crystalline, metal-poor, plagioclase-rich breccias which do not seem to be represented even in the clasts of the polymict breccias (67915,82 and 65095,53). They are substantially lower than the coarse crystalline breccias in alkalis, U and Au, and may be divided into a dense fine-grained variety (porosity 1–2%) and a porous fine to medium-grained type.

(a) DENSE TYPE. These are a mixed bag. Breccia 67702,16-5 is almost identical in trace element composition to the fine-grained breccia 63502,13-2. Both resemble rocks 63335 and 68415 in Rb, Cs, and U. The rocks contain a Group 4 ancient meteoritic component, whereas the breccias belong to Group 2. Yet all have similar Ir/Au. There is a possibility that Ge may redistribute during metamorphism, or be lost by volatilization, and 68415 may represent a recrystallized example of the source material of the soil separates.

Fine crystalline breccia 67602,14-3 contains a unique meteoritic component (Fig. 1) characterized by a very high ratio of refractory siderophiles to Au and Ni (Ir/Au = 6.4; Re/Au = 7.0; Ir/Ni = 3.2; Re/Ni = 3.5 relative to Cl chondrites). Ratios of this order would be expected for a high-temperature condensate (Grossman, 1972) and may be representative of a refractory-rich moon (Wänke et al., 1973; Krähenbühl et al., 1973c; Ganapathy and Anders, 1974).

*67602,14-5: Very fine-gained medium-gray fragments with white inclusions. A dense, fine-grained breccia. The porosity is 1–2%, mostly as cracks. There is about 70% plagioclase and 30% pyroxene. Tiny sparsely disseminated metal grains amount to 0.002% or less and are difficult to distinguish from sulfide because of the small size.

67702,16-5: Fine-grained medium-gray breccia. A dense, fine-grained breccia, similar to 67602,14-3.

(b) POROUS TYPE. Two of the porous breccias, 67602,14-7 and 67702,16-6, apparently represent a substantial component in soils in the North Ray Crater ejecta blanket. They are very similar to each other and to 63502,13-1 and 63502,13-8, in alkalis, siderophiles and meteoritic component (Group 6). Among the large rocks only 60017 contains this component, and compositionally is the closest relative to these porous breccias.

A third porous breccia, 67602,14-4, apparently containing a Group 3 meteoritic component, differs considerably in composition from the coarse crystalline breccias having the same component. On an AuIrGe classification diagram (Fig. 1) 67602,14-4 is clearly separated from the coarse breccias, plotting at the low Ge extremity of Group 3. A composition in this region of the diagram could arise from a mixture of Group 1 or 2 with Group 5 or 6 material.

67702,16-6: Fine-grained medium- to light-gray breccia. A fine- to medium-grained rock. It resembles 67702,16-5 except for a more evenly distributed porosity of ~3%. Metal (~0.006%) is present as tiny, evenly disseminated grains.

67602,14-7: White chalky matrix with dark vitreous inclusions. A fairly porous breccia. The matrix has 3–5% porosity, mostly as cracks around and between grains. Crystal and rock fragments include a dense fine-grained breccia which contains most of the metal as tiny grains. The metal content is perhaps 0.005% overall. There is considerable ilmenite (~0.2%).

67602,14-4: Medium- to fine-grained light- to medium-gray breccia. A moderately dense rock. The granular texture suggests moderate temperature (subsolidus) recrystallization. Porosity is ~3%, as cracks and vugs. Metal (0.01%) is present as tiny grains in matrix.

Polymict breccias

These are common to both soils and rocks (65095,53 and 67915,82) and are generally rather poor in metal (<0.01% for the soil fragments and <0.2% for the rock sections). The distribution of metal varies greatly from one fragment or clast to another.

67702,16-7: Chalky white matrix with dark clasts. A medium-grained basaltic or gabbroic fragment. The grains are broken, but the texture is still partly intact. Traces of metal (probably <0.001%) are present in some olivine grains.

67602,14-5: Chalky white feldspar-rich fragments. A complex breccia fragment. The interior is porous and friable but the outer 0.5 mm or so is progressively less porous and more coherent. Granular textured gabbroic rock fragments are common. The metal content is about 0.0002% or less.

67602,14-9: Polymict breccia with white chalky matrix. A fine-grained, porous matrix (5% porosity) which contains metagabbro and dense metabreccia fragments. The metal is mostly restricted to the fragments, and is generally about 0.002 percent of the total but up to 0.1% within individual clasts.

Anorthositic breccias

Except for possible fragments in the polymict breccias, this type of anorthositic breccia is unrepresented in the rock sections studied.

67702,16-8: Fine-grained light-gray breccia. A porous breccia of about 80–90% plagioclase. About 10% porosity evenly disseminated. Fragments are noticeably subround and weakly sorted. Maybe a crystal tuff. Metal less than 0.0002% as tiny grains.

Miscellaneous samples

The glass (67602,14-6) and variolitic basalt (67602,14-10) have counterparts in the rocks 68815,147 and 60017,124, respectively.

67602,14-6: Medium- to dark-brown splashed glass with adhering fragments. Vesicular glass with irregular crystal fragments. There is about 30–40% porosity. Rare metal (about 0.0003% or less) is present as tiny spheres.

67602,14-10: Miscellaneous material. (a) An intersertal to variolitic basalt. Radiating sheaves of plagioclase and pyroxene have interstitial devitrified glass. Minor metal (about 0.003%) is present as tiny interstitial grains. Some interstitial voids are primary (diktytaxitic). The rock contains about 70% plagioclase.

(b) A large olivine crystal rimmed by fine-grained breccia. Olivine has two tiny metal inclusions and one small glass inclusion. A second fragment consists of a granular mosaic of medium-grained plagioclase and olivine with 0.01% metal as tiny interstitial grains associated with sulfide.

Rock Petrography

60016,88. A fractured microsubophitic metabreccia. The outlines of fragments are indistinct. There are some vuggy, diktytaxitic patches and many large blobs of metal up to 0.2 mm diameter. There is ~ 1% metal, ~ 50% pyroxene, and ~ 5% olivine.

60017,124. A vesicular microvariolitic basalt. Radiating plagioclase laths are up to 0.2 mm long. Irregular fragments of cherty anorthosite with chill margins are present and rare (0.02%) metal occurs as irregular blobs up to 0.1 mm diameter. The metal/sulfide ratio is ~ 1.

60315,65. A poikilitic diabase (described in detail by Bence *et al.*, 1973; Hodges and Kushiro, 1973; and Walker *et al.*, 1973). There are large, irregular, poikilitic to subophitic blobs of Fe metal up to 100 μ in between oikocrysts and as blobs up to 1 mm diameter in vugs. The metal contains 5.0% Ni (Nagata *et al.*, 1973). Sulfide is associated with Fe in vuggy regions which are interstitial to oikocrysts. There is ~ 10% metal visible in this thin section. Magnetic measurements suggest that the metal content is very variable (1.36% Brecher *et al.*, 1973; 4.45% Nagata *et al.*, 1973).

61016,215. A compound rock.

(A) 100% maskelynite. No pyroxene, olivine, or metal is present except for crystallization of basaltic glass along cracks.

(B) An ophitic diabase. It consists chiefly of plagioclase (mostly as an isotropic maskelynite), clinopyroxene, and perhaps olivine. Interstitial devitrified glass is apparently not shock produced. The clinopyroxene and olivine are badly cracked. There appears to be ~0.3% metal. The metal/troilite ratio is 10:1, mostly as subcircular blobs up to 50 μ diameter.

62295,76. A variolitic and intersertal spinel olivine basalt. [Described in detail as an allivalite (Agrell *et al.*, 1973) or spinel troctolite (Brown *et al.*, 1973; Nord *et al.*, 1973; Walker *et al.*, 1973).] Irregular metal grains, up to 50 μ diameter, occur sometimes as inclusions in olivine, but mostly interstitially associated with devitrified glass. There appears to be ~ 0.01% visible metal, in contrast to 0.3% from magnetic measurements on a different sample (Brecher *et al.*, 1973). The metal to sulfide ratio is about 1:1.

65015,90. A poikilitic diabase. (Described in detail by Albee *et al.*, 1973). Sulfide oikocrysts are interstitial to orthopyroxene oikocrysts and are associated with void space. There are plagioclase and breccia fragments up to 0.3 mm. Iron occurs as oikocrysts and as blobs ~ 0.1 mm diameter, especially in vesicles. There appears to be 0.1% metal and a metal to sulfide ratio of 2:1. Thin section 65015,83 is richer in metal containing 1.1% (with 4.6% Ni) and 0.3% troilite (Albee *et al.*, 1973).

65095,53. A porous, polymict, multigeneration breccia. The porosity is ~ 15% occurring mostly at the fractures and fragment boundary cracks. There is ~ 5% of glass as irregular fragments and devitrified spheres. The largest and most abundant clasts are of gray mottled metabreccia with as much as 0.5% metal as irregular blobs up to 60 μ across. There is some alteration of Fe to gray reflective goethite. Large clods of metal may be intergranular or independent in matrix. There is ~ 0.2% metal and the metal to sulfide is ~ 3:1.

66095,79. A subophitic metabreccia (also described by Taylor *et al.*, 1973). Porosity is 5%, occurring as microcracks and microvesicles up to 0.05 mm diameter. Oikocrysts of orthopyroxene are up to 0.1 mm long. Plagioclase occurs as euhedral to subhedral laths up to 0.05 mm long in orthopyroxene. Plagioclase and anorthositic clasts are irregular or subangular. Metal and sulfide is present in polygranular aggregates which are irregular to subophitic between plagioclase. There is occasionally some alteration of metal and sulfide to gray reflective goethite. The visible metal content is ~ 0.2%, with a metal to sulfide ratio of ~ 3:1. Nagata *et al.* (1973) report 1.21% metal (6% Ni in kamacite) and Pearce *et al.* (1973) 1.44% by magnetic measurements.

67915,82. A fractured polymict breccia. The clast population has been studied by Roedder and Weiblen (1974). The rock consists of irregular subrounded mineral and rock fragments set in a brownish occult matrix. The plagioclase fragments show undulating to cherty extinction. There are granular pyroxene drops in a "mottled" plagioclase matrix. Fragments of ilmenite are irregular. There are a few fragments of quartz and perhaps of Fe-rich pyroxene. A trace of magnetite has ilmenite lamellae. Metal (<0.05%) is up to 40 μ across.

*68415,153. A coarse-grained plagioclase basalt. It has been described in detail by Helz and Applemann (1973), Nord *et al.* (1973), and Walker *et al.* (1973). It consists of ~75% sub-euhedral plagioclase, ~25% subophitic orthopyroxene interstitial to plagioclase, ~1% ilmenite, and ~2% residuum. There is <0.1% interstitial Fe especially with residuum. The Fe is variable in amount [0.3% Nagata *et al.* (1973), 0.9% Pearce *et al.* (1973)] and the Ni distribution of kamacite is bimodal, mostly being 12% Ni with a smaller group at 4–9% (Gancarz *et al.*, 1972; Nagata *et al.*, 1973).

68815,147. A glassy, polymict metabreccia. This rock has been described in detail by Brown *et al.* (1973). It is made up mostly of devitrified glass in streamers, with some mottled clasts. Some clasts are very rich in sulfide, and there are trace SiO_2 clasts. Metal occurs as tiny grains < 0.05 mm diameter, the metal to sulfide ratio is ~ 1:2. There is 0.6% metal containing 6% Ni in kamacite (Nagata *et al.*, 1973).

APOLLO 16 STRATIGRAPHY

In our previous Apollo 16 paper (Krähenbühl *et al.*, 1973c), we briefly examined the stratigraphy of the Apollo 16 site, attempting to use trace siderophile and alkali elements as horizon "markers." At North Ray Crater there appeared to be a close association between rock type, "marker" chemistry and the crater stratigraphy suggested by the AFGIT (1973) and Ulrich (1973a). More recently, the stratigraphy of North Ray Crater has been discussed from a petrological viewpoint by Delano *et al.* (1973) and Taylor *et al.* (1973).

North Ray samples were collected by the astronauts from areas very near the crater (less than 2 crater radii) and should therefore represent mainly deep ejecta. South Ray Crater was approached only to within 10 crater radii, and hence the distant ejecta collected should be largely derived from the upper strata. Proximate samples were taken of several older craters, but the latter were either small (Buster, Flag) or so old that few of their ejected rocks have survived.

North Ray Crater

The geology of North Ray Crater has been discussed by the AFGIT (1973), Ulrich (1973a), more recently by Hodges *et al.* (1973) and in detail by Ulrich (1973b). In our previous discussion of North Ray Crater (Krähenbühl *et al.*, 1973c), we tentatively associated the following rock types with the three stratigraphic units suggested by the AFGIT (1973).

(1) The uppermost unit of light matrix breccias is probably represented mainly by B2 breccias (e.g. 60016 and 65095). These are relatively high in Au (> 3 ppb) and Rb (> 1 ppm) and have the Ir/Au characteristic of Imbrium ejecta (Group 1 in our later classification, Morgan *et al.*, 1974).

(2) The light friable unit, underlying the light matrix breccias, may consist of cataclastic anorthosites, which are 10^2 times lower in Au and Rb. Since these rocks appear to be entirely free of any meteoritic material, they probably are deep ejecta from a nearby basin, most probably Mare Nectaris.

(3) The lowest dark unit may be represented by samples from Outhouse Rock 67915 which have moderate levels of Au (1.1 and 1.9 ppb) and Rb (0.8–1.1 ppm), and a characteristically high Ir/Au (Group 5 in our current classification scheme). These may be shallow ejecta from Nectaris, which should also contain debris from older basins. Let us see how the new data affect the picture.

The trace elements show little correlation with petrographic breccia type (Wilshire *et al.*, 1973). Since the texture of a rock may be changed drastically by shock melting and vitrification, the chemical composition may be a more fundamental criterion. Consequently, the trace elements will be given major weight in the following discussion. Of the 6 North Ray rocks analyzed by us, only one appears to represent the uppermost stratum: 63355, which has a Group 1 (= Imbrium) meteoritic component. The 2 North Ray stations are probably too close to the crater rim to have abundant upper level ejecta. In addition, since North Ray Crater is on a ridge, the upper strata may have been attenuated by mass wasting. Imbrium-type Group 1 material is well represented among soil separates from both North Ray stations.

The meteorite-free, low-alkali cataclastic anorthosites which we have interpreted as deep Nectaris ejecta (Krähenbühl *et al.*, 1973c) are not widely represented among North Ray rocks. The extremely low-alkali anorthosites analyzed by us to date (60025, 61016, 68115) are all South Ray ejecta judging from exposure ages (Lightner and Marti, 1974; Stettler *et al.*, 1973; Drozd *et al.*, 1974). The most similar North Ray rock is 67075 which appears to be free of meteoritic material (1 ppm Ni) but is higher than 60025 in Rb (0.6–0.8 ppm) and rare earth elements (LSPET, 1973; Haskin *et al.*, 1973).

The presence of shallow Nectaris material in the third unit (Krähenbühl *et al.*, 1973c) is confirmed by additional samples (67955 and several soil separates) containing Group 5 meteoritic material, and resembling 67915 (Outhouse Rock) in lithophile elements (Fig. 2) and absolute siderophile abundance (Fig. 3). The Group 5 meteoritic component is also present in an anorthositic soil separate of similar Rb content at Luna 20, as shown in Fig. 1 (Ganapathy *et al.*, 1974).

In addition to the three units discussed above, North Ray ejecta may contain pre-Nectaris material. Rock 60017 and 4 soil separates (63502,13-1, 63502,13-8, 67602,14-7, and 67702,16-6) appear to have the lowest Au and Rb contents of any North Ray ejecta. [Although collected at Station 13, 2 crater radii from the rim, the 63502 breccias are almost certainly of North Ray origin, since anorthositic separates of similar alkali content from this soil have an exposure age of 42 m.y. (Schaeffer and Husain, 1973)]. These 5 samples are the only Apollo 16 material which contain a Group 6 meteoritic component. This component has been tentatively assigned to Humorum (Morgan *et al.*, 1974). From the basin age sequence of Hartmann and Wood (1971), 60017 and the associated soil fractions should be derived from the pre-Nectaris basement. Crisium ejecta, which should lie between those from Nectaris and Humorum (Fig. 1), are not represented among the major rocks, but may be present in those soil fractions containing a Group 3 meteoritic component.

South Ray Crater

There are 2, or just possibly 3 units exposed in South Ray Crater (Muehlberger *et al.*, 1972; Hodges *et al.*, 1973). A bench 50–60 m from the top shows as an albedo contrast and may mark a stratigraphic discontinuity. A third stratum might be inferred from the dark annulus at the margin of the crater floor, but that could be the result of mass wasting (Hodges *et al.*, 1973). The second layer, which is the source of the high albedo material forming the prominent ejecta blanket, is probably composed of cataclastic anorthosites (60025 and 68115). These are devoid of meteoritic material, and probably represent deep ejecta from a basin-forming event. By analogy with North Ray, we interpret these rocks as deep Nectaris ejecta.

The meteoritic components in South Ray Crater ejecta (exposure age ≤ 7 m.y.) belong largely to the low Ir/Au groups, 1 and 4, and probably represent Imbrium ejecta mantling the Apollo 16 site.

Volatile enrichment. At South Ray Crater, Tl, Br, and Cd are enriched, often spectacularly, in several rocks, notably "rusty rock" 66095 (Krähenbühl *et al.*, 1973b,c). Our more recent analyses have confirmed this. There are also sporadic enrichments of Ag, Bi, and Zn. The elevated volatile content appears to be the result of a redistribution of the indigenous lunar complement of these elements, rather than simple addition by cometary impact as proposed by El Goresy *et al.* (1973), since 66095 is substantially enriched in Pb of typical lunar isotopic abundance (Nunes and Tatsumoto, 1973).

North Ray and the remaining samples fall in mutually exclusive regions on a plot of Tl and Cd (Fig. 4). Not all non-North Ray rocks are uniformly enriched in these elements, since there appears to be a group of samples showing a complementary depletion. Three of these (60315, 65015, and 69935) are significantly depleted in Tl relative to alkalis; mean $Tl/Cs = (6.2 \pm 1.7) \times 10^{-4}$ in contrast to 1×10^{-2} for average lunar rocks (Krähenbühl *et al.*, 1973c). A fourth, 68415, has normal Tl but is depleted in Cd (and Bi). (The rock 69955 plots in the low-Tl area, but this sample is alkali-poor and $Tl/Cs \sim 6 \times 10^{-2}$ indicates slight volatile *enrichment*.) Three of the four volatile-depleted samples are in Group 4, and the fourth plots at the extreme low-Ge end of Group 1. The implication is that Group 4 may not be a distinct component, but contains Group 1 samples which have lost Ge, as suggested by Morgan *et al.* (1974). Only 60315 and 69935 have South Ray exposure ages. Rock 68415 has an age of 100 m.y. and 65015 has an age of 385 m.y.

Orientale Ejecta?

Group 1 ancient meteoritic material probably represents Imbrium ejecta (Morgan *et al.*, 1974); however, there are indications that this group is composite. At Apollo 14, Group 1 material generally has Re/Ir (Cl normalized) < 1.5, whereas at Apollo 16, and particularly at South Ray Crater, many Group 1 samples have Re/Ir > 1.5 (the R group of Krähenbühl *et al.*, 1973c). The high Re/Ir Group 1

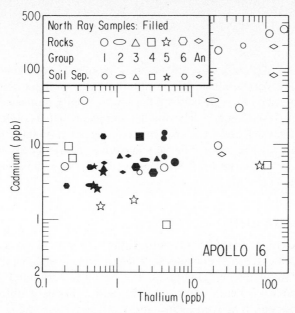

Fig. 4. North Ray ejecta separate from other Apollo 16 rocks in a Tl–Cd diagram. Three volatile-depleted rocks (open symbols between 0.2 and 0.3 ppb Tl, 4 and 10 ppb Cd) are significantly depleted in Tl relative to alkalis (Krähenbühl *et al.*, 1973c). A fourth (open symbol at 4 ppb Tl, 0.8 ppb Cd) has normal Tl but is depleted in Cd (and Br). Three of the 4 volatile-depleted rocks are Group 4 and may have lost Ge (see Fig. 1).

[hereafter called subgroup 1(R)] may be a local Apollo 16 variant of Imbrian ejecta. Alternatively, the Cayley Plains unit may be Orientale ejecta (Chao *et al.*, 1973; Hodges *et al.*, 1973; for a contrary view see Oberbeck *et al.*, 1974; Grudewicz, 1974). If subgroup 1(R) is Orientale ejecta it should have been present at Apollo 14, since, on the average, we might expect 4 m of Orientale material to overlie 130 m of Imbrium ejecta (McGetchin *et al.*, 1973). The Apollo 14 samples in which we found Group 1 meteoritic material were predominantly derived from Cone Crater and unfortunately, probably represent largely Imbrium material. Orientale material should be more apparent at stations more distant from Cone Crater.

In soil separates at Apollo 16 the 1(R) material is best seen in magnetic separates with visible metal (e.g. 64502,4a). At Fra Mauro, a single metal sphere, 14258,36,14, does indeed have very high Re/Ir. Antimony and Ge are very high but this is not uncommon in Apollo 16 1(R) material (e.g. 60016). The Ir/Au ratio in 14258,36,14 is among the lowest measured (0.23 Cl-normalized), and resembles that seen in 60315 (0.18). These very low ratios, if real, may provide a useful inclusive criterion, but not an exclusive one, for the 1(R) subgroup. In other words, such low ratios seem to occur *only* in the 1(R) subgroup, although not all 1(R) samples have them.

Glasses

Spheres

Two large glass spheres were collected at Apollo 16; 60095 (47 g) and 65016 (21.0 g). These are very similar to each other in trace element content (Table 1). They are characterized by high siderophiles (Au = 7 ppb), moderate alkalis and U, and low volatiles (Figs. 2 and 3). The smaller sample, 65016, is hollow and appears to have been outgassed since it is systematically lower in volatile trace elements. This is true even for the more volatile siderophile elements Sb and Ge, though here the effect is not large (~ 30%). Both spheres have a Group 5 meteoritic component (Fig. 1).

Glass-coated basalt 64455

The glass coating 64455,25 resembles the spheres in having a Group 5 meteoritic component, but is considerably higher in siderophiles (Au = 12.7 ppb). It is also higher in alkalis and U (Fig. 2). The basalt is very much lower than the glass in siderophile elements (Au = 1.6 ppb), and contains a different Group 4, meteoritic component. It is even higher, however, in alkalis and U. The coating is clearly splashed-on glass, and not simply a fusion product of the original rock surface, a conclusion also reached petrographically by Grieve and Plant (1973).

Our sample of glass coating is too high in alkalis to be the pure outer glass (Grieve and Plant, 1973). Perhaps our sample has assimilated some basalt substrate by partial melting or alternatively some basalt mechanically adhered to our glass sample. The basalt is a "rusty rock" and is highly enriched in T1 and Br (but not in Cd). Corresponding enrichments are found in the glass.

Origin of glasses

Apparently the glasses have a common origin, since all have the same Group 5 meteoritic component, and similar compositions except for some differences which are readily explained. At least one of the glasses, 60095, has been exposed to micrometeorites for a very short time, judging from microcraters (Neukum *et al.*, 1973). South Ray is the youngest crater at Apollo 16, and seems the most likely local source for the glasses, especially considering their location, and the volatile-rich substrate of 64455.

There seem to be three possible explanations for the Group 5 component in the glasses. Firstly, that the glasses themselves are of Nectaris origin. In view of the prevalent recrystallization in Apollo 16 rocks, it seems unlikely that the glasses would survive since the Nectaris event. Secondly, that the glasses are Nectaris material with a Group 5 component, shock-melted by the South Ray projectile, which was itself chemically inconspicuous. If our stratigraphy is correct, there are two objections to this explanation. Homogeneous glasses are formed near the region of maximum shock pressure, i.e. in excess of 600 kbar (Dence *et al.*, 1970). This is more likely to occur in the upper one-third of South Ray Crater than at the depth of the shallow Nectaris ejecta layer. In addition, the absolute abundance of the Group 5 component in the glasses is much higher than in Group 5 material at North Ray Crater.

The third alternative is that the glasses were formed by a local impact by a Type IIIA iron or perhaps a mesosiderite. On a ternary AuIrGe plot (Fig. 7 of Morgan *et al.*, 1974), IIIA irons lie in a narrow band across the lower part of the diagram and there is a small but finite probability (perhaps 5–10%) of a member of this group falling in the field of the Group 5 ancient meteoritic component.

SOILS

Micrometeorite content

The cosmogenic Ne^{21} exposure ages of Apollo 16 soils have been discussed by Walton *et al.* (1973), using their own data and those of Kirsten *et al.* (1973) and Bogard *et al.* (1973). The ages range from 50 m.y. (or perhaps as low as 30 m.y.) for soils from the rim of North Ray Crater, to more than 300 m.y. at Stations 4 and 5. The content of micrometeorites of primitive (Cl chondrite-like) composition in the soils should correlate with exposure age, and both siderophile (Ir, Re, Au, and Ni) and volatile (Ag, Bi, Br, Cd, Ge, Pb, Sb, Se, Te, and Zn) elements should be correspondingly enriched (Keays *et al.*, 1970 and later papers summarized by Anders *et al.*, 1973). At Apollo 16, however, the siderophiles are from ancient meteoritic sources, and the large volatile enrichments near South Ray may also mask the micrometeorite component. Of our elements, Bi, Te, and Se seem least likely to be affected by these factors. The Bi data for 9 soils (Fig. 5) are strongly correlated with exposure age ($r = 0.902$), and the slope corresponds to an addition of $2.4 \times 10^{-3}\%$ Cl chondrite material per million years. The correlations are less good for Se and Te. Of the 9 soils, 7 give good age correlation with Te ($r = 0.949$) and 6 with Se ($r = 0.996$). Low Te- and Se-points may reflect losses by volatilization as hydride by interaction with hydrogen of solar wind or cometary origin. (61221 is unusually high in volatile compounds and hydrogen; Gibson and Moore, 1973). Soil 68841 is too high in Se, possibly because of high indigenous or ancient meteoritic contribution. The regression slopes for Te and Se correspond to the addition of 2.2×10^{-3} and $2.3 \times 10^{-3}\%$ Cl material per million years. The intercepts at zero age, corresponding to 0.35 ppb Bi, 8.2 ppb Te, and 150 ppb Se, represent the indigenous lunar contribution, plus an addition from volatile-poor ancient meteoritic material.

The saturation exposure ages of mature 11, 12, 14, and 15 soils are all about 400 m.y., corresponding to a uniform mixing rate of 2 mm m.y.$^{-1}$ (Kirsten *et al.*, 1972). Taking the mean value of $2.3 \times 10^{-3}\%$ Cl per million years, we arrive at a saturated micrometeorite component of 0.92% Cl at Apollo 16. This is in good agreement with values from other sites: Apollo 11, 1.14; Apollo 12, 1.28; Apollo 15, 0.96 (Anders *et al.*, 1973).

The remaining volatile elements in the soil are largely indigenous and cannot be used for the calculation of micrometeorite influx. Bromine is quite constant in

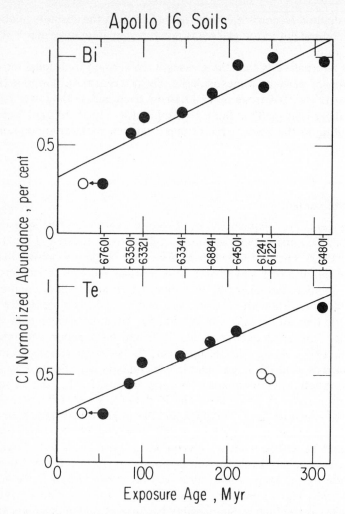

Fig. 5. In Apollo 16 < 1 mm soils, Bi, Te (and Se) abundances tend to correlate with Ne_c^{21} exposure age and correspond to an addition of 2.3×10^{-5} g Cl chondrite material g^{-1} soil m.y.$^{-1}$. Samples denoted by open symbols were not used in the regression analysis. The two Station 1 soils may have lost Te.

all soils, whereas Cd varies greatly and is particularly enriched in the Station 1 soils. The indigenous volatiles show good correlations with each other, e.g. Tl with Br ($r = 0.908$) and Ag with Zn (0.970).

Ancient meteorite component

Almost all lunar soils, even from mare areas, contain an ancient meteoritic component (Anders *et al.*, 1973). At many landing sites this is difficult to characterize except in fresh crater rim soils or in coarse lithic fragments, because of the micrometeorite component. At Apollo 16 an ancient component appears to be prevalent even in the soils, and since Bi seems to be a reliable indicator of micrometeorite material, we can make a small correction for this minor contribution. The net ancient component may then give some indication of the dominant types at each sampling site. The results for 9 Apollo 16 soils are given in Fig. 6. The corrected points are represented by filled symbols.

Three soils (61221, 63321, and 64501) are closely grouped in the lower left, and plot very closely to rock 60315. This cluster lies outside the range of the established groups and cannot be explained simply on the basis of mixing of other components. Possibly all four samples are slightly contaminated with Au, but this seems unlikely, since it would require fine adjustment of the contamination levels to the individual absolute abundances. Note also that the Ir/Au ratios are extremely similar to that of 14258 metal, although the metal is much richer in Ge. The suspicion lingers that this may be Orientale material.

Three soils (63501, 64801, and 68841) fall into Group 1. In the case of 63501, several soil separates were picked from the corresponding 1–2 mm sieve fraction

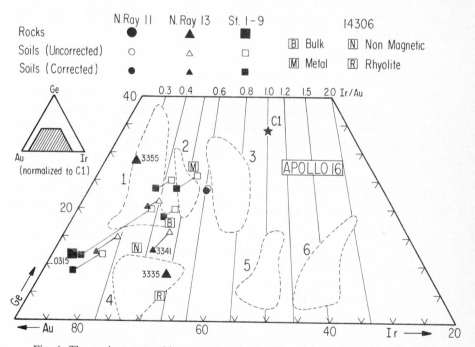

Fig. 6. The ancient meteoritic component in Apollo 16 soils may be estimated by correcting for the micrometeorite contribution, and characterized by Ir, Au, Ge. Three soils in the lower left (61221, 63321, 64501) have very low Ir/Au, and, together with rock 60315, may represent a new meteoritic group. Shadowed soil 63341 appears to contain a mixture of debris from Shadow Rock (rocks 63335 and 63355). Soils 61241 and 67601 fall between groups and apparently are a complex mixture. In breccia 14306, bulk sample, rhyolite and metal fall into different groups but have very similar Ir/Au, perhaps indicating Ge redistribution during metamorphism.

63502. Two fractions, 63502,12-10 and 63502,12-11, have high Au and a Group 1 meteoritic component, and material similar to these probably dominates the average siderophile content.

Shadowed soil 63341 was collected from beneath Shadow Rock from which rocks 60017, 63355, and 63355 were sampled. It falls between Groups 1 and 4, but close to the mixing line between our samples of 63355 and 63335. It is possible that this soil consists largely of material eroded from the boulder similar to 63335 and 63355 in composition.

The two remaining soils (61241, 67601) fall between groups and clearly represent a complex mixture of a number of components. In the case of 67601, the 1–2 mm fraction of the same soil contains components from Groups 1, 3, 5(?), 6 and beyond (67602,14-3).

Bulk soils can be interpreted explicitly only in very favorable cases, but may provide valuable information concerning the dominant meteoritic component in the regolith. A good example is the low Ir/Au–Ge/Au material, which is apparently common in soils (3 out of 9), but found in only one rock. This may be a result of Orientale material arriving at the Apollo 16 site in ballistic trajectory. In the secondary crater formation the primary ejectum is comminuted (Oberbeck *et al.*, 1974) and reejected as fine debris.

BRECCIA 14306

This rock is a multigeneration metamorphosed breccia, described in detail by Anderson *et al.* (1972). Its indurated texture, and small clast size, prevented the extraction of clasts for analysis. A sample was taken of a metal-bearing rhyolite clast, and magnetic and nonmagnetic fractions of the bulk rock were separated. In addition, a chip of the whole rock was analyzed. The distribution of trace elements is shown in Fig. 7.

Many of our elements (11 or 12) appear to be enriched by factors of 10 or more in the metal relative to the bulk sample. The archetype siderophiles (Ir, Re, Au,

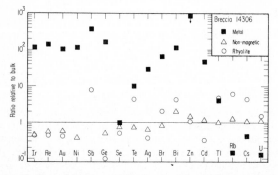

Fig. 7. "Archetype" siderophile elements Ir, Re, Au, Ni, and Ge are enriched 10^2 times in 14306 metal over the bulk rock. The elements Te, Ag, Br, Bi, and Cd are also enriched between 10 and 10^2 times and hence must be strongly siderophile in lunar breccias.

Ni, Sb, and Ge) are rather uniformly enriched by a factor of $\sim 10^2$. Other elements not generally considered to be siderophile (Te, Ag, Br, Bi, and Cd) are also enriched in the metal fraction, by factors of between 10 and 10^2. It appears that at least 50% of the siderophile elements reside in the metal phase in 14306, since they are depleted in the nonmagnetic fraction by a factor of 2. The remaining siderophiles may reside in metal which was too fine to separate, rather than in the silicate itself. In other elements the nonmagnetic fraction closely resembles the bulk rock. The many stages of grinding and acetone washing greatly increase the likelihood of contamination, hence the very high Zn in the metal and high Sb in the nonmagnetic fraction. The rhyolite is similar to the nonmagnetic fraction in Ir, Re, and Au, but is substantially depleted in Ge. The high Sb could possibly be spurious. The rhyolite contains 4–6 times as much alkalis and Tl as the bulk rock, yet, surprisingly, it is not greatly enriched in U.

The meteoritic elements present an interesting picture (Fig. 6). The bulk sample is nominally Group 1, though it is at the extreme high Ir/Au limit. The rhyolite and metal fall into Groups 4 and 2, respectively. Yet all three samples have very similar Ir/Ar. This strongly suggests that there has been some internal redistribution of Ge during metamorphism, and that the rhyolite and metal are both Group 2 material of possible Serenitatis origin (Morgan *et al.*, 1974). In clasts from breccia 14321, there appears to have been a similar redistribution. Three of 4 KREEP-rich clasts fall into Group 1 but the fourth occupies the low end of Group 4. In our AuIrGe ternary classification (Fig. 1) the redistribution of Ge has the effect of moving the point along a line of constant Ir/Au. This effect seems to have influenced only Groups 1, 2, and 4. There is no low-Ge group corresponding to Group 3, and there are no high-Ge equivalents of Groups 5 and 6 (overlap between Ir/Au for Groups 3 and 5 is very marginal). The redistribution of Ge in breccias is not a frequent occurrence. We have analyzed 2 or more samples from each of 13 breccias, and find reasonably firm evidence of Ge movement in only 2: 14306 and 14321. Both belong to Warner's (1972) Type 4 breccias, that is, they show medium-grade metamorphism with partly annealed matrix. In the other breccias, samples either belong to the same ancient meteoritic group, or to groups which cannot be related to each other by simple Ge transfer.

The 14306 nonmagnetic fraction falls slightly above Group 4, but the Ir/Au ratio is substantially lower than in the other 3 samples from this breccia, being more typical of Group 1. It is possible that the lower Ir/Au ratio may be the result of Au contamination during the many stages of separation. Nevertheless, it is also conceivable that the multigeneration breccia 14306 (Anderson *et al.*, 1972) may contain at least 2 types of ancient meteoritic components: Group 1 and Group 2, in some cases modified to Group 4 by Ge redistribution.

Acknowledgments—We thank Dr. Urs Krähenbühl for the magnetic separation of breccia 14306 and for experimental assistance. This work was supported in part by NASA grant NGL 14-001-167. One of us (EA) thanks the John Simon Guggenheim Memorial Foundation for a fellowship.

REFERENCES

Agrell S. O., Agrell J. E., Arnold A. R., and Long J. V. P. (1973) Some observations on rock 62295 (abstract). In *Lunar Science—IV*, pp. 15–17. The Lunar Science Institute, Houston.

Albee A. L., Gancarz A. J., and Chodos A. A. (1973) Metamorphism of Apollo 16 and 17 and Luna 20 metaclastic rocks at about 3.95 AE: Samples 61156, 64423,14-2, 65015, 67483,15-2, 76055, 22006, and 22007. *Proc. Fourth Lunar Sci. Conf., Geochim. Cosmochim. Acta*, Suppl. 4, Vol. 1, pp. 569–595. Pergamon.

Anders E., Ganapathy R., Krähenbühl U., and Morgan J. W. (1973) Meteoritic material on the Moon. *The Moon* **8**, 3–24.

Anderson A. T., Braziunas T. F., Jacoby J., and Smith J. V. (1972) Thermal and mechanical history of breccias 14306, 14063, 14270, and 14321. *Proc. Third Lunar Sci. Conf., Geochim. Cosmochim. Acta*, Suppl. 3, Vol. 1, pp. 819–835. MIT Press.

AFGIT (Apollo Field Geology Investigation Team) (1973) Apollo 16 exploration of Descartes: A geologic summary. *Science* **179**, 62–68.

Behrmann C., Crozaz G., Drozd R., Hohenberg C., Ralston C., Walker R., and Yuhas D. (1973) Cosmic-ray exposure history of North Ray and South Ray material. *Proc. Fourth Lunar Sci. Conf., Geochim. Cosmochim. Acta*, Suppl. 4, Vol. 2, pp. 1957–1974. Pergamon.

Bence A. E., Papike J. J., Sueno S., and Delano J. W. (1973) Pyroxene poikiloblastic rocks from the lunar highlands. *Proc. Fourth Lunar Sci. Conf., Geochim. Cosmochim. Acta*, Suppl. 4, Vol. 1, pp. 597–611. Pergamon.

Bogard D. D., Nyquist L. E., Hirsch W. C., and Moore D. R. (1973) Trapped noble gas abundances in surface and sub-surface fines from Apollo 15 and 16 (abstract). In *Lunar Science—IV*, pp. 79–81. The Lunar Science Institute, Houston.

Brecher A., Vaughan D. J., Burns R. G., and Morash K. R. (1973) Magnetic and Mössbauer studies of Apollo 16 rock chips 60315,51 and 62295,27. *Proc. Fourth Lunar Sci. Conf., Geochim. Cosmochim. Acta*, Suppl. 4, Vol. 3, pp. 2991–3001. Pergamon.

Brown G. M., Peckett A., Phillips R., and Emeleus C. H. (1973) Mineral-chemical variations in the Apollo 16 magnesio-feldspathic highland rocks. *Proc. Fourth Lunar Sci. Conf., Geochim. Cosmochim. Acta*, Suppl. 4, Vol. 1, pp. 505–518. Pergamon.

Chao E. C. T., Soderblom L. A., Boyce J. M., Wilhelms D. E., and Hodges C. A. (1973) Lunar light plains deposits (Cayley Formation)—A reinterpretation of origin (abstract). In *Lunar Science—IV*, pp. 127–129. The Lunar Science Institute, Houston.

Delano J. W., Bence A. E., Papike J. J., and Cameron K. L. (1973) Petrology of the 2–4 mm soil fraction from the Descartes region of the moon and stratigraphic implications. *Proc. Fourth Lunar Sci. Conf., Geochim. Cosmochim. Acta*, Suppl. 4, Vol. 1, pp. 537–551. Pergamon.

Dence M. R., Douglas J. A. V., Plant A. G., and Traill R. J. (1970) Petrology, mineralogy and deformation of Apollo 11 samples. *Proc. Apollo 11 Lunar Sci. Conf., Geochim. Cosmochim. Acta*, Suppl. 1, Vol. 1, pp. 315–340. Pergamon.

Drozd R. J., Hohenberg C. M., Morgan C. J., and Ralston C. E. (1974) Cosmic-ray exposure history at the Apollo 16 and other lunar sites: Lunar surface dynamics. *Proc. Fifth Lunar Sci. Conf., Geochim. Cosmochim. Acta*. This volume.

El Goresy A., Ramdohr P., and Medenbach O. (1973) Lunar samples from Descartes site: Opaque mineralogy and geochemistry. *Proc. Fourth Lunar Sci. Conf., Geochim. Cosmochim. Acta*, Suppl. 4, Vol. 1, pp. 733–750. Pergamon.

Ganapathy R. and Anders E. (1974) Bulk compositions of the Moon and Earth, estimated from meteorites. *Proc. Fifth Lunar Sci. Conf., Geochim. Cosmochim. Acta*. This volume.

Ganapathy R., Morgan J. W., Krähenbühl U., and Anders E. (1973) Ancient meteoritic components in lunar highland rocks: Clues from trace elements in Apollo 15 and 16 samples. *Proc. Fourth Lunar Sci. Conf., Geochim. Cosmochim. Acta*, Suppl. 4, Vol. 2, pp. 1239–1261. Pergamon.

Gancarz A. J., Albee A. L., and Chodos A. A. (1972) Comparative petrology of Apollo 16 sample 68415 and Apollo 14 samples 14276 and 14310. *Earth Planet. Sci. Lett.* **16**, 307–330.

Gibson E. K., Jr. and Moore G. W. (1973) Volatile-rich lunar soil: Evidence of possible cometary impact. *Science* **179**, 69–71.

Grieve R. A. F. and Plant A. G. (1973) Partial melting on the lunar surface, as observed in glass coated

Apollo 16 samples. *Proc. Fourth Lunar Sci. Conf., Geochim. Cosmochim. Acta*, Suppl. 4, Vol. 1, pp. 667–679. Pergamon.

Grossman L. (1972) Condensation in the primitive solar nebula. *Geochim. Cosmochim. Acta* **36**, 597–619.

Grudewicz E. B. (1974) Lunar upland plains relative age determinations and their bearing on the provenance of the Cayley Formation (abstract). In *Lunar Science—V*, pp. 301–303. The Lunar Science Institute, Houston.

Hartmann W. K. and Wood C. A. (1971) Moon: Origin and evolution of multi-ring basins. *The Moon* **3**, 3–78.

Haskin L. A., Helmke P. A., Blanchard D. P., Jacobs J. W., and Telander K. (1973) Major and trace element abundances in samples from the lunar highlands. *Proc. Fourth Lunar Sci. Conf., Geochim. Cosmochim. Acta*, Suppl. 4, Vol. 2, pp. 1275–1296. Pergamon.

Helz R. T. and Appleman D. E. (1973) Mineralogy, petrology, and crystallization history of Apollo 16 rock 68415. *Proc. Fourth Lunar Sci. Conf., Geochim. Cosmochim. Acta*, Suppl. 4, Vol. 1, pp. 643–659. Pergamon.

Hodges C. A., Muehlberger W. R., and Ulrich G. E. (1973) Geologic setting of Apollo 16. *Proc. Fourth Lunar Sci. Conf., Geochim. Cosmochim. Acta*, Suppl. 4, Vol. 1, pp. 1–25. Pergamon.

Hodges F. N. and Kushiro I. (1973) Petrology of Apollo 16 lunar highland rocks. *Proc. Fourth Lunar Sci. Conf., Geochim. Cosmochim. Acta*, Suppl. 4, Vol. 1, pp. 1033–1048. Pergamon.

Huneke J. C., Jessberger E. K., Podosek F. A., and Wasserburg G. J. (1973) $^{40}Ar/^{39}Ar$ measurements in Apollo 16 and 17 samples and the chronology of metamorphic and volcanic activity in the Taurus-Littrow region. *Proc. Fourth Lunar Sci. Conf., Geochim. Cosmochim. Acta*, Suppl. 4, Vol. 2, pp. 1725–1756. Pergamon.

Keays R. R., Ganapathy R., Laul J. C., Anders E., Herzog G. F., and Jeffery P. M. (1970) Trace elements and radioactivity in lunar rocks: Implications for meteorite infall, solar-wind flux, and formation conditions of Moon. *Science* **167**, 490–493.

Keays R. R., Ganapathy R., Laul J. C., Krähenbühl U., and Morgan J. W. (1974) The simultaneous determination of 20 trace elements in terrestrial, lunar and meteoritic material by radiochemical neutron activation analysis. *Anal. Chim. Acta*. In press.

Kirsten T., Deubner J., Horn P., Kaneoka I., Kiko J., Schaeffer O. A., and Thio S. K. (1972) The rare gas record of Apollo 14 and 15 samples. *Proc. Third Lunar Sci. Conf., Geochim. Cosmochim. Acta*, Suppl. 3, Vol. 2, pp. 1865–1889. MIT Press.

Kirsten T., Horn P., and Kiko J. (1973) $^{39}Ar–^{40}Ar$ dating and rare gas analysis of Apollo 16 rocks and soils. *Proc. Fourth Lunar Sci. Conf., Geochim. Cosmochim. Acta*, Suppl. 4, Vol. 2, pp. 1757–1784. Pergamon.

Krähenbühl U., Morgan J. W., Ganapathy R., and Anders E. (1973a) Abundance of 17 trace elements in carbonaceous chondrites. *Geochim. Cosmochim. Acta* **37**, 1353–1370.

Krähenbühl U., Ganapathy R., Morgan J. W., and Anders E. (1973b) Volatile elements in Apollo 16 samples: Possible evidence for outgassing of the Moon. *Science* **180**, 858–861.

Krähenbühl U., Ganapathy R., Morgan J. W., and Anders E. (1973c) Volatile elements in Apollo 16 samples: Implications for highland volcanism and accretion history of the Moon. *Proc. Fourth Lunar Sci. Conf., Geochim. Cosmochim. Acta*, Suppl. 4, Vol. 2, pp. 1325–1348. Pergamon.

Lightner B. D. and Marti K. (1974) Lunar trapped xenon (abstract). In *Lunar Science—V*, pp. 447–449. The Lunar Science Institute, Houston.

LSPET (1973) The Apollo 16 lunar samples: A petrographic and chemical description. *Science* **179**, 23–24.

Marti K., Lightner B. D., and Osborn T. W. (1973) Krypton and xenon in some lunar samples and the age of North Ray Crater. *Proc. Fourth Lunar Sci. Conf., Geochim. Cosmochim. Acta*, Suppl. 4, Vol. 2, pp. 2037–2048. Pergamon.

McGetchin T. R., Settle M., and Head J. W. (1973) Radial thickness variation in impact crater ejecta: Implications for lunar basin deposits. *Earth Planet. Sci. Lett.* **20**, 226–236.

Morgan J. W., Laul J. C., Krähenbühl U., Ganapathy R., and Anders E. (1972) Major impacts on the moon: Characterization from trace elements in Apollo 12 and 14 samples. *Proc. Third Lunar Sci. Conf., Geochim. Cosmochim. Acta*, Suppl. 3, Vol. 2, pp. 1377–1395. MIT Press.

Morgan J. W., Ganapathy R., Higuchi H., Krähenbühl U., and Anders E. (1974) Lunar basins: Tentative characterization of projectiles, from meteoritic elements in Apollo 17 boulders. *Proc. Fifth Lunar Sci. Conf., Geochim. Cosmochim. Acta.* This volume.

Muehlberger W. R., Batson R. M., Boudette E. L., Duke C. M., Eggleton R. E., Elston D. P., England A. W., Freeman V. L., Hait M. H., Hall T. A., Head J. W., Hodges C. A., Holt H. E., Jackson E. D., Jordan J. A., Larson K. B., Milton D. J., Reed V. S., Rennilson J. J., Schaber G. G., Schafer J. P., Silver L. T., Stuart-Alexander D. E., Sutton R. L., Swann G. A., Tyner R. L., Ulrich G. E., Wilshire H. G., Wolfe E. W., and Young J. W. (1972) Preliminary geologic investigation of the Apollo 16 landing site. *Apollo 16 Preliminary Science Report*, NASA SP-315, pp. 6-1 to 6-81.

Nagata T., Fisher R. M., Schwerer F. C., Fuller M. D., and Dunn J. R. (1973) Magnetic properties and natural remanent magnetization of Apollo 15 and 16 lunar material. *Proc. Fourth Lunar Sci. Conf., Geochim. Cosmochim. Acta*, Suppl. 4, Vol. 3, pp. 3019–3043. Pergamon.

Neukum G., Hörz F., Morrison D. A., and Hartung J. B. (1973) Crater populations on lunar rocks. *Proc. Fourth Lunar Sci. Conf., Geochim. Cosmochim. Acta*, Suppl. 4, Vol. 3, pp. 3255–3276. Pergamon.

Nord G. L., Jr., Lally J. S., Heuer A. H., Christie J. M., Radcliffe S. V., Griggs D. T., and Fisher R. M. (1973) Petrologic study of igneous and metaigneous rocks from Apollo 15 and 16 using high voltage transmission electron microscopy. *Proc. Fourth Lunar Sci. Conf., Geochim. Cosmochim. Acta*, Suppl. 4, Vol. 1, pp. 953–970. Pergamon.

Nunes P. D. and Tatsumoto M. (1973) Excess lead in "Rusty Rock" 66095 and implications for an early lunar differentiation. *Science* **182**, 916–920.

Oberbeck V. R., Morrison R. H., Hörz F., Quaide W. L., and Gault D. E. (1974) Smooth plains and continuous deposits of craters and basins. *Proc. Fifth Lunar Sci. Conf., Geochim. Cosmochim. Acta.* Volume 1.

Pearce G. W., Gose W. A., and Strangway D. W. (1973) Magnetic studies on Apollo 15 and 16 lunar samples. *Proc. Fourth Lunar Sci. Conf., Geochim. Cosmochim. Acta*, Suppl. 4, Vol. 3, pp. 3045–3076. Pergamon.

Pearce G. W. and Simonds C. H. (1974) Magnetic properties of Apollo 16 samples and implications for their mode of formation. *J. Geophys. Res.* In press.

Roedder E. and Weiblen P. W. (1974) Petrology of clasts in breccia 67915 (abstract). In *Lunar Science—V*, pp. 642–644. The Lunar Science Institute, Houston.

Schaeffer O. A. and Husain L. (1973) Early lunar history: Ages of 2 to 4 mm soil fragments from the lunar highlands. *Proc. Fourth Lunar Sci. Conf., Geochim. Cosmochim. Acta*, Suppl. 4, Vol. 2, pp. 1847–1863. Pergamon.

Steele I. M. and Smith J. V. (1973) Mineralogy and petrology of some Apollo 16 rocks and fines: General petrologic model of moon. *Proc. Fourth Lunar Sci. Conf., Geochim. Cosmochim. Acta*, Suppl. 4, Vol. 1, pp. 519–536. Pergamon.

Stettler A., Eberhardt P., Geiss J., Grögler N., and Maurer P. (1973) Ar^{39}–Ar^{40} ages and Ar^{37}–Ar^{38} exposure ages of lunar rocks. *Proc. Fourth Lunar Sci. Conf., Geochim. Cosmochim. Acta*, Suppl. 4, Vol. 2, pp. 1865–1888. Pergamon.

Taylor G. J., Drake M. J., Hallam M. E., Marvin U. B., and Wood J. A. (1973) Apollo 16 stratigraphy: The ANT hills, the Cayley Plains, and a pre-Imbrian regolith. *Proc. Fourth Lunar Sci. Conf., Geochim. Cosmochim. Acta*, Suppl. 4, Vol. 1, pp. 553–568. Pergamon.

Taylor L. A., Mao H. K. and Bell P. M. (1973) "Rust" in the Apollo 16 rocks. *Proc. Fourth Lunar Sci. Conf., Geochim. Cosmochim. Acta*, Suppl. 4, Vol. 1, pp. 829–839. Pergamon.

Turner G., Cadogan P. H., and Yonge C. J. (1973) Argon selenochronology. *Proc. Fourth Lunar Sci. Conf., Geochim. Cosmochim. Acta*, Suppl. 4, Vol. 2, pp. 1889–1914. Pergamon.

Ulrich G. E. (1973a) A geologic model for North Ray Crater (abstract). In *Lunar Science—IV*, pp. 745–747. The Lunar Science Institute, Houston.

Ulrich G. E. (1973b) A geologic model for North Ray Crater and stratigraphic implications for the Descartes region. *Proc. Fourth Lunar Sci. Conf., Geochim. Cosmochim. Acta*, Suppl. 4, Vol. 1, pp. 27–39. Pergamon.

Walker D., Longhi J., Grove T. L., Stolper E., and Hays J. F. (1973) Experimental petrology and origin of rocks from the Descartes highlands. *Proc. Fourth Lunar Sci. Conf., Geochim. Cosmochim. Acta*, Suppl. 4, Vol. 1, pp. 1013–1032. Pergamon.

Walton J. R., Lakatos S., and Heymann D. (1973) Distribution of inert gases in fines from the Cayley–Descartes region. *Proc. Fourth Lunar Sci. Conf., Geochim. Cosmochim. Acta*, Suppl. 4, Vol. 2, pp. 2079–2095. Pergamon.

Wänke H., Baddenhausen H., Dreibus G., Jagoutz E., Kruse H., Palme H., Spettel B., and Teschke F. (1973) Multielement analyses of Apollo 15, 16, and 17 samples and the bulk composition of the moon. *Proc. Fourth Lunar Sci. Conf., Geochim. Cosmochim. Acta*, Suppl. 4, Vol. 2, pp. 1461–1481. Pergamon.

Warner J. L., Simonds C. H., Phinney W. C. (1973) Apollo 16 rocks: Classification and petrogenetic model. *Proc. Fourth Lunar Sci. Conf., Geochim. Cosmochim. Acta*, Suppl. 4, Vol. 1, pp. 481–504. Pergamon.

Wilshire H. G., Stuart-Alexander D. E., and Jackson E. D. (1973) Apollo 16 rocks: Petrology and classification. *J. Geophys. Res.* **78**, 2379–2391.

Proceedings of the Fifth Lunar Conference
(Supplement 5, Geochimica et Cosmochimica Acta)
Vol. 2 pp. 1685–1701 (1974)
Printed in the United States of America

Labile and nonlabile element relationships among Apollo 17 samples*

STANKA JOVANOVIC and GEORGE W. REED, JR.

Chemistry Division, Argonne National Laboratory, Argonne, Illinois 60439

Abstract—The trace element—F, Cl, Br, I, Li, U, Ru, Os, Hg, and Te—contents of a suite of Apollo 17 soils and rocks are reported. Unusually large concentrations of the halogens are found in the orange and associated gray soils and unusually low concentrations are found in dunite and troctolitic granulite. There is a high degree of coherence among the halogens; this suggests that they evolved under closed system conditions on the moon.

Ion microprobe studies of orange glass spherules have yielded F and Cl concentration ratios consistent with those from the bulk sample.

For most samples, F and Cl are associated with P_2O_5 as apatites. In a few samples, controls other than apatite stoichiometry determined the correlation with P_2O_5. These controls were probably either very late crystallization or very early partial melting and they also appear to have determined the correlation of Li with Cl and possibly that found among other lithophile trace elements. The lack of specificity in terms of sample type suggests that these controls are independent of parental rock or magma.

Ru and Os exhibit several correlation trends indicating chemical processing on the moon. The samples which have meteoritic ratios are interpreted as being primitive and not extensively recycled.

Local mixing of soils from different geologic units is evident in the U data.

Evidence for lunar atmospheric Hg is provided by the Hg found in the sealed sample return containers.

INTRODUCTION

THE APOLLO 17 ASTRONAUTS returned samples from the Taurus-Littrow region of the moon. The terrain sampled was the most geologically diverse of any of the Apollo landing sites. The LM landed in a valley, flooded with mare basalts, flanked by massifs of an early lunar crust uplifted by the Serenitatis Basin forming event and by a lower lying hilly region of possibly distinct origin. The formations sampled were the valley floor, down-wasted boulders and soils from North and South Massifs and Sculptured Hills, and a land slide at South Massif. A wide variety of samples, ranging from possible subfloor basalts to possibly exotic orange and black soils, were collected from the valley floor.

We have studied the halogens and Hg as volatiles, U and Li as nonvolatile lithophilic elements and Ru and Os as both siderophile and possible volatile elements, in a fairly representative suite of samples by activation analysis. Samples were subjected to leaching for ten minutes with hot water and, in the case of Hg, to volatilization by stepwise heating. The term labile in this paper refers to either readily volatile, i.e. appreciable vapor pressure at 100–200°C, or to hot

---

*Work performed under the auspices of the USAEC and NASA.

1685

water soluble elements or compounds. The nonlabile fraction is that remaining after such mild treatment.

In this report, concentrations, inter-element relations, and some inferences regarding the Apollo 17 site, are presented. Correlations among $F-Cl-P_2O_5$, $Cl-Br-I$, $Ru-Os-Ni$, the coherence between Li and Cl and chemical evidence via U for local soil admixture are discussed. Additional evidence for lunar atmospheric Hg has been obtained from the Apollo 16 and 17 sample return containers. Hg data for Apollo 17 samples are being evaluated, and the sample results given here should be considered as preliminary. The data are given in Tables 1 and 2. The concentrations in the leach (l) and residue (r) are listed separately as is the Hg volatilized below 130°C and total Hg. The U contents, obtained independently in different sets of experiments are listed in the respective tables. Neutron and photon irradiations followed by radiochemical separations were used to determine the various elements. Sample handling, prior to sealing the samples in fused silica vials for irradiation, was restricted to a N_2-dry box. Duplicate samples were not determined although different phases or parts, i.e. interior and exterior, of the same sample may have been determined. The uncertainty in the results is 10% or less except where noted, and is based on counting statistics.

RESULTS

$F-Cl-P_2O_5$

The F and Cl contents of Apollo 17 samples are similar to those from other sites. In general, concentrations range from 30–60 ppm for F and from 15–35 ppm for Cl in soils and breccias. Igneous rocks analyzed have Cl contents from 1–15 ppm. The exceptional samples are the orange soil 74220 and the associated gray soil 74241 which have high halogen contents; the black soil, from a layer below the orange soil, has 40 ppm Cl but very low F in spite of a major element composition similar to the orange soil; and the troctolitic granulite 76535 which is highly depleted in Cl (and Br) relative to F, and in fact, relative to Cl and Br in all lunar samples except anorthositic samples 15415 and 15418.

The F/Cl_r ratio for most Apollo 17 samples measured, 5 soils and 6 rocks, is 4.0 ± 0.7. South Massif samples 73141, 73235 and possibly 73275 have lower ratios. High F/Cl_r ratios are found in two rocks of apparent deep-seated origin, dunite 72417 and granulite 76535 as well as in basalt 74275. This observation and low vesicularity (LSPET, 1973) suggests that 74275 may have been derived from deep in the lunar crust. The relatively high F/Cl_r ratios parallel the terrestrial trend for ultrabasic rocks from depth. The F/Cl_r weight ratio of ~4 relates the eleven samples noted above to those from other sites.

The F/P_2O_5 and Cl_r/P_2O_5 correlations have been discussed previously (Reed and Jovanovic, 1972; Jovanovic and Reed, 1973). The F data have been fragmentary so far and detailed correlation studies have been directed primarily toward Cl. The Apollo 17 basaltic samples tend to fall on or near the F/P_2O_5 correlation line of Apollo 11 soil and Apollo 15 basalts, Fig. 1. The Apollo 17

Table 1. Halogens, lithium, uranium and tellurium in Apollo 17 samples.*

Sample	F ppm	Cl r† ppm	Cl l† ppm	Br r ppb	Br l ppb	I‡ ppb	Li[a] ppm	U[a] ppm
Soils								
70181,17	52	14	19	840	43	1.2	7.2	0.22
70006,11[b] 94 cm		13	5	190	60	3	9.4	0.29
70005,11[b] 135 cm		14	3.8	130	60	3	7.2	0.24
70002,11[b] 256 cm		21	7.9	190	90	9	9.8	0.51
71501,34		5.7	2.1	86*	26*	2	11	0.17
72701,34	55	12	1.8	80	35	1.2	8	0.79
73141,20	17	10	6.5	210	130	2	9.3	
74220,34[b]	102	23	49	380	1200	14	9.7	0.17
74220,111*	25, 36[c]	82	21	120	300	13	11	0.17
74241,64[b]	230	27	32	420	380	13	13	0.14
74001,12	<2	11	29	95	85	5.9	11	0.15
75080,8	58	12	22	140	230	5.1	7.3	0.25
76281,7*	<60	11	3.5	100	100	2.4	8.2	
78501,52	38	12	3.9	180	200	2.2	5.6	
Rocks								
72275,110[d] ext.	117[e]	28	1.6	94	30	3.3	12	1.6
72395,43 int.	41	7.4	1.0	42*	14*	1.7	—	0.59
72417,1,1 int.	154[e]	5.9	0.79	18	10*	0.9	2.3	0.002
73235,48 int. white clast		17	4.8	58*	100*	4.4	8.1	0.48
73235,48 int. dark matrix	30[f]	17	3.0	58*	57*	1.9	5.5	1.2
73275,27	30	11	0.89	71	44	0.9	9.4	1.1
74275,61	43	2.8	0.64	8*	3*	0.9	9	0.17
75075,24[d] ext.	39	12	3.1	10*	<2	0.8	8.7	0.13
76315,69[d] ext.		7.8	1.3	52*	24*	1.3	15	1.0
76315,70 int.	49	15	1.4	68*	10*	1.2	13	1.1
76535,23	9	0.22	0.77	6	<3	1.1	—	0.03
77035,82	39	9.6	1.3	15*	12*	1	7.1	0.83

*The counting statistical errors are usually 10% or less. Errors in Br are up to 30% in indicated samples. Tellurium at greater than ppb level detected only in 74220,34 = 10^4 ppb; 74220,111 = 16 ± 6 ppb and 76281,7 = 26 ± 12 ppb.

†r = residue after leach. l = leach solution.

‡I detected in leach only.

[a]U and Li are <10% leachable.

[b]Jovanovic *et al.* (1973).

[c]pH 5 leach.

[d]Exterior surface with patina.

[e]Values for the interior sample 72275,66, and exterior 72417,1,7.

[f]Sample 73235,44 exterior, dark, has 14 ppm F.

massif rocks 73235, 76315, 73275, and soil 73141 appear to form a new correlation which includes 66095. South Massif samples 72275 and 72701 fall close to the correlation line for Apollo 12, 14, and 15 soils; this is consistent with their KREEP-like nature. Samples 72417, 75075, 78501, and 74241 have more F and Cl than can be stoichiometrically accommodated as apatite.

Table 2. Mercury, Ruthenium, Osmium, and Uranium in Apollo 17 Samples.*

Sample		$Hg^a_{\leqslant130°C}$ ppb	Hg^a_{total} ppb	Ru ppb	Os ppb	U ppm
70181,17	scoop, 5 cm	2.3	26	14 ± 4	1.5 ± 0.2	0.16
72275,66	interior	0.11	3.8	$\leqslant 3$	1.5 ± 0.1	0.98
72395,43	interior	0.26	4.4	17	9	1.3
72417,1,7	exterior			$\leqslant 3$	1.2 ± 0.7	0.002
72417,1,1	interior	0.35	2.2		$\leqslant 0.8$	
73141,20	trench, 15 cm	0.3	5.3	10 ± 2	18	0.79
73235,48	int. dark	2.3	4.4	7.6	12	0.61
73235,44	exterior			5.3	14	0.42
73235,48	int. white	0.9	3.2			
74220,111	trench, 8 cm	n.d.	5.7	$\leqslant 1$	0.7 ± 0.3	0.13
74241,18		n.d.	3.2	3 ± 1	0.8 ± 0.2	0.36
74001,12	core, 69.5 cm	5.3	13	$\leqslant 1$	20	
74275,61	chip of rock	6.4	61			
76241,17[b]	scoop, 5 cm	0.2	2.2	17		0.68
76241,17[b]	scoop, 5 cm			193	12	1.38
76241,17[b]	scoop, 5 cm			12	28	
76261,17	skimm, 2 cm	0.3	3	19	35	
76281,7	scoop, 5 cm	0.8	6.3	46		0.86
76315,69	patina			6.7 ± 1.5	4.7	
76315,70	interior			6.3	12	
76535,23	interior	0.5	1.1	0.5 ± 0.4	0.5 ± 0.2	0.02
78501,52	scoop, 5 cm			18	37	0.53

*The counting statistical errors are about 10%, or as indicated.
[a]Hg data are preliminary and may contain errors up to 50%.
[b]Permanently shadowed sample.
n.d. = not detected.

The Cl_r/P_2O_5 ratios for many Apollo 17 samples coincide with those already reported for samples from other sites, Fig. 2. Basalt 74275 falls on the Apollo 11, 12, 15 basalt correlation line but so do massif samples 72395, 73275, and 76315. KREEP-like sample 72275 and both dark matrix and white clast of 73235 and soils 72701, 73141, and 71501 fall along the curve for Apollo 11, 12, 14, and 15 KREEP and front samples. A group of samples which may be unique to the Apollo 17 suite includes 74241, 74220, 70181, 78501, 75075, and dunite 72417 and granulite 76535. The line through these samples intercepts the P_2O_5 axis at 0.03 wt.%. Since over one-half of these samples contain halogens in excess of that required for apatite, the correlation with P_2O_5 is either fictitious or is the result of controls other than apatite stoichiometry. The P_2O_5 contents with which the Cl_r contents of these

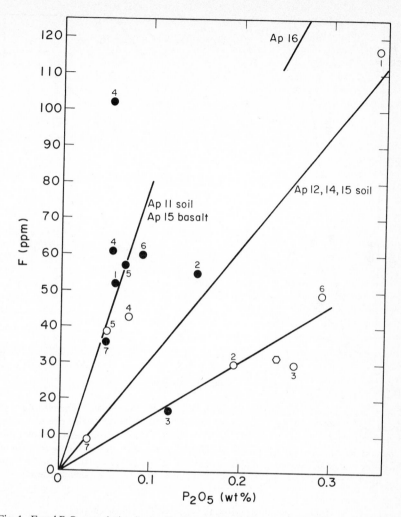

Fig. 1. F and P_2O_5 correlation in Apollo 17 samples. Symbols—open = rocks, 1–7: 72275, 73235, 73275, 74275, 75075, 76315, 76535; closed = soil, 1–7: 70181, 72701, 73141, 74220, 75080, 76281, 78501; Basaltic samples fall along with Apollo 11 soil and Apollo 15 basalts; Station 2 samples are close to Apollo 11, 12, and 15 soils correlation lines. Station 3 and Station 6 samples establish a new correlation that includes 66095 (hexagonal symbol); other Apollo 16 samples fall along the indicated line. P_2O_5 data are from our laboratory, LSPET (1973), Duncan *et al.* (1974) and Rhodes *et al.* (1974).

samples are correlated are very low, 0.00–0.02 wt.%. This P_2O_5 may be associated with a residual P-bearing phase remaining after most of the phosphates have crystallized; or these samples may contain an early partial melt that extracted halogens but very little P_2O_5.

The Cl_r/P_2O_5 ratios associated with various suites of lunar samples are summarized in Table 3.

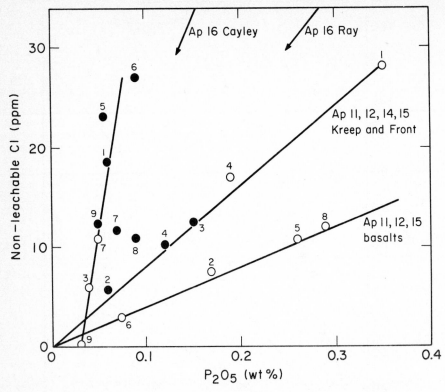

Fig. 2. Nonleachable Cl and P_2O_5 correlation in Apollo 17 samples. Symbols—open = rocks, 1–9, 72275, 72395, 72417, 73235, 73275, 74275, 75075, 76315, 76535; closed = soils, 1–9: 70181, 71501, 72701, 73141, 74220, 74241, 75080, 76281, 78501. Massif samples fall along established correlations for samples from other sites. A possible new correlation including orange soil, dunite, granulite and basaltic samples is indicated. Data for P_2O_5 are given in Fig. 1; P_2O_5 for 72395 obtained from A. L. Albee, private communication.

Table 3. Residual and total Cl versus P_2O_5 in lunar samples.

Samples	Cl_r/P_2O_5	Cl_t/P_2O_5
Mare basalts (8) 12033, 14301, 15418	0.003 ± 0.001	0.007 ± 0.003
Apollo 11, 12, 14, 15 soils and breccias (11); 10017, 10044, 63501 Apollo 17—South and North Massif (5); 71501	0.008 ± 0.001	0.011 ± 0.002
Apollo 17—Plains and Sculptured Hills (5) Apollo 16—Cayley (4) Luna 16, Luna 20	0.022 ± 0.004	0.035 ± 0.012
Apollo 16—Ray (5)	0.012 ± 0.001	0.020 ± 0.004

Cl–Br–I

Rather striking correlations are obtained from the data on the leach and residual halogens found in Apollo 17 soils and rocks. What makes these correlations so surprising is that Br and I are considered to be dispersed elements, whereas Cl has been established to be associated with a phosphate phase (Reed and Jovanovic, 1973). The fact that Br is excluded from terrestrial and meteoritic apatites makes the Br correlation with lunar Cl of additional interest.

Nonleachable Cl, Br in rocks. The average Cl_r/Br_r ratio of Apollo 17 rocks is ~300, Fig. 3. Some of the rocks tend to have a Cl_r/Br_r ratio of about one-half this amount. The only other suite for which we have sufficient data is Apollo 16 where the Cl_r/Br_r ratio is ~46. These ratios may be compared with the cosmic abundance ratio of 100–200 based on C-I and C-II chondrites, a ratio of ~100 for basaltic achondrites, and ratios of 20–100 for terrestrial basalts and granites. A fractionation of Cl and Br seems to have occurred in lunar rocks.

Nonleachable Cl, Br in soils. Most soils have Cl_r/Br_r ratios between 64 and 110. A correlation is not obvious. Two-thirds of the soils have Cl_r concentrations within a 20% spread whereas Br_r varies by greater than 50%. Soils have larger Br contents than rocks. Since soil consists primarily of comminuted local rock it appears that Br from another source is present, either incorporated into the rock detritus or as a nonleachable soil component. Since the Cl content is fairly constant, a Br source relatively free of Cl is required.

Leachable Cl and Br in soils. A source of the excess nonleachable Br in the soils discussed above could be the leachable Br associated with soils. For most samples the labile Br is positively correlated with labile Cl, however, a few samples have an excess of Br_l, Fig. 4b. Removal of the ~50 ppb excess Br from the three samples in which it is found makes these samples colinear with most other soils. The Cl_l/Br_l ratio for soils is ~83.

Leachable Cl, Br in rocks. Most Apollo 17 rocks measured fall on one of two Cl_l versus Br_l correlation lines, Fig. 4a. Both intercept the Cl_l axis at ~0.3 ppm. When the excess Cl_l is subtracted, the Cl_l/Br_l ratios are 46 and 93. These ratios are low relative to Cl_r/Br_r and are consistent with Cl being present in a phosphate which is not water soluble.

Leachable I–Cl–Br. All comparisons for this set of elements are for the leachable fraction since this is the only I we detect in lunar samples.

No correlation is found for I_l and Cl_l in soils. Two trends are noted for I_l and Cl_l in rocks, Fig. 4c. One of the curves intercepts the Cl_l axis at the same concentration of ~0.3 ppm Cl_l as do the Br_l versus Cl_l correlation trends, Fig. 4a. The second curve intercepts the I axis at ~0.5 ppb I, i.e. an excess of I_l over Cl_l.

The I_l–Br_l correlation, Fig. 4d, further supports the I_l–Cl_l correlation. Two curves may be drawn based on the same samples which defined the I_l–Cl_l curves. A 0.75 ppb excess I_l relative to Br_l is found; a deficiency of labile Br with respect to both labile, Cl and I may be indicated.

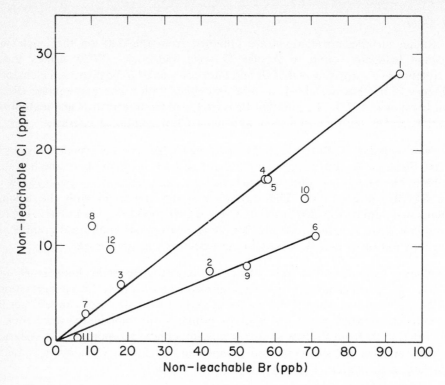

Fig. 3. Nonleachable Cl and Br correlation in Apollo 17 rocks. Symbols and sample numbers are as in Fig. 4.

Cl–Br–I conclusions. We summarize here the correlations observed and their possible inferences.

1. In Apollo 17 rocks correlations exist for leachable and residual Cl and Br and leachable Cl, Br, and I.

2. In Apollo 17 soils leachable Cl and Br are correlated but residual Cl and Br are not.

3. The high degree of coherence between these three elements in rocks suggest that (a) they are indigenous, (b) lunar geochemical processes occurred in closed systems, i.e. due to their significantly different chemistries and volatilities strong fractionations would be anticipated had the parent liquids effervesced in an open system, and (c) halogens were, however, exsolved from crystalline phases as soluble salts; all Apollo 17 rocks analyzed fall into one of two patterns even though the extent of this "exsolution" varied somewhat.

Ion microprobe study of orange spherules

The very high surface F concentrations observed by Surveyor VII at Tyco (Patterson *et al.*, 1970) has stimulated considerable interest. In a number of experiments we have attempted to measure water leachable F in lunar samples

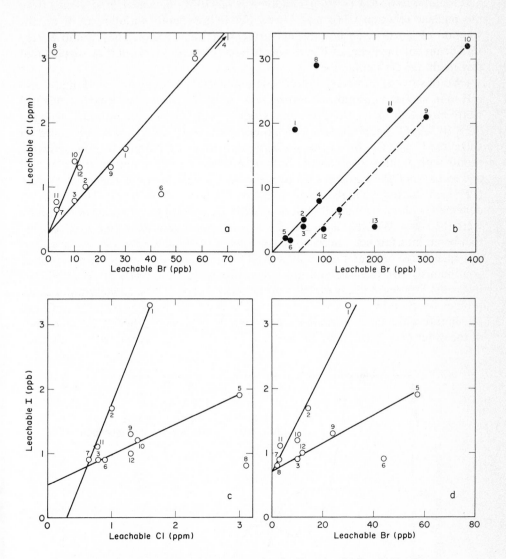

Fig. 4(a) Leachable Cl and Br correlation in Apollo 17 rocks. (b) Leachable Cl and Br correlation in Apollo 17 soils. Samples along the dashed line have an apparent excess of leachable Br. (c) Leachable I and Cl correlation in Apollo 17 rocks. One set of samples appears to have excess I and the other excess Cl. (d) Leachable I and Br correlation in Apollo 17 rocks. Two trends occur with apparently same amount of excess I. Symbols—open = rocks, 1–12: 72275, 72395, 72417, 73235 white, 73235 dark, 73275, 74275, 75075, 76315 exterior, 76315 interior, 76535, 77035; closed = soil, 1–13: 70181, 70006, 70005, 70002, 71501, 72701, 73141, 74001, 74220, 74241, 75080, 76281, 78501.

and found it to be low (<10%). Goldberg *et al.* (1974) have measured γ-rays from the nuclear reaction $^{19}F(p, \alpha\gamma)$ ^{16}O to obtain the depth variation of F on the fraction of a μm scale in exterior surfaces of lunar rocks. They found a very significant enhancement of F in the outermost few tenths μm but they suspect that this might be contamination.

We observed, however, appreciable amounts of F leachable with dilute acid (pH 5) in an 100 mg sample of orange soil 74220,111. A few spherules of this soil were mounted on a Cu disc and studied with an ion microprobe using a 5.3×10^{-3} μa O$^-$ beam. In scanning experiments, F, Cl, and Br were observed on surfaces; F and Cl coated the surfaces uniformly, Br did not. A concentration profile with depth was obtained. To suppress the number of mass peaks detected and to enhance the halogen signals, negative secondaries were collected. Wexler (1973) conducted a similar experiment on a glass sphere from 14163 soil. He collected positive secondaries but otherwise the operating conditions of the probe were the same. We have used the rate of excavation determined by Wexler using two beam interference microscopy since we were unable to relocate our crater. The qualitative nature of our results must be emphasized. The sphere was probed continuously for 1520 sec and mass measurements were made at about 300 sec intervals. The depth excavated was of the order of 500 Å. The concentration depth profiles for F and Cl are given in Fig. 5. The concentrations of both F and Cl fall off sharply with depth. Assuming equal detection efficiency, the F/Cl atom ratio is of the order of 1, hence a wt. ratio of 0.5. This may be compared with a bulk F/Cl

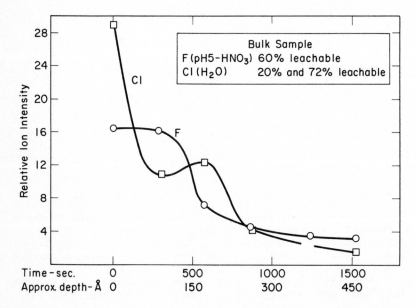

Fig. 5. F and Cl depth profile in an orange glass sphere from 74220 via ion microprobe. The readily leachable, probably surficial, halogens in the bulk sample are given for comparison.

wt. ratio of 1.4 and 0.6 for 74220,34 or of 1.7 and 0.3 for 74220,111 for leachable and residual F and Cl, respectively. All such bulk measurements contain uncertainties since different sample aliquants are required for F and Cl measurements. However, the agreement between the probe and bulk analyses is satisfactory and consistent with the possibility that F in both the surface and bulk sample is lunar.

Li

The Li contents of Apollo 17 samples range from 2.3 to 15 ppm. The low value is in dunite 72417 and is one of the lowest so far observed in lunar samples. The only significant enrichments of Li in lunar samples observed so far are in Apollo 14 and 15 KREEP-rich samples with concentrations up to ~60 ppm. In general, mare basalts have lower (~6–10 ppm) Li than anorthositic samples, including anorthosite 15415 with 15 ppm. This suggests that in lunar rocks Li may be associated with an Al-bearing host phase. Since Li appears to be enriched in volatile-rich KREEP samples, we have compared Li with Cl_r in Apollo 17 samples. Two trends are observed, Fig. 6. One set of samples contains sympathetically varying Li and Cl_r. They lie along a correlation line with a mean Cl_r/Li ratio of 2.3 ± 0.28. This set of samples includes five basaltic soils and the Sculptured Hills soil which contains ~70% mare basalt (see U discussion).

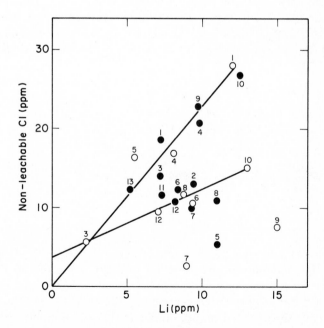

Fig. 6. Nonleachable Cl and Li correlation in Apollo 17 samples. Symbols and sample numbers are as in Fig. 4.

However, two South Massif boulder rocks 72417, 72275, and a Station 3 rock 73235 also lie along the curve.

The second group of samples form a cluster at 7–11 ppm Li and 10–13 ppm Cl_r. However, if rocks only are considered, a correlation appears. This correlation line intercepts the Cl_r axis at about 4 ppm Cl_r. If this excess Cl_r is subtracted from the samples, a $\Delta Cl_r/Li$ ratio of 0.86 ± 0.05 is obtained. The soils cluster along this line. The lack of specificity with regards to sample type occurs here also.

For both groups of samples the colinearity of soils and rocks would seem to preclude a mixing process to account for the soil correlation. These sets of samples may represent progressively later fluids proportionately enriched in Li and Cl_r or successive partial melts into which proportionate amounts of Li and Cl_r were extracted. The lack of specificity regarding sample type suggests that these correlations are dependent on factors not controlled by the parental magmas. This may illustrate a problem in studying correlations among trace elements which cannot be identified stoichiometrically with discrete phases.

U

U data determined in separate experiments via fission-I and fission-Ru are listed in Tables 1 and 2, respectively. The two dominant geologic units at the Apollo 17 site, the massifs and the valley are also characterized by the U contents of the returned samples. Massif boulder samples have U concentrations of $\sim$1.15 ppm, the valley soils and basalts have U contents of $\sim$0.17 ppm. A local admixture of soils of one type with the other has apparently occurred. We estimate the fractions of massif soil based on U content and compare these with the estimated fractions based on soil petrography by Heiken and McKay (1974):

Station	2	2a	6	8
% Massif via U	65	63	79	37
% Massif soil	67	60	49	28

Our North Massif sample gives a higher massif content than Heiken's estimate. The trend noted here has also been commented on by Silver (1974).

Ru–Os

The Ru and Os data are given in Table 2. The concentrations are compared in Fig. 7a and the relationships between Os and Ni are given in Fig. 7b. The basaltic samples measured have very low Os ($\sim$1 ppb) with the exception of black soil 74001 which has 20 ppb Os but $\sim$1 ppb Ru. Very low Os and Ru are found in rocks of possible deep seated origin, dunite 72417 and granulite 76535.

The depletion of Os in the exterior of 76315 relative to that in the interior is consistent with the pattern previously noted in three Apollo 12 basalts (Reed and

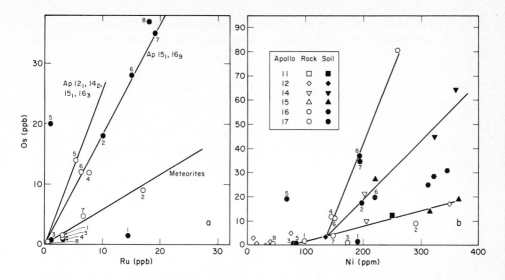

Fig. 7(a) Osmium and ruthenium correlation in Apollo 17 samples. Data fall along correlation lines established by samples from other sites. Subscript indicates the number of samples. (b) Osmium and nickel correlation in lunar samples. Three trends converging at ~0.4 ppb Os, ~130 ppm Ni (12033) are observed. Symbols—open = rocks, 1–8: 72275, 72395, 72417, 73235 interior, 73235 exterior, 76315 interior, 76315 exterior, 76535; closed = soil, 1–8: 70181, 73141, 74220, 74241, 74001, 76241, 76261, 78501. Ni data are from Warner (1974).

Jovanovic, 1971). In two other cases 73235 and 72417, the interior and exterior have the same Os contents within the errors.

For the first time we have observed several samples with meteoritic Ru/Os ratios: 72275, 72395, 72417, and 74220 give a ratio of 1.7 ± 0.3. Stations 2a, 6, and 8 samples have the Ru/Os ratio of 0.50 ± 0.04 found in Apollo 16 samples which we have designated as Cayley (Jovanovic and Reed, 1973). The meteoritic ratio may merely indicate that the samples are relatively unreworked primitive materials that have preserved their original complements of these Pt-metals. This is in agreement with the interpretation of the petrological observations on boulder # 2 Station 2, samples by Albee *et al.* (1974).

Os–Ni. Two Os versus Ni correlation lines involving Apollo 11–14 samples have been suggested previously (Reed *et al.*, 1972). Apollo 15–17 data support these trends and help to better define the curves. There also appears to be a third correlation through Apollo 17 samples from Stations 6 and 8 and possibly including 66095.

All three curves converge at ~4 ppb Os and 140 ppm Ni. This is the same Os concentration, previously arrived at on the basis of three other observations, that we suggested as being indigenous lunar.

Ru, Os and the meteoritic component. Soils from North Massif and Sculptured Hills have higher Ru and Os contents than the 73141 soil from Station 2a at South

Massif and basaltic soils from the valley. Contamination by valley ejecta would, at least on the basis of present data, tend to dilute Ru and Os, hence the North Massif and Sculptured Hills soil values may be lower limits. Thus, soils blanketing the massifs may be inherently more Pt-metal rich. This possibility is not supported by the approximately constant Ir contents at Stations 2 and 8 reported by Baedecker *et al.* (1974) and by Morgan *et al.* (1974).

Some further insight into the question of lunar versus extralunar sources for Pt-metals can be arrived at by comparing Ru and Os with Ir, presumably the "par excellence" extralunar component. Ir data, used here are from Morgan *et al.* (1974). Breccias 72275, 73235, and 76315 and soil 76241 give Ru/Ir ratios of 1.2–1.7; the Os/Ir ratio is 2.7 excluding the 0.6 for 72275. The meteoritic ratios for Ru/Ir and Os/Ir are 2.8 and 1.7, respectively. Soil 76261 has a meteoritic Ru/Ir ratio but a nonmeteoritic Os/Ir ratio of 5.4. The granulite 76535 and the black soil 74001 show extreme depletion in Ir relative to Ru and Os. The Apollo 17 samples, therefore provide further evidence that at least two of the Pt-metals, Ru and Os, have been chemically processed on the moon.

We have assumed that the Ni–Os correlation found in Apollo 11 through 14 samples may have been due to at least two populations of extralunar objects. We wonder now if the evidence is not sufficiently convincing to conclude that most if not all of the trends are due to fractionation of Ru and Os indigenous to the moon. As already noted, when a meteoritic Ru/Os ratio occurs, it may merely be an indication that the sample is primitive lunar material and has not been extensively recycled.

Lunar atmospheric Hg

We have established by stepwise heating experiments that a significant fraction of the Hg found in returned lunar samples is labile at daytime temperatures on the lunar surface. This labile Hg exhibits a concentration gradient in the upper several centimeters of the regolith consistent with the thermal gradient. This may be considered atmospheric Hg. Another assessment of its presence was achieved by measuring Hg in the Apollo lunar sample return containers (ALSRC). This experiment was made possible through the cooperation and assistance of R. Stoenner and R. Davis who tapped the sealed ALSRC's and cryogenically pumped and trapped the gases present. A stainless steel tube containing fine gold mesh, supported by stainless steel screens near each end, was placed between the ALSRC and the charcoal trap. Each collector consisted of a stack of three layers of Au mesh. The system was evacuated to $\sim 10^{-6}$ torr and baked at 400°C. The down-stream foil was originally introduced to serve as a blank and the data are treated as though this was the case. However, the collection efficiency of the Au mesh is not known, hence the results may be lower limits. Stoenner and Davis (1974) report that $\sim 90\%$ of the gas in the ALSRC was pumped out. The Au mesh was sealed in fused silica vials and irradiated with neutrons.

Radioactive Hg was extracted by stepwise heating. Only the Hg released below 500°C, which was >85% of the total Hg, was considered to be lunar.

We assume that the mean depth of the soil gathered is ~3 cm. At this depth we estimate a lunar daytime temperature of 27°C. The Hg extracted from the ALSRC's from Apollo 16 and 17 and the approximate weights of soil are given in Table 4. Based on the amount of Hg extracted previously from various soil samples at 130°C and in a few cases at 50°C and 75°C the amount found in the ALSRC's is reasonable and is probably a lower limit. Clearly, if this amount of Hg can be removed at 23°C from soil present in the ALSRC, then it should have been lost from soil surfaces on the moon unless it was in dynamic equilibrium with a source, namely the lunar atmosphere (Reed *et al.*, 1971)

Table 4. Hg collected from the Apollo 16 and 17 lunar sample return containers (ALSRC).

ALSRC	Hg–Au 1 10^{-6} g	Hg–Au 2 10^{-6} g	Fines g	Hg g/g Fines
APOLLO 16	0.25	0.07	5770	3.0×10^{-11}
APOLLO 17	0.88	0.63	7898	3.1×10^{-11}

Hg released at ≤130°C in lunar samples   ~1×10^{-9}

CONCLUSIONS

(1) For Apollo 17 samples, the halogens exhibit a remarkable degree of coherence in view of their apparent diverse geochemistry, i.e. Cl bound as apatite versus dispersed elements Br and I. The residual Cl/Br ratios for rocks and soils range from 60 to about 300. Thus they span the cosmic abundance ratio of 100–200. Lunar chemical processing may account for the fractionation observed.

(2) The F–Cl–P_2O_5 coherence, previously noted, persists among Apollo 17 samples. Li and residual Cl also exhibit a good correlation.

(3) The nonselectivity in terms of sample type in various inter-element correlations suggests that the coherence among these elements is controlled by factors other than the composition of the parent magma or rock.

(4) Both U and Ru–Os support lateral mixing of regolith on a local scale.

(5) The meteoritic Ru/Os ratios found in a few samples are attributed to primitive lunar material that has not been extensively recycled.

(6) The labile Hg pumped from sealed lunar sample return containers must have been in dynamic equilibrium with Hg in the lunar atmosphere.

Acknowledgments—We thank the Chemistry Department and the operating staff of the High Flux Reactor, Brookhaven National Laboratory, for making their facilities available. We are especially grateful to J. Hudis and Mrs. E. Rowland for their assistance. We thank D. Steidl for his assistance in the ion microprobe experiment. The cooperation of the Argonne National Laboratory LINAC staff is greatly appreciated. Association of G. W. Reed with and the use of facilities at the Enrico Fermi Institute, University of Chicago, is gratefully acknowledged.

REFERENCES

Albee A. L., Chodos A. A., Dymek R. F., Gancarz A. J., and Goldman D. S. (1974) Preliminary investigation of boulders 2 and 3, Apollo 17, Station 2: Petrology and Rb–Sr Model Ages (abstract). In *Lunar Science—V*, pp. 6–8. The Lunar Science Institute, Houston.

Baedecker P. A., Chou C. L., Grudewicz E. B., Sundberg L. L., and Wasson J. T. (1974) Extralunar materials in lunar soils and rocks (abstract). In *Lunar Science—V*, pp. 28–30. The Lunar Science Institute, Houston.

Duncan A. R., Erlank A. J., Willis P. J., and Sher M. K. (1974) Compositional characteristics of the Apollo 17 regolith (abstract). In *Lunar Science—V*, pp. 184–186. The Lunar Science Institute, Houston.

Goldberg R. H., Leich D. A., Burnett D. S., and Tombrello T. A. (1974) Surface fluorine on lunar samples (abstract). In *Lunar Science—V*, pp. 277–279. The Lunar Science Institute, Houston.

Heiken G. and McKay D. A. (1974) Petrography of Apollo 17 soils (abstract). In *Lunar Science—V*, pp. 319–321. The Lunar Science Institute, Houston.

Jovanovic S. and Reed G. W., Jr. (1973) Volatile trace elements and the characterization of the Cayley formation and the primitive lunar crust. *Proc. Fourth Lunar Sci. Conf., Geochim. Cosmochim. Acta*, Suppl. 4, Vol. 2, pp. 1313–1324. Pergamon.

Jovanovic S., Jensen K., and Reed G. W., Jr. (1973) The halogens, U, Li, Te, and P_2O_5 in five Apollo 17 soil samples. *Trans. Am. Geophys. Union* **54**, 595.

LSPET (Lunar Science Preliminary Examination Team) (1973) Apollo 17 lunar samples: Chemical and petrographic description. *Science* **182**, 659–672.

Morgan J. W., Ganapathy R., Higuchi H., Krähenbühl U., and Anders E. (1974) Lunar basins: Tentative characterization of projectiles from meteoritic elements in Apollo 17 boulders (abstract). In *Lunar Science—V*, pp. 526–528. The Lunar Science Institute, Houston.

Patterson J. A., Turkevich A. L., Franzgrote E. J., Economou T. E., and Sowinski K. P. (1970) Chemical composition of the lunar surface in the terra region near the crater Tycho. *Science* **168**, 825.

Reed G. W. and Jovanovic S. (1971) The halogens and other trace elements in Apollo 12 samples and the implications of halides, platinum metals, and mercury on surfaces. *Proc. Second Lunar Sci. Conf., Geochim. Cosmochim. Acta*, Suppl. 2, Vol. 2, pp. 1261–1276. MIT Press.

Reed G. W., Jr. and Jovanovic S. (1973) Fluorine in lunar samples: Implications concerning lunar fluorapatite. *Geochim. Cosmochim. Acta* **37**, 1457–1462.

Reed G. W., Goleb J. A., and Jovanovic S. (1971) Surface-related mercury in lunar samples. *Science* **172**, 258.

Reed G. W., Jr., Jovanovic S., and Fuchs L. H. (1972) Trace element relations between Apollo 14 and 15 and other lunar samples, and the implications of a moon-wide Cl-KREEP coherence and Pt-metal non-coherence. *Proc. Third Lunar Sci. Conf., Geochim. Cosmochim. Acta*, Suppl. 3, Vol. 2, pp. 1989–2001. MIT Press.

Rhodes J. M., Rogers K. V., Shih C., Bansal B. M., Nyquist L. E., and Wiesmann H. (1974) The relationship between geology and soil chemistry at the Apollo 17 landing site (abstract). In *Lunar Science—V*, pp. 630–632. The Lunar Science Institute, Houston.

Silver L. T. (1974) Patterns of U–Th–Pb distributions and isotope relations in Apollo 17 soils (abstract). In *Lunar Science—V*, pp. 706–708. The Lunar Science Institute, Houston.

Stoenner R. W., Davis R., Jr., and Bauer M. (1973) Radioactive rare gases and tritium in the sample return container and the ^{37}Ar and ^{39}Ar depth profile in the Apollo 16 drill stem. Unpublished report, Brookhaven National Laboratory, New York.

Warner J. L. (1973) *Lunar Science Data Base.* L. B. J. Space Center, Houston.

Wexler, S. (1974) Private communication.

Proceedings of the Fifth Lunar Conference
(Supplement 5, Geochimica et Cosmochimica Acta)
Vol. 2 pp. 1703–1736 (1974)
Printed in the United States of America

Lunar basins: Tentative characterization of projectiles, from meteoritic elements in Apollo 17 boulders

John W. Morgan, R. Ganapathy, Hideo Higuchi,
Urs Krähenbühl, and Edward Anders

Enrico Fermi Institute and Department of Chemistry, University of Chicago, Chicago,
Illinois 60637

Abstract—Thirty-one Apollo 17 samples were analyzed by neutron activation analysis for Ag, Au, Bi, Br, Cd, Cs, Ge, Ir, Ni, Rb, Re, Sb, Se, Te, Tl, U, and Zn. Like most other highland samples, the breccias contain ancient meteoritic debris, judging from the 10^2–10^3-fold enrichment of siderophile elements over their levels in crystalline rocks.

A review of data for 74 highland samples from 5 sites shows that 6 discrete ancient meteoritic components are present, differing in their proportions of the less volatile to the more volatile siderophiles (Ir and Re versus Au and especially Ge, Sb). These components do not match any of the known meteorite classes, but seem to represent an extinct population of bodies (planetesimals, moonlets?) that produced the mare basins. On the basis of their stratigraphy and geographic distribution, five of the six groups are tentatively assigned to specific mare basins: Imbrium, Serenitatis, Crisium, Nectaris, and Humorum or Nubium.

A few properties of the basin-forming objects are inferred from the trace element data. They were independently formed bodies of roughly chondritic composition, not castoffs of a larger body. They were internally undifferentiated, had radii less than 100 km, contained between ~15% and ~40% iron, and struck the moon at velocities generally less than 8 km/sec. None match the earth or moon in bulk composition; in fact, several are complementary, being lower in refractories and higher in volatiles than the earth or moon. Possibly this implies a genetic link between these bodies, the earth, and the moon.

An attempt is made to reconstruct the bombardment history of the moon from the observation that only $\sim 80^{+80}_{-40}$ basin-forming objects, totaling $\sim 2 \times 10^{-3}$ $M_{\mathbb{D}}$, fell on the moon after crustal differentiation ~4.5 AE ago, and that only one such body, of $\sim 5 \times 10^{-5}$ $M_{\mathbb{D}}$, was left 3.9 AE ago. The apparent half-life of basin-forming bodies, ~100 m.y., is close to the calculated value for earth-crossing planetesimals. According to this picture, a gap in radiometric ages is expected between the Imbrium (3.9 AE) and Nectaris (4.2 AE) impacts, because all 7 basins formed in this interval lie on the farside or east limb.

Several Apollo 17 samples are discussed in detail: boulder samples 72255, 72275, 76315, 77075, and 77135; troctolite 76535; and orange soil 74220. Station 6 and 7 boulders contain mainly the Group 2 meteoritic component assigned to Serenitatis, while the Station 2 boulder contains a distinctive, Group 3 component which we attribute to Crisium.

Introduction

Ever since Apollo 14, we have tried to understand the ancient meteoritic debris in highland breccias (Morgan *et al.*, 1972). Though it was clear from the beginning that several discrete components were present, their number and source were hard to establish unequivocally, even after 120 analyses of highland rocks (Ganapathy *et al.*, 1973). The present picture looks rather more hopeful, not only because we now use a different graphic representation, but also because the latest

1703

set of 34 samples has greatly sharpened previous compositional trends. The Apollo 17 boulders (this paper) and a suite of Apollo 16 samples (Ganapathy *et al.*, 1974) were especially helpful in this regard.

The major part of this paper is devoted to the ancient meteoritic components. The final section discusses the remaining data on Apollo 17 rocks and soils.

Experimental

Samples

Rock samples were interior chips without sawn surfaces. Those distributed by boulder consortia are described in reports by the consortium leaders E. C. T. Chao, W. A. Phinney, and John A. Wood (Chao and Minkin, 1974; Heiken *et al.*, 1973; Consortium Indomitabile, 1974).

One aliquot of the orange soil 74220,54 was analyzed as received. The other (0.46 g) was cleaned ultrasonically in acetone and separated by hand-picking into orange and dark glass fractions (mass 51.8 and 42.4 mg; estimated compositions >60% orange glass + ~10% light material, and >80% dark glass + <5% orange glass).

Method

Samples were irradiated at the NBS reactor, Gaithersburg, Md., at a fluence of 6 to 8 × 10^{19} neutrons. Our radiochemical procedure has been described by Keays *et al.* (1974).

Results

The data are given in Table 1. Accuracy and precision have been discussed by Keays *et al.* (1974).

Ancient Meteoritic Components

Indigenous or meteoritic?

Two lines of evidence suggest that the siderophile elements in highland breccias are extraneous.

First, the siderophiles are greatly overabundant in *breccias* compared to *genuine* igneous* rocks. This is true for all highland rock types: anorthosites, norites, and granites. Figure 1 shows an alkali-poor and an alkali-rich highland breccia (black circles and black squares), each matched up with a crystalline rock of similar bulk composition and Rb, Cs, U content (open symbols). Most of the 16 elements have comparable abundances in the breccia and its crystalline counterpart, except for the first 5, siderophile elements which are enhanced in the breccias by factors of 10^2 to 10^3. Nickel, not plotted here, behaves similarly.

---

*As opposed to *pseudo*-igneous rocks such as 14310 or 68415. These rocks, though crystalline in texture, match the breccias in siderophile content and age, and are therefore best interpreted as impact melts.

Table 1. Trace elements in Apollo 17 rocks and soils (ppb, except Ni, Rb, Zn, ppm).

Sample	Classification*	Mass mg	Ir	Re	Au	Ni	Sb	Ge	Se	Te	Ag	Br	Bi	Zn	Cd	Tl	Rb	Cs	U
Rocks																			
70215,64	Mare Basalt	94	0.003	0.0015	0.026	1	0.18	1.66	176	2.1	1.1	8.1	0.099	2.1	1.8	0.16	0.36	15	118
72255,42,9001	GB B&W Ct	22	0.0040	0.0068	0.008	4	0.26	61	280	14.3	0.76	15.3	0.30	4.5	5.8	0.30	1.27	67	240
72255,52,9004	GB Mx	34	5.28	0.498	2.00	227	0.77	174	77	4.7	0.57	101	0.21	2.8	8.1	1.18	5.8	240	1790
72275,57,9002	GB Mx	75	2.26	0.225	0.82	97	1.17	406	34	4.14	0.74	48	0.11	2.7	13	0.71	5.9	260	1500
72275,80,9004	GB Ctl bl rim	29	2.54	0.233	1.16	122	0.94	137	63	3.46	0.93	290	0.14	2.8	15	0.71	11.3	480	3100
72275,83,9005	GB Ct2 G Aph	57	3.44	0.334	1.30	147	1.06	178	52	2.74	0.56	95	0.12	2.4	26	0.62	5.4	260	1840
72275,91,9006	GB Ct5 Basalt	43	0.023	0.0066	0.045	43	2.87	1290	230	7.8	0.58	44	0.14	2.7	8.3	0.58	8.0	360	1060
73235,45	BGB	83	3.71	0.385	2.31	144	1.14	230	53	4.3	1.0	90	0.69	9.4	27	2.1	4.7	198	1060
73275,23	Breccia	96	5.71	0.494	3.34	182	1.19	265	92	5.5	0.74	73	0.16	2.5	4.1	1.60	6.9	270	1360
75035,35	Mare Basalt	104	<0.0007	0.007	0.0084	1	0.04	1.27	156	1.5	0.62	8.0	0.043	2.3	1.1	0.29	0.79	29	153
76315,73	BGB Mx	92	5.42	0.507	3.21	256	1.49	346	100	4.04	0.84	48	0.098	3.1	5.0	0.31	5.91	250	1540
76315,74	BGB C3	71	5.97	0.575	3.48	260	1.54	354	107	5.1	0.88	44	0.28	3.4	6.4	0.34	5.9	250	1490
76535,20	Troctolite	151	0.0054	0.0012	0.0025	44	0.014	1.70	4.1	0.28	0.12	3.2	0.037	1.2	0.60	0.012	0.20	14	19.4
77017,48	An Ol Gabbro	83	17.0	1.73	5.65	443	0.72	110	68	1.9	0.87	35	0.22	2.5	9.0	0.77	1.34	61	137
77075,19	GGB bl dike	28	8.89	0.781	5.09	286	1.92	532	112	6.3	1.2	81	0.34	2.8	7.5	2.4	6.4	270	1450
77135,10	GGB Ctl/Mx?	82	3.78	0.485	3.57	205	1.21	295	137	3.6	1.1	47	0.18	2.9	10.5	2.6	6.5	270	1390
77135,50	GGB Ct2:TrAn	28	7.20	0.662	1.46	174	0.58	50	11.3	1.32	0.38	11.6	0.17	2.6	6.8	0.48	1.80	74	260
77135,62	GGB Ctl:rTrB	70	15.1	1.42	4.74	412	0.47	78	33	1.1	0.58	17.6	0.14	2.4	3.7	0.58	2.6	73	450
77135,69	GGB Nonves Mx	61	10.5	1.06	6.45	438	2.16	618	144	8.84	1.2	45	0.23	3.3	3.5	2.3	6.1	250	1380
78155,30	Catacl An	132	3.32	0.278	0.66	68	20.4	27	49	3.2	1.0	65	0.29	2.3	63	5.9	1.76	84	250
79035,19	Soil B	105	7.50	0.629	2.39	162	0.89	278	300	18.6	19	117	0.70	40	71	2.2	1.69	72	310
79155,26,9001	bl Glass	17	2.40	0.143	0.81	79	2.45	24	205	<1	5.1	14.3	0.32	2.6	3.5	0.60	0.84	42	178
Soil Separates, <1 mm																			
74220,54,1	Dark Glass	41	0.114	0.0135	0.23	70	1.00	41	129	10.0	320	15	0.50	45	92	1.60	0.66	30	130
74220,54,2	Orange Glass	27	0.214	0.0553	1.07	72	25.3	191	460	49	75	88	1.53	141	260	9.9	0.77	44	115
Soils, <1 mm																			
72321,1	Shadowed	84	8.87	1.07	6.03	550	1.81	625	240	24	6.5	78	0.65	18	37	1.51	4.1	170	900
74001,5	Below Orange	39	0.021	0.213	0.705	68	1.16	105	350	38	72	210	0.67	148	25	4.0	0.76	37	141
74220,54	Orange	122	0.411	0.052	0.99	67	0.65	250	640	62	111	520	1.43	230	320	20	0.95	53	168
74241,30	Above Orange	122	2.78	0.296	1.01	64	0.55	155	340	24	25	610	0.75	86	210	9.1	2.3	107	330
75081,33	Nr Camelot	116	5.36	0.470	1.70	113	0.67	190	250	10	9.9	100	0.54	27	32	1.3	1.2	47	240
76241,12	Shadowed	104	8.57	0.820	3.81	220	1.34	420	240	17.0	9.4	88	0.67	25	82	1.95	2.8	133	550
76261,13	Control	130	6.46	0.671	2.52	160	1.06	300	210	13.6	7.9	65	0.59	23	39	1.59	2.7	115	490

*An = anorthite, anorthosite   B = breccia   bl = black   G = gray   Mx = matrix   Pl = plagioclase   Tr = troctolite
Aph = aphanitic   BG = blue gray   Ct = clast   GG = green gray   Ol = olivine   r = recrystallized

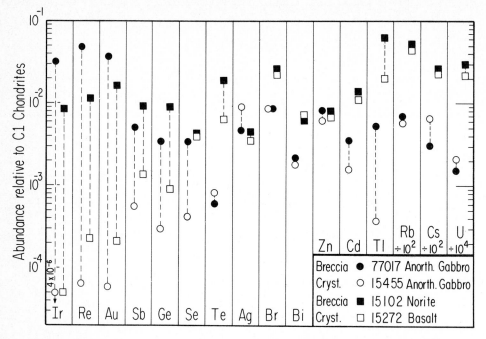

Fig. 1. Highland breccias (solid symbols) are strongly enriched in the first 5, siderophile elements, compared to crystalline rocks of the same Rb, Cs, and U content (open symbols). The enrichment is due to an ancient meteoritic component, dating from the era of intense bombardment.

Second, the lowest abundances of siderophiles in highland rocks, e.g. <0.05 ppb Au (Fig. 2), are consistent with measured metal–silicate distribution coefficients, which range from 10^4 to 10^6 (Kimura *et al.*, 1974). It appears that these abundances represent true indigenous levels. On the other hand, the more common, high abundances around 1–10 ppb (Fig. 2) exceed equilibrium solubilities in silicates by several orders of magnitude. The extraneous nature of these siderophiles is shown by the fact that they reside largely in discrete metal grains of micron-size and larger (Wlotzka *et al.*, 1972, 1973; Ganapathy *et al.*, 1974), which comprise 0.1–1% of highland breccias and impact melts (Dickey, 1970; Vinogradov, 1973). A suspension of metal droplets in a silicate melt is dynamically unstable on time scales comparable to the freezing time of basaltic flows (Fish *et al.*, 1960; Provost and Bottinga, 1972).

For these reasons, it appears that high siderophile abundances in highland rocks are diagnostic of a meteoritic component. Because this component occurs in breccias that have remained closed systems for at least 3.9 AE, it can properly be called an ancient meteoritic component.

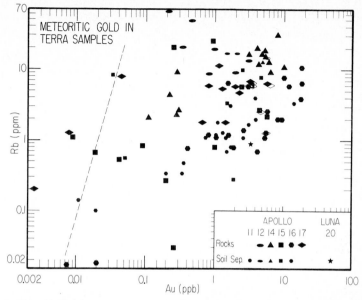

Fig. 2. Most highland samples are enriched in meteoritic gold above indigenous levels (approximated by dashed line). Enrichment is nearly independent of Rb content and rock type. Only anorthosites of <0.2 ppm Rb tend to be free of meteoritic gold. Perhaps they come from deeper crustal levels, not penetrated by meteoritic projectiles.

Classification

The 6 diagnostic elements differ greatly in condensation temperature from a solar gas. Re and Ir condense a few hundred degrees above iron, Sb and Ge condense a few hundred degrees below iron, while Ni and Au condense almost concurrently with iron, slightly before and after, respectively (Grossman, 1972 and unpublished data; Larimer, 1967 and unpublished data). Presumably for this reason these elements show large abundance variations in meteorites, which have been used as the basis for classification (Wasson and Schaudy, 1971 and earlier papers; Baedecker, 1971; Scott, 1972).

Similar variations in Ir/Au and Ge/Au ratios have been seen in lunar samples, suggesting the presence of more than one kind of meteoritic component (Morgan *et al.*, 1972; Ganapathy *et al.*, 1973). The trends show up more clearly on ternary plots. Figures 3 and 4 show results for 74 highland samples from 5 sites. The data are normalized to Cl chondrite abundances (Krähenbühl *et al.*, 1973b), to prevent the plots from being dominated by the element of highest absolute abundance. We have again applied corrections for indigenous contributions, using average composition of meteorite-free, alkali-poor, and alkali-rich highland rocks, and assuming that the correction for rocks of intermediate composition varies linearly with Rb content (Ganapathy *et al.*, 1973). The corrections for Sb, not given in the earlier paper, were 0.064 and 0.25 ppb for the two rock types.

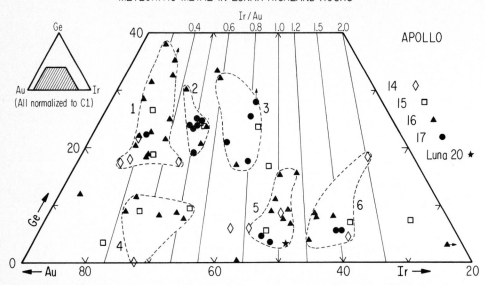

Fig. 3. Six distinct types of ancient meteoritic component seem to be present, differing in their proportions of Ir, Au, and Ge. Both the clustering of the points and their uneven geographic distribution suggest that this material is derived mainly from a few large objects—probably the projectiles that excavated the mare basins.

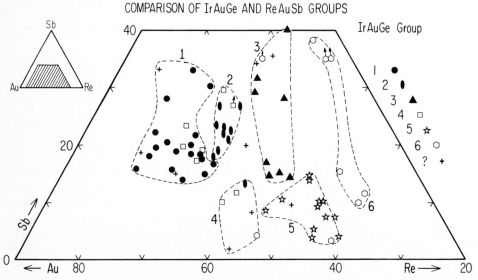

Fig. 4. Six analogous groupings may be recognized on the basis of Re, Au, and Sb content. Most of the points belonging to a given Ir–Au–Ge group fall in a similarly situated cluster on an Re–Au–Sb plot. The principal exception is Group 4, which is largely dispersed on this plot.

In the past, we considered only samples of ⩾0.8 or ⩾1 ppb Au, for fear that the signal-to-noise ratio might become too small at lower Au contents. However, upon examining the data, we found that samples with Au contents down to 0.30 ppb still gave reasonable results, fitting into the general pattern. In fact, even in the next interval (0.2–0.3 ppb Au), 4 of 8 samples still fell into existing fields, while 4 others either had large corrections or fell outside the fields (mainly in the Ir-rich corner). We therefore extended the cutoff to 0.2 ppb Au, but discarded the last 4 samples.

In order to avoid samples of mixed parentage, we limited ourselves to rocks and to lithic fragments from coarse soils. Agglutinates, magnetic separates, soil breccias, and glasses from coarse soils were excluded.

Six groups seem to be present, as indicated by the dashed lines (Fig. 3). The elongated shape of the groups reflects the fact that Ge and Sb are less reliable indicator elements than are Ir, Re, and Au. Their indigenous contributions are larger and more variable, and their volatility makes them more prone to metamorphic redistribution. Consequently, we used Ir/Au and Re/Au ratios as our prime criteria in defining the groups.

The reality of these groups is supported by the mutual consistency of the two plots (each of which has a volatile element at the apex and a refractory element in the right corner). Most points belonging to a given Ir–Au–Ge group fall in a similarly situated, compact cluster on the Re–Au–Sb plot. The principal exceptions are Group 4, and a few samples of sporadically high indigenous Sb or Ge content.

There is a rough correspondence between the new groups and the old ones, as follows:

$$1 \approx LN1 + R \quad 2 \approx LN2 \quad 3 \approx NR \quad 4 \approx DN\,(Ir/Au < 0.6)$$
$$5 \approx DN\,(Ir/Au > 0.6)\text{ and A }(Ir/Au < 1.1) \quad 6 \approx A\,(Ir/Au > 1.1)$$

The definition of these groups involves some subjectivity, and we do not claim that the present 6 groups are final. We intend to study the problem by cluster analysis, once high-quality data for chondrites are available for control purposes (see next section). In general, we have tried to be conservative, creating as few groups as possible. The separation between some of the groups is not as sharp as one might like, e.g. 1 and 2 or 5 and 6. However, in both cases there were reasons not to combine them. Group 2 contains a compact cluster of Apollo 17 samples, which would stand out as a distinctive subgroup in a combined Group 1 + 2. Groups 5 and 6 also contained distinctive subsets of samples; moreover, if combined, these 2 groups would cover a wider range in Ir/Au ratio than any other.

Comparison with meteorites

The ancient lunar meteoritic bodies show little resemblance to present-day meteorites (Fig. 5–7). Most meteorite groups avoid the lunar fields. Only a few (E4, E6, H-chondrites, and I and IIIA irons) show as much as a marginal overlap. It appears that the ancient meteoritic bodies represent a distinct population, different from present-day meteorites.

One interesting point that emerges from Figs. 5 and 6 is that the chondrite groups are not much more compact than the lunar meteorite groups. Probably the

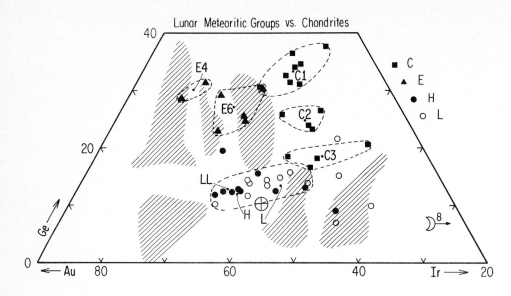

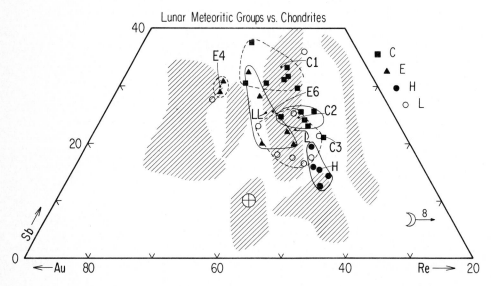

Figs. 5 and 6. Ancient lunar meteoritic components show little resemblance to known
chondrite classes.

best indication of the "natural width" of a chondrite group comes from Cl
chondrites (Figs. 5 and 6) and H, L chondrites (Fig. 6) where at least 5 samples
were analyzed for all 3 elements in the same laboratory (Krähenbühl *et al.*, 1973b;
Fouché and Smales, 1967a, b). For most other classes, data were few and from
different sources.

It would be useful to obtain more high-quality data on chondrites, to see how

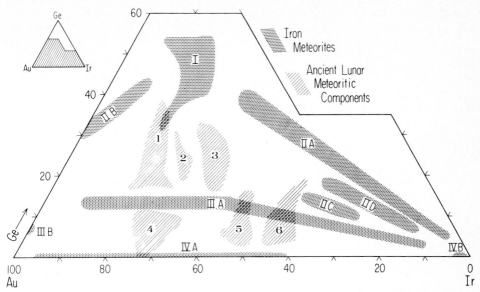

Fig. 7. Ancient lunar meteorite components do not resemble iron meteorites.

well the diagnostic elements used can resolve individual chondrite groups. Note that the H, L, and LL groups are not resolved in Fig. 5, presumably because of the poor quality of the data.

Basins or craters?

Granted that the ancient meteoritic material dates from the early intense bombardment, does it come mainly from a few large basins or from a multitude of smaller craters? The clustering into 6 groups alone does not prove that only 6 bodies were involved. Six families of crater-forming bodies, each with hundreds of members, would lead to the same result. A more conclusive argument comes from the size distribution of lunar craters and basins.

Contribution of basins. Various authors have estimated the amount of ejecta in the lunar highlands, and their results consistently show that the larger objects will contribute the lion's share. This is a consequence of the small slope of the size-distribution function. Öpik (1971), using a single size-distribution law (Baldwin, 1964), estimated that 56% of the highland ejecta will be contributed by craters (actually basins) larger than 412 km in diameter, and an additional 25% by craters between 206 and 412 km. Short and Forman (1972) determined the crater contribution separately, by integrating the size-distribution function to 400 km, and then adding the basins individually. Their results indicate a basin contribution of 30–60%, depending on the degree of slumping assumed. The true value must be higher, because several more basins have been discovered since 1972 (e.g. El Baz, 1973), and the radius of the Imbrian basin seems to be 485 rather than 335 km

(McGetchin *et al.*, 1973). Moreover, Short and Forman used a relation for the crater depth/diameter ratio which gave preposterously small volumes for the larger basins—comparable to the projectile volume. Thus the basin contribution is likely to be substantially higher than 30–60%, probably close to Öpik's 81%. In terms of projectile debris, the basins' share may be still greater, if the basin-forming objects had systematically lower impact velocities than the crater-forming objects (Hartmann and Wood, 1971).

Shallow ejecta only? Even though the basins may account for most of the ejecta, not all authors agree that their projectile debris will dominate at distant sites. Dence and Plant (1972) pointed out that the Pre-Imbrian surface was covered with mixed ejecta from many highland craters, and argued, by analogy with Ries Crater "bunte Breccia," that it was this surface layer which contributed the bulk of the ejecta at Fra Mauro and other distant sites. Because of its shallow origin and gentle shock history, it would not contain any Imbrium projectile debris. Deeper, more heavily shocked material analogous to suevite would be found only at closer range, e.g. at the Apollo 15 site. Other authors have also used the Ries analogy to argue for a shallow origin and short range of lunar crater ejecta (Stöffler *et al.*, 1974). However, these arguments neglect the difference in gravity between earth and moon. The range R of a fragment ejected with velocity v at an angle β is given by

$$R = (v^2/g)\, 2 \sin \beta \cos \beta$$

where g is the acceleration due to gravity. Other things being equal, a fragment on the moon thus will fly 6 times farther than on the earth. Indeed, several rocks of deep-seated origin have been found on the lunar surface: a troctolite from 15 to 40 km depth (Gooley *et al.*, 1974) and a number of dunite clasts, presumably from below the 65 km crust (e.g. Albee *et al.*, 1974).

Secondary craters? Oberbeck *et al.* (1974) have raised a different objection. They note that any basin ejecta reaching Apollo landing sites would strike at high enough velocities to excavate a secondary crater of several times their own mass. Thus Imbrium or Orientale material reaching the Apollo 16 site would be diluted 4- or 7-fold with local rock. They therefore contend that the ancient meteoritic component is derived mainly from many small, local craters, not a few large, distant basins.

Oberbeck's mechanism certainly applies to large blocks, tens or hundreds of meters in diameter. But mass distribution functions show that such blocks comprise only a minor part of the distant ejecta. Moreover, their very size precludes any admixture of projectile debris. The great bulk of the distant ejecta seems to have been of millimeter to decimeter size, judging from both the grain and clast size of lunar breccias, and the uniform dispersion of the metal on a millimeter scale.

Now, when the first 10 cm of 100 m of such fine-grained basin ejecta arrive at a given site, they will indeed mix with the local regolith, to a depth on the order of the diameter of the largest fragments, i.e. ~10 cm. Successive 10-cm layers will mix not with pristine local regolith, but with material containing progressively larger amounts of basin ejecta, and so, before as much as a meter of material has been deposited, the process has turned into orderly sedimentation, without appreciable mixing. (The situation is analogous to that in the mare regolith, where lateral deposition greatly predominates over vertical mixing. A spectacular demonstration is the integrity of coarse-grained, Bi- and Cd-rich layer in core 12028; Laul et al., 1971.) Greater mixing depths will be attained at the sites of secondary craters, but because these craters account for only a minor part of the ejecta, the resulting polymict breccias will not be very common.

This is confirmed by trace element and electron microprobe data. Thirteen out of fourteen breccias studied by us contain only one kind of meteoritic metal (see section "Petrographic Relations"). The Co and Ni contents of metal grains in a given breccia also tend to cluster strongly, at least in cases where the Ni content is low enough for all the metal to be present as kamacite (see, for example, Taylor et al., 1973).

Finally, we note that local impacts can never become dominant over basins if basins contribute the major part of the ejecta. At best local impacts can mobilize and remobilize older basin ejecta. But if basins contribute 80–90% of the projectile debris on the moon, they will dominate the meteoritic component in second-, third-, and nth generation breccias. The only secular trend of such repeated mixing will be eventual homogenization of the metal. Our data on breccias, and Figs. 3, 4, and 10 show that very little such homogenization has taken place.

Thus it seems safe to conclude that the ancient meteoritic debris comes mainly from a few large events, superimposed on a continuum of smaller events. This is consistent with the clustering of the points in Fig. 3: all but 8 of the 74 points fall within the six groups.

Assignment to Basins

Ejecta thickness

We can try to assign the groups to individual basins, by comparing their frequency of occurrence at each landing site (Fig. 3) with expected average contributions of ejecta from each basin, according to McGetchin et al. (1973). We have recalculated their data, using the basin coordinates and age sequence of Hartmann and Wood (1971) rather than Stuart-Alexander and Howard (1970), and including the Luna 20 site (Table 2).

Pre-basin overlay?

In principle, these trends can be blurred by two-stage transport, i.e. reejection of older ejecta during formation of a basin. To assess the importance of this effect, we have calculated the average overlay thickness at the center of each pre-basin surface, using the equations of McGetchin et al. (1973). It turns out that the old overlay comprises only some 10–200 m, or 1–5% of the new ejecta, hardly a significant fraction.

Table 2. Ejecta depth in meters.*

Contributing basin	Crater density†	Apollo 11	Apollo 12	Apollo 14	Apollo 15	Apollo 16	Apollo 17	Luna 16	Luna 20
Orientale	2.4 ± 0.2	1.1	4	4	1.9	1.3	1.1	0.6	0.6
Imbrium	2.5 ± 0.2	59	133	128	560	54	66	19	19
Nectaris	16 ± 2	232	8	10	13	211	35	78	64
Crisium	17 ± 4	42	4	5	13	20	100	482	1134
Humorum	25 ± 2	4	68	47	5	7	3	1.4	1.8
Smythii	27 ± 7	2	0.4	0.5	0.9	1.4	3	18	18
Serenitatis	28	144	20	26	693	57	1089	47	40
Tranquillitatis W.	30	121	0.7	0.9	4	1.8	54	3	3
Tranquillitatis E.		9	0.2	0.3	1.0	1.9	32	5	6
Nubium	30	3	44	63	2	6	1.1	0.6	0.5
Fecunditatis	30	2	0.1	0.2	0.3	1.1	1.8	258	80
Australe		17	6	6	10	15	15	46	39

*Calculated according to McGetchin *et al.* (1973).

†Relative to average mare = 1. These crater densities refer to ejecta blanket of flooded basins or to floors of unflooded basins, and should be proportional to age (Hartmann and Wood, 1971).

Highland craters, on the other hand, contribute ~1 km of overlay (Short and Forman, 1972), and thus may pose a more serious problem. A large crater contribution could manifest itself in two ways, depending on whether or not the crater-forming population was chemically distinct from the basin-forming population. If it was, it would fill the continuum between the groups in Fig. 3. If it was not, it would merely blur the geographic trends, all groups being equally common at all landing sites. Neither of these is the case: note, for example, the scarcity of Apollo 17 samples in Groups 1 and 4, and their dominance in Groups 2 and 3. We thus conclude that the crater contribution was small compared to the basin contribution. Let us therefore try to assign each group to an individual basin.

Tentative assignments

Imbrium. Group 1 (Fig. 3) has all attributes expected for an Imbrian origin, according to Table 2. It is prominent at all sites except Luna 20 which is the most distant from the Imbrium Basin, and Apollo 17 which is partly shielded by South Massif from Imbrian ejecta. It seems to comprise the topmost stratigraphic unit everywhere, judging from its prominence in intercrater areas and lower abundance on crater rims (e.g. North Ray Crater at Apollo 16), where deep-seated material is expected to dominate.

Serenitatis. According to Table 2, material from this basin should be dominant at Apollo 17. This clue leads us to Group 2, six of whose ten members come from Apollo 17. Not only is this the largest grouping of Apollo 17 samples, but it also includes all samples of the oldest breccia type at this site, the blue–gray breccias (Fig. 8). Two of them (76315) come from the Station 6 boulder, which originated 1600 m below the crest of North Massif. This suggests a position fairly low in the stratigraphic sequence: Serenitatis or older (Table 2).

Crisium. Group 3 is most likely derived from Crisium. It is represented mainly by the gray, foliated boulder from Station 2 of Apollo 17 (Fig. 8), which apparently comes from near the top of South Massif (Schmitt, 1973) and hence must represent one of the younger basins. With Imbrium already assigned, the main possibilities are Nectaris or Crisium. If it came from Nectaris, then Group 3 should be rather common at Apollo 16 (Table 2); yet only 2 of 37 samples from that site fall in Group 3. On the other hand, if it came from Crisium, then Group 3 should be more prominent at Apollo 17 than at other sites—and indeed, Apollo 17 accounts for 4 of the 7 samples in this group.

Nectaris. Either Group 5 or 6 may be derived from the Nectaris body. At Apollo 16, both are represented mainly by a distinctive type of alkali-poor, dark-matrix breccia from Stations 11 and 13, near the rim of North Ray Crater (Ganapathy *et al.*, 1974). Group 5 also includes three shocked glasses from other stations (Fig. 9). The dark-matrix breccias comprise the third stratigraphic unit at that site, being overlain by nearly 300 m of light-matrix breccias and cataclastic anorthosites (AFGIT, 1973; Hodges *et al.*, 1973). The latter contain no meteoritic component and have therefore been interpreted as deep ejecta from the Nectaris basin (Krähenbühl *et al.*, 1973a; Ganapathy *et al.*, 1974). Thus, Groups 5 and 6 must represent basins of Nectarian or older age. We tentatively assign Group 5 to the Nectaris body, because of its presence at Luna 20 (Fig. 9). Both Nectaris and Crisium ejecta should be prominent at that site, but by analogy to the stratigraphy of Apollo 16, all but the bottom portion of the thick Crisium blanket at Luna 20 should consist of deep ejecta, uncontaminated by projectile debris. It must be

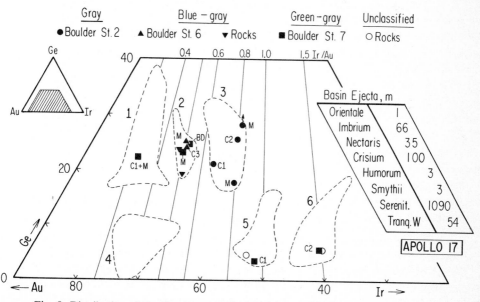

Fig. 8. Distribution of Apollo 17 samples among lunar meteoritic groups. C = clast, M = matrix, BD = black dike.

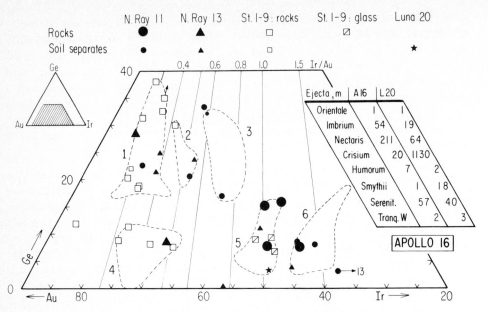

Fig. 9. Distribution of Apollo 16 and Luna 20 samples among lunar meteoritic groups.

overlain by a thin blanket of meteorite-bearing Nectaris material. The source crater of Luna 20 soil, Apollonius C (Vinogradov, 1973; Crozaz *et al.*, 1973), should have ejected mainly such Nectaris material.

Humorum(?). Group 6 must be pre-Imbrian in age. It is found mainly in rocks from crater rims which tend to come from lower stratigraphic units: 60017 from Outhouse Rock at North Ray Crater (Apollo 16), and 14063 from one of the White Rocks at Cone Crater (Apollo 14). With Nectaris, Crisium, and Serenitatis already assigned, the principal remaining possibility is the Humorum body. Alternatives are Nubium or the South Imbrium basin at 10°N, 16°W (Wilhelms and McCauley, 1971). Perhaps a further study of Apollo 14 samples will shed some light on this question, because material from all three basins should be prominent at that site.

Unassigned. This leaves Group 4 unassigned. It is the least well-defined of the 6 groups (note the dispersal of its members in Fig. 4), and contains a high proportion of impact melts. Perhaps it is not a true group, but merely a chance cluster of samples from Groups 1 and 2 that lost some volatiles in impact melting or metamorphism.

Orientale. We also do not have a strong candidate for Orientale material. Early estimates (Hodges *et al.*, 1973; Chao *et al.*, 1973) suggested that some tens of meters of Orientale material might be present at Apollo 16 and other sites, but the calculations of McGetchin *et al.* (1973) suggest much smaller thicknesses, 1–4 m (Table 2). Moreover, because of the great ejection distances, such material probably fragmented on impact and became dispersed throughout the local regolith (Oberbeck *et al.*, 1974). Two highly speculative candidates are a subgroup of Group 1, distinguished by a high Re/Ir ratio (*R* group of Ganapathy *et al.*, 1973) or some samples of very low Ir/Au ratio (Ganapathy *et al.*, 1974).

Tests of proposed assignments

Table 3 summarizes our tentative assignments, first for the age sequence of Hartmann and Wood (1971) used above, and then for the sequence of Stuart-Alexander and Howard (1970): Orientale, Imbrium, Crisium, Humorum, Nectaris, Serenitatis, Fecunditatis, and Tranquillitatis W. The latter set, which reverses the assignments of Crisium and Nectaris, is slightly less self-consistent, but not by a decisive margin.

It is obviously of interest to check whether our assignments are consistent with radiometric ages or relative ages inferred from petrographic or field geological relationships. To facilitate such tests, we have listed our assignments (for rocks only) in Table 5 in the Appendix.

Radiometric ages. Nothing definite can yet be said about the radiometric ages. Only a few of our samples have been dated thus far, and although the available ages show a slight trend in the right direction (P. Horn, private communication, 1974), better statistics will be needed before any firm conclusions are drawn. In any case, the trend may never be completely clearcut. The majority of highland rock ages fall between 3.8 and 4.0 AE (Podosek *et al.*, 1973; Huneke *et al.*, 1973), with only a few stragglers up to 4.3 AE (Schaeffer and Husain, 1973, 1974; Kirsten *et al.*, 1973; Kirsten and Horn, 1974; Stettler *et al.*, 1974). Apparently many of these ages were reset by thermal metamorphism or impact melting in the Imbrian event, and because similar rejuvenation may have happened in earlier impacts, a 1:1 correspondence between age and basin assignment should not be expected. It seems more realistic to look for trends in maximum or average ages of these groups.

Petrographic relations. According to the petrographic tenet that clasts are always older than matrix, they should never contain a younger component than the matrix. Most of our data are consistent with this principle, but this does not

Table 3. Tentative basin assignments.

| Group | Ir/Au | Ge | Mare Basin* | |
			HW	SH
1	0.30–0.40	high	Imbrium	Imbrium
2	0.40–0.55	high	Serenitatis	Serenitatis
3	0.55–0.85	high	Crisium	Nectaris
4	0.30–0.50	low	??	??
5	0.85–1.20	low	Nectaris	Crisium
6	1.50–2.0	low	Humorum or Nubium	Humorum or Nubium

*HW = Basin age sequence of Hartmann and Wood (1971).
SH = Basin age sequence of Stuart-Alexander and Howard (1970).

strengthen our model; it merely fails to weaken it. Seven clasts from six breccias contain *no* meteoritic component, or too little to characterize, though the matrix does have meteoritic material in each case (15265, 15405, 15418, 15455, 15459, 72255, 72275). Seven other clasts from seven rocks contain the same meteoritic component as the matrix, to within the accuracy of the data (14063, 14321, 61016, 67915, 72255, 72275, and 76315).

On the other hand, one rock (77135 from the Station 7 boulder) has clasts containing *younger* components than the matrix. Matrix and black dike material fall in Group 2, while two clasts fall in Groups 5 and 6 (Fig. 8).* We are convinced that these differences in meteoritic pattern are real, because they are paralleled by differences in petrography and nonmeteoritic trace elements. Both clasts are troctolites, both have low alkali and uranium contents, and both have counterparts among the Apollo 17 rocks that match them in meteoritic pattern (open circles in Fig. 8; the two rocks matching clasts 1 and 2 are 77017 and 78155, respectively), and, to a fair degree, even in Rb, Cs, and U content (Fig. 10).

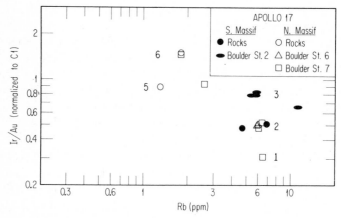

Fig. 10. At Apollo 17 site, ancient meteoritic components show some correlation with rock type (as expressed by Rb content). Low-alkali, troctolitic rocks and clasts (all from North Massif or Sculptured Hills) belong to Groups 5 or 6. High-alkali rocks are found only in Groups 1 to 3.

According to conventional petrographic reasoning, matrix (which falls in Group 2) is younger than the clasts (Groups 5 and 6), while our assignments imply that it is older. We cannot repair the situation by assuming that Group 2 belongs to a basin younger than Serenitatis. Imbrium is ruled out, because Group 2 is not

---

*A third sample, 77135,10, falls in Group 1, but both its nature and purity are in doubt. Unlike the other samples from this boulder, it was prepared by the curatorial staff rather than by the consortium leader, E. C. T. Chao. Though ostensibly a sample of clast 1, it has a much higher alkali content than authentic clast 1 (77135,62)—in fact, the same as matrix (77135,69). Perhaps it represents still another clast type, but we have trouble suppressing the wicked thought that it may be ordinary matrix with a small amount of gold contamination. One ppb of Au, i.e. 10^{-10} g Au in a 0.1 g sample, would suffice to move matrix from Group 2 to Group 1.

common enough at other sites. Nectaris, Crisium, Humorum, and Smythii escape this objection, but encounter two others: how could these distant basins furnish the bulk of the material of North and South Massifs? (As we saw earlier, Group 2 seems to comprise the most common rock type of these massifs.) And where then is the ~1 km of Serenitatis ejecta expected at this site?

A more attractive solution is to assume a Pre-Serenitatis age for Groups 5 and 6. This would imply that material from old basins finds its way to the surface more often than is thought. It would also require new identifications for Nectaris and Humorum—perhaps unrecognized subgroupings within the 6 groups of Fig. 3. This is not out of the question. For example, Apollo 16 samples fall mainly in the upper parts of Groups 5 and 6, while samples from other sites fall in the lower parts.

A third, and to us, most satisfactory alternative is that the green–gray breccias are Serenitatis material that was remobilized during the Imbrium event, incorporating clasts of Imbrian age or older. Indeed, Schmitt (1973) proposed that the tan–gray* breccias are younger than the blue–gray breccias, having intruded them in partially molten state. Head (1974) suggested that the Taurus-Littrow area was tectonically reactivated by the Imbrium event.

From our results, it seems rather likely that the blue–gray breccias actually were the principal *parent rock* of the green–gray breccias. All blue–gray breccias analyzed thus far fall in Group 2, as do the matrices of green–gray breccias, and all have virtually the same Rb, Cs, and U contents (Figs. 8 and 9; also 4 samples of the green–gray boulder 2 from Station 2, analyzed by Laul and Schmitt, 1974). If one breccia type is indeed derived from the other, then very little extraneous material, differing in KREEP or siderophile element content, can have been added in the process.

Field geology. Chao and Minkin (1974) have pointed out that the massifs at Taurus-Littrow are uplifted crustal blocks, not fallback breccias. Because the blocks were uplifted intact during the Serenitatis event, any meteoritic component contained in them must be of Pre-Serenitatis age. (In this view, the clasts would be older still.)

Though we agree that the massifs originated by crustal uplift, we note that this uplift must have been slow compared to ballistic flight of ejecta. Thus, at the time the massifs rose, the overlying ground was already covered with a thick layer of Serenitatis ejecta. These ejecta would cap the massifs.

The average thickness of ejecta should have been about 1100 m, according to McGetchin *et al.* (1973). This is consistent with the relief of the Littrow Sculptured Hills and the Vitruvius Plateau, both of which are believed to be Serenitatis ejecta (Head, 1974). It seems unlikely that the pre-massif area remained bare, while kilometer-thick ejecta deposits were laid down in the immediate vicinity.

---

*Tan–gray on the lunar surface, green–gray in the laboratory.

Properties of Basin-Forming Objects

Let us try to infer the properties of the basin-forming objects from the clues available.

Primary or secondary bodies?

The high siderophile element content of highland breccias (several percent of solar abundance) shows that the basin-forming objects were independently formed primary bodies, not secondary castoffs of a larger body that had undergone a "planetary" segregation of metal and silicate. Such segregation typically depletes siderophiles by factors of 10^{-4} to 10^{-6}. Hence the basin-forming objects cannot represent material spun off the earth after core formation (Wise, 1963; O'Keefe, 1970, 1972), condensates or volatilization residues from a hot earth (Ringwood, 1970), or fragments of a differentiated proto-moon disrupted during capture (Urey and Macdonald, 1971; Öpik, 1972; Smith, 1974; Wood and Mitler, 1974).

Internal structure

The ubiquity of metal in highland breccias (Fig. 2) suggests that the basin-forming bodies were undifferentiated, chondrite-like, with the metal phase uniformly dispersed throughout. Otherwise, if the metal had segregated into a core, the projectile would have to mix with the target rock to an incredibly uniform degree, to produce the distribution in Fig. 2.

Cooling rates of iron meteorites (Goldstein and Short, 1967) support this conclusion. Fricker *et al.* (1970) have compared the observed cooling rates with thermal models of asteroids, and conclude that the IVA irons, with a wide range of cooling rates, came from a 100-km body whose metal phase did not coalesce into a single core but formed scattered pockets throughout the body. Smaller bodies should be still less completely differentiated. According to published estimates, even the Imbrium body, largest of the basin-forming objects, was smaller than 100 km ($R = 95$ km, Baldwin, 1963; $R = 68$ km, Urey, 1968).

Impact velocity

The consistently high siderophile element content of highland breccias may also be a significant clue to the impact velocities of the bodies. According to cratering theory, the ratio of eroded mass, M, to projectile mass, μ, is a function of impact velocity, w. Thus, if the *mean* value of M/μ is known for a given impact, the velocity can be calculated.

In practice, the true mean value is hard to determine, because projectile and target rock do not mix uniformly. At Meteor Crater, Arizona, rim material and nearby ejecta are very low in meteoritic elements (Morgan *et al.*, unpublished work), while some of the subfloor material is high in Ni (Nininger, 1956). Unless these effects exactly cancel each other, distant ejecta will contain either more or less than their share of projectile material. For large basins, the situation is simpler, inasmuch as even rim material (e.g. Apollo 15, 17) typically had flight distances of $\sim 10^2$ km, and thus must have been strongly shocked and meteorite-bearing. Consequently, all our samples are distant ejecta in the above sense, and so any bias should be the same for all groups. Thus the M/μ values, and the velocities derived therefrom, should be reliable at least in a relative sense.

We have estimated M/μ in three ways: from the Au, Ni, and metal content of lunar breccias. To determine M/μ from the mean Au content, Au_B, we need to know the Au content of the projectile, Au_P. This can be estimated via the iron content of the projectile, Fe_P:

$$(M + \mu)/\mu \approx M/\mu = Au_P/Au_B = (Au_P/Fe_P)(Fe_P/Au_B)$$

Because Au and Fe condense together from a solar gas and scarcely fractionate from each other in chondrites (Larimer and Anders, 1970), it is reasonable to assume that Au_P/Fe_P equaled the Cl chondrite ratio, 8.0×10^{-7}. This leaves only Fe_P to be determined. We have assumed two extreme values: "lunar" at 9% and "terrestrial" at 36% Fe. They bracket all known chondrite classes, which range from 18% to 35% Fe.

About 19% of the highland breccias in Fig. 2 contain no meteoritic material, or too little to characterize. In order to make our M/μ values representative of the whole of the ejecta, we have corrected the mean Au contents of each group for a 19% contribution of meteorite-poor material.

A second set of M/μ values was calculated in an analogous manner from the Ni contents, using the Ni/Fe ratio of Cl chondrites, 5.67×10^{-2}. The third set was based on the metal contents of the breccias, on the assumption that all the iron in the projectiles was in the metallic state. (This may not be such a bad assumption, because the Ni content of the metal is about 6%, close to the value of 5.3% expected for completely reduced Cl chondrite material.) The three sets of M/μ values, and a weighted average, are given in Table 4. Only the values for "lunar" iron content (9%) are shown. Those for terrestrial iron content (36%) are a factor of 4 higher.

It is interesting to compare these values with our original estimates from Apollo 14 alone (Ganapathy et al., 1972). Both Au_B and M/μ have changed very little, as we went from 7 samples at one site to 102 samples at 7 sites (Fig. 2). Our original Au_B value, based mainly on Group 1 samples, was 4.6 ppb, which gives $M/\mu = 16$ for the parameters used in Table 4. The current value for Group 1 is 12.

These M/μ values may be converted to impact velocity by a suitable cratering relation. There still is some controversy whether cratering is governed by the momentum or kinetic energy of the projectile, and we have therefore used both kinds of relation.

$$Momentum: \quad M/\mu = 5.67 \times 10^{-5} \, kw \text{ (Öpik, 1958, 1961)}$$
$$Energy: \quad M/\mu = 1.58 \times 10^{-10} \, w^{2.117} \text{ (Gault, 1973)}$$

The k in Öpik's equation is a velocity-dependent parameter, varying from 2 to 4.7 in the range 0 to 35 km/sec.

Table 4. Target/Projectile mass ratio.

M/μ based on*	M/μ value†					
	Group 1	Group 2	Group 3	Group 4	Group 5	Group 6
Au	10.1 (23)	25.3 (12)	47.6 (8)	36.2 (4)	21.3 (13)	153 (9)
Ni	12.7 (15)	26.0 (10)	41.8 (6)	58.7 (3)	16.4 (10)	96 (6)
Metal‡	15.5 (9)	25.2 (5)	24.0 (3)	19.9 (2)	33.9 (3)	69 (4)
Weighted mean	12.0	25.6	41.4	40.0	20.9	117

*For a "lunar" iron content of the projectile, i.e. 9 wt.%. For a "terrestrial" iron content of 36%, the M/μ values will be 4 times larger.

†Number of samples is given in parentheses. The observed ratios, all based on samples with >0.2–0.3 ppb Au, were divided by 0.81, to allow for 19% highland samples with lower siderophile element contents.

‡Banerjee (1974), Brecher et al. (1973), Gose et al. (1972), Nagata et al. (1973), Nagata et al. (1974), Pearce et al. (1973), and Pearce et al. (1974).

The two relations are plotted in Fig. 11, along with M/μ values for lunar and terrestrial iron contents (solid and open circles).

The velocities are consistently low; indeed, for a lunar iron content, 6 of the 12 points fall below lunar escape velocity (2.4 km/sec), for either cratering relation. To the extent that these M/μ values are representative, this would seem to suggest that the projectiles had Fe contents higher than 9%. But even an iron content as high as 36% (open circles) gives quite low velocities: less than 8 km/sec in 9 of 12 cases. (In this M/μ range, Öpik's relation predicts much higher velocities than Gault's. It is in fact responsible for all 3 cases above 8 km/sec.)

These velocities may not be accurate in an absolute sense, because they are based on uncertain cratering relations and the uncertain assumption that M/μ in the ejecta was the same as for the whole impact. However, these uncertainties cancel out when we compare M/μ in ancient breccias and in ejecta from more recent craters. Although we have looked hard for contributions from recent craters, they are at best barely detectable against the background of the ancient component (Anders et al., 1973). It seems difficult to escape the conclusion that most of the ancient projectiles had *systematically lower velocities* than the present crater-forming population (~15 km/sec for asteroidal material; more for cometary material). Only Group 6 comes close to the present-day range.

Chemical composition

We can try to characterize the basin-forming bodies more fully, using data on volatile elements. Though we have thus far confined ourselves to the 5 siderophile elements Ir, Re, Au, Sb, and Ge, 8 of the remaining elements measured by us also are in part meteoritic. Figure 12 shows a rather favorable case, with a high content of meteoritic material. The indigenous corrections (black bars) are larger for volatiles than for siderophiles, but remain manageably small for most elements except Br, Zn, and Cd.

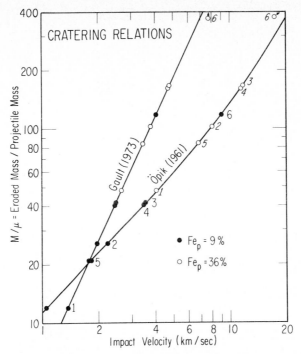

Fig. 11. Nominal impact velocity of basin-forming objects, estimated from cratering relations of Gault and Öpik. The two sets of points represent extreme values of iron content of the projectiles. Except for No. 6, these bodies seem to have had impact velocities well below the mean geocentric velocity for present-day meteorites, ~15 km/sec.

We have derived the abundance patterns of the 6 groups in this manner, using 3–4 samples of high siderophile element content from each group (Fig. 13). The patterns were very consistent for the first 6 elements; less so for the last 4, Se to Bi. For those elements, we gave greatest weight to the lowest values.

Several trends are apparent. (1) Groups 1 and 2 have almost identical patterns, except for the higher relative Au abundance in Group 1. (2) In Groups 1 to 3, Sb and Ge are equally abundant, while in Groups 4 to 6, Ge is less abundant than Sb. Such differences have also been noted between carbonaceous chondrites and ordinary chondrites, where they were attributed to a higher effective temperature of chondrule formation (Case *et al.*, 1973; Krähenbühl *et al.*, 1973b; Anders, 1968). (3) In all groups except 6, highly volatile Te and Bi are less abundant than the preceding 4, moderately volatile elements. This trend is also observed in ordinary chondrites (Larimer and Anders, 1967; Laul *et al.*, 1973; Larimer, 1973), where it has been attributed to accretion in the range of partial condensation of these elements. In terms of this explanation, the flat pattern of Group 6 would imply a lower accretion temperature and hence a more distant origin. This is an interesting thought, in view of the high M/μ value which implies a high impact velocity (Fig. 11) and hence an eccentric orbit. However, as expressed in the high M/μ, the samples of this group have a low content of meteoritic material, and hence the

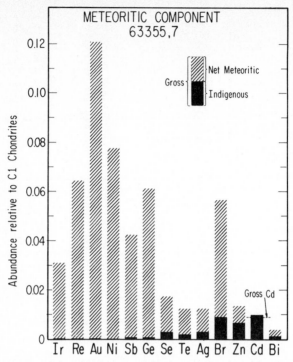

Fig. 12. Even some volatile elements are largely meteoritic in highland samples of high siderophile element abundance.

abundance pattern is poorly defined. (4) Group 3, which plots closest to Cl chondrites in Figs. 5 and 6, is highest in volatiles.

We can extend the comparison to the bulk earth and moon (Fig. 14), using the compositional model of Ganapathy and Anders (1974). None of the 6 groups is a perfect match for the earth or moon, and perhaps none should be, if these bodies accreted from planetesimals of varied compositions. The earth's composition could perhaps be approximated by a combination of Groups 3 and 5, or possibly others. The moon's composition requires a more refractory-rich material than even Group 6. Only one sample approaching this composition has been found thus far: a breccia fraction from Apollo 16 soil (67602,14-3; off-scale in lower right of Fig. 3). However, a mixture of Group 5 with an Ir, Re-rich early condensate would give a pretty good match.

Interestingly, Groups 1 and 2 are complementary to the earth and moon: poorer in refractories (Ir, Re), richer in at least two well-determined volatiles (Sb and Ge). This is consistent with accretion from a cooling nebula (Larimer and Anders, 1970; Anders, 1971; Grossman, 1972; Grossman and Clark, 1973), where refractories concentrate in the large, early-formed bodies, and volatiles, in the small, late-formed bodies. This complementarity may signify a genetic link between earth and moon on the one hand and basin-forming bodies on the other. In retrospect, it makes sense that the last bodies to fall on the moon were complementary to it in composition.

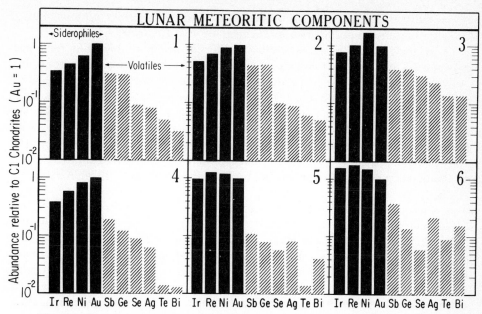

Fig. 13. Abundance patterns of 6 ancient meteoritic groups, corrected for indigenous contribution.

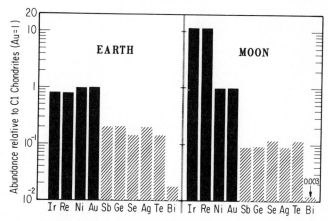

Fig. 14. Abundance patterns of bulk earth and moon, according to model of Ganapathy and Anders (1974). Compare with Fig. 13. Groups 1 and 2 are complementary to earth and moon, being poorer in refractories (Ir, Re) and richer in volatiles (Ge, Sb, etc.).

ORIGIN OF BASIN-FORMING OBJECTS

Clues from the bombardment history of the moon

An important clue is the rather young age of the Imbrium Basin, 3.9 AE. This is not a statistical fluke, because at least one body (Orientale) fell still later.

Apparently, the basin-forming objects were stored in orbits of long enough collision lifetime to permit two of them to survive for ≥700 m.y.

A second, more ambiguous, clue is the clustering of highland rock ages between 3.8 and 4.0 AE (Podosek *et al.*, 1973, and references cited therein). It is not yet clear whether this clustering implies a genuine cataclysm, i.e. a peak in the bombardment rate (Tera *et al.*, 1973, 1974), or merely resetting of a continuum of older ages by the Imbrium impact. In terms of the cataclysm hypothesis, all basins formed within 100–200 m.y. of each other; any older ages are due to earlier magmatism, or metamorphism unrelated to basin formation. The opposite, "steady bombardment" hypothesis interprets older ages, running up to ~4.3 AE, as the actual impact dates of basins (Schaeffer and Husain, 1974; Kirsten and Horn, 1974).

A third clue is the existence of meteorite-free crustal rocks (Fig. 2). It limits the amount of meteoritic material that could have fallen on the moon after crustal differentiation (~4.5 AE ago? Nunes and Tatsumoto, 1973), and thus fixes a point on the bombardment curve of the moon. The crust, comprising $0.1\,M_{\jmath}$, probably contains no more than 2% of meteoritic material on the average, or $0.002\,M_{\jmath}$ [Baldwin, 1971, arrived at a figure of $0.003\,M_{\jmath}$ from crater counts and an extrapolation of meteorite fluxes. His latest value (private communication, 1974),

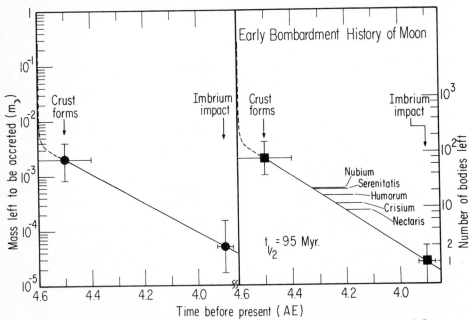

Fig. 15. Two points on the bombardment curve of the moon are fixed by the mass influx after formation of crust ($\sim 2 \times 10^{-3}\,M_{\jmath}$ or 80^{+80}_{-40} basin-forming objects) and the population surviving after the Imbrium impact 3.9 AE ago (Orientale, $\sim 5 \times 10^{-5}\,M_{\jmath}$). Horizontal bars at right give expected formation times of major basins. The 300 m.y. gap between Nectaris and Imbrium impacts reflects the fact that all 7 basins formed in this interval are on the farside or east limb.

is 0.002 $M_{\rangle}$.] Another point on the bombardment curve is fixed by the Imbrium impact 3.9 AE ago. Only the Orientale body, of estimated mass $5 \times 10^{-5} M_{\rangle}$, was yet to fall after that time. These two points are shown in the left-hand portion of Fig. 15. The right-hand portion shows the same data, but with the number rather than mass of bodies as the ordinate. The number of basin-forming objects 4.5 AE ago is taken as 80^{+80}_{-40}, about twice the number actually known (Hartmann and Wood, 1971). This is an attempt to allow for additional discoveries (El Baz, 1973; Scott, 1974).

It is not certain that the flux in this interval was exponential. However, if we make this assumption, we can estimate the probable formation times of the basins from their relative age. The horizontal bars in Fig. 15 correspond to the age sequence of Hartmann and Wood (1971), omitting basins $\leqslant 300$ km diameter. (The age sequence of Stuart-Alexander and Howard gives fairly similar results, except that Crisium falls at 4.1 AE.)

All of the ages fall between 4.2 and 4.35 AE. The absolute values of these ages should not be taken seriously because they depend critically on the initial number of basin-forming bodies 4.5 AE ago. However, the time gap between 3.9 and 4.2 AE is an interesting and significant feature. It reflects the fact that all 7 of the basins formed in the interval between Nectaris and Imbrium lie on the farside or east limb, and hence contribute negligible material to the lunar landing sites on the nearside. This fact, rather than a special cataclysm, may be responsible for the scarcity of ages in this interval.

Origins

Four origins may be considered for the basin-forming objects (Hartmann and Wood, 1971; Morgan *et al.*, 1972):

(1) Planetesimals in earth-crossing orbits
(2) Planetesimals in earth-grazing orbits
(3) Earth satellites, swept up by the moon during tidal recession or capture
(4) Stray asteroids, perturbed by Mars or Jupiter into earth-crossing orbits.

Of these 4 types of bodies, moonlets in earth orbit (Ruskol, 1962, 1971, 1972) most readily meet the constraints of survival to 3.9 AE, rapid extinction thereafter, chemical complementarity, and low impact velocity. Planetesimals in earth-crossing orbits are adequate in most respects, provided their half-life against planetary collision is as long as ~ 100 m.y. (Fig. 15). The original Monte Carlo calculations by Arnold (1965) and Wetherill (1968) gave a half-life of only ~ 30 m.y. (after the initial, rapid sweep-up of some 90% of the bodies). However, the actual lifetime may be longer, because secular variations in eccentricity, inclination, and node may put the object out of the earth's range for part of the time. Unpublished calculations by Mellick and Anders that make some allowance for these effects do in fact give $t_{1/2} \approx 80$ m.y., close to the required value.

The other two types of bodies are too long-lived to be the main source. However, they may have been a subsidiary source, contributing a few objects of unusual composition or impact velocity (e.g. the Group 6 body).

MISCELLANEOUS OBSERVATIONS ON APOLLO 17 SAMPLES

Highland rocks

Most Apollo 17 highland samples have high and remarkably uniform Rb, Cs, and U contents, well in the KREEP range (Rb 6 ppm, Cs 260 ppb, U 1.5 ppm). More so than at other sites, there is a tendency for meteoritic and non-meteoritic elements to correlate. The alkali-rich samples fall into Groups 1 to 3 whereas the alkali-poor samples, all from North Massif or the Sculptured Hills, fall into Groups 5 and 6 (Fig. 10).

Igneous rocks. Three samples contain no meteoritic component, judging from the very low abundance of siderophiles. Two of these are clasts in the No. 1 boulder from Station 2.

Sample 72275,91 is a pigeonite basalt (Stoeser *et al.*, 1974), possibly ancestral to KREEP-rich breccias. In its overall trace element pattern, it is very similar to mesostasis-rich basalt 15272,9,22 (Morgan *et al.*, 1973), except for a 50-fold higher Ge abundance (1290 ppb), and lesser enrichments of Sb and Se. This is by far the highest Ge content seen in a meteorite-free lunar rock thus far. Igneous rocks usually contain only a few ppb to a few tens of ppb. Soils and breccias contain up to 2000 ppb, but their Ge is invariably accompanied by high Ir, Au, Ni, etc., and thus appears to be of meteoritic origin.

The black-and-white ("Civet Cat") clast 72255,42 is low in alkalis and U, and has essentially no meteoritic component. This suggests a deep-seated origin, below the ancient highland regolith. Indeed, Stoeser *et al.* (1974) have argued on textural grounds that this clast represents a plutonic rock. Compston and Gray (1974) have reported a Rb–Sr age of 4.18 ± 0.04 AE.

Troctolite 76535 also lacks a meteoritic component, and is moreover exceedingly low in U, alkalis, and volatiles (Table 1). This is consistent with the deep-seated origin (15–40 km) proposed by Gooley *et al.* (1974).

Volatile enrichment. In contrast to Apollo 16, only one anorthosite, 78155 from Sculptured Hills, shows appreciable enrichment in Tl, Cd, and Sb, which we attributed to fumarolic volcanism (Krähenbühl *et al.*, 1973a). If such activity took place at Apollo 17, it must have been less widespread than at Apollo 16.

Mare rocks

The two mare basalts show no striking features. The unusually low-Ni content noted by the AFGIT (1973) is confirmed, and turns out to be paralleled by similarly low abundances of other siderophiles (Table 1).

Soils, soil breccias, and glasses

Orange soil. Our samples of 74220 are enriched in Ag, Br, Zn, Cd, and Tl, compared to ordinary lunar rocks or soils, but are quite low in siderophile meteoritic elements. Judging from the data on separates (Table 1), most of the enrichment resides in the orange glass. We do not confirm the high-Te content of 1.2×10^4 ppb found by Jovanovic et al. (1973).

Dark glass 79155, proposed as a source material of orange soil (Mao et al., 1974), differs from it in being much lower in volatiles and higher in siderophiles. We therefore believe it to be an ordinary impact glass, not the parent material of orange soil.

Our data place some constraints on the source of the volatiles. The low siderophile element content precludes a direct meteoritic source. All known meteorites rich in volatiles also are rich in siderophiles, and from our knowledge of cosmochemical processes, it seems likely that this trend will always hold. An internal source seems to be required.

An important constraint on this source comes from the isotopic composition of the lead in orange soil (Tatsumoto et al., 1973). Because this lead is much less radiogenic than any other lunar lead, it must have come from an isolated source that did not equilibrate with the lunar crust: either a volatile-rich, undifferentiated lunar interior as proposed by Urey (1972), or a volatile-rich projectile at the bottom of a mare basin. From the abundance pattern in Fig. 13, it seems that two of the projectiles (Nos. 3 and 6) were relatively rich in volatiles, but they happen to be assigned to the wrong basins, Crisium and Humorum. The No. 2 body assigned to Serenitatis was not notably volatile-rich. Perhaps an older, unrecognized projectile was responsible. It is interesting in this connection that Scott (1974) has found a second mascon in the northwestern part of Mare Serenitatis.

Shadowed soils. The two shadowed soils 72321 and 76241 do not appear to be significantly enriched in volatile elements. (A control soil, 76261, was measured for only one of them, however.) Only Cd shows an appreciable enrichment (by a factor of 2), as first noted by Chou et al. (1974). This supports their suggestion of an atmophile character for this element. Other shadowed soils (Krähenbühl et al., 1973a; Laul and Schmitt, 1974) do not show consistent Cd enrichments, however, and so some further work may be needed to settle the matter.

Acknowledgments—This work was supported in part by NASA Grant NGL 14-001-167. One of us (EA) thanks the John Simon Guggenheim Memorial Foundation for a fellowship.

References

AFGIT (Apollo Field Geology Investigation Team) (1973) Apollo 16 exploration of Descartes: A geologic summary. *Science* **179**, 62–68.

Albee A. L., Chodos A. A., Dymek R. F., Gancarz A. J., Goldman D. S., Papanastassiou D. A., and Wasserburg G. J. (1974) Dunite from the lunar highlands: Petrography, deformation history, Rb–Sr age (abstract). In *Lunar Science—V*, pp. 3–5. The Lunar Science Institute, Houston.

Anders E. (1968) Chemical processes in the early solar system, as inferred from meteorites. *Acc. Chem. Res.* **1**, 289–298.

Anders E. (1971) Meteorites and the early solar system. *Ann. Rev. Astron. Astrophys.* **9**, 1–34.

Anders E., Ganapathy R., Krähenbühl U., and Morgan J. W. (1973) Meteoritic material on the Moon. *The Moon* **8**, 3–24.

Arnold J. R. (1965) The origin of meteorites as small bodies. III. General considerations. *Astrophys. J.* **141**, 1548–1556.

Baedecker P. A. (1971) Iridium (77) In *Handbook of Elemental Abundances in Meteorites* (editor B. Mason), pp. 463–472. Gordon and Breach, New York.

Baldwin R. B. (1963) *The Measure of the Moon.* University of Chicago Press, Chicago. 488 pp.

Baldwin R. B. (1964) Lunar crater counts. *Astron. J.* **69**, 377–392.

Baldwin R. B. (1971) On the history of lunar impact cratering: The absolute time scale and the origin of planetesimals. *Icarus* **14**, 36–52.

Banerjee S. K. (1974) A preliminary report on the magnetic measurements of samples 72275 and 72255. In *Consortium Indomitabile* (q.v.), pp. 153–155.

Brecher A., Vaughan D. J., Burns R. G., and Morash K. R. (1973) Magnetic and Mössbauer studies of Apollo 16 rock chips 60315,51 and 62295,27. *Proc. Fourth Lunar Sci. Conf., Geochim. Cosmochim. Acta,* Suppl. 4, Vol. 3, pp. 2991–3001. Pergamon.

Case D. R., Laul J. C., Pelly I. Z., Wechter M. A., Schmidt-Bleek F., and Lipschutz M. E. (1973) Abundance patterns of thirteen trace elements in primitive carbonaceous and unequilibrated ordinary chondrites. *Geochim. Cosmochim. Acta* **37**, 19–33.

Chao E. C. T. and Minkin J. A. (1974) The petrogenesis of 77135, a fragment-laden pigeonite feldspathic basalt—a major highland rock type (abstract). In *Lunar Science—V,* pp. 112–114. The Lunar Science Institute, Houston. Report in preparation.

Chao E. C. T., Soderblom L. A., Boyce J. M., Wilhelms D. E., and Hodges C. A. (1973) Lunar light plains deposits (Cayley Formation)—a reinterpretation of origin (abstract). In *Lunar Science—V,* pp. 127–128. The Lunar Science Institute, Houston.

Chou C.-L., Baedecker P. A., Bild R. W., Robinson K. L., and Wasson J. T. (1974) Volatile elements in lunar soils (abstract). In *Lunar Science—V,* pp. 115–117. The Lunar Science Institute, Houston.

Compston W. and Gray C. M. (1974) Rb–Sr age of the Civet Cat Clast, 72255. In *Consortium Indomitabile* (q.v.), pp. 139–143.

Consortium Indomitabile (1974) *Interdisciplinary Studies of Samples from Boulder 1, Station 2, Apollo 17,* Vol. 1. Smithsonian Astrophysical Observatory, Cambridge. 191 pp.

Crozaz G., Walker R., and Zimmerman D. (1973) Fossil track and thermoluminescence studies of Luna 20 material. *Geochim. Cosmochim. Acta* **37**, 825–830.

Dence M. R. and Plant A. G. (1972) Analysis of Fra Mauro samples and the origin of the Imbrium Basin. *Proc. Third Lunar Sci. Conf., Geochim. Cosmochim. Acta,* Suppl. 3, Vol. 1, pp. 379–399. MIT Press.

Dickey J. S., Jr. (1970) Nickel-iron in lunar anorthosites. *Earth Planet. Sci. Lett.* **8**, 387–392.

El Baz F. (1973) Al-Khwarizmi: A new-found basin on the lunar far side. *Science* **180**, 1173–1176.

Fish R. A., Goles G. G., and Anders E. (1960) The record in the meteorites—III. On the development of meteorites in asteroidal bodies. *Astrophys. J.* **132**, 243–258.

Fouché K. F. and Smales A. A. (1967a) The distribution of trace elements in chondritic meteorites. I. Gallium, germanium and indium. *Chem. Geol.* **2**, 5–33.

Fouché K. F. and Smales A. A. (1967b) The distribution of trace elements in chondritic meteorites. II. Antimony, arsenic, gold, palladium, and rhenium. *Chem. Geol.* **2**, 105–134.

Fricker P. E., Goldstein J. I., and Summers A. L. (1970) Cooling rates and thermal histories of iron and stony-iron meteorites. *Geochim. Cosmochim. Acta* **34**, 475–491.

Ganapathy R. and Anders E. (1974) Bulk composition of the moon and earth, estimated from meteorites. *Proc. Fifth Lunar Sci. Conf., Geochim. Cosmochim. Acta,* Suppl. 5. This volume.

Ganapathy R., Laul J. C., Morgan J. W., and Anders E. (1972) Moon: Possible nature of the body that produced the Imbrian Basin, from the composition of Apollo 14 samples. *Science* **175**, 55–59.

Ganapathy R., Morgan J. W., Krähenbühl U., and Anders E. (1973) Ancient meteoritic components in lunar highland rocks: Clues from trace elements in Apollo 15 and 16 samples. *Proc. Fourth Lunar Sci. Conf., Geochim. Cosmochim. Acta,* Suppl. 4, Vol. 2, pp. 1239–1261. Pergamon.

Ganapathy R., Morgan J. W., Higuchi H., Anders E., and Anderson A. T. (1974) Meteoritic and volatile elements in Apollo 16 rocks and in separated phases from 14306. *Proc. Fifth Lunar Sci. Conf., Geochim. Cosmochim. Acta,* Suppl. 5. This volume.

Gault D. E. (1973) Displaced mass, depth, diameter, and effects of oblique trajectories for impact craters formed in dense crystalline rocks. *The Moon* **6**, 32–44.

Goldstein J. I. and Short J. M. (1967) The iron meteorites, their thermal history, and parent bodies. *Geochim. Cosmochim. Acta* **31**, 1733–1770.

Gooley R., Brett R., Warner J., and Smyth J. R. (1974) Sample 76535, a deep lunar crustal rock. *Proc. Fifth Lunar Sci. Conf.*, *Geochim. Cosmochim. Acta*, Suppl. 5. This volume.

Gose W. A., Pearce G. W., Strangway D. W., and Larson E. E. (1972) Magnetic properties of Apollo 14 breccias and their correlation with metamorphism. *Proc. Third Lunar Sci. Conf.*, *Geochim. Cosmochim. Acta*, Suppl. 3, Vol. 3, pp. 2387–2395. MIT Press.

Grossman L. (1972) Condensation in the primitive solar nebula. *Geochim. Cosmochim. Acta* **36**, 597–619.

Grossman L. and Clark S. P., Jr. (1973) High-temperature condensates in chondrites and the environment in which they formed. *Geochim. Cosmochim. Acta* **37**, 635–649.

Hartmann W. K. (1972) Paleocratering of the Moon: Review of post-Apollo data. *Astrophys. Space Sci.*: **17**, 48–64.

Hartmann W. K. and Wood C. A. (1971) Moon: Origin and evolution of multi-ring basins. *The Moon* **3**, 3–78.

Head J. W. (1974) Morphology and structure of the Taurus-Littrow highlands (Apollo 17): Evidence for their origin and evolution. *The Moon* **9**, 355–395.

Heiken G. H., Butler P., Jr., Simonds C. H., Phinney W. C., Warner J., Schmitt H. H., Bogard D. D., and Pearce W. G. (1973) Preliminary data on boulders at Station 6, Apollo 17 landing site. NASA TM X-58116. 56 pp.

Hodges C. A., Muehlberger W. R., and Ulrich G. E. (1973) Geologic setting of Apollo 16. *Proc. Fourth Lunar Sci. Conf.*, *Geochim. Cosmochim. Acta*, Suppl. 4, Vol. 1, pp. 1–25. Pergamon.

Huneke J. C., Jessberger E. K., Podosek F. A., and Wasserburg G. J. (1973) $^{40}Ar/^{39}Ar$ measurements in Apollo 16 and 17 samples and the chronology of metamorphic and volcanic activity in the Taurus-Littrow region. *Proc. Fourth Lunar Sci. Conf.*, *Geochim. Cosmochim. Acta*, Suppl. 4, Vol. 2, pp. 1725–1756. Pergamon.

Jovanovic S., Jensen K., and Reed G. W., Jr. (1973) The halogens, U, Li, Te, and P_2O_5 in five Apollo 17 soil samples (abstract). *EOS* **54**, 595.

Keays R. R., Ganapathy R., Laul J. C., Krähenbühl U., and Morgan J. W. (1974) The simultaneous determination of 20 trace elements in terrestrial, lunar and meteoritic material by radiochemical neutron activation analysis. *Anal. Chim. Acta*. In press.

Kimura K., Lewis R. S., and Anders E. (1974) Distribution of gold and rhenium between nickel–iron and silicate melts: Implications for the abundance of siderophile elements on the Earth and Moon. *Geochim. Cosmochim. Acta* **38**, 683–701.

Kirsten T. and Horn P. (1974) ^{39}Ar–^{40}Ar-chronology of the Taurus-Littrow region II: A 4.29 b.y. old troctolite and ages of basalts and highland breccias (abstract). In *Lunar Science—V*, pp. 419–421. The Lunar Science Institute, Houston.

Kirsten T., Horn P., and Kiko J. (1973) ^{39}Ar–^{40}Ar dating and rare gas analysis of Apollo 16 rocks and soils. *Proc. Fourth Lunar Sci. Conf.*, *Geochim. Cosmochim. Acta*, Suppl. 4, Vol. 2, pp. 1757–1784. Pergamon.

Krähenbühl U., Ganapathy R., Morgan J. W., and Anders E. (1973a) Volatile elements in Apollo 16 samples: Implications for highland volcanism and accretion history of the moon. *Proc. Fourth Lunar Sci. Conf.*, *Geochim. Cosmochim. Acta*; Suppl. 4, Vol. 2, pp. 1325–1348. Pergamon.

Krähenbühl U., Morgan J. W., Ganapathy R., and Anders E. (1973b) Abundance of 17 trace elements in carbonaceous chondrites. *Geochim. Cosmochim. Acta* **37**, 1353–1370.

Larimer J. W. (1967) Chemical fractionations in meteorites—I. Condensation of the elements. *Geochim. Cosmochim. Acta* **31**, 1215–1238.

Larimer J. W. (1973) Chemical fractionations in meteorites—VII. Cosmothermometry and cosmobarometry. *Geochim. Cosmochim. Acta* **37**, 1603–1623.

Larimer J. W. and Anders E. (1967) Chemical fractionations in meteorites—II. Abundance patterns and their interpretation. *Geochim. Cosmochim. Acta* **31**, 1239–1270.

Larimer J. W. and Anders E. (1970) Chemical fractionations in meteorites—III. Major element fractionations in chondrites. *Geochim. Cosmochim. Acta* **34**, 367–388.

Laul J. C. and Schmitt R. A. (1974) Chemical composition of Boulder 2 rocks and soils, Apollo 17, Station 2. *Earth Planet. Sci. Lett.* In press.

Laul J. C., Morgan J. W., Ganapathy R., and Anders E. (1971) Meteoritic material in lunar samples: Characterization from trace elements. *Proc. Second Lunar Sci. Conf., Geochim. Cosmochim. Acta,* Suppl. 2, Vol. 2, pp. 1139–1158. MIT Press.

Laul J. C., Ganapathy R., Anders E., and Morgan J. W. (1973) Chemical fractionations in meteorites— VI. Accretion temperatures of H-, LL-, and E-chondrites, from abundances of volatile trace elements. *Geochim. Cosmochim. Acta* **37**, 329–357.

McGetchin T. R., Settle M., and Head J. W. (1973) Radial thickness variation in impact crater ejecta: Implications for lunar basin deposits. *Earth Planet. Sci. Lett.* **20**, 226–236.

Mao H. K., El Goresy A., and Bell P. M. (1974) Orange glasses: Reaction of molten liquids with Apollo 17 soil breccia (70019) and gabbro (79155) (abstract). In *Lunar Science—V*, p. 489. The Lunar Science Institute, Houston.

Morgan J. W., Laul J. C., Krähenbühl U., Ganapathy R., and Anders E. (1972) Major impacts on the moon: Characterization from trace elements in Apollo 12 and 14 samples. *Proc. Third Lunar Sci. Conf., Geochim. Cosmochim. Acta,* Suppl. 3, Vol. 2, pp. 1377–1395. MIT Press.

Morgan J. W., Krähenbühl U., Ganapathy R., Anders E., and Marvin U. B. (1973) Trace element abundances and petrology of separates from Apollo 15 soils. *Proc. Fourth Lunar Sci. Conf., Geochim. Cosmochim. Acta,* Suppl. 4, Vol. 2, pp. 1379–1398. Pergamon.

Nagata T., Fisher R. M., Schwerer F. C., Fuller M. D., and Dunn J. R. (1973) Magnetic properties and natural remanent magnetization of Apollo 15 and 16 lunar material. *Proc. Fourth Lunar Sci. Conf., Geochim. Cosmochim. Acta,* Suppl. 4, Vol. 3, pp. 3019–3043. Pergamon.

Nagata T., Sugiura N., Fisher R. M., Schwerer F. C., Fuller M. D., and Dunn J. R. (1974) Magnetic properties and natural remanent magnetization of Apollo 16 and 17 lunar material (abstract). In *Lunar Science—V*, pp. 540–542. The Lunar Science Institute, Houston.

Nininger H. H. (1956) *Arizona's Meteorite Crater: Past, Present, Future.* American Meteorite Museum, Sedona, Arizona. 232 pp.

Nunes P. D. and Tatsumoto M. (1973) Excess lead in "Rusty Rock" 66095 and implications for an early lunar differentiation. *Science* **182**, 916–920.

Oberbeck V. R., Morrison R. H., Hörz F., Quaide W. L., and Gault D. E. (1974) Smooth plains and continuous deposits of craters and basins. *Proc. Fifth Lunar Sci. Conf., Geochim. Cosmochim. Acta.* This volume.

O'Keefe J. A. (1970) The origin of the Moon. *J. Geophys. Res.* **75**, 6565–6574.

O'Keefe J. A. (1972) The origin of the Moon: Theories involving joint formation with the Earth. *Astrophys. Space Sci.* **16**, 201–211.

Öpik E. J. (1958) Meteorite impact on solid surface. *Irish Astron. J.* **5**, 14–33.

Öpik E. J. (1961) Notes on the theory of impact craters. In *Proc. Geophys. Lab.-Lawrence Radiation Lab. Cratering Symp.*, Vol. 2, Paper S. URCL Report 6438, pp. 1–28.

Öpik E. J. (1971) Cratering and the Moon's surface. In *Advan. Astron. Astrophys.* (editor Z. Kopal), Vol. 8, pp. 107–337. Academic Press, New York.

Öpik E. J. (1972) Comments on lunar origin. *Irish Astron. J.* **10**, 190–238.

Pearce G. W., Gose W. A., and Strangway D. W. (1973) Magnetic studies on Apollo 15 and 16 lunar samples. *Proc. Fourth Lunar Sci. Conf., Geochim. Cosmochim. Acta,* Suppl. 4, Vol. 3, pp. 3045–3076. Pergamon.

Pearce G. W., Gose W. A., and Strangway D. W. (1974) Magnetism of the Apollo 17 samples (abstract). In *Lunar Science—V*, pp. 590–592. The Lunar Science Institute, Houston.

Podosek F. A., Huneke J. C., Gancarz A. J., and Wasserburg G. J. (1973) The age and petrography of two Luna 20 fragments and inferences for widespread lunar metamorphism. *Geochim. Cosmochim. Acta* **37**, 887–904.

Provost A. and Bottinga Y. (1972) Rates of solidification of Apollo 11 basalt and Hawaiian tholeiite. *Earth Planet. Sci. Lett.* **15**, 325–337.

Ringwood A. E. (1970) Origin of the Moon: The precipitation hypothesis. *Earth Planet. Sci. Lett.* **8**, 131–140.

Ruskol E. L. (1962) The origin of the Moon. In *The Moon* (editors Z. Kopal and Z. Mikhailov), pp. 149–155. Academic Press, New York.

Ruskol E. L. (1971) On the origin of the Moon. III. Some aspects of the dynamics of the circumterrestrial swarm. *Astron. J.* **48**, 819–829.

Ruskol E. L. (1972) On the initial distance of the Moon forming in the circumterrestrial swarm. In *The Moon* (editors S. K. Runcorn and H. C. Urey, D. Reidel), pp. 402–404. Dordrecht.

Schaeffer O. A. and Husain L. (1973) Early lunar history: Ages of 2 to 4 mm soil fragments from the lunar highlands. *Proc. Fourth Lunar Sci. Conf., Geochim. Cosmochim. Acta*, Suppl. 4, Vol. 2, pp. 1847–1863. Pergamon.

Schaeffer O. A. and Husain L. (1974) Chronology of lunar basin formation and ages of lunar anorthositic rocks (abstract). In *Lunar Science—V*, pp. 663–665. The Lunar Science Institute, Houston.

Schmitt H. H. (1973) Apollo 17 report on the valley of Taurus-Littrow. *Science* **182**, 681–690.

Scott D. H. (1974) The geologic significance of some lunar gravity anomalies (abstract). In *Lunar Science—V*, pp. 693–694. The Lunar Science Institute, Houston.

Scott E. (1972) Chemical fractionation in iron meteorites and its interpretation. *Geochim. Cosmochim. Acta* **36**, 1205–1236.

Short N. M. and Forman M. L. (1972) Thickness of impact crater ejecta on the lunar surface. *Mod. Geol.* **3**, 69–91.

Smith J. V. (1974) Origin of Moon by disintegrative capture with chemical differentiation followed by sequential accretion (abstract). In *Lunar Science—V*, pp. 718–720. The Lunar Science Institute, Houston.

Stettler A., Eberhardt P., Geiss J., Grögler N., and Maurer P. (1974) Sequence of terra rock formation and basaltic lava flows on the Moon (abstract). In *Lunar Science—V*, pp. 738–740. The Lunar Science Institute, Houston.

Stoeser D. B., Wolfe R. W., Wood J. A., and Bower J. F. (1974) Petrology. In *Consortium Indomitabile* (q.v.), pp. 35–109.

Stöffler D., Dence M. R., Abadian M., and Graup G. (1974) Ejecta formations and pre-impact stratigraphy of lunar and terrestrial craters: Possible implications for the ancient lunar crust (abstract). In *Lunar Science—V*, pp. 746–748. The Lunar Science Institute, Houston.

Stuart-Alexander D. E. and Howard K. A. (1970) Lunar maria and circular basins—A review. *Icarus* **12**, 440–456.

Tatsumoto M., Nunes P. D., Knight R. J., Hedge C. E., and Unruh D. M. (1973) U–Th–Pb, Rb–Sr, and K measurements of two Apollo 17 samples (abstract). *EOS* **54**, 614.

Taylor L. A., Mao H. K., and Bell P. M. (1973) "Rust" in the Apollo 16 rocks. *Proc. Fourth Lunar Sci. Conf., Geochim. Cosmochim. Acta*, Suppl. 4, Vol. 1, pp. 829–839. Pergamon.

Tera F., Papanastassiou D. A., and Wasserburg G. J. (1973) A lunar cataclysm at ~3.95 AE and the structure of the lunar crust (abstract). In *Lunar Science—IV*, pp. 723–725. The Lunar Science Institute, Houston.

Tera F., Papanastassiou D. A., and Wasserburg G. J. (1974) The lunar time scale and a summary of isotopic evidence for a terminal lunar cataclysm (abstract). In *Lunar Science—V*, pp. 792–794. The Lunar Science Institute, Houston.

Urey H. C. (1968) Mascons and the history of the Moon. *Science* **162**, 1408–1410.

Urey H. C. (1972) Evidence for objects of lunar mass in the early solar system. *The Moon* **4**, 383–389.

Urey H. C. and MacDonald G. J. E. (1971) Origin and history of the Moon. In *Physics and Astronomy of the Moon* (editor Z. Kopal), pp. 213–289. Academic Press, New York.

Vinogradov A. P. (1973) Preliminary data on lunar soil collected by the Luna 20 unmanned spacecraft. *Geochim. Cosmochim. Acta* **37**, 721–729.

Wasson J. T. and Schaudy R. (1971) The chemical classification of iron meteorites—V. Groups IIIC and IIID and other irons with germanium concentrations between 1 and 25 ppm. *Icarus* **14**, 59–70.

Wetherill G. W. (1968) Dynamical studies of asteroidal and cometary orbits and their relation to the origin of meteorites. In *Origin and Distribution of the Elements* (editor L. H. Ahrens), pp. 423–443. Pergamon Press, Oxford.

Wilhelms D. E. and McCauley J. F. (1971) Geologic map of the near side of the Moon. In *Geologic Atlas of the Moon*, No. I-703, U.S. Geological Survey, Washington, D.C.

Wise D. U. (1963) An origin of the Moon by rotational fission during formation of the Earth's core. *J. Geophys. Res.* **68**, 1547–1554.

Wlotzka F., Jagoutz E., Spettel B., Baddenhausen H., Balacescu A., and Wänke H. (1972) On lunar metallic particles and their contribution to the trace element content of Apollo 14 and 15 soils. *Proc. Third Lunar Sci. Conf., Geochim. Cosmochim. Acta*, Suppl. 3, Vol. 1, pp. 1077–1084. MIT Press.

Wlotzka F., Spettel B., and Wänke H. (1973) On the composition of metal from Apollo 16 fines and the meteoritic component. *Proc. Fourth Lunar Sci. Conf., Geochim. Cosmochim. Acta*, Suppl. 4, Vol. 2, pp. 1483–1491. Pergamon.

Wood J. A. and Mitler H. E. (1974) Origin of the Moon by a modified capture mechanism, *or* half a loaf is better than a whole one (abstract). In *Lunar Science—V*, pp. 851–853. The Lunar Science Institute, Houston.

APPENDIX

Table 5. Highland rocks: Tentative assignments to basins.

Sample	Description*	Ir Au Ge	Re Au Sb	Best estimate	Au ppb	Comments
			Group†			
Group 1: Imbrium						
14306,35,9	B	1	1	(1)	5.3	Ir/Au nearly high enough for 2
14321,184,14A	MB	4	1	1?	6.06	
14321,184,15	Dk Ct(MB)	1	1	1	8.08	
14321,184,16A	MB	1	1	1	9.96	
14321,184,19A	MB	1	1	1	6.41	
15455,38	B Dk Ves	1	1	1	5.8	
15455,183	B Dk Dense	1	1	1	4.6	
60016,6	B2	1	1	(1)	5.91	Ge high, Re/Ir high
60315,79	C2	?	(1)	1?	18.3	Ir/Au too low for 1, Re/Au lowest in group
61016,132a	C1	1	1	1	9.55	
61016,132b	C1	1	(1)	(1)	5.60	Re/Ir high
62295,40	C2	1?	1	(1)	3.10	Ge too high
63355,7	B4	1	1	1	18.4	
65015,51	C2	4	1	1?	10.2	
65095,30	B2	1	1	1	5.45	
66095,55	C2	1	1	1	17.9	
69935,8	B4	4	1	1?	11.9	
69955,11	C1	1	1	(1)	0.307	Low Au
77135,10	GGB Ct + Mx?	1	(1)	(1)	3.57	Re/Au nearly high enough for Group 2

Table 5. (*Continued*).

Sample	Description*	Ir Au Ge	Re Au Sb	Best estimate	Au ppb	Comments
Group 2: Serenitatis						
15265,13,5	B Mx	2	2	2	3.46	
15405,5,A5	B Mx	4	2	2?	0.93	
68815,124	B5	2	2	(2)	8.32	Ir/Au lowest in group
73235,45	BGB	2	2	2	2.31	
73275,23	B	2	2	2	3.34	
76315,73	BGB Mx	2	2	2	3.21	
76315,74	BGB Ct	2	2	2	3.48	
77075,19	GGB bl Dike	2	2	2	5.09	
77135,69	GGB Non-Ves Mx	2	2	2	6.45	
Group 3: Crisium						
14310,119	IM	3?	3	3?	4.31	Ge too low
15059,13,1	B Mx	3	3	(3)	2.45	Mare soil admixture?
72255,52	GB Mx	3	3	3	2.00	
72275,57	GB Mx	3	3	3	0.82	High Sb
72275,80	GB Ctl bl rim	3	3	3	1.16	
72275,83	GB Ct2 G Aph	3	3	3	1.30	
Group 4: ???						
63335,17	B5	4	2?	4?	0.81	Sb much too high
64455,27	C2	4	4	4	1.56	
68415,67	C1	4	4	(4)	2.65	Re/Ir lowest in group
Group 5: Nectaris						
14063,37,A10	Fine Mx	5	?	5?	0.22	Re contamination; low Au, high Sb
22006,3	Anorthosite	5	6	(5)	3.37	Re contamination, as in all USSR samples
60095,5	Glass	5	5	5	7.11	
64455,25	Glass	5	5	5	12.7	
65016,7	Glass	5	5	5	7.19	
67915,63a	B4 Mx	5	5	5	1.90	
67915,63b	B4 Ct	5	5	5	1.06	
67955,20	C2(B1?)	5	(6)	(5)	1.60	Re–Au–Sb Groups 5 + 6 not well separated
77017,48	An Ol Gabbro	5	5	5	5.65	
77135,62	GGB Ctl:Troct	5	5	5	4.74	

Table 5. (*Continued*).

Sample	Description*	Ir Au Ge	Group† Re Au Sb	Best estimate	Au ppb	Comments
Group 6: Humorum?						
14063,37,A11	Coarse Mx	6	3?	6?	0.28	Low Au, high Sb
14063,37,A12	Dk Ct	6	6	(6)	0.28	Low Au
15418,30,05D	B Ves Mx	6	6	6	1.00	
60017,8	B4	6	(6)	(6)	0.41	Low Au; Sb too high
77135,50	GGB Ct2:Troct	6	6	6	1.46	
78155,30	Catacl. An	6	6	6	0.66	Very high Sb

*An = anorthite; -osite   Ct = clast   MB = microbreccia
 Aph = aphanitic   Dk = dark   Mx = matrix   Alphanumeric symbols refer to
 B = breccia   G = gray   Ol = olivine   classification of Wilshire
 BG = blue–gray   GG = green–gray   Pl = plagioclase   *et al.* (1973)
 C = crystalline   IM = impact melt   Ves = vesicular
†Less reliable assignments are indicated by parentheses (= fair) or question marks (= poor).

Proceedings of the Fifth Lunar Conference
(Supplement 5, Geochimica et Cosmochimica Acta)
Vol. 2 pp. 1737–1746 (1974)
Printed in the United States of America

Composition of the gases associated with the magmas that produced rocks 15016 and 15065

Colin Barker

Chemistry Department, University of Tulsa, Tulsa, Oklahoma 74104

Abstract—Gases may be trapped in crystallizing minerals in many different ways but gas analyses can provide geologically meaningful information only if the gases can be related to their specific locations in the host mineral. The gases trapped in glass inclusions can be representative samples of the gases associated with the magma adjacent to the mineral at the time it crystallized. Analyses of the gases in glass inclusions in gabbro 15065,44 and basalt 15016,48 show that in both cases the gases in the parent magma started with a composition close to 46% oxygen, 42% carbon and 12% hydrogen and became more water-rich as crystallization progressed. This enrichment is probably due to a selective removal of the less soluble carbon compounds into the vapor phase. Some of the volatile components may react with the rock as temperature falls to give siderite and possibly hydrous phases.

Introduction

MANY LABORATORY EXPERIMENTS have demonstrated the importance of volatiles in controlling the viscosity and freezing range of silicate melts. The volatile components of melts also have an important influence on the sequence of crystallization and on the types of minerals which form (obviously hydroxyl-bearing minerals such as micas and amphiboles cannot crystallize from an anhydrous melt). The absence, or rare occurrence, of micas, amphiboles, clays, carbonates, late-stage alteration products or fluid inclusions in lunar rocks all suggest that lunar magmas were much poorer in volatiles than terrestrial magmas. The lunar volatiles also appear to be more reducing than the volatiles associated with terrestrial magmas. This conclusion is supported by the oxidation state of iron and by the rare-earth (lanthanide) distribution pattern. This shows a marked depletion in europium which is the only rare earth that can be reduced to the +II state (Helmke *et al.*, 1972; and others). More direct evidence for the reducing nature of lunar gases is provided by studies of equilibria involving oxide minerals since these lead to values for the fugacity of oxygen which are lower than those obtained for corresponding terrestrial systems (Sato *et al.*, 1973; Nash and Hausel, 1973).

On earth direct analysis of the gases associated with magmas is possible in volcanic areas but the data can be obtained for only one point in geologic time and can provide no information about the composition of volcanic gases released in the past. On the moon the collection and analysis of volcanic gases has not yet been achieved although the results from the Suprathermal Ion Detector Experiment (SIDE) have been interpreted as showing that water vapor is still being degassed from the lunar surface (Freeman *et al.*, 1972). An alternative approach to the problem of establishing the composition of the gases associated with magmas is to study rocks which crystallized, since the minerals can trap small samples of

Table 1. A classification of possible locations for gases in minerals.

Primary gas I	Representative I.1	Fluid inclusions (primary) Glass inclusions (primary)
	Nonrepresentative I.2	Fluid inclusions (primary) Glass inclusions (primary) Mineral structure Structure sites Structure holes ("clathrates") Vesicles Grain boundaries
Secondary gas II		Fluid inclusions (secondary) Glass inclusions (secondary) Decay products Mineral alteration products Solar wind implants Surface adsorption

the gases associated with the melt. Gases may be trapped in many different ways and these are summarized in Table 1.

Any gas which is associated with the melt and subsequently becomes incorporated into a mineral is called "Primary" (Type I). Any gas introduced into the mineral after it is formed is called "Secondary" (Type II). The locations which contain primary gases are divided into two subgroups, Types I.1 and I.2. If the composition of the gas in the location in the mineral is the same as it was in the parent liquid adjacent to the growing crystal, the gas composition is "representative" (Type I.1). However, many locations contain gases derived from the melt but which have a different composition from the gases in the melt because minerals have selectively removed one or more of the volatile compounds during crystallization. For example, water is selectively removed during the crystallization of hydrous minerals. Also, if a magma develops into two or more immiscible phases, the gases will be distributed between the phases on the basis of solubilities. When a melt inclusion forms, it may trap the various phases in proportions different from those in the magma and therefore will contain a gas sample which does not have the same composition as the total gases associated with the magma. Locations of this type are classified as "nonrepresentative" (Type I.2). This classification of gas locations will be presented in detail elsewhere. For the study of lunar samples, the most important locations are those that contain primary, representative gases (Type I.1). Because lunar minerals crystallized from silicate melts containing low concentrations of volatiles, fluid inclusions are not formed. This leaves glass inclusions as the only locations in a mineral which may contain representative samples of the gases associated with the parent melt.

Gas can be released from minerals by either heating or crushing. Crushing has several disadvantages. The new surfaces formed as the mineral is shattered are chemically very active and strongly adsorb some of the gases that are released (Piperov and Penchev, 1973; Barker and Torkelson, 1974). Because different gases are adsorbed to different extents crushing changes both the amount and composition of the evolved gases. Crushing may also release nonrepresentative gases from representative (I.1) locations, such as glass inclusions. This situation arises because a contraction bubble frequently forms in glass inclusions as the mineral starts to cool (Roedder and Weiblen, 1972). This bubble is likely to be enriched in the gases which are least soluble in the melt. As temperature falls further the melt solidifies and the bubble-glass distribution of gas may be frozen in. If such an inclusion is shattered by crushing, only the vapor content of the bubble is released and this contains a gas sample enriched in the components which are least soluble in the silicate melt.

The alternative to crushing is heating, but it is obvious that fusing a rock gives a mixture of gases coming from a variety of different locations in the minerals. Fortunately, the different gas locations have their own characteristic temperatures for gas loss so that on sequential heating the contents of different locations are released at different temperatures. This approach was used in the current study. Similar experimental procedures have been described by Wachi et al. (1972), Gibson and Johnson (1972), and Gibson and Moore (1972a).

Experimental

A calibrated E.A.I. QUAD 1110 mass spectrometer was used to monitor the release of gases from heated rock samples. The stainless-steel vacuum system was essentially the same as that described by Barker and Sommer (1973) except that the ion pump was replaced with a 175l/sec diffusion pump. Mullite sample tubes and boats were used because they are chemically inert at high temperatures. All new tubes and boats were given an initial outgassing at 1400°C under vacuum for several hours. In order to keep hot metal out of contact with the evolved gases the thermocouples which controlled the furnace and monitored the sample temperature were both outside of the sample tube. This probably results in the indicated temperatures being slightly too high, particularly at low temperatures.

Small pieces of the rock samples, weighing approximately 60 mg, were placed in the side arm of one of the sample tubes (Fig. 1). After evacuation, the sample tube and boat were heated to ~1000°C and cooled in order to remove gases adsorbed on surfaces. The lunar sample was then pushed into the sample boat by moving the steel slug with a magnet. A linear temperature programmer controlled the input to the furnace to give a temperature increase of 5.7°C/min. As the sample temperature increased the mass spectrometer analyzed the evolved gases by scanning the mass spectrum from mass 2 to mass 50 every 137 sec, which is equivalent to a data point every 13°C for any given gas. The carbon monoxide and nitrogen peaks at mass 28 could not be resolved and the fragment peaks at masses 12 and 14 (corrected for any methane contribution) were used in calculating the amounts of these two gases. Samples were heated to 1400°C and were completely fused onto the walls of the sample boat when cooled and removed from the vacuum system.

Lunar samples were received in nitrogen-filled containers and all sample handling was carried out in a nitrogen dry box to avoid adsorption of terrestrial gases, especially water. Both samples were from the interiors of rocks and therefore should not contain solar wind implanted ions (Leich et al., 1973).

Results and Discussion

The evolution of water, carbon dioxide and carbon monoxide from gabbro 15065,44 as a function of temperature is shown in Fig. 2. Analytical data for sulfur

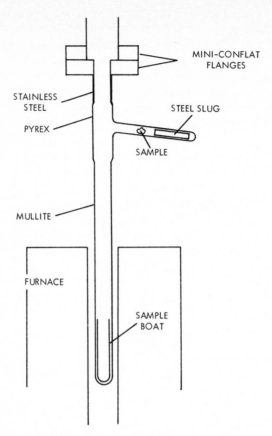

Fig. 1. Sample heating system. The rock sample is kept in the side arm while the tube and
boat are outgassed. Overall tube length is approximately 35 cm.

gases are considered unreliable and are not reported because the vacuum system
was assembled using copper gaskets and these may react with sulfur compounds.
In the temperature range from 100°C to about 450°C water predominates, but as
temperature continues to rise carbon dioxide evolution increases and reaches a
maximum at about 550°C. For terrestrial samples, most of the gases evolved in
this temperature range can be attributed to surface adsorption, rupturing of fluid
inclusions, or to the decomposition of carbonates and hydrous minerals. Fluid
inclusions are not important in lunar samples, the effects of adsorbed gases are
minimized by the sample handling procedures but small amounts of carbonates
and clays may be present. This is discussed below. Gibson and Moore (1972b) also
analyzed a piece of rock 15065 but failed to find the carbon dioxide release in the
range 450–750°C. This probably reflects sample inhomogeneity rather than
differences in sample handling procedures.

Water reaches its maximum value in the interval from 950°C to 1050°C. This
range also marks the beginning of carbon monoxide evolution. The final peak in
gas evolution occurs close to the melting point with carbon monoxide as the major

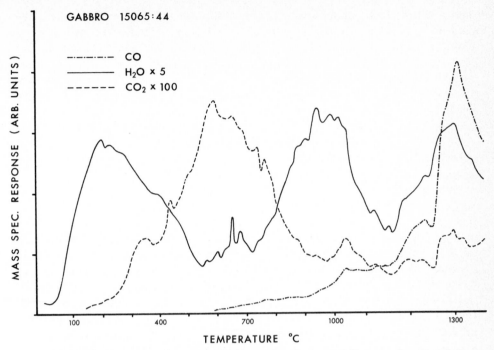

Fig. 2. Release of water, carbon dioxide, and carbon monoxide as a function of temperature for gabbro 15065,44. The abscissa is derived from the heating time which only approximates a linear relationship with temperature. Heating rate was 5.7°C/min.

component. This is in marked contrast to terrestrial basalts for which carbon dioxide always greatly exceeds carbon monoxide while both are much less abundant than water. The high-temperature gas release from the gabbro is attributed to gas evolving from glass inclusions, although it is possible that some of the carbon monoxide is produced by reaction of carbon compounds with silicates particularly at temperatures close to the melting point. Glass inclusions normally contain a contraction bubble (see photomicrographs in Roedder and Weiblen, 1972). As the host mineral is heated the glass in the inclusion melts and expands and will exactly fill the cavity at the temperature of mineral formation if no vapor phase was present when the melt was originally trapped. The continued expansion of the melt as temperature rises will cause some of the melt inclusions to rupture their host mineral and allow the gases dissolved in the melt to boil into the vacuum system. It is apparent from this sequence of events that the glass inclusions in different minerals can be expected to rupture at various temperatures depending on the temperature of mineral crystallization, the difference in coefficient of expansion between the glass and the mineral and the strength of the mineral. For the glass inclusions that do rupture, the major control is probably the temperature of mineral formation. Thus, the first mineral that formed from the original magma crystallized at the highest temperature so that any glass inclusions in it will be the last to rupture during the laboratory heating. Figure 2 clearly shows that the high-temperature gas release is made up of several episodes. The gas released at 1300°C comes from the glass inclusions in the first-formed minerals and corresponds to the composition of the early magmatic gas composition. The peak at 1225°C is produced by gases released from the glass inclusions in one of the later-formed minerals and so corresponds to the composition of the gases in the magma at some later stage in its evolution. In the last stages of magma crystallization, some glass may form between mineral grains. Gases will be released from this as soon as it softens because it does not have to rupture a host mineral to come into communication with the surrounding vacuum. The peak at 1030°C probably corresponds to the gases released in this way. These gases will be a sample of those associated with the later stages of magma crystallization.

It is possible that some gases are present in substitutional sites in the mineral lattice and are released at high temperatures. Barker and Sommer (1974) showed that water in the quartz lattice is not released until the structure is destroyed by melting. They also showed that the quantity of water in the quartz lattice depends on the partial pressure of water in the growth medium. From this, it would be expected that this source of gas would be insignificant in the volatile-poor lunar rocks which crystallized at low pressures. Further confirmation was obtained by analyzing a plagioclase phenocryst from a Hawaiian lava flow. This sample crystallized at high pressure but contained virtually no glass inclusions. Thus, it provided an upper limit for the amount of gas expected from the lattice. On melting, it produced only a barely detectable high-temperature gas release showing that the lattice contribution is minor. This result also supports the contention that glass inclusions are the major source of gas released in the few hundred degrees up to the melting point.

The gas release patterns for the 15016,48 basalt were broadly similar to those obtained for the gabbro (Fig. 3). Again the gases released in the temperature range 1200–1300°C are thought to come from the glass inclusions and so give the best estimate of the composition of the gases associated with the magma. One striking difference between the basalt and the gabbro is the sharp peak in the release of carbon dioxide which occurs at 490°C for the basalt. No comparable increase occurs for water, carbon monoxide, or other component in this temperature range. The carbon dioxide release is attributed to the thermal decomposition of a carbonate. The evolution of carbon dioxide as a function of temperature for selected pure carbonates has been determined under the same experimental conditions as those used for the lunar samples. All the carbonates studied showed a single maximum in the carbon dioxide release curve. The temperature of the peak maximum is characteristic of the carbonate and selected values are given in Table 2. The peak observed for basalt 15016 is consistent with the thermal decomposition of siderite and the amount of carbon dioxide released requires that approximately 5 ppm of siderite are present in the basalt. The siderite is not an artifact of the sample handling procedures because all handling was carried out under dry nitrogen. Also, the basalt and the gabbro were treated in exactly the same way but only the basalt showed evidence of siderite. Gibson and Chang (1974) found a peak at ~500°C for the thermal release of carbon dioxide from rock

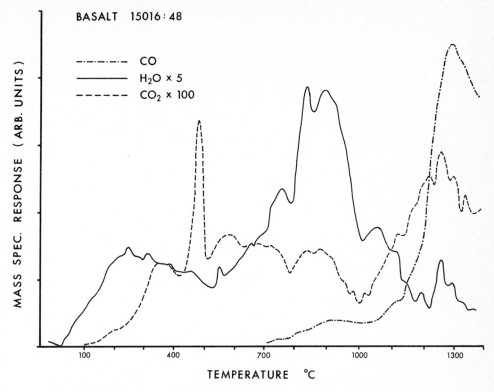

Fig. 3. Release of water, carbon dioxide and carbon monoxide as a function of temperature for gabbro 15016,48. The abscissa is derived from the heating time which only approximates a linear relationship with temperature. Heating rate was 5.7°C/min.

Table 2. Temperature of the peak maximum in the carbon dioxide evolution curves for selected carbonates.

Carbonate	Temperature, °C
Calcite	690
Dolomite 1	640
Dolomite 2	630
Siderite 1	515
Siderite 2	490
Rhodochrosite	465
Cerussite	320

67016. They also attributed this to carbonate decomposition. Isotopic data for carbon and oxygen on this sample were not consistent with a terrestrial source for the carbonate (Gibson and Chang, 1974).

So far, gas analyses have been discussed in terms of individual compounds—water, carbon monoxide, etc. This is probably not strictly valid because gases react very rapidly with one another at high temperatures and reequilibrations will occur. In the immediate vicinity of the sample, gas composition is probably buffered by the minerals present in the rock. Because of these reactions, compositions are better expressed in terms of the number of atoms of each element if all the reacting species remain in the vapor phase. Carbon, hydrogen, and oxygen account for the bulk of the volatiles evolved from the lunar samples, so compositions can be conveniently plotted on carbon–hydrogen–oxygen ternary diagrams after normalizing these three elements to 100% (Barker and Sommer, 1973). The composition of the gases from different generations of glass inclusions (Figs. 2 and 3) in basalt 15016 and gabbro 15065 are shown plotted on a C–H–O ternary diagram in Fig. 4. Certain trends are apparent. For the gabbro, the gases associated with the magma in its early stages of crystallization (as sampled by the glass inclusions in the first-formed minerals) are richest in the carbon component. The melt became richer in water as crystallization progressed, as indicated by the

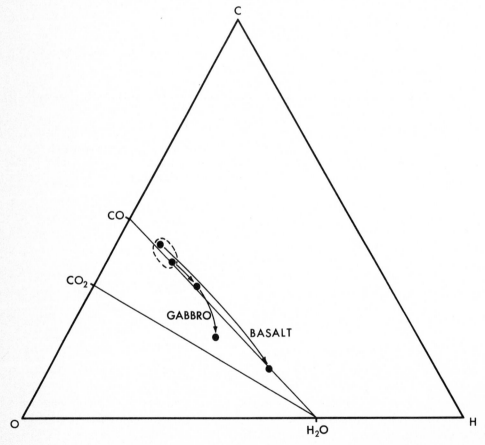

Fig. 4. Carbon–hydrogen–oxygen ternary diagram showing the composition of the gases in various locations in gabbro 15065,44 and basalt 15016,48 (see text for discussion).

composition of the volatiles in melt inclusions in later-formed minerals and in interstitial glass. The same trend holds for the basalt. The composition of the gases in glass inclusions show that the gases associated with the magmas that produced both gabbro 15065 and basalt 15016 started with very similar gas compositions, approximately 46% oxygen, 42% carbon and 12% hydrogen (Fig. 4).

If the melt gets richer in the water component as crystallization progresses, then either water is being added to the melt or, more likely, the carbon component is being removed from it. Since carbon dioxide is about an order of magnitude less soluble than water in silicate melts (Khitarov and Kadik, 1973), a gas phase would be much enriched in carbon gases at the expense of the water component. If vesicles were formed when the pressure dropped due to the extrusion of basalt on the lunar surface, then these would contain the gases enriched in carbon compounds. As temperature fell, carbonates would become stable and would be formed by reaction of the carbon dioxide with the inner walls of the vesicles. Small quantities of clay minerals may be formed in a similar way by reactions involving water vapor (Drever, 1970). So far SEM studies of vesicle walls have not lead to the identification of carbonates.

The formation of glass inclusions during the crystallization of the lunar minerals is fortuitous so that the total quantity of gas released from these inclusions has no significance unless the volume of glass is known. These data were not available for the samples analyzed.

Acknowledgments—I am grateful to NASA for financial support (grant NGR 37-008-002). My thanks are also due to M. A. Sommer for help in analyzing the lunar samples and the Hawaiian plagioclase, and to P. Wagenhofer for analyzing the carbonates.

References

Barker C. and Sommer M. A. (1973) Mass spectrometric analysis of the volatiles released by heating or crushing rocks. In *Analytical Methods Developed for Application to Lunar Samples Analyses*. A.S.T.M., STP 539, pp. 56–70.

Barker C. and Sommer M. A. (1974) Water content of the quartz lattice: A potential method of geobarometry. *Nature*. In press.

Barker C. and Torkelson B. E. (1974) Gas adsorption on crushed quartz and basalt. *Geochim. Cosmochim. Acta*. In press.

Drever J. I., Fitzgerald R. W., Liang S. S., and Arrhenius G. (1970) Phyllosilicates in Apollo 11 samples. *Proc. Apollo 11 Lunar Sci. Conf., Geochim. Cosmochim. Acta*, Suppl. 1, Vol. 1, pp. 341–345. Pergamon.

Freeman J. W., Jr., Hills H. K., and Vondrak R. R. (1972) Water vapor, whence comest thou? *Proc. Third Lunar Sci. Conf., Geochim. Cosmochim. Acta*, Suppl. 3, Vol. 3, pp. 2217–2230. MIT Press.

Gibson E. K. and Johnson S. M. (1972) Thermogravimetric-quadrupole mass-spectrometric analysis of geochemical samples. *Thermochim. Acta* 4, 49–56.

Gibson E. K. and Moore G. W. (1972a) Inorganic gas release and thermal analysis study of Apollo 14 and 15 soils. *Proc. Third Lunar Sci. Conf., Geochim. Cosmochim. Acta*, Suppl. 3, Vol. 2, pp. 2029–2040. MIT Press.

Gibson E. K. and Moore G. W. (1972b) Thermal analysis—inorganic gas release studies on Apollo 14, 15 and 16 lunar samples. In *The Apollo 15 Lunar Samples*, pp. 307–310. The Lunar Science Institute, Houston.

Gibson E. K. and Chang S. (1974) Abundance and isotopic composition of carbon in lunar rock 67016: suggestions of a carbonate-like phase (abstract). In *Lunar Science—V*, pp. 264–266. The Lunar Science Institute, Houston.

Helmke P. A., Haskin L. A., Korotev R. L., and Ziege K. E. (1972) Rare earth and other trace elements in Apollo 14 samples. *Proc. Third Lunar Sci. Conf., Geochim. Cosmochim. Acta*, Suppl. 3, Vol. 2, pp. 1275–1292. MIT Press.

Khitarov N. I. and Kadik A. A. (1973) Water and carbon dioxide in magmatic melts and peculiarities of the melting process. *Contr. Mineral. and Petrol.* **41**, 205–215.

Leich D. A., Tombrello T. A., and Burnett D. S. (1973) The depth distribution of hydrogen in lunar materials. *Earth Planet. Sci. Lett.* **19**, 305–314.

Nash W. P. and Hausel W. D. (1973) Partial pressures of oxygen, phosphorus and fluorine in some lunar lavas. *Earth Planet. Sci. Lett.* **20**, 13–27.

Piperov N. B. and Penchev N. P. (1973) A study on gas inclusions in minerals. Analysis of the gases from micro-inclusions in allanite. *Geochim. Cosmochim. Acta* **37**, 2075–2097.

Roedder E. and Weiblen P. W. (1972) Petrographic features and petrologic significance of melt inclusions in Apollo 14 and 15 rocks. *Proc. Third Lunar Sci. Conf., Geochim. Cosmochim. Acta*, Suppl. 3, Vol. 1, pp. 251–279. MIT Press.

Sato O., Hickling N. L., and McLane J. E. (1973) Oxygen fugacity values of Apollo 12, 14 and 15 lunar samples and reduced state of lunar magmas. *Proc. Fourth Lunar Sci. Conf., Geochim. Cosmochim. Acta*, Suppl. 4, Vol. 1, pp. 1061–1079. Pergamon.

Wachi F. M., Gilmartin D. E., Oro J., and Updegrove W. S. (1971) Differential thermal analysis and gas release studies of Apollo 11 samples. Air Force Report No. SAMSO-TR-71-313.

Proceedings of the Fifth Lunar Conference
(Supplement 5, Geochimica et Cosmochimica Acta)
Vol. 2 pp. 1747–1762 (1974)
Printed in the United States of America

Simulation of lunar carbon chemistry: I. Solar wind contribution

J. P. Bibring,[1] A. L. Burlingame,[2] J. Chaumont,[1]
Y. Langevin,[1] M. Maurette,[1] and P. C. Wszolek[2]

[1]Laboratoire René Bernas, 91406, Orsay, France.
[2]Space Sciences Laboratory, University of California, Berkeley, California 94720.

Abstract—Various mineral targets prepared either as large single crystals or as micron-sized fragments were exposed to high fluxes ($\geqslant 10^{15}$ ions/cm²) of low-energy ^{13}C, ^{15}N, and D ions. Species possibly synthesized during implantation were searched for in large crystals by using high-resolution mass spectrometry and ionic analyzer techniques. Ultramicroscopic features induced in micron-sized fragments were analyzed with a high-voltage electron microscope. Results for the targets were compared with those obtained by applying the same techniques to the study of size fractions and mineral separates from the mature 10086 sample. The main conclusions of these investigations are: (1) The implantation of "solar wind" C, D, and N ions in solids synthesizes small molecules that can be released into vacuum either by ion sputtering or by heating. (2) This "ion implantation synthesis" is highly specific when compared to other processes so far proposed to explain the formation of molecules in the solar nebula and in interstellar space. (3) In the crystalline fraction of mature lunar soil samples the carbon injected by the solar wind should reach a saturation concentration of ~200 ppm. (4) The carbon chemistry of the crystalline components of mature soil samples is dominated by solar wind implantation effects. (5) The reduction of metal cations by hydrogen is not a prerequisite for the formation of carbides in lunar minerals.

INTRODUCTION

THE TOP SUPERFICIAL LAYER of the lunar soil, from depths of $\sim 100\,\mu$ to 1 mm, is very active during the maturation of the lunar regolith. In this lunar "skin" the position and orientation of a dust grain at an initial depth d changes frequently with time: a nearby impact may bury the grain under an ejecta blanket of thickness σ; a direct impact may excavate a grain if the overlying crater has a depth $D > \sigma$; or the grain may be incorporated into a breccia or a glassy agglutinate and perhaps released by some subsequent impact. During its integrated residence time on the top surface of the regolith the grain is irradiated with low-energy nuclear particle fluxes (including solar wind nuclei as well as ions set in ballistic motion from the lunar atmosphere by various processes), exposed to various types of "vapor" phases, bombarded with small grains that become attached to it, etc. In this paper and in our companion paper (Bibring *et al.*, this volume) we present in detail some simulation experiments intended to aid in understanding the role of the solar and lunar winds in the evolution of the lunar "skin"; we more specifically apply the results of such experiments in discussing the contribution of these two different types of winds to lunar carbon chemistry.

Our study of simulated solar wind implantation synthesis differs from previous ones in several important aspects: (1) The targets were bombarded with a mixture of low-energy ^{13}C, ^{15}N, and D_2 ions; the species artificially synthesized during the ion implantation (IIS species) were searched for in a great variety of targets including metals with known but different characteristics for radiation damage defects (Al, Au) as well as solids either found in the lunar regolith (glass, pyroxenes, feldspars, ilmenite) or proposed as plausible models for cosmic dust grains (olivine, MgO, SiC, graphite, magnetite). (2) The IIS species were analyzed by using, in addition to the mass spectrometric techniques already applied to lunar carbon studies (acid dissolution and thermal evolution "pyrolysis" techniques), an ionic analyzer which gives the implantation depth profiles of all species released upon sputtering from the targets with a good mass resolution (~ 7000) in the mass range $M \leq 50$. (3) The physics of ion implantation phenomena was taken into consideration as follows: (i) micron-sized grains were obtained by crushing a chunk of each nonmetallic single crystal so far investigated, dispersed on electron microscope substrates and finally irradiated under the same conditions as the targets. For the metals the high-voltage electron microscope (HVEM) thin samples were obtained by using known dissolution techniques. Then after irradiation the samples were observed with the HVEM both before and after thermal treatments where they were "pyrolyzed" similarly to the samples analyzed with the mass spectrometer; (ii) the sputtering rate of each target was determined as described elsewhere (Bibring, 1972; Bibring *et al.*, 1974a) and used in combination with a recent range-energy theory (Winterbon *et al.*, 1970) to define more carefully the irradiation conditions for implanting a mixture of ions that can sputter each other away.

Experimental

The minerals to be irradiated for use in the acid dissolution and thermal evolution experiments were wrapped with thin platinum wire to prevent surface charging effects and were cleaned by sonication with methanol and dichloromethane plus heating under vacuum: olivine (room temperature), plagioclase (750°C), and ilmenite (950°C).

The plagioclase, An_{66}, has the following composition: CaO, 13.75%; Na_2O, 3.74%; FeO, 0.413%; K_2O, 0.121%; MgO, 0.115%; Al_2O_3, 29.31%; SiO_2, 51.60% (D. S. Burnett, private communication). The olivine has an estimated composition of Fo_{92} (J. T. O'Connor, private communication). The ilmenite was obtained from Ward's Natural Science Establishment, Inc.

The ionic implanter used in the present work has been described (Chaumont *et al.*, 1971). The dose rate was always smaller than 20 μamps/cm^2 and was measured by monitoring the total current leaking from the sample holder. The beam section of about 10 cm^2 was large enough to simultaneously irradiate all the samples when identical conditions were required.

^{13}C only was implanted into plagioclase at an energy of 13 keV (1 keV/amu) and a total dose of 10^{16} ions/cm^2. The ^{13}C, ^{15}N, and D_2 mixture was implanted into the various targets in the following way: 3 keV/amu ^{13}C atoms (40 keV) were first injected up to a total dose of 10^{16} ions/cm^2; then 3 keV/amu ^{15}N atoms (45 keV) were mixed into the carbon layer with the same dose of 10^{16} ions/cm^2; finally deuterium ions (D_2^+) were implanted at 3 separate energies ($E = 3$, 4.5, and 6 keV/amu) with a total dose of 10^{17} ions/cm^2 and 10^{18} ions/cm^2 for one olivine experiment.

Mineral targets of about 0.5 cm^2 surface area were heated under vacuum in a quartz tube at a linear temperature program rate of 25°C/min from ambient to 1400°C. The furnace used to obtain these temperatures has been described (Simoneit *et al.*, 1973b). A Pt–Pt/Rh thermocouple is used to control the linear temperature program with excellent reproducibility. The temperature gradient between the thermocouple and the sample itself is not yet known, but we estimate the difference to be less than 10%. All gaseous species released by heating were continuously monitored using a high-sensitivity, high-resolution (set at 10,000) MS-902 mass spectrometer coupled to an XDS Sigma-7 computer. Perfluorokerosene is used as an external mass standard. Details for quantitating gaseous species and the calibration procedure have been reported (Simoneit *et al.*, 1973b). Reproducibility is ±10% for this technique (Simoneit *et al.*, 1973a). Corrections were made for background levels of ^{13}CO and $^{13}CO_2$ in the targets which amounted to about 8% and 20%, respectively, of the total ^{13}CO and $^{13}CO_2$ released by the irradiated plagioclase.

Mineral targets of 1.5–2.0 cm^2 surface area were treated with either 40% HF or 40% DF under vacuum. The gases released were collected as described by Holland *et al.* (1972a, 1972b) and analyzed using an MS-902 mass spectrometer set at 10,000 resolution. The gas mixture was introduced into a heated, all glass inlet system. The system was calibrated for quantitation using methane or a standard mixture of methane and C$_2$, C$_3$ hydrocarbons, neon, and argon in helium. The reproducibility of the acid dissolution technique is within ±25%. Background levels for all labeled species quantitated in these experiments were assessed with unirradiated minerals and were found to be negligible except for $^{13}CO_2$. Corrections made for background $^{13}CO_2$ amounted to 15–35% of the total $^{13}CO_2$ detected. The ionic analyzer used in this work is an instrument at ONERA (Office National d'Etudes et de Recherches Aerospatiales) (Hernandez *et al.*, 1972). A 10 keV argon ion beam was used for sputtering.

ION IMPLANTATION SYNTHESIS (IIS)

Results

The following high-resolution mass spectral data were obtained by acid etch of irradiated targets: (1) In plagioclase, ilmenite, and olivine implanted with ^{13}C, ^{15}N, and D$_2$ ions, mixtures of simple molecules are released by HF including deuterocarbons, hydrocarbons, carbon dioxide, nitrogen, hydrogen cyanide, and deuterium cyanide. The most abundant of these molecules are listed in Table 1, columns 2–5, along with their yields based on the original amount of ^{13}C and ^{15}N implanted. About 20–30% of the carbon and about 12% of the nitrogen are released in these gaseous species. The deutero- and hydrocarbons represent

Table 1. Species released by acid hydrolysis of minerals irradiated with solar wind type ions (% of implanted ^{13}C and ^{15}N).

Mineral Ions	Plagioclase ^{13}C $D/^{13}C = 0$	Plagioclase ^{13}C, ^{15}N, D$_2$ $D/^{13}C = 20$	Ilmenite ^{13}C, ^{15}N, D$_2$ $D/^{13}C = 20$	Olivine ^{13}C, ^{15}N, D$_2$ $D/^{13}C = 20$	Olivine ^{13}C, ^{15}N, D$_2$ $D/^{13}C = 200$
Etching agent	DF	HF	HF	HF	HF
Species					
$^{13}C_2H_2$	—	0.53	0.12	trace	trace
$^{13}C_2D_2$	3.0	—	—	—	—
$^{13}CH_4$	—	—	0.18	0.15	0.72
$^{13}CD_4$	—	0.67	1.27	0.78	12.9
$^{13}C_2D_4$	—	1.04	0.24	trace	0.72
$^{13}C_2D_6$	—	—	—	trace	0.59
$^{13}CO_2$	26.8	17.9	25.2	18.6	13.8
$^{15}N_2$	—	11.0	3.70	10.6	3.6
$H^{13}C^{15}N + D^{13}C^{15}N$	—	1.10	1.32	1.07	2.21
$\dfrac{H^{13}C^{15}N + D^{13}C^{15}N}{^{13}CD_4}$	—	1.60	1.04	1.37	0.17

species synthesized by implanted ions and reaction products of HF with carbide-like materials, respectively. Small amounts of $^{13}C_3D_6$ and $^{13}C_3D_8$ have also been detected in some of the experiments. Neither $^{15}NO_2$ nor ^{15}NO was detected. In the case of ^{15}NO, there is interference from the CF^+ ion deriving from the per-fluorokerosene mass standard. ^{13}CO was not detected above background levels plus the $^{13}CO^+$ fragment from $^{13}CO_2$. Deuterium gas is almost certainly formed, but we have not analyzed for this species. Other species, e.g. $^{15}ND_3$, may be synthesized but would be retained in the reaction solution under our current experimental conditions. (2) For the experiment in which ^{13}C only was implanted into plagioclase (Table 1, column 1), DF released deuteroacetylene and carbon dioxide only. The deuteroacetylene should derive from material behaving like calcium and/or sodium carbides. This experiment shows that the yield of at least some carbide-like materials does not seem to depend on the simultaneous irradiation of the targets with fluxes of reducing D ions. (3) The yields for the IIS products, except for the hydrocarbons which derive from carbide-like material, appear to be independent of the mineralogy of the substrate when one compares the results for plagioclase and olivine (columns 2 and 4) irradiated under the same conditions. However, the yields of some of the IIS products in ilmenite (column 3) are higher. This difference may be due to smaller or negligible sputtering losses of carbon from ilmenite compared with the other two minerals. As we discuss below, the sputtering rate for ilmenite, experimentally determined with helium, is about ten times lower than for feldspar, pyroxene, and glasses. (4) The type of carbides and thus the hydrocarbons detected do depend on the mineral irradiated and its chemical composition. Plagioclase yields acetylene but no methane, whereas olivine and ilmenite yield methane which presumably derives from material behaving like iron carbide. Some acetylene was detected in the ilmenite experiment (Table 1, column 3), and may be due to impurities in the mineral. We estimate that about 5–10% could be silicates (J. T. O'Connor, private communication). (5) The yields of all the IIS products in olivine clearly depend on the $D/^{13}C$ ratio which was ten times higher for the second irradiation (column 5) than for the first irradiation (column 4). The amounts of carbon dioxide and nitrogen decreased while the yields of methane, the deuterocarbons, and hydrogen cyanide plus deuterium cyanide increased.

The major species released by heating of plagioclase irradiated with ^{13}C, ^{15}N, and D_2 are ^{13}CO, $^{15}N_2$, and $^{13}CO_2$. Thermal release profiles for these species and for HOD, which is released in small amounts are shown in Fig. 1. Figure 2 shows the profiles for ^{13}CO, $^{15}N_2$, $^{13}CO_2$, and HOD for *unirradiated* plagioclase which was cleaned and exposed to the vacuum system of the ion implanter. In the irradiated plagioclase, most of the ^{13}CO and $^{15}N_2$ are evolved below 1000°C while most of the $^{13}CO_2$ is released above 1100°C. HOD evolves mostly from 400°C to 700°C. No $D^{13}C^{15}N$ nor $H^{13}C^{15}N$ was detected in this experiment although $H^{13}CN$ was detected in small amounts during heating of irradiated plagioclase in a previous experiment (Wszolek *et al.*, 1974a). Neither ^{15}NO nor $^{15}NO_2$ was detected. Thirty-seven percent and 8% of the implanted carbon were recovered as ^{13}CO and $^{13}CO_2$, respectively; 26% of the implanted nitrogen was recovered as $^{15}N_2$. When

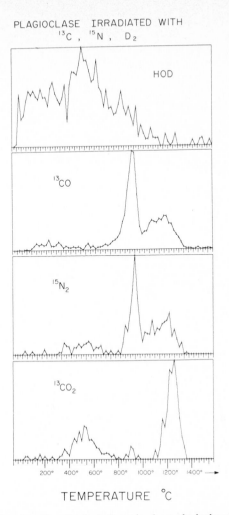

PLAGIOCLASE IRRADIATED WITH
^{13}C, ^{15}N, D_2

HOD

^{13}CO

$^{15}N_2$

$^{13}CO_2$

200°   400°   600°   800°   1000°   1200°   1400° ⟶

TEMPERATURE °C

Fig. 1. Thermal release profiles of selected species from plagioclase irradiated with ^{13}C, ^{15}N (10^{16} atoms/cm^2; 3 keV/amu) and D_2 (10^{17} atoms/cm^2; 3, 4.5, and 6 keV/amu). The intensity scale is different for each species. The linear temperature program rate is 25°C/min.

ilmenite irradiated with ^{13}C, ^{15}N, and D_2 was heated, the major species evolved were $^{13}CO_2$ and $^{15}N_2$ which were released as broad peaks from about 800°C to 1350°C. Very little ^{13}CO was detected.

Our experiments with the artificially irradiated plagioclase and olivine have accounted for about 30% (dissolution) and 45% (pyrolysis) of the implanted carbon. These amounts may represent a lower limit, however, particularly if the mineral targets have reached saturation at a dose less than 10^{16} ^{13}C/cm^2. We have not yet determined the saturation flux value for carbon for these minerals, but we have some pertinent preliminary data. In an experiment with plagioclase ir-

UNIRRADIATED   PLAGIOCLASE

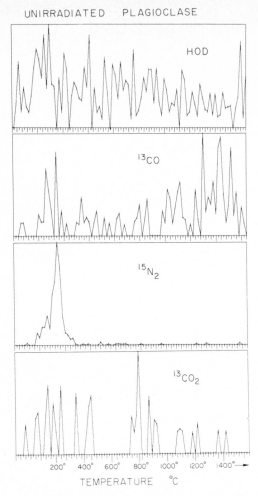

Fig. 2. Thermal release profiles from unirradiated plagioclase. The intensity scale is
different for each species and different from Fig. 1.

radiated with 3×10^{15} $^{13}C/cm^2$, 70% of the ^{13}C was recovered during pyrolysis. This
preliminary result suggests that 3×10^{15} $^{13}C/cm^2$ may be closer to the saturation
value for plagioclase than 10^{16} $^{13}C/cm^2$.

The mass spectra obtained with the ionic analyzer are very complex since both
gaseous and nongaseous species released from the targets are detected. Furthermore, absolute yields can't be determined at the present time because the
ionization efficiency of each species during sputtering is not yet known. For this
reason the abundances of the IIS species have been only qualitatively analyzed in
the mass range $M < 40$. Such species have been classified in 4 distinct groups
(deuterocarbons, oxygen species, nitrogen compounds, carbide, and deuteride-
like material) and we present some of our preliminary results for 3 "oxygen-poor"

(Al, Au, graphite) and 2 "oxygen-rich" (olivine, plagioclase) targets: (1) Deuterocarbon lines. In Al single crystals ^{13}CD, $^{13}CD_2$, $^{13}CD_3$, and $^{13}C_2D$ lines were detected in a very reproducible way. Furthermore, the implantation depth profiles of all $^{13}CD_n$ species were strikingly similar to that corresponding to the ^{13}C line. On the other hand, with the polycrystalline Al foil the relative yield of these species was extremely variable with some target areas only showing the ^{13}CD line. In all the other targets including gold, only lighter molecular fragments were identified—^{13}CD, $^{13}CD_2$ in Au and graphite; ^{13}CD in plagioclase and olivine. However, in plagioclase which was pre-irradiated with a high dose ($\sim 10^{12}$ cm^{-2}) of 3 MeV/amu krypton ions, $^{13}CD_2$ and $^{13}C_2D$ were also detected. (2) Oxygen species. Strong OD lines were clearly detected in the oxygen-rich targets, with olivine showing the most intense line. ^{13}CO was detected in olivine, but in plagioclase there was a very intense interference from ^{27}AlD. No ^{15}NO was observed in plagioclase and we have not yet searched for this species in olivine. (3) Nitrogen compounds. ND and weak ^{13}CN and D^{13}CN lines were detected in Al single crystals and in the pre-irradiated plagioclase. (4) Carbide- and deuteride-like material. A rich variety of D-metal and ^{13}C-metal lines (Na–D, Mg–D, Al–D; Na–^{13}C, Mg–^{13}C, Al–^{13}C) have been found in the targets where these metals are present.

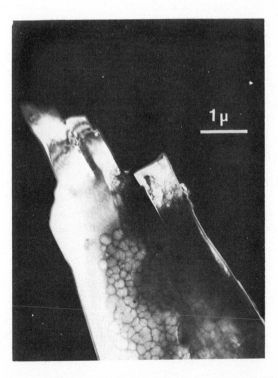

Fig. 3. Darkfield micrograph (HVEM) of a feldspar grain irradiated with 1 keV/amu ions (10^{16}/cm^{-2}).

The solid state radiation damage chemistry of the IIS process in silicates

It could be argued that during the acid or ion etching of the targets, "active sites" are created on the surface that trigger the synthesis of the IIS species. However, this interpretation seems unlikely because these two very different etching techniques give about the same highly specific mixture of species and in the following sections we will assume that IIS species can be synthesized *in situ* in solids. To definitively check the occurrence of the IIS process in solids, A. Bernas is currently trying to detect IIS species in KBr and quartz by using infrared techniques.

We already reported (Bibring, 1972, Bibring *et al.*, 1974a) that above a critical flux value, ϕ_c, that depends both on the target and on the projectile ion, all the mineral grains with the exception of graphite are both severely rounded by ion sputtering and coated with ultrathin amorphous layers of radiation damaged materials (Fig. 3). In metals no amorphous coating is observed, and the main results of ion implantation, in aluminum for example (Jouffrey *et al.*, 1972, 1973), are first to increase the density of dislocations up to maximum values $\sim 10^{11}$ cm^{-2} and to produce interstitial clusters, gas bubbles, graphite inclusions, etc.; then for higher doses the density of dislocations slightly decreases as a result of the reorganization of the dislocation network that triggers the formation of new features such as geometrical voids, etc.

In fact, after implantation at flux values $\phi \geq \phi_c$, all the minerals can be considered as being coated with a "giant" nuclear particle track which is spread all over the surface. In such an extremely reactive material the chemical reactions between the implanted D, C, and N atoms could occur more readily for the following reasons: (1) Diffusion processes are generally considered to be much enhanced (Matzke and Whitton, 1966) and the free energy of the system is much higher. (2) By using the model of Kinchin and Pease (1955) we deduce that each incident 40 keV ^{13}C ion produces a cascade of 800 recoils in the silicate targets and therefore the IIS process involves *"hot" atom chemistry* which usually gives a great variety of reaction channels.

On the other hand, the following causes could limit the IIS yields: (1) During implantation the sputtering rate of the targets, S, leads to saturation values in the concentrations of most* implanted ions. The saturation flux values, ϕ_s, are roughly proportional both to S^{-1} and to the thickness of the implanted layers, and they should roughly be given by the ϕ_c values that correspond to the removal of

---

*Exceptions to the saturation theory have been reported for alkaline elements. In particular, the concentration of Na in aluminum increases linearly with the dose up to values that are ~ 1000 times higher than the expected "sputtering" values. Such a drastic effect seems to be due to a very rapid volume diffusion that redistributes Na deeper inside the target. This could occur on the moon as well, with the result of enriching the finest size fraction of the regolith in very "mobile" elements such as the alkalis that are possibly injected by the solar and lunar winds. So far we have not conducted exact measurements of the saturation values for the D, C, and N atoms to check if they could be enriched over the ϕ_s values. However, the similarity of their implantation depth profiles as measured with the ionic analyzer argues against a strong redistribution for any one of them, and therefore for such elements we will assume that $\phi_s \sim \phi_c$.

about one thickness of amorphous coating on the target. Thus, on this basis alone, we expect that ilmenite will be a much more efficient matrix than plagioclase on the moon (S ilmenite ~0.1 S plagioclase), both for the trapping of implanted solar wind ions and the synthesis of IIS species.

(2) After implantation the D, C, and N "reactants" are quickly redistributed *inside the amorphous coating* by "damage" diffusion, (Matzke and Whitton, 1966), and this is illustrated by our pyrolysis runs of targets implanted with low- and high-energy ^{13}C ions (0.3 and 1 keV/amu, respectively) that are discussed in our companion paper (Bibring *et al.*, this volume). During this process they could precipitate at "sinks" that discriminate between the D, C, and N atoms or they could be incorporated in large clusters of defects. Then it could be argued either that they are lost for the IIS reactions or that the "sinks" or clusters act in fact as active sites for such reactions (see below). However, we did not detect large-sized precipitates ($\geqslant 50$ Å) such as graphite inclusions and gas bubbles in the minerals so far investigated. Therefore, we have now undertaken the difficult task of searching for weakly contrasted and small-sized precipitates embedded in the amorphous coatings, to check the possibility of a fractional precipitation of the D, C, and N atoms in the mineral targets.

(3) Many different reaction channels are opened for the implanted ions. This is evidenced by the great number of species detected ranging from Mg–^{13}C species up to ^{13}C$_3$D$_8$ molecules. The rich variety of carbide-like fragments that we observed with the ionic analyzer supports the view that the active sites for the formation of the C-metal species are carbon-rich inclusions into which knocked-on "hot" target atoms are injected. It is more difficult to identify the active sites involved in the formation of the IIS species. Since such sites should show no steric factor hindrance for the synthesis of a rigid chain molecule such as ^{13}C$_3$D$_8$, they could be voids containing an excess of deuterium and with a size at least comparable to the length of the ^{13}C$_3$D$_8$ molecule (~20 Å). Then both the nature and the pre-implantation history of the targets should play an important role in the IIS process, and this conclusion is supported by the results obtained with the ionic analyzer first for the plagioclase pre-irradiated with krypton tracks and, second, by the very different IIS yields observed for aluminum prepared either as single crystals or as a polycrystalline foil.

Discussion

Specificity of the IIS process. The IIS process is highly specific when compared to other processes so far proposed to explain the formation of molecules in space which either yield very rich mixtures of compounds containing up to 18 carbon atoms (Anders *et al.*, 1973) or about which it has been said "the only molecules everyone agrees must be formed in interstellar grains are H$_2$" (Metz, 1973). In contrast, the IIS process in the targets and lunar fines gives measurable amounts of only small molecules with up to 3 carbon atoms, the HCN/CH$_4$ ratio measured during acid dissolution is about 0.2, the CO/CNH ratio evolved during pyrolysis is about 25 or higher and the yield of nitrogen oxides is very small. Furthermore, our ionic analyzer runs show quite clearly that the ion sputtering of amorphous coatings loaded with IIS species releases a rich variety of small species including OD, which are very difficult to synthesize by other processes. Therefore we believe that the IIS process could contribute to the formation of some of the small molecules that are found in interstellar clouds (Bibring *et al.*, 1974a).

Limitations of the simulation experiments. All implantation experiments so far conducted to simulate the injection of solar wind species in the regolith, including ours, have severe inherent limitations: (1) After the pioneering work of Zeller *et al.* (1966), the first low energy ion implantations intended specifically for lunar research were reported by Pillinger *et al.* (1972). This experiment has two major drawbacks: (i) Contrary to solar wind species that have energies of about 1 keV/amu (same speed), the ^{13}C and D ion energies were very different (1 keV and 2.5 keV/amu, respectively). From range-energy theories (Winterbon *et al.*, 1970) it can be shown that the ^{13}C and D layers were not completely superimposed; (ii) Furthermore, from our comparative study of the sputtering rates of various targets for solar wind type ions (Bibring, 1972; Bibring *et al.*, 1974a) we deduce that the flux of deuterium ions used by Pillinger *et al.* ($\sim 2 \times 10^{18}$ ions/cm^2) was high enough to remove most of the carbon that was first implanted in the targets. Therefore, when this experiment was first reported, it was difficult to understand why IIS species were formed, unless a small proportion ($\sim$ a few percent) of the implanted ^{13}C atoms were in fact reimplanted during knock-on collisions with the deuterium ions. Now from our low and high energy ^{13}C implantations in plagioclase (Bibring *et al.*, this volume) we believe that in the experiment of Pillinger *et al.* (1972) a high proportion of the ^{13}C atoms were redistributed in the thick amorphous coatings subsequently produced by the deuterium ions.

(2) The second series of simulation experiments was then reported by Chang *et al.* (1973a, 1973b) and ourselves (Wszolek *et al.*, 1973). In Chang's experiment, only one ion species was implanted in a given target with the correct solar wind energy. Therefore, this experiment cannot be subjected to the criticism that the first ion layer was sputtered away during the subsequent implantations. However, this experiment did not come close to simulating the mixture of ions present in the solar wind, and the flux of ^{13}C ions was 600 times greater than that actually injected by the solar wind in lunar samples (see below), with the possible consequence of a very different type of "hot" atom chemistry in the targets.

(3) Our experiment has limitations also. In particular, we did play the dangerous game of injecting a mixture of ions in the targets. But we were as careful as possible, using recent range-energy theories (Winterbon *et al.*, 1970) as well as our directly measured sputtering rate to define irradiation conditions which superimpose various layers of ions without noticeably sputtering away any one of them. For these reasons, the ^{13}C, ^{15}N, and D_2 mixture was implanted in the following way: 3 keV/amu ^{13}C atoms (40 keV) were first injected, up to a flux value of about 10^{16} ions/cm^2; then 3 keV/amu (45 keV) ^{15}N atoms were mixed into the

carbon layer with the ratio $^{15}N/^{13}C = 1$; finally deuterium ions were implanted at 3 separate energies ($E = 3, 4.5, 6$ keV/amu), up to a total dose of 10^{17} ions/cm^2 and 10^{18} ions/cm^2 for one olivine experiment. Unfortunately, such conditions ruled out the possibility of injecting the right C/N/D "solar wind" ratios in the targets. We estimate below that in crystalline particles in the regolith the saturation concentration of carbon should be 10 times lower ($\sim 10^{15}$ ions/cm^2) but that the D/C ratio should be 5–50 times greater than the values used in the present experiment. As indicated by the olivine yields reported in Table 1 (columns 4 and 5), a higher D/^{13}C ratio gives heavier IIS species, and therefore we suspect that the present simulation experiments do not perfectly match the solar wind synthesis on the moon. However, we expect very soon to decrease the ^{13}C flux down to $10^{15} \cdot$ cm^{-2} in order to have a better solar wind simulation.

(4) A final limitation of all simulation experiments concerns the differences between the targets and the lunar dust grains that have suffered a complex maturation in the regolith. In particular, during HVEM runs we discovered that the finest lunar dust grains were both loaded with a very high density of nuclear particle tracks ($>10^{11}$ tracks/cm^2) and aggregated as "nano-breccias." As a first attempt to subject the targets to a maturation procedure, we irradiated plagioclase crystals with a high dose of tracks ($\sim 3 \times 10^{11}$ tracks/cm^2) produced with 3 MeV/amu krypton ions. The heavier species detected in such crystals with the ionic analyzer then strongly suggest that the pre-implantation history of the targets should also play an important role in the IIS yields. Another attempt to better simulate lunar conditions was to use pellets of lunar dust grains as targets. Chang *et al.* (1973a, 1973b) applied 10,000 psi pressure to make pellets and their approach is probably correct, although it could be argued that the high sintering pressure has likely produced changes in the texture of the grains. On the other hand, Pillinger *et al.* (1972) made pellets of lunar dust by using 30% of KBr as a binding agent. No report is made in their paper of the IIS yield of a pure KBr target, but it could be that such crystals are in fact efficient matrices for the IIS process since they are known to contain a high density of microscopic voids. Then the hydrolysis of the lunar dust pellets could yield a complex mixture of IIS species, released by both KBr and lunar dust grains. Therefore it is quite clear that more work is needed to define realistic IIS yields that are produced by solar wind nuclei in lunar dust grains.

Saturation flux values in the contemporary solar wind and amount of total solar wind carbon in the regolith. On the moon the saturation concentrations of solar wind species that are not deeply redistributed into the grains by volume diffusion depend on the sputtering rate of solar wind α-particles, which are the most efficient "rounding + coating" ions in space (Bibring *et al.*, 1974a). Almost complete saturation will be observed at a flux value ϕ_c (He) which removes about one implanted layer—or one amorphous coating—from the grains. For glasses, pyroxenes, and feldspars on the moon, ϕ_c (He) $\sim 5 \times 10^{16}$ ions/cm^2 and this value is about 10 times higher for ilmenites. The similarity in the implantation depth profiles of the D, C, and N ions indicates that their saturation fluxes are likely to be given by the product of ϕ_c (He) by the ratios H/He ~ 25, C/He $\sim 2 \times 10^{-2}$, N/He $\sim 10^{-2}$, respectively (these solar wind ratios can be deduced from the data compiled by Bames (1972)). Therefore, if we neglect the contribution of ilmenite grains, the sputtering saturation concentration of solar wind carbon injected into the crystalline component of a mature soil sample, c(C), will be approximately given by: c(C) ppm $\sim S_{eff} \times \phi_c$ (He) $\times$ (C/He)$_{sw} \times 12 \times N_0^{-1}$, where S_{eff} is the effective area of the sample as determined from gas adsorption isotherms (Holmes *et al.*, 1973) and which is about 1 m^2/g for mature soils, ϕ_c (He) $\sim 5 \times 10^{16}$ α/cm^2, C/He $\sim 2 \times 10^{-2}$, and N_0 the Avogadro's number. With these values we get c(C) ~ 200 ppm.

Some implications of the present results for lunar carbon studies

Even though we have not achieved a perfect simulation of lunar solar wind conditions, many similarities are found when comparing the targets with lunar material. Acid hydrolysis of the mineral targets (Table 1) releases a suite of small molecules similar to those detected for lunar fines (Holland *et al.*, 1972b). The simulation experiments further show that CO_2 and N_2 deriving from solar wind ions can be released when the mineral matrix is dissolved with acid. Although there have been some tentative reports that these two species are released by acid hydrolysis of lunar fines (Sakai *et al.*, 1972; Kerridge *et al.*, 1974), their presence has been difficult to demonstrate because of experimental difficulties with terrestrial atmospheric contaminants and background levels of CO_2 released by DF etch of glass reaction vessels.

While it is perhaps premature to compare absolute yields of the IIS products we can consider some relative yields. In particular, the ratio of H^{13}C^{15}N + D^{13}C^{15}N/^{13}CD$_4$ obtained for olivine with the higher and more favorable D/^{13}C ratio (Table 1, column 5) compares very favorably with the corresponding figure for lunar soils: 0.17 versus about 0.1–0.2 for most bulk mature and nonmature soils (Holland *et al.*, 1972b) and 0.12 for the crystal-rich 5–10 μ fraction of 10086 (Bibring *et al.*, this volume). The molar ratio of ^{13}CD$_4$/^{13}C$_2$D$_6$ for the same olivine experiment is about 43 compared with a corresponding ratio of about 30–100 for bulk lunar fines (Holland *et al.*, 1972b).

The thermal release profiles obtained for ^{13}CO and ^{15}N$_2$ from irradiated plagioclase (Fig. 1) show the characteristic bimodal evolution found for most lunar fines. Most of the ^{13}CO and ^{15}N$_2$ is released below 1000°C for these mineral targets, and identical behavior is observed for crystal-rich fractions from lunar soil 10086, especially the finest size fractions ($<5 \mu$ and 5–10 μ) and a pure plagioclase separate (Wszolek *et al.*, 1974b; see Fig. 5 in Bibring *et al.*, this volume). We believe that the release of CO and N$_2$ below 1000°C (maximum ~ 800°C) from the minerals and lunar soils is due to reactions occurring during recrystallization of the amorphous coatings on the surface of solar wind irradiated grains. The maximum release temperature for ^{13}CO did not change substantially when the energy of the implanted ^{13}C was varied from 0.3 up to 3 keV/amu (compare Fig. 1, this paper, with Fig. 2 of Bibring *et al.*, this volume). The evolution of $^{13}CO_2$ derived from plagioclase implanted with ^{13}C at solar wind energy occurs at a temperature lower than that corresponding to a phase transformation, 400–750°C (Bibring et al., this volume). Unlike that of ^{13}CO, the maximum release temperature for $^{13}CO_2$ did vary with the energy and depth of implanted ^{13}C (Fig. 1, this paper, and Fig. 2 of Bibring et al., this volume). One interpretation of this behavior is that diffusion may be the rate controlling process in the release of $^{13}CO_2$ from plagioclase.

It is not yet clear from the simulation experiments what parameters are important for the formation of carbides in lunar fines. Our data show that calcium and/or sodium carbides can be produced without reduction of the plagioclase with solar wind hydrogen. Instead, the amounts of these carbides appear to decrease when hydrogen is added (Table 1, column 2 versus column 1) but the interpretation of the data is complicated by the fact that carbon was implanted at two different energies in these plagioclase experiments. On the other hand, the yield of iron carbide produced by solar wind ions may increase with an increase in the amount of hydrogen as indicated by our data for olivine (Table 1, columns 4 and 5). The present results suggest that two or more mechanisms may control the amount of carbides on the moon.

The (carbide/methane)$_{IIS}$ ratios (0.19 and 0.06) for the olivine dissolution experiments (Table 1, columns 4 and 5) are much lower than the corresponding (carbide/methane)$_{lunar fines}$ ratio ($\sim$3–4), even taking into account the low concentration ($\sim$8%) of iron silicate in the olivine. This discrepancy between the simulations and lunar conditions is probably due mostly to selective loss of CH_4 in the lunar thermal cycle and during formation of glassy agglutinates and other complex particles (Cadogan et al., 1973; Des Marais et al., 1973). An alternative explanation is that iron carbide is added from meteorites or synthesized during meteoritic impact (Pillinger et al., 1972; Cadogan et al., 1973), and it is expected that this additional carbide would be concentrated in glassy agglutinates formed by micrometeorites (Cadogan et al., 1973). These complex particles do in fact exhibit a very high CD_4/CH_4 ratio of about 20 (Cadogan et al., 1973; Bibring et al., this volume). However, from preliminary data it appears unlikely that significant amounts of carbide have been added to or synthesized in the glassy agglutinates since the percentage of carbide (detected as CD_4) versus total carbon (measured as CO) is about the same or higher for the 5–10 μ fraction of 10086 as for the glassy agglutinates in the 37–105 μ fraction of 10086 (Bibring et al., this volume). Another possible explanation for the differences in the carbide/methane ratios is that solar wind carbon is implanted into a hydrogen-reduced layer on the surface of lunar dust grains (Gammage and Becker, 1971), and carbide may be formed directly with reduced metal. This condition is lacking in our simulation experiments.

CONCLUSIONS

After five years of lunar sample studies it is quite clear that a good knowledge of the processes occurring in the most active superficial layer of the lunar regolith, at depths of about 100 μ to 1 mm, is a prerequisite to the understanding of many fields of lunar science. To tackle such difficult problems it is first necessary to conduct meaningful simulation experiments concerning the interactions between the lunar surface and its space environment, and to combine very different analytical approaches. The work reported in this paper is a preliminary attempt to satisfy these various requirements. We have combined the fields of light element chemistry and radiation damage effects to analyze the contribution of the solar wind in the active "skin" of the regolith; in our companion paper we applied the same philosophy to simulations of lunar winds.

The main results of our solar wind simulation experiments are: (1) The implantation of solar wind H, C, and N ions in silicates synthesizes small molecules that can be released in vacuum either by ion sputtering or by heating. (2) This ion implantation synthesis is highly specific when compared to other processes so far proposed to explain the formation of molecules in the solar nebula or in interstellar space. (3) The ultra-thin amorphous coatings of radiation damaged material that are produced during the ion implantations have peculiar physicochemical properties that dominate the thermal release of various carbon species. (4) In the crystalline component of mature soils, the carbon injected by the solar wind should reach a saturation concentration of about 200 ppm. (5) A comparison of the targets and various size fractions and mineral separates from the mature *10086 sample* shows that in the "crystalline" fraction of mature soils (10μ-residues that contain very few glasses; feldspar separates extracted from coarser size fraction), the carbon chemistry is clearly dominated by solar wind implantation effects, when no evidence of track metamorphism is observed with the HVEM. In particular, the HCN/CH_4 ratio evolved during dissolutions as well as the CO and N_2 pyrolysis curves are very similar in the "crystalline" lunar components and in the targets. (6) The reduction by hydrogen of metal cations in minerals is not a prerequisite for the formation of carbides in lunar samples.

Acknowledgments—One of us (M.M.) is deeply indebted to R. Klapish and B. Jouffrey for their enthusiastic support and interest and to R. M. Walker and D. S. Burnett for many helpful suggestions and criticisms. This work would have been impossible without the superb technical assistance of C. Jouret during the HVEM observations and of F. C. Walls with mass spectrometry runs. We are also very grateful to E. El-Gammal and B. Vidal for the use of the ionic analyzer at ONERA and we are indebted to G. Slodzian for his constant advice and interest during the ionic analyzer runs. We thank D. S. Burnett and J. T. O'Connor for the plagioclase and olivine, respectively. We are indebted to J. T. O'Connor for help in preparing mineral and size fractions. This work was supported in France by CNRS and in the United States by the National Aeronautics and Space Administration grant NGR 05-003-435.

References

Anders E., Hayatsu R., and Studier M. (1973) Organic compounds in meteorites. *Science* **182**, 781.

Bames S. J. (1972) Spacecraft observations of the solar wind composition. NASA Special Publication SP-308, p. 335.

Bibring J. P. (1972) Effets d'implantation ionique dans l'espace. These de Doctorat 3e cycle, Faculté des Science d'Orsay, France.

Bibring J. P., Langevin Y., Maurette M., Meunier R., Jouffrey B., and Jouret C. (1974a) Ion implantation effects in "cosmic" dust grains. *Earth Planet. Sci. Lett.* **22**, 205–214.

Bibring J. P., Burlingame A. L., Langevin Y., Maurette M., and Wszolek P. C. (1974b) Simulation of lunar carbon chemistry. II. Lunar wind contribution. *Proc. Fifth Lunar Sci. Conf., Geochim. Cosmochim. Acta.* This volume.

Cadogan P. H., Eglinton G., Gowar A. P., Jull A. J. T., Maxwell J. R., and Pillinger C. T. (1973) Location of methane and carbide in Apollo 11 and 16 lunar fines. *Proc. Fourth Lunar Sci. Conf., Geochim. Cosmochim. Acta*, Suppl. 4, Vol. 2, pp. 1493–1508. Pergamon.

Chang S., Kvenvolden K. A., and Gibson E. K., Jr. (1973a) Simulated solar wind implantation of carbon and nitrogen containing ions into an analogue of lunar fines (abstract). In *Lunar Science—IV.* pp. 124–126. The Lunar Science Institute, Houston.

Chang S., Mack R., Gibson E. K. Jr., and Moore G. W. (1973b) Simulated solar wind implantation of carbon and nitrogen ions into terrestrial basalt and lunar fines. *Proc. Fourth Lunar Sci. Conf., Geochim. Cosmochim. Acta*, Suppl. 4, Vol. 2, pp. 1509–1522. Pergamon.

Chaumont J., Baran-Marzak M., Camplan J., Meunier R., and Sarrouy J. L. (1971) The Ion Implantation Facility at Orsay. *Le Zide* 152, 105–112.

DesMarais D. J., Hayes J. M., and Meinschein W. G. (1973) Accumulation of carbon in lunar soils. *Nature (Phys. Sci.)* 246, 65–68.

Gammage R. B. and Becker K. (1971) Exoelectron emission and surface characteristics of lunar materials. *Proc. Second Lunar Sci. Conf., Geochim. Cosmochim. Acta*, Suppl. 2, Vol. 3, pp. 2057–2067. MIT Press.

Hernandez R., Lanusse P., Slodzian G., and Vidal G. (1972) Spectrometrie de masse avec source à emission ionique secondaire. *La Recherche Aerospatiale* 6, 313–324.

Holland P. T., Simoneit B. R., Wszolek P. C., and Burlingame A. L. (1972a) Study of carbon compounds in Apollo 12 and 14 lunar samples. *Space Life Sciences* 3, 551–561.

Holland P. T., Simoneit B. R., Wszolek P. C., and Burlingame A. L. (1972b) Compounds of carbon and other volatile elements in Apollo 14 and 15 samples. *Proc. Third Lunar Sci. Conf., Geochim. Cosmochim. Acta*, Suppl. 3, Vol. 2, pp. 2131–2147. MIT Press.

Holmes H. F., Fuller E. L., Jr., and Gammage R. B. (1973) Interaction of gases with lunar materials: Apollo 12, 14, and 16 samples. *Proc. Fourth Lunar Sci. Conf., Geochim. Cosmochim. Acta*, Suppl. 4, Vol. 3, pp. 2413–2423. Pergamon.

Jouffrey B., Joyes P., and Ruault M. O. (1972) Etude de defauts crees dans une matrice d'aluminium, par bombardement d'ions Al^+. *Phil. Mag* 25, 833–851.

Jouffrey B., Chaumont J., Bernas H., and Ruault, M. O. (1973) Etude en microscopie electronique de defauts dans les metaux par implantation ionique. *J. Phys.* (France) 34, Suppl. 11, 12, pp. 21–25.

Kerridge J. F., Petrowski C., and Kaplan I. R. (1974) Sulfur, methane and metallic iron in some Apollo 17 fines (abstract). In *Lunar Science—V*, pp. 411–412. The Lunar Science Institute, Houston.

Kinchin G. H. and Pease R. S. (1955) The displacement of atoms in solids by radiation. *Progress in Physics* 18, 1.

Matzke H. J. and Whitton J. L. (1966) Ion-bombardment-induced radiation damage in some ceramics and ionic crystals. *Can. J. Physics* 44, 995–1010.

Metz W. D. (1973) Interstellar molecules: New theory of formation from gases. *Science* 182, 466.

Pillinger C. T., Cadogan P. H., Eglinton G., Maxwell J. R., Mays B. J., Grant W. A., and Nobes M. J. (1972) Simulation study of lunar carbon chemistry. *Nature (Phys. Sci.)* 235, 108–109.

Sakai H., Chang S., Petrowski C., Smith J., and Kaplan I. R. (1972) Distribution of carbon and sulfur in hydrolyzed Apollo 15 lunar fines. In *The Apollo 15 Lunar Samples*, pp. 319–323. The Lunar Science Institute, Houston.

Simoneit B. R., Wszolek P. C., Christiansen P., Jackson R. F., and Burlingame A. L. (1973a) Carbon chemistry of Luna 16 and Luna 20 samples. *Geochim. Cosmochim. Acta* 37, 1063–1074.

Simoneit B. R., Christiansen P. C., and Burlingame A. L. (1973b) Volatile element chemistry of selected lunar, meteoritic, and terrestrial samples. *Proc. Fourth Lunar Sci. Conf., Geochim. Cosmochim. Acta*, Suppl. 4, Vol. 2, pp. 1635–1650. Pergamon.

Winterbon K. B., Sigmund P., and Sanders J. P. (1970) Spatial distribution of energy deposited by atomic particles in elastic collisions. *Math. Fys. Medd.* 37, 14.

Wszolek P. C., Jackson R. F., Burlingame A. L., and Maurette M. (1973) Carbon chemistry of Apollo

15 and 16 samples and solar wind ion implantation studies (abstract). In *Lunar Science—IV*, pp. 801–803. The Lunar Science Institute, Houston.

Wszolek P. C., Walls F. C., and Burlingame A. L. (1974a) Simulation of lunar carbon chemistry in plagioclase (abstract). In *Lunar Science—V*, pp. 854–856. The Lunar Science Institute, Houston.

Wszolek P. C., O'Connor J. T., Walls F. C., and Burlingame A. L. (1974b) Thermal release profiles and the distribution of carbon and nitrogen among minerals and aggregate particles separated from lunar soil (abstract). In *Lunar Science—V*, pp. 857–859. The Lunar Science Institute, Houston.

Zeller E. J., Ronca L. B., and Levy P. W. (1966) Proton-induced hydroxyl formation on the lunar surface. *J. Geophys. Res.* **71**, 4855–4860.

Proceedings of the Fifth Lunar Conference
(Supplement 5, Geochimica et Cosmochimica Acta)
Vol. 2 pp. 1763–1784 (1974)
Printed in the United States of America

Simulation of lunar carbon chemistry: II. Lunar winds contribution

J. P. Bibring,[1] A. L. Burlingame,[2] Y. Langevin,[1]
M. Maurette,[1] and P. C. Wszolek[2]

[1]Laboratoire René Bernas, 91406, Orsay, France
[2]Space Sciences Laboratory, University of California, Berkeley, California 94720

Abstract—To further identify the major processes responsible for both the synthesis and the redistribution of carbon compounds in the lunar regolith we investigated the effects of different types of lunar "winds" that are possibly active in the lunar atmosphere. First we analyzed both by high-voltage electron microscopy and high-resolution mass spectrometry various samples including: thin vapor phase deposits obtained from the Murchison meteorite and from lunar rock 15065; silicate targets artificially irradiated with low-energy carbon nuclei and then subjected to thermal treatments; and finally glassy agglutinates isolated from mature 10086 lunar fines. Then we computed the concentration of total carbon in the lunar soil resulting from the implantation of lunar winds components injected in the lunar atmosphere. These studies suggest that: (1) The directly condensed lunar winds phases so far examined are strongly depleted in carbon and sulfur with respect to the parent material, and should recrystallize quickly in the lunar thermal wave. (2) The coating of radiation damaged material produced on grains during ion implantation has physicochemical properties which prevent identification of low-energy species implanted from the lunar atmosphere and therefore one has to rely on computations to evaluate the contribution of the lunar winds to lunar carbon chemistry. (3) Such computations reveal that the implanted lunar winds carbon originates both from the vapor phases injected into the lunar atmosphere during the thermal metamorphism of mature lunar soil grains ($\sim$15 ppm) and from the direct volatilization of the impacting micrometeorites ($\sim$10 ppm). However, these values are still much smaller than the saturation concentration of carbon ($\sim$200 ppm) expected from the solar wind. (4) The crystalline component which makes up about 90% of the glassy agglutinates has not been strongly heated during "agglutination."

INTRODUCTION

There are possibly various types of vapor phases that can either inject carbon compounds in the lunar atmosphere or directly condense on the surface of lunar dust grains to form carbon-rich deposits. Let us first classify these vapor phases into three different types of lunar "winds" and estimate the *maximum amount* of carbon contamination expected from such winds over a period of $\sim$4 b.y., in a thoroughly mixed regolith of $\sim$5 m in thickness, by supposing that the vapor phases are totally recondensed on the moon:

(1) Lunar wind I (LWI). Rehfus (1972) was the first to define this type of lunar wind which is constituted of expanding gas clouds that are produced during the volatilization of micrometeorites impacting on the moon. From Gault *et al.* (1972) the flux of micrometeorites responsible for LWI is $\sim$2 × 10^{-3} g/cm^2/10^6 yr. If we assume that the micrometeorites are completely volatilized upon impact and have a composition similar to type I carbonaceous chondrite ($\leqslant$3.5% of carbon), the maximum amount of carbon which can be expected from LWI in our "test"

1763

regolith ($\sim$300 ppm) is then comparable to the calculated "saturation" concentration of solar wind carbon ($\sim$200 ppm) in mature soil (see Bibring *et al.*, this volume). In this estimate we have neglected the contribution of the 4.7×10^{-3} g/cm^2/10^6 yr of lunar material volatilized during impact since this matter is much poorer in carbon ($\leqslant$200 ppm) than carbonaceous chondrites.

(2) Lunar wind II (LWII). The tiny crystallites found in rock vesicles by McKay *et al.* (1972), as well as some of the microscopic surface features reported by Blandford *et al.* (1974) could result from the direct condensation of other types of vapors, such as those produced when lunar material is "thermally metamorphosed" into glass at a temperature of about 1500°C. We will define such vapor phases as lunar wind II (LWII). An upper limit for the amount of carbon injected in our 5 m test regolith can be obtained by making the following additional "maximizing" assumptions: the regolith contains 50% glass; the glass was formed from lunar dust grains saturated in solar wind carbon ($\sim$200 ppm); and the dust grains were completely degassed during melting. This upper limit of $\sim$100 ppm is about $\frac{1}{3}$ the value expected from LWI, and is comparable to the calculated saturation concentration of solar wind carbon.

(3) Lunar wind III (LWIII). The atoms sputtered away from the surface of lunar dust grains by the solar wind constitute a third type of lunar wind (LWIII). From our recent estimates of sputtering erosion rates of about 0.5 Å/yr in lunar conditions (Borg *et al.*, 1974), we deduce that the total amount of material sputtered away in 4 b.y. is about $20 \, \text{g} \cdot \text{cm}^{-2}$. This material is quite peculiar because it is likely to be the amorphous coating of solar wind radiation damaged material on the grains (Bibring *et al.*, this volume) containing an atomic concentration of carbon of $\sim$4 $\times$ 10^{-3} ($\sim$10^{15} C $\cdot$ cm^{-2}, up to a depth of about 500 Å). Therefore we deduce a maximum carbon contamination due to LWIII of about 30 ppm. However, since early studies (Wehner *et al.*, 1963) concerning the speed distribution of sputtered away atoms on the moon strongly suggest that a major part of the sputtered species are lost from the moon, we will neglect the contribution of LWIII to carbon chemistry in this paper.

Several qualitative arguments have already been proposed to decrease the contribution of LWI. In particular, carbonaceous chondrites contain large concentrations of volatile elements such as sulfur, bismuth, mercury, etc. The heavy volatiles (Bi, Hg) are generally enriched in the lunar soil (10 to 100 fold) with respect to the concentration measured in lunar igneous rocks (Morgan *et al.*, 1972) thus indicating a carbonaceous chondrite contamination. On the other hand, sulfur and water (volatiles normally associated with carbonaceous chondrites) are depleted in the lunar soil, and this observation was used (Gibson and Johnson, 1971) to argue that a preferential gravitational escape of light species injected in the lunar atmosphere via the lunar wind occurs because light species should acquire speeds greater than the escape velocity, $V_e \sim 2.4$ km/sec. Therefore, it was concluded that carbon as well as sulfur in the lunar wind should be lost from the moon. However, such arguments are probably not correct for the following reasons: (1) In the more realistic model of LWI that we will use later, the speed distribution of the constituent species is such that an important proportion (20%) of all species, *even the light ones*, have speeds $< V_e$. (2) The composition of LWI is certainly dominated by CO species since the initial temperature of the expanding gas cloud is $\sim 10^{40}$ K. But the LWII mixture can possibly contain heavier species (CO_2) similar to those that we obtain from the vacuum pyrolysis of silicate targets artificially exposed to a simulated solar wind (Bibring *et al.*, this volume). (3) As suggested by Hayes (1972), carbon species injected into the lunar atmosphere could be reimplanted in the regolith via photoionization and acceleration in the solar wind electric field.

In this paper we present a preliminary attempt to quantitatively evaluate the contribution of the lunar winds to lunar carbon chemistry by specifically answering the following questions: (1) What is the fate of the *directly condensed* lunar winds phases as maturation proceeds? (2) Are such direct condensates strongly depleted in carbon and sulfur, or not? (3) Using thermal release profiles, is it possible to differentiate the carbon reimplanted from the lunar atmosphere from that directly implanted from the solar wind at much higher energy? (4) If it is not possible to detect experimentally the reimplanted carbon species, then could the concentration of the reimplanted species be estimated in a reasonable way by using both the theory already applied to ^{40}Ar (Manka and Michel, 1971) and a realistic model of lunar winds? (5) Can the glassy agglutinates when observed with our combined high-voltage electron microscope (HVEM) and mass spectrometer techniques give clues concerning both meteoritic contamination and the redistribution of carbon during maturation?

EXPERIMENTAL

Vacuum sublimations of Murchison meteorite and lunar rock 15065 were carried out in an evacuated ($\sim 10^{-6}$ torr) bell jar using a 126115 Sloan electron gun (Model 4B) as a heat source. Deposits of 500 and 5000 Å were obtained on lunar dust grains for HVEM observations; deposits of 1000 Å of Murchison meteorite were obtained on quartz plates cooled at 10°C for pyrolysis. The thickness of the deposits was determined with a quartz balance. The quartz plates had been cleaned with methanol and dichloromethane followed by heating to 1300°C under vacuum. No contaminants were observed during pyrolysis of a blank quartz plate.

Details concerning ion implantation and our acid dissolution and thermal evolution (pyrolysis) techniques are given in our companion paper (Bibring *et al.*, this volume) and references cited therein. One plagioclase target (~ 0.6 cm^2 surface area) was irradiated with 4 keV ^{13}C at 10^{16} atoms/cm^2. Two other plagioclase targets (~ 0.4 cm^2 surface area) were first irradiated with 10^{17} α-particles/cm^2 at 2 keV/nucleon. Then one of these targets was subsequently irradiated with 13 keV ^{13}C at 10^{16} atoms/cm^2, and the other one with 4 keV ^{13}C at 10^{16} atoms/cm^2.

The separation of size fractions, mineral separates, and glassy agglutinates from 10086 D lunar fines is reported elsewhere (Wszolek *et al.*, 1974a). Contamination of these materials during the separation scheme appears to be confined to atmospheric CO_2, Teflon from the coated isodynamic separator, and possibly some hydrocarbons in the finest size fractions. These contaminants are all released below 600°C during pyrolysis. Total carbon was determined as the sum of CO and CO_2 released by each fraction. The glassy agglutinates with $2.73 < \rho < 3.21$ were crushed and then dispersed on a microscope grid for HVEM studies.

THE FATE OF THE DIRECTLY CONDENSED LUNAR WINDS VAPOR PHASES

To get further insight into the fate of the directly condensed phases we conducted the following experiments. (1) A variety of micron-sized grains dispersed on electron microscope substrates were observed with the HVEM both before and after thermal treatments similar to those used during pyrolysis runs. The grains were either extracted from lunar soil samples with different indices of maturity or obtained by crushing an internal chunk of lunar igneous rock 15065. The crushed grains were then subjected to artificial vapor phases obtained from rock 15065 or from the Murchison meteorite, either by vacuum sublimation with an electron gun or by ion sputtering. (2) $5\,\mu$-residues from the same lunar dust samples as well as Murchison vapor phases condensed on quartz plates at 10°C and plagioclase targets exposed to a simulated solar wind were subjected to pyrolysis runs up to 1400°C. Some of the electron microscope preparations with the vapor deposits were then treated through similar thermal runs, up to temperatures corresponding to the various peaks in the pyrolysis release profiles, with a view to possibly correlating textural changes in the grains to the release peaks.

The main conclusions of such combined experiments are: (1) The artificial "lunar wind" coatings (Fig. 1a, b) recrystallize at a low temperature ($\sim$300°C) and are clearly separated from the host crystals (Fig. 1c) at about 500°C. On the other hand, the amorphous coatings observed on the lunar dust grains are not recrystallized by 700°C (Fig. 1d), but at about 800°C they begin to give tiny crystallites that are firmly embedded in the host crystal, and such annealing features are exactly like those observed both for the fossil nuclear particle tracks registered in the same grains and for the amorphous coatings produced during artificial solar wind implantations. (2) Therefore, we suggest that any initial lunar wind deposit when subjected to the lunar thermal cycle should recrystallize and add to the regolith a new generation of tiny "lunar wind" grains that probably have sizes similar to the thickness of the deposit (mostly $<$1000 Å). However, we still have to definitively check the validity of this hypothesis by making sure that a long term annealing at 130°C produces the same results that a short term ($\sim$2 hours) at 500°C does.

The ultimate fate of the lunar wind grains is very difficult to define at the present time. In fact we already presented evidence that indicates that the smallest lunar dust grains are preferentially "agitated" on the top surface of the regolith (Borg et al., 1974). Therefore, one possibility could be that they are preferentially sputtered away by the solar wind. Then the early studies by Wehner et al. (1963) concerning the speed distribution of sputtered away atoms under "lunar" conditions suggest that a major part of the sputtered species are lost from the moon. Another possibility is that they are incorporated into glassy agglutinates during micrometeoritic impact. Still another possibility would be that the lunar wind grains are welded to the surface of coarser grains during low-speed collisions. Then they could be responsible for the apparent surface enrichment of the lunar dust grains in carbon, which is generally deduced from either size

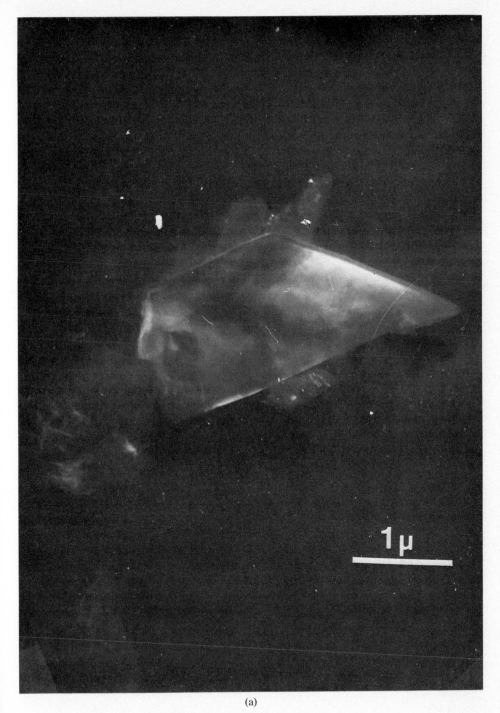

(a)

Fig. 1. Darkfield micrographs (HVEM) showing: (a) grain with artificial lunar wind deposit (500 Å thick); (b) grain with artificial lunar wind deposit (5000 Å thick); (c) the 500 Å thick lunar wind deposit after 2 hours of annealing at 500°C; crystallites and deposit are separating from the grain; (d) no changes in a grain with a solar wind amorphous coating after 2 hours of annealing at 700°C.

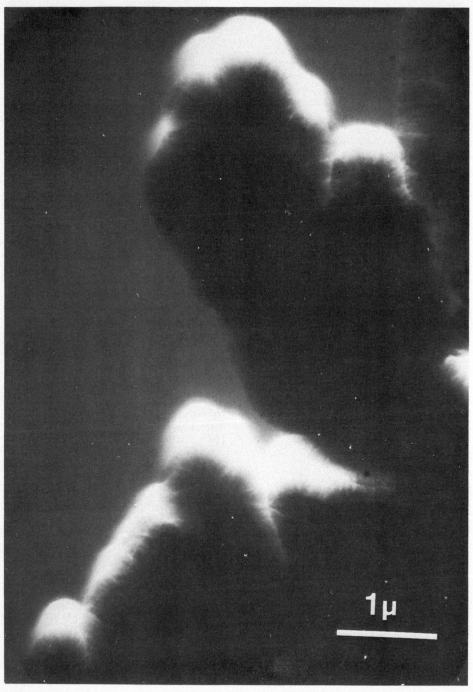

(b)

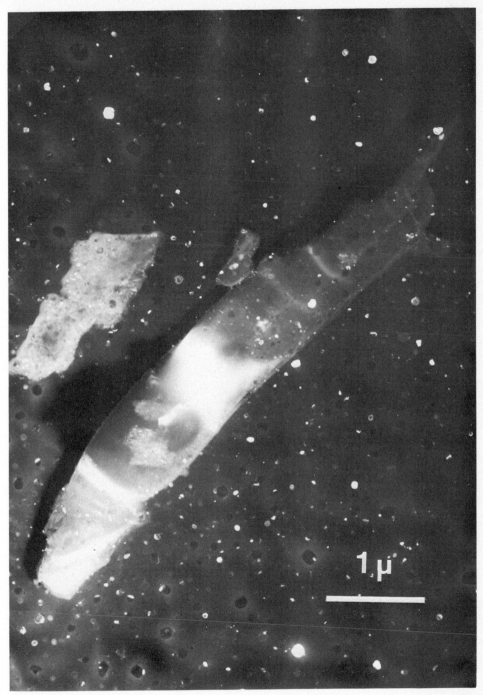

(c)

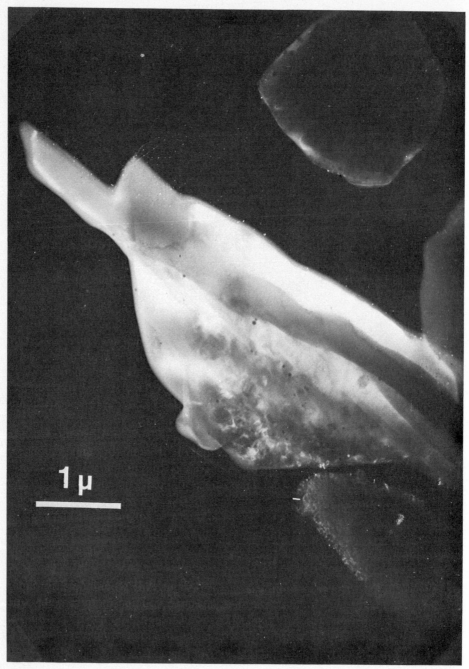

(d)

fraction effects (Cadogan *et al.*, 1972; Holland *et al.*, 1972a, 1972b) or differential etching experiments (Cadogan *et al.*, 1971), and which is generally considered as being highly specific for implanted and/or reimplanted species. However, our preliminary pyrolysis results show that the "Murchison" vapor deposit is strongly depleted both in carbon and sulfur relative to the parent material. While the parent Murchison meteorite releases a suite of carbon- and sulfur-containing compounds when heated (Simoneit *et al.*, 1973), only CO was detected when the "Murchison" vapor deposit, collected on quartz plates, was heated under similar conditions. This CO was evolved at about 1200°C, and the amount detected corresponds to roughly 700 ppm carbon in the deposit. Since Murchison contains ~2% carbon and 3% sulfur (Lovering *et al.*, 1971), the vapor deposit appears to be depleted more than 20 times with respect to both of these elements. Nevertheless, these experiments need to be confirmed and extended. In particular, it will be of interest to determine if carbon and sulfur are retained to a greater extent when the vapor deposition is carried out under conditions simulating the lunar night. We believe that whatever the ultimate fate of the lunar wind crystallites is, their contribution to carbon and sulfur compounds on the moon may be negligible and we will now present two different attempts to evaluate the amount of lunar wind carbon implanted from the lunar atmosphere in lunar soil.

PHYSICOCHEMICAL PROPERTIES OF AMORPHOUS COATINGS AND THE
EXPERIMENTAL SEARCH FOR IMPLANTED LUNAR WIND CARBON

The release of $^{13}CO_2$ by acid dissolution of mineral targets implanted with low-energy carbon ions (Bibring *et al.*, this volume) suggests that this species could be synthesized by the IIS process. Whether the $^{13}CO_2$ is synthesized *in situ* or generated during subsequent treatment or heatings, its release during pyrolysis experiments seems to be controlled by diffusion processes for two reasons. We observed that in plagioclase targets exposed to simulated solar wind ($\leqslant 13$ keV ^{13}C), $^{13}CO_2$ evolves over a wide range of *low* temperatures before the occurrence of the first-phase transformation in the targets which corresponds to the recrystallization of the amorphous coatings. In addition, the temperature at which the $^{13}CO_2$ is released depends on the energy of the implanted ^{13}C and/or the thickness of the amorphous coating into which the ^{13}C is implanted (see below and Bibring *et al.*, this volume). Then since carbon ions implanted from the lunar atmosphere are expected to have ranges (< 100 Å) much shorter than those of solar wind nuclei (~ 500 Å), the CO_2 deriving from such ions could be released at a lower temperature than that expected from solar wind carbon, if it undergoes a diffusion process which is dominated by "surface proximity" (Lewis *et al.*, 1964). Therefore, from thermal release experiments it should be possible to separate the CO_2 molecules originating from lunar wind carbon from those resulting from the direct implantation of solar wind nuclei. However, on the moon both the solar and the low-energy lunar wind species are implanted into an amorphous coating of radiation damaged material. In these coatings the distribution of the implanted species as well as their release during pyrolysis could be dominated by two very different mechanisms: (1) As already suggested by earlier work (Matzke and Whitton, 1966) in solids that are *severely damaged* during ion implantation, the diffusion of implanted species is much faster than in nondamaged crystals ("damage" diffusion). Therefore, carbon ions reimplanted from the lunar atmosphere could be quickly redistributed in the solar wind coatings, matching the distribution of the more energetic solar wind carbon. Thus, they would lose the memory of their low initial energy and would be released at the same temperature as solar wind species. (2) On the other hand, the work of Kelly *et al.* (1968) indicates that in solids that *are not damaged* during ion implantation, the species implanted during high dose experiments are trapped in clusters of defects that are much more difficult to move by volume diffusion. We also note that Ducati *et al.* (1972) suggest that such a "trapping" diffusion should be responsible for the low-temperature features observed in the release pattern of rare gases in lunar glasses.

In order to determine if it is possible to detect any reimplanted species from thermal release profiles as well as to try to identify the diffusion mechanism which may be responsible for the low-temperature release of carbon species, we conducted the following experiments: (1) We irradiated one plagioclase crystal with low-energy (0.3 keV/amu) ^{13}C ions only. Two other plagioclase crystals were first irradiated with a high dose ($\sim 10^{17}$ ions/cm^2) of 2 keV/amu α-particles to produce a thick amorphous coating on the targets. Then one of these crystals was subsequently irradiated with 1 keV/amu "solar wind" ^{13}C ions and the other one with 0.3 keV/amu "lunar wind" ^{13}C ions. In Fig. 2 it can be seen that the low- and high-energy carbon atoms reimplanted into an amorphous coating release $^{13}CO_2$ at the same temperature, in contrast with the lower $^{13}CO_2$ release temperature observed for the target implanted only with the low-energy carbon. (2) We irradiated plagioclase targets with ^{13}C ions of 40 keV, with doses of 3×10^{15} and $10^{16} \cdot cm^{-2}$, and we did not detect the marked changes in the position of the low-temperature $^{13}CO_2$ peaks which could be expected from a "trapping" diffusion process. (3) Most of the ^{13}C is released as ^{13}CO at higher temperatures, and the thermal release profiles for ^{13}CO are nearly identical for all the targets (Fig. 2).

From these preliminary experiments we deduce the following conclusions: (1) The dose-independent release of the CO_2 species at low-temperature suggests that, contrary to rare gas atoms, their release during pyrolysis is not dominated by a "trapping diffusion" mechanism (see the work of Kelly *et al.*, 1968, for the reasoning). (2) This conclusion is further supported by the observation that low-energy reimplanted C ions cannot be differentiated from solar wind carbon when they are injected into a solar wind amorphous coating where they should be quickly redistributed by "damage diffusion." Therefore, it is not possible to detect the contribution of the lunar winds reimplanted species from pyrolysis experiments alone; this remark also applies to the "orphan" ^{40}Ar found in lunar soil samples (Heymann and Yaniv, 1970). Thus, it appears that the only way to estimate the contribution of the lunar wind to lunar carbon chemistry is to compute the quantity of total carbon implanted from the lunar atmosphere.

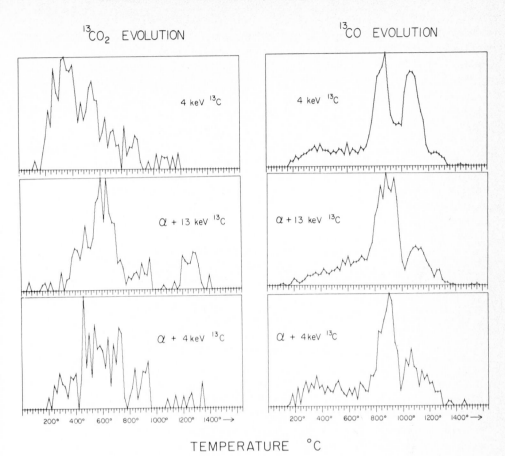

$^{13}CO_2$ EVOLUTION ^{13}CO EVOLUTION

4 keV ^{13}C 4 keV ^{13}C

α + 13 keV ^{13}C α + 13 keV ^{13}C

α + 4 keV ^{13}C α + 4 keV ^{13}C

200° 400° 600° 800° 1000° 1200° 1400° →   200° 400° 600° 800° 1000° 1200° 1400° →

TEMPERATURE   °C

Fig. 2. Thermal release profiles of $^{13}CO_2$ and ^{13}CO from plagioclase irradiated with: (top) 4 keV ^{13}C at 10^{16} atoms/cm²; (middle) 2 keV/nucleon α-particles at 10^{17} atoms/cm² plus 13 keV ^{13}C at 10^{16} atoms/cm²; (bottom) 2 keV/nucleon α-particles at 10^{17} atoms/cm² plus 4 keV ^{13}C at 10^{16} atoms/cm². The linear temperature program rate is 25°C/minute.

COMPUTATION OF THE AMOUNT OF IMPLANTED LUNAR WINDS CARBON

To evaluate the quantity of total carbon implanted from the lunar atmosphere, we have adapted the ^{40}Ar theory of Manka and Michel (1971) to the carbon species injected in the lunar atmosphere by using a realistic model of lunar wind I that derives from that first presented by Rehfus (1972). We note here that the previous qualitative evaluations of Hayes (1972) based on the amount of ^{40}Ar implanted in the regolith were not exact because the speed distribution of the lunar wind species, which is very different from that of ^{40}Ar atoms leaking through the lunar soil, was not taken into consideration. We summarize below the main steps of our computations, discussing each time the validity of our assumptions.

Mass yields. In the introduction we evaluated the total mass yields of carbon expected from lunar wind I (~300 ppm) and lunar wind II (~100 ppm) by considering that the lunar wind carbon is completely mixed *without losses* into a mature regolith of about 5 m in thickness. In the following section we will evaluate the proportion of this carbon, which is indeed firmly implanted into the lunar soil grains at the end of the atmospheric "cycle." We note here that the 100 ppm value for LWII is an upper limit for the following reasons: not all soil samples contain 50% glass; before their transformation into glass the dust grains were perhaps not all loaded with saturation concentrations (~200 ppm) of solar wind carbon; the glasses were probably not completely degassed at the time of formation. On the other hand, the value of ~300 ppm for LWI could be higher if the micrometeorite fluxes were more intense in the past or if the carbon concentration of the impacting particles is much higher than 3.5%.

Chemical composition of the lunar winds vapor phases. (1) Lunar wind I. From the kinetics of the reaction network ($CO_2 + C \rightleftharpoons 2CO$; $CO_2 \rightleftharpoons CO + \frac{1}{2}O_2$; $CO \rightleftharpoons C + \frac{1}{2}O_2$; $CS_2 + 3O_2 \rightleftharpoons CO_2 + 2SO_2$) at the very high temperature ($\sim 10^{4\circ}K$) of the gas cloud during the first 10^{-8} sec of the expansion, we deduce that CO is the major carbon compound in LWI. (2) Lunar wind II. It is likely that at 1200–1500°C the composition of LWII is intermediate between that of LWI (CO) and that of the mixture of compounds released during the pyrolysis of plagioclase targets exposed to simulated solar wind: CO (80%) $< CO_2$ (20%) (see Bibring *et al.*, this volume).

Speed distribution of the lunar winds species. (1) Lunar wind I. The model we use is derived from that of Rehfus (1972): the gas cloud expands adiabatically as an homogeneous hemisphere. We deduce that about 20% of the species will be injected into the lunar atmosphere with $V < V_e$, whatever their mass may be. In particular, we insist that there is no "critical" mass below which the species escape the moon. (2) Lunar wind II. Here we have considered a Maxwellian speed distribution for species emitted at 1500°C and we deduce that 14%, 0.6%, and 0.002% of the C, CO, CO_2 species, respectively, will have $V > V_e$.

Atmospheric distribution of the lunar winds species. All species with $V < V_e$ are thermalized very quickly by colliding frequently (once every ~100 sec) with the lunar surface at temperature T_L. Therefore in this "thermodynamic" model their distribution in the lunar atmosphere is scaled as $\exp-(MGh/kT_L)$, where M and h are the mass and the altitude of the species, and G and k the gravitational and Boltzmann constants, respectively. To take into account all possible situations on the moon we have also considered a very different "ballistic" model which scales the species as:

$$\frac{1}{\sqrt{z}} \int_{\sqrt{z}}^{\infty} e^{-x^2} \, dx$$

where $z = MGh/kT_L$.

Ionization and photodissociation lifetimes (τ_i; τ_d). (1) Ionization. All the species with $V < V_e$ are ionized either by photoionization or by charge exchange with solar wind protons. The τ_i values against charge exchange are of about a few months (Bernstein *et al.*, 1963); the τ_i values of C, O, S, Ar, CO, CO_2 species against photoionization are of about 6 days, 2.5 months, 1.5 days, 1.5 months, 2 months, and 1 month, respectively. Therefore, with the exception of the C and S species both processes occur with about the same probability. (2) Photodissociation. It is difficult at the present time to infer with certainty if τ_i is shorter than τ_d. Accurate measurements of photoionization cross sections down to $\lambda \sim 300\,\text{Å}$ indicate that at least for CO, $\tau_i < \tau_d$ (Sun and Weissler, 1955a, 1955b; Wilkinson and Johnston, 1950; Wainpan *et al.*, 1955). On the other hand, for molecules like CO_2 and H_2O, $\tau_i > \tau_d$ (Rank *et al.*, 1971). However, these uncertainties for the τ_d values are likely to play a minor role in the evaluation of lunar wind carbon in the regolith. In fact, low-energy molecules ($E \geqslant 50\,\text{eV}$) are dissociated any way into their constituent atoms as soon as they penetrate into the target, and the only possible distinct effect of photodissociation is to slightly increase the sticking coefficient of carbon on the lunar dust grains (see below).

Interactions of the ionized species with the solar wind electromagnetic field. The results of this computation can be summarized as follows: (1) About 50% of the ionized species with $V < V_e$ do impact the moon in periods that range from days for C and S to months for the other species. (2) The more massive species are accelerated for a shorter distance in the lunar atmosphere and they have a lower speed than the light ones. Thus their penetration and therefore their sticking coefficient, η, in the soil grains is smaller on the average than those estimated for the light species unless they break down by photodissociation.

Sticking of the implanted species in the lunar dust grains. This part of our analysis is the most uncertain since very few results concerning the sticking coefficients, $\eta(E)$, of low-energy ions in solids are known. As a first estimate of η for species implanted into an amorphous coating of radiation damaged material we used the values reported for rare gases in glass (Carter and Colligon, 1968). Then from these values we infer a model where η is mass independent and varies from 0 to a plateau value of 1, when the energy of the implanted species varies from 0 to $E \sim 1\,\text{keV}$. The final results show that the η value for the larger species will be smaller on the average than those estimated for the light species. Fortunately, the differences are not very large, and we deduced that the proportions of ionized species with $V < V_e$ that are effectively trapped (trapping efficiency, K) are 28%, 18%, 14%, and 14% for C, CO, Ar, and CO_2 species, respectively. Furthermore, by using the very different "ballistic" model to scale the distribution of the lunar winds species in the lunar atmosphere, we get K values that are only 1.5 times lower than those deduced from the pure thermodynamic model. Since the real situation on the moon is probably intermediate between these extreme models, we will adopt an average K value of about 15% for the last part of our computation.

The main conclusions of these computations are: (1) Amount of lunar winds carbon in the regolith. About 0.2×0.15 and 0.9×0.15 of the "total" carbon injected in the lunar atmosphere by lunar wind I and II, respectively, is firmly implanted into the lunar dust grains. From our evaluation of "total" lunar winds carbon given in the introduction, we then estimate that the carbon contamination produced in 4 b.y. in a mature regolith of 5 m by lunar wind I and II is about 10 and 15 ppm, respectively. Therefore, we deduce that the contribution of solar wind carbon (~ 200 ppm) dominates lunar carbon chemistry. (2) Limitations of the computations. The trapping efficiencies ($K \sim 0.15$) are quite insensitive to both the composition of the lunar winds and the models used to scale the distribution of the lunar wind species in the lunar atmosphere; furthermore they are only slightly affected by photodissociation. On the other hand, the computations strongly depend on the values of the sticking coefficients as well as important parameters in the model of lunar wind I, such as the proportion of the energy of the impacting micrometeorite which is transformed into internal energy of the expanding gas cloud.

The Magnetic Glassy Agglutinates

Magnetic glassy agglutinates are apparently formed when glasses produced during micrometeoritic impacts splash on the lunar surface and are impregnated with lunar dust grains (McKay et al., 1972a, 1972b). Therefore, since they represent one of the end products of maturation and since they are known to contain high concentrations of carbon compounds (Cadogan et al., 1973; Des Marais et al., 1973), they should be good tracers to study the possible alteration of solar wind carbon during maturation, as well as to detect a possible micrometeorite carbon contamination in the regolith.

Acid dissolution experiments (Cadogan et al., 1973) indicated a much higher concentration of carbide-like material in the glassy agglutinates than in crystal-rich soil fractions in the same size range. In addition, from Mossbauer studies (Housley et al., 1970) it was inferred that a high concentration of tiny iron particles, with sizes up to ~ 100 Å, are trapped in the agglutinates. Several 100 keV electron microscope observations of tiny crystallites in the glassy matrix of the agglutinates were even reported to be due to such iron particles (Housley et al., 1973), although it is well known that great care should be exercised when interpreting the 100 keV electron micrographs of insulator grains.* Furthermore, no electron diffraction pattern was reported to justify this interpretation. Therefore, since the present status of lunar carbon chemistry in the agglutinates still appeared very confusing, we decided to apply our HVEM-mass spectrometer approach to this problem by using a separate extracted from mature soil 10086 (Wszolek et al., 1974a).

The main results of our HVEM observations are: (1) The glassy agglutinates

---

*Such grains degrade very quickly under the beam of a 100 keV electron microscope as a result of beam ionization effects that are at least 20 times greater at 100 keV than at 1 MeV, because of both higher electron ionization losses and higher beam current. Therefore, insulator grains have to be observed with a HVEM but even in this case several micrographs of the same object have to be taken at different beam intensities to insure that the features under study do not "grow" in situ under the beam.

are a two-component mixture made up of $\sim 90\%$ crystals that do not seem to be attached to glass ("isolated crystals") (Fig. 3) and of $\sim 10\%$ glassy matrix. (2) The "isolated crystals" in the agglutinates are loaded with high track densities $(\rho \sim 10^{11} \cdot cm^{-2})$, and they do not contain crystallites indicative of track annealing (Dran *et al.*, 1972). (3) The glassy matrix shows very complex features. It does in fact contain tiny crystallites (see feature "a" in Fig. 4) that are much smaller than those observed on 100 keV electron micrographs (Housley, 1973). However, these particles do not give the electron diffraction pattern expected from pure iron and upon heating they grow like crystallites nucleated by tracks. In addition, the glass is impregnated with an abundance of tiny crystals that show very angular "broken"-like habits and that are not always well "wet" to the glass (feature "b" in Fig. 4). (4) During pyrolysis experiments the crystallites dispersed in the glassy matrix increase in size and those expected from track annealing appear in the crystals at about 800°C. At 1050°C the glassy agglutinates have completely recrystallized and observations with a petrographic microscope indicate that they are darker than the starting material. Finally, we note that unlike crystal separates extracted from the same size fraction which start forming pellets of sintered dust at a low temperature, the glassy agglutinates do not aggregate before 1100°C.

From these observations we deduce the following conclusions: (1) The "isolated crystals" have been irradiated as lunar dust grains before their incorporation into the agglutinates because they contain track densities that are much higher than those observed in feldspar grains from the same size fractions.

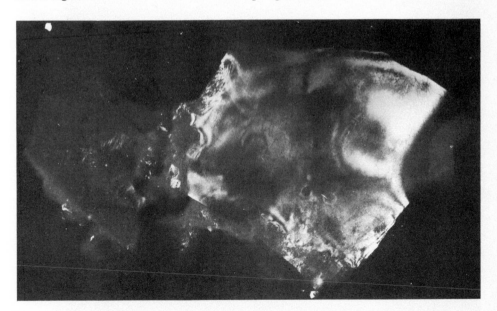

Fig. 3. Darkfield micrograph (HVEM) of a crystal ($\sim 5\ \mu$) isolated from crushed glassy agglutinates from the 37–105 μ fraction of 10086 D fines. This crystal contains a very high density ($\sim 10^{11} \cdot cm^{-2}$) of tracks that appear as lines of dark contrast, but no crystallites indicative of track metamorphism during thermal annealing can be observed.

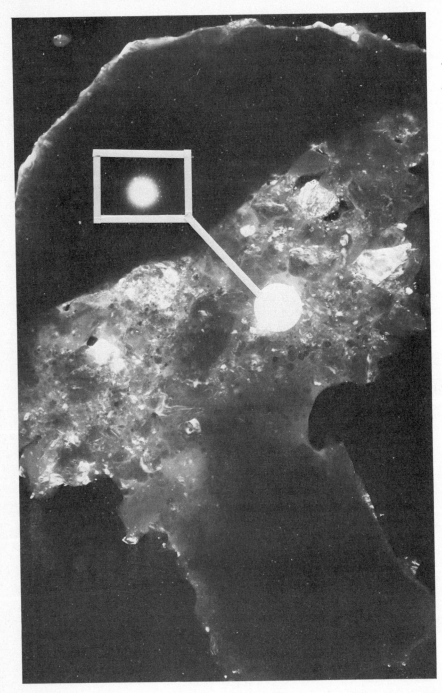

Fig. 4. Darkfield micrograph (HVEM) of the glassy matrix from crushed glassy agglutinates from the 37–105 μ fraction of 10086 D fines. Entire field of view is about 10 μ. The matrix is very complex and contains tiny crystallites (feature "a") and a high proportion of small angular crystals as well as voids and crystals that are not completely "wet" to the glass (feature "b").

(2) The average sizes of these crystals should be quite small (10–20 μ), due to the high probability of finding a micron-sized broken crystal with high ρ values. (3) The process of agglutination does not involve strong metamorphism, since the "pre-agglutinate" radiation damage features are not annealed. We estimate that the crystals have not been exposed to temperatures exceeding 700°C.

The results of our mass spectrometry experiments with the glassy agglutinates complete the HVEM study as follows. Thermal evolution "pyrolysis" experiments give a bimodal release pattern for CO and N_2 (Wszolek et al., 1974a) with the low-temperature peak corresponding to the track annealing temperature (~ 800°C). However, unlike both the solar wind simulation targets (Wszolek et al., 1974b; Bibring et al., this volume) and the crystal-rich fractions from 10086 [e.g. 5 μ-residues, 5–10 μ fraction, <37 μ fraction and feldspar separate (Wszolek et al., 1974a)], the high-temperature peak (~ 1100°C) is much more intense than the low-temperature one (Fig. 5). It is interesting to note that the thermal release profiles for the *bulk* 10086 sample (Gibson and Johnson, 1971) show an almost equal distribution of CO and N_2 between the two temperature regions, with an apparent slight predominance of the low-temperature peak. We do not expect the temperature shift for the glassy agglutinates to be due to a selective loss of most of the CO and N_2 evolved at lower temperatures if the crystals incorporated into the agglutinates have not experienced temperatures >700°C. Thermal volatilization studies by Gibson and Moore (1973) have shown that heating lunar fines even at 750°C under vacuum for 24 hours does not release large amounts of CO, but that 12–30% of the carbon is lost, mostly as CH_4 and CO_2. Nevertheless, a comparison of the total CO contents of the glassy agglutinates (~ 130 ppm, Table 1) versus the crystal-rich 5–10 μ fraction of 10086 (258 ppm, Table 1) and the <37 μ fraction of 10086 (174 ppm; unpublished results, this laboratory) suggests that significant amounts (50% and 25%) of CO could be lost from the fine dust grains incorporated into glassy agglutinates. Unfortunately, we have not separated a 10–20 μ fraction from 10086 which might be more representative of the incorporated fines.

The difference in the pyrolysis behavior of the glassy agglutinates versus crystal-rich fractions could be due to one or more other factors besides the loss of some CO: (1) chemical alteration (Chang et al., 1973; Wszolek et al., 1974a) of carbon during agglutination; (2) addition of meteoritic carbon, e.g. incorporation of tiny lunar wind grains deriving from vapor deposits similar to the Murchison deposit which released CO at 1200°C (see above); (3) a physical change which somehow delays the release of the major part of the carbon in the incorporated fines simultaneous with the release of carbon trapped in the glassy matrix.

Acid dissolution data for the glassy agglutinates show a high concentration of carbide-like material and very high CD_4/CH_4 ratios (Table 1; Cadogan et al., 1973). However, the proportion of carbide-like material, accounted for by CD_4, relative to total carbon or relative to total CO (to neglect probable terrestrial CO_2 contamination during separation) is about the same for the crystal-rich 5–10 μ fraction as for the glassy agglutinates (Table 1). These limited data suggest that there has been no significant synthesis of carbide nor any significant addition of

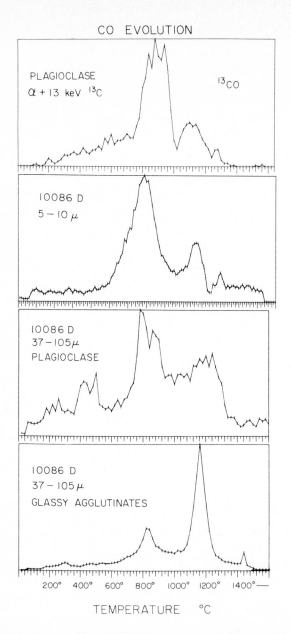

Fig. 5. Thermal release profiles of CO from three crystal-rich materials (plagioclase artificially irradiated with α-particles plus 13 keV ^{13}C; the 5–10 μ fraction of 10086D fines; and plagioclase from the 37–105 μ fraction of 10086D fines). These three patterns are very similar and are different from the one obtained for the glassy agglutinates from the 37–105 μ fraction of 10086D fines (bottom). The intensity scale is different for each profile.

Table 1. Yields of gases released by DF acid dissolution (nanomoles/g).

	5–10 μ	glassy agglutinates 37–105 μ	
		$\rho < 2.73$	$2.73 < \rho < 3.21$
$CD_4 + CHD_3$	2610	1074	1022
CH_4	388	55.6	79.6
^{20}Ne	172	8.4	28.9
^{36}Ar	32.3	10.1	11.2
C_2H_6	73.0	53.8	65.2
C_2D_2	*	*	*
C_2D_4	272	204	114
C_2D_6	80.2	46.2	68.2
C_3 deuterocarbons	45.5	47.2	34.8
C_4 deuterocarbons	*	6.5	*
HCN + DCN	48.5	95.7	134
CD_4/CH_4 molar ratio	6.7	19.4	12.9
Total carbon (ppm)	382	165	166
Total CO (ppm)	258	123	132
Total nitrogen (ppm)	157	69	63

(Table header: Sample 10086 D)

*Not detected.

meteoritic carbide during the agglutination process. The higher CD_4/CH_4 ratio for the agglutinates could then be simply explained as being due to diffusive loss of CH_4 from fine grains when they are incorporated into the agglutinates (Cadogan *et al.*, 1973; Des Marais *et al.*, 1973). Depletion of ^{20}Ne and ^{36}Ar is also observed for the agglutinates relative to the 5–10 μ fraction (Table 1) whereas C_2H_6 appears to be well retained. The amount of HCN + DCN is about five times greater in the glassy agglutinates relative to total carbon or total CO than in the 5–10 μ fraction. These data may indicate the addition of cyanide species or their synthesis during agglutination from pre-existing carbon and nitrogen. In summary, it is clear that more study of the glassy agglutinates is required before conclusions can be drawn regarding the extent of redistribution or alteration of carbon species and possible meteoritic contamination during the agglutination process.

The only other possible "exotic" contributors to lunar carbon chemistry that we have not investigated in the present work are the micro-glass splashes that seem to be found in abundance on the external surface of lunar rocks that contain a high density of micro-craters (R. M. Walker, personal communication; Blandford *et al.*, 1974). However, the following lines of evidence indicate that the disc-shaped glass splashes do not play a dominating role in modifying the initial solar wind implantation yields: (1) On the surface of individual soil grains the abundance of the micro-glass splashes seems to be much smaller than that observed on rock surface (Blandford *et al.*, 1974). (2) By crushing grains with a diameter of about 40 microns (400 mesh fraction of mature soil samples) we have observed with the HVEM that the external surface of such grains was very similar to that of a micron-sized dust grain, in only showing an amorphous coating of solar wind radiation damage and small particles (mostly crystalline) that are firmly welded to the coatings. Therefore we deduce that the external surface of the coarser soil grains is unlikely to be "glass coated." (3) The temperature release profiles of both the 10 μ-residue and the feldspar separates from sample 10086 D look very similar to that of the silicate targets exposed to simulated solar wind (Fig. 5). Furthermore, the study of solar wind rare gases trapped in pyroxene, feldspar, and ilmenite separates from mature soil samples clearly indicates that the rare gas distributions vary markedly from one mineral species to the other. It is unlikely that either the small particles or the glass splashes attached to the grain modify solar wind "chemistry" in the lunar dust grains.

Conclusions

Our simulation experiments, computations, and detailed examination of glassy agglutinates provide the following evaluation of the lunar winds contribution: (1) The directly condensed lunar winds vapor phase is probably strongly depleted in carbon and sulfur compounds and may recrystallize quickly in the lunar thermal cycle and separate from the host crystals. It could add a new generation of tiny grains to the lunar regolith. (2) The ultra-thin coatings of radiation damaged materially produced on grains during ion implantation has peculiar physicochemical properties that prevent the experimental identification of the very low-energy species reimplanted from the lunar atmosphere. (3) The contamination of a mature regolith due to lunar wind I and II, which is of about 10 and 15 ppm, respectively, is much smaller than the contribution expected from the implantation of solar wind carbon (~200 ppm). The computations of these estimates are quite insensitive to both the composition of the lunar winds and to the models used to scale the distribution of the carbon species in the lunar atmosphere. Furthermore, they are only very slightly affected by photodissociation but they strongly depend on the models of lunar wind I as well as the values of the sticking coefficients. Therefore, we do not know if these estimates represent upper limits or not, although we maximized all the other parameters. (4) The extent of carbon losses, meteoritic carbon contamination, and alteration of carbon during the formation of glassy agglutinates is not yet clear. (5) The micro-glass splashes as well as the tiny crystalline grains that are possibly attached to the surface of the coarser grains probably do not affect the characteristics of solar wind carbon chemistry in the lunar soil.

Acknowledgments—One of us (MM) is deeply indebted to R. Klapish and B. Jouffrey for their enthusiastic support and interest and to R. M. Walker, D. S. Burnett, and one of our referees (S.

Chang) for many helpful suggestions and criticisms. This work would have been impossible without the superb technical assistance of C. Jouret during the HVEM observations and of F. C. Walls with high-resolution mass spectrometry data. We thank J. T. O'Connor for help in preparing mineral separates from lunar sample 10086. This work was supported in France by CNRS and in the United States by National Aeronautics and Space Administration grant NGR 05-003-435.

REFERENCES

Berstein W., Fredricks R. W., Vogl J. L., and Fowler W. A. (1963) The lunar atmosphere and the solar wind. *Icarus* **2**, 233–248.

Bibring J. P., Burlingame A. L., Chaumont J., Langevin Y., Maurette M., and Wszolek P. C. (1974) Simulation of lunar carbon chemistry. I. Solar wind contribution. *Proc. Fifth Lunar Sci. Conf., Geochim. Cosmochim. Acta.* This volume.

Blandford G., McKay D., and Morrison D. (1974) Accretionary particles and microcraters (abstract). In *Lunar Science—V*, pp. 67–69. The Lunar Science Institute, Houston.

Borg J., Comstock G. M., Langevin Y., Maurette M., and Thibaut C. (1974) Lunar soil maturation, Part I: Microscopic and macroscopic dynamic processes in the regolith (abstract). In *Lunar Science—V*, pp. 79–81. The Lunar Science Institute, Houston.

Cadogan P. H., Eglinton G., Maxwell J. R., and Pillinger C. T. (1971) Carbon chemistry of the lunar surface. *Nature* **231**, 29–31.

Cadogan P. H., Eglinton G., Firth J. N. M., Maxwell J. R., Mays B. J, and Pillinger C. T. (1972) Survey of carbon compounds II: The carbon chemistry of Apollo 11, 12, 14 and 15 samples. *Proc. Third Lunar Sci. Conf., Geochim. Cosmochim. Acta*, Suppl. 3, Vol. 2, pp. 2069–2090. MIT Press.

Cadogan P. H., Eglinton G., Maxwell J. R., and Pillinger C. T. (1973) Distribution of methane and carbide in Apollo 11 fines. *Nature (Phys. Sci.)* **241**, 81–83.

Carter G. and Colligon J. S. (1968) In *Ion Bombardments of Solids*, pp. 363–367. Heinemann Educational Books.

Chang S., Mack R., Gibson E. K., Jr., and Moore G. W. (1973) Simulated solar wind implantation of carbon and nitrogen ions into terrestrial basalt and lunar fines. *Proc. Fourth Lunar Sci. Conf., Geochim. Cosmochim. Acta*, Suppl. 4, Vol. 2, pp. 1509–1522. Pergamon.

Des Marais D. J., Hayes J. M., and Meinschein W. G. (1973) Accumulation of carbon in lunar soils. *Nature (Phys. Sci.)* **246**, 65–68.

Dran J. C., Duraud J. P., Maurette M., Durrieu L., Jouret C., and Legressus C. (1972) Track metamorphism in extraterrestrial breccias. *Proc. Third Lunar Sci. Conf., Geochim. Cosmochim. Acta*, Suppl. 3, Vol. 3, pp. 2883–2903. MIT Press.

Ducati H., Kalbitzer S., Kiko J., Kirsten T., and Müller H. W. (1974) Rare gas diffusion studies in individual lunar soil particles and in artificially implanted glasses. Preprint.

Gault D. E., Hörz F., and Hartung J. B. (1972) Effects of microcratering on the lunar surface. *Proc. Third Lunar Sci. Conf., Geochim. Cosmochim. Acta*, Suppl. 3, Vol. 3, pp. 2713–2734. MIT Press.

Gibson E. K. and Johnson S. M. (1971) Thermal analysis—inorganic gas release studies of lunar samples. *Proc. Second Lunar Sci. Conf., Geochim. Cosmochim. Acta*, Suppl. 2, Vol. 2, pp. 1351–1366. MIT Press.

Gibson E. K. and Moore G. W. (1973) Carbon and sulfur distributions and abundances in lunar fines. *Proc. Fourth Lunar Sci. Conf., Geochim. Cosmochim. Acta*, Suppl. 4, Vol. 2, pp. 1577–1586. Pergamon.

Hayes J. M. (1972) Extralunar sources for carbon on the moon. *Space Life Sciences* **3**, 474–483.

Heymann D. and Yaniv A. (1970) [40]Ar anomaly in lunar sample from Apollo 11. *Proc. Apollo 11 Lunar Sci. Conf., Geochim. Cosmochim. Acta*, Suppl. 1, Vol. 2, pp. 1261–1267. Pergamon.

Holland P. T., Simoneit B. R., Wszolek P. C., and Burlingame A. L. (1972a) Study of carbon compounds in Apollo 12 and 14 lunar samples. *Space Life Sciences* **3**, 551–561.

Holland P. T., Simoneit B. R., Wszolek P. C., and Burlingame A. L. (1972b) Compounds of carbon and other volatile elements in Apollo 14 and 15 samples. *Proc. Third Lunar Sci. Conf., Geochim. Cosmochim. Acta*, Suppl. 3, Vol. 2, pp. 2131–2147. MIT Press.

Housley R. M., Blander M., Abdel-Gawad M., Grant R. W., and Muir A. H. Jr. (1970) Mössbauer spectroscopy of Apollo 11 samples. *Proc. Apollo 11 Lunar Sci. Conf., Geochim. Cosmochim. Acta,* Suppl. 1, Vol. 3, pp. 2251–2268. Pergamon.

Housley R. M., Grant R. W., and Paton N. E. (1973) Origin and characteristics of excess Fe metal in lunar glass welded aggregates. *Proc. Fourth Lunar Sci. Conf., Geochim. Cosmochim. Acta,* Suppl. 4, Vol. 3, pp. 2737–2749. Pergamon.

Kelly R., Jech C., and Matzke H. J. (1968) Diffusion in inert-gas bombarded KCl, KBr, and KI. *Phys. Stat. Sol.* 25, 641–653.

Lewis W. B., MacEwan J. R., Stevens W. H., and Hart R. G. (1964) Fission-gas behaviour in UO_2 fuel. Paper presented by Canada to the Third U.N. International Conf. on the Peaceful Uses of Atomic Energy, Geneva. Report AECL-2019.

Manka R. H. and Michel F. C. (1971) Lunar atmosphere as a source of lunar surface elements. *Proc. Second Lunar Sci. Conf., Geochim. Cosmochim. Acta,* Suppl. 2, Vol. 2, pp. 1717–1728. MIT Press.

Matzke H. J. and Whitton J. L. (1966) Ion-bombardment-induced radiation damage in some ceramics and ionic crystals. *Can. J. Physics* 44, 995–1010.

McKay D. S., Clanton U. S., Morrison, D. A., and Ladle G. H. (1972a) Vapor phase crystallization in Apollo 14 breccia. *Proc. Third Lunar Sci. Conf., Geochim. Cosmochim. Acta,* Suppl. 3, Vol. 1, pp. 739–752. MIT Press.

McKay D. S., Heiken G. H., Taylor R. M., Clanton U. S., Morrison D. A., and Ladle G. H. (1972b) Apollo 14 soils: Size distribution and particle types. *Proc. Third Lunar Sci. Conf., Geochim. Cosmochim. Acta,* Suppl. 3, Vol. 1, pp. 983–994. MIT Press.

Morgan J. W., Krähenbühl U., Ganapathy R., and Anders E. (1972) Trace elements in Apollo 15 samples: Implications for meteorite influx and volatile depletion on the moon. *Proc. Third Lunar Sci. Conf., Geochim. Cosmochim. Acta,* Suppl. 3, Vol. 2, pp. 1361–1376. MIT Press.

Rank D. M., Townes, C. H., and Welch W. J. (1971) Interstellar molecules and dense clouds. *Science* 174, 1083.

Rehfus D. E. (1972) Lunar winds. *J. Geophys. Res.* 77, 6303.

Simoneit B. R., Christiansen P. C., and Burlingame A. L. (1973) Volatile element chemistry of selected lunar, meteoritic, and terrestrial samples. *Proc. Fourth Lunar Sci. Conf., Geochim. Cosmochim. Acta,* Suppl. 4, Vol. 2, pp. 1635–1650. Pergamon.

Sun H. and Weissler G. L. (1955a) Absorption coefficients of nitric oxide in the vacuum ultraviolet. *J. Chem. Phys.* 23, 1372.

Sun H. and Weissler G. L. (1955b) Absorption cross sections of carbon dioxide and carbon monoxide in the vacuum ultraviolet. *J. Chem. Phys.* 23, 1625.

Wainpan N., Walker W. C., and Weissler G. L. (1955) Photoionization efficiencies and cross sections in O_2, N_2, CO_2, A, H_2O, H_2 and CH_4. *Phys. Rev.* 99, 542.

Wehner G. K., Kenknight C., and Rosenberg D. L. (1963) Sputtering rates under solar-wind bombardment. *Planet. Space Sci.* 11, 885–895.

Wilkinson P. G. and Johnston H. L. (1950) The absorption spectra of methane, carbon dioxide, water vapor, and ethylene in the vacuum ultraviolet. *J. Chem. Phys.* 18, 190.

Wszolek P. C., O'Connor J. T., Walls F. C., and Burlingame A. L. (1974a) Thermal release profiles and the distribution of carbon and nitrogen among minerals and aggregate particles separated from lunar soil (abstract). In *Lunar Science—V,* pp. 857–859. The Lunar Science Institute, Houston.

Wszolek P. C., Walls F. C., and Burlingame A. L. (1974b) Simulation of lunar carbon chemistry in plagioclase (abstract). In *Lunar Science—V,* pp. 854–856. The Lunar Science Institute, Houston.

Proceedings of the Fifth Lunar Conference
(Supplement 5, Geochimica et Cosmochimica Acta)
Vol. 2 pp. 1785–1800 (1974)
Printed in the United States of America

Abundances of C, N, H, He, and S in Apollo 17 soils from Stations 3 and 4: Implications for solar wind exposure ages and regolith evolution

S. Chang and K. Lennon

NASA Ames Research Center, Moffett Field, California 94035

E. K. Gibson, Jr.

NASA Johnson Space Center, Houston, Texas 77058

Abstract—The abundances and isotopic compositions of C and S and the abundances of N, H, He, and metallic Fe have been measured in bulk samples of Apollo 17 soils from Stations 3 and 4. The amounts of solar wind derived elements in these samples indicate lunar surface exposures typical of immature and submature soils. In general, their S contents are consistent with the mixing model proposed by Gibson and Moore (1974). The orange soil, 74220,84, however, appears to be S-depleted despite its basaltic composition. Furthermore, it exhibited unusual heterogeneity both in its S content and isotopic composition. Comparison of the elemental abundance data for C, N, H, and He in these soils shows that the amounts of CO released by pyrolysis from 600–1200°C and the quantity of N_2 released from 150–1200°C are more reliable than total carbon, solar wind H_2 and He, and acid released CH_4 as indices of solar wind exposure. With CO and N_2 as criteria, the following decreasing order of solar wind exposure was established: $73261 > 73221 > 73281 \gg 73241 > 74240 > 74260 > 74220$. Correlations of CO and N_2 abundances with agglutinate contents of these soils are excellent, but no anticorrelation with mean grain size was observed. Regolith evolution at Station 3 is discussed in the context of solar wind exposure times and stratigraphic inversion associated with the Ballet Crater event. Partial grain size sorting or particle density sorting or both may have occurred during deposition of the light mantle at Station 3.

Introduction

ANALYSES OF LUNAR FINES have demonstrated that carbon, nitrogen, hydrogen, and rare gases are located primarily near the surfaces of grains, thus reflecting a preponderant solar wind origin and relatively minor contributions of these elements from parent lunar rocks. Abundances of carbon species and hydrogen have been correlated with the duration of lunar surface exposure experienced by the samples (Cadogan *et al.*, 1972; Des Marais *et al.*, 1973, 1974; Epstein and Taylor, 1970; Holland *et al.*, 1972), and nitrogen abundances have recently been correlated with agglutinate content, an index of surface exposure (Goel *et al.*, 1974), and noble gas content (Müller, 1973). Unlike solar wind rare gases, solar wind carbon and nitrogen are expected to be reactive species and, therefore, retained in soil particles by chemical as well as physical binding forces. Thus, retention and, conversely, loss by diffusion of carbon and nitrogen are not expected to be as sensitive to particle mineralogy and moderate thermal regimes as they appear to be for some of the rare gases (Bauer *et al.*, 1972; Kirsten *et al.*, 1971). Solar wind carbon (Chang *et al.*, 1973) and nitrogen (Müller, 1973) appear to exist in particles primarily in chemically bound form and are released by vacuum pyrolysis primarily as CO and N_2 above 500°C (Chang *et al.*, 1973; Wszolek *et al.*, 1974). The state of trapped solar wind hydrogen is not clear; it is released in molecular form by vacuum pyrolysis at relatively low temperatures (Gibson and Johnson, 1971; Epstein and Taylor, 1970) and, if it is trapped in that form, it must be subject to severe loss by diffusion. Taking these factors into consideration, we undertook a comparative study of carbon, nitrogen, hydrogen, and helium abundances in a number of bulk Apollo 17 soil samples as measures of relative solar wind exposure. Solar wind exposure ages together with agglutinate content, particle track, and noble gas spallation ages, and other indices of exposure applicable to various depths in the regolith serve to elucidate the record of the timing and duration of surface events and processes that is preserved in regolith stratigraphy.

In addition to C, N, H, and He, data on S and metallic iron contents are obtained in our analyses. The distribution and abundance of sulfur in soils at the Apollo 17 site can be accounted for by a simple mixing model involving sulfur-rich mare-type basalts and various sulfur-poor massif materials (Gibson and Moore, 1974). For that reason, S abundances may be important indicators of lateral particle transport and volatile element mobilization on the lunar surface. Metallic iron is comprised of lunar and meteoritic components, with an additional contribution resulting, presumably, from reduction of ferrous minerals accompanying solar wind hydrogen and micrometeorite bombardment. The relevance of metallic iron measurements to processes of regolith evolution have been discussed by Housley *et al.* (1974), Kerridge *et al.* (1974), and Pillinger *et al.* (1973).

Methods and Samples

Analytical procedures are similar to those reported previously (Chang *et al.*, 1971; Kaplan *et al.*, 1970). However, recent modifications permit extraction and measurement from the same sample of the abundances and isotopic compositions of a variety of carbon-, nitrogen-, and sulfur-containing gases along with the abundances of helium and hydrogen. Details of the procedure will be described elsewhere (Sakai *et al.*, 1974). In brief, by means of stepwise vacuum pyrolysis to 1200°C followed by combustion at that temperature, C, N, H, and He are quantitatively extracted as gases—carbon as CH_4, CO, and CO_2; nitrogen as N_2; and hydrogen as H_2—and measured manometrically. Acid hydrolysis in $6N$ H_2SO_4 at 120°C for at least 150 hours frees troilitic sulfur as H_2S and carbide-like carbon as hydrocarbons, of which we measure only CH_4. As in pyrolysis experiments, gases of interest are measured manometerically. Indigenous CH_4 is also freed by acid and is analyzed in combination with CH_4 derived from carbide-like carbon. Helium abundances obtained by both acid treatment and vacuum pyrolysis are in excellent agreement. H_2 released by acid treatment includes both solar wind H_2 and H_2 evolved as a result of metallic iron dissolution. The amount of solar wind H_2 released by pyrolysis of a sample at 600°C is subtracted from the H_2 released by acid in order to assess the amount of H_2 arising only from metallic iron dissolution. Gas release data of Gibson and Johnson (1971) indicate that solar wind H_2 is essentially completely released from lunar fines by vacuum pyrolysis at 600°C. For isotopic measurements, CH_4 and CO are converted to CO_2, and H_2S is converted to SO_2. Most $\delta^{13}C_{PDB}$ and $\sigma^{34}S_{CD}$ values were determined by conventional means and supplied by Dr. I. R. Kaplan. Measurements of carbon isotopic compositions in samples containing 1 to 4 μg C as CO_2 were supplied by Dr. J. M. Hayes by means of a computer controlled ion-counting mass spectrometer (Schoeller and Hayes, 1974). Prior to all analyses, lunar samples are outgassed at 140–160°C under vacuum (less than 10^{-4} mm Hg) for at least three hours to remove adsorbed terrestrial N_2 and CO_2. Some indigenous volatile gases may be lost under these outgassing conditions. Holmes *et al.* (1974) report that N_2 and CO_2 are not chemisorbed by lunar fines, but rather are reversibly physically adsorbed.

Analyses have been performed on two suites of Apollo 17 soils. One suite of samples is a trench sequence sampled from just outside the rim of Ballet Crater at Station 3. These include: 73221,7, a skim sample of a thin 0.5 cm layer of medium gray soil; 73241,5, a light gray soil from the upper 5 cm of the trench which overlies a marbled zone; and 73261,8 and 73281,6, which represent the medium gray and light gray portions of the marbled zone, respectively. The second suite of samples includes 74220,84, the orange soil, and its companion gray soils 74240,16 and 74260,7.

RESULTS AND DISCUSSION

1. *Elemental abundances*

Elemental abundance data obtained by vacuum pyrolysis-combustion and acid hydrolysis of bulk soil samples are presented in Tables 1 and 2, respectively. The distribution of carbon between CH_4, CO, and CO_2 (Table 1) is typical of results obtained by pyrolysis of Apollo 15 (Chang *et al.*, 1974) and Apollo 16 (Kerridge *et al.*, 1974) soils. Up to 600°C CO_2 is the major carbon species, while at 1200°C CO predominates. Except for N the very small amounts of all the elements released from pyrolysis residues by combustion at 1200°C indicate their virtually complete extraction by pyrolysis at 1200°C. Typically, pyrolyses at 1200°C were carried out for 2–3 hours. The range of elemental abundances is characteristic of immature and submature soils. The observation of very low contents of light element volatiles in 74220,84 parallels some of the results obtained by others (see Gibson and Moore (1973) for a review of orange soil data including variations in light element content) and is consistent with extremely low solar wind exposure. The total carbon contents for 74240 and 74260 are in excellent agreement with values of 55 and 45 ppm C, respectively, reported by Moore *et al.* (1974). Large inexplicable discrepancies exist between the values of 155 and 170 ppm C reported by the same authors during PET for 73221 and 73261, respectively, and the corresponding carbon abundances of 81 and 88 ppm in Table 1. However, our total carbon values when compared with the other elemental data in Tables 1 and 2 present an internally consistent picture (see below). The nitrogen content of 18.3 ppm N for 74240,16 agrees well with the 23 ppm reported by Müller (1974) for a submillimeter sample of 74241, especially since the exclusion of 1 mm particles from his sample represents an effective enrichment in finer particles and, therefore, in total solar wind derived N.

Amounts of CH_4 released by acid treatment (Table 2) also reflect the immaturity of these soils. The origin of small amounts of CO released by acid is not clear; terrestrial contamination is not likely. Reimplanted atmospheric carbon and/or solar wind origins are possible, and indigenous CO trapped in vesicles cannot be ruled out. Relatively large amounts of CO_2 are also evolved during acid treatment. Significantly, the amounts of CO_2 released from the orange soil by both low-temperature pyrolysis and acid treatment are consistent and constitute the

Table 1. Abundances of C, N, He, and H released by vacuum pyrolysis-combustion of soils from Stations 3 and 4.[a]

			Gases released						
			CH_4	CO	CO_2	Total C	N_2	He	H_2
Sample	(wt. g)	T(°C)	ppm C	ppm C	ppm C	ppm C	ppm	ppm	ppm
73221,7	(0.384)	600	—[b]	3	16		—	—	20
		1200	1	53	7	81	44	8	—
		1200 + O_2	—	—	1		0	0	—
73241,5	(1.040)	600	—	1.7	9.8		—	—	5.2
		1200	0.2	26	3.2	42	21.9	3.8	—
		1200 + O_2	—	—	0.7		1.5	0	—
73261,8	(0.995)	600	—	1.6	11.3		—	—	22
		1200	0.6	61.4	11.6	88	51.2	9.2	—
		1200 + O_2	—	—	1.0		3.4	0	—
73281,6	(0.969)	600	—	1.8	11.9		—	—	14.8
		1200	0.4	44.3	21.4	81	39.8	NM[c]	—
		1200 + O_2	—	—	0.7		0	NM	—
74220,84	(1.265)	600	—	0.4	6.5		—	—	0.6
		1200	0	2.4	0.9	11	7.3	2	—
		1200 + O_2	—	—	1.2		0	0.3	—
74240,16	(1.488)	600	—	3.5	16		—	—	5.5
		1200	0.6	21.6	6	50	16.4	19.4	—
		1200 + O_2	—	—	1.9		1.9	0	—
74260,7	(1.266)	600	—	2.1	11.3		—	—	7.8
		1200	0.4	19.6	7.4	42	13.1	24	—
		1200 + O_2	—	—	1.5		5.0	0	—

[a]Detection limits are as follows: CH_4, CO, CO_2, 0.2 μgC; N_2, 0.4 μg; He, 0.06 μg; H_2, 0.03 μg. System blanks for gases are: CH_4, <0.5 μgC; CO, <0.2 μgC; CO_2, <0.2 μgC; CO_2 from direct combustion, <0.7 μgC; N_2, <0.4 μg; He, <0.06 μg; H_2, <0.06 μg.

[b]Signifies that measurements were not performed at the indicated temperature. For CH_4, He, and N_2, gases evolved at 600°C were stored until after the 1200°C pyrolysis for combined measurements. For solar wind H_2, gas evolution is essentially complete at 600°C.

[c]NM indicates no measurement due to loss of gas sample.

bulk of the total carbon. Except for 73221,7 and 73241,5, all the CO_2 released by acid can be accounted for during pyrolysis. The two exceptions may possibly reflect several different sources of CO_2, all of which may be heterogeneously distributed among soils. (We will return to this point in Section 2 below.)

Sulfur abundances for the Station 3 soils cover a range (300 to 677 ppm S) previously observed for highland soils (Gibson and Moore, 1974; Kerridge *et al.*, 1974; Rees and Thode, 1974), whereas the higher abundances of the gray Station 4 soils are characteristic of soils containing relatively high amounts of mare basalt components (Gibson and Moore, 1974). Analyses of two additional aliquots of 74220,84 gave widely differing results: 212 and 441 ppm. Sulfur determination by combustion of a fourth aliquot released 182 ppm. In this last case, there is reason to believe that the yield was not quantitative. By comparison, an aliquot of

Table 2. Abundances of C, S, H, and He released by acid hydrolysis of soils from Stations 3 and 4.[a]

Sample	(wt. g)	CH_4[b] ppm C	CO ppm C	CO_2 ppm C	H_2S ppm S	He ppm	H_2[c] ppm	Fe^{od} wt.%
		Gases released						
73221,7	(0.111)	7	6	28	310	3	211	0.53
73241,5	(0.958)	3.5	1.2	15.8	606	3.7	124	0.33
73261,8	(0.951)	7.1	2.0	16.2	677	7.0	184	0.45
73281,6	(0.992)	3.4	5.8	10.8	578	6.0	191	0.49
74220,84	(1.142)	0	0	5.8	564	1.5	13.3	0.035
74240,16	(0.957)	1.3	1.9	15.8	1199	22	58	0.15
74260,7	(0.931)	1.3	1.8	16.5	1015	21	41	0.18

[a]Detection limits are as follows: CH_4, CO, CO_2, He, H_2, see footnote a, Table 1; H_2S, 0.5 μgS. Analytical blanks for CH_4, CO, CO_2, He, H_2, see footnote a, Table 1; H_2S, <0.5 μgS.

[b]CH_4 represents indigenous methane plus methane derived from hydrolysis of carbide-like carbon.

[c]H_2 represents sum of solar wind hydrogen and hydrogen released by acid dissolution of metallic iron.

[d]Metallic iron is calculated from H_2 after correction for solar wind hydrogen from Table 1.

74220,84 examined by Gibson and Moore (1974) yielded 750 ppm, while replicate samples of 74220,5 afforded 550 ± 20 ppm. Either sulfur is very heterogeneously distributed in the orange soil, or substantial amounts of sulfur occur in forms other than troilite, or a combination of both. The orange soil 74220,84 yields a unique trimodal gas release profile for SO_2 during vacuum pyrolysis which consists of significant SO_2 evolution at 150–350°C, 500–800°C, and 950–1150°C (Gibson and Moore, 1973). Typical lunar soils release SO_2 only above 900°C through reaction of troilite with other mineral phases. In any case the depletion of S in the orange soil despite its basaltic composition is indicative of an unusual high-temperature history. The sulfur abundance for 74260,7 (1015 ppm) agrees well with that reported by Gibson *et al.* (1974) for 74260,5 (1080 ± 20 ppm).

2. *Carbon and sulfur isotopic composition*

Results of isotopic measurements on gases released by vacuum pyrolysis and acid hydrolysis are presented in Tables 3 and 4, respectively. The soils are listed in order of decreasing solar wind exposure as determined in Section 3 below. Carbon released at 600°C consists primarily of CO_2 and is uniformly isotopically light by comparison with that released largely as CO between 600°C and 1200°C. The isotopic difference reflects different origins for the carbon released in the two temperature ranges. Solar wind carbon accounts for the bulk of the carbon at the high temperature range. Because indigenous lunar carbon in rocks is isotopically

Table 3. Isotopic composition of carbon released by vacuum pyrolysis of soils from Stations 3 and 4.

Sample	$\delta^{13}C_{PDB}$ (per mil)[a]	
	150–600°C[d]	600–1200°C[e]
73261,8	NM	+17
73221,7	−2	+19
73281,6	−4[b]	+19[c]
73241,5	−7	+16
74240,16	−18	+5
74260,7	−7	+18
74220,84	NM	+10[c]

[a]Weighted average of CO and CO_2 contributions unless otherwise indicated. ±1 per mil. CO was converted to CO_2 for isotopic measurements.
[b]Isotopic composition of CO_2 component only.
[c]Isotopic composition of CO component only.
[d]CO_2/CO = 5 to 7, except for 74220,84 for which ratio is 15.
[e]CO/CO_2 = 3 to 8; Station 4 soils at low end, Station 3 soils at high end.

Table 4. Abundances and isotopic compositions of carbon in CH_4 and CO_2 and S in H_2S released by acid hydrolysis of soils from Stations 3 and 4.[a]

	CH_4		CO_2		H_2S	
	ppm C	$\delta^{12}C_{PDB}$	ppm C	$^{13}C_{PDB}$	ppm S	$\delta^{34}S_{CD}$
73261,6	7.1	—	16	+6	677	+5.0
73221,7	7	—	28	−1	310	+2.5
73281,6	3.4	+23	11	+4	578	+3.3
73241,5	3.5	+16	16	−3	606	+3.7
74240,16	1.3	—	16	+7	1199	+2.7
74260,7	1.3	—	17	+8	1015	+4.2
74220,84	0	—	6	−15	564	−3.1
74220,84	—	—	—	—	442	−1.1
74220,84	—	—	—	—	212	+3.3

[a]CH_4 was converted to CO_2 for isotopic measurements; ±1 per mil for $\delta^{13}C_{PDB}$. H_2S was converted to SO_2 for isotopic measurements; ±0.1 per mil for $\delta^{34}S_{CD}$.

light, about −25 per mil, the high enrichment of ^{13}C in the carbon released at high temperatures indicates that only a minor amount of rock carbon is retained in these soils. That is not to say, however, that solar wind carbon is intrinsically isotopically heavy, since potential isotopic fractionation processes associated with solar wind hydrogen-stripping, sputtering, and fractional volatilization by micrometeorite impacts may alter the isotopic composition of initially implanted solar wind carbon (see Chang *et al.*, 1974; Kerridge *et al.*, 1974a, and references therein). The isotopic composition of the small amount of carbon released from the orange soil from 600–1200°C is consistent with a solar wind carbon component originating from a 5–10% contamination by the closely associated gray soils. The origin of the low-temperature carbon is not clear; possibilities include trace amounts of terrestrial contamination, CO and CO_2 adsorbed on particle surfaces from transient atmospheres associated with impact events (Gibson and Moore, 1973a) or CO_2 from decomposition of carbonate-like phases derived from parent rocks (Gibson and Chang, 1974). Data in Table 3 show no apparent correlation of isotopic content with solar wind exposure, although there is reason to believe that the ^{13}C enrichment of soil carbon relative to rock carbon is a result of isotopic fractionation by processes acting at the lunar surface (see above).

The high enrichment of ^{13}C in carbon released as CH_4 by acid hydrolysis (Table 4) is consistent with its solar wind origin and is similar to earlier results for Apollo 12 (Chang *et al.*, 1971), Apollo 15 (Chang *et al.*, 1974), and Apollo 16 soils (Kerridge *et al.*, 1974).

Abundances and isotopic compositions of CO_2 released by acid (Table 4) are variable and show no clear correlation with solar wind exposure. On the other hand, the general enrichment in ^{13}C (except for 74220,84) relative to rock carbon and carbon released by pyrolysis at 600°C argue that some solar wind derived carbon is released as CO_2 by acid treatment. In support of this view, Wszolek *et al.* reported at the Fifth Lunar Science Conference that a small fraction of carbon injected into plagioclase targets under simulated solar wind conditions is released as CO_2 by acid. In addition, adsorbed gases (of lunar, terrestrial, or meteoritic origin) may be desorbed, carbonates may decompose to CO_2, and indigenous CO_2 trapped in vesicles may be freed by acid breakdown of the mineral matrix. Therefore, the CO_2 freed by acid is likely to be a complex mixture derived from as many as four different sources, each possibly having its own characteristic isotopic composition. For soils in general, neither vacuum pyrolysis nor acid hydrolysis can be considered specific means for distinguishing any single source of the CO_2 from any other. On the other hand, the orange soil may be an exception to this generalization. Its high degree of compositional homogeneity, its putative origin in a high temperature event, and its subsequent shielding from solar wind and micrometeorite bombardment could have effectively reduced the sources of CO_2 to gas trapped within the grains and/or adsorbed on the grains during its formation. If this is the case, then the value of −15 per mil may be close to the isotopic composition of CO_2 in the gas phase associated with the formation of the orange soil. It should be noted that carbon in orange soil, like sulfur, appears to be heterogeneously distributed (Gibson and Moore, 1973). The processes which led to such distributions may be the same, and the CO_2 and the hypothesized nontroilitic sulfur component may have the same origin. Additional measurements are needed to assess the isotopic homogeneity of CO_2 released both by acid and by low-temperature pyrolysis of orange soil.

The isotopic composition of sulfur released as H_2S by acid from the soils (with the exception of 74220,84) spans a range from 2.5 to 5.0 per mil, well within the values obtained from soils from other landing sites (see Chang *et al.*, 1974; Kaplan, 1972; Kerridge *et al.*, 1974; Rees and Thode, 1974; and references therein). Although there is reason to believe that enrichment of [34]S in lunar fines is correlatable with surface exposure (Kerridge *et al.*, 1974; Rees and Thode, 1974), there is no discernible correlation between either sulfur abundance and isotopic composition or sulfur isotopic composition and surface exposure for the soils in Table 4.

The isotopic composition of sulfur in three aliquots of 74220,84 shows unusual variations from −3.1 to 3.3 per mil with [34]S enrichment increasing with decreasing sulfur abundance. In a fourth sample examined by pyrolysis-combustion, 182 ppm S was found with an isotopic composition of 4.7 per mil. A weighted average of the first three samples yields a value of −1.3 per mil. Such depletions in [34]S as these have not been observed in any other soil examined thus far. Values of −2.2 per mil and −0.2 per mil, however, have been measured in breccia 67015 (Kerridge *et al.*, 1974) and basalt 12022 (Kaplan and Petrowski, 1971). The reason for the wide isotopic variation between aliquots of the same sample is not clear. These results may reflect heterogeneous distribution of troilite phases having different isotopic compositions. Incomplete hydrolysis of troilite in some samples is not consistent with the inverse relationship between sulfur abundance and isotopic composition. Furthermore, prolonged rehydrolysis of a previously treated sample yielded no additional sulfur. Further study of the isotopic composition of carbon, sulfur, and other volatile elements in the orange soil is clearly desirable.

3. *Solar wind exposure ages*

For purposes of establishing the order of bulk solar wind exposure experienced by the Station 3 and 4 soils, we chose from among our data the abundances of CO released by pyrolysis from 600–1200°C and N_2 released over the entire pyrolysis range as the best indices. The reasons for this choice are as follows. Lunar soil irradiated with ^{13}C ions under simulated solar wind conditions released greater than 90% of the implanted carbon as [13]CO in the 600–1150°C range. Furthermore, there is good reason to believe that reimplanted atmospheric carbon (Hayes, 1972) of lunar or extralunar origin is released as CO by pyrolysis over the same temperature range (Chang *et al.*, 1974a). In simulation studies with ^{15}N ions, the implanted nitrogen is released as [15]N_2 from 500–1150°C.

In Fig. 1a and b, elemental abundance data are plotted for each soil according to solar wind exposure, the order decreasing from left to right as determined by the CO and N_2 criteria. The He data in Fig. 1b are taken from Table 2.

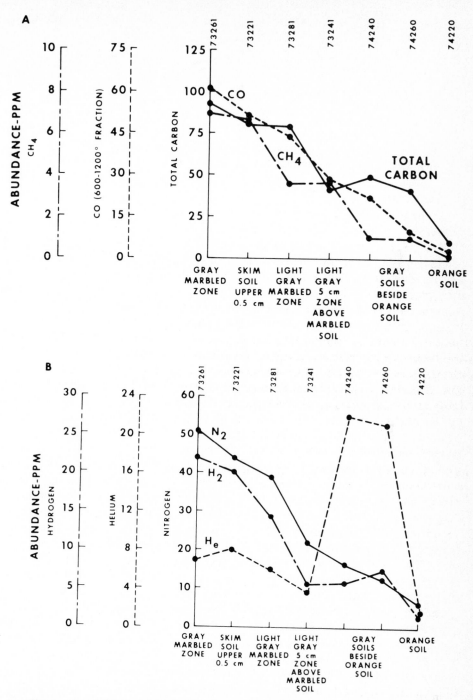

Fig. 1. (a) and (b) Variations in abundances of solar wind derived elements among Apollo 17 soils from Stations 3 and 4.

C

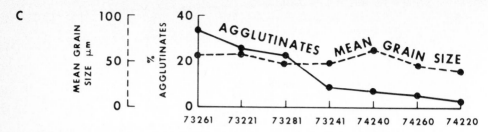

Fig. 1. (c) Variations in agglutinate contents and mean grain sizes among Apollo 17 soils from Stations 3 and 4. Data supplied by personal communications from G. Heiken and D. S. McKay.

At the time of this report, the only other available information relevant to the surface exposure of all these soils consisted of the agglutinate data of Heiken and McKay (1974) and the grain size analyses of McKay *et al.* (1974). For purposes of comparison the mean grain size and agglutinate contents of the soils are shown in Fig. 1c. (Data from Heiken and McKay, personal communication.) All grain size data were obtained with submillimeter fines, and agglutinate data were derived from the 90–150 micron size fraction. Figure 1c shows that an expected anticorrelation of grain size with agglutinate content does not hold for these soils. Even if the mean grain size from analysis of subcentimeter samples of 74240 and 74260 were included (107 and 157 microns, respectively, personal communication from Dr. D. S. McKay), a clear anticorrelation is lacking. This is especially noticeable for 73241, 73261, and 74220. The case of the orange soil is understandable on the basis of its particle size homogeneity and possible origin in lava fountaining (McKay and Heiken, 1973), combined with its lack of surface exposure. On the other hand, the agglutinate contents of 73261 and 73241 show them to be the most mature and the least mature soils, respectively, in the Station 3 suite; yet 73241 is finer grained (49.7 microns) than 73261 (54.5 microns). McKay (personal communication) has suggested that the presence of larger amounts of fine-grained orange and black glass in 73241 than in 73261 may account for the grain size discrepancy. Although size parameters permit division of soils into fields representing immature, submature, and mature groupings, there are wide variations in the size parameters within each field (McKay *et al.*, 1974). Moreover, the presence of significant populations of rather homogeneously sized components such as orange glass may inordinately bias the grain size analysis. For these reasons judgments of relative soil maturity between individual samples on the basis of grain size parameters may not always be valid. Indeed, it seems likely that at the present time no single criterion for surface exposure of soils is altogether unambiguous and convincing. However, several criteria in combination may provide a valid basis for quantitative assessments of degrees of surface exposure.

Data depicted in Fig. 1 show strikingly close parallels between the abundances of CO, N_2, and agglutinate particles, especially between the last two. If, as generally believed, the agglutinate content of a soil is a valid index of surface exposure in the uppermost grain layers of the regolith, then the parallels provide compelling evidence that CO and N_2 contents are also indices of comparable validity. A less satisfactory correlation with surface exposure is given by the amount of CH_4 released by acid. Quantitative measurements of small amounts of CH_4 by our methods is difficult, and other methods (Cadogan et al., 1972) may reveal a better correlation with agglutinate content or other exposure criteria. However, the CH_4 abundance comprises only a small fraction of the solar wind carbon input for a soil and may not be sufficiently representative. H_2 and He abundances show their largest departures from the CO, N_2, and agglutinate trend in going from the Station 3 to the Station 4 soils. The dramatic increase in He undoubtedly reflects substantially higher abundances of mare-derived soil components with gas-retentive titanium-rich phases in the Station 4 soils (Rhodes et al., 1974). The departure of H_2 abundances from the trend in Station 4 soils may have a similar explanation, but it is clear that the retention of H_2 is far less sensitive to differences in soil mineralogy than is that of He. The departure of the total carbon contents from the agglutinate trend reflects the fact that the amounts of CO_2 released by pyrolysis from the Station 4 soils and from 73281,6 represent substantially larger fractions of the total carbon than do those from the other Station 3 soils, which in turn indicates that carbon derived from nonsolar wind sources is relatively more plentiful in the former group of soils. Apparently from the limited data available, He, H_2, and total carbon contents are sensitive to variations in soil mineralogy while CO, N_2, and agglutinate contents are relatively insensitive. Des Marais et al. (1974) have found that the decreasing order of overall retention of solar wind implanted volatile elements in mature soils is $N > C \gg {}^{84}K > {}^{36}Ar > {}^{20}Ne \gg {}^{4}He > H$. It should be pointed out that the overall order of retention and the order of sensitivity to mineralogy need not be the same, as is shown by our results for He and H_2.

The chemical alteration and redistribution of solar wind carbon (and nitrogen, presumably) in soil particles during exposure on the lunar surface are complex processes (Cadogan et al., 1973; Chang et al., 1973; Des Marais et al., 1973; Wszolek et al., 1974). Nevertheless, the correlations obtained here based simply on determining abundances of CO and N_2, which are most representative of solar wind carbon and nitrogen inputs, indicate that these kinds of measurements can provide valid indices of surface exposure. Clearly, in order to place the correlations on as secure a footing as possible, similar analyses of more soils is desirable.

Intense churning of a 0.5 to 1 mm lunar surface layer by micrometeorite bombardment should provide the primary means for exposure of fresh grain surfaces to solar wind irradiation (Gault et al., 1974). Therefore, although the lunar surface layer in which agglutinate formation occurs is expected to extend to greater depth than the penetration range of solar wind particles, the time scales for accumulation of solar wind species and formation of agglutinates should be similar or closely related. On this basis the time required for a soil to attain its equilibrium concentrations of solar wind carbon and nitrogen and agglutinate particles may also be similar; these equilibration times set limits on the time spans over which these measurements may serve as useful indices of surface exposure (see McKay and Heiken, 1973a, for example).

4. Regolith evolution at Station 3

The origin of the light mantle has been tentatively attributed to deposition through a single dynamic event such as an avalanche or fluidized flow of South Massif talus (Schmitt, 1973). At all localities observed by Schmitt on the light mantle, a 5 to 10 cm thick layer of medium gray soil overlies a layer of light gray soil which extends to a meter or more in depth. Possibly, the medium gray surface layer resulted from progressive darkening of the initially light gray deposit by agglutinate formation, which appears to be an effective age-darkening mechanism on the lunar surface (Adams and McCord, 1973). This possibility is supported by the observation that petrographic differences between the light gray and medium gray soils lie mainly in their agglutinate content (Heiken and McKay, 1974), their component compositions being otherwise very similar.

On the basis of his observations on the moon, Schmitt (1973) suggested that the Ballet Crater event interrupted the normal development of the medium gray regolith and produced a distension and inversion of pre-existing soil strata. Unlike elsewhere on the light mantle, the medium gray surface soil at Station 3 is only 0.5 cm thick and presumably represents new regolith developed since the formation of Ballet Crater. The resulting strata sampled are: the 0.5 cm thick layer of medium gray surface soil, 73221,7; a 3 cm thick layer of subsurface light gray material, 73241,5; at least 15 cm of medium gray soil, 73261,8, into which light gray material, 73281,6 is irregularly marbled. If this trench sequence is a result of stratigraphic inversion, then the two light gray soils should be related to each other by common origin. Moreover, the processes responsible for development of both medium gray layers should have been the same since the development of the upper 0.5 cm amounted to resumption of processes which were momentarily interrupted by the cratering event.

From the CO, N_2, and agglutinate data in Fig. 1, it is clear that the solar wind exposure of the light gray soil in the marbled zone, 73281,6, is twice that of the subsurface light gray soil, 73241,5. Also, the medium gray soil in the marbled region, 73261,8 has undergone slightly longer solar wind exposure (about 15%) than the medium gray surface soil, 73221,7. If the two light gray soils are related, several explanations for the difference in solar wind exposure are possible which are consistent with deposition of the light mantle in a single dynamic event. (1) Deposition was accompanied by vertical size sorting of pre-irradiated particles which resulted in enrichment of the lower region of the deposit in coarse particles and the upper in finer particles, thus producing apparently higher concentrations of solar wind species in the latter. Evidence of vertical size sorting of rocks in the light mantle has been reported by Schmitt (1973). (2) Deposition was accompanied by density sorting of pre-irradiated particles which enriched the lower region of the deposit in more dense particles and the upper region in less dense particles. Analysis of carbon species in lunar soils separated according to density (Cadogan et al., 1973) revealed strong enrichment (about a factor of 10) of solar wind derived carbon in the least dense particles. These particles are glassy agglutinates which have been shown to contain the highest total concentrations of solar wind carbon among various particle types (Des Marais et al., 1973); similar enrichment of solar wind nitrogen in the agglutinate is expected (Goel *et al.*, 1974). (3) Deposition was accompanied by a combination of (1) and (2). Presumably, deposition of the light gray mantle deposit was followed by development of a medium gray surface layer from what was originally light gray material through age-darkening by means of agglutinate formation. In this context it should be noted that low density agglutinates occur widely distributed over particle sizes. Of the two sorting processes, the density sorting is supported by the relative agglutinate contents of 73241 (8.4%) and 73281 (24.6%), while size sorting is not obvious in the mean grain sizes of the two soils, 49.7 microns and 48.0 microns, respectively. In any case the impact event, first of all, would have caused inversion of the strata, placing the layers of the light gray soil least enriched in solar wind species at the very surface and, second, marbling, the latter involving mainly the light gray soil layers most enriched in solar wind species which originally interfaced with the medium gray soil. Development of the 0.5 cm surface gray layer then proceeded.

The preliminary particle track data of Crozaz *et al.* (1974) are consistent with the exposure sequence of Station 3 soils given in Fig. 1. However, they provide no additional evidence for the proposed stratigraphic inversion. Their data are also consistent with sequential layering of strata at different times without inversion, but if such were the case, it is not easy to understand the marbling of the soil strata and the very thin medium gray surface layer in the vicinity of Ballet Crater as compared to thicker deposits elsewhere on the light mantle.

The small difference in solar wind exposure of samples 73221,7 and 73261,8 suggests that age-darkening of the light mantle has been proceeding at a rate which differs little from that which was operative prior to the cratering event. If the event can be dated, detailed study of the upper 0.5 cm soil layer may provide a basis for determining the rates of some of the processes operating on and in the regolith at this locality.

One way to arrive at a crude estimate of the date is to make use of the agglutinate data. At the Fifth Lunar Science Conference, McKay estimated that agglutinate formation at the Apollo 17 site occurred at a rate of about 2.5% per million years. If 73221 is simply reworked material that was originally the uppermost layer of 73241, the inverted light gray material, then the agglutinate content formed since the time of the cratering event can be roughly estimated as 18% by subtracting the agglutinate content of 73241 (8.4%) from that of 73221 (26%). About seven million years would be required to produce the net 18% agglutinate content. An estimate of this type must be viewed with appropriate caution, however (McKay and Heiken, 1973a). Some particle track data of Crozaz *et al.* (1974) may also be relevant in this context. First of all, the estimated age since deposition of the first 5 cm of soil at Station 3 is 5–20 m.y. In addition, although it is hazardous to associate rock 73275 with the Ballet Crater event, it is interesting that its track exposure age is 4.7 ± 1.0 m.y. Obviously, the similarity in these ages may well be fortuitous. Firm conclusions may be drawn about the timing of events at Station 3 when further analyses of rocks, soils, and nearby drive tube samples are completed.

One final aspect of the Station 3 soils concerns their sulfur abundances. Whereas the subsurface soils contain from 580 to 680 ppm S, the surface soil, 73221,7, is strikingly depleted in S at the 310 ppm level (Table 4). If indeed the Ballet Crater event was responsible for the existing stratigraphy at Station 3, and if the surface soil layer has developed since that event, the sulfur data suggest that either the parent rocks of the light mantle soils are sulfur-poor and recent lateral mixing of light mantle soils with more sulfur-rich mare components has been less extensive than before the cratering event, or the parent rocks contain moderate amounts of sulfur and recent lateral mixing has primarily involved even more severely sulfur-depleted anorthositic highland material. The latter view is supported by low abundances (1–7%) of mare derived lithic and mineral fragments in the massif and light mantle soils (Heiken and McKay, 1974). An unknown factor may be responsible for the sulfur data, however. Detailed study of volatile elements such as lead and the halogens in the Station 3 soils would be instructive.

REFERENCES

Adams J. B. and McCord T. B. (1973) Vitrification darkening in the lunar highlands and identification of Descartes material at the Apollo 16 site. *Proc. Fourth Lunar Sci. Conf., Geochim. Cosmochim. Acta*, Suppl. 4, Vol. 1, pp. 163–177. Pergamon.

Baur H., Frick U., Funk H., Schultz L., and Signer P. (1972) Thermal release of helium, neon, and argon from lunar fines and minerals. *Proc. Third Lunar Sci. Conf., Geochim. Cosmochim. Acta*, Suppl. 3, Vol. 2, pp. 1947–1966. MIT Press.

Cadogan P. H., Eglinton G., Firth J. N. M., Maxwell J. R., Mays B. J., and Pillinger C. T. (1972) Survey of lunar carbon compounds, II: The carbon chemistry of Apollo 11, 12, 14, and 15 samples. *Proc. Third Lunar Sci. Conf., Geochim. Cosmochim. Acta*, Suppl. 3, Vol. 2, pp. 2069–2090. MIT Press.

Cadogan P. H., Eglinton G., Gowar A. P., Jull A. J. T., Maxwell J. R., and Pillinger C. T. (1973) Location of methane and carbide in Apollo 11 and 16 lunar fines. *Proc. Fourth Lunar Sci. Conf., Geochim. Cosmochim. Acta*, Suppl. 4, Vol. 2, pp. 1493–1508. Pergamon.

Chang S., Kvenvolden K., Lawless J., Ponnamperuma C., and Kaplan I. R. (1971) Carbon, carbides, and methane in an Apollo 12 sample. *Science* 171, 474–477.

Chang S., Mack R., Gibson E. K., Jr., and Moore G. W. (1973) Simulated solar wind implantation of carbon and nitrogen ions into terrestrial basalt and lunar fines. *Proc. Fourth Lunar Sci. Conf., Geochim. Cosmochim. Acta*, Suppl. 4, Vol. 2, pp. 1509–1522. Pergamon.

Chang S., Lawless J., Romiez M., Kaplan I. R., Petrowski C., Sakai H., and Smith J. W. (1974) Carbon, nitrogen, and sulfur in lunar fines 15012 and 15013: Abundances, distributions, and isotopic compositions. *Geochim. Cosmochim. Acta.* In press.

Chang S., Lennon K., and Gibson E. K., Jr. (1974a) Abundances of C, N, H, He, and S in Apollo 17 soils from Stations 3 and 4: Implications for solar wind exposure ages and regolith evolution (abstract). In *Lunar Science—V*, pp. 106–108. The Lunar Science Institute, Houston.

Crozaz G., Drozd R., Hohenberg C., Morgan C., Ralston C., Walker R., and Yuhas D. (1974) Lunar surface dynamics: Some general conclusions and new results from Apollo 16 and 17 (abstract). In *Lunar Science—V*, pp. 157–159. The Lunar Science Institute, Houston.

Des Marais D. J., Hayes J. M., and Meinschein W. G. (1973) The distribution in lunar soil of carbon released by pyrolysis. *Proc. Fourth Lunar Sci. Conf., Geochim. Cosmochim. Acta*, Suppl. 4, Vol. 2, pp. 1543–1558. Pergamon.

Des Marais D. J., Hayes J. M., and Meinschein W. G. (1974) Retention of solar wind-implanted elements in lunar soils (abstract). In *Lunar Science—V*, pp. 168–170. The Lunar Science Institute, Houston.

Des Marais D. J., Hayes J. M., and Meinschein W. G. (1974a) The distribution in lunar soil of hydrogen released by pyrolysis. *Proc. Fifth Lunar Sci. Conf., Geochim. Cosmochim. Acta*, Suppl. 5. This volume.

Epstein S. and Taylor H. P., Jr. (1970) The concentration and isotopic composition of hydrogen, carbon, and silicon in Apollo 11 lunar rocks and minerals. *Proc. Apollo 11 Lunar Sci. Conf., Geochim. Cosmochim. Acta*, Suppl. 1, Vol. 2, pp. 1085–1096. Pergamon.

Gault D. E., Hörz F., Brownlee D. E., and Hartung J. B. (1974) Mixing of the lunar regolith (abstract). In *Lunar Science—V*, pp. 260–262. The Lunar Science Institute, Houston.

Gibson E. K., Jr. and Chang S. (1974) Abundance and isotopic composition of carbon in lunar rock 67016: Suggestions of a carbonate-like phase (abstract). In *Lunar Science—V*, pp. 264–266. The Lunar Science Institute, Houston.

Gibson E. K., Jr. and Johnson S. M. (1971) Thermal analysis–inorganic gas release studies of lunar samples. *Proc. Second Lunar Sci. Conf., Geochim. Cosmochim. Acta*, Suppl. 2, Vol. 2, pp. 1351–1366. MIT Press.

Gibson E. K., Jr. and Moore C. B. (1973) Variable contents of lunar soil 74220. *Earth Planet. Sci. Lett.* **20**, 404–408.

Gibson E. K., Jr. and Moore G. W. (1973a) Volatile rich lunar soil: Evidence of possible cometary impact. *Science* **179**, 69–71.

Gibson E. K., Jr. and Moore G. W. (1974) Total sulfur abundances and distributions in the valley of Taurus-Littrow: Evidence of mixing (abstract). In *Lunar Science—V*, pp. 267–269. The Lunar Science Institute, Houston.

Goel P. S., Shukla P. N., Kothari B. K., and Garg A. N. (1974) Solar wind as source of nitrogen in lunar fines (abstract). In *Lunar Science—V*, pp. 270–272. The Lunar Science Institute, Houston.

Hayes J. M. (1972) Extralunar sources for carbon on the moon. *Space Life Sci.* **3**, 474–483.

Heiken G. and McKay D. S. (1974) Petrography of Apollo 17 soils (abstract). In *Lunar Science—V*, pp. 319–321. The Lunar Science Institute, Houston.

Holland P. T., Simoneit B. R., Wszolek P. C., and Burlingame A. L. (1972) Compounds of carbon and other volatile elements in Apollo 14 and 15 samples. *Proc. Third Lunar Sci. Conf., Geochim. Cosmochim. Acta*, Suppl. 3, Vol. 2, pp. 2131–2147. MIT Press.

Holmes H. F., Fuller E. L., Jr., and Gammage R. B. (1974) Some surface properties of Apollo 17 soils (abstract). In *Lunar Science—V*, pp. 349–351. The Lunar Science Institute, Houston.

Housley R. M., Cirlin E. H., and Grant R. W. (1974) Solar wind and micrometeorite alteration of the lunar regolith (abstract). In *Lunar Science—V*, pp. 360–362. The Lunar Science Institute, Houston.

Kaplan I. R. (1972) Distribution and isotopic abundance of biogenic elements in lunar samples. *Space Life Sci.* **3**, 383–403.

Kaplan I. R. and Petrowski C. (1971) Carbon and sulfur isotope studies on Apollo 12 lunar samples. *Proc. Second Lunar Sci. Conf., Geochim. Cosmochim. Acta*, Suppl. 2, Vol. 2, pp. 1397–1406. MIT Press.

Kaplan I. R., Smith J. W., and Ruth E. (1970) Carbon and sulfur concentration and isotopic composition in Apollo 11 lunar samples. *Proc. Apollo 11 Lunar Sci. Conf., Geochim. Cosmochim. Acta*, Suppl. 1, Vol. 2, pp. 1317–1329. Pergamon.

Kerridge J. F., Kaplan I. R., Petrowski C., and Chang S. (1974) Light element geochemistry of the Apollo 16 site. *Geochim. Cosmochim. Acta.* Submitted.

Kerridge J. F., Kaplan I. R., and Lesley F. D. (1974a) Accumulation and isotopic evolution of carbon on the lunar surface (abstract). In *Lunar Science—V*, pp. 408–410. The Lunar Science Institute, Houston.

Kirsten T., Steinbrunn F., and Zähringer J. (1971) Location and variation of trapped rare gases in Apollo 12 lunar samples. *Proc. Second Lunar Sci. Conf., Geochim. Cosmochim. Acta*, Suppl. 2, Vol. 2, pp. 1651–1669. MIT Press.

McKay D. S. and Heiken G. H. (1973) Petrography and scanning electron microscope study of Apollo 17 orange and black glass. *Trans. Am. Geophys. Union* **54**, 599.

McKay D. S. and Heiken G. H. (1973a) The South Ray Crater age paradox. *Proc. Fourth Lunar Sci. Conf., Geochim. Cosmochim. Acta*, Suppl. 4, Vol. 1, pp. 41–47. Pergamon.

McKay D. S., Fruland R. M., and Heiken G. H. (1974) Grain size distribution as an indicator of the maturity of lunar soils (abstract). In *Lunar Science—V*, pp. 480–482. The Lunar Science Institute, Houston.

Moore C. B., Lewis C. F., Cripe J. D., and Volk M. (1974) Total carbon and sulfur contents of Apollo 17 lunar samples (abstract). In *Lunar Science—V*, pp. 520–522. The Lunar Science Institute, Houston.

Müller O. (1973) Chemically bound nitrogen contents of Apollo 16 and Apollo 15 lunar fines. *Proc. Fourth Lunar Sci. Conf., Geochim. Cosmochim. Acta*, Suppl. 4, Vol. 2, pp. 1625–1634. Pergamon.

Müller O. (1974) Solar wind and indigenous nitrogen in Apollo 17 lunar samples (abstract). In *Lunar Science—V*, pp. 534–536. The Lunar Science Institute, Houston.

Pillinger C. T., Batts B. D., Eglinton G., Gowar A. P., Jull A. J. T., and Maxwell J. R. (1973) Formation of lunar carbide from lunar iron silicates. *Nature Phys. Sci.* **245**, 3–5.

Rees C. E. and Thode H. G. (1974) Sulfur concentrations and isotope ratios in Apollo 16 and Apollo 17 samples (abstract). In *Lunar Science—V*, pp. 621–623. The Lunar Science Institute, Houston.

Rhodes J. M., Rodgers K. V., Shih C., Bansal B. M., Nyquist L. E., and Wiesmann H. (1974) Relationship between geology and soil chemistry at the Apollo 17 landing site (abstract). In *Lunar Science—V*, pp. 630–632. The Lunar Science Institute, Houston.

Sakai H., Kaplan I. R., Petrowski C., and Smith J. W. (1974) Analyses of C, N, S, He, H, and metallic Fe in lunar samples by pyrolysis and hydrolysis. In preparation.

Schmitt H. H. (1973) Apollo 17 report on the valley of Taurus-Littrow. *Science* **182**, 681–690.

Schoeller D. A. and Hayes J. M. (1974) Computer controlled ion-counting isotope ratio mass spectrometer. *Rev. Sci. Instrum.* Submitted.

Wszolek P. C., Walls F. C., and Burlingame A. L. (1974) Simulation of lunar carbon chemistry in plagioclase (abstract). In *Lunar Science—V*, pp. 854–856. The Lunar Science Institute, Houston.

Proceedings of the Fifth Lunar Conference
(Supplement 5, Geochimica et Cosmochimica Acta)
Vol. 2 pp. 1801–1809 (1974)
Printed in the United States of America

Loss of oxygen, silicon, sulfur, and potassium from the lunar regolith

Robert N. Clayton,* Toshiko K. Mayeda, and Julie M. Hurd

Enrico Fermi Institute, University of Chicago

Abstract—The processes of formation and maturation of lunar soils lead to enrichments in the heavy stable isotopes of oxygen, silicon, sulfur, and potassium. The isotopic enrichment implies substantial losses of these elements from the moon. Vaporization by micrometeorite impact and by ion sputtering have removed at least 1% of the mass of the regolith. The losses of sulfur and potassium amount to at least 20–30% of their original abundance in the regolith.

Introduction

Variations in isotopic compositions in lunar soils have been reported for several light elements: H, C, N, O, Si, S, and K. The elemental abundances and isotopic compositions of the first three—hydrogen, carbon, and nitrogen—are dominated by the addition of atoms of extralunar origin, predominantly from the solar wind. Detailed interpretation of their isotopic variations is difficult due to lack of knowledge of the isotopic composition of the incoming material, uncertainties in the fluxes of solar wind, meteoritic and cometary matter, and inadequate understanding of the mechanisms for trapping the extralunar matter in the regolith. Of the other four elements, oxygen and silicon are major elements, sulfur and potassium are minor elements, and all exhibit the same range of concentrations in the soil as in crystalline rocks. Their abundances are sufficiently high that contributions to the soil from extralunar sources cannot exceed a few percent of the indigenous complement of these elements. Thus the observed variations in isotopic composition are attributable to processes of formation and maturation of the regolith, and interpretation of the variations should enhance our understanding of these processes.

Results and Discussion

Processes leading to variation in isotopic abundance may be broadly classified as either nuclear or chemical, the latter encompassing all mass-dependent phenomena, including diffusion. For elements with three or more stable isotopes, an experimental distinction between nuclear and chemical processes can be made by comparing the relative magnitudes of isotope ratio variations for two or more

---

*Also from Departments of Chemistry and Geophysical Sciences, University of Chicago.

different isotope ratios. Rees and Thode (1972) measured S^{33}/S^{32}, S^{34}/S^{32}, and S^{36}/S^{32} in lunar rocks and soils, and found the variations to be in the proportions $1:2:4$ as expected for chemical processes. Epstein and Taylor (1971) reported that variations in Si^{29}/Si^{28} were one-half as great as the corresponding variations in Si^{30}/Si^{28}. A similar test can be made, in principle, for potassium, but the low abundance of K^{40} makes this difficult (Barnes *et al.*, 1973). Partial fluorination reactions similar to those first described by Epstein and Taylor (1971) have been carried out on soils 14163 and 76501, and on soil breccia 70019. The resulting δO^{17} and δO^{18} values are shown in Fig. 1. For each sample, the first-reacting oxygen, representing $<0.6\%$ of the whole sample, is greatly enriched in the heavy isotopes, and successive fractions approach the whole-soil value ($\delta^{18} \approx 6\%_0$) monotonically. The points conform well to a chemical fractionation line of slope $\frac{1}{2}$.

Isotopic compositions of oxygen, silicon, sulfur, and potassium are very uniform in lunar igneous rocks (Epstein and Taylor, 1972; Clayton *et al.*, 1973; Rees and Thode, 1972; Barnes *et al.*, 1973). It can therefore be concluded that the isotopic variations observed in lunar soils result from chemical fractionation processes associated with formation of soil from rocks, and with maturation of the soil on exposure at the surface.

The isotopic variations of the four elements under discussion have several features in common, which suggest a common origin:

(a) All show enrichments in the heavy isotopes in soils relative to rocks.
(b) The magnitudes of the heavy isotope enrichments in oxygen, silicon, and sulfur are correlated with measures of soil maturity, such as noble gas content, metallic iron, particle tracks, etc. (Epstein and Taylor, 1971, 1972; Taylor and Epstein, 1973; Rees and Thode, 1974).
(c) There is some evidence for surface correlation for the effects in oxygen and silicon (Epstein and Taylor, 1971; Epstein and Taylor, 1972; Taylor and Epstein, 1973; Clayton *et al.*, 1971) and sulfur (Rees and Thode, 1974).

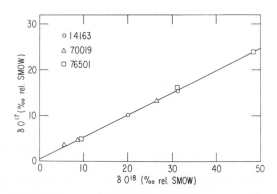

Fig. 1. Results of partial fluorination reactions of lunar soils (14163, 76501) and a soil breccia (70019). The data points fall along a line with slope $\frac{1}{2}$, showing that the isotopic variations are the result of mass-dependent processes.

Taken together, these observations imply that the heavy-isotope enrichment processes operate continuously or repeatedly within the regolith, and lead to a net enrichment of heavy isotopes relative to the source materials. No complementary heavy-isotope depleted materials are known among the lunar samples. Material balance thus requires that the heavy-isotope depleted component has been removed from the moon (Epstein and Taylor, 1971; Gibson et al., 1973).

Removal of matter from the moon may be visualized in two steps: (1) transport from the surface or interior of a solid grain to the lunar "atmosphere" and (2) loss from the atmosphere to interplanetary space. These may be parts of the same event, as in a hypervelocity micrometeorite impact (Gault et al., 1972), or may be two discrete processes, such as diffusive loss of volatile compounds from grains (Kaplan et al., 1970; Housley et al., 1974) followed by photoionization and removal by the solar wind (Manka and Michel, 1971). More complex models can be constructed, involving volatile transport, deposition on grain surfaces, and subsequent removal from the moon by sputtering due to solar ion bombardment (R. M. Housley, personal communication, 1974).

Although it is possible that different combinations of processes are predominant for different elements, the parallelism in behavior of the four elements under consideration suggests that a common mechanism may govern all four. This conclusion was also reached by Epstein and Taylor (1974). These elements are very disparate in their pertinent chemical properties, such as ionization potentials, volatility as elements, stability and volatility of hydrides, etc. In particular, potassium forms no compounds volatile at lunar surface temperatures, so that removal of a substantial fraction of potassium from the interiors of grains, as implied by the large isotope effects, requires volatilization at high temperatures, probably in excess of $1000°C$ (Gibson et al., 1973). Under these conditions, extensive decomposition of sulfides and loss of sulfur by vaporization also occurs (Gibson and Moore, 1973). Loss of oxygen and silicon by volatilization in micrometeorite impacts is also feasible, as shown by the vapor pressure measurements of DeMaria et al. (1971).

There is good evidence that brief exposure of liquid droplets of lunar magma to the interplanetary vacuum does not result in significant loss of these elements by vaporization. Neither the Apollo 15 green glass spheres nor the Apollo 17 orange glass show any measurable heavy-isotope enrichments in oxygen (Clayton et al., 1973; Taylor and Epstein, 1973), silicon (Taylor and Epstein, 1973), or potassium (Barnes et al., 1973).

We have also examined glassy agglutinates from lunar soils in order to determine whether the glasses are enriched in O^{18} relative to the remainder of the soil. The results are shown in Table 1. No measurable differences were found between the agglutinates and the whole soil. Thus, although the O^{18}-enrichment in the soil is correlated with the agglutinate content, the glass itself does not have an exceptional isotopic enrichment.

By means of a simple model, it is possible to set lower limits on the amounts of material which must be removed from the moon in order to account for the heavy-isotope enrichments found in the regolith. If isotopic fractionation takes

Table 1. Oxygen isotopic compositions of soils and glassy agglutinates.

Sample no.	δO^{18} soil (‰)	δO^{18} agglutinates (‰)
14163	6.14	6.01
15270	5.96	6.00
66081	6.07	6.12

place either by diffusive loss of the elements from a grain or liquid drop, or by diffusive separation in the lunar atmosphere, a fractionation factor equal to the inverse square root of the atomic masses is a good approximation. For a given fractionation factor, the maximum isotope effect in the residual portion will result if a Rayleigh process is assumed, i.e. the residue within the grain, liquid drop or atmosphere remains well mixed. Then the isotopic enrichment in the residue can be calculated as a function of the fraction of the element remaining in the residue. Results are shown in Fig. 2 for the four elements under consideration. The shaded regions along the curves for sulfur and potassium cover the ranges observed for isotopic compositions of whole soil samples: 4–10‰ for sulfur (Kaplan et al., 1970; Kaplan and Petrowski, 1971; Rees and Thode, 1972; Smith et al., 1973), and 5–8‰ for potassium (Barnes et al., 1973). These isotopic compositions imply that, of the original potassium and sulfur which were present in the rock from which the regolith was formed, some 20–30% has been lost from the moon. The "whole-soil" heavy-isotope enrichments for oxygen are only about 0.5‰ (Onuma et al., 1970; Epstein and Taylor, 1970; O'Neil and Adami, 1970) and for silicon are about 0.3‰ (Epstein and Taylor, 1970), implying losses of approximately 1% of these elements from the moon. Since oxygen and silicon are the two most abundant elements in lunar rocks, it can be concluded that the moon has lost at least 1% of the mass of the regolith through processes of micrometeorite bombardment and ion sputtering.

It has been shown by partial fluorination reactions (Epstein and Taylor, 1971) that the small O^{18} and Si^{30} enrichments in the lunar soil are the consequence of very large enrichments localized in a small fraction of the soil, which is especially reactive with fluorine. These large enrichments are indicated schematically in Fig. 2 by hash-marks along the fractionation curves for oxygen and silicon. In the extreme cases, these samples represent residues from which over half of the original oxygen and silicon has been removed. These can hardly be in situ residues of thermal volatilization, nor is it feasible to produce them by sputtering of solid particles at low temperatures, since no mechanism exists to maintain the Rayleigh condition of an homogenized residue. It is more likely that the extreme isotope enrichments are in vapor deposits derived from matter volatilized by impact and fractionated in the atmosphere. Large enrichments in S^{34} are found in the finest size-fractions of the soils (Rees and Thode, 1974) which could result from a similar process occuring for sulfur. Tests for surface correlation of the potassium isotope effects have not yet been reported.

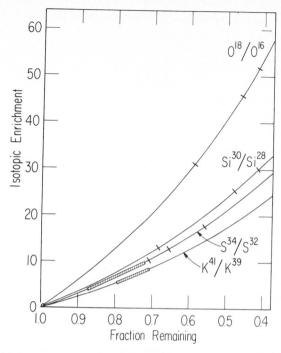

Fig. 2. Rayleigh fractionation curves showing heavy-isotope enrichment (in permil, ‰) as a function of the residual fraction of each element. The cross-hatched regions along the sulfur and potassium curves, and near the origin on the oxygen and silicon curves indicate the ranges of isotopic compositions observed for whole soils. Larger isotopic enrichments, found in surface or reactive sites, are indicated by hash-marks along the curves. Potassium data are from Barnes *et al.* (1973); sulfur data are from Kaplan *et al.* (1970), Kaplan and Petrowski (1971), Rees and Thode (1972, 1974), Smith *et al.* (1973), Kerridge *et al.* (1974); data on silicon and oxygen (whole-soil and stripping reactions) are from Epstein and Taylor (1971, 1972, 1974) and Taylor and Epstein (1973); additional whole-soil oxygen data are from Onuma *et al.* (1970), Clayton *et al.* (1971, 1972, 1973).

Could losses of sulfur and potassium of 20–30% have gone undetected by ordinary elemental analyses? The range of sulfur contents of lunar surface rocks is broader than that of lunar soils. At the Apollo 12, 14, 15, and 16 sites, the vast majority of rocks *and* soils analyzed contain 500–800 ppm sulfur (Kaplan and Petrowski, 1971; Moore *et al.*, 1972; Rees and Thode, 1972, 1974; Gibson and Moore, 1973; Smith *et al.*, 1973; Cripe and Moore, 1974). The basalts at the Apollo 11 and 17 sites are much richer in sulfur (1500–2800 ppm), whereas soils in these areas are only moderately richer (600–1300 ppm) (Kaplan *et al.*, 1970; Gibson and Moore, 1974; Chang *et al.*, 1974; Kerridge *et al.*, 1974; Rees and Thode, 1974; Moore *et al.*, 1974). Thus, although the coverage is too sparse to permit good averaging over the lunar surface, the sulfur analyses are entirely compatible with the hypotheses of considerable volatile transport and of loss of a large fraction from the moon. At any particular site on the moon, local variations of sulfur content of soils is dominated by mixing in various proportions of sulfur-rich and sulfur-poor components (Cripe and Moore, 1974; Gibson and Moore, 1974).

Comparison between igneous rock analyses and soil analyses is even more difficult in the case of potassium, due to the ubiquitous admixture of a KREEP component in the soils, bringing in large amounts of potassium from a source not represented among the major igneous rock types sampled. Thus the substantial net loss of potassium from the regolith could not have been observed on the basis of elemental analyses alone. The increase of potassium content with decreasing grain size (and hence increasing specific surface area) in several soils (Evenson *et al.*, 1973), has been cited as evidence for vapor transport and redeposition, but has been considered primarily in terms of addition rather than removal of potassium from the regolith.

It has long been recognized that potassium–argon dating of lunar soils is seriously complicated by the presence of large amounts of reimplanted Ar^{40} (Heymann and Yaniv, 1970). It is evident from the discussion above that the regolith system is open with respect to potassium, as well as argon, so that whole-soil "ages" do not measure single lunar events (Pepin *et al.*, 1972, 1974).

We may also examine the implication of the observation of major losses of potassium to abundance of the related element, rubidium. Gibson *et al.* (1973) measured the loss of potassium and rubidium from lunar rocks and soils by heating in vacuum, and found rubidium to be lost preferentially. Their experiments were designed to simulate conditions in a hot ejecta blanket, and thus involved heating at temperatures near 1000°C for many hours. Under these conditions, the relative volatilities of potassium and rubidium are probably determined by their relative equilibrium vapor pressures. However, in the rapid heating in a lunar micrometeorite impact, the relative volatile loss is more likely to be governed by transport processes, favoring potassium over rubidium. Furthermore, after vaporization, removal from the moon is easier for potassium than for rubidium due to its lower mass. Hence, large losses of potassium do not necessarily imply large losses of rubidium. Major rubidium losses would be inconsistent with the common observations of Rb–Sr model ages near 4.6 b.y. However, Nyquist *et al.* (1973) have shown that rubidium losses from soils are not entirely negligible, leading to a correlation between increased model ages (Rb loss) and soil maturity, as measured by the content of glassy agglutinates.

CONCLUSIONS

Several recent papers have noted that stable isotope variations of some light elements imply significant degrees of volatilization and vapor phase transport (Epstein and Taylor, 1971; Barnes *et al.*, 1973; Gibson *et al.*, 1973; Housley *et al.*, 1974). In this paper we have summarized the data concerning oxygen, silicon, sulfur, and potassium, and have emphasized the fact that material balance requires removal of substantial amounts of these elements from the moon. The data for oxygen and silicon imply a mass loss from the moon of at least 1% of the mass of the regolith. The data for potassium and sulfur imply removal of at least

20–30% of the original complement of these elements from the regolith and thence from the moon. These four elements were selected for discussion because of the wealth of isotopic evidence available. It is clear that similar conclusions about loss from the moon must also hold for many other major and minor elements for which similar isotopic evidence is not available.

Thermal volatilization by micrometeorite impact is the important first step leading to loss of volatiles. Some material may reach escape velocity in the impact process, the remainder is redeposited on solid surfaces where it is susceptible to removal by ion sputtering. It is not evident that chemical interactions with implanted solar wind hydrogen or processes specifically involving radiation-damaged surfaces play any significant role in removal of these elements from the regolith and from the moon.

Acknowledgments—This research was supported by NASA grant NGL-14-001-169.

REFERENCES

Barnes I. L., Garner E. L., Gramlich J. W., Machlan L. A., Moody J. R., Moore L. J., Murphy T. J., and Shields W. R. Isotopic abundance ratios and concentrations of selected elements in some Apollo 15 and Apollo 16 samples. *Proc. Fourth Lunar Sci. Conf., Geochim. Cosmochim. Acta*, Suppl. 4, Vol. 2, pp. 1197–1207. Pergamon.

Chang S., Lennon K., and Gibson E. K., Jr. (1974). Abundances of C, N, H, He, and S in Apollo 17 soils from Stations 3 and 4: Implications for solar wind exposure ages and regolith evolution (abstract). In *Lunar Science—V*, pp. 106–108. The Lunar Science Institute, Houston.

Clayton R. N., Onuma N., and Mayeda T. K. (1971). Oxygen isotope fractionation in Apollo 12 rocks and soils. *Proc. Second Lunar Sci. Conf., Geochim. Cosmochim. Acta*, Suppl. 2, Vol. 2, pp. 1417–1420. MIT Press.

Clayton R. N., Hurd J. M., and Mayeda T. K. (1972). Oxygen isotopic compositions and oxygen concentrations of Apollo 14 and Apollo 15 rocks and soils. *Proc. Third Lunar Sci. Conf., Geochim. Cosmochim. Acta*, Suppl. 3, Vol. 2, pp. 1455–1463. MIT Press.

Clayton R. N., Hurd J. M., and Mayeda T. K. (1973). Oxygen isotopic compositions of Apollo 15, 16, and 17 samples, and their bearing on lunar origin and petrogenesis. *Proc. Fourth Lunar Sci. Conf., Geochim. Cosmochim. Acta*, Suppl. 4, Vol. 2, pp. 1535–1542. Pergamon.

Cripe J. D. and Moore C. B. (1974). Total sulfur contents of Apollo 15 and Apollo 16 lunar samples (abstract). In *Lunar Science—V*, pp. 523–525. The Lunar Science Institute, Houston.

DeMaria G., Balducci G., Guido M., and Piacente V. (1971). Mass spectrometric investigation of the vaporization process of Apollo 12 lunar samples. *Proc. Second Lunar Sci. Conf., Geochim. Cosmochim. Acta*, Suppl. 2, Vol. 2, pp. 1367–1380. MIT Press.

Epstein S. and Taylor H. P., Jr. (1970). The concentration and isotopic composition of hydrogen, carbon and silicon in Apollo 11 lunar rocks and minerals. *Proc. Apollo 11 Lunar Sci. Conf., Geochim. Cosmochim. Acta*, Suppl. 1, Vol. 2, pp. 1085–1096. Pergamon.

Epstein S. and Taylor H. P., Jr. (1971). O^{18}/O^{16}, Si^{30}/Si^{28}, D/H and C^{13}/C^{12} ratios in lunar samples. *Proc. Second Lunar Sci. Conf., Geochim. Cosmochim. Acta*, Suppl. 2, Vol. 2, pp. 1421–1441. MIT Press.

Epstein S. and Taylor H. P., Jr. (1972). O^{18}/O^{16}, Si^{30}/Si^{28}, C^{13}/C^{12}, and D/H studies of Apollo 14 and 15 samples. *Proc. Third Lunar Sci. Conf., Geochim. Cosmochim. Acta*, Suppl. 3, Vol. 2, pp. 1429–1454. MIT Press.

Epstein S. and Taylor H. P., Jr. (1974). Oxygen, silicon, carbon and hydrogen isotope fractionation processes in lunar surface materials (abstract). In *Lunar Science—V*, pp. 212–214. The Lunar Science Institute, Houston.

Evenson N. M., Rama Murthy V., and Coscio M. R., Jr. (1973). Rb–Sr ages of some mare basalts and the isotopic and trace element systematics in lunar fines. *Proc. Third Lunar Sci. Conf., Geochim. Cosmochim. Acta*, Suppl. 3, Vol. 2, pp. 1707–1724. MIT Press.

Gault D. E., Hörz F., and Hartung J. B. (1972). Effects of microcratering on the lunar surface. *Proc. Third Lunar Sci. Conf., Geochim. Cosmochim. Acta*, Suppl. 3, Vol. 3, pp. 2713–2734. MIT Press.

Gibson E. K., Jr. and Moore G. W. (1973). Carbon and sulfur distributions and abundances in lunar fines. *Proc. Fourth Lunar Sci. Conf., Geochim. Cosmochim. Acta*, Suppl. 4, Vol. 2, pp. 1577–1586. Pergamon.

Gibson E. K., Jr. and Moore G. W. (1974). Total sulfur abundances in the valley of Taurus-Littrow: Evidence of mixing (abstract). In *Lunar Science—V*, pp. 267–269. The Lunar Science Institute, Houston.

Gibson E. K., Jr., Hubbard N. J., Wiesmann H., Bansal B. M., and Moore G. W. (1973). How to lose Rb, K, and change the K/Rb ratio: An experimental study. *Proc. Fourth Lunar Sci. Conf., Geochim. Cosmochim. Acta*, Suppl. 4, Vol. 2, pp. 1263–1273. Pergamon.

Heymann D. and Yaniv A. (1970). Inert gases in the fines from the Sea of Tranquillity. *Proc. Apollo 11 Lunar Sci. Conf., Geochim. Cosmochim. Acta*, Suppl. 1, Vol. 2, pp. 1247–1259. Pergamon.

Housley R. M., Cirlin E. H., and Grant R. W. (1974). Solar wind and micrometeorite alteration of the lunar regolith (abstract). In *Lunar Science—V*, pp. 360–362, The Lunar Science Institute, Houston.

Kaplan I. R., Smith J. W., and Ruth E. (1970). Carbon and sulfur concentration and isotopic composition in Apollo 11 lunar samples. *Proc. Apollo 11 Lunar Sci. Conf., Geochim. Cosmochim. Acta*, Suppl. 1, Vol. 2, pp. 1317–1329. Pergamon.

Kaplan, I. R. and Petrowski C. (1971). Carbon and sulfur isotope studies on Apollo 12 lunar samples. *Proc. Second Lunar Sci. Conf., Geochim. Cosmochim. Acta*, Suppl. 2, Vol. 2, pp. 1397–1406. MIT Press.

Kerridge J. F., Petrowski C., and Kaplan I. R. (1974). Sulfur, methane and metallic iron in some Apollo 17 fines (abstract). In *Lunar Science—V*. pp. 411–413. The Lunar Science Institute, Houston.

Manka R. H. and Michel F. C. (1971). Lunar atmosphere as a source of lunar surface elements. *Proc. Second Lunar Sci. Conf., Geochim. Cosmochim. Acta*, Suppl. 2, Vol. 2, pp. 1717–1728. MIT Press.

Moore C. B., Lewis C. F., Cripe J., Delles F. M., Kelley W. R., and Gibson E. K. (1972). Total carbon, nitrogen, and sulfur in Apollo 14 lunar samples. *Proc. Third Lunar Sci. Conf., Geochim. Cosmochim. Acta*, Suppl. 3, Vol. 2, pp. 2051–2058. MIT Press.

Moore C. B., Lewis C. F., Cripe J., and Volk M. (1974). Total carbon and sulfur contents of Apollo 17 lunar samples (abstract). In *Lunar Science—V*, pp. 520–522. The Lunar Science Institute, Houston.

Nyquist L. E., Hubbard N. J., Gast P. W., Bansal B. M., Wiesmann H., and Jahn B. (1973). Rb–Sr systematics for chemically defined Apollo 15 and 16 materials. *Proc. Fourth Lunar Sci. Conf., Geochim. Cosmochim. Acta*, Suppl. 4, Vol. 2, pp. 1823–1846. Pergamon.

O'Neil J. R. and Adami L. H. (1970). Oxygen isotope analyses of selected Apollo 11 materials. *Proc. Apollo 11 Lunar Sci. Conf., Geochim. Cosmochim. Acta*, Suppl. 1, Vol. 2, pp. 1425–1427. Pergamon.

Onuma N., Clayton R. N., and Mayeda T. K. (1970). Apollo 11 rocks: Oxygen isotope fractionation between minerals, and an estimate of the temperature of formation. *Proc. Apollo 11 Lunar Sci. Conf., Geochim. Cosmochim. Acta*, Suppl. 1, Vol. 2, pp. 1429–1434. Pergamon.

Pepin R. O., Bradley J. G., Dragon J. C., and Nyquist L. E. (1972). K–Ar dating of lunar fines: Apollo 12, Apollo 14 and Luna 16. *Proc. Third Lunar Sci. Conf., Geochim. Cosmochim. Acta*, Suppl. 3, Vol. 2, pp. 1569–1588. MIT Press.

Pepin R. O., Basford J. R., and Dragon J. C. (1974). K–Ar ages and depositional chronologies of Apollo 15 drill core fines (abstract). In *Lunar Science—V*, pp. 593–595. The Lunar Science Institute, Houston.

Rees C. E. and Thode H. G. (1972). Sulphur concentrations and isotope ratios in lunar samples. *Proc. Third Lunar Sci. Conf., Geochim. Cosmochim. Acta*, Suppl. 3, Vol. 2, pp. 1479–1485. MIT Press.

Rees C. E. and Thode H. G. (1974). Sulphur concentrations and isotope ratios in Apollo 16 and 17 samples (abstract). In *Lunar Science—V*, pp. 621–623. The Lunar Science Institute, Houston.

Smith J. W., Kaplan I. R., and Petrowski C. (1973). Carbon, nitrogen, sulfur, helium, hydrogen and metallic iron in Apollo 15 drill stem fines. *Proc. Fourth Lunar Sci. Conf., Geochim. Cosmochim. Acta*, Suppl. 4, Vol. 2, pp. 1651–1656. Pergamon.

Taylor H. P. and Epstein S. (1973). O^{18}/O^{16} and Si^{30}/Si^{28} studies of some Apollo 15, 16 and 17 samples. *Proc. Fourth Lunar Sci. Conf., Geochim. Cosmochim. Acta*, Suppl. 4, Vol. 2, pp. 1657–1679. Pergamon.

Proceedings of the Fifth Lunar Conference
(Supplement 5, Geochimica et Cosmochimica Acta)
Vol. 2 pp. 1811–1822 (1974)
Printed in the United States of America

The distribution in lunar soil of hydrogen released by pyrolysis

David J. DesMarais, J. M. Hayes,* and W. G. Meinschein

Department of Geology, Indiana University, Bloomington, Indiana 47401

Abstract—The hydrogen contents of various Apollo 15 rocks and Apollo 15 and 16 soil particle types and sieve fractions have been determined by the pyrolysis method. The Apollo 15 crystalline rocks analyzed release very little hydrogen (approximately 6 μmole H/g sample). The mineral, glass, and light-colored breccia fragments in the soils contain approximately 5, 30, and 10 μmole H/g sample, respectively. The dark breccias and glassy agglutinates examined have much higher hydrogen contents (60 and 110 μmole H/g sample).

For soil particle diameters less than 105 μm, hydrogen content increases linearly with particle surface area per particle mass. The calculated grain surface average concentrations of hydrogen in the Apollo 15 and 16 soils fall short of experimentally determined saturation values by a factor of 20. The volume-correlated component of hydrogen in a soil increases with the maturity of the soil.

Introduction

It is generally accepted that most of the hydrogen observed in lunar soils derives from the solar wind. The concentration of hydrogen in lunar crystalline rocks is much lower than that observed in soils (Fireman *et al.*, 1970; Stoenner *et al.*, 1970; Epstein and Taylor, 1970; Friedman *et al.*, 1970). Analyses of soil and rock surfaces (e.g. Leich *et al.*, 1973) indicate a pronounced enrichment of hydrogen which has presumably been implanted as solar wind protons. Furthermore, the $^1H/^4He$ ratio in soils is within a factor of two of the presumed average solar wind value (Hanneman *et al.*, 1970; Stoenner *et al.*, 1970; Hintenberger *et al.*, 1970).

Isotopic studies of lunar soil hydrogen have been employed to determine the origin of water detected in the fines. Friedman *et al.* (1970) proposed that the observed water derived from the interaction of solar wind protons and the oxygen contained in the fines materials. Epstein and Taylor (1970, 1972), citing hydrogen and oxygen isotopic data as well as the low-temperature release of the water, argued that the water in the soils is terrestrial in origin. Zeller (1966) has shown that impacting high-energy protons can form hydroxyl groups in silicates. Any water formed in lunar soils by this process may be lost prior to sampling by the astronauts (see Housley *et al.*, 1973, 1974).

Solar wind hydrogen apparently contains little or no deuterium (Epstein and Taylor, 1972), an observation supported by numerous isotopic measurements of lunar fines (e.g. Epstein and Taylor, 1972; Friedman *et al.*, 1970; Hintenberger *et al.*, 1970). The presence of deuterium in some breccias and in hydrogen released at high pyrolysis temperatures may be at least partly due to cosmic ray-induced spallation reactions in the soil (Friedman *et al.*, 1971; Epstein and Taylor, 1972).

---

*Also at Department of Chemistry.

The complex assemblage of particles comprising lunar soil has derived from the reworking of original lunar material by extralunar bombardment. Soil maturity, as defined by McKay *et al.* (1972), is measured by the abundance of reworked particles (e.g. glassy agglutinates, and soil breccias), the mean particle size, and other properties of the fines material which are influenced by bombardment. In general, mature soils are characterized by high contents of agglutinates and soil breccias, small mean particle diameters, high cosmic ray track densities, and high contents of solar wind-derived elements. Due to the continuous transport of material by meteorite impacts at the lunar surface, some soils may contain recognizable subpopulations of particles with distinct degrees of maturity (McKay *et al.*, 1974). For example, sample 68501 is a highly mature soil which may have received more than one input of regolith material (DesMarais *et al.*, 1973a; McKay *et al.*, 1974). Consequently, the assemblage of particles in 68501 reflects not only its particular exposure history, but also the histories of its component grain populations.

Previous studies of carbon contents in soil size fractions and grain types have helped to elucidate how solar wind-implanted carbon accumulates in lunar fines material (DesMarais *et al.*, 1973a, 1973b; Cadogan *et al.*, 1973). In general, carbon which has built up on grain surfaces is continually redistributed into the interiors of agglutinates and breccias. Therefore, a mature soil contains surface correlated carbon and a sizable volume correlated component of carbon which is at least partly of solar wind origin. It is logical to assume that other solar wind-derived elements, including hydrogen, have accumulated by a similar mechanism.

Due to the high abundance of protons in the solar wind, any lunar soil or rock surface exposed to the solar wind flux for a few years should reach steady state with respect to its surface correlated hydrogen content (for relevant discussions, see Housley *et al.*, 1974; Chang *et al.*, 1973). This "saturation" effect, the volatility of H_2 gas, and the reactivity of H with other elements, will contribute to a distribution of this element different from those of the other solar wind-implanted species. The present paper reports hydrogen abundances in soil size fractions and particle types. Such elemental distribution data should indicate how the physical and chemical properties of solar wind-implanted elements influence their observed abundances in the lunar soils.

Experimental

The methods employed for sample handling, pyrolyses, and gas chromatographic analyses have been discussed elsewhere (DesMarais *et al.*, 1972, 1973a, 1973c). Briefly, each lunar soil sample was pyrolyzed in a quartz tube filled with helium at 2 atm pressure. The sample was heated at $10°C\ min^{-1}$ to 1300°C, and the evolved gas was injected into a gas chromatograph at sample temperatures of 150°C, 450°C, 750°C, 1050°C, and 1300°C. At 1300°C, three additional gas samples were taken in 30 min intervals. The helium ionization detector in the gas chromatograph was calibrated using pure hydrogen and the technique of exponential dilution (Lovelock, 1960). The total hydrogen content of each sample was obtained by summing the hydrogen evolved from ambient temperature to 1300°C. Generally, two analyses of each sample were made. The analysis of variance used took advantage of the large number of measured samples. This method yielded a more accurate estimate of the overall experimental precision than an independent analysis of variance for each individual sample. Comparisons to the

Table 1. Abundances of hydrogen and carbon released from lunar materials.

Sample	Comments, Station	C^a μmoles C/g	H^b μmoles H/g	H^c μmoles H/g
15001,166	Core, 214 cm depth	5.8 ± 0.45^d	42 ± 2	
15002,308	Core, 186 cm depth	7.8 ± 0.55	52 ± 2.5	
15004,164	Core, 113 cm depth	8.2 ± 0.55	$92 \pm 4.$	
15005,232	Core, 54 cm depth	7.7 ± 0.55	$62 \pm 3.$	
15012,7	SESC fines, 6	6.2 ± 0.45	$65 \pm 3.$	
15012,8	SESC fines, 6	6.2 ± 0.45	72 ± 3.5	70 (1)
15013,8	Fines under LM	8.6 ± 0.6	68 ± 3.5	
15058,73	Porphyritic basalt, ALSEP		12 ± 0.5	
15080,5	Mare fines, 1	7.3 ± 0.5	62 ± 3	
15100,9	Apennine front fines, 2	6.8 ± 0.5	45 ± 2.5	36 (2)
15301,24	Apennine front fines, 7	5.8 ± 0.45	50 ± 2.5	52.2 (3)
15401,13	"Fillet", Apennine front, 6a	1.8 ± 0.25	13 ± 0.5	
15415,44	Anorthosite, 7	0.3 ± 0.2	2 ± 0.1	
15426,31	Green glass clod, 7	1.9 ± 0.25	12 ± 0.5	
15556,56	Vesicular basalt, 9a		4 ± 0.1	
61221,3	White fines, trench bottom, 1	4.4 ± 0.5	$35 \pm 4.$	7.8 (4)
61241,19	Gray fines, trench top, 1	5.5 ± 0.55	65 ± 5.5	
62240,6	Buster crater fines, 2	7.7 ± 0.65	$79 \pm 6.$	
63320,10	Shadowed fines, 13	5.5 ± 0.55	$63 \pm 5.$	
63340,10	Buried fines, 13	5.3 ± 0.55	50 ± 4.5	
63500,6	Control fines, 13	5.8 ± 0.55	49 ± 4.5	
67481,24	Fines, 11	4.3 ± 0.5	$31 \pm 4.$	
68501,37	Fines, 8	6.8 ± 0.6	74 ± 5.5	34 (5)

[a]DesMarais et al. (1973a).

[b]This work.

[c]Other analyses: (1) Sakai et al., 1972; (2) Friedman et al., 1972; (3) Epstein and Taylor, 1972; (4) Epstein and Taylor, 1973; (5) Merlivat et al., 1974.

[d]Uncertainties are one standard deviation.

hydrogen analyses of other authors are presented in Table 1. In soils, most of the hydrogen is released below 1000°C, but a small increase in hydrogen evolution from the sample above 1050°C may be caused by the melting of the lunar material.

The liquid argon wet-sieving procedure previously described by DesMarais et al. (1972) and subsequently modified by DesMarais et al. (1973a) was employed for the present work. Individual minerals were isolated from the 149–250 μm size fractions as described by DesMarais et al. (1973a).

RESULTS AND DISCUSSION

Bulk analyses

Table 1 shows experimentally determined hydrogen and carbon contents of the Apollo 15 and 16 samples analyzed in this laboratory. Results of available analyses from other laboratories are also shown. Apollo 15 crystalline basalts contain less than 12 μmoles H/g. Virtually all of the hydrogen in these basalts is released at pyrolysis temperatures above 1050°C. Sample 15058,73 (12 μmole H/g) is a porphyritic basalt with a very low abundance of vugs or other cavities. In contrast, sample 15556,56 (4 μmole H/g) is a vesicular basalt which bears abundant evidence of outgassing and which apparently formed at or near the top of a lava flow (Lofgren, 1971). This comparison is supported by Moore *et al.* (1972), who found a lower carbon content in 15556 than in 15058.

Both 15401,13 and 15426,31 have hydrogen contents which are very low for soils. Approximately two-thirds of the hydrogen in these soils is released between pyrolysis temperatures of 450–1050°C. The low hydrogen content of 15401,13 supports the observation by Fleischer and Hart (1972) and DesMarais *et al.* (1973a) that soil 15401 has a relatively short exposure history.

Mature Apollo 15 soils typically contain between 40 and 70 μmoles H/g. The hydrogen in these soils is released essentially between the pyrolysis temperatures of 150–1050°C, with a strong evolution in the vicinity of 600°C. Mature soils sampled from the Mare generally have higher contents than those sampled nearer the Apennine Front. Core fines 15001,166; 15002,308; 15004,164; and 15005,232 do not show the strong decrease in hydrogen content with increasing depth reported by Smith *et al.* (1973) on the basis of three other samples in the same core. Instead, our analyses reflect a random pattern of hydrogen contents with depth. Heiken *et al.* (1973a) report a similar pattern of soil exposure parameters along the Apollo 15 core. Apparently, the rare gas (Bogard and Nyquist, 1973) and hydrogen contents of the core samples reflect the exposure histories of the various strata.

The mature Apollo 16 soils analyzed released between 50 and 80 μmoles H/g. The rather similar hydrogen contents of 63320,10; 63340,10; and 63500,6 do not indicate that these soils had radically different exposure histories. Apparently, the large boulder at Station 13 which was shielding soils 63320 and 63340 from solar wind bombardment has had a minimal effect on the solar wind-derived hydrogen content of these soils. Soil 61221,3 is unusual because it combines the physical characteristics of an immature soil with a hydrogen content within a factor of two of a typical mature soil value. Although anomalously high carbon contents were also found (Moore *et al.*, 1973; Epstein and Taylor, 1973; DesMarais *et al.*, 1973a), the hydrogen content reported here is not supported by Epstein and Taylor (1973), who measured a very low hydrogen content for 61221. The hydrogen content of 67481,24 indicates a submature classification for that soil, a conclusion confirmed by its rare gas (Walton *et al.*, 1973) and agglutinate contents (Heiken *et al.*, 1973b).

Figure 1 depicts the relationship between the total hydrogen and total carbon contents of the Apollo 15 and 16 soils analyzed. The percentages of agglutinates in the soils are provided as a measure of soil maturity (McKay *et al.*, 1972). As expected, the hydrogen and carbon contents correlate with each other and also with soil maturity.

The measurements reported here cannot reveal the actual chemical state of the hydrogen in the lunar material. The chemical state of hydrogen, soil composition, and sites of entrapment can all affect the observed release of hydrogen as a function of temperature.

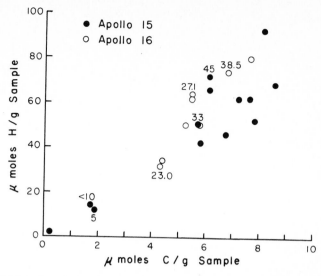

Fig. 1. Total hydrogen contents (μmoles H/g sample) versus total carbon contents (μmole C/g sample) in Apollo 15 and 16 samples. Numbers adjacent to data points indicate the percentages of agglutinates in the 105–149 μm size fractions of the Apollo 15 soils, and the 90–150 μm size fractions of the Apollo 16 soils. Apollo 16 agglutinate data from Heiken *et al.* (1973b).

Hydrogen contents of lunar soil particles

In general, the different particle types have different hydrogen contents (Table 2, Fig. 2) which parallel the carbon contents reported by DesMarais *et al.* (1973a). The hydrogen contents of the plagioclase grains from soils 61221,3 and 68501,37 are low because these grains probably derive directly from crystalline lunar rocks, which as a rule, have very low hydrogen contents. The glasses and white breccias analyzed contained more hydrogen than the plagioclase. Virtually all of this hydrogen was found to evolve from the samples above 1050°C, suggesting that

Table 2. Hydrogen contents of individual particle types.

Sample	Size fraction (μm)	Glass	Plagioclase crystals	White breccia	Dark breccia	Glassy agglutinates
			Hydrogen content, μmoles H/g sample			
15012,7	149–250	36 ± 4.0		19 ± 3.0	100 ± 7.0	106 ± 7.5
15301,24	149–250	19 ± 3.0		15 ± 3.0	65[a] ± 5.0	
15401,13	149–250	9 ± 2.5		19 ± 3.0	35[a] ± 4.0	
15426,31	149–1000	10 ± 2.5				
61221,3	250–1000		4 ± 2.5	12 ± 2.5	30 ± 3.5	
68501,37	149–250		7 ± 2.5	33 ± 3.5	60 ± 5.0	117 ± 7.5

[a]Indicates composite fractions containing glassy agglutinates, dark breccia, and basalt.

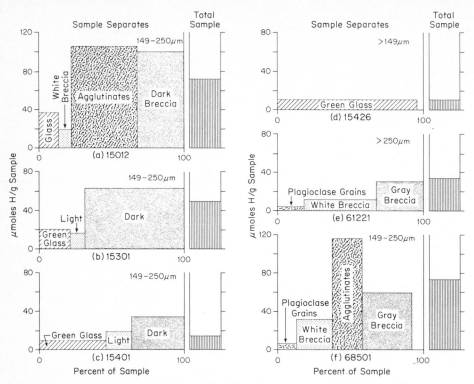

Fig. 2. Total hydrogen contents (μmoles H/g sample) and abundances of particle types in the coarse sieve fractions of fines (a) 15012,7; (b) 15301,24; (c) 15401,13; (d) 15426,31; (e) 61221,3; and (f) 68501,37. In the "sample separates" portion of each diagram, the horizontal axis depicts the relative percentages of the soil particle types.

hydrogen is liberated when the samples melt. DesMarais *et al.* (1973a) observed that the glass fractions of soils 15012,7; 15301,24; and 15401,13 all had similar carbon contents. Given this uniformity, it is difficult to explain why these glass fractions have substantially different hydrogen contents.

All agglutinates and the Apollo 15 dark breccias analyzed released substantial amounts of hydrogen during pyrolysis between 450 and 750°C. The Apollo 15 breccias analyzed were low grade. Warner (1972) proposes that low-grade breccias were formed under relatively mild thermal conditions. It therefore seems possible that these breccias retained hydrogen which might have been lost during a more intense thermal event. For example, 68501,37 has a medium-grade breccia which has been more extensively remelted, and most of the hydrogen in this breccia was released above 1050°C. DesMarais *et al.* (1973a) reported carbon contents for light and dark breccia fractions which were about equal in most of the soils analyzed. The hydrogen contents of these same light and dark breccia fractions are noticeably different, reflecting an enrichment of hydrogen in the dark breccias. This observation indicates that hydrogen retention in lunar fines depends somewhat on the chemical nature of the soil material. Housley *et al.* (1973) noted that the darkened soil fractions contained iron apparently reduced by solar proton bombardment. It is possible that the associated high iron contents of these fractions are due to the prolonged exposure of these particular fractions to proton bombardment. Alternatively, all of the fraction in a given soil could have a similar exposure history, with hydrogen being better retained by mineral phases having a high iron content. Pillinger et al. (1974) report that the carbide content of a lunar soil depends upon the amount of indigenous Fe^{II} available for exposure-induced reduction to Fe^{o} in the silicates. Perhaps the hydrogen content of a soil displays a similar dependence.

Glass agglutinates contained the most hydrogen of any particle type, an observation consistent with that of DesMarais et al. (1973a) for carbon.

The distribution of hydrogen with respect to particle size

DesMarais et al. (1973a) have recognized surface- and volume-correlated components of carbon in lunar fines by means of analyses of the distribution of carbon with respect to particle size. The equation used to resolve these two components can be rewritten in terms of hydrogen:

$$[H]_r = [H]_v + \frac{3S_H}{r\rho} \tag{1}$$

where $[H]_r$ is the total abundance of H(μmoles H/g sample) in some size fraction of particle radius, r; $[H]_v$ is the volume-correlated component (μmoles H/g sample); S_H is the "surface concentration"(μmoles H/cm^2 of particle surface); r is the average particle radius (cm) in the size fraction;* and ρ is the average soil grain density (arbitrarily set at 3 g/cm^3 for this paper). The "surface concentrations" are not genuine because they are calculated assuming the soil particles are smooth spheres. Nevertheless, because most soils contain particles with equivalent surface roughness, this particular assumption is not troublesome where comparisons of S_H values between soils are involved.

Figure 3 and Table 3 show $[H]_r$ versus r^{-1} relationships for five Apollo 15 and 16 soils. For particle size fractions with radii less than 50 μm ($r^{-1} > 200$ cm^{-1}) the linear relationship appears valid for all soils. The coarsest size fractions analyzed of soils 15401, 15012, and 61221 show an upswing in hydrogen content. Similar upswings noted for carbon by DesMarais et al. (1973) in soils 15012, 15301, and 15401 were attributed to an excess content of carbon-rich agglutinates in these size fractions. Since the agglutinates of these soils are also rich in hydrogen, the upswing in the $[H]_r$ versus r^{-1} at small r^{-1} values is presumably caused by an excess of hydrogen-rich agglutinates. Soil 61221,3 did not display such an upswing for carbon at small r^{-1}, yet an upswing in hydrogen content is present. Analyses of

---

*The mean radius used for plotting each size fraction corresponds to the radius of a sphere whose surface area is the average of the surface areas of the largest and smallest spheres which could be present in the size fraction. For soil 15301,24, these calculated radii were found to agree closely with radii empirically determined using Eq. (4) given by Eberhardt et al. (1965).

Table 3. Hydrogen, total carbon, and methane contents of soil size fractions.

Sample	Size fraction (μm)	Inverse mean radius (cm^{-1})	H (μmoles H/g)	C (μmoles C/g)	CH$_4$ (μmoles C/g)
15012,7	<37	760	96 ± 7	6.6 ± 0.6	0.14 ± 0.02
	37–53	440	68 ± 5.5	4.8 ± 0.5	0.073 ± 0.012
	53–74	310	56 ± 5	4.1 ± 0.5	0.064 ± 0.011
	74–105	220	58 ± 5	4.1 ± 0.5	0.051 ± 0.01
	105–149	150	51 ± 4.5	5.2 ± 0.55	0.025 ± 0.006
	149–250	98	34 ± 4	4.7 ± 0.5	0.033 ± 0.006
	>250	36	39 ± 4	6.2 ± 0.6	0.10 ± 0.015
15080,5	<37	760	111 ± 8	7.9 ± 0.65	0.27 ± 0.036
	37–53	440	73 ± 5.5	5.2 ± 0.55	0.092 ± 0.014
	53–74	310	70 ± 5.5	5.3 ± 0.55	0.039 ± 0.008
	74–105	220	53 ± 4.5	5.4 ± 0.55	0.053 ± 0.01
	105–149	150	57 ± 5	3.8 ± 0.45	0.036 ± 0.008
	>149	38	30 ± 3.5	3.3 ± 0.45	0.037 ± 0.008
15301,24	<37	760		7.4 ± 0.5	0.28 ± 0.036
	37–53	440		6.2 ± 0.45	0.15 ± 0.021
	53–105	240		4.0 ± 0.35	0.10 ± 0.015
	105–420	69		3.6 ± 0.35	0.043 ± 0.009
	>420	26		2.6 ± 0.3	0.063 ± 0.011
15401,13	<37	760	39 ± 4	4.3 ± 0.5	0.049 ± 0.009
	37–53	440	27 ± 3.5	2.6 ± 0.4	0.013 ± 0.005
	53–74	310	17 ± 3	1.7 ± 0.35	0.009 ± 0.004
	74–105	220	16 ± 3	1.8 ± 0.35	0.024 ± 0.006
	105–149	150	29 ± 3.5	2.4 ± 0.4	0.008 ± 0.004
61221,3	<20	1410	54 ± 5	7.6 ± 0.65	0.055 ± 0.01
	20–30	780	36 ± 4	5.4 ± 0.55	0.026 ± 0.007
	30–37	600	25 ± 3.5	4.3 ± 0.4	0.030 ± 0.007
	37–53	440	30 ± 3.5	3.3 ± 0.35	0.020 ± 0.006
	53–74	310	25 ± 3.5	2.4 ± 0.3	0.011 ± 0.005
	74–105	220	20 ± 3.0	1.8 ± 0.3	0.010 ± 0.004
	105–149	150	40 ± 4.0	2.2 ± 0.3	0.014 ± 0.005
	149–250	98	28 ± 3.5	2.0 ± 0.3	0.008 ± 0.004
68501,37	<20	1410	132 ± 9	15.3 ± 1.2	0.28 ± 0.036
	20–30	780	91 ± 6.5	11.3 ± 0.85	0.20 ± 0.026
	30–37	600	92 ± 6.5	9.1 ± 0.75	0.12 ± 0.018
	37–53	440	86 ± 6.5	6.3 ± 0.6	0.12 ± 0.018
	53–74	310	65 ± 5.5	5.7 ± 0.55	0.045 ± 0.009
	74–105	220	50 ± 4.5	3.3 ± 0.45	0.035 ± 0.008
	105–149	150	41 ± 4.0	3.0 ± 0.45	0.048 ± 0.009

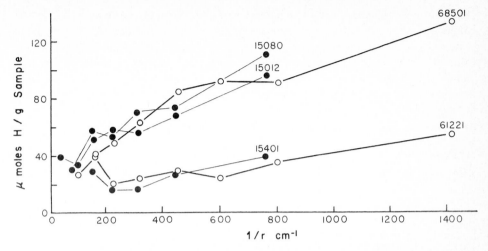

Fig. 3. Total hydrogen concentrations of sieved lunar fines 15012,7, 15080,5, 15401,13, 61221,3, and 68501,37.

grain types in the $>250\ \mu$m size fraction of 61221,3 reveal that the dark breccia contains much more hydrogen than the light breccia, whereas the carbon contents of the two breccias are equivalent. Apparently, the excess hydrogen observed at small r^{-1} is due to the presence of dark breccia material with a 30 μmole H/g hydrogen content.

Table 4 lists S_H and $[H]_v$ values which were obtained by the regression of $[H]_r$ on r^{-1} for six soil samples. The uncertainties accompanying these values are the standard deviations as determined from the regression analyses. The values shown for S_H display considerable variations which do not correlate well with the total hydrogen contents of the respective soils. In contrast, carbon surface concentrations (S_C) reported by DesMarais *et al.* (1973b) for the *same* four soils, appear to be largely constant. Two other hydrogen–carbon contrasts can be cited: First, comparing experimentally determined nitrogen surface concentrations

Table 4. Surface and volume-correlated hydrogen abundances in the $<105\ \mu$m size fraction of lunar soils.

Samples	S_H, nmoles H/cm^2	$[H]_v$, μmoles H/g	Total H, μmoles H/g
15012,7	76 ± 12[a]	37 ± 6	72 ± 3.5
15080,5	102 ± 13	33 ± 6	62 ± 3
15401,13	45 ± 5	5 ± 3	13 ± 0.5
15426,31	42 ± 14	2 ± 5	12 ± 0.5
61221,3	27 ± 4	15 ± 3	35 ± 4
68501,37	62 ± 10	47 ± 7	74 ± 5.5

[a]Indicated uncertainties are one standard deviation.

(S_N), S_C, and S_H values with expected solar wind values, Hayes *et al.* (1974) calculated relative surface retentions of carbon and hydrogen. Assuming complete retention of nitrogen on soil grain surfaces, Hayes *et al.* found that hydrogen losses from grain surfaces were at least 1000 times as great as carbon losses. Second, the present work indicates that dark-colored (Fe°-rich?) breccias were enriched in hydrogen. Such contrasts may indicate that the observed S_H values are controlled by loss effects involving hydrogen on grain surfaces. These values may suggest that hydrogen, more so than carbon, displays an affinity for phases with high iron contents.

Lord (1968) irradiated samples of forsterite (Mg_2SiO_4) with a simulated solar wind flux of 2 keV protons. He observed that the forsterite surface became saturated with hydrogen in a layer 500 Å thick when the atoms attained a concentration of 5×10^{17} atoms/cm^{-2} or 830 nmoles/cm^2. Values of S_H (Table 4) found for six Apollo soils range from 14 to 51 nmoles/cm^2. The S_H values are calculated assuming the soil grains are smooth spheres. A correction factor allowing for the rough grain surfaces would decrease the calculated S_H values. Furthermore, hydrogen atoms are observed to depths greater than 1000 Å in lunar soil grain surfaces (Leich *et al.*, 1973); therefore the complete chemical saturation of lunar material by the solar wind may occur at a value larger than 5×10^{17} atoms/cm^{-2} as measured by Lord. Consequently, the actual average surface concentrations of hydrogen in lunar soils are at least a factor of 20 below the saturation concentration. It is therefore required that a substantial percentage of the lunar soil grain surfaces are not *saturated* with respect to hydrogen. The observed lunar surface concentrations of hydrogen must therefore represent a kinetically controlled steady state in which the solar wind hydrogen input is balanced by losses due to diffusive escape, sputtering, etc. (see Hayes *et al.*, 1974).

The $[H]_v$ values listed in Table 4 correlate very well with the total hydrogen contents of the respective soils. Observing a similar correlation for carbon, DesMarais *et al.* (1973a, b) proposed that the carbon volume-correlated component, $[C]_v$, reflected a soil's indigenous lunar carbon content plus solar wind carbon which had been redistributed into the interiors of agglutinates and breccias by the energetic events of soil cycling. A soil's $[C]_v$ value thus increases with its agglutinate content, hence its maturity (McKay *et al.*, 1972). The correlation of $[H]_v$ with total hydrogen, and the high hydrogen content of the agglutinates, indicates that $[H]_v$ also increases with soil maturity and constitutes approximately half the hydrogen found in a mature soil.

Acknowledgments—We thank the graduate school of Indiana University for a fellowship for D. J. D. and appreciate the conscientious technical assistance of Kaye Fichman, Robert Yount, John Ling, Joseph Stellevato, and William Keesom. This work was supported by NASA grant NGL 15-003-115.

References

Bogard D. D. and Nyquist L. E. (1973) $^{40}Ar/^{36}Ar$ variations in Apollo 15 and 16 regolith. *Proc. Fourth Lunar Sci. Conf., Geochim. Cosmochim. Acta*, Suppl. 4, Vol. 2, pp. 1975–1985. Pergamon.

Cadogan P. H., Eglinton G., Maxwell J. R., and Pillinger C. T. (1973) Distribution of methane and carbide in Apollo 11 fines. *Nature Phys. Sci.* **241**, 81–82.

Chang S., Mack R., Gibson E. K., Jr., and Moore G. W. (1973) Simulated solar wind implantation of carbon and nitrogen ions into terrestrial basalt and lunar fines. *Proc. Fourth Lunar Sci. Conf., Geochim. Cosmochim. Acta*, Suppl. 4, Vol. 2, pp. 1509–1522. Pergamon.

DesMarais D. J., Hayes J. M., and Meinschein W. G. (1972) Pyrolysis study of carbon in lunar fines and rocks. In *The Apollo 15 Lunar Samples*, pp. 294–297. The Lunar Science Institute, Houston.

DesMarais D. J., Hayes J. M., and Meinschein W. G. (1973a) The distribution in lunar soil of carbon released by pyrolysis. *Proc. Fourth Lunar Sci. Conf., Geochim. Cosmochim. Acta*, Suppl. 4, Vol. 2, pp. 1543–1558. Pergamon.

DesMarais D. J., Hayes J. M., and Meinschein W. G. (1973b) Accumulation of carbon in lunar soils. *Nature Phys. Sci.* **246**, 65–68.

DesMarais D. J., Hayes J. M., and Meinschein W. G. (1973c) Techniques for the analysis of gases sequentially released from lunar samples. *Analytical Methods Developed for Application to Lunar Samples Analysis*, ASTM STP 539, 71–79. American Society for Testing and Materials.

Eberhardt P., Geiss J., and Grögler N. (1965) Ueber die Verteilung der Uredelgase im Meteoriten Khor Temiki. *Tschermaks Mineral. Petrogr. Mitt.* **10**, 535–551.

Epstein S. and Taylor H. P., Jr. (1970) The concentration and isotopic composition of hydrogen, carbon and silicon in Apollo 11 lunar rocks and minerals. *Proc. Apollo 11 Lunar Sci. Conf., Geochim. Cosmochim. Acta*, Suppl. 1, Vol. 2, pp. 1085–1096. Pergamon.

Epstein S. and Taylor H. P., Jr. (1972) $^{18}O/^{16}O$, $^{30}Si/^{28}Si$, $^{13}C/^{12}C$, and D/H studies of Apollo 14 and 15 samples. *Proc. Third Lunar Sci. Conf., Geochim. Cosmochim. Acta*, Suppl. 3, Vol. 2, pp. 1429–1454. MIT Press.

Epstein S. and Taylor H. P., Jr. (1973) The isotopic composition and concentration of water, hydrogen, and carbon in some Apollo 15 and 16 orange soils and in the Apollo 17 orange soil. *Proc. Fourth Lunar Sci. Conf., Geochim. Cosmochim. Acta*, Suppl. 4, Vol. 2, pp. 1559–1575. Pergamon.

Fireman E. L., D'Amico J. C., and DeFelice J. C. (1970) Tritium and argon radioactivities in lunar material. *Science* **167**, 566–568.

Fleischer R. L. and Hart H. R., Jr. (1972) Particle track record of Apollo 15 green soil and rock. In *The Apollo 15 Lunar Samples*, pp. 368–370. The Lunar Science Institute, Houston.

Friedman I., Gleason J. D., and Hardcastle K. G. (1970) Water, hydrogen, deuterium, carbon and ^{13}C content of selected lunar material. *Proc. Apollo 11 Lunar Sci. Conf., Geochim. Cosmochim. Acta*, Suppl. 1, Vol. 2, pp. 1103–1109. Pergamon.

Friedman I., O'Neil J. R., Gleason J. D., and Hardcastle K. G. (1971) The carbon and hydrogen content and isotopic composition of some Apollo 12 materials. *Proc. Second Lunar Sci. Conf., Geochim. Cosmochim. Acta*, Suppl. 2, Vol. 2, pp. 1407–1415. MIT Press.

Friedman I., Hardcastle K. G., and Gleason J. D. (1972) Isotopic composition of carbon and hydrogen in some Apollo 14 and 15 samples. In *The Apollo 15 Lunar Samples*, pp. 302–306. The Lunar Science Institute, Houston.

Hanneman R. E. (1970) Thermal and gas evolution behavior of Apollo 11 samples. *Proc. Apollo 11 Lunar Sci. Conf., Geochim. Cosmochim. Acta*, Suppl. 1, Vol. 2, pp. 1207–1211. Pergamon.

Hayes J. M., DesMarais D. J., and Meinschein W. G. (1974) Retention of solar wind-implanted elements in lunar soils. Submitted to *Geochim. Cosmochim. Acta*.

Heiken G., Duke M., McKay D. S., Clanton U. S., Fryxell R., Nagle J. S., Scott R., and Sellers G. A. (1973a) Preliminary stratigraphy of the Apollo 15 drill core. *Proc. Fourth Lunar Sci. Conf., Geochim. Cosmochim. Acta*, Suppl. 4, Vol. 1, pp. 191–213. Pergamon.

Heiken G. H., McKay D. S., and Fruland R. M. (1973b) Apollo 16 soils: Grain size analyses and petrography. *Proc. Fourth Lunar Sci. Conf., Geochim. Cosmochim. Acta*, Suppl. 4, Vol. 1, pp. 251–265. Pergamon.

Hintenberger H., Weber H. W., Voshage H., Wanke H., Begeman F., and Wlotzka F. (1970) Concentrations and isotopic abundances of the rare gases, hydrogen, and nitrogen in Apollo 11 lunar matter. *Proc. Apollo 11 Lunar Sci. Conf., Geochim. Cosmochim. Acta*, Suppl. 1, Vol. 2, pp. 1269–1282. Pergamon.

Housley R. M., Grant R. W., and Paton N. E. (1973) Origin and characteristics of excess Fe metal in lunar glass welded aggregates. *Proc. Fourth Lunar Sci. Conf., Geochim. Cosmochim. Acta*, Suppl. 4, Vol. 3, pp. 2737–2749. Pergamon.

Housley R. M., Cirlin E. H., and Grant R. W. (1974) Solar wind and micrometeorite alteration of the lunar regolith (abstract). In *Lunar Science—V*, pp. 360–362. The Lunar Science Institute, Houston.

Leich D. A., Tombrello T. A., and Brunett D. S. (1973) The depth distribution of hydrogen and fluorine in lunar samples. *Proc. Fourth Lunar Sci. Conf., Geochim. Cosmochim. Acta*, Suppl. 4, Vol. 2, pp. 1597–1612. Pergamon.

Lofgren G. E. (1971) *Lunar Sample Information Catalog, Apollo 15*, NASA-MSC 03209, p. 263.

Lord H. D. (1968) Hydrogen and helium ion implantation into olivine and enstatite: Retention coefficients, saturation concentrations, and temperature release profiles. *J. Geophys. Res.* **73**, 5271–5280.

Lovelock J. E. (1960) *Gas Chromatography 1960* (editor R. P. W. Scott), p. 26. Butterworth.

McKay D. S., Heiken G. H., Taylor R. M., Clanton U. S., Morrison D. A., and Ladle G. H. (1972) Apollo 14 soils: Size distribution and particle types. *Proc. Third Lunar Sci. Conf., Geochim. Cosmochim. Acta*, Suppl. 3, Vol. 1, pp. 983–994. MIT Press.

McKay D. S., Fruland R. M., Heiken G. H. (1974) Grain size distribution as an indicator of the maturity of lunar soils (abstract). In *Lunar Science—V*, pp. 480–482. The Lunar Science Institute, Houston.

Merlivat L., Lelu M., Nief G., and Roth E. (1974) Deuterium content of lunar material (abstract). In *Lunar Science—V*, pp. 498–500. The Lunar Science Institute, Houston.

Moore C. B., Lewis C. F., and Gibson E. K., Jr. (1972) Carbon and nitrogen in Apollo 15 lunar samples. In *The Apollo 15 Lunar Samples*, pp. 316–318. The Lunar Science Institute, Houston.

Moore C. B., Lewis C. F., and Gibson E. K., Jr. (1973) Total carbon contents of Apollo 15 and 16 lunar samples. *Proc. Fourth Lunar Sci. Conf., Geochim. Cosmochim. Acta*, Suppl. 4, Vol. 2, pp. 1613–1623. Pergamon.

Pillinger C. T., Davis P. R., Eglinton G., Gowar A. P., Jull A. J. T., Maxwell J. R., Housley R. M., and Cirlin E. H. (1974) The association between carbide and finely-divided metallic iron in lunar fines. *Proc. Fifth Lunar Sci. Conf., Geochim. Cosmochim. Acta*. This volume.

Sakai H., Chang S., Petrowski C., Smith J., and Kaplan I. R. (1972) Distribution of carbon and sulfur in hydrolyzed Apollo 15 lunar fines. In *The Apollo 15 Lunar Samples*, pp. 319–323. The Lunar Science Institute, Houston.

Smith J. W., Kaplan I. R., and Petrowski C. (1973) Carbon, nitrogen, sulfur, helium, hydrogen, and metallic iron in the Apollo 15 drill stem fines. *Proc. Fourth Lunar Sci. Conf., Geochim. Cosmochim. Acta*, Suppl. 4, Vol. 2, pp. 1651–1656. Pergamon.

Stoenner R. W., Lyman W. J., and Davis R., Jr. (1970) Cosmic ray production of rare gas radioactivities and tritium in lunar material. *Science* **167**, 553–555.

Walton J. R., Lakatos S., and Heymann D. (1973) Distribution of inert gases in fines from the Cayley–Descartes region. *Proc. Fourth Lunar Sci. Conf., Geochim. Cosmochim. Acta*, Suppl. 4, Vol. 2, pp. 2079–2095. Pergamon.

Warner J. L. (1972) Metamorphism of Apollo 14 breccias. *Proc. Third Lunar Sci. Conf., Geochim. Cosmochim. Acta*, Suppl. 3, Vol. 1, pp. 623–643. MIT Press.

Zeller E. J., Ronca L. B., and Levy P. W. (1966) Proton induced hydroxyl formation on the lunar surface. *J. Geophys. Res.* **71**, 4855–4860.

Proceedings of the Fifth Lunar Conference
(Supplement 5, Geochimica et Cosmochimica Acta)
Vol. 2 pp. 1823–1837 (1974)
Printed in the United States of America

Sulfur abundances and distributions in the valley of Taurus-Littrow

EVERETT K. GIBSON, JR.

TN7, Geochemistry, NASA Johnson Space Center, Houston, Texas 77058

GARY W. MOORE

Lockheed Electronics Corp., Houston, Texas 77058

Abstract—Total sulfur abundances have been determined for 36 Apollo 17 soil, breccia and crystalline rock samples. Sulfur concentrations range from 550 to 1300 μgS/g for soil samples with the orange soil containing the lowest amount of sulfur. Noritic breccias contain between 720 and 950 μgS/g while the anorthositic rocks have sulfur contents of 270 and 368 μgS/g. The dunite 72415 contained the lowest sulfur content (44 μgS/g) of any Apollo 17 sample studied.

Apollo 17 basalts have unusually high sulfur contents (1580–2770 μgS/g) as compared to Apollo 12 and 15 basalts and terrestrial basalts. Sulfur abundances for the Apollo 17 basalts are almost identical to those from Apollo 11 basalts. A negative correlation between percent metallic iron and total sulfur for the Apollo 17 and 15 basalts was found and suggests that a portion of the metallic iron in lunar basalts may result from desulfuration of the melt prior to crystallization from the lunar magma.

INTRODUCTION

THE TAURUS-LITTROW LANDING SITE presents a unique opportunity to examine in detail several of the geochemical trends suggested by studies on returned samples from previous lunar mare and highland locations. The valley's subfloor basalts permit studies on mare basalts in contrast to the lunar highland materials of the North and South Massifs. The Apollo 17 site is very suitable for study of some regolith processes, i.e. mixing of highlands materials with mare materials. The Apollo 17 crew extensively sampled both geologic units.

We wish to report on our investigations of total sulfur abundances and distributions for 36 Apollo 17 soils, breccias and basaltic rock samples. Sulfur is usually present in lunar basalts and breccias as the accessory sulfide mineral troilite, which is commonly associated with metallic iron. Within the lunar magmas from which the mare basalts were derived, sulfur would occur as sulfide species dissolved in the magma and later separated immiscibly from the silicate magma as an Fe–FeS melt (Skinner, 1970). The source regions for lunar basalts must have been enriched in sulfur as compared to those of terrestrial basalts because of the greater sulfur content of lunar basalts and the abundance of troilite. Fresh Hawaiian basalts typically contain between 13 and 264 μgS/g (Gibson, unpublished results). Sulfur is usually found in the forms of pyrrhotite or chalopyrite in Hawaiian basalts (Skinner and Peck, 1969). Lunar basalts examined from Apollo 11 through 16 range in sulfur content from 600 to 2200 μgS/g (Kaplan *et al.*, 1970; Cripe and Moore, 1974; Rees and Thode, 1974; Cripe, 1973; Moore *et al.*, 1974).

The total sulfur content of lunar fines from Apollo 11, 12, 14, 15, and 16 has been accounted for by mixing models based upon the sulfur contribution from individual components found in the fines at a particular sample site (Moore *et al.*, 1972, 1973; Cripe and Moore, 1974). Addition of sulfur to the lunar regolith from meteorite infall is not significant in contributing to the observed sulfur abundances for lunar fines. Carbonaceous chondrites (Type I) contain up to 6% sulfur (Gibson and Moore, 1974b). For any lunar fines mixing model, large amounts of sulfur cannot have been added to the lunar surface from meteorites unless some major loss mechanism has operated. Gibson and Moore (1973) have shown that between 12 and 30% of the sulfur in lunar fines can be volatilized at temperatures as low as 750°C during heating under vacuum and such regolith processes as microcratering, vapor transport, and lunar outgassing may move small amounts of sulfur around on the lunar surface and a steady-state situation may exist for sulfur addition and loss. The 750°C temperature is well below the FeS solidus and if large amounts of sulfur are lost at this temperature, it implies that a fraction of the sulfur is surface correlated rather than volume correlated. Rees and Thode (1974) reported that sulfur content of lunar fines increased with decreasing grain size. Their results indicate that a major fraction of the sulfur in lunar fines is found on the surfaces of particles. Rees and Thode (1974) have also reported that sulfur present on the surfaces of lunar particles is highly fractionated ($\delta^{34}S = +21.7‰$) as compared to the inner portions of the particles ($\delta^{34}S = +5.2‰$). Thus, loss of a small amount of sulfur is believed to have occurred during regolith processes (Rees and Thode, 1974; Clayton *et al.*, 1974; Moore *et al.*, 1974).

EXPERIMENTAL

Total sulfur measurements have been made using a LECO IR-32 total sulfur analyzer. Sample sizes ranging from 20 to 50 mg were combusted in oxygen at 1600°C, and the resulting SO_2 was detected using an infrared Luft cell detector. The detection limit of the analyzer was 1 μgS. The accuracy of the sulfur analysis was checked against N.B.S. standard steel 55E (S = 110 ± 10 μgS/g) and U.S.G.S. standard reference basalt BCR-1 (S = 464 ± 10 μgS/g). Experimentally obtained values were within ±5% of the previously reported values. Sample powders were prepared using a tool steel percussion mortar followed by hand grinding in an agate mortar and pestle until 100 mesh size was reached. Lunar soil samples were not crushed. Powders used for 17 of the samples analyzed were the same splits used for the Apollo 17 LSPET studies (LSPET, 1973).

DISCUSSION OF RESULTS

The analytical values obtained for sulfur abundances on Apollo 17 samples are given in Table 1. The experimental results in Table 1 are mean values of duplicate or triplicate analyses. They are similar to those reported by Cripe and Moore (1974); Rees and Thode (1974); Kerridge *et al.* (1974); Rhodes *et al.* (1974), and Duncan *et al.* (1974) for Apollo 17. Apollo 17 basalts have unusually high sulfur contents (1580–2770 μgS/g) as compared to Apollo 12 and 15 basalts and

Table 1. Apollo 17 total sulfur abundances.

Basalts	Description	Mean μgS/g
70035,1	Basalt	1580 ± 40
70215,2	Basalt	2210 ± 30
70215,54	Basalt	2210 ± 20
74275,56	Basalt	1650 ± 20
75035,37	Basalt	2770 ± 40
75055,6	Basalt	2210 ± 20
Massif Samples, Breccias and Miscellaneous		
72275,2	Noritic breccia	890 ± 20
72415,2	Dunite clast	44 ± 10
72435,1	Noritic breccia	945 ± 20
73275,24	Green-gray breccia	927 ± 10
76015,32	Poikilitic rock-matrix	895 ± 40
76015,58	Poikilitic rock-matrix	641 ± 30
76015,59	Poikilitic rock-matrix	834 ± 30
76015,65	Poikilitic rock-matrix	679 ± 30
76015,70	Porous clast	552 ± 30
76055,5	Noritic breccia	720 ± 40
76230,4	Anorthosite	270 ± 20
76315,2	Noritic breccia	755 ± 40
76315,33	Dark-gray phase of noritic breccia	950 ± 30
76315,65	Blue-gray phase of noritic breccia	785 ± 20
77017,2	Anorthositic gabbro with glass	955 ± 20
77135,2	Noritic breccia	800 ± 30
78155,2	Anorthosite-cataclasite	368 ± 15
79135,1	Dark matrix breccia	1020 ± 20
Soils		
70011,19	S.E.S.C. soil	1300 ± 30
73141,8	LRV stop between ST 2-3	630 ± 30
74220,5	Orange soil	560 ± 20
74220,84	Orange soil	750 ± 20
74260,4	Gray soil	1080 ± 20
75111,5	LRV stop between ST 5-6	1260 ± 30
75121,6	LRV stop between ST 5-6	1140 ± 20
76240,9	Soil beside Station 6 boulder	850 ± 10
76260,3	Soil beside Station 6 boulder	795 ± 20
76280,6	Soil beside Station 6 boulder	822 ± 10
76501,18	Rake reference soil	665 ± 40
78501,20	Rake reference soil	1125 ± 20

terrestrial basalts. However, Apollo 17 basalt sulfur concentrations (mean value = 2105 μgS/g) are almost identical with those reported for the Apollo 11 basalts (mean value = 2200 μgS/g; Kaplan *et al.*, 1970). Apollo 17 basalt 75035 contains the greatest sulfur (2770 μgS/g) of any lunar sample analyzed to date. Cripe and Moore (1974) reported that a separate split of 75035 contained 3140 μgS/g, similar to the high sulfur concentration we found for the rock. Apollo 17 basalts have sulfur contents 10–100 times greater than some fresh Hawaiian basalts. Enrichment in sulfur for Apollo 17 and 11 titanium-rich basalts as compared to those from Apollo 12 and 15 mare sites suggest a higher sulfur content in the lunar magmas which generated these basalts.

Sulfur contents for the three major rock types (basalts, noritic breccias, and anorthositic rocks) found at the Apollo 17 location vary with their major element chemistry (LSPET, 1973). Figure 1 shows the correlation between FeO and total sulfur content. Basalts with FeO contents between 18 and 20% have sulfur concentrations between 1580 and 2770 µgS/g. Two anorthositic rocks have sulfur contents of 270 and 368 µgS/g while their FeO contents are 5.14 and 6.19%, respectively. Anorthositic gabbro 77017, which has been severely shocked resulting in glass injection throughout the fractures, has a sulfur content of 955 µgS/g. The anomalous sulfur content suggests addition of sulfur during the cataclastic event which introduced the melt into the fractures of the sample.

Noritic breccias which have FeO values between 8.7 and 11.58% have sulfur contents between 552 and 950 µgS/g. These abundances are intermediate between

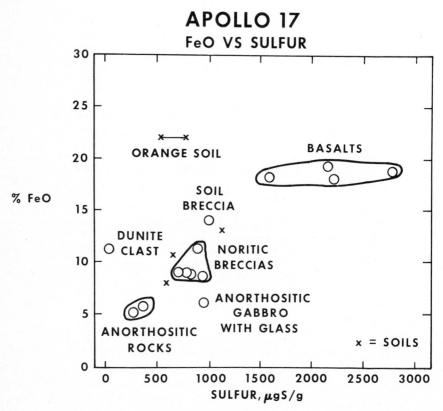

Fig. 1. Total sulfur and FeO concentrations for Apollo 17 rock samples. The FeO values were taken from LSPET (1973).

the Apollo 17 anorthositic and basaltic rocks (Fig. 1). Three separate samples from the noritic breccia 76315 produced sulfur values of 755, 950, and 785 μgS/g for (1) composite sample, (2) dark-gray phase, and (3) blue-gray phase respectively. For this fragmental rock the observed concentration range indicates that for the three separate samples analyzed, sulfur distribution is moderately similar throughout the sample. However, differing results were obtained from the analysis of five samples from the poikilitic rock 76015. Four matrix samples contained sulfur values of 895, 641, 834, and 679 μgS/g respectively, while a porous clast from 76015 contained only 552 μgS/g. The variations in sulfur content within the matrix of 76015 are important observations. Interpretations of data from a single analysis point for fragmental rocks of this nature should be made with extreme caution. In contrast, two separate samples of the fine grained basalt 70215 gave sulfur values of 2210 and 2210 μgS/g indicating a high degree of sample homogeneity.

Dunite clast 72415 contained the lowest sulfur content (44 ± 10 μgS/g) of any Apollo 17 sample examined. The low sulfur abundance was similar to those measured previously for the white portion of the black and white breccias from Apollo 16. However, these two samples differ widely in their chemical compositions. Both the anorthositic materials and the dunite clast are low in their sulfur contents. Albee *et al.* (1974) have suggested that 72415 may represent a very early differentiate derived from the upper lunar mantle. The sulfur depletion in 72415 raises several interesting questions concerning the behavior of sulfur in processes occurring within the interior of the moon. For example, what is the role of sulfur during partial melting processes? We might expect that with small degrees of partial melting the melt or liquid would be enriched in sulfur while the residue would be depleted in sulfur. If 72415 represents a very early differentiate, sulfur-bearing components were excluded from it during its crystallization or its source material was depleted in sulfur. At the present time it is difficult to understand the role of sulfur during partial melting processes. Investigations into the sulfur distribution between different layered rocks in terrestrial intrusive bodies such as the Skaergaard will perhaps answer some of these questions.

Soil breccia 79315 contains a sulfur value (1020 μgS/g) intermediate between the massif samples and subfloor basalts. The difference in sulfur content between the massif samples (275–950 μgS/g) and mare basalts (1580–2770 μgS/g) indicates either that highland materials were initially low in sulfur or they have been efficiently outgassed and depleted in sulfur relative to mare basalts.

Orange soil 74220 has a chemical composition similar to the Apollo 17 titanium-rich basalts (LSPET, 1973). However, two separate sample allocations gave sulfur contents of 550 and 750 μgS/g. The low sulfur content, in light of the basaltic composition of the soil (Fig. 1), shows that whatever source or origin chosen for the orange soil (fire fountaining or impact derived (Heiken *et al.*, in press)), the process was very efficient in removing sulfur from the sample, or we are looking at material from a different source with the same major element chemistry. Low sulfur content along with associated surface correlated volatile elements such as Pb, halogens, and Zn and surface related compounds (e.g. Gibson and Moore, 1973; Heiken *et al.*, 1974) is consistent with the fire fountaining

Table 2. Apollo 17 mixing model for sulfur in lunar soils.

Rock Type	µgS/g	LM Relative Abundance (%)	LM Sulfur Concentration µgS/g	Station 4 Relative Abundance (%)	Station 4 Sulfur Concentration µgS/g	Station 5 Relative Abundance (%)	Station 5 Sulfur Concentration µgS/g	LRV-3 Relative Abundance (%)	LRV-3 Sulfur Concentration µgS/g
Basalt	2100	54	1134	41.7	876	66.7	1400	27.4	575
Orange Glass	650	16.8	109	21.9	142	12.5	81	26.2	170
Noritic breccia	850	12.6	107	20.3	173	10.8	92	17.9	152
Anorthosite	320	16.7	53	16.1	52	10.0	32	28.5	91
Total Calculated fines			1403		1191		1605		988
Experimentally measured			1300		1080		1260		990
% Difference			7.3%		9.3%		21.5%		0.2%

Rock Type	µgS/g	Station 9 Relative Abundance (%)	Station 9 Sulfur Concentration µgS/g	Station 2 Relative Abundance (%)	Station 2 Sulfur Concentration µgS/g	LRV-6 Relative Abundance (%)	LRV-6 Sulfur Concentration µgS/g
Basalt	2100	35.9	754	1.8	38	11	231
Orange Glass	650	19.4	126	6.7	43	8	52
Noritic breccia	850	13.5	115	47.9	407	37.5	319
Anorthosite	320	31.2	100	42.9	138	43.5	139
Total Calculated fines			1095		626		741
Experimentally measured			1160		750		890
% Difference			5.6%		16.5%		16.7%

	Station 6		Station 8		LRV-2	
Basalt	2100	15.9	34.0	714	23.3	489
Orange Glass	650	11.7	14.1	92	19.4	126
Noritic breccia	850	24	13.1	111	21.2	180
Anorthosite	320	48.3	38.5	123	36.1	116
Total Calculated fines		769		1040		911
Experimentally measured		795		1125		950
% Difference		3.2%		7.5%		4%

Relative abundance data from Rhodes *et al.* (1974).

hypothesis because terrestrial pyroclastic materials with similar morphologies are also depleted in sulfur relative to their source materials (Gibson and Moore, unpublished data). The depletion in sulfur indicates severe outgassing of the material during the pyroclastic event and associated elevated temperatures. The extremely low vesicularity of the orange glass supports the hypothesis that the samples were completely "degassed" (Epstein and Taylor, 1973).

Twelve Apollo 17 soils have sulfur concentrations ranging from 550 to 1300 μgS/g (Table 1). Those soils collected at both the North and South Massifs have sulfur concentrations below 1000 μgS/g with the exception of soil 78501 (1125 μgS/g) collected at the Sculptured Hills Station. Soils collected on the Taurus-Littrow valley floor, which were associated with the subfloor basalts, have sulfur concentrations greater than 1000 μgS/g with the exception of the orange soil.

MIXING OF APOLLO 17 SOILS

Various mixing models for the Apollo 17 soils have been proposed by numerous investigators (Schonfeld, 1974; LSPET, 1973; Rhodes et al., 1974). Moore et al. (1972), Cripe (1972), and Cripe and Moore (1974) have noted that the total sulfur abundances for lunar soils could be accounted for by a mixing model based upon the sulfur contribution from individual components found in the soils at a particular site. Their model explained the observed sulfur abundances for the Apollo 11, 12, 14, 15, and 16 soils. Mixing models of the Moore et al. type for the Apollo 17 site are lacking because soil survey data of the type done by Reid et al. (1972) are not yet available for the Apollo 17 site. However, Rhodes et al. (1974) have grouped the Apollo 17 soils into three major categories: (1) Valley Floor type soils; (2) South Massif type soils; and (3) North Massif type soils. They noted that the chemical variations, based upon major element analyses, within and between these three groups can be accounted for by means of chemical mixing models in terms of lateral transport and mixing of prevailing local rock types, such as high-titanium basalts, KREEP-like noritic breccias, anorthositic gabbro breccias and orange glass. Rhodes et al. (1974) have used their chemical mixing model to determine the inferred proportions of locally derived components (i.e. major rock types) which comprise each of the soils samples at the Apollo 17 site. Using the mixing models of Moore et al. (1972) and Cripe and Moore (1974) along with the inferred proportions of major rock types found in the Apollo 17 soils by Rhodes et al. (1974), we can calculate the sulfur content expected for each of the samples studied. Table 2 illustrates the calculated total sulfur balance for ten Apollo 17 soils based on Rhodes et al. (1974) inferred rock components in the soils. The calculated sulfur abundances are given along with the experimentally measured sulfur values from Table 1 and Moore et al. (1974). Sulfur values chosen for the four components (basalts, orange glass, noritic breccias, and anorthositic breccias) are the mean values for the four major components. The calculated sulfur abundances differ from the experimentally determined values by 0.2–21.5% with a mean difference of 9.2%. Only three samples differ by more than 10% and the calculated values are within the sampling errors reported by Moore et al. (1972) and Cripe and Moore (1974).

Data presented in Table 2 indicate that the sulfur content of the Apollo 17 soils can be accounted for by a mixing of mare basalt which is enriched in sulfur with massif materials which are depleted in sulfur. Mixing models constructed by Schonfeld (1974) for the Apollo 17 soils indicate that around 1–2% meteoritic component (based upon the observed Ni content) is present in the Apollo 17 soils. Moore *et al.* (1972) have shown that sulfur addition to the lunar fines from meteorite infall (based upon C-1 composition of $S = 6.0\%$) is not significant in contribution to the observed sulfur abundances for lunar fines. Their mixing model does not take into account sulfur from the solar wind (only a minor contributor) or recycled meteoritic sulfur, but it does indicate that large amounts of sulfur cannot have been added to the lunar surface from meteorites unless some major loss mechanism has operated.

CORRELATION BETWEEN SULFUR AND ZINC IN APOLLO 17 SOILS

Zinc has been shown to be concentrated on the surface of the orange soil at the Apollo 17 site (G. Reed, personal communication). The orange soil has the greatest zinc concentration of any Apollo 17 sample (292 μgZn/g) (Rhodes *et al.*, 1974). Most Apollo 17 soils have zinc abundances between 18 and 50 μgZn/g (Rhodes *et al.*, 1974). Rees and Thode (1974) have recently shown that sulfur abundances increase with decreasing grain size for two Apollo 17 soils. Their data imply that the major portion of the sulfur found in the lunar soils is surface correlated. The S/Zn ratio for soils can provide a useful indicator for identifying samples which are enriched in a surface deposited Zn phase. For example, the orange soil 74220 has a S/Zn ratio of 1.9. Soils 71061, 74260, and 74240 which contain 6.3, 7.7, and 4.0% orange glass (Heiken and McKay, 1974) have S/Zn ratios of 7.9, 9.9, and 12.5, respectively. These three soils along with the orange soil have the lowest S/Zn ratios of the 22 Apollo 17 soils examined. Generally the Apollo 17 soils have S/Zn ratios between 20 and 40.

A comparison between the S/Zn ratio and percent basaltic component is seen in Fig. 2. The soils collected near the North and South Massifs (Stations 2, 3, 6, and 7) are separated from the mare soils (Stations 1, 4, 5, 9, and LM) because of the increased basaltic component found in the mare soils. The separation between the North and South Massif data in Fig. 3 undoubtedly results from the location of the North Massif sampling sites. The three large craters Henry, Shakespeare, and Cochise located at the base of the North Massif have thrown valley floor mare material onto the lower slopes of the North Massif. The resulting dilution or highland soils with mare soils would cause the observed separation between the two massifs for their S/Zn ratio versus percent basaltic components as seen in Fig. 3. The soils collected from the mare sites (Stations 1, 4, 5, 9, and LM) generally have an increasing S/Zn ratio as the percent basaltic component increases.

The four soils (74220, 71061, 74260, 74240) with apparent large quantities of surface zinc and possessing very low S/Zn ratios lie along a straight line (Fig. 2). The straight line correlation is to be expected if both sulfur and zinc were surface correlated elements for these lunar fines. The fractionation behavior of zinc and sulfur, both volatile elements, should be geochemically related. In terrestrial

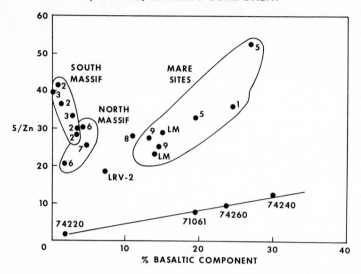

Fig. 2. S/Zn ratios and percent basaltic component in Apollo 17 soils. The Zn data is from Rhodes *et al.* (1974) and the percent basaltic component data from Heiken and McKay (1974).

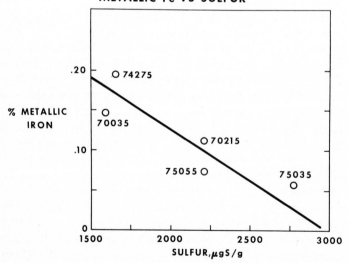

Fig. 3. Metallic iron and total sulfur contents for the Apollo 17 basalts. The metallic iron data is from Pearce *et al.* (1974).

systems zinc commonly behaves as a chalcophile element. The surface correlation between zinc and sulfur probably results from the condensation of both volatile elements on the surfaces of the orange glass during the event which formed Shorty Crater. In addition to enrichment of zinc and sulfur on the surfaces of the orange glass, other volatile elements such as chlorine are also enriched (Reed and Jovanovic, 1973). There is some evidence that a small portion of the total sulfur in the orange soil may be released at low temperatures (Gibson and Moore, 1973). Gas release studies have shown that around 15 ppm SO_2 is evolved below 250°C from orange soil.

Metallic Iron and Total Sulfur in Lunar Basalts

Skinner (1970) noted that the ubiquitous occurrence of intergrown troilite and metallic iron in three Apollo 11 basalts was suggestive of an initially homogeneous sulfide liquid that separated immiscibly from the parent magma. Skinner (1970) ruled out a post-magmatic desulfuration of pyrrhotite grains to produce the metallic iron and troilite intergrowths observed. The Apollo 17 titanium-rich basalts are chemically and petrographically similar to those from Apollo 11. We have reexamined the possibility that a portion of the metallic iron found in lunar basalts may be derived from the desulfuration of the magma before crystallization occurs. After the extrusion of a lunar lava onto the surface of the moon, the high temperature (greater than 1200°C) of the lava along with its low viscosity should result in mixing during flowage with exposure to the lunar vacuum. During this period desulfuration of troilite could occur at elevated temperatures resulting in the formation of metallic iron and subsequent loss of S_2. The iron and FeS are late stage phases occurring in the interstices between earlier crystallized phases.

If desulfuration is an operative process during crystallization of lunar basalts, we should find an inverse correlation between the metallic iron content of a lunar basalt and its total sulfur content. Such a relationship can be seen in Fig. 3 for five Apollo 17 basalts. The metallic iron contents were measured magnetically by Pearce *et al.* (1974). The inverse correlation between Fe° and S supports the hypothesis that desulfuration has occurred within the Apollo 17 basalts. Brown *et al.* (1974) noted that the Apollo 17 basalt 75035 (Group II in their classification scheme) had a lower melting temperature than did other Apollo 17 basalts. However, this does not necessarily imply it was therefore cooler when extruded. Sample 75035 contained the greatest total sulfur content (2770 μgS/g) and the least amount of metallic iron for Apollo 17 basalts studied. This further suggests that sample 75035 has been subjected to lesser degrees of heating or exposure to lunar vacuum before crystallization than basalts such as 74275 and 70035. In order to have further evidence supporting the desulfuration mechanism operating in lunar basalts, measurements on the isotopic fractionation of sulfur in lunar basalts, are required. Samples which have undergone greater degrees of desulfuration should have the greatest loss of the light isotope. Unfortunately at the present time, no isotopic data exists to test the model. It might be noted that the inverse correlation

APOLLO 15 BASALTS
METALLIC Fe° VS SULFUR

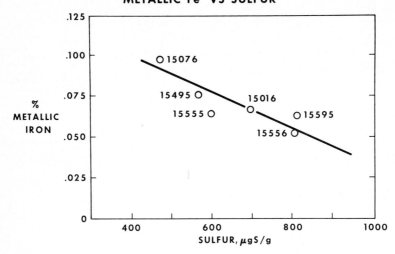

Fig. 4. Metallic iron and total sulfur contents for Apollo 15 basalts. Metallic iron data is from Pearce (1973).

between percent metallic iron and total sulfur content observed for the Apollo 17 basalts also holds for the Apollo 15 basalts (Fig. 4).

An alternative explanation for the inverse correlation between metallic iron abundances and total sulfur content for the lunar basalts may result from variations in the activity of iron and sulfur content of the melts from which the lunar basalts crystallized. Should the sulfur content be relatively high in the melt, available iron would tend to combine with sulfur, therefore less metallic iron would form during the crystallization of the lunar basalt. Alternatively, should the sulfur abundances be low in the melt, the iron activity would be greater leading to the formation of larger amounts of metallic iron. Thus the inverse correlation between metallic iron and total sulfur for lunar basalts could possibly result from initial sulfur abundance in the melt along with the activity of iron before crystallization. We believe that when the sulfur isotopic fractionation data becomes available for lunar basalts we can distinguish between the two possibilities.

Comparison of Sulfur Abundances in Lunar Rocks with Terrestrial Basalts

Cripe (1973) in his extensive study of sulfur abundances in fresh basaltic rocks found that the sulfur content was dependent on the environment of deposition. Moore and Fabbi (1971) had previously noted that subaerial basalts reflected a loss of sulfur by outgassing during eruption and cooling, whereas submarine basalts retained sulfur because of hydrostatic pressure. Flood plateau basalts were found to be intermediate in their sulfur abundances (Cripe, 1973). Recent analyses (Gibson, 1974, unpublished) of 80 Hawaiian basalts ranging from prehistoric flow to recent Mauna Ulu material found that the total sulfur abundances ranged from 13 to 264 μgS/g with a median value of 115 μgS/g. These 80 basalts had previously been studied by Wright (1971). A summary of sulfur abundances for terrestrial basalts is given in Fig. 5.

Lunar basalts from Apollo 11 and 17 range in sulfur content from 1500 to 2770 μgS/g. Their sulfur abundance is greater than twice the value observed for terrestrial submarine basalts which have retained sulfur because of hydrostatic pressure. The question could now be asked, "What was the initial sulfur abundance of lunar basalts before extrusion on the lunar surface?" Once on the lunar surface, the lava would have an even lower confining pressure than in the source reservoir. Flowage and outgassing of the lunar basalt should outgas the lava even more. Loss of sulfur from lunar basalts has been documented by Skinner (1970). McKay et al. (1972) found vapor phase deposits of sulfides within vugs and fractures of lunar breccias suggesting movement of sulfur-rich vapors.

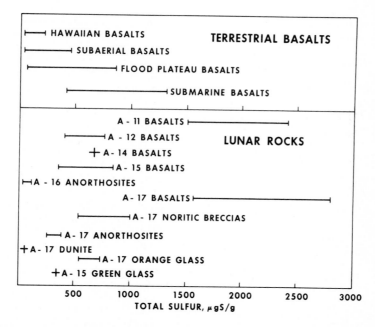

Fig. 5. Total sulfur abundances for terrestrial basalts and lunar crystalline rocks. The Hawaiian basalt data is from Gibson (1974, unpublished). The subaerial basalts, flood plateau basalts and submarine basalts data is from Cripe (1973). The Apollo 11, 12, 14, and 15 sulfur abundance data is from Compston et al. (1970); Kaplan and Smith (1970); Maxwell et al. (1970); Kaplan and Petrowski (1971); Compston et al. (1971); Thode and Rees (1972); Cripe (1973), and Cripe and Moore (1974). All other data is from Gibson and Moore (1973, 1974a).

Additional evidence of sulfur loss in lunar soils can be seen from the isotopic fractionation data of Rees and Thode (1972). It is quite obvious that the enrichment in sulfur for lunar basalts as compared to terrestrial basalts requires a source region richer in sulfur than normally found for terrestrial magmas.

The sulfur abundance data (Fig. 5) for the lunar highland anorthositic material (sulfur less than 400 $\mu gS/g$) as compared to mare samples indicates either that highland materials were initially low in sulfur or they have been efficiently outgassed and depleted in sulfur relative to mare basalts. Sulfur could be removed from the lower lunar crust or source region for the mare basalts by formation of an immiscible Fe–FeS liquid which would sink to form a core (Brett, 1973; Taylor and Jakes, 1974). This would effectively remove the other siderophile elements along with sulfur.

Acknowledgments—The authors wish to thank Robin Brett, David S. McKay, Richard J. Williams, G. W. Pearce, Carleton B. Moore, Sherwood Chang, Jerry D. Cripe, and Thomas Cobleigh for comments and suggestions during various stages of this study. The assistance of the technical support staff of Lockheed Electronics Company is appreciated in keeping the equipment operating, especially the efforts of Tom Cobleigh.

References

Albee A. L., Chodos A. A., Dymek R. F., Gancarz A. J., Goldman D. S., Papanastassiou D. A., and Wasserburg G. J. (1974) Dunite from the lunar highlands: Petrography, deformational history, Rb–Sr age (abstract). In *Lunar Science—V*, pp. 3–5. The Lunar Science Institute, Houston.

Brett R. (1973) A lunar core of Fe–Ni–S. *Geochim. Cosmochim. Acta* **37**, 165–170.

Brown G. M., Peckett A., Emeleus C. H., and Phillips R. (1974) Mineral-chemical properties of Apollo 17 mare basalts and terra fragments (abstract). In *Lunar Science—V*, pp. 89–91. The Lunar Science Institute, Houston.

Clayton R. N., Mayeda T. K., and Hurd J. M. (1974) Loss of O, Si, S and K from the lunar regolith (abstract). In *Lunar Science—V*, pp. 129–131. The Lunar Science Institute, Houston.

Compston W., Berry H., Vernon M. J., Chappell B. W., and Kaye M. J. (1971) Rubidium-strontium chronology and chemistry of lunar material from the Ocean of Storms. *Proc. Second Lunar Sci. Conf., Geochim. Cosmochim. Acta*, Suppl. 2, Vol. 2, pp. 1471–1485. MIT Press.

Cripe J. D. (1973) The total sulfur content of lunar samples and terrestrial basalts. M.S. Thesis, Arizona State Univ., Tempe, Ariz.

Cripe J. D. and Moore C. B. (1974) Total sulfur contents of Apollo 15 and Apollo 16 lunar samples (abstract). In *Lunar Science—V*, pp. 523–525. The Lunar Science Institute, Houston.

Duncan A. R., Erlank A. J., Willis J. P., and Sher M. K. (1974) Compositional characteristics of the Apollo 17 regolith (abstract). In *Lunar Science—V*, pp. 184–186. The Lunar Science Institute, Houston.

Epstein S. and Taylor H. (1973) The isotopic composition and concentration of water, hydrogen, and carbon in some Apollo 15 and 16 soils and in the Apollo 17 orange soil. *Proc. Fourth Lunar Sci. Conf., Geochim. Cosmochim. Acta*, Suppl. 4, Vol. 2, pp. 1559–1575. Pergamon.

Gibson E. K., Jr. and Moore G. W. (1973) Carbon and sulfur distributions and abundances in lunar fines. *Proc. Fourth Lunar Sci. Conf., Geochim. Cosmochim. Acta*, Suppl. 3, Vol. 2, pp. 1577–1586. Pergamon.

Gibson E. K., Jr. and Moore G. W. (1974a) Total sulfur abundances and distributions in the valley of Taurus-Littrow: Evidence of mixing (abstract). In *Lunar Science—V*, pp. 267–269. The Lunar Science Institute, Houston.

Gibson E. K., Jr. and Moore G. W. (1974b) Total sulfur abundances in carbonaceous chondrites. *Trans. Amer. Geophys. Union* **55**, 333.

Heiken G. and McKay D. S. (1974) Petrography of Apollo 17 soils (abstract). In *Lunar Science—V*, pp. 319–321. The Lunar Science Institute, Houston.

Heiken G., McKay D. S., and Brown R. W. (1974) Lunar deposits of possible pyroclastic origin. *Geochim. Cosmochim. Acta.* In press.

Jovanovic S., Jensen K., and Reed G. W., Jr. (1973) The halogens, U, Li, Te and P_2O_5 in five Apollo 17 soil samples. *Trans. Amer. Geophys. Union* **54**, 595.

Jovanovic S. and Reed G. W., Jr. (1974) Labile trace elements in Apollo 17 samples (abstract). In *Lunar Science—V*, pp. 391–393. The Lunar Science Institute, Houston.

Kaplan I. R., Smith J. W., and Ruth E. (1970) Carbon and sulfur concentration and isotopic composition in Apollo 11 lunar samples. *Proc. Apollo 11 Lunar Sci. Conf., Geochim. Cosmochim. Acta,* Suppl. 1, Vol. 2, pp. 1317–1329. Pergamon.

Kaplan I. R. and Petrowski C. (1971) Carbon and sulfur isotope studies on Apollo 12 lunar samples. *Proc. Second Lunar Sci. Conf., Geochim. Cosmochim. Acta,* Suppl. 2, Vol. 2, pp. 1397–1406. MIT Press.

Kerridge J. F., Petrowski C., and Kaplan I. R. (1974) Sulfur, methane and metallic iron in some Apollo 17 fines (abstract). In *Lunar Science—V*, pp. 411–413. The Lunar Science Institute, Houston.

LSPET (1973) Apollo 17 lunar samples: Chemical and petrographic description. *Science* **182**, 659–672.

McKay D. S., Clanton U. S., and Morrison D. A. (1972) Vapor phase crystallization in Apollo 14 breccia. *Proc. Third Lunar Sci. Conf., Geochim. Cosmochim. Acta,* Suppl. 3, Vol. 1, pp. 739–752. MIT Press.

Maxwell J. A., Peck L. C., and Wiik H. B. (1970) Chemical composition of Apollo 11 lunar samples 10017, 10020, 10072 and 10084. *Proc. Apollo 11 Lunar Sci. Conf., Geochim. Cosmochim. Acta,* Suppl. 1, Vol. 2, pp. 1369–1374. Pergamon.

Moore C. B., Lewis C. F., Cripe J., Delles F. M., Kelly W. R., and Gibson E. K., Jr. (1972) Total carbon, nitrogen and sulfur in Apollo 14 lunar samples. *Proc. Third Lunar Sci. Conf., Geochim. Cosmochim. Acta,* Suppl. 3, Vol. 2, pp. 2051–2058. MIT Press.

Moore C. B., Lewis C. F., Cripe J. D., and Volk M. (1974) Total carbon and sulfur contents of Apollo 17 lunar samples (abstract). In *Lunar Science—V*, pp. 520–522. The Lunar Science Institute, Houston.

Moore J. G. and Fabbi B. P. (1971) An estimate of the juvenile sulfur content of basalt. *Contr. Mineral. Petrol.* **33**, 118–127.

Pearce G. W. (1973) Magnetism and lunar surface samples. Ph.D. Dissertation, Univ. of Toronto, Toronto, Canada.

Pearce G. W., Gose W. A., and Strangway D. W. (1974) Magnetism of the Apollo 17 samples (abstract). In *Lunar Science—V*, pp. 590–592. The Lunar Science Institute, Houston.

Rees C. E. and Thode H. G. (1972) Sulfur concentrations and isotope ratios in lunar samples. *Proc. Third Lunar Sci. Conf., Geochim. Cosmochim. Acta,* Suppl. 3, Vol. 2, pp. 1479–1485. MIT Press.

Rees C. E. and Thode H. G. (1974) Sulphur concentrations and isotopic ratios in Apollo 16 and 17 samples (abstract). In *Lunar Science—V*, pp. 621–623. The Lunar Science Institute, Houston.

Reid A. M., Warner J., Ridley W. I., and Brown R. W. (1972) Major element composition of glasses in three Apollo 15 soils. *Meteoritics* **7**, 395–415.

Rhodes J. M., Rodgers K. V., Shih C., Bansal B. M., Nyquist L. E., and Wiesmann H. (1974) The relationship between geology and soil chemistry at the Apollo 17 landing site. *Proc. Fifth Lunar Sci. Conf., Geochim. Cosmochim. Acta,* Suppl. 5. This volume.

Schonfeld E. (1974) Component abundance and evolution of regoliths and breccias: Interpretation by mixing models (abstract). In *Lunar Science—V*, pp. 669–671. The Lunar Science Institute, Houston.

Skinner B. J. and Peck D. L. (1969) An immiscible sulfide melt from Hawaii. *Economic Geology Monograph* **4**, 310–322.

Skinner B. J. (1970) High crystallization temperatures indicated for igneous rocks from Tranquillity Base. *Proc. Apollo 11 Lunar Sci. Conf., Geochim. Cosmochim. Acta,* Suppl. 1, Vol. 1, pp. 891–895. Pergamon.

Taylor S. R. and Jakes P. (1974) Geochemical zoning in the moon (abstract). In *Lunar Science—V*, pp. 786–788. The Lunar Science Institute, Houston.

Thode H. G. and Rees C. E. (1972) Sulfur concentrations and isotope ratios in Apollo 14 and 15 samples. In *The Apollo 15 Lunar Samples*, pp. 402–404. The Lunar Science Institute, Houston.

Wright T. L. (1971) Chemistry of Kilauea and Mauna Loa lava in space and time. U.S.G.S. Prof. paper 735, 40 pp.

Proceedings of the Fifth Lunar Conference
(Supplement 5, Geochimica et Cosmochimica Acta)
Vol. 2 pp. 1839–1854 (1974)
Printed in the United States of America

D/H and $^{18}O/^{16}O$ ratios of H_2O in the "rusty" breccia 66095 and the origin of "lunar water"*

Samuel Epstein and Hugh P. Taylor, Jr.

Division of Geological and Planetary Sciences, California Institute of Technology, Pasadena, California 91109

Abstract—Stepwise vacuum extractions of H_2O at temperatures ranging from 25°C to 700°C were carried out on lunar "rusty rock" 66095. This rock contains larger amounts of H_2O (20–40 μmoles/g) than any other lunar rock or soil sample yet analyzed. About 20% of the evolved water is liberated at 25°C, and nearly all the water is removed by the time the rock is heated to 200°C. Thus, the H_2O in 66095 is even more loosely bound than the adsorbed H_2O in lunar soils, which ordinarily do not undergo significant dehydration below 200–300°C. The average δD and $\delta^{18}O$ values of H_2O evolved from 66095 are about −100 and −15, respectively. These values are remarkably similar to those in terrestrial atmospheric water vapor. The stepwise isotopic release patterns for 66095 show no appreciable variation in either δD or $\delta^{18}O$, indicating that there is only one major type of H_2O present in the sample. Similar stepwise heating experiments were also carried out on samples of rehydrated 66095, and on two forms of FeOOH, a terrestrial goethite and a synthetic akaganeite. Except for the $\delta^{18}O$ values in the synthetic akaganeite, the amounts and isotopic compositions of H_2O obtained by stepwise heating of all these samples are practically identical to that obtained for 66095. The abnormally heavy $\delta^{18}O$ value (mean = +4) of the akaganeite probably can be accounted for by nonequilibrium evaporation of H_2O from the water-rich akaganeite. The similarity in H_2O release patterns and isotopic compositions between 66095 and the terrestrial samples formed from meteoric waters indicates that the H_2O extracted from 66095 is a terrestrial contaminant. The rusty coating could have formed by oxidation–hydration of $FeCl_2$ during exposure to terrestrial meteoric H_2O vapor.

Introduction

EXAMINATION OF THE ROCKS brought back from the Apollo 16 mission showed that several had a faint "rusty" coating, and in one sample, 66095, this "rusty" coating was particularly prominent. It was suggested by the Preliminary Examination Team that this coating might be hydrated iron oxide and that it may have formed on the moon, because it was observed during the initial examination of the Apollo 16 samples at the Lunar Receiving Laboratory.

Because rock 66095 is the first lunar *rock* that showed promise of containing appreciable amounts of H_2O, it was obviously important to analyze the D/H and $^{18}O/^{16}O$ ratios of this water. After several requests we were finally able to obtain a 5 g sample of 66095, and the isotopic analyses of the H_2O extracted from this sample are discussed below.

Previous Work

D/H and $^{18}O/^{16}O$ analyses of H_2O are the two most important geochemical tracers that can be used to distinguish waters of varying origins. On the earth,

*Contribution No. 2481, Publications of the Division of Geological and Planetary Sciences, California Institute of Technology, Pasadena, California 91109.

waters of various origins have distinctly different isotopic compositions, and in many circumstances waters of meteoric, magmatic, metamorphic, and oceanic origin can all be distinguished one from the other by their $\delta^{18}O$ and δD values. Water from carbonaceous chondrite meteorities is also isotopically distinct. Thus, although we have no *a priori* way of knowing what truly indigenous lunar water should look like isotopically, it is likely that isotopic analyses can at least be used to evaluate the degree of contamination from terrestrial sources. Also, if H_2O from a lunar sample should have an isotopic composition distinct from any terrestrial water, a lunar origin would be indicated.

Small amounts of water have been found in all the lunar soils returned from every Apollo mission (e.g. see Epstein and Taylor, 1970, 1971, 1972; Friedman *et al.*, 1970). This water is typically present in concentrations of only about 7–20 μmoles/g (i.e. 125–360 ppm). However, even though H_2O is present in only small amounts in the lunar samples, it is very important to understand its origin and nature because of its possible importance in the original differentiation of the moon and because of the well-known importance of water in almost all geochemical and petrological processes.

The isotopic and concentration data obtained for the H_2O from the lunar soils strongly suggest that terrestrial atmospheric water vapor (meteoric H_2O) is the dominant source of this "lunar water," and the latter is therefore largely a result of terrestrial contamination (e.g. see Epstein and Taylor, 1970, 1971, 1972, 1973). This evidence can be briefly summarized as follows: (1) Almost all the H_2O in the lunar soils escapes much more readily during heating than any of the solar wind gases, including the very loosely bound helium. (2) Making allowances for cross contamination with the coexisting solar wind H_2, the estimated δD values of this H_2O are very similar to those of temperate zone atmospheric water vapor. (3) Most of the water in the lunar samples can, in a matter of a few hours, be isotopically exchanged with H_2O vapor at about 300°C. (4) The grain surfaces in the lunar fines have all undergone a great deal of radiation, probably of sufficient magnitude to produce chemically active surfaces that would be expected to readily adsorb H_2O. (5) Much of this H_2O is known to reside on the very outermost grain surfaces of the lunar fines, because it is largely "stripped off" during the initial partial fluorinations of the lunar soils, even though only 20–40 μmoles/g of silicate oxygen are removed (Taylor and Epstein, 1973). (6) The $\delta^{18}O$ values of the "lunar H_2O" in 3 lunar samples, determined by direct and indirect methods are very low ($\delta^{18}O \approx -20$ to -5), even though in some cases these values have probably been increased by exchange with the adjacent lunar silicates during the extraction procedure; the only common waters that have such low $\delta^{18}O$ values are terrestrial meteoric waters. (7) No OH-bearing mineral of clear-cut lunar origin has yet been found in any lunar sample. Lunar vulcanism and igneous activity clearly took place under conditions of very low P_{H_2O}. Because of abundant vesicles in many lunar basalts, there was a gas phase present, but it need not have contained significant H_2O. (8) There are significant amounts of essentially deuterium-free solar wind H_2 present in all lunar soils. However, the bulk of the "lunar H_2O" cannot have been produced from this solar wind H_2 by some kind of oxidation process on the lunar surface because this water is much too rich in deuterium.

Previous work on rock 66095 concerning the problem of origin of the hydrated iron oxides in this sample can be divided into two categories: (1) mineralogical studies on the nature of the iron hydroxide material; (2) hydrogen and oxygen isotopic studies by Friedman et al. (1974).

Friedman et al. (1974) observed δD values of about -75 to -200 and $\delta^{18}O$ of -1 to $+20$ in the H_2O liberated from rock 66095. Even though these δD values are essentially identical to those of common terrestrial meteoric waters as well as to the extrapolated δD values of the adsorbed H_2O on previously analyzed lunar soils (Epstein and Taylor, 1970, 1971, 1972, 1973), Friedman et al. (1974) concluded that the H_2O in rock 66095 was probably of lunar, not terrestrial, origin. To a large degree this conclusion was based on the very positive $\delta^{18}O$ values obtained on the H_2O. The latter are much higher than in terrestrial meteoric H_2O or in the H_2O obtained from terrestrial rust by Friedman et al. (1974).

Taylor et al. (1973), on the other hand, concluded from mineralogical studies on 66095 that the iron hydroxide in this lunar sample probably formed by hydration–oxidation of the mineral lawrencite ($FeCl_2$). Significant amounts of Cl were found in the rusty coating, and it is well-known that in a terrestrial environment lawrencite will hydrate and oxidize very rapidly. Recently, Taylor (1974) was able to demonstrate the definite presence of the metastable mineral akaganeite (β FeOOH) in the rusty coating of 66095. This mineral is known to rapidly form when $FeCl_2$ is exposed to a terrestrial oxidizing atmosphere. However, it will not form simply through exposure to H_2O in a reducing environment, because there is an oxidation involved in going from lawrencite to akaganeite. Other materials such as amorphous iron hydroxide or goethite may also be present in the rusty coating. The above results led Taylor et al. (1973) to conclude that the rusty coating on 66095 is a result of exposure to terrestrial contamination, either during sample return in the Apollo capsule or during handling here on Earth.

EXPERIMENTAL PROCEDURES AND RESULTS

Our method of extraction of H_2O from lunar samples involves a stepwise heating of the sample in vacuum by means of an RF induction furnace and the collection and purification of the water; the water is then quantitatively converted to H_2 or CO_2 for δD and $\delta^{18}O$ analysis. The same basic procedure described in Epstein and Taylor (1970), with some modifications, was used in the present study in analyzing the ferric hydroxides in rock 66095.

At the time our experiments were initiated, akaganeite had not been identified in 66095 and the exact nature of the "goethite" in 66095 was not known. In addition to true goethite or some other iron hydroxide mineral, it could have contained amorphous ferric hydroxide, in which some of the H_2O was chemisorbed or adsorbed on the surfaces of the grains. Consequently, we modified our analytical procedure utilizing certain precautions, so that as little

H_2O as possible was lost from sample 66095 prior to placing it in our analytical apparatus. Since partial evaporation can be accompanied by isotopic fractionations, we felt it was necessary to avoid excessive vacuum pumping on the sample or excessive exposure of the sample to the dry nitrogen atmosphere in the dry box.

Briefly our experimental procedure was as follows. Rock 66095 in its air-tight stainless steel O-ring container was placed in a P_2O_5-dried, N_2-filled, gloved dry box. Then 2–3 chips (each about 0.5–1.0 g) were quickly removed and transferred into a platinum crucible which had been previously heated to high temperature in vacuum; the crucible was then placed into a degassed glass vessel, thus isolating the lunar sample from the room atmosphere when it was taken out of the dry box. The glass vessel was then connected to the vacuum line and we proceeded with the stepwise extraction of the water. The transfer of the sample from the steel container to the glass vessel took less than 30 sec. The evacuation of the dry nitrogen from the vessel containing the sample of rock 66095 was also done very quickly, and the pressure in the glass envelope was not allowed to reach a value much below several hundred microns. Any further vacuum pumping on the sample was done after first assuring that any water given off by the sample would be trapped in a liquid nitrogen-cooled trap.

In our usual procedure of extraction of gases from lunar samples, we extracted the total amount of gas given off in a temperature interval and then separated and purified these gases for volume and isotopic analyses. In our extraction of H_2O from rock 66095, our aim was to get as many successive increments of H_2O as possible. We wanted to obtain samples of about 10–20 μmoles, and to analyze each of these separately.

The data in Table 1 and Fig. 1 show the amounts of water, isotopic analyses, and range of temperatures for each successive increment of H_2O sample. The δD and $\delta^{18}O$ values are defined in the usual way as the difference in per mil of the D/H or $^{18}O/^{16}O$ ratio relative to the well-known SMOW standard. As these data show, in some cases we were able to obtain several increments of H_2O from the sample at a single temperature. However, in other cases it was necessary to raise the temperature to obtain the 10–20 μmoles of H_2O desired for analysis.

Two different aliquots of rusty rock 66095 were used for extracting water for the δD analyses and for the $\delta^{18}O$ analysis. As Table 1 and Fig. 1 indicate, the δD and the $\delta^{18}O$ values of consecutive increments of extracted H_2O are very uniform; they are also not very different from the values in terrestrial meteoric waters. Note that appreciable quantities of H_2O were obtained by simple pumping at 25°C, and that the isotopic composition of this easily removed H_2O is virtually indistinguishable from the values obtained for the high-temperature H_2O.

To better understand the isotopic and concentration data obtained on rock 66095, we decided that it would be useful to carry out a number of stepwise pyrolysis experiments on terrestrial ferric hydroxide compounds of various types. Also, the aliquot of 66095 from which we had previously extracted water for $\delta^{18}O$ analysis was rehydrated for 5 hours at room temperature with water vapor that was in contact with a large reservoir of liquid water of known $\delta^{18}O$ and δD. It was

Table 1. The δD and $\delta^{18}O$ values of water extracted from lunar sample 66095.*

Temp. range	D/H extraction (1.627 g)		$^{18}O/^{16}O$ extraction (1.775 g)	
	μmoles H_2O/g	δD (‰)	μmoles H_2O/g	$\delta^{18}O$ (‰)
25°C	2.7	−92		
	3.1	−104	3.5	−17.8
	2.1	−146		
25–125°C	3.4	−98		
	4.1	−89		
			12.6	−17.2
125–200°C	6.0	−73		
	4.7	−94		
	5.1	−83		
200–700°C	4.2	−97		
	3.2	−136	2.7	−13.3
	1.4	−96		
Totals	40.0	−97	18.8	−16.6

*Stepwise pyrolysis extractions of H_2O utilizing small incremental increases of temperature by R. F. induction heating. The temperatures other than 25°C are only approximate in this table and in Tables 2, 3, and 4.

hoped that the rehydration process would form hydrous ferric material similar to that present initially in 66095 prior to its dehydration, and thus provide information on the δD and $\delta^{18}O$ of the water that might have produced the iron hydroxides in 66095.

Table 2 gives the quantities and the δD and $\delta^{18}O$ values for the water extracted from the rehydrated sample of 66095. It was necessary to rehydrate twice, once to obtain water for the δD values and then again to obtain hydration water for the $\delta^{18}O$ analyses. Note that about three times as much water was available after the first hydration than after the second hydration, even though the latter exposure time was longer (17 hours). The number of possible hydration sites apparently decreases after heating, perhaps because of annealing.

Table 3 gives the yields and isotope data for water obtained by the stepwise dehydration of powdered crystalline goethite, a terrestrial sample from Wisconsin. This crystalline goethite is a pseudomorph after pyrite, and contains some carbonate impurities.

Table 4 gives the H_2O yields and isotope data for water obtained by the stepwise dehydration of akaganeite (β FeOOH) made by L. A. Taylor in his laboratory at the University of Tennessee and shipped to us for isotope analysis. Note that, compared to the large δD variations typically observed during stepwise heating of lunar soils, the δD analyses given in Tables 2, 3, and 4 in each case are quite uniform, irrespective of the temperature interval of dehydration. However,

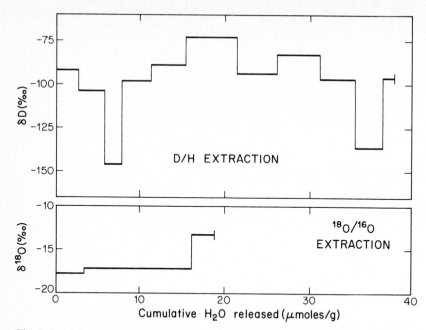

Fig. 1. Isotopic release patterns in H₂O during two dehydration experiments involving stepwise heating of 0.5–1.0 g chips of lunar sample 66095 in vacuum. The cumulative amounts of H₂O differ in the two extractions because different aliquots of sample were used and the H₂O is heterogeneously distributed in the sample. Details of the extraction procedure are given in the text and in Table 1.

Table 2. The δD and $\delta^{18}O$ values of water extracted from a sample of 66095 which had previously been hydrated with H_2O vapor using a meteoric H_2O sample having $\delta D = -39$ and $\delta^{18}O = -5.8$.

Temp. range	D/H extraction†		$^{18}O/^{16}O$ extraction†	
	μmoles H_2O/g	δD (‰)	μmoles H_2O/g	$\delta^{18}O$ (‰)
25°C	13.7	−30	4.0	−10.8
25–300°C	17.7	+13	5.6	−7.8
300–700°C	—	—	1.6	+2.6
Totals	31.4	−6	11.2	−7.4
			(9.6)*	(−9.0)*

*These values exclude the 1.6 μmoles/g of H_2O obtained at high temperatures because of the probable oxygen isotopic exchange with the silicate.

†For both extractions, the weight of the sample (2 fragments) was 1.775 g.

the $\delta^{18}O$ values are less systematic, and apparently are more dependent upon the detailed history of hydration and dehydration that each sample has undergone. This feature is discussed in more detail below.

Table 3. The δD and $\delta^{18}O$ values of water extracted from a terrestrial goethite sample from Wisconsin.

Temp. range	D/H extraction (22.6 mg)		$^{18}O/^{16}O$ extraction (48.0 mg)	
	μmoles H_2O/g	δD (‰)	μmoles H_2O/g	$\delta^{18}O$ (‰)
25–100°C	970	−113	630	−18.9
100–250°C	1370	−132	1670	−15.2
	1900	−154	1670	−19.7
250–750°C	660	−152	330	−5.2
Totals	4900	−139	4300	−16.7

DISCUSSION OF THE ANALYTICAL DATA

Concentration of H_2O in 66095

The data in Table 1 show that our sample of 66095 contains 20–40 μmoles of water per gram of rock. Two different portions of 66095 gave significantly different amounts of H_2O, indicating that the water is heterogeneously distributed in the rock. It is obvious from the large quantities of H_2O evolved during simple pumping in vacuum at 25°C ($\sim$20% of the total) that the H_2O concentrations must be to some extent dependent on the handling history of the sample. For example, comparison of our data with those of Friedman et al. (1974) indicates that they obtained even larger amounts of H_2O from 66095 at room temperature than we did (70 μmoles/g versus 8 μmoles/g). The order of magnitude higher concentration observed by Friedman et al. (1974) is probably due to the fact that they exposed their sample of 66095 to the terrestrial atmosphere while ours was transferred in a N_2-filled dry box. Note, however, that at temperatures higher than 25°C our measured H_2O concentrations in 66095 are similar to those of Friedman et al. (1974).

The total concentration of H_2O in our sample of 66095 is far in excess of that found in any other lunar basalt or rock breccia, and it is also somewhat higher than that in any other lunar soil or soil breccia. The release pattern of H_2O during vacuum heating of 66095 is also markedly different from the patterns obtained on lunar soils. Most of the H_2O extracted from lunar soils does not evolve until temperatures of about 200°C are reached, and in some cases as much as one-third of the water is not given off until the soil is heated to about 500°C. In contrast, about 20% of the water in 66095 is released at 25°C, and *most* of the water was removed by the time the temperature of the sample had been raised to approximately 200°C (Table 1). This dehydration pattern is somewhat similar to the pattern obtained for the terrestrial goethite (Table 3) and akaganeite (Table 4), but is not at all like that shown by the typical lunar soil. Obviously, the bulk of the

Table 4. The δD and $\delta^{18}O$ values of water extracted from a synthetic akaganeite.

| Temp. range | D/H extraction | | $^{18}O/^{16}O$ extraction | |
	μmoles H_2O/g	δD (‰)	μmoles H_2O/g	$\delta^{18}O$ (‰)
Sample A—(Divided into aliquots of 5.8 mg for D/H and 3.8 mg for $^{18}O/^{16}O$)				
25–75°C	2590	−86	1790	+7.7
75–150°C	2070	−91	2370	+13.5
150–350°C	3620	−113	2890	+3.7
350–700°C	2410	−110	3000	−5.3
Totals	10690	−102	10050	+4.1
Sample B—(An aliquot of 13.2 mg used for D/H extraction)				
	410	−72		
25°C	230	−69		
	780	−73		
	650	−76		
25–75°C	330	−71		
	650	−108		
	710	−101		
	920	−55		
75–150°C	1180	−86		
	2100	−128		
150–350°C	1080	−153		
350–600°C	540	—		
Totals	9580	−100		

H_2O in the lunar soils cannot be present as iron hydroxides, but must be bound very tightly to active sites on the grain surfaces in the lunar fines.

The presence of H_2O that can be so easily removed from rock 66095 by simple pumping at 25°C or even at 125°C suggests that such water cannot have been present in the rock during its original residence on the lunar surface. Much higher temperatures are attained in the moon's surface during the two-week lunar day, and considering the lunar vacuum and the lengths of time involved, it is inconceivable that such large amounts of low-temperature water could have been retained by the rock. Note that during our laboratory experiments these large amounts of H_2O were evolved from single fragments of 66095, not from the crushed or powdered rock. Much of this water must have been present as a surface coating on the rock chip, or present along accessible fractures.

Isotopic composition of H_2O in 66095

As shown in Fig. 1, the δD and $\delta^{18}O$ values of the H_2O obtained by stepwise heating of 66095 are fairly uniform. The $\delta^{18}O$ values show a slight tendency to become richer in ^{18}O in the higher temperature fractions, but this is perhaps to be expected either because of high-temperature oxygen isotope exchange with the adjacent silicate oxygen, or if the lighter isotopic species $H_2^{16}O$ (mass 18) diffuses away preferentially relative to $H_2^{18}O$ (mass 20). Note that such mass-dependent kinetic effects would probably be less important for D/H because the masses of HDO and $H_2^{16}O$ are only 19 and 18, respectively; also any such effects would tend to be masked by the much greater chemical fractionations exhibited by the hydrogen isotopes (where per mil variations in nature are typically an order of magnitude higher than for oxygen isotopes).

The uniform isotopic release pattern shown in Fig. 1 is extremely significant with regard to the origin of the H_2O in 66095, because it indicates that the higher temperature H_2O probably has the same origin as the low-temperature H_2O, and we have previously shown above that the latter could not have been present on the lunar surface. Thus, the isotopic evidence strongly favors the idea that essentially *all* of this water was added to 66095 subsequent to the arrival of the Apollo 16 astronauts on the moon.

Inasmuch as the only logical source of such H_2O contamination is terrestrial meteoric water, in Fig. 2 we plot the 66095 data on a δD–$\delta^{18}O$ diagram that also shows the meteoric water line. The latter is the locus of essentially all rain, snow, lake, river, and atmospheric water on the earth. Note that the different temperature fractions of H_2O from 66095 straddle the terrestrial meteoric H_2O line, and the average isotopic composition of the H_2O from 66095 plots only slightly above the line. Excluding some remarkable coincidence, these isotopic correlations *prove* that the bulk of the water in 66095 must be a result of terrestrial contamination.

Rehydration of 66095

In order to test the reversibility and the isotopic fractionations involved in the hydration process, the 66095 sample that had been previously dehydrated to obtain the $\delta^{18}O$ release pattern shown in Fig. 1 was rehydrated twice with a terrestrial meteoric water. As shown in Table 2 and Fig. 2 rehydration does indeed take place, but the subsequent H_2O release patterns behave in a different manner than did the original sample of 66095. Instead of having uniform δD and $\delta^{18}O$ values in the stepwise increments, the evolved H_2O shows a systematic progressive enrichment of 43‰ in D/H and 13‰ in $^{18}O/^{16}O$. This probably means that the hydrous compounds are different than in the original sample. It may be that original FeOOH in the 66095 sample, once dehydrated at temperatures up to 700°C, does not simply reform during later hydration. Instead, the dehydrated FeOOH may simply readsorb H_2O on active surfaces, or perhaps form amorphous iron hydroxides. It is plausible that such adsorbed H_2O would progressively

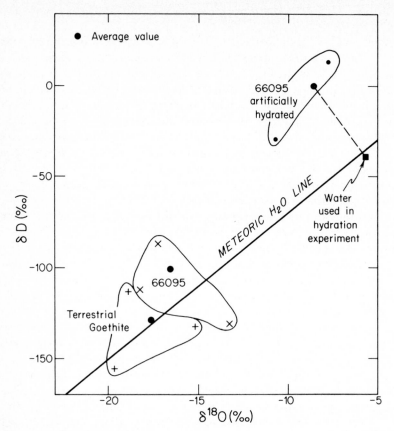

Fig. 2. δD–$\delta^{18}O$ diagram comparing the isotopic compositions of H_2O extracted during stepwise heating of sample 66095 with analogous data obtained from a terrestrial goethite sample formed from meteoric waters. Also shown are data obtained from a previously dehydrated sample of 66095 that was subsequently exposed at 25°C to water vapor in equilibrium with a large reservoir of liquid water having the isotopic composition shown by the solid square.

fractionate during heating, with the lighter isotopic species being evolved preferentially at the beginning; such effects might be smaller during dehydration of hydroxyl-bearing compounds.

It is clear that high-temperature heating changes 66095 in such a way as to make it less susceptible to rehydration, because even though it was exposed to 100% humidity it contains nowhere near as much H_2O as did the sample that Friedman *et al.* (1974) exposed to the terrestrial atmosphere. Also, our second rehydration experiment produced a factor of 3 lower adsorption of H_2O than was observed in the first rehydration. In addition, as shown in Fig. 2, the H_2O evolved from the artificially hydrated 66095 sample plots somewhat above the meteoric water line, whereas the H_2O from the original sample lies closer to the meteoric water line.

Dehydration of terrestrial goethite and akaganeite

A hydroxyl-bearing iron compound has been definitely identified in rock 66095 (Taylor *et al.*, 1973; P. M. Bell, personal communication). At first this was thought to be goethite, but it has recently been established that the mineral akaganeite (β FeOOH) is present. Akaganeite is only known to form by hydration–oxidation of $FeCl_2$. It is metastable at room temperature, and Cl or some other halogen is probably necessary for it to be preserved (G. Rossman, personal communication).

Our stepwise dehydration experiments on a sample of terrestrial goethite formed from meteoric waters (Table 3) show a pattern that is almost identical to that shown by 66095. Namely, the stepwise δD and $\delta^{18}O$ values are both very uniform, although the H_2O fraction obtained at 250–750°C is enriched in ^{18}O by about 10 per mil. This latter effect may be due to high-temperature exchange with the other $\frac{3}{4}$ of the goethite oxygen that remains behind after complete dehydration. If we exclude this high-temperature data on the $\delta D–\delta^{18}O$ diagram of Fig. 2, the different fractions of H_2O evolved from the goethite plot in the neighborhood of the meteoric water line in a fashion that is practically identical to the 66095 data. Note that the concentrations of H_2O measured on two goethite samples are 4900 and 4300 μmoles/g. These are lower than the theoretical yield of 5630 μmoles/g from stoichiometric goethite, presumably because of impurities in our sample.

The stepwise dehydration experiments on a sample of synthetic akaganeite made in Tennessee (Table 4) show a uniform δD pattern similar to that shown by both 66095 and by the Wisconsin goethite. The average δD value of the water in the akaganeite sample is identical to that in 66095. The $\delta^{18}O$ release pattern is, however, totally different; instead of the low $\delta^{18}O$ values characteristic of pristine meteoric H_2O the average $\delta^{18}O$ is +4, and even higher $\delta^{18}O$ values are observed in the low-temperature fractions. We do not understand the reasons for the peculiarly heavy $\delta^{18}O$ values, but one possibility is that after it was synthesized the mineral has "dried out" to a certain extent, with an accompanying preferential loss of $H_2^{16}O$ by nonequilibrium evaporation. Such a fractionation process would be expected to more strongly affect the loosely bound water, perhaps explaining why the highest temperature fraction in Table 4 has the most negative $\delta^{18}O$.

The analyzed akaganeite sample is a "fluffy" material with a high surface area; it contains about twice as much H_2O as does stoichiometric FeOOH. Also, the akaganeite crystal structure is made up of tubes that are known to contain loosely bound water (G. Rossman, personal communication). Such excess adsorbed water might fractionate ^{18}O more strongly during evaporation than does the hydroxyl in the akaganeite structure. Note that our $\delta^{18}O$ values on the synthetic akaganeite are very similar to those obtained by Friedman *et al.* (1974) on a sample of 66095 that outgassed a very large amount of H_2O at room temperature. Clearly, in view of its presence in the terrestrial akaganeite, the existence of heavy $\delta^{18}O$ values in H_2O from such samples cannot be considered diagnostic of a lunar or extraterrestrial origin; such heavy $\delta^{18}O$ values probably can be produced by the vagaries of the sample handling procedures.

Another possibility, however, is that the heavy $\delta^{18}O$ values shown in Table 4 may in fact be characteristic of akaganeite, and that the 66095 sample obtained by Friedman *et al.* (1974) contained more akaganeite than did ours. This is plausible because Friedman *et al.* (1974) obtained their sample a full year before we did, and during such a time interval the metastable akaganeite may have partially transformed to goethite (the stable form of FeOOH). More experiments will be necessary to finally work out these details.

CONCLUSIONS

Our δD analyses of H_2O liberated from 66095 are compared with the δD values obtained on the small amounts of "lunar water" from Apollo 11, 12, 14, 15, 16, and 17 regolith samples in Fig. 3. Note that both the high-temperature and low-temperature H_2O from 66095 are identical in δD to the δD "plateau" previously established by us for H_2O from the lunar soils. This "plateau" at high values of H_2O/H_2 represents the maximum δD value of H_2O observed in the lunar soils (Epstein and Taylor, 1973). Water extracted from soils in which there is appreciable deuterium-free solar wind hydrogen is always depleted in deuterium to some extent because of cross-contamination exchange during the high labora-

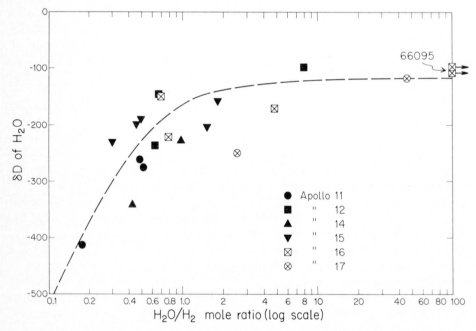

Fig. 3. Modification of Fig. 2 of Epstein and Taylor (1973) showing all of our isotopic data on lunar samples, with the addition of data points on 66095 (this work) and on some other lunar samples (S. Epstein and H. P. Taylor, unpublished data). The data points for 66095 lie well off to the right of the diagram; the lower point represents the average H_2O evolved at 25°C and the upper point represents the average H_2O evolved in the interval 25–700°C.

tory temperatures that are necessary to dehydrate the lunar soils. Inasmuch as sample 66095 has by far the highest H_2O/H_2 ratio of any lunar sample (≥ 400), the δD of its H_2O cannot have been isotopically affected by the laboratory cross-contamination.

It is clear from the systematics shown in Fig. 3 that the bulk of the H_2O in 66095 and the adsorbed water in the lunar soils almost certainly must have identical sources. Inasmuch as we have previously argued that the small amounts of water in the lunar soils are a terrestrial contaminant, Fig. 3 suggests a similar origin for the 66095 water.

The 66095 isotopic data are plotted on a δD–$\delta^{18}O$ diagram in Fig. 4 and compared with our previous isotopic data on water from lunar soils. Only a few $\delta^{18}O$ analyses have been done on H_2O from lunar soils, so the number of data

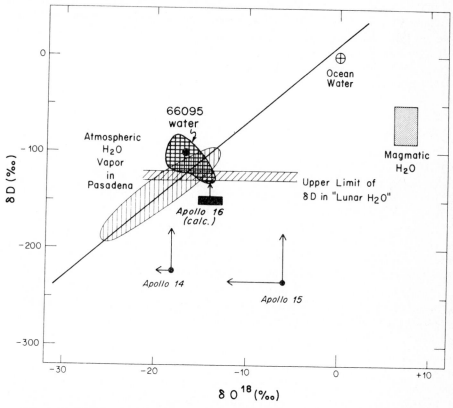

Fig. 4. Modification of Fig. 3 of Epstein and Taylor (1973) with the addition of the field obtained for sample 66095 from Fig. 2 of the present paper. The upper limit of δD in "lunar H_2O" is defined by the "plateau" in Fig. 3. The arrows indicate the directions in which the isotopic compositions of the other lunar H_2O samples should be corrected to take into account certain isotopic fractionations that accompany the laboratory extraction procedures (see text). The positions of the terrestrial meteoric water line, typical magmatic water, and ocean water are also shown for comparison.

points is much smaller than shown in Fig. 3. The comparisons in Fig. 4 further demonstrate the isotopic similarities between the 66095 H_2O and the adsorbed H_2O in lunar soils. Note that the H_2O data points from the lunar soils must be corrected as shown by the arrows in Fig. 4 to take into account (1) the D/H cross-contamination referred to above, and (2) oxygen isotope exchange between the H_2O and the adjacent high-^{18}O silicate oxygen during the high-temperature extraction. When these effects are taken into account, there is a very good correspondence in the isotopic compositions of all the "lunar water" samples.

The data in the present paper thus further substantiate our previous conclusion that most of the H_2O so far identified in lunar samples is a minor terrestrial contaminant. It is clear that the lunar surface samples have undergone a long exposure history involving exposure (1) to the lunar surface vacuum, (2) to the high surface temperatures produced by solar heating during lunar daylight, and (3) to a massive particle and radiation bombardment. These effects, combined with the absence of any yet identifiable source of indigenous lunar water, have apparently produced materials so totally dehydrated and with so many surface-active sites that they almost instantaneously adsorb any H_2O with which they come into contact. Even so-called "perfectly sealed samples" brought back from the moon must have been exposed to some H_2O outgassed from the astronauts backpacks and from the other Apollo paraphernalia.

Note that even the orange soil, 74220, the least irradiated and most carefully handled lunar soil sample, contained as much as 4.6 μmoles H_2O/g. Although this is the lowest concentration of H_2O yet reported from a lunar soil, even this carefully handled sample must have adsorbed some H_2O (Epstein and Taylor, 1973). Because of its very high H_2O/H_2 ratio the δD value of 74220 plots near the right-hand side of the "plateau" shown on Fig. 3. Thus the lunar sample with the highest H_2O concentration (66095) contains H_2O with almost exactly the same δD value as that shown by the lunar soil with the *lowest* H_2O concentration.

What is the explanation for the high concentrations of terrestrial meteoric H_2O in 66095? This problem has been dealt with by Taylor *et al.* (1973) and we basically agree with their conclusion. An easily hydrated chemical compound of some type *must* have been present in 66095 at the time the Apollo 16 astronauts landed on the moon. In view of the identification of Cl in the rust, the most likely candidate is lawrencite, $FeCl_2$, which in contact with even small amounts of H_2O and O_2 quickly hydrates to akaganeite and/or amorphous iron hydroxides, with attendant formation of HCl. Other "volatile" elements such as Zn, Pb, Cd, Bi, Tl, etc. are also abnormally enriched in 66095 (Nunes and Tatsumoto, 1973; El Goresy *et al.*, 1973; Krahenbuhl *et al.*, 1973), so it is reasonable that whatever event was responsible for the high concentrations of these "volatile" elements was also responsible for the addition of abnormal amounts of Cl to the rock.

Some workers have suggested that these "volatile" elements were added to 66095 by a lunar or meteoritic hydrothermal event that also produced the rusty coatings. We cannot totally rule out such a phenomenon, but if it happened, any hydrous minerals that were formed must have subsequently largely dehydrated on the moon, only to be rehydrated again upon coming into contact with H_2O vapor of terrestrial origin. It seems more likely to us that the addition of the "volatile" elements and formation of lawrencite in 66095 was essentially an anhydrous event.

There is some evidence in our data on synthetic akaganeite and in the data of Friedman *et al.* (1974) on a particularly H_2O-rich sample of 66095 that there exist isotopic fractionations of the type expected during nonequilibrium evaporation of water. For such fractionations the ^{18}O effects are significantly larger than the deuterium effects. However, there is a lack of such fractionation effects in both the well-crystallized goethite and in the 66095 sample that we have analyzed. The $\delta^{18}O$ values obtained from our sample of akaganeite are somewhat erratic and in a recent attempt to extract H_2O from another sample of akaganeite we ran into difficulties probably because of the presence of ferrous chloride hydrate and the possible production of HCl. It is important to do additional experiments to determine the relationship between the δD and $\delta^{18}O$ dehydration patterns of the water extracted in a stepwise fashion from various forms of FeOOH to ascertain the significance of the $\delta^{18}O$ patterns observed. Only then will we really understand all the implications of the isotopic values found for water extracted from these minerals.

Although we have argued both here and in previous publications that the water that we have extracted from lunar materials is primarily of terrestrial origin, nevertheless, we recognize that certainly some true lunar water may exist on the moon. For example, the oxidation of solar wind or solar flare H_2 will produce H_2O that has a very low deuterium concentration. However, the D/H ratio of small amounts of any such water formed on the moon could fluctuate wildly because of the presence of possible oxidized D-rich spallation products that vary in amount from rock to rock. We also recognize the possibility that lunar water may exist that just by coincidence happens to have a D/H and $^{18}O/^{16}O$ ratio similar to terrestrial meteoric H_2O. Such a lunar H_2O cannot be proven *a priori* to be entirely absent by any stable isotope technique. Our main point is that such lunar water has as yet not been observed nor its isotopic composition measured in any mineral or fluid inclusion that is definitely of lunar origin. Certainly it is most difficult to defend a conclusion that the water extractable from rock 66095 is lunar in origin unless some unlikely coincidences have taken place.

Acknowledgments—We wish to thank Paul Yanagisawa for carrying out some of the laboratory work, L. A. Taylor for supplying the samples of synthesized and identified akaganeite, C. Yapp for supplying the goethite sample, and I. Friedman for permission to discuss his unpublished data on 66095. We are also grateful for helpful discussions with L. A. Taylor, P. M. Bell, G. R. Rossman, M. N. Bass, and L. T. Silver. This research was supported by the National Aeronautics and Space Administration Grant No. NGL-05-002-190.

REFERENCES

El Goresy P., Ramdahr P., Pavićević M., Medenbach O., Müller O., and Gentner W. (1973) Zinc, lead, chlorine and FeOOH-bearing assemblages in the Apollo 16 sample 66095: Origin by impact of a comet or a carbonaceous chondrite? *Earth Planet. Sci. Lett.* **18**, 411–419.

Epstein S. and Taylor H. P., Jr. (1970) The concentration and isotopic composition of hydrogen,

carbon, and silicon in Apollo 11 lunar rocks and minerals. *Proc. Apollo 11 Lunar Sci. Conf.,
Geochim. Cosmochim. Acta*, Suppl. 1, Vol. 2, pp. 1085–1096. Pergamon.

Epstein S. and Taylor H. P., Jr. (1971) O^{18}/O^{16}, Si^{30}/Si^{28}, D/H, and C^{13}/C^{12} ratios in lunar samples. *Proc.
Second Lunar Sci. Conf., Geochim. Cosmochim. Acta*, Suppl. 2, Vol. 2, pp. 1421–1441. MIT Press.

Epstein S. and Taylor H. P., Jr. (1972) O^{18}/O^{16}, Si^{30}/Si^{28}, C^{13}/C^{12}, and D/H studies of Apollo 14 and 15
samples. *Proc. Third Lunar Sci. Conf., Geochim. Cosmochim. Acta*, Suppl. 3, Vol. 2, pp. 1429–1454.
MIT Press.

Epstein S. and Taylor H. P., Jr. (1973) The isotopic composition and concentration of water, hydrogen,
and carbon in some Apollo 15 and 16 soils and in the Apollo 17 orange soil. *Proc. Fourth Lunar Sci.
Conf., Geochim. Cosmochim. Acta*, Suppl. 4, Vol. 2, pp. 1559–1575. Pergamon.

Friedman I., Gleason J. D., and Hardcastle K. G. (1970) Water, hydrogen, deuterium carbon, and C^{13}
content of selected lunar material. *Proc. Apollo 11 Lunar Sci. Conf., Geochim. Cosmochim. Acta*,
Suppl. 1, Vol. 2, pp. 1103–1109. Pergamon.

Friedman I., Hardcastle K. G., and Gleason J. D. (1974) Water and carbon in rusty lunar rock 66095.
Science. In press.

Krahenbuhl U., Ganapathy R., Morgan J. W., and Anders E. (1973) Volatile elements in Apollo 16
samples: Possible evidence for outgassing of the Moon. *Science* **180**, 858–861.

Nunes P. D. and Tatsumoto M. (1973) Excess lead in "rusty rock" 66095 and implications for an early
lunar differentiation. *Science* **182**, 916–920.

Taylor H. P., Jr. and Epstein S. (1973) O^{18}/O^{16}, Si^{30}/Si^{28} studies of some Apollo 15, 16, and 17 samples.
Proc. Fourth Lunar Sci. Conf., Geochim. Cosmochim. Acta, Suppl. 4, Vol. 2, pp. 1657–1679.
Pergamon.

Taylor L. A., Mao H. K., and Bell P. M. (1973) "Rust" in the Apollo 16 rocks. *Proc. Fourth Lunar Sci.
Conf., Geochim. Cosmochim. Acta*, Suppl. 4, Vol. 1, pp. 829–839. Pergamon.

Proceedings of the Fifth Lunar Conference
(Supplement 5, Geochimica et Cosmochimica Acta)
Vol. 2 pp. 1855–1868 (1974)
Printed in the United States of America

Accumulation and isotopic evolution of carbon on the lunar surface

John F. Kerridge and I. R. Kaplan

Institute of Geophysics, University of California, Los Angeles, California 90024

F. David Lesley

Department of Mathematics, San Diego State University, San Diego, California 92115

Abstract—The processes of hydrogen stripping, agglutination and impact comminution, acting upon a regolith containing both indigenous and solar wind-implanted carbon, provide a quantitative explanation for the relationship between carbon abundance and isotopic composition observed for Apollo 16 soils.

Introduction

With the discovery that Apollo 11 soils were substantially enriched in carbon over their ancestral rocks, came the realization that most of the C on the lunar surface derived from extralunar sources. These sources were readily identified with, principally, the solar wind (Epstein and Taylor, 1970; Kaplan *et al.*, 1970) and, secondarily, meteorites, cometary particles and the lunar atmosphere (for review, see Hayes, 1972). The fact that soil C was found to be considerably augmented in ^{13}C with respect to rock C (Kaplan *et al.*, 1970) was attributed either to "heavy" solar wind or to fractionating processes on the lunar surface (Kaplan *et al.*, 1970; Epstein and Taylor, 1972; Moore *et al.*, 1973). However, a quantitative treatment of both accumulation and fractionation processes has been lacking and this paper represents a progress report on the construction of a model describing the abundance and isotopic composition of lunar C in terms of processes known, or believed, to be active on the surface of the moon. A model describing accumulation alone has been recently presented by DesMarais *et al.* (1973a, 1973b) and its relationship to our model will be discussed below.

Our approach has consisted of a number of steps which may be briefly summarized. First, sources and processes most responsible for lunar C dynamics were identified. Second, differential equations were written to describe the effects of each source and process on the abundance of both total C and ^{13}C. Third, these equations were combined and integrated to yield expressions for total C and ^{13}C as functions of time. Fourth, since a reliable measure of surface exposure age is not available for each sample, this term was eliminated by expressing C abundance in terms of ^{13}C. This is equivalent to defining the rate of the accumulation process in terms of the rate of the isotope fractionation process. Finally, the predicted relationship between C abundance and ^{13}C content was compared with appropriate experimental data for lunar soils. We conclude that our model provides an acceptable approximation to reality, although some details need to be elaborated

1855

further and a statistically firmer base of experimental data is required for refinement of imprecisely known parameters.

Derivation of the Model

Regolith C consists of two components, one correlated with surface area (S) and the other with volume (V). It is convenient to treat S in terms of the surficial density of carbon, $D\,\mu g\,C\,cm^{-2}$, and the geometrical surface area, $A\,cm^2\,g^{-1}$. Fresh ejecta from an impact event, the starting material of the regolith, are characterized by values of S, V, and A equal to 0, $V(0)$, and $A(0)$, respectively, which increase after time, t, to $S(t)$, $V(t)$, and $A(t)$. Our definition of t is the average, cumulative residence time of a constituent grain on the very surface. It should be noted that, during the period of grain size evolution, described below, our measure of sample "age," its exposure to solar wind bombardment, is not linearly related to ages determined by such parameters as cosmic ray exposure.

Sources of carbon

For this model, we consider only the solar wind as the source of extralunar C. Clearly, this is an oversimplification since the existence of a solid extralunar component, of either meteoritic or cometary origin, or both, has been well established, via trace element analysis, in lunar soil (e.g. Ganapathy *et al.*, 1970; Baedecker *et al.*, 1971). Because at least a major part of this component appears to have the composition of C1 carbonaceous meteorites (for review, see Anders *et al.*, 1973), it is reasonable to suppose that some C is reaching the surface of the moon in solid form. However, our ignorance of the way in which a potentially volatile species, such as C, behaves during hypervelocity impact precludes quantitative treatment at this time. The possible existence of a small C component due to reimplantation of atmospheric species was considered by Hayes (1972) who concluded that it could supply at most 10 ppm C to a mature soil. Consequently, a component resulting from implantation of photoionized species is ignored at the present time.

The magnitude of the solar wind input of C was calculated from the cosmic abundance tables of Cameron (1973) and the proton flux observed by spacecraft in the solar wind (Wolfe, 1972). An atomic C/H ratio of 3.7×10^{-4} and a proton flux of $2 \times 10^8\,cm^{-2}\,sec^{-1}$ lead to a C flux of $5 \times 10^{-5}\,\mu g\,cm^{-2}\,y^{-1}$. Uncertainty in the abundance of C relative to H could be as high as a factor of four, compared with which the uncertainty in the proton flux is trivial.

The abundance of indigenous C varies significantly from site to site. We shall argue later that samples from the Apollo 16 landing site serve as the most sensitive assay for extralunar surface processes. Although local igneous rocks at this site are characterized by very low C abundances (Moore *et al.*, 1973), the C contribution from all lithic materials which predate the present regolith is actually quite high. We have elsewhere calculated it to be about 40 ppm and reasonably uniform throughout the landing site (Kerridge *et al.*, 1974). It seems probable that much of this C is of extralunar origin which became incorporated into the moon during the epoch which culminated in formation of the breccias which dominate the lithology at the Apollo 16 site.

Because the isotopic composition of the solar wind is not initially known, we have assumed an arbitrary value for the solar wind $^{13}C/^{12}C$ ratio. We have then calculated fractionations relative to this ratio and finally attempted to relate this arbitrary scale to the standard conventionally employed in the laboratory for isotope ratios expressed as $\delta^{13}C$. A flux of ^{13}C in the solar wind equal to $5 \times 10^{-7}\,\mu g\,cm^{-2}\,y^{-1}$ was assumed, i.e. a $^{13}C/^{12}C$ ratio of 0.01. The isotopic composition of indigenous C is known at the outset only in terms of the experimental scale and not relative to the solar wind value. However, it turns out that the magnitude of the indigenous C isotope ratio has a negligible effect on the isotopic values eventually achieved by the lunar soils and so, as a first approximation, indigenous C is assigned the same $^{13}C/^{12}C$ ratio as the solar wind.

Lunar surface processes

Implantation. The fact that lunar soil grains contain an implanted solar wind component has been established via a number of lines of evidence (e.g. Eberhardt *et al.*, 1970; Borg *et al.*, 1971; Leich *et al.*,

1973). Although the actual depth of penetration is somewhat uncertain it is sufficiently small, less than 2000 Å, that this component, at least initially, may be regarded as purely surficial. Retention efficiencies for implantation of different elements at solar wind energies are generally unknown. However, C is generally believed to be efficiently retained (DesMarais *et al.*, 1974) and so we have assumed 100% sticking efficiency. Because only one side of a grain is irradiated at a time, the solar wind flux, $F \mu g \, C \, cm^{-2} \, y^{-1}$, is reduced by a shape factor, q, to which we have assigned a value of 0.25. Thus, the increase in density of surficial carbon, D, due to solar wind implantation in time, dt, is

$$F \cdot q \cdot dt \tag{1}$$

Increase in surface area. The mean size of grains in the lunar regolith tends initially to decrease with time resulting in a corresponding increase in specific surface area, if particle shapes are assumed to remain the same. An equilibrium grain size is finally achieved, which has been interpreted as representing a balance between the competitive effects of impact comminution and growth by agglutination (McKay *et al.*, 1972). The exact form of the surface area evolution curve is not known. A plot of mean grain size against $^{21}Ne_c$ age (Kerridge *et al.*, 1974) suggests an exponential decay curve and we have assumed an expression of the form:

$$A(t) = A(0) + G(1 - e^{-Bt}) \tag{2}$$

Although the empirical rate constant, B, in Eq. (2) could apparently be calculated from the $^{21}Ne_c$ data, this would relate to measurement of t on a different scale from the one defined earlier. The $^{21}Ne_c$ age is a measure of the time which the sample has spent at depths less than one or two meters, whereas the time scale used here is based upon exposure to radiation with a penetration depth of about 2000 Å. The problem of evaluating B indirectly will be considered below.

Estimation of values for the other empirical constants, $A(0)$ and G, also presents problems, as few literature data exist on the geometrical surface areas of lunar soils. Fortunately, data do exist for measurements of mean grain size of several soils of varying maturity (e.g. Heiken *et al.*, 1973; Butler *et al.*, 1973) and from these one may conclude that the equilibrium grain size is about 0.4 times that of pristine ejecta, i.e. maturation increases the specific surface area of a soil over its initial value by a factor of about 2.5. Mature soils are characterized by a geometrical surface area of about 250 cm² g⁻¹ (Fanale *et al.*, 1971) so that values of 100 cm² g⁻¹ and 150 cm² g⁻¹ were assigned to $A(0)$ and G, respectively, by considering limiting cases for Eq. (2).

Hydrogen stripping. It has been suggested that solar wind H bombarding the surface of lunar soil grains will tend to react with C, already implanted there, to form CH_4 (Abell *et al.*, 1970), much of which will diffuse out of the grain and be lost into the lunar atmosphere and ultimately into space (Kaplan *et al.*, 1970). Certainly, evidence for the existence of CH_4 attributed to *in situ* hydrogenation of C in lunar soils has been advanced by Cadogan *et al.* (1972); however, neither the rate of reaction nor the rate of diffusive loss is known *ab initio*. It is worth mentioning that all the surficial C appears to reside within range of solar wind H, evidence for which has been found to a depth of about 4000 Å (Leich *et al.*, 1973). Evidence for the reactivity of solar wind H comes from the observation of metallic Fe in lunar fines, apparently produced by *in situ* reduction of lunar Fe^{++} (Gammage and Becker, 1971; Housley *et al.*, 1973; Kerridge *et al.*, 1974). The change in D as a result of H stripping in time dt is

$$-K \cdot D(t) \cdot dt \tag{3}$$

The constant, K, incorporates both the rate of reaction between C and H and the rate of loss of CH_4 by thermal diffusion. Assignment of a value to K will be considered below.

The mass dependence of the diffusion process results in $^{12}CH_4$ being lost preferentially to $^{13}CH_4$. Assuming that the rates of diffusion are inversely proportional to the square roots of the respective molecular weights, and that the rate of diffusion is the dominant part of K, it follows that a value of K for $^{13}CH_4$ will be equal to 0.97 times the value for $^{12}CH_4$.

Agglutination. Aggregates of grains welded together by glass, apparently formed during micrometeorite impact (McKay *et al.*, 1972), are a common feature of lunar soils. This agglutination process has three consequences for the C chemistry of the lunar surface. First, some originally surficial C is translated into a volume-correlated component (DesMarais *et al.*, 1973a). Second, C is lost from the molten fraction by volatilization. This loss is assumed to be effectively complete and to involve no isotopic fractionation. Third, volume-correlated C produced by burial of implanted solar wind C is no longer vulnerable to H stripping and is therefore "frozen in" both with respect to its abundance and to its isotopic composition. The change in D due to agglutination in time dt is:

$$-P \cdot D(t) \cdot dt \tag{4}$$

The constant, P, is equal to the rate of agglutination as fractional increase in y^{-1}, and appropriate numerical values are considered below, together with those for B and K.

Surficial carbon

Total incremental change in density of surficial C is given by adding expressions (1), (3), and (4):

$$dD = F \cdot q \cdot dt - K \cdot D(t) \cdot dt - P \cdot D(t) \cdot dt \tag{5}$$

Integration of this expression gives:

$$D(t) = \frac{Fq}{P+K} (1 - e^{-(P+K)t})$$

Thus, the magnitude of the surface-correlated component is given by:

$$S(t) = \frac{Fq}{P+K} (1 - e^{-(P+K)t}) [A(0) + G(1 - e^{-Bt})] \tag{6}$$

Volume carbon

Similarly, the change in V through addition of formerly surficial C in time dt is

$$(1-y) \cdot P \cdot D(t) \cdot A(t) \cdot dt \quad \text{or} \quad (1-y) \cdot P \cdot S(t) \cdot dt \tag{7}$$

where y is the fraction of each agglutinate which was completely melted and consequentially decarbonized. Microscopic examination of agglutinates suggests a value of about 0.5 for y, although Bibring *et al.* (1974) have reported that electron microscopy and diffraction reveal only about 10% glass in agglutinates. This question should be resolved. Volume-correlated C is considered to be lost from molten agglutinate material together with its surficial C. The corresponding change in V in time dt is:

$$-y \cdot P \cdot V(t) \cdot dt \tag{8}$$

Total incremental change in volume C is given by addition of expressions (7) and (8):

$$dV = (1-y) \cdot P \cdot S(t) \cdot dt - y \cdot P \cdot V(t) \cdot dt \tag{9}$$

Substituting for $S(t)$ and rearranging we obtain:

$$\frac{dV}{dt} + y \cdot P \cdot V(t) = \frac{Fq(1-y)P}{P+K} (1 - e^{-(P+K)t}) [A(0) + G(1 - e^{-Bt})]$$

Multiplying by the integrating factor we have:

$$e^{yPt} \cdot \frac{dV}{dt} + yPe^{yPt} \cdot V(t) = e^{yPt} \frac{Fq(1-y)P}{P+K} (1 - e^{-(P+K)t}) [A(0) + G(1 - e^{-Bt})]$$

so that

$$\frac{d}{dt} [e^{yPt} \cdot V(t)] = e^{yPt} \frac{Fq(1-y)P}{P+K} (1 - e^{-(P+K)t}) [A(0) + G(1 - e^{-Bt})]$$

Integrating we obtain:

$$e^{yPt} \cdot V(t) = \frac{Fq(1-y)P}{P+K} \int_0^t e^{yPt}(1 - e^{-(P+K)t})[A(0) + G(1 - e^{-Bt})] \cdot dt + V(0).$$

Whence:

$$V(t) = \frac{Fq(1-y)P}{P+K} \left\{ \frac{[A(0)+G](1-e^{-yPt})}{yP} + \frac{G(e^{-yPt}-e^{-Bt})}{yP-B} \right.$$

$$\left. + \frac{[A(0)+G](e^{-(P+K)t}-e^{-yPt})}{(1-y)P+K} + \frac{G(e^{-yPt}-e^{-(P+K+B)t})}{(1-y)P+K+B} \right\}$$

$$+ V(0) \cdot e^{-yPt}$$

$$(10)$$

Finally, adding Eqs. (6) and (10), and taking $V(0) = C(0)$, the expression for total C content at time t is

$$C(t) = \frac{Fq}{P+K} \left\{ (1-e^{-(P+K)t})[A(0)+G(1-e^{-Bt})] + \frac{(1-y)[A(0)+G](1-e^{-yPt})}{y} \right.$$

$$+ \frac{P(1-y)G(e^{-yPt}-e^{-Bt})}{yP-B} + \frac{P(1-y)[A(0)+G](e^{-(P+K)t}-e^{-yPt})}{(1-y)P+K}$$

$$\left. + \frac{P(1-y)G(e^{-yPt}-e^{-(P+K+B)t})}{(1-y)P+K+B} \right\} + C(0) \cdot e^{-yPt}$$

$$(11)$$

This expression may be applied to either ^{12}C or ^{13}C by employing the appropriate values for F, $C(0)$, and K. Isotopic compositions are expressed in terms of $\delta^{13}C$, which is defined as

$$\delta^{13}C = \frac{^{13}C/^{12}C - (^{13}C/^{12}C)_{std}}{(^{13}C/^{12}C)_{std}} \times 1000$$

$$(12)$$

The standard in the case of our calculated functions is the arbitrary value, 0.01, assigned to the solar wind.

Equation (11) may therefore be used, if the various constants are known, to calculate C and $\delta^{13}C$ in terms of t, and hence in terms of each other.

Because the magnitudes of the parameters K, P, and B are currently unknown, a trial-and-error procedure was adopted in order to derive tentative values. However, first it is necessary to consider the experimental data with which the model is to be compared.

EXPERIMENTAL RESULTS

In order to follow secular changes in the chemistry of the lunar regolith it is desirable to use soil samples with a range of exposure ages but with basically the same major element composition. Such samples may be found at the Apollo 16 landing site at Cayley–Descartes, in which several relatively recent craters (AFGIT, 1973) penetrate a regolith which is fairly homogeneous with respect to the major rock-forming elements (LSPET, 1972). Thus, the total Fe contents of the soils which we studied showed a standard deviation of only $\pm 14\%$. Although the exact mechanisms of C retention are not clear, Pillinger et al. (1973) have shown that carbide production is strongly influenced by Fe content and it seems likely that such major element effects are important in controlling the retention of chemically active species such as C. Our results for the abundances and isotopic compositions of a number of light elements for fourteen soils and five rocks from this site are published elsewhere (Kerridge et al., 1974). In Fig. 1 we illustrate the observed relationship between total C and $\delta^{13}C$ for soils, employing three data points from Epstein and Taylor (1973) in addition to our own results. We conclude that a significant trend is apparent with total C increasing in a nonlinear fashion with $\delta^{13}C$. This increase also matches a number of rather crude estimates of increasing

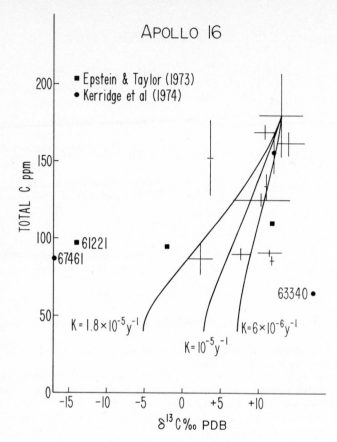

Fig. 1. Relationship between C and δ^{13}C observed for Apollo 16 soils compared with predictions of the present model. The effect of varying the rate of H stripping, K, is illustrated, the optimum value being 10^{-5} y^{-1}. Coefficients P and B in Eq. (11) were held constant at their optimum values of 8.1×10^{-6} y^{-1} and 9×10^{-6} y^{-1}, respectively.

surface exposure age based upon CH_4 released during acid hydrolysis (Cadogan *et al.*, 1972), solar wind helium, and Fe^0 formed by *in situ* reduction of lunar Fe^{++} by solar wind H (Kerridge *et al.*, 1974).

Two anomalous points in Fig. 1 should be noted. Sample 61221 is believed to have acquired a significant proportion of its C, and other volatile elements, from the impact of a comet (Gibson and Moore, 1973) or carbonaceous meteorite (Epstein and Taylor, 1973) so that its C systematics are unlikely to match the normal solar wind-based pattern. On the other hand, sample 67461 contains C of probably solar wind origin which has experienced little, if any, of the isotopic fractionation which characterizes other soils. We have attributed this phenomenon to the fact that 67461 is a fillet soil in which most of the grains were irradiated while still contained within the host boulder (Behrmann *et al.*, 1973) before being deposited into an environment in which they experienced virtually no turnover

(Kerridge *et al.*, 1974). We also point out that the $\delta^{13}C$ value of this sample ($-17\%_o$ PDB) may represent the best measurement to date of the isotopic composition of solar wind C. However, because of the somewhat speculative nature of the arguments involved, we do *not* start with this value in our present calculations, the results of which, in fact, suggest a value closer to zero on the PDB scale.

A third point, that due to sample 63340, also lies well away from the main sequence in Fig. 1, but for this we have no reasonable explanation.

Shown also in Fig. 1 are some curves derived from Eq. (11) utilizing tentative values of the rate constants K, P, and B as described in the next section. Carbon and $\delta^{13}C$ model data for constructing these curves were calculated using a WATF IV program on the IBM 360 computers at SDSU and UCLA.

EVALUATION OF RATE CONSTANTS

The experimentally derived relationships between C, $\delta^{13}C$, mean grain size and agglutinate content can be used to impose useful constraints on the values which may be assigned to K, P, and B. Acquisition of data on further samples will improve the statistical weight of the observations and permit yet more rigorous definition of the rate constants. For simplicity we give the values calculated for K, P, and B which give the best fit to the data and show how deviations from these values perturb the model relationships away from the experimental data. The optimum values are $K = 10^{-5}\,y^{-1}$, $P = 8.1 \times 10^{-6}\,y^{-1}$ and $B = 9 \times 10^{-6}\,y^{-1}$.

Figure 1 compares the experimental data, for C and $\delta^{13}C$, with model curves calculated for the optimum values given above and also for deviations of K by a factor of about 1.8 either greater or smaller. The isotopic fractionations predicted by the model are relative to the $^{13}C/^{12}C$ ratio assumed for the solar wind. This arbitrary scale is brought into coincidence with experimental scale, relative to PDB, by assuming that the abundance value of 180 ppm C corresponds to the isotopic value of $+13\%_o$ PDB for both calculation and experiment. It is apparent that the lower value of K fails to yield an acceptable degree of isotopic fractionation. The higher value of K is also unacceptable, although this is not obvious from Fig. 1, because the surface residence time required to build up the observed levels of C is probably incompatible with the age of the regolith. An approximate upper limit to such a residence time may be inferred from drill stem data. Uniform irradiation of grains to a depth of at least 2.4 m (Price *et al.*, 1973) in a regolith whose total age is about 4 G.y. (Tera *et al.*, 1973; Schaeffer and Husain, 1973) constrains each grain of mean diameter 60 μm (Heiken *et al.*, 1973) to a surface residence of about 10^5 y.

The relationship between C and $\delta^{13}C$, illustrated in Fig. 1, is not strongly affected by changes in P and B. In order to constrain these factors more closely, measured quantities such as mean grain size and agglutinate content may be used. In fact, the evolution of grain size within the regolith, of which Eq. (2) is an idealized representation, is controlled by the balance between the breaking down of grains by impact and the building up of welded aggregates by agglutination. Thus, the rate at which equilibrium grain size is reached is at least partially

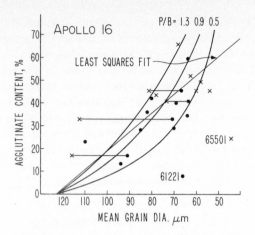

Fig. 2. The rate of agglutination, P, and the rate at which mean grain size approaches equilibrium, B, are closely related, as shown by data for Apollo 16 soils. This relationship can be used to constrain values of the ratio P/B to within 0.4 of the optimum value of 0.9. Mean grain size data, for 0–1 mm fines, from Heiken *et al.* (1973) and Butler *et al.* 1973. Agglutinate contents from Heiken *et al.* (1973), Adams and McCord (1973) and Kerridge *et al.* (1974). As these three groups studied different grain size fractions, which revealed systematic differences in agglutinate content, these data have been normalized to those of Heiken *et al.*

dependent on the rate of agglutination, i.e. B and P are closely related and this is borne out by Fig. 2 in which agglutinate content is plotted against mean grain size, M_z, for a number of Apollo 16 soils. The data correlate strongly with an intercept at zero agglutinate content of 123 μm, which we take as the mean grain size of fresh ejecta. Shown also in Fig. 2 are curves predicted by the model, assuming that geometrical surface area is inversely proportional to mean grain size and that the value taken for $A(0)$ corresponds to an M_z value of 123 μm, as found above. A ratio of P/B equal to 0.9 is found to give the optimum fit to the data, with virtually all the data being contained within uncertainty limits of ±0.4. Note, however, that the agglutinate data relate to the 90–150 μm grain size fraction and this may not be representative of the soil.

The effect of varying B, while maintaining a P/B ratio of 0.9, is illustrated in Fig. 3 in which total C is plotted against mean grain size for a number of Apollo 16 soils. Model curves are drawn for the optimum value of B and for upper and lower limits constrained by the data. These values correspond to a factor of about 1.4 greater or smaller than the optimum value of $9 \times 10^{-6} \, y^{-1}$. Consequently we can give the following tentative values to the rate constants in Eq. (11): $K = 1.0^{+0.8}_{-0.4} \times 10^{-5} \, y^{-1}$; $B = 9.0 \pm 4.0 \times 10^{-6} \, y^{-1}$; $P = B \times 0.9 \pm 0.4$.

DISCUSSION

Work on this model is not complete; however, it is already capable of predicting the salient features of the relationships observed to exist between total

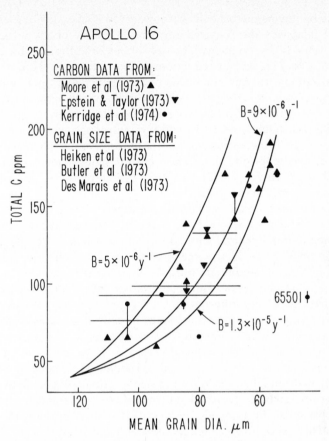

Fig. 3. Relationship between C and mean grain size observed for Apollo 16 soils can be used to constrain permissible values of the coefficient B. An optimum value of 9×10^{-6} y^{-1} is found, with limits of ± 4.0. The ratio P/B was held constant at 0.9 and the coefficient K was maintained at 10^{-5} y^{-1}.

C abundance, δ^{13}C, mean grain size and agglutinate content. In particular, both the general nature of the C systematics, as shown by Figs. 1 and 3, and the absolute abundances and isotopic fractionations observed are approximated without recourse to any unknown sources or processes.

Limitations of the model

One of the most serious drawbacks of the model in its present form is not apparent from the figures. The time taken to achieve the observed abundance and isotopic composition (taken at 180 ppm and +13‰ PDB) is quite long. Typically, for optimum values of K, P, and B, an exposure time of 1.2×10^5 y is required which corresponds to an actual residence time of 5×10^5 y when lunar nighttime and shielding by the magnetosheath are considered (Geiss, 1973). Increasing K to $1.8 \times 10^{-5}\,y^{-1}$ increases the required residence time by about a factor of two; reducing K to $6 \times 10^{-6}\,y^{-1}$ reduces it by only about 20%. Permissible changes in P and B do not effect significant reductions in exposure time. However, this time scale for exposure is strongly influenced by the value taken for F, the flux of C in the solar wind. Not only is this currently uncertain within a factor of at least two, but there are good grounds for believing that the solar wind was significantly more intense in the past (Geiss, 1973). In addition, introduction of some C from any other sources, such as meteorites, comets or the lunar atmosphere, would serve to reduce the required exposure time. Thus, our preliminary results tend to support the conclusion of Hayes (1972) that a contribution of C from some other extralunar source may be required.

A number of other limitations may be mentioned. It is possible that some C is lost, as species which are less volatile than CH_4, at the relatively high temperatures produced during hypervelocity impact and that this loss could contribute to isotopic fractionation. In fact, some loss of C may occur by micrometeorite-induced turnover while grains are buried below the surface layer. Furthermore, the actual size distribution of a lunar soil is not fully represented by the mean grain size and soils of very different physical properties, but the same mean grain size, may exist. In particular, the role played by ultra-fine grains in masking larger grains from the solar wind is ignored for lack of quantitative information. Moore *et al.* (1973) have pointed out that local variations in lunar topography can cause gravity sorting in soils and hence local variations in C content because C is nonuniformly distributed among different size fractions. We agree that this factor should be considered but, again, lack information for a quantitative treatment. We have tacitly assumed in applying Eq. (11) that C(0) is uniform throughout the landing site. There is some evidence that this may not be correct for the Apollo 16 site, but the issue is not fully clarified at the present time. On the one hand, Moore *et al.* (1973) have shown, by means of the distribution of C among different grain size fractions, that some samples contain significantly less indigenous C than others. On the other hand, Kerridge *et al.* (1974) noted that CH_4 produced by hydrolysis correlates very strongly with total C for soils from Apollo 16. Such CH_4 probably originates from solar wind C (Cadogan *et al.*, 1972) and is virtually absent from rocks. Thus, by extrapolation to zero CH_4, a value for indigenous C could be obtained. For Apollo 16 this was found to be 44 ppm, in remarkable agreement with the average value of 41 ppm found by us for five Apollo 16 rocks (including breccias). It is disturbing that the level of indigenous C calculated by Kerridge *et al.* is so much higher than the 2–6 ppm inferred by Moore *et al.* for the same site, especially as agreement between the two groups on individual samples is quite good. Clearly, more work needs to be done to resolve this problem.

Finally, we have assumed that the rate of reaction with H is the same for both isotopes of C and that fractionation results only from differences in diffusion rate. This may not be correct and we may be underestimating the efficiency with which H stripping fractionates implanted C. Laboratory simulation might enable this question, at least, to be answered.

Comparison with other models

Several more-or-less qualitative ideas concerning lunar C systematics have been put forward by various workers. Thus, Hayes (1972), based on the relative cosmic abundances of C and the noble gases, and the noble gas content of lunar soils, derived a minimum value of about 50 ppm for lunar C of solar wind origin. We have elsewhere compared data for noble gas isotopes of solar wind origin with other exposure age criteria and found no significant correlations between them for the Apollo 16 soil samples (Kerridge et al., 1974). We therefore concluded that most, if not all, soil samples have lost substantial fractions of their implanted noble gases, giving misleadingly low exposure ages.

Epstein and Taylor (1972) suggested three possible explanations for the observed abundance and isotope patterns. First, a two-component mixing of indigenous C with solar wind of variable $\delta^{13}C$ between $+10\%_0$ and $+30\%_0$ PDB. Second, two-component mixing of indigenous C with solar wind, or meteoritic, C with $\delta^{13}C$ in the range $-10\%_0$ to $+10\%_0$ PDB, with subsequent enrichment in ^{13}C by H stripping. Third, three-component mixing of indigenous C, "light" meteoritic C and "heavy" solar wind C. The discovery of some soils enriched in C with very low or even negative $\delta^{13}C$ has made the first model less likely. The second picture is clearly akin to our present model, while the possibility of a third, probably meteoritic, component cannot be ruled out at present, though it does not seem necessary as an explanation of low $\delta^{13}C$ values.

Moore et al. (1973) considered that surficial C would reach a steady-state concentration as loss through sputtering equaled solar wind input. They pointed out that total C could still increase beyond this point by incorporation of surficial C into agglutinates, as in the present model. Preliminary calculations suggest that sputtering by ^4He, the most effective species, will have only a small, and indirect, effect on lunar C chemistry. The sputter rate is sufficiently small, 0.043 Å y^{-1} (McDonnell and Flavill, 1974) and the depth of solar wind implantation is sufficiently great, 200–1000 Å, that the main effect of sputtering is to erode the mineral matrix, thus increasing straggling of the implanted solar wind atoms with a possible decrease in the rate of H stripping of the C. However, the effect of sputtering on implanted solar wind species should certainly be studied further and it is fortunate that this interaction should be amenable to laboratory simulation.

Incorporation of surficial C into agglutinates also forms the basis of a model advanced by DesMarais et al. (1973a, 1973b), with which we shall deal in more detail. Based on analyses of grain size separates from a number of soils, they concluded that most soils show constant surficial C density but varying amounts of volume-correlated C. This they interpreted to mean that redistribution of C on the lunar surface, by agglutination and erosion, occurs more rapidly than implantation of extralunar C, of mainly solar wind origin. Their model is based upon plots which show the distribution of C among different size fractions of a sample, expressed as C content versus reciprocal grain size. From such a plot, the slope yields the surficial density, apparently constant among samples of differing degrees of maturity, and the intercept on the ordinate, when corrected for indigenous C, gives a measure of the sample exposure age.

Clearly, the two models have much in common. Their differences stem mainly from different rates attributed to the processes of implantation and agglutination and from the inclusion in our model of H stripping as an important process. Detailed comparison of the models is not possible at present. However, it is worth noting that if grain size evolution proceeds much more rapidly than C implantation, as suggested by their model, the curve in Fig. 3 relating C content to mean grain size should lie well to the right of our optimum model curve and outside the limits imposed by the experimental data. In addition, plots of the type used by DesMarais *et al.* (1973a, 1973b) have, in fact, been drawn by Moore *et al.* (1973) for two soils from Apollo 16 (see their Fig. 1) and actually revealed significantly different slopes, greater for the more mature 65701 than for 67701, in contrast to the common slopes predicted by DesMarais *et al.*

Studies, such as those of DesMarais *et al.* (1973a, b) of the distribution of C species within individual samples are clearly complementary to our approach and we anticipate that discrepancies between the models will be shortly resolved.

Acknowledgments—Work at UCLA supported by NASA grant 05-007-221.

REFERENCES

Abell P. I., Draffan G. H., Eglinton G., Hayes J. M., Maxwell J. R., and Pillinger C. T. (1970) Organic analysis of the returned lunar sample. *Science* **167**, 757–759.

Adams J. B. and McCord T. B. (1973) Vitrification darkening in the lunar highlands and identification of Descartes material at the Apollo 16 site. *Proc. Fourth Lunar Sci. Conf., Geochim. Cosmochim. Acta*, Suppl. 4, Vol. 1, pp. 163–177. Pergamon.

AFGIT (Apollo Field Geology Investigation Team) (1973) Apollo 16 exploration of Descartes; a geologic summary. *Science* **179**, 62–69.

Anders E., Ganapathy R., Krähenbühl U., and Morgan J. W. (1973) Meteoritic material on the moon. *The Moon* **8**, 3–24.

Baedecker P. A., Schaudy R., Elzie J. L., Kimberlin J., and Wasson J. T. (1971) Trace element studies of rocks and soils from Oceanus Procellarum and Mare Tranquillitatis. *Proc. Second Lunar Sci. Conf., Geochim. Cosmochim. Acta*, Suppl. 2, Vol. 2, pp. 1037–1061. MIT Press.

Behrmann C., Crozaz G., Drozd R., Hohenburg C., Ralston C., Walker R., and Yuhas D. (1973) Cosmic-ray exposure history of North Ray and South Ray material. *Proc. Fourth Lunar Sci. Conf., Geochim. Cosmochim. Acta*, Suppl. 4, Vol. 2, pp. 1957–1974. Pergamon.

Bibring J. P., Burlingame A. L., Maurette M., O'Connor J. T., and Wszolek P. (1974) *Proc. Fifth Lunar Sci. Conf., Geochim. Cosmochim. Acta.* This volume.

Borg J., Maurette M., Durrieu L., and Jouret C. (1971) Ultra-microscopic features in micron-sized lunar dust grains and cosmophysics. *Proc. Second Lunar Sci. Conf., Geochim. Cosmochim. Acta*, Suppl. 2, Vol. 2, pp. 2027–2040. MIT Press.

Butler J. C., Greene G. M., and King E. A., Jr. (1973) Grain size frequency distributions and modal analyses of Apollo 16 fines. *Proc. Fourth Lunar Sci. Conf., Geochim. Cosmochim. Acta*, Suppl. 4, Vol. 1, pp. 267–278. Pergamon.

Cadogan P. H., Eglinton G., Firth J. N. M., Maxwell J. R., Mays B. J., and Pillinger C. T. (1972) Survey of lunar carbon compounds: II. The carbon chemistry of Apollo 11, 12, 14 and 15 samples. *Proc. Third Lunar Sci. Conf., Geochim. Cosmochim. Acta*, Suppl. 3, Vol. 2, pp. 2069–2090. MIT Press.

Cameron A. G. W. (1973) Abundances of the elements in the solar system. *Space Sci. Rev.* **15**, 121–146.

DesMarais D. J., Hayes J. M., and Meinschein W. G. (1973a) The distribution in lunar soil of carbon released by pyrolysis. *Proc. Fourth Lunar Sci. Conf., Geochim. Cosmochim. Acta*, Suppl. 4, Vol. 2, pp. 1543–1558. Pergamon.

DesMarais D. J., Hayes J. M., and Meinschein W. G. (1973b) Accumulation of carbon in lunar soils. *Nature Phys. Sci.* **246**, 65–68.

DesMarais D. J., Hayes J. M., and Meinschein W. G. (1974) Retention of solar wind-implanted elements in lunar soils (abstract). In *Lunar Science—V*, pp. 168–170. The Lunar Science Institute, Houston.

Eberhardt P., Geiss J., Graf H., Grögler N., Krähenbühl U., Schwaller H., Schwarzmüller J., and Stettler A. (1970) Trapped solar wind noble gases, exposure age and K/Ar-age in Apollo 11 lunar fine material. *Proc. Apollo 11 Lunar Sci. Conf., Geochim. Cosmochim. Acta*, Suppl. 1, Vol. 2, pp. 1037–1070. Pergamon.

Epstein S. and Taylor H. P., Jr. (1970) $^{18}O/^{16}O$, $^{30}Si/^{28}Si$, D/H and $^{13}C/^{12}C$ studies of lunar rocks and minerals. *Science* **167**, 533–535.

Epstein S. and Taylor H. P., Jr. (1972) O^{18}/O^{16}, Si^{30}/Si^{28}, C^{13}/C^{12}, D/H studies of Apollo 14 and 15 samples. *Proc. Third Lunar Sci. Conf., Geochim. Cosmochim. Acta*, Suppl. 3, Vol. 2, pp. 1429–1454. MIT Press.

Epstein S. and Taylor H. P., Jr. (1973) The isotopic composition and concentration of water, hydrogen and carbon in some Apollo 15 and 16 soils and in the Apollo 17 orange soil. *Proc. Fourth Lunar Sci. Conf., Geochim. Cosmochim. Acta*, Suppl. 4, Vol. 2, pp. 1559–1575. Pergamon.

Fanale F. P., Nash D. B., and Cannon W. A. (1971) Lunar fines and terrestrial rock powders: relative surface areas and heats of adsorption. *J. Geophys. Res.* **76**, 6459–6461.

Gammage R. B. and Becker K. (1971) Exoelectron emission and surface characteristics of lunar materials. *Proc. Second Lunar Sci. Conf., Geochim. Cosmochim. Acta*, Suppl. 2, Vol. 2, pp. 2057–2067. MIT Press.

Ganapathy R., Keays R. R., Laul J. C., and Anders E. (1970) Trace elements in Apollo 11 lunar rocks: implications for meteorite influx and origin of Moon. *Proc. Apollo 11 Lunar Sci. Conf., Geochim. Cosmochim. Acta*, Suppl. 1, Vol. 2, pp. 1117–1142. Pergamon.

Geiss J. (1973) Solar wind composition and implications about the history of the solar system. Paper presented at 13th International Cosmic Ray Conference, Denver.

Gibson E. K., Jr. and Moore G. W. (1973) Volatile-rich lunar soil: Evidence of possible cometary impact. *Science* **179**, 69–71.

Hayes J. M. (1972) Extralunar sources for carbon on the moon. *Space Life Sci.* **3**, 474–483.

Heiken G., McKay D. S., and Fruland R. M. (1973) Apollo 16 soils: Grain size analysis and petrography. *Proc. Fourth Lunar Sci. Conf., Geochim. Cosmochim. Acta*, Suppl. 4, Vol. 1, pp. 251–265. Pergamon.

Housley R. M., Cirlin E. H., and Grant R. W. (1973) Characterization of fines from the Apollo 16 site. *Proc. Fourth Lunar Sci. Conf., Geochim. Cosmochim. Acta*, Suppl. 4, Vol. 3, pp. 2729–2735. Pergamon.

Kaplan I. R., Smith J. W., and Ruth E. (1970) Carbon and sulfur concentration and isotopic composition in Apollo 11 lunar samples. *Proc. Apollo 11 Lunar Sci. Conf., Geochim. Cosmochim. Acta*, Suppl. 1, Vol. 2, pp. 1317–1329. Pergamon.

Kerridge J. F., Kaplan I. R., Petrowski C., and Chang S. (1974) Light element geochemistry of the Apollo 16 site. In press *Geochim. Cosmochim. Acta.*

Leich D. A., Tombrello T. A., and Burnett D. S. (1973) The depth distribution of hydrogen and fluorine in lunar samples. *Proc. Fourth Lunar Sci. Conf., Geochim. Cosmochim. Acta*, Suppl. 4, Vol. 2, pp. 1597–1612. Pergamon.

LSPET (Lunar Sample Preliminary Examination Team) (1972) Preliminary examination of lunar samples. *Apollo 16 Preliminary Science Report*, NASA SP-315.

McDonnell J. A. M. and Flavill R. P. (1974) Sputter erosion on the lunar surface: measurements and features under simulated solar He^+ bombardment (abstract). In *Lunar Science—V*, pp. 478–479. The Lunar Science Institute, Houston.

McKay D. S., Heiken G. H., Taylor R. M., Clanton U. S., Morrison D. A., and Ladle G. H. (1972) Apollo 14 soils: Size distribution and particle types. *Proc. Third Lunar Sci. Conf., Geochim. Cosmochim. Acta*, Suppl. 3, Vol. 1, pp. 983–984. MIT Press.

Moore C. B., Lewis C. F., and Gibson E. K., Jr. (1973) Total carbon contents of Apollo 15 and 16 lunar samples. *Proc. Fourth Lunar Sci. Conf., Geochim. Cosmochim. Acta*, Suppl. 4, Vol. 2, pp. 1613–1623. Pergamon.

Pillinger C. T., Batts B. D., Eglinton G., Gowar A. P., Jull A. J. T., and Maxwell J. R. (1973) Formation of lunar carbide from lunar iron silicates. *Nature Phys. Sci.* **245**, 3–5.

Price P. B., Chan J. H., Hutcheon I. D., MacDougall D., Rajan R. S., Shirk E. K., and Sullivan J. D. (1973) Low energy heavy ions in the solar system. *Proc. Fourth Lunar Sci. Conf., Geochim. Cosmochim. Acta*, Suppl. 4, Vol. 3, pp. 2347–2361. Pergamon.

Schaeffer O. A. and Husain L. (1973) Early lunar history: Ages of 2 to 4 mm soil fragments from the lunar highlands. *Proc. Fourth Lunar Sci. Conf., Geochim. Cosmochim. Acta*, Suppl. 4, Vol. 2, pp. 1847–1863. Pergamon.

Tera F., Papanastassiou D. A., and Wasserburg G. J. (1973) A lunar cataclysm at 3.95 AE and the structure of the lunar crust (abstract). In *Lunar Science—IV*, pp. 723–725. The Lunar Science Institute, Houston.

Wolfe J. H. (1972) The large-scale structure of the solar wind. In *Solar Wind*, pp. 170–196, NASA SP-308.

Proceedings of the Fifth Lunar Conference
(Supplement 5, Geochimica et Cosmochimica Acta)
Vol. 2 pp. 1869–1884 (1974)
Printed in the United States of America

Hydrogen and fluorine in the surfaces of lunar samples*

D. A. Leich,† R. H. Goldberg, D. S. Burnett, and T. A. Tombrello

California Institute of Technology, Pasadena, California 91109

Abstract—The resonant nuclear reaction $^{19}F(p, \alpha\gamma)^{16}O$ has been used to perform depth-sensitive analyses for both fluorine and hydrogen in lunar samples. The resonance at 0.83 MeV (center-of-mass) in this reaction has been applied to the measurement of the distribution of trapped solar protons in lunar samples to depths up to 0.45 μm. These results are interpreted in terms of terrestrial H_2O surface contamination and of a redistribution of the implanted solar H which has been influenced by heavy radiation damage in the surface region. Results are also presented for an experiment to test the penetration of H_2O into laboratory glass samples which have been irradiated with ^{16}O to simulate the radiation damaged surfaces of lunar glasses. Fluorine determinations have been performed in a 1 μm surface layer on lunar samples using the same $^{19}F(p, \alpha\gamma)^{16}O$ resonance. The data are discussed from the standpoint that observed fluorine concentrations are a mixture of true lunar fluorine and Teflon contamination.

I. Introduction

Solar wind ion implantation has been shown to be an important source of both high rare gas contents (see, for example, Eberhardt *et al.*, 1970) and extreme radiation damage (Borg *et al.*, 1971) within one micron of the surfaces of lunar soil grains. Significant surface-correlated enrichments of H, C, and N in lunar soils, presumably due to solar wind implantation, have also been identified (see, for example, Kaplan *et al.*, 1970; Moore *et al.*, 1970; Goel and Kothari 1972). Also, chemical and physical alterations of grain surfaces can be caused by solar wind related processes, including atmospheric reimplantation and ion sputtering, as well as unrelated mechanisms such as diffusive loss at lunar daytime temperatures, impact volatilization, condensation of impact-generated volatiles, and, possibly, reaction with volcanic emanations.

In our continuing study, we have used a nuclear technique to investigate the effects of this complex environment on the depth distribution of hydrogen and fluorine in the outer micron of 2–5 mm lunar soil fragments and in chips from lunar rocks. Observed surface concentrations of hydrogen (other than contamination) can be interpreted in terms of an expected solar wind source; however, the solar wind is not likely to be responsible for appreciable surface concentrations of fluorine. Consequently, measurements of the fluorine distributions in the outer micron of lunar sample surfaces may yield information concerning other processes such as the mobilization of volatiles on the moon or reaction of the sample surfaces with fluorine-bearing volcanic emanations (Turkevich 1973), provided the extent of fluorine contamination can be correctly assessed.

---

*Supported in part by the National Science Foundation [GP-28027] and the National Aeronautics and Space Administration [NGR 05-002-333].

†Present address: Physics Department, University of California, Berkeley, California 94720.

II. Experimental Technique

The basic technique has been described in detail elsewhere (Leich and Tombrello 1973), but a few points bear repeating. Both hydrogen and fluorine measurements are performed using the narrow (5 keV wide) resonance at 0.83 MeV center-of-mass energy in the nuclear reaction $^{19}F(p, \alpha\gamma)^{16}O$. A proton beam is directed onto the lunar sample for analysis of the fluorine distribution. Due to the sharp resonant nature of this reaction, the gamma-ray counting rate is proportional to the fluorine content at a particular depth determined by the choice of the incident proton energy. Higher energy protons penetrate deeper before being slowed down to the resonant energy. An analogous situation holds for measurement of the hydrogen distributions, in which a ^{19}F beam is used to induce the reaction on the hydrogen contained in the lunar sample. In either case, the gamma-ray counting rate is measured at a number of incident beam energies in the vicinity of the resonance, and these data are converted directly into a depth profile using proportionalities between beam energy and depth and between counting rate and concentration of hydrogen or fluorine. Measurements have been performed over the depth range for which counts from only one resonance need be considered: 0–0.45 μm for hydrogen, with a depth resolution (given by the depth corresponding to half the resonance width) of ~0.02 μm, and 0–1.0 μm for fluorine, with a resolution of ~0.05 μm.

A new scattering chamber has been used for some of the measurements reported in this paper, enabling the following improvements over the previous configuration described by Leich and Tombrello (1973):

(A) The base pressure of the scattering chamber has been lowered from ~1×10^{-9} torr to $\leqslant 1 \times 10^{-10}$ torr.

(B) The gamma-ray detection efficiency, determined by bombarding a set of analyzed standards (Belvidere Mountain chlorite: 1.63% H, Durango apatite: 3.53% F, and reagent CaF_2), has been increased from ~0.02 to ~0.06.

(C) Precise visual positioning of the beam (~2×2 mm) on the target surface is now possible. Many samples fluoresce under bombardment, making the positioning quite accurate, within a few tenths of a millimeter.

The following points should also be emphasized:

(1) Both all-metal scattering chambers are vacuum baked and are pumped by clean, getter-ion pumps with a liquid-nitrogen-cooled baffle at the interface with the accelerator vacuum (~10^{-6} torr) to keep contamination to an absolute minimum. We have never seen a hydrogen buildup on any sample. Such buildup commonly occurs in poor vacuum as a result of beam-induced polymerization of residual hydrocarbon gases from the vacuum system.

(2) Both in simulation experiments (discussed below) and in exposure of lunar samples to atmospheric humidity we have not observed an increase in the hydrogen content below a depth of ~500 Å. Nevertheless, many of the samples used in this study were returned from the lunar surface in the vacuum-sealed rock boxes and have subsequently been carefully protected from atmospheric exposure by storing and handling them entirely in dry nitrogen gas (20–50 ppm H_2O in SSPL; P_2O_5 gettered in our lab).

(3) No attempt has been made to neutralize the charging of the sample due to bombardment of targets by the proton or $^{19}F^{4+}$ ion beams, other than to collect continuously the charge from the aluminum target holder. By noting the shift in proton beam energy necessary to induce resonant gamma-ray production on a 1.6-cm diameter fused silica target coated with a thin film of CaF_2, we have verified that surface potentials as high as ~20 kV can be acquired by silicate targets under the bombardment conditions used in our analysis. For the lunar samples, a correction to the depth scale for such a surface potential can be applied by measuring the shift in the proton beam energy corresponding to the $^{27}Al(p, \gamma)^{28}Si$ resonance at 0.992 MeV proton energy, acting on the aluminum content (assumed to be relatively uniform) of the sample. In this manner, surface potentials of 2–14 kV have been observed on these samples during illumination with the proton beam. While surface potentials of this magnitude are significant for the fluorine distribution data, in that they correspond to apparent shifts of up to 0.25 μm in depth, potentials of a few tens of kilovolts have no significant effect on the hydrogen depth distribution measurements (Leich *et al.*, 1973b).

(4) We have never observed any indication (e.g. an increased hydrogen or fluorine content in repeated analyses) that the relatively small amounts ($\sim 10^{15}$ atoms) of hydrogen and fluorine implanted in the course of our analyses contribute to the measured concentrations. This observation is in agreement with the expected result based on the deep penetration of the energetic beam ions ($\sim 10 \mu$m) and the negligible diffusion of the implanted ions over the times between analyses (a few days). On the other hand, beam-induced perturbations of the H distribution may make H-profile data less meaningful if taken after the sample has been analyzed for F. Since the F distributions are quite stable under bombardment, the information gained from a particular sample can be optimized by analyzing for H first and then for F. Except for samples 66044 and 68124, all analyses for hydrogen were performed first.

III. Hydrogen Depth Distributions

We have previously reported results of hydrogen profile measurements on a number of Apollo 11, Apollo 15, and Apollo 16 samples (Leich *et al.* 1973a and 1973b), noting that the hydrogen profiles measured on these samples fell into two distinct classes. One of these classes is characterized by a surface concentration of hydrogen, not more than a few hundred angstroms thick (e.g. the glass sphere profile in Fig. 1). The location of this hydrogen concentration appears to be reasonably consistent with expected solar wind proton penetration depths. However, small amounts of surface adsorbed hydrogen ($\sim 2 \times 10^{15}$ atoms/cm^2, equivalent to one monolayer of H_2O) are routinely observed on *interior* rock samples which have been exposed only to dry nitrogen gas. The source of this adsorbed hydrogen is undoubtedly terrestrial (most likely the small residual H_2O content of the "dry" nitrogen), leading to the obvious conclusion that the similar features observed on the lunar exterior surfaces of some of the same samples, including sealed rock box samples, are due primarily to terrestrial contamination.

Figure 2 shows the results of a simulation experiment performed to determine the effects of extreme radiation damage on the adsorption and penetration of H_2O contamination. Fused silica targets were irradiated with 86 keV $^{16}O^+$ ions at doses up to 1.4×10^{17} ions/cm^2 in order to produce heavy radiation damage down to depths of ~ 3000 Å. Subsequent exposure of some of these targets to water, in either liquid or vapor form, resulted in adsorption of measurable amounts of hydrogen surface contamination. Observed surface (within hundreds of angstroms) hydrogen concentrations were consistently a factor of two to three higher on the radiation damaged surfaces than on undamaged surfaces. However, the hydrogen concentrations exhibited only a slight dependence on the total dose between 2×10^{16} and 1.4×10^{17} ions/cm^2. Most important, it is evident from Fig. 2 that very little hydrogen contamination has penetrated to depths of 1000 Å or greater, in spite of the fact that radiation damage should be at saturation levels in the region from the surface to ~ 3000 Å deep (Winterbon *et al.*, 1970). From a chemical point of view, O^+ ions might not be the optimum choice for simulating the solar wind radiation damage, since the excess oxygen may affect the penetration of H_2O into the sample in a way that would not be applicable for the reducing conditions present on the surfaces of lunar samples (Epstein and Taylor, 1971). Nevertheless, the similarity of the hydrogen profiles obtained on H_2O contaminated radiation damaged quartz glass samples with many of the lunar

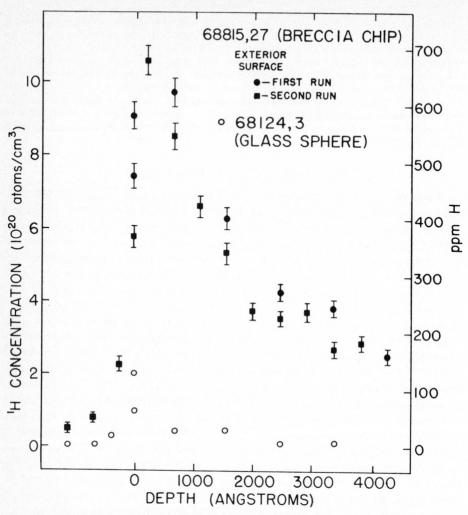

Fig. 1. Hydrogen concentration versus depth for two samples: breccia chip 68815,27 and glass spherule 68124,3. The apparently nonzero H content at negative depth (i.e. in vacuum) is due to the finite resolution (~ 200 Å) of the measurement technique. 68815,27 and 68124,3 are sealed rock box samples.

sample hydrogen profiles (e.g. 68124,3; see also Leich *et al.*, 1973a, 1973b) argues strongly for a similar origin of the observed hydrogen in these cases, namely terrestrial H_2O contamination preferentially bound on radiation damaged surfaces.

In addition to the surface adsorbed hydrogen, a small hydrogen content with a uniform distribution between ~ 1000 and 4000 Å deep (the limit of our measurements) has been observed in most of the interior rock samples, suggesting that this hydrogen component (normally between 20 and 50 ppm) is probably representative of a small volume content of lunar hydrogen in these rocks. An alternative explanation is that the deep hydrogen corresponds to H adsorbed on reentrant surfaces or along microfractures.

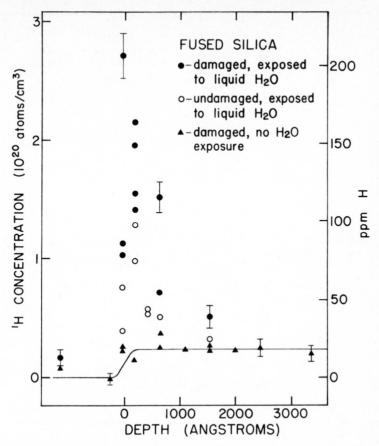

Fig. 2. Fused silica simulation experiment results. Data shown are representative of a set of samples subjected to radiation damage and/or to H₂O exposure tests. Two of the samples were damaged by irradiating them with 86-keV $^{16}O^{+}$ ions for 4 hours to a total dose of 1.4×10^{17} ions/cm². One of these (solid circles) was subsequently exposed to H₂O in both liquid (submerged in distilled water for 24 hours) and vapor (laboratory atmosphere for one week) form, while the other (solid triangles) was exposed only to dry N₂ gas for 2 hours. A third sample (open circles) was not radiation damaged but was given the same H₂O exposure as the first sample. Only sample error bars are shown on the data points obtained during the subsequent H analysis, performed to determine the extent of H₂O penetration. The solid curve represents typical results for a clean fused silica sample with a normal (for this batch) volume H content of ~20 ppm.

A second type of profile is characterized by an additional component with a maximum hydrogen content near 1000 Å deep and concentrated mainly within ~2000 Å of the surface (the full width at half maximum ranges from 1800 to 2700 Å with a mean of 2200 Å) and extending to depths greater than 4000 Å. Figures 1 and 3 show two examples of this type of distribution. While extensive penetration of a terrestrial contaminant cannot be completely ruled out as a possible origin for this hydrogen component, we believe we are measuring solar hydrogen because: (1) Exposure of artificially radiation-damaged fused silica

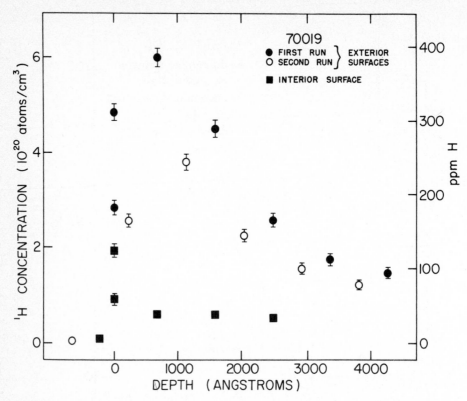

Fig. 3. Hydrogen concentration versus depth for 70019,17, a glass coated soil-breccia
chip (sealed rock box sample). The exterior surface was glass; the interior surface, soil
breccia. During bombardment the H is relatively mobile, as indicated by the difference in
the first and second run.

surfaces to H_2O, even in liquid form, produced no penetration of H_2O to depths
greater than 1000 Å, even for surfaces which were heavily damaged with ^{16}O ions
prior to H_2O exposure (Fig. 2). (2) It appears doubtful that the exterior surfaces of
68815 and 70019, which had never been exposed to the atmosphere, could have
adsorbed such large quantities of H_2O while the exterior surfaces of other
samples, such as 68124,3 (the glass sphere, Fig. 1), returned in the same sealed
rock box as 68815, showed only small quantities of surface adsorbed hydrogen.
Given the known tendency of glasses to hydrate, the 68124,3 glass sphere would
have been expected to adsorb much more H_2O than the surface of a crystalline
rock sample such as 68815. (3) The large amounts of hydrogen observed in these
samples are much greater than those seen in the damaged, H_2O-exposed glasses.

Our measurements of Apollo 17 samples, which consist of surface chips from
two very different rocks, show similar hydrogen depth profiles. Sample 70019,17,
shown in Fig. 3, is a glass-coated soil breccia which has a maximum hydrogen
content near 1000 Å deep, closely paralleling 68815,27. Samples 75075,2 and
75075,18 are from opposite sides of a heavily patinated basalt; both were exposed

to the solar wind. Although 75075 is very different physically from other samples we have measured, the hydrogen depth profiles are similar to 68815 and 70019 (see Fig. 4). Both 75075 and 70019 are sealed rock box samples and thus have not been exposed to atmospheric H_2O contamination. The hydrogen content of the deeper regions (100–200 ppm) is typical of lunar soil samples and this would be consistent with a model of patina formation by deposition of nearby soil without significant depletion of hydrogen from depths greater than 2000–3000 Å (as opposed to splattering of a coating from the host rock itself—see Blanford *et al.*, 1974). However, the similarity of the overall profile to that found on unpatinated rock surfaces is suggestive of a patina formed from material which had not been exposed to solar wind prior to deposition or which was outgassed during

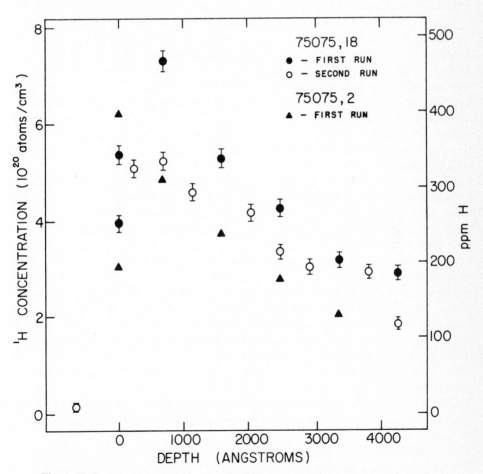

Fig. 4. Hydrogen concentration versus depth for two samples (sealed rock box) from basalt 75075, both patinated surface chips. Points for first and second runs on sample 75075,18 are shown to demonstrate the mobility of the hydrogen distribution, typical of other samples we have measured. Error bars for 75075,2 have been omitted for purposes of clarity.

deposition. In any case the similarity of the 75075 profiles to those observed on unpatinated rocks or glass samples is best interpreted as indicating that the patina on these two surfaces has been deposited in layers at least 0.4μm thick. Alternatively, it is more difficult to rule out H_2O contamination as an origin for the 75075 hydrogen because our argument that contamination H_2O does not penetrate to depths $\geqslant 1000$ Å may not be valid for the porous patina surface. However, the low hydrogen concentrations observed for the interior, soil-breccia surface of 70019 (Fig. 3), which is similar material, argues against the 75075 profiles being due to contamination.

We now wish to consider in more detail the origin of the solar hydrogen at depths between 500 and 4000 Å in rocks such as 68815 and 70019. Although implantation of solar wind protons is the most likely original source, the observed hydrogen profiles are significantly more penetrating than would be derived from the direct implantation of solar wind protons, in agreement with conclusions based on chemical etching experiments for implanted rare gases (Eberhardt *et al.*, 1970; Kirsten *et al.*, 1970; Hintenberger *et al.*, 1970). If solar wind is the source of this hydrogen component, extensive modification by diffusion and trapping of hydrogen atoms is implied. If diffusion rates for hydrogen in terrestrial silicates (Brückner 1971) are applicable to the lunar samples, it appears that bulk volume diffusion would be too rapid to result in the observed profiles without some sort of trapping to slow down the diffusion process (Ducati *et al.*, 1973). A hypothesis in which implanted solar wind hydrogen diffuses rapidly into (and out of) the samples with a small remnant of the implanted dose being retained in radiation damage traps seems plausible. The radiation damage is evidently so heavy in the outer 500 Å that no isolated traps remain (Borg *et al.*, 1971). Beneath this depth, relatively intense radiation damage (but below saturation) may persist to a depth of ~2000 Å, corresponding closely with the radiation-damage range of He ions with velocities near those of frequent high-velocity (up to 800 km/sec) solar wind streams observed by satellites (Wolfe 1972). The population of isolated radiation-damage traps by diffusing solar wind atoms may then result in a hydrogen depth profile which reflects the distribution of radiation damage. A discontinuity in the radiation damage gradient near 2000 Å deep may account for the characteristic bend observed in the measured hydrogen profiles, with the tail of the hydrogen distribution (below 2000 Å deep) representing diffusion of hydrogen out of the region of high hydrogen concentration into a region in which the radiation damage (due to solar flare and suprathermal ions) is much less intense. The slope of the hydrogen profile in the 2000–4000 Å depth region may reflect population of traps in a radiation-damage gradient, or it may represent a dynamic profile of inward diffusion with only weak trapping.

An alternative explanation for the observed hydrogen profiles is the direct implantation of "suprathermal" (10–100 keV) protons. If this hypothesis is correct, the measured hydrogen profiles provide information about the energy spectrum of the incident protons. A reasonably good fit to the initial set of hydrogen profile data for 68815,27 is shown by the solid curve in Fig. 5. This curve was obtained from range-energy relations (Lindhard *et al.*, 1963) using the energy

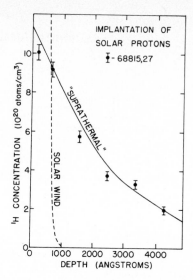

Fig. 5. Implantation of solar protons in lunar samples. The data points are from sample 68815,27 (Fig. 1). The solid curve is the erosional equilibrium distribution resulting from the flux spectrum indicated by the solid lines in Fig. 6, assuming an atomic erosion rate of 0.5 Å/yr, and calculated from the expected flux, incidence angle, irradiation time, and projected range of suprathermal protons (Leich, 1973). The spectrum was chosen to give a rough fit to the data, using a proton range-energy relation derived from Lindhard *et al.* (1963) and neglecting range straggling and diffusion. The dashed curve indicates the limit of penetration of the present-day solar wind protons including the effects of range straggling. With no diffusive losses, but allowing for erosion, the peak hydrogen content due to solar wind protons at the surface would be greater than 1 mole/cm³, more than three orders of magnitude higher than the observed hydrogen content near the surface of sample 68815,27, and nearly as dense as the sample itself!

spectrum shown in Fig. 6, assuming that no post-implantation diffusion occurs and taking into account the effects of erosion and of incidence from a solar direction. A two-component energy spectrum is necessary to produce the characteristic bend in the profile near ~2000 Å deep. An atomic erosion rate of 0.5 Å/year (Wehner *et al.*, 1963) and an erosional equilibrium profile has been assumed. The attainment of equilibrium requires an exposure time $\geq 10^4$ yr. The required flux of suprathermal protons is $\sim 10^{13}$ cm^{-2} y^{-1}, within a factor of about 3 of the flux for an event yielding the data shown in Fig. 6 taken from satellite observations (Frank, 1970). Since the long-term flux of protons in this energy range is likely to be a few orders of magnitude lower than the flux during such an event, it appears unlikely that the average flux has been high enough to account for the measured hydrogen profiles by direct implantation. However, little data have been obtained in this energy range, and it is possible that long-term fluxes may have been high enough to account for a significant portion of the hydrogen distributions, or at least to account for a significant radiation-damage gradient in the 2000–4000 Å depth region.

However, if the suprathermal hypothesis is adopted, it is still necessary to

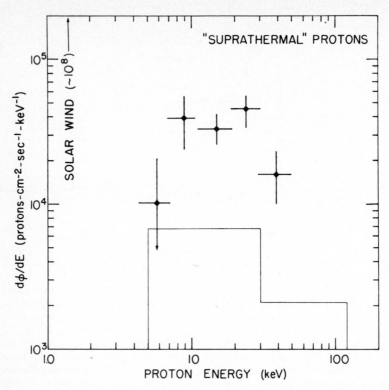

Fig. 6. "Suprathermal" proton flux spectrum $d\phi/dE$ versus proton energy E. Data points with associated error bars are taken from satellite observations reported by Frank (1970) for a single event. The solid line is the flux spectrum, assumed to be uniform in time, which gives a rough fit to the observed hydrogen distribution in 68815,27, assuming an atomic erosion rate of 0.5 Å/yr.

invoke radiation-damage hindrance to explain why the direct implantation profile has not been modified by diffusion. The profile may or may not be controlled by the radiation-damage gradient, depending on the strength and capacity of the trapping site. Postulating a suprathermal flux may be unnecessary. In any case, the important point is that radiation damage is the dominant mechanism for localizing hydrogen near the surfaces of lunar samples.

IV. FLUORINE DEPTH DISTRIBUTION

Previously we have reported fluorine surface concentration (up to 1000 ppm) on several Apollo 16 samples. These F concentrations (Table 1) were much larger than those of bulk analyses of rocks (<50 ppm) or soils (40–100 ppm) as reported by Jovanovic and Reed (1973); however, the high (~2500 ppm) surface fluorine concentration measured by the *in situ* Surveyor VII analysis (Patterson *et al.*, 1970) agree with our data—if taken at face value. The critical question, which we were unable to answer previously, was the level of fluorine contamination. This is

Table 1. Fluorine concentration data in lunar samples. The two regions listed are surface 0.5 μm deep, and 0.5–1.0 μm deep.

Sample	Surface-averaged F content (ppm)	
	0–0.5 μm	0.5–1.0 μm
65315,6	1000	480
65315,6 interior	100	50
*68124,3-A	410	50
*68124,3-B	850	120
*68124,10-A	260	130
*68124,10-B	220	100
*66044,8-A	820	540
*66044,8-B	1900	1400
*66044,8 interior	75	40
†70019,17	235	60
†70019,17 interior	180	130
†75075,2	975	550
†75075,18	330	150

*Sealed rock box sample.
†Sealed rock box sample; not Teflon bagged.

of particular concern because fluorine-rich materials (Teflon, Freon, etc.) have been used extensively both in the mission and post-mission handling of lunar samples. Our recent work has indicated that fluorine contamination is present— making the study of surface fluorine in returned lunar samples difficult.

There are two features of our data which are not easily explained by contamination: (1) High fluorine concentrations are observed even at depths of 1 μm; this is illustrated in Fig. 7 (bottom)—for the cases of 66044,8 and 75075,2. The former is a 5 mm crystalline anorthosite fragment which has the highest surface-averaged fluorine concentration we have measured. Our measurements of 75075,2 are on a heavily patinated surface which shows a fluorine content similar to that of 66044,8. On 66044,8, the fluorine concentrations are high on two surfaces, and, along with 75075,2, are uniform at depths from 0.2 to at least 1 μm. Table 1 indicates the range of fluorine contents in the depth intervals 0–0.5 μm and 0.5–1.0 μm for some of the samples we have studied. While these high levels persist to 1 μm, our experience in hydrogen depth studies has shown the hydrogen contamination on lunar samples is usually manifested as a surface film which is less than ~0.1 μm thick. (2) Fluorine concentrations are usually much higher on the lunar exterior surface of rock chips than on the interior surface. The samples studied to date were allocated primarily for the purpose of hydrogen studies; in order to provide a dry N_2 atmosphere and to protect against surface abrasion during shipping, the samples were sealed in Teflon bags, excepting our Apollo 17 samples which were wrapped in aluminum foil prior to bagging. The interior surfaces produced by chipping in Houston provide a control on fluorine contamination from the packaging process. However, all rocks returned on Apollo

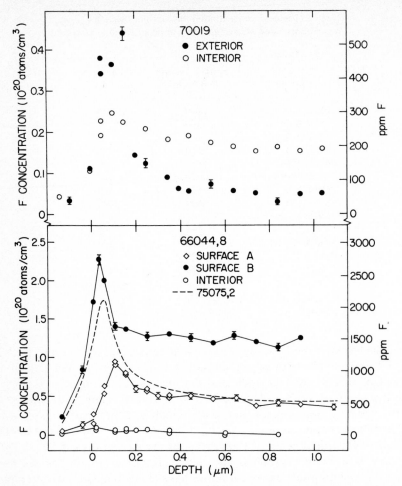

Fig. 7. Top: Fluorine concentration versus depth for sample 70019,17 (sealed rock box sample). Exterior points are from measurements on the glass coating; interior points are from measurements of an interior soil-breccia surface, freshly exposed in our laboratory. Bottom: Fluorine concentration versus depth for anorthosite coarse fine 66044,8, and patinated breccia 75075,2, both sealed rock box samples. Shown are data from two surfaces of 66044,8, and from an interior surface freshly exposed in our laboratory. The smoothed dashed curve is drawn through the data points of 75075,2 for clarity in the figure. The position of the origin of the depth scale has *not* been corrected for charging effects (see text) on any of these profiles (top and bottom), but measured surface potentials are consistent with the location of the peaks in the F profiles on the surfaces (zero-depth) of the samples.

15–17 were contained in fluorocarbon bags cleaned in Freon, which could provide additional fluorine contamination on lunar exterior surfaces, especially in view of the vibration associated with the return trip and of the possibility that the radiation-damaged exterior surfaces may have an enhanced affinity, compared to a fresh, undamaged surface, for fluorine as well as for other contaminants.

In order to measure the contamination due to Teflon packaging, we cleaned and baked quartz glass disks and found, after this procedure, a fluorine level of <20 ppm. Two of these disks were transported to the curatorial facility where they were heat sealed in Teflon bags in the same way as lunar samples. Subsequent measurements showed a surface fluorine peak (~200 ppm) which at a depth of 1 μm had not yet reached zero (<20 ppm) concentration (Fig. 8). The disks which remained in our laboratory as controls had <20 ppm, as before. Figure 8 shows that readily measurable amounts of fluorine were produced either by the heat sealing or by abrasion; however, the amounts are much lower (by a factor of 5–10) than on lunar exterior surfaces. Conceivably, the rough surface of a rock is much more susceptible to contamination than the disks. However, the lunar interior samples packaged in Teflon tend to show concentrations no higher

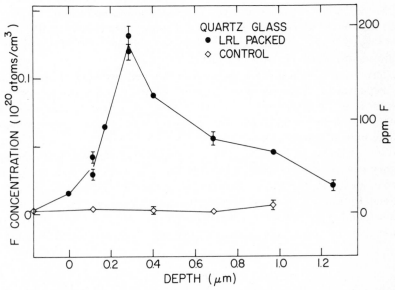

Fig. 8. Fluorine concentration versus depth for quartz glass disks: solid points correspond to one that was packaged in Teflon in the curatorial facility; open points to an identical disk that served as a control. The apparent location of the peak F content between 0.3 and 0.4 μm deep instead of on the surface is probably due to charging of the sample by the proton beam (~20 kV).

than those found on the disks. Also, this difference could be due to higher levels of contamination on the lunar samples resulting from any of the factors mentioned above (longer storage time, strong vibrations while in contact with fluorocarbon bags, and the possible effects of radiation damage). Further, Fig. 8 shows that contamination is present to depths as great as 1 μm, implying that the high fluorine at this depth in lunar samples (see point [1]) could also be contamination.

If the surface fluorine contents of the lunar samples represent contamination from the mission packaging materials, it is somewhat surprising that the most contaminated sample should be a coarse-fine fragment, because 66044,8 was transported to earth as part of a soil sample and, statistically, should have been protected. To check that 66044,8 was not an anomalously fluorine-rich lunar rock it was cleaved in half in our laboratory and analyzed for fluorine on the interior surface. As shown in Fig. 7 (bottom), the average fluorine content at 0.5–1.0 μm is ≤40 ppm, consistent with bulk fluorine measurements (Jovanovic and Reed, 1973).

Although the consistently higher fluorine on exterior relative to interior surfaces suggests a lunar origin for some of the surface fluorine, we now have some experimental evidence that the original arguments against contamination may not be valid. It may be that no lunar sample is sufficiently uncontaminated for the purposes of our experiment and perhaps other surface property experiments (e.g. carbon) as well. The best candidates for resolving this issue would be samples from surface indentations, e.g. vesicles, which have been protected from abrasion, and some SESC (Surface Environment Sample Container) or core samples that have never been exposed to Teflon. Assuming that all fluorine we have observed is due to Teflon, the associated carbon contamination would be insignificant in most bulk carbon chemistry studies.

We have so far limited the discussion to the measurement of surface fluorine layers because of their possible connection with condensation or reaction processes. It is clear that the interpretation of these results is clouded by the possibility of Teflon contamination; however, the use of our technique to determine bulk fluorine content is *not* open to question (see Fig. 7, bottom and Fig. 8). For example, an interior surface of 70019 that was freshly exposed in our laboratory has not been contaminated: the flat fluorine depth distribution and the absence of a large surface peak (see the top part of Fig. 7) support this conclusion. The concentration observed (~200 ppm) is approximately a factor of two higher than that found in most Apollo 16 samples (Jovanovic and Reed, 1973); thus, we conclude that the soil breccia part of 70019 is a fluorine-rich lunar sample.

V. Summary

A significant portion of the hydrogen contents of lunar samples appears to reside within a few thousand angstroms of the sample surfaces. Besides surface contamination, in quantities corresponding to a monolayer ($\sim 2 \times 10^{15}$ H atoms/cm^2) or so of H$_2$O, and perhaps a small (normally 20–50 ppm) volume content of lunar hydrogen, an additional component has been observed in several of the lunar samples analyzed in this and previous studies. Although apparently of solar origin, the distribution of this component over characteristic depths of 1000–3000 Å shows that a simple picture of direct solar wind implantation is not adequate. Instead, the observed hydrogen depth distributions appear to result from some combination of a redistribution of solar wind hydrogen by diffusion and trapping in radiation damage sites and the direct implantation of "suprathermal" (10–100 keV) protons.

Although we are convinced that solar hydrogen has been measured in at least four of our samples [10085,31-12 (Leich *et al.*, 1973a), 68815, 75075, and 70019], we do not understand why *only* these four show high concentrations. The 10085 and 70019 samples are glasses but eight other glass samples analyzed (both coarse fines and rock surface chips) show low concentrations (Leich, 1973). The 75075 (a patinated rock surface) sample is unique. However, five other rock samples which should be comparable to 68815 show only small concentrations (Leich, 1973). Although some of these samples may not have been exposed or may have been abraded, this variability is puzzling and suggests that the nature of the "trapping" mechanism we postulate may be quite complex in detail.

In contrast to hydrogen, the fluorine contents of lunar sample surfaces are not likely to be strongly influenced by solar particle implantation. On the other hand, if volcanic emanations are present on the lunar surface, they may, by analogy with terrestrial volcanic emanations, contain significant quantities of fluorine and other halogens. Thus, detection of surface enrichments of fluorine in lunar samples could indicate the presence of such emanations from the lunar interior. While consistently higher fluorine concentrations have been found on the lunar exterior surfaces than on the interior surfaces of the samples analyzed in this study, it is now certain that significant fluorine contamination has been introduced in the course of sample return and processing, and the possibility that the exterior surfaces were merely more thoroughly contaminated than the interior surfaces cannot, at this time, be ruled out. Hence, although tempted, we are unable to defend a lunar origin of observed surface enrichments of fluorine with presently measured samples.

Acknowledgments—We thank W. Russell for his help with the experimental measurements, S. Epstein for the use of his facilities for sample handling, and P. Eberhardt for helpful comments.

References

Blanford G., McKay D., and Morrison D. (1974) Accretionary particles and microcraters (abstract). In *Lunar Science—V*, pp. 67–69. The Lunar Science Institute, Houston.

Borg J., Maurette M., Durrieu L., and Jouret C. (1971) Ultramicroscopic features in micron-sized lunar dust grains and cosmophysics. *Proc. Second Lunar Sci. Conf., Geochim. Cosmochim. Acta*, Suppl. 2, Vol. 3, pp. 2027–2040. MIT Press.

Brückner R. (1971) Properties and structure of vitreous silica. II. *J. Non-Crystalline Solids* 5, 177–216.

Ducati H., Kalbitzer S., Kiko J., Kirsten T., and Müller H. W. (1973) Rare gas diffusion studies in individual lunar soil particles and in artificially implanted glasses. *The Moon* 8, 210–227.

Eberhardt P., Geiss J., Graf H., Grögler N., Krähenbühl U., Schwaller H., Schwarzmuller J., and Stettler A. (1970) Trapped solar wind noble gases, exposure age and K/Ar-age in Apollo 11 lunar fine material. *Proc. Apollo 11 Lunar Sci. Conf., Geochim. Cosmochim. Acta*, Suppl. 1, Vol. 2, pp. 1037–1070. Pergamon.

Epstein S. and Taylor H. P. (1971) $^{18}O/^{16}O$, $^{30}Si/^{28}Si$, D/H, and $^{13}C/^{12}C$ ratios in lunar samples. *Proc. Second Lunar Sci. Conf., Geochim. Cosmochim. Acta*, Suppl. 2, Vol. 2, pp. 1421–1441. MIT Press.

Frank L. A. (1970) On the presence of low-energy protons ($5 \leqslant E \leqslant 50$ keV) in the interplanetary medium. *J. Geophys. Res.* 75, 707–716.

Goel P. S. and Kothari B. K. (1972) Total nitrogen contents of some Apollo 14 lunar samples by neutron activation analysis. *Proc. Third Lunar Sci. Conf., Geochim. Cosmochim Acta*, Suppl. 3, Vol. 2, pp. 2041–2050. MIT Press.

Hintenberger H., Weber H. W., Voshage H., Wänke H., Begemann F., and Wlotzka F. (1970) Concentrations and isotopic abundances of the rare gases, hydrogen and nitrogen in Apollo 11 lunar matter. *Proc. Apollo 11 Lunar Sci. Conf., Geochim. Cosmochim. Acta,* Suppl. 1, Vol. 2, pp. 1269–1282. Pergamon.

Jovanovic S. and Reed G. W., Jr. (1973) Volatile trace elements and the characterization of the Cayley Formation and the primitive lunar crust. *Proc. Fourth Lunar Sci. Conf., Geochim. Cosmochim. Acta,* Suppl. 4, Vol. 2, pp. 1313–1324. Pergamon.

Kaplan I. R., Smith J. W., and Ruth E. (1970) Carbon and sulfur concentration and isotopic composition of Apollo 11 lunar samples. *Proc. Apollo 11 Lunar Sci. Conf., Geochim. Cosmochim. Acta,* Suppl. 1, Vol. 2, pp. 1317–1329. Pergamon.

Kirsten T., Müller O., Steinbrunn F., and Zähringer J. (1970) Study of distribution and variations of rare gases in lunar material by a microprobe technique. *Proc. Apollo 11 Lunar Sci. Conf., Geochim. Cosmochim. Acta,* Suppl. 1, Vol. 2, pp. 1331–1343. Pergamon.

Leich D. A. (1973) Applications of a nuclear technique for depth-sensitive hydrogen analysis: trapped H in lunar samples and the hydration of terrestrial obsidian (thesis). Unpublished.

Leich D. A. and Tombrello T. A. (1973) A technique for measuring hydrogen concentration versus depth in solid samples. *Nuclear Instruments and Methods* **108,** 67–71.

Leich D. A., Tombrello T. A., and Burnett D. S. (1973a) The depth distribution of hydrogen in lunar materials. *Earth Planet. Sci. Lett.* **19,** 305–314.

Leich D. A., Tombrello T. A., and Burnett D. S. (1973b) The depth distribution of hydrogen and fluorine in lunar samples. *Proc. Fourth Lunar Sci. Conf., Geochim. Cosmochim. Acta,* Suppl. 4, Vol. 2, pp. 1597–1612. Pergamon.

Lindhard J., Scharff M., and Schiøtt H. E. (1963) Range concepts and heavy ion ranges (notes on atomic collisions, II). *Mat. Fys. Medd. Dan. Videnskab. Selskab* **33**(14), 1–42.

Moore D. B., Gibson E. K., Larimer J. W., Lewis C. F., and Nichiporuk W. (1970) Total carbon and nitrogen abundances in Apollo 11 lunar samples and selected achondrites and basalts. *Proc. Apollo 11 Lunar Sci. Conf., Geochim. Cosmochim. Acta,* Suppl. 1, Vol. 2, pp. 1375–1382. Pergamon.

Patterson J. H., Turkevich A. L., Franzgrote E. J., Economou T. E., and Sowinski K. P. (1970) Chemical composition of the lunar surface in a terra region near the crater Tycho. *Science* **168,** 825–828.

Turkevich A. L. (1973) Private communication.

Wehner G. K., Kenknight C., and Rosenberg D. L. (1963) Sputtering rates under solar-wind bombardment. *Planet. Space Sci.* **11,** 885–895.

Winterbon K. B., Sigmund P., and Sanders J. B. (1970) Spatial distribution of energy deposited by atomic particles in elastic collisions. *Mat. Fys. Dan. Videnskab. Selskab* **37**(14), 1–73.

Wolfe J. H. (1972) The large scale structure of the solar wind. *Solar Wind,* NASA SP-308, pp. 170–196.

Proceedings of the Fifth Lunar Conference
(Supplement 5, Geochimica et Cosmochimica Acta)
Vol. 2 pp. 1885–1895 (1974)
Printed in the United States of America

Deuterium, hydrogen, and water content of lunar material

L. Merlivat, M. Lelu, G. Nief, and E. Roth

D.R.A. Centre d'Etudes Nucléaires de Saclay, B.P. No. 2, 91190 GIF-sur-YVETTE, France

Abstract—Hydrogen, water, and deuterium content of Apollo 15, 16, and 17 samples have been measured. A new extraction line based on a flow system operated at atmospheric pressure has been built.

Mean values of hydrogen concentrations and D/H ratios measured in fines and rock breccia samples spread from 14 to 26 μmoles/g and from 28×10^{-6} ($\delta D = -820\permil$) to 47×10^{-6} ($\delta D = -700\permil$), respectively. The isotopic content of water is very similar to terrestrial values and argues in favor of a contamination origin for this water. Experimental evidence of a partial isotopic exchange between hydrogen and water during extraction is given.

Amounts of hydrogen measured in two Apollo 17 rock samples (70215 and 75035) are about 25 times less than in fines samples. Relative values of D/H ratio of hydrogen and water preclude an isotopic exchange with terrestrial material to explain high values measured in hydrogen (D/H $= 187 \times 10^{-6}$ or $\delta D = +200\permil$). Deuterium formed by nuclear reaction on the lunar surface has been measured. Estimated lower and upper limits lie between 1×10^{-11} and 1.6×10^{-10} mole of deuterium per gram. Comparison between experimental and calculated values is made. A very encouraging agreement is shown.

INTRODUCTION

THE POSSIBLE SOURCES of elemental hydrogen in lunar material are solar wind, solar flares, and cosmic ray particles (Reedy and Arnold, 1972), while water can be either originally lunar or formed later by reduction of oxides by solar wind hydrogen or added by terrestrial contamination.

Deuterium is found in hydrogen as well as in water molecules. In the latter case, measuring its concentration enables to trace the origin of water by comparing the value obtained for lunar material with known terrestrial values. Also, as deuterium is formed under nuclear reactions induced by impinging solar and galactic cosmic ray particles on the lunar surface, its abundance and distribution in lunar samples provide information on their exposure and radiation history.

Deuterium, hydrogen, and water were measured in lunar samples since the beginning of Apollo missions (Epstein and Taylor, 1970, 1971, 1972, 1973, 1974; Friedman *et al.*, 1970, 1971, 1972; Merlivat *et al.*, 1972). We concluded from our results (Merlivat *et al.*, 1972) that the original content of water, hydrogen, and deuterium in lunar samples could not be directly measured experimentally with the vacuum extraction technique used. The two main reasons for that are:

- hydrogen and water can undergo an isotopic exchange,
- oxidation of hydrogen to water takes place at extraction temperature higher than 600°C.

We have tried to overcome these difficulties by changing the kinetic conditions of extraction. For that purpose we have built a flow system operated at atmospheric pressure. The experimental arrangement is described in the first part of this paper. Results pertaining to fine and rock samples brought back to earth by Apollo 15, 16, and 17 missions follow.

Experimental Procedure

Lunar samples are handled in a glove box under dry air in order to preclude exposure to atmospheric moisture. The box is dried by forcing the air circulation over vessels containing P_2O_5 and offering about a 0.1 m^2 apparent surface. Though we do not make a measurement of the remaining moisture content, we check that when a vessel filled with fresh P_2O_5 ($\sim$20 g) is introduced it does not only increase in weight after one hour exposure but we observe a decrease of the order of 1–2 mg of the weight due to the drying of the walls of the vessel by the main P_2O_5 reservoirs. All the vessels are dried by standing in the glove box for at least 24 hours. The handling of a lunar sample requires about 30 min. Samples are wrapped in a platinum foil in the glove box (Merlivat *et al.*, 1974). They are subsequently introduced in a vessel closed with a ground joint for weighing outside of the box. Back in the dry box, the samples are loaded in a quartz container which is sealed afterwards on the extraction line. Figure 1 shows the quartz container in which the samples are kept under dry air. One end is closed by a stopcock. The other by a break seal to prevent the presence of grease upstream of the furnace point, preliminary experiments having shown that it leads to an hydrogen contamination due to cracking of grease to methane by the hot vanadium. We have found no difficulty with grease downstream of the furnace in blank experiments. The lunar samples are loaded in the horizontal left arm of the container through the ground joint. The iron bar close to the sample is used during the extraction procedure, to push the sample to the bottom of the U-tube after a preliminary degassing of the quartz at 1200°C.

Figure 2 is a drawing of the extraction line. Extracted gases are carried by helium consecutively through two traps kept at liquid nitrogen temperature. Condensable gases are retained in the first one filled with quartz wool. Other gases (H_2, rare gases except He) are held back in the second one, filled

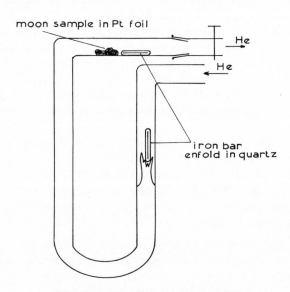

Fig. 1. Quartz holder of the lunar sample.

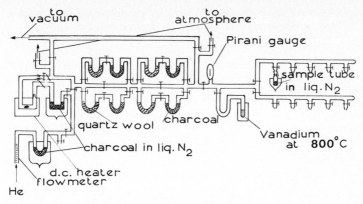

Fig. 2. Extraction experimental line.

with charcoal. Four separate sets of such double traps are used for different ranges of extraction temperature between 175°C and 1200°C. A constant heating rate of 4°C by minute is adopted. Helium is purified by flowing also through two other traps both filled respectively with charcoal and kept at liquid nitrogen temperature. The flow value is 50 cc per minute.

Blanks are run before each experiment at the flow rate value of the helium carrier gas quoted above. Amounts of hydrogen and water collected are always smaller than 0.01 μmole.

Extraction experiments start by raising the temperature of the vessel to about 175°C and degassing during 90 min. Gases evolved at that temperature cannot be surely considered as representative of lunar conditions as the surface of the moon reaches in the daytime temperature as high as 150°C. They are therefore pumped out. The gases are then collected and the heating procedure ends by a 30 min collection period during which the sample is heated at 1200°C. Blanks are run after breaking the seal, under the same condition of flow rate as the experiment, the furnace being at 1200°C, temperature at which pollution by carried over grease is most to be feared. The sample is still in the initial position in the quartz container.

Values in Tables 1 and 2 show that baking the sample longer does not significantly change the amount collected. Helium is pumped out of the whole line. Gases are then released from the traps to react with vanadium at 800°C. At this temperature, vanadium acts as a chemical trap for all compounds except rare gases and hydrogen (Merlivat *et al.*, 1972). Water is reduced to hydrogen. Hydrogen samples are transferred to ampoules filled with charcoal, kept at liquid nitrogen temperature. The samples are ready for analysis by mass spectrometry.

RESULTS AND DISCUSSION

All results obtained with this extraction technique are listed in Tables 1 to 3. Isotopic ratios are given in parts per million (ppm).*

---

*Deuterium concentrations found in the literature are either expressed by the ratio $R = (D/H)$ in ppm or by their relative value δ compared to the standard S.M.O.W., according to:

$$\delta\%_0 = \left(\frac{R}{R_{\text{S.M.O.W.}}} - 1\right) \times 1000.$$

The absolute value of $R_{\text{S.M.O.W.}}$ is 155.76 ppm (Hagemann *et al.*, 1970).

Table 1. Hydrogen and water content and isotopic ratios measured in fines and breccias samples for different intervals of temperature extraction.

Sample	H_2 μmoles/g	$(D/H)_{H_2}$ ppm	H_2O μmoles/g	$(D/H)_{H_2O}$ ppm
15600, m = 0.588 g—fines				
(a) 200–450°C	3.9	39.3	2.1	144.3
(b) 450–725°C	12.8	27.0	0.5	155.6
(c) 725–990°C	1.0	122.5	<0.01	—
(d) 990–1200°C	0.2	112.6	<0.01	—
whole sample	17.9	36.0	2.6	146.5
15600, m = 0.536 g—fines				
a* (a₁) 160–380°C	2.9 {0.9	44.0 {67.6	3.2 {3.2	140.4 {140.4
(a₂) 380–460°C	{2.0	{33.3	{<0.01	{—
(b) 460–720°C	10.3	27.5	0.6	143.6
(c) 720–1000°C	0.7	124.3	<0.01	—
whole sample	13.9	35.8	3.8	140.9
68501, m = 0.546 g—fines				
(a) 160–460°C	3.2	43.7	4.1	140.3
(b) 460–735°C	12.0	36.9	1.1	129.2
(c) 735–985°C	1.6	95.5	<0.01	—
(d) 985–1200°C	0.3	87.5	<0.01	—
whole sample	17.1	44.5	5.2	137.9
72501, m = 0.189 g—fines				
(a) 175–460°C	5.3	42.7	4.2	141.9
(b) 460–725°C	16.7	22.2	0.8	144.5
(c) 725–1000°C	0.8	60.3	<0.01	—
(d) 1000–1200°C	<0.01	—	<0.01	—
whole sample	22.8	28.3	5.0	142.0
72501, m = 0.141 g—fines				
(a) 175–460°C	5.9	40.4	4.2	144.2
(b) 460–725°C	18.4	23.9	<0.01	—
(c) 725–1000°C	1.8	77.5	0.4	135.6
(d) 1000–1200°C	0.2	129.0	<0.01	—
whole sample	26.3	32.0	4.6	143.0
78501, m = 0.221 g—fines				
(a) 175–475°C	3.4	43.8	4.4	142.4
(b) 475–725°C	12.1	27.6	1.4	150.9
(c) 725–1000°C	0.8	111.0	<0.02	—
(d) 1000–1200°C	0.06	—	<0.01	—
whole sample	16.4	35.4	5.8	144.5
15299, m = 0.755 g—rock breccia				
(a) 195–455°C	3.9	71.5	8.5	151.8
(b) 455–690°C	16.8	42.7	3.5	144.6
(c) 690–996°C	1.3	70.7	0.3	120.3
(d) 996–1200°C	0.2	105.4	0.4	109.4
whole sample	19.4	46.8	12.7	147.7

Table 1. (*Continued*).

| Sample | H₂ | | H₂O | |
	μmoles/g	(D/H)$_{H_2}$ ppm	μmoles/g	(D/H)$_{H_2O}$ ppm
15299, m = 0.827 g—rock breccia				
(a) 200–455°C	1.1	59.5	n.d.	n.d.
(b) 455–725°C	14.8	37.2	n.d.	n.d.
(c) 725–990°C	1.2	71.2	0.3	119.6
(d) 990–1200°C	0.03	110.8	0.1	116.4
whole sample	17.1	41.2	—	—

*The usual (a) sample has been treated in two separate fractions in this experiment.

Table 2. Hydrogen and water content and isotopic ratios measured in rock samples for different intervals of temperature extraction.

| Sample | H₂ | | H₂O | |
	μmoles/g	(D/H)$_{H_2}$ ppm	μmoles/g	(D/H)$_{H_2O}$ ppm
70215, m = 0.585 g—basalt				
(a) 175–470°C	<0.01	—	2.3	124.8
(b) 470–725°C	0.75	187.4	<0.01	—
(c) 725–995°C	<0.01	—	0.02	—
(d) 995–1200°C	<0.01	—	0.4	162.4
whole sample	0.75	187.4	2.7	130.0
75035, m = 0.653 g—basalt				
(a) 170–470°C	0.04	120.3	2.22	140.2
(b) 470–725°C	0.19	170.5	0.26	151.7
(c) 725–995°C	0.07	258.5	<0.01	—
(d) 995–1200°C	0.02	129.6	<0.01	—
whole sample	0.32	180.9	2.48	141.4
75035, m = 0.862 g—basalt				
(a) 170–470°C	0.22	98.3	2.31	116.6
(b) 470–725°C	0.69	130.5	0.52	115.9
(c) 725–995°C	0.28	164.4	<0.01	—
(d) 995–1200°C	<0.01	—	<0.01	—
whole sample	1.2	132.7	2.8	116.5

Fines and breccia samples

Abundance and isotopic content of water. The range of our determinations of water content and isotopic concentration extends respectively from 2.6 to 12.7 μmoles/g and 138 to 148 ppm. Those isotopic variations differ from the values obtained with our previous vacuum extraction technique which spread from 96 to 115 ppm (Merlivat *et al.*, 1974). Epstein and Taylor (1973) give a value of about $\delta = -125‰$ (137 ppm) for the isotopic composition of water in lunar samples.

The lower values we found in the vacuum technique are explained by the

Table 3. Columns 1 and 4: Total amounts of hydrogen and water extracted. Columns 2 and 5: Mean isotopic ratios measured in hydrogen and water. Columns 3 and 6: Total amounts of deuterium measured in the hydrogen and water phases.

	q_{H_2} moles/g	$(D/H)_{H_2}$ ppm	q_{D_2} (H_2) moles/g	q_{H_2O} moles/g	$(D/H)_{H_2O}$ ppm	q_{D_2} (H_2O) moles/g
	1	2	3	4	5	6
Fines						
15600	17.9×10^{-6}	36.0	6.4×10^{-10}	2.6×10^{-6}	146.5	3.8×10^{-10}
15600	13.9×10^{-6}	35.8	5.0×10^{-10}	3.8×10^{-6}	149.9	5.3×10^{-10}
68501	17.1×10^{-6}	44.5	7.6×10^{-10}	5.2×10^{-6}	137.9	7.2×10^{-10}
72501	22.8×10^{-6}	28.3	6.4×10^{-10}	5.1×10^{-6}	142.0	7.2×10^{-10}
72501	26.3×10^{-6}	32.0	8.4×10^{-10}	4.6×10^{-6}	143.0	6.6×10^{-10}
78501	16.4×10^{-6}	35.4	5.8×10^{-10}	5.8×10^{-6}	144.5	8.4×10^{-10}
Rock breccia						
15299	19.4×10^{-6}	46.8	9.1×10^{-10}	12.7×10^{-6}	147.7	18.7×10^{-10}
15299	17.1×10^{-6}	41.2	7.0×10^{-10}	—	—	—
Rock (basalt)						
70215	0.75×10^{-6}	187.4	1.4×10^{-10}	2.7×10^{-6}	130.0	3.5×10^{-10}
75035	0.32×10^{-6}	180.9	0.6×10^{-10}	2.5×10^{-6}	141.5	3.5×10^{-10}
75035	1.2×10^{-6}	132.7	1.6×10^{-10}	2.8×10^{-6}	116.5	3.2×10^{-10}

contribution of water formed by oxidation of low deuterium content hydrogen to the total amount of water collected using the vacuum technique. Moreover, recovery of water in the flow system at temperatures higher than 700°C is always small, the largest part being extracted below 500°C. Significative amounts of water extracted at temperature higher than 700°C have been measured only for two samples (15299 and 72501). They represent 6% and 9% of the total amount of extracted water. This remark leads to assume an upper limit to the contribution of oxidized hydrogen to the total water sample, since we have observed that a temperature of about 600°C is needed to start the oxidation reaction under vacuum (Merlivat *et al.*, 1972).

Consequently, measured isotopic compositions appear to be representative of the water contained in the samples. The spread of mean values between 138 and 148 ppm is very comparable to that of terrestrial vapor and liquid water in mid latitudes regions. This is a strong argument in favor of a contamination origin for the water extracted from the sample. This observation has already been pointed out by other authors for several years (Epstein and Taylor, 1971, 1972, 1973, 1974). This contamination could have taken place either on the lunar surface as leakage through the astronauts' suits or in the Apollo cabin or in subsequent handling about which we possess no details. Several deuterium content values of separate water fraction (cf. Table 1) are rather high. Figures close to 156 ppm have been obtained. An explanation for those high values could be an isotopic enrichment during the adsorption and (or) desorption of water on the lunar material.

Abundance and isotopic content of hydrogen. The amount of hydrogen extracted with the flow system (cf. Table 3) is systematically higher than that using the vacuum technique (Merlivat *et al.*, 1974). This corroborates the observation that no noticeable oxidation of hydrogen takes place at atmospheric pressure under our working conditions. However, the total amount of hydrogen extracted under the form of H_2 and H_2O should be constant and independent of the extraction method. This is not the case as a comparison of the values for 15600 shows

	$H_2O + H_2$	μmol/g
	individual extractions	average
Vacuum extractions	24.6; 26.6	25.6
He flow extractions	20.5; 17.7	19.1

We have no explanation for this difference. That should be investigated further. We have checked that the efficiency of degassing at 175°C is the same under vacuum and with a flow of helium. Incomplete extraction using the flow system is not plausible as the amount of gas extracted at the highest temperature is very small (5< % of the total). Incomplete recovery from vanadium cannot be the cause of a discrepancy, the same technique being used in both cases.

Results from duplicate extractions of separate aliquots of the same sample differ significantly. It has been shown (Heymann and Yaniv, 1970; Muller, 1974) that the volumes of solar wind gases measured in fines material are grain size dependent. This could explain why duplication of hydrogen extraction is not very good, aliquots of samples being chosen without reference to their grain size.

The deuterium content of hydrogen fractions extracted follows always the same pattern during the rise of temperature up to 1200°C (cf. Table 1). The first extracted fraction ($t < 460$°C) is always enriched in deuterium with respect to the second one. We assign the comparatively higher value observed for the low-temperature fraction results to a partial isotopic exchange between hydrogen and extracted water which is richer in deuterium. Figure 3 represents the variation of [D/H] ratio in H_2 versus the molar ratio H_2O/H_2 for the low-temperature fraction (a). The positive correlation between the two variables supports our assumption that adsorbed water is the main factor influencing the isotopic composition of the low-temperature extracted hydrogen.

The largest part of hydrogen is extracted within the second temperature interval. Like for the (a) fraction, its origin is mainly solar wind. Its deuterium content should therefore be very close to 0 ppm (Epstein and Taylor, 1972). We interpret our measured isotopic values in the (b) fractions as the result, first, of an isotopic partial exchange with adsorbed water, as for the (a) fraction, and second, as the contribution of deuterium formed by nuclear reaction of high-energy particles (solar and galactic cosmic rays) with the target elements of the lunar samples. This last source of deuterium is also responsible for the fractions (c) and (d) results, when the samples are large enough to be analytically significant.

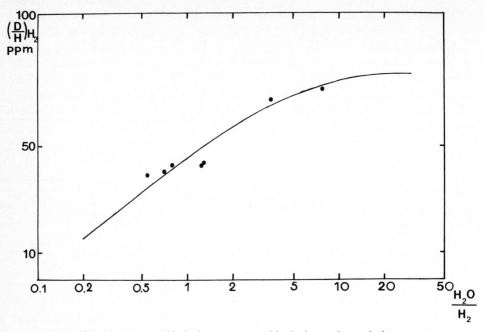

Fig. 3. D/H ratios measured in hydrogen extracted in the lowest interval of temperature
($t < 460°$) versus water and hydrogen amounts ratios extracted in the same temperature
range.

This interpretation of the high deuterium content of hydrogen extracted at high temperature from fines and breccia rests largely on results of rock sample analysis given below. Our analysis of hydrogen and deuterium from fines and breccia do not allow quantitative estimations either of the deuterium content of solar wind hydrogen or of the absolute amount of deuterium formed on the lunar surface by nuclear reaction since the experimental values reflect also, but for an unspecifiable amount, the contribution of adsorbed water which is most likely of terrestrial origin.

Rock samples

Sample 70215. The amount of hydrogen extracted from this basalt sample is about 25 times less than in fines and breccia samples, due to the smaller contribution of the solar wind component (cf. Table 2). The gas has been released in a single temperature interval. Its deuterium content is higher than that of all natural terrestrial hydrogen compounds. This excludes that this high deuterium content could result from an isotopic exchange with lower deuterium terrestrial material. On the contrary, if some partial exchange could have taken place, it would have lowered the D/H ratio of hydrogen. The measured deuterium abundance in rock 70215 is 1.4×10^{-10} moles/g (cf. Table 3). A lower limit of the fraction which is certainly nonterrestrial is given by the excess of deuterium present over the quantity corresponding to terrestrial material. The isotopic composition of the latter being about 150 ppm and the ratio of the sample 187 ppm. This shows a 20% excess of deuterium giving 3×10^{-11} moles/g lower limit for the amount of deuterium formed by nuclear reaction.

A D/H ratio of 212 ppm has already been measured on an Apollo 12 basalt by Friedman *et al.* (1971) in a sample containing 3 μmoles/g of H_2 leading to a 6.4×10^{-10} moles/g for the deuterium amount.

The amount of water extracted in the (a) fraction is smaller than for fines and breccia samples except for the first 15600 extraction. Its low isotopic content is not clearly understood. Either adsorption–desorption kinetics or a mixing of adsorbed and solar wind hydrogen oxidized water could be considered to explain it.

The water extracted in the (d) fraction (995–1200°C) is of special interest. Two possible explanations are proposed for this isotopic rich water:

first, it can be oxidized hydrogen with high D/H ratio due to nuclear processes; second, it could be indigenous occluded water, its deuterium value, 162.4 ppm, being either primordial or already modified by exchange with hydrogen.

Further experimental evidence of high temperature extracted water from other rock samples is needed to enable choosing between various hypotheses.

Sample 75035. We have analyzed two aliquots of the rock 75035. There is a difference between the amount of hydrogen found, 0.35× and 1.2×10^{-6} moles/g (cf. Table 3). The first experiment run with rock 75035 leads to a lower limit of 1×10^{-11} moles/g for deuterium formed on the lunar surface. We think that this difference has to be related to the position of the two aliquots in our piece of rock, mainly with respect to surface, and therefore to exposure. A larger amount of hydrogen combined with a smaller D/H ratio argues in favor of a major contribution of solar wind hydrogen for one of the two aliquots. The same explanation is valid to interpret the difference between the isotopic ratio of extracted water samples.

No water is extracted at high temperature for this sample.

Deuterium produced by nuclear reaction. Our results show that lower and upper limits of the amount of deuterium formed by nuclear reaction in rock samples is experimentally attainable. The values of upper limits given in Table 3 for rocks lie from 0.6× to 1.6×10^{-10} mole of deuterium per gram. The determined exposure ages of these two samples are 100×10^6 yr for rock 70215 (Kirsten and Horn, 1974) and $(83 \pm 18) \times 10^6$ yr for rock 75035 (Crozaz *et al.*, 1974).

Calculations of deuterium production by solar and galactic cosmic rays have been made by Yokoyama (1974). Results, using different types of excitation function are listed in Table 4.

As in our experiments, we did not characterize the position of the aliquots analyzed within the sample and as we cannot assign a position with respect to the lunar surface, the agreement between measured upper limits and calculated values is very encouraging.

Table 4. Calculated amounts of deuterium formed by action of solar and galactic cosmic rays on exposed lunar rocks at different depths using two types of excitation function.

Depth (cm)	0	1	2	7
D_2 (10^{-10} mole/g/10^8 yr)				
model I	1.5	0.6	0.5	0.35
model II	1.2	0.4	0.3	0.2

It will be interesting to measure depth profiles in well documented, oriented rocks and compare the deuterium distribution to other trace elements repartition formed by nuclear reaction on the surface of the moon, as has previously been suggested by Epstein and Taylor (1972).

Acknowledgment—We thank the reviewers for many helpful criticisms and suggestions.

REFERENCES

Crozaz G., Drozd R., Hohenberg C., Morgan C., Ralston C., Walker R., and Yuhas D. (1974) Lunar surface dynamics: Some general conclusions and new results from Apollo 16 and 17 (abstract). In *Lunar Science—V*, pp. 157–159. The Lunar Science Institute, Houston.

Epstein S. and Taylor H. P., Jr. (1970) The concentration and isotopic composition of hydrogen, carbon and silicon in Apollo 11 lunar rocks and minerals. *Proc. Apollo 11 Lunar Sci. Conf., Geochim. Cosmochim. Acta,* Suppl. 1, Vol. 2, pp. 1085–1096. Pergamon.

Epstein S. and Taylor H. P., Jr. (1971) O^{18}/O^{16}, Si^{30}/Si^{28}, D/H, and C^{13}/C^{12} ratios in lunar samples. *Proc. Second Lunar Sci. Conf., Geochim. Cosmochim. Acta,* Suppl. 2, Vol. 2, pp. 1421–1441. MIT Press.

Epstein S. and Taylor H. P., Jr. (1972) O^{18}/O^{16}, Si^{30}/Si^{28}, C^{13}/C^{12}, and D/H studies of Apollo 14 and 15 samples. *Proc. Third Lunar Sci. Conf., Geochim. Cosmochim. Acta,* Suppl. 3, Vol. 2, pp. 1429–1454. MIT Press.

Epstein S. and Taylor H. P., Jr. (1973) The isotopic composition and concentration of water, hydrogen, and carbon in some Apollo 15 and 16 soils and in the Apollo 17 orange soil. *Proc. Fourth Lunar Sci. Conf., Geochim. Cosmochim. Acta,* Suppl. 4, Vol. 2, pp. 1559–1575. Pergamon.

Epstein S. and Taylor H. P., Jr. (1974) Oxygen, silicon, carbon, and hydrogen isotope fractionation processes in lunar surface materials (abstract). In *Lunar Science—V*, pp. 212–214. The Lunar Science Institute, Houston.

Friedman I., Gleason J. D., and Hardcastle K. G. (1970) Water, hydrogen, deuterium, carbon, and C^{13} content of selected lunar material. *Proc. Apollo 11 Lunar Sci. Conf., Geochim. Cosmochim. Acta,* Suppl. 1, Vol. 2, pp. 1103–1109. Pergamon.

Friedman I., O'Neil J. R., Gleason J. D., and Hardcastle K. (1971) The carbon and hydrogen content and isotopic composition of some Apollo 12 materials. *Proc. Second Lunar Sci. Conf., Geochim. Cosmochim. Acta,* Suppl. 2, Vol. 2, pp. 1407–1415. MIT Press.

Friedman I., Hardcastle K. G., and Gleason J. D. (1972) Isotopic composition of carbon and hydrogen in some Apollo 14 and 15 samples. In *The Apollo 15 Lunar Samples*, pp. 302–306. The Lunar Science Institute, Houston.

Hagemann R., Nief G., and Roth E. (1970) Absolute isotopic scale for deuterium analysis of natural water. *Tellus* 22, 712–715.

Heymann D. and Yaniv A. (1970) Inert gases in the fines from the Sea of Tranquillity. *Proc. Apollo 11 Lunar Sci. Conf., Geochim. Cosmochim. Acta,* Suppl. 1, Vol. 2, pp. 1247–1259. Pergamon.

Kirsten T. and Horn P. (1974) ^{39}Ar–^{40}Ar chronology of the Taurus-Littrow region II: A 4.28 b.y. old troctolite and ages of basalts and highland breccias (abstract). In *Lunar Science—V*, pp. 419–421. The Lunar Science Institute, Houston.

Merlivat L., Nief G., and Roth E. (1972) Deuterium content of lunar material. *Proc. Third Lunar Sci. Conf., Geochim. Cosmochim. Acta*, Suppl. 3, Vol. 2, pp. 1473–1477. MIT Press.

Merlivat L., Lelu M., Nief G., and Roth E. (1974) Deuterium content of lunar material (abstract). In *Lunar Science—V*, pp. 498–500. The Lunar Science Institute, Houston.

Muller O. (1974) Solar wind and indigenous nitrogen in Apollo 17 lunar samples (abstract). In *Lunar Science—V*, pp. 534–536. The Lunar Science Institute, Houston.

Reddy R. C. and Arnold J. R. (1972) Interaction of solar and galactic cosmic-ray particles with the moon. *J. Geophys. Res.* **77**(4), 537–555.

Yokoyama Y. (1974) Personal communication.

Proceedings of the Fifth Lunar Conference
(Supplement 5, Geochimica et Cosmochimica Acta)
Vol. 2 pp. 1897–1906 (1974)
Printed in the United States of America

Total carbon and sulfur contents of Apollo 17 lunar samples

Carleton B. Moore, Charles F. Lewis, and Jerry D. Cripe

Arizona State University, Tempe, Arizona 85281

Abstract—The total carbon contents of Apollo 17 fines range from 5 to 200 $\mu g/g$. Only samples from Shorty Crater and Steno Crater have values less than 100 $\mu g/g$ total carbon. Soil breccias have total carbon contents similar to dark fines. Massif (Highland) breccias have total carbon contents in the range 21–105 $\mu g/g$. Apollo 17 basalts have from 31 to 64 $\mu g/g$ total carbon and 1860 to 3140 $\mu g/g$ total sulfur. These values are similar to those measured for basalts from Apollo 11.

Apollo 17 fines have total sulfur contents ranging from 710 to 1330 $\mu g/g$. The total sulfur contents in the fines reflects the nature of local underlying rocks diluted by material moved laterally from other lunar sources. Highland breccias have total sulfur contents ranging from 380 to 910 $\mu g/g$. Sulfur analyses of sieved fines indicate that some have surface correlated sulfur.

Introduction

Total carbon and sulfur contents were determined in lunar fines, breccias and fragmental and igneous rocks from the Apollo 17 mission. The analytical techniques utilized were the same as those used previously. These analytical methods are outlined and reviewed by Moore *et al.* (1971, 1972).

Investigations of total carbon in the Apollo 11, 12, 14, 15, and 16 samples indicated that the lunar fines had higher carbon contents than the lunar rocks. This was primarily attributed by several investigators to the solar wind addition of carbon to the surfaces of fines particles (Abell *et al.*, 1970; Kaplan and Smith, 1970; Moore *et al.*, 1970a, 1970b, 1971, 1973; Moore, 1972; Cadogan *et al.*, 1973; Chang *et al.*, 1973; Des Marais *et al.*, 1973; Epstein and Taylor, 1973; Simoneit *et al.*, 1973; Smith *et al.*, 1973). Lunar breccia samples and some fines samples have total carbon contents that may be as high as the typical fines or as low as lunar rocks. Breccias from Apollo 11 and 12 were similar in total carbon content to the typical fines and were concluded to be derived from high carbon fines material. The majority of breccias from Apollo 14, however, were lower in total carbon than those from Apollo 11 and 12. They appeared to be complex, primary breccias produced during the Imbrium or other large impact events. Although the term breccia is widely used for lunar sample classification purposes, it is important to realize that it can include both compacted indurated fines and fragmental rocks that have not been exposed to lunar surface processes. Compacted fines or soil breccias have high total carbon contents analogous to the lunar fines, while the fragmental rock type breccias do not reflect the addition of extralunar carbon, unless thermal metamorphism has removed it. Lunar samples collected on Apollo 15, 16, and 17 include breccias of both of these types. The total carbon content is a useful indicator of a lunar breccia's history and classification.

Our carbon results from earlier missions have been substantiated by other investigators especially those investigating carbon isotopes (Kaplan and Smith, 1970; Kaplan and Petrowski, 1971; Epstein and Taylor, 1970, 1971, 1972, 1973). Studies of the chemical species released from the lunar materials by pyrolysis, leaching, and crushing by several investigators indicated that the carbon was mainly present as inorganic carbon. A review of these studies for Apollo 14 and earlier missions has been compiled by Gibson and Moore (1972) and Kaplan (1972). More recent studies of this type include Des Marais *et al.* (1973), Simoneit *et al.* (1973), and Wszolek and Burlingame (1973). Investigators using these sample opening techniques have sometimes estimated total carbon contents by totaling the concentrations of species detected. Often such totals do not approach the carbon contents determined by complete combustion.

Investigations by the authors and other investigators indicated that the finest lunar regolith material had significantly higher contents than the coarser lunar fines (Moore *et al.*, 1970b; Kaplan and Smith, 1970). Des Marais *et al.* (1973) also emphasize that coarser fines from mature soils usually contain high carbon agglutinate particles.

Based on the distribution of total carbon in the Apollo 11, 12, 14, 15, and 16 samples, the authors developed qualitative models which derived the carbon present in fines from indigenous lunar carbon and solar wind implanted carbon. It was assumed in these models that essentially all meteoritic carbon was not retained by the moon during impacts. Hayes (1972) has reviewed the extra lunar sources for carbon on the moon and indicated that in addition to the solar wind, meteorites and comets may make a significant contribution.

Des Marais *et al.* (1973) and Moore *et al.* (1973) have shown that as a soil matures the volume component of total carbon increases due to agglutinate formation. The agglutinate particles contain fine grained particles with high concentrations of surface correlated carbon. The authors (Moore *et al.*, 1973) have estimated that the surface component distribution of total carbon remains essentially constant after cosmic ray exposure ages of roughly 20–30 m.y.

Total sulfur contents of the lunar samples appeared to be unrelated to either solar wind or meteorite effects. The distribution of sulfur in the lunar fines may be explained by mixing models of indigenous lunar rock types (Moore *et al.*, 1972; Cripe and Moore, 1974). The sulfur contents of rocks appear to be specific for samples collected from each mission. Basalts generally have higher total sulfur values than do fragmental rocks. Crystalline anorthosites have the lowest total sulfur contents measured. The results of the authors are generally consistent with those found by using X-ray fluorescence techniques (Rhodes and Hubbard, 1973; Rhodes *et al.*, 1974), and Gibson and Moore (1973) using a combustion technique similar to ours. Refinements on the mixing models are apparently needed in order to explain the isotopic variation of sulfur in fines samples as measured by Rees and Thode (1972) and Sakai *et al.* (1972).

Results

The results from the Apollo 17 total carbon and sulfur analyses are given in Table 1 for those samples provided during regular sample allocation. The total

Table 1. Total carbon and sulfur in Apollo 17 samples.

Sample No.	Station	Sample Type	Total μg/g C	μg/g S
70011,17	LM	SESC	120	1200
70215,53	SEP-LM	Fine basalt	31	2040
71055,29	1A	Medium basalt	54	1860
72141,17	LRV 2	<1 mm fines	155	950
72161,12	LRV 3	<1 mm fines	200	990
72275,71	2	Layered light gray breccia	23	860
72395,42	2	Green–gray breccia	105	
72441,8	2	<1 mm fines	135	750
72701,30	2	<1 mm fines	125	790
73121,12	2a	<1 mm fines	120	710
73141,14	2	<1 mm fines	120	730
73235,46	3	Blue gray breccia	54	500
74121,14	LRV 6	<1 mm fines	140	890
74220,84	4	unsieved fines	5	—
75035,42	5	Medium basalt	64	3140
75121,10	LRV 7	<1 mm fines	145	1120
76240,12	6	unsieved fines	125	870
76260,6	6	unsieved fines	100	880
76321,12	6	<1 mm fines	140	770
77017,47	7	Brecciated olivine gabbro	80	910
78121,8	LRV 11	<1 mm fines	125	970
78155,28	8	Gabbroic breccia	21	380
78221,9	8	<1 mm fines	190	950
78421,32	8	<1 mm fines	165	890
78441,17	8	<1 mm fines	155	890
78461,13	8	<1 mm fines	180	890
78481,29	8	<1 mm fines	180	950
79135,27	9	Dark matrix breccia	150	1110
79221,20	9	<1 mm fines	150	1280
79241,20	9	<1 mm fines	140	1160
79261,26	9	<1 mm fines	110	1330

carbon values determined by the authors during LSPET are given in Table 2. No attempt was made to homogenize or mechanically split the lunar fines samples. This minimized the possibility of contamination but increased the possibility of sampling differences. Differences between LSPET analyses and later analyses may be attributed to sampling inhomogeneities due to grain size distribution differences or exotic particles, although no detailed studies have been made concerning this possible mechanism. It has not been uncommon for rock samples provided for the LSPET analyses to contain fines particles adhering to the rock surface. The higher LSPET total carbon of the gabbroic breccia 78155 as compared to the regular allocation of the same sample may be attributed to contamination by fines in the former.

Rock samples were crushed with a single stroke in a clean diamond mortar. The crushed material was not sieved. Samples ranging from 100 to 400 mg were taken for an individual analysis. The analytical precision for all lunar samples is a

Table 2. Total carbon in Apollo 17 P.E.T. samples.

Sample No.	Station	Sample Type	$\mu g/g$ C
70161,2	ALSEP	<1 mm fines	150
70181,2	ALSEP	<1 mm fines	165
71041,2	1A	<1 mm fines	90
71061,2	1A	<1 mm fines	40
71501,2	1A	<1 mm fines	75
72501,3	2	<1 mm fines	125
72701,3	2	<1 mm fines	140
73221,2	3	<1 mm fines	155
73261,2	3	<1 mm fines	170
74220,4	4	unsieved fines	100
74240,4	4	unsieved fines	55
74260,3	4	unsieved fines	45
75081,2	5	<1 mm fines	115
76501,3	6	<1 mm fines	120
77531,2	7	<1 mm fines	180
78155,3	8	Gabbroic breccia	85
78501,3	8	<1 mm fines	170
79221,3	9	<1 mm fines	160
79261,3	9	<1 mm fines	145

function of the total carbon found and the sample size, but generally averages about 10% of the concentration given.

The analytical results for total carbon and sulfur follow the same pattern as those of the earlier Apollo missions. Fines and dark soil-breccia samples generally have higher total carbon contents than do the fragmental rock-breccias and igneous rocks. It is of importance to note that the total carbon contents of the Apollo 17 basalts are similar to those found for the Apollo 11 basalts. For the Apollo missions 11 through 16 the trend for total carbon in rocks was constantly downward dropping from 60 to 70 $\mu g/g$ in the Apollo 11 basalts to lows of less than 5 $\mu g/g$ in the Apollo 16 anorthosites. The return to higher values for the Apollo 17 rocks indicates that the trend was a real one and not due to improved contamination control in the sample handling procedure. Likewise, the total sulfur contents in the Apollo 17 basalts and associated fines returned to the higher values, in the greater than 1000 $\mu g/g$ range, found in the related Apollo 11 samples.

Most of the Apollo 17 fines samples are gray to black in color and are similar in total carbon content to the normal mature lunar fines of earlier Apollo missions. The total carbon values for separate samples at different traverse stations are shown in Fig. 1. It may be seen that individual samples at each station generally show lower ranges in total carbon content than do samples from different stations. Samples from Station 1A near Steno Crater and Station 4 at Shorty Crater are both lower in total carbon than the mature fines from the other sampling stations. Following the models put forward in our reports on earlier lunar missions, it appears that the lower values at Steno Crater may be attributed to the fact that the fines samples are in part composed of rock fragments from the cratering event

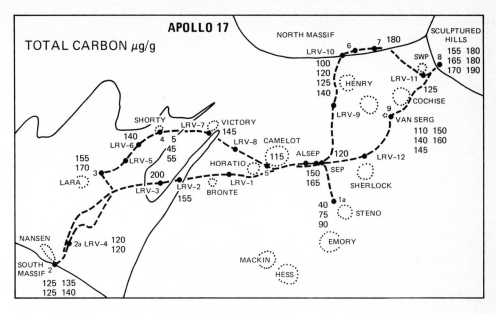

Fig. 1. Total carbon contents of individual samples from different Apollo 17 traverse stations at the Taurus-Littrow landing site.

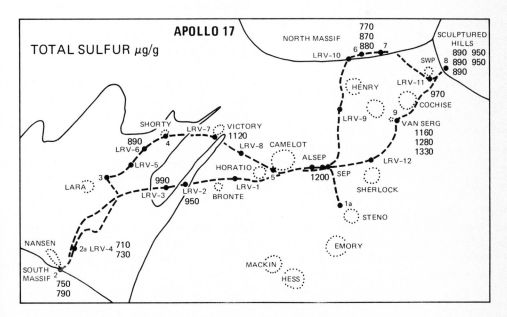

Fig. 2. Total sulfur contents of individual samples from different Apollo traverse stations at the Taurus-Littrow landing site.

that have not had surface exposure to the solar wind. In other words, they are not mature fines (McKay *et al.*, 1974). The samples from Station 4 include the orange soil sample and adjacent samples of darker fines. The orange soil appears to have a low total carbon content of about 5 μg/g and the adjacent darker fines carbon contents in the 50 μg/g range. The 5 μg/g carbon value is preferred to the LSPET value of 100 μg/g in Table 2. A detailed discussion of the variable carbon contents of lunar soil 74220 has been published by Gibson and Moore (1973). Measured total carbon variations between mature soils are attributed to grain size differences and agglutinate percentage differences at the sampling stations in the Taurus-Littrow landing site. Inspection of both of these parameters for the LSPET data shows direct correlations between them and the total carbon content. The fines samples collected at Station 8 from different depths in a trench show no significant differences in either total carbon or sulfur content.

Total sulfur contents of Apollo 17 fines at different sampling stations are plotted on Fig. 2. The stations located on the valley floor have significantly higher total sulfur contents than those near the massifs and landslide deposits from the massifs. The higher sulfur fines appear thus to have a higher proportion of the high sulfur basaltic material in their makeup. It may be seen in Table 1 that the Apollo 17 basalts range from 1860–3140 μg/g in total sulfur while the highland type breccias have total sulfur contents in the 380–910 μg/g range.

DISCUSSION

As indicated above, the total carbon and sulfur abundances for the Apollo 17 samples are generally similar in their gross characteristics to those for the earlier Apollo missions. The total carbon content of mature fines appears to be independent of the underlying rock types while the total sulfur is partially dependent upon the local rock types. In our earlier papers we attributed the higher carbon content in the fines to the addition of carbon by the solar wind. Based upon mass balance models and the distribution of total sulfur, we concluded that meteoritic carbon and sulfur have made little if any contribution to the observed abundances.

A major question still remains with respect to the exact proportions of carbon contributed by comminuted lunar rocks, solar wind and meteorites to that observed in the fines. A critical discussion of this problem has been prepared by Hayes (1972). In this review he concludes that in addition to the solar wind, meteorites and comets may also contribute to the lunar carbon via a cycle of carbon through the lunar atmosphere. Such a contribution is supported by the work of Heymann and Yaniv (1970), who showed an implantation on the lunar surface of Ar-40 from a tenuous lunar atmosphere. Evidence from carbon isotope studies also indicates difficulty in developing a simple model for the extralunar sources of fines carbon. The carbon fines (Epstein and Taylor, 1970, 1971, 1972; Kaplan and Smith, 1970; Kaplan and Petrowski, 1971) are richer in C-13 than the carbon in lunar rocks and carbonaceous chondrites indicating either a C-13 enriched solar wind or more likely an extensive carbon exchange mechanism on fines particle surfaces.

Des Marais *et al.* (1973) and Moore *et al.* (1973) as well as other investigators have shown that the analyses of sieved splits of the lunar samples are important in understanding the origin and history of carbon (and sulfur) in the lunar fines. Such studies allow surface and volume components of total carbon to be differentiated. They indicate that the surface component reaches a steady-state maturity within a period of about 20–30 m.y., while the volume component continues to rise slowly due to the formation of carbon-containing agglutinate particles. The total carbon measured in lunar fines is thus a function of the indigenous lunar carbon plus carbon added by solar wind and to a minor amount by meteorite impact. Removal of this surface carbon by proton stripping together with carbon addition from the solar wind causes its total concentration to reach a steady-state level. At the same time, the loss of volatile carbon species from the surface causes the isotopic distribution of this carbon to become heavier. Some of the extra lunar, surface related, carbon is stabilized by the formation of agglutinate particles by meteorite impact. Superimposed on this stabilization may be an erosion process by surface sputtering that causes changes in the grain size and distribution of the fines and most notably the loss by comminution of the finest particles. Such events are recorded in the chemistry of different particle sizes of sieved samples.

Although generalized models of lunar regolith evolution may be developed, it is also important to remember that each lunar sample has had a unique history. This is particularly true with respect to the effect of relatively recent meteorite impacts in the vicinity of the sampling localities (Gault *et al.*, 1974). The effect of such impacts may be seen in the lighter colored low carbon fines from Stations 1A and 4. In these cases, relatively large impacts exposed coarse-grained rock fragments at the lunar surface. Short-term gardening effects did not have enough time to mature these soils before they were collected.

Although the general model for regolith evolution indicates that the sulfur contents of the lunar fines is primarily due to indigenous lunar sulfur, analysis of sieved samples indicates that in some samples sulfur is surface correlated. Figure 3 illustrates the distribution of total sulfur in several lunar fines samples. The distribution may vary from samples with no surface sulfur component such as 14298 to those with a distinctive surface sulfur related component such as 12032. We attribute the surface sulfur to meteoritic sulfur and lunar sulfur deposited on particle surfaces from the vapor cloud produced by a hypervelocity meteorite impact. Proton stripping will remove this surface sulfur as individual particles are exposed to the solar wind. Sample 14298 is a mature dark fines sample with a total carbon content of 150 μg/g, while sample 12032 is an immature light colored fines sample with a total carbon content of 25 to 60 μg/g. Samples 15012, 65701, and 67701 show pattern intermediate to the two end members. Sample 15012, a mature dark fines, shows a possible surface component only on the finest material. The Apollo 16 samples 67501 and 67701 show chemical anomalies in the coarser sized fragments due to poor sampling statistics. The high sulfur value in the coarse 65701 material may be due to a high sulfur agglutinate fragment while the low sulfur value in the coarse size fragments in 67701 may be due to a low sulfur rock fragment. As with carbon, surface distribution studies of sulfur are best made with particles in the 40–250 μm size range.

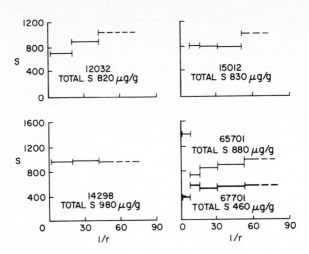

Fig. 3. Reciprocal radius versus total sulfur for sieved fractions of lunar fines 12032, 14298, 15012, 65701, and 67701.

Further size distribution studies for carbon, sulfur, and nitrogen in the Apollo 17 samples will give us a better understanding of the chemical evolution of the lunar regolith.

Acknowledgments—This research was supported in part by NASA grants NGL-03-001-001 and NGR-03-001-057. E. K. Gibson, Jr. kindly helped by transporting the total carbon LSPET samples from Houston to Tempe and in the preparation of the total carbon section of the Apollo 17 LSPET report. This article was written while C. B. Moore was a NSF-NRC Postdoctoral Fellow in the Chemical Evolution Branch of the NASA Ames Research Center. The interest and help of the personnel of that Branch is appreciated.

REFERENCES

Abell P. T., Draffan G. H., Eglinton G., Hayes J. M., Maxwell J. R., and Pillinger C. T. (1970) Organic analysis of the returned lunar sample. *Science* **167**, 757–759.

Cadogan P. H., Eglinton G., Gowar A. P., Jull A. J. T., Maxwell J. R., and Pillinger C. T. (1973) Location of Methane and Carbide in Apollo 11 and 16 Lunar Fines. *Proc. Fourth Lunar Sci. Conf.*, *Geochim. Cosmochim. Acta*, Suppl. 4, Vol. 1, pp. 1493–1508. Pergamon.

Chang S., Mack R., Gibson E. K., Jr., and Moore G. W. (1973) Simulated solar wind implantation of carbon and nitrogen ions into terrestrial basalt and lunar fines. *Proc. Fourth Lunar Sci. Conf.*, *Geochim. Cosmochim. Acta*, Suppl. 4, Vol. 1, pp. 1509–1522. Pergamon.

Cripe J. D. and Moore C. B. (1974) Total sulfur contents of Apollo 15 and Apollo 16 lunar samples (abstract). In *Lunar Science—V*, pp. 523–525. The Lunar Science Institute, Houston.

Des Marais D. J., Hayes J. M., and Meinschein W. G. (1973) The distribution in lunar soil of carbon released by pyrolysis. *Proc. Fourth Lunar Sci. Conf.*, *Geochim. Cosmochim. Acta*, Suppl. 4, Vol. 1, pp. 1543–1558. Pergamon.

Epstein S. and Taylor H. P., Jr. (1970) The concentration and isotopic composition of hydrogen, carbon, and silicon in Apollo 11 lunar rocks and minerals. *Proc. Apollo 11 Lunar Sci. Conf.*, *Geochim. Cosmochim. Acta*, Suppl. 1, Vol. 2, pp. 1085–1096. Pergamon.

Epstein S. and Taylor H. P., Jr. (1971) O^{18}/O^{16}, Si^{30}/Si^{28}, D/H, and C^{13}/C^{12} ratios in lunar samples. *Proc. Second Lunar Sci. Conf.*, *Geochim. Cosmochim. Acta*, Suppl. 2, Vol. 2, pp. 1421–1441. MIT Press.

Epstein S. and Taylor H. P., Jr. (1972) O^{18}/O^{16}, Si^{30}/Si^{28}, C^{13}/C^{12}, and D/H studies of Apollo 14 and 15 samples. *Proc. Third Lunar Sci. Conf., Geochim. Cosmochim. Acta*, Suppl. 2, Vol. 2, pp. 1429–1454. MIT Press.

Epstein S. and Taylor H. P., Jr. (1973) The isotopic composition and concentration of water, hydrogen and carbon in some Apollo 15 and 16 soils and in the Apollo 17 orange soil. *Proc. Fourth Lunar Sci. Conf., Geochim. Cosmochim. Acta*, Suppl. 4, Vol. 1, pp. 1559–1575. Pergamon.

Gault D. E., Hörz F., Brownlee D. E., and Hartung J. B. (1974) Mixing of the lunar regolith (abstract). In *Lunar Science—V*, pp. 260–262. The Lunar Science Institute, Houston.

Gibson E. K., Jr. and Moore G. W. (1973) Carbon and Sulfur distributions and abundances in lunar fines. *Proc. Fourth Lunar Sci. Conf., Geochim. Cosmochim. Acta*, Suppl. 4, Vol. 1, pp. 1577–1586. Pergamon.

Gibson E. K., Jr. and Moore C. B. (1973) Variable carbon contents of lunar soil 74220. *Earth Planet. Sci. Lett.*, **20**, 404–408. North-Holland, Amsterdam.

Gibson E. K. and Moore C. B. (1972) Compounds of the organogenic elements in Apollo 11 and 12 lunar samples: A review. *Space Life Sciences* **3**, 404–414.

Hayes J. M. (1972) Extralunar sources for carbon on the Moon. *Space Life Sciences* **3**, 474–483.

Heymann D. and Yaniv A. (1970) Ar^{40} anomaly in lunar samples from Apollo 11. *Proc. Apollo 11 Lunar Sci. Conf., Geochim. Cosmochim. Acta*, Suppl. 1, Vol. 2, pp. 1261–1267. Pergamon.

Kaplan I. R. (1972) Distribution and isotopic abundance of biogenic elements in lunar samples. *Space Life Sciences* **11**, 111–222.

Kaplan I. R. and Petrowski C. (1971) Carbon and sulfur isotopic studies on Apollo 12 lunar samples. *Proc. Second Lunar Sci. Conf., Geochim. Cosmochim. Acta*, Suppl. 2, Vol. 2, pp. 1397–1406. MIT Press.

Kaplan I. R. and Smith J. W. (1970) Concentration and isotopic composition of carbon and sulfur in Apollo 11 lunar samples. *Science* **167**, 541–543.

McKay D. S., Fruland R. M., and Heiken G. H. (1974) Grain size distribution as an indicator of the maturity of lunar soils (abstract). In *Lunar Science—V*, pp. 480–482. The Lunar Science Institute, Houston.

Moore C. B. (1972) The distribution of carbon in lunar samples from Apollo 11, 12, and 14. In *International Geological Congress, Twenty-fourth Session, Section 15*, Planetology, (Editor J. E. Gill), Montreal, pp. 69–74.

Moore C. B., Lewis C. F., Gibson E. K., and Nichiporuk W. (1970a) Total carbon and nitrogen abundances in lunar samples. *Science* **167**, 496–497.

Moore C. B., Gibson E. K., Larimer J. W., Lewis C. F., and Nichiporuk W. (1970b) Total carbon and nitrogen abundances in Apollo 11 lunar samples and selected achondrites and basalts. *Proc. Apollo 11 Lunar Sci. Conf., Geochim. Cosmochim. Acta*, Suppl. 1, Vol. 2, pp. 1375–1382. Pergamon.

Moore C. B., Lewis C. F., Larimer J. W., Delles F. M., Gooley R., and Nichiporuk W. (1971) Total carbon and nitrogen abundances in Apollo 12 lunar samples. *Proc. Second Lunar Sci. Conf., Geochim. Cosmochim. Acta*, Suppl. 2, Vol. 2, pp. 1343–1350. MIT Press.

Moore C. B., Lewis C. F., Cripe J. D., Delles F. M., Kelly W. R., and Gibson E. K., Jr. (1972) Total carbon, nitrogen and sulfur in Apollo 14 lunar samples. *Proc. Third Lunar Sci. Conf., Geochim. Cosmochim. Acta*, Suppl. 3, Vol. 2, pp. 2051–2058. MIT Press.

Moore C. B., Lewis C. F., and Gibson E. K., Jr. (1973) Total carbon contents of Apollo 15 and 16 lunar samples. *Proc. Fourth Lunar Sci. Conf., Geochim. Cosmochim. Acta*, Suppl. 4, Vol. 2, pp. 1613–1623. Pergamon.

Rees C. E. and Thode H. G. (1972) Sulfur concentrations and isotope ratios in lunar samples. *Proc. Third Lunar Sci. Conf., Geochim. Cosmochim. Acta*, Suppl. 3, Vol. 2, pp. 1479–1485. MIT Press.

Rhodes J. M. and Hubbard N. J. (1973) Chemistry, classification and petrogenesis of Apollo 15 mare basalts. *Proc. Fourth Lunar Sci. Conf., Geochim. Cosmochim. Acta*, Suppl. 4, Vol. 1, pp. 1127–1148. Pergamon.

Rhodes J. M., Rodgers K. V., Shih C., Bansal B. M., Nyquist L. E., and Wiesmann H. (1974) The relationship between geology and soil chemistry at the Apollo 17 landing site (abstract). In *Lunar Science—V*, pp. 630–633. The Lunar Science Institute, Houston.

Sakai H., Petrowski C., Goldhaber M. B., and Kaplan I. R. (1972) Distribution of carbon, sulfur and nitrogen in Apollo 14 and 15 material (abstract). In *Lunar Science—III*, pp. 672–674. The Lunar Science Institute, Houston.

Simoneit B. R., Christiansen P. C., and Burlingame A. L. (1973) Volatile element chemistry of selected lunar, meteoritic, and terrestrial samples. *Proc. Fourth Lunar Sci. Conf., Geochim. Cosmochim. Acta*, Suppl. 4, Vol. 1, pp. 1635–1650. Pergamon.

Smith J. W., Kaplan I. R., and Petrowski C. (1973) Carbon, nitrogen, sulfur, helium, hydrogen, and metallic iron in Apollo 15 drill stem fines. *Proc. Fourth Lunar Sci. Conf., Geochim. Cosmochim. Acta*, Suppl. 4, Vol. 1, pp. 1651–1656. Pergamon.

Wszolek P. C. and Burlingame A. L. (1973) Carbon chemistry of the Apollo 15 and 16 deep drill cores. *Proc. Fourth Lunar Sci. Conf., Geochim. Cosmochim. Acta*, Suppl. 4, Vol. 1, pp. 1681–1692. Pergamon.

Proceedings of the Fifth Lunar Conference
(Supplement 5, Geochimica et Cosmochimica Acta)
Vol. 2 pp. 1907–1918 (1974)
Printed in the United States of America

Solar wind nitrogen and indigenous nitrogen in Apollo 17 lunar samples

O. Müller

Max-Planck-Institut für Kernphysik, Heidelberg, Germany

Abstract—Chemically bound nitrogen contents have been determined in Apollo 17 fines <1 mm and grain size fractions, in Apollo 17 rocks, and in some stone meteorites using the Kjeldahl method. Rocks contain, in general, a factor of about ten less nitrogen than fines. The lunar samples results show, as in our previous work, that the main amount of nitrogen is present in a chemically bound state. Leach-experiments on Apollo 15 and Apollo 17 fines indicate two chemical forms of nitrogen: ammonium- and nitride-nitrogen. Nitrogen in fines is concentrated on grain surfaces and is derived most likely from solar wind implantation. Correlation of trapped ^{36}Ar with nitrogen provides further evidence for solar wind origin of nitrogen in fines. Implanted nitrogen is better retained in fine-grained material than the light noble gases and carbon, and is a useful reference in terms of solar elemental abundances. Carbon to nitrogen ratios of lunar fines are interpreted as lower limits for the abundance ratio of these elements in the sun.

Introduction

THE ABUNDANCE and isotopic composition of the light elements hydrogen, helium, carbon, nitrogen, and neon in fine-grained regolith samples of the moon demonstrate that the major portion of these elements is derived from implantation by the solar wind. The extralunar contribution is easy to detect, because lunar rocks, which are the principal source material of lunar fines, are, in general, strongly depleted in light volatile elements. A further important aspect of solar wind-implanted chemically active species is their ability to form compounds in the surface layers of mineral grains, e.g. methane and carbides, and probably ammonia and nitrides. Hydrogen compounds of carbon and nitrogen may be relevant precursors of more complex organic molecules and compounds of biological significance on planetary surfaces.

This report is on chemically bound nitrogen concentrations in various Apollo 17 lunar samples from the Taurus Littrow landing site. The work continues our earlier investigations of bound nitrogen in lunar material with the Kjeldahl method (Müller, 1972, 1973). The Kjeldahl technique detects nitrogen only in a bound state and discriminates against molecular nitrogen N_2, which may be either derived from atmospheric contamination or be indigenous to lunar samples. This method also discriminates against nitrate- and nitrite-nitrogen. We have analyzed eight Apollo 17 fines of grain size <1 mm from different stations. Two of them have been separated into several grain size fractions to study the grain size dependence of nitrogen implanted by the solar wind. These are fines 74241, a gray soil adjacent to the orange soil at the rim of Shorty Crater, and fines 78501 collected from sloping terrain at the base of Sculptured Hills. Three rock samples, a crushed anorthositic gabbro, a homogeneous gabbro, and a fine-grained basalt, have been analyzed to determine bound nitrogen indigenous to Apollo 17 rocks. Furthermore, some stone meteorite samples have been investigated: the light and dark, noble gas-rich portions of the achondrite Kapoeta to study possible contribution of solar wind-implanted nitrogen in the dark portion, and the carbonaceous chondrites Allende, Murchison and Murray.

The aims of the present investigation have been: (1) to compare chemically bound nitrogen abundances of Apollo 17 lunar samples with data for total nitrogen obtained by other techniques; (2) to study the dependence of bound nitrogen on surface area in grain size fractions of lunar fines to demonstrate the solar wind origin of nitrogen; (3) to prove the relation of trapped ^{36}Ar with bound nitrogen as additional evidence for the solar wind origin of nitrogen; (4) to show the high retention of solar wind-implanted nitrogen by comparing ^{4}He/N, ^{20}Ne/N, and ^{36}Ar/N ratios of lunar fines with the abundance ratio of these elements in the sun; (5) to learn more about the chemical forms of nitrogen in fines by performing leach-experiments; and (6) to compare C/N ratios of fines with those of the sun.

ANALYTICAL PROCEDURE

The technique for the chemically bound nitrogen determinations (Kjeldahl method) in lunar and meteorite samples was the same as that previously described (Müller, 1972, 1973). Leach-experiments were performed on Apollo 15 and Apollo 17 fines of grain size <1 mm and <24 μm, and on some synthetic materials as follows. Samples of about a 100 mg were weighed in teflon centrifuge tubes, 10 ml 2N H_2SO_4 were added, and the lunar fines <1 mm and synthetic materials were shaken for 1 hour at 20°C. The fines <24 μm were treated for 15 min. The material was centrifuged and the supernatant leach-solution transferred to the Kjeldahl apparatus for ammonia distillation. The residue was decomposed with highly pure 40% HF and 96% H_2SO_4 in a platinum crucible, and the non-leachable nitrogen determined by the Kjeldahl technique. Prior to the ammonia distillations, an aliquot of the leach- and residue-portion, respectively, was pipetted to determine the elements Na, Mg, Al, Ca, and Fe by atomic absorption spectrophotometry with a Perkin–Elmer model 403 spectrophotometer.

RESULTS AND DISCUSSION

Bound nitrogen contents

The results of the bound nitrogen determinations for Apollo 17 lunar fines <1 mm and grain size fractions, and for lunar igneous rocks are compiled in Table 1. Also included are the results of replicate runs for Apollo 15 soils 15012 and 15013 from the SESC, and for some stone meteorites. These Apollo 15 fines were also analyzed in replicates for total nitrogen by Kothari and Goel (1972), and Chang *et al.* (1973a) using (*n, p*)-activation and pyrolysis-combustion, respectively, and can, therefore, be used for interlaboratory comparison. The data of the three laboratories agree within 10%, and no marked difference is observed between total nitrogen and chemically bound nitrogen contents. Kothari and Goel found 98 and 110 ppm N in 15012,11 and 15013,14, respectively. Chang *et al.*

Table 1. Chemically bound nitrogen concentrations in Apollo 17 fines <1 mm and grain size fractions, in Apollo 17 rocks, in Apollo 15 SESC-fines, and in stone meteorites. Nitrogen determinations as ammonia by the Kjeldahl method and Nessler's reagent. Error: ±6%. SESC = Special Environment Sample Container. A.S.U. = Arizona State University.

Sample No. Apollo 17	Station and Landmark	Sample Type	Chem. Bound N (ppm)
70011,18	LM, ALSEP Valley floor	fines <1 mm SESC	73
71501,23	Sta. 1a Valley floor	fines <1 mm rake soil	51
72501,27	Sta. 2, Mountain slopes South Massif	fines <1 mm rake soil	70
72701,27	Sta. 2, Mountain slopes Upslope from the valley at Nansen Crater	fines <1 mm rake soil	73
74241,11	Sta. 4, Valley floor Near Shorty Crater	fines <1 mm gray soil beside orange soil	23
Grain size fractions (μm)		3–12 μm	117
		12–24	82
		24–48	19
		48–60	19
		60–109	13
75061,23	Sta. 5, Valley floor Near Camelot Crater	fines <1 mm Skim-top of flat boulder	42
76501,44	Sta. 6, Mountain slopes North Massif	fines <1 mm rake soil	63
78501,26	Sta. 8, Mountain slopes Base of Sculptured Hills	fines <1 mm rake soil	73
Grain size fractions (μm)		3–12 μm	107
		<24	109
		12–24	91
		24–48	68
		48–60	48
		60–109	42
		109–272	40
77017,32	Sta. 7 North Massif	Crushed anorthositic gabbro, brecciated and invaded by glass	<8
79155,24	Sta. 9, Rim of Van Serg Crater	Homogeneous gabbro, partially coated with glass	<8
70215,21	Near LM	Fine-grained basalt	<8
Apollo 15			
15012,14 SESC-1	Sta. 6, east of Spur Crater	fines bottom of trench	108 (111, 101, 113)*
15013,11 SESC	Sta. 8, between LM and ALSEP	fines bottom of trench	110 (108, 117, 104)*

Table 1. (*Continued*).

Sample No. Apollo 17	Station and Landmark	Sample Type	Chem. Bound N (ppm)
Stone meteorites			
Kapoeta	light portion	Achondrite	86
MPI No. 137	dark, gas-rich portion	(Howardite)	82
Allende		Carb. Chondrite	15
A.S.U. No. 8185		C-3	(16, 14)*
Murchison		C-2	226
A.S.U. No. 828			(221, 231)*
Murray		C-2	302
A.S.U. No. 635.2			(313, 291)*

*Replicate analyses.

reported 96 and 110 ppm N in 15012,16 and 15013,3, respectively. Our mean values are 108 and 110 ppm N in 15012,14 and 15013,11, respectively.

Fines of the Apollo 17 sampling areas range in bound nitrogen contents from 23 to 73 ppm N. The Apollo 17 fines contain, in the mean, the lowest amounts of bound nitrogen among all Apollo fines analyzed by us so far. This is illustrated in Fig. 1. The varying nitrogen contents are functions of the exposure age and the maturity of the soil samples. Indeed, nitrogen contents of individual Apollo 17 soils nicely correlate with their state of maturity. McKay *et al.* (1974) divided the Apollo 17 soil samples into mature, submature and immature groups by plotting graphic mean grain size versus graphic standard deviation. The average agglutinate content increases from immature to submature and mature soils being 16.3, 36.5, 46.7%, respectively. The mature soils 72501, 72701, 76501, and 78501 have nitrogen contents of 70, 73, 63, and 73 ppm N, respectively, the submature soil 71501 51 ppm N, and the immature soils 74241 and 75061 23 and 42 ppm N, respectively, showing strong relation of nitrogen content of soils with the state of maturity. Similarly, nitrogen contents of Apollo 16 fines are positively correlated with agglutinate contents as shown by Goel *et al.* (1974).

Comparison of bound nitrogen data for Apollo 17 fines with those of Petrowski *et al.* (1974) who used vacuum pyrolysis-technique for the detection of total nitrogen shows satisfactory agreement, yet our numbers are systematically lower. Five soils can be compared; the numbers in parentheses are own values: 70011 77 ppm N (73 ppm N), 71501 60 (51), 72701 91 (73), 75061 49 (42), 76501 73 (63). This indicates that the Kjeldahl method detects the main amount of nitrogen present in lunar fines. Goel *et al.* (1974) using (n, p)-activation found distinctly higher nitrogen values than ours. The largest discrepancy is for soil 78501 for which Goel *et al.* measured 125 ppm N. Chang *et al.* (1974) obtained 18.3 ppm N for the gray soil 74240 by using vacuum pyrolysis-combustion; we found 23 ppm N for 74241 in fairly good agreement.

To learn more about bound nitrogen indigenous to lunar rocks we have

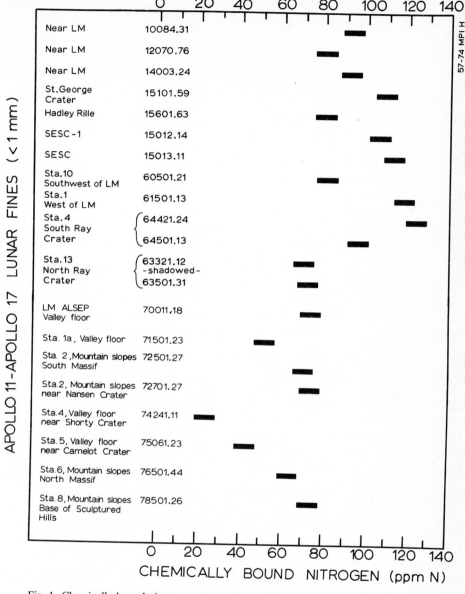

Fig. 1. Chemically bound nitrogen contents (in ppm N) in Apollo 11 through Apollo 17 lunar fines <1 mm with error bars.

analyzed rock samples 77017, 79155, and 70215. They contain less than 8 ppm N (Table 1). Their nitrogen contents are similar to those of igneous rocks 12063, 12075, and 15556, previously analyzed (Müller, 1972). The low nitrogen contents of igneous rocks, illustrated in Fig. 2 together with two breccias, suggest that indigenous nitrogen was low in abundance at the time of mineral formation

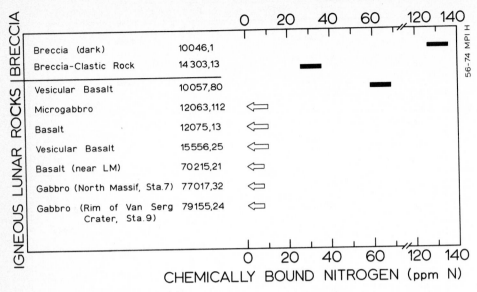

Fig. 2. Chemically bound nitrogen contents (in ppm N) in various lunar igneous rocks and breccias with error bars. The crushed, brecciated gabbro 77017 is not strictly "igneous."

billions of years ago, and demonstrate once more the severe loss of volatile elements in the early history of the moon.

Bound nitrogen contents of four stone meteorites are shown in Table 1. The data of the light and dark, noble gas-rich portion of the achondrite Kapoeta reveal that there was no significant contribution of solar wind-derived nitrogen to the dark portion. This result is surprising, because, besides the trapped noble gases, carbon is enriched in the dark portion by a factor of almost two relative to the light portion (Müller and Zähringer, 1966). The carbonaceous chondrites Allende, Murchison, and Murray have mean values of 15, 226, and 302 ppm bound nitrogen, respectively. The number for Allende agrees well with data of Moore and Cripe (1974) who found (13 ± 3) ppm N in replicate runs. Our data for Murchison and Murray are, however, much lower than previous values for C2-chondrites (Moore, 1971). It appears that C2-chondrites contain substantial amounts of nitrate- and/or nitrite-nitrogen, which are not detected with the Kjeldahl technique. Heterogeneity from sample to sample has also to be taken into account for the differences. Recently, Kaplan and Injerd (1974) measured around 850 ppm N for Murchison by sealed-tube Kjeldahl digestion. The discrepancy deserves further investigation.

Grain size analysis of lunar fines

The bound nitrogen results of grain size fraction analyses of Apollo 17 fines 74241 and 78501 are given in Table 1 and illustrated in Fig. 3. Previous data on Apollo 14, 15, and 16 fines are included for comparison in this graph (Müller, 1972,

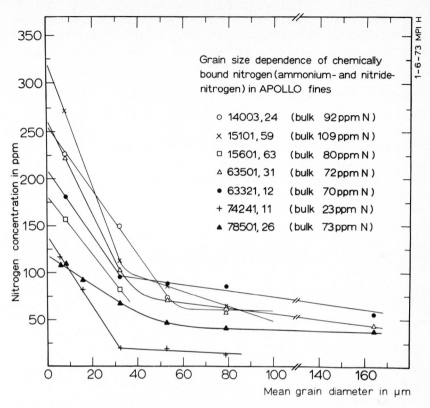

Fig. 3. Grain size dependence of chemically bound nitrogen in sieve fractions of Apollo 14, 15, 16, and 17 lunar fines. The "bulk" nitrogen contents refer to those of fines <1 mm.

1973). The nitrogen concentrations of sieve fractions are plotted versus the mean grain diameter. As approximate mean grain diameter of the fractions 3–12, <24, 12–24, 24–48, 48–60, 60–109, and 109–272 μm, we have used 6, 8, 16, 32, 53, 79, and 164 μm, respectively, in this plot. The means for grain sizes have been calculated by adding one third of the size range to the lower limit of each grain size fraction. The data of fines 74241 show, as in our previous studies, a distinct increase of nitrogen with decreasing grain size. It is now commonly accepted that the surface-correlated nitrogen in lunar fines is mainly derived from solar wind implantation. The grain size fraction curve for 74241 is distinctly different from those of other lunar fines. Two features are remarkable: the nitrogen contents for very small and very large grain sizes are relatively low at about 140 ppm and 13–19 ppm N, respectively. This is probably due to the low agglutinate content (8%) and coarse graphic mean grain size (115 μm) of immature soil 74241 (McKay et al., 1974). The grain size dependence of bound nitrogen for mature soil 78501 is also unique. The slope at smaller grain sizes is distinctly less than that of other fines, and the nitrogen content at a very small grain size is the lowest found by us so far (about 120 ppm N).

From the fines shown in Fig. 3 only 15101 and 63501 lie nearly on a straight line in a log–log plot of nitrogen content versus reciprocal grain diameter, whereas data points of the other fines show larger scatter. The gradients of the lines are rather similar, except soil 74241, for which the gradient is much steeper.

Nitrogen and trapped noble gases in fines

Besides grain size analysis, correlation of nitrogen with trapped noble gases can provide further evidence for solar wind origin of nitrogen in fines. In Fig. 4 we plotted trapped ^{36}Ar versus bound nitrogen for seven Apollo 17 fines. Most of the fines are well correlated; fines 78501 strongly deviate from the dashed line due to a smaller ^{36}Ar/N ratio compared to other fines. Trapped helium and neon are not well related with bound nitrogen. Noble gas data are from Hübner *et al.*, (1974, and this volume).

In previous reports we have shown by means of the ^{4}He/N and ^{20}Ne/N ratios in Apollo 14 and Apollo 16 fines <24 μm that implanted nitrogen is more tightly retained because of its chemically active nature than are the noble gases (Müller, 1972, 1973). This is also true for carbon, yet its retention is probably somewhat less than that of nitrogen as discussed below. Nitrogen contents of lunar fines are, therefore, a useful measure of exposure time to solar wind. Moreover, amounts of solar wind-implanted elements normalized to nitrogen demonstrate their depletion relative to solar abundances. In Table 2 the abundance of bound nitrogen in two fine grain size fractions of soils 74241 and 78501 is compared with the contents of trapped noble gases ^{4}He, ^{20}Ne, and ^{36}Ar. The calculated ^{4}He/N, ^{20}Ne/N, and ^{36}Ar/N ratios are distinctly lower than the corresponding solar abundance ratios (Cameron, 1973) showing the loss of noble gases in lunar fines. The best agreement with the solar ratio is for the ^{36}Ar/N ratio of fines 78501, 3–12 μm, which differs by a

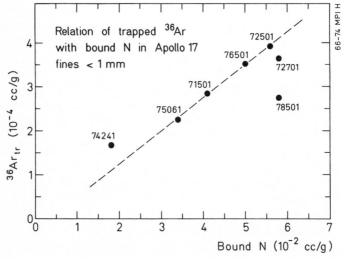

Fig. 4. Correlation of trapped ^{36}Ar with chemically bound nitrogen in Apollo 17 fines <1 mm. Contents are given in cm³ STP/g.

Table 2. Relation of chemically bound nitrogen with trapped noble gases in fine sieve fractions of Apollo 17 soils. Noble gas data: Hübner *et al.* (1974). Solar abundance ratios: Cameron (1973).

Lunar fines Sieve fractions	^{4}He/N	^{20}Ne/N $\times 10^{-2}$	^{36}Ar/N $\times 10^{-3}$
74241,11			
~3–12 μm	4.4	4.0	3.9
12–24 μm	2.1	2.1	1.8
78501,26			
~3–12 μm	2.7	4.2	7.6
12–24 μm	1.9	3.2	5.4
Solar abundance ratio	590	82	26

factor of 3.4. The finest sieve fractions, 3–12 μm, of the two fines have consistently higher noble gas to nitrogen ratios than the 12–24 μm fractions. The lighter noble gases appear to be more strongly surface-correlated than nitrogen.

Leach-experiments on lunar fines

Leach-experiments on fines were performed to study the chemical nature of bound nitrogen. The results compiled in Table 3 show that a substantial amount of bound nitrogen is easily leached in the form of ammonia by treating soils with 2N H_2SO_4 at 20°C. This indicates that NH_3 molecules and/or NH_4^+ cations are liberated during breakup of grain surfaces, being possibly originally present in minerals and glasses of fines. These species may have been formed by the interaction of solar wind nitrogen with hydrogen. It cannot be excluded that ammonia is derived from easily hydrolyzable nitrides of lunar fines. Indication of ammonia was found by Chang *et al.* (1971) for fines 12023, and by Holland *et al.* (1972) for fines 15261. Bound nitrogen measured in the residue after leaching lunar fines is interpreted as nitride-nitrogen, which is hydrolyzable only with difficulty.

Test experiments were performed on powdered samples of a sodium–calcium–silica glass containing 2.1 wt.% bound nitrogen, silicon nitride and titanium nitride. In the glass sample, 1.1% of the bound nitrogen was detected in the leach-fraction; the nitrides showed no nitrogen in the leach. Nitrogen in the glass is present in the form of nitride-, NH_2- and NH-groups. A NH-vibration band at about 3 μm was detected by infrared spectrometry (Mulfinger and Franz, 1965). The low amount of leachable nitrogen in the glass indicates that NH- and NH_2-nitrogen is practically not attacked in our leaching procedure.

In aliquots of both the leach-solution and residue-solution five elements were determined, which were mainly indicative for the olivine (Mg, Fe) and plagioclase (Na, Al, Ca) contents of fines. The results in Table 3 show that about 30–37% of each element were dissolved with 2N H_2SO_4 at 20°C, meaning that all mineral components are broken up about equally. A similar result was found by

Table 3. Leach-experiments on lunar fines and some test materials. Agent:
2N H_2SO_4, 20°C, 1 hour; <24 μm: 15 min. The numbers for leach- and
residue-portion sum up to approximately 100%.

Sample N Content	Bound Nitrogen Leach	Residue	Elements Determined Na, Mg, Al, Ca, Fe Leach	Residue
	rel. percent		rel. percent for each element	
15101 < 1 mm 109 ppm N	61	46	~37	~66
15101 < 24 μm 272 ppm N	41	66	~33	~57
78501 < 1 mm 73 ppm N	37	50	~36	~71
78501 < 24 μm 109 ppm N	27	74	~30	~73
72501 < 1 mm 70 ppm N	86	14	~31	~68
Na–Ca–silica– Glass 2.1 wt.% N	1.1	99	Na: 4 Ca: 0	96 100
Si_3N_4 40 wt.% N	0	100	—	—
TiN 23 wt.% N	0	100	—	—

Begemann *et al.* (1970) when treating Apollo 11 fines with cold HNO_3 (1:3). In
summary, the leach-experiments indicate that nitrogen and carbon chemistry in
lunar fines appear to be analogous in respect to the hydrogen compounds
ammonia and methane, and the presence of nitrides and carbides. Indigenous
methane and carbides were detected in Apollo 11 and Apollo 12 fines by Abell *et
al.* (1970) and Chang *et al.* (1971).

Solar wind carbon and nitrogen

Because of its chemically active character, solar wind-implanted carbon has
also a high retention on grain surfaces, which is comparable to that of nitrogen.
DesMarais *et al.* (1974) calculated the overall retention for carbon to be 0.65
relative to 1.0 for nitrogen. The carbon to nitrogen atomic ratios of Apollo 17 fines
listed in Table 4 range from 1.68 to 2.80, and are in the average distinctly higher
than those previously calculated for Apollo 14, 15, and 16 fines, which vary
between 1.35 and 1.75 (Müller, 1973). Carbon data of Apollo 17 fines are from
Moore *et al.* (1974). When comparing C/N ratios of lunar fines with the ratio of 4.2
of the solar photosphere found by Lambert (1968) from a reliable analysis of
atomic and molecular spectra, it turns out that C/N ratios of fines are consistently
lower than that of the sun. Volatile carbon compounds such as methane appear,
therefore, to be preferentially lost from fine-grained lunar material by diffusion

Table 4. Total carbon and chemically bound nitrogen data, and C/N atomic ratios in Apollo 17 lunar fines and the solar photosphere. Carbon data: Moore et al. (1974).

Fines <1 mm	70011	71501	72501	72701	74240† 74241	76501	78501	Solar Photosphere* (C = 100)
C (ppm)	120	75	125	140	55	120	170	
C (10^{18} at./g)	6.0	3.7	6.3	7.0	2.8	6.0	8.5	100
N (ppm)	73	51	70	73	23	63	73	
N (10^{18} at./g)	3.1	2.2	3.0	3.1	1.0	2.7	3.1	24
C/N (at./at.)	1.94	1.68	2.10	2.26	2.80	2.22	2.74	4.2

*Lambert (1968).
†Unsieved.

and possibly during reworking of the regolith. If ammonia were actually present in fines, a comparable loss of nitrogen should be expected. This is apparently not the case on the basis of C/N ratios. We interpret, therefore, the substantial amounts of leachable nitrogen in fines (Table 3) as being present as NH_4^+ cations and/or easily hydrolyzable nitrides, the diffusion rate of which is much smaller than that of ammonia.

Solar wind implantation, trapping efficiencies, loss mechanisms, molecule formation and isotopic alteration of light elements on the lunar surface are, at present, not fully understood. Simulated solar wind implantation experiments with the labeled ions ^2H, ^{13}C, ^{15}N have provided some insight into the complex processes on the very surface of the moon (Pillinger et al., 1972; Chang et al., 1973b; Wszolek et al., 1974). Moreover, model calculations for the accumulation of carbon in the lunar regolith as a function of time and the corresponding evolution of its carbon-13 content (Kerridge et al., 1974) are a promising approach for better understanding the light element chemistry on the moon's surface.

Acknowledgments—We thank the National Aeronautics and Space Administration for providing the lunar samples. Valuable discussions with Prof. W. Gentner, Dr. P. Horn, W. Hübner, and Dr. T. Kirsten are highly appreciated. I thank D. Kaether for his skillful assistance, and R. Schwan for making the grain size fractions and for sample preparation. Prof. C. B. Moore, Arizona State University, Tempe, made the carbonaceous chondrites available to us, which is gratefully acknowledged.

REFERENCES

Abell P. I., Eglinton G., Maxwell J. R., Pillinger C. T., and Hayes J. M. (1970) Indigenous lunar methane and ethane. Nature 226, 251–252.
Begemann F., Vilcsek E., Rieder R., Born W., and Wänke H. (1970) Cosmic-ray produced radioisotopes in lunar samples from the Sea of Tranquillity (Apollo 11). Proc. Apollo 11 Lunar Sci. Conf., Geochim. Cosmochim. Acta, Suppl. 1, Vol. 2, pp. 995–1005. Pergamon.
Cameron A. G. W. (1973) Abundances of the elements in the solar system. Space Sci. Rev. 15, 121–146.
Chang S., Kvenvolden K., Lawless J., Ponnamperuma C., and Kaplan I. R. (1971) Carbon, carbides, and methane in an Apollo 12 sample. Science 171, 474–477.

Chang S., Lawless J., Romiez M., Kaplan I. R., Petrowski C., Sakai H., and Smith J. W. (1973a) Carbon, nitrogen and sulfur in lunar fines 15012 and 15013: Abundances, distributions and isotopic compositions. Preprint.

Chang S., Mack R., Gibson E. K., Jr., and Moore G. W. (1973b) Simulated solar wind implantation of carbon and nitrogen ions into terrestrial basalt and lunar fines. *Proc. Fourth Lunar Sci. Conf.*, *Geochim. Cosmochim. Acta*, Suppl. 4, Vol. 2, pp. 1509–1522. Pergamon.

Chang S., Lennon K., and Gibson E. K., Jr. (1974) Abundances of C, N, H, He, and S in Apollo 17 soils from stations 3 and 4: Implications for solar wind exposure ages and regolith evolution (abstract). In *Lunar Science—V*, pp. 106–108. The Lunar Science Institute, Houston.

DesMarais D. J., Hayes J. M., and Meinschein W. G. (1974) Retention of solar wind-implanted elements in lunar soils (abstract). In *Lunar Science—V*, pp. 168–170. The Lunar Science Institute, Houston.

Goel P. S., Shukla P. N., Kothari B. K., and Garg A. N. (1974) Solar wind as source of nitrogen in lunar fines (abstract). In *Lunar Science—V*, pp. 270–272. The Lunar Science Institute, Houston.

Holland P. T., Simoneit B. R., Wszolek P. C., and Burlingame A. L. (1972) Compounds of carbon and other volatile elements in Apollo 14 and 15 samples. *Proc. Third Lunar Sci. Conf.*, *Geochim. Cosmochim. Acta*, Suppl. 3, Vol. 2, pp. 2131–2147. MIT Press.

Hübner W., Kirsten T., and Kiko J. (1974) Rare gases in Apollo 17 soils with emphasis on analysis of size and mineral fractions of soil 74241 (abstract). In *Lunar Science—V*, pp. 369–371. The Lunar Science Institute, Houston, and this volume.

Kaplan I. R. and Injerd W. (1974). Personal communication.

Kerridge J. F., Kaplan I. R., and Lesley F. D. (1974). Accumulation and isotopic evolution of carbon on the lunar surface (abstract). In *Lunar Science—V*, pp. 408–410. The Lunar Science Institute, Houston.

Kothari B. K. and Goel P. S. (1972). Total nitrogen abundances in five Apollo 15 samples (Hadley–Apennine region) by neutron activation analysis. In *The Apollo 15 Lunar Samples*, pp. 282–283. The Lunar Science Institute, Houston.

Lambert D. L. (1968). The abundances of the elements in the solar photosphere—I. Carbon, nitrogen, and oxygen. *Monthly Notices Roy. Astron. Soc.* **138**, 143–179.

McKay D. S., Fruland R. M., and Heiken G. H. (1974). Grain size distribution as an indicator of the maturity of lunar soils (abstract). In *Lunar Science—V*, pp. 480–482. The Lunar Science Institute, Houston.

Moore C. B. (1971) Nitrogen. In *Handbook of Elemental Abundances in Meteorites* (editor B. Mason), pp. 93–98. Gordon and Breach.

Moore C. B., Lewis C. F., Cripe J. D., and Volk M. (1974) Total carbon and sulfur contents of Apollo 17 lunar samples (abstract). In *Lunar Science—V*, pp. 520–522. The Lunar Science Institute, Houston.

Moore C. B. and Cripe J. D. (1974) Written communication.

Müller O. (1972) Chemically bound nitrogen abundances in lunar samples, and active gases released by heating at lower temperatures (250 to 500°C). *Proc. Third Lunar Sci. Conf.*, *Geochim. Cosmochim. Acta*, Suppl. 3, Vol. 2, pp. 2059–2068. MIT Press.

Müller O. (1973) Chemically bound nitrogen contents of Apollo 16 and Apollo 15 lunar fines. *Proc. Fourth Lunar Sci. Conf.*, *Geochim. Cosmochim. Acta*, Suppl. 4, Vol. 2, pp. 1625–1634. Pergamon.

Müller O. and Zähringer J. (1966) Chemische Unterschiede bei uredelgashaltigen Steinmeteoriten. *Earth Planet. Sci. Lett.* **1**, 25–29.

Mulfinger H. O. and Franz H. (1965) Über den Einbau von chemisch gelöstem Stickstoff in oxydischen Glasschmelzen. *Glastechn. Berichte* **38**, 235–242.

Petrowski C., Kerridge J. F., and Kaplan I. R. (1974) Light element geochemistry of the Apollo 17 site. This volume.

Pillinger C. T., Cadogan P. H., Eglinton G., Maxwell J. R., Mays B. T., Grant W. A., and Nobes M. J. (1972) Simulation study of lunar carbon chemistry. *Nature, Phys. Sci.* **235**, 108–109.

Wszolek P. C., Walls F. C. and Burlingame A. L. (1974). Simulation of lunar carbon chemistry in plagioclase (abstract). In *Lunar Science—V*, pp. 854–856. The Lunar Science Institute, Houston.

Proceedings of the Fifth Lunar Conference
(Supplement 5, Geochimica et Cosmochimica Acta)
Vol. 2 pp. 1919–1937 (1974)
Printed in the United States of America

Concentration-versus-depth profiles of hydrogen, carbon, and fluorine in lunar rock surfaces

G. M. Padawer, E. A. Kamykowski, M. C. Stauber, and M. D. D'Agostino

Research Department, Grumman Aerospace Corporation, Bethpage, New York 11714

W. Brandt

Physics Department, New York University, New York, New York 10003

Abstract—Nuclear microanalysis assays for hydrogen, carbon, and fluorine in surfaces of lunar rocks are described. Each of the hydrogen depth profiles manifests a strong ($\sim$400 ppm) surface prominence, attributable primarily to terrestrial contamination; between 1 and 2 μm the profiles are nearly flat ($\sim$80 ppm), possibly indicative of indigenous lunar hydrogen. Assays for carbon and fluorine were performed on the lunar exterior and interior surfaces of 62255,27. The carbon profiles are each characterized by a prominent surface (-contaminant) peak ($\sim$1 μg-cm^{-2} for each surface initially) that subsides gradually within 4.5 μm of the surface, to approximately constant values of 347^{+42}_{-94} and 216^{+39}_{-93} ppm in the lunar exterior and interior subsurfaces, respectively. The observed fluorine concentrations, 70–270 ppm, within the depth range of our measurements, 0–1 μm, are ascribed to contamination that may have been introduced during Teflon-bagging procedures.

Introduction

THE ENRICHMENT with various light elemental species, such as hydrogen and carbon, of lunar material surfaces exposed to the lunar exterior has been ascribed to the implantation of particulate radiation, primarily of solar wind origin (e.g. Moore *et al.*, 1970; DesMarais *et al.*, 1973). The possible enriching of the lunar atmosphere by a meteorite impact cloud (Pillinger *et al.*, 1972) has recently been considered as an additional factor in the surface chemistry of lunar material. Since the penetration depth of the bombarding atomic particles is correlated with their impingement energy, a measurement of the concentration-versus-depth profile of the atomic species in question should provide clues as to the energy distribution of the bombarding particles, and to the exposure time of the lunar specimen, provided the magnitude of the relevant flux component can be independently determined, and if concentration saturation conditions have not been attained (Chang *et al.*, 1973). For geologically derivative material (e.g. soil particles, breccia) the implantation record for a particular species, such as carbon, would be superimposed on a concentration level prevailing in the parent material. The interpretation of observed concentration profiles, however, is influenced by several factors. These include:

(a) Ablation of the surface through sputtering by the total impinging atomic particle flux, as well as through micrometeoroid impact. Atomic sputtering alone, if assumed to ablate a surface exposed to the lunar exterior at the rate of

0.4 Å-y^{-1} (Wehner *et al.*, 1963), would limit the age of the surviving implantation record in a 1-μm-deep region to about 2.5×10^4 y.

(b) Diffusion of implanted species out of the material as well as into greater depth regions, a process that may be significantly accelerated by the surface heating induced during the lunar day. Quantitative assessment of the mobility of implanted particles on the basis of existing bulk diffusivity data is difficult, since such data do not reflect the complex microstructure of most lunar material. In addition, the radiation damage introduced into the surface regions by the bombarding atomic particle fluxes, spatially distributed with a steep gradient into the material surface, may be assumed to affect strongly the diffusion characteristics of the medium, as well as the saturation concentrations of implanted atoms that the medium can locally support.

(c) Likelihood of terrestrial contamination of the sample surfaces with the particle species of interest. This problem has been cited by various workers involved with the analysis of hydrogen, carbon, and fluorine in lunar samples (Leich *et al.*, 1973; Epstein and Taylor, 1974; Goldberg *et al.*, 1974), and becomes particularly acute when the analysis seeks to address low concentrations of these elements in surface regions. The possibilities for contamination are numerous, for which reason extreme care in the handling of lunar samples is called for. Even the contents of the sealed SRC's (sample return containers) are susceptible to potential contamination by the outgassed fluorocarbons from Teflon packaging material, by the residual water vapor in the dry nitrogen used as storage medium, and by hydrocarbons in the vacuum chamber in which the experimental analysis is performed.

Despite the possible limitation imposed by the factors cited above, data on the depth distributions of light elements, such as hydrogen, carbon, and fluorine, in lunar material surfaces, can, if carefully interpreted, furnish detailed information on the lunar surface constituency of these elements. Microprobe techniques utilizing nuclear reactions induced by fast charged particles are uniquely suited for *in situ* mapping of depth distributions of such elements, since they can provide good depth resolution and sensitivity. Moreover, these techniques produce little perturbation of the element distribution being assayed, provided the beam-induced temperature excursions in the target are minimized.

Descriptions of Nuclear Microanalysis Techniques

Hydrogen

We again employed the lithium nuclear microprobe (LNM) technique to measure hydrogen concentrations in the surfaces of lunar samples. In essence, the method, described in detail elsewhere (Stauber *et al.*, 1973), utilizes fast lithium-7 ions to interact with hydrogen-1 nuclei (protons) via the ^{1}H(^{7}Li, γ)^{8}Be reaction. Signature gamma rays at 14.7 and 17.6 MeV are detected, and their intensities are related to the yields from calibration standards having known hydrogen composition in order that the hydrogen concentration in the unknown sample can be inferred.

Carbon

To perform nuclear microanalysis for carbon we utilized 1.5-MeV deuterons to stimulate the (nonresonant) $^{12}C(d, p_0)^{13}C$ reaction ($Q = 2.72$ MeV) in the lunar exterior and lunar interior surfaces of 62255,27. This technique is capable of measuring the carbon concentration within the first ~ 6 microns of the surface of lunar rock. The depth range is limited by the onset of $^{16}O(d, p_0)^{17}O$ ($Q = 1.92$ MeV) protons produced in host oxygen residing near the surface of the specimen. (Our earlier attempts to perform microanalysis for carbon via the $^{12}C(p, \gamma)^{13}N$ reaction had been unsuccessful because of severe competition from gamma-ray background attributable primarily to the $^{27}Al(p, \gamma)^{28}Si$ reaction stimulated in the host rock matrix.) The deuteron enters the target with energy E_d^0, and is slowed by atomic collision processes as it penetrates to a depth, x, where it interacts with kinetic energy E_d, $E_d < E_d^0$, with a carbon-12 nucleus via the $^{12}C(d, p_0)^{13}C$ reaction. The proton from this reaction, observed at the angle θ with respect to the incident beam direction, similarly sustains energy loss as it traverses the skew distance $x/|\cos \theta|$. The kinetic energy of this reaction proton at its birth, E_p^0, is given by reaction kinematics

$$E_p^0 = E_p^0(E_d, \theta)$$

The proton arrives at the detector with energy E_p:

$$E_p = E_p\left(E_p^0, \frac{x}{|\cos \theta|}, S_p\right)$$

where S_p is the effective stopping power of the target material for protons, and

$$E_d = E_d(E_d^0, x, S_d)$$

where S_d is the effective stopping power of the target material for deuterons.

In this way the abscissa of the differential pulse-height spectrum for the emergent protons as sensed in the detector may be related to the depth at which the protons of each particular energy originated. The practicable depth resolution is the result of the combined influences of several factors, the most significant of which are

  * Dispersion-in-energy as a function of detection angle, which results from the geometry of the detector as well as from the finite beam spot
  * Intrinsic energy resolution of the detector
  * Straggling of deuterons in the target, and the straggling of protons in the target and in the foil intervening between the detector and the target
  * Energy dispersion of the incident beam
  * Compositional and geometric irregularities in the (surface of the) target.

The beam-dependent background, attributable to constituents in the host rock other than carbon, is inferred by bombardment of pure targets of suitable candidate materials, chiefly aluminum and silicon, present in the lunar rock matrix in reasonably well known concentrations.

Fluorine

We applied the $^{19}F(p, \alpha\gamma)^{16}O$ reaction to probe the fluorine in the lunar exterior and interior surfaces of 62255,27. The reaction is strongly resonant for protons at 0.874 MeV, and is marked by the copious emission of 6.1-MeV gamma rays emanating from the second excited state in ^{16}O, the residual nucleus. The gamma-ray yield at precisely the resonant energy derives from fluorine atoms located at the very surface. As the bombarding energy is raised, the incident protons need to traverse a specific thickness of target material in order for their energy to be reduced sufficiently to initiate the reaction in question. The slowing characteristics of protons in matter being well understood, one can infer from the energy loss the depth at which the gamma rays originate. One obtains the relationship between gamma-ray yield and absolute fluorine concentration by bombarding a target of known fluorine constituency and observing the emitted gamma rays. Gamma-ray yield as a function of bombarding energy may then be converted to a concentration-versus-depth profile. The technique is described in greater detail elsewhere (Padawer, 1970).

Experimental Procedure

The LNM analysis of hydrogen in 61016,135, 61016,171, and 68815,25 was carried out at the High Voltage Laboratory at Oak Ridge National Laboratory, where an HVEC Model CN accelerator delivered a beam of singly ionized ^7Li projectiles between 2.7 and 4.8 MeV. The beam-tube extension, terminated by a carousel target chamber having a capacity for eight specimens per loading, incorporated a liquid-nitrogen in-line cold finger, a 0.25-cm-d collimator, and a quartz viewer for alignment adjustments. A suppressor section held at -300 V immediately preceded the chamber and served to prevent escape of secondary electrons generated at the site of bombardment. A silicone-fluid diffusion pump serviced the extension and maintained a $\sim 10^{-6}$ torr vacuum during the experiments. The detector consisted of a 20.3-cm-d $\times$ 15.2-cm-h NaI(Tl) crystal encased in a 5.1-cm-thick anticoincidence (A-C) plastic scintillator which served as a cosmic-ray rejector. The detection system was positioned at $90°$ with respect to the beam and was enclosed in a 10.2-cm-thick lead house. This structure in turn was surrounded by a 20.3-cm-thick wall of borated paraffin, introduced to reduce the beam-related neutron flux admitted to the detector. This final configuration placed the face of the NaI(Tl) 28 cm from the beam impingement site. The output pulses from both the NaI(Tl) and the A-C shield were processed to generate timing pulses for examination by a coincidence-anticoincidence module. A third signal obtained from a pile-up inspector subcircuit associated with the NaI(Tl) was also presented to this unit. A linear signal from the NaI(Tl) was thereby eliminated from further analysis if coincident with an A-C pulse or with a pile-up event. Data acquisition was carried out simultaneously in two modes, one of which made use of a multichannel analyzer (MCA) for full spectral processing of the NaI(Tl) pulses; the other utilized a single-channel analyzer (SCA) with upper and lower levels set to accept pulses corresponding only to the gamma-ray energy region of interest (13–18 MeV). The SCA window was established during a bombardment of a highly enriched hydrogen-in-titanium target. We measured the intensity of the beam-independent background to be very nearly 0.020 count-sec^{-1}. Measurements performed on hydrogen-in-titanium NBS standards and on Kapton targets of known hydrogen constituency provided calibration data whereby raw counts could be related to hydrogen concentration in the lunar specimens. A least-squares-analysis fitting to a straight line yielded a calibration constant of $(5.5 \pm 0.5) \times 10^{-18}$ cm^2-eV-$(\mu C)^{-1}$-(ppm H)$^{-1}$.

LNM microanalysis for hydrogen in 73235,28 was attempted with ^{7}Li ions delivered by the tandem Van de Graaff accelerator facility at SUNY, Stony Brook, New York. The higher bombarding energies obtainable there enabled us to probe deeper into the lunar sample than had been possible at ORNL or, earlier (Stauber et al., 1973), at the State University of Iowa (SUI). For signal detection we utilized SUNY's large volume, high sensitivity, shielded NaI(Tl) crystal, which in physical configuration did not differ significantly from the array employed at ORNL.

The Grumman (3 MV) Van de Graaff accelerator provided analyzed proton and deuteron beams required for the microprobe analysis of carbon and fluorine. The customary collimator system preceding the target defined the beam axis and spot size. Samples under study were mounted in the target chamber cited above. Actual mounting of the lunar samples took place in a dry nitrogen atmosphere; a gate valve affixed to the chamber sealed it for transfer between the glove box and the experimental area, thereby precluding the exposing of the specimens to atmospheric contamination. Liquid-nitrogen trapping in the roughing line and in the beam line extension afforded rapid evacuation and served to reduce possible contamination introduced by the vacuum pumping system. Typical pressures were on the order of 10^{-6} torr.

We applied the $^{12}C(d, p_0)^{13}C$ ($Q = 2.72$ MeV) reaction to the analysis for carbon in lunar samples. A circular, 0.8-cm-d 1000-μm-deep silicon surface-barrier transmission detector, affixed to the inside of the target chamber, detected the reaction protons at 130° with respect to the beam direction. A 25-micron-thick aluminum foil overlaying the detector arrested all back-scattered deuterons, but permitted the more highly penetrating protons to be sensed. The deuteron bombarding energy in the analysis for ^{12}C in the lunar sample was maintained constant at 1.5 MeV, the beam uniformly illuminating a target spot defined by a 0.64-cm-d circular collimator aperture. For each of the two surfaces the first few runs were as brief as practicable, each corresponding to the accumulation of only 0.8 μC of charge, and lasting some 25–30 sec, that we might monitor any early perturbative effects of the bombardments. This precautionary procedure was motivated by anticipation of: (1) the accretion of surface carbon resulting from the cracking of vacuum hydrocarbons; (2) subsequent diffusing of carbon into the subsurface; and (3) the implanting of deuterons, which would contribute spurious signals attributable to the $^2H(d, p)^3H$ reaction.

Bombardments of a thick graphite target offered a means of establishing a concentration standard; these runs were conducted with very low beam currents to obviate electronics response problems due to high count rates. Pure aluminum and pure silica were bombarded for the purpose of empirically ascertaining the contribution of these materials to the background.

The investigations for fluorine were implemented with the resonant nuclear reaction $^{19}F(p, \alpha\gamma)^{16}O$ ($E_{resonance} = 0.874$ MeV), which yielded a 6.1-MeV signature gamma ray from the decay of the excited ^{16}O residual nucleus. The detection system for this study comprised a 7.6-cm-d × 7.6-cm-h NaI(Tl) crystal surrounded by a nominal 2.5-cm-thick liquid-scintillator cosmic-ray rejector. The entire unit, enclosed in a 5.1-cm-thick walled lead house, open to the target, was positioned 15.2 cm from the target at 0° with respect to the beam direction. The beam current ranged between 0.2 and 0.5 μA when the 0.64-cm-d collimator was used and 0.2 μA for those measurements that required a 0.32-cm-d collimation aperture. A surface-aluminized LiF crystal served as the concentration standard.

DESCRIPTION OF SAMPLES

61016,135 and 61016,171 are two chips removed from two different sites on the exterior surface of a football-sized anorthosite that was returned loose (unbagged) in the BSLSS. Small impact craters mark the lunar exterior surface of both samples.

68815,25 is a chip from the top of a large (1 m) block, a dense breccia. It was returned in a sealed sample-return container, SRC2, and was maintained in a dry nitrogen atmosphere up to the time of our first measurements of the hydrogen constituency of its exterior surface (Stauber et al., 1973); subsequent to these measurements it was exposed to the laboratory atmosphere. We report on our assay of the vuggy interior surface of the sample.

62255, a breccia found perched on the lunar regolith, was returned in SRC1, a sample-return container that failed to seal. 62255,27 was chipped from a larger fragment that, when removed from the sample bag, was found to have been detached from the parent rock. The catalog photograph and the chipping schematic identify its sugary surface as the true lunar exterior; however, we perceived no micrometeoroid-impact craters anywhere on the sample. For shipment to our laboratory the chip had been packaged in a Teflon-bagged, stoppered plastic vial.

73235,28 is a chip sawn from the lunar exterior of a metabreccia.

RESULTS

Hydrogen

In the LNM measurements performed at ORNL in June 1973, we found the three probed samples, 61016,135 (exterior), 61016,171 (exterior), and 68815,25 (interior) to contain very nearly identical hydrogen concentration profiles. As shown in Fig. 1, each of these distributions is characterized by a surface prominence within the first 0.5 micron or so, having ~ 400 ppm peak intensity ($\sim 6 \times 10^{20}$ H atoms-cm^{-3}). Integrated over depth, this is equivalent to 30×10^{15}-atoms-cm^{-2}. That a ~ 400 ppm peak intensity was observed in the lunar interior surface of 68815,25, a chip that was stored exposed to laboratory atmosphere following our original measurements at SUI in February 1973 (Stauber et al., 1973) provides strong evidence that the surface prominences for all three samples are chiefly due to contamination. This contaminant hydrogen, equivalent to ~ 15 monolayers of water, obscures any true lunar hydrogen within ~ 1 μm of the surface.

Between 1 and 2 μm, the maximum depth limit of our measurements, the profiles are quite flat; the observed mean concentration within this depth range is ~ 80 ppm. Although the hydrogen constituency of lunar fines has been extensively reported, no data are available for comparison with others' findings for Apollo 16 rock interiors at depths greater than 1 μm. We note, however, that our results are in accord with the bulk hydrogen values measured by D'Amico et al. (1970) for Apollo 11 rock interiors. Their values of 0.5–1.4 STP cm^3 H$_2$ per gram sample translate to 45–126 ppm hydrogen. (See also Epstein and Taylor, 1970, and Epstein and Taylor, 1971.)

Employing the ^{7}Li beam of the Van de Graaff accelerator at SUNY, Stony Brook, New York, we attempted LNM analysis of 73235,28 at 5.2, 5.6, 6.0, and 6.4 MeV, corresponding to 3.8, 4.6, 5.4, and 6.2 μm, respectively, depths beyond the regions previously accessible to us with the Model CN single stage Van de Graaff accelerators at SUI and ORNL. The sensitivity of the SUNY measurements was restricted by the intrusion of strong beam-dependent background in the gamma-ray pulse-height region of our interest. This competitive interference was

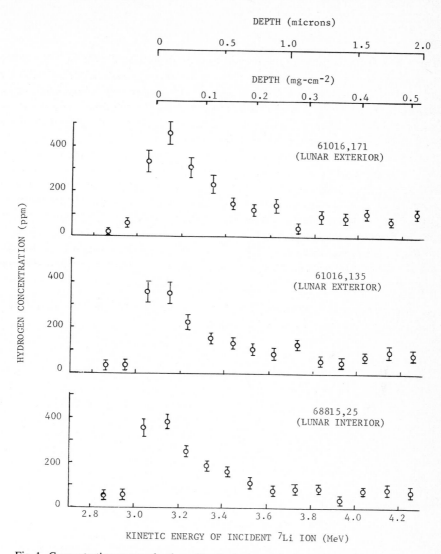

Fig. 1. Concentration-versus-depth profiles for hydrogen in the surface regions of three lunar samples. In each case the surface prominence subsides near 0.25 mg-cm^{-2} (1 μm), and is attributable to terrestrial contamination. The mean concentration averaged over the 0.25–0.50 mg-cm^{-2} (1–2 μm) range, ~80 ppm for all three samples, is in accord with measured values for bulk hydrogen in Apollo 11 rock interiors. To convert to atomic density, multiply ordinate by 156×10^{22}.

too intense to permit us to unfold the hydrogen concentration in the rock at these greater depths.

Carbon

The yield from a pure graphite calibration target is shown in Fig. 2. For comparison, the energy dependence of the differential cross section, $d\sigma/d\Omega$, for the $^{12}C(d, p_0)^{13}C$ reaction at $\theta = 130°$ (Huez *et al.*, 1972) is shown in Fig. 3, and serves to illustrate the similarity between $d\sigma/d\Omega$-versus-E_d and the thick-target differential pulse-height spectrum associated with carbon that is homogeneously distributed in the near-surface region of a specimen.

Figure 4a is the differential pulse-height spectrum accumulated during the (early) bombardment of the lunar exterior surface of 62255,27, newly exposed to the 1.5-MeV deuteron beam. (All error bars appearing in this report represent one standard deviation in counts statistics.) These data were obtained immediately after the first five short (0.8 μC) doses were delivered to the target. Figure 4b is the spectrum accumulated immediately after a deuteron dose of 132 μC had been delivered to the same surface of the target. Similar spectra obtain for bombardments of the lunar interior surface as well. In all cases the differential pulse-height distributions are characterized by three salient features:

(1) A proton prominence between MCA channels 93 and 102, the signature of carbon residing within 1.7 μm of the surface;

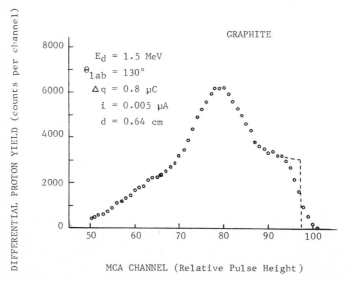

Fig. 2. Differential pulse-height spectrum for protons during 1.5-MeV deuteron bombardment of thick graphite target. The dispersion with respect to the ideal (dashed-line) cutoff near channel 97 is a measure of the practicable depth resolution, ~1.0 μm for this work.

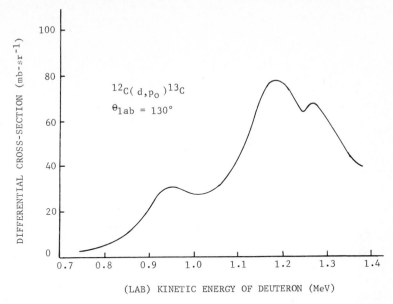

Fig. 3. Differential cross section for $^{12}C(d, p_0)^{13}C$ reaction (Huez *et al.*, 1972). The prominence near 1.2 MeV corresponds to the yield near channel 80, Fig. 2.

(2) A broad, somewhat less energetic spectral subgroup, bounded by channels 77 and 93, corresponding to protons emanating from carbon residing at depths between 1.7 and 6.2 μm;

(3) A broad, intense, proton prominence, below channel 76, deriving from deuterons reacting with oxygen in the host material via the $^{16}O(d, p_0)^{17}O$ ($Q = 1.92$ MeV) reaction.

To monitor the early perturbative effects of the bombardments we performed the initial assays for carbon in the lunar exterior and interior surfaces of 62255,27 in short steps, each consisting of a beam charge accumulation of 0.8 μC; the current was held at ~ 30 nA during these initial runs. The yield was, statistically, reasonably constant for these first few short steps (see inserts, Fig. 5), conforming well with the linear extrapolations fitted to the data of the subsequent runs. We conclude that any drastic perturbation of the initial carbon constituency of the sample, as received by us, would have had to have been induced before, say, $\frac{1}{4} \times 0.8$ μC $\cong 200$ nC had accrued, i.e. before ~ 8 sec has elapsed into the first run.

A slow, monotonic increase in yield per unit charge with increasing bombardment duration, shown in Figs. 5a and b, gives evidence for the accretion of carbon at both bombarded surfaces and within both subsurface regions. The observed growth is attributed to the condensing of vacuum-system hydrocarbons and the subsequent "cracking" of these molecules at the beam impingement site; the vacuum was maintained at $3(\pm 1) \times 10^{-6}$ torr. Figure 5 illustrates that although a fourfold variation in the current density and a twofold variation in the vacuum

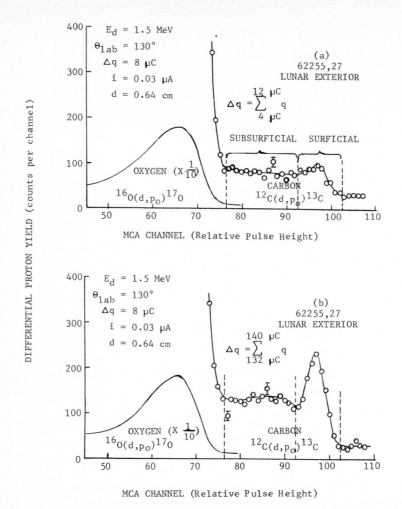

Fig. 4. Differential pulse-height spectrum for protons during 1.5-MeV deuteron bombardment of exterior surface of 62255,27 (a) at the outset of the measurements and (b) at
the conclusion of the measurements. The intense broad prominence below channel 75 is
attributed to oxygen in the host rock matrix (compare Fig. 6a). The peak centered at
channel 97 corresponds to protons stimulated in surface carbon.

quality were experienced in the course of the measurements, the *rate* of carbon
buildup remained very nearly constant.

The empirically derived background yields that were obtained by our bombarding targets of pure silica (Fig. 6a) and pure aluminum (Fig. 6b) were weighted
with respect to sample composition and were further corrected to compensate for
the fact that those targets had not been mounted in the carousel in the same
configuration as the lunar sample.

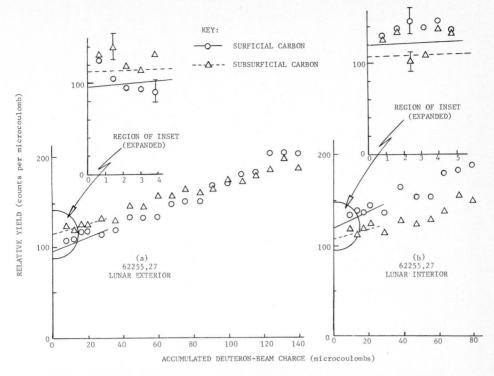

Fig. 5. Rate of buildup of carbon upon and within bombarded surfaces of 62255,27. Data shown in insets imply that Fig. 4a is very likely an accurate representation of the carbon constituency that preexisted in the sample immediately prior to the initiating of the first bombardment.

Figure 7a illustrates the net yield associated with an early bombardment of the exterior surface of the sample after the relevant raw spectrum (Fig. 4a) had been corrected for background attributable to oxygen, aluminum, and silicon. From the yield above channel 105 we infer $\sim 7.5~(\mu C)^{-1}$-channel^{-1} to be a probable maximum background contribution from other elements. This last correction, if not applied, would result in an overestimating of the true initial carbon concentration in the subsurficial regions by some 30%.

Figure 7b is the difference between Figs. 4a and b, and shows the net accrued surface carbon as well as the carbon that, under the influence of the prolonged exposure to the beam, had been caused to diffuse into the subsurface region.

Figures 8a and b are the extracted concentration profiles for carbon in subsurficial regions of the lunar exterior and interior faces, respectively, of the sample. The surface prominences, initially $\sim 1~\mu g\text{-cm}^{-2}$, or $\sim 45 \times 10^{15}$ C atoms-cm^{-2}, are not depicted. The profiles subside to approximately constant values of 347^{+42}_{-94} and 216^{+39}_{-93} ppm in the lunar exterior and interior subsurfaces, respectively. Our results are inordinately higher than those reported by other workers, e.g. Moore *et al.* (1973), who found, by pyrolysis measurements, no more than 59 ppm

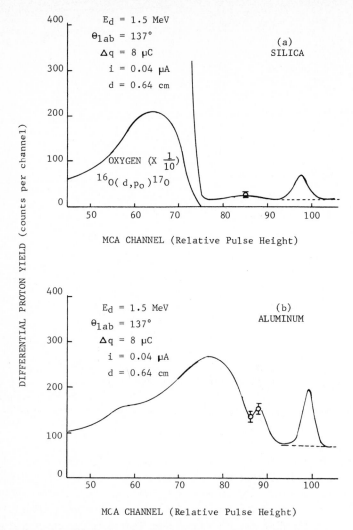

Fig. 6. Pulse-height distributions for 1.5-MeV deuteron bombardments of (a) silica and (b) aluminum. These spectra, appropriately weighted, afforded background corrections with respect to oxygen, silicon, and aluminum.

in Apollo 16 rocks. This discrepancy, together with the fact that similar carbon profiles obtained for two surfaces of 62255,27 that had different lunar exposure histories, leads us to conclude that the carbon observed is primarily attributable to contamination introduced after sampling. A comparison with our fluorine measurements (see below) shows that fully twenty times more carbon is present than would have been observed were the carbon to be associated exclusively with $(CF_2)_x$-like fluorocarbons (e.g. Teflon, Freon). The observed carbon might be a constituent of plastic that had abraded from the inner wall of the vial in which the

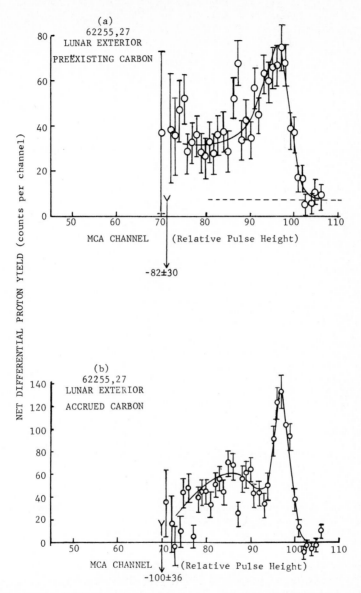

Fig. 7. (a) Data of Fig. 4a corrected with respect to background attributable to oxygen, silicon, and aluminum; the dashed line represents the estimated residual (uncompensated) background contribution from elements other than C, O, Si, and Al. (b) Difference between spectra of Figs. 4b and a, showing net carbon accrual during entire series of bombardments of exterior surface of 62255,27.

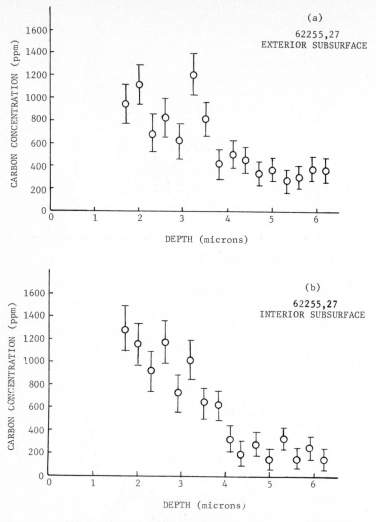

Fig. 8. Concentration-versus-depth profiles for carbon in 1.7–6.2 μm depth range in (a) exterior and (b) interior subsurfaces of 62255,27. Specific profiles for shallower depths could not be unfolded. To convert to atomic density, multiply ordinate by 13.0×10^{22}.

sample was shipped. Unfortunately, we did not perform microanalysis for hydrogen in this same sample (62255,27); the results of such measurements, when compared with the carbon results, might have served to help identify the source of contamination.

Fluorine

Figure 9a illustrates the results of our probe of the fluorine composition in the exterior surface of 62255,27. The onset of the reaction of interest is displaced

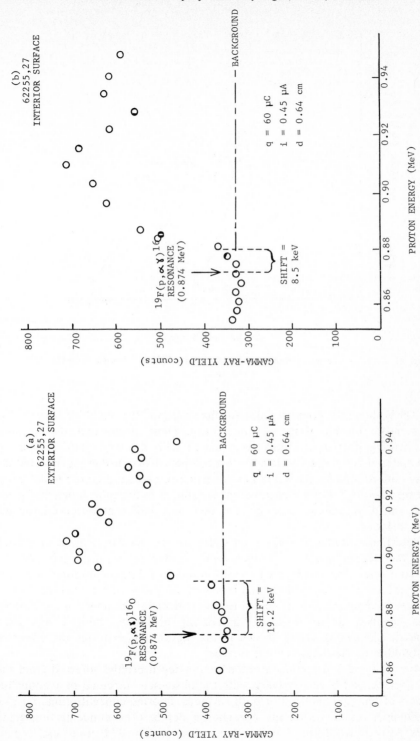

Fig. 9. Unreduced data from nuclear microprobe analysis for fluorine in (a) exterior and (b) interior surfaces of 62255,27.

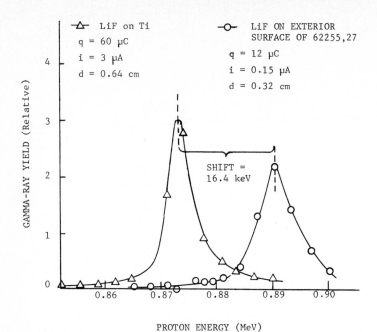

Fig. 10. Effect of charge buildup on apparent energy shift of $^{19}F(p, \alpha\gamma)^{16}O$ 0.874 MeV resonance.

some 19 keV above the nominal value for the reaction "threshold," 874 keV. One might surmise, on the basis of this measurement alone, that the intrinsic fluorine-bearing host matrix had been coated with a ~ 0.4-μm-thick layer of surface material devoid of fluorine. To test this possibility further we inverted the specimen and probed similarly the interior fracture surface. Figure 9b illustrates that the results of this latter measurement are similar to the data shown in Fig. 9a, although the displacement is somewhat less than half that observed for the obverse surface.

To determine whether the observed shifts are attributable to the decelerating effect of the electrostatic charge concurrently implanted into the highly nonconductive specimen, we evaporated a thin coating of lithium fluoride onto the exterior surface of our specimen, and repeated our probe of this surface. Figure 10 presents a comparison of these last results with those obtained for a grounded metallic substrate upon which a thin coating of lithium fluoride had been evaporated. The displacement shown in this figure demonstrates the effect of charge buildup in the lunar sample surface.

Figures 11a and b are concentration-versus-depth profiles inferred from the data of Figs. 9a and b, respectively, with the abscissae corrected to account for the effects of charge buildup. For both surfaces, fluorine concentrations ranging up to 270 ppm reside within the one-micron depth. (The concentration-depth integrals derived from our data are equivalent to 2×10^{15} F atoms-cm^{-2}.) Our

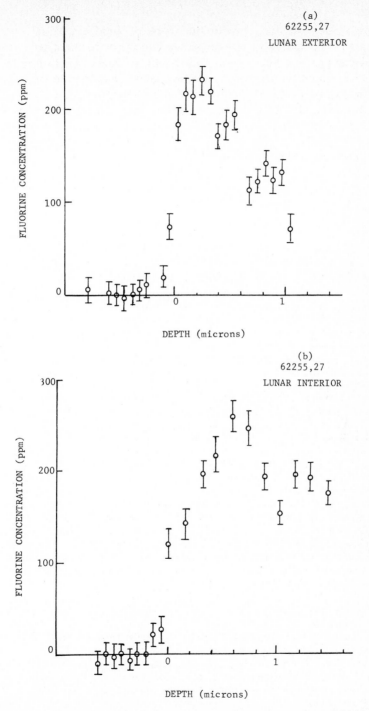

Fig. 11. Concentration-versus-depth profiles for fluorine in (a) exterior and (b) interior surfaces of 62255,27. To convert to atomic density, multiply ordinate by 8.2×10^{22}.

results are in general accord with the work of Goldberg *et al.* (1974), who demonstrated the culpability, with respect to fluorine contamination, of the LRL's Teflon-bagging process. Such intense levels of contamination obscure indigenous lunar fluorine, which, by neutron activation analysis, has been found to be present in bulk concentrations ranging between 18 and 41 ppm in various Apollo 16 rocks (Jovanovic and Reed, 1973). We offer, further, that differences between the two profiles that we measured in 62255,27 might merely reflect differences in the absorptive properties of these morphologically dissimilar surfaces, vis-à-vis a contaminant fluorine environment.

Acknowledgment—This research is supported in part by the National Aeronautics and Space Administration under NASA contract NAS 9-12655.

References

Chang S., Mack R., Gibson E. K., Jr., and Moore G. W. (1973) Simulated solar wind implantation of carbon and nitrogen ions into terrestrial basalt and lunar fines. *Proc. Fourth Lunar Sci. Conf., Geochim. Cosmochim. Acta*, Suppl. 4, Vol. 2, pp. 1509–1522. Pergamon.

D'Amico J., DeFelice J., and Fireman E. L. (1970) The cosmic-ray and solar-flare bombardment of the moon. *Proc. Apollo 11 Lunar Sci. Conf., Geochim. Cosmochim. Acta*, Suppl. 1, Vol. 2, pp. 1029–1035. Pergamon.

DesMarais D. J., Hayes J. M., and Meinschein W. G. (1973) The distribution in lunar soil of carbon released by pyrolysis. *Proc. Fourth Lunar Sci. Conf., Geochim. Cosmochim. Acta*, Suppl. 4, Vol. 2, pp. 1543–1558. Pergamon.

Epstein S. and Taylor H. P., Jr. (1970) The concentration and isotopic composition of hydrogen, carbon, and silicon in Apollo 11 lunar rocks and minerals. *Proc. Apollo 11 Lunar Sci. Conf., Geochim. Cosmochim. Acta*, Suppl. 1, Vol. 2, pp. 1085–1096. Pergamon.

Epstein S. and Taylor H. P., Jr. (1971) $^{18}O/^{16}O$, $^{30}Si/^{28}Si$, D/H, and $^{13}C/^{12}C$ ratios in lunar samples. *Proc. Second Lunar Sci. Conf., Geochim. Cosmochim. Acta*, Suppl. 2, Vol. 2, pp. 1421–1441. MIT Press.

Epstein S. and Taylor H. P., Jr. (1974) Oxygen, silicon, carbon, and hydrogen isotope fractionation processes in lunar surface materials (abstract). In *Lunar Science—V*, pp. 212–214. The Lunar Science Institute, Houston.

Goldberg R. H., Leich D. A., Burnett D. S., and Tombrello T. A. (1974) Surface fluorine on lunar samples (abstract). In *Lunar Science—V*, pp. 277–279. The Lunar Science Institute, Houston.

Huez M., Quaglia L., and Weber G. (1972) Fonction d'excitation de la réaction $^{12}C(d, p_0)^{13}C$ entre 400 et 1350 keV distributions angulaires. *Nucl. Instr. and Meth.* **105**, 197–203.

Jovanovic S. and Reed G. W., Jr. (1973) Volatile trace elements and the characterization of the Cayley Formation and the primitive lunar crust. *Proc. Fourth Lunar Sci. Conf., Geochim. Cosmochim. Acta*, Suppl. 4, Vol. 2, pp. 1313–1324. Pergamon.

Leich D. A., Tombrello T. A., and Burnett D. S. (1973) The depth distribution of hydrogen and fluorine in lunar samples. *Proc. Fourth Lunar Sci. Conf., Geochim. Cosmochim. Acta*, Suppl. 4, Vol. 2, pp. 1597–1612. Pergamon.

Moore C. B., Gibson E. K., Larimer J. W., Lewis C. F., and Nichiporuk W. (1970) Total carbon and nitrogen abundances in Apollo 11 lunar samples and selected achondrites and basalts. *Proc. Apollo 11 Lunar Sci. Conf., Geochim. Cosmochim. Acta*, Suppl. 1, Vol. 2, pp. 1375–1382. Pergamon.

Moore C. B., Lewis C. F., and Gibson E. K., Jr. (1973) Total carbon contents of Apollo 15 and 16 lunar samples. *Proc. Fourth Lunar Sci. Conf., Geochim. Cosmochim. Acta*, Suppl. 4, Vol. 2, pp. 1613–1623. Pergamon.

Padawer G. M. (1970) Proton microprobe analysis of the surface of stranded wire in the lunar module. *Nuclear Applications* **9**, 856–860.

Pillinger C. T., Cadogan P. H., Eglinton G., Maxwell J. R., Mays B. J., Grant W. A., and Nobes M. J. (1972) Simulation study of lunar carbon chemistry. *Nature Phys. Sci.* **235**, 108–109.

Stauber M. C., Padawer G. M., Brandt W., D'Agostino M. D., Kamykowski E. A., and Young D. A. (1973) Nuclear microprobe analysis of solar proton implantation profiles in lunar rock surfaces. *Proc. Fourth Lunar Sci. Conf., Geochim. Cosmochim. Acta*, Suppl. 4, Vol. 2, pp. 2189–2201. Pergamon.

Wehner G. K., Kenknight C., and Rosenberg D. L. (1963) Sputtering rates under solar-wind bombardment. *Planet. Space Sci.*, **11**, 885–895.

Proceedings of the Fifth Lunar Conference
(Supplement 5, Geochimica et Cosmochimica Acta)
Vol. 2 pp. 1939–1948 (1974)
Printed in the United States of America

Light element geochemistry of the Apollo 17 site*

CHARI PETROWSKI, JOHN F. KERRIDGE, and I. R. KAPLAN

Institute of Geophysics and Department of Geology, UCLA, Los Angeles, California 90024

Abstract—Abundances of C, N, S, He, H, and Fe° at Taurus-Littrow are typical of a mixed mare and highland environment. Isotopic compositions of C, N, and S also generally conform to patterns observed elsewhere on the moon. Further evidence is presented for a complex lunar S cycle.

INTRODUCTION

OUR STUDY of the lunar distribution of stable light elements is continued here with a report on the analysis of six soils, three rock breccias, five basalts and an agglutinate from the Apollo 17 landing site at Taurus-Littrow. These samples were analyzed to give the abundance and isotopic composition of carbon, sulfur, and nitrogen, the abundance of helium and hydrogen, and the content of metallic iron. Agglutinate contents of soil fractions were also determined. Results are generally in line with those found in our laboratory for earlier missions (Kaplan *et al.*, 1970; Kaplan and Petrowski 1971; Smith *et al.*, 1973; Chang *et al.*, 1974; Kerridge *et al.*, 1974a). It is well known that the lunar geochemistry of these elements is most strongly influenced by lunar surface processes so that interpretation of the data relate primarily to evolution of the regolith. However, the major element chemistry of the local regolith has a significant effect on certain of these surface processes (thus the titanium content controls He retention to a considerable extent) and consequently chemical inhomogenities, such as those found at Taurus-Littrow, serve to confuse the record of secular developments in local light element chemistry. Interpretation is further hindered by the statistically small number of samples analyzed. It is particularly noticeable that most of the strong interelement, or interisotope, correlations found for Apollo 16 soils (Kerridge *et al.*, 1974a) are much weaker, or even absent, for Apollo 17.

EXPERIMENTAL

Carbon, N, He, and H were determined by pyrolysis-combustion and S, He, and Fe° by hydrolysis in $6 N H_2SO_4$, which also generated CH_4 comprising those fractions termed "methane" and "carbide" by other workers. Sulfur evolved during pyrolysis–combustion and N produced during hydrolysis were also measured, but have been found to be unreliable estimates of these elements so they are not reported here. Further experimental details may be found in the publications cited above.

---

*Publication No. 1336. Institute of Geophysics and Planetary Physics, University of California, Los Angeles.

Results

Tables 1 through 6 list the abundances and isotopic values measured in this study together with such comparative data as may be found in the literature. Agreement with other laboratories is generally good and detailed comments follow for each element.

Carbon

Results for C are summarized in Table 1. During combustion of soil 78421 the sample furnace cracked so that only pyrolysis data were obtained. The strong correlation between pyrolysis C and combustion C for the other soils enabled us to estimate the probable combustion yield of 78421 as 7 ± 1 ppm which was added to the pyrolysis measurement of 135 to give 142 ppm total C.

Values for total C in the soils range from 94 to 149 ppm, characteristic of mature regolith of mixed mare and highland origin. Similarly, values of $\delta^{13}C$ correspond, with one exception, to the high values found for most mature soils

Table 1. Abundance and isotopic composition of carbon in Apollo 17 samples.

Sample	C total ppm	$\delta^{13}C$ ‰ PDB	CH_4(hyd) as C ppm	C total ppm (Moore *et al.*, 1974)
Fines				
70011,27	131	+ 16.3	21.3	120
71501,22	133	+ 1.8	16.2	75
	141	+ 4.5		
72701,15	149	+ 14.7	13.5	133
75061,33	94	+ 13.7	14.5	
76501,20	99	+ 21.2	12.8	120
78421,28	142*	+ 16.1	20.8	165
Breccias				
77017,41	25	− 17.4	2.0	80
77035,63	12		0	
77135,75	26	− 30.1	0	
Basalts				
70017,64	22	− 22.1	0	
70215,52	36	− 39.3	0	31
74275,54	18	− 28.2	0	
75035,41	23	− 28.5	0	64
75075,55	16	− 25.4	0	
Agglutinate				
70019,23	142	+ 11.6	17.3	

*Estimated value, see text.

from earlier missions. The exception 71501, gave replicate values of $+1.8$ and $+4.5‰$ PDB, which is unusually low for a soil containing 137 ppm C. This fact, and the measurement of only 75 ppm C in another split of the same sample by Moore *et al.* (1974), cause us to suspect contamination in the present case. (The observation of McKay *et al.* (1974) that 71501 is submature constitutes an additional reason for preferring the lower C abundance value.) Agreement between our results and those of Moore *et al.*, is otherwise quite good, giving an average deviation from the mean of 6.8% for the other four soils studied in common, with no systematic discrepancy. There are no isotope data with which ours may be compared. It is noticeable that the marked trend for $\delta^{13}C$ to increase with increasing C content which we observed for Apollo 16 soils (Kerridge *et al.*, 1974a) and which we discuss elsewhere in this volume (Kerridge *et al.*, 1974b) is not apparent within these limited Apollo 17 data.

Abundances and isotopic compositions of C for both basaltic and highland rocks from Taurus-Littrow conform to patterns observed at other landing sites. Total C values range from 12 to 36 ppm with $\delta^{13}C$ values ranging from -17.4 to $-39.3‰$ PDB. System blanks were at or below the detection limit which was less than 1 ppm C for all rocks and most soil samples. Agreement with literature values was markedly worse for rocks than for soils. For two of the three rocks studied in common, Moore *et al.* found approximately three times as much C as we did, although agreement for the third sample was reasonable.

The agglutinate sample, 70019, gave C data which were indistinguishable from those of a mature soil, including 17.3 ppm C as CH_4 by acid hydrolysis. Values of CH_4 found for soils ranged from 12.8 to 21.3 ppm C but failed to show the strong correlation with total C found for Apollo 16 samples (Kerridge *et al.*, 1974a).

Nitrogen

Table 2 gives N abundance data for soils, and the agglutinate, which are typical of a mature regolith, ranging from 49 to 101 ppm. These data correlate with C abundances though, again, much more weakly than for Apollo 16 samples (Kerridge *et al.*, 1974a). Comparison with the data of Müller (1974) reveals that his values for the abundance of chemically bound N are systematically 14% below the present data, with a generally small dispersion, suggesting that the difference between the different techniques represents that N which is not chemically bound but is physically trapped as molecular N_2, possibly in microvesicles. With only two samples studied in common, comparison with the data of Goel *et al.* (1974) is not statistically meaningful, however both values lay 10–20% above ours. Two samples were studied in duplicate, or triplicate, and gave a standard deviation in abundance of about 1%. System blanks were below the detection limit of about 2 ppm.

Values of $\delta^{15}N$ are similar to those reported for Apollo 16 (Kerridge *et al.*, 1974a) both in magnitude and in their failure to correlate with other parameters. As is the case for the earlier mission, however, there is a suggestion of an anti-correlation with $\delta^{13}C$, but this is not statistically significant at the present

Table 2. Abundance and isotopic composition of nitrogen in Apollo 17 samples.

Sample	N ppm This work	Literature	δ^{15}N ‰ AIR
Fines			
70011,27	77	73(1), 92(2)	+ 27.1
71501,22	59, 60, 61	51(1), 68(2)	+ 26.0, + 24.7
72701,15	91	73(1)	+ 59.0
75061,33	49	42(1)	+ 36.5
76501,20	73	63(1)	+ 12.0*
78421,28	100, 102		
Agglutinate			
70019,23	70		− 8.5

References: (1) Müller (1974), (2) Goel *et al.* (1974).
*Possible air contamination.

time. Particularly noticeable is the fact that δ^{15}N of the agglutinate sample (70019) is much lighter than the N in soils and is, in fact, one of the lightest lunar samples that we have measured.

Sulfur

Abundance and isotopic data for sulfur are given in Table 3 together with comparative data from the literature. Included in Table 3 are results from the total combustion of two soils as this technique apparently yields more reliable sulfur data than the combined pyrolysis–combustion procedure. For these two cases, hydrolysis and combustion data agree to within about 5%, providing a rough measure of reproducibility. System blanks for the hydrolysis line were below the detection limit of approximately 3 ppm.

Comparison with literature data reveals some interesting trends. Agreement with other investigators is generally reasonable but significantly worse than that which we found, with the same investigators, for Apollo 16 samples. In that case, the average deviation from the mean of our results and those in the literature was only 6% with no systematic discrepancy. In the present case, the agreement varies considerably from group to group, depending upon the analysis technique used, with some groups showing a substantial systematic discrepancy. Thus, the abundance data of Gibson *et al.*, (1974), using a combustion procedure and an infrared Luft cell detector, are systematically 30% higher than ours, with a 20% dispersion. Similarly, Moore *et al.* (1974), using combustion followed by a titration procedure, give abundances which are systematically 29% above the present data, also with about 20% dispersion. Rees and Thode (1974), on the other hand, using hydrolysis with hydrochloric acid, found sulfur abundances in good agreement with ours, showing an average deviation from the mean of better than 4% with a small systematic discrepancy toward lower values. Finally, the X-ray fluorescence

Table 3. Abundance and isotopic composition of sulfur in Apollo 17 samples.

Sample	S ppm This work	Literature	$\delta^{34}S$ ‰ CDT This work	Literature
Fines				
70011,27	968	878(1), 1200(2), 1300(3)	+ 7.4	+ 7.4(1)
71501,22	1075	996(1), 1200(4)	+ 6.4	+ 5.3(1)
71501,22(C)*	1136		+ 7.3	
72701,15	617	500(5), 700(6), 790(2)	+ 10.1	
75061,33	1151	1300(4)	+ 5.3	
76501,20	598	620(1), 665(3), 700(6)	+ 8.0	+ 8.9(1)
78421,28	804	890(2)		
78421,28(C)	828		+ 11.7	
Breccias				
77017,14	700	910(2), 955(3)	− 5.2	
77035,63	733		+ 2.1	
77135,75	634	700(6), 800(3)	+ 1.7	
Basalts				
70017,64	2283	1500(6)	+ 1.4	
70215,52	1689	1581(1), 1700(6), 2210(3)	+ 1.5	+ 0.6(1)
74275,54	1397	1223(1), 1650(3)	+ 2.0	+ 0.6(1)
75035,41	1849	2770(3), 3140(2)	+ 1.7	
75075,55	1708	1600(6)	+ 1.8	
Agglutinate				
70019,23	999	1000(6)	+ 7.6	

*(C) indicates data obtained by combustion in oxygen and not by hydrolysis.
References: (1) Rees and Thode (1974), (2) Moore et al. (1974), (3) Gibson and Moore (1974), (4) LSPET (1973), (5) Scoon (1974), (6) Rhodes et al. (1974).

data of LSPET (1973) and Rhodes et al. (1974) are in good agreement with our results, giving an average deviation from the mean of 6% with no systematic discrepancy. However, X-ray fluorescence is a relatively imprecise technique for analyzing S in the presence of silicon, as in lunar samples, and these data carry less weight than the others quoted above.

The fact that hydrolysis-based techniques agree with combustion-based techniques for Apollo 16 samples but not for those from Apollo 17 suggests a difference in the type of chemical combination experienced by S at the two landing sites. Possibly a fraction of the S at Apollo 17 is in the form of a sulfide, or some other mineral, which does not dissolve quantitatively in either of the acids used for hydrolysis.

Values for $\delta^{34}S$ are similar to those found at other landing sites, with soils giving consistently higher enrichment in ^{34}S than rocks. The correlation of $\delta^{34}S$ with S abundance found for Apollo 16 soils (Kerridge et al., 1974a) is absent at Apollo 17, however, the correlation between $\delta^{34}S$ and surface exposure age is reflected in correlations with both N abundance and agglutinate content. With one exception, S from rocks, of both mare and highland origin, shows the same isotopic uniformity found for Apollo 16 highland rocks. The exception is the brecciated gabbro, 77017 which gave by far the lightest $\delta^{34}S$ we have yet found for a lunar sample. We believe that this value is real as S from pyrolysis–combustion gave a similarly negative $\delta^{34}S$, although the yield of S was less than 50%. Either this breccia incorporated some fractionated S of extralunar origin or indigenous S is less isotopically homogeneous than was apparent from other analyses. Gibson and Moore (1974) have suggested that S was added to this rock during the cataclastic event responsible for its pronounced shock features.

Our isotopic data for soils are in good agreement with those of Rees and Thode (1974); however, the results for rock S show the same intergroup discrepancy which we observed for Apollo 16, with Rees and Thode reporting values about 1.0–1.5‰ lighter than ours.

Metallic iron

During acid hydrolysis, H_2 is released quantitatively by dissolution of $Fe°$. This metal may be in the form of grains crystallized during lunar igneous activity, fragments of meteoritic debris or the product of solar wind reduction of indigenous lunar Fe^{++} (Gammage and Becker, 1971; Housley *et al.*, 1973; Kerridge *et al.*, 1974a). The first component is predominant in basalts, the second in highland rocks and all three occur in soils. Because of the absence of basalts from the Apollo 16 site, it was possible to disregard the first component and thus to distinguish between the second and third (Kerridge *et al.*, 1974a). In the present case, however, the intimate mixture of basalt and highland rock at Taurus-Littrow precludes such treatment and, although Apollo 17 soils, including the agglutinate sample, give consistently higher values than do rocks, it is not possible to separate the contribution due to *in situ* reduction of Fe^{++}.

The data in Table 4, which have been corrected for implanted H, determined by pyrolysis, are compared with values obtained by a variety of other techniques. The agreement with determinations made by means of saturation magnetization (Pearce *et al.*, 1974; Nagata *et al.*, 1974), ferromagnetic resonance (Griscom, 1974) and Mössbauer spectroscopy (Housley *et al.*, 1974) is generally good, particularly for magnetic measurements on rocks. Note that the data of Housley *et al.* pertain to < 45 μm size fractions of the soils, whereas our measurements were made on 0–1 mm fines. The data of other investigators are directly comparable.

Helium

Soils from Apollo 17 are extremely rich in He, see Table 5, as expected for a mature regolith containing a substantial basaltic component. Intersample differences in He content reflect primarily mineralogical variations, in particular, differences in ilmenite content. Thus, 72701, which contains only 1.52% TiO_2, has a He content of only 14.7 ppm, whereas 75061, which is significantly less mature

Table 4. Content of metallic iron in Apollo 17
samples.

Sample	Fe° wt.% This work	Literature values
Fines		
70011,27	0.70	
71501,22	0.60	0.69(1)
72701,15	0.64	
75061,33	0.51	
76501,20	0.70	0.60(1)
78421,28	0.84	0.75(2)
Breccias		
77017,41	0.26	0.15(3); 0.19(4)
77035,63	0.61	
77135,75	0.41	0.51(4)
Basalts		
70017,64	0.28	0.17(3)
70215,52	0.08	0.11(4)
74275,54	0.17	0.19(4)
75035,41	0.06	0.06(4)
75075,55	0.55	
Agglutinate		
70019,23	0.70	

References: (1) Housley *et al.* (1974), (2) Griscom (1974), (3) Nagata *et al.* (1974), (4) Pearce *et al.* (1974).

Table 5. Abundance of helium and hydrogen in Apollo 17 samples.

Sample	He ppm This work	Literature	H ppm This work
Fines			
70011,27	36.2, 37.0(*H*)	39(1)	55.1
71501,22	36.0, 37.2, 37.4, 38.3(*H*)	41(1)	59.6
72701,15	14.7, 14.7(*H*)	13(2), 14(3), 16(1)	53.9
75061,33	27.1, 31.3(*H*)	19(3), 29(1)	43.0
76501,20	22.3, 23.5(*H*)	18(2), 23(1)	43.0
78421,28	30.6, 30.7, 31.2(*H*)		67.0
Agglutinate			
70019,23	31.5(*H*), 34.0		39.5

(*H*) signifies data from acid hydrolysis.
References: (1) Hübner *et al.* (1974), (2) Bogard *et al.* (1974), (3) Hintenberger *et al.* (1974).

than 72701 but which contains 10.31% TiO_2, has twice as much He. Similar behavior can be deduced for 71501 (9.52% TiO_2; 37.2 ppm He) and 76501 (3.15% TiO_2; 22.9 ppm He). (Titanium data from LSPET, 1973.) Clearly, He abundance data cannot yield any useful information concerning solar wind exposure ages of Apollo 17 soils.

Agreement with the results of other investigators is good, showing an average deviation from the mean of 5%, significantly better than the agreement among the other investigators. Precision, based on differences between pyrolysis and hydrolysis values and on replicate analyses, was 2.5% average deviation from the mean. System blanks were below the detection limits of about 0.5 ppm He.

Agglutinate content

Grain mounts of 90–150 μm size fractions of the soil samples were examined in the optical microscope in order to determine their contents of agglutinates. The results of this survey are given in Table 6 and compared with the results of LSPET (1973) on different splits of the same soils. Agreement is excellent for two samples and bad for two others. As numerous workers have pointed out, agglutinate content is a good measure of soil maturity, and in the present case our values correlate very well with abundance of N and reasonably well with abundance of C. They also correlate strongly with $\delta^{34}S$, confirming our earlier suggestion that $\delta^{34}S$ increases with increasing surface exposure (Kerridge *et al.*, 1974a).

DISCUSSION

Interpretation of the Apollo 17 data in terms of lunar surface processes is complicated for reasons given above. We restrict our discussion here to some comments concerning S systematics. Several attempts have been made to explain abundances of S in lunar soils by means of mixing models with major local rock types as end members (e.g. Moore *et al.*, 1972; Gibson and Moore, 1974). In particular, Gibson and Moore (1974) have interpreted Apollo 17 soil data in terms of simple mixing of a basaltic component with anorthositic and/or noritic material.

Table 6. Agglutinate contents of Apollo 17 fines (90–150 μm size fraction).

Sample	% Agglutinates This work	LSPET, 1973
70011,27	37	
71501,22	34	35
72701,15	38	
75061,33	25	24
76501,20	32	47
78421,28	48	63

However, such a simple mixing model offers no explanation for the systematic enrichment of soils in ^{34}S. If, as has been suggested by Clayton *et al.* (1974), impact induced volatilization has caused preferential loss of ^{32}S, enhancing the residual S in ^{34}S by Rayleigh fractionation, the soil data should be displaced relative to the true mixing line by about 15–30% of the predicted S content. Such a displacement is not incompatible with the data of Gibson and Moore, or with our results. However, a consequence of the model of Clayton *et al.* is that soils which have lost the most S should show the largest isotopic fractionations and this is not borne out by our, admittedly scanty, data. In fact, for Apollo 16 soils, we have demonstrated a strong positive correlation between S content and δ^{34}S, on the basis of which we outlined a tentative suggestion for a relatively complex S cycle on the lunar surface (Kerridge *et al.*, 1974a).

Acknowledgments—We thank J. Cline, J. De Grosse, and D. Winter for assistance. Work supported by NASA grant no. 05-007-221.

REFERENCES

Bogard D. D., Nyquist L. E., and Hirsch W. C. (1974). Noble gases in Apollo 17 boulders and soils (abstract). In *Lunar Science—V*, pp. 73–75. The Lunar Science Institute, Houston.

Chang S., Lawless J., Romiez M., Kaplan I. R., Petrowski C., Sakai H., and Smith J. W. (1974). Carbon, nitrogen and sulfur in lunar fines 15012 and 15013: Abundances, distributions and isotopic compositions. *Geochim. Cosmochim. Acta.* **38**, 853–872.

Clayton R. N., Mayeda T. K., and Hurd J. M. (1974). Loss of O, Si, S and K from the lunar regolith (abstract). In *Lunar Science—V*, pp. 129–131. The Lunar Science Institute, Houston.

Gammage R. B. and Becker K. (1971). Exoelectron emission and surface characteristics of lunar materials. *Proc. Second Lunar Sci. Conf., Geochim. Cosmochim. Acta*, Suppl. 2, Vol. 3, pp. 2057–2067. MIT Press.

Gibson E. K., Jr. and Moore G. W. (1974). Total sulfur abundances and distributions in the valley of Taurus-Littrow: Evidence of mixing (abstract). In *Lunar Science—V*, pp. 267–269. The Lunar Science Institute, Houston.

Goel P. S., Shukla P. N., Kothari B. K., and Garg A. N. (1974). Solar wind as source of nitrogen in lunar fines. Paper presented at Fifth Lunar Science Conference, Houston.

Griscom D. L. (1974). On the nature and distribution of ferromagnetic phases in the Taurus-Littrow valley: Comparison with other regions of the moon (abstract). In *Lunar Science—V*, pp. 296–298. The Lunar Science Institute, Houston.

Hintenberger H., Weber H. W., and Schultz L. (1974). Solar, spallogenic and radiogenic rare gases in Apollo 17 soils and breccias (abstract). In *Lunar Science—V*, pp. 334–336. The Lunar Science Institute, Houston.

Housley R. M., Cirlin E. H., and Grant R. W. (1973). Characterization of fines from the Apollo 16 site. *Proc. Fourth Lunar Sci. Conf., Geochim. Cosmochim. Acta.*, Suppl. 4, Vol. 3, pp. 2729–2735. Pergamon.

Housley R. M., Cirlin E. H., and Grant R. W. (1974). Solar wind and micrometeorite alteration of the lunar regolith (abstract). In *Lunar Science—V*, pp. 360–362. The Lunar Science Institute, Houston.

Hübner W., Kirsten T., and Kiko J. (1974). Rare gases in Apollo 17 soils with emphasis on analysis of size and mineral fractions of soil 74241 (abstract). In *Lunar Science—V*, pp. 369–371. The Lunar Science Institute, Houston.

Kaplan I. R. and Petrowski C. (1971). Carbon and sulfur isotope studies on Apollo 12 lunar samples. *Proc. Second Lunar Sci. Conf., Geochim. Cosmochim. Acta.*, Suppl. 2, Vol. 2, pp. 1397–1406. MIT Press.

Kaplan I. R., Smith J. W., and Ruth E. (1970). Carbon and sulfur concentration and isotopic composition in Apollo 11 lunar samples. *Proc. Apollo 11 Lunar Sci. Conf.*, *Geochim. Cosmochim. Acta.*, Suppl. 1, pp. 1317–1329. Pergamon.

Kerridge J. F., Kaplan I. R., Petrowski C., and Chang S. (1974). Light element geochemistry of the Apollo 16 site. *Geochim. Cosmochim. Acta.* In press.

Kerridge J. F., Kaplan I. R., and Lesley F. D. (1974). Accumulation and isotopic evolution of carbon on the lunar surface. This volume.

LSPET (Lunar Sample Preliminary Examination Team) (1973). Apollo 17 lunar sample information catalog, NASA MSC 03211.

Moore C. B., Lewis C. F., Cripe J., Delles F. M., Kelly W. R., and Gibson E. K., Jr. (1972). Total carbon, nitrogen and sulfur in Apollo 14 lunar samples. *Proc. Third Lunar Sci. Conf.*, *Geochim. Cosmochim. Acta.*, Suppl. 3, Vol. 3, pp. 2051–2058. MIT Press.

Moore C. B., Lewis C. F., Cripe J., and Volk M. (1974). Total carbon and sulfur contents of Apollo 17 lunar samples (abstract). In *Lunar Science—V*, pp. 520–522. The Lunar Science Institute, Houston.

Müller O. (1974). Solar wind- and indigenous nitrogen in Apollo 17 lunar samples (abstract). In *Lunar Science—V*, pp. 534–536. The Lunar Science Institute, Houston.

Nagata T., Sugiura N., Fisher R. M., Schwerer F. C., Fuller M. D., and Dunn J. R. (1974). Magnetic properties and natural remanent magnetization of Apollo 16 and 17 lunar materials (abstract). In *Lunar Science—V*, pp. 540–542. The Lunar Science Institute, Houston.

Pearce G. W., Gose W. A., and Strangway D. W. (1974). Magnetism of the Apollo 17 samples (abstract). In *Lunar Science—V*, pp. 590–592. The Lunar Science Institute, Houston.

Rees C. E. and Thode H. G. (1974). Sulphur concentrations and isotope ratios in Apollo 16 and 17 samples. Paper presented at Fifth Lunar Science Conference, Houston.

Rhodes J. M., Rodgers K. V., Shih C., Bansal B. M., Nyquist L. E., and Wiesmann H. (1974). The relationship between geology and soil chemistry at the Apollo 17 site (abstract). In *Lunar Science—V*, pp. 630–632. The Lunar Science Institute, Houston.

Scoon J. H. (1974). Chemical analysis of lunar samples from the Apollo 16 and 17 collections (abstract). In *Lunar Science—V*, pp. 690–692. The Lunar Science Institute, Houston.

Smith J. W., Kaplan I. R., and Petrowski C. (1973). Carbon, nitrogen, sulfur, helium, hydrogen and metallic iron in Apollo 15 drill stem fines. *Proc. Fourth Lunar Sci. Conf.*, *Geochim. Cosmochim. Acta.*, Suppl. 4, Vol. 2, pp. 1651–1656. Pergamon.

Proceedings of the Fifth Lunar Conference
(Supplement 5, Geochimica et Cosmochimica Acta)
Vol. 2 pp. 1949–1961 (1974)
Printed in the United States of America

The association between carbide and finely divided metallic iron in lunar fines

C. T. Pillinger, P. R. Davis, G. Eglinton, A. P. Gowar, A. J. T. Jull, and J. R. Maxwell

School of Chemistry, University of Bristol, Bristol BS8 1TS, U.K.

R. M. Housley and E. H. Cirlin

Rockwell International Science Center, Thousand Oaks, California 91360

Abstract—The carbide concentration in a variety of samples of lunar fines measured as the CD_4 evolved on deuterated acid dissolution is directly related to finely divided iron $Fe°$ (<about 1 μm) as measured by ferromagnetic resonance spectroscopy. Both carbide and this free metallic iron, when normalized to ^{36}Ar content as an exposure parameter, are directly related to the bulk Fe^{II} content of the soil. These data indicate that carbide synthesis occurring at the lunar surface depends on the amount of indigenous Fe^{II} in silicates available for exposure-induced reduction to $Fe°$.

Introduction

The concentration of acid-hydrolysable carbide, measured as CD_4 by DCl dissolution of a variety of Apollo 11, 12, and 14 bulk fines, correlates (Abell *et al.*, 1971; Cadogan *et al.*, 1971, 1972; Holland *et al.*, 1972) with a number of surface exposure parameters, including ^{36}Ar content and radiation damage (the percentage of grains with very high densities of solar flare tracks and the percentage of fine grains with amorphous coatings). Thus, samples with the greatest surface exposure have the highest carbide concentrations, although there is a significant degree of scatter from a linear relationship.

Indirect evidence suggests that the major carbide species in the bulk fines is associated with metallic iron and is a form of iron carbide. Thus, separation studies on 10086 and 60501 fines have shown that carbide is concentrated in very fine grains (<10 μm) and glassy agglutinates and microbreccias with high magnetic susceptibilities (Cadogan *et al.*, 1973a, b). Agglutinates are known to contain high concentrations of finely divided metallic iron (Housley *et al.*, 1972, 1973a, b).

The distribution and relative abundances of the C_1–C_3 gaseous species released from bulk fines by deuterated or undeuterated acid are similar to those released by cohenite (Fe, Ni)$_3$C (Chang *et al.*, 1970; Abell *et al.*, 1971), with methane being by far the most abundant species. A notable exception is that acetylene is released from lunar fines but not from cohenite. A C_2D_2 releasing species has been shown to be concentrated in plagioclase-rich fractions of highland soils (Wszolek and Burlingame, 1973; Wszolek *et al.*, 1973) and detected as $^{13}C_2D_2$ among the species released from plagioclase implanted with ^{13}C ions, and may result from dissolution of an ionic carbide such as CaC_2 (Wszolek *et al.*, 1974;

Bibring *et al.*, 1974). The relationship between iron carbide and lunar surface exposure taken together with the observation that high concentrations of CD_4 are released by very fine grains, glassy agglutinates, and microbreccias, suggests that synthesis of iron carbide on the lunar surface is brought about by exposure-induced reactions involving meteorite impact, solar wind implantation, or both (Cadogan *et al.*, 1972, 1973b). Only small amounts of CD_4 are released by glasses, mineral grains, or iron-rich grains presumed to be of meteorite origin (Cadogan *et al.*, 1973a, b; Wszolek *et al.*, 1973). The carbon released by pyrolysis follows similar, though less well-defined trends with particle type (DesMarais *et al.*, 1973a). These findings support the hypothesis that a primary contribution to lunar carbide from meteorites, in the form of debris or as vapor phase deposits, can only be minor (Cadogan *et al.*, 1972, 1973b; Wszolek *et al.*, 1973). In a preliminary paper (Pillinger *et al.*, 1973) it was suggested that lunar iron carbide (as measured by CD_4 released by DCl dissolution) had been largely formed from lunar iron silicates. Carbide concentrations normalized to constant exposure (i.e. $CD_4/^{36}Ar$) were found to correlate with indigenous iron (Fe^{II} as FeO). Thus, for samples collected from missions up to Apollo 16, the generation of carbide was interpreted as depending not only on surface exposure, but also on the availability of indigenous lunar Fe^{II} in silicates for reduction to $Fe°$. On the basis of limited data available from low temperature Mössbauer measurements it was concluded that a metallic $Fe°$ phase involved was superparamagnetic iron ($Fe°sp$) up to about 100 Å in diameter. As a result of saturation effects encountered in the low temperature Mössbauer (Housley *et al.*, 1974b) measurements, sufficient consistent data on $Fe°sp$ concentrations are not yet available to investigate this relationship further. The preliminary study also showed that the $Fe°$ phase (0.01–25 μm) measured by room temperature Mössbauer spectroscopy is not related to carbide concentration. In this present study we have examined the relationship between carbide and $Fe°$ in further detail and have shown that a metallic phase apparently associated with carbide is the finely divided iron (0.01–1 μm) which is detected by ferromagnetic resonance spectroscopy. It appears that the particle size range of $Fe°$ associated with carbide is restricted.

Experimental

The methods used for sample handling and determination of carbide were identical to those previously described (Cadogan *et al.*, 1972, 1973b). Ferromagnetic resonance spectroscopic methods and the data obtained are discussed in detail separately (Cirlin *et al.*, 1974; Housley *et al.*, 1974b); other ferromagnetic resonance data were obtained from Griscom *et al.* (1973), Griscom (1974), and Weeks (1973a, b).

^{36}Ar data were obtained from the laboratories of Bogard, Geiss, Heymann, Hintenberger, Kirsten, Marti, and Pepin (see references cited in Abell *et al.*, 1971; Cadogan *et al.*, 1971; Cadogan *et al.*, 1972; Pillinger *et al.*, 1973; plus Bogard *et al.*, 1974; Hübner *et al.*, 1974; Heymann *et al.*, 1974; Jordan *et al.*, 1974). Fe^{II} (as FeO) data were obtained from the laboratories of Brunfelt, Compston, Hubbard, Juan, Laul, Philpotts, Rose, and Wänke (references cited in Pillinger *et al.*, 1973; plus LSPET, 1973; Wänke *et al.*, 1973; Laul and Schmitt, 1973).

Table 1. Concentration of CD_4 and CH_4 released by DCl
dissolution of Apollo 17 fines.

Sample*	CD_4 (μg/g)	CH_4 (μg/g)	CD_4/CH_4 ratio	Total C† ppm
71501	15.1	7.6	2.0	75
72141	21.1	7.7	2.8	155
72501	10.5	4.3	2.5	125
72701	8.3	3.1	2.7	140
73221	6.0	2.3	2.6	—
73281	5.1	2.2	2.4	—
74220	0.1	0.1	—	5
75080	24.6	5.7	4.3	115
76501	7.3	4.4	1.6	120

*Bulk soils, except 74220 (<45 μm).
†Data from Moore *et al.* (1974).

RESULTS

Table 1 lists the concentrations of CH_4 and CD_4 released by DCl dissolution of a number of Apollo 17 soils; the carbide data for the other Apollo missions have been reported elsewhere (Abell *et al.*, 1971; Cadogan *et al.*, 1972; Pillinger *et al.*, 1974).

Carbide and surface exposure

Figure 1 shows a plot of CD_4 released by DCl dissolution of bulk fines from Apollo 11 to Apollo 17 against measurements of ^{36}Ar concentrations in the same samples. The relationship between carbide and surface exposure indicated by data obtained from Apollo 11, 12, and 14 samples (Cadogan *et al.*, 1972) is no longer apparent when samples from all missions are considered. A similar situation arises when CD_4 concentration is plotted against exposure as measured by the proportion of grains with very high track densities. These observations indicate that surface exposure is not the only factor controlling carbide synthesis and Fig. 1 suggests that a landing site dependence is also involved as previously observed for Apollo 11, 12, 14, and 16 samples (Pillinger *et al.*, 1973); Apollo 15 and 17 samples show extensive scatter. The apparent landing site dependence and the scatter can be accounted for when bulk chemistry is considered, as carbide would be expected to show some proportionality to Fe^{II} content if Fe^{II} were the source of the associated free metal (Pillinger *et al.*, 1973). The plot (Fig. 2) of $CD_4/^{36}$Ar against Fe^{II} for samples from all Apollo missions, supports this hypothesis.

Separate relationships are observed for the two groups of rare gas investigators, presumably as a result of systematic differences in the quantitation of ^{36}Ar.

Normalizing to constant ^{36}Ar content will not provide an absolute correction for surface exposure. ^{36}Ar may diffuse more readily than carbon species because it

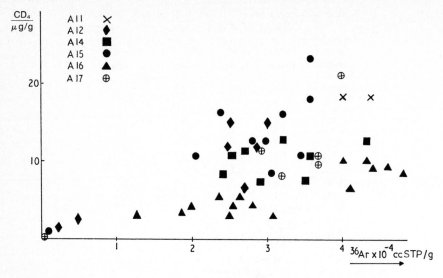

Fig. 1. Carbon as CD_4 versus ^{36}Ar concentration in Apollo 11–17 fines (data from all rare gas investigators). A similar degree of scatter is observed when systematic errors are minimized by plotting the data from each rare gas investigator separately.

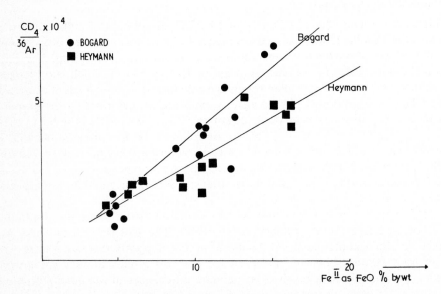

Fig. 2. Carbon as CD_4 normalized to constant ^{36}Ar concentration in Apollo 11–17 fines versus total iron content (as FeO). Only the rare gas data from the groups of Bogard and Heymann are plotted to minimize the effects of systematic errors, while allowing measurements for a wide range of samples.

is not chemically bound in the matrix and major losses may occur during impact-induced surface melting. Nevertheless, ^{36}Ar is thought to represent a more reliable measure of solar wind exposure than lighter rare gases such as ^{20}Ne because the latter have lower, and variable retentivities in certain minerals (e.g. Hintenberger and Weber, 1973).

Carbide and free iron

Carbide in lunar fines is associated with magnetic phases, presumably free iron (Cadogan *et al.*, 1973a, b) and a correlation between carbide and some form of elemental iron would therefore be expected (Pillinger *et al.*, 1973). From limited data it was suggested (Pillinger *et al.*, 1973) that carbide concentration might be directly related to Fe°sp concentration estimated from the excess area measurements of low temperature Mössbauer spectroscopy (Housley *et al.*, 1971, 1972). The Fe°sp is thought to derive from Fe^{II} by a reaction induced during lunar surface exposure (Grant *et al.*, 1973). At present, the Mössbauer data are not internally consistent (Housley *et al.*, 1974b) and the hypothesis still awaits substantiation. However, ferromagnetic resonance data (for a more complete discussion, see Housley *et al.*, 1974b; Cirlin *et al.*, 1974) are now available for twelve samples of fines for which the carbide concentrations have been measured. There is a direct relationship between CD$_4$ released by acid dissolution and ferromagnetic resonance intensity, measured for the same samples (Fig. 3). Similar correlations are observed between carbide and the ferromagnetic resonance data of two other groups of investigators (Griscom *et al.*, 1973; Griscom, 1974; Weeks, 1972, 1973a, b).

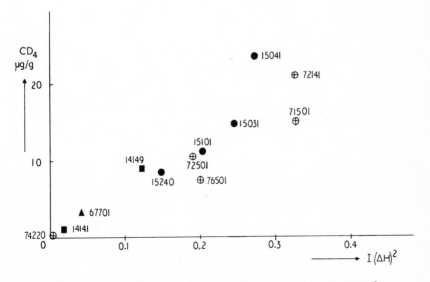

Fig. 3. Carbon as CD$_4$ versus ferromagnetic resonance intensity [I(ΔH)^2].

Some controversy exists over the nature of the species giving rise to the ferromagnetic resonance (see below). However, there is strong evidence (Tsay *et al.*, 1971; Cirlin *et al.*, 1974) to support the interpretation of Manatt *et al.* (1970), who suggested that small, spherical, single crystals of Fe in the size range 0.01–1 μm were responsible for the characteristic resonance. This fine-grained metal (Fe°_{fmr}) may have derived by exposure-induced reduction processes (Cirlin *et al.*, 1974; Housley *et al.*, 1973b). There is no direct relation between the ferromagnetic resonance intensity and the exposure parameter ^{36}Ar (Fig. 4). A relationship between Fe°_{fmr} and exposure is revealed, however, by taking into consideration the Fe^{II} content of the silicate. Thus, analogous to the carbide, there is a correlation, when Fe°_{fmr} is normalized to constant exposure and plotted as $Fe^{\circ}_{fmr}/^{36}Ar$ against Fe^{II} (Fig. 5). The correlation also holds when the data of two other groups of investigators (Griscom *et al.*, 1973; Griscom, 1974; Weeks, 1973a, b) are considered independently.

DISCUSSION

The occurrence in the lunar regolith of carbide (giving rise to CD_4 by DCl dissolution) and the occurrence of finely divided iron species (giving rise to ferromagnetic resonance) appear to be intimately associated. Several measures exist for iron species in lunar samples. Among the different techniques used there is some overlap in the particle size range of Fe° which it is assumed that each technique measures (Table 2). In general, there is good numerical agreement for the same measurements made by different investigators, but there is some

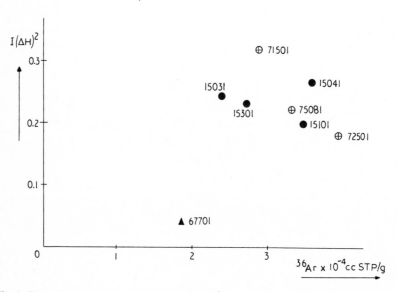

Fig. 4. Ferromagnetic resonance intensity $[I(\Delta H)^2]$ versus ^{36}Ar concentration for various soil samples. Rare gas data of Heymann and Kirsten.

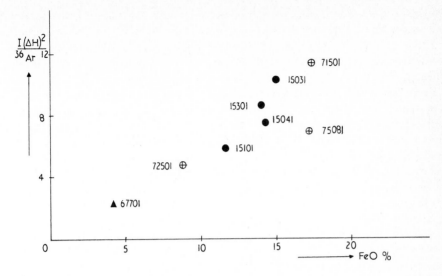

Fig. 5. Ferromagnetic resonance intensity $[I(\Delta H)^2]$ normalized to constant ^{36}Ar concentration versus total iron content (as FeO) for various soil samples. Rare gas data from Heymann and Kirsten.

Table 2. Measurement of metallic iron in lunar fines and soil breccias.

Technique	Iron species	Size range*	Investigators
Mössbauer, ambient temperature	Ferromagnetic metal	0.01–25 μm	Housley
Mössbauer, low temperature	Isolated Fe° atoms (Fe°sp)	<100 Å	Housley
Ferromagnetic resonance	Metallic iron† (Fe°$_{fmr}$)	0.01–1 μm	Housley, Griscom, Weeks
Magnetic hysteresis	Superparamagnetic Single domain Multi domain	<150 Å 150–300 Å >300 Å	Pearce, Fuller

*Size ranges listed are approximate, and are given as quoted by the various investigators (references in text).

†Griscom and Weeks and their colleagues suggest this may include magnetite.

disagreement as to the identities of the species measured. Cirlin *et al.* (1974) maintain that the ferromagnetic species are almost entirely finely divided Fe°, whereas Griscom *et al.* (1973, 1974) have pointed out that the resonance due to Fe° and ferric oxides such as magnetite are almost identical for very small grains. Thus, it has been suggested (Weeks, 1973a; Weeks and Prestel, 1974; Griscom, 1974) that up to about 20% of the characteristic resonance observed for lunar fines is attributable to the presence of magnetite-like material, because the line width (ΔH) is less than that expected for pure Fe°. However, since carbide is associated with the iron, the $Fe°$ species cannot be pure. An investigation of iron doped with light elements such as carbon, nitrogen, etc. may help to resolve the apparent anomalies in observed line width. The relationship involving ferromagnetic resonance intensity and carbide and Fe^{II} is consistent with most, if not all, of the ferromagnetic resonance measurement being attributable to $Fe°$.

From the results presented above it seems that finely divided metal up to about 1 μm diameter is produced by an exposure process. In the fines, large grains of multidomain iron (>1 μm) may be contributed additionally as meteorite fragments (e.g. Goldstein and Axon, 1973). Some finely divided iron may derive from micrometeorites but such iron should not exhibit a direct relationship with the total iron ($Fe°$ and Fe^{II}) content of the sample (Pillinger et al., 1973) and hence, its contribution to total $Fe°$ must be correspondingly small. Although both CD_4 and ferromagnetic resonance data correlate well with exposure, neither parameter correlates with Mössbauer room temperature data available at present (Pillinger et al., 1973; Cirlin et al., 1974).

It has been suggested by several investigators that the finely divided $Fe°$ in the lunar fines arises by reduction of Fe^{II} in silicates during meteorite impact heating. The reducing agent could be implanted solar wind hydrogen (Carter and McKay, 1972; Grant et al., 1973; Housley et al., 1973b) or a component of the gas phase of the impact cloud, for example, CO (Pearce et al., 1972). The necessary thermal conditions could arise in an ejecta blanket (Pearce et al., 1972) or by shock (Cisowski et al., 1973). Thermal treatment ($>1100°C$) of iron-containing minerals such as ilmenite and ulvöspinel at low oxygen fugacity results in formation of $Fe°$ from $FeTiO_3$ (Taylor et al., 1972); the required temperature can be reached during simulated meteorite impacts on terrestrial ilmenite (Sclar et al., 1973). Alternatively, solar wind implantation alone could bring about the reduction by cleavage of Fe–O bonds or by ion sputtering (Yin et al., 1972; Kelly and Lam, 1973). Any superparamagnetic and single domain $Fe°$ formed by the reduction of Fe^{II} can give rise to larger grains of micron sized $Fe°$ under thermal conditions (Pearce et al., 1972; Housley et al., 1973b; Usselman and Pearce, 1974). Thus, the $Fe°$ metal grain size would be expected to be a continuum from single atoms to micron sized species.

We have previously proposed (Cadogan et al., 1973b) that lunar iron carbide is synthesized by one or more of the following mechanisms:

(1) Reaction with $Fe°$ of carbon present from any source but particularly of solar wind origin. The process could involve incorporation of carbon into $Fe°$ during formation of larger grains. The migration of carbon as volatile species such as CH_4 cannot be overlooked.

(2) Implantation of solar wind carbon into $Fe°$, shown to be possible by simulation (Pillinger et al., 1972).

(3) Carburization of exposed $Fe°$ by carbon-containing species in a gas cloud created by meteorite decomposition and soil melting.

As an interpretation of the plot CD_4 versus $1/r$ (Cadogan et al., 1972), DesMarais et al. (1973b) have suggested that the CD_4 released from DCl dissolution experiments represents only recently synthesized and unaltered carbide as might be produced by (2) and (3) above. However, this is not a likely possibility, because complex recycled particles such as agglutinates and micro-breccias have carbide contents which are undoubtedly volume related (Cadogan *et al.*, 1972; Cadogan *et al.*, 1973a, b). In view of the limited data available (Cadogan *et al.*, 1972) it may be an oversimplification to extrapolate the CD_4 versus $1/r$ plot to infinitely large particle radius.

As a result of the reduction processes which lead to the generation of metallic iron, both surface exposure and ferrous iron content should be controlling factors in carbide synthesis. Other species which are exposure-dependent may also be related to the occurrence of free iron. For example, the approximately linear relationships between exposure parameters and the concentration of CH_4 in fines from Apollo 11, 12, and 14 (Cadogan *et al.*, 1972) show considerably greater scatter when samples from later Apollo missions are included. When $CH_4/^{36}Ar$ is plotted against FeO content the scatter is reduced suggesting that CH_4 may be involved in some way with iron. In analogy to carbon, lunar nitrogen is thought to be present partly as nitride (Müller, 1972, 1973; Holland *et al.*, 1972). Thus, nitrogen could also be involved with Fe° and possible relationships with iron in lunar fines should be sought.

Exposure processes may cause valency state changes for elements other than iron, or alter the chemical environments of some ions. The formation of a species, presumed to be calcium carbide, by implantation of $^{13}C^+$ into plagioclase is indicated by the $^{13}C_2D_2$ released as a major species on deuterated acid treatment (Wszolek *et al.*, 1974; Bibring *et al.*, 1974). Since CaC_2 is an ionic carbide, $^{13}C_2D_2$ is presumably formed from carbon atoms in a Ca^{2+}-rich environment. The release of C_2D_2 from lunar soils, particularly those of high-plagioclase content (Wszolek and Burlingame, 1973; Wszolek *et al.*, 1973) indicates that similar material may have resulted from solar wind implantation into lunar grains. Thus, C_2D_2 may be a significant indicator of exposure processes in the lunar highlands. However, the contribution to total carbide from the acetylene-producing species is not great and is probably lower than the relative yields of CD_4 and C_2D_2 on acid dissolution would suggest. Typical mare soils, e.g. fines from the Apollo 15 deep drill core, have CD_4/C_2D_2 ratios of 20–100 (Wszolek and Burlingame, 1973). However, an ionic carbide such as CaC_2 normally has a much higher yield of deuterocarbons on acid dissolution than does covalent iron carbide. CD_4 represents by far the major proportion of lunar carbide and the species thought to be iron carbide is the largest fraction of lunar carbon chemically characterized to date. Even so, only up to about 15% of carbon in lunar fines is actually released as deuterocarbon and hydrocarbon gases on acid dissolution; however, the deuterocarbon yield of the reaction is unknown and the actual percentages of carbide present are likely to be higher than those quoted.

CONCLUSIONS

This work provides further circumstantial evidence that iron carbide is the major source of CD_4 released by deuterated acid treatment of lunar fines.

Correlations between sets of analytical data do not constitute *proof* that this carbide is predominantly associated with sub-micron metal grains. However, correlations of this type allow working hypotheses to be formulated, indicate where further work should be directed, and stimulate further analytical studies. A definitive characterization of the species giving rise to deuterocarbons by deuterated acid dissolution will only be forthcoming from:

(1) Quantitative estimation of the carbon as carbide (this is by no means a trivial problem).

(2) A knowledge of the precise location of surface-located carbide species, i.e. whether they are within the depth of penetration of the solar wind hydrogen (e.g. 2000 ± 500 Å, Leich *et al.*, 1973) or at the *very* surface of lunar grains (Batts *et al.*, 1974).

(3) A study of the mechanisms involved in the formation of both iron and carbide (Batts *et al.*, 1974).

Carbide formation has been shown to be dependent on both exposure and bulk chemistry. It may be necessary to further clarify these factors by subdividing "exposure" into solar wind implantation and micrometeorite bombardment and by considering the bulk chemistry in terms of the abundances of different iron-containing minerals and glasses in the fines. The proportion of glass-welded agglutinates which is known to be a measure of soil maturity (McKay *et al.*, 1974) may be a valuable guide to the role of impact processes.

At present carbide may only be used as a measure of exposure differences between samples of very similar Fe^{II} content. Further work, including the simulations mentioned above, may allow a "carbide-derived exposure scale" to be established. However, before such a scale can be derived, the reaction mechanisms for the Fe^{II} to $Fe°$ reduction and carbide formation need to be better understood.

Unlike many other elements, carbon cannot be considered without regard to its chemical state. It is the chemistry of lunar carbon we are attempting to clarify and it is this very chemistry which makes the problem a complex one.

REFERENCES

Abell P. I., Cadogan P. H., Eglinton G., Maxwell J. R., and Pillinger C. T. (1971) Survey of lunar carbon compounds, I: The presence of indigenous gases and hydrolysable carbon compounds in Apollo 11 and Apollo 12 samples. *Proc. Second Lunar Sci. Conf., Geochim. Cosmochim. Acta,* Suppl. 2, Vol. 2, pp. 1843–1863. MIT Press.

Batts B. D., Biggar G. M., Billetop M. C. J., Davis P. R., Eglinton G., Erents S. K., Gardiner L. R., Gowar A. P., Housley R. M., Humphries D. J., Jull A. J. T., Maxwell J. R., Mays B. J., McCracken G. M., and Pillinger C. T. (1974) The origin of lunar carbide (abstract). In *Lunar Science—V,* pp. 47–49. The Lunar Science Institute, Houston.

Bibring J. P., Burlingame A., Chaumont J., Maurette M., Slodzian G., and Wszolek P. (1974) Lunar soil maturation, II: Synthesis of carbon compounds in the regolith (abstract). In *Lunar Science—V,* pp. 57–59. The Lunar Science Institute, Houston.

Bogard D. D., Nyquist L. E., and Hirsch W. C. (1974) Noble gases in Apollo 17 boulders and soils (abstract). In *Lunar Science—V,* pp. 73–75. The Lunar Science Institute, Houston.

Cadogan P. H., Eglinton G., Maxwell J. R., and Pillinger C. T. (1971) Carbon chemistry of the lunar surface. *Nature* **231**, 29–31.

Cadogan P. H., Eglinton G., Firth J. N. M., Maxwell J. R., Mays B. J., and Pillinger C. T. (1972) Survey of lunar carbon compounds, II: The carbon chemistry of Apollo 11, 12, 14, and 15 samples. *Proc. Third Lunar Sci. Conf., Geochim. Cosmochim. Acta*, Suppl. 3, Vol. 2, pp. 2069–2090. MIT Press.

Cadogan P. H., Eglinton G., Maxwell J. R., and Pillinger C. T. (1973a) Distribution of methane and carbide in Apollo 11 fines. *Nature Phys. Sci.* **241**, 81–82.

Cadogan P. H., Eglinton G., Gowar A. P., Jull A. J. T., Maxwell J. R., and Pillinger C. T. (1973b) Location of methane and carbide in Apollo 11 and 16 lunar fines. *Proc. Fourth Lunar Sci. Conf., Geochim. Cosmochim. Acta*, Suppl. 4, Vol. 2, pp. 1493–1508. Pergamon.

Carter J. L. and McKay D. S. (1972) Metallic mounds produced by reduction of material of simulated lunar composition and implications on the origins of metallic mounds on lunar glasses. *Proc. Third Lunar Sci. Conf., Geochim. Cosmochim. Acta*, Suppl. 3, Vol. 1, pp. 953–970. MIT Press.

Chang S., Smith J. W., Kaplan I., Lawless J., Kvenvolden K. A., and Ponnamperuma C. (1970) Carbon compounds in lunar fines from Mare Tranquillitatis, IV: Evidence for oxides and carbides. *Proc. Apollo 11 Lunar Sci. Conf., Geochim. Cosmochim. Acta*, Suppl. 1, Vol. 2, pp. 1857–1869. Pergamon.

Cirlin E. H., Housley R. M., Goldberg I. B., and Paton N. E. (1974) Ferromagnetic resonance as a method for studying regolith dynamics and breccia formation (abstract). In *Lunar Science—V*, pp. 121–122. The Lunar Science Institute, Houston.

Cisowski C. S., Fuller M., Rose M. F., and Wasilewski P. J. (1973) Magnetic effects of experimental shocking of lunar soil. *Proc. Fourth Lunar Sci. Conf., Geochim. Cosmochim. Acta*, Suppl. 4, Vol. 3, pp. 3003–3017. Pergamon.

DesMarais D. J., Hayes J. M., and Meinschein W. G. (1973a) The distribution in lunar soil of carbon released by pyrolysis. *Proc. Fourth Lunar Sci. Conf., Geochim. Cosmochim. Acta*, Suppl. 4, Vol. 2, pp. 1543–1558. Pergamon.

DesMarais D. J., Hayes J. M., and Meinschein W. G. (1973b) Accumulation of carbon in lunar soils. *Nature Phys. Sci.* **246**, 65–68.

Goldstein J. I. and Axon H. J. (1973) Metallic particles from 3 Apollo 16 soils (abstract). In *Lunar Science—IV*, pp. 299–301. The Lunar Science Institute, Houston.

Grant R. W., Housley R. M., and Paton N. E. (1973) Origin and characteristics of excess Fe metal in lunar glass welded aggregates. *Proc. Fourth Lunar Sci. Conf., Geochim. Cosmochim. Acta*, Suppl. 4, Vol. 3, pp. 2737–2749. Pergamon.

Griscom D. L., Friebele E. J., and Marquardt C. L. (1973) Evidence for a ubiquitous, sub-microscopic "magnetite-like" constituent in the lunar soils. *Proc. Fourth Lunar Sci. Conf., Geochim. Cosmochim. Acta*, Suppl. 4, Vol. 3, pp. 2709–2727. Pergamon.

Griscom D. L. (1974) On the nature and distribution of ferromagnetic phases in the Taurus-Littrow Valley: Comparison with other regions of the moon (abstract). In *Lunar Science—V*, pp. 296–297. The Lunar Science Institute, Houston.

Griscom D. L., Marquardt C. L., and Friebele E. J. (1974) Ferromagnetic resonance of fine grained iron and magnetite precipitates in simulated lunar glasses: Comparison with lunar soils (abstract). In *Lunar Science—V*, pp. 293–295. The Lunar Science Institute, Houston.

Heymann D., Jordan J. L., Walton J. R., and Lakatos S. (1974) An inert gas "borscht" from the Taurus-Littrow site (abstract). In *Lunar Science—V*, pp. 331–333. The Lunar Science Institute, Houston.

Hintenberger H. and Weber H. W. (1973) Trapped rare gases in lunar fines and breccias. *Proc. Fourth Lunar Sci. Conf., Geochim. Cosmochim. Acta*, Suppl. 4, Vol. 2, pp. 2003–2019. Pergamon.

Holland P. T., Simoneit B. R., Wszolek P. C., and Burlingame A. L. (1972) Compounds of carbon and other volatile elements in Apollo 14 and 15 samples. *Proc. Third Lunar Sci. Conf., Geochim. Cosmochim. Acta*, Suppl. 3, Vol. 2, pp. 2131–2147. MIT Press.

Housley R. M., Grant R. W., Muir A. H., Blander M., and Abdel-Gawad M. (1971) Mössbauer studies of Apollo 12 samples. *Proc. Second Lunar Sci. Conf., Geochim. Cosmochim. Acta*, Suppl. 2, Vol. 3, pp. 2125–2136. MIT Press.

Housley R. M., Grant R. W., and Abdel-Gawad M. (1972) Study of excess Fe metal in the lunar fines by magnetic separation, Mössbauer spectroscopy, and microscopic examination. *Proc. Third Lunar Sci. Conf., Geochim. Cosmochim. Acta*, Suppl. 3, Vol. 1, pp. 1065–1076. MIT Press.

Housley R. M., Cirlin E. H., and Grant R. W. (1973a) Characterization of fines from the Apollo 16 site. *Proc. Fourth Lunar Sci. Conf., Geochim. Cosmochim. Acta,* Suppl. 4, Vol. 3, pp. 2729–2735. Pergamon.

Housley R. M., Grant R. W., and Paton N. E. (1973b) Origin and characteristics of excess Fe metal in lunar glass welded aggregates. *Proc. Fourth Lunar Sci. Conf., Geochim. Cosmochim. Acta,* Suppl. 4, Vol. 3, pp. 2737–2749. Pergamon.

Housley R. M., Cirlin E. H., and Grant R. W. (1974a) Solar wind and micrometeorite alteration of the lunar regolith (abstract). In *Lunar Science—V,* pp. 360–362. The Lunar Science Institute, Houston.

Housley R. M., Cirlin E. H., and Grant R. W. (1974b) Solar wind and micrometeorite alteration of the lunar regolith. *Proc. Fifth Lunar Sci. Conf., Geochim. Cosmochim. Acta,* Suppl. 5. This volume.

Hübner W., Kirsten T., and Kiko J. (1974) Rare gases in Apollo 17 soils with emphasis on analysis of size and mineral fractions of soil 74241 (abstract). In *Lunar Science—V,* pp. 369–371. The Lunar Science Institute, Houston.

Jordan J. L., Heymann D., and Lakatos S. (1974) Inert gas patterns in the regolith at the Apollo 15 landing site. *Geochim. Cosmochim. Acta* **38,** 65–78.

Kelly R. and Lam N. Q. (1973) The sputtering of oxides. *Radiation effects* **19,** 39–47.

Laul J. C. and Schmitt R. A. (1973) Chemical composition of Apollo 15, 16, and 17 samples. *Proc. Fourth Lunar Sci. Conf., Geochim. Cosmochim. Acta,* Suppl. 4, Vol. 2, pp. 1349–1367. Pergamon.

Leich D. A., Tombrello T. A., and Burnett D. S. (1973) The depth distribution of hydrogen in lunar materials. *Earth Planet. Sci. Lett.* **19,** 305–314.

LSPET (1973) Preliminary examination of lunar samples. *Apollo 17 Preliminary Science Report,* NASA SP-330, p. 7-1. NASA.

Manatt S. L., Elleman D. D., Vaughan R. W., Chan S. I., Tsay F. D., and Huntress W. D., Jr. (1970) Magnetic resonance studies of lunar samples. *Proc. Apollo 11 Lunar Sci. Conf., Geochim. Cosmochim. Acta,* Suppl. 1, Vol. 3, pp. 2321–2323. Pergamon.

McKay D. S., Fruland R. M., and Heiken G. H. (1974) Grain size distribution as an indicator of the maturity of lunar soils (abstract). In *Lunar Science—V,* pp. 480–482. The Lunar Science Institute, Houston.

Moore C. B., Lewis C. F., Cripe J. D., and Volk M. (1974) Total carbon and sulfur contents of Apollo 17 lunar soils (abstract). In *Lunar Science—V,* pp. 520–522. The Lunar Science Institute, Houston.

Müller O. (1972) Chemically bound nitrogen abundances in lunar samples and active gases released by heating at lower temperatures (250 to 500°C). *Proc. Third Lunar Sci. Conf., Geochim. Cosmochim. Acta,* Suppl. 3, Vol. 2, pp. 2059–2068. MIT Press.

Müller O. (1973) Chemically bound nitrogen contents of Apollo 16 and Apollo 15 lunar fines. *Proc. Fourth Lunar Sci. Conf., Geochim. Cosmochim. Acta,* Suppl. 4, Vol. 2, pp. 1625–1634. Pergamon.

Pearce G. W., Williams R. J., and McKay D. S. (1972) The magnetic properties and morphology of metallic iron produced by subsolidus reduction of synthetic Apollo 11 composition glasses. *Earth Planet. Sci. Lett.* **17,** 95–104.

Pillinger C. T., Cadogan P. H., Eglinton G., Maxwell J. R., Mays B. J., Grant W. A., and Nobes M. J. (1972) Simulation study of lunar carbon chemistry. *Nature* **235,** 108–109.

Pillinger C. T., Batts B. D., Eglinton G., Gowar A. P., Jull A. J. T., and Maxwell J. R. (1973) Formation of lunar carbide from lunar iron silicates. *Nature Phys. Sci.* **245,** 3–5.

Pillinger C. T., Eglinton G., Gowar A. P., Jull A. J. T., and Maxwell J. R. (1974) The exposure history of the Apollo 16 site determined by carbon chemistry measurements. *Proc. of the Joint Soviet-American Conference on Cosmochemistry of the Moon and Planets* (abstract).

Sclar C. B., Bauer J. F., Ribbe P. H., Phillips M. W., Pickart S. J., and Alperin H. A. (1973) Shock effects in lunar minerals and their experimentally shocked counterparts (abstract). In *Lunar Science—IV,* pp. 666–668. The Lunar Science Institute, Houston.

Taylor L. A., Williams R. J., and McCallister R. H. (1972) Stability relations of ilmenite and ulvöspinel in the Fe–Ti–O system and application of these data to lunar material assemblages. *Earth Planet. Sci. Lett.* **16,** 282–288.

Tsay F. D., Chan S. I., and Manatt S. L. (1971) Ferromagnetic resonance of lunar samples. *Geochim. Cosmochim. Acta* **35,** 865–875.

Usselman T. M. and Pearce G. W. (1974) Grain growth of iron: Implications for the thermal conditions in a lunar ejecta blanket (abstract). In *Lunar Science—V*, pp. 809–811. The Lunar Science Institute, Houston.

Wänke H., Baddenhausen H., Dreibus G., Jagoutz E., Kruse H., Palme H., Spettel B., and Teschke F. (1973) Multielement analyses of Apollo 15, 16, and 17 samples and the bulk composition of the moon. *Proc. Fourth Lunar Sci. Conf., Geochim. Cosmochim. Acta*, Suppl. 4, Vol. 2, pp. 1461–1481. Pergamon.

Weeks R. A. (1972) Magnetic phases in lunar material and their electron magnetic resonance spectra: Apollo 14. *Proc. Third Lunar Sci. Conf., Geochim. Cosmochim. Acta*, Suppl. 3, Vol. 3, pp. 2503–2517. MIT Press.

Weeks R. A. (1973a) Ferromagnetic phases of lunar fines and breccias: Electron magnetic resonance spectra of Apollo 16 samples. *Proc. Fourth Lunar Sci. Conf., Geochim. Cosmochim. Acta*, Suppl. 4, Vol. 3, pp. 2763–2781. Pergamon.

Weeks R. A. (1973b) Personal communication.

Weeks R. A. and Prestel D. (1974) Ferromagnetic resonance properties of some Apollo 16 and 17 fines and comparison to those of terrestrial analogues (abstract). In *Lunar Science—V*, pp. 833–835. The Lunar Science Institute, Houston.

Wszolek P. C. and Burlingame A. L. (1973) Carbon chemistry of the Apollo 15 and 16 deep drill cores. *Proc. Fourth Lunar Sci. Conf., Geochim. Cosmochim. Acta*, Suppl. 4, Vol. 2, pp. 1681–1692. Pergamon.

Wszolek P. C., Simoneit B. R., and Burlingame A. L. (1973) Studies of magnetic fines and volatile-rich soils: Possible meteoritic and volcanic contributions to lunar carbon and light element chemistry. *Proc. Fourth Lunar Sci. Conf., Geochim. Cosmochim. Acta*, Suppl. 4, Vol. 2, pp. 1693–1706. Pergamon.

Wszolek P. C., Walls F. C., and Burlingame A. L. (1974) Simulation of lunar carbon chemistry in plagioclase (abstract). In *Lunar Science—V*, pp. 854–856. The Lunar Science Institute, Houston.

Yin L. I., Ghose S., and Adler I. (1972) Investigation of possible solar wind darkening of the lunar surface by photoelectron spectroscopy. *J. Geophys. Res.* **77**, 1360–1367.

Proceedings of the Fifth Lunar Conference
(Supplement 5, Geochimica et Cosmochimica Acta)
Vol. 2 pp. 1963–1973 (1974)
Printed in the United States of America

Sulfur concentrations and isotope ratios in Apollo 16 and 17 samples

C. E. Rees and H. G. Thode

Department of Chemistry, McMaster University, Hamilton, Ontario, Canada L8S 4K1

Abstract—Sulfur concentrations and isotope ratios have been measured for Apollo 16 and 17 samples. The results are in accord with those from other Apollo sites. Grain size separations on two Apollo 17 soils of different maturities show increasing sulfur concentration and heavy-isotope enrichment with decreasing grain size. These findings suggest that the original fragmentation of a rock into soil leads to preferential enrichment of sulfur in the finer grained particles. Subsequent reworking and comminution of the soil causes sulfur loss by an isotopically competitive mechanism which leads to the heavy-isotope enrichment in the residual sulfur.

Introduction

Previous studies of sulfur concentrations and $\delta^{34}S$* values for lunar samples have revealed significant variations between rocks, breccias and soil (see, for example, Rees and Thode (1972) and references therein; Smith *et al.* (1973)). In general it has been found that $\delta^{34}S$ values for rocks fall close to zero while for soils values have been found between $+5‰$ and $\sim+10‰$. In addition, measurements of $^{33}S/^{32}S$ and $^{36}S/^{32}S$ (Thode and Rees, 1971; Rees and Thode, 1972) show that the sulfur isotope abundance variations are mass-dependent.

We report here measurements of sulfur concentrations and $\delta^{34}S$ values for Apollo 16 and 17 samples. In addition, we report on determinations of these same parameters as a function of grain size for two Apollo 17 samples of different maturities.

Analytical Procedures

Sulfur is released as hydrogen sulfide from fines, and crushed rock samples, by treatment with hydrochloric acid. Successive conversions are made from hydrogen sulfide to cadmium sulfide to silver sulfide to sulfur hexafluoride, the latter compound being the gas used for isotopic analysis. Sulfur concentrations are determined gravimetrically at the silver sulfide step. Details of these procedures and of the mass spectrometry have been given by Thode and Rees (1971). Reported sulfur concentrations have an estimated error of ±3% except, as discussed below, in the case of the grain size separations. The standard deviation for an individual determination of $\delta^{34}S$ is estimated to be ±0.07‰.

Grain size separations of soils 72501 and 75081 were performed in an acetone medium with Buckbee Mears stainless steel sieves of 3-in. diameter. Samples of ~4.5 g were used in order to obtain sufficient sulfur for analysis from each grain size fraction. Losses of material were greater for the fine fractions and total losses amounted to ~2%. Because of the small amounts of sulfur in the grain size fractions, the errors in reported sulfur concentrations may be up to ±10%. The precision of the $\delta^{34}S$

---

*$\delta^{34}S$, ‰ = $\{[(^{34}S/^{32}S)_{sample}/(^{34}S/^{32}S)_{standard}] - 1\} \times 1000$, where the standard reference material is sulfur from the troilite of the Canyon Diablo meteorite.

determinations was not affected by the smallness of the available samples, except in one instance, which is noted in the results section below.

RESULTS

Sulfur concentrations and $\delta^{34}S$ values

Data for Apollo 16 and 17 samples are shown in Table 1. The results show the same pattern as for samples from previous Apollo missions, with $\delta^{34}S$ values for rocks lying close to zero and the values for soils lying in the range +5‰ to +10‰. The Apollo 16 anorthosite 61016 is the first sample for which we have found a negative $\delta^{34}S$ value. Breccias from Apollo 16 and 17 span a $\delta^{34}S$ range from +0.4‰ to +9.4‰, this latter value being for the soil breccia 79135. It should be noted that the data for 79135 and 75081 were inadvertently interchanged in Table 2 of Rees and Thode (1974).

There appear to be some discrepancies between our $\delta^{34}S$ data and those of the U.C.L.A. group (Kerridge *et al.*, 1974; and J. Kerridge, personal communication) that lie outside the range of variations expected within samples. The reasons for these discrepancies are being investigated. In addition, in cases where splits of the same samples have been analyzed, we obtain significantly lower values for sulfur

Table 1. Apollo 16 and 17 sulfur contents and isotopic compositions.

Sample No.	Description	Sulfur content ppm	$\delta^{34}S$, ‰
61016,137	anorthosite	518	−0.1
68815,101	breccia	500	+0.4
68502,20	1–2 mm fines		
I	nonmagnetic fraction	509	+1.4
II	magnetic fraction	742	+1.9
III	fine residue	646	+5.8
68501,47	<1 mm fines	581	+8.6
61501,18	<1 mm fines	589	+9.6
64501,21	<1 mm fines	500	+9.7
60601,16	<1 mm fines	663	+9.9
70215,80	surface chips fine basalt	1581	+0.6
74275,62	fine basalt	1223	+0.6
73235,89	breccia	400	+1.5
79135,36	breccia	828	+9.4
71501,42	<1 mm fines	996	+5.3
75081,5	<1 mm fines	1014	+5.5
70011,22	<1 mm fines	878	+7.4
76501,47	<1 mm fines	620	+8.9
72501,42	<1 mm fines	540	+9.9

concentrations than do Gibson and Moore (1974), Moore *et al.* (1974), and Cripe and Moore (1974). Again, the reasons for this are being investigated.

Magnetic separation

The separation of soil sample 68502 (1–2 mm) into nonmagnetic and magnetic fractions was made by placing the sample on a white teflon sheet and selecting those particles which adhered to an alnico magnet. The third fraction, consisting of a small amount of fine residue left on the teflon sheet after carrying out the above separation procedure, was also analyzed. The separation was obviously a crude one; however, it is clear from Table 1 that the more magnetic particles have higher sulfur contents and $\delta^{34}S$ values. The relatively low $\delta^{34}S$ values for Fractions 1 and 2 suggest that the coarse particles are more rock-like than the fine residue which has a $\delta^{34}S$ value more typical of soil. The ^{34}S enrichment of the original coarse fines sample is $+1.93\%o$ (weighted average).

Figure 1 shows all of our data on <1 mm soil samples. It should be noted that the tendency, reported earlier (Thode and Rees, 1972), for Apollo 15 soils to show increasing $\delta^{34}S$ with increasing sulfur concentration is reversed in the case of Apollo 17 soils.

Apollo 17 grain size separations

The data for the grain size fractions are shown in Tables 2 and 3.

The two samples, 72501 and 75081, fall at extremes of the field represented in Fig. 1. The former soil has low sulfur and high ^{34}S enrichment, while the latter has high sulfur and low ^{34}S enrichment. Both soils show a marked increase of $\delta^{34}S$ value with decreasing grain size. There is also a tendency, less marked in the case of 72501, for the sulfur concentration to pass through a minimum and then to increase with decreasing grain size. Unfortunately, the comparison of the grain

Table 2. Grain size fraction data for Apollo Sample No. 72501,42.

Sample	Description	% of mass in fraction	Sulfur content ppm	$\delta^{34}S$, %o
bulk	<1 mm fines	—	540	+9.9
IA	1000–53 μ	50	580	+5.3
IB	53–45 μ	4	436	+7.8
II	45–30 μ	11	580	+8.9
III	30–20 μ	10	560	+10.4
IV	20–10 μ	16	693	+12.9
V	10–5 μ	8	745	+17.7
VI	<5 μ	~1	~1164	—*

*Sample too small for isotopic analysis.

Table 3. Grain size fraction data for Apollo Sample No. 75081,5.

Sample	Description	% of mass in fraction	Sulfur content ppm	$\delta^{34}S$, ‰
Bulk	<1 mm fines	—	1014	+5.5
2A	500–1000 μ	7	1129	+2.1
2B	200–500 μ	16	970	+2.3
2C	90–200 μ	22	670	+3.6
2D	60–90 μ	11	765	+4.3
2E	45–60 μ	8	846	+4.8
2F	30–45 μ	9	840	+5.0
2G	20–30 μ	7	1029	+5.3
2H	10–20 μ	13	1543	+6.0
2I	5–10 μ	6	1717	+9.1
2J	0–5 μ	~1	~1717	+10.1*

*Very small sample. Uncertainty ±0.2‰.

size distributions is hampered because different sets of sieves were used. However, it can be seen that 75081 has markedly more coarse-grained material than 72501.

DISCUSSION

The values found for $^{33}S/^{32}S$, $^{34}S/^{32}S$, and $^{36}S/^{32}S$ in lunar samples (Thode and Rees, 1971; Rees and Thode, 1972) showed that the sulfur isotope abundance variations were caused by mass-dependent processes. Clayton *et al.* (1974) discuss the isotope abundance variations of oxygen, silicon, sulphur, and potassium in lunar samples. They suggest that the ^{34}S enrichment in lunar soils relative to the rocks implies that ^{32}S enriched material has been lost from the moon. They

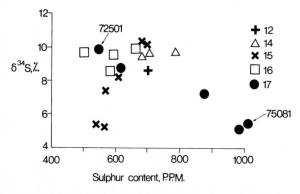

Fig. 1. $\delta^{34}S$ values and sulfur concentrations for soil samples analyzed in this laboratory. The samples for which grain size separations were made are indicated by number.

propose that estimates of the minimum fractional losses of material can be made by assuming a Rayleigh process where the isotope fractionation is given by the inverse square root of the mass ratio for ^{32}S and ^{34}S. On this basis, the enrichment of ^{34}S in soils relative to rocks by from 5‰ to 10‰ implies a depletion of total sulfur from the soils of from 15% to 30%. Previously reported data on sulfur concentrations in rocks and the wide scatter of concentrations and $\delta^{34}S$ values for soils, shown in Fig. 1, make it clear that it is difficult to separate out such depletion from whatever other effects are controlling the sulfur concentrations in particular samples.

The variability of $\delta^{34}S$ between roughly +5‰ and +10‰ for individual soils indicates that different extents of sulfur depletion have occurred. In addition, the results of the grain size separation, shown in Tables 2 and 3, show that different extents of depletion have occurred for the coarse and fine fractions of these soils. The $\delta^{34}S$ values indicate that more depletion of sulfur has occurred in the fine fractions than in the coarse. It is necessary to reconcile this observation with the additional observation that for both samples the sulfur concentration is highest for the finest grain sizes.

The partial fluorination experiments of Epstein and Taylor (see, for example, Taylor and Epstein, 1973) indicate that the heavy-isotope enrichments of oxygen and silicon in soils are surface correlated and that the interiors of individual soil grains are isotopically unaltered from rock-like values. Because of its low concentration it is not feasible to perform such stripping experiments for sulfur. However, the examination of various grain size fractions, with different ratios of surface to bulk material, yield somewhat similar information. We believe that the most basic piece of information yielded by the grain size data is that at least a part of the $\delta^{34}S$ enrichment of soil is a *bulk* property extending throughout individual grains.

Consider a sphere of radius r which has an outer shell of depth a and a spherical core of radius $(r - a)$. The fraction f, of material in the outer shell, is given by

$$f = \frac{r^3 - (r - a)^3}{r^3}.$$

Figure 2 shows the grain size data for 72501. The $\delta^{34}S$ value of each size fraction is plotted against the value of f calculated for the average grain size in each fraction. This has been done for three shell thicknesses: 1 μ, 3 μ, and 5 μ. All three choices of shell thickness yield variations of $\delta^{34}S$ with f which are monotonic. In particular, a choice of 3 μ for the shell thickness yields a linear variation which can be interpreted as a mixing line for two $\delta^{34}S$ components. One component, the core component, corresponds to $f = 0$ and is at ~5.2‰; the other, the shell component, corresponds to $f = 1$ and is at ~+21.2‰.

The deficiencies of this simple model are many and obvious. For example; the mean grain size of a fraction is not accurately represented by the mean value of its grain size extremes. The true mean grain size depends upon the distribution of sizes within the individual fractions. This means that the simple function f does

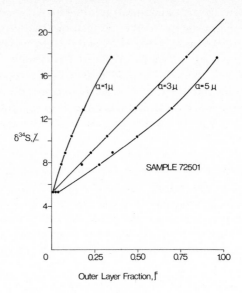

Fig. 2. Plots of δ^{34}S versus the outer layer fraction f (see text) for three choices of shell thickness.

not accurately represent the outer layer fraction. In addition, the idea of a sharply defined change from one isotopic composition to another at a particular depth is probably an oversimplification. The alteration of isotope ratios by isotopically competitive processes depends, in a nonlinear manner, on the degree of depletion of the original material. A continuous variation of sulfur depletion with depth would give rise to continuous rather than sharp variation of δ^{34}S.

Regardless of these and other objections, we believe that variation of δ^{34}S with f for 72501 represents, at least qualitatively, the combination of sulfur components of different isotopic compositions producing the observed δ^{34}S values for each grain size fraction. The values +5.2‰ and +21.2‰ may not be exactly correct, but will not be grossly in error, while the suggested shell depth of $\sim$3 μ requires further investigation. It should be pointed out that the depth to which O and Si isotope effects are present is much smaller. Typically these effects persist to $\sim$1% yield of total oxygen. Rearranging the previous equation for f gives

$$a = r[1 - (1 - f)^{1/3}].$$

Assigning a mean grain size of $\sim$100 μ and a value of 0.01 for f gives

$$a \, \Omega \, 0.3 \, \mu.$$

In an earlier publication (Rees and Thode, 1972), we reported the results of a chemical leaching-experiment on an Apollo 14 soil sample. In view of the present results on grain size fractions, we now believe our interpretation of the leaching-experiment to have been incorrect. In the leaching-experiment a sample of 14163 was treated with HCl and the H_2S generated was collected in three successive fractions. The first fraction was 2.4‰ light, the second 0.1‰ heavy and the third 2.0‰ heavy with respect to the bulk. We interpreted this as showing that the HCl had attacked the outer layers of the grains first, so that the sulfur in the outer layers was isotopically light relative to the bulk sulfur. It is now necessary to amend that interpretation and to say that the leaching-experiment indicates that the sulfur *most readily accessible to attack by HCl* is isotopically light with respect to the bulk. This sulfur is not necessarily that closest to the surface of grains but is more likely that in the least altered or melted crystalline fragments in the soil. This alternative interpretation, which is more in accord with the grain size data, renders invalid the mechanism we suggested on p. 1484 of our previous report.

If the shell model approach is used for the data on 75081, there is no value of shell thickness for which a linear relation holds between $\delta^{34}S$ and f. Figure 3 shows plots of $\delta^{34}S$ versus f for both 72501 and 75081 for a shell thickness of 3 μ. In order to explain the difference between 72501 and 75081, it is necessary to consider information regarding them from other sources.

McKay *et al.* (1974) have classified 72501 as a mature soil and 75081 as an submature one. This classification was made on the basis of the grain size distribution of these and other soils. According to McKay *et al.*, the process of soil maturation is one in which immature soil is formed by a rock fragmentation event which is then followed by successive comminution events and by possible mixing with other soils of differing degrees of maturity. The initial fragmentation leads to a particle size distribution which is skewed toward coarse grains. Subsequent comminution alters this distribution so that the proportion of coarse grains drops

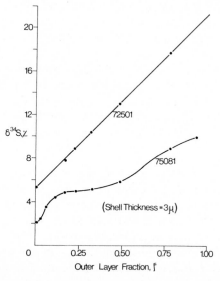

Fig. 3. Plots of $\delta^{34}S$‰ versus outer layer fraction f for samples 72501 and 75081. The chosen shell thickness, 3 μ, is that which gives a linear variation in the case of 72501.

while the proportion of fine grains rises. This process may be complicated by the admixture of soil components of differing maturity.

Evensen *et al.* (1973) have performed grain size separations on 75081 and other soils and have determined various trace elements in the size fractions. The trends that they note for K, Rb, Sr, Ba, and ^{87}Rb/^{86}Sr parallel the trends we have observed for sulfur. The variation of each of these quantities with decreasing grain size generally shows an initial fall followed by an increase. One exception is that in Apollo 14 soils, Murthy *et al.* (1972) found no consistent trends in either elemental abundances or ^{87}Rb/^{86}Sr as a function of grain size. Our crude sieve experiment with 14163, however, showed slightly increasing sulfur concentration and δ^{34}S value with decreasing grain size. Evensen *et al.* (1973, 1974) identify Apollo 14 soils as representing the "exotic" component which is found to varying degrees in other soils. Their "exotic" component is "apparently similar in chemical and isotopic character to KREEP." The trace element data on soils other than Apollo 14 are explained by assuming that, concurrent with mixing of local and Apollo 14 type material, there are differential comminution effects resulting in trace element enrichments in the finer fractions of the locally derived component.

The above considerations, together with our size fraction data, make it possible to put forward an interpretation of the fragmentation-comminution-mixing process from the point of view of sulfur concentrations and isotopic abundances.

(1) Initial fragmentation leads to pulverized, but essentially unaltered, rock with a grain size distribution skewed toward coarse grains. It is this initial fragmentation that preferentially concentrates sulfur-rich material in the fine grains, possibly because the occurrence of sulfide in small grains in the rock favors its subsequent appearance in the finer fractions of the fresh soil. Possibly at this stage there is a small loss of sulfur from the bulk material leading to a small overall enrichment of δ^{34}S.

(2) Reworking and comminution of the dust now leads to the shifting of its grain size distribution toward small grain sizes. It also leads to or involves processes causing sulfur loss and isotopic alteration. These processes are most effective on the smaller grains because of their large relative surface areas.

(3) The sequence (1) and (2) may be complicated by the mixing of soils of different maturity.

We can apply this scheme to a qualitative explanation of the differences found between 72501 and 75081. With reference to Fig. 3, we hypothesize that a δ^{34} versus f plot for a totally immature soil would be a horizontal straight line passing through a value of δ^{34}S which is close to zero. The size distribution of this immature soil would favor coarse grains. We further take 72501 to be typical of a mature soil that has suffered fragmentation, comminution, sulfur loss and δ^{34}S enrichment as described above, without significant dilution of these effects by admixture with any less mature component. We now suggest that the δ^{34}S versus f plot for the submature soil 75081 represents a combination of the mature and the immature plots.

If the mature and immature components of the mixture had identical grain size distributions and sulfur content variations, differing only in $\delta^{34}S$ variation with particle size, then the $\delta^{34}S$ versus f plot for the mixture would be a straight line with intermediate slope and intercepts. However, the enhancement of coarse grains in the immature component pulls the $\delta^{34}S$ value down for values of f close to zero. For smaller grains (increasing f) the relative abundance of the mature component increases and pushes $\delta^{34}S$ upward toward the mature line. At the very finest grain sizes (f approaching unity), the high $\delta^{34}S$ and enhanced fine grain abundances of the mature component may or may not dominate the high sulfur concentration but low $\delta^{34}S$ and low fine grain abundance of the immature component.

This argument is conjectural and needs more evidence before it can be taken very seriously. It does, however, offer a qualitative explanation of how the fine grain size fractions of soils can have high $\delta^{34}S$ values, implying loss of sulfur, while at the same time having higher sulfur concentrations than the coarser grained fractions. Regardless of the precise mechanisms giving rise to these patterns, it seems clear that the $\delta^{34}S$ values for soils are closely linked to maturity and to the mixing of different soil components.

It is not clear at this time to what extent our discussion of the sulfur isotopic evolution of soils is in accord with the findings of Evensen *et al.* (1973, 1974) and Murthy *et al.* (1972). As mentioned above, they found no systematic variation of trace element concentrations as a function of grain size for Apollo 14 soils. Our single crude sieving experiment on 14163 indicates an increase of sulfur concentration and $\delta^{34}S$ value with decreasing grain size. The bulk $\delta^{34}S$ for this sample, $\sim +9.5‰$, suggests that it is similar in its sulfur systematics to the mature soil, 72501, that we have examined in detail. The sample, 15531, examined by Evensen *et al.* (1973) has trace element concentrations that increase markedly with decreasing grain size. This is attributed to either mixing of a local soil with a significant proportion of exotic material or to the effects of comminution on a single component soil. They present arguments based on Rb–Sr systematics to justify the multicomponent scheme. Our analysis of a bulk sample of 15531 gave a low sulfur concentration and a low $\delta^{34}S$ value ($+5.4‰$). According to our interpretation, in the absence of any grain size fraction data for sulfur, this soil should be similar to the submature sample 75081 and could indeed be a mixture of a (local?) immature component with a mature one.

Our consideration of fragmentation-comminution-mixing for soils has to include processes involving the loss of sulfur by an isotopically competitive mechanism, in order to explain the high $\delta^{34}S$ values of soils relative to rocks. It should be noted that measurements by Barnes *et al.* (1973, 1974) on K isotope ratios suggest that for at least one of the trace elements measured by Evensen *et al.* (1973) loss may be a significant factor in determining abundance in soils.

The loss of sulfur must also be borne in mind when considering attempts to 'manufacture' soils from rock mixtures, e.g. Gibson and Moore (1974). Before it can be accepted that a soil's sulfur concentration results from the mixing of sulfur-rich components such as basalt with sulfur-poor materials like anorthosite, it must be shown that enough sulfur loss can have occurred from the mixture to allow the observed $\delta^{34}S$ enrichment from the near zero value of each starting component.

Acknowledgments—This work was supported by the National Research Council of Canada. We thank C. McEwing for his assistance in the mass spectrometry.

Note added in proof In the results section above we refer to the discrepancies in sulfur concentrations determined by ourselves and the U.C.L.A. group, using acid hydrolysis and gravimetric methods, and by the Arizona and Texas groups, using fusion and SO_2 titration. The acid hydrolysis methods give consistently lower values than the fusion methods. We became concerned that possibly not all the sulfur in a sample is accessible to acid hydrolysis, either because some of it is dissolved in glass or is present as troilite but totally surrounded by a hard-to-attack mineral such as ilmenite. The Apollo 17 rock sample 70215 has been reported as having sulfur concentration values of: 1581 ppm (this paper); 1689 ppm (J. F. Kerridge, personal communication); 2040 ppm (Moore *et al.*, 1974) and 2210 ppm (Gibson and Moore, 1974). In an effort to resolve these discrepancies we have performed two further determinations on this rock (Monster *et al.*, in preparation). Using 85% phosphoric acid we obtained a value of 1600 ppm. Using fusion at 900°C with sodium carbonate + potassium chlorate we obtained a value of 1630 ppm. We are reassured by the closeness of the values we obtain using three different methods.

REFERENCES

Barnes I. L., Garner E. L., Gramlich J. W., Machlan L. A., Moody J. R., Moore L. J., Murphy T. J., and Shields W. R. (1973) Isotopic abundance ratios and concentrations of selected elements in some Apollo 15 and Apollo 16 samples (abstract). *Proc. Fourth Lunar Sci. Conf., Geochim. Cosmochim. Acta*, Suppl. 4, Vol. 2, pp. 1197–1207. Pergamon.

Barnes I. L., Garner E. L., Gramlich J. W., Machlan L. A., Moody J. R., Moore L. J., Murphy T. J., and Shields W. R. (1974) Isotopic abundance ratios and concentrations of selected elements in Apollo 17 samples. In *Lunar Science—V*, pp. 38–40. The Lunar Science Institute, Houston.

Clayton R. N., Mayeda T. K., and Hurd J. M. (1974) Loss of O, Si, S and K from the lunar regolith (abstract). In *Lunar Science—V*, pp. 129–131. The Lunar Science Institute, Houston.

Cripe J. D. and Moore C. B. (1974) Total sulfur contents of Apollo 15 and Apollo 16 lunar samples (abstract). In *Lunar Science—V*, pp. 523–525. The Lunar Science Institute, Houston.

Evensen N. M., Murthy V. R., and Coscio M. R., Jr. (1973) Rb–Sr ages of some mare basalts and the isotopic and trace element systematics in lunar fines. *Proc. Fourth Lunar Sci. Conf., Geochim. Cosmochim. Acta*, Suppl. 4, Vol. 2, pp. 1707–1724. Pergamon.

Evensen N. M., Murthy V. R., and Coscio M. R., Jr. (1974) Episodic Lunacy—V: Origin of the exotic component (abstract). In *Lunar Science—V*, pp. 220–221. The Lunar Science Institute, Houston.

Gibson E. K. and Moore G. W. (1974) Total sulfur abundances and distributions in the valley of Taurus-Littrow: Evidence of mixing (abstract). In *Lunar Science—V*, pp. 267–269. The Lunar Science Institute, Houston.

Kaplan I. R., Smith J. W., and Ruth E. (1970) Carbon and sulfur concentration and isotopic composition of Apollo 11 lunar samples. *Proc. Apollo 11 Lunar Sci. Conf., Geochim. Cosmochim. Acta*, Suppl. 1, Vol. 2, pp. 1317–1330. Pergamon.

Kerridge J. F., Petrowski C., and Kaplan I. R. (1974) Sulfur, methane and metallic iron in some Apollo 17 fines (abstract). In *Lunar Science—V*, pp. 411–412. The Lunar Science Institute, Houston.

McKay D. S., Fruland R. M., and Heiken G. H. (1974) Grain size distribution as an indicator of the maturity of lunar soils (abstract). In *Lunar Science—V*, pp. 480–482. The Lunar Science Institute, Houston.

Moore C. B., Lewis C. F., Cripe J. D., and Volk M. (1974) Total carbon and sulfur contents of Apollo 17 lunar samples (abstract). In *Lunar Science—V*, pp. 520–522. The Lunar Science Institute, Houston.

Murthy V. R., Evensen N. M., Jahn B., and Coscio M. R., Jr. (1972) Apollo 14 and 15 samples: Rb–Sr ages, trace elements, and lunar evolution. *Proc. Third Lunar Sci. Conf., Geochim. Cosmochim. Acta*, Suppl. 3, Vol. 2, pp. 1503–1514. MIT Press.

Rees C. E. and Thode H. G. (1972) Sulphur concentrations and isotope ratios in lunar samples. *Proc. Third Lunar Sci. Conf., Geochim. Cosmochim. Acta*, Suppl. 3, Vol. 2, pp. 1479–1485. MIT Press.

Rees C. E. and Thode H. G. (1974) Sulphur concentrations and isotope ratios in Apollo 16 and 17 samples. In *Lunar Science—V*, pp. 621–623. The Lunar Science Institute, Houston.

Smith J. W., Kaplan I. R., and Petrowski C. (1973) Carbon, nitrogen, sulfur, helium, hydrogen, and metallic iron in Apollo 15 drill stem fines. *Proc. Fourth Lunar Sci. Conf., Geochim. Cosmochim. Acta*, Suppl. 4, Vol. 2, pp. 1651–1656. Pergamon.

Taylor H. P., Jr. and Epstein S. (1973) O^{18}/O^{16} and Si^{30}/Si^{28} studies of some Apollo 15, 16, and 17 samples. *Proc. Fourth Lunar Sci. Conf., Geochim. Cosmochim. Acta*, Suppl. 4, Vol. 2, pp. 1657–1679. Pergamon.

Thode H. G. and Rees C. E. (1971) Measurement of sulphur concentrations and the isotope ratios $^{33}S/^{32}S$, $^{34}S/^{32}S$, and $^{36}S/^{32}S$ in Apollo 12 samples. *Earth Planet. Sci. Lett.* **12**, 434–438.

Thode H. G. and Rees C. E. (1972) Sulphur concentrations and isotope ratios in Apollo 14 and 15 samples. In *The Apollo 15 Lunar Samples*, pp. 402–403. The Lunar Science Institute, Houston.

Proceedings of the Fifth Lunar Conference
(Supplement 5, Geochimica et Cosmochimica Acta)
Vol. 2 pp. 1975–2003 (1974)
Printed in the United States of America

Noble gases in Apollo 17 fines: Mass fractionation effects in trapped Xe and Kr

D. D. BOGARD, W. C. HIRSCH,* and L. E. NYQUIST

NASA Johnson Space Center, Houston, Texas 77058

Abstract—Isotopic abundances of He, Ne, Ar, Kr, and Xe have been measured in eight Apollo 17 bulk soils, in grain size separates of three soils, and in a stepwise temperature release of 76321. Cosmic ray exposure ages for Apollo 17, Station 6 soils are $\sim$230–330 $\times 10^6$ yr, and the K–^{40}Ar gas retention age of $>$30 μ size fractions of 76321 is 4.0 $\times 10^9$ yr. Significant variations were observed among bulk soils and grain size separates for ratios of trapped ^{4}He/^{22}Ne, ^{22}Ne/^{36}Ar, and isotopic ratios of Ne and Ar. The isotopic composition of trapped Xe and Kr determined for soils 15271, 76501, and 76321 by ordinate–intercept plots are similar to previous determinations on other lunar soils. Ratios of trapped ^{129}Xe/^{136}Xe, ^{132}Xe/^{136}Xe, and ^{134}Xe/^{136}Xe in a large number of Apollo 15 and 16 deep drill core and surface fines show total variations of approximately 1% per mass. Similar isotopic variations have been observed previously in the composition of various trapped Xe and Kr components. A stepwise temperature release of Xe and Kr in $<$20 μ fraction of 76321 showed a progressive change in isotopic ratios as a function of gas release which closely resembles the isotopic variations in different trapped components. The relative changes in isotopic ratios closely follow mass fractionation trends. It is argued that processes acting on the lunar surface which produce mass fractionation effects are the major cause of the observed isotopic variations in trapped Xe and Kr, and that the "true" solar wind composition has not yet been identified.

INTRODUCTION

NOBLE GAS MEASUREMENTS on lunar samples yield diverse information ranging from gas retention ages and cosmic ray exposure ages to the nature of the implanted solar wind species and regolith surface processes. We have performed measurements of the abundances and isotopic compositions of He, Ne, Ar, Kr, and Xe in Apollo 17 soil and rock samples with an emphasis on Station 6 samples. Preliminary results on Station 6 boulders have already been presented in abstract form (Bogard *et al.*, 1974). In this paper we present noble gas measurements made on eight bulk soils (four from Station 6), grain size separates of soil 76321 and 76501, and a stepwise temperature release of 76321. Measurements were also made on grain size separates of 15271. We shall discuss these data largely in terms of the effects of mass fractionation processes, and we shall be particularly concerned with mass fractionation effects on the isotopic compositions of "trapped" Xe and Kr.

He, Ne AND Ar RESULTS

The analyses were performed mass spectrometrically by standard procedures (Bogard *et al.*, 1973). Data are presented in Tables 1–3.

---

*Also Northrop Services, Inc. Houston, Texas 77058.

D. D. BOGARD *et al.*

Table 1. Abundances and isotopic ratios of He, Ne, and Ar in lunar soils. Abundances are in units of cm³ STP/g and possess estimated uncertainties of ±5–10%. Uncertainties given for isotopic ratios are derived from one sigma of multiple measurements plus, for a few measurements, uncertainties in small applied blank corrections. Relative values of ^{4}He/^{3}He possess uncertainties of approximately ±3%, and absolute ratios, ±5%. Small blank corrections have been applied, which in most cases were considerably less than 1%.

Sample	Weight (mg)	^{4}He (10^{-2})	^{4}He/^{3}He	^{22}Ne (10^{-6})	^{20}Ne/^{22}Ne	^{22}Ne/^{21}Ne	^{36}Ar (10^{-6})	^{36}Ar/^{38}Ar	^{40}Ar/^{36}Ar
Bulk soils									
72501,14	5.87	7.78	2700	119	12.84 ± 0.03	29.65 ± 0.35	292	5.40 ± 0.01	1.109 ± 0.011
72701,14	4.57	7.30	2720	111	12.88 ± 0.03	29.45 ± 0.11	323	5.40 ± 0.01	1.770 ± 0.234
75081,32	5.59	16.2	2820	139	12.92 ± 0.03	28.79 ± 0.07	268	5.38 ± 0.01	0.793 ± 0.009
76501,12	6.21	10.0	2715	127	12.84 ± 0.03	28.92 ± 0.09	371	5.39 ± 0.01	0.897 ± 0.010
76261,23	7.05	8.99	2730	103	12.91 ± 0.03	28.64 ± 0.09	307	5.39 ± 0.01	0.982 ± 0.007
76321,2	5.77	12.5	2710	146	12.85 ± 0.03	29.30 ± 0.09	351	5.38 ± 0.01	0.952 ± 0.025
76281,2	6.14	9.87	2760	107	12.87 ± 0.04	28.66 ± 0.06	286	5.39 ± 0.01	1.133 ± 0.009
70051,17	6.24	15.1	2870	182	12.81 ± 0.03	28.62 ± 0.16	352	5.36 ± 0.01	1.429 ± 0.005
76321,2 grain sizes									
1000–250 μ	24.92	1.27	2380	20.8	12.61 ± 0.03	24.60 ± 0.05	64.2	5.34 ± 0.01	1.809 ± 0.005
250–125 μ	13.40	2.11	2395	36.6	12.62 ± 0.03	25.48 ± 0.10	113	5.32 ± 0.01	1.359 ± 0.005
125–63 μ	13.30	3.43	2470	52.9	12.61 ± 0.07	26.63 ± 0.18	152	5.35 ± 0.01	1.365 ± 0.005
125–63 μ	7.27	3.06	2440	54.7	12.64 ± 0.03	26.62 ± 0.05	164	5.33 ± 0.01	1.209 ± 0.005
63–44 μ	10.44	4.10	2570	66.8	12.65 ± 0.03	27.38 ± 0.05	177	5.35 ± 0.01	1.030 ± 0.005
63–44 μ	10.67	4.41	2450	70.0	12.73 ± 0.06	27.06 ± 0.20	181	5.33 ± 0.01	1.062 ± 0.007
44–30 μ	8.24	5.58	2450	90.3	12.69 ± 0.03	28.01 ± 0.06	226	5.35 ± 0.01	0.950 ± 0.005
30–20 μ	4.57	7.62	2490	125	12.63 ± 0.03	29.16 ± 0.09	229	5.36 ± 0.01	1.049 ± 0.005
<20 μ	4.04	27.1	2770	331	12.86 ± 0.03	30.10 ± 0.09	827	5.40 ± 0.01	0.801 ± 0.005
76501,12 grain sizes									
1000–250 μ	19.27	1.81	2495	25.4	12.79 ± 0.03	23.53 ± 0.05	82.7	5.36 ± 0.01	1.388 ± 0.005
250–125 μ	21.03	2.23	2560	32.9	12.67 ± 0.03	24.07 ± 0.06	116	5.32 ± 0.01	1.276 ± 0.005
125–63 μ	10.30	3.28	2390	49.5	12.67 ± 0.03	25.18 ± 0.06	126	5.32 ± 0.01	1.149 ± 0.005
63–44 μ	5.43	4.33	2925	61.3	12.67 ± 0.04	26.38 ± 0.10	178	5.34 ± 0.01	1.020 ± 0.005
44–30 μ	4.17	5.87	2795	87.4	12.74 ± 0.03	27.44 ± 0.07	250	5.36 ± 0.01	.955 ± 0.005
30–20 μ	5.19	7.98	2680	110	12.78 ± 0.03	28.23 ± 0.07	297	5.38 ± 0.01	0.876 ± 0.005
<20 μ	5.46	19.7	3000	216	12.96 ± 0.04	29.64 ± 0.18	722	5.40 ± 0.01	0.788 ± 0.005
15271,64 agglutinate and nonagglutinate grain sizes									
250–125 μ									
Agglutinate	15.78	2.92	2480	59.0	12.67 ± 0.03	25.72 ± 0.18	223	5.36 ± 0.01	1.184 ± 0.020
Nonagglutinate	22.45	1.16	2065	20.8	12.33 ± 0.14	17.29 ± 0.15	58.9	5.19 ± 0.01	2.012 ± 0.013
125–63 μ									
Agglutinate	10.68	3.77	2605	74.8	12.63 ± 0.03	26.70 ± 0.07	290	5.34 ± 0.01	1.103 ± 0.010
Nonagglutinate	15.48	1.71	2190	34.8	12.54 ± 0.03	20.33 ± 0.05	82.4	5.27 ± 0.01	1.569 ± 0.015
63–44 μ	4.97	4.15	2330	79.4	12.71 ± 0.03	26.07 ± 0.09	200	5.35 ± 0.01	1.138 ± 0.019
63–44 μ	6.64	3.95	2305	80.1	12.64 ± 0.03	26.25 ± 0.39	207	5.36 ± 0.01	1.150 ± 0.005
<44 μ	4.14	12.2	2970	199	12.82 ± 0.03	29.35 ± 0.05	426	5.40 ± 0.01	1.068 ± 0.017
44–30 μ	5.59	3.44	2395	63.9	12.60 ± 0.03	24.50 ± 0.22	159	5.35 ± 0.01	1.186 ± 0.018
44–30 μ	8.89	3.29	2285	69.5	12.56 ± 0.03	24.90 ± 0.24	149	5.31 ± 0.01	1.187 ± 0.006
30–20 μ	7.39	4.36	2415	88.6	12.65 ± 0.03	26.25 ± 0.06	218	5.34 ± 0.01	1.117 ± 0.009
30–20 μ	4.14	4.25	2355	91.3	12.60 ± 0.03	26.54 ± 0.06	207	5.34 ± 0.01	1.129 ± 0.013
<20 μ	4.88	11.6	2730	193	12.82 ± 0.03	28.91 ± 0.13	537	5.40 ± 0.01	1.000 ± 0.007

The He and Ne contents of Apollo 17 soils (Table 1) vary by a factor of two, whereas the Ar, Kr, and Xe contents are much less variable. The ^{4}He contents of the eight bulk fines (Table 1) show a strong positive correlation with Ti concentrations (Rhodes *et al.*, 1974), which is probably due to the higher retentivity of ilmenite for He and Ne (Eberhardt *et al.*, 1970). However, other

Table 2. Abundance (in 10^{-9} cm^3 STP/g) and isotopic composition (relative to ^{86}Kr = 1.00) of Kr in lunar soils. Abundances possess estimated uncertainties of ±10%. Small hydrocarbon corrections have been applied to ^{78}Kr and ^{80}Kr based on mass 79 and ratios of background peaks. Even smaller $(^{40}$Ar$)_2^+$ corrections have been applied to ^{80}Kr based on ^{40}Ar$^+$ signals. Uncertainties given for isotopic ratios are one sigma of the average of multiple measurements plus one-half the magnitude of the above corrections in the case of ^{78}Kr and ^{80}Kr. Absolute ratios have an additional uncertainty of ±0.1%/mass in the discrimination correction applied.

Sample	^{84}Kr (10^{-9})	^{78}Kr/^{86}Kr	^{80}Kr/^{86}Kr	^{82}Kr/^{86}Kr	^{83}Kr/^{86}Kr	^{84}Kr/^{86}Kr
Bulk soils						
72501,14	184	0.0232	0.1401	0.6901	0.6823	3.338
		±0.005	±0.0010	±0.0015	±0.0027	±0.010
72701,14	192	0.0237	0.1317	0.6632	0.6622	3.268
		±0.0007	±0.0020	±0.0030	±0.0033	±0.007
75081,32	148	0.0227	0.1337	0.6676	0.6664	3.268
		±0.0008	±0.0014	±0.0032	±0.0040	±0.010
76501,12	196	0.0219	0.1317	0.6632	0.6619	3.263
		±0.0007	±0.0015	±0.0014	±0.0038	±0.006
76261,23	164	0.0203	0.1315	0.6627	0.6635	3.264
		±0.0014	±0.0016	±0.0013	±0.0030	±0.008
76321,2	211	0.0224	0.1310	0.6623	0.6615	3.263
		±0.0005	±0.0014	±0.0060	±0.0028	±0.014
76281,2	148	0.0239	0.1333	0.6660	0.6660	3.269
		±0.0003	±0.0020	±0.0033	±0.0027	±0.011
70051,17	185	0.0229	0.1344	0.6669	0.6689	3.270
		±0.0010	±0.0008	±0.0029	±0.0028	±0.013
76321,2 grain sizes						
1000–250 μ	41.6	0.0261	0.1373	0.6706	0.6724	3.268
		±0.0013	±0.0015	±0.0029	±0.0044	±0.015
250–125 μ	68.4	0.0286	0.1344	0.6660	0.6684	3.263
		±0.0020	±0.0019	±0.0038	±0.0024	±0.014
125–63 μ	82.1	0.0280	0.1325	0.6598	0.6606	3.253
		±0.0016	±0.0021	±0.0036	±0.0036	±0.012
125–63 μ	99.2	0.0326	0.1332	0.6639	0.6668	3.265
		±0.0031	±0.0026	±0.0049	±0.0045	±0.012
63–44 μ	106	0.0267	0.1321	0.6603	0.6626	3.255
		±0.0009	±0.0010	±0.0040	±0.0048	±0.012
63–44 μ	110	0.0278	0.1334	0.6646	0.6640	3.265
		±0.0043	±0.0015	±0.0054	±0.0043	±0.021
44–30 μ	142	0.0241	0.1321	0.6601	0.6621	3.252
		±0.0018	±0.0011	±0.0037	±0.0020	±0.006
30–20 μ	187	0.0320	0.1315	0.6630	0.6631	3.265
		±0.0036	±0.0016	±0.0061	±0.0026	±0.014
<20 μ	461	0.0218	0.1290	0.6559	0.6568	3.261
		±0.0012	±0.0015	±0.0032	±0.0029	±0.012
76321,2, <20 μ stepwise temperature release						
170	0.0953	0.30	0.163	0.672	0.666	2.98
		±0.11	±0.40	±0.102	±0.111	±0.29
320	0.186	0.23	0.131	0.694	0.671	3.35
		±0.16	±0.037	±0.044	±0.048	±0.17

Table 2. (*Continued*).

Sample	^{84}Kr (10^{-9})	$^{78}Kr/^{86}Kr$	$^{80}Kr/^{86}Kr$	$^{82}Kr/^{86}Kr$	$^{83}Kr/^{86}Kr$	$^{84}Kr/^{86}Kr$
500	1.36	0.0416	0.1310	0.675	0.681	3.302
		±0.0122	±0.0100	±0.019	±0.038	±0.060
700	24.2	0.02233	0.1380	0.6884	0.6781	3.337
		±0.00029	±0.0014	±0.0024	±0.0026	±0.008
800	126	0.02064	0.1319	0.6676	0.6639	3.293
		±0.00006	±0.0004	±0.0007	±0.0008	±0.002
900	145	0.02003	0.1305	0.6582	0.6576	3.262
		±0.00004	±0.0008	±0.0012	±0.0008	±0.003
1000	39.3	0.01997	0.1286	0.6510	0.6543	3.242
		±0.00010	±0.0005	±0.0010	±0.0011	±0.003
1100	33.7	0.02031	0.1279	0.6500	0.6530	3.237
		±0.00035	±0.0004	±0.0014	±0.0013	±0.004
1200	11.5	0.02096	0.1283	0.6498	0.6545	3.237
		±0.00054	±0.0010	±0.0011	±0.0022	±0.005
1400	7.38	0.02114	0.1293	0.6483	0.6554	3.215
		±0.00071	±0.0022	±0.0039	±0.0028	±0.019
1620	.156	0.189	0.139	0.643	0.645	3.21
		±0.027	±0.027	±0.064	±0.047	±0.18
76501,12 grain sizes						
1000–250 μ	39.0	0.0266	0.1351	0.6816	0.6825	3.290
		±0.0012	±0.0075	±0.0044	±0.0042	±0.012
250–125 μ	67.4	0.0244	0.1370	0.6696	0.6719	3.267
		±0.0005	±0.0015	±0.0026	±0.0027	±0.009
125–63 μ	77.8	0.0251	0.1353	0.6633	0.6701	3.263
		±0.0002	±0.0012	±0.0031	±0.0042	±0.009
63–44 μ	99.4	0.0235	0.1311	0.6571	0.6645	3.187
		±0.0009	±0.0039	±0.0073	±0.0115	±0.063
44–30 μ	137	0.0248	0.1333	0.6636	0.6660	3.261
		±0.0015	±0.0021	±0.0041	±0.0038	±0.013
30–20 μ	156	0.0213	0.1371	0.6625	0.6641	3.261
		±0.0018	±0.0030	±0.0037	±0.0033	±0.010
<20 μ	379	0.0215	0.1312	0.6594	0.6593	3.263
		±0.0003	±0.0008	±0.0021	±0.0022	±0.010
15271,64 agglutinate and nonagglutinate grain sizes						
250–125 μ						
Agglutinate	152	0.0245	0.1361	0.6692	0.6705	3.270
		±0.0004	±0.0011	±0.0026	±0.0027	±0.009
Nonagglutinate	41.6	0.0334	0.1569	0.6694	0.7128	3.276
		±0.0005	±0.0035	±0.0071	±0.0064	±0.024
125–63 μ						
Agglutinate	174	0.0236	0.1349	0.6670	0.6688	3.263
		±0.0005	±0.0014	±0.0024	±0.0023	±0.008
Nonagglutinate	51.7	0.0319	0.1510	0.6908	0.6987	3.280
		±0.0011	±0.0017	±0.0017	±0.0025	±0.007
63–44 μ	133	0.0261	0.1388	0.6700	0.6747	3.273
		±0.0015	±0.0022	±0.0035	±0.0033	±0.018
63–44 μ	126	0.0285	0.1380	0.6732	0.6747	3.271

Table 2. (*Continued*).

Sample	^{84}Kr (10^{-9})	^{78}Kr/^{86}Kr	^{80}Kr/^{86}Kr	^{82}Kr/^{86}Kr	^{83}Kr/^{86}Kr	^{84}Kr/^{86}Kr
<44 μ	309	±0.0041 0.0216	±0.0025 0.1346	±0.0049 0.6660	±0.0056 0.6649	±0.015 3.277
44–30 μ	93.8	±0.0006 0.0232	±0.0012 0.1402	±0.0017 0.6764	±0.0046 0.6803	±0.005 3.271
44–30 μ	94.1	±0.0032 0.0289	±0.0038 0.1414	±0.0029 0.6735	±0.0040 0.6803	±0.014 3.271
30–20 μ	127	±0.0009 0.0258	±0.0014 0.1380	±0.0029 0.6680	±0.0044 0.6757	±0.012 3.260
30–20 μ	133	±0.0011 0.0295	±0.0020 0.1366	±0.0030 0.6677	±0.0028 0.6719	±0.017 3.264
<20 μ	289	±0.0008 0.0232 ±0.0007	±0.0019 0.1329 ±0.0024	±0.0044 0.6649 ±0.0021	±0.0039 0.6677 ±0.0028	±0.012 3.266 ±0.008

Table 3. Abundances (in 10^{-9} cm^3 STP/g) and isotopic composition (relative to ^{136}Xe = 1.00) of Xe in lunar soils. Abundances possess estimated uncertainties of ±5–10%. Uncertainties given for isotopic ratios are one sigma of multiple measurements. Absolute ratios have an additional uncertainty of ±0.1%/mass in the discrimination correction applied.

Sample	^{132}Xe (10^{-9})	^{124}Xe/ ^{136}Xe	^{126}Xe/ ^{136}Xe	^{128}Xe/ ^{136}Xe	^{129}Xe/ ^{136}Xe	^{130}Xe/ ^{136}Xe	^{131}Xe/ ^{136}Xe	^{132}Xe/ ^{136}Xe	^{134}Xe/ ^{136}Xe
Bulk soils									
72501,14	26.1	0.0203 ±0.0016	0.0222 ±0.0010	0.2949 ±0.0054	3.575 ±0.032	0.5610 ±0.0078	2.798 ±0.024	3.301 ±0.023	1.239 ±0.012
72701,14	26.6	0.0183 ±0.0017	0.0222 ±0.0021	0.2849 ±0.0056	3.445 ±0.031	0.5487 ±0.0078	2.743 ±0.024	3.318 ±0.027	1.233 ±0.016
75081,32	22.7	0.0199 ±0.0012	0.0211 ±0.0012	0.2859 ±0.0033	3.477 ±0.033	0.5525 ±0.0060	2.760 ±0.016	3.331 ±0.014	1.232 ±0.012
76501,12	27.5	0.0185 ±0.0012	0.0208 ±0.0021	0.2831 ±0.0048	3.464 ±0.030	0.5491 ±0.0062	2.749 ±0.023	3.318 ±0.026	1.231 ±0.014
76261,23	22.7	0.0196 ±0.0017	0.0214 ±0.0016	0.2846 ±0.0033	3.463 ±0.028	0.5486 ±0.0084	2.741 ±0.024	3.321 ±0.024	1.224 ±0.012
76321,2	30.7	0.0179 ±0.0007	0.0169 ±0.0009	0.2828 ±0.0041	3.454 ±0.027	0.5474 ±0.0074	2.735 ±0.022	3.319 ±0.022	1.234 ±0.012
76281,2	22.2	0.0191 ±0.0017	0.0218 ±0.0005	0.2868 ±0.0037	3.458 ±0.038	0.5494 ±0.0075	2.741 ±0.032	3.310 ±0.034	1.231 ±0.015
70051,17	31.1	0.0175 ±0.0008	0.0189 ±0.0005	0.2808 ±0.0047	3.496 ±0.019	0.5439 ±0.0051	2.702 ±0.014	3.304 ±0.015	1.226 ±0.010
76321,2 grain sizes									
1000–250 μ	5.40	0.0267 ±0.0022	0.0363 ±0.0020	0.3080 ±0.0029	3.465 ±0.027	0.5631 ±0.056	2.818 ±0.022	3.319 ±0.021	1.227 ±0.010
250–125 μ	8.38	0.0215 ±0.0018	0.0194 ±0.0016	0.2921 ±0.0079	3.449 ±0.032	0.5526 ±0.0066	2.769 ±0.030	3.308 ±0.028	1.226 ±0.014
125–63 μ	11.1	0.0202 ±0.0014	0.0240 ±0.0012	0.2897 ±0.0038	3.449 ±0.033	0.5504 ±0.0051	2.744 ±0.019	3.315 ±0.021	1.229 ±0.010
125–63 μ	13.8	0.0175 ±0.0043	0.0176 ±0.0020	0.2774 ±0.0059	3.370 ±0.037	0.5373 ±0.0047	2.700 ±0.028	3.261 ±0.027	1.214 ±0.015
63–44 μ	13.3	0.0207 ±0.0011	0.0229 ±0.0021	0.2878 ±0.0047	3.444 ±0.023	0.5495 ±0.0033	2.746 ±0.018	3.310 ±0.027	1.224 ±0.009

D. D. BOGARD et al.

Table 3. (Continued).

Sample	^{132}Xe (10^{-9})	^{124}Xe/ ^{136}Xe	^{126}Xe/ ^{136}Xe	^{128}Xe/ ^{136}Xe	^{129}Xe/ ^{136}Xe	^{130}Xe/ ^{136}Xe	^{131}Xe/ ^{136}Xe	^{132}Xe/ ^{136}Xe	^{134}Xe/ ^{136}Xe
63–44 μ	13.7	0.0195	0.0230	0.2859	3.415	0.5455	2.731	3.298	1.221
		±0.0006	±0.0008	±0.0037	±0.030	±0.0063	±0.029	±0.019	±0.011
44–30 μ	17.3	0.0198	0.0225	0.2881	3.442	0.5490	2.746	3.319	1.230
		±0.0024	±0.0018	±0.0042	±0.039	±0.0061	±0.031	0.031	±0.015
30–20 μ	21.7	0.0184	0.0197	0.2838	3.411	0.5420	2.720	3.294	1.220
		±0.0029	±0.0051	±0.0038	±0.030	±0.0086	±0.025	±0.038	±0.011
<20 μ	55.8	0.0174	0.0167	0.2780	3.439	0.5454	2.719	3.321	1.228
		±0.0009	±0.0009	±0.0042	±0.015	±0.0035	±0.012	±0.013	±0.006

76321,2, <20 μ stepwise temperature release

Sample	^{132}Xe (10^{-9})	^{124}Xe/ ^{136}Xe	^{126}Xe/ ^{136}Xe	^{128}Xe/ ^{136}Xe	^{129}Xe/ ^{136}Xe	^{130}Xe/ ^{136}Xe	^{131}Xe/ ^{136}Xe	^{132}Xe/ ^{136}Xe	^{134}Xe/ ^{136}Xe
170	0.0191	—	—	—	3.15	0.506	2.58	3.11	1.18
					±0.52	±0.091	±0.41	±0.49	±0.20
320	0.0721	—	—	—	3.06	0.390	2.44	3.03	1.12
					±0.27	±0.074	±0.24	±0.13	±0.13
500	0.0949	—	—	—	3.45	0.521	2.73	3.34	1.30
					±0.23	±0.050	±0.21	±0.21	±0.13
700	0.451	0.0142	0.0137	0.2835	3.535	0.553	2.761	3.364	1.238
		±0.0055	±0.0080	±0.0073	±0.059	±0.015	±0.034	±0.037	±0.028
800	5.76	0.01617	0.01432	0.2855	3.550	0.5604	2.781	3.377	1.239
		±0.00044	±0.00075	±0.0059	±0.022	±0.0052	±0.017	±0.018	±0.009
900	12.3	0.01615	0.01517	0.2841	3.517	0.5558	2.765	3.364	1.238
		±0.00044	±0.00034	±0.0018	±0.009	±0.0014	±0.008	±0.005	±0.004
1000	9.49	0.01739	0.01802	0.2776	3.413	0.5442	2.714	3.300	1.224
		±0.00079	±0.00063	±0.0014	±0.013	±0.0024	±0.011	±0.009	±0.005
1100	9.18	0.01680	0.01726	0.2746	3.382	0.5379	2.692	3.281	1.219
		±0.00069	±0.00033	±0.0014	±0.015	±0.0027	±0.013	±0.012	±0.005
1200	2.70	0.01619	0.0155	0.2701	3.315	0.5318	2.660	3.251	1.210
		±0.00095	±0.0016	±0.0034	±0.023	±0.0057	±0.020	±0.021	±0.009
1400	2.02	0.0187	0.0195	0.2745	3.311	0.5335	2.663	3.247	1.216
		±0.0013	±0.0011	±0.0030	±0.020	±0.0045	±0.033	±0.015	±0.009
1620	.0535	—	—	0.239	3.17	0.503	2.568	3.183	1.222
				±0.010	±0.12	±0.022	±0.085	±0.084	±0.041

76501,12 grain sizes

Sample	^{132}Xe (10^{-9})	^{124}Xe/ ^{136}Xe	^{126}Xe/ ^{136}Xe	^{128}Xe/ ^{136}Xe	^{129}Xe/ ^{136}Xe	^{130}Xe/ ^{136}Xe	^{131}Xe/ ^{136}Xe	^{132}Xe/ ^{136}Xe	^{134}Xe/ ^{136}Xe
1000–250 μ	4.92	0.0320	0.0416	0.3219	—	0.577	2.877	3.370	1.235
		±0.0014	±0.0036	±0.0097		±0.011	±0.040	±0.039	±0.016
250–125 μ	9.52	0.0242	0.0319	0.3020	3.487	0.5594	2.799	3.321	1.229
		±0.0011	±0.0012	±0.0034	±0.020	±0.0062	±0.020	±0.016	±0.009
125–63 μ	10.9	0.0245	0.0218	0.2925	3.434	0.5455	2.751	3.294	1.224
		±0.0028	±0.0022	±0.0103	±0.046	±0.0092	±0.037	±0.035	±0.016
63–44 μ	14.6	0.0205	0.0235	0.2943	3.437	0.5495	2.758	3.305	1.221
		±0.0020	±0.0031	±0.0054	±0.042	±0.0065	±0.031	±0.027	±0.017
44–30 μ	20.7	0.0183	0.0184	0.2907	3.444	0.5483	2.747	3.325	1.229
		±0.0025	±0.0045	±0.0054	±0.039	±0.0084	±0.028	±0.0295	±0.012
30–20 μ	21.8	0.0200	0.0217	0.2875	3.461	0.5499	2.745	3.304	1.230
		±0.0021	±0.0022	±0.0066	±0.050	±0.0097	±0.035	±0.037	±0.025
<20 μ	51.0	0.0186	0.0176	0.2845	3.469	0.5502	2.750	3.331	1.229
		±0.0012	±0.001	±0.0061	±0.045	±0.0076	±0.035	±0.041	±0.016

15271,64 agglutinate and nonagglutinate grain sizes
250–125 μ

Sample	^{132}Xe (10^{-9})	^{124}Xe/ ^{136}Xe	^{126}Xe/ ^{136}Xe	^{128}Xe/ ^{136}Xe	^{129}Xe/ ^{136}Xe	^{130}Xe/ ^{136}Xe	^{131}Xe/ ^{136}Xe	^{132}Xe/ ^{136}Xe	^{134}Xe/ ^{136}Xe
Agglutinate	22.8	0.0278	0.0371	0.3088	3.463	0.5590	2.801	3.274	1.223
		±0.0006	±0.0012	±0.0065	±0.069	±0.0097	±0.054	±0.045	±0.022

Table 3. (*Continued*).

Sample	^{132}Xe (10^{-9})	^{124}Xe/ ^{136}Xe	^{126}Xe/ ^{136}Xe	^{128}Xe/ ^{136}Xe	^{129}Xe/ ^{136}Xe	^{130}Xe/ ^{136}Xe	^{131}Xe/ ^{136}Xe	^{132}Xe/ ^{136}Xe	^{134}Xe/ ^{136}Xe
Nonagglutinate	7.02	0.0605	0.0965	0.3932	3.498	0.6025	3.066	3.323	1.223
		±0.0031	±0.0038	±0.0069	±0.026	±0.0106	±0.024	±0.021	±0.011
125–63 μ									
Agglutinate	26.9	0.0262	0.0336	0.3024	3.468	0.5604	2.797	3.310	1.226
		±0.0010	±0.0006	±0.0031	±0.014	±0.0037	±0.011	±0.008	±0.007
Nonagglutinate	7.64	0.0529	0.0822	0.3734	3.542	0.6080	3.020	3.347	1.226
		±0.0020	±0.0026	±0.0050	±0.029	±0.0081	±0.028	±0.022	±0.017
63–44 μ	21.3	0.0280	0.0374	0.2945	3.358	0.5447	2.763	3.229	1.217
		±0.0028	±0.0036	±0.0090	±0.073	±0.013	±0.083	±0.060	±0.028
63–44 μ	16.7	0.0305	0.0411	0.3154	3.461	0.5634	2.818	3.304	1.221
		±0.0021	±0.0028	±0.0047	±0.035	±0.0069	±0.030	±0.026	±0.014
<44 μ	51.2	0.0209	0.0238	0.2777	3.371	0.5350	2.690	3.260	1.048
		±0.0005	±0.0007	±0.0035	±0.033	±0.0056	±0.022	±0.023	±0.011
44–30 μ	14.0	0.0345	0.0495	0.3279	3.482	0.5769	2.856	3.324	1.225
		±0.0015	±0.0022	±0.0074	±0.033	±0.0064	±0.022	±0.022	±0.015
44–30 μ	12.0	0.0361	0.0482	0.3332	3.507	0.5813	2.887	3.329	1.225
		±0.0028	±0.0021	±0.0064	±0.031	±0.0097	±0.029	±0.017	±0.023
30–20 μ	20.0	0.0305	0.0397	0.3096	3.455	0.5619	2.809	3.306	1.226
		±0.0009	±0.0017	±0.0041	±0.030	±0.0069	±0.021	±0.011	±0.011
30–20 μ	16.6	0.0308	0.0405	0.3112	3.449	0.5681	2.811	3.280	1.224
		±0.0021	±0.0019	±0.0066	±0.059	±0.0134	±0.047	±0.051	±0.023
<20 μ	43.4	0.0217	0.0272	0.2972	3.450	0.5559	2.762	3.306	1.224
		±0.0028	±0.0022	±0.0039	±0.031	±0.0055	±0.024	±0.027	±0.014

factors must also affect relative abundances. Figure 1 shows a tendency for bulk soils from a given area and particularly grain size separates to form groups of characteristic ^{4}He/^{22}Ne, but variable ^{22}Ne/^{36}Ar. The two Station 2 soils, which have the lowest Ti and He contents, have identical ^{4}He/^{22}Ne of 650, but different ^{22}Ne/^{36}Ar. The four Station 6 soils have ^{4}He/^{22}He of 850 ± 60, but show variations

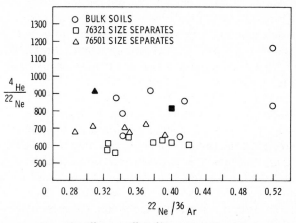

Fig. 1. Elemental ratios of ^{4}He/^{22}Ne and ^{22}Ne/^{36}Ar for Apollo 17 bulk soils and grain size separates. Solid points are the <20 μ fractions.

in ^{22}Ne/^{36}Ar of 0.335–0.415. Mare soil 75081 shows the highest value for both of these ratios. Coarser grain size separates (20–1000 microns) of 76321 and 76501 show nearly constant values for ^{4}He/^{22}Ne of 600 ± 40 and 690 ± 40, respectively, but appreciably wider ranges in their ^{22}Ne/^{36}Ar ratios. However, the <20 micron grain sizes for both soils (which comprise ~24% of the total sample weight, but contain roughly half of the total trapped gas content) show ^{4}He/^{22}Ne which are 30% higher. Although it is conceiveable that the higher ^{4}He/^{22}Ne in the <20 μ fractions are due to a preferential concentration of ilmenite in the finest fraction, comparison of neon isotopic data in bulk soils to that in the grain size fractions argues against this explanation. Whereas ^{20}Ne/^{22}Ne ratios for all eight bulk soils, which incorporate a wide variation in ilmenite concentration, fall in the narrow range 12.81–12.92, the ^{20}Ne/^{22}Ne for the <20 μ fractions of 76321 and 76501 are appreciably higher than the ^{20}Ne/^{22}Ne for the coarser fractions (Table 1). Corrections for cosmogenic Ne have only a small effect on this trend (<1%, except for the coarsest size fractions), as can be verified by constructing a three isotope diagram of the Ne data. The smaller grain sizes thus contain neon which is closer in composition to that measured in the solar wind composition experiments (Geiss *et al.*, 1972) than the larger ones. In fact, both ^{20}Ne/^{22}Ne and ^{36}Ar/^{38}Ar in these bulk soil analyses are slightly higher than commonly has been measured in previous lunar soils (c.f. Eberhardt *et al.*, 1970, 1972; Hintenberger *et al.*, 1971; Bogard *et al.*, 1973). On the other hand, the ^{84}Kr/^{36}Ar and ^{132}Xe/^{36}Ar for bulk soils and nearly all grain sizes fall in the narrow range of $5.3–6.6 \times 10^{-4}$ and $7.4–9.0 \times 10^{-5}$, respectively, which is typical of many lunar soils.

Eberhardt *et al.* (1972) also noted appreciable variations in Ne/Ar (but not He/Ne, Ar/Kr, or Kr/Xe) with grain size for three ilmenite separates from Apollo 11 and 12 samples. These authors discussed two possible models to explain such a trend; a volume correlated Ar, Kr, and Xe component, and surface correlated gases which have lost constant amounts of He and Ne (i.e. a fractionation process). These Apollo 17 data also support our earlier observations from Apollo 15 and 16 soils (Bogard *et al.*, 1973) that while He/Ne and Ne/Ar ratios are both sensitive to sample chemistry, the Ne/Ar ratio appears to be particularly sensitive to elemental fractionation effects which are apparently unrelated to bulk sample chemistry. Mass fractionation effects also occur in the heavy noble gases, as will be argued later from Kr and Xe isotopic data.

Grain size analyses of soils 76321, 76501, and 15271 show an inverse dependence of trapped noble gas concentrations with grain size. This relationship, which is created by a surface correlated trapped component and a volume correlated cosmogenic component, can be represented by the equation (Eberhardt *et al.*, 1970), $C = kd^{-n}$, where C is gas concentration and d is grain diameter. The values of $-n$ are similar for the two Apollo 17 soils, and have the following average values: ^{4}He = 0.59, ^{22}Ne = 0.57, ^{36}Ar = 0.49, ^{84}Kr = 0.45, ^{132}Xe = 0.45. Through use of the ordinate–intercept technique (Eberhardt *et al.*, 1970), these grain size analyses also permit more precise determinations of trapped isotopic ratios and cosmogenic and radiogenic nuclide abundances than can be determined from bulk soil analyses alone. Trapped Ne and Ar isotopic ratios and concentrations of cosmogenic ^{21}Ne, cosmogenic ^{38}Ar, and radiogenic ^{40}Ar derived in this manner are summarized in Table 4. Two values of trapped ^{20}Ne/^{22}Ne are listed for 76321 and 76501 and correspond to the coarser grain sizes (20–1000 μ) and to the <20 μ grain size, respectively. Soils 76501 and 76321 (collected from the sloping side of an ~5 m boulder) possess noble gas concentrations and isotopic compositions, which are also similar to surface (76261) and subsurface (76281) soils collected from a partially shielded position in a cleft between boulders. The cosmogenic ^{21}Ne for 76321 and 76501 (Table 4) can be combined with chemical determinations (Rhodes *et al.*, 1974) and relative elemental production rates (Bogard and Cressy, 1973) to calculate a ^{21}Ne production rate of 0.12 × 10^{-8} cm^3/g 10^6 yr and cosmic ray exposure ages of 250 and 330 m.y. respectively. Similar ages of 230 and 240 m.y. are obtained for cosmogenic ^{21}Ne calculated from the bulk analyses of soils 76261 and 76281. These soil ages, while not precise, are at least an order of magnitude older than exposure ages for the large boulders at Station 6 (Bogard *et al.*, 1974), and indicate that these boulders have not contributed the major portion of even the 250–1000 μ grain sizes of the boulder soil 76321 (probably <25%).

The radiogenic ^{40}Ar contents derived from ordinate–intercept plots for soils 76321 and 76501 (Table 4) can be combined with bulk K abundances (Rhodes *et al.*, 1974) to give "apparent" K–^{40}Ar gas retention ages of ~4.3 × 10^9 yr for both soils. These ages, however, are inconsistent with the fact that most of the K in these soils is contributed by the noritic breccia component (Rhodes *et al.*, 1974; and this volume) which has been dated by ^{39}Ar/^{40}Ar and by ^{87}Rb/^{87}Sr methods at this site at 4.0 × 10^9 yr (cf. Huneke *et al.*, 1973; Nyquist *et al.*, 1974). For this reason the K concentrations of several grain sizes of 76501 were measured by isotope dilution (Bansal and Wiesmann, unpublished data) and were found to vary from 785 (63–125 μ) to 1015 ppm (<20 μ). Figure 2 shows a modified ordinate–intercept plot where ^{40}Ar/^{36}Ar is plotted against K/^{36}Ar for 76501 (Basford, 1974). Four out of five grain sizes above 30 μ define a straight line with a trapped ^{40}Ar/^{36}Ar = 0.80 and a K–^{40}Ar gas-retention age of 4.0 × 10^9 yr, which is in good agreement with the reported ages of local breccias. The <20 μ grain size (and consequently the bulk soil also) fall below this linear trend, indicating a lower

Table 4. Summary of trapped Ne and Ar ratios and cosmogenic ^{21}Ne and ^{38}Ar and radiogenic ^{40}Ar abundances obtained from ordinate–intercept plots of grain size separates of three soils (Table 1). Grain sizes used in deriving these values are also indicated. Uncertainties in trapped ratios are one sigma of the intercept value; uncertainties in abundances are derived from one sigma of the slope and uncertainties in measured abundances.

Soil	Trapped ^{22}Ne/^{21}Ne	Trapped ^{20}Ne/^{22}Ne	Cosmogenic ^{21}Ne	Trapped ^{36}Ar/^{38}Ar	Trapped ^{40}Ar/^{36}Ar	Cosmogenic ^{38}Ar	Radiogenic ^{40}Ar
76321,2	31.0 ± 0.2 (<125 μ)	12.65–12.85	30.2 ± 3.8 (<125 μ)	5.40 ± 0.015 (<250 μ)	0.766 ± 0.035 (<1000 μ)	43.0 ± 3.0 (<250 μ)	6850 ± 650 (<1000 μ)
76501,12	31.4 ± 0.2 (<125 μ)	12.75–12.95	39.5 ± 4.1 (<125 μ)	5.41 ± 0.01 (<250 μ)	0.752 ± 0.025 (<1000 μ)	42.2 ± 6.5 (<250 μ)	5770 ± 540 (<1000 μ)
15271,64	32.1 ± 0.4 (<63 μ)	12.8	63 ± 7 (<63 μ)	5.43 ± 0.012 (<63 μ)	0.964 ± 0.017 (<63 μ)	66 ± 13 (<63 μ)	3480 ± 450 (<1000 μ)

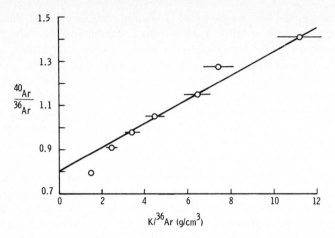

Fig. 2. Ordinate–intercept plot of $^{40}Ar/^{36}Ar$ against $K/^{36}Ar$ for grain sizes of 76501. Grain sizes above 30 μ indicate a $K-^{40}Ar$ age of 4.0×10^9 yr and a trapped $^{40}Ar/^{36}Ar = 0.80$. The <20 μ size fraction apparently possesses a lower trapped $^{40}Ar/^{36}Ar$.

trapped $^{40}Ar/^{36}Ar$. If the <20 μ size fraction has an ^{40}Ar age of 4.0×10^9 yr its trapped $^{40}Ar/^{36}Ar$ would be 0.71. An upper limit of trapped $^{40}Ar/^{36}Ar$ in the <20 μ fraction of 0.80 is established for the unlikely situation that the ^{40}Ar gas retention age is zero. Thus, as was the case with trapped $^{20}Ne/^{22}Ne$ and He/Ne, trapped $^{40}Ar/^{36}Ar$ in the finest grain sizes of 76501 differs from the rest of the soil.

XENON AND KRYPTON RESULTS

The ordinate–intercept technique was used to derive the isotopic composition of the trapped solar wind Kr and Xe in the grain size separates of 76501, 76321, and 15271 (Tables 5 and 6). As discussed by Eberhardt *et al.* (1970) and Basford *et al.* (1973), the use of this technique assumes that for all grain sizes the cosmogenic gas concentrations and the isotopic composition of the trapped component remain constant. Constancy of the isotopic composition of the trapped component has

Table 5. Isotopic composition of surface-correlated trapped Xe obtained from ordinate–intercept plots (relative to $^{130}Xe = 1.00$). Uncertainties given are two sigma of the least-squares fit to the data. The entire range of grain sizes has been used, except for a few individual ratios which were not consistent with a linear trend.

Sample	124	126	128	129	131	132	134	136
76321,2	0.0291	0.0257	0.5083	6.334	5.001	6.126	2.269	1.845
	±0.0026	±0.0045	±0.0062	±0.020	±0.020	±0.018	±0.010	±0.016
76501,12	0.0296	0.0269	0.5140	6.320	5.014	6.095	2.263	1.845
	±0.0024	±0.0060	±0.0056	±0.038	±0.022	±0.040	±0.020	±0.022
15271,64	0.0298	0.0277	0.5094	6.340	4.987	6.087	2.264	1.857
	±0.0030	±0.0052	±0.0078	±0.022	±0.020	±0.034	±0.020	±0.018

Table 6. Isotopic composition of surface-correlated trapped Kr obtained from ordinate–intercept plots (relative to ^{86}Kr = 1.00). Uncertainties given are two sigma of the least-squares fit to the data. The entire range of grain size has been used, except for a few individual ratios which were not consistent with a linear trend.

Sample	78	80	82	83	84
76321,2	0.021	0.1293	0.6567	0.6575	3.257
	±0.002	±0.0010	±0.0034	±0.0032	±0.008
76501,12	0.020	0.1309	0.6566	0.6571	3.263
	±0.002	±0.0042	±0.0034	±0.0016	±0.007
15271,64	0.021	0.1301	0.6580	0.6572	3.270
	±0.002	±0.0010	±0.0028	±0.0022	±0.016

always been assumed, and more will be said on this later. It has been shown that biased results occur if the major target elements for production of cosmogenic Kr and Xe exhibit a uniform change in concentration with grain size, but that corrections for these target element variations can be made in a manner analogous to Fig. 2 (Eberhardt *et al.*, 1972). However, for soils 76321 and 76501 concentrations of cosmogenic Kr and Xe are unusually low in proportion to the trapped gas abundances. Ordinate–intercept plots for several Kr and Xe isotopic ratios for 76321 (Figs. 3 and 4) demonstrate this point. Analogous plots for 76501 are similar,

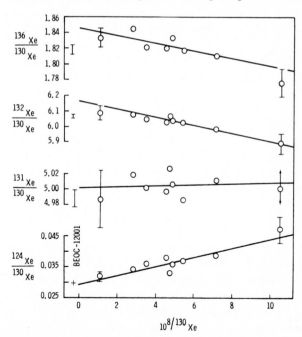

Fig. 3. Ordinate–intercept plots for four isotopic ratios of Xe for grain size separates of 76321. Trapped ratios for BEOC 12001 (Eberhardt *et al.*, 1972) are shown on the left for comparison. Uncertainties in isotopic ratios (Table 3) are indicated for two grain sizes.

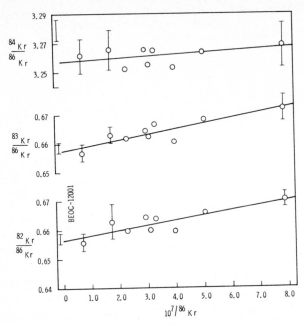

Fig. 4. Ordinate–intercept plots for three isotopic ratios of Kr for grain size separates of 76321. Trapped ratios for BEOC 12001 (Eberhardt *et al.*, 1972) are shown on the left for comparison. Uncertainties in isotopic ratios (Table 2) are indicated for three grain sizes.

whereas 15271 shows greater relative cosmogenic enrichments. For the grain size range of <20–1000 μ the total change in abundances of ^{84}Kr, ^{83}Kr, and ^{82}Kr relative to ^{86}Kr and of ^{134}Xe, ^{132}Xe, ^{131}Xe, ^{130}Xe, and ^{129}Xe relative to ^{136}Xe due to the cosmogenic component is less than 5% for both Apollo 17 soils. Considerably larger cosmogenic effects occur, however, for ^{78}Kr, ^{80}Kr, ^{124}Xe, and ^{126}Xe. When similar grain sizes of 76321 are compared to soil 12001 used by Eberhardt *et al.* (1972) to derive the BEOC-12001 trapped Kr and Xe, 76321 shows cosmogenic Kr and Xe enrichments of only 1/2 and 1/3, respectively, that of 12001. In fact, the measured ratios of ^{82}Kr/^{86}Kr, ^{83}Kr/^{86}Kr, ^{84}Kr/^{86}Kr, ^{134}Xe/^{136}Xe, ^{132}Xe/^{136}Xe, ^{131}Xe/^{136}Xe, ^{130}Xe/^{136}Xe, ^{129}Xe/^{136}Xe, and ^{128}Xe/^{136}Xe in the <20 μ fractions of both 76321 and 76501 are as low or lower than the derived BEOC-12001 trapped values (where target element corrections were made). In other words, the smallest grain size fractions measured on both 76321 and 76501 contain no apparent cosmogenic enrichments for the above ratios compared to BEOC-12001. This means that the derived trapped values for the above ratios for 76321 and 76501 are only very slightly affected by the cosmogenic component, and by possible small variations in target element concentrations. To further confirm this, [Ba] was measured by isotope dilution (Bansal and Wiesmann, unpublished data) in the <20, 30–44, 63–125, and 250–1000 μ grain sizes of 76501, and was found to vary by only 20% (123, 106, 103, and 124 ppm, respectively). If corrections for these [Ba] differences are inserted into the ordinate–intercept plot for 76501, the derived trapped Xe composition changes by amounts which are considerably smaller than the uncertainties given in Table 5.

Because of the very low relative abundances of cosmogenic Kr and Xe in the $<20\ \mu$ fraction of 76321, the data from the stepwise temperature releases (Tables 2 and 3) are little affected by the cosmogenic component. In the case of Xe, most isotopic ratios show a nearly smooth, mass-dependant trend across essentially the entire temperature release, with preferential enrichments of the lighter isotopes over the heavier at low temperature. This trend is demonstrated for Xe by Fig. 5, where the $^{132}Xe/^{136}Xe$ is plotted against $^{134}Xe/^{136}Xe$, $^{129}Xe/^{136}Xe$, and $^{130}Xe/^{136}Xe$ for various release temperatures. The solid lines in the three isotope correlation plots are fitted to the five temperature releases between 800°C and 1200°C, which contained 93.6% of the total Xe and possessed the smallest uncertainties. Slopes of these lines, as well as slopes for analogous plots of $^{128}Xe/^{136}Xe$ and $^{131}Xe/^{136}Xe$, are given in Table 7. These isotopic trends shown by Fig. 5 and Table 7 cannot be produced by a two component mixing of trapped plus cosmogenic Xe. If this were so, the slopes in Table 7 would be the spectrum relative to ^{132}Xe of this cosmogenic Xe; however, cosmogenic Xe with extreme ratios of $128/132 = 0.119$, $130/132 = 0.22$, and $131/132 = 0.93$ has never been observed. Also, for such a two component system the trapped Xe composition would have to deviate from the bulk composition and previous measurements of trapped Xe and would have to lie on the mixing line at or beyond one extreme of the temperature data. This second objection would also apply to a mixture of trapped plus excess fissiogenic Xe. Supposition of a three component Xe model with the trapped composition near the bulk analyses also cannot explain the data and would meet with the same objection concerning cosmogenic Xe, plus the additional objections that the third component would have to lie colinear with the first two components on the isotope–correlation diagrams, and both nonsolar components would have to comprise a large proportion of the total Xe.

The trends of Xe isotopic fractionation which are linear in mass are shown in Fig. 5 as dotted lines through $^{132}Xe/^{136}Xe = 3.321$ (the measured value in the $<20\ \mu$ grain size) and the corresponding value of $(^{i}Xe/^{136}Xe)_0$ on the least-squares fit to

Table 7. Slopes of $^{i}Xe/^{136}Xe$ relative to $^{132}Xe/^{136}Xe$ and of $^{i}Kr/^{86}Kr$ relative to $^{83}Kr/^{86}Kr$ as defined by the temperature release data of 76321 and by linear mass fractionation. Slopes were obtained from a least-squares fit to the 800–1200°C data for Xe and the 700–1200°C data for Kr. Uncertainties are two sigma of the data fit.

	128/136	129/136	130/136	131/136	134/136	80/86	82/86	84/86
Slope of least-squares fit to temperature data	0.119 ±0.014	1.788 ±0.13	0.220 ±0.022	0.928 ±0.076	0.229 ±0.028	—	—	—
Slope corrected for cosmogenic Xe and Kr for >1000°C	0.154 ±0.010	1.790 ±0.13	0.243 ±0.012	1.026 ±0.022	0.224 ±0.030	0.415 ±0.048	1.485 ±0.070	3.67 ±0.40
Slope of linear mass fractionation	0.168	1.81	0.247	1.03	0.185	0.397	1.34	3.28

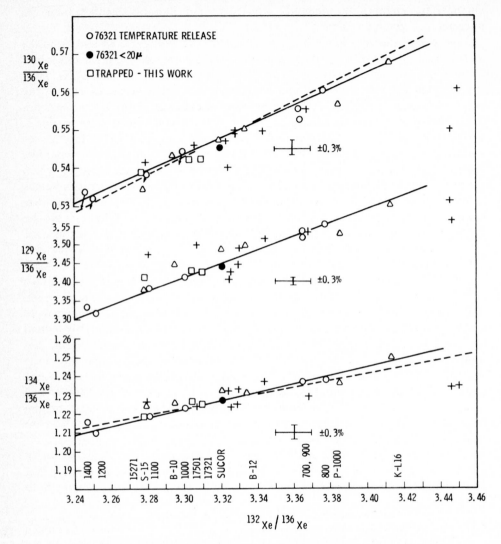

Fig. 5. Isotope correlation plots for five isotopes of Xe from the stepwise temperature release of 76321, <20 μ. Temperatures are indicated in C°. The solid lines represent linear mass fractionation trends normalized to $^{132}Xe/^{136}Xe = 3.321$ as measured in the <20 μ size fraction. (The mass fractionation trend for $^{129}Xe/^{136}Xe$ is graphically indistinguishable from the solid line.) Corrections to the 1000–1400°C data for cosmogenic Xe are shown in the $^{130}Xe/^{136}Xe$ plot as short arrows. A number of trapped Xe compositions (including those in Table 5) are also shown and are identified as follows: crosses (+), various soils from five lunar sites, Basford *et al.* (1973); triangles: SUCOR, Podosek *et al.* (1971); B-10 = BEOC 10084, Eberhardt *et al.* (1970); B-12 = BEOC 12001, Eberhardt *et al.* (1972); P-1000, the 1000°C extraction of Pesyanoe, Marti (1969); K-L16 = Luna 16 fines, Kaiser (1972); S-15 = 15601, Srinivasan *et al.* (1972). A typical uncertainty in trapped isotopic ratios of ±0.3% is shown; some reported uncertainties are less, some are greater.

the data. The mass fractionation trend for $^{129}Xe/^{136}Xe$ versus $^{132}Xe/^{136}Xe$ coincides with the solid-line fit to the temperature data. Slopes of these trends were calculated according to the formula:

$$\frac{^iXe}{^{136}Xe} = \left(\frac{^iXe}{^{136}Xe}\right)_0\left[1-\frac{136-i}{4}\right] + \frac{(^iXe/^{136}Xe)_0}{3.321}\left(\frac{136-i}{4}\right)\left(\frac{^{132}Xe}{^{136}Xe}\right)$$

and are given in Table 7. As can be seen from Fig. 5 and Table 7, the calculated linear mass fractionation and the trend established by the temperature release data for 76321 are very similar for masses 128–136.

Plots analogous to Fig. 5 for $^{124}Xe/^{136}Xe$ and $^{126}Xe/^{136}Xe$ do not show smooth decreases in these ratios with increasing extraction temperatures above 900°C. This fact is undoubtedly caused by the release of significant amounts of cosmogenic ^{124}Xe and ^{126}Xe at higher temperatures. These two isotopes are the most sensitive to the addition of small amounts of cosmogenic Xe, as can be seen by comparing the derived trapped composition (Table 5) with the Xe composition measured in the <20 μ grain size (Table 3) for 76321 and 76501. We have utilized this cosmogenic ^{124}Xe present in the temperature-release data to make small cosmogenic corrections to the other Xe isotopes in the following manner. We assume that, like the heavier Xe isotopes, trapped $^{124}Xe/^{136}Xe$ released by the temperature extraction can be represented by a linear mass fractionation line which passes through the trapped compositions determined for 76321 and 12001 (Eberhardt et al., 1972). The 800°C and 900°C data are in fair agreement with this line, whereas the 1000–1400°C data plot far above the line in the direction of excess ^{124}Xe. We take the excess ^{124}Xe relative to the fractionation line to be cosmogenic ^{124}Xe, and calculate the equivalent amounts of cosmogenic ^{128}Xe, ^{129}Xe, ^{130}Xe, ^{131}Xe, and ^{132}Xe, assuming a typical spallation Xe spectrum of 124/128/129/130/131/132 = 1.0/2.7/2.7/1.8/9.1/1.4 (Bogard et al., 1971; Basford et al., 1973; Eberhardt et al., 1974). We then apply these small corrections to the 1000–1400°C data. The magnitude of these corrections are shown in Table 7 as "corrected slopes" and in Fig. 5 as small arrows on the plotted points for $^{130}Xe/^{136}Xe$ and $^{132}Xe/^{136}Xe$. Except for $^{128}Xe/^{136}Xe$, these cosmogenic corrections are generally less than the 1 σ uncertainty in the measured ratios. However, in all cases the cosmogenic corrections produce better agreement between slopes defined by the temperature data and slopes defined by linear mass fractionation (Table 7). The small disagreement in slopes remaining for $^{128}Xe/^{136}Xe$ is probably due to the uncertainty in the relatively much larger cosmogenic corrections applied to ^{128}Xe. The small disagreement in slopes remaining for $^{134}Xe/^{136}Xe$ cannot be due to a cosmogenic effect, but could conceiveably be due to small amounts of Xe from the spontaneous fission of ^{238}U if this U is contained in phases which degas preferentially at lower extraction temperatures. If this is the case, the change in $^{132}Xe/^{136}Xe$ at 800°C and 900°C due to in situ fission Xe would be only of the order of 0.1%, but this amount could essentially account for the differences in slopes in Table 7. Therefore, we consider the Xe isotopic data from the temperature release of soil 76321, <20 μ to be completely consistent with a linear mass fraction of trapped solar wind Xe, with only small effects due to cosmogenic and fissiogenic Xe.

The relative abundances of ^{82}Kr, ^{83}Kr, ^{84}Kr, and ^{86}Kr from the temperature release of 76321 follow a smooth fractionation trend through approximately 3/4 of the total Kr release (900°C), then begin to show deviations (Fig. 6). The ^{80}Kr/^{86}Kr data (not plotted) show an essentially linear trend against ^{83}Kr/^{86}Kr, but after a low value of ^{80}Kr/^{86}Kr = 0.1279 at 1100°C, ^{80}Kr/^{86}Kr and ^{83}Kr/^{86}Kr increase again while still maintaining their constant ratio. Values of ^{78}Kr/^{86}Kr decrease for a time, then increase. For all temperatures above 900°C, the ^{80}Kr/^{86}Kr, ^{82}Kr/^{86}Kr, ^{83}Kr/^{86}Kr, and ^{84}Kr/^{86}Kr ratios fall between the trapped values of BEOC-10084 (Eberhardt *et al.*, 1970) and SUCOR (Podosek *et al.*, 1971).

As was the case with ^{124}Xe, we assume that ^{78}Kr/^{86}Kr, the ratio most sensitive to cosmogenic Kr, can be represented by a linear mass fractionation line passing through the low-temperature (800°C and 900°C) data. We take the excess ^{78}Kr relative to this fractionation line to be cosmogenic ^{78}Kr, and calculate the equivalent amounts of cosmogenic ^{80}Kr, ^{82}Kr, ^{83}Kr, and ^{84}Kr assuming a spallation Kr spectrum of 78/80/82/83/84 = 0.2/0.5/0.8/1.0/0.4 (Bogard *et al.*, 1971; Basford *et al.*, 1973; Eberhardt *et al.*, 1974). We then apply these cosmogenic corrections to the 1000–1400°C data. Unfortunately, the ^{78}Kr/^{86}Kr and ^{84}Kr/^{86}Kr cosmogenic ratios are quite sensitive to both shielding and target chemistry, and the magnitude of the corrections applied to Kr are therefore subject to greater uncertainties than those applied to Xe. Nevertheless, the corrected Kr isotopic ratios show reasonably good agreement with a process of linear mass fractionation, as shown by Fig. 6 and Table 7. The fact that ^{80}Kr/^{86}Kr and ^{83}Kr/^{86}Kr maintain a nearly constant ratio to each other while individual ratios decrease to 1100°C, then increase, occurs because the mass fractionation trend and the trapped plus cosmogenic Kr trend are nearly colinear. That is, on a plot of ^{80}Kr/^{86}Kr versus ^{83}Kr/^{86}Kr the temperature data and the grain size data define lines of similar slopes. The observed trends in the Kr isotopic ratios for the temperature release are not likely due solely to a mixture of trapped plus cosmogenic Kr by the same reasons given earlier for Xe. Addition of the above average cosmogenic Kr to trapped Kr with the composition of SUCOR (Fig. 6), which falls below the temperature data on all Kr correlation plots, would not generate the same linear relationships observed, and would be in contradiction to the composition of trapped Kr in 76321 determined by ordinate–intercept plots. However, isotopic trends similar to those of Fig. 6 have been previously noted for low temperature extractions (<900°C) of soils 10084 (Pepin *et al.*, 1970a) and 14163 (Bogard and Nyquist, 1972). Both investigations attributed the low-temperature data to the presence of a second, cosmogenic component. In the case of 14163 the quite large isotopic enhancement at low temperatures (up to 0.80 for ^{83}Kr/^{86}Kr) led Bogard and Nyquist to suggest low-energy nuclear reactions on Rb as a possible source of this cosmogenic Kr. If one assumes that trapped Kr in 76321 has a composition similar to SUCOR (Fig. 6), then the stepwise temperature release observed for Kr in this soil could conceiveably be produced by a two-component mixture of this trapped composition and the postulated low-temperature cosmogenic Kr component. Such an explanation, however, is in contradiction with the trapped Kr composition determined by the ordinate–intercept technique (Table 6) and the

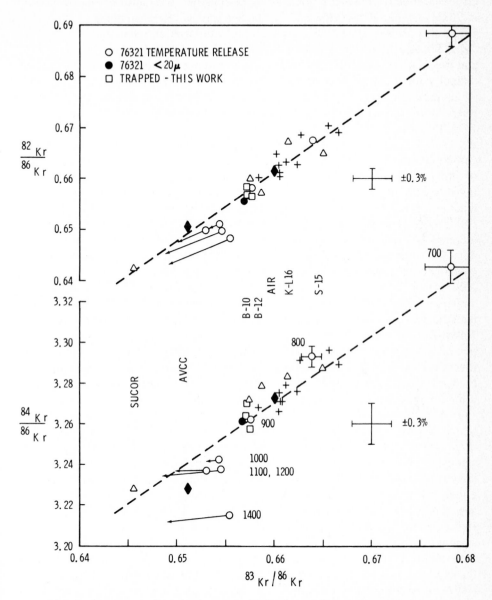

Fig. 6. Isotope correlation plots for four isotopes of Kr from the stepwise temperature release of 76321, <20 μ. Temperatures are indicated in C°. Corrections to the 1000–1400°C data for cosmogenic Kr are indicated by the arrows. The dotted lines represent linear mass fractionation trends normalized to the 900°C data. The compositions of Kr for the atmosphere (Nief, 1960), average carbonaceous chondrites (AVCC— Eugster et al., 1967), and trapped Kr in lunar fines are also shown. See Fig. 5 for identification and references of trapped compositions.

clear evidence for mass fractionation in Xe. Further, the factor of six lower [Rb] in 76321 compared to 14163 and the higher trapped Kr/cosmogenic Kr ratio in 76321 also argues against this explanation. The large enrichments of ^{82}Kr/^{86}Kr, ^{83}Kr/^{86}Kr, and ^{84}Kr/^{86}Kr over typical trapped values seen in 14163 (up to ~25%) and the lack of analogous low-temperature enrichments for Xe presents a major difficulty for a mass fractionation explanation of the 14163 Kr data.

Isotopic Xe and Kr data have been previously determined for the stepwise temperature release of several lunar fines (Hohenberg *et al.*, 1970; Marti *et al.*, 1970; Pepin *et al.*, 1970b; Kaiser, 1972; Bogard and Nyquist, 1972; Srinivasan *et al.*, 1972; Basford *et al.*, 1973). The concept of mass fractionation as a probable cause of isotopic variations of trapped Ne, Ar, Kr, and Xe during stepwise release of lunar soils has been suggested and discussed by a number of the above investigators and is not original with this work. However, the general trend for most of these determinations was for isotopic ratios such as ^{132}Xe/^{136}Xe and ^{83}Kr/^{86}Kr to remain nearly constant with increasing extraction temperature or to increase due to a cosmogenic component. In this regard the isotopic Kr and Xe data for stepwise temperature release of 76321 is much less affected by the cosmogenic component than were these previous analyses, and the Xe data for 76321 (Fig. 5) essentially monitors the behavior of only trapped Xe during stepwise release of this soil. Hohenberg *et al.* (1970) noted that variations in the isotopic composition of trapped He, Ne, and Ar during temperature release of soil 10084 was greater than that expected from single-stage fractionation, but appeared to be approximately equal to the mass ratio of the isotopes involved. These investigators suggested that isotopic variations for these gases could be explained by deeper burial of heavier isotopes during solar wind bombardment, followed by diffusive separation of isotopes during thermal release in the laboratory. Srinivasan (1973) reported Ne, Ar, and Kr data from a stepwise heating of breccia 14318, which showed appreciable evidence of isotopic fractionation and gas loss ($[^{22}$Ne$] = 4.5 \times 10^{-6}$ cm^3/g; trapped ^{20}Ne/^{22}Ne $= 11.2$). Srinivasan suggested that significant variations with extraction temperature also existed in the isotopic composition of trapped Kr, and that a mass dependent fractionation process either during solar wind implantation or subsequently was responsible for the apparent variation in trapped Ne, Ar, and Kr during temperature release. However, to obtain the trapped Kr composition it was necessary to subtract an assumed constant cosmogenic Kr component from all measured temperature data, and to assume that the trapped Kr composition lay on a fractionation line through previous determinations of trapped Kr. For soil 15601, Srinivasan *et al.* (1972) noted a tendency for lighter Xe and Kr isotopes to be released preferentially over heavier isotopes at low-extraction temperatures, but this effect tended to be masked by cosmogenic gases at intermediate and higher temperatures. They suggested a mass fractionation source for this effect.

ISOTOPIC VARIATIONS IN TRAPPED Xe AND Kr

The isotopic compositions of *trapped* solar wind Xe and Kr in lunar soils as determined by several laboratories are also plotted in Figs. 5 and 6. These data include determinations on lunar soils from Apollo 11, 12, 14, 15, and 17, Luna-16 and the 1000°C Xe release of the Pesyanoe gas-rich meteorite. Most of the trapped compositions were derived by ordinate–intercept plots such as Figs. 3 and 4, but several were derived from three-isotope correlation diagrams where one trapped isotopic ratio (usually $^{126}Xe/^{130}Xe$ and $^{78}Kr/^{82}Kr$) had to be assumed. The total spread in isotopic ratios among various *trapped* Xe components is appreciable, amounting to 5.1% for $^{132}Xe/^{136}Xe$, with similar variations for other ratios (Fig. 5). The total spread in trapped $^{83}Kr/^{86}Kr$ is 3.2%. This is to be contrasted with typical analytical uncertainties of about ±0.3% in the isotopic ratios of the various trapped components. Most trapped Xe and Kr compositions plotted in Figs. 5 and 6 also agree within their uncertainties with the trend established by the temperature release of 76321, which has been shown to be one consistent with linear mass fractionation. The appreciable variations in most trapped ratios are not likely due to the undercorrection or overcorrection of cosmogenic gases by the same arguments given earlier that the 76321 temperature release data cannot be produced by a mixture of trapped plus cosmogenic Xe. Trapped ^{124}Xe and ^{126}Xe variations among soils are more erratic and may be in part due to the large cosmogenic corrections which are applied and to the considerably larger relative uncertainties assigned to these isotopes. However, Basford *et al.* (1973) note an apparent correlation between trapped $^{124}Xe/^{130}Xe$ and $^{126}Xe/^{130}Xe$ values. Several investigations have considered trapped solar wind-implanted Xe and Kr to possess an essentially unique (although not necessarily solar) composition, which would be best determined by the most carefully designed experiment, and have attempted to relate this composition to Xe and Kr compositions of the earth's atmosphere and of carbonaceous chondrites (Eberhardt *et al.*, 1970, 1972; Kaiser, 1972; Podosek *et al.*, 1971; Srinivasan *et al.*, 1972). Other investigators have commented on the possibility that trapped Xe and Kr compositions are variable among lunar soils due to mass fractionation and other effects (Burnett *et al.*, 1971; Basford *et al.*, 1973). However, until recently there were insufficient determinations of the composition of trapped Xe and Kr in lunar soils to permit a detailed analysis of data trends. Basford *et al.* (1973) determined trapped Xe and Kr compositions for a number of fines and noted that significant variations existed. These investigators considered mass fractionation effects as a possible cause of the isotopic variations in trapped Kr, but favored addition of excess Xe components such as those trapped due to fission (from ^{238}U, ^{244}Pu, and other possible sources), radioactive decay (from extinct ^{129}I), and neutron capture as the major cause of the isotopic variation observed in trapped Xe in lunar soils.

Additional evidence of appreciable variations in trapped Xe isotopic composition is given by Table 8 and Fig. 7, where $^{129}Xe/^{136}Xe$ and $^{134}Xe/^{136}Xe$ ratios are plotted against $^{132}Xe/^{136}Xe$ for a large number of Apollo 15, 16, and 17 deep drill core and surface soils. (Additional noble gas data for these samples were reported by Bogard *et al.*, 1973.) The data in Fig. 7 have been corrected for a cosmogenic component assuming the composition used earlier for corrections to the 76321 temperature data. These four Xe isotopes are only slightly affected by corrections made for cosmogenic Xe, and the basic data trends are unchanged even if one allows extreme variations in cosmogenic ratios of ±50%. The cosmogenic

Table 8. Xe isotopic ratios (normalized to ^{136}Xe = 1.00) for Apollo 15 and 16 deep drill core and surface fines. Corrected ratios have been corrected for cosmogenic Xe assuming trapped 126/136 = 0.0144 (SUCOR) and assuming a cosmogenic Xe spectrum of 126/129/132/134/136 = 1.0/1.5/0.75/0.05/0. Uncertainties listed are one sigma of multiple measurements. Absolute ratios have an additional uncertainty of ±0.1% mass in the discrimination correction applied.

| Sample | 126/136 | 129/136 | 132/136 | 134/136 | Corrected | | |
					129/136	132/136	134/136
Apollo 16 deep drill core							
60007,111	0.0218	3.489	3.331	1.227	3.478	3.325	1.227
	±0.0007	±0.035	±0.032	±0.012			
60007,118	0.0215	3.390	3.279	1.224	3.379	3.274	1.224
	±0.0086	±0.028	±0.019	±0.022			
60006,11	0.0235	3.517	3.348	1.241	3.503	3.341	1.241
	±0.0021	±0.065	±0.055	±0.021			
60004,26	0.0274	3.543	3.352	1.236	3.523	3.342	1.236
	±0.0012	±0.029	±0.022	±0.009			
60004,17	0.0244	3.494	3.327	1.235	3.479	3.320	1.235
	±0.0010	±0.027	±0.020	±0.009			
60003,13	0.0239	3.421	3.294	1.228	3.407	3.287	1.228
	±0.0014	±0.023	±0.019	±0.008			
60001,11	0.0249	3.542	3.335	1.245	3.526	3.327	1.245
	±0.0008	±0.022	±0.014	±0.008			
Apollo 16 surface fines							
63321,3	0.0110	3.339	3.232	1.209	3.339	3.232	1.209
	±0.0004	±0.026	±0.025	±0.012			
65511,1	0.0239	3.434	3.288	1.220	3.420	3.281	1.220
	±0.0010	±0.049	±0.044	±0.017			
64421,16	0.0208	3.398	3.269	1.219	3.388	3.264	1.219
	±0.0007	±0.029	±0.023	±0.011			
65501,4	0.0270	3.444	3.278	1.221	3.425	3.269	1.221
	±0.0020	±0.025	±0.022	±0.016			
63341,2	0.0105	3.338	3.242	1.217	3.338	3.242	1.217
	±0.0002	±0.031	±0.027	±0.012			
60501,15	0.0205	3.414	3.281	1.222	3.405	3.276	1.222
	±0.006	±0.012	±0.011	±0.005			
Apollo 15 deep drill core							
15006,20	0.0452	3.451	3.304	1.233	3.405	3.281	1.231
	±0.0025	±0.042	±0.036	±0.023			
15005,19	0.0454	3.490	3.314	1.224	3.443	3.291	1.222
	±0.0027	±0.038	±0.023	±0.015			
15004,20	0.0405	3.512	3.333	1.232	3.473	3.313	1.231
	±0.0024	±0.040	±0.033	±0.015			
15003,22	0.0421	3.438	3.276	1.220	3.396	3.255	1.219
	±0.0026	±0.041	±0.028	±0.013			
15002,20	0.0356	3.441	3.268	1.218	3.409	3.252	1.217
	±0.0009	±0.032	±0.021	±0.011			
15001,24	0.0528	3.518	3.328	1.225	3.460	3.299	1.223
	±0.0018	±0.032	±0.020	±0.010			

Table 8. (*Continued*).

Sample	126/136	129/136	132/136	134/136	Corrected 129/136	Corrected 132/136	Corrected 134/136
15005,185	0.0509 ±0.0021	3.488 ±0.035	3.310 ±0.025	1.232 ±0.011	3.433	3.283	1.230
15006,174	0.0340 ±0.0010	3.477 ±0.029	3.321 ±0.021	1.224 ±0.010	3.448	3.306	1.223
15006,167	0.0321 ±0.0022	3.404 ±0.019	3.283 ±0.014	1.222 ±0.006	3.377	3.270	1.221
15006,149	0.0423 ±0.0014	3.464 ±0.032	3.318 ±0.029	1.226 ±0.012	3.422	3.297	1.225
15006,145	0.0371 ±0.0023	3.454 ±0.021	3.316 ±0.016	1.227 ±0.009	3.420	3.299	1.226
15006,138	0.0514 ±0.0026	3.510 ±0.026	3.333 ±0.022	1.232 ±0.010	3.455	3.305	1.230
15005,177	0.0574 ±0.0017	3.555 ±0.025	3.364 ±0.019	1.245 ±0.013	3.491	3.332	1.243
15005,180	0.0712 ±0.0032	3.561 ±0.055	3.329 ±0.040	1.228 ±0.020	3.476	3.286	1.225
15005,227	0.0529 ±0.0017	3.536 ±0.025	3.338 ±0.013	1.228 ±0.006	3.478	3.309	1.226
15005,239	0.0526 ±0.0021	3.535 ±0.031	3.343 ±0.024	1.239 ±0.012	3.478	3.314	1.237
15005,255	0.0468 ±0.0176	3.480 ±0.029	3.317 ±0.023	1.226 ±0.011	3.431	3.293	1.224
15005,267	0.0393 ±0.0011	3.397 ±0.025	3.255 ±0.018	1.220 ±0.099	3.360	3.236	1.219
15005,283	0.0512 ±0.0023	3.520 ±0.043	3.327 ±0.033	1.230 ±0.018	3.465	3.299	1.228
Apollo 15 bulk surface fines							
15041,80	0.0325 ±0.0015	3.438 ±0.037	3.324 ±0.026	1.226 ±0.015	3.411	3.310	1.225
15031,70	0.0391 ±0.0014	3.564 ±0.014	3.372 ±0.011	1.244 ±0.007	3.527	3.353	1.243
15301,1	0.0241 ±0.0013	3.466 ±0.032	3.301 ±0.021	1.226 ±0.011	3.451	3.294	1.226
15021,4	0.0356 ±0.0008	3.574 ±0.041	3.388 ±0.036	1.243 ±0.015	3.542	3.372	1.242
15101,2	0.0347 ±0.0010	3.509 ±0.034	3.339 ±0.031	1.233 ±0.012	3.479	3.324	1.232
15601,3	0.0325 ±0.0023	3.484 ±0.031	3.320 ±0.023	1.224 ±0.012	3.457	3.306	1.223
15251,49	0.0325 ±0.0022	3.416 ±0.045	3.287 ±0.042	1.220 ±0.016	3.389	3.273	1.219
15291,34	0.0345 ±0.0012	3.392 ±0.022	3.276 ±0.019	1.215 ±0.011	3.362	3.261	1.214
15471,50	0.0336 ±0.0015	3.552 ±0.033	3.374 ±0.022	1.239 ±0.013	3.523	3.360	1.238
15271,64	0.0326 ±0.0022	3.449 ±0.024	3.309 ±0.016	1.230 ±0.011	3.422	3.295	1.229

Table 8. (*Continued*).

Sample	126/136	129/136	132/136	134/136	Corrected 129/136	Corrected 132/136	Corrected 134/136
15271,64 agglutinate and nonagglutinate grain sizes							
1000–500 μ							
Agglutinate	0.0387	3.388	3.262	1.211	3.352	3.244	1.210
	±0.0010	±0.034	±0.0230	±0.013			
Nonagglutinate	0.1607	3.578	3.360	1.222	3.358	3.250	1.215
500–250 μ	±0.0079	±0.0418	±0.034	±0.013			
Agglutinate	0.0403	3.465	3.308	1.222	3.426	3.288	1.221
	±0.0011	±0.019	±0.016	±0.008			
Nonagglutinate	0.0957	3.495	3.313	1.222	3.373	3.252	1.218
250–125 μ	±0.0014	±0.037	±0.027	±0.012			
Agglutinate	0.0371	3.463	3.274	1.223	3.429	3.257	1.222
	±0.0012	±0.069	±0.045	±0.022			
Nonagglutinate	0.0965	3.498	3.323	1.223	3.375	3.261	1.219
125–63 μ	±0.0038	±0.026	±0.021	±0.011			
Agglutinate	0.0336	3.468	3.310	1.226	3.439	3.296	1.225
	±0.0006	±0.013	±0.008	±0.007			
Nonagglutinate	0.0822	3.542	3.347	1.226	3.440	3.296	1.223
	±0.0026	±0.028	±0.022	±0.017			

corrections were considerably less than the total spread among data for nearly all plotted points (Table 8), especially for Apollo 16 samples. The total spread in $^{132}Xe/^{136}Xe$ is 4.2%, and the three trapped ratios plotted are consistent with a linear mass fractionation trend. The trapped values plotted in Fig. 7 are similar to those in Fig. 5, except that a few of the bulk samples show relatively greater fractionation trends (lower $^{i}Xe/^{136}Xe$). There is no apparent relationship between trapped $^{132}Xe/^{136}Xe$ and depth in the Apollo 15 and 16 deep drill cores. However, the Apollo 16 surface fines show lower $^{132}Xe/^{136}Xe$ than the Apollo 16 core samples, and the two soils collected near North Ray Crater (63321 and 63341) show the lowest $^{132}Xe/^{136}Xe$. There is no apparent correlation between $^{132}Xe/^{136}Xe$ and $^{20}Ne/^{22}Ne$ for these data. *Variations in isotopic composition of up to several percent for trapped Xe in lunar soils are apparently common*, and suggest a possible problem in derivations of trapped Xe by ordinate–intercept plots. One of the basic assumptions in the use of these plots is that the isotopic composition of the trapped component does not vary with grain size. However, the variations in trapped Xe shown by Table 8, suggest that this assumption may not always be true. Small variations of trapped Xe (and Kr) with grain size may have produced some of the scatter in Figs. 5 and 6 and analogous ratio plots. Although a few of the various determined trapped components do not fall along the fractionation lines in Figs. 5 and 7, most do, and strongly suggest that mass fractionation has played an important role in modifying the composition of "trapped" Xe.

The trends shown by Fig. 6 and Table 7 strongly suggest that the various trapped Kr components in soils are also all related by a linear mass fractionation

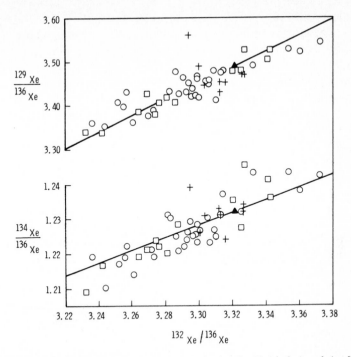

Fig. 7. Isotope correlation plots for trapped Xe in Apollo 15 (circles) and Apollo 16 (squares) deep drill cores and surface soils and in Apollo 17 surface soils (crosses). Apollo 15 and 16 data are the cosmogenic-corrected ratios of Table 8, except that the 15271 nonagglutinate data is not plotted. Solid lines indicate linear mass fractionation trends normalized to SUCOR (dark triangle).

process. Some controversy has existed as to whether "trapped" Kr is similar to atmospheric Kr (BEOC-10084 and 12001) or to Kr in carbonaceous chondrites (SUCOR). However, deviations of various trapped Kr determinations (including SUCOR) from the fractionation line are essentially all within their experimental uncertainties. The SUCOR trapped Kr composition appears to be highly fractionated, whereas all other trapped components trend along the fractionation line within ±1% of atmospheric Kr (Fig. 6). There is no apparent requirement to invoke excess, fissiogenic ^{86}Kr to explain any of these trapped Kr components. Basford *et al.* (1973) also recognized that several trapped Kr components showed small, mass-dependent differences relative to atmospheric Kr, and stated that they possessed no definitive evidence of the presence of fissiogenic ^{86}Kr. Isotopic compositions of "trapped" components have been commonly expressed in terms of δ-plots relative to a reference isotopic composition, usually that of air or AVCC (e.g. Basford *et al.*, 1973). We have not illustrated the data in Figs. 5–7 in this manner because there is no unique δ-plot representation corresponding to the single trend lines in the figures. For example, to calculate δ-plots for trapped Kr relative to air, data from the upper right portion of the Fig. 6 trend line will yield linear δ-plots showing relative enrichment of the light isotopes; data from the lower left portion of the trend line will yield δ-plots showing relative enrichment of the heavy isotopes. Thus, the single trend line in Fig. 6 is transferred into a family of δ-plots with both positive and negative slopes relative to air Kr. If certain preferred positions along the fractionation line exist, "characteristic" δ-patterns will occur, as is suggested by Basford *et al.* (1973) to be the case for Kr. However, it is not clear that this suggestion is supported by the totality of the data shown in Fig. 6.

It is informative to compare the relative extent of Kr and Xe isotopic fractionation for specific samples. Trapped Xe in the two Luna 16 samples measured by Basford *et al.* (1973) have unusually high values for trapped $^{132}Xe/^{136}Xe$ ($\sim$3.45) but fall significantly below the mass fractionation trends for other isotopic ratios. Trapped Xe in Luna 16 determined by Kaiser (1972) also possesses a high $^{132}Xe/^{136}Xe$, but falls on the mass fractionation trend for all isotope correlation plots. Trapped Kr in all three Luna 16 samples do not show extreme fractionation effects relative to other lunar soils. SUCOR trapped Kr (Podosek *et al.*, 1971) appears highly fractionated in favor of heavier isotopes compared to other trapped Kr, whereas SUCOR trapped Xe does not deviate significantly from other trapped Xe. The SUCOR components were not derived directly from ordinate–intercept plots, but rather by the assumption of common components among three different Apollo 11 fines samples, which may not have been the case for both Kr and Xe. For most other trapped components there is generally good agreement between the fractionation trends of Kr and Xe. For example, BEOC-10084 and the three trapped components derived in this work show Kr and Xe which are isotopically heavier relative to several fines reported by Basford *et al.* (1973) which show Kr and Xe which are isotopically lighter.

The severely fractionated nature of Kaiser's (1972) Luna 16 Xe determination relative to most trapped compositions (Fig. 5) appears readily explainable. This trapped Xe was determined from a temperature extraction, and because of the cosmogenic Xe present at higher temperatures, Kaiser used only the 550°C and 700°C extractions to obtain relative abundances of masses 130–136. As our temperature data on 76321 indicate, low temperature extractions are enriched in lighter isotopes. On the other hand, trapped Kr in Kaiser's Luna 16 sample was apparently derived using all of the temperature data, and thus a strong mass fractionation in Kr isotopic ratios does not result. The same type of temperature effect is also apparent between trapped Kr isotopic compositions derived for 10084 for the 600–900°C releases and the 1000–1300°C releases (Basford *et al.*, 1973). The effect presumably exists also in previous use of specific temperature data for trapped Xe and Kr compositions (Marti, 1969; Pepin *et al.*, 1970; Hohenberg *et al.*, 1970). This situation points out a disadvantage in using temperature extractions directly to derive the isotopic compositions of various components (including three isotope correlation plots). If the mass fractionation trend observed in 76321 Kr and Xe was produced during laboratory gas release, then even the assumption of a mixture of two components during temperature release does not necessarily yield the exact isotopic compositions of the two components. For example, trapped Xe released at low temperatures may be enriched in lighter isotopes due to fractionation, while cosmogenic Xe released at high temperatures may be enriched in heavier isotopes.

We believe that the data contained in Figs. 5–7, and analogous ratio plots present a good case for processes of linear mass fractionation on the lunar surface having produced much of the observed variations in trapped Xe and Kr isotopic compositions. Mass fractionation processes are known to affect the lighter noble gases in ratios such as He/Ne, Ne/Ar, ^{20}Ne/^{22}Ne and probably also in ^{3}He/^{4}He, ^{36}Ar/^{38}Ar, Kr/Ar and Xe/Ar (e.g. Eberhardt *et al.*, 1970, 1972; Hohenberg *et al.*, 1970; Pepin *et al.*, 1970; Hintenberger *et al.*, 1971; Kirsten *et al.*, 1971; Geiss *et al.*, 1972; Srinivasan *et al.*, 1972; Bogard *et al.*, 1973; Srinivasan, 1973). Thus, it is not unreasonable to expect mass fractionation effects in Xe and Kr isotopic compositions as well. A number of such processes can be envisioned: gas loss during ion implantation, gas loss during micrometeorite impact melting, implantation energies and depths which are mass-dependent, lunar atmosphere-implanted components, real temporal variations in the solar wind composition, etc., and all of these have been suggested in some context by one or more of the above investigators. Unfortunately, both empirical data and theoretical considerations concerning these processes are sparse. Huneke (1973) has presented a model of diffusive fractionation of surface implanted gases during and subsequent to ion implantation. In this model the concentration (C) of a particular gaseous species is parameterized by $\tau_0 D/R^2$, where τ_0 is a generalized time parameter for irradiation on the lunar surface, D the diffusion coefficient, and R the range of implantation. Huneke's model predicts that for large values of $\tau_0 D/R^2$ the isotopic composition of gases in lunar soils will be fractionated by the factor $R_i \sqrt{D_j}/R_j \sqrt{D_i}$ for isotopes i and j. Utilizing the ranges observed for 24 keV ^{22}Na and ^{24}Na ions incident on A1 and assuming $D \propto M^{-1/2}$ (M = mass of ion) for ^{20}Ne and ^{22}Ne, he calculated $R_{20}\sqrt{D_{22}}/R_{22}\sqrt{D_{20}} = 0.91$ in good agreement with the observed ^{20}Ne/^{22}Ne fractionation in many lunar soils with respect to the solar wind compositions. The model also predicts that initial gas released in stepwise temperature analyses will be biased in favor of lighter isotopes relative to the solar wind composition for low diffusion susceptibility (low $\tau_0 D/R^2$), but that this bias disappears as diffusion susceptibility increases, i.e. the ^{20}Ne/^{22}Ne in the initial gas release approaches the solar wind value. If we assume that this model is applicable to trapped Xe and Kr; specifically, if we assume by analogy to Ne that $R_i \sqrt{D_j}/R_j \sqrt{D_i} \approx M_i/M_j$ for Kr and Xe also, then the following anticipations result. If lunar soils possess intermediate to high susceptibility to Xe and Kr diffusion, trapped gas compositions ought to be biased in favor of heavier isotopes relative to the solar wind (up to a factor of ~1.007/mass for Xe and ~1.012/mass for Kr). If diffusion susceptibility for Xe and Kr is high (as is the case for Ne), trapped Xe and Kr compositions in most soils should group near this limiting fractionation value; if diffusion susceptibility is intermediate and highly variable, trapped compositions should show a spread between the solar wind value and a composition which is "heavier" by the above limiting fractionation factors. A spread in the isotopic compositions is definitely indicated by Figs. 5–7. The total spread in trapped compositions is ~1.016/mass for Xe and ~1.011/mass for Kr; omitting the Luna 16 and Pesyanoe Xe and the SUCOR Kr compositions, the spread becomes 1.011/mass for Xe and 1.005/mass for Kr. Huneke's (1973) model under the above assumptions of extension to Xe and Kr, would anticipate the following results for stepwise temperature release of trapped Xe and Kr. For soils with high diffusion susceptibility, the trapped gas composition released at low temperatures should approximate the solar-implanted composition, but at higher temperatures the composition released should quickly approach the limiting, fractionated value. For soils with low diffusion susceptibility, the composition released at low-extraction temperatures could be enriched in the lighter isotopes relative to the solar-implanted values, and with increasing extraction temperature the composition of the gas released would become progressively "heavier" as it approached toward limiting fractionation values. As is the apparent case with bulk soils, the Xe and Kr data from stepwise release of 76321 suggests that the model case of low diffusion susceptibility is more likely. However, such conclusions can be only very tentative because we do not accurately know the diffusion susceptibility of lunar soils for Xe and Kr under these model conditions, and because we have not considered other processes listed above which may produce isotope fractionation (lunar atmosphere-implanted components, gas loss during impact melting, etc.). Therefore, we cannot accurately deduce the composition of solar wind-implanted Xe and Kr relative to measured data trends (Figs. 5–7). However, it does appear reasonable to conclude that such compositions lie within or very near the range of measured trapped compositions, and generally related to them by mass fractionation.

As stated previously, Basford *et al.* (1973) accept mass fractionation as a possible cause of variations in trapped Kr, but believe excess Xe components (from fission, ^{129}I decay, and neutron capture) have caused much of the variations in trapped Xe in their samples. These authors have discussed in detail the isotopic effects such excess components are expected to produce. Unfortunately, for isotope correlation plots such as Fig. 5 the addition of fission Xe lowers the isotope ratios in such a manner as to produce a trend similar to the mass fractionation trend and not easily distinguishable from it. Trapped Xe for two Luna 16 samples (Basford *et al.*, 1973) are not completely consistent with either a mass fractionation or a trapped plus fissiogenic Xe trend, and to a lesser degree this is also true for some isotopic ratios of other samples. Excess Xe components from fission and radiogenic decay cannot be completely ruled out, especially for certain samples such as Apollo 14 soils where some local breccias apparently contain these excess Xe components (e.g. Crozaz *et al.*, 1972). However, it should be emphasized that to produce the observed variations in trapped Xe by addition of *excess, surface-correlated* fission Xe to a common trapped component requires that trapped component to fall at the extreme right of Fig. 5; i.e. to possess values of ^{132}Xe/^{136}Xe, ^{134}Xe/^{136}Xe, etc. at least as large as the highest trapped Xe value. This concept would also require trapped Xe in most lunar soils to contain an amount of excess fissiogenic ^{136}Xe which is several percent of the total ^{136}Xe present. Similarly, to explain the total observed variations in trapped ^{129}Xe and ^{128}Xe by addition of excess, surface-correlated Xe to a common trapped component (Basford *et al.*, 1973) would require amounts of ^{129}Xe from ^{129}I decay and ^{128}Xe from neutron capture which are a few percent of the total ^{129}Xe and ^{128}Xe in soils. In the case of the mare soils, which are presumably composed largely of basalt fragments with crystallization ages similar to mare basalts (Papanastassiou and Wasserburg, 1973), there is also the problem of storing large excesses of ^{129}Xe from ^{129}I and fissiogenic Xe from ^{244}Pu for times of ~ 0.7–1×10^9 yr, before implanting it in grain surfaces. Also, contrary to what might be expected from the excess Xe concept, no apparent relationship exists between trapped Xe isotopic ratios (Figs. 5 and 7) and either crystallization ages assigned to various lunar sites, or to measured ^{40}Ar/^{36}Ar in the soils.

In conclusion, we emphasize that appreciable variations in the isotopic compositions of trapped Xe and Kr occur in lunar soils. Much of these variations can apparently be explained by processes on the lunar surface which produce mass fractionation without resorting to large amounts of excess components. However, the occurrence of appreciable concentrations of certain excess Xe components in some soils as argued by Basford *et al.* (1973) cannot be ruled out. It is necessary to understand the exact causes of isotopic variations in Kr and Xe in order to determine the true composition of the solar wind component. Also, several investigators (Eberhardt *et al.*, 1970, 1972; Podosek *et al.*, 1971; Kaiser, 1972; Basford *et al.*, 1973; Srinivasan *et al.*, 1972) have attempted to relate AVCC and atmospheric Kr and Xe components to the "solar wind" composition by various processes, especially mass fractionation. If trapped Kr and Xe components in lunar soils are all related by mass fractionation processes, such comparisons are probably valid. However, if appreciable excess Xe components exist in lunar soils, the "trapped" component will not be representative of the solar wind composition, and such comparisons may be misleading.

REFERENCES

Basford J. R. (1974) Potassium-argon analysis of Apollo 11 regolith (abstract). In *Lunar Science—V*, pp. 41–43. The Lunar Science Institute, Houston.

Basford J. R., Dragon J. C., and Pepin R. O. (1973) Krypton and xenon in lunar fines. *Proc. Fourth Lunar Sci. Conf., Geochim. Cosmochim. Acta*, Suppl. 4, Vol. 2, pp. 1915–1955. Pergamon.

Bogard D. D. and Cressy P. J., Jr. (1973) Spallation production of ^{3}He, ^{21}Ne, and ^{38}Ar from target elements in the Bruderheim chondrite. *Geochim. Cosmochim. Acta* **37**, 527–546.

Bogard D. D., Funkhouser J. G., Schaeffer O. A., and Zahringer J. (1971) Noble gas abundances in lunar material: Cosmic-ray spallation products and radiation ages from the Sea of Tranquillity and the Ocean of Storms. *J. Geophys. Res.* **76**, 2757–2799.

Bogard D. D. and Nyquist L. E. (1972) Noble gas studies on regolith materials from Apollo 14 and 15. *Proc. Third Lunar Sci. Conf., Geochim. Cosmochim. Acta*, Suppl. 3, Vol. 2, pp. 1797–1820. MIT Press.

Bogard D. D., Nyquist L. E., Hirsch W. C., and Moore D. R. (1973) Trapped solar and cosmogenic noble gas abundances in Apollo 15 and 16 deep drill samples. *Earth Planet. Sci. Lett.* **21**, 52–69.

Bogard D. D., Nyquist L. E., and Hirsch W. C. (1974) Noble gases in Apollo 17 boulders and soils (abstract). In *Lunar Science—V*, pp. 73–75. The Lunar Science Institute, Houston.

Burnett D. S., Huneke J. C., Podosek F. A., Russ G. P., III, and Wasserburg G. J. (1971) The irradiation history of lunar samples. *Proc. Second Lunar Sci. Conf., Geochim. Cosmochim. Acta*, Suppl. 2, Vol. 2, pp. 1671–1679. MIT Press.

Eberhardt P., Geiss J., Graf H., Grögler N., Krähenbühl U., Schwaller H., Schwarzmüller J., and Stettler A. (1970) Trapped solar wind noble gases, exposure age and K/Ar-age in Apollo 11 fine material. *Proc. Apollo 11 Lunar Sci. Conf., Geochim. Cosmochim. Acta*, Suppl. 1, Vol. 1, pp. 1037–1070. Pergamon.

Eberhardt P., Geiss J., Graf H., Grögler N., Mendia M. D., Mörgeli M., Schwaller H., and Stettler A. (1972) Trapped solar wind gases in Apollo 12 lunar fines 12001 and Apollo 11 breccia 10046. *Proc. Second Lunar Sci. Conf., Geochim. Cosmochim. Acta*, Suppl. 3, Vol. 2, pp. 1821–1856. MIT Press.

Eberhardt P., Geiss J., Graf N., Grögler U., Krähenbühl H., Schwaller H., and Stettler A. (1974) Noble gas investigations of lunar rocks 10017 and 10071. *Geochim. Cosmochim. Acta* **38**, 97–120.

Eugster O., Eberhardt P., and Geiss J. (1967) Kr and Xe isotopic composition in three carbonaceous chondrites. *Earth Planet. Sci. Lett.* **3**, 249–257.

Geiss J., Buehler F., Cerutti H., Eberhardt P., and Filleaux Ch. (1972) Solar wind composition experiment. *Apollo 16 Preliminary Science Report*, NASA SP-315, Chapter 14.

Hintenberger H., Weber H. W., and Takaoka N. (1971) Concentrations and isotopic abundances of the rare gases in lunar matter. *Proc. Second Lunar Sci. Conf., Geochim. Cosmochim. Acta*, Suppl. 2, Vol. 2, pp. 1607–1625. MIT Press.

Hohenberg C. M., Davis P. K., Kaiser W. A., Lewis R. S., and Reynolds J. H. (1970) Trapped and cosmogenic rare gases from stepwise heating of Apollo 11 samples. *Proc. Apollo 11 Lunar Sci. Conf., Geochim. Cosmochim. Acta*, Suppl. 1, Vol. 2, pp. 1283–1310. Pergamon.

Huneke J. C. (1973) Diffusive fractionation of surface implanted gases. *Earth Planet. Sci. Lett.* **21**, 35–44.

Huneke J. C., Jessberger E. K., Podosek F. A., and Wasserburg G. J. (1973) $^{40}Ar/^{39}Ar$ measurements in Apollo 16 and 17 samples and the chronology of metamorphic and volcanic activity in the Taurus-Littrow region. *Proc. Fourth Lunar Sci. Conf., Geochim. Cosmochim. Acta*, Suppl. 4, Vol. 2, pp. 1725–1726. Pergamon.

Kaiser W. A. (1972) Rare gas studies in Luna-16-G-7 fines by stepwise heating technique. A low fission solar wind Xe. *Earth Planet. Sci. Lett.* **13**, 387–399.

Kirsten T., Steinbrum F., and Zähringer J. (1971) Location and variation of trapped rare gases in Apollo 12 lunar samples. *Proc. Second Lunar Sci. Conf., Geochim. Cosmochim. Acta*, Suppl. 2, Vol. 2, pp. 1651–1669. MIT Press.

Marti K. (1969) Solar-type xenon: A new isotopic composition of xenon in the Pesyanoe meteorite. *Science* **166**, 1263–1265.

Marti K., Lugmair G. W., and Urey H. C. (1970) Solar wind gases, cosmic ray spallation products and the irradiation history of Apollo 11 samples. *Proc. Apollo 11 Lunar Sci. Conf., Geochim. Cosmochim. Acta*, Suppl. 1, Vol. 2, pp. 1357–1368. Pergamon.

Nief G. (1960) NBS Tech. Note 51 (editor F. Mohler).

Nyquist L. E., Bansal B. M., Wiesmann H., and Jahn B. M. (1974) Taurus-Littrow chronology: Some constraints on early lunar crustal development. *Proc. Fifth Lunar Sci. Conf., Geochim. Cosmochim. Acta*. This volume.

Papanastassiou D. A. and Wasserburg G. J. (1973) Rb–Sr ages and initial strontium in basalts from Apollo 15. *Earth Planet. Sci. Lett.* **17**, 324–337.

Pepin R. O., Nyquist L. E., Phinney D., and Black D. C. (1970a) Isotopic composition of rare gases in lunar samples. *Science* **167**, 550–553.

Pepin R. O., Nyquist L. E., Phinney D., and Black D. C. (1970b) Rare gases in Apollo 11 lunar material. *Proc. Apollo 11 Lunar Sci. Conf., Geochim. Cosmochim. Acta*, Suppl. 1, Vol. 2, pp. 1435–1444. Pergamon.

Podosek F. A., Huneke J. C., Burnett D. S., and Wasserburg G. J. (1971) Isotopic composition of xenon and krypton in the lunar soil and in the solar wind. *Earth Planet. Sci. Lett.* **10**, 199–216.

Rhodes J. M., Rodgers K. V., Shih C., Bansal C. M., Nyquist L. E., and Wiesmann H. (1974) The relationship between geology and soil chemistry at the Apollo 17 landing site (abstract). In *Lunar Science—V*, pp. 630–632. The Lunar Science Institute, Houston.

Srinivasan B. (1973) Variations in the isotopic composition of trapped rare gases in lunar sample 14318. *Proc. Third Lunar Sci. Conf., Geochim. Cosmochim. Acta*, Suppl. 3, Vol. 2, pp. 2049–2064. MIT Press.
Srinivasan B., Hennecke E. W., Sinclair D. E., and Manuel O. K. (1972) A comparison of noble gases released from lunar fines (# 1560164) with noble gases in meteorites and in the earth. *Proc. Third Lunar Sci. Conf., Geochim. Cosmochim. Acta*, Suppl. 3, Vol. 2, pp. 1927–1946. MIT Press.

Proceedings of the Fifth Lunar Conference
(Supplement 5, Geochimica et Cosmochimica Acta)
Vol. 2 pp. 2005–2022 (1974)
Printed in the United States of America

Solar, spallogenic, and radiogenic rare gases in Apollo 17 soils and breccias

H. Hintenberger, H. W. Weber, and L. Schultz

Max-Planck-Institut für Chemie (Otto-Hahn-Institut), Mainz, Germany

Abstract—Concentration and isotopic composition of helium, neon and argon as well as ^{84}Kr and ^{132}Xe concentrations have been measured in Apollo 17 fines and breccias 72701, 74260, 75061, 75081, 79035, and 79135. From all samples, at least eight grain size fractions were analyzed and surface and volume correlated components determined. Ilmenite separates have been measured from the 35–54 μm fraction of 72701, 74241, 75081, and 79035.

^{21}Ne exposure ages for soils range from 27 m.y. (74220) to 230 m.y. (75081). The breccias, however, have exposure ages of about 600 m.y. (79035) and about 800 m.y. (79135). K–Ar ages for four soils lie between 2.9 and 3.9 b.y.

The trapped ^{4}He and ^{20}Ne concentrations in various soils are determined to a large degree by their ilmenite content. In ilmenite-rich material (TiO$_2$ > 6%) correlations were found between the ^{4}He/^{36}Ar and ^{20}Ne/^{36}Ar ratios and the ^{36}Ar concentration, which is used as an indicator of the solar wind exposure. These ratios decrease with increasing solar wind exposure.

The concentrations of trapped ^{20}Ne and ^{36}Ar are correlated with the spallogenic ^{21}Ne. Furthermore, the trapped (^{4}He/^{3}He) ratios and the surface correlated (^{40}Ar/^{36}Ar) ratios appear to be correlated.

Characteristic differences in the rare gas abundance pattern have been found between breccias and soils.

(a) The concentrations of trapped gases as well as of spallogenic ^{21}Ne are higher in breccias than in soils.
(b) The ratio ^{132}Xe/^{36}Ar of the trapped gases in breccias is higher than in soils, indicating an excess of ^{132}Xe or a loss of ^{36}Ar and ^{84}Kr in breccias relative to soil materials.

In ilmenite the trapped He and Ne concentrations are higher than in bulk samples; the ^{36}Ar concentration, however, is for the three soils analyzed smaller in ilmenite than in the three bulk fines analyzed.

INTRODUCTION

THE CONCENTRATION PATTERN of surface correlated rare gas nuclides reflects—with the exception of ^{40}Ar—the abundance distribution of these isotopes in the solar wind. This pattern is, however, modified by various secondary processes acting on the lunar material such as diffusion losses, saturation effects, shock influences, or erosion phenomena. It is the aim of this investigation to elucidate the influence of these secondary processes on the elemental and isotopic composition of rare gases in lunar materials. Therefrom, information on the solar wind composition, on the history of soils and breccias, and on the mechanisms of the incorporation of rare gas nuclides into lunar materials can be obtained.

This paper reports on concentration and isotopic composition of He, Ne, and Ar as well as the concentration of ^{84}Kr and ^{132}Xe in bulk material, in grain size fractions of bulk material, and to some extent in ilmenite separates from soils and breccias of the Apollo 17 mission. It continues our earlier investigations on lunar material from the Apollo 11 to Apollo 16 missions (Hintenberger *et al.*, 1970a, 1970b, 1971; Hintenberger and Weber, 1973).

The samples investigated are listed in Table 1. They cover all major regolith types of the Apollo 17 landing site including the unique "orange" soil 74220 and the two gray soils collected adjacent to 74220. Furthermore, two dark matrix breccias have been analyzed. To increase the number of data points in the graphs and to verify that the relations found for Apollo 17 samples also hold for samples from other lunar missions, additional data from Hintenberger *et al.* (1970b, 1971) and Hintenberger and Weber (1971) are included in some figures. These samples and their abbreviations are listed in a footnote of Table 1.

EXPERIMENTAL PROCEDURES

The experimental details of the noble gas analyses have been described in earlier publications (Hintenberger *et al.*, 1970b, 1971; Schultz and Hintenberger, 1967; Weber, 1973).

At least eight grain size fractions from each of the six Apollo 17 soils and two breccias listed in Table 1 have been prepared by dry sieving. Stainless steel sieves of 20, 25, 35.5, 54, 75, 120, 200, and 300 μm, respectively, were used. In some cases, three smaller grain size fractions have been obtained using nylon sieves of 5 and 10 μm, respectively.

The ilmenite fractions were separated from bulk grain size fractions by Clerici solution in special glass vessels. The purity of these fractions was estimated from optical inspections to be better than 90% with the exception of the very small sample 72701 whose ilmenite content is only about 70%.

Table 1. Sample description[f].

Sample No. and description			K ppm	U ppm	Th ppm	TiO$_2$ %
7a	72701	light mantle surface soil	1330[a]	0.808[b]	2.962[b]	1.52[a]
7b	74220	"orange" soil	665[a]	0.161[c]	0.556[c]	8.81[a]
7c	74241 ⎱ gray soils adjacent	1000[a]	0.14[d]	—	8.61[a]
7d	74260 ⎰ to 74220	1000[a]		—	7.68[a]
7e	75061 ⎱	630[a]	0.35[a]	0.89[a]	10.31[a]
7f	75081 ⎰ dark mantle surface soils	670[a]	0.25[d]	—	9.41[a]
7G*	79035	dark matrix breccia (Van Serg Crater ejecta?)	—	0.31[e]	—	—
7H	79135	dark matrix breccia (interior of a boulder)	830[a]	—	—	5.15[a]

[a]LSPET (1973).
[b]Nunes *et al.* (1974).
[c]Tatsumoto *et al.* (1973).
[d]Jovanovic and Reed (1974).
[e]Morgan *et al.* (1974).
[f]In the figures other lunar samples are indicated by the following numbers:
10021 = 1A; 10060 = 1B; 10061 = 1C;
10084 = 1d; 10087 = 1e; 12070 = 2a;
14163 = 4a; 15471 = 5a; 68501 = 6a
*Note that numbers for breccias have a capital letter.

RESULTS

Bulk samples

Table 2 shows the results on bulk samples of Apollo 17 soils and breccias. Due to the heterogeneity of the samples and the small sample weights used the measured absolute concentrations vary far outside the analytical errors, which are in the order of ±3% for the concentrations of the He, Ne, and Ar isotopes and about ±10% for ^{84}Kr and ^{132}Xe. The precision of the data obtained have been tested by measurements on the Bruderheim standard. The concentrations given in Table 2 are mean values for the results obtained on different aliquots of the same sample. The number of aliquots measured and the maximum and minimum values observed for the ^{4}He contents are given in the last three lines. Table 2 includes also some elemental and isotopic abundance ratios.

Grain size fractions

The gas concentrations as well as the atomic and isotopic abundance ratios in grain size fractions of 72701, 74260, 75061, 75081, 79035, and 79135 are compiled in the Tables 3 to 8. The corresponding data for the orange soil 74220 and the

Table 2. Rare gas concentrations in cm³ STP/g in bulk material of fines and breccias from Apollo 17 mission. Errors in isotope ratios 2%, errors in concentrations of He, Ne, and Ar about 3%, of Kr and Xe about 10%. Sample weights between 0.5–2.1 mg.

Nuclides and Ratios	72701,25	74220,47	74241,24	74260,9	75061,21	75081,72	79035,15	79135,32
^{3}He	2.81×10^{-5}	4.85×10^{-6}	4.91×10^{-5}	4.63×10^{-5}	4.14×10^{-5}	6.82×10^{-5}	6.68×10^{-5}	4.33×10^{-5}
^{4}He	8.07×10^{-2}	1.426×10^{-2}	1.573×10^{-1}	1.420×10^{-1}	1.086×10^{-1}	1.845×10^{-1}	1.882×10^{-1}	1.295×10^{-1}
^{20}Ne	1.664×10^{-3}	1.566×10^{-4}	1.542×10^{-3}	1.594×10^{-3}	1.248×10^{-3}	2.33×10^{-3}	3.10×10^{-3}	2.75×10^{-3}
^{21}Ne	4.27×10^{-6}	4.35×10^{-7}	3.98×10^{-6}	4.18×10^{-6}	3.26×10^{-6}	5.97×10^{-6}	8.48×10^{-6}	7.48×10^{-6}
^{22}Ne	1.292×10^{-4}	1.213×10^{-5}	1.228×10^{-4}	1.270×10^{-4}	9.47×10^{-5}	1.812×10^{-4}	2.42×10^{-4}	2.13×10^{-4}
^{36}Ar	3.81×10^{-4}	1.653×10^{-6}	1.593×10^{-4}	1.708×10^{-4}	1.676×10^{-4}	3.72×10^{-4}	4.10×10^{-4}	4.47×10^{-4}
^{38}Ar	7.24×10^{-5}	3.14×10^{-6}	3.06×10^{-5}	3.27×10^{-5}	3.17×10^{-5}	7.02×10^{-5}	7.84×10^{-5}	8.51×10^{-5}
^{40}Ar	4.15×10^{-4}	1.067×10^{-4}	1.197×10^{-3}	1.232×10^{-3}	1.565×10^{-4}	3.05×10^{-4}	9.19×10^{-4}	1.217×10^{-3}
^{84}Kr	2.0×10^{-7}	8.5×10^{-9}	7.2×10^{-8}	7.1×10^{-8}	9.4×10^{-8}	1.9×10^{-7}	1.9×10^{-7}	1.7×10^{-7}
^{132}Xe	2.2×10^{-8}	1.8×10^{-9}	1.5×10^{-8}	1.2×10^{-8}	1.2×10^{-8}	2.6×10^{-8}	4.4×10^{-8}	5.0×10^{-8}
^{4}He/^{3}He	2870	2940	3200	3070	2620	2710	2820	2990
^{20}Ne/^{22}Ne	12.88	12.91	12.56	12.55	13.18	12.86	12.92	12.91
^{22}Ne/^{21}Ne	30.3	27.9	30.9	30.4	29.0	30.4	28.5	28.5
^{36}Ar/^{38}Ar	5.26	5.26	5.21	5.22	5.29	5.30	5.22	5.24
^{40}Ar/^{36}Ar	1.089	6.45	7.51	7.21	0.934	0.820	2.24	2.73
^{4}He/^{20}Ne	48.5	91.1	102.0	89.1	87.0	79.2	60.7	47.1
^{20}Ne/^{36}Ar	4.37	9.47	9.68	9.33	7.45	6.26	7.56	6.17
^{36}Ar/^{84}Kr	1910	1940	2220	2410	1780	1960	2160	2620
^{84}Kr/^{132}Xe	9.1	4.7	4.7	5.9	8.1	7.3	4.3	3.4
number of aliquots	5	3	3	4	3	4	3	3
^{4}He-content max.	8.96×10^{-2}	1.458×10^{-2}	1.615×10^{-1}	1.442×10^{-1}	1.408×10^{-1}	2.02×10^{-1}	2.04×10^{-1}	1.712×10^{-1}
min.	7.03×10^{-2}	1.397×10^{-2}	1.530×10^{-1}	1.403×10^{-1}	7.10×10^{-2}	1.566×10^{-1}	1.630×10^{-1}	1.064×10^{-1}

Table 3. Rare gas concentrations in cm³ STP/g in grain size fractions of fines from Apollo 17 sample 72701,25. Errors in isotope ratios 2%, errors in concentrations of He, Ne, and Ar about 3%, of Kr and Xe about 10%. Sample weights between 0.8–3.7 mg.

	<20 μm	20–25 μm	25–35 μm	35–54 μm	54–75 μm	75–120 μm	120–200 μm	200–300 μm
^{3}He	4.15×10^{-5}	1.789×10^{-5}	1.169×10^{-5}	7.23×10^{-6}	5.10×10^{-6}	4.02×10^{-6}	2.57×10^{-6}	3.00×10^{-6}
^{4}He	1.158×10^{-1}	5.08×10^{-2}	3.24×10^{-2}	1.967×10^{-2}	1.165×10^{-2}	9.33×10^{-3}	5.32×10^{-3}	6.66×10^{-3}
^{20}Ne	2.51×10^{-3}	1.240×10^{-3}	7.44×10^{-4}	4.77×10^{-4}	2.53×10^{-4}	2.09×10^{-4}	1.395×10^{-4}	1.839×10^{-4}
^{21}Ne	6.28×10^{-6}	3.32×10^{-6}	2.11×10^{-6}	1.497×10^{-6}	1.055×10^{-6}	8.45×10^{-7}	6.26×10^{-7}	7.37×10^{-7}
^{22}Ne	1.957×10^{-4}	9.60×10^{-5}	5.90×10^{-5}	3.83×10^{-5}	2.03×10^{-5}	1.695×10^{-5}	1.114×10^{-5}	1.482×10^{-5}
^{36}Ar	5.36×10^{-4}	2.54×10^{-4}	1.741×10^{-4}	1.115×10^{-4}	3.43×10^{-5}	4.65×10^{-5}	3.12×10^{-5}	4.65×10^{-5}
^{38}Ar	9.85×10^{-5}	4.83×10^{-5}	3.28×10^{-5}	2.14×10^{-5}	6.83×10^{-6}	8.91×10^{-6}	6.09×10^{-6}	8.99×10^{-6}
^{40}Ar	5.26×10^{-4}	2.96×10^{-4}	2.11×10^{-4}	1.601×10^{-4}	8.45×10^{-5}	1.130×10^{-4}	9.54×10^{-5}	1.315×10^{-4}
^{84}Kr	3.12×10^{-7}	1.46×10^{-7}	9.33×10^{-8}	6.17×10^{-8}	1.91×10^{-8}	2.48×10^{-8}	1.75×10^{-8}	2.59×10^{-8}
^{132}Xe	4.1×10^{-8}	1.92×10^{-8}	1.26×10^{-8}	9.5×10^{-9}	4.6×10^{-9}	4.1×10^{-9}	2.8×10^{-9}	4.3×10^{-9}
^{4}He/^{3}He	2790	2840	2770	2720	2280	2320	2070	2220
^{20}Ne/^{22}Ne	12.83	12.92	12.61	12.45	12.46	12.33	12.52	12.41
^{22}Ne/^{21}Ne	31.2	28.9	28.0	25.6	19.24	20.1	17.80	20.1
^{36}Ar/^{38}Ar	5.44	5.26	5.31	5.21	5.02	5.22	5.12	5.17
^{40}Ar/^{36}Ar	0.981	1.165	1.212	1.436	2.46	2.43	3.06	2.83
^{4}He/^{20}Ne	46.1	41.0	43.5	41.2	46.0	44.6	38.1	36.2
^{20}Ne/^{36}Ar	4.68	4.88	4.27	4.28	7.38	4.49	4.47	3.95
^{36}Ar/^{84}Kr	1720	1740	1870	1860	1800	1880	1780	1800
^{84}Kr/^{132}Xe	7.6	7.6	7.4	6.5	4.2	6.0	6.3	6.0

Table 4. Rare gas concentrations in cm³ STP/g in grain size fractions of fines from Apollo 17 sample 74260,9. Errors in isotope ratios 2%, errors in concentrations of He, Ne, and Ar about 3%, of Kr and Xe about 10%. Sample weights between 0.5–2.9 mg.

	<20 μm	<20 μm	20–25 μm	25–35 μm	35–54 μm	75–120 μm	120–200 μm	200–300 μm
^{3}He	5.88×10^{-5}	5.80×10^{-5}	1.580×10^{-5}	1.364×10^{-5}	7.48×10^{-6}	5.93×10^{-6}	2.46×10^{-6}	6.91×10^{-6}
^{4}He	1.759×10^{-1}	1.745×10^{-1}	4.42×10^{-2}	3.86×10^{-2}	1.840×10^{-2}	1.454×10^{-2}	4.76×10^{-3}	1.997×10^{-2}
^{20}Ne	2.05×10^{-3}	2.10×10^{-3}	6.04×10^{-4}	4.84×10^{-4}	2.69×10^{-4}	1.626×10^{-4}	7.10×10^{-5}	2.22×10^{-4}
^{21}Ne	5.45×10^{-6}	5.40×10^{-6}	1.811×10^{-6}	1.531×10^{-6}	1.004×10^{-6}	7.24×10^{-7}	5.12×10^{-7}	8.20×10^{-7}
^{22}Ne	1.643×10^{-4}	1.688×10^{-4}	4.98×10^{-5}	4.04×10^{-5}	2.24×10^{-5}	1.364×10^{-5}	6.30×10^{-6}	1.905×10^{-5}
^{36}Ar	2.05×10^{-4}	2.25×10^{-4}	4.95×10^{-5}	4.66×10^{-5}	2.59×10^{-5}	1.727×10^{-5}	1.107×10^{-5}	2.90×10^{-5}
^{38}Ar	3.87×10^{-5}	4.34×10^{-5}	9.68×10^{-6}	9.16×10^{-6}	5.23×10^{-6}	3.55×10^{-6}	2.37×10^{-6}	5.66×10^{-6}
^{40}Ar	1.367×10^{-3}	1.588×10^{-3}	3.03×10^{-4}	3.08×10^{-4}	1.698×10^{-4}	1.353×10^{-4}	8.72×10^{-5}	2.29×10^{-4}
^{84}Kr	8.7×10^{-8}	8.8×10^{-8}	2.0×10^{-8}	1.9×10^{-8}	1.0×10^{-8}	6.4×10^{-9}	4.2×10^{-9}	1.3×10^{-8}
^{132}Xe	1.5×10^{-8}	1.7×10^{-8}	3.7×10^{-9}	2.3×10^{-9}	2.1×10^{-9}	1.2×10^{-9}	1.1×10^{-9}	2.2×10^{-9}
^{4}He/^{3}He	2990	3010	2800	2830	2460	2450	1935	2890
^{20}Ne/^{22}Ne	12.48	12.44	12.13	11.98	12.01	11.92	11.27	11.65
^{22}Ne/^{21}Ne	30.1	31.3	27.5	26.4	22.3	18.84	12.30	23.2
^{36}Ar/^{38}Ar	5.30	5.18	5.11	5.09	4.95	4.86	4.67	5.12
^{40}Ar/^{36}Ar	6.67	7.06	6.12	6.61	6.56	7.83	7.88	7.90
^{4}He/^{20}Ne	85.8	83.1	73.2	79.8	68.4	89.4	67.0	90.0
^{20}Ne/^{36}Ar	10.00	9.33	12.20	10.39	10.39	9.42	6.41	7.66
^{36}Ar/^{84}Kr	2350	2550	2540	2410	2560	2720	2640	2250
^{84}Kr/^{132}Xe	6.0	5.2	5.3	8.2	4.9	5.3	3.9	5.8

Table 5. Rare gas concentrations in cm³ STP/g in grain size fractions of fines from Apollo 17 sample 75061,21. Errors in isotope ratios 2%, errors in concentrations of He, Ne, and Ar about 3%, of Kr and Xe about 10%. Sample weights between 0.75–2.8 mg.

	<20 μm	20–25 μm	25–35 μm	35–54 μm	54–75 μm	75–120 μm	120–200 μm	200–300 μm
^{3}He	5.55×10^{-5}	1.842×10^{-5}	1.443×10^{-5}	1.411×10^{-5}	6.46×10^{-6}	6.47×10^{-6}	8.19×10^{-6}	6.05×10^{-6}
^{4}He	1.532×10^{-1}	4.97×10^{-2}	3.94×10^{-2}	3.58×10^{-2}	1.416×10^{-2}	1.429×10^{-2}	1.940×10^{-2}	1.338×10^{-2}
^{20}Ne	2.07×10^{-3}	7.91×10^{-4}	6.60×10^{-4}	4.70×10^{-4}	2.80×10^{-4}	2.76×10^{-4}	2.62×10^{-4}	2.23×10^{-4}
^{21}Ne	5.44×10^{-6}	2.31×10^{-6}	2.04×10^{-6}	1.445×10^{-6}	1.083×10^{-6}	1.102×10^{-6}	8.79×10^{-7}	8.61×10^{-7}
^{22}Ne	1.657×10^{-4}	6.31×10^{-5}	5.45×10^{-5}	3.72×10^{-5}	2.23×10^{-5}	2.24×10^{-5}	2.11×10^{-5}	1.824×10^{-5}
^{36}Ar	2.80×10^{-4}	7.60×10^{-5}	6.78×10^{-5}	5.16×10^{-5}	2.59×10^{-5}	2.45×10^{-5}	3.47×10^{-5}	2.90×10^{-5}
^{38}Ar	5.33×10^{-5}	1.452×10^{-5}	1.320×10^{-5}	1.010×10^{-5}	5.02×10^{-6}	4.78×10^{-6}	6.71×10^{-6}	5.74×10^{-6}
^{40}Ar	2.62×10^{-4}	7.44×10^{-5}	7.26×10^{-5}	6.21×10^{-5}	3.73×10^{-5}	3.68×10^{-5}	5.41×10^{-5}	5.24×10^{-5}
^{84}Kr	1.64×10^{-7}	5.04×10^{-8}	4.13×10^{-8}	3.14×10^{-8}	1.33×10^{-8}	1.34×10^{-8}	2.01×10^{-8}	1.60×10^{-8}
^{132}Xe	2.3×10^{-8}	8.5×10^{-9}	1.3×10^{-8}	5.8×10^{-9}	2.2×10^{-9}	2.0×10^{-9}	2.9×10^{-9}	2.7×10^{-9}
^{4}He/^{3}He	2760	2700	2730	2540	2190	2210	2370	2210
^{20}Ne/^{22}Ne	12.49	12.54	12.11	12.63	12.56	12.32	12.42	12.23
^{22}Ne/^{21}Ne	30.5	27.3	26.7	25.7	20.6	20.3	24.0	21.2
^{36}Ar/^{38}Ar	5.25	5.23	5.14	5.11	5.16	5.13	5.17	5.05
^{40}Ar/^{36}Ar	0.936	0.979	1.071	1.203	1.440	1.502	1.559	1.807
^{4}He/^{20}Ne	74.0	62.8	59.7	76.2	50.6	51.8	74.0	60.0
^{20}Ne/^{36}Ar	7.39	10.41	9.73	9.11	10.81	11.27	7.55	7.69
^{36}Ar/^{84}Kr	1710	1510	1640	1640	1950	1830	1730	1810
^{84}Kr/^{132}Xe	7.1	5.9	3.1	5.4	6.0	6.7	6.9	5.9

Table 6. Rare gas concentrations in cm³ STP/g in grain size fractions of fines from Apollo 17 samples 75081,72. Errors in isotope ratios 2%, errors in concentrations of He, Ne, and Ar about 3%, of Kr and Xe about 10%. Sample weights between 0.9–3.0 mg.

	<20 μm	20–25 μm	25–35 μm	35–54 μm	54–75 μm	75–120 μm	120–200 μm	200–300 μm
^{3}He	7.98×10^{-5}	2.62×10^{-5}	2.22×10^{-5}	1.876×10^{-5}	9.69×10^{-6}	1.008×10^{-5}	9.59×10^{-6}	8.06×10^{-6}
^{4}He	2.15×10^{-1}	7.25×10^{-2}	5.79×10^{-2}	5.01×10^{-2}	2.29×10^{-2}	2.40×10^{-2}	2.24×10^{-2}	1.907×10^{-2}
^{20}Ne	2.72×10^{-3}	1.035×10^{-3}	8.49×10^{-4}	6.81×10^{-4}	3.45×10^{-4}	3.04×10^{-4}	2.90×10^{-4}	2.42×10^{-4}
^{21}Ne	6.89×10^{-6}	3.00×10^{-6}	2.45×10^{-6}	2.05×10^{-6}	1.266×10^{-6}	1.107×10^{-6}	1.017×10^{-6}	9.00×10^{-7}
^{22}Ne	2.18×10^{-4}	8.12×10^{-5}	6.79×10^{-5}	5.39×10^{-5}	2.78×10^{-5}	2.43×10^{-5}	2.33×10^{-5}	1.946×10^{-5}
^{36}Ar	4.38×10^{-4}	1.247×10^{-4}	1.101×10^{-4}	9.90×10^{-5}	3.76×10^{-5}	3.49×10^{-5}	3.87×10^{-5}	3.78×10^{-5}
^{38}Ar	8.24×10^{-5}	2.43×10^{-5}	2.12×10^{-5}	1.935×10^{-5}	7.34×10^{-6}	7.11×10^{-6}	7.54×10^{-6}	7.53×10^{-6}
^{40}Ar	3.46×10^{-4}	1.155×10^{-4}	9.97×10^{-5}	1.039×10^{-4}	4.36×10^{-5}	4.08×10^{-5}	4.40×10^{-5}	5.42×10^{-5}
^{84}Kr	2.2×10^{-7}	7.8×10^{-8}	5.8×10^{-8}	6.2×10^{-8}	2.2×10^{-8}	2.1×10^{-8}	2.1×10^{-8}	2.3×10^{-8}
^{132}Xe	2.8×10^{-8}	1.0×10^{-8}	6.5×10^{-9}	8.4×10^{-9}	2.9×10^{-9}	3.2×10^{-9}	2.7×10^{-9}	2.8×10^{-9}
^{4}He/^{3}He	2690	2770	2610	2670	2360	2380	2340	2370
^{20}Ne/^{22}Ne	12.48	12.74	12.50	12.63	12.41	12.51	12.45	12.44
^{22}Ne/^{21}Ne	31.6	27.1	27.7	26.3	22.0	22.0	22.9	21.6
^{36}Ar/^{38}Ar	5.32	5.13	5.19	5.12	5.12	4.91	5.13	5.02
^{40}Ar/^{36}Ar	0.790	0.926	0.906	1.049	1.160	1.169	1.137	1.434
^{4}He/^{20}Ne	79.0	70.0	68.2	73.6	66.4	78.9	77.2	78.8
^{20}Ne/^{36}Ar	6.21	8.30	7.71	6.88	9.18	8.71	7.49	6.40
^{36}Ar/^{84}Kr	1960	1600	1900	1600	1710	1660	1840	1640
^{84}Kr/^{132}Xe	8.1	7.8	8.9	7.4	7.6	6.6	7.8	8.2

Table 7. Rare gas concentrations in cm³ STP/g in grain size fractions of fines from Apollo 17 sample 79035,15. Errors in isotope ratios 2%, errors in concentrations of He, Ne, and Ar about 3%, of Kr and Xe about 10%. Sample weights between 0.6–3 mg.

	<20 μm	20–25 μm	25–35 μm	35–54 μm	54–75 μm	75–120 μm	120–200 μm	200–300 μm
^{3}He	9.11×10^{-5}	3.51×10^{-5}	2.57×10^{-5}	1.705×10^{-5}	1.346×10^{-5}	1.202×10^{-5}	9.15×10^{-6}	7.84×10^{-6}
^{4}He	2.71×10^{-1}	9.84×10^{-2}	7.21×10^{-2}	4.33×10^{-2}	3.18×10^{-2}	2.85×10^{-2}	2.05×10^{-2}	1.840×10^{-2}
^{20}Ne	4.33×10^{-3}	1.706×10^{-3}	1.206×10^{-3}	8.10×10^{-4}	5.78×10^{-4}	4.99×10^{-4}	4.14×10^{-4}	3.18×10^{-4}
^{21}Ne	1.136×10^{-5}	5.19×10^{-6}	3.90×10^{-6}	3.01×10^{-6}	2.56×10^{-6}	2.24×10^{-6}	1.970×10^{-6}	1.598×10^{-6}
^{22}Ne	3.48×10^{-4}	1.378×10^{-4}	9.79×10^{-5}	6.55×10^{-5}	4.83×10^{-5}	4.14×10^{-5}	3.41×10^{-5}	2.67×10^{-5}
^{36}Ar	5.40×10^{-4}	1.882×10^{-4}	1.269×10^{-4}	9.39×10^{-5}	5.89×10^{-5}	5.99×10^{-5}	6.04×10^{-5}	5.03×10^{-5}
^{38}Ar	1.015×10^{-4}	3.65×10^{-5}	2.49×10^{-5}	1.861×10^{-5}	1.187×10^{-5}	1.196×10^{-5}	1.231×10^{-5}	1.033×10^{-5}
^{40}Ar	1.193×10^{-3}	4.13×10^{-4}	2.78×10^{-4}	2.22×10^{-4}	1.333×10^{-4}	1.518×10^{-4}	1.599×10^{-4}	1.323×10^{-4}
^{84}Kr	2.51×10^{-7}	8.58×10^{-8}	6.15×10^{-8}	4.30×10^{-8}	3.02×10^{-8}	2.81×10^{-8}	2.94×10^{-8}	2.92×10^{-8}
^{132}Xe	6.4×10^{-8}	2.2×10^{-8}	1.47×10^{-8}	1.16×10^{-8}	7.12×10^{-9}	7.7×10^{-9}	7.8×10^{-9}	7.3×10^{-9}
^{4}He/^{3}He	2970	2800	2810	2540	2360	2370	2240	2350
^{20}Ne/^{22}Ne	12.44	12.38	12.32	12.37	11.97	12.05	12.14	11.91
^{22}Ne/^{21}Ne	30.6	26.6	25.1	21.8	18.87	18.48	17.31	16.71
^{36}Ar/^{38}Ar	5.32	5.16	5.10	5.05	4.96	5.01	4.91	4.87
^{40}Ar/^{36}Ar	2.21	2.19	2.19	2.36	2.26	2.53	2.65	2.63
^{4}He/^{20}Ne	62.6	57.7	59.8	53.5	55.0	57.1	49.5	57.9
^{20}Ne/^{36}Ar	8.02	9.06	9.50	8.63	9.81	8.33	6.85	6.32
^{36}Ar/^{84}Kr	2150	2190	2060	2180	1950	2130	2050	1720
^{84}Kr/^{132}Xe	3.9	3.9	4.2	3.7	4.3	3.6	3.8	4.0

reference soil 74241 have been published recently (Hintenberger and Weber, 1973, Tables 3 and 4). Some bulk samples have isotopic and elemental noble gas ratios outside the spread of values obtained in grain size fractions. Especially the ^{4}He/^{3}He, ^{20}Ne/^{22}Ne, and ^{4}He/^{20}Ne ratios tend to be higher in bulk samples. This effect is larger than the experimental error and could be attributable to gas losses or connected with the loss of very fine grained material during sieving. Experiments are planed to clarify this discrepancy.

Mineral separates

More detailed information can be anticipated from the investigation of mineral separates of known grain sizes. These investigations are in progress and therefore only a few preliminary results can be included in this paper. Up to now, we have measured the rare gases in the 35.5–54 μm ilmenite grain size fractions of 3 soils and a breccia from the Apollo 17 missions. These preliminary results are listed in Table 9.

DISCUSSION

In order to compare the concentration of trapped noble gases in different soils from Apollo 17 the isotopes ^{4}He, ^{20}Ne, ^{36}Ar, ^{84}Kr, and ^{132}Xe are plotted in Fig. 1. The samples are arranged from the top to bottom with decreasing ^{36}Ar-

Table 8. Rare gas concentrations in cm³ STP/g in grain size fractions of breccia from Apollo 17 sample 79135,32. Errors in isotope ratios 2%, errors in concentrations of He, Ne, and Ar about 3%, of Kr and Xe about 10%. Sample weights between 0.6–2.1 mg.

	$<5\ \mu m$	$5-10\ \mu m$	$10-20\ \mu m$	$<20\ \mu m$	$20-25\ \mu m$	$25-35\ \mu m$	$35-54\ \mu m$	$54-75\ \mu m$	$75-120\ \mu m$	$120-200\ \mu m$
^{3}He	9.98×10^{-5}	4.96×10^{-5}	5.64×10^{-5}	8.70×10^{-5}	3.98×10^{-5}	3.29×10^{-5}	3.21×10^{-5}	1.203×10^{-5}	1.683×10^{-5}	3.38×10^{-5}
^{4}He	3.04×10^{-1}	1.463×10^{-1}	1.699×10^{-1}	2.63×10^{-1}	1.166×10^{-1}	9.52×10^{-2}	9.13×10^{-2}	2.67×10^{-2}	4.32×10^{-2}	9.77×10^{-2}
^{20}Ne	6.31×10^{-3}	3.31×10^{-3}	3.65×10^{-3}	5.36×10^{-3}	2.61×10^{-3}	2.20×10^{-3}	2.15×10^{-3}	6.76×10^{-4}	9.32×10^{-4}	2.35×10^{-3}
^{21}Ne	1.664×10^{-5}	9.03×10^{-6}	9.89×10^{-6}	1.430×10^{-5}	7.49×10^{-6}	6.68×10^{-6}	6.52×10^{-6}	3.07×10^{-6}	3.53×10^{-6}	6.82×10^{-6}
^{22}Ne	4.98×10^{-4}	2.66×10^{-4}	2.87×10^{-4}	4.26×10^{-4}	2.07×10^{-4}	1.764×10^{-4}	1.712×10^{-4}	5.53×10^{-5}	7.52×10^{-5}	1.863×10^{-4}
^{36}Ar	8.86×10^{-4}	4.27×10^{-4}	5.16×10^{-4}	8.76×10^{-4}	3.39×10^{-4}	3.42×10^{-4}	2.99×10^{-4}	8.68×10^{-5}	1.291×10^{-4}	3.89×10^{-4}
^{38}Ar	1.671×10^{-4}	8.09×10^{-5}	9.75×10^{-5}	1.692×10^{-4}	6.43×10^{-5}	6.62×10^{-5}	5.77×10^{-5}	1.748×10^{-5}	2.54×10^{-5}	7.43×10^{-5}
^{40}Ar	2.14×10^{-3}	1.007×10^{-3}	1.236×10^{-3}	2.19×10^{-3}	8.12×10^{-4}	8.53×10^{-4}	7.46×10^{-4}	2.23×10^{-4}	3.34×10^{-4}	9.56×10^{-4}
^{84}Kr	4.05×10^{-7}	1.94×10^{-7}	2.27×10^{-7}	3.79×10^{-7}	1.54×10^{-7}	1.45×10^{-7}	1.33×10^{-7}	4.43×10^{-8}	5.77×10^{-8}	1.90×10^{-7}
^{132}Xe	1.3×10^{-7}	6.5×10^{-8}	6.7×10^{-8}	1.1×10^{-7}	4.9×10^{-8}	3.7×10^{-8}	4.0×10^{-8}	1.2×10^{-8}	1.7×10^{-8}	5.2×10^{-8}
^{4}He/^{3}He	3050	2950	3010	3020	2930	2890	2840	2220	2570	2890
^{20}Ne/^{22}Ne	12.67	12.44	12.72	12.58	12.61	12.47	12.56	12.22	12.39	12.61
^{22}Ne/^{21}Ne	29.9	29.5	29.0	29.8	27.6	26.4	26.3	18.01	21.3	27.3
^{36}Ar/^{38}Ar	5.30	5.28	5.29	5.18	5.27	5.17	5.18	4.97	5.08	5.24
^{40}Ar/^{36}Ar	2.42	2.36	2.40	2.50	2.40	2.49	2.49	2.57	2.59	2.46
^{4}He/^{20}Ne	48.2	44.2	46.5	49.1	44.7	43.3	42.5	39.5	46.4	41.6
^{20}Ne/^{36}Ar	7.12	7.75	7.07	6.12	7.78	6.43	7.19	7.79	7.22	6.04
^{36}Ar/^{84}Kr	2190	2200	2240	2310	2200	2360	2300	1960	2240	2050
^{84}Kr/^{132}Xe	3.2	3.0	3.4	3.6	3.1	3.9	3.3	3.8	3.4	3.7

Table 9. Rare gas content and elemental ratios in bulk and ilmenite grain size fractions (35–54 μm) of Apollo 17 soils and breccias (cc STP/g^{-1}).

	Weight (mg)		^{4}He	^{20}Ne	^{36}Ar	^{4}He/^{36}Ar	^{20}Ne/^{36}Ar
	0.20	Ilm.	1.110×10^{-1}	5.75×10^{-4}	4.29×10^{-5}	2590	13.4
72701,25		Bulk	1.967×10^{-2}	4.77×10^{-4}	1.115×10^{-4}	176.4	4.28
		I/B	5.64	1.21	0.385	14.6	3.1
	0.56	Ilm.	9.61×10^{-2}	3.41×10^{-4}	1.284×10^{-5}	7480	26.6
74241,24		Bulk	2.18×10^{-2}	2.66×10^{-4}	2.38×10^{-5}	916	11.2
		I/B	4.41	1.28	0.539	8.18	2.4
	1.20	Ilm.	1.419×10^{-1}	6.67×10^{-4}	3.09×10^{-5}	4590	21.6
75081,72		Bulk	5.01×10^{-2}	6.81×10^{-4}	9.90×10^{-5}	506	6.88
		I/B	2.83	0.979	0.312	9.07	3.1
	1.00	Ilm.	1.224×10^{-1}	9.91×10^{-4}	8.61×10^{-5}	1420	11.5
79035,15		Bulk	4.33×10^{-2}	8.10×10^{-4}	9.39×10^{-5}	461	8.63
		I/B	2.83	1.22	0.917	3.09	1.33

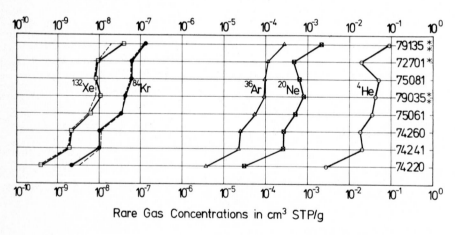

Rare Gas Concentrations in cm^3 STP/g

Breccias ✳✳

Grain Size
35μ < d < 54μ

Grain Size Fractions
of
Apollo 17 Soils and
Breccias

Ilmenite poor
Soil ✳

Fig. 1. Concentration of ^{4}He, ^{20}Ne, ^{36}Ar, ^{84}Kr, and ^{132}Xe in 35–54 μm grain size fractions of Apollo 17 soils and breccias. The dashed lines are parallel to the ^{36}Ar line. The low concentration of ^{4}He and ^{20}Ne in 72701 is explained by its low ilmenite content.

concentration. Hintenberger and Weber (1973) have used a similar diagram to compare trapped noble gas concentrations from different landing sites. Figure 1 is, however, modified with respect to the following points:

(1) The influence of different grain size distributions in the various soils has been eliminated by using only 35–54 μm grain size fractions.
(2) ^{36}Ar was taken as an indicator of the solar wind irradiation (i.e. the samples are arranged after decreasing ^{36}Ar concentration) because ^{36}Ar is normally more precisely determined than ^{84}Kr.

In Fig. 1 the lines for the heavy rare gases ^{36}Ar, ^{84}Kr, and ^{132}Xe of soils are parallel to each other, indicating constant ratios of ^{36}Ar/^{84}Kr and ^{36}Ar/^{132}Xe of 2100 ± 300 and 13400 ± 2000, respectively. The concentrations of trapped ^{36}Ar, ^{84}Kr, and ^{132}Xe in lunar samples can be taken—at least as an approximation—as indicators of solar wind exposure. Therefore, in Fig. 1 the samples suffered from top to bottom less and less solar irradiation. This sequence is the same for each of the three isotopes ^{36}Ar, ^{84}Kr, and ^{132}Xe. The ^{4}He and ^{20}Ne concentrations, however, would define completely different sequences, an effect which is understood in terms of diffusive losses of the light trapped rare gases. To become independent as far as possible from these losses, ^{132}Xe would be the best rare gas isotope as indicator of solar wind exposure. However, in Fig. 1 we prefered ^{36}Ar because of its smaller experimental error.

As noted by Eugster *et al.* (1972) and Hintenberger and Weber (1973), high ^{4}He/^{20}Ne and ^{20}Ne/^{36}Ar ratios in lunar soil samples are connected with high ilmenite contents. This holds also for Apollo 17 samples. The ^{20}Ne and the ^{4}He lines show a marked minimum for the light mantle soil 72701 which has the lowest ilmenite content (1.5% TiO_2) of all the samples in this diagram.

This indicates again a loss of the light rare gases ^{4}He and ^{20}Ne in ilmenite poor soils. Figure 2 demonstrates the correlation of the ratios ^{4}He/^{20}Ne and ^{20}Ne/^{36}Ar of the trapped rare gas nuclides with the concentrations of TiO_2 as a measure for the ilmenite content in the Apollo 17 soils and breccias investigated in this paper.

A more detailed inspection of Fig. 1 shows two maxima in the ^{132}Xe line for the two breccias 79035 and 79135 which indicate a ^{132}Xe excess as compared to ^{132}Xe calculated from the ^{36}Ar content. In the diagram of our Apollo 11 to Apollo 16 data (Hintenberger and Weber, 1973, Fig. 1) also the breccias 10021 and 10046 show distinct humps in the ^{132}Xe line. This feature is more clearly shown in Fig. 3, where the ^{132}Xe/^{36}Ar ratios versus the ^{36}Ar concentrations are plotted. We have included in this diagram also our data from the samples of earlier missions. Most soil samples show distinctly lower ^{132}Xe/^{36}Ar ratios than the breccias. Thus, the formation of a breccia is at least in many cases connected either with an incorporation of ^{132}Xe or with a loss of ^{36}Ar. Figure 4 shows the corresponding diagram for the ^{84}Kr/^{36}Ar ratio. An argon loss without a krypton loss would influence this ratio in a similar way as the ^{132}Xe/^{36}Ar ratio, but no systematic increase of the ^{84}Kr/^{36}Ar ratio in breccias is observed. Therefore, the high

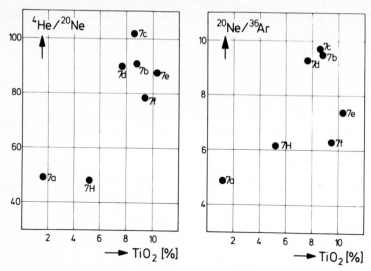

Fig. 2. Correlation between the TiO₂ contents and trapped rare gas ratios in Apollo 17 soils and breccias. TiO₂ is a measure of the ilmenite content of the sample. Numbers on points are explained in Table 1.

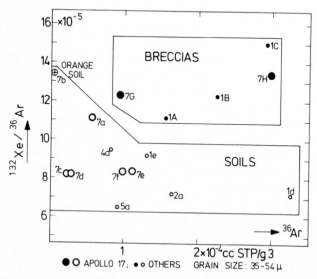

Fig. 3. $^{132}Xe/^{36}Ar$ as a function of trapped ^{36}Ar in Apollo 17 soils (large open circles) and breccias (large closed circles). For comparison measurements from other lunar missions are included. Grain size effects are excluded by taking for all samples the analyses of the same grain size fractions (35–54 μm). Numbers on points are explained in Table 1.

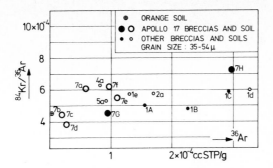

Fig. 4. $^{84}Kr/^{36}Ar$ as a function of ^{36}Ar in soils (open circles) and breccias (closed circles).

$^{132}Xe/^{36}Ar$ ratios in these breccias must be caused by an excess of trapped Xe, or a simultaneous loss of ^{36}Ar and ^{84}Kr.

The influence of the dose of solar wind irradiation on the $^{4}He/^{36}Ar$ and $^{20}Ne/^{36}Ar$ ratios can be seen from Fig. 5 and 6. The concentration of ^{36}Ar is taken as a measure of the total solar exposure. To reduce the influence of differences in the gas retentivities we have compared only the data obtained on ilmenite rich samples ($TiO_2 > 6\%$). Both ratios, $^{4}He/^{36}Ar$ as well as $^{20}Ne/^{36}Ar$, decrease systematically with increasing ^{36}Ar concentrations. This can be explained by an enhanced diffusive loss of ^{4}He and ^{20}Ne due to larger radiation damages in highly irradiated crystals. However, the same pattern would be expected if saturation effects as mentioned by Eberhardt *et al.* (1970) are responsible for the observed noble gas pattern in lunar soils. More detailed information on this subject can be

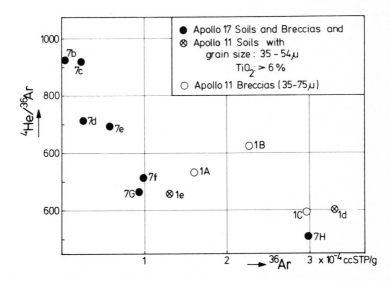

Fig. 5. $^{4}He/^{36}Ar$ versus ^{36}Ar in lunar soils and breccias. Only materials with TiO_2 contents larger than 6% are considered. Numbers on points are explained in Table 1.

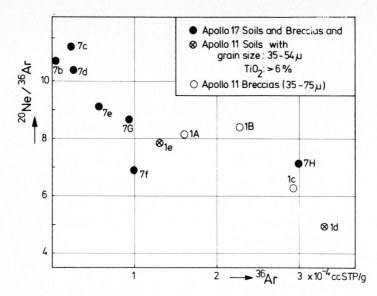

Fig. 6. ^{20}Ne/^{36}Ar versus ^{36}Ar in lunar soils and breccias.

obtained by the investigation of mineral fractions, first of all from the rare gas retentive ilmenite.

Figure 7 shows the concentrations of ^{4}He, ^{20}Ne, and ^{36}Ar in ilmenite normalized to the concentration of the same isotope in bulk material of the same grain size.

^{4}He and ^{20}Ne have higher concentrations in ilmenite. This reflects the high retentivity of implanted light noble gases in ilmenite (Eberhardt *et al.*, 1970). All three soils contain, however, more ^{36}Ar in the bulk material. This may imply different irradiation histories for bulk and ilmenite of the same soil. Eberhardt *et*

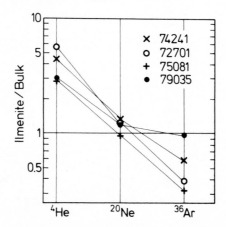

Fig. 7. Comparison of concentrations of ^{4}He, ^{20}Ne, and ^{36}Ar in ilmenite and bulk samples (grain size: 35–54 μm).

al. (1972) have shown that in 10084 and 12001 the cosmic ray exposure age of ilmenite is smaller than the exposure age of the bulk soil. Measurements on ilmenite of different grain sizes will be necessary to prove whether this explanation is also true for the Apollo 17 soils investigated here. It may be noted that the breccia 79035 has about the same concentration of trapped ^{36}Ar in ilmenite and bulk samples.

Besides the surface correlated trapped component, lunar soils and breccias contain volume correlated spallogenic and radiogenic noble gases. A separation of these components is possible if grain size fractions are analyzed (Eberhardt *et al.*, 1970; Heymann and Yaniv, 1970; Hintenberger *et al.*, 1970b, 1971; Hintenberger and Weber, 1973). This method has been used to determine the radiogenic ^{40}Ar concentration. Figure 8 shows ^{40}Ar correlation plots for Apollo 17 soils and breccias. The radiogenic ^{40}Ar is given by the ordinate intercept; the ratio of the surface correlated argon components $(^{40}Ar/^{36}Ar)_{sc}$ is given by the slope of the line. The method assumes that radiogenic ^{40}Ar is the same in all grain size fractions.

Table 10 contains the results for radiogenic ^{40}Ar and K–Ar ages, calculated with K values from Table 1. For the soils 72701 and 74220, and to some extent also for 75061 and 75081, the correlation lines of Fig. 8 can be used to determine the radiogenic and the surface correlated ^{40}Ar components with relatively small errors. The two reference soils 74241 and 74260, however, as well as the breccias 79035 and 79135 contain very high amounts of surface correlated ^{40}Ar. In such cases, correlation plots cannot resolve the surface and volume correlated components. Therefore, we can give only upper limits for the K–Ar ages of these soils. These limits are higher than our recently published results for 74241 (Hintenberger and Weber, 1973) due to a computational error. The difference in age between the orange soil and the adjacent gray soil has become smaller and—considering the difficulties in resolving the two components—does not appear significant.

Table 10 also includes isotopic ratios of surface correlated gases and concentrations of volume correlated spallogenic nuclides (^{3}He, ^{21}Ne, ^{38}Ar).

The ratio $^{40}Ar/^{36}Ar$ of the surface correlated argon show a large range from 0.77 in 75081 to 7.6 in 74241. The ratio $^{4}He/^{3}He$ of the surface correlated helium isotopes shows a range from 2740 to 3260 in the corresponding samples. Geiss (1973) pointed out that these two ratios seem to be correlated. This is supported by our new results on the Apollo 17 samples (see Fig. 9). Geiss attempted to explain the correlation by a decrease of solar wind $^{4}He/^{3}He$ ratio with time combined with an enhanced ^{40}Ar retrapping from the moon's atmosphere at earlier times (Yaniv and Heymann, 1972). Such a correlation could, however, also be obtained if radiogenic ^{4}He and ^{40}Ar released from the moon's interior is adsorbed and partly incorporated into the highly damaged grain surfaces.

For the lunar regolith with a rather continuous mixing, a correlation between surface correlated gases in the grains incorporated at the very surface of the regolith and spallogenic gases produced in the top layer (~ 1 m) is expected. Figures 10 and 11 show the concentrations of trapped $(^{20}Ne)_{tr}$ and $(^{36}Ar)_{tr}$ versus spallogenic $(^{21}Ne)_{sp}$ in soils and breccias. Indeed, a correlation between trapped

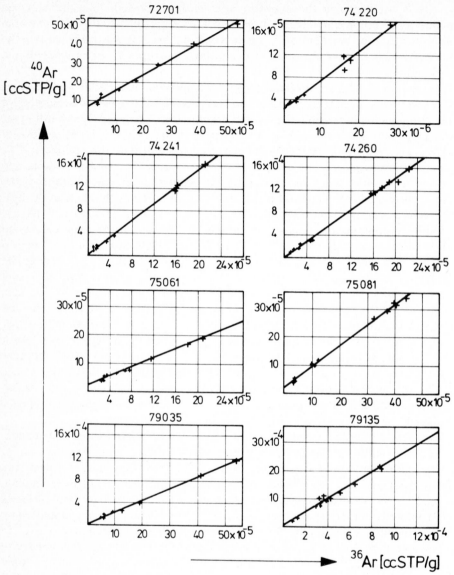

Fig. 8. ^{40}Ar versus ^{36}Ar in grain size fractions of Apollo 17 samples. The ordinate intercept determines the amount of radiogenic ^{40}Ar, the slope of the line the surface correlated ^{40}Ar/^{36}Ar ratio (see Table 10). The radiogenic ^{40}Ar cannot be deduced from such diagrams in cases of high surface correlated ^{40}Ar concentrations (74241, 74260, 79035, 79135). Similar diagrams have been used to compute the spallogenic isotopes given in Table 10.

Table 10. Concentrations (cm³ STP/g) and isotope ratios of trapped, spallogenic and radiogenic rare gas nuclides of lunar fines and breccias.

	$\left(\frac{^4He}{^3He}\right)_{tr}$	$\left(\frac{^{20}Ne}{^{21}Ne}\right)_{tr}$	$\left(\frac{^{22}Ne}{^{21}Ne}\right)_{tr}$	$\left(\frac{^{36}Ar}{^{38}Ar}\right)_{tr}$	$\left(\frac{^{40}Ar}{^{36}Ar}\right)_{tr}$	$^3He_{sp}$	$^{21}Ne_{sp}$	$^{38}Ar_{sp}$	$^{40}Ar_{rad}$	$^{21}Ne^*$ Exposure age 10^6 yr	K–Ar† age 10^9 yr
						10^{-8} cm³ STP/g					
72701,25	2900	417	32.4	5.38	0.89	68	34	50	6700	210	3.85
	±40	±9	±0.7	±.04	±.05	±15	±4	±30	±1000	±20	±.25
74220,47	2970	399	30.8	5.26	5.2	15	4.3	2.6	2600	27	3.44
	±40	±4	±0.3	±0.06	±0.2	±2	±0.2	±1.3	±300	±1	±.17
74241,24	3260	411	32.8	5.20	7.4	120	25	24	≤2100	160	—
	±50	±4	±0.4	±0.04	±0.3	±21	±2	±5		±10	
74260,9	3090	411	32.7	5.27	7.0	100	32	30	≤2100	200	—
	±40	±5	±0.4	±0.03	±0.2	±30	±3	±3		±20	
75061,21	2770	413	32.1	5.29	0.85	120	33	22	1800	210	3.0
	±50	±12	±.8	±.03	±.04	±30	±6	±8	±400	±40	±.3
75081,72	2760	413	32.3	5.33	0.78	110	37	53	1800	230	2.9
	±30	±5	±0.5	±0.02	±0.02	±30	±4	±11	±300	±20	±.2
79035,15	2980	409	32.5	5.32	2.18	220	96	100	1700	600	2.5
	±60	±8	±.7	±.03	±.05	±50	±8	±15	±700	±50	±.5
79135,32	3170	423	33.0	5.29	2.5	305	130	100	—	810	—
	±20	±8	±.5	±.03	±.1	±25	±10	±40		±60	

*Calculated with a production rate of 0.16×10^{-8} cc STP g⁻¹ m.y.⁻¹.
†K values from Table 1. For 79035 a K concentration of 830 ppm was used.

and spallogenic rare gases is observed. The following relations hold for all concentrations obtained on 35–54 μm fractions of soils and breccias listed in Table 1, two additional Apollo 17 soils (74220 and 74241) as well as three Apollo 11 breccias 10021, 10046, 10061, and five soils from earlier missions analyzed in our

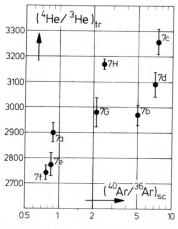

Fig. 9. Correlation of trapped $^4He/^3He$ and surface correlated $^{40}Ar/^{36}Ar$ in Apollo 17 materials. Numbers on points are explained in Table 1.

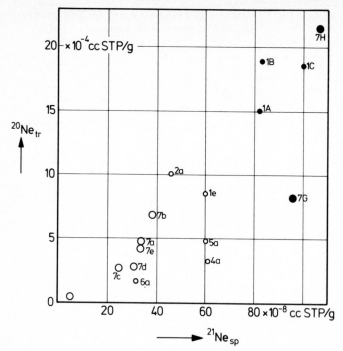

Fig. 10. Correlation of trapped ^{20}Ne in the 35–54 μm grain size fraction with spallogenic ^{21}Ne in Apollo 17 soils (large open circles) and breccias (large filled circles). For comparison data obtained in our laboratory from other lunar missions are included. The abbreviations used are explained in Table 1.

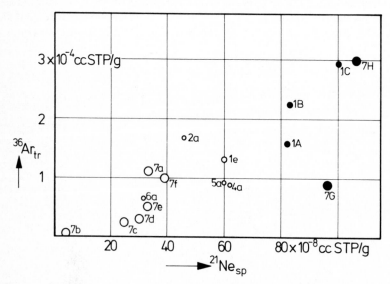

Fig. 11. Trapped ^{36}Ar versus spallogenic ^{21}Ne (see caption of Fig. 10). Numbers on points are explained in Table 1.

laboratory:

$$(^{20}Ne)_{tr} \leqslant 2.3 \times 10^3 \times (^{21}Ne)_{sp}$$
$$(^{36}Ar)_{tr} \leqslant 380 \times (^{21}Ne)_{sp}$$

In a very simplified model, one may explain these relations as valid for soils with continuous mixing. The deviations of data from this correlation may be due to a diffusion loss of ^{20}Ne or ^{36}Ar, or by an interruption of the solar wind exposure by an inhomogeneous mixing or by formation of a breccia.

An explanation for the extremely low $(^{20}Ne)_{tr}$ and $(^{36}Ar)_{tr}$ concentration together with the relatively high $(^{21}Ne)_{sp}$ content of the breccia 79035 (Figs. 10 and 11) seems to be that about one-half of the spallogenic gases is produced before compaction and the other half after excavation of this rock. The use of such a plot could help to decipher the complex irradiation history of some lunar breccias. However, a larger number of different breccias and soils have to be analyzed to demonstrate that it is really possible to draw conclusions on the exposure time of lunar breccias from correlation diagrams like Figs. 10 and 11.

Acknowledgments—We are grateful to NASA for generously providing the lunar samples. We thank Mrs. Chr. Reitz for technical assistance.

REFERENCES

Baur H., Frick U., Funk H., Schultz L., and Signer P. (1972) On the question of retrapped ^{40}Ar in lunar fines (abstract). In *Lunar Science—III*, pp. 47–49. The Lunar Science Institute, Houston.

Eberhardt P., Geiss J., Graf H., Grögler N., Krähenbühl U., Schwaller H., Schwarzmüller J., and Stettler A. (1970) Trapped solar wind noble gases, exposure age and K/Ar-age in Apollo 11 lunar fine material. *Proc. Apollo 11 Lunar Sci. Conf., Geochim. Cosmochim. Acta*, Suppl. 1, Vol. 2, pp. 1037–1070. Pergamon.

Eberhardt P., Geiss J., Graf H., Grögler N., Mendia M. D., Mörgeli M., Schwaller H., and Stettler A. (1972) Trapped solar wind noble gases in Apollo 12 lunar fines 12001 and Apollo 11 breccia 10046. *Proc. Third Lunar Sci. Conf., Geochim. Cosmochim. Acta*, Suppl. 3, Vol. 2, pp. 1821–1856. MIT Press.

Eugster O., Grögler N., Mendia M. D., Eberhardt P., and Geiss J. (1973) Trapped solar wind noble gases and exposure age of Luna 16 lunar fines. *Geochim. Cosmochim. Acta* **37**, pp. 1991–2003.

Frick U., Baur H., Funk H., Phinney D., Schäfer Chr., Schultz L., and Signer P. (1973) Diffusion properties of light noble gases in lunar fines. *Proc. Fourth Lunar Sci. Conf., Geochim. Cosmochim. Acta*, Suppl. 4, Vol. 2, pp. 1987–2002. Pergamon.

Geiss J. (1973) Solar wind composition and implications about the history of the solar system. Preprint.

Heymann D. and Yaniv A. (1970) Inert gases in the fines from the Sea of Tranquillity. *Proc. Apollo 11 Lunar Sci. Conf., Geochim. Cosmochim. Acta*, Suppl. 1, Vol. 2, pp. 1247–1259. Pergamon.

Hintenberger H. and Weber H. W. (1973) Trapped rare gases in lunar fines and breccias. *Proc. Fourth Lunar Sci. Conf., Geochim. Cosmochim. Acta*, Suppl. 4, Vol. 2, pp. 2003–2019. Pergamon.

Hintenberger H., Weber H. W., Voshage H., Wänke H., Begemann F., and Wlotzka F. (1970) Concentrations and isotopic abundances of the rare gases, hydrogen, and nitrogen in Apollo 11 lunar matter. *Proc. Apollo 11 Lunar Sci. Conf., Geochim. Cosmochim. Acta*, Suppl. 1, Vol. 2, pp. 1269–1282. Pergamon.

Hintenberger H., Weber H. W., and Takaoka N. (1971) Concentrations and isotopic abundance of the rare gases in lunar matter. *Proc. Second Lunar Sci. Conf., Geochim. Cosmochim. Acta*, Suppl. 2, Vol. 2, pp. 1607–1625. MIT Press.

Hoffman J. H., Hodges R. R., and Johnson F. S. (1973) Lunar atmospheric composition results from Apollo 17. *Proc. Fourth Lunar Sci. Conf., Geochim. Cosmochim. Acta,* Suppl. 4, Vol. 3, pp. 2865–2875. Pergamon.

Jovanovic S. and Reed G. W., Jr. (1974) Labile trace elements in Apollo 17 samples (abstract). In *Lunar Science—V,* pp. 391–393. The Lunar Science Institute, Houston.

LSPET (Lunar Sample Preliminary Examination Team) (1973) Apollo 17 Lunar Samples: Chemical and Petrographic Description. *Science* **182,** 659–671.

Morgan J. W., Ganapathy R., Higuchi H., Krähenbühl U., and Anders E. (1974) Lunar basins: Tentative characterization of projectiles, from meteorite elements in Apollo 17 boulders (abstract). In *Lunar Science—V,* pp. 526–528. The Lunar Science Institute, Houston.

Nunes P. D., Tatsumoto M., and Unruh D. M. (1974) U–Th–Pb systematics of some Apollo 17 samples (abstract). In *Lunar Science—V,* pp. 562–564. The Lunar Science Institute, Houston.

Schultz L. and Hintenberger H. (1967) Edelgasmessungen an Eisenmeteoriten. *Zs. Naturf.* **22a,** pp. 773–779.

Tatsumoto M., Nunes P. D., Knight R. J., Hedge C. E., and Unruh D. M. (1973) U–Th–Pb, Rb–Sr, and K measurements of two Apollo 17 samples. *EOS* **54,** 614–615.

Weber H. (1973) Thesis. Johannes Gutenberg-Universität, Mainz.

Yaniv A. and Heymann D. (1972) Atmospheric Ar^{40} in lunar fines. *Proc. Third Lunar Sci. Conf., Geochim. Cosmochim. Acta,* Suppl. 3, Vol. 2, pp. 1967–1980. MIT Press.

Proceedings of the Fifth Lunar Conference
(Supplement 5, Geochimica et Cosmochimica Acta)
Vol. 2 pp. 2023–2031 (1974)
Printed in the United States of America

Lunar trapped xenon

B. D. LIGHTNER and K. MARTI

Department of Chemistry, University of California, San Diego, La Jolla, California 92037

Abstract—Evidence is presented for the presence in Apollo 16 light matrix breccias of trapped xenon which is isotopically similar to terrestrial xenon. This trapped component is predominantly released above 900°C, together with cosmic ray produced spallation gases. The relative abundances of Ar, Kr, and Xe are distinct from both solar and terrestrial patterns. Terrestrial-type xenon has so far been observed in both Apollo 14 and 16 breccias.

INTRODUCTION

A DETAILED STUDY of breccia 14321 in this laboratory (Marti *et al.*, 1973a) revealed that trapped xenon in this breccia is distinct in isotopic composition from solar xenon. This result called for an investigation of indigenous lunar xenon, a component which is generally masked by abundant spallation and fission gases. Such a study is expected to be most successful when it is carried out on samples with (1) short exposure ages and (2) low abundances of U, Ba and the REE, the elements responsible for fission and spallation xenon components. We have investigated the anorthositic or light matrix breccias 60025, 62255, and 67075. The first two belong to a group of rocks with 2 m.y. exposure ages (data will be published elsewhere) and, therefore, may represent ejecta from South Ray Crater; 67075 is an ejecta from North Ray Crater with a 49 m.y. exposure age (Marti *et al.*, 1973b).

Although the present study will not be complete until all possible origins of this distinctive xenon component have been investigated, we feel that a report at this time is desirable, partly because a knowledge of this component is important in attempts to unravel lunar xenon spectra which are mixtures of several components.

EXPERIMENTAL TECHNIQUES

Sample 60025,83, a collection of several small chips from the interior of a cataclastic anorthosite, was wrapped in aluminum foil for analysis. Sample 62255,17, two interior chips from a white breccia, and sample 67075,8, four chips from the interior of an anorthositic rock from North Ray Crater, were run without foil containers. All samples were stored in vacuum at 80°C for several days in the side tube of the extraction system preceding analysis. Gases were released stepwise at several crucible temperatures. Extraction times were 1 hour, excepting the 1700°C steps, which were limited to 25 min. Reextractions at 1800°C for 25 min indicated complete gas extraction.

Before any samples were run, the extraction system was thoroughly degassed at 500°C. Crucible extraction blanks yielded the following amounts of ^{132}Xe in units of 10^{-14} cm^3 STP: 1.1, 1.2, 2.0, and 6.8 for the 700°C, 900°C, 1300°C, and 1700°C fractions, respectively. A blank using an aluminum foil container identical in weight (31 mg) to the 60025 sample container gave a factor of 2 higher 700°C blank. At higher temperatures, the foil blank amounts were similar to those given above. An extraction blank was measured before and after each sample. These blanks include gas released by vapor deposition on and subsequent induced heating of the extraction system walls.

Reactive gases were removed by reaction on hot Ti–Pd getters, and He–Ne, Ar, Kr, and Xe were separated into four fractions by adsorption on charcoal. Each fraction was run separately using an automated magnetic field stepping mass spectrometer. Peak heights were recorded digitally on paper tape. Data reduction was done off-line on a TI980A minicomputer system. Ratios were extrapolated to the time of inlet. Errors given in Table 1 represent 95% confidence intervals. They were derived from the statistical fluctuation of the measured ratios and include uncertainties in mass discrimination and blank corrections. Errors reflect the extremely small gas amounts released by each temperature step. ^{132}Xe from individual steps varied from a low of 8.4×10^{-13} cc STP to a high of 9.5×10^{-12} cc STP. Xenon concentrations were determined by peak height comparison with air standards. Total xenon isotopic compositions given in Table 1 are the sums of the temperature fractions.

ISOTOPIC CORRELATIONS

The isotopic compositions and concentrations of xenon in the individual temperature fractions are set out in Table 1. The spallation components in samples 60025,83 and 62255,17 are very small, reflecting the low concentrations of large ion lithophilic elements in these anorthosites (Laul and Schmitt, 1973; Krähenbühl *et al.*, 1973), as well as their short cosmic ray exposure ages.

Table 1. Isotopic abundances of xenon released by stepwise heating of Apollo 16 light matrix breccias, normalized to ^{132}Xe = 100.

Sample	Temperature °C	^{132}Xe $\times 10^{-12}$ cm^3 STP g^{-1}	^{124}Xe	^{126}Xe	^{128}Xe	^{129}Xe	^{130}Xe	^{131}Xe	^{132}Xe	^{134}Xe	^{136}Xe
60025,83	700	11.8	0.40	0.32	7.22	99.8	15.37	80.0	100	38.5	33.4
(0.244 g)		±1.8	±.06	.11	.33	1.0	.27	.6		.8	.6
	900	4.6	0.37	0.42	7.3	97.9	15.16	79.2	100	37.8	32.8
		±.7	±.12	.12	.6	2.1	.33	1.6		.7	.6
	1300	39	0.392	0.371	7.26	98.1	15.16	78.73	100	39.00	33.07
		±6	±.022	.027	.15	.6	.18	.31		.29	.28
	1700	10.1	0.37	0.45	7.23	98.0	14.83	79.6	100	38.7	32.7
		±1.5	±.07	.08	.24	1.7	.27	.8		.7	.4
	Total	66	0.389	0.376	7.25	98.4	15.15	79.13	100	38.77	33.05
		±9	±.021	.029	.12	.5	.13	.27		.25	.21
62255,17	700	7.6	0.32	0.30	7.0	96.2	14.9	79.8	100	38.5	32.9
(0.228 g)		±1.1	±.07	.10	.6	2.6	.5	1.2		.5	.4
	900	3.7	0.45	0.49	7.5	99.8	16.2	79.1	100	38.8	32.9
		±.6	±.16	.12	.7	2.4	.7	2.3		.5	.5
	1300	21.4	0.38	0.422	7.44	99.0	15.23	79.5	100	38.8	32.9
		±3.2	±.06	.034	.32	.8	.34	.9		.5	.5
	1700	5.4	0.41	0.42	7.4	96.6	15.3	78.6	100	40.1	33.8
		±.8	±.10	.09	.5	2.0	.6	1.0		.8	.8
	Total	38	0.38	0.404	7.33	98.2	15.27	79.4	100	38.96	32.97
		±6	.04	.033	.23	.8	.24	.24		.31	.31
67075,8	900	4.1	0.57	1.11	8.5	98.7	16.0	81.5	100	38.9	32.4
(0.214 g)		±.6	±.15	.25	.6	3.7	.8	2.1		1.9	.8
	1700	6.3	2.47	3.68	12.1	95.8	17.99	86.5	100	38.6	32.8
		±.9	±.20	.36	.6	1.9	.34	3.0		.5	1.2
	Total	10.4	1.72	2.67	10.7	96.9	17.22	84.5	100	38.7	32.6
		±1.5	±.14	.24	.4	1.9	.36	2.0		.8	.8

We discuss now the isotopic abundances of trapped xenon in the investigated samples by making use of isotopic correlations. Figure 1 is an example of the correlations which are obtained from three fission-shielded xenon isotopes. It is analogous to Fig. 1 of Marti *et al.* (1973a). The plotted isotope ratios represent a variable mixture of two components: trapped and spallation xenon. For reference, a line is drawn between AIR and pure spallation xenon (Marti *et al.*, 1973a). Similar correlations are obtained for all the other fission-shielded isotopes. Figure 2 shows a correlation of $^{136}Xe/^{130}Xe$ with $^{128}Xe/^{130}Xe$, similar to the one exhibited by Marti *et al.* (1973a). The 60025 and 62255 data points cluster near AIR xenon, far from the solar-xenon data points. Addition of cosmic ray spallation products shifts points along the indicated line. As can be seen in both Figs. 1 and 2, the contribution from spallation xenon is very small for 60025 and 62255. A displacement in Fig. 2 in the ordinate direction would indicate the addition of fission produced $^{136}Xe_f$. The Apollo 16 samples, as expected from the anorthositic character of these rocks, show no evidence of fission xenon. In fact, the data indicate that no more than 40 ppb uranium is present in either 60025 or 62255, given a 4 b.y. rock age. The measured U concentrations in 60025 are much lower (Laul and Schmitt, 1973; Krähenbühl *et al.*, 1973).

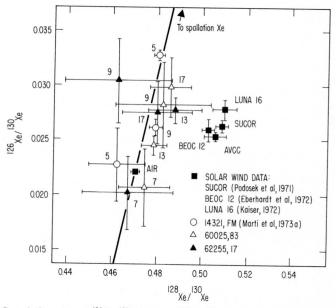

Fig. 1. Correlation plot of $^{126}Xe/^{130}Xe$ versus $^{128}Xe/^{130}Xe$ for several Apollo 14 and 16 samples. One sigma error bars are given if larger than the symbols. Numbers by the symbols are stepwise heating extraction temperatures in hundreds of degrees Celsius. The 95% confidence errors reported for the Apollo 16 points (Table 1) have been decreased by a factor of two for the plot in order to reduce congestion and to approximate the 1 sigma errors reported for the other samples. AIR is from Nier (1950). For reference, a line has been drawn between AIR and pure spallation xenon (Marti *et al.*, 1973a).

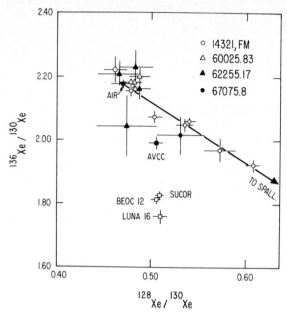

Fig. 2. Correlation of $^{136}Xe/^{130}Xe$ with $^{128}Xe/^{130}Xe$ from Apollo 14 and 16 samples in the region near trapped xenon. Apollo 14 points are from Marti *et al.* (1973a). See Fig. 1 for definition of reference points. Reported Apollo 16 errors have been halved, as discussed in Fig. 1. For reference, a solid line is drawn between AIR and pure spallation xenon.

It is not possible to determine the mixing endpoints from such correlation plots. The exact composition of trapped xenon cannot be obtained without an additional datum. Nevertheless, it is possible to eliminate the data point cluster representing either the solar xenon or AVCC composition, as one endpoint, since these points lie off the correlation lines. Most data points cluster near the AIR point, and it appears that the actual "trapped" data point cannot be much different. If we fix one of the spallation sensitive isotope ratios to the AIR value, then all other ratios of trapped component are determined from the correlations. Figure 3 shows the results obtained if the $^{126}Xe/^{130}Xe$ ratio is taken as the AIR value. In this case, Apollo 14 soil fragment 14160,8 (Marti *et al.*, 1973b) was used to approximate the spallation xenon yields. However, since the corrections for spallation are very small, the choice of these yields is not critical. Actually, the only xenon isotopes noticeably affected by this small spallation correction are masses 124 and 126. More specifically, a 20% uncertainty in the choice of value for $^{126}Xe/^{130}Xe$ affects the relative abundance of ^{124}Xe by 13%, ^{128}Xe by 1.0% and all the other isotopes by less than 0.5%. The isotopic similarity of the trapped component to terrestrial xenon is striking, including the distinctive fine structure at mass 129.

As will be reported elsewhere, the isotopic abundances of trapped krypton are similar to both solar (BEOC 12: Eberhardt *et al.*, 1972) and terrestrial krypton, within experimental uncertainty.

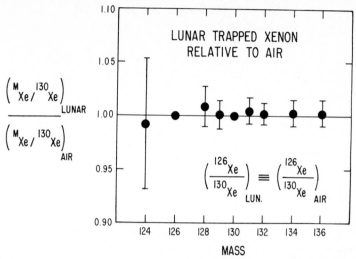

Fig. 3. Isotopic composition of trapped xenon in samples 60025 and 62255 relative to terrestrial atmospheric xenon. (See text for discussion.)

DISCUSSION OF ORIGIN

We believe that with the experimental procedures employed in this study, we have eliminated the possibility of a contamination. A duplicate analysis of an unwrapped chip of 60025 in a different extraction system shows a similar trapped Xe component, but unfortunately shows a detectable memory of spallation-rich gases consistent with those of a previously analyzed sample. The isotopic information is, therefore, of limited value and is not discussed here. We conclude that the investigated xenon was in fact trapped in the samples. Although trapped xenon in the studied lunar rocks is similar to terrestrial atmospheric xenon isotopically, we consider it unlikely to be of terrestrial origin because of the following observations.

If the xenon extracted from these samples was adsorbed terrestrial gas, the observed enrichment of Xe relative to Kr and Ar (Fig. 5) would be expected. A rough estimate of the amount of xenon adsorbed on Apollo 14 soils and breccias, $0.3 \, \text{m}^2/\text{g}$ effective surface area (Cadenhead et al., 1972), can be derived by using the xenon adsorption data of Fanale and Cannon (1971) from a terrestrial shale of $19.4 \, \text{m}^2/\text{g}$ effective surface area. If we assume that at a fixed xenon partial pressure, the amount of xenon adsorbed per gram is directly proportional to the effective surface area per gram, then a lunar breccia exposed to air at 25°C should contain approximately $5 \times 10^{-10} \, \text{cm}^3$ STP/g adsorbed ^{132}Xe. This calculated concentration is about a factor of ten larger than that measured in 60025 and 62255. For comparison, the other breccia known to contain this component, 14321, contained trapped ^{132}Xe varying between 1 and $5 \times 10^{-10} \, \text{cm}^3$ STP/g (Marti et al., 1973a); an effective surface area of $0.34 \, \text{m}^2/\text{g}$ for 14321 was measured by Cadenhead et al. (1972).

However, these samples were stored at 80°C under high vacuum for several days before analysis. After initial pump-down, during the two weeks preceding analysis, the maximum xenon partial pressure that the 60025 sample was exposed to was less than 10^{-12} torr. The partial pressure of xenon in air is 7×10^{-5} torr. Clearly, if adsorbed xenon on this sample had achieved equilibrium with the ambient gas in the extraction system at 80°C, adsorbed xenon would have been absolutely undetectable.

For an adsorbed gas component to remain locked into the sample under the conditions prevailing in the storage arm of the extraction system, a complete closing of the individual pores in the samples by some unknown process, such as a chemical alteration, would be required. This mechanism is open to experimental study and work along these lines is in progress in our laboratory.

Nevertheless, it appears improbable that the gas is adsorbed and "locked-in" atmospheric xenon, because temperatures in excess of 900°C are required for release. As can be seen in Fig. 4, trapped xenon is released predominantly at the higher extraction temperatures. The release pattern of the trapped gas more closely follows that of the tightly bound *in situ* generated constituents such as spallation produced noble gases. A *direct* measure of the degree of association between *in situ* produced and trapped noble gas is the ratio $^{81}Kr/^{86}Kr$. ^{81}Kr is produced by spallation reactions exclusively; ^{86}Kr is practically pure trapped gas.

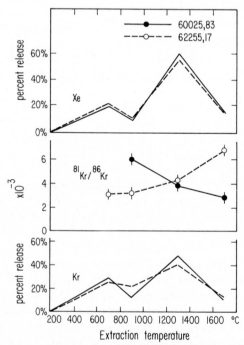

Fig. 4. Trapped krypton and xenon released from samples 60025 and 62255 as a function of extraction temperature. Also plotted is the ratio $^{81}Kr/^{86}Kr$ versus extraction temperature. (See text for discussion.)

Their ratio is plotted against extraction temperature in Fig. 4. The spallation to trapped krypton ratio for 62255 increases slightly during extraction; that ratio from 60025 actually decreases with higher activation energy. This difference in slope is consistent with the evidence discussed below (Fig. 5) for the additional presence of solar-type noble gas in 62255. Of course, this line of argument assumes trapped xenon and trapped krypton share a common origin. A similar correlation of trapped xenon with spallation xenon is evident in ^{124}Xe and ^{126}Xe but cannot be precisely quantified.

Figure 5 gives the elemental abundances of the noble gases relative to ^{36}Ar. Included for reference is solar xenon (Pesyanoe: Marti, 1969; Lunar soil 10084: Marti *et al.*, 1970), meteoritic xenon and terrestrial atmospheric xenon. Only the trapped noble gas component is plotted for the lunar samples. The relative abundances of trapped argon, krypton and xenon are clearly different than those of both the terrestrial and solar noble gas reservoirs. It should be noted, however, that a significant fraction of the earth's xenon may be trapped in sedimentary rock, as first pointed out by Canales *et al.* (1968); therefore, atmospheric noble gas abundances may not be representative of terrestrial abundances. In Fig. 5, the relative abundances of Ar, Kr, and Xe in 62255 may be derived from 60025 trapped gas by the admixture of a solar-type component. Such a mixture would

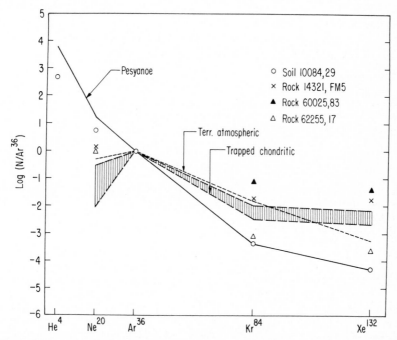

Fig. 5. Elemental abundances of trapped noble gases relative to ^{36}Ar from Apollo 14 and 16 samples. Terrestrial atmospheric, trapped chondritic and solar (Pesyanoe: Marti, 1969; Lunar soil 10084; Marti *et al.*, 1970) noble gas abundances are also given for reference.

contain approximately 10% solar xenon. The 62255 xenon isotopic data (Fig. 1) do not conflict with this suggestion.

So far, lunar trapped xenon discussed above has been observed in the various components of the Apollo 14 breccia 14321 and in the reported Apollo 16 anorthositic breccias 60025 and 62255. These samples differ considerably in texture and mineralogy. The observed ^{132}Xe concentrations range from 35 to 400×10^{-12} cc STP/g. On the other hand, three other anorthositic breccias which we have studied, 67015, 67075, and 67915, contain less than 5, 10 and 3×10^{-12} cc STP/g of trapped ^{132}Xe, respectively (Lightner and Marti, 1974) of unknown isotopic composition. None of the investigated samples are known to contain "rust" (Taylor et al., 1973).

It is interesting to compare the trapped xenon of Apollo 16 anorthositic breccias with abundances of other volatile elements. Thallium data are available for 60025 and 67915 (Krähenbühl et al., 1973) and show Tl enrichment by a factor of 50 in 60025 if compared to 67915. The corresponding trapped Xe ratio is 25. Krähenbühl et al. (1973) have discussed in detail the Tl abundances in Apollo 16 anorthosites and conclude that the generally large Tl enrichments are probably due to lunar fumaroles. On the other hand, El Goresy et al. (1973) favor an impact origin for some of the lunar volatiles at the Apollo 16 site. Nunes and Tatsumoto (1973) observed a large excess of lead in the anorthositic breccia 66095 which is not supported by U and Th in the rock but which has to be of lunar origin.

We expect that additional studies which are in progress may help to clarify some of the remaining questions and remote alternatives regarding the origin of lunar trapped xenon. The possibility of an association with other volatiles is intriguing.

Acknowledgments—We acknowledge the capable assistance of K. R. Goldman. We have benefited from discussions with E. Anders, G. W. Lugmair, J. H. Reynolds, and H. C. Urey. This work was supported by NASA grant NGR 05-009-150.

References

Cadenhead D. A., Wagner N. J., Jones B. R., and Stetter J. R. (1972) Some surface characteristics and gas interactions of Apollo 14 fines and rock fragments. *Proc. Third Lunar Sci. Conf., Geochim. Cosmochim. Acta*, Suppl. 3, Vol. 3, pp. 2243–2257. MIT Press.

Canales R. A., Alexander E. C., Jr., and Manuel O. K. (1968) Terrestrial abundances of noble gases. *J. Geophys. Res.* **73**, 3331.

Eberhardt P., Geiss J., Graf H., Grögler N., Mendia M. D., Mögeli M., Schwaller H., Stettler A., Krähenbühl U., and von Gunten H. R. (1972) Trapped solar wind noble gases in Apollo 12 lunar fines 12001 and Apollo 11 breccia 10046. *Proc. Third Lunar Sci. Conf., Geochim. Cosmochim. Acta*, Suppl. 3, Vol. 2, pp. 1821–1856. MIT Press.

El Goresy A., Ramdohr P., Pavicevic M., Medenbach O., Müller O., and Gentner W. (1973) Zinc, lead, chlorine and FeOOH-bearing assemblages in the Apollo 16 sample 66095: Origin by impact of a comet or a carbonaceous chondrite? *Earth Planet. Sci. Lett.* **18**, 411.

Fanale F. P. and Cannon W. A. (1971) Physical adsorption of rare gases on terrigenous sediments. *Earth Planet. Sci. Lett.* **11**, 362–368.

Kaiser W. A. (1972) Rare gas studies in Luna-16-G-7 fines by stepwise heating technique. A low fission solar wind Xe. *Earth Planet. Sci. Lett.* **13**, 387–399.

Krähenbühl U., Ganapathy R., Morgan J., and Anders E. (1973) Volatile elements in Apollo 16 samples: Implications for highland volcanism and accretion history of the moon. *Proc. Fourth Lunar Sci. Conf., Geochim. Cosmochim. Acta*, Suppl. 4, Vol. 2, pp. 1325–1348. Pergamon.

Laul J. C. and Schmitt R. A. (1973) Chemical composition of Apollo 15, 16 and 17 samples. *Proc. Fourth Lunar Sci. Conf., Geochim. Cosmochim. Acta*, Suppl. 4, Vol. 2, pp. 1349–1367. Pergamon.

Lightner B. D. and Marti K. (1974) Lunar trapped xenon (abstract). In *Lunar Science—V*, pp. 447–449. The Lunar Science Institute, Houston.

Marti K. (1969) Solar-type xenon: A new isotopic composition of xenon in the Pesyanoe meteorite. *Science* **166**, 1263–1265.

Marti K., Lugmair G. W., and Urey H. C. (1970) Solar wind gases, cosmic-ray spallation products and the irradiation history of Apollo 11 samples. *Proc. Apollo 11 Lunar Sci. Conf., Geochim. Cosmochim. Acta*, Suppl. 1, Vol. 2, pp. 1357–1367. Pergamon.

Marti K., Lightner B. D., and Lugmair G. W. (1973a) On ^{244}Pu in lunar rocks from Fra Mauro and implications regarding their origin. *The Moon* **8**, 241–250.

Marti K., Lightner B. D., and Osborn T. W. (1973b) Krypton and xenon in some lunar samples and the age of North Ray crater. *Proc. Fourth Lunar Sci. Conf., Geochim. Cosmochim. Acta*, Suppl. 4, Vol. 2, pp. 2037–2048. Pergamon.

Nier A. O. (1950) A redetermination of the relative abundances of the isotopes of neon, krypton, rubidium, xenon and mercury. *Phys. Rev.* **79**, 450–454.

Nunes P. D. and Tatsumoto M. (1973) Excess lead in "rusty rock" 66095 and implications for an early lunar differentiation. *Science* **182**, 916–920.

Podosek F. A., Huneke J. C., Burnett D. S., and Wasserburg G. J. (1971) Isotopic composition of xenon and krypton in the lunar soil and in the solar wind. *Earth Planet. Sci. Lett.* **10**, 199–216.

Taylor L. A., Mao H. D., and Bell P. M. (1973) "Rust" in the Apollo 16 rocks. *Proc. Fourth Lunar Sci. Conf., Geochim. Cosmochim. Acta*, Suppl. 4, Vol. 1, pp. 829–839. Pergamon.

Proceedings of the Fifth Lunar Conference
(Supplement 5, Geochimica et Cosmochimica Acta)
Vol. 2 pp. 2033–2044 (1974)
Printed in the United States of America

Lunar breccia 14066: ^{81}Kr–^{83}Kr exposure age, evidence for fissiogenic xenon from ^{244}Pu, and rate of production of spallogenic ^{126}Xe

B. Srinivasan*

Department of Physics, University of California, Berkeley, California 94720

Abstract—Krypton and xenon from two samples of lunar breccia 14066 were analyzed mass spectrometrically. Xenon in this rock apparently contains fissiogenic components from decay *in situ* of ^{238}U and ^{244}Pu. The initial ^{244}Pu/^{238}U ratio (4.56 b.y. ago) appears to be nearly equal to or greater than Podosek's value (0.013) for the bulk St. Severin meteorite. The trapped xenon in sample 14066 is probably similar to "Sucor" in composition and certainly not of atmospheric composition. The spallogenic spectra for krypton show high values for $(^{78}$Kr$/^{83}$Kr$)_{sp}$ relative to other Apollo 14 breccias of similar target element concentrations. The ^{81}Kr–^{83}Kr exposure age is ≈ 27 m.y. The spallogenic spectra for Xe exhibit very low values for the ratio $(^{131}$Xe$/^{126}$Xe$)_{sp}$. The above three lines of evidence indicate a simple irradiation history for 14066—that is, burial at great depth until excavation to the lunar surface at the time of the Cone Crater event, 27 m.y. ago. The rate of production of spallogenic ^{126}Xe from Ba and REE (rare-earth elements) has been inferred to be

$$[(1.20 \pm 0.09) \times g\ Ba/g + (0.73 \pm 0.05) \times g\ REE/g] \times 10^{-9}\ cc\ STP/(g)(m.y.)$$

where $REE = La + Ce + Pr$.

INTRODUCTION

FROM THE BEGINNING of the lunar sample analysis program, rare gas mass spectroscopists were searching for xenon anomalies resulting from the decay of now extinct radioactivities ^{129}I and ^{244}Pu. Such observations are expected to play an important role in deciphering the very early history of the moon. The first positive evidence for the occurrences of radiogenic ^{129}Xe from the decay of ^{129}I and fissiogenic $^{131-136}$Xe from the decay of ^{244}Pu, in lunar breccia 14301, was reported by Crozaz *et al.* (1972). Fissiogenic xenon isotopes have been identified in breccia 14318 (Behrmann *et al.*, 1973; Reynolds *et al.*, 1974) and in breccia 14313 (Behrmann *et al.*, 1973). In these three breccias, the products of extinct radioactivities appear to be "parentless." In other words, an *in situ* origin seems to be excluded. To explain the "parentless" components, Reynolds *et al.* (1974) have examined four different models, all of which are unsatisfactory in some respects. Additional studies are required for a complete understanding of this problem.

In contrast to the above mentioned samples, breccia 14321 contains fissiogenic xenon which was apparently produced by decay of ^{244}Pu *in situ* (Marti *et al.*, 1973). The inferred ^{244}Pu$/^{238}$U in this sample is consistent with values observed in

---

*Present address: Enrico Fermi Institute and Department of Chemistry, University of Chicago, Chicago, Illinois 60637.

St. Severin meteorite. Thus, four lunar samples from the Apollo 14 mission have shown evidence for the existence of now extinct radioactivities in the early history of the moon. More recently, Haines *et al.* (1974) have found excess fission tracks attributable to spontaneous fission of ^{244}Pu in Apollo 15 Kreep basalts. Another lunar breccia, 14066, appears to contain fissiogenic xenon from decay of ^{244}Pu *in situ*, similar to 14321. In this report, the results of mass spectrometric analyses of krypton and xenon in two samples of 14066 are described. The work of Alexander and Davis (1974) and this work constitute a survey of rare gases in the consortium rock 14066 headed by Reynolds. Some of the results obtained by Kaiser (1972) in an earlier study of this sample are worth recapitulating.

Kaiser (1972) deduced a very unusual spallogenic spectrum for xenon in 14066 by assuming the observed xenon to be comprised of (1) isotopes 131–136 from spontaneous fission of ^{238}U, (2) a trapped component of atmospheric composition and (3) isotopes 124–134 from spallation. No allowance was made for a fissiogenic component from ^{244}Pu. The spallogenic spectrum obtained by Kaiser (1972) was characterized by very low values for $(^{130}\text{Xe}/^{126}\text{Xe})_{sp}$ and $(^{131}\text{Xe}/^{126}\text{Xe})_{sp}$ ratios (subscript sp refers to spallation). In order to verify the unusual spallogenic yields in 14066, experiments were carried out with aliquots of samples for which trace element concentrations (U, Ba, rare-earth elements, etc.) had already been determined. This particular approach is useful in correctly evaluating the fissiogenic component from ^{238}U and to look for additional fissiogenic contribution from ^{244}Pu. Further, the use of aliquots for rare gases and trace elements analyses is useful in deducing the rate of production of spallogenic ^{126}Xe from the observed concentrations of ^{126}Xe$_{sp}$ and the ^{81}Kr–^{83}Kr exposure age.

EXPERIMENTAL PROCEDURE

Samples

Lunar sample 14066 is a recrystallized or metamorphosed breccia and belongs to group 7 in the scheme of Warner (1972). The following two samples from the breccia were analyzed for rare gases: (1) an aliquot of saw dust, sample 14066,31,1 weighing 0.2536 g; (2) an aliquot of homogenized sample 14066,21,2.01 obtained from clast 17 (Mark *et al.*, 1973) weighing 0.2796 g.

Laul *et al.* (1972) and Maxwell (1972) have determined the concentrations of trace elements on separate aliquots of the above samples. The pertinent results are shown in Table 1.

Krypton and xenon analyses

Rare gases were extracted by heating the samples to 1650°C in a tungsten crucible. The various techniques involved in the purification, separation of krypton from xenon, and mass spectrometric analyses of krypton and xenon are described elsewhere (Hohenberg *et al.*, 1970; Srinivasan, 1973). The methods for the reduction of data and the calculation of errors are the same as reported by Srinivasan (1973).

The completeness of extraction of rare gases was checked for sample 14066,31,1. In the second heating to 1650°C, the observed abundances of ^{84}Kr and ^{132}Xe were reduced by factors of 60 and 130, respectively, in comparison to the observed abundances in the first heating.

Table 1. Trace element concentrations in two samples from lunar breccia 14066.

Element (ppm)	14066,31,3 saw dust (Laul *et al.*, 1972)	14066,31,1 saw dust (Maxwell, private communication)	140661,21,2.01 breccia clast (Laul *et al.*, 1972)
Sr		170	
Y		190	
Zr	950 ± 86	1000	550 ± 110
Ba	920 ± 41	760	800 ± 70
La	79 ± 2	73	61 ± 3
Ce	178 ± 3		180 ± 24
U	4.0 ± 0.5		2.7 ± 0.3

Hot blanks

Prior to the analysis of the samples, the empty crucible was heated to 1650°C and the rare gases were analyzed. These measurements monitored the cleanliness of the sample system. No corrections were made for the hot blanks either in the results or in the subsequent calculations. The results and conclusions of this paper are not altered, significantly, by the presence of hot blanks. The hot blanks are listed separately in the footnotes for the tables. Instrument blanks characteristic of the mass spectrometer were monitored before every analysis and found to be insignificant.

RESULTS AND DISCUSSION

Krypton

The observed isotopic compositions and abundances of krypton in the two samples of 14066 are set out in Table 2. Also given in the table are the spallogenic spectra for krypton. The spallation spectra were obtained by assuming that the observed krypton is a mixture of the following components:

(1) isotopes 83–86 from spontaneous fission of ^{238}U,
(2) trapped krypton of atmospheric composition, and
(3) isotopes 78–84 from spallation.

The observed krypton was corrected consecutively for components (1) and (2) by the use of the following information: (a) ^{40}Ar–^{39}Ar age of (3.93 ± 0.03) b.y. for breccia 14066 (Alexander and Davis, 1974), (b) the concentration of U in the samples (Table 1), (c) the α-decay half-life of 4.51×10^9 yr and spontaneous fission half-life of 1×10^{16} yr, for ^{238}U, (d) the spontaneous fission yields for krypton isotopes (Wetherill, 1953), and (e) the isotopic composition of atmospheric krypton (Eugster *et al.*, 1967).

The contributions to ^{86}Kr from component (1) are 0.3% and 0.9% of the observed ^{86}Kr, in samples 14066,31,1 and 14066,21,2.01, respectively. The corrections for ^{84}Kr and ^{83}Kr are negligible.

Table 2. Isotopic compositions of observed krypton and spallogenic krypton.

Isotope	14066,31,1 (0.2536 g)		14066,21,2.01 (0.2796 g)	
	observed	spallation	observed	spallation
78	0.02863 ±.00024	0.2306 ±.0035	0.09393 ±.00061	0.2341 ±.0018
80	0.09039 ±.00039	0.5262 ±.0074	0.2363 ±.0012	0.5304 ±.0041
81	0.000687 ±.000027	0.00698* ±.00029	0.002640 ±.000047	0.00699 ±.00013
82	0.27267 ±.00070	0.767 ±.014	0.4769 ±.0024	0.7759 ±.0080
83	0.29508 ±.00072	≡1.000	0.5610 ±.0023	≡1.000
84	≡1.0000	0.250 ±.036	≡1.0000	0.240 ±.010
86	0.29902 ±.00095	—	0.2805 ±.0011	—
Abundance†	999 ± 53	98.3 ± 5.4	264 ± 15	99.6 ± 5.6

Notes: A 1650°C blank measured immediately before the analyses of the samples and under identical procedure, yielded 2.3×10^{-12} cc STP of ^{84}Kr; the isotopic composition was atmospheric. Mass discrimination and sensitivity values were obtained from air standards.

*A correction was made for ^{81}Kr based on the abundance at mass 79. Half the correction is added as an error.

†Abundances in units of 10^{-12} cc STP/g relate to isotopes used for normalization.

The trapped component is assumed to be of atmospheric composition. The other possibility is "solar krypton" approximated by surface correlated krypton in lunar fines (Podosek *et al.*, 1971; Eberhardt *et al.*, 1972.) However, the differences between atmospheric krypton and surface correlated krypton in lunar fines are very small. Either of the assumptions for trapped krypton give rise to spallation spectra which are indistinguishable from one another within the limits of experimental error.

^{81}Kr–^{83}Kr exposure age:

The cosmic ray exposure age (t_{exp}) can be calculated from the data for the radioactive isotope ^{81}Kr (Marti, 1967). The equations pertinent to the calculations are given below.

$$t_{exp} = (^{83}Kr/^{81}Kr)_{sp} \times (P_{81}/P_{83}) \times 1/\lambda_{81} \tag{1}$$

$$P_{81}/P_{83} = 0.95 \times [(^{80}Kr + {}^{82}Kr)/2 \times {}^{83}Kr]_{sp} \tag{2}$$

where λ_{81} is the decay constant $(3.3 \times 10^{-6}\,\mathrm{yr}^{-1})$ and P_{81}/P_{83} is the relative production rate of ^{81}Kr with respect to ^{83}Kr. Equation (2) provides a way to obtain P_{81}/P_{83} from the observed spallation yields. Using the data in Table 1, the exposure ages are calculated to be (26.7 ± 1.2) m.y. and (26.9 ± 0.5) m.y. for 14066,31,1 and 14066,21,2.01, respectively. The reported errors do not include uncertainties in λ_{81} and in P_{81}/P_{83}. These exposure ages are in excellent agreement with the ages obtained for other Apollo 14 samples which are believed to be ejecta from Cone Crater event (Turner et al., 1971; Lugmair and Marti, 1972).

Lugmair and Marti (1972) observe that for Apollo 14 samples containing excesses of ^{80}Kr and ^{82}Kr from neutron capture reactions on bromine, Eq. (2) cannot be employed for obtaining P_{81}/P_{83}. Instead, they suggest the use of an empirical equation which has been obtained through studies of samples 10017, 10057, 10071, and 14321, which contain similar abundances of target elements (Sr/Zr) and are noted for the absence of neutron capture products. The equation is

$$P_{81}/P_{83} = 0.850 \times (^{78}\mathrm{Kr}/^{83}\mathrm{Kr})_{\mathrm{sp}} + 0.442 \qquad (3)$$

Use of Eq. (3) yields $t_{\mathrm{exp}} = (27.7 \pm 1.2)$ m.y. for sample 14066,31,1 and $t_{\mathrm{exp}} = (27.8 \pm 0.5)$ m.y. for 14066,21,2.01. These values are nearly identical to the exposure ages given in the preceding paragraph. The agreement suggests the absence of neutron capture anomalies in krypton in breccia 14066. This conclusion is consistent with the low abundance of bromine, 37 ppb (Alexander and Davis, 1974), and to minimal exposure to neutrons as deduced from the $(^{131}\mathrm{Xe}/^{126}\mathrm{Xe})_{\mathrm{sp}}$ ratio (see spallogenic xenon).

Xenon

The observed isotopic compositions and abundances of xenon in the two samples of 14066 are shown in Table 3. Also included in the table are the spallation spectra derived with the help of the following assumptions. The observed xenon is assumed to be comprised of

(1) isotopes 131–136 from spontaneous fission of ^{238}U,

(2) isotopes 131–136 from spontaneous fission of ^{244}Pu,

(3) trapped xenon of "Sucor" composition (Podosek et al., 1971), and

(4) isotopes 124–132 from spallation. The reaction $^{130}\mathrm{Ba}(n, \gamma) \rightarrow {}^{131}\mathrm{Xe}$ has been included in this component.

Trapped xenon

It is essential to define the nature of the trapped component before attempting to unravel the observed spectra. As for krypton, the two choices for this component are (a) atmospheric xenon and (b) "solar xenon" as approximated by surface correlated xenon in lunar fines. Unlike krypton, the isotopic compositions of xenon in these two reservoirs are distinctly different. Further, there are small variations in the isotopic compositions of xenon between different samples of lunar fines (Podosek et al., 1971; Eberhardt et al., 1972). Among the Apollo 14

Table 3. Isotopic compositions of observed xenon and spallogenic xenon.

Isotope	14066,31,1 (0.2536 g)		14066,21,2.01 (0.2796 g)	
	observed	spallation	observed	spallation
124	0.08093 ±.00092	0.571 ±.014	0.2033 ±.0027	0.577 ±.020
126	0.1383 ±.0013	≡1.000	0.3494 ±.0034	≡1.000
128	0.2680 ±.0017	1.462 ±.049	0.5710 ±.0041	1.479 ±.054
129	1.0913 ±.0047	1.48 ±.41	1.2265 ±.0067	1.42 ±.25
130	0.2608 ±.0012	0.898 ±.066	0.4146 ±.0016	0.864 ±.050
131	1.0978 ±.0028	2.87 ±.33	1.5547 ±.0091	2.75 ±.22
132	≡1.0000	0.68 ±.41	≡1.0000	0.54 ±.25
134	0.3844 ±.0014	—	0.3899 ±.0017	—
136	0.3343 ±.0019	—	0.3579 ±.0020	—
Abundance†	233 ± 12	31.3 ± 1.8	92.9 ± 4.8	32.2 ± 2.1

Notes: A 1650°C blank measured immediately before the analyses of the samples and under identical procedure, yielded 2.6×10^{-13} cc STP of ^{132}Xe; the isotopic composition was atmospheric. Mass discrimination and sensitivity values were obtained from air standards (Nier, 1950).

†Abundances in units of 10^{-12} cc STP/g relate to isotopes used for normalization.

breccias reported to contain fissiogenic xenon from ^{244}Pu, samples 14301, 14313, and 14318 contain "solar" xenon for the trapped component, whereas sample 14321 includes a trapped component of atmospheric composition (Marti *et al.*, 1973). For trapped xenon in 14066, the above two possibilities are considered, immediately below.

Let us assume, tentatively, that the isotopes ^{134}Xe and ^{136}Xe are a simple mixture of xenon from spontaneous fission of ^{238}U and trapped xenon either of atmospheric composition (Nier, 1950) or of "Sucor" composition (Podosek *et al.*, 1971). The abundance of ^{136}Xe$_f$ can be calculated for the two different sets of assumptions, from the following equation.

$$^{136}\text{Xe}_f = \frac{(^{134}\text{Xe}/^{136}\text{Xe})_{obs} - (^{134}\text{Xe}/^{136}\text{Xe})_t}{(^{134}\text{Xe}/^{136}\text{Xe})_f - (^{134}\text{Xe}/^{136}\text{Xe})_t} \times {}^{136}\text{Xe}_{obs} \qquad (4)$$

where the subscripts obs, t, and f stand for observed, trapped and fissiogenic values, respectively. The fissiogenic yields of xenon isotopes from spontaneous fission of ^{238}U were obtained from Wetherill (1953). The results of the calculation are set out in Table 4. For comparison the amounts of ^{136}Xe$_f$ expected from decay

Table 4. Comparison between abundances of ^{136}Xe$_f$ calculated for two different
assumptions of trapped component.

| Sample | ^{136}Xe$_f$ in 10^{-12} cc STP/g | | |
	Atmospheric trapped	Sucor trapped	From uranium*
14066,31,1	6.1 ± 2.0	16.0 ± 1.4	8.3 ± 0.8
14066,21,2.01	8.4 ± 1.0	11.9 ± 0.8	5.6 ± 0.6

*Calculated for decay *in situ* of ^{238}U (see text).

of ^{238}U *in situ* for a period of 3.93 ± 0.03 b.y. have also been calculated and shown in the last column. The uranium concentrations in Table 1, the decay constant of ^{238}U given earlier and a fissiogenic yield of 6% for ^{136}Xe were used in the calculations.

The large error in the abundance of ^{136}Xe$_f$ in column 2 for sample 14066,31,1 is a direct consequence of the proximity of ^{134}Xe/^{136}Xe ratios in the sample and in the atmosphere. Confining our attention to the case of atmospheric xenon as trapped component, the following points are evident from a comparison of results between columns 2 and 4: (a) Sample 14066,31,1 appears to have lost some of the fissiogenic xenon originating from ^{238}U and (b) sample 14066,21,2.01 demands an additional source for fissiogenic xenon apart from ^{238}U. On the other hand if "Sucor" xenon is used as a trapped component (column 3), both samples require an additional source for fissiogenic xenon. Preferring a self-consistent explanation for both samples rather than an *ad hoc* explanation for each sample, I postulate that the trapped component is of "Sucor" composition and the excess fissiogenic xenon arises from the decay of ^{244}Pu *in situ*. However, the exact nature of the trapped component is likely to be a modification of "Sucor" xenon. Detailed stepwise heating experiments might provide a decisive answer to this question. Assumption of a trapped component other than "Sucor" xenon, say fractionated "Sucor" xenon, would yield different amounts of "plutonogenic" xenon than the one calculated here (see fissiogenic xenon).

Fissiogenic xenon

Xenon from spontaneous fission of ^{238}U was calculated by the same method as for krypton. After subtracting this component from the observed xenon, the contributions from ^{244}Pu were calculated by the use of an equation similar to Eq. (4).

$$^{136}\text{Xe}_f = \frac{(^{134}\text{Xe}/^{136}\text{Xe})_{corr} - (^{134}\text{Xe}/^{136}\text{Xe})_{sucor}}{(^{134}\text{Xe}/^{136}\text{Xe})_f - (^{134}\text{Xe}/^{136}\text{Xe})_{sucor}} \times {}^{136}\text{Xe}_{corr} \qquad (5)$$

where subscript corr refers to values corrected for spontaneous fission of ^{238}U and subscript f refers to fissiogenic yields for xenon isotopes from ^{244}Pu (Alexander *et al.*, 1971). The calculated abundances of ^{136}Xe$_f$ from ^{238}U as well as ^{244}Pu are set out in Table 5.

Assuming that both ^{238}U and ^{244}Pu have decayed *in situ*, the ratios of ^{244}Pu/^{238}U

Table 5. Abundance of ^{136}Xe from spontaneous fission of ^{238}U and ^{244}Pu; inferred ^{244}Pu/^{238}U ratios.

Sample	^{136}Xe$_f \times 10^{-12}$ cc STP/g from		^{244}Pu/^{238}U at	
	^{238}U	^{244}Pu	$t = 3.93$ b.y.	$t = 4.56$ b.y.
14066,31,1	8.3 ± 0.8	9.9 ± 4.1	$(2.0 \pm 0.9) \times 10^{-4}$	0.04 ± 0.02
14066,21,2.01	5.6 ± 0.6	8.1 ± 2.7	$(2.5 \pm 0.9) \times 10^{-4}$	0.05 ± 0.02

were calculated at $t = 3.93$ b.y. and $t = 4.56$ b.y. In these calculations, the following parameters were used for ^{244}Pu in addition to the data given earlier for ^{238}U: α-decay half-life of 8.18×10^7 yr, spontaneous fission half-life of 6.5×10^{10} yr and yield of 6% for ^{136}Xe$_f$. The initial ^{244}Pu/^{238}U ratio ($t = 4.56$ b.y.), although critically dependent on the age difference (4.56–3.93) and on the nature of the trapped component assumed, appears to be in the range of 0.02–0.07. This may be compared with the initial ^{244}Pu/^{238}U ratios of 0.013 in bulk St. Severin meteorite (Podosek, 1970) and 0.025 in lunar breccia 14321 (Marti et al., 1973). It is also worthwhile to note that the initial ratios may be altered by chemical fractionation of U and Pu. In the lunar sample 15272 studied by Haines et al. (1974) large variations are observed in the Pu/U ratios, in the different grains from the same basalt fragment. Nevertheless, the above exercise provides qualitative evidence for the contemporaneity of moon and meteorites.

Spallogenic xenon

The subtraction of fissiogenic and trapped components from the observed spectra yield the isotopic compositions of spallogenic xenon. The results are set out in Table 3. The ratio $(^{131}$Xe/^{126}Xe$)_{sp} = 2.75$ is one of the lowest values for lunar samples and consistent with minimal exposure to secondary particles, mainly neutrons. Bogard et al. (1974) have found the same value (2.75) for the ratio $(^{131}$Xe/^{126}Xe$)_{sp}$ in sample 76315. The low $(^{131}$Xe/^{126}Xe$)_{sp}$ and high $(^{78}$Kr/^{83}Kr$)_{sp}$ have already been recognized as correlated indicators of minimal shielding during cosmic ray exposure (Marti et al., 1973a) and reveal a simple irradiation history for 14066—that is, burial at great depth until transport to the lunar surface at the time of the Cone Crater event.

Rate of production of spallogenic ^{126}Xe

The simple history of cosmic ray exposure and the well-defined ^{81}Kr–^{83}Kr exposure ages are profitably used to deduce the average rate of production of ^{126}Xe$_{sp}$ from the two samples (which agree with one another within experimental error). The rate of production of ^{126}Xe$_{sp}$ in units of 10^{-9} cc STP/(g)(m.y.) is

$$(1.20 \pm 0.09) \times (g\ Ba/g) + (0.73 \pm 0.05) \times (g\ REE/g) \qquad (6)$$

where REE = La + Ce + Pr. The relative production rates from Ba and REE have been inferred from Rudstam's systematics (1966) as used by Hohenberg *et al.* (1967) for the Pasamonte meteorite, since Ba and REE appear to be present in nearly the same proportions in Pasamonte and in breccia 14066. Further, the Pr contents in 14066 were calculated using the Pr/Ce ratio in Pasamonte and Ce concentrations in the samples of 14066. Since two independent determinations were available for Ba and La concentrations in the saw dust sample, the following average values were used: Ba = (840 ± 80) ppm; La = (76 ± 3) ppm. The usefulness of Eq. (6) for determining the "surface" exposure ages of lunar samples is discussed below.

Exposure ages

In this section the ^{81}Kr–^{83}Kr ages of various lunar samples are compared with the ^{126}Xe$_{sp}$ ages calculated from Eq. (6). The results are set out in Table 6. The samples represent a broad spectrum of exposure ages from 25 to 500 m.y. The samples are also characterized by nearly sixfold variation in the abundances of Ba and REE. For example, the Ba contents are 110 ppm in 10044 and 628 ppm in 14321 with corresponding variations in the REE contents.

The experimentally determined concentrations of ^{126}Xe$_{sp}$, Ba and REE were substituted in Eq. (6) to calculate the ^{126}Xe$_{sp}$ ages. The concentrations of ^{126}Xe$_{sp}$

Table 6. Comparison of ^{81}Kr–^{83}Kr exposure ages with the spallogenic ^{126}Xe exposure ages.

	Exposure age		References for
Lunar sample	^{81}Kr–^{83}Kr	^{126}Xe*	concentrations of Ba, REE
10017	509 ± 29 (a)	308 ± 47 (a)	(b, h, i, k)
	480 ± 25 (b)	371 ± 80 (b)	
10044	70 ± 17 (c)	72 ± 13 (c)	(i, l)
10057	34 ± 5 (c)	54 ± 9 (c)	(j)
	47 ± 2 (a)	34 ± 6 (a)	
10071	372 ± 22 (a)	212 ± 31 (a)	(b, h)
	350 ± 15 (b)	300 ± 71 (b)	
14310	259 ± 7 (d)	176 ± 31 (f)	(m, n, o, p, q)
	268 ± 14 (e)	239 ± 24 (e)	
14321 FM5	27.2 ± 0.5 (d)	25 ± 5 (g)	(p, r)
14321 FM1 + 2	23.8 ± 0.6 (d)	17 ± 3 (g)	(p, r)

*^{126}Xe exposure ages were calculated from Eq. (6). The concentrations of spallogenic ^{126}Xe, Ba and REE were obtained from references corresponding to the letter(s) in parentheses: (a) Marti *et al.* (1970); (b) Eberhardt *et al.* (1974); (c) Hohenberg *et al.* (1970); (d) Lugmair and Marti (1972); (e) Kaiser (1972); (f) Marti *et al.* (1973a); (g) Marti *et al.* (1973); (h) Gast *et al.* (1970); (i) Goles *et al.* (1970); (j) Morrison *et al.* (1970); (k) Philpotts and Schnetzler (1970); (l) Tera *et al.* (1970); (m) Brunfelt *et al.* (1972); (n) Helmke *et al.* (1972); (o) Hubbard *et al.* (1972); (p) Masuda *et al.* (1972); (q) Philpotts *et al.* (1972); (r) Strasheim *et al.* (1972).

were obtained from the same experiments that were used to determine the $^{81}Kr-^{83}Kr$ ages. The concentrations of Ba and REE were simple averages of the various determinations reported in the literature. The average method gave rise to errors in the ranges of 5–10% for Ba and 2–5% for REE. The errors in the $^{126}Xe_{sp}$ exposure ages (Table 6) include uncertainties in the concentrations of $^{126}Xe_{sp}$, Ba, REE, and the uncertainties in Eq. (6).

An inspection of Table 6 reveals that although there is a general agreement between the $^{81}Kr-^{83}Kr$ ages and $^{126}Xe_{sp}$ ages, there is a tendency for the xenon ages to be lower than the krypton ages except for sample 10057 analyzed by Hohenberg *et al.* (1970). The lower xenon ages are more pronounced for samples 10017, 10057, 10071, and 14310 analyzed by Marti and co-workers (1970, 1973a). The following three causes are examined for an explanation of the lower $^{126}Xe_{sp}$ ages in these samples.

(1) The measured concentration of $^{126}Xe_{sp}$ in the samples may be low. This is likely to result from (a) diffusive loss of xenon from the samples and/or (b) underestimation of the abundances of xenon in the samples, due to experimental error.

(2) An overestimation of the abundances of xenon in samples of 14066 would result in the overestimation of the rate of production of $^{126}Xe_{sp}$ from Ba and REE. In that event, the calculated $^{126}Xe_{sp}$ exposure ages would be low. However, this possibility appears to be unlikely from our results on a split sample of Berkeley Bruderheim interlaboratory standard; the analysis of Bruderheim meteorite was carried out immediately following the analyses of samples from 14066. The abundances of $^{84}Kr = (70 \pm 4) \times 10^{-12}$ cc STP/g and $^{132}Xe = (91 \pm 7) \times 10^{-12}$ cc STP/g in our sample of Bruderheim are in agreement with those from other laboratories (Reynolds, private communication 1972). This is taken as evidence that the concentrations of rare gases in samples of 14066 are correctly estimated.

(3) The $^{81}Kr-^{83}Kr$ age might have been overestimated. This could result if the isotope ^{81}Kr has not reached radioactive equilibrium between production from cosmic rays and radioactive decay. If this were the case, application of Eqs. (1) and (2) to calculate t_{exp} would result in high ages. Since all of the samples under consideration, have sufficiently long exposure ages compared to the half-life of ^{81}Kr, this explanation is also untenable.

In conclusion, I suggest that other lunar samples with simple irradiation history be examined by experimental methods analogous to the one followed here (for 14066). Such studies would yield a statistically meaningful equation for the production of $^{126}Xe_{sp}$ from Ba and REE. Pending the availability of such studies, Eq. (6) presented in this paper is expected to be useful for the calculation of "surface" exposure ages of lunar samples.

Acknowledgments—The author is grateful to J. H. Reynolds for valuable criticism and advice. Thanks are due to P. K. Davis for help with computer programs and G. A. McCrory for maintenance of the mass spectrometer system. The work was supported by a research grant from NASA.

REFERENCES

Alexander E. C., Jr., Lewis R. S., Reynolds J. H., and Michel M. C. (1971) Plutonium-244; Confirmation as an extinct radioactivity. *Science* **172**, 837–840.

Alexander E. C., Jr., and Davis P. K. (1974) ^{40}Ar–^{39}Ar ages and trace element contents of Apollo 14 breccias; an interlaboratory cross-calibration of ^{40}Ar–^{39}Ar standards. *Geochim. Cosmochim. Acta* **38**, 911–928.

Behrmann C. J., Drozd R. J., and Hohenberg C. M. (1973) Extinct Lunar radioactivities: Xenon from ^{244}Pu and ^{129}I in Apollo 14 breccias. *Earth Planet. Sci. Lett.* **17**, 446–455.

Bogard D. D., Nyquist L. E., and Hirsch W. C. (1974) Noble gases in Apollo 17 boulders and soils (abstract). In *Lunar Science—V*, pp. 73–75. The Lunar Science Institute, Houston.

Brunfelt A. O., Heier, K. S., Nilssen B., Sundvoll B., and Steinnes E. (1972) Distribution of elements between different phases of Apollo 14 rocks and soils. *Proc. Third Lunar Sci. Conf., Geochim. Cosmochim. Acta*, Suppl. 3, Vol. 2, pp. 1133–1147. MIT Press.

Crozaz G., Drozd R., Graf H., Hohenberg C. M., Monin M., Ragan D., Seitz M., Shirck J., Walker R. M., and Zimmerman J. (1972) Evidence for extinct ^{244}Pu: Implications of the age of the Pre-Imbrium Crust (abstract). In *Lunar Science—III*, pp. 164–166. The Lunar Science Institute, Houston.

Eberhardt P., Geiss J., Graf H., Grögler N., Mendia M. D., Mörgeli M., Schwaller H., Stettler A., Krähenbühl U., and von Gunten H. R. (1972) Trapped solar wind gases in Apollo 12 lunar fines 12001 and Apollo 11 breccia 10046. *Proc. Third Lunar Sci. Conf., Geochim. Cosmochim. Acta*, Suppl. 3, Vol. 2, pp. 1821–1856. MIT Press.

Eberhardt P., Geiss J., Graf H., Grögler N., Krähenbühl U., Schwaller H., and Stettler A. (1974) Noble gas investigations of lunar rocks 10017 and 10071. *Geochim. Cosmochim. Acta* **38**, 97–120.

Eugster O., Eberhardt P. and Geiss J. (1967) The isotopic composition of krypton in unequilibrated and gas rich chondrites. *Earth Planet. Sci. Lett.* **2**, 385–393.

Gast P. W., Hubbard N. J., and Wiesmann H. (1970) Chemical composition and petrogenesis of basalts from Tranquillity base. *Proc. Apollo 11 Lunar Sci. Conf., Geochim. Cosmochim. Acta*, Suppl. 1, Vol. 2, pp. 1143–1163. Pergamon.

Goles G. G., Randle K., Osawa M., Lindstrom D. J., Jerome D. Y., Steinborn T. L., Beyer R. L., Martin M. R., and McKay S. M. (1970) Interpretations and speculations on elemental abundances in lunar samples. *Proc. Apollo 11 Lunar Sci. Conf., Geochim. Cosmochim. Acta*, Suppl. 1, Vol. 2, pp. 1177–1194. Pergamon.

Haines E. L., Hutcheon I. D., and Weiss J. R. (1974) Excess fission tracks in Apennine front kreep basalts (abstract). In *Lunar Science—V*, pp. 304–306. The Lunar Science Institute, Houston.

Helmke P. A., Haskin L. A., Korotev R. L., and Karen E. Z. (1972) Rare earths and other trace elements in Apollo 14 samples. *Proc. Third Lunar Sci. Conf., Geochim. Cosmochim. Acta*, Suppl. 3, Vol. 2 pp. 1275–1292. MIT Press.

Hohenberg, C. M., Munk M. N., and Reynolds J. H. (1967) Spallation and fissiogenic xenon and krypton from stepwise heating of the Pasamonte achondrite; the case for extinct plutonium 244 in meteorites; relative ages of chondrites and achondrites. *J. Geophys. Res.* **72**, 3139–3177.

Hohenberg, C. M., Davis P. K., Kaiser W. A., Lewis R. S., and Reynolds J. H. (1970) Trapped and cosmogenic rare gases from stepwise heating of Apollo 11 samples. *Proc. Apollo 11 Lunar Sci. Conf., Geochim. Cosmochim. Acta*, Suppl. 1, Vol. 2, pp. 1283–1309. Pergamon.

Hubbard N. J., Gast P. W., Rhodes J. M., Bansal B. M., Wiesmann H., and Church S. E. (1972) Nonmare basalts: Part II. *Proc. Third Lunar Sci. Conf., Geochim. Cosmochim. Acta*, Suppl. 3, Vol. 2, pp. 1161–1179. MIT Press.

Kaiser W. A. (1972) Kr and Xe in three Apollo 14 samples by stepwise heating technique. Unpublished manuscript available in mimeographed form.

Laul J. C., Wakita H., Showalter D. L., Boynton W. V., and Schmitt R. A. (1972) Bulk, rare earth, and other trace elements in Apollo 14, 15 and Luna 16 samples. *Proc. Third Lunar Sci. Conf., Geochim. Cosmochim. Acta*, Suppl. 3, Vol. 2, pp. 1181–1200. MIT Press.

Lugmair G. W., and Marti K. (1972) Exposure ages and neutron capture record in lunar samples from Fra Mauro. *Proc. Third Lunar Sci. Conf., Geochim. Cosmochim. Acta*, Suppl. 3, Vol. 2, pp. 1891–1897. MIT Press.

Mark R. K., Cliff R. A., Lee-Hu C., and Wetherill G. W. (1973) Rb–Sr studies of lunar breccias and soils. *Proc. Fourth Lunar Sci. Conf., Geochim. Cosmochim. Acta*, Suppl. 4, Vol. 2, pp. 1785–1795. Pergamon.

Marti K. (1967) Mass spectrometric detection of cosmic-ray produced ^{81}Kr in meteorites and the possibility of Kr–Kr dating. *Phys. Rev. Lett.* **18**, 264–266.

Marti K., Lugmair G. W., and Urey H. C. (1970) Solar wind gases, cosmic-ray spallation products and the irradiation history of Apollo 11 samples. *Proc. Apollo 11 Lunar Sci. Conf., Geochim. Cosmochim. Acta*, Suppl. 1, Vol. 2, pp. 1357–1367. Pergamon.

Marti K., Lightner B. D., and Lugmair G. W. (1973) On ^{244}Pu in lunar rocks from Fra Mauro and implications regarding their origin. *The Moon* **8**, 241–250.

Marti, K., Lightner B. D., and Osborn T. W. (1973a) Krypton and xenon in some lunar samples and the age of North Ray Crater. *Proc. Fourth Lunar Sci. Conf., Geochim. Cosmochim. Acta*, Suppl. 4, Vol. 2, pp. 2037–2048. Pergamon.

Masuda A., Nakamura N., Kurasawa H., and Tanaka T. (1972). Precise determination of rare earth elements in the Apollo 14 and 15 samples. *Proc. Third Lunar Sci. Conf., Geochim. Cosmochim. Acta*, Suppl. 3, Vol. 2, pp. 1307–1313. MIT Press.

Maxwell J. A. (1972) Private communication to J. H. Reynolds on consortium rock 14066.

Morrison G. H., Gerard J. T., Kashuba A. T., Gangadharam E. V., Rothenberg Ann M., Potter N. H., and Miller G. W. (1970) Elemental abundances of lunar soils and rocks. *Proc. Apollo 11 Lunar Sci. Conf., Geochim. Cosmochim. Acta*, Suppl. 1, Vol. 2, pp. 1383–1392. Pergamon.

Nier A. O. (1950) A redetermination of the relative abundances of the isotopes of neon, krypton, rubidium, xenon and mercury. *Phys. Rev.* **79**, 450–454.

Philpotts J. A. and Schnetzler C. C. (1970) Apollo 11 lunar samples: K, Rb, Sr, Ba and rare earth concentrations in some rocks and separated phases. *Proc. Apollo 11 Lunar Sci. Conf., Geochim. Cosmochim. Acta*, Suppl. 1, Vol. 2, pp. 1471–1486. Pergamon.

Philpotts J. A., Schnetzler C. C., Nava D. F., Bottino M. L., Fullagar P. D., Thomas H. H., Schuhmann S., and Kouns C. W. (1972) Apollo 14: Some geochemical aspects. *Proc. Third Lunar Sci. Conf., Geochim. Cosmochim. Acta*, Suppl. 3, Vol. 2, pp. 1293–1305. MIT Press.

Podosek F. A. (1970) The abundance of ^{244}Pu in the early solar system. *Earth Planet. Sci. Lett.* **8**, 183–187.

Podosek F. A., Huneke J. C., Burnett D. S., and Wasserburg G. J. (1971) Isotopic composition of xenon and krypton in the lunar soil and in the solar wind. *Earth Planet. Sci. Lett.* **10**, 199–216.

Reynolds J. H., Alexander E. C., Jr., Davis P. K., and Srinivasan B. (1974) Studies of K–Ar dating and xenon from extinct radioactivities in breccia 14318; implications for early lunar history. *Geochim. Cosmochim. Acta* **38**, 401–417.

Rudstam G. (1966) Systematics of spallation yields. *Z. Naturforsch.* **21a**, 1027–1041.

Srinivasan B. (1973) Variations in the isotopic composition of trapped rare gases in lunar sample 14318. *Proc. Fourth Lunar Sci. Conf., Geochim. Cosmochim. Acta*, Suppl. 4, Vol. 2, pp. 2049–2064. Pergamon.

Strasheim A., Jackson P. F. S., Coetzee J. H. J., Strelow F. W. E., Wybenga F. T., Gricus A. J., Kokot M. L., and Scott R. H. (1972) Analysis of lunar samples 14163, 14259 and 14321 with isotopic data for ^{7}Li/^{6}Li. *Proc. Third Lunar Sci. Conf., Geochim. Cosmochim. Acta*, Suppl. 3, Vol. 2, pp. 1337–1342. MIT Press.

Tera F., Eugster O., Burnett D. S., and Wasserburg G. J. (1970) Comparative study of Li, Na, K, Rb, Cs, Ca, Sr and Ba abundances in achondrites and in Apollo 11 lunar samples. *Proc. Apollo 11 Lunar Sci. Conf., Geochim. Cosmochim. Acta*, Suppl. 1, Vol. 2, pp. 1637–1657. Pergamon.

Turner G., Huneke J. C., Podosek F. A., and Wasserburg G. J. (1971) ^{40}Ar–^{39}Ar ages and cosmic-ray exposure ages of Apollo 14 samples. *Earth Planet. Sci. Lett.* **12**, 19–35.

Warner J. L. (1972) Metamorphism of Apollo 14 breccias. *Proc. Third Lunar Sci. Conf., Geochim. Cosmochim. Acta*, Suppl. 3, Vol. 1, pp. 623–643. MIT Press.

Wetherill G. W. (1953) Spontaneous fission yields from uranium and thorium. *Phys. Rev.* **92**, 907–912.

Proceedings of the Fifth Lunar Conference
(Supplement 5, Geochimica et Cosmochimica Acta)
Vol. 2 pp. 2045–2060 (1974)
Printed in the United States of America

Evidence for solar cosmic ray proton-produced neon in fines 67701 from the rim of North Ray Crater

J. R. Walton, D. Heymann, J. L. Jordan, and A. Yaniv*

Department of Space Physics and Astronomy and Department of Geology, Rice University, Houston, Texas 77001

Abstract—Fines 67701 from the rim of North Ray Crater (Apollo 16) contain cosmogenic Ne with $(Ne^{21}/Ne^{22})_C = 0.4 \pm 0.1$, significantly smaller than the ratio of about 1 produced by galactic cosmic rays. Calculations of neon compositions produced by solar cosmic rays in a soil cover on the lunar surface (using WALTON'S cross sections of Ne produced by ~10–45 MeV protons in Mg, Al, and Si) show that fines 67701 may contain detectable amounts of solar cosmic ray produced neon. The 4.5×10^{-8} cm³ STP/g of Ne_C^{21} in 67701 could have been produced in 50 m.y., the age of North Ray Crater, by an average integral (4π) flux, $F(E > 10 \text{ MeV})$ of about 60 protons/cm² sec. Neon in gas-rich meteorites could also be a mixture of trapped solar wind neon, SCR and GCR-produced neon. However, the quantitative agreement with Black's data is still unsatisfactory.

INTRODUCTION

THE DISCOVERY of substantial quantities of He³ in an iron meteorite by Paneth *et al.* (1952) marked the beginning of studies of stable and radioactive nuclides produced by galactic cosmic rays (GCR) in extraterrestrial matter. However, investigations into the possible effects of solar cosmic rays (SCR) on extraterrestrial matter have been much more limited, mainly because the SCR record is restricted to the first few centimeters below the pre-atmospheric surfaces of meteorites, which are strongly affected by ablation losses during the meteoroids' passage through our atmosphere.

Lunar surface materials, however, show abundant evidence of exposure to SCR, namely from particle track records (e.g. Crozaz *et al.*, 1970; Fleischer *et al.*, 1970) and from radionuclide records (e.g. Shedlovsky *et al.*, 1970; D'Amico *et al.*, 1970). Yaniv *et al.* (1971) suggested that stable Ne²¹ and Ar³⁸ in single particles from Apollo 11 and Apollo 12 fines could have been *in part* produced by SCR, but their conclusions were not firm because they did not know the cross sections of neon and argon in lunar materials by the relatively low-energy SCR protons.

One of us (JRW) has recently completed measurements of the cross sections of stable Ne and Ar isotopes and radioactive Na²², Ar³⁷, and Ar³⁹ in Mg, Al, Si, and Ca targets bombarded with ~10–45 MeV protons. Production rates calculated with these cross sections (Walton *et al.*, 1974) provide—for the first time—a quantitative basis for the discussion of SCR effects in stable inert gases.

In the present paper we will discuss the case of fines 67701 collected at Station 11 on the rim of North Ray Crater (Apollo 16) and show, briefly, how the results

---

*Present Address: Department of Physics and Astronomy, Tel Aviv University, Ramat Aviv, Israel.

can be used as a basis for discussing other systems, such as the gas-rich meteorites.

INERT GAS CONTENTS OF SIZE FRACTIONS AND SINGLE PARTICLES FROM 67701

Unsieved soil 67700, from which 67701 was derived, was collected on the rim of North Ray Crater, approximately halfway between the "white breccia boulder" and "house rock." According to the crew's comments (Muehlberger *et al.*, 1972), the soil cover at the collection site is uncommonly thin, i.e. only one centimeter or less. Thin soil covers are favorable cases for detecting SCR produced inert gases, because—as will be shown in the following sections—SCR production of Ne is probably only significant in the first few centimeters below the surface of the regolith.

We have determined the inert gases in a bulk sample, in sieve fractions, and in nine single particles from the 500–1000 μm sieve fraction by mass-spectrometric techniques which have been described elsewhere (Heymann and Yaniv, 1970). The results are given in Table 1. The effective diameter, D_{eff}, of a given sieve fraction was calculated from the grain size distribution of 67701 (Heiken *et al.*, 1973; and D. S. McKay, private communication), such that one gram of spherules with diameter D_{eff} has the same surface area and volume as one gram of the real sample of the sieve fraction.

Fines 67701 are fairly "normal" in that:

(1) Their inert gas contents are anticorrelated with grain diameter; when the concentration is expressed as $C_D = S(D/D_0)^{-n}$, the n-values are approximately 0.7 for the principal "trapped" species He^4, Ne^{20}, Ar^{36}, Kr^{84}, and Xe^{132}.

(2) The amounts of surface correlated trapped gases (atoms/cm^2) are less than those in more mature fines from the landing area, such as 61241 (Walton *et al.*, 1973) but only by some 20–30%.

(3) The He^4/Ne^{20} ratios of ~40 are typical of plagioclase-rich, TiO_2-poor fines (Walton *et al.*, 1973).

In the three isotope representation of Ne, i.e. the plot of Ne^{20}/Ne^{22} versus Ne^{21}/Ne^{22} any measured composition can be analyzed in terms of two or more components (cf. Black, 1972). We shall restrict our calculations to a two-component mixture. One component is trapped Ne_T. The other component is proton-produced Ne_{PR}. Normally, lunar fines are discussed with a spallation component, Ne_{SP} or Ne_C, which is Ne produced by GCR. Since detectable amounts of SCR-produced Ne may be present in 67701, we treat the combined GCR + SCR produced Ne as a "virtual" component, i.e. Ne_{PR}.

For the analysis of our size fractions and bulk data from 67701 we have used Eq. (2) from Eberhardt *et al.* (1970):

$$(Ne^{22}/Ne^{21}) = \frac{Ne^{21}_{PR}}{Ne^{21}} [(Ne^{22}/Ne^{21})_{PR} - (Ne^{22}/Ne^{21})_T] + (Ne^{22}/Ne^{21})_T \qquad (1)$$

Table 1. Inert gas contents (cm³ STP/g), isotopic composition, and Ne_{PR}^{21} of Apollo 16 bulk fines 67701,20; of its size fractions; and of single particles (500–1000 µm) from 67701,20.

Sample and D_{eff} (µm)[a]	Weight (µg)	He^4 (10^{-3})	$\dfrac{He^4}{He^3}$	Ne^{20} (10^{-5})	$\dfrac{Ne^{20}}{Ne^{22}}$	$\dfrac{Ne^{21}}{Ne^{22}}$	Ar^{36} (10^{-5})	$\dfrac{Ar^{36}}{Ar^{38}}$	$\dfrac{Ar^{40}}{Ar^{36}}$	Kr^{84} (10^{-8})	Xe^{132} (10^{-8})	Ne_{PR}^{21} (10^{-8})
67701,20												
Bulk[b]	2322	20.6	2810	50.5	13.1	.0338	18.6	5.35	0.0966	6.37	1.05	6.4±3.0
11 µm	1063	41.9	2920	94.2	13.0	.0327	36.9	5.30	0.912	12.2	2.47	3.6±5.2
56	939	10.5	3000	25.4	12.8	.0349	10.0	5.26	1.22	3.43	0.59	5.4±3.1
88	1355	6.83	2730	17.6	12.7	.0350	8.24	5.25	1.31	2.98	0.48	3.9±1.6
157	3127	4.98	2750	12.6	12.7	.0364	5.35	5.29	1.28	2.01	0.32	4.2±1.2
347	3791	3.10	2600	8.57	12.8	.0388	3.53	5.23	2.12	1.36	0.22	4.5±0.6
680[c]	3896	1.40	2620	4.22	12.6	.0423	1.71	5.27	1.81	0.55	0.15	3.4±0.6
											Avg.	4.5±2.2

Single particles	Weight (µg)	He^4 (10^{-4})	$\dfrac{He^4}{He^3}$	Ne^{20} (10^{-5})	$\dfrac{Ne^{20}}{Ne^{22}}$	$\dfrac{Ne^{21}}{Ne^{22}}$	Ar^{36} (10^{-6})	$\dfrac{Ar^{36}}{Ar^{38}}$	$\dfrac{Ar^{40}}{Ar^{36}}$			Ne_{PR}^{21} (10^{-8})
1	1422	4.19	1910	1.42	12.2	.066	6.2	5.39	.560			4.0±1.3
		(2%)	(16%)	(2%)	(3%)	(16%)	(12%)	(2%)	(8%)			
2	1438	6.61	2070	1.33	12.1	.058	3.55	5.00	3.38			3.0±1.6
		(2%)	(8%)	(2%)	(4%)	(23%)	(5%)	(3%)	(4%)			
3	545	2.98	2380	.536	13.0	.062	1.15	4.9	20.2			1.3±1.5
		(4%)	(23%)	(3%)	(14%)	(52%)	(4%)	(8%)	(3%)			
4	1208	2.52	1650	.506	11.5	.15	1.13	4.64	11.1			5.3±1.1
		(3%)	(20%)	(2%)	(6%)	(15%)	(5%)	(4%)	(3%)			
5	1395	.054	615	.156	9.3	.20	2.22	4.85	5.41			2.8±0.5
		(60%)	(74%)	(5%)	(10%)	(12%)	(4%)	(4%)	(3%)			
6	1055	13.1	2220	3.83	12.5	.054	12.	5.11	0.17			6.9±1.2
		(2%)	(6%)	(2%)	(2%)	(6%)	(27%)	(2%)	(19%)			
7	700	64.2	2480	11.5	12.9	.039	92.	5.17	4.07			5.7±1.9
		(2%)	(4%)	(2%)	(2%)	(5%)	(3%)	(2%)	(2%)			
8	900	17.8	2800	1.91	12.8	.055	3.41	5.02	28.0			3.5±1.0
		(2%)	(6%)	(4%)	(4%)	(10%)	(3%)	(2%)	(2%)			
9	768	20.3	2250	4.26	13.0	.043	11.4	5.35	2.92			3.6±2.9
		(2%)	(5%)	(2%)	(3%)	(20%)	(3%)	(2%)	(2%)			
											Avg.	4.0±1.4

[a]D_{eff} is the effective grain diameter for the following sieve fractions: ≤44, 44–74, 74–105, 105–250, 250–500, 500–1000 µm, respectively.

[b]From Walton et al. (1973).

[c]This sample consisted of nine particles. Errors for size fractions. He^4, Ne^{20}, $Ar^{36} = \pm3\%$ or less; $Kr^{84} = \pm10\%$ or less; and $Xe^{132} = \pm20\%$ or less; $He^4/He^3 = \pm4\%$ or less; $Ne^{20}/Ne^{22} = \pm2\%$ or less; $Ne^{21}/Ne^{22} = \pm3\%$ or less; $Ar^{36}/Ar^{38} = \pm2\%$ or less; $Ar^{40}/Ar^{36} = \pm4\%$ or less. For single particles errors are given in parenthesis below the entries in the table.

and its companion equation:

$$(Ne^{20}/Ne^{21}) = \frac{Ne_{PR}^{21}}{Ne^{21}}[(Ne^{20}/Ne^{21})_{PR} - (Ne^{20}/Ne^{21})_T] + (Ne^{20}/Ne^{21})_T \qquad (2)$$

Least-square fits (Eberhard plots) of (Ne^{22}/Ne^{21}) versus $1/Ne^{21}$ and (Ne^{20}/Ne^{21}) versus $1/Ne^{21}$ yield two slopes and the two trapped ratios. Implicit in this treatment is the assumption that all samples contain the same amount of Ne_{PR}^{21}.

The slope values represent two equations with three unknowns:

$$\text{Slope 1} = Ne_{PR}^{21}[(Ne^{22}/Ne^2)_{PR} - (Ne^{22}/Ne^{21})_T] \qquad (3)$$

$$\text{Slope 2} = Ne_{PR}^{21}[(Ne^{20}/Ne^{21})_{PR} - (Ne^{20}/Ne^{21})_T] \qquad (4)$$

which we reduce to two unknowns, namely Ne_{PR}^{21} and $(Ne^{22}/Ne^{21})_{PR}$ by the additional good assumption that $(Ne^{20}/Ne^{21})_{PR} \ll (Ne^{20}/Ne^{21})_T$ (i.e. $\sim 1 \ll \sim 400$; see also Nyquist and Pepin, 1971).

We have analyzed the data with these equations for a number of combinations: A = bulk, 11 and 56 μm fractions; B = bulk, 11, 56, 88, 157, and 347 μm fractions; C = all seven data points; D = 11, 56, 88, and 157 μm fractions: E = same as D plus 347 μm fraction; and F = all size fractions, no bulk.

The range of $(Ne^{21}/Ne^{22})_T$ values calculated is 0.0320 to 0.0329 with an average of 0.032 ± 0.001. The range of $(Ne^{20}/Ne^{21})_T$ values is 392 to 408 with an average of 400 ± 8. The range of $(Ne^{20}/Ne^{22})_T$ values is 12.9–13.1 with an average of 13.0 ± 0.1. The range of $(Ne^{21}/Ne^{22})_{PR}$ values is 0.18–0.48 with an average of 0.4 ± 0.1.

The trapped neon ratios in 67701 fall within the range of values reported for lunar fines. However, the $(Ne^{21}/Ne^{22})_{PR}$ ratio is much smaller than the value of about 0.8–1.0 produced by GCR in stony meteorites and lunar rocks. The questions before us are: Is the relatively small $(Ne^{21}/Ne^{22})_{PR}$ ratio in 67701 real, and is it evidence for SCR-produced neon in these fines?

As for the first question, we note that the $(Ne^{21}/Ne^{22})_{PR}$ ratios (calculated in the same manner) of 61221 and 61241 (our unpublished data) lie between 0.8 and 1.0. In addition, we have fitted our data by regression calculations to straight lines in the three isotope plot: $(Ne^{21}/Ne^{22}) = a(Ne^{20}/Ne^{22}) + b$. The intercept value b, i.e. for $(Ne^{20}/Ne^{22}) = 0$, represents a *maximum* value for $(Ne^{21}/Ne^{22})_{PR}$. For the size fractions only $b = 0.3 \pm 0.1$; for size fractions, bulk, and single particles combined $b = 0.6 \pm 0.2$. In this respect, it is important to note that a single particle need not fall on or near the line of the sieve fractions, either because the particle has had an unusual irradiation history, or because its composition is significantly different from that of the bulk soil. The same treatment for 61221 and 61241 yields the same maximum values of $b = 0.8 \pm 0.2$. Although the evidence is not wholly conclusive, we feel that there are strong indications that the $(Ne^{21}/Ne^{22})_{PR}$ ratio of 67701 lies in the range 0.3–0.6, with 0.4 ± 0.1 the most probable value. The chemical composition of 61221 (Rose *et al.*, 1973) is very similar to that of 67701 (Compston *et al.*, 1973); hence the difference in the ratios is apparently not caused by chemical differences alone.

We have calculated Ne_{PR}^{21} by the equation:

$$Ne_{PR}^{21} = [(31.0 \pm 0.5)Ne^{21} - Ne^{22}]/(28.5 \pm 1.1)$$

The results are given in Table 1. If Ne_C^{21} (cosmogenic) is calculated in the normal manner (using $(Ne^{22}/Ne^{21})_C = 1.1$, instead of $(Ne^{22}/Ne^{21})_{PR} = 2.5$), the resulting values of Ne_C^{21} are about 5% smaller than the corresponding Ne_{PR}^{21} values. The Ne_C^{21} contents of bulk, all sieve fractions and eight of the nine single particles agree with the average of $(4.5 \pm 2.2) \times 10^{-8}$ cm³ STP/g of the sieve fractions, indicating that 67701 is a fairly homogeneous soil with a relatively simple and short irradiation history.

When Ne_C^{21} exposure ages are calculated with the Ne_C^{21} production rate of 0.17×10^{-8} cm³ STP/g m.y. (production rate equation from Yaniv *et al.*, 1971; chemical composition from Compston *et al.*, 1973) the ages of the size fractions range from 20 to 38 m.y. with an average of 27 m.y.; the ages of the single particles range from 8 to 41 m.y. with an average of 24 m.y. These "ages" are systematically younger than the 50 m.y. Kr^{81}–Kr ages (Marti et al., 1973; Crozaz et al., 1974) of rocks from the rim of North Ray Crater, which have been interpreted as the time of formation of the crater and its associated ejecta blanket. The most likely explanation for the age-discordancy is the following. Wright et al. (1973) have shown that the GCR-production of Ne_C^{21} at the surface of the Keyes chondrite is only about 70% of that at a depth of some 16 cm. Likewise, Schultz et al. (1973) have found a similar effect in the St. Séverin meteorite. The surface production rate of Ne_C^{21} in the lunar case is probably less than 70% of that at about 50 g/cm^2 depth, because of its 2π geometry. We infer from this that Ne_C^{21} production rate equations cannot be used for soil covers such as at the site of 67701 without corrections, and that the Ne_C^{21} "ages" of 25 m.y. do *not* date the North Ray Crater event. The most likely age of the crater is therefore 50 m.y.

Neon Produced by "Low-Energy" SCR Protons

The detailed results on cross sections and production rates will be published elsewhere (Walton, 1974). A summary of the data has been reported in the conference abstracts (Walton et al., 1974). For the discussion at hand, we present the information in Fig. 1, where we have plotted the *calculated isotopic compositions* of neon produced in Mg, Al, and Si:

(1) By solar flare protons, using the two spectral distributions A and B given in the Appendix.

(2) At different depths (g/cm^2) in a lunar soil cover (details in Walton, 1974).

(3) For two cases: (a) integration up to proton energies of 45 MeV and (b) integration up to 200 MeV. The two cases are different in that the excitation functions of Ne^{20}, Ne^{21}, Ne^{22}, and Na^{22} are now well established up to 45 MeV (Walton, 1974), but the first three are not known from 45 to 540 MeV (Bieri and Rutsch, 1962). The 200 MeV curves have been approximated by interpolated excitation functions, which we have obtained by using published neon cross sections at higher energies. Because of this, we consider the 200 MeV curves less reliable than the 45 MeV curves. From an analysis of all the errors, we have concluded that the uncertainties of the 45 MeV curves are about ±20% in both ratios. No reliable errors can be given for the 200 MeV curves, but they are probably not less than those of the 45 MeV curves.

The salient features of Fig. 1 are:

(1) The two spectra yield results which are the same within the errors of the calculations.

(2) The relative positions of the Mg, Al, and Si curves, in conjunction with the absolute production rates (Walton et al., 1974) show that SCR produced Ne in Al-rich, Mg-poor systems (e.g. plagioclase feldspars) has distinctly smaller Ne^{21}/Ne^{22} ratios than Ne produced in Mg-rich systems (e.g. olivines and pyrox

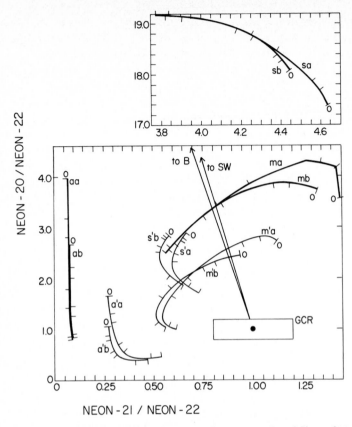

Fig. 1. Neon compositions produced by SCR protons in Mg, Al, and Si as a function of depth in a layer of soil. Isotropic proton flux and 2π irradiation geometry have been assumed. Curve aa represents Al, Spectrum A (see Appendix), integration to 45 MeV. Curve ab: Al, Spectrum B (Appendix), 45 MeV. Curves ma and mb: same for Mg. Curves sa and sb: same for Si. Note large values of Ne^{20}/Ne^{22} and Ne^{21}/Ne^{22} in the upper portion of the figure. Curves $a'a$, $a'b$, $m'a$, $m'b$, $s'a$, and $s'b$ differ from the unprimed set in that the integrations were carried out to 200 MeV. The tickmarks along the curves indicate depths starting at 0 g/cm², which is shown, and 0.5, 1, 2, 5, 10, and 25 g/cm². The box marked "GCR" represents neon compositions produced by galactic cosmic rays in stony meteorites and in the interior of lunar rocks. The lines labeled "to B" and "to SW" connect GCR neon with trapped solar wind Neon B (Black, 1972) and the average solar wind composition (Geiss *et al.*, 1972) respectively. Note the relative positions of the Al, Mg, and Si curves with respect to GCR neon. Neon produced by low-energy protons in plagioclase feldspar should have distinctly smaller Ne^{21}/Ne^{22} ratios than neon produced in ferromagnesian minerals such as olivine or pyroxene.

enes). This feature reflects the difficulty of producing Ne^{21} from Al^{27} by protons <50 MeV.

(3) Integration to 200 MeV shifts all curves in the direction of the GCR composition near $Ne^{20}:Ne^{21}:Ne^{22} = 1:1:1$.

(4) The 45 MeV curves for Si show extraordinarily large Ne^{20}/Ne^{22} (17–19) and

Ne^{21}/Ne^{22} (3.8–4.6) ratios. The principal reason is that proton initiated reactions on Si^{28} leading to Ne^{22} and Na^{22} (p, Be^{7}; p, $He^{4}He^{3}$; p, He^{4} $3p$) have large Q-values of -26.6, -26.4, and -30.5 MeV respectively, whereas the (p, $2He^{4}$) reaction to Ne^{21} has a Q-value of -16.8 MeV and the (p, $2He^{4}p$) reaction to Ne^{20} has a Q-value of -19.3 MeV.

(5) The Mg curves all trend from relatively large Ne^{20}/Ne^{22} and Ne^{21}/Ne^{22} ratios at the surface (0 g/cm^2) to smaller ratios at depth (25 g/cm^2). This reflects the energy dependency of these ratios as produced in Mg. At 40 MeV, Ne^{20}/Ne^{22} = 0.93, and Ne^{21}/Ne^{22} = 0.27.

(6) The Ne^{21}/Ne^{22} minima in the 200 MeV curves of Mg, and Si have been "forced" upon these curves by our demand that $\sigma_{20}:\sigma_{21}:\sigma_{22} = 1:1:1$ at 200 MeV.

Figure 1 shows unequivocally that we cannot represent proton-produced neon compositions in complex silicate systems by a single component. Instead, we must discuss neon systematics in terms of a wide range of possible compositions. Only for those cases where samples have always been irradiated at depths where SCR production becomes undetectable relative to GCR production can we represent proton-produced Ne with a single, GCR "component."

GCR- AND SCR-PRODUCED NEON IN 67701

We can now calculate the neon compositions produced by the exposure of a soil cover consisting of 67701 fines to GCR and SCR protons simultaneously. Since the chemical composition of 67701 was not available at the time of the calculations, we have used the composition of fines 67461 as given by Wänke *et al.* (1973), i.e. Mg = 2.38%, Al = 16.0%, and Si = 21.1% (wt.%).

We have made the following assumptions:

(1) For the Ne^{21}_{GCR} production rate we have adopted a value of 0.17×10^{-8} cm^3 STP/g m.y. at 20–25 g/cm^2 depth. Between 0 and 20 g/cm^2 we have *approximated* the profiles of Wright *et al.* (1973) and Schultz *et al.* (1973) by two straight-line portions, with the production rate at 10 g/cm^2 set at 0.136×10^{-8} cm^3 STP/g m.y., and at 0 g/cm^2 at 0.068×10^{-8} cm^3 STP/g m.y. In this way, the average Ne^{21}_{GCR} production rate in the first 10 g/cm^2 is only 0.6 that assumed at 20 g/cm^2 which, if applied to our Ne^{21}_{PR} data, would bring our Ne^{21}_{PR} ages in line with the published Kr^{81}–Kr ages of North Ray Crater. We have further adopted $(Ne^{22}/Ne^{21})_{GCR} = 1.15$ and $(Ne^{20}/Ne^{21})_{GCR} = 1.00$. These ratios are known to vary somewhat (by approximately 20%).

(2) The average flux of solar flare protons during the last 50 m.y. can be represented by Spectrum B (Appendix). This assumption may not be valid. Spectrum B represents the average flux of three high years (1956, 1959, and 1960) of solar cycle 19 (More and Tiffany, 1962). The average flux since the formation of the North Ray Crater ejecta blanket may have been greater or smaller. We have therefore calculated three cases: (a) average SCR flux = Spectrum B, no GCR (by disregarding GCR production); (b) average SCR flux = Spectrum B, GCR; (c)

average SCR flux = 0.1 Spectrum B, with no change in the spectral shape, GCR. For cases (b) and (c) we have kept the GCR-profile constant.

For practical purposes, we are more interested in the *bulk* neon composition produced in a soil cover of a given thickness (say, between 0 and 1 g/cm^2) than in the compositions produced at a given depth, because the first kind of information is a more realistic case for data obtained on the rake fines 67701. The results of our calculations are shown by the curves in the lower part of Fig. 2, again for the two cases of integration to 45 MeV and 200 MeV. The salient features of the curves are:

(1) A wide range of neon compositions can be produced in soil covers with the composition of 67701 ranging in thickness from 0.1 to 25 g/cm^2. Ne^{20}/Ne^{22} lies between about 0.8 and 3.3; Ne^{21}/Ne^{22} between about 0.4 and 0.9.

(2) Five of the six curves show Ne^{21}/Ne^{22} minima for thicknesses between 0.5 and 2.0 g/cm^2. The thickness of the present soil cover at the collection site is 1.5 g/cm^2, clearly near the "minimum" values.

(3) A decrease of the SCR proton flux relative to the GCR flux results in a shifting of the curves toward the GCR composition. The smallest Ne^{21}/Ne^{22} values for cases (a), (b), and (c) are 0.395, 0.436, and 0.540, respectively.

In the upper part of Fig. 2 we have shown the positions of trapped neon in 67701 and neon as measured in bulk 67701, in size fractions of 67701 and in seven of the nine single particles. We have seen earlier that the Eberhardt analysis gives $(Ne^{21}/Ne^{22})_{PR} = 0.4 \pm 0.1$. We do not know the value of $(Ne^{20}/Ne^{22})_{PR}$, but from the calculated proton-produced compositions (lower part of Fig. 2), we conclude that $0.8 < (Ne^{20}/Ne^{22})_{PR} < 3.0$. Lines I and II are two-component mixing lines of trapped neon (20/22 = 13.0; 21/22 = 0.032) with two proton produced components; 20/22 = 0.8; 21/22 = 0.3 and 20/22 = 3.0; 21/22 = 0.5, respectively. Lines I and II represent lower and upper bounds of mixing lines compatible with the error of $(Ne^{21}/Ne^{22})_{PR}$ and the range of values of $(Ne^{20}/Ne^{22})_{PR}$. Line III is the mixing line of trapped neon and GCR-produced neon.

The salient features of Fig. 2 are:

(1) Within errors, the data points of the sieve fractions (and bulk) fall within the bounds of lines I and II. Only one sieve fraction ($D_{eff} = 347 \, \mu m$) agrees, within errors, with mixing line III.

(2) Within errors, all single particles, except 4 and 5 (not shown) fall within the bounds of lines I and II, but also agree with mixing line III. Particle 4 (20/22 = 11.5; 21/22 = 0.15) falls above line III, but within the errors agrees with this line. Particle 5 (20/22 = 9.3; 21/22 = 0.20) falls very close to line II, but does not agree with line III.

(3) Five of the six calculated 67701 curves show values in the region between lines I and II and two (S' and S'G) fall entirely within this region.

From this we conclude that the relatively small $(Ne^{21}/Ne^{22})_{PR}$ value, as deduced from the sieve fractions of 67701 could well be due to SCR-produced neon in the

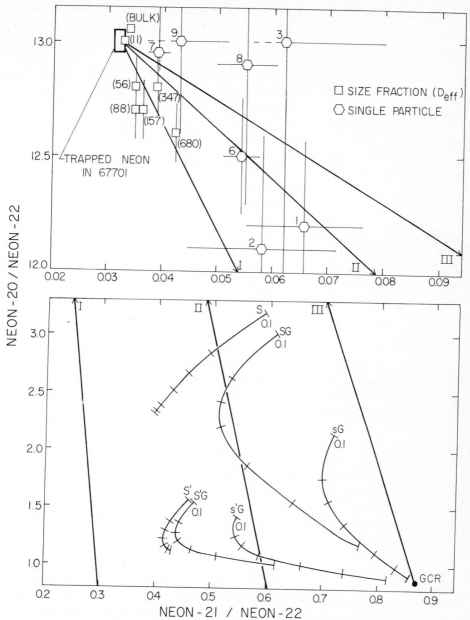

Fig. 2. Neon compositions in size fractions and single particles from 67701 (upper) and calculated neon compositions in soil covers of increasing thickness (0.1, 0.5, 1, 2, 5, 10, 25 g/cm²), consisting of 67701 material (lower). The chemical composition of 67461 (Wänke *et al.*, 1973) has been used in the calculations. Lines labeled I, II, and III are explained in the text. Curve labeled S represents neon compositions produced only by SCR in 67701 (spectrum B); SG: neon produced by SCR (Spectrum B) plus galactic cosmic rays in 67701; SG: neon produced by SCR (Spectrum B, 1/10 proton flux) plus GCR in 67701; in all three cases the integration is carried to 45 MeV. Curves S′, S′G, and s′G: same for integration to 200 MeV. Point labeled "GCR" is GCR neon composition used in the calculations.

thin soil cover at North Ray Crater. The single particles do not support or contradict this conclusion, except particle 5, whose Ne composition cannot be explained in the two component Ne_T–Ne_{GCR} system, but requires the presence of Ne with a relatively small $(Ne^{21}/Ne^{22})_{PR}$ value.

Let us assume that the average Ne_C^{21} content of the size fractions, 4.5×10^{-8} cm^3 STP/g (Table 1), represents the *maximum* amount of SCR-produced Ne^{21} in a 67701-type soil cover of 1.5 g/cm^2 thickness *during 50 m.y.* Our calculations for Spectrum B yield 9.3×10^{-8} cm^3 STP/g for integration to 45 MeV and 33×10^{-8} cm^3 STP/g for integration to 200 MeV. These numbers suggest that the average proton flux during the last 50 m.y. was no greater than one-half, and possibly no greater than one-tenth that of Spectrum B, with the same spectral shape. The latter conclusion means that the average integral (4π) flux F $(E > 10$ MeV) was no greater than 60 protons/cm^2 sec. This number increases if the effective time of exposure to SCR protons has been less than 50 m.y. The number of 60 protons/cm^2 is similar to the average value (40–80) over the last 5000 yr as given by Hoyt *et al.* (1973), and the values (80–120) as given by Wahlen *et al.* (1972) and Rancitelli *et al.* (1972). However, the spectral parameters used in this paper and in the others are all different, hence it is difficult to conclude that the agreement is as strong as suggested by these data.

THE GAS-RICH METEORITES

We wish to consider briefly the implications for the gas-rich meteorites. The gas-rich, dark portions of these meteorites are probably indurated soil covers from their parent bodies. Black (1972) has proposed that the dark portions contain Neon C, a Ne component with $(Ne^{20}/Ne^{22}) = 10.6 \pm 0.3$ and $(Ne^{21}/Ne^{22}) = 0.042 \pm 0.003$. According to Black, Neon C represents directly implanted solar flare ions.

In Fig. 3 we show the Ne compositions which result from the addition of trapped solar wind Ne (Neon B from Black, 1972) to the proton-produced (SCR + GCR) compositions in 67701 from the preceding section. We have only considered calculations for the 200 MeV case and flare Spectrum B. For the average solar wind Ne_{SW}^{20} flux we have adopted 1.7×10^4 ions/cm^2 sec from Geiss *et al.* (1972). As before we consider only *bulk* Ne compositions in soil covers of increasing thickness.

From inert gas studies of lunar fines it is known that only a fraction of all implanted solar wind Ne_{SW}^{20} ions is ultimately retained as trapped Ne_T^{20}. The trapping efficiency in 67701 can be estimated as follows. With the present-day Ne_{SW}^{20} flux as given above, and an implantation time of 50 m.y., about 1.0 cm^3 STP of Ne_{SW}^{20} has been implanted per square meter of the soil cover. For a soil cover of 1.5 g/cm^2 thickness this corresponds to a Ne_T^{20} content of 0.67 cm^3 STP/g, for 100% trapping efficiency. The actually observed Ne_T^{20} of 67701 (bulk) is 5.05×10^{-4} cm^3 STP/g, which implies an apparent trapping efficiency of only 7.5×10^{-4}. (The Ne_T^{20} content of the unsieved soil 67700 is not known, but judging from the total weight of the >1 mm fraction as given in the Apollo 16 Lunar Sample

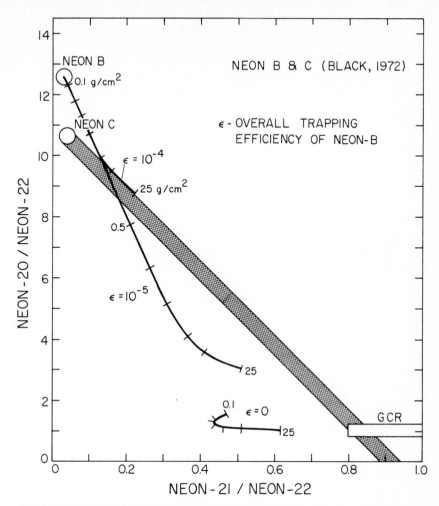

Fig. 3. Calculated neon compositions in soil covers of increasing thickness (0.1, 0.5, 1, 2, 5, 10, 25 g/cm²) consisting of 67701 material, which result from adding trapped solar wind neon (represented by Neon B from Black, 1972) to SCR + GCR produced neon. For the solar wind Ne^{20} flux we have assumed 1.7×10^4 ions/cm² sec (Geiss *et al.*, 1972). The parameter ϵ is explained in the text. Neon C and the shaded band are from Black (1972). The calculations show that either very low Ne trapping efficiencies, or very large SCR proton fluxes are required to produce neon compositions below the shaded band. This conclusion is wholly independent of the amounts of trapped Neon B present in the material.

Information Catalog, the Ne_T^{20} content of the unsieved soil could be at most about 20% smaller than that of 67701; hence the trapping efficiency as calculated above could also be at most 20% smaller.)

This apparent trapping efficiency can be interpreted in a number of ways:

(1) The *real* trapping efficiency of Ne^{20} in 67701 material is as small as 7.5×10^{-4}.

(2) The real trapping efficiency is greater than 7.5×10^{-4}, and:

(a) The average Ne^{20}_{SW} flux during the last 50 m.y. has been smaller than 1.7×10^4 ions/cm^2 sec.

(b) The surface from which 67700 was collected had been an *erosional* surface; i.e. more material had been transported away from the collection site than had been deposited there, during the last 50 m.y.

(c) The soil cover had evolved on an essentially soil-free surface by meteorite impacts in the underlying rocks; the median grain size had been fairly coarse (say $>100 \mu$m) during the buildup of the soil cover in 50 m.y.; hence, the soil cover became relatively quickly saturated with Ne_T^{20}.

(d) Soil 67700 had *recently* become "contaminated" with substantial quantities of material that had not (or only briefly) been exposed to solar wind implantation.

Our results on the single particles seem to rule out 2(d), but cannot help us to decide between the other possibilities. Hence, in order to account for the possible effects combined under 1 and 2(a–c) we shall use the parameter $\alpha \equiv Ne_T^{20}/Ne^{20}_{SW}$; i.e. the ratio of the total number of Ne_T^{20} atoms present in a column of 1 cm^2 cross section of a soil cover to the total number of Ne^{20}_{SW} ions which have struck 1 cm^2 of this cover.

In the two-component system Ne_T–Ne_{GCR} all points in the three isotope plot fall along a straight line between the two end-members. The degree to which points in the system Ne_T–Ne_{GCR}–Ne_{SCR} deviate from this line depends on the relative (to Ne_T and Ne_{GCR}) amounts of Ne_{SCR} produced in the soil cover. In the present calculations we have kept the Ne_{GCR} production rates constant. In order to account for possible variations in the Ne^{20}_{SW}/SCR (proton) fluxes we use the parameter $\beta \equiv$ SCR flux of Spectrum B/actual SCR flux. The curves in Fig. 3 have been constructed; however, for different values of the parameter $\epsilon \equiv \alpha \times \beta$. Epsilon, then, accounts for any combination of the possible effects discussed above.

Let us now return to Fig. 3. For $\epsilon \geqslant 10^{-3}$ all compositions (0.1–25 g/cm^2) fall within the circle about the composition of Neon B; i.e. the bulk Ne compositions are overwhelmingly determined by the composition of trapped Ne. For $\epsilon = 10^{-4}$ and 10^{-5} the points begin to differ substantially from the trapped composition. The "shaded band" in Fig. 3 is the high-temperature trend region from Black (1972). Black has pointed out that few, if any data points from gas-rich meteorites or lunar fines fall below the shaded region, despite the fact that the amounts of trapped Neon B vary in these systems by at least one order of magnitude. Figure 3 shows that one requires extraordinary conditions to move the Ne composition of the soil cover at North Ray Crater below the shaded region. Either the Ne_T retention must be $<10^{-4}$, or very large SCR fluxes (10^4 times Spectrum B at least) or a combination of the two are required. This conclusion is absolutely independent of the *amounts* of trapped Neon B present in the soil, because the Ne compositions arise from the coupling of solar wind Ne implantation with SCR and GCR production of Ne in a soil cover.

The trend of the two curves $\epsilon = 10^{-4}$ and 10^{-5} can be easily understood. For very thin soil covers ($<0.1 \, g/cm^2$) the bulk composition is again overwhelmingly determined by Ne_T. As the thickness of the (always well-mixed) soil cover increases, Ne_{SCR} produced in the Al-rich, Mg-poor 67700 material begins to move the compositions away from Neon B, but along a trend toward Ne^{21}/Ne^{22} values substantially smaller than the GCR value. For soil covers between 2 and 15 g/cm^2 Ne_{GCR} becomes clearly detectable and the curve bends over to the right in the direction of Ne_{GCR}.

Our discussion, then, shows that the Ne compositions in (bulk) gas-rich meteorites can be explained in the Ne_T–Ne_{GCR}–Ne_{SCR} system; and that it is not necessary to invoke the Neon C component, for which there is no direct experimental evidence at present. This does, however, not imply that directly implanted solar flare Ne is not present in gas-rich meteorites or lunar fines. Solar flare iron tracks have been detected in these systems (see Crozaz *et al.*, 1970; Fleischer *et al.*, 1970; and Pellas *et al.*, 1969). However, if the isotopic composition of directly trapped solar flare Ne is closely similar to that of trapped solar wind Neon B, the two kinds of trapped Ne cannot be distinguished by mass-spectrometric techniques. Clearly, our discussion allows the two kinds of trapped Ne to have *identical* isotopic compositions.

Our own dilemma is illustrated in Fig. 4, where we show the Ne compositions produced by SCR protons (Spectrum B) in olivine (Fa_{20}) for soil covers of increasing thickness. The olivine curve is a very close approximation for (ordinary) chondrites. The dashed lines (I to VI) have been taken from Black (1972). They connect Neon B with bulk Ne compositions from Kapoeta, Fayetteville and Holman Island. Only the Holman Island lines (V and VI) intersect the olivine curve (200 MeV). Lines III and IV (Kapoeta and Fayetteville D_2) are sufficiently close to the olivine curve, such that, within the error of the Ne measurements in the meteorites and the errors of the calculated Ne compositions, there is marginal agreement. However, the remaining two lines (I and II) trend toward Ne^{21}/Ne^{22} values far below the predicted one. The line farthest to the left (I) represents the $<600°C$ Ne fraction released by Kapoeta. It seems possible that the low-temperature release from this meteorite was principally from plagioclase feldspar, which would then explain its trend toward very small Ne^{21}/Ne^{22} values. However, the case of the bulk sample of Fayetteville D_2 (line II) remains essentially unexplained.

We note, however, that the Ne^{21}/Ne^{22} ratio produced by protons in Mg decreases rapidly above 30 MeV (Walton, 1974). If the materials of the dark portions of Fayetteville had been exposed to a much "flatter" energy distribution than Spectrum B; one in which the relative abundance of protons <40 MeV had been *much* smaller than in Spectrum B, then the olivine curves may become displaced towards smaller Ne^{20}/Ne^{22} and Ne^{21}/Ne^{22} values. Neon cross sections in the 50–200 MeV range are urgently needed to confirm or disprove this suggestion.

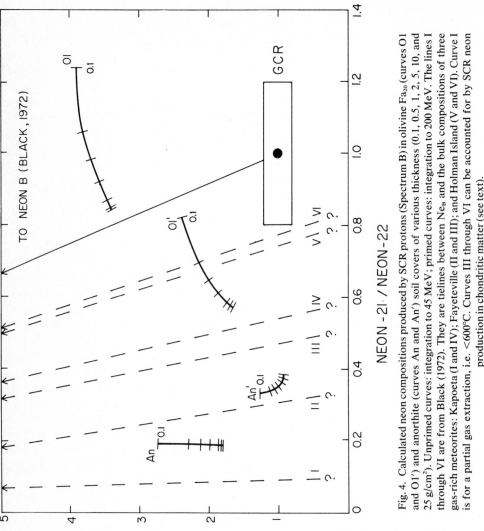

Fig. 4. Calculated neon compositions produced by SCR protons (Spectrum B) in olivine Fa_{20} (curves O1 and O1′) and anorthite (curves An and An′) soil covers of various thickness (0.1, 0.5, 1, 2, 5, 10, and 25 g/cm²). Unprimed curves: integration to 45 MeV; primed curves: integration to 200 MeV. The lines I through VI are from Black (1972). They are tielines between Ne$_B$ and the bulk compositions of three gas-rich meteorites: Kapoeta (I and IV); Fayeteville (II and III); and Holman Island (V and VI). Curve I is for a partial gas extraction, i.e. <600°C. Curves III through VI can be accounted for by SCR neon production in chondritic matter (see text).

Acknowledgments—We thank Dr. Marvin Rowe and Mr. Dennis Edgerley for their help in determining the cross sections upon which the calculations of this paper are based. Dr. Steven Lakatos has given valuable assistance in the measurements of the single particles and the preparation of the size fractions. Mrs. Penelope Bennett has prepared the manuscript. The research was supported in part by the Atomic Energy Commission Contract No. AT-(40-1)-3815, the National Science Foundation grant No. Ga-4042, and NASA grant NGL44-006-127.

REFERENCES

Bieri R. H. and Rutsch W. (1962) Erzeugungsquerschnitte für Edelgase aus Mg, Al, Fe, Ni, Cu, und Ag bei Bestrahlung mit 540 MeV protonen. *Helv. Phys. Acta* **35**, 553–554.

Black D. C. (1972) On the origins of trapped helium, neon, and argon isotopic variations in meteorites—I. Gas-rich meteorites, lunar soil and breccia. *Geochim. Cosmochim. Acta* **36**, 347–375.

Compston W., Vernon M. J., Chappell Bruce W., and Freeman R. (1973) Rb–Sr model ages and chemical composition of nine Apollo 16 soils (abstract). In *Lunar Science—IV*, pp. 158, 158a, 158b. The Lunar Science Institute, Houston.

Crozaz G., Haack U., Hair M., Maurette M., Walker R., and Woolum D. (1970) Nuclear track studies of ancient solar radiations and dynamic lunar surface processes. *Proc. Apollo 11 Lunar Sci. Conf., Geochim. Cosmochim. Acta*, Suppl. 1, Vol. 3, pp. 2051–2080. Pergamon.

Crozaz G., Drozd R., Hohenberg C. C., Morgan C., Ralston C., Walker R., and Yuhas D. (1974) Lunar surface dynamics: Some general conclusions and new results from Apollo 16 and 17 (abstract). In *Lunar Science—V*, pp. 157–159. The Lunar Science Institute, Houston.

D'Amico J., DeFelice J., and Fireman E. L. (1970) The cosmic-ray solar-flare bombardment of the moon. *Proc. Apollo 11 Lunar Sci. Conf., Geochim. Cosmochim. Acta*, Suppl. 1, Vol. 2, pp. 1029–1036. Pergamon.

Eberhardt P., Geiss J., Graf H., Grögler N., Krähenbühl U., Schwaller H., Schwarzmüller J., and Stettler A. (1970) Trapped solar wind noble gases, exposure age and K/Ar-age in Apollo 11 lunar fine material. *Proc. Apollo 11 Lunar Sci. Conf., Geochim. Cosmochim. Acta*, Suppl. 1, Vol. 2, pp. 1037–1070. Pergamon.

Fleischer R. L., Haines E. L., Hart H. R., Jr., Woods R. T., and Comstock G. M. (1970) The particle track record of the Sea of Tranquillity. *Proc. Apollo 11 Lunar Sci. Conf., Geochim. Cosmochim. Acta*, Suppl. 1, Vol. 3, pp. 2103–2120. Pergamon.

Geiss J., Buehler F., Cerutti H., Eberhardt P., and Fillieux Ch. (1972) Solar wind composition experiment. *Apollo 16 Preliminary Science Report*, NASA SP-315, Section 14, pp. 1–10.

Heiken G. H., McKay D. S., and Fruland R. M. (1973) Apollo 16 soils: Grain size analyses and petrography. *Proc. Fourth Lunar Sci. Conf., Geochim. Cosmochim. Acta*, Suppl. 4, Vol. 1, pp. 251–265. Pergamon.

Heymann D. and Yaniv A. (1970) Inert gases in the fines from the Sea of Tranquillity. *Proc. Apollo 11 Lunar Sci. Conf., Geochim. Cosmochim. Acta*, Suppl. 1, Vol. 2, pp. 1247–1259. Pergamon.

Hoyt H. P., Jr., Walker R. M., and Zimmerman D. W. (1973) Solar flare proton spectrum averaged over the last 5×10^3 years. *Proc. Fourth Lunar Sci. Conf., Geochim. Cosmochim. Acta*, Suppl. 4, Vol. 3, pp. 2489–2502. Pergamon.

Marti K., Lightner B. D., and Osborn T. W. (1973) Krypton and xenon in some lunar samples and the age of North Ray Crater. *Proc. Fourth Lunar Sci. Conf., Geochim. Cosmochim. Acta*, Suppl. 4, Vol. 2, pp. 2037–2048. Pergamon.

More K. A. and Tiffany O. L. (1962) Comparison of Monte Carlo and ionization calculations for spacecraft shielding. In *Proc. of the Symposium on the protection against radiation hazards in space*, pp. 682–685. Oak Ridge National Laboratory TID-7652.

Muehlberger W. R., Bailey N. G., Batson R. M., Freeman V. L., Hait M. H., Hodges C. A., Jackson E. D., Larson K. B., Reed V. S., Schaber G. G., Stuart-Alexander D. E., Sutton P. L., Swann G. A., Tyner R. L., Ulrich G. E., Wilshire H. G., and Wolfe E. W. (1972) Documentation and environment of the Apollo 16 samples: A preliminary report. U.S. Department of the Interior, Geological Survey.

Nyquist L. E. and Pepin R. O. (1971) Rare gases in Apollo 12 surface and subsurface fine materials. Unpublished proceedings of the Second Lunar Science Conference, Houston.

Paneth F. A., Reasbeck P., and Mayne K. I. (1952) Helium 3 contents and age of meteorites. *Geochim. Cosmochim. Acta* **2**, 300–303.

Pellas P., Poupeau G., Lorin J. C., Reeves H., and Audouze J. (1969) Primitive low-energy particle irradiation of meteoritic crystals. *Nature* **223**, 272.

Rancitelli L. A., Perkins R. W., Felix W. D., and Wogman N. A. (1972) Lunar surface processes and cosmic ray characterization from Apollo 12–15 lunar sample analyses. *Proc. Third Lunar Sci. Conf., Geochim. Cosmochim. Acta,* Suppl. 3, Vol. 2, pp. 1681–1691. MIT Press.

Rose H. J., Cuttitta Frank, Berman Sol, Carron M. K., Christian R. P., Dwornik E. J., Greenland L. P., and Ligon D. T., Jr. (1973) Compositional data for twenty-two Apollo 16 samples. *Proc. Fourth Lunar Sci. Conf., Geochim. Cosmochim. Acta,* Suppl. 4, Vol. 2, pp. 2037–2048. Pergamon.

Schultz L., Phinney D., and Signer P. (1973) Depth dependence of spallogenic noble gases in the St. Séverin chondrite. *Meteoritics* **8**, 435–436.

Shedlovsky J. P., Honda M., Reedy R. C., Evans Jr. J. C., Lal D., Lindstrom R. M., Delany A. C., Arnold J. R., Loosli H. H., Fruchter J. S., and Finkel R. C. (1970) Pattern of bombardment-produced radionuclides in rocks 10017 and in lunar soil. *Proc. Apollo 11 Lunar Sci. Conf., Geochim. Cosmochim. Acta,* Suppl. 1, Vol. 2, pp. 1503–1532. Pergamon.

Wahlen M., Honda M., Imamura M., Fruchter J. S., Finkel R. C., Kohl D. P., Arnold J. R., and Reedy R. C. (1972) Cosmogenic nuclides in football-sized rocks. *Proc. Third Lunar Sci. Conf., Geochim. Cosmochim. Acta,* Suppl. 3, Vol. 2, pp. 1719–1732. MIT Press.

Wänke H., Baddenhausen H., Dreibus G., Jagoutz E., Kruse H., Palme H., Spettel B., and Teschke F. (1973) Multielement analyses of Apollo 15, 16, and 17 samples and the bulk composition of the Moon. *Proc. Fourth Lunar Sci. Conf., Geochim. Cosmochim. Acta,* Suppl. 4, Vol. 2, pp. 1461–1482. Pergamon.

Walton J. R., Lakatos S., and Heymann D. (1973) Distribution of inert gases in fines from the Cayley-Descartes region. *Proc. Fourth Lunar Sci. Conf., Geochim. Cosmochim. Acta,* Suppl. 4, Vol. 2, pp. 2079–2096. Pergamon.

Walton J. R. (1974) Production of He, Ne, and Ar in lunar material by solar cosmic ray protons. Ph.D. thesis, Rice University, Houston, Texas.

Walton J. R., Heymann D., Yaniv A., Edgerley D., and Rowe M. W. (1974) Production of He, Ne, Ar, and U-236 in lunar material by solar cosmic ray protons (abstract). In *Lunar Science—V*, pp. 817–819. The Lunar Science Institute, Houston.

Webber W. R. (1967) Sunspot number and solar cosmic ray productions for cycle 20 (1965–1975) with preliminary estimates for cycle 21. The Boeing Company, Report D2-113522-1.

Wright R. J., Simms L. A., Reynolds M. A., and Bogard D. D. (1973) Depth variation of cosmogenic noble gases in the ~120-kg Keyes chondrite. *J. Geophys. Res.* **78**, 1308–1318.

Yaniv A., Taylor G. J., Allen S., and Heymann D. (1971) Stable rare gas isotopes produced by solar flares in single particles of Apollo 11 and Apollo 12 fines. *Proc. Second Lunar Sci. Conf., Geochim. Cosmochim. Acta,* Suppl. 2, Vol. 2, pp. 1705–1716. MIT Press.

APPENDIX

Spectrum A is an energy-power-law best fit of data for the years 1959 and 1960 presented in Webber (1967) to $dF/dE = CE^{-\gamma}$, with $C = 4.5 \times 10^{11}$ protons/cm^2 yr MeV (4π) and $\gamma = 2.55$.

Spectrum B is the spectrum for the average of 1956, 1959, and 1960, as given by More and Tiffany (1962). Their curve was fitted to a fifth order polynomial of the form $\log F = \Sigma_{i=1}^{5} C_i (\log E)^{-i} + C_6$. The integral flux ($4\pi$) of protons/cm^2 sec F ($E > 10$ MeV) is 520 for Spectrum A and 630 for Spectrum B.

Proceedings of the Fifth Lunar Conference
(Supplement 5, Geochimica et Cosmochimica Acta)
Vol. 2 pp. 2061–2074 (1974)
Printed in the United States of America

Lunar neutron capture as a tracer for regolith dynamics*

D. S. Burnett

Division of Geological and Planetary Sciences, California Institute of Technology,
Pasadena, California 91109

Dorothy S. Woolum

California Institute of Technology and Physics Department, California State University,
Fullerton, Fullerton, California 92631

Abstract—The Apollo 17 Lunar Neutron Probe Experiment measured both the boron-10 neutron capture rate and the uranium-235 neutron-induced fission rate as a function of depth. Cd absorption gave a measure of the neutron energy spectrum. Comparisons of the results are made with theory, and good agreement is obtained for the magnitudes and depth dependences of the capture rates. While the low-energy neutron spectrum at depth agrees with theory, the spectrum near the peak of the flux profile is harder than predicted. In light of these results, several alternatives for interpreting the magnitude and uniformity of the neutron capture data from lunar surface soil samples are outlined. While none of the alternatives can be unquestionably defended or discarded, a surface layer mixing model is discussed in detail.

INTRODUCTION

ALL THE LUNAR SAMPLES available to us for study come from the upper few meters of the lunar surface. An overall assessment of particle exposure data indicates that most lunar samples represent material which has resided at more than one depth during its near surface residence time. Because of the strong depth dependence of neutron capture rates and because the neutron flux penetrates to comparatively large (meter) depths, isotopic variations produced by neutron capture reactions, as first reported by Eugster *et al.* (1970), are the most important data for unraveling these complex irradiation histories. However, a necessary precondition for the interpretation of the isotopic variations is knowledge of absolute rates of neutron capture and how these rates vary with depth. The Lunar Neutron Probe Experiment (LNPE) was flown on Apollo 17 in order to provide these data as well as to provide some information on the neutron energy spectrum. The energy spectrum is important because different nuclei capture neutrons in different energy ranges and some knowledge of the neutron energy distribution is required to convert the capture rate of one nucleus into that of another.

Previously we have published the results of our ^{235}U fission detectors (Woolum and Burnett, 1974a; hereafter referred to as WBI). In this paper we present the results of the measurements of the ^{10}B neutron capture rate and of the Cd absorption measurements and discuss their implications for current lunar prob-

---

*Division of Geological and Planetary Sciences Contribution No. 2477.

lems. A more complete report on the LNPE ^{10}B data, focussed on the documentation of the accuracy of the experiment, will be published separately.

Experimental

A description of the neutron probe was published in the Apollo 17 Preliminary Science Report
(Woolum et al., 1973). The probe was deployed in the hole drilled for the deep core sample, and data
were obtained down to depths of 210 cm. Exposure of the detector materials occurred only when the
probe was inserted into the lunar surface. Two target-detector systems were employed, both using
particle track detectors: (1) Foils of ^{235}U were placed at 8 discrete depths and fission fragments were
recorded in mica detectors. (2) An essentially continuous series of 23 targets of ^{10}B was exposed to
cellulose triacetate (Triafol TN) plastic which recorded tracks from the ^{10}B(n, α) ^{7}Li reaction.
Cylindrial Cd absorbers were placed over the ^{10}B targets and detectors at two depths, 180 and
370 g/cm^2. Cd strongly absorbs neutrons of energies below about 0.5 eV; thus the B-plastic detector
enclosed by the Cd records only the fraction of the neutron captures which occur above 0.5 eV.

Results

The analysis of the ^{235}U-mica detectors from the LNPE [WBI] has shown that:
(A) the depth profile of the ^{235}U neutron-induced fission rate rises steeply from the
surface reaching a broad maximum from 100 to 160 g/cm^2 and then decreases at
larger depths. The trend of the experimental data is in agreement with the
theoretical profile calculated by Lingenfelter, Canfield, and Hampel (1972),
hereafter referred to as LCH. (B) The overall comparison between theory and
experiment (Fig. 4 of WBI) for all depths indicated that the experimental fission
rates were about $11 \pm 11\%$ lower than those calculated. The 11% error figure is
our best present estimate of the accuracy of the experimental ^{235}U fission rate,
allowing for all known sources of error. This error, as with all errors quoted in this
paper, is calculated as one standard deviation.

The uncertainty in the overall normalization of the theoretical calculations was
estimated to be $\pm 30\%$ by LCH; consequently there is excellent agreement
between experiment and theory in the absolute magnitude as well as the depth
dependence of the fission rate.

Figure 1 shows the measured track density in the plastic detectors plotted as a
function of depth beneath the lunar surface. The depth scale has been calculated
using the measured densities for the Apollo 17 deep core sample (Apollo 17
Catalog, 1973). For reference the theoretical capture rate profile, as calculated
by LCH, has been normalized to the data point at ~ 140 g/cm^2. Background track
densities from all sources are estimated to be small (100–200 cm^{-2}) and no
corrections have been applied to the data on Fig. 1. The error bars on the
individual data points on Fig. 1 are based only on counting statistics (typically
3%). However, allowing for all known sources of error, the overall precision of a
single measurement is estimated to be $\pm 7\%$. Within these limits there is good
agreement between the observed and theoretical capture rate profile. In fact, the
experimental data are much smoother and correspond to the theoretical profile
much better than would be expected for a precision of $\pm 7\%$. The data in Fig. 1

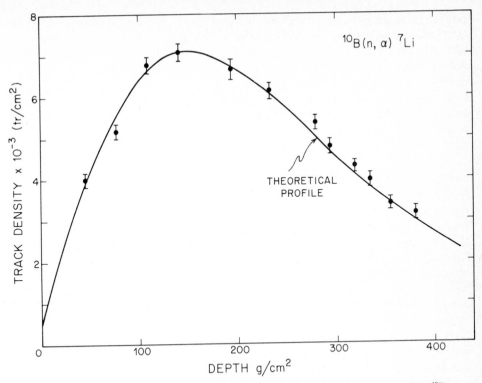

Fig. 1. Data are measured track densities in cellulose triacetate plastic exposed to ^{10}B targets at various depths. Allowing for all sources of error, the overall precision for a single measurement should be about twice the size of the error bars shown, which are based solely on counting statistics. For comparison, the theoretical capture rate profile of Lingenfelter *et al.* (1972) has been normalized to the data point at 140 g/cm^2. The data points and the theoretical profile correspond to well within experimental errors.

show somewhat less scatter than those given in our 5th Lunar Science Conference abstract (Woolum and Burnett, 1974b) because of additional measurements. The data points in several cases are the average of replicate measurements of a single sample. No measurements have been excluded.

The relationship between the measured track density, ρ, and the ^{10}B capture rate, n, in captures (g ^{10}B)$^{-1}$ (sec)$^{-1}$, is

$$f\rho = n\epsilon T$$

where f is a self-shielding correction required because the ^{10}B targets are strong neutron absorbers and thus produce a depression of the neutron flux which is being measured, T is the total exposure time of the LNPE, and ϵ (in g/cm^2) is a measure of the overall efficiency of the probe. For an ideal track detector placed in contact with an infinitely thick ^{10}B target, $\epsilon = \frac{1}{4}(R\alpha + R_{Li})$, where $R\alpha$ and R_{Li} are the alpha and ^{7}Li recoil ranges in boron metal. The efficiency, ϵ, was determined on a unit of the neutron probe constructed with natural B targets. For natural B, setting $f = 1$ introduces no significant error. See WBI for a discussion of the calibration irradiations. The self-absorption factor was determined, in the same manner as for the U-mica detectors [WBI], to be $1.37 \pm .16$. As with the U-mica detectors, the error in f is the largest source of error. Unless a specific assumption about the neutron energy spectrum is made, it is only possible to place bounds on f [see WBI for details]. In order to keep our experimental results independent of any theoretical calculations, we have been unwilling to make such an assumption; however, as a consequence we must accept larger error limits on the absolute capture rates.

Using the above calibration factors, we obtain an absolute capture rate of $467 \pm 74 \, (g \, ^{10}B)^{-1} \, (sec)^{-1}$ at a depth of $150 \, g/cm^2$. This value can be used to normalize the capture rate profile shown in Fig. 1. The experimental capture rates quoted above should be essentially final. However, we are assuming that the track registration efficiency is not pressure dependent below 1 mm pressure and that there is no systematic scanning bias between the lunar samples (track densities $\sim 5 \times 10^3 \, cm^{-2}$) and the calibration samples (track densities $\sim 5 \times 10^5 \, cm^{-2}$). Both effects are subject to further evaluation, but we now believe the pressure effect to be small. For comparison, the theoretical capture rate from LCH at this depth is $575 \, (g \, ^{10}B)^{-1} \, sec^{-1}$. The theoretical value has been adjusted to correspond to the cosmic ray intensity at the time of the Apollo 17 mission according to the methods given in [WBI]. Again, considering the estimated $\pm 30\%$ uncertainty in the normalization of the LCH theory, there is quite satisfactory agreement between the experimental and theoretical values.

Independent estimates of neutron capture rates are available from measured concentrations of radioactive nuclei produced by neutron capture in lunar samples. Figure 2 compares the ratio of the experimental rate to that calculated by LCH for all available data. For the ^{235}U fission, ^{10}B, ^{60}Co, and ^{37}Ar, the comparison between experiment and theory was made using the data points at the depths corresponding to the peaks at their respective depth profiles. Similar results would be obtained for comparisons at other depths.

For ^{37}Ar production, where the neutron capture occurs in the MeV range, the theoretical capture rate has been calculated by us: We adopted the energy spectrum of neutrons in the MeV range, as calculated by Armstrong et al. (1973) for the earth's atmosphere. The Mev neutron energy spectrum used by LCH, when applied to the atmosphere, is known to be too steep, whereas that of Armstrong et al. fits data on MeV neutrons in the atmosphere well. (This is not important for the capture rates at low energies, where all the other neutron captures occur, since they are insensitive to the shape of the MeV neutron spectrum.) We then normalized the Armstrong et al. spectrum to the total flux of neutrons greater than 1 MeV, as given by LCH, and used the resulting energy distribution to calculate the rate for $^{40}Ca \, (n, \alpha) \, ^{37}Ar$.

The only serious discrepancy between experiment and theory in Fig. 2 is for the case of ^{236}U (Fields et al., 1973). A detailed comparison of the ^{236}U data and our data for $^{235}U \, (n, f)$ was presented in WBI. The experimental ^{236}U activities, although measured in surface samples, are at least twice those calculated by LCH

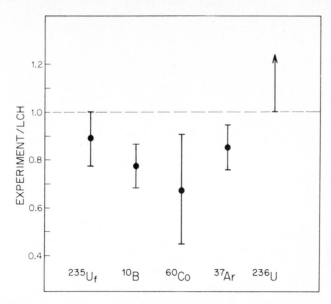

Fig. 2. The figure displays the ratio of the experimental neutron reaction rate to that calculated theoretically (LCH). For the LNPE data (^{235}U fission and ^{10}B) the symbols denote the targets for the reaction, whereas for lunar sample data the induced radioactivity is denoted by the symbol: ^{60}Co (Wahlen *et al.*, 1973), ^{37}Ar (Kornblum *et al.*, 1973), ^{236}U (Fields *et al.*, 1973). The arrow in the ^{236}U position indicates that the measured ^{236}U decay rate is at least twice that calculated theoretically. The agreement between theory and experiment is good overall.

for the *peak* of the capture rate profile. It is very important that the ^{236}U data be confirmed, because the ^{236}U activity depends on the cosmic ray intensity over the last 50×10^6 yr, whereas the other radioactive nuclei shown have shorter half-lives and, thus, measure the present day cosmic ray intensity.

Omitting the ^{236}U, the data in Fig. 2 are best interpreted as indicating that, overall, agreement between experiment and the LCH theory is excellent but that the normalization of the LCH neutron flux should be lowered by 15–20%. Except for the detailed shape of the spectrum in the MeV region discussed above, there is no evidence for systematic differences in the total flux in any neutron energy region compared to that calculated by LCH. (The ^{235}U and ^{10}B reactions are primarily due to neutrons below 1 eV, whereas the ^{60}Co and ^{37}Ar are produced by a resonance at 132 eV and by MeV neutrons respectively).

Figure 3 shows the relative track density profile as a function of position underneath the Cd absorber at 180 g/cm^2. For comparison the profile for the exposure of the probe to a well-thermalized neutron flux is also shown. The nonthermal character of the low-energy lunar neutron flux is responsible for the large difference in the two profiles. Table 1 compares the experimental and theoretical Cd ratios, defined as the ratio of the total neutron density to the density of neutrons greater than 0.5 eV. The experimental data have been corrected for differences in self-absorption between the bare and the Cd-covered detectors,

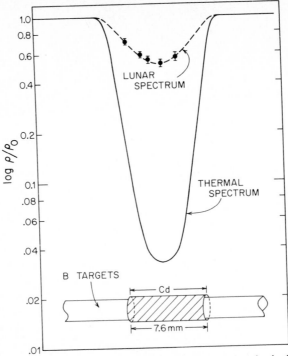

Fig. 3. The data points are the measured track densities in the plastic detectors as a function of position under the Cd absorber at 180 g/cm² relative to the track density outside the Cd. For comparison the solid curve is the measured profile obtained by exposure of the probe to a well-thermalized neutron spectrum. The difference in the two curves graphically illustrates the nonthermal character of the lunar neutron energy spectrum.

Table 1. Cd ratios for ^{10}B-plastic detectors.

$$\text{Cd ratio} = \frac{\text{Total neutron density}}{\text{Neutron density} > 0.5 \text{ eV}}$$

Depth (g/cm)²	LNPE (Expt)	LCH (Theory)
180	2.11 + .16	2.7
370	2.9 ± 0.3	2.9

leakage of neutrons from the open ends of the Cd cylinders, a small background track density, and for the relative efficiencies of the specific target positions used to calculate the Cd ratios. Considering the sensitivity of the Cd ratio to the energy spectrum (Fig. 3), the agreement between the theoretical and experimental Cd ratio is reasonable; however, the experimental value at 180 g/cm² is distinctly lower than that calculated. This ratio has been independently confirmed by two different scanners. The measurement at 370 g/cm² is much more difficult because the track density beneath the Cd absorber drops to 1500 cm⁻². Consequently, the quoted errors are twice as large; nevertheless, the Cd ratio at 370 g/cm^2 appears to be significantly larger than that at 180 g/cm^2. Thus, we conclude that at great depths the LCH low-energy neutron spectrum is correct but that there is more variation in the spectrum with depth than is predicted theoretically. The differences between experiment and theory at 180 g/cm^2 indicate that the LCH spectrum has too many low-energy neutrons, in agreement with the conclusion drawn previously from a comparison of Sm, Gd, and Eu isotopic variations in lunar samples (Russ *et al.*, 1971; Russ, 1973a, 1973b).

DISCUSSION

The overall good agreement between the LNPE results and the LCH calculations shows that lunar neutron capture processes are relatively well understood and that previous interpretations of the ^{157}Gd neutron data obtained from lunar samples, based on the LCH calculations, should be reliable (Eugster *et al.*, 1970; Burnett *et al.*, 1971; Marti and Lugmair, 1971; Russ *et al.*, 1971; Russ *et al.*, 1972; Russ, 1973a, 1973b). However, because of differences in the actual and theoretical energy spectra at depths where neutron capture is most important, additional consideration must be given to the uncertainties in the critical ^{157}Gd capture rate. Because the capture occurs in a resonance at 0.03 eV, the ^{157}Gd rate is sensitive to the shape of the spectrum below 0.2 eV. It is still possible that the LCH capture rate for ^{157}Gd may be high by roughly 40%, although even a change of this amount would not drastically alter previous conclusions. In contrast the calculations of Kornblum and Fireman (Fireman, 1974), give ^{157}Gd capture rates which are *higher* than LCH by factors ranging from 1 to 1.5, depending on the choice of input parameters. It is important that these differences be understood. However, because a detailed comparison with experimental data has been possible and good agreement obtained, we shall at present adopt the LCH ^{157}Gd capture rate.

Fluences in surface soils

Even with the first measurement of neutron fluences from Gd isotopic variations in Apollo 11 soils (Eugster *et al.*, 1970), it was clear that it was not possible to reconcile the measured fluences with the simple idea of a uniformly mixed regolith using a regolith depth inferred from field observations. Using the LCH ^{157}Gd capture rates, the observed fluences are a factor of 2–3 lower than predicted from a uniformly mixed regolith (Russ, 1973a, 1973b). The LNPE data indicate that this factor of 2–3 cannot be ascribed to inaccuracies in the magnitude or depth dependence of the LCH capture rate. Using the Kornblum–Fireman capture rates, the discrepancies would be even larger. (The above discussion of course assumes, supported by data on lunar rocks with minimal cosmic ray exposure [Eugster *et al.*, 1970; Russ, 1973b], that the isotopic composition of Gd in unexposed lunar material is identical to that of terrestrial Gd.)

The nature of this "neutron deficit" can be most simply seen by considering the Apollo 11 case as a numerical example, where, from observations of benches and flat-bottomed craters, a regolith depth of 4–5 m has been estimated (Shoemaker *et al.*, 1969). Assuming that, on a 10^9 yr time scale, the regolith has been uniformly mixed and that the total mixing time is equal to 3.7×10^9 yr (the age of the local basalts; Papanastassiou *et al.*, 1970), the predicted neutron fluence (time-integrated flux) is just the average neutron flux in the upper 5 m times 3.7×10^9 yr. The calculated value is $6.3 \times 10^{16}\ n/\mathrm{cm}^2$, whereas the observed fluence in 10084 is $2.5 \times 10^{16}\ n/\mathrm{cm}^2$.

A variety of explanations can be postulated [WBI]: (1) The regolith is not well mixed; preferentially less irradiated material occupies the upper portions of the regolith with more highly irradiated material at larger depths. (2) The regolith depths estimated from field or seismic data (see Watkins and Kovach, 1973 for a summary) are too low by factors of 2–3. (3) The cosmic ray flux averaged over the past $3–4 \times 10^9$ yr was less than at present by factors of 2–3 [WBI]. (4) Large quantities of irradiated material have been ejected from the moon by small impacts (Fireman, 1974). (5) The "neutron deficit" is not significant; an accumulation of small effects is creating an apparent discrepancy.

None of these alternatives can be totally ruled out; however, in this paper we wish only to comment further on alternatives (5), (4), and (1).

Significance of neutron deficit

In discussing (5) we first wish to emphasize that the observed smooth fluence profiles in the Apollo 15 and 16 deep cores (Russ *et al.*, 1972; Russ, 1973a, 1973b) are not incompatible with the concept of a uniformly mixed regolith. The distinction is one of time scale. The profiles observed for the cores reflect the history of that spot on the lunar surface in the last 500 m.y. Except for their burial in the last 500 m.y., the core samples are generally indistinguishable from surface soil samples; in particular they were all preirradiated with neutrons prior to deposition. These predepositional fluences, or equivalently, the fluences in the surface soils are the subject of the present discussion, and they presumably reflect exposure on a 10^9 yr time scale or longer. On the other hand, the core data tell us that the surface soil samples have not been mixed to meter depths in the last 500 m.y. and thus have been effectively "shut off" from neutron exposure for this period of time, because the neutron capture rate is low at the surface. In other words, it is probably somewhat unfair to compare predictions of a uniform mixing model with surface soil fluences when the available data for Apollo 15 and 16 tell us that the average fluence over the upper 2 m is about 1.3 times the surface fluence. If this factor of 1.3 is combined with our estimate above, that the LCH ^{157}Gd capture rate could be high by a factor of 1.4 (stretching all error bars to the limit, this factor could be 1.6), then the differences between the "observed" fluences and those calculated for a well-mixed regolith could be cut by almost a factor of 2. This would almost, but not entirely, eliminate any discrepancy between the observed fluences and a uniformly mixed regolith. Given the validity of the above factors, small contributions from alternatives (2), (3), and (4) could resolve any remaining discrepancy.

Lunar mass loss

The mean loss rate $\dot{M}$ required to explain the neutron deficit entirely by alternative (4) can be crudely estimated by assuming that the regolith is mixed rapidly throughout a depth that has not varied with time over the last $2-3 \times 10^9$ yr. In this case, irradiated lunar material which escapes from the lunar surface is replenished by previously unirradiated material and a steady-state fluence is attained which is equal to $(1/\dot{M}) \int_0^L \phi(x) \, dx$, where $\phi(x)$ is the flux of $0-0.2$ eV neutrons at depth x. For typical soil fluences $\dot{M} \sim 3$ m/10^9 yr is obtained. Fireman's (1974) estimate was 0.5 m/10^9 yr. Either estimate is much higher than expected from laboratory cratering studies: Gault *et al.* (1963) present curves (Fig. 11 of their paper) which indicate that, for a nominal orbital velocity of 20 km/sec, a mass of ejecta about 4 times the projected mass is accelerated to velocities greater than the lunar escape velocity. Combining this with the estimate of Gault *et al.* (1974) that the present-day mass influx rate to the lunar surface is 1.4 g/cm^2-10^9 yr, a mass loss rate of about 5 cm/10^9 yr is obtained. The above estimates are for cratering in solid rock. Cratering in regolith materials should lead to even lower mass loss rates according to the cumulative mass versus range curves of Gault *et al.* (1974).

Further, although any spot on the lunar surface is the site of both impact erosion and accretion, the Gd data on the Apollo 15 and Apollo 16 deep drill core samples (Russ *et al.*, 1972; Russ, 1973a) are most simply interpreted as these spots being sites of *net* accretion over the last 10^9 yr. In particular, it is difficult to account for the approximately monotonic rise in the fluence with depth for the Apollo 16 core if there is an appreciable impact mass loss from the regolith.

For these reasons, it appears more difficult to understand the neutron deficit in terms of mass loss than by means of the other alternatives.

Surface layer mixing

Previously (WBI: Woolum and Burnett, 1974b) we outlined a model, falling into the general class of alternative (1) which we feel should be seriously considered and wish to discuss in more detail here. This model is strongly suggested by considering the nature of the samples analyzed from Apollo 12.

The concentrations of many trace elements, in particular the rare earths, are much higher in the Apollo 12 soils than in the local basaltic rocks indicating that these elements are concentrated in a material which is foreign to at least the local neighborhood of the Apollo 12 landing site. The foreign component has been identified in 1–4 mm particles as a light-colored glassy material, "KREEP," and is believed by many to be represented in rock 12013 (see, for example, Meyer *et al.*, 1971). On two KREEP concentrates from 12033, Eberhardt *et al.* (1973) obtained a well-defined intermediate temperature ^{40}Ar/^{39}Ar plateau corresponding to an age of 800 ± 40 m.y. Argon released at higher temperatures gave ratios characteristic of higher ages. The straightforward interpretation of the data—as given by the authors—is that 800 m.y. ago an impact melted the source material of the KREEP glass, partially outgassing the radiogenic ^{40}Ar, and deposited the fragments at the

Apollo 12 site. From a detailed decomposition of bulk Ar and K data in Apollo 12 soil samples, Pepin *et al.* (1972) estimated a K–Ar age of 950 ± 50 m.y. in rough agreement with the $^{40}Ar/^{39}Ar$ age. Pb–U isotopic data on Apollo 12 soils and breccias are also suggestive of a U-rich component having an age of 800–1000 m.y. (Silver, 1971; Tatsumoto *et al.*, 1971).

Material balance calculations (Meyer *et al.*, 1971) indicate that essentially all of the Gd in soils 12033, 12070, and 12042 (for which Gd isotopic data are available) (Burnett *et al.*, 1971; Russ, 1973b) should be concentrated in KREEP. Consequently, the neutron fluences reported for the Apollo 12 soils should be associated with KREEP.

For our purposes it is not necessary to assume that all KREEP was deposited in a single event, e.g. by the Copernican impact. The conclusions drawn below would require modification only if most of the KREEP were deposited at the Apollo 12 site at times prior to $1–2 \ 10^9$ yr. However, there is no evidence for such early deposition of KREEP fragments.

In the following discussion we shall assume that the KREEP material was unirradiated at the time of deposition. The spallation ^{21}Ne and ^{38}Ar contents of the KREEP fragments are uniform and are low compared to other lunar soils (Eberhardt *et al.*, 1973); however, it is likely that any accumulation of these gases prior to 800 m.y. would have been outgassed at the time of melting. For the same reason the anomolously low concentrations of particle tracks (Crozaz *et al.*, 1971), solar wind ^{36}Ar (Yaniv and Heymann, 1972), solar wind hydrogen (Epstein and Taylor, 1971) and carbon (Epstein and Taylor, 1971) in 12033, although suggestive, cannot be used as an argument either for or against preirradiation. (On the other hand, because of the refractory nature of Gd, the record of prior neutron irradiation would survive the melting at 800 m.y.) Possible evidence supporting the absence of preirradiation is the relatively low concentrations of "meteoritic" siderophile elements in 12033 (Wasson and Baedecker, 1972). Like Gd, these elements should not be easily depleted by heating. However, we do not believe that the origin(s) of the siderophile elements is sufficiently well understood to be used as an argument to support the absence of preirradiation, and, further, even if their extralunar character is accepted, it is easy to construct models in which the KREEP source material would be preirradiated with neutrons but not with meteoritic siderophile elements (e.g. impact into an exposed outcrop).

If rock 12013 were deposited at the same time as the soil KREEP fragments, then the low fluences measured in both light and dark portions of 12013 (Albee *et al.*, 1971; Russ, 1973b) are evidence that the KREEP fragments were unirradiated at the time of deposition. The low fluences can be readily explained by local burial to a depth $\geqslant 3$ m for 800 m.y. and then emplacement near the lunar surface by a small recent impact. This irradiation history is consistent with available data on 12013.

In summary our assumption of no preirradiation of Apollo 12 KREEP before deposition is consistent with available data but not firmly established. But, if the fluences were acquired subsequent to deposition, then the approximate mixing time for the Apollo 12 soils is 800 m.y. rather than the 3.3×10^9 yr age of the local basalts (Papanastassiou and Wasserburg, 1970). In this case, the observed fluences can be accounted for by "surface mixing" of a relatively recently deposited layer(s), i.e. a ray, rather than deep mixing of the entire regolith.

Thus the Apollo 12 soil fluences can be explained by utilizing special properties of the Apollo 12 site. In addition to the obvious question about other sites, we must also ask whether such an *ad hoc* explanation is compatible with the striking uniformity of the fluences among different soil samples. When corrected for differences in chemical composition, the fluences in most lunar surface soils from Apollos 11–16 are constant to within ±30% (Russ, 1973a, 1973b). (A similar uniformity, but with a wider spread, may exist for spallation rare gas contents in soils; however, we have not yet thoroughly analyzed the literature data on this point.) Soils from unique, particularly young, sites do not fall into this group. For example, 14141 shows a very low fluence (Russ, 1973a, 1973b) because it is composed primarily of recent Cone Crater ejecta. However, for surface soils not associated with young craters there appears to be a relatively well-defined average lunar soil neutron fluence.

The surface mixing model can also account for the uniformity of lunar soil fluences. We assume that at each site a relatively recent ($T \sim 10^9$ yr) event(s) has deposited unirradiated, preferably Gd-rich, material. The time of deposition, T, will vary from site to site. Subsequent to deposition smaller impacts will mix the fresh material into the older regolith. Moreover, there will be a correlation between the depth of mixing, L, and the time of deposition; older deposits will have been mixed deeper, etc. The neutron fluence, Ψ, in such a mixed layer is given by: $\Psi = T/L \int_0^L \phi(x)\, dx$, where $\phi(x)$ is the flux of neutrons ≤ 0.2 eV at depth, x. Provided that $L \geq 2$ m, the integral will be constant and the fluence will be inversely proportional to L/T which is the average surface mixing rate down to depth L, and which is the average lunar property measured by the soil fluences. This possibility has been noted previously (Burnett *et al.*, 1971). The measured soil fluences correspond to a mixing rate of about 3 m/10^9 yr which should apply to a depth of 2–3 m. Mixing rates at shallower depths will be much faster and rates to larger depths much slower.

There is some evidence that our surface layer model would work for Apollo 11 and 15 because the soils have a "magic component" which is rich in K, Rb, and probably rare earths (Papanastassiou *et al.*, 1970; Evensen *et al.*, 1974). Rays from the craters Aristillus and Autolycus pass through the Apollo 15 site; these are plausible sources for Gd-rich material, and the craters are probably about the right age for our model. On the other hand, as pointed out to us by Wänke, there is no large excess in the concentrations of Gd in the Apollo 11 soil compared to either type of local basalt nor is there any distinction in the rare earth abundance patterns. Also, there is no chemical or petrological evidence that would support this model for Apollo 14 and 16. Nevertheless, the model is so compelling for Apollo 12 that it should be seriously considered for the other sites.

There are other objections to the surface mixing model. For one, the above surface mixing rate of 3 m/10^9 yr is significantly higher than the turnover times calculated by Gault *et al.*, (1974). This may mean that the association of L/T with a turnover time is incorrect. On the other hand, there are other suggestions that the Gault *et al.* turnover times are somewhat low. The Gd data for the Apollo 15 and 16 deep cores (Russ *et al.*, 1972; Russ, 1973a) indicate that in the period around 500–1000 m.y. ago both sites were areas of deposition. The cores differ in that the deposition ceased 500 m.y. ago for Apollo 15 and continued until recent times for Apollo 16. The simplest interpretation is that there was a cratering event of at least 2 m depth at both sites around 10^9 yr ago which was then refilled. The probability of this is quite low in the Gault *et al.* calculations.

It is also not clear at present whether the spallation rare gas data on surface soils can be understood with the surface layer model proposed to explain the fluences. Because Ba is geochemically correlated with rare earths and because Ba is the primary target element for ^{126}Xe spallation, ^{126}Xe contents should be low compared to those calculated assuming a uniformly mixed regolith. Calculations by Burnett *et al.* (1971) and Geiss (1973) indicate that this is true. On the other hand, spallation Ne, Ar, and Kr isotopes should be less affected by the deposition of a rare-earth, Ba-rich surface layer. Thus, soil "exposure ages" calculated from ^{126}Xe should be lower than those calculated from other rare gases. There is no consistent trend in the literature of this type; however because, relative to a uniformly mixed regolith, we are concerned with a factor of 2–3 discrepancy, a detailed review of spallation rare gas data and production rates should be made before drawing final conclusions. Use of solar wind derived elements to investigate regolith mixing problems is even more difficult because of large diffusion losses and of uncertainties in both the present-day (particularly for heavy elements) and long-term fluxes.

Summary

We have discussed interpretations of the low neutron fluences observed in lunar surface soils. One model discussed at length was based on mixing of a relatively recently deposited surface layer for which there was independent evidence from Apollo 12. Such a model can also explain the uniformity of the fluences in all surface soils if the mixing rate is about $3 \text{ m}/10^9$ yr. In addition, several objections have been discussed: (a) there is no independent evidence for such depositions for Apollo 14 and 16, (b) the required mixing rate is high, and (c) it may not be possible to understand spallation rare gas data with the same model. If these objections prove to be valid, then an alternative explanation will have to be adopted.

It should be emphasized that any explanation which reconciles the measured fluences with those calculated assuming a uniformly mixed regolith (e.g. alternatives (2) and (5), or, as in alternative (1)), postulates a less than average fluence for the upper few meters of the regolith, probably cannot be reconciled with the presently accepted lunar cratering history. Given a steep decrease in the lunar surface cratering rate with time and the corresponding increase in regolith turnover times (see, for example, Gault *et al.*, 1974), and assuming the constancy of galactic cosmic rays, one would not expect a uniform mixing, and in fact one would predict a higher than average fluence in the upper few meters, not a lower one.

In general, it would be more satisfying if the neutron fluences and spallation rare gas contents can be explained from some general statistical mixing model for the regolith rather than on an *ad hoc* basis as we have done here. In any case, the list of alternative interpretations for the soil fluences given above shows that understanding the levels of galactic cosmic ray exposure in lunar soils is an important lunar science problem and should receive more discussion than it has hitherto. We have recently found that, by adopting semi-empirical low energy neutron spectra which fit both the observed Sm/Gd capture rate ratios in lunar samples and the LNPE Cd ratio at 180 g/cm², the calculated absolute ^{157}Gd capture rates for these spectra are about 0.5 of those calculated by LCH. This is a larger effect than we estimated above and removes much, but not necessarily all, of the neutron deficit. A more detailed account of these estimates is being prepared.

Acknowledgments—Critical contributions to the LNPE design and data processing were made by Marian Furst, J. R. Weiss, and C. A. Bauman. We have benefited in all aspects of our work from consultation with G. P. Russ. Our ideas on regolith mixing have been strongly influenced by discussions with G. J. Wasserburg, J. Huneke, and J. R. Arnold. Clearly, this experiment could not have been performed without close cooperation from various branches in NASA; in particular we acknowledge the JSC Science and Applications Division, the Crew Training Division and, most of all, the Apollo 17 crew for the flawless execution of the experiment. This research was supported by NASA contract number NAS-9-12585.

REFERENCES

Albee A., Burnett D. S., Chodos A. A., Haines E. L., Huneke J. C., Papanastassiou D. A., Podosek F. A., Russ G. P., and Wasserburg G. J. (1970) Mineralogic and isotopic investigations on 12013. *Earth Planet. Sci. Lett.* **9**, 137–163.

Apollo 17 Lunar Sample Information Catalog (1973) MSC 03211, 61.

Armstrong T. W., Chandler K. L., and Barish J. (1973) Calculations of neutron flux spectra induced in the earth's atmosphere. *J. Geophys. Res.* **78**, 2715–2726.

Burnett D. S., Huneke J. C., Podosek F. A., Russ G. P., III, and Wasserburg G. J. (1971) The irradiation history of lunar samples. *Proc. Second Lunar Sci. Conf., Geochim. Cosmochim. Acta*, Suppl. 2, Vol. 2, pp. 1671–1697. MIT Press.

Crozaz G., Walker R., and Woolum D. S. (1971) Nuclear track studies of dynamic surface processes on the moon and the constancy of solar activity. *Proc. Second Lunar Sci. Conf., Geochim. Cosmochim. Acta*, Suppl. 2, Vol. 2, pp. 2543–2558. MIT Press.

Eberhardt P., Geiss J., Grögler N., and Stettler A. (1973) How old is the crater, Copernicus? University of Bern. Preprint.

Epstein S. and Taylor H. P. (1971) ^{18}O/^{16}O, ^{30}Si/^{28}Si, D/H and ^{13}C/^{12}C ratios in lunar samples. *Proc. Second Lunar Sci. Conf., Geochim. Cosmochim. Acta*, Suppl. 2, Vol. 2, pp. 1421–1441. MIT Press.

Eugster O., Tera F., Burnett D. S., and Wasserburg G. J. (1970) The isotopic composition of Gd and the neutron capture effects in samples from Apollo 11. *Earth Planet. Sci. Lett.* **8**, 20–30.

Evensen N. M., Murthy Rama V., and Coscio M. R. (1974) Episodic lunacy V: Origin of the exotic component (abstract). In *Lunar Science—V*, pp. 220–222. The Lunar Science Institute, Houston.

Fields P. R., Diamond H., Metta D. N., and Rokop D. J. (1973) Reaction products of lunar uranium and cosmic rays. *Proc. Fourth Lunar Sci. Conf., Geochim. Cosmochim. Acta*, Suppl. 4, Vol. 2, pp. 2123–2130. Pergamon.

Fireman E. L. (1974) History of the lunar regolith from neutrons (abstract). In *Lunar Science—V*, pp. 230–232. The Lunar Science Institute, Houston.

Gault D. E., Shoemaker E. M., and Moore H. J. (1963) Spray ejected from the lunar surface by meteoroid impact. NASA TN D-1767.

Gault D. E., Hörz F., Brownlee D. E., and Hartung J. B. (1974) Mixing of the lunar regolith (abstract).

In *Lunar Science—V*, pp. 260–262. The Lunar Science Institute, Houston; also see paper in these proceedings, Vol. 3.

Geiss J. (1973) Solar wind composition and implications about the history of the solar system. Proc. 13th Cosmic Ray Conf., Denver, Colorado. To be published.

Kornblum J. J., Fireman E. L., Levine M., and Aronson A. (1973) Neutrons in the moon. *Proc. Fourth Lunar Sci. Conf., Geochim. Cosmochim. Acta*, Suppl. 4, Vol. 2, pp. 2171–2182. Pergamon.

Lingenfelter R. E., Canfield E. H., and Hampel V. E. (1972) The lunar neutron flux revisited. *Earth Planet. Sci. Lett.* **16**, 355–369.

Marti K. and Lugmair G. W. (1971) Kr^{81}–Kr and K–Ar^{40} ages, cosmic-ray spallation products, and neutron capture effects in lunar samples from Oceanus Procellarum. *Proc. Second Lunar Sci. Conf., Geochim. Cosmochim. Acta*, Suppl. 2, Vol. 2, pp. 1591–1605. MIT Press.

Meyer C., Brett R., Hubbard N. J., Morrison D. A., McKay D. S., Aitken F. K., Takeda H., and Schonfeld E. (1971) Mineralogy, chemistry and origin of the KREEP in soil samples from the Ocean of Storms. *Proc. Second Lunar Sci. Conf., Geochim. Cosmochim. Acta*, Suppl. 2, Vol. 1, pp. 393–411. MIT Press.

Papanastassiou D. A., Wasserburg G. J., and Burnett D. S. (1970) Rb–Sr ages of lunar rocks from the Sea of Tranquillity. *Earth Planet. Sci. Lett.* **8**, 1–9.

Papanastassiou D. A. and Wasserburg G. J. (1970) Rb–Sr ages from the Ocean of Storms. *Earth Planet. Sci. Lett.* **8**, 269–278.

Pepin R. O., Bradley J. G., Dragon J. C., and Nyquist L. E. (1972) K–Ar dating of lunar fines: Apollo 12, Apollo 14, and Luna 16. *Proc. Third Lunar Conf., Geochim. Cosmochim. Acta*, Suppl. 3, Vol. 2, pp. 1569–1588. MIT Press.

Russ G. P., III (1973a) Apollo 16 neutron stratigraphy. *Earth Planet. Sci. Lett.* **16**, 275–289.

Russ G. P., III (1973b) Thesis, California Institute of Technology.

Russ G. P., III, Burnett D. S., Lingenfelter R. E., and Wasserburg G. J. (1971) Neutron capture on ^{149}Sm in lunar samples. *Earth Planet. Sci. Lett.* **13**, 53–60.

Russ G. P., III, Burnett D. S., and Wasserburg G. J. (1972) Lunar neutron stratigraphy. *Earth Planet. Sci. Lett.* **15**, 172–180.

Shoemaker E. M., Baily N. G., Batson R. M., Dahlem D. H., Foss T. H., Grolier M. J., Goddard E. N., Hait M. H., Holt H. E., Larson R. B., Rennilson J. J., Schaber G. G., Schleicher D. L., Schmitt H. H., Sutton R. L., Swann G. A., Waters A. C., and West M. N. (1969) Geologic setting of the lunar samples returned by the Apollo 11 mission. In *Apollo 11 Preliminary Science Report*. NASA SP-214, pp. 41–84.

Silver L. T. (1971) U–Th–Pb isotope systems in Apollo 11 and 12 regolith materials and a possible age for the Copernicus impact event. *EOS, Trans. Amer. Geophys. Union* **52**, 534; and unpublished Proc. Second Lunar Sci. Conf. (abstract).

Stoenner R. W. and Davis R. (1974) The fast neutron production of ^{37}Ar in the Apollo 17 deep drill string (abstract). In *Lunar Science—V*, pp. 741–743. The Lunar Science Institute, Houston.

Tatsumoto M., Knight R. J., and Doe B. R. (1971) U–Th–Pb systematics of Apollo 12 lunar samples. *Proc. Second Lunar Sci. Conf., Geochim. Cosmochim. Acta*, Suppl. 2, Vol. 2, pp. 1521–1546. MIT Press.

Wahlen M., Finkel R. L., Imamura M., Kohl C. P., and Arnold J. R. (1973) ^{60}Co in lunar samples. *Earth Planet. Sci. Lett.* **19**, 315–320.

Wasson J. I. and Baedecker P. A. (1972) Provenance of Apollo 12 KREEP. *Proc. Third Lunar Sci. Conf., Geochim. Cosmochim. Acta*, Suppl. 3, Vol. 2, pp. 1315–1326. MIT Press.

Watkins J. S. and Kovach R. L. (1973) Seismic investigation of the lunar regolith. *Proc. Fourth Lunar Sci. Conf., Geochim. Cosmochim. Acta*, Suppl. 4, Vol. 3, pp. 2561–2574. Pergamon.

Woolum D. S. and Burnett D. S. (WBI) (1974a) In-situ measurement of the rate of ^{235}U fission induced by lunar neutrons. *Earth Planet. Sci. Lett.* **21**, 153–163.

Woolum D. S. and Burnett D. S. (1974b) Neutron capture and the surface mixing of the regolith (abstract). In *Lunar Science—V*, pp. 848–850. The Lunar Science Institute, Houston.

Woolum D. S., Burnett D. S., and Bauman C. A. (1973) The Apollo 17 lunar neutron probe experiment. In *Apollo 17 Preliminary Science Report*. NASA SP-330, pp. 18-1.

Yaniv A. and Heymann D. (1972) Atmospheric ^{40}Ar in lunar fines. *Proc. Third Lunar Sci. Conf., Geochim. Cosmochim. Acta*, Suppl. 3, Vol. 2, pp. 1967–1980. MIT Press.

Proceedings of the Fifth Lunar Conference
(Supplement 5, Geochimica et Cosmochimica Acta)
Vol. 2 pp. 2075–2092 (1974)
Printed in the United States of America

Regolith history from cosmic-ray-produced nuclides

E. L. Fireman

Center for Astrophysics, Harvard College Observatory and Smithsonian Astrophysical Observatory,
Cambridge, Massachusetts 02138

Abstract—A model is given for regolith development by which the meteoroid impact parameters (time of last major impact, soil escape rate, and soil turnover depths) are determined from the Gd isotope data for the Apollo 15 and 16 drill stems. The time of the last major impact, cosmic-ray-produced-nuclide-clock reset time, at the Apollo 15 site is 0.6–1.3 aeon. The reset time at the Apollo 16 site is between 1.54 ± 0.14 and 3.3–4.0 aeon. The average soil escape rate from the moon corresponding to the 1.54 ± 0.14 aeon reset time is between 70 and 110 cm/aeon and corresponding to the 3.3 and 4.0 aeon reset time is between 36 and 56 cm/aeon. The soil turnover depth for the 1.54 ± 0.14 aeon reset time is 250 cm and for the 3.3–4.0 aeon reset time is 750 cm. The Gd data restrict the change with time of the meteoroid flux during the past 1.4 aeon.

INTRODUCTION

THE LUNAR REGOLITH is very complicated. Each gram of soil contains millions of separate grains, each with a distinct cosmic-ray irradiation history. The histories of a few of the larger grains (rock chips) have been traced by isotope analysis; however, it would be impossible to trace the separate histories of all the grains that have been recovered. Some similarities should exist in the grain histories, because the number of impacting cosmic rays below 10^{10} GeV energy per square meter is uniform over the moon during one-tenth of an aeon or longer and the number of impacting meteoroids below one gram mass per square meter is also uniform over the moon during one-tenth of an aeon or longer according to the meteoroid mass distribution (Fechtig, 1971; Gault *et al.*, 1972). From these two uniformity conditions and from neutron isotope data with a model of soil development described by an integral equation with rigid boundary conditions, we found that the moon has been losing mass during the past aeon at a rate of approximately 80 g/cm² aeon (Fireman, 1974). We shall examine the consequences of these two uniformity conditions with an equivalent differential equation, impose less rigid boundary conditions, and consider how the consequences change when the meteoroid flux increased in the past.

Cosmic-ray-produced isotopes have been measured as a function of depth to almost 3 m in lunar soil. All soil samples contain large amounts of solar wind which contain the stable rare-gas isotopes normally used for exposure age estimates. In spite of this difficulty, Bogard *et al.* (1973), Hubner *et al.* (1973), and Pepin *et al.* (1974) have measured the concentrations of spallation-produced stable isotopes as a function of depth. The depth variation of the neutron-produced stable isotopes of Gd and Sm were measured (Russ *et al.* 1972; Russ, 1973) with high accuracy.

The depth variations of the spallation stable rare-gas isotopes and the neutron-produced isotopes of Gd and Sm are not as would be expected if there were no soil movement during the irradiation. Russ *et al.* (1972), Russ (1973), Hubner *et al.* (1973), and Bogard *et al.* (1973) have proposed accretionary-type soil models that account for the depth variations. Accretionary models necessitate that a large amount of irradiated soil be buried below 3 m, if the moon has been bombarded by cosmic rays at the current intensity for 4.0 aeons. The amount required is at least a factor of 2 larger than the soil thickness estimates from crater data (Shoemaker *et al.*, 1970) and from active seismic investigations (Watkins and Kovach, 1973), which range between 3 and 12 m. If the ancient soil has escaped from the moon, then a regolith thickness even as small as 250 cm is consistent with a constant cosmic-ray irradiation for 4.0 aeons (Fireman, 1974). Accretionary-type soil models are also inconsistent with the type of soil motion expected for meteoroid impacts. Material removed from one site and deposited at another by meteoroid impacts is generally deposited at shallower depths. The irradiated material would be enhanced near the surface if soil movement is dominated by impacts of meteoroids of less than 50 cm diameter. We mention three other aspects of the statistical model: (1) A statistical model relates isotope data at different sites. The accretionary models consider each site to be so unique that there is little possibility for testing these models. (2) With the statistical model, the average mass loss rate of the moon can be estimated. (3) With the statistical model, isotope data can be related to meteor impact parameters.

A GENERAL COSMIC-RAY REQUIREMENT FOR SOIL MODELS

The stable cosmic-ray-produced isotopes at all six Apollo sites require that irradiated soil* be diluted by unirradiated soil during times of the order of aeons. Such dilution is greater than the implantation of meteors with low exposure ages would have caused. The chemical composition of the soil limits the amount of meteoric material to less than 2% (Laul *et al.*, 1971); the dilution process alters this estimate.*

In a restricted sense, only large impacts that excavate more than 3 m of material and deposit unirradiated (lightly irradiated) grains cause dilution. In a general sense, however, when mass escapes from the moon, small impacts also enhance the dilution. The irradiated grains near the surface are subjected to many more impacts and have a higher escape probability than the unirradiated grains at large depths. Small impacts thereby increase the dilution rate. The probability for a large impact at a site during a time period, such as 0.1 aeon, is very small even though many large impacts, that contributed unirradiated grains to the site, have occurred on the moon. During the same time period many small impacts directly on the site excavate irradiated grains. Even if only a small fraction of the

---

*A discussion of the exposure ages of lunar rocks and of the meteoric influx in the context of our soil model is given later.

excavated mass escapes from the moon, a general dilution essentially caused by small impacts occurs for the moon. On the basis of impact studies with high-speed projectiles, Gault *et al.* (1972) estimated that approximately 10^4 times the projectile mass is the amount of soil excavated by an impact and that approximately one-tenth of one percent of the excavated soil escapes from the moon. Furthermore, the fraction that escapes is independent of the projectile mass. Fechtig *et al.* (1973) and Gault *et al.* (1974) conclude that most of the impacting mass is in particles between 10^{-8} and 10 g masses; therefore, in a nonrestricted sense, small impacts strongly influence the dilution process.

In our formulation, the cosmic-ray exposure age at a particular depth is expressed in terms of the dilution rate at the same depth. The dilution rate decreases with increasing depth because the influence of small impacts diminishes. There is an inverse relation between exposure age and the dilution rate that causes the exposure age to increase with depth. The dilution rate is related to the number of times the soil at a depth is turned over. Turnover times have been calculated on the basis of cratering data by Gault *et al.* (1974). The dilution rate is zero for depths that have not turned over once; the smallest depth at which the dilution rate is zero is called the turnover depth.

Cosmic-ray-produced isotope data are most relevant for times between 0.1 and 4 aeons and depths between 10 and 300 cm. On the other hand, solar wind and solar flare track data are relevant for much smaller times and depths. Soil grains were saturated with the solar wind in approximately a thousand years, and a significant fraction of grains were exposed to solar flares for one million years. Models for soil development from cosmic-ray data can be made consistent with these solar wind and flare data by requiring the soil grains to spend a short time during their history near the surface.

Occasionally, soil layers approximately 1 cm thick have a somewhat different color, elemental abundance, and texture than the soil above and below. The most distinctive of these layers is at 13 cm depth for Apollo 12. According to Laul *et al.* (1971), this layer was undisturbed for more than 10 m.y. Models for soil development from cosmic-ray data should therefore permit a 1 cm layer of material at approximately 13 cm depth to persist undisturbed for more than 10 m.y.

From the point of view of small meteoroid impacts and cosmic rays, there is nothing unique about the six Apollo sites. If irradiated soil is diluted by unirradiated soil at these sites, it is natural to expect irradiated material to be diluted everywhere on the moon. If the cosmic-ray bombardment rate determined from thick samples of lunar material is uniform over the moon, then the rate of dilution of irradiated material integrated over the depth is also uniform. We impose this requirement on our soil model for times since the last major impact at a site; the upper limit for the major impact time is the solidification age of the rocks at the site. The time of the last major impact, T_s, is defined to be the most recent time when more than 3 m of soil were deposited or removed from a site. So few major impacts have occurred at the Apollo sites, during the past 3 aeons, that major impacts at the Apollo sites cannot be treated statistically. The number of major impacts for the entire moon, on the other hand, is very large so that for the entire moon major impacts can be treated statistically and contribute to the dilution process.

The uniformity condition provides a relation between the dilution rate $q(x)$ and the escape rate of soil from the moon, E:

$$E = \int_0^\infty q(x)\, dx \text{ (cm aeon}^{-1})$$

(1)

where $q(x)$ is the fraction of mass at depth x per aeon replaced by unirradiated material. If all the irradiated grains that leave a site return to the moon, then the lunar soil could not be diluted everywhere. Relation (1) requires that the soil escape rate caused by small meteoroid impacts be uniform over the moon. Relation (1) does not require that the dilution rate, $q(x)$, be uniform over the moon. At a young site, one with a recent major impact time, T_s, the time-averaged dilution rate, $\overline{q(x)}$, is not zero only for shallow depths. At a mature site, one with an ancient major impact time, T_s, $\overline{q(x)}$ does not become zero until large depths.

The dilution of irradiated soil affects the relation between the cosmic-ray isotope exposure age and the isotope production rate. The modification in the exposure age is discussed in the next section.

THE SOIL MODEL

Lunar soil is thought to have developed by meteoroid impacts. Theoretically, the meteoroid impact parameters can be calculated from the soil movement and soil escape rate if they are known. The cosmic-ray-produced isotopes in the soil and, especially, their depth variations also depend on the soil movement and the soil escape rate. Since some of these isotope measurements are quite accurate, it is desirable to relate them to the soil movement and the soil escape rate for the moon. Our model formulates such relations.

Cosmic-ray-produced isotopes have been measured to depths of nearly 3 m and show approximately smooth variations over this depth range. An irregularity exists in the Gd isotope measurements of the Apollo 16 drill stem in the neighborhood of 200 cm depth (Russ, 1973), but it can be attributed to a difficulty that occurred during the sample collection. Even if this irregularity is real, our statistical model would encompass it. In fact, if the turnover depth were 250 cm for 1.5 aeon, irregularities at ~250 cm depth would be quite frequent at sites irradiated for 1.5 aeon. An intermediate-sized impact could also cause an irregularity. The turnover depth, h, is defined to be the depth where the dilution rate becomes zero, i.e. less than 0.01.

Meteoroid impacts affect the soil by removing and replacing surface material and by mixing surface material to depth. Soil layers can persist for 10^7 yr even when the soil is uniformly mixed to a depth of 3 m in an aeon. If the meteoroid flux were constant in our model, soil would then be uniformly mixed to 3 cm depth in 10 m.y., and a layer at 13 cm depth would be preserved for 40 m.y.

A statistical soil model was mathematically formulated by an integral equation with stringent boundary conditions (Fireman, 1974). Here, we describe the model in terms of an equivalent differential equation and relax the boundary conditions.

The rate of change of the isotope content at depth x' and time t, $I(x', t)$, is

$$\frac{dI(x', t)}{dt} = R(x') - q(x')I(x', t) \tag{2}$$

where $R(x')$ is the production rate and $q(x')$ is the dilution rate. Because of soil escape, the depth, x', differs from the present depth, x; $x' = x + \bar{E}t$, where $\bar{E}$ is the time averaged soil escape rate. If we consider the grains presently at depth x back in time, then we can denote the isotope content of these grains by $I(x)$ where

$$\int_0^{R(x)T(x)} \frac{dI(x)}{\widetilde{R(x)} - \widetilde{q(x)}I(x)} = \int_{-T_s}^0 dt \tag{3}$$

At large depths, $x > h$, there is no soil mixing so that: $\widetilde{R(x)} = R(x + \bar{E}t)$. At shallow depths, $x < h$, dilution and mixing occur, so that $\widetilde{R(x)}$ is the average production rate in these grains and $\widetilde{q(x)}$ is the rate of replacement with unirradiated grains. Dilution occurs near the surface; soil mixing then moves the less irradiate grains to other depths.

The number of turnovers for the soil at a particular depth has been calculated from cratering theory by Gault (1974); h is the depth of the material that has been turned over once during the time, T_s, and $q(x)$ is related to the number of turnovers experienced by the material presently at x during the time, T_s. The integration of Eq. (3) can be approximately carried out for two regions: (1) $x < h$, where the soil is mixed to the surface so that

$$\overline{R(x)}T(x) = \frac{\bar{R}}{q(x)}[1 - e^{-\overline{q(x)}T_s}] \tag{4}$$

and

$$\bar{R} = \left(\frac{1}{T_s}\right)\int_{-T_s}^0 \widetilde{R(x)}\, dT \cong \frac{1}{h}\int_0^h R(x)\, dx; \qquad \overline{q(x)} = \left(\frac{1}{T_s}\right)\int_{-T_s}^0 \widetilde{q(x)}\, dT$$

and (2) $x > h$, where there is no soil mixing so that

$$\overline{R(x)}T(x) = R\left(x + \frac{\bar{E}T_s}{2}\right)T_s \tag{5}$$

and

$$q(x) = 0$$

At $x = h$ there is a small transition region where some soil mixing and no dilution occur.

The function $R(x)$ is known from target measurements, *in situ* measurements, and also theoretical calculations. The soil escape rate is obtained from the dilution rates by Eq. (1).

The product of the spallation production rate and exposure age, $\overline{R_{sp}(x)}T(x)$, is the spallation cosmic-ray-produced isotope content (e.g. ${}^3\text{He}_{sp}$, ${}^{21}\text{Ne}_{sp}$, ${}^{38}\text{Ar}_{sp}$)

presently at x since the content is zero at time T_s. According to Eugster *et al.* (1970), if the initial isotope content is not known but the initial isotope ratio is known (e.g. $^{158}Gd/^{157}Gd$), then

$$[\overline{R_{158}(x) T(x)}] = \frac{\dfrac{^{158}Gd}{^{157}Gd} - \left(\dfrac{^{158}Gd}{^{157}Gd}\right)_i}{1 + \dfrac{^{158}Gd}{^{157}Gd}} \tag{6}$$

where $R_{158}(x)$ is the production rate per ^{157}Gd atom.

We calculate the average dilution rates, $\overline{q(x)}$, and the turnover depth, h, from the Gd data at the Apollo 16 site with minimal and maximal values for T_{16}. We then examine whether the Gd ratios calculated with the corresponding $\overline{q(x)}$ and h but with a different impact time, T_{15}, fit the Gd measurements (Russ *et al.*, 1972) for the Apollo 15 site. The initial isotope ratio, $(^{158}Gd/^{157}Gd)_i$, is assumed to be the terrestrial value, 1.587, and the lowest lunar value, 1.592 (Russ *et al.*, 1972); these ratios are reasonable upper and lower limits. A lower limit of 1.587 for $(^{158}Gd/^{157}Gd)_i$ is also derived from our model with a 4.0 aeon impact time. Cosmic-ray-produced nuclides and their depth variations at sites are calculable when $\overline{q(x)}$, h, and T_s are known. The Gd data at the Apollo 16 site place limits on $\overline{q(x)}$, h, and T_{16}. The limits are narrowed by the Gd data at the Apollo 15 site. The fact that the Gd data at two sites can be fit with the corresponding dilution and mixing rates is a test of our model; however, more stringent tests will be possible when the contents of other cosmic-ray-produced isotopes and their depth variations are measured and when Gd depth measurements are made for the Apollo 17 drill stem.

If h were less than 250 cm for the Apollo 16 site, then Eq. (5) requires that the $^{158}Gd/^{157}Gd$ ratios decrease with increasing depth at approximately 250 cm depth. Such a decrease is not observed (Russ *et al.*, 1973); so that $h \geqslant 250$ cm. Equation (4) for Gd at the Apollo 16 site, when the initial Gd ratio is 1.592, becomes

$$\frac{\dfrac{^{158}Gd}{^{157}Gd} - 1.592}{1 + \dfrac{^{158}Gd}{^{157}Gd}} = \frac{\bar{R}_{158}}{\bar{q}(x)} [1 - e^{-\overline{q(x)}T_{16}}] \tag{7}$$

where

$$\bar{R}_{158} = \frac{1}{h} \int_0^{h \geqslant 250 cm} R_{158}(x)\, dx$$

For $h > 250$ cm, $\bar{R}_{158}$ decreases with increasing h: $\bar{R}_{158} = 0.68 \times 10^{-2}$ for $h = 250$ cm and $\bar{R}_{158} = 0.30 \times 10^{-2}$ for $h = 750$ cm. The production rates, $R_{158}(x)$, calculated by Lingenfelter *et al.* (1972) and by Kornblum and Fireman (1974), agree to better than a factor of 2. The values used here are those of Lingenfelter *et al.* (1972). The maximum T_{16} is 4.0 aeons, the solidification age of the rocks at the Apollo 16 site. For each value of h, there is a minimum value for T_{16} obtained by setting $q(h) = 0$ in Eq. (7). For $h = 250$ cm the minimum T_{16} is 1.54 ± 0.14 aeon; for $h = 750$ cm the minimum T_{16} is 3.3–4.0 aeon. Since T_{16} cannot be larger than

4.0 aeon, h is less than 750 cm. Gault (1974) calculated probable times for soil turnover on the basis of crater data. The probable turnover time for soil at depths larger than 250 cm was between 3.8 and 4.0 aeon. The limits imposed on the turnover depth h, by the Apollo 16 Gd data encompass the probable values deduced from cratering theory by Gault *et al.* (1974) only for T_{16} close to 4.0 aeon.

Table 1 gives the maximum and minimum soil escape and dilution rates and the major impact times, T_{16}, obtained for the Apollo 16 drill stem with the two initial Gd ratios.

The minimum impact time, 1.4 aeons, gives the maximum average soil escape rate, 110 cm/aeon. The maximum impact time, 4.0 aeon, gives the minimum average soil escape rate, 36 cm/aeon. Figures 1 and 2 compare the Gd ratios measured in the Apollo 16 drill stem with the calculated ratios. The ratios

406-053

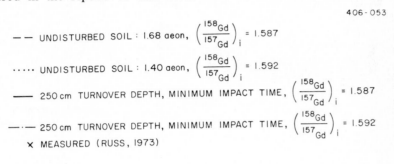

— — UNDISTURBED SOIL : 1.68 aeon, $\left(\dfrac{^{158}Gd}{^{157}Gd}\right)_i = 1.587$

····· UNDISTURBED SOIL : 1.40 aeon, $\left(\dfrac{^{158}Gd}{^{157}Gd}\right)_i = 1.592$

——— 250 cm TURNOVER DEPTH, MINIMUM IMPACT TIME, $\left(\dfrac{^{158}Gd}{^{157}Gd}\right)_i = 1.587$

—·— 250 cm TURNOVER DEPTH, MINIMUM IMPACT TIME, $\left(\dfrac{^{158}Gd}{^{157}Gd}\right)_i = 1.592$

× MEASURED (RUSS, 1973)

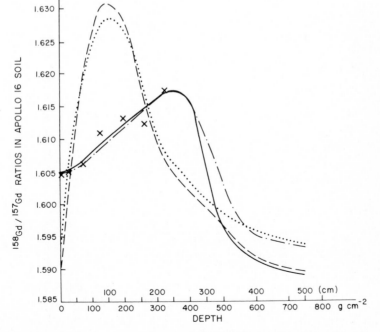

Fig. 1. $^{158}Gd/^{157}Gd$ ratio versus depth in Apollo 16 soil for undisturbed soil and for soil model with $(^{158}Gd/^{157}Gd)_i = 1.587$ and 1.592 and corresponding minimum impact times; $T_{16} = 1.68$ and 1.40 aeon, respectively.

Table 1. Average dilution rates, $\overline{q(x)}$, and soil escape rates $\overline{E}$, determined from Apollo 16 soil Gd^{158}/Gd^{157} data (Russ, 1973).

Depth (cm)	$R_{158}(x)$ $\left(\frac{\text{captures}}{^{157}\text{Gd-aeon}}\right)$	A $\frac{Gd^{158}}{Gd^{157}} - 1.592$	$A/[1 + (Gd^{158}/Gd^{157})]$ $\left(\frac{1}{250}\int_0^{250} R_{158}(x)\,dx\right)$ (aeon)	$\overline{q(x)}$† (aeon⁻¹)	B $\frac{Gd^{158}}{Gd^{157}} - 1.587$	$B/[1 + (Gd^{158}/Gd^{157})]$ $\left(\frac{1}{250}\int_0^{250} R_{158}(x)\,dx\right)$ (aeon)	$\overline{q(x)}$‡ (aeon⁻¹)
0	~0.1 × 10⁻²	0.0125	0.70, 1.58*	1.15, 0.51*	0.0175	0.98, 2.22*	0.71, 0.33*
35	0.60 × 10⁻²	0.014	0.79, 1.79*	0.90, 0.42*	0.019	1.07, 2.42*	0.58, 0.27*
70	0.87 × 10⁻²	0.017	0.95, 2.15*	0.60, 0.28*	0.022	1.23, 2.78*	0.41, 0.19*
100	1.00 × 10⁻²	0.019	1.07, 2.42*	0.41, 0.20*	0.024	1.35, 3.06*	0.28, 0.14*
135	0.95 × 10⁻²	0.020	1.13, 2.56*	0.32, 0.16*	0.025	1.41, 3.20*	0.22, 0.12*
170	0.75 × 10⁻²	0.021	1.18, 2.67*	0.25, 0.13*	0.026	1.46, 3.30*	0.17, 0.10*
200	0.52 × 10⁻²	0.022	1.26, 2.85*	0.16, 0.09*	0.027	1.55, 3.51*	0.09, 0.07*
250	0.38 × 10⁻²	0.025	T_{16} = 1.40, 3.18*	0, 0.03*	0.030	T_{16} = 1.68, 3.80*	0, 0.025*
300	0.27 × 10⁻²	—	—, —	0, —	—	—, —	0, —
400	~0.12 × 10⁻²	—	—, —	0, —	—	—, —	0, —
500	~0.06 × 10⁻²	—	T_{16}^* = —, 3.30*	0, ~0.01*	—	T_{16}^* = —, 4.00*	0, 0.01*

$\frac{1}{250}\int_0^{250} R_{158}(x)\,dx = 0.68 \times 10^{-2}$.

$\overline{E} = \int_0^{250} q(x)\,dx \left(\frac{\text{cm}}{\text{aeon}}\right) = 110$

$†\left(\frac{^{158}Gd}{^{157}Gd}\right)_i = 1.592.$

$\overline{E}^* = \int_0^{750} q(x)\,dx \left(\frac{\text{cm}}{\text{aeon}}\right) = 56$

$‡\left(\frac{^{158}Gd}{^{157}Gd}\right)_i = 1.587$

$\frac{1}{250}\int_0^{250} R_{158}(x)\,dx = 70$

$\frac{1}{750}\int_0^{750} R_{158}(x)\,dx = 36$

*Soil escape and dilution rates with 750 cm turnover depth,

$$\frac{1}{750}\int_0^{750} R_{158}(x)\,dx = 0.30 \times 10^{-2},$$

are approximately half and impact times are approximately twice those with 250 cm turnover depth.

406-053

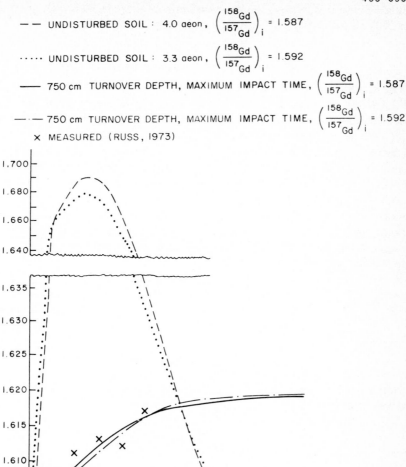

Fig. 2. ^{158}Gd/^{157}Gd ratio versus depth in Apollo 16 soil for undisturbed soil and for soil model with $(^{158}$Gd/^{157}Gd$)_i = 1.587$ and 1.592 and the corresponding maximum impact times, $T_{16} = 4.0$ and 3.3 aeon, respectively.

calculated with minimum and maximum impact times are identical for depths less than 250 cm but differ markedly at depths beyond 250 cm.

SOIL MODEL APPLIED AT THE APOLLO 15 SITE

The solidification ages of the rocks at the Apollo 15 site cluster between 3.0 and 3.3 aeons (Husain, 1972). A natural question to ask is: Can our statistical model account for the Apollo 15 and 16 Gd data without major impacts at the sites (i.e. $T_{15} = 3$ aeons and $T_{16} = 4$ aeons)? According to cratering data the meteoroid flux increased markedly in the ancient past, i.e. between 3 and 4 aeons ago. How the meteoroid flux increased during the previous 3 aeons is uncertain. According to Fechtig et al. (1971), the meteoroid flux increased at all times in the past as $e^{-2.6t}$, where t is in negative aeons. On the other hand according to Gault et al. (1974) the meteoroid flux was approximately constant in the recent past and increased markedly in the ancient past. In our model the soil escape rate by small impacts is the same over the entire moon, and the soil escape rate is proportional to the meteoroid flux.

If the meteoroid flux increased exponentially in the past as $e^{-2.6t}$ and $T_{16} = 4.0$ aeons, the soil is essentially undisturbed during the past 3 aeons at the Apollo 15 site. The Gd ratios produced in 3 aeons for undisturbed soil are higher than observed. Even if the meteoroid flux were approximately constant for 3 aeons and increased as $e^{-2.6t}$ between 3 and 4 aeons ago, the Gd ratios at the Apollo 15 site would be higher than observed. The model cannot account for the Apollo 15 Gd data without having a major impact at the site.

In the Apollo 15 drill stem, Russ et al. (1972) measured the ^{158}Gd/^{157}Gd ratio to be 1.596 at 250 cm depth. The major impact time,

$$T_{15} = \left[\frac{^{158}\text{Gd}/^{157}\text{Gd} - (^{158}\text{Gd}/^{157}\text{Gd})_i}{R_{158}(250)[2.596]} \right]_{250\text{cm}}$$

which is 0.58 aeon with an initial ratio of 1.592 and is 1.25 aeon with an initial ratio of 1.587 based on no soil escape. Soil escape affects these impact times very little. The effect is negligible if the meteoroid flux increased in the past. If the meteoroid flux was constant, the major impact times are 0.6 and 1.3 aeon for the two initial ratios.

Relations (8), (9), and (10) are based on a constant meteoroid flux, an initial Gd ratio of 1.592, and the minimum impact time:

$$\bar{E} = 110 \text{ cm/aeon} = E_{15} = E_{16} \tag{8}$$

$$h_{15} = \left(\frac{T_{15}}{T_{16}}\right) h_{16} = \left(\frac{0.6}{1.4}\right)(250 \text{ cm}) = 111 \text{ cm} \tag{9}$$

and

$$\overline{q_{15}(x)} = \left(\frac{h_{16}}{h_{15}}\right) \overline{q_{16}(x)} = 2.25\overline{q_{16}(x)} \qquad \text{for } x < 111 \text{ cm} \tag{10}$$

$$\overline{q_{15}(x)} = 0 \qquad \text{for } x > 111 \text{ cm}$$

because

$$E = \int_0^{h_{15}} q_{15}(x) \, dx = \int_0^{h_{16}} q_{16}(x) \, dx$$

For the maximum impact time, and the same initial Gd ratio, the $\bar{E}$ is 56 cm/aeon, h_{15} is 0.6/3.3 (750 cm) or 136 cm, and $\overline{q_{15}(x)} = 5.5\,\overline{q_{16}(x)}$ for $x < 136$ cm. The Gd ratios as a function of depth for the Apollo 15 site are calculated with Eq. (11), which results from Eqs. (7) and (4) for $x < 111$ cm or 136 cm and with Eq. (12), which results from Eqs. (6) and (5) for $x > 111$ cm or 136 cm and with $T_{15} = 0.6$ aeon.

$$\frac{\bar{R}_{158}}{q(x)}[1 - h^{0.6\overline{q(x)}}] = \frac{\dfrac{^{158}Gd}{^{157}Gd} - 1.592}{1 + \dfrac{^{158}Gd}{^{157}Gd}} \tag{11}$$

where

$$\bar{R}_{158} = \left(\frac{1}{111}\right) \int_0^{111} R_{158}(x) \, dx$$

for $x < 111$ cm, and for $x > 111$ cm

$$[R_{158}(x + 33 \text{ cm})]0.6 \text{ aeon} = \frac{\dfrac{^{158}Gd}{^{157}Gd} - 1.592}{1 + \dfrac{^{158}Gd}{^{157}Gd}} \tag{12}$$

For the other turnover rate,

$$\bar{R}_{158} \text{ is } \left(\frac{1}{136}\right) \int_0^{136} R(x) \, dx \quad \text{and} \quad R\left(x + \frac{\bar{E}}{2} T_{15}\right)$$

is $R(x + 17)$. The ^{158}Gd production rate, R_{158}, for Apollo 15 soil is 0.70 times that of the Apollo 16 soil because the elemental abundances in the soils differ (Lingenfelter *et al.*, 1972). Table 2 gives the ^{158}Gd production rates, the dilution rates, and the calculated Gd ratios versus depth for the Apollo 15 site with a constant meteoroid flux and $(^{158}Gd/^{157}Gd)_i = 1.592$. The calculated Gd ratios are compared with the experimentally measured ratios (Russ *et al.*, 1972) in Fig. 3. A similar calculation was carried out with an initial Gd ratio of 1.587 in Table 3; the calculated and measured ratios are compared in Fig. 4. The calculated and measured Gd ratios at every depth agree except for a meteoroid flux that is constant for 3.8 aeon. The depth variation of the stable spallation rare-gas contents for the Apollo 15 site may also differ from their production rates at depths less than 150 cm because soil movement and soil dilution occurred.

If the meteoroid flux increased exponentially as proposed by Fechtig *et al.* (1971), then the average soil escape rates, the turnover depth, and the dilution

Table 2. Calculated $\dfrac{^{158}\text{Gd}}{^{157}\text{Gd}}$ ratios for Apollo 15 soil $\left[T_{15} = 0.6 \text{ aeon}, \left(\dfrac{^{158}\text{Gd}}{^{157}\text{Gd}}\right)_i = 1.592\right]$ dilution and turnover rates obtained from Table 1.

Depth (cm)	$R_{158}(x)$ $\left(\dfrac{\text{captures}}{^{157}\text{Gd-aeon}}\right)$	$\overline{R(x)}^*$ $\left(\dfrac{\text{captures}}{^{157}\text{Gd-aeon}}\right)$	$\overline{q(x)}^*$ aeon^{-1}	$\left(\dfrac{^{158}\text{Gd}}{^{157}\text{Gd}}\right)^*$	$\overline{R(x)}^{**}$ $\left(\dfrac{\text{captures}}{^{157}\text{Gd-aeon}}\right)$	$\overline{q(x)}^{**}$ aeon^{-1}	$\left(\dfrac{^{158}\text{Gd}}{^{157}\text{Gd}}\right)^{**}$
0	$\sim 0.07 \times 10^{-2}$	0.47×10^{-2}	2.6	1.5961	0.49×10^{-2}	2.80	1.5963
35	0.42×10^{-2}	0.47×10^{-2}	2.0	1.5970	0.49×10^{-2}	2.30	1.5967
70	0.61×10^{-2}	0.47×10^{-2}	1.35	1.5979	0.49×10^{-2}	1.54	1.5978
100	0.70×10^{-2}	0.47×10^{-2}	0.92	1.5987	0.49×10^{-2}	1.10	1.5985
135	0.67×10^{-2}	0.53×10^{-2}	0	1.6004	0.49×10^{-2}	0.88	1.5998
170	0.527×10^{-2}	0.37×10^{-2}	0	1.5981	0.47×10^{-2}	0	1.5993
200	0.365×10^{-2}	0.33×10^{-2}	0	1.5972	0.35×10^{-2}	0	1.5975
250	0.276×10^{-2}	0.23×10^{-2}	0	1.5957	0.26×10^{-2}	0	1.5960

*250 cm turnover depth for Apollo 16 soil and meteoroid flux constant

$$\overline{R(x)}^* = \frac{1}{111} \int_0^{111} R(x)\,dx = 0.47 \times 10^{-2} \text{ for } x < 111 \text{ cm}$$

$$\overline{R(x)}^* = R(x + 33) \text{ for } x > 111 \text{ cm}$$

$$\overline{q(x)}^* = 2.25 q(x)^\dagger \text{ which is in Table 1.}$$

$$\overline{q(x)}^* = 0 \text{ for } x > 111 \text{ cm}$$

**750 cm turnover depth for Apollo 16 soil and meteoroid flux constant

$$\overline{R(x)}^{**} = \frac{1}{136} \int_0^{136} R(x)\,dx = 0.49 \times 10^{-2} \text{ for } x < 136 \text{ cm}$$

$$\overline{R(x)}^{**} = R(x + 17) \text{ for } x > 136 \text{ cm}$$

$$\overline{q(x)}^{**} = 5.5 \text{ times the starred values of } q(x)^\dagger \text{ of Table 1}$$

$$\overline{q(x)}^{**} = 0 \text{ for } x > 136 \text{ cm}$$

rates at the sites are related by

$$\bar{E}_{15} = \left(\frac{e^{2.6 T_{15}} - 1}{e^{2.6 T_{16}} - 1}\right) \bar{E}_{16} \tag{13}$$

$$h_{15} = \left(\frac{e^{2.6 T_{15}} - 1}{e^{2.6 T_{16}} - 1}\right)\left(\frac{T_{15}}{T_{16}}\right) h_{16} \tag{14}$$

$$h_{15}\, \overline{q_{15}(x)} = h_{16}\, \overline{q_{16}(x)} \tag{15}$$

With $T_{15} = 0.6$ aeon and $T_{16} = 1.4$ aeon, $\bar{E}_{15} = 11$ cm/aeon and $h_{15} = 0.043 h_{16}$ or 10.7 cm. For $T_{15} = 0.6$ and $T_{16} = 3.3$ aeon, E_{15} is less than 0.04 cm/aeon and h_{15} is 0.01 cm. The depth variation of all cosmic-ray-produced isotopes would then be practically the same as the production rates with no soil dilution or soil mixing. Figure 3 shows that the Gd ratios calculated with the exponentially increasing meteoroid flux during the past and $^{158}\text{Gd}/^{157}\text{Gd}_i = 1.592$ do not agree with the

406-053

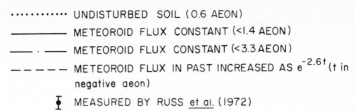

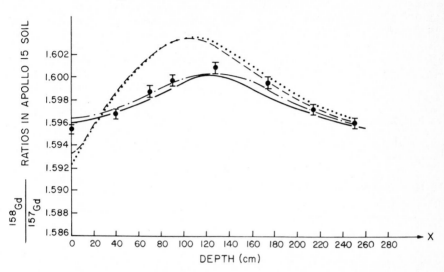

Fig. 3. Comparison of measured and calculated ratios, $^{158}Gd/^{157}Gd$, in Apollo 15 soil with $(^{158}Gd/^{157}Gd)_i = 1.592$. The last major impact at the site is then 0.6 aeon.

measurements as well as those calculated with a constant meteoroid flux. With the initial Gd ratio of 1.587, $\bar{E}_{15} = 26$ cm/aeon and $h_{15} = 70$ cm for the 250 cm Apollo 16 turnover depth. For this initial ratio, $\bar{E}_{15} = 0.05$ cm/aeon, $h_{15} = 0.04$ cm and the soil is essentially undisturbed during 1.3 aeon when the Apollo 16 turnover depth is 750 cm. Figure 4 shows the results. The Gd ratios that agree best with the measurements are those calculated with a meteoroid flux that is constant only during the past 1.68 aeon.

Gault *et al.* (1974) did not give an analytical expression for the meteoroid flux in the past. We therefore did not calculate the depth variation of the Gd isotopes at the Apollo 15 site with their meteoroid flux. Better accord with the Gd isotope data would probably be achieved with their estimate of the meteoroid flux in the past than with $e^{-2.6t}$ because the Gd isotope data are in better accord with the constant meteoroid flux during the recent past than with the exponentially increasing flux.

For simplicity we have assumed a constant cosmic-ray flux in the past. If the cosmic-ray flux increased in the past, it would be more difficult to achieve accord with the isotope data. If the cosmic-ray flux decreased in the past, it would be simpler to achieve accordance.

Table 3. Calculated $\frac{^{158}Gd}{^{157}Gd}$ ratios for Apollo 15 soil $\left[T_{15} = 1.3 \text{ aeon}, \left(\frac{^{158}Gd}{^{157}Gd}\right)_i = 1.587\right]$ dilution and turnover rates obtained from Table 1.

Depth (cm)	$R_{158}(x)$ $\left(\dfrac{\text{captures}}{^{157}\text{Gd-aeon}}\right)$	$\overline{R(x)}^*$ $\left(\dfrac{\text{captures}}{^{157}\text{Gd-aeon}}\right)$	$\overline{q(x)}^*$ aeon^{-1}	$\left(\dfrac{^{158}Gd}{^{157}Gd}\right)^*$	$\overline{R(x)}^{**}$ $\left(\dfrac{\text{captures}}{^{157}\text{Gd-aeon}}\right)$	$\overline{q(x)}^{**}$ aeon^{-1}	$\left(\dfrac{^{158}Gd}{^{157}Gd}\right)^{**}$
0	$\sim 0.07 \times 10^{-2}$	0.48×10^{-2}	0.91	1.5961	0.47×10^{-2}	1.01	1.5958
35	0.42×10^{-2}	0.48×10^{-2}	0.75	1.5970	0.47×10^{-2}	0.79	1.5968
70	0.61×10^{-2}	0.48×10^{-2}	0.53	1.5988	0.47×10^{-2}	0.56	1.5983
100	0.70×10^{-2}	0.48×10^{-2}	0.364	1.6000	0.47×10^{-2}	0.41	1.5992
135	0.67×10^{-2}	0.48×10^{-2}	0.286	1.6005	0.47×10^{-2}	0.35	1.6000
170	0.527×10^{-2}	0.48×10^{-2}	0.220	1.6010	0.47×10^{-2}	0.29	1.6002
200	0.365×10^{-2}	0.33×10^{-2}	0	1.5972	0.47×10^{-2}	0.205	1.6007
250	0.276×10^{-2}	0.23×10^{-2}	0	1.5957	0.47×10^{-2}	0.075	1.6016

*250 cm turnover depth for Apollo 16 soil, meteoroid flux constant

$$\left\{\begin{array}{l} \overline{R(x)}^* = \dfrac{1}{193}\displaystyle\int_0^{193} R(x)\,dx = 0.48 \times 10^{-2} \text{ for } x > 193 \text{ cm} \\[2ex] \overline{R(x)}^* = R(x + 45) \text{ for } x > 193 \text{ cm} \\[2ex] \overline{q(x)}^* = 1.30 q(x)\ddagger \text{ which is in Table 1} \\[2ex] \overline{q(x)}^* = 0 \text{ for } x > 193 \text{ cm} \end{array}\right.$$

**750 cm turnover depth for Apollo 16 soil, meteoroid flux constant

$$\left\{\begin{array}{l} \overline{R(x)}^{**} = \dfrac{1}{256}\displaystyle\int_0^{256} R(x)\,dx = 0.47 \times 10^{-2} \text{ for } x < 256 \text{ cm} \\[2ex] \overline{q(x)}^{**} = 2.93 \text{ times the starred values of } q(x)\ddagger \text{ of Table 1} \\[2ex] \overline{q(x)}^{**} = 0 \text{ for } x > 256 \text{ cm} \end{array}\right.$$

The soil model presented here is equivalent to the more abstract statistical soil model presented previously (Fireman, 1974; Fireman and Kornblum, 1974) when the statistical noise in the abstract model is caused by meteoroid impacts, the time of statistical applicability is the time of the last major impact, and the rate of change of the fraction of soil grains that have been irradiated for the time of statistical applicability is the dilution rate. The abstract model is a more general model since statistical noise in soil encompasses both meteoroid impacts and moonquakes. The integral equation which described the abstract model was solved by a more mathematically rigorous procedure than used for the solution of the differential equation presented here; however, the differential equation is more pictorial.

EXPOSURE AGES OF ROCKS

The exposure ages of the lunar rocks at a site differ from that of the soil for three reasons: (1) rocks are not diluted; (2) the time-averaged depth of rocks

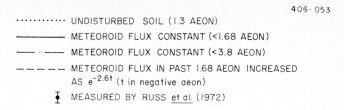

406-053

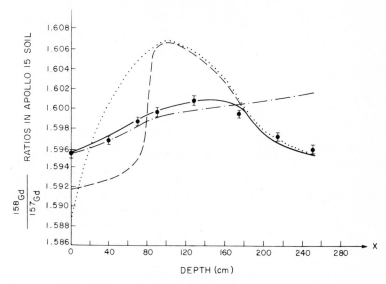

Fig. 4. Comparison of measured and calculated ratios, $^{158}Gd/^{157}Gd$, in Apollo 15 soil with $(^{158}Gd/^{157}Gd)_i = 1.587$. The last major impact at the site is then 1.2 aeons.

buried in undisturbed soil is the same as that of the adjacent soil, but rocks that were buried in the soil that has turned over are carried to the surface and then remain on the surface; (3) rocks on the surface are eroded into soil by small impacts. In the context of our model the rocks at a site suddenly become exposed to cosmic rays at the time of the last major impact. Because of soil escape the depth of a rock in the soil gradually becomes smaller. At approximately the time when the turnover depth of the soil equals the depth of the rock, soil motion carries the rock to the surface. The spallation content, (e.g. ^3He), of a rock presently at a depth, x, greater than h is

$$^3He \cong T_s R_3 \left(x + \frac{\bar{E}T_s}{2} \right), \quad \text{where} \quad R_3 \left(x + \frac{\bar{E}T_s}{2} \right) \quad (16)$$

is the average ^3He production rate and T_s is the time of the last major impact at the site. The rocks presently on the surface have a distribution of exposure ages which depends upon their burial at the time, T_s. A rock implanted on the surface by the last major impact has the maximum exposure, T_s. A rock recently brought

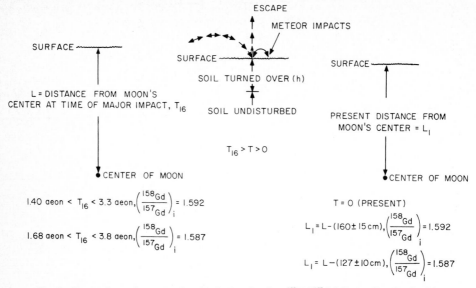

Fig. 5. Comparison of measured and calculated ratios, ^{158}Gd/^{157}Gd, in Apollo 15 soil with (^{158}Gd/^{157}Gd)$_i$ = 1.587. The last major impact at the site is then 1.2 aeons.

to the surface by soil turnover has the minimum spallation age,

$$^3\text{He}/R_3 \left(h + \frac{E T_s}{2} \right)$$

Rock erosion modifies the spallation ages of some of the rocks on the surface.

METEOROID INFLUX

The meteoroid influx estimated from the abundances of meteoritic-type elements in the soil is modified because of soil escape and soil dilution. The most direct method to determine the average meteoroid influx is from the concentration of the meteor-type elements as a function of depth in the Apollo drill stems. Since these data are not available, we estimate the influx rate from the soil escape rate.

Small meteoroids are implanted near the surface and then mixed to other depths in a fashion similar to that of the diluting soil grains. The average meteoroid influx rate since the last major impact is

$$\overline{\text{influx}} = (\text{const.}) \int_0^h \overline{q(x)\, dx} \tag{17}$$

where the (const.) is the ratio of impacting mass to the soil escape mass. According to Gault *et al.* (1972), this ratio is approximately 0.1 so that the average meteoroid influx rate is approximately 0.1 times the average soil escape rate. With a soil density of 1.5 g/cm^3, the meteoroid influx rate is approximately

15 g/cm^2 aeon. This estimate is twice the value estimated from a 2% abundance of meteoric material in 10 m of soil.

Summary

Two general consequences result from the application of our soil model to the Gd data. (1) The initial (^{158}Gd/^{157}Gd) ratio for the moon is between 1.587 and 1.592. The condition that the Apollo 16 soil be irradiated for less than 4.0 aeons gives the 1.587 limit. The condition that the soil at all sites have the same initial Gd isotope ratio gives the 1.592 limit. (2) The soil escape rate for the moon is between 36 and 110 cm/aeon, if the meteoroid flux was constant in the past.

Figure 5 is an illustration of how the Apollo 16 site developed according to our model.

Acknowledgments—This work was supported in part by grant NGR 09-015-145 from the National Aeronautics and Space Administration. The author gratefully acknowledges helpful discussion with Dr. F. Steinbrunn, Professor R. O. Pepin, and J. R. Basford.

References

Bogard D. D., Nyquist L. E., Hirsch H. C., and Moore D. R. (1973) Trapped solar and cosmogenic noble gas abundances in Apollo 15 and 16 deep drill samples. *Earth Planet. Sci. Lett.* **21**, 52–69.

Eugster O., Tera F., Burnett D. S., and Wasserburg G. J. (1970) The isotopic composition of Gd and the neutron capture effects in samples from Apollo 11. *Earth Planet. Sci. Lett.* **8**, 20–30.

Fechtig H. (1971) Cosmic dust in the atmosphere and in the interplanetary space at 1 AU today and in the early solar system. Presented at IAU Colloqu. No. 13, Albany, N.Y.; proceedings in press.

Fireman E. L. (1974) History of the lunar regolith from neutrons (abstract). In *Lunar Science—V*, pp. 230–232. The Lunar Science Institute, Houston.

Gault D. E., Horz F., and Hartung J. B. (1972) Effects of microcratering on the lunar surface. *Proc. Third Lunar Sci. Conf., Geochim. Cosmochim. Acta*, Suppl. 3, Vol. 3, pp. 2713–2734. MIT Press.

Gault D. E., Horz F., Brownlee D. E., and Hartung J. B. (1974) Mixing of the lunar regolith. *Abstracts Fifth Lunar Conf.* March 18–22, 1974, Part 1, pp. 260–262.

Hubner W. D., Heymann D., Kirsten T. (1973) Inert gas stratigraphy of Apollo 15 drill core sections 15001 and 15003. *Proc. Fourth Lunar Sci. Conf., Geochim. Cosmochim. Acta*, Suppl. 4, Vol. 2, pp. 2021–2036. Pergamon.

Husain L. (1972) The ^{40}Ar–^{39}Ar and cosmic ray exposure ages of Apollo 15 crystalline rocks, breccias, and glasses. In *The Apollo 15 Lunar Samples*, pp. 374–377. The Lunar Science Institute, Houston.

Kornblum J. J. and Fireman E. L. (1974) Neutrons and the moon. Submitted to *Earth Planet. Sci. Lett.*

Laul J. C., Morgan J. W., Ganapathy R., and Anders E. (1971) Meteoritic material in lunar samples: Characterization from trace elements. *Proc. Second Lunar Sci. Conf., Geochim. Cosmochim. Acta*, Suppl. 2, Vol. 2, pp. 1139–1158. MIT Press.

Lingenfelter R. E., Canfield E. H., and Hampel V. E. (1972) The lunar neutron flux revisited. *Earth Planet. Sci. Lett.* **16**, 335–369.

Pepin R. O., Basford J. R., Dragon J. C., Coscio M. R., Jr., and Murthy V. R. (1974) K–Ar ages and depositional chronologies of Apollo 15 drill core (abstract). In *Lunar Science—V*, pp. 593–595. The Lunar Science Institute, Houston.

Russ G. P., III (1973) Apollo 16 neutron stratigraphy. *Earth Planet. Sci. Lett.* **19**, 275–289.

Russ G. P., III, Burnett D. S., and Wasserburg G. J. (1972) Lunar neutron stratigraphy. *Earth Planet. Sci. Lett.* **15**, 172–186.

Shoemaker E. M., Hait M. H., Swann G. A., Schleicher D. L., Schaber G. G., Sutton R. L., Dahlem D. H., Goddard E. N., and Waters A. C. (1970) Origin of the lunar regolith at Tranquillity Base. *Proc. Apollo 11 Lunar Sci. Conf., Geochim. Cosmochim. Acta*, Suppl. 1, Vol. 3, pp. 2399–2412. Pergamon.

Watkins J. S. and Kovach R. L. (1973) Seismic investigation of the lunar regolith. *Proc. Fourth Lunar Sci. Conf., Geochim. Cosmochim. Acta*, Suppl. 4, Vol. 3, pp. 2561–2574. Pergamon.

Proceedings of the Fifth Lunar Conference
(Supplement 5, Geochimica et Cosmochimica Acta)
Vol. 2 pp. 2093–2103 (1974)
Printed in the United States of America

Depth profiles of ^{53}Mn in lunar rocks and soils

M. IMAMURA, K. NISHIIZUMI, and M. HONDA

University of Tokyo, Tokyo, Japan

R. C. FINKEL, J. R. ARNOLD, and C. P. KOHL

University of California, San Diego, La Jolla, California

Abstract—Cosmic ray produced ^{53}Mn has been measured in a series of samples down the length of the Apollo 16 deep drill core. The results indicate that the lunar regolith has been unmixed, on a meter scale, for the past 5 m.y. at the location of this core. The data are in agreement with earlier ^{53}Mn measurements on the Apollo 15 drill core (Imamura *et al.*, 1973). The ^{53}Mn activity profile in rock 14310 has been measured from the surface to a depth of 8 mm. This profile in comparison with profiles from 12002 and 14321 indicates, in contradiction to the track exposure age (1–3 m.y.) obtained for 14310 by Crozaz *et al.* (1972), that all three rocks have probably been on the lunar surface long enough to saturate their solar cosmic ray produced ^{53}Mn ($t_{1/2} = 3.7 \times 10^6$ yr) activity. The best fit to the data is obtained using an SCR flux with an R_0 of 100 MV and a J (> 10 MeV) of 100 protons/cm^2 sec (4π). Differences in the rock profiles can be explained by differences in erosion rates; 0.5 mm/10^6 yr for 12002 and 14310 and 2.0 mm/10^6 yr for 14321. ^{53}Mn activities are also reported for the Apollo 16 so-called permanently shadowed soil and soils from an Apollo 16 trench.

INTRODUCTION

^{53}MN IS A WELL STUDIED COSMOGENIC RADIONUCLIDE. Wahlen *et al.* (1972) have given a detailed experimental and theoretical treatment of the solar cosmic ray (SCR) production of ^{53}Mn in near-surface samples of lunar rocks. Imamura *et al.* (1973) have extended the study to include the galactic cosmic ray (GCR) production of the nuclide to a depth of 416 g/cm^2 in the Apollo 15 drill core. The full range of the ^{53}Mn production profile is delineated by a combination of the SCR and GCR profiles. Now that the total profile is understood theoretically, experimental ^{53}Mn profiles can be interpreted in terms of the specific histories of the individual samples. ^{53}Mn is useful as a probe for the rates and depths of lunar surface processes. The time scale observed is limited by the ^{53}Mn half-life (3.7×10^6 yr) to the past 5–8 m.y., an interesting region with regard to expected mixing rates of regolith material and expected erosion rates and exposure times of rocks. This nuclide is produced in the moon almost totally from iron, eliminating complications due to differences in sample chemistry. It can be determined in fairly small samples (typically 200 mg).

Cores from all the Apollo missions have distinct layering indicating complex depositional history at any given location. Theoretical models have been constructed for the mixing of the regolith by impact processes (Gault *et al.*, 1974; J. R. Arnold, unpublished). Experimental evidence obtained from stable isotope studies (Russ, 1973), rare gas work (Hübner *et al.*, 1973), and studies of particle track densities (Bhandari *et al.*, 1973), can be used to estimate the mixing in individual cores. For recent mixing a radioactive tracer such as ^{53}Mn has an advantage over the methods listed above because it looks at a time interval which is approximated by its mean life and not at a cumulative record. ^{53}Mn studies on suitable samples should provide a good test for any mixing model. The core samples are obvious candidates for such a study. We present here results for seven samples (0–364 g/cm^2) from the Apollo 16 drill core and for the trench soils from Plum Crater which were taken from two distinct layers.

Cosmogenic radionuclides have been used to establish exposure ages of relatively young lunar samples (Rancitelli *et al.*, 1972). Crozaz *et al.* (1972) have determined from GCR track studies that rock 14310 has been exposed on the lunar surface for only 1.5–3 m.y. ^{53}Mn measurements on near-surface samples of this rock were undertaken in an effort to confirm the short exposure.

Experimental

Samples

The first seven samples listed in Table 1 are drill core samples from the ends of the separate sections of the deep drill core taken at the Apollo 16 ALSEP site. When the core was examined after return from the moon, section 60005 and part of section 60006 were nearly empty. This gap makes the absolute depths of the four deep samples uncertain. The depths given in Table 1 are those of Russ (1973), calculated for a loss of 180 g from the void area. If a negligible amount had been lost these

Table 1. ^{53}Mn in Apollo 16 soil samples and in Rock 14310 (all count rates are corrected to Nov. 19, 1973).[a]

Samples (depth)	Sample description			Sample counted after irradiation			^{54}Mn (cpm/ mg Mn)	^{54}Mn (cpm/ mg Mn) after $(n, 2n)$, (n, p) corrections	^{53}Mn[e] (dpm/ kg Fe)
	Fe[b] (%)	Mn[c] (ppm)	I.P.[d]	Fe (μg)	Mn[c] (μg)	^{54}Mn (cpm)			
Apollo 16 drill core samples[f]									
60007,105 (1.0 g/cm^2)	4.04	538	A	0.2	79.0	1.401 ± .052	17.73 ± .75	15.19 ± .75	417 ± 35
60007,112 (32.4 g/cm^2)	2.99	385	A	0.2	62.5	0.693 ± .030	11.08 ± .53	8.53 ± .54	226 ± 21
60006,5 (83.1 g/cm^2)	3.98	523	A	0.2	75.9	0.793 ± .038	10.44 ± .54	7.90 ± .55	214 ± 20
60004,20 (161.5 g/cm^2)	4.38	574	A	0.0	85.2	0.786 ± .037	9.22 ± .48	6.71 ± .49	181 ± 18
60004,11 (223.5 g/cm^2)	4.42	583	A	0.0	89.8	0.641 ± .031	7.14 ± .37	4.63 ± .38	126 ± 13
60003,7 (289.5 g/cm^2)	4.78	584	A	0.2	87.8	0.519 ± .030	5.91 ± .36	3.37 ± .37	84.9 ± 10.9
60001,5 (363.6 g/cm^2)	4.31	565	A	0.2	86.4	0.380 ± .023	4.39 ± .28	1.85 ± .29	50.0 ± 8.5
Trench soils									
61241,7-1	3.97	525	B	0.3	86.2	1.383 ± .060	16.04 ± .76	13.60 ± .77	372 ± 35
61241,7-2 (≤ 2 g/cm^2)			A	0.7	78.4	1.370 ± .060	17.48 ± .84	14.86 ± .84	405 ± 35
61221,21-1	3.41	450	A	0.5	76.4	0.905 ± .048	11.85 ± .67	9.26 ± .67	252 ± 25
61221,21-2 (~ 50 g/cm^2)			B	0.3	71.7	0.824 ± .048	11.49 ± .71	9.04 ± .72	247 ± 27

Table 1. (*Continued*).

Samples (depth)	Sample description Fe^b (%)	Mn^c (ppm)	I.P.^{d}	Sample counted after irradiation Fe (μg)	Mn^c (μg)	^{54}Mn (cpm)	^{54}Mn (cpm/ mg Mn)	^{54}Mn (cpm/ mg Mn) after (n, $2n$), (n, p) corrections	^{53}Mn^e (dpm/ kg Fe)
permanently shadowed soil									
63321,2-1	3.62	487	A	0.2	76.8	1.191 ± .049	15.50 ± .71	12.96 ± .71	359 ± 31
63321,2-2			B	0.5	74.4	1.131 ± .060	15.20 ± .86	12.72 ± .86	354 ± 36
Surface soil near "Shadow Rock"									
63501,14-1	3.58	484	A	0.4	80.6	1.484 ± .067	18.41 ± .91	15.84 ± .91	442 ± 39
63501,14-2			B	0.4	84.7	1.623 ± .055	19.16 ± .75	16.70 ± .76	467 ± 41
14310,152									
Ia (0–2 mm)			C	0.2	152.8	2.87 ± .10	18.81 ± .76	16.52 ± .76	518 ± 44
Ib (0–2 mm)			C	0.1	139.0	2.61 ± .10	18.76 ± .82	16.47 ± .82	517 ± 45
IIa (2–4 mm)	6.38	887	C	0.1	165.8	2.66 ± .07	16.06 ± .53	13.77 ± .54	432 ± 35
IIb (2–4 mm)			C	0.1	179.3	2.82 ± .11	15.75 ± .69	13.46 ± .69	422 ± 37
IIIa (4–8 mm)			C	0.4	267	3.75 ± .15	14.06 ± .63	11.76 ± .63	369 ± 33
IIIb (4–8 mm)			C	0.2	267	3.80 ± .12	14.23 ± .53	11.94 ± .53	375 ± 31
Reference samples									
14321 a	10.5	1280	B	0.3	235	2.54 ± .10	10.82 ± .48	8.40 ± .49	212 ± 20
14321 b (40 g/cm^2)			B	0.2	115.6	1.303 ± .062	11.27 ± .58	8.85 ± .59	223 ± 22
15031 a	12.0	1560	B	0.4	231	2.57 ± .12	11.12 ± .57	8.70 ± .58	234 ± 24
15031 b ($\sim$ 60 g/cm^2)			B	0.4	123.2	1.348 ± .059	10.94 ± .53	8.50 ± .54	229 ± 23
Standard terrestrial basalt									
JB-1	6.37	1135	B	0.2	111.8	0.246 ± .019	2.20 ± .18	− 0.22 ± .20	
Standard terrestrial granodiorite									
JB-1	1.47	487	A	0.3	72.7	0.174 ± .022	2.40 ± .31	− 0.16 ± .32	

Standards	^{53}Mn(dpm)	Mn(μg)		Fe (μg)	Mn^c (μg)	^{54}Mn (cpm)	^{54}Mn (cpm/ mg Mn)	after (n, $2n$), (n, p) corrections	cpm ^{54}Mn^g/ dpm ^{53}Mn
^{53}Mn-1	0.00804	144	B	0.1	123.1	3.63 ± .12	29.5 ± 1.2	27.1 ± 1.2	483 ± 32
^{53}Mn-2	0.0110	196	A	0.2	164	4.88 ± .09	29.7 ± 0.8 ⎫	27.2 ± 0.7	485 ± 27
^{53}Mn-4	0.00643	115	A	0.1	87.6	2.60 ± .08	29.7 ± 1.1 ⎭		
^{53}Mn-3	0.0195	348	C	0.2	320	8.68 ± .21	27.1 ± 0.9	24.8 ± 0.9	443 ± 27
Mn-1		2785	B	0.2	2430	5.84 ± .18	2.40 ± .09 cpm ^{54}Mn/mg Mn		
Mn-2		1932	C	0.1	1685	3.84 ± .10	2.28 ± .08 cpm ^{54}Mn/mg Mn		
Mn-3		1600	A	0.1	1341	3.36 ± .09	2.51 ± .08 cpm ^{54}Mn/mg Mn		

Standards	Fe (mg)	Mn carrier(μg)		Fe (μg)		^{54}Mn (cpm)			
Fe-1	1.497	474	A	410		15.3 ± 1.1	0.012 ± 0.001 cpm ^{54}Mn/μg Fe		
Fe-2	1.682	448	C	383		16.3 ± 1.2	0.011 ± 0.001 cpm ^{54}Mn/μg Fe		

^{a}November 19, 1973 is an arbitrary date about $1\frac{1}{2}$ months after the end of the irradiation.

^{b}3% uncertainty.

^{c}2% uncertainty.

^{d}Irradiation position of the sample (see text).

^{e}Errors are quadratically added; a 2 σ counting error, a 3% and 2% error for the Fe and Mn determinations respectively, and the error given for the amplification factor (cpm ^{54}Mn/dpm ^{53}Mn). The amplification factor error includes a ± 5% (systematic) error for the original ^{53}Mn standard.

^{f}Depths given for the Apollo 16 core samples are those of Russ (1973) assuming a 180 g loss.

^{g}Excluding the 5% error for the original standard would decrease the errors in these amplification factors to 21, 12, and 15 for positions B, A, and C, respectively.

depths would decrease by 55.1 g/cm². The total region of the core sampled in each of the seven cases is about 1 g/cm² thick. (M. B. Duke, personal communication)

The next two samples listed are from a trench dug on the rim of Plum Crater. 61241,7 is the gray surface soil which extended to a depth of about 1 cm and 61221,21 is the lighter colored material from about 30–35 cm depth (ALGIT, 1972). It is difficult to assign errors to these depths.

Sample 63321,2 is a supposedly permanently shadowed soil from a cavity under the large "Shadow Rock" at Station 13. Sample 63501,14 is a surface soil taken about 10 m to the west of the boulder and is used here as a reference for the activity in the shadowed soil (Muehlberger *et al.*, 1972).

The sample of rock 14310 is a 6 mm by 15 mm slab extending from the surface to a depth of about 8 mm. This slab was divided into three depth fractions (0–2, 2–4, and 4–8 mm) by the grinding technique of Finkel *et al.* (1971). The uncertainty in these depths is estimated to be 0.3 mm. The slab is from the top surface of the rock about 1 cm to the "south" (LRL notation) of the sample used by the Walker group for track studies. Figure 1 in the paper by Yuhas *et al.* (1972) shows the location of the Walker sample.

Chemical procedures and counting

The samples, generally about 200 mg, were analyzed for ^{53}Mn activity by neutron activation using the reaction ^{53}Mn$(n, \gamma)^{54}$Mn (Millard, 1965). The chemical and irradiation procedures used in this work have been described by Imamura *et al.* (1973).

The samples were irradiated in the isotope tray of the Argonne CP-5 reactor for 104 days for a total flux of thermal neutrons of 5.4×10^{19} n/cm² as estimated from the reaction ^{59}Co$(n, \gamma)^{60}$Co, $\sigma = 37$b. The samples and standards were divided into three groups A, B, and C depending on their positions in the quartz ampule during the irradiation. There was a small difference in the irradiation of the standards in the three sections; therefore the results for the samples in each group were calculated using the standard irradiated with that group. Two terrestrial rock samples and two previously measured lunar samples (Imamura *et al.*, 1973 and Wahlen *et al.*, 1972) were included in the irradiation as additional checks on the consistency of the results. The activities obtained for the lunar samples agree within the errors with the previous measurements.

Following the post-irradiation chemistry an aliquot was taken to determine the final chemical yield of Mn. The purified MnCl₂ was dried under a heat lamp on to the bottom of a small plastic vial for counting. The ^{54}Mn activity was determined using a 100 cc Ge(Li) detector and a Nuclear Data 2200 multichannel analyzer. The detector head was shielded with 2 cm of oxygen-free Cu and the assembly was operated in a horizontal position inside a 15 cm thick Pb shield (10 cm on the bottom). The efficiency for the ^{54}Mn $(t_{1/2} = 312$ d) peak was 4% (835 ± 3 keV) and the background for the peak was 0.07 cpm.

RESULTS

The detailed results for the Apollo 16 and 14310 samples are presented in Table 1. The amplification factor for this irradiation varied from 485 ± 27 to 443 ± 27 cpm ^{54}Mn/dpm ^{53}Mn depending on the position of the sample during irradiation. This factor is larger than the one obtained for the Apollo 15 samples by Imamura *et al.* (1973) because of the longer irradiation time which was needed due to the lower Fe content of the samples presently being studied. The contribution of the ^{54}Fe$(n, p)^{54}$Mn reaction to the total ^{54}Mn activity was less than 0.7%. The ^{55}Mn$(n, 2n)^{54}$Mn contribution ranged from 13% to 55%. The cosmic ray produced ^{54}Mn present in these samples contributes a negligible activity. The errors given in Table 1 for the ^{54}Mn activities are 2 σ counting errors only. The errors in the ^{53}Mn activities given in the last column contain all known sources of error (see footnote e to Table 1).

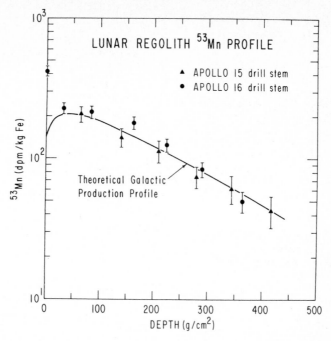

Fig. 1. ^{53}Mn activity depth profile in the lunar regolith. The data for the Apollo 15 core are from Imamura *et al.* (1973). The error bars on the experimental points include all known sources of error. The line is a theoretical estimate of the ^{53}Mn activity due to GCR only (see text). It is normalized to the deepest experimental point in the Apollo 15 core.

The ^{53}Mn activities in Table 1 and in Figs. 1 and 2 are given in units of dpm/kg Fe. Because Fe is the only significant target element for the production of ^{53}Mn on the moon, the use of these units allows direct comparisons to be made between samples of different chemistries. The chemical compositions of the Apollo 16 core samples are seen to be more variable than those obtained by Imamura *et al.* (1973) for the Apollo 15 drill core.

DISCUSSION

Apollo 16 soil samples

The ^{53}Mn results for the Apollo 16 drill core are displayed graphically in Fig. 1. Also plotted in the figure are the points for the Apollo 15 drill core. The theoretical GCR production profile in the figure is the same line as was given in the earlier paper by Imamura *et al.* (1973) and the curve is normalized to the same Apollo 15 point. This curve was calculated using the model of Reedy and Arnold (1972). The agreement between the Apollo 15 and 16 data and between the experimental points and the theoretical profile indicates that the cores have experienced similar bombardment histories for the last 5 m.y. and that the Reedy–Arnold model is a good estimate of the GCR production profile for both of them. The profile for the

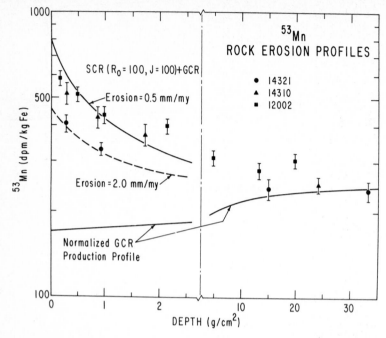

Fig. 2. Measured ^{53}Mn activity depth profile in rocks 12002, 14310, and 14321. Data for 12002, 14321 and the deep sample of 14310 are from Wahlen *et al.* (1973). The experimental points are plotted at the average depths for the samples. The lower line is the theoretical GCR activity profile normalized to go through the deepest points in rocks 14310 and 14321. The upper curves represent a sum of the lower GCR line and a theoretical estimate of the SCR production. J is the 4π integral flux above 10 MeV in units of protons/cm^2 sec.

four deep Apollo 16 samples might be somewhat steeper than that obtained from the Apollo 15 core samples, but the uncertainties in the activities preclude a definite conclusion.

The SCR contribution to the ^{53}Mn activity in near-surface samples has been discussed in detail by Wahlen *et al.* (1972). It is interesting to compare the surface sample from the Apollo 16 core with the surface values obtained for other lunar samples. Figure 2, which will be discussed in the next section, presents the experimental ^{53}Mn activity profiles for three lunar rocks. Core sample 60007,105 was obtained from about the same depth as the third sample of 12002 (~ 1 g/cm^2) and the activities almost overlap.

The ^{53}Mn profiles for the Apollo 15 and 16 drill cores are quite similar indicating that the same conclusions with regard to mixing times and depths should apply to both collection sites. Imamura *et al.* (1973) found the Apollo 15 core to have been unmixed on a meter scale over the time interval accessible to ^{53}Mn investigations (the past 5 m.y. or so). This conclusion is in accord with other core studies—the Gd isotope work of Russ *et al.* (1972) and Russ (1973) and the rare gas measurements made by Bogard *et al.* (1973). These studies using stable nuclides investigate much longer time scales than are possible using ^{53}Mn. They set an upper limit for the time since the last mixing of the upper 100 g/cm^2 for the individual cores. The ^{53}Mn results indicate no disturbance for the last 5 m.y. and give a lower limit for the time of regolith mixing at the locations sampled.

It is not surprising that the ^{53}Mn results show the cores to have been unmixed on a meter depth scale. The theoretical calculations of Gault *et al.* (1974) indicate that impact processes would not be expected to mix the lunar regolith to more than a depth of 1 or 2 cm on the time scale of around 5 m.y. Monte Carlo calculations by J. R. Arnold (unpublished) indicate a somewhat greater mean gardening depth, about 3–5 cm, on this time scale. However, both models are consistent with the observations. Samples from shallow depths in cores should show evidence of this mixing.

The ^{53}Mn activities for the trench soils 61241 and 61221 (Table 1) can be compared with the core results in Fig. 1 and are seen to agree fairly well with the experimental points for the cores. The surface and deep trench samples are quite distinct in appearance and most probably originate from different events. The main conclusion to be drawn from the trench data is that, since both layers seem to have the saturated ^{53}Mn activities expected for samples at their present depths, the events which deposited them must both have occurred more than 5 m.y. ago. The top layer has been interpreted by Gibson and Moore (1973) as being ejecta from South Ray Crater. The ^{53}Mn results indicate that this sample cannot be pure South Ray material since the age of that crater appears to be only 2 m.y. (Behrmann *et al.*, 1973; Lightner and Marti, 1974). This conclusion is consistent with the one reached by Walton *et al.* (1973) who obtained ^{21}Ne exposure ages of 250 m.y. for both the top and the bottom of the trench.

The trench is located on the rim of Plum Crater and it seems reasonable to speculate that the surface layer was deposited during its formation. If it was, then Plum Crater must be > 5 m.y. old.

When the "shadowed" soil sample 63321,2 was collected on the moon it was thought that this soil had been completely shielded from SCR bombardment due to its location in a cavity under a large rock. However, the ^{53}Mn activity for this sample is considerably higher than the GCR activity expected for a sample at a depth estimated from the size of the overlying boulder. If the ^{53}Mn in 63321 was all produced when this sample was in its shadowed location the cavity must have received about 75% of the SCR flux seen by the nearby exposed surface soil 63501,14. The shadowed soil undoubtedly saw some fraction of the sky, but 75% seems too high to reconcile with the observations made by the astronauts and with the photographic documentation (ALGIT, 1972).

The high ^{53}Mn activity in 63321 could be due to lateral transport of freshly irradiated material into the area sampled, either by nearby impacts or by electrostatic effects, as described by Gold and Williams (1974). The difficulty with this interpretation is that the activity in the shadowed soil is quite high and would require the surface material to be laterally well-mixed on a time scale of only several million years.

The most likely possibility is that Shadow Rock was emplaced in its present location a short time ago. An upper limit for the time the rock began shadowing the soil can be calculated, for the case of perfect SCR shielding assuming an initial ^{53}Mn concentration for the soil and assuming an average depth for the soil with regard to the overlying mass of the rock. If the average depth of rock over the sample is taken to be 1 m or 300 g/cm^2 and the original activity is assumed to be equal to the activity in the surface sample of rock 12002, 583 dpm/kg Fe (the highest experimental value obtained by this group), then the length of time the rock has been in its present location cannot be greater than one ^{53}Mn half-life or 3.7 m.y. Another approach is to use the activity for the nearby soil 63501 as the initial activity in the shadowed soil. Using this lower value reduces the derived age of the boulder to 2 m.y. Clark and Keith (1973) have measured ^{26}Al ($t_{1/2} = 7.4 \times 10^5$ yr) in chips from Shadow Rock and find values which are quite ordinary and indicate saturation. Thus the rock could not have a surface exposure time of much less than 2 m.y. It is interesting that 2 m.y. is the estimated age of South Ray Crater. It may be that Shadow Rock originated in that event and that the soil from the cavity may have been permanently shadowed since that time. However, the ^{21}Ne exposure age of the rock, 25 m.y., measured by Bogard *et al.* (1973) indicates that South Ray is an unlikely source for the boulder.

Rock 14310

The ^{53}Mn results for 14310 are plotted in Fig. 2. The experimental profiles for 12002 and 14321 are presented for comparison. Rocks 14310 and 12002 are basaltic rocks with fairly constant compositions and densities. For the figure the densities of 12002 and 14310 have been taken to be 3.3 and 2.9 g/cm^2, respectively. Rock 14321 is a complex fragmental rock. Wahlen *et al.* (1972) have used a density of 3.09 g/cm^2 for 14321, but the value seems high compared to 2.4 g/cm^2 measured by Chung *et al.* (1972). The density used in this work is 2.65 g/cm^2, which was calculated from the total weight of the rock, 8996 g, and the volume of a plaster model. The 14321 depths are plotted assuming no attrition, i.e. surface loss after collection. The attrition estimated by Wahlen *et al.* (1972) for this rock, 0.1 g/cm^2, would not significantly change the conclusions. The GCR production profile at the bottom of the figure is the same as that in Fig. 1 except that the normalization has been raised by 15% over that in Fig. 1 allowing the line to go through the experimental points for the deep samples of 14310 and 14321.

Samples from a rock have quite different cosmic ray irradiation geometries than samples from comparable depths in a core. The net effect of rock geometry for a vertical column, down to a depth of ~ 50 g/cm^2 would be to increase the activities in the deeper rock samples relative to core samples from the same depths and thus to flatten the observed rock profile. The flattening would be expected to be more noticeable for smaller rocks. 12002 is the smallest rock in Fig. 2 and 14321 is the largest. The geometry of a sample from an irregular rock such as 14321 is very difficult to interpret and some of the differences in the experimental profiles for the three rocks might be due to differences in their geometries.

For depths down to about 3 g/cm^2 the GCR production profile is substantially constant, while the SCR production is a steeply falling function of depth. Wahlen *et al.* (1972) have discussed in detail the SCR ^{53}Mn profile with reference to the 12002 and 14321 samples. The upper curves in Fig. 2 represent the sum of the lower renormalized GCR profile and the Wahlen *et al.* (1972) theoretical estimates for the SCR production rates assuming saturation and an infinite plane geometry. These two curves agree fairly well with the experimental points on the assumption that all three rocks are saturated, have experienced similar cosmic ray irradiations and have undergone various erosion rates, 0.5 mm/m.y. for 14310 and 12002, 2.0 mm/m.y. for 14321. The friable rock 14321 is expected to have a higher erosion rate than the others, but according to their observed physical properties, 12002 and 14310 should have similar erosion rates. There are many parameters involved in the theoretical SCR calculation and other curves can be obtained which fit the data for a particular rock. However, because the profiles have been measured for three rocks, there are additional constraints on the possible combinations of theoretical parameters. In particular, it seems reasonable to require the same hardness and flux for SCR irradiations of all three rocks at least as a first approximation.

Total exposure ages obtained from spallation rare gases and GCR tracks for 12002 (94 ± 6 m.y., Marti and Lugmair, 1971; 55 m.y., Fleischer *et al.*, 1971), 14310 (259 ± 7 m.y., Lugmair and Marti, 1972; 300 m.y., Burnett *et al.*, 1972) and 14321 (27 m.y., Lugmair and Marti, 1972; 24 ± 2 m.y., Burnett *et al.*, 1972) are all quite long compared to the half-life of ^{53}Mn. However, these exposure times give little information about detailed exposure history. Estimates of surface dwell-times are more useful for interpreting the ^{53}Mn profiles. Alexander (1971) in an extensive study of rare gases in 12002, estimated a surface exposure time of 2–5 m.y. in agreement with the 4.4 ± 2.2 m.y. ^{3}He–^{3}H surface exposure age of D'Amico *et al.* (1971) and the 2.2 m.y. suntan age of Bhandari *et al.* (1972). 14310 has been extensively studied by track workers and a surface exposure age of 1.5–3 m.y. (Yuhas *et al.*, 1972; Crozaz *et al.*, 1972) is generally accepted. Rancitelli *et al.* (1972) measured ^{26}Al–^{22}Na and found a saturated value for ^{26}Al ($t_{1/2} = 7.4 \times 10^5$ yr) in 14310, indicating a surface exposure of at least 2–3 m.y. Bhandari *et al.* (1972) obtained a suntan age for 14321 of 4.2 m.y. and Hart *et al.* (1972) calculated a maximum surface exposure time of 8.2 m.y. for this rock.

An attempt was made, for the case of a short exposure time, to obtain theoretical ^{53}Mn profiles which would agree with the experimental points for the rocks. Using our model, an R_0 of 100 MV and an exposure time of 2 m.y. agreement can be obtained only if 12002 and 14310 have experienced an erosion rate >2 mm/m.y. and a flux of solar protons >200 protons/cm^2 sec (4π) (>10 MeV). In light of earlier studies (Wahlen *et al.*, 1972; Finkel *et al.*, 1971), we do not find such an erosion rate or flux satisfactory. The lack of agreement with the track results is not understood at the present time.

Acknowledgments—The authors are grateful to R. M. Walker for suggesting the 14310 ^{53}Mn measurements and to G. P. Russ, III for valuable discussions. Norman Fong and Florence Kirchner furnished essential support to the work. This work was supported by NASA grant NGL 05–009–148.

References

Alexander E. C., Jr. (1971) Spallogenic Ne, Kr, and Xe from a depth study of 12002. *Proc. Second Lunar Sci. Conf., Geochim. Cosmochim. Acta*, Suppl. 2, Vol. 2, pp. 1643–1650. MIT Press.

(ALGIT) Apollo Lunar Geology Investigation Team (1972) Documentation and environment of the Apollo 16 sample, A preliminary report. *Interagency Report: Astrogeology* **51**, U.S. Geological Survey.

Behrmann C., Crozaz G., Drozd R., Hohenberg C., Ralston C., Walker R., and Yuhas D. (1973) Cosmic-ray exposure history of North Ray and South Ray material. *Proc. Fourth Lunar Sci. Conf., Geochim. Cosmochim. Acta*, Suppl. 4, Vol. 2, pp. 1957–1974. Pergamon.

Bhandari N., Goswami J. N., Gupta S. K., Lal D., Tamhane A. S., and Venkatavaradan V. S. (1972) Collision controlled radiation history of the lunar regolith. *Proc. Third Lunar Sci. Conf., Geochim. Cosmochim. Acta*, Suppl. 3, Vol. 3, pp. 2811–2829. MIT Press.

Bhandari N., Goswami J., and Lal D. (1973) Surface irradiation and evolution of the lunar regolith. *Proc. Fourth Lunar Sci. Conf., Geochim. Cosmochim. Acta*, Suppl. 4, Vol. 3, pp. 2275–2290. Pergamon.

Bogard D. D., Nyquist L. E., Hirsch W. C., and Moore D. R. (1973) Trapped solar and cosmogenic noble gas abundances in Apollo 15 and 16 deep drill samples. *Earth Planet. Sci. Lett.* **21**, 52–69.

Burnett D. S., Huneke J. C., Podosek F. A., Russ G. P., III, Turner G., and Wasserburg G. J. (1972) The irradiation history of lunar samples (abstract). In *Lunar Science—III*, pp. 105–107. The Lunar Science Institute, Houston.

Chung D. H., Westphal W. B., and Olhoeft G. R. (1972) Dielectric properties of Apollo 14 lunar samples. *Proc. Third Lunar Sci. Conf., Geochim. Cosmochim. Acta*, Suppl. 3, Vol. 3, pp. 3161–3172. MIT Press.

Clark R. S. and Keith J. E. (1973) Determination of natural and cosmic ray induced radionuclides in Apollo 16 lunar samples. *Proc. Fourth Lunar Sci. Conf., Geochim. Cosmochim. Acta*, Suppl. 4, Vol. 2, pp. 2105–2113. Pergamon.

Crozaz G., Drozd R., Hohenberg C. M., Hoyt H. P., Jr., Ragan D., Walker R. M., and Yuhas D. (1972) Solar flare and galactic cosmic ray studies of Apollo 14 and 15 samples. *Proc. Third Lunar Sci. Conf., Geochim. Cosmochim. Acta*, Suppl. 3, Vol. 3, pp. 2917–2931. MIT Press.

D'Amico J., DeFelice J., Fireman E. L., Jones C., and Spannagel G. (1971) Tritium and argon radioactivities and their depth variations in Apollo 12 samples. *Proc. Second Lunar Sci. Conf., Geochim. Cosmochim. Acta*, Suppl. 2, Vol. 2, pp. 1825–1839. MIT Press.

Finkel R. C., Arnold J. R., Imamura M., Reedy R. C., Fruchter J. S., Loosli H. H., Evans J. C., and Delany A. C. (1971) Depth variation of cosmogenic nuclides in a lunar surface rock and lunar soil. *Proc. Second Lunar Sci. Conf., Geochim. Cosmochim. Acta*, Suppl. 2, Vol. 2, pp. 1773–1789. MIT Press.

Fleischer R. L., Hart H. R., Jr., Comstock G. M., and Evwaraye A. O. (1971) The particle track record of the Ocean of Storms. *Proc. Second Lunar Sci. Conf., Geochim. Cosmochim. Acta*, Suppl. 2, Vol. 3, pp. 2559–2568. MIT Press.

Gault D. E., Hörz F., Brownlee D. E., and Hartung J. B. (1974) Mixing of the lunar regolith (abstract). In *Lunar Science—V*, pp. 260–262. The Lunar Science Institute, Houston.

Gibson E. K., Jr. and Moore G. W. (1973) Volatile-rich lunar soil: Evidence of possible cometary impact. *Science* **179**, 69–71.

Gold T. and Williams G. J. (1974) Electrostatic effects on the lunar surface (abstract). In *Lunar Science—V*, pp. 273–274. The Lunar Science Institute, Houston.

Hart H. R., Jr., Comstock G. M., and Fleischer R. L. (1972) The particle track record at Fra Mauro. *Proc. Third Lunar Sci. Conf., Geochim. Cosmochim. Acta*, Suppl. 3, Vol. 3, pp. 2831–2844. MIT Press.

Hübner W., Heymann D., and Kirsten T. (1973) Inert gas stratigraphy of Apollo 15 drill core sections 15001 and 15003. *Proc. Fourth Lunar Sci. Conf., Geochim. Cosmochim. Acta*, Suppl. 4, Vol. 2, pp. 2021–2036. Pergamon.

Imamura M., Finkel R. C., and Wahlen M. (1973) Depth profile of ^{53}Mn in the lunar surface. *Earth Planet. Sci. Lett.* **20**, 107–112.

Lightner B. D. and Marti K. (1974) Lunar trapped xenon (abstract). In *Lunar Science—V*, pp. 447–449. The Lunar Science Institute, Houston.

Lugmair G. W. and Marti K. (1972) Exposure ages and neutron capture record in lunar samples from Fra Mauro. *Proc. Third Lunar Sci. Conf., Geochim. Cosmochim. Acta,* Suppl. 3, Vol. 2, pp. 1891–1897. MIT Press.

Marti K. and Lugmair G. W. (1971) ^{81}Kr–Kr and K–^{40}Ar ages, cosmic-ray spallation products, and neutron effects in lunar samples from Oceanus Procellarum. *Proc. Second Lunar Sci. Conf., Geochim. Cosmochim. Acta,* Suppl. 2, Vol. 2, pp. 1591–1605. MIT Press.

Millard H. T. (1965) Thermal neutron activation: Measurement of cross section for Manganese-53. *Science* 147, 503–504.

Muehlberger W. R., Batson R. M., Boudette E. L., Duke C. M., Eggleton R. E., Elston D. P., England A. W., Freeman V. L., Hait M. H., Hall T. A., Head J. W., Hodges C. A., Holt H. E., Jackson E. D., Jordan J. A., Larson K. B., Milton D. J., Reed V. S., Rennilson J. J., Schaber G. G., Schafer J. P., Silver L. T., Stuart-Alexander D., Sutton R. L., Swann G. A., Tyner R. L., Ulrich G. E., Wilshire H. G., Wolfe E. W., and Young J. W. (1972) Preliminary geologic investigation of the Apollo 16 landing site. *Apollo 16 Preliminary Science Report,* NASA SP-315, 6, 1–81.

Rancitelli L. A., Perkins R. W., Felix W. D., and Wogman N. A. (1972) Lunar surface processes and cosmic ray characterization from Apollo 12–15 lunar sample analyses. *Proc. Third Lunar Sci. Conf., Geochim. Cosmochim. Acta,* Suppl. 3, Vol. 2, pp. 1681–1691. MIT Press.

Reedy R. C. and Arnold J. R. (1972) Interaction of solar and galactic cosmic-ray particles with the moon. *J. Geophys. Res.* 77, 537–555.

Russ G. P., III (1973) Apollo 16 neutron stratigraphy. *Earth Planet. Sci. Lett.* 19, 275–289.

Russ G. P., III, Burnett D. S., and Wasserburg G. J. (1972) Lunar neutron stratigraphy. *Earth Planet. Sci. Lett.* 15, 172–186.

Wahlen M., Honda M., Imamura M., Fruchter J. S., Finkel R. C., Kohl C. P., Arnold J. R., and Reedy R. C. (1972) Cosmogenic nuclides in football-sized rocks. *Proc. Third Lunar Sci. Conf., Geochim. Cosmochim. Acta,* Suppl. 3, Vol. 2, pp. 1719–1732. MIT Press.

Walton J. R., Lakatos S., and Heymann D. (1973) Distribution of inert gases in fines from the Cayley–Descartes region. *Proc. Fourth Lunar Sci. Conf., Geochim. Cosmochim. Acta,* Suppl. 4, Vol. 2, pp. 2079–2095. Pergamon.

Yuhas D. E., Walker R. M., Reeves H., Poupeau G., Pellas P., Lorin J. C., Chetrit G. C., Berdot J. L., Price P. B., Hutcheon I. D., Hart H. R., Jr., Fleischer R. L., Comstock G. M., Lal D., Goswami J. N., and Bhandari N. (1972) Track consortium report on rock 14310. *Proc. Third Lunar Sci. Conf., Geochim. Cosmochim. Acta,* Suppl. 3, Vol. 3, pp. 2941–2947. MIT Press.

Proceedings of the Fifth Lunar Conference
(Supplement 5, Geochimica et Cosmochimica Acta)
Vol. 2 pp. 2105–2119 (1974)
Printed in the United States of America

The saturated activity of ^{26}Al in lunar samples as a function of chemical composition and the exposure ages of some lunar samples

J. E. KEITH and R. S. CLARK

NASA Johnson Space Center, Houston, Texas 77058

"You haven't told me yet," said Lady Nuttal, "what it is your fiancé does for a living."

"He's a statistician," replied Lamia, with an annoying sense of being on the defensive.

Lady Nuttal was obviously taken aback. It had not occurred to her that statisticians entered into normal social relationships. The species, she would have surmised, was perpetuated in some collateral manner, like mules.

"But Aunt Sara, it's a very interesting profession," said Lamia warmly.

"I don't doubt it," said her aunt, who obviously doubted it very much. "To express anything important in mere figures is so plainly impossible that there must be endless scope for well-paid advice on how to do it. But don't you think that life with a statistician would be rather, shall we say, humdrum?"

Lamia was silent. She felt reluctant to discuss the surprising depth of emotional possibility which she had discovered below Edward's numerical veneer.

"It's not the figures themselves," she said finally, "it's what you do with them that matters."

K. A. C. Manderville, *The Undoing of Lamia Gurdleneck.*
Preface to Volume 2, *The Advanced Theory of Statistics.* M. G. Kendall and A. Stuart, Hafner, N.Y. (1961).

Quoted with the permission of the author.

Abstract—The saturated specific activity of ^{26}Al in lunar samples is shown to be a function of the silica, alumina, and magnesia contents as well as the reciprocal of the cube root of the sample weight, by linear multivariate regression. The sample weight term is introduced to remove the dependence of the specific activity on the surface to volume ratio of the sample. Since only whole lunar rocks were considered, the resulting expression is only applicable to such samples. For those 28 samples whose activities and compositions are known, 21 are shown to be saturated and all those in the set previously thought to be unsaturated—12034, 12064, and 14301—are shown to be unsaturated by internal consistency arguments. In addition, 10003, 15265, 15415, and marginally, 60315 are shown to be unsaturated. The expression representing the specific activity of the saturated set is:

$$^{26}\text{Al(dpm/kg)} = 0.652[\text{SiO}_2] + 2.50[\text{Al}_2\text{O}_3] + 0.560[\text{Al}_2\text{O}_3]w^{-1/3} + 2.28[\text{MgO}]w^{-1/3}$$

where $[\text{M}_x\text{O}_y]$ is the weight percent oxide as percent and w is the sample weight in kilograms. The residuals from the regression are examined and found to be well-behaved. The exposure ages of the seven unsaturated samples are calculated and the probability of picking up an unsaturated lunar rock from the lunar surface is shown to be roughly 20%.

Introduction

Since Apollo 12, some lunar samples have been described as unsaturated in [26]Al. The identification of these samples was always somewhat uncertain since no systematic method had been evolved, and usually involved comparison of the observed [26]Al activity with that predicted for meteorites, with the observed [22]Na activity, or with calculations based on some model of thick-target reactions. Nevertheless, there is an extensive list of samples thought to be unsaturated. O'Kelley *et al.* (1971, 1972) have found that 12054, 12062, and 12064, as well as 15475 and 15495 show evidence of low exposure. Rancitelli *et al.* (1972) have shown that the boulder from which 15205 and 15206 were taken was on the surface for a time comparable to the half-life of [26]Al, and that the soil breccia 15501,2 was also unsaturated. In addition, Keith *et al.* (1972) have found that 14045, 15085, 15086, 15426, and 15431 had lower [26]Al specific activities than might be expected at saturation.

Samples unsaturated in [26]Al are of interest to those doing other lunar studies, such as track counting, rare gas mass spectroscopy, pit counting, etc., and the distribution of unsaturated samples itself should yield information about the recent gardening of the lunar regolith. In this work we attempt to find the dependence of the saturation activity of [26]Al in lunar samples on chemical composition, as others have for meteorites (see Fuse and Anders, 1969; Heyman and Anders, 1967; and Cressy, 1971).

Method

We will attempt to elucidate the conditions of irradiation of various lunar samples by removing the variation in their specific activities due to variations in their chemical composition and other factors that may suggest themselves, by multivariate linear regression, and allow only the conditions of irradiation and the errors in the measurements of the specific activities to vary randomly. In this paper we shall define "saturated in [26]Al" to mean the attainment of secular equilibrium in the [26]Al decay rate as the result of having been on the lunar surface, exposed to both solar flare accelerated protons and cosmic rays for more than the last few million years. We recognize at the outset that the model must take into account the strong dependence on depth (Rancitelli *et al.*, 1971) as a result of the solar proton bombardment of sample surfaces. If this "surface effect" is to be allowed to vary randomly then only samples can be selected whose cosmogenic radionuclides were determined before a significant man-made alteration of the original shape took place. In addition, only those samples whose shape could be expected to remain essentially constant over the last few million years can be included. This consideration would exclude fines, since Rancitelli *et al.* (1971) have shown that extensive gardening of the lunar surface takes place over times comparable to an [26]Al half-life. Consequently, the literature has been searched to find all those lunar samples that meet the following three conditions:

(1) That the [26]Al specific activities were determined on the intact rock sample, or at most one from which only a small chip had been taken.

(2) That the sample be sufficiently coherent as to make the assumption of constant shape reasonable.

(3) That the chemical analysis for major elements be known. As a result of the search, 28 such cases were found and are shown in Table 1.

We will use the standard model for multivariate linear regression:

$$y_i = \Sigma_j \beta_j x_{ij} + \epsilon_i$$

where y_i is the observed [26]Al specific activity of the ith sample ($i = 1, 2, 3, \ldots, N$).

Table 1. Data used in regressions in this paper.

Lunar Sample	Weight (g)	^{26}Al (dpm/kg)	$\pm s$	SiO$_2$ (%)	Al$_2$O$_3$ (%)	MgO (%)	Reference
10003	213	74	8	38.54	10.32	7.34	a, b
10017	971	73	8	42.01	7.80	7.74	a, b
10018	211.5	108	16	42.57	12.49	8.18	a, b
10019	234	101	15	40.64	12.81	7.37	a, b
10021	157	110	15	42.84	12.83	8.29	a, b
10057	897	75	8	41.66	8.65	6.93	a, b
10072	399	73	8	40.64	8.01	7.48	a, b
12002	1529	75	6	43.56	7.87	14.88	c, b
12004	502	89.8	5.3	44.3	8.5	10.9	c + q, e
12010	288.7	83	19	46.27	10.04	8.29	f, g
12013	80	115	16	61	12	6	c, i
12034	155	45	5	45.8	15.7	9.6	c, e
12053	879	81	12	46.21	10.14	8.17	c, d
12063	2360	78	2	43.48	9.27	9.56	h, d
12064	1214	51	5	46.41	10.50	6.38	c, i
12065	2088	82	2	47.1	9.8	9.2	h, j
12073	405	110	10	41	15	11	c, i
14053	251.3	101	4	46.4	13.6	8.48	f, k
14066	510	103	6	51	15	9.5	f, l
14301	1370	62	18	47.6	15.9	10.4	f, k
15256	201	97	6	44.93	8.89	9.08	f, m
15265	314.2	72	8	46.94	16.71	9.95	f, m
15415	269.4	116	9	44.08	35.49	0.09	f, m
15418	1127.5	120	5	44.97	26.73	5.38	f, m
15558	1333.3	84	5	46.31	12.40	10.51	f, m
60315,0	787.7	92	8	45.61	17.18	13.15	n, o
60335,0	311.0	140	8	46.19	25.27	8.14	p, o
62235,0	317.7	137	8	47.04	18.69	10.14	p, o
$\bar{x}$	692.0	90.99		45.22	13.84	8.65	
S_x	608.3	23.40		4.11	6.39	2.64	

The first reference in the right hand column refers to the radiochemical measurement, and the second to the chemical analysis. References: (a) O'Kelley *et al.* (1970); (b) Warner (1971); (c) O'Kelley *et al.* (1971); (d) Willis *et al.* (1971); (e) Wahita and Schmitt (1971); (f) Keith *et al.* (1972); (g) Compston *et al.* (1971); (h) Rancitelli *et al.* (1971); (i) LSPET (1970); (j) Smales *et al.* (1971); (k) Hubbard *et al.* (1972); (l) LSPET (1971); (m) LSPET (1972); (n) Eldridge *et al.* (1973); (o) LSPET (1973); (p) Clark and Keith (1973); (q) Wrigley (1971). Where more than one reference appears for a measurement, the weighted average of the measurements found in the references, and its standard deviation, are listed.

x_{ij} is the jth independent variable (SiO$_2$ content, etc.) of the ith sample ($j = 1, 2, 3, \ldots, p$).

ϵ_i is the true random error and is normally distributed with expected mean 0 and expected variance σ^2.

$\hat{y}_i$ is the ^{26}Al specific activity predicted for the ith sample by the model under discussion. $\hat{y}_i = \Sigma_j b_j x_{ij}$.

b_j is the estimate of β_j obtained by minimizing the sum $\Sigma_i (y_i - \hat{y}_i)^2$, and is called the jth regression coefficient.

$y_i - \hat{y}_i$ is the ith residual and estimates ϵ_i, if there is no bias in the model.

$S_{y/x}^2$ is defined as $(N - p)^{-1} \Sigma_i (y_i - \hat{y}_i)^2$, where N is the number of observations and p is the number of regression coefficients in the model. $S_{y/x}$ is called the standard error of estimate, and estimates σ if there is no bias in the model.

$r(u, v)$ is defined as $\Sigma_i(u_i - \bar{u})(v_i - \bar{v})[\Sigma_i(u_i - \bar{u})^2\Sigma_i(v_i - \bar{v})^2]^{-1/2}$; $r(u, v)$ is called the coefficient of correlation of u and v and can take on values between -1 and 1.

$\mathbf{X}_k$ is the vector containing the independent variables relating to the kth sample $(x_{k1}, x_{k2}, x_{k3}, \ldots, x_{kp})$.

$\mathbf{X}$ is the $N \times p$ matrix assembled from the $\mathbf{X}_k$ vectors.

In addition to requiring that $\Sigma_i(y_i - \hat{y}_i)^2$ be a minimum, it is desirable that neither the observed values, y_i, the predicted values, $\hat{y}_i$, nor any of the independent variables, x_{ij}, exhibit a significant correlation with the residuals, $y_i - \hat{y}_i$. (It might be noted that requiring the correlation of $\hat{y}$ and $y - \hat{y}$ to be small is equivalent to requiring the average of the residuals to be small, since $\Sigma_i\hat{y}_i(y_i - \hat{y}_i) = 0$ for any regression, and the average of the observed values is not small.) We will choose to exclude the constant term from our model, i.e. if the model be written $y_i = b_0 + \Sigma_j b_j x_{ij}$ we will set $b_0 = 0$, as both Cressy (1971) and Fuse and Anders (1969) did, and for the same reason; our belief that all the contributors to the specific activity of ^{26}Al have been accounted for.

Physical intuition suggests that a proper regression should reveal two sorts of residuals; the majority clustered about zero with a variance larger than that due to the uncertainty in the measurements, since the variation in the irradiation conditions would also be included, and a smaller group lying below them, representing those samples whose irradiation was anomalously low, due to burial, shielding, ablation, etc. (Note that this expectation prevents the use of a weighted model, where the weights are determined solely by the accuracy of the determination of the dependent variable. We anticipate that the variations in ^{26}Al activity due to variations in exposure are at least as large as the uncertainties of measurement.) Anomalously high residuals would be very rare if the model were correct.

A stepwise regression program, STPREG, (Smith, 1971) was selected from the Johnson Space Center library and applied to the data in Table 1. Longley (1967) has published a study of the numerical accuracy of various linear regression programs on a wide range of computers. For one of his criteria of accuracy he took the first regression coefficient, b_1, of a specified problem. Two of the stepwise regression programs tested by Longley had errors in the value of b_1 of 82% and 340%. For the same problems, the value of b_1 obtained from STPREG is in error by 1.3×10^{-5}%. By the examination of the results of the application of STPREG to all the observations in Table 1, we will select a subset of observations whose regression fulfills the requirements set forth above, and which we find from internal consistency considerations to consist wholly of samples saturated in ^{26}Al, and then compare all doubtful samples, or those thought to be unsaturated, with the model arising from the saturated set.

RESULTS

Preliminary investigations showed that the effect of correcting the ^{26}Al activity for iron and calcium content, using either Cressy's (1971) or Fuse and Ander's (1969) results produced corrections that were small compared to the uncertainties in the measurements, and resulted in no improvement in early regressions. In addition, there was a serious theoretical objection to their inclusion; ^{26}Al could only be made from iron and calcium by high energy reactions induced by galactic cosmic rays, and the fraction of the total activity induced by them was uncertain, but surely not near one.

In Table 2 we show the results of a few of the many regressions of the observed ^{26}Al activities on the chemical compositions and other variables that we have performed, using the observations shown in Table 1. All the regressions are unweighted, and are the result of stepwise regression with the criterion for acceptance of a candidate term being that the "partial" or "sequential" F statistics be greater than or equal to one (see Draper and Smith, 1966, pp. 67–72 and 171–172).

Table 2. Results of various regressions.

Regression Number	Number of observations	$S_{y/x}$	$r(y, y - \hat{y})$	$x_1 = [SiO_2]$ $b_1 \pm s_1$	$x_2 = [Al_2O_3]$ $b_2 \pm s_2$	$x_3 = [MgO]$ $b_3 \pm s_3$	$b_4 \pm s_4$	$b_5 \pm s_5$	$b_6 \pm s_6$
1	25	18.98	0.82	1.50 ± .19	1.47 ± .62	0			
2	21	11.00	0.60	1.28 ± .29	2.00 ± .46	1.19 ± 1.03			
3	21	7.93	0.52	0.44 ± .29	2.15 ± .33	2.24 ± .78	20.36 ± 4.34		
4	19	4.88	0.19	0.23 ± .18	2.93 ± .27	1.84 ± .51	24.12 ± 3.16		
5	22	7.45	0.29	0.26 ± .25	3.02 ± .31	1.38 ± .70	25.60 ± 4.78		
6	21	5.78	0.24	0	3.28 ± .22	2.33 ± .36	26.32 ± 2.60		
7	21	5.38	0.20	1.64 ± .30	0	−1.66 ± 1.53	−0.602 ± .172	2.43 ± .20	2.86 ± 1.11
8	21	5.54	0.13	0.652 ± .141	2.50 ± .48	0		0.562 ± .330	2.28 ± .502
Fuse and Anders	34	4.86		1.45 ± .05	2.52 ± .29				
Cressy	8	1.91		1.14 ± .14	5.98 ± 1.01	0.17 ± .18			

The column headed by $S_{y/x}$ contains the standard error of estimate, and that headed by $r(y, y - \hat{y})$ contains the coefficient of correlation of the observed value of the specific activity of ^{26}Al, y, and the residuals, $y - \hat{y}$, from the model corresponding to that line. Under $b_j \pm s_j$ one finds the regression coefficients and their standard deviations relating to the corresponding independent variables, x_j, in units determined by the units of y (dpm/kg), and those of the independent variable.

All observations used to produce these coefficients are to be found in Table 1. If a zero appears in a position in the table, the stepwise regression rejected this variable ($F < 1.0$); if it is a blank, the corresponding independent variable was not included in the model. The variables x_1, x_2, and x_3 remain the same in all regressions; the identity of the others changes as indicated below:

(1) All Apollo 11, 12, 14, and 15 observations in Table 1 are included. (2) Observations pertaining to 12034, 12064, 14301, and 15265 have been omitted. (3) Same set of observations as in line 2, $x_4 = w^{-1/3}$ (w is in kg). (4) Observations pertaining to 10003 and 15415 have been removed from the set used in line 3, $x_4 = w^{-1/3}$. (5) Add the observations concerning 60315, 60335, and 62235 to the set used in line 4, $x_4 = w^{-1/3}$. (6) The observation pertaining to 60315 has been removed from the set used in line 5, $x_4 = w^{-1/3}$. (7) The same set of observations as in line 6; $x_4 = [SiO_2]w^{-1/3}$, $x_5 = [Al_2O_3]w^{-1/3}$, $x_6 = [MgO]w^{-1/3}$. (8) The same set of observations as line 6, $x_4 = [SiO_2]w^{-1/3}$ removed from the model. Fuse and Anders—A weighted (by the reciprocal of the variance of ^{26}Al measurement) regression of the ^{26}Al specific activity of various meteorites on the Si, Al, and S contents as independent variables. The contributions of Mg, Ca, and Fe + Ni were fixed at 0, 7, and 2.2 dpm/kg target element, respectively. The results shown in the table have been converted to our units. Cressy—A weighted (by the reciprocal of the variance of the ^{26}Al measurement) regression of the ^{26}Al specific activity found in various mineral-separate fractions of the Bruderheim meteorite on Si, Al, Mg, and S as the independent variables. The Ca and Fe + Ni were fixed at 24 and 2.2 dpm/kg target element, respectively. The results shown in the table have been converted to our units.

When this work was started no Apollo 16 measurements existed, and none are included in the first four lines of Table 2. In line 1 of this table we show the results of a regression of the ^{26}Al specific activity (in dpm/kg) on the major element composition with $x_1 = [SiO_2]$, $x_2 = [Al_2O_3]$, and $x_3 = [MgO]$, where the concentrations are in weight percent. The assignment of x_1, x_2, x_3 remains the same for the remainder of this discussion. Figure 1 shows the results of this regression. The appearance of this figure confirms our previously mentioned expectation about the existence of a saturated main set and a smaller set located below it. The members of the lower set are, reading from the lowest to the highest:

12034—which was dug up from a trench (LSPET, 1970);
12064—previously suggested to be unsaturated (O'Kelley et al., 1971);

14301—shown to have been heavily dust covered (R. L. Sutton, private communication, 1972);

15265—about which no previous suggestion of unsaturation had been made.

Note the relatively large value of the standard error of estimate ($S_{y/x}$) and the significant coefficient of correlation of the residuals and the observed values, $r(y, y - \hat{y})$. Both must diminish if we are to produce a useful regression.

Line 2 of Table 2 shows the results of removing the four low observations from consideration, but keeping the model the same. Magnesia (x_3) now attains a nonzero regression coefficient, which gratifies our physical intuition, but the coefficient of correlation of the residuals with the observed values is still significantly (at the 95% level) greater than zero. We shall hypothesize that part of this dependence of the residuals on the observed values is due to the dependence of the observed values upon the solar flare proton irradiation, of which no account has yet been made. Specifically, one would expect that the specific ^{26}Al activity would vary directly with the surface to volume ratio. To estimate this ratio several variables were constructed, such as the ratio of the surface of the smallest rectangular parallelepiped that would contain the sample (as determined by measurements of the models of the samples) to the surface of a cube of equal volume, or of a volume equal to the sample's weight divided by an arbitrary but constant density, 2.75 g/cc. None of these variables worked well, but it was found that part of the surface to volume ratio due to change in the size of the sample could be accounted for quite well by assuming that the shape of the samples did not vary widely from some norm (the preceding negative results suggest this) and that their densities could also be considered constant, so that the cube root of the reciprocal of the sample weight was a sufficient measure of the variation in the surface to volume ratio due to variation in size. This variable (x_4 in line 3 of Table 2) entered the regression with a high degree of significance. Two more samples, 10003 and 15415, now appear to be unsaturated, and were eliminated from the saturated set (line 4), with a marked improvement in $r(y, y - \hat{y})$.

At this point, three observations from Apollo 16 (60315, 60335, and 62235) became available and were of considerable interest due to their larger specific activities and higher alumina contents. They were added to the remaining 19 in the saturated set after assuring ourselves that their inclusion did not invalidate previous steps and a new regression was performed (line 5). One of the newcomers was found to be marginally unsaturated (60315) and was removed, lowering $S_{y/x}$ and $r(y, y - \hat{y})$.

However, at this point the model contains one term related to the solar proton irradiation, and it is independent of the chemical composition, which seems unlikely. Therefore we proposed a model in which $x_4 = w^{-1/3}$ was replaced by three new variables, $x_4 = [SiO_2]w^{-1/3}$, $x_5 = [Al_2O_3]w^{-1/3}$, and $x_6 = [MgO]w^{-1/3}$, where w is the sample weight in kilograms. The results using this new model are shown in line 7, and are disappointing due to the emergence of two negative coefficients and a zero value for the coefficient of the alumina content. This difficulty was overcome by eliminating x_4 from the regression on the grounds that silica is the target least likely to produce ²⁶Al from solar protons, on pure energetic considerations. The resulting regression appears in line 8, and recommends itself not only by the nonnegativity of its coefficients, but also by the gratifying drop in the value of $r(y, y - \hat{y})$, and we will accept this regression as the best predictor of the saturated activity of ²⁶Al in lunar samples from other easily observed intensive variables. The sequence of events in lines 1 through 8 are also depicted in Figs. 1 and 2.

We have examined the residuals for correlation with the independent variables and have found no significant correlation at the 90% level, nor any suggestion of more complex dependence in scatter plots of the residuals versus the independent

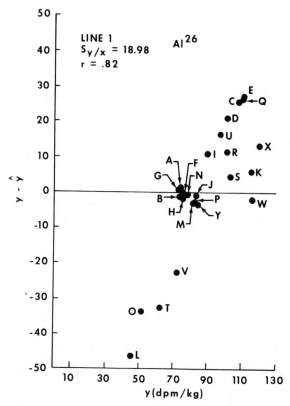

Fig. 1. Scatter plot of residuals for initial model. This is a scatter plot of the residuals versus the observed values derived from the model in line 1, Table 2. All Apollo 11, 12, 14, and 15 samples from Table 1 were considered in this regression—hence the large scatter. The four points forming the group corresponding to the four samples—12034, 12064, 14301, and 15265—are omitted in the next regression. The points identified by letters correspond to the following lunar samples:

A—10003, B—10017, C—10018, D—10019, E—10021, F—10057, G—10072, H—12002, I—12004, J—12010, K—12013, L—12034, M—12053, N—12063, O—12064, P—12065, Q—12073, R—14053, S—14066, T—14301, U—15256, V—15265, W—15415, X—15418, Y—15558.

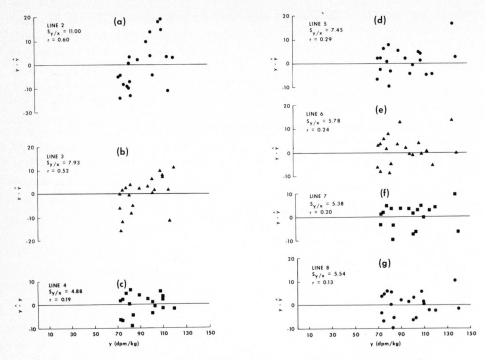

Fig. 2. Scatter plots for succeeding models. These are scatter plots of the residuals as a function of the observed values for the regressions whose results are shown in lines 2 through 8 of Table 2. Figure 2a corresponds to line 2, 2b to line 3, etc.

variables, or y, or $\hat{y}$. The residuals also do not show any evidence of a non-normal distribution, as one can see by examination of the cumulative distribution of the residuals displayed in Fig. 3. A Kolmogorov–Smirnov one-sample test on the residuals shows that there is no reason to reject the hypothesis that this sample was drawn from a normal population, even at very generous risk levels.

An examination of the distributions of the independent variables reveals that they are not all uniformly distributed over their ranges; only the magnesia-related variables (x_3 and x_6) are found to be without significant clustering. Silica concentrations (x_1) have a small scatter compared to their average (see Table 1) varying from 38% to 51% with one outlier at 61% (12013). This means that the observation corresponding to 12013 can have a disproportionately large effect on the value of b_1. However, we have no reason to reject this observation as invalid, and omitting the observation from the calculation of the coefficients in line 8 of Table 2 does not change the value of b_1 radically. The alumina-related variables (x_2 and x_5) also show outliers due to the inclusion of Apollo 16 samples, but in neither case is the gap between the outliers and the main group of values so large compared with the spread of the latter, and hence the maldistribution is potentially less damaging.

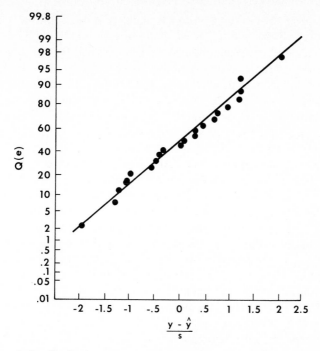

Fig. 3. Cumulative distribution of normalized residuals. The residuals in this figure are derived from the expression for the saturated ^{26}Al activity whose constants can be found in line 8 of Table 2. The variable plotted on the abscissa is the normalized residual, $e_i = (y_i - \hat{y}_i)/s$, where s is the standard deviation of the residuals ($s = S_{y/x}[(N-p)/N]^{1/2}$). The variable plotted on the ordinate is the cumulative frequency (expressed as percent), corrected for continuity, $Q(e) = [T(e_i) - \frac{1}{2}]/N$, where $T(e)$ is the number of residuals having values less than or equal to e. The straight line is derived from the unit normal distribution ($\mu = 0, \sigma = 1$). The ordinate is in the usual cumulative normal scale.

Discussion

Below line 8 in Table 2 we have appended the regression coefficients (for the 4π case) of Cressy (1971) and Fuse and Anders (1969), converted to our units. Where comparisons can be made, that is for the high energy variables x_1, x_2, and x_3, the results are similar. The solar proton components were not considered in the case of meteorites, due to the expectation that the ablation losses suffered during entry into the atmosphere would prevent the recovery of any solar flare activated material, and no provision is made for them in either regression. In both of these regressions, however, the effect of higher energy protons on higher Z targets has been estimated, found to be small, and subtracted previous to the regression (or the mathematical equivalent of this operation). Such corrections, as mentioned previously, would make very little difference in the regressions involving lunar samples.

The coefficients from the regressions such as those found in Cressy (1971), Fuse and Anders (1969), and this work, have often been identified as estimates of the "production rate" of ^{26}Al by the following argument. By "production rate" we mean the product of the cosmic ray and solar proton flux as a function of depth, energy, secondary particle flux, etc. and the excitation function for a particular product isotope and target element integrated over energy, depth, and other appropriate variables. Let us suppose that we were able to measure the ^{26}Al activity in a large number of saturated (in ^{26}Al) lunar samples of differing composition but with the same shape and exposure geometry. Then one could perform a regression of the observed ^{26}Al activities (and it would be permissible to weight them by their uncertainties of measurement, since all other uncertainties have been eliminated) against the compositions x_i, and obtain the regression $\hat{y}_j = \Sigma_i b_i x_{ij}$. Both the weighted sum of the production rates, P, and the regression, $\Sigma_i b_i x_{ij}$ estimate the ^{26}Al content: $\Sigma_i P_i x_{ij} = \Sigma_i b_i x_{ij}$, for any composition. Therefore, $P_i = b_i$. Notice that this argument is dependent on the assumption that the forms of $\Sigma_i P_i x_{ij}$ and $\Sigma_i b_i x_{ij}$ are identical. If a term is added or omitted from either, the conclusion follows for all the terms common to both linear combinations if and only if each column vector, X_i, which is made up of all the concentrations of the ith oxide in the X matrix, is orthogonal to every other such column vector X_k in X (see Draper and Smith, 1966, pp. 69–71). The orthogonality condition does not apply to this work (the off-diagonal terms in Table 3 are not zero). The regression coefficients are dependent on each other. In addition, the samples are not identical in shape and exposure geometry, so that even if either the orthogonality or the identical form condition applied, the b_j's would estimate some sort of average of a large number of integrations over different conditions.

The nonorthogonality condition has consequences when we attempt to calculate the variance of the estimate of the saturated ^{26}Al activity for a new sample whose composition vector is X_k. The expression for this variance is $Var(\hat{y}_k) =$

Table 3. Coefficients of correlation and inverse matrix for line 8, Table 2.

	$[SiO_2]$	$[Al_2O_3]$	$[Al_2O_3]w^{-1/3}$	$[MgO]w^{-1/3}$
b_1	6.487	−.6763	.4939	−.8601
b_2	−14.68	72.64	−.8643	.5650
b_5	7.481	−43.81	35.37	−.6367
b_6	−19.90	43.74	−34.40	82.52

This table refers to the model shown in line 8 of Table 2. The entries above the slanted line are the coefficients of correlation of the regression coefficients, $r(b_j, b_k)$, and those below the slanting line are the elements of the inverse matrix $(X'X)^{-1}$, each multiplied by 10^4. Since both matrices are symmetric, both may be recovered in full from this table. The inverse matrix $(X'X)^{-1}$ is necessary for calculation of the variance of any new prediction of the saturated ^{26}Al activity from the model whose constants are shown in line 8 of Table 2.

$S^2_{y/x}[1 + \mathbf{X}'_k(\mathbf{X}'\mathbf{X})^{-1}\mathbf{X}_k]$, where $(\mathbf{X}'\mathbf{X})^{-1}$ is the inverse matrix (see Table 3) arising from the regression. Since the standard deviations of the predictions of the saturated ^{26}Al activity for all the observations rejected as unsaturated are of interest, we calculate them by this method and show them in Table 4. In the right hand column of Table 4 one finds the ratio of the residual to the standard deviation of the predicted values. As one can see, none of these values would lead one to suspect that they should really be included as members of the saturated set. The largest negative value of this ratio for a member of the saturated set is −1.6.

We also show in Table 4 the variance of the predicted values for these samples calculated under the false assumption that they arise from an orthogonal data set. The variance of the predicted value for the kth new observation under this assumption would be the same as that for a linear combination: $\text{Var}(\hat{y}_k) = \Sigma_j x^2_{kj} S^2_j$, where S^2_j is the variance of the jth coefficient, b_j. As one can see, there is a considerable difference between this result and the correct method of calculation of the variance.

It also behoves us to examine the results obtained by applying the regression equation that we have selected to those samples (more than 50) which we originally excluded from consideration; i.e. fines and parts of rocks. Let us calculate the predicted ^{26}Al specific activity using the chemical analysis and the weight of the whole, undisturbed sample from the regression equations represented by line 8, Table 3, and subtract the predicted value from the observed value for the part of the sample measured. If we compare these pseudoresiduals with the observed and predicted values, we find results consistent with our hypotheses: the pseudoresiduals of the parts of rocks have a large scatter and no correlation with either the observed or the predicted values of the ^{26}Al specific activity. There is general good behavior however; all pieces known to have been

Table 4. Comparison of rejected samples with predictions of model.

Sample	Observed ^{26}Al value $y \pm S$	Residual $y - \hat{y}$	Standard deviation from matrix $S_{\hat{y}}$	"Standard deviation" as $\sqrt{\Sigma x_j^2 S_j^2}$	$\dfrac{y - \hat{y}}{S_{\hat{y}}}$
10003	74 ± 8	−14.80	5.73	11.14	−2.58
12034	45 ± 5	−81.48	6.28	16.45	−13.0
12064	51 ± 5	−24.76	6.13	9.33	−4.04
14301	62 ± 18	−38.30	6.33	15.47	−6.40
15265	72 ± 8	−26.84	5.86	15.03	−4.58
15415	116 ± 9	−32.83	11.09	25.45	−2.96
60315	92 ± 9	−23.77	6.30	14.01	−3.77

The residuals are the differences between the observed values and the values predicted by the expression whose constants appear in line 8, Table 2. The "standard deviation from the matrix" is the standard deviation of the predicted value for a sample whose composition vector is $\mathbf{X}_k$: $S_{y/x}[1 + \mathbf{X}'_k(\mathbf{X}'\mathbf{X})^{-1}\mathbf{X}_k]^{1/2}$. The matrix $(\mathbf{X}'\mathbf{X})^{-1}$ may be recovered from Table 3.

near the top of sectioned rocks and pieces consisting mostly of exposed surfaces all have large positive pseudoresiduals. Those known to be well shielded all have negative pseudoresiduals. The pseudoresiduals of the fines samples also show a large scatter and no correlation with the ^{26}Al values predicted as described above. However, there is a strong positive correlation (significant at >95% confidence level) between the pseudoresiduals of the fines and the observed values. The relationships described above are those one would expect if the model chosen were at least approximately correct, and they are also consistent with our exclusion of these samples from the set used in the regressions.

CONCLUSIONS

(1) We have developed a method for predicting the saturated ^{26}Al specific activity of lunar rock samples from various readily observable variables. The results of this method show internal consistency, within the limitations imposed by the number of observations available, and the limitations and possible sources of bias in the method have been explored.

(2) It is possible by postulating a one-step elevation to the surface to predict the most probable length of time that some lunar samples have been exposed to cosmic rays and solar protons. These "exposure ages" or "most recent surface residence times" are shown in Table 5. These ages should be regarded as upper limits, since it is assumed that the sample has been lifted in one step from a level of shielding that had a negligible ^{26}Al production rate. Lunar sample 12034 is a good illustration of this problem, since its exposure age may be presumed to be zero; it was dug up. Yet it was not deeply enough buried to have a negligible ^{26}Al activity, and its fictitious age is included in the table as a matter of interest. It is also assumed that all the samples retain about the same amount of their solar proton induced ^{26}Al, but the greater friability of 15415 suggests that increased erosion may be the cause of its unsaturation, and hence that its true exposure age may be much larger than calculated.

(3) Of the 27 lunar samples in the set of observations found on the lunar surface, six have been found to be very probably unsaturated. Thus, the probability of picking up a lunar rock whose specific ^{26}Al is unsaturated is $0.22^{+.18}_{-.11}$ (90% confidence limits assuming a binomial distribution). This affords us a new, quantitative estimate of the recent deep gardening of the lunar regolith.

(4) The small number of observations clearly imposes limitations on the conclusions that can be drawn from this regression analysis and we have attempted to point them out. Apollo 14 and 16 samples are underrepresented in the set of observations, and the addition of Apollo 16 samples would be especially valuable since this would increase the accuracy of our knowledge of the alumina-related coefficients (b_2 and b_5). It has been assumed implicitly that there is no effect on the ^{26}Al activity that is related to the mission on which they were recovered, after compositional effects have been removed. Although this is a plausible assumption, there is no way to test it with the present limited set of observations.

Table 5. Exposure ages of some lunar samples.

Sample	Exposure age (m.y.)† $t_{-\sigma}^{+\sigma}$
10003	$1.91_{-.52}^{+1.1}$
12034*	$0.47_{-.09}^{+.07}$
12064	$1.19_{-.25}^{+.32}$
14301	$1.03_{-.42}^{+.70}$
15265	$0.97_{-.33}^{+.48}$
15415	$1.61_{-.34}^{+.51}$
60315	$1.69_{-.38}^{+.61}$

*The age listed here is that calculated from the predicted and observed ^{26}Al activities as if this sample had been found on the surface. The exposure age of this rock is zero; it was dug up. See text.

†The error limits shown here are derived from the usual truncated Taylor's theorem approximation for the variance of a function. However, since the distribution of the ratio of two normally-distributed variables is not normally-distributed, these error limits must be interpreted with discretion.

The exposure ages in this table are calculated from the data in Table 4 on the assumption of a one-step elevation to the surface from depths deep enough to reduce the ^{26}Al production rate there to a negligible level. The one standard deviation values shown include the uncertainties reported for both the observed and predicted values. Note that exposure ages so calculated represent the most recent surface residence time, and are not cumulative over the entire history of the rock as gas ages are.

Finally, in using our results to predict the saturated ^{26}Al activities one ought to be careful when using compositions which lie near or beyond the edge of the set of observations used in the regression. Although the variance of the predicted value, $\hat{y}$, will be correctly calculated by the expression given in the text, it assumes that there is no bias in the model, and the consequence of some small amount of bias in the model is likely to be far more severe near the edge of the set than well within it.

Acknowledgments—Technical support in the operation and maintenance of the equipment in our laboratory was provided by Warren R. Portenier and Marshall K. Robbins of Northrup Services, Inc. Linda Bennett assisted us in the data reduction. We wish to thank A. H. Feiveson of NASA-JSC and R. B. Harrist of the U. of Texas School of Public Health at Houston for many helpful suggestions and discussions.

REFERENCES

Armstrong T. W. and Alsmiller R. G., Jr. (1971) Calculation of cosmogenic radionuclides in the Moon and comparison with Apollo measurements. *Proc. Second Lunar Sci. Conf., Geochim. Cosmochim. Acta,* Suppl. 2, Vol. 2, pp. 1729–1745. MIT Press.

Clark R. S. and Keith J. E. (1973) Determination of natural and cosmic ray induced radionuclides in Apollo 16 lunar samples. *Proc. Fourth Lunar Sci. Conf., Geochim. Cosmochim. Acta*, Suppl. 4, Vol. 2, pp. 2105–2113. Pergamon.

Compston W., Berry H., Vernon M. J., Chappell B. W., and Kaye M. J. (1971) Rubidium–strontium chronology and chemistry of lunar material from the Ocean of Storms. *Proc. Second Lunar Sci. Conf., Geochim. Cosmochim. Acta*, Suppl. 2, Vol. 2, pp. 1471–1485. MIT Press.

Cressy P. J., Jr. (1971) The production rate of ^{26}Al from target elements in the Bruderheim chondrite. *Geochim. Cosmochim. Acta* **35**, 1283–1296.

Draper N. R. and Smith H. (1966) *Applied Regression Analysis*. Wiley, New York.

Eldridge J. S., O'Kelley G. D., and Northcutt K. J. (1973) Radionuclide concentrations in Apollo 16 samples. *Proc. Fourth Lunar Sci. Conf., Geochim. Cosmochim. Acta*, Suppl. 4, Vol. 2, pp. 2115–2122. Pergamon.

Fuse K. and Anders E. (1969) Aluminum–26 in meteorites, VI: Achondrites. *Geochim. Cosmochim. Acta* **33**, 653–670.

Heyman D. and Anders E. (1967) Meteorites with short cosmic ray exposure ages. *Geochim. Cosmochim. Acta* **31**, 1793–1809.

Keith J. E., Clark R. S. and Richardson K. A. (1972) Gamma-ray measurements of Apollo 12, 14 and 15 lunar samples. *Proc. Third Lunar Sci. Conf., Geochim. Cosmochim. Acta*, Suppl. 3, Vol. 2, pp. 1671–1680. MIT Press.

Longley J. W. (1967) An appraisal of least squares programs from the point of view of the user. *J. Amer. Stat. Assoc.* **62**, 819.

LSPET (Lunar Sample Preliminary Examination Team) (1970) Preliminary examination of lunar samples from Apollo 12. *Science* **167**, 1325–1339.

LSPET (Lunar Sample Preliminary Examination Team) (1971) Preliminary examination of lunar samples from Apollo 14. *Science* **173**, 681–693.

LSPET (Apollo 15 Preliminary Examination Team) (1972) The Apollo 15 lunar samples. A preliminary description. *Science* **175**, 363–375.

LSPET (Apollo 16 Preliminary Examination Team) (1973) The Apollo 16 lunar samples. *Science* **179**, 23–34.

O'Kelley G. D., Eldridge J. S., Schonfeld E., and Bell P. R. (1970) Primordial radionuclide abundances, solar proton and cosmic-ray effects and ages of Apollo lunar samples by nondestructive gamma-ray spectroscopy. *Proc. Apollo 11 Lunar Sci. Conf., Geochim. Cosmochim. Acta*, Suppl. 1, Vol. 2, pp. 1407–1423. Pergamon.

O'Kelley G. D., Eldrige J. S., Schonfeld E., and Bell P. R. (1971) Cosmogenic radionuclide concentrations and exposure ages of lunar samples from Apollo 12. *Proc. Second Lunar Sci. Conf., Geochim. Cosmochim. Acta*, Suppl. 2, Vol. 2, pp. 1747–1755. MIT Press.

O'Kelley G. D., Eldridge J. S., Northcutt K. J., and Schonfeld E. (1972) Primordial radioelements and cosmogenic radionuclides in lunar samples from Apollo 15. *Proc. Third Lunar Sci. Conf., Geochim. Cosmochim. Acta*, Suppl. 3, Vol. 2, pp. 1659–1670. MIT Press.

Perkins R. W., Rancitelli L. A., Cooper J. A., Kaye J. H., and Wogman N. A. (1970) Cosmogenic and primordial radionuclide measurements in Apollo 11 lunar samples by nondestructive analysis. *Proc. Apollo 11 Lunar Sci. Conf., Geochim. Cosmochim. Acta*, Suppl. 1, Vol. 2, pp. 1455–1469. Pergamon.

Rancitelli L. A., Perkins R. W., Felix W. D., and Wogman N. A. (1971) Erosion and mixing of the lunar surface from cosmogenic and primordial radionuclide measurements in Apollo 12 lunar samples. *Proc. Second Lunar Sci. Conf., Geochim. Cosmochim. Acta*, Suppl. 2, Vol. 2, pp. 1757–1772. MIT Press.

Rancitelli L. A., Perkins R. W., Felix W. D., and Wogman N. A. (1972) Lunar surface processes and cosmic ray characterization from Apollo 12–15 lunar sample analysis. *Proc. Third Lunar Sci. Conf., Geochim. Cosmochim. Acta*, Suppl. 3, Vol. 2, pp. 1681–1691. MIT Press.

Smith E. O. (1971) STPREG—Stepwise regression with restricted least squares, Lockheed Electronics Co., Inc. NASA Contract NAS 9-12200.

Wahita H. and Schmitt R. A. (1971) Bulk chemical composition of Apollo 12 samples: Five igneous and one breccia rocks and four soils. *Proc. Second Lunar Sci. Conf., Geochim. Cosmochim. Acta*, Suppl. 2, Vol. 2, pp. 1231–1236. MIT Press.

Warner J. (Compiler) (1971) A summary of Apollo 11 chemical, age, and model data, extracted from *Proc. Apollo 11 Lunar Sci. Conf., Geochim. Cosmochim. Acta*, Suppl. 1. Curator's Office, Johnson Space Center.

Willis J. P., Ahrens L. H., Danchin R. V., Erlank A. J., Gurney J. J., Hofmeyr P. K., McCarthy T. S., and Orren J. J. (1971) Some interelement relationships between lunar rocks and fines, and stony meteorites. *Proc. Second Lunar Sci. Conf., Geochim. Cosmochim. Acta*, Suppl. 2, Vol. 2, pp. 1123–1138. MIT Press.

Wrigley R. C. (1972) Some cosmogenic and primordial radionuclides in Apollo 12 lunar surface materials. *Proc. Second Lunar Sci. Conf., Geochim. Cosmochim. Acta*, Suppl. 2, Vol. 2, pp. 1791–1796. MIT Press.

Proceedings of the Fifth Lunar Conference
(Supplement 5, Geochimica et Cosmochimica Acta)
Vol. 2 pp. 2121–2138 (1974)
Printed in the United States of America

Determination of natural and cosmic ray induced radionuclides in Apollo 17 lunar samples

J. E. Keith, R. S. Clark, and L. J. Bennett

NASA Johnson Space Center, Houston, Texas 77058

Abstract—Natural and cosmic ray induced radionuclides have been determined in 15 rocks and 7 soils from the Apollo 17 mission by nondestructive gamma-ray spectroscopy. The Th and K contents of these rocks and soils are shown to be linear functions of the U contents. Large amounts of ^{56}Co, ^{46}Sc, and ^{54}Mn were found in these samples due to the large proton-accelerating flares which occurred between August 2 and 6, 1972. The behavior of the ^{56}Co activity as a function of the ^{54}Mn activity seems to be the result of variation in surface to volume ratios among the samples and exposure angle. It is not necessary to invoke an asymetric solar flare flux to explain this behavior.

Introduction

THE NONDESTRUCTIVE GAMMA-RAY SPECTROSCOPY of lunar samples has enabled many properties of solar protons and cosmic rays and their interactions with the lunar regolith to be determined. The measurements have been used to determine the exposure ages of lunar samples, the average of the cosmic ray flux for the last few million years, the flux of solar flares, the erosion rates of lunar rocks, the gardening of the lunar regolith, and the movement of samples on the lunar surface (O'Kelley *et al.*, 1970, 1971b, 1972; Eldridge *et al.*, 1972, 1973; Rancitelli *et al.*, 1971, 1972, 1973; Perkins *et al.*, 1970; Wrigley, 1971, 1973; Wrigley and Quaide, 1970; Keith *et al.*, 1972; Clark and Keith, 1973).

The extremely large and energetic proton-accelerating solar flares which occurred between August 2 and 6, 1972 (Bostrom *et al.*, 1972) made possible the measurement of the largest amounts of solar proton induced radionuclides in any set of lunar samples and the identification of several isotopes not seen previously in lunar samples (LSPET, 1974). In this paper we present the results of the measurement of the gamma-ray spectra of 15 rocks and 7 soils from the Apollo 17 mission.

Procedure

The dual parameter gamma-ray spectrometer data acquisition system in the Radiation Counting Laboratory of the Johnson Space Center has been previously described by its developers (O'Kelley *et al.*, 1970, 1971a, 1971b) and the location of the facility has been described by McLane *et al.* (1967). The procedures and containers are the same as were described in Keith *et al.* (1972). The data are reduced by a computer stripping program which has been previously described (Keith *et al.*, 1972; Clark and Keith, 1973).

Results and Discussion

In Table 1 we present the results of gamma-ray spectrometry on 15 rock samples from Apollo 17. The Preliminary Examination (LSPET, 1974) has

Table 1. Gamma-ray analysis of Apollo 17 rocks.

Type*	Basalts					Breccia		
Subtype	Fine		Medium		Coarse	Dr. Matrix	Agglutinate	Grn–gray matrix
Sample	70255	71155	70275	78135	78505	70175	70019	76215
Weight (g)	224.9	25.80	171.4	133.93	499.6	338.8	128.0	642.8
Comments		Boulder chip			With soil sample	Friable	from crater floor	boulder spall
Th (ppm)	.31 ± .03	.31 ± .06	.43 ± .04	.26 ± .05	.39 ± .05	.40 ± .04	1.03 ± .10	4.6 ± .2
U (ppm)	.107 ± .008	.118 ± .017	.120 ± .013	.107 ± .012	.135 ± .006	.105 ± .007	.23 ± .02	1.27 ± .06
K (%)	.048 ± .004	.040 ± .003	.043 ± .006	.0525 ± .0018	.0508 ± .008	.055 ± .002	.0692 ± .0012	.215 ± .014
^{26}Al (dpm/kg)	49 ± 6	105 ± 8	91 ± 5	42 ± 4	72 ± 10	42 ± 5	45 ± 3	56 ± 3
^{22}Na (dpm/kg)	72 ± 7	119 ± 11	84 ± 5	74 ± 5	67 ± 8	76 ± 18	110 ± 8	60 ± 4
^{54}Mn (dpm/kg)	137 ± 15	160 ± 20	180 ± 30	180 ± 20	100 ± 6	156 ± 9	166 ± 12	22 ± 17
^{56}Co (dpm/kg)	211 ± 19	310 ± 50	220 ± 20	240 ± 20	59 ± 13	300 ± 70	240 ± 30	45 ± 6
^{46}Sc (dpm/kg)	63 ± 6	81 ± 7	83 ± 20	76 ± 5	45 ± 4	39 ± 5	59 ± 5	5 ± 3
^{48}V (dpm/kg)	<30		32 ± 15	18 ± 5		17 ± 5		
Th/U	2.9	2.6	3.6	2.4	2.9	3.8	4.5	3.6
K/U	4500	3400	3600	4900	3800	5200	3000	1690
Group composition†	70215				75055		79135	77135
SiO$_2$ (%)	37.19				41.27		42.29	46.13
TiO$_2$ (%)	13.14				10.17		5.15	1.54
Al$_2$O$_3$ (%)	8.76				9.75		15.08	18.01
FeO (%)	19.62				18.24		14.01	9.11
MgO (%)	8.52				6.84		10.42	12.63
CaO (%)	10.43				12.30		11.44	11.03

Type*	Breccia	Brecciated or shocked anorthosites to gabbros			Miscellaneous		
Subtype	Lt.-gray banded matrix	Norite	Coarse norite		Poikilitic clasts		Brecciated dunite
Sample	72255	76535	78235	78255,1	72315,17	72355	72415
Weight (g)	402.57	145.57	128.8	31.01	84.82	367.4	29.79
Comments	boulder chip	rake sample	boulder chip	boulder chip	boulder clast	boulder chip	boulder clast
Th (ppm)	4.4 ± .4	.19 ± .02	.59 ± .08	.83 ± .06	4.80 ± .5	5.3 ± .3	<.15
U (ppm)	1.20 ± .15	.054 ± .01	.196 ± .016	.227 ± .016	1.53 ± .15	1.39 ± .15	<.06
K (%)	.184 ± .008	.0235 ± .0008	.0490 ± .0015	.059 ± .003	.284 ± .015	.253 ± .3	.007 ± .002
^{26}Al (dpm/kg)	78 ± 6	94 ± 5	77 ± 7	65 ± 6	62 ± 8	84 ± 6	77 ± 6
^{22}Na (dpm/kg)	61 ± 5	119 ± 13	111 ± 8	50 ± 5	73 ± 5	87 ± 6	290 ± 30
^{54}Mn (dpm/kg)	41 ± 6	24 ± 6	55 ± 8	10 ± 5	90 ± 9	66 ± 7	77 ± 16
^{56}Co (dpm/kg)	35 ± 15	8 ± 5	52 ± 9	30 ± 20		58 ± 13	150 ± 30
^{46}Sc (dpm/kg)	6 ± 6	<1.5	1.4 ± .9	<15		12 ± 3	8 ± 3
^{48}V (dpm/kg)			<12				
Th/U	3.7	3.5	3.0	3.7	3.1	3.8	
K/U	1500	4400	2500	2600	1900	1820	>1170
Group composition†	77135	76535‡	78155				72415
SiO$_2$ (%)	47.54	43.4	45.75				39.93
TiO$_2$ (%)	.91	.03	.27				.03
Al$_2$O$_3$ (%)	17.01	20.7	25.94				1.53
FeO (%)	11.58	4.99	5.82				11.34
MgO (%)	9.35	19.2	6.33				43.61
CaO (%)	11.71	11.0	15.18				1.14

*As classified by LSPET (1973).

†Taken from rocks in the same group (LSPET, 1973).

Shortlived activities are corrected to time of lunar liftoff.

‡Cooley *et al.*, In press.

Table 2. Gamma-ray analysis of Apollo 17 fines.

Sample* Desc.	72221,0 boulder overhang fillet	72241,0 boulder fillet	72441,0 under boulder	74220,92 orange soil	75061,5 boulder top skim	76240,2 shadowed soil	76261,1 skim reference
Approx. depth (cm)	0–3	0–5	70†	5–8	0–1	0–5	0–2
Th (ppm)	3.6 ± .4	3.6 ± .5	3.5 ± .4	.65 ± .09	.91 ± .13	2.5 ± .3	2.1 ± .3
U (ppm)	.89 ± .03	.94 ± .03	.83 ± .03	.164 ± .010	.248 ± .015	.618 ± .018	.49 ± .02
K (%)	.142 ± .004	.144 ± .004	.141 ± .004	.068 ± .002	.066 ± .002	.118 ± .004	.102 ± .003
^{26}Al (dpm/kg)	132 ± 12	187 ± 17	65 ± 6	45 ± 4	180 ± 16	156 ± 14	182 ± 17
^{22}Na (dpm/kg)	63 ± 4	143 ± 8	47 ± 3	51 ± 3	187 ± 10	41 ± 3	148 ± 8
^{54}Mn (dpm/kg)	48 ± 11	78 ± 13	38 ± 11	50 ± 3	200 ± 14	28 ± 6	93 ± 7
^{56}Co (dpm/kg)	<200	130 ± 90	<300	31 ± 6	490 ± 30	28 ± 17	240 ± 20
^{46}Sc (dpm/kg)	5 ± 2	11 ± 3	6 ± 2	19.1 ± 1.6	86 ± 5	7.2 ± .8	23 ± 2
^{48}V (dpm/kg)				13 ± 14	47 ± 12	7.0 ± 1.4	18 ± 12
Th/U	4.0	3.8	4.2	4.0	3.7	4.0	4.3
K/U	1600	1500	1700	4100	2700	1900	2100

*All Sample weights were approximately 100 g.
†Approximate thickness of the boulder.

resulted in the division of the rock samples into 8 groups, 7 of which are represented in our sample set, but only 1 of which has had chemical analysis. Where direct comparison is not possible, the composition of other samples in the same group are listed with the radiochemical data. In Table 2 we present the data for the lunar soils.

Natural Radioactivities

The analyses of the rocks in Table 1 are divided into two groups by their K, U, and Th contents, with the breccias in the ranges $1750 \leqslant [K] \leqslant 3000$, $1.2 \leqslant [U] \leqslant 2$, and $4 \leqslant [Th] \leqslant 8$ ppm, and the basalts and gabbros in the ranges $270 \leqslant [K] \leqslant 590$, $0.1 \leqslant [U] \leqslant 0.23$, and $0.26 \leqslant [Th] \leqslant 0.83$ ppm. The basalts are strongly clustered about $[K] = 500$ ppm, $[U] = 0.1$ ppm, and $[Th] = 0.4$ ppm, and the gabbros are in the higher part of the gabbro-basalt range. This clustering of components strongly suggests that the basalts are derived from a single source.

In Fig. 1 we show the Th content of Apollo 17 samples as a function of their U content, using data from this work and LSPET (1974). As in the case of the Apollo 16 samples, the dependence of the Th content on the U content can be represented as a linear function using the method of York (1966), with an insignificant constant term:

for rocks:

$$[Th] = -0.052 \pm 0.023 + (3.85 \pm 0.11)[U], (\chi_\nu^2 = 1.23), \text{ and}$$

for soils:

$$[Th] = +0.052 \pm 0.041 + (3.64 \pm 0.11)[U], (\chi_\nu^2 = 0.71).$$

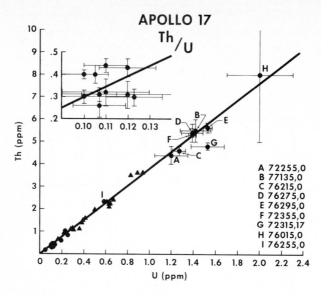

Fig. 1. Th as a function of U for Apollo 17 samples. The circles represent data from rocks; the triangles represent data from soils. The anomalous (see Fig. 2) point from rock 76255 falls on the line (shown for rocks only) found by the regression method of York (1966).

In Fig. 2 we show the K and U data for the same samples. The dependence of K on U is not as well represented by a linear function, even after the exceptional point due to the breccia 76255 is omitted. Moreover, the constant terms are no longer insignificant:

for rocks:

$$[K] = 250 \pm 30 + (1570 \pm 110)[U], (\chi_\nu^2 = 8.23), \text{ and}$$

for soils:

$$[K] = 426 \pm 29 + (1130 \pm 64)[U], (\chi_\nu^2 = 2.11).$$

The brecciated dunite boulder clast (72415) has as low a content of K, U, and Th as any sample thus far measured. The K, U, and Th contents found in this sample are comparable to those found in the anorthositic samples 15415 and 15418 (Keith *et al.*, 1972).

Scatter plots of both K as a function of U and Th as a function of U have been examined for each Apollo mission for all known determinations of these elements by gamma-ray spectrometry. The data were fitted to the straight lines $[Th] = a + b[U]$ and $[K] = a + b[U]$ by the method of York (1966). The results are summarized in Table 3. For all the Th versus U data, the y intercept a is never significantly different from zero (see t_a) and the values of the reduced chi-square (whose expectation is 1) suggest that all the scatter about the various lines can be accounted for by the errors of measurement.

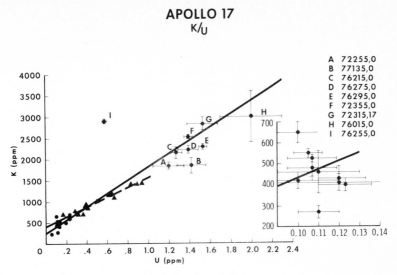

Fig. 2. K as a function of U for Apollo 17 samples. The circles represent data from rocks; the triangles represent data from soils. The solid line represents the expression found by the method of York (1966) for the variation of K as a function of U for rocks (76255 being omitted from the calculation), and the dashed line that of the soils.

The case of K as a function of U is different however, as Table 3 shows. It is clear that the behavior of K versus U is similar in all missions except Apollo 11, which is quite different; it is the only one to have a negative y intercept, and the slope is about $2\frac{1}{2}$ times as great as the other slopes. The result for Apollo 11 is strongly dependent on the results for 10003, and three other low-K Apollo 11 samples 10045, 10050, and 10062 lie near 10003, although lack of the knowledge of the analytical uncertainties of these three samples prevents their inclusion in the set used for the regression. Other smaller differences can be noted: Apollo 16 and 17 have the only large, strongly significant positive y intercepts, and the reduced chi-squares suggest the existence of real scatter, unlike the Th versus U case. Thus, the results are consistent with the view that K has been differentiated from U while Th has not.

Cosmic Ray and Solar Proton Induced Radioactivities

The integral fluxes of protons from solar flares have been measured during the last two solar cycles at >10, >30, and >60 MeV (Webber, 1963; Kinzey and McDonald, private communication to R. C. Reedy, 1969; Bostrom et al., 1967–1972). In Fig. 3 we show the relative amounts of ^{22}Na and ^{54}Mn produced in a target of constant but arbitrary chemical composition and exposure by solar protons of >30 MeV since 1956 as a function of time. We shall call this the "relative decayed flux," in order to distinguish it from observed activities. Contributions to the relative decayed flux previous to 1956 may be neglected since

Table 3. The regression coefficients for [Th] = a + b[U] and
for [K] = a + b[U] for each mission, using the method of
York (1966).

Mission	n	a	σ_a	b	σ_b	χ_ν^2	t_a
		[Th] = a + b[U]					
11	10	.03	.13	3.9	.24	.39	.22
12	23	−.01	.02	3.7	.06	.44	−.27
14	24	.21	.14	3.6	.06	.34	1.5
15	35	.04	.01	3.9	.06	1.2	−3.0
16	94	−.01	.02	3.7	.05	1.2	−.34
17	43	−.02	.02	3.8	.07	.99	−1.1
		[K] = a + b[U]					
11	9	−579	195	3672	377	1.2	−3.0
12	21	166	34	1428	107	7.5	4.9
14	24	109	135	1363	60	2.9	0.8
15	35	89	16	1517	69	8.4	5.6
16	94	217	21	1289	44	7.1	10.3
17	41	310	22	1424	65	4.3	14.1

The standard deviations of the regression coefficients (σ_a and σ_b) are given here, and include correction for scatter. Student's t statistic ($t_a = a/\sigma_a$) for the significance of the y intercept is also shown, and if $|t_a| \gtrsim 2$, a cannot be considered to be zero. The reduced chi-square, χ_ν^2, is the ratio of the weighted (σ^{-2}) sum of the squares of the x and y residuals to the number of degrees of freedom ($n - 2$). Values of this statistic greater than 2 or 3 (depending on the number of degrees of freedom) suggest the existence of scatter in the data not accounted for by the assigned errors.

even for ^{22}Na they would be about 3% of their original value. In Fig. 4 we show the relative decayed fluxes varying with time with the half-lives of ^{22}Na, ^{54}Mn, and ^{56}Co or ^{46}Sc due to solar protons with energies >10, >30, and >60 MeV during the period of the Apollo 11 through 17 missions. We have selected a set of representative samples: 10003, 12004, 14053, 15256, 68815, and 70275 to be as similar in weight and major element composition as the scarcity of reliable determinations of the short-lived activities and the paucity of major element analyses would allow (the analysis of 70275 is inferred from a similar sample, 70215). The observed activities for these samples and their uncertainties were corrected for target composition using the Fe concentration for ^{54}Mn and ^{56}Co, the Ti concentration for ^{46}Sc and the Si, Al, and Mg concentration (see Yokoyama et al., 1973) for ^{22}Na. A galactic component corresponding to the target composition was also subtracted. They were also reduced to a constant surface-to-volume ratio by multiplication by the cube root of their weight (see Keith and Clark, 1974). These adjusted observed values were then normalized to the relative decayed flux at the time of Apollo 11 and are shown together as a function of time in Fig. 4. Considering that no attempt has been made to account for the errors in the flux

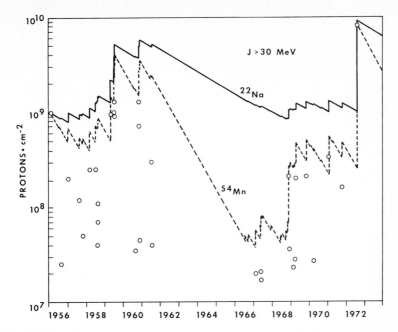

Fig. 3. Relative amounts of ^{22}Na and ^{54}Mn (relative decayed flux) produced in a target of constant but arbitrary composition (see text) and exposure by solar protons of >30 MeV since 1956. Relative contributions by individual flares to the particular isotope at a given time may be estimated by extrapolating the data point from an individual flare to the time in question by the slope determined by the half-life. The open circles show the time-integrated fluxes ($J > 30$ MeV) (Bostrom *et al.*, 1967–1972; Weber, 1963) at the dates shown, and have the units protons/cm^2. The units of the relative decayed flux are, of course, arbitrary.

determinations (consistent biases in these have been "normalized out," however) or for the secular variation in the galactic cosmic ray flux, the agreement between the observed and calculated value is reasonably good.

In Table 2 we present the results of the measurement of the gamma-ray spectra of 7 lunar soil samples from Apollo 17. The fillet under the overhanging boulder at Station 2, 72221, shows evidence (low ^{22}Na and ^{54}Mn) of considerable shadowing as might be expected, but the sample was made available too late to obtain anything but an upper limit for the ^{56}Co activity. If the chemical composition of 72441 is similar to that of 75201 and 72701, the activities listed in the table for it are consistent with burial under a 70 cm boulder.

The ^{26}Al activities found in 72221, 72241, and 76240 are all considerably higher than might be expected. Two explanations seem possible: that the boulders rolled over these soils and started shadowing them some time within the last million years, or, more reasonably, that the filleting has been the result of movement of material from the top few millimeters of the regolith under or against the boulder as a result of small meteorite bombardment. The suggestion that sample 70175

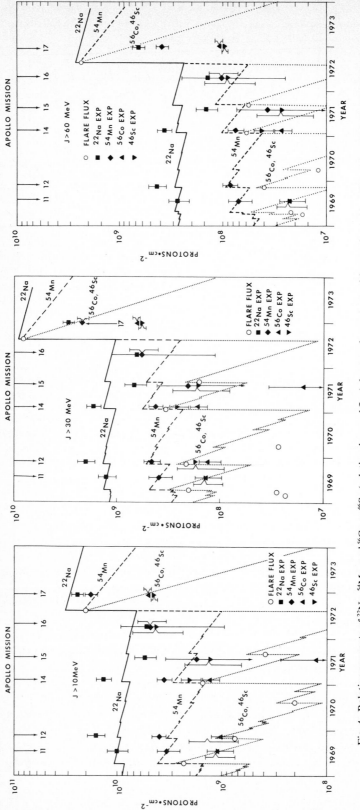

Fig. 4. Relative amounts of ^{22}Na, ^{54}Mn, and ^{56}Co or ^{46}Sc (relative decayed flux) due to solar protons with energies > 10, > 30, and > 60 MeV during the period of the Apollo 11 through Apollo 17 missions. Experimental results of each of the isotopes for each mission (normalized to Apollo 11) are also shown. As in Fig. 3, the units of the relative decayed flux (and the normalized experimental results) are relative, and those of the time-integrated flare fluxes (open circles) in protons/cm^2.

(ALGIT, 1973) might have been buried is not consistent with its large ^{56}Co activity (contrast it with that of the rake sample, 78505).

Two rock samples, 78135 and 70255, appear to be unsaturated in ^{26}Al. Using the chemical analyses of other samples of the same rock types and the method of calculating the saturated ^{26}Al activity discussed by the authors elsewhere (Keith and Clark, 1974), it is estimated that the surface exposure ages of these samples are about $\frac{3}{4}$ million years. Two other samples, 70175 and 70019, also appear to be unsaturated in ^{26}Al, probably because of their great friability, as in the case of 15415 (Keith and Clark, 1974).

The ^{56}Co, ^{54}Mn, and ^{46}Sc Behavior in Lunar Samples

Rancitelli *et al.* (1974) have reported finding evidence for a spatial asymmetry in the proton flux arising from the solar flares of August 2–6, 1972. This evidence consists of the change in the ratio ^{56}Co/^{54}Mn in their boulder chip samples as a function of the angle between the normal to the boulder face where the chip was taken and the local lunar zenith. Their explanation of this variation requires that the solar flare flux have a variable rigidity that was directly proportional to the angle between the direction of propagation of the protons and the local lunar zenith at the Apollo 17 site.

The difficulty of proposing a geophysical mechanism for such an asymmetric flux led us to examine all the data from Apollo 17 relating to induced activities, and we will show that for soils, rocks, and boulder chips the behavior of ^{56}Co as a function of ^{54}Mn, and also in the case of soils, ^{46}Sc, is the same regardless of orientation. This behavior is one of approximate linear dependence. There is no theoretical reason that the dependence should be linear, but we believe that inspection of Figs. 5–9 will show that there is little, if any, reason to force additional terms on the model. By examining the case of the ^{56}Co behavior versus ^{54}Mn and ^{56}Sc in soils we also will show that the most probable cause for variation in the activities of these three isotopes is the variation in the irradiation conditions: thickness of sample and degree of shielding or degree of exposure.

In order to describe the behavior of ^{56}Co as a function of ^{54}Mn and ^{46}Sc as quantitatively as possible, we will fit it to the model $y = a + bx$ by a method of weighted least squares where the weights are determined by the uncertainties in both variables; the method due to York (1966). Such weighting is required since the uncertainties in all the variables are comparable. From this treatment we will also obtain a quantitative measure, the reduced chi-square, of the degree to which the departures from the model can be accounted for by the uncertainties of measurement.

In Figs. 5 and 6 we present the ^{56}Co activity in Apollo 17 soils as a function of those of the ^{54}Mn and the ^{46}Sc activities, respectively in Apollo 17 soils. Each activity in Figs. 5 and 6 has been divided by the concentration of its probable target element (expressed as the oxide), and if that concentration were unknown, it is taken to be that of the average of the soils from the same station. The data for these figures were drawn from LSPET (1974), Eldridge *et al.* (1974) and this work.

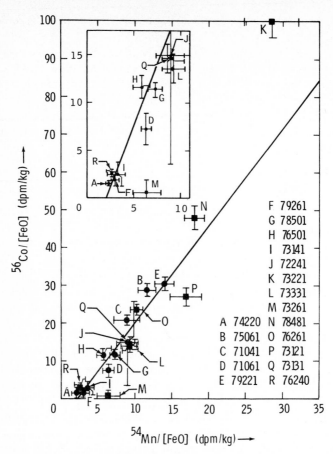

Fig. 5. Scatter diagram for ^{56}Co/[FeO] versus ^{54}Mn/[FeO] for activities from Apollo 17 soils. The origin and treatment of the data shown here are described in the text. The chemical analyses are from LSPET (1974). Only the points from samples where the radiochemical and chemical analyses were made on the same sample (solid circles) were considered in calculating the least squares (York, 1966) expression:

$$^{56}\text{Co/[FeO]} = -5.40 \pm 1.01 + (2.60 \pm 0.208)^{54}\text{Mn/[FeO]}, \chi_\nu^2 = 0.934.$$

The solid squares represent samples whose chemical analyses were inferred from other soils with the station number. The activities, i.e. ^{56}Co and ^{54}Mn are in dpm/kg, as of lunar lift off, and the concentration of ferrous oxide [FeO] in weight percent.

The most current value from each laboratory was used and where two or more laboratories reported values for the same sample, the weighted (by the inverse of the variances of the measurements) average was taken. The uncertainties shown have been adjusted for disagreement where it existed by multiplication by $\sqrt{\chi_\nu^2}$ if this statistic were greater than 1.

As one can see in Figs. 5 and 6, ^{56}Co tends to be a linear function of ^{54}Mn and the ratio ^{56}Co/[FeO] tends to be a linear function of the ratio ^{46}Sc/[TiO$_2$]. Straight

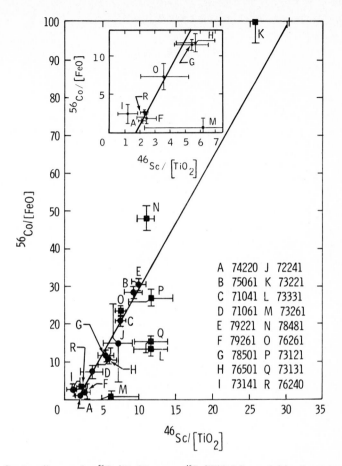

Fig. 6. Scatter diagram for ^{56}Co/[FeO] versus ^{46}Sc/[TiO$_2$] for activities from Apollo 17 soils. The origin and treatment of the data shown here are described in the text. The chemical analyses are from LSPET (1974). Only the points from samples where the radiochemical and chemical analyses were made on the same sample (solid circles) were considered in calculating the least squares (York, 1966) expression:

$$^{56}\text{Co/[FeO]} = -6.12 \pm 0.834 + (3.60 \pm 0.181)^{46}\text{Sc/[TiO}_2\text{]}, \chi_\nu^2 = 0.486.$$

The solid squares represent samples whose chemical analyses were inferred from other soils with the same station number. The activities of ^{56}Co and ^{46}Sc are in dpm/kg as of lunar liftoff, and the concentrations of the oxides in weight percent.

lines have been fitted to these data by the method of York (1966), with ^{56}Co/[FeO] as the dependent variable, using only the data for which a chemical analysis was available. In both cases, the data are well represented by the line, and the y intercept is negative. Presumably, the ^{54}Mn and ^{46}Sc activities near the x intercepts represent the values of the ^{54}Mn and ^{46}Sc due to irradiation by galactic cosmic rays and a small amount from previous solar flares. (We can deduce from Fig. 4 (>30 MeV) that the amount of ^{54}Mn present due to previous solar flares is about 3% of that due to the August 1972 flares; the corresponding ratio for ^{56}Co and ^{46}Sc is about 0.2%.)

If we arrange the data in order of decreasing distance from the x intercept, as we have in Table 4, we see that the same ordering, within the uncertainties where clusters occur, is achieved whether the ^{56}Co/[FeO] versus ^{54}Mn/[FeO] or ^{46}Sc/[TiO$_2$] plot is considered, and that this order is intuitively consistent with the comments concerning exposure and depth of sampling made by ALGIT (1973). This suggests very strongly that this joint variation of the ^{56}Co with the two higher energy products is the result of the combined effects of the average depth of sampling and the exposure geometry. If we examine the behavior of ^{56}Co and ^{54}Mn in whole rocks picked up on the lunar surface, we find the same behavior as in the soil samples (see Fig. 7).

In Fig. 8 we show the variation of ^{56}Co activity as a function of ^{54}Mn activity in all boulder chips for which these two activities have been reported. As one can see, the dependence of ^{56}Co activity on the ^{54}Mn activity is the same as seen previously in the whole-rock and soil samples, which have little or no dependence on exposure angle: a linear dependence with a negative y intercept. In Fig. 9 we show the data reported by Rancitelli *et al.* (1974) who have corrected this data for

Table 4. Ordering of Apollo 17 soils by ^{56}Co versus ^{54}Mn and ^{56}Co versus ^{46}Sc.

Sample number	Order by ^{56}Co and ^{54}Mn	Sampling depth (cm)	Comments	Order by ^{56}Co and ^{46}Sc
73221	1	0–0.5	Skim soil from top of trench	1
78481	2	0–0.5 to 1	Skim soil from top of trench	2
79221	⌈3	0–2	Top of trench at Station 9	3⌉
75021	4	0–1	Soil from top of boulder	5
73121	⌊5		Upper few cm from trench at Station 2a	4⌋
76261	6	0–2	Skim soil	6
71041	7	0–1 to 2	Soil taken in shadow of boulder	7
72241	⌈8	0–5	Boulder fillet	9⌉
73131	9		Middle of trench at Station 2a	8
73331	⌊10	0–5		10⌋
76501	⌈11		Rake soil Station 6	12⌉
78501	⌊12		Upper few cm	11⌋
71061	⌈13	0–5 to 6	Soil taken in shadow of boulder	13
73141	14	ca. 15	Soil from bottom of 15 cm trench	15⌉
76240	15		Soil from underneath overhang of boulder	14
74220	16	5–8	Orange soil	17
79261	17	ca. 17	Soil from bottom of trench at Station 9	16
73261	⌊18	5–10	Trench sample	18⌋

The samples were ordered from 1 to 18 by their distance from the x intercept in Fig. 5 to the place on the line closest to data point for the sample, starting with the greatest distance. This ordering is shown in the second column from the left. The square brackets enclose order numbers indistinguishable within the errors. The samples were ordered in the same manner using Fig. 6 and assigned the numbers in column 2. As inspection shows, the two orderings are identical within the uncertainties of measurement.

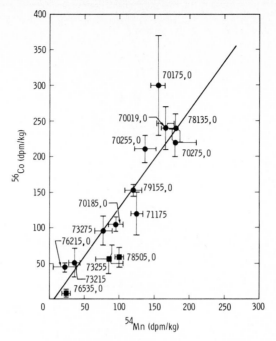

Fig. 7. Scatter diagram of the ^{56}Co and ^{54}Mn activities found in whole rocks from Apollo 17. The origin and treatment of the data shown here are described in the text. The solid circles represent data from samples not known to have been buried, and are the only data considered in calculating the least squares (York, 1966) expression:

$$^{56}\text{Co} = -7.14 \pm 16.9 + (1.37 \pm 0.153)^{54}\text{Mn}, \chi_\nu^2 = 0.836.$$

The solid squares represent data from samples known to have been or suspected to have been buried when recovered. The activities of ^{56}Co and ^{54}Mn are in dpm/kg as of lunar liftoff.

galactic production in both nuclides, and report the remainder extrapolated back to the end of the great flares (August 7, 1972). Even so, the linearity (with a different slope, of course) and negative intercept previously found persist.

In Table 5 we show the results of all the regressions of ^{56}Co activities as a function of other activities for various sets of Apollo 17 samples. The constants for the whole-rock set seem least like the others, probably because of the dissimilarity of their shapes to those of boulder chips or the scoops of fines. We have excluded those rock samples suspected of complete burial to make the comparison to boulder chips more direct. If these three samples were included in the data set the resulting slope and y intercept would be more like those of the other sets. Nevertheless, certain strong similarities remain. All the y intercepts are negative, usually significantly so, and the slopes are all positive and significantly so. The values in parentheses have been derived from the regression constants immediately above them in order to correct for the relative decay of ^{56}Co and ^{54}Mn from the zero time (August 7, 1972) used by Rancitelli *et al.* (1974)

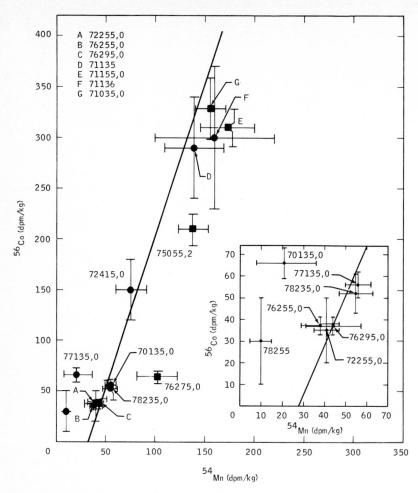

Fig. 8. Scatter diagram of the ^{56}Co and ^{54}Mn activities in boulder chips from Apollo 17. The origin and treatment of the data shown here are described in the text. The solid squares represent data from samples also recently discussed by Rancitelli *et al.* (1974), but have no corrections for galactic contributions. All the data shown was considered in calculating the least squares (York, 1966) expression:

$$^{56}\text{Co} = -62.1 \pm 18.6 + (2.26 \pm 0.270)^{54}\text{Mn}, \; \chi_\nu^2 = 1.34.$$

The activities of ^{56}Co and ^{54}Mn are in dpm/kg as of lunar liftoff.

to that (lunar liftoff, December 14, 1972) used in the other data sets. This is only an approximate correction, and the corrections made by Rancitelli *et al.* (1974) for the galactic cosmic ray contribution to these activities have not been accounted for. Nevertheless, it is still apparent that the behavior of the activities of the ^{56}Co and ^{54}Mn in both the boulder chips and soil samples cited by Rancitelli *et al.* (1974) as strongly suggesting an asymmetric solar flare flux bear a strong similarity to other data sets which do not have any dependence on exposure angles.

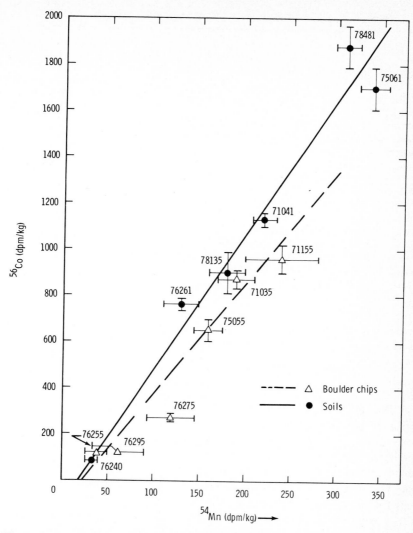

Fig. 9. Scatter diagram of the activities of ^{56}Co and ^{54}Mn in some Apollo 17 soils and boulder chips. The data shown were taken from Rancitelli *et al.* (1974) without alteration. The lines represent the following expressions which were calculated by the method of York (1966):

$$\text{Soils: } {}^{56}\text{Co} = -106 \pm 54.7(5.84 \pm 0.346){}^{54}\text{Mn}, \chi_\nu^2 = 1.37$$
$$\text{Chips: } {}^{56}\text{Co} = -110 \pm 71.3 + (4.82 \pm 0.585){}^{54}\text{Mn}, \chi_\nu^2 = 0.968.$$

If $y = a + bx$, and $a < 0$, then y/x will have a nonlinear dependence on x near the origin. If x (the ^{54}Mn activity) is now also caused to vary by changes in exposure geometry, or angle, then the ratio ^{56}Co/^{54}Mn $= y/x$ will have the variation with angle or exposure geometry shown by the boulder chip samples reported by Rancitelli *et al.* (1974), without any asymmetry in the bombarding flux,

Table 5. Comparison of behavior of ^{56}Co, ^{54}Mn, and ^{46}Sc in Apollo 17 samples.

	Figure	$a \pm \sigma_a$	t_a	$b \pm \sigma_b$	t_b	χ_ν^2
Soils ^{56}Co/[FeO] versus ^{54}Mn/[FeO]	5	-5.40 ± 1.01	$-.535$	$2.60 \pm .208$	12.5	.934
Soils ^{56}Co/[FeO] versus ^{46}Sc/[TiO$_2$]	6	$-6.12 \pm .834$	-7.34	$3.60 \pm .181$	19.9	.486
Soils ^{56}Co versus ^{54}Mn	9	-106 ± 54.7	-1.94	$5.84 \pm .346$	16.9	1.37
(Rancitelli abstract)		(-45 ± 23)		$(2.47 \pm .146)$		
Whole rocks, ^{56}Co versus ^{54}Mn	7	-7.14 ± 16.9	$-.422$	$1.37 \pm .153$	8.95	.836
Boulder chips ^{56}Co versus ^{54}Mn	8	-62.1 ± 18.6	-3.34	$2.26 \pm .270$	8.37	1.34
Boulder chips ^{56}Co versus ^{54}Mn	9	-110 ± 71.3	-1.55	$4.82 \pm .585$	8.24	.968
(Rancitelli abstract)		(-46 ± 30)		$(2.04 \pm .247)$		

The constant determined by the weighted least-squares method of York (1966), on the model $y = a + bx$ where the weights are determined by both the uncertainty in x and in y are shown here for various sets of Apollo 17 data. The number of the corresponding figure is also shown. The values for the slope and y intercept in parentheses have been reduced to the lunar liftoff date from the date (August 7, 1972) used in Rancitelli *et al.* (1974). The reduced chi-square (whose expectation is one) is a measure of how well the data is represented by the line and is defined as the weighted sum of the residuals in both x and y divided by the number of degrees of freedom. Student's t statistic for the y intercept (t_a) and the slope (t_b) are also shown.

and we believe this to be the explanation of the dependence of the ratio ^{56}Co/^{54}Mn on exposure angle found by these workers.

Summary

(1) The Th contents of both Apollo 17 rock and soil samples are a linear function of the U contents with approximately zero y intercept. The experimental errors seem to account for all the scatter. The K contents are also linear functions of the U content, but the y intercepts are significantly different from zero, and the experimental errors do not account for all the scatter, especially in the case of the rocks. These relationships also hold for the samples from previous missions, except for the atypical behavior of the K content of Apollo 11 rocks.

(2) The variation in the solar proton induced activities in short-lived ($t_{1/2} < 3$ yr) radionuclides has been shown to be broadly consistent with the pattern and fluences of solar flare accelerated protons since 1956.

(3) The behavior of ^{56}Co activity in Apollo 17 samples as a function of ^{54}Mn activity seems to be the result of variations in the surface to volume ratios of the samples and in their exposure. It is not necessary to invoke an asymmetric solar proton flux to explain this behavior.

References

ALGIT (Apollo Lunar Geology Investigation Team USGS) (1973) Documentation and environment of the Apollo 17 samples: A preliminary report. *U.S. Geological Survey Interagency Report: Astrogeology* **71**, NASA Contract T-5874A.

Bostrom C. O., Williams D. J., and Arens J. F. (1967–1972) Solar proton monitor experiment. *Solar Geophysical Data*, Vols. 282–317, ESSA.

Clark R. S. and Keith J. E. (1973) Determination of natural and cosmic ray induced radionuclides in Apollo 16 lunar samples. *Proc. Fourth Lunar Sci. Conf., Geochim. Cosmochim. Acta*, Suppl. 4, Vol. 2, pp. 2105–2113. Pergamon.

Eldridge J. S., O'Kelley G. D., and Northcutt K. J. (1972) Abundances of primordial and cosmogenic radionuclides in Apollo 14 rocks and fines. *Proc. Third Lunar Sci. Conf., Geochim. Cosmochim. Acta*, Suppl. 3, Vol. 2, pp. 1651–1658. MIT Press.

Eldridge J. S., O'Kelley G. D., and Northcutt K. J. (1973) Radionuclide concentrations in Apollo 16 samples (abstract). In *Lunar Science—IV*, pp. 219–221. The Lunar Science Institute, Houston.

Eldridge J. S., O'Kelley G. D., and Northcutt K. J. (1974) Radionuclide concentrations in Apollo 16 lunar samples determined by non-destructive gamma-ray spectrometry. *Proc. Fourth Lunar Sci. Conf., Geochim. Cosmochim. Acta*, Suppl. 4, Vol. 2, pp. 2115–2122. Pergamon.

Keith J. E., Clark R. S., and Richardson K. A. (1972) Gamma-ray measurements of Apollo 12, 14 and 15 lunar samples. *Proc. Third Lunar Sci. Conf., Geochim. Cosmochim. Acta*, Suppl. 3, Vol. 2, pp. 1671–1680. MIT Press.

LSPET (Lunar Sample Preliminary Examination Team) (1974) Preliminary examination of lunar samples. In *Apollo 17 Preliminary Science Report*, NASA SP-330, pp. 7-1 to 7-46.

McLane J. C., King E. A., Flory D. A., Richardson K. A., Dawson J. P., Kemmerer W. W., and Wooley B. C. (1967) Lunar receiving laboratory. *Science* **155**, 525–529.

O'Kelley G. D., Eldridge J. S., Schonfeld E., and Bell P. R. (1970) Primordial radionuclide abundance, solar proton and cosmic-ray effects and ages of Apollo 11 lunar samples by non-destructive gamma-ray spectrometry. *Proc. Apollo 11 Lunar Sci. Conf., Geochim. Cosmochim. Acta*, Suppl. 1, Vol. 2, pp. 1407–1423. Pergamon.

O'Kelley G. D., Eldridge J. S., Schonfeld E., and Bell P. R. (1971a) Abundances of the primordial radionuclides K, Th, and U in Apollo 12 lunar samples by nondestructive gamma-ray spectrometry: Implications for origin of lunar soils. *Proc. Second Lunar Sci. Conf., Geochim. Cosmochim. Acta*, Suppl. 2, Vol. 2, pp. 1159–1168. MIT Press.

O'Kelley G. D., Eldridge J. S., Schonfeld E., and Bell P. R. (1971b) Cosmogenic radionuclide concentrations and exposure ages of lunar samples from Apollo 12. *Proc. Second Lunar Sci. Conf., Geochim. Cosmochim. Acta*, Suppl. 2, Vol. 2, pp. 1747–1755. MIT Press.

O'Kelley G. D., Eldridge J. S., Northcutt K. J., and Schonfeld E. (1972) Primordial radioelements and cosmogenic radionuclides in lunar samples from Apollo 15. *Proc. Third Lunar Sci. Conf., Geochim. Cosmochim. Acta*, Suppl. 3, Vol. 2, pp. 1659–1670. MIT Press.

Perkins R. W., Rancitelli L. A., Cooper J. A., Kaye J. H., and Wogman N. A. (1970) Cosmogenic and primordial radionuclide measurements in Apollo 11 lunar samples by nondestructive analysis. *Proc. Apollo 11 Lunar Sci. Conf., Geochim. Cosmochim. Acta*, Suppl. 1, Vol. 2, pp. 1455–1471. Pergamon.

Rancitelli L. A., Perkins R. W., Felix W. D., and Wogman N. A. (1971) Erosion and mixing of the lunar surface from cosmogenic and primordial and radionuclide measurements in Apollo 12 lunar samples. *Proc. Second Lunar Sci. Conf., Geochim. Cosmochim. Acta*, Suppl. 2, Vol. 2, pp. 1757–1772. MIT Press.

Rancitelli L. A., Perkins R. W., Felix W. D., and Wogman N. A. (1972) Lunar surface processes and cosmic ray characterization from Apollo 12–15 lunar sample analyses. *Proc. Third Lunar Sci. Conf., Geochim. Cosmochim. Acta*, Suppl. 3, Vol. 2, pp. 1681–1691. MIT Press.

Rancitelli L. A., Perkins R. W., Felix W. D., and Wogman N. A. (1973) Primordial radionuclides in soils and rocks from the Apollo 16 site (abstract). In *Lunar Science—IV*, pp. 615–617. The Lunar Science Institute, Houston.

Rancitelli L. A., Perkins R. W., Felix W. D., and Wogman N. A. (1974) Anisotrope of the August 4–7 solar flares at the Apollo 17 site (abstract). In *Lunar Science—V*, pp. 618–620. The Lunar Science Institute, Houston.

Webber W. R. (1963) A summary of solar cosmic ray events. In *Solar Proton Manual* (editor Frank McDonald), NASA TR R-169, p. 1.

Wrigley R. C. (1971) Some cosmogenic and primordial radionuclides in Apollo 12 lunar surface materials. *Proc. Second Lunar Sci. Conf., Geochim. Cosmochim. Acta*, Suppl. 2, Vol. 2, pp. 1791–1796. MIT Press.

Wrigley R. C. (1973) Radionuclides at Descartes (abstract). In *Lunar Science—IV*, pp. 799–800. The Lunar Science Institute, Houston.

Wrigley R. C. and Quaide W. L. (1970) [26]Al and [22]Na in lunar surface materials: Implications for depth distribution studies. *Proc. Apollo 11 Lunar Sci. Conf., Geochim. Cosmochim. Acta*, Suppl. 1, Vol. 2, pp. 1751–1757. Pergamon.

Yokoyama Y., Sato J., Reyes J. L., and Guichard F. (1973) Variation of solar cosmic-ray flux deduced from [22]Na–[26]Al data in lunar samples. *Proc. Fourth Lunar Sci. Conf., Geochim. Cosmochim. Acta*, Suppl. 4, Vol. 2, pp. 2209–2227. Pergamon.

York D. (1966) Least-squares fitting of a straight line. *Can. J. Phys.* **44**, 1079–1086.

Proceedings of the Fifth Lunar Conference
(Supplement 5, Geochimica et Cosmochimica Acta)
Vol. 2 pp. 2139–2147 (1974)
Printed in the United States of America

Cosmogenic radionuclides in samples from Taurus-Littrow: Effects of the solar flare of August 1972*

G. Davis O'Kelley, James S. Eldridge, and K. J. Northcutt

Oak Ridge National Laboratory, Oak Ridge, Tennessee 37830

Abstract—Cosmogenic radionuclide concentrations in a suite of Apollo 17 samples were determined nondestructively by use of gamma-ray spectrometers with high sensitivity and low background. Samples investigated were rocks 70135, 70185, 71135, 71136, 71175, 71566, 73215, 73255, 73275, 76295, 76597, and 79155; and <1 mm fines from soils 71131, 71501, 73121, 73131, 73141, 73221, 73241, 73261, 73281, 76501, 78501, 79221, and 79261. Cosmogenic radionuclides determined in this study were ^{22}Na, ^{26}Al, ^{46}Sc, ^{48}V, ^{54}Mn, ^{56}Co, and ^{7}Be.

The pattern of radionuclide concentrations observed in these Apollo 17 samples is quite distinct from that of any previous lunar sampling mission, due to proton bombardment by the intense solar flares of August 4–9, 1972. This intense proton irradiation made it possible for us to identify ^{7}Be in a lunar sample for the first time.

Data on the ^{56}Co and ^{54}Mn contents of thin surface samples were used to calculate the proton flux and rigidity of the August, 1972 solar flare. We determined the integrated proton flux $J_i(> 10 \text{ MeV}) = 1.9 \times 10^{10} \text{ cm}^{-2}$ and for an energy spectrum expressed as the function $E^{-\alpha}$, the shape parameter α was found to be $1.9^{+0.2}_{-0.1}$. Thus, the solar cosmic ray event of August 1972 exhibited a much higher average proton energy than other flares of cycle 20, which had characteristic values of $\alpha \approx 3.0$.

Introduction

THE APOLLO 17 voyage of exploration to the valley of Taurus-Littrow yielded the greatest scientific benefits of the six manned lunar landing missions. As a result of experience on previous Apollo flights, planning for Apollo 17 was characterized by a carefully selected landing site of great importance, a high degree of astronaut mobility, rapid scientific assessment both on the moon and on the earth, and a collection of well-documented samples representative of the geological complexities of the area. In addition, samples from the Taurus-Littrow Valley proved to be of great interest because, prior to collection, they had been subjected to proton bombardment by the intense sequence of solar flares of August 4–9, 1972. Data on radionuclide concentrations in thin surface samples offered the opportunity to determine the flux and energy spectrum of the solar flare protons, while samples from depth could be used to measure the influence of the galactic cosmic ray component.

As in our studies on samples from previous Apollo flights, we applied the methods of nondestructive gamma-ray spectrometry to the determination of radionuclides in samples from Taurus-Littrow. Measurements of cosmogenic radionuclides in 25 samples from Apollo 17 are reported below. Because several

*Research carried out under Union Carbide's contract with the U.S. Atomic Energy Commission through interagency agreements with the National Aeronautics and Space Administration.

radionuclides have short half-lives, it was desirable to begin measurements as soon as possible. Our first samples were received six days after splashdown, and additional measurements were carried out during the preliminary examination. An early account of the results on some of these samples was given in LSPET (1973b), and a more complete account in O'Kelley *et al.* (1974).

Nondestructive gamma-ray spectrometry has proved to be valuable as an accurate and rapid method for scanning a large number of samples to determine concentrations of the primordial radioelements K, Th, and U in lunar samples. Data on the K, Th, and U contents of 26 samples from Taurus-Littrow and correlations with local terrain features are reported by Eldridge *et al.* (1974) in a companion paper of this volume.

EXPERIMENTAL PROCEDURES

Two gamma-ray scintillation spectrometers were used for these analyses. The first was a Ge(Li) detector enclosed in a passive lead shield of 8 cm wall thickness. The detection efficiency at 1332 keV was 16% that of a 7.6×7.6 cm NaI(Tl) detector at a source-to-detector distance of 25 cm. Energy resolution expressed as full width at half-maximum counting rate at 1332 was 2.5 keV. Pulse-height distributions were recorded on a 4096-channel analyzer.

Most of the measurements were carried out with a scintillation gamma-ray spectrometer of low background, which contained two NaI(Tl) detectors, each 23 cm in diameter and 13 cm long, with 10 cm pure NaI light guides. The detectors were operated in singles and coincidence modes with a large plastic scintillator in anticoincidence. Further background reduction was achieved with a massive lead shield 20 cm thick. The data acquisition system included a coincidence–anticoincidence logic circuit interfaced to a 4096-channel analyzer. Singles data from each NaI(Tl) detector were recorded, and coincidence events were sorted in a 63×63-channel matrix. The mechanical design and performance of this system have been discussed in detail by Eldridge *et al.* (1973).

Least-squares fitting of the gamma-ray spectra was performed on an IBM 360/91 computer with ALPHA-M, a program written by Schonfeld (1967). Standard gamma-ray spectrum libraries were acquired by measurement of replicas containing radionuclide standards under the same experimental conditions as for the lunar samples. Replica fabrication, least-squares data analysis, and results on analyses of radionuclide test mixtures were discussed by O'Kelley *et al.* (1970) and by Eldridge *et al.* (1973). Errors listed in the tables below are overall estimates including counting statistics and uncertainties in the calibrations.

RESULTS AND DISCUSSION

Cosmogenic radionuclide concentrations

Concentrations of cosmogenic radionuclides determined in this study are given in Table 1. Values for short-lived nuclides have been corrected for decay to 2300 GMT December 14, 1972, the approximate time of departure of the lunar module's ascent stage from the moon.

The general pattern of radionuclide concentrations in Table 1 is profoundly different from that observed for samples from previous Apollo missions since, prior to collection, the Taurus-Littrow samples had been subjected to bombardment by the intense series of solar flares of August 4–9, 1972. Large enhancements in the yields of ^{22}Na, ^{46}Sc, ^{54}Mn, and ^{56}Co are due to the solar event of August 1972.

Table 1. Concentrations (dpm/kg) of cosmogenic radionuclides in Apollo 17 samples. Decay corrected to 2300 GMT, December 14, 1972.

Sample number	Type[a]	Mass (g)	^{26}Al	^{22}Na	^{54}Mn	^{56}Co	^{46}Sc	^{48}V	^{7}Be
					Rocks				
70135	CB[b]	446	38 ± 2	33 ± 3	56 ± 6	56 ± 6	32 ± 3	10 ± 5	
70185	CB	449	70 ± 4	50 ± 4	95 ± 10	105 ± 10	47 ± 5		
71135	FB[b]	36.2	80 ± 6	95 ± 7	140 ± 15	290 ± 50	70 ± 30		
71136	FB[b]	23.8	90 ± 8	93 ± 9	160 ± 60	300 ± 70	70 ± 30		
71175	MB	188	60 ± 3	68 ± 4	125 ± 8	120 ± 30	43 ± 12		
71566	CB	415	50 ± 2	49 ± 3	95 ± 8	—	—		
73215	BR	1057	85 ± 5	59 ± 5	36 ± 8	51 ± 20	3 ± 7		
73255	BR	392	75 ± 4	88 ± 6	86 ± 20	56 ± 20	—		
73275	BR	430	107 ± 5	99 ± 6	78 ± 12	96 ± 20	8 ± 5		
76295	BR[b]	261	67 ± 5	54 ± 4	38 ± 15	41 ± 7	5 ± 2		
78597	FB	319	48 ± 4	33 ± 4	80 ± 10	80 ± 20	25 ± 10		
79155	CB	316	70 ± 3	63 ± 3	120 ± 12	153 ± 8	65 ± 3		
					Soils (< 1 mm fines)				
71131	BS	50.1	69 ± 4	62 ± 6	102 ± 13	—	—		
71501	BS	100	73 ± 4	89 ± 6	135 ± 15	—	—		
73121	TT	100	189 ± 10	189 ± 10	137 ± 20	218 ± 20	15 ± 5		
73131	BS	100	54 ± 3	126 ± 5	75 ± 10	119 ± 12	15 ± 3		
73141	TB	101	62 ± 3	49 ± 5	26 ± 8	20 ± 10	10 ± 5		
73221	SS	46.3	197 ± 10	310 ± 15	230 ± 30	810 ± 40	33 ± 6		450 ± 350
73241	TT	100	92 ± 5	110 ± 5	80 ± 8	95 ± 10	10 ± 3		
73261	TB	100	57 ± 4	42 ± 4	52 ± 12	5 ± 10	8 ± 5		
73281	TB	50.3	46 ± 5	42 ± 9	50 ± 40	—	—		
76501	RS	97.9	90 ± 9	90 ± 9	60 ± 10	120 ± 12	18 ± 4	15 ± 10	
78501	RS	113	90 ± 9	105 ± 10	96 ± 10	150 ± 10	30 ± 6	10 ± 8	
79221	TT	100	130 ± 7	165 ± 10	215 ± 20	470 ± 25	65 ± 7	30 ± 20	
79261	TB	100	45 ± 4	43 ± 4	44 ± 6	26 ± 10	15 ± 4		

[a]CB = coarse basalt; FB = fine basalt; MB = medium basalt; BR = breccia; BS = bulk soil; TT = trench top; TM = trench middle; TB = trench bottom; SS = skim soil; RS = rake soil.
[b]Fragment or chip sampled from large boulder.

The high intensity and energy of the flare made it possible for several groups to identify ^{7}Be in a lunar sample for the first time. During the preliminary examination of the Apollo 17 samples O'Kelley et al., Rancitelli et al., and Schonfeld, all obtained results on ^{7}Be concentrations of some samples by nondestructive gamma-ray spectrometry, and these data were published in LSPET (1973b). In addition, Finkel et al. (1973) reported measurements of ^{7}Be abundance in Apollo 17 soils by use of destructive radiochemical analysis methods of high sensitivity.

Because chemical analysis data are lacking for most of the samples reported here, detailed interpretations cannot be made in some cases. However, the chemical analyses carried out during the preliminary examination of the Apollo 17 samples and reported by LSPET (1973a, 1973b) showed that the major element compositions of rocks and soils at Taurus-Littrow are similar to those at the

Apollo 15 site. Further, the variation in major element concentrations of specific types of geologic samples appears to be sufficiently regular to permit estimation of target element concentrations where needed.

Except for the Fra Mauro materials, surface samples from previous Apollo sites typically showed ^{26}Al/^{22}Na concentration ratios $\geqslant 2$ for samples in which the ^{26}Al concentration had reached its saturation value. Such concentration ratios have proved to be useful guides in searching for rocks with low exposure, since the ^{22}Na concentration measures short-term exposure and the ^{26}Al concentration determines cosmic ray exposure on a scale of a few million years. However, the ratio ^{26}Al/^{22}Na in most Apollo 17 samples is close to unity because the intense solar flare bombardment in August 1972 more than doubled the amount of ^{22}Na already present before the flare occurred. In addition, uncertainties in chemical composition of the samples make it difficult to identify samples of low exposure with respect to the 0.74 m.y. half-life of ^{26}Al.

The elevated yields of ^{56}Co and ^{54}Mn over samples examined previously are due to solar proton bombardment of iron. As will be shown later, the ^{56}Co concentration is especially useful in measuring the total proton flux, while ^{54}Mn is a product of higher energy proton bombardment and is helpful in determining the rigidity of the proton energy distribution.

Similarly, the high yields of ^{46}Sc in Apollo 17 surface samples arise from solar proton irradiation of titanium. Concentrations of ^{46}Sc, corrected for galactic cosmic ray production in iron, correlate well with estimates and measurements of titanium target element concentrations in these samples. Although sixteen-day ^{48}V is produced in solar flares by the reaction ^{48}Ti$(p, n)^{48}$V, over four months had elapsed between the August, 1972 solar flare and collection of the samples at Taurus-Littrow. Because of this long decay period following the solar flare, the concentrations of ^{48}V reported in Table 1 correspond to the steady-state concentration of ^{48}V expected (O'Kelley *et al.*, 1972) for the bombardment of iron by high-energy, galactic protons.

Concentration patterns for some of the radionuclides discussed above suggest that some of the rocks listed in Table 1 were shielded at least partially from the most recent solar flare. Three possibilities are 73215, for which no orientation exists, 70185, which may have been buried, and 78597, a rake sample. Additionally, 70135 and 76295 are boulder chips which either were shadowed by overhanging boulder material as partial shielding, or were irradiated by a solar proton flux whose average energy was a function of the incident angle. The latter explanation based on solar proton anisotropy has been applied by Rancitelli *et al.* (1974) to explain the low ratio of ^{56}Co to ^{54}Mn in boulder chip 76295.

Characterization of the August 1972 solar flare

Chemical analysis data are lacking for many of the samples which might be used to characterize the solar flare of August, 1972. For this reason, ^{56}Co and ^{54}Mn concentrations may be used to advantage, since both are produced by solar proton bombardment of iron, and concentrations of iron in the Apollo 17 samples concerned may be estimated by analogy to samples of known composition to an accuracy of about ± 10%. Excitation functions for production of ^{56}Co and ^{54}Mn are shown in Fig. 1 from the work of Brodzinski *et al.* (1971) and references therein. As shown in Fig. 1 the excitation function for production of ^{56}Co from iron has an effective threshold of about 6 MeV; hence, the ^{56}Co concentrations in thin surface samples is especially useful as a monitor of total proton flux. Information on the rigidity of the flare can be derived from the yields of ^{54}Mn, a product of higher energy proton bombardment, with a threshold of about 25 MeV.

Data on solar flare production of ^{54}Mn and ^{56}Co in five thin surface samples are compiled in Table 2. Concentrations shown earlier in Table 1 were corrected for galactic cosmic ray production by methods discussed by O'Kelley *et al.* (1971), and then the solar production component was corrected for decay to August 7, 1972. The low-energy product ^{56}Co exhibits very high concentration gradients at

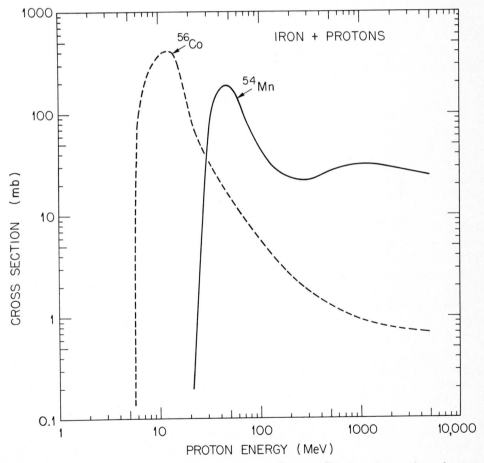

Fig. 1. Excitation functions for the production of ^{54}Mn and ^{56}Co from the reactions of protons with natural iron.

Table 2. Solar flare production of ^{54}Mn and ^{56}Co in thin surface samples.*

Sample	Type	^{54}Mn (dpm/kg)	^{56}Co (dpm/kg)	$\dfrac{^{56}\text{Co}}{^{54}\text{Mn}}$
71135	Chip	155 ± 40	890 ± 160	5.7
71136	Chip	182 ± 80	920 ± 220	5.1
73121	Trench Top	152 ± 25	660 ± 70	4.3
73221	Skim Soil	275 ± 30	2540 ± 130	9.2
79221	Trench Top	255 ± 30	1460 ± 80	5.7

*Decay corrected to August 7, 1972. Concentrations of ^{54}Mn and ^{56}Co shown in Table 1 were corrected for galactic cosmic ray production.

exposed surfaces, so the concentrations of ^{56}Co determined for the thin surface samples of Table 2 are very high with respect to the high-energy product ^{54}Mn. This is especially striking in the case of skim soil 73221, which has an effective thickness of only about 0.6 g/cm^2 and a concentration ratio ^{56}Co/^{54}Mn of 9.2.

The data were analyzed in terms of a model which for simplicity assumed the solar proton spectrum to be of the form

$$\frac{dJ}{dE} = kE^{-\alpha}$$

where J is the proton flux, E is the proton energy in MeV, k is a constant related to the flare intensity, and α is a shape parameter which measures the average energy of the flare. The model assumed the irradiated surface to be an infinite plane. The proton energy spectrum and relative flux were calculated as a function of depth, assuming that the number of secondary particles was negligible. Production of ^{54}Mn and ^{56}Co within each differential of thickness was obtained by integrating the product of the number of target atoms and the excitation function for the appropriate reaction over the attenuated energy spectrum. Details of more extensive calculations of this type have been discussed by Reedy and Arnold (1972) and by Rancitelli *et al.* (1971).

Results of the analysis are given in Table 3 and are compared with a report by Rancitelli *et al.* (1974) and a summary of satellite measurements compiled by King (1973). Agreement between all groups is quite satisfactory, even though the first two measurements of Table 3 were carried out on lunar surface samples and the data compiled by King (1973) were obtained from satellite measurements in a relatively unshielded region of space. The very high value obtained for the integrated proton flux $J_i(>10\text{ MeV}) = 1.9 \times 10^{10}$ protons/cm^2 verified that the solar flare event of August 4–9, 1972 was the largest of solar cycle 20. In fact, the solar proton fluences observed for the period August 4–9, 1972 constituted about 70% of the proton fluence above 10 MeV for the entire cycle (King, 1973). A shape parameter $\alpha = 1.9$ also implies that the average energy, or "rigidity," of the protons from the August 1972 flare is much higher than the average energy of the present solar cycle, which is characterized by a shape parameter $\alpha \simeq 3.0$.

Table 3. Integrated proton flux $J_i(>10\ \text{MeV})$ and spectrum data for the August 4–9, 1972 solar flare.

$J_i(>10\ \text{MeV})$, protons/cm^2	Shape parameter,* α	Reference
1.9×10^{10}	$1.9^{+0.2}_{-0.1}$	Present work
1.58×10^{10}	2.0 ± 0.1	Rancitelli et al. (1974)
2.25×10^{10}	—	King (1973)

*The parameter α in the energy distribution $dJ/dE = kE^{-\alpha}$.

Trench samples

Of thirteen soil samples from the valley of Taurus-Littrow which we have analyzed, more than half were associated with trenching operations. Because samples from a trench give a rather low-resolution view of the variations in radionuclide concentration with depth, as compared with that obtained from segments of a core in the lunar regolith, our discussion will be rather qualitative. However, trench samples are of much more interest for their information on color, texture, mineralogy, and chemistry (Eldridge et al., 1974).

Sample 73121 was discussed earlier, and sample 73141 was taken from a trench near 73121 at a depth of 15 cm in the light mantle. The radionuclide concentrations found at 15 cm depth and shown in Table 1 are those expected for galactic cosmic ray production. Not related to this trench, as might be expected from its number, is 73131. This sample was removed from the wall of a 2 m diameter crater, 15 cm down the crater wall and may contain particles from the comminution of a soil clod or soft breccia removed simultaneously with the soil. Sample 73131 has received an intense, recent, solar-flare irradiation; however, the concentration of ^{26}Al appears to be well below the surface saturation value. Although speculative in the absence of more detailed information concerning the sampling procedures used, the low ^{26}Al content of 73131 suggests that the small 2 m crater at Station 2a was formed within the last million years.

From Station 3 we obtained a skim soil 73221, light-gray soil 73241 from 0 to 5 cm depth of a trench, a medium-gray soil 73261 from 5 to 10 cm depth, and a light-gray soil 73281 from an adjacent zone in the trench, also from about 5 to 10 cm depth. Again, the expected depth dependence was found, as shown in Table 1. It will be noted that for our present purpose, 73261 and 73281 are duplicate samples and exhibit essentially the same radionuclide concentrations.

At Station 9, on the southeast flank of the Van Serg Crater ejecta blanket, sample 79221 was taken from the uppermost 2 cm of a trench. Data from this sample were used earlier in Table 2. Sample 79261, from a sampling depth of 7 to 17 cm, was removed from a light-gray or white layer. The expected decrease in radionuclide concentrations from surface to depth again was noted. The concentrations of ^{26}Al, ^{22}Na, ^{54}Mn in 79261 agree well with calculations of galactic cosmic ray production rates (cf. Reedy and Arnold, 1972), but concentrations of ^{56}Co and ^{46}Sc are somewhat higher than expected.

Radionuclides produced by solar proton bombardment, especially ^{56}Co, are sensitive indicators of the mixing of surface soil into deeper trench samples. Data on ^{56}Co presented in Table 1 demonstrate that the trenching discussed in this section was carried out under excellent control. Only 79261 may show evidence for material mixed from above.

Acknowledgments—The authors gratefully acknowledge the assistance and advice provided by the Lunar Sample Analysis Planning Team. We wish to thank the staff of the Curator's Office, NASA Johnson Space Center, especially J. O. Annexstad, R. S. Clark, and M. A. Reynolds, for their aid in rapidly and efficiently processing the samples for this study.

REFERENCES

Brodzinski R. L., Rancitelli L. A., Cooper J. A., and Wogman N. A. (1971) High-energy proton spallation of iron. *Phys. Rev.* **C4**, 1257–1265.

Eldridge J. S., O'Kelley G. D., Northcutt K. J., and Schonfeld E. (1973) Nondestructive determination of radionuclides in lunar samples using a large, low-background gamma-ray spectrometer and a novel application of least-squares fitting. *Nucl. Instrum. Methods* **112**, 319–322.

Eldridge J. S., O'Kelley G. D., and Northcutt K. J. (1974) Primordial radioelement concentrations in rocks and soils from Taurus-Littrow. *Proc. Fifth Lunar Sci. Conf., Geochim. Cosmochim. Acta.* This volume.

Finkel R. C., Wahlen M., Arnold J. R., Kohl C. P., and Immamura M. (1973) The gradient of cosmogenic radionuclides to a depth of 400 g/cm^2 in the moon (abstract). In *Lunar Science—IV*, pp. 242–243. The Lunar Science Institute, Houston.

King J. H. (1973) Solar proton fluences as observed during 1966–1972 and as predicted for 1977–1983 space missions. Goddard Spaceflight Center report X-601-73-324. Submitted for publication in the *Journal of Spacecraft and Rockets.*

LSPET (Lunar Sample Preliminary Examination Team) (1973a) Apollo 17 lunar samples: Chemical and petrographic description. *Science* **182**, 659–672.

LSPET (Lunar Sample Preliminary Examination Team) (1973b) Preliminary examination of lunar samples. In *Apollo 17 Preliminary Science Report*. NASA Johnson Space Center Report NASA SP-330, Chapter 7.

O'Kelley G. D., Eldridge J. S., Schonfeld E., and Bell P. R. (1970) Primordial radionuclide abundances, solar proton and cosmic-ray effects, and ages of Apollo 11 lunar samples by nondestructive gamma-ray spectrometry. *Proc. Apollo 11 Lunar Sci. Conf., Geochim. Cosmochim. Acta*, Suppl. 1, Vol. 2, pp. 1407–1423. Pergamon.

O'Kelley G. D., Eldridge J. S., Schonfeld E., and Bell P. R. (1971) Cosmogenic radionuclide concentrations and exposure ages of lunar samples from Apollo 12. *Proc. Second Lunar Sci. Conf., Geochim. Cosmochim. Acta*, Suppl. 2, Vol. 2, pp. 1747–1755. MIT Press.

O'Kelley G. D., Eldridge J. S., Northcutt K. J., and Schonfeld E. (1972) Primordial radioelements and cosmogenic radionuclides in lunar samples from Apollo 15. *Proc. Third Lunar Sci. Conf., Geochim. Cosmochim. Acta*, Suppl. 3, Vol. 2, pp. 1659–1670. MIT Press.

O'Kelley G. D., Eldridge J. S., and Northcutt K. J. (1974) Concentrations of cosmogenic radionuclides in Apollo 17 samples: Effects of the solar flare of August, 1972 (abstract). In *Lunar Science—V*, pp. 577–579. The Lunar Science Institute, Houston.

Rancitelli L. A., Perkins R. W., Felix W. D., and Wogman N. A. (1971) Erosion and mixing of the lunar surface from cosmogenic and primordial radionuclide measurements in Apollo 12 lunar samples. *Proc. Second Lunar Sci. Conf., Geochim. Cosmochim. Acta*, Suppl. 2, Vol. 2, pp. 1757–1772. MIT Press.

Rancitelli L. A., Perkins R. W., Felix W. D., and Wogman N. A. (1974) Anisotropy of the August 4–7, 1972 solar flares at the Apollo 17 site (abstract). In *Lunar Science—V*, pp. 618–620. The Lunar Science Institute, Houston; also, verbal report.

Reedy R. C. and Arnold J. R. (1972) Interaction of solar and galactic cosmic-ray particles with the moon. *J. Geophys. Res.* **77**, 537–555.

Schonfeld E. (1967) ALPHA-M—An improved computer program for determining radioisotopes by least-squares resolution of the gamma-ray spectra. *Nucl. Instrum. Methods* **52**, 177–178.

Proceedings of the Fifth Lunar Conference
(Supplement 5, Geochimica et Cosmochimica Acta)
Vol. 2 pp. 2149–2184 (1974)
Printed in the United States of America

Rare gases and trace elements in Apollo 15 drill core fines: Depositional chronologies and K–Ar ages, and production rates of spallation-produced ^{3}He, ^{21}Ne, and ^{38}Ar versus depth

R. O. Pepin, J. R. Basford, and J. C. Dragon

School of Physics and Astronomy, University of Minnesota, Minneapolis, Minnesota 55455

M. R. Coscio, Jr. and V. R. Murthy

Department of Geology and Geophysics, University of Minnesota, Minneapolis, Minnesota 55455

Abstract—Concentrations of spallation-produced ^{3}He, ^{21}Ne, ^{38}Ar, ^{78,83}Kr, and ^{126,131}Xe were derived by applying correlation techniques to rare gas and chemical data from six sized soil samples evenly spaced down the length of the Apollo 15 drill core. Profiles defined by these spallation nuclides are all smooth functions of depth, and are therefore in agreement with the conclusion of Russ *et al.* (1972), based on the neutron fluence versus depth profile, that the depositional history of the drill core section is coherent and relatively simple.

A rapid deposition model for the core section, in which ^{38}Ar$_{sp}$ production with depth is assumed to have the same relative shape as the ^{37,39}Ar production rate profiles calculated by Reedy and Arnold (1972), yields best-fit parameters virtually identical to those derived from the neutron fluence data by Russ *et al.*: rapid deposition of uniformly preirradiated material from at or below the base of the drill core to a level ~45 g/cm² below the present regolith surface, followed by *in situ* cosmic-ray irradiation of the section for 450 m.y. and recent deposition of the upper surface layer.

If the Reedy and Arnold production rate profiles for ^{3}H and ^{22}Na are assumed to apply to ^{3}He$_{sp}$ and ^{21}Ne$_{sp}$ production with depth, there exist no simple model fits to the observed spallation ^{3}He distribution, and the rapid deposition model is incompatible with the ^{21}Ne$_{sp}$ data. Continuously accreting model fits to the Ne, Ar, and neutron fluence data yield discordant depositional parameters. The ^{3}He$_{sp}$ profile strongly suggests an unrecognized low-energy neutron capture reaction for either ^{3}H or ^{3}He production. Extension of the rapid deposition model to include a production term which follows the low-energy neutron flux profile yields quantitative matches of the predicted ^{3}He$_{sp}$ and ^{21}Ne$_{sp}$ profiles to the measured profiles. Spallation Kr and Xe distributions with depth are consistent with their identification as high-energy spallation products. In the absence of calculated production rate versus depth profiles for these nuclides, they provide only general constraints on depositional history.

Production rates of spallogenic ^{3}He, ^{21}Ne, and ^{38}Ar versus depth are calculated from rapid deposition model fits to the measured profiles. Production rates for ^{3}He$_{sp}$ and ^{21}Ne$_{sp}$ are significantly lower than the values derived by Bogard *et al.* (1971). These production rates yield concordant ^{81}Kr–Kr, ^{21}Ne and ^{38}Ar exposure chronologies for specific shielding depths and irradiation history models of rock samples from the Apollo 14, 15, and 16 sites.

K–Ar ages of these soils from isochron correlations are close to 3 AE. There is marginal evidence in the data that K–Ar ages may increase uniformly down the cored section, from ~2.7 AE at 65 g/cm² to ~3.2 AE at 412 g/cm². If correct, this nonspallogenic depth effect effectively rules out instantaneous deposition of the whole section. With K–Ar ages spanning the interval around 2.9 AE, the drill core samples thus join soils from several other lunar sites in showing evidence for major thermal activity on the lunar surface ~2.9 AE ago.

Introduction

The discovery by Russ *et al.* (1972) that neutron fluences vary smoothly with depth in the Apollo 15 deep drill core implies a relatively simple irradiation history for this regolith section. Depositional histories ranging from rapid accretion of the present section from uniformly preirradiated regolith 450 m.y. ago, to continuous accretion for ~400 m.y., ending ~500 m.y. ago, are compatible with the neutron fluence data (Russ *et al.*, 1972). These models require the specific concentrations (abundances/target element abundances) of all cosmogenic nuclei to be smooth functions of depth, and predict within narrow limits the functional forms for those nuclei whose production rates versus depth are known. Examination of the depth dependence of the cosmic-ray spallation components of all five rare gases in the Apollo 15 drill core samples might therefore provide enough information to assess both the validity of the neutron fluence models and the applicability to the rare gases of production rate versus depth relationships calculated for certain cosmogenic radionuclides.

Studies of concentration versus depth profiles for certain spallogenic rare gas nuclides in Apollo 15 drill core fines have been carried out by Bogard *et al.* (1973) and Hübner *et al.* (1973). Both investigations led to similar conclusions: (1) that the depth distribution of $^{21}Ne_{sp}$ could on the average be consistent with either of the extreme neutron fluence models—rapid accumulation of preirradiated regolith or uniform accumulation of unirradiated or weakly preirradiated regolith, both followed by several hundred million years of *in situ* irradiation—with roughly comparable accumulation and/or *in situ* irradiation times; and (2) that the concentration of $^{21}Ne_{sp}$ is not a smooth function of depth over short vertical distances. Concentration fluctuations of 10–30% or larger occur in samples a few centimeters apart, not only in Ne but in $^{80}Kr_{sp}$ and $^{126}Xe_{sp}$ as well (Bogard *et al.*, 1973); in both studies these are attributed to probable variations in the near-surface radiation dosages experienced by these materials prior to or during accumulation of the present core section.

Both Bogard *et al.* (1973) and Hübner *et al.* (1973) calculated spallogenic rare gas concentrations by subtracting an assumed "solar" (nonspallogenic) component from measured isotopic abundances in bulk soil samples. Since the method must assume that a particular "solar" component isotopic composition is uniformly applicable to the drill core samples, and since the spallation gas concentrations in these samples are relatively small fractions of the total gases and are therefore sensitive to the assumed "solar" compositions, there are substantial uncertainties associated with spallation gas concentrations calculated in this way; this point was explicitly recognized and discussed by both Bogard *et al.* and Hübner *et al.* It is possible that a significant fraction of their observed departures from smooth concentration versus depth relations are generated by the analytical method. For this reason, we chose to carry out rare gas and trace element chemical analyses on separated grain-sized fractions of the drill core fines. In most cases this procedure, utilizing correlation techniques, yields a check on the assumption of a two-component gas system, a direct determination of the isotopic composition of the nonspallogenic (*surface-correlated* or *trapped*) gas component, and a measure of the spallation gas composition, all of which permit calculation of spallation gas concentrations with a minimum of assumptions (Eberhardt *et al.*, 1970, 1972; Basford *et al.*, 1973). Applied to the argon isotopes, the grain-sized method can also yield the bulk K–Ar ages of lunar soils from isochron analysis (Pepin *et al.*, 1972; Venkatesan *et al.*, 1974).

We have measured all stable isotopes of He, Ne, Ar, Kr, and Xe, and the concentrations of K, Rb, Sr, and Ba, in four grain-sized fractions of each of six samples of $<250 \mu$ Apollo 15 drill core stem fines, spaced at $\sim$40 cm intervals down the length of the core. These samples were from materials extracted from the base of each of the six drill stems and distributed as "early allocation" samples. The Minnesota and CIT samples were taken from the same level in each stem, and therefore the neutron fluence data reported by Russ *et al.* (1972) and the rare gas and chemical data presented in this report were obtained on at least approximate aliquots of the same materials.

Experimental Procedure and Results

Grain-sized separates of each of the $\sim$25 mg drill core samples were prepared by sieving in acetone. Grain size intervals, and masses of the aliquots used for rare gas analysis, are given in Table 1. Masses of the aliquots taken for chemical determinations were generally in the range 1–2 mg. Sample preparation procedures, and details of both rare gas and isotope dilution mass spectrometry, were unchanged from those described by Basford *et al.* (1973) and Evensen *et al.* (1973).

Concentrations and isotopic compositions of He, Ne, Ar, Kr, and Xe are set out in Tables 1–3, together with concentrations of K, Sr, and Ba; for Rb concentrations, and additional details and discussion of the chemical measurements, see Evensen *et al.* (1974). Isotopic compositions in Tables 1–3 are referenced to an internal laboratory standard for He, to Eberhardt *et al.*'s (1965) composition ($^{20}Ne/^{22}Ne = 9.800 \pm 0.080$, $^{21}Ne/^{22}Ne = 0.02899 \pm 0.00025$) of atmospheric Ne, to Nier's (1950) composition ($^{40}Ar/^{36}Ar = 296.0 \pm 0.5$, $^{38}Ar/^{36}Ar = 0.1880 \pm 0.0005$) of atmospheric Ar, and to the atmospheric compositions given in Tables 5 and 6 (Basford *et al.*, 1973) for Kr and Xe. System blank corrections were always $<1\%$ for ^{3}He and negligible for ^{4}He; $<2\%$ in three samples and $<1\%$ otherwise for ^{22}Ne, where the dominant correction is for CO_2^{++}, and negligible for ^{21}Ne and ^{20}Ne; $\leqslant 3\%$ for ^{40}Ar and negligible for ^{36}Ar and ^{38}Ar; and $\leqslant 2\%$ for ^{84}Kr and ^{132}Xe. Uncertainties in isotopic compositions in Tables 1–3 are 1 σ statistical errors which include uncertainties in blank and mass discrimination corrections as well as statistical errors in the measured ratios; the 1 σ errors in absolute concentrations include both systematic components originating in the gas calibration system and statistical errors in peak height measurements and instrumental nonlinearity corrections.

Table 1. Helium, neon, argon, and potassium in grain-sized separates of $<250\ \mu$ fines from the Apollo 15 drill core stems.

Sample	Mass (mg)	^{4}He	^{22}Ne	^{36}Ar	K (ppm)	^{3}He/^{4}He ($\times 10^{-4}$)	^{20}Ne/^{22}Ne	^{21}Ne/^{22}Ne ($\times 10^{-2}$)	^{38}Ar/^{36}Ar ($\times 10^{-1}$)	^{40}Ar/^{36}Ar
		(units of 10^{-5} cc STP/g)								
15006,26										
105–250 μ	2.894	919.	2.334	6.68	1899	6.93	11.880	6.692	1.975	1.539
		±16	±.047	±.13	±95	±.12	±.010	±.007	±.007	±.016
37–105 μ	7.219	1876.	4.33	10.17	2020	5.57	12.038	5.313	1.950	1.204
		±43	±.11	±.15	±100	±.14	±.007	±.009	±.006	±.006
16–37 μ	3.423	4185.	8.60	17.08	2440	4.58	12.262	4.255	1.930	1.020
		±96	±.28	±.24	±120	±.11	±.010	±.006	±.006	±.004
$<16\ \mu$	4.296	18610.	25.83	56.5	1750	4.31	12.538	3.428	1.904	0.929
		±480	±.72	±1.5	±88	±.13	±.008	±.006	±.006	±.003
15005,25										
105–250 μ	3.345	499.	1.281	3.271	2120	10.04	11.270	10.552	2.058	3.036
		±16	±.027	±.049	±110	±.20	±.014	±.023	±.008	±.010
37–105 μ	6.134	1673.	3.029	6.255	1778	6.48	12.023	6.616	1.981	2.099
		±57	±.094	±.093	±89	±.13	±.024	±.017	±.006	±.007
16–37 μ	3.687	3653.	6.28	11.98	1875	4.87	12.272	4.718	1.923	1.635
		±99	±.19	±.18	±94	±.10	±.021	±.012	±.006	±.005
$<16\ \mu$	5.396	13930.	18.26	40.9	2120	4.249	12.594	3.614	1.883	1.520
		±490	±.60	±1.3	±110	±.093	±.026	±.016	±.006	±.005
15004,26										
74–250 μ	6.465	1738.	3.416	7.37	1851	5.80	11.932	5.897	1.966	1.706
		±40	±.085	±.11	±93	±.12	±.011	±.009	±.006	±.009
37–74 μ	3.462	2411.	4.64	9.30	1422	5.18	12.190	5.179	1.948	1.460
		±55	±.13	±.16	±71	±.23	±.012	±.009	±.006	±.011
16–37 μ	3.446	4670.	9.06	15.49	1703	4.45	12.201	4.212	1.925	1.487
		±110	±.24	±.24	±85	±.12	±.011	±.005	±.006	±.008
$<16\ \mu$	4.176	17650.	26.27	50.2	1809	4.08	12.471	3.461	1.905	1.140
		±460	±.74	±1.3	±90	±.18	±.020	±.008	±.006	±.005
15003,28										
74–250 μ	5.392	1409.	2.764	5.36	1815	5.89	11.837	6.388	1.990	2.222
		±34	±.083	±.10	±91	±.16	±.027	±.024	±.007	±.020
37–74 μ	3.896	1746.	3.591	7.01	1887	5.79	12.076	5.406	1.964	1.798
		±40	±.090	±.12	±94	±.23	±.040	±.012	±.007	±.013
16–37 μ	3.535	3443.	7.13	12.26	1896	4.60	12.133	4.365	1.936	1.591
		±79	±.18	±.19	±95	±.13	±.036	±.007	±.006	±.009
$<16\ \mu$	5.208	12500.	19.30	40.9	2060	4.25	12.523	3.501	1.908	1.243
		±320	±.54	±1.0	±100	±.13	±.008	±.006	±.006	±.005
15002,26										
74–250 μ	5.400	1324.	2.345	5.798	2400	5.69	12.065	5.915	1.959	2.218
		±30	±.059	±.090	±120	±.24	±.011	±.012	±.007	±.012
37–74 μ	3.544	2325.	4.99	8.76	1987	4.92	11.984	4.819	1.950	1.597
		±53	±.13	±.15	±99	±.18	±.014	±.009	±.006	±.011
16–37 μ	3.767	4690.	12.76	15.32	1892	4.25	11.874	3.960	1.930	1.318
		±120	±.34	±.23	±95	±.42	±.012	±.006	±.006	±.007
$<16\ \mu$	4.335	17440.	26.99	48.4	2290	4.22	12.391	3.383	1.906	1.128
		±450	±.78	±1.2	±110	±.16	±.006	±.003	±.006	±.005
15001,30										
74–250 μ	6.222	817.	1.548	3.273	2390	5.32	11.641	7.318	.2.005	3.929
		±19	±.043	±.046	±120	±.32	±.056	±.031	±.007	±.013
37–74 μ	2.124	1542.	2.842	5.206	2500	4.98	11.994	5.710	1.980	2.960
		±26	±.060	±.076	±120	±.12	±.016	±.009	±.007	±.013
16–37 μ	2.365	2480.	4.75	8.48	2300	4.58	12.135	4.660	1.952	2.092
		±57	±.11	±.12	±120	±.19	±.006	±.005	±.006	±.008
$<16\ \mu$	3.687	9440.	15.41	34.49	3930	4.18	12.413	3.521	1.894	1.725
		±250	±.43	±.79	±200	±.23	±.009	±.005	±.006	±.008

Table 2. Krypton and strontium in grain-sized separates of <250 μ fines from the Apollo 15 drill core stems.

Sample	^{84}Kr ($\times 10^{-8}$ cc STP/g)	Sr (ppm)	^{78}Kr	^{80}Kr	^{82}Kr	^{83}Kr	^{86}Kr
					relative to ^{84}Kr = 100		
15006,26							
105–250 μ	4.66	129	0.9995	5.020	21.47	21.86	30.24
	±.22	±6	±.0070	±.030	±.08	±.08	±.09
37–105 μ	7.48	*132	0.8853	4.719	21.25	21.45	30.40
	±.24	±13	±.0042	±.017	±.05	±.06	±.06
16–37 μ	11.98	133	0.8015	4.466	21.00	21.06	30.38
	±.46	±7	±.0046	±.018	±.04	±.05	±.07
<16 μ	37.5	135	0.6721	4.160	20.48	20.48	30.42
	±1.2	±7	±.0025	±.010	±.04	±.03	±.05
15005,25							
105–250 μ	2.20	119	1.481	6.495	23.38	24.31	30.03
	±.12	±6	±.013	±.037	±.09	±.10	±.11
	4.17	*125	1.121	5.453	21.99	22.51	30.16
	±.13	±12	±.006	±.028	±.07	±.06	±.09
16–37 μ	7.81	133	0.8856	4.717	21.18	21.33	29.98
	±.32	±7	±.0061	±.024	±.10	±.07	±.10
<16 μ	27.3	149	0.7124	4.297	20.74	20.68	30.29
	±1.1	±7	±.0025	±.014	±.04	±.05	±.07
15004,26							
74–250 μ	5.10	130	0.9201	4.929	21.45	21.71	30.24
	±.23	±6	±.0051	±.021	±.05	±.06	±.07
37–74 μ	6.22	117	0.8646	4.710	21.21	21.40	30.22
	±.21	±6	±.0056	±.024	±.06	±.05	±.07
16–37 μ	10.88	125	0.7834	4.495	20.96	21.08	30.36
	±.41	±6	±.0040	±.021	±.05	±.05	±.08
<16 μ	33.1	138	0.6771	4.192	20.56	20.50	30.41
	±1.4	±7	±.0026	±.012	±.04	±.04	±.05
15003,28							
74–250 μ	3.37	*120	1.030	5.212	21.82	22.20	30.13
	±.11	±12	±.009	±.031	±.07	±.08	±.09
37–74 μ	4.59	116	0.9105	4.855	21.31	21.63	30.12
	±.29	±6	±.0061	±.027	±.06	±.06	±.09
16–37 μ	8.44	130	0.8127	4.552	20.97	21.21	30.22
	±.53	±6	±.0048	±.017	±.06	±.05	±.08
<16 μ	24.00	153	0.6752	4.214	20.58	20.58	30.43
	±.77	±8	±.0030	±.012	±.04	±.04	±.06
15002,26							
74–250 μ	3.88	120	0.9832	5.093	21.59	21.96	30.23
	±.12	±6	±.0058	±.020	±.06	±.06	±.08
37–74 μ	5.58	116	0.8773	4.739	21.21	21.49	30.07
	±.18	±6	±.0063	±.026	±.08	±.07	±.08
16–37 μ	10.01	126	0.7696	4.437	20.81	20.96	30.40
	±.35	±6	±.0040	±.018	±.05	±.05	±.07
<16 μ	31.6	155	0.6727	4.177	20.52	20.44	30.37
	±1.1	±8	±.0030	±.012	±.04	±.04	±.06

Table 2. (*Continued*).

Sample	^{84}Kr ($\times 10^{-8}$ cc STP/g)	Sr (ppm)	^{78}Kr	^{80}Kr	^{82}Kr	^{83}Kr	^{86}Kr
			relative to ^{84}Kr = 100				
15001,30							
74–250 μ	2.15	139	1.242	5.894	22.71	23.48	30.03
	±.07	±7	±.010	±.035	±.07	±.05	±.09
37–74 μ	3.43	128	1.020	5.131	21.69	21.95	30.28
	±.11	±6	±.008	±.043	±.09	±.10	±.14
16–37 μ	5.24	*140	0.8855	4.764	21.25	21.49	30.22
	±.17	±14	±.0074	±.037	±.08	±.08	±.11
<16 μ	20.76	153	0.7025	4.264	20.66	20.59	30.31
	±.66	±8	±.0036	±.016	±.05	±.04	±.06

*Sr fraction lost: concentration estimated from adjacent or similar size fractions.

Table 3. Xenon and barium in grain-sized separates of <250 μ fines from the Apollo 15 drill core stems.

Sample	^{132}Xe ($\times 10^{-9}$ cc STP/g)	Ba (ppm)	^{124}Xe	^{126}Xe	^{128}Xe	^{129}Xe	^{130}Xe	^{131}Xe	^{134}Xe	^{136}Xe
			relative to ^{132}Xe = 100							
15006,26										
105–250 μ	6.28	272	2.127	3.098	12.14	106.9	18.41	92.89	36.71	29.54
	±.24	±14	±.028	±.064	±.18	±.7	±.20	±.53	±.25	±.34
37–105 μ	9.53	278	1.556	2.368	11.27	105.8	17.78	90.42	36.72	29.62
	±.41	±14	±.013	±.021	±.09	±.4	±.07	±.36	±.25	±.11
16–37 μ	14.07	270	1.224	1.630	10.25	105.8	17.33	87.00	36.55	29.83
	±.39	±14	±.013	±.020	±.09	±.4	±.10	±.34	±.18	±.17
<16 μ	45.0	302	0.7357	0.917	9.054	104.7	16.85	84.12	36.89	30.10
	±1.1	±15	±.0063	±.011	±.053	±.3	±.06	±.19	±.12	±.08
15005,25										
105–250 μ	3.51	340	3.406	6.004	16.11	109.0	20.62	104.99	35.61	29.14
	±.11	±17	±.052	±.071	±.15	±.6	±.24	±.78	±.36	±.35
37–105 μ	6.09	253	2.119	3.309	12.57	107.8	19.01	95.40	36.75	29.52
	±.23	±13	±.026	±.034	±.13	±.5	±.17	±.45	±.28	±.17
16–37 μ	11.22	279	1.410	2.055	10.72	107.1	17.70	89.41	37.05	29.64
	±.30	±14	±.019	±.024	±.12	±.5	±.14	±.60	±.26	±.15
<16 μ	34.9	316	0.8406	1.085	9.451	106.2	17.09	85.09	36.97	30.00
	±.9	±16	±.0078	±.008	±.055	±.3	±.06	±.20	±.14	±.10
15004,26										
74–250 μ	6.62	*252	1.548	2.467	11.48	107.0	17.74	92.78	36.79	29.70
	±.17	±25	±.023	±.018	±.11	±.4	±.21	±.47	±.23	±.13
37–74 μ	8.06	248	1.421	2.090	11.05	105.5	17.74	90.32	36.36	30.34
	±.21	±12	±.023	±.034	±.14	±.6	±.18	±.60	±.22	±.20
16–37 μ	12.62	240	1.104	1.562	10.11	105.6	17.36	87.75	37.16	30.15
	±.33	±12	±.014	±.022	±.09	±.4	±.12	±.32	±.17	±.15
<16 μ	40.7	268	0.7266	0.8898	9.043	105.4	16.79	84.51	36.90	29.97
	±1.1	±13	±.0075	±.0062	±.045	±.2	±.08	±.19	±.11	±.10

Table 3. (*Continued*).

Sample	^{132}Xe ($\times 10^{-9}$ cc STP/g)	Ba (ppm)	^{124}Xe	^{126}Xe	^{128}Xe	^{129}Xe	^{130}Xe	^{131}Xe	^{134}Xe	^{136}Xe
			relative to ^{132}Xe = 100							
15003,28										
74–250 μ	5.75	*270	1.927	3.115	12.34	106.8	18.37	97.06	36.26	29.47
	±.16	±27	±.019	±.043	±.17	±.7	±.20	±.55	±.26	±.26
37–74 μ	6.30	245	1.446	2.304	10.91	105.3	17.65	91.48	36.73	29.30
	±.17	±12	±.034	±.030	±.06	±.7	±.16	±.41	±.28	±.31
16–37 μ	11.15	263	1.136	1.635	10.04	105.2	17.52	87.76	36.91	29.87
	±.28	±13	±.019	±.022	±.13	±.3	±.13	±.39	±.17	±.21
<16 μ	35.3	297	0.7326	0.8996	9.067	105.5	16.86	84.77	36.85	29.86
	±.9	±15	±.0090	±.0071	±.039	±.3	±.07	±.19	±.10	±.12
15002,26										
74–250 μ	5.20	337	1.934	3.056	12.26	106.2	18.83	97.35	36.53	29.70
	±.15	±17	±.029	±.030	±.14	±.5	±.18	±.51	±.16	±.15
37–74 μ	7.49	280	1.442	2.083	10.66	105.4	17.53	90.13	36.64	29.54
	±.20	±14	±.025	±.027	±.13	±.6	±.19	±.60	±.23	±.37
16–37 μ	12.98	292	1.052	1.442	9.72	105.1	17.35	87.43	37.20	29.96
	±.35	±15	±.018	±.021	±.12	±.5	±.14	±.34	±.19	±.15
<16 μ	39.6	356	0.6869	0.8344	8.992	105.2	16.83	84.17	36.86	30.06
	±1.0	±18	±.0065	±.0094	±.049	±.3	±.07	±.21	±.10	±.08
15001,30										
74–250 μ	3.31	377	2.672	4.296	14.10	107.1	19.93	104.82	36.46	29.56
	±.09	±19	±.025	±.056	±.15	±.6	±.21	±.44	±.36	±.22
37–74 μ	5.15	344	1.652	2.967	11.84	106.2	18.19	95.07	37.39	29.67
	±.17	±17	±.039	±.062	±.20	±.9	±.28	±.74	±.38	±.36
16–37 μ	8.55	342	1.273	1.857	10.38	106.0	17.99	90.38	37.15	30.35
	±.23	±17	±.021	±.026	±.18	±.5	±.19	±.51	±.34	±.26
<16 μ	30.91	383	0.7927	0.987	9.285	106.6	17.00	85.47	36.87	29.91
	±.77	±19	±.0088	±.011	±.050	±.2	±.07	±.26	±.10	±.12

*Ba fraction lost: concentration estimated from adjacent or similar size fractions.

DISCUSSION

Correlation systematics

The two-component structure of rare gases in lunar fines may be represented by the equations $(^M X) = (^M X)_{sc} + (^M X)_{vc}$ and $(^{M'} X) = (^{M'} X)_{sc} + (^{M'} X)_{vc}$, where M and M' are any two isotopes of gas X, and sc and vc designate the surface and volume correlated components respectively. With M and M' representing isotopic abundances, the ratio of these equations may be expressed as

$$\left(\frac{M'}{M}\right) = \frac{(M)_{vc}}{M}\left[\left(\frac{M'}{M}\right)_{vc} - \left(\frac{M'}{M}\right)_{sc}\right] + \left(\frac{M'}{M}\right)_{sc} \tag{1}$$

or

$$\left(\frac{M'}{M}\right) = \frac{[C]}{M}\left[\frac{(M)_{vc}}{[C]}\left[\left(\frac{M'}{M}\right)_{vc} - \left(\frac{M'}{M}\right)_{sc}\right]\right] + \left(\frac{M'}{M}\right)_{sc} \tag{2}$$

In Eq. (2), $[C]$ is the concentration of chemical elements associated with the volume correlated gases, e.g. spallation target nuclides or parent radionuclides. With M'' representing a third isotopic abundance, the two-component equations also yield the three-isotope correlation equations

$$(M'/M) = (M''/M)\left[\frac{(M'/M)_{vc} - (M'/M)_{sc}}{(M''/M)_{vc} - (M''/M)_{sc}}\right]$$
$$+ \left[(M'/M)_{sc} - (M''/M)_{sc}\left(\frac{(M'/M)_{vc} - (M'/M)_{sc}}{(M''/M)_{vc} - (M''/M)_{sc}}\right)\right] \qquad (3)$$

which in the special case $(M''/M)_{vc} = 0$ reduce to

$$(M'/M) = (M''/M)\left[\frac{(M'/M)_{sc} - (M'/M)_{vc}}{(M''/M)_{sc}}\right] + (M'/M)_{vc} \qquad (4)$$

If the measured quantities $1/M$, $[C]/M$ or M''/M are plotted versus M'/M for a suite of several samples, linear correlations result if and only if the isotopic compositions $(M'/M)_{sc}$, $(M'/M)_{vc}$, $(M''/M)_{sc}$ and $(M''/M)_{vc}$ are identical in all samples, i.e. if the assumption that rare gases in the samples are variable mixtures of isotopically uniform surface and volume correlated components is correct. In addition, Eqs. (1) and (2), the ordinate intercept relations developed by Eberhardt *et al.* (1970, 1972), yield correlations only if the concentrations $(M)_{vc}$ or specific concentrations $(M)_{vc}/[C]$ are the same in all samples. Surface correlated rare gases in lunar fines are dominated by the solar wind, and are relatively constant in isotopic composition although variations apparently due to lunar surface and atmospheric transport and fractionation mechanisms are known to occur (Basford *et al.*, 1973). Volume correlated gases are chiefly spallogenic except for radiogenic components in ^{4}He and ^{40}Ar—discussed below—and in the heavy Xe isotopes (Basford *et al.*, 1973). Spallogenic nuclide concentrations $(M)_{vc} = (M)_{sp}$ depend directly on the energy spectrum and duration of cosmic-ray irradiation and on target element abundances. Since ordinate intercept analysis by Eq. (1) requires identical $(M)_{sp}$ in all samples, this condition is best satisfied by a sample suite which maximizes the chance for identical cosmic ray exposure histories, i.e. grain-sized separates of single soil samples. Variations in target element abundances over the grain-sized suite are accommodated through Eq. (2) correlations, which require only that specific concentrations $(M)_{sp}/[C]$ be constant.

The ordinate intercept correlations, of form $(M'/M) = (1/M)\lambda + b$ or $(M'/M) = ([C]/M)\lambda + b$, yield the surface correlated or trapped isotopic ratios $(M'/M)_{sc} = (M'/M)_{tr}$ as ordinate intercepts, and, from the measured slopes λ,

$$(M)_{sp} = \lambda/((M'/M)_{sp} - (M'/M)_{tr}) \qquad (5)$$

or

$$(M)_{sp}/[C] = \lambda/((M'/M)_{sp} - (M'/M)_{tr}) \qquad (6)$$

The only unknowns in Eqs. (5) and (6) are the spallation ratios $(M'/M)_{sp}$. Concentrations of ^{3}He$_{sp}$, ^{21}Ne$_{sp}$, and ^{38}Ar$_{sp}$ calculated from Eq. (5) are quite insensitive to the isotopic composition of the spallation component, and literature values may be used without significant error. For Kr and Xe, the reasonably good assumption that $^{86}Kr_{sp}$ and $^{136}Xe_{sp} = 0$ permits direct determination of $(M'/M)_{sp}$ from Eq. (4).

It is useful to rewrite the ordinate intercept relations (1) and (2) to accommodate explicitly the special case of volume correlated rare gases which consist of mixtures of spallogenic and radiogenic components. For $(M)_{vc} = (M)_{sp} + (M)_{rad}$, Eq. (1) becomes

$$\left(\frac{M'}{M}\right)\frac{1}{\gamma} = \frac{(M)_{sp}}{M\gamma}\left[\left(\frac{M'}{M}\right)_{sp} - \left(\frac{M'}{M}\right)_{tr}\right] + \left(\frac{M'}{M}\right)_{tr}, \quad \text{with } \gamma \equiv (1 - (M)_{rad}/M). \quad (7)$$

For $(M')_{vc} = (M')_{sp} + (M')_{rad}$, Eq. (2) becomes

$$\left(\frac{M'}{M}\right)\frac{1}{\gamma} = \frac{[C]}{M\gamma}\left[\frac{(M')_{rad} + (M')_{sp}}{[C]}\right] + \left(\frac{M'}{M}\right)_{tr}, \quad \text{with } \gamma \equiv (1 - (M)_{sp}/M). \quad (8)$$

These forms apply in calculations of spallogenic ^3He concentrations and of K–Ar ages.

Core deposition models and depositional chronologies

We consider two of the deposition models introduced and discussed by Russ *et al.* (1972) in connection with neutron fluence data from the drill core fines: (1) rapid deposition of the present section T years ago, with no subsequent significant vertical mixing; and (2) uniform accretion of the core material over a period of T' years, followed by T years as a stable regolith section. To assess these models in terms of the measured concentrations versus depth of the spallogenic rare gases, we represent the production rate of a spallogenic nuclide as a function of depth h in the regolith as $P(h) = P_s f(h)$, where P_s is the *surface* production rate. Then for model (1), the concentration of a spallogenic nuclide $(M)_{sp}$ at present depth d in the core section is simply $M_{sp}(d) = P_s Tf(d)$ if the source materials were initially unirradiated, and $M_{sp}(d) = (M_{sp})_0 + P_s Tf(d)$ if the materials had experienced uniform preirradiation resulting in an initial spallation concentration $(M_{sp})_0$ in the accreted section. For model (2), during the uniform accretion stage, the differential production $dM_{sp}(h)$ at depth h for time dt is $dM_{sp}(h) = P_s f(h)dt$; if ρ is the accretion rate, then $dt = dh/\rho$, $dM_{sp}(h) = P_s(h)dh/\rho$, and the integrated production in a sample now at regolith depth d is

$$M_{sp}(d) = \frac{P_s}{\rho}\int_0^d f(h)dh$$

for initially unirradiated material. As in model (1), production during the subsequent *in situ* irradiation stage is $P_s Tf(d)$, and therefore the total spallation production of isotope M is

$$M_{sp}(d) = \frac{P_s}{\rho}\int_0^d f(h)dh + P_s Tf(d);$$

the accretion time T' of the section from depth d to the surface is given by $T' = d/\rho$, and T is the *in situ* irradiation time.

Reedy and Arnold (1972) have calculated production rates by galactic cosmic-ray particles as a function of depth in the lunar regolith for a number of radionuclides, among them ^{3}H, ^{22}Na, and ^{37}Ar and ^{39}Ar. In the following comparisons of measured spallation rare gas concentrations versus depth with model predictions, we assume that the *relative* production rates $P(h)/P_s$ deduced from data given by Reedy and Arnold for these four radionuclides are applicable to the production of ^{3}He$_{sp}$, ^{21}He$_{sp}$, and ^{38}Ar, respectively. Reedy and Arnold's absolute production rates do not apply directly to the rare gases—roughly half of the spallogenic ^{3}He, for example, is produced directly, with the balance *via* ^{3}H.

Spallation rare gases versus depth: Comparison with model predictions

Argon. Ordinate intercept analysis of the Ar data in Table 1, with $M = {}^{36}$Ar and $M' = {}^{38}$Ar in Eq. (1), yields the trapped ratios $({}^{38}\text{Ar}/{}^{36}\text{Ar})_{tr}$ and the concentrations of ^{38}Ar$_{sp}$ given in Table 4 for the six drill stem samples. To calculate the latter quantity, Eq. (5) was rewritten as $(38)_{sp} = (36)_{sp}(38/36)_{sp} = \lambda/[1 - (38/36)_{tr}/(38/36)_{sp}]$, with $(38/36)_{sp}$ taken as 1.52 (Huneke *et al.*, 1972). Correlations are satisfactory, with data points within $\pm 1\ \sigma$ of the correlation lines, except for 15001 where the $<16\ \mu$ point appeared anomalous and was dropped in calculating $(38)_{sp}$; it is possible that the concentration of Ca, the principal target element, is significantly variable over the grain-sized separates of this sample. In addition to possible variation within 15001, bulk CaO concentrations are known to vary over the sample suite 15001–15006 (Rhodes, 1972); ^{38}Ar$_{sp}$ concentrations normalized to the CaO concentration of 10% in 15001 are given in a Table 4 footnote. These normalized concentrations are plotted versus depth in Figs. 1 and 2.

It is clear from Figs. 1 and 2 that spallation Ar varies smoothly with depth in the core samples, with a concentration peak near ~ 100 g/cm^2. For comparison, we have plotted in Fig. 2 the relative production rate $P(d)/P_s$ versus depth for ^{37}Ar (choosing the "surface" at a depth of 40 g/cm^2 in the present core for reasons discussed below), calculated from data taken from Fig. 7 of Reedy and Arnold (1972). The relative depth dependence of ^{39}Ar production is very similar. The close correspondence in shape between the spallogenic ^{38}Ar and the ^{37,39}Ar production rate distributions with depth suggests that a dominant fraction of the spallogenic ^{38}Ar was produced in the core materials during *in situ* irradiation of the accreted core section. However, it is clear that in this model, because of the ~ 40 g/cm^2 offset between peak ^{38}Ar$_{sp}$ concentration and peak production rate, a surface slab of thickness $t_s \sim 40$ g/cm^2 must have been absent during most of the *in situ* irradiation time.

For a variation of the rapid deposition model (1) of Russ *et al.* (1972) in which specific allowance is made for the recent deposition of material of thickness t_s on top of the original regolith section, the predicted concentration of ^{38}Ar$_{sp}$ as a function of depth d below the *present* surface is ^{38}Ar$_{sp}(d \geq t_s) =$

Table 4. Spallogenic concentrations and trapped (surface-correlated) isotopic compositions of helium, neon and argon in $<250~\mu$ fines from the Apollo 15 drill core stems.

Sample	*Depth (g/cm²)	†³He$_{sp}$ ²¹Ne$_{sp}$ ‡³⁸Ar$_{sp}$ (units of 10⁻⁷ cc STP/g)			(⁴He/³He)$_{tr}$	(²⁰Ne/²²Ne)$_{tr}$	(²¹Ne/²²Ne)$_{tr}$ (×10⁻²)	(³⁸Ar/³⁶Ar)$_{tr}$ (×10⁻¹)	(⁴⁰Ar/³⁶Ar)$_{tr}$
15006,26	65	30.0	9.39	6.12	2459.	12.585	3.088	1.896	0.780
		±1.7	±.63	±.29	±64	±.075	±.056	±.002	±.079
15005,25	138	37.0	10.33	7.28	2640.	12.612	3.083	1.871	1.321
		±3.7	±.44	±.49	±110	±.058	±.064	±.006	±.061
15004,26	208	39.1	10.09	5.98	2667.	12.47	3.095	1.893	1.110
		±2.6	±.24	±.31	±73	±.11	±.020	±.002	±.063
15003,28	276	31.8	9.34	5.67	2530.	12.634	3.038	1.895	1.074
		±3.7	±.44	±.18	±100	±.041	±.045	±.002	±.076
15002,26	341	25.7	—	4.09	2466	—	—	1.903	0.960
		±2.3		±.73	±59			±.007	±.026
15001,30	412	17.7	7.60	§3.42	2410.	12.499	3.052	1.894	1.216
		±1.9	±.42	±.89	±76	±.041	±.057	±.014	±.077

*Taken from Russ *et al.* (1972).

†From data correlations corrected for radiogenic ⁴He (see text). Without these corrections, the concentrations are, for 15006–15001, respectively: 26.4 ± 1.8, 32.2 ± 4.1, 34.8 ± 2.3, 27.8 ± 3.8, 20.9 ± 2.1, 11.4 ± 2.5.

‡CaO varies by ~10% among these samples (Rhodes, 1972). When normalized to CaO = 10.0% in 15001, the specific concentrations are, for 15006–15001, respectively: 6.15, 7.33, 5.91, 5.31, 4.23, 3.42.

§Poor correlation. The $<16~\mu$ point lies well off the correlation defined by the three coarser grain sizes, and is omitted in calculating this concentration.

$(^{38}\mathrm{Ar_{sp}})_0 + P_s^{38}Tf(d - t_s)$. Assuming that the function f derived from Reedy and Arnold's ³⁷Ar data applies to ³⁸Ar$_{sp}$ production, there are three adjustable parameters in the model: t_s; $(^{38}\mathrm{Ar_{sp}})_0$, arising from irradiation of the drill core fines prior to accretion and assumed uniform with depth; and $P_s^{38}T$, the product of the ³⁸Ar$_{sp}$ surface production rate and the *in situ* irradiation time. The procedure in assessing the compatibility of this model with the measured ³⁸Ar$_{sp}$ concentrations is then to seek values for these parameters which give a reasonable fit of the predicted and measured profiles.

The profiles calculated from the model converge strongly to the measured profile for the case $t_s = 45 \pm 5$ g/cm², $(^{38}\mathrm{Ar_{sp}})_0 = (15 \pm 2) \times 10^{-8}$ cc STP/g, and $P_s^{38}T = (28 \pm 2) \times 10^{-8}$ cc STP/g, where the errors are estimated from deterioration in the goodness of fit. The calculated fit is shown as the solid curve in Fig. 1. There are no fits for any case in which $t_s = 0$ and/or $(^{38}\mathrm{Ar_{sp}})_0 = 0$, i.e. the data require both recent addition of surficial material at the core site and preirradiated source material. There is no criterion in the fitting procedure for separate determination of P_s^{38} and T, so evaluation of the *in situ* irradiation time T depends on independent evaluation of the surface production rate of ³⁸Ar$_{sp}$. P_s^{38} was estimated by assuming that a typical sampled lunar soil has been irradiated at or above a regolith depth of ~1 m; the average ³⁸Ar$_{sp}$ production rate $\langle P^{38} \rangle$ is then $1.93P_s^{38}$, deduced by integration over the ³⁷Ar production rate profile to this depth. The calculation is insensitive, within comparatively wide limits, to the choice of

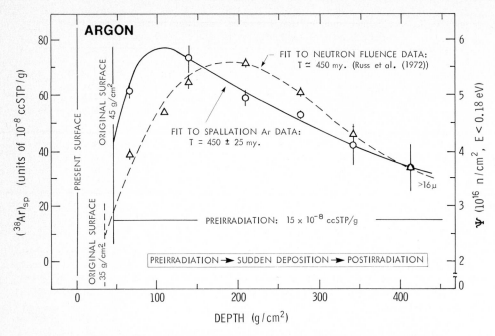

Fig. 1. Spallation ^{38}Ar concentrations, normalized to 10% CaO, and neutron fluences (Russ *et al.*, 1972) versus depth. The solid and dashed curves are sudden deposition model fits to the Ar and neutron data, respectively. The concentration of spallogenic Ar produced in uniform preirradiation of the section according to this model is represented as a horizontal line; a preirradiation neutron fluence is also required by the model, but is not shown.

maximum irradiation depth: $\langle P^{38} \rangle$ is within 10% of $1.93 P_s^{38}$ for maximum depths ranging from 30 to 150 cm. $\langle P^{38} \rangle$ was taken equal to 1.69×10^{-8} cc STP/gCa-m.y., the average measured production rate in ~50 Apollo 11 and 12 crystalline rock samples (Bogard *et al.*, 1971). P_s^{38} is then 8.8×10^{-9} cc STP/gCa-m.y. = 6.25×10^{-10} cc STP/g-m.y. for 10% CaO in the drill core samples. With this choice, the *in situ* irradiation time for the lower ~370 g/cm^2 of the drill core section is $T = 450 \pm 25$ m.y., where the error is calculated from the parameter errors given above; it does not include uncertainty in P_s^{38}, which is probably larger.

The concentration of ^{38}Ar$_{sp}$ versus depth predicted by the continuous accretion model (2) of Russ *et al.* (1972), again allowing for recent deposition of a surface slab of thickness t_s, is

$$^{38}\text{Ar}_{sp}(d \geq t_s) = \frac{P_s^{38}}{\rho} \int_0^{d-t_s} f(h)dh + P_s^{38} T f(d - t_s)$$

The adjustable parameters are t_s; P_s^{38}/ρ, where ρ is the accretion rate; and $P_s^{38}T$. An excellent fit to the measured data, shown as the solid curve in Fig. 2, is obtained for the case $t_s = 40 \pm 5$ g/cm^2, $P_s^{38}/\rho = (2.3 \pm 0.1) \times 10^{-10}$ cc STP/g-(g/cm^2), and $P_s^{38}T = (33 \pm 2) \times 10^{-8}$ cc STP/g. Using the value of P_s^{38} estimated

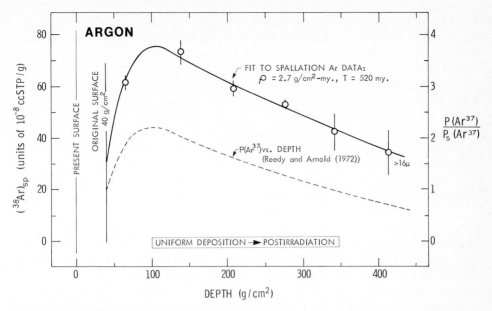

Fig. 2. Uniform deposition model fit to the spallation Ar versus depth profile. The dashed curve, referred to a surface at present depth 40 g/cm², is the relative production function $P^{37}(d)/P_s^{37}$ calculated from Reedy and Arnold's (1972) data.

above, this model fit yields an accretion rate $\rho = 2.7$ g/cm²-m.y. and thus an accretion time T' for the lower ~ 370 g/cm² of the core section of ~ 140 m.y., and subsequent *in situ* irradiation time $T \simeq 520$ m.y. The thickness of the recently deposited surface slab is roughly the same as in the model (1) fit, but model (2) does not require preirradiation of the accreting regolith.

The depositional chronology derived for the drill core fines by means of the first of these two model fits to the spallation Ar data agrees very well with the conclusions reached by Russ *et al.* (1972) from their neutron fluence data. The sudden accretion model fit to the neutron fluence depth profile is shown as the dashed curve in Fig. 1. As far as the *in situ* irradiation times, the existence and thickness of a recent surface layer, and the requirement for preirradiated source materials are concerned, the results are essentially identical even though the depth dependencies of the two measured profiles are quite different. The continuously accreting model fits to the two sets of data, however, yield disparate results. For the Ar data, the excellent model fit (Fig. 2) yields $t_s \simeq 40$ g/cm², $\rho \simeq 2.7$ g/cm²-m.y., $T' \simeq 140$ m.y. and $T \simeq 520$ m.y.; the less convincing fit to the neutron fluence data gives $t_s \sim 0$, $\rho \sim 1.2$ g/cm²-m.y., $T' \sim 350$ m.y., and $T \sim 600$ m.y. (Russ *et al.*, Fig. 6A). The apparent disagreements in t_s and ρ may prove to be grounds for rejecting this model. In addition, these relatively rapid accumulation rates may be incompatible with heavy ion track data from the drill core soils, as pointed out by Russ *et al.* The constraint on accumulation rates imposed by more recent track measurements is unclear, however: Bhandari *et al.* (1972) propose an average sedimentation rate of ≤0.5 g/cm²-m.y. based on their study of fines from drill stem 15002, while Fleischer and Hart (1972) and Fleischer *et al.* (1974) find evidence that materials in drill stems 15005 and 15003 accumulated at rates of ≥0.8 g/cm²-m.y. and ≥0.6 g/cm²-m.y., respectively.

A final comment on the model fits to the spallation Ar data is appropriate in view of a recent report of directly measured ^{37}Ar and ^{39}Ar depth profiles in the Apollo 16 and 17 drill stems by Stoenner *et al.* (1974). For both ^{37}Ar and ^{39}Ar, Reedy and Arnold's (1972) theoretical profiles and Stoenner *et al.*'s (1974) measured profiles peak at ~65 g/cm² depth. Below the peak production depth, the theoretical ^{37}Ar production rate decreases with exponential mean free paths of 300 g/cm² to a depth of ~250 g/cm², and 200 g/cm² for $d > 250$ g/cm²; the measured ^{37}Ar profile decreases from the peak with an exponential mean free path of ~240 g/cm². The rapid deposition model fit to the spallation ^{38}Ar data, based on the theoretical profile, is compatible with the measured profile as well. Both relative production rates for $d \geq 65$ g/cm² are shown in Fig. 3, arbitrarily normalized to 1.0 (theoretical) and 0.5 (measured) at 65 g/cm², and compared to the normalized ^{38}Ar$_{sp}$ concentrations produced *in situ* (Table 4 concentrations *minus* the constant preirradiation contribution of 1.5×10^{-7} cc STP/g yielded by the rapid deposition model) in samples 15001–15005. The data points follow both profiles reasonably well.

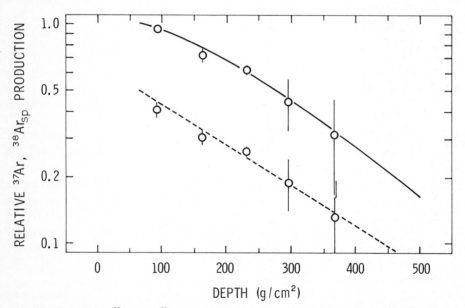

Fig. 3. Relative ^{37}Ar and ^{38}Ar$_{sp}$ production versus depth. The upper, solid curve, normalized to 1.0 at 65 g/cm², is the Reedy and Arnold (1972) theoretical production rate profile for ^{37}Ar. The lower, dashed line, normalized to 0.5 at 65 g/cm², is the ^{37}Ar profile measured in Apollo 16 and 17 drill core samples by Stoenner *et al.* (1974). The data points, normalized to best fit to the two profiles, are ^{38}Ar$_{sp}$ concentrations produced *in situ* in the Apollo 15 drill core according to the rapid deposition model. The ^{38}Ar$_{sp}$ production rate versus depth is reasonably consistent with either ^{37}Ar profile.

The rapid deposition model parameters derived above are *not* compatible with Stoenner *et al.*'s measured ^{39}Ar profile, which falls off below the production peak with an exponential mean free path of ~ 170 g/cm^2. However, recalculation of model fits to the measured ^{38}Ar$_{sp}$ data, with the relative production function $f(h)$ now representing the measured ^{39}Ar profile rather than the theoretical ^{37,39}Ar profile, yields a close, least error fit for t_s and $P_s^{38}T$ within error of their previous values of 45 ± 5 g/cm^2 and $(28 \pm 2) \times 10^{-8}$ cc STP/g, and therefore a similar *in situ* irradiation time T. The preirradiation concentration $(^{38}$Ar$_{sp})_0$ is now required to be 20×10^{-8} cc STP/g, about $\frac{1}{3}$ higher. The principal difference imposed by the adoption of the measured ^{39}Ar profile, however, is that it is now necessary to postulate an additional, relatively small production of ^{38}Ar$_{sp}$ which closely follows the low-energy neutron flux profile (Russ *et al.*, 1972; Woolum and Burnett, 1974). This type of "spallation" component, which in smaller relative amount also optimizes the fit of the rapid deposition model, utilizing the *measured* ^{37}Ar profile, to the spallation Ar data, is discussed in the following sections on the ^{3}He$_{sp}$ and ^{21}Ne$_{sp}$ profiles, where it appears to play a more dominant role.

Helium. (^{4}He/^{3}He) trapped ratios and ^{3}He$_{sp}$ concentrations (Table 4) were calculated by ordinate intercept analysis of He data in Table 1, utilizing Eq. (7) with $M' = {}^3$He and $M = {}^4$He to explicitly remove a probable radiogenic ^{4}He component from the volume correlated ^{4}He concentrations. In calculating $\gamma \equiv (1 - (^4$He$)_{rad}/^4$He) for each grain-sized fraction, we have used U and Th abundances measured by Silver (1972) in bulk (unseparated) samples located within a few vertical millimeters of the Minnesota samples, and assumed that Ba/U is roughly constant in the size separates and that the U, Th–He ages and K–Ar ages of these samples (Fig. 10) are equal. The correction for ^{4}He$_{rad}$ is relatively small ($\gamma > 0.9$) except in the coarsest sized fractions of 15005 ($\gamma = 0.76$) and 15001 ($\gamma = 0.81$), which are poor in trapped gases. From Eq. (5), ^{3}He$_{sp} = {}^4$He$_{sp}/(^4$He$/^3$He$)_{sp} = \lambda/[1 - (4/3)_{sp}/(4/3)_{tr}]$. $(^4$He$/^3$He$)_{sp}$ is assumed $\simeq 4$; results are insensitive to this choice. The corrected concentrations of ^{3}He$_{sp}$ are plotted versus depth in Fig. 4. The *uncorrected* concentrations are given in a footnote to Table 4; the effect of the correction is to shift the profile to higher concentrations, while essentially preserving its shape.

The ^{3}He$_{sp}$ depth profile in Fig. 4 is a smooth function of depth, but its shape is completely unexpected. If the relative production rate function $P^3(d)/P_s^3$ for ^{3}H, derived from data taken from Fig. 8 of Reedy and Arnold (1972) and plotted in Fig. 4, is assumed to represent the relative production of ^{3}He$_{sp}$ with depth, then both depositional models discussed above are incompatible with the measured profile for any combination of parameters. Short of throwing out the models (which seems somewhat drastic in view of the experimental support for the sudden deposition model from the Ar and neutron fluence data), there appear to be at least two other possibilities: (1) that there exists an unrecognized production reaction for either ^{3}He or ^{3}H, with peak production in the depth range ~ 150–200 g/cm^2; or (2) that the pattern in Fig. 4 represents a concentration profile resulting from diffusional equilibrium of ^{3}H, ^{3}He, or both. We have not evaluated this latter

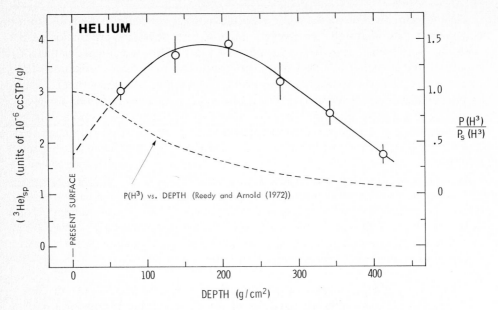

Fig. 4. Spallation ^{3}He concentrations versus depth. The solid curve is a visual fit to the data point distribution. The Reedy and Arnold (1972) relative production rate profile for ^{3}H is shown for comparison.

model in detail, but we note that in this case one would expect the ^{3}He to be both surface correlated and very loosely bound. The ^{3}He$_{sp}$ in Table 4 and Fig. 4 could be loosely bound—although this is not typical of lunar soils in general—but the correlation techniques used in deriving ^{3}He$_{sp}$ concentrations ensure that this component is *volume* correlated. It is possible that the falloff in ^{3}He$_{sp}$ concentration from peak to surface could be due to diffusive loss from a production profile resembling the Reedy and Arnold function. The magnitude of the implied loss, however, is a large fraction of the integrated production; it is an open question as to whether ^{3}H (or ^{3}He) can be quickly and efficiently mobilized from production sites in grain interiors in the subsurface thermal environment.

In reference to the first possibility—a production mechanism for helium and/or tritium in addition to that represented by the Reedy and Arnold spallation function—the shape of the depth profile in Fig. 4 is suggestive: it appears qualitatively to be a superposition of the ^{3}H production profile (Fig. 4), a production function which varies with depth roughly as the neutron fluence profile (Fig. 1), and preirradiation, all referenced to an original surface at present depth ~50 g/cm^2. This observation strongly suggests that the ^{3}H$_{sp}$ profile could be in at least approximate agreement with the sudden deposition model if the unknown production reaction is a low-energy (<0.2 eV) neutron capture producing either ^{3}H or ^{3}He.

It is straightforward to assess this hypothesis quantitatively in terms of model fits to the measured ^{3}He$_{sp}$ profile. In a rapid deposition model extended to include the postulated low-energy capture reaction, the predicted concentration of $^3He_{sp}$ as a function of depth d below the present surface is $^3He_{sp}(d \geqslant t_s) = (^3He_{sp})_0 + P_s^3 Tf(d - t_s) + N_s^3 Tg(d - t_s)$, where $(^3He_{sp})_0$ is the concentration arising from uniform preirradiation of the drill core materials, P_s^3 and N_s^3 are the surface production rates of $^3H + {^3He}$ by spallation and neutron capture respectively, $f(d - t_s) \equiv P^3(d - t_s)/P_s^3$ is the relative spallation production rate versus depth from Reedy and Arnold, and $g(d - t_s) \equiv N^3(d - t_s)/N_s^3 = \phi(d - t_s)/\phi_s$ is the relative neutron capture rate versus depth, which is simply the neutron fluence profile given by Russ *et al.* (1972) after subtraction of the preirradiation fluence and normalization to the surface *in situ* fluence; ϕ is the neutron flux, and *in situ* fluence is $\psi = \phi T$. Chemical compositions are assumed constant with depth.

Free parameters in the model are t_s, $(^3He_{sp})_0$, $P_s^3 T$, and $N_s^3 T$. Minimum error fits to the measured profile occur for the parameter values $t_s = 50 \pm 10$ g/cm^2, $(^3He_{sp})_0 = (3 \pm 1) \times 10^{-7}$ cc STP/g, $P_s^3 T = (20 \pm 3) \times 10^{-7}$ cc STP/g, and $N_s^3 T = (2.9 \pm 0.2) \times 10^{-7}$ cc STP/g. This range of fits is compared to the measured profile in Fig. 5, where the three separate functions which add to form the gross median spallation concentration versus depth are also shown. The fits are in quite satisfactory agreement with the measured concentrations; in addition, the fact

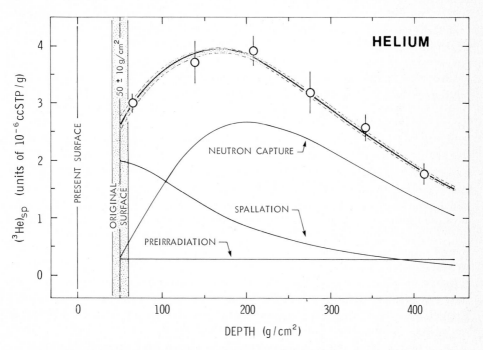

Fig. 5. Rapid deposition model fits to the observed $^3He_{sp}$ profile in the Apollo 15 drill core. The stippled areas represent the ranges of fits to the data within the uncertainties in the best-fit model parameters given in the text. The "neutron capture," "spallation," and "preirradiation" curves represent the depth profiles of the separate $^3He_{sp}$ components in the model; their sum at any depth matches the observed concentration profile.

that the optimum fits occur for t_s near the value $45 \pm 5 \, \text{g/cm}^2$ derived from the rapid deposition model fit to the Ar data suggests that the model descriptions of the two spallation profiles are coherent, as of course they must be if the model is valid.

We have no explanation for the apparent generation of such relatively large amounts of tritium and/or helium in low-energy neutron capture reactions. The one obvious candidate, $^6\text{Li}(n, \alpha)^3\text{H}$, fails to account for the large concentrations indicated by the "neutron capture" profile in Fig. 5. The surface production rate by neutron capture in an element with cross section σ and concentration C is $N_s^3 \cong C\sigma\phi_s$, and the integrated production in time T is $N_s^3 T$; from the parameters above, $N_s^3 T \cong 2.9 \times 10^{-7} \, \text{cc STP/g} = 7.8 \times 10^{12} \, \text{atoms/g}$. Therefore $C \cong 7.8 \times 10^{12}/\sigma\phi_s T$. Woolum and Burnett (1974) give the maximum low-energy neutron flux $\phi_{\text{max}} \cong 1.4 \, \text{n/cm}^2\text{-sec}$; from their measured profile, or from the "neutron capture" profile in Fig. 5, the surface neutron flux ϕ_s is then $\sim 0.15 \, \text{n/cm}^2\text{-sec}$. With $T = 450 \, \text{m.y.}$, $C \cong 3.7 \times 10^{21}/\sigma$ atoms/g for σ in barns. For the $^6\text{Li}(n, \alpha)^3\text{H}$ slow neutron cross section $\sigma_{n\alpha} = 950b$, and $^6\text{Li}/\text{Li} = 0.074$, the lithium concentration in the drill core fines required to account for all of the neutron capture component is $\sim 530 \, \text{ppm}$, about 35 times greater than the 15 ppm concentration reported by Reed *et al.* (1972) for sample 15041, taken from Station 8 near the drill core site and probably representative for drill core Li. It appears that the explanation must lie elsewhere. Accurate measurements of ^3H versus depth in the Apollo 15 drill core are now of particular interest.

Neon. Trapped ratios and $^{21}\text{Ne}_{\text{sp}}$ concentrations listed in Table 4 were calculated from Eq. (1) correlations of Table 1 data, with $M' = {}^{21}\text{Ne}$, ^{20}Ne and $M = {}^{22}\text{Ne}$, and from Eq. (5) in the form $^{21}\text{Ne}_{\text{sp}} = {}^{22}\text{Ne}_{\text{sp}}(^{21}\text{Ne}/^{22}\text{Ne})_{\text{sp}} = \lambda/[1 - (21/22)_{\text{tr}}/(21/22)_{\text{sp}}]$ with $(^{21}\text{Ne}/^{22}\text{Ne})_{\text{sp}} = 0.95$. Ne data from sample 15002,26 do not define linear correlations in ordinate intercept analysis, and therefore either the data in Table 1 are suspect (although a careful review failed to reveal experimental anomalies) or Ne in this sample contains a significant admixture of a third, isotopically distinct component. Spallation ^{21}Ne concentrations for the five remaining samples are plotted versus depth in Fig. 6, along with the relative production rate function $P^{22}(d)/P_s^{22}$ for ^{22}Na from Reedy and Arnold (1972).

The data points in Fig. 6 define a smooth and remarkably flat depth profile with a broad concentration peak near $\sim 150 \, \text{g/cm}^2$. In contrast, $P^{22}(d)/P_s^{22}$ maximizes at $d \simeq 30 \, \text{g/cm}^2$, so even if the surface of the regolith section was between 35 and 55 g/cm^2 for most of the *in situ* irradiation period, as required by model fits to the neutron fluence, Ar and He data, the peak in the $^{21}\text{Ne}_{\text{sp}}$ concentration profile predicted by the sudden deposition model is too shallow by $\sim 75 \, \text{g/cm}^2$. Subtracting various assumed values of $(^{21}\text{Ne}_{\text{sp}})_0$, arising from uniform preirradiation of the core materials, from the measured concentrations in Fig. 6 results in concentration profiles due to *in situ* production which fall off more rapidly than the measured profile on both sides of the peak, but in which the peak concentration depth of $\sim 150 \, \text{g/cm}^2$ is unchanged. Therefore the combination of the measured $^{21}\text{Ne}_{\text{sp}}$ concentration profile and the assumption that the ^{22}Na relative production

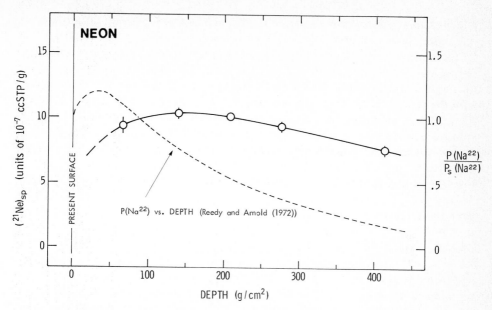

Fig. 6. Spallation ^{21}Ne concentrations versus depth. The solid curve is a visual fit to the data point distribution. The Reedy and Arnold (1972) relative production rate profile for ^{22}Na is shown for comparison.

function correctly describes the relative production of ^{21}Ne$_{sp}$ with depth is incompatible with the sudden deposition model.

A continuously accreting model (model (2)) with parameter values $t_s = 55$ g/cm^2, $P_s^{21}T = 8.0 \times 10^{-7}$ cc STP/g, and $P_s^{21}/\rho = 2.5 \times 10^{-9}$ cc STP/g-(g/cm^2) fits the observed profile in Fig. 6 barely within error. A procedure identical to that employed earlier in calculating P_s^{38} from $\langle P^{38} \rangle$, using production rate data from Bogard *et al.* (1971) and chemical data from Rhodes and Hubbard (1972), yields $P_s^{21} \simeq 1.63 \times 10^{-9}$ cc STP/g-m.y. for the drill core samples, and therefore $T \simeq 490$ m.y., $\rho \simeq 0.64$ g/cm^2-m.y., and $T' \simeq 560$ m.y. These accretion and *in situ* irradiation times are in very good agreement with the values $T \sim 430$ m.y. and $T' \sim 570$ m.y. estimated by Bogard *et al.* (1973) from a best fit of the same model to much more erratic spallation Ne data, but the agreement would seem to be largely fortuitous since their *model* fit yields $t_s = 0$ and a ^{21}Ne$_{sp}$ concentration peak near ~ 40 g/cm^2 depth, both incompatible with the Fig. 6 profile. Bogard *et al.*'s highest ^{21}Ne$_{sp}$ concentrations occur near ~ 130 g/cm^2 depth, in rough agreement with the Fig. 6 profile, but these concentrations are inconsistent with their model fit.

The surface slab thickness t_s deduced from the continuous accretion model fit to the Fig. 6 data is significantly larger than that required to fit the same model to the spallation Ar data (Fig. 2), but a more profound discrepancy is the factor 4 difference in the sedimentation rates, from ρ(Ar) = 2.7 g/cm^2-m.y. to ρ(Ne) = 0.64 g/cm^2-m.y. The neutron fluence data yields $\rho(\psi) \sim 1.2$ g/cm^2-m.y., and in addition requires $t_s \sim 0$. Therefore the detailed depositional histories derived from the appropriate Ar, Ne and neutron fluence fit parameters are discordant, and the continuous accretion model is consequently suspect.

In contrast, the sudden deposition model yields concordant depositional and chronological histories when applied to the measured spallation Ar and neutron fluence drill core profiles. As discussed above, this model could also be consistent with the spallation He profile, but although the fit of model to data is impressive, the *ad hoc* postulation of a component generated by an unspecified low-energy capture reaction does not in itself constitute a convincing argument. The evidence for low-energy capture might be strengthened if it could be shown that the rapid deposition model which fits the He data, involving capture production following the same depth profile, is also compatible with the spallation Ne data in Fig. 6. This, in fact, is the case; the requirement imposed by both the He and Ne data is for a production reaction which effectively shifts the peak of the total production profile to greater depths than those predicted by Reedy and Arnold's spallation profiles. For Ne, the rapid deposition model gives $^{21}Ne_{sp}$ $(d \geqslant t_s) = (^{21}Ne_{sp})_0 + P_s{}^{21}Tf(d - t_s) + N_s{}^{21}Tg(d - t_s)$; here f is Reedy and Arnold's relative spallation production profile for ^{22}Na (Fig. 6), and g, as before, is the surface-normalized low-energy neutron flux profile. Extremely close minimum error fits to the measured $^{21}Ne_{sp}$ profile, shown in Fig. 7, occur for the parameter values

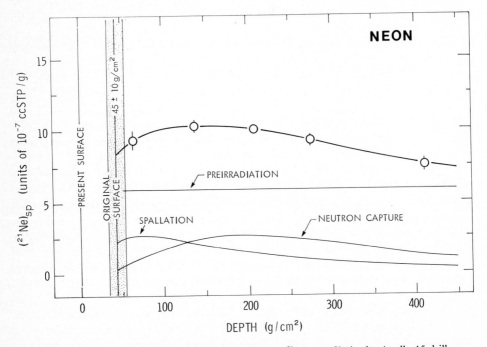

Fig. 7. Rapid deposition model fits to the observed $^{21}Ne_{sp}$ profile in the Apollo 15 drill core. The "neutron capture," "spallation," and "preirradiation" curves represent the depth profiles of the separate $^{21}Ne_{sp}$ components in the model; their sum at any depth matches the observed concentration profile.

$t_s = 45 \pm 10 \text{ g/cm}^2$, $({}^{21}\text{Ne}_{\text{sp}})_0 = (5.9 \pm 0.1) \times 10^{-7} \text{ cc STP/g}$, $P_s^{21}T = (2.2 \pm 0.3) \times 10^{-7} \text{ cc STP/g}$, and $N_s^{21}T = (0.30 \pm 0.03) \times 10^{-7} \text{ cc STP/g}$; the individual components of the total profile are also plotted in Fig. 7.

A strong argument for the validity of the rapid deposition model is the relatively invariant value of t_s, $45 \pm 10 \text{ g/cm}^2$, required for model fits to *all* measured spallation profiles. The case for the existence of a production mechanism involving low-energy neutron capture is strengthened by the Ne results, since it is at least unlikely that diffusional effects in He and Ne, considering the large mass difference, could fortuitously alter both measured profiles in exactly the way required to resemble the same, spurious production profile. As pointed out earlier, low-energy capture production may also contribute to spallation Ar; for the rapid deposition model utilizing the relative spallation function f based on Stoenner *et al.*'s (1974) measured ${}^{39}\text{Ar}$ profile, minimum error fits of model to data occur for the parameter values $t_s = 45 \pm 10 \text{ g/cm}^2$, $({}^{38}\text{Ar}_{\text{sp}})_0 = (20 \pm 1) \times 10^{-8} \text{ cc STP/g}$, $P_s^{38}T = (26 \pm 2) \times 10^{-8} \text{ cc STP/g}$, and $N_s^{38}T = (1.8 \pm 0.3) \times 10^{-8} \text{ cc STP/g}$; Stoenner *et al.*'s measured ${}^{37}\text{Ar}$ profile requires $t_s = 45 \pm 10 \text{ g/cm}^2$, $({}^{38}\text{Ar}_{\text{sp}})_0 = (14 \pm 1) \times 10^{-8} \text{ cc STP/g}$, $P_s^{38}T = (26 \pm 2) \times 10^{-8} \text{ cc STP/g}$, and $N_s^{38}T = (1.4 \pm 0.5) \times 10^{-8} \text{ cc STP/g}$.

Krypton and xenon. Trapped isotopic compositions in Tables 5 and 6 are defined by the intercepts of Eq. (2) correlations of data in Tables 2 and 3 for each Kr and Xe isotope, with $M = {}^{82}\text{Kr}$ and $[C] = \text{Sr}$ for Kr and $M = {}^{130}\text{Xe}$ and $[C] = \text{Ba}$ for Xe. The specific spallation concentrations ${}^{78}\text{Kr}_{\text{sp}}/\text{Sr}$ and ${}^{83}\text{Kr}_{\text{sp}}/\text{Sr}$ in Table 5, and ${}^{126}\text{Xe}_{\text{sp}}/\text{Ba}$ and ${}^{131}\text{Xe}_{\text{sp}}/\text{Ba}$ in Table 6, were calculated from Eq. (6) using Eq. (2) correlation slopes, with $M = {}^{78}\text{Kr}$ and ${}^{83}\text{Kr}$ and $[C] = \text{Sr}$ for Kr and $M = {}^{126}\text{Xe}$ and ${}^{131}\text{Xe}$ and $[C] = \text{Ba}$ for Xe; in most cases several different Kr and Xe isotopes were used for M' in separate correlations, and the Eq. (6) results averaged to obtain the specific concentrations listed in Tables 5 and 6. The various values of $(M'/M)_{\text{sp}}$ required in Eq. (6) calculations were determined directly as the intercepts of Eq. (4) three-isotope correlations, with $M'' = {}^{86}\text{Kr}$ and ${}^{136}\text{Xe}$ and the assumption that $({}^{86}\text{Kr})_{\text{vc}} \equiv ({}^{86}\text{Kr})_{\text{sp}} \equiv 0$ and $({}^{136}\text{Xe})_{\text{vc}} \equiv ({}^{136}\text{Xe})_{\text{sp}} \equiv 0$. The assumption that $({}^{136}\text{Xe})_{\text{vc}} \equiv ({}^{136}\text{Xe})_{\text{sp}}$ is known to be invalid for certain lunar soils because of the presence of volume correlated fission Xe (Basford *et al.*, 1973); however, there is no evidence in the spallation Xe isotopic compositions for a significant admixture of this component in the drill core fines.

The specific concentrations ${}^{78}\text{Kr}_{\text{sp}}/\text{Sr}$ and ${}^{83}\text{Kr}_{\text{sp}}/\text{Sr}$ given in Table 5 vary somewhat erratically with depth: sample 15004 in particular is low compared to 15003 and 15005. The effect is chemical rather than irradiational, and is due to the fact that the tacit assumption made in expressing specific concentrations as $\text{Kr}_{\text{sp}}/\text{Sr}$—that the Sr concentration is representative of the concentration of all major Kr spallation target elements—is not correct. About half of the ${}^{83}\text{Kr}_{\text{sp}}$ production—and more than half of the ${}^{78}\text{Kr}_{\text{sp}}$ production—is from Zr, and Sr and Zr do *not* correlate well in the drill core fines: Zr/Sr ratios from data given by Philpotts (1972) range from 3.88 in 15001 to a minimum of 3.13 in 15004. The only completely accurate way to resolve this difficulty is to measure Zr as well as Sr

Table 5. Spallogenic specific concentrations and trapped (surface-correlated) isotopic compositions of krypton in <250 μ fines from the Apollo 15 drill core stems.

Sample	Depth (g/cm^2)	*^{78}Kr$_{sp}$/Sr	*^{83}Kr$_{sp}$/Sr	^{78}Kr	^{80}Kr	^{83}Kr	^{84}Kr	^{86}Kr
		(units of 10^{-7} cc STP/gSr)		Trapped ratios: relative to ^{82}Kr = 100				
15006,26	65	15.2 (4.57)	68.8 (27.7)	3.050	19.819	99.64	491.1	149.49
		±1.4	±5.4	±.031	±.024	±.15	±2.7	±.74
15005,25	138	16.0 (4.86)	77.4 (31.4)	3.059	19.80	99.18	489.7	148.34
		±.5	±2.9	±.016	±.13	±.16	±1.6	±.13
15004,26	208	12.8 (4.06)	68.3 (28.7)	3.073	19.865	99.50	490.0	149.26
		±1.2	±5.1	±.035	±.063	±.16	±1.3	±.39
15003,28	276	13.5 (3.68)	68.2 (25.3)	2.918	19.669	99.69	492.3	150.16
		±1.3	±5.4	±.046	±.053	±.32	±0.6	±.29
15002,26	341	12.6 (3.43)	65.9 (24.4)	3.015	19.711	99.34	492.4	149.84
		±.5	±4.9	±.017	±.013	±.26	±0.6	±.62
15001,30	412	10.1 (2.74)	55.8 (20.6)	3.088	19.873	99.14	490.6	148.89
		±.6	±5.7	±.032	±.050	±.19	±0.8	±.32
†ATMOSPHERE				3.011	19.587	99.60	494.63	150.98
				±.010	±.009	±.10	±.10	±.12

*Specific concentrations in parentheses are ^{78}Kr$_{sp}$/[Sr(1 + 0.69(Zr/Sr))] and ^{83}Kr$_{sp}$/[Sr(1 + 0.44(Zr/Sr))]. See text.

†Basford *et al.* (1973).

concentrations in each grain size fraction; failing this, the specific concentrations may be corrected to first order by including the known Zr/Sr ratios for each bulk sample and assuming uniformity with grain size. Marti *et al.* (1966) give the dependence of ^{83}Kr$_{sp}$ on target element abundances as ^{83}Kr$_{sp} \propto$ [Sr + 0.59Y + 0.32Zr] = Sr[1 + 0.59(Y/Sr) + 0.32(Zr/Sr)]. Y and Zr correlate reasonably well in lunar material; with (Y/Zr) $\cong$ 0.21 (Duncan *et al.*, 1973), the spallogenic ^{83}Kr specific concentrations may be rewritten as ^{83}Kr$_{sp}$/[Sr(1 + 0.44(Zr/Sr))]. For ^{78}Kr$_{sp}$, the dependence on target element abundances calculated according to the method of Marti *et al.* (1966) is ^{78}Kr$_{sp} \propto$ [Sr + 0.78Y + 0.53Zr] = Sr[1 + 0.78(Y/Sr) + 0.53(Zr/Sr)], and the spallogenic ^{78}Kr specific concentrations are ^{78}Kr$_{sp}$/[Sr(1 + 0.69(Zr/Sr))]. These revised specific concentrations are given in parentheses in Table 5 and plotted versus depth in Fig. 8; in both cases Zr/Sr for each of the drill core samples (except 15002) is from Philpotts' (1972) data.

The profiles in Fig. 8 are relatively smooth functions of depth. The one deviant sample, 15002 at 341 g/cm^2, is the only one in which Zr was not determined; Zr/Sr was taken as the average of adjacent samples. Current cross-section data on Kr and Xe target elements are inadequate to allow reliable calculation of the theoretical production rate versus depth profiles (Reedy and Arnold, 1972), but the relatively shallow ^{78}Kr$_{sp}$ concentration peak—particularly referenced to an original surface at a present depth of ~35–55 g/cm^2—seems roughly consistent with expected production by high-energy components of the galactic particle flux. An isotope effect appears fairly clearly in Fig. 8: ^{83}Kr$_{sp}$, closer in mass to the abundant

Table 6. Spallogenic specific concentrations and trapped (surface-correlated) isotopic compositions of xenon in <250 μ fines from the Apollo 15 drill core stems.

Sample	Depth	$^{126}Xe_{sp}/Ba$	$^{131}Xe_{sp}/Ba$	^{124}Xe	^{126}Xe	^{128}Xe	^{129}Xe	^{131}Xe	^{132}Xe	^{134}Xe	^{136}Xe
	(g/cm²)	(units of 10^{-7} cc STP/gBa)				Trapped ratios: relative to $^{130}Xe = 100$					
15006,26	65	6.10	31.4	2.911	3.05	50.95	630.5	497.1	603.7	222.75	182.30
		±.62	±2.6	±.057	±.16	±.55	±2.2	±2.5	±0.8	±.49	±.35
15005,25	138	6.15	29.3	3.22	3.20	52.32	634.3	497.0	599.1	222.3	180.2
		±.99	±2.9	±.31	±.36	±.46	±5.8	±1.6	±5.8	±2.1	±2.0
15004,26	208	5.12	27.8	3.259	3.355	51.419	634.2	499.4	603.3	223.3	180.72
		±.34	±4.8	±.072	±.095	±.030	±5.4	±3.3	±3.5	±1.7	±.80
15003,28	276	4.97	29.0	3.08	2.91	51.08	635.0	496.3	602.7	222.8	181.7
		±.80	±3.8	±.19	±.26	±.70	±5.2	±5.4	±5.1	±2.0	±2.1
15002,26	341	4.26	26.7	2.83	2.78	51.19	635.8	496.9	605.5	223.9	182.5
		±.51	±1.8	±.17	±.17	±.35	±3.6	±1.5	±3.5	±1.3	±1.0
15001,30	412	3.39	19.7	3.44	3.42	52.18	639.0	498.7	600.2	221.6	181.0
		±.41	±1.4	±.11	±.22	±.51	±4.3	±2.9	±3.6	±1.8	±2.1
†ATMOSPHERE				2.337	2.180	47.147	649.56	521.25	660.68	256.30	217.64
				±.007	±.011	±.047	±.58	±.59	±.53	±.37	±.22

†Basford *et al.* (1973).

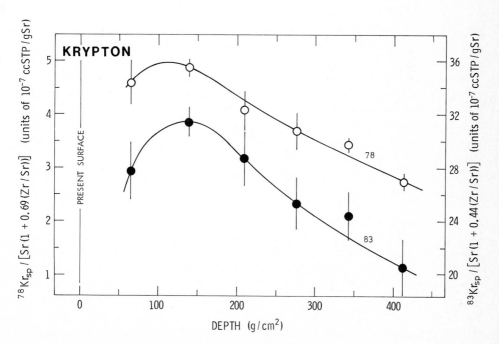

Fig. 8. Spallation ^{78}Kr (left scale) and ^{83}Kr (right scale) specific concentrations versus depth. Solid curves are visual fits to the data point distributions.

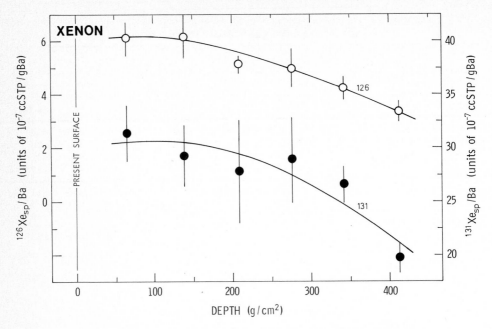

Fig. 9. Spallation ^{126}Xe (left scale) and ^{131}Xe (right scale) specific concentrations versus depth. Solid curves are visual fits to the data point distributions.

target nuclides, has a depth profile which peaks significantly deeper and decreases with depth $\sim$20% less rapidly than ^{78}Kr$_{sp}$, both suggesting production by lower energy particles.

Spallogenic Xe specific concentrations from Table 6 are plotted versus depth in Fig. 9. The flat ^{126}Xe$_{sp}$ profile in the upper part of the regolith section convincingly demonstrates its origin as a high-energy spallation product. The ^{131}Xe$_{sp}$ data are too imprecise to support more than the general observation that its profile is similar to that of ^{126}Xe$_{sp}$ but may drop off somewhat more slowly with depth.

Spallation He, Ne, and Ar production rates versus depth

Effective production rate profiles. The weight of the evidence discussed above appears to favor a rapid depositional model for the drill core section sufficiently strongly to justify adopting it as a working hypothesis. Application of the model to measured neutron fluence and spallation Ar depth distributions yields concordant stable stratification ages of $\sim$450 m.y. The effective *in situ* production rate of a particular spallogenic nuclide ^{A}X is $R^{A}(d) = [^{A}X_{sp}(d) - (^{A}X_{sp})_0]/T$, and the surface production rate is $R_s^{A} = R^{A}(0) = [^{A}X_{sp}(0) - (^{A}X_{sp})_0]/T$, where $^{A}X_{sp}(d)$ is the measured concentration profile, $(^{A}X_{sp})_0$ is the uniform preirradiation concentration yielded by the model fit, and T is the *in situ* irradiation time. Using the rapid deposition model equations with the best-fit parameters given above, we have calculated the *relative* production function $F^A(d) \equiv R^A(d)/R_s^A$ versus depth in the drill core fines for $^3\mathrm{He_{sp}}$, $^{21}\mathrm{Ne_{sp}}$, and $^{38}\mathrm{Ar_{sp}}$, and tabulated the results in Table 7; for the present, we have used for Ar the model shown in Fig. 1, which is based on Reedy and Arnold's theoretical $^{37}\mathrm{Ar}$ production rate profile. Total surface production rates are $R_s^A = [P_s^A T f(0) + N_s^A T g(0)]/T$, and with $f(0) = g(0) = 1$ and $T = 450$ m.y., these are $R_s^3 = 0.509 \times 10^{-8}$ cc STP/g-m.y., $R_s^{21} = 0.0554 \times 10^{-8}$ cc STP/g-m.y., and $R_s^{38} = 0.0625 \times 10^{-8}$ cc STP/g-m.y. for the drill core fines. A rudimentary chemical coefficient may be incorporated into these production rates by adopting Bogard *et al.*'s (1971) empirical determination that the mean $^{21}\mathrm{Ne_{sp}}$ and $^{38}\mathrm{Ar_{sp}}$ production rates in ~50 Apollo 11 and 12 crystalline rocks are roughly proportional to concentrations of (Mg + Si) and (Ca) respectively. Average (Mg + Si) = 0.279 g/g, and average (Ca) = 0.0715 g/g in the Apollo 15 drill core samples (Rhodes, 1972). The effective production rates versus depth $R^A(d) = R_s^A F^A(d)$ of these three light rare gas spallogenic nuclides are therefore

$$R^3(d) = 0.509 F^3(d) \times 10^{-8} \text{ cc STP/g-m.y.}$$
$$R^{21}(d) = 0.199 F^{21}(d) \times 10^{-8} \text{ cc STP/g(Mg + Si)-m.y.}$$
$$R^{38}(d) = 0.874 F^{38}(d) \times 10^{-8} \text{ cc STP/g(Ca)-m.y.}$$

Calculated errors in the numerical coefficients of these equations, arising from

Table 7. Relative production functions $F^A \equiv R^A/R_s^A$ versus depth for $^3\mathrm{He_{sp}}$, $^{21}\mathrm{Ne_{sp}}$, and $^{38}\mathrm{Ar_{sp}}$.

Depth (d) (g/cm²)	F^3	F^{21}	F^{38}	Depth (d) (g/cm²)	F^3	F^{21}	F^{38}
0	1.00	1.00	1.00	190	1.41	1.56	1.50
5	1.07	1.13	1.21	200	1.37	1.52	1.45
10	1.13	1.24	1.41	225	1.26	1.40	1.33
15	1.18	1.32	1.58	250	1.15	1.27	1.21
20	1.23	1.40	1.73	275	1.03	1.14	1.10
25	1.26	1.45	1.88	300	0.91	1.01	0.97
30	1.30	1.51	1.99	325	0.80	0.89	0.86
40	1.36	1.59	2.11	350	0.70	0.78	0.76
50	1.43	1.67	2.17	375	0.61	0.68	0.67
60	1.48	1.72	2.19	400	0.53	0.59	0.60
70	1.51	1.75	2.18	425	0.46	0.51	0.53
80	1.54	1.77	2.15	450	0.40	0.45	0.47
90	1.57	1.79	2.09	475	0.35	0.39	0.41
100	1.58	1.80	2.02	500	0.30	0.34	0.37
110	1.59	1.80	1.96	525	0.26	0.30	0.32
120	1.59	1.79	1.90	550	0.23	0.26	0.29
130	1.58	1.77	1.83	575	0.20	0.22	0.25
140	1.57	1.76	1.78	600	0.17	0.20	0.22
150	1.55	1.73	1.72	625	0.15	0.17	0.20
160	1.52	1.69	1.66	650	0.13	0.15	0.18
170	1.49	1.65	1.61	675	0.11	0.13	0.15
180	1.45	1.61	1.56	700	0.10	0.11	0.14

uncertainties in the model best-fit parameters, are $\sim 15\%$. Values of F^A for $d = 0-700\,\text{g/cm}^2$ are given in Table 7.

These effective production rate versus depth expressions are at best relatively crude approximations to the correct functions. It is clear, for example, that the dependence on target chemistry must itself be a function of depth since the variety of different nuclear reactions involving different target elements which contribute to the total spallation yield in general are characterized by strongly energy-dependent (and therefore depth-dependent) excitation functions. It is equally clear that these expressions are totally model-dependent, and are valid only to the extent that the rapid deposition model for the drill core fines is itself valid.

It is informative to compare these production rate expressions to those derived by Bogard *et al.* (1971): $\langle P^3 \rangle \equiv 1 \times 10^{-8}\,\text{cc STP/g}$, $\langle P^{21} \rangle = 0.564 \times 10^{-8}\,\text{cc STP/g(Mg + Si)-m.y.}$, and $\langle P^{38} \rangle = 1.69 \times 10^{-8}\,\text{cc STP/g(Ca)-m.y.}$ We have *assumed*, in our initial discussion of model fits to the Ar data, that these production rates are averages for samples cycling through roughly the upper meter of the lunar regolith: i.e. that

$$\langle P^A \rangle \simeq (1/D) \int_0^D R^A(x)dx, \quad \text{with } D \sim 165\,\text{g/cm}^2.$$

Numerical integration of F^A in Table 7 from 0 to $165\,\text{g/cm}^2$ yields $\langle F^3 \rangle = 1.45$, $\langle F^{21} \rangle = 1.65$, and $\langle F^{38} \rangle = 1.93$, and therefore $\langle R^3 \rangle = 0.74 \times 10^{-8}\,\text{cc STP/g-m.y.}$, $\langle R^{21} \rangle = 0.328 \times 10^{-8}\,\text{cc STP/g(Mg + Si)-m.y.}$, and $\langle R^{38} \rangle = 1.69 \times 10^{-8}\,\text{cc STP/g(Ca)-}$ m.y. Our integrated production rate for $^{38}\text{Ar}_{sp}$ agrees with Bogard *et al.*'s value, as it must since we assumed their value in evaluating T from the best-fit model parameters to the Ar data. But Bogard *et al.*'s mean production rates for $^3\text{He}_{sp}$ and $^{21}\text{Ne}_{sp}$ exceed our integrated values by $\sim 35\%$ and 70%, respectively. It is possible that the disagreement in the He production rates is due to diffusive loss of spallogenic ^3H or ^3He from the drill core fines, but this is by no means certain for reasons outlined earlier. Diffusive loss of such a large fraction ($\sim 40\%$) of spallogenic ^{21}Ne from the drill core fines seems definitely out of the question: one of the two spallation Ne production rate expressions therefore appears to be incorrect for other reasons. This discrepancy is addressed in detail in the following section.

Application to rock exposure ages. Drozd *et al.* (1974) have measured Ne, Ar, and Kr abundances in a number of Apollo 14, 15, and 16 rock samples. Their calculated $^{81}\text{Kr–Kr}$, ^{21}Ne and ^{38}Ar exposure ages, the latter two based on Bogard *et al.*'s (1971) production rates, are tabulated in Table 8. With few exceptions, the ^{21}Ne ages are strikingly lower than the $^{81}\text{Kr–Kr}$ ages; ^{38}Ar ages are either comparable to, or lower than, the $^{81}\text{Kr–Kr}$ ages.

Also shown in Table 8 are ^{21}Ne and ^{38}Ar exposure ages calculated for several of these samples from the effective production rate versus depth expressions derived in the previous section. Ages are given by the equations $T^{21}(d) = {}^{21}\text{Ne}_{sp}/R^{21}(d)$ and $T^{38}(d) = {}^{38}\text{Ar}_{sp}/R^{38}(d)$, where $^{21}\text{Ne}_{sp}$ and $^{38}\text{Ar}_{sp}$ are the measured

Table 8. Cosmic-ray exposure ages in millions of years (m.y.) for Apollo 14, 15, and 16 rock samples.

Sample		$*^{81}$Kr–Kr	$*^{21}$Ne	$*^{38}$Ar	$T^{21}(d)$	$T^{38}(d)$	Depth d (g/cm²)
67455		50.3	17.3	38.	42.6	59.0	6.3
		±3.1	±4.1	±13.	±7.1	±7.6	
67915	North Ray	50.6	21.0	26.	60.3	41.8	2.8
		±3.0	±4.9	±10.	±9.7	±5.6	
67955		50.1	17.9	32.	45.2	55.0	4.8
		±3.2	±4.2	±12.	±7.6	±7.1	
68115		2.08	1.75	1.63			
		±.32	±.41	±.67			
68815		2.04	1.21	2.18	1.97	1.97	80
	South Ray	±.20	±.29	±.98	±.32	±.26	±60
69935		1.99	1.40	4.0			
		±.37	±.33	±1.7	See text		
69955		4.23	2.13	—			
		±.41	±.51				
68415		92.5	32.5	113.			
		±13.3	±7.8	±42.			
62235		153.3	104.	163.	164.	153.	90
		±6.5	±24.	±54.	±27.	±20.	±20
15475		473.	336.	543.	529.	529.	105
		±20.	±79.	±183.	±88.	±72.	±50
15595		110.	112.	105.			
		±17.	±26.	±34.			
14310		268.	113.	173.	275.	263.	6.0
		±20.	±26.	±55.	±45.	±35.	

*Data from Drozd et al. (1974). ^{21}Ne and ^{38}Ar exposure ages calculated using production rates given by Bogard et al. (1971).

concentrations (Drozd, personal communication); chemical compositions were taken from the references cited by Drozd et al. (1974) for chemistry. Clearly, both "ages" are now functions of depth. Since ^{81}Kr–Kr ages are relatively insensitive to shielding effects—except for neutron capture effects (Drozd et al., 1974)—the relevant questions are: (1) whether there exists an irradiation depth d for which T^{Kr-Kr}, $T^{21}(d)$ and $T^{38}(d)$ are equal; and (2) whether, in view of constraints imposed by geologic or regolith dynamics arguments, the particular sample under consideration could have resided at effectively this shielding depth during its history of exposure to the galactic radiation.

The three exposure ages for the North Ray ejecta samples 67455, 67915, and 67955 are roughly concordant within error for shielding depths ranging from ~ 3 to ~6 g/cm². The host boulders were apparently ejected from well-shielded to surface exposure locations in a single-stage impact event (Drozd et al., 1974); the small shielding depths are fully consistent with samples taken from near the lightly eroded surfaces of the boulder chips.

The South Ray samples, in contrast, show evidence in both the ^{81}Kr–Kr ages (samples 69935 and 69955) and the rare gas ages (samples 68815, 69935 and 69955) for presurface irradiation. The occurrence of a neutron-generated Kr component in 69935 and the discrepant Kr–Kr ages of 69935 and 69955—samples from, respectively, the top and bottom of the same boulder—are consistent with a presurface and surface exposure history for the parent boulder in which the boulder, previously unirradiated, was buried with its base ~350 g/cm^2 (and its top ~170 g/cm^2) below the regolith surface for ~1–6 m.y. prior to its ejection to the surface, probably by a South Ray secondary (Drozd *et al.*, 1974). The Drozd *et al.* model requires a 180° orientation flip during ejection, and a post-ejection surface residence time of 2 m.y., during which the sample from what is now the boulder top (69935) was essentially unshielded and the bottom sample (69955) was shielded by the ~180 g/cm^2 of overlying rock.

This irradiation model, which accounts fully for both the observed low-energy neutron-produced Kr component in 69935 and the discordant ^{81}Kr–Kr exposure ages in 69935 and 69955, provides a significant test (although not a definitive test, since the model is not necessarily unique) of the spallation Ne production rate profile derived from the drill core samples. The model predicts the following spallation Ne concentrations in these samples: $(^{21}\text{Ne}_{sp})_{69935} = R^{21}(d_b)T_b + R^{21}(d_s)T_s$, and $(^{21}\text{Ne}_{sp})_{69955} = R^{21}(D_b)T_b + R^{21}(D_s)T_s$, where T_b and T_s are the subsurface and surface residence times for the boulder, and d_b, D_b and d_s, D_s are the subsurface and surface shielding depths for the two samples. We adopt the values $d_b = 350$ g/cm^2, $D_b = 170$ g/cm^2, $d_s = 0$, and $D_s = 180$ g/cm^2 from the model, and assume $T_s = 2$ m.y. since this number seems relatively well established from other samples (Table 8). Using the Ne production rate equation given above, with F^{21} values for the appropriate depths from Table 7, and spallation Ne concentrations of 2.0×10^{-9} cc STP/g and 3.0×10^{-9} cc STP/g for 69935 and 69955, respectively (Drozd, personal communication), we calculate $T_b = 3.2$ m.y. and $T_b = 2.1$ m.y., or on the average $T_b = 2.6 \pm 0.6$ m.y. This subsurface residence time for the Station 9 boulder is close to the nominal value of 2.1 m.y. given by Drozd *et al.* (1974), and well within their rather wide limits of ~1–6 m.y. The Ne production rate profile is therefore entirely consistent with the Kr–Kr chronology of these samples. It is interesting to note that if certain obvious refinements are made in the shielding depths assumed in the calculation, i.e. a reduction of d_b to ~300 g/cm^2 and of D_s to ~150 g/cm^2 to compensate roughly for the nonhemispherical shape of the boulder, and an increase in d_s to ~5 g/cm^2 since the samples are taken from nonzero depths in the rock chips—then the calculations yield $T_b = 2.1 \pm 0.1$ m.y.

Two other Apollo 16 rock samples, 68815 and 62235, show definite evidence in the spallogenic rare gases for a history of subsurface irradiation. The $T^{21}(d)$ and $T^{38}(d)$ functions for 68815 intersect at 1.97 m.y.—very close to the ^{81}Kr–Kr age of 2.04 m.y.—at a depth of ~80 g/cm^2; the depth is uncertain since it is near the peak of both production rate profiles, where production rates are insensitive functions of depth. An immediate difficulty with this single-stage irradiation history is that the sample was collected *on the surface*, and moreover had probably been on the surface for ~2 m.y. if we accept Drozd *et al.*'s arguments that secondary impacts from the South Ray event excavated many of the Station 8 and Station 9 boulders, presumably including this one. We therefore seek an irradiation model, consistent with the spallation rare gas concentrations, in which the 68815 parent boulder resided on the surface for the past ~ 2 m.y. In a simple two-stage model, similar to that applied to the 69935/55 boulder, $^{21}Ne_{sp} = R^{21}(d_b)T_b + R^{21}(d_s)T_s$, and $^{38}Ar_{sp} = R^{38}(d_b)T_b + R^{38}(d_s)T_s$. With $T_s = 2$ m.y., $d_s \sim 5$ g/cm^2, $^{21}Ne_{sp} = 1.7 \times 10^{-9}$ cc STP/g, and $^{38}Ar_{sp} = 4.1 \times 10^{-9}$ cc STP/g, the equation pair yields solutions for T_b only at depths where the production function ratio $F^{38}(d_b)/F^{21}(d_b)$ is $\simeq 1.5$. This ratio is higher than any value in the depth range 0–700 g/cm^2 covered by Table 7, but is reached at $d_b \sim 1050$ g/cm^2. At $d_b \sim 1050$ g/cm^2, $T_b \sim 70$ m.y.; therefore, an irradiation history for 68815 of ~ 70 m.y. at ~ 6.5 m depth followed by 2 m.y. on the surface is consistent with spallation Ne and Ar concentrations. We cannot take the nominal values for d_b and T_b too seriously, since they require extrapolation of the F-values in Table 7 to about three times the maximum depth of the measured data on which they are based. However, it is clear that in this irradiation model, d_b must be large.

Sample 62235 is a small rock fragment collected on the rim of Buster Crater at Station 2. Neon and argon exposure ages are in excellent agreement with the Kr–Kr age if the effective irradiation depth was ~ 90 g/cm^2 (Table 8). Although this sample was also collected from the surface, and there is some evidence that its surface residence time was also a few million years (see Drozd et al., 1974), most of its spallation gases must have been acquired at depth since its total exposure age is so comparatively large. The absence of a "soft" component in Kr does not rule against prolonged exposure near 90 g/cm^2, since according to Drozd et al. slow neutron effects are expected to be observable only at considerably greater shielding depths.

Of the remaining samples in Table 8, 15475 appears to have resided at an effective depth of ~ 100 g/cm^2 for more than 500 m.y. In contrast, 14310 has apparently been on or very close to a regolith or rock surface during its entire exposure history: the three exposure ages are in striking agreement for $d \simeq 6$ g/cm^2. Two samples, 68115 and 68415, yield no concordant exposure chronologies for any shielding depth. The final sample, 15595, is the only one which yields *concordant* exposure ages using Bogard et al.'s (1971) rare gas production rates, and *discordant* ages using those derived from the drill core data. Although in general the evidence from rock exposure ages, discussed above and summarized in Table 8, strongly supports the validity of these production rate profiles, and therefore the validity of the rapid deposition model for the Apollo 15 drill core section, there are clearly still questions to be answered.

Trapped gases

In general, isotopic compositions of the trapped rare gases in the drill core fines (Tables 4–6) are similar to those deduced in similar studies of other lunar fines by many other investigators. A few specific comments are applicable to previous rare gas studies of the drill core materials, and to current work with these and other lunar fines:

(1) While the $(^4He/^3He)_{tr}$ ratios are all fairly close to the average value of 2500, there is a suggestion in the Table 4 data of a rough (and unexplained) correlation between the trapped ratio and the $^3He_{sp}$ concentration.

(2) There is no evidence in our single samples from drill stems 15001 and 15003 for the high $(^4He/^3He)_{tr}$ ratios of ~2700 and ~3000 inferred by Hübner *et al.* (1973) in their study of several bulk samples from these two stems.

(3) The ratio $(^{22}Ne/^{21}Ne)_{tr}$ varies from 32.3 to 32.9 in the five samples where it was measured; all are within error of the average value of 32.6. This is within error of the value 32 ± 1 determined by Hübner *et al.* (1973) and used by them in calculating $^{21}Ne_{sp}$ concentrations in bulk samples from stems 15001 and 15003, but does not agree with the number 31 ± 0.5 assumed and used by Bogard *et al.* (1973) for the same purpose. We therefore suspect that spallation Ne concentrations in bulk drill core samples calculated by Bogard *et al.* are both statistically erratic and systematically low.

(4) Most rare gas mass spectrometer systems, including ours, use atmospheric Ar for instrumental calibration. Since the atmospheric $^{40}Ar/^{36}Ar$ ratio is ~100–300 times larger than that typical of lunar soils, spectrometer memory effects can seriously hamper accurate measurement of instrumental mass discrimination of Ar and thus limit the accuracy of sample isotopic compositions. Over the past year we have devoted some considerable effort to obtaining accurate discrimination corrections, and recent measurements of $(^{38}Ar/^{36}Ar)_{tr}$ in a variety of lunar fines by ordinate intercept analysis (Table 4; Venkatesan *et al.*, 1974; Basford, 1974) suggest a value of 0.1890 ± 0.0010 $(^{36}Ar/^{38}Ar = 5.29 \pm 0.03)$ as a current best estimate for average trapped lunar Ar. As is the case for other trapped lunar gases, however, there are probably real variations in this ratio from sample to sample.

(5) The isotopic compositions of trapped Kr in Table 5 vary slightly from sample to sample, probably due in part to small sample masses, particularly in the coarse grain size fractions. However, the only strikingly anomalous value is the low $(^{78}Kr/^{82}Kr)_{tr}$ ratio in 15003. If all other ratios are averaged over the six samples and then compared to the composition of atmospheric Kr (Table 5), the result is a familiar isotopic pattern: within error, a linear fractionation with respect to the terrestrial isotopic composition by ~-0.25%/mass number (Basford *et al.*, 1973).

(6) Trapped $^{124}Xe/^{130}Xe$ and $^{126}Xe/^{130}Xe$ ratios in Table 6 are both imprecise and highly erratic. Within error, the remaining isotopic ratios fall near or within the ranges in isotopic composition characteristic of trapped lunar Xe (Basford *et al.*, 1973).

K–Ar ages

Equation (8), with $M = {}^{36}Ar$, $M' = {}^{40}Ar$, $[C] =$ potassium concentration K, and the excellent assumption that $({}^{40}Ar)_{sp} \ll ({}^{40}Ar)_{rad}$, may be written as $({}^{40}Ar/{}^{36}Ar)(1/\gamma) = (K/({}^{36}Ar)\gamma)({}^{40}Ar_r/K) + ({}^{40}Ar/{}^{36}Ar)_{tr}$, where $\gamma = (1 - ({}^{36}Ar)_{sp}/{}^{36}Ar)$ is a correction factor for spallogenic ^{36}Ar. The ratio of parentless, surface correlated ^{40}Ar to solar wind-derived ^{36}Ar is represented as $(^{40}Ar/^{36}Ar)_{tr}$. If the grain size separates of a single soil sample contain trapped Ar of uniform isotopic composition and the same ratio of volume correlated radiogenic ^{40}Ar to K, then a plot of $(^{40}Ar/^{36}Ar)(1/\gamma)$ versus $K/(^{36}Ar)\gamma$ for these fractions defines a linear correlation of intercept $(^{40}Ar/^{36}Ar)_{tr}$ and slope $^{40}Ar_r/K$; the K–Ar age of the soil is calculated directly from the slope by means of the standard K–Ar dating equation. It is important to point out that the method is essentially a bulk K–Ar dating technique, in which the correlation permits resolution of the radiogenic and parentless ^{40}Ar components; as such, the ages are precise measures of event chronologies only to the extent that inheritance and diffusive loss of radiogenic Ar are minor effects.

Correlations and slope ages resulting from the application of this technique to the drill core fines are shown in Fig. 10, and the slope ages are plotted versus depth in Fig. 11. $(^{40}Ar/^{36}Ar)_{tr}$ ratios are tabulated in Table 4. Because of the coarse vertical scales in Fig. 10, all data point errors in this direction are smaller than the symbol size and the correlations appear better than they actually are; the 1 σ age errors given in the figure are a more valid measure of the correlation fits. The relative imprecision of the ages calculated from these correlations means that we cannot rule convincingly between the two alternatives shown in Fig. 11: (1) that the soil ages are roughly uniform at 2.9 ± 0.2 AE for section between 65 and 340 g/cm^2, with the deep sample 15001 probably older at $\sim$3.3 AE; or (2) that the core material is progressively older with increasing depth, with the age increasing by $\sim$450 m.y. down the lower 350 g/cm^2 of the core. The latter possibility is more consistent with the data. If it is correct, it constitutes a constraint on the core accretion models discussed above. A truly sudden, single event deposition could not reasonably be expected either to accidentally create a coherent age versus depth profile by mixing source materials of different ages, or to preserve a preexisting profile in the source materials. On the other hand, a rapid deposition of the core section—over a time perhaps as long as 100 m.y.—as a superposition of thin layers derived from preirradiated materials of progressively younger K–Ar age could still be compatible with the "sudden" deposition model fits to the neutron fluence and Ar data, and with the presence of the large numbers of sedimentary horizons which exist in the drill core (Heiken et al., 1972).

Independent of the questions of which age profile in Fig. 11 may be correct, and of the resultant constraints imposed on the stratigraphy and history of the source materials for the drill core section, it seems clear that the fine materials now in the section were outgassed in a major thermal event $\sim$2.9 AE ago, or perhaps in a series of such events spanning a few hundred million years around 2.9 AE. We have noted previously that K–Ar ages of lunar soils show a tendency to group around $\sim$2.8 and $\sim$4.0 AE (Pepin et al., 1972). All K–Ar ages measured on <1 mm fines in our laboratory, plus one age derived from grain size data reported by Yaniv and Heymann (1972), are plotted on the Fig. 12 histogram. The earlier conclusion is still valid: of the eight sampled lunar sites, soil samples from five are represented in a grouping between $\sim$2.7 and 3.2 AE, and four in a peak near 4 AE. The one Apollo 16 soil examined by the isochron technique described

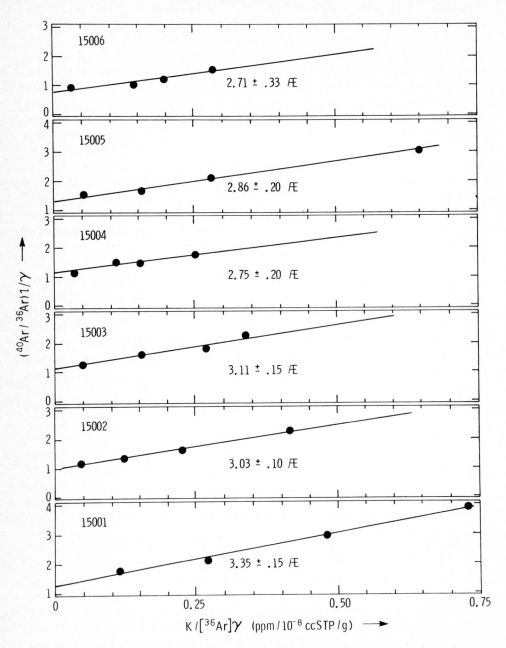

Fig. 10. K–Ar isochron correlations (equation (8)) for the drill core fines. Values of $\gamma = 1 - (^{36}Ar_{sp}/^{36}Ar)$ are ≥ 0.99 for all separates and do not affect the correlations in a substantive way. Vertical errors are smaller than data points; horizontal errors are uniformly $\sim 5\%$. Numbers are slope K–Ar ages $\pm 1\,\sigma$ errors. Ordinate intercepts are $(^{40}Ar/^{36}Ar)_{tr}$ ratios, given in Table 4.

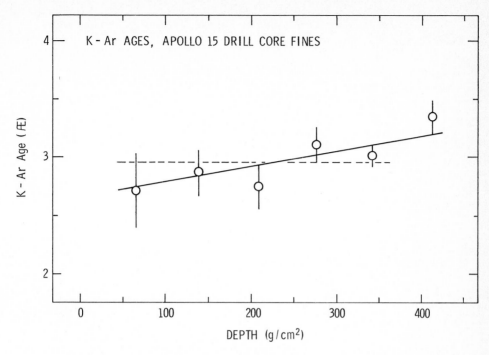

Fig. 11. K–Ar ages (from Fig. 10) versus depth. The horizontal dashed line at ~2.9 AE is a possible fit to the depth profile defined by the upper five data points. The solid line is a best fit to the complete profile.

above failed to yield an interpretable correlation, and we have not studied material from Luna 20. We briefly summarize the Fig. 12 data site by site: (1) Apollo 11. It is not uncommon for different soils from the same site to show widely variant ages, e.g. Apollo 12 and 15 samples in Fig. 12. The two Apollo 11 data points, however, refer to ilmenite (~2.8 AE) and plagioclase (~4.0 AE) separates from the same sample, 10084 (Basford, 1974). Basford proposes a mixing model in which 10084 consists of ~70% locally derived products and ~30% highlands throw-in material, represented chronologically by the ilmenite and plagioclase ages, respectively. The data suggest an intense local weathering of the ilmenite-rich mare basalts ~2.8 AE ago. (2) Apollo 12. The ages in Fig. 12 were derived from a multicomponent mixing model of ten Apollo 12 soils (Pepin *et al.*, 1972). The ~1 AE age refers to the KREEP endpoint of the mixing curves and is close to the Copernican impact age. The ~2.8 AE age occurs at the mare basalt endpoint of the model and was originally thought to indicate Ar loss during comminution of the ~3.2 AE old local basalts, but the recurrence of this age at sites where the local rocks are considerably older, notably at the Apollo 11 and 14 sites, casts doubt on this explanation. Evidence for the ~4 AE age is rather indirect. (3) Apollo 14. The samples are 14259 (~2.9 AE) and 14149 (~4.1 AE). The ages have been recalculated using Eq. (8), but are only marginally different from the values originally obtained by another method (Pepin *et al.*, 1972). (4) Apollo 15. The drill

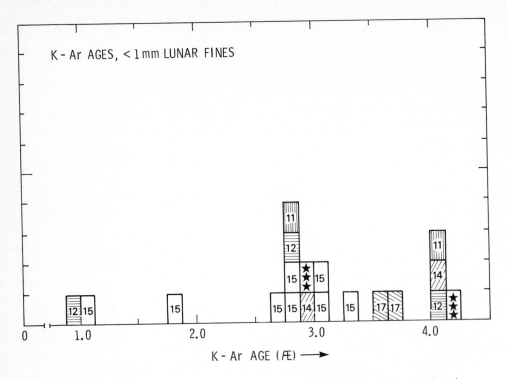

Fig. 12. K–Ar ages of lunar fines, as determined in the Minnesota laboratory. Numbered rectangles designate samples from Apollo 11–17; starred rectangles are Luna 16 samples. Ar data for Apollo 15 sample at ~1.8 AE (15601) from Yaniv and Heymann (1972).

core samples are grouped within ~300 m.y. of 2.9 AE. The two other Apollo 15 ages in Fig. 12 are conspicuous because they fall below 2.5 AE in an otherwise almost empty region of the histogram. Sample 15601 (Ar data from Yaniv and Heymann, 1972) falls at ~1.8 AE, and a rather imprecise but interestingly low value, within ~300 m.y. of 1 AE, was found for 15531. Both 15601 and 15531 were collected from near the edge of Hadley Rille. (5) Luna 16. The ages of the shallow (~2.9 AE) and deep (~4.2 AE) layers of the Luna 16 core are represented by the starred rectangles in Fig. 12. These recalculated ages are slightly higher than those originally reported by Pepin *et al.* (1972). K was not measured in the individual grain size fractions of these soils. (6) Apollo 17. The samples are 75081 and 71501, both "dark-mantle" soils; the ages are close to 3.6 AE (Venkatesan *et al.*, 1974), somewhat lower than the 3.8 AE local basalts, and are interesting in that they are almost alone in the gap between ~3.2 and 4 AE.

The peak in Fig. 12 near 4.0 AE is coincident with the apparent cessation of saturation bombardment of the lunar surface. These soils, outgassed in events near the end of that active early epoch, have experienced no significant subsequent heating. With regard to the grouping around 2.9 AE, we reiterate an earlier conclusion (Pepin *et al.*, 1972): there appear to have been widespread, large-scale inputs of thermal energy into the lunar regolith for a period perhaps as long as

~500 m.y. centered around 2.9 AE. The evidence discussed here suggests a late episode of relatively intense meteoritic bombardment as the probable mechanism.

Acknowledgments—Comments on the manuscript by Dr. D. D. Bogard and Dr. Raymond Davis, Jr. were very helpful. This research was supported by NASA grants NGL 24-005-225 and NGR 24-005-223.

REFERENCES

Basford J. R., Dragon J. C., Pepin R. O., Coscio M. R., and Murthy V. R. (1973) Krypton and xenon in lunar fines. *Proc. Fourth Lunar Sci. Conf., Geochim. Cosmochim. Acta,* Suppl. 4, Vol. 2, pp. 1915–1955. Pergamon.

Basford J. R. (1974) K–Ar analysis of Apollo 11 fines 10084. *Proc. Fifth Lunar Sci. Conf., Geochim. Cosmochim. Acta.* This volume.

Bhandari J. N., Goswami J. N., and Lal D. (1972) Apollo 15 regolith: a predominantly accretion or mixing model? In *The Apollo 15 Lunar Samples,* pp. 336–341. The Lunar Science Institute, Houston.

Bogard D. D., Funkhouser J. G., Schaeffer O. A., and Zähringer J. (1971) Noble gas abundances in lunar material—cosmic ray spallation products and radiation ages from the Sea of Tranquillity and Ocean of Storms. *J. Geophys. Res.* **76,** 2757–2779.

Bogard D. D., Nyquist L. E., Hirsch W. C., and Moore D. R. (1973) Trapped solar and cosmogenic noble gas abundances in Apollo 15 and 16 deep drill samples. *Earth Planet. Sci. Lett.* **21,** 52–69.

Drozd R. J., Hohenberg C. M., Morgan C. J., and Ralston C. E. (1974) Cosmic-ray exposure history at the Apollo 16 and other lunar sites: lunar surface dynamics. *Geochim. Cosmochim. Acta.* In press.

Duncan A. R., Erlank A. J., Willis J. P., and Ahrens L. H. (1973) Composition and inter-relationships of some Apollo 16 samples. *Proc. Fourth Lunar Sci. Conf., Geochim. Cosmochim. Acta,* Suppl. 4, Vol. 2, pp. 1097–1113. Pergamon.

Eberhardt P., Eugster O., and Marti K. (1965) A redetermination of the isotopic composition of atmospheric neon. *Z. Naturforsch.* **21a,** 623–624.

Eberhardt P., Geiss J., Graf H., Grögler N., Krähenbühl U., Schwaller H., Schwarzmüller J., and Stettler A. (1970) Trapped solar wind noble gases, exposure age and K/Ar-age in Apollo 11 lunar fine material. *Proc. Apollo 11 Lunar Sci. Conf., Geochim. Cosmochim. Acta,* Suppl. 1, Vol. 2, pp. 1037–1070. Pergamon.

Eberhardt P., Geiss J., Graf H., Grögler N., Mendia M. D., Mörgeli M., Schwaller H., and Stettler A. (1972) Trapped solar wind noble gases in Apollo 12 lunar fines 12001 and Apollo 11 breccia 10046. *Proc. Third Lunar Sci. Conf., Geochim. Cosmochim. Acta,* Suppl. 3, Vol. 2, pp. 1821–1856. MIT Press.

Evensen N. M., Murthy V. R., and Coscio M. R. (1973) Rb–Sm ages of some mare basalts and the isotopic and trace element systematics in lunar fines. *Proc. Fourth Lunar Sci. Conf., Geochim. Cosmochim. Acta,* Suppl. 4, Vol. 2, pp. 1707–1724. Pergamon.

Evensen N. M., Murthy V. R. and Coscio M. R. (1974) Provenance of KREEP and the exotic component: Elemental and isotopic studies of grain size fractions in lunar soils. *Proc. Fifth Lunar Sci. Conf., Geochim. Cosmochim. Acta.* This volume.

Fleischer R. L. and Hart H. R., Jr. (1972) Particle track record in Apollo 15 deep drill core from 54 to 80 cm depths. In *The Apollo 15 Lunar Samples,* pp. 371–373. The Lunar Science Institute, Houston.

Fleischer R. L., Hart H. R., Jr., and Giard W. R. (1974) Surface history of lunar soil and soil columns. *Geochim. Cosmochim. Acta* **38,** 365–380.

Heiken G., Duke M., Fryxell R., Nagle J. S., Scott R., and Sellers G. A. (1972) Stratigraphy of the Apollo 15 drill core, NASA Technical Memorandum X-58101, November.

Hübner W., Heymann D., and Kirsten T. (1973) Inert gas stratigraphy of Apollo 15 drill core sections 15001 and 15003. *Proc. Fourth Lunar Sci. Conf., Geochim. Cosmochim. Acta,* Suppl. 4, Vol. 2, pp. 2021–2036. Pergamon.

Huneke J. C., Podosek F. A., Burnett D. S., and Wasserberg G. J. (1972) Rare gas studies of the galactic cosmic ray irradiation history of lunar rock. *Geochim. Cosmochim. Acta* **36,** 269–301.

Marti K., Eberhardt P., and Geiss J. (1966) Spallation, fission, and neutron capture anomalies in meteoritic krypton and xenon. *Z. Naturforschg.* **21a**, 398–413.

Nier A. O. (1950) A redetermination of the relative abundances of the isotopes of carbon, nitrogen, oxygen, argon and potassium. *Phys. Rev.* **77**, 789–793.

Pepin R. O., Bradley J. C., Dragon J. C., and Nyquist L. E. (1972) K–Ar dating of lunar fines: Apollo 12, Apollo 14, and Luna 16. *Proc. Third Lunar Sci. Conf., Geochim. Cosmochim. Acta*, Suppl. 3, Vol. 2, pp. 1569–1588. MIT Press.

Philpotts J. A. (1972) Apollo 15 early allocation results (communication to principal investigators from L. A. Haskin) 22 March 1972.

Reed G. W., Jr., Jovanovic S., and Fuchs L. H. (1972) Trace element relations between Apollo 14 and Apollo 15 and other lunar samples and the implications of a moon-wide C1-KREEP coherence and Pt-metal non-coherence. *Proc. Third Lunar Sci. Conf., Geochim. Cosmochim. Acta*, Suppl. 3. Vol. 2, pp. 1989–2001. MIT Press.

Reedy R. C. and Arnold J. R. (1972) Interaction of solar and galactic cosmic-ray particles with the moon. *J. Geophys. Res.* **77**, 537–555.

Rhodes J. M. (1972) Apollo 15 early allocation results (communication to principal investigators from L. A. Haskin) 22 March 1972.

Rhodes J. M. and Hubbard N. J. (1972) Apollo 15 early allocation results (communication to principal investigators from L. A. Haskin) 22 March 1972.

Russ G. P., III, Burnett D. S., and Wasserberg G. J. (1972) Lunar neutron stratigraphy. *Earth Planet. Sci. Lett.* **15**, 172–186.

Silver L. T. (1972) Apollo 15 early allocation results (communication to principal investigators from L. A. Haskin) 22 March 1972.

Stoenner R. W., Davis R., Jr., Norton E., and Bauer M. (1974) Radioactive rare gases and tritium in the sample return container, and the ^{37}Ar, ^{39}Ar, tritium, hydrogen and helium depth profiles in the Apollo 16 and 17 drill stems. *Proc. Fifth Lunar Sci. Conf., Geochim. Cosmochim. Acta.* This volume.

Venkatesan T. R., Johnson N. L., Pepin R. O., Evensen N., Coscio M. R., and Murthy V. R. (1974) K–Ar dating of lunar fines: the Apollo 17 dark mantle. *Earth Planet. Sci. Lett.* In press.

Woolum D. S. and Burnett D. (1974) In-situ measurement of the rate of ^{235}U fission induced by lunar neutrons. *Earth Planet. Sci. Lett.* **31**, 153–163.

Yaniv A. and Heymann D. (1972) Atmospheric Ar^{40} in lunar fines. *Proc. Third Lunar Sci. Conf., Geochim. Cosmochim Acta*, Suppl. 3, Vol. 2, pp. 1967–1980. MIT Press.

Proceedings of the Fifth Lunar Conference
(Supplement 5, Geochimica et Cosmochimica Acta)
Vol. 2 pp. 2185–2203 (1974)
Printed in the United States of America

Solar flare and lunar surface process characterization at the Apollo 17 site*

L. A. Rancitelli, R. W. Perkins, W. D. Felix, and N. A. Wogman

Battelle, Pacific Northwest Laboratories, Richland, Washington 99352

Abstract—The Apollo 17 lunar samples, of which 13 are considered here, have been analyzed for a wide spectrum of cosmogenic and primordial radionuclides. The intense solar flares of August 4–7, 1972 produced unusually high concentrations of the short-lived radionuclides; ^{48}V (16d), ^{51}Cr (28d), ^{7}Be (53d), ^{56}Co (78d), ^{46}Sc (84d), and substantially increased the concentration of ^{54}Mn (312d), ^{22}Na (2.60 yr) and ^{60}Co (5.24 yr). Based on very careful and accurate measurements of these flare-produced radionuclides, the intensity and energy spectrum of the August 1972 flare at the Apollo 17 site has been established. The time integrated proton flux above 10 MeV was 7.9×10^{9} protons/cm². This flux was much harder than flares during the past five years with 36% of the proton flux above 30 MeV and 20% above 60 MeV. Expressed as a power function, $E^{-\alpha}$, the value of α was 1.9. This is a substantially more rigid spectrum than the average for the past million years which is estimated to have an α value of 3.1. Measurements of high- and low-energy reaction products in lunar samples exposed at different angles on the moon provide strong evidence for substantial angular anisotropy of the solar flare protons. The hardness of the solar flare proton flux increases dramatically as the incident angle moves from normal to low angles.

Soil which was permanently shadowed by a large boulder at Station 6 allowed the residence time of that boulder to be established at about one-half million years. The range in primordial radionuclide ratios was greater in samples at the Apollo 17 site than at any of the previous landing sites. Potassium to uranium ratios as high as 2860 and 6500 were observed for lunar soil and basalt, respectively, while many of the rock samples showed thorium to uranium ratios of less than 3.

Introduction

THE APOLLO 17 LUNAR MATERIAL provides a unique and precious set of samples not only because of the interesting geographical area which they represent, but also because of the spectrum of short-lived radionuclides which they contain. These radionuclides were produced by the two giant solar flares of August 4–7, 1972. This resulted in a very high proton flux on the lunar surface at the Apollo 17 site which was at that time 5 days into the lunar night. The relatively high concentrations of the short-lived cosmogenic radionuclides allowed their measurement with a high degree of accuracy which was sufficient to establish the energy spectrum of the incident proton flux and to provide an indication of its angular anisotropy.

Analyses of lunar samples from previous Apollo missions have already provided a great deal of information concerning the history and processes of the moon and sun (Perkins *et al.*, 1970; Rancitelli *et al.*, 1971, 1972; Finkel *et al.*, 1971). The cosmogenic radionuclides reflect both the recent and long-term character of the solar and galactic cosmic ray flux, while primordial radionuclides help to

*This paper is based on work supported by the National Aeronautics and Space Administration, Johnson Space Center, Houston, Texas under Contract NAS 9-11712.

describe the magmatic differentiation processes at the lunar surface. Apollo 11 sample studies provided our first observation of cosmogenic radionuclide production rates on the moon and confirmed the fact that the top centimeter of the lunar surface received a relatively intense periodic solar bombardment with a much smaller continuous galactic cosmic ray contribution (Perkins *et al.*, 1970; Shedlovsky *et al.*, 1970; O'Kelley *et al.*, 1970). Measurements on core samples and other soil and lunar rock specimens from the Apollo 12 mission provided our first determination of erosion rate of lunar rocks, mixing of the lunar soil, together with an indication of the long-term average galactic cosmic ray flux and that of the recent solar flare (Rancitelli *et al.*, 1971; Finkel *et al.*, 1971). Apollo 15 samples provided several examples of relatively short exposure times, on the order of one-half million years and less. Samples from Apollo 16 were of a very unusual chemistry resulting in very different cosmogenic radionuclide ratios as a result of differences in target element abundance. A quiet sun for a long period preceding this mission allowed the measurement of ambient levels of galactic cosmic ray-produced short-lived radionuclides, while the friable soil clods collected by rake sampling established the fact that these materials with short lunar surface age—on the order of 10^5 yr—are evidently common.

Receipt of Apollo 17 samples for analysis at our laboratory began only 6 days after splashdown and this has permitted their measurement with the minimum of decay loss. The very high degree of accuracy and precision obtained in our earlier measurements (Perkins *et al.*, 1970; Rancitelli *et al.*, 1971, 1972) was maintained in these analyses and has allowed the observation of subtle differences in radionuclide concentrations. This high precision and accuracy has been exceedingly helpful as it has permitted small, but very meaningful variations of the cosmogenic radionuclides in soils and rocks to be observed.

Procedure

The multiple gamma coincidence counting techniques (Perkins *et al.*, 1970) described in our earlier work were used to analyze the Apollo 17 samples. Briefly, the technique involves counting the sample for 8000–12,000 min in anticoincidence shielded multidimensional NaI(Tl) gamma-ray spectrometers. A comparison of the observed concentrations after background subtraction with the count rate of a sample mockup containing known radionuclide additions provides the basis for determining the radionuclide concentrations.

The mockups are prepared from a mixture of casting resin, iron powder, and aluminum oxide that contains a precisely known radionuclide addition. These mockups reproduce the precise shape, size, physical and electron densities of the samples. In most cases, the uncertainties associated with counting statistics were on the order 1–2% for ^{22}Na, ^{26}Al, K, Th, and U while the absolute errors for these radionuclides based on all analytical uncertainties including the error in radioisotope standards ranged from 3% to 8%. Precise mockups were not available for all the Apollo 17 samples analyzed. In these cases, the counting efficiencies and Compton corrections were estimated from our library of over 100 lunar mockups of various sizes and shapes by the methods used by O'Kelley *et al.* (1970). We have assigned appropriately larger uncertainties to these determinations. These values will be refined when sample models become available. Because of the relatively high concentrations of the short-lived radionuclides ^{46}Sc, ^{54}Mn, and ^{56}Co, it was possible to achieve similar accuracies. For the measurement of the short-lived cosmogenic radionuclides, a large Ge(Li) diode with an annular sodium-iodide anticoincidence shield (Cooper and Perkins, 1972) proved very useful. Spectra of lunar rocks 75055 recorded with this analyzer are shown in Fig. 1. Two spectra are actually recorded simultaneously. The one, called a coincidence spectrum, is recorded where events are seen in the sodium-iodide crystal at the same time as an event in the Ge(Li) detector. This, of course, requires the simultaneous emission of two or more gammas. The anticoincidence spectrum is accumulated where only a single gamma ray is emitted by the radionuclide. The value of this technique for measuring short-lived radionuclides such as ^{51}Cr, ^{7}Be and ^{54}Mn which emit only one gamma ray is illustrated in Fig. 2 for rock 75055. The upper spectra show the results which would be achieved with a normal Ge(Li) diode, while the two lower sets of spectra are amplified portions of the anticoincidence spectra shown in Fig. 1. The improvement in ability to measure all of these radionuclides is obvious, and in the case of ^{7}Be and ^{51}Cr, their measurement without the anticoincidence technique would not be nearly as reliable.

RESULTS AND DISCUSSION

Cosmogenic radionuclide concentrations

The cosmogenic radionuclide concentrations observed in the Apollo 17 lunar samples are summarized in Table 1. In Table 2 the concentrations of the cosmogenic radionuclides in a soil sample from each of the Apollo missions is compared. Excluding the analysis from Apollo 16, which had a very unusual chemistry, it is evident that the cosmogenic radionuclide concentrations of the Apollo 17 samples are strikingly different. One of the most obvious differences is the ^{26}Al to ^{22}Na ratio which was relatively constant at approximately two for the first four missions, yet is approximately one for the Apollo 17 samples. The target elements from which these radionuclides are produced have approximately the same concentrations in Apollo 17 soils as that in Apollo 11 and 15. It is thus evident that the solar flare of August 1972 produced approximately as much additional ^{22}Na in the lunar surface as was present just prior to the flare. The relatively high concentrations of ^{46}Sc result from a proton reaction on titanium, which itself was at a very high concentration in Apollo 17 soil. The cosmogenic radionuclides ^{54}Mn, ^{56}Co, ^{57}Co, and ^{58}Co result almost entirely from proton and alpha-particle spallation reactions on iron. Among the Apollo 17 samples (see Table 3) there is approximately a factor of 2 variation in both the iron and aluminum concentrations; therefore, an interpretation of the ^{26}Al content relative to the cosmogenic radionuclides resulting from spallation reactions on iron must be based on chemical composition. Since the radionuclides ^{56}Co and ^{54}Mn are produced almost entirely by spallation reactions on iron, the ratios of these radionuclides in lunar samples are essentially independent of chemical composition of the specimen. Because of their relatively high concentrations and the associated ease of accurate measurements, they serve here as important indicators in interpreting the characteristics of the solar flare.

Lunar surface processes

Of the 23 Apollo 17 lunar samples which have been received at our laboratory at this writing, the analyses of 13 are sufficiently well analyzed to include in this discussion. These include 3 samples from Station 1; 2 from Station 5; 5 from

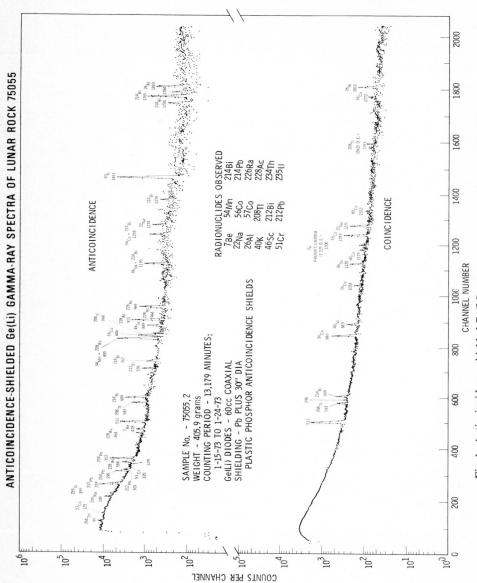

Fig. 1. Anticoincidence shielded Ge(Li) gamma-ray spectra of lunar rock 75055.

^{7}Be, ^{51}Cr, AND ^{54}Mn REGIONS OF ANTI-COINCIDENCE SHIELDED Ge(Li) GAMMA RAY SPECTRA OF APOLLO 17 ROCK, 75055, 1

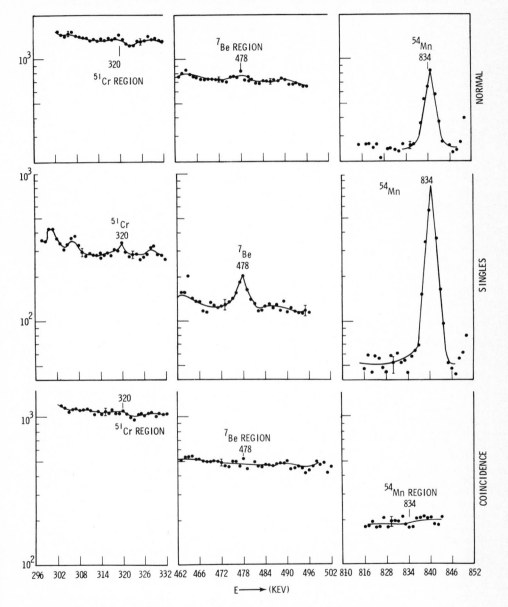

Fig. 2. ^{7}Be, ^{51}Cr, and ^{54}Mn regions of anticoincidence shielded Ge(Li) gamma-ray spectra of Apollo 17 Rock, 75055,1.

Table 1. Cosmogenic* and primordial radionuclides in Apollo 17 materials.

Radionuclides	71035,0	71041,4	71155,0	75055,2	75061,5	76240,2
^{7}Be (dpm/kg)	—	—	—	140 ± 20	350 ± 120	—
^{22}Na (dpm/kg)	92 ± 4	123 ± 4	112 ± 4	85 ± 5	171 ± 5	42 ± 2
^{26}Al (dpm/kg)	79 ± 3	123 ± 4	105 ± 4	69 ± 7	174 ± 5	151 ± 6
^{46}Sc (dpm/kg)	87 ± 5	75 ± 3	80 ± 4	62 ± 7	112 ± 5	8 ± 4
^{48}V (dpm/kg)	—	27 ± 15	<60	<22	26 ± 13	2.6
^{51}Cr (dpm/kg)	—	—	—	75 ± 37	—	<200
^{54}Mn (dpm/kg)	164 ± 15	198 ± 10	227 ± 30	139 ± 15	286 ± 12	21 ± 8
^{56}Co (dpm/kg)	279 ± 14	379 ± 10	310 ± 20	210 ± 15	548 ± 30	27 ± 3
^{57}Co (dpm/kg)	—	—	—	7.4 ± 1.5	18 ± 7	<12
^{58}Co (dpm/kg)	—	—	—	6.9 ± 3	21 ± 14	<12
^{60}Co (dpm/kg)	<4.6	<2.2	<4.4	<3	<2.8	0.8 ± 0.4
K (ppm)	200 ± 20	630 ± 40	<450	650 ± 50	600 ± 30	1100 ± 50
Th (ppm)	0.36 ± 0.03	0.863 ± 0.080	0.29 ± 0.05	0.40 ± 0.02	0.87 ± 0.03	2.30 ± 0.11
U (ppm)	0.096 ± 0.011	0.25 ± 0.01	0.13 ± 0.02	0.10 ± 0.01	0.22 ± 0.01	0.60 ± 0.03
K/U	2450	2520	<3460	6500	2730	1830
Th/U	4.00	3.93	2.23	4.00	3.95	3.83
Sample weight (grams)	144.11	111.1	25.8	405.9	100.0	104.98
Sample type†	VFCB	Soil	VFGB	Vesicular Basalt	Soil, top of Boulder	Soil, permanent Shadow

*Decay corrected to December 17, 1972.

Radionuclides	76255,0	76261,1	76275,0	76295,0	77135,0	78421,1	78481,1
^{7}Be (dpm/kg)	—	—	—	—	—	—	370 ± 90
^{22}Na (dpm/kg)	71 ± 4	143 ± 4	100 ± 3	64 ± 3	100 ± 5	39 ± 2	244 ± 12
^{26}Al (dpm/kg)	79 ± 4	171 ± 5	110 ± 3	71 ± 4	111 ± 6	55 ± 2	257 ± 12
^{46}Sc (dpm/kg)	3.9 ± 1.2	27 ± 3	7 ± 2	6.4 ± 2.6	7.2 ± 2.2	9.3 ± 2.5	59 ± 3
^{48}V (dpm/kg)	<4	<29	—	—	<4	—	34 ± 15
^{51}Cr (dpm/kg)	—	—	—	—	—	—	<340
^{54}Mn (dpm/kg)	38 ± 9	106 ± 14	103 ± 20	69 ± 26	21 ± 15	<42	264 ± 20
^{56}Co (dpm/kg)	37 ± 4	246 ± 8	86 ± 9	35 ± 5	66 ± 4	<6	606 ± 30
^{57}Co (dpm/kg)	—	—	—	—	—	—	18 ± 3
^{58}Co (dpm/kg)	—	—	—	—	—	—	18 ± 12
^{60}Co (dpm/kg)	2.5 ± 0.5	<2.5	<1.1	<1.1	3.4 ± 1.5	1.6 ± 0.8	—
K (ppm)	2900 ± 140	970 ± 40	2250 ± 90	2300 ± 80	1850 ± 90	840 ± 30	950 ± 40
Th (ppm)	2.33 ± 0.10	1.92 ± 0.04	5.69 ± 11	5.76 ± 0.17	5.51 ± 0.28	1.58 ± 0.07	1.49 ± 0.08
U (ppm)	0.58 ± 0.02	0.52 ± 0.02	1.40 ± 0.03	1.55 ± 0.05	1.42 ± 0.07	0.41 ± 0.02	0.39 ± 0.02
K/U	5000	1865	1610	1480	1300	2050	2440
Th/U	4.01	3.69	4.06	3.72	3.88	3.86	3.82
Sample weight (grams)	393.2	100.7	55.93	260.7	316.7	94.51	101.27
Sample type†	Breccia	Soil, 2 cm skim	PRB	PRB	VHB	Soil, 15–25 cm	Soil, 1 cm skim

†VFGB—Very fine-grained basalt.
 PRB—Partially recrystallized breccia.
 VHB—Vesicular hemfelsic breccia.

Table 2. Cosmogenic radionuclide content of Apollo soils.

			(dpm/kg)			
	10084,41	12070,13	14163,0	15041,14	69941,1	71041,4
^{22}Na	64 ± 3	80 ± 2	44 ± 2	65 ± 2	39 ± 2	123 ± 4
^{26}Al	131 ± 4	165 ± 4	89 ± 4	123 ± 4	147 ± 5	123 ± 4
^{26}Al/^{22}Na	2.05 ± 0.11	2.06 ± 0.07	2.02 ± 0.13	1.89 ± 0.08	3.77 ± 0.23	1.00 ± 0.04
^{46}Sc	11 ± 1	5.9 ± 1.6	<5.3	2.9 ± 1.6	<6.9	75 ± 3
^{48}V	<10	—	—	<10	—	27 ± 15
^{54}Mn	24 ± 3	21 ± 8	<38	14 ± 8	37 ± 14	198 ± 10
^{56}Co	53 ± 10	57 ± 7	<13	27 ± 6	<23	379 ± 10

Station 6; 1 from Station 7; and 2 from Station 8, consisting of 3 basalt samples, 4 breccia, and 6 soils. The ^{26}Al to ^{22}Na ratio for most of the surface samples is approximately 1, and significant deviations from this ratio may suggest a short exposure age relative to the half-life of ^{26}Al (0.74 m.y.), or partial shielding relative to 2π irradiation geometry. Three lunar samples were analyzed from Station 1. They include chips (71035,0 and 71155,0) from the top surface of boulders approximately one meter apart and one soil sample (71041,4). The ^{26}Al to ^{22}Na ratios in each of these were essentially 1, indicating a lunar surface exposure long compared with the half-life of ^{26}Al. Samples from Station 6 present a very interesting picture, since they indicate the boulders in this area have attained their present position within the past million years. Photographic documentation of this area clearly showed tracks of 3 large boulders which they had left in rolling

Table 3. Elemental composition of Apollo 17 specimens.*

Sample No.	Si	Al	Ca	Mg	Ti	Fe	Na	Mn	K	S	P	O[a]
71041,4	18.58	5.72	7.66	5.86	5.74	13.78	0.26	0.19	0.07	0.13	0.03	41.98
71155,0[b]	17.38	4.59	7.45	5.14	7.88	15.25	0.24	0.22	0.03	0.18	0.04	41.60
75055,2	19.29	5.16	8.79	4.13	6.10	14.18	0.33	0.22	0.07	0.19	0.03	41.51
75061,5	18.38	5.51	7.66	5.75	6.18	14.14	0.24	0.19	0.07	0.13	0.03	41.72
76215,0[c]	21.56	9.53	7.88	7.62	0.92	7.08	0.39	0.10	0.25	0.80	0.12	43.75
76240,2[e]	20.29	9.86	8.78	6.68	1.89	8.02	0.26	0.11	0.08	0.07	0.03	43.93
76261,1[e]	20.29	9.86	8.78	6.68	1.89	8.02	0.26	0.11	0.08	0.07	0.03	43.93
76255,0[d]	21.42	9.53	7.90	7.48	0.88	6.95	0.42	0.09	0.22	0.08	0.13	44.90
76275,0[d]	21.42	9.53	7.90	7.48	0.88	6.95	0.42	0.09	0.22	0.08	0.13	44.90
76295,0[d]	21.42	9.53	7.90	7.48	0.88	6.95	0.42	0.09	0.22	0.08	0.13	44.90
77135,0	21.56	9.53	7.88	7.62	0.92	7.08	0.39	0.10	0.25	0.80	0.12	43.75
78481[f]	19.95	8.33	8.41	5.98	3.28	10.22	0.26	0.14	0.07	0.10	0.02	43.24

*Lunar Sample Information Catalog, Apollo 17 (1973).
[a]By difference.
[b]Composition of sample 70215 assumed. See text.
[c]Composition of sample 77135 assumed. See text.
[d]Composition of sample 76315 assumed. See text.
[e]Composition of sample 76501,2 assumed. See text.
[f]Composition of sample 78501,2 assumed. See text.

down an adjacent hillside. The lunar soil samples 76261,2 and 76240,2 both indicate shielding and undersaturation with respect to ^{26}Al. Soil sample 76240,2 was collected beneath an overhang on a very large boulder in a permanently shielded area, while sample 76261,1 was collected about one meter from the boulder. Based on the target element concentrations of 76261,1 a saturation concentration for ^{26}Al would be approximately 250 dpm/kg. The much lower observed concentration of 171 dpm/kg is attributed to a 60% shielding relative to the 2π irradiation that it would have been exposed to if it were on a flat plain for the past several million years. The highly shadowed lunar sample, 76240, is shielded from about 85% of the 2π radiation it would have been exposed to were it on a flat plain, as indicated by the ^{22}Na concentration of only 42 dpm/kg. A similar analysis of the ^{26}Al excess for its present shielding indicates that it has been shielded by the adjacent boulder for a period of approximately one-half million years. This interpretation has also been suggested by Keith *et al.* (1973) for the high ^{26}Al to ^{22}Na ratio for soil sample 76240.

Sample 78421,1 is a trench soil sample from Station 8. Its depth, which is greater than 15 cm, shielded it almost completely from the August 1972 solar flare and the concentrations are therefore what one would predict from the galactic cosmic ray component (Rancitelli *et al.*, 1971).

The cosmogenic radionuclides ^{56}Co and ^{54}Mn are spallation products of iron and their concentrations are thus dependent on chemical composition as well as the geometry of the sample on the lunar surface. As we will consider in the following discussion, the ratios of these radionuclides can provide direct information on the rigidity of the solar flare and the angular anisotropy; however, since the radionuclide ratio is also a function of the angle of the exposure surface of lunar samples, it can provide direct information on the sample geometry as well.

In our previous work (Rancitelli *et al.*, 1971), we described a model for calculating the depth dependence of ^{26}Al and ^{22}Na production from a representative incident solar proton spectrum, the appropriate excitation functions and the sample's target element composition. This model employs the kinetic power law form:

$$\frac{dJ}{dE} = kE^{-\alpha} \tag{1}$$

to describe the incident proton spectrum, where J is the proton flux, E is the proton energy, α is a shape function, and k a constant related to the particle intensity. Using this model, we have calculated the best fit for the observed ^{26}Al gradients in solar and rock samples to be 3.1. The concentration gradient for ^{22}Na in Apollo 12 core samples and rock sections also showed an alpha equal to 3.1. Thus, it appeared that the solar flares which had produced ^{22}Na in lunar material had the same energy shape function as the average for the past few million years. An excellent pair of radionuclides for determining the rigidity of the August 1972 solar flare is ^{56}Co and ^{54}Mn. Both of these are produced almost entirely from iron, yet ^{56}Co is a very low-energy product, whereas ^{54}Mn is a relatively high-energy product with a threshold of 25 MeV (see Fig. 3). Other radionuclides, including

EXCITATION FUNCTIONS FOR PRODUCTION OF
^{7}Be, ^{46}Sc, ^{54}Mn, ^{56}Co FROM MAJOR TARGET ELEMENTS

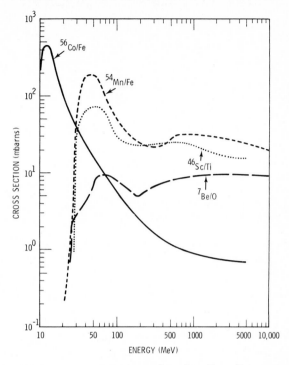

Fig. 3. Excitation functions for production of ^{7}Be, ^{46}Sc, ^{54}Mn, ^{56}Co from major target elements.

^{46}Sc and ^{7}Be, can serve to verify the energy spectrum estimate based on the ^{56}Co to ^{54}Mn ratio. In Fig. 4 the calculated ratios of ^{56}Co to ^{54}Mn are plotted as a function of sample thickness for alpha values ranging from 1.2 to 4. Also we have indicated our observed values for 4 Apollo 17 lunar soil samples, for rock sample, 75055, and for the Apollo 11 soil sample 10084. For this analysis the contribution from galactic cosmic rays to ^{54}Mn and ^{56}Co production was subtracted using galactic cosmic ray production rates from our earlier work (Perkins *et al.*, 1970). From the soil samples, we have established the best fit for alpha as 1.9 for the August 1972 solar flare. Thus, the proton energy spectrum for the August solar flare was much more rigid than that of both previous recent flares and of the average for the past million years. The average rigidity of the flares which preceded the Apollo 11 mission was about 3.2 as indicated by plotting our value for the Apollo 11 soil 10089 (Perkins *et al.*, 1970) on Fig. 4.

In Figs. 5, 6, and 7 the expected concentrations of ^{7}Be, ^{22}Na, ^{46}Sc, ^{51}Cr, ^{54}Mn, and ^{56}Co have been calculated as a function of depth for an alpha value of 1.9 for the lunar soil samples 75061 and 78481 and rock chip sample 75055. We have also plotted the observed concentrations of these radionuclides on the curves. The

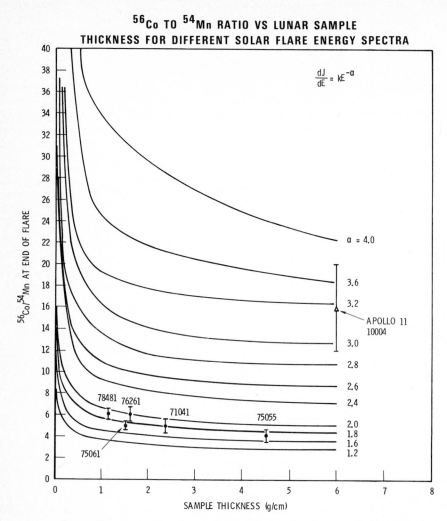

Fig. 4. ^{56}Co to ^{54}Mn ratio versus lunar sample thickness for different solar flare energy spectra.

observed radionuclide concentrations in the two soil samples fit the calculated data very well, indicating good agreement with the alpha value of 1.9. For rock sample 75055, the concentrations were all low by approximately 60%. For the comparison, we have therefore normalized the curves so that the observed ^{46}Sc, ^{54}Mn, ^{56}Co, and ^{22}Na best fit to the calculated lines. The relatively low concentrations of radionuclides in this rock sample are apparently due to the fact the sample was taken from the southeast face of the boulder which was inclined about 45° to the horizon, and thus received a substantially lower integrated exposure than the relatively flat lunar soil samples.

From Eq. (1) and the radionuclide concentrations in the lunar soil, it is possible

COMPARISON OF CALCULATED TO OBSERVED
RADIONUCLIDE CONTENT OF APOLLO 17 SOIL 75061

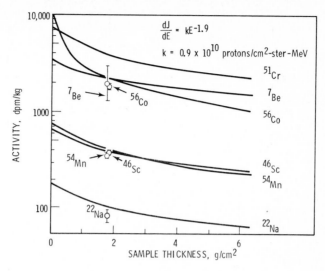

Fig. 5. Comparison of calculated to observed radionuclide content of Apollo 17 soil
75061.

COMPARISON OF CALCULATED TO OBSERVED
RADIONUCLIDE CONTENT OF APOLLO 17 SOIL 78481

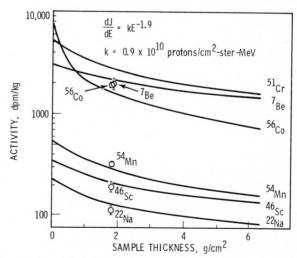

Fig. 6. Comparison of calculated to observed radionuclide content of Apollo 17 soil
78481.

COMPARISON OF CALCULATED TO OBSERVED
RADIONUCLIDE CONTENT OF APOLLO 17 ROCK 75055

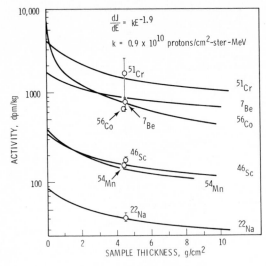

Fig. 7. Comparison of calculated to observed radionuclide content of Apollo 17 rock
75055.

to calculate the total proton flux from the August flare. This flux and the
associated energy shape function are included in Fig. 8. Included in Fig. 9 are
similar data by King (1973) back through 1965. As indicated in the figures, the
integral proton flux above 10 MeV from the August solar flare of 7.8×10^9 protons/cm^2 far exceeds the intensity and rigidity of the recent solar flares.
This can be compared to the value of 1.4×10^9 protons from the January 1971 flare
(Rancitelli *et al.*, 1971). Integration of energy areas about 30 and 60 MeV,
respectively, show 36% and 20% of the protons in these areas. In Table 4 we have
compared our integrated proton fluxes with those reported by King (1972) and
Bostrum *et al.* (1972) from satellite measurements. Therefore, in Table 5 we have
summarized all of our observed ^{56}Co to ^{54}Mn ratios for the Apollo 17 samples and
the approximate angle of the surface of the rock samples relative to the horizontal
are presented. While the surface angle of these samples is only an estimate from

Table 4. Integrated proton flux for the August 4–7, 1972 solar flares.
J: (10^8 protons/cm^2-ster.)

Energy interval (MeV)	This work	King (1972)	Bostrum *et al.* (1972)
>10	12.6	11	16
>30	4.6	6.1	6.4
>60	2.5	2.0	1.9

COMPARISON OF TIME INTEGRATED FLUX OF AUGUST 4-7, 1972
SOLAR FLARE WITH PREVIOUSLY REPORTED SOLAR FLARES

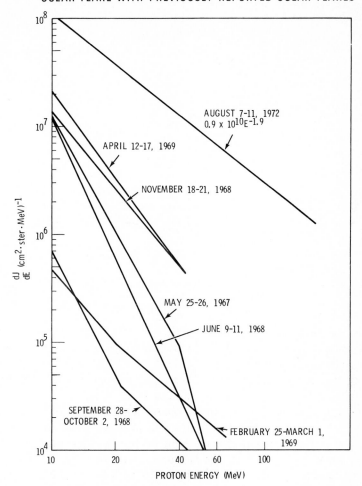

Fig. 8. Comparison of time integrated flux of August 4–7, 1972 solar flare with previously
reported solar flares (King 1973).

the lunar sample catalog, it is sufficient to indicate the extreme dependence of the
ratio of these isotopes on the surface sample's orientation.

From our analysis of the proton energy spectrum (Fig. 4) we selected mainly
samples which had been exposed on a flat plain (i.e. 4 soils). Here we observed
^{56}Co to ^{54}Mn ratios of 5 to 6. Examination of the ^{56}Co to ^{54}Mn ratios in other
samples indicated that information on angular anisotropy could also be obtained.
The samples that had their surface at the greatest angle relative to the horizontal
plain showed the lowest ^{56}Co to ^{54}Mn ratio. This ratio is, of course, very sensitive
to the hardness of the incident solar flare energy spectrum (see Fig. 4). For

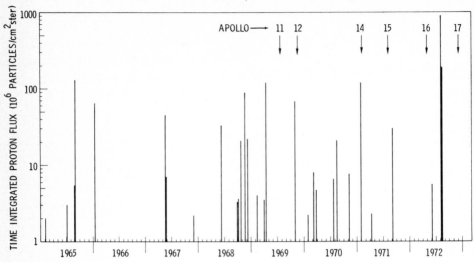

Fig. 9. Major solar flare intensities from 1965 to 1972 (King, 1972).

example, the 3 boulder chips 76255, 76275, and 76295, which had high surface angles relative to the horizon, had very low ^{56}Co to ^{54}Mn ratios. The implication of these observations is that there is a much lower ratio of low- to high-energy protons arriving at the lunar surface at low angles. Thus, a high degree of anisotropy exists which is a function of the incident angle. Sample 76295,0 was taken from an essentially vertical surface and had one of the lowest observed ^{56}Co to ^{54}Mn ratios, 1.85. This would be comparable to an alpha value of less than 1 (see Fig. 4) thus an extremely rigid spectrum. Anisotropy is also indicated in soil sample 76240 taken from beneath the overhang of the very large boulder at Station 6. Here, the ^{56}Co to ^{54}Mn ratio of 2.54 is much lower than the value of 6 in the adjacent solar sample 76261,1. Since only protons of low incident angle could strike soil sample 76240,2, this is substantiating evidence of anisotropy favoring high-energy protons at low angles. While all of the boulder chip samples were taken from southeast or east faces of lunar surface boulders, soil sample 76240 was shielded except from the north. Thus, angular anisotropy appears to be independent of compass direction.

The energy spectrum which impinged at different angles of lunar surface orientation can be quantified by integrating the solar proton flux over several energy intervals and employing the proton spectrum hardness factor alpha from Eq. (1) for that angle. When this is done over the energy interval 30–100 MeV and compared with the total number of protons present in the flare (between 10 and 100 MeV), this defines the relative number of protons with energies higher than the ^{54}Mn production threshold. Results of these calculations are summarized in Table 6 for lunar surface orientation between 0° and 90°. The alpha value

Table 5. Relationship of solar cosmic ray-induced ^{56}Co and ^{54}Mn to lunar surface orientation.*

Sample No. (Type)	^{56}Co (dpm/kg)	^{54}Mn (dpm/kg)	^{56}Co/^{54}Mn	Lunar surface† orientation
71035,0 (Boulder chip)	870 ± 40	190 ± 20	4.58	SE, 15°
71041,4 (Soil)	1130 ± 30	220 ± 13	5.13	0°
71155,0 (Boulder chip)	960 ± 60	240 ± 40	4.00	SE, 30°
75055,2 (Boulder chip)	650 ± 47	160 ± 16	4.06	SE, 45°
75061,5 (Soil)	1700 ± 90	340 ± 16	5.00	0°
76240,2 (Soil)	84 ± 9	33 ± 7	2.55	N, 0°
76255,0 (Boulder chip)	120 ± 12	38 ± 12	3.16	SE, 60°
76261,1 (Soil)	760 ± 25	130 ± 19	5.97	0°
76275,0 (Boulder chip)	270 ± 19	120 ± 26	2.25	SE, 60°
76295,0 (Boulder chip)	120 ± 16	62 ± 29	1.94	SE, 90°
78135,0 (Rock)	900 ± 90	180 ± 20	5.00	0°
78481,1 (Soil)	1880 ± 90	310 ± 13	6.06	0°

*Decay corrected to August 7, 1972.

†Lunar compass direction of sample surface and angle of surface to horizontal.

Table 6. Relationship of lunar sample surface orientation at the Apollo 17 site to the proton energy spectrum of the August 1972 solar flare.

Lunar surface orientation (degrees)*	α	$\dfrac{J_i\,(30 > Ep > 100\ \mathrm{MeV})}{J_i\,(10 > Ep > 100\ \mathrm{MeV})}$
0°	2.0 ± 0.1	0.26 ± 0.02
15°	1.8 ± 0.2	0.29
30°	1.7 ± 0.1	0.33
45°	1.7 ± 0.2	0.31
60°	1.4 ± 0.2	0.41
90°	0.9 ± 0.3	0.55
0° (Apollo 11)	3.2 ± 0.4	0.08 ± 0.04
0° (m.y. Average)	3.0 ± 0.2	0.10 ± 0.02

*Angle to the horizontal.

decreases from 2.0 for sample surfaces at 0° to 0.9 for sample surfaces at 90°. The fraction with energies above 30 MeV is approximately 26% at 0° and increases to 55% for orientations of 90°. As an indication of the extreme hardness of August 1972 flares these proton ratios are compared with those calculated for the April 1969 flare which preceded the Apollo 11 landing. This 1969 flare had an average alpha of 3.2 with only 8% of the proton having energies in excess of 30 MeV. These alpha values can be compared with the long-term average value of 3.0 which we previously determined from ^{26}Al gradients in soil and rock samples. Thus for the last one or two million years, 10% of the protons have energies of greater than 30 MeV.

Primordial radionuclides

The primordial radionuclide content of lunar samples provides basic information on the moon's geochemistry, on local transport and mixing of soil, and on the moon's relationship to other bodies in the solar system. Where sufficient and properly selected samples are studied, their primordial radionuclide content may also serve to define stratigraphic relationships through core sample analysis (Rancitelli *et al.*, 1970) and also the relationship of breccia and crystalline rocks to their surrounding and underlying soil.

Analysis of samples from the Apollo 17 site show far greater differentiation of primordial radionuclide ratios than has been observed in samples from any of the other previous missions. The primoridal radionuclide concentrations and their ratios are summarized in Table 1 along with the cosmogenic radionuclides. For comparison, the potassium to uranium content of the Apollo 17 samples analyzed at our laboratory are plotted in Fig. 10 along with most of the large sample analyses from previous missions (Perkins *et al.*, 1970; O'Kelley *et al.*, 1970, 1971; Rancitelli *et al.*, 1971, 1972). Table 7 compares the primordial radionuclide concentrations in lunar soil samples for all of the Apollo missions. From the comparisons in Table 7, it is evident that higher potassium to uranium ratios were observed in soil samples from the Apollo 17 site than at any of the other sites. These soil samples were taken from Stations 1 and 5 where potassium to uranium ratios of 2860 and 2730, respectively, were observed. One of the more spectacular observations was the very high potassium to uranium ratio of 6500 in the basalt

Table 7. Comparison of primordial radionuclide concentrations in lunar soils from the Apollo mission.

	K (ppm)	Th (ppm)	U (ppm)	K/U
Apollo 11	1100	2.3	0.55	2000
Apollo 12	2000	6.7	1.7	1180
Apollo 14	4400	14.6	3.6	1220
Apollo 15	1200–1700	3.8–4.7	0.94–1.3	1200–1500
Apollo 16	740–1200	1.8–2.7	0.41–0.74	1470–2070
Apollo 17	600–1100	0.86–2.3	0.22–0.60	1830–2860

POTASSIUM vs URANIUM CONTENT
OF LUNAR MATERIALS

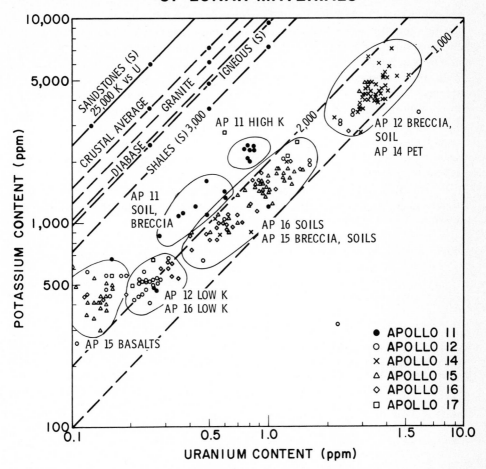

Fig. 10. Comparison of K versus U content of Apollo 17 samples with previous Apollo samples and terrestrial materials.

sample 75055. This is considerably higher than has been observed in any of the previous Apollo samples, thus substantiating an unusual differentiation process at this site. Another indication of major differentiation is evidenced by the very low thorium to uranium ratios in several of the Apollo 17 rock samples. The lowest observed ratio is 2.3 for the basalt rock chip 71155; however, values of three and less were observed in six other rock samples studied by the preliminary examination team (Lunar Sample Information Catalog, Apollo 17). Lunar rock samples from previous Apollo missions had shown an atomic ratio for ^{232}Th: ^{238}U of about 3.8, which is the average for our present-day solar system (Fowler and Hoyle, 1960). The average thorium to uranium atomic ratio for the Apollo 17 soils was 3.95. As has been stated earlier (Perkins *et al.*, 1970), the soil values are in some question. This is due to the fact that radon loss could produce a negative bias in the uranium value, thus giving an artificially high thorium to uranium ratio. We believe that this bias is less in the more recent Apollo missions because of the Curator's practice of sealing the samples in double teflon bags which should help minimize radon loss. The lowest value of potassium observed in the Apollo 17 samples was 200 ppm in the basalt chip 71035. Except for a few samples from the Apollo 16 mission, and rock 12005 from the Apollo 12 mission, this is the lowest observed value for a lunar sample. From the sites where we have thus far analyzed both soil and rock samples, there was a correlation between their relative primordial radionuclide contents. For example, Station 1 showed low concentrations in both the soil and rock samples, while Station 6 showed much higher concentrations in both soil and rock samples. Station 5 showed intermediate values for the soil and rocks.

However, in all cases the primordial radionuclide ratios are substantially different in the rocks than in the surrounding soils, thus making the relationship of the soil to the rocks a difficult problem.

Acknowledgments—We wish to thank R. M. Campbell, D. R. Edwards, J. G. Pratt, and J. H. Reeves of this Laboratory for their aid in standards preparation and in data acquisition. The unique and sensitive instrumentation that made this work possible was developed during the past decade under sponsorship of the United States Atomic Energy Commission, Division of Biomedical and Environmental Research.

REFERENCES

Bostrom C. O., Kohl J. W., McEntire R. W., and Williams D. J. (1972) The solar proton flux—August 2–12. Preprint—Applied Physics Laboratory (November 1972).

Cooper J. A. and Perkins R. W. (1972) A versatile Ge(Li)–NaI(Tl) coincidence—anticoincidence gamma-ray spectrometer for environmental and biological problems. *Nucl. Instr. & Methods* **99**(1), 125.

Finkel R. C., Arnold J. R., Imamura M., Reedy R. C., Fruchter J. S., Loosli H. H., Evans J. C., Delany A. C., and Shedlovsky J. P. (1971) Depth variation of cosmogenic nuclides in a lunar surface rock and lunar soil. *Proc. Second Lunar Sci. Conf., Geochim. Cosmochim. Acta*, Suppl. 2, Vol. 2, pp. 1773–1789. MIT Press.

Fowler W. A. and Hoyle F. (1960) Nuclear cosmochronology. *Ann. Phys.* **10**, 280.

Keith J. E. and Clark R. S. (1973) Personal communication.

King J. H. (1972) Personal communication (September 29).

King J. H. (1972) Study of mutual consistency of IMP 4 solar proton data. NSSDC 72-14 (October).

Lunar Sample Information Catalog, Apollo 17 (1973). National Aeronautics and Space Administration MSC 03211.

O'Kelley G. D., Eldridge J. S., Schonfeld E., and Bell P. R. (1970) Primordial radionuclide abundances, solar proton and cosmic-ray effects, and ages of Apollo lunar samples by nondestructive gamma ray spectrometry. *Proc. Apollo 11 Lunar Sci. Conf., Geochim. Cosmochim. Acta*, Suppl. 1, Vol. 2, pp. 1407–1423. Pergamon.

O'Kelley G. D., Eldridge J. S., Schonfeld E., and Bell P. R. (1971) Cosmogenic radionuclide concentrations and exposure ages of lunar samples from Apollo 12. *Proc. Second Lunar Sci. Conf., Geochim. Cosmochim. Acta*, Suppl. 2, Vol. 2, pp. 1747–1755. MIT Press.

Perkins R. W., Rancitelli L. A., Cooper J. A., Kaye J. H., and Wogman N. A. (1970) Cosmogenic and primordial radionuclide measurements in Apollo 11 lunar samples by nondestructive analysis. *Proc. Apollo 11 Lunar Sci. Conf., Geochim. Cosmochim. Acta*, Suppl. 1, Vol. 2, pp. 1455–1469. Pergamon.

Rancitelli L. A., Perkins R. W., Felix W. D., and Wogman N. A. (1971) Erosion and mixing of the lunar surface from cosmogenic and primordial radionuclide measurements in Apollo 12 lunar samples. *Proc. Second Lunar Sci. Conf., Geochim. Cosmochim. Acta*, Suppl. 2, Vol. 2, pp. 1757–1772. MIT Press.

Rancitelli L. A., Perkins R. W., Felix W. D., and Wogman N. A. (1972) Lunar surface processes and cosmic ray characterization from Apollo 12–15 lunar sample analyses. *Proc. Third Lunar Sci. Conf., Geochim. Cosmochim. Acta*, Suppl. 3, Vol. 2, pp. 1681–1691. MIT Press.

Shedlovsky J. P., Honda M., Reedy R. C., Evans J. C., Jr., Lal D., Lindstrom R. M., Delany A. C., Arnold J. R., Loosli H., Fruchter J. S., and Finkel R. C. (1970) Pattern of bombardment-produced radionuclides in rock 10017 and in lunar soil. *Proc. Apollo 11 Lunar Sci. Conf., Geochim. Cosmochim. Acta*, Suppl. 1, Vol. 2, pp. 1503–1532. Pergamon.

Proceedings of the Fifth Lunar Conference
(Supplement 5, Geochimica et Cosmochimica Acta)
Vol. 2 pp. 2205–2209 (1974)
Printed in the United States of America

Titanium spallation cross sections between 30 and 584 MeV and Ar^{39} activities on the moon

F. STEINBRUNN and E. L. FIREMAN

Center for Astrophysics, Harvard College Observatory and Smithsonian Astrophysical Observatory, Cambridge, Massachusetts 02138

Abstract—The production cross sections of Ar^{39} for Ti spallation at 45-, 319-, 433-, and 584-MeV proton energies were measured to be 0.37 ± 0.09, 12.4 ± 3.7, 9.1 ± 2.7, and 17.8 ± 6.2 mb, respectively. Normalized Ar^{39} production rates and activities are also derived for protons above 40 MeV and for three differential proton spectra of the type $\sim E^{-\alpha}$. It is concluded that, even for samples of high-Ti content, Ti spallation by solar protons below 200-MeV energy does not contribute significantly to their Ar^{39} radioactivity.

INTRODUCTION

THE Ar^{39} ACTIVITIES were measured in lunar surface and drill-core samples from all Apollo missions (Fireman *et al.*, 1970, 1972, 1973; D'Amico *et al.*, 1970, 1971; Stoenner *et al.*, 1970, 1971). The Ar^{39} activity in near-surface samples is produced by solar- and cosmic ray-proton spallation of Ti and Fe; therefore, it should be possible to obtain information on the average proton intensity and rigidity over the past ~ 1000 yr from Ar^{39} data. If cross sections extrapolated to low energies are used, an unusually high solar-proton intensity is indicated by Ar^{39} data (Fireman, 1973). The Ti content of some samples at the Apollo 11 and 17 sites is quite high and can reach about 50% of the Fe content (Rose *et al.*, 1974). It is necessary to know the Ti spallation cross sections produced by 45- to 600-MeV protons in order to calculate the contribution of Ti to the production of Ar^{39} by solar protons. The Ar^{39} activity in six proton-irradiated Ti foils was measured (Steinbrunn and Fireman, 1974), and the cross sections are reported together with Ar^{39} production rates for Ti bombarded by solar protons of $E^{-\alpha}$ differential energy spectra.

EXPERIMENTAL PROCEDURE

Table 1 summarizes data on the six Ti foils including proton fluxes and energies obtained by R. Perkins (private communication, 1973). Previously, a number of other radioisotopes were measured in the same foils by nondestructive γ-spectrometry (Brodzinski *et al.*, 1971). Since it is nearly 5 yr since the irradiation, no interference from the Ar^{37} activity ($T_{1/2} = 35$ days) was observed.

The Ar^{39} extraction procedure was the same as that used earlier in our laboratory for the extraction of Ar from magnetic fractions of meteorites (Fireman and Spannagel, 1971). The Ar was released by dissolving the foils in hot sulfuric acid of $9n$ concentration. The Ti foils were cut in ~ 100 pieces and placed inside a three-neck flask of the extraction system. Under vacuum conditions, 400 cm³ of sulfuric acid was dripped onto the Ti chips. Air had been removed from the acid by bubbling He through it for 15 min. The system was filled with He at a pressure of ~ 90 mm Hg above atmospheric, and the He flow was started at an estimated rate of 80–100 cm³/min. The He bubbled through the solution and then passed through a condenser, a silica-gel trap, a charcoal trap at liquid-nitrogen temperature, and, finally,

Table 1. Target and irradiation summary.

Target	Area (cm²)	Weight (g)	Proton dose (×10¹³)	Proton energy (MeV)	Date of irradiation
8600-40-Ti	23 (round)	2.600	304.6	15	3/13/69
8600-41-Ti	23 (round)	2.588	448.9	30	3/13/69
8600-42-Ti	23 (round)	2.583	333.9	45	3/13/69
Ti-A (front foil)	40.3 (square)	13.9	7.582	319	7/12/68
8600-8-Ti	40.3 (square)	14.05	1.983	433	7/12/68
8600-9-Ti	41.5 (square)	14.4	5.405	584	7/12/68

through a flow meter into the room. Argon carrier of 0.5-cm³ volume was mixed into the He flow. The flow was continued (4–14 hour) until the Ti was completely dissolved. The Ar yield was $100 \pm 5\%$ of the initial carrier. The Ar, together with 10% methane, was placed into a small proportional counter at 1.4 atm pressure for counting.

RESULTS

The 15-MeV proton energy is below the threshold of any nuclear reaction of protons with any of the stable Ti isotopes leading to Ar^{39}. For protons of 30 MeV, only the reaction $Ti^{47}(p; 2He^4, p)Ar^{39}$ contributes, if a Coulomb barrier of 5.5 MeV is added. The Ar^{39} activities in both 15- and 30-MeV foils were within the counter background activity. The cross sections derived for these energies are smaller than $\sim 5 \mu b$. The Ar^{39} activities of the remaining four foils were corrected for counter efficiency and carrier yield. The derived cross sections are the following:

Proton energy (MeV)	45	319	433	584
Cross section (mb)	0.37 ± 0.09	12.4 ± 3.7	9.1 ± 2.7	17.8 ± 6.2

The errors include the statistical counting error (less than 5% at 2σ throughout), 5% for the extraction efficiency, and 10% for the counter efficiency. Errors assumed for the proton fluxes are 5% at 45 MeV, 10% at 319 and 433 MeV, and 15% at 584 MeV, according to Brodzinski et al. (1971) and Perkins (private communication, 1973).

In Fig. 1, the measured cross sections are shown together with a yield curve calculated after Rudstam's (1966) CDMD formula. At high energies, the cross sections agree well with the semiempirical formula; however, at 45 MeV, the cross section does not agree, because it was deduced from data at higher energies. Our cross section at 584 MeV of 17.8 ± 6.2 mb agrees with the 600-MeV value of 15.7 mb measured by Stoenner et al. (1970).

AR³⁹ PRODUCTION RATES AND ACTIVITIES

In order to estimate the Ar^{39} production from Ti by solar-flare proton bombardment, three differential proton spectra of the form $\sim E^{-2}$, $\sim E^{-2.5}$, and

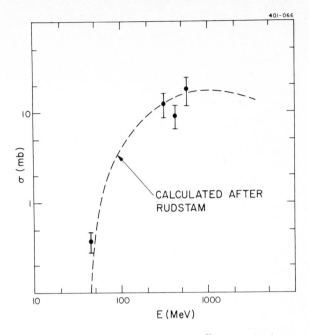

Fig. 1. Proton spallation cross section of ^{39}Ar from titanium.

$\sim E^{-3}$ were assumed. Normalized production rates,

$$Ar^{39} = \int_{40}^{\infty} F(E)x\sigma(E)\,dE,$$

were calculated in 17 steps for each proton spectrum. The cross sections as a function of energy listed in Table 2 were taken graphically from a yield curve similar in shape to the Rudstam curve in Fig. 1, but the curve connected our measured data points.

The calculated production rates and activities, normalized to an integrated flux

Table 2. Cross sections used for the calculation of Ar^{39}
production rates in Table 3.

Energy interval (MeV)	σ (mb)	Energy interval (MeV)	σ (mb)
40–50	0.37	130–140	6.25
50–60	1.25	140–150	6.7
60–70	2.1	150–160	7.15
70–80	2.8	160–170	7.6
80–90	3.4	170–180	8.05
90–100	4.1	180–190	8.45
100–110	4.65	190–200	8.9
110–120	5.2	200–∞	12.5
120–130	5.75		

Table 3. Normalized production rates and equilibrium activities
of Ar^{39} for three differential proton spectra.

Differential proton spectrum ($E > 40$ MeV)	Ar^{39} production* (10^{-27} Ar^{39} atoms)	Ar^{39} activity† (10^{-4} dis/min kg)
$\sim E^{-2}$	4.8 ± 1.2	1.0 ± 0.25
$\sim E^{-2.5}$	3.3 ± 0.8	0.69 ± 0.17
$\sim E^{-3}$	2.4 ± 0.6	0.50 ± 0.12

*Ar^{39} production normalized to an integral proton flux of
1 proton/cm² above 40 MeV and to a target of 1 Ti atom/cm².
†Ar^{39} activity normalized to an integral proton flux of
1 proton/cm² sec above 40 MeV and 1% by weight Ti content.

of 1 proton/cm² and a target of 1 Ti atom/cm² and to 1% by weight Ti content, are given in Table 3. The data represent a "thin target" case and apply only to surface samples of ~ 1 mm thickness.

For a sample with a 10% Ti content, only 0.1 Ar^{39} dis/min kg is produced by a flux of 100 protons/cm² sec above 40 MeV with an E^{-2} differential spectrum. From the Ar^{39} activities calculated with our cross sections, we conclude that solar protons of relatively low energies and of the assumed spectral distribution do not contribute significantly to lunar Ar^{39} activities via Ti spallation. For near-surface samples of lunar material, Ar^{39} activities are generally of the order of 10 dis/min kg. In order to account for the high Ar^{39} activities in lunar samples, the solar-flare protons during the past 1000 yr need to have higher energies than has been assumed here, or the Ar^{39} production rate from Ca has been underestimated. Measurements of cross sections for Ar^{39} production from Ca are needed to clarify this point.

Acknowledgment—This work was supported in part by grant NGR 09-015-145 from the National Aeronautics and Space Administration.

REFERENCES

Brodzinski R. L., Rancitelli L. A., Cooper J. A., and Wogman N. A. (1971) High-energy proton spallation of titanium. *Phys. Rev. C*, 3rd series, **4**, 1250–1257.

D'Amico J., DeFelice J., and Fireman E. L. (1970) The cosmic-ray and solar flare bombardment of the moon. *Proc. Apollo 11 Lunar Sci. Conf., Geochim. Cosmochim. Acta*, Suppl. 1, Vol. 2, pp. 1029–1036. Pergamon.

D'Amico J., DeFelice J., Fireman E. L., Jones C., and Spannagel G. (1971) Tritium and argon radioactivities and their depth variations in Apollo 12 samples. *Proc. Second Lunar Sci. Conf., Geochim. Cosmochim. Acta*, Suppl. 2, Vol. 2, pp. 1825–1839. MIT Press.

Fireman E. L. (1973) Solar flares during the past 1000 years as revealed by lunar studies. *Proc. First European Astron. Meeting*, Athens, September 4–9, 1972, Vol. 1. Springer-Verlag.

Fireman E. L., D'Amico J., and DeFelice J. (1970) Tritium and argon radioactivities in lunar material. *Science* **167**, 566–568.

Fireman E. L., D'Amico J., and DeFelice J. (1973) Radioactivities vs. depth in Apollo 16 and 17 soil. *Proc. Fourth Lunar Sci. Conf., Geochim. Cosmochim. Acta*, Suppl. 4, Vol. 2, pp. 2131–2143. Pergamon.

Fireman E. L., D'Amico J., DeFelice J., and Spannagel G. (1972) Radioactivities in returned lunar material. *Proc. Third Lunar Sci. Conf., Geochim. Cosmochim. Acta*, Suppl. 3, Vol. 2, pp. 1747–1761. MIT Press.

Fireman E. L. and Spannagel G. (1971) Radial gradient of cosmic rays from the Lost City Meteorite. *J. Geophys. Res.* **76**, 4127–4134.

Rose H. J., Jr., Brown F. W., Carron M. K., Christian R. P., Cuttitta F., Dwornik E. J., and Ligon D. T., Jr. (1974) Composition of some Apollo 17 samples (abstract). In *Lunar Science—V*, pp. 645–647. The Lunar Science Institute, Houston.

Rudstam G. (1966) Systematics of spallation yields. *Z. Naturforsch.* **21a**, 1027–1041.

Steinbrunn F. and Fireman E. L. (1974) ^{39}Ar production cross sections in Ti for solar-proton effects in lunar surface samples (abstract). In *Lunar Science—V*, pp. 732–734. The Lunar Science Institute, Houston.

Stoenner R. W., Lyman W. J., and Davis R., Jr. (1970) Cosmic-ray production of rare-gas radioactivities and tritium in lunar material. *Proc. Apollo 11 Lunar Sci. Conf., Geochim. Cosmochim. Acta*, Suppl. 1, Vol. 2, pp. 1583–1594. Pergamon.

Stoenner R. W., Lyman W. J., and Davis R., Jr. (1971) Radioactive rare gases and tritium in lunar rocks and in the sample return container. *Proc. Second Lunar Sci. Conf., Geochim. Cosmochim. Acta*, Suppl. 2, Vol. 2, pp. 1813–1823. MIT Press.

Proceedings of the Fifth Lunar Conference
(Supplement 5, Geochimica et Cosmochimica Acta)
Vol. 2 pp. 2211–2229 (1974)
Printed in the United States of America

Radioactive rare gases, tritium, hydrogen, and helium in the sample return container, and in the Apollo 16 and 17 drill stems

R. W. Stoenner, Raymond Davis, Jr., Elinor Norton, and Michael Bauer

Brookhaven National Laboratory, Upton, New York 11973

Abstract—The gas was extracted from the sample return container from the Apollo 16 and 17 missions by adsorption on charcoal and activated vanadium metal. The hydrogen, argon, and radon were separated and counted to give the tritium, ^{37}Ar, ^{39}Ar, and ^{222}Rn activities. The tritium and argon activities observed could be explained by diffusive losses of these gases from the fine material in the container. There was *no excess* tritium present in the Apollo 17 containers that could be attributed to solar tritons remaining from the intense flare of August 4, 1972. The ^{222}Rn observed in the sample return container was interpreted as an emanation product from lunar fines and an emanation yield of 1×10^{-4} was calculated. This yield is consistent with the low radon content observed in the lunar atmosphere.

The tritium, ^{37}Ar, ^{39}Ar, and ^{222}Rn activities and the K, Ca, Ti, Fe, Mn, hydrogen and helium contents were measured on a set of samples from the Apollo 16 and 17 deep drill stems at depths from 49 to 472 g/cm^2. The ^{37}Ar and ^{39}Ar activities combined with similar measurements at more shallow depths by Fireman and associates (SAO) in the Apollo 16 drill stems give the complete activity profile in the lunar regolith. Since ^{37}Ar is produced mainly by the ^{40}Ca$(n, \alpha)^{37}$Ar reaction it is possible to determine the neutron production rate in the regolith as a function of the depth. The ^{222}Rn extracted from the samples by vacuum melting was found to be lower than expected in most of the samples based upon their uranium contents, but in two samples excess ^{222}Rn was observed. The hydrogen and helium contents of the drill stem samples were measured. The helium contents were found to be relatively uniform with depth in the Apollo 16 soil, but increased a factor of two with increasing depth in the Apollo 17 drill stem. The H/He atom ratio was higher than the accepted solar wind value by a factor of two, probably due to water contamination.

Introduction

This report will present two separate series of measurements on lunar material from the Apollo 16 and 17 missions. The gas in the sample return container (SRC) has been collected in all Apollo missions with the exception of Apollo 11 and analyzed for the radioactive rare gases, ^{37}Ar, ^{39}Ar, and ^{222}Rn, and tritium. In Part I, we will give the new results for the Apollo 16 and 17 missions and a discussion summarizing all of the results and their relation to the lunar atmosphere. In Part II, we will present the results of measurements of the radioactive gases ^{37}Ar, ^{39}Ar, ^{222}Rn, and tritium evolved from vacuum melting of samples from the Apollo 16 and 17 missions. Of special interest is a series of measurements on the Apollo 16 and 17 deep drill strings. The ^{37}Ar produced in the lunar soil by ^{40}Ca$(n, \alpha)^{37}$Ar reaction can be used to determine the fast neutron production as a function of the depth by galactic cosmic rays. These observations are of interest in theoretical modeling of the neutron flux as a function of the depth. An understanding of the neutron production and moderation in the lunar soil is essential to derive the gardening history of lunar soil from gadolinium and samarium isotope data (e.g. Russ III, 1973) and to derive the mass loss from the moon from impacts (Fireman, 1974).

In addition to the radioactivity measurements, we also determined the hydrogen, helium, K, Ca, Ti, Fe, Mn and uranium concentrations in these samples. The chemical analyses for K, Ca, Ti, and Fe are essential for interpreting the ^{37}Ar and ^{39}Ar measurements. The uranium concentrations will be compared to the ^{222}Rn activities observed from the melts to test whether the ^{226}Ra is in equilibrium with the ^{238}U. The results will be reported, though this question remains in a rather unsatisfactory state on lunar soil samples. The hydrogen and helium contents are useful in understanding the solar wind composition and flux in the past. The helium content of the Apollo 17 deep drill string shows a dramatic increase with depth whereas the Apollo 16 samples are lower and relatively constant.

Part I   Radioactive Rare Gases and Tritium from the Sample Return Container

The sample return container (SRC) was closed on the surface of the moon by the astronauts and remains sealed until it is opened at the Lunar Receiving Laboratory (LRL). The container is provided with an O-ring and an indium metal seal and therefore should remain vacuum tight during transit. The fine lunar material in the container is essentially in its pristine condition as it existed on the lunar surface. Our interest was in observing the emanation of ^{222}Rn (3.8 day) from ^{226}Ra decay, and the diffusive loss of the radioactive isotopes, ^{37}Ar (35 day), ^{39}Ar (269 year), and tritium (12.3 year) from lunar fines. Our results on the Apollo 12, 14, and 15 missions were reported (Stoenner *et al.*, 1972), and we will give the results from one box from the Apollo 16 mission (SRC no. 2) and both SRC from the Apollo 17 mission.

Experimental

The gas was removed by adsorption on charcoal at liquid-nitrogen temperatures (77°K) using a stainless steel system located directly under the LRL cabinets used for opening and unloading the sample containers. This system was pumped initially with a cryopump, and finally a high vacuum (10^{-6} torr) was achieved with a vac-ion pump. The system consisted of bakeable high-vacuum valves (2.5 cm) leading to a charcoal-filled finger (2.5 cm dia.) containing 10 g of petroleum-base charcoal (low in radium), and a second finger containing 10 g of activated vanadium metal. A thermocouple gauge was used to measure the pressure in the system and to follow the pressure in the SRC as the gas was removed. The probe was attached through an O-ring seal to the outside surface of the SRC at a point where the wall was thinned to 0.5 mm. Prior to extracting the gas, the system was evacuated and baked at 400°C. The SRC was penetrated with a sharp beveled rod 6 mm in diameter and the gas was collected over a 60–90 min period. During this time the pressure in the SRC drops to about 5% of its initial value. The charcoal adsorption technique effectively removes the rare gases (Ar to Rn), other gaseous compounds, and partially adsorbs hydrogen. The activated vanadium metal is an effective adsorber for hydrogen gas at room temperature and was used in an effort to determine the chemical form of tritium. Following the extraction, the charcoal and vanadium fingers were closed off with valves and brought to Brookhaven for analysis.

The charcoal finger was connected to a glass vacuum system containing a 700 cc Toepler pump that removes the gas from the finger, and places it in a tube containing in series a small charcoal trap (0.1 g C) and a quartz tube containing activated vanadium metal. The system that was used for vacuum melting and extracting gases from lunar samples will be described later. The collected gas volume was measured, an aliquot was taken for mass spectrometer analysis, and then the gas was heated to 1000°C over vanadium metal to getter all reactive species. The vanadium metal was cooled to room temperature to retain hydrogen, and the rare gases were removed in sequence by adjusting the temperature of the charcoal. With the charcoal at 77°K, helium was removed. The volume of helium in all cases was too small to measure accurately, and was less than 10^{-3} cm^3 STP. The charcoal temperature was raised to -76°C. Argon carrier with 10% CH$_4$ was introduced and removed into a small proportional counter. Finally, the charcoal temperature was raised to 250°C. Argon carrier with 10% CH$_4$ was introduced once again and removed together with radon into a second proportional counter. After the rare gases were removed the vanadium was heated to 1000°C to release hydrogen. The hydrogen was passed through a palladium thimble, collected with a Toepler pump, its volume measured, and placed in a small proportional counter for tritium measurements.

The stainless steel finger containing the vanadium absorber was treated separately. The vanadium metal was in a quartz tube with a taper joint so that it could be quickly removed from the finger and plugged into the processing system. It was evacuated, and then heated to 1000°C to release hydrogen. The hydrogen was passed through a palladium thimble, collected with a Toepler pump, its volume measured, and finally introduced together with argon–CH$_4$ into a small proportional counter for tritium measurement.

RESULTS AND DISCUSSION

One Apollo 16 box and both Apollo 17 sample return containers were analyzed. The first SRC from the Apollo 16 mission was not sealed properly on the lunar surface and consequently had a pressure of one atmosphere and thus could not be analyzed. The volumes collected by the charcoal adsorber plus the vanadium metal are given in Table 1. The volume of gas collected can be used to estimate the pressure in the SRC prior to puncturing. The total volume of the box is 23 liters, the rocks occupy approximately 5 liters of this space leaving a void volume of 18 liters. The corresponding pressure in the box is given in the last column of Table 1, and it can be seen that the calculated pressure agrees reasonably well with the thermocouple gauge reading. An aliquot of the gas recovered from the charcoal finger was analyzed with a mass spectrometer and found to contain predominantly N$_2$ and a small amount of water. Thus most of the pressure in the sample return container is air, and probably resulted from a small leak in its cover seal.

The volumes of hydrogen ultimately recovered from the charcoal and vanadium adsorbers are listed in Table 1. The volume of hydrogen obtained from the charcoal absorber would include water, hydrogen, and any other hydrogen containing compound that existed as gas in the SRC. It can be seen that the volumes were quite

Table 1. Gases collected from the sample return containers.

Mission-SRC	SRC* pressure (torr)	Volume gas collected (cm^3 STP)	Volume H$_2$ on vanadium (cm^3 STP)	Volume H$_2$ on charcoal (cm^3 STP)	SRC pressure calculated from recovered gas (torr)
16-2	—	3.93	0.35	0.05	0.17
17-1	0.170	4.79	0.87	0.90	0.20
17-2	0.025	0.277	0.12	0.07	0.012

*Observed on thermocouple gauge on SRC.

low for containers 16-2 and 17-2, comparable to the system blank, but much higher for 17-1. The vanadium absorber will not react with water or hydrocarbons at room temperature but does adsorb hydrogen efficiently. The volumes of hydrogen from the vanadium absorber are a fair measure of the actual volume of hydrogen in the SRC gas. However, there is a system blank in the range of 0.05–0.10 cm^3 H$_2$ at STP which is small compared to the volume of hydrogen from 17-1, but comparable to the volume of hydrogen from 17-2.

It will be shown later that lunar fines contain about 1.0–1.4 cm^3 of H$_2$ at STP per gram. It is interesting to estimate the fraction of hydrogen present in fine material that diffuses out into the gas phase during the 8–21 days the material was present in the SRC. The weights of fine material in these three SRC's are given in Table 2. From these weights one estimates that the fraction of hydrogen released was 8×10^{-5}, 7×10^{-4}, and 2×10^{-5}, respectively from the individual containers 16-2, 17-1, and 17-2. These values will be compared to the corresponding ones for tritium release.

The radioactive ^{37}Ar, ^{39}Ar, ^{222}Rn, and tritium observed in the Apollo 16 and 17 SRC are listed in Table 2. We have also included in the table the reported values obtained from earlier missions. The argon activities and tritium observed in the SRC gas can be attributed to the diffusion of these radioactive products out of the lunar fines (Stoenner *et al.*, 1971, 1972).

Tritium. The tritium activities in the Apollo 16 and 17 SRC are very low; the uncertainties listed are counting statistics (1 σ). The lowest activity observed (17-1), is approximately twice counter background. It may be noticed that the tritium activity was approximately equally divided between the charcoal and vanadium adsorbers. The hydrogen from the Apollo 16 mission had the highest specific activity in the sample collected on the charcoal adsorber (0.90 dpm T/cm^3 H$_2$ STP), and a somewhat lower specific activity on the vanadium absorber (0.13 dpm T/cm^3 H$_2$ STP). The tritium specific activities from the Apollo 17 mission were low and were approximately the same in the samples collected on charcoal and on vanadium (0.02–0.07 dpm T/cm^3 H$_2$ STP). It is interesting to compare these specific tritium activities with those obtained from vacuum melting of soil, see discussion later on. Although there is a large variation in the specific tritium activity from soil, 0.1–0.4 dpm T/cm^3 H$_2$ STP, the average specific activity is 0.23 dpm T/cm^3 H$_2$ STP. The soil value is higher than the specific tritium activity from the Apollo 17 sample return containers, but much lower than the value of 0.90 dpm T/cm^3 H$_2$ STP obtained from the charcoal adsorber from the Apollo 16 sample return container. One might expect that soil samples would be contaminated with some atmospheric water, and therefore the specific tritium activity from soil samples may be lower than actual lunar hydrogen. It is difficult to assess the hydrogen contamination present in the sample return container. The hydrogen volumes recovered were small, but contamination from astronaut effluent water vapor is very likely. The high specific tritium activity observed in the Apollo 16 mission may be attributed to a solar flare on April 19, two days before the lunar landing. Tritium in solar flares will be discussed later.

Table 2. Radioactivities from the sample return container.

Mission	Date	SRC pressure (torr)	Wt. fines in SRC (kg)	Observed activity* (dpm)				
				^{222}Rn	^{37}Ar	^{39}Ar	T from Charcoal	T from Vanadium
12	Nov. 19–20, 1969	0.030	2.7	4.9 ± 0.3	0.040 ± 0.002	0.0023 ± 0.0010	—	—
14	Feb. 5–6, 1971	0.070	0.6	1.74 ± 0.02	0.040 ± 0.002	0.0008 ± 0.0005	0.280 ± 0.020	—
15	July 30–Aug. 2, 1971	0.032	2.3	0.65 ± 0.04	0.106 ± 0.003	0.0030 ± 0.0012	0.279 ± 0.005	0.275 ± 0.007
16	April 21–24, 1972	—	4.3	0.116 ± 0.007	0.155 ± 0.005	<0.002	0.045 ± 0.002	0.045 ± 0.002
17-1	Dec. 11–14, 1972	0.170	2.0	0.181 ± 0.012	0.0052 ± 0.0014	<0.002	0.034 ± 0.002	0.051 ± 0.002
17-2	Dec. 11–14, 1972	0.025	7.7	0.131 ± 0.007	0.0087 ± 0.0014	<0.002	0.005 ± 0.001	0.006 ± 0.001

*Errors listed are the 1 σ statistical counting errors.

The average tritium activity we observed in Apollo 16 samples was approximately 200 dpm/kg soil, see Part II. It would follow that the fraction of the tritium that diffused out of the soil was about 10^{-4}. This fraction compares well with the fraction of hydrogen gas that diffused out of the soil, and is about a factor of ten lower than was observed in the Apollo 15 mission. These diffusion processes were discussed in more detail in our earlier report (Stoenner *et al.*, 1972).

Tritium was observed in the solar flare particles from the flare of November 12, 1960 by Fireman, DeFelice, and Tilles (1961). This flare had an intensity of 10^9 protons/cm^2 with energy above 30 MeV. In our analysis of lunar samples we have been interested in obtaining further evidence for tritium from the sun. Very high amounts (20–200 dpm) of tritium were observed lightly adsorbed on two Apollo 12 samples (12063 and 12065), and searches for sources of tritium contamination within LRL and the spacecraft were made (Stoenner *et al.*, 1971, 1972). Searches for adsorbed tritium on rock 12065 made six months later by D'Amico *et al.* (1971) revealed that less than 0.01 dpm T was present. There still remains a question whether the high tritium level observed on the two Apollo 12 rocks was from the sun or from some unknown source of contamination. If there is a high concentration of tritium present in solar flare particles or solar wind, one might expect to observe it in the gas in the SRC, particularly after a solar flare event. Tritium measurements were not carried out on the Apollo 12 because of the difficulties with biological quarantine procedures, but measurements were made on succeeding missions. Very low tritium activities were observed, and these could be explained by diffusive losses of hydrogen from the fine material in the container. There were no flares prior to the Apollo 15 mission, but there were flares prior to the other missions. The flare dates, days before mission, and proton fluxes (>50 MeV protons/cm^2) were as follows (see Fireman, 1972): (12) November 2, 1969, 21 days, $(3.4 \pm 0.9) \times 10^7$; (14) January 24, 1971, 12 days, $(3.7 \pm 0.6) \times 10^7$; (16) April 19, 1972, 2 days, $(0.7 \pm 0.3) \times 10^7$. The intensity of these flares prior to the Apollo 12, 14, and 16 missions were about two orders of magnitude less than that of November 12, 1960 that Fireman and his associates found to contain tritons. However, the flare of August 4–9, 1972, 124 days prior to the Apollo 17 mission was very intense, and the total number of protons/cm^2 with energy above 60 MeV was 2.3×10^{10} (Solar Geophysical Data, 1973). However, we found very low tritium levels in both SRC's from this mission. We have concluded that the tritium observed in the sample return containers from the Apollo 14, 16, and 17 missions could be accounted for by diffusion of cosmic ray produced tritium from the fine material present. There is no evidence for a triton component in the solar flare particles from these experiments. The fact that lightly adsorbed tritium was not observed from the Apollo 17 SRC's can perhaps be explained by noting that the lunar surface at the collection site was exposed to two months of direct solar radiation, and all easily evolved hydrogen was lost. Recent results from IMP-5 and IMP-6 show that the T/H ratio in solar flares is in the range $(6–13) \times 10^{-5}$ (Anglin *et al.*, 1973). Flare triton contents as low as this, if present in the August 4–9, 1972 flare, would not be observable in the SRC from the Apollo 17 mission, especially if allowances are made for diffusive losses.

^{37}Ar and ^{39}Ar. The ^{39}Ar activities are all very low and are at the limit of the sensitivity of our counters. The ^{37}Ar activities are considerably higher, well above counter backgrounds, and in all cases a clearly resolved ^{37}Ar peak at 2.8 keV was observed (see Stoenner et al., 1970, for counting technique). The ^{37}Ar is produced in lunar materials principally by the ^{40}Ca$(n, \alpha)^{37}$Ar reaction from galactic and solar cosmic rays. The amount of ^{37}Ar present in lunar fines, therefore, depends upon the calcium content and the cosmic ray intensity. Since ^{37}Ar has a 35-day half-life, it is particularly sensitive to periodic solar flares. However, flare particles increase the ^{37}Ar levels only at shallow depths (<10 g/cm^2) in the lunar soil (Fireman et al., 1973). The high amount of ^{37}Ar observed in the SRC from the Apollo 16 mission is the result of a large amount of fine material in the container (4.3 kg), with high calcium content, and there was a low intensity solar flare two days before the sample was recovered. The ^{37}Ar activity in both of the Apollo 17 containers is extremely low, whereas one might expect a high level resulting from the large solar flare of August 4, 1972, 124 days prior to the Apollo 17 mission. Our ^{37}Ar measurements on surface samples collected during this mission are as high as 120 dpm/kg ($\sim$900 dpm/g on August 4), but samples at depths greater than 50 g/cm^2 were approximately a factor of two lower than samples from the corresponding depth from the Apollo 16 mission, because of lower calcium content. The low ^{37}Ar values obtained from the Apollo 17 containers cannot be attributed to incomplete collection of the gas from the container, because the drop in pressure was carefully observed during the collection of the gas, and the quantity of gas recovered corresponds closely to the initial pressure in the container. Perhaps the diffusion of ^{37}Ar from the material in these containers was unusually low. We have not studied the rate of ^{37}Ar diffusion from Apollo 17 samples to test this possibility.

Radon. The loss of radon from the lunar soil and into the lunar atmosphere has been the subject of many investigations. Of special interest are the spacecraft observations that have revealed variations in the alpha particle emission rates from the lunar surface that are attributed to variations in radon emission (Gorenstein and Bjorkholm, 1973; Bjorkholm et al., 1973; Bjorkholm and Gorenstein, 1974). However, in general the amount of radon in the lunar atmosphere is low, and this may be attributed to the low emanating power of lunar soil. We have been particularly interested in observing ^{222}Rn in the sample return containers because they contain a large mass of lunar fines, and the physical and chemical characteristics of the soil in the container were essentially the same as they were when present in the high-vacuum, dry environment of the lunar surface. We have improved our system for recovering radon by using charcoal low in radium, and have eliminated grease-lubricated stopcocks from our system. The sample recovered is counted in a small proportional counter with a well-measured counting efficiency for radon and its daughters. Thus, by using a large mass of lunar sample and a counter with a very low background, we can measure very low emanation rates. The measured radon activities at the time of puncturing the box, corrected for the radioactive growth factor for the period of time the container was closed are given in Table 2. From

Table 3. Radon emanation rates in sample return containers.

Mission, SRC	Average U content (ppm)	Total ^{238}U disintegration rate in fines (dpm)*	Fraction emanated ($\times 10^3$)
12	0.9	1800	2.7
14	3.4	1500	1.2
15	1.4	2400	0.3
16-2	0.64	2000	0.06
17-1	0.25	370	0.5
17-2	0.25	1400	0.09

*0.74 dpm ^{226}Ra/μg U.

these saturation ^{222}Rn production rates, the weight, and average uranium content of the fine material in each container, we have calculated the fraction of ^{222}Rn emanated. The values are listed in Table 3. The emanation rates obtained are considerably lower than the values reported by Adams *et al.* (1972), and in the same range as the values reported by Yaniv and Heyman (1972). The relation between these emanation measurements and the lunar atmosphere was discussed in our earlier report (Stoenner *et al.*, 1972). Low emanation rates observed here are consistent with the observed low alpha particle emission rates from the surface of the moon.

Part II   Radioactive ^{37}Ar, ^{39}Ar, Tritium, and ^{222}Rn, and the Hydrogen and Helium Concentrations in the Apollo 16 and 17 Drill Stems

The primary interest in these investigations was to measure the ^{37}Ar (35.1 day) production rate in the lunar soil by cosmic rays. This isotope is of particular interest because it is produced primarily by the fast neutron reaction ^{40}Ca$(n, \alpha)^{37}$Ar, and can therefore be used to measure the fast neutron production rate as a function of the depth. The slow neutron intensity was measured to a depth of 2 meters by the Apollo 17 neutron probe experiment (Woolum *et al.*, 1974). These slow and fast neutron fluxes are of special interest in understanding the long-term development of the lunar regolith as revealed by isotopic variations in gadolinium and samarium (Russ III *et al.*, 1972; Russ III, 1973; Fireman, 1974).

The Apollo 16 drill stem samples were measured in two laboratories. Fireman and his associates (1973) at the Smithsonian Astrophysical Observatory measured samples from the surface to a depth of 27 g/cm^2, and we measured samples from 83 to 342 g/cm^2 depth (set 60001 to 60006). In addition to the drill string samples, we made measurements on a sample scooped from the surface (69941,16), and a soil sample from under a 1 meter diameter boulder (69961,17). A set of samples from the Apollo 17 drill string was measured; these ranged in depth from 49 g/cm^2 (70008,9) to 472 g/cm^2 (70002,6). Samples from the top section of the Apollo 17 drill string were not available, so it was not possible to determine the ^{37}Ar activity profile from 49 g/cm^2 depth to the surface. Instead, this depth range was studied using a set of secondary samples whose depths were not accurately known. These samples were: a scooped surface sample 3 meters from the deep core site at the ALSEP station (70181,6), a scooped surface sample from Station 5 near Camelot crater 4.9 km distant (75081,41-43), and the bottom 15–25 cm (78421,13) and sides 6–15 cm (78441,2) of a trench dug at Station 8, 2.3 km from the ALSEP station.

In addition to ^{37}Ar, we measured three other radioactive products, ^{39}Ar (269 year), tritium, and ^{222}Rn. Argon-39 is produced as a spallation product from titanium and iron, and by (n, p) reaction on potassium. Radon-222 arises from ^{226}Ra decay and it is of interest to test whether the equilibrium amount is present in the lunar regolith that would be expected from the decay of ^{238}U. Our vacuum melting and gas separation techniques also permitted measuring the hydrogen and helium concentrations in the lunar soil as a function of the depth. These measurements are of interest in understanding the solar wind fluxes and H/He ratios in the past. Chemical analyses of these same samples are very important in the interpretation of the results, and for this reason 25 mg aliquots of the samples were analyzed for K, Ca, Ti, Fe, Mn, and U.

EXPERIMENTAL

Extraction. The vacuum melting procedure and counting techniques were described in our earlier report (Stoenner *et al.*, 1971). However, there were some important improvements that will be described. Radon can readily dissolve in grease, and our glass system used prior to the Apollo 16 utilized grease-lubricated stopcocks. For the measurements reported herein our glass system was completely rebuilt and high-vacuum Teflon plunger type valves with viton O-ring seals were used. The system is similar to that diagrammed in our earlier paper, but without quarantine restrictions it could be combined into a single system (Stoenner *et al.*, 1970).

The sample (1.5–5 g) was weighed into a platinum crucible that fitted into a molybdenum crucible with a close-fitting alumina protective liner. The crucible was heated by induction inside a quartz tube that was attached to an automatic Toepler pump to collect the evolved gases. The purification, separation, and subsequent treatment of He, Ar, Rn, and H$_2$ utilizes the same equipment and procedures as described in the previous section.

The system was evacuated, the crucible baked out, and the line flamed prior to each sample to eliminate water contamination. The sample was ground to 80 mesh, a 25 mg aliquot taken for chemical analysis, and the remainder placed in the platinum crucible. The system was evacuated for 4 min reaching a vacuum of 10^{-4} torr. With the sample present in the system for 16–72 hours, the line was flamed and the accumulated gas collected and carried through the procedure for hydrogen purification. This procedure tests the line blank for hydrogen, and tests for the presence of lightly adsorbed tritium. The line hydrogen blank was in the range 0.5–0.15 cm^3 STP, and less than 0.020 dpm T/g was observed. Following this tritium–hydrogen blank measurement, argon and hydrogen (0.15 cm^3 STP each) were added to the system for carriers, and the sample was melted. The sample was maintained at 1300°C for 1 hour, and finally the gases were collected with the Toepler pump while the sample was held at the maximum temperature of 1400°C.

The gases were collected with the Toepler pump and placed in a tube containing a small charcoal U-trap and activated vanadium metal in a silica glass tube at a temperature of 900°C. The vanadium metal was then cooled to room temperature, adsorbing the hydrogen and the charcoal trap was cooled with liquid nitrogen to adsorb argon and radon. Helium was then removed from the tube with the Toepler pump and its volume measured. The temperature of the charcoal was set to $-78°C$ (dry-ice) and the argon removed with the Toepler pump, its volume measured, and placed in a small proportional counter. The charcoal was then heated to 250°C and radon removed. An argon–methane (P-10 counting gas) mixture was admitted to the charcoal–vanadium tube to serve as a carrier gas to assure quantitative removal of radon. This gas was then placed in a small proportional counter for measuring ^{222}Rn activity. Finally, the vanadium metal was heated to 900°C to evolve hydrogen. The hydrogen was allowed to diffuse through a paladium thimble at 850°C and the purified hydrogen collected with a Teopler pump. The hydrogen was counted in a small proportional counter using an argon(90%)–methane(10%) counting gas. The fraction of hydrogen in the counter varied from 2–85%.

Argon counting. The counting procedures used were identical to those described earlier, except for the ^{37}Ar measurements on samples 69941,16 and 69961,17. These samples were received very late and it was necessary to use a rise-time analysis of the pulses to resolve ^{37}Ar decay events from ^{39}Ar events. A counting system was used that measures both the rise-time of the pulse, and the pulse height for each event (Davis *et al.*, 1972). The events are plotted on a two-dimensional plot, the rise-time (called amplitude of the differentiated pulse) on the ordinate, and the pulse height (energy) on the abscissa. The 2.8-keV Auger electron from an ^{37}Ar decay event has a very short range in the counter gas (~0.05 mm), and therefore the ion pairs created are localized in the counter gas. These ions are collected at the center-wire as a bunch and give a fast-rising pulse. On the other hand, the energetic betas from ^{39}Ar decays, in passing through the counter gas, produce ion pairs along the track of the beta. If in an ^{39}Ar decay the same number of ion pairs are produced in the counting gas as in an ^{37}Ar decay, these ion pairs will be more widely separated in the gas, consequently collected over a longer period of time, and result in a slower rising pulse. The energy scale and region of fast events was calibrated with an ^{55}Fe X-ray source. Figure 1 shows a plot of counting events recorded for sample 69941,16 during a 36.7-day period. The region of fast events is defined by the heavy lines, and the energy region for ^{37}Ar decays centered about 2.8 keV is indicated by the vertical lines (fwhm). The cluster of ^{37}Ar events (146) can be clearly observed in the region defined by the correct rise-time and energy for ^{37}Ar Auger electrons. The multitude of ^{39}Ar events (971) occur well outside of the region of fast events.

Radon. To test our procedures for measuring radon a sample of a standard rock (W-1) containing 0.58 μg U per gram was melted following the identical procedures used with lunar material. The ^{222}Rn observed for two aliquots of this rock was 0.41 and 0.43 dpm/g. These values can be compared to the equilibrium rate of 0.43 dpm $^{222}Rn/g$ corresponding to the uranium content of the rock. The ^{222}Rn counting efficiency of the small proportional counters used was determined by counting the ^{222}Rn from a National Bureau of Standards standard ^{226}Ra solution. The ^{222}Rn was collected from the solution by a helium purge and placed in the proportional counter with the usual argon–10% methane filling gas.

Results and Discussion

The radioactive argon and tritium activities observed in the Apollo 16 and 17 drill stems are given in Table 4 and the elemental compositions are listed in Table 5. The errors noted are statistical (1 σ). The depth of each sample is given, based on the information given in the Apollo 16 and 17 information catalogs. It was assumed that although section 60005 was returned with only 76 g of material compared to approximately 200 g contained in the other sections, it was in fact full at the time of sampling. If there was truly a void at the level of 60005, the depth of the four lowest data points would be decreased by about 40 g/cm^2.

Argon-37. Argon-37 is produced primarily by the $^{40}Ca(n, \alpha)^{37}Ar$ reaction, though some ^{37}Ar is produced by spallation on titanium and iron and by $^{40}Ca(p, \alpha)^{37}K$ (β^+ decay)^{37}Ar. The (p, α) production mechanism is unimportant at

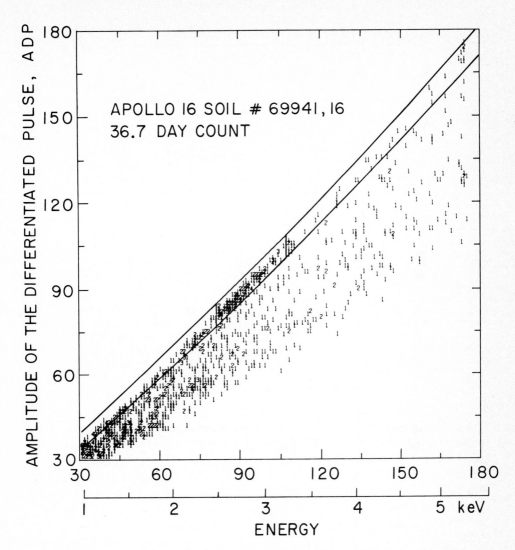

Fig. 1. Pulse rise-time (ADP) versus pulse height (Energy) plot for Apollo 16 soil sample. Fast-rising ^{37}Ar pulses should fall in fast region defined by heavy lines, and half-width (fwhm) defined by channels 80–107 on energy scale.

depth, but is an important process for the production of ^{37}Ar by solar flare protons in surface samples. The spallation production of ^{37}Ar is significant and we can make an estimate of this effect by comparing the activity levels in lunar soil of ^{37}Ar with those of ^{39}Ar, a product produced mostly by spallation. The ^{39}Ar activity levels are about a factor of 7.5 lower than the ^{37}Ar activities. The spallation cross sections for producing ^{37}Ar and ^{39}Ar at 600 MeV are about the same for iron (5.6 and 6.3 mb, respectively) and titanium (11.6 and 15.7 mb, respectively), see Stoenner *et al.* (1970). It follows that not more than 12% of the ^{37}Ar was produced by spallation of

Table 4. Radioactive argon and tritium in the Apollo 16 and 17 soil.

Sample No.	Depth (g/cm^2)	Activity (dpm/Kg)			dpm ^{37}Ar/gCa
		^{37}Ar	^{39}Ar	T	
60006,12	83	62.8 ± 2.4	11.5 ± 0.6	130 ± 4	0.537 ± 0.020
60004,18	143	59.5 ± 2.0	8.0 ± 0.5	442 ± 10	0.518 ± 0.017
60004,10	205	39.4 ± 1.9	6.6 ± 0.5	166 ± 2	0.345 ± 0.017
60003,6	272	31.7 ± 1.7	3.5 ± 0.5	151 ± 5	0.271 ± 0.015
60001,13	342	20.8 ± 1.2	2.7 ± 0.4	227 ± 6	0.178 ± 0.010
69941,16	0–4.5	56 ± 4	8.2 ± 0.4	246 ± 5	0.495 ± 0.035
69961,17	100	97 ± 6	10.7 ± 0.5	283 ± 4	0.932 ± 0.058
70181,6	0–4.5	115 ± 4	9.3 ± 0.4	261 ± 4	1.474 ± 0.051
75081,41-43	0–4.5	81 ± 3	10.2 ± 0.4	186 ± 3	1.025 ± 0.038
78441,2	11–27	48.2 ± 5.6	8.9 ± 0.4	235 ± 2	0.560 ± 0.065
78421,13	27–45	41.7 ± 5.4	10.9 ± 0.4	221 ± 4	0.485 ± 0.063
70008,9	49	54.5 ± 3.3	10.6 ± 0.5	97 ± 3	0.717 ± 0.043
70008,7	72	49.4 ± 2.7	12.7 ± 0.6	208 ± 4	0.650 ± 0.035
70008,5	99	43.5 ± 3.1	10.9 ± 0.6	196 ± 4	0.551 ± 0.039
70008,3	124	37.4 ± 2.4	7.2 ± 0.5	403 ± 6	0.492 ± 0.032
70006,6	180	28.6 ± 2.0	6.3 ± 0.6	1561 ± 22	0.362 ± 0.025
70005,6	252	19.5 ± 1.5	5.6 ± 0.4	118 ± 3	0.240 ± 0.019
70004,6	325	21.7 ± 1.7	—	—	0.271 ± 0.021
70003,6	399	13.6 ± 1.1	2.3 ± 0.3	134 ± 4	0.170 ± 0.014
70002,6	472	9.2 ± 1.2	2.6 ± 0.8	376 ± 6	0.114 ± 0.015

iron and titanium. Since this is true, we can then compare the results for the Apollo 16 and 17 missions on the basis of the calcium content of the two samples. In Table 4 we have listed the dpm ^{37}Ar per gram of calcium, and this data is plotted in Fig. 2. We have included in Fig. 2 four points measured by Fireman *et al.* (1973) at the Smithsonian Astrophysical Observatory (SAO). These values from SAO define the shape of the production curve at shallow depths for galactic cosmic radiation. The two surface samples from the Apollo 17 mission contained large amounts of ^{37}Ar produced by the August 4–9, 1972 flare, but the two samples from the trench appeared to fit reasonably well on the curve. Unfortunately, as mentioned earlier, the upper portion of the Apollo 17 drill stem was not made available, so that the detailed structure of the depth curve could not be obtained. We regret very much missing the data from this most unusual event, namely a very large solar flare superimposed on a normal galactic spectrum. It may be seen from Fig. 2 that the values for dpm ^{37}Ar per gram Ca versus depth from the two missions fit together rather well. We have made a least-squares fit of the data starting from a depth of $49 \, g/cm^2$ presuming that the activity falls off exponentially with depth. This analysis yields an average exponential mean-free path of $238 \pm 21 \, g/cm^2$. The corresponding values for the individual missions are: Apollo 16, $230 \pm 28 \, g/cm^2$ and Apollo 17, $248 \pm 31 \, g/cm^2$. This general shape and magnitude agrees well with the calculations of Reedy and Arnold (1972), only their calculated mean-free path of

Table 5. Chemical composition of samples from Apollo 16 and 17.

Sample No.	Percent composition					U ppm	H_2 (cm³ STP/g)	He (cm³ STP/g)
	K	Ca	Ti	Fe	Mn			
60006,12	0.078*	11.7	0.33	3.9	0.063	—	1.0	0.055
60004,18	0.109*	11.5	0.41	4.8	0.069	—	1.3	0.054
60004,10		11.4	0.39	4.8	0.065	—	1.1	0.047
60003,6	0.106*	11.7	0.39	5.0	0.068	—	1.4	0.056
60001,13	0.111*	11.7	0.37	4.7	0.067	—	1.1	0.025
69941,16	0.094	11.3	0.36	4.4	—	—	0.8	0.041
69961,17	0.096	10.4	0.37	4.7	—	—	0.6	0.033
70181,6	0.077	7.8	4.93	14.1	—	0.28	0.85	0.168
75081,41-43	0.069	7.9	5.76	14.0	—	0.24	0.66	0.146
78441,2	0.084	8.6	2.50	9.7	—	0.41	1.24	0.149
78421,13	0.092	8.6	2.49	9.6	—	—	1.13	0.150
70008,9	0.073	7.6	5.71	14.8	—	0.20	0.80	0.069
70008,7	0.083	7.6	5.60	14.8	—	0.22	0.85	0.066
70008,5	0.080	7.9	5.58	14.9	—	0.21	0.80	0.063
70008,3	0.090	7.6	4.29	13.5	—	0.29	1.07	0.110
70006,6	0.102	7.9	5.01	13.9	—	0.30	0.83	0.101
70005,6	0.094	8.1	3.47	12.2	—	0.41	1.18	0.176
70004,6	0.090	8.0	3.60	12.6	—	0.41	1.19	0.144
70003,6	0.111	8.0	3.37	12.3	—	0.48	1.41	0.176
70002,6	0.168	8.1	3.56	12.1	—	0.66	1.63	0.136

*K from Philpotts *et al.* (1973).

300 g/cm² is somewhat greater than we observed. Lingenfelter *et al.* (1972) and Kornblum *et al.* (1973) use an exponential mean-free path of 165 and 155 g/cm² respectively, for the neutron source function in their calculations of neutron moderation in the lunar regolith. This value is considerably lower than the value reported here, and will therefore change their calculated slow neutron depth profile. One may integrate the depth curve shown in Fig. 2 to obtain the total ^{37}Ar production rate in lunar soil from galactic cosmic rays, namely 190 ± 17 dpm ^{37}Ar/gCa · cm².

The two Apollo 17 surface samples were extremely high, 1.48 and 1.04 cpm ^{37}Ar/g Ca, compared to 0.34 dpm ^{37}Ar/g Ca from the SAO data from the Apollo 16 drill string. One can attribute this large increase to the flare of August 4–9, 1972. It is interesting to calculate the ^{37}Ar activity that was present at the end of the flare event. At that time there was 8–14 dpm ^{37}Ar/g Ca in lunar surface material. Since only surface material had high ^{37}Ar activity, it is clear that the ^{37}Ar was not produced by neutrons. One presumes that activity was formed by the decay of ^{37}K produced by the ^{40}Ca$(p, \alpha)^{37}$K reaction, threshold of 5.6 MeV. It will be interesting to know the excitation function for this reaction and make an estimate of the ^{37}Ar production from the solar flare proton spectrum. Measurement of this excitation function is now in progress at Rice (Heymann, private communication, 1974).

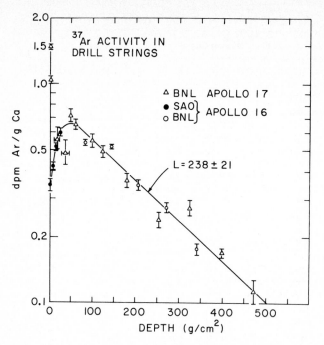

Fig. 2. Argon-37 as a function of the depth in the Apollo 16 and 17 deep drill strings.

Argon-39 and tritium. The ^{39}Ar and tritium activities are listed in Table 4. The ^{39}Ar activities from the Apollo 17 mission will be counted again to obtain more accurate results, so for the present we will leave these out of the discussion.

The ^{39}Ar activity as a function of the depth for the Apollo 16 mission is shown in Fig. 3. It may be observed that the production rate increases from surface values by a factor of approximately 1.6 and thereafter falls off with depth with an exponential mean-free path of 166 g/cm^2. Comparing this behavior with the Reedy and Arnold (1972) calculations, we find the measured ^{39}Ar activities are about 1.7 times higher than calculated (they used 16.8% Fe and 1.5% Ti) and the calculated exponential mean-free path is greater (280 g/cm^2).

Tritium values appear to vary considerably. There is an apparent decrease with depth but a reliable value of L cannot be derived. Tritium is a high-energy evaporation product, and one would expect it to be produced primarily from oxygen and aluminum. It is difficult to understand why the tritium results are so erratic, particularly since the fines appear to retain hydrogen reasonably well.

Radon. The ^{222}Rn values are a direct measurement of the ^{226}Ra present in the samples. One would expect the ^{226}Ra in these samples to be in equilibrium with the ^{238}U present. As mentioned earlier, our extraction system was tested by melting samples of the standard rock W-1 and the quantity of ^{222}Rn was obtained that corresponded to the uranium content. Uranium was not measured on Apollo 16 samples, but one can compare our ^{222}Rn measurements with the uranium

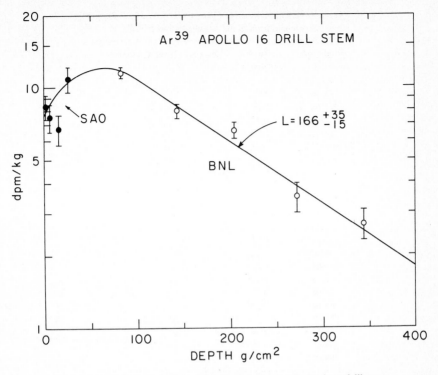

Fig. 3. Argon-39 as a function of the depth in the Apollo 16 deep drill stem.

concentrations of the drill stem measured by Silver (1973). The radon extracted from the drill stem samples and the two Station 9 samples are listed in Table 6, and compared with the equilibrium ^{222}Rn disintegration rate calculated from Silver's uranium contents. Our ^{222}Rn activities are about a factor of two low in the top two samples of the drill stem series, and in the remaining samples the agreement is better, though far from satisfactory. The lack of agreement possibly is due to the fact that Silver's uranium measurements may not be applicable to our specific samples, and to resolve this question we measured the uranium content on an aliquot of each particular sample melted in the Apollo 17 drill stem.

The uranium contents for the Apollo 17 samples are given in Table 5, and the calculated ^{222}Rn equilibrium values are compared with the observed ^{222}Rn from the melt in Table 6. It may be observed from the comparison, that the observed ^{222}Rn is usually low by 15–25%, but two samples are 27% and 38% high. It is very difficult to explain the high values. One sample was remelted to see if one could reproduce the evaluation of ^{222}Rn. However, on remelting, less ^{222}Rn was evolved than in the first melt. This low result could be explained by volatilization of the parent ^{226}Ra during the initial melting. To test this possibility, the material from the furnace wall was removed by an acid leach, placed in a flask. After allowing the ^{222}Rn to grow back in equilibrium, the radon was removed from the liquid by helium purge, trapped in charcoal, purified, and counted. It was found that 24% of the original ^{222}Rn was

Table 6. Comparison of ^{222}Rn activity and uranium concentrations in the Apollo 16 and 17 soil.

Sample No.	^{222}Rn* (dpm/g)	^{222}Rn calc. from U content (dpm/g)
60006,12	0.26	0.53 (6)†
60004,18	0.29	0.47 (21)†
60004,10	0.38	0.50 (12)†
60003,6	0.54	0.48 (8)†
60001,13	0.61	0.61 (6)†
69941,16	0.58	0.56 (22)†
69961,17	0.43	0.53 (25)†
70181,6	0.21	0.21
75081,41-43	0.15	0.18
788441,2	0.32	0.30
78421,13	0.31	—
70008,9	0.11	0.15
70008,7	0.088	0.16
70008,5	0.11	0.15
70008,3	0.16	0.21
70006,6	0.28	0.22
70005,6	0.26	0.30
70004,6	0.24	0.30
70003,6	0.30	0.36
70002,6	0.62	0.45

*Counting errors ± 3%.

†U from Silver (1973), number in parentheses is the fraction number of the same parent sample. ^{222}Rn in dpm/g = 0.74 × (ppm U).

indeed present on the walls of the furnace. This experiment showed that ^{226}Ra does vaporize at the temperature of 1500°C used in melting procedures.

It is clear that more effort must be placed on these radon measurements to resolve the questions raised by this work. The high ^{222}Rn values reported for two soil samples should be explored further, and the volatilization properties of radium should be studied in detail. It is conceivable that there is a problem here that would be of interest in interpreting lead isotope dating.

Hydrogen and helium. Since rather large samples were required for the radioactivity measurements, it was possible to measure the hydrogen and helium recovered, and compare our volumes with those measured by other investigators. The hydrogen and helium contents of all samples are given in Table 5, estimated errors are 10% for hydrogen and 5% for helium. The hydrogen reported includes hydrogen present in the sample as water. It may be observed that the hydrogen content of the lunar soil from all samples is relatively constant, about 1.2 ± 0.2 cm^3 STP per gram. These results compare well with total hydrogen reported for soil samples by other investigators (e.g. Epstein and Taylor, 1972; and Friedman *et al.*, 1972), and our earlier results (Stoenner *et al.*, 1970).

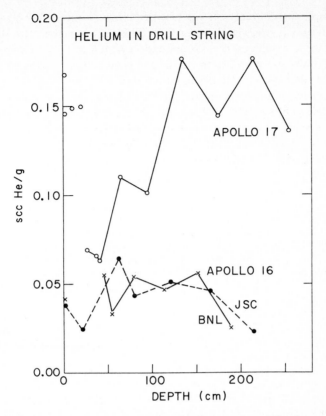

Fig. 4. The helium content in the Apollo 16 and 17 deep drill strings.

The helium content of the Apollo 16 drill stem is relatively constant except for the deepest sample. It is especially interesting to compare these helium contents of the Apollo 16 drill stem with those measured by Bogard *et al.* (1973). A comparison of the data is given in Fig. 4, and it may be observed that the agreement is satisfactory. The helium contents of the Apollo 17 drill string is entirely different, exhibiting a dramatic increase with depth, see Fig. 4. It is clear from these results that the depth variation of the hydrogen and helium contents of the lunar regolith varies with the location. The hydrogen and helium are derived from the solar wind and one might hope to observe long period variations in the solar wind flux by studying the depth variation of these gases in the lunar soil. The increase with depth in the Apollo 17 drill string would indicate an increased solar wind flux in the past.

The H/He ratio in the sun is a very important quantity, and plays an important role in astrophysics. This ratio has been measured spectroscopically in solar prominences, and the chromosphere, in solar cosmic rays, and the solar wind (see review of Hirshberg, 1973). The spectroscopic values are considered by Hirshberg to be the most accurate and give a H/He atom ratio of 12 to 20. Solar wind H/He ratios give a value of 7 to 10, and solar cosmic rays indicate a H/He ratio of about 17 at the flare site. The H/He atom ratio we observed in the Apollo 16 drill stem and the

Station 9 samples is relatively constant (with the exception of the bottom of the drill stem), and give an average H/He ratio of 49. This ratio is higher than the solar wind value, and can be explained by contamination with terrestrial water. Epstein and Taylor (1972) finds about 50% of the hydrogen in lunar fines is present as water. Our samples were also probably contaminated with terrestrial water, and if so, this would explain the high H/He ratios. Studies of drill stem hydrogen and helium contents coupled with exposure age measurements should be a valuable method for studying the history of the solar wind flux, and H/He ratios. It is possible that the sun's luminosity has varied in the past few million years (see recent review by Cameron, 1973), and it would be interesting to search for possible changes in solar wind and solar cosmic ray intensities in lunar material.

Acknowledgments—We would like to acknowledge the assistance of Dr. Donald Bogard and Mr. Lewis Sims in setting up the apparatus and collecting the gas from the sample return container. We are indebted to LSAPT, the Curator's office and the staff at the Lunar Receiving Laboratory for their help in making early allocations of samples from the drill stems. Dr. Don Burnett stimulated our interest in making the drill stem measurements, and assisted us on several technical problems. We would like to acknowledge many helpful discussions with Dr. John C. Evans during the progress of this work.

REFERENCES

Adams J. A. S., Barretto P. M., Clark R. B., and Duval J. S. (1971) Radon-222 emanation and the high apparent lead isotope ages in lunar dust. *Nature* **231**, 174–175.

Anglin J. D., Dietrich W. F., and Simpson J. A. (1973) Deuterium and tritium from solar flares of 10 MeV per nucleon, *5th Int. Seminar*, Leningrad, June 26–29, 1973.

Bogard D., Nyquist L. E., Hirsch W., and Moore D. (1973) Trapped solar and cosmogenic noble gas abundances in Apollo 15 and 16 deep drill samples. *Earth Planet. Sci. Lett.* **21**, 52.

Bjorkholm P., Golub L., Gorenstein P. (1973) Detection of a nonuniform distribution of Polonium-210 on the moon with the Apollo 16 particle spectrometer. *Science* **180**, 957–959.

Cameron A. G. W. (1973) Major variations in solar luminosity. *Rev. of Geophysics and Space Physics* **11**, 505–510.

D'Amico J., De Felice J., Fireman E. L., Jones C., and Spannagel G. (1971) Tritium and argon radioactivities and their depth variations in Apollo 12 samples. *Proc. Second Lunar Sci. Conf.*, *Geochim. Cosmochim. Acta*, Suppl. 2, Vol. 2, pp. 1825–1839. MIT Press.

Davis R., Jr., Evans J. C., Radeka V., and Rogers L. C., Report on the Brookhaven solar neutrino experiment. *Proc. Conf. Neutrino '72, Europhysics Conf.*, Balatonfured, Hungary, June 1972, Vol. 1, pp. 5–22.

Davis R., Jr. and Evans J. C. (1973) Experimental limits on extraterrestrial sources of neutrinos. *13th Int. Conf. on Cosmic Rays*, Vol. 3, pp. 2001–2006; revised as BNL 17945 (report).

Epstein S. and Taylor H. P., Jr. (1973) $^{18}O/^{16}O$, $^{30}Si/^{28}Si$, $^{13}C/^{12}C$, and D/H studies of Apollo 14 and 15 samples. *Proc. Third Lunar Sci. Conf.*, *Geochim. Cosmochim. Acta*, Suppl. 3, Vol. 2, pp. 1429–1454. MIT Press.

Fireman E. L., De Felice J., and Tilles D. (1961) Solar flare tritium in a recovered satellite. *Phys. Rev.* **123**, 1935–1936.

Fireman E. L. (1972) In *Apollo 15 Lunar Samples.* The Lunar Science Institute, Houston, pp. 364–367.

Fireman E. L., D'Amico J., and De Felice J. (1973) Radioactivities vs. depth in Apollo 16 and 17 soil. *Proc. Fourth Lunar Sci. Conf.*, *Geochim. Cosmochim. Acta*, Suppl. 4, Vol. 2, pp. 2131–2143. Pergamon.

Fireman E. L. (1974) History of the lunar regolith from neutrons. *Proc. Fifth Lunar Sci. Conf.*, *Geochim. Cosmochim. Acta.* This volume.

Friedman I., Hardcastle K. G., and Geason J. D. (1972) Isotopic composition of carbon and hydrogen in some Apollo 14 and 15 samples. In *The Apollo 15 Samples*. The Lunar Science Institute, Houston.

Gorenstein P. and Bjorkholm P. (1973) Detection of radon emanation from the crater Aristarchus by the Apollo 15 particle spectrometer. *Science* **179**, 792–794.

Heymann D. (1974) Private communication.

Hirshberg J. (1973) Helium abundance in the sun. *Rev. Geophysics and Space Physics* **11**, 115–131.

Kornblum J. J., Fireman E. L., Levine M., and Aronson A. (1973) Neutrons in the moon. *Proc. Fourth Lunar Sci. Conf., Geochim. Cosmochim. Acta*, Suppl. 4, Vol. 2, pp. 2171–2182. Pergamon.

Lingenfelter R. E., Caufield E. H., and Hampel V. E. (1972) The lunar neutron flux revisited. *Earth Planet. Sci. Lett.* **16**, 355–369.

Philpotts J. A., Schumann L., Korins C. W., Lum R. K. L., Bickel A. L., and Schnetzer C. C. (1973) Apollo 16 returned lunar samples: lithophile trace-element abundances. *Proc. Fourth Lunar Sci. Conf., Geochim. Cosmochim. Acta*, Suppl. 4, Vol. 2, pp. 1427–1436. Pergamon.

Reedy R. C. and Arnold J. R. (1972) Interaction of solar and galactic cosmic-ray particles with the moon. *J. Geophys. Res.* **77**, 537–555.

Russ III G. P., Burnett D. S., and Wasserburg G. J. (1972) Lunar neutron stratigraphy. *Earth Planet. Sci. Lett.* **15**, 172–186.

Russ III G. P. (1973) Apollo 16 neutron stratigraphy. *Earth Planet. Sci. Lett.* **19**, 275–289.

Silver L. T. (1973) Uranium–thorium–lead isotopic characteristics (abstract). In *Lunar Science—IV*, p. 673. The Lunar Science Institute, Houston.

Smith J. W., Kaplan I. R., and Petrowski C. (1973) Carbon, nitrogen, sulfur, helium, hydrogen, and metallic iron in Apollo 15 drill stem fines. *Proc. Fourth Lunar Sci. Conf., Geochim. Cosmochim. Acta*, Suppl. 4, Vol. 2, pp. 1651–1656. Pergamon.

Solar Geophysical Data (1973), CRPL-FB 135, pp. 86–91, February 1973. U.S. Department of Commerce (Boulder, Colorado, USA 80303).

Stoenner R. W., Lyman W. J., and Davis R., Jr. (1970) Cosmic-ray production of rare-gas radioactivities and tritium in lunar material. *Proc. Apollo 11 Lunar Sci. Conf., Geochim. Cosmochim. Acta*, Suppl. 1, Vol. 2, pp. 1583–1594. Pergamon.

Stoenner R. W., Lyman W., and Davis R., Jr. (1971) Radioactive rare gases and tritium in lunar rocks and in the sample return container. *Proc. Second Lunar Sci. Conf., Geochim. Cosmochim. Acta*, Suppl. 2, Vol. 2, pp. 1813–1823. MIT Press.

Stoenner R. W., Lindstrom R. M., Lyman W., Davis R., Jr. (1972) Argon, radon, and tritium radioactivities in the sample return container and the lunar surface. *Proc. Third Lunar Sci. Conf., Geochim. Cosmochim. Acta*, Suppl. 3, Vol. 2, pp. 1703–1717. MIT Press.

Woolum D. S. and Burnett D. S. (1974a) In-situ measurement of the rate of ^{235}U fission induced by lunar neutrons. *Earth Planet. Sci. Lett.* In press.

Woolum D. S. and Burnett D. S. (1974b) Lunar neutron capture rates and surface mixing of the regolith. *Proc. Fifth Lunar Sci. Conf., Geochim. Cosmochim. Acta*. This volume.

Woolum D. S., Burnett D. S., and Bauman C. A. (1974) Lunar neutron probe experiment. *Apollo 17 Prelim. Sci. Report* NASA SP 330, pp. 18-1 to 18-12.

Yaniv A. and Heymann D. (1972) Radon emanation from Apollo 11, 12, and 14 fines. *Earth Planet. Sci. Lett.* **15**, 95–99.

Proceedings of the Fifth Lunar Conference
(Supplement 5, Geochimica et Cosmochimica Acta)
Vol. 2 pp. 2231–2247 (1974)
Printed in the United States of America

²²Na–²⁶Al chronology of lunar surface processes

Y. YOKOYAMA, J. L. REYSS, and F. GUICHARD

Centre des Faibles Radioactivités, Laboratoire Mixte CEA-CNRS, 91190-Gif-sur-Yvette, France

Abstract—²²Na–²⁶Al determinations of 155 rocks returned from Apollo 11 to 17 missions are compiled, normalized for chemical composition and classified by ²²Na–²⁶Al correlation diagram in regard to their saturation or nonsaturation in ²⁶Al activity: 84 of them are saturated, 35 unsaturated, and 36 undecided.

Clusters of unsaturated samples are found at LM-ALSEP-SEP Station of the Apollo 17 mission and on the rims of North Ray Crater and of Buster Crater of the Apollo 16 mission.

Anorthosite 15415 and its pedestal 15431 are unsaturated probably because of a rapid erosion.

A correlation was found between the saturated (and normalized) activity of ²⁶Al and the sample weight, which is in agreement with the theoretically calculated formula:

$$^{26}\text{Al (dpm/kg)} = 55 + 38 \; W^{-1/3} \qquad \text{for } W \text{ (weight in kg)} > 0.1 \text{ kg}$$

INTRODUCTION

THE METHOD is based on the fact that ²²Na and ²⁶Al were produced on the lunar surface through similar reactions, but with different half-lives: the activity of ²²Na ($T_{1/2} = 2.6$ yr) in a sample can be saturated in several years, whereas the saturation takes about two to three million years for ²⁶Al ($T_{1/2} = 7.16 \times 10^5$ yr).

Up to the present, more than three hundred determinations of the pair ²²Na–²⁶Al have been made in many laboratories on the samples returned from Apollo 11 to 17 missions. It was found that the ²⁶Al/²²Na ratios of about thirty samples were unusually low, and the low ²⁶Al activities of these samples were attributed to the undersaturation due to short exposure ages, but with a certain reserve because very unusual chemical compositions could also explain the unusual ²⁶Al/²²Na ratios.

In the present work, by using the method proposed in a previous paper (Yokoyama *et al.*, 1973), all data available up to the present were normalized for their chemical compositions, and compared with the general distribution pattern of the pair ²²Na–²⁶Al, in order to see if the samples are really saturated in ²⁶Al activity.

NORMALIZATION FOR CHEMICAL COMPOSITION

Activities of ²⁶Al (and ²²Na) are complex functions of target-element concentration, of cosmic ray-exposure geometry, and of exposure time:

$$A_{\text{obs}} = f(\text{Si, Al, Mg, Na; geometry; time}) \qquad (1)$$

where A_{obs} is the observed activity in a sample, Si, Al, Mg, and Na represent the concentrations of these elements in the sample.

In the previous paper, we showed that the ratio of the production rate from a target element, for example Al, to that from Si is practically independent of sample size, so that the above-mentioned function can be separated as a product of three factors:

$$A_{obs} = f_1(\text{Si, Al, Mg, Na}) \cdot f_2(\text{geometry}) \cdot f_3(\text{time}) \qquad (2)$$

Observed activities can therefore be normalized to a reference target composition, and then normalized activities A_{nor} become free from the function due to the variation in chemical composition:

$$A_{nor} = f_2(\text{geometry}) \cdot f_3(\text{time}) \qquad (3)$$

The following equations were proposed for the normalization:

$$A_{nor} = A_{obs} \frac{(\text{Si} + 2.7 \text{ Al} + 0.05 \text{ Mg})_{ref}}{(\text{Si} + 2.7 \text{ Al} + 0.05 \text{ Mg})_{sample}} \qquad \text{for } ^{26}\text{Al} \qquad (4)$$

$$A_{nor} = A_{obs} \frac{(\text{Si} + 1.8 \text{ Al} + 4.0 \text{ Mg} + 10 \text{ Na})_{ref}}{(\text{Si} + 1.8 \text{ Al} + 4.0 \text{ Mg} + 10 \text{ Na})_{sample}} \qquad \text{for } ^{22}\text{Na} \qquad (5)$$

where Si, Al, Mg, and Na represent their weight percent in the reference (suffixed as ref) and in the sample to be normalized. The mean chemical composition of Apollo 11 to 15 soils is adopted as the reference, since it represents nearly the middle composition of variety: $\text{Si} = 21.7\%$, $\text{Al} = 7.7\%$, $\text{Mg} = 5.8\%$, and $\text{Na} = 0.37\%$.

The method of estimation of the numerical constants applied in the equations and their validity were discussed in detail in the previous paper, and the uncertainties of the numerical constants were estimated to be about ±15%.

For about half of the samples studied here, no chemical analyses have been made up to the present. For these samples, the chemical compositions were estimated from their rock type and from their U, Th, and K contents (details are given in Appendix 1).

All errors introduced by the normalization, namely uncertainties of the numerical constants in the formulae and uncertainties of chemical compositions (including the errors of the estimation of Si, Al, and Mg contents from K, Th, and U contents) are considered. The propagation of these errors is calculated by differentiating Eqs. (4) and (5) and the results are quadratically added to the errors of radioactive measurement which were expressed on the basis of one standard deviation including all known sources of error.

DISTRIBUTION PATTERN OF THE PAIR ^{22}Na–^{26}Al

All data available up to the present were compiled. The references were given in the previous paper (Yokoyama *et al.*, 1973). Some additional or corrected data on the Apollo 16 and 17 samples were obtained from the papers of Clark and Keith (1973), Eldridge *et al.* (1973, 1974), Keith *et al.* (1974), LSPET (1973), O'Kelley *et al.* (1974), Rancitelli *et al.* (1974), and Wrigley (1973).

The measurement of two rocks from Apollo 17 was carried out in the present study by nondestructive gamma–gamma coincidence spectrometry. The result is given in Appendix 2.

The crude data were normalized for their elemental composition, and normalized activities of the pair ^{22}Na–^{26}Al were plotted for each Apollo mission and it was shown that the distribution pattern of the pair ^{22}Na–^{26}Al of saturated samples lies along a theoretically calculated line which is characterized by the mean solar proton flux for each mission. An example of the ^{22}Na–^{26}Al distribution pattern is given in Fig. 1 for Apollo 17 samples. Theoretically calculated lines in the figure represent the saturation activities for the sum of the production due to GCR (galactic cosmic rays) and that due to SCR (solar cosmic rays) and correspond to a mean solar proton flux of $J(E_p > 10 \text{ MeV}) = 260 \pm 50$ protons/cm^2 sec 4π for the production of ^{22}Na with $R_0 = 100$ MV or $J(>10 \text{ MeV}) = 165 \pm 30$ p/cm^2 sec 4π with $R_0 = 150$ MV. These fluxes were obtained, with a method described in the previous paper, from the data of saturated soils. They correspond to the integrated proton fluxes for the August 1972 flare of $J(>10 \text{ MeV}) = (2.7 \pm 0.7) \times 10^{10}$ and $(1.7 \pm 0.4) \times 10^{10}$ p/cm^2, respectively. These fluxes may be compared with the flux estimated by O'Kelley *et al.* (1974), 1.9×10^{10} p/cm^2, and with the result of a compilation made by King (1973) for the proton flux integration for 4–9 August 1972, 2.25×10^{10} p/cm^2.

The rock samples of which ^{22}Na–^{26}Al data are distributed within the error margin of the saturation line are considered to be saturated in ^{26}Al activity.

In Fig. 1, the deviation of a sample far from the saturation line to the lower ^{26}Al activity side means that the sample has an evidently short exposure age and is unsaturated in ^{26}Al activity, which is the case of 70019, 70175, 70255, 78135, and 78235 (in the present paper, the complete comma number is omitted if it is xxxxx,0). Keith and Clark (1974) proposed a method for calculating the saturated ^{26}Al activity from the chemical composition and the weight of a sample and showed that two samples, 70255 and 78135, are unsaturated in ^{26}Al activity. They suggested that two other samples, 70019 and 70175, are also unsaturated in ^{26}Al activity. Our conclusion drawn from the ^{22}Na–^{26}Al method for these samples is therefore consistent with their results. We also found that two other samples, 72315,17 and 78235, are unsaturated in ^{26}Al. Since we compare the activities of the two radionuclides formed through similar reactions, the ^{22}Na–^{26}Al method can be applied for a part of sample, as long as its result constrains only the part in question (because a part does not necessarily represent the whole). Nevertheless, if a part of a rock is saturated in SCR-produced ^{26}Al, it implies that the rock has been stayed at the lunar surface for more than two million years, if not the saturation of the part itself is impossible.

The data of three rocks 70135, 76215, and 76295 are situated near the GCR point which represents the almost pure production by galactic cosmic rays. This suggests that these samples have been partially (or almost completely) shielded from solar flare protons by their parent boulders. The activities of other short-lived nuclides such as ^{54}Mn and ^{56}Co (expressed in the unit of dpm/kg target) are also low in these samples. It also supports the partial shielding of the samples

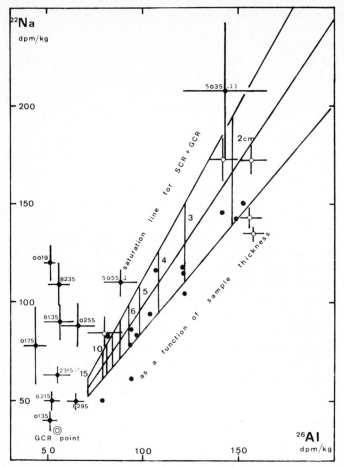

Fig. 1. ^{22}Na–^{26}Al correlation diagram for Apollo 17 samples. Open circles with error bars represent soils, solid circles represent rocks saturated in ^{26}Al, and solid circles with errors bars represent rocks either unsaturated in ^{26}Al or uncertain. The first digit (7) of the LRL number of sample is omitted. Saturation lines correspond to the production of ^{22}Na and ^{26}Al by GCR + SCR with SCR parameters of $J = 260 \pm 50$ protons $(E_p > 10 \text{ MeV})/\text{cm}^2\text{-sec-}4\pi$ and $R_0 = 100 \text{ MV}$ for ^{22}Na and $J = 70$ protons $(E_p > 10 \text{ MeV})/\text{cm}^2\text{-sec-}4\pi$ with $R_0 = 150 \text{ MV}$ for ^{26}Al. The determination of these SCR parameters was given in the previous paper (Yokoyama *et al.*, 1973). The approximate sample thickness corresponding to the saturated activity is also indicated along the saturation line.

from solar protons. Keith *et al.* (1974) and O'Kelley *et al.* (1974) suggest the same conclusion.

Two samples, 75035,22 and 75055,2, are situated near the saturation line in Fig. 1. They can be saturated in ^{26}Al activity within the margin of error, but a slight degree of undersaturation is also possible, and therefore, these samples are classified by this method as undecided samples.

The results of the study of Apollo 17 rocks by ^{22}Na–^{26}Al method are summarized in Table 1. Among twenty-nine samples studied, six are unsaturated and half of them are clustered in Station LM-ALSEP-SEP, which is located near the center of the Taurus-Littrow Valley. Apollo Lunar Geology Investigation Team (ALGIT, 1973) reported that the abundance of surface rocks in this area is higher than the regional average for the valley floor. In this connection, a relatively young exposure age suggested by ^{22}Na–^{26}Al chronology is interesting.

Samples taken at Station 1, which is located about 1 km of the LM-ALSEP-SEP but also near the center of the Taurus-Littrow Valley, exhibit a striking contrast: five samples studied are all saturated in ^{26}Al activity.

Rock chips collected from three boulders at Station 2 (located at the foot of South Massif) indicate that two of them are saturated and another unsaturated in ^{26}Al activity. The unsaturated rock, 72315,17, has a normalized activity in ^{26}Al of 55 ± 7 dpm/kg. Since the saturation activity for GCR is 55 dpm/kg, the SCR-produced ^{26}Al activity of this sample is 0 ± 7 dpm/kg which indicates an extremely

Table 1. Classification of Apollo 17 rocks by saturation or nonsaturation in ^{26}Al activity.

Station	Total	Saturated	Unsaturated	Undecided or shielded	
LM-ALSEP	7	$3 \begin{pmatrix} 0017, 96 \\ 0185; 0275 \end{pmatrix}$	$3 \begin{pmatrix} 0019, 0175 \\ 0255 \end{pmatrix}$	$1 (0135^\mathrm{S})$	Central Cluster
1	5	$5 \begin{pmatrix} 1035; 1135 \\ 1136; 1155 \\ 1175 \end{pmatrix}$			Steno Crater
2	3	2 (2255; 2415)	1 (2315,17)		South Massif
3	3	$3 \begin{pmatrix} 3215; 3255 \\ 3275 \end{pmatrix}$			Light Mantle
5	2			$2 \begin{pmatrix} 5035,22 \\ 5055,2 \end{pmatrix}$	Camelot Crater
6, 7	5	2 (6275; 7135)		$3 \begin{pmatrix} 6215^\mathrm{S}; 6255^\mathrm{S} \\ 6295^\mathrm{S} \end{pmatrix}$	North Massif
8	3	$1 (8505^\mathrm{R})$	2 (8135; 8235)		Sculptured Hills
9	1	1 (9155)			Van Serg Crater

The first digit (7) of the LRL number is omitted for brevity.
The complete comma number is omitted when it is xxxx,0.
$^\mathrm{R}$ Rake sample.
$^\mathrm{S}$ Shielded sample from solar flare particles by its parent boulder.

short exposure age for SCR of this sample (equal or less than an order of 10^5 yr). Hutcheon *et al.* (1974) studied track profiles and micrometeorite craters of this rock, and found about 10^5 yr by track method and 1.2×10^5 yr by microcrater counts. Our result from ^{22}Na–^{26}Al method agrees well with their results. This rock was proposed by Hutcheon *et al.* as a new standard for solar flare particle studies.

Two rocks samples at Station 8 located near the base of the Sculptured Hills are unsaturated in ^{26}Al activity. This is consistent with the fact that most of the rocks at this area have poorly developed fillets (ALGIT, 1973).

In contrast to the Apollo 17 samples, the saturated ^{22}Na activities of the Apollo 15 and 16 samples are very low (Fig. 2). It is due to the fact that there were no major flares during the period preceding the Apollo 15 and 16 missions, whereas the largest flares of the twentieth solar cycle occurred during 4–7 August 1972 enhanced about three to four times the production of ^{22}Na for samples returned from the Apollo 17 mission (long-lived ^{26}Al activity is not influenced by short term variation of solar activity).

The results of the ^{22}Na–^{26}Al method for the Apollo 16 samples are summarized in Table 2.

A striking feature seen in Table 2 is that there are more unsaturated samples than saturated ones among the samples taken on the rim of North Ray Crater (at Station 11) and also on the rim of Buster Crater (at Station 2), whereas saturated samples dominate at the other stations.

North Ray Crater is a large young crater (900–950 m across) which was, for the first time in lunar exploration, investigated along its rim, and several samples were taken from large boulders for the age determination. The formation of North Ray Crater was dated to be 30–50 m.y. by rare gas and track studies (Behrmann *et al.*, 1973; Kirsten *et al.*, 1973; Schaeffer and Husain 1973). On South Boulder located next to House Rock, an impact-produced spall zone surrounding a percussion

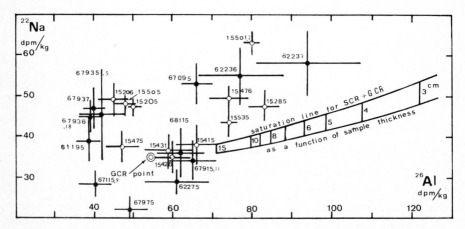

Fig. 2. ^{22}Na–^{26}Al correlation diagram for Apollo 15 and 16 rocks. Only unsaturated samples are shown (saturated samples are given in Figs. 5, 6 of the previous paper (Yokoyama *et al.*, 1973)).

Table 2. Saturation or nonsaturation of Apollo 16 rocks in ^{26}Al activity.

Station	Number of sample				
	Total	Saturated	Unsaturated	Undecided or shielded	
LM, 10	6	4 (0135; 0255 / 0275; 0335)		2 (0115,1 / 0315)	
1	7	4 (1155; 1156 / 1175 / 1295,18)	1 (1195)	2 (1016,120 / 1016,173 / 1135,1)	Plum Crater
2	5	1 (2235)	2 (2236, 2237)	2 (2275^s; 2295)	Buster Crater
4, 5, 6	9	7 (4476; 5035 / 5056; 5095 / 5075,1–2 / 6035; 6075)		2 (5055 / 6095,13)	
8, 9	7	4 (8415,1; 8416 / 9935; 9955)		3 (8035; 8115^s / 8815,2)	South Ray Crater
11, 13	15	4 (7055; 7435,7 / 7475; 7955,1)	6 (7095; 7115,9 / 7935,5 / 7936,18 / 7937; 7975)	5 (3335,5; 3355,1 / 7035,17 / 7455,56 / 7915,11^s)	North Ray Crater

The first digit (6) of the LRL number is omitted.

The complete comma number is omitted when it is xxxx,0.

^{s} Shielded sample from solar flare particles.

cone was found and four chips were sampled within a freshly broken spall zone. The depth of the spall zone (i.e. the thickness of the material removed) is approximately 2–3 cm (LSPET, 1972). Since the penetration depth of solar cosmic rays is about 2–3 cm, just the material containing the SCR products should be removed by this impact. Indeed, three of the four chips, 67935,5, 67936,18 and 67937, were reported as unsaturated in ^{26}Al by Eldridge *et al.* (1973) and Clark and Keith (1973). It was confirmed by the present work. The other chip, 67955,1, taken from the same zone is saturated in ^{26}Al activity. Another chip, 67915,11, was taken from an area about 2 m away from the spall zone, but from the same boulder. The sample seems to have been shielded from solar flare protons. These facts suggest a rather complex history of this 5 m boulder.

Two chips taken from a white breccia show that one is saturated (67475), and the other is probably not (67455,56), which suggests a probably complex history of this boulder. Among six small breccias studied, three (67095, 67115,9, and 67975) were unsaturated in ^{26}Al. The fact that about half of the samples are unsaturated in ^{26}Al activity suggests that either this region was recently bombarded with an intense flux of relatively small (and probably secondary) projectiles, or these unsaturated fragments have the same origin.

Buster Crater (90 m in diameter) is a fresh and probably primary crater (*Apollo 16 Preliminary Science Report*, 1972). Six rock samples which can be associated with the Buster ejecta blanket with high confidence were collected at its rim. Five of them were measured and two are unsaturated in ^{26}Al. The dating by rare gas and track methods is not yet reported for these samples (a soil sample taken on the rim, 62241, shows an exposure age of 56 m.y., Behrmann *et al.*, 1973). The short exposure age (about one million years) deduced by ^{22}Na–^{26}Al method may be the age of a post-cratering recycle event rather than the age of Buster Crater itself, because a rock sample, 62235, rested completely saturated in SCR-produced ^{26}Al on the rim of this crater, that is improbable if the crater formation is so recent.

The results of the ^{22}Na–^{26}Al method for Apollo 15 rocks are summarized in Table 3.

Table 3. Saturation or nonsaturation of Apollo 15 rocks in ^{26}Al activity.

Station	Number of sample				
	Total	Saturated	Unsaturated	Undecided	
1	4	2 (065; 085)	1 (086)	1 (076)	Elbow Crater
2	2		2 (205; 206)		St. George Crater
3	1	1 (016)			Mare surface
4	6	$2\left(\begin{array}{c}486\\499,9\end{array}\right)$	$3\left(\begin{array}{c}475;476\\495\end{array}\right)$	1 (498)	Dune Crater
6	4	2 (255; 256)	2 (265; 285)		Apennine Front
7	9	$3\left(\begin{array}{c}418;459^P\\465\end{array}\right)$	$3\left(\begin{array}{c}415;426^P\\431\end{array}\right)$	$3\left(\begin{array}{c}445;455\\466\end{array}\right)$	Spur Crater
8 LM-ALSEP	2	$2\left(\begin{array}{c}058,29\\558\end{array}\right)$			Ray-mantled mare surface
9	2		$2\left(\begin{array}{c}501,2^C\\505\end{array}\right)$		Very fresh surface
9a	5	$4\left(\begin{array}{c}545;556\\557;597\end{array}\right)$	1 (535)		Hadley Rille

The first two digits (15) of the LRL number are omitted.
^{C}Soil clod.
^{P}Partial sample.

Among the Apollo 15 rocks, the exposure history of anorthosite 15415 seems interesting. The ^{26}Al activity of 116 ± 9 dpm/kg and the ^{26}Al/^{22}Na ratio of 3.2 ± 0.5 measured by Keith *et al.* (1972) are apparently in the range of saturation, but after the normalization for the chemical composition, the ^{26}Al activity of 68 ± 6 dpm/kg and the ^{26}Al/^{22}Na ratio of 1.8 ± 0.3 were obtained. Sample 15415 was taken from the top of a poorly indurated breccia, 15431–7. From its position, sample 15415 should be well exposed for SCR. The expected saturated ^{26}Al activity of the sample having 4–5 cm thickness is about 110 dpm/kg. Since the rare gas exposure age of this sample is of the order of 100 m.y. (Stettler *et al.*, 1973), the sample should be saturated for GCR-produced ^{26}Al (55 dpm/kg). The SCR-produced ^{26}Al of 13 ± 6 dpm/kg is about a quarter of the saturation of SCR-produced ^{26}Al (about 55 dpm/kg). An exposure age of 0.4 ± 0.2 m.y. can be deduced from this data.

On the other hand, the breccia 15431, from which 15415 was picked up, is also unsaturated in ^{26}Al and shows the same age. Keith *et al.* (1972) suggested that the unsaturation of the very friable breccia 15431 is due to a rapid erosion. In this connection, the unsaturation of anorthosite 15415 deduced by the present study is consistent with the suggestion made by Keith *et al.* It is therefore possible to conclude that the anorthosite, 15415, was incorporated in the breccia, but "weathered out" very recently, arriving at the "come and sample me" situation on top of rock 15431–7. It was very fortunate for the astronauts and the lunar sample investigators. A friable green clod, 15426 taken at the same Station 7 as that of 15415 is also unsaturated in ^{26}Al, indicating also rapid erosion.

Four samples were collected at Station 9 located at a 15 m diameter crater of very fresh appearance. The ejecta blanket of this crater probably represents the freshest surface of any significant extent that has been sampled by the Apollo missions (at least before Apollo 16), according to the description of the Apollo 15 Preliminary Science Report (1972).

Two samples studied, 15501,2 and 15505, are both unsaturated in ^{26}Al. Rancitelli *et al.* (1972) estimated an exposure age of a few hundred thousand years for 15505. Our estimation, less than 2×10^5 yr, agrees with theirs. For the soil clod 15501,2 Rancitelli *et al.* suggested that either the sample has an exposure age of three-fourths million years or this very soft soil clod was eroded at a high rate of greater than 3 cm per million years. Our estimation of the SCR exposure age is 0.4 ± 0.2 m.y., which was deduced from the normalized ^{26}Al activity of 80 ± 3 dpm/kg: subtracting the galactic contribution of 55 dpm/kg, we obtained a SCR-produced ^{26}Al activity of 25 ± 3 dpm/kg in contrast to the expected SCR production of 90 ± 20 dpm/kg for this sample. Thus, the possibility that these two samples have the same exposure age of about 0.2 m.y. can not be excluded. Alternatively, the soft soil clod 15501,2 was eroded rapidly as suggested by Rancitelli *et al.*

The results of the ^{22}Na–^{26}Al method for Apollo 14 rocks are summarized in Table 4 and in Fig. 3, and those of Apollo 12 in Table 5 and in Fig. 4.

For Apollo 11 samples, we found that all rocks studied are saturated in ^{26}Al activity: 10003, 10017, 10018,1, 10019,1, 10021,1, 10057, 10061,21, and 10072,1.

Table 4. Saturation or nonsaturation of Apollo 14 rocks in ^{26}Al activity.

Sample area	Number of sample			
	Total	Saturated	Unsaturated	Undecided
Comprehensive sample area	5	1 (271)	3 (265; 272; 273)	1 (169)
Others	10	7 ⎛053; 066 305,18; 310^T 315; 318; 321^T⎞		3 ⎛045; 082 301⎞

The first two digits (14) of the LRL number are omitted.
^{T}Top slices are saturated in ^{26}Al.

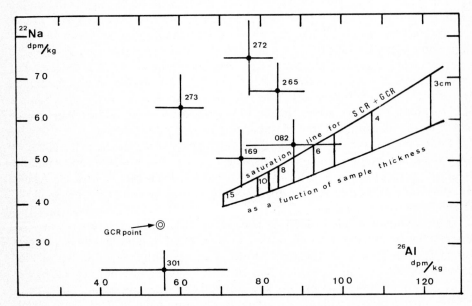

Fig. 3. ^{22}Na–^{26}Al correlation diagram for Apollo 14 rocks. The first two digits (14) of the LRL number are omitted. Only unsaturated samples are shown (saturated samples are given in Fig. 4 of the previous paper (Yokoyama *et al.*, 1973).

RELATION BETWEEN NORMALIZED ^{26}Al ACTIVITY AND WEIGHT FOR SATURATED SAMPLES

As it was discussed at the beginning of the present paper, we have shown that the function of chemical composition and the function of the exposure geometry can be separated and the normalized activity of a saturated sample becomes uniquely a function of exposure geometry:

$$A_{nor} = f_2(\text{geometry}) \text{ for a saturated sample} \tag{6}$$

Table 5. Saturation or nonsaturation of Apollo 12 rocks in ^{26}Al activity.

Sample area	Number of sample			
	Total	Saturated	Unsaturated	Undecided
Selected samples[a]	6	$3 \begin{pmatrix} 02^{\mathrm{T}}; 04,1 \\ 16 \end{pmatrix}$		3 (05; 10; 13)
Contingency sample[b]	2	1 (75,5)		1 (73)
Head Crater	2	1 (52,1)		1 (31)
Bench Crater	3	2 (39; 53^{T})	1 (40,30)	
Surveyor Crater	6	$4 \begin{pmatrix} 51,1\text{–}3 \\ 62; 63; 65 \end{pmatrix}$	2 (54; 64)	

The first three digits (120) of the LRL number are omitted.
[a]Sampled near Middle Crescent Crater.
[b]Sampled near LM.
[T]Top slices are saturated in ^{26}Al.

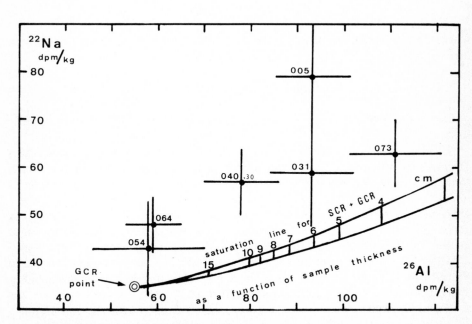

Fig. 4. ^{22}Na–^{26}Al correlation diagram for Apollo 12 (unsaturated) rocks. Saturated samples are given in Fig. 3 of the previous paper (Yokoyama *et al.*, 1973). The first two digits (12) of the LRL number are omitted.

In order to determine the form of the function, we calculated theoretically the ^{26}Al activity of a sample as a function of the weight of the sample, W (expressed in the unit of kg). The calculation was made in the following way: first, the mean production rate of ^{26}Al was calculated as a function of sample thickness for the infinite semi-plane geometry, by the method described in the previous paper (Yokoyama *et al.*, 1973). The calculated activity is approximately expressed with the equation,

$$A_{nor}(\text{dpm/kg}) = 55 + 195(1 - e^{-1.26/h}) \frac{\lambda}{\lambda + \alpha R} \qquad (7)$$

where h is the sample thickness, λ is the decay constant of the ^{26}Al $(1.0 \times 10^{-6}\,\text{y}^{-1})$, R is the erosion rate (in cm y^{-1}) and α is the mean attenuation constant of the ^{26}Al activity when this latter is expressed as an exponentially decreasing function of the depth $\alpha = 2\,\text{cm}^{-1}$). The first term of the right hand is independent of the sample thickness and corresponds approximately to the production of ^{26}Al by GCR (55 dpm/kg). The second term increases with the decrease of the sample thickness and corresponds to the additional production of ^{26}Al by SCR near the surface of the sample.

Then, the mean thickness, $\bar{h}$, of the sample is expressed as a function of the sample weight by assuming that the form of the sample is a cube or a sphere ($\bar{h}$ (cm) $= 7\,W^{1/3}$ for density $= 3$) or a hemisphere ($\bar{h} = 4.5\,W^{1/3}$). By replacing h by W in Eq. (7), we obtain

$$A_{nor}(\text{dpm/kg}) = 55 + 195(1 - e^{-kW^{-1/3}}) \frac{\lambda}{\lambda + \alpha R} \qquad (8)$$

where k is a constant (0.18 for the cubic or spheric form and 0.28 for the hemispheric form). For $W > 0.1$ kg, the relation can be simplified to

$$A_{nor}(\text{dpm/kg}) = 55 + k'W^{-1/3} \frac{\lambda}{\lambda + \alpha R} \qquad (9)$$

where k' is a constant (31 for the cubic or spheric form and 45 for the hemispheric form).

Equation (9) can be derived from a simpler model: the total activity of a sample is the sum of the GCR-produced ^{26}Al which is approximately proportional to the sample weight and the SCR-produced ^{26}Al which is approximately proportional to the exposed area, S. Total activity $= k_1 W + k_2 S$, where k_1 and k_2 are proportional constants. From the theoretical calculation, k_1 is estimated to be about 55 dpm/kg, and k_2 is estimated to be $0.66\,\lambda/\lambda + \alpha R$ dpm/cm^2. Since W (kg) $= \rho S\bar{h}/1000$ (ρ is the density), the specific activity (dpm/kg) $= k_1 + 1000\,k_2/\rho\bar{h}$. By replacing $\bar{h}$ by $7\,W^{1/3}$ or $4.5\,W^{1/3}$), an equation comparable to Eq. (9) can be obtained, with nearly the same values for k' (31 for the cubic or spheric form and 49 for the hemispheric form).

These theoretical functions are compared, in Fig. 5, with the experimental results: the observed ^{26}Al activities of Keith–Clark's selected samples (Keith and Clark, 1974) are normalized for chemical composition by our Eq. (4), and then, thus obtained normalized activities are plotted against the inverse cubic root of

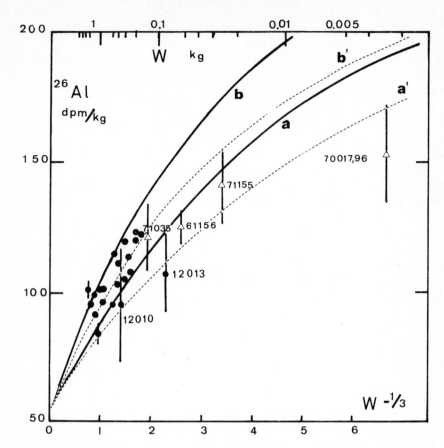

Fig. 5. Correlation diagram between the saturated (and normalized) activity in ^{26}Al and the sample weight. Lines "a" and "a'" are theoretically calculated lines for the cubic (or spheric) sample form respectively with an erosion rate of zero and of 1 mm/10^6 yr. Lines "b" and "b'" are theoretical lines for the hemispheric form respectively with an erosion rate of zero and of 1 mm/10^6 yr. Solid circles represent Keith–Clark's selected samples (for some samples, error bars and LRL sample number are given; 12010 and 12013 seem to be slightly unsaturated in ^{26}Al). Triangles represent additional samples selected by the present study (70017,96 may be slightly shielded from solar flare protons).

the sample weight, $W^{-1/3}$. Four light-weight samples which were judged as saturated in ^{26}Al by the ^{22}Na–^{26}Al method are also plotted in the figure in order to see the tendency of the relation for a lighter mass. In Fig. 5, the normalized ^{26}Al activities are distributed within the area surrounded by two theoretical curves, "a" (cubic or spheric form) and "b" (hemispheric form). Most of the samples can therefore by approximated by the spheric or hemispheric form.

For a weight less than 100 g, the experimental results agree better with the theoretical curves "a'" and "b'" for which an erosion rate of 1 mm/10^6 yr is assumed, than the curves "a" and "b" corresponding to the no-erosion cases.

The linear function of $W^{-1/3}$, Eq. (9) can be a good approximation for the weight between 100 g and 10 kg. The smallest sample of Keith–Clark's set was 80 g and the largest was less than 2500 g.

From the combination of the weight-function Eq. (9) with the normalization Eq. (4), we obtained an equation which predicts the saturated ^{26}Al activity of a sample having a chemical composition other than that of the reference:

$$A_{obs}(dpm/kg) = \frac{Si + 2.7\ Al + 0.05\ Mg}{42.8}(55 + 38\ W^{-1/3}) \tag{10}$$

where the denominater, 42.8, corresponds to $(Si + 2.7\ Al + 0.05\ Mg)_{ref}$ of Eq. (4), and the coefficient of the term including $W^{-1/3}$ of Eq. (9) was simplified by assuming an erosion rate of $1\ mm/10^6\ yr$ and by adopting the value of k' for the hemispheric form. After converting a net element percent to the oxide basis and converting the term of MgO to the term of MgO $W^{-1/3}$ (because the production of ^{26}Al from Mg is due to SCR), we obtained an equation,

$$A_{obs}(dpm/kg) = 0.60[SiO_2] + 1.8[Al_2O_3] + 0.42[SiO_2]\ W^{-1/3}$$
$$+ 1.27[Al_2O_3]\ W^{-1/3} + 0.03[MgO]\ W^{-1/3} \tag{11}$$

where $[SiO_2]$, $[Al_2O_3]$, and $[MgO]$ are the percent oxides. This theoretical equation can be compared with an experimental Eq. (12), which was deduced by Keith and Clark (1974) from the regression analysis of twenty-one samples from the Apollo 11 to 16 missions:

$$^{26}Al(dpm/kg) = 0.652[SiO_2] + 2.50[Al_2O_3] + 0.560[Al_2O_3]\ W^{-1/3}$$
$$+ 2.28[MgO]\ W^{-1/3} \tag{12}$$

The first two terms of the right hand of these two equations are in good agreement. A considerable disagreement, however, is seen between the terms including $W^{-1/3}$: Keith–Clark's equation lacks the term of $SiO_2\ W^{-1/3}$ and has a much larger coefficient for the term of MgO $W^{-1/3}$. We calculated the predicted ^{26}Al activities of Keith–Clark's samples by their Eq. (12), and found that the smallest production of ^{26}Al from Mg was 12 dpm/kg (rock 15418) and the largest was 38 dpm/kg (rock 12010). From our theoretical calculation, however, the ^{26}Al production from Mg was an order of 1 dpm/kg or less. For the present, we point out the difference between the two equations, but without arguing which is better. Nevertheless, since the theoretical method and the experimental one should not be opposed but should cooperate, we suggest that the regression analysis can be made with a minimum number, 3, of the unknown variables; as our theoretical calculation shows on the one hand that the ratio of the ^{26}Al production from Al to that from Si is practically constant independently of sample size, and on the other hand, that the ^{26}Al production from Mg is probably negligible, the regression analysis can be made by the following equation:

$$^{26}Al(dpm/kg) = ([SiO_2] + C_1[Al_2O_3]) \cdot (C_2 + C_3\ W^{-1/3}) \tag{13}$$

where C_1, C_2, and C_3 are the regression coefficients to be determined. We also suggest the elimination, from the set, of rock 12013 which is probably unsaturated in ^{26}Al.

In conclusion, we propose the application of two methods in order to know whether the sample is saturated in ^{26}Al activity or not, further to see the degree of undersaturation, and finally to estimate the exposure age: the first method is the ^{22}Na–^{26}Al correlation diagram after normalization for chemical composition by Eqs. (4) and (5); in this method, the ^{22}Na activity indicates the exposure conditions of the sample for SCR. The second one is the ^{26}Al–$W^{-1/3}$ correlation diagram always after normalization for chemical composition by Eq. (4). Simultaneous and comparative application of the two methods gives more precision.

Acknowledgments—We are grateful to Professor J. Labeyrie for his constant encouragement. We wish to thank Dr. J. Sato, Tokyo, for his participation in the early part of this work.

REFERENCES

ALGIT (Apollo Lunar Geology Investigation Team) (1973) Documentation and environment of the Apollo 17 samples: a preliminary report. *U.S. Geological Survey. Interagency Report: Astrogeology* **71**.

Apollo 15 Preliminary Science Report NASA SP-289 (1972).

Apollo 16 Preliminary Science Report NASA SP-315 (1972).

Apollo 17 Preliminary Science Report NASA SP-330 (1973).

Behrmann C., Crozaz G., Drozd R., Hohenberg C., Ralston C., Walker R., and Yuhas D. (1973) Cosmic-ray exposure history of North Ray and South Ray material. *Proc. Fourth Lunar Sci. Conf., Geochim. Cosmochim. Acta*, Suppl. 4, Vol. 2, pp. 1957–1974. Pergamon.

Clark R. S. and Keith J. E. (1973) Determination of natural and cosmic ray induced radionuclides in Apollo 16 lunar samples. *Proc. Fourth Lunar Sci. Conf., Geochim. Cosmochim. Acta*, Suppl. 4, Vol. 2, pp. 2105–2113. Pergamon.

Eldridge J. S., O'Kelley G. D., and Northcutt K. J. (1973) Radionuclides concentrations in Apollo 16 lunar samples determined by nondestructive gamma-ray spectrometry. *Proc. Fourth Lunar Sci. Conf., Geochim. Cosmochim. Acta*, Suppl. 4, Vol. 2, pp. 2115–2122. Pergamon.

Eldridge J. S., O'Kelley G. D., and Northcutt K. J. (1974) Primordial radioelement concentrations in rocks and soils from Taurus-Littrow (abstract). In *Lunar Science—V*, pp. 206–208. The Lunar Science Institute, Houston.

Hutcheon I. D., Macdougall D., Price P. B., Hörz F., Morrison D., and Schneider E. (1974) Rock 72315: A new lunar standard for solar flare and micro-meteorite exposure (abstract). In *Lunar Science—V*, pp. 378–380. The Lunar Science Institute, Houston.

Keith J. E. and Clark R. S. (1974) The saturation activity of ^{26}Al in lunar samples as a function of chemical composition and the exposure ages of some lunar samples (abstract). In *Lunar Science—V*, pp. 405–407. The Lunar Science Institute, Houston.

Keith J. E., Clark R. S., and Richardson K. A. (1972) Gamma-ray measurements of Apollo 12, 14, and 15 lunar samples. *Proc. Third Lunar Sci. Conf., Geochim. Cosmochim. Acta*, Suppl. 3, Vol. 2, pp. 1671–1680. MIT Press.

Keith J. E., Clark R. S., and Bennett L. J. (1974) Determination of natural and cosmic ray induced radionuclides in Apollo 17 lunar samples (abstract). In *Lunar Science—V*, pp. 402–404. The Lunar Science Institute, Houston.

King J. H. (1973) Solar proton fluences as observed during 1966–1972 and as predicted for 1977–1983 space missions. Goddard Space Flight Center Preprint X-601-73-324.

Kirsten T., Horn P., and Kiko J. (1973) ^{39}Ar–^{40}Ar dating and rare gas analysis of Apollo 16 rocks and soils. *Proc. Fourth Lunar Sci. Conf., Geochim. Cosmochim. Acta*, Suppl. 4, Vol. 2, pp. 1757–1784. Pergamon.

LSPET (Lunar Sample Preliminary Examination Team) (1972) *Apollo 16 Preliminary Science Report*, NASA SP-315.

LSPET (Lunar Sample Preliminary Examination Team) (1973) *Apollo 17 Preliminary Science Report*, NASA SP-330.

O'Kelley G. D., Eldridge J. S., and Northcutt K. J. (1974) Concentrations of cosmogenic radionuclides in Apollo 17 samples: Effects of solar flare of August 1972 (abstract). In *Lunar Science—V*, pp. 577–579. The Lunar Science Institute, Houston.

Rancitelli L. A., Perkins R. W., Felix W. D., and Wogman N. A. (1972) Lunar surface processes and cosmic ray characterization from Apollo 12–15 lunar sample analyses. *Proc. Third Lunar Sci. Conf.*, *Geochim. Cosmochim. Acta*, Suppl. 3, Vol. 2, pp. 1681–1691. MIT Press.

Rancitelli L. A., Perkins R. W., Felix W. D., and Wogman N. A. (1974) Anisotropy of the August 4–7, 1972 solar flares at the Apollo 17 site (abstract). In *Lunar Science—V*, pp. 618–620. The Lunar Science Institute, Houston.

Schaeffer O. A. and Husain L. (1973) Early lunar history: ages of 2 to 4 mm soil fragments from the lunar highlands. *Proc. Fourth Lunar Sci. Conf.*, *Geochim. Cosmochim. Acta*, Suppl. 4, Vol. 2, pp. 1847–1863. Pergamon.

Stettler A., Eberhardt P., Geiss J., Grögler N., and Maurer P. (1973) Ar^{39}–Ar^{40} ages and Ar^{37}–Ar^{38} exposure ages of lunar rocks. *Proc. Fourth Lunar Sci. Conf.*, *Geochim. Cosmochim. Acta*, Suppl. 4, Vol. 2, pp. 1865–1888. Pergamon.

Wrigley R. C. (1973) Radionuclides at Descartes in the central highlands. *Proc. Fourth Lunar Sci. Conf.*, *Geochim. Cosmochim. Acta*, Suppl. 4, Vol. 2, pp. 2203–2208. Pergamon.

Yokoyama Y., Sato J., Reyss J. L., and Guichard F. (1973) Variation of solar cosmic-ray flux deduced from ^{22}Na–^{26}Al data in lunar samples. *Proc. Fourth Lunar Sci. Conf.*, *Geochim. Cosmochim. Acta*, Suppl. 4, Vol. 2, pp. 2209–2227. Pergamon.

APPENDIX 1

The main target elements for the production of ^{22}Na and ^{26}Al are Si, Al, Mg, and Na. Since the lunar samples contain less than one percent Na, the production of ^{22}Na from Na can be neglected. On the other hand, the production of ^{26}Al from Mg is not important and can be neglected as it was seen in Fig. 1 of the previous paper (Yokoyama *et al.*, 1973). The Si content of lunar samples is practically constant. The factors which determine the production of ^{26}Al are therefore uniquely the Al content, and the factors for ^{22}Na are the Al and Mg contents.

Since the Th, U, and K contents are simultaneously determined with ^{22}Na and ^{26}Al activities by gamma-spectrometry, we made an attempt to see if there is a correlation between the Al and Mg contents and the Th, U, and K contents. For this purpose, the Al_2O_3 and MgO contents are plotted as a function of the Th (or U) content. An example of such plotting is given in Fig. 6. The mare basalts are distributed in the figure within a narrow area both in the Al_2O_3 content and in the Th content. The Al_2O_3 content of unanalyzed sample can therefore be estimated with a good precision for mare basalts. For breccias, the dispersion of data is more important. Nevertheless, the distribution range of the Al_2O_3 contents can be estimated for each Apollo mission as a function of the Th content. For example, for the Apollo 16 breccias, a following correlation was obtained from the unweighted regression analysis of 16 samples,

$$Al_2O_3(\%) = (31.4 \pm 1.2) - (2.3 \pm 0.6) \text{ Th (ppm)} \qquad (14)$$

The correlation coefficient was -0.70 and $\Sigma(\Delta S)^2/N\text{-}2$ was 7.3, where ΔS is the difference between calculated and measured Al_2O_3 contents for a given sample and N is the number of samples (if we take into account two Apollo 16 breccias which situate near the Apollo 14 KREEP breccias in Fig. 6, we obtain the following correlation: $Al_2O_3 = (30.1 \pm 0.9) - (1.47 \pm 0.26)$ Th, with a correlation coefficient of -0.85 and $\Sigma(\Delta S)^2/N - 2 = 7.5$). For the normalization, we adopted an error of $\pm 4.0\%$ Al_2O_3 which is considerably larger than that calculated from the above-mentioned figure.

The MgO content is estimated in the same manner.

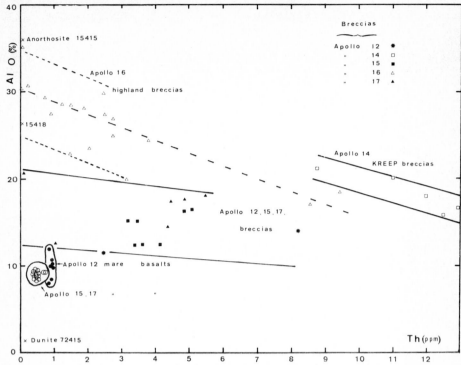

Fig. 6. Correlation diagram between Al_2O_3 contents and Th contents of lunar samples.

APPENDIX 2

Natural and cosmogenic radionuclides of two basalts from Apollo 17, 70017,96 and 75035,22, were measured with a gamma–gamma coincidence spectrometer consisting of two 12.7×12.7 cm NaI (Tl) scintillators shielded successively by 1.5 cm thick Hg, 5 cm thick plastic anticosmic scintillator, 2 cm thick Cu, 12 cm thick Fe, and 15 cm thick Pb. The background counting rate was 0.007 cpm with a counting efficiency of 2.3% for the coincidence peak of 0.511–2.3 MeV of ^{26}Al. The data reduction was made by the successive peeling off method based on the standard spectra of Th, U, ^{26}Al, and ^{22}Na sources. The results are given in Table 6.

Table 6. Cosmogenic and natural radionuclides in two basalts of Apollo 17.

Sample	Weight (g)	^{22}Na (dpm/kg)	^{26}Al (dpm/kg)	Th (ppm)	U (ppm)
70017,96 (Top)[a]	3.38	130 ± 13	110 ± 14	0.58 ± 0.31	0.11 ± 0.11
75035,22 (Top)[b]	2.19	170 ± 25	107 ± 15	0.65 ± 0.39	0.22 ± 0.22

[a]Mean thickness of about 8 mm.
[b]0–9 mm deep.
Top refers to the probable lunar surface orientation.
The errors include all known sources of error on the basis of one standard deviation.

Proceedings of the Fifth Lunar Conference
(Supplement 5, Geochimica et Cosmochimica Acta)
Vol. 2 pp. 2249–2250 (1974)
Printed in the United States of America

Examination of Apollo 17 surface fines for porphyrins and aromatic hydrocarbons

J. H. RHO

Jet Propulsion Laboratory, California Institute of Technology, Pasadena, California 91103

Abstract—An Apollo 17 surface fines sample 75081,78, collected from Station 5, was extracted and examined fluorometrically. No porphyrins were found with fluorometric methods capable of detecting 2×10^{-14} moles/g of Ni-mesoporphyrin IX in the sample. Also aromatic hydrocarbons were undetected.

ANALYTICAL PROCEDURES

A surface fine sample from Apollo 17 mission was examined for aromatic hydrocarbons and porphyrins. The sample 75081,78 (10 g) was collected unsieved from Station 5. The sample was extracted in a Soxhlet extractor under an atmosphere of argon, with benzene: methanol (3/2 v/v) for a period of 24 hours. The extracts were concentrated to a volume of 0.3 ml under a stream of nitrogen and their fluorescence properties examined for free base porphyrins and aromatic hydrocarbons. The extracts were then carried through a demetallation procedure (Rho *et al.*, 1971) and examined fluorometrically for metalloporphyrins. A procedural blank in which 200-mesh optical quartz was substituted for the sample and a reagent blank were carried through the procedure and compared to a solution made by demetallating 0.1 ng of Ni-mesoporphyrin IX. Fluorescence spectra were obtained with an Aminco–Bowman spectrophotofluorometer.

RESULTS AND DISCUSSION

Fluorescence spectra of the lunar sample extracts were obtained on an Aminco–Bowman spectrophotofluorometer at excitation wavelengths, ranging from 350 to 450 nm. No fluorescence attributable to either porphyrins or aromatic hydrocarbons was found, even with computer-enhancement of the data (Rho *et al.*, 1972). The averages of 16 scans of fluorescence from 550 to 750 nm were recorded on a signal analyzer at 5 sec per scan, as shown in Fig. 1. Individual scans were corrected for variation of scan rate before averaging. One hundred and eighty data points in this range were processed by the method of least squares. Plots of deviations from computed background for the Apollo 17 surface sample and the optical quartz blank indicate that there are some coincidences between the spectral features in the range from 570 to 650 nm for the quartz blank and the lunar samples. These peaks are instrumental artifacts from undispersed light from the xenon lamp, as noted previously (Rho *et al.*, 1972). When the deviations from the blank were subtracted from those of the lunar sample extracts, no features attributable to porphyrins remained, as shown in Fig. 1.

With the method used one could conservatively detect 0.1 ng of Ni-mesoporphyrin IX (2×10^{-13} moles), a typical metalloporphyrin. The lunar sam-

FLUORESCENCE SPECTRA OF
APOLLO SAMPLE EXTRACTS

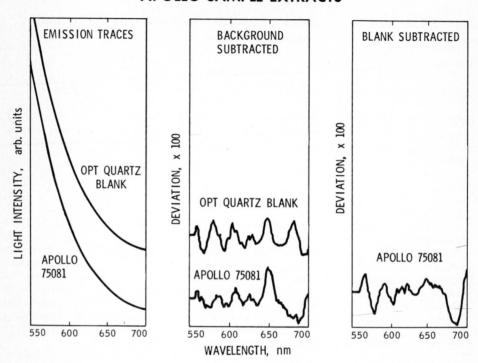

Fig. 1. Averaged traces of fluorescence spectra of the Apollo samples and the deviations of the traces from best fitting six-term polynomials. Fluorescence excitation was set at 400 nm.

ples examined in this study weighed 10 g. Therefore, 2×10^{-14} moles/g on Ni-mesoporphyrin IX or equivalent could have been detected in the surface sample. This work indicates that no porphyrins or aromatic hydrocarbons are present in these lunar samples in any significant amounts.

Acknowledgment—The author thanks J. R. Thompson for his technical assistance. This work represents one phase of research carried out at the Jet Propulsion Laboratory, California Institute of Technology, under Contract No. NAS 7-100, sponsored by the National Aeronautics and Space Administration.

REFERENCES

Rho J. H., Bauman A. J., Yen T. F., and Bonner J. (1971) Absence of porphyrins in an Apollo 12 lunar surface sample. *Proc. Second Lunar Sci, Conf., Geochim. Cosmochim. Acta*, Suppl. 2, Vol. 2, pp. 1875–1877. MIT Press.

Rho J. H., Cohen E. A., and Bauman A. J. (1972) Spectrofluorometric search for porphyrins in Apollo 14 surface fines. *Proc. Third Lunar Sci. Conf., Geochim. Cosmochim. Acta*, Suppl. 3, Vol. 2, pp. 2149–2155. MIT Press.

Lunar Sample Index

Pages   1–973:   Volume 1, Mineralogy and Petrology
Pages 975–2250: Volume 2, Chemical and Isotope Analyses/Organic Chemistry
Pages 2251–3134: Volume 3, Physical Properties

(An index entry refers to the opening page of an article.)

Subject Index

Author Index
(Volume 2)